GENERAL CHEMISTRY

TENTH EDITION

DARRELL D. EBBING

Wayne State University, Emeritus

STEVEN D. GAMMON

Western Washington University

BROOKS/COLE
CENGAGE Learning™

Australia • Brazil • Japan • Korea • Mexico • Singapore • Spain • United Kingdom • United States

BROOKS/COLE
CENGAGE Learning™

General Chemistry, Tenth Edition
International Edition
Darrell D. Ebbing, Steven D. Gammon

Publisher: Mary Finch

Executive Editor: Lisa Lockwood

Developmental Editor: Alyssa White

Assistant Editor: Elizabeth Woods

Editorial Assistant: Krista Mastroianni

Senior Media Editor: Lisa Weber

Media Editor: Stephanie VanCamp

Marketing Manager: Nicole Hamm

Marketing Program Manager: Jing Hu

Marketing Communications Manager:
    Darlene Macanan

Content Project Manager: Teresa L. Trego

Art Director: Maria Epes

Manufacturing Planner: Karen Hunt

Rights Acquisitions Specialist: Tom McDonough

Production Service: PreMediaGlobal

Photo Researcher: Bill Smith Group

Text Researcher: Isabel Saraiva

Copy Editor: PreMediaGlobal

OWL Producers: Stephen Battisti, Cindy Stein,
    David Hart (Center for Educational Software
    Development, University of Massachusetts,
    Amherst)

Text Designer: Tani hasagawa

Cover Designer: Natalie Hill

Cover Image: ©Charles D. Winters

Compositor: PreMediaGlobal

International Edition:

ISBN-13: 978-1-111-98949-1

ISBN-10: 1-111-98949-4

Cengage Learning International Offices

Asia
www.cengageasia.com
tel: (65) 6410 1200

Australia/New Zealand
www.cengage.com.au
tel: (61) 3 9685 4111

Brazil
www.cengage.com.br
tel: (55) 11 3665 9900

India
www.cengage.co.in
tel: (91) 11 4364 1111

Latin America
www.cengage.com.mx
tel: (52) 55 1500 6000

UK/Europe/Middle East/Africa
www.cengage.co.uk
tel: (44) 0 1264 332 424

Represented in Canada by
Nelson Education, Ltd.
www.nelson.com
tel: (416) 752 9100 / (800) 668 0671

Cengage Learning is a leading provider of customized learning solutions with office locations around the globe, including Singapore, the United Kingdom, Australia, Mexico, Brazil, and Japan. Locate your local office at: **www.cengage.com/global**

For product information and free companion resources:
www.cengage.com/international
Visit your local office: **www.cengage.com/global**
Visit our corporate website: **www.cengage.com**

Printed in Canada
1 2 3 4 5 6 7 15 14 13 12

Brief Contents

Contents

1 Chemistry and Measurement   1

2 Atoms, Molecules, and Ions   41

3 Calculations with Chemical Formulas and Equations 88

4 Chemical Reactions 126

5 The Gaseous State   178

6 Thermochemistry   226

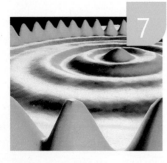

7 Quantum Theory of the Atom   268

8 Electron Configurations and Periodicity 299

9 Ionic and Covalent Bonding 334

12 Solutions 487

13 Rates of Reaction 533

14 Chemical Equilibrium   590

15 Acids and Bases   634

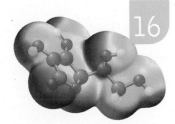

16 Acid–Base Equilibria   663

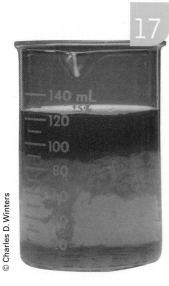

© Charles D. Winters

17 Solubility and Complex-Ion Equilibria **709**

© Charles D. Winters

18 Thermodynamics and Equilibrium **741**

19 Electrochemistry 778

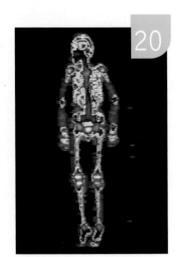

20 Nuclear Chemistry 827

21 Chemistry of the Main-Group Elements 873

22 The Transition Elements and Coordination Compounds 936

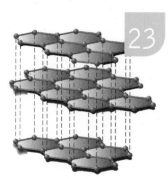

23 Organic Chemistry 975

Essays

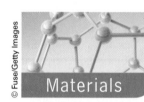

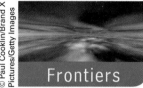

Preface

In the preface to the first edition, we wrote, "Scientists delve into the molecular machinery of the biological cell and examine bits of material from the planets of the solar system. The challenge for the instructors of introductory chemistry is to capture the excitement of these discoveries [of chemistry] while giving students a solid understanding of the basic principles and facts. The challenge for the students is to be receptive to a new way of thinking, which will allow them to be caught up in the excitement of discovery." From the very first edition of this text, our aims have always been to help instructors capture the excitement of chemistry and to teach students to "think chemistry." Here are some of the features of the text that we feel are especially important in achieving these goals.

Clear, Lucid Explanations of Chemical Concepts

We have always placed the highest priority on writing clear, lucid explanations of chemical concepts. We have strived to relate abstract concepts to specific real-world events and have presented topics in a logical, yet flexible, order. With succeeding editions we have refined the writing, incorporating suggestions from instructors and students.

Coherent Problem-Solving Approach

With the first edition, we presented a coherent problem-solving approach that involved worked-out *Examples* coupled with in-chapter *Exercises* and corresponding end-of-chapter *Problems*. This approach received an enormously positive response, and we have continued to refine the pedagogical and conceptual elements in each subsequent edition.

In the previous edition, we revised every Example, dividing the problem-solving process into a *Problem Strategy*, a *Solution*, and an *Answer Check*. By doing this, we hoped to help students develop their problem-solving skills: *think* how to proceed, *solve* the problem, and *check* the answer. This last step is one that is often overlooked by students, but it is critical if one is to obtain consistently reliable results.

For this tenth edition, we have added yet another level of support for students in this problem-solving process. In every Example, we have added what we call the *Gaining Mastery Toolbox*. We based this Toolbox on how we as instructors might help a student who is having trouble with a particular problem. We imagine a student coming to our office because of difficulty with a particular problem. We begin the help session by pointing out to the student the "big idea" that one needs to solve the problem. We call this the *Critical Concept*. But suppose the student is still having difficulty with the problem. We now ask the student about his or her knowledge of prior topics that will be needed to approach the problem. We call these needed prior topics the *Solution Essentials*. Each Gaining Mastery Toolbox that we have added to an Example begins by pointing out the Critical Concept involved in solving the problem posed in that Example. Then, under the heading of Solution Essentials, we list the topics the student needs to have mastered to solve this problem. We hope the Gaining Mastery Toolbox helps the student in much the way that an individual office visit can. Over several Examples, these Toolboxes should help the student develop the habit of focusing on the Critical Concept and the Solution Essentials while engaged in general problem solving.

While we believe in the importance of this coherent example/exercise approach, we also think it is necessary to have students expand their understanding of the concepts. For this purpose, we have a second type of in-chapter problem, *Concept Checks*. We have written these to force students to think about the concepts involved, rather than to focus on the final result or numerical answer—or to try to fit the problem to a memorized algorithm. We want students to begin each problem by asking, "What are the chemical concepts that apply here?" Many of these problems involve visualizing a molecular situation, since visualization is such a critical part of learning and understanding modern chemistry. Similar types of end-of-chapter problems, the *Conceptual Problems,* are provided for additional practice.

Expanded Conceptual Focus

A primary goal of recent editions has been to strengthen the conceptual focus of the text. To that end we have added three new types of end-of-chapter problems, *Concept Explorations*, *Strategy Problems*, and *Self-Assessment Questions*. While we have included them in the end-of-chapter material, *Concept Explorations* are unlike any of the other end-of-chapter problems. These multipart, multistep problems are structured activities developed to help students explore important chemical concepts—the key ideas in general chemistry—and confront common misconceptions or gaps in learning. Often deceptively simple, *Concept Explorations* ask probing questions to test student's understanding. Because we feel strongly that in order to develop a lasting conceptual understanding, students must think about the question without jumping quickly to formulas or algorithms (or even a calculator); we have purposely not included their answers in the *Student Solutions Manual.* As *Concept Explorations* are ideally used in an interactive classroom situation, we have reformatted them into workbook style in-class handouts with space for written answers and drawings to facilitate their use in small groups. In the *Instructor's Resource Manual* we provide additional background on the literature and theories behind their development, information on how Steve Gammon has implemented them into his classroom and suggestions for integration, and a listing of the concepts (and common misconceptions thereof) that each *Concept Exploration* addresses.

We recognize a need to challenge students to build a conceptual understanding rather than simply memorizing the algorithm from the matched pair and then applying it to a similar problem to get a solution. The *Strategy Problems* were written to extend students' problem-solving skills beyond those developed in the *Practice* and *General Problems.* With this edition, we have nearly doubled the number of these problems. To work a *Strategy Problem,* students will need to think about the problem (which might involve several concepts or problem-solving skills from the chapter), then solve it on their own without a similar problem from which to model their answer. For this reason, we have explicitly chosen not to include their answers in the *Student Solutions Manual.*

On the basis of student feedback, we developed conceptually focused multiple-choice questions to provide students with a quick opportunity for self-assessment. As they are intended primarily for self-study, these questions have been included with the Review Questions, in the re-titled *Self-Assessment and Review Questions* section. As an instructor, you know that a student may answer a multiple-choice question correctly but still use incorrect reasoning to arrive at the answer. You would certainly like to know whether the student has used correct reasoning. In this edition, we have explored using *two-tier questions* to address whether the student's learning of a concept has depth or is superficial. The first tier of a question might be fairly straightforward. For example, we might begin a question by listing a number of formulas of compounds and ask the student to classify each one as an ionic compound or a molecular compound. The student might give correct answers, but we want to draw him or her out as to the reasoning used by adding a further question (the second tier), such as "Which of the following is the best statement regarding molecular compounds?" By seeing how the student answers the second tier of a two-tier question, we can learn whether he or she may have a misconception of the material. In other words, we learn whether the student has a complete and correct understanding of an important concept.

An Illustration Program with an Emphasis on Molecular Concepts

Most of us (and our students) are highly visual in our learning. When we see something, we tend to remember it. As in the previous edition, we went over each piece of art, asking how it might be improved or where art could be added to improve student comprehension. We continue to focus on the presentation of chemistry at the molecular level. The molecular "story" starts in Chapter 1, and by Chapter 2, we have developed the molecular view and have integrated it into the problem-solving apparatus as well as into the text discussions. The following chapters continue to use the molecular view to strengthen chemical concepts. We have introduced electrostatic potential maps where pedagogically relevant to show how electron density changes across a molecule. This is especially helpful for visually demonstrating such things as bond and molecular polarity and acid–base behavior.

Chapter Essays Showcasing Chemistry as a Modern, Applicable Science

We continue our *A Chemist Looks at . . .* essays, which cover up-to-date issues of science and technology. We have chosen topics that will engage students' interest while at the same time highlight the chemistry involved. Icons are used to describe the content area (materials, environment, daily life, frontiers, and life science) being discussed. The essays show students that chemistry is a vibrant, constantly changing science that has relevance for our modern world. The essay "Gecko Toes, Sticky But Not Tacky," for example, describes the van der Waals forces used by gecko toes and their possible applications to the development of infinitely reusable tape or robots that can climb walls!

Also, with this edition, we continue our *Instrumental Methods* essays. These essays demonstrate the importance of sophisticated instruments for modern chemistry by focusing on an instrumental method used by research chemists, such as mass spectroscopy or nuclear magnetic resonance. Although short, these essays provide students with a level of detail to pique the students' interest in this subject.

We recognize that classroom and study times are very limited and that it can be difficult to integrate this material into the course. For that reason, we include end-of-chapter essay questions based on each *A Chemist Looks at . . .* and *Instrumental Methods* essay. These questions promote the development of scientific writing skills, another area that often gets neglected in packed general chemistry courses. It is our hope that having brief essay questions ready to assign will allow both students and instructors to value the importance of this content and make it easier to incorporate into their curriculums.

Additions and Changes Made in This Edition

- A beautiful new book design that is user friendly but that also conveys the sense of excitement surrounding chemistry!
- A heavily revised art and photography program that helps highlight the excitement of chemistry while explaining key concepts. New talking labels have been added to the art and photos. These talk students through each figure and explain what they are seeing—what they need to know.
- The rewriting of the Instrumental Methods essay "Mass Spectroscopy and Molecular Formula."
- The updating of several "A Chemist Looks At" essays, including those on "Lasers and Compact Disc Players" (rewritten), "Carbon Dioxide Gas and the Greenhouse Effect" (updated figure), and "Stratospheric Ozone" (updated figure).
- The complete rewriting of Sections 6.2 and 6.3 (in the chapter titled "Thermochemistry") to include an explicit discussion of the first law of thermodynamics.
- The rewriting in Chapter 8 of the section on atomic radius to enlarge the discussion.
- The redefining in Chapter 8 of electron affinity with a positive sign for stable negative ions. Although many general chemistry texts use the opposite sign, a

positive sign convention is that used by researchers in the field, as well as by inorganic and physical chemistry texts, and we think that it can be defined quite naturally. All values of electron affinity have been updated.

- An explicit defining in Chapter 9 of Coulomb's law.
- The rewriting in Chapter 9 of the section on bond length and covalent radii to dovetail with the earlier discussion of atomic radius in Chapter 8.
- The changing in Chapter 9 of the term bond energy to bond enthalpy to conform to preferred terminology.
- The addition in Chapter 10 of a subsection to include the geometry of large molecules. The discussion of dipole moment has been rewritten.
- The rewriting in Chapter 18 of the first section since Chapter 6 now includes a full discussion of the first law of thermodynamics, which had previously been discussed in Chapter 18.
- The making in Chapter 21 of a number of small changes to update the discussions of contemporary applications of main-group elements and compounds.
- The rewriting in Chapter 24 of a paragraph on protein biosynthesis to clarify the discussion.
- The doubling of the number of Strategy Problems in all chapters, except for two descriptive chapters.

Complete Instructional Package

For the Instructor

A complete suite of customizable teaching tools accompanies *General Chemistry* tenth edition. Whether available in print, online, or via CD, these integrated resources are designed to save you time, and help make class preparation, presentation, assessment, and course management more efficient and effective.

Supporting Materials

OWL for General Chemistry

Instant Access OWL with Cengage YouBook (6 months) ISBN: 978-1-133-04506-9
Instant Access OWL with Cengage YouBook (24 months) ISBN: 978-1-4282-7225-5

OWL
Online Web
Learning

By Roberta Day and Beatrice Botch of the University of Massachusetts, Amherst, and William Vining of the State University of New York at Oneonta.

OWL Online Web Learning offers more assignable, gradable content (including end-of-chapter questions specific to this textbook) and more reliability and flexibility than any other system. OWL's powerful course management tools allow instructors to control due dates, number of attempts, and whether students see answers or receive feedback on how to solve problems. OWL includes the Cengage YouBook, an interactive, customizable eBook with a text edit tool that allows instructors to modify the textbook narrative as needed. Instructors can publish web links, quickly reorder entire sections and chapters, and hide any content they don't teach to create an eBook that perfectly matches their syllabus. The Cengage YouBook includes animated figures, video clips, highlighting, notes, and more.

Developed by chemistry instructors for teaching chemistry, OWL is the only system specifically designed to support **mastery learning**, where students work as long as they need to master each chemical concept and skill. OWL has already helped hundreds of thousands of students master chemistry through a wide range of assignment types, including tutorials, interactive simulations, and algorithmically generated homework questions that provide instant, answer-specific feedback. OWL is continually enhanced with online learning tools to address the various learning styles of today's students such as:

- **Quick Prep** review courses that help students learn essential skills to succeed in General and Organic Chemistry
- **Jmol** molecular visualization program for rotating molecules and measuring bond distances and angles
- **Go Chemistry**® mini video lectures on key concepts that students can play on their computers or download to their video iPods, smart phones, or personal video players

In addition, when you become an OWL user, you can expect service that goes far beyond the ordinary. For more information or to see a demo, please contact your Cengage Learning representative or visit us at **www.cengage.com/owl**.

For the Instructor

Instructor Companion Site

Supporting materials are available to qualified adopters. Please consult your local Cengage Learning sales representative for details. Go to **login.cengage.com**, find this textbook, and choose Instructor Companion Site to see samples of these materials, request a desk copy, locate your sales representative, and download the following assets:

- **Complete Solutions Manual**, revised by David Shinn of the U.S. Merchant Marine Academy, contains complete and detailed solutions to all exercises and problems.

- **Instructor's Resource Manual**, written by Darrell Ebbing of Wayne State University and Steven Gammon of Western Washington University, offers information about chapter essays, suggestions for alternative sequencing of topics, short chapter descriptions, a master list of learning objectives, correlation or cumulative skills problems with text topics, alternative examples for lectures, and suggestions for lecture demonstrations.

- **PowerPoint Lectures**, revised by Mary Frances Barber of Wayne State University, written for this text so that instructors can customize by importing their own lecture slides or other materials. These lectures feature new application questions to help the student understand and master each chemical concept.

- **Concept Exploration Worksheets** are print-ready PDF versions of the *Concept Exploration* problems from the text, reformatted in worksheet style for use as in-class handouts or to facilitate group discussions.

- **ExamView Testing Software** enables you to create, deliver, and customize tests using the more than 500 test bank questions written specifically for this text by Nathanael Fackler of the Nebraska Wesleyan University. Digital files of the test banks in both Microsoft Word and PDF format are also included for your convenience.

- **Image libraries** contain digital files for figures, photographs, and numbered tables from the text, as well as multimedia animations in a variety of digital formats. Use these files to print transparencies, create your own PowerPoint slides, and supplement your lectures.

- **Instructor's Resource Manual for the Lab Manual,** by Barbara Munk of Wayne State University, provides instructions with possible sequences of experiments and alternatives, notes and materials for preparing the labs, and sample results to all pre- and post-lab activities in *Experiments in General Chemistry*. Also available on the companion website.

Chemistry CourseMate

Instant Access (two semester) ISBN: 978-1-133-50691-1

This book includes Chemistry CourseMate, a complement to your textbook. Chemistry CourseMate includes an interactive eBook, interactive teaching and learning tools such as quizzes, flashcards, videos, and more. Chemistry CourseMate also includes the Engagement Tracker, a first-of-its-kind tool that monitors student engagement in the course. Go to **login.cengage.com** to access these resources. Look for the CourseMate icon, which denotes a resource available within CourseMate.

For the Student

Visit CengageBrain.com

To access these and additional course materials, please visit **www.cengagebrain.com**. At the CengageBrain.com home page, search for this textbook's ISBN (from the back cover of your book). This will take you to the product page where these resources can be found. (Instructors can log in at **login.cengage.com**.)

Quick Prep for General Chemistry

Instant Access (90 days) ISBN: 978-0-495-56029-6

Quick Prep is a self-paced online short course that helps students succeed in general chemistry. Students who completed Quick Prep through an organized class or self-study averaged almost a full letter grade higher in their subsequent general chemistry course than those who did not. Intended to be taken prior to the start of the semester, Quick Prep is appropriate for both underprepared students and for students who seek a review of basic skills and concepts. Quick Prep features an assessment quiz to focus students on the concepts they need to study to be prepared for general chemistry. Quick Prep is approximately 20 hours of instruction delivered through OWL with no textbook required and can be completed at any time in the student's schedule. Professors can package a printed access card for Quick Prep with the textbook or students can purchase instant access at **www.cengagebrain.com**. To view an OWL Quick Prep demonstration and for more information, visit **www.cengage.com/chemistry/quickprep**.

Chemistry CourseMate

Instant Access (two semester) ISBN: 978-1-133-50691-1

This book includes Chemistry CourseMate, which helps you make the grade. Chemistry CourseMate includes an interactive eBook with highlighting, note taking and search capabilities, as well as interactive learning tools such as quizzes, flashcards, videos, and more. Go to **login.cengage.com** to access these resources, and look for the CourseMate icon to find resources related to your text in Chemistry CourseMate.

Instant Access Go Chemistry® for General Chemistry (27-video set)

ISBN: 978-0-495-38228-7

Pressed for time? Miss a lecture? Need more review? Go Chemistry for General Chemistry is a set of 27 downloadable mini video lectures, accessible via the printed access card packaged with your textbook or available for purchase separately. Developed by one of this book's authors, Go Chemistry helps you quickly review essential topics—whenever and wherever you want! Each video contains animations and problems and can be downloaded to your computer desktop or portable video player (like iPod or iPhone) for convenient self study and exam review. Selected Go Chemistry videos have e-Flashcards to briefly introduce a key concept and then test student understanding with a series of questions. The Cengage YouBook in OWL contains Go Chemistry. Students can enter the ISBN above at **www.cengagebrain.com** to download two free videos or to purchase instant access to the 27-video set or individual videos.

Essential Math for Chemistry Students, Second Edition

By David W. Ball, Cleveland State University. This short book is intended to help you gain confidence and competency in the essential math skills you need to succeed in general chemistry. Each chapter focuses on a specific type of skill and has worked-out examples to show how these skills translate to chemical problem solving. The book includes references to the OWL learning system where you can access online algebra skills exercises. ISBN: 978-0-495-01327-3

Survival Guide for General Chemistry with Math Review, Second Edition

By Charles H. Atwood, University of Georgia. Intended to help you practice for exams, this survival guide shows you how to solve difficult problems by dissecting them into manageable chunks. The guide includes three levels of proficiency questions—A, B, and minimal—to quickly build confidence as you master the knowledge you need to succeed in your course. ISBN: 978-0-495-38751-0

For the Laboratory

Experiments in General Chemistry

By Barbara Munk, Wayne State University. Forty-one traditional experiments parallel the material found in the textbook. Each lab exercise has a prelab assignment, background information, clear instructions for performing the experiment, and a convenient section for reporting results and observations. New to this edition are ten Inquiries with Limited Guidance. Following the conceptual focus of the text, these new experiments allow students to work at their own intellectual levels, design their own experiments, and analyze the data from those experiments without help or prompting form the manual. ISBN: 978-1-111-98942-2

Cengage Learning Brooks/Cole Lab Manuals

Cengage Learning offers a variety of printed manuals to meet all general chemistry laboratory needs. Visit **www.cengage.com/chemistry** for a full listing and description of these laboratory manuals and laboratory notebooks. All of our lab manuals can be customized for your specific needs.

Signature Labs . . . for the Customized Laboratory

Signature Labs is Cengage Learning's digital library of tried-and-true labs that help you take the guesswork out of running your chemistry laboratory. Select just the experiments you want from hundreds of options and approaches. Provide your students with only the experiments they will conduct and know you will get the results you seek. Visit **www.signaturelabs.com** to begin building your manual today.

Acknowledgments

The preparation of a general chemistry textbook and its ancillary materials, even a revision, is a complex project involving many people because the accuracy of the text, as well as the problems presented, and the solutions of those problems, is of extreme importance. The initial planning for this revision began with discussions in the Boston offices of Cengage Learning that included Charlie Hartford, publisher; Lisa Lockwood, executive editor; Alyssa White, development editor; and Lisa Weber and Stephanie VanCamp, media editors. These people took a keen interest in this revision from its inception. Alyssa worked with us on a day-to-day basis and deserves our special thanks for her many contributions that have added so much to this revision. In particular, she completely reworked the art program, which is so important to a general chemistry book, and she also set up the format of such features as the Gaining Mastery Toolboxes (added in this revision) and the end-of-chapter Learning Objectives. We also want to thank Lisa Lockwood for assembling an outstanding group of people to work on this project and for her infectious enthusiasm for this revision. Our thanks go to Teresa Trego, content project manager, who was directly responsible for the production phase of this work. Greg Johnson, our project manager at PreMediaGlobal, kept the electronic files moving on schedule, and Scott Rosen of Bill Smith Studio Group was our photo researcher, who given a sketchy idea found the perfect picture. We thank Greg and Scott for carrying out these important tasks with such professional skill. We give our special thanks to David Shinn of the U.S. Merchant Marine Academy for his prodigious and precise work on this project. Liz Woods, assistant editor, had the task of overseeing the revision of the ancillary materials, including the Solutions Manual and Instructor's Resource Manual. Our many thanks go to her for this important job. Our thanks also go to our media editors, Lisa Weber and Stephanie VanCamp, with Lisa working on the OWL (Online Web Learning) products and Stephanie working on the electronic version of our book (YouBook). Finally, Nicole Hamm, marketing manager, was instrumental in launching the tenth edition of *General Chemistry*. It was a pleasure working with her.

Darrell wishes to thank his wife Jean and their children, Julie, Linda, and Russell, for their continued support and encouragement over many years of writing. Steve thanks his wife Jodi and their two children, Katie and Andrew, and his parents, Judy and Dick, for their support and for helping him keep a perspective on the important things in life.

Reviewers

The development of any revision would be impossible without the help of many reviewers. We are enormously grateful to the following people for giving their time and ideas to this, the tenth edition of *General Chemistry*, as well as to the many reviewers who have helped shape the book over the course of previous editions.

Reviewers of the Tenth Edition

Bryan Breyfogle, *Missouri State University*
Michelle Coach, *Asnuntuck Community College*
Jared DeCoste, *Ball State University*
Michael Golde, *University of Pittsburgh*
Margie Haak, *Oregon State University*
Patrick Lloyd, *Kingsborough Community College*
Catherine Malele, *Binghamton University*
Eric Malina, *University of Nebraska, Lincoln*
David Manke, *University of Massachusetts, Dartmouth*
Michael Melcer, *U.S. Merchant Marine Academy*
Pedro Patino, *University of Central Florida*
Gregory Raner, *University of North Carolina at Greensboro*
Steven Rowley, *Middlesex County College*
Laura Sessions, *Valencia Community College*
Clarissa Sorensen-Unruh, *Central New Mexico Community College*

Reviewers of the Ninth Edition

Edwin H. Abbott, *Montana State University*
Zerihun Assefa, *North Carolina A&T State University*
Maryfran Barber, *Wayne State University*
Mufeed M. Basti, *North Carolina A&T State University*
Alan H. Bates, *University of Massachusetts, Dartmouth*
Eric R. Bittner, *University of Houston*
Gary L. Blackmer, *Western Michigan University*
Simon Bott, *University of Houston*
J. J. Breen, *Providence College*
David R. Burgess, *Rivier College*
Jerry Burns, *Pellissippi State Technical Community College*
Robert F. Cozzens, *George Mason University*
Paul M. Dickson, *Schoolcraft College*
William Donovan, *University of Akron*

Cheryl B. French, *University of Central Oklahoma*
Luther Giddings, *Salt Lake Community College*
Thomas Grow, *Pensacola Junior College*
Michael A. Hauser, *St. Louis Community College*
Sherell Hickman, *Brevard Community College*
Dr. Richard H. Hoff, *U.S. Military Academy*
Songping D. Huang, *Kent State University*
Dr. T. Fred Johnson, *Brevard Community College*
Rebecca M. Jones, *Austin Peay State University*
Myung-Hoon Kim, *Georgia Perimeter College*
Richard H. Langley, *Stephen F. Austin State University*
Mohammad Mahroof-Tahir, *St. Cloud State University*
Jamie L. Manson, PhD, *Eastern Washington University*
Debbie McClinton, *Brevard Community College*
Lauren E. H. McMills, *Ohio University*
Randy M. Miller, *California State University, Chico*
Richard L. Nafshun, *Oregon State University*
Thomas Neils, *Grand Rapids Community College*
Emmanual Ojadi, *University of Massachusetts, Dartmouth*
Eugene Pinkhassik, *University of Memphis*
Jeffrey J. Rack, *Ohio University*
Robert Sharp, *University of Michigan*
Yury Skorik, *University of Pittsburgh*
Cheryl A. Snyder, *Schoolcraft College*
Shujun Su, *Missouri State University*
Kurt Teets, *Okaloosa-Walton College*
P. Gregory Van Patten, *Ohio University*
Ramaiyer Venkatraman, *Jackson State University*
Victor Vilchiz, *Virginia State University*
James A. Zimmerman, *Missouri State University*
Eric J. A Zückerman, *Augusta State University*
Yuegang Zuo, *University of Massachusetts, Dartmouth*
Lisa A. Zuraw, *The Citadel*
Tatiana Zuvich, *Brevard Community College*

Student Focus Group Participants

We are grateful to the Boston College students who shared with our team all their thoughts on the content and design of the text, supplements, and technology. It's our hope that their experience studying from the eighth edition and their ideas for improvement will help all of the future students who use our text.

Brendan Dailey, Class of 2009
Jessica DeLuca, Class of 2009
Jon Durante, Class of 2009
Mykael Garcia, Class of 2009
Christina Murphy, Class of 2009
Kristen Pfau, Class of 2009
Katie Phillips, Class of 2009
Johnny Stratigis, Class of 2009

Reviewers of the Eighth Edition

Mufeed Basti, *North Carolina A&T State University*
Kenneth Brown, *Old Dominion University*
P. J. Brucat, *University of Florida*
Joe Casalnuovo, *California State Polytechnic University, Pomona*
Edward Case, *Clemson University*
David Chitharanjan, *University of Wisconsin, Stevens Point*
Kevin Crawford, *The Citadel*
Thomas Dowd, *William Rainey Harper College*
Jack Gill, *Texas Woman's University*
John Hardee, *Henderson State University*
Daniel Haworth, *Marquette University*
David Herrick, *University of Oregon*
Linda Hobart, *Finger Lakes Community College*
Donna Hobbs, *Augusta State University*
Kirk Kawagoe, *Fresno City College*
Alvin Kennedy, *Morgan State University*
Cathy MacGowan, *Armstrong Atlantic State University*
Deborah McClinton, *Brevard Community College*
Abdul Mohammed, *North Carolina A&T State University*
Ray Mohseni, *East Tennessee State University*
Gary Mort, *Lane Community College*
Patricia Pieper, *Anoka-Ramsey Community College*
John Pollard, *University of Arizona*
Dennis Sardella, *Boston College*
John Thompson, *Lane Community College*
Mike Van Stipdonk, *Wichita State University*
Carmen Works, *Sonoma State University*
Tim Zauche, *University of Wisconsin, Platteville*

Reviewers of the Seventh Edition

Carey Bissonnette, *University of Waterloo*
Bob Belford, *West Virginia University*
Conrad Bergo, *East Stroudsburg University*
Aaron Brown, *Ventura College*
Tim Champion, *Johnson C. Smith University*
Paul Cohen, *College of New Jersey*
Lee Coombs, *California Polytechnic State University*
Jack Cummins, *Metro State College*
William M. Davis, *The University of Texas, Brownsville*
Earline F. Dikeman, *Kansas State University*
Evelyn S. Erenrich, *Rutgers University*
Greg Ferrance, *Illinois State University*
Renee Gittler, *Penn State Lehigh University*
Brian Glaser, *Black Hawk College*
David Grainger, *Colorado State University*
Christopher Grayce, *University of California, Irvine*
John M. Halpin, *New York University*
Carol Handy, *Portland Community College*
Daniel Haworth, *Marquette University*
Gregory Kent Haynes, *Morgan State University*
Robert Henry, *Tarrant County College*
Grant Holder, *Appalachian State University*
Andrew Jorgensen, *University of Toledo*
Kirk Kawagoe, *Fresno City College*
David Kort, *Mississippi State University*
Charles Kosky, *Borough of Manhattan Community College*
Jeffrey Kovac, *University of Tennessee at Knoxville*
Art Landis, *Emporia State University*
Richard Langley, *Stephen F. Austin State University*
Robert Mentore, *Ramapo College*
Joyce Miller, *San Jacinto College (South)*
Bob Morris, *Ball State University*
John Nash, *Purdue University*
Deborah Nycz, *Broward Community College*
Michael A. Quinlan, *University of Southern California*
Joe Rorke, *College of DuPage*
John Schaumloffel, *University of Massachusetts, Dartmouth*
Vernon Thielmann, *Southwest Missouri State University*
Jennifer Travers, *Oregon State University*
Gershon Vincow, *Syracuse University*
Donald Wirz, *University of California, Riverside*
Pete Witt, *Midlands Technical College*
Kim Woodrum, *University of Kentucky*

Reviewers of the Sixth Edition

Robert Balahura, *University of Guelph*

Kenneth Brooks, *New Mexico State University*

Barbara Burke, *California State Polytechnic University, Pomona*

Ernest Davidson, *Indiana University*

Janice Ems-Wilson, *Valencia Community College*

Louis Farrugia, *The University of Glasgow*

Mike Herman, *Tulane University*

Sharon Hutchinson, *University of Idaho*

C. Frederick Jury, *Collin County Community College*

Wolter Kaper, *Faculeit Der Scheikunde, Amsterdam*

Anne Loeb, *College of Lake County*

Stephen Loeb, *University of Windsor*

Adrienne Loh, *University of Wisconsin, La Crosse*

David Miller, *California State University, Northridge*

Robert Morris, *Ball State University*

Ates Tanin, *University of Toronto*

Donald Wirz, *Crafton Hills College*

Robert Zellmer, *The Ohio State University*

Darrell D. Ebbing
Steven D. Gammon

A Note to Students

Having studied and taught chemistry for some years, we are well aware of the problems students encounter. We also know that students don't always read the Preface, so we wanted to remind you of all the resources available to help you master general chemistry.

Read the book

Each individual learns in a different way. We have incorporated a number of features into the text to help you tailor a study program that meets your particular needs and learning style.

Practice, practice, practice

Problem solving is an important part of chemistry, and it only becomes easier with practice. We worked hard to create a consistent three-part problem-solving approach (*Problem Strategy, Solution,* and *Answer Check*) in each in-chapter *Example*. Try the related *Exercise* on your own, and use the corresponding end-of-chapter *Practice Problems* to gain mastery of your problem-solving skills.

In every Example, we have also added what we call the *Gaining Mastery Toolbox*. We based this Toolbox on how we as instructors might help a student who is having trouble with a particular problem. We imagine a student coming to our office because of difficulty with a particular problem. We begin the help session by pointing out to the student the "big idea" that one needs to solve the problem. We call this the *Critical Concept*. But suppose the student is still having difficulty with the problem. We now ask the student about his or her knowledge of prior topics that will be needed to approach the problem. We call these needed prior topics the *Solution Essentials*. Each Gaining Mastery Toolbox that we have added to an Example begins by pointing out the Critical Concept involved in solving the problem posed in that Example. Then, under the heading of Solution Essentials, we list the topics the student needs to have mastered to solve this problem. We hope the Gaining Mastery Toolbox helps the student in much the way that an individual office visit can. Over several Examples, these Toolboxes should help the student develop the habit of focusing on the Critical Concept and the Solution Essentials while engaged in general problem solving.

Get help when you need it

Don't hesitate to ask your instructor or teaching assistant for help. You can also take advantage of the following helpful aids:

- The **Student Solutions Manual** contains detailed solutions to textbook problems.
- The **Study Guide** reinforces concepts and further builds problem-solving skills.

We have put a lot of time and thought into how to help you succeed. We hope you take advantage of all the technology and resources available with *General Chemistry,* Tenth Edition. Best of luck in your study!

Darrell D. Ebbing
Steven D. Gammon

About the Authors

DARRELL D. EBBING

Darrell Ebbing became interested in chemistry at a young age when he tried his hand at doing magic tricks for friends, such as turning water to wine and having (nitrated) handkerchiefs disappear in a poof. Soon, however, his interests turned to the chemistry behind the magic and he even set up a home laboratory. After briefly becoming interested in botany in high school (having gathered several hundred plant specimens), his interest in chemistry was especially piqued when he managed to isolate white crystals of caffeine from tea. From that point, he knew he would go on to major in chemistry. During college, he helped pay his expenses by working at the USDA lab in Peoria, Illinois, as an assistant to a carbohydrate chemist, where he worked on derivatives of starch. As a graduate student at Indiana University, his interests gravitated to the theoretical—to understand the basis of chemistry—and he pursued a PhD in physical chemistry in the area of quantum chemistry.

Professor Ebbing began his professional career at Wayne State University where he taught courses at the undergraduate and graduate level and was for several years the Head of the Physical Chemistry Division. He soon became especially involved in teaching general chemistry, taking the position of Coordinator of General Chemistry. In his teaching, he used his knowledge of "chemical magic" to do frequent lecture demonstrations. He has written a book for introductory chemistry as well as this one for general chemistry (where you will see many of those lecture demonstrations). Although retired from active teaching, he retains a keen interest in frontier topics of science and in the history and philosophy of physical science, interests he hopes to turn into another book.

Having grown up in farm country, surrounded by fields and woods, Professor Ebbing has always maintained a strong interest in the great outdoors. He enjoys seeing nature up close through hiking and birding. His interests also include concerts and theater, as well as world travel.

STEVEN D. GAMMON

Steve Gammon started on his path to becoming a chemist and science educator in high school where he was captivated by a great instructor. After receiving a PhD in inorganic chemistry and chemical education from the University of Illinois-Urbana, he worked for two years at the University of Wisconsin-Madison, serving as the General Chemistry Laboratory Coordinator and becoming immersed in the field of chemical education. Professor Gammon then went on to join the faculty at the University of Idaho as the Coordinator of General Chemistry. In this role, he taught thousands of students, published instructional software, directed federally funded projects involving K-12 teachers, and began his work on *General Chemistry* (then going into its sixth edition). During his 11 years at the University of Idaho he was honored with both university and national (Carnegie Foundation) teaching awards.

In his current appointment at Western Washington University in Bellingham, Washington, Professor Gammon's focus has expanded from chemical education into science education, where he has the opportunity to work with a wide variety of students including undergraduate science majors, preservice K-12 teachers, and practicing K-12 teachers. Some of his current work has been funded by grants from the National Science Foundation. Throughout all of these changes, Professor Gammon has maintained his keen interest in the learning and teaching of introductory chemistry; his science education experiences have only made him better at understanding and addressing the needs of general chemistry students. Professor Gammon's motivation in working with all groups is to create materials and methods that inspire his students to be excited about learning chemistry and science.

When Professor Gammon isn't thinking about the teaching and learning of chemistry, he enjoys doing a variety of activities with his family, including outdoor pursuits such as hiking, biking, camping, gold panning, and fishing. Scattered throughout the text you might find some examples of where his passion for these activities is used to make connections between chemistry and everyday living.

About the Cover

The cover photo depicts the "volcano reaction," in which ammonium dichromate (orange crystals), $(NH_4)_2Cr_2O_7$, decomposes exothermically after it has been ignited. A plume of the dark green chromium(III) oxide, Cr_2O_3, forms as the solid is forced upward by the other products (steam and nitrogen gas).

1 Chemistry and Measurement

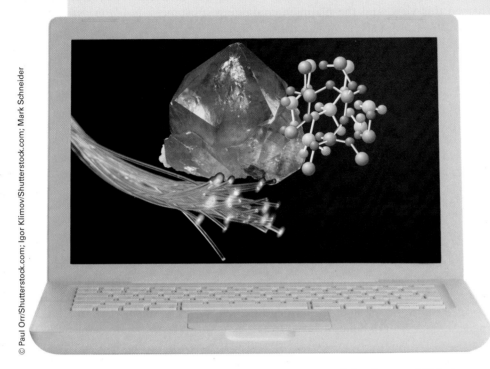

One of the forms of SiO_2 in nature is the quartz crystal. Optical fibers that employ light for data transmission use ultrapure SiO_2 that is produced synthetically.

Contents and Concepts

An Introduction to Chemistry

We start by defining the science called chemistry and introducing some fundamental concepts.

1.1 Modern Chemistry: A Brief Glimpse

1.2 Experiment and Explanation

1.3 Law of Conservation of Mass

1.4 Matter: Physical State and Chemical Constitution

Physical Measurements

Making and recording measurements of the properties and chemical behavior of matter is the foundation of chemistry.

1.5 Measurement and Significant Figures

1.6 SI Units

1.7 Derived Units

1.8 Units and Dimensional Analysis (Factor-Label Method)

© Paul Orr/Shutterstock.com; Igor Klimov/Shutterstock.com; Mark Schneider

In 1964 Barnett Rosenberg and his coworkers at Michigan State University were studying the effects of electricity on bacterial growth. They inserted platinum wire electrodes into a live bacterial culture and allowed an electric current to pass. After 1 to 2 hours, they noted that cell division in the bacteria stopped. The researchers were very surprised by this result, but even more surprised by the explanation. They were able to show that cell division was inhibited by a substance containing platinum, produced from the platinum electrodes by the electric current. A substance such as this one, the researchers thought, might be useful as an anticancer drug, because cancer involves runaway cell division. Later research confirmed this view, and the platinum-containing substances, *cisplatin, carboplatin,* and *oxaliplatin* are all current anticancer drugs. (Figure 1.1).

This story illustrates three significant reasons to study chemistry. First, chemistry has important practical applications. The development of lifesaving drugs is one, and a complete list would touch upon most areas of modern technology.

Second, chemistry is an intellectual enterprise, a way of explaining our material world. When Rosenberg and his coworkers saw that cell division in the bacteria had ceased, they systematically looked for the chemical substance that caused it to cease. They sought a chemical explanation for the occurrence.

Finally, chemistry figures prominently in other fields. Rosenberg's experiment began as a problem in biology; through the application of chemistry, it led to an advance in medicine. Whatever your career plans, you will find that your knowledge of chemistry is a useful intellectual tool for making important decisions.

An Introduction to Chemistry

All of the objects around you—this book, your pen or pencil, and the things of nature such as rocks, water, and plant and animal substances—constitute the *matter* of the universe. Each of the particular kinds of matter, such as a certain kind of paper or plastic or metal, is referred to as a *material*. We can define chemistry as the science of the composition and structure of materials and of the changes that materials undergo.

One chemist may hope that by understanding certain materials, he or she will be able to find a cure for a disease or a solution for an environmental ill. Another chemist may simply want to understand a phenomenon. Because chemistry deals with all materials, it is a subject of enormous breadth. It would be difficult to exaggerate the influence of chemistry on modern science and technology or on our ideas about our planet and the universe. In the section that follows, we will take a brief glimpse at modern chemistry and see some of the ways it has influenced technology, science, and modern thought.

Figure 1.1 ▶

Barnett Rosenberg and Cisplatin

© National Geographic/Getty Images; Science Source/Photoresearchers

Photo of the discoverer of the anti-cancer activity of Cisplatin.

Photo of Cisplatin crystals as seen with a microscope.

1.1 Modern Chemistry: A Brief Glimpse

For thousands of years, human beings have fashioned natural materials into useful products. Modern chemistry certainly has its roots in this endeavor. After the discovery of fire, people began to notice changes in certain rocks and minerals exposed to high temperatures. From these observations came the development of ceramics, glass, and metals, which today are among our most useful materials. Dyes and medicines were other early products obtained from natural substances. For example, the ancient Phoenicians extracted a bright purple dye, known as Tyrian purple, from a species of sea snail. One ounce of Tyrian purple required over 200,000 snails. Because of its brilliant hue and scarcity, the dye became the choice of royalty.

Although chemistry has its roots in early technology, chemistry as a field of study based on scientific principles came into being only in the latter part of the eighteenth century. Chemists began to look at the precise quantities of substances they used in their experiments. From this work came the central principle of modern chemistry: the materials around us are composed of exceedingly small particles called *atoms,* and the precise arrangement of these atoms into *molecules* or more complicated structures accounts for the many different characteristics of materials. Once chemists understood this central principle, they could begin to fashion molecules to order. They could *synthesize* molecules; that is, they could build large molecules from small ones. Tyrian purple, for example, was eventually synthesized from the simpler molecule aniline; see Figure 1.2. Chemists could also correlate molecular structure with the characteristics of materials and so begin to fashion materials with special characteristics.

The liquid-crystal displays (LCDs) that are used on everything from watches and cell phones to computer monitors and televisions are an example of an application that depends on the special characteristics of materials (Figure 1.3). The liquid crystals used in these displays are a form of matter intermediate in characteristics between those of liquids and those of solid crystals—hence the name. Many of these liquid crystals are composed of rodlike molecules that tend to align themselves something like the wood matches in a matchbox. The liquid crystals are held in alignment in layers by plates that have microscopic grooves. The molecules are attached to small electrodes or transistors. When the molecules are subjected to an electric charge from the transistor or electrode, they change alignment to point in a new direction. When they change direction, they change how light passes through their layer. When the liquid-crystal layer is combined with a light source and color filters, incremental changes of alignment of the molecules throughout the display allow for images that have high contrast and millions of colors. ▶ Figure 1.4 shows a model of one of the molecules that

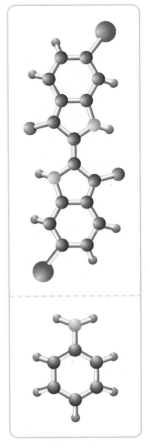

Figure 1.2 ▲

Molecular models of Tyrian purple and aniline Tyrian purple (*top*) is a dye that was obtained by the early Phoenicians from a species of sea snail. The dye was eventually synthesized from aniline (*bottom*). In molecular models the balls represent atoms; each element is represented by a particular color. The lines between the balls indicate that there is a connection holding the atoms together.

Liquid crystals and liquid-crystal displays are described in the essay at the end of Section 11.7.

Figure 1.3 ▶

An iPad© that uses a liquid-crystal display These liquid-crystal displays are used in a variety of electronic devices.

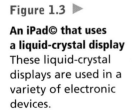

© PSL Images/Alamy

Figure 1.4 ▶

Model of a molecule that forms a liquid crystal Note that the molecule has a rodlike shape. In this model each atom in the structure is represented by a colored shape. Atoms that are touching are connected to each other.

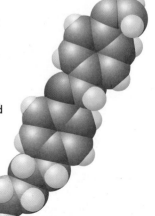

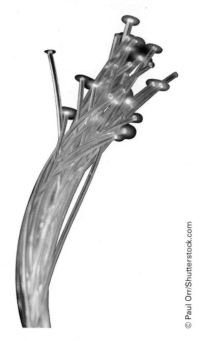

© Paul Orr/Shutterstock.com

Figure 1.5 ▲

Optical fibers A bundle of optical fibers that can be used to transmit data via pulses of light.

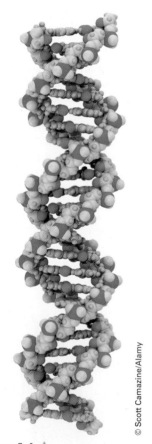

© Scott Camazine/Alamy

Figure 1.6 ▲

A side view of a fragment of a DNA molecule DNA contains the hereditary information of an organism that is passed on from one generation to the next.

forms a liquid crystal; note the rodlike shape of the molecule. Chemists have designed many similar molecules for liquid-crystal applications.

Chemists continue to develop new materials and to discover new properties of old ones. Electronics and communications, for example, have been completely transformed by technological advances in materials. Optical-fiber cables have replaced long-distance telephone cables made of copper wire. Optical fibers are fine threads of extremely pure glass. Because of their purity, these fibers can transmit laser light pulses for miles compared with only a few inches in ordinary glass. Not only is optical-fiber cable cheaper and less bulky than copper cable carrying the same information, but through the use of different colors of light, optical-fiber cable can carry voice, data, and video information at the same time (Figure 1.5). At the ends of an optical-fiber cable, devices using other new materials convert the light pulses to electrical signals and back, while computer chips constructed from still other materials process the signals.

Chemistry has also affected the way we think of the world around us. For example, biochemists and molecular biologists—scientists who study the molecular basis of living organisms—have made a remarkable finding: all forms of life appear to share many of the same molecules and molecular processes. Consider the information of inheritance, the genetic information that is passed on from one generation of organism to the next. Individual organisms, whether bacteria or human beings, store this information in a particular kind of molecule called deoxyribonucleic acid, or DNA (Figure 1.6).

DNA consists of two intertwined molecular chains; each chain consists of links of four different types of molecular pieces, or bases. Just as you record information on a page by stringing together characters (letters, numbers, spaces, and so on), an organism stores the information for reproducing itself in the order of these bases in its DNA. In a multicellular organism, such as a human being, every cell contains the same DNA.

One of our first projects will be to look at this central concept of chemistry, the atomic theory of matter. We will do that in the next chapter, but first we must lay the groundwork for this discussion. We will need some basic vocabulary to talk about science and to describe materials; then we will need to discuss measurement and units, because measurement is critical for quantitative work.

1.2 Experiment and Explanation

Experiment and explanation are the heart of chemical research. A chemist makes observations under circumstances in which variables, such as temperature and amounts of substances, can be controlled. An **experiment** is *an observation of natural phenomena carried out in a controlled manner so that the results can be duplicated and rational conclusions made.* In the chapter opening it was mentioned that Rosenberg studied the effects of electricity on bacterial growth. Temperature and amounts of nutrients in a given volume of bacterial medium are important variables in such experiments. Unless these variables are controlled, the work cannot be duplicated, nor can any reasonable conclusion be drawn.

After a series of experiments, perhaps a researcher sees some relationship or regularity in the results. For instance, Rosenberg noted that in each experiment in which an electric current was passed through a bacterial culture by means of platinum wire electrodes, the bacteria ceased dividing. If the regularity or relationship is fundamental and we can state it simply, we call it a law. A **law** is *a concise statement or mathematical equation about a fundamental relationship or regularity of nature.* An example is the law of conservation of mass, which says that the mass, or quantity of matter, remains constant during any chemical change.

At some point in a research project, a scientist tries to make sense of the results by devising an explanation. Explanations help us organize knowledge and predict future events. A **hypothesis** is *a tentative explanation of some regularity of nature.*

The Birth of the Post-it Note®

Have you ever used a Post-it and wondered where the idea for those little sticky notes came from? You have a chemist to thank for their invention. The story of the Post-it Note illustrates how the creativity and insights of a scientist can result in a product that is as common in the office as the stapler or pen.

In the early 1970s, Art Fry, a 3M scientist, was standing in the choir at his church trying to keep track of all the little bits of paper that marked the music selections for the service. During the service, a number of the markers fell out of the music, making him lose his place. While standing in front of the congregation, he realized that he needed a bookmark that would stick to the book, wouldn't hurt the book, and could be easily detached. To make his plan work, he required an adhesive that would not *permanently* stick things together. Finding the appropriate adhesive was not as simple as it may seem, because most adhesives at that time were created to stick things together permanently.

Still thinking about his problem the next day, Fry consulted a colleague, Spencer Silver, who was studying adhesives at the 3M research labs. That study consisted of conducting a series of tests on a range of adhesives to determine the strength of the bond they formed. One of the adhesives that Silver created for the study was an adhesive that always remained sticky. Fry recognized that this adhesive was just what he needed for his bookmark. His first bookmark, invented the day after the initial idea, consisted of a strip of Silver's tacky adhesive applied to the edge of a piece of paper.

Part of Fry's job description at 3M was to spend time working on creative ideas such as his bookmark. As a result, he continued to experiment with the bookmark to improve its properties of sticking, detaching, and not hurting the surface to which it was attached. One day, while doing some paperwork, he wrote a question on one of his experimental strips of paper and sent it to his boss stuck to the top of a file folder. His boss then answered the question on the note and returned it attached to some other documents. During a later discussion over coffee, Fry and his boss realized that they had invented a new way for people to communicate: the Post-it Note was born. Today the Post-it Note is one of the top-selling office products in the United States.

■ See Problems 1.143 and 1.144.

Having seen that bacteria ceased to divide when an electric current from platinum wire electrodes passed through the culture, Rosenberg was eventually able to propose the hypothesis that certain platinum compounds were responsible. If a hypothesis is to be useful, it should suggest new experiments that become tests of the hypothesis. Rosenberg could test his hypothesis by looking for the platinum compound and testing for its ability to inhibit cell division.

If a hypothesis successfully passes many tests, it becomes known as a theory. A **theory** is *a tested explanation of basic natural phenomena.* An example is the molecular theory of gases—the theory that all gases are composed of very small particles called molecules. This theory has withstood many tests and has been fruitful in suggesting many experiments. Note that we cannot prove a theory absolutely. It is always possible that further experiments will show the theory to be limited or that someone will develop a better theory. For example, the physics of the motion of objects devised by Isaac Newton withstood experimental tests for more than two centuries, until physicists discovered that the equations do not hold for objects moving near the speed of light. Later physicists showed that very small objects also do not follow Newton's equations. Both discoveries resulted in revolutionary developments in physics. The first led to the theory of relativity, the second to quantum mechanics, which has had an immense impact on chemistry.

The two aspects of science, experiment and explanation, are closely related. A scientist performs experiments and observes some regularity; someone explains this regularity and proposes more experiments; and so on. From his experiments, Rosenberg explained that certain platinum compounds inhibit cell division. This explanation led him to do new experiments on the anticancer activity of these compounds.

Figure 1.7 ▶

A representation of the scientific method This flow diagram shows the general steps in the scientific method. At the right, Rosenberg's work on the development of an anticancer drug illustrates the steps.

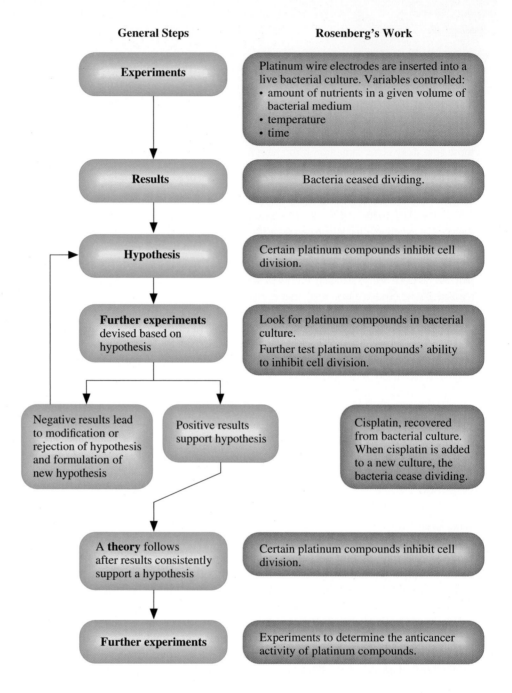

General Steps

Experiments

Results

Hypothesis

Further experiments devised based on hypothesis

Negative results lead to modification or rejection of hypothesis and formulation of new hypothesis

Positive results support hypothesis

A **theory** follows after results consistently support a hypothesis

Further experiments

Rosenberg's Work

Platinum wire electrodes are inserted into a live bacterial culture. Variables controlled:
- amount of nutrients in a given volume of bacterial medium
- temperature
- time

Bacteria ceased dividing.

Certain platinum compounds inhibit cell division.

Look for platinum compounds in bacterial culture.
Further test platinum compounds' ability to inhibit cell division.

Cisplatin, recovered from bacterial culture. When cisplatin is added to a new culture, the bacteria cease dividing.

Certain platinum compounds inhibit cell division.

Experiments to determine the anticancer activity of platinum compounds.

© Martin Shields/Alamy

Figure 1.8 ▲

Laboratory balance A modern single-pan balance. The mass of the material on the pan appears on the digital readout.

The *general* process of advancing scientific knowledge through observation; the framing of laws, hypotheses, or theories; and the conducting of more experiments is called the *scientific method* (Figure 1.7). It is not a method for carrying out a *specific* research program, because the design of experiments and the explanation of results draw on the creativity and individuality of a researcher.

1.3 Law of Conservation of Mass

Modern chemistry emerged in the eighteenth century, when chemists began to use the balance systematically as a tool in research. Balances measure **mass,** which is *the quantity of matter in a material* (Figure 1.8). **Matter** is the general term for the material things around us; we can define it as *whatever occupies space and can be perceived by our senses.*

Figure 1.9 ▶

Heating mercury metal in air Mercury metal reacts with oxygen to yield mercury(II) oxide. The color of the oxide varies from red to yellow, depending on the particle size.

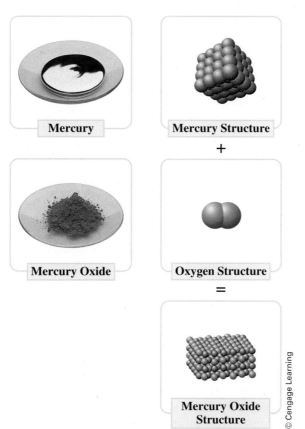

Mercury

Mercury Structure

+

Mercury Oxide

Oxygen Structure

=

Mercury Oxide Structure

© Cengage Learning

© Cengage Learning

Figure 1.10 ▲

Heated mercury(II) oxide When you heat mercury(II) oxide, it decomposes to mercury and oxygen gas.

Antoine Lavoisier (1743–1794), a French chemist, was one of the first to insist on the use of the balance in chemical research. By weighing substances before and after chemical change, he demonstrated the **law of conservation of mass,** which states that *the total mass remains constant during a chemical change (chemical reaction).* ▶

In a series of experiments, Lavoisier applied the law of conservation of mass to clarify the phenomenon of burning, or combustion. He showed that when a material burns, a component of air (which he called oxygen) combines chemically with the material. For example, when the liquid metal mercury is heated in air, it burns or combines with oxygen to give a red-orange substance, whose modern name is mercury(II) oxide. We can represent the chemical change as follows:

Mercury + oxygen ⟶ mercury(II) oxide

The arrow means "is changed to." See Figure 1.9.

By strongly heating the red-orange substance, Lavoisier was able to decompose it to yield the original mercury and oxygen gas (Figure 1.10). The following example illustrates how the law of conservation of mass can be used to study this reaction.

Chemical reactions may involve a gain or loss of heat and other forms of energy. According to Einstein, mass and energy are equivalent. Thus, when energy is lost as heat, mass is also lost, but such small changes in mass in chemical reactions (billionths of a gram) are too small to detect.

⦾WL Interactive Example 1.1 **Using the Law of Conservation of Mass**

Gaining Mastery Toolbox

Critical Concept 1.1
The law of conservation of mass applies to chemical reactions. Whenever a chemical reaction occurs, the mass of the substances before the reaction (reactants) is identical to the mass of the newly formed substances after the reaction (products).

Solution Essentials:
• Definition of mass
• Definition of matter
• Law of conservation of mass

You heat 2.53 grams of metallic mercury in air, which produces 2.73 grams of a red-orange residue. Assume that the chemical change is the reaction of the metal with oxygen in air.

Mercury + oxygen ⟶ red-orange residue

What is the mass of oxygen that reacts? When you strongly heat the red-orange residue, it decomposes to give back the mercury and release the oxygen, which you collect. What is the mass of oxygen you collect?

(continued)

(continued)

Problem Strategy You apply the law of conservation of mass to the reaction. According to this law, the total mass remains constant during a chemical reaction; that is,

$$\text{Mass of substances before reaction} = \text{mass of substances after reaction}$$

Solution From the law of conservation of mass,

$$\text{Mass of mercury} + \text{mass of oxygen} = \text{mass of red-orange residue}$$

Substituting, you obtain

$$2.53 \text{ grams} + \text{mass of oxygen} = 2.73 \text{ grams}$$

or

$$\text{Mass of oxygen} = (2.73 - 2.53) \text{ grams} = \mathbf{0.20 \text{ grams}}$$

The mass of oxygen collected when the red-orange residue decomposes equals the mass of oxygen that originally reacted **(0.20 grams).**

Answer Check Arithmetic errors account for many mistakes. You should always check your arithmetic, either by carefully redoing the calculation or, if possible, by doing the arithmetic in a slightly different way. Here, you obtained the answer by subtracting numbers. You can check the result by addition: the sum of the masses of mercury and oxygen, 2.53 + 0.20 grams, should equal the mass of the residue, 2.73 grams.

Exercise 1.1 You place 1.85 grams of wood in a vessel with 9.45 grams of air and seal the vessel. Then you heat the vessel strongly so that the wood burns. In burning, the wood yields ash and gases. After the experiment, you weigh the ash and find that its mass is 0.28 gram. What is the mass of the gases in the vessel at the end of the experiment?

■ See Problems 1.37, 1.38, 1.39, and 1.40.

Lavoisier set out his views on chemistry in his *Traité Élémentaire de Chimie* (*Basic Treatise on Chemistry*) in 1789. The book was very influential, especially among younger chemists, and set the stage for modern chemistry.

Before leaving this section, you should note the distinction between the terms *mass* and *weight* in precise usage. The weight of an object is the force of gravity exerted on it. The weight is proportional to the mass of the object divided by the square of the distance between the center of mass of the object and that of the earth. ◀ Because the earth is slightly flattened at the poles, an object weighs more at the North Pole, where it is closer to the center of the earth, than at the equator. The mass of an object is the same wherever it is measured.

The force of gravity F between objects whose masses are m_1 and m_2 is Gm_1m_2/r^2, where G is the gravitational constant and r is the distance between the centers of mass of the two objects.

1.4 Matter: Physical State and Chemical Constitution

We describe iron as a silvery-colored metal that melts at 1535°C (2795°F). Once we have collected enough descriptive information about many different kinds of matter, patterns emerge that suggest ways of classifying it. There are two principal ways of classifying matter: by its physical state as a solid, liquid, or gas, and by its chemical constitution as an element, compound, or mixture.

Solids, Liquids, and Gases

Commonly, a given kind of matter exists in different physical forms under different conditions. Water, for example, exists as ice (solid water), as liquid water, and as steam (gaseous water) (Figure 1.11). The main identifying characteristic of solids is their rigidity: they tend to maintain their shapes when subjected to outside forces. Liquids and gases, however, are *fluids;* that is, they flow easily and change their shapes in response to slight outside forces.

What distinguishes a gas from a liquid is the characteristic of *compressibility* (and its opposite, *expansibility*). A gas is easily compressible, whereas a liquid is not. You can put more and more air into a tire, which increases only slightly in volume. In fact, a given quantity of gas can fill a container of almost any size. A small quantity would expand to fill the container; a larger quantity could be compressed to fill the same space. By contrast, if you were to try to force more liquid water into a closed glass bottle that was already full of water, it would burst.

These two characteristics, rigidity (or fluidity) and compressibility (or expansibility), can be used to frame definitions of the three common states of matter:

solid *the form of matter characterized by rigidity;* a solid is relatively incompressible and has fixed shape and volume.

liquid *the form of matter that is a relatively incompressible fluid;* a liquid has a fixed volume but no fixed shape.

gas *the form of matter that is an easily compressible fluid;* a given quantity of gas will fit into a container of almost any size and shape.

The term *vapor* is often used to refer to the gaseous state of any kind of matter that normally exists as a liquid or a solid.

These three forms of matter—solid, liquid, gas—comprise the common **states of matter.**

Elements, Compounds, and Mixtures

To understand how matter is classified by its chemical constitution, we must first distinguish between physical and chemical changes and between physical and chemical properties. A **physical change** is *a change in the form of matter but not in its chemical identity*. Changes of physical state are examples of physical changes. The process of dissolving one material in another is a further example of a physical change. For instance, you can dissolve sodium chloride (table salt) in water. The result is a clear liquid, like pure water, though many of its other characteristics are different from those of pure water. The water and sodium chloride in this liquid retain their chemical identities and can be separated by some method that depends on physical changes.

Distillation is one way to separate the sodium chloride and water components of this liquid. You place the liquid in a flask to which a device called a *condenser* is attached (see Figure 1.12). The liquid in the flask is heated to bring it to a boil. (Boiling entails the formation of bubbles of the vapor in the body of the liquid.) Water vapor forms and passes from the flask into the cooled condenser, where the vapor changes back to liquid water. The liquid water is collected in another flask, called a *receiver*. The original flask now contains the solid sodium chloride. Thus, by means of physical changes (the change of liquid water to vapor and back to liquid), you have separated the sodium chloride and water that you had earlier mixed together.

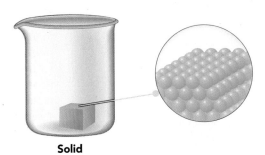

Solid

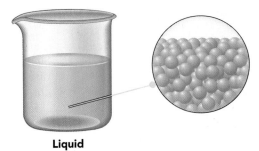

Liquid

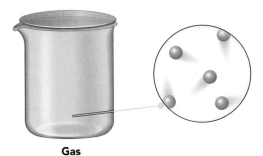

Gas

Figure 1.11 ▲

Molecular representations of solid, liquid, and gas The top beaker contains a solid with a molecular view of the solid; the molecular view depicts the closely packed, immobile atoms that make up the solid structure. The middle beaker contains a liquid with a molecular view of the liquid; the molecular view depicts atoms that are close together but moving freely. The bottom beaker contains a gas with a molecular view of the gas; the molecular view depicts atoms that are far apart and moving freely.

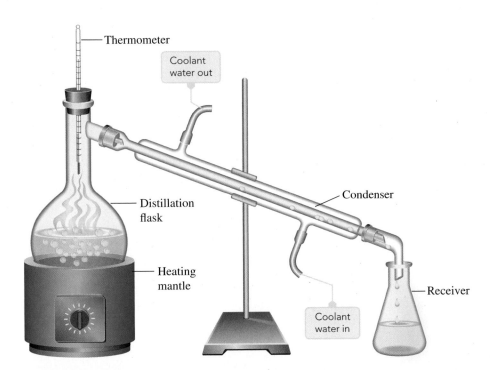

Thermometer

Coolant water out

Distillation flask

Condenser

Heating mantle

Receiver

Coolant water in

Figure 1.12 ◀

Separation by distillation You can separate an easily vaporized liquid from another substance by distillation.

© Cengage Learning

Figure 1.13 ▲

Reaction of sodium with water Sodium metal flits around the water surface as it reacts briskly, giving off hydrogen gas. The other product is sodium hydroxide, which changes a substance added to the water (phenolphthalein) from colorless to pink.

A **chemical change,** or **chemical reaction,** is *a change in which one or more kinds of matter are transformed into a new kind of matter or several new kinds of matter.* The rusting of iron, during which iron combines with oxygen in the air to form a new material called rust, is a chemical change. The original materials (iron and oxygen) combine chemically and cannot be separated by any physical means. To recover the iron and oxygen from rust requires a chemical change or a series of chemical changes.

We characterize or identify a material by its various properties, which may be either physical or chemical. A **physical property** is *a characteristic that can be observed for a material without changing its chemical identity.* Examples are physical state (solid, liquid, or gas), melting point, and color. A **chemical property** is *a characteristic of a material involving its chemical change.* A chemical property of iron is its ability to react with oxygen to produce rust.

Substances The various materials we see around us are either substances or mixtures of substances. A **substance** is *a kind of matter that cannot be separated into other kinds of matter by any physical process.* Earlier you saw that when sodium chloride is dissolved in water, it is possible to separate the sodium chloride from the water by the physical process of distillation. However, sodium chloride is itself a substance and cannot be separated by physical processes into new materials. Similarly, pure water is a substance.

No matter what its source, a substance always has the same characteristic properties. Sodium is a solid metal having a melting point of 98°C. The metal also reacts vigorously with water (Figure 1.13). No matter how sodium is prepared, it always has these properties. Similarly, whether sodium chloride is obtained by burning sodium in chlorine or from seawater, it is a white solid melting at 801°C.

Exercise 1.2 Potassium is a soft, silvery-colored metal that melts at 64°C. It reacts vigorously with water, with oxygen, and with chlorine. Identify all of the physical properties given in this description. Identify all of the chemical properties given.

■ See Problems 1.47, 1.48, 1.49, and 1.50.

In Chapter 2, we will redefine an element in terms of atoms.

Elements Millions of substances have been characterized by chemists. Of these, a very small number are known as elements, from which all other substances are made. Lavoisier was the first to establish an experimentally useful definition of an element. He defined an **element** as *a substance that cannot be decomposed by any chemical reaction into simpler substances.* In 1789 Lavoisier listed 33 substances as elements, of which more than 20 are still so regarded. Today 118 elements are known. Some elements are shown in Figure 1.14. ◄

Compounds Most substances are compounds. A **compound** is *a substance composed of two or more elements chemically combined.* By the end of the eighteenth century, Lavoisier and others had examined many compounds and showed that all of them were composed of the elements in definite proportions by mass. Joseph Louis Proust (1754–1826), by his painstaking work, convinced the majority of chemists of the general validity of the **law of definite proportions** (also known as the **law of constant composition**): *a pure compound, whatever its source, always contains definite or constant proportions of the elements by mass.* For example, 1.0000 gram of sodium chloride always contains 0.3934 gram of sodium and 0.6066 gram of chlorine, chemically combined. Sodium chloride has definite proportions of sodium and chlorine; that is, it has constant or definite composition. ◄

It is now known that some compounds do not follow the law of definite proportions. These nonstoichiometric compounds, as they are called, are described briefly in Chapter 11.

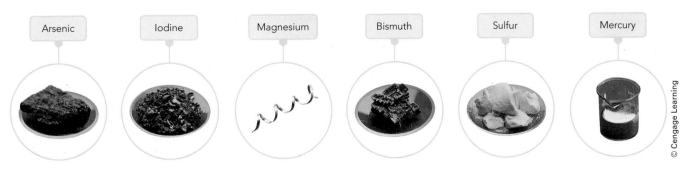

Arsenic   Iodine   Magnesium   Bismuth   Sulfur   Mercury

© Cengage Learning

Figure 1.14 ▲

Some elements

Mixtures Most of the materials around us are mixtures. A **mixture** is *a material that can be separated by physical means into two or more substances.* Unlike a pure compound, a mixture has variable composition. When you dissolve sodium chloride in water, you obtain a mixture; its composition depends on the relative amount of sodium chloride dissolved. You can separate the mixture by the physical process of distillation. ▶

Mixtures are classified into two types. A **heterogeneous mixture** is *a mixture that consists of physically distinct parts, each with different properties.* Figure 1.15 shows a heterogeneous mixture of potassium dichromate and iron filings. Another example is salt and sugar that have been stirred together. If you were to look closely, you would see the separate crystals of sugar and salt. A **homogeneous mixture** (also known as a **solution**) is *a mixture that is uniform in its properties throughout given samples.* When sodium chloride is dissolved in water, you obtain a homogeneous mixture, or solution. Air is a gaseous solution, principally of two elementary substances, nitrogen and oxygen, which are physically mixed but not chemically combined.

A **phase** is *one of several different homogeneous materials present in the portion of matter under study.* A heterogeneous mixture of salt and sugar is said to be composed of two different phases: one of the phases is salt; the other is sugar. Similarly, ice cubes in water are said to be composed of two phases: one phase is ice; the other is liquid water. Ice floating in a solution of sodium chloride in water also consists of two phases, ice and the liquid solution. Note that a phase

Chromatography, another example of a physical method used to separate mixtures, is described in the essay at the end of this section.

Figure 1.15 ▼

A heterogeneous mixture

© Cengage Learning

The mixture on the watch glass consists of potassium dichromate (orange crystals) and iron filings.

A magnet separates the iron filings from the mixture.

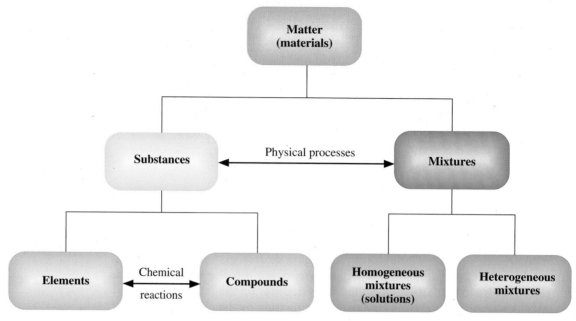

Figure 1.16 ▲

Relationships among elements, compounds, and mixtures Mixtures can be separated by physical processes into substances, and substances can be combined physically into mixtures. Compounds can be separated by chemical reactions into their elements, and elements can be combined chemically to form compounds.

may be either a pure substance in a particular state or a solution in a particular state (solid, liquid, or gaseous). Also, the portion of matter under consideration may consist of several phases of the same substance or several phases of different substances.

Figure 1.16 summarizes the relationships among elements, compounds, and mixtures. Materials are either substances or mixtures. Substances can be mixed by physical processes, and other physical processes can be used to separate the mixtures into substances. Substances are either elements or compounds. Elements may react chemically to yield compounds, and compounds may be decomposed by chemical reactions into elements.

CONCEPT CHECK 1.1

Matter can be represented as being composed of individual units. For example, the smallest individual unit of matter can be represented as a single circle, •, and chemical combinations of these units of matter as connected circles, ••, with each element represented by a different color. Using this model, place the appropriate label—element, compound, or mixture—below each representation.

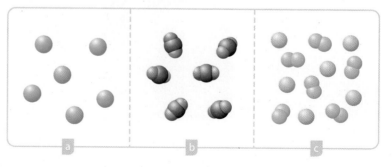

Separation of Mixtures by Chromatography

Chromatography is a group of similar separation techniques. Each depends on how fast a substance moves, in a stream of gas or liquid, past a stationary phase to which the substance may be slightly attracted. An example is provided by a simple experiment in *paper chromatography* (see Figure 1.17). In this experiment, a line of ink is drawn near one edge of a sheet of paper, and the paper is placed upright with this edge in a solution of methanol and water. As the solution creeps up the paper, the ink moves upward, separating into a series of different-colored bands that correspond to the different dyes in the ink. All the dyes are attracted to the wet paper fibers, but with different strengths of attraction. As the solution moves upward, the dyes less strongly attracted to the paper fibers move more rapidly.

The Russian botanist Mikhail Tswett was the first to understand the basis of chromatography and to apply it systematically as a method of separation. In 1906 Tswett separated pigments in plant leaves by *column chromatography.* He first dissolved the pigments from the leaves in petroleum ether, a liquid similar to gasoline. After packing a glass tube or column with powdered chalk, he poured the solution of pigments into the top of the column (see Figure 1.18). When he washed the column by pouring in more petroleum ether, it began to show distinct yellow and green bands. These bands, each containing a pure pigment, became well separated as they moved down the column, so that the pure pigments could be obtained. The name *chromatography* originates from this early separation of colored substances (the stem *chromato-* means "color"), although the technique is not limited to colored substances.

Gas chromatography (GC) is a more recent separation method. Here the moving stream is a gaseous mixture of vaporized substances plus a gas such as helium, which is called the *carrier.* The stationary material is either a solid or a liquid adhering to a solid, packed in a column. As the gas passes through the column, substances in the mixture are attracted differently to the stationary column packing and thus are separated. Gas chromatography is a rapid, small-scale method of separating mixtures. It is also important in the analysis of mixtures because the time it takes for a substance at a given temperature to travel through the column to a detector (called the *retention time*) is fixed. You can therefore use retention times to help identify substances. Figure 1.19 shows a gas chromatograph and a portion of a computer plot (*chromatogram*). Each peak on the chromatogram corresponds to a specific substance. The peaks were automatically recorded by the instrument as the different substances in the mixture passed the detector. Chemists have analyzed complicated mixtures by gas chromatography. Analysis of chocolate, for example, shows that it contains over 800 flavor compounds.

Figure 1.17 ▷

An illustration of paper chromatography A line of ink has been drawn along the lower edge of a sheet of paper. The dyes in the ink separate as a solution of methanol and water creeps up the paper.

© Cengage Learning

(continued)

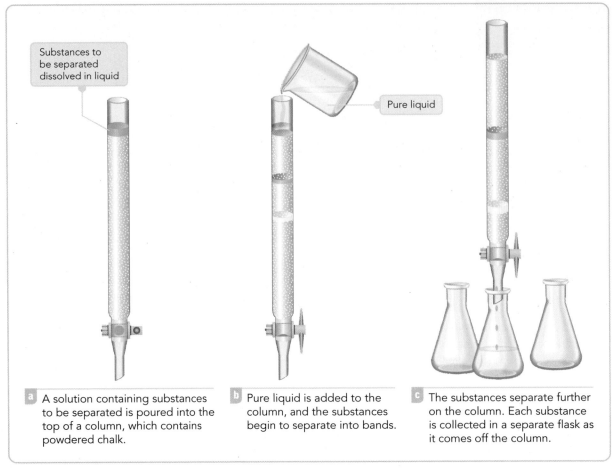

a A solution containing substances to be separated is poured into the top of a column, which contains powdered chalk.

b Pure liquid is added to the column, and the substances begin to separate into bands.

c The substances separate further on the column. Each substance is collected in a separate flask as it comes off the column.

Figure 1.18 ▲

Column chromatography

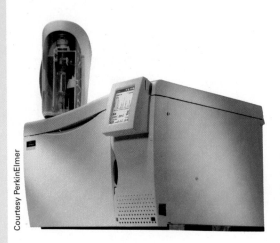

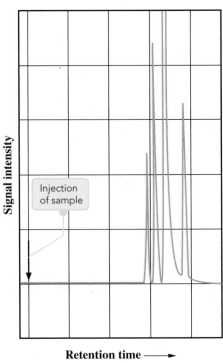

Figure 1.19 ◄

Gas chromatography *Left:* A modern gas chromatograph. *Right:* This is a chromatogram of a hexane mixture, showing its separation into four different substances. Such hexane mixtures occur in gasoline; hexane is also used as a solvent to extract the oil from certain vegetable seeds.

■ See Problems 1.145 and 1.146.

Physical Measurements

Chemists characterize and identify substances by their particular properties. To determine many of these properties requires physical measurements. Cisplatin, the substance featured in the chapter opening, is a yellow substance whose solubility in water (the amount that dissolves in a given quantity of water) is 0.252 gram in 100 grams of water. You can use the solubility, as well as other physical properties, to identify cisplatin. In a modern chemical laboratory, measurements often are complex, but many experiments begin with simple measurements of mass, volume, time, and so forth. In the next few sections, we will look at the measurement process. We first consider the concept of *precision* of a measurement.

1.5 Measurement and Significant Figures

Measurement is the comparison of a physical quantity to be measured with a **unit** of measurement—that is, with *a fixed standard of measurement*. On a centimeter scale, the centimeter unit is the standard of comparison. ▶ A steel rod that measures 9.12 times the centimeter unit has a length of 9.12 centimeters. To record the measurement, you must be careful to give both the *measured number* (9.12) and the *unit* (centimeters).

2.54 centimeters = 1 inch

If you repeat a particular measurement, you usually do not obtain precisely the same result, because each measurement is subject to experimental error. The measured values vary slightly from one another. Suppose you perform a series of identical measurements of a quantity. The term **precision** refers to *the closeness of the set of values obtained from similar measurements of an identical quantity.* **Accuracy** is a related term; it refers to *the closeness of a single measurement to its true value.* To illustrate the idea of precision, consider a simple measuring device, the centimeter ruler. In Figure 1.20, a steel rod has been placed near a ruler subdivided into tenths of a centimeter. You can see that the rod measures just over 9.1 cm (cm = centimeter). With care, it is possible to estimate by eye to hundredths of a centimeter. Here you might give the measurement as 9.12 cm. Suppose you measure the length of this rod twice more. You find the values to be 9.11 cm and 9.13 cm. Thus, you record the length of the rod as being somewhere between 9.11 cm and 9.13 cm. The spread of values indicates the precision with which a measurement can be made by this centimeter ruler. ▶

To indicate the precision of a measured number (or result of calculations on measured numbers), we often use the concept of significant figures. **Significant figures** are *those digits in a measured number (or in the result of a calculation with measured numbers) that include all certain digits plus a final digit having some uncertainty.* When you measured the rod, you obtained the values 9.12 cm, 9.11 cm, and 9.13 cm. You could report the result as the average, 9.12 cm. The first two digits (9.1) are certain; the next digit (2) is estimated, so it has some uncertainty. It would be incorrect to write 9.120 cm for the length of the rod. This would say that the last digit (0) has some uncertainty but that the other digits (9.12) are certain, which is not true.

Measurements that are of high precision are usually accurate. It is possible, however, to have a systematic error in a measurement. Suppose that, in calibrating a ruler, the first centimeter is made too small by 0.1 cm. Then, although the measurements of length on this ruler are still precise to 0.01 cm, they are accurate to only 0.1 cm.

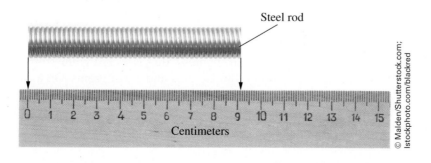
Steel rod

Centimeters

© Malden/Shutterstock.com;
Istockphoto.com/blackred

Figure 1.20 ◀

Precision of measurement with a centimeter ruler The length of the rod is just over 9.1 cm. On successive measurements, we estimate the length by eye at 9.12, 9.11, and 9.13 cm. We record the length as between 9.11 cm and 9.13 cm.

Number of Significant Figures

The term **number of significant figures** refers to *the number of digits reported for the value of a measured or calculated quantity, indicating the precision of the value.* Thus, there are three significant figures in 9.12 cm, whereas 9.123 cm has four. To count the number of significant figures in a given measured quantity, you observe the following rules:

1. All digits are significant except zeros at the beginning of the number and possibly terminal zeros (one or more zeros at the end of a number). Thus, 9.12 cm, 0.912 cm, and 0.00912 cm all contain three significant figures.

2. Terminal zeros ending at the right of the decimal point are significant. Each of the following has three significant figures: 9.00 cm, 9.10 cm, 90.0 cm.

3. Terminal zeros in a number without an explicit decimal point may or may not be significant. If someone gives a measurement as 900 cm, you do not know whether one, two, or three significant figures are intended. If the person writes 900. cm (note the decimal point), the zeros are significant. More generally, you can remove any uncertainty in such cases by expressing the measurement in scientific notation.

Scientific notation is *the representation of a number in the form $A \times 10^n$, where A is a number with a single nonzero digit to the left of the decimal point and n is an integer, or whole number.* In scientific notation, the measurement 900 cm precise to two significant figures is written 9.0×10^2 cm. If precise to three significant figures, it is written 9.00×10^2 cm. Scientific notation is also convenient for expressing very large or very small quantities. It is much easier (and simplifies calculations) to write the speed of light as 3.00×10^8 (rather than as 300,000,000) meters per second. ◀

See Appendix A for a review of scientific notation.

Figure 1.21 ▲

Significant figures and calculators
Not all of the figures that appear on a calculator display are significant. In performing the calculation 100.0 × 0.0634 ÷ 25.31, the calculator display shows 0.250493875. We would report the answer as 0.250, however, because the factor 0.0634 has the least number of significant figures (three).

© Cengage Learning

Significant Figures in Calculations

Measurements are often used in calculations. How do you determine the correct number of significant figures to report for the answer to a calculation? The following are two rules that we use:

1. **Multiplication and division.** When multiplying or dividing measured quantities, give as many significant figures in the answer as there are in the measurement with *the least number of significant figures.*

2. **Addition and subtraction.** When adding or subtracting measured quantities, give the same number of decimal places in the answer as there are in the measurement with *the least number of decimal places.*

Let us see how you apply these rules. Suppose you have a substance believed to be cisplatin, and, in an effort to establish its identity, you measure its solubility (the amount that dissolves in a given quantity of water). You find that 0.0634 gram of the substance dissolves in 25.31 grams of water. The amount dissolving in 100.0 grams is

$$100.0 \text{ grams of water} \times \frac{0.0634 \text{ gram cisplatin}}{25.31 \text{ grams of water}}$$

Performing the arithmetic on a pocket calculator (Figure 1.21), you get 0.250493875 for the numerical part of the answer (100.0 × 0.0634 ÷ 25.31). It would be incorrect to give this number as the final answer. The measurement 0.0634 gram has the least number of significant figures (three). Therefore, you report the answer to three significant figures—that is, 0.250 gram.

Now consider the addition of 184.2 grams and 2.324 grams. On a calculator, you find that 184.2 + 2.324 = 186.524. But because the quantity 184.2 grams has the least number of decimal places—one, whereas 2.324 grams has three—the answer is 186.5 grams.

Exact Numbers

So far we have discussed only numbers that involve uncertainties. However, you will also encounter exact numbers. An **exact number** is *a number that arises when you count items or sometimes when you define a unit.* For example, when you say there are 9 coins in a bottle, you mean exactly 9, not 8.9 or 9.1. Also, when you say there are 12 inches to a foot, you mean exactly 12. Similarly, the inch is defined to be exactly 2.54 centimeters. The conventions of significant figures do not apply to exact numbers. Thus, the 2.54 in the expression "1 inch equals 2.54 centimeters" should not be interpreted as a measured number with three significant figures. In effect, the 2.54 has an infinite number of significant figures, but of course it would be impossible to write out an infinite number of digits. You should make a mental note of any numbers in a calculation that are exact, because they have no effect on the number of significant figures in a calculation. The number of significant figures in a calculation result depends only on the numbers of significant figures in quantities having uncertainties. For example, suppose you want the total mass of 9 coins when each coin has a mass of 3.0 grams. The calculation is

$$3.0 \text{ grams} \times 9 = 27 \text{ grams}$$

You report the answer to two significant figures because 3.0 grams has two significant figures. The number 9 is exact and does not determine the number of significant figures.

Rounding

In reporting the solubility of your substance in 100.0 grams of water as 0.250 gram, you *rounded* the number you read off the calculator (0.2504938). **Rounding** is *the procedure of dropping nonsignificant digits in a calculation result and adjusting the last digit reported.* The general procedure is as follows: Look at the leftmost digit to be dropped. Then . . .

1. If this digit is 5 or greater, add 1 to the last digit to be retained and drop all digits farther to the right. Thus, rounding 1.2151 to three significant figures gives 1.22.

2. If this digit is less than 5, simply drop it and all digits farther to the right. Rounding 1.2143 to three significant figures gives 1.21.

In doing a calculation of two or more steps, it is desirable to retain nonsignificant digits for intermediate answers. This ensures that accumulated small errors from rounding do not appear in the final result. If you use a calculator, you can simply enter numbers one after the other, performing each arithmetic operation and rounding only the final answer. ▶ To keep track of the correct number of significant figures, you may want to record intermediate answers with a line under the last significant figure, as shown in the solution to part d of the following example.

The answers for Exercises and Problems will follow this practice. For most Example Problems where intermediate answers are reported, all answers, intermediate and final, will be rounded.

Example 1.2 **Using Significant Figures in Calculations**

Gaining Mastery Toolbox

Critical Concept 1.2
The final result of any calculation must be reported to the correct number of significant figures. When performing numerical operations (addition, subtraction, multiplication, and division), rules must be applied to determine the correct reporting of the final answer. Rounding should *only* be applied to the final result.

Solution Essentials:
- Determination of the number of significant figures in a measured value
- Tracking the number of significant figures of each measured and calculated value throughout the calculation process
- Reporting a calculated value (often the "answer") to the correct number of significant figures by applying the appropriate rules for the operations (multiplication/division and addition/subtraction) and the rule for rounding

Perform the following calculations and round the answers to the correct number of significant figures (units of measurement have been omitted).

a $\dfrac{2.568 \times 5.8}{4.186}$

b $5.41 - 0.398$

c $3.38 - 3.01$

d $4.18 - 58.16 \times (3.38 - 3.01)$

Problem Strategy Parts a, b, and c are relatively straightforward; use the rule for significant figures that applies in each case. In a multistep calculation as in d, proceed step by step, applying the relevant rule to each step.

(continued)

(continued)

First, do the operations within parentheses. By convention, you perform multiplications and divisions before additions and subtractions. Do not round intermediate answers, but do keep track of the least significant digit—say, by underlining it.

Solution

a The factor 5.8 has the fewest significant figures; therefore, the answer should be reported to two significant figures. Round the answer to **3.6**. **b** The number with the least number of decimal places is 5.41. Therefore, round the answer to two decimal places, to **5.01**. **c** The answer is **0.37**. Note how you have lost one significant figure in the subtraction. **d** You first do the subtraction within parentheses, underlining the least significant digits.

$$4.1\underline{8} - 58.1\underline{6} \times (3.3\underline{8} - 3.0\underline{1}) = 4.1\underline{8} - 58.1\underline{6} \times 0.3\underline{7}$$

Following convention, you do the multiplication before the subtraction.

$$4.1\underline{8} - 58.1\underline{6} \times 0.3\underline{7} = 4.1\underline{8} - 2\underline{1}.5192 = -1\underline{7}.3392$$

The final answer is **−17**.

Answer Check When performing any calculation, you should always report answers to the correct number of significant figures. If a set of calculations involves *only* multiplication and division, it is not necessary to track the significant figures through intermediate calculations; you just need to use the measurement with the least number of significant figures as the guide. However, when addition or subtraction is involved, make sure that you have noted the correct number of significant figures in each calculated step in order to ensure that your answer is correctly reported.

Exercise 1.3 Give answers to the following arithmetic setups. Round to the correct number of significant figures.

a $\dfrac{5.61 \times 7.891}{9.1}$ **b** $8.91 - 6.435$

c $6.81 - 6.730$ **d** $38.91 \times (6.81 - 6.730)$

■ See Problems 1.61 and 1.62.

CONCEPT CHECK 1.2

a When you report your weight to someone, how many significant figures do you typically use?

b What is your weight with two significant figures?

c Indicate your weight and the number of significant figures you would obtain if you weighed yourself on a truck scale that can measure in 50 kg or 100 lb increments.

1.6 SI Units

The first measurements were probably based on the human body (the length of the foot, for example). In time, fixed standards developed, but these varied from place to place. Each country or government (and often each trade) adopted its own units. As science became more quantitative in the seventeenth and eighteenth centuries, scientists found that the lack of standard units was a problem. ◄ They began to seek a simple, international system of measurement. In 1791 a study committee of the French Academy of Sciences devised such a system. Called the *metric system,* it became the official system of measurement for France and was soon used by scientists throughout the world. Most nations have since adopted the metric system or, at least, have set a schedule for changing to it.

SI Base Units and SI Prefixes

In 1960 the General Conference of Weights and Measures adopted the **International System of units** (or **SI**, after the French *le Système International d'Unités*), which is *a particular choice of metric units.* This system has seven **SI base units,** *the SI units from which all others can be derived.* Table 1.1 lists these base units and the symbols used to represent them. In this chapter, we will discuss four base quantities: length, mass, time, and temperature. ◄

One advantage of any metric system is that it is a decimal system. In SI, a larger or smaller unit for a physical quantity is indicated by an **SI prefix,** which is *a prefix used in the International System to indicate a power of 10.* For example,

In the system of units that Lavoisier used in the eighteenth century, there were 9216 grains to the pound (the livre). English chemists of the same period used a system in which there were 7000 grains to the pound—unless they were trained as apothecaries, in which case there were 5760 grains to the pound!

The amount of substance is discussed in Chapter 3, and the electric current (ampere) is introduced in Chapter 19. Luminous intensity will not be used in this book.

Table 1.1	SI Base Units	
Quantity	Unit	Symbol
Length	meter	m
Mass	kilogram	kg
Time	second	s
Temperature	kelvin	K
Amount of substance	mole	mol
Electric current	ampere	A
Luminous intensity	candela	cd

Table 1.2	Selected SI Prefixes	
Prefix	Multiple	Symbol
mega	10^6	M
kilo	10^3	k
deci	10^{-1}	d
centi	10^{-2}	c
milli	10^{-3}	m
micro	10^{-6}	μ*
nano	10^{-9}	n
pico	10^{-12}	p

*Greek letter mu, pronounced "mew."

the base unit of length in SI is the meter (somewhat longer than a yard), and 10^{-2} meter is called a *centi*meter. Thus 2.54 centimeters equals 2.54×10^{-2} meters. The SI prefixes used in this book are presented in Table 1.2.

Length, Mass, and Time

The **meter (m)** is *the SI base unit of length.* ▶ By combining it with one of the SI prefixes, you can get a unit of appropriate size for any length measurement. For the very small lengths used in chemistry, the nanometer (nm; 1 nanometer = 10^{-9} m) or the picometer (pm; 1 picometer = 10^{-12} m) is an acceptable SI unit. *A non-SI unit of length* traditionally used by chemists is the **angstrom (Å),** which equals 10^{-10} m. (An oxygen atom, one of the minute particles of which the substance oxygen is composed, has a diameter of about 1.3 Å. If you could place oxygen atoms adjacent to one another, you could line up over 75 million of them in 1 cm.)

The **kilogram (kg)** is *the SI base unit of mass,* equal to about 2.2 pounds. ▶ This is an unusual base unit in that it contains a prefix. In forming other SI mass units, prefixes are added to the word *gram* (g) to give units such as the *milli*gram (mg; 1 mg = 10^{-3} g).

The **second (s)** is *the SI base unit of time* (Figure 1.22). Combining this unit with prefixes such as *milli-, micro-, nano-,* and *pico-,* you create units appropriate for measuring very rapid events. The time required for the fastest chemical processes is about a picosecond, which is the current limit of timekeeping technology. When you measure times much longer than a few hundred seconds, you revert to *minutes* and *hours,* an obvious exception to the prefix–base format of the International System.

The meter was originally defined in terms of a standard platinum–iridium bar kept at Sèvres, France. In 1983, the meter was defined as the distance traveled by light in a vacuum in 1/299,792,458 seconds.

The present standard of mass is the platinum–iridium kilogram mass kept at the International Bureau of Weights and Measures in Sèvres, France.

Exercise 1.4 Express the following quantities using an SI prefix and a base unit. For instance, 1.6×10^{-6} m = 1.6 μm. A quantity such as 0.000168 g could be written 0.168 mg or 168 μg.

a 1.84×10^{-9} m   b 5.67×10^{-12} s   c 7.85×10^{-3} g

d 9.7×10^3 m   e 0.000732 s   f 0.000000000154 m

■ See Problems 1.65 and 1.66.

Temperature

Temperature is difficult to define precisely, but we all have an intuitive idea of what we mean by it. It is a measure of "hotness." A hot object placed next to a cold one becomes cooler, while the cold object becomes hotter. Heat energy passes from a hot object to a cold one, and the quantity of heat passed between the objects

Figure 1.22 ▲

NIST F-1 cesium fountain clock First put into service in 1999 at the National Institute of Standards and Technology (NIST) in Boulder, Colorado, this clock represents a new generation of super-accurate atomic clocks. The clock is so precise that it loses only one second every 20 million years.

depends on the difference in temperature between the two. Therefore, temperature and heat are different, but related, concepts.

A thermometer is a device for measuring temperature. The common type consists of a glass capillary containing a column of liquid whose length varies with temperature. A scale alongside the capillary gives a measure of the temperature. The **Celsius scale** (formerly the centigrade scale) is *the temperature scale in general scientific use.* On this scale, the freezing point of water is 0°C and the boiling point of water at normal barometric pressure is 100°C. However, *the SI base unit of temperature* is the **kelvin (K),** a unit on an *absolute temperature* scale. (See the first margin note on the next page.) On any absolute scale, the lowest temperature that can be attained theoretically is zero. The Celsius and the Kelvin scales have equal-size units (that is, a change of 1°C is equivalent to a change of 1 K), where 0°C is a temperature equivalent to 273.15 K. Thus, it is easy to convert from one scale to the other, using the formula

$$T_K = \left(t_C \times \frac{1\ K}{1°C}\right) + 273.15\ K$$

where T_K is the temperature in kelvins and t_C is the temperature in degrees Celsius. A temperature of 20°C (about room temperature) equals 293 K.

The Fahrenheit scale is at present the common temperature scale in the United States. Figure 1.23 compares Kelvin, Celsius, and Fahrenheit scales. As the figure shows, 0°C is the same as 32°F (both exact), and 100°C corresponds to 212°F

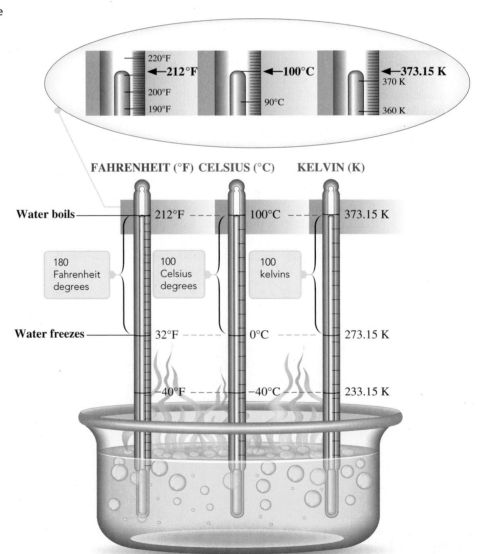

Figure 1.23 ▷

Comparison of temperature scales Room temperature is about 68°F, 20°C, and 293 K. Water freezes at 32°F, 0°C, and 273.15 K. Water boils under normal pressure at 212°F, 100°C, and 373.15 K.

(both exact). Therefore, there are 212 − 32 = 180 Fahrenheit degrees in the range of 100 Celsius degrees. That is, there are exactly 9 Fahrenheit degrees for every 5 Celsius degrees. Knowing this, and knowing that 0°C equals 32°F, we can derive a formula to convert degrees Celsius to degrees Fahrenheit. ▶

$$t_F = \left(t_C \times \frac{9°F}{5°C}\right) + 32°F$$

where t_F is the temperature in Fahrenheit and t_C is the temperature in Celsius. By rearranging this, we can obtain a formula for converting degrees Fahrenheit to degrees Celsius:

$$t_C = \frac{5°C}{9°F} \times (t_F - 32°F)$$

Note that the degree sign (°) is not used with the Kelvin scale, and the unit of temperature is simply the kelvin (not capitalized).

These two temperature conversion formulas are often written K = °C + 273.15 and °F = (1.8 × °C) + 32. Although these yield the correct number, they do not take into account the units.

Example 1.3 **Converting from One Temperature Scale to Another**

The hottest place on record in North America is Death Valley in California. It reached a temperature of 134°F in 1913. What is this temperature reading in degrees Celsius? in kelvins?

Problem Strategy This calculation involves conversion of a temperature in degrees Fahrenheit (°F) to degrees Celsius (°C) and kelvins (K). This is a case where using formulas is the most reliable method for arriving at the answer.

Solution Substituting, we find that

$$t_C = \frac{5°C}{9°F} \times (t_F - 32°F) = \frac{5°C}{9°F} \times (134°F - 32°F) = \mathbf{56.7°C}$$

In kelvins,

$$T_K = \left(t_C \times \frac{1\ K}{1°C}\right) + 273.15\ K = \left(56.7°C \times \frac{1\ K}{1°C}\right) + 273.15\ K = \mathbf{329.9\ K}$$

Answer Check An easy way to check temperature conversions is to use the appropriate equation and convert your answer back to the original temperature to see whether it matches.

Exercise 1.5 a A person with a fever has a temperature of 102.5°F. What is this temperature in degrees Celsius? b A cooling mixture of dry ice and isopropyl alcohol has a temperature of −78°C. What is this temperature in kelvins?

See Problems 1.69, 1.70, 1.71, and 1.72.

Gaining Mastery Toolbox

Critical Concept 1.3
There are several commonly used temperature scales: Celsius, Fahrenheit, and Kelvin. The Celsius temperature scale is used most frequently in the laboratory. Using mathematical relationships, temperature values can be converted between the Celsius, Kelvin, and Fahrenheit scales.

Solution Essentials:
- Relationship (conversion formula) between Kelvin and Celsius temperatures
- Relationship (conversion formula) between Celsius and Fahrenheit temperatures

CONCEPT CHECK 1.3

a Estimate and express the length of your leg in an appropriate metric unit.

b What would be a reasonable height, in meters, of a three-story building?

c How would you be feeling if your body temperature was 39°C?

d Would you be comfortable sitting in a room at 23°C in a short-sleeved shirt?

1.7 Derived Units

Once base units have been defined for a system of measurement, you can derive other units from them. You do this by using the base units in equations that define other physical quantities. For example, *area* is defined as length times length. Therefore,

$$\text{SI unit of area} = (\text{SI unit of length}) \times (\text{SI unit of length})$$

Table 1.3	Derived Units	
Quantity	**Definition of Quantity**	**SI Unit**
Area	Length squared	m^2
Volume	Length cubed	m^3
Density	Mass per unit volume	kg/m^3
Speed	Distance traveled per unit time	m/s
Acceleration	Speed changed per unit time	m/s^2
Force	Mass times acceleration of object	$kg \cdot m/s^2$ (= newton, N)
Pressure	Force per unit area	$kg/(m \cdot s^2)$ (= pascal, Pa)
Energy	Force times distance traveled	$kg \cdot m^2/s^2$ (= joule, J)

From this, you see that the SI unit of area is meter × meter, or m^2. Similarly, *speed* is defined as the rate of change of distance with time; that is, speed = distance/time. Consequently,

$$\text{SI unit of speed} = \frac{\text{SI unit of distance}}{\text{SI unit of time}}$$

The SI unit of speed is meters per second (that is, meters divided by seconds). The unit is symbolized m/s or $m \cdot s^{-1}$. The unit of speed is an example of an **SI derived unit,** which is *a unit derived by combining SI base units.* Table 1.3 defines a number of derived units. Volume and density are discussed in this section; pressure and energy are discussed later (in Sections 5.1 and 6.1, respectively).

Volume

Volume is defined as length cubed and has the SI unit of cubic meter (m^3). This is too large a unit for normal laboratory work, so we use either cubic decimeters (dm^3) or cubic centimeters (cm^3, also written cc). Traditionally, chemists have used the **liter (L),** which is *a unit of volume equal to a cubic decimeter* (approximately one quart). In fact, most laboratory glassware (Figure 1.24) is calibrated in liters or milliliters (1000 mL = 1 L). Because 1 dm equals 10 cm, a cubic decimeter, or one liter, equals $(10 \text{ cm})^3 = 1000 \text{ cm}^3$. Therefore, a milliliter equals a cubic centimeter. In summary

$$1 \text{ L} = 1 \text{ dm}^3 \qquad \text{and} \qquad 1 \text{ mL} = 1 \text{ cm}^3$$

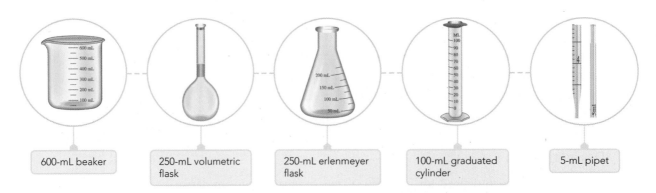

600-mL beaker   250-mL volumetric flask   250-mL erlenmeyer flask   100-mL graduated cylinder   5-mL pipet

Figure 1.24 ▲

Some laboratory glassware

Figure 1.25 ▶

The relative densities of copper and mercury Copper floats on mercury because the density of copper (8.95 g/cm³) is less than the density of mercury (13.5 g/cm³). (This is a copper penny; recently minted pennies are copper-clad zinc.)

Figure 1.26 ▲

Relative densities of some liquids Shown are three liquids (dyed so that they will show up clearly): *ortho*-xylene, water, and 1,1,1-trichloroethane. Water (blue layer) is less dense than 1,1,1-trichloroethane (colorless) and floats on it. *Ortho*-xylene (top, golden layer) is less dense than water and floats on it.

Density

The **density** of an object is its *mass per unit volume.* You can express this as

$$d = \frac{m}{V}$$

where d is the density, m is the mass, and V is the volume. Suppose an object has a mass of 15.0 g and a volume of 10.0 cm³. Substituting, you find that

$$d = \frac{15.0 \text{ g}}{10.0 \text{ cm}^3} = 1.50 \text{ g/cm}^3$$

The density of the object is 1.50 g/cm³ (or 1.50 g·cm⁻³).

Density is an important physical property of a material. Water, for example, has a density of 1.000 g/cm³ at 4°C and a density of 0.998 g/cm³ at 20°C. Lead has a density of 11.3 g/cm³ at 20°C. (Figures 1.25 and 1.26 dramatically show some relative densities.) ▶ Oxygen gas has a density of 1.33×10^{-3} g/cm³ at normal pressure and 20°C. (Like other gases under normal conditions, oxygen has a density that is about 1000 times smaller than those of liquids and solids.) Because the density is characteristic of a substance, it can be helpful in identifying it. Example 1.4 illustrates this point. Density can also be useful in determining whether a substance is pure. Consider a gold bar whose purity is questioned. The metals likely to be mixed with gold, such as silver or copper, have lower densities than gold. Therefore, an adulterated (impure) gold bar can be expected to be far less dense than pure gold.

The density of solid materials on earth ranges from about 1 g/cm³ to 22.5 g/cm³ (osmium metal). In the interior of certain stars, the density of matter is truly staggering. Black neutron stars—stars composed of neutrons, or atomic cores compressed by gravity to a superdense state—have densities of about 10¹⁵ g/cm³.

Example 1.4   **Calculating the Density of a Substance**

Gaining Mastery Toolbox

Critical Concept 1.4
Density is the relationship of the mass to volume (m/V) of a substance. Unlike mass, density is a characteristic (physical property) that can be used to identify and distinguish pure substances from each other.

Solution Essentials:
• Density of a substance is calculated by dividing the mass of substance (m) by the volume (V) that the mass of substance occupies
• Rules for significant figures and rounding
• The metric system

A colorless liquid, used as a solvent (a liquid that dissolves other substances), is believed to be one of the following:

Substance	Density (in g/mL)
n-butyl alcohol	0.810
ethylene glycol	1.114
isopropyl alcohol	0.785
toluene	0.866

(continued)

(continued)

To identify the substance, a chemist determined its density. By pouring a sample of the liquid into a graduated cylinder, she found that the volume was 35.1 mL. She also found that the sample weighed 30.5 g. What was the density of the liquid? What was the substance?

Problem Strategy The solution to this problem lies in finding the density of the unknown substance. Once the density of the unknown substance is known, you can compare it to the list of known substances presented in the problem and look for a match. Density is the relationship of the mass of a substance per volume of that substance. Expressed as an equation, density is the mass divided by the volume: $d = m/V$.

Solution You substitute 30.5 g for the mass and 35.1 mL for the volume into the equation.

$$d = \frac{m}{V} = \frac{30.5 \text{ g}}{35.1 \text{ mL}} = \textbf{0.869 g/mL}$$

The density of the liquid equals that of **toluene** (within experimental error).

Answer Check Always be sure to report the density in the units used when performing the calculation. Density is not always reported in units of g/mL or g/cm^3, for example; gases are often reported with the units of g/L.

Exercise 1.6 A piece of metal wire has a volume of 20.2 cm^3 and a mass of 159 g. What is the density of the metal? We know that the metal is manganese, iron, or nickel, and these have densities of 7.21 g/cm^3, 7.87 g/cm^3, and 8.90 g/cm^3, respectively. From which metal is the wire made?

■ See Problems 1.73, 1.74, 1.75, and 1.76.

In addition to characterizing a substance, the density provides a useful relationship between mass and volume. For example, suppose an experiment calls for a certain mass of liquid. Rather than weigh the liquid on a balance, you might instead measure out the corresponding volume. Example 1.5 illustrates this idea.

⚙WL Interactive Example 1.5 Using the Density to Relate Mass and Volume

Gaining Mastery Toolbox

Critical Concept 1.5
Density is a mathematical equation expressed as $d = m/V$. Because this relationship is an equation with three variables, it can be manipulated to solve for one variable when two of the variables are known.

Solution Essentials:
- Identify the unknown quantity in the $d = m/V$ equation.
- Rearrange the equation to solve for the unknown quantity
- Substitute known values into the rearranged equation to solve for unknown quantity
- The metric system
- Rules for significant figures and rounding

An experiment requires 43.7 g of isopropyl alcohol. Instead of measuring out the sample on a balance, a chemist dispenses the liquid into a graduated cylinder. The density of isopropyl alcohol is 0.785 g/mL. What volume of isopropyl alcohol should he use?

Problem Strategy The formula that defines density provides a relationship between the density of a substance and the mass and volume of the substance: $d = m/V$. Because this relationship is an equation, it can be used to solve for one variable when the other two are known. Examining this problem reveals that the mass (m) and

density (d) of the substance are known, so the equation can be used to solve for the volume (V) of the liquid.

Solution You rearrange the formula defining the density to obtain the volume.

$$V = \frac{m}{d}$$

Then you substitute into this formula:

$$V = \frac{43.7 \text{ g}}{0.785 \text{ g/mL}} = \textbf{55.7 mL}$$

Answer Check Note that the density of the alcohol is close to 1 g/mL, which is common for many substances. In cases like this, you can perform a very quick check of your answer, because the numerical value for the volume of the liquid should not be much different from the numerical value of the mass of the liquid.

Exercise 1.7 Ethanol (grain alcohol) has a density of 0.789 g/cm^3. What volume of ethanol must be poured into a graduated cylinder to equal 30.3 g?

■ See Problems 1.77, 1.78, 1.79, and 1.80.

CONCEPT CHECK 1.4

You are working in the office of a precious metals buyer. A miner brings you a nugget of metal that he claims is gold. You suspect that the metal is a form of "fool's gold," called marcasite, that is composed of iron and sulfur. In the back of your office, you have a chunk of pure gold. What simple experiments could you perform to determine whether the miner's nugget is gold?

1.8 Units and Dimensional Analysis (Factor-Label Method)

In performing numerical calculations with physical quantities, it is good practice to enter each quantity as a number with its associated unit. Both the numbers and the units are then carried through the indicated algebraic operations. The advantages of this are twofold:

1. The units for the answer will come out of the calculations.
2. If you make an error in arranging factors in the calculation (for example, if you use the wrong formula), this will become apparent because the final units will be nonsense.

Dimensional analysis (or the **factor-label method**) is *the method of calculation in which one carries along the units for quantities.* As an illustration, suppose you want to find the volume V of a cube, given s, the length of a side of the cube. Because $V = s^3$, if $s = 5.00$ cm, you find that $V = (5.00 \text{ cm})^3 = 5.00^3 \text{ cm}^3$. There is no guesswork about the unit of volume here; it is cubic centimeters (cm^3).

Suppose, however, that you wish to express the volume in liters (L), a metric unit that equals 10^3 cubic centimeters (approximately one quart). You can write this equality as

$$1 \text{ L} = 10^3 \text{ cm}^3$$

If you divide both sides of the equality by the right-hand quantity, you get

$$\frac{1 \text{ L}}{10^3 \text{ cm}^3} = \frac{10^3 \cancel{\text{cm}^3}}{10^3 \cancel{\text{cm}^3}} = 1$$

Observe that you treat units in the same way as algebraic quantities. Note too that the right-hand side now equals 1 and no units are associated with it. Because it is always possible to multiply any quantity by 1 without changing that quantity, you can multiply the previous expression for the volume by the factor $1 \text{ L}/10^3 \text{ cm}^3$ without changing the actual volume. You are changing only the way you express this volume:

$$V = 5.00^3 \cancel{\text{cm}^3} \times \underbrace{\frac{1 \text{ L}}{10^3 \cancel{\text{cm}^3}}}_{\substack{\text{converts} \\ \text{cm}^3 \text{ to L}}} = 125 \times 10^{-3} \text{ L} = 0.125 \text{ L}$$

In order to solve problems using dimensional analysis, you may find it helpful to diagram a solution pathway using units as a guide. For example, in the problem above:

Starting Units → Goal Units

cm³ ⟹ L

In this diagram the ⟹ represents the ratio $1 \text{ L}/10^3 \text{ cm}^3$, which we use to convert from units of cm^3 to L.

The ratio $1 \text{ L}/10^3 \text{ cm}^3$ is called a **conversion factor** because it is *a factor equal to 1 that converts a quantity expressed in one unit to a quantity expressed in another unit.* ▶ Note that the numbers in this conversion factor are *exact*, because 1 L equals exactly 1000 cm^3. Such exact conversion factors do not affect the number of significant figures in an arithmetic result. In the previous calculation, the

It takes more room to explain conversion factors than it does to use them. With practice, you will be able to write the final conversion step without the intermediate algebraic manipulations outlined here.

quantity 5.00 cm (the measured length of the side) does determine or limit the number of significant figures.

The next two examples illustrate how the conversion-factor method may be used to convert one metric unit to another.

Example 1.6

Converting Units: Metric Unit to Metric Unit

Nitrogen gas is the major component of air. A sample of nitrogen gas in a glass bulb weighs 243 mg. What is this mass in SI base units of mass (kilograms)?

Problem Strategy This problem requires converting a mass in milligrams to kilograms. Finding the correct relationship for performing such a conversion directly in one step might be more difficult than doing the conversion in multiple steps. In this case, conversion factors for converting milligrams to grams and from grams to kilograms are relatively easy to derive, so it makes sense to do this conversion in two steps. First convert milligrams to grams; then convert grams to kilograms. To convert from milligrams to grams, note that the prefix *milli-* means 10^{-3}. To convert from grams to kilograms, note that the prefix *kilo-* means 10^3.

Diagramming the solution clearly shows how two conversion factors ($2 \Rightarrow$) are used to solve the problem.

Starting Units ... mg ⟹ g ⟹ kg ... Goal Units

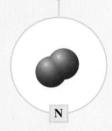

Nitrogen molecular model

N

Solution Since 1 mg = 10^{-3} g, you can write

$$243 \text{ mg} \times \frac{10^{-3} \text{ g}}{1 \text{ mg}} = 2.43 \times 10^{-1} \text{ g}$$

Then, because the prefix *kilo-* means 10^3, you write

$$1 \text{ kg} = 10^3 \text{ g}$$

and

$$2.43 \times 10^{-1} \text{ g} \times \frac{1 \text{ kg}}{10^3 \text{ g}} = \mathbf{2.43 \times 10^{-4} \text{ kg}}$$

Note, however, that you can combine the two conversion steps as follows:

$$243 \text{ mg} \times \underbrace{\frac{10^{-3} \text{ g}}{1 \text{ mg}}}_{\substack{\text{converts} \\ \text{mg to g}}} \times \underbrace{\frac{1 \text{ kg}}{10^3 \text{ g}}}_{\substack{\text{converts} \\ \text{g to kg}}} = 2.43 \times 10^{-4} \text{ kg}$$

Answer Check Often, mistakes lead to answers that are easily detected by giving them a *reasonableness check*. For example, if while solving this problem you mistakenly write the conversion factors in inverted order (although this should not happen if you use units in your calculation), the final answer will be very large. Or perhaps you inadvertently copy the exponent from your calculator with the wrong sign, giving the answer as 2.43×10^4 kg. A large answer is not reasonable, given that you are converting from small mass units (mg) to relatively large mass units (kg). Both of these errors can be caught simply by taking the time to make sure that your answer is reasonable. After completing any calculations, be sure to check for reasonableness in your answers.

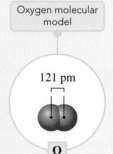

Oxygen molecular model

121 pm

O

Exercise 1.8 The oxygen molecule (the smallest particle of oxygen gas) consists of two oxygen atoms a distance of 121 pm apart. How many millimeters is this distance?

■ See Problems 1.81, 1.82, 1.83, and 1.84.

| **Example** 1.7 | **Converting Units: Metric Volume to Metric Volume** |

Critical Concept 1.7
Dimensional analysis can be used to convert between volumes. Volume measurements typically involve cubed conversion factors.

Solution Essentials:
• Derive cubic (volume) conversion factors from metric conversion factors
• Metric conversion factors
• The metric system
• Rules for significant figures and rounding

The world's oceans contain approximately 1.35×10^9 km^3 of water. What is this volume in liters?

Problem Strategy This problem requires several conversions. A relationship that will prove helpful here is the definition of the liter: 1 dm^3 = 1 L. This provides a clue that if we can convert cubic kilometers (km^3) to cubic decimeters (dm^3), we can use this relationship to convert to liters. When solving problems that involve cubic units such as this one, it is often helpful to think of the conversion factors that you would use if the units were not cubed. In this case, convert from kilometers to meters and then from meters to decimeters. Next, think about how these conversion factors would look if they were cubed. For example, 10^3 m = 1 km, so $(10^3\,\text{m})^3 = (1\,\text{km})^3$ is the cubic relationship. Likewise, 10^{-1} m = 1 dm, so $(10^{-1}\,\text{m})^3 = (1\,\text{dm})^3$ is the cubic relationship. Now these two cubic conversion factors can be combined into three steps to solve the problem.

Solution

$$1.35 \times 10^9\ \text{km}^3 \times \underbrace{\left(\frac{10^3\ \text{m}}{1\ \text{km}}\right)^3}_{\substack{\text{converts} \\ \text{km}^3\ \text{to m}^3}} \times \underbrace{\left(\frac{1\ \text{dm}}{10^{-1}\ \text{m}}\right)^3}_{\substack{\text{converts} \\ \text{m}^3\ \text{to dm}^3}} = 1.35 \times 10^{21}\ \text{dm}^3$$

Because a cubic decimeter is equal to a liter, the volume of the oceans is

$$\mathbf{1.35 \times 10^{21}\ L}$$

Answer Check One of the most common errors made with this type of problem is a failure to cube both the numerator and the denominator in the conversion factors. For example, $(10^3\ \text{m}/1\ \text{km})^3 = 10^9\ \text{m}^3/1\ \text{km}^3$.

Exercise 1.9   A large crystal is constructed by stacking small, identical pieces of crystal, much as you construct a brick wall by stacking bricks. A unit cell is the smallest such piece from which a crystal can be made. The unit cell of a crystal of gold metal has a volume of 67.6 Å^3. What is this volume in cubic decimeters?

■ See Problems 1.85 and 1.86.

The conversion-factor method can be used to convert any unit to another unit, provided a conversion equation exists between the two units. Relationships between certain U.S. units and metric units are given in Table 1.4. You can use these to convert between U.S. and metric units. Suppose you wish to convert 0.547 lb to grams. From Table 1.4, note that 1 lb = 0.4536 kg, or 1 lb = 453.6 g, so the conversion factor from pounds to grams is 453.6 g/1 lb. Therefore,

$$0.547\ \text{lb} \times \frac{453.6\ \text{g}}{1\ \text{lb}} = 248\ \text{g}$$

The next example illustrates a conversion requiring several steps.

Table 1.4	**Relationships of Some U.S. and Metric Units**	
Length	**Mass**	**Volume**
1 in. = 2.54 cm (exact)	1 lb = 0.4536 kg	1 qt = 0.9464 L
1 yd = 0.9144 m (exact)	1 lb = 16 oz (exact)	4 qt = 1 gal (exact)
1 mi = 1.609 km	1 oz = 28.35 g	
1 mi = 5280 ft (exact)		

Example **1.8**   **Converting Units: Any Unit to Another Unit**

Critical Concept 1.8
Dimensional analysis can be used to perform any type of unit conversion. As long as conversion factors between units exist, dimensional analysis can be applied to solve the problem.

Solution Essentials:
• U.S. to metric conversion factors
• Metric conversion factors
• The metric system
• Rules for significant figures and rounding

© Spencer Grant/Photo Edit

How many centimeters are there in 6.51 miles?

Problem Strategy This problem involves converting from U.S. to metric units: specifically, from miles to centimeters. One path to a solution is using three conversions: miles to feet, feet to inches, and inches to centimeters. Table 1.4 has the information you need to develop the conversion factors for U.S. to metric units.

Starting Units   mi   ft   in   Goal Units   cm

Solution From the definitions, you obtain the following conversion factors:

$$1 = \frac{5280 \text{ ft}}{1 \text{ mi}} \qquad 1 = \frac{12 \text{ in.}}{1 \text{ ft}} \qquad 1 = \frac{2.54 \text{ cm}}{1 \text{ in.}}$$

Then,

$$6.51 \text{ mi} \times \underbrace{\frac{5280 \text{ ft}}{1 \text{ mi}}}_{\substack{\text{converts} \\ \text{mi to ft}}} \times \underbrace{\frac{12 \text{ in.}}{1 \text{ ft}}}_{\substack{\text{converts} \\ \text{ft to in.}}} \times \underbrace{\frac{2.54 \text{ cm}}{1 \text{ in.}}}_{\substack{\text{converts} \\ \text{in. to cm}}} = \mathbf{1.05 \times 10^6 \text{ cm}}$$

All of the conversion factors used in this example are exact, so the number of significant figures in the result is determined by the number of significant figures in 6.51 mi.

Answer Check Because miles are a relatively large unit of length and centimeters are a relatively small unit of length, one would expect that converting from miles to centimeters would yield a very large number. This is the case here, so the answer is reasonable.

Exercise 1.10 Using the definitions 1 in. = 2.54 cm and 1 yd = 36 in. (both exact), obtain the conversion factor for yards to meters. How many meters are there in 3.54 yd?

■ See Problems 1.87, 1.88, 1.89, and 1.90.

A Checklist for Review

ⓦWL and **GOChemistry** Sign in at **www.cengage.com/owl** to:
• View tutorials and simulations, develop problem-solving skills, and complete online homework assigned by your professor.
• For quick review and exam prep, download Go Chemistry mini lecture modules from OWL or purchase them at **www.cengagebrain.com**.

Summary of Facts and Concepts

Chemistry is an experimental science in that the facts of chemistry are obtained by *experiment.* These facts are systematized and explained by *theory,* and theory suggests more experiments. The *scientific method* involves this interplay, in which the body of accepted knowledge grows as it is tested by experiment.

Chemistry emerged as a *quantitative science* with the work of the eighteenth-century French chemist Antoine Lavoisier. He made use of the idea that the mass, or quantity of matter, remains constant during a chemical reaction *(law of conservation of mass).*

Matter may be classified by its physical state as a *gas, liquid,* or *solid.* Matter may also be classified by its chemical constitution as an *element, compound,* or *mixture.* Materials are either substances or mixtures of substances. Substances are either elements or compounds, which are composed of two or more elements. Mixtures can be separated into substances by physical processes, but compounds can be separated into elements only by chemical reactions.

A quantitative science requires the making of measurements. Any measurement has limited *precision,* which you convey by writing the measured number to a certain number of *significant figures*. There are many different systems of measurement, but scientists generally use the metric system. The International System (SI) uses a particular selection of metric units. It employs seven *base units* combined with *prefixes* to obtain units of various size. Units for other quantities are derived from these. To obtain a *derived unit* in SI for a quantity such as the volume or density, you merely substitute base units into a defining equation for the quantity.

Dimensional analysis is a technique of calculating with physical quantities in which units are included and treated in the same way as numbers. You can thus obtain the *conversion factor* needed to express a quantity in new units.

Learning Objectives

1.1 Modern Chemistry: A Brief Glimpse
- Provide examples of the contributions of chemistry to humanity.

1.2 Experiment and Explanation
- Describe how chemistry is an experimental science.
- Understand how the scientific method is an approach to performing science.

1.3 Law of Conservation of Mass
- Explain the law of conservation of mass.
- Apply the law of the conservation of mass. Example 1.1

1.4 Matter: Physical State and Chemical Constitution
- Compare and contrast the three common states of matter: solid, liquid, and gas.
- Describe the classifications of matter: elements, compounds, and mixtures (heterogeneous and homogeneous).
- Understand the difference between chemical changes (chemical reactions) and physical changes.
- Distinguish between chemical properties and physical properties.

1.5 Measurement and Significant Figures
- Define and use the terms *precision* and *accuracy* when describing measured quantities.
- Learn the rules for determining significant figures in reported measurements.
- Know how to represent numbers using scientific notation.
- Apply the rules of significant figures to reporting calculated values.
- Be able to recognize exact numbers.
- Know when and how to apply the rules for rounding.
- Use significant figures in calculations. Example 1.2

Important Terms

1.2: experiment, law, hypothesis, theory

1.3: mass, matter, law of conservation of mass

1.4: solid, liquid, gas, states of matter, physical change, chemical change (chemical reaction), physical property, chemical property, substance, element, compound, law of definite proportions (law of constant composition), mixture, heterogeneous mixture, homogeneous mixture (solution), phase

1.5: unit, precision, accuracy, significant figures, number of significant figures, scientific notation, exact number, rounding

1.6 SI Units

- Become familiar with the SI (metric) system of units, including the SI prefixes.
- Convert from one temperature scale to another. Example 1.3

International System of units (SI)
SI base units
SI prefix
meter (m)
angstrom (Å)
kilogram (kg)
second (s)
Celsius scale
kelvin (K)

1.7 Derived Units

- Define and provide examples of derived units.
- Calculate the density of a substance. Example 1.4
- Use density to relate mass and volume. Example 1.5

SI derived unit
liter (L)
density

1.8 Units and Dimensional Analysis (Factor-Label Method)

- Apply dimensional analysis to solving numerical problems.
- Convert from one metric unit to another metric unit. Example 1.6
- Convert from one metric volume to another metric volume. Example 1.7
- Convert from any unit to another unit. Example 1.8

dimensional analysis (factor-label method)
conversion factor

Key Equations

$$T_K = \left(t_C \times \frac{1\ K}{1°C} \right) + 273.15\ K$$

$$t_C = \frac{5°C}{9°F} \times (t_F - 32°F)$$

$$d = m/V$$

Questions and Problems

⏾WL Interactive versions of these problems may be assigned in OWL.

Self-Assessment and Review Questions

Key: These questions test your understanding of the ideas you worked with in the chapter. These problems vary in difficulty and often can be used for the basis of discussion.

1.1 Discuss some ways in which chemistry has changed technology. Give one or more examples of how chemistry has affected another science.

1.2 Define the terms *experiment* and *theory*. How are theory and experiment related? What is a hypothesis?

1.3 Illustrate the steps in the scientific method using Rosenberg's discovery of the anticancer activity of cisplatin.

1.4 Define the terms *matter* and *mass*. What is the difference between mass and weight?

1.5 State the law of conservation of mass. Describe how you might demonstrate this law.

1.6 A chemical reaction is often accompanied by definite changes in appearance. For example, heat and light may be emitted, and there may be a change of color of the substances. Figure 1.9 shows the reactions of the metal mercury with oxygen in air. Describe the changes that occur.

1.7 Characterize gases, liquids, and solids in terms of compressibility and fluidity.

1.8 Choose a substance and give several of its physical properties and several of its chemical properties.

1.9 Give examples of an element, a compound, a heterogeneous mixture, and a homogeneous mixture.

1.10 What distinguishes an element from a compound? Can a compound also be an element?

1.11 What phases or states of matter are present in a glass of bubbling carbonated beverage that contains ice cubes?

1.12 What is meant by the precision of a measurement? How is it indicated?

1.13 Two rules are used to decide how to round the result of a calculation to the correct number of significant figures. Use a calculation to illustrate each rule. Explain how you obtained the number of significant figures in the answers.

1.14 Why should units be carried along with numbers in a calculation?

1.15 Distinguish between a measured number and an exact number. Give examples of each.

1.16 How does the International System (SI) obtain units of different size from a given unit? How does the International System obtain units for all possible physical quantities from only seven base units?

1.17 What is an absolute temperature scale? How are degrees Celsius related to kelvins?

1.18 Define *density*. Describe some uses of density.

1.19 When the quantity 12.9 g is added to 2×10^{-02} g, how many significant figures should be reported in the answer?
- a. one
- b. two
- c. three
- d. four
- e. five

1.20 You perform an experiment in the lab and determine that there are 36.3 inches in a meter. Using this experimental value, how many millimeters are there in 1.34 feet?
- a. 4.05×10^8 mm
- b. 4.43×10^2 mm
- c. 4.05×10^2 mm
- d. 44.3 mm
- e. 4.43×10^5 mm

1.21 A 75.0-g sample of a pure liquid, liquid A, with a density of 3.00 g/mL is mixed with a 50.0-mL sample of a pure liquid, liquid B, with a density of 2.00 g/mL. What is the total volume of the mixture? (Assume there is no reaction upon the mixing of A and B and volumes are additive.)
- a. 275 mL
- b. 175 mL
- c. 125 mL
- d. 100. mL
- e. 75 mL

1.22 Which of the following represents the *smallest* mass?
- a. 0.23 mg
- b. 23 cg
- c. 2.3×10^{-2} kg
- d. 2.3×10^3 μg
- e. 0.23 g

Concept Explorations

Key: Concept explorations are comprehensive problems that provide a framework that will enable you to explore and learn many of the critical concepts and ideas in each chapter. If you master the concepts associated with these explorations, you will have a better understanding of many important chemistry ideas and will be more successful in solving all types of chemistry problems. These problems are well suited for group work and for use as in-class activities.

1.23 Physical and Chemical Changes

Say you are presented with two beakers, beaker A and beaker B, each containing a white, powdery compound.

- a. From your initial observations, you suspect that the two beakers contain the same compound. Describe, in general terms, some experiments in a laboratory that you could do to help prove or disprove that the beakers contain the same compound.

- b. Would it be easier to prove that the compounds are the same or to prove that they are different? Explain your reasoning.

- c. Which of the experiments that you listed above are the most convincing in determining whether the compounds are the same? Justify your answer.

- d. A friend states that the best experiment for determining whether the compounds are the same is to see if they both dissolve in water. He proceeds to take 10.0 g of each compound and places them in separate beakers, each containing 100 mL of water. Both compounds completely dissolve. He then states, "Since the same amount of both substances dissolved in the same volume of water, they must both have the same chemical composition." Is he justified in making this claim? Why or why not?

1.24 Significant Figures
Part 1:

- a. Consider three masses that you wish to add together: 3 g, 1.4 g, and 3.3 g. These numbers represent measured values. Add the numbers together and report your answer to the correct number of significant figures.

- b. Now perform the addition in a stepwise fashion in the following manner. Add 3 g and 1.4 g, reporting this sum to the correct number of significant figures. Next, take the number from the first step and add it to 3.3 g, reporting this sum to the correct number of significant figures.

- c. Compare your answers from performing the addition in the two distinct ways presented in parts a and b. Does one of the answers represent a "better" way of reporting the results of the addition? If your answer is yes, explain why your choice is better.

- d. A student performs the calculation $(5.0 \times 5.143 \text{ g}) + 2.80 \text{ g}$ and, being mindful of significant figures, reports an answer of 29 g. Is this the correct answer? If not, what might this student have done incorrectly?

- e. Another student performs the calculation $(5 \times 5.143 \text{ g}) + 2.80$ and reports an answer of 29 g. Is this the correct answer? If not, what might this student have done incorrectly?

- f. Yet another student performs the calculation $(5.00 \times 5.143 \text{ g}) + 2.80$ and reports an answer of 28.5 g. Is this the correct answer? If not, what did this student probably do incorrectly?

- g. Referring to the calculations above, outline a procedure or rule(s) that will always enable you to report answers using the correct number of significant figures.

Part 2:

a   A student wants to determine the volume of 27.2 g of a substance. He looks up the density of the material in a reference book, where it is reported to be 2.4451 g/cm³. He performs the calculation in the following manner:

$$27.2 \text{ g} \times 1.0 \text{ cm}^3/2.4 \text{ g} = 11.3 \text{ cm}^3$$

Is the calculated answer correct? If not, explain why it is not correct.

b   Another student performs the calculation in the following manner:

$$27.2 \text{ g} \times 1.00 \text{ cm}^3/2.45 \text{ g} = 11.1 \text{ cm}^3$$

Is this a "better" answer than that of the first student? Is this the "best" answer, or could it be "improved"? Explain.

c   Say that you have ten ball bearings, each having a mass of 1.234 g and a density of 3.1569 g/cm³. Calculate the volume of these ten ball bearings. In performing the calculation, present your work as unit conversions, and report your answer to the correct number of significant figures.

d   Explain how the answer that you calculated in part c is the "best" answer to the problem?

Conceptual Problems

Key: These problems are designed to check your understanding of the concepts associated with some of the main topics presented in each chapter. A strong conceptual understanding of chemistry is the foundation for both applying chemical knowledge and solving chemical problems. These problems vary in level of difficulty and often can be used as a basis for group discussion.

1.25 A material is believed to be a compound. Suppose you have several samples of this material obtained from various places around the world. Comment on what you would expect to find upon observing the melting point and color for each sample. What would you expect to find upon determining the elemental composition for each sample?

1.26

a   Sodium metal is partially melted. What are the two phases present?

b   A sample of sand is composed of granules of quartz (silicon dioxide) and seashells (calcium carbonate). The sand is mixed with water. What phases are present?

1.27 You need a thermometer that is accurate to ±5°C to conduct some experiments in the temperature range of 0°C to 100°C. You find one in your lab drawer that has lost its markings.

a   What experiments could you do to make sure your thermometer is suitable for your experiments?

b   Assuming that the thermometer works, what procedure could you follow to put a scale on your thermometer that has the desired accuracy?

1.28 Imagine that you get the chance to shoot five arrows at each of the targets depicted below. On each of the targets, indicate the pattern that the five arrows make when

a   You have poor accuracy and good precision.

b   You have poor accuracy and poor precision.

c   You have good accuracy and good precision.

1.29 Say you live in a climate where the temperature ranges from −100°F to 20°F and you want to define a new temperature scale, YS (YS is the "Your Scale" temperature scale), which defines this range as 0.0°YS to 100.0°YS.

a   Come up with an equation that would allow you to convert between °F and °YS.

b   Using your equation, what would be the temperature in °F if it were 66°YS?

1.30 You have two identical boxes with interior dimensions of 8.0 cm × 8.0 cm × 8.0 cm. You completely fill one of the boxes with wooden spheres that are 1.6 cm in diameter. The other box gets filled with wooden cubes that are 1.6 cm on each edge. After putting the lid on the filled boxes, you then measure the density of each. Which one is more dense?

1.31 You are presented with a piece of metal in a jar. It is your job to determine what the metal is. What are some physical properties that you could measure in order to determine the type of metal? You suspect that the metal might be sodium; what are some chemical properties that you could investigate? (See Section 1.4 for some ideas.)

1.32 Consider the following compounds and their densities.

Substance	Density (g/mL)	Substance	Density (g/mL)
Isopropyl alcohol	0.785	Toluene	0.866
n-Butyl alcohol	0.810	Ethylene glycol	1.114

You create a column of the liquids in a glass cylinder with the most dense material on the bottom layer and the least dense on the top. You do not allow the liquids to mix.

a   First you drop a plastic bead that has a density of 0.24 g/cm³ into the column. What do you expect to observe?

b   Next you drop a different plastic bead that has a volume of 0.043 mL and a mass of 3.92 × 10⁻² g into the column. What would you expect to observe in this case?

c   You drop another bead into the column and observe that it makes it all the way to the bottom of the column. What can you conclude about the density of this bead?

1.33

a   Which of the following items have a mass of about 1 g?
a grain of sand
a paper clip
a nickel
a 5.0-gallon bucket of water
a brick
a car

b   What is the approximate mass (using SI mass units) of each of the items in part a?

1.34 What is the length of the nail reported to the correct number of significant figures?

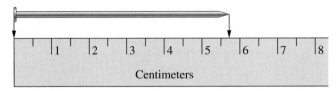

Centimeters

1.35 For these questions, be sure to apply the rules for significant figures.

    a You are conducting an experiment where you need the volume of a box; you take the length, height, and width measurements and then multiply the values together to find the volume. You report the volume of the box as 0.310 m³. If two of your measurements were 0.7120 m and 0.52145 m, what was the other measurement?

    b If you were to add the two measurements from the first part of the problem to a third length measurement with the reported result of 1.509 m, what was the value of the third measurement?

1.36 You are teaching a class of second graders some chemistry, and they are confused about an object's mass versus its density. Keeping in mind that most second graders don't understand fractions, how would you explain these two ideas and illustrate how they differ? What examples would you use as a part of your explanation?

Practice Problems

Key: These problems are for practice in applying problem-solving skills. They are divided by topic, and some are keyed to exercises (see the ends of the exercises). The problems are arranged in matching pairs; the odd-numbered problem of each pair is listed first, and its answer is given in the back of the book.

Conservation of Mass

1.37 Some iron wire weighing 5.6 g is placed in a beaker and covered with 15.0 g of dilute hydrochloric acid. The acid reacts with the metal and gives off hydrogen gas, which escapes into the surrounding air. After reaction, the contents of the beaker weighs 20.4 g. What is the mass of hydrogen gas produced by the reaction?

1.38 A 15.9-g sample of sodium carbonate is added to a solution of acetic acid weighing 20.0 g. The two substances react, releasing carbon dioxide gas to the atmosphere. After reaction, the contents of the reaction vessel weighs 29.3 g. What is the mass of carbon dioxide given off during the reaction?

1.39 Zinc metal reacts with yellow crystals of sulfur in a fiery reaction to produce a white powder of zinc sulfide. A chemist determines that 65.4 g of zinc reacts with 32.1 g of sulfur. How many grams of zinc sulfide could be produced from 20.0 g of zinc metal?

1.40 Aluminum metal reacts with bromine, a red-brown liquid with a noxious odor. The reaction is vigorous and produces aluminum bromide, a white crystalline substance. A sample of 27.0 g of aluminum yields 266.7 g of aluminum bromide. How many grams of bromine react with 18.1 g of aluminum?

Solids, Liquids, and Gases

1.41 Give the normal state (solid, liquid, or gas) of each of the following.

    a potassium hydrogen tartrate (cream of tartar)

    b tungsten

    c carbon (graphite)

    d mercury

1.42 Give the normal state (solid, liquid, or gas) of each of the following.

    a sodium hydrogen carbonate (baking soda)

    b isopropyl alcohol (rubbing alcohol)

    c carbon monoxide

    d lead

Chemical and Physical Changes; Properties of Substances

1.43 Which of the following are physical changes and which are chemical changes?

    a melting of sodium chloride

    b pulverizing of rock salt

    c burning of sulfur

    d dissolving of salt in water

1.44 For each of the following, decide whether a physical or a chemical change is involved.

    a dissolving of sugar in water

    b rusting of iron

    c burning of wood

    d evaporation of alcohol

1.45 Solid iodine, contaminated with salt, was heated until the iodine vaporized. The violet vapor of iodine was then cooled to yield the pure solid. Solid iodine and zinc metal powder were mixed and ignited to give a white powder. Identify each physical change and each chemical change.

1.46 A sample of mercury(II) oxide was heated to produce mercury metal and oxygen gas. Then the liquid mercury was cooled to −40°C, where it solidified. A glowing wood splint was thrust into the oxygen, and the splint burst into flame. Identify each physical change and each chemical change.

1.47 The following are properties of substances. Decide whether each is a physical property or a chemical property.

    a Chlorine gas liquefies at −35°C under normal pressure.

    b Hydrogen burns in chlorine gas.

    c Bromine melts at −7.2°C.

    d Lithium is a soft, silvery-colored metal.

    e Iron rusts in an atmosphere of moist air.

1.48 Decide whether each of the following is a physical property or a chemical property of the substance.

a Salt substitute, potassium chloride, dissolves in water.

b Seashells, calcium carbonate, fizz when immersed in vinegar.

c The gas hydrogen sulfide smells like rotten eggs.

d Fine steel wool (Fe) can be burned in air.

e Pure water freezes at 0°C.

1.49 Mercury(II) oxide is an orange-red solid with a density of 11.1 g/cm^3. It decomposes when heated to give mercury and oxygen. The compound is insoluble in water (does not dissolve in water). Identify the physical and the chemical properties of mercury(II) oxide that are cited.

1.50 Iodine is a solid having somewhat lustrous, blue-black crystals. The crystals vaporize readily to a violet-colored gas. Iodine combines with many metals. For example, aluminum combines with iodine to give aluminum iodide. Identify the physical and the chemical properties of iodine that are cited.

Elements, Compounds, and Mixtures

1.51 Consider the following separations of materials. State whether a physical process or a chemical reaction is involved in each separation.

a Sodium chloride is obtained from seawater by evaporation of the water.

b Mercury is obtained by heating the substance mercury(II) oxide; oxygen is also obtained.

c Pure water is obtained from ocean water by evaporating the water, then condensing it.

d Iron is produced from an iron ore that contains the substance iron(III) oxide.

e Gold is obtained from river sand by panning (allowing the heavy metal to settle in flowing water).

1.52 All of the following processes involve a separation of either a mixture into substances or a compound into elements. For each, decide whether a physical process or a chemical reaction is required.

a Sodium metal is obtained from the substance sodium chloride.

b Iron filings are separated from sand by using a magnet.

c Sugar crystals are separated from a sugar syrup by evaporation of water.

d Fine crystals of silver chloride are separated from a suspension of the crystals in water.

e Copper is produced when zinc metal is placed in a solution of copper(II) sulfate, a compound.

1.53 Indicate whether each of the following materials is a substance, a heterogeneous mixture, or a solution.

a milk   b bromine

c gasoline   d aluminum

1.54 Label each of the following as a substance, a heterogeneous mixture, or a solution.

a seawater   b sulfur

c fluorine   d beach sand

1.55 Which of the following are pure substances and which are mixtures? For each, list all of the different phases present.

a bromine liquid and its vapor

b paint, containing a liquid solution and a dispersed solid pigment

c partially molten iron

d baking powder containing sodium hydrogen carbonate and potassium hydrogen tartrate

1.56 Which of the following are pure substances and which are mixtures? For each, list all of the different phases present.

a a sugar solution with sugar crystals at the bottom

b ink containing a liquid solution with fine particles of carbon

c a sand containing quartz (silicon dioxide) and calcite (calcium carbonate)

d liquid water and steam at 100°C

Significant Figures

1.57 How many significant figures are there in each of the following measurements?

a 57.00 g   b 0.0503 kg

c 6.310 J   d 0.80090 m

e 5.06×10^{-7} cm   f 2.010 s

1.58 How many significant figures are there in each of the following measurements?

a 4.0100 mg   b 0.05930 g

c 0.035 mm   d 3.100 s

e 8.91×10^1 L   f 9.100×10^4 cm

1.59 The circumference of the earth at the equator is 40,000 km. This value is precise to four significant figures. Write this in scientific notation to express correctly the number of significant figures.

1.60 The astronomical unit equals the mean distance between the earth and the sun. This distance is 150,000,000 km, which is precise to three significant figures. Express this in scientific notation to the correct number of significant figures.

1.61 Assuming all numbers are measured quantities, do the indicated arithmetic and give the answer to the correct number of significant figures.

a $\dfrac{8.71 \times 0.0301}{0.031}$

b $0.71 + 89.3$

c $934 \times 0.00435 + 107$

d $(847.89 - 847.73) \times 14673$

1.62 Assuming all numbers are measured quantities, do the indicated arithmetic and give the answer to the correct number of significant figures.

a $\dfrac{08.71 \times 0.57}{5.871}$

b $8.937 - 8.930$

c $8.937 + 8.930$

d $0.00015 \times 54.6 + 1.002$

1.63 One sphere has a radius of 5.15 cm; another has a radius of 5.00 cm. What is the difference in volume (in cubic centimeters) between the two spheres? Give the answer to the correct number of significant figures. The volume of a sphere is $(4/3)\,\pi r^3$, where $\pi = 3.1416$ and r is the radius.

1.64 A solid circular cylinder of iron with a radius of 1.500 cm has a ruler etched along its length. What is the volume of iron contained between the marks labeled 3.20 cm and 3.50 cm? The volume of a circular cylinder is $\pi r^2 l$, where $\pi = 3.1416$, r is the radius, and l is the length.

SI Units

1.65 Write the following measurements, without scientific notation, using the appropriate SI prefix.
- a 4.851×10^{-6} g
- b 3.16×10^{-2} m
- c 2.591×10^{-9} s
- d 8.93×10^{-12} g

1.66 Write the following measurements, without scientific notation, using the appropriate SI prefix.
- a 5.89×10^{-12} s
- b 0.2010 m
- c 2.560×10^{-9} g
- d 6.05×10^3 m

1.67 Using scientific notation, convert:
- a 6.15 ps to s
- b 3.781 μm to m
- c 1.546 Å to m
- d 9.7 mg to g

1.68 Using scientific notation, convert:
- a 6.20 km to m
- b 1.98 ns to s
- c 2.54 cm to m
- d 5.23 μg to g

Temperature Conversion

1.69 Convert:
- a 68°F to degrees Celsius
- b −23°F to degrees Celsius
- c 26°C to degrees Fahrenheit
- d −81°C to degrees Fahrenheit

1.70 Convert:
- a 51°F to degrees Celsius
- b −11°F to degrees Celsius
- c −41°C to degrees Fahrenheit
- d 22°C to degrees Fahrenheit

1.71 The reaction of oxygen and hydrogen is used to propel rockets. Liquid oxygen has a boiling point of –222.7°C. What is this temperature in degrees Fahrenheit?

1.72 Salt and ice are stirred together to give a mixture to freeze ice cream. The temperature of the mixture is −20.0°C. What is this temperature in degrees Fahrenheit?

Density

1.73 A certain sample of the mineral galena (lead sulfide) weighs 12.4 g and has a volume of 1.64 cm³. What is the density of galena?

1.74 A flask contains a 30.0 mL sample of acetone (nail polish remover) that weighs 23.6 grams. What is the density of the acetone?

1.75 A mineral sample has a mass of 5.94 g and a volume of 0.73 cm³. The mineral is either sphalerite (density = 4.0 g/cm³), cassiterite (density = 6.99 g/cm³), or cinnabar (density = 8.10 g/cm³). Which is it?

1.76 A liquid with a volume of 8.5 mL has a mass of 6.71 g. The liquid is either octane, ethanol, or benzene, the densities of which are 0.702 g/cm³, 0.789 g/cm³, and 0.879 g/cm³, respectively. What is the identity of the liquid?

1.77 Platinum has a density of 21.4 g/cm³. What is the mass of 5.9 cm³ of this metal?

1.78 What is the mass of a 43.8-mL sample of gasoline, which has a density of 0.70 g/cm³?

1.79 Ethanol has a density of 0.789 g/cm³. What volume must be poured into a graduated cylinder to give 19.8 g of alcohol?

1.80 Bromine is a red-brown liquid with a density of 3.10 g/mL. A sample of bromine weighing 88.5 g occupies what volume?

Unit Conversions

1.81 Sodium hydrogen carbonate, known commercially as baking soda, reacts with acidic materials such as vinegar to release carbon dioxide gas. An experiment calls for 0.480 kg of sodium hydrogen carbonate. Express this mass in milligrams.

1.82 The acidic constituent in vinegar is acetic acid. A 10.0-mL sample of a certain vinegar contains 611 mg of acetic acid. What is this mass of acetic acid expressed in micrograms?

1.83 Water consists of molecules (groups of atoms). A water molecule has two hydrogen atoms, each connected to an oxygen atom. The distance between any one hydrogen atom and the oxygen atom is 0.96 Å. What is this distance in millimeters?

1.84 The different colors of light have different wavelengths. The human eye is most sensitive to light whose wavelength is 555 nm (greenish-yellow). What is this wavelength in centimeters?

1.85 A submicroscopic particle suspended in a solution has a volume of 1.3 μm³. What is this volume in liters?

1.86 The total amount of fresh water on earth is estimated to be 3.73×10^8 km³. What is this volume in cubic meters? in liters?

1.87 The calorie, the Btu (British thermal unit), and the joule are units of energy; 1 calorie = 4.184 joules (exact), and 1 Btu = 252.0 calories. Convert 3.15 Btu to joules.

1.88 How many grams are there in 3.58 short tons? Note that 1 g = 0.03527 oz (ounces avoirdupois), 1 lb (pound) = 16 oz, and 1 short ton = 2000 lb. (These relations are exact.)

1.89 The first measurement of sea depth was made in 1840 in the central South Atlantic, where a plummet was lowered 2425 fathoms. What is this depth in meters? Note that 1 fathom = 6 ft, 1 ft = 12 in., and 1 in. = 2.54×10^{-2} m. (These relations are exact.)

1.90 The estimated amount of recoverable oil from the field at Prudhoe Bay in Alaska is 1.3×10^{10} barrels. What is this amount of oil in cubic meters? One barrel = 42 gal (exact), 1 gal = 4 qt (exact), and 1 qt = 9.46×10^{-4} m^3.

1.91 A fish tank is 20.0 in. long, 20.0 in. deep, and 10.0 in. high. What is the maximum volume of water, in liters, that the fish tank can hold?

1.92 The population density of worms in a particular field is 33 worms per cubic meter of soil. How many worms would there be in the top meter of soil in a field that has dimensions of 1.00 km by 2.0 km?

General Problems

Key: These problems provide more practice but are not divided by topic or keyed to exercises. Each section ends with essay questions, each of which is color coded to refer to the *A Chemist Looks at Daily Life* (orange) or *Instrumental Methods* (brown) chapter essay on which it is based. Odd-numbered problems and the even-numbered problems that follow are similar; answers to all odd-numbered problems except the essay questions are given in the back of the book.

1.93 Sodium metal reacts vigorously with water. A piece of sodium weighing 19.70 g was added to a beaker containing 126.22 g of water. During reaction, hydrogen gas was produced and bubbled from the solution. The solution, containing sodium hydroxide, weighed 145.06 g. How many grams of hydrogen gas were produced?

1.94 An antacid tablet weighing 0.853 g contained calcium carbonate as the active ingredient, in addition to an inert binder. When an acid solution weighing 56.519 g was added to the tablet, carbon dioxide gas was released, producing a fizz. The resulting solution weighed 57.152 g. How many grams of carbon dioxide were produced?

1.95 When a mixture of aluminum powder and iron(III) oxide is ignited, it produces molten iron and aluminum oxide. In an experiment, 5.40 g of aluminum was mixed with 18.50 g of iron(III) oxide. At the end of the reaction, the mixture contained 11.17 g of iron, 10.20 g of aluminum oxide, and an undetermined amount of unreacted iron(III) oxide. No aluminum was left. What is the mass of the iron(III) oxide?

1.96 When chlorine gas is bubbled into a solution of sodium bromide, the sodium bromide reacts to give bromine, a red-brown liquid, and sodium chloride (ordinary table salt). A solution was made by dissolving 20.6 g of sodium bromide in 100.0 g of water. After passing chlorine through the solution, investigators analyzed the mixture. It contained 16.0 g of bromine and 11.7 g of sodium chloride. How many grams of chlorine reacted?

1.97 A graduated cylinder weighed 66.5 g. To the cylinder was added 58.2 g of water and 5.279 g of sodium chloride. What was the total mass of the cylinder and the solution? Express the answer to the correct number of significant figures.

1.98 A beaker weighed 53.10 g. To the beaker was added 5.348 g of iron pellets and 56.1 g of hydrochloric acid. What was the total mass of the beaker and the mixture (before reaction)? Express the answer to the correct number of significant figures.

1.99 Describe each of the following as a physical or chemical property of each listed chemical substance.
a baking soda reacts with vinegar
b ice melts at 0°C
c graphite is a soft, black solid
d hydrogen burns in air

1.100 Describe each of the following as a physical or chemical property of each listed chemical substance.
a chlorine is a green gas
b zinc reacts with acids
c magnesium has a density of 1.74 g/cm^3
d iron can rust

1.101 A red-orange solid contains only mercury and oxygen. Analyses of three different samples gave the following results. Are the data consistent with the hypothesis that the material is a compound?

	Mass of Sample	Mass of Mercury	Mass of Oxygen
Sample A	1.0410 g	0.9641g	0.0769 g
Sample B	1.5434 g	1.4293 g	0.1141 g
Sample C	1.2183 g	1.1283 g	0.0900 g

1.102 Analyses of several samples of a material containing only iron and oxygen gave the following results. Could this material be a compound?

	Mass of Sample	Mass of Iron	Mass of Oxygen
Sample A	1.518 g	1.094 g	0.424 g
Sample B	2.056 g	1.449 g	0.607 g
Sample C	1.873 g	1.335 g	0.538 g

1.103 A cubic box measures 39.3 cm on an edge. What is the volume of the box in cubic centimeters? Express the answer to the correct number of significant figures.

1.104 A cylinder with circular cross section has a radius of 2.56 cm and a height of 56.32 cm. What is the volume of the cylinder? Express the answer to the correct number of significant figures.

1.105 A spherical tank has a radius of 175.0 in. Calculate the volume of the tank in cubic inches; then convert this to Imperial gallons. The volume of a sphere is $(4/3)\pi r^3$, where *r* is the radius. One Imperial gallon equals 277.4 in^3.

1.106 An aquarium has a rectangular cross section that is 47.8 in. by 12.5 in.; it is 19.5 in. high. How many U.S.

gallons does the aquarium contain? One U.S. gallon equals exactly 231 in^3.

1.107 Obtain the difference in volume between two spheres, one of radius 5.61 cm, the other of radius 5.85 cm. The volume V of a sphere is $(4/3)\,\pi r^3$, where r is the radius. Give the result to the correct number of significant figures.

1.108 What is the difference in surface area between two circles, one of radius 7.98 cm, the other of radius 8.50 cm? The surface area of a circle of radius r is πr^2. Obtain the result to the correct number of significant figures.

1.109 Perform the following arithmetic setups and express the answers to the correct number of significant figures.

a $\dfrac{56.1 - 51.1}{6.58}$ b $\dfrac{56.1 + 51.1}{6.58}$

c $0.0065 \times 3.21 + 0.0911$ d $(9.1 + 8.6) \times 26.91$

1.110 Assuming all of the numbers are measured quantities, perform the following arithmetic setups and report the answers to the correct number of significant figures.

a $\dfrac{9.345 + 9.005}{9.811}$ b $\dfrac{9.345 - 9.005}{9.811}$

c $(7.50 + 7.53) \times 3.71$ d $0.71 \times 0.36 + 17.36$

1.111 For each of the following, write the measurement in terms of an appropriate prefix and base unit.

a The mass of magnesium per milliliter in a sample of blood serum is 0.0186 g.

b The radius of a carbon atom is about 0.000000000077 m.

c The hemoglobin molecule, a component of red blood cells, is 0.0000000065 m in diameter.

d The wavelength of a certain infrared radiation is 0.00000085 m.

1.112 For each of the following, write the measurement in terms of an appropriate prefix and base unit.

a The mass of calcium per milliliter in a sample of blood serum is 0.0912 g.

b The radius of an oxygen atom is about 0.000000000066 m.

c A particular red blood cell measures 0.0000071 m.

d The wavelength of a certain ultraviolet radiation is 0.000000056 m.

1.113 Write each of the following in terms of the SI base unit (that is, express the prefix as the power of 10).

a 1.07 ps   b 5.8 μm
c 319 nm   d 15.3 ms

1.114 Write each of the following in terms of the SI base unit (that is, express the prefix as the power of 10).

a 6.6 mK   b 275 pm
c 22.1 ms   d 45 μm

1.115 Titanium metal is used in aerospace alloys to add strength and corrosion resistance. Titanium melts at 1677°C. What is this temperature in degrees Fahrenheit?

1.116 Tungsten metal, which is used in lightbulb filaments, has the highest melting point of any metal (3410°C). What is this melting point in degrees Fahrenheit?

1.117 Calcium carbonate, a white powder used in toothpastes, antacids, and other preparations, decomposes when heated to about 825°C. What is this temperature in degrees Fahrenheit?

1.118 Sodium hydrogen carbonate (baking soda) starts to decompose to sodium carbonate (soda ash) at about 50°C. What is this temperature in degrees Fahrenheit?

1.119 Mercury metal is liquid at normal temperatures but freezes at −38.9°C. What is this temperature in kelvins? in degrees Fahrenheit?

1.120 Gallium metal can be melted by the heat of one's hand. Its melting point is 29.8°C. What is this temperature in kelvins? in degrees Fahrenheit?

1.121 Zinc metal can be purified by distillation (transforming the liquid metal to vapor, then condensing the vapor back to liquid). The metal boils at normal atmospheric pressure at 833°F. What is this temperature in degrees Celsius? in kelvins?

1.122 Iodine is a bluish-black solid. It forms a violet-colored vapor when heated. The solid melts at 236°F. What is this temperature in degrees Celsius? in kelvins?

1.123 An aluminum alloy used in the construction of aircraft wings has a density of 2.70 g/cm^3. Express this density in SI units (kg/m^3).

1.124 Vanadium metal is added to steel to impart strength. The density of vanadium is 5.96 g/cm^3. Express this in SI units (kg/m^3).

1.125 Hematite (iron ore) weighing 70.7 g was placed in a flask whose volume was 53.2 mL. The flask with hematite was then carefully filled with water and weighed. The hematite and water weighed 109.3 g. The density of the water was 0.997 g/cm^3. What was the density of the hematite?

1.126 The density of quartz mineral was determined by adding a weighed piece to a graduated cylinder containing 51.2 mL water. After the quartz was submerged, the water level was 65.7 mL. The quartz piece weighed 38.4 g. What was the density of the quartz?

1.127 Some bottles of colorless liquids were being labeled when the technicians accidentally mixed them up and lost track of their contents. A 30.0-mL sample withdrawn from one bottle weighed 44.6 g. The technicians knew that the liquid was either acetone, benzene, chloroform, or carbon tetrachloride (which have densities of 0.792 g/cm^3, 0.899 g/cm^3, 1.489 g/cm^3, and 1.595 g/cm^3, respectively). What was the identity of the liquid?

1.128 Providing no reaction occurs, a solid will float on any liquid that is more dense than it is. The volume of a piece of calcite weighing 35.6 g is 12.9 cm^3. On which of the following liquids will the calcite float: carbon tetrachloride (density = 1.60 g/cm^3), methylene bromide (density = 2.50 g/cm^3), tetrabromoethane (density = 2.96 g/cm^3), or methylene iodide (density = 3.33 g/cm^3)?

1.129 Ultrapure silicon is used to make solid-state devices, such as computer chips. What is the mass of a circular cylinder of silicon that is 12.40 cm long and has a radius of 4.00 cm? The density of silicon is 2.33 g/cm^3.

1.130 Platinum metal is used in jewelry; it is also used in automobile catalytic converters. What is the mass of a cube of platinum 4.40 cm on an edge? The density of platinum is 21.4 g/cm^3.

1.131 Vinegar contains acetic acid (about 5% by mass). Pure acetic acid has a strong vinegar smell but is corrosive to the skin. What volume of pure acetic acid has a mass of 35.00 g? The density of acetic acid is 1.053 g/mL.

1.132 Ethyl acetate has a characteristic fruity odor and is used as a solvent in paint lacquers and perfumes. An experiment requires 0.070 kg of ethyl acetate. What volume is this (in liters)? The density of ethyl acetate is 0.902 g/mL.

1.133 Convert:
a 37.1 mm to centimeters
b 318 μs to milliseconds
c 8.45 kg to micrograms
d 93 km to nanometers

1.134 Convert:
a 127 Å to micrometers
b 21.0 kg to milligrams
c 1.09 cm to millimeters
d 4.6 ns to microseconds

1.135 Convert:
a 5.91 kg of chrome yellow to milligrams
b 753 mg of vitamin A to micrograms
c 90.1 MHz (megahertz), the wavelength of an FM signal, to kilohertz
d 498 mJ (the joule, J, is a unit of energy) to kilojoules

1.136 Convert:
a 104 pm, the radius of a sulfur atom, to angstroms
b 7.19 μg of cyanocobalamin (vitamin B$_{12}$) to milligrams

c 0.0605 kPa (the pascal, Pa, is a unit of pressure) to centipascals
d 0.010 mm, the diameter of a typical blood capillary, to centimeters

1.137 The largest of the Great Lakes is Lake Superior, which has a volume of 12,230 km^3. What is this volume in liters?

1.138 The average flow of the Niagara River is 3.50 km^3 per week. What is this volume in liters?

1.139 A room measures 10.0 ft × 11.0 ft and is 9.0 ft high. What is its volume in liters?

1.140 A cylindrical settling tank is 5.0 ft deep and has a radius of 15.0 ft. What is the volume of the tank in liters?

1.141 The masses of diamonds and gems are measured in carats. A carat is defined as 200 mg. If a jeweler has 275 carats of diamonds, how many grams does she have?

1.142 One year of world production of gold was 49.6 × 10^6 troy ounces. One troy ounce equals 31.10 g. What was the world production of gold in metric tons (10^6 g) for that year?

■ **1.143** What are some characteristics of the adhesive used for Post-it Notes?

■ **1.144** All good experiments start with a scientific question. What was the scientific question that Art Fry was trying to answer when he embarked on finding the adhesive for the Post-it Note?

■ **1.145** Describe how gas chromatography works.

■ **1.146** What do the various chromatographic separation techniques have in common?

Strategy Problems

Key: As noted earlier, all of the practice and general problems are matched pairs. This section is a selection of problems that are not in matched-pair format. These challenging problems require that you employ many of the concepts and strategies that were developed in the chapter. In some cases, you will have to integrate several concepts and operational skills in order to solve the problem successfully.

1.147 When the quantity 5 × 10^{-2} mg is subtracted from 4.7 mg, how many significant figures should be reported in the answer?

1.148 A 33.0-g sample of an unknown liquid at 20.0°C is heated to 120°C. During this heating, the density of the liquid changes from 0.854 g/cm^3 to 0.797 g/cm^3. What volume would this sample occupy at 120°C?

1.149 A 175-g sample of a pure liquid, liquid A, with a density of 3.00 g/mL is mixed with a 50.0-mL sample of a pure liquid, liquid B, with a density of 2.00 g/mL. What is the total volume of the mixture? (Assume there is no reaction upon the mixing of A and B, and volumes are additive.)

1.150 On a long trip you travel 832 miles in 21 hours. During this trip, you use 31 gallons of gasoline.

a Calculate your average speed in miles per hour.
b Calculate your average speed in kilometers per hour.
c Calculate your gas mileage in kilometers per liter.

1.151 The density of lead at 20°C is 11.3 g/cm^3. Rank the volumes of the following quantities of lead from least to greatest: 50.0 mL, 0.100 qt, 50.0 g, and 0.0250 lb.

1.152 A 2010 oil spill in the Gulf of Mexico was estimated to dump as much as 7.6 × 10^8 L of crude oil into the water. Calculate the volume of this spill in cubic feet.

1.153 The density of liquid water at 80°C is 972 kg/m^3 and at 20°C is 998 kg/m^3. If you have 200.0 mL of water at 20°C, what volume (mL) will the water occupy at 80°C? Which will contain more water molecules, 1.0 L of water at 80°C or 1.0 L of water at 20°C?

1.154 Catalytic converters are used in automobiles to convert harmful gaseous substances in the exhaust to more benign substances, for example, carbon monoxide to carbon dioxide. Beads used in these converters have extremely large surface areas. For instance, a single 3.0-mm porous bead can have a surface area of 1.0 × 10^6 cm^2. If a car has two catalytic converters, each with 5.0 × 10^3 beads, calculate the surface area of the two catalytic converters in km^2.

1.155 A 55.0-cm³ sample of ocean water has a density of 1025 kg/m³. Assuming all of the solids stay in solution, what will be the density of this water sample in units of g/cm³ after 5.0 mL of water has evaporated?

1.156 At 20°C liquid gasoline gas has a density of 0.75 g/cm³. If a 5.5-mL sample of gasoline is placed into a sealed 3.00-L container at 50°C and allowed to completely evaporate, what will be the density of the gasoline vapor in the container?

1.157 The figures below represent a gas trapped in containers. The orange balls represent individual gas atoms. Container A on the left has a volume that is one-half the volume of container B on the right.

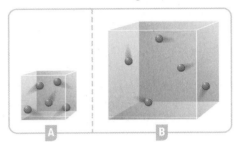

a How does the mass of gas in one container compare with the mass of gas in the other?

b Which container has the greater density of gas?

c If the volume of container A were increased to the same volume as container B, how would the density of the gas in container A change?

1.158 An ice cube measures 3.50 cm on each edge and weighs 39.45 g.

a Calculate the density of ice.

b Calculate the mass of 400.4 mL of water in an ice cube.

1.159 The total length of all the DNA molecules contained in a human body is 1×10^{10} miles. The population of the United States is about 300 million. What is the total length of all the DNA of the U.S. population in lightyears? (A lightyear is the distance that light travels in a year and is 9.46×10^{15} m.)

1.160 Prospectors are considering searching for gold on a plot of land that contains 1.31 g of gold per bucket of soil. If the volume of the bucket is 4.67 L, how many grams of gold are there likely to be in 2.38×10^3 cubic feet of soil?

1.161 Water and saline (salt) solution have in common that they are both homogeneous. How do these materials differ? Be specific and use chemical terms to describe the two systems.

1.162 A solution is prepared by dissolving table salt, sodium chloride, in water at room temperature.

a Assuming there is no significant change in the volume of water during the preparation of the solution, how would the density of the solution compare to that of pure water?

b If you were to boil the solution for several minutes and then allow it to cool to room temperature, how would the density of the solution compare to the density in part a?

c If you took the solution prepared in part a and added more water, how would this affect the density of the solution?

Cumulative-Skills Problems

Key: The problems under this heading combine skills introduced in the previous chapter with those given in the current one.

1.163 When 10.0 g of marble chips (calcium carbonate) is treated with 50.0 mL of hydrochloric acid (density 1.096 g/mL), the marble dissolves, giving a solution and releasing carbon dioxide gas. The solution weighs 60.4 g. How many liters of carbon dioxide gas are released? The density of the gas is 1.798 g/L.

1.164 A steel sphere has a radius of 1.58 in. If this steel has a density of 7.88 g/cm³, what is the mass of the sphere in grams?

1.165 Zinc ore (zinc sulfide) is treated with sulfuric acid, leaving a solution with some undissolved bits of material and releasing hydrogen sulfide gas. If 10.8 g of zinc ore is treated with 50.0 mL of sulfuric acid (density 1.153 g/mL), 65.1 g of solution and undissolved material remains. In addition, hydrogen sulfide (density 1.393 g/L) is evolved. What is the volume (in liters) of this gas?

1.166 A weather balloon filled with helium has a diameter of 3.50 ft. What is the mass in grams of the helium in the balloon at 21°C and normal pressure? The density of helium under these conditions is 0.166 g/L.

1.167 The land area of Greenland is 840,000 mi², with only 132,000 mi² free of perpetual ice. The average thickness of this ice is 5000 ft. Estimate the mass of the ice (assume two significant figures). The density of ice is 0.917 g/cm³.

1.168 Antarctica, almost completely covered in ice, has an area of 5,500,000 mi² with an average height of 7500 ft. Without the ice, the height would be only 1500 ft. Estimate the mass of this ice (two significant figures). The density of ice is 0.917 g/cm³.

1.169 A sample of an ethanol–water solution has a volume of 54.2 cm³ and a mass of 49.6 g. What is the percentage of ethanol (by mass) in the solution? (Assume that there is no change in volume when the pure compounds are mixed.) The density of ethanol is 0.789 g/cm³ and that of water is 0.998 g/cm³. Alcoholic beverages are rated in *proof,* which is a measure of the relative amount of ethanol in the beverage. Pure ethanol is exactly 200 proof; a solution that is 50% ethanol by volume is exactly 100 proof. What is the proof of the given ethanol–water solution?

1.170 You have a piece of gold jewelry weighing 9.35 g. Its volume is 0.654 cm³. Assume that the metal is an alloy (mixture) of gold and silver, which have densities of 19.3 g/cm³ and 10.5 g/cm³, respectively. Also assume that there is no change in volume when the pure metals are mixed. Calculate the percentage of gold (by mass) in the alloy. The relative amount of gold in an alloy is measured in *karats.* Pure gold is 24 karats; an alloy of 50% gold is 12 karats. State the proportion of gold in the jewelry in karats.

1.171 A sample of a bright blue mineral was weighed in air, then weighed again while suspended in water. An object is buoyed up by the mass of the fluid displaced by the object. In air, the mineral weighed 7.35 g; in water, it weighed 5.40 g. The densities of air and water are 1.205 g/L and 0.9982 g/cm^3, respectively. What is the density of the mineral?

1.172 A sample of vermilion-colored mineral was weighed in air, then weighed again while suspended in water. An object is buoyed up by the mass of the fluid displaced by the object. In air, the mineral weighed 18.49 g; in water, it weighed 16.21 g. The densities of air and water are 1.205 g/L and 0.9982 g/cm^3, respectively. What is the density of the mineral?

1.173 A student gently drops an object weighing 15.8 g into an open vessel that is full of ethanol, so that a volume of ethanol spills out equal to the volume of the object. The experimenter now finds that the vessel and its contents weigh 10.5 g more than the vessel full of ethanol only. The density of ethanol is 0.789 g/cm^3. What is the density of the object?

1.174 An experimenter places a piece of a solid metal weighing 255 g into a graduated cylinder, which she then fills with liquid mercury. After weighing the cylinder and its contents, she removes the solid metal and fills the cylinder with mercury. She now finds that the cylinder and its contents weigh 101 g less than before. The density of mercury is 13.6 g/cm^3. What is the density of the solid metal?

2

Atoms, Molecules, and Ions

The probe of a scanning tunneling microscope (as sharp as a single atom!) was used to arrange iron atoms in an oval shape on a surface layer of copper atoms and collect data to produce this image.

Contents and Concepts

Atomic Theory and Atomic Structure

The key concept in chemistry is that all matter is composed of very small particles called atoms. We look at atomic theory, discuss atomic structure, and finally describe the periodic table, which organizes the elements.

2.1 Atomic Theory of Matter
2.2 The Structure of the Atom
2.3 Nuclear Structure; Isotopes
2.4 Atomic Masses
2.5 Periodic Table of the Elements

Chemical Substances: Formulas and Names

We explore how atoms combine in various ways to yield the millions of known substances.

2.6 Chemical Formulas; Molecular and Ionic Substances
2.7 Organic Compounds
2.8 Naming Simple Compounds

Chemical Reactions: Equations

2.9 Writing Chemical Equations
2.10 Balancing Chemical Equations

◑WL Sign in to OWL at www.cengage.com/owl to view tutorials and simulations, develop problem-solving skills, and complete online homework assigned by your professor.

go Chemistry Download mini lecture videos for key concept review and exam prep from OWL or purchase them from www.cengagebrain.com.

odium is a soft, silvery metal. This metal cannot be handled with bare fingers, because it reacts with any moisture on the skin, causing a burn. Chlorine is a poisonous, greenish yellow gas with a choking odor. When molten sodium is placed into an atmosphere of chlorine, a dramatic reaction occurs: the sodium bursts into flame and a white, crystalline powder forms (see Figure 2.1). What is particularly interesting about this reaction is that this chemical combination of two toxic substances produces sodium chloride, the substance that we commonly call table salt.

Sodium metal and chlorine gas are particular forms of matter. In burning, they undergo a chemical change—a chemical reaction—in which these forms of matter change to a form of matter with different chemical and physical properties. How do we explain the differences in properties of different forms of matter? And how do we explain chemical reactions such as the burning of sodium metal in chlorine gas? This chapter and the next take an introductory look at these basic questions in chemistry. In later chapters we will develop the concepts introduced here.

Atomic Theory and Atomic Structure

As noted in Chapter 1, all matter is composed of atoms. In this first part of Chapter 2, we explore the atomic theory of matter and the structure of atoms. We also examine the *periodic table,* which is an organizational chart designed to highlight similarities among the elements.

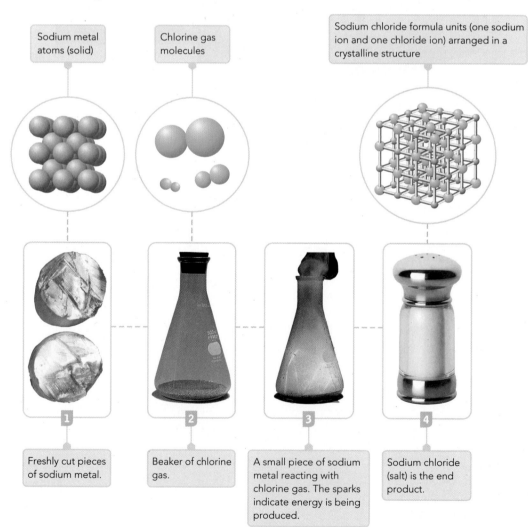

Sodium metal atoms (solid)

Chlorine gas molecules

Sodium chloride formula units (one sodium ion and one chloride ion) arranged in a crystalline structure

1 Freshly cut pieces of sodium metal.

2 Beaker of chlorine gas.

3 A small piece of sodium metal reacting with chlorine gas. The sparks indicate energy is being produced.

4 Sodium chloride (salt) is the end product.

© Cengage Learning; Andrew Lambert Photography/Photo Researchers, Inc.; Chimpinski/Shutterstock.com

Figure 2.1 ▲

Reaction of sodium and chlorine *Left to Right:* Sodium metal, chlorine gas, a small piece of sodium metal reacting with chlorine, the sodium chloride product of the reaction. The sparks in the flask of the reaction indicate that energy is produced.

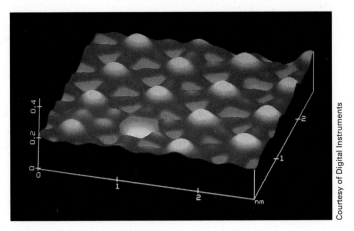

Courtesy of Digital Instruments

Figure 2.2 ◄

Image of iodine atoms on a platinum metal surface This computer-screen image was obtained by a scanning tunneling microscope (discussed in Chapter 7); the color was added to the image by computer. Iodine atoms are the large peaks with pink tops. Note the "vacancy" in the array of iodine atoms. A scale shows the size of the atoms in nanometers.

2.1 Atomic Theory of Matter

As we noted in Chapter 1, Lavoisier laid the experimental foundation of modern chemistry. But the British chemist John Dalton (1766–1844) provided the basic theory: all matter—whether element, compound, or mixture—is composed of small particles called atoms. The postulates, or basic statements, of Dalton's theory are presented in this section. Note that the terms *element, compound,* and *chemical reaction,* which were defined in Chapter 1 in terms of matter as we normally see it, are redefined here by the postulates of Dalton's theory in terms of atoms.

Postulates of Dalton's Atomic Theory

The main points of Dalton's **atomic theory,** *an explanation of the structure of matter in terms of different combinations of very small particles,* are given by the following postulates:

1. All matter is composed of indivisible atoms. An **atom** is *an extremely small particle of matter that retains its identity during chemical reactions.* (See Figure 2.2.)

2. An **element** is *a type of matter composed of only one kind of atom,* each atom of a given kind having the same properties. Mass is one such property. Thus, the atoms of a given element have a characteristic mass. (We will need to revise this definition of element in Section 2.3 so that it is stated in modern terms.) ►

3. A **compound** is *a type of matter composed of atoms of two or more elements chemically combined in fixed proportions.* The relative numbers of any two kinds of atoms in a compound occur in simple ratios. Water, for example, a compound of the elements hydrogen and oxygen, consists of hydrogen and oxygen atoms in the ratio of 2 to 1.

4. A **chemical reaction** consists of *the rearrangement of the atoms present in the reacting substances to give new chemical combinations present in the substances formed by the reaction.* Atoms are not created, destroyed, or broken into smaller particles by any chemical reaction.

Today we know that atoms are not truly indivisible; they are themselves made up of particles. Nevertheless, Dalton's postulates are essentially correct.

Atomic Symbols and Models

It is convenient to use symbols for the atoms of the different elements. An **atomic symbol** is *a one- or two-letter notation used to represent an atom corresponding to a particular element.* ► Typically, the atomic symbol consists of the first letter, capitalized, from the name of the element, sometimes with an additional letter from the name in lowercase. For example, chlorine has the symbol Cl. Other symbols are derived from a name in another language (usually Latin). Sodium is given the symbol Na from its Latin name, *natrium.* Symbols of selected elements are listed in Table 2.1.

Early in the development of his atomic theory, Dalton built spheres to represent atoms and used combinations of these spheres to represent compounds. Chemists continue to use this idea of representing atoms by three-dimensional models, but because we now know so much more about atoms and molecules than Dalton did, these models have become more refined. Section 2.6 describes these models in more detail.

As you will see in Section 2.4, it is the average mass of an atom that is characteristic of each element on earth.

The atomic names (temporary) given to newly discovered elements are derived from the atomic number (discussed in Section 2.3) using the numerical roots: nil (0), un (1), bi (2), tri (3), quad (4), pent (5), hex (6), sept (7), oct (8), and enn (9) with the ending -ium. For example, the element with atomic number 116 is ununhexium; its atomic symbol is Uuh.

Table 2.1	Some Common Elements	
Name of Element	**Atomic Symbol**	**Physical Appearance of Element***
Aluminum	Al	Silvery-white metal
Barium	Ba	Silvery-white metal
Bromine	Br	Reddish-brown liquid
Calcium	Ca	Silvery-white metal
Carbon	C	
Graphite		Soft, black solid
Diamond		Hard, colorless crystal
Chlorine	Cl	Greenish-yellow gas
Chromium	Cr	Silvery-white metal
Cobalt	Co	Silvery-white metal
Copper	Cu (from *cuprum*)	Reddish metal
Fluorine	F	Pale yellow gas
Gold	Au (from *aurum*)	Pale yellow metal
Helium	He	Colorless gas
Hydrogen	H	Colorless gas
Iodine	I	Bluish-black solid
Iron	Fe (from *ferrum*)	Silvery-white metal
Lead	Pb (from *plumbum*)	Bluish-white metal
Magnesium	Mg	Silvery-white metal
Manganese	Mn	Gray-white metal
Mercury	Hg (from *hydrargyrum*)	Silvery-white liquid metal
Neon	Ne	Colorless gas
Nickel	Ni	Silvery-white metal
Nitrogen	N	Colorless gas
Oxygen	O	Colorless gas
Phosphorus (white)	P	Yellowish-white, waxy solid
Potassium	K (from *kalium*)	Soft, silvery-white metal
Silicon	Si	Gray, lustrous solid
Silver	Ag (from *argentum*)	Silvery-white metal
Sodium	Na (from *natrium*)	Soft, silvery-white metal
Sulfur	S	Yellow solid
Tin	Sn (from *stannum*)	Silvery-white metal
Zinc	Zn	Bluish-white metal

*Common form of the element under normal conditions.

Deductions from Dalton's Atomic Theory

Note how atomic theory explains the difference between an element and a compound. Atomic theory also explains two laws we considered earlier. One of these is the law of conservation of mass, which states that the total mass remains constant during a chemical reaction. By postulate 2, every atom has a definite mass. Because a chemical reaction only rearranges the chemical combinations of atoms (postulate 4), the mass must remain constant. The other law explained by atomic theory is the law of definite proportions (constant composition). Postulate 3 defines a compound as a type of matter containing the atoms of two or more elements in definite proportions. Because the atoms have definite mass, compounds must have the elements in definite proportions by mass.

A good theory should not only explain known facts and laws but also predict new ones. The **law of multiple proportions,** deduced by Dalton from his atomic theory, states that *when two elements form more than one compound, the masses of one element in these compounds for a fixed mass of the other element are in ratios of small whole numbers.* To illustrate this law, consider the following. If we take a fixed mass of carbon, 1.000 gram, and react it with oxygen, we end up with two compounds: one that contains 1.3321 grams of oxygen for each 1.000 gram of carbon, and one that contains 2.6642 grams of oxygen per 1.000 gram of carbon. Note that the ratio of the amounts of oxygen in the two compounds is 2 to 1 (2.6642 ÷ 1.3321), which indicates that there must be twice as much oxygen in the second compound. Applying atomic theory, if we assume that the compound that has 1.3321 grams of oxygen to 1.000 gram of carbon is CO (the combination of one carbon atom and one oxygen atom), then the compound that contains twice as much oxygen per 1.000 gram of carbon must be CO_2 (the combination of one carbon atom and two oxygen atoms). The deduction of the law of multiple proportions from atomic theory was important in convincing chemists of the validity of the theory.

CONCEPT CHECK 2.1

Like Dalton, chemists continue to model atoms using spheres. Modern models are usually drawn with a computer and use different colors to represent atoms of different elements. Which of the models below represents CO_2?

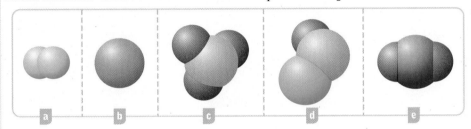

a   b   c   d   e

2.2 The Structure of the Atom

Although Dalton had postulated that atoms were indivisible particles, experiments conducted around the beginning of the last century showed that atoms themselves consist of particles. These experiments showed that an atom consists of two kinds of particles: a **nucleus,** *the atom's central core, which is positively charged and contains most of the atom's mass,* and one or more electrons. An **electron** is *a very light, negatively charged particle that exists in the region around the atom's positively charged nucleus.*

Discovery of the Electron

In 1897 the British physicist J. J. Thomson (Figure 2.3) conducted a series of experiments that showed that atoms were not indivisible particles. Figure 2.4 shows an experimental apparatus similar to the one used by Thomson. In this apparatus, two electrodes from a high-voltage source are sealed into a glass tube from which the air has been evacuated. The negative electrode is called the *cathode;* the positive one, the *anode.* When the high-voltage current is turned on, the glass tube emits a greenish light. Experiments showed that this greenish light is caused by the interaction of the glass with *cathode rays,* which are rays that originate from the cathode. ▶

After the cathode rays leave the negative electrode, they move toward the anode, where some rays pass through a hole to form a beam (Figure 2.4). This beam bends away from the negatively charged plate and toward the positively charged plate. (Cathode rays are not directly visible, but do cause certain materials such as zinc sulfide to glow so you can observe them.) Figure 2.5 shows a similar experiment, in which cathode rays are seen to bend when a magnet is brought toward them. Thomson showed that the characteristics of cathode rays are independent of the material making up the cathode. From such evidence, he concluded that a cathode ray consists of a beam of negatively charged particles (or electrons) and that electrons are constituents of all matter.

From his experiments, Thomson could also calculate the ratio of the electron's mass, m_e, to its electric charge, e. He could not obtain either the mass or the charge

Figure 2.3 ▲

Joseph John Thomson (1856–1940) J. J. Thomson's scientific ability was recognized early with his appointment as professor of physics in the Cavendish Laboratory at Cambridge University when he was not quite 28 years old. Soon after this appointment, Thomson began research on the discharge of electricity through gases. This work culminated in 1897 with the discovery of the electron. Thomson was awarded the Nobel Prize in physics in 1906.

A television tube uses the deflection of cathode rays by magnetic fields. A beam of cathode rays is directed toward a coated screen on the front of the tube, where by varying the magnetism generated by electromagnetic coils, the beam traces a luminescent image. Television tubes are becoming much less common with the advent of flat screen displays. Flat screen displays use a much different technology.

© The Cavendish Laboratory

Figure 2.4 ▶

Formation of cathode rays Cathode rays leave the cathode, or negative electrode, and are accelerated toward the anode, or positive electrode. Some of the rays pass through the hole in the anode to form a beam, which then is bent by the electric plates in the tube.

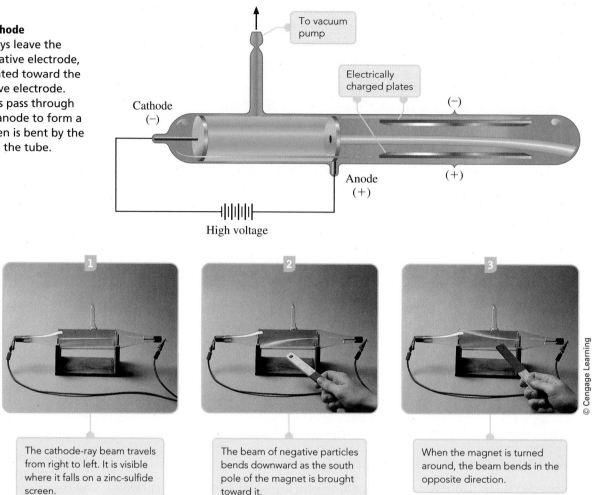

Figure 2.5 ▲

Bending of cathode rays by a magnet

separately, however. In 1909 the U.S. physicist Robert Millikan performed a series of ingenious experiments in which he obtained the charge on the electron by observing how a charged drop of oil falls in the presence and in the absence of an electric field (Figure 2.6). From this type of experiment, the charge on the electron is found to be 1.602×10^{-19} coulombs (the *coulomb,* abbreviated C, is a unit of electric charge). If you use this charge with the most recent value of the mass-to-charge ratio of the electron, you obtain an electron mass of 9.109×10^{-31} kg, which is more than 1800 times smaller than the mass of the lightest atom (hydrogen). This shows quite clearly that the electron is indeed a subatomic particle.

The Nuclear Model of the Atom

Ernest Rutherford (1871–1937), a British physicist, put forth the idea of the *nuclear model* of the atom in 1911, based on experiments done in his laboratory by Hans Geiger and Ernest Marsden. These scientists observed the effect of bombarding thin gold foil (and other metal foils) with alpha radiation from radioactive substances such as uranium (Figure 2.7). Rutherford had already shown that alpha rays consist of positively charged particles. Geiger and Marsden found that most of the alpha particles passed through a metal foil as though nothing were there, but a few (about 1 in 8000) were scattered at large angles and sometimes almost backward.

According to Rutherford's model, most of the mass of the atom (99.95% or more) is concentrated in a positively charged center, or nucleus, around which the negatively charged electrons move. Although most of the mass of an atom is in its nucleus, the nucleus occupies only a very small portion of the space of the atom. Nuclei have diameters of about 10^{-15} m (10^{-3} pm), whereas atomic diameters are about 10^{-10} m (100 pm), a hundred thousand times larger. If you were to use a golf ball to represent the nucleus, the atom would be about 3 miles in diameter!

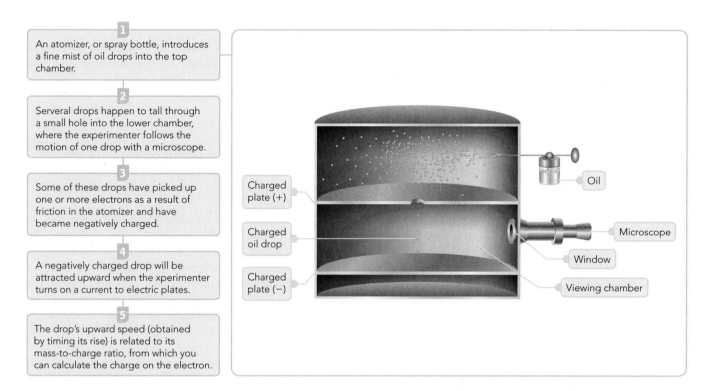

1. An atomizer, or spray bottle, introduces a fine mist of oil drops into the top chamber.

2. Serveral drops happen to fall through a small hole into the lower chamber, where the experimenter follows the motion of one drop with a microscope.

3. Some of these drops have picked up one or more electrons as a result of friction in the atomizer and have became negatively charged.

4. A negatively charged drop will be attracted upward when the xperimenter turns on a current to electric plates.

5. The drop's upward speed (obtained by timing its rise) is related to its mass-to-charge ratio, from which you can calculate the charge on the electron.

Charged plate (+)

Charged oil drop

Charged plate (−)

Oil

Microscope

Window

Viewing chamber

The nuclear model easily explains the results of bombarding gold and other metal foils with alpha particles. Alpha particles are much lighter than gold atoms. (Alpha particles are helium nuclei.) Most of the alpha particles pass through the metal atoms of the foil, undeflected by the lightweight electrons. When an alpha particle does happen to hit a metal-atom nucleus, however, it is scattered at a wide angle because it is deflected by the massive, positively charged nucleus (Figure 2.8).

Figure 2.6 ▲

Millikan's oil-drop experiment

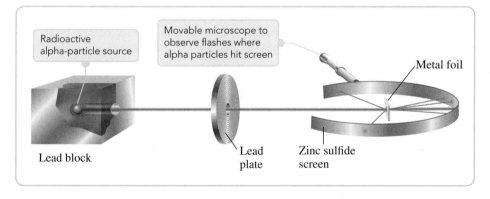

Radioactive alpha-particle source

Movable microscope to observe flashes where alpha particles hit screen

Metal foil

Lead block

Lead plate

Zinc sulfide screen

Figure 2.7 ◄

Alpha-particle scattering from metal foils Alpha radiation is produced by a radioactive source and formed into a beam by a lead plate with a hole in it. (Lead absorbs the radiation.) Scattered alpha particles are made visible by a zinc sulfide screen, which emits tiny flashes where particles strike it. A movable microscope is used for viewing the flashes.

Some alpha particles are scattered

Beam of alpha particles

Nucleus

Electrons

Most alpha particles pass straight through

Figure 2.8 ◄

Representation of the scattering of alpha particles by a gold foil Most of the alpha particles pass through the foil barely deflected. A few, however, collide with gold nuclei and are deflected at large angles. (The relative sizes of nuclei are smaller than can be drawn here.)

What would be a feasible model for the atom if Geiger and Marsden had found that 7999 out of 8000 alpha particles were deflected back at the alpha-particle source?

2.3 Nuclear Structure; Isotopes

The nucleus of an atom also has a structure; the nucleus is composed of two different kinds of particles, protons and neutrons. The type of alpha-particle scattering experiment that led to the nuclear model of the atom was also instrumental in clarifying this structure of the nucleus.

An important property of the nucleus is its electric charge. One way to determine the value of the positive charge of a nucleus is by analyzing the distribution of alpha particles scattered from a metal foil; other experiments also provide the nuclear charge. ◄ From such experiments, researchers discovered that each element has a unique nuclear charge that is an integer multiple of the magnitude of the electron charge. This integer, which is characteristic of an element, is called the *atomic number* (Z). A hydrogen atom nucleus, whose magnitude of charge equals that of the electron, has the smallest atomic number, which is 1.

The simplest way to obtain the nuclear charge is by analyzing the x rays emitted by the element when irradiated with cathode rays. This is discussed in the essay at the end of Section 8.2.

In 1919, Rutherford discovered that hydrogen nuclei, or what we now call protons, form when alpha particles strike some of the lighter elements, such as nitrogen. A **proton** is *a nuclear particle having a positive charge equal in magnitude to that of the electron and a mass more than 1800 times that of the electron.* When an alpha particle collides with a nitrogen atom, a proton may be knocked out of the nitrogen nucleus. The protons in a nucleus give the nucleus its positive charge.

The **atomic number** (Z) is therefore *the number of protons in the nucleus of an atom.* Because you can determine the atomic number experimentally, you can determine unambiguously whether or not a sample is a pure element or if perhaps you have discovered a new element. We can now state the definition of an element with more precision than we could in Section 2.1. An **element** is *a substance whose atoms all have the same atomic number.* The inside back cover of the book lists the elements and their atomic numbers (see the Table of Atomic Numbers and Atomic Masses).

The neutron was also discovered by alpha-particle scattering experiments. When beryllium metal is irradiated with alpha rays, a strongly penetrating radiation is obtained from the metal. In 1932 the British physicist James Chadwick (1891–1974) showed that this penetrating radiation consists of neutral particles, or neutrons. The **neutron** is *a nuclear particle having a mass almost identical to that of the proton but no electric charge.* When beryllium nuclei are struck by alpha particles, neutrons are knocked out. Table 2.2 compares the masses and charges of the electron and the two nuclear particles, the proton and the neutron.

Now consider the nucleus of some atom, say of the naturally occurring sodium atom. The nucleus contains 11 protons and 12 neutrons. Thus, the charge on the

Table 2.2	Properties of the Electron, Proton, and Neutron			
Particle	Mass (kg)	Charge (C)	Mass (amu)*	Charge (e)
Electron	9.10938×10^{-31}	-1.60218×10^{-19}	0.00055	-1
Proton	1.67262×10^{-27}	$+1.60218 \times 10^{-19}$	1.00728	$+1$
Neutron	1.67493×10^{-27}	0	1.00866	0

*The atomic mass unit (amu) equals 1.66054×10^{-27} kg; it is defined in Section 2.4.

sodium nucleus is +11e, which we usually write as simply +11, meaning +11 units of electron charge e. Similarly, the nucleus of a naturally occurring aluminum atom contains 13 protons and 14 neutrons; the charge on the nucleus is +13.

We characterize a nucleus by its atomic number (Z) and its mass number (A). The **mass number (A)** is *the total number of protons and neutrons in a nucleus.* The nucleus of the naturally occurring sodium atom has an atomic number of 11 and a mass number of 23 (11 + 12).

A **nuclide** is *an atom characterized by an atomic number and mass number.* The shorthand notation for any nuclide consists of the symbol of the element with the atomic number written as a subscript on the left and the mass number as a superscript on the left. You write the *nuclide symbol* for the naturally occurring sodium nuclide as follows:

$$\text{Mass number} \longrightarrow {}^{23}_{11}\text{Na} \longleftarrow \text{Atomic number}$$

An atom is normally electrically neutral, so it has as many electrons about its nucleus as the nucleus has protons; that is, the number of electrons in a neutral atom equals its atomic number. A sodium atom has a nucleus of charge +11, and around this nucleus are 11 electrons (with a charge of −11, giving the atom a charge of 0).

All nuclei of the atoms of a particular element have the same atomic number, but the nuclei may have different mass numbers. **Isotopes** are *atoms whose nuclei have the same atomic number but different mass numbers; that is, the nuclei have the same number of protons but different numbers of neutrons.* Sodium has only one naturally occurring isotope, which we denote by the same symbol we used for the nuclide (${}^{23}_{11}\text{Na}$); we also call the isotope sodium-23. Naturally occurring oxygen is a mixture of isotopes; it contains 99.759% oxygen-16, 0.037% oxygen-17, and 0.204% oxygen-18. ▶

Figure 2.9 may help you to visualize the relationship among the different subatomic particles in the isotopes of an element. The size of the nucleus as represented in the figure is very much exaggerated in order to show the numbers of protons and neutrons. In fact, the space taken up by the nucleus is minuscule compared with the region occupied by electrons. Electrons, although extremely light particles, move throughout relatively large diffuse regions, or "shells," about the nucleus of an atom.

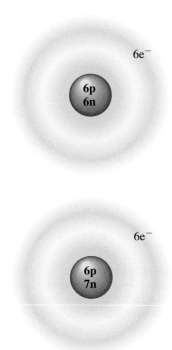

Figure 2.9 ▲

A representation of two isotopes of carbon The drawing shows the basic particles making up the carbon-12 and carbon-13 isotopes. (The relative sizes of the nuclei are much exaggerated here.)

Isotopes were first suspected in about 1912 when chemically identical elements with different atomic masses were found in radioactive materials. The most convincing evidence, however, came from the mass spectrometer, discussed in Section 2.4.

| **Example** 2.1 | **Writing Nuclide Symbols** |

What is the nuclide symbol for a nucleus that contains 38 protons and 50 neutrons?

Problem Strategy To solve this problem, we need to keep in mind that the number of protons in the nucleus (the atomic number) is what uniquely identifies an element and that the mass number is the sum of the protons and neutrons.

Solution From the Table of Atomic Numbers and Atomic Masses on the inside back cover of this book, you will note that the element with atomic number 38 is strontium, symbol Sr. The mass number is 38 + 50 = 88. The symbol is ${}^{88}_{38}\text{Sr}$.

Answer Check When writing elemental symbols, as you have done here, make sure that you always capitalize *only* the first letter of the symbol.

Exercise 2.1 A nucleus consists of 17 protons and 18 neutrons. What is its nuclide symbol?

■ See Problems 2.47 and 2.48.

2.4 Atomic Masses

A central feature of Dalton's atomic theory was the idea that an atom of an element has a characteristic mass. Now we know that a naturally occurring element may be a mixture of isotopes, each isotope having its own characteristic mass. However, the percentages of the different isotopes in most naturally occurring elements have remained essentially constant over time and in most cases are independent of the origin of the element. Thus, what Dalton actually calculated were *average* atomic masses (actually average *relative* masses, as we will discuss). Since you normally work with naturally occurring mixtures of elements (whether in pure form or as compounds), it is indeed these average atomic masses that you need in ordinary chemical work. ◄

Some variation of isotopic composition occurs in a number of elements, and this limits the significant figures in their average atomic masses.

Relative Atomic Masses

Dalton could not weigh individual atoms. What he could do was find the average mass of one atom relative to the average mass of another. We will refer to these relative atomic masses as *atomic masses.* To see how Dalton obtained his atomic masses, imagine that you burn hydrogen gas in oxygen. The product of this reaction is water, a compound of hydrogen and oxygen. By experiment, you find that 1.0000 gram of hydrogen reacts with 7.9367 g of oxygen to form water. To obtain the atomic mass of oxygen (relative to hydrogen), you need to know the relative numbers of hydrogen atoms and oxygen atoms in water. As you might imagine, Dalton had difficulty with this. Today we know that water contains two atoms of hydrogen for every atom of oxygen. So the atomic mass of oxygen is $2 \times 7.9367 = 15.873$ times that of the mass of the average hydrogen atom. ◄

Dalton had assumed the formula for water to be HO. Using this formula and accurate data, he would have obtained 7.9367 for the relative atomic weight of oxygen.

© G. Tompkinson

Figure 2.10 ▲

A mass spectrometer This instrument measures the masses of atoms and molecules.

Atomic Mass Units

Dalton's hydrogen-based atomic mass scale was eventually replaced by a scale based on oxygen and then, in 1961, by the present carbon-12 mass scale. This scale depends on measurements of atomic mass by an instrument called a *mass spectrometer* (Figure 2.10), which we will describe briefly later in this section. You make accurate measurements of mass on this instrument by comparing the mass of an atom to the mass of a particular atom chosen as a standard. On the present atomic mass scale, the carbon-12 isotope is chosen as the standard and is arbitrarily assigned a mass of exactly 12 atomic mass units. One **atomic mass unit (amu)** is, therefore, *a mass unit equal to exactly one-twelfth the mass of a carbon-12 atom.*

On this modern scale, the **atomic mass** of an element is *the average atomic mass for the naturally occurring element, expressed in atomic mass units.* A complete table of atomic masses appears on the inside back cover of this book. (Atomic mass is sometimes referred to as atomic weight. Although the use of the term *weight* in *atomic weight* is not strictly correct, it is sanctioned by convention to mean "average atomic mass.")

Mass Spectrometry and Atomic Masses

Earlier we noted that you use a mass spectrometer to obtain accurate masses of atoms. The earliest mass spectrometers measured the mass-to-charge ratios of positively charged atoms using the same ideas that Thomson used to study electrons. Modern instruments may use other techniques, but all measure the mass-to-charge ratios of positively charged atoms (and molecules as well). Figure 2.11 shows a simplified sketch of a mass spectrometer running a neon sample (see the essay on page 98 for a discussion of modern mass spectrometry).

Mass spectrometers produce a *mass spectrum,* which shows the relative numbers of atoms for various masses. If a sample of neon, which has the three isotopes depicted in Figure 2.12, is introduced into a mass spectrometer, it produces the mass spectrum shown in Figure 2.13. This mass spectrum gives us the information we need to calculate the atomic mass of neon: all of the isotopic masses (neon-20, neon-21, and neon-22) and the relative number, or fractional abundance, of each isotope. The **fractional abundance** of an isotope is *the fraction*

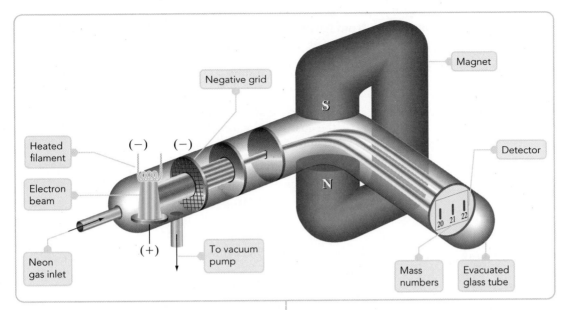

Labels (clockwise): Negative grid · Magnet · S · N · Heated filament · (−) · (−) · Electron beam · Detector · Neon gas inlet · (+) · To vacuum pump · Mass numbers · Evacuated glass tube · 20 21 22

1 Neon gas enters an evacuated chamber, where neon atoms from positive ions when they collide with electrons.

2 Positively charged neon atoms, Ne$^+$, are accelerated from this region by the negative grid and pass between the poles of a magnet.

3 The beam of positively charged atoms is split into three beams by the magnetic feild according to the mass-to-charge ratios.

4 The three beams then travel to a detector at a end of the tube. (The detector is shown here as a photographic plate; in modern spectrometers, the detector is electronic, and the mass positions are recorded on a computer.)

Figure 2.11 ▲

Diagram of a simple mass spectrometer, showing the separation of neon isotopes

of the total number of atoms that is composed of a particular isotope. The fractional abundances of the neon isotopes in naturally occurring neon are neon-20, 0.9051; neon-21, 0.0027; and neon-22, 0.0922.

You calculate the atomic mass of an element by multiplying each isotopic mass by its fractional abundance and summing the values. If you do that for neon using the data given here, you will obtain 20.179 amu. The next example illustrates the calculation in full for chromium.

Figure 2.12 ▼

Representations of the three naturally occurring isotopes of Ne *Top to bottom:* Ne-20, Ne-21, and Ne-22. Ne-20 is the most abundant isotope. Representations are not drawn to scale.

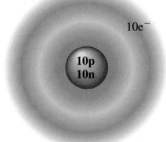

10e$^-$

10p
10n

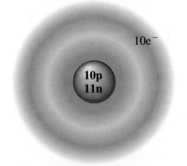

10e$^-$

10p
11n

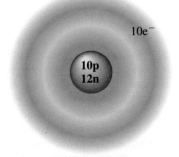

10e$^-$

10p
12n

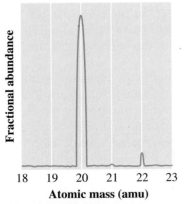

Fractional abundance

18 19 20 21 22 23

Atomic mass (amu)

Figure 2.13 ◄

The mass spectrum of neon Neon is separated into its isotopes Ne-20, Ne-21, and Ne-22. The height at each mass peak is proportional to the fraction of that isotope in the element.

Example 2.2   Determining Atomic Mass from Isotopic Masses and Fractional Abundances

Critical Concept 2.2
The atomic masses of the elements are calculated. Typically each element has more than one naturally occurring isotope. Therefore, the mass of each element must be calculated based on the mass and relative abundance of each isotope; it is the average atomic mass.

Solution Essentials:
• Isotope
• Atomic mass unit
• Fractional abundance
• Rules for significant figures and rounding

Chromium, Cr, has the following isotopic masses and fractional abundances:

Mass Number	Isotopic Mass (amu)	Fractional Abundance
50	49.9461	0.0435
52	51.9405	0.8379
53	52.9407	0.0950
54	53.9389	0.0236

What is the atomic mass of chromium?

Problem Strategy The type of average used to calculate the atomic mass of an element is similar to the "average" an instructor might use to obtain a student's final grade in a class. Suppose the student has a total exam grade of 76 and a total laboratory grade of 84. The instructor decides to give a weight of 70% to the exams and 30% to the laboratory. How would the instructor calculate the final grade? He or she would multiply each type of grade by its weight factor and add the results:

$$(76 \times 0.70) + (84 \times 0.30) = 78$$

The final grade is closer to the exam grade, because the instructor chose to give the exam grade greater weight.

Solution Multiply each isotopic mass by its fractional abundance, then sum:

$$49.9461 \text{ amu} \times 0.0435 = 2.17 \text{ amu}$$
$$51.9405 \text{ amu} \times 0.8379 = 43.52 \text{ amu}$$
$$52.9407 \text{ amu} \times 0.0950 = 5.03 \text{ amu}$$
$$53.9389 \text{ amu} \times 0.0236 = \underline{1.27 \text{ amu}}$$
$$51.99 \text{ amu}$$

The atomic mass of chromium is **51.99 amu.**

Answer Check The average mass (atomic mass) should be near the mass of the isotope with greatest abundance: in this case, 51.9405 amu with fractional abundance of 0.8379. This provides a quick check on your answer to this type of problem; any "answer" that is far from this will be incorrect.

Exercise 2.2 Chlorine consists of the following isotopes:

Isotope	Isotopic Mass (amu)	Fractional Abundance
Chlorine-35	34.96885	0.75771
Chlorine-37	36.96590	0.24229

What is the atomic mass of chlorine?

■ See Problems 2.51, 2.52, 2.53, and 2.54.

2.5 Periodic Table of the Elements

In 1869 the Russian chemist Dmitri Mendeleev (1834–1907) (Figure 2.14) and the German chemist J. Lothar Meyer (1830–1895), working independently, made similar discoveries. They found that when they arranged the elements in order of atomic mass, they could place them in horizontal rows, one row under the other, so that the elements in each vertical column have similar properties. *A tabular arrangement of elements in rows and columns, highlighting the regular repetition of properties of the elements,* is called a **periodic table.**

Eventually, more accurate determinations of atomic masses revealed discrepancies in this ordering of the elements. However, in the early part of the last century, it was shown that the elements are characterized by their atomic numbers, rather than atomic masses. When the elements in the periodic table are ordered by atomic number, such discrepancies vanish.

A modern version of the periodic table, with the elements arranged by atomic number, is shown in Figure 2.15 (see also inside front cover). Each entry lists the atomic number, atomic symbol, and atomic mass of an element. This is a convenient way of tabulating such information, and you should become familiar with using the periodic table for that purpose. As we develop the subject matter of chemistry throughout the text, you will see how useful the periodic table is.

© Edgar Fahs Smith Collection, University of Pennsylvania

Figure 2.14 ▲

Dmitri Ivanovich Mendeleev (1834–1907) Mendeleev constructed a periodic table as part of his effort to organize chemistry. He received many international honors for his work, but his reception in czarist Russia was mixed. He had pushed for political reforms and made many enemies as a result.

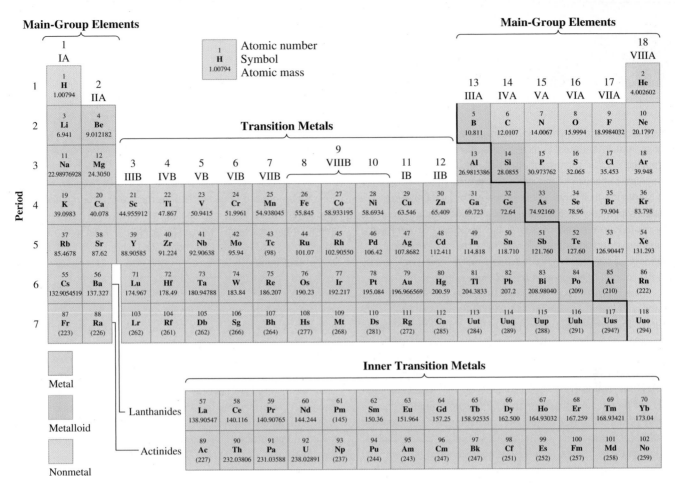

Figure 2.15 ▲

A modern form of the periodic table This table is also given on the inside front cover of the book.

Periods and Groups

The basic structure of the periodic table is its division into rows and columns, or periods and groups. A **period** consists of *the elements in any one horizontal row of the periodic table.* A **group** consists of *the elements in any one column of the periodic table.*

The first period of elements consists of only hydrogen (H) and helium (He). The second period has 8 elements, beginning with lithium (Li) and ending with neon (Ne). There is then another period of 8 elements, and this is followed by a period having 18 elements, beginning with potassium (K) and ending with krypton (Kr). The fifth period also has 18 elements. The sixth period actually consists of 32 elements, but in order for the row to fit on a page, part of it appears at the bottom of the table. Otherwise the table would have to be expanded, with the additional elements placed after barium (Ba, atomic number 56). The seventh period, though not complete, also has some of its elements placed as a row at the bottom of the table.

The groups are usually numbered. The numbering frequently seen in North America labels the groups with Roman numerals and A's and B's. In Europe a similar convention has been used, but some columns have the A's and B's interchanged. To eliminate this confusion, the International Union of Pure and Applied Chemistry (IUPAC) suggested a convention in which the columns are numbered 1 to 18. Figure 2.15 shows the traditional North American and the IUPAC conventions. When we refer to an element by its periodic group, we will use the traditional North American convention. The A groups are called *main-group* (or *representative*) *elements;* the B groups are called *transition elements.* The two rows of elements at the bottom of the table are called *inner transition elements* (the first row is referred to as the *lanthanides;* the second row, as the *actinides*).

Thirty Seconds on the Island of Stability

It appears that scientists have now been on the "Island of Stability"—for at least 30 seconds. In recent years, the island of stability has attracted nuclear scientists the way Mount Everest has attracted mountain climbers. Like mountain climbers, scientists have wanted to surmount their own tallest peak—discovery of the relatively stable, superheavy nuclide at the peak of stability that is surrounded by foothills composed of less stable nuclides. Discovering the superheavy nuclide at the peak would also mean discovering a new element—a task that has had its allure since Lavoisier gave currency to the concept of elements in the late eighteenth century.

Theoretical physicists predicted the existence of the island of stability, centered at element 114 with mass number 298, in the 1960s. The term *stability* refers here to that of the atomic nucleus. An unstable nucleus tends to fall apart by radioactive decay—a piece spontaneously flies off the nucleus, leaving a different one behind. Approximately 275 nuclides are completely stable, or non-radioactive. All of these nuclides have atomic numbers (or numbers of protons) no greater than 83 (the atomic number for the element bismuth). Beyond bismuth, all elements are radioactive and become increasingly unstable. In fact, none of the original, primeval elements past uranium (element 92)—the *transuranium*

elements—exists any longer; they have long since vanished by radioactive decay. Scientists have made transuranium elements in the laboratory, however.

In February 1996, for example, scientists at Darmstadt, Germany, produced a few atoms of element 112 by bombarding a lead target with zinc atoms. Each atom of element 112 lasted only about 240 microseconds. We can represent the result by a nuclear equation, which is similar to a chemical equation (described at the end of the chapter).

$$^{70}_{30}\text{Zn} + ^{208}_{82}\text{Pb} \longrightarrow ^{277}_{112}\text{Uub} + ^{1}_{0}\text{n}$$

We introduced the nuclide symbols used here in Section 2.3. The equation is a summary of the following. A high-speed zinc-70 nucleus (as an ion, which is an electrically charged atom) smashes into a lead-208 target where, as the arrow denotes, it yields new products, a nucleus of element 112 (mass number 277) and a neutron, denoted $^{1}_{0}\text{n}$. Element 112 has the provisional name of ununbium until several other laboratories can reproduce the work, after which the discoverer may name it. Figure 2.16 shows the ion accelerator (UNILAC) that produced the zinc ions for this experiment.

As noted earlier, the elements in any one group have similar properties. For example, the elements in Group IA, often known as the *alkali metals,* are soft metals that react easily with water. (Hydrogen, a gas, is an exception and might better be put in a group by itself.) Sodium is an alkali metal. So is potassium. The Group VIIA elements, known as *halogens,* are also reactive elements. Chlorine is a halogen. We have already noted its vigorous reaction with sodium. Bromine, which is a red-brown liquid, is another halogen. It too reacts vigorously with sodium.

Metals, Nonmetals, and Metalloids

The elements of the periodic table in Figure 2.15 are divided by a heavy "staircase" line into metals on the left and nonmetals on the right. A **metal** is *a substance that has a characteristic luster, or shine, and is generally a good conductor of heat and electricity.* Except for mercury, the metallic elements are solids at room temperature (about 20°C). They are more or less *malleable* (can be hammered into sheets) and *ductile* (can be drawn into wire).

A **nonmetal** is *an element that does not exhibit the characteristics of a metal.* Most of the nonmetals are gases (for example, chlorine and oxygen) or solids (for example, phosphorus and sulfur). The solid nonmetals are usually hard, brittle substances. Bromine is the only liquid nonmetal.

Most of the elements bordering the staircase line in the periodic table (Figure 2.15) are metalloids, or semimetals. A **metalloid,** or **semimetal,** is *an element having both metallic and nonmetallic properties.* These elements, such as silicon (Si) and germanium (Ge), are usually good *semiconductors*—elements that, when pure, are poor conductors of electricity at room temperature but become moderately good conductors at higher temperatures. ◀

When these pure semiconductor elements have small amounts of certain other elements added to them (a process called doping), they become very good conductors of electricity. Semiconductors are the critical materials in solid-state electronic devices.

Scientists in Dubna, north of Moscow in Russia, and their American collaborators from Lawrence Livermore Laboratory in California produced one atom of element 114 (provisional name ununquadium) in December 1998, placing scientists on the shore of the island of stability. In this experiment, scientists bombarded a plutonium-244 target with calcium-48 atoms.

$$^{48}_{20}\text{Ca} + {}^{244}_{94}\text{Pu} \longrightarrow {}^{289}_{114}\text{Uuq} + 3{}^{1}_{0}\text{n}$$

The atom of element 114 lasted about 30 seconds. While a 30-second lifetime may seem brief, it is many times longer than that seen in the previously discovered element 112. In 2001 this same group of scientists reported the discovery of element 116 (ununhexium).

As a rather bizarre footnote to this story, the "discovery" of element 118 was reported in 1999 by scientists at Lawrence Berkeley National Laboratory in California. Two years later, these scientists reported that they were unable to reproduce their results. The laboratory fired one of the scientists, who they alleged had faked the data. Fortunately, fraud is rather unusual in chemistry; when it does occur, it is discovered, as in this case, when the work cannot be reproduced. Notably, element 118 was rediscovered in 2006 by the same Russian and American team who discovered elements 114 and 116, bringing scientists one step closer to the island of stability.

■ See Problems 2.131 and 2.132.

Figure 2.16 ▲

Within the UNILAC ion accelerator The accelerator consists of a series of electrodes arranged one after the other in a copper-plated steel container. Each electrode accelerates the ion. Eventually, the ion gains sufficient speed, and therefore energy, to cause its nucleus to fuse on collision with a target atom nucleus giving a new nucleus.

© A. Zschau/GSI

Exercise 2.3 By referring to the periodic table (Figure 2.15 or inside front cover), identify the group and period to which each of the following elements belongs. Then decide whether the element is a metal, nonmetal, or metalloid.

a Se   b Cs   c Fe   d Cu   e Br

■ See Problems 2.57 and 2.58.

CONCEPT CHECK 2.3

Consider the elements He, Ne, and Ar. Can you come up with a reason why they are in the same group in the periodic table?

Chemical Substances: Formulas and Names

Atomic theory has developed steadily since Dalton's time and has become the cornerstone of chemistry. It results in an enormous simplification: all of the millions of compounds we know today are composed of the atoms of just a few elements. Now we look more closely at how we describe the composition and structure of chemical substances in terms of atoms.

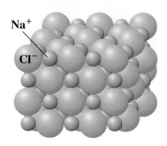

Figure 2.17 ▲

Sodium chloride model This model of a sodium chloride crystal illustrates the 1 : 1 packing of the Na^+ and Cl^- ions.

Another way to understand the large numbers of molecules involved in relatively small quantities of matter is to consider 1 g of water (about one-fifth teaspoon). It contains 3.3×10^{22} water molecules. If you had a penny for every molecule in this quantity of water, the height of your stack of pennies would be about 300 million times the distance from the earth to the sun.

2.6 Chemical Formulas; Molecular and Ionic Substances

The **chemical formula** of a substance is *a notation that uses atomic symbols with numerical subscripts to convey the relative proportions of atoms of the different elements in the substance.* Consider the formula of aluminum oxide, Al_2O_3. This means that the compound is composed of aluminum atoms and oxygen atoms in the ratio 2 : 3. Consider the formula for sodium chloride, NaCl. When no subscript is written for a symbol, it is assumed to be 1. Therefore, the formula NaCl means that the compound is composed of sodium atoms and chlorine atoms in the ratio 1 : 1 (Figure 2.17).

Additional information may be conveyed by different kinds of chemical formulas. To understand this, we need to look briefly at two main types of substances: molecular and ionic.

Molecular Substances

A **molecule** is *a definite group of atoms that are chemically bonded together—that is, tightly connected by attractive forces.* The nature of these strong forces is discussed in Chapters 9 and 10. A *molecular substance* is a substance that is composed of molecules, all of which are alike. The molecules in such a substance are so small that even extremely minute samples contain tremendous numbers of them. One billionth (10^{-9}) of a drop of water, for example, contains about 2 trillion (2×10^{12}) water molecules. ◄

A **molecular formula** *gives the exact number of different atoms of an element in a molecule.* The hydrogen peroxide molecule contains two hydrogen atoms and two oxygen atoms chemically bonded. Therefore, its molecular formula is H_2O_2. Other simple molecular substances are water, H_2O; ammonia, NH_3; carbon dioxide, CO_2; and ethanol (ethyl alcohol), C_2H_6O. In Chapter 3, you will see how we determine such formulas.

The atoms in a molecule are not simply piled together randomly. Rather, the atoms are chemically bonded in a definite way. A *structural formula* is a chemical formula that shows how the atoms are bonded to one another in a molecule. For example, it is known that each of the hydrogen atoms in the water molecule is bonded to the oxygen atom. Thus, the structural formula of water is H—O—H. A line joining two atomic symbols in such a formula represents the chemical bond connecting the atoms. Figure 2.18 shows some structural formulas. Structural formulas are sometimes condensed in writing. For example, the structural formula of ethanol may be written CH_3CH_2OH or C_2H_5OH, depending on the detail you want to convey.

The atoms in a molecule are not only connected in definite ways but exhibit definite spatial arrangements as well. Chemists often construct molecular models as an aid in visualizing the shapes and sizes of molecules. Figure 2.18 shows molecular models for several compounds. While the ball-and-stick type of model shows the bonds and bond angles clearly, the space-filling type gives a more realistic feeling of the space occupied by the atoms. Chemists also use computer-generated models of molecules, which can be produced in a variety of forms in addition to those shown here.

Some elements are molecular substances and are represented by molecular formulas. Chlorine, for example, is a molecular substance and has the formula Cl_2, each molecule being composed of two chlorine atoms bonded together. Sulfur consists of molecules composed of eight atoms; its molecular formula is S_8. Helium and neon are composed of isolated atoms; their formulas are He and Ne, respectively. Other elements, such as carbon (in the form of graphite or diamond), do not have a simple molecular structure but consist of a very large, indefinite number of atoms bonded together. These elements are represented simply by their atomic symbols. (An important exception is a form of carbon called buckminsterfullerene, which was discovered in 1985 and has the molecular formula C_{60}.) Models of some elementary substances are shown in Figure 2.19.

	WATER	**AMMONIA**	**ETHANOL**
Molecular formula	H_2O	NH_3	C_2H_6O

Structural formula	H–O–H	H–N–H
		    |
		    H
			H H
			| |
			H–C–C–O–H
			| |
			H H

Molecular model (ball-and-stick type)	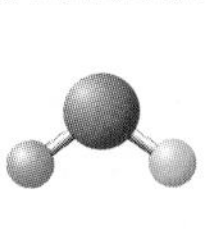	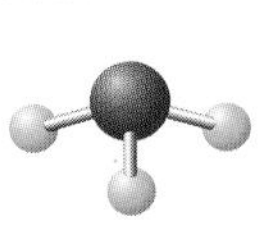	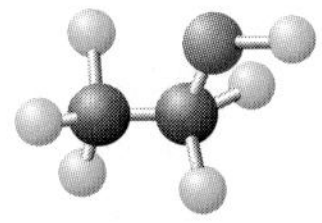
Molecular model (space-filling type)			
Electrostatic potential map	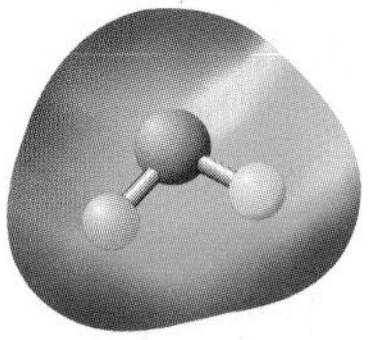		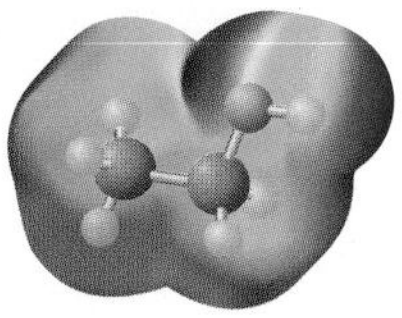

Figure 2.18 ◄

Examples of molecular and structural formulas, molecular models, and electrostatic potential maps Three common molecules—water, ammonia, and ethanol—are shown. The electrostatic potential map representation at the bottom of the figure illustrates the distribution of electrons in the molecule using a color spectrum. Colors range from red (relatively high electron density) all the way to blue (low electron density).

An important class of molecular substances is the polymers. **Polymers** are *very large molecules that are made up of a number of smaller molecules repeatedly linked together.* **Monomers** are *the small molecules that are linked together to form the polymer.* A good analogy for the formation of polymers from monomers is that of making a chain of paper clips. Say you have boxes of red, blue, and yellow paper clips. Each paper clip represents a monomer. You can make a large chain of paper clips, representing the polymer, with a variety of patterns and repeating units. One chain might involve a repeating pattern of two red, three yellow, and one blue. Another chain might consist of only blue paper clips. Just like the paper clips, the creation of a particular polymer in a laboratory involves controlling both the monomers that are being linked together and the linkage pattern.

Polymers are both natural and synthetic. Hard plastic soda bottles are made from chemically linking two different monomers in an alternating pattern. Wool and silk are natural polymers of amino acids linked by peptide bonds. Nylon® for fabrics, Kevlar® for bulletproof vests, and Nomex® for flame-retardant clothing all contain a CONH linkage. Plastics and rubbers are also polymers that are made from carbon- and hydrogen-containing monomers. Even the Teflon® coating on cookware is a polymer that is the result of linking CF_2CF_2 monomers (Figure 2.20). From these examples, it is obvious that polymers are very important molecular materials that we use every day in a wide variety of applications. For more on the chemistry of polymers, refer to Chapter 24.

Ionic Substances

Although many substances are molecular, others are composed of ions (pronounced "eye'-ons"). An **ion** is *an electrically charged particle obtained from an atom or chemically bonded group of atoms by adding or removing electrons.* Sodium chloride is a substance made up of ions.

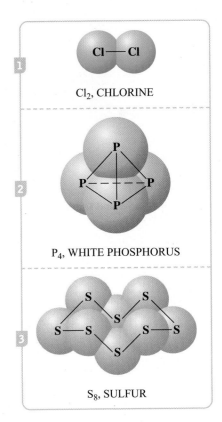

Figure 2.19 ▲

Molecular models of some elementary substances

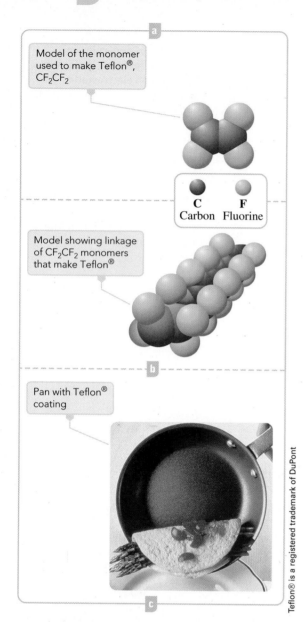

Figure 2.20 ▲

F₂C═CF₂ monomer and Teflon®

Although isolated atoms are normally electrically neutral and therefore contain equal numbers of positive and negative charges, during the formation of certain compounds atoms can become ions. Metal atoms tend to lose electrons, whereas nonmetals tend to gain electrons. When a metal atom such as sodium and a non-metal atom such as chlorine approach one another, an electron can transfer from the metal atom to the nonmetal atom to produce ions.

An atom that picks up an extra electron becomes *a negatively charged ion,* called an **anion** (pronounced "an'-ion"). An atom that loses an electron becomes *a positively charged ion,* called a **cation** ("cat'-ion"). A sodium atom, for example, can lose an electron to form a sodium cation (denoted Na^+). A chlorine atom can gain an electron to form a chloride anion (denoted Cl^-). A calcium atom can lose two electrons to form a calcium cation, denoted Ca^{2+}. Note that the positive-two charge on the ion is indicated by a superscript 2+.

Some ions consist of two or more atoms chemically bonded but having an excess or deficiency of electrons so that the unit has an electric charge. An example is the sulfate ion, SO_4^{2-}. The superscript 2− indicates an excess of two electrons on the group of atoms.

An **ionic compound** is *a compound composed of cations and anions.* Sodium chloride consists of equal numbers of sodium ions, Na^+, and chloride ions, Cl^-. The strong attraction between positive and negative charges holds the ions together in a regular arrangement in space. For example, in sodium chloride, each Na^+ ion is surrounded by six Cl^- ions, and each Cl^- ion is surrounded by six Na^+ ions. The result is a *crystal,* which is a kind of solid having a regular three-dimensional arrangement of atoms, molecules, or (as in the case of sodium chloride) ions. Figure 2.21 shows sodium chloride crystals and two types of models used to depict the arrangement of the ions in the crystal. The number of ions in an individual sodium chloride crystal determines the size of the crystal.

The formula of an ionic compound is written by giving the smallest possible integer number of different ions in the substance, except that the charges on the ions are omitted so that the formulas merely indicate the atoms involved. For example, sodium chloride contains equal numbers of Na^+ and Cl^- ions. The formula is written NaCl (not Na^+Cl^-). Iron(III) sulfate is a compound consisting of iron(III) ions, Fe^{3+}, and sulfate ions, SO_4^{2-}, in the ratio 2 : 3. The formula is written $Fe_2(SO_4)_3$, in which parentheses enclose the formula of an ion composed of more than one atom (again, omitting ion charges); parentheses are used only when there are two or more such ions.

Although ionic substances do not contain molecules, we can speak of the smallest unit of such a substance. The **formula unit** of a substance is *the group of atoms or ions explicitly symbolized in the formula.* For example, the formula unit of water, H_2O, is the H_2O molecule. The formula unit of iron(III) sulfate, $Fe_2(SO_4)_3$, consists of two Fe^{3+} ions and three SO_4^{2-} ions. The formula unit is the smallest neutral unit of such substances.

All substances, including ionic compounds, are electrically neutral. You can use this fact to obtain the formula of an ionic compound, given the formulas of the ions. This is illustrated in the following example.

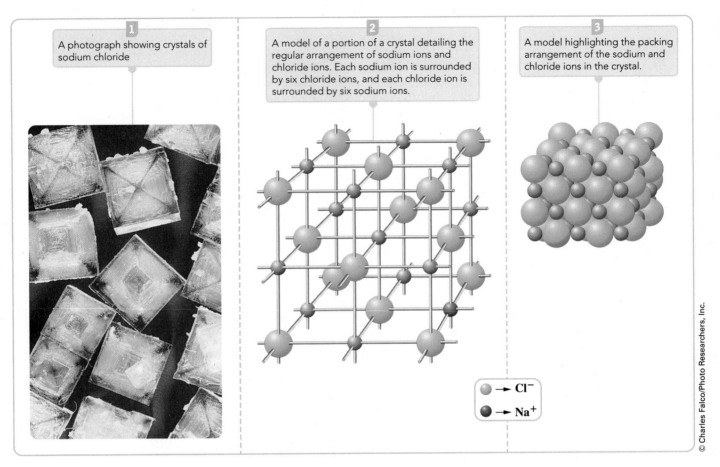

1 A photograph showing crystals of sodium chloride

2 A model of a portion of a crystal detailing the regular arrangement of sodium ions and chloride ions. Each sodium ion is surrounded by six chloride ions, and each chloride ion is surrounded by six sodium ions.

3 A model highlighting the packing arrangement of the sodium and chloride ions in the crystal.

○ → Cl⁻

● → Na⁺

© Charles Falco/Photo Researchers, Inc.

Figure 2.21 ▲

The sodium chloride crystal

| **Example** 2.3 | **Writing an Ionic Formula, Given the Ions** |

Gaining Mastery Toolbox

Critical Concept 2.3
Ionic compounds are formed by the combination of positive and negatively charged particles. When the charged particles (ions) combine, they combine in a ratio that yields a compound with an overall charge of zero. Ionic compounds are often combinations of metal and nonmetal atoms.

Solution Essentials:
• Compound
• Ion, anion, and cation
• Ionic compound
• Formula unit

a Chromium(III) oxide is used as a green paint pigment (Figure 2.22). It is a compound composed of Cr^{3+} and O^{2-} ions. What is the formula of chromium(III) oxide?

b Strontium oxide is a compound composed of Sr^{2+} and O^{2-} ions. Write the formula of this compound.

Problem Strategy Because a compound is neutral, the sum of positive and negative charges equals zero. Consider the ionic compound $CaCl_2$, which consists of one Ca^{2+} ion and two Cl^- ions. The sum of the charges is

$$\underbrace{1 \times (+2)}_{\text{Positive charge}} + \underbrace{2 \times (-1)}_{\text{Negative charge}} = 0$$

Note that the number of calcium ions in $CaCl_2$ equals the magnitude of charge on the chloride ion (1), whereas the number of chloride ions in $CaCl_2$ equals the magnitude of charge on the calcium ion (2). In general, you use the magnitude of charge on one ion to obtain the subscript for the other ion. You may need to simplify the formula you obtain this way so that it expresses the simplest ratio of ions.

© Cengage Learning

Figure 2.22 ▲

Chromium(III) oxide The compound is used as a green paint pigment.

(*continued*)

(*continued*)

Solution

a You can achieve electrical neutrality by taking as many cations as there are units of charge on the anion and as many anions as there are units of charge on the cation. The unit of charge on the O^{2-} anion is 2, and the unit of charge on the Cr^{3+} cation is 3. Therefore, we predict that the ratio of Cr^{3+} to O^{2-} is $2 : 3$. The correct formula of an ionic compound must have a net charge of zero when the charges of all the ions are combined; two Cr^{3+} ions have a total charge of 6^{+}, and three O^{2-} ions have a total charge of 6^{-}. The $2 : 3$ ratio of Cr^{3+} to O^{2-} is the simplest ratio, and the formula is $\mathbf{Cr_2O_3}$. The solution to this problem can be diagrammed in the following manner. However, keep in mind that one additional step of reducing the ratio might be required.

$$Cr^{③+} \quad O^{②-} \qquad \text{or} \qquad Cr_2O_3$$

b You see that equal numbers of Sr^{2+} and O^{2-} ions will give a neutral compound. Thus, the formula is SrO. If you use the units of charge to find the subscripts, you get

$$Sr^{②+} \quad O^{②-} \qquad \text{or} \qquad Sr_2O_2$$

The final formula is **SrO**, because this gives the simplest ratio of ions.

Answer Check When writing ionic formulas, always make certain that the formula that you write reflects the smallest whole-number ratio of the ions. For example, using the technique presented in this example, Pb^{4+} and O^{2-} combine to give Pb_2O_4; however, the correct formula is PbO_2, which reflects the simplest whole-number ratio of ions.

Exercise 2.4 Potassium chromate is an important compound of chromium (Figure 2.23). It is composed of K^{+} and $CrO_4{}^{2-}$ ions. Write the formula of the compound.

■ See Problems 2.75 and 2.76.

© Cengage Learning

Figure 2.23 ▲

Potassium chromate
Many compounds of chromium have bright colors, which is the origin of the name of the element. It comes from the Greek word *chroma*, meaning "color."

CONCEPT CHECK 2.4

Classify each of the following as either an ionic or molecular compound.

CH_4 Br_2 $CaCl_2$ KNO_3 CH_4O LiF

Which is the best statement regarding molecular compounds?

a Molecular compounds always contain at least two elements. For example, H_2O is a molecular compound because it contains two elements: hydrogen and oxygen.

b Molecular compounds usually contain carbon.

c Molecular compounds consist of cations and anions.

d A molecular compound is a combination of nonmetal atoms.

e Molecular compounds are gases at room temperature.

2.7 Organic Compounds

An important class of *molecular substances that contain carbon combined with other elements, such as hydrogen, oxygen, and nitrogen,* is **organic compounds.** Organic chemistry is the area of chemistry that is concerned with these compounds (Chapter 23 is devoted to this topic). Historically, organic compounds were restricted to those that could be produced only from living entities and were thought to contain a "vital force" based on their natural origin. The concept of the vital force was disproved in 1828 when a German chemist, Friedrick Wöhler, synthesized urea (a molecular compound in human urine, CH_4N_2O, Figure 2.24) from the molecular compounds ammonia (NH_3) and cyanic acid (HNCO). His work clearly demonstrated

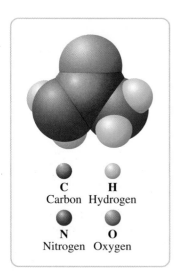

C Carbon   **H** Hydrogen

N Nitrogen   **O** Oxygen

Figure 2.24 ▲

Molecular model of urea (CH_4N_2O) Urea was the first organic molecule deliberately synthesized by a chemist from non-organic compounds.

that a given compound is exactly the same whether it comes from a living entity or is synthesized.

Organic compounds make up the majority of all known compounds. Since 1957, more than 13 million (60%) of the recorded substances in an international materials registry have been listed as organic. You encounter organic compounds in both living and nonliving materials every day. The proteins, amino acids, enzymes, and DNA that make up your body are all either individual organic molecules or contain organic molecules. Table sugar, peanut oil, antibiotic medicines, and methanol (windshield washer) are all examples of organic molecules as well. Organic chemistry and the compounds produced by the reactions of organic molecules are probably responsible for the majority of the materials that currently surround you as you read this page of text.

The simplest organic compounds are hydrocarbons. **Hydrocarbons** are *those compounds containing only hydrogen and carbon.* Common examples include methane (CH_4), ethane (C_2H_6), propane (C_3H_8), acetylene (C_2H_2), and benzene (C_6H_6). Hydrocarbons are used extensively as sources of energy for heating our homes, for powering internal combustion engines, and for generating electricity. They also are the starting materials for most plastics. Much of the mobility and comfort of our current civilization is built on the low cost and availability of hydrocarbons.

The chemistry of organic molecules is often determined by groups of atoms in the molecule that have characteristic chemical properties. A **functional group** is a *reactive portion of a molecule that undergoes predictable reactions.* When you use the term *alcohol* when referring to a molecular compound, you actually are indicating a molecule that contains an —OH functional group. Methyl alcohol has the chemical formula CH_3OH. The term *ether* indicates that an organic molecule contains an oxygen atom between two carbon atoms, as in diethyl ether ($CH_3CH_2OCH_2CH_3$). Table 2.3 contains a few examples of organic functional groups along with example compounds.

Chapter 23 contains a much more extensive treatment of organic chemistry. Due to its importance and its relation to living organisms, a substantial part of your future chemistry experience will likely be in the field of organic chemistry.

2.8 Naming Simple Compounds

Before the structural basis of chemical substances became established, compounds were named after people, places, or particular characteristics. Examples are Glauber's salt (sodium sulfate, discovered by J. R. Glauber), sal ammoniac (ammonium chloride, named after the ancient Egyptian deity Ammon from the temple near which the substance was made), and washing soda (sodium carbonate, used for softening wash water). Today several million compounds are known and thousands of new ones are discovered every year. Without a system for naming compounds, coping with this multitude of substances would be a hopeless task. **Chemical nomenclature** is *the systematic naming of chemical compounds.*

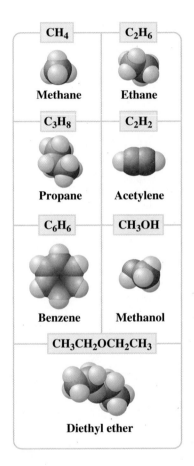

CH_4 — Methane

C_2H_6 — Ethane

C_3H_8 — Propane

C_2H_2 — Acetylene

C_6H_6 — Benzene

CH_3OH — Methanol

$CH_3CH_2OCH_2CH_3$ — Diethyl ether

Table 2.3	Examples of Organic Functional Groups		
Functional Group	**Name of Functional Group**	**Example Molecule**	**Common Use**
—OH	Alcohol	Methyl alcohol (CH_3OH)	Windshield washer
—O—	Ether	Dimethyl ether (CH_3OCH_3)	Solvent
—COOH	Carboxylic acid	Acetic acid (CH_3COOH)	Acid in vinegar

CONCEPT CHECK 2.5

Identify the following compounds as being a hydrocarbon, an alcohol, an ether, or a carboxylic acid.

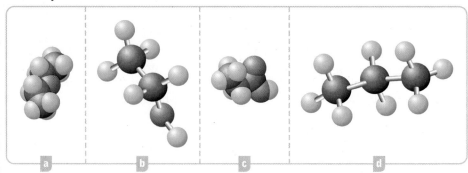

If a compound is not classified as organic, as discussed in Section 2.7, it must be inorganic. **Inorganic compounds** are *composed of elements other than carbon.* A few notable exceptions to this classification scheme include carbon monoxide, carbon dioxide, carbonates, and the cyanides; all contain carbon and yet are generally considered to be inorganic.

In this section, we discuss the nomenclature of some simple inorganic compounds. We first look at the naming of ionic compounds. Then, we look at the naming of some simple molecular compounds, including binary molecular compounds (molecular compounds of two elements) and acids. Finally, we look at hydrates of ionic compounds. These substances contain water molecules in loose association with ionic compounds.

Ionic Compounds

Ionic compounds, as we saw in the previous section, are substances composed of ions. Most ionic compounds contain metal and nonmetal atoms—for example, NaCl. (The ammonium salts, such as NH_4Cl, are a prominent exception.) You name an ionic compound by giving the name of the cation followed by the name of the anion. For example,

$$\underbrace{\text{potassium}}_{\substack{\text{cation} \\ \text{name}}} \underbrace{\text{sulfate}}_{\substack{\text{anion} \\ \text{name}}}$$

Before you can name ionic compounds, you need to be able to write and name ions.

The simplest ions are monatomic. A **monatomic ion** is *an ion formed from a single atom.* Table 2.4 lists common monatomic ions of the main-group elements. You may want to look at the table while you read first the rules for predicting the charges on such ions and then the rules for naming the monatomic ions.

Table 2.4	Common Monatomic Ions of the Main-Group Elements*						
	IA	IIA	IIIA	IVA	VA	VIA	VIIA
Period 1							H^-
Period 2	Li^+	Be^{2+}	B	C	N^{3-}	O^{2-}	F^-
Period 3	Na^+	Mg^{2+}	Al^{3+}	Si	P	S^{2-}	Cl^-
Period 4	K^+	Ca^{2+}	Ga^{3+}	Ge	As	Se^{2-}	Br^-
Period 5	Rb^+	Sr^{2+}	In^{3+}	Sn^{2+}	Sb	Te^{2-}	I^-
Period 6	Cs^+	Ba^{2+}	Tl^+, Tl^{3+}	Pb^{2+}	Bi^{3+}		

*Elements shown in color do not normally form compounds having monatomic ions.

Rules for Predicting the Charges on Monatomic Ions

1. Most of the main-group metallic elements have one monatomic cation with a charge equal to the group number in the periodic table (the Roman numeral). Example: aluminum, in Group IIIA, has a monatomic ion Al^{3+}.

2. Some metallic elements of high atomic number are exceptions to the previous rule; they have more than one cation. These elements have common cations with a charge equal to the group number minus 2, in addition to having a cation with a charge equal to the group number. Example: The common ion of lead is Pb^{2+}. (The group number is 4; the charge is $4 - 2$.) In addition to compounds containing Pb^{2+}, some lead compounds contain Pb^{4+}.

3. Most transition elements form more than one monatomic cation, each with a different charge. Most of these elements have one ion with a charge of $+2$. Example: Iron has common cations Fe^{2+} and Fe^{3+}. Copper has common cations Cu^+ and Cu^{2+}.

4. The charge on a monatomic anion for a nonmetallic main-group element equals the group number minus 8. Example: Oxygen has the monatomic anion O^{2-}. (The group number is 6; the charge is $6 - 8$.)

Rules for Naming Monatomic Ions

1. Monatomic cations are named after the element if there is only one such ion. Example: Al^{3+} is called aluminum ion; Na^+ is called sodium ion.

2. If there is more than one monatomic cation of an element, Rule 1 is not sufficient. The *Stock system* of nomenclature names the cations after the element, as in Rule 1, but follows this by a Roman numeral in parentheses denoting the charge on the ion. Example: Fe^{2+} is called iron(II) ion and Fe^{3+} is called iron(III) ion. ▶

   In an older system of nomenclature, such ions are named by adding the suffixes *-ous* and *-ic* to a stem name of the element (which may be from the Latin) to indicate the ions of lower and higher charge, respectively. Example: Fe^{2+} is called ferrous ion; Fe^{3+}, ferric ion. Cu^+ is called cuprous ion; Cu^{2+}, cupric ion.

   Table 2.5 lists some common cations of the transition elements. Most of these elements have more than one ion, so require the Stock nomenclature system or the older suffix system. A few, such as zinc, have only a single ion that is normally encountered, and you usually name them by just the metal name. You would not be wrong, however, if, for example, you named Zn^{2+} as zinc(II) ion.

3. The names of the monatomic anions are obtained from a stem name of the element followed by the suffix *-ide*. Example: Br^- is called bromide ion, from the stem name *brom-* for bromine and the suffix *-ide*.

   A **polyatomic ion** is *an ion consisting of two or more atoms chemically bonded together and carrying a net electric charge*. Table 2.6 lists some common polyatomic ions. The first two are cations (Hg_2^{2+} and NH_4^+); the rest are anions. There are

The Roman numeral actually denotes the oxidation state, or oxidation number, of the atom in the compound. For a monatomic ion, the oxidation state equals the charge. Otherwise, the oxidation state is a hypothetical charge assigned in accordance with certain rules; see Section 4.5.

Table 2.5	**Common Cations of the Transition Elements**					
Ion	**Ion Name**	**Ion**	**Ion Name**	**Ion**	**Ion Name**
Cr^{3+}	Chromium(III) or chromic	Co^{2+}	Cobalt(II) or cobaltous	Zn^{2+}	Zinc
Mn^{2+}	Manganese(II) or manganous	Ni^{2+}	Nickel(II) or nickel	Ag^+	Silver
Fe^{2+}	Iron(II) or ferrous	Cu^+	Copper(I) or cuprous	Cd^{2+}	Cadmium
Fe^{3+}	Iron(III) or ferric	Cu^{2+}	Copper(II) or cupric	Hg^{2+}	Mercury(II) or mercuric

Table 2.6	Some Common Polyatomic Ions		
Name	**Formula**	**Name**	**Formula**
Mercury(I) or mercurous	Hg_2^{2+}	Permanganate	MnO_4^-
Ammonium	NH_4^+	Nitrite	NO_2^-
Cyanide	CN^-	Nitrate	NO_3^-
Carbonate	CO_3^{2-}	Hydroxide	OH^-
Hydrogen carbonate (or bicarbonate)	HCO_3^-	Peroxide	O_2^{2-}
Acetate	$C_2H_3O_2^-$	Phosphate	PO_4^{3-}
Oxalate	$C_2O_4^{2-}$	Monohydrogen phosphate	HPO_4^{2-}
Hypochlorite	ClO^-	Dihydrogen phosphate	$H_2PO_4^-$
Chlorite	ClO_2^-	Sulfite	SO_3^{2-}
Chlorate	ClO_3^-	Sulfate	SO_4^{2-}
Perchlorate	ClO_4^-	Hydrogen sulfite (or bisulfite)	HSO_3^-
Chromate	CrO_4^{2-}	Hydrogen sulfate (or bisulfate)	HSO_4^-
Dichromate	$Cr_2O_7^{2-}$	Thiosulfate	$S_2O_3^{2-}$

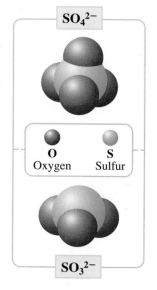

Figure 2.25 ▲

Sulfate and sulfite oxoanions Molecular models of the sulfate *(top)* and sulfite *(bottom)* oxoanions.

no simple rules for writing the formulas of such ions, although you will find it helpful to note a few points.

Most of the ions in Table 2.6 are *oxoanions* (also called *oxyanions*), which consist of oxygen with another element (called the characteristic or central element). Sulfur, for example, forms the oxoanions sulfate ion, SO_4^{2-}, and sulfite ion, SO_3^{2-} (Figure 2.25). The oxoanions in the table are grouped by the characteristic element. For instance, the sulfur ions occur as a group at the end of the table.

Note that the names of the oxoanions have a stem name from the characteristic element, plus a suffix *-ate* or *-ite*. These suffixes denote the relative number of oxygen atoms in the oxoanions of a given characteristic element. The name of the oxoanion with the greater number of oxygen atoms has the suffix *-ate;* the name of the oxoanion with the lesser number of oxygen atoms has the suffix *-ite.* The sulfite ion, SO_3^{2-}, and sulfate ion, SO_4^{2-}, are examples. Another example are the two oxoanions of nitrogen listed in Table 2.6:

$$NO_2^- \qquad \text{nitr\underline{ite} ion}$$
$$NO_3^- \qquad \text{nitr\underline{ate} ion}$$

Unfortunately, the suffixes do not tell you the actual number of oxygen atoms in the oxoanion, only the relative number. However, if you were given the formulas of the two anions, you could name them.

The two suffixes *-ite* and *-ate* are not enough when there are more than two oxoanions of a given characteristic element. The oxoanions of chlorine are an example. Table 2.6 lists four oxoanions of chlorine: ClO^-, ClO_2^-, ClO_3^-, and ClO_4^-. In such cases, two prefixes, *hypo-* and *per-,* are used in addition to the two suffixes. The two oxoanions with the least number of oxygen atoms (ClO^- and ClO_2^-) are named using the suffix *-ite,* and the prefix *hypo-* is added to the one of these two ions with the fewer oxygen atoms:

$$ClO^- \qquad \text{hypochlor\underline{ite} ion}$$
$$ClO_2^- \qquad \text{chlor\underline{ite} ion}$$

The two oxoanions with the greatest number of oxygen atoms (ClO_3^- and ClO_4^-) are named using the suffix *-ate,* and the prefix *per-* is added to the one of these two ions with the greater number of oxygen atoms:

$$ClO_3^- \qquad \text{chlor\underline{ate} ion}$$
$$ClO_4^- \qquad \text{perchlor\underline{ate} ion}$$

Some of the polyatomic ions in Table 2.6 are oxoanions bonded to one or more hydrogen ions (H^+). They are sometimes referred to as *acid anions,* because acids are substances that provide H^+ ions. As an example, monohydrogen phosphate ion, HPO_4^{2-}, is essentially a phosphate ion (PO_4^{3-}) to which a hydrogen ion (H^+) has bonded. The prefix *mono-,* from the Greek, means "one." Similarly, *di-* is a prefix from the Greek meaning "two," so dihydrogen phosphate ion is a phosphate ion to which two hydrogen ions have bonded. In an older terminology, ions such as hydrogen carbonate and hydrogen sulfate were called bicarbonate and bisulfate, respectively.

The last anion in the table is thiosulfate ion, $S_2O_3^{2-}$. The prefix *thio-* means that an oxygen atom in the root ion name (sulfate, SO_4^{2-}) has been replaced by a sulfur atom.

There are only a few common polyatomic cations. Those listed in Table 2.6 are the mercury(I) ion (also called mercurous ion) and the ammonium ion. Mercury(I) ion is one of the few common metal ions that is not monatomic; its formula is Hg_2^{2+}. The ion charge indicated in parentheses is the charge per metal atom. The ammonium ion, NH_4^+, is one of the few common cations composed of only nonmetal atoms.

Now that we have discussed the naming of ions, let us look at the naming of ionic compounds. Example 2.4 illustrates how you name an ionic compound given its formula.

◉WL Interactive Example | **2.4** | **Naming an Ionic Compound from Its Formula**

Gaining Mastery Toolbox

Critical Concept 2.4
There are rules that can be applied to naming ionic compounds. In order to be successful in naming ionic compounds, you have to know both the names and charges of the cation and anion that form the compound. Ionic compounds are often metal-nonmetal combinations of atoms.

Solution Essentials:
- Monatomic ions and charges
- Polyatomic ions and charges
- Cation names (monatomic and polyatomic)
- Anion names (monatomic and polyatomic)
- The naming rules of three possible combinations of ions: monatomic ions, monatomic and polyatomic ions, and polyatomic ions.
- Metal and nonmetal
- Main group and transition metal

Name the following: a Mg_3N_2, b $CrSO_4$.

Problem Strategy First, you need to identify the type of compound that you are dealing with. Because each of these compounds contains both metal and non-metal atoms, we expect them to be ionic, so we need to apply the rules for naming ionic compounds. When naming ionic compounds, start by writing the formulas of the cations and anions in the compound, then name the ions. Look first for any monatomic cations and anions whose charges are predictable. Use the rules given in the text to name the ions. For those metals having more than one cation, you will need to deduce the charge on the metal ion (say the ion of Cr in $CrSO_4$) from the charge on the anion. You will need

to memorize or have available the formulas and names of the polyatomic anions to be able to write them from the formula of the compound.

Solution

a Magnesium, a Group IIA metal, is expected to form only a 2+ ion (Mg^{2+}, the magnesium ion). Nitrogen (Group VA) is expected to form an anion of charge equal to the group number minus 8 (N^{3-}, the nitride ion). You can check that these ions would give the formula Mg_3N_2. The name of the compound is **magnesium nitride** (the name of the cation followed by the name of the anion).

b Chromium is a transition element and, like most such elements, has more than one monatomic ion. You can find the charge on the Cr ion if you know the formula of the anion. From Table 2.6, you see that the SO_4 in $CrSO_4$ refers to the anion SO_4^{2-} (the sulfate ion). Therefore, the Cr cation must be Cr^{2+} to give electrical neutrality. The name of Cr^{2+} is chromium(II) ion, so the name of the compound is **chromium(II) sulfate.**

Answer Check Whenever you have to name a compound, you can check your answer to see if the name will lead you back to the correct formula.

Exercise 2.5 Write the names of the following compounds: a CaO, b $PbCrO_4$.

■ See Problems 2.77 and 2.78.

Example 2.5   Writing the Formula from the Name of an Ionic Compound

Critical Concept 2.5
The name of the ionic compound provides the names of the ions that form the compound. In the formula the cation name will always be listed first, followed by the anion name. For metals that can have more than one charge, Roman numerals following the cation name indicate the charge.

Solution Essentials:
- Monatomic ions and charges
- Polyatomic ions and charges
- Cation names (monatomic and polyatomic)
- Anion names (monatomic and polyatomic)
- Ionic compound
- Example 2.3
- Metal and nonmetal
- Main group elements and transition metal

Write formulas for the following compounds: **a** iron(II) phosphate, **b** titanium(IV) oxide.

Problem Strategy As in all nomenclature problems, first determine the type of compound you are working with. The compounds presented in this problem contain both metal and nonmetal atoms, so they are ionic. Use the name of the compound, obtain the name of the ions, and then write the formulas of the ions. Finally, use the method of Example 2.3 to obtain the formula of the compound from the formulas of the ions.

Solution

a Iron(II) phosphate contains the iron(II) ion, Fe^{2+}, and the phosphate ion, PO_4^{3-}. (You will need to have memorized the formula of the phosphate ion, or have access to Table 2.6.) Once you know the formulas of the ions, you use the method of Example 2.3 to obtain the formula. The formula is $\mathbf{Fe_3(PO_4)_2}$.

b Titanium(IV) oxide is composed of titanium(IV) ions, Ti^{4+}, and oxide ions, O^{2-}. The formula of titanium(IV) oxide is $\mathbf{TiO_2}$.

Answer Check Whenever you have to write the formula of a compound from its name, make sure that the formula you have written will result in the correct name.

Exercise 2.6   A compound has the name thallium(III) nitrate. What is its formula? The symbol of thallium is Tl.

■ See Problems 2.79 and 2.80.

CONCEPT CHECK 2.6

Which of the following are polyatomic ions?

$$SO_3 \quad SO_4^{2-} \quad NO_2^- \quad NO_2 \quad I_2 \quad I_3^-$$

Which of the following is a false statement regarding polyatomic ions?

a Polyatomic ions must contain two or more atoms in their chemical formula.

b Polyatomic ions must have either a positive or negative charge.

c Polyatomic ions must have at least two different elements chemically bonded together.

d Polyatomic ions can be combinations of metals and nonmetals.

e Polyatomic ions can be oxoanions.

Binary Molecular Compounds

A **binary compound** is *a compound composed of only two elements.* Binary compounds composed of a metal and a nonmetal are usually ionic and are named as ionic compounds, as we have just discussed. (For example, NaCl, $MgBr_2$, and Al_2N_3 are all binary ionic compounds.) Binary compounds composed of two nonmetals or metalloids are usually molecular and are named using a prefix system. Examples of binary molecular compounds are H_2O, NH_3, and CCl_4. Using this prefix system, you name the two elements using the order given by the formula of the compound.

Order of Elements in the Formula The order of elements in the formula of a binary molecular compound is established by convention. By this convention, the nonmetal or metalloid occurring first in the following sequence is written first in the formula of the compound.

Element B Si C Sb As P N H Te Se S I Br Cl O F
Group IIIA IVA VA VIA VIIA

You can reproduce this order easily if you have a periodic table available. You arrange the nonmetals and metalloids in the order of the groups of the periodic table, listing the elements from the bottom of the group upward. Then you place H between Groups VA and VIA, and move O so that it is just before F.

This order places the nonmetals and metalloids approximately in order of increasing nonmetallic character. Thus, in the formula, the first element is the more metallic and the second element is the more nonmetallic, similar to the situation with binary ionic compounds. For example, the compound whose molecule contains three fluorine atoms and one nitrogen atom is written NF_3, not F_3N.

Rules for Naming Binary Molecular Compounds Now let us look at the rules for naming binary molecular compounds by the *prefix system*.

1. The name of the compound usually has the elements in the order given in the formula.
2. You name the first element using the exact element name.
3. You name the second element by writing the stem name of the element with the suffix *-ide* (as if the element occurred as the anion).
4. You add a prefix, derived from the Greek, to each element name to denote the subscript of the element in the formula. (The Greek prefixes are listed in Table 2.7.) Generally, the prefix *mono-* is not used, unless it is needed to distinguish two compounds of the same two elements.

Table 2.7	Greek Prefixes for Naming Compounds
Number	**Prefix**
1	mono-
2	di-
3	tri-
4	tetra-
5	penta-
6	hexa-
7	hepta-
8	octa-
9	nona-
10	deca-

Consider N_2O_3. This is a binary molecular compound, as you would predict because N and O are nonmetals. According to Rule 1, you name it after the elements, N before O, following the order of elements in the formula. By Rule 2, the N is named exactly as the element (nitrogen), and by Rule 3, the O is named as the anion (oxide). The compound is a nitrogen oxide. Finally, you add prefixes to denote the subscripts in the formula (Rule 4). As Table 2.7 shows, the prefix for two is *di-*, and the prefix for three is *tri-*. The name of the compound is dinitrogen trioxide.

Here are some examples to illustrate how the prefix *mono-* is used. There is only one compound of hydrogen and chlorine: HCl. It is called hydrogen chloride, not monohydrogen monochloride, since the prefix *mono-* is not generally written. On the other hand, there are two common compounds of carbon and oxygen: CO and CO_2. They are both carbon oxides. To distinguish one from the other, we name them carbon monoxide and carbon dioxide, respectively.

Here are some other examples of prefix names for binary molecular compounds.

SF_4	sulfur tetrafluoride	SF_6	sulfur hexafluoride
ClO_2	chlorine dioxide	Cl_2O_7	dichlorine hept(a)oxide

The final vowel in a prefix is often dropped before a vowel in a stem name, for ease in pronunciation. Tetraoxide, for instance, becomes tetroxide.

A few compounds have older, well-established names. The compound H_2S would be named dihydrogen sulfide by the prefix system, but is commonly called hydrogen sulfide. NO is still called nitric oxide, although its prefix name is nitrogen monoxide. The names water and ammonia are used for H_2O and NH_3, respectively.

Example 2.6   Naming a Binary Compound from Its Formula

Critical Concept 2.6
There are rules that can be applied to naming binary molecular compounds. These rules are specific to binary molecular compounds and do not apply to ionic compounds. Molecular compounds are combinations of non-metal atoms.

Solution Essentials:
- Greek prefixes
- Binary molecular compound naming rules
- Naming order of elements
- Binary compound
- Periodic table
- Nonmetal

Name the following compounds: a N_2O_4, b P_4O_6.

Problem Strategy First, you need to determine the type of compound that you are dealing with: molecular or ionic. In this case the compound contains only nonmetals, so it is molecular. Molecular compounds

require that you use the Greek prefixes (Table 2.7), naming the elements in the same order as in the formulas.

Solution
a N_2O_4 contains two nitrogen atoms and four oxygen atoms. Consulting Table 2.7 reveals that the corresponding prefixes are *di-* and *tetra-*, so the name is **_di_nitrogen _tetr_oxide.**

b P_4O_6 contains four phosphorus atoms and six oxygen atoms. Consulting Table 2.7 reveals that the corresponding prefixes are *tetra-* and *hexa-*, so the name is **_tetra_phosphorus _hex_oxide.**

Answer Check Check to see whether the name leads back to the correct formula.

Exercise 2.7 Name the following compounds: a Cl_2O_6, b PCl_3, c PCl_5.

■ See Problems 2.83 and 2.84.

Example 2.7   Writing the Formula from the Name of a Binary Compound

Critical Concept 2.7
Greek prefixes are used to determine the correct formula of binary molecular compounds. The elements in the formula are written in the same order as they are presented in the name.

Solution Essentials:
- Greek prefixes
- Binary molecular compound naming rules
- Element symbols
- Periodic table
- Nonmetal

Give the formulas of the following compounds: a disulfur dichloride, b tetraphosphorus trisulfide.

Problem Strategy As always, first determine whether you are working with a molecular or ionic compound.

Because these compounds contain only nonmetals, they must be molecular. Given that they are molecular, you will need to use the Greek prefixes in Table 2.7 to give you the corresponding names for each subscript.

Solution You change the names of the elements to symbols and translate the prefixes to subscripts. The formulas are a S_2Cl_2 and b P_4S_3.

Answer Check Check to see whether the formula you have written will result in the correct name.

Exercise 2.8 Give formulas for the following compounds: a carbon disulfide, b sulfur trioxide.

■ See Problems 2.85 and 2.86.

Example 2.8   Naming a Binary Chemical Compound from Its Molecular Model

Critical Concept 2.8
Models of molecular compounds depict the number and identity of each element in a single molecule. The molecular formula can be written directly from the information contained in the molecular model.

Solution Essentials:
- Element symbols
- Naming order of elements
- Example 2.6
- Molecular models

Name the chemical compounds shown in the margin at the top of the next page.

Problem Strategy After recognizing that these are molecular compounds, start by writing down the elements using subscripts to denote how many atoms of each element occur in the compound. Next, determine which element should be written first in the chemical formula. (The more metallic element is written first.)

(continued)

(*continued*)

Solution

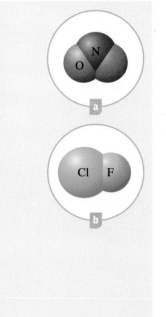

a The elements in the compound are O and N. There are two O atoms for each N atom, so the chemical formula could be O_2N. Applying the rule for writing the more metallic element first, the formula is rearranged to yield the correct formula, NO_2. Using Greek prefixes and naming the elements in the order in which they appear, we get the correct chemical formula: **nitrogen dioxide.**

b Following the same procedure as part a, we get **chlorine monofluoride.**

Answer Check Make sure that the answer reflects the correct number of each of the elements given in the molecular structure.

Exercise 2.9 Using the molecular models, name the following chemical compounds:

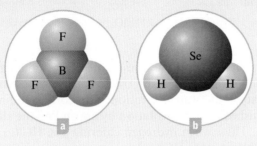

■ See Problems 2.87 and 2.88.

Acids and Corresponding Anions

Acids are an important class of compounds, and we will look closely at them in the next chapter. Here we want merely to see how we name these compounds and how they are related to the anions that we encounter in some ionic compounds. For our present purposes, an *acid* is a molecular compound that yields hydrogen ions, H^+, and an anion for each acid molecule when the acid dissolves in water. An example is nitric acid, HNO_3. The HNO_3 molecule yields one H^+ ion and one nitrate ion, NO_3^-, in aqueous (water) solution.

Nitric acid, HNO_3, is an oxoacid, or oxyacid (Figure 2.26). An **oxoacid** is *an acid containing hydrogen, oxygen, and another element (often called the central atom)*. In water the oxoacid molecule yields one or more hydrogen ions, H^+, and an oxoanion. The names of the oxoacids are related to the names of the corresponding oxoanions. If you know the name of the oxoanion, you can obtain the name of the corresponding acid by replacing the suffix as follows:

Anion Suffix	Acid Suffix
-ate	*-ic*
-ite	*-ous*

The following diagram illustrates how to apply this information to naming oxoacids.

Acid	Contains	Name
HNO_3 ➡	nitrate anion therefore nitric acid	
	ate to ic	
HNO_2 ➡	nitrite anion therefore nitrous acid	
	ite to ous	

Table 2.8 lists some oxoanions and their corresponding oxoacids.

Some binary compounds of hydrogen and nonmetals yield acidic solutions when dissolved in water. These *solutions* are named like compounds by using the

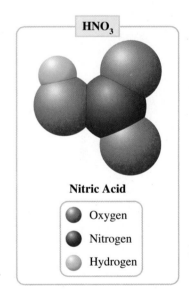

HNO₃

Nitric Acid

● Oxygen
● Nitrogen
○ Hydrogen

Figure 2.26 ▲

Molecular model of nitric acid

Table 2.8   Some Oxoanions and Their Corresponding Oxoacids

Oxoanion		Oxoacid	
CO_3^{2-}	Carbon*ate ion*	H_2CO_3	Carbon*ic acid*
NO_2^-	Nit*rite ion*	HNO_2	Nitr*ous acid*
NO_3^-	Nitr*ate ion*	HNO_3	Nitr*ic acid*
PO_4^{3-}	Phosph*ate ion*	H_3PO_4	Phosphor*ic acid*
SO_3^{2-}	Sulf*ite ion*	H_2SO_3	Sulfur*ous acid*
SO_4^{2-}	Sulf*ate ion*	H_2SO_4	Sulfur*ic acid*
ClO^-	*Hypo*chlor*ite ion*	$HClO$	*Hypo*chlor*ous acid*
ClO_2^-	Chlor*ite ion*	$HClO_2$	Chlor*ous acid*
ClO_3^-	Chlor*ate ion*	$HClO_3$	Chlor*ic acid*
ClO_4^-	*Per*chlor*ate ion*	$HClO_4$	*Per*chlor*ic acid*

prefix *hydro-* and the suffix *-ic* with the stem name of the nonmetal, followed by the word *acid*. We denote the solution by the formula of the binary compound followed by (*aq*) for aqueous (water) solution. The corresponding binary compound can be distinguished from the solution by appending the state of the compound to the formula. Thus, when hydrogen chloride gas, HCl(*g*), is dissolved in water, it forms hydrochloric acid, HCl(*aq*).

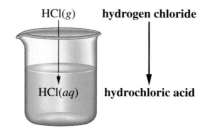

Here are some other examples:

Binary Compound	*Acid Solution*
HBr(*g*), hydrogen bromide	*hydro*brom*ic acid,* HBr(*aq*)
HF(*g*), hydrogen fluoride	*hydro*fluor*ic acid,* HF(*aq*)

Example 2.9   Writing the Name and Formula of an Anion from the Acid

Gaining Mastery Toolbox

Critical Concept 2.9
There are rules that can be applied to naming acids. Depending on the type of acid, there are two sets of rules: one for oxoacids and one for acids that are binary compounds of hydrogen and a nonmetal.

Solution Essentials:
• Recognition of the chemical formula of an acid
• Oxoacid
• Naming oxoacid from oxoanion name
• Polyatomic anion names

Selenium has an oxoacid, H_2SeO_4, called selenic acid. What is the formula and name of the corresponding anion?

Problem Strategy Knowing that H_2SeO_4 is an oxyacid, we must first determine the oxoanion name from

the examples provided in Table 2.8. From the oxoanion name you determine whether a prefix (*per-* or *hypo-*) is necessary and whether the proper suffix is *-ite* or *-ate*. You then name the anion.

Solution When you remove two H^+ ions from H_2SeO_4, you obtain the SeO_4^{2-} ion. You name the ion from the acid by replacing *-ic* with *-ate*. The anion is called the **selenate ion.**

Answer Check Check to see whether the name you have written leads back to the correct formula.

Exercise 2.10 What are the name and formula of the anion corresponding to perbromic acid, $HBrO_4$?

■ See Problems 2.89 and 2.90.

Hydrates

A **hydrate** is *a compound that contains water molecules weakly bound in its crystals.* These substances are often obtained by evaporating an aqueous solution of the compound. Consider copper(II) sulfate. When an aqueous solution of this substance is evaporated, blue crystals form in which each formula unit of copper(II) sulfate, $CuSO_4$, is associated with five molecules of water. The formula of the hydrate is written $CuSO_4 \cdot 5H_2O$, where a centered dot separates $CuSO_4$ and $5H_2O$. When the blue crystals of the hydrate are heated, the water is driven off, leaving behind white crystals of copper(II) sulfate without associated water, a substance called *anhydrous* copper(II) sulfate (see Figure 2.27).

Hydrates are named from the anhydrous compound, followed by the word *hydrate* with a prefix to indicate the number of water molecules per formula unit of the compound. For example, $CuSO_4 \cdot 5H_2O$ is known as copper(II) sulfate pentahydrate.

$CuSO_4 \cdot 5H_2O$

Hydrate

Figure 2.27 ◄

Copper(II) sulfate The hydrate $CuSO_4 \cdot 5H_2O$ is blue; the anhydrous compound, $CuSO_4$, is white.

$CuSO_4$

Anhydrous

© Cengage Learning

Example 2.10   **Naming a Hydrate from Its Formula**

Gaining Mastery Toolbox

Critical Concept 2.10
Hydrates are ionic compounds that contain water molecules as part of the chemical formula. Naming hydrates requires that you first apply the rules of naming ionic compounds and then account for the number of water molecules in the chemical formula.

Solution Essentials:
- Recognition of the chemical formula of a hydrate
- Rules for naming hydrates
- Rules for naming ionic compounds
- Greek prefixes

Epsom salts has the formula $MgSO_4 \cdot 7H_2O$. What is the chemical name of the substance?

Problem Strategy You need to identify the type of compound that you are dealing with, ionic or molecular. Because this compound contains both metals

and nonmetals, it is ionic. However, this compound has the additional feature that it contains water molecules. Therefore, you need to apply ionic naming rules and then use Greek prefixes (Table 2.7) to indicate the number of water molecules that are part of the formula. The word *hydrate* will be used to represent water in the name.

Solution $MgSO_4$ is magnesium sulfate. $MgSO_4 \cdot 7H_2O$ is **magnesium sulfate heptahydrate.**

Answer Check Make sure that the name leads back to the formula that you started with.

Exercise 2.11   Washing soda has the formula $Na_2CO_3 \cdot 10H_2O$. What is the chemical name of this substance?

■ See Problems 2.91 and 2.92.

Example 2.11   **Writing the Formula from the Name of a Hydrate**

Gaining Mastery Toolbox

Critical Concept 2.11
A hydrate formula has two parts: the formula of the ionic compound and the number of water molecules bound to the ionic compound. A centered dot is used to separate the two parts, and Greek prefixes are used to indicate the number of water molecules.

Solution Essentials:
- Rules for naming hydrates
- Rules for naming ionic compounds
- Greek prefixes

The mineral gypsum has the chemical name calcium sulfate dihydrate. What is the chemical formula of this substance?

Problem Strategy You need to identify the type of compound that you are dealing with, ionic or molecular. Because this compound contains both metals and nonmetals, it is ionic. However, this compound has the additional feature that it contains water molecules. Therefore, we need to apply ionic naming rules and then use Greek prefixes (Table 2.7) to indicate the number of water molecules that are part of the formula. The word *hydrate* will be used to represent water in the name.

Solution Calcium sulfate is composed of calcium ions (Ca^{2+}) and sulfate ions (SO_4^{2-}), so the formula of the anhydrous compound is $CaSO_4$. Since the mineral is a *di*hydrate, the formula of the compound is **$CaSO_4 \cdot 2H_2O$.**

(continued)

(*continued*)

Answer Check Make sure that the formula will result in the correct name that you started with.

Exercise 2.12 Photographers' hypo, used to fix negatives during the development process, is sodium thiosulfate

pentahydrate. What is the chemical formula of this compound?

■ See Problems 2.93 and 2.94.

CONCEPT CHECK 2.7

You take a job with the U.S. Environmental Protection Agency (EPA) inspecting college chemistry laboratories. On the first day of the job while inspecting Generic University you encounter a bottle with the formula Al_2Q_3 that was used as an unknown compound in an experiment. Before you send the compound off to the EPA lab for analysis, you want to narrow down the possibilities for element Q. What are the likely (real element) candidates for element Q?

Chemical Reactions: Equations

An important feature of atomic theory is its explanation of a chemical reaction as a rearrangement of the atoms of substances. The reaction of sodium metal with chlorine gas described in the chapter opening involves the rearrangement of the atoms of sodium and chlorine to give the new combination of atoms in sodium chloride. Such a rearrangement of atoms is conveniently represented by a chemical equation, which uses chemical formulas.

2.9 Writing Chemical Equations

A **chemical equation** is *the symbolic representation of a chemical reaction in terms of chemical formulas.* For example, the burning of sodium in chlorine to produce sodium chloride is written

$$2Na + Cl_2 \longrightarrow 2NaCl$$

The formulas on the left side of an equation (before the arrow) represent the reactants; a **reactant** is *a starting substance in a chemical reaction.* The arrow means "react to form" or "yield." The formulas on the right side represent the products; a **product** is *a substance that results from a reaction.* Note the *coefficient* of 2 in front of the formula NaCl. Coefficients in front of the formula give the relative number of molecules or formula units involved in the reaction. When no coefficient is written, it is understood to be 1.

In many cases, it is useful to indicate the states or phases of the substances in an equation. You do this by placing appropriate labels indicating the phases within parentheses following the formulas of the substances. You use the following phase labels:

(g) = gas, (l) = liquid, (s) = solid, (aq) = aqueous (water) solution

When you use these labels, the previous equation becomes

$$2Na(s) + Cl_2(g) \longrightarrow 2NaCl(s)$$

You can also indicate in an equation the conditions under which a reaction takes place. If the reactants are heated to make the reaction go, you can indicate this by putting the symbol Δ (capital Greek delta) over the arrow. For example, the equation

$$2NaNO_3(s) \xrightarrow{\Delta} 2NaNO_2(s) + O_2(g)$$

indicates that solid sodium nitrate, $NaNO_3$, decomposes when heated to give solid sodium nitrite, $NaNO_2$, and oxygen gas, O_2.

When an aqueous solution of hydrogen peroxide, H_2O_2, comes in contact with platinum metal, Pt, the hydrogen peroxide decomposes into water and oxygen gas. The platinum acts as a *catalyst,* a substance that speeds up a reaction without undergoing any net change itself. You write the equation for this reaction as follows. The catalyst, Pt, is written over the arrow. ▶

$$2H_2O_2\,(aq) \xrightarrow{\text{Pt}} 2H_2O(l) + O_2(g)$$

You also can combine the symbol Δ with the formula of a catalyst, placing one above the arrow and the other below the arrow.

Platinum is a silvery-white metal used for jewelry. It is also valuable as a catalyst for many reactions, including those that occur in the catalytic converters of automobiles.

2.10 Balancing Chemical Equations

When the coefficients in a chemical equation are correctly given, the numbers of atoms of each element are equal on both sides of the arrow. The equation is then said to be *balanced*. That a chemical equation should be balanced follows from atomic theory. A chemical reaction involves simply a recombination of the atoms; none are destroyed and none are created. Consider the burning of natural gas, which is composed mostly of methane, CH_4. Using atomic theory, you describe this as the chemical reaction of one molecule of methane, CH_4, with two molecules of oxygen, O_2, to form one molecule of carbon dioxide, CO_2, and two molecules of water, H_2O.

CH_4	+	$2O_2$	$\longrightarrow$	CO_2	+	$2H_2O$
one molecule of methane	+	two molecules of oxygen	react to form	one molecule of carbon dioxide	+	two molecules of water

Figure 2.28 represents the reaction in terms of molecular models.

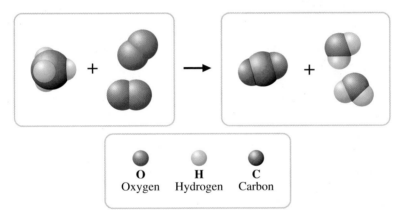

Figure 2.28 ◀

Representation of the reaction of methane with oxygen Molecular models represent the reaction of CH_4 with O_2 to give CO_2 and H_2O.

Before you can write a balanced chemical equation for a reaction, you must determine *by experiment* those substances that are reactants and those that are products. You must also determine the formulas of each substance. Once you know these things, you can write the balanced chemical equation.

As an example, consider the burning of propane gas (Figure 2.29). By experiment, you determine that propane reacts with oxygen in air to give carbon dioxide and water. You also determine from experiment that the formulas of propane, oxygen, carbon dioxide, and water are C_3H_8, O_2, CO_2, and H_2O, respectively. Then you can write

$$C_3H_8 + O_2 \longrightarrow CO_2 + H_2O$$

This equation is not balanced because the coefficients that give the relative number of molecules involved have not yet been determined. *To balance the equation, you select coefficients that will make the numbers of atoms of each element equal on both sides of the equation.*

Because there are three carbon atoms on the left side of the equation (C_3H_8), you must have three carbon atoms on the right. This is achieved by writing 3 for the coefficient of CO_2.

$$\underline{1}C_3H_8 + O_2 \longrightarrow \underline{3}CO_2 + H_2O$$

(We have written the coefficient for C_3H_8 to show that it is now determined; normally when the coefficient is 1, it is omitted.) Similarly, there are eight hydrogen atoms on the left (in C_3H_8), so you write 4 for the coefficient of H_2O.

$$\underline{1}C_3H_8 + O_2 \longrightarrow \underline{3}CO_2 + \underline{4}H_2O$$

The coefficients on the right are now determined, and there are ten oxygen atoms on this side of the equation: 6 from the three CO_2 molecules and 4 from the four

Figure 2.29 ▶

The burning of propane gas
Propane gas, C_3H_8, from the tank burns by reacting with oxygen, O_2, in air to give carbon dioxide, CO_2, and water, H_2O.

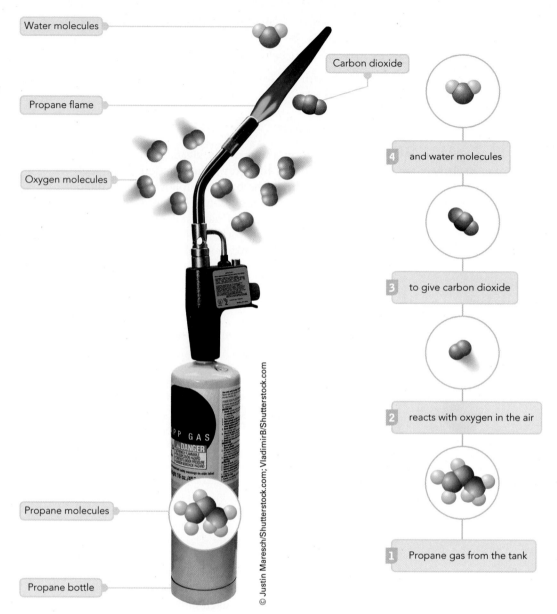

Water molecules

Carbon dioxide

Propane flame

Oxygen molecules

4 and water molecules

3 to give carbon dioxide

2 reacts with oxygen in the air

Propane molecules

1 Propane gas from the tank

Propane bottle

© Justin Maresch/Shutterstock.com; VladimirB/Shutterstock.com

H_2O molecules. You now write 5 for the coefficient of O_2. We drop the coefficient 1 from C_3H_8 and have the balanced equation.

$$C_3H_8 + 5O_2 \longrightarrow 3CO_2 + 4H_2O$$

It is a good idea to check your work by counting the number of atoms of each element on the left side and then on the right side of the equation.

The combustion of propane can be equally well represented by

$$2C_3H_8 + 10O_2 \longrightarrow 6CO_2 + 8H_2O$$

in which the coefficients of the previous equation have been doubled. Usually, however, it is preferable to *write the coefficients so that they are the smallest whole numbers possible.* When we refer to a balanced equation, we will assume that this is the case unless otherwise specified.

The method we have outlined—called *balancing by inspection*—is essentially a trial-and-error method. ◄ It can be made easier, however, by observing the following rule:

Balancing equations this way is relatively quick for all but the most complicated equations. A special method for such equations is described in Chapter 19.

Balance first the atoms for elements that occur in only one substance on each side of the equation.

The usefulness of this rule is illustrated in the next example. Remember that in any formula, such as $Fe_2(SO_4)_3$, a subscript to the right of the parentheses multiplies *each* subscript within the parentheses. Thus, $Fe_2(SO_4)_3$ represents 4 × 3 oxygen atoms. Remember also that you cannot change a subscript in any formula; only the coefficients can be altered to balance an equation.

OWL Interactive Example 2.12 Balancing Simple Equations

Gaining Mastery Toolbox

Critical Concept 2.12
When a chemical reaction occurs, elements are neither created nor destroyed; they are rearranged. A balanced chemical equation is one that correctly represents the substances (elements and compounds) before and after a reaction. When the equation is balanced, the starting numbers of atoms of elements (reactants) equal those after the reaction (products).

Solution Essentials:
- Balancing chemical equations by inspection
- Balanced chemical equation
- Chemical equation
- Reactants
- Products

Balance the following equations.

a $H_3PO_3 \longrightarrow H_3PO_4 + PH_3$ b $Ca + H_2O \longrightarrow Ca(OH)_2 + H_2$

c $Fe_2(SO_4)_3 + NH_3 + H_2O \longrightarrow Fe(OH)_3 + (NH_4)_2SO_4$

Problem Strategy Look at the chemical equation for atoms of elements that occur in only one substance on each side of the equation. Begin by balancing the equation in one of these atoms. Use the number of atoms on the left side of the arrow as the coefficient of the substance containing that element on the right side, and vice versa. After balancing one element in an equation, the rest become easier.

Solution

a Oxygen occurs in just one of the products (H_3PO_4). It is therefore easiest to balance O atoms first. To do this, note that H_3PO_3 has three O atoms; use 3 as the coefficient of H_3PO_4 on the right side. There are four O atoms in H_3PO_4 on the right side; use 4 as the coefficient of H_3PO_3 on the left side.

$$\underline{4}H_3PO_3 \longrightarrow \underline{3}H_3PO_4 + PH_3$$

This equation is now also balanced in P and H atoms. Thus, the balanced equation is

$$\mathbf{4H_3PO_3 \longrightarrow 3H_3PO_4 + PH_3}$$

b The equation is balanced in Ca atoms as it stands.

$$\underline{1}Ca + H_2O \longrightarrow \underline{1}Ca(OH)_2 + H_2$$

O atoms occur in only one reactant and in only one product, so they are balanced next.

$$\underline{1}Ca + \underline{2}H_2O \longrightarrow \underline{1}Ca(OH)_2 + H_2$$

The equation is now also balanced in H atoms. Thus, the answer is

$$\mathbf{Ca + 2H_2O \longrightarrow Ca(OH)_2 + H_2}$$

c To balance the equation in Fe atoms, you write

$$\underline{1}Fe_2(SO_4)_3 + NH_3 + H_2O \longrightarrow \underline{2}Fe(OH)_3 + (NH_4)_2SO_4$$

You balance the S atoms by placing the coefficient 3 for $(NH_4)_2SO_4$.

$$\underline{1}Fe_2(SO_4)_3 + \underline{6}NH_3 + H_2O \longrightarrow \underline{2}Fe(OH)_3 + \underline{3}(NH_4)_2SO_4$$

Now you balance the N atoms.

$$\underline{1}Fe_2(SO_4)_3 + \underline{6}NH_3 + H_2O \longrightarrow \underline{2}Fe(OH)_3 + \underline{3}(NH_4)_2SO_4$$

To balance the O atoms, you first count the number of O atoms on the right (18). Then you count the number of O atoms in substances on the left with known coefficients. There are 12 O's in $Fe_2(SO_4)_3$; hence, the number of remaining O's (in H_2O) must be 18 − 12 = 6.

$$\underline{1}Fe_2(SO_4)_3 + \underline{6}NH_3 + \underline{6}H_2O \longrightarrow \underline{2}Fe(OH)_3 + \underline{3}(NH_4)_2SO_4$$

Finally, note that the equation is now balanced in H atoms. The answer is

$$\mathbf{Fe_2(SO_4)_3 + 6NH_3 + 6H_2O \longrightarrow 2Fe(OH)_3 + 3(NH_4)_2SO_4}$$

(*continued*)

(*continued*)

Answer Check As a final step when balancing chemical equations, always count the number of atoms of each element on both sides of the equation to make certain that they are equal.

Exercise 2.13 Find the coefficients that balance the following equations.

a $O_2 + PCl_3 \longrightarrow POCl_3$

b $P_4 + N_2O \longrightarrow P_4O_6 + N_2$

c $As_2S_3 + O_2 \longrightarrow As_2O_3 + SO_2$

d $Ca_3(PO_4)_2 + H_3PO_4 \longrightarrow Ca(H_2PO_4)_2$

■ See Problems 2.97 and 2.98.

A Checklist for Review

Summary of Facts and Concepts

Atomic theory is central to chemistry. According to this theory, all matter is composed of small particles, or *atoms.* Each *element* is composed of the same kind of atom, and a *compound* is composed of two or more elements chemically combined in fixed proportions. A *chemical reaction* consists of the rearrangement of the atoms present in the reacting substances to give new chemical combinations present in the substances formed by the reaction. Although Dalton considered atoms to be the ultimate particles of matter, we now know that atoms themselves have structure. An atom has a *nucleus* and *electrons.* These atomic particles were discovered by J. J. Thomson and Ernest Rutherford.

Thomson established that cathode rays consist of negatively charged particles, or electrons, that are constituents of all atoms. Thomson measured the mass-to-charge ratio of the electron, and later Millikan measured its charge. From these measurements, the electron was found to be more than 1800 times lighter than the lightest atom.

Rutherford proposed the nuclear model of the atom to account for the results of experiments in which alpha particles were scattered from metal foils. According to this model, the atom consists of a central core, or nucleus, around which the electrons exist. The nucleus has most of the mass of the atom and consists of *protons* (with a positive charge) and *neutrons* (with no charge). Each chemically distinct atom has a nucleus with a specific number of protons (*atomic number*), and around the nucleus in the neutral atom are an equal number of electrons. The number of protons plus neutrons in a nucleus equals the *mass number.* Atoms whose nuclei have the same number of protons but different numbers of neutrons are called *isotopes.*

The *atomic mass* can be calculated from the isotopic masses and *fractional abundances* of the isotopes in a naturally occurring element. These data can be determined by the use of a mass spectrometer, which separates ions according to their mass-to-charge ratios.

The elements can be arranged in rows and columns by atomic number to form the *periodic table.* Elements in a given *group* (column) have similar properties. (A *period* is a row in the periodic table.) Elements on the left and at the center of the table are *metals;* those on the right are *nonmetals.*

A *chemical formula* is a notation used to convey the relative proportions of the atoms of the different elements in a substance. If the substance is molecular, the formula gives the precise number of each kind of atom in the molecule. If the substance is ionic, the formula also gives the relative number of different ions in the compound.

Chemical nomenclature is the systematic naming of compounds based on their formulas or structures. Rules are given for naming *ionic compounds, binary molecular compounds, acids,* and *hydrates.*

A chemical reaction occurs when the atoms in substances rearrange and combine into new substances. We represent a reaction by a *chemical equation,* writing a chemical formula for each reactant and product. The coefficients in the equation indicate the relative numbers of reactant and product molecules or formula units. Once the reactants and products and their formulas have been determined by experiment, we determine the coefficients by *balancing* the numbers of each kind of atom on both sides of the equation.

Learning Objectives	Important Terms
2.1 Atomic Theory of Matter	
▪ List the postulates of atomic theory. ▪ Define *element, compound,* and *chemical reaction* in the context of these postulates. ▪ Recognize the atomic symbols of the elements. ▪ Explain the significance of the law of multiple proportions.	**atomic theory** **atom** **element** **compound** **chemical reaction** **atomic symbol** **law of multiple proportions**
2.2 The Structure of the Atom	
▪ Describe Thomson's experiment in which he discovered the electron. ▪ Describe Rutherford's experiment that led to the nuclear model of the atom.	**nucleus** **electron**
2.3 Nuclear Structure; Isotopes	
▪ Name and describe the nuclear particles making up the nucleus of the atom. ▪ Define *atomic number, mass number,* and *nuclide.* ▪ Write the nuclide symbol for a given nuclide. ▪ Define and provide examples of *isotopes* of an element. ▪ Write the nuclide symbol of an element. Example 2.1	**proton** **atomic number (Z)** **neutron** **mass number (A)** **nuclide** **isotope**
2.4 Atomic Masses	
▪ Define *atomic mass unit* and *atomic mass.* ▪ Describe how a *mass spectrometer* can be used to determine the *fractional abundance* of the isotopes of an element. ▪ Determine the atomic mass of an element from the isotopic masses and fractional abundances. Example 2.2	**atomic mass unit (amu)** **atomic mass** **fractional abundance**
2.5 Periodic Table of the Elements	
▪ Identify periods and groups on the periodic table. ▪ Find the *main-group* and *transition* elements on the periodic table. ▪ Locate the *alkali metal* and *halogen* groups on the periodic table. ▪ Recognize the portions of the periodic table that contain the *metals, nonmetals,* and *metalloids* (*semimetals*).	**periodic table** **period (of periodic table)** **group (of periodic table)** **metal** **nonmetal** **metalloid (semimetal)**
2.6 Chemical Formulas; Molecular and Ionic Substances	
▪ Determine when the *chemical formula* of a compound represents a *molecule.* ▪ Determine whether a chemical formula is also a *molecular formula.* ▪ Define *ion, cation,* and *anion.* ▪ Classify compounds as *ionic* or *molecular.* ▪ Define and provide examples for the term *formula unit.* ▪ Specify the charge on all substances, ionic and molecular. ▪ Write an ionic formula, given the ions. Example 2.3	**chemical formula** **molecule** **molecular formula** **polymer** **monomer** **ion** **anion** **cation** **ionic compound** **formula unit**

2.7 Organic Compounds

- List the attributes of molecular substances that make them *organic compounds.*
- Explain what makes a molecule a *hydrocarbon.*
- Recognize some *functional groups* of organic molecules.

organic compound
hydrocarbon
functional group

2.8 Naming Simple Compounds

- Recognize *inorganic compounds.*
- Learn the rules for predicting the charges of *monatomic ions* in ionic compounds.
- Apply the rules for naming monatomic ions.
- Learn the names and charges of common *polyatomic ions.*
- Name an ionic compound from its formula. Example 2.4
- Write the formula of an ionic compound from its name. Example 2.5
- Determine the order of elements in a *binary (molecular) compound.*
- Learn the rules for naming binary molecular compounds, including the Greek prefixes.
- Name a binary compound from its formula. Example 2.6
- Write the formula of a binary compound from its name. Example 2.7
- Name a binary molecular compound from its molecular model. Example 2.8
- Recognize molecular compounds that are acids.
- Determine whether an acid is an *oxoacid.*
- Learn the approach for naming binary acids and oxoacids.
- Write the name and formula of an anion from the acid. Example 2.9
- Recognize compounds that are *hydrates.*
- Learn the rules for naming hydrates.
- Name a hydrate from its formula. Example 2.10
- Write the formula of a hydrate from its name. Example 2.11

chemical nomenclature
inorganic compound
monatomic ion
polyatomic ion
binary compound
oxoacid
hydrate

2.9 Writing Chemical Equations

- Identify the *reactants* and *products* in a chemical equation.
- Write chemical equations using appropriate phase labels, symbols of reaction conditions, and the presence of a catalyst.

chemical equation
reactant
product

2.10 Balancing Chemical Equations

- Determine if a chemical reaction is balanced.
- Master the techniques for balancing chemical equations. Example 2.12

Questions and Problems

Self-Assessment and Review Questions

Key: These questions test your understanding of the ideas you worked with in the chapter. These problems vary in difficulty and often can be used for the basis of discussion.

2.1 Describe atomic theory and discuss how it explains the great variety of different substances. How does it explain chemical reactions?

2.2 Two compounds of iron and chlorine, A and B, contain 1.270 g and 1.904 g of chlorine, respectively, for each gram of iron. Show that these amounts are in the ratio 2:3. Is this consistent with the law of multiple proportions? Explain.

2.3 Explain the operation of a cathode-ray tube. Describe the deflection of cathode rays by electrically charged plates placed within the cathode-ray tube. What does this imply about cathode rays?

2.4 Explain Millikan's oil-drop experiment.

2.5 What are the different kinds of particles in the atom's nucleus? Compare their properties with each other and with those of an electron.

2.6 Describe the nuclear model of the atom. How does this model explain the results of alpha-particle scattering from metal foils?

2.7 Describe how protons and neutrons were discovered to be constituents of nuclei.

2.8 Oxygen consists of three different _____, each having eight protons but different numbers of neutrons.

2.9 Describe how Dalton obtained relative atomic masses.

2.10 Define the term *atomic mass*. Why might the values of atomic masses on a planet elsewhere in the universe be different from those on earth?

2.11 Briefly explain how a mass spectrometer works. What kinds of information does one obtain from the instrument?

2.12 What is the name of the element in Group IVA and Period 5?

2.13 Cite some properties that are characteristic of a metal.

2.14 What is the difference between a molecular formula and a structural formula?

2.15 Ethane consists of molecules with two atoms of carbon and six atoms of hydrogen. Write the molecular formula for ethane.

2.16 What is the fundamental difference between an organic substance and an inorganic substance? Write chemical formulas of three inorganic molecules that contain carbon.

2.17 Give an example of a binary compound that is ionic. Give an example of a binary compound that is molecular.

2.18 Which of the following models represent a(n):

 a element
 b compound
 c mixture
 d ionic solid
 e gas made up of an element and a compound
 f mixture of elements
 g solid element
 h solid
 i liquid

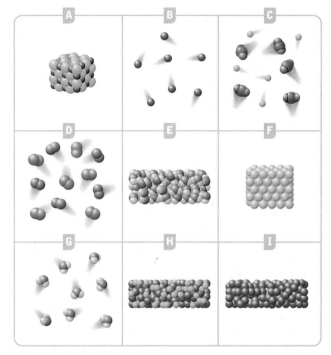

2.19 Explain what is meant by the term *balanced chemical equation.*

2.20 The compounds CuCl and $CuCl_2$ were formerly called cuprous chloride and cupric chloride, respectively. What are their names using the Stock system of nomenclature? What are the advantages of the Stock system of nomenclature over the former one?

2.21 How many protons, neutrons, and electrons are in $^{119}Sn^{2+}$?

 a 119 p, 50 n, 69 e$^-$ d 50 p, 69 n, 50 e$^-$
 b 50 p, 119 n, 52 e$^-$ e 50 p, 69 n, 48 e$^-$
 c 69 p, 50 n, 69 e

2.22 The atomic mass of Ga is 69.72 amu. There are only two naturally occurring isotopes of gallium: ^{69}Ga, with a mass of 69.0 amu, and ^{71}Ga, with a mass of 71.0 amu. The natural abundance of the ^{69}Ga isotope is approximately:

 a 15%   b 30%   c 50%   d 65%   e 80%

2.23 In which of the following are the name and formula correctly paired?

 a sodium sulfite: Na_2S
 b calcium carbonate: $Ca(CO_3)_2$
 c magnesium hydroxide: $Mg(OH)_2$
 d nitrite: NO_2
 e iron (III) oxide: FeO

2.24 A chunk of an unidentified element (let's call it element "X") is reacted with sulfur to form an ionic compound with the chemical formula X_2S. Which of the following elements is the most likely identity of X?

 a C   b Al   c Cl   d Mg   e Li

Concept Explorations

Key: Concept explorations are comprehensive problems that provide a framework that will enable you to explore and learn many of the critical concepts and ideas in each chapter. If you master the concepts associated with these explorations, you will have a better understanding of many important chemistry ideas and will be more successful in solving all types of chemistry problems. These problems are well suited for group work and for use as in-class activities.

2.25 Average Atomic Mass

Part 1: Consider the four identical spheres below, each with a mass of 2.00 g.

Calculate the average mass of a sphere in this sample.

Part 2: Now consider a sample that consists of four spheres, each with a different mass: blue mass is 2.00 g, red mass is 1.75 g, green mass is 3.00 g, and yellow mass is 1.25 g.

 a Calculate the average mass of a sphere in this sample.

 b How does the average mass for a sphere in this sample compare with the average mass of the sample that consisted just of the blue spheres? How can such different samples have their averages turn out the way they did?

Part 3: Consider two jars. One jar contains 100 blue spheres, and the other jar contains 25 each of red, blue, green, and yellow colors mixed together.

 a If you were to remove 50 blue spheres from the jar containing just the blue spheres, what would be the total mass of spheres left in the jar? (Note that the masses of the spheres are given in Part 2.)

 b If you were to remove 50 spheres from the jar containing the mixture (assume you get a representative distribution of colors), what would be the total mass of spheres left in the jar?

 c In the case of the mixture of spheres, does the average mass of the spheres *necessarily* represent the mass of an individual sphere in the sample?

 d If you had 80.0 grams of spheres from the blue sample, how many spheres would you have?

 e If you had 60.0 grams of spheres from the mixed-color sample, how many spheres would you have? What assumption did you make about your sample when performing this calculation?

Part 4: Consider a sample that consists of three green spheres and one blue sphere. The green mass is 3.00 g, and the blue mass is 1.00 g.

 a Calculate the *fractional abundance* of each sphere in the sample.

 b Use the fractional abundance to calculate the average mass of the spheres in this sample.

 c How are the ideas developed in this Concept Exploration related to the atomic masses of the elements?

2.26 Model of the Atom

Consider the following depictions of two atoms, which have been greatly enlarged so you can see the subatomic particles.

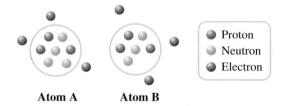

Atom A   Atom B

 a How many protons are present in atom A?

 b What is the significance of the number of protons depicted in atom A or any atom?

 c Can you identify the real element represented by the drawing of atom A? If so, what element does it represent?

 d What is the charge on atom A? Explain how you arrived at your answer.

 e Write the nuclide symbol of atom A.

 f Write the atomic symbol and the atomic number of atom B.

 g What is the mass number of atom B? How does this mass number compare with that of atom A?

 h What is the charge on atom B?

 i Write the nuclide symbol of atom B.

 j Draw pictures like those above of $^6_3Li^+$ and $^6_3Li^-$ atoms. What are the mass number and atomic number of each of these atoms?

 k Consider the two atoms depicted in this problem and the two that you just drew. What is the total number of lithium isotopes depicted? How did you make your decision?

 l Is the mass number of an isotope of an atom equal to the mass of the isotope of the atom? Be sure to explain your answer.

Conceptual Problems

Key: These problems are designed to check your understanding of the concepts associated with some of the main topics presented in each chapter. A strong conceptual understanding of chemistry is the foundation for both applying chemical knowledge and solving chemical problems.

These problems vary in level of difficulty and often can be used as a basis for group discussion.

2.27 A friend is trying to balance the following equation:

$$N_2 + H_2 \longrightarrow NH_3$$

He presents you with his version of the "balanced" equation:

$$N + H_3 \longrightarrow NH_3$$

You immediately recognize that he has committed a serious error; however, he argues that there is nothing wrong, since the equation is balanced. What reason can you give to convince him that his "method" of balancing the equation is flawed?

2.28 One of the early models of the atom proposed that atoms were wispy balls of positive charge with the electrons evenly distributed throughout. What would you expect to observe if you conducted Rutherford's experiment and the atom had this structure?

2.29 Given that the periodic table is an organizational scheme for the elements, what might be some other logical ways in which to group the elements that would provide meaningful chemical information in a periodic table of your own devising?

2.30 You have the mythical metal element "X" that can exist as X^+, X^{2+}, and X^{5+} ions.

a What would be the chemical formula for compounds formed from the combination of each of the X ions and $SO_4{}^{2-}$?

b If the name of the element X is exy, what would be the names of each of the compounds from part a of this problem?

2.31 You discover a new set of polyatomic anions that has the newly discovered element "X" combined with oxygen. Since you made the discovery, you get to choose the names for these new polyatomic ions. In developing the names, you want to pay attention to convention. A colleague in your lab has come up with a name that she really likes for one of the ions. How would you name the following in a way consistent with her name?

Formula	Name
$XO_4{}^{2-}$	
$XO_3{}^{2-}$	
$XO_2{}^{2-}$	excite
XO^{2-}	

2.32 Match the molecular model with the correct chemical formula: CH_3OH, NH_3, KCl, H_2O.

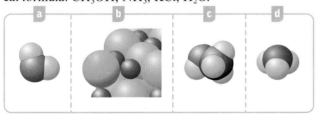

2.33 Currently, the atomic mass unit (amu) is based on being exactly one-twelfth the mass of a carbon-12 atom and is equal to 1.66×10^{-27} kg.

a If the amu were based on sodium-23 with a mass equal to exactly 1/23 of the mass of a sodium-23 atom, would the mass of the amu be different?

b If the new mass of the amu based on sodium-23 is 1.67×10^{-27} kg, how would the mass of a hydrogen atom, in amu, compare with the current mass of a hydrogen atom in amu?

2.34 Consider a hypothetical case in which the charge on a proton is twice that of an electron. Using this hypothetical case, and the fact that atoms maintain a charge of 0, how many protons, neutrons, and electrons would a potassium-39 atom contain?

2.35 For each of the following chemical reactions, write the correct element and/or compound symbols, formulas, and coefficients needed to produce complete, balanced equations. In all cases, the reactants are elements and the products are binary compounds.

a	____	+	____	$\longrightarrow$ LiCl
b	Na	+	S	$\longrightarrow$
c	Al	+	I_2	$\longrightarrow$
d	Ba	+	N_2	$\longrightarrow$
e	____	+	____	$\longrightarrow$ V_3P_5

2.36 You perform a chemical reaction using the hypothetical elements A and B. These elements are represented by their molecular models shown below:

A B_2

The product of the reaction represented by molecular models is

a Using the molecular models and the boxes, present a balanced chemical equation for the reaction of elements A and B.

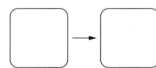

b Using the symbols A and B_2 for the chemical reaction, write a balanced chemical equation.

c What are some real-element possibilities for element B?

Practice Problems

Key: These problems are for practice in applying problem-solving skills. They are divided by topic, and some are keyed to exercises (see the ends of the exercises). The problems are arranged in matching pairs; the odd-numbered problem of each pair is listed first, and its answer is given in the back of the book.

2.37 What is the name of the element represented by each of the following atomic symbols?

a Ar   b Zn   c Ag   d Mg

2.38 For each atomic symbol, give the name of the element.

a Be   b Al   c Si   d C

2.39 Give the atomic symbol for each of the following elements.

a potassium   b sulfur   c iron   d manganese

2.40 Give the atomic symbol for each of the following elements.

a mercury   b calcium   c tin   d copper

Electrons, Protons, and Neutrons

2.41 The mass-to-charge ratio for the positive ion F^+ is 1.97×10^{-7} kg/C. Using the value of 1.602×10^{-19} C for the charge on the ion, calculate the mass of the fluorine atom. (The mass of the electron is negligible compared with that of the ion, so the ion mass is essentially the atomic mass.)

2.42 A student has determined the mass-to-charge ratio for an electron to be 5.64×10^{-12} kg/C. In another experiment, using Millikan's oil-drop apparatus, he found the charge on the electron to be 1.605×10^{-19} C. What would be the mass of the electron, according to these data?

2.43 The following table gives the number of protons and neutrons in the nuclei of various atoms. Which atom is the isotope of atom A? Which atom has the same mass number as atom A?

	Protons	Neutrons
Atom A	18	19
Atom B	16	19
Atom C	18	18
Atom D	17	20

2.44 The following table gives the number of protons and neutrons in the nuclei of various atoms. Which atom is the isotope of atom A? Which atom has the same mass number as atom A?

	Protons	Neutrons
Atom A	32	39
Atom B	33	38
Atom C	38	50
Atom D	32	38

2.45 Naturally occurring chlorine is a mixture of the isotopes Cl-35 and Cl-37. How many protons and how many neutrons are there in each isotope? How many electrons are there in the neutral atoms?

2.46 Naturally occurring nitrogen is a mixture of ^{14}N and ^{15}N. Give the number of protons, neutrons, and electrons in the neutral atom of each isotope.

2.47 An atom contains 34 protons and 45 neutrons. What is the nuclide symbol for the nucleus?

2.48 What is the nuclide symbol for the nucleus that contains 14 protons and 14 neutrons?

Atomic Masses

2.49 Ammonia is a gas with a characteristic pungent odor. It is sold as a water solution for use in household cleaning. The gas is a compound of nitrogen and hydrogen in the atomic ratio 1 : 3. A sample of ammonia contains 7.933 g N and 1.712 g H. What is the atomic mass of N relative to H?

2.50 Hydrogen sulfide is a gas with the odor of rotten eggs. The gas can sometimes be detected in automobile exhaust. It is a compound of hydrogen and sulfur in the atomic ratio 2 : 1. A sample of hydrogen sulfide contains 0.587 g H and 9.330 g S. What is the atomic mass of S relative to H?

2.51 An element has two naturally occurring isotopes with the following masses and abundances:

Isotopic Mass (amu)	Fractional Abundance
49.9472	2.500×10^{-3}
50.9440	0.9975

What is the atomic mass of this element? What is the identity of the element?

2.52 Calculate the atomic mass of an element with two naturally occurring isotopes, from the following data:

Isotope	Isotopic Mass (amu)	Fractional Abundance
X-63	62.930	0.6909
X-65	64.928	0.3091

What is the identity of element X?

2.53 An element has three naturally occurring isotopes with the following masses and abundances:

Isotopic Mass (amu)	Fractional Abundance
38.964	0.9326
39.964	1.000×10^{-4}
40.962	0.0673

Calculate the atomic mass of this element. What is the identity of the element?

2.54 An element has three naturally occurring isotopes with the following masses and abundances:

Isotopic Mass (amu)	Fractional Abundance
27.977	0.9221
28.976	0.0470
29.974	0.0309

Calculate the atomic mass of this element. What is the identity of the element?

2.55 While traveling to a distant universe, you discover the hypothetical element "X." You obtain a representative sample of the element and discover that it is made up of two isotopes, X-23 and X-25. To help your science team calculate the atomic mass of the substance, you send the following drawing of your sample with your report.

In the report, you also inform the science team that the brown atoms are X-23, which have an isotopic mass of 23.02 amu, and the green atoms are X-25, which have an isotopic mass of 25.147 amu. What is the atomic mass of element X?

2.56 While roaming a parallel universe, you discover the hypothetical element "Z." You obtain a representative sample of the element and discover that it is made up of two isotopes, Z-47 and Z-51. To help your science team calculate the atomic mass of the substance, you send the following drawing of your sample with your report.

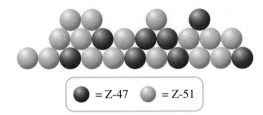

= Z-47   = Z-51

In the report, you also inform the science team that the blue atoms are Z-47, which have an isotopic mass of 47.621 amu, and the orange atoms are Z-51, which have an isotopic mass of 51.217 amu. What is the atomic mass of element Z?

Periodic Table

2.57 Refer to the periodic table (Figure 2.15 or inside front cover) and answer the following questions.
  a What Group VIA element is a metalloid?
  b What is the Group IIIA element in Period 3?

2.58 Refer to the periodic table (Figure 2.15 or inside front cover) and answer the following questions.
  a What Group VA element is a metal?
  b What is the Group IIA element in Period 2?

2.59 Identify the group and period for each of the following. Refer to the periodic table (Figure 2.15 or inside front cover). Label each as a metal, nonmetal, or metalloid.
  a C   b Po   c Cr   d Mg   e B

2.60 Refer to the periodic table (Figure 2.15 or inside front cover) and obtain the group and period for each of the following elements. Also determine whether the element is a metal, nonmetal, or metalloid.
  a V   b Rb   c B   d I   e He

2.61 Give one example (atomic symbol and name) for each of the following.
  a a main-group (representative) element in the second period
  b a halogen
  c a transition element in the fourth period
  d a lanthanide element

2.62 Give one example (atomic symbol and name) for each of the following.
  a an actinide element
  b a transition element in the fourth period
  c a main-group (representative) element in the third period
  d an alkali metal

Molecular and Ionic Substances

2.63 The normal form of the element sulfur is a brittle, yellow solid. This is a molecular substance, S_8. If this solid is vaporized, it first forms S_8 molecules; but at high temperature, S_2 molecules are formed. How do the molecules of the solid sulfur and of the hot vapor differ? How are the molecules alike?

2.64 White phosphorus is available in sticks, which have a waxy appearance. This is a molecular substance, P_4. When this solid is vaporized, it first forms P_4 molecules; but at high temperature, P_2 molecules are formed. How do the molecules of white phosphorus and those of the hot vapor differ? How are the molecules alike?

2.65 A 1.50-g sample of nitrous oxide (an anesthetic, sometimes called laughing gas) contains 2.05×10^{22} N_2O molecules. How many nitrogen atoms are in this sample? How many nitrogen atoms are in 44.0 g of nitrous oxide?

2.66 Nitric acid is composed of HNO_3 molecules. A sample weighing 4.50 g contains 4.30×10^{22} HNO_3 molecules. How many nitrogen atoms are in this sample? How many oxygen atoms are in 61.0 g of nitric acid?

2.67 A sample of ethanol (ethyl alcohol), C_2H_5OH, contains 4.2×10^{23} hydrogen atoms. How many C_2H_5OH molecules are in this sample?

2.68 A sample of ammonia, NH_3, contains 3.3×10^{21} hydrogen atoms. How many NH_3 molecules are in this sample?

2.69 Give the molecular formula for each of the following structural formulas.

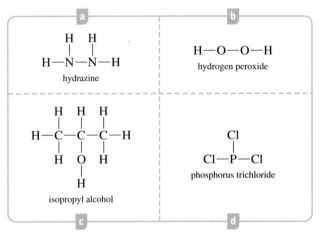

2.70 What molecular formula corresponds to each of the following structural formulas?

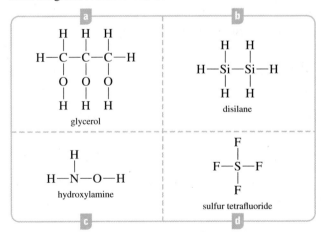

2.71 Write the molecular formula for each of the following compounds represented by molecular models.

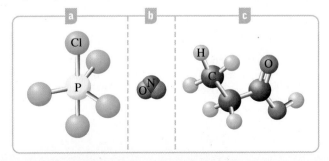

2.72 Write the molecular formula for each of the following compounds represented by molecular models.

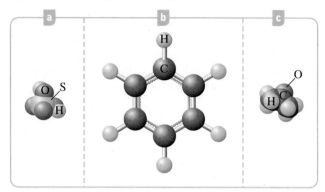

2.73 Ammonium phosphate, $(NH_4)_3PO_4$, has how many oxygen atoms for each nitrogen atom?

2.74 Iron(II) nitrate has the formula $Fe(NO_3)_2$. What is the ratio of iron atoms to oxygen atoms in this compound?

2.75 Write the formula for the compound of each of the following pairs of ions.
- a Fe^{3+} and CN^-
- b K^+ and CO_3^{2-}
- c Li^+ and N^{3-}
- d Ca^{2+} and P^{3-}

2.76 For each of the following pairs of ions, write the formula of the corresponding compound.
- a Co^{2+} and N^{3-}
- b NH_4^+ and PO_4^{3-}
- c Na^+ and SO_3^{2-}
- d Fe^{3+} and OH^-

Chemical Nomenclature

2.77 Name the following compounds.
- a Na_2SO_4
- b Na_3N
- c $CuCl$
- d Cr_2O_3

2.78 Name the following compounds.
- a CaO
- b Mn_2O_3
- c NH_4HCO_3
- d $Cu(NO_3)_2$

2.79 Write the formulas of:
- a lead(II) permanganate
- b barium hydrogen carbonate
- c cesium sulfide
- d iron(III) acetate

2.80 Write the formulas of:
- a sodium thiosulfate
- b copper(II) hydroxide
- c calcium hydrogen carbonate
- d chromium(III) phosphide

2.81 For each of the following binary compounds, decide whether the compound is expected to be ionic or molecular.
- a SeF_4
- b $LiBr$
- c SiF_4
- d Cs_2O

2.82 For each of the following binary compounds, decide whether the compound is expected to be ionic or molecular.
- a H_2O
- b As_4O_6
- c ClF_3
- d Fe_2O_3

2.83 Give systematic names to the following binary compounds.
- a N_2O
- b P_4O_{10}
- c $AsCl_3$
- d Cl_2O_7

2.84 Give systematic names to the following binary compounds.
- a N_2F_2
- b CF_4
- c N_2O_5
- d As_4O_6

2.85 Write the formulas of the following compounds.
- a diphosphorus pentoxide
- b nitrogen dioxide
- c dinitrogen tetrafluoride
- d boron trifluoride

2.86 Write the formulas of the following compounds.
- a nitrogen tribromide
- b xenon hexafluoride
- c carbon monoxide
- d dichlorine pentoxide

2.87 Write the systematic name for each of the following compounds represented by a molecular model.

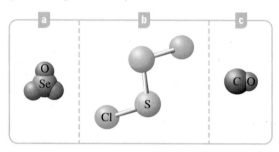

2.88 Write the systematic name for each of the following molecules represented by a molecular model.

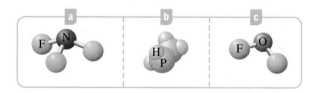

2.89 Give the name and formula of the acid corresponding to each of the following oxoanions.
- a sulfite ion, SO_3^{2-}
- b hyponitrite ion, $N_2O_2^{2-}$
- c disulfite ion, $S_2O_5^{2-}$
- d arsenate ion, AsO_4^{3-}

2.90 Give the name and formula of the acid corresponding to each of the following oxoanions.
- a hypoiodite ion, IO^-
- b chlorite ion, ClO_2^-
- c selenite ion, SeO_3^{2-}
- d nitrate ion, NO_3^-

2.91 Glauber's salt has the formula $Na_2SO_4 \cdot 10H_2O$. What is the chemical name of this substance?

2.92 Emerald-green crystals of the substance $NiSO_4 \cdot 6H_2O$ are used in nickel plating. What is the chemical name of this compound?

2.93 Iron(II) sulfate heptahydrate is a blue-green, crystalline compound used to prepare other iron compounds. What is the formula of iron(II) sulfate heptahydrate?

2.94 Cobalt(II) chloride hexahydrate has a pink color. It loses water on heating and changes to a blue-colored compound. What is the formula of cobalt(II) chloride hexahydrate?

Chemical Equations

2.95 In the equation $2PbS + O_2 \longrightarrow 2PbO + 2SO_2$, how many oxygen atoms are there on the right side? Is the equation balanced as written?

2.96 For the balanced chemical equation $Pb(NO_3)_2 + K_2CO_3 \longrightarrow PbCO_3 + 2KNO_3$, how many oxygen atoms are on the right side?

2.97 Balance the following equations.
a $Sn + NaOH \longrightarrow Na_2SnO_2 + H_2$
b $Al + Fe_3O_4 \longrightarrow Al_2O_3 + Fe$
c $CH_3OH + O_2 \longrightarrow CO_2 + H_2O$
d $P_4O_{10} + H_2O \longrightarrow H_3PO_4$
e $PCl_5 + H_2O \longrightarrow H_3PO_4 + HCl$

2.98 Balance the following equations.
a $NO_2 + H_2O \longrightarrow HNO_3 + NO$
b $MnO_2 + HCl \longrightarrow MnCl_2 + Cl_2 + H_2O$
c $Na_2S_2O_3 + I_2 \longrightarrow NaI + Na_2S_4O_6$

d $Al_4C_3 + H_2O \longrightarrow Al(OH)_3 + CH_4$
e $Ca_3(PO_4)_2 + H_3PO_4 \longrightarrow Ca(H_2PO_4)_2$

2.99 Solid calcium phosphate and aqueous sulfuric acid solution react to give calcium sulfate, which comes out of the solution as a solid. The other product is phosphoric acid, which remains in solution. Write a balanced equation for the reaction using complete formulas for the compounds with phase labels.

2.100 Solid sodium metal reacts with water, giving a solution of sodium hydroxide and releasing hydrogen gas. Write a balanced equation for the reaction using complete formulas for the compounds with phase labels.

2.101 Lead metal is produced by heating solid lead(II) sulfide with solid lead(II) sulfate, resulting in liquid lead and sulfur dioxide gas. Write a balanced equation for the reaction using complete formulas for the compounds with phase labels; indicate that the reactants are heated.

2.102 An aqueous solution of ammonium chloride and barium hydroxide is heated, and the compounds react to give off ammonia gas. Barium chloride solution and water are also products. Write a balanced equation for the reaction using complete formulas for the compounds with phase labels; indicate that the reactants are heated.

General Problems

Key: These problems provide more practice but are not divided by topic or keyed to exercises. Each section ends with essay questions, each of which is color coded to refer to the *A Chemist Looks at Frontiers* (purple) chapter essay on which it is based. Odd-numbered problems and the even-numbered problems that follow are similar; answers to all odd-numbered problems except the essay questions are given in the back of the book.

2.103 Two samples of different compounds of sulfur and oxygen have the following compositions. Show that the compounds follow the law of multiple proportions. What is the ratio of oxygen in the two compounds for a fixed amount of sulfur?

	Amount S	*Amount O*
Compound A	1.210 g	1.811 g
Compound B	1.783 g	1.779 g

2.104 Two samples of different compounds of nitrogen and oxygen have the following compositions. Show that the compounds follow the law of multiple proportions. What is the ratio of oxygen in the two compounds for a fixed amount of nitrogen?

	Amount N	*Amount O*
Compound A	1.206 g	2.755 g
Compound B	1.651 g	4.714 g

2.105 In a series of oil-drop experiments, the charges measured on the oil drops were -3.20×10^{-19} C, -6.40×10^{-19} C, -9.60×10^{-19} C, and -1.12×10^{-18} C. What is the smallest difference in charge between any two drops? If this is assumed to be the charge on the electron, how many excess electrons are there on each drop?

2.106 In a hypothetical universe, an oil-drop experiment gave the following measurements of charges on oil drops: -5.55×10^{-19} C, -9.25×10^{-19} C, -1.11×10^{-18} C, and -1.48×10^{-18} C. Assume that the smallest difference in charge equals the unit of negative charge in this universe. What is the value of this unit of charge? How many units of excess negative charge are there on each oil drop?

2.107 Compounds of europium, Eu, are used to make color television screens. The europium nucleus has a charge of $+63$. How many electrons are there in the neutral atom? in the Eu^{3+} ion?

2.108 Cesium, Cs, is used in photoelectric cells ("electric eyes"). The cesium nucleus has a charge of $+55$. What is the number of electrons in the neutral atom? in the Cs^+ ion?

2.109 One isotope of a metallic element has mass number 74 and has 51 neutrons in the nucleus. An atomic ion has 18 electrons. Write the symbol for this ion (give the symbol for the nucleus and give the ionic charge as a right superscript).

2.110 A nucleus of mass number 81 contains 46 neutrons. An atomic ion of this element has 36 electrons in it. Write the symbol for this atomic ion (give the symbol for the nucleus and give the ionic charge as a right superscript).

2.111 Obtain the fractional abundances for the two naturally occurring isotopes of copper. The masses of the isotopes are $^{63}_{29}Cu$, 62.9298 amu; $^{65}_{29}Cu$, 64.9278 amu. The atomic mass is 63.546 amu.

2.112 Silver has two naturally occurring isotopes, one of mass 106.91 amu and the other of mass 108.90 amu. Find the fractional abundances for these two isotopes. The atomic mass is 107.87 amu.

2.113 Identify the following elements, giving their name and atomic symbol.

 a a transition element in Group VB, Period 5
 b the halogen in Period 2
 c a nonmetal that is normally a liquid
 d a normally gaseous element in Group IA

2.114 Identify the following elements, giving their name and atomic symbol.

 a a normally liquid element in Group VIIA
 b a metal that is normally a liquid
 c a main-group element in Group IIIA, Period 3
 d the alkali metal in Period 4

2.115 Give the names of the following ions.
 a Mn^{2+} b Ni^{2+} c Co^{2+} d Fe^{3+}

2.116 Give the names of the following ions.
 a Cr^{3+} b Pb^{4+} c Ti^{2+} d Cu^{2+}

2.117 Write formulas for all the ionic compounds that can be formed by combinations of these ions: Na^+, Co^{2+}, SO_4^{2-}, and Cl^-.

2.118 Write formulas for all the ionic compounds that can be formed by combinations of these ions: Mg^{2+}, Cr^{3+}, S^{2-}, and NO^{3-}.

2.119 Name the following compounds.
 a $Sn_3(PO_4)_2$ b NH_4NO_2
 c $Mg(OH)_2$ d $NiSO_3$

2.120 Name the following compounds.
 a $Cu(NO_2)_2$ b $(NH_4)_3P$
 c K_2SO_3 d Hg_3N_2

2.121 Give the formulas for the following compounds.
 a mercury(I) sulfide
 b cobalt(III) sulfite
 c ammonium dichromate
 d aluminum fluoride

2.122 Give the formulas for the following compounds.
 a magnesium phosphate
 b calcium carbonate
 c lead(IV) phosphide
 d hydrogen peroxide

2.123 Name the following molecular compounds.
 a $AsBr_3$ b H_2Te c P_2O_5 d SiO_2

2.124 Name the following molecular compounds.
 a ClF_4 b CS_2 c PF_3 d SF_6

2.125 Balance the following equations.
 a $NaOH + H_2CO_3 \longrightarrow Na_2CO_3 + H_2O$
 b $SiCl_4 + H_2O \longrightarrow SiO_2 + HCl$
 c $Ca_3(PO_4)_2 + C \longrightarrow Ca_3P_2 + CO$
 d $H_2S + O_2 \longrightarrow SO_2 + H_2O$
 e $N_2O_5 \longrightarrow NO_2 + O_2$

2.126 Balance the following equations.
 a $C_2H_6 + O_2 \longrightarrow CO_2 + H_2O$
 b $P_4O_6 + H_2O \longrightarrow H_3PO_3$
 c $KClO_3 \longrightarrow KCl + KClO_4$
 d $(NH_4)_2SO_4 + NaOH \longrightarrow NH_3 + H_2O + Na_2SO_4$
 e $NBr_3 + NaOH \longrightarrow N_2 + NaBr + HOBr$

2.127 A monatomic ion has a charge of +2. The nucleus of the ion has a mass number of 62. The number of neutrons in the nucleus is 1.21 times that of the number of protons. How many electrons are in the ion? What is the name of the element?

2.128 A monatomic ion has a charge of +1. The nucleus of the ion has a mass number of 85. The number of neutrons in the nucleus is 1.30 times that of the number of protons. How many electrons are in the ion? What is the name of the element?

2.129 Natural carbon, which has an atomic mass of 12.011 amu, consists of carbon-12 and carbon-13 isotopes. Given that the mass of carbon-13 is 13.00335 amu, what would be the average atomic mass (in amu) of a carbon sample prepared by mixing equal numbers of carbon atoms from a sample of natural carbon and a sample of pure carbon-13?

2.130 Natural chlorine, which has an atomic mass of 35.4527 amu, consists of chlorine-35 and chlorine-37 isotopes. Given that the mass of chlorine-35 is 34.96885 amu, what is the average atomic mass (in amu) of a chlorine sample prepared by mixing equal numbers of chlorine atoms from a sample of natural chlorine and a sample of pure chlorine-35?

■ **2.131** Write the equation for the nuclear reaction in which element 112 was first produced.

■ **2.132** Describe the island of stability. What nuclide is predicted to be most stable?

Strategy Problems

Key: As noted earlier, all of the practice and general problems are matched pairs. This section is a selection of problems that are not in matched-pair format. These challenging problems require that you employ many of the concepts and strategies that were developed in the chapter. In some cases, you will have to integrate several concepts and operational skills in order to solve the problem successfully.

2.133 Correct any mistakes in the naming of the following compounds or ions.

   SO_3: sulfite

   NO_2: nitrate

   PO_4^{3-}: phosphite ion

   N_2: nitride

   $Mg(OH)_2$: maganese dihydroxide

2.134 Aluminum is reacted with oxygen to form a binary compound. Write the balanced chemical reaction and write the name of the compound formed by the reaction.

2.135 An unknown metal (let's call it "M") is reacted with sulfur to produce a compound with the chemical formula MS. What is the charge on the metal in the compound MS? Name a metal that could be metal M.

2.136 A hypothetical element X is found to have an atomic mass of 37.45 amu. Element X has only two isotopes, X-37 and X-38. The X-37 isotope has a fractional abundance of 0.7721 and an isotopic mass of 37.24. What is the isotopic mass of the other isotope?

2.137 Ammonia gas reacts with molecular oxygen gas to form nitrogen monoxide gas and liquid water. Write the complete balanced reaction with all proper state symbols.

2.138 A monotomic ion has a charge of $+3$. The nucleus of the ion has a mass number of 27. The number of neutrons in the nucleus is equal to the number of electrons in a S^{2+} ion. Identify the element and indicate the number of protons, neutrons, and electrons.

2.139 A small crystal of $CaCl_2$ that weighs 0.12 g contains 6.5×10^{20} formula units of $CaCl_2$. What is the total number of ions (cations and anions) that make up this crystal?

2.140 Write the formulas and names for all the ionic compounds that can form by combinations of the following ions: Mg^{2+}, Cr^{3+}, the carbonate anion, and the nitride anion.

2.141 The IO_3^- anion is called iodate. There are three related ions: IO^-, IO_2^-, and IO_4^-. Using what you have learned about similar groups of anions, write the name for each of the following compounds:

HIO_3

$NaIO_4$

$Mg(IO_2)_2$

$Fe(IO_2)_3$

2.142 Name the following compounds:

SO_3

HNO_2

Mg_3N_3

$HI(aq)$

$Cu_3(PO_4)_2$

$CuSO_4 \cdot 5H_2O$

2.143 Write out in words the following chemical equations:

 a $PbCl_2(aq) + Na_2S(aq) \longrightarrow PbS(s) + 2NaCl(aq)$

 b $H_2O(l) + SO_3(g) \longrightarrow H_2SO_4(aq)$

 c $C(s) + O_2(g) \longrightarrow CO_2(g)$

 d $H_2(g) + I_2(g) \longrightarrow 2HI(g)$

2.144 From the following written description, write the balanced chemical equation for the reaction including state symbols. A diatomic gaseous molecule that contains 17 protons per atom is reacted with a solid element that has an atomic number of 19 to yield an ionic compound.

2.145 An 18-mL sample of water contains 6.0×10^{23} water molecules. Assuming that this sample consists only of the Hydrogen-1 and Oxygen-16 isotopes, calculate the number of neutrons, protons, and electrons in this water sample.

2.146 Name the following compounds:

 a $HCl(g)$

 b $HF(g)$

 c $HBr(aq)$

 d $HNO_3(aq)$

2.147 During nuclear decay a ^{238}U atom can break apart into a helium-4 atom and one other atom. Assuming that no subatomic particles are destroyed during this decay process, what is the other element produced?

2.148 Write the balanced chemical equation for the reaction of an aqueous oxacid that contains the phosphate anion with solid magnesium hydroxide. The reaction produces liquid dihydrogen monoxide and a solid ionic compound formed from the combination of magnesium cations and phosphate anions.

Cumulative-Skills Problems

Key: The problems under this heading combine skills introduced in the previous chapter with those given in the current one.

2.149 There are 2.619×10^{22} atoms in 1.000 g of sodium. Assume that sodium atoms are spheres of radius 1.86 Å and that they are lined up side by side. How many miles in length is the line of sodium atoms?

2.150 There are 1.699×10^{22} atoms in 1.000 g of chlorine. Assume that chlorine atoms are spheres of radius 0.99 Å and that they are lined up side by side in a 0.5-g sample. How many miles in length is the line of chlorine atoms in the sample?

2.151 A sample of green crystals of nickel(II) sulfate heptahydrate was heated carefully to produce the bluish green nickel(II) sulfate hexahydrate. What are the formulas of the hydrates? If 8.753 g of the heptahydrate produces 8.192 g of the hexahydrate, how many grams of anhydrous nickel(II) sulfate could be obtained?

2.152 Cobalt(II) sulfate heptahydrate has pink-colored crystals. When heated carefully, it produces cobalt(II) sulfate monohydrate, which has red crystals. What are the formulas of these hydrates? If 3.548 g of the heptahydrate yields 2.184 g of the monohydrate, how many grams of the anhydrous cobalt(II) sulfate could be obtained?

2.153 A sample of metallic element X, weighing 4.315 g, combines with 0.4810 L of Cl_2 gas (at normal pressure and 20.0°C) to form the metal chloride with the formula XCl. If the density of Cl_2 gas under these conditions is 2.948 g/L, what is the mass of the chlorine? The atomic mass of chlorine is 35.453 amu. What is the atomic mass of X? What is the identity of X?

2.154 A sample of metallic element X, weighing 3.177 g, combines with 0.6015 L of O_2 gas (at normal pressure and 20.0°C) to form the metal oxide with the formula XO. If the density of O_2 gas under these conditions is 1.330 g/L, what is the mass of this oxygen? The atomic mass of oxygen is 15.9994 amu. What is the atomic mass of X? What is the identity of X?

Calculations with Chemical Formulas and Equations

© Cengage Learning; Handout/Reuters/Corbis

Zinc and iodine react to produce zinc iodide, a solution used to find flaws in composite airplane wings. Zinc iodide solution sprayed on a wing will fill a crack, if present, which shows up on an x ray as an opaque area (from the absorption of x rays by iodine atoms).

Contents and Concepts

Mass and Moles of Substance
Here we will establish a critical relationship between the mass of a chemical substance and the quantity of that substance (in moles).

3.1 Molecular Mass and Formula Mass

3.2 The Mole Concept

Determining Chemical Formulas
Explore how the percentage composition and mass percentage of the elements in a chemical substance can be used to determine the chemical formula.

3.3 Mass Percentages from the Formula

3.4 Elemental Analysis: Percentages of Carbon, Hydrogen, and Oxygen

3.5 Determining Formulas

Stoichiometry: Quantitative Relations in Chemical Reactions
Develop a molar interpretation of chemical equations, which then allows for calculation of the quantities of reactants and products.

3.6 Molar Interpretation of a Chemical Equation

3.7 Amounts of Substances in a Chemical Reaction

3.8 Limiting Reactant; Theoretical and Percentage Yields

OWL Sign in to OWL at **www.cengage.com/owl** to view tutorials and simulations, develop problem-solving skills, and complete online homework assigned by your professor.

GoChemistry Download mini lecture videos for key concept review and exam prep from OWL or purchase them from **www.cengagebrain.com**.

Acetic acid (ah-see'-tik acid) is a colorless liquid with a sharp, vinegary odor. In fact, vinegar contains acetic acid, which accounts for vinegar's odor and sour taste. The name *vinegar* derives from the French word *vinaigre,* meaning "sour wine." Vinegar results from the fermentation of wine or cider by certain bacteria. These bacteria require oxygen, and the overall chemical change is the reaction of ethanol (alcohol) in wine with oxygen to give acetic acid (Figure 3.1).

Laboratory preparation of acetic acid may also start from ethanol, which reacts with oxygen in two steps. First, ethanol reacts with oxygen to yield a compound called acetaldehyde, in addition to water. In the second step, the acetaldehyde reacts with more oxygen to produce acetic acid. (The human body also produces acetaldehyde and then acetic acid from alcohol, as it attempts to eliminate alcohol from the system.)

This chapter focuses on two basic questions, which we can illustrate using these compounds: How do you determine the chemical formula of a substance such as acetic acid (or acetaldehyde)? How much acetic acid can you prepare from a given quantity of ethanol (or a given quantity of acetaldehyde)? These types of questions are very important in chemistry. You must know the formulas of all the substances involved in a reaction before you can write the chemical equation, and you need the balanced chemical equation to determine the quantitative relationships among the different substances in the reaction. We begin by discussing how you relate number of atoms or molecules to grams of substance, because this is the key to answering both questions.

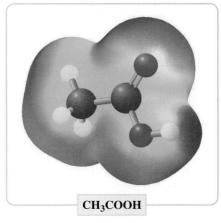

CH_3COOH

Figure 3.1 ▲

Electrostatic potential map of acetic acid molecule (CH_3COOH) Acetic acid is the compound responsible for the sour taste and smell of vinegar.

Mass and Moles of Substance

You buy a quantity of groceries in several ways. You often purchase items such as oranges and lemons by counting out a particular number. Some things, such as eggs and soda, can be purchased in a "package" that represents a known quantity—for example, a dozen or a case. Bulk foods, such as peanuts or hard candy, are usually purchased by mass, because it is too tedious to count them out, and because we know that a given mass yields a certain quantity of the item. All three of these methods are used by chemists to determine the quantity of matter: counting, using a package that represents a quantity, and measuring the mass. For a chemist, it is a relatively easy matter to weigh a substance to obtain the mass. However, the number of atoms or molecules in even a seemingly minute amount is much too large to count. (You may recall from Section 2.6 that a billionth of a drop of water contains 2×10^{12} H_2O molecules.) Nevertheless, chemists are interested in knowing such numbers. How many atoms of carbon are there in one molecule of acetic acid? How many molecules of acetic acid can be obtained from one molecule of ethanol? Before we look at how chemists solve this problem of measuring numbers of atoms, molecules, and ions, we must look at the concept of molecular mass (or molecular weight) and formula mass (or formula weight) and introduce the package chemists call the mole.

3.1 Molecular Mass and Formula Mass

In Chapter 2 (Section 2.4), we discussed the concept of atomic mass. We can easily extend this idea to include molecular mass. The **molecular mass** (MM) of a substance is *the sum of the atomic masses of all the atoms in a molecule of the substance.* It is, therefore, the average mass of a molecule of that substance, expressed in atomic mass units. For example, the molecular mass of water, H_2O, is 18.0 amu (2×1.0 amu from two H atoms plus 16.0 amu from one O atom). If the molecular formula for the substance is not known, you can determine the molecular mass experimentally by means of a mass spectrometer. In later chapters we will discuss simple, inexpensive methods for determining molecular mass.

The **formula mass** (FM) of a substance is *the sum of the atomic masses of all atoms in a formula unit of the compound,* whether molecular or not. Sodium chloride, with the formula unit NaCl, has a formula mass of 58.44 amu (22.99 amu from Na plus 35.45 amu from Cl). NaCl is ionic, so strictly speaking the expression "molecular mass of NaCl" has no meaning. On the other hand, the molecular mass and the formula mass calculated from the molecular formula of a substance are identical. ◄

Some chemists use the term molecular *mass in a less strict sense for ionic as well as molecular compounds.*

Example 3.1

Calculating the Formula Mass from a Formula

Gaining Mastery Toolbox

Critical Concept 3.1
The formula mass of a compound is based on the number and type of each element in the compound. The formula mass is calculated by summing the masses of all atoms in the chemical formula.

Solution Essentials:
• Formula mass
• Atomic mass
• Atomic mass unit
• Periodic table
• Rules for significant figures and rounding

Calculate the formula mass of each of the following to three significant figures, using a table of atomic masses (AM): a chloroform, $CHCl_3$; b iron(III) sulfate, $Fe_2(SO_4)_3$.

Problem Strategy Identify the number and type of atoms in the chemical formula. Use the periodic table to obtain the atomic mass of each of the elements present in the compounds. Taking into account the number of each atom present in the formula, sum the masses. If atoms in the formula are enclosed within parentheses as in part b, the number of each element within the parentheses should be multiplied by the subscript that follows the final, or closing, parenthesis.

Solution

a The calculation is

$$
\begin{array}{ll}
1 \times \text{AM of C} = & 12.0 \text{ amu} \\
1 \times \text{AM of H} = & 1.0 \text{ amu} \\
3 \times \text{AM of Cl} = 3 \times 35.45 \text{ amu} = & \underline{106.4 \text{ amu}} \\
\text{FM of } CHCl_3 = & 119.4 \text{ amu}
\end{array}
$$

The answer rounded to three significant figures is **119 amu.**

b The calculation is

$$
\begin{array}{lll}
2 \times \text{AM of Fe} = & 2 \times 55.8 \text{ amu} = & 111.6 \text{ amu} \\
3 \times \text{AM of S} = & 3 \times 32.1 \text{ amu} = & 96.3 \text{ amu} \\
3 \times 4 \times \text{AM of O} = & 12 \times 16.00 \text{ amu} = & \underline{192.0 \text{ amu}} \\
\text{FM of } Fe_2(SO_4)_3 = & & 399.9 \text{ amu}
\end{array}
$$

The answer rounded to three significant figures is **4.00×10^2 amu.**

Answer Check The most common error when calculating formula masses is not correctly accounting for those atoms that are enclosed in parentheses.

Exercise 3.1   Calculate the formula masses of the following compounds, using a table of atomic masses. Give the answers to three significant figures. a nitrogen dioxide, NO_2; b glucose, $C_6H_{12}O_6$; c sodium hydroxide, NaOH; d magnesium hydroxide, $Mg(OH)_2$.

■ See Problems 3.27 and 3.28.

Example 3.2

Calculating the Formula Mass from Molecular Models

Gaining Mastery Toolbox

Critical Concept 3.2
A molecular model depicts the number and type of each atom present in the molecular compound. Molecular models can be used to determine the chemical formula of a molecular compound.

Solution Essentials:
• Formula mass   • Periodic table
• Atomic mass   • Molecular model
• Atomic mass unit   • Rules for significant figures and rounding

For the following two compounds, write the molecular formula and calculate the formula mass to four significant figures:

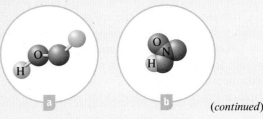

(continued)

(continued)

Problem Strategy In order to calculate the formula mass, you need to have the total number of each type of atom present in the structure, so look at the structure and keep a tally of all the atoms present.

Solution

a This molecular model is of a molecule that is composed of two O and two H atoms. For inorganic compounds, the elements in a chemical formula are written in order such that the most metallic element is listed first. (Even though H is not metallic, it is positioned in the periodic table in such a way that it is considered to be a metal when writing formulas.) Hence, the chemical formula is H_2O_2. Using the same approach as Example 3.1, calculating the formula mass yields **34.02 amu.**

b This molecular model represents a molecule made up of one N atom, three O atoms, and one H atom. The chemical formula is then HNO_3. The formula mass is **63.01 amu.**

Answer Check Always make sure that your answer is reasonable. If you obtain a molecular mass of more than 200 amu for a simple molecular compound (this is possible but not typical), it is advisable to check your work closely.

Exercise 3.2 For the following compounds, write the molecular formula and calculate the formula mass to three significant figures.

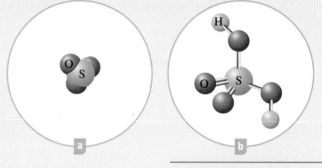

■ See Problems 3.29 and 3.30.

3.2 The Mole Concept

When we prepare a compound industrially or even study a reaction in the laboratory, we deal with tremendous numbers of molecules or ions. Suppose you wish to prepare acetic acid, starting from 10.0 g of ethanol. This small sample (less than 3 teaspoonsful) contains 1.31×10^{23} molecules, a truly staggering number. Imagine a device that counts molecules at the rate of one million per second. It would take more than four billion years—nearly the age of the earth—for this device to count that many molecules! Chemists have adopted the *mole concept* as a convenient way to deal with the enormous numbers of molecules or ions in the samples they work with.

Definition of Mole and Molar Mass

A **mole** (symbol **mol**) is defined as *the quantity of a given substance that contains as many molecules or formula units as the number of atoms in exactly 12 g of carbon-12.* One mole of ethanol, for example, contains the same number of ethanol molecules as there are carbon atoms in 12 g of carbon-12.

The number of atoms in a 12-g sample of carbon-12 is called **Avogadro's number** (to which we give the symbol N_A). Recent measurements of this number give the value 6.0221367×10^{23}, which to three significant figures is 6.02×10^{23}.

A mole of a substance contains Avogadro's number (6.02×10^{23}) of molecules (or formula units). The term *mole,* like a dozen or a gross, thus refers to a particular number of things. A dozen eggs equals 12 eggs, a gross of pencils equals 144 pencils, and a mole of ethanol equals 6.02×10^{23} ethanol molecules.

In using the term *mole* for ionic substances, we mean the number of formula units of the substance. For example, a mole of sodium carbonate, Na_2CO_3, is a quantity containing 6.02×10^{23} Na_2CO_3 units. But each formula unit of Na_2CO_3 contains two Na^+ ions and one CO_3^{2-} ion. Therefore, a mole of Na_2CO_3 also contains $2 \times 6.02 \times 10^{23}$ Na^+ ions and $1 \times 6.02 \times 10^{23}$ CO_3^{2-} ions. ▶

When using the term *mole,* it is important to specify the formula of the unit to avoid any misunderstanding. For example, a mole of oxygen atoms (with the formula O) contains 6.02×10^{23} O atoms. A mole of oxygen molecules (formula O_2) contains 6.02×10^{23} O_2 molecules—that is, $2 \times 6.02 \times 10^{23}$ O atoms.

The **molar mass** of a substance is *the mass of one mole of the substance.* Carbon-12 has a molar mass of exactly 12 g/mol, by definition.

Sodium carbonate, Na_2CO_3, is a white, crystalline solid known commercially as soda ash. Large amounts of soda ash are used in the manufacture of glass. The hydrated compound, $Na_2CO_3 \cdot 10H_2O$, is known as washing soda.

Figure 3.2 ▲

One mole each of various substances

For all substances, the molar mass in grams per mole is numerically equal to the formula mass in atomic mass units.

Ethanol, whose molecular formula is C_2H_6O (frequently written as the condensed structural formula C_2H_5OH), has a molecular mass of 46.1 amu and a molar mass of 46.1 g/mol. Figure 3.2 shows molar amounts of different substances.

| **Example** 3.3 | **Calculating the Mass of an Atom or Molecule** |

Gaining Mastery Toolbox

Critical Concept 3.3
The mass of one mole
(6.022×10^{23} units) of a substance is the molar mass of a substance. The units of molar mass are grams per mole (g/mol). The molar mass of each element can be found on the periodic table and the inside back cover of the text.

Solution Essentials:
- Mole (mol)
- Avogadro's number (6.022×10^{23})
- Molar mass
- Periodic table
- Rules for significant figures and rounding

a What is the mass in grams of a chlorine atom, Cl?
b What is the mass in grams of a hydrogen chloride molecule, HCl?

Problem Strategy In order to solve this type of problem, we need to consider using molar mass and the relationship between the number of atoms or molecules and the molar mass.

Solution

a The atomic mass of Cl is 35.5 amu, so the molar mass of Cl is 35.5 g/mol. Dividing 35.5 g (per mole) by 6.02×10^{23} (Avogadro's number) gives the mass of one atom.

$$\text{Mass of a Cl atom} = \frac{35.5 \text{ g}}{6.02 \times 10^{23}} = \mathbf{5.90 \times 10^{-23} \text{ g}}$$

b The molecular mass of HCl equals the AM of H plus the AM of Cl, or 1.01 amu + 35.5 amu = 36.5 amu. Therefore, 1 mol HCl contains 36.5 g HCl and

$$\text{Mass of an HCl molecule} = \frac{36.5 \text{ g}}{6.02 \times 10^{23}} = \mathbf{6.06 \times 10^{-23} \text{ g}}$$

Answer Check As you know, individual atoms and molecules are incredibly small. Therefore, whenever you are asked to calculate the mass of a few atoms or molecules, you should expect a very small mass.

Exercise 3.3 a What is the mass in grams of a calcium atom, Ca? b What is the mass in grams of an ethanol molecule, C_2H_5OH?

■ See Problems 3.33, 3.34, 3.35, and 3.36.

Mole Calculations

Now that you know how to find the mass of one mole of a substance, there are two important questions to ask. First, how much does a given number of moles of a substance weigh? Second, how many moles of a given formula unit does a given mass of substance contain? Both questions are easily answered using dimensional analysis, or the conversion-factor method. ▶

To illustrate, consider the conversion of grams of ethanol, C_2H_5OH, to moles of ethanol. The molar mass of ethanol is 46.1 g/mol, so we write

$$1 \text{ mol } C_2H_5OH = 46.1 \text{ g } C_2H_5OH$$

Thus, the factor converting grams of ethanol to moles of ethanol is 1 mol C_2H_5OH/46.1 g C_2H_5OH. To convert moles of ethanol to grams of ethanol, we simply invert the conversion factor (46.1 g C_2H_5OH/1 mol C_2H_5OH). Note that the unit you are converting *from* is on the bottom of the conversion factor; the unit you are converting *to* is on the top.

Again, suppose you are going to prepare acetic acid from 10.0 g of ethanol, C_2H_5OH. How many moles of C_2H_5OH is this? You convert 10.0 g C_2H_5OH to moles C_2H_5OH by multiplying by the appropriate conversion factor.

$$10.0 \text{ g } C_2H_5OH \times \frac{1 \text{ mol } C_2H_5OH}{46.1 \text{ g } C_2H_5OH} = 0.217 \text{ mol } C_2H_5OH$$

Diagramming this solution (recall that the ⇒ indicates a conversion factor).

Start		Goal
g C_2H_5OH	⟹	mol C_2H_5OH

The following examples further illustrate this conversion-factor technique.

Alternatively, because the molar mass is the mass per mole, you can relate mass and moles by means of the formula

Molar mass = mass/moles

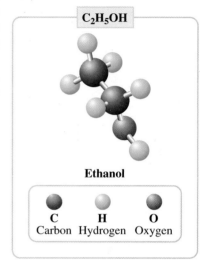

C_2H_5OH

Ethanol

C	H	O
Carbon	Hydrogen	Oxygen

Example 3.4   Converting Moles of Substance to Grams

Gaining Mastery Toolbox

Critical Concept 3.4
Molar mass is a conversion factor. This conversion factor can be used with dimensional analysis to convert from grams of substance to moles of substance.

Solution Essentials:
• Molar mass
• Mole (mol)
• Conversion factor
• Dimensional analysis
• Rules for significant figures and rounding

Zinc iodide, ZnI_2, can be prepared by the direct combination of elements (Figure 3.3). A chemist determines from the amounts of elements that 0.0654 mol ZnI_2 can form. How many grams of zinc iodide is this?

Problem Strategy Use the molar mass to write the factor that converts from mol ZnI_2 to g ZnI_2. Note that the unit you are converting from (mol ZnI_2) is on the bottom of the conversion factor, and the unit you are converting to (g ZnI_2) is on the top.

The problem strategy is diagrammed as

Start		Goal
mol ZnI_2	⟹	g ZnI_2

Solution The molar mass of ZnI_2 is 319 g/mol. (The formula mass is 319 amu, which is obtained by summing the atomic masses in the formula.) Therefore,

$$0.0654 \text{ mol } ZnI_2 \times \frac{319 \text{ g } ZnI_2}{1 \text{ mol } ZnI_2} = \textbf{20.9 g } ZnI_2$$

Answer Check Whenever you solve a problem of this type, be sure to write all units, making certain that they will cancel. This "built-in" feature of dimensional analysis ensures that you are correctly using the conversion factors.

Exercise 3.4 Hydrogen peroxide, H_2O_2, is a colorless liquid. A concentrated solution of it is used as a source of oxygen for rocket propellant fuels. Dilute aqueous solutions are used as a bleach. Analysis of a solution shows that it contains 0.909 mol H_2O_2 in 1.00 L of solution. What is the mass of hydrogen peroxide in this volume of solution?

Figure 3.3 ◀

Reaction of zinc and iodine Heat from the reaction of the elements causes some iodine to vaporize (violet vapor).

© Cengage Learning

■ See Problems 3.37, 3.38, 3.39, and 3.40.

ⓌWL Interactive Example 3.5   Converting Grams of Substance to Moles

Gaining Mastery Toolbox

Critical Concept 3.5
Molar mass is a conversion factor. This conversion factor can be used with dimensional analysis to convert from moles of substance to mass of substance.

Solution Essentials:
• Molar mass
• Mole (mol)
• Conversion factor
• Dimensional analysis
• Rules for significant figures and rounding

Lead(II) chromate, $PbCrO_4$, is a yellow paint pigment (called chrome yellow) prepared by a precipitation reaction (Figure 3.4). In a preparation, 45.6 g of lead(II) chromate is obtained as a precipitate. How many moles of $PbCrO_4$ is this?

Problem Strategy Since we are starting with a mass of $PbCrO_4$, we need the conversion factor for grams of $PbCrO_4$ to moles of $PbCrO_4$. The molar mass of $PbCrO_4$ will provide this information.
The problem strategy is diagrammed as

Start   Goal
g $PbCrO_4$ ⟹   mol $PbCrO_4$

Solution The molar mass of $PbCrO_4$ is 323 g/mol. That is,

$$1 \text{ mol } PbCrO_4 = 323 \text{ g } PbCrO_4$$

Therefore,

$$45.6 \text{ g } \cancel{PbCrO_4} \times \frac{1 \text{ mol } PbCrO_4}{323 \text{ g } \cancel{PbCrO_4}} = \textbf{0.141 mol } PbCrO_4$$

Answer Check Note that the given amount of material in this problem (45.6 g $PbCrO_4$) is much less than its molar mass (323 g/mol). Therefore, we would expect the number of moles of $PbCrO_4$ to be much less than 1, which is the case here. Quick, alert comparisons such as this can be very valuable in checking for calculation errors.

Exercise 3.5   Nitric acid, HNO_3, is a colorless, corrosive liquid used in the manufacture of nitrogen fertilizers and explosives. In an experiment to develop new explosives for mining operations, a sample containing 28.5 g of nitric acid was poured into a beaker. How many moles of HNO_3 are there in this sample of nitric acid?

© Cengage Learning

Figure 3.4 ▲

Preparation of lead(II) chromate When lead(II) nitrate solution (colorless) is added to potassium chromate solution (clear yellow), bright yellow solid lead(II) chromate forms (giving a cloudlike formation of fine crystals).

■ See Problems 3.41 and 3.42.

Example 3.6   Calculating the Number of Molecules in a Given Mass

Gaining Mastery Toolbox

Critical Concept 3.6
Avogadro's number is a conversion factor. Whenever you are required to know "how many" atoms, elements, molecules, etc., use the conversion factor 1 mol = 6.022×10^{23}.

Solution Essentials:
• Molar mass
• Avogadro's number (6.022×10^{23})
• Mole (mol)
• Conversion factor
• Dimensional analysis
• Rules for significant figures and rounding

How many molecules are there in a 3.46-g sample of hydrogen chloride, HCl?

Problem Strategy The number of molecules in a sample is related to moles of compound (1 mol HCl = 6.02×10^{23} HCl molecules). Therefore, if you first convert grams HCl to moles, then you can convert moles to number of molecules.
This problem strategy that requires two conversion factors is diagrammed as

Start   Goal
g HCl   ⟹   mol HCl   ⟹   molecules HCl

Solution Here is the calculation:

$$3.46 \text{ g } \cancel{HCl} \times \frac{1 \text{ mol } \cancel{HCl}}{36.5 \text{ g } \cancel{HCl}} \times \frac{6.02 \times 10^{23} \text{ HCl molecules}}{1 \text{ mol } \cancel{HCl}}$$

$$= \textbf{5.71} \times \textbf{10}^{22} \textbf{ HCl molecules}$$

(continued)

(continued)

Note how the units in the numerator of a factor are canceled by the units in the denominator of the following factor.

Answer Check A very common mistake made when solving this type of problem is to use an incorrect conversion factor, such as 1 mol HCl = 6.02×10^{23} g HCl. If this statement were true, the mass of HCl contained in a mole would be far more than the mass of all the matter in the Milky Way galaxy! Therefore, if you end up with a gigantic mass for an answer, or get stuck on how to use the quantity 6.02×10^{23} as a conversion factor, improper unit assignment is a likely culprit.

Exercise 3.6 Hydrogen cyanide, HCN, is a volatile, colorless liquid with the odor of certain fruit pits (such as peach and cherry pits). The compound is highly poisonous. How many molecules are there in 56 mg HCN, the average toxic dose?

■ See Problems 3.45, 3.46, 3.47, and 3.48.

CONCEPT CHECK 3.1

You have 1.5 moles of tricycles.

a How many moles of seats do you have?

b How many moles of tires do you have?

c How could you use parts a and b as an analogy to teach a friend about the number of moles of OH^- ions in 1.5 moles of $Mg(OH)_2$?

Determining Chemical Formulas

When a chemist has discovered a new compound, the first question to answer is, What is the formula? To answer, you begin by analyzing the compound to determine amounts of the elements for a given amount of compound. This is conveniently expressed as **percentage composition**—that is, as *the mass percentages of each element in the compound.* You then determine the formula from this percentage composition. If the compound is a molecular substance, you must also find the molecular mass of the compound in order to determine the molecular formula.

The next section describes the calculation of mass percentages. Then, in two following sections, we describe how to determine a chemical formula.

3.3 Mass Percentages from the Formula

Suppose that A is a part of something—that is, part of a whole. It could be an element in a compound or one substance in a mixture. We define the **mass percentage** of A as the parts of A *per hundred parts of the total, by mass.* That is,

$$\text{Mass \% } A = \frac{\text{mass of } A \text{ in the whole}}{\text{mass of the whole}} \times 100\%$$

You can look at the mass percentage of A as the number of grams of A in 100 g of the whole.

The next example will provide practice with the concept of mass percentage. In this example we will start with a compound (formaldehyde, CH_2O) whose formula is given and obtain the percentage composition.

Example **3.7** **Calculating the Percentage Composition from the Formula**

Critical Concept 3.7
You assume that you have one mole of compound. Making this assumption, you then know the total mass of the compound (molar mass), and from the chemical formula, the number of moles of each element in the compound.

Solution Essentials:
• Percentage composition
• Mass percentage
• Molar mass
• Rules for significant figures and rounding

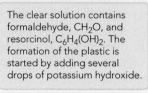

The clear solution contains formaldehyde, CH_2O, and resorcinol, $C_6H_4(OH)_2$. The formation of the plastic is started by adding several drops of potassium hydroxide.

Formaldehyde, CH_2O, is a toxic gas with a pungent odor. Large quantities are consumed in the manufacture of plastics (Figure 3.5), and a water solution of the compound is used to preserve biological specimens. Calculate the mass percentages of the elements in formaldehyde (give answers to three significant figures).

Problem Strategy To calculate mass percentage, you need the mass of an element in a given mass of compound. You can get this information by interpreting the formula in molar terms and then converting moles to masses, using a table of atomic masses. Thus, 1 mol CH_2O has a mass of 30.0 g and contains 1 mol C (12.0 g), 2 mol H (2 × 1.01 g), and 1 mol O (16.0 g). You divide each mass of element by the molar mass, then multiply by 100, to obtain the mass percentage.

Solution Here are the calculations:

$$\% \text{ C} = \frac{12.0 \text{ g}}{30.0 \text{ g}} \times 100\% = \textbf{40.0\%}$$

$$\% \text{ H} = \frac{2 \times 1.01 \text{ g}}{30.0 \text{ g}} \times 100\% = \textbf{6.73\%}$$

You can calculate the percentage of O in the same way, but it can also be found by subtracting the percentages of C and H from 100%:

$$\% \text{ O} = 100\% - (40.0\% + 6.73\%) = \textbf{53.3\%}$$

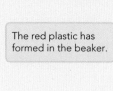

The red plastic has formed in the beaker.

© Cengage Learning

Figure 3.5 ▲

Preparing resorcinol-formaldehyde plastic

Answer Check A typical mistake when working a mass percentage problem is to forget to account for the number of moles of each element given in the chemical formula. An example would be to answer this question as though the formula were CHO instead of CH_2O.

Exercise 3.7 Ammonium nitrate, NH_4NO_3, which is prepared from nitric acid, is used as a nitrogen fertilizer. Calculate the mass percentages of the elements in ammonium nitrate (to three significant figures).

■ See Problems 3.57, 3.58, 3.59, and 3.60.

Example **3.8** **Calculating the Mass of an Element in a Given Mass of Compound**

Critical Concept 3.8
The percentage composition of an element is the *mass* percentage of the element. The mass percentage of an element can be used to calculate the mass of that element in a sample.

Solution Essentials:
• Percentage composition
• Mass percentage
• Rules for significant figures and rounding

How many grams of carbon are there in 83.5 g of formaldehyde, CH_2O? Use the percentage composition obtained in the previous example (40.0% C, 6.73% H, 53.3% O).

Problem Strategy Solving this type of problem requires that

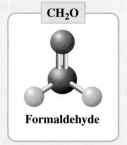

CH_2O

Formaldehyde

(continued)

(*continued*)

you first calculate the mass percentage of the element of interest, carbon in this case (as noted in the problem, this was done in Example 3.7). Multiplying of this percentage, expressed as a decimal, times the mass of formaldehyde in the sample, will yield the mass of carbon present.

Solution CH_2O is 40.0% C, so the mass of carbon in 83.5 g CH_2O is

$$83.5 \text{ g} \times 0.400 = \textbf{33.4 g}$$

Answer Check Make sure that you have converted the percent to a decimal prior to multiplication. An

answer that is more than the starting mass of material is an indication that you probably made this mistake.

Exercise 3.8   How many grams of nitrogen, N, are there in a fertilizer containing 48.5 g of ammonium nitrate and no other nitrogen-containing compound? See Exercise 3.7 for the percentage composition of NH_4NO_3.

■ See Problems 3.61 and 3.62.

3.4 Elemental Analysis: Percentages of Carbon, Hydrogen, and Oxygen

Suppose you have a newly discovered compound whose formula you wish to determine. The first step is to obtain its percentage composition. As an example, consider the determination of the percentages of carbon, hydrogen, and oxygen in compounds containing only these three elements. The basic idea is this: You burn a sample of the compound of known mass and get CO_2 and H_2O. Next you relate the masses of CO_2 and H_2O to the masses of carbon and hydrogen. Then you calculate the mass percentages of C and H. You find the mass percentage of O by subtracting the mass percentages of C and H from 100.

Figure 3.6 shows an apparatus used to find the amount of carbon and hydrogen in a compound. The compound is burned in a stream of oxygen gas. The vapor of the compound and its combustion products pass over copper pellets coated with copper(II) oxide, CuO, which supplies additional oxygen and ensures that the compound is completely burned. As a result of the combustion, every mole of carbon (C) in the compound ends up as a mole of carbon dioxide (CO_2), and every mole of hydrogen (H) ends up as one-half mole of water (H_2O). The water is collected by a drying agent, a substance that has a strong affinity for water. The carbon dioxide is collected by chemical reaction with sodium hydroxide, NaOH.

Figure 3.6 ▼

Combustion method for determining the percentages of carbon and hydrogen in a compound

Vapor of the compound burns in O_2 in the presence of CuO pellets, giving CO_2 and H_2O. The water vapor is collected by a drying agent, and CO_2 combines with the sodium hydroxide. Amounts of CO_2 and H_2O are obtained by weighing the U-tubes before and after combustion.

Oxygen in

Furnace

The compound is placed in the sample dish and is heated by the furnace.

Copper(II) oxide (CuO) pellets

Drying agent (traps H_2O)

Sodium hydroxide (traps CO_2)

Excess oxygen out

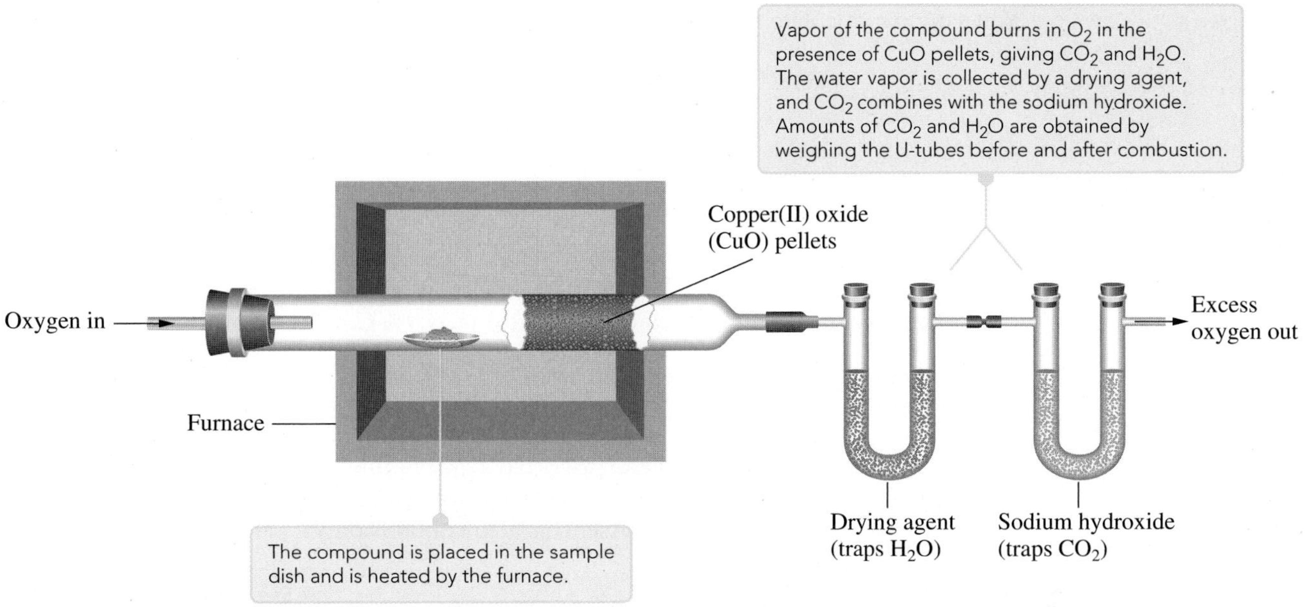

Sodium hydroxide reacts with carbon dioxide according to the following equations:

$$NaOH + CO_2 \longrightarrow NaHCO_3$$
$$2NaOH + CO_2 \longrightarrow Na_2CO_3 + H_2O$$

◀ By weighing the U-tubes containing the drying agent and the sodium hydroxide before and after combustion, it is possible to determine the masses of water and carbon dioxide produced. From these data, you can calculate the percentage composition of the compound.

The chapter opened with a discussion of acetic acid. The next example shows how to determine the percentage composition of this substance from combustion data. We will use this percentage composition later in Example 3.12 to determine the formula of acetic acid.

☉WL Interactive Example 3.9   Calculating the Percentages of C and H by Combustion

Gaining Mastery Toolbox

Critical Concept 3.9
The mass percentages of carbon and hydrogen can be calculated from the masses of CO_2 and H_2O. During the reaction *all* of the carbon atoms in the sample end up in the CO_2 and *all* of the hydrogen atoms end up in the H_2O. Therefore, the mass of carbon in the CO_2 is equal to the mass of carbon in the sample, and the mass of hydrogen in the H_2O is equal to the mass of hydrogen in the sample. We *cannot* make the statement that the mass of oxygen in the H_2O and CO_2 is equal to the mass of oxygen in the sample because oxygen is both in the sample and is a reactant.

Solution Essentials:
• Elemental analysis
• Percentage composition
• Mass percentage
• Molar mass
• Dimensional analysis
• Rules for significant figures and rounding

Acetic acid contains only C, H, and O. A 4.24-mg sample of acetic acid is completely burned. It gives 6.21 mg of carbon dioxide and 2.54 mg of water. What is the mass percentage of each element in acetic acid?

Problem Strategy Note that in the products of the combustion, all of the carbon from the sample ends up in the CO_2, all of the hydrogen ends up in the H_2O, and the oxygen is in both compounds. Because of this, you should concentrate first on the carbon and hydrogen and worry about the oxygen last. If we can determine the mass of carbon and hydrogen in the original sample, we should then be able to determine the percentage of each of these elements present in the compound. Once we know the mass percentages of carbon and hydrogen in the original compound, the remaining mass percentage (100% total) must be due to oxygen. Let's start by determining the mass of carbon that was originally contained in the compound. You first convert the mass of CO_2 to moles of CO_2. Then you convert this to moles of C, noting that 1 mol C produces 1 mol CO_2. Finally, you convert to mass of C. Similarly, for hydrogen, you convert the mass of H_2O to mol H_2O, then to mol H, and finally to mass of H. (Remember that 1 mol H_2O produces 2 mol H.) Once you have the masses of C and H, you can calculate the mass percentages. Subtract from 100% to get % O.

Solution Following is the calculation of grams C:

$$6.21 \times 10^{-3} \text{ g } CO_2 \times \frac{1 \text{ mol } CO_2}{44.0 \text{ g } CO_2} \times \frac{1 \text{ mol C}}{1 \text{ mol } CO_2} \times \frac{12.0 \text{ g C}}{1 \text{ mol C}}$$
$$= 1.69 \times 10^{-3} \text{ g C (or 1.69 mg C)}$$

For hydrogen, you note that 1 mol H_2O yields 2 mol H, so you write

$$2.54 \times 10^{-3} \text{ g } H_2O \times \frac{1 \text{ mol } H_2O}{18.0 \text{ g } H_2O} \times \frac{2 \text{ mol H}}{1 \text{ mol } H_2O} \times \frac{1.01 \text{ g H}}{1 \text{ mol H}}$$
$$= 2.85 \times 10^{-4} \text{ g H (or 0.285 mg H)}$$

You can now calculate the mass percentages of C and H in acetic acid.

$$\text{Mass \% C} = \frac{1.69 \text{ mg}}{4.24 \text{ mg}} \times 100\% = 39.9\%$$

$$\text{Mass \% H} = \frac{0.285 \text{ mg}}{4.24 \text{ mg}} \times 100\% = 6.72\%$$

You find the mass percentage of oxygen by subtracting the sum of these percentages from 100%:

$$\text{Mass \% O} = 100\% - (39.9\% + 6.72\%) = 53.4\%$$

Thus, the percentage composition of acetic acid is **39.9% C, 6.7% H, and 53.4% O.**

Answer Check The most common error for this type of problem is not taking into account the fact that each mole of water contains two moles of hydrogen.

Exercise 3.9   A 3.87-mg sample of ascorbic acid (vitamin C) gives 5.80 mg CO_2 and 1.58 mg H_2O on combustion. What is the percentage composition of this compound (the mass percentage of each element)? Ascorbic acid contains only C, H, and O.

■ See Problems 3.63 and 3.64.

CONCEPT CHECK 3.2

You perform combustion analysis on a compound that contains only C and H.

a Considering the fact that the combustion products CO_2 and H_2O are colorless, how can you tell if some of the product got trapped in the CuO pellets (see Figure 3.6)?

b Would your calculated results of mass percentage of C and H be affected if some of the combustion products got trapped in the CuO pellets? If your answer is yes, how might your results differ from the expected values for the compound?

3.5 Determining Formulas

The percentage composition of a compound leads directly to its empirical formula. An **empirical formula** (or **simplest formula**) for a compound is *the formula of a substance written with the smallest integer (whole number) subscripts.* For most ionic substances, the empirical formula is the formula of the compound. ▶ This is often not the case for molecular substances. For example, hydrogen peroxide has the molecular formula H_2O_2. The molecular formula, you may recall, tells you the precise number of atoms of different elements in a molecule of the substance. The empirical formula, however, merely tells you the ratio of numbers of atoms in the compound. The empirical formula of hydrogen peroxide (H_2O_2) is HO (Figure 3.7).

Compounds with different molecular formulas can have the same empirical formula, and such substances will have the same percentage composition. An example is acetylene, C_2H_2, and benzene, C_6H_6. Acetylene is a gas used as a fuel and also in welding. Benzene, in contrast, is a liquid that is used in the manufacture of plastics and is a component of gasoline. Table 3.1 illustrates how these two compounds, with the same empirical formula, but different molecular formulas, also have different chemical structures. Because the empirical formulas of acetylene and benzene are the same, they have the same percentage composition: 92.3% C and 7.7% H, by mass.

To obtain the molecular formula of a substance, you need two pieces of information: (1) as in the previous section, the percentage composition, from which the empirical formula can be determined; and (2) the molecular mass. The molecular mass allows you to choose the correct multiple of the empirical formula for the molecular formula. We will illustrate these steps in the next three examples.

Empirical Formula from the Composition

The empirical formula of a compound shows the ratios of numbers of atoms in the compound. You can find this formula from the composition of the compound by converting masses of the elements to moles. The next two examples show these calculations in detail.

The formula of sodium peroxide, an ionic compound of Na^+ and O_2^{2-}, is Na_2O_2. Its empirical formula is NaO.

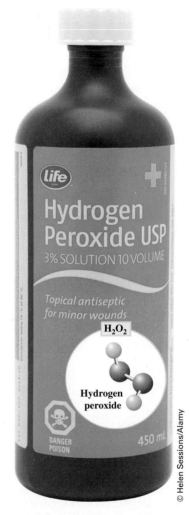

Figure 3.7 ▲

Molecular model of hydrogen peroxide (H_2O_2) Hydrogen peroxide has the empirical formula HO and the molecular formula H_2O_2. In order to arrive at the correct structure of molecular compounds like H_2O_2 shown here, chemists must have the molecular formula.

	Table 1.3	**Molecular Models of Two Compounds That Have the Empirical Formula CH**

Although benzene and acetylene have the same empirical formula, they do not have the same molecular formula or structure.

Compound	Empirical Formula	Molecular Formula	Molecular Model
Acetylene	CH	C_2H_2	
Benzene	CH	C_6H_6	

Mass Spectrometry and Molecular Formula

Some very sophisticated instruments have become indispensable to modern chemical research. One such instrument is the **mass spectrometer,** which measures the masses of positive ions produced from an extremely small sample of a substance, displaying the data as a *mass spectrum*. Figure 3.8 shows the mass spectrum of methylene chloride, CH_2Cl_2. Each of the positive ions produced in the spectrometer from the substance is shown as a bar or line (called a *peak*) at the mass of the ion, in amu. (Strictly speaking, each peak appears at the mass-to-charge ratio of the ion, but the ion charge is usually one.) The height of each peak is proportional to the relative abundance of the ion. The mass spectrum of a substance can be used to identify it or, if the substance is a new one, to obtain its molecular formula. The spectrum may also give information about the structure of the molecule.

Figure 2.11 on page 51 shows a simplified diagram of a mass spectrometer, used there to separate atomic neon into its three isotopic ions. Neon gas (which consists of neon atoms) enters the spectrometer, where the atoms are ionized by bombarding them with a beam of high-energy electrons. (An electron from the beam strikes a neon atom, knocking out one of neon's electrons.) The positive ions are accelerated by a negative plate into the field of a magnet, which bends the path of each ion depending on its mass-to-charge ratio. The less massive ions are bent more strongly than the heavier ones for the same charge. In most modern spectrometers, the ions fall on an electronic detector, whose current is recorded and displayed as the mass spectrum on a computer screen and printed on chart paper.

The spectrum of a molecular substance is more complicated than that of an atom. Consider methylene chloride, CH_2Cl_2. When its molecule is struck by an electron of sufficient energy,

it may lose one of its own electrons (now giving two free electrons), leaving behind a positive ion:

$$CH_2Cl_2 + e^- \longrightarrow CH_2Cl_2^+ + 2e^-$$

The ion gains a great deal of energy from the collision of the molecule with an electron, and it may lose this energy by breaking into smaller pieces. For example:

$$CH_2Cl_2^+ \longrightarrow CH_2Cl^+ + Cl$$

Thus, the original molecule, even one as simple as CH_2Cl_2, can give rise to a number of ions. In addition, each ion may consist of atoms having more than one natural isotope, so that each ion may occur in the spectrum as a group of several peaks. Note that the mass spectrum of methylene chloride shown in Figure 3.8 displays four prominent peaks and many smaller ones. A larger molecule can give an even more complicated mass spectrum, oftentimes yielding a "fingerprint" useful in identification. Only methylene chloride has exactly the spectrum shown in Figure 3.8. Thus, by comparing the mass spectrum of an unknown substance with those in a catalog of mass spectra of known compounds, you can determine its identity. The more information you have about the compound, the shorter the search through the catalog of spectra.

The mass spectrum itself contains a wealth of information about molecular structure. Some experience is needed to analyze the spectrum of a compound, but you can get an idea of how it is done by looking at Figure 3.8. Suppose you do not know the identity of the compound. The most intense peaks at the greatest mass usually correspond to the ion from the original molecule and give you the molecular mass. Thus, you would expect the peaks at mass 84 and mass 86 to be from the original molecular ion, so the molecular mass is approximately 84 to 86.

Example 3.10 Determining the Empirical Formula from Masses of Elements (Binary Compound)

Critical Concept 3.10
The mole ratio of elements in the compound is the same ratio as the element subscripts in the empirical formula. The subscripts of the compound must be written as whole (integer) numbers. In some cases the empirical formula is the same as the molecular formula.

Solution Essentials:
- Empirical formula
- Molar mass
- Dimensional analysis
- Rules for significant figures and rounding

A compound of nitrogen and oxygen is analyzed, and a sample weighing 1.587 g is found to contain 0.483 g N and 1.104 g O. What is the empirical formula of the compound?

Problem Strategy This compound has the formula N_xO_y, where x and y are whole numbers that we need to determine. Masses of N and O are given in the problem. If we convert the masses of each of these elements to moles, they will be proportional to the subscripts in the empirical formula. The formula must be the smallest integer ratio of the elements. To obtain the smallest integers from the moles, we divide each by the smallest one. If the results are all whole numbers, they

(continued)

Figure 3.8 ▶

Mass spectrum of methylene chloride, CH_2Cl_2 The lines (*peaks*) at higher mass correspond to the ions $CH_2Cl_2^+$ with different isotopes, produced by the loss of a single electron from the original CH_2Cl_2 molecule. (The spectrum is from the NIST Chemistry WebBook, http://webbook.nist.gov/chemistry/.)

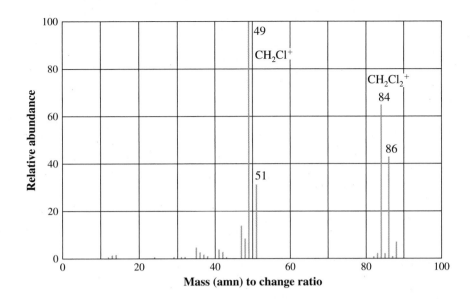

An elemental analysis is also possible. The two most intense peaks in Figure 3.8 are at 84 amu and 49 amu. They differ by 35 amu. Perhaps the original molecule that gives the peak at 84 amu contains a chlorine-35 atom, which is lost to give the peak at 49 amu. If this is true, you should expect another peak at mass 86, corresponding to the original molecular ion with chlorine-37 in place of a chlorine-35 atom. This is indeed what you see.

The relative heights of the peaks, which depend on the natural abundances of atoms, are also important, because they give you additional information about the elements present. The relative heights can also tell you how many atoms of a given element are in the original molecule. Naturally occurring chlorine is 75.8% chlorine-35 and 24.2% chlorine-37. If the original molecule contained only one Cl atom, the peaks at 84 amu and 86 amu would be in the ratio 0.758 : 0.242. That is, the peak at 86 amu would be about one-third the height of the one at 84 amu. In fact, the relative height is twice this value. This means that the molecular ion contains two chlorine atoms, because the chance that any such ion contains one chlorine-37 atom is then twice as great. Thus, simply by comparing relative peak heights, you can confirm the presence of particular elements.

The spectrum in Figure 3.8 shows the peaks at nearest whole-number masses. There are mass spectrometers, however, that can obtain the masses of moderate-sized molecules to perhaps four decimal places. With this high resolution, you can in principle obtain the molecular formula of a substance. For example, the CH_2Cl_2 molecule composed of the most abundant isotopes has a whole-number molecular mass of 84 amu (the largest peak at the right in Figure 3.8), but its mass to four decimals is 83.9534 amu. There are several other molecules with whole-number masses of 84 amu. One of these is hexene, C_6H_{12} (the species composed of the more abundant isotopes); but, its more exact mass is 84.0939 amu. In general, molecules with different molecular formulas will have different masses, so you can turn this problem around. Given a molecular mass, you can use special computer programs to obtain the molecular formula.

■ See Problems 3.115 and 3.116.

(*continued*)

will be the subscripts in the formula. (Otherwise, you will need to multiply by some factor, as illustrated in Example 3.11.)

Solution You convert the masses to moles:

$$0.483 \text{ g N} \times \frac{1 \text{ mol N}}{14.0 \text{ g N}} = 0.0345 \text{ mol N}$$

$$1.104 \text{ g O} \times \frac{1 \text{ mol O}}{16.00 \text{ g O}} = 0.06900 \text{ mol O}$$

In order to obtain the smallest integers, you divide each mole number by the smaller one (0.0345 mol). For N, you get 1.00; for O, you get 2.00. Thus, the ratio of the number of N atoms to the number of O atoms is 1 to 2. Hence, the empirical formula is NO_2.

Answer Check Keep in mind that the empirical formula determined by this type of calculation is not necessarily the molecular formula of the compound. The only information that it provides is the smallest whole-number ratio of the elements.

Exercise 3.10 A sample of compound weighing 83.5 g contains 33.4 g of sulfur. The rest is oxygen. What is the empirical formula?

■ See Problems 3.65 and 3.66.

Example 3.11   Determining the Empirical Formula from Percentage Composition (General)

Gaining Mastery Toolbox

Critical Concept 3.11
You assume that you are starting with 100 g of substance. By making this assumption, it simplifies the process of determining the relative masses of each element in the substance.

Solution Essentials:
• Empirical formula
• Percentage composition
• Molar mass
• Dimensional analysis
• Rules for significant figures and rounding

Chromium forms compounds of various colors. (The word *chromium* comes from the Greek *khroma,* meaning "color"; see Figure 3.9.) Sodium dichromate is the most important commercial chromium compound, from which many other chromium compounds are manufactured. It is a bright orange, crystalline substance. An analysis of sodium dichromate gives the following mass percentages: 17.5% Na, 39.7% Cr, and 42.8% O. What is the empirical formula of this compound? (Sodium dichromate is ionic, so it has no molecular formula.)

Problem Strategy Here we need to determine the whole-number values for *x, y,* and *z* in the compound $Na_xCr_yO_z$. The first task is to use the percent composition data to determine the moles of Na, Cr, and O in the compound. Then we can use mole ratios of these elements to determine the empirical formula. Assume for the purposes of this calculation that you have 100.0 g of substance. Then the mass of each element in the sample equals the numerical value of the percentage. For example, the quantity of sodium in 100 g of sodium dichromate is 17.5 g, since the substance is 17.5% Na. Now convert the masses to moles and divide each mole number by the smallest. In this example, you do not obtain a series of integers, or whole numbers, from this division. You will need to find a whole-number factor to multiply these results by to obtain integers. Normally this factor will be 2 or 3, though it might be larger.

Solution Of the 100.0 g of sodium dichromate, 17.5 g is Na, 39.7 g is Cr, and 42.8 g is O. You convert these amounts to moles.

$$17.5 \text{ g Na} \times \frac{1 \text{ mol Na}}{23.0 \text{ g Na}} = 0.761 \text{ mol Na}$$

$$39.7 \text{ g Cr} \times \frac{1 \text{ mol Cr}}{52.0 \text{ g Cr}} = 0.763 \text{ mol Cr}$$

$$42.8 \text{ g O} \times \frac{1 \text{ mol O}}{16.0 \text{ g O}} = 2.68 \text{ mol O}$$

Now you divide all the mole numbers by the smallest one.

$$\text{For Na:} \frac{0.761 \text{ mol}}{0.761 \text{ mol}} = 1.00$$

$$\text{For Cr:} \frac{0.763 \text{ mol}}{0.761 \text{ mol}} = 1.00$$

$$\text{For O:} \frac{2.68 \text{ mol}}{0.761 \text{ mol}} = 3.52$$

You must decide whether these numbers are integers (whole numbers), within experimental error. If you round off the last digit, which is subject to experimental error, you get $Na_{1.0}Cr_{1.0}O_{3.5}$. In this case, the subscripts are not all integers. However, they can be made into integers by multiplying each one by 2; you get $Na_{2.0}Cr_{2.0}O_{7.0}$. Thus, the empirical formula is $\mathbf{Na_2Cr_2O_7}$.

Answer Check If you are working this type of problem and are not arriving at an answer that is readily converted to integers, it might be that you did not use enough significant figures throughout the calculation and/or that you rounded intermediate calculations.

Figure 3.9 ▼

Chromium compounds of different colors

Exercise 3.11 Benzoic acid is a white, crystalline powder used as a food preservative. The compound contains 68.8% C, 5.0% H, and 26.2% O, by mass. What is its empirical formula?

Potassium chromate, K_2CrO_4

Chromium(VI) oxide, CrO_3

Chromium(III) sulfate, $Cr_2(SO_4)_3$

Chromium metal

Potassium dichromate, $K_2Cr_2O_7$

chromium(III) oxide, Cr_2O_3

■ See Problems 3.67, 3.68, 3.69, and 3.70.

Molecular Formula from Empirical Formula

The molecular formula of a compound is a multiple of its empirical formula. For example, the molecular formula of acetylene, C_2H_2, is equivalent to $(CH)_2$, and the molecular formula of benzene, C_6H_6, is equivalent to $(CH)_6$. Therefore, the molecular mass is some multiple of the empirical formula mass, which is obtained by summing the atomic masses of the atoms in the empirical formula. For any molecular compound, you can write

$$\text{Molecular mass} = n \times \text{empirical formula mass}$$

where n is the number of empirical formula units in the molecule. You get the molecular formula by multiplying the subscripts of the empirical formula by n, which you calculate from the equation

$$n = \frac{\text{molecular mass}}{\text{empirical formula mass}}$$

Once you determine the empirical formula for a compound, you can calculate its empirical formula mass. If you have an experimental determination of its molecular mass, you can calculate n and then the molecular formula. The next example illustrates how you use percentage composition and molecular mass to determine the molecular formula of acetic acid.

Example 3.12

Determining the Molecular Formula from Percentage Composition and Molecular Mass

Gaining Mastery Toolbox

Critical Concept 3.12
The molecular mass of a compound is a multiple of the empirical formula mass. The multiple is an integer and can be used to determine the molecular formula of the compound from the empirical formula.

Solution Essentials:
• Empirical formula
• Empirical formula mass
• Molecular mass
• Percentage composition
• Dimensional analysis
• Rules for significant figures and rounding

In Example 3.9, we found the percentage composition of acetic acid to be 39.9% C, 6.7% H, and 53.4% O. Determine the empirical formula. The molecular mass of acetic acid was determined by experiment to be 60.0 amu. What is its molecular formula?

Problem Strategy This problem employs the same general strategy as Example 3.11, with a few additional steps. In this case the compound is composed of C, H, and O. After determining the empirical formula of the compound, you calculate the empirical formula mass. You then determine how many times more massive the molecular compound is than the empirical compound by dividing the molecular mass by the empirical formula mass. You now have a factor by which the subscripts of the empirical formula need to be multiplied to arrive at the molecular formula.

Solution A sample of 100.0 g of acetic acid contains 39.9 g C, 6.7 g H, and 53.4 g O. Converting these masses to moles gives 3.33 mol C, 6.6 mol H, and 3.34 mol O. Dividing the mole numbers by the smallest one gives 1.00 for C, 2.0 for H, and 1.00 for O. **The empirical formula of acetic acid is CH_2O.** (You may have noted that the percentage composition of acetic acid is, within experimental error, the same as that of formaldehyde—see Example 3.7—so they must have the same empirical formula.) The empirical formula mass is 30.0 amu. Dividing the empirical formula mass into the molecular mass gives the number by which the subscripts in CH_2O must be multiplied.

$$n = \frac{\text{molecular mass}}{\text{empirical formula mass}} = \frac{60.0 \text{ amu}}{30.0 \text{ amu}} = 2.00$$

$C_2H_4O_2$

The molecular formula of acetic acid is $(CH_2O)_2$, or $C_2H_4O_2$. (A molecular model of acetic acid is shown in Figure 3.10.)

Answer Check Calculate the molecular mass of your answer. It should agree with the molecular mass of the compound given in the problem.

Exercise 3.12 The percentage composition of acetaldehyde is 54.5% C, 9.2% H, and 36.3% O, and its molecular mass is 44 amu. Obtain the molecular formula of acetaldehyde.

Figure 3.10 ▲
Molecular model of acetic acid

■ See Problems 3.73, 3.74, 3.75, and 3.76.

The condensed structural formula for acetaldehyde is CH_3CHO. Its structural formula is

$$\begin{array}{c} H \quad O \\ | \quad \| \\ H-C-C-H \\ | \\ H \end{array}$$

Determining this structural formula requires additional information.

The formula of acetic acid is often written $HC_2H_3O_2$ to indicate that one of the hydrogen atoms is acidic (lost easily), while the other three are not (Figure 3.10). Now that you know the formulas of acetic acid and acetaldehyde (determined from the data in Exercise 3.12 to be C_2H_4O), you can write the equations for the preparation of acetic acid described in the chapter opener. ◀ The first step consists of reacting ethanol with oxygen to obtain acetaldehyde and water. If you write the chemical equation and balance it, you obtain

$$2C_2H_5OH + O_2 \longrightarrow 2C_2H_4O + 2H_2O$$
$$\text{ethanol} \qquad\qquad\qquad \text{acetaldehyde}$$

In practice, the reaction is carried out with the reactants as gases (gas phase) at about 400°C using silver as a catalyst.

The second step consists of reacting acetaldehyde with oxygen to obtain acetic acid. Acetaldehyde liquid is mixed with a catalyst—manganese(II) acetate—and air is bubbled through it. The balanced equation is

$$2C_2H_4O + O_2 \longrightarrow 2HC_2H_3O_2$$
$$\text{acetaldehyde} \qquad\qquad \text{acetic acid}$$

Once you have this balanced equation, you are in a position to answer quantitative questions such as, How much acetic acid can you obtain from a 10.0-g sample of acetaldehyde? You will see how to answer such questions in the next sections.

CONCEPT CHECK 3.3

A friend has some questions about empirical formulas and molecular formulas. You can assume that he is good at performing the calculations.

a For a problem that asked him to determine the empirical formula, he came up with the answer $C_2H_8O_2$. Is this a possible answer to the problem? If not, what guidance would you offer your friend?

b For another problem he came up with the answer $C_{1.5}H_4$ as the empirical formula. Is this answer correct? Once again, if it isn't correct, what could you do to help your friend?

c Since you have been a big help, your friend asks one more question. He completed a problem of the same type as Example 3.12. His answers indicate that the compound had an empirical formula of C_3H_8O and the molecular formula C_3H_8O. Is this result possible?

Stoichiometry: Quantitative Relations in Chemical Reactions

In Chapter 2, we described a chemical equation as a representation of what occurs when molecules react. We will now study chemical equations more closely to answer questions about the stoichiometry of reactions. **Stoichiometry** (pronounced "stoy-key-om'-e-tree") is *the calculation of the quantities of reactants and products involved in a chemical reaction.* It is based on the chemical equation and on the relationship between mass and moles. Such calculations are fundamental to most quantitative work in chemistry. In the next sections, we will use the industrial Haber process for the production of ammonia to illustrate stoichiometric calculations.

3.6 Molar Interpretation of a Chemical Equation

In the Haber process for producing ammonia, NH_3, nitrogen (from the atmosphere) reacts with hydrogen at high temperature and pressure.

$$N_2(g) + 3H_2(g) \longrightarrow 2NH_3(g)$$

Hydrogen is usually obtained from natural gas or petroleum and so is relatively expensive. For this reason, the price of hydrogen partly determines the price of ammonia. Thus, an important question to answer is, How much hydrogen is required to give a particular quantity of ammonia? For example, how much hydrogen would be needed to produce one ton (907 kg) of ammonia? Similar kinds of questions arise throughout chemical research and industry.

To answer such quantitative questions, you must first look at the balanced chemical equation: one N_2 molecule and three H_2 molecules react to produce two NH_3 molecules (see models above). A similar statement involving multiples of these numbers of molecules is also correct. For example, 6.02×10^{23} N_2 molecules react with $3 \times 6.02 \times 10^{23}$ H_2 molecules, giving $2 \times 6.02 \times 10^{23}$ NH_3 molecules. This last statement can be put in molar terminology: one mole of N_2 reacts with three moles of H_2 to give two moles of NH_3.

> You may interpret a chemical equation either in terms of numbers of molecules (or ions or formula units) or in terms of numbers of moles, depending on your needs.

Because moles can be converted to mass, you can also give a mass interpretation of a chemical equation. The molar masses of N_2, H_2, and NH_3 are 28.0, 2.02, and 17.0 g/mol, respectively. Therefore, 28.0 g of N_2 reacts with 3×2.02 g of H_2 to yield 2×17.0 g of NH_3.

We summarize these three interpretations as follows:

$$N_2 \quad + \quad 3H_2 \quad \longrightarrow \quad 2NH_3$$

1 molecule N_2 + 3 molecules $H_2 \longrightarrow$ 2 molecules NH_3	(molecular interpretation)
1 mol N_2 + 3 mol $H_2 \longrightarrow$ 2 mol NH_3	(molar interpretation)
28.0 g N_2 + 3×2.02 g $H_2 \longrightarrow 2 \times 17.0$ g NH_3	(mass interpretation)

Suppose you ask how many grams of atmospheric nitrogen will react with 6.06 g (3×2.02 g) of hydrogen. You see from the last equation that the answer is 28.0 g N_2. We formulated this question for one mole of atmospheric nitrogen. Recalling the question posed earlier, you may ask how much hydrogen (in kg) is needed to yield 907 kg of ammonia in the Haber process. The solution to this problem depends on the fact that *the number of moles involved in a reaction is proportional to the coefficients in the balanced chemical equation.* In the next section, we will describe a procedure for solving such problems.

Exercise 3.13   In an industrial process, hydrogen chloride, HCl, is prepared by burning hydrogen gas, H_2, in an atmosphere of chlorine, Cl_2. Write the chemical equation for the reaction. Below the equation, give the molecular, molar, and mass interpretations.

■ See Problems 3.77 and 3.78.

3.7 Amounts of Substances in a Chemical Reaction

You see from the preceding discussion that a balanced chemical equation relates the amounts of substances in a reaction. The coefficients in the equation can be given a molar interpretation, and using this interpretation you can, for example, calculate the moles of product obtained from any given moles of reactant. Also, you can extend this type of calculation to answer questions about masses of reactants and products.

Again consider the Haber process for producing ammonia gas. Suppose you have a mixture of H_2 and N_2, and 4.8 mol H_2 in this mixture reacts with N_2 to produce NH_3. How many moles of NH_3 can you produce from this quantity of H_2?

In asking questions like this, you assume that the mixture contains a sufficient quantity of the other reactant (N_2, in this case).

The balanced chemical equation for the Haber process tells you that 3 mol H_2 produce 2 mol NH_3. You can express this as a conversion factor: ◀

$$\underbrace{\frac{2 \text{ mol } NH_3}{3 \text{ mol } H_2}}_{\substack{\text{Converts from} \\ \text{mol } H_2 \text{ to mol } NH_3}}$$

Multiplying any quantity of H_2 by this conversion factor mathematically converts that quantity of H_2 to the quantity of NH_3 as specified by the balanced chemical equation. Note that you write the conversion factor with the quantity you are converting *from* on the bottom (3 mol H_2), and the quantity you are converting *to* on the top (2 mol NH_3).

To calculate the quantity of NH_3 produced from 4.8 mol H_2, you write 4.8 mol H_2 and multiply this by the preceding conversion factor:

$$4.8 \text{ mol } H_2 \times \frac{2 \text{ mol } NH_3}{3 \text{ mol } H_2} = 3.2 \text{ mol } NH_3$$

Note how the unit mol H_2 cancels to give the answer in mol NH_3.

In this calculation, you converted moles of reactant to moles of product.

$$\text{Moles reactant} \longrightarrow \text{moles product}$$

It is just as easy to calculate the moles of reactant needed to obtain the specified moles of product. You mathematically convert moles of product to moles of reactant.

$$\text{Moles product} \longrightarrow \text{moles reactant}$$

To set up the conversion factor, you refer to the balanced chemical equation and place the quantity you are converting from on the bottom and the quantity you are converting to on the top.

$$\underbrace{\frac{3 \text{ mol } H_2}{2 \text{ mol } NH_3}}_{\substack{\text{Converts from} \\ \text{mol } NH_3 \text{ to mol } H_2}}$$

Now consider the problem asked in the previous section: How much hydrogen (in kg) is needed to yield 907 kg of ammonia by the Haber process? The balanced chemical equation directly relates moles of substances, not masses. Therefore, you must first convert the mass of ammonia to moles of ammonia, then convert moles of ammonia to moles of hydrogen. Finally, you convert moles of hydrogen to mass of hydrogen.

$$\text{Mass } NH_3 \longrightarrow \text{mol } NH_3 \longrightarrow \text{mol } H_2 \longrightarrow \text{mass } H_2$$

You learned how to convert mass to moles, and vice versa, in Section 3.2.

The calculation to convert 907 kg NH_3, or 9.07×10^5 g NH_3, to mol NH_3 is as follows:

$$9.07 \times 10^5 \text{ g } NH_3 \times \underbrace{\frac{1 \text{ mol } NH_3}{17.0 \text{ g } NH_3}}_{\substack{\text{Converts from} \\ \text{g } NH_3 \text{ to mol } NH_3}} = 5.34 \times 10^4 \text{ mol } NH_3$$

Now you convert from moles NH_3 to moles H_2.

$$5.34 \times 10^4 \text{ mol } NH_3 \times \underbrace{\frac{3 \text{ mol } H_2}{2 \text{ mol } NH_3}}_{\substack{\text{Converts from} \\ \text{mol } NH_3 \text{ to mol } H_2}} = 8.01 \times 10^4 \text{ mol } H_2$$

Finally, you convert moles H_2 to grams H_2.

$$8.01 \times 10^4 \text{ mol } H_2 \times \underbrace{\frac{2.02 \text{ g } H_2}{1 \text{ mol } H_2}}_{\substack{\text{Converts from} \\ \text{mol } H_2 \text{ to g } H_2}} = 1.62 \times 10^5 \text{ g } H_2$$

The result says that to produce 907 kg NH_3, you need 1.62×10^5 g H_2, or 162 kg H_2.

Once you feel comfortable with the individual conversions, you can do this type of calculation in a single step by multiplying successively by conversion factors, as follows:

$$9.07 \times 10^5 \text{ g } NH_3 \times \frac{1 \text{ mol } NH_3}{17.0 \text{ g } NH_3} \times \frac{3 \text{ mol } H_2}{2 \text{ mol } NH_3} \times \frac{2.02 \text{ g } H_2}{1 \text{ mol } H_2} =$$

$$1.62 \times 10^5 \text{ g } H_2 \text{ (or 162 kg } H_2\text{)}$$

Note that the unit in the denominator of each conversion factor cancels the unit in the numerator of the preceding factor. Figure 3.11 illustrates this calculation diagrammatically.

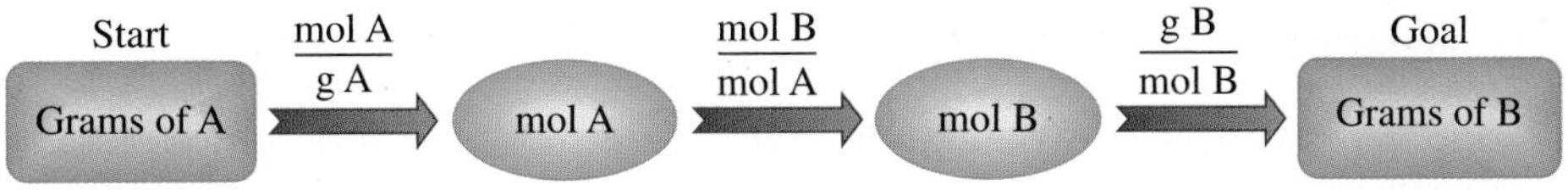

Figure 3.11 ◀

Steps in a stoichiometric calculation You convert the mass of substance A in a reaction to moles of substance A, then to moles of another substance B, and finally to mass of substance B.

The following two examples illustrate additional variations of this type of calculation.

◗WL Interactive Example 3.13 **Relating the Quantity of Reactant to Quantity of Product**

Gaining Mastery Toolbox

Critical Concept 3.13
The coefficients in a balanced chemical equation represent molar relationships between the reactants and products. These relationships can be used as conversion factors when performing calculations.

Solution Essentials:
• Molar interpretation of a balanced chemical equation.
• Molar mass
• Balanced chemical equation
• Dimensional analysis
• Rules for significant figures and rounding

Hematite, Fe_2O_3, is an important ore of iron; see Figure 3.12. (An ore is a natural substance from which the metal can be profitably obtained.) The free metal is obtained by reacting hematite with carbon monoxide, CO, in a blast furnace. Carbon monoxide is formed in the furnace by partial combustion of carbon. The reaction is

$$Fe_2O_3(s) + 3CO(g) \longrightarrow 2Fe(s) + 3CO_2(g)$$

How many grams of iron can be produced from 1.00 kg Fe_2O_3?

Problem Strategy This calculation involves the conversion of a quantity of Fe_2O_3 to a quantity of Fe. In performing this calculation, we assume that there is enough CO for the complete reaction of the Fe_2O_3. An essential feature of this type of calculation is that you must use the information from the balanced chemical equation to convert from the moles of a given substance to the moles of another substance. Therefore, you first convert the mass of Fe_2O_3 (1.00 kg $Fe_2O_3 = 1.0 \times 10^3$ g Fe_2O_3) to moles of Fe_2O_3. Then, using the relationship from the balanced chemical equation, you convert moles of Fe_2O_3 to moles of Fe. Finally, you convert the moles of Fe to grams of Fe.

The problem strategy is diagrammed as

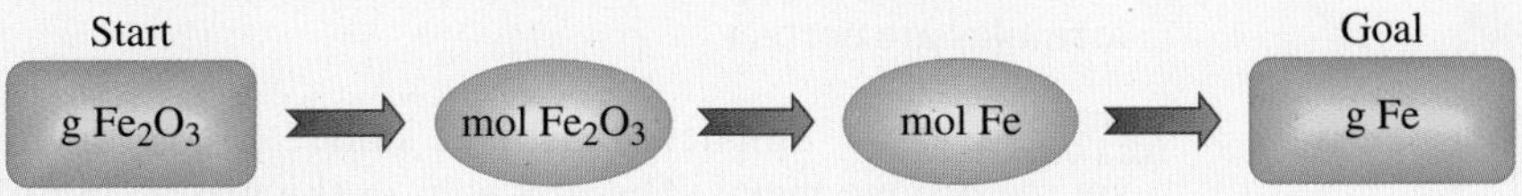

Solution The calculation is as follows:

$$1.00 \times 10^3 \text{ g } Fe_2O_3 \times \frac{1 \text{ mol } Fe_2O_3}{160 \text{ g } Fe_2O_3} \times \frac{2 \text{ mol } Fe}{1 \text{ mol } Fe_2O_3} \times \frac{55.8 \text{ g } Fe}{1 \text{ mol } Fe} = \textbf{698 g Fe}$$

(continued)

(*continued*)

Figure 3.12 ▲

Hematite The name of this iron mineral stems from the Greek word for blood, which alludes to the color of certain forms of the mineral.

© Cengage Learning

Answer Check When calculating the mass of product produced in a chemical reaction such as this, keep in mind that the total mass of reactants must equal the total mass of products. Because of this, you should be suspicious of any calculation where you find that a few grams of reactant produces a large mass of product.

Exercise 3.14   Sodium is a soft, reactive metal that instantly reacts with water to give hydrogen gas and a solution of sodium hydroxide, NaOH. How many grams of sodium metal are needed to give 7.81 g of hydrogen by this reaction? (Remember to write the balanced equation first.)

■ See Problems 3.83, 3.84, 3.85, and 3.86.

⚙WL Interactive Example 3.14   **Relating the Quantities of Two Reactants (or Two Products)**

Today chlorine is prepared from sodium chloride by electrochemical decomposition. Formerly chlorine was produced by heating hydrochloric acid with pyrolusite (manganese dioxide, MnO_2), a common manganese ore. Small amounts of chlorine may be prepared in the laboratory by the same reaction (see Figure 3.13):

$$4HCl(aq) + MnO_2(s) \longrightarrow 2H_2O(l) + MnCl_2(aq) + Cl_2(g)$$

How many grams of HCl react with 5.00 g of manganese dioxide, according to this equation?

Problem Strategy When starting a problem like this, determine and note which quantities in the balanced chemical equation are being related. Here you are looking at the two reactants, and the relationship is that 4 mol HCl reacts with 1 mol MnO_2. Answer this problem following the same strategy as outlined in Example 3.13, only this time use the relationship between the HCl and MnO_2.

The problem strategy is diagrammed as

Start   g MnO_2 ⟹ mol MnO_2 ⟹ mol HCl ⟹ g HCl   Goal

Solution You write what is given (5.00 g MnO_2) and convert this to moles, then to moles of what is desired (mol HCl). Finally you convert this to mass (g HCl). The calculation is:

$$5.00 \text{ g } MnO_2 \times \frac{1 \text{ mol } MnO_2}{86.9 \text{ g } MnO_2} \times \frac{4 \text{ mol HCl}}{1 \text{ mol } MnO_2} \times \frac{36.5 \text{ g HCl}}{1 \text{ mol HCl}} = \textbf{8.40 g HCl}$$

Answer Check To avoid incorrect answers, before starting any stoichiometry problem, make sure that the chemical equation is complete and balanced and that the chemical formulas are correct.

Exercise 3.15   Sphalerite is a zinc sulfide (ZnS) mineral and an important commercial source of zinc metal. The first step in the processing of the ore consists of heating the sulfide with oxygen to give zinc oxide, ZnO, and sulfur dioxide, SO_2. How many kilograms of oxygen gas combine with 5.00×10^3 g of zinc sulfide in this reaction? (You must first write the balanced chemical equation.)

— Cl_2

© Cengage Learning

Figure 3.13 ◄

Preparation of chlorine Concentrated hydrochloric acid was added to manganese dioxide in the beaker. Note the formation of yellowish green gas (chlorine), which is depicted by molecular models.

■ See Problems 3.87 and 3.88.

Exercise 3.16   The British chemist Joseph Priestley prepared oxygen in 1774 by heating mercury(II) oxide, HgO. Mercury metal is the other product. If 6.47 g of oxygen is collected, how many grams of mercury metal are also produced?

■ See Problems 3.89 and 3.90.

CONCEPT CHECK 3.4

The main reaction of a charcoal grill is $C(s) + O_2(g) \longrightarrow CO_2(g)$. Which of the statements below are incorrect? Why?

a 1 atom of carbon reacts with 1 molecule of oxygen to produce 1 molecule of CO_2.

b 1 g of C reacts with 1 g of O_2 to produce 2 grams of CO_2.

c 1 g of C reacts with 0.5 g of O_2 to produce 1 g of CO_2.

d 12 g of C reacts with 32 g of O_2 to produce 44 g of CO_2.

e 1 mol of C reacts with 1 mol of O_2 to produce 1 mol of CO_2.

f 1 mol of C reacts with 0.5 mol of O_2 to produce 1 mol of CO_2.

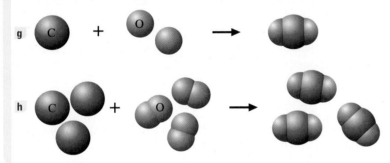

3.8 Limiting Reactant; Theoretical and Percentage Yields

Often reactants are added to a reaction vessel in amounts different from the molar proportions given by the chemical equation. In such cases, only one of the reactants may be completely consumed at the end of the reaction, whereas some amounts of other reactants will remain unreacted. The **limiting reactant** (or **limiting reagent**) is *the reactant that is entirely consumed when a reaction goes to completion.* A reactant that is not completely consumed is often referred to as an *excess reactant.* Once one of the reactants is used up, the reaction stops. This means that:

> The moles of product are always determined by the starting moles of limiting reactant.

A couple of analogies may help you understand the limiting reactant problem. Suppose you want to make some cheese sandwiches. Each is made from two slices of bread and a slice of cheese. Let's write that in the form of a chemical equation:

$$2 \text{ slices bread} + 1 \text{ slice cheese} \longrightarrow 1 \text{ sandwich}$$

You look in the kitchen and see that you have six slices of bread and two slices of cheese. The six slices of bread would be enough to make three sandwiches if you had enough cheese. The two slices of cheese would be enough to make two sandwiches if you had enough bread. How many sandwiches can you make?

Figure 3.14 ▶

Limiting reactant analogy using cheese sandwiches Start with six slices of bread, two slices of cheese, and the sandwich-making equation. Even though you have extra bread, you are limited to making two sandwiches by the amount of cheese you have on hand. Cheese is the limiting reactant.

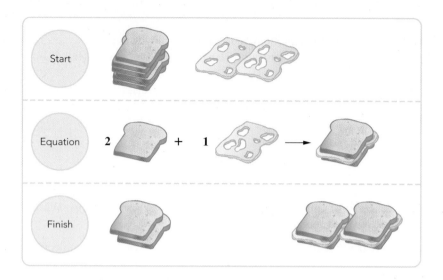

What you will find is that once you have made two sandwiches, you will be out of cheese (Figure 3.14). Cheese is the "limiting reactant" in the language of chemistry. If you look at each "reactant," and ask how much "product" you can make from it, the reactant that limits your amount of product is called the limiting reactant.

Here is another analogy. Suppose you are supervising the assembly of automobiles. Your plant has in stock 300 steering wheels and 900 tires, plus an excess of every other needed component. How many autos can you assemble from this stock? Here is the "balanced equation" for the auto assembly:

1 steering wheel + 4 tires + other components $\longrightarrow$ 1 auto

One way to solve this problem is to calculate the number of autos that you could assemble from each component. In this case, you only need to concentrate on the effects of the steering wheels and tires since production will not be limited by the other, excess components. Looking at the equation, you can determine that from 300 steering wheels, you could assemble 300 autos; from 900 tires you could assemble $900 \div 4 = 225$ autos.

How many autos can you produce, 225 or 300? Note that by the time you have assembled 225 autos, you will have exhausted your stock of tires, so no more autos can be assembled. Tires are the "limiting reactant," since they limit the number of automobiles that you can actually assemble (225).

You can set up these calculations in the same way for a chemical reaction. First, you calculate the numbers of autos you could assemble from the number of steering wheels and from the number of tires available, and then you compare the results. From the balanced equation, you can see that one steering wheel is equivalent to one auto, and four tires are equivalent to one auto. Therefore:

$$300 \text{ steering wheels} \times \frac{1 \text{ auto}}{1 \text{ steering wheel}} = 300 \text{ autos}$$

$$900 \text{ tires} \times \frac{1 \text{ auto}}{4 \text{ tires}} = 225 \text{ autos}$$

Comparing the numbers of autos produced (225 autos versus 300 autos), you conclude that tires are the limiting component. (The component producing the fewer autos is the limiting component.) Note that apart from the units (or factor labels), the calculation is identical with what was done previously.

Now consider a chemical reaction, the burning of hydrogen in oxygen.

$$2H_2(g) + O_2(g) \longrightarrow 2H_2O(g)$$

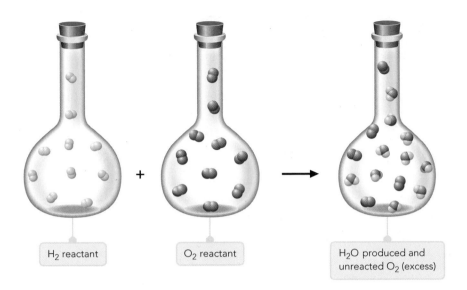

Figure 3.15 ◀

Molecular view of 2H$_2$ + O$_2$ ⟶ 2H$_2$O reaction with H$_2$ as the limiting reactant When an equal number of moles of H$_2$ and O$_2$ are reacted according to the equation 2H$_2$ + O$_2$ ⟶ 2H$_2$O, all of the H$_2$ completely reacts, whereas only half of the O$_2$ is consumed. In this case, the H$_2$ is the limiting reactant and the O$_2$ is the excess reactant.

H$_2$ reactant   O$_2$ reactant   H$_2$O produced and unreacted O$_2$ (excess)

Suppose you put 1 mol H$_2$ and 1 mol O$_2$ into a reaction vessel. How many moles of H$_2$O will be produced? First, you note that 2 mol H$_2$ produces 2 mol H$_2$O and that 1 mol O$_2$ produces 2 mol H$_2$O. Now you calculate the moles of H$_2$O that you could produce from each quantity of reactant, assuming that there is sufficient other reactant.

$$1 \text{ mol } H_2 \times \frac{2 \text{ mol } H_2O}{2 \text{ mol } H_2} = 1 \text{ mol } H_2O$$

$$1 \text{ mol } O_2 \times \frac{2 \text{ mol } H_2O}{1 \text{ mol } O_2} = 2 \text{ mol } H_2O$$

Comparing these results, you see that hydrogen, H$_2$, yields the least amount of product, so it must be the limiting reactant. By the time 1 mol H$_2$O is produced, all of the hydrogen is used up; the reaction stops. Oxygen, O$_2$, is the excess reactant. Figure 3.15 depicts this reaction from a molecular viewpoint.

We can summarize the limiting-reactant problem as follows. Suppose you are given the amounts of reactants added to a vessel, and you wish to calculate the amount of product obtained when the reaction is complete. Unless you know that the reactants have been added in the molar proportions given by the chemical equation, the problem is twofold: (1) you must first identify the limiting reactant; (2) you then calculate the amount of product from the amount of limiting reactant. The next examples illustrate the steps.

CONCEPT CHECK 3.5

Solid ReO$_3$ is a material that is an extremely good conductor of electricity. It can be formed by the chemical reaction Re(s) + 3Re$_2$O$_7$(s) ⟶ 7ReO$_3$(s). If 3 mol of Re and 3 mol of Re$_2$O$_7$ are allowed to react, 7 mol of ReO$_3$ are produced. Which of the following statements is the best reason that 7 mol of ReO$_3$ are produced during the chemical reaction?

a According to the balanced chemical equation, 7 mol of ReO$_3$ will always be formed regardless of the starting amounts of the reactants.

b Rhenium is the limiting reactant, so 7 mol of product will form.

c When 3 mol of Re$_2$O$_7$ undergoes reaction, 7 mol of ReO$_3$ is produced.

d Re$_2$O$_7$ is the excess reactant, and therefore 7 mol of ReO$_3$ are produced.

e The excess rhenium completely reacts producing 7 mol of ReO$_3$.

OWL Interactive Example 3.15   Calculating with a Limiting Reactant (Involving Moles)

Critical Concept 3.15
During a chemical reaction, after any one of the reactants is completely consumed, no additional products are formed. In order to calculate the amount of product formed during a chemical reaction, one must first identify the reactant—the limiting reactant— that would be used up prior to the other reactants.

Solution Essentials:
• Limiting reactant
• Molar interpretation of a balanced chemical equation.
• Molar mass
• Balanced chemical equation
• Dimensional analysis
• Rules for significant figures and rounding

Zinc metal reacts with hydrochloric acid by the following reaction:

$$Zn(s) + 2HCl(aq) \longrightarrow ZnCl_2(aq) + H_2(g)$$

If 0.30 mol Zn is added to hydrochloric acid containing 0.52 mol HCl, how many moles of H_2 are produced?

Problem Strategy

Step 1: Which is the limiting reactant? To answer this, using the relationship from the balanced chemical equation, you take each reactant in turn and ask how much product (H_2) would be obtained if each were totally consumed. The reactant that gives the smaller amount of product is the limiting reactant. (Remember how you obtained the limiting component in the auto-assembly analogy.)

Step 2: You obtain the amount of product actually obtained from the amount of limiting reactant.

Solution

Step 1:

$$0.30 \; \cancel{\text{mol Zn}} \times \frac{1 \; \text{mol } H_2}{1 \; \cancel{\text{mol Zn}}} = 0.30 \; \text{mol } H_2$$

$$0.52 \; \cancel{\text{mol HCl}} \times \frac{1 \; \text{mol } H_2}{2 \; \cancel{\text{mol HCl}}} = 0.26 \; \text{mol } H_2$$

You see that hydrochloric acid must be the limiting reactant and that some zinc must be left unconsumed. (Zinc is the excess reactant.)

Step 2: Since HCl is the limiting reactant, the amount of H_2 produced must be **0.26 mol.**

Answer Check A common assumption that often leads to errors in solving a problem of this type is to assume that whichever reactant is present in the least quantity (in mass or moles of material) is automatically the limiting reactant. Note how that assumption would have led to an incorrect answer in this problem. Always be sure to account for the stoichiometry of the balanced chemical equation.

Exercise 3.17   Aluminum chloride, $AlCl_3$, is used as a catalyst in various industrial reactions. It is prepared from hydrogen chloride gas and aluminum metal shavings.

$$2Al(s) + 6HCl(g) \longrightarrow 2AlCl_3(s) + 3H_2(g)$$

Suppose a reaction vessel contains 0.15 mol Al and 0.35 mol HCl. How many moles of $AlCl_3$ can be prepared from this mixture?

■ See Problems 3.91 and 3.92.

OWL Interactive Example 3.16   Calculating with a Limiting Reactant (Involving Masses)

Critical Concept 3.16
The amount (mol) of each reactant decreases during a chemical reaction. After the reaction is complete, the limiting reactant is completely consumed, and the other reactants are partially consumed.

Solution Essentials:
• Limiting reactant
• Molar interpretation of a balanced chemical equation.
• Molar mass
• Balanced chemical equation
• Dimensional analysis
• Rules for significant figures and rounding

In a process for producing acetic acid, oxygen gas is bubbled into acetaldehyde, CH_3CHO, containing manganese(II) acetate (catalyst) under pressure at 60°C.

$$2CH_3CHO(l) + O_2(g) \longrightarrow 2HC_2H_3O_2(l)$$

In a laboratory test of this reaction, 20.0 g CH_3CHO and 10.0 g O_2 were put into a reaction vessel. a How many grams of acetic acid can be produced by this reaction from these amounts of reactants? b How many grams of the excess reactant remain after the reaction is complete?

Problem Strategy

a This part is similar to the preceding example, but now you must convert grams of each reactant (acetaldehyde and oxygen) to moles of product (acetic acid). From these results, you decide which is the limiting reactant and the moles of product obtained, which you convert to grams of product.

b In order to calculate the amount of excess reactant remaining after the reaction is complete, you need to know the identity of the excess reactant and how much of this

(continued)

(continued)

excess reactant was needed for the reaction. The result from Step 1 provides the identity of the limiting reactant, so you will know which reactant was in excess. From the moles of product produced by the limiting reactant, you calculate the grams of the excess reactant needed in the reaction. You now know how much of this excess reactant was consumed, so subtracting the amount consumed from the starting amount will yield the amount of excess reactant that remains.

Solution

a How much acetic acid is produced?

Step 1: To determine which reactant is limiting, you convert grams of each reactant (20.0 g CH_3CHO and 10.0 g O_2) to moles of product, $HC_2H_3O_2$. Acetaldehyde has a molar mass of 44.1 g/mol, and oxygen has a molar mass of 32.0 g/mol.

$$20.0 \text{ g } CH_3CHO \times \frac{1 \text{ mol } CH_3CHO}{44.1 \text{ g } CH_3CHO} \times \frac{2 \text{ mol } HC_2H_3O_2}{2 \text{ mol } CH_3CHO} = 0.454 \text{ mol } HC_2H_3O_2$$

$$10.0 \text{ g } O_2 \times \frac{1 \text{ mol } O_2}{32.0 \text{ g } O_2} \times \frac{2 \text{ mol } HC_2H_3O_2}{1 \text{ mol } O_2} = 0.625 \text{ mol } HC_2H_3O_2$$

Thus, acetaldehyde, CH_3CHO, is the limiting reactant, so 0.454 mol $HC_2H_3O_2$ was produced.

Step 2: You convert 0.454 mol $HC_2H_3O_2$ to grams of $HC_2H_3O_2$.

$$0.454 \text{ mol } HC_2H_3O_2 \times \frac{60.1 \text{ g } HC_2H_3O_2}{1 \text{ mol } HC_2H_3O_2} = \mathbf{27.3 \text{ g } HC_2H_3O_2}$$

b How much of the excess reactant (oxygen) was left over? You convert the moles of acetic acid to grams of oxygen (the quantity of oxygen needed to produce this amount of acetic acid).

$$0.454 \text{ mol } HC_2H_3O_2 \times \frac{1 \text{ mol } O_2}{2 \text{ mol } HC_2H_3O_2} \times \frac{32.0 \text{ g } O_2}{1 \text{ mol } O_2} = 7.26 \text{ g } O_2$$

You started with 10.0 g O_2, so the quantity remaining is

$$(10.0 - 7.26) \text{ g } O_2 = \mathbf{2.7 \text{ g } O_2} \quad \text{(mass remaining)}$$

Answer Check Whenever you are confronted with a stoichiometry problem you should always determine if you are going to have to solve a limiting reactant problem like this one, or a problem like Example 3.13 that involves a single reactant and one reactant in excess. A good rule of thumb is that when two or more reactant quantities are specified, you should approach the problem as was done here.

© Cengage Learning

Figure 3.16 ▲

Reaction of zinc and sulfur When a hot nail is stuck into a pile of zinc and sulfur, a fiery reaction occurs and zinc sulfide forms.

Exercise 3.18 In an experiment, 7.36 g of zinc was heated with 6.45 g of sulfur (Figure 3.16). Assume that these substances react according to the equation

$$8Zn + S_8 \longrightarrow 8ZnS$$

What amount of zinc sulfide was produced?

■ See Problems 3.93 and 3.94.

The **theoretical yield** of product is *the maximum amount of product that can be obtained by a reaction from given amounts of reactants.* It is the amount that you calculate from the stoichiometry based on the limiting reactant. In Example 3.16, the theoretical yield of acetic acid is 27.3 g. In practice, the *actual yield* of a product may be much less for several possible reasons. First, some product may be lost during the process of separating it from the final reaction mixture. Second, there may be other, competing reactions that occur simultaneously with the reactant on which the theoretical yield is based. Finally, many reactions appear to stop before they reach completion; they give mixtures of reactants and products. ◄

It is important to know the actual yield from a reaction in order to make economic decisions about a preparation method. The reactants for a given method may not be too costly per kilogram, but if the actual yield is very low, the final cost can be very high. The **percentage yield** of product is *the actual yield (experimentally determined) expressed as a percentage of the theoretical yield (calculated).*

Such reactions reach chemical equilibrium. We will discuss equilibrium quantitatively in Chapter 14.

$$\text{Pecentage yield} = \frac{\text{actual yield}}{\text{theoretical yield}} \times 100\%$$

To illustrate the calculation of percentage yield, recall that the theoretical yield of acetic acid calculated in Example 3.16 was 27.3 g. If the actual yield of acetic acid obtained in an experiment, using the amounts of reactants given in Example 3.16, is 23.8 g, then

$$\text{Percentage yield of HC}_2\text{H}_3\text{O}_2 = \frac{23.8 \text{ g}}{27.3 \text{ g}} \times 100\% = 87.2\%$$

Exercise 3.19 New industrial plants for acetic acid react liquid methanol with carbon monoxide in the presence of a catalyst.

$$\text{CH}_3\text{OH}(l) + \text{CO}(g) \longrightarrow \text{HC}_2\text{H}_3\text{O}_2(l)$$

In an experiment, 15.0 g of methanol and 10.0 g of carbon monoxide were placed in a reaction vessel. What is the theoretical yield of acetic acid? If the actual yield is 19.1 g, what is the percentage yield?

■ See Problems 3.97 and 3.98.

CONCEPT CHECK 3.6

You perform the hypothetical reaction of an element, $X_2(g)$, with another element, $Y(g)$, to produce $XY(g)$.

a Write the balanced chemical equation for the reaction.

b If X_2 and Y were mixed in the quantities shown in the container on the left below and allowed to react, which of the three options is the correct representation of the contents of the container after the reaction has occurred?

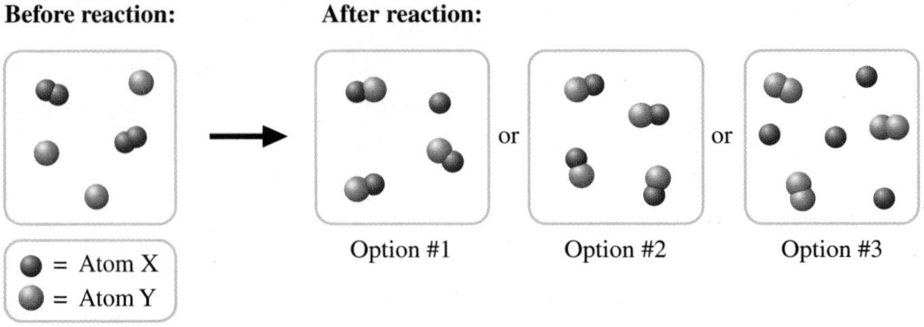

c Using the information presented in part b, identify the limiting reactant.

A Checklist for Review

⊙WL and 🔲 Chemistry Sign in at **www.cengage.com/owl** to:
- View tutorials and simulations, develop problem-solving skills, and complete online homework assigned by your professor.
- For quick review and exam prep, download Go Chemistry mini lecture modules from OWL or purchase them at **www.cengagebrain.com**.

Summary of Facts and Concepts

A *formula mass* equals the sum of the atomic masses of the atoms in the formula of a compound. If the formula corresponds to that of a molecule, this sum of atomic masses equals the *molecular mass* of the compound. The mass of *Avogadro's number* (6.02×10^{23}) of formula units—that is, the mass of one *mole* of substance—equals the mass in grams that corresponds to the numerical value of the formula mass in amu. This mass is called the *molar mass*.

The *empirical formula (simplest formula)* of a compound is obtained from the *percentage composition* of the substance, which is expressed as *mass percentages* of the elements. To calculate the empirical formula, you convert mass percentages to ratios of moles, which, when expressed in smallest whole numbers, give the subscripts in the formula. A molecular formula is a multiple of the empirical formula; this multiple is determined from the experimental value of the molecular mass.

A chemical equation may be interpreted in terms of moles of reactants and products, as well as in terms of molecules. Using this *molar interpretation*, you can convert from the mass of one substance in a chemical equation to the mass of another. The maximum amount of product from a reaction is determined by the *limiting reactant,* the reactant that is completely used up; the other reactants are in excess.

Learning Objectives

3.1 Molecular Mass and Formula Mass

- Define the terms *molecular mass* and *formula mass* of a substance.
- Calculate the formula mass from a formula. Example 3.1
- Calculate the formula mass from molecular models. Example 3.2

3.2 The Mole Concept

- Define the quantity called the *mole*.
- Learn *Avogadro's number.*
- Understand how the *molar mass* is related to the formula mass of a substance.
- Calculate the mass of atoms and molecules. Example 3.3
- Perform calculations using the mole.
- Convert from moles of substance to grams of substance. Example 3.4
- Convert from grams of substance to moles of substance. Example 3.5
- Calculate the number of molecules in a given mass of substance. Example 3.6

3.3 Mass Percentages from the Formula

- Define *mass percentage.*
- Calculate the percentage composition of the elements in a compound. Example 3.7
- Calculate the mass of an element in a given mass of compound. Example 3.8

3.4 Elemental Analysis: Percentage of Carbon, Hydrogen, and Oxygen

- Describe how C, H, and O combustion analysis is performed.
- Calculate the percentage of C, H, and O from combustion data. Example 3.9

Important Terms

molecular mass
formula mass

mole (mol)
Avogadro's number (N_A)
molar mass

percentage composition
mass percentage

3.5 Determining Formulas

- Define *empirical formula*.
- Determine the empirical formula of a binary compound from the masses of its elements. Example 3.10
- Determine the empirical formula from the percentage composition. Example 3.11
- Understand the relationship between the molecular mass of a substance and its *empirical formula mass*.
- Determine the molecular formula from the percentage composition and molecular mass. Example 3.12

empirical (simplest) formula

3.6 Molar Interpretation of a Chemical Equation

- Relate the coefficients in a balanced chemical equation to the number of molecules or moles (*molar interpretation*).

stoichiometry

3.7 Amounts of Substances in a Chemical Reaction

- Use the coefficients in a chemical reaction to perform calculations.
- Relate the quantities of reactant to the quantity of product. Example 3.13
- Relate the quantities of two reactants or two products. Example 3.14

3.8 Limiting Reactant; Theoretical and Percentage Yields

- Understand how a *limiting reactant* or *limiting reagent* determines the moles of product formed during a chemical reaction and how much *excess reactant* remains.
- Calculate with a limiting reactant involving moles. Example 3.15
- Calculate with a limiting reactant involving masses. Example 3.16
- Define and calculate the *theoretical yield* of chemical reactions.
- Determine the *percentage yield* of a chemical reaction.

limiting reactant (reagent)
theoretical yield
percentage yield

Key Equations

$$\text{Mass\% } A = \frac{\text{mass of } A \text{ in the whole}}{\text{mass of the whole}} \times 100\% \qquad \text{Percentage yield} = \frac{\text{actual yield}}{\text{theoretical yield}} \times 100\%$$

$$n = \frac{\text{molecular mass}}{\text{empirical formula mass}}$$

Questions and Problems

⊽WL Interactive versions of these problems may be assigned in OWL.

Self-Assessment and Review Questions

Key: These questions test your understanding of the ideas you worked with in the chapter. These problems vary in difficulty and often can be used for the basis of discussion.

3.1 Describe in words how to obtain the formula mass of a compound from the formula.

3.2 What is the difference between a formula mass and a molecular mass? Could a given substance have both a formula mass and a molecular mass?

3.3 One mole of N_2 contains how many N_2 molecules? How many N atoms are there in one mole of N_2? One mole of iron(III) sulfate, $Fe_2(SO_4)_3$, contains how many moles of SO_4^{2-} ions? How many moles of O atoms?

3.4 Explain what is involved in determining the composition of a compound of C, H, and O by combustion.

3.5 A substance has the molecular formula $C_6H_{12}O_2$. What is its empirical formula?

3.6 Explain what is involved in obtaining the empirical formula from the percentage composition.

3.7 Hydrogen peroxide has the empirical formula HO and an empirical formula weight of 17.0 amu. If the molecular mass is 34.0 amu, what is the molecular formula?

3.8 Describe in words the meaning of the equation

$$CH_4 + 2O_2 \longrightarrow CO_2 + 2H_2O$$

using a molecular, a molar, and then a mass interpretation.

3.9 Explain how a chemical equation can be used to relate the masses of different substances involved in a reaction.

3.10 What is a limiting reactant in a reaction mixture? Explain how it determines the amount of product.

3.11 Come up with some examples of limiting reactants that use the concept but don't involve chemical reactions.

3.12 Explain why it is impossible to have a theoretical yield of more than 100%.

3.13 How many grams of NH_3 will have the same number of molecules as 15.0 g of C_6H_6?

- a 15.0
- b 3.27
- c 17.0
- d 1.92
- e 14.2

3.14 Which of the following has the largest number of molecules?

- a 1 g of benzene, C_6H_6
- b 1 g of formaldehyde, CH_2O
- c 1 g of TNT, $C_7H_5N_3O_6$
- d 1 g of naphthalene, $C_{10}H_8$
- e 1 g of glucose, $C_6H_{12}O_6$

3.15 When 2.56 g of a compound containing only carbon, hydrogen, and oxygen is burned completely, 3.84 g of CO_2 and 1.05 g of H_2O are produced. What is the empirical formula of the compound?

- a $C_3H_4O_3$
- b $C_5H_6O_4$
- c $C_5H_6O_5$
- d $C_4H_4O_3$
- e $C_4H_6O_3$

3.16 How many atoms are present in 123 g of magnesium cyanide?

- a 9.7×10^{23}
- b 2.91×10^{24}
- c 2.83×10^{28}
- d 4.85×10^{24}
- e 5.65×10^{27}

Concept Explorations

Key: Concept explorations are comprehensive problems that provide a framework that will enable you to explore and learn many of the critical concepts and ideas in each chapter. If you master the concepts associated with these explorations, you will have a better understanding of many important chemistry ideas and will be more successful in solving all types of chemistry problems. These problems are well suited for group work and for use as in-class activities.

3.17 Moles and Molar Mass

Part 1: The mole provides a convenient package where we can make a connection between the mass of a substance and the number (count) of that substance. This is a familiar concept if you have ever bought pieces of bulk hard candy, where you purchase candy by mass rather than count. Typically, there is a scale provided for weighing the candy. For example, a notice placed above the candy bin might read, "For the candy in the bin below, there are 500 pieces of candy per kg." Using this conversion factor, perform the following calculations.

- a How many candies would you have if you had 0.2 kg?
- b If you had 10 dozen candies, what would be their mass?
- c What is the mass of one candy piece?
- d What is the mass of 2.0 moles of candies?

Part 2: The periodic table provides information about each element that serves somewhat the same purpose as the label on the nail bin described in Part 1, only in this case, the mass (molar mass) of each element is the number of grams of the element that contain 6.02×10^{23} atoms or molecules of the element. As you are aware, the quantity 6.02×10^{23} is called the mole.

- a If you had 0.2 kg of helium, how many helium atoms would you have?
- b If you had 10 dozen helium atoms, what would be their mass?
- c What is the mass of one helium atom?
- d What is the mass of 2.0 moles of helium atoms?

Part 3: Say there is a newly defined "package" called the binkle. One binkle is defined as being exactly 3×10^{12}.

- a If you had 1.0 kg of nails and 1.0 kg of helium atoms, would you expect them to have the same number of binkles? Using complete sentences, explain your answer.
- b If you had 3.5 binkles of nails and 3.5 binkles of helium atoms, which quantity would have more (count) and which would have more mass? Using complete sentences, explain your answers.
- c Which would contain more atoms, 3.5 g of helium or 3.5 g of lithium? Using complete sentences, explain your answer.

3.18 Moles Within Moles and Molar Mass

Part 1:

- a How many hydrogen and oxygen atoms are present in 1 molecule of H_2O?
- b How many moles of hydrogen and oxygen atoms are present in 1 mol H_2O?
- c What are the masses of hydrogen and oxygen in 1.0 mol H_2O?
- d What is the mass of 1.0 mol H_2O?

Part 2: Two hypothetical ionic compounds are discovered with the chemical formulas XCl_2 and YCl_2, where X and Y represent symbols of the imaginary elements. Chemical analysis of the two compounds reveals that 0.25 mol XCl_2 has a mass of 100.0 g and 0.50 mol YCl_2 has a mass of 125.0 g.

a What are the molar masses of XCl_2 and YCl_2?

b If you had 1.0-mol samples of XCl_2 and YCl_2, how would the number of chloride ions compare?

c If you had 1.0-mol samples of XCl_2 and YCl_2, how would the masses of elements X and Y compare?

d What is the mass of chloride ions present in 1.0 mol XCl_2 and 1.0 mol YCl_2?

e What are the molar masses of elements X and Y?

f How many moles of X ions and chloride ions would be present in a 200.0-g sample of XCl_2?

g How many grams of Y ions would be present in a 250.0-g sample of YCl_2?

h What would be the molar mass of the compound YBr_3?

Part 3: A minute sample of $AlCl_3$ is analyzed for chlorine. The analysis reveals that there are 12 chloride ions present in the sample. How many aluminum ions must be present in the sample?

a What is the total mass of $AlCl_3$ in this sample?

b How many moles of $AlCl_3$ are in this sample?

Conceptual Problems

Key: These problems are designed to check your understanding of the concepts associated with some of the main topics presented in each chapter. A strong conceptual understanding of chemistry is the foundation for both applying chemical knowledge and solving chemical problems. These problems vary in level of difficulty and often can be used as a basis for group discussion.

3.19 You react nitrogen and hydrogen in a container to produce ammonia, $NH_3(g)$. The following figure depicts the contents of the container after the reaction is complete.

= NH_3
= N_2

a Write a balanced chemical equation for the reaction.
b What is the limiting reactant?
c How many molecules of the limiting reactant would you need to add to the container in order to have a complete reaction (convert all reactants to products)?

3.20 Propane, C_3H_8, is the fuel of choice in a gas barbecue. When burning, the balanced equation is

$$C_3H_8 + 5O_2 \longrightarrow 3CO_2 + 4H_2O$$

a What is the limiting reactant in cooking with a gas grill?
b If the grill will not light and you know that you have an ample flow of propane to the burner, what is the limiting reactant?
c When using a gas grill you can sometimes turn the gas up to the point at which the flame becomes yellow and smokey. In terms of the chemical reaction, what is happening?

3.21 A critical point to master in becoming proficient at solving problems is evaluating whether or not your answer is reasonable. A friend asks you to look over her homework to see if she has done the calculations correctly. Shown below are descriptions of some of her answers. Without using your calculator or doing calculations on paper, see if you can judge the answers below as being reasonable or ones that will require her to go back and work the problems again.

a 0.33 mol of an element has a mass of 1.0×10^{-3} g.
b The mass of one molecule of water is 1.80×10^{-10} g.
c There are 3.01×10^{23} atoms of Na in 0.500 mol of Na.
d The molar mass of CO_2 is 44.0 kg/mol.

3.22 High cost and limited availability of a reactant often dictate which reactant is limiting in a particular process. Identify the limiting reactant when the reactions below are run, and come up with a reason to support your decision.

a Burning charcoal on a grill:

$$C(s) + O_2(g) \longrightarrow CO_2(g)$$

b Burning a chunk of Mg in water:

$$Mg(s) + 2H_2O(l) \longrightarrow Mg(OH)_2(aq) + H_2(g)$$

c The Haber process of ammonia production:

$$3H_2(g) + N_2(g) \longrightarrow 2NH_3(g)$$

3.23 An exciting, and often loud, chemical demonstration involves the simple reaction of hydrogen gas and oxygen gas to produce water vapor:

$$2H_2(g) + O_2(g) \longrightarrow 2H_2O(g)$$

The reaction is carried out in soap bubbles or balloons that are filled with the reactant gases. We get the reaction to proceed by igniting the bubbles or balloons. The more H_2O that is formed during the reaction, the bigger the bang. Explain the following observations.

a A bubble containing just H_2 makes a quiet "ffffft" sound when ignited.
b When a bubble containing equal amounts of H_2 and O_2 is ignited, a sizable bang results.
c When a bubble containing a ratio of 2 to 1 in the amounts of H_2 and O_2 is ignited, the loudest bang results.
d When a bubble containing just O_2 is ignited, virtually no sound is made.

3.24 A few hydrogen and oxygen molecules are introduced into a container in the quantities depicted in the following drawing. The gases are then ignited by a spark, causing them to react and form H_2O.

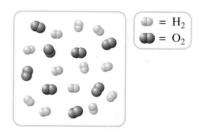
= H_2
= O_2

What is the maximum number of water molecules that can be formed in the chemical reaction?

Draw a molecular level representation of the container's contents after the chemical reaction.

3.25 A friend asks if you would be willing to check several homework problems to see if she is on the right track. Following are the problems and her proposed solutions. When you identify the problem with her work, make the appropriate correction.

Calculate the number of moles of calcium in 27.0 g of Ca.

$$27.0 \text{ g Ca} \times \frac{1 \text{ mol Ca}}{6.022 \times 10^{23} \text{ g Ca}} = ?$$

Calculate the number of potassium ions in 2.5 mol of K_2SO_4.

$$2.5 \text{ mol } K_2SO_4 \times \frac{1 \text{ mol } K^+ \text{ions}}{1 \text{ mol } K_2SO_4}$$
$$\times \frac{6.022 \times 10^{23} \text{ } K^+\text{ions}}{1 \text{ mol } K^+ \text{ ions}} = ?$$

Sodium reacts with water according to the following chemical equation.

$$2Na + 2H_2O \longrightarrow H_2 + 2NaOH$$

Assuming complete reaction, calculate the number of moles of water required to react with 0.50 mol of Na.

$$0.50 \text{ mol Na} \times \frac{1 \text{ mol } H_2O}{2 \text{ mol Na}} = ?$$

3.26 A friend is doing his chemistry homework and is working with the following chemical reaction.

$$2C_2H_2(g) + 5O_2(g) \longrightarrow 4CO_2(g) + 2H_2O(g)$$

He tells you that if he reacts 2 moles of C_2H_2 with 4 moles of O_2, then the C_2H_2 is the limiting reactant since there are fewer moles of C_2H_2 than O_2.

How would you explain to him where he went wrong with his reasoning (what concept is he missing)?

After providing your friend with the explanation from part a, he still doesn't believe you because he *had* a homework problem where 2 moles of calcium were reacted with 4 moles of sulfur and he needed to determine the limiting reactant. The reaction is

$$Ca(s) + S(s) \longrightarrow CaS(s)$$

He obtained the correct answer, Ca, by reasoning that since there were fewer moles of calcium reacting, calcium *had* to be the limiting reactant. How would you explain his reasoning flaw and why he got "lucky" in choosing the answer that he did?

Practice Problems

Key: These problems are for practice in applying problem-solving skills. They are divided by topic, and some are keyed to exercises (see the ends of the exercises). The problems are arranged in matching pairs; the odd-numbered problem of each pair is listed first, and its answer is given in the back of the book.

Formula Masses and Mole Calculations

3.27 Find the formula masses of the following substances to three significant figures.

methanol, CH_3OH

nitrogen trioxide, NO_3

potassium carbonate, K_2CO_3

nickel(II) phosphate, $Ni_3(PO_4)_2$

3.28 Find the formula masses of the following substances to three significant figures.

sulfuric acid, H_2SO_4

phosphorus pentachloride, PCl_5

ammonium chloride, NH_4Cl

calcium hydroxide, $Ca(OH)_2$

3.29 Calculate the formula mass of the following molecules to three significant figures.

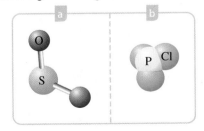

3.30 Calculate the formula mass of the following molecules to three significant figures.

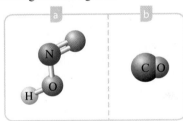

3.31 Phosphoric acid, H_3PO_4, is used to make phosphate fertilizers and detergents and is also used in carbonated beverages. What is the molar mass of H_3PO_4?

3.32 Ammonium nitrate, NH_4NO_3, is used as a nitrogen fertilizer and in explosives. What is the molar mass of NH_4NO_3?

3.33 Calculate the mass (in grams) of each of the following species.

Na atom   N atom

CH_3Cl molecule $Hg(NO_3)_2$ formula unit

3.34 Calculate the mass (in grams) of each of the following species.

PBr_3 molecule   Te atom

$Fe(OH)_2$ formula unit   Br atom

3.35 Diethyl ether, $(C_2H_5)_2O$, commonly known as ether, is used as an anesthetic. What is the mass in grams of a molecule of diethyl ether?

3.36 Glycerol, $C_3H_8O_3$, is used as a moistening agent for candy and is the starting material for nitroglycerin. Calculate the mass of a glycerol molecule in grams.

3.37 Calculate the mass in grams of the following.
- a 0.331 mol Fe
- b 2.1 mol F
- c 0.034 mol CO_2
- d 1.89 mol K_2CrO_4

3.38 Calculate the mass in grams of the following.
- a 0.15 mol Na
- b 0.594 mol S
- c 2.78 mol CH_2Cl_2
- d 38 mol $(NH_4)_2S$

3.39 Boric acid, H_3BO_3, is a mild antiseptic and is often used as an eyewash. A sample contains 0.543 mol H_3BO_3. What is the mass of boric acid in the sample?

3.40 Carbon disulfide, CS_2, is a colorless, highly flammable liquid used in the manufacture of rayon and cellophane. A sample contains 0.0116 mol CS_2. Calculate the mass of carbon disulfide in the sample.

3.41 Obtain the moles of substance in the following.
- a 76 g C_4H_{10}
- b 7.05 g Cl_2
- c 26.2 g $Al_2(CO_3)_3$
- d 2.86 g C

3.42 Obtain the moles of substance in the following.
- a 2.57 g As
- b 7.83 g S_8
- c 33.8 g N_2H_4
- d 227 g $Al_2(SO_4)_3$

3.43 A 1.547-g sample of blue copper(II) sulfate pentahydrate, $CuSO_4 \cdot 5H_2O$, is heated carefully to drive off the water. The white crystals of $CuSO_4$ that are left behind have a mass of 0.989 g. How many moles of H_2O were in the original sample? Show that the relative molar amounts of $CuSO_4$ and H_2O agree with the formula of the hydrate.

3.44 Calcium sulfate, $CaSO_4$, is a white, crystalline powder. Gypsum is a mineral, or natural substance, that is a hydrate of calcium sulfate. A 1.000-g sample of gypsum contains 0.791 g $CaSO_4$. How many moles of $CaSO_4$ are there in this sample? Assuming that the rest of the sample is water, how many moles of H_2O are there in the sample? Show that the result is consistent with the formula $CaSO_4 \cdot 2H_2O$.

3.45 Calculate the following.
- a number of atoms in 8.21 g Li
- b number of atoms in 32.0 g Br_2
- c number of molecules in 45 g NH_3
- d number of formula units in 201 g $PbCrO_4$
- e number of SO_4^{2-} ions in 14.3 g $Cr_2(SO_4)_3$

3.46 Calculate the following.
- a number of Ca^{2+} ions in 4.71 g $Ca_3(PO_4)_2$
- b number of molecules in 14.9 g N_2O_5
- c number of atoms in 25.7 g Al
- d number of formula units in 2.99 g $NaClO_4$
- e number of atoms in 5.66 g I_2

3.47 Carbon tetrachloride is a colorless liquid used in the manufacture of fluorocarbons and as an industrial solvent. How many molecules are there in 7.58 mg of carbon tetrachloride?

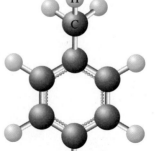

3.48 Chlorine trifluoride is a colorless, reactive gas used in nuclear fuel reprocessing. How many molecules are there in a 8.55-mg sample of chlorine trifluoride?

Mass Percentage

3.49 A 6.01-g aqueous solution of isopropyl alcohol contains 3.67 g of isopropyl alcohol. What is the mass percentage of isopropyl alcohol in the solution?

3.50 A 1.836-g sample of coal contains 1.584 g C. Calculate the mass percentage of C in the coal.

3.51 Phosphorus oxychloride is the starting compound for preparing substances used as flame retardants for plastics. An 8.53-mg sample of phosphorus oxychloride contains 1.72 mg of phosphorus. What is the mass percentage of phosphorus in the compound?

3.52 Ethyl mercaptan is an odorous substance added to natural gas to make leaks easily detectable. A sample of ethyl mercaptan weighing 3.52 mg contains 1.64 mg of sulfur. What is the mass percentage of sulfur in the substance?

3.53 Seawater contains 0.0065% (by mass) of bromine. How many grams of bromine are there in 2.50 L of seawater? The density of seawater is 1.025 g/cm^3.

3.54 A fertilizer is advertised as containing 14.0% nitrogen (by mass). How much nitrogen is there in 4.15 kg of fertilizer?

3.55 A sample of an alloy of aluminum contains 0.0898 mol Al and 0.0381 mol Mg. What are the mass percentages of Al and Mg in the alloy?

3.56 A sample of gas mixture from a neon sign contains 0.0856 mol Ne and 0.0254 mol Kr. What are the mass percentages of Ne and Kr in the gas mixture?

Chemical Formulas

3.57 Calculate the percentage composition for each of the following compounds (three significant figures).
- a CO
- b $Co(NO_3)_2$
- c CO_2
- d NaH_2PO_4

3.58 Calculate the percentage composition for each of the following compounds (three significant figures).
- a NO_2
- b H_2N_2
- c $KClO_4$
- d $Mg(NO_3)_2$

3.59 Calculate the mass percentage of each element in toluene, represented by the following molecular model.

3.60 Calculate the mass percentage of each element in 2-propanol, represented by the following molecular model.

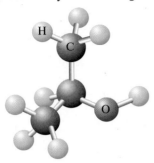

3.61 Which contains more sulfur, 40.8 g of calcium sulfate, $CaSO_4$, or 38.8 g of sodium sulfite, Na_2SO_3?

3.62 Which contains more carbon, 6.01 g of glucose, $C_6H_{12}O_6$, or 5.85 g of ethanol, C_2H_6O?

3.63 Ethylene glycol is used as an automobile antifreeze and in the manufacture of polyester fibers. The name glycol stems from the sweet taste of this poisonous compound. Combustion of 6.38 mg of ethylene glycol gives 9.06 mg CO_2 and 5.58 mg H_2O. The compound contains only C, H, and O. What are the mass percentages of the elements in ethylene glycol?

3.64 Phenol, commonly known as carbolic acid, was used by Joseph Lister as an antiseptic for surgery in 1865. Its principal use today is in the manufacture of phenolic resins and plastics. Combustion of 5.23 mg of phenol yields 14.67 mg CO_2 and 3.01 mg H_2O. Phenol contains only C, H, and O. What is the percentage of each element in this substance?

3.65 An oxide of osmium (symbol Os) is a pale yellow solid. If 2.89 g of the compound contains 2.16 g of osmium, what is its empirical formula?

3.66 An oxide of tungsten (symbol W) is a bright yellow solid. If 5.34 g of the compound contains 4.23 g of tungsten, what is its empirical formula?

3.67 Hydroquinone, used as a photographic developer, is 65.4% C, 5.5% H, and 29.1% O, by mass. What is the empirical formula of hydroquinone?

3.68 Potassium manganate is a dark green, crystalline substance whose composition is 39.6% K, 27.9% Mn, and 32.5% O, by mass. What is its empirical formula?

3.69 Acrylic acid, used in the manufacture of acrylic plastics, has the composition 50.0% C, 5.6% H, and 44.4% O. What is its empirical formula?

3.70 Malonic acid is used in the manufacture of barbiturates (sleeping pills). The composition of the acid is 34.6% C, 3.9% H, and 61.5% O. What is malonic acid's empirical formula?

3.71 Two compounds have the same composition: 92.25% C and 7.75% H.
- **a** Obtain the empirical formula corresponding to this composition.
- **b** One of the compounds has a molecular mass of 52.03 amu; the other, of 78.05 amu. Obtain the molecular formulas of both compounds.

3.72 Two compounds have the same composition: 85.62% C and 14.38% H.
- **a** Obtain the empirical formula corresponding to this composition.
- **b** One of the compounds has a molecular mass of 28.03 amu; the other, of 56.06 amu. Obtain the molecular formulas of both compounds.

3.73 Compounds of boron with hydrogen are called boranes. One of these boranes has the empirical formula BH_3 and a molecular mass of 28 amu. What is its molecular formula?

3.74 Putrescine, a substance produced by decaying animals, has the empirical formula C_2H_6N. Several determinations of molecular mass give values in the range of 87 to 90 amu. Find the molecular formula of putrescine.

3.75 Oxalic acid is a toxic substance used by laundries to remove rust stains. Its composition is 26.7% C, 2.2% H, and 71.1% O (by mass), and its molecular mass is 90 amu. What is its molecular formula?

3.76 Adipic acid is used in the manufacture of nylon. The composition of the acid is 49.3% C, 6.9% H, and 43.8% O (by mass), and the molecular mass is 146 amu. What is the molecular formula?

Stoichiometry: Quantitative Relations in Reactions

3.77 Ethylene, C_2H_4, burns in oxygen to give carbon dioxide, CO_2, and water. Write the equation for the reaction, giving molecular, molar, and mass interpretations below the equation.

3.78 Hydrogen sulfide gas, H_2S, burns in oxygen to give sulfur dioxide, SO_2, and water. Write the equation for the reaction, giving molecular, molar, and mass interpretations below the equation.

3.79 Butane, C_4H_{10}, burns with the oxygen in air to give carbon dioxide and water.

$$2C_4H_{10}(g) + 13O_2(g) \longrightarrow 8CO_2(g) + 10H_2O(g)$$

What is the amount (in moles) of carbon dioxide produced from 0.60 mol C_4H_{10}?

3.80 Ethanol, C_2H_5OH, burns with the oxygen in air to give carbon dioxide and water.

$$C_2H_5OH(l) + 3O_2(g) \longrightarrow 2CO_2(g) + 3H_2O(l)$$

What is the amount (in moles) of water produced from 0.77 mol C_2H_5OH?

3.81 Iron in the form of fine wire burns in oxygen to form iron(III) oxide.

$$4Fe(s) + 3O_2(g) \longrightarrow 2Fe_2O_3(s)$$

How many moles of O_2 are needed to produce 7.82 mol Fe_2O_3?

3.82 Nickel(II) chloride reacts with sodium phosphate to precipitate nickel(II) phosphate.

$$3NiCl_2(aq) + 2Na_3PO_4(aq) \longrightarrow Ni_3(PO_4)_2(s) + 6NaCl(aq)$$

How many moles of nickel(II) chloride are needed to produce 0.715 mol nickel(II) phosphate?

3.83 White phosphorus, P_4, is prepared by fusing calcium phosphate, $Ca_3(PO_4)_2$, with carbon, C, and sand, SiO_2, in an electric furnace.

$$2Ca_3(PO_4)_2(s) + 6SiO_2(s) + 10C(s) \longrightarrow$$
$$P_4(g) + 6CaSiO_3(l) + 10CO(g)$$

How many grams of calcium phosphate are required to give 30.0 g of phosphorus?

3.84 Nitric acid, HNO_3, is manufactured by the Ostwald process, in which nitrogen dioxide, NO_2, reacts with water.

$$3NO_2(g) + H_2O(l) \longrightarrow 2HNO_3(aq) + NO(g)$$

How many grams of nitrogen dioxide are required in this reaction to produce 7.50 g HNO_3?

3.85 Tungsten metal, W, is used to make incandescent bulb filaments. The metal is produced from the yellow tungsten(VI) oxide, WO_3, by reaction with hydrogen.

$$WO_3(s) + 3H_2(g) \longrightarrow W(s) + 3H_2O(g)$$

How many grams of tungsten can be obtained from 4.81 kg of hydrogen with excess tungsten(VI) oxide?

3.86 Acrylonitrile, C_3H_3N, is the starting material for the production of a kind of synthetic fiber (acrylics). It can be made from propylene, C_3H_6, by reaction with nitric oxide, NO.

$$4C_3H_6(g) + 6NO(g) \longrightarrow 4C_3H_3N(g) + 6H_2O(g) + N_2(g)$$

How many grams of acrylonitrile are obtained from 452 kg of propylene and excess NO?

3.87 Using the following reaction (depicted using molecular models), large quantities of ammonia are burned in the presence of a platinum catalyst to give nitric oxide, as the first step in the preparation of nitric acid.

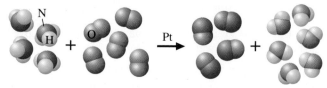

Suppose a vessel contains 5.5 g of NH_3. How many grams of O_2 are needed for a complete reaction?

3.88 The following reaction, depicted using molecular models, is used to make carbon tetrachloride, CCl_4, a solvent and starting material for the manufacture of fluorocarbon refrigerants and aerosol propellants.

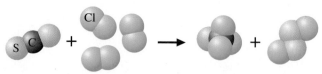

Calculate the number of grams of carbon disulfide, CS_2, needed for a laboratory-scale reaction with 62.7 g of chlorine, Cl_2.

3.89 When dinitrogen pentoxide, N_2O_5, a white solid, is heated, it decomposes to nitrogen dioxide and oxygen.

$$2N_2O_5(s) \overset{\Delta}{\longrightarrow} 4NO_2(g) + O_2(g)$$

If a sample of N_2O_5 produces 1.381 g O_2, how many grams of NO_2 are formed?

3.90 Copper metal reacts with nitric acid. Assume that the reaction is

$$3Cu(s) + 8HNO_3(aq) \longrightarrow$$
$$3Cu(NO_3)_2(aq) + 2NO(g) + 4H_2O(l)$$

If 5.58 g $Cu(NO_3)_2$ is eventually obtained, how many grams of nitrogen monoxide, NO, would have formed?

Limiting Reactant; Theoretical and Percentage Yields

3.91 Solutions of sodium hypochlorite, NaClO, are sold as a bleach (such as Clorox). They are prepared by the reaction of chlorine with sodium hydroxide.

$$2NaOH(aq) + Cl_2(g) \longrightarrow$$
$$NaCl(aq) + NaClO(aq) + H_2O(l)$$

If you have 1.44 mol of NaOH in solution and 1.47 mol of Cl_2 gas available to react, which is the limiting reactant? How many moles of NaClO(aq) could be obtained?

3.92 Potassium superoxide, KO_2, is used in rebreathing gas masks to generate oxygen.

$$4KO_2(s) + 2H_2O(l) \longrightarrow 4KOH(s) + 3O_2(g)$$

If a reaction vessel contains 0.25 mol KO_2 and 0.15 mol H_2O, what is the limiting reactant? How many moles of oxygen can be produced?

3.93 Methanol, CH_3OH, is prepared industrially from the gas-phase catalytic balanced reaction that has been depicted here using molecular models.

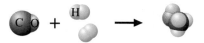

In a laboratory test, a reaction vessel was filled with 35.4 g CO and 10.2 g H_2. How many grams of methanol would be produced in a complete reaction? Which reactant remains unconsumed at the end of the reaction? How many grams of it remain?

3.94 Carbon disulfide, CS_2, burns in oxygen. Complete combustion gives the balanced reaction that has been depicted here using molecular models.

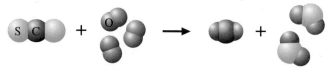

Calculate the grams of sulfur dioxide, SO_2, produced when a mixture of 35.0 g of carbon disulfide and 35.0 g of oxygen reacts. Which reactant remains unconsumed at the end of the combustion? How many grams remain?

3.95 Titanium, which is used to make airplane engines and frames, can be obtained from titanium(IV) chloride, which in turn is obtained from titanium(IV) oxide by the following process:

$$3TiO_2(s) + 4C(s) + 6Cl_2(g) \longrightarrow$$
$$3TiCl_4(g) + 2CO_2(g) + 2CO(g)$$

A vessel contains 4.15 g TiO_2, 5.67 g C, and 6.78 g Cl_2. Suppose the reaction goes to completion as written. How many grams of titanium(IV) chloride can be produced?

3.96 Hydrogen cyanide, HCN, is prepared from ammonia, air, and natural gas (CH_4) by the following process:

$$2NH_3(g) + 3O_2(g) + 2CH_4(g) \xrightarrow{Pt}$$
$$2HCN(g) + 6H_2O(g)$$

Hydrogen cyanide is used to prepare sodium cyanide, which is used in part to obtain gold from gold-containing rock. If a reaction vessel contains 11.5 g NH_3, 12.0 g O_2, and 10.5 g CH_4, what is the maximum mass in grams of hydrogen cyanide that could be made, assuming the reaction goes to completion as written?

3.97 Methyl salicylate (oil of wintergreen) is prepared by heating salicylic acid, $C_7H_6O_3$, with methanol, CH_3OH.

$$C_7H_6O_3 + CH_3OH \longrightarrow C_8H_8O_3 + H_2O$$

In an experiment, 2.50 g of salicylic acid is reacted with 10.31 g of methanol. The yield of methyl salicylate, $C_8H_8O_3$, is 1.27 g. What is the percentage yield?

3.98 Aspirin (acetylsalicylic acid) is prepared by heating salicylic acid, $C_7H_6O_3$, with acetic anhydride, $C_4H_6O_3$. The other product is acetic acid, $C_2H_4O_2$.

$$C_7H_6O_3 + C_4H_6O_3 \longrightarrow C_9H_8O_4 + C_2H_4O_2$$

What is the theoretical yield (in grams) of aspirin, $C_9H_8O_4$, when 2.00 g of salicylic acid is heated with 4.00 g of acetic anhydride? If the actual yield of aspirin is 1.86 g, what is the percentage yield?

General Problems

Key: These problems provide more practice but are not divided by topic or keyed to exercises. Each section ends with essay questions, each of which is color coded to refer to the *Instrumental Methods* (brown) chapter essay on which it is based. Odd-numbered problems and the even-numbered problems that follow are similar; answers to all odd-numbered problems except the essay questions are given in the back of the book.

3.99 Morphine, a narcotic substance obtained from opium, has the molecular formula $C_{17}H_{19}NO_3$. What is the mass percentage of each element in morphine (to three significant figures)?

3.100 Caffeine, the stimulant in coffee and tea, has the molecular formula $C_8H_{10}N_4O_2$. Calculate the mass percentage of each element in the substance. Give the answers to three significant figures.

3.101 A moth repellent, *para*-dichlorobenzene, has the composition 49.1% C, 2.7% H, and 48.2% Cl. Its molecular mass is 147 amu. What is its molecular formula?

3.102 Sorbic acid is added to food as a mold inhibitor. Its composition is 64.3% C, 7.2% H, and 28.5% O, and its molecular mass is 112 amu. What is its molecular formula?

3.103 Thiophene is a liquid compound of the elements C, H, and S. A sample of thiophene weighing 7.96 mg was burned in oxygen, giving 16.65 mg CO_2. Another sample was subjected to a series of reactions that transformed all of the sulfur in the compound to barium sulfate. If 4.31 mg of thiophene gave 11.96 mg of barium sulfate, what is the empirical formula of thiophene? Its molecular mass is 84 amu. What is its molecular formula?

3.104 Aniline, a starting compound for urethane plastic foams, consists of C, H, and N. Combustion of such compounds yields CO_2, H_2O, and N_2 as products. If the combustion of 9.71 mg of aniline yields 6.63 mg H_2O and 1.46 mg N_2, what is its empirical formula? The molecular mass of aniline is 93 amu. What is its molecular formula?

3.105 A sample of limestone (containing calcium carbonate, $CaCO_3$) weighing 438 mg is treated with oxalic acid, $H_2C_2O_4$, to give calcium oxalate, CaC_2O_4.

$$CaCO_3(s) + H_2C_2O_4(aq) \longrightarrow$$
$$CaC_2O_4(s) + H_2O(l) + CO_2(g)$$

The mass of the calcium oxalate produced is 469 mg. What is the mass percentage of calcium carbonate in this limestone?

3.106 A titanium ore contains rutile (TiO_2) plus some iron oxide and silica. When it is heated with carbon in the presence of chlorine, titanium tetrachloride, $TiCl_4$, is formed.

$$TiO_2(s) + C(s) + 2Cl_2(g) \longrightarrow TiCl_4(g) + CO_2(g)$$

Titanium tetrachloride, a liquid, can be distilled from the mixture. If 35.4 g of titanium tetrachloride is recovered from 18.1 g of crude ore, what is the mass percentage of TiO_2 in the ore (assuming all TiO_2 reacts)?

3.107 Nitrobenzene, $C_6H_5NO_2$, an important raw material for the dye industry, is prepared from benzene, C_6H_6, and nitric acid, HNO_3.

$$C_6H_6(l) + HNO_3(l) \longrightarrow C_6H_5NO_2(l) + H_2O(l)$$

When 21.6 g of benzene and an excess of HNO_3 are used, what is the theoretical yield of nitrobenzene? If 30.0 g of nitrobenzene is recovered, what is the percentage yield?

3.108 Ethylene oxide, C_2H_4O, is made by the oxidation of ethylene, C_2H_4.

$$2C_2H_4(g) + O_2(g) \longrightarrow 2C_2H_4O(g)$$

Ethylene oxide is used to make ethylene glycol for automobile antifreeze. In a pilot study, 10.6 g of ethylene gave 9.91 g of ethylene oxide. What is the percentage yield of ethylene oxide?

3.109 Zinc metal can be obtained from zinc oxide, ZnO, by reaction at high temperature with carbon monoxide, CO.

$$ZnO(s) + CO(g) \longrightarrow Zn(s) + CO_2(g)$$

The carbon monoxide is obtained from carbon.

$$2C(s) + O_2(g) \longrightarrow 2CO(g)$$

What is the maximum amount of zinc that can be obtained from 75.0 g of zinc oxide and 50.0 g of carbon?

3.110 Hydrogen cyanide, HCN, can be made by a two-step process. First, ammonia is reacted with O_2 to give nitric oxide, NO.

$$4NH_3(g) + 5O_2(g) \longrightarrow 4NO(g) + 6H_2O(g)$$

Then nitric oxide is reacted with methane, CH_4.

$$2NO(g) + 2CH_4(g) \longrightarrow 2HCN(g) + 2H_2O(g) + H_2(g)$$

When 24.2 g of ammonia and 25.1 g of methane are used, how many grams of hydrogen cyanide can be produced?

3.111 A mixture consisting of 11.9 g of calcium fluoride, CaF_2, and 12.1 g of sulfuric acid, H_2SO_4, is heated to drive off hydrogen fluoride, HF.

$$CaF_2(s) + H_2SO_4(l) \longrightarrow 2HF(g) + CaSO_4(s)$$

What is the maximum number of grams of hydrogen fluoride that can be obtained?

3.112 Calcium carbide, CaC_2, used to produce acetylene, C_2H_2, is prepared by heating calcium oxide, CaO, and carbon, C, to high temperature.

$$CaO(s) + 3C(s) \longrightarrow CaC_2(s) + CO(g)$$

If a mixture contains 2.60 kg of each reactant, how many grams of calcium carbide can be prepared?

3.113 Alloys, or metallic mixtures, of mercury with another metal are called amalgams. Sodium in sodium amalgam reacts with water. (Mercury does not.)

$$2Na(s) + 2H_2O(l) \longrightarrow 2NaOH(aq) + H_2(g)$$

If a 15.23-g sample of sodium amalgam evolves 0.108 g of hydrogen, what is the percentage of sodium in the amalgam?

3.114 A sample of sandstone consists of silica, SiO_2, and calcite, $CaCO_3$. When the sandstone is heated, calcium carbonate, $CaCO_3$, decomposes into calcium oxide, CaO, and carbon dioxide.

$$CaCO_3(s) \longrightarrow CaO(s) + CO_2(g)$$

What is the percentage of silica in the sandstone if 18.7 mg of the rock yields 3.95 mg of carbon dioxide?

■ **3.115** What type of information can you obtain from a compound using a mass spectrometer?

■ **3.116** Why is the mass spectrum of a molecule much more complicated than that of an atom?

Strategy Problems

Key: As noted earlier, all of the practice and general problems are matched pairs. This section is a selection of problems that are not in matched-pair format. These challenging problems require that you employ many of the concepts and strategies that were developed in the chapter. In some cases, you will have to integrate several concepts and operational skills in order to solve the problem successfully.

3.117 Exactly 4.0 g of hydrogen gas combines with 32 g of oxygen gas according to the following reaction.

$$2H_2 + O_2 \longrightarrow 2H_2O$$

a How many hydrogen molecules are required to completely react with 48 oxygen molecules?

b If you conducted the reaction, and it produced 5.0 mol H_2O, how many moles of both O_2 and H_2 did you start with?

c If you started with 37.5 g O_2, how many grams of H_2 did you start with to have a complete reaction?

d How many grams of O_2 and H_2 were reacted to produce 30.0 g H_2O?

3.118 Aluminum metal reacts with iron(III) oxide to produce aluminum oxide and iron metal.

a How many moles of Fe_2O_3 are required to completely react with 44 g Al?

b How many moles of Fe are produced by the reaction of 3.14 mol Fe_2O_3 and 99.1 g Al?

c How many atoms of Al are required to produce 7.0 g Fe?

3.119 Consider the equation

$$2KOH + H_2SO_4 \longrightarrow K_2SO_4 + 2H_2O$$

a If 25 g H_2SO_4 is reacted with 7.7 g KOH, how many grams of K_2SO_4 are produced?

b For part a of this problem, identify the limiting reactant and calculate the mass of excess reactant that remains after the reaction is completed.

c Calculate the theoretical yield of the reaction. How many grams of material would you expect to obtain if the reaction has a 67.1% yield?

3.120 You perform a combustion analysis on a 255 mg sample of a substance that contains only C, H, and O, and you find that 561 mg CO_2 is produced, along with 306 mg H_2O.

a If the substance contains only C, H, and O, what is the empirical formula?

b If the molar mass of the compound is 180 g/mol, what is the molecular formula of the compound?

3.121 A 3.0-L sample of paint that has a density of 4.65 g/mL is found to contain 33.1 g $Pb_3N_2(s)$. How many grams of lead were in the paint sample?

3.122 When ammonia and oxygen are reacted, they produce nitric oxide and water. When 8.5 g of ammonia is allowed to react with an excess of O_2, the reaction produces 12.0 g of nitrogen monoxide. What is the percentage yield of the reaction?

3.123 A 12.1-g sample of Na_2SO_3 is mixed with a 14.6-g sample of $MgSO_4$. What is the total mass of oxygen present in the mixture?

3.124 Potassium superoxide, KO_2, is employed in a self-contained breathing apparatus used by emergency personnel as a source of oxygen. The reaction is

$$4KO_2(s) + 2H_2O(l) \longrightarrow 4KOH(s) + 3O_2(g)$$

If a self-contained breathing apparatus is charged with 750 g KO_2 and then is used to produce 188 g of oxygen, was all of the KO_2 consumed in this reaction? If the KO_2 wasn't all consumed, how much is left over and what mass of additional O_2 could be produced?

3.125 Calcium carbonate is a common ingredient in stomach antacids. If an antacid tablet has 68.4 mg of calcium carbonate, how many moles of calcium carbonate are there in 175 tablets?

3.126 While cleaning out your closet, you find a jar labeled "2.21 moles lead nitrite." Since Stock convention was not used, you do not know the oxidation number of the lead. You weigh the contents, and find a mass of 6.61×10^5 mg. What is the percentage composition of nitrite?

3.127 Sulfuric acid can be produced by the following sequence of reactions.

$$S_8 + 8O_2 \longrightarrow 8SO_2$$
$$2SO_2 + O_2 \longrightarrow 2SO_3$$
$$SO_3 + H_2O \longrightarrow H_2SO_4$$

You find a source of sulfur that is 65.0% sulfur by mass. How many grams of this starting material would be required to produce 87.0 g of sulfuric acid?

3.128 A sample of methane gas, $CH_4(g)$, is reacted with oxygen gas to produce carbon dioxide and water. If 20.0 L of methane (density = 1.82 kg/m^3) and 30.0 L of oxygen gas (density = 1.31 kg/m^3) are placed into a container and allowed to react, how many kg of carbon dioxide will be produced by the reaction?

3.129 Copper reacts with nitric acid according to the following reaction.

$$3Cu(s) + 8HNO_3(aq) \longrightarrow$$
$$3Cu(NO_3)_2(aq) + 2NO(g) + 4H_2O(l)$$

If 2.40 g of Cu is added to a container with 2.00 mL of concentrated nitric acid (70% by mass HNO_3; density =

1.42 g/cm^3), what mass of nitrogen monoxide gas will be produced?

3.130 A sample containing only boron and fluorine was decomposed yielding 4.75 mg of boron and 17.5 mL of fluorine gas (density = 1.43 g/L). What is the empirical formula of the sample compound?

3.131 Sodium azide, NaN_3, undergoes the reaction $NaN_3(s) \longrightarrow 2Na(s) + 3N_2(g)$. Because this reaction is very fast and produces nitrogen gas, NaN_3 is used to inflate airplane escape chutes. Sodium azide can be produced through two reaction steps.

$$2Na(s) + 2NH_3(g) \longrightarrow 2NaNH_2(s) + H_2(g)$$
$$2NaNH_2(s) + N_2O(g) \longrightarrow$$
$$NaN_3(s) + NaOH(s) + NH_3(g)$$

Starting with 1.0 kg of Na, 6.0 kg of NH_3, and 1.0 kg of N_2O, what is the maximum mass (kg) of sodium azide that can be produced?

3.132 Dinotrogen monoxide, commonly known as laughing gas, can be obtained by cautiously warming ammonium nitrate according to the equation

$$NH_4NO_3(s) \longrightarrow N_2O(g) + 2H_2O(g)$$

If the reaction has a 75% yield, what mass of ammonium nitrate must be used to produce 685 mg of N_2O?

Cumulative-Skills Problems

Key: The problems under this heading combine skills introduced in previous chapters with those given in the current one.

3.133 A 0.500-g mixture of Cu_2O and CuO contains 0.425 g Cu. What is the mass of CuO in the mixture?

3.134 A mixture of Fe_2O_3 and FeO was found to contain 72.00% Fe by mass. What is the mass of Fe_2O_3 in 0.750 g of this mixture?

3.135 Hemoglobin is the oxygen-carrying molecule of red blood cells, consisting of a protein and a nonprotein substance. The nonprotein substance is called heme. A sample of heme weighing 35.2 mg contains 3.19 mg of iron. If a heme molecule contains one atom of iron, what is the molecular mass of heme?

3.136 Penicillin V was treated chemically to convert sulfur to barium sulfate, $BaSO_4$. An 8.19-mg sample of penicillin V gave 5.46 mg $BaSO_4$. What is the percentage of

sulfur in penicillin V? If there is one sulfur atom in the molecule, what is the molecular mass?

3.137 A 3.41-g sample of a metallic element, M, reacts completely with 0.0158 mol of a gas, X_2, to form 4.52 g MX. What are the identities of M and X?

3.138 1.92 g M^{2+} ion reacts with 0.158 mol X^- ion to produce a compound, MX_2, which is 86.8% X by mass. What are the identities of M^+ and X^-?

3.139 An alloy of iron (71.0%), cobalt (12.0%), and molybdenum (17.0%) has a density of 8.20 g/cm^3. How many cobalt atoms are there in a cylinder with a radius of 2.50 cm and a length of 10.0 cm?

3.140 An alloy of iron (54.7%), nickel (45.0%), and manganese (0.3%) has a density of 8.17 g/cm^3. How many iron atoms are there in a block of alloy measuring 10.0 cm $\times$ 20.0 cm $\times$ 15.0 cm?

4

Chemical Reactions

© Cengage Learning; Siede Preis/Getty Images

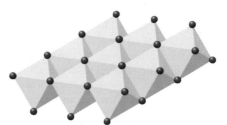

Adding lead(II) nitrate to potassium iodide gives a yellow precipitate of lead(II) iodide. Precipitation reactions are important in biological organisms—for example, in the production of bone (calcium phosphate) and seashell (calcium carbonate).

Contents and Concepts

Ions in Aqueous Solution
Explore how molecular and ionic substances behave when they dissolve in water to form *solutions*.

4.1 Ionic Theory of Solutions and Solubility Rules
4.2 Molecular and Ionic Equations

Types of Chemical Reactions
Investigate several important types of reactions that typically occur in aqueous solution: *precipitation reactions, acid–base reactions,* and *oxidation–reduction reactions.*

4.3 Precipitation Reactions
4.4 Acid–Base Reactions
4.5 Oxidation–Reduction Reactions
4.6 Balancing Simple Oxidation–Reduction Equations

Working with Solutions
Now that we have looked at how substances behave in solution, it is time to quantitatively describe these solutions using *concentration.*

4.7 Molar Concentration
4.8 Diluting Solutions

Quantitative Analysis
Using chemical reactions in aqueous solution, determine the amount of substance or species present in materials

4.9 Gravimetric Analysis
4.10 Volumetric Analysis

OWL Sign in to OWL at **www.cengage.com/owl** to view tutorials and simulations, develop problem-solving skills, and complete online homework assigned by your professor.

go Chemistry Download mini lecture videos for key concept review and exam prep from OWL or purchase them from **www.cengagebrain.com**.

Chemical reactions are the heart of chemistry. Some reactions, such as those accompanying a forest fire or the explosion of dynamite, are quite dramatic. Others are much less obvious, although all chemical reactions must involve detectable change. A chemical reaction involves a change from reactant substances to product substances, and the product substances will have physical and chemical properties different from those of the reactants.

Figure 4.1 shows an experimenter adding a colorless solution of potassium iodide, KI, to a colorless solution of lead(II) nitrate, $Pb(NO_3)_2$. What you see is the formation of a cloud of bright yellow crystals where the two solutions have come into contact, clear evidence of a chemical reaction. The bright yellow crystals are lead(II) iodide, PbI_2, one of the reaction products. We call a solid that forms during a chemical reaction in solution a *precipitate;* the reaction is a precipitation reaction.

In this chapter, we will discuss the major types of chemical reactions, including precipitation reactions. Some of the most important reactions we will describe involve ions in aqueous (water) solution. Therefore, we will first look at these ions and see how we represent by chemical equations the reactions involving ions in aqueous solution.

Some questions we will answer are: What is the evidence for ions in solution? How do we write chemical equations for reactions involving ions? How can we classify and describe the many reactions we observe so that we can begin to understand them? What is the quantitative description of solutions and reactions in solution?

Figure 4.1 ▲

Reaction of potassium iodide solution and lead(II) nitrate solution The reactant solutions are colorless, but one of the products, lead(II) iodide, forms as a yellow precipitate.

Ions in Aqueous Solution

You probably have heard that you should not operate electrical equipment while standing in water. And you may have read a murder mystery in which the victim was electrocuted when an electrical appliance "accidentally" fell into his or her bath water. Actually, if the water were truly pure, the person would be safe from electrocution, because pure water is a nonconductor of electricity. Bath water, or water as it flows from the faucet, however, is a *solution* of water with small amounts of dissolved substances in it, and these dissolved substances make the solution an electrical conductor. This allows an electric current to flow from an electrical appliance to the human body. Let us look at the nature of such solutions.

4.1 Ionic Theory of Solutions and Solubility Rules

Chemists began studying the electrical behavior of substances in the early nineteenth century, and they knew that you could make pure water electrically conducting by dissolving certain substances in it. In 1884, the young Swedish chemist Svante Arrhenius proposed the *ionic theory of solutions* to account for this conductivity. He said that certain substances produce freely moving ions when they dissolve in water, and these ions conduct an electric current in an aqueous solution. ▶

Suppose you dissolve sodium chloride, NaCl, in water. From our discussion in Section 2.6, you may remember that sodium chloride is an ionic solid consisting of sodium ions, Na^+, and chloride ions, Cl^-, held in a regular, fixed array. When you dissolve solid sodium chloride in water, the Na^+ and Cl^- ions go into solution as freely moving ions. Now suppose you dip electric wires that are connected to the poles of a battery into a solution of sodium chloride. The wire that connects to the positive pole of the battery attracts the negatively charged chloride ions in solution, because of their opposite charges. Similarly, the wire connected to the negative pole of the battery attracts the positively charged sodium ions in

Arrhenius submitted his ionic theory as part of his doctoral dissertation to the faculty at Uppsala, Sweden, in 1884. It was not well received and he barely passed. In 1903, however, he was awarded the Nobel Prize in chemistry for this theory.

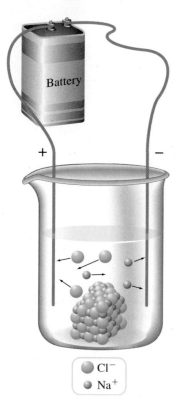

Figure 4.2 ▲

Motion of ions in solution Ions are in fixed positions in a crystal. During the solution process, however, ions leave the crystal and become freely moving. Note that Na^+ ions (small gray spheres) are attracted to the negative wire, whereas Cl^- ions (large green spheres) are attracted to the positive wire.

solution (Figure 4.2). Thus, the ions in the solution begin to move, and these moving charges form the electric current in the solution. (In a wire, it is moving electrons that constitute the electric current.)

Now consider pure water. Water consists of molecules, each of which is electrically neutral. Since each molecule carries no net electric charge, it carries no overall electric charge when it moves. Thus, pure water is a nonconductor of electricity.

In summary, although water is itself nonconducting, it has the ability to dissolve various substances, some of which go into solution as freely moving ions. An aqueous solution of ions is electrically conducting.

Electrolytes and Nonelectrolytes

We can divide the substances that dissolve in water into two broad classes, electrolytes and nonelectrolytes. An **electrolyte** is *a substance that dissolves in water to give an electrically conducting solution.* Sodium chloride, table salt, is an example of an electrolyte. When most ionic substances dissolve in water, ions that were in fixed sites in the crystalline solid go into the surrounding aqueous solution, where they are free to move about. The resulting solution is conducting because the moving ions form an electric current. Thus, in general, *ionic solids that dissolve in water are electrolytes.*

Not all electrolytes are ionic substances. Certain molecular substances dissolve in water to form ions. The resulting solution is electrically conducting, and so we say that the molecular substance is an electrolyte. An example is hydrogen chloride gas, $HCl(g)$, which is a molecular substance. Hydrogen chloride gas dissolves in water, giving $HCl(aq)$, which in turn produces hydrogen ions, H^+, and chloride ions, Cl^-, in aqueous solution. (The solution of H^+ and Cl^- ions is called hydrochloric acid.)

$$HCl(aq) \xrightarrow{H_2O} H^+(aq) + Cl^-(aq)$$

We will look more closely at molecular electrolytes, such as HCl, at the end of this section.

A **nonelectrolyte** is *a substance that dissolves in water to give a nonconducting or very poorly conducting solution.* A common example is sucrose, $C_{12}H_{22}O_{11}$, which is ordinary table sugar. Another example is methanol, CH_3OH, a compound used in car window washer solution. Both of these are molecular substances. The solution process occurs because molecules of the substance mix with molecules of water. Molecules are electrically neutral and cannot carry an electric current, so the solution is electrically nonconducting.

Observing the Electrical Conductivity of a Solution

Figure 4.3 shows a simple apparatus that allows you to observe the ability of a solution to conduct an electric current. The apparatus has two electrodes; here they are flat metal plates, dipping into the solution in a beaker. One electrode connects directly to a battery through a wire. The other electrode connects by a wire to a light bulb that connects with another wire to the other side of the battery. For an electric current to flow from the battery, there must be a complete circuit, which allows the current to flow from the positive pole of the battery through the circuit to the negative pole of the battery. To have a complete circuit, the solution in the beaker must conduct electricity, as the wires do. If the solution is conducting, the circuit is complete and the bulb lights. If the solution is nonconducting, the circuit is incomplete and the bulb does not light.

The beaker shown on the left side of Figure 4.3 contains pure water. Because the bulb is not lit, we conclude that pure water is a nonconductor (or very poor conductor) of electricity, which is what we expect from our earlier discussion. The beaker shown on the right side of Figure 4.3 contains a solution of sodium chloride

Figure 4.3 ◄

Testing the electrical conductivity of a solution

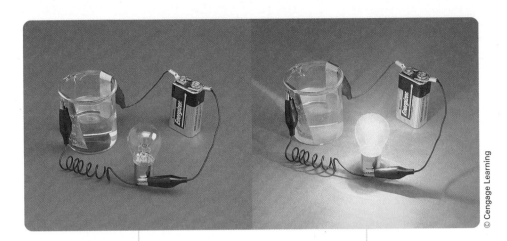

© Cengage Learning

Pure water does not conduct; therefore, the bulb does not light.

A solution of sodium chloride in which moving ions are present allows the current to pass through it, and the bulb lights.

in water. In this case, the bulb burns brightly, showing that the solution is a very good conductor of electricity, due to the movement of ions in the solution.

How brightly the bulb lights tells you whether the solution is a very good conductor (contains a "strong" electrolyte) or only a moderately good conductor (contains a "weak" electrolyte). The solution of sodium chloride burns brightly, so we conclude that sodium chloride is a strong electrolyte. Let us look more closely at such strong and weak electrolytes.

Strong and Weak Electrolytes

When electrolytes dissolve in water they produce ions, but to varying extents. A **strong electrolyte** is *an electrolyte that exists in solution almost entirely as ions.* Most ionic solids that dissolve in water do so by going into the solution almost completely as ions, so they are strong electrolytes. An example is sodium chloride. We can represent the dissolution of sodium chloride in water by the following equation:

$$NaCl(s) \xrightarrow{H_2O} Na^+(aq) + Cl^-(aq)$$

A **weak electrolyte** is *an electrolyte that dissolves in water to give a relatively small percentage of ions.* These are generally molecular substances. Ammonia, NH_3, is an example. Pure ammonia is a gas that readily dissolves in water and goes into solution as ammonia molecules, $NH_3(aq)$. When you buy "ammonia" in the grocery store, you are buying an aqueous solution of ammonia. Ammonia molecules react with water to form ammonium ions, NH_4^+, and hydroxide ions, OH^-.

$$NH_3(aq) + H_2O(l) \longrightarrow NH_4^+(aq) + OH^-(aq)$$

However, these ions, $NH_4^+ + OH^-$, react with each other to give back ammonia molecules and water molecules.

$$NH_4^+(aq) + OH^-(aq) \longrightarrow NH_3(aq) + H_2O(l)$$

Both reactions, the original one and its reverse, occur constantly and simultaneously. We denote this situation by writing a single equation with a double arrow:

$$NH_3(aq) + H_2O(l) \rightleftharpoons NH_4^+(aq) + OH^-(aq)$$

As a result of this forward and reverse reaction, just a small percentage of the NH_3 molecules (about 3%) have reacted at any given moment to form ions. Thus, ammonia is a weak electrolyte. The electrical conductivities of a strong and a weak electrolyte are contrasted in Figure 4.4.

Figure 4.4 ▶

Comparing strong and weak electrolytes

© Cengage Learning

The apparatus is similar to that in Figure 4.3, but this time strong and weak electrolytes are compared. The solution on the left is of HCl (a strong electrolyte) and that on the right is of NH_3 (a weak electrolyte). Note how much more brightly the bulb on the left burns compared with that on the right.

Most soluble molecular substances are either nonelectrolytes or weak electrolytes. An exception is hydrogen chloride gas, $HCl(g)$, which dissolves in water to produce hydrogen ions and chloride ions. We represent its reaction with water by an equation with a single arrow:

$$HCl(aq) \longrightarrow H^+(aq) + Cl^-(aq)$$

Since hydrogen chloride dissolves to give almost entirely ions, hydrogen chloride (or hydrochloric acid) is a strong electrolyte.

Solubility Rules

It is clear from the preceding discussion that substances vary widely in their *solubility,* or ability to dissolve, in water. Some compounds, such as sodium chloride and ethyl alcohol (CH_3CH_2OH), dissolve readily and are said to be *soluble.* Others, such as calcium carbonate (which occurs naturally as limestone and marble) and benzene (C_6H_6), have quite limited solubilities and are thus said to be *insoluble.*

Soluble ionic compounds form solutions that contain many ions and therefore are strong electrolytes. To predict the solubility of ionic compounds, chemists have developed solubility rules. Table 4.1 lists eight solubility rules for ionic compounds. These rules apply to most of the common ionic compounds that we will discuss in this course. Example 4.1 illustrates how to use the rules.

CONCEPT CHECK 4.1

Which of the following would you expect to be strong electrolytes when placed in water?

NH_4Cl $MgBr_2$ H_2O HCl $Ca_3(PO_4)_2$ CH_3OH

Which of the statements below support the answers you chose above? (Pick as many as apply.)

a   Ionic compounds always produce a significant number of ions in aqueous solution.

b   A molecular compound can be either a weak electrolyte or a nonelectrolyte.

c   Soluble ionic compounds are strong electrolytes.

d   Water conducts electricity so that whenever water is present, we would expect a strong electrolyte.

e   Some molecular substances completely ionize in aqueous solution.

Table 4.1		Solubility Rules for Ionic Compounds	
Rule	Applies to	Statement	Exceptions
1	Li^+, Na^+, K^+, NH_4^+	Group IA and ammonium compounds are soluble.	—
2	$C_2H_3O_2^-, NO_3^-$	Acetates and nitrates are soluble.	—
3	Cl^-, Br^-, I^-	Most chlorides, bromides, and iodides are soluble.	$AgCl, Hg_2Cl_2, PbCl_2, AgBr, HgBr_2, Hg_2Br_2, PbBr_2, AgI, HgI_2, Hg_2I_2, PbI_2$
4	SO_4^{2-}	Most sulfates are soluble.	$CaSO_4, SrSO_4, BaSO_4, Ag_2SO_4, Hg_2SO_4, PbSO_4$
5	CO_3^{2-}	Most carbonates are insoluble.	Group IA carbonates, $(NH_4)_2CO_3$
6	PO_4^{3-}	Most phosphates are insoluble.	Group IA phosphates, $(NH_4)_3PO_4$
7	S^{2-}	Most sulfides are insoluble.	Group IA sulfides, $(NH_4)_2S$
8	OH^-	Most hydroxides are insoluble.	Group IA hydroxides, $Ca(OH)_2, Sr(OH)_2, Ba(OH)_2$

Example 4.1   Using the Solubility Rules

Gaining Mastery Toolbox

Critical Concept 4.1
The solubility rules are used to predict if an ionic compound will dissolve in water. Soluble ionic compounds are strong electrolytes; therefore, they exist as ions in solution. Insoluble ionic compounds do not appreciably dissolve in water.

Solution Essentials:
• Solubility rules
• Ionic compound
• Ions

Determine whether the following compounds are soluble or insoluble in water.

a Hg_2Cl_2 **b** KI   **c** lead(II) nitrate

Problem Strategy You should refer to Table 4.1 for the solubility rules of ionic compounds to see which compounds are soluble in water. The soluble ionic compounds are those that readily dissolve in water. The insoluble compounds hardly dissolve at all.

Solution

a According to Rule 3 in Table 4.1, most compounds that contain chloride, Cl^-, are soluble. However,

Hg_2Cl_2 is listed as one of the exceptions to this rule, so it does not dissolve in water. Therefore, **Hg_2Cl_2 is not soluble in water**.

b According to both Rule 1, Group IA compounds are soluble, and Rule 3, most iodides are soluble, KI is expected to be soluble. Therefore, **KI is soluble in water**.

c According to Rule 2, compounds containing nitrates, NO_3^-, are soluble. Since there are no exceptions to this rule, **lead(II) nitrate, $Pb(NO_3)_2$, is soluble in water**.

Answer Check Most incorrect determinations of solubility occur when the exceptions are overlooked, so always be sure to check the exceptions column of Table 4.1.

Exercise 4.1 Determine whether the following compounds are soluble or insoluble in water.

a NaBr **b** $Ba(OH)_2$ **c** calcium carbonate

■ See Problems 4.29 and 4.30.

Let us summarize the main points in this section. Compounds that dissolve in water are soluble; those that dissolve little, or not at all, are insoluble. Soluble substances are either electrolytes or nonelectrolytes. Nonelectrolytes form nonconducting aqueous solutions because they dissolve completely as molecules. Electrolytes form electrically conducting solutions in water because they dissolve to give ions in solution. Electrolytes can be strong or weak. Almost all soluble ionic substances are strong electrolytes. Soluble molecular substances usually are nonelectrolytes or weak electrolytes; the latter solution consists primarily of molecules, but has a small percentage of ions. Ammonia, NH_3, is an example of a molecular substance that is a weak electrolyte. A few molecular substances (such as HCl) dissolve almost entirely as ions in the solution and are therefore strong electrolytes. The solubility rules can be used to predict the solubility of ionic compounds in water.

LiI(*s*) and CH$_3$OH(*l*) are introduced into separate beakers containing water. Using the drawings shown here, label each beaker with the appropriate compound and indicate whether you would expect each substance to be a strong electrolyte, a weak electrolyte, or a nonelectrolyte.

© Cengage Learning

4.2 Molecular and Ionic Equations

We use chemical equations to help us describe chemical reactions. For a reaction involving ions, we have a choice of chemical equations, depending on the kind of information we want to convey. We can represent such a reaction by a *molecular equation,* a *complete ionic equation,* or a *net ionic equation.*

To illustrate these different kinds of equations, consider the preparation of precipitated calcium carbonate, CaCO$_3$. This white, fine, powdery compound is used as a paper filler to brighten and retain ink, as an antacid (as in the trade-named Tums), and as a mild abrasive in toothpastes. One way to prepare this compound is to react calcium hydroxide, Ca(OH)$_2$, with sodium carbonate, Na$_2$CO$_3$. Let us look at the different ways to write the equation for this reaction.

Molecular Equations

You could write the equation for this reaction as follows:

$$Ca(OH)_2(aq) + Na_2CO_3(aq) \longrightarrow CaCO_3(s) + 2NaOH(aq)$$

We call this a **molecular equation**, which is *a chemical equation in which the reactants and products are written as if they were molecular substances, even though they may actually exist in solution as ions.* The molecular equation is useful because it is explicit about what the reactant solutions are and what products you obtain. The equation says that you add aqueous solutions of calcium hydroxide and sodium carbonate to the reaction vessel. As a result of the reaction, the insoluble, white calcium carbonate solid forms in the solution; that is, calcium carbonate precipitates. After you remove the precipitate, you are left with a solution of sodium hydroxide. The molecular equation closely describes what you actually do in the laboratory or in an industrial process.

Complete Ionic Equations

Although a molecular equation is useful in describing the actual reactant and product substances, it does not tell you what is happening at the level of ions. That is, it does not give you an ionic-theory interpretation of the reaction. Because this kind of information is useful, you often need to rewrite the molecular equation as an ionic equation.

Again, consider the reaction of calcium hydroxide, $Ca(OH)_2$, and sodium carbonate, Na_2CO_3. Both are soluble ionic substances and therefore strong electrolytes; when they dissolve in water, they go into solution as ions. Each formula unit of $Ca(OH)_2$ forms one Ca^{2+} ion and two OH^- ions in solution. If you want to emphasize that the solution contains freely moving ions, it would be better to write $Ca^{2+}(aq) + 2OH^-(aq)$ in place of $Ca(OH)_2(aq)$. Similarly, each formula unit of Na_2CO_3 forms two Na^+ ions and one CO_3^{2-} ion in solution, and you would emphasize this by writing $2Na^+(aq) + CO_3^{2-}(aq)$ in place of $Na_2CO_3(aq)$. The reactant side of the equation becomes

$$Ca^{2+}(aq) + 2OH^-(aq) + 2Na^+(aq) + CO_3^{2-}(aq) \longrightarrow$$

Thus, the reaction mixture begins as a solution of four different kinds of ions.

Now let us look at the product side of the equation. One product is the precipitate $CaCO_3(s)$. According to the solubility rules, this is an insoluble ionic compound, so it will exist in water as a solid. We leave the formula as $CaCO_3(s)$ to convey this information in the equation. On the other hand, NaOH is a soluble ionic substance and therefore a strong electrolyte; it dissolves in aqueous solution to give the freely moving ions, which we denote by writing $Na^+(aq) + OH^-(aq)$. The complete equation is

$$Ca^{2+}(aq) + 2OH^-(aq) + 2Na^+(aq) + CO_3^{2-}(aq) \longrightarrow$$
$$CaCO_3(s) + 2Na^+(aq) + 2OH^-(aq)$$

The purpose of such a *complete ionic equation* is to represent each substance by its predominant form in the reaction mixture. For example, if the substance is a soluble ionic compound, it probably dissolves as individual ions (so it is a strong electrolyte). In a complete ionic equation, you represent the compound as separate ions. If the substance is a weak electrolyte, it is present in solution primarily as molecules, so you represent it by its molecular formula. If the substance is an insoluble ionic compound, you represent it by the formula of the compound, not by the formulas of the separate ions in solution.

Thus, a **complete ionic equation** is *a chemical equation in which strong electrolytes (such as soluble ionic compounds) are written as separate ions in the solution.* You represent other reactants and products by the formulas of the compounds, indicating any soluble substance by *(aq)* after its formula and any insoluble solid substance by *(s)* after its formula.

Net Ionic Equations

In the complete ionic equation representing the reaction of calcium hydroxide and sodium carbonate, some ions (OH^- and Na^+) appear on both sides of the equation. This means that nothing happens to these ions as the reaction occurs. They are called spectator ions. A **spectator ion** is *an ion in an ionic equation that does not take part in the reaction.* You can cancel such ions from both sides to express the essential reaction that occurs.

$$Ca^{2+}(aq) + 2\cancel{OH^-(aq)} + 2\cancel{Na^+(aq)} + CO_3^{2-}(aq) \longrightarrow$$
$$CaCO_3(s) + 2\cancel{Na^+(aq)} + 2\cancel{OH^-(aq)}$$

The resulting equation is

$$Ca^{2+}(aq) + CO_3^{2-}(aq) \longrightarrow CaCO_3(s)$$

This is the **net ionic equation**, *an ionic equation from which spectator ions have been canceled.* It shows that the reaction that actually occurs at the ionic level is between calcium ions and carbonate ions to form solid calcium carbonate.

From the net ionic equation, you can see that mixing any solution of calcium ion with any solution of carbonate ion will give you this same reaction. For example, the strong electrolyte calcium nitrate, $Ca(NO_3)$, dissolves readily in water to provide a source of calcium ions. (According to the solubility rules, calcium nitrate is soluble, so it goes into solution as Ca^{2+} and NO_3^- ions.) Similarly, the strong electrolyte potassium carbonate, K_2CO_3, dissolves readily in water to provide a source of carbonate ions. (According to the solubility rules, potassium carbonate is soluble, so it goes into solution as K^+ and CO_3^{2-} ions.) When you mix solutions of these two compounds, you obtain a solution of calcium ions and carbonate ions, which react to form the insoluble calcium carbonate. The other product is potassium nitrate, a soluble ionic compound, and therefore a strong electrolyte. The molecular equation representing the reaction is

$$Ca(NO_3)_2(aq) + K_2CO_3(aq) \longrightarrow CaCO_3(s) + 2KNO_3(aq)$$

You obtain the complete ionic equation from this molecular equation by rewriting each of the soluble ionic compounds as ions, but retaining the formula for the precipitate $CaCO_3(s)$:

$$Ca^{2+}(aq) + 2NO_3^-(aq) + 2K^+(aq) + CO_3^{2-}(aq) \longrightarrow$$
$$CaCO_3(s) + 2K^+(aq) + 2NO_3^-(aq)$$

The net ionic equation is

$$Ca^{2+}(aq) + CO_3^{2-}(aq) \longrightarrow CaCO_3(s)$$

Note that the net ionic equation is identical to the one obtained from the reaction of $Ca(OH)_2$ and Na_2CO_3. The essential reaction is the same whether you mix solutions of calcium hydroxide and sodium carbonate or solutions of calcium nitrate and potassium carbonate.

The value of the net ionic equation is its generality. For example, seawater contains Ca^{2+} and CO_3^{2-} ions from various sources. Whatever the sources of these ions, you expect them to react to form a precipitate of calcium carbonate. In seawater, this precipitate results in sediments of calcium carbonate, which eventually form limestone (Figure 4.5).

© Royalty-Free/Corbis

Figure 4.5 ▲

Limestone formations It is believed that most limestone formed as a precipitate of calcium carbonate (and other carbonates) from seawater. The photograph shows limestone formations at Bryce Point, Bryce Canyon National Park, Utah. More than 60 million years ago, this area was covered by seawater.

| Example | 4.2 | **Writing Net Ionic Equations** |

Gaining Mastery Toolbox

Critical Concept 4.2
The net ionic equation depicts the reaction that is occurring. In the net ionic equation, soluble ionic compounds are represented as aqueous ions, whereas nonelectrolytes and weak electrolytes are written with their molecular formula (not as ions). You expect a reaction to occur when a non-electrolyte, weak electrolyte, or insoluble solid (precipitate) is formed as a product. If no reaction occurs, then there will be no net ionic equation.

Solution Essentials:
• Net ionic equation
• Spectator ion
• Complete ionic equation
• Molecular equation
• Solubility rules
• Nonelectrolyte
• Weak electrolyte

Write a net ionic equation for each of the following molecular equations.

a $2HClO_4(aq) + Ca(OH)_2(aq) \longrightarrow Ca(ClO_4)_2(aq) + 2H_2O(l)$
Perchloric acid, $HClO_4$, is a strong electrolyte, forming H^+ and ClO_4^- ions in solution. $Ca(ClO_4)_2$ is a soluble ionic compound.

b $HC_2H_3O_2(aq) + NaOH(aq) \longrightarrow NaC_2H_3O_2(aq) + H_2O(l)$
Acetic acid, $HC_2H_3O_2$, is a molecular substance and a weak electrolyte.

Problem Strategy You will need to convert the molecular equation to the complete ionic equation, then cancel spectator ions to obtain the net ionic equation. For each ionic compound in the reaction, use the solubility rules to determine if the compound will be soluble (in the solution as ions) or insoluble (present as an undissolved solid). For the complete ionic equation, represent all of the strong electrolytes by their separate ions in solution, adding (aq) after the formula of each. Retain the formulas of the other compounds. An ionic compound should have (aq) after its formula if it is soluble or (s) if it is insoluble.

Solution

a According to the solubility rules presented in Table 4.1 and the problem statement, $Ca(OH)_2$ and $Ca(ClO_4)_2$ are soluble ionic compounds, so they are strong electrolytes. The problem statement notes that $HClO_4$ is also a strong electrolyte. You write each strong electrolyte in the form of separate ions. Water, H_2O, is a nonelectrolyte (or very

(continued)

(*continued*)

weak electrolyte), so you retain its molecular formula. The complete ionic equation is

$$2H^+(aq) + 2ClO_4^-(aq) + Ca^{2+}(aq) + 2OH^-(aq) \longrightarrow$$
$$Ca^{2+}(aq) + 2ClO_4^-(aq) + 2H_2O(l)$$

After canceling spectator ions and dividing by 2, you get the following net ionic equation:

$$\mathbf{H^+(aq) + OH^-(aq) \longrightarrow H_2O(l)}$$

b According to the solubility rules, NaOH and $NaC_2H_3O_2$ are soluble ionic compounds, so they are strong electrolytes. The problem statement notes that $HC_2H_3O_2$ is a weak electrolyte, which you write by its molecular formula. Water, H_2O, is a nonelectrolyte, so you retain its molecular formula also. The complete ionic equation is

$$HC_2H_3O_2(aq) + Na^+(aq) + OH^-(aq) \longrightarrow Na^+(aq) + C_2H_3O_2^-(aq) + H_2O(l)$$

and the net ionic equation is

$$\mathbf{HC_2H_3O_2(aq) + OH^-(aq) \longrightarrow C_2H_3O_2^-(aq) + H_2O(l)}$$

Answer Check If both of the reactants are strong electrolytes and a reaction occurs, your correct net ionic equation should not have any ions as products of the reaction. In this case, because the reaction is with a weak electrolyte, ions in the net ionic equation are possible.

Exercise 4.2 Write complete ionic and net ionic equations for each of the following molecular equations.

a $2HNO_3(aq) + Mg(OH)_2(s) \longrightarrow 2H_2O(l) + Mg(NO_3)_2(aq)$
Nitric acid, HNO_3, is a strong electrolyte.

b $Pb(NO_3)_2(aq) + Na_2SO_4(aq) \longrightarrow PbSO_4(s) + 2NaNO_3(aq)$

■ See Problems 4.33 and 4.34.

Types of Chemical Reactions

Among the several million known substances, many millions of chemical reactions are possible. Beginning students are often bewildered by the possibilities. How can I know whether two substances will react when they are mixed? How can I predict the products? Although it is not possible to give completely general answers to these questions, it is possible to make sense of chemical reactions. Most of the reactions we will study belong to one of three types:

1. Precipitation reactions. In these reactions, you mix solutions of two ionic substances, and a solid ionic substance (a precipitate) forms.

2. Acid–base reactions. An acid substance reacts with a substance called a base. Such reactions involve the transfer of a proton between reactants.

3. Oxidation–reduction reactions. These involve the transfer of electrons between reactants.

We will look at each of these types of reactions. By the time you finish this chapter, you should feel much more comfortable with the descriptions of chemical reactions that you will encounter in this course.

4.3 Precipitation Reactions

In the previous section, we used a precipitation reaction to illustrate how to convert a molecular equation to an ionic equation. A precipitation reaction occurs in aqueous solution because one product is insoluble. A **precipitate** is *an insoluble solid compound formed during a chemical reaction in solution.* To predict whether a precipitate will form when you mix two solutions of ionic compounds, you need to know whether any of the potential products that might form are insoluble. This is another application of the solubility rules (Section 4.1).

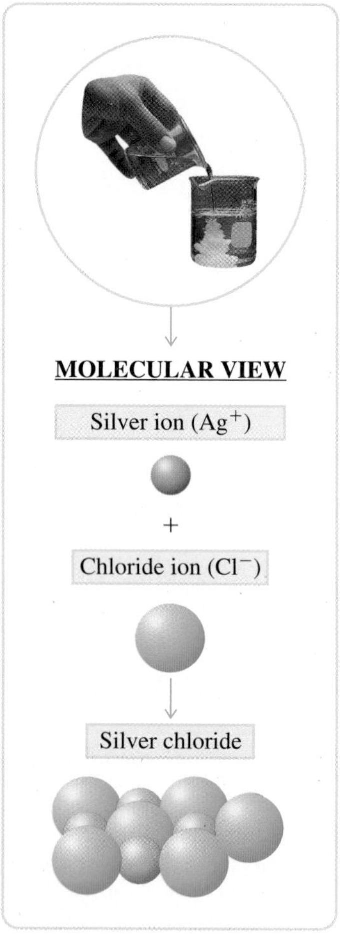

Predicting Precipitation Reactions

Now let us see how you would go about predicting whether a precipitation reaction will occur. Suppose you mix together solutions of magnesium chloride, $MgCl_2$, and silver nitrate, $AgNO_3$. You can write the potential reactants as follows:

$$MgCl_2 + AgNO_3 \longrightarrow$$

How can you tell if a reaction will occur, and if it does, what products to expect?

When you write a precipitation reaction as a molecular equation, the reaction has the form of an exchange reaction. An **exchange** (or **metathesis**) **reaction** is *a reaction between compounds that, when written as a molecular equation, appears to involve the exchange of parts between the two reactants.* In a precipitation reaction, the anions exchange between the two cations (or vice versa).

Let us momentarily assume that a reaction does occur between magnesium chloride and silver nitrate. If you exchange the anions, you get silver chloride and magnesium nitrate. Once you figure out the formulas of these potential products, you can write the molecular equation. The formulas are $AgCl$ and $Mg(NO_3)_2$. (If you do not recall how to write the formula of an ionic compound, given the ions, turn back to Example 2.3.) The balanced equation, assuming there *is* a reaction, is

$$MgCl_2 + 2AgNO_3 \longrightarrow 2AgCl + Mg(NO_3)_2$$

Let us verify that $MgCl_2$ and $AgNO_3$ are soluble and then check the solubilities of the products. Rule 3 in Table 4.1 says that chlorides are soluble, with certain exceptions, which do not include magnesium chloride. Thus, we predict that magnesium chloride is soluble. Rule 2 indicates that nitrates are soluble, so $AgNO_3$ is soluble as well. The potential products are silver chloride and magnesium nitrate. According to Rule 3, silver chloride is one of the exceptions to the general solubility of chlorides. Therefore, we predict that the silver chloride is insoluble. Magnesium nitrate is soluble according to Rule 2.

Now we can append the appropriate phase labels to the compounds in the preceding equation.

$$MgCl_2(aq) + 2AgNO_3(aq) \longrightarrow 2AgCl(s) + Mg(NO_3)_2(aq)$$

We predict that reaction occurs because silver chloride is insoluble, and precipitates from the reaction mixture. Figure 4.6 shows the formation of the white silver chloride from this reaction and depicts the net ionic equation using molecular models. If you separate the precipitate from the solution by pouring it through filter paper, the solution that passes through (the filtrate) contains magnesium nitrate, which you could obtain by evaporating the water. The molecular equation is a summary of the actual reactants and products in the reaction. ◀

To see the reaction that occurs on an ionic level, you need to rewrite the molecular equation as a net ionic equation. You first write the strong electrolytes (here soluble ionic compounds) in the form of ions, leaving the formula of the precipitate unchanged.

$$Mg^{2+}(aq) + 2Cl^-(aq) + 2Ag^+(aq) + 2NO_3^-(aq) \longrightarrow$$
$$2AgCl(s) + Mg^{2+}(aq) + 2NO_3^-(aq)$$

After canceling spectator ions and reducing the coefficients to the smallest whole numbers, you obtain the net ionic equation:

$$Ag^+(aq) + Cl^-(aq) \longrightarrow AgCl(s)$$

MOLECULAR VIEW

Silver ion (Ag^+)

+

Chloride ion (Cl^-)

Silver chloride

Figure 4.6 ▲

Reaction of magnesium chloride and silver nitrate Magnesium chloride solution is added to a beaker of silver nitrate solution. A white precipitate of silver chloride forms.

The reactants $MgCl_2$ and $AgNO_3$ must be added in correct amounts; otherwise, the excess reactant will remain along with the product $Mg(NO_3)_2$.

This equation represents the essential reaction that occurs: Ag^+ ions and Cl^- ions in aqueous solution react to form solid silver chloride.

If silver chloride were soluble, a reaction would not have occurred. When you first mix solutions of $MgCl_2(aq)$ and $AgNO_3(aq)$, you obtain a solution of four ions: $Mg^{2+}(aq)$, $Cl^-(aq)$, $Ag^+(aq)$, and $NO_3^-(aq)$. If no precipitate formed, you would end up simply with a solution of these four ions. However, because Ag^+ and Cl^- react to give the precipitate AgCl, the effect is to remove these ions from the reaction mixture as an insoluble compound and leave behind a solution of $Mg(NO_3)_2(aq)$.

⚙WL Interactive Example 4.3   Deciding Whether a Precipitation Reaction Occurs

Gaining Mastery Toolbox

Critical Concept 4.3
Use the solubility rules to predict if a precipitate will form during the reaction. Whenever one of the products of an aqueous reaction between two soluble salts is an insoluble compound, then a precipitation reaction will occur. Otherwise, there will be no reaction, and we predict that the solution will consist only of dissolved ions.

Solution Essentials
- Exchange (metathesis) reaction
- Precipitate
- Net ionic equation
- Spectator ion
- Complete ionic equation
- Molecular equation
- Solubility rules

For each of the following, decide whether a precipitation reaction occurs. If it does, write the balanced molecular equation and then the net ionic equation. If no reaction occurs, write the compounds followed by an arrow and then *NR* (no reaction).

a Aqueous solutions of sodium chloride and iron(II) nitrate are mixed.
b Aqueous solutions of aluminum sulfate and sodium hydroxide are mixed.

Problem Strategy When aqueous solutions containing soluble salts are mixed, a precipitation reaction occurs if an insoluble compound (precipitate) forms. Start by writing the formulas of the compounds that are mixed. (If you have trouble with this, see Examples 2.3 and 2.4.) Then, assuming momentarily that the compounds do react, write the exchange reaction. Make sure that you write the correct formulas of the products. Using the solubility rules, append phase labels to each formula in the equation consulting the solubility rules. If one of the products forms an insoluble compound (precipitate), reaction occurs; otherwise, no reaction occurs.

Solution

a The formulas of the compounds are NaCl and $Fe(NO_3)_2$. Exchanging anions, you get sodium nitrate, $NaNO_3$, and iron(II) chloride, $FeCl_2$. The equation for the exchange reaction is

$$NaCl + Fe(NO_3)_2 \longrightarrow NaNO_3 + FeCl_2 \qquad \text{(not balanced)}$$

Referring to Table 4.1, note that NaCl and $NaNO_3$ are soluble (Rule 1). Also, iron(II) nitrate is soluble (Rule 2), and iron(II) chloride is soluble. (Rule 3 says that chlorides are soluble with some exceptions, none of which include $FeCl_2$.) Since there is no precipitate, no reaction occurs. You obtain simply an aqueous solution of the four different ions (Na^+, Cl^-, Fe^{2+}, and NO_3^-). For the answer, we write

$$\textbf{NaCl}(aq) + \textbf{Fe(NO}_3)_2(aq) \longrightarrow \textbf{NR}$$

b The formulas of the compounds are $Al_2(SO_4)_3$ and NaOH. Exchanging anions, you get aluminum hydroxide, $Al(OH)_3$, and sodium sulfate, Na_2SO_4. The equation for the exchange reaction is

$$Al_2(SO_4)_3 + NaOH \longrightarrow Al(OH)_3 + Na_2SO_4 \qquad \text{(not balanced)}$$

From Table 4.1, you see that $Al_2(SO_4)_3$ is soluble (Rule 4), NaOH and Na_2SO_4 are soluble (Rule 1), and $Al(OH)_3$ is insoluble (Rule 8). Thus, aluminum hydroxide precipitates. The balanced molecular equation with phase labels is

$$\textbf{Al}_2(\textbf{SO}_4)_3(aq) + \textbf{6NaOH}(aq) \longrightarrow \textbf{2Al(OH)}_3(s) + \textbf{3Na}_2\textbf{SO}_4(aq)$$

To get the net ionic equation, you write the strong electrolytes (here, soluble ionic compounds) as ions in aqueous solution and cancel spectator ions.

$$2Al^{3+}(aq) + 3\cancel{SO_4^{2-}(aq)} + 6\cancel{Na^+(aq)} + 6OH^-(aq) \longrightarrow$$
$$2Al(OH)_3(s) + 6\cancel{Na^+(aq)} + 3\cancel{SO_4^{2-}(aq)}$$

The net ionic equation is

$$\textbf{Al}^{3+}(aq) + \textbf{3OH}^-(aq) \longrightarrow \textbf{Al(OH)}_3(s)$$

Thus, aluminum ion reacts with hydroxide ion to precipitate aluminum hydroxide.

(continued)

(continued)

Answer Check One of the most critical aspects of getting this type of problem right is making certain that you are using the correct chemical formulas for the reactants and products.

Exercise 4.3 You mix aqueous solutions of sodium iodide and lead(II) acetate. If a reaction occurs, write the balanced molecular equation and the net ionic equation. If no reaction occurs, indicate this by writing the formulas of the compounds and an arrow followed by *NR*.

■ See Problems 4.37, 4.38, 4.39, and 4.40.

CONCEPT CHECK 4.3

Your lab partner tells you that she mixed two solutions that contain ions. You analyze the solution and find that it contains the ions and precipitate shown in the beaker.

 a Write the molecular equation for the reaction.

 b Write the complete ionic equation for the reaction.

 c Write the net ionic equation for the reaction.

$Na^+(aq)$

$C_2H_3O_2^-(aq)$

$SrSO_4(s)$

4.4 Acid–Base Reactions

Acids and bases are some of the most important electrolytes. You can recognize acids and bases by some simple properties. Acids have a sour taste. Solutions of bases, on the other hand, have a bitter taste and a soapy feel. (Of course, you should *never* taste laboratory chemicals.) Some examples of acids are acetic acid, present in vinegar; citric acid, a constituent of lemon juice; and hydrochloric acid, found in the digestive fluid of the stomach. An example of a base is aqueous ammonia, often used as a household cleaner. Table 4.2 lists further examples; see also Figure 4.7.

Another simple property of acids and bases is their ability to cause color changes in certain dyes. An **acid–base indicator** is *a dye used to distinguish between acidic and basic solutions by means of the color changes it undergoes in these solutions.* Such dyes are common in natural materials. The amber color of tea, for example, is lightened by the addition of lemon juice (citric acid). Red cabbage

Table 4.2	Common Acids and Bases	
Name	**Formula**	**Remarks**
Acids		
Acetic acid	$HC_2H_3O_2$	Found in vinegar
Acetylsalicylic acid	$HC_9H_7O_4$	Aspirin
Ascorbic acid	$H_2C_6H_6O_6$	Vitamin C
Citric acid	$H_3C_6H_5O_7$	Found in lemon juice
Hydrochloric acid	HCl	Found in gastric juice (digestive fluid in stomach)
Sulfuric acid	H_2SO_4	Battery acid
Bases		
Ammonia	NH_3	Aqueous solution used as a household cleaner
Calcium hydroxide	$Ca(OH)_2$	Slaked lime (used in mortar for building construction)
Magnesium hydroxide	$Mg(OH)_2$	Milk of magnesia (antacid and laxative)
Sodium hydroxide	$NaOH$	Drain cleaners, oven cleaners

juice changes to green and then yellow when a base is added (Figure 4.8). The green and yellow colors change back to red when an acid is added. Litmus is a common laboratory acid–base indicator. This dye, produced from certain species of lichens, turns red in acidic solution and blue in basic solution. Phenolphthalein (fee' nol thay' leen), another laboratory acid–base indicator, is colorless in acidic solution and pink in basic solution.

An important chemical characteristic of acids and bases is the way they react with one another. To understand these acid–base reactions, we need to have precise definitions of the terms *acid* and *base*.

Definitions of Acid and Base

When Arrhenius developed his ionic theory of solutions, he also gave the classic definitions of acids and bases. According to Arrhenius, an **acid** is *a substance that produces hydrogen ions, H^+, when it dissolves in water.* An example is nitric acid, HNO_3, a molecular substance that dissolves in water to give H^+ and NO_3^-.

$$HNO_3(aq) \xrightarrow{H_2O} H^+(aq) + NO_3^-(aq)$$

An Arrhenius **base** is *a substance that produces hydroxide ions, OH^-, when it dissolves in water.* For example, sodium hydroxide is a base.

$$NaOH(s) \xrightarrow{H_2O} Na^+(aq) + OH^-(aq)$$

The molecular substance ammonia, NH_3, is also a base because it yields hydroxide ions when it reacts with water. ▶

$$NH_3(aq) + H_2O(l) \rightleftharpoons NH_4^+(aq) + OH^-(aq)$$

Although the Arrhenius concept of acids and bases is useful, it is somewhat limited. For example, it tends to single out the OH^- ion as the source of base character, when other ions or molecules can play a similar role. In 1923, Johannes N. Brønsted and Thomas M. Lowry independently noted that many reactions involve nothing more than the transfer of a proton (H^+) between reactants, and they realized that they could use this idea to expand the definitions of acids and bases

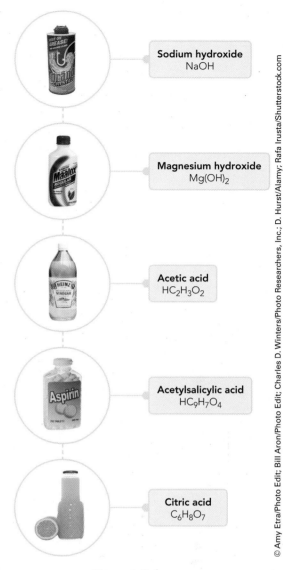

Sodium hydroxide
NaOH

Magnesium hydroxide
$Mg(OH)_2$

Acetic acid
$HC_2H_3O_2$

Acetylsalicylic acid
$HC_9H_7O_4$

Citric acid
$C_6H_8O_7$

Figure 4.7 ▲

Household acids and bases Shown are a variety of household products that are either acids or bases.

Solutions of ammonia are sometimes called ammonium hydroxide and given the formula $NH_4OH(aq)$, based on analogy with NaOH. No NH_4OH molecule or compound has ever been found, however.

Preparation of red cabbage juice. The beaker (green solution) contains red cabbage juice and sodium bicarbonate (baking soda), which is a basic solution. The flask in the center contains red cabbage juice in an acidic solution.

Red cabbage juice has been added from a pipet to solutions in the beakers. These solutions vary in acidity from highly acidic on the left to highly basic on the right.

Figure 4.8 ◀

Red cabbage juice as an acid–base indicator

to describe a large class of chemical reactions. In this view, acid–base reactions are *proton-transfer reactions.*

Consider the preceding reaction. It involves the transfer of a proton from the water molecule, H_2O, to the ammonia molecule, NH_3.

$$\overset{\curvearrowleft H^+}{NH_3(aq) + H_2O(l)} \rightleftharpoons NH_4^+(aq) + OH^-(aq)$$

Once the proton, H^+, has left the H_2O molecule, it leaves behind an OH^- ion. When the H^+ adds to NH_3, an NH_4^+ ion results. H_2O is said to *donate* a proton to NH_3, and NH_3 is said to *accept* a proton from H_2O.

Brønsted and Lowry defined an **acid** as *the species (molecule or ion) that donates a proton to another species in a proton-transfer reaction.* They defined a **base** as *the species (molecule or ion) that accepts a proton in a proton-transfer reaction.* In the reaction of ammonia with water, the H_2O molecule is the acid, because it donates a proton. The NH_3 molecule is a base, because it accepts a proton.

$$\underset{\text{base} \qquad \text{acid}}{\overset{\curvearrowleft H^+}{NH_3(aq) + H_2O(l)}} \rightleftharpoons NH_4^+(aq) + OH^-(aq)$$

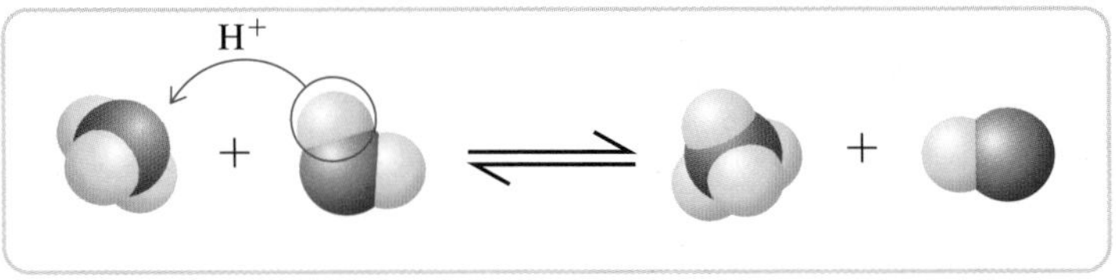

The dissolution of nitric acid, HNO_3, in water is actually a proton-transfer reaction, although the following equation, which we used as an illustration of an Arrhenius acid, does not spell that out.

$$HNO_3(aq) \overset{H_2O}{\longrightarrow} H^+(aq) + NO_3^-(aq)$$

To see that this is a proton-transfer reaction, we need to clarify the structure of the hydrogen ion, $H^+(aq)$. This ion consists of a proton (H^+) in association with water molecules, which is what (*aq*) means. This is not a weak association, however, because the proton (or hydrogen nucleus) would be expected to attract electrons strongly to itself. In fact, the $H^+(aq)$ ion might be better thought of as a proton chemically bonded to a water molecule to give the H_3O^+ ion, with other water molecules less strongly associated with this ion, which we represent by the phase-labeled formula $H_3O^+(aq)$. Written in this form, we usually call this the *hydronium ion.* It is important to understand that the hydrogen ion, $H^+(aq)$, and the hydronium ion, $H_3O^+(aq)$, represent precisely the same physical ion. For simplicity, we often write the formula for this ion as $H^+(aq)$, but when we want to be explicit about the proton-transfer aspect of a reaction, we write the formula as $H_3O^+(aq)$. ◄

The H^+(aq) ion has variable structure in solution. In solids, there is evidence for [H(H₂O)ₙ]⁺, where n is 1, 2, 3, 4, or 6.

Now let us rewrite the preceding equation by replacing $H^+(aq)$ by $H_3O^+(aq)$. To maintain a balanced equation, we will also need to add $H_2O(l)$ to the left side.

$$HNO_3(aq) + H_2O(l) \longrightarrow NO_3^-(aq) + H_3O^+(aq)$$

Note that the reaction involves simply a transfer of a proton (H^+) from HNO_3 to H_2O:

$$\underset{\text{acid} \qquad \text{base}}{\overset{\curvearrowright H^+}{HNO_3(aq) + H_2O(l)}} \longrightarrow NO_3^-(aq) + H_3O^+(aq)$$

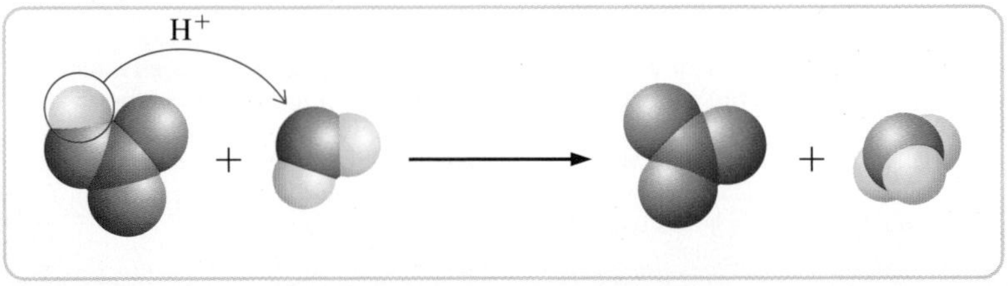

The HNO$_3$ molecule is an acid (proton donor) and H$_2$O is a base (proton acceptor). Note that H$_2$O may function as an acid or a base, depending on the other reactant.

The Arrhenius definitions and those of Brønsted and Lowry are essentially equivalent for aqueous solutions, although their points of view are different. For instance, sodium hydroxide and ammonia are bases in the Arrhenius view because they increase the percentage of OH$^-$ ion in the aqueous solution. They are bases in the Brønsted–Lowry view because they provide species (OH$^-$ from the strong electrolyte sodium hydroxide and NH$_3$ from ammonia) that can accept protons. ▶

We will discuss the Brønsted–Lowry concept of acids and bases more thoroughly in Chapter 15.

Strong and Weak Acids and Bases

Acids and bases are classified as strong or weak, depending on whether they are strong or weak electrolytes. A **strong acid** is *an acid that ionizes completely in water; it is a strong electrolyte.* Hydrochloric acid, HCl(*aq*), and nitric acid, HNO$_3$(*aq*), are examples of strong acids. Using the hydronium ion notation, we write the respective equations as follows:

$$HCl(aq) + H_2O(l) \longrightarrow H_3O^+(aq) + Cl^-(aq)$$

$$HNO_3(aq) + H_2O(l) \longrightarrow H_3O^+(aq) + NO_3^-(aq)$$

Table 4.3 lists six common strong acids. Most of the other acids we will discuss are weak acids.

A **weak acid** is *an acid that only partly ionizes in water; it is a weak electrolyte.* Examples of weak acids are hydrocyanic acid, HCN(*aq*), and hydrofluoric acid, HF(*aq*). These molecules react with water to produce a small percentage of ions in solution.

$$HCN(aq) + H_2O(l) \rightleftharpoons H_3O^+(aq) + CN^-(aq)$$

$$HF(aq) + H_2O(l) \rightleftharpoons H_3O^+(aq) + F^-(aq)$$

Figure 4.9 presents molecular views of the strong acid HCl(*aq*) and the weak acid HF(*aq*). Note how both the strong and weak acid undergo the same reaction with water (proton donors); however, in the case of the weak acid such as HF, only a small portion of the acid molecules actually undergoes reaction, leaving the majority of the acid molecules unreacted. On the other hand, strong acids like HCl and HNO$_3$ undergo complete reaction with water. The result is that weak acids can produce little H$_3$O$^+$ in aqueous solution, whereas strong acids can produce large amounts. Hence, when chemists refer to a weak acid, they are often thinking about a substance that produces a limited amount of H$_3$O$^+$(*aq*), whereas when chemists refer to a strong acid, they are thinking about a substance that can produce large amounts of H$_3$O$^+$(*aq*).

Table 4.3	Common Strong Acids and Bases	
Strong Acids	**Strong Bases**	
HClO$_4$	LiOH	
H$_2$SO$_4$	NaOH	
HI	KOH	
HBr	Ca(OH)$_2$	
HCl	Sr(OH)$_2$	
HNO$_3$	Ba(OH)$_2$	

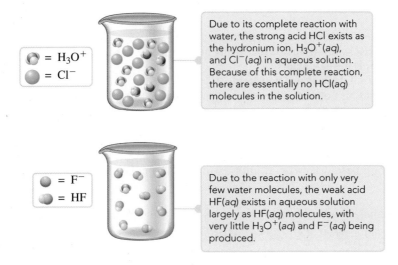

Figure 4.9 ◀

Molecular views comparing the strong acid HCl and the weak acid HF in water (H$_2$O molecules are omitted for clarity)

A **strong base** is *a base that is present in aqueous solution entirely as ions, one of which is OH$^-$; it is a strong electrolyte.* The ionic compound sodium hydroxide, NaOH, is an example of a strong base. It dissolves in water as Na$^+$ and OH$^-$.

$$NaOH(s) \xrightarrow{\text{H}_2\text{O}} Na^+(aq) + OH^-(aq)$$

The hydroxides of Groups IA and IIA elements, except for beryllium hydroxide, are strong bases (see Table 4.3).

A **weak base** is *a base that is only partly ionized in water; it is a weak electrolyte.* Ammonia, NH$_3$, is an example.

$$NH_3(aq) + H_2O(l) \rightleftharpoons NH_4^+(aq) + OH^-(aq)$$

You will find it important to be able to identify an acid or base as strong or weak. When you write an ionic equation, you represent strong acids and bases by the ions they form and weak acids and bases by the formulas of the compounds. ◄ The next example will give you some practice identifying acids and bases as strong or weak.

In many cases, when chemists are thinking about strong versus weak bases in aqueous solution, they consider a weak base to be a substance that can produce only limited amounts of OH$^-$(aq) and a strong base to be a substance that can produce large amounts of OH$^-$(aq).

CONCEPT CHECK 4.4

Complete and balance the two chemical equations.

$$HClO_4(aq) + H_2O(l) \longrightarrow$$

$$HBr(aq) + H_2O(l) \longrightarrow$$

Which of the following statements is true regarding the solutions formed after the two chemical reactions have occurred?

  a   The HBr is partially ionized in water.

  b   The HClO$_4$ is completely ionized in water.

  c   In the solution of HClO$_4$, you would expect to find very few HClO$_4$(aq) molecules relative to ClO$_4^-$(aq) ions.

  d   In the HBr solution there are significant quantities of HBr(aq) molecules.

  e   None of the other statements (a–d) is true.

Example   **4.4**   **Classifying Acids and Bases as Strong or Weak**

Gaining Mastery Toolbox

Critical Concept 4.4
Assume that if an acid or base is not listed in Table 4.3 ("Common Strong Acids and Bases"), it is a weak acid or a weak base. This assumption works very well for most acids and bases typically encountered in the laboratory.

Solution Essentials:
• Weak acid
• Weak base
• Strong acid
• Strong base

Identify each of the following compounds as a strong or weak acid or base:

  a LiOH   b HC$_2$H$_3$O$_2$ c HBr   d HNO$_2$

Problem Strategy Refer to Table 4.3 for the common strong acids and bases. You can assume that other acids and bases are weak.

Solution

  a As noted in Table 4.3, **LiOH is a strong base**.

  b Acetic acid, HC$_2$H$_3$O$_2$, is not one of the strong acids listed in Table 4.3; therefore, we assume **HC$_2$H$_3$O$_2$ is a weak acid**.

  c As noted in Table 4.3, **HBr is a strong acid**.

  d Nitrous acid, HNO$_2$, is not one of the strong acids listed in Table 4.3; therefore, we assume **HNO$_2$ is a weak acid**.

Answer Check Do not fall into the trap of assuming that if a compound has OH in the formula, it must be a base. For example, methyl alcohol, CH$_3$OH, has OH in the formula, but it is not a strong base. Likewise, do not assume that a compound is an acid just because it has H.

Exercise 4.4   Label each of the following as a strong or weak acid or base:

  a H$_3$PO$_4$ b HClO   c HClO$_4$ d Sr(OH)$_2$

■ See Problems 4.41 and 4.42.

Neutralization Reactions

One of the chemical properties of acids and bases is that they neutralize one another. A **neutralization reaction** is *a reaction of an acid and a base that results in an ionic compound and possibly water.* When a base is added to an acid solution, the acid is said to be neutralized. *The ionic compound that is a product of a neutralization reaction* is called a **salt**. Most ionic compounds other than hydroxides and oxides are salts. Salts can be obtained from neutralization reactions such as

$$2HCl(aq) + Ca(OH)_2(aq) \longrightarrow CaCl_2(aq) + 2H_2O(l)$$
$$\text{acid} \qquad\qquad \text{base} \qquad\qquad \text{salt}$$

$$HCN(aq) + KOH(aq) \longrightarrow KCN(aq) + H_2O(l)$$
$$\text{acid} \qquad\quad \text{base} \qquad\quad \text{salt}$$

The salt formed in a neutralization reaction consists of cations obtained from the base and anions obtained from the acid. In the first example, the base is $Ca(OH)_2$, which supplies Ca^{2+} cations; the acid is HCl, which supplies Cl^- anions. The salt contains Ca^{2+} and Cl^- ions ($CaCl_2$).

We wrote these reactions as molecular equations. Written in this form, the equations make explicit the reactant compounds and the salts produced. However, to discuss the essential reactions that occur, you need to write the net ionic equations. You will see that the net ionic equation for each acid–base reaction involves the transfer of a proton.

The first reaction involves a strong acid, HCl(aq), and a strong base, $Ca(OH)_2(aq)$. Both $Ca(OH)_2$ and the product $CaCl_2$, being soluble ionic compounds, are strong electrolytes. Writing these strong electrolytes in the form of ions gives the following complete ionic equation:

$$2H^+(aq) + 2Cl^-(aq) + Ca^{2+}(aq) + 2OH^-(aq) \longrightarrow$$
$$Ca^{2+}(aq) + 2Cl^-(aq) + 2H_2O(l)$$

Canceling spectator ions and dividing by 2 gives the net ionic equation:

$$H^+(aq) + OH^-(aq) \longrightarrow H_2O(l)$$

Note the transfer of a proton from the hydrogen ion (or hydronium ion, H_3O^+) to the hydroxide ion.

The second reaction involves HCN(aq), a weak acid, and KOH(aq), a strong base; the product is KCN, a strong electrolyte. The net ionic equation is

$$HCN(aq) + OH^-(aq) \longrightarrow CN^-(aq) + H_2O(l)$$

Note the proton transfer, characteristic of an acid–base reaction.

In each of these examples, hydroxide ions latch strongly onto protons to form water. Because water is a very stable substance, it effectively provides the driving force of the reaction.

Although water is one of the products in most neutralization reactions, the reaction of an acid with the base ammonia provides a prominent exception. Consider the reaction of sulfuric acid with ammonia:

$$H_2SO_4(aq) + 2NH_3(aq) \longrightarrow (NH_4)_2SO_4(aq)$$
$$\text{acid} \qquad\quad \text{base} \qquad\qquad \text{salt}$$

The net ionic equation is

$$H^+(aq) + NH_3(aq) \longrightarrow NH_4^+(aq)$$

Note the proton transfer, a hallmark of an acid–base reaction. In this case, NH_3 molecules latch onto protons and form relatively stable NH_4^+ ions.

Example 4.5

Critical Concept 4.5
One product of a neutralization reaction will be a salt. The salt is an ionic compound formed from the acid anions and the base cations. Water is often an additional product of neutralization reactions.

Solution Essentials:
• Neutralization reaction
• Salt
• Strong and weak acid
• Net ionic equation
• Spectator ion
• Strong and weak electrolyte

Writing an Equation for a Neutralization

Write the molecular equation and then the net ionic equation for the neutralization of nitrous acid, HNO_2, by sodium hydroxide, $NaOH$, both in aqueous solution. Use an arrow with H^+ over it to show the proton transfer.

Problem Strategy Because this is a neutralization reaction, you know that the reactants are an acid and a base and that there will be a salt and possibly water as a product. Therefore, start by writing the acid and base reactants and the salt product. Recall that the salt consists of cations from the base and anions from the acid. Note also that water is frequently a product; if it is, you will not be able to balance the equation without it. Once you have the balanced molecular equation, write the complete ionic equation. Do that by representing any strong electrolytes by the formulas of their ions. Finally, write the net ionic equation by canceling any spectator ions, and from it note the proton transfer.

Solution The salt consists of the cation from the base (Na^+) and the anion from the acid (NO_2^-); its formula is $NaNO_2$. You will need to add H_2O as a product to complete and balance the molecular equation:

$$HNO_2(aq) + NaOH(aq) \longrightarrow NaNO_2(aq) + H_2O(l) \quad \text{(molecular equation)}$$

Note that $NaOH$ (a strong base) and $NaNO_2$ (a soluble ionic substance) are strong electrolytes; HNO_2 is a weak electrolyte (it is not one of the strong acids in Table 4.3). You write both $NaOH$ and $NaNO_2$ in the form of ions. The complete ionic equation is

$$HNO_2(aq) + Na^+(aq) + OH^-(aq) \longrightarrow Na^+(aq) + NO_2^-(aq) + H_2O(l)$$

The net ionic equation is

$$HNO_2(aq) + OH^-(aq) \longrightarrow NO_2^-(aq) + H_2O(l)$$

Note that a proton is transferred from HNO_2 to the OH^- ion to yield the products:

$$\overset{\overset{\displaystyle H^+}{\frown}}{HNO_2(aq) + OH^-(aq)} \longrightarrow NO_2^-(aq) + H_2O(l) \quad \text{(net ionic equation)}$$

Answer Check Remember that correct answers include phase labels and that ions must have correct charges.

Exercise 4.5   Write the molecular equation and the net ionic equation for the neutralization of hydrocyanic acid, HCN, by lithium hydroxide, $LiOH$, both in aqueous solution.

■ See Problems 4.43, 4.44, 4.45, and 4.46.

Acids such as HCl and HNO_3 that have only one acidic hydrogen atom per acid molecule are called *monoprotic* acids. A **polyprotic acid** is *an acid that yields two or more acidic hydrogens per molecule.* Phosphoric acid is an example of a triprotic acid. By reacting this acid with different amounts of a base, you can obtain a series of salts:

$$H_3PO_4(aq) + NaOH(aq) \longrightarrow NaH_2PO_4(aq) + H_2O(l)$$

$$H_3PO_4(aq) + 2NaOH(aq) \longrightarrow Na_2HPO_4(aq) + 2H_2O(l)$$

$$H_3PO_4(aq) + 3NaOH(aq) \longrightarrow Na_3PO_4(aq) + 3H_2O(l)$$

Salts such as NaH_2PO_4 and Na_2HPO_4 that have acidic hydrogen atoms and can undergo neutralization with bases are called *acid salts.*

Exercise 4.6   Write molecular and net ionic equations for the successive neutralizations of each of the acidic hydrogens of sulfuric acid with potassium hydroxide. (That is, write equations for the reaction of sulfuric acid with KOH to give the acid salt and for the reaction of the acid salt with more KOH to give potassium sulfate.)

■ See Problems 4.47, 4.48, 4.49, and 4.50.

Acid–Base Reactions with Gas Formation

Certain salts, notably carbonates, sulfites, and sulfides, react with acids to form a gaseous product. Consider the reaction of sodium carbonate with hydrochloric acid. The molecular equation for the reaction is

$$\underset{\text{carbonate}}{Na_2CO_3(aq)} + \underset{\text{acid}}{2HCl(aq)} \longrightarrow \underset{\text{salt}}{2NaCl(aq)} + H_2O(l) + CO_2(g)$$

Here, a carbonate (sodium carbonate) reacts with an acid (hydrochloric acid) to produce a salt (sodium chloride), water, and carbon dioxide gas. A similar reaction is shown in Figure 4.10, in which baking soda (sodium hydrogen carbonate) reacts with the acetic acid in vinegar. Note the bubbles of carbon dioxide gas that evolve during the reaction. This reaction of a carbonate with an acid is the basis of a simple test for carbonate minerals. When you treat a carbonate mineral or rock, such as limestone, with hydrochloric acid, the material fizzes, as bubbles of odorless carbon dioxide form.

It is useful to consider the preceding reaction as an exchange, or metathesis, reaction. Recall that in an exchange reaction, you obtain the products from the reactants by exchanging the anions (or the cations) between the two reactants. In this case, we interchange the carbonate ion with the chloride ion and we obtain the following:

$$Na_2CO_3(aq) + 2HCl(aq) \longrightarrow 2NaCl(aq) + H_2CO_3(aq)$$

The last product shown in the equation is carbonic acid, H_2CO_3. (You obtain the formula of carbonic acid by noting that you need to associate two H^+ ions with the carbonate ion, CO_3^{2-}, to obtain a neutral compound.) Carbonic acid is unstable and decomposes to water and carbon dioxide gas. The overall result is the equation we wrote earlier:

$$Na_2CO_3(aq) + 2HCl(aq) \longrightarrow 2NaCl(aq) + \underbrace{H_2O(l) + CO_2(g)}_{H_2CO_3(aq)}$$

The net ionic equation for this reaction is

$$CO_3^{2-}(aq) + 2H^+(aq) \longrightarrow H_2O(l) + CO_2(g)$$

The carbonate ion reacts with hydrogen ion from the acid. If you write the hydrogen ion as H_3O^+ (hydronium ion), you can see also that the reaction involves a proton transfer:

$$\overset{\displaystyle 2H^+}{\overset{\frown}{CO_3^{2-}(aq)}} + 2H_3O^+(aq) \longrightarrow H_2CO_3(aq) + 2H_2O(l) \longrightarrow 3H_2O(l) + CO_2(g)$$

From the broader perspective of the Brønsted–Lowry view, this is an acid–base reaction.

Sulfites behave similarly to carbonates. When a sulfite, such as sodium sulfite (Na_2SO_3), reacts with an acid, sulfur dioxide (SO_2) is a product. Sulfides, such as sodium sulfide (Na_2S), react with acids to produce hydrogen sulfide gas. The reaction can be viewed as a simple exchange reaction. The reactions are summarized in Table 4.4. ▶

+

=

© Charles D. Winters

Figure 4.10 ▲

Reaction of a carbonate with an acid Baking soda (sodium hydrogen carbonate) reacts with acetic acid in vinegar to evolve bubbles of carbon dioxide.

CO_3^{2-}, S^{2-}, and SO_3^{2-} are bases. Thus the net ionic equations involve a basic anion reacting with H^+, an acid.

Table 4.4	Some Ionic Compounds That Evolve Gases When Treated with Acids	
Ionic Compound	**Gas**	**Example**
Carbonate (CO_3^{2-})	CO_2	$Na_2CO_3 + 2HCl \longrightarrow 2NaCl + H_2O + CO_2$
Sulfite (SO_3^{2-})	SO_2	$Na_2SO_3 + 2HCl \longrightarrow 2NaCl + H_2O + SO_2$
Sulfide (S^{2-})	H_2S	$Na_2S + H_2SO_4 \longrightarrow Na_2SO_4 + H_2S$

ⓄWL Interactive Example **4.6** **Writing an Equation for a Reaction with Gas Formation**

Critical Concept 4.6
Ionic compounds that contain the carbonate, sulfite, or sulfide anions will produce a gas when reacted with an acid. In addition to the gas, these reactions will produce a salt.

Solution Essentials:
• Acid
• Salt
• Exchange reaction
• Solubility rules
• Carbonate, sulfite, and sulfide anions

Write the molecular equation and the net ionic equation for the reaction of zinc sulfide with hydrochloric acid.

Problem Strategy Start by writing the reactants, using the solubility rules to determine the phase labels. Next think about the ways in which this reaction can occur; first try to write it as an exchange reaction. Examine the products to see whether an insoluble compound forms and/or there is gas formation; either indicates a reaction.

Solution First write the reactants, noting from Table 4.1 that most sulfides (except those of the group IA metals and $(NH_4)_2S$) are insoluble.

$$ZnS(s) + HCl(aq) \longrightarrow$$

If you look at this as an exchange reaction, the products are $ZnCl_2$ and H_2S. Since chlorides are soluble (with some exceptions, not including $ZnCl_2$), we write the following balanced equation.

$$\mathbf{ZnS}(s) + \mathbf{2HCl}(aq) \longrightarrow \mathbf{ZnCl_2}(aq) + \mathbf{H_2S}(g)$$

The complete ionic equation is

$$ZnS(s) + 2H^+(aq) + 2Cl^-(aq) \longrightarrow Zn^{2+}(aq) + 2Cl^-(aq) + H_2S(g)$$

and the net ionic equation is

$$\mathbf{ZnS}(s) + \mathbf{2H^+}(aq) \longrightarrow \mathbf{Zn^{2+}}(aq) + \mathbf{H_2S}(g)$$

Answer Check When you are trying to solve problems of this type, if starting with an exchange reaction doesn't yield the desired results, use the examples in Table 4.4 as a guide.

Exercise 4.7 Write the molecular equation and the net ionic equation for the reaction of calcium carbonate with nitric acid.

■ See Problems 4.51, 4.52, 4.53, and 4.54.

CONCEPT CHECK 4.5

At times, we want to generalize the formula of certain important chemical substances; acids and bases fall into this category. Given the following reactions, try to identify the acids, bases, and some examples of what the general symbols (M and A^-) represent.

a $MOH(s) \longrightarrow M^+(aq) + OH^-(aq)$

b $HA(aq) + H_2O(l) \rightleftharpoons H_3O^+(aq) + A^-(aq)$

c $H_2A(aq) + H_2O(l) \rightleftharpoons H_3O^+(aq) + HA^-(aq)$

d For parts a through c, provide real examples for M and A.

4.5 Oxidation–Reduction Reactions

In the two preceding sections, we described precipitation reactions (reactions producing a precipitate) and acid–base reactions (reactions involving proton transfer). Here we discuss the third major class of reactions, oxidation–reduction reactions, which are reactions involving a transfer of electrons from one species to another.

As a simple example of an oxidation–reduction reaction, let us look at what happens when you dip an iron nail into a blue solution of copper(II) sulfate

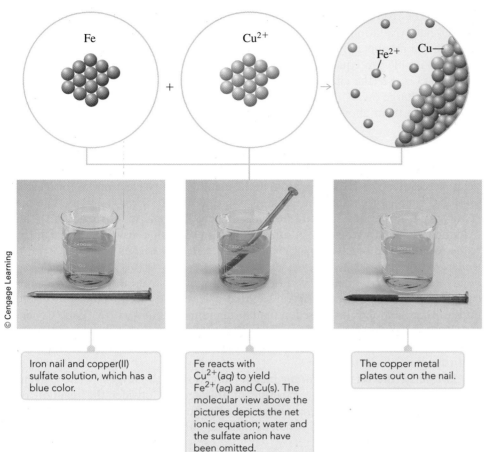

Figure 4.11 ◀

Reaction of iron with Cu²⁺(aq)

Iron nail and copper(II) sulfate solution, which has a blue color.

Fe reacts with $Cu^{2+}(aq)$ to yield $Fe^{2+}(aq)$ and $Cu(s)$. The molecular view above the pictures depicts the net ionic equation; water and the sulfate anion have been omitted.

The copper metal plates out on the nail.

(Figure 4.11). What you see is that the iron nail becomes coated with a reddish-brown tinge of metallic copper. The molecular equation for this reaction is

$$Fe(s) + CuSO_4(aq) \longrightarrow FeSO_4(aq) + Cu(s)$$

The net ionic equation is

$$Fe(s) + Cu^{2+}(aq) \longrightarrow Fe^{2+}(aq) + Cu(s)$$

The electron-transfer aspect of the reaction is apparent from this equation. Note that each iron atom in the metal loses two electrons to form an iron(II) ion, and each copper(II) ion gains two electrons to form a copper atom in the metal. The net effect is that two electrons are transferred from each iron atom in the metal to each copper(II) ion.

The concept of *oxidation numbers* was developed as a simple way of keeping track of electrons in a reaction. Using oxidation numbers, you can determine whether electrons have been transferred from one atom to another. If electrons have been transferred, an oxidation–reduction reaction has occurred.

Oxidation Numbers

We define the **oxidation number** (or **oxidation state**) of an atom in a substance as *the actual charge of the atom if it exists as a monatomic ion, or a hypothetical charge assigned to the atom in the substance by simple rules.* An oxidation–reduction reaction is one in which one or more atoms change oxidation number, implying that there has been a transfer of electrons.

Consider the combustion of calcium metal in oxygen gas (Figure 4.12).

$$2Ca(s) + O_2(g) \longrightarrow 2CaO(s)$$

This is an oxidation–reduction reaction. To see this, you assign oxidation numbers to the atoms in the equation and then note that the atoms change oxidation number during the reaction.

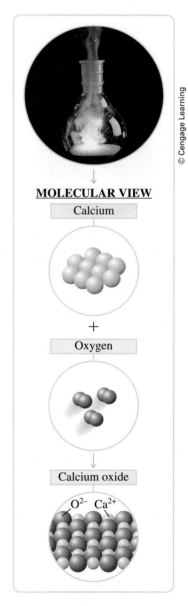

MOLECULAR VIEW

Calcium

+

Oxygen

Calcium oxide

O^{2-} Ca^{2+}

Figure 4.12 ▲

The burning of calcium metal in oxygen The burning calcium emits a red-orange flame.

Since the oxidation number of an atom in an element is always zero, Ca and O in O_2 have oxidation numbers of zero. Another rule follows from the definition of oxidation number: The oxidation number of an atom that exists in a substance as a monatomic ion equals the charge on that ion. So the oxidation number of Ca in CaO is +2 (the charge on Ca^{2+}), and the oxidation number of O in CaO is −2 (the charge on O^{2-}). To emphasize these oxidation numbers in an equation, we will write them above the atomic symbols in the formulas.

$$2\overset{0}{Ca}(s) + \overset{0}{O_2}(g) \longrightarrow 2\overset{+2\,-2}{CaO}(s)$$

From this, you see that the Ca and O atoms change in oxidation number during the reaction. In effect, each calcium atom in the metal loses two electrons to form Ca^{2+} ions, and each oxygen atom in O_2 gains two electrons to form O^{2-} ions. The net result is a transfer of electrons from calcium to oxygen, so this reaction is an oxidation–reduction reaction. In other words, an **oxidation–reduction reaction** (or **redox reaction**) is *a reaction in which electrons are transferred between species or in which atoms change oxidation number.*

Note that calcium has gained in oxidation number from 0 to +2. (Each calcium atom loses two electrons.) We say that calcium has been *oxidized*. Oxygen, on the other hand, has decreased in oxidation number from 0 to −2. (Each oxygen atom gains two electrons.) We say that oxygen has been *reduced*. An oxidation–reduction reaction always involves both oxidation (the loss of electrons) and reduction (the gain of electrons).

Formerly, the term *oxidation* meant "reaction with oxygen." The current definition greatly enlarges the meaning of this term. Consider the reaction of calcium metal with chlorine gas (Figure 4.13); the reaction looks similar to the burning of calcium in oxygen. The chemical equation is

$$\overset{0}{Ca}(s) + \overset{0}{Cl_2}(g) \longrightarrow \overset{+2\,-1}{CaCl_2}(s)$$

In this reaction, the calcium atom is oxidized, because it increases in oxidation number (from 0 to +2, as in the previous equation). Chlorine is reduced; it decreases in oxidation number from 0 to −1. This is clearly an oxidation–reduction reaction that does not involve oxygen.

© Cengage Learning

Figure 4.13 ▲

The burning of calcium metal in chlorine The reaction appears similar to the burning of calcium in oxygen.

Oxidation-Number Rules

So far, we have used two rules for obtaining oxidation numbers: (1) the oxidation number of an atom in an element is zero, and (2) the oxidation number of an atom in a monatomic ion equals the charge on the ion. These and several other rules for assigning oxidation numbers are given in Table 4.5.

Table 4.5	Rules for Assigning Oxidation Numbers	
Rule	**Applies to**	**Statement**
1	Elements	The oxidation number of an atom in an element is zero.
2	Monatomic ions	The oxidation number of an atom in a monatomic ion equals the charge on the ion.
3	Oxygen	The oxidation number of oxygen is −2 in most of its compounds. (An exception is O in H_2O_2 and other peroxides, where the oxidation number is −1.)
4	Hydrogen	The oxidation number of hydrogen is +1 in most of its compounds. (The oxidation number of hydrogen is −1 in binary compounds with a metal, such as CaH_2.)
5	Halogens	The oxidation number of fluorine is −1 in all of its compounds. Each of the other halogens (Cl, Br, I) has an oxidation number of −1 in binary compounds, except when the other element is another halogen above it in the periodic table or the other element is oxygen.
6	Compounds and ions	The sum of the oxidation numbers of the atoms in a compound is zero. The sum of the oxidation numbers of the atoms in a polyatomic ion equals the charge on the ion.

In molecular substances, we use these rules to give the *approximate* charges on the atoms. Consider the molecule SO_2. Oxygen atoms tend to attract electrons, pulling them from other atoms (sulfur in the case of SO_2). As a result, an oxygen atom in SO_2 takes on a negative charge relative to the sulfur atom. The magnitude of the charge on an oxygen atom in a molecule is not a full -2 charge as in the O^{2-} ion. However, it is convenient to assign an oxidation number of -2 to oxygen in SO_2 (and in most other compounds of oxygen) to help us express the approximate charge distribution in the molecule. Rule 3 in Table 4.5 says that an oxygen atom has an oxidation number of -2 in most of its compounds.

Rules 4 and 5 are similar in that they tell you what to expect for the oxidation number of certain elements in their compounds. Rule 4, for instance, says that hydrogen has an oxidation number of $+1$ in most of its compounds.

Rule 6 states that the sum of the oxidation numbers of the atoms in a compound is zero. This rule follows from the interpretation of oxidation numbers as (hypothetical) charges on the atoms. Because any compound is electrically neutral, the sum of the charges on its atoms must be zero. This rule is easily extended to ions: the sum of the oxidation numbers (hypothetical charges) of the atoms in a polyatomic ion equals the charge on the ion.

You can use Rule 6 to obtain the oxidation number of one atom in a compound or ion, if you know the oxidation numbers of the other atoms in the compound or ion. Consider the SO_2 molecule. According to Rule 6,

$$\text{(Oxidation number of S)} + 2 \times \text{(oxidation number of O)} = 0$$

or

$$\text{(Oxidation number of S)} + 2 \times (-2) = 0.$$

Therefore,

$$\text{Oxidation number of S (in } SO_2) = -2 \times (-2) = +4$$

The next example illustrates how to use the rules in Table 4.5 to assign oxidation numbers.

Example 4.7

Assigning Oxidation Numbers

Critical Concept 4.7
The sum of all the oxidation numbers in a compound or polyatomic ion is equal to the overall charge on the compound or ion. This concept is very helpful in assigning the oxidation number of elements that are not covered by the rules (Table 4.5).

Solution Essentials:
• Oxidation number
• Rules for assigning oxidation numbers (Table 4.5)

Use the rules from Table 4.5 to obtain the oxidation number of the chlorine atom in each of the following: $HClO_4$ (perchloric acid), b ClO_3^- (chlorate ion).

Problem Strategy We need to apply the rules in Table 4.5. Rule 6 is the best place to start. Therefore, in each case, write the expression for the sum of the oxidation numbers, equating this to zero for a compound or to the charge for an ion. Now, use Rules 2 to 5 to substitute oxidation numbers for particular atoms, such as -2 for oxygen and $+1$ for hydrogen, and solve for the unknown oxidation number (Cl in this example).

Solution
a For perchloric acid, Rule 6 gives the equation

$$\text{(Oxidation number of H)} + \text{(oxidation number of Cl)} + 4 \times \text{(oxidation number of O)} = 0$$

Using Rules 3 and 4, you obtain

$$(+1) + \text{(oxidation number of Cl)} + 4 \times (-2) = 0$$

(continued)

(continued)

Therefore,

$$\text{Oxidation number of Cl (in HClO}_4) = -(+1) - 4 \times (-2) = \textbf{+7}$$

b For the chlorate ion, Rule 6 gives the equation

$$(\text{Oxidation number of Cl}) + 3 \times (\text{oxidation number of O}) = -1$$

Using Rule 3, you obtain

$$(\text{Oxidation number of Cl}) + 3 \times (-2) = -1$$

Therefore,

$$\text{Oxidation number of Cl (in ClO}_3^-) = -1 - 3 \times (-2) = \textbf{+5}$$

Answer Check Because most compounds do not have elements with very large positive or very large negative oxidation numbers, you should always be on the alert for a *possible* assignment mistake when you find oxidation states greater than +6 or less than −4. (From this example you see that a +7 oxidation state is possible; however, it occurs in only a limited number of cases.)

Exercise 4.8   Obtain the oxidation numbers of the atoms in each of the following: **a** potassium dichromate, $K_2Cr_2O_7$, **b** permanganate ion, MnO_4^-.

■ See Problems 4.55, 4.56, 4.57, and 4.58.

Describing Oxidation–Reduction Reactions

We use special terminology to describe oxidation–reduction reactions. To illustrate this, we will look again at the reaction of iron with copper(II) sulfate. The net ionic equation is

$$\overset{0}{\text{Fe}}(s) + \overset{+2}{\text{Cu}^{2+}}(aq) \longrightarrow \overset{+2}{\text{Fe}^{2+}}(aq) + \overset{0}{\text{Cu}}(s)$$

We can write this reaction in terms of two half-reactions. A **half-reaction** is *one of two parts of an oxidation–reduction reaction, one part of which involves a loss of electrons (or increase of oxidation number) and the other a gain of electrons (or decrease of oxidation number).* The half-reactions for the preceding equation are

$$\overset{0}{\text{Fe}}(s) \longrightarrow \overset{+2}{\text{Fe}^{2+}}(aq) + 2e^- \qquad \text{(electrons lost by Fe)}$$
$$\overset{+2}{\text{Cu}^{2+}}(aq) + 2e^- \longrightarrow \overset{0}{\text{Cu}}(s) \qquad \text{(electrons gained by Cu}^{2+})$$

Oxidation is *the half-reaction in which there is a loss of electrons by a species (or an increase of oxidation number of an atom).* **Reduction** is *the half-reaction in which there is a gain of electrons by a species (or a decrease in the oxidation number of an atom).* Thus, the equation $\text{Fe}(s) \longrightarrow \text{Fe}^{2+}(aq) + 2e^-$ represents the oxidation half-reaction, and the equation $\text{Cu}^{2+}(aq) + 2e^- \longrightarrow \text{Cu}(s)$ represents the reduction half-reaction.

Recall that a species that is *oxidized* loses electrons (or contains an atom that increases in oxidation number) and a species that is *reduced* gains electrons (or contains an atom that decreases in oxidation number). An **oxidizing agent** is *a species that oxidizes another species; it is itself reduced.* Similarly, a **reducing agent** is *a species that reduces another species; it is itself oxidized.* In our example reaction, the copper(II) ion is the oxidizing agent, whereas iron metal is the reducing agent.

The relationships among these terms are shown in the following diagram for the reaction of iron with copper(II) ion.

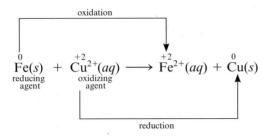

Some Common Oxidation–Reduction Reactions

Many oxidation–reduction reactions can be described as one of the following:

1. Combination reaction
2. Decomposition reaction
3. Displacement reaction
4. Combustion reaction

We will describe examples of each of these in this section.

Combination Reactions A **combination reaction** is *a reaction in which two substances combine to form a third substance.* Note that not all combination reactions are oxidation–reduction reactions. However, the simplest cases are those in which two elements react to form a compound; these are clearly oxidation–reduction reactions. In Chapter 2, we discussed the reaction of sodium metal and chlorine gas, which is a redox reaction (Figure 4.14).

$$2Na(s) + Cl_2(g) \longrightarrow 2NaCl(s)$$

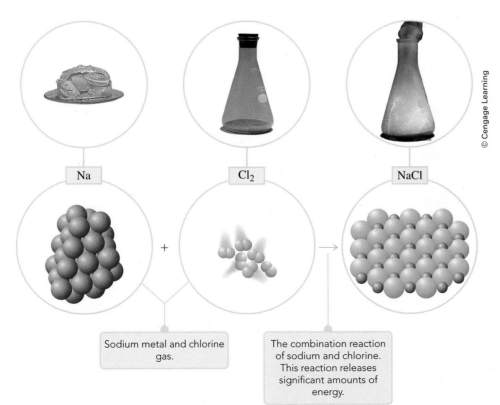

Figure 4.14 ◄
Combination reaction

© Cengage Learning

Na

Cl$_2$

NaCl

+

Sodium metal and chlorine gas.

The combination reaction of sodium and chlorine. This reaction releases significant amounts of energy.

Figure 4.15 ▲

Decomposition reaction The decomposition reaction of mercury(II) oxide into its elements, mercury and oxygen.

Antimony and chlorine also combine in a fiery reaction.

$$2Sb + 3Cl_2 \longrightarrow 2SbCl_3$$

Some combination reactions involve compounds as reactants and are not oxidation–reduction reactions. For example,

$$CaO(s) + SO_2(g) \longrightarrow CaSO_3(s)$$

(If you check the oxidation numbers, you will see that this is not an oxidation–reduction reaction.)

Decomposition Reactions   A **decomposition reaction** is *a reaction in which a single compound reacts to give two or more substances.* Often these reactions occur when the temperature is raised. In Chapter 1, we described the decomposition of mercury(II) oxide into its elements when the compound is heated (Figure 4.15). This is an oxidation–reduction reaction.

$$2HgO(s) \xrightarrow{\Delta} 2Hg(l) + O_2(g)$$

Another example is the preparation of oxygen by heating potassium chlorate with manganese(IV) oxide as a catalyst.

$$2KClO_3(s) \xrightarrow[MnO_2]{\Delta} 2KCl(s) + 3O_2(g)$$

In this reaction, a compound decomposes into another compound and an element; it also is an oxidation–reduction reaction.

Not all decomposition reactions are of the oxidation–reduction type. For example, calcium carbonate at high temperatures decomposes into calcium oxide and carbon dioxide.

$$CaCO_3(s) \xrightarrow{\Delta} CaO(s) + CO_2(g)$$

Is there a change in oxidation numbers? If not, this confirms that this is not an oxidation–reduction reaction.

Displacement Reactions   A **displacement reaction** (also called a **single-replacement reaction**) is *a reaction in which an element reacts with a compound, displacing another element from it.* Since these reactions involve an element and one of its compounds, these must be oxidation–reduction reactions. An example is the reaction that occurs when you dip a copper metal strip into a solution of silver nitrate.

$$Cu(s) + 2AgNO_3(aq) \longrightarrow Cu(NO_3)_2(aq) + 2Ag(s)$$

From the molecular equation, it appears that copper displaces silver in silver nitrate, producing crystals of silver metal and a greenish-blue solution of copper(II) nitrate. The net ionic equation, however, shows that the reaction involves the transfer of electrons from copper metal to silver ion:

$$Cu(s) + 2Ag^+(aq) \longrightarrow Cu^{2+}(aq) + 2Ag(s)$$

When you dip a zinc metal strip into an acid solution, bubbles of hydrogen form on the metal and escape from the solution (Figure 4.16).

$$Zn(s) + 2HCl(aq) \longrightarrow ZnCl_2(aq) + H_2(g)$$

Figure 4.16 ▲

Displacement reaction Displacement reaction of zinc metal and hydrochloric acid. Hydrogen gas formed in the reaction bubbles from the submerged metal surface.

Table 4.6	Activity Series of the Elements
React vigorously with acidic solutions and water to give H_2	Li, K, Ba, Ca, Na
React with acids to give H_2	Mg, Al, Zn, Cr, Fe, Cd, Co, Ni, Sn, Pb
Do not react with acids to give H_2*	H_2, Cu, Hg, Ag, Au

*Cu, Hg, and Ag react with HNO_3 but do not produce H_2. In these reactions, the metal is oxidized to the metal ion, and NO_3^- ion is reduced to NO_2 or other nitrogen species.

Zinc displaces hydrogen in the acid, producing zinc chloride solution and hydrogen gas. The net ionic equation is

$$Zn(s) + 2H^+(aq) \rightarrow Zn^{2+}(aq) + H_2(g)$$

Whether a reaction occurs between a given element and a monatomic ion depends on the relative ease with which the two species gain or lose electrons. Table 4.6 shows the activity series of the elements, a listing of the elements in decreasing order of their ease of losing electrons during reactions in aqueous solution. The metals listed at the top are the strongest reducing agents (they lose electrons easily); those at the bottom, the weakest. A free element reacts with the monatomic ion of another element if the free element is above the other element in the activity series. The highlighted elements react slowly with liquid water, but readily with steam, to give H_2.

Consider this reaction:

$$2K(s) + 2H^+(aq) \longrightarrow 2K^+(aq) + H_2(g)$$

You would expect this reaction to proceed as written, because potassium metal (K) is well above hydrogen in the activity series. In fact, potassium metal reacts violently with water, which contains only a very small percentage of H^+ ions. Imagine the reaction of potassium metal with a strong acid like HCl!

Combustion Reactions A **combustion reaction** is *a reaction in which a substance reacts with oxygen, usually with the rapid release of heat to produce a flame.* The products include one or more oxides. Oxygen changes oxidation number from 0 to −2, so combustions are oxidation–reduction reactions.

Organic compounds usually burn in oxygen or air to yield carbon dioxide. If the compound contains hydrogen (as most do), water is also a product. For instance, butane (C_4H_{10}) burns in air as follows:

$$2C_4H_{10}(g) + 13O_2(g) \longrightarrow 8CO_2(g) + 10H_2O(g)$$

Many metals burn in air as well. Although chunks of iron do not burn readily in air, iron wool, which consists of fine strands of iron, does (Figure 4.17). The increased surface area of the metal in iron wool allows oxygen from air to react quickly with it.

$$4Fe(s) + 3O_2(g) \longrightarrow 2Fe_2O_3(s)$$

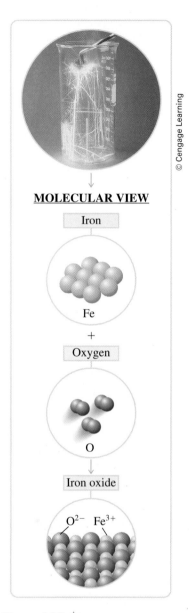

Figure 4.17 ▲

The combustion of iron wool Iron reacts with oxygen in the air to produce iron(III) oxide, Fe_2O_3. The reaction is similar to the rusting of iron but is much faster.

4.6 Balancing Simple Oxidation–Reduction Equations

Oxidation–reduction reactions can often be quite difficult to balance. ▶ Some are so complex in fact that chemists have written computer programs to accomplish the

Balancing equations was discussed in Section 2.10.

task. In this section, we will develop a method for balancing simple oxidation–reduction reactions that can later be generalized for far more complex reactions. One of the advantages of using this technique for even simple reactions is that you focus on what makes oxidation–reduction reactions different from other reaction types. See Chapter 19 for a comprehensive treatment of this topic.

At first glance, the equation representing the reaction of zinc metal with silver(I) ions in solution might appear to be balanced.

$$Zn(s) + Ag^+(aq) \longrightarrow Zn^{2+}(aq) + Ag(s)$$

However, because a balanced chemical equation must have a charge balance as well as a mass balance, this equation is not balanced: it has a total charge of $+1$ for the reactants and $+2$ for the products. Let us apply the half-reaction method for balancing this equation.

Half-Reaction Method Applied to Simple Oxidation–Reduction Equations

The *half-reaction method* consists of first separating the equation into two half-reactions, one for oxidation, the other for reduction. You balance each half-reaction, then combine them to obtain a balanced oxidation–reduction reaction. Here is an illustration of the process. First we identify the species being oxidized and reduced and assign the appropriate oxidation states.

$$\overset{0}{Zn}(s) + \overset{+1}{Ag}^+(aq) \longrightarrow \overset{+2}{Zn}^{2+}(aq) + \overset{0}{Ag}(s)$$

Next, write the half-reactions in an unbalanced form.

$$Zn \longrightarrow Zn^{2+} \qquad \text{(oxidation)}$$
$$Ag^+ \longrightarrow Ag \qquad \text{(reduction)}$$

Next, balance the charge in each equation by adding electrons to the more positive side to create balanced half-reactions. Following this procedure, the balanced half-reactions are:

$$Zn \longrightarrow Zn^{2+} + 2e^- \qquad \text{(oxidation half-reaction)}$$
$$Ag^+ + e^- \longrightarrow Ag \qquad \text{(reduction half-reaction)}$$

Note that the number of electrons that Zn loses during the oxidation process (two) exceeds the number of electrons gained by Ag^+ during the reduction (one). Since, according to the reduction half-reaction, each Ag^+ is capable of gaining only one electron, we need to double the amount of Ag^+ in order for it to accept all of the electrons produced by Zn during oxidation. To meet this goal and obtain the balanced oxidation–reduction reaction, we multiply each half-reaction by a factor (integer) so that when we add them together, the electrons cancel. We multiply the first equation by 1 (the number of electrons in the second half-reaction) and multiply the second equation by 2 (the number of electrons in the first half-reaction).

$$1 \times (Zn \longrightarrow Zn^{2+} + 2e^-)$$

$$Zn + 2Ag^+ + 2e^- -$$

The electrons cancel, which finally yields the balanced oxidation–reduction equation:

$$Zn(s) + 2Ag^+(aq) \longrightarrow Zn^{2+}(aq) + 2Ag(s)$$

Example 4.8 further illustrates this technique.

Example 4.8

Gaining Mastery Toolbox

Critical Concept 4.8
A balanced oxidation-reduction reaction has both mass balance and charge balance. Using the half-reaction method of balancing oxidation-reduction reactions assures that there is charge balance.

Solution Essentials:
- Half-reaction (oxidation and reduction)
- Oxidation
- Reduction
- Oxidation-reduction reaction
- Rules for assigning oxidation numbers (Table 4.5)

Balancing Simple Oxidation–Reduction Reactions by the Half-Reaction Method

Consider a more difficult problem, the combination (oxidation–reduction) reaction of magnesium metal and nitrogen gas:

$$Mg(s) + N_2(g) \longrightarrow Mg_3N_2(s)$$

Apply the half-reaction method to balance this equation.

Problem Strategy Start by identifying the species undergoing oxidation and reduction and assigning oxidation numbers. Then write the two balanced half-reactions, keeping in mind that you add the electrons to the more positive side of the reaction. Next, multiply each of the half-reactions by a whole number that will cancel the electrons on each side of the equation. Finally, add the half-reactions together to yield the balanced equation.

Solution Identify the oxidation states of the elements:

$$\overset{0}{Mg}(s) + \overset{0}{N_2}(g) \longrightarrow \overset{+2\;-3}{Mg_3N_2}(s)$$

In this problem, a molecular compound, nitrogen (N_2), is undergoing reduction. When a species undergoing reduction or oxidation is a molecule, write the formula of the molecule (do not split it up) in the half-reaction. Also, make sure that both the mass and the charge are balanced. (Note the $6e^-$ required to balance the charge due to the $2N^{3-}$.)

$$Mg \longrightarrow Mg^{2+} + 2e^- \qquad \text{(balanced oxidation half-reaction)}$$
$$N_2 + 6e^- \longrightarrow 2N^{3-} \qquad \text{(balanced reduction half-reaction)}$$

We now need to multiply each half-reaction by a factor that will cancel the electrons.

$$3 \times (Mg \longrightarrow Mg^{2+} + 2e^-)$$
$$\underline{1 \times (N_2 + 6e^- \longrightarrow 2N^{3-})}$$
$$3Mg + N_2 + \cancel{6e^-} \longrightarrow 3Mg^{2+} + 2N^{3-} + \cancel{6e^-}$$

Therefore, the balanced combination (oxidation–reduction) reaction is

$$\mathbf{3Mg + N_2 \longrightarrow 3Mg^{2+} + 2N^{3-}}$$

From inspecting the coefficients in this reaction, looking at the original equation, and knowing that Mg^{2+} and N^{3-} will combine to form an ionic compound (Mg_3N_2), we can rewrite the equation in the following form:

$$3Mg(s) + N_2(g) \longrightarrow Mg_3N_2(s)$$

Answer Check Do a final check for balance by counting the number of atoms of each element on both sides of the equation.

Exercise 4.9 Use the half-reaction method to balance the equation $Ca(s) + Cl_2(g) \longrightarrow CaCl_2(s)$.

■ See Problems 4.65 and 4.66.

Working with Solutions

The majority of the chemical reactions discussed in this chapter take place in solution. This is because the reaction between two solid reactants often proceeds very slowly or not at all. In a solid, the molecules or ions in a crystal tend to occupy approximately fixed positions, so the chance of two molecules or ions coming together to react is small. In liquid solutions, reactant molecules are free to move throughout the liquid; therefore, reaction is much faster. When you run reactions in liquid solutions, it is convenient to dispense the amounts of reactants by measuring out volumes of reactant solutions. In the next two sections, we will discuss calculations involved in making up solutions, and in Section 4.10 we will describe stoichiometric calculations involving such solutions.

4.7 Molar Concentration

When we dissolve a substance in a liquid, we call the substance the *solute* and the liquid the *solvent.* Consider ammonia solutions. Ammonia gas dissolves readily in water, and aqueous ammonia solutions are often used in the laboratory. In such solutions, ammonia gas is the solute and water is the solvent.

The general term *concentration* refers to the quantity of solute in a standard quantity of solution. Qualitatively, we say that a solution is *dilute* when the solute concentration is low and *concentrated* when the solute concentration is high. Usually these terms are used in a comparative sense and do not refer to a specific concentration. We say that one solution is more dilute, or less concentrated, than another. However, for commercially available solutions, the term *concentrated* refers to the maximum, or near maximum, concentration available. For example, concentrated aqueous ammonia contains about 28% NH_3 by mass.

In this example, we expressed the concentration quantitatively by giving the mass percentage of solute—that is, the mass of solute in 100 g of solution. However, we need a unit of concentration that is convenient for dispensing reactants in solution, such as one that specifies moles of solute per solution volume. **Molar concentration**, or **molarity (*M*)**, is defined as *the moles of solute dissolved in one liter (cubic decimeter) of solution.*

$$\text{Molarity } (M) = \frac{\text{moles of solute}}{\text{liters of solution}}$$

An aqueous solution that is 0.15 *M* NH_3 (read this as "0.15 molar NH_3") contains 0.15 mol NH_3 per liter of solution. If you want to prepare a solution that is, for example, 0.200 *M* $CuSO_4$, you place 0.200 mol $CuSO_4$ in a 1.000-L volumetric flask, or a proportional amount in a flask of a different size (Figure 4.18). You add a small quantity of water to dissolve the $CuSO_4$. Then you fill the flask with additional water to the mark on the neck and mix the solution. The following example shows how to calculate the molarity of a solution given the mass of solute and the volume of solution.

This is the last instance where dimensional analysis and the rules for significant figures and rounding will be listed in the Solution Essentials. Any time that you perform a calculation and report an answer, you must consider the rules for rounding and significant figures. Dimensional analysis is a powerful tool that is applicable to most problem solving in which molar quantities are involved, as they are in this example.

Example 4.9   **Calculating Molarity from Mass and Volume**

Critical Concept 4.9
Molarity is a common way for chemists to express solution concentration. Molarity is a ratio of the moles of dissolved substance (solute) to the volume of solution in liters: mol/L. The concentration of a solution is always independent of the amount of solution. ▲

Solution Essentials:
- Molarity (molar concentration)
- Mole
- Molar mass
- Dimensional analysis
- Rules for significant figures and rounding

A sample of $NaNO_3$ weighing 0.38 g is placed in a 50.0 mL volumetric flask. The flask is then filled with water to the mark on the neck, dissolving the solid. What is the molarity of the resulting solution?

Problem Strategy To calculate the molarity, you need the moles of solute. Therefore, you first convert grams $NaNO_3$ to moles. The molarity equals the moles of solute divided by the liters of solution.

(continued)

0.0500 mol CuSO₄·5H₂O (12.48 g) is weighed on a platform balance.

The copper(II) sulfate pentahydrate is transferred carefully to the volumetric flask.

Water is added to bring the solution level to the mark on the neck of the 250-mL volumetric flask. The molarity is 0.0500 mol/0.250 L = 0.200 *M*.

© Cengage Learning

Figure 4.18 ▲

Preparing a 0.200 *M* CuSO₄ solution

(*continued*)

Solution You find that 0.38 g NaNO₃ is $4.\underline{47} \times 10^{-3}$ mol NaNO₃; the last significant figure is underlined. The volume of solution is 50.0 mL, or 50.0×10^{-3} L, so the molarity is

$$\text{Molarity} = \frac{4.\underline{47} \times 10^{-3} \text{ mol NaNO}_3}{50.0 \times 10^{-3} \text{ L soln}} = \textbf{0.089 } \textbf{\textit{M}} \textbf{ NaNO}_3$$

Answer Check Although very dilute solutions are possible, there is a limit as to how concentrated solutions can be. Therefore, any answer that leads to solution concentrations that are in excess of 20 *M* should be suspect.

Exercise 4.10 A sample of sodium chloride, NaCl, weighing 0.0678 g is placed in a 25.0-mL volumetric flask. Enough water is added to dissolve the NaCl, and then the flask is filled to the mark with water and carefully shaken to mix the contents. What is the molarity of the resulting solution?

■ See Problems 4.67, 4.68, 4.69, and 4.70.

The advantage of molarity as a concentration unit is that the amount of solute is related to the volume of solution. Rather than having to weigh out a specified mass of substance, you can instead measure out a definite volume of solution of the substance, which is usually easier. As the following example illustrates, molarity can be used as a factor for converting from moles of solute to liters of solution, and vice versa.

Example 4.10   **Using Molarity as a Conversion Factor**

Gaining Mastery Toolbox

Critical Concept 4.10
Molarity is a conversion factor. As a conversion factor it can be used to convert between moles of solute and the volume of solution or from volume of solution to moles of solute.

Solution Essentials:
• Molarity (molar concentration)
• Mole
• Molar mass
• Conversion factor

An experiment calls for the addition to a reaction vessel of 0.184 g of sodium hydroxide, NaOH, in aqueous solution. How many milliliters of 0.150 *M* NaOH should be added?

Problem Strategy Because molarity relates moles of solute to volume of solution, you first need to convert grams of NaOH to moles of NaOH. Then, you can convert moles NaOH to liters of solution, using the molarity as a conversion factor. Here, 0.150 *M* means that 1 L of solution contains 0.150 moles of solute, so the conversion factor is

$$\frac{1 \text{ L soln}}{0.150 \text{ mol NaOH}}$$

Converts
mol NaOH to L soln

(*continued*)

(continued)

Solution Here is the calculation. (The molar mass of NaOH is 40.0 g/mol.)

$$0.184 \text{ g NaOH} \times \frac{1 \text{ mol NaOH}}{40.0 \text{ g NaOH}} \times \frac{1 \text{ L soln}}{0.150 \text{ mol NaOH}} = 3.07 \times 10^{-2} \text{ L soln (or 30.7 mL)}$$

You need to add **30.7 mL** of 0.150 *M* NaOH solution to the reaction vessel.

Answer Check Note that the mass of NaOH required for the experiment is relatively small. Because of this, you should not expect that a large volume of the 0.150 *M* NaOH solution will be required.

Exercise 4.11 How many milliliters of 0.163 *M* NaCl are required to give 0.0958 g of sodium chloride?

■ See Problems 4.71, 4.72, 4.73, and 4.74.

Exercise 4.12 How many moles of sodium chloride should be put in a 50.0-mL volumetric flask to give a 0.15 *M* NaCl solution when the flask is filled to the mark with water? How many grams of NaCl is this?

■ See Problems 4.75, 4.76, 4.77, and 4.78.

4.8 Diluting Solutions

Commercially available aqueous ammonia (28.0% NH_3) is 14.8 *M* NH_3. Suppose, however, that you want a solution that is 1.00 *M* NH_3. You need to dilute the concentrated solution with a definite quantity of water. For this purpose, you must know the relationship between the molarity of the solution before dilution (the *initial molarity*) and that after dilution (the *final molarity*).

To obtain this relationship, first recall the equation defining molarity:

$$\text{Molarity} = \frac{\text{moles of solute}}{\text{liters of solution}}$$

You can rearrange this to give

$$\text{Moles of solute} = \text{molarity} \times \text{liters of solution}$$

The product of molarity and the volume (in liters) gives the moles of solute in the solution. Writing M_i for the initial molar concentration and V_i for the initial volume of solution, you get

$$\text{Moles of solute} = M_i \times V_i$$

When the solution is diluted by adding more water, the concentration and volume change to M_f (the final molar concentration) and V_f (the final volume), and the moles of solute equals

$$\text{Moles of solute} = M_f \times V_f$$

Because the moles of solute has not changed during the dilution (Figure 4.19),

$$M_i \times V_i = M_f \times V_f$$

(Note: You can use any volume units, but both V_i and V_f must be in the same units.)

The next example illustrates how you can use this formula to find the volume of a concentrated solution needed to prepare a given volume of dilute solution.

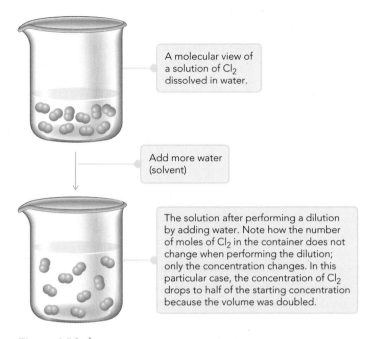

A molecular view of a solution of Cl_2 dissolved in water.

Add more water (solvent)

The solution after performing a dilution by adding water. Note how the number of moles of Cl_2 in the container does not change when performing the dilution; only the concentration changes. In this particular case, the concentration of Cl_2 drops to half of the starting concentration because the volume was doubled.

Figure 4.19 ▲

Molecular view of the dilution process

© Cengage Learning

Figure 4.20 ▲

Preparing a solution by diluting a concentrated one A volume of concentrated ammonia, similar to the one in the beaker, has been added to the volumetric flask. Here water is being added from the plastic squeeze bottle to bring the volume up to the mark on the flask.

Example 4.11   Diluting a Solution

Critical Concept 4.11
The product of a solution's molar concentration and the volume of the solution gives the moles of solute in a solution. This relationship can be expressed mathematically as $(M_{solute})(V_{solute}) = mol_{solute}$. This relationship can be applied to derive the dilution equation $(M_i \times V_i = M_f \times V_f)$.

Solution Essentials:
• Dilution
• Dilution equation $(M_i \times V_i = M_f \times V_f)$
• Molarity

You are given a solution of 14.8 M NH_3. How many milliliters of this solution do you require to give 100.0 mL of 1.00 M NH_3 when diluted (Figure 4.20)?

Problem Strategy Because this problem is a dilution, you can use the dilution formula. In this case you want to know the initial volume (V_i) of the solution that you are going to dilute; all of the other quantities needed for the formula are given in the problem.

Solution You know the final volume (100.0 mL), final concentration (1.00 M), and initial concentration (14.8 M). You write the dilution formula and rearrange it to give the initial volume.

$$M_i V_i = M_f V_f$$

$$V_i = \frac{M_f V_f}{M_i}$$

Now you substitute the known values into the right side of the equation.

$$V_i = \frac{1.00\ M \times 100.0\ \text{mL}}{14.8\ M} = \textbf{6.76 mL}$$

Answer Check When you perform a dilution, the volume of the more concentrated solution should always be less than the volume of the final solution, as it is in the photo. Related to this concept is the fact that the initial concentration of a solution is always greater than the final concentration after dilution. These two concepts will enable you to check the reasonableness of your calculated quantities whenever you use the dilution equation.

Exercise 4.13 You have a solution that is 1.5 M H_2SO_4 (sulfuric acid). How many milliliters of this acid do you need to prepare 100.0 mL of 0.18 M H_2SO_4?

■ See Problems 4.79 and 4.80.

CONCEPT CHECK 4.6

Consider the following beakers. Each contains a solution of the hypothetical atom X.

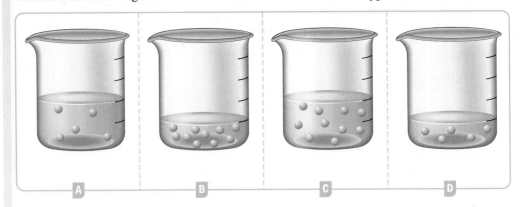

a Arrange the beakers in order of increasing concentration of X.

b Without adding or removing X, what specific things could you do to make the concentrations of X equal in each beaker? (*Hint:* Think about dilutions.)

Quantitative Analysis

Analytical chemistry deals with the determination of composition of materials—that is, the analysis of materials. The materials that one might analyze include air, water, food, hair, body fluids, pharmaceutical preparations, and so forth. The analysis of materials is divided into qualitative and quantitative analysis. *Qualitative analysis* involves the identification of substances or species present in a material. For instance, you might determine that a sample of water contains lead(II) ion. **Quantitative analysis**, which we will discuss in the last sections of this chapter, involves *the determination of the amount of a substance or species present in a material.* In a quantitative analysis, you might determine that the amount of lead(II) ion in a sample of water is 0.067 mg/L.

4.9 Gravimetric Analysis

Gravimetric analysis is *a type of quantitative analysis in which the amount of a species in a material is determined by converting the species to a product that can be isolated completely and weighed.* Precipitation reactions are frequently used in gravimetric analyses. In these reactions, you determine the amount of an ionic species by precipitating it from solution. The precipitate, or solid formed in the reaction, is then filtered from the solution, dried, and weighed. The advantages of a gravimetric analysis are its simplicity (at least in theory) and its accuracy. The chief disadvantage is that it requires meticulous, time-consuming work. Because of this, whenever possible, chemists use modern instrumental methods.

As an example of a gravimetric analysis, consider the problem of determining the amount of lead in a sample of drinking water. Lead, if it occurs in the water, probably exists as the lead(II) ion, Pb^{2+}. Lead(II) sulfate is a very insoluble compound of lead(II) ion. When sodium sulfate, Na_2SO_4, is added to a solution containing Pb^{2+}, lead(II) sulfate precipitates (that is, $PbSO_4$ comes out of the solution as a fine, crystalline solid). If you assume that the lead is present in solution as lead(II) nitrate, you can write the following equation for the reaction:

$$Na_2SO_4(aq) + Pb(NO_3)_2(aq) \longrightarrow 2NaNO_3(aq) + PbSO_4(s)$$

You can separate the white precipitate of lead(II) sulfate from the solution by filtration. Then you dry and weigh the precipitate. Figure 4.21 shows the laboratory setup used in a similar analysis.

Figure 4.21 ◀

Gravimetric analysis for barium ion

© Cengage Learning

a A solution of potassium chromate (yellow) is poured down a stirring rod into a solution containing an unknown amount of barium ion, Ba^{2+}. The yellow precipitate that forms is barium chromate, $BaCrO_4$.

b The solution is filtered by pouring it into a crucible containing a porous glass partition. Afterward, the crucible is heated to dry the barium chromate. By weighing the crucible before and afterward, you can determine the mass of precipitate.

⚭WL Interactive Example 4.12 Determining the Amount of a Species by Gravimetric Analysis

Gaining Mastery Toolbox

Critical Concept 4.12
The precipitate will contain the species (element, ion, etc.) of interest. If you know the mass and chemical formula of the precipitate, you can determine the mass of each element in the precipitate as outlined in Example 3.7.

Solution Essentials:
• Mass percentage
• Molar mass
• Example 3.7

A 1.000-L sample of polluted water was analyzed for lead(II) ion, Pb^{2+}, by adding an excess of sodium sulfate to it. The mass of lead(II) sulfate that precipitated was 229.8 mg. What is the mass of lead in a liter of the water? Give the answer as milligrams of lead per liter of solution.

Problem Strategy Because an excess of sodium sulfate was added to the solution, you can expect that all of the lead is precipitated as lead(II) sulfate, $PbSO_4$. If you determine the percentage of lead in the $PbSO_4$ precipitate, you can calculate the quantity of lead in the water sample.

Solution Following Example 3.7, you obtain the mass percentage of Pb in $PbSO_4$ by dividing the molar mass of Pb by the molar mass of $PbSO_4$, then multiplying by 100%:

$$\% \, Pb = \frac{207.2 \text{ g/mol}}{303.3 \text{ g/mol}} \times 100\% = 68.32\%$$

Therefore, the 1.000-L sample of water contains

Amount Pb in sample = 229.8 mg $PbSO_4$ × 0.6832 = 157.0 mg Pb

The water sample contains **157.0 mg Pb per liter**.

Answer Check Check to make sure that the mass of the element of interest, 157.0 mg Pb in this case, is less than the total mass of the precipitate, 229.8 mg in this case. If you do not find this to be true, you have made an error.

Exercise 4.14 You are given a sample of limestone, which is mostly $CaCO_3$, to determine the mass percentage of Ca in the rock. You dissolve the limestone in hydrochloric acid, which gives a solution of calcium chloride. Then you precipitate the calcium ion in solution by adding sodium oxalate, $Na_2C_2O_4$. The precipitate is calcium oxalate, CaC_2O_4. You find that a sample of limestone weighing 128.3 mg gives 140.2 mg of CaC_2O_4. What is the mass percentage of calcium in the limestone?

■ See Problems 4.83 and 4.84.

4.10 Volumetric Analysis

As you saw earlier, you can use molarity as a conversion factor, and in this way you can calculate the volume of solution that is equivalent to a given mass of solute (see Example 4.10). This means that you can replace mass measurements in solution reactions by volume measurements. In the next example, we look at the volumes of solutions involved in a given reaction.

♨WL Interactive Example 4.13 **Calculating the Volume of Reactant Solution Needed**

Gaining Mastery Toolbox

Critical Concept 4.13
You need to start with a balanced chemical equation. From the balanced chemical equation, you can determine the molar relationship (conversion factor) between the reacting quantities. This conversion factor is critical for correctly solving this type of problem. Do *not* use the dilution equation ($M_i \times V_i = M_f \times V_f$) to solve problems that involve chemical reactions in solution.

Solution Essentials:
• Molarity (molar concentration)
• Molar interpretation of a balanced chemical equation.
• Stoichiometry

Consider the reaction of sulfuric acid, H_2SO_4, with sodium hydroxide, NaOH.

$$H_2SO_4(aq) + 2NaOH(aq) \longrightarrow 2H_2O(l) + Na_2SO_4(aq)$$

Suppose a beaker contains 35.0 mL of 0.175 M H_2SO_4. How many milliliters of 0.250 M NaOH must be added to react completely with the sulfuric acid?

Problem Strategy This is a stoichiometry problem that involves the reaction of sulfuric acid and sodium hydroxide. You have a known volume and concentration of H_2SO_4 in the beaker, and you want to determine what volume of a known concentration of NaOH is required for complete reaction. If you can determine the number of moles of H_2SO_4 that are contained in the beaker, you can then use the balanced chemical reaction to determine the number of moles of NaOH required to react completely with the H_2SO_4. Finally, you can use the concentration of the NaOH to determine the required volume of NaOH. Following this strategy, you convert from 35.0 mL (or 35.0×10^{-3} L) H_2SO_4 solution to moles H_2SO_4 (using the molarity of H_2SO_4), then to moles NaOH (from the chemical equation). Finally, you convert this to volume of NaOH solution (using the molarity of NaOH).

The problem strategy is diagrammed as

Start

$$\boxed{\text{L } H_2SO_4} \Longrightarrow \boxed{\text{mol } H_2SO_4} \Longrightarrow \boxed{\text{mol NaOH}} \Longrightarrow \boxed{\text{L NaOH}}$$

Goal

Solution The calculation is as follows:

$$35.0 \times 10^{-3} \text{ L } H_2SO_4 \text{ soln} \times \frac{0.175 \text{ mol } H_2SO_4}{1 \text{ L } H_2SO_4 \text{ soln}} \times \frac{2 \text{ mol NaOH}}{1 \text{ mol } H_2SO_4} \times$$

$$\frac{1 \text{ L NaOH soln}}{0.250 \text{ mol NaOH}} = 4.90 \times 10^{-2} \text{ L NaOH soln (or 49.0 mL NaOH soln)}$$

Thus, 35.0 mL of 0.175 M sulfuric acid solution reacts with exactly **49.0 mL** of 0.250 M sodium hydroxide solution.

Answer Check Whenever you perform a titration calculation, be sure that you have taken into account the stoichiometry of the reaction between the acid and base (use the balanced chemical equation). In this case, two moles of NaOH are required to neutralize each mole of acid. Furthermore, when performing titration calculations, do not be tempted to apply the dilution equation to solve the problem. If you were to take such an approach here, you would arrive at an incorrect result since the dilution equation fails to take into account the stoichiometry of the reaction.

Exercise 4.15 Nickel sulfate, $NiSO_4$, reacts with sodium phosphate, Na_3PO_4, to give a pale yellow-green precipitate of nickel phosphate, $Ni_3(PO_4)_2$, and a solution of sodium sulfate, Na_2SO_4.

$$3NiSO_4(aq) + 2Na_3PO_4(aq) \longrightarrow Ni_3(PO_4)_2(s) + 3Na_2SO_4(aq)$$

How many milliliters of 0.375 M $NiSO_4$ will react with 45.7 mL of 0.265 M Na_3PO_4?

■ See Problems 4.89 and 4.90.

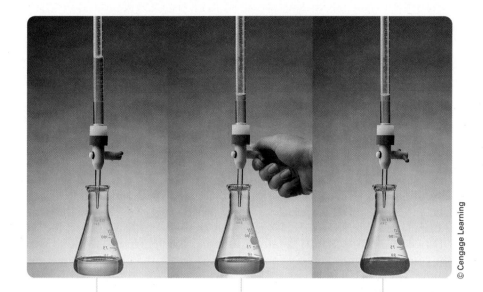

The flask contains HCl(*aq*) and a few drops of phenolphthalein indicator; the buret contains 0.207 *M* NaOH(*aq*) (the buret reading is 44.97 mL).

NaOH(*aq*) was added to the solution in the flask until a persistent faint pink color was reached, marking the endpoint of the titration (the buret reading is 49.44 mL). The amount of HCl can be determined from the volume of NaOH(*aq*) used (4.47 mL); see Example 4.14.

The addition of several drops of NaOH(*aq*) solution beyond the endpoint gives a deep pink color.

Figure 4.22 ▲

Titration of an unknown amount of HCl with NaOH

An important method for determining the amount of a particular substance is based on measuring the volume of reactant solution. Suppose substance *A* reacts in solution with substance *B*. If you know the volume and concentration of a solution of *B* that just reacts with substance *A* in a sample, you can determine the amount of *A*. **Titration** is *a procedure for determining the amount of substance* A *by adding a carefully measured volume of a solution with known concentration of* B *until the reaction of* A *and* B *is just complete.* **Volumetric analysis** is *a method of analysis based on titration.*

Figure 4.22 shows a flask containing hydrochloric acid with an unknown amount of HCl being titrated with sodium hydroxide solution, NaOH, of known molarity. The reaction is

$$NaOH(aq) + HCl(aq) \longrightarrow NaCl(aq) + H_2O(l)$$

To the HCl solution are added a few drops of phenolphthalein indicator. ▶ Phenolphthalein is colorless in the hydrochloric acid but turns pink at the completion of the reaction of NaOH with HCl. Sodium hydroxide with a concentration of 0.207 *M* is contained in a *buret,* a glass tube graduated to measure the volume of liquid delivered from the stopcock. The solution in the buret is added to the HCl in the flask until the phenolphthalein just changes from colorless to pink. At this point, the reaction is complete and the volume of NaOH that reacts with the HCl is read from the buret. This volume is then used to obtain the mass of HCl in the original solution.

An indicator is a substance that undergoes a color change when a reaction approaches completion. See Section 4.4.

☉WL Interactive Example 4.14   Calculating the Quantity of Substance in a Titrated Solution

Gaining Mastery Toolbox

Critical Concept 4.14
You need to start with a balanced chemical equation for the acid-base neutralization reaction. Just as in Example 4.13, a balanced chemical equation is essential for correctly solving the problem. In most instances the titration calculations will be based on acid-base neutralization reactions.

Solution Essentials:
• Molarity
• Acid-base neutralization reaction
• Molar interpretation of a balanced chemical equation.
• Stoichimetry

A flask contains a solution with an unknown amount of HCl. This solution is titrated with 0.207 M NaOH. It takes 4.47 mL of the NaOH solution to complete the reaction. What is the mass of the HCl?

Problem Strategy First, in order to determine the stoichiometry of the reaction, you start by writing the balanced chemical equation.

$$\text{NaOH}(aq) + \text{HCl}(aq) \longrightarrow \text{NaCl}(aq) + \text{H}_2\text{O}(l)$$

If you use the numerical information from the problem to determine the moles of NaOH added to the solution, you then can use the stoichiometry of the reaction to determine the moles of HCl that reacted. Once you know the moles of HCl, you can use the molar mass of HCl to calculate the mass of HCl. Employing this strategy, you convert the volume of NaOH (4.47×10^{-3} L NaOH solution) to moles NaOH (from the molarity of NaOH). Then you convert moles NaOH to moles HCl (from the chemical equation). Finally, you convert moles HCl to grams HCl.

The problem strategy is diagrammed as

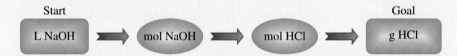

Solution The calculation is as follows:

$$4.47 \times 10^{-3}\,\text{L NaOH soln} \times \frac{0.207\,\text{mol NaOH}}{1\,\text{L NaOH soln}} \times \frac{1\,\text{mol HCl}}{1\,\text{mol NaOH}} \times \frac{36.5\,\text{g HCl}}{1\,\text{mol HCl}}$$

$$= \textbf{0.0338 g HCl}$$

Answer Check Before you perform the titration calculations, always write down the balanced chemical equation.

Exercise 4.16   A 5.00-g sample of vinegar is titrated with 0.108 M NaOH. If the vinegar requires 39.1 mL of the NaOH solution for complete reaction, what is the mass percentage of acetic acid, $\text{HC}_2\text{H}_3\text{O}_2$, in the vinegar? The reaction is

$$\text{HC}_2\text{H}_3\text{O}_2(aq) + \text{NaOH}(aq) \longrightarrow \text{NaC}_2\text{H}_3\text{O}_2(aq) + \text{H}_2\text{O}(l)$$

■ See Problems 4.91 and 4.92.

CONCEPT CHECK 4.7

Consider three flasks, each containing 0.10 mol of acid. You need to learn something about the acids in each of the flasks, so you perform titration using a NaOH solution. Here are the results of the experiment:

Flask A	10 mL of NaOH required for neutralization
Flask B	20 mL of NaOH required for neutralization
Flask C	30 mL of NaOH required for neutralization

a What have you learned about each of these acids from performing the experiment?

b Could you use the results of this experiment to determine the concentration of the NaOH? If not, what assumption about the molecular formulas of the acids would allow you to make the concentration determination?

A Checklist for Review

Summary of Facts and Concepts

Reactions often involve ions in aqueous solution. Many of the compounds in such reactions are *electrolytes,* which are substances that dissolve in water to give ions. Electrolytes that exist in solution almost entirely as ions are called *strong electrolytes.* Electrolytes that dissolve in water to give a relatively small percentage of ions are called *weak electrolytes.* The *solubility rules* can be used to predict the extent to which an ionic compound will dissolve in water. Most soluble ionic compounds are strong electrolytes.

We can represent a reaction involving ions in one of three different ways, depending on what information we want to convey. A *molecular equation* is one in which substances are written as if they were molecular, even though they are ionic. This equation closely describes what you actually do in the laboratory. However, this equation does not describe what is happening at the level of ions and molecules. For that purpose, we rewrite the molecular equation as a *complete ionic equation* by replacing the formulas for strong electrolytes by their ion formulas. If you cancel *spectator ions* from the complete ionic equation, you obtain the *net ionic equation.*

Most of the important reactions we consider in this course can be divided into three major classes: (1) precipitation reactions, (2) acid–base reactions, and (3) oxidation–reduction reactions. A *precipitation reaction* occurs in aqueous solution because one product is insoluble. You can decide whether two ionic compounds will result in a precipitation reaction, if you know from solubility rules that one of the potential products is insoluble.

Acids are substances that yield hydrogen ions in aqueous solution or donate protons. *Bases* are substances that yield hydroxide ions in aqueous solution or accept protons. These *acid–base reactions* are proton-transfer reactions. In this chapter, we covered neutralization reactions (reactions of acids and bases to yield salts) and reactions of certain salts with acids to yield a gas.

Oxidation–reduction reactions are reactions involving a transfer of electrons from one species to another or a change in the oxidation number of atoms. The concept of *oxidation numbers* helps us describe this type of reaction. The atom that increases in oxidation number is said to undergo *oxidation;* the atom that decreases in oxidation number is said to undergo *reduction.* Oxidation and reduction must occur together in a reaction. Many oxidation–reduction reactions fall into the following categories: combination reactions, decomposition reactions, displacement reactions, and combustion reactions. Oxidation–reduction reactions can be balanced by the *half-reaction method.*

Molar concentration, or *molarity,* is the moles of solute in a liter of solution. Knowing the molarity allows you to calculate the amount of solute in any volume of solution. Because the moles of solute are constant during the *dilution of a solution,* you can determine to what volume to dilute a concentrated solution to give one of desired molarity.

Quantitative analysis involves the determination of the amount of a species in a material. In *gravimetric analysis,* you determine the amount of a species by converting it to a product you can weigh. In *volumetric analysis,* you determine the amount of a species by titration. *Titration* is a method of chemical analysis in which you measure the volume of solution of known molarity that reacts with a compound of unknown amount. You determine the amount of the compound from this volume of solution.

Learning Objectives	Important Terms
4.1 Ionic Theory of Solutions and Solubility Rules	
■ Describe how an ionic substance can form ions in aqueous solution. ■ Explain how an *electrolyte* makes a solution electrically conductive. ■ Give examples of substances that are electrolytes. ■ Define *nonelectrolyte,* and provide an example of a molecular substance that is a nonelectrolyte. ■ Compare the properties of solutions that contain *strong electrolytes* and *weak electrolytes.* ■ Learn the *solubility rules* for ionic compounds. ■ Use the solubility rules. Example 4.1	**electrolyte** **nonelectrolyte** **strong electrolyte** **weak electrolyte**
4.2 Molecular and Ionic Equations	
■ Write the *molecular equation* of a chemical reaction. ■ From the molecular equation of both strong electrolytes and weak electrolytes, determine the *complete ionic equation.* ■ From the complete ionic equation, write the *net ionic equation.* ■ Write net ionic equations. Example 4.2	**molecular equation** **complete ionic equation** **spectator ion** **net ionic equation**

4.3 Precipitation Reactions

- Recognize *precipitation* (*exchange*) reactions.
- Write molecular, complete ionic, and net ionic equations for precipitation reactions.
- Decide whether a precipitation reaction will occur. Example 4.3
- Determine the product of a precipitation reaction.

precipitate
exchange (metathesis) reaction

4.4 Acid–Base Reactions

- Understand how an *acid–base* indicator is used to determine whether a solution is acidic or basic.
- Define *Arrhenius acid* and *Arrhenius base*.
- Write the chemical equation of an Arrhenius base or acid in aqueous solution.
- Define *Brønsted–Lowry acid* and *Brønsted–Lowry base*.
- Write the chemical equation of a Brønsted–Lowry base or acid in aqueous solution.
- Write the chemical equation of an acid in aqueous solution using a *hydronium ion*.
- Learn the common *strong acids* and *strong bases*.
- Distinguish between a strong acid and a *weak acid* and the solutions they form.
- Distinguish between a strong base and a *weak base* and the solutions they form.
- Classify acids and bases as strong or weak. Example 4.4
- Recognize *neutralization reactions*.
- Write an equation for a neutralization reaction. Example 4.5
- Write the reactions for a *polyprotic acid* in aqueous solution.
- Recognize acid–base reactions that lead to gas formation.
- Write an equation for a reaction with gas formation. Example 4.6

acid–base indicator
acid (Arrhenius)
base (Arrhenius)
acid (Brønsted–Lowry)
base (Brønsted–Lowry)
strong acid
weak acid
strong base
weak base
neutralization reaction
salt
polyprotic acid

4.5 Oxidation–Reduction Reactions

- Define *oxidation–reduction* reaction.
- Learn the oxidation-number rules.
- Assign oxidation numbers. Example 4.7
- Write the *half-reactions* of an oxidation–reduction reaction.
- Determine the species undergoing *oxidation* and *reduction*.
- Recognize *combination reactions, decomposition reactions, displacement reactions,* and *combustion reactions*.
- Use the activity series to predict when displacement reactions will occur.

oxidation number (oxidation state)
oxidation–reduction reaction (redox reaction)
half-reaction
oxidation
reduction
oxidizing agent
reducing agent
combination reaction
decomposition reaction
displacement reaction (single-replacement reaction)
combustion reaction

4.6 Balancing Simple Oxidation–Reduction Equations

- Balance simple oxidation–reduction reactions by the half-reaction method. Example 4.8

4.7 Molar Concentration

- Define *molarity* or *molar concentration* of a solution.
- Calculate the molarity from mass and volume. Example 4.9
- Use molarity as a conversion factor. Example 4.10

molar concentration (molarity) (M)

4.8 Diluting Solutions	
■ Describe what happens to the concentration of a solution when it is diluted. ■ Perform calculations associated with dilution. ■ Diluting a solution. Example 4.11	

4.9 Gravimetric Analysis	
■ Determine the amount of a species by *gravimetric analysis.* Example 4.12	**quantitative analysis** **gravimetric analysis**

4.10 Volumetric Analysis	
■ Calculate the volume of reactant solution needed to perform a reaction. Example 4.13 ■ Understand how to perform a *titration.* ■ Calculate the quantity of substance in a titrated solution. Example 4.14	**titration** **volumetric analysis**

Key Equations

$$\text{Molarity } (M) = \frac{\text{moles of solute}}{\text{liters of solution}}$$

$$M_i \times V_i = M_f \times V_f$$

Questions and Problems

⦿WL Interactive versions of these problems may be assigned in OWL.

Self-Assessment and Review Questions

Key: These questions test your understanding of the ideas you worked with in the chapter. These problems vary in difficulty and often can be used for the basis of discussion.

4.1 Define the terms *strong electrolyte* and *weak electrolyte.* Give an example of each.

4.2 Explain why some electrolyte solutions are strongly conducting, whereas others are weakly conducting.

4.3 Explain the terms *soluble* and *insoluble.* Use the solubility rules to write the formula of an insoluble ionic compound.

4.4 What are the advantages and disadvantages of using a molecular equation to represent an ionic reaction?

4.5 What is a net ionic equation? What is the value in using a net ionic equation? Give an example.

4.6 What is a *spectator ion*? Illustrate with a complete ionic reaction.

4.7 What are the major types of chemical reactions? Give a brief description and an example of each.

4.8 Describe in words how you would prepare pure crystalline $AgCl$ and $NaNO_3$ from solid $AgNO_3$ and solid $NaCl$.

4.9 Give an example of a neutralization reaction. Label the acid, base, and salt.

4.10 Give an example of a polyprotic acid and write equations for the successive neutralizations of the acidic hydrogen atoms of the acid molecule to produce a series of salts.

4.11 Why must oxidation and reduction occur together in a reaction?

4.12 Give an example of a displacement reaction. What is the oxidizing agent? What is the reducing agent?

4.13 Why is the product of molar concentration and volume constant for a dilution problem?

4.14 Describe how the amount of sodium hydroxide in a mixture can be determined by titration with hydrochloric acid of known molarity.

4.15 What is the net ionic equation for the following molecular equation?

$$HF(aq) + KOH(aq) \longrightarrow KF(aq) + H_2O(l)$$

Hydrofluoric acid, HF, is a molecular substance and a weak electrolyte.

ⓐ $HF(aq) + KOH(aq) \longrightarrow K^+(aq) + F^-(aq)$
ⓑ $H^+(aq) + OH^-(aq) \longrightarrow H_2O(l)$
ⓒ $H^+(aq) + KOH(aq) \longrightarrow K^+(aq) + H_2O(l)$
ⓓ $HF(aq) + OH^-(aq) \longrightarrow F^-(aq) + H_2O(l)$
ⓔ $HF(aq) + K^+(aq) + OH^-(aq) \longrightarrow KF(aq) + H_2O(l)$

4.16 An aqueous sodium hydroxide solution mixed with an aqueous magnesium nitrate solution yields which of the following products?

ⓐ magnesium hydroxide(*aq*)
ⓑ magnesium dihydroxide(*s*)
ⓒ magnesium hydroxide(*s*)
ⓓ dimagnesium hydroxide(*s*)
ⓔ sodium nitrate(*l*)

4.17 Which of the following compounds would produce the highest concentration of Cl⁻ ions when 0.10 mol of each is placed in separate beakers containing equal volumes of water?

a NaCl   b PbCl₂   c HClO₄
d MgCl₂   e HCl

4.18 In an aqueous 0.10 M HNO₂ solution (HNO₂ is a weak electrolyte), which of the following would you expect to find in the highest concentration?

a HNO₂   b OH⁻   c H₃O⁺
d NO₂⁻   e H⁺

Concept Explorations

Key: Concept explorations are comprehensive problems that provide a framework that will enable you to explore and learn many of the critical concepts and ideas in each chapter. If you master the concepts associated with these explorations, you will have a better understanding of many important chemistry ideas and will be more successful in solving all types of chemistry problems. These problems are well suited for group work and for use as in-class activities.

4.19 The Behavior of Substances in Water

Part 1:

a Ammonia, NH₃, is a weak electrolyte. It forms ions in solution by *reacting* with water molecules to form the ammonium ion and hydroxide ion. Write the balanced chemical reaction for this process, including state symbols.

b From everyday experience you are probably aware that table sugar (sucrose), C₁₂H₂₂O₁₁, is soluble in water. When sucrose dissolves in water, it doesn't form ions through any reaction with water. It just *dissolves* without forming ions, so it is a nonelectrolyte. Write the chemical equation for the dissolving of sucrose in water.

c Both NH₃ and C₁₂H₂₂O₁₁ are soluble molecular compounds, yet they behave differently in aqueous solution. Briefly explain why one is a weak electrolyte and the other is a nonelectrolyte.

d Hydrochloric acid, HCl, is a molecular compound that is a strong electrolyte. Write the chemical reaction of HCl with water.

e Compare the ammonia reaction with that of hydrochloric acid. Why are both of these substances considered electrolytes?

f Explain why HCl is a strong electrolyte and ammonia is a weak electrolyte.

g Classify each of the following substances as either ionic or molecular.

- KCl
- NH₃
- CO₂
- MgBr₂
- HCl
- Ca(OH)₂
- PbS
- HC₂H₃O₂

h For those compounds above that you classified as ionic, use the solubility rules to determine which are soluble.

i The majority of ionic substances are solids at room temperature. Describe what you would observe if you placed a soluble ionic compound and an insoluble ionic compound in separate beakers of water.

j Write the chemical equation(s), including state symbols, for what happens when each soluble ionic compound that you identified above is placed in water.

Are these substances reacting with water when they are added to water?

k How would you classify the soluble ionic compounds: strong electrolyte, weak electrolyte, or nonelectrolyte? Explain your answer.

l Sodium chloride, NaCl, is a strong electrolyte, as is hydroiodic acid, HI. Write the chemical equations for what happens when these substances are added to water.

m Are NaCl and HI strong electrolytes because they have similar behavior in aqueous solution? If not, describe, using words and equations, the different chemical process that takes place in each case.

Part 2: You have two hypothetical molecular compounds, AX and AY. AX is a strong electrolyte and AY is a weak electrolyte. The compounds undergo the following chemical reactions when added to water.

$$AX(aq) + H_2O(l) \longrightarrow AH_2O^+(aq) + X^-(aq)$$

$$AY(aq) + H_2O(l) \longrightarrow AH_2O^+(aq) + Y^-(aq)$$

a Explain how the relative amounts of AX(aq) and AY(aq) would compare if you had a beaker of water with AX and a beaker of water with AY.

b How would the relative amounts of X⁻(aq) and Y⁻(aq) in the two beakers compare? Be sure to explain your answer.

4.20 Working with Concentration (Molarity Concepts)

Note: You should be able to answer all of the following questions without using a calculator.

Part 1:

a Both NaCl and MgCl₂ are soluble ionic compounds. Write the balanced chemical equations for these two substances dissolving in water.

b Consider the pictures below. These pictures represent 1.0-L solutions of 1.0 M NaCl(aq) and 1.0 M MgCl₂(aq). The representations of the ions in solution are the correct relative amounts. Water molecules have been omitted for clarity. Correctly label each of the beakers, provide a key to help identify the ions, and give a brief explanation of how you made your assignments.

Keeping in mind that the pictures represent the relative amounts of ions in the solution and that the numerical information about these solutions is presented above, answer the following questions c through f.

c How many moles of NaCl and MgCl₂ are in each beaker?

d How many moles of chloride ions are in each beaker? How did you arrive at this answer?

e What is the concentration of chloride ions in each beaker? Without using mathematical equations, briefly explain how you obtained your answer.

f Explain how it is that the concentrations of chloride ions in these beakers are different even though the concentrations of each substance (compound) are the same.

Part 2: Say you were to dump out half of the $MgCl_2$ solution from the beaker above.

a What would be the concentration of the $MgCl_2(aq)$ and of the chloride ions in the remaining solution?

b How many moles of the $MgCl_2$ and of the chloride ions would remain in the beaker?

c Explain why the concentration of $MgCl_2(aq)$ would not change, whereas the number of moles of $MgCl_2$ would change when solution was removed from the beaker. As part of your answer, you are encouraged to use pictures.

Part 3: Consider the beaker containing 1.0 L of the 1.0 *M* NaCl(*aq*) solution. You now add 1.0 L of water to this beaker.

a What is the concentration of this NaCl(*aq*) solution?

b How many moles of NaCl are present in the 2.0 L of NaCl(*aq*) solution?

c Explain why the concentration of NaCl(*aq*) *does* change with the addition of water, whereas the number of moles *does not* change. Here again, you are encouraged to use pictures to help answer the question.

Conceptual Problems

Key: These problems are designed to check your understanding of the concepts associated with some of the main topics presented in each chapter. A strong conceptual understanding of chemistry is the foundation for both applying chemical knowledge and solving chemical problems. These problems vary in level of difficulty and often can be used as a basis for group discussion.

4.21 You need to perform gravimetric analysis of a water sample in order to determine the amount of Ag^+ present.

a List three aqueous solutions that would be suitable for mixing with the sample to perform the analysis.

b Would adding $KNO_3(aq)$ allow you to perform the analysis?

c Assume you have performed the analysis and the silver solid that formed is moderately soluble. How might this affect your analysis results?

4.22 In this problem you need to draw two pictures of solutions in beakers at different points in time. Time zero (*t* = 0) will be the *hypothetical* instant at which the reactants dissolve in the solution (*if* they dissolve) *before* they react. Time after mixing (*t* > 0) will be the time required to allow sufficient interaction of the materials. For now, we assume that insoluble solids have no ions in solution and do not worry about representing the stoichiometric amounts of the dissolved ions. Here is an example: Solid NaCl and solid $AgNO_3$ are added to a beaker containing 250 mL of water.

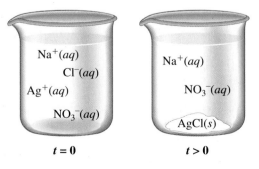

Note that we are not showing the H_2O, and we are representing only the ions and solids in solution. Using the same conditions as the example (adding the solids to H_2O), draw pictures of the following:

a Solid lead(II) nitrate and solid ammonium chloride at *t* = 0 and *t* > 0

b FeS(*s*) and $NaNO_3(s)$ at *t* = 0 and *t* > 0

c Solid lithium iodide and solid sodium carbonate at *t* = 0 and *t* > 0

4.23 Three acid samples are prepared for titration by 0.01 *M* NaOH:

1. Sample 1 is prepared by dissolving 0.01 mol of HCl in 50 mL of water.

2. Sample 2 is prepared by dissolving 0.01 mol of HCl in 60 mL of water.

3. Sample 3 is prepared by dissolving 0.01 mol of HCl in 70 mL of water.

a Without performing a formal calculation, compare the concentrations of the three acid samples (rank them from highest to lowest).

b When the titration is performed, which sample, if any, will require the largest volume of the 0.01 *M* NaOH for neutralization?

4.24 You come across a beaker that contains water, aqueous ammonium acetate, and a precipitate of calcium phosphate.

a Write the balanced molecular equation for a reaction between two solutions containing ions that could produce this solution.

b Write the complete ionic equation for the reaction in part a.

c Write the net ionic equation for the reaction in part a.

4.25 Would you expect a precipitation reaction between an ionic compound that is an electrolyte and an ionic compound that is a nonelectrolyte? Justify your answer.

4.26 Equal quantities of the hypothetical strong acid HX, weak acid HA, and weak base BZ are added to separate beakers of water, producing the solutions depicted in the drawings. In the drawings, the relative amounts of each substance present in the solution (neglecting the water) are shown. Identify the acid or base that was used to produce each of the solutions (HX, HA, or BZ).

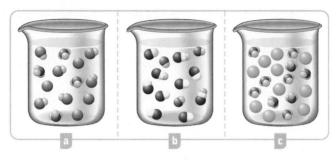

$\bigcirc$ = H_3O^+

$\bullet$ = OH^-

4.27 If one mole of the following compounds were each placed into separate beakers containing the same amount of water, rank the $Cl^-(aq)$ concentrations from highest to lowest (some may be equivalent): KCl, $AlCl_3$, $PbCl_2$, NaCl, HCl, NH_3, KOH, and HCN.

4.28 Try and answer the following questions without using a calculator.

a A solution is made by mixing 1.0 L of 0.5 *M* NaCl and 0.5 L of 1.0 *M* $CaCl_2$. Which ion is at the highest concentration in the solution?

b Another solution is made by mixing 0.50 L of 1.0 *M* KBr and 0.50 L of 1.0 *M* K_3PO_4. What is the concentration of each ion in the solution?

Practice Problems

Key: These problems are for practice in applying problem-solving skills. They are divided by topic, and some are keyed to exercises (see the ends of the exercises). The problems are arranged in matching pairs; the odd-numbered problem of each pair is listed first, and its answer is given in the back of the book.

Solubility Rules

4.29 Using solubility rules, predict the solubility in water of the following ionic compounds.

a $Al(OH)_3$ b Ca_3N_2
c NH_4Cl d KOH

4.30 Using solubility rules, predict the solubility in water of the following ionic compounds.

a PbS   b $AgNO_3$
c Na_2CO_3 d CaI_2

4.31 Using solubility rules, decide whether the following ionic solids are soluble or insoluble in water. If they are soluble, write the chemical equation for dissolving in water and indicate what ions you would expect to be present in solution.

a AgBr   b Li_2SO_4
c $Ca_3(PO_4)_2$ d Na_2CO_3

4.32 Using solubility rules, decide whether the following ionic solids are soluble or insoluble in water. If they are soluble, write the chemical equation for dissolving in water and indicate what ions you would expect to be present in solution.

a $(NH_4)_2SO_4$ b $BaCO_3$
c $Pb(NO_3)_2$ d $Ca(OH)_2$

Ionic Equations

4.33 Write net ionic equations for the following molecular equations. HBr is a strong electrolyte.

a $AgNO_3(aq) + NaBr(aq) \longrightarrow AgBr(s) + NaNO_3(aq)$

b $NaOH(aq) + NH_4Br(aq) \longrightarrow$
$NaBr(aq) + NH_3(g) + H_2O(l)$

c $HBr(aq) + KOH(aq) \longrightarrow KBr(aq) + H_2O(l)$

d $K_2S(aq) + 2HBr(aq) \longrightarrow 2KBr(aq) + H_2S(g)$

4.34 Write net ionic equations for the following molecular equations. HBr is a strong electrolyte.

a $HBr(aq) + NH_3(aq) \longrightarrow NH_4Br(aq)$

b $2HCl(aq) + Ba(OH)_2(aq) \longrightarrow 2H_2O(l) + BaCl_2(aq)$

c $Pb(NO_3)_2(aq) + 2NaBr(aq) \longrightarrow$
$PbBr_2(s) + 2NaNO_3(aq)$

d $MgCO_3(s) + H_2SO_4(aq) \longrightarrow$
$MgSO_4(aq) + H_2O(l) + CO_2(g)$

4.35 Lithium carbonate solution reacts with aqueous hydrobromic acid to give a solution of lithium bromide, carbon dioxide gas, and water. Write the molecular equation and the net ionic equation for the reaction.

4.36 Lead(II) nitrate solution and sodium sulfate solution are mixed. Crystals of lead(II) sulfate come out of solution, leaving a solution of sodium nitrate. Write the molecular equation and the net ionic equation for the reaction.

Precipitation

4.37 Write the molecular equation and the net ionic equation for each of the following aqueous reactions. If no reaction occurs, write *NR* after the arrow.

a $FeSO_4 + NaCl \longrightarrow$
b $Na_2CO_3 + MgBr_2 \longrightarrow$
c $MgSO_4 + NaOH \longrightarrow$
d $NiCl_2 + NaBr \longrightarrow$

4.38 Write the molecular equation and the net ionic equation for each of the following aqueous reactions. If no reaction occurs, write *NR* after the arrow.

a $LiCl + Al(NO_3)_3 \longrightarrow$
b $AgNO_3 + NaI \longrightarrow$
c $Ba(NO_3)_2 + K_2SO_4 \longrightarrow$
d $NH_4NO_3 + K_2SO_4 \longrightarrow$

4.39 For each of the following, write molecular and net ionic equations for any precipitation reaction that occurs. If no reaction occurs, indicate this.

a Solutions of barium nitrate and lithium sulfate are mixed.

b Solutions of sodium bromide and calcium nitrate are mixed.

c Solutions of aluminum sulfate and sodium hydroxide are mixed.

d Solutions of calcium bromide and sodium phosphate are mixed.

4.40 For each of the following, write molecular and net ionic equations for any precipitation reaction that occurs. If no reaction occurs, indicate this.

a Zinc chloride and sodium sulfide are dissolved in water.

b Sodium sulfide and calcium chloride are dissolved in water.

c Magnesium sulfate and potassium bromide are dissolved in water.

d Magnesium sulfate and potassium carbonate are dissolved in water.

Strong and Weak Acids and Bases

4.41 Classify each of the following as a strong or weak acid or base.

a NH_3 **b** HCNO   **c** $Mg(OH)_2$ **d** $HClO_3$

4.42 Classify each of the following as a strong or weak acid or base.

a HF   **b** KOH   **c** $HClO_4$ **d** HIO

Neutralization Reactions

4.43 Complete and balance each of the following molecular equations (in aqueous solution); include phase labels. Then, for each, write the net ionic equation.

a $NaOH + HNO_3 \longrightarrow$

b $HCl + Ba(OH)_2 \longrightarrow$

c $HC_2H_3O_2 + Ca(OH)_2 \longrightarrow$

d $NH_3 + HNO_3 \longrightarrow$

4.44 Complete and balance each of the following molecular equations (in aqueous solution); include phase labels. Then, for each, write the net ionic equation.

a $Ba(OH)_2 + HC_2H_3O_2 \longrightarrow$

b $H_2SO_4 + KOH \longrightarrow$

c $Al(OH)_3 + HCl \longrightarrow$

d $HClO + Sr(OH)_2 \longrightarrow$

4.45 For each of the following, write the molecular equation, including phase labels. Then write the net ionic equation. Note that the salts formed in these reactions are soluble.

a the neutralization of hydrobromic acid with calcium hydroxide solution

b the reaction of solid aluminum hydroxide with nitric acid

c the reaction of aqueous hydrogen cyanide with calcium hydroxide solution

d the neutralization of lithium hydroxide solution by aqueous hydrogen cyanide

4.46 For each of the following, write the molecular equation, including phase labels. Then write the net ionic equation. Note that the salts formed in these reactions are soluble.

a the neutralization of lithium hydroxide solution by aqueous chloric acid

b the reaction of barium hydroxide solution and aqueous nitrous acid

c the reaction of sodium hydroxide solution and aqueous nitrous acid

d the neutralization of aqueous hydrogen cyanide by aqueous strontium hydroxide

4.47 Complete the right side of each of the following molecular equations. Then write the net ionic equations. Assume all salts formed are soluble. Acid salts are possible.

a $Ca(OH)_2(aq) + 2H_2SO_4(aq) \longrightarrow$

b $2H_3PO_4(aq) + Ca(OH)_2(aq) \longrightarrow$

c $NaOH(aq) + H_2SO_4(aq) \longrightarrow$

d $Sr(OH)_2(aq) + 2H_2CO_3(aq) \longrightarrow$

4.48 Complete the right side of each of the following molecular equations. Then write the net ionic equations. Assume all salts formed are soluble. Acid salts are possible.

a $2KOH(aq) + H_3PO_4(aq) \longrightarrow$

b $3H_2SO_4(aq) + 2Al(OH)_3(s) \longrightarrow$

c $2HC_2H_3O_2(aq) + Ca(OH)_2(aq) \longrightarrow$

d $H_2SO_3(aq) + NaOH(aq) \longrightarrow$

4.49 Write molecular and net ionic equations for the successive neutralizations of each acidic hydrogen of sulfurous acid by aqueous calcium hydroxide. $CaSO_3$ is insoluble; the acid salt is soluble.

4.50 Write molecular and net ionic equations for the successive neutralizations of each acidic hydrogen of phosphoric acid by calcium hydroxide solution. $Ca_3(PO_4)_2$ is insoluble; assume that the acid salts are soluble.

Reactions Evolving a Gas

4.51 The following reactions occur in aqueous solution. Complete and balance the molecular equations using phase labels. Then write the net ionic equations.

a $BaCO_3 + HNO_3 \longrightarrow$

b $K_2S + HCl \longrightarrow$

c $CaSO_3(s) + HI \longrightarrow$

4.52 The following reactions occur in aqueous solution. Complete and balance the molecular equations using phase labels. Then write the net ionic equations.

a $CaS + HBr \longrightarrow$ **b** $MgCO_3 + HNO_3 \longrightarrow$

c $K_2SO_3 + H_2SO_4 \longrightarrow$

4.53 Write the molecular equation and the net ionic equation for the reaction of solid iron(II) sulfide and hydrochloric acid. Add phase labels.

4.54 Write the molecular equation and the net ionic equation for the reaction of solid barium carbonate and hydrogen bromide in aqueous solution. Add phase labels.

Oxidation Numbers

4.55 Obtain the oxidation number for the element noted in each of the following.

a Mn in K_2MnO_4 **b** Br in $KBrO_4$

c Nb in NbO_2 **d** Ga in Ga_2O_3

4.56 Obtain the oxidation number for the element noted in each of the following.

a Cr in CrO_3 **b** Hg in Hg_2Cl_2

c Ga in $Ga(OH)_3$ **d** P in Na_3PO_4

4.57 Obtain the oxidation number for the element noted in each of the following.

a N in NH_2^-
b I in IO_3^-
c H in H_2
d Cl in $HClO_4$

4.58 Obtain the oxidation number for the element noted in each of the following.

a N in N_2
b Cr in CrO_4^{2-}
c Zn in $Zn(OH)_4^{2-}$
d As in $H_2AsO_3^-$

4.59 Determine the oxidation numbers of all the elements in each of the following compounds. (*Hint:* Look at the ions present.)

a $Hg_2(ClO_4)_2$
b $Cr_2(SO_4)_3$
c $CoSeO_4$
d $Pb(OH)_2$

4.60 Determine the oxidation numbers of all the elements in each of the following compounds. (*Hint:* Look at the ions present.)

a $Mn(ClO_2)_2$
b $Fe_2(CrO_4)_3$
c $HgCr_2O_7$
d $Co_3(PO_4)_2$

Describing Oxidation–Reduction Reactions

4.61 In the following reactions, label the oxidizing agent and the reducing agent.

a $Co(s) + Cl_2(g) \longrightarrow CoCl_2(s)$
b $P_4(s) + 5O_2(g) \longrightarrow P_4O_{10}(s)$

4.62 In the following reactions, label the oxidizing agent and the reducing agent.

a $ZnO(s) + C(s) \longrightarrow Zn(g) + CO(g)$
b $8Fe(s) + S_8(s) \longrightarrow 8FeS(s)$

4.63 In the following reactions, label the oxidizing agent and the reducing agent.

a $2Al(s) + 3F_2(g) \longrightarrow 2AlF_3(s)$
b $Hg^{2+}(aq) + NO_2^-(aq) + H_2O(l) \longrightarrow$
$Hg(s) + 2H^+(aq) + NO_3^-(aq)$

4.64 In the following reactions, label the oxidizing agent and the reducing agent.

a $PbS(s) + 4H_2O_2(aq) \longrightarrow PbSO_4(s) + 4H_2O(l)$
b $Fe_2O_3(s) + 3CO(g) \longrightarrow 2Fe(s) + 3CO_2(g)$

Balancing Oxidation–Reduction Reactions

4.65 Balance the following oxidation–reduction reactions by the half-reaction method.

a $CuCl_2(aq) + Al(s) \longrightarrow AlCl_3(aq) + Cu(s)$
b $Cr^{3+}(aq) + Zn(s) \longrightarrow Cr(s) + Zn^{2+}(aq)$

4.66 Balance the following oxidation–reduction reactions by the half-reaction method.

a $FeI_3(aq) + Mg(s) \longrightarrow Fe(s) + MgI_2(aq)$
b $H_2(g) + Ag^+(aq) \longrightarrow Ag(s) + H^+(aq)$

Molarity

4.67 A sample of 0.0512 mol of iron(III) chloride, $FeCl_3$, was dissolved in water to give 50.0 mL of solution. What is the molarity of the solution?

4.68 A 50.0-mL volume of $AgNO_3$ solution contains 0.0345 mol $AgNO_3$ (silver nitrate). What is the molarity of the solution?

4.69 An aqueous solution is made from 0.798 g of potassium permanganate, $KMnO_4$. If the volume of solution is 50.0 mL, what is the molarity of $KMnO_4$ in the solution?

4.70 A sample of oxalic acid, $H_2C_2O_4$, weighing 1.200 g is placed in a 100.0-mL volumetric flask, which is then filled to the mark with water. What is the molarity of the solution?

4.71 What volume of 0.120 M $CuSO_4$ is required to give 0.250 mol of copper(II) sulfate, $CuSO_4$?

4.72 How many milliliters of 0.126 M $HClO_4$ (perchloric acid) are required to give 0.150 mol $HClO_4$?

4.73 An experiment calls for 0.0353 g of potassium hydroxide, KOH. How many milliliters of 0.0176 M KOH are required?

4.74 What is the volume (in milliliters) of 0.100 M H_2SO_4 (sulfuric acid) containing 0.949 g H_2SO_4?

4.75 Heme, obtained from red blood cells, binds oxygen, O_2. How many moles of heme are there in 150 mL of 0.0019 M heme solution?

4.76 Insulin is a hormone that controls the use of glucose in the body. How many moles of insulin are required to make up 28 mL of 0.0048 M insulin solution?

4.77 Describe how you would prepare 2.50×10^2 mL of 0.50 M Na_2SO_4. What mass (in grams) of sodium sulfate, Na_2SO_4, is needed?

4.78 How many grams of sodium dichromate, $Na_2Cr_2O_7$, should be added to a 100.0-mL volumetric flask to prepare 0.025 M $Na_2Cr_2O_7$ when the flask is filled to the mark with water?

4.79 You wish to prepare 0.12 M HNO_3 from a stock solution of nitric acid that is 15.8 M. How many milliliters of the stock solution do you require to make up 1.00 L of 0.12 M HNO_3?

4.80 A chemist wants to prepare 0.75 M HCl. Commercial hydrochloric acid is 12.4 M. How many milliliters of the commercial acid does the chemist require to make up 1.50 L of the dilute acid?

4.81 Calculate the concentrations of each ion present in a solution that results from mixing 50.0 mL of a 0.20 M $NaClO_3(aq)$ solution with 25.0 mL of a 0.20 M $Na_2SO_4(aq)$ solution. Assume that the volumes are additive.

4.82 A 4.00 g sample of KCl is dissolved in 10.0 mL of water. The resulting solution is then added to 60.0 mL of a 0.500 M $CaCl_2(aq)$ solution. Assuming that the volumes are additive, calculate the concentrations of each ion present in the final solution.

Gravimetric Analysis

4.83 A chemist added an excess of sodium sulfate to a solution of a soluble barium compound to precipitate all of the barium ion as barium sulfate, $BaSO_4$. How many grams of barium ion are in a 458-mg sample of the barium compound if a solution of the sample gave 513 mg $BaSO_4$ precipitate? What is the mass percentage of barium in the compound?

4.84 A soluble iodide was dissolved in water. Then an excess of silver nitrate, $AgNO_3$, was added to precipitate all of the iodide ion as silver iodide, AgI. If 1.545 g of the soluble iodide gave 2.185 g of silver iodide, how many grams of iodine are in the sample of soluble iodide? What is the mass percentage of iodine, I, in the compound?

4.85 Gold has compounds containing gold(I) ion or gold(III) ion. A compound of gold and chlorine was treated with a solution of silver nitrate, $AgNO_3$, to convert the chloride ion in the compound to a precipitate of $AgCl$. A 162.7-mg sample of the gold compound gave 100.3 mg $AgCl$.
  - a Calculate the percentage of the chlorine in the gold compound.
  - b Decide whether the formula of the compound is $AuCl$ or $AuCl_3$.

4.86 Copper has compounds with copper(I) ion or copper(II) ion. A compound of copper and chlorine was treated with a solution of silver nitrate, $AgNO_3$, to convert the chloride ion in the compound to a precipitate of $AgCl$. A 59.40-mg sample of the copper compound gave 86.00 mg $AgCl$.
  - a Calculate the percentage of chlorine in the copper compound.
  - b Decide whether the formula of the compound is $CuCl$ or $CuCl_2$.

4.87 A compound of iron and chlorine is soluble in water. An excess of silver nitrate was added to precipitate the chloride ion as silver chloride. If a 134.8-mg sample of the compound gave 304.8 mg $AgCl$, what is the formula of the compound?

4.88 A 1.345-g sample of a compound of barium and oxygen was dissolved in hydrochloric acid to give a solution of barium ion, which was then precipitated with an excess of potassium chromate to give 2.012 g of barium chromate, $BaCrO_4$. What is the formula of the compound?

Volumetric Analysis

4.89 A flask contains 49.8 mL of 0.150 M $Ca(OH)_2$ (calcium hydroxide). How many milliliters of 0.500 M Na_2CO_3 (sodium carbonate) are required to react completely with the calcium hydroxide in the following reaction?

$$Na_2CO_3(aq) + Ca(OH)_2(aq) \longrightarrow$$
$$CaCO_3(s) + 2NaOH(aq)$$

4.90 What volume of 0.250 M HNO_3 (nitric acid) reacts with 44.8 mL of 0.150 M Na_2CO_3 (sodium carbonate) in the following reaction?

$$2HNO_3(aq) + Na_2CO_3(aq) \longrightarrow$$
$$2NaNO_3(aq) + H_2O(l) + CO_2(g)$$

4.91 How many milliliters of 0.150 M H_2SO_4 (sulfuric acid) are required to react with 8.20 g of sodium hydrogen carbonate, $NaHCO_3$, according to the following equation?

$$H_2SO_4(aq) + 2NaHCO_3(aq) \longrightarrow$$
$$Na_2SO_4(aq) + 2H_2O(l) + 2CO_2(g)$$

4.92 How many milliliters of 0.250 M $KMnO_4$ are needed to react with 3.55 g of iron(II) sulfate, $FeSO_4$? The reaction is as follows:

$$10FeSO_4(aq) + 2KMnO_4(aq) + 8H_2SO_4(aq) \longrightarrow$$
$$5Fe_2(SO_4)_3(aq) + 2MnSO_4(aq) + K_2SO_4(aq) +$$
$$8H_2O(l)$$

4.93 A 3.75-g sample of iron ore is transformed to a solution of iron(II) sulfate, $FeSO_4$, and this solution is titrated with 0.150 M $K_2Cr_2O_7$ (potassium dichromate). If it requires 43.7 mL of potassium dichromate solution to titrate the iron(II) sulfate solution, what is the percentage of iron in the ore? The reaction is

$$6FeSO_4(aq) + K_2Cr_2O_7(aq) + 7H_2SO_4(aq) \longrightarrow$$
$$3Fe_2(SO_4)_3(aq) + Cr_2(SO_4)_3(aq) + 7H_2O(l) +$$
$$K_2SO_4(aq)$$

4.94 A solution of hydrogen peroxide, H_2O_2, is titrated with a solution of potassium permanganate, $KMnO_4$. The reaction is

$$5H_2O_2(aq) + 2KMnO_4(aq) + 3H_2SO_4(aq) \longrightarrow$$
$$5O_2(g) + 2MnSO_4(aq) + K_2SO_4(aq) + 8H_2O(l)$$

It requires 51.7 mL of 0.145 M $KMnO_4$ to titrate 20.0 g of the solution of hydrogen peroxide. What is the mass percentage of H_2O_2 in the solution?

General Problems

Key: These problems provide more practice but are not divided by topic or keyed to exercises. Each section ends with essay questions. Odd-numbered problems and the even-numbered problems that follow are similar; answers to all odd-numbered problems except the essay questions are given in the back of the book.

4.95 Aluminum metal reacts with perchloric acid to produce hydrogen gas and a solution of aluminum perchlorate. Write the molecular equation for this reaction. Then write the corresponding net ionic equation.

4.96 Magnesium metal reacts with hydrobromic acid to produce hydrogen gas and a solution of magnesium bromide. Write the molecular equation for this reaction. Then write the corresponding net ionic equation.

4.97 Potassium sulfate solution reacts with barium bromide solution to produce a precipitate of barium sulfate and a solution of potassium bromide. Write the molecular equation for this reaction. Then write the corresponding net ionic equation.

4.98 Nickel(II) sulfate solution reacts with sodium hydroxide solution to produce a precipitate of nickel(II) hydroxide and a solution of sodium sulfate. Write the molecular equation for this reaction. Then write the corresponding net ionic equation.

4.99 Decide whether a reaction occurs for each of the following. If it does not, write *NR* after the arrow. If it does, write the balanced molecular equation; then write the net ionic equation.

a $Al(OH)_3 + HNO_3 \longrightarrow$
b $NaBr + HClO_4 \longrightarrow$
c $CaCl_2 + NaNO_3 \longrightarrow$
d $MgSO_4 + Ba(NO_3)_2 \longrightarrow$

4.100 Decide whether a reaction occurs for each of the following. If it does not, write *NR* after the arrow. If it does, write the balanced molecular equation; then write the net ionic equation.

a $LiOH + HCN \longrightarrow$
b $Li_2CO_3 + HNO_3 \longrightarrow$
c $LiCl + AgNO_3 \longrightarrow$
d $NaCl + MgSO_4 \longrightarrow$

4.101 Complete and balance each of the following molecular equations, including phase labels, if a reaction occurs. Then write the net ionic equation. If no reaction occurs, write *NR* after the arrow.

a $Sr(OH)_2 + HC_2H_3O_2 \longrightarrow$
b $NH_4I + CsCl \longrightarrow$
c $NaNO_3 + CsCl \longrightarrow$
d $NH_4I + AgNO_3 \longrightarrow$

4.102 Complete and balance each of the following molecular equations, including phase labels, if a reaction occurs. Then write the net ionic equation. If no reaction occurs, write *NR* after the arrow.

a $K_3PO_4 + MgCl_2 \longrightarrow$
b $H_2CO_3 + Sr(OH)_2 \longrightarrow$
c $FeSO_4 + MgCl_2 \longrightarrow$
d $HClO_4 + BaCO_3 \longrightarrow$

4.103 Describe in words how you would do each of the following preparations. Then give the molecular equation for each preparation.

a $MgCl_2(s)$ from $MgCO_3(s)$
b $NaNO_3(s)$ from $NaCl(s)$
c $Al(OH)_3(s)$ from $Al(NO_3)_3(s)$
d $HCl(aq)$ from $H_2SO_4(aq)$

4.104 Describe in words how you would do each of the following preparations. Then give the molecular equation for each preparation.

a $CuCl_2(s)$ from $CuSO_4(s)$
b $Ca(C_2H_3O_2)_2(s)$ from $CaCO_3(s)$
c $NaNO_3(s)$ from $Na_2SO_3(s)$
d $MgCl_2(s)$ from $Mg(OH)_2(s)$

4.105 Classify each of the following reactions as a combination reaction, decomposition reaction, displacement reaction, or combustion reaction.

a When they are heated, ammonium dichromate crystals, $(NH_4)_2Cr_2O_7$, decompose to give nitrogen, water vapor, and solid chromium(III) oxide, Cr_2O_3.
b When aqueous ammonium nitrite, NH_4NO_2, is heated, it gives nitrogen and water vapor.

c When gaseous ammonia, NH_3, reacts with hydrogen chloride gas, HCl, fine crystals of ammonium chloride, NH_4Cl, are formed.
d Aluminum added to an aqueous solution of sulfuric acid, H_2SO_4, forms a solution of aluminum sulfate, $Al_2(SO_4)_3$. Hydrogen gas is released.

4.106 Classify each of the following reactions as a combination reaction, decomposition reaction, displacement reaction, or combustion reaction.

a When solid calcium oxide, CaO, is exposed to gaseous sulfur trioxide, SO_3, solid calcium sulfate, $CaSO_4$, is formed.
b Magnesium reacts with bromine to give magnesium bromide, $MgBr_2$.
c Calcium metal (solid) reacts with water to produce a solution of calcium hydroxide, $Ca(OH)_2$, and hydrogen gas.
d When solid sodium hydrogen sulfite, $NaHSO_3$, is heated, solid sodium sulfite, Na_2SO_3, sulfur dioxide gas, SO_2, and water vapor are formed.

4.107 Consider the reaction of all pairs of the following compounds in water solution: $Ba(OH)_2$, $Pb(NO_3)_2$, H_2SO_4, $NaNO_3$, $MgSO_4$.

a Which pair (or pairs) forms one insoluble compound and one soluble compound (not water)?
b Which pair (or pairs) forms two insoluble compounds?
c Which pair (or pairs) forms one insoluble compound and water?

4.108 Consider the reaction of all pairs of the following compounds in water solution: $Sr(OH)_2$, $AgNO_3$, H_3PO_4, KNO_3, $CuSO_4$.

a Which pair (or pairs) forms one insoluble compound and one soluble compound (not water)?
b Which pair (or pairs) forms two insoluble compounds?
c Which pair (or pairs) forms one insoluble compound and water?

4.109 An aqueous solution contains 3.75 g of iron(III) sulfate, $Fe_2(SO_4)_3$, per liter. What is the molarity of $Fe_2(SO_4)_3$? When the compound dissolves in water, the Fe^{3+} ions and SO_4^{2-} ions in the crystal go into the solution. What is the molar concentration of each ion in the solution?

4.110 An aqueous solution contains 5.00 g of calcium chloride, $CaCl_2$, per liter. What is the molarity of $CaCl_2$? When calcium chloride dissolves in water, the calcium ions, Ca^{2+}, and chloride ions, Cl^-, in the crystal go into the solution. What is the molarity of each ion in the solution?

4.111 A stock solution of potassium dichromate, $K_2Cr_2O_7$, is made by dissolving 89.3 g of the compound in 1.00 L of solution. How many milliliters of this solution are required to prepare 1.00 L of 0.100 M $K_2Cr_2O_7$?

4.112 A 71.2-g sample of oxalic acid, $H_2C_2O_4$, was dissolved in 1.00 L of solution. How would you prepare 2.50 L of 0.150 M $H_2C_2O_4$ from this solution?

4.113 A solution contains 6.00% (by mass) NaBr (sodium bromide). The density of the solution is 1.046 g/cm^3. What is the molarity of NaBr?

4.114 An aqueous solution contains 3.75% NH_3 (ammonia) by mass. The density of the aqueous ammonia is 0.979 g/mL. What is the molarity of NH_3 in the solution?

4.115 Bone was dissolved in hydrochloric acid, giving 50.0 mL of solution containing calcium chloride, $CaCl_2$. To precipitate the calcium ion from the resulting solution, an excess of potassium oxalate was added. The precipitate of calcium oxalate, CaC_2O_4, weighed 1.437 g. What was the molarity of $CaCl_2$ in the solution?

4.116 A barium mineral was dissolved in hydrochloric acid to give a solution of barium ion. An excess of potassium sulfate was added to 50.0 mL of the solution, and 1.128 g of barium sulfate precipitate formed. Assume that the original solution was barium chloride. What was the molarity of $BaCl_2$ in this solution?

4.117 You have a sample of a rat poison whose active ingredient is thallium(I) sulfate. You analyze this sample for the mass percentage of active ingredient by adding potassium iodide to precipitate yellow thallium(I) iodide. If the sample of rat poison weighed 759.0 mg and you obtained 212.2 mg of the dry precipitate, what is the mass percentage of the thallium(I) sulfate in the rat poison?

4.118 An antacid tablet has calcium carbonate as the active ingredient; other ingredients include a starch binder. You dissolve the tablet in hydrochloric acid and filter off insoluble material. You add potassium oxalate to the filtrate (containing calcium ion) to precipitate calcium oxalate. If a tablet weighing 0.750 g gave 0.629 g of calcium oxalate, what is the mass percentage of active ingredient in the tablet?

4.119 A sample of $CuSO_4 \cdot 5H_2O$ was heated to 110°C, where it lost water and gave another hydrate of copper(II) ion that contains 32.50% Cu. A 98.77-mg sample of this new hydrate gave 116.66 mg of barium sulfate precipitate when treated with a barium nitrate solution. What is the formula of the new hydrate?

4.120 A sample of $CuSO_4 \cdot 5H_2O$ was heated to 100°C, where it lost water and gave another hydrate of copper(II) ion that contained 29.76% Cu. An 85.42-mg sample of this new hydrate gave 93.33 mg of barium sulfate precipitate when treated with a barium nitrate solution. What is the formula of the new hydrate?

4.121 A solution of scandium chloride was treated with silver nitrate. The chlorine in the scandium compound was converted to silver chloride, AgCl. A 58.9-mg sample of scandium chloride gave 167.4 mg of silver chloride. What are the mass percentages of Sc and Cl in scandium chloride? What is its empirical formula?

4.122 A water-soluble compound of gold and chlorine is treated with silver nitrate to convert the chlorine completely to silver chloride, AgCl. In an experiment, 328 mg of the compound gave 464 mg of silver chloride. Calculate the percentage of Cl in the compound. What is its empirical formula?

4.123 A 0.608-g sample of fertilizer contained nitrogen as ammonium sulfate, $(NH_4)_2SO_4$. It was analyzed for nitrogen by heating with sodium hydroxide.

$$(NH_4)_2SO_4(s) + 2NaOH(aq) \longrightarrow$$
$$Na_2SO_4(aq) + 2H_2O(l) + 2NH_3(g)$$

The ammonia was collected in 46.3 mL of 0.213 M HCl (hydrochloric acid), with which it reacted.

$$NH_3(g) + HCl(aq) \longrightarrow NH_4Cl(aq)$$

This solution was titrated for excess hydrochloric acid with 44.3 mL of 0.128 M NaOH.

$$NaOH(aq) + HCl(aq) \longrightarrow NaCl(aq) + H_2O(l)$$

What is the percentage of nitrogen in the fertilizer?

4.124 An antacid tablet contains sodium hydrogen carbonate, $NaHCO_3$, and inert ingredients. A 0.500-g sample of powdered tablet was mixed with 50.0 mL of 0.190 M HCl (hydrochloric acid). The mixture was allowed to stand until it reacted.

$$NaHCO_3(s) + HCl(aq) \longrightarrow$$
$$NaCl(aq) + H_2O(l) + CO_2(g)$$

The excess hydrochloric acid was titrated with 45.3 mL of 0.128 M NaOH (sodium hydroxide).

$$HCl(aq) + NaOH(aq) \longrightarrow NaCl(aq) + H_2O(l)$$

What is the percentage of sodium hydrogen carbonate in the antacid?

Strategy Problems

Key: As noted earlier, all of the practice and general problems are matched pairs. This section is a selection of problems that are not in matched-pair format. These challenging problems require that you employ many of the concepts and strategies that were developed in the chapter. In some cases, you will have to integrate several concepts and operational skills in order to solve the problem successfully.

4.125 You order a glass of juice in a restaurant, only to discover that it is warm and too sweet. The sugar concentration of the juice is 3.47 M, but you would like it reduced to a concentration of 1.78 M. How many grams of ice should you add to 100 mL of juice, knowing that only a third of the ice will melt before you take the first sip? (The density of water is 1.00 g/mL.)

4.126 If 38.2 mL of 0.248 M aluminum sulfate solution is diluted with deionized water to a total volume of 0.639 L, how many grams of aluminum ion are present in the diluted solution?

4.127 If 45.1 mL of a solution containing 8.30 g of silver nitrate is added to 31.3 mL of 0.511 M sodium carbonate solution, calculate the molarity of silver ion in the resulting solution. (Assume volumes are additive.)

4.128 An aluminum nitrate solution is labeled 0.256 M. If 31.6 mL of this solution is diluted to a total of 65.1 mL, calculate the molarity of nitrate ion in the resulting solution.

4.129 Zinc acetate is sometimes prescribed by physicians for the treatment of Wilson's disease, which is a genetically caused condition wherein copper accumulates to toxic levels in the body. If you were to analyze a sample of zinc acetate and find that it contains 3.33×10^{23} acetate ions, how many grams of zinc acetate must be present in the sample?

4.130 Arsenic acid, H_3AsO_4, is a poisonous acid that has been used in the treatment of wood to prevent insect damage. Arsenic acid has three acidic protons. Say you take a 25.00-mL sample of arsenic acid and prepare it for titration with NaOH by adding 25.00 mL of water. The complete neutralization of this solution requires the addition of 53.07 mL of 0.6441 M NaOH solution. Write the balanced chemical reaction for the titration, and calculate the molarity of the arsenic acid sample.

4.131 When the following equation is balanced by the half-reaction method using the smallest set of whole-number stoichiometric coefficients possible, how many electrons are canceled when the two half-reactions are added together?

$$K(s) + N_2(g) \longrightarrow K_3N(s)$$

4.132 Identify each of the following reactions as being a neutralization, precipitation, or oxidation–reduction reaction.

a $Mg(NO_3)_2(g) + Na_2S(aq) \longrightarrow MgS(s) + 2NaNO_3(aq)$

b $CsOH(aq) + HClO_4(aq) \longrightarrow$
$$Cs^+(aq) + 2H_2O(l) + ClO_4^-(aq)$$

c $Na_2SO_4(aq) + Hg(NO_3)_2(aq) \longrightarrow$
$$HgSO_4(s) + 2NaNO_3(aq)$$

d $Fe_2O_3(s) + 3CO(g) \longrightarrow 2Fe(s) + 3CO_2(g)$

4.133 A 414-mL sample of 0.196 M $MgBr_2$ solution is prepared in a large flask. A 43.0-mL portion of the solution is then placed into an empty 100.0-mL beaker. What is the concentration of the solution in the beaker?

4.134 Three 1.0-g samples of $PbCl_2$, KCl, and $CaCl_2$ are placed in separate 500-mL beakers. In each case, enough 25°C water is added to bring the total volume of the mixture to 250 mL. Each of the mixtures is then stirred for five minutes. Which of the mixtures will have the highest concentration of chloride (Cl^-) ion?

4.135 What is the molarity of pure water with a density of 1.00 g/mL?

4.136 A 25-mL sample of 0.50 M NaOH is combined with a 75-mL sample of 0.50 M NaOH. What is the concentration of the resulting NaOH solution?

4.137 Nitric acid can be reacted with zinc according to the following chemical equation.

$$4HNO_3(aq) + Zn(s) \longrightarrow Zn(NO_3)_2(aq) + 2H_2O(l) + 2NO_2(g)$$

If 3.75 g of Zn is added to 175 mL of 0.500 M HNO_3, what mass of NO_2 would be produced by the chemical reaction?

4.138 You are asked to prepare 0.250 L of a solution that is 0.500 M in nitrate ion. Your only source of nitrate ion is a bottle of 1.00 M calcium nitrate. What volume (mL) of the calcium nitrate solution must you use?

4.139 How many grams of precipitate are formed if 175 mL of a 0.750 M aluminum sulfate solution and 375 mL of a 1.15 M sodium hydroxide solution are mixed together?

4.140 Potassium hydrogen phthalate (abbreviated as KHP) has the molecular formula $KHC_8H_4O_4$ and a molar mass of 204.22 g/mol. KHP has one acidic hydrogen. A solid sample of KHP is dissolved in 50 mL of water and titrated to the equivalence point with 22.90 mL of a 0.5010 M NaOH solution. How many grams of KHP were used in the titration?

Cumulative-Skills Problems

Key: The problems under this heading combine skills introduced in previous chapters with those given in the current one.

4.141 Silver nitrate reacts with strontium chloride in an aqueous precipitation reaction. What are the formulas of silver nitrate and strontium chloride? Write the molecular equation and net ionic equation for the reaction. What are the names of the products? Give the molecular equation for another reaction that produces the same precipitate.

4.142 Lead(II) nitrate reacts with cesium sulfate in an aqueous precipitation reaction. What are the formulas of lead(II) nitrate and cesium sulfate? Write the molecular equation and net ionic equation for the reaction. What are the names of the products? Give the molecular equation for another reaction that produces the same precipitate.

4.143 Elemental bromine is the source of bromine compounds. The element is produced from certain brine solutions that occur naturally. These brines are essentially solutions of calcium bromide that, when treated with chlorine gas, yield bromine in a displacement reaction. What are the molecular equation and net ionic equation for the reaction? A solution containing 40.0 g of calcium bromide requires 14.2 g of chlorine to react completely with it, and 22.2 g of calcium chloride is produced in addition to whatever bromine is obtained. How many grams of calcium bromide are required to produce 10.0 pounds of bromine?

4.144 Barium carbonate is the source of barium compounds. It is produced in an aqueous precipitation reaction from barium sulfide and sodium carbonate. (Barium sulfide is a soluble compound obtained by heating the mineral barite, which is barium sulfate, with carbon.) What are the molecular equation and net ionic equation for the precipitation reaction? A solution containing 33.9 g of barium sulfide requires 21.2 g of sodium carbonate to react completely with it, and 15.6 g of sodium sulfide is produced in addition to whatever barium carbonate is obtained. How many grams of barium sulfide are required to produce 5.00 tons of barium carbonate? (One ton equals 2000 pounds.)

4.145 Mercury(II) nitrate is treated with hydrogen sulfide, H_2S, forming a precipitate and a solution. Write the molecular equation and the net ionic equation for the reaction. An acid is formed; is it strong or weak? Name each of the products. If 81.15 g of mercury(II) nitrate and 8.52 g of hydrogen sulfide are mixed in 550.0 g of water to form 58.16 g of precipitate, what is the mass of the solution after the reaction?

4.146 Mercury(II) nitrate is treated with hydrogen sulfide, H_2S, forming a precipitate and a solution. Write the molecular equation and the net ionic equation for the reaction. An acid is formed; is it strong or weak? Name each of the products. If 65.65 g of mercury(II) nitrate and 4.26 g of hydrogen sulfide are mixed in 395.0 g of water to form 54.16 g of precipitate, what is the mass of the solution after the reaction?

4.147 A transition metal X forms an oxide of formula X_2O_3. It is found that only 50% of X atoms in this compound are in the +3 oxidation state. The only other stable oxidation states of X are +2 and +5. What percentage of X atoms are in the +2 oxidation state in this compound?

4.148 Iron forms a sulfide with the approximate formula Fe_7S_8. Assume that the oxidation state of sulfur is −2 and that iron atoms exist in both +2 and +3 oxidation states. What is the ratio of Fe(II) atoms to Fe(III) atoms in this compound?

4.149 What volume of a solution of ethanol, C_2H_6O, that is 94.0% ethanol by mass contains 0.200 mol C_2H_6O? The density of the solution is 0.807 g/mL.

4.150 What volume of a solution of ethylene glycol, $C_2H_6O_2$, that is 56.0% ethylene glycol by mass contains 0.350 mol $C_2H_6O_2$? The density of the solution is 1.072 g/mL.

4.151 A 10.0-mL sample of potassium iodide solution was analyzed by adding an excess of silver nitrate solution to produce silver iodide crystals, which were filtered from the solution.

$$KI(aq) + AgNO_3(aq) \longrightarrow KNO_3(aq) + AgI(s)$$

If 2.183 g of silver iodide was obtained, what was the molarity of the original KI solution?

4.152 A 25.0-mL sample of sodium sulfate solution was analyzed by adding an excess of barium chloride solution to produce barium sulfate crystals, which were filtered from the solution.

$$Na_2SO_4(aq) + BaCl_2(aq) \longrightarrow 2NaCl(aq) + BaSO_4(s)$$

If 5.719 g of barium sulfate was obtained, what was the molarity of the original Na_2SO_4 solution?

4.153 A metal, M, was converted to the chloride MCl_2. Then a solution of the chloride was treated with silver nitrate to give silver chloride crystals, which were filtered from the solution.

$$MCl_2(aq) + 2AgNO_3(aq) \longrightarrow$$
$$M(NO_3)_2(aq) + 2AgCl(s)$$

If 2.434 g of the metal gave 7.964 g of silver chloride, what is the atomic weight of the metal? What is the metal?

4.154 A metal, M, was converted to the sulfate, $M_2(SO_4)_3$. Then a solution of the sulfate was treated with barium chloride to give barium sulfate crystals, which were filtered off.

$$M_2(SO_4)_3(aq) + 3BaCl_2(aq) \longrightarrow$$
$$2MCl_3(aq) + 3BaSO_4(s)$$

If 1.200 g of the metal gave 6.026 g of barium sulfate, what is the atomic weight of the metal? What is the metal?

4.155 Phosphoric acid is prepared by dissolving phosphorus(V) oxide, P_4O_{10}, in water. What is the balanced equation for this reaction? How many grams of P_4O_{10} are required to make 1.50 L of aqueous solution containing 5.00% phosphoric acid by mass? The density of the solution is 1.025 g/mL.

4.156 Iron(III) chloride can be prepared by reacting iron metal with chlorine. What is the balanced equation for this reaction? How many grams of iron are required to make 3.00 L of aqueous solution containing 9.00% iron(III) chloride by mass? The density of the solution is 1.067 g/mL.

4.157 An alloy of iron and carbon was treated with sulfuric acid, in which only iron reacts.

$$2Fe(s) + 3H_2SO_4(aq) \longrightarrow Fe_2(SO_4)_3(aq) + 3H_2(g)$$

If a sample of alloy weighing 2.358 g gave 0.1067 g of hydrogen, what is the percentage of iron in the alloy?

4.158 An alloy of aluminum and magnesium was treated with sodium hydroxide solution, in which only aluminum reacts.

$$2Al(s) + 2NaOH(aq) + 6H_2O(l) \longrightarrow$$
$$2NaAl(OH)_4(aq) + 3H_2(g)$$

If a sample of alloy weighing 1.118 g gave 0.1068 g of hydrogen, what is the percentage of aluminum in the alloy?

4.159 Determine the volume of sulfuric acid solution needed to prepare 37.4 g of aluminum sulfate, $Al_2(SO_4)_3$, by the reaction

$$2Al(s) + 3H_2SO_4(aq) \longrightarrow Al_2(SO_4)_3(aq) + 3H_2(g)$$

The sulfuric acid solution, whose density is 1.104 g/mL, contains 15.0% H_2SO_4 by mass.

4.160 Determine the volume of sodium hydroxide solution needed to prepare 26.2 g sodium phosphate, Na_3PO_4, by the reaction

$$3NaOH(aq) + H_3PO_4(aq) \longrightarrow Na_3PO_4(aq) + 3H_2O(l)$$

The sodium hydroxide solution, whose density is 1.133 g/mL, contains 12.0% NaOH by mass.

4.161 The active ingredients in an antacid tablet contained only calcium carbonate and magnesium carbonate. Complete reaction of a sample of the active ingredients required 39.20 mL of 0.08750 *M* hydrochloric acid. The chloride salts from the reaction were obtained by evaporation of the filtrate from this titration; they weighed 0.1900 g. What was the percentage by mass of the calcium carbonate in the active ingredients of the antacid tablet?

4.162 The active ingredients of an antacid tablet contained only magnesium hydroxide and aluminum hydroxide. Complete neutralization of a sample of the active ingredients required 48.5 mL of 0.187 *M* hydrochloric acid. The chloride salts from this neutralization were obtained by evaporation of the filtrate from the titration; they weighed 0.4200 g. What was the percentage by mass of magnesium hydroxide in the active ingredients of the antacid tablet?

5

The Gaseous State

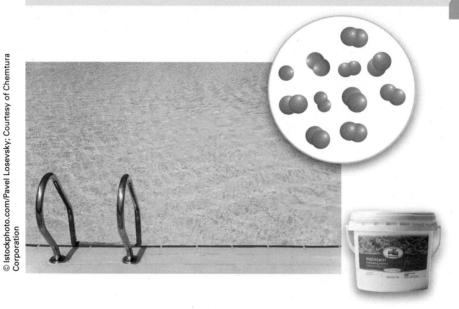

© Istockphoto.com/Pavel Losevsky; Courtesy of Chemtura Corporation

Bromine evaporates at room temperature to produce a dense gas. Compounds that contain bromine are used to disinfect swimming pools and hot tubs.

ases have several characteristics that distinguish them from liquids and solids. You can compress gases into smaller and smaller volumes. Anyone who has transported gases such as oxygen compressed in steel cylinders can appreciate this (Figure 5.1). Also, for gases, unlike for liquids and solids, you can relate pressure, volume, temperature, and molar amount of substance with fair accuracy by one simple equation, the ideal gas law. Suppose you wish to determine the amount of oxygen in a cylinder of compressed gas. You measure the pressure (with a gauge), the temperature, and the volume of gas (volume of the tank). Then, using the ideal gas law, you can calculate the amount of gas in the tank.

Kinetic-molecular theory describes a gas as composed of molecules in constant motion. This theory helps explain the simple relationship that exists among the pressure, volume, temperature, and amount of a gas. Kinetic theory has also enhanced our understanding of the flow of fluids, the transmission of sound, and the conduction of heat.

This chapter introduces the empirical gas laws, such as the ideal gas law, and the kinetic-molecular theory that explains these laws. After finishing the chapter, you will be able to answer questions such as, How many grams of oxygen are there in a 50.0-L gas cylinder at 21°C when the oxygen pressure is 15.7 atmospheres? (An atmosphere is a unit of pressure equal to that of the normal atmosphere.) You will also be able to calculate the average speed of an oxygen molecule in this tank (479 m/s, or 1071 mi/hr).

Figure 5.1 ▲

Cylinders of gas Gases such as oxygen and nitrogen can be transported as compressed gases in steel cylinders. Large volumes of gas at normal pressures can be compressed into a small volume. Note the pressure gauges.

Gas Laws

Most substances composed of small molecules are gases under normal conditions or else are easily vaporized liquids. Table 5.1 lists selected gaseous substances and some of their properties.

In the first part of this chapter, we will examine the quantitative relationships, or empirical laws, governing gases. We will first consider the concept of pressure.

5.1 Gas Pressure and Its Measurement

Pressure is defined as *the force exerted per unit area of surface.* A coin resting on a table exerts a force, and therefore a pressure, downward on the table as a consequence of gravity. The air above the table exerts an additional pressure on the table, because the air is also being pulled downward by gravity.

To obtain the SI unit of pressure and a feeling for its size, let us calculate the pressure on a table from a perfectly flat coin with a radius and mass equal to that of a new penny (9.3 mm in radius and 2.5 g). The force exerted by the coin from gravity equals the mass of the coin times the constant acceleration of gravity.

Table 5.1	Properties of Selected Gases			
Name	**Formula**	**Color**	**Odor**	**Toxicity**
Ammonia	NH_3	Colorless	Penetrating	Toxic
Carbon dioxide	CO_2	Colorless	Odorless	Nontoxic
Carbon monoxide	CO	Colorless	Odorless	Very toxic
Chlorine	Cl_2	Pale green	Irritating	Very toxic
Hydrogen	H_2	Colorless	Odorless	Nontoxic
Hydrogen sulfide	H_2S	Colorless	Foul	Very toxic
Methane	CH_4	Colorless	Odorless	Nontoxic
Nitrogen dioxide	NO_2	Red-brown	Irritating	Very toxic

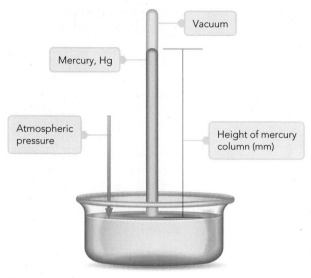

Figure 5.2 ▲

A mercury barometer The height *h* is proportional to the barometric pressure. For that reason, the pressure is often given as the height of the mercury column, in units of millimeters of mercury, mmHg.

Atmospheric, or barometric, pressure depends not only on altitude but also on weather conditions. The "highs" and "lows" of weather reports refer to high- and low-pressure air masses. A high is associated with fair weather; a low brings unsettled weather, or storms.

Acceleration is the change of speed per unit time, so the SI unit of acceleration is meters per second per second, abbreviated m/s². The constant acceleration of gravity is 9.81 m/s², and the force on the coin due to gravity is

Force = mas × constant acceleration of gravity
= $(2.5 \times 10^{-3}$ kg$) \times (9.81$ m/s²$) = 2.5 \times 10^{-2}$ kg·m/s²

The cross-sectional area of the coin is $\pi \times$ (radius)² = 3.14 × $(9.3 \times 10^{-3}$ m$)^2 = 2.7 \times 10^{-4}$ m². Therefore,

$$\text{Pressure} = \frac{\text{force}}{\text{area}} = \frac{2.5 \times 10^{-2} \text{ kg·m/s}^2}{2.7 \times 10^{-4} \text{ m}^2} = 93 \text{ kg/(m·s}^2)$$

The SI unit of pressure, kg/(m·s²), is given the name **pascal (Pa),** after the French physicist Blaise Pascal (1623–1662), who studied fluid pressure. Note that the pressure exerted by a coin the size and mass of a penny is approximately 100 Pa. The pressure exerted by the atmosphere is about 1000 times larger, or about 100,000 Pa. Thus, the pascal is an extremely small unit.

Chemists have traditionally used two other units of pressure, based on the mercury barometer. A **barometer** is *a device for measuring the pressure of the atmosphere.* The mercury barometer consists of a glass tube about one meter long, filled with mercury and inverted in a dish of mercury (Figure 5.2). At sea level the mercury in the tube falls to a height of about 760 mm above the level in the dish. This height is a direct measure of the atmospheric pressure (Figure 5.3). ◄ Air pressure downward on the surface of the mercury in the dish is transmitted through the liquid and exerts a pressure upward at the base of the mercury column, supporting it. A mercury column placed in a sealed flask, as in Figure 5.4, measures the gas pressure in the flask. It acts as a **manometer,** *a device that measures the pressure of a gas or liquid in a vessel.*

The unit **millimeters of mercury (mmHg),** also called the **torr** (after Evangelista Torricelli, who invented the mercury barometer in 1643), is *a unit of pressure equal to that exerted by a column of mercury 1 mm high at 0.00°C.* The **atmosphere (atm)** is a related *unit of pressure equal to exactly 760 mmHg.*

The general relationship between the pressure *P* and the height *h* of a liquid column in a barometer or manometer is

$$P = gdh$$

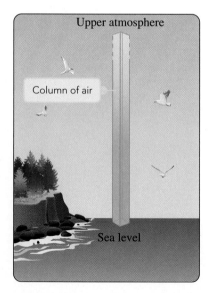

Figure 5.3 ◄

Atmospheric pressure from air mass The force of gravity on the mass of a column of air exerts a pressure at the earth's surface. This pressure varies slightly with weather, but is approximately 101 kPa (760 mmHg, or 1 atm).

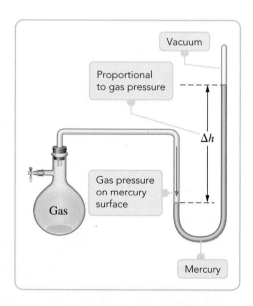

Figure 5.4 ▶

A flask equipped with a closed-tube manometer The gas pressure in the flask is proportional to the difference in heights between the liquid levels in the manometer, Δ*h*.

Table 5.2	Important Units of Pressure
Unit	**Relationship or Definition**
Pascal (Pa)	$kg/(m \cdot s^2)$
Atmosphere (atm)	$1 \text{ atm} = 1.01325 \times 10^5 \text{ Pa} \simeq 101 \text{ kPa}$
mmHg, or torr	$760 \text{ mmHg} = 1 \text{ atm}$
Bar	$1.01325 \text{ bar} = 1 \text{ atm}$

Here g is the constant acceleration of gravity (9.81 m/s^2) and d is the density of the liquid in the manometer. If g, d, and h are in SI units, the pressure is given in pascals. Table 5.2 summarizes the relationships among the various units of pressure. ▶

*In 1982, the International Union of Pure and Applied Chemistry (IUPAC) recommended that standard pressure be defined as equal to 1 **bar**, a unit of pressure equal to 1×10^5 Pa, slightly less than 1 atm. Despite this recommendation, 1 atm is still in common use by chemists as the standard pressure. Until the bar is more widely adopted by chemists as the standard unit of pressure, we will continue to use the atmosphere.*

Example 5.1   Converting Units of Pressure

Gaining Mastery Toolbox

Critical Concept 5.1
Many different units are used to express gas pressure. Of greatest importance to chemists are the units: pascal (Pa), atmosphere (atm), mmHg, or bar. There are conversion factors listed in Table 5.2 that allow conversion between these units.

Solution Essentials:
• Units of pressure (Table 5.2)
• Pressure conversion factors (Table 5.2)

The pressure of a gas in a flask is measured to be 797.7 mmHg using a mercury-filled manometer as in Figure 5.4. What is this pressure in pascals and atmospheres?

Problem Strategy This is a unit conversion problem. To convert to the appropriate units, use the conversion factors presented in Table 5.2 and the techniques presented in Chapter 1. Keep in mind that the conversion 760 mmHg = 1 atm is exact.

Solution Conversion to pascals:

$$797.7 \text{ mmHg} \times \frac{1.01325 \times 10^5 \text{ Pa}}{760 \text{ mmHg}} = \mathbf{1.064 \times 10^5 \text{ Pa}}$$

Conversion to atmospheres:

$$797.7 \text{ mmHg} \times \frac{1 \text{ atm}}{760 \text{ mmHg}} = \mathbf{1.050 \text{ atm}}$$

Answer Check Pressure conversions between the units of atm and mmHg are very common, so it is helpful to remember the relationship 1 atm = 760 mmHg in order to check your answers quickly.

Exercise 5.1 A gas in a container had a measured pressure of 57 kPa. Calculate the pressure in units of atm and mmHg.

■ See Problems 5.37 and 5.38.

CONCEPT CHECK 5.1
Suppose that you set up two barometers like the one shown in Figure 5.2. In one of the barometers you use mercury, and in the other you use water. Which of the barometers would have a higher column of liquid, the one with Hg or H_2O? Explain your answer.

5.2 Empirical Gas Laws

All gases under moderate conditions behave quite simply with respect to pressure, temperature, volume, and molar amount. By holding any two of these physical properties constant, it is possible to show a simple relationship between the other two. The discovery of these quantitative relationships, the empirical gas laws, occurred from the mid-seventeenth to the mid-nineteenth century.

Boyle's Law: Relating Volume and Pressure

One characteristic property of a gas is its *compressibility*—its ability to be squeezed into a smaller volume by the application of pressure. By comparison, liquids

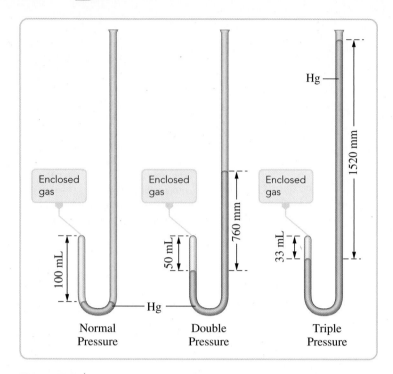

Table 5.3	Pressure–Volume Data for 1.000 g O₂ at 0°C		
	P (atm)	*V* (L)	*PV*
	0.2500	2.801	0.7002
	0.5000	1.400	0.7000
	0.7500	0.9333	0.7000
	1.000	0.6998	0.6998
	2.000	0.3495	0.6990
	3.000	0.2328	0.6984
	4.000	0.1744	0.6976
	5.000	0.1394	0.6970

Figure 5.5 ▲

Boyle's experiment The volume of the gas at normal atmospheric pressure (760 mmHg) is 100 mL. When the pressure is doubled by adding 760 mm of mercury, the volume is halved (to 50 mL). Tripling the pressure decreases the volume to one-third of the original (to 33 mL).

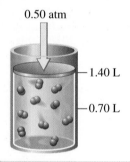

a When a 1.000-g sample of O₂ gas at 0°C is placed in a container at a pressure of 0.50 atm, it occupies a volume of 1.40 L.

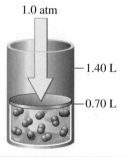

b When the pressure on the O₂ sample is doubled to 1.0 atm, the volume is reduced to 0.70 L, which is half the original volume.

Figure 5.6 ▲

Model of gas pressure–volume relationship at a constant temperature

and solids are relatively incompressible. The compressibility of gases was first studied quantitatively by Robert Boyle in 1661. When he poured mercury into the open end of a J-shaped tube, the volume of the enclosed gas decreased (Figure 5.5). Each addition of mercury increased the pressure on the gas, decreasing its volume. From such experiments, he formulated the law now known by his name. According to **Boyle's law,** *the volume of a sample of gas at a given temperature varies inversely with the applied pressure.* That is, $V \propto 1/P$, where V is the volume, P is the pressure, and $\propto$ means "is proportional to." Thus, if the pressure is doubled, the volume is halved.

Boyle's law can also be expressed in the form of an equation. Putting pressure and volume on the same side of the equation, you can write

> Boyle's law:
> $PV = \text{constant}$ (for a given amount of gas at fixed temperature)

That is, for a given amount of gas at a fixed temperature, the pressure times the volume equals a constant. Table 5.3 gives some pressure and volume data for 1.000 g O₂ at 0°C. Figure 5.6 presents a molecular view of the pressure–volume relationship for two data points in Table 5.3. Note that the product of the pressure and volume for each data pair is nearly constant. By plotting the volume of the oxygen at different pressures (as shown in Figure 5.7), you obtain a graph showing the inverse relationship of P and V.

You can use Boyle's law to calculate the volume occupied by a gas when the pressure changes. Consider the 50.0-L gas cylinder of oxygen mentioned in the chapter opening. The pressure of gas in the cylinder is 15.7 atm at 21°C. What volume of oxygen can you get from the cylinder at 21°C if the atmospheric pressure is 1.00 atm? You can write P_i and V_i for the initial pressure (15.7 atm) and initial volume (50.0 L), and P_f and V_f for the final pressure (1.00 atm) and final volume (to be determined). Because the temperature does not change, the product of the pressure and volume remains constant. Thus, you can write

$$P_f V_f = P_i V_i$$

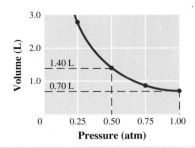

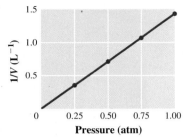

Figure 5.7 ◀

Gas pressure–volume relationship

a Plot of volume vs. pressure for a sample of oxygen. The volume (of 1.000 g O_2 at 0°C) decreases with increasing pressure. When the pressure is doubled (from 0.50 atm to 1.00 atm), the volume is halved (from 1.40 L to 0.70 L).

b Plot of $1/V$ vs. pressure (at constant temperature) for the same sample. The straight line indicates that volume varies inversely with pressure.

Dividing both sides of the equation by P_f gives

$$V_f = V_i \times \frac{P_i}{P_f}$$

Substituting into this equation yields

$$V_f = 50.0 \text{ L} \times \frac{15.7 \text{ atm}}{1.00 \text{ atm}} = 785 \text{ L}$$

Note that the initial volume is multiplied by a ratio of pressures. You know the oxygen gas is changing to a lower pressure and will therefore expand, so this ratio will be greater than 1. The final volume, 785 L, is that occupied by all of the gas at 1.00 atm pressure. However, because the cylinder holds 50.0 L, only this volume of gas remains in the cylinder. The volume that escapes is $(785 - 50.0)$ L $= 735$ L.

Example 5.2   Using Boyle's Law

Gaining Mastery Toolbox

Critical Concept 5.2
At constant temperature there is an inverse relationship between the volume (V) and pressure (P) of a gas. Because it is an inverse relationship between V and P, if one of the two variables (V or P) increases, the other variable must decrease by a proportional amount. For example, you can use this relationship to predict that the volume of a gas sample will decrease by a factor of 2 when the pressure on the gas is doubled.

Solution Essentials:
• Boyle's law (PV = constant)
• Units of pressure (Table 5.2)

A volume of air occupying 12.0 dm³ at 98.9 kPa is compressed to a pressure of 119.0 kPa. The temperature remains constant. What is the new volume?

Problem Strategy Putting the data for the problem in tabular form, you see what data you have and what you must find. This will suggest the method of solution.

$V_i = 12.0 \text{ dm}^3 \quad P_i = 98.9 \text{ kPa}$ ⎫ Temperature and
$V_f = \quad ? \quad\quad\quad P_f = 119.0 \text{ kPa}$ ⎬ moles remain
⎭ constant.

Because P and V vary but temperature and moles are constant, you use Boyle's law.

Solution Application of Boyle's law gives

$$V_f = V_i \times \frac{P_i}{P_f} = 12.0 \text{ dm}^3 \times \frac{98.9 \text{ kPa}}{119.0 \text{ kPa}} = \mathbf{9.97 \text{ dm}^3}$$

Answer Check In general, when the pressure on a gas increases, the gas is compressed and therefore occupies a smaller volume. When the pressure on a gas decreases, the gas expands—it will occupy a larger volume. Use these concepts to check your answers. In this example, since the pressure increases, the smaller final volume of the answer is what we would expect. (Note that you get the answer to this type of problem by multiplying the volume by a ratio of pressures. This ratio should be less than 1 if the volume decreases, but more than 1 if the volume increases.)

Exercise 5.2 A volume of carbon dioxide gas, CO_2, equal to 20.0 L was collected at 23°C and 1.00 atm pressure. What would be the volume of carbon dioxide collected at 23°C and 0.830 atm?

■ See Problems 5.39, 5.40, 5.41, and 5.42.

Figure 5.8 ▶

Effect of temperature on volume of a gas

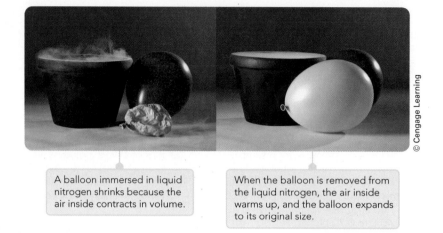

A balloon immersed in liquid nitrogen shrinks because the air inside contracts in volume.

When the balloon is removed from the liquid nitrogen, the air inside warms up, and the balloon expands to its original size.

© Cengage Learning

Before leaving the subject of Boyle's law, we should note that the pressure–volume product for a gas is not precisely constant. You can see this from the PV data given in Table 5.3 for oxygen. In fact, all gases follow Boyle's law at low to moderate pressures but deviate from this law at high pressures. The extent of deviation depends on the gas. We will return to this point at the end of the chapter.

Charles's Law: Relating Volume and Temperature

Temperature also affects gas volume. When you immerse an air-filled balloon in liquid nitrogen (−196°C), the balloon shrinks (Figure 5.8). After the balloon is removed from the liquid nitrogen, it returns to its original size. A gas contracts when cooled and expands when heated.

One of the first quantitative observations of gases at different temperatures was made by Jacques Alexandre Charles in 1787. Charles was a French physicist and a pioneer in hot-air and hydrogen-filled balloons. ◀ Later, John Dalton (in 1801) and Joseph Louis Gay-Lussac (in 1802) continued these kinds of experiments, which showed that a sample of gas at a fixed pressure increases in volume *linearly* with temperature. By "linearly," we mean that if we plot the volume occupied by a given sample of gas at various temperatures, we get a straight line (Figure 5.9).

When you extend the straight lines in Figure 5.9 from the last experimental point toward lower temperatures—that is, when you *extrapolate* the straight lines backward—you find that they all intersect at a common point. This point occurs at a temperature of −273.15°C, where the graph indicates a volume of zero. This seems to say that if the substances remain gaseous, the volumes occupied will be zero at −273.15°C. This could not happen, however; all gases liquefy before they reach this temperature, and Charles's law does not apply to liquids. These extrapolations do show that we can express the volume variation of a gas with temperature more simply by choosing a different thermometer scale. Let us see how to do this.

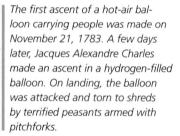

The first ascent of a hot-air balloon carrying people was made on November 21, 1783. A few days later, Jacques Alexandre Charles made an ascent in a hydrogen-filled balloon. On landing, the balloon was attacked and torn to shreds by terrified peasants armed with pitchforks.

Figure 5.9 ▶

Linear relationship of gas volume and temperature at constant pressure The graph shows gas volume versus temperature for a given mass of gas at 1.00 atm pressure. This linear relationship is independent of amount or type of gas. Note that all lines extrapolate to −273°C at zero volume.

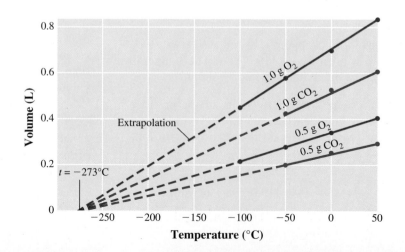

The fact that the volume occupied by a gas varies linearly with degrees Celsius can be expressed mathematically by the following equation:

$$V = bt + a$$

where t is the temperature in degrees Celsius and a and b are constants that determine the straight line. ▶ You can eliminate the constant a by observing that $V = 0$ at $t = -273.15$ for any gas. Substituting into the preceding equation, you get

$$0 = b(-273.15) + a \quad \text{or} \quad a = 273.15b$$

The equation for the volume can now be rewritten:

$$V = bt + 273.15b = b(t + 273.15)$$

Suppose you use a temperature scale equal to degrees Celsius plus 273.15, which you may recognize as the *Kelvin scale* (an absolute scale). ▶

$$K = {}^{\circ}C + 273.15$$

If you write T for the temperature on the Kelvin scale, you obtain

$$V = bT$$

This is **Charles's law,** which we can state as follows: *the volume occupied by any sample of gas at a constant pressure is directly proportional to the absolute temperature.* Thus, doubling the absolute temperature of a gas doubles its volume (Figure 5.10).

Charles's law can be rearranged into a form that is very useful for computation.

> Charle's law:
> $$\frac{V}{T} = \text{constant} \quad \text{(for a given amount of gas at a fixed pressure)}$$

This equation says that the volume divided by the absolute temperature, for a sample of gas at a fixed pressure, remains constant.

Consider a sample of gas at a fixed pressure, and suppose the temperature changes from its initial value T_i to a final value T_f. How does the volume change? Because the volume divided by absolute temperature is constant, you can write

$$\frac{V_f}{T_f} = \frac{V_i}{T_i}$$

Or, rearranging slightly,

$$V_f = V_i \times \frac{T_f}{T_i}$$

Note that to obtain the final volume, the initial volume is multiplied by a ratio of absolute temperatures. The next example illustrates a calculation involving a change in the temperature of a gas.

The mathematical equation of a straight line, $y = mx + b$, is discussed in Appendix A.

The Kelvin temperature scale and a more formal equation for temperature conversion were discussed in Section 1.6.

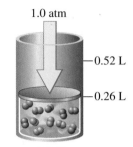

a A 1.0-g sample of O_2 at 100 K and 1.0 atm of pressure occupies a volume of 0.26 L.

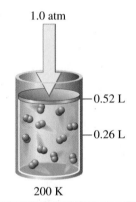

b When the absolute temperature of the gas is doubled to 200 K, the volume of O_2 doubles to 0.52 L.

Figure 5.10 ▲

Model of the temperature–volume relationship for a gas at a fixed volume

Example 5.3   Using Charles's Law

Critical Concept 5.3
At constant pressure there is a direct proportional relationship between the volume (V) and temperature (T) of a gas. Because it is a direct proportional relationship between V and T, if one of the two variables (V or T) increases, the other variable must increase by a proportional amount. For example, you can use this relationship to predict that the volume of a gas sample will triple when the Kelvin temperature is increased by a factor of 3.

Solution Essentials:
• Charles's law (V/T = constant)
• Units of pressure (Table 5.2)
• Kelvin temperature

Earlier we found that the total volume of oxygen that can be obtained from a particular tank at 1.00 atm and 21°C is 785 L (including the volume remaining in the tank). What would be this volume of oxygen if the temperature had been 28°C?

Problem Strategy You must first convert temperatures to the Kelvin scale. Then, list the data in tabular form so you can easily see what varies and what remains constant. You apply Charles's law when V and T vary and P is constant.

(continued)

(continued)

Solution $T_i = (21 + 273)K = 294$ K

$T_f = (28 + 273)K = 301$ K

Following is the data table:

$V_i = 785$ L $P_i = 1.00$ atm $T_i = 294$ K

$V_f = ?$ $P_f = 1.00$ atm $T_f = 301$ K

Note that T varies and P remains constant, so V must change. These are the conditions needed to apply Charles's law.

$$V_f = V_i \times \frac{T_f}{T_i} = 785 \text{ L} \times \frac{301 \text{ K}}{294 \text{ K}} = \mathbf{804\ L}$$

Answer Check   In general, when the temperature increases, the volume of gas increases, and when the temperature decreases, the volume of gas decreases. Applying these concepts to check the answer, we verify that when the temperature was increased, the final volume is greater than the starting volume. (Note that you get the answer to this type of problem by multiplying the volume by a ratio of absolute temperatures. This ratio should be more than 1 if the volume increases, but less than 1 if the volume decreases.)

Exercise 5.3   If you expect a chemical reaction to produce 4.38 dm³ of oxygen, O_2, at 19°C and 101 kPa, what will be the volume at 25°C and 101 kPa?

■ See Problems 5.45, 5.46, 5.47, and 5.48.

Although most gases follow Charles's law fairly well, they deviate from it at high pressures and low temperatures.

Combined Gas Law: Relating Volume, Temperature, and Pressure

Boyle's law ($V \propto 1/P$) and Charles's law ($V \propto T$) can be combined and expressed in a single statement: the volume occupied by a given amount of gas is proportional to the absolute temperature divided by the pressure ($V \propto T/P$). You can write this as an equation.

$$V = \text{constant} \times \frac{T}{P} \quad \text{or} \quad \frac{PV}{T} = \text{constant} \quad \text{(for a given amount of gas)}$$

Consider a problem in which you wish to calculate the final volume of a gas when the pressure and temperature are changed. PV/T is constant for a given amount of gas, so you can write

$$\frac{P_f V_f}{T_f} = \frac{P_i V_i}{T_i}$$

which rearranges to

$$V_f = V_i \times \frac{P_i}{P_f} \times \frac{T_f}{T_i}$$

Thus, the final volume is obtained by multiplying the initial volume by ratios of pressures and absolute temperatures.

♥WL Interactive Example 5.4   Using the Combined Gas Law

Gaining Mastery Toolbox

Critical Concept 5.4
The relationships between the pressure (P), volume (V), and temperature (T) of a gas can be combined into a single expression. This expression encompasses Boyle's and Charles's laws.

Solution Essentials:
- Combined gas law $\frac{P_f V_f}{T_f} = \frac{P_i V_i}{T_i}$
- Units of pressure (Table 5.2)
- Kelvin temperature

Modern determination of %N in an organic compound is an automated version of one developed by the French chemist Jean-Baptiste Dumas in 1830. The Dumas method uses hot copper(II) oxide to oxidize C and H in the compound to CO_2 and H_2O (both are trapped chemically) and to convert N in the compound to N_2 gas (Figure 5.11). Using the Dumas method, 39.8 mg of caffeine gives 10.1 cm³ of nitrogen gas at 23°C and 746 mmHg. What is the volume of nitrogen at 0°C and 760 mmHg?

(continued)

(*continued*)

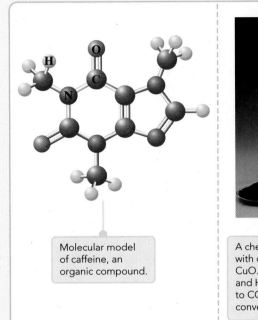

Molecular model of caffeine, an organic compound.

A chemist heats caffeine with copper(II) oxide, CuO. CuO oxidizes C and H in the compound to CO_2 and H_2O, and converts N to N_2.

Copper metal produced by the reaction.

Figure 5.11 ◀

Oxidation of an organic compound with copper(II) oxide

© Cengage Learning

Problem Strategy As always, you will need to convert temperature to the Kelvin scale and put the data in tabular form to determine the law(s) that apply. In this case, since both pressure and temperature are changing, we should expect to use more than one law to arrive at an answer.

Solution You first express the temperatures in kelvins:

$$T_i = (23 + 273) \text{ K} = 296 \text{ K}$$

$$T_f = (0 + 273) \text{ K} = 273 \text{ K}$$

Now you put the data for the problem in tabular form:

$V_i = 10.1 \text{ cm}^3$ $P_i = 746 \text{ mmHg}$ $T_i = 296 \text{ K}$
$V_f = ?$ $P_f = 760 \text{ mmHg}$ $T_f = 273 \text{ K}$

From the data, you see that you need to use Boyle's and Charles's laws combined to find how V varies as P and T change. Note that the increase in pressure decreases the volume, making the pressure ratio less than 1.

The decrease in temperature also decreases the volume, and the ratio of absolute temperatures is less than 1.

$$V_f = V_i \times \frac{P_i}{P_f} \times \frac{T_f}{T_i} = 10.1 \text{ cm}^3 \times \frac{746 \text{ mmHg}}{760 \text{ mmHg}} \times$$
$$\frac{273 \text{ K}}{296 \text{ K}} = \textbf{9.14 cm}^3$$

Answer Check This gas sample is undergoing a temperature decrease and a pressure increase. These two changes will have opposite effects on the final gas volume. Considering the magnitude of the changes in temperature and pressure in this problem, we can conclude that the final volume of the gas should not be greatly different from its initial volume.

Exercise 5.4   A balloon contains 5.41 dm^3 of helium, He, at 24°C and 101.5 kPa. Suppose the gas in the balloon is heated to 35°C. If the helium pressure is now 102.8 kPa, what is the volume of the gas?

■ See Problems 5.51 and 5.52.

CONCEPT CHECK 5.2

To conduct some experiments, a 10.0-L flask equipped with a movable plunger, as illustrated here, is filled with enough H_2 gas to come to a pressure of 20 atm.

a In the first experiment, we decrease the temperature in the flask by 10°C and then increase the volume. Predict how the pressure in the flask changes during each of these events and, if possible, how the final pressure compares to your starting pressure.

b Once again we start with the pressure in the flask at 20 atm. The flask is then heated 10°C, followed by a volume decrease. Predict how the pressure in the flask changes during each of these events and, if possible, how the final pressure compares to your starting pressure.

Figure 5.12 ▶

The molar volume of a gas The box has a volume of 22.4 L, equal to the molar volume of a gas at STP. The basketball is shown for comparison.

© Cengage Learning

Avogadro's Law: Relating Volume and Amount

In 1808 the French chemist Joseph Louis Gay-Lussac (1778–1850) concluded from experiments on gas reactions that the volumes of reactant gases at the same pressure and temperature are in ratios of small whole numbers (*the law of combining volumes*). For example, two volumes of hydrogen react with one volume of oxygen gas to produce water.

$$2H_2(g) \ + \ O_2(g) \ \longrightarrow \ 2H_2O(g)$$
2 volumes   1 volume

Three years later, the Italian chemist Amedeo Avogadro (1776–1856) interpreted the law of combining volumes in terms of what we now call **Avogadro's law:** *equal volumes of any two gases at the same temperature and pressure contain the same number of molecules.* Thus, two volumes of hydrogen contain twice the number of molecules as in one volume of oxygen, in agreement with the chemical equation for the reaction.

One mole of any gas contains the same number of molecules (Avogadro's number = 6.02×10^{23}) and by Avogadro's law must occupy the same volume at a given temperature and pressure. This *volume of one mole of gas* is called the **molar gas volume, V_m**. Volumes of gases are often compared at **standard temperature and pressure (STP),** *the reference conditions for gases chosen by convention to be 0°C and 1 atm pressure.* At STP, the molar gas volume is found to be 22.4 L/mol (Figure 5.12).

We can re-express Avogadro's law as follows: the molar gas volume at a given temperature and pressure is a specific constant independent of the nature of the gas.

> Avogadro's law:
> V_m = specific constant (= 22.4 L/mol at STP)
> (depending on T and P but independent of the gas)

Table 5.4 lists the molar volumes of several gases. The values agree within about 2% of the value expected from the empirical gas laws (the "ideal gas" value). We will discuss the reason for the deviations from this value at the end of this chapter, in the section on real gases.

5.3 The Ideal Gas Law

In the previous section, we discussed the empirical gas laws. Here we will show that these laws can be combined into one equation, called the *ideal gas equation.* Earlier we combined Boyle's law and Charles's law into the equation

$$V = \text{constant} \times \frac{T}{P} \qquad \text{(for a given amount of gas)}$$

This "constant" is independent of the temperature and pressure but does depend on the amount of gas. For one mole, the constant will have a specific value, which we will denote as R. The molar volume, V_m, is

$$V_m = R \times \frac{T}{P}$$

Table 5.4	Molar Volumes of Several Gases at 0.0°C and 1.00 atm
Gas	**Molar Volume (L)**
He	22.40
H_2	22.43
O_2	22.39
CO_2	22.29
NH_3	22.09
Ideal gas*	22.41

*An ideal gas follows the empirical gas laws.

Nitrogen Monoxide Gas and Biological Signaling

In 1998, the Nobel committee awarded its prize in physiology or medicine to three scientists for the astounding discovery that nitrogen monoxide (nitric oxide) gas, NO, functions as the signaling agent between biological cells in a wide variety of chemical processes. Until this discovery, biochemists had thought that the major chemical reactions in a cell always involved very large molecules. Now they discovered that a simple gas, NO, could play a central role in cell chemistry.

Prizewinners Robert Furchgott and Louis Ignarro, independently, unraveled the role of nitrogen monoxide in blood-pressure regulation. Cells in the lining of arteries detect increased blood pressure and respond by producing nitrogen monoxide. NO rapidly diffuses through the artery wall to cells in the surrounding muscle tissue. In response, the muscle tissue relaxes, the blood vessel expands, and the blood pressure drops.

In a related discovery, another prizewinner, Ferid Murad, explained how nitroglycerin works to alleviate the intense chest pain of an angina attack, which results from reduced blood flow to the heart muscle as a result of partial blockage of arteries by plaque. Physicians have prescribed nitroglycerin for angina for more than a century, knowing only that it works. Murad found that nitroglycerin breaks down in the body to form nitrogen monoxide, which relaxes the arteries, allowing greater blood flow to the heart. Alfred Nobel, were he alive today, would no doubt be stunned by this news. Nobel, who established the prizes bearing his name, made his fortune in the nineteenth century from his invention of dynamite, a mixture of nitroglycerin with clay that tamed the otherwise hazardous explosive. When Nobel had heart trouble, his physician recommended that he eat a small quantity of nitroglycerin; he refused. In a letter, he wrote, "It is ironical that I am now ordered by my physician to eat nitroglycerin." Today,

a patient may either take nitroglycerin pills (containing tenths of a milligram of compound in a stabilized mixture) for occasional use or apply a chest patch to dispense nitroglycerin continuously to the skin, where it is absorbed (Figure 5.13).

Research papers on the biological role of nitrogen monoxide now number in the tens of thousands. For example, scientists have found that white blood cells use nitrogen monoxide in a kind of chemical warfare. These cells emit concentrated clouds of NO that surround bacterial or tumor cells, killing them by interfering with certain cell processes. Researchers have also discovered that nitrogen monoxide plays a role in penile erection. Pharmacologists found that the drug Viagra (Figure 5.14) assists in the action of NO in dilating arteries, leading to erection.

Courtesy Schwarz Pharma

Figure 5.13 ▲

Nitroglycerin patch Nitroglycerin has been used for more than a hundred years to treat and prevent angina attacks. Here a patient wears a patch that dispenses nitroglycerin steadily over a period of time.

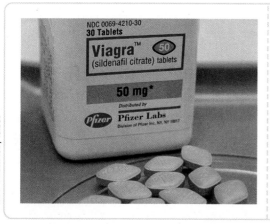

© AP Photo/Toby Talbot

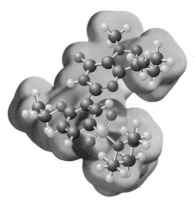

Figure 5.14 ◄

Viagra Viagra is marketed as an aid to erectile dysfunction. Sexual stimulation produces nitrogen monoxide, which in turn aids an enzyme in the production of a substance called cGMP that relaxes arterial muscles, allowing blood to flow to the penis. Viagra enhances the action of NO by inhibiting an enzyme that degrades cGMP.

■ See Problems 5.131 and 5.132.

Table 5.5	Molar Gas Constant in Various Units

Value of R

0.082058 L·atm/(K·mol)

8.3145 J/(K·mol)*

8.3145 kg·m²/(s²·K·mol)

8.3145 kPa·dm³/(K·mol)

1.9872 cal/(K·mol)*

*The units of pressure times volume are the units of energy—for example, joules (J) or calories (cal).

According to Avogadro's law, the molar volume at a specific value of T and P is a constant independent of the nature of the gas, and this implies that R is a constant independent of the gas. The **molar gas constant, R,** is *the constant of proportionality relating the molar volume of a gas to T/P.* Values of R in various units are given in Table 5.5.

The preceding equation can be written for n moles of gas by multiplying both sides by n.

$$\underbrace{nV_m}_{V} = \frac{nRT}{P} \quad \text{or} \quad PV = nRT$$

Because V_m is the volume per mole, nV_m is the total volume V. *The equation PV = nRT, which combines all of the gas laws,* is called the **ideal gas law.**

> Ideal gas law:
> $PV = nRT$

The ideal gas law includes all the information contained in Boyle's, Charles's, and Avogadro's laws. In fact, starting with the ideal gas law, you can derive any of the other gas laws.

Example 5.5 Deriving Empirical Gas Laws from the Ideal Gas Law

Gaining Mastery Toolbox

Critical Concept 5.5
The ideal gas law can be used to derive relationships between n, P, T, and V. Interpreting the equation $PV = nRT$ can lead you to the relationships between each of the variables. For example, for a sample of gas the pressure (P) is directly proportional to the number of moles (n) of gas and the temperature (T) of the gas.

Solution Essentials:
- Ideal gas law ($PV = nRT$)
- Units of pressure (Table 5.2)
- Kelvin temperature

Prove the following statement: the pressure of a given amount of gas at a fixed volume is proportional to the absolute temperature. This is sometimes called *Amontons's law.* In 1702 Guillaume Amontons constructed a thermometer based on measurement of the pressure of a fixed volume of air. The principle he employed is now used in special gas thermometers to establish the Kelvin scale.

Problem Strategy Because we need a relationship between pressure and temperature, the ideal gas law is a logical starting place.

Solution From the ideal gas law,
$$PV = nRT$$
Solving for P, you get
$$P = \left(\frac{nR}{V}\right)T$$
Note that everything in parentheses in this equation is constant. Therefore, you can write
$$P = \text{constant} \times T$$
Or, expressing this as a proportion,
$$P \propto T$$
Boyle's law and Charles's law follow from the ideal gas law by a similar derivation.

Answer Check We can apply common experience to qualitatively verify this relationship. Think about what happens to the pressure in a tire when the temperature is increased: the hotter the tire gets, the more the pressure increases.

Exercise 5.5 Show that the moles of gas are proportional to the pressure for constant volume and temperature.

■ See Problems 5.55 and 5.56.

No gas is "ideal." But the ideal gas law is very useful even though it is only an approximation. The behavior of real gases is described at the end of this chapter.

The limitations that apply to Boyle's, Charles's, and Avogadro's laws also apply to the ideal gas law. That is, the ideal gas law is most accurate for low to moderate pressures and for temperatures that are not too low. ◄

Calculations Using the Ideal Gas Law

The type of problem to which Boyle's and Charles's laws are applied involves a change in conditions (P, V, or T) of a gas. The ideal gas law allows us to solve another type of problem: given any three of the quantities P, V, n, and T, calculate the unknown quantity. Example 5.6 illustrates such a problem.

Example 5.6   Using the Ideal Gas Law

Critical Concept 5.6
The ideal gas law ($PV = nRT$) is applied to systems that are not changing. Using the ideal gas law you can determine the value of one of the variables (P, V, n, or T) while the remaining values represented in the equation are known.

Solution Essentials:
- Ideal gas law ($PV = nRT$)
- Units of pressure (Table 5.2)
- Kelvin temperature

Answer the question asked in the chapter opening: how many grams of oxygen, O_2, are there in a 50.0-L gas cylinder at 21°C when the oxygen pressure is 15.7 atm?

Problem Strategy In asking for the mass of oxygen, we are in effect asking for moles of gas, n, because mass and moles are directly related. The problem gives P, V, and T, so you can use the ideal gas law to solve for **n**. The proper value to use for R depends on the units of P and V. Here they are in atmospheres and liters, respectively, so you use $R = 0.0821$ L·atm/(K·mol) from Table 5.5. Once you have solved for n, convert to mass by using the molar mass of oxygen.

Solution The data given in the problem are

Variable	Value
P	15.7 atm
V	50.0 L
T	(21 + 273) K = 294 K
n	?

Solving the ideal gas law for n gives

$$n = \frac{PV}{RT}$$

Substituting the data gives

$$n = \frac{15.7 \text{ atm} \times 50.0 \text{ L}}{0.0821 \text{ L·atm/(K·mol)} \times 294 \text{ K}} = 32.5 \text{ mol}$$

and converting moles to mass of oxygen yields

$$32.5 \text{ mol } O_2 \times \frac{32.0 \text{ g } O_2}{1 \text{mol } O_2} = \mathbf{1.04 \times 10^3 g\ O_2}$$

Answer Check If you were to forget that oxygen is a diatomic gas with the formula O_2, you would not get this problem right. Keep in mind, too, that the other gaseous diatomic elements are H_2, N_2, F_2, and Cl_2.

Exercise 5.6 What is the pressure in a 50.0-L gas cylinder that contains 3.03 kg of oxygen, O_2, at 23°C?

■ See Problems 5.57, 5.58, 5.59, 5.60, 5.61, and 5.62.

Gas Density; Molecular-Mass Determination

The density of a substance is its mass divided by its volume, and because the volume of a gas varies with temperature and pressure, the density of a gas also varies with temperature and pressure (Figure 5.15). In the next example, we use the ideal gas law to calculate the density of a gas at any temperature and pressure. We do this by calculating the moles, and from this the mass, in a liter of gas at the given temperature and pressure; the mass per liter of gas is its density. At the end of this section, we will give an alternative method of solving this problem, using a formula relating density to temperature, pressure, and molecular weight.

Example 5.7   Calculating Gas Density

Critical Concept 5.7
Gas density varies with temperature and pressure. In a fixed volume, at a given temperature and pressure, the density of the gas will depend on the mass of gas in that volume (e.g., units of g/L).

Solution Essentials:
- Ideal gas law ($PV = nRT$)
- Units of pressure (Table 5.2)
- Kelvin temperature

What is the density of oxygen, O_2, in grams per liter at 25°C and 0.850 atm?

Problem Strategy The density of a gas is often expressed in g/L. Using the ideal gas law, you can calculate the moles of O_2 in 1 L of O_2. Next, you can convert the moles of O_2 to a mass of O_2, keeping in mind that this mass is the amount of O_2 per liter of O_2 (g/1 L), which is the density.

(continued)

(continued)

Solution The data given are

Variable	Value
P	0.850 atm
V	1 L (exact value)
T	(25 + 273) K = 298 K
n	?

Therefore,

$$n = \frac{PV}{RT} = \frac{0.850 \text{ atm} \times 1 \text{ L}}{0.0821 \text{ L·atm/(K·mol)} \times 298 \text{ K}} = 0.0347 \text{ mol}$$

Now convert mol O_2 to grams:

$$0.0347 \text{ mol } O_2 \times \frac{32.0 \text{ g } O_2}{1 \text{ mol } O_2} = 1.11 \text{ g } O_2$$

Therefore, the density of O_2 at 25°C and 0.850 atm is **1.11 g/L.**

Suppose you wanted the density of Cl_2, instead of O_2, at this T and P. Only the conversion of moles to mass is affected. The previous calculation becomes

$$0.0347 \text{ mol } Cl_2 \times \frac{70.9 \text{ g } Cl_2}{1 \text{ mol } Cl_2} = 2.46 \text{ g } Cl_2$$

The density of Cl_2 at 25°C and 0.850 atm is 2.46 g/L. Note that the density of a gas is directly proportional to its molecular mass.

Answer Check Because gases consist of molecules spread very far apart, you should expect any reasonable answer for gas density to be very low when compared to values for liquids and solids. Therefore, gas densities should be on the order of this answer.

Figure 5.15 ▲

Hot-air ballooning A propane gas burner on board a balloon heats the air. The heated air expands, occupying a larger volume; it therefore has a lower density than the surrounding air. The hot air and balloon rise.

Exercise 5.7 Calculate the density of helium, He, in grams per liter at 21°C and 752 mmHg. The density of air under these conditions is 1.188 g/L. What is the difference in mass between 1 liter of air and 1 liter of helium? (This mass difference is equivalent to the buoyant, or lifting, force of helium per liter.)

■ See Problems 5.63, 5.64, 5.65, and 5.66.

Figure 5.16 ▲

A gas whose density is greater than that of air The reddish-brown gas being poured from the flask is bromine. Note how the gas, which is denser than air, hugs the bottom of the beaker.

The previous example demonstrates that the density of a gas is directly proportional to its molecular mass. Bromine, whose molecular mass is five times that of oxygen, is therefore more than five times as dense as air. Figure 5.16 shows bromine gas being poured into a beaker where it displaces the air and fills the beaker.

The relation between density and molecular mass of a gas suggests that you could use a measurement of gas density to determine its molecular mass. In fact, gas-density (vapor-density) measurements provided one of the first methods of determining molecular mass. The method was worked out by the French chemist Jean-Baptiste André Dumas in 1826. It can be applied to any substance that can be vaporized without decomposing.

As an illustration, consider the determination of the molecular mass of halothane, an inhalation anesthetic. The density of halothane vapor at 71°C (344 K) and 768 mmHg (1.01 atm) is 7.05 g/L. To obtain the molecular mass, you calculate the moles of vapor in a given volume. The molar mass equals mass divided by the moles in the same volume. From the density, you see that one liter of vapor has a mass of 7.05 g. The moles in this volume are obtained from the ideal gas law.

$$n = \frac{PV}{RT} = \frac{1.01 \text{ atm} \times 1 \text{ L}}{0.0821 \text{ L·atm}/(\text{K·mol}) \times 344 \text{ K}} = 0.0358 \text{ mol}$$

Therefore, the molar mass, M_m, is

$$M_m = \frac{m}{n} = \frac{7.05 \text{ g}}{0.0358 \text{ mol}} = 197 \text{ g/mol}$$

Thus, the molecular mass is 197 amu.

If you look at what we have done, you will see that all you need to have is the mass of vapor in *any* given volume (for a given T and P). An explicit determination of density is not necessary. You determine the moles of vapor from the volume using the ideal gas law, then determine the molar mass by dividing the mass by moles for this volume. The next example illustrates this. At the end of this section, we will derive an explicit formula for the molecular mass in terms of the density of a gas, which will provide an alternative way to solve this problem.

⊘WL Interactive Example 5.8   Determining the Molecular Mass of a Vapor

Gaining Mastery Toolbox

Critical Concept 5.8
The molar mass of a substance is the grams of substance per mol of substance (g/mol). The ideal gas law can be used to determine the moles of a gas sample when the pressure (P), temperature (T), and volume (V) are known.

Solution Essentials:
• Ideal gas law (PV = nRT)
• Units of pressure (Table 5.2)
• Molar mass
• Kelvin temperature

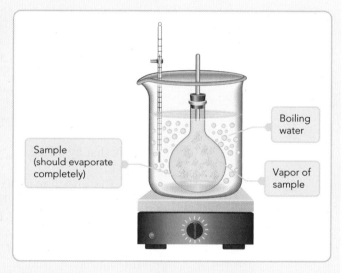

Using the apparatus described in Figure 5.17, a 200.0-mL flask at 99°C and a pressure of 733 mmHg is filled with the vapor of a volatile (easily vaporized) liquid. The mass of the substance in the flask is 0.970 g. What is the molecular mass of the liquid?

Problem Strategy You can calculate the moles of vapor from the ideal gas law and then calculate the molar mass by dividing mass by moles. Note that the temperature of the vapor is the same as the temperature of the boiling water and that the pressure of the vapor equals the barometric pressure.

Solution You have the following data for the vapor.

Figure 5.17 ▲

Finding the vapor density of a substance After the flask fills with the vapor of the substance at the temperature of the boiling water, you allow the flask to cool so that the vapor condenses. You obtain the mass of the substance (which was vapor) by weighing the flask and substance and then subtracting the mass of the empty flask. The pressure of the vapor equals the barometric pressure, and the temperature equals that of the boiling water.

Variable	Value
P	$733 \text{ mmHg} \times \dfrac{1 \text{ atm}}{760 \text{ mmHg}} = 0.964 \text{ atm}$
V	$200.0 \text{ mL} = 0.2000 \text{ L}$
T	$(99 + 273) \text{ K} = 372 \text{ K}$
n	?

From the ideal gas law, $PV = nRT$, you obtain

$$n = \frac{PV}{RT} = \frac{0.964 \text{ atm} \times 0.2000 \text{ L}}{0.0821 \text{ L·atm}/(\text{K·mol}) \times 372 \text{ K}}$$

$$n = 0.00631 \text{ mol}$$

Dividing the mass of the vapor by moles gives you the mass per mole (the molar mass).

$$\text{Molar mass} = \frac{\text{grams vapor}}{\text{moles vapor}} = \frac{0.970 \text{ g}}{0.00631 \text{ mol}}$$
$$= 154 \text{ g/mol}$$

Thus, the molecular mass is **154 amu.**

(continued)

(*continued*)

Answer Check No gas has a smaller molar mass than H_2 at 2 g/mol. Therefore, if you obtain an answer that is smaller than this, it must be incorrect. The upper limit on molar masses that you would expect is less definite, but solids and liquids that readily vaporize will have molar masses not much larger than a few hundred. An answer much larger than this should be regarded with suspicion.

Exercise 5.8 A sample of a gaseous substance at 25°C and 0.862 atm has a density of 2.26 g/L. What is the molecular mass of the substance?

■ See Problems 5.67, 5.68, 5.69, and 5.70.

From the ideal gas law, you can obtain an explicit relationship between the molecular mass and density of a gas. Recall that the molar mass ($M_m = m/n$) when expressed in grams per mole is numerically equal to the molecular mass. If you substitute $n = m/M_m$ into the ideal gas law, $PV = nRT$, you obtain

$$PV = \frac{m}{M_m} RT \quad \text{or} \quad PM_m = \frac{m}{V} RT$$

But m/V equals the density, d. Substituting this gives

$$PM_m = dRT$$

We can illustrate the use of this equation by solving Examples 5.7 and 5.8 again. In Example 5.7, we asked, "What is the density of oxygen, O_2, in grams per liter at 25°C (298 K) and 0.850 atm?" You rearrange the previous equation to give an explicit formula for the density and then substitute into this formula.

$$d = \frac{PM_m}{RT} = \frac{0.850 \text{ atm} \times 32.0 \text{ g/mol}}{0.0821 \text{ L·atm/(K·mol)} \times 298 \text{ K}} = 1.11 \text{ g/L}$$

Note that by giving M_m in g/mol and R in L·atm/(K·mol), you obtain the density in g/L.

The question in Example 5.8 is equivalent to the following. What is the molecular mass of a substance weighing 0.970 g whose vapor occupies 200.0 mL (0.2000 L) at 99°C (372 K) and 733 mmHg (0.964 atm)? The density of the vapor is 0.970 g/0.2000 L = 4.85 g/L. To obtain the molecular mass directly, you rearrange the equation $PM_m = dRT$ to give an explicit formula for M_m, then substitute into it.

$$M_m = \frac{dRT}{P} = \frac{4.85 \text{ g/L} \times 0.0821 \text{ L·atm/(K·mol)} \times 372 \text{ K}}{0.964 \text{ atm}} = 154 \text{ g/mol}$$

CONCEPT CHECK 5.3

Three 3.0-L flasks, each at a pressure of 878 mmHg, are in a room. The flasks contain He, Ar, and Xe, respectively.

a Which of the flasks contains the most atoms of gas?

b Which of the flasks has the greatest density of gas?

c If the He flask were heated and the Ar flask cooled, which of the three flasks would be at the highest pressure?

d Say the temperature of the He was lowered while that of the Xe was raised. Which of the three flasks would have the greatest number of moles of gas?

5.4 Stoichiometry Problems Involving Gas Volumes

In Chapter 3 you learned how to find the mass of one substance in a chemical reaction from the mass of another substance in the reaction. Now that you know how to use the ideal gas law, we can extend these types of problems to include gas volumes.

Consider the following reaction, which is often used to generate small quantities of oxygen gas:

$$2KClO_3(s) \xrightarrow[MnO_2]{\Delta} 2KCl(s) + 3O_2(g)$$

Suppose you heat 0.0100 mol of potassium chlorate, $KClO_3$, in a test tube. How many liters of oxygen can you produce at 298 K and 1.02 atm?

You solve such a problem by breaking it into two problems, one involving stoichiometry and the other involving the ideal gas law. You note that 2 mol $KClO_3$ yields 3 mol O_2. Therefore,

$$0.0100 \; \cancel{\text{mol } KClO_3} \times \frac{3 \text{ mol } O_2}{2 \; \cancel{\text{mol } KClO_3}} = 0.0150 \text{ mol } O_2$$

Now that you have the moles of oxygen produced, you can use the ideal gas law to calculate the volume of oxygen under the conditions given. You rearrange the ideal gas law, $PV = nRT$, and solve for the volume.

$$V = \frac{nRT}{P}$$

Then you substitute for n, T, and P. Because the pressure is given in units of atmospheres, you choose the value of R in units of L·atm/(K·mol). The answer comes out in liters.

$$V = \frac{0.0150 \; \cancel{\text{mol}} \times 0.0821 \text{ L} \cdot \cancel{\text{atm}}/(\cancel{K} \cdot \cancel{\text{mol}}) \times 298 \; \cancel{K}}{1.02 \; \cancel{\text{atm}}} = 0.360 \text{ L}$$

The next example further illustrates this method.

⒪WL Interactive Example 5.9   Solving Stoichiometry Problems Involving Gas Volumes

Gaining Mastery Toolbox

Critical Concept 5.9
Balanced chemical equations provide molar conversion factors between the reactants and products. The ideal gas law can be used to determine the number of moles of reactant prior to performing a stoichiometric calculation.

Solution Essentials:
- Ideal gas law ($PV = nRT$)
- Units of pressure (Table 5.2)
- Kelvin temperature

Automobiles are equipped with air bags that inflate on collision to protect the occupants from injury (Figure 5.18). Many such air bags are inflated with nitrogen, N_2, using the rapid reaction of sodium azide, NaN_3, and iron(III) oxide, Fe_2O_3, which is initiated by a spark. The overall reaction is

$$6NaN_3(s) + Fe_2O_3(s) \longrightarrow 3Na_2O(s) + 2Fe(s) + 9N_2(g)$$

How many grams of sodium azide would be required to provide 75.0 L of nitrogen gas at 25°C and 748 mmHg?

Problem Strategy This is a stoichiometry problem and a gas-law problem. The chemical equation relates the moles of NaN_3 to moles of N_2. You can use the ideal gas law to relate the volume of N_2 to moles of N_2 and hence moles of NaN_3. Then you can obtain the grams of NaN_3 from the moles of NaN_3.

The strategy for the stoichiometry portion of the problem is diagrammed as

Solution Here are the available data:

Variable	Value
P	$748 \; \cancel{\text{mmHg}} \times \dfrac{1 \text{ atm}}{760 \; \cancel{\text{mmHg}}} = 0.984$ atm
V	75.0 L
T	$(25 + 273)$ K $= 298$ K
n	?

(continued)

(*continued*)

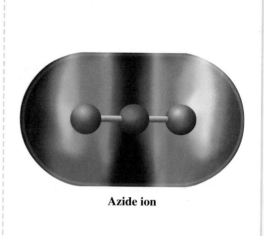

Azide ion

Figure 5.18 ▲

Automobile air bag Automobiles are equipped with air bags that inflate during collisions. Most air bags are inflated with nitrogen obtained from the rapid reaction of sodium azide, NaN_3 (the azide ion is N_3^-).

Note that the pressure was converted from mmHg to atmospheres and the temperature from degrees Celsius to kelvins. Rearrange the ideal gas law to obtain n.

$$n = \frac{PV}{RT}$$

Substituting from the data gives you the moles of N_2.

$$n = \frac{0.984 \text{ atm} \times 75.0 \text{ L}}{0.0821 \text{ L} \cdot \text{atm}/(\text{K} \cdot \text{mol}) \times 298 \text{ K}} = 3.02 \text{ mol}$$

From the moles of N_2, you use the chemical equation to obtain moles of NaN_3.

$$3.02 \text{ mol N}_2 \times \frac{6 \text{ mol NaN}_3}{9 \text{ mol N}_2} = 2.01 \text{ mol NaN}_3$$

Now you can calculate the grams of NaN_3. The molar mass of NaN_3 is 65.01 g/mol.

$$2.01 \text{ mol NaN}_3 \times \frac{65.01 \text{ g NaN}_3}{1 \text{ mol NaN}_3} = \textbf{131 g NaN}_3$$

Answer Check As with any problem, see whether the context can help you evaluate your answer. Here we determined the mass of a substance used to inflate an airbag. Using this problem context, we would expect the amount of NaN_3 not to be many kilograms; otherwise, it wouldn't fit in a steering wheel. Also, it shouldn't be some minuscule amount (1 mol of gas will occupy a volume of about 22 L at STP), because the airbag has a large volume to fill (75.0 L in this case).

Exercise 5.9 How many liters of chlorine gas, Cl_2, can be obtained at 40°C and 787 mmHg from 9.41 g of hydrogen chloride, HCl, according to the following equation?

$$2KMnO_4(s) + 16HCl(aq) \longrightarrow 8H_2O(l) + 2KCl(aq) + 2MnCl_2(aq) + 5Cl_2(g)$$

■ See Problems 5.73, 5.74, 5.75, 5.76, 5.77, and 5.78.

5.5 Gas Mixtures; Law of Partial Pressures

While studying the composition of air, John Dalton concluded in 1801 that each gas in a mixture of unreactive gases acts, as far as its pressure is concerned, as though it were the only gas in the mixture. To illustrate, consider two 1-L flasks, one filled with helium to a pressure of 152 mmHg at a given temperature and the other

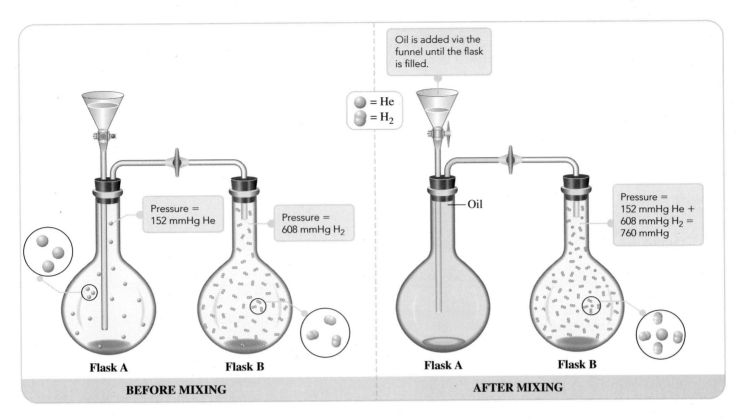

Oil is added via the funnel until the flask is filled.

= He
= H₂

Pressure = 152 mmHg He

Pressure = 608 mmHg H₂

Oil

Pressure = 152 mmHg He + 608 mmHg H₂ = 760 mmHg

Flask A   **Flask B**

BEFORE MIXING

Flask A   **Flask B**

AFTER MIXING

Figure 5.19 ▲

An illustration of Dalton's law of partial pressures The valve connecting the flasks and the valve at the funnel are opened so that flask A fills with mineral oil. The helium flows into flask B (having the same volume as flask A), where it mixes with hydrogen. Each gas exerts the pressure it would exert if the other were not there.

filled with hydrogen to a pressure of 608 mmHg at the same temperature. Suppose all of the helium in the one flask is put in with the hydrogen in the other flask (see Figure 5.19). After the gases are mixed in one flask, each gas occupies a volume of one liter, just as before, and has the same temperature.

According to Dalton, each gas exerts the same pressure it would exert if it were the only gas in the flask. Thus, the pressure exerted by helium in the mixture is 152 mmHg, and the pressure exerted by hydrogen in the mixture is 608 mmHg. The total pressure exerted by the gases in the mixture is 152 mmHg + 608 mmHg = 760 mmHg.

Partial Pressures and Mole Fractions

The pressure exerted by a particular gas in a mixture is the **partial pressure** of that gas. The partial pressure of helium in the preceding mixture is 152 mmHg; the partial pressure of hydrogen in the mixture is 608 mmHg. According to **Dalton's law of partial pressures,** *the sum of the partial pressures of all the different gases in a mixture is equal to the total pressure of the mixture.*

If you let P be the total pressure and P_A, P_B, P_C, . . . be the partial pressures of the component gases in a mixture, the law of partial pressures can be written as

Dalton's law of partial pressures:
$$P = P_A + P_B + P_C + \cdots$$

The individual partial pressures follow the ideal gas law. For component A,
$$P_A V = n_A RT$$
where n_A is the number of moles of component A.

The composition of a gas mixture is often described in terms of the mole fractions of component gases. The **mole fraction** of a component gas is *the fraction* of moles of that component in the total moles of gas mixture (or the fraction of molecules that are component molecules). Because the pressure of a gas is proportional to moles, for fixed volume and temperature ($P = nRT/V \propto n$), the mole fraction also equals the partial pressure divided by total pressure.

$$\text{Mole fraction of } A = \frac{n_A}{n} = \frac{P_A}{P}$$

Mole percent equals mole fraction $\times$ 100. Mole percent is equivalent to the percentage of the molecules that are component molecules.

Example 5.10   Calculating Partial Pressures and Mole Fractions of a Gas in a Mixture

Gaining Mastery Toolbox

Critical Concept 5.10
Each gas in a mixture of gases can be treated independently. This concept is based on ideal gases, with each gas in the mixture having no interaction with the other gases.

Solution Essentials:
- Dalton's law of partial pressure
- Partial pressure
- Mole fraction
- Ideal gas law ($PV = nRT$)
- Units of pressure (Table 5.2)
- Kelvin temperature

A 1.00-L sample of dry air at 25°C and 786 mmHg contains 0.925 g N_2, plus other gases including oxygen, argon, and carbon dioxide.

a What is the partial pressure (in mmHg) of N_2 in the air sample?

b What is the mole fraction and mole percent of N_2 in the mixture?

Problem Strategy The key to understanding how to solve this problem is to assume that each gas in the container is independent of other gases; that is, it occupies the entire volume of the sample. Therefore, each gas in the mixture follows the ideal gas law. Using the ideal gas law, we can solve for the pressure of just the N_2, which is its partial pressure. The mole fraction of N_2 can be calculated from the partial pressure of N_2 and the total pressure of all the gases in the sample.

Solution a Each gas in a mixture follows the ideal gas law. To calculate the partial pressure of N_2, you convert 0.925 g N_2 to moles N_2.

$$0.925 \text{ g } N_2 \times \frac{1 \text{ mol } N_2}{28.0 \text{ g } N_2} = 0.0330 \text{ mol } N_2$$

You substitute into the ideal gas law (noting that 25°C is 298 K).

$$P_{N_2} = \frac{n_{N_2}RT}{V}$$

$$= \frac{0.0330 \text{ mol} \times 0.0821 \text{ L·atm/(K·mol)} \times 298 \text{ K}}{1.00 \text{ L}}$$

$$= 0.807 \text{ atm } (= \mathbf{613 \text{ mmHg}})$$

b The mole fraction of N_2 in air is

$$\text{Mole fraction of } N_2 = \frac{P_{N_2}}{P} = \frac{613 \text{ mmHg}}{786 \text{ mmHg}} = \mathbf{0.780}$$

Air contains **78.0 mole percent** of N_2.

Answer Check Whenever you are solving for the mole fraction of a substance in a mixture, the number must always be less than 1 (as it was in this problem). Therefore, if you obtain a value greater than 1, it is a sure sign that you have made an error.

Exercise 5.10   A 10.0-L flask contains 1.031 g O_2 and 0.572 g CO_2 at 18°C. What are the partial pressures of oxygen and carbon dioxide? What is the total pressure? What is the mole fraction of oxygen in the mixture?

■ See Problems 5.81, 5.82, 5.83, and 5.84.

CONCEPT CHECK 5.4

A flask equipped with a valve contains 3.0 mol of H_2 gas. You introduce 3.0 mol of Ar gas into the flask via the valve and then seal the flask.

a What happens to the pressure of just the H_2 gas in the flask after the introduction of the Ar? If it changes, by what factor does it do so?

b How do the pressures of the Ar and the H_2 in the flask compare?

c How does the total pressure in the flask relate to the pressures of the two gases?

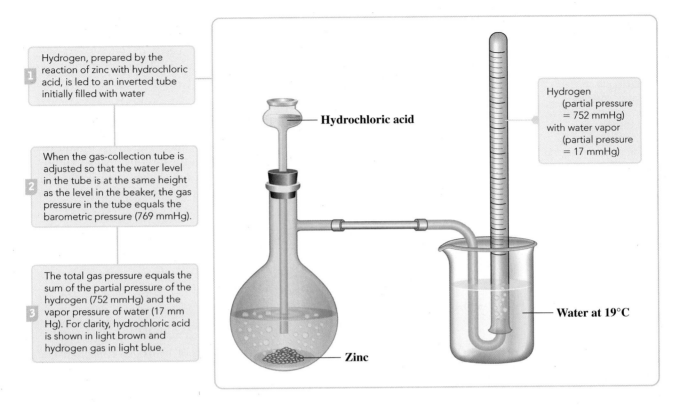

Hydrogen, prepared by the reaction of zinc with hydrochloric acid, is led to an inverted tube initially filled with water

When the gas-collection tube is adjusted so that the water level in the tube is at the same height as the level in the beaker, the gas pressure in the tube equals the barometric pressure (769 mmHg).

The total gas pressure equals the sum of the partial pressure of the hydrogen (752 mmHg) and the vapor pressure of water (17 mm Hg). For clarity, hydrochloric acid is shown in light brown and hydrogen gas in light blue.

Hydrochloric acid

Hydrogen (partial pressure = 752 mmHg) with water vapor (partial pressure = 17 mmHg)

Water at 19°C

Zinc

Figure 5.20 ▲

Collection of gas over water

Collecting Gases over Water

A useful application of the law of partial pressures arises when you collect gases over water (a method used for gases that do not dissolve appreciably in water). Figure 5.20 shows how a gas, produced by chemical reaction in the flask, is collected by leading it to an inverted tube, where it displaces water. As gas bubbles through the water, the gas picks up molecules of water vapor that mix with it. The partial pressure of water vapor in the gas mixture in the collection tube depends only on the temperature. This partial pressure of water vapor is called the *vapor pressure* of water. ▶ Values of the vapor pressure of water at various temperatures are listed in Table 5.6 (see Appendix B for a more complete table). The following example shows how to find the partial pressure and then the mass of the collected gas.

Vapor pressure is the maximum partial pressure of the vapor in the presence of the liquid. It is defined more precisely in Chapter 11.

Table 5.6	Vapor Pressure of Water at Various Temperatures*
Temperature (°C)	**Pressure (mmHg)**
0	4.6
10	9.2
15	12.8
17	14.5
19	16.5
21	18.7
23	21.1
25	23.8
27	26.7
30	31.8
40	55.3
60	149.4
80	355.1
100	760.0

*Appendix B contains a more complete table.

⦿WL Interactive Example `5.11`   **Calculating the Amount of Gas Collected over Water**

Hydrogen gas is produced by the reaction of hydrochloric acid, HCl, on zinc metal.

$$2HCl(aq) + Zn(s) \longrightarrow ZnCl_2(aq) + H_2(g)$$

The gas is collected over water. If 156 mL of gas is collected at 19°C (two significant figures) and 769 mmHg total pressure, what is the mass of hydrogen collected?

Problem Strategy   The gas collected is hydrogen mixed with water vapor. To obtain the amount of hydrogen, you must first find its partial pressure in the mixture, using Dalton's law (Step 1). Then you can calculate the moles of hydrogen from the ideal gas law (Step 2). Finally, you can obtain the mass of hydrogen from the moles of hydrogen (Step 3).

Solution

Step 1:   The vapor pressure of water at 19°C is 16.5 mmHg. From Dalton's law of partial pressures, you know that the total gas pressure equals the partial pressure of hydrogen, P_{H_2}, plus the partial pressure of water, P_{H_2O}.

$$P = P_{H_2} + P_{H_2O}$$

Substituting and solving for the partial pressure of hydrogen, you get

$$P_{H_2} = P - P_{H_2O} = (769 - 16.5) \text{ mmHg} = 752 \text{ mmHg}$$

Step 2:   Now you can use the ideal gas law to find the moles of hydrogen collected. The data are

Variable	Value
P	$752 \text{ mmHg} \times \dfrac{1 \text{ atm}}{760 \text{ mmHg}} = 0.989 \text{ atm}$
V	$156 \text{ mL} = 0.156 \text{ L}$
T	$(19 + 273) \text{ K} = 292 \text{ K}$
n	?

From the ideal gas law, $PV = nRT$, you have

$$n = \frac{PV}{RT} = \frac{0.989 \text{ atm} \times 0.156 \text{ L}}{0.0821 \text{ L·atm/(K·mol)} \times 293 \text{ K}} = 0.00641 \text{ mol}$$

Step 3:   You convert moles of H_2 to grams of H_2.

$$0.00664 \text{ mol } H_2 \times \frac{2.02 \text{ g } H_2}{1 \text{ mol } H_2} = \textbf{0.0130 g } H_2$$

Answer Check   Note that this problem assumes that the water level in the collection tube is the same as the water level in the beaker (Figure 5.20). This enables us to equate the gas pressure in the collection tube to atmospheric pressure.

Exercise 5.11   Oxygen can be prepared by heating potassium chlorate, $KClO_3$, with manganese dioxide as a catalyst. The reaction is

$$2KClO_3(s) \xrightarrow[MnO_2]{\Delta} 2KCl(s) + 3O_2(g)$$

How many moles of O_2 would be obtained from 1.300 g $KClO_3$? If this amount of O_2 were collected over water at 23°C and at a total pressure of 745 mmHg, what volume would it occupy?

■ See Problems 5.87 and 5.88.

Kinetic-Molecular Theory

In the following sections, you will see how the interpretation of a gas in terms of the **kinetic-molecular theory** (or simply **kinetic theory**) leads to the ideal gas law. According to this theory, *a gas consists of molecules in constant random motion.*

The word *kinetic* describes something in motion. Kinetic energy, E_k, is the energy associated with the motion of an object of mass m. From physics,

$$E_k = \tfrac{1}{2}m \times (\text{speed})^2$$

We will use the concept of kinetic energy in describing kinetic theory.

5.6 Kinetic Theory of an Ideal Gas

Our present explanation of gas pressure is that it results from the continual bombardment of the container walls by constantly moving molecules (Figure 5.21). This kinetic interpretation of gas pressure was first put forth in 1676 by Robert Hooke, who had been an assistant of Boyle. Hooke did not pursue this idea, however, so the generally accepted interpretation of gas pressure remained the one given by Isaac Newton, a contemporary of Hooke.

According to Newton, the pressure of a gas was due to the mutual repulsions of the gas particles (molecules). These repulsions pushed the molecules against the walls of the gas container, much as coiled springs packed tightly in a box would push against the walls of the box. This interpretation continued to be the dominant view of gas pressure until the mid-nineteenth century.

Despite the dominance of Newton's view, some people followed the kinetic interpretation. In 1738, Daniel Bernoulli, a Swiss mathematician and physicist, gave a quantitative explanation of Boyle's law using the kinetic interpretation. He even suggested that molecules move faster at higher temperatures, in order to explain Amontons's experiments on the temperature dependence of gas volume and pressure. However, Bernoulli's paper attracted little notice. A similar kinetic interpretation of gases was submitted for publication to the Royal Society of London in 1848 by John James Waterston. His paper was rejected as "nothing but nonsense." ▶

Soon afterward, the scientific climate for the kinetic view improved. The kinetic theory of gases was developed by a number of influential physicists, including James Clerk Maxwell (1859) and Ludwig Boltzmann (in the 1870s). Throughout the last half of the nineteenth century, research continued on the kinetic theory, making it a cornerstone of our present view of molecular substances.

Figure 5.21 ▲

Kinetic-theory model of gas pressure According to kinetic theory, gas pressure is the result of the bombardment of the container walls by constantly moving molecules.

Waterston's paper, with the reviewers' comments, was discovered in the Royal Society's files by Lord Rayleigh in 1892. (Rayleigh codiscovered argon and received the 1904 Nobel Prize in physics.)

Postulates of Kinetic Theory

Physical theories are often given in terms of *postulates:* the basic statements from which all conclusions or predictions of a theory are deduced. The postulates are accepted as long as the predictions from the theory agree with experiment. If a particular prediction did not agree with experiment, we would limit the area to which the theory applies, modify the postulates, or start over with a new theory.

The kinetic theory of an *ideal gas* (a gas that follows the ideal gas law) is based on five postulates.

Postulate 1: *Gases are composed of molecules whose size is negligible compared with the average distance between them.* Most of the volume occupied by a gas is empty space. This means that you can usually ignore the volume occupied by the molecules.

Postulate 2: *Molecules move randomly in straight lines in all directions and at various speeds.* This means that properties of a gas that depend on the motion of molecules, such as pressure, will be the same in all directions.

Postulate 3: *The forces of attraction or repulsion between two molecules (intermolecular forces) in a gas are very weak or negligible, except when they collide.* ▶ This means that a molecule will continue moving in a straight line with undiminished speed until it collides with another gas molecule or with the walls of the container.

Postulate 4: *When molecules collide with one another, the collisions are elastic.* In an elastic collision, the total kinetic energy remains constant; no kinetic energy is lost. To understand the difference between an elastic and an inelastic collision, compare the collision of two hard steel spheres with

The statements that there are no intermolecular forces and that the volume of molecules is negligible are simplifications that lead to the ideal gas law. But intermolecular forces are needed to explain how we get the liquid state from the gaseous state; it is intermolecular forces that hold molecules together in the liquid state.

Figure 5.22 ▶

Elastic collision of steel balls

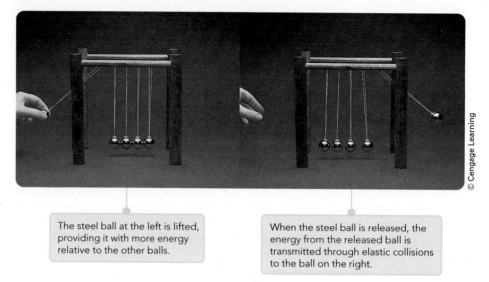

The steel ball at the left is lifted, providing it with more energy relative to the other balls.

When the steel ball is released, the energy from the released ball is transmitted through elastic collisions to the ball on the right.

© Cengage Learning

the collision of two masses of putty. The collision of steel spheres is nearly elastic (that is, the spheres bounce off each other and continue moving), but that of putty is not (Figure 5.22). Postulate 4 says that unless the kinetic energy of molecules is removed from a gas—for example, as heat—the molecules will forever move with the same average kinetic energy per molecule.

Postulate 5: *The average kinetic energy of a molecule is proportional to the absolute temperature.* This postulate establishes what we mean by temperature from a molecular point of view: the higher the temperature, the greater the molecular kinetic energy.

Qualitative Interpretation of the Gas Laws

According to the kinetic theory, the pressure of a gas results from the bombardment of container walls by molecules. Both the concentration of molecules (number per unit volume) and the average speed of the molecules are factors in determining this pressure. Molecular concentration and average speed determine the frequency of collisions with the wall. Average molecular speed determines the average force of a collision.

Let's consider how these terms apply to Avogadro's law, which relates the volume of gas to the moles of gas at constant (fixed) temperature and pressure. Constant temperature means that the average kinetic energy of a molecule is constant (Postulate 5). Therefore, the average molecular speed and thus the average molecular force from collision remain constant. Gas pressure depends on the force and frequency of molecular collisions with container walls. Consider the container depicted in Figure 5.23*A*, where the pressure of the gas molecules in the container equals atmospheric pressure. What happens when you introduce more gas molecules into the container? More moles of gas increase the number of molecules per unit volume (concentration) and therefore increase the frequency of collisions with the container walls. If the volume of the container were fixed, the pressure would increase (Figure 5.23*B*). However, Avogadro's law requires that the pressure remain constant, so the container volume expands until the concentration and, therefore, the frequency of collisions with container walls decrease (Figure 5.23*C*). The result is that when you increase the number of moles of gas in a container at a constant pressure and temperature, the volume increases.

The same concepts can be used to discuss Boyle's law. Suppose you increase the volume of a gas while holding both the temperature and number of moles of gas constant (Figure 5.24). This decreases the concentration of molecules and the frequency of collisions per unit wall area; the pressure must decrease. Thus, when you increase the volume of a gas while keeping the temperature constant, the pressure decreases.

Now consider Charles's law. If you raise the temperature, you increase the average molecular speed. The average force of a collision increases. If all other

a A container where the gas molecules are at atmospheric pressure ($P_{inside} = P_{atm}$). The pressure in the container is due to the force and frequency of molecular collisions with the container walls.

b The container after increasing the number of gas molecules while not allowing the piston to move. Due to the greater concentration of gas (more moles of gas in the same volume), the frequency of collisions of the gas molecules with the walls of the container has increased, causing the pressure to increase ($P_{inside} = P_{atm}$).

c Container after the molecules are allowed to move the piston and increase the volume. The concentration of gas molecules and frequency of collisions with the container walls have decreased, and the pressure of the gas molecules inside the container is again equal to atmospheric pressure ($P_{inside} = P_{atm}$).

Figure 5.23 ▲

Molecular description of Avogadro's law

factors remained fixed, the pressure would increase. For the pressure to remain constant as it does in Charles's law, it is necessary for the volume to increase so that the number of molecules per unit volume decreases and the frequency of collisions decreases. Thus, when you raise the temperature of a gas while keeping the pressure constant, the volume increases.

a A fixed number of moles of gas in a container at room temperature. The pressure in the container is due to the force and frequency of molecular collisions with the container walls.

b The same number of moles of gas at the same temperature as A, but in a container of greater volume. Because of the larger container volume, the gas concentration is now lower, which leads to a decrease in the frequency of molecular collisions per unit of wall area, resulting in a lower pressure.

Figure 5.24 ▲

Molecular description of Boyle's law

CONCEPT CHECK 5.5

Consider a sealed glass bottle of helium gas at room temperature. If you immerse the bottle in an ice water bath, how will this immersion affect the pressure of the gas?

   I. Pressure increase

  II. Pressure decrease

 III. No pressure change

The best explanation for your answer is?

a The force of the collisions of the helium atoms with the container walls increases with decreasing temperature.

b The helium atoms have significantly greater volumes at lower temperatures.

c The frequency and force of the collisions of the helium atoms with the container walls decrease with decreasing temperature.

d The average kinetic energy of the helium atoms remains constant when the temperature decreases.

e The concentration of helium decreases, resulting in a decrease in the frequency of the collisions with the container walls.

The Ideal Gas Law from Kinetic Theory

One of the most important features of kinetic theory is its explanation of the ideal gas law. To show how you can get the ideal gas law from kinetic theory, we will first find an expression for the pressure of a gas. ◀

According to kinetic theory, the pressure of a gas, P, will be proportional to the frequency of molecular collisions with a surface and to the average force exerted by a molecule in collision.

$$P \propto \text{frequency of collisions} \times \text{average force}$$

The average force exerted by a molecule during a collision depends on its mass m and its average speed u—that is, on its average momentum mu. ◀ In other words, the greater the mass of the molecule and the faster it is moving, the greater the force exerted during collision. The frequency of collisions is also proportional to the average speed u, because the faster a molecule is moving, the more often it strikes the container walls. Frequency of collisions is inversely proportional to the gas volume V, because the larger the volume, the less often a given molecule strikes the container walls. Finally, the frequency of collisions is proportional to the number of molecules N in the gas volume. Putting these factors together gives

$$P \propto \left(u \times \frac{1}{V} \times N \right) \times mu$$

Bringing the volume to the left side, you get

$$PV \propto Nmu^2$$

Because the average kinetic energy of a molecule of mass m and average speed u is $\frac{1}{2}mu^2$, PV is proportional to the average kinetic energy of a molecule. ◀ Moreover, the average kinetic energy is proportional to the absolute temperature (Postulate 5). Noting that the number of molecules, N, is proportional to the moles of molecules, n, you have

$$PV \propto nT$$

You can write this as an equation by inserting a constant of proportionality, R, which you can identify as the molar gas constant.

$$PV = nRT$$

The next two sections give additional deductions from kinetic theory.

This derivation is essentially the one given by Daniel Bernoulli.

Consistent with kinetic theory, we assume that the duration of collision of a molecule with a container wall is the same, regardless of the type of molecule.

Recall that kinetic energy is defined as $\frac{1}{2}m$ multiplied by (speed)^2.

5.7 Molecular Speeds; Diffusion and Effusion

The principal tenet of kinetic theory is that molecules are in constant random motion. Now we will look at the speeds of molecules and at some conclusions of kinetic theory regarding molecular speeds.

Molecular Speeds

According to kinetic theory, the speeds of molecules in a gas vary over a range of values. The British physicist James Clerk Maxwell (1831–1879) showed theoretically—and it has since been demonstrated experimentally—that molecular speeds are distributed as shown in Figure 5.25. This distribution of speeds depends on the temperature. At any temperature, the molecular speeds vary widely, but most are close to the average speed, which is close to the speed corresponding to the maximum in the distribution curve. As the temperature increases, the average speed increases.

The **root-mean-square (rms) molecular speed,** u, is *a type of average molecular speed, equal to the speed of a molecule having the average molecular kinetic energy.* It is given by the following formula:

$$u = \sqrt{\frac{3RT}{M_m}} = \left(\frac{3RT}{M_m} \right)^{\frac{1}{2}}$$

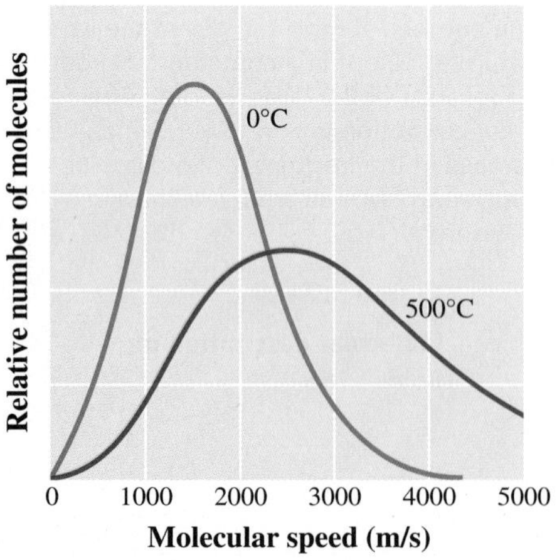

Figure 5.25 ▲

Maxwell's distribution of molecular speeds The distributions of speeds of H_2 molecules are shown for 0°C and 500°C. Note that the speed corresponding to the maximum in the curve (the most probable speed) increases with temperature.

where R is the molar gas constant, T is the absolute temperature, and M_m is the molar mass for the gas. This result follows from Postulate 5 of kinetic theory. ▶ Note that of two gases, the one with the higher molar mass will have the lower rms speed.

In applying this equation, *care must be taken to use consistent units.* If SI units are used for R ($= 8.31$ kg·m²/(s²·K·mol)), T (K), and M_m (kg/mol), as in the following example, the rms speed will be in meters per second. Note that in these units, H_2, whose molecular weight is 2.02 amu, has a molar mass of 2.02×10^{-3} kg/mol.

Values of the rms speed calculated from this formula indicate that molecular speeds are astonishingly high. For example, the rms speed of H_2 molecules at 20°C is 1.90×10^3 m/s (over 4000 mi/hr).

> According to kinetic theory, the total kinetic energy of a mole of any gas equals $\frac{3}{2}RT$. The average kinetic energy of a molecule, which by definition is $\frac{1}{2}mu^2$, is obtained by dividing $\frac{3}{2}RT$ by Avogadro's number, N_A. Therefore, the kinetic energy is $\frac{1}{2}mu^2 = 3RT/(2N_A)$. Hence $u^2 = 3RT/(N_Am)$. Or, noting that N_Am equals the molar mass M_m, you get $u^2 = 3RT/M_m$, from which you get the text equation.

Example 5.12

Calculating the rms Speed of Gas Molecules

Calculate the rms speed of O_2 molecules in a cylinder at 21°C and 15.7 atm. See Table 5.5 for the appropriate value of R.

Problem Strategy The rms speed of a gas can be determined by substituting the appropriate values into the preceding equation.

Solution The rms molecular speed is independent of pressure but does depend on the absolute temperature, which is $(21 + 273)$ K $= 294$ K. To calculate u, it is best to use SI units throughout. In these units, the molar mass of O_2 is 32.0×10^{-3} kg/mol, and $R = 8.31$ kg·m²/(s²·K·mol). Hence,

$$u = \left(\frac{3 \times 8.31 \text{ kg} \cdot \text{m}^2/(\text{s}^2 \cdot \text{K} \cdot \text{mol}) \times 294 \text{ K}}{32.0 \times 10^{-3} \text{ kg/mol}} \right)^{\frac{1}{2}} = \textbf{479 m/s}$$

Answer Check When performing calculations that involve this formula, be sure to use the appropriate value of R, $(8.31$ kg·m²)/(s²·K·mol), and to express the molar mass in units of kg/mol, and the temperature in kelvins.

Exercise 5.12   What is the rms speed (in m/s) of a carbon tetrachloride molecule at 22°C?

■ See Problems 5.89, 5.90, 5.91, and 5.92.

Exercise 5.13   At what temperature do hydrogen molecules, H_2, have the same rms speed as nitrogen molecules, N_2, at 455°C? At what temperature do hydrogen molecules have the same average kinetic energy?

■ See Problems 5.93 and 5.94.

Gaining Mastery Toolbox

Critical Concept 5.12
The rms (root-mean-square) speed of a gas depends on the temperature and molar mass of the gas. The rms speed is directly proportional to the square root of the kelvin temperature (greater temperature equals greater rms speed) and inversely proportional to the square root of the molar mass (more mass equals less rms speed).

Solution Essentials:
- Root-mean-square (rms) molecular speed
- rms equation
$$u = \sqrt{\frac{3RT}{M_m}} = \left(\frac{3RT}{M_m}\right)^{\frac{1}{2}}$$
- Kinetic molecular theory
- Molar mass

Diffusion and Effusion

The pleasant odor of apple pie baking in the oven quickly draws people to the kitchen. The spread of an odor is easily explained by kinetic theory. All molecules are in constant, chaotic motion, and eventually a cluster of molecules of a particular substance will spread out to occupy a larger and larger space. Gaseous **diffusion** is *the process whereby a gas spreads out through another gas to occupy the space uniformly.* Figure 5.26 is a kinetic-theory explanation of the diffusion of two gases. An individual molecule moves chaotically as it is buffeted by other molecules. After sufficient time, this constant chaotic motion of molecules results in a complete mixing of the gases. Figure 5.27 demonstrates the diffusion of ammonia gas through air.

When you think about the kinetic-theory calculations of molecular speed, you might ask why diffusion is not even much faster than it is. Why does it take minutes for the gas to diffuse throughout a room when the molecules are moving at perhaps a thousand miles per hour? This was, in fact, one of the first criticisms of kinetic theory. The answer is simply that a molecule never travels very far in one direction (at ordinary pressures) before it collides with

Figure 5.26 ▲

Kinetic-theory model of gas diffusion
A vessel contains two different gases. For clarity, the figure shows the motion of only one molecule. The path of the molecule is chaotic because of the constant collisions with other molecules. After some time, the "green" molecule has by its random motion mixed with the "red" gas. Similar motions of other molecules eventually result in the complete mixing of the two gases.

Figure 5.27 ▶

Gaseous diffusion

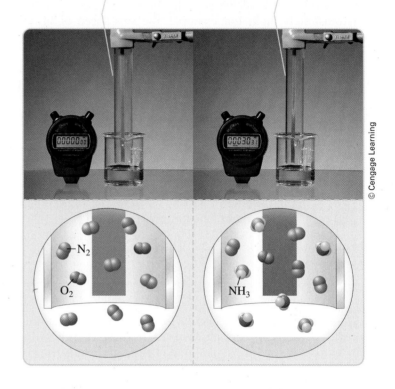

Concentrated aqueous ammonia in the beaker evolves ammonia gas into the glass tube, which contains a strip of wet indicator paper. In the molecular view there are only nitrogen and oxygen molecules (air) surrounding the indicator paper.

The indicator changes color as the ammonia gas diffuses upward through the air in the tube. The molecular view depicts the appearance of the NH_3 molecules in the tube due to diffusion.

© Cengage Learning

Figure 5.28 ▼

Model of gaseous effusion
According to kinetic theory, gas molecules are in constant random motion. When a molecule in the container happens to hit the small opening, it passes (or effuses) to the outside. The rate of effusion depends on the speed of the molecules—the faster the molecules move, the more likely they are to encounter the opening and pass from the inside of the box to the outside.

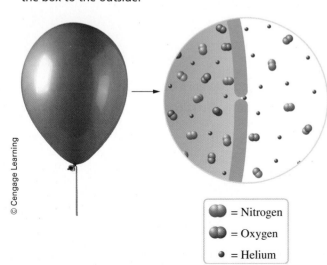

© Cengage Learning

= Nitrogen
= Oxygen
= Helium

another molecule and moves off in another direction. If you could trace out the path of an individual molecule, it would be a zigzagging trail. For a molecule to cross a room, it has to travel many times the straight-line distance.

Although the rate of diffusion certainly depends in part on the average molecular speed, the effect of molecular collisions makes the theoretical picture a bit complicated. *Effusion,* like diffusion, is a process involving the flow of a gas but is theoretically much simpler.

If you place a container of gas in a vacuum and then make a very small hole in the container, the gas molecules escape through the hole at the same speed they had in the container (Figure 5.28). *The process in which a gas flows through a small hole in a container* is called **effusion.** It was first studied by Thomas Graham, who discovered in 1846 that the rate of effusion of a gas is inversely proportional to the square root of its density. Today we usually state **Graham's law of effusion** in terms of molecular mass: *the rate of effusion of gas molecules from a particular hole is inversely proportional to the square root of the molecular mass of the gas at constant temperature and pressure.*

Let us consider a kinetic-theory analysis of an effusion experiment. Suppose the hole in the container is made small enough so that the gas molecules continue to move randomly (rather than moving together as they would in a wind). When a molecule happens to encounter the hole, it leaves the container. The collection of molecules leaving the container by chance encounters with the hole constitutes the effusing gas. All you have to consider for effusion is the rate at which molecules encounter the hole in the container.

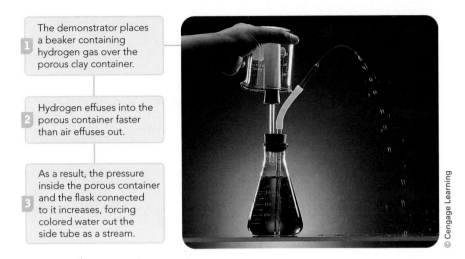

1 The demonstrator places a beaker containing hydrogen gas over the porous clay container.

2 Hydrogen effuses into the porous container faster than air effuses out.

3 As a result, the pressure inside the porous container and the flask connected to it increases, forcing colored water out the side tube as a stream.

© Cengage Learning

Figure 5.29 ▲

The hydrogen fountain

The rate of effusion of molecules from a container depends on three factors: the cross-sectional area of the hole (the larger it is, the more likely molecules are to escape); the number of molecules per unit volume (the more crowded the molecules are, the more likely they are to encounter the hole); and the average molecular speed (the faster the molecules are moving, the sooner they will find the hole and escape).

If you compare the effusion of different gases from the same container, at the same temperature and pressure, the first two factors will be the same. The average molecular speeds will be different, however.

Because the average molecular speed essentially equals $\sqrt{3RT/M_m}$, where M_m is the molar mass, the rate of effusion is proportional to $\sqrt{M_m}$. That is, the rate of effusion is inversely proportional to the square root of the molar mass (or molecular weight), as Graham's law states. The derivation of Graham's law from kinetic theory was considered a triumph of the theory and greatly strengthened confidence in its validity.

Graham's law of effusion:

$$\text{Rate of effusion of molecules} \propto \frac{1}{\sqrt{M_m}} \quad \text{(for the same container at constant } T \text{ and } P\text{)}$$

The hydrogen fountain, shown in Figure 5.29, is dependent on the differences in rates of effusion of gases. (Can you explain why a helium-filled balloon loses pressure after a few hours?)

Example 5.13   Calculating the Ratio of Effusion Rates of Gases

Gaining Mastery Toolbox

Critical Concept 5.13
For gases in a container at a constant temperature and pressure, the effusion rate of the gas *only* depends on the molar mass of the gas. Gases with greater molar mass will effuse more slowly than those with lower molar mass.

Solution Essentials:
• Effusion
• Graham's law of effusion: rate of effusion of molecules $\propto \dfrac{1}{\sqrt{M_m}}$ (for the same container at constant T and P)
• Kinetic molecular theory
• Molar mass

Calculate the ratio of effusion rates of molecules of carbon dioxide, CO_2, and sulfur dioxide, SO_2, from the same container and at the same temperature and pressure.

Problem Strategy Here we need to compare the rates of effusion of two gases having different molar masses. We can use Graham's law of effusion for the two gases to set up a relationship

(continued)

(*continued*)

$$\frac{\text{Rate of effusion of CO}_2}{\text{Rate of effusion of SO}_2} = \frac{\frac{1}{\sqrt{M_m(\text{CO}_2)}}}{\frac{1}{\sqrt{M_m(\text{SO}_2)}}}$$

where $M_m(\text{SO}_2)$ is the molar mass of SO_2 (64.1 g/mol) and $M_m(\text{CO}_2)$ is the molar mass of CO_2 (44.0 g/mol). This equation can be simplified to

$$\frac{\text{Rate of effusion of CO}_2}{\text{Rate of effusion of SO}_2} = \sqrt{\frac{M_m(\text{SO}_2)}{M_m(\text{CO}_2)}}$$

(Note that the two rates of effusion are inversely proportional to the square roots of the molar masses.)

Solution Substituting these molar masses into the formula gives

$$\frac{\text{Rate of effusion of CO}_2}{\text{Rate of effusion of SO}_2} = \sqrt{\frac{64.1 \text{ g/mol}}{44.0 \text{ g/mol}}} = \mathbf{1.21}$$

In other words, carbon dioxide effuses 1.21 times faster than sulfur dioxide (because CO_2 molecules move 1.21 times faster on average than SO_2 molecules).

Answer Check Analyze your answer to make sure that the gas with the lower molar mass has the greater rate of effusion. In this case, the CO_2 would be expected to have the greater effusion rate, so we should expect the effusion rate ratio of $\mathbf{CO_2}$ to $\mathbf{SO_2}$ to be greater than 1.

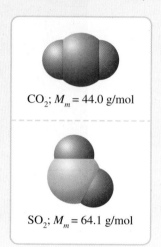

CO_2; M_m = 44.0 g/mol

SO_2; M_m = 64.1 g/mol

Exercise 5.14 If it takes 3.52 s for 10.0 mL of helium to effuse through a hole in a container at a particular temperature and pressure, how long would it take for 10.0 mL of oxygen, O_2, to effuse from the same container at the same temperature and pressure? (Note that the rate of effusion can be given in terms of volume of gas effused per second.)

■ See Problems 5.95, 5.96, 5.97, and 5.98.

Exercise 5.15 If it takes 4.67 times as long for a particular gas to effuse as it takes hydrogen under the same conditions, what is the molecular mass of the gas? (Note that the rate of effusion is inversely proportional to the time it takes for a gas to effuse.)

■ See Problems 5.99 and 5.100.

Graham's law has practical application in the preparation of fuel rods for nuclear fission reactors. Such reactors depend on the fact that the uranium-235 nucleus undergoes fission (splits) when bombarded with neutrons. When the nucleus splits, several neutrons are emitted and a large amount of energy is liberated. These neutrons bombard more uranium-235 nuclei, and the process continues with the evolution of more energy. However, natural uranium consists of 99.27% uranium-238 (which does not undergo fission) and only 0.72% uranium-235 (which does undergo fission). A uranium fuel rod must contain about 3% uranium-235 to sustain the nuclear reaction.

To increase the percentage of uranium-235 in a sample of uranium (a process called *enrichment*), one first prepares uranium hexafluoride, UF_6, a white, crystalline solid that is easily vaporized. ◀ Uranium hexafluoride vapor is allowed to pass through a series of porous membranes. Each membrane has many small holes through which the vapor can effuse. Because the UF_6 molecules with the lighter isotope of uranium travel about 0.4% faster than the UF_6 molecules with the heavier isotope, the gas that passes through first is somewhat richer in uranium-235. When this vapor passes through another membrane, the uranium-235 vapor becomes further concentrated. It takes many effusion stages to reach the necessary enrichment.

Uranium hexafluoride vapor has a density of 13.0 g/L at 57°C (the temperature at which the solid sublimes). This is more than 12 times as dense as air at the same temperature.

CONCEPT CHECK 5.6

Consider the experimental apparatus shown. In this setup, each round flask contains a gas, and the long tube contains no gas (that is, it is a vacuum).

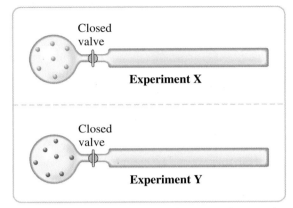

a We use 1.0 mol of He for experiment X and 1.0 mol of Ar for experiment Y. If both valves are opened at the same time, which gas would you expect to reach the end of the long tube first?

b If you wanted the Ar to reach the end of the long tube at the same time as the He, what experimental condition (that is, you cannot change the equipment) could you change to make this happen?

5.8 Real Gases

In Section 5.2, we found that, contrary to Boyle's law, the pressure–volume product for O_2 at 0°C was not quite constant, particularly at high pressures (see PV column in Table 5.3). Experiments show that the ideal gas law describes the behavior of real gases quite well at low pressures and moderate temperatures, but not at high pressures and low temperatures.

Figure 5.30 shows the behavior of the pressure–volume product at various pressures for several gases. The gases deviate noticeably from Boyle's law (ideal gas) behavior at high pressures. Also, the deviations differ for each kind of gas. We can explain why this is so by examining the postulates of kinetic theory, from which the ideal gas law can be derived.

Postulate 1 of kinetic theory says that the volume of space occupied by molecules is negligible compared with the total gas volume. To derive the ideal gas law from kinetic theory, we must assume that each molecule is free to move throughout the entire gas volume, V. At low pressures, where the volume of individual molecules is negligible compared with the total volume available (Figure 5.31*A*), the ideal gas law is a good approximation. At higher pressures, where the volume of individual molecules becomes important (Figure 5.31*B*), the space through which a molecule can move is significantly different from V.

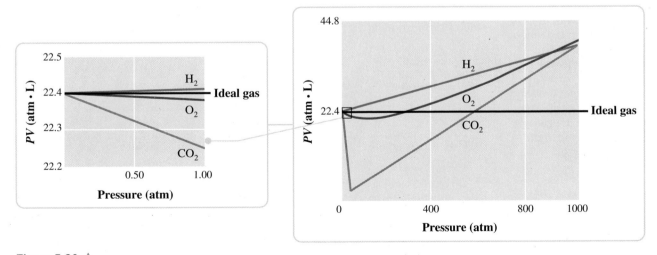

Figure 5.30 ▲

Pressure–volume product of gases at different pressures *Right:* The pressure–volume product of one mole of various gases at 0°C and at different pressures. *Left:* Values at low pressure. For an ideal gas, the pressure–volume product is constant.

a At low pressures, the volume of molecules is a small fraction of the total volume and can be neglected, as in the ideal gas low

b At high pressures, the volume of molecules is a significant fraction of the total volume and cannot be neglected. The ideal gas law is no longer a good approximation.

Figure 5.31 ▲

Effect of molecular volume at high pressure

Postulate 3 says that the forces of attraction between molecules (intermolecular forces) in a gas are very weak or negligible. This is a good approximation at low pressures, where molecules tend to be far apart, because these forces diminish rapidly as the distance between molecules increases. Intermolecular forces become significant at higher pressures, though, because the molecules tend to be close together. Because of these intermolecular forces, the actual pressure of a gas is less than that predicted by ideal gas behavior. As a molecule begins to collide with a wall surface, neighboring molecules pull this colliding molecule slightly away from the wall, giving a reduced pressure (Figure 5.32).

The Dutch physicist J. D. van der Waals (1837–1923) was the first to account for these deviations of a real gas from ideal gas behavior. The **van der Waals equation** is *an equation similar to the ideal gas law, but includes two constants,* a *and* b, *to account for deviations from ideal behavior.*

$$\left(P + \frac{n^2 a}{V^2}\right)(V - nb) = nRT$$

The constants *a* and *b* are chosen to fit experiment as closely as possible. Table 5.7 gives values of van der Waals constants for several gases.

You can obtain the van der Waals equation from the ideal gas law, $PV = nRT$, by making the following replacements:

Ideal Gas Equation		*van der Waals Equation*
V	becomes	$V - nb$
P	becomes	$P + n^2 a/V^2$

To obtain these substitutions, van der Waals reasoned as follows. The volume available for a molecule to move in equals the gas volume, V, minus the volume occupied by molecules. So he replaced V in the ideal gas law with $V - nb$, where nb represents the volume occupied by n moles of molecules. Then he noted that the total force of attraction on any molecule about to hit a wall is proportional to the concentration of neighboring molecules, n/V. However, the number of molecules about to hit the wall per unit wall area is also proportional to the concentration, n/V. Therefore, the force per unit wall area, or pressure, is reduced from that assumed in the ideal gas law by a factor proportional to n^2/V^2. Letting a be the proportionality constant, we can write

$$P(\text{actual}) = P(\text{ideal}) - n^2 a/V^2$$

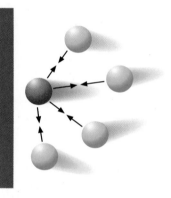

Figure 5.32 ▲

Effect of intermolecular attractions on gas pressure Gas pressure on a wall is due to molecules colliding with it. Here a molecule (shown in red) about to collide with the wall is attracted away from the wall by the weak attractive forces of neighboring molecules (intermolecular forces). As a result, the pressure exerted by the gas is less than that expected in the absence of those forces.

Table 5.7	van der Waals Constants for Some Gases	
	a	*b*
Gas	**L²·atm/mol²**	**L/mol**
CO_2	3.658	0.04286
C_2H_6	5.570	0.06499
C_2H_5OH	12.56	0.08710
He	0.0346	0.0238
H_2	0.2453	0.02651
O_2	1.382	0.03186
SO_2	6.865	0.05679
H_2O	5.537	0.03049

or

$$P(\text{ideal}) = P(\text{actual}) + n^2a/V^2$$

In other words, you replace P in the ideal gas law by $P + n^2a/V^2$.

The next example illustrates the use of the van der Waals equation to calculate the pressure exerted by a gas under given conditions.

Example 5.14	**Using the van der Waals Equation**

Critical Concept 5.14
Real gases do not always behave as ideal gases. Gases behave most ideally under low pressure-high temperature conditions. At higher pressures intermolecular forces (attractions between molecules) and the space occupied by the gas particles must be taken into account when making predictions about gas behavior. At low temperatures intermolecular forces must be taken into account.

Solution Essentials:
• Van der Waals equation
$$\left(P + \frac{n^2a}{V^2}\right)(V - nb) = nRT$$
• Real gas
• Kinetic-molecular theory
• Ideal gas

If sulfur dioxide were an ideal gas, the pressure at 0.0°C exerted by 1.000 mole occupying 22.41 L would be 1.000 atm (22.41 L is the molar volume of an ideal gas at STP). Use the van der Waals equation to estimate the pressure of this volume of 1.000 mol SO_2 at 0.0°C. See Table 5.7 for values of a and b.

Problem Strategy You first rearrange the van der Waals equation to give P in terms of the other variables.

$$P = \frac{nRT}{V - nb} - \frac{n^2a}{V^2}$$

Now you substitute $R = 0.08206$ L·atm/(K·mol), $T = 273.2$ K, $V = 22.41$ L, $a = 6.865$ L^2·atm/mol^2, and $b = 0.05679$ L/mol.

Solution

$$P = \frac{1.000 \text{ mol} \times 0.08206 \text{ L·atm/(K·mol)} \times 273.2 \text{ K}}{22.41 \text{ L} - (1.000 \text{ mol} \times 0.05679 \text{ L/mol})} -$$
$$\frac{(1.000 \text{ mol})^2 \times 6.865 \text{ L}^2\text{·atm/mol}^2}{(22.41 \text{ L})^2}$$

$$= 1.003 \text{ atm} - 0.014 \text{ atm} = \mathbf{0.989 \text{ atm}}$$

Note that the first term, $nRT/(V - nb)$, gives the pressure corrected for molecular volume. This pressure, 1.003 atm, is slightly higher than the ideal value, 1.000 atm, because the volume through which a molecule is free to move, $V - nb$, is smaller than what is assumed for an ideal gas (V). Therefore, the concentration of molecules in this volume is greater, and the number of collisions per unit surface area is expected to be greater. However, the last term, n^2a/V^2, corrects for intermolecular forces, which reduce the force of the collisions, and this reduces the pressure to 0.989 atm.

Answer Check Make sure that you used the appropriate value for the ideal gas constant and that the units of volume and pressure match those given for the constants a and b.

Exercise 5.16   Use the van der Waals equation to calculate the pressure of 1.000 mol ethane, C_2H_6, that has a volume of 22.41 L at 0.0°C. Compare the result with the value predicted by the ideal gas law.

■ See Problems 5.101 and 5.102.

CONCEPT CHECK 5.7

A 1.00-L container is filled with an ideal gas and the recorded pressure is 350 atm. We then put the same amount of a real gas into the container and measure the pressure.

a If the real gas molecules occupy a relatively small volume and have large intermolecular attractions, how would you expect the pressures of the two gases to compare?

b If the real gas molecules occupy a relatively large volume and there are negligible intermolecular attractions, how would you expect the pressures of the two gases to compare?

c If the real gas molecules occupy a relatively large volume and have large intermolecular attractions, how would you expect the pressures of the two gases to compare?

Carbon Dioxide Gas and the Greenhouse Effect

Yes, I know, I shouldn't have left that chocolate bar on the seat of my car, which I parked on a sunny street near the university. You and I know that the interior of such a car can become quite hot. Have you ever wondered why? If the sun's heat energy can get into the car through its window glass, why can't it leave equally well?

Parked cars are similar to greenhouses, which people have used for centuries to protect plants from cold weather. The Swedish chemist Svante Arrhenius, whom you may remember from the previous chapter (for his theory of ionic solutions), realized that carbon dioxide in the atmosphere acts like the glass in a greenhouse. His calculations, published in 1898, were the first

to show how sensitive the temperature of the earth might be to the percentage of carbon dioxide in the atmosphere. Arrhenius's reference to the earth's atmospheric envelope, with its carbon dioxide, as a "hothouse" has evolved into our present-day term *greenhouse effect*.

Here is the explanation of the greenhouse effect. The principal gases in the atmosphere are oxygen, O_2, and nitrogen, N_2. These gases are transparent to visible light from the sun, and when this light reaches the surface of the earth, it is absorbed and converted to heat. This heat causes atoms in the earth's surface to vibrate, which then radiate the heat energy as infrared radiation, or heat rays (Figure 5.33). Neither oxygen nor nitrogen absorb infrared

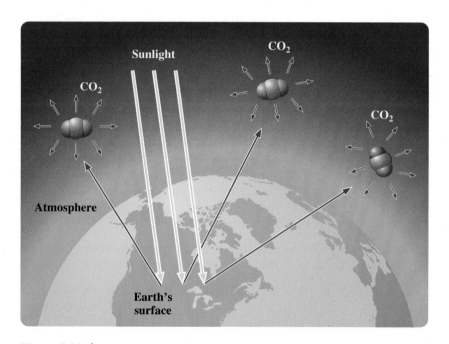

Figure 5.33 ▲

Greenhouse effect of certain gases in the atmosphere The atmosphere is transparent to sunlight, except for radiation in the far ultraviolet. When sunlight reaches the earth's surface, it is absorbed. The heated surface then radiates the energy back as infrared radiation. The major components of the atmosphere, O_2 and N_2, are transparent to this radiation, but gases such as CO_2, H_2O, CH_4, and chlorofluorocarbons, do absorb infrared rays. These substances in the atmosphere then reradiate the infrared rays, with the result that a significant fraction of the radiation returns to earth. In effect, these gases trap the radiation, acting like the glass on a greenhouse.

radiation, so if the earth's atmosphere contained only these gases, the infrared radiation, or heat rays, would simply escape into outer space; there would be no greenhouse effect. However, other gases in the atmosphere, especially carbon dioxide, do absorb infrared radiation, and it is this absorption that warms the atmosphere and eventually also the earth's surface, giving us a greenhouse effect. The greater the percentage of carbon dioxide in the atmosphere, the warmer the earth should be. (Glass, acting like carbon dioxide, allows visible light to pass into a greenhouse, but absorbs infrared radiation, effectively trapping the heat.)

The greenhouse effect has become associated with the concept of *global warming*. Records kept since the nineteenth century clearly show that the amount of carbon dioxide in the atmosphere has increased dramatically, from about 284 ppm (parts per million) in 1830 to about 370 ppm in 2000 (Figure 5.34). (Parts per million is equivalent to mole percent $\times$ 10^{-4}.) Scientists believe that our burning of fossil fuels (coal, oil, and natural gas) is responsible for this change. Moreover, temperature measurements made at the surface of the earth during the past 100 years indicate an average temperature increase of between 0.4°C and 0.8°C. Thus, many climate scientists believe that human activities are at least partially responsible for global warming, and hence, they predict drastic changes in climate.

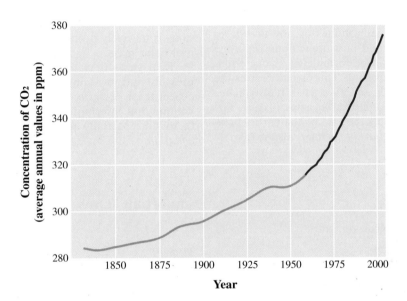

Figure 5.34 ▲

Concentrations of carbon dioxide in the atmosphere (average annual values) Concentrations of carbon dioxide in the atmosphere have been increasing steadily since about 1830. (*Source of blue curve*: D. M. Etheridge, L. P. Steele, R. L. Langenfelds, R. J. Francey, J. M. Barnola, and V. I. Morgan. 1998. Historical CO_2 records from the Law Dome DE08, DE08-2, and DSS ice cores. *Source of the purple curve:* R. F. Keeling, S. C. Piper, A. F. Bollenbacker and J. S. Walker. 2009. Atmospheric CO_2 records from sites in the SIO air sampling network. Both sets of data are from *Trends Online: A Compendium of Data on Global Change*. Carbon Dioxide Information Analysis Center, Oak Ridge National Laboratory, U. S. Department of Energy, Oak Ridge, TN, U.S.A.)

■ See Problems 5.133 and 5.134.

A Checklist for Review

Summary of Facts and Concepts

The *pressure* of a gas equals the force exerted per unit area of surface. It is measured by a *manometer* in units of *pascals, millimeters of mercury,* or *atmospheres.*

Gases at low to moderate pressures and moderate temperatures follow the same simple relationships, or gas laws. Thus, for a given amount of gas at constant temperature, the volume varies inversely with pressure *(Boyle's law)*. Also, for a given amount of gas at constant pressure, the volume is directly proportional to the absolute temperature *(Charles's law)*. These two laws, together with *Avogadro's law* (equal volumes of any two gases at the same temperature and pressure contain the same number of molecules), can be formulated as one equation, *PV = nRT (ideal gas law)*. This equation gives the relationship among *P, V, n,* and *T* for a gas. It also relates these quantities for each component in a gas mixture. The total gas pressure, *P*, equals the sum of the partial pressures of each component *(law of partial pressures)*.

The ideal gas law can be explained by the *kinetic-molecular theory*. We define an ideal gas as consisting of molecules with negligible volume that are in constant random motion. In the ideal gas model, there are no intermolecular forces between molecules, and the average kinetic energy of molecules is proportional to the absolute temperature. From kinetic theory, one can show that the *rms molecular speed* equals $\sqrt{3RT/M_m}$. Given this result, one can derive *Graham's law of effusion:* the rate of effusion of gas molecules under identical conditions is inversely proportional to the square root of the molecular mass.

Real gases deviate from the ideal gas law at high pressure and low temperature. From kinetic theory, we expect these deviations, because real molecules do have volume and intermolecular forces do exist. These two factors are partially accounted for in the *van der Waals equation*.

Learning Objectives

Important Terms

5.1 Gas Pressure and Its Measurement

- Define *pressure* and its units.
- Convert units of pressure. Example 5.1

pressure
pascal (Pa)
barometer
manometer
millimeters of mercury (mmHg or torr)
atmosphere (atm)
bar

5.2 Empirical Gas Laws

- Express *Boyle's law* in words and as an equation.
- Use Boyle's law. Example 5.2
- Express *Charles's law* in words and as an equation.
- Use Charles's law. Example 5.3
- Express the *combined gas law* as an equation.
- Use the combined gas law. Example 5.4
- State *Avogadro's law*.
- Define *standard temperature and pressure (STP)*.

Boyle's law
Charles's law
Avogadro's law
molar gas volume (V_m)
standard temperature and pressure (STP)

5.3 The Ideal Gas Law

- State what makes a gas an *ideal gas*.
- Learn the *ideal gas law* equation.
- Derive the empirical gas laws from the ideal gas law. Example 5.5
- Use the ideal gas law. Example 5.6
- Calculate gas density. Example 5.7
- Determine the molecular mass of a vapor. Example 5.8
- Use an equation to calculate gas density.

molar gas constant (R)
ideal gas law

5.4 Stoichiometry Problems Involving Gas Volumes

- Solve stoichiometry problems involving gas volumes. Example 5.9

5.5 Gas Mixture; Law of Partial Pressures

- Learn the equation for *Dalton's law of partial pressures*.
- Define the *mole fraction* of a gas.
- Calculate the partial pressure and mole fractions of a gas in a mixture. Example 5.10
- Describe how gases are collected over water and how to determine the *vapor pressure* of water.
- Calculate the amount of gas collected over water. Example 5.11

partial pressure
Dalton's law of partial pressures
mole fraction

5.6 Kinetic Theory of an Ideal Gas

- List the five postulates of the *kinetic theory*.
- Provide a qualitative description of the gas laws based on the kinetic theory.

kinetic-molecular theory of gases (kinetic theory)

5.7 Molecular Speeds; Diffusion and Effusion

- Describe how the *root-mean-square (rms) molecular speed* of gas molecules varies with temperature.
- Describe the molecular-speed distribution of gas molecules at different temperatures.
- Calculate the rms speed of gas molecules. Example 5.12
- Define *effusion* and *diffusion*.
- Describe how individual gas molecules move undergoing diffusion.
- Calculate the ratio of effusion rates of gases. Example 5.13

root-mean-square (rms) molecular speed
diffusion
effusion
Graham's law of effusion

5.8 Real Gases

- Explain how and why a *real gas* is different from an ideal gas.
- Use the van der Waals equation. Example 5.14

van der Waals equation

Key Equations

$PV = \text{constant}$ (constant n, T)

$\dfrac{V}{T} = \text{constant}$ (constant n, P)

$V_m = \text{specific constant}$
(depending on T, P; independent of gas)

$PV = nRT$

$PM_m = dRT$

$P = P_A + P_B + P_C + \cdots$

Mole fraction of $A = \dfrac{n_A}{n} = \dfrac{P_A}{P}$

$u = \sqrt{\dfrac{3RT}{M_m}} = \left(\dfrac{3RT}{M_m}\right)^{\frac{1}{2}}$

$\text{Rate of effusion} \propto \dfrac{1}{\sqrt{M_m}}$

(same container at constant T, P)

$\left(P + \dfrac{n^2 a}{V^2}\right)(V - nb) = nRT$

Questions and Problems

Self-Assessment and Review Questions

Key: These questions test your understanding of the ideas you worked with in the chapter. These problems vary in difficulty and often can be used for the basis of discussion.

5.1 Define *pressure*. From the definition, obtain the SI unit of pressure in terms of SI base units.

5.2 For what purpose is a manometer used? How does it work?

5.3 What variables determine the height of the liquid in a manometer?

5.4 Starting with Boyle's law (stated as an equation), obtain an equation for the final volume occupied by a gas from the initial volume when the pressure is changed at constant temperature.

5.5 Explain how you would set up an absolute temperature scale based on the Fahrenheit scale, knowing that absolute zero is $-273.15°C$.

5.6 The volume occupied by a gas depends linearly on degrees Celsius at constant pressure, but it is not directly proportional to degrees Celsius. However, it is directly proportional to kelvins. What is the difference between a linear relationship and a direct proportion?

5.7 Starting with Charles's law (stated as an equation), obtain an equation for the final volume of a gas from its initial volume when the temperature is changed at constant pressure.

5.8 State Avogadro's law in words. How does this law explain the law of combining volumes? Use the gas reaction

$$N_2 + 3H_2 \longrightarrow 2NH_3$$

as an example in your explanation.

5.9 What are the standard conditions for comparing gas volumes?

5.10 What does the term *molar gas volume* mean? What is the molar gas volume (in liters) at STP for an ideal gas?

5.11 Starting from Boyles's, Charles's, and Avogadro's laws, obtain the ideal gas law, $PV = nRT$.

5.12 What is the value of R in units of $L \cdot mmHg/(K \cdot mol)$?

5.13 What variables are needed to describe a gas that obeys the ideal gas law? What are the SI units for each variable?

5.14 The ideal gas law relates four variables. An empirical gas law relates two variables, assuming the other two are constant. How many empirical gas laws can be obtained? Give statements of each.

5.15 Give the postulates of kinetic theory and state any evidence that supports them.

5.16 What is the origin of gas pressure, according to kinetic theory?

5.17 Explain Boyle's law in terms of the kinetic theory.

5.18 How does the rms molecular speed depend on absolute temperature? on molar volume?

5.19 Explain why a gas appears to diffuse more slowly than average molecular speeds might suggest.

5.20 What is effusion? Why does a gas whose molecules have smaller mass effuse faster than one whose molecules have larger mass?

5.21 Under what conditions does the behavior of a real gas begin to differ significantly from the ideal gas law?

5.22 What is the physical meaning of the *a* and *b* constants in the van der Waals equation?

5.23 Which of the following is *not* part of the kinetic molecular theory?

> a   Gas molecules have no volume.
> b   Gas particles are in constant motion.
> c   The greater the volume occupied by a given amount of gas, the higher the intermolecular forces.
> d   Total kinetic energy is conserved during the collision.
> e   The average kinetic energy is proportional to the absolute temperature.

5.24 A sample of nitrogen gas is placed into a container at 200 K. If the temperature of the container is reduced by 100°C and the measured pressure decreases by 50%, how does the volume of the container change?

> a   The container volume increases by less than 50%.
> b   The container volume increases by more than 50%.
> c   The container volume doesn't change.
> d   The container volume increases by 50%.
> e   The container volume decreases by 50%.

5.25 A 1-liter container is filled with 2.0 mol Ar, 2.0 mol H_2, and 4.0 mol Kr. Which of the following statements about these gases is *false*?

> a   The Kr is the densest of the three gases.
> b   The mole fraction of Ar in the flask is 0.25.
> c   The total pressure in the flask is four times the pressure of the Ar.
> d   The Ar atoms hit the walls of the flask with the greatest force of the three gases.
> e H_2 has the highest rms velocity of the three gases.

5.26 At standard temperature and pressure, a 1.00-mol sample of argon gas is vented into a 22.4-L rigid box that already contains 1.00 mol of nitrogen gas. We would expect the argon gas to

> a   decrease the total gas pressure in the box by 50%.
> b   occupy the entire 22.4-L volume of the box.
> c   increase the total gas pressure in the box by less than 50%.
> d   spread out into the box, but the actual volume occupied by the gas cannot be known without pressure information.
> e   occupy only a volume of 11.2 L of the box, since there is already 1.0 mol of nitrogen present in the container.

Concept Explorations

Key: Concept explorations are comprehensive problems that provide a framework that will enable you to explore and learn many of the critical concepts and ideas in each chapter. If you master the concepts associated with these explorations, you will have a better understanding of many important chemistry ideas and will be more successful in solving all types of chemistry problems. These problems are well suited for group work and for use as in-class activities.

5.27 Gas Laws and Kinetic Theory of Gases I

Shown below are two identical containers labeled A and B. Container A contains a molecule of an ideal gas, and container B contains two molecules of an ideal gas. Both containers are at the same temperature. (Note that small numbers of molecules and atoms are being represented in these examples in order that you can easily compare the amounts. Real containers with so few molecules and atoms would be unlikely.)

 a How do the pressures in the two containers compare? Be sure to explain your answer.

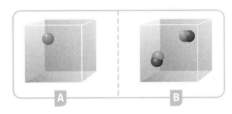

 b Shown below are four different containers (C, D, E, and F), each with the same volume and at the same temperature. How do the pressures of the gases in the containers compare?

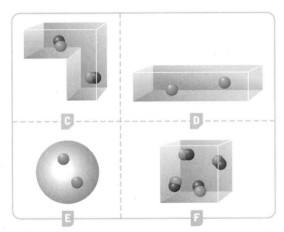

 c Container H below has twice the volume of container G. How will the pressure in the containers compare? Explain your reasoning.

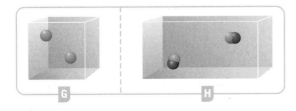

 d How will the pressure of containers G and H compare if you add two more gas molecules to container H?

 e Consider containers I and J below. Container J has twice the volume of container I. Container I is at a temperature of 100 K, and container J is at 200 K. How does the pressure in container I compare with that in container J? Include an explanation as part of your answer.

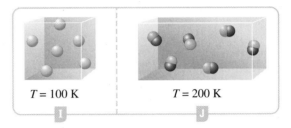

T = 100 K   *T* = 200 K

5.28 Gas Laws and Kinetic Theory of Gases II

Consider the box below that contains a single atom of an ideal gas.

 a Assuming that this gas atom is moving, describe how it creates pressure inside the container.

 b Now consider the two containers below, each at the same temperature. If we were to measure the gas pressure in each container, how would they compare? Explain your answer.

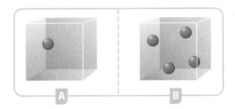

 c Consider the two containers below, labeled C and D, one with an ideal gas atom and one with ideal gas molecules, each at the same temperature. The gas molecule has more mass than the gas atom. How do the pressures of the two containers compare? Why?

 d For the containers C and D above, which would have the higher root-mean-square (rms) molecular speed? How does the difference in rms speeds support or contradict the answer you obtained above when you compared the pressures?

e   Consider containers E and F below. How do the pressures in these containers compare? Be sure to explain your answer.

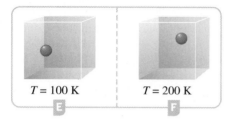

g   Now think about the containers below, I and J. How do the pressures in these containers compare? Be sure to justify your answer.

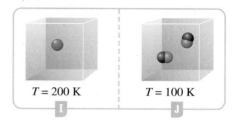

f   Consider containers G and H below, one with a gas atom and one with a gas molecule. As before, the gas molecule has more mass than the gas atom. Explain how the pressures in these containers compare.

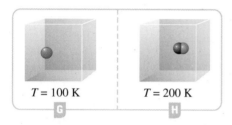

h   Finally, how do the pressures in containers K and L compare?

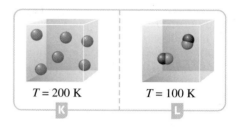

Conceptual Problems

Key: These problems are designed to check your understanding of the concepts associated with some of the main topics presented in each chapter. A strong conceptual understanding of chemistry is the foundation for both applying chemical knowledge and solving chemical problems. These problems vary in level of difficulty and often can be used as a basis for group discussion.

5.29 Using the concepts developed in this chapter, explain the following observations.

a   The lid of a water bottle pops off when the bottle sits in the sun.

b   You are not supposed to dispose of aerosol cans in a fire.

c   Automobile tires are flatter on cold days.

d   A balloon pops when you squeeze it.

5.30 You have three identical flasks, each containing equal amounts of N_2, O_2, and He. The volume of the N_2 flask is doubled, the O_2 flask volume is halved, and the He flask volume is reduced to one-third of the original volume. Rank the flasks from highest to lowest pressure both before and after the volume is changed and indicate by what factor the pressure has changed.

5.31 Consider the following gas container equipped with a movable piston.

a   By what factor (increase by 1, decrease by 1.5, etc.) would you change the pressure if you wanted the volume to change from volume C to volume D?

b   If the piston were moved from volume C to volume A, by what factor would the pressure change?

c   By what factor would you change the temperature in order to change from volume C to volume B?

d   If you increased the number of moles of gas in the container by a factor of 2, by what factors would the pressure and the volume change?

5.32 A 3.00-L flask containing 2.0 mol of O_2 and 1.0 mol of N_2 is in a room that is at 22.0°C.

a   How much (what fraction) of the total pressure in the flask is due to the N_2?

b   The flask is cooled and the pressure drops. What happens, if anything, to the mole fraction of the O_2 at the lower temperature?

c   1.0 L of liquid water is introduced into the flask containing both gases. The pressure is then measured about 45 minutes later. Would you expect the measured pressure to be higher or lower?

d   Given the information in this problem and the conditions in part c, would it be possible to calculate the pressure in the flask after the introduction of the water? If it is not possible with the given information, what further information would you need to accomplish this task?

5.33 Consider the following setup, which shows identical containers connected by a tube with a valve that is presently closed. The container on the left has 1.0 mol of H_2 gas; the container on the right has 1.0 mol of O_2.

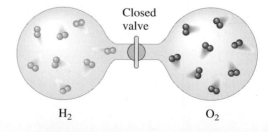

Note: Acceptable answers to some of these questions might be "both" or "neither one."

a Which container has molecules that are moving at a faster average molecular speed?

b 2.0 mol of Ar is added to the system with the valve open. What fraction of the total pressure will be due to the H_2?

c If the valve is opened, will the pressure in each of the containers change? If it does, how will it change (increase, decrease, or no change)?

d Which container has the greatest density of gas?

e Which container has more molecules?

5.34 Two identical He-filled balloons, each with a volume of 20 L, are allowed to rise into the atmosphere. One rises to an altitude of 3000 m while the other rises to 6000 m.

a Assuming that the balloons are at the same temperature, which balloon has the greater volume?

b What information would you need in order to calculate the volume of each of the balloons at their respective heights?

5.35 Three 25.0-L flasks are placed next to each other on a shelf in a chemistry stockroom. The first flask contains He at a pressure of 1.0 atm, the second contains Xe at 1.50 atm, and the third contains F_2 and has a label that says 2.0 mol F_2. Consider the following questions about these flasks.

a Which flask has the greatest number of moles of gas?

b If you wanted each of the flasks to be at the same pressure as the He flask, what general things could you do to the other two containers to make this happen?

5.36 You have a balloon that contains O_2. What could you do to the balloon in order to double the volume? Be specific in your answers; for example, you could increase the number of moles of O_2 by a factor of 2.

Practice Problems

Key: These problems are for practice in applying problem-solving skills. They are divided by topic, and some are keyed to exercises (see the ends of the exercises). The problems are arranged in matching pairs; the odd-numbered problem of each pair is listed first, and its answer is given in the back of the book.

Note: In these problems, the final zeros given in temperatures and pressures (for example, 20°C, 760 mmHg) are significant figures.

Units of Pressure

5.37 A gas in a closed-tube manometer (Figure 5.4) has a measured pressure of 0.094 atm. Calculate the pressure in mmHg.

5.38 The barometric pressure measured outside an airplane at 9 km (30,000 ft) was 266 torr. Calculate the pressure in kPa.

Empirical Gas Laws

5.39 You fill a balloon with helium gas to a volume of 2.68 L at 23°C and 789 mmHg. Now you release the balloon. What would be the volume of helium if its pressure changed to 499 mmHg but the temperature were unchanged?

5.39 Suppose you had a 3.15-L sample of neon gas at 21°C and a pressure of 0.951 atm. What would be the volume of this gas if the pressure were increased to 1.292 atm while the temperature remained constant?

5.41 You have a cylinder of argon gas at 19.8 atm pressure at 19°C. The volume of argon in the cylinder is 25.0 L. What would be the volume of this gas if you allowed it to expand to the pressure of the surrounding air (0.974 atm)? Assume the temperature remains constant.

5.42 A diving bell is a container open at the bottom. As the bell descends, the water level inside changes so that the pressure inside equals the pressure outside. Initially, the volume of air is 8.58 m^3 at 1.020 atm and 20°C. What is the volume at 1.212 atm and 20°C?

5.43 A McLeod gauge measures low gas pressures by compressing a known volume of the gas at constant temperature. If 315 cm^3 of gas is compressed to a volume of 0.0457 cm^3 under a pressure of 2.51 kPa, what was the original gas pressure?

5.44 If 456 dm^3 of krypton at 101 kPa and 21°C is compressed into a 30.1-dm^3 tank at the same temperature, what is the pressure of krypton in the tank?

5.45 A sample of nitrogen gas at 18°C and 760 mmHg has a volume of 3.92 mL. What is the volume at 0°C and 1 atm of pressure?

5.46 A mole of gas at 0°C and 760 mmHg occupies 22.41 L. What is the volume at 20°C and 760 mmHg?

5.47 Helium gas, He, at 22°C and 1.00 atm occupied a vessel whose volume was 5.08 L. What volume would this gas occupy if it were cooled to liquid-nitrogen temperature (−197°C)?

5.48 An experiment called for 4.83 L of sulfur dioxide, SO_2, at 0°C and 1.00 atm. What would be the volume of this gas at 29°C and 1.00 atm?

5.49 A sample of 62.3 cm^3 of argon gas at 18°C was contained at a pressure of 155 kPa in a J-shaped tube with mercury, as in Figure 5.5. Later the temperature changed. When the mercury level was adjusted to give the same pressure of argon, the gas volume changed to 61.2 cm^3. What was the final temperature of the argon?

5.50 A vessel containing 39.5 cm^3 of helium gas at 25°C and 106 kPa was inverted and placed in cold ethanol. As the gas contracted, ethanol was forced into the vessel to maintain the same pressure of helium. If this required 7.5 cm^3 of ethanol, what was the final temperature of the helium?

5.51 A bacterial culture isolated from sewage produced 35.5 mL of methane, CH_4, at 31°C and 753 mmHg. What is the volume of this methane at standard temperature and pressure (0°C, 760 mmHg)?

5.52 Pantothenic acid is a B vitamin. Using the Dumas method, you find that a sample weighing 71.6 mg gives 3.84 mL of nitrogen gas at 23°C and 785 mmHg. What is the volume of nitrogen at STP?

5.53 Methanol, CH_3OH, can be produced in industrial plants by reacting carbon dioxide with hydrogen in the presence of a catalyst. Water is the other product. How many volumes of hydrogen are required for each volume of carbon dioxide when each gas is at the same temperature and pressure?

5.54 In the presence of a platinum catalyst, ammonia, NH_3, burns in oxygen, O_2, to give nitric oxide, NO, and water vapor. How many volumes of nitric oxide are obtained from one volume of ammonia, assuming each gas is at the same temperature and pressure?

Ideal Gas Law

5.55 Starting from the ideal gas law, prove that the volume of a mole of gas is inversely proportional to the pressure at constant temperature (Boyle's law).

5.56 Starting from the ideal gas law, prove that the volume of a mole of gas is directly proportional to the absolute temperature at constant pressure (Charles's law).

5.57 In an experiment, you fill a heavy-walled 6.00-L flask with methane gas, CH_4. If the flask contains 7.13 g of methane at 19°C, what is the gas pressure?

5.58 A cylinder of oxygen gas contains 91.3 g O_2. If the volume of the cylinder is 8.58 L, what is the pressure of the O_2 if the gas temperature is 21°C?

5.59 An experiment calls for 3.50 mol of chlorine, Cl_2. What volume will this be if the gas volume is measured at 34°C and 4.00 atm?

5.60 According to your calculations, a reaction should yield 5.67 g of oxygen, O_2. What do you expect the volume to be at 23°C and 0.894 atm?

5.61 The maximum safe pressure that a certain 4.00-L vessel can hold is 3.50 atm. If the vessel contains 0.410 mol of gas, what is the maximum temperature (in degrees Celsius) to which this vessel can be subjected?

5.62 A 2.50-L flask was used to collect a 5.65-g sample of propane gas, C_3H_8. After the sample was collected, the gas pressure was found to be 741 mmHg. What was the temperature of the propane in the flask?

5.63 Calculate the density of hydrogen sulfide gas, H_2S, at 49°C and 967 mmHg. Obtain the density in grams per liter.

5.64 What is the density of ammonia gas, NH_3, at 31°C and 751 mmHg? Obtain the density in grams per liter.

5.65 Butane, C_4H_{10}, is an easily liquefied gaseous fuel. Calculate the density of butane gas at 0.897 atm and 24°C. Give the answer in grams per liter.

5.66 Chloroform, $CHCl_3$, is a volatile (easily vaporized) liquid solvent. Calculate the density of chloroform vapor at 98°C and 797 mmHg. Give the answer in grams per liter.

5.67 A chemist vaporized a liquid compound and determined its density. If the density of the vapor at 90°C and 753 mmHg is 1.585 g/L, what is the molecular mass of the compound?

5.68 You vaporize a liquid substance at 100°C and 755 mmHg. The volume of 0.548 g of vapor is 237 mL. What is the molecular mass of the substance?

5.69 A 2.30-g sample of white solid was vaporized in a 345-mL vessel. If the vapor has a pressure of 985 mmHg at 148°C, what is the molecular mass of the solid?

5.70 A 2.56-g sample of a colorless liquid was vaporized in a 250-mL flask at 121°C and 786 mmHg. What is the molecular mass of this substance?

5.71 Ammonium chloride, NH_4Cl, is a white solid. When heated to 325°C, it gives a vapor that is a mixture of ammonia and hydrogen chloride.

$$NH_4Cl(s) \longrightarrow NH_3(g) + HCl(g)$$

Suppose someone contends that the vapor consists of NH_4Cl molecules rather than a mixture of NH_3 and HCl. Could you decide between these alternative views on the basis of gas-density measurements? Explain.

5.72 Phosphorus pentachloride, PCl_5, is a white solid that sublimes (vaporizes without melting) at about 100°C. At higher temperatures, the PCl_5 vapor decomposes to give phosphorus trichloride, PCl_3, and chlorine, Cl_2.

$$PCl_5(g) \longrightarrow PCl_3(g) + Cl_2(g)$$

How could gas-density measurements help to establish that PCl_5 vapor is decomposing?

Stoichiometry with Gas Volumes

5.73 Magnesium metal reacts with hydrochloric acid to produce hydrogen gas, H_2.

$$Mg(s) + 2HCl(aq) \longrightarrow MgCl_2(aq) + H_2(g)$$

Calculate the volume (in liters) of hydrogen produced at 33°C and 665 mmHg from 0.0840 mol Mg and excess HCl.

5.74 Calcium carbide reacts with water to produce acetylene gas, C_2H_2.

$$CaC_2(s) + 2H_2O(l) \longrightarrow Ca(OH)_2(aq) + C_2H_2(g)$$

Calculate the volume (in liters) of acetylene produced at 26°C and 684 mmHg from 0.075 mol CaC_2 and excess H_2O.

5.75 Lithium hydroxide, LiOH, is used in spacecraft to recondition the air by absorbing the carbon dioxide exhaled by astronauts. The reaction is

$$2LiOH(s) + CO_2(g) \longrightarrow Li_2CO_3(s) + H_2O(l)$$

What volume of carbon dioxide gas at 21°C and 781 mmHg could be absorbed by 327 g of lithium hydroxide?

5.76 Magnesium burns in air to produce magnesium oxide, MgO, and magnesium nitride, Mg_3N_2. Magnesium nitride reacts with water to give ammonia.

$$Mg_3N_2(s) + 6H_2O(l) \longrightarrow 3Mg(OH)_2(s) + 2NH_3(g)$$

What volume of ammonia gas at 24°C and 753 mmHg will be produced from 3.93 g of magnesium nitride?

5.77 Urea, NH_2CONH_2, is a nitrogen fertilizer that is manufactured from ammonia and carbon dioxide.

$$2NH_3(g) + CO_2(g) \longrightarrow NH_2CONH_2(aq) + H_2O(l)$$

What volume of ammonia at 25°C and 3.00 atm is needed to produce 908 g (2 lb) of urea?

5.78 Nitric acid is produced from nitrogen monoxide, NO, which in turn is prepared from ammonia by the Ostwald process:

$$4NH_3(g) + 5O_2(g) \longrightarrow 4NO(g) + 6H_2O(g)$$

What volume of oxygen at 35°C and 2.15 atm is needed to produce 100.0 g of nitrogen monoxide?

5.79 Sodium hydrogen carbonate is also known as baking soda. When this compound is heated, it decomposes to sodium carbonate, carbon dioxide, and water vapor. Write the balanced equation for this reaction. What volume (in liters) of carbon dioxide gas at 77°C and 756 mmHg will be produced from 26.8 of sodium hydrogen carbonate?

5.80 Ammonium sulfate is used as a nitrogen and sulfur fertilizer. It is produced by reacting ammonia with sulfuric acid. Write the balanced equation for the reaction of gaseous ammonia with sulfuric acid solution. What volume (in liters) of ammonia at 15°C and 1.15 atm is required to produce 150.0 g of ammonium sulfate?

Gas Mixtures

5.81 Calculate the total pressure (in atm) of a mixture of 0.0300 mol of helium, He, and 0.0400 mol of oxygen, O_2, in a 4.00-L flask at 20°C. Assume ideal gas behavior.

5.82 Calculate the total pressure (in atm) of a mixture of 0.0200 mol of helium, He, and 0.0100 mol of hydrogen, H_2, in a 2.50-L flask at 10°C. Assume ideal gas behavior.

5.83 A 200.0-mL flask contains 1.03 mg O_2 and 0.56 mg He at 15°C. Calculate the partial pressures of oxygen and of helium in the flask. What is the total pressure?

5.84 The atmosphere in a sealed diving bell contained oxygen and helium. If the gas mixture has 0.200 atm of oxygen and a total pressure of 3.00 atm, calculate the mass of helium in 10.0 L of the gas mixture at 40°C.

5.85 In a series of experiments, the U.S. Navy developed an undersea habitat. In one experiment, the mole percent composition of the atmosphere in the undersea habitat was 79.0% He, 17.0% N_2, and 4.0% O_2. What will the partial pressure of each gas be when the habitat is 58.8 m below sea level, where the pressure is 6.91 atm?

5.86 The gas from a certain volcano had the following composition in mole percent (that is, mole fraction × 100): 65.0% CO_2, 25.0% H_2, 5.4% HCl, 2.8% HF, 1.7% SO_2, and 0.1% H_2S. What would be the partial pressure of each of these gases if the total pressure of volcanic gas were 760 mmHg?

5.87 Formic acid, $HCHO_2$, is a convenient source of small quantities of carbon monoxide. When warmed with sulfuric acid, formic acid decomposes to give CO gas.

$$HCHO_2(l) \longrightarrow H_2O(l) + CO(g)$$

If 3.85 L of carbon monoxide was collected over water at 25°C and 689 mmHg, how many grams of formic acid were consumed?

5.88 An aqueous solution of ammonium nitrite, NH_4NO_2, decomposes when heated to give off nitrogen, N_2.

$$NH_4NO_2(s) \longrightarrow 2H_2O(g) + N_2(g)$$

This reaction may be used to prepare pure nitrogen. How many grams of ammonium nitrite must have reacted if 4.00 dm^3 of nitrogen gas was collected over water at 26°C and 97.8 kPa?

Molecular Speeds; Effusion

5.89 Calculate the rms speeds of N_2 molecules at 25°C and at 125°C. Sketch approximate curves of the molecular speed distributions of N_2 at 25°C and at 125°C.

5.90 Calculate the rms speed of Br_2 molecules at 23°C and 1.00 atm. What is the rms speed of Br_2 at 23°C and 2.00 atm?

5.91 Uranium hexafluoride, UF_6, is a white solid that sublimes (vaporizes without melting) at 57°C under normal atmospheric pressure. The compound is used to separate uranium isotopes by effusion. What is the rms speed (in m/s) of a uranium hexafluoride molecule at 57°C?

5.92 For a spacecraft or a molecule to leave the moon, it must reach the escape velocity (speed) of the moon, which is 2.37 km/s. The average daytime temperature of the moon's surface is 365 K. What is the rms speed (in m/s) of a hydrogen molecule at this temperature? How does this compare with the escape velocity?

5.93 At what temperature would CO_2 molecules have an rms speed equal to that of H_2 molecules at 25°C?

5.94 At what temperature does the rms speed of O_2 molecules equal 475. m/s?

5.95 Obtain the ratio of rates of effusion of H_2 and H_2Se under the same conditions.

5.96 What is the ratio of rates of effusion of N_2 and O_2 under the same conditions?

5.97 If 0.10 mol of I_2 vapor can effuse from an opening in a heated vessel in 39 s, how long will it take 0.10 mol H_2 to effuse under the same conditions?

5.98 If it takes 11.2 hours for 1.00 L of nitrogen, N_2, to effuse through the pores in a balloon, how long would it take for 1.00 L of helium, He, to effuse under the same conditions?

5.99 If 4.83 mL of an unknown gas effuses through a hole in a plate in the same time it takes 9.23 mL of argon, Ar, to effuse through the same hole under the same conditions, what is the molecular mass of the unknown gas?

5.100 A given volume of nitrogen, N_2, required 68.3 s to effuse from a hole in a chamber. Under the same conditions, another gas required 85.6 s for the same volume to effuse. What is the molecular mass of this gas?

van der Waals Equation

5.101 Calculate the pressure of water vapor at 120.0°C if 1.000 mol of water vapor occupies 32.50 L. Use the van der Waals equation (see Table 5.7 for data). Compare with the result from the ideal gas law.

5.102 Calculate the pressure of ethanol vapor, $C_2H_5OH(g)$, at 82.0°C if 1.000 mol $C_2H_5OH(g)$ occupies 30.00 L. Use the van der Waals equation (see Table 5.7 for data). Compare with the result from the ideal gas law.

5.103 Calculate the molar volume of ethane at 1.00 atm and 0°C and at 10.0 atm and 0°C, using the van der Waals equation. The van der Waals constants are given in Table 5.7. To simplify, note that the term n^2a/V^2 is small compared with P. Hence, it may be approximated with negligible error by substituting nRT/P from the ideal gas law for V in this term. Then the van der Waals equation can be solved for the volume. Compare the results with the values predicted by the ideal gas law.

5.104 Calculate the molar volume of oxygen at 1.00 atm and 0°C and at 10.0 atm and 0°C, using the van der Waals equation. The van der Waals constants are given in Table 5.7. (See the note on solving the equation given in Problem 5.103.) Compare the results with the values predicted by the ideal gas law. Also compare with the values obtained from Table 5.3.

General Problems

Key: These problems provide more practice but are not divided by topic or keyed to exercises. Each section ends with essay questions, each of which is color coded to refer to the *A Chemist Looks at Life Science* (pink) or *A Chemist Looks at Environment* (green) chapter essay on which it is based. Odd-numbered problems and the even-numbered problems that follow are similar; answers to all odd-numbered problems except the essay questions are given in the back of the book.

5.105 A glass tumbler containing 243 cm³ of air at 1.00×10^2 kPa (the barometric pressure) and 20°C is turned upside down and immersed in a body of water to a depth of 20.5 m. The air in the glass is compressed by the weight of water above it. Calculate the volume of air in the glass, assuming the temperature and barometric pressure have not changed.

5.106 The density of air at 20°C and 1.00 atm is 1.205 g/L. If this air were compressed at the same temperature to equal the pressure at 50.0 m below sea level, what would be its density? Assume the barometric pressure is constant at 1.00 atm. The density of seawater is 1.025 g/cm³.

5.107 A flask contains 201 mL of argon at 21°C and 738 mmHg. What is the volume of gas, corrected to STP?

5.108 A steel bottle contains 24.0 L of a gas at 11.0 atm and 20°C. What is the volume of gas at STP?

5.109 A balloon containing 5.0 dm³ of gas at 14°C and 100.0 kPa rises to an altitude of 2000. m, where the temperature is 20°C. The pressure of gas in the balloon is now 79.0 kPa. What is the volume of gas in the balloon?

5.110 A volume of air is taken from the earth's surface, at 19°C and 1.00 atm, to the stratosphere, where the temperature is −21°C and the pressure is 1.00×10^{-3} atm. By what factor is the volume increased?

5.111 A radioactive metal atom decays (goes to another kind of atom) by emitting an alpha particle (He^{2+} ion). The alpha particles are collected as helium gas. A sample of helium with a volume of 12.05 mL was obtained at 765 mmHg and 23°C. How many atoms decayed during the period of the experiment?

5.112 The combustion method used to analyze for carbon and hydrogen can be adapted to give percentage N by collecting the nitrogen from combustion of the compound as N_2. A sample of a compound weighing 8.75 mg gave 1.59 mL N_2 at 25°C and 749 mmHg. What is the percentage N in the compound?

5.113 Dry air at STP has a density of 1.2929 g/L. Calculate the average molecular weight of air from the density.

5.114 A hydrocarbon gas has a density of 1.22 g/L at 20°C and 1.00 atm. An analysis gives 80.0% C and 20.0% H. What is the molecular formula?

5.115 Pyruvic acid, $HC_3H_3O_3$, is involved in cell metabolism. It can be assayed for (that is, the amount of it determined) by using a yeast enzyme. The enzyme makes the following reaction go to completion:

$$HC_3H_3O_3(aq) \longrightarrow C_2H_4O(aq) + CO_2(g)$$

If a sample containing pyruvic acid gives 21.2 mL of carbon dioxide gas, CO_2, at 349 mmHg and 30°C, how many grams of pyruvic acid are there in the sample?

5.116 A person exhales about 5.8×10^2 L of carbon dioxide per day (at STP). The carbon dioxide exhaled by an astronaut is absorbed from the air of a space capsule by reaction with lithium hydroxide, LiOH.

$$2LiOH(s) + CO_2(g) \longrightarrow Li_2CO_3(s) + H_2O(l)$$

How many grams of lithium hydroxide are required per astronaut per day?

5.117 Liquid oxygen was first prepared by heating potassium chlorate, $KClO_3$, in a closed vessel to obtain oxygen at high pressure. The oxygen was cooled until it liquefied.

$$2KClO_3(s) \longrightarrow 2KCl(s) + 3O_2(g)$$

If 170.0 g of potassium chlorate reacts in a 2.50-L vessel, which was initially evacuated, what pressure of oxygen will be attained when the temperature is finally cooled to 25°C? Use the preceding chemical equation and ignore the volume of solid product.

5.118 Raoul Pictet, the Swiss physicist who first liquefied oxygen, attempted to liquefy hydrogen. He heated potassium formate, $KCHO_2$, with KOH in a closed 2.50-L vessel.

$$KCHO_2(s) + KOH(s) \longrightarrow K_2CO_3(s) + H_2(g)$$

If 75.0 g of potassium formate reacts in a 2.50-L vessel, which was initially evacuated, what pressure of hydrogen will be attained when the temperature is finally cooled to 25°C? Use the preceding chemical equation and ignore the volume of solid product.

5.119 A 24.9-mL volume of hydrochloric acid reacts completely with 55.0 mL of aqueous Na_2CO_3. The reaction is

$$2HCl(aq) + Na_2CO_3(aq) \longrightarrow$$
$$CO_2(g) + H_2O(l) + 2NaCl(aq)$$

The volume of CO_2 formed is 141 mL at 27°C and 727 mmHg. What is the molarity of the HCl solution?

5.120 A 21.4-mL volume of hydrochloric acid reacts completely with a solid sample of $MgCO_3$. The reaction is

$$2HCl(aq) + MgCO_3(s) \longrightarrow$$
$$CO_2(g) + H_2O(l) + MgCl_2(aq)$$

The volume of CO_2 formed is 159 mL at 23°C and 731 mmHg. What is the molarity of the HCl solution?

5.121 A 48.90-mL sample of a 0.2040 M acid reacts with an excess of Na_2CO_3 to form 125.0 mL CO_2 at 722 mmHg and 17°C. If the acid is either HCl or H_2SO_4, which is it?

5.122 A 41.41-mL sample of a 0.1250 M acid reacts with an excess of Na_2CO_3 to form 150.0 mL CO_2 at 646 mmHg and 27°C. If the acid is either HCl or H_2SO_4, which is it?

5.123 If the rms speed of NH_3 molecules is found to be 0.600 km/s, what is the temperature (in degrees Celsius)?

5.124 If the rms speed of He atoms in the exosphere (highest region of the atmosphere) is 3.53×10^3 m/s, what is the temperature (in kelvins)?

5.125 Calculate the ratio of rates of effusion of $^{235}UF_6$ and $^{238}UF_6$, where ^{235}U and ^{238}U are isotopes of uranium. The atomic masses are ^{235}U, 235.04 amu; ^{238}U, 238.05 amu; ^{19}F (the only naturally occurring isotope), 18.998 amu. Carry five significant figures in the calculation.

5.126 Hydrogen has two stable isotopes, ^1H and ^2H, with atomic masses of 1.0078 amu and 2.0141 amu, respectively. Ordinary hydrogen gas, H_2, is a mixture consisting mostly of $^1H_2$ and $^1H^2H$. Calculate the ratio of rates of effusion of $^1H_2$ and $^1H^2H$ under the same conditions.

5.127 A 1.000-g sample of an unknown gas at 0°C gives the following data:

P (atm)	V (L)
0.2500	3.1908
0.5000	1.5928
0.7500	1.0601
1.0000	0.7930

Use these data to calculate the value of the molar mass at each of the given pressures from the ideal gas law (we will call this the "apparent molar mass" at this pressure). Plot the apparent molar masses against pressure and extrapolate to find the molar mass at zero pressure. Because the ideal gas law is most accurate at low pressures, this extrapolation will give an accurate value for the molar mass. What is the accurate molar mass?

5.128 Plot the data given in Table 5.3 for oxygen at 0°C to obtain an accurate molar mass for O_2. To do this, calculate a value of the molar mass at each of the given pressures from the ideal gas law (we will call this the "apparent molar mass" at this pressure). On a graph show the apparent molar mass versus the pressure and extrapolate to find the molar mass at zero pressure. Because the ideal gas law is most accurate at low pressures, this extrapolation will give an accurate value for the molar mass. What is the accurate molar mass?

5.129 Carbon monoxide, CO, and oxygen, O_2, react according to

$$2CO(g) + O_2(g) \longrightarrow 2CO_2(g)$$

Assuming that the reaction takes place and goes to completion, determine what substances remain and what their partial pressures are after the valve is opened in the apparatus represented in the accompanying figure. Also assume that the temperature is fixed at 300 K.

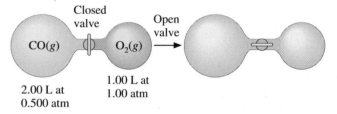

5.130 Suppose the apparatus shown in the figure accompanying Problem 5.129 contains H_2 at 0.500 atm in the left vessel separated from O_2 at 1.00 atm in the other vessel. The valve is then opened. If H_2 and O_2 react to give H_2O when the temperature is fixed at 300 K, what substances remain and what are their partial pressures after reaction?

■ **5.131** How does nitroglycerine alleviate the pain of an angina attack?

■ **5.132** How does nitrogen monoxide, NO, function in the body to regulate blood pressure?

■ **5.133** Explain the greenhouse effect.

■ **5.134** What famous chemist first predicted the greenhouse effect of carbon dioxide gas?

Strategy Problems

Key: As noted earlier, all of the practice and general problems are matched pairs. This section is a selection of problems that are not in matched-pair format. These challenging problems require that you employ many of the concepts and strategies that were developed in the chapter. In some cases, you will have to integrate several concepts and operational skills in order to solve the problem successfully.

5.135 A 19.9-mL volume of a hydrochloric acid solution reacts completely with a solid sample of magnesium carbonate, producing 183 mL of CO_2 that is collected over water at 24.0°C and 738 torr total pressure. The reaction is

$$2HCl(aq) + MgCO_3(s) \longrightarrow$$
$$CO_2(g) + H_2O(l) + MgCl_2(aq)$$

What is the molarity of the HCl solution?

5.136 The graph below represents the distribution of molecular speeds of hydrogen and neon at 200 K.

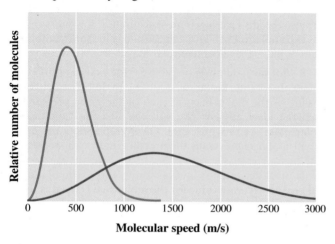

Relative number of molecules (y-axis)

Molecular speed (m/s) (x-axis: 0, 500, 1000, 1500, 2000, 2500, 3000)

a Match each curve to the appropriate gas.
b Calculate the rms speed (in m/s) for each of the gases at 200 K.
c Which of the gases would you expect to have the greater effusion rate at 200 K? Justify your answer.
d Calculate the temperature at which the rms speed of the hydrogen gas would equal the rms speed of the neon at 200 K.

5.137 A submersible balloon is sent to the bottom of the ocean. On shore, the balloon had a capacity of 162 L when it was filled at 21.0°C and standard pressure. When it reaches the ocean floor, which is at 5.92°C, the balloon occupies 18.8 L of space. What is the pressure on the ocean floor?

5.138 A given mass of gas occupies a volume of 870 mL at 28°C and 740 mmHg. What will be the new volume at STP?

5.139 A container is filled with 16.0 g of O_2 and 14.0 g of N_2.
a What is the volume of the container at STP?
b What is the partial pressure of the O_2 gas?
c What are the mole fraction and the mole percent of the N_2 in the mixture.

5.140 Sulfur-containing compounds give skunks their potent smell. One of the principal smelly compounds in skunk spray is (E)-2-butene-1-thiol, C_4H_7S.
a What is the root-mean-square (rms) molecular speed of a gas molecule of this compound at 25°C?
b Using the value from part a, calculate how long it would take a molecule of C_4H_7S to reach your nose if you were 150 m from the skunk.
c Does the calculation that you performed in part b provide an accurate estimate for the length of time it would take for the molecule to travel 150 m, or is there something that was overlooked in performing the calculation?

5.141 Sulfur hexafluoride, SF_6, is an extremely dense gas. How does its density compare with the density of air? Use a molar mass for air of 29.0 g/mol.

5.142 A rigid 1.0-L container at 75°C is fitted with a gas pressure gauge. A 1.0-mol sample of ideal gas is introduced into the container.
a What would the pressure gauge in the container be reading in mmHg?
b Describe the interactions in the container that are causing the pressure.

c Say the temperature in the container were increased to 150°C. Describe the effect this would have on the pressure, and, in terms of kinetic theory, explain why this change occurred.

5.143 The reaction $8H_2(g) + S_8(l) \longrightarrow 8H_2S(g)$ is run at 125°C and a constant pressure of 12.0 atm. Assuming complete reaction, what mass of S_8 would be required to produce 5.00×10^2 mL of H_2S gas under these conditions?

5.144 Shown below are three containers of an ideal gas (A, B, and C), each equipped with a movable piston (assume that atmospheric pressure is 1.0 atm).
a How do the pressures in these containers compare?
b Are all the gases at the same temperature? If not, compare the temperatures.
c If you cooled each of the containers in an ice-water bath to 0.0°C, describe how the volumes and pressures of the gases in these containers would compare.

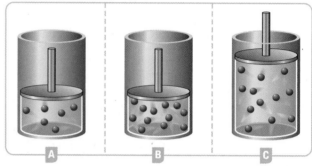

5.145 A 275-mL sample of CO gas is collected over water at 31°C and 755 mmHg. If the temperature of the gas collection apparatus rises to 39°C, what is the new volume of the sample? Assume that the barometric pressure does not change.

5.146 Ethanol, the alcohol used in automobile fuels, is produced by the fermentation of sugars present in plants. Corn is often used as the sugar source. The following equation represents the fermentation of glucose, the sugar in corn, by yeast to produce ethanol and carbon dioxide.

$$C_6H_{12}O_6(aq) \longrightarrow 2C_2H_5OH(aq) + 2CO_2(g)$$

Ethanol is combusted in an automobile engine according to the equation

$$C_2H_5OH(l) + 3O_2(g) \longrightarrow 2CO_2(g) + 3H_2O(g)$$

What would be the *total* volume of CO_2 gas formed at STP when 3.00 kg of sugar is fermented and the ethanol is then combusted in an automobile engine?

5.147 Silicon nitride, Si_3N_4, is a material that is used in computer chips as an insulator. Silicon nitride can be prepared according to the chemical equation

$$3SiH_4(g) + 4NH_3(g) \longrightarrow Si_3N_4(g) + 12H_2(g)$$

If you wanted to prepare a surface film of $Si_3N_4(s)$ that had an area of 9.0 mm² and was 4.0×10^{-5} mm thick, what volume of SiH_4 gas would you need to use at 1.0×10^{-5} torr and 775 K? The density of Si_3N_2 is 3.29 g/cm³.

5.148 A sample of neon gas is placed into a 22.4-L rigid container that already contains 1.00 mol of nitrogen gas. Indicate which statement is correct in describing the change that occurs when the neon gas is added to the container. Assume ideal gas behavior.

a The partial pressure of the nitrogen gas in the container would decrease by 1/2 (50%).

b The partial pressure of the nitrogen gas in the container would double.

c The partial pressure of the nitrogen gas in the container would decrease by some amount, but we cannot know the exact amount without more information.

d The partial pressure of the nitrogen gas would not change.

e The partial pressure of the nitrogen gas would increase by some amount, but we cannot know the exact amount without more information.

5.149 If you have a 150-L cylinder filled with chlorine gas to a density of 2.8 g/L, how many moles of chlorine would you need to add to the cylinder? If you were to double the temperature of the cylinder, would the gas density change?

5.150 A sample of carbon dioxide gas is placed in a container. The volume of the container is reduced to 1/3 of its original volume while the pressure is observed to double. In this system did the temperature change? Explain your answer.

Cumulative-Skills Problems

Key: The problems under this heading combine skills introduced in previous chapters with those given in the current one.

5.151 A sample of a breathing mixture for divers contained 34.3% helium, He; 51.7% nitrogen, N_2; and 14.0% oxygen, O_2 (by mass). What is the density of this mixture at 22°C and 775 mmHg?

5.152 A sample of natural gas is 85.2% methane, CH_4, and 14.8% ethane, C_2H_6, by mass. What is the density of this mixture at 18°C and 771 mmHg?

5.153 A sample of sodium peroxide, Na_2O_2, was reacted with an excess of water.

$$2Na_2O_2(s) + 2H_2O(l) \longrightarrow 4NaOH(aq) + O_2(g)$$

All of the sodium peroxide reacted, and the oxygen was collected over water at 21°C. The barometric pressure was 771 mmHg. The apparatus was similar to that shown in Figure 5.20. However, the level of water inside the tube was 25.0 cm above the level of water outside the tube. If the volume of gas in the tube is 31.0 mL, how many grams of sodium peroxide were in the sample?

5.154 A sample of zinc metal was reacted with an excess of hydrochloric acid.

$$Zn(s) + 2HCl(aq) \longrightarrow ZnCl_2(aq) + H_2(g)$$

All of the zinc reacted, and the hydrogen gas was collected over water at 17°C; the barometric pressure was 751 mmHg. The apparatus was similar to that shown in Figure 5.20, but the level of water inside the tube was 29.7 cm above the level outside the tube. If the volume of gas in the tube is 22.1 mL, how many grams of zinc were there in the sample?

5.155 A mixture contained calcium carbonate, $CaCO_3$, and magnesium carbonate, $MgCO_3$. A sample of this mixture weighing 7.85 g was reacted with excess hydrochloric acid. The reactions are

$$CaCO_3(g) + 2HCl(aq) \longrightarrow$$
$$CaCl_2(aq) + H_2O(l) + CO_2(g)$$
$$MgCO_3(s) + 2HCl(aq) \longrightarrow$$
$$MgCl_2(aq) + H_2O(l) + CO_2(g)$$

If the sample reacted completely and produced 1.94 L of carbon dioxide, CO_2, at 25°C and 785 mmHg, what were the percentages of $CaCO_3$ and $MgCO_3$ in the mixture?

5.156 A mixture contained zinc sulfide, ZnS, and lead sulfide, PbS. A sample of the mixture weighing 6.12 g was reacted with an excess of hydrochloric acid. The reactions are

$$ZnS(s) + 2HCl(aq) \longrightarrow ZnCl_2(aq) + H_2S(g)$$
$$PbS(s) + 2HCl(aq) \longrightarrow PbCl_2(aq) + H_2S(g)$$

If the sample reacted completely and produced 1.049 L of hydrogen sulfide, H_2S, at 23°C and 762 mmHg, what were the percentages of ZnS and PbS in the mixture?

5.157 A mixture of N_2 and Ne contains equal moles of each gas and has a total mass of 10.0 g. What is the density of this gas mixture at 500 K and 15.00 atm? Assume ideal gas behavior.

5.158 A mixture of Ne and Ar gases at 350 K contains twice as many moles of Ne as of Ar and has a total mass of 50.0 g. If the density of the mixture is 4.00 g/L, what is the partial pressure of Ne?

5.159 An ideal gas with a density of 3.00 g/L has a pressure of 675 mmHg at 25°C. What is the root-mean-square speed of the molecules of this gas?

5.160 The root-mean-square speed of the molecules of an ideal gas at 25°C and a pressure of 2.50 atm is 5.00×10^2 m/s. What is the density of this gas?

6 Thermochemistry

© Joseph P. Sinnot/Fundamental Photographs, NYC;
Harry Taylor/Getty Images

Ammonium dichromate decomposes in a fiery reaction to give a "volcano" of chromium(III) oxide that spews out N_2 gas and H_2O vapor. The color of ruby is due to the chromium(III) ion.

Contents and Concepts

Understanding Heats of Reaction

The first part of the chapter lays the groundwork for understanding what we mean by heats of reaction.

6.1 Energy and Its Units

6.2 First Law of Thermodynamics; Work and Heat

6.3 Heat of Reaction; Enthalpy of Reaction

6.4 Thermochemical Equations

6.5 Applying Stoichiometry to Heats of Reaction

6.6 Measuring Heats of Reaction

Using Heats of Reaction

Now that we understand the basic properties of heats of reaction and how to measure them, we can explore how to use them.

6.7 Hess's Law

6.8 Standard Enthalpies of Formation

6.9 Fuels—Foods, Commercial Fuels, and Rocket Fuels

OWL Sign in to OWL at www.cengage.com/owl to view tutorials and simulations, develop problem-solving skills, and complete online homework assigned by your professor.

Chemistry Download mini lecture videos for key concept review and exam prep from OWL or purchase them from www.cengagebrain.com.

early all chemical reactions involve either the release or the absorption of heat, a form of energy. The burning of coal and gasoline are dramatic examples of chemical reactions in which a great deal of heat is released. Such reactions are important sources of warmth and energy. Chemical reactions that absorb heat are usually less dramatic. The reaction of barium hydroxide with an ammonium salt is an exception. If crystals of barium hydroxide octahydrate, $Ba(OH)_2 \cdot 8H_2O$, are mixed with crystals of ammonium nitrate, NH_4NO_3, in a flask, the solids form first a slush, then a liquid. Because the reaction mixture absorbs heat from the surroundings, the flask feels cool. It soon becomes so cold that if it is set in a puddle of water on a board, the water freezes; the board can then be inverted with the flask frozen to it (Figure 6.1).

In this chapter, we will be concerned with the quantity of heat released or absorbed in a chemical reaction. We will address several questions: How do you measure the quantity of heat released or absorbed by a chemical reaction? To what extent can you relate the quantity of heat involved in a given reaction to the quantities of heat in other reactions? And how can you use this information?

Understanding Heats of Reaction

Thermodynamics is the science of the relationships between heat and other forms of energy. *Thermochemistry* is one area of thermodynamics. It concerns the study of the quantity of heat absorbed or evolved (given off) by chemical reactions. An example of a heat-evolving reaction is the burning of fuel. There may be practical reasons why you want to know the quantity of heat evolved during the burning of a fuel: you could calculate the cost of the fuel per unit of heat energy produced; you could calculate the quantity of heat obtained per unit mass of rocket fuel; and so forth. But there are also theoretical reasons for wanting to know the quantity of heat involved in a reaction. For example, knowing such values, you are able to calculate the amount of energy needed to break a particular kind of chemical bond and so learn something about the strength of that bond. Heat measurements also provide data needed to determine whether a particular chemical reaction occurs and, if so, to what extent. ▶

These questions concern chemical equilibrium and will be discussed in Chapter 18.

We have just used the terms *energy* and *heat,* assuming you have some idea of what they mean. But to proceed, you will need precise definitions of these and other terms. In the following section, we will define energy and its different forms and introduce units of energy.

Figure 6.1 ◀

A reaction that absorbs heat Two crystalline substances, barium hydroxide octahydrate and an ammonium salt, are mixed thoroughly in a flask. Then the flask, which feels quite cold to the touch, is set in a puddle of water on a board. In a couple of minutes, the flask and board are frozen solidly together. The board can then be inverted with the flask frozen to it.

© Cengage Learning

6.1 Energy and Its Units

We can define **energy** briefly as *the potential or capacity to move matter*. According to this definition, energy is not a material thing but rather a property of matter. Energy exists in different forms that can be interconverted. You can see the relationship of a given form of energy to the motion of matter by following its interconversions into different forms.

Consider the interconversions of energy in a steam-driven electrical generator. A fuel is burned to heat water and generate steam. The steam expands against a piston (or turbine), which is connected to a drive shaft that turns an electrical coil in a magnetic field. Electricity is generated in the coil. The fuel contains chemical energy, which is converted to heat. Part of the heat is then converted to motion of the drive shaft, and this motion is converted to electrical energy. The electrical energy could be used to run a motor, transforming the electrical energy back to the energy of motion. Or you could send the electricity into a lightbulb, converting electrical energy to heat energy and light energy. Photovoltaic cells can convert light back to electricity, which could be used to run a motor that can move matter (Figure 6.2). These examples show that energy can exist in different forms, including heat, light, and electrical energy, and these different forms can be interconverted. You also see their relationship to the energy of motion.

In this chapter, we will be especially concerned with the energy of substances, or chemical energy, and its transformation during chemical reaction into heat energy. To prepare for this, we will first explore the quantitative meaning of the energy of motion (*kinetic energy*). Then we will look at the concepts of *potential energy* and of the *internal energy* of substances, which is defined in terms of the kinetic and potential energies of the particles making up the substance.

Figure 6.2 ▲

Conversion of light energy to kinetic energy
Solar-powered vehicles use panels of photovoltaic cells.

© Reuters/Corbis

Kinetic Energy; Units of Energy

Kinetic energy is *the energy associated with an object by virtue of its motion*. An object of mass m and speed or velocity v has kinetic energy E_k equal to

$$E_k = \tfrac{1}{2}mv^2$$

This formula shows that the kinetic energy of an object depends on both its mass and its speed. A heavy object can move more slowly than a light object and still have the same kinetic energy. ◄

Consider the kinetic energy of a person whose mass is 59.0 kg and whose speed is 26.8 m/s. (This is equivalent to a person with a mass of 130 lb traveling in an automobile going 60 miles per hour.) You substitute the mass and speed into the formula.

$$E_k = \tfrac{1}{2} \times (59.0 \text{ kg}) \times (26.8 \text{ m/s})^2 = 2.12 \times 10^4 \text{ kg·m}^2/\text{s}^2$$

In the previous chapter, we used the symbol u for average molecular speed. Here v is the speed of an individual object or particle.

Note that the unit of energy comes out of the calculation. Because you substituted SI units of mass and speed, you obtain the SI unit of energy. *The SI unit of energy, $kg·m^2/s^2$*, is given the name **joule (J)** (pronounced "jewl") after the English physicist James Prescott Joule (1818–1889), who studied the energy concept. You see that a person weighing 130 lb and traveling 60 miles per hour has a kinetic energy equal to 2.12×10^4 J, or 21.2 kJ (21.2 kilojoules).

The joule is an extremely small unit. To appreciate its size, note that the *watt* is a measure of the quantity of energy used per unit time and equals 1 joule per second. A 100-watt bulb, for example, uses 100 joules of energy every second. A kilowatt-hour, the unit by which electric energy is sold, equals 3600 kilowatt-seconds (there are 3600 seconds in 1 hour), or 3.6 million joules. A household might use something like 1000 kilowatt-hours (3.6 billion joules) of electricity in a month.

The **calorie (cal)** is *a non-SI unit of energy commonly used by chemists, originally defined as the amount of energy required to raise the temperature of one gram of water by one degree Celsius.* This is only an approximate definition, however, because we now know that the energy needed to heat water depends slightly on the temperature of the water. In 1925 the calorie was defined in terms of the joule:

$$1 \text{ cal} = 4.184 \text{ J} \qquad \text{(exact definition)}$$

A person weighing 130 lb and traveling 60 miles per hour has a kinetic energy of

$$2.12 \times 10^4 \text{ J} \times \frac{1 \text{ cal}}{4.184 \text{ J}} = 5.07 \times 10^3 \text{ cal (5.07 kcal)}$$

Example 6.1   Calculating Kinetic Energy

Gaining Mastery Toolbox

Critical Concept 6.1
Kinetic energy is the energy associated with an object's motion and depends on its mass and speed. When performing calculations, express quantities in SI units to obtain the energy in joules.

Solution Essentials:
• Kinetic energy: $E_k = \frac{1}{2}mv^2$
• Energy

A good pitcher can throw a baseball so that it travels between 60 and 90 miles per hour (although speeds in excess of 100 miles per hour have been recorded). A regulation baseball weighing 143 g (0.143 kg) travels 75 miles per hour (33.5 m/s). What is the kinetic energy of this baseball in joules? in calories?

Problem Strategy Kinetic energy is defined by the formula $E_k = \frac{1}{2}mv^2$. You simply substitute into this formula, making certain that m and v are expressed in the same system of units.

Solution Substitute into the defining equation for the kinetic energy, using SI units.

$$E_k = \frac{1}{2} \times 0.143 \text{ kg} \times (33.5 \text{ m/s})^2 = \textbf{80.2 J}$$

Using the conversion factor 1 cal/4.184 J we obtain

$$80.2 \text{ J} \times \frac{1 \text{ cal}}{4.184 \text{ J}} = \textbf{19.2 cal}$$

Answer Check Make sure that the values for m and v are in the same system of units; SI units are generally appropriate. The SI unit of mass is kg, and the SI unit of speed is m/s. The answer then comes out in SI units (joules in this case).

Exercise 6.1   An electron, whose mass is 9.11×10^{-31} kg, is accelerated by a positive charge to a speed of 5.0×10^6 m/s. What is the kinetic energy of the electron in joules? in calories?

■ See Problems 6.45, 6.46, 6.47, and 6.48.

Potential Energy

Potential energy is *the energy an object has by virtue of its position in a field of force.* For example, water at the top of a dam has potential energy (in addition to whatever kinetic energy it may possess), because the water is at a relatively high position in the gravitational force field of the earth. You can calculate this potential energy of the water from the formula $E_p = mgh$. Here E_p is the potential energy of a quantity of water at the top of the dam, m is the mass of the water, g is the constant acceleration of gravity (equal to 9.807 m/s²), and h is the height of the water measured from some standard level. The choice of this standard level is arbitrary, because only *differences* of potential energy are important in any physical situation. It is convenient to choose the standard level to be the surface of the earth. As a quantity of water falls over the dam, its potential energy decreases from mgh at the top of the dam to zero at the earth's surface.

Figure 6.3 ▲

Potential energy and kinetic energy Water at the top of the dam has potential energy. As the water falls over the dam, this potential energy is converted to kinetic energy.

The potential energy of the water at the top of the dam is converted to kinetic energy when the water falls to a lower level. As the water falls, it moves more quickly. The potential energy decreases and the kinetic energy increases. Figure 6.3 shows the potential energy of water being converted to kinetic energy as the water falls over a dam.

Internal Energy

Consider the total energy of a quantity of water as it moves over the dam. This water as a whole has kinetic energy and potential energy. However, we know that water is made up of molecules, which are made up of smaller particles, electrons and nuclei. Each of these particles also has kinetic energy and potential energy. *The sum of the kinetic and potential energies of the particles making up a substance* is referred to as the **internal energy, U,** of the substance. Therefore, the total energy, E_{tot}, of a quantity of water equals the sum of its kinetic and potential energies as a whole ($E_k + E_p$) plus its internal energy.

$$E_{tot} = E_k + E_p + U$$

Normally when you study a substance in the laboratory, the substance is at rest in a vessel. Its kinetic energy as a whole is zero. Moreover, its potential energy as a whole is constant, and you can take it to be zero. In this case, the total energy of the substance equals its internal energy, U.

Law of Conservation of Energy

We have discussed situations in which one form of energy can be converted into another form of energy. For example, when water falls over a dam, potential energy is converted into kinetic energy. Some of the kinetic energy of the water may also be converted into random molecular motion—that is, into internal energy of the water. The total energy, E_{tot}, of the water, however, remains constant, equal to the sum of the kinetic energy, E_k, the potential energy, E_p, and the internal energy, U, of the water.

This result can be stated more generally as the **law of conservation of energy**: *energy may be converted from one form to another, but the total quantity of energy remains constant.* We say energy is conserved. The *first law of thermodynamics* is a specific statement of the conservation of energy, especially useful in discussing chemical reactions. We describe this law in the next section.

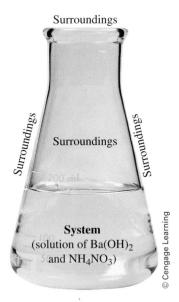

Surroundings

Surroundings

Surroundings

Surroundings

System (solution of $Ba(OH)_2$ and NH_4NO_3)

Figure 6.4 ▲

Illustration of a thermodynamic system The *system* consists of the portion of the universe that we choose to study; in this case, it is a solution of $Ba(OH)_2$ and NH_4NO_3 and the flask. Everything else constitutes the *surroundings*.

CONCEPT CHECK 6.1

A solar-powered water pump has photovoltaic cells that protrude from top panels. These cells collect energy from sunlight, storing it momentarily in a battery, which later runs an electric motor that pumps water up to a storage tank on a hill. What energy conversions are involved in using sunlight to pump water into the storage tank?

6.2 First Law of Thermodynamics; Work and Heat

The first law of thermodynamics relates the change in internal energy of a physical or chemical change taking place in a container to the flows (transfers) of energy into or out of the container. These energy transfers are of two kinds: work and heat. Although you may have some idea of what these terms mean, to be quantitative we will need to define them precisely, which in turn will require us to first define what we mean by a thermodynamic system and its surroundings.

Thermodynamic System and Its Surroundings

In the chapter opening, we looked at the reaction between crystals of barium hydroxide octahydrate, $Ba(OH)_2 \cdot 8H_2O$, and crystals of ammonium nitrate, NH_4NO_3, taking place in a flask. Suppose we decide to study the change of internal energy of the chemical substances in this flask as the reaction occurs. We refer to these chemical substances as the thermodynamic system and everything around the system as the surroundings (Figure 6.4). (We might include the flask with the chemical substances as our system, or not, depending on what we decide to include in our study.) In general, *the substance or mixture of substances that we single out for study (perhaps including the vessel) in which a change occurs* is called the **thermodynamic system**, or simply the **system**. *Everything in the vicinity of the thermodynamic system (its environment)*, we refer to as the **surroundings.**

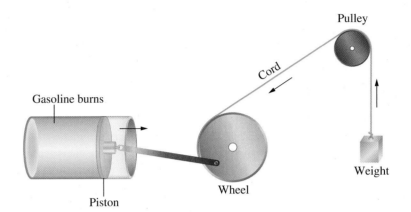

Figure 6.5 ◄

Work done by a system Gasoline burning in a cylinder produces gases that expand against the piston attached to a wheel and pulley. As the piston moves outward, the cord wraps around the wheel, pulling the weight upward.

Definition of Work

Work is *an energy transfer (or energy flow) into or out of a thermodynamic system whose effect on the surroundings is equivalent to moving an object through a field of force.* Imagine a system consisting of a cylinder in which gasoline burns to produce gases (Figure 6.5). These gases push against the surroundings, consisting of a piston attached to a wheel and pulley, which in turn lifts a weight. As the weight moves upward against gravity (a field of force), it gains potential energy, while the chemical substances within the engine lose an equivalent quantity of energy (in order to conserve energy—neglecting any losses due to friction). ► Thus, energy flows from the engine to its surroundings (to the weight). We say the engine (the system) does work on its surroundings.

You can also have the surroundings do work on the thermodynamic system. Consider a piston-type air pump with its air outlet closed off (Figure 6.6). Imagine that you hang a weight across the pump handle so that the handle is pulled down and air within the pump is compressed. As the weight falls downward, it loses potential energy, and the system (air within the pump) gains energy. We say that the falling weight has done work on the system.

We denote *work* by the symbol *w*. The algebraic sign of *w* is chosen to be positive when work is done on the thermodynamic system and negative when work is done by the thermodynamic system. (Notice that in this convention *w* has been defined from the perspective of the system.) You can remember the signs this way: when you do work on a system, you add (+) energy to it; when the system does work on the surroundings, energy is subtracted (−) from it.

The potential energy, E_p, of a mass m lifted by a height h equals mgh, where g is the constant acceleration due to gravity (g = 9.81 m/s²).

Definition of Heat

Heat is *an energy transfer (energy flow) into or out of a thermodynamic system that results from a temperature difference between the system and its surroundings.* As long as the system and its surroundings are in thermal contact (not insulated from one another), energy (heat) flows between them to establish equal temperatures, or thermal equilibrium. Heat flows from a region of higher temperature to one of lower temperature; once the temperatures become equal, heat flow stops. Note that when heat flows into a system, it appears in the system as an increase in the system's internal energy. We do not say that the system has heat because heat is only an energy flow; you can obtain the same increase in internal energy by doing work on the system as by adding heat to it.

We can explain this flow of energy between two regions of different temperatures in terms of kinetic-molecular theory. Imagine two vessels in contact,

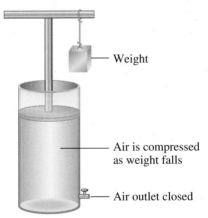

Weight

Air is compressed as weight falls

Air outlet closed

Figure 6.6 ▲

Work done on a system A weight hangs across the handle of a piston-type air pump (with the air outlet closed). As the weight falls downward, it loses potential energy and the system (air within the cylinder) gains energy.

Figure 6.7 ▲

A kinetic-theory explanation of heat The vessel on the left contains oxygen molecules at a higher temperature than the oxygen molecules on the right. Molecules collide with the vessel walls, thereby losing or gaining energy. The faster molecules tend to slow down, and the slower molecules tend to speed up. The net result is that energy is transferred through the vessel walls from the hot gas to the cold gas. We call this energy transfer *heat*.

each containing oxygen gas and the one on the left being hotter (Figure 6.7). According to kinetic theory, the average speed of molecules in the hotter gas is greater than that of molecules in the colder gas. But as the molecules in their random motions collide with the vessel walls, they lose energy to or gain energy from the walls. The faster molecules tend to slow down, while the slower molecules tend to speed up. Eventually, the average speeds of the molecules in the two vessels (and therefore the temperatures of the two gases) become equal. The net result is that energy is transferred through the vessel walls from the hot gas to the cold gas; that is, heat has flowed from the hotter vessel to the cooler one.

Heat and temperature are sometimes confused. The distinction is clear in the kinetic picture of a gas. According to kinetic theory, the absolute temperature of a gas is directly proportional to the average kinetic energy of the molecules. When you add heat to a gas, you increase its internal energy—and therefore its total kinetic energy. This increase in kinetic energy will be distributed over the molecules in the sample. Therefore, the increase in average kinetic energy per molecule (and thus the increase in temperature) depends on the size of the gas sample. A given quantity of heat will raise the temperature of a sample more if the sample is small.

Heat is denoted by the symbol q. The algebraic sign of q is chosen to be positive when the thermodynamic system absorbs heat and negative when the thermodynamic system evolves heat. (This convention for q, like the one for w, is from the perspective of the system.) You can remember the signs this way: when a system absorbs heat, energy is added (+) to it; when a system evolves heat, energy is subtracted (−) from it.

Change of Internal Energy

As we mentioned earlier, the internal energy, U, of a thermodynamic system is the sum of the kinetic and potential energies of all the particles of the system. We generally do not need to know the total internal energy of a thermodynamic system; what interests us is the *change of internal energy* when the system changes in some way.

For example, suppose you have 1.00 g of water at 0.00°C and 1.00 atm pressure. Let's consider this quantity of water to be our thermodynamic system and denote its internal energy as U_i (the initial internal energy). Suppose you change the system by adding 4.184 J (1.00 cal) of heat to it. You now have 1.00 g of water at 1.00°C and 1.00 atm. Let's denote the internal energy of this system as U_f (the final internal energy). The change in internal energy in going from the initial to final conditions is

$$\Delta U = U_f - U_i$$

where we have used the symbol Δ to mean "change in." (You read ΔU as "delta U" or "change in U.") When you add 4.184 J of heat to 1.00 g of water, the change in internal energy of the water, ΔU, is 4.184 J.

Internal energy is an *extensive property*, that is, it depends on the amount of substances in the system. If you double the amount of substances in the system under given conditions, you double the system's quantity of internal energy. Other examples of extensive properties are mass and volume.

Internal energy is also a state function. A **state function** is *a property of a system that depends only on its present state, which is determined by variables such as temperature and pressure and which is independent of any previous history of the system.* This means that any change in internal energy of a system depends only on the initial state and the final state of the system and not on how the change was made. If you start with 1.00 g of water at 0.00°C and 1.00 atm pressure and if you add 4.184 J (1.00 cal) of energy to it, you will obtain 1.00 g of water at 1.00°C and 1.00 atm. It doesn't matter how the energy is added, whether as heat or as work, or some combination of both.

An analogy might clarify the concept of a state function. Suppose you are hiking in mountainous terrain. You start walking from campsite A, which is at an altitude of 1200 ft above sea level according to your pocket altimeter, and you take the curving graded path upward to campsite B, which you find has an altitude of 4800 ft (Figure 6.8). You could take a more direct but more difficult route to campsite B. The distances traveled are different; the distance by the graded path is longer than by the more direct route. However, the difference in altitudes between campsite A and campsite B is independent of how you go from A to B (independent of any history). It is 3600 ft by whatever path you take. Altitude is analogous to a state function, whereas distance traveled is not.

First Law of Thermodynamics

Thermodynamics is described in terms of three laws. The first law of thermodynamics, as we noted, is essentially the law of conservation of energy applied to a thermodynamic system and relates the change in internal energy of the system, ΔU, to transfers of energy (heat, q, and work, w) into or out of the system. We can express the first law as follows:

$$\Delta U = q + w$$

Thus, the **first law of thermodynamics** states that *the change in internal energy of a system, ΔU, equals $q + w$ (heat plus work)*. We will discuss the second and third laws of thermodynamics in Chapter 18.

Figure 6.9 illustrates the addition of heat to a system that consists of a gas in a cylindrical vessel fitted with a piston. On top of the piston is a weight, which we will consider to be part of the surroundings. We momentarily fix the position of the piston so it does not move, and then raise the temperature of the surroundings above that of the system so that heat passes from the surroundings to the system. As you add heat to the system, its temperature rises until it equals that of the surroundings. Suppose the total heat added to the system is 165 J. Using the sign convention that q is positive when heat is added to (or absorbed by) the system, we write $q = +165$ J.

As the temperature of the gas in the vessel increases, the gas pressure increases at fixed volume (Boyle's law). We now allow the piston to move so that the gas expands and lifts the piston and the weight on top of it. In lifting the weight, the system does work. The energy gained by the weight equals the force of gravity on the weight times the height to which the weight was raised (the distance it was moved). Suppose this energy is 92 J. Because the surroundings, which include the weight, have gained 92 J of energy, the system must have lost 92 J by the conservation of energy. The system has done work on the surroundings. Adhering to the sign convention that work done by the system is negative (energy is subtracted from the system), we write $w = -92$ J.

The system in Figure 6.9 gains internal energy from the heat absorbed and loses internal energy via the work done. The net change of internal energy according to the first law is

$$\Delta U = q + w = (+165\text{ J}) + (-92\text{ J}) = +73\text{ J}$$

Figure 6.8 ▲

An analogy to illustrate a state function The two campsites differ by 3600 ft in altitude. This difference in altitude is independent of the path taken between the campsites. The distance traveled and the things encountered along the way, however, do depend on the path taken. Altitude here is analogous to a thermodynamic state function.

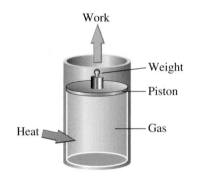

Figure 6.9 ▲

Exchanges of heat and work with the surroundings A gas is enclosed in a vessel with a piston. Heat flows into the vessel from the surroundings, which are at a higher temperature. As the temperature of the gas increases, the gas expands, lifting the weight (doing work).

Exercise 6.2   A gas is enclosed in a system similar to that shown in Figure 6.9. More weights are added to the piston, giving a total mass of 2.20 kg. As a result, the gas is compressed, and the weights are lowered 0.250 m. At the same time, 1.50 J of heat evolves from the system. What is the change in internal energy of the system, ΔU? The force of gravity on a mass m is mg, where g is the constant acceleration of gravity ($g = 9.81$ m/s^2).

■ See Problems 6.49 and 6.50.

CONCEPT CHECK 6.2

In closed breathing systems, such as in space ships, it is necessary to remove carbon dioxide exhaled during breathing from the air as well as to replenish oxygen. One way to remove exhaled carbon dioxide is to circulate the air through a solution of LiOH with which CO_2 reacts:

$$2LiOH(aq) + CO_2(g) \longrightarrow Li_2CO_3(s) + H_2O(l)$$

Imagine a setup in which a piston and weight (the surroundings) enclose a cylinder containing CO_2 gas above a solution of LiOH (the system).

    As CO_2 reacts with LiOH, the gas volume decreases, and the piston and weight moves downward. When 1.0 mol CO_2 reacts, 89.1 kJ of heat evolves, and the piston and weight do 2.5 kJ of work on the system. The change of internal energy is (choose one):

   a   +91.6 kJ   **b**   −91.6 kJ   **c**   +86.6 kJ   **d**   −86.6 kJ

Explain how you arrived at your answer; in particular, what signs would you attach to the heat of reaction and work done on the system?

6.3 Heat of Reaction; Enthalpy of Reaction

Generally when a chemical reaction occurs, the temperature of the reaction vessel or system begins to change as a result of the reaction. It may start to increase or perhaps decrease. If we imagine the system in thermal contact with its surroundings, however, heat will flow either into or out of the system. For example, if the system temperature begins to rise as a result of chemical reaction, heat flows from the system to the surroundings (heat evolves from the system). Or if the system temperature begins to fall, heat flows from the surroundings to the system (the system absorbs heat).

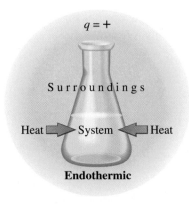

Heat of Reaction

The **heat of reaction** (at a given temperature) is *the heat, q, absorbed or evolved from a reaction system to retain a fixed temperature of the system under the conditions specified for the reaction (such as a fixed pressure).* Chemical reactions or physical changes are classified as exothermic or endothermic, depending on the sign of q. An **exothermic process** is *a chemical reaction or a physical change in which heat is evolved (q is negative);* see Figure 6.10. An **endothermic process** is *a chemical reaction or a physical change in which heat is absorbed (q is positive).* Experimentally, you note that in the exothermic reaction, the reaction flask initially warms; in the endothermic reaction, the reaction flask cools. We can summarize as follows:

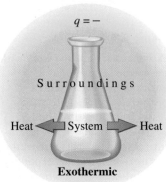

Type of Reaction	Experimental Effect Noted	Result on System	Sign of q
Endothermic	Reaction vessel cools (heat is absorbed)	Energy added	+
Exothermic	Reaction vessel warms (heat is evolved)	Energy subtracted	−

Suppose that in an experiment 1 mol of methane burns in oxygen and evolves 890 kJ of heat: $CH_4(g) + 2O_2(g) \longrightarrow CO_2(g) + 2H_2O(l)$. The reaction is exothermic. Therefore the heat of reaction, q, is −890 kJ.

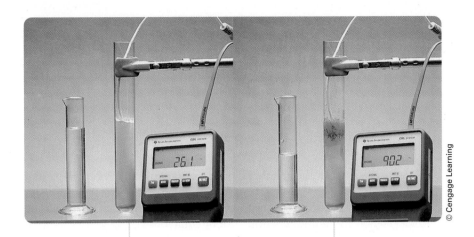

Figure 6.10 ◀

An exothermic process

The test tube contains anhydrous copper(II) sulfate (white crystals) and a thermometer that registers 26.1°C. The graduated cylinder contains water.

Water from the graduated cylinder has been added to the test tube, where the copper(II) sulfate forms hydrated copper(II) ion (blue color). The thermometer now registers 90.2°C, because the hydration process is exothermic.

Or consider the reaction described in the chapter opening, in which crystals of barium hydroxide octahydrate, $Ba(OH)_2 \cdot 8H_2O$, react with crystals of ammonium nitrate, NH_4NO_3.

$$Ba(OH)_2 \cdot 8H_2O(s) + 2NH_4NO_3(s) \longrightarrow 2NH_3(g) + 10H_2O(l) + Ba(NO_3)_2(aq)$$

When 1 mol $Ba(OH)_2 \cdot 8H_2O$ reacts with 2 mol NH_4NO_3, the reaction mixture absorbs 170.8 kJ of heat. The reaction is endothermic. Therefore the heat of reaction, q, is +170.8 kJ.

> **Exercise 6.3**   Ammonia burns in the presence of a platinum catalyst to give nitric oxide, NO.
>
> $$4NH_3(g) + 5O_2(g) \xrightarrow{\text{Pt}} 4NO(g) + 6H_2O(l)$$
>
> In an experiment, 4 mol NH_3 is burned and evolves 1170 kJ of heat. Is the reaction endothermic or exothermic? What is the value of q?
>
> ■ See Problems 6.51, 6.52, 6.53, and 6.54.

Pressure-Volume Work

In general, a reaction system may involve work as well as heat, depending on the conditions under which the reaction occurs. Often we carry out a reaction in a beaker or flask open to the atmosphere, in which case the reaction occurs at the fixed pressure of the atmosphere, P. If the reaction system changes volume, ΔV, pressure-volume work is either done by the system (if it expands against the surrounding atmosphere) or pressure-volume work is done on the system (as the system contracts from the pressure of the surrounding atmosphere). We can show that this **pressure-volume work**, w, equals *the negative of the pressure times the change in volume of the system.*

$$w = -P\Delta V$$

(Note that the negative sign results from the convention that w is defined from the perspective of the system. When ΔV is positive, the system expands, doing work on the surroundings, and thus the system loses energy, hence the negative sign.)

Consider the reaction of zinc metal with hydrochloric acid in a beaker open to the atmosphere at pressure P.

$$Zn(s) + 2H_3O^+(aq) \longrightarrow Zn^{2+}(aq) + 2H_2O(l) + H_2(g)$$

This is an exothermic reaction, with $q = -152.4$ kJ per mole of zinc consumed. Also, pressure-volume work is done: as hydrogen gas evolves, the system expands,

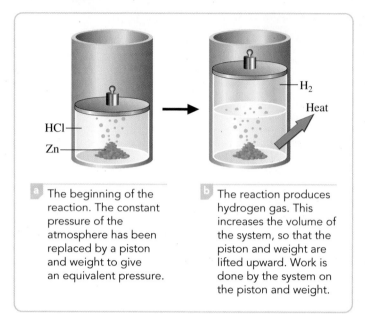

a The beginning of the reaction. The constant pressure of the atmosphere has been replaced by a piston and weight to give an equivalent pressure.

b The reaction produces hydrogen gas. This increases the volume of the system, so that the piston and weight are lifted upward. Work is done by the system on the piston and weight.

Figure 6.11 ▲

Reaction of zinc metal with hydrochloric acid at constant pressure

pushing back the surrounding atmosphere. The effect is to lift a mass of air upward against gravity. (See Figure 6.11 in which the atmosphere has been replaced by an equivalent weight to make it clear that a mass has been lifted.) The work done by the system equals the force of gravity, F, on the system times the height, h, that the mass of air is moved upward. This work is energy lost by the system so that w is negative:

$$w = -F \times h$$

The atmospheric pressure, P, equals the force, F, exerted per cross-sectional area, A, of the beaker: $P = F/A$. Therefore, the force F equals PA. The change in volume of the reaction system equals the cross-sectional area, A, times the height it is raised, h. Thus, $\Delta V = A \times h$, and so $h = \Delta V/A$. Substituting for F and h in the formula for work gives

$$w = -F \times h = -(PA) \times (\Delta V/A) = -P\Delta V$$

We can use this result to calculate the pressure-volume work for the reaction of zinc with hydrochloric acid. When 1.00 mol Zn reacts with excess HCl, the reaction produces 1.00 mol H_2, which at 25°C and 1.00 atm (= 1.01×10^5 Pa) occupies 24.5 L (= 24.5×10^{-3} m^3). The work done by the chemical system in pushing back the atmosphere is

$$w = -P\Delta V = -(1.01 \times 10^5 \text{ Pa}) \times (24.5 \times 10^{-3} \text{ m}^3)$$
$$= -2.47 \times 10^3 \text{ J, or } -2.47 \text{ kJ}$$

Note that we used SI units throughout the calculation to obtain the work in SI units (joules).

Suppose we would like to know the change of internal energy for this reaction. We apply the first law of thermodynamics to the system: The internal energy change for the system, ΔU, equals the heat absorbed, q (= -152.4 kJ), plus the work done, w (which equals $-P\Delta V = -2.47$ kJ):

$$\Delta U = q - P\Delta V$$
$$= -152.4 \text{ kJ} - 2.47 \text{ kJ} = -154.9 \text{ kJ}$$

In summary, when 1.00 mol Zn reacts with excess hydrochloric acid, the internal energy changes as kinetic and potential energies of electrons and nuclei change in going from reactants to products; the internal energy change is -154.9 kJ. Energy leaves the system mostly as heat ($q = -152.4$ kJ) but also partly as expansion work ($w = -2.47$ kJ).

Exercise 6.4 ▲ Consider the combustion (burning) of methane, CH_4, in oxygen.

$$CH_4(g) + 2O_2(g) \longrightarrow CO_2(g) + 2H_2O(l)$$

The heat of reaction at 77°C and 1.00 atm is -885.5 kJ. What is the change in volume when 1.00 mol CH_4 reacts with 2.00 mol O_2? (You can ignore the volume of liquid water, which is insignificant compared with volumes of gases.) What is w for this change? Calculate ΔU for the change indicated by the chemical equation.

■ See Problems 6.55 and 6.56.

Enthalpy and Enthalpy of Reaction

We began this chapter by discussing the heat of reaction, noting that we have a special interest in heat. How much heat can we obtain from a fuel? What can the quantity of heat tell us about a chemical reaction? As you will see, we can often directly measure this heat. For reactions in which the volume does not change, this heat equals the change of internal energy, ΔU. In general, though, the reaction system also involves work, in which case the heat of reaction is not immediately related to ΔU.

When the reaction occurs against a constant external pressure (say the atmosphere), the heat of reaction instead equals the change of another thermodynamic quantity, this one called the enthalpy (pronounced en-THAL-py) and denoted by the symbol H. The **enthalpy** of a thermodynamic system is defined as *its internal energy,* U, *plus pressure,* P, *times volume,* V.

$$H = U + PV$$

Because U, P, and V are state functions, H is also a state function. This means that for a given temperature and pressure, a given amount of substance has a definite enthalpy. In addition, enthalpy is an extensive property (it depends on the amount of substance). And like the internal energy, we are normally interested only in changes of the property enthalpy. This change of enthalpy, ΔH, equals the final enthalpy H_f minus the initial enthalpy H_i,

$$\Delta H = H_f - H_i$$

Let's now show that the heat for a reaction occurring at constant pressure (and involving only pressure-volume work) equals the enthalpy change. To begin, we write the first law of thermodynamics for this reaction,

$$\Delta U = q + w = q - P\Delta V$$

and solve for the heat of reaction, q.

$$q = \Delta U + P\Delta V = (U_f - U_i) + P(V_f - V_i)$$

Now we regroup the terms:

$$q = (U_f + PV_f) - (U_i + PV_i) = H_f - H_i$$

The result is the central equation for this chapter: The heat of reaction at a fixed pressure and a given temperature is

$$q = \Delta H \text{ (At fixed pressure and a given temperature)}$$

We refer to *the change in enthalpy,* ΔH, *for a reaction at a given temperature and fixed pressure* (for which there is only pressure-volume work) as the **enthalpy of reaction.**

As an illustration of the ideas introduced in this section, consider the reaction at 25°C of sodium metal and water, carried out in a beaker open to the atmosphere at 1.00 atm pressure.

$$2Na(s) + 2H_2O(l) \longrightarrow 2NaOH(aq) + H_2(g)$$

The metal and water react vigorously and evolve heat. Experiments have been done to show that 2 moles of sodium metal reacts with 2 moles of water to evolve 368.6 kJ of heat. Therefore, the enthalpy of reaction (or enthalpy change for the reaction) is $\Delta H = -368.6$ kJ. Figure 6.12 shows this change as an *enthalpy diagram* in which the reactants $2Na(s) + 2H_2O(aq)$ are shown at higher enthalpy than the products $2NaOH(aq) + H_2(g)$. When the reaction occurs, the enthalpy changes, as shown by the downward arrow, whose length equals the heat of reaction.

What is the change of internal energy in this sodium-water reaction? We know that this change is related to the enthalpy of reaction, but there is a small correction due to the pressure-volume work. Let's begin by writing the first law of thermodynamics, then substituting ΔH for q and $-P\Delta V$ for w:

$$\Delta U = q + w = \Delta H - P\Delta V$$

As we noted in the preceding paragraph, the enthalpy of reaction is -368.6 kJ; earlier we showed that the pressure-volume work done when a system increases in volume by 1 mole of gas at 25°C and 1.00 atm is -2.47 kJ. So, the change of internal energy of the reaction system is

$$\Delta U = -368.6 \text{ kJ} - 2.47 \text{ kJ} = -371.1 \text{ kJ}$$

6.4 Thermochemical Equations

We will often find it convenient to write the enthalpy of reaction, ΔH, with the chemical equation. A **thermochemical equation** is *the chemical equation for a reaction (including phase labels) in which the equation is given a molar interpretation, and*

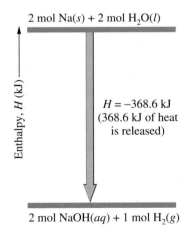

2 mol Na(s) + 2 mol H$_2$O(l)

Enthalpy, H (kJ)

$H = -368.6$ kJ
(368.6 kJ of heat is released)

2 mol NaOH(aq) + 1 mol H$_2$(g)

Figure 6.12 ▲

An enthalpy diagram When 2 mol Na(s) and 2 mol H$_2$O(l) react to give 2 mol NaOH(aq) and 1 mol H$_2$(g), 368.6 kJ of heat is released, and the enthalpy of the system decreases by 368.6 kJ.

the enthalpy of reaction for these molar amounts is written directly after the equation. For the reaction of sodium and water, you would write

$$2Na(s) + 2H_2O(l) \longrightarrow 2NaOH(aq) + H_2(g); \Delta H = -368.6 \text{ kJ}$$

This equation says that 2 mol of sodium reacts with 2 mol of water to produce 2 mol of sodium hydroxide and 1 mol of hydrogen gas, and 368.6 kJ of heat evolves.

Note that the thermochemical equation includes phase labels. This is because the enthalpy change, ΔH, depends on the phases of the substances. Consider the reaction of hydrogen and oxygen to produce water. If the product is water vapor, 2 mol of H_2 burn to release 483.7 kJ of heat.

$$2H_2(g) + O_2(g) \longrightarrow 2H_2O(g); \Delta H = -483.7 \text{ kJ}$$

On the other hand, if the product is liquid water, the heat released is 571.7 kJ.

$$2H_2(g) + O_2(g) \longrightarrow 2H_2O(l); \Delta H = -571.7 \text{ kJ}$$

| *It takes 44.0 kJ of heat to vaporize 1 mol of liquid water at 25°C.*

In this case, additional heat is released when water vapor condenses to liquid. ◀

⌾WL **Interactive Example** 6.2   **Writing Thermochemical Equations**

Aqueous sodium hydrogen carbonate solution (baking soda solution) reacts with hydrochloric acid to produce aqueous sodium chloride, water, and carbon dioxide gas. The reaction absorbs 12.7 kJ of heat at constant pressure for each mole of sodium hydrogen carbonate. Write the thermochemical equation for the reaction.

Problem Strategy Remember that the thermochemical equation is the balanced equation for the reaction, given a molar interpretation, with the enthalpy of reaction written directly after it. You need to translate the names of substances given in the problem statement into their formulas and then balance the equation. (This is an acid–base reaction; refer to Section 4.4 for a discussion.) The problem statement gives you the quantity of heat absorbed (heat is added to the system), so the enthalpy change is a positive quantity. Note too that this is the heat absorbed for each mole of sodium hydrogen carbonate.

Solution You first write the balanced chemical equation.

$$NaHCO_3(aq) + HCl(aq) \longrightarrow NaCl(aq) + H_2O(l) + CO_2(g)$$

The equation is for 1 mol $NaHCO_3$, with the absorption of 12.7 kJ of heat. The corresponding ΔH is $+12.7$ kJ. The thermochemical equation is

$$\mathbf{NaHCO_3(aq) + HCl(aq) \longrightarrow NaCl(aq) + H_2O(l) + CO_2(g); \Delta H = +12.7 \text{ kJ}}$$

Answer Check Check that the formulas of the substances in the equation are correct and that the equation is balanced. Also check that the enthalpy of reaction has the correct sign and that its value is for the correct amount of reactant given in the problem statement (1 mol $NaHCO_3$).

Exercise 6.5 A propellant for rockets is obtained by mixing the liquids hydrazine, N_2H_4, and dinitrogen tetroxide, N_2O_4. These compounds react to give gaseous nitrogen, N_2, and water vapor, evolving 1049 kJ of heat at constant pressure when 1 mol N_2O_4 reacts. Write the thermochemical equation for this reaction.

■ See Problems 6.57 and 6.58.

The following are two important rules for manipulating thermochemical equations:

1. When a thermochemical equation is multiplied by any factor, the value of ΔH for the new equation is obtained by multiplying the value of ΔH in the original equation by that same factor.

2. When a chemical equation is reversed, the value of ΔH is reversed in sign.

Consider the thermochemical equation for the synthesis of ammonia.

$$N_2(g) + 3H_2(g) \longrightarrow 2NH_3(g); \Delta H = -91.8 \text{ kJ}$$

Suppose you want the thermochemical equation to show what happens when twice as many moles of nitrogen and hydrogen react to produce ammonia. Because double the amounts of substances are present, the enthalpy of reaction is doubled (enthalpy is an extensive quantity). Doubling the previous equation, you obtain

$$2N_2(g) + 6H_2(g) \longrightarrow 4NH_3(g); \Delta H = -184 \text{ kJ}$$

Suppose you reverse the first equation we wrote for the synthesis of ammonia. Then the reaction is the dissociation of 2 mol of ammonia into its elements. The thermochemical equation is

$$2NH_3(g) \longrightarrow N_2(g) + 3H_2(g); \Delta H = +91.8 \text{ kJ}$$

If you want to express this in terms of 1 mol of ammonia, you simply multiply this equation by a factor of $\frac{1}{2}$. ▶

The next example further illustrates the use of these rules of thermochemical equations.

If you use the molar interpretation of a chemical equation, there is nothing unreasonable about using such coefficients as $\frac{1}{2}$ and $\frac{3}{2}$.

Example 6.3   Manipulating Thermochemical Equations

Gaining Mastery Toolbox

Critical Concept 6.3
The rules for manipulating thermochemical equations are: (1) When multiplying an equation by any factor, multiply ΔH by the same factor; (2) When reversing an equation, change the sign of ΔH.

Solution Essentials:
- Rules for manipulating thermochemical equations
- Thermochemical equation
- Enthalpy of reaction, ΔH
- Heat of reaction, q

When 2 mol $H_2(g)$ and 1 mol $O_2(g)$ react to give liquid water, 572 kJ of heat evolves.

$$2H_2(g) + O_2(g) \longrightarrow 2H_2O(l); \Delta H = -572 \text{ kJ}$$

Write this equation for 1 mol of liquid water. Give the reverse equation, in which 1 mol of liquid water dissociates into hydrogen and oxygen.

Problem Strategy Recall the rules for manipulating thermochemical equations. When you multiply an

equation by a factor, you also multiply ΔH by the same factor. When you reverse an equation, you multiply ΔH by -1.

Solution You multiply the coefficients and ΔH by $\frac{1}{2}$:

$$H_2(g) + \tfrac{1}{2}O_2(g) \longrightarrow H_2O(l); \Delta H = -286 \text{ kJ}$$

Reversing the equation, you get

$$H_2O(l) \longrightarrow H_2(g) + \tfrac{1}{2}O_2(g); \Delta H = +286 \text{ kJ}$$

Answer Check If you multiply an equation by a factor, check that every formula and ΔH have been multiplied by that factor.

Exercise 6.6 ⓐ Write the thermochemical equation for the reaction described in Exercise 6.5 for the case involving 1 mol N_2H_4. ⓑ Write the thermochemical equation for the reverse of the reaction described in Exercise 6.5.

■ See Problems 6.59, 6.60, 6.61, and 6.62.

CONCEPT CHECK 6.3

Natural gas consists primarily of methane, CH_4. It is used in a process called *steam reforming* to prepare a gaseous mixture of carbon monoxide and hydrogen for industrial use.

$$CH_4(g) + H_2O(g) \longrightarrow CO(g) + 3H_2(g); \Delta H = 206 \text{ kJ}$$

The reverse reaction, the reaction of carbon monoxide and hydrogen, has been explored as a way to prepare methane (synthetic natural gas). Which of the following are exothermic? Of these, which one is the most exothermic?

a $CH_4(g) + H_2O(g) \longrightarrow CO(g) + 3H_2(g)$

b $2CH_4(g) + 2H_2O(g) \longrightarrow 2CO(g) + 6H_2(g)$

c $CO(g) + 3H_2(g) \longrightarrow CH_4(g) + H_2O(g)$

d $2CO(g) + 6H_2(g) \longrightarrow 2CH_4(g) + 2H_2O(g)$

Lucifers and Other Matches

Samuel Jones, an Englishman, patented one of the first kinds of matches in 1828. It consisted of a glass bead containing sulfuric acid surrounded by a coating of sugar with an oxidizing agent. You ignited the match by breaking the bead using a pair of pliers or, if you were more daring, your teeth. This action released the acid, which ignited an exothermic reaction in the surrounding combustible materials.

Later Jones began to market a friction match discovered, but not patented, by John Walker. Walker, who had been experimenting with explosives, discovered this match one day when he tried to remove a small glob of a dried mixture of antimony sulfide and potassium chlorate from a stick. He rubbed the stick on the floor and was surprised when it burst into flame. Jones called his matches "Lucifers." They were well named; when lighted, they gave off a shower of sparks and smoky fumes with the acrid odor of sulfur dioxide. Jones had every box inscribed with the warning "Persons whose lungs are delicate should by no means use Lucifers."

A few years later, a Frenchman, Charles Sauria, invented the white phosphorus match, which became an immediate success. When rubbed on a rough surface, the match lighted easily, without hazardous sparks, and smelled better than Lucifers. The match head contained white phosphorus, an oxidizing agent, and glue. White phosphorus is a yellowish-white, waxy substance, often sold in the form of sticks looking something like fat crayons. Unlike crayons, though, white phosphorus ignites spontaneously in air (so it is generally stored under water). The glue in the match mixture had two purposes: it protected the white phosphorus from air, and it held the match mixture firmly together. The white phosphorus match had one serious drawback. White phosphorus is quite poisonous. Workers in match factories often began to show the agonizing symptoms of "phossy jaw," from white phosphorus poisoning, in which the jawbone disintegrates. The manufacture of white phosphorus matches was outlawed in the early 1900s.

The head of the "strike-anywhere" match contains the relatively nontoxic tetraphosphorus trisulfide, P_4S_3, and potassium chlorate, $KClO_3$ (Figure 6.13). By rubbing the match head against a surface, you create enough heat to ignite the match. Tetraphosphorus trisulfide then burns in a very exothermic reaction.

$$P_4S_3(s) + \tfrac{16}{3}KClO_3(s) \longrightarrow$$
$$P_4O_{10}(s) + 3SO_2(g) + \tfrac{16}{3}KCl(s); \Delta H = -3884 \text{ kJ}$$

Safety matches have a head containing mostly an oxidizing agent and require a striking surface containing nonpoisonous red phosphorus.

| The head of the match contains tetraphosphorus trisulfide and potassium chlorate. | The substances in the match head react when frictional heat ignites the mixture. |

Figure 6.13 ▲

Strike-anywhere matches

■ See Problems 6.127 and 6.128.

6.5 Applying Stoichiometry to Heats of Reaction

As you might expect, the quantity of heat obtained from a reaction will depend on the amount of reactants. We can extend the method used to solve stoichiometry problems, described in Chapter 3, to problems involving the quantity of heat.

Consider the reaction of methane, CH_4 (the principal constituent of natural gas), burning in oxygen at constant pressure. How much heat could you obtain from 10.0 g of methane, assuming you had an excess of oxygen? You can answer this question if you know the enthalpy change for the reaction of 1 mol of methane. The thermochemical equation is

$$CH_4(g) + 2O_2(g) \longrightarrow CO_2(g) + 2H_2O(l); \Delta H = -890.3 \text{ kJ}$$

Starting Units

Goal Units

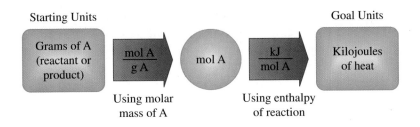

Figure 6.14 ◀

Calculating the heat obtained from a reaction Convert the grams of a reactant or product to moles (using the molar mass). Then convert these moles to kilojoules of heat (using the enthalpy of reaction).

The calculation involves the following conversions:

$$\text{Grams of CH}_4 \longrightarrow \text{moles of CH}_4 \longrightarrow \text{kilojoules of heat}$$

$$10.0 \text{ g CH}_4 \times \frac{1 \text{ mol CH}_4}{16.0 \text{ g CH}_4} \times \frac{-890.3 \text{ kJ}}{1 \text{ mol CH}_4} = -556 \text{ kJ}$$

Figure 6.14 illustrates the general calculation. ▶

If you wanted the result in kilocalories, you could convert this answer as follows:

$$556 \text{ kJ} \times \frac{1 \text{ kcal}}{4.184 \text{ kJ}} = 133 \text{ kcal}$$

◉WL Interactive Example **6.4**   **Calculating the Heat of Reaction from the Stoichiometry**

Gaining Mastery Toolbox

Critical Concept 6.4
Stoichiometry can be extended to include the heat of reaction; starting from the thermochemical equation, you write the conversion factor from moles of substance to heat of reaction.

Solution Essentials:
- Conversion of amount of substance to quantity of heat
- Enthalpy of reaction, ΔH
- Heat of reaction, q
- Stoichiometry

How much heat is evolved when 9.07×10^5 g of ammonia is produced according to the following equation? (Assume that the reaction occurs at constant pressure.)

$$N_2(g) + 3H_2(g) \longrightarrow 2NH_3(g); \Delta H = -91.8 \text{ kJ}$$

Problem Strategy The calculation involves converting grams of NH_3 to moles of NH_3 and then to kilojoules of heat.

Starting Units

Goal Units

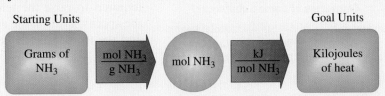

You obtain the conversion factor for the second step from the thermochemical equation, which says that the production of 2 mol NH_3 is accompanied by $q = -91.8$ kJ.

Solution $9.07 \times 10^5 \text{ g NH}_3 \times \frac{1 \text{ mol NH}_3}{17.0 \text{ g NH}_3} \times \frac{-91.8 \text{ kJ}}{2 \text{ mol NH}_3} = -2.45 \times 10^6 \text{ kJ}$

Thus, **2.45×10^6 kJ of heat evolves.**

Answer Check Make sure that the units in the calculation cancel to give the units of heat. Also note the sign of ΔH to decide whether heat is evolved or absorbed.

Exercise 6.7   How much heat evolves when 10.0 g of hydrazine reacts according to the reaction described in Exercise 6.5?

■ See Problems 6.63, 6.64, 6.65, and 6.66.

6.6 Measuring Heats of Reaction

So far we have introduced the concept of heat of reaction and have related it to the enthalpy change. We also showed how to display the enthalpy change in a thermochemical equation, and from this equation how to calculate the heat of reaction for any given amount of substance. Now that you have a firm idea of what heats of reaction are, how would you measure them?

First, we need to look at the heat required to raise the temperature of a substance, because a thermochemical measurement is based on the relationship between heat and temperature change. The heat required to raise the temperature of a substance is called its *heat capacity.*

Heat Capacity and Specific Heat

Heat is required to raise the temperature of a given amount of substance, and the quantity of heat depends on the temperature change. The **heat capacity** (C) of a sample of substance is *the quantity of heat needed to raise the temperature of the sample of substance one degree Celsius (or one kelvin).* Changing the temperature of the sample from an initial temperature t_i to a final temperature t_f requires heat equal to

$$q = C\Delta t$$

where Δt is the change of temperature and equals $t_f - t_i$. ◄

Suppose a piece of iron requires 6.70 J of heat to raise the temperature by one degree Celsius. Its heat capacity is therefore 6.70 J/°C. The quantity of heat required to raise the temperature of the piece of iron from 25.0°C to 35.0°C is

$$q = C\Delta t = (6.70 \text{ J/°C}) \times (35.0°C - 25.0°C) = 67.0 \text{ J}$$

Heat capacity is directly proportional to the amount of substance. Often heat capacities are listed for molar amounts of substances. The *molar heat capacity* of a substance is its heat capacity for one mole of substance.

Heat capacities are also compared for one-gram amounts of substances. The **specific heat capacity** (or simply **specific heat**) is *the quantity of heat required to raise the temperature of one gram of a substance by one degree Celsius (or one kelvin) at constant pressure.* To find the heat q required to raise the temperature of a sample, you multiply the specific heat of the substance, s, by the mass in grams, m, and the temperature change, Δt.

$$q = s \times m \times \Delta t$$

The specific heats and molar heat capacities of a few substances are listed in Table 6.1. Values depend somewhat on temperature, and those listed are for 25°C. Water has a specific heat of 4.18 J/(g·°C)—that is, 4.18 joules per gram per degree Celsius. In terms of calories, the specific heat of water is 1.00 cal/(g·°C). Example 6.5 illustrates the use of the preceding equation.

> The heat capacity will depend on whether the process is constant-pressure or constant-volume. We will assume a constant-pressure process unless otherwise stated.

Table 6.1	Specific Heats and Molar Heat Capacities of Some Substances*	
Substance	**Specific Heat** J/(g·°C)	**Molar Heat Capacity** J/(mol·°C)
Aluminum, Al	0.901	24.3
Copper, Cu	0.384	24.4
Ethanol, C_2H_5OH	2.43	112.2
Iron, Fe	0.449	25.1
Water, H_2O	4.18	75.3

*Values are for 25°C.

Example 6.5

Relating Heat and Specific Heat

Gaining Mastery Toolbox

Critical Concept 6.5
The heat q required to raise the temperature Δt of a given mass m of a substance depends on the substance's specific heat s; given any three of these quantities, the fourth one can be calculated.

Solution Essentials:
- Equation relating heat to specific heat ($q = s \times m \times \Delta t$)
- Specific heat, s
- Heat capacity

Calculate the heat absorbed by 15.0 g of water to raise its temperature from 20.0°C to 50.0°C (at constant pressure). The specific heat of water is 4.18 J/(g·°C).

Problem Strategy Note that specific heat is the heat needed to raise the temperature of one gram of substance one degree Celsius. To get the total heat, multiply specific heat by the grams of substance and by the change in degrees Celsius.

Solution You substitute into the equation

$$q = s \times m \times \Delta t$$

The temperature change is

$$\Delta t = t_f - t_i = 50.0°C - 20.0°C = +30.0°C$$

(continued)

(*continued*)

Therefore,

$$q = 4.18 \text{ J/(g·°C)} \times 15.0 \text{ g} \times (+30.0°C) = \textbf{1.88} \times \textbf{10}^\textbf{3} \textbf{ J}$$

Answer Check Be sure to put units on quantities in your calculation and to note whether units cancel to give units of heat.

Exercise 6.8   Iron metal has a specific heat of 0.449 J/(g·°C). How much heat is transferred to a 5.00-g piece of iron, initially at 20.0°C, when it is placed in a pot of boiling water? Assume that the temperature of the water is 100.0°C and that the water remains at this temperature, which is the final temperature of the iron.

■ See Problems 6.69 and 6.70.

Measurement of Heat of Reaction

You measure the heat of reaction in a **calorimeter,** *a device used to measure the heat absorbed or evolved during a physical or chemical change.* The device can be as simple as the apparatus sketched in Figure 6.15, which consists of an insulated container (for example, a pair of polystyrene coffee cups) with a thermometer. More elaborate calorimeters are employed when precise measurements are needed for research, although the basic idea remains the same—to measure temperature changes under controlled circumstances and relate these temperature changes to heat.

The coffee-cup calorimeter shown in Figure 6.15 is a constant-pressure calorimeter. The heat of the reaction is calculated from the temperature change caused by the reaction, and since this is a constant-pressure process, the heat can be directly related to the enthalpy change, ΔH. Research versions of a constant-pressure calorimeter are available, and these are used when gases are not involved.

For reactions involving gases, a *bomb calorimeter* is generally used (Figure 6.16). Consider the heat of combustion of graphite, the form of carbon used in the "lead" of a pencil. To measure the heat released when graphite burns in oxygen, a sample of graphite is placed in a small cup in the calorimeter. The graphite is surrounded by oxygen, and the graphite and oxygen are sealed in a steel vessel, or bomb. An electrical circuit is activated to start the burning of the graphite. The bomb is surrounded by water in an insulated container, and the heat of reaction is calculated from the temperature change of the calorimeter caused by the reaction.

Because the reaction in a bomb calorimeter occurs in a closed vessel, the pressure does not generally remain constant. Rather, the volume remains constant. In this case, the heat of reaction equals the change in internal energy, ΔU, and to obtain ΔH, a small correction is needed. This correction is normally small and is negligible when the reaction does not involve gases or when the moles of reactant gases equals the moles of product gases, as in the combustion of graphite to carbon dioxide.

The next example describes the calculations needed to obtain the heat of reaction from calorimetric measurements. Note that the heat of reaction in a constant-pressure calorimeter equals the enthalpy of reaction, ΔH, whereas the heat of reaction in a bomb (constant volume) calorimeter equals ΔH (approximately) only when the moles of gas remain constant.

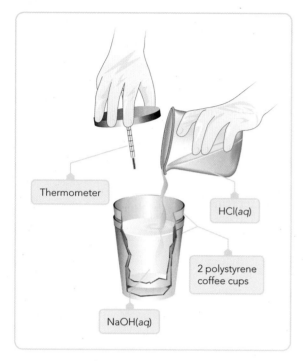

Figure 6.15 ▲

A simple coffee-cup calorimeter This calorimeter is made of two nested polystyrene coffee cups. The outer cup helps to insulate the reaction mixture from its surroundings. After the reactants are added to the inner cup, the calorimeter is covered to reduce heat loss by evaporation and convection. The heat of reaction is determined by noting the temperature rise or fall.

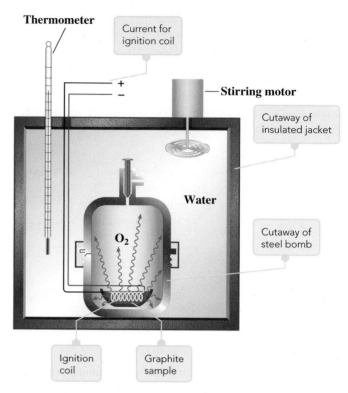

Figure 6.16 ▲

A bomb calorimeter This type of calorimeter can be used to determine the heat of combustion of a substance. The figure shows a sample of graphite being burned in oxygen within a vessel called the bomb. An electric current passing through the ignition coil starts the graphite burning. The temperature of the water surrounding the bomb rises as heat flows from the exothermic reaction of graphite and oxygen. Example 6.6 illustrates the calculations involved in determining the heat of combustion in such an experiment.

Example 6.6

Calculating ΔH from Calorimetric Data

Gaining Mastery Toolbox

Critical Concept 6.6
By running an exothermic reaction in a calorimeter (whose heat capacity is known and whose temperature rise is noted), you can deduce the heat released by the reaction, and from the mass of reactant, you can calculate the heat of reaction (equal to ΔH, assuming the moles of gas remain constant).

Solution Essentials:
- Equation relating heat to specific heat ($q = s \times m \times \Delta t$)
- Specific heat, s
- Heat capacity
- Conversion of amount of substance to quantity of heat
- Thermochemical equation
- Enthalpy of reaction, ΔH
- Heat of reaction, q
- Stoichiometry

Suppose 0.562 g of graphite is placed in a bomb calorimeter with an excess of oxygen at 25.00°C and 1 atm pressure (Figure 6.16). Excess O_2 ensures that all carbon burns to form CO_2. The graphite is ignited, and it burns according to the equation

$$C(\text{graphite}) + O_2(g) \longrightarrow CO_2(g)$$

On reaction, the calorimeter temperature rises from 25.00°C to 25.89°C. The heat capacity of the calorimeter and its contents was determined in a separate experiment to be 20.7 kJ/°C. What is the heat of reaction at 25.00°C and 1 atm pressure? Express the answer as a thermochemical equation. (Note that moles of gas remain constant, so the heat of reaction is essentially the enthalpy of reaction.)

Problem Strategy First, you need to obtain the heat absorbed by the calorimeter when the sample burns, using the calorimeter's temperature rise and its heat capacity. Note that the heat released by reaction of the sample (a negative quantity) equals the heat absorbed by the calorimeter (a positive quantity), except for this sign change. Then, you need to convert this heat to the heat released per mole to obtain ΔH.

Step 1: To obtain heat absorbed by the calorimeter, you reason as follows. The heat released by the reaction is absorbed by the calorimeter and its contents. Let q_{rxn} be the quantity of heat from the reaction mixture, and let C_{cal} be the heat capacity of the calorimeter and contents. The quantity of heat absorbed by the calorimeter is $C_{cal}\Delta t$. This will have the same magnitude as q_{rxn}, but the opposite

(continued)

(continued)

sign: $q_{rxn} = -C_{cal}\Delta t$. Substituting values of C_{cal} and Δt, you obtain the heat q_{rxn} from the sample of C (0.562 g). The factor $(q_{rxn}/0.562\ \text{g C})$ converts

$$\text{Grams C} \longrightarrow \text{kJ heat}$$

Step 2: What is the heat released from 1 mol C (equal to ΔH)? Convert this amount of C (mathematically) to grams C. Then convert this to kJ heat, using the conversion factor you obtained in Step 1 $(q_{rxn}/0.562\ \text{g C})$. The conversion steps are

$$\text{Moles C} \longrightarrow \text{grams C} \longrightarrow \text{kJ heat }(\Delta H)$$

Solution

Step 1: The heat from the graphite sample is

$$q_{rxn} = -C_{cal}\Delta t = -20.7\ \text{kJ/°C} \times (25.89°C - 25.00°C)$$
$$= -20.7\ \text{kJ/°C} \times 0.89°C = -18.4\ \text{kJ}$$

An extra figure is retained in q_{rxn} for further computation. The negative sign indicates that the reaction is exothermic, as expected for a combustion. The factor to convert grams C to kJ heat is $-18.4\ \text{kJ}/0.562\ \text{g C}$.

Step 2: The conversion of 1 mol C to kJ heat for 1 mol (ΔH) is

$$1\ \cancel{\text{mol C}} \times \frac{12\ \cancel{\text{g C}}}{1\ \cancel{\text{mol C}}} \times \frac{-18.4\ \text{kJ}}{0.562\ \cancel{\text{g C}}} = -3.9 \times 10^2\ \text{kJ}$$

(The final answer has been rounded to two significant figures.) When 1 mol of carbon burns, 3.9×10^2 kJ of heat is released. You can summarize the results by the thermochemical equation:

$$\textbf{C(graphite)} + \textbf{O}_2\textbf{(g)} \longrightarrow \textbf{CO}_2\textbf{(g); } \Delta H = -3.9 \times 10^2\ \textbf{kJ}$$

Answer Check Go over your calculations carefully. Check that the units within each step cancel properly. Also check that you have the right sign for ΔH; an exothermic reaction will have a negative sign.

Exercise 6.9 Suppose 33 mL of 1.20 *M* HCl is added to 42 mL of a solution containing excess sodium hydroxide, NaOH, in a coffee-cup calorimeter. The solution temperature, originally 25.0°C, rises to 31.8°C. Give the enthalpy change, ΔH, for the reaction:

$$\text{HCl}(aq) + \text{NaOH}(aq) \longrightarrow \text{NaCl}(aq) + \text{H}_2\text{O}(l)$$

Express the answer as a thermochemical equation. For simplicity, assume that the heat capacity and the density of the final solution in the cup are those of water. (In more accurate work, these values must be determined.) Also assume that the total volume of the solution equals the sum of the volumes of HCl(*aq*) and NaOH(*aq*).

■ See Problems 6.73, 6.74, 6.75, and 6.76.

Using Heats of Reaction

In the first part of this chapter, we looked at the basic properties of the heat of reaction and how to measure it. Now we want to find how heats of reaction can be used. We will see that the ΔH for one reaction can be obtained from the ΔH's of other reactions. This means that we can tabulate a small number of values and use them to calculate others.

6.7 Hess's Law

Enthalpy, you may recall, is a state function (Section 6.3). This means that the enthalpy change for a chemical reaction is independent of the path by which the products are obtained. In 1840, the Russian chemist Germain Henri Hess, a professor at the University of St. Petersburg, discovered this result by experiment. **Hess's law of heat summation** states that *for a chemical equation that can be written as the sum of two or more steps, the enthalpy change for the overall equation equals the sum of the enthalpy changes for the individual steps.* In other words, no matter how you go from given reactants to products (whether in one step or several), the enthalpy change for the overall chemical change is the same.

To understand Hess's law fully and to see how you can use it, consider a simple example. Suppose you would like to find the enthalpy change for the combustion of graphite (carbon) to carbon monoxide.

$$2C(graphite) + O_2(g) \longrightarrow 2CO(g)$$

The direct determination of this enthalpy change is very difficult, because once carbon monoxide forms it reacts further with oxygen to yield carbon dioxide. If you do the experiment in an excess of oxygen, you obtain only carbon dioxide, and the enthalpy change is the heat of complete combustion of graphite. On the other hand, if you do the experiment in a limited quantity of oxygen, you obtain a mixture of carbon monoxide and carbon dioxide, and the heat of reaction is a value appropriate for a mixture of these products. How can you obtain the enthalpy change for the preparation of pure carbon monoxide from graphite and oxygen?

The answer is to apply Hess's law. To do this, imagine that the combustion of graphite to carbon monoxide takes place in two separate steps:

$$2C(graphite) + 2O_2(g) \longrightarrow 2CO_2(g) \qquad \text{(first step)}$$

$$2CO_2(g) \longrightarrow 2CO(g) + O_2(g) \qquad \text{(second step)}$$

In the first step, you burn 2 mol of graphite in 2 mol of oxygen to produce 2 mol of carbon dioxide. In the second step, you decompose this carbon dioxide to give 2 mol of carbon monoxide and 1 mol of oxygen. The net result is the combustion of 2 mol of graphite in 1 mol of oxygen to give 2 mol of carbon monoxide. You can obtain this result by adding the two steps, canceling out 2 mol CO_2 and 1 mol O_2 on both sides of the equation.

$$2C(graphite) + \cancel{2}O_2(g) \longrightarrow \cancel{2CO_2(g)}$$
$$\underline{\cancel{2CO_2(g)} \qquad\qquad\qquad \longrightarrow 2CO(g) + \cancel{O_2(g)}}$$
$$2C(graphite) + O_2(g) \longrightarrow 2CO(g)$$

According to Hess's law, the enthalpy change for the overall equation (which is the equation you want) equals the sum of the enthalpy changes for the two steps. Now you need to determine the enthalpy changes for the separate steps.

You can determine the enthalpy change for the first step by simply burning graphite in an excess of oxygen, as described in Example 6.6. The result is $\Delta H = -393.5$ kJ per mole of CO_2 formed. For 2 mol CO_2, you multiply by 2.

$$2C(graphite) + 2O_2(g) \longrightarrow 2CO_2(g); \Delta H = (-393.5 \text{ kJ}) \times (2)$$

The second step, the decomposition of carbon dioxide, is not an easy experiment. However, the reverse of this decomposition is simply the combustion of carbon monoxide. You could determine the ΔH for that combustion by burning carbon monoxide in an excess of oxygen. The experiment is similar to the one for the combustion of graphite to carbon dioxide.

$$2CO(g) + O_2(g) \longrightarrow 2CO_2(g); \Delta H = -566.0 \text{ kJ}$$

From the properties of thermochemical equations (Section 6.4), you know that the enthalpy change for the reverse reaction is simply (-1) times the original reaction.

$$2CO_2(g) \longrightarrow 2CO(g) + O_2(g); \Delta H = (-566.0 \text{ kJ}) \times (-1)$$

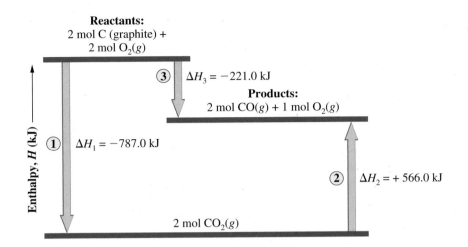

Figure 6.17 ◀

Enthalpy diagram illustrating Hess's law The diagram shows two different ways to go from graphite and oxygen (reactants) to carbon monoxide (products). Going by way of reactions 1 and 2 is equivalent to the direct reaction 3.

If you now add these two steps and add their enthalpy changes, you obtain the chemical equation and the enthalpy change for the combustion of carbon to carbon monoxide, which is what you wanted.

$$2C(graphite) + 2O_2(g) \longrightarrow 2CO_2(g) \qquad \Delta H_1 = (-393.5 \text{ kJ}) \times (2)$$
$$\underline{2CO_2(g) \qquad\qquad \longrightarrow 2CO(g) + O_2(g) \qquad \Delta H_2 = (-566.0 \text{ kJ}) \times (-1)}$$
$$2C(graphite) + O_2(g) \longrightarrow 2CO(g) \qquad \Delta H_3 = -221.0 \text{ kJ}$$

You see that the combustion of 2 mol of graphite to give 2 mol of carbon monoxide has an enthalpy change of -221.0 kJ. Figure 6.17 gives an enthalpy diagram showing the relationships among the enthalpy changes for this calculation.

This calculation illustrates how you can use Hess's law to obtain the enthalpy change for a reaction that is difficult to determine by direct experiment. Hess's law is more generally useful, however, in that it allows you to calculate the enthalpy change for one reaction from the values for others, whatever their source. Suppose you are given the following data: ▶

$$S(s) + O_2(g) \longrightarrow SO_2(g); \Delta H = -297 \text{ kJ} \qquad\qquad (1)$$

$$2SO_3(g) \longrightarrow 2SO_2(g) + O_2(g); \Delta H = 198 \text{ kJ} \qquad\qquad (2)$$

How could you use these data to obtain the enthalpy change for the following equation?

$$2S(s) + 3O_2(g) \longrightarrow 2SO_3(g) \qquad\qquad (3)$$

You need to multiply Equations 1 and 2 by factors (perhaps reversing one or both equations) so that when you add them together you obtain Equation 3. You can usually guess what you need to do to the first two equations to obtain the third one. Note that Equation 3 has a coefficient of 2 for S(s). This suggests that you should multiply Equation 1 by 2 (and multiply the ΔH by 2). Note also that $SO_3(g)$ in Equation 3 is on the right-hand side. This suggests that you should reverse Equation 2 (and multiply the ΔH by -1). Here is the calculation:

$$2S(s) + 2O_2(g) \longrightarrow 2SO_2(g) \qquad \Delta H = (-297 \text{ kJ}) \times (2)$$
$$\underline{2SO_2(g) + O_2(g) \longrightarrow 2SO_3(g) \qquad \Delta H = (198 \text{ kJ}) \times (-1)}$$
$$2S(s) + 3O_2(g) \longrightarrow 2SO_3(g) \qquad \Delta H = -792 \text{ kJ}$$

> We have followed common convention in writing S for the formula of sulfur in Equations 1 and 3, although the molecular formula is S_8. Doing this gives much simpler equations. You might try to do the calculations again, rewriting the equations using S_8 for sulfur. Remember to multiply the value of ΔH in each case by 8.

The next example gives another illustration of how Hess's law can be used to calculate the enthalpy change for a reaction from the enthalpy values for other reactions. In this case, the problem involves three equations from which we obtain a fourth. Although this problem is somewhat more complicated than the one we just did, the basic procedure is the same: Compare the coefficients in the equations. See what factors you need to multiply the equations whose ΔH's you know to obtain the equation you want. Check your results to be sure that the equations (and the ΔH's) add up properly.

Example 6.7 Applying Hess's Law

What is the enthalpy of reaction, ΔH, for the formation of tungsten carbide, WC, from the elements? (Tungsten carbide is very hard and is used to make cutting tools and rock drills.)

$$W(s) + C(graphite) \longrightarrow WC(s)$$

The enthalpy change for this reaction is difficult to measure directly, because the reaction occurs at 1400°C. However, the heats of combustion of the elements and of tungsten carbide can be measured easily:

$$2W(s) + 3O_2(g) \longrightarrow 2WO_3(s); \Delta H = -1685.8 \text{ kJ} \qquad (1)$$

$$C(graphite) + O_2(g) \longrightarrow CO_2(g); \Delta H = -393.5 \text{ kJ} \qquad (2)$$

$$2WC(s) + 5O_2(g) \longrightarrow 2WO_3(s) + 2CO_2(g); \Delta H = -2391.8 \text{ kJ} \qquad (3)$$

Problem Strategy Note that Equations 1, 2, and 3 involve the reactants and products for the desired reaction (as well as their enthalpy changes). You apply Hess's law to these equations, perhaps reversing an equation or multiplying it by a factor, so that when you add all of them together, you obtain the desired equation.

First, look to see whether any equation needs to be reversed to get a reactant or product on the correct side. If you do reverse an equation, remember to change the sign of its ΔH. Second, to obtain the multiplicative factor for each equation, compare that equation with the desired one. For instance, note that Equation 1 has $2W(s)$ on the left side, whereas the desired equation has $W(s)$ on the left. Therefore, you divide Equation 1 (and its ΔH) by 2, or multiply by the factor $\frac{1}{2}$.

Solution Do any equations need to be reversed? Note that you want to end up with $W(s)$ and C(graphite) on the left side and with WC(s) on the right side of the final equation (the formation of WC from its elements). Equation 1 has $W(s)$ on the left, as required after the addition of the equations. Similarly, Equation 2 has C(graphite) on the left, as required. However, Equation 3 has WC(s) on the left, but it should be on the right. Therefore, you reverse Equation 3 (and multiply its ΔH by -1).

$$2WO_3(s) + 2CO_2(g) \longrightarrow 2WC(s) + 5O_2(g); \Delta H = 2391.8 \text{ kJ} \qquad (3')$$

To obtain the multiplicative factor for each equation, you compare Equations 1, 2, and 3', in turn, with the desired equation. We did this earlier (at the end of the Problem Strategy) for Equation 1. Multiplying it by $\frac{1}{2}$, you obtain

$$W(s) + \tfrac{3}{2}O_2(g) \longrightarrow WO_3(s); \Delta H = \tfrac{1}{2} \times (-1685.8 \text{ kJ}) = -842.9 \text{ kJ}$$

Compare Equation 2 with the desired equation. Both have C(graphite) on the left side; therefore, you leave Equation 2 as it is.

$$C(graphite) + O_2(g) \longrightarrow CO_2(g); \Delta H = -393.5 \text{ kJ}$$

Now, compare Equation 3' with the desired equation. Equation 3' has $2WC(s)$ on the right side, whereas the desired equation has WC(s) on the right side. Therefore, you multiply Equation 3' (and its ΔH) by $\frac{1}{2}$.

$$WO_3(s) + CO_2(g) \longrightarrow WC(s) + \tfrac{5}{2}O_2(g); \Delta H = \tfrac{1}{2} \times (2391.8 \text{ kJ}) = 1195.9 \text{ kJ}$$

Now add these last three equations and their corresponding ΔH's.

$$W(s) + \tfrac{3}{2}O_2(g) \longrightarrow WO_3(s) \qquad \Delta H = -842.9 \text{ kJ}$$
$$C(graphite) + O_2(g) \longrightarrow CO_2(g) \qquad \Delta H = -393.5 \text{ kJ}$$
$$\underline{WO_3(s) + CO_2(g) \longrightarrow WC(s) + \tfrac{5}{2}O_2(g) \qquad \Delta H = 1195.9 \text{ kJ}}$$
$$W(s) + C(graphite) \longrightarrow WC(s) \qquad \Delta H = \mathbf{-40.5 \text{ kJ}}$$

Answer Check Check that when you add the three equations substances that do not appear in the final one cancel correctly. For example, note that $\frac{3}{2}O_2(g)$ and $O_2(g)$ on the left sides of the equations cancel $\frac{5}{2}O_2(g)$ on the right side. Then check that the final equation is the one asked for.

(continued)

(continued)

Exercise 6.10   Manganese metal can be obtained by reaction of manganese dioxide with aluminum.

$$4Al(s) + 3MnO_2(s) \longrightarrow 2Al_2O_3(s) + 3Mn(s)$$

What is ΔH for this reaction? Use the following data:

$$2Al(s) + \tfrac{3}{2}O_2(g) \longrightarrow Al_2O_3(s); \Delta H = -1676 \text{ kJ}$$

$$Mn(s) + O_2(g) \longrightarrow MnO_2(s); \Delta H = -520 \text{ kJ}$$

■ See Problems 6.77, 6.78, 6.79, and 6.80.

CONCEPT CHECK 6.4

The heat of fusion (also called heat of melting), ΔH_{fus}, of ice is the enthalpy change for

$$H_2O(s) \longrightarrow H_2O(l); \Delta H_{fus}$$

Similarly, the heat of vaporization, ΔH_{vap}, of liquid water is the enthalpy change for

$$H_2O(l) \longrightarrow H_2O(g); \Delta H_{vap}$$

How is the heat of sublimation, ΔH_{sub}, the enthalpy change for the reaction

$$H_2O(s) \longrightarrow H_2O(g); \Delta H_{sub}$$

related to ΔH_{fus} and ΔH_{vap}?

6.8 Standard Enthalpies of Formation

Because Hess's law relates the enthalpy changes of some reactions to the enthalpy changes of others, we only need to tabulate the enthalpy changes of certain types of reactions. We also generally list enthalpy changes only for certain standard thermodynamic conditions (which are not identical to the standard conditions for gases, STP).

The term **standard state** refers to *the standard thermodynamic conditions chosen for substances when listing or comparing thermodynamic data: 1 atm pressure and the specified temperature (usually 25°C).* ▶ These standard conditions are indicated by a superscript degree sign (°). The enthalpy change for a reaction in which reactants in their standard states yield products in their standard states is denoted $\Delta H°$ ("delta H degree," but often read as "delta H zero"). The quantity $\Delta H°$ is called the *standard enthalpy of reaction.*

As we will show, it is sufficient to tabulate just the enthalpy changes for formation reactions—that is, for reactions in which compounds are formed from their elements. To specify the formation reaction precisely, however, we must specify the exact form of each element.

Some elements exist in the same physical state (gas, liquid, or solid) in two or more distinct forms. For example, oxygen in any of the physical states occurs both as dioxygen (commonly called simply oxygen), with O_2 molecules, and as ozone, with O_3 molecules. Dioxygen gas is odorless; ozone gas has a characteristic pungent odor. Solid carbon has two principal crystalline forms: graphite and diamond. Graphite is a soft, black, crystalline substance; diamond is a hard, usually colorless crystal. The elements oxygen and carbon are said to exist in different allotropic forms. An **allotrope** is *one of two or more distinct forms of an element in the same physical state* (Figure 6.18).

The **reference form** of an element for the purpose of specifying the formation reaction is usually *the stablest form (physical state and allotrope) of the element under standard thermodynamic conditions.* The reference form of oxygen at 25°C is $O_2(g)$; the reference form of carbon at 25°C is graphite. ▶

Table 6.2 lists standard enthalpies of formation of substances and ions (a longer table is given in Appendix C). The **standard enthalpy of formation** (also called the **standard heat of formation**) of a substance, denoted $\Delta H_f°$, is *the enthalpy change for the formation of one mole of the substance in its standard state from its elements in their reference form and in their standard states.*

The International Union of Pure and Applied Chemistry (IUPAC) recommends that the standard pressure be 1 bar (1 × 10⁵ Pa). Thermodynamic tables are becoming available for 1 bar pressure, and in the future such tables will probably replace those for 1 atm.

Although the reference form is usually the stablest allotrope of an element, the choice is essentially arbitrary as long as one is consistent.

Figure 6.18 ▶

Allotropes of sulfur *Left:* An evaporating dish contains rhombic sulfur, the stable form of the element at room temperature. *Right:* When this sulfur is melted, then cooled, it forms long needles of monoclinic sulfur, another allotrope. At room temperature, monoclinic sulfur will slowly change back to rhombic sulfur. Both forms contain the molecule S_8, depicted by the model.

© Cengage Learning

Table 6.2	Standard Enthalpies of Formation (at 25°C)*				
Substance or Ion	ΔH_f° (kJ/mol)	**Substance or Ion**	ΔH_f° (kJ/mol)	**Substance or Ion**	ΔH_f° (kJ/mol)
$e^-(g)$	0	$CH_3CHO(g)$	−166.1	$NH_3(g)$	−45.90
Bromine		$CH_3CHO(l)$	−191.8	$NH_4^+(aq)$	−132.5
$Br(g)$	111.9	**Chlorine**		$NO(g)$	90.29
$Br^-(aq)$	−121.5	$Cl(g)$	121.3	$NO_2(g)$	33.10
$Br^-(g)$	−219.0	$Cl^-(aq)$	−167.2	$HNO_3(aq)$	−207.4
$Br_2(g)$	30.91	$Cl^-(g)$	−234.0	**Oxygen**	
$Br_2(l)$	0	$Cl_2(g)$	0	$O(g)$	249.2
$HBr(g)$	−36.44	$HCl(g)$	−92.31	$O_2(g)$	0
Calcium		**Fluorine**		$O_3(g)$	142.7
$Ca(s)$	0	$F(g)$	79.39	**Silicon**	
$Ca^{2+}(aq)$	−542.8	$F^-(g)$	−255.1	$Si(s)$	0
$CaCO_3(s, \text{calcite})$	−1206.9	$F^-(aq)$	−332.6	$SiCl_4(l)$	−687.0
$CaO(s)$	−635.1	$F_2(g)$	0	$SiF_4(g)$	−1614.9
Carbon		$HF(g)$	−272.5	$SiO_2(s, \text{quartz})$	−910.9
$C(g)$	716.7	**Hydrogen**		**Silver**	
$C(s, \text{diamond})$	1.897	$H(g)$	218.0	$Ag(s)$	0
$C(s, \text{graphite})$	0	$H^+(aq)$	0	$Ag^+(aq)$	105.6
$CCl_4(g)$	−95.98	$H^+(g)$	1536.2	$AgBr(s)$	−100.4
$CCl_4(l)$	−135.4	$H_2(g)$	0	$AgCl(s)$	−127.1
$CO(g)$	−110.5	$H_2O(g)$	−241.8	$AgF(s)$	−204.6
$CO_2(g)$	−393.5	$H_2O(l)$	−285.8	$AgI(s)$	−61.84
$CO_3^{2-}(aq)$	−677.1	$OH^-(aq)$	−230.0	**Sodium**	
$CS_2(g)$	116.9	**Iodine**		$Na(g)$	107.3
$CS_2(l)$	89.70	$I(g)$	106.8	$Na(s)$	0
$HCN(g)$	135.1	$I^-(aq)$	−55.19	$Na^+(aq)$	−240.1
$HCN(l)$	108.9	$I^-(g)$	−194.6	$Na^+(g)$	609.3
$HCO_3^-(aq)$	−692.0	$I_2(s)$	0	$Na_2CO_3(s)$	−1130.8
Hydrocarbons		$HI(g)$	26.36	$NaCl(s)$	−411.1
$CH_4(g)$	−74.87	**Lead**		$NaHCO_3(s)$	−950.8
$C_2H_4(g)$	52.47	$Pb(s)$	0	**Sulfur**	
$C_2H_6(g)$	−84.68	$Pb^{2+}(aq)$	−1.7	$S(g)$	277.0
$C_6H_6(l)$	49.0	$PbO(s)$	−219.4	$S(s, \text{monoclinic})$	0.360
Alcohols		$PbS(s)$	−98.32	$S(s, \text{rhombic})$	0
$CH_3OH(l)$	−238.7	**Nitrogen**		$S_2(g)$	128.6
$C_2H_5OH(l)$	−277.7	$N(g)$	472.7	$SO_2(g)$	−296.8
Aldehydes		$N_2(g)$	0	$H_2S(g)$	−20.50
$HCHO(g)$	−117				

*See Appendix C for additional values.

To understand this definition, consider the standard enthalpy of formation of liquid water. Note that the stablest forms of hydrogen and oxygen at 1 atm and 25°C are $H_2(g)$ and $O_2(g)$, respectively. These are therefore the reference forms of the elements. You write the formation reaction for 1 mol of liquid water as follows:

$$H_2(g) + \tfrac{1}{2}O_2(g) \longrightarrow H_2O(l)$$

The standard enthalpy change for this reaction is -285.8 kJ per mole of H_2O. Therefore, the thermochemical equation is

$$H_2(g) + \tfrac{1}{2}O_2(g) \longrightarrow H_2O(l); \Delta H_f^\circ = -285.8 \text{ kJ}$$

The values of standard enthalpies of formation listed in Table 6.2 and in other tables are determined by direct measurement in some cases and by applying Hess's law in others. Oxides, such as water, can often be determined by direct calorimetric measurement of the combustion reaction. If you look back at Example 6.7, you will see an illustration of how Hess's law can be used to obtain the enthalpy of formation of tungsten carbide, WC.

Note that the standard enthalpy of formation of an element will depend on the form of the element. For example, the ΔH_f° for diamond equals the enthalpy change from the stablest form of carbon (graphite) to diamond. The thermochemical equation is

$$C(\text{graphite}) \longrightarrow C(\text{diamond}); \Delta H_f^\circ = 1.9 \text{ kJ}$$

On the other hand, the ΔH_f° for graphite equals zero. Note the values of ΔH_f° for the elements listed in Table 6.2; the reference forms have zero values.

Now let us see how to use standard enthalpies of formation (listed in Table 6.2) to find the standard enthalpy change for a reaction. We will first look at this problem from the point of view of Hess's law. But when we are finished, we will note a pattern in the result, which will allow us to state a simple formula for solving this type of problem.

Consider the equation

$$CH_4(g) + 4Cl_2(g) \longrightarrow CCl_4(l) + 4HCl(g); \Delta H^\circ = ?$$

From Table 6.2 you find the enthalpies of formation for $CH_4(g)$, $CCl_4(l)$, and $HCl(g)$. You can then write the following thermochemical equations:

$$C(\text{graphite}) + 2H_2(g) \longrightarrow CH_4(g); \Delta H_f^\circ = -74.9 \text{ kJ} \qquad \textbf{(1)}$$

$$C(\text{graphite}) + 2Cl_2(g) \longrightarrow CCl_4(l); \Delta H_f^\circ = -135.4 \text{ kJ} \qquad \textbf{(2)}$$

$$\tfrac{1}{2}H_2(g) + \tfrac{1}{2}Cl_2(g) \longrightarrow HCl(g); \Delta H_f^\circ = -92.3 \text{ kJ} \qquad \textbf{(3)}$$

You now apply Hess's law. Since you want CH_4 to appear on the left, and CCl_4 and 4HCl on the right, you reverse Equation 1 and add Equation 2 and $4 \times$ Equation 3.

$CH_4(g) \longrightarrow \cancel{C(\text{graphite})} + \cancel{2H_2(g)}$	$(-74.9 \text{ kJ}) \times (-1)$
$\cancel{C(\text{graphite})} + 2Cl_2(g) \longrightarrow CCl_4(l)$	$(-135.4 \text{ kJ}) \times (1)$
$\underline{\cancel{2H_2(g)} + 2Cl_2(g) \longrightarrow 4HCl(g)}$	$\underline{(-92.3 \text{ kJ}) \times (4)}$
$CH_4(g) + 4Cl_2(g) \longrightarrow CCl_4(l) + 4HCl(g)$	$\Delta H^\circ = -429.7 \text{ kJ}$

The setup of this calculation can be greatly simplified once you closely examine what you are doing. Note that the ΔH_f° for each compound is multiplied by its coefficient in the chemical equation whose ΔH° you are calculating. Moreover, the ΔH_f° for each reactant is multiplied by a negative sign. You can symbolize the enthalpy of formation of a substance by writing the formula in parentheses following ΔH_f°. Then your calculation can be written as follows:

$$\Delta H^\circ = [\Delta H_f^\circ(CCl_4) + 4\,\Delta H_f^\circ(HCl)] - [\Delta H_f^\circ(CH_4) + 4\,\Delta H_f^\circ(Cl_2)]$$
$$= [(-135.4) + 4(-92.3)]\,\text{kJ} - [(-74.9) + 4(0)]\,\text{kJ} = -429.7 \text{ kJ}$$

In general, you can calculate the ΔH° for a reaction by the equation

$$\Delta H^\circ = \Sigma\, n\Delta H_f^\circ(\text{products}) - \Sigma\, m\Delta H_f^\circ(\text{reactants})$$

Here Σ is the mathematical symbol meaning "the sum of," and m and n are the coefficients of the substances in the chemical equation.

The next two examples illustrate the calculation of enthalpies of reaction from standard enthalpies of formation.

Example 6.8

CS$_2$

Calculating the Heat of Phase Transition from Standard Enthalpies of Formation

Use values of ΔH°_f to calculate the heat of vaporization, ΔH°_{vap}, of carbon disulfide at 25°C. The vaporization process is

$$CS_2(l) \longrightarrow CS_2(g)$$

Problem Strategy The vaporization process can be treated like a chemical reaction, with the liquid as the "reactant" and the vapor as the "product." It is convenient to record the values of ΔH°_f under the formulas in the equation, multiplying them by the coefficients in the equation. You calculate ΔH°_{vap} by subtracting values for the "reactant" from values for the "product."

Solution Here is the equation for the vaporization, with values of ΔH°_f multiplied by coefficients (here, all 1's).

$$CS_2(l) \longrightarrow CS_2(g)$$
$$\underset{1(89.7)}{} \qquad \underset{1(116.9)\ (kJ)}{}$$

The calculation is

$$\Delta H^\circ_{vap} = \Sigma\, n\, \Delta H^\circ_f(\text{products}) - \Sigma\, m\, \Delta H^\circ_f(\text{reactants})$$
$$= \Delta H^\circ_f[CS_2(g)] - \Delta H^\circ_f[CS_2(l)]$$
$$= (116.9 - 89.7)\ kJ = \mathbf{27.2\ kJ}$$

Answer Check Review your arithmetic carefully. Note that vaporization requires adding heat, so the enthalpy change should be positive.

Exercise 6.11 Calculate the heat of vaporization, ΔH°_{vap}, of water, using standard enthalpies of formation (Table 6.2).

■ See Problems 6.83 and 6.84.

Example 6.9

Calculating the Enthalpy of Reaction from Standard Enthalpies of Formation

Large quantities of ammonia are used to prepare nitric acid. The first step consists of the catalytic oxidation of ammonia to nitric oxide, NO.

$$4NH_3(g) + 5O_2(g) \xrightarrow{\text{Pt}} 4NO(g) + 6H_2O(g)$$

What is the standard enthalpy change for this reaction? Use Table 6.2 for data.

Problem Strategy You record the values of ΔH°_f under the formulas in the equation, multiplying them by the coefficients in the equation. You calculate ΔH° by subtracting values for the reactants from values for the products.

Solution Here is the equation with the ΔH°_f's recorded beneath it:

$$4NH_3(g) + 5O_2(g) \longrightarrow 4NO(g) + 6H_2O(g)$$
$$\underset{4(-45.9)}{} \quad \underset{5(0)}{} \qquad\qquad \underset{4(90.3)}{} \quad \underset{6(-241.8)\ (kJ)}{}$$

The calculation is

$$\Delta H^\circ = \Sigma\, n\, \Delta H^\circ_f(\text{products}) - \Sigma\, m\, \Delta H^\circ_f(\text{reactants})$$
$$= [4\, \Delta H^\circ_f(NO) + 6\, \Delta H^\circ_f(H_2O)] - [4\, \Delta H^\circ_f(NH_3) + 5\, \Delta H^\circ_f(O_2)]$$
$$= [4(90.3) + 6(-241.8)]\ kJ - [4(-45.9) + 5(0)]\ kJ$$
$$= \mathbf{-906\ kJ}$$

(continued)

(*continued*)

Answer Check Be very careful of arithmetical signs—they are a likely source of mistakes. Also pay particular attention to the state of each substance. Here, for example, you must use the ΔH_f° for $H_2O(g)$, not for $H_2O(l)$.

> **Exercise 6.12** Calculate the enthalpy change for the following reaction:
>
> $$3NO_2(g) + H_2O(l) \longrightarrow 2HNO_3(aq) + NO(g)$$
>
> Use standard enthalpies of formation.

■ See Problems 6.85, 6.86, 6.87, and 6.88.

Enthalpies of formation can also be defined for ions. It is not possible to make thermal measurements on individual ions, so we must arbitrarily define the standard enthalpy of formation of one ion as zero. Then values for all other ions can be deduced from calorimetric data. The standard enthalpy of formation of $H^+(aq)$ is taken as zero. Values of ΔH_f° for some ions are given in Table 6.2.

> **Exercise 6.13** Calculate the standard enthalpy change for the reaction of an aqueous solution of barium hydroxide, $Ba(OH)_2$, with an aqueous solution of ammonium nitrate, NH_4NO_3, at 25°C. (Figure 6.1 illustrated this reaction using solids instead of solutions.) The complete ionic equation is
>
> $$Ba^{2+}(aq) + 2OH^-(aq) + 2NH_4^+(aq) + 2NO_3^-(aq) \longrightarrow 2NH_3(g) + 2H_2O(l) + Ba^{2+}(aq) + 2NO_3^-(aq)$$

■ See Problems 6.89 and 6.90.

6.9 Fuels—Foods, Commercial Fuels, and Rocket Fuels

A *fuel* is any substance that is burned or similarly reacted to provide heat and other forms of energy. The earliest use of fuels for heat came with the control of fire, which was achieved perhaps 400,000 years ago. This major advance allowed the human species to migrate from tropical savannas and eventually to inhabit most of the earth. Through cooking, fire also increased the variety of edible food supplies and provided some protection of the food from bacterial decay. During the mid-eighteenth century, the discovery of the steam engine, which converts the chemical energy latent in fuels to mechanical energy, ushered in the Industrial Revolution. Today fuels not only heat our homes and move our cars but are absolutely necessary for every facet of modern technology. For example, fuels generate the electricity required for our modern computing and communications technologies, and they propel the rocket engines that make possible our explorations of outer space. In this section we will look at foods as fuels; at fossil fuels (which include gas, oil, and coal); at coal gasification and liquefaction; and at rocket fuels.

Foods as Fuels

Foods fill three needs of the body: they supply substances for the growth and repair of tissue, they supply substances for the synthesis of compounds used in the regulation of body processes, and they supply energy. About 80% of the energy we need is for heat. The rest is used for muscular action, chemical processes, and other body processes. ▶

The human body requires about as much energy in a day as does a 100-watt lightbulb.

The body generates energy from food by the same overall process as combustion, so the overall enthalpy change is the same as the heat of combustion, which can be determined in a calorimeter. You can get some idea of the energy available from carbohydrate foods by looking at a typical one, glucose ($C_6H_{12}O_6$). The thermochemical equation for the combustion of glucose is

$$C_6H_{12}O_6(s) + 6O_2(g) \longrightarrow 6CO_2(g) + 6H_2O(l); \Delta H^\circ = -2803 \text{ kJ}$$

One gram of glucose yields 15.6 kJ (3.73 kcal) of heat when burned.

A representative fat is glyceryl trimyristate, $C_{45}H_{86}O_6$. The equation for its combustion is

$$C_{45}H_{86}O_6(s) + \tfrac{127}{2}O_2(g) \longrightarrow 45CO_2(g) + 43H_2O(l);\ \Delta H^\circ = -27{,}820\ kJ$$

One gram of this fat yields 38.5 kJ (9.20 kcal) of heat when burned. The average values quoted for carbohydrates and fats are 4.0 kcal/g and 9.0 kcal/g, respectively. ◀

In the popular literature of nutrition, the kilocalorie is referred to as the Calorie (capital C). Thus, these values are given as 4.0 Calories and 9.0 Calories.

Note that fats contain more than twice the fuel value per gram as do carbohydrates. Thus, by storing its fuel as fat, the body can store more fuel for a given mass of body tissue.

Fossil Fuels

All of the fossil fuels in existence today were created millions of years ago when aquatic plants and animals were buried and compressed by layers of sediment at the bottoms of swamps and seas. Over time this organic matter was converted by bacterial decay and pressure to petroleum (oil), gas, and coal. Figure 6.19 gives the percentages of the total energy consumed in the United States from various sources. Fossil fuels account for over 80% of the total.

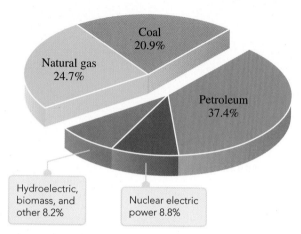

Figure 6.19 ▲

Sources of energy consumed in the United States (2009) Data are from *Monthly Energy Review On-line*, http//www.eia.doe.gov.

Anthracite, or hard coal, the oldest variety of coal, was laid down as long as 250 million years ago and may contain over 80% carbon. Bituminous coal, a younger variety of coal, has between 45% and 65% carbon. Fuel values of coals are rated in Btu's (British thermal units) per pound, which are essentially heats of combustion per pound of coal. A typical value is 13,200 Btu/lb. A Btu equals 1054 J, so 1 Btu/lb equals 2.32 J/g. Therefore, the combustion of coal in oxygen yields about 30.6 kJ/g. You can compare this value with the heat of combustion of pure carbon (graphite).

$$C(graphite) + O_2(g) \longrightarrow CO_2(g);\ \Delta H^\circ = -393.5\ kJ$$

The value given in the equation is for 1 mol (12.0 g) of carbon. Per gram, you get 32.8 kJ/g, which is comparable with the values obtained for coal.

Natural gas and petroleum together account for nearly two-thirds of the fossil fuels consumed per year. They are very convenient fluid fuels, being easily transportable and having no ash. Purified natural gas is primarily methane, CH_4, but it also contains small amounts of ethane, C_2H_6; propane, C_3H_8; and butane, C_4H_{10}. We would expect the fuel values of natural gas to be close to that for the heat of combustion of methane:

$$CH_4(g) + 2O_2(g) \longrightarrow CO_2(g) + 2H_2O(g);\ \Delta H^\circ = -802\ kJ$$

This value of ΔH° is equivalent to 50.1 kJ per gram of fuel.

Petroleum is a very complicated mixture of compounds. Gasoline, which is obtained from petroleum by chemical and physical processes, contains many different hydrocarbons (compounds of carbon and hydrogen). One such hydrocarbon is octane, C_8H_{18}. The combustion of octane evolves 5074 kJ of heat per mole.

$$C_8H_{18}(l) + \tfrac{25}{2}O_2(g) \longrightarrow 8CO_2(g) + 9H_2O(g);\ \Delta H^\circ = -5074\ kJ$$

This value of ΔH° is equivalent to 44.4 kJ/g. These combustion values indicate another reason why the fluid fossil fuels are popular: they release more heat per gram than coal does.

Coal Gasification and Liquefaction

The major problem with petroleum and natural gas as fuels is their relative short supply. It has been estimated that petroleum supplies will be 80% depleted by about the year 2030. Natural-gas supplies may be depleted even sooner.

Coal supplies, on the other hand, are sufficient to last several more centuries. This abundance has spurred much research into developing commercial methods for converting coal to the more easily handled liquid and gaseous fuels. Most of these methods begin by converting coal to carbon monoxide, CO. One way involves the water–gas reaction.

$$C(s) + H_2O(g) \longrightarrow CO(g) + H_2(g)$$

In this reaction, steam is passed over hot coal. Once a mixture of carbon monoxide and hydrogen is obtained, it can be transformed by various reactions into useful products. For example, in the methanation reaction (discussed at length in Chapter 14), this mixture is reacted over a catalyst to give methane.

$$CO(g) + 3H_2(g) \longrightarrow CH_4(g) + H_2O(g)$$

Different catalysts and different reaction conditions result in liquid fuels. An added advantage of coal gasification and coal liquefaction is that sulfur, normally present in fossil fuels, can be removed during the process. The burning of sulfur-containing coal is a major source of air pollution and acid rain.

Rocket Fuels

Rockets are self-contained missiles propelled by the ejection of gases from an orifice. Usually these are hot gases propelled from the rocket by the reaction of a fuel with an oxidizer. Rockets are believed to have originated with the Chinese—perhaps before the thirteenth century, which is when they began to appear in Europe. However, it was not until the twentieth century that rocket propulsion began to be studied seriously, and since World War II rockets have become major weapons. Space exploration with satellites propelled by rocket engines began in 1957 with the Russian satellite *Sputnik I*. Today weather and communications satellites are regularly put into orbit about the earth using rocket engines.

One of the factors determining which fuel and oxidizer to use is the mass of the fuel and oxidizer required. We have already seen that natural gas and gasoline have higher fuel values per gram than does coal. The difference is caused by the higher hydrogen content of natural gas and gasoline. Hydrogen is the element of lowest density, and at the same time it reacts exothermically with oxygen to give water. You might expect hydrogen and oxygen to be an ideal fuel–oxidizer combination. The thermochemical equation for the combustion of hydrogen is

$$H_2(g) + \tfrac{1}{2}O_2(g) \longrightarrow H_2O(g); \Delta H° = -242 \text{ kJ}$$

This value of $\Delta H°$ is equivalent to 120 kJ/g of fuel (H_2) compared with 50 kJ/g of methane. The second and third stages of the *Saturn V* launch vehicle that sent a three-man Apollo crew to the moon used a hydrogen/oxygen system. The launch vehicle contained liquid hydrogen (boiling at $-253°C$) and liquid oxygen, or LOX (boiling at $-183°C$).

The first stage of liftoff used kerosene and oxygen, and an unbelievable 550 metric tons (550×10^3 kg) of kerosene were burned in 2.5 minutes! It is interesting to calculate the average rate of energy production in this 2.5-minute interval. Kerosene is approximately $C_{12}H_{26}$. The thermochemical equation is

$$C_{12}H_{26}(l) + \tfrac{37}{2}O_2(g) \longrightarrow 12CO_2(g) + 13H_2O(g); \Delta H° = -7513 \text{ kJ}$$

This value of $\Delta H°$ is equivalent to 44.1 kJ/g. Thus, 550×10^6 g of fuel generated 2.43×10^{10} kJ in 150 s (2.5 min). Each second, the average energy produced was 1.62×10^{11} J. This is equivalent to 1.62×10^{11} watts, or 217 million horsepower (1 horsepower equals 745.7 watts, or J/s).

The landing module for the Apollo mission used a fuel made of hydrazine, N_2H_4, and a derivative of hydrazine. The oxidizer was dinitrogen tetroxide, N_2O_4. These substances are normally liquids and are therefore easier to store than liquid hydrogen and oxygen. The reaction of the oxidizer with hydrazine is

$$2N_2H_4(l) + N_2O_4(l) \longrightarrow 3N_2(g) + 4H_2O(g); \Delta H° = -1049 \text{ kJ}$$

Solid propellants are also used as rocket fuels. The mixture used in the booster rockets of the space shuttles (Figure 6.20) is a fuel containing aluminum metal powder. An oxidizer of ammonium perchlorate, NH_4ClO_4, is mixed with the fuel.

© NASA

Figure 6.20 ▲

The launching of a space shuttle The solid fuel for the booster rockets is a mixture of aluminum metal powder and other materials with ammonium perchlorate as the oxidizer. A cloud of aluminum oxide forms as the rockets burn.

A Checklist for Review

Summary of Facts and Concepts

Energy exists in various forms, including *kinetic energy* and *potential energy*. The SI unit of energy is the *joule* (1 calorie = 4.184 joules). The *internal energy* of a substance is the sum of the kinetic energies and potential energies of the particles making up the substance. According to the *law of conservation of energy*, the total quantity of energy remains constant.

Reactions absorb or evolve definite quantities of heat under given conditions. At constant pressure, this *heat of reaction* is the *enthalpy of reaction*, ΔH. The chemical equation plus ΔH for molar amounts of reactants is referred to as the *thermochemical equation*. With it, you can calculate the heat for any amount of

substance by *applying stoichiometry to heats of reaction*. One measures the heat of reaction in a *calorimeter*. Direct calorimetric determination of the heat of reaction requires a reaction that goes to completion without other reactions occurring at the same time. Otherwise, the heat or enthalpy of reaction is determined indirectly from other enthalpies of reaction by using *Hess's law of heat summation*. Thermochemical data are conveniently tabulated as *enthalpies of formation*. If you know the values for each substance in an equation, you can easily compute the enthalpy of reaction. As an application of thermochemistry, the last section of the chapter discusses fuels.

Learning Objectives

6.1 Energy and Its Units

- Define *energy, kinetic energy, potential energy,* and *internal energy.*
- Define the SI unit of energy *joule,* as well as the common unit of energy *calorie.*
- Calculate the kinetic energy of a moving object. Example 6.1
- State the law of conservation of energy.

6.2 First Law of Thermodynamics; Work and Heat

- Define a thermodynamic system and its surroundings.
- Define *work* and *heat.*
- Define the change of internal energy of a system.
- Express the *first law of thermodynamics* verbally and mathematically.

6.3 Heat of Reaction; Enthalpy of Reaction

- Define the heat of reaction.
- Distinguish between an exothermic process and an endothermic process.
- Describe *pressure-volume work* verbally and mathematically.
- Define enthalpy and enthalpy of reaction.
- Relate the heat of reaction at constant pressure and the enthalpy of reaction.

6.4 Thermochemical Equations

- Define a *thermochemical equation.*
- Write a thermochemical equation given pertinent information. Example 6.2
- Learn the two rules for manipulating (reversing and multiplying) thermochemical equations.
- Manipulate a thermochemical equation using these rules. Example 6.3

Important Terms

energy
kinetic energy
joule (J)
calorie (cal)
potential energy
internal energy
law of conservation of energy

thermodynamic system (or system)
surroundings
work
heat
state function
first law of thermodynamics

heat of reaction
exothermic process
endothermic process
pressure-volume work
enthalpy
enthalpy of reaction

thermochemical equation

6.5 Applying Stoichiometry to Heats of Reaction

■ Calculate the heat absorbed or evolved from a reaction given its enthalpy of reaction and the mass of a reactant or product. Example 6.4

6.6 Measuring Heats of Reaction

■ Define *heat capacity* and *specific heat*.
■ Relate the heat absorbed or evolved to the specific heat, mass, and temperature change.
■ Calculate the heat involved in changing the temperature of a material, given its specific heat. Example 6.5
■ Define *calorimeter*.
■ Calculate the enthalpy of reaction from calorimetric data (its temperature change and heat capacity). Example 6.6

heat capacity
specific heat capacity (specific heat)
calorimeter

6.7 Hess's Law

■ State Hess's law of heat summation.
■ Apply Hess's law to obtain the enthalpy change for one reaction from the enthalpy changes of a number of other reactions. Example 6.7

Hess's law of heat summation

6.8 Standard Enthalpies of Formation

■ Define *standard state* and *reference form*.
■ Define *standard enthalpy of formation*.
■ Calculate the heat of a phase transition using standard enthalpies of formation for the different phases. Example 6.8
■ Calculate the heat (enthalpy) of reaction from the standard enthalpies of formation of the substances in the reaction. Example 6.9

standard state
allotrope
reference form
standard enthalpy of formation (standard heat of formation)

6.9 Fuels—Foods, Commercial Fuels, and Rocket Fuels

■ Define *fuel*.
■ Describe the three needs of the body that are fulfilled by foods.
■ Give the approximate average values quoted (per gram) for the heat values (heats of combustion) for fats and for carbohydrates.
■ List the three major fossil fuels.
■ Describe the processes of coal gasification and coal liquefaction.
■ Describe some fuel–oxidizer systems used in rockets.

Key Equations

$E_k = \frac{1}{2}mv^2$

$1 \text{ cal} = 4.184 \text{ J}$

$\Delta U = q + w$

$w = -P\Delta V$

$H = U + PV$

$q = \Delta H$ (At fixed pressure and a given temperature)

$q = C\,\Delta t$

$q = s \times m \times \Delta t$

$\Delta H° = \Sigma\, n\, \Delta H_f°(\text{products}) - \Sigma\, m\, \Delta H_f°(\text{reactants})$

Questions and Problems

Self-Assessment and Review Questions

Key: These questions test your understanding of the ideas you worked with in the chapter. These problems vary in difficulty and often can be used for the basis of discussion.

6.1 Define *energy, kinetic energy, potential energy,* and *internal energy.*

6.2 Describe the interconversions of potential and kinetic energy in a moving pendulum. A moving pendulum eventually comes to rest. Has the energy been lost? If not, what has happened to it?

6.3 Define the joule in terms of SI base units.

6.4 What is the original definition of the calorie? What is the present definition?

6.5 Suppose heat flows into a vessel containing a gas. As the heat flows into the gas, what happens to the gas molecules? What happens to the internal energy of the gas?

6.6 The internal energy of a substance is a state function. What does this mean?

6.7 Define an *exothermic* reaction and an *endothermic* reaction. Give an example of each.

6.8 Under what condition is the enthalpy change equal to the heat of reaction?

6.9 How does the enthalpy change for an endothermic reaction occurring at constant pressure?

6.10 Why is it important to give the states of the reactants and products when giving an equation for ΔH?

6.11 If an equation for a reaction is doubled and then reversed, how is the value of ΔH changed?

6.12 Consider the reaction of methane, CH_4, with oxygen, O_2, discussed in Section 6.5. How would you set up the calculation if the problem had been to compute the heat if 10.0 g H_2O were produced (instead of 10.0 g CH_4 reacted)?

6.13 Define the heat capacity of a substance. Define the specific heat of a substance.

6.14 Describe a simple calorimeter. What measurements are needed to determine the heat of reaction?

6.15 You discover that you cannot carry out a particular reaction for which you would like the enthalpy change. Does this mean that you will be unable to obtain this enthalpy change? Explain.

6.16 What property of enthalpy provides the basis of Hess's law? Explain.

6.17 What is meant by the thermodynamic standard state?

6.18 What is meant by the reference form of an element? What is the standard enthalpy of formation of an element in its reference form?

6.19 What is meant by the standard enthalpy of formation of a substance?

6.20 Write the chemical equation for the formation reaction of $H_2S(g)$.

6.21 Is the following reaction the appropriate one to use in determining the enthalpy of formation of methane, $CH_4(g)$? Why or why not?

$$C(g) + 4H(g) \longrightarrow CH_4(g)$$

6.22 What is a fuel? What are the fossil fuels?

6.23 List some rocket fuels and corresponding oxidizers. Give thermochemical equations for the exothermic reactions of these fuels with the oxidizers.

6.24 Give chemical equations for the conversion of carbon in coal to methane, CH_4.

6.25 The equation for the combustion of 2 mol of butane can be written

$$2C_4H_{10}(g) + O_2(g) \longrightarrow 8CO_2(g) + 10H_2O(g); \Delta H < O$$

Which of the following produces the *least* heat?

a   Burning enough butane to produce 1 mol of water.

b   All of the above reactions (a, b, c, and d) produce the same amount of heat.

c   Burning enough butane to produce 1 mol of carbon dioxide.

d   Burning 1 mol of butane.

e   Reacting 1 mol of oxygen with excess butane.

6.26 A 5.0-g sample of water starting at 60.0°C loses 418 J of energy in the form of heat. What is the *final* temperature of the water after this heat loss?

a   20.°C   b   40.°C

c   50.°C   d   60.°C

e   80.°C

6.27 Hypothetical elements A_2 and B_2 react according to the following equation, forming the compound AB.

$$A_2(aq) + B_2(aq) \longrightarrow 2AB(aq); \Delta H° = +271 \text{ kJ/mol}$$

If solutions $A_2(aq)$ and $B_2(aq)$, starting at the same temperature, are mixed in a coffee-cup calorimeter, the reaction that occurs is

a   exothermic, and the temperature of the resulting solution falls.

b   exothermic or endothermic, depending on the original and final temperatures.

c   endothermic, and the temperature of the resulting solution falls.

d   exothermic, and the temperature of the resulting solution rises.

e   endothermic, and the temperature of the resulting solution rises.

6.28 Consider the following specific heats of metals.

Metal	Specific Heat
copper	0.385 J/(g·°C)
magnesium	1.02 J/(g·°C)
mercury	0.138 J/(g·°C)
silver	0.237 J/(g·°C)

Four 25-g samples, one of each metal, and four insulated containers with identical water volumes, all start out at room temperature. Now suppose you add exactly the same quantity of heat to each metal sample. Then you place the hot metal samples in different containers of water (that all have the same volume of water). Which of the answers below is true?

a   The water with the copper will be the hottest.

b   The water with the magnesium will be the hottest.

c   The water with the mercury will be the hottest.

d   The water with the silver will be the hottest.

e   The temperature of the water will be the same in all the cups.

Concept Explorations

Key: Concept explorations are comprehensive problems that provide a framework that will enable you to explore and learn many of the critical concepts and ideas in each chapter. If you master the concepts associated with these explorations, you will have a better understanding of many important chemistry ideas and will be more successful in solving all types of chemistry problems. These problems are well suited for group work and for use as in-class activities.

6.29 Thermal Interactions

Part 1: In an insulated container, you mix 200. g of water at 80°C with 100. g of water at 20°C. After mixing, the temperature of the water is 60°C.

  a How much did the temperature of the hot water change? How much did the temperature of the cold water change? Compare the magnitudes (positive values) of these changes.

  b During the mixing, how did the heat transfer occur: from hot water to cold, or from cold water to hot?

  c What quantity of heat was transferred from one sample to the other?

  d How does the quantity of heat transferred to or from the hot-water sample compare with the quantity of heat transferred to or from the cold-water sample?

  e Knowing these relative quantities of heat, why is the temperature change of the cold water greater than the magnitude of the temperature change of the hot water.

  f A sample of hot water is mixed with a sample of cold water that has twice its mass. Predict the temperature change of each of the samples.

  g You mix two samples of water, and one increases by 20°C, while the other drops by 60°C. Which of the samples has less mass? How do the masses of the two water samples compare?

  h A 7-g sample of hot water is mixed with a 3-g sample of cold water. How do the temperature changes of the two water samples compare?

Part 2: A sample of water is heated from 10°C to 50°C. Can you calculate the amount of heat added to the water sample that caused this temperature change? If not, what information do you need to perform this calculation?

Part 3: Two samples of water are heated from 20°C to 60°C. One of the samples requires twice as much heat to bring about this temperature change as the other. How do the masses of the two water samples compare? Explain your reasoning.

6.30 Enthalpy

  a A 100.-g sample of water is placed in an insulated container and allowed to come to room temperature at 21°C. To heat the water sample to 41°C, how much heat must you add to it?

  b Consider the hypothetical reaction,

$$2X(aq) + Y(l) \longrightarrow X_2Y(aq)$$

being run in an insulated container that contains 100. g of solution. If the temperature of the solution changes from 21°C to 31°C, how much heat does the chemical reaction produce? How does this answer compare with that in part a? (You can assume that this solution is so dilute that it has the same heat capacity as pure water.)

  c If you wanted the temperature of 100. g of this solution to increase from 21°C to 51°C, how much heat would you have to add to it? (Try to answer this question without using a formula.)

  d If you had added 0.02 mol of X and 0.01 mol of Y to form the solution in part b, how many moles of X and Y would you need to bring about the temperature change described in part c.

  e Judging on the basis of your answers so far, what is the enthalpy of the reaction $2X(aq) + Y(l) \longrightarrow X_2Y(aq)$?

Conceptual Problems

Key: These problems are designed to check your understanding of the concepts associated with some of the main topics presented in each chapter. A strong conceptual understanding of chemistry is the foundation for both applying chemical knowledge and solving chemical problems. These problems vary in level of difficulty and often can be used as a basis for group discussion.

6.31 Chemical reactions are run in each of the beakers depicted below (labeled A, B, and C). The magnitude and direction of heat and work for each reaction are represented as arrows, with the length of an arrow depicting the relative magnitude of the heat or work.

  a Is the reaction endothermic or exothermic?
  b What is the sign (+ or −) of the work?
  c What is the sign (+ or −) of the enthalpy of each reaction?
  d Is there an increase or decrease in internal energy?
  e What is the temperature of the reaction mixture immediately after the reaction when compared to room temperature?

6.32 Shown below is a diagram depicting the enthalpy change of a chemical reaction run at constant pressure.

Products

Reactants

Enthalpy, *H*

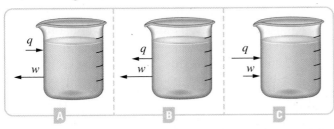

For the reaction in each beaker, answer the following and explain your reasoning:

a Is the reaction exothermic or endothermic?

b What is the sign of ΔH?

c What is the sign of q?

d If the reaction does no work, what is the sign of ΔE for this process?

6.33 A small car is traveling at twice the speed of a larger car, which has twice the mass of the smaller car. Which car has the greater kinetic energy? (Or do they both have the same kinetic energy?)

6.34 The equation for the combustion of butane, C_4H_{10}, is

$$C_4H_{10}(g) + \tfrac{13}{2}O_2(g) \longrightarrow 4CO_2(g) + 5H_2O(g)$$

Which one of the following generates the least heat? Why?

a producing one mole of carbon dioxide by burning butane

b reacting one mole of oxygen with excess butane

c producing one mole of water by burning butane

d burning one mole of butane

6.35 A 250-g sample of water at 20.0°C is placed in a freezer that is held at a constant temperature of −20.0°C. Considering the water as the "system," answer the following questions:

a what is the sign of q_{sys} for the water after it is placed in the freezer?

b After a few hours, what will be the state of the water?

c How will the initial enthalpy for the water compare with the final enthalpy of the water after it has spent several hours in the freezer?

d What will the temperature of the water be after several hours in the freezer?

6.36 A 20.0-g block of iron at 50.0°C and a 20.0 g block of aluminum at 45°C are placed in contact with each other. Assume that heat is only transferred between the two blocks.

a Draw an arrow indicating the heat flow between the blocks.

b What is the sign of q_{sys} for the aluminum when the blocks first come into contact?

c What will you observe when q_{sys} for the iron is zero?

d Estimate the temperature of the Al and Fe blocks when q_{sys} of the iron equals q_{sys} of the aluminum.

6.37 A block of aluminum and a block of iron, both having the same mass, are removed from a freezer and placed outside on a warm day. When the same quantity of heat has flowed into each block, which block will be warmer? Assume that neither block has yet reached the outside temperature. (See Table 6.1 for the specific heats of the metals.)

6.38 What is the enthalpy change for the preparation of one mole of liquid water from the elements, given the following equations?

$$H_2(g) + \tfrac{1}{2}O_2(g) \longrightarrow H_2O(g); \Delta H_f$$

$$H_2O(l) \longrightarrow H_2O(g); \Delta H_{vap}$$

6.39 You have two samples of different metals, metal A and metal B, each having the same mass. You heat both metals to 95°C and then place each one into separate beakers containing the same quantity of water at 25°C.

a You measure the temperatures of the water in the two beakers when each metal has cooled by 10°C and find that the temperature of the water with metal A is higher than the temperature of the water with metal B. Which metal has the greater specific heat? Explain.

b After waiting a period of time, the temperature of the water in each beaker rises to a maximum value. In which beaker does the water temperature rise to the higher value, the one with metal A or the one with metal B? Explain.

6.40 Consider the reactions of silver metal, Ag(s), with each of the halogens: fluorine, $F_2(g)$, chlorine, $Cl_2(g)$, and bromine, $Br_2(l)$. What chapter data could you use to decide which reaction is most exothermic? Which reaction is that?

6.41 Tetraphosphorus trisulfide, P_4S_3, burns in excess oxygen to give tetraphosphorus decoxide, P_4O_{10}, and sulfur dioxide, SO_2. Suppose you have measured the enthalpy change for this reaction. How could you use it to obtain the enthalpy of formation of P_4S_3? What other data do you need?

6.42 A soluble salt, MX_2, is added to water in a beaker. The equation for the dissolving of the salt is:

$$MX_2(s) \longrightarrow M^{2+}(aq) + 2X^-(aq); \quad \Delta H > 0$$

a Immediately after the salt dissolves, is the solution warmer or colder?

b Indicate the direction of heat flow, in or out of the beaker, while the salt dissolves.

c After the salt dissolves and the water returns to room temperature, what is the value of q for the system?

Practice Problems

Key: These problems are for practice in applying problem-solving skills. They are divided by topic, and some are keyed to exercises (see the ends of the exercises). The problems are arranged in matching pairs; the odd-numbered problem of each pair is listed first, and its answer is given in the back of the book.

Energy and Its Units

6.43 Hydrogen sulfide, H_2S, is produced during decomposition of organic matter. When 0.5000 mol H_2S burns to produce $SO_2(g)$ and $H_2O(l)$, −281.0 kJ of heat is released. What is this heat in kilocalories?

6.44 Methane, CH_4, is a major component of marsh gas. When 0.5000 mol methane burns to produce carbon dioxide and liquid water, −445.1 kJ of heat is released. What is this heat in kilocalories?

6.45 A car whose mass is 4.85×10^3 lb is traveling at a speed of 57 miles per hour. What is the kinetic energy of the car in joules? in calories? See Table 1.4 for conversion factors.

6.46 A bullet weighing 235 grains is moving at a speed of 2.52×10^3 ft/s. Calculate the kinetic energy of the bullet in joules and in calories. One grain equals 0.0648 g.

6.47 Chlorine dioxide, ClO_2, is a reddish yellow gas used in bleaching paper pulp. The average speed of a ClO_2 molecule at 25°C is 306 m/s. What is the kinetic energy (in joules) of a ClO_2 molecule moving at this speed?

6.48 Nitrous oxide, N_2O, has been used as a dental anesthetic. The average speed of an N_2O molecule at 25°C is

379 m/s. Calculate the kinetic energy (in joules) of an N_2O molecule traveling at this speed.

First Law of Thermodynamics

6.49 An ideal gas expands isothermally (at constant temperature). The internal energy of an ideal gas remains constant during an isothermal change. If q is -76 J, what are ΔU and w?

6.50 A gas is cooled and loses 82 J of heat. The gas contracts as it cools, and work done on the system equal to 29 J is exchanged with the surroundings. What are q, w, and ΔU?

Heat of Reaction

6.51 The decomposition of ozone, O_3, to oxygen, O_2, is an exothermic reaction. What is the sign of q? If you were to touch a flask in which ozone is decomposing to oxygen, would you expect the flask to feel warm or cool?

6.52 The process of dissolving ammonium nitrate, NH_4NO_3, in water is an endothermic process. What is the sign of q? If you were to add some ammonium nitrate to water in a flask, would you expect the flask to feel warm or cool?

6.53 Nitric acid, a source of many nitrogen compounds, is produced from nitrogen dioxide. An old process for making nitrogen dioxide employed nitrogen and oxygen.

$$N_2(g) + 2O_2(g) \longrightarrow 2NO_2(g)$$

The reaction absorbs 66.2 kJ of heat per 2 mol NO_2 produced. Is the reaction endothermic or exothermic? What is the value of q?

6.54 Hydrogen cyanide is used in the manufacture of clear plastics such as Lucite and Plexiglas. It is prepared from ammonia and natural gas (CH_4).

$$2NH_3(g) + 3O_2(g) + 2CH_4(g) \longrightarrow 2HCN(g) + 6H_2O(g)$$

The reaction evolves 469 kJ of heat per mol of HCN formed. Is the reaction endothermic or exothermic? What is the value of q when 2 mol HCN forms?

6.55 What is ΔU when 1.00 mol of liquid water vaporizes at 100°C? The heat of vaporization, ΔH°_{vap}, of water at 100°C is 40.66 kJ/mol.

6.56 What is ΔU for the following reaction at 25°C?

$$2H_2(g) + O_2(g) \longrightarrow 2H_2O(l)$$

Thermochemical Equations

6.57 When 2 mol of potassium chlorate crystals decompose to potassium chloride crystals and oxygen gas at constant temperature and pressure, 78.0 kJ of heat is given off. Write a thermochemical equation for this reaction.

6.58 When 1 mol of iron metal reacts with hydrochloric acid at constant temperature and pressure to produce hydrogen gas and aqueous iron(II) chloride, 89.1 kJ of heat evolves. Write a thermochemical equation for this reaction.

6.59 When white phosphorus burns in air, it produces phosphorus(V) oxide.

$$P_4(s) + 5O_2(g) \longrightarrow P_4O_{10}(s); \Delta H = -3010 \text{ kJ}$$

What is ΔH for the following equation?

$$P_4O_{10}(s) \longrightarrow P_4(s) + 5O_2(g)$$

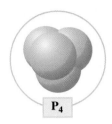

P_4

6.60 Carbon disulfide burns in air, producing carbon dioxide and sulfur dioxide.

$$CS_2(l) + 3O_2(g) \longrightarrow CO_2(g) + 2SO_2(g); \Delta H = -1077 \text{ kJ}$$

What is ΔH for the following equation?

$$\tfrac{1}{2}CO_2(g) + SO_2(g) \longrightarrow \tfrac{1}{2}CS_2(l) + \tfrac{3}{2}O_2(g)$$

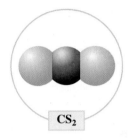

CS_2

6.61 With a platinum catalyst, ammonia will burn in oxygen to give nitric oxide, NO.

$$4NH_3(g) + 5O_2(g) \longrightarrow 4NO(g) + 6H_2O(g);$$
$$\Delta H = -906 \text{ kJ}$$

What is the enthalpy change for the following reaction?

$$NO(g) + \tfrac{3}{2}H_2O(g) \longrightarrow NH_3(g) + \tfrac{5}{4}O_2(g)$$

6.62 Phosphoric acid, H_3PO_4, can be prepared by the reaction of phosphorus(V) oxide, P_4O_{10}, with water.

$$\tfrac{1}{4}P_4O_{10}(s) + \tfrac{3}{2}H_2O(l) \longrightarrow H_3PO_4(aq); \Delta H = -96.2 \text{ kJ}$$

What is ΔH for the reaction involving 1 mol of P_4O_{10}?

$$P_4O_{10}(s) + 6H_2O(l) \longrightarrow 4H_3PO_4(aq)$$

Stoichiometry of Reaction Heats

6.63 Colorless nitric oxide, NO, combines with oxygen to form nitrogen dioxide, NO_2, a brown gas.

$$2NO(g) + O_2(g) \longrightarrow 2NO_2(g); \Delta H = -114 \text{ kJ}$$

What is the enthalpy change per gram of nitric oxide?

6.64 Hydrogen, H_2, is used as a rocket fuel. The hydrogen is burned in oxygen to produce water vapor.

$$2H_2(g) + O_2(g) \longrightarrow 2H_2O(g); \Delta H = -484 \text{ kJ}$$

What is the enthalpy change per kilogram of hydrogen?

6.65 Ammonia burns in the presence of a copper catalyst to form nitrogen gas.

$$4NH_3(g) + 3O_2(g) \longrightarrow 2N_2(g) + 6H_2O(g);$$
$$\Delta H = -1267 \text{ kJ}$$

What is the enthalpy change to burn 35.8 g of ammonia?

6.66 Hydrogen sulfide, H_2S, is a foul-smelling gas. It burns to form sulfur dioxide.

$$2H_2S(g) + 3O_2(g) \longrightarrow 2SO_2(g) + 2H_2O(g);$$
$$\Delta H = -1036 \text{ kJ}$$

Calculate the enthalpy change to burn 27.4 g of hydrogen sulfide.

6.67 Propane, C_3H_8, is a common fuel gas. Use the following to calculate the grams of propane you would need to provide 369 kJ of heat.

$$C_3H_8(g) + 5O_2(g) \longrightarrow 3CO_2(g) + 4H_2O(g);$$
$$\Delta H = -2043 \text{ kJ}$$

6.68 Ethanol, C_2H_5OH, is mixed with gasoline and sold as gasohol. Use the following to calculate the grams of ethanol needed to provide 429 kJ of heat.

$$C_2H_5OH(l) + 3O_2(g) \longrightarrow 2CO_2(g) + 3H_2O(g);$$
$$\Delta H = -1235 \text{ kJ}$$

Heat Capacity and Calorimetry

6.69 An iron skillet weighing 1.63 kg is heated on a stove to 178°C. Suppose the skillet is cooled to room temperature, 21°C. How much heat energy (in joules) must be removed to affect this cooling? The specific heat of iron is 0.449 J/(g·°C).

6.70 You wish to heat water to make coffee. How much heat (in joules) must be used to raise the temperature of 0.180 kg of tap water (enough for one cup of coffee) from 19°C to 96°C (near the ideal brewing temperature)? Assume the specific heat is that of pure water, 4.18 J/(g·°C).

6.71 When steam condenses to liquid water, 2.26 kJ of heat is released per gram. The heat from 168 g of steam is used to heat a room containing 6.44×10^4 g of air (20 ft × 12 ft × 8 ft). The specific heat of air at normal pressure is 1.015 J/(g·°C). What is the change in air temperature, assuming the heat from the steam is all absorbed by air?

6.72 When ice at 0°C melts to liquid water at 0°C, it absorbs 0.334 kJ of heat per gram. Suppose the heat needed to melt 38.0 g of ice is absorbed from the water contained in a glass. If this water has a mass of 0.210 kg and a temperature of 21.0°C, what is the final temperature of the water? (Note that you will also have 38.0 g of water at 0°C from the ice.)

6.73 When 23.6 g of calcium chloride, $CaCl_2$, was dissolved in water in a constant-pressure calorimeter, the temperature rose from 25.0°C to 38.7°C. If the heat capacity of the solution and the calorimeter is 1258 J/°C, what is the enthalpy change when 1.20 mol of calcium chloride dissolves in water? The solution process is

$$CaCl_2(s) \longrightarrow Ca^{2+}(aq) + 2Cl^-(aq)$$

6.74 When 15.3 g of sodium nitrate, $NaNO_3$, was dissolved in water in a constant-pressure calorimeter, the temperature fell from 25.00°C to 21.56°C. If the heat capacity of the solution and the calorimeter is 1071 J/°C, what is the enthalpy change when 1 mol of sodium nitrate dissolves in water? The solution process is

$$NaNO_3(s) \longrightarrow Na^+(aq) + NO_3^-(aq); \Delta H = ?$$

6.75 A sample of ethanol, C_2H_5OH, weighing 2.84 g was burned in an excess of oxygen in a bomb calorimeter. The temperature of the calorimeter rose from 25.00°C to 33.73°C. If the heat capacity of the calorimeter and contents was 9.63 kJ/°C, what is the value of q for burning 1.00 mol of ethanol at constant volume and 25.00°C? The reaction is

$$C_2H_5OH(l) + 3O_2(g) \longrightarrow 2CO_2(g) + 3H_2O(l)$$

Is q equal to ΔU or ΔH?

6.76 A sample of benzene, C_6H_6, weighing 3.51 g was burned in an excess of oxygen in a bomb calorimeter. The temperature of the calorimeter rose from 25.00°C to 37.18°C. If the heat capacity of the calorimeter and contents was 12.05 kJ/°C, what is the value of q for burning 1.00 mol of benzene at constant volume and 25.00°C? The reaction is

$$C_6H_6(l) + \tfrac{15}{2}O_2(g) \longrightarrow 6CO_2(g) + 3H_2O(l)$$

Is q equal to ΔU or ΔH?

Hess's Law

6.77 Hydrazine, N_2H_4, is a colorless liquid used as a rocket fuel. What is the enthalpy change for the process in which hydrazine is formed from its elements?

$$N_2(g) + 2H_2(g) \longrightarrow N_2H_4(l)$$

Use the following reactions and enthalpy changes:

$$N_2H_4(l) + O_2(g) \longrightarrow N_2(g) + 2H_2O(l);$$
$$\Delta H = -622.2 \text{ kJ}$$

$$H_2(g) + \tfrac{1}{2}O_2(g) \longrightarrow H_2O(l); \Delta H = -285.8 \text{ kJ}$$

6.78 Hydrogen peroxide, H_2O_2, is a colorless liquid whose solutions are used as a bleach and an antiseptic. H_2O_2 can be prepared in a process whose overall change is

$$H_2(g) + O_2(g) \longrightarrow H_2O_2(l)$$

Calculate the enthalpy change using the following data:

$$H_2O_2(l) \longrightarrow H_2O(l) + \tfrac{1}{2}O_2(g); \Delta H = -98.0 \text{ kJ}$$

$$2H_2(g) + O_2(g) \longrightarrow 2H_2O(l); \Delta H = -571.6 \text{ kJ}$$

6.79 Hydrogen cyanide is a highly poisonous, volatile liquid. It can be prepared by the reaction

$$CH_4(g) + NH_3(g) \longrightarrow HCN(g) + 3H_2(g)$$

What is the heat of reaction at constant pressure? Use the following thermochemical equations:

$$N_2(g) + 3H_2(g) \longrightarrow 2NH_3(g); \Delta H = -91.8 \text{ kJ}$$

$$C(\text{graphite}) + 2H_2(g) \longrightarrow CH_4(g); \Delta H = -74.9 \text{ kJ}$$

$$\tfrac{1}{2}H_2(g) + C(\text{graphite}) + \tfrac{1}{2}N_2(g) \longrightarrow HCN(g);$$
$$\Delta H = 135.1 \text{ kJ}$$

6.80 Ammonia will burn in the presence of a platinum catalyst to produce nitric oxide, NO.

$$4NH_3(g) + 5O_2(g) \longrightarrow 4NO(g) + 6H_2O(g)$$

What is the heat of reaction at constant pressure? Use the following thermochemical equations:

$$N_2(g) + O_2(g) \longrightarrow 2NO(g); \Delta H = 180.6 \text{ kJ}$$

$$N_2(g) + 3H_2(g) \longrightarrow 2NH_3(g); \Delta H = -91.8 \text{ kJ}$$

$$2H_2(g) + O_2(g) \longrightarrow 2H_2O(g); \Delta H = -483.7 \text{ kJ}$$

6.81 Compounds with carbon–carbon double bonds, such as ethylene, C_2H_4, add hydrogen in a reaction called hydrogenation.

$$C_2H_4(g) + H_2(g) \longrightarrow C_2H_6(g)$$

Calculate the enthalpy change for this reaction, using the following combustion data:

$$C_2H_4(g) + 3O_2(g) \longrightarrow 2CO_2(g) + 2H_2O(l);$$
$$\Delta H = -1411 \text{ kJ}$$

$$C_2H_6(g) + \frac{7}{2}O_2(g) \longrightarrow 2CO_2(g) + 3H_2O(l);$$
$$\Delta H = -1560 \text{ kJ}$$

$$H_2(g) + \frac{1}{2}O_2(g) \longrightarrow H_2O(l); \Delta H = -286 \text{ kJ}$$

6.82 Acetic acid, CH_3COOH, is contained in vinegar. Suppose acetic acid was formed from its elements, according to the following equation:

$$2C(\text{graphite}) + 2H_2(g) + O_2(g) \longrightarrow CH_3COOH(l)$$

Find the enthalpy change, ΔH, for this reaction, using the following data:

$$CH_3COOH(l) + 2O_2(g) \longrightarrow 2CO_2(g) + 2H_2O(l);$$
$$\Delta H = -874 \text{ kJ}$$

$$C(\text{graphite}) + O_2(g) \longrightarrow CO_2(g); \Delta H = -394 \text{ kJ}$$
$$H_2(g) + \frac{1}{2}O_2(g) \longrightarrow H_2O(l); \Delta H = -286 \text{ kJ}$$

Standard Enthalpies of Formation

6.83 Carbon tetrachloride, CCl_4, is a liquid used as an industrial solvent and in the preparation of fluorocarbons. What is the heat of vaporization of carbon tetrachloride?

$$CCl_4(l) \longrightarrow CCl_4(g); \Delta H° = ?$$

Use standard enthalpies of formation (Table 6.2).

6.84 The cooling effect of alcohol on the skin is due to its evaporation. Calculate the heat of vaporization of ethanol (ethyl alcohol), C_2H_5OH.

$$C_2H_5OH(l) \longrightarrow C_2H_5OH(g); \Delta H° = ?$$

The standard enthalpy of formation of $C_2H_5OH(l)$ is -277.7 kJ/mol and that of $C_2H_5OH(g)$ is -235.1 kJ/mol.

6.85 Hydrogen sulfide gas is a poisonous gas with the odor of rotten eggs. It occurs in natural gas and is produced during the decay of organic matter, which contains sulfur. The gas burns in oxygen as follows:

$$2H_2S(g) + 3O_2(g) \longrightarrow 2H_2O(l) + 2SO_2(g)$$

Calculate the standard enthalpy change for this reaction using standard enthalpies of formation.

6.86 Carbon disulfide is a colorless liquid. When pure, it is nearly odorless, but the commercial product smells vile.

Carbon disulfide is used in the manufacture of rayon and cellophane. The liquid burns as follows:

$$CS_2(l) + 3O_2(g) \longrightarrow CO_2(g) + 2SO_2(g)$$

Calculate the standard enthalpy change for this reaction using standard enthalpies of formation.

6.87 Iron is obtained from iron ore by reduction with carbon monoxide. The overall reaction is

$$Fe_2O_3(s) + 3CO(g) \longrightarrow 2Fe(s) + 3CO_2(g)$$

Calculate the standard enthalpy change for this equation. See Appendix C for data.

6.88 The first step in the preparation of lead from its ore (galena, PbS) consists of roasting the ore.

$$PbS(s) + \frac{3}{2}O_2(g) \longrightarrow SO_2(g) + PbO(s)$$

Calculate the standard enthalpy change for this reaction, using enthalpies of formation (see Appendix C).

6.89 Hydrogen chloride gas dissolves in water to form hydrochloric acid (an ionic solution).

$$HCl(g) \xrightarrow{H_2O} H^+(aq) + Cl^-(aq)$$

Find $\Delta H°$ for the above reaction. The data are given in Table 6.2.

6.90 Carbon dioxide from the atmosphere "weathers," or dissolves, limestone ($CaCO_3$) by the reaction

$$CaCO_3(s) + CO_2(g) + H_2O(l) \longrightarrow$$
$$Ca^{2+}(aq) + 2HCO_3^-(aq)$$

Obtain $\Delta H°$ for this reaction. See Table 6.2 for the data.

6.91 The Group IIA carbonates decompose when heated. For example,

$$BaCO_3(s) \longrightarrow BaO(s) + CO_2(g)$$

Use enthalpies of formation (see Appendix C) and calculate the heat required to decompose 15.0 g of barium carbonate.

6.92 The Group IIA carbonates decompose when heated. For example,

$$MgCO_3(s) \longrightarrow MgO(s) + CO_2(g)$$

Use enthalpies of formation (see Appendix C) and calculate the heat required to decompose 10.0 g of magnesium carbonate.

General Problems

Key: These problems provide more practice but are not divided by topic or keyed to exercises. Each section ends with essay questions, each of which is color coded to refer to the *A Chemist Looks at Daily Life* (orange) chapter essay on which it is based. Odd-numbered problems and the even-numbered problems that follow are similar; answers to all odd-numbered problems except the essay questions are given in the back of the book.

6.93 The potential energy of an object in the gravitational field of the earth is $E_p = mgh$. What must be the SI unit of g if this equation is to be consistent with the SI unit of energy for E_p?

6.94 The energy, E, needed to move an object a distance d by applying a force F is $E = F \times d$. What must be the SI unit of force if this equation is to be consistent with the SI unit of energy for E?

6.95 Liquid hydrogen peroxide has been used as a propellant for rockets. Hydrogen peroxide decomposes into oxygen and water, giving off heat energy equal to 686 Btu per pound of propellant. What is this energy in joules per gram of hydrogen peroxide? (1 Btu = 252 cal; see also Table 1.4.)

6.96 Hydrogen is an ideal fuel in many respects; for example, the product of its combustion, water, is nonpolluting. The heat given off in burning hydrogen to gaseous water is 5.16×10^4 Btu per pound. What is this heat energy in joules per gram? (1 Btu = 252 cal; see also Table 1.4.)

6.97 Niagara Falls has a height of 167 ft (American Falls). What is the potential energy in joules of 1.00 lb of water at the top of the falls if we take water at the bottom to have a potential energy of zero? What would be the speed of this water at the bottom of the falls if we neglect friction during the descent of the water?

6.98 Any object, be it a space satellite or a molecule, must attain an initial upward velocity of at least 11.2 km/s in order to escape the gravitational attraction of the earth. What would be the kinetic energy in joules of a satellite weighing 2458 lb that has the speed equal to this escape velocity of 11.2 km/s?

6.99 When calcium carbonate, $CaCO_3$ (the major constituent of limestone and seashells), is heated, it decomposes to calcium oxide (quicklime).

$$CaCO_3(s) \longrightarrow CaO(s) + CO_2(g); \Delta H = 177.9 \text{ kJ}$$

How much heat is required to decompose 21.3 g of calcium carbonate?

6.100 Calcium oxide (quicklime) reacts with water to produce calcium hydroxide (slaked lime).

$$CaO(s) + H_2O(l) \longrightarrow Ca(OH)_2(s); \Delta H = -65.2 \text{ kJ}$$

The heat released by this reaction is sufficient to ignite paper. How much heat is released when 28.6 g of calcium oxide reacts?

6.101 Acetic acid, $HC_2H_3O_2$, is the sour constituent of vinegar (*acetum* is Latin for "vinegar"). In an experiment, 3.58 g of acetic acid was burned.

$$HC_2H_3O_2(l) + 2O_2(g) \longrightarrow 2CO_2(g) + 2H_2O(l)$$

If 52.0 kJ of heat evolved, what is ΔH per mole of acetic acid?

6.102 Formic acid, $HCHO_2$, was first discovered in ants (*formica* is Latin for "ant"). In an experiment, 5.48 g of formic acid was burned at constant pressure.

$$2HCHO_2(l) + O_2(g) \longrightarrow 2CO_2(g) + 2H_2O(l)$$

If 30.3 kJ of heat evolved, what is ΔH per mole of formic acid?

6.103 Suppose you mix 21.0 g of water at 52.7°C with 54.9 g of water at 31.5°C in an insulated cup. What is the maximum temperature of the water after mixing?

6.104 Suppose you mix 23.6 g of water at 66.2°C with 45.4 g of water at 35.7°C in an insulated cup. What is the maximum temperature of the solution after mixing?

6.105 A piece of lead of mass 60.8 g was heated by an electrical coil. From the resistance of the coil, the current, and the time the current flowed, it was calculated that 235 J of heat was added to the lead. The temperature of the lead rose from 20.4°C to 35.5°C. What is the specific heat of the lead?

6.106 The specific heat of copper metal was determined by putting a piece of the metal weighing 35.4 g in hot water. The quantity of heat absorbed by the metal was calculated to be 47.0 J from the temperature drop of the water. What was the specific heat of the metal if the temperature of the metal rose 3.45°C?

6.107 A 50.0-g sample of water at 100.00°C was placed in an insulated cup. Then 25.3 g of zinc metal at 25.00°C was added to the water. The temperature of the water dropped to 96.68°C. What is the specific heat of zinc?

6.108 A 19.6-g sample of a metal was heated to 61.67°C. When the metal was placed into 26.7 g of water in a calorimeter, the temperature of the water increased from 25.00°C to 30.61°C. What is the specific heat of the metal?

6.109 A 29.1-mL sample of 1.05 *M* KOH is mixed with 20.9 mL of 1.07 *M* HBr in a coffee-cup calorimeter (see Section 6.6 of your text for a description of a coffee-cup calorimeter). The enthalpy of the reaction, written with the lowest whole-number coefficients, is −55.8 kJ. Both solutions are at 21.8°C prior to mixing and reacting. What is the final temperature of the reaction mixture? When solving this problem, assume that no heat is lost from the calorimeter to the surroundings, the density of all solutions is 1.00 g/mL, and volumes are additive.

6.110 A 14.1-mL sample of 0.996 *M* NaOH is mixed with 32.3 mL of 0.905 *M* HCl in a coffee-cup calorimeter (see Section 6.6 of your text for a description of a coffee-cup calorimeter). The enthalpy of the reaction, written with the lowest whole-number coefficients, is −55.8 kJ. Both solutions are at 21.6°C prior to mixing and reacting. What is the final temperature of the reaction mixture? When solving this problem, assume that no heat is lost from the calorimeter to the surroundings, the density of all solutions is 1.00 g/mL, the specific heat of all solutions is the same as that of water, and volumes are additive.

6.111 In a calorimetric experiment, 6.48 g of lithium hydroxide, LiOH, was dissolved in water. The temperature of the calorimeter rose from 25.00°C to 36.66°C. What is ΔH for the solution process?

$$LiOH(s) \longrightarrow Li^+(aq) + OH^-(aq)$$

The heat capacity of the calorimeter and its contents is 547 J/°C.

6.112 When 21.45 g of potassium nitrate, KNO_3, was dissolved in water in a calorimeter, the temperature fell from 25.00°C to 14.14°C. What is the ΔH for the solution process?

$$KNO_3(s) \longrightarrow K^+(aq) + NO_3^-(aq)$$

The heat capacity of the calorimeter and its contents is 682 J/°C.

6.113 A 10.00-g sample of acetic acid, $HC_2H_3O_2$, was burned in a bomb calorimeter in an excess of oxygen.

$$HC_2H_3O_2(l) + 2O_2(g) \longrightarrow 2CO_2(g) + 2H_2O(l)$$

The temperature of the calorimeter rose from 25.00°C to 35.84°C. If the heat capacity of the calorimeter and its contents is 13.43 kJ/°C, what is the enthalpy change for the reaction?

6.114 The sugar arabinose, $C_5H_{10}O_5$, is burned completely in oxygen in a calorimeter.

$$C_5H_{10}O_5(s) + 5O_2(g) \longrightarrow 5CO_2(g) + 5H_2O(l)$$

Burning a 0.548-g sample caused the temperature to rise from 20.00°C to 20.54°C. The heat capacity of the calorimeter and its contents is 15.8 kJ/°C. Calculate ΔH for the combustion reaction per mole of arabinose.

6.115 Ethylene glycol, $HOCH_2CH_2OH$, is used as antifreeze. It is produced from ethylene oxide, C_2H_4O, by the reaction

$$C_2H_4O(g) + H_2O(l) \longrightarrow HOCH_2CH_2OH(l)$$

Use Hess's law to obtain the enthalpy change for this reaction from the following enthalpy changes:

$$2C_2H_4O(g) + 5O_2(g) \longrightarrow 4CO_2(g) + 4H_2O(l);$$
$$\Delta H = -2612.2 \text{ kJ}$$

$$HOCH_2CH_2OH(l) + \tfrac{5}{2}O_2(g) \longrightarrow 2CO_2(g) + 3H_2O(l);$$
$$\Delta H = -1189.8 \text{ kJ}$$

6.116 Hydrogen sulfide, H_2S, is a poisonous gas with the odor of rotten eggs. The reaction for the formation of H_2S from the elements is

$$H_2(g) + \tfrac{1}{8}S_8(\text{rhombic}) \longrightarrow H_2S(g)$$

Use Hess's law to obtain the enthalpy change for this reaction from the following enthalpy changes:

$$H_2S(g) + \tfrac{3}{2}O_2(g) \longrightarrow H_2O(g) + SO_2(g); \Delta H = -518 \text{ kJ}$$
$$H_2(g) + \tfrac{1}{2}O_2(g) \longrightarrow H_2O(g); \Delta H = -242 \text{ kJ}$$
$$\tfrac{1}{8}S_8(\text{rhombic}) + O_2(g) \longrightarrow SO_2(g); \Delta H = -297 \text{ kJ}$$

6.117 Hydrogen, H_2, is prepared by *steam reforming,* in which hydrocarbons are reacted with steam. For CH_4,

$$CH_4(g) + H_2O(g) \longrightarrow CO(g) + 3H_2(g)$$

Calculate the enthalpy change $\Delta H°$ for this reaction, using standard enthalpies of formation.

6.118 Hydrogen is prepared from natural gas (mainly methane, CH_4) by partial oxidation.

$$CH_4(g) + \tfrac{1}{2}O_2(g) \longrightarrow CO(g) + 2H_2(g)$$

Calculate the enthalpy change $\Delta H°$ for this reaction, using standard enthalpies of formation.

6.119 Calcium oxide, CaO, is prepared by heating calcium carbonate (from limestone and seashells).

$$CaCO_3(s) \longrightarrow CaO(s) + CO_2(g)$$

Calculate the standard enthalpy of reaction, using enthalpies of formation. The $\Delta H_f°$ of CaO(s) is -635 kJ/mol. Other values are given in Table 6.2.

6.120 Sodium carbonate, Na_2CO_3, is used to manufacture glass. It is obtained from sodium hydrogen carbonate, $NaHCO_3$, by heating.

$$2NaHCO_3(s) \longrightarrow Na_2CO_3(s) + H_2O(g) + CO_2(g)$$

Calculate the standard enthalpy of reaction, using enthalpies of formation (Table 6.2).

6.121 Calculate the heat released when 5.000 L Cl_2 with a density of 2.46 g/L at 25°C reacts with an excess of sodium metal to form solid sodium chloride at 25°C.

6.122 Calculate the heat released when 2.000 L O_2 with a density of 1.11 g/L at 25°C reacts with an excess of hydrogen to form liquid water at 25°C.

6.123 Sucrose, $C_{12}H_{22}O_{11}$, is common table sugar. The enthalpy change at 25°C and 1 atm for the complete burning of 1 mol of sucrose in oxygen to give $CO_2(g)$ and $H_2O(l)$ is -5641 kJ. From this and from data given in Table 6.2, calculate the standard enthalpy of formation of sucrose.

6.124 Acetone, CH_3COCH_3, is a liquid solvent. The enthalpy change at 25°C and 1 atm for the complete burning of 1 mol of acetone in oxygen to give $CO_2(g)$ and $H_2O(l)$ is -1791 kJ. From this and from data given in Table 6.2, calculate the standard enthalpy of formation of acetone.

6.125 The thermite reaction is a very exothermic reaction; it has been used to produce liquid iron for welding. A mixture of 2 mol of powdered aluminum metal and 1 mol of iron(III) oxide yields liquid iron and solid aluminum oxide. How many grams of the mixture are needed to produce 348 kJ of heat? See Appendix C for data.

6.126 Ammonium nitrate is an oxidizing agent and can give rise to explosive mixtures. A mixture of 2.00 mol of powdered aluminum and 3.00 mol of ammonium nitrate crystals reacts exothermically, yielding nitrogen gas, water vapor, and aluminum oxide. How many grams of the mixture are required to provide 245 kJ of heat? See Appendix C for data.

■ **6.127** Describe the physical characteristics of white phosphorus. Is it found in any modern matches? Why or why not?

■ **6.128** What is the phosphorus compound used in "strike anywhere" matches. What is the chemical equation for the burning of this compound in air?

Strategy Problems

Key: As noted earlier, all of the practice and general problems are matched pairs. This section is a selection of problems that are not in matched-pair format. These challenging problems require that you employ many of the concepts and strategies that were developed in the chapter. In some cases, you will have to integrate several concepts and operational skills in order to solve the problem successfully.

6.129 The maximum mass allowed for a bowling ball in ten-pin bowling is 16 lb. The average speed of the ball when released during a game is between about 17 to 19 miles per hour (mph). Imagine that a 16.0-lb bowling ball is flung up an incline at 18.0 mph. The ball rolls up the incline, reaching a certain height h above where it started before rolling backward down the slope. What is the maximum height h (in meters) attained by the bowling ball? Neglect any loss of energy through friction with the incline or with air.

6.130 How fast (in meters per second) must a spherical iron ball with a diameter of 4.00 cm be traveling in order to have a kinetic energy of 94.7 J? The density of iron is 7.87 g/cm³.

6.131 An experimenter puts 3.00 mol H_2 gas and 1.00 mol O_2 gas, both at 1.00 atm and 25°C, in a cylinder with a piston (the setup is similar to that in Figure 6.9 but with the weight replaced by the weight of the atmosphere). Suppose the gases are then ignited so that they react to form liquid water, $H_2O(l)$. When the reaction is complete (at 1.00 atm and 25°C), what is the work done by an external atmosphere of 1.00 atm on this system. (You can ignore the volume of liquid water.) What is the net change of internal energy from this reaction? Note that 285.8 kJ of heat evolves when 1.00 mol liquid water forms from the gaseous elements at 1.00 atm and 25°C.

6.132 A researcher places an aqueous solution containing 16.8 g $NaHCO_3$ and another aqueous solution containing 18.2 g HCl into a stout container that she closes with a sturdy lid. The reaction that occurs is

$$NaHCO_3(aq) + HCl(aq) \longrightarrow NaCl(aq) + H_2O(l) + CO_2(g)$$

She determines that the reaction mixture evolves 3.04 kJ of heat. What is the work done by the system (the reaction mixture) for these amounts of reactants? What is the change of internal energy, ΔU, for this reaction per mole of $NaHCO_3$?

6.133 Potassium superoxide, KO_2, is used in some rebreathing gas masks (in which the exhaled breath is recycled in a closed space). Water vapor in the exhaled air reacts with potassium superoxide to produce oxygen gas:

$$4KO_2(s) + 2H_2O(l) \longrightarrow 4KOH(s) + 3O_2(g)$$

In an experiment at 25°C and 1.00 atm in which water vapor reacts with a quantity of potassium superoxide, it was found that 2.70 kJ of heat was absorbed. The oxygen gas produced was then collected at 23.0°C and 789 mmHg. What volume (in liters) of oxygen was collected?

6.134 Dry ice is solid carbon dioxide; it vaporizes at room temperature and normal pressures to the gas. Suppose you put 21.5 g of dry ice in a vessel fitted with a piston (similar to the one in Figure 6.9 but with the weight replaced by the atmosphere), and it vaporizes completely to the gas, pushing the piston upward until its pressure and temperature equal those of the surrounding atmosphere at 24.0°C and 751 mmHg. Calculate the work done by the gas in expanding against the atmosphere. Neglect the volume of the solid carbon dioxide, which is very small in comparison to the volume of the gas phase.

6.135 Sulfur forms a number of allotropes; two important ones are rhombic and monoclinic (see Figure 6.18). Suppose you start with a 10.0-g sample of rhombic sulfur at 25.0°C and 1.00 atm and heat the sample until it melts at 113°C. The liquid is then quickly cooled to 25.0°C and 1.000 atm at which it freezes to form crystalline needles of monoclinic sulfur. After a time, the monoclinic crystals form rhombic sulfur at 25.0°C and 1.000 atm. What is the total change of enthalpy, $\Delta H°$, for this process in going from rhombic sulfur to liquid sulfur to monoclinic sulfur and then to rhombic sulfur?

6.136 Sulfur dioxide gas reacts with oxygen, $O_2(g)$, to produce $SO_3(g)$. This reaction releases 99.0 kJ of heat (at constant pressure) for each mole of sulfur dioxide that reacts. Write the thermochemical equation for the reaction of 2 mol of sulfur dioxide, and then also for the decomposition of 3 mol of sulfur trioxide gas into oxygen gas and sulfur dioxide gas. Do you need any other information to answer either question?

6.137 When solid iron burns in oxygen gas (at constant pressure) to produce $Fe_2O_3(s)$, 1651 kJ of heat is released for every 4 mol of iron burned. How much heat is released when 10.3 g $Fe_2O_3(s)$ is produced (at constant pressure)? What additional information would you need to calculate the heat released to produce this much $Fe_2O_3(s)$ if you burned iron in ozone gas, $O_3(g)$, instead of $O_2(g)$?

6.138 Hydrogen is burned in oxygen to release heat (see equation below). How many grams of hydrogen gas must be burned to release enough heat to warm a 50.0-g block of iron from 21°C to 225°C?

$$2H_2(g) + O_2(g) \longrightarrow 2H_2O(g); \Delta H = -484 \text{ kJ}$$

Iron has a specific heat of 0.449 J/(g·°C).

6.139 Calculate the grams of oxygen gas required to produce 7.60 kJ of heat when hydrogen gas burns at constant pressure to give liquid water, given the following:

$$2H_2(g) + O_2(g) \longrightarrow 2H_2O(g); \Delta H = -484 \text{ kJ}$$

Liquid water has a heat of vaporization of 44.0 kJ per mole at 25°C.

6.140 Sodium metal reacts with water to produce hydrogen gas in an exothermic process.

$$2Na(s) + 2H_2O(l) \longrightarrow 2NaOH(aq) + H_2(g)$$

The reaction was carried out at 25.0°C and 760.0 mmHg, and the hydrogen gas generated was then collected over water at 23.0°C and 785 mmHg, yielding 23.1 L of gas. Calculate the heat evolved; use data from Appendix C.

6.141 A mixture of hydrogen gas and oxygen gas at 25.0°C and 1.00 atm with a total mass of 2.500 g was burned to yield liquid water and any excess reactant. If this reaction generated 28.6 kJ of heat, what was the composition of the initial gas mixture? Is there any ambiguity in the answer? If so, describe.

6.142 Water gas is a mixture of equal molar amounts of H_2 and CO gases produced by passing steam through hot coke (C):

$$H_2O(g) + C(s) \longrightarrow H_2(g) + CO(g)$$

Water gas was originally burned as a fuel gas, though it is now used to produce hydrogen. How much heat would you obtain by completely burning 1.00 L of water gas at 25.0°C and 1.00 atm? The products are liquid water and gaseous carbon dioxide.

6.143 A piece of iron was heated to 95.4°C and dropped into a constant-pressure calorimeter containing 284 g of water at 32.2°C. The final temperature of the water and iron was 51.9°C. Assuming that the calorimeter itself absorbs a negligible amount of heat, what was the mass (in grams) of the piece of iron? The specific heat of iron is 0.449 J/(g·°C), and the specific heat of water is 4.18 J/(g·°C).

6.144 You heat 1.000 quart of water from 25.0°C to its normal boiling point by burning a quantity of methane gas, CH_4. What volume of methane at 23.0°C and 745 mmHg would you require to heat this quantity of water, assuming that the methane is completely burned? The products are liquid water and gaseous carbon dioxide.

6.145 The enthalpy of combustion, ΔH, for benzoic acid, C_6H_5COOH, is −3226 kJ/mol. When a sample of benzoic acid was burned in a calorimeter (at constant pressure), the temperature of the calorimeter and contents rose from 23.44°C to 27.65°C. The heat capacity of the calorimeter and contents was 12.41 kJ/°C. What mass of benzoic acid was burned?

6.146 Given the following (hypothetical) thermochemical equations:

$$A + B \longrightarrow 2C; \Delta H = -447 \text{ kJ}$$
$$A + 3D \longrightarrow 2E; \Delta H = -484 \text{ kJ}$$
$$2D + B \longrightarrow 2F; \Delta H = -429 \text{ kJ}$$

Calculate ΔH, in kJ, for the equation

$$4E + 5B \longrightarrow 4C + 6F$$

6.147 The head of a "strike anywhere" match contains tetraphosphorus trisulfide, P_4S_3. In an experiment, a student burned this compound in an excess of oxygen and found that it evolved 3651 kJ of heat per mole of P_4S_3 at a constant pressure of 1 atm. She wrote the following thermochemical equation:

$$P_4S_3(s) + 8O_2(g) \longrightarrow P_4O_{10}(s) + 3SO_2(g);$$
$$\Delta H° = -3651 \text{ kJ}$$

Calculate the standard enthalpy of formation of P_4S_3, using this student's result and the following standard enthalpies of formation: $P_4O_{10}(s)$, -3009.9 kJ/mol; $SO_2(g)$, -296.8 kJ/mol. How does this value compare with the value given in Appendix C?

6.148 Toluene, $C_6H_5CH_3$, has an enthalpy of combustion of -3908 kJ/mol. Using data from Appendix C, calculate the enthalpy of formation of toluene.

Cumulative-Skills Problems

Key: The problems under this heading combine skills introduced in previous chapters with those given in the current one.

6.149 What will be the final temperature of a mixture made from 25.0 g of water at 15.0°C, 45.0 g of water at 50.0°C, and 15.0 g of water at 37.0°C?

6.150 What will be the final temperature of a mixture made from equal masses of the following: water at 25.0°C, ethanol at 35.5°C, and iron at 95°C?

6.151 Graphite is burned in oxygen to give carbon monoxide and carbon dioxide. If the product mixture is 33% CO and 67% CO_2 by mass, what is the heat from the combustion of 1.00 g of graphite?

6.152 A sample of natural gas is 80.0% CH_4 and 20.0% C_2H_6 by mass. What is the heat from the combustion of 1.00 g of this mixture? Assume the products are $CO_2(g)$ and $H_2O(l)$.

6.153 A 10.0-g sample of a mixture of CH_4 and C_2H_4 reacts with oxygen at 25°C and 1 atm to produce $CO_2(g)$ and $H_2O(l)$. If the reaction produces 520 kJ of heat, what is the mass percentage of CH_4 in the mixture?

6.154 A sample containing 2.00 mol of graphite reacts completely with a limited quantity of oxygen at 25°C and 1.0 atm pressure, producing 481 kJ of heat and a mixture of CO and CO_2. Calculate the masses of CO and CO_2 produced.

6.155 How much heat is released when a mixture containing 10.0 g NH_3 and 20.0 g O_2 reacts by the following equation?

$$4NH_3(g) + 5O_2(g) \longrightarrow 4NO(g) + 6H_2O(g);$$
$$\Delta H° = -906 \text{ kJ}$$

6.156 How much heat is released when a mixture containing 10.0 g CS_2 and 10.0 g Cl_2 reacts by the following equation?

$$CS_2(g) + 3Cl_2(g) \longrightarrow S_2Cl_2(g) + CCl_4(g);$$
$$\Delta H° = -230 \text{ kJ}$$

6.157 Consider the Haber process:

$$N_2(g) + 3H_2(g) \longrightarrow 2NH_3(g); \Delta H° = -91.8 \text{ kJ}$$

The density of ammonia at 25°C and 1.00 atm is 0.696 g/L. The density of nitrogen, N_2, is 1.145 g/L, and the molar heat capacity is 29.12 J/(mol·°C). (a) How much heat is evolved in the production of 1.00 L of ammonia at 25°C and 1.00 atm? (b) What percentage of this heat is required to heat the nitrogen required for this reaction (0.500 L) from 25°C to 400°C, the temperature at which the Haber process is run?

6.158 An industrial process for manufacturing sulfuric acid, H_2SO_4, uses hydrogen sulfide, H_2S, from the purification of natural gas. In the first step of this process, the hydrogen sulfide is burned to obtain sulfur dioxide, SO_2.

$$2H_2S(g) + 3O_2(g) \longrightarrow 2H_2O(l) + 2SO_2(g);$$
$$\Delta H° = -1124 \text{ kJ}$$

The density of sulfur dioxide at 25°C and 1.00 atm is 2.62 g/L, and the molar heat capacity is 30.2 J/(mol·°C). (a) How much heat would be evolved in producing 1.00 L of SO_2 at 25°C and 1.00 atm? (b) Suppose heat from this reaction is used to heat 1.00 L of the SO_2 from 25°C to 500°C for its use in the next step of the process. What percentage of the heat evolved is required for this?

6.159 A rebreathing gas mask contains potassium superoxide, KO_2, which reacts with moisture in the breath to give oxygen.

$$4KO_2(s) + 2H_2O(l) \longrightarrow 4KOH(s) + 3O_2(g)$$

Estimate the grams of potassium superoxide required to supply a person's oxygen needs for one hour. Assume a person requires 1.00×10^2 kcal of energy for this time period. Further assume that this energy can be equated to the heat of combustion of a quantity of glucose, $C_6H_{12}O_6$, to $CO_2(g)$ and $H_2O(l)$. From the amount of glucose required to give 1.00×10^2 kcal of heat, calculate the amount of oxygen consumed and hence the amount of KO_2 required. The $\Delta H_f°$ for glucose(s) is -1273 kJ/mol.

6.160 The carbon dioxide exhaled in the breath of astronauts is often removed from the spacecraft by reaction with lithium hydroxide.

$$2LiOH(s) + CO_2(g) \longrightarrow Li_2CO_3(s) + H_2O(l)$$

Estimate the grams of lithium hydroxide required per astronaut per day. Assume that each astronaut requires 2.50×10^3 kcal of energy per day. Further assume that this energy can be equated to the heat of combustion of a quantity of glucose, $C_6H_{12}O_6$, to $CO_2(g)$ and $H_2O(l)$. From the amount of glucose required to give 2.50×10^3 kcal of heat, calculate the amount of CO_2 produced and hence the amount of LiOH required. The $\Delta H_f°$ for glucose(s) is -1273 kJ/mol.

Quantum Theory of the Atom

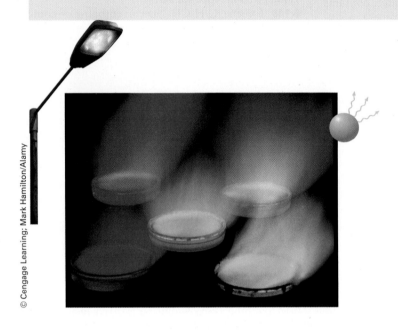

Colored flames from several metallic compounds; the visible light is emitted from the metal atoms. The yellow emission from sodium atoms is used in sodium street lamps.

Contents and Concepts

Light Waves, Photons, and the Bohr Theory

To understand the formation of chemical bonds, you need to know something about the electronic structure of atoms. Because light gives us information about this structure, we begin by discussing the nature of light. Then we look at the Bohr theory of the simplest atom, hydrogen.

Quantum Mechanics and Quantum Numbers

The Bohr theory firmly establishes the concept of energy levels but fails to account for the details of atomic structure. Here we discuss some basic notions of quantum mechanics, which is the theory currently applied to extremely small particles, such as electrons in atoms.

OWL Sign in to OWL at **www.cengage.com/owl** to view tutorials and simulations, develop problem-solving skills, and complete online homework assigned by your professor.

Chemistry Download mini lecture videos for key concept review and exam prep from OWL or purchase them from **www.cengagebrain.com**.

ccording to Rutherford's model (Section 2.2), an atom consists of a nucleus many times smaller than the atom itself, with electrons occupying the remaining space. How are the electrons distributed in this space? Or, we might ask, what are the electrons doing in the atom?

The answer was to come from an unexpected area: the study of colored flames. When metallic compounds burn in a flame, they emit bright colors (Figure 7.1). The spectacular colors of fireworks are due to the burning of metal compounds. Lithium and strontium compounds give a deep red color; barium compounds, a green color; and copper compounds, a bluish green color.

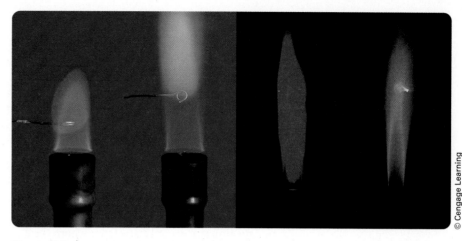

© Cengage Learning

Figure 7.1 ▲

Flame tests of Groups IA and IIA elements A wire loop containing a sample of a metallic compound is placed in a flame. *Left to right:* flames of lithium (red), sodium (yellow), strontium (red), and calcium (orange).

Although the red flames of lithium and strontium appear similar, the light from each can be resolved (separated) by means of a prism into distinctly different colors. This resolution easily distinguishes the two elements. A prism disperses the colors of white light just as small raindrops spread the colors of sunlight into a rainbow or spectrum. But the light from a flame, when passed through a prism, reveals something other than a rainbow. Instead of the continuous range of color from red to yellow to violet, the spectrum from a strontium flame, for example, shows a cluster of red lines and blue lines against a black background. The spectrum of lithium is different, showing a red line, a yellow line, and two blue lines against a black background. (See Figure 7.2.)

Each element, in fact, has a characteristic line spectrum because of the emission of light from atoms in the hot gas. The spectra can be used to identify elements. How is it that each atom emits particular colors of light? What does a line spectrum tell us about the structure of an atom? If you know something about the structures of atoms, can you explain the formation of ions and molecules? We will answer these questions in this and the next few chapters.

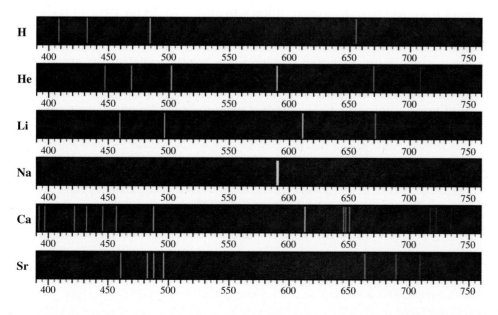

Figure 7.2 ◄

Emission (line) spectra of some elements The lines correspond to visible light emitted by atoms. (Wavelengths of lines are given in nanometers.)

Light Waves, Photons, and the Bohr Theory

In Chapter 2 we looked at the *basic structure* of atoms and we introduced the concept of a chemical bond. To understand the formation of a chemical bond between atoms, however, you need to know something about the *electronic structure* of atoms. The present theory of the electronic structure of atoms started with an explanation of the colored light produced in hot gases and flames. Before we can discuss this, we need to describe the nature of light.

7.1 The Wave Nature of Light

If you drop a stone into one end of a quiet pond, the impact of the stone with the water starts an up-and-down motion of the water surface. This up-and-down motion travels outward from where the stone hit; it is a familiar example of a wave. A *wave* is a continuously repeating change or oscillation in matter or in a physical field. Light is also a wave. It consists of oscillations in electric and magnetic fields that can travel through space. Visible light, x rays, and radio waves are all forms of *electromagnetic radiation*.

You characterize a wave by its wavelength and frequency. The **wavelength,** denoted by the Greek letter λ (lambda), is *the distance between any two adjacent identical points of a wave.* Thus, the wavelength is the distance between two adjacent peaks or troughs of a wave. Figure 7.3 shows a cross section of a water wave at a given moment, with the wavelength (λ) identified. Radio waves have wavelengths from approximately 100 mm to several hundred meters. Visible light has much shorter wavelengths, about 10^{-6} m. Wavelengths of visible light are often given in nanometers (1 nm = 10^{-9} m). For example, light of wavelength 5.55×10^{-7} m, the greenish yellow light to which the human eye is most sensitive, equals 555 nm.

The **frequency** of a wave is *the number of wavelengths of that wave that pass a fixed point in one unit of time* (usually one second). For example, imagine you are anchored in a small boat on a pond when a stone is dropped into the water. Waves travel outward from this point and move past your boat. The number of wavelengths that pass you in one second is the frequency of that wave. Frequency is denoted by the Greek letter ν (nu, pronounced "new"). The unit of frequency is /s, or s^{-1}, also called the *hertz* (Hz).

The wavelength and frequency of a wave are related to each other. Figure 7.4 shows two waves, each traveling from left to right at the same speed; that is, each wave moves the same total length in 1 s. The top wave, however, has a wavelength twice that of the bottom wave. In 1 s, two complete wavelengths of the top wave move left to right from the origin. It has a frequency of 2/s, or 2 Hz. In the same time, four complete wavelengths of the bottom wave move left to right from the origin. It has a frequency of 4/s, or 4 Hz. Note that for two waves traveling with a given speed, wavelength and frequency are inversely related: the greater the wavelength, the lower the frequency, and vice versa. In general, with a wave of frequency ν and wavelength λ, there are ν wavelengths, each of length λ, that pass a fixed point every second. The product $\nu\lambda$ is the total length of the wave that

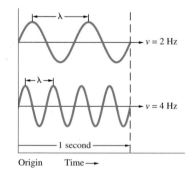

Figure 7.4 ▲

Relation between wavelength and frequency Both waves are traveling at the same speed. The top wave has a wavelength twice that of the bottom wave. The bottom wave, however, has twice the frequency of the top wave.

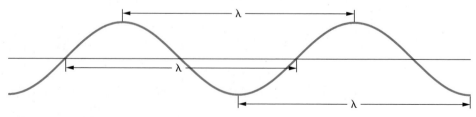

Figure 7.3 ▲

Water wave (ripple) The wavelength (λ) is the distance between any two adjacent identical points of a wave, such as two adjacent peaks or two adjacent troughs.

has passed the point in 1s. This length of wave per second is the speed of the wave. For light of speed c,

$$c = \nu\lambda$$

The speed of light waves in a vacuum is a constant and is independent of wavelength or frequency. This speed is 3.00×10^8 m/s, which is the value for c that we use in the following examples. ▶

To understand how fast the speed of light is, it might help to realize that it takes only 2.6 s for radar waves (which travel at the speed of light) to leave Earth, bounce off the moon, and return—a total distance of 478,000 miles.

Example 7.1

Gaining Mastery Toolbox

Critical Concept 7.1
A wave is characterized by its wavelength, λ, and its frequency, ν, which are related by the speed of the wave (which for light waves is denoted by the symbol c). You need to be able to manipulate this relationship to solve for wavelength in terms of frequency.

Solution Essentials:
• Equation relating frequency and wavelength: $c = \nu\lambda$

Obtaining the Wavelength of Light from Its Frequency

What is the wavelength of the yellow sodium emission, which has a frequency of 5.09×10^{14}/s?

Problem Strategy Note that frequency and wavelength are related (their product equals the speed of light). Write this as an equation, and then rearrange it to give the wavelength on the left and the frequency and speed of light on the right.

Solution The frequency and wavelength are related by the formula $c = \nu\lambda$. You rearrange this formula to give

$$\lambda = \frac{c}{\nu}$$

in which c is the speed of light (3.00×10^8 m/s). Substituting yields

$$\lambda = \frac{3.00 \times 10^8 \, \text{m/s}}{5.09 \times 10^{14}/\text{s}} = 5.89 \times 10^{-7} \text{m, or } \textbf{589 nm}$$

Answer Check Make sure you use the same units for c and ν (we used SI units in the solution). Check that the units cancel on the right to give the units of wavelength. As a check on your arithmetic, note that the wavelength for the answer should be in the visible range (400 nm to 750 nm).

Exercise 7.1 The frequency of the strong red line in the spectrum of potassium is 3.91×10^{14}/s. What is the wavelength of this light in nanometers?

■ See Problems 7.35 and 7.36.

♥WL InteractiveExample 7.2   Obtaining the Frequency of Light from Its Wavelength

Gaining Mastery Toolbox

Critical Concept 7.2
A wave is characterized by its wavelength, λ, and its frequency, ν, which are related by the speed of the wave (which for light waves is denoted by the symbol c). You need to be able to manipulate this relationship to solve for frequency in terms of wavelength.

Solution Essentials:
• Equation relating frequency and wavelength: $c = \nu\lambda$

What is the frequency of violet light with a wavelength of 408 nm?

Problem Strategy Rearrange the equation relating frequency and wavelength so that the frequency is on the left and the wavelength and speed of light are on the right.

Solution You rearrange the equation relating frequency and wavelength to give

$$\nu = \frac{c}{\lambda}$$

Substituting for λ (408 nm = 408×10^{-9} m) gives

$$\nu = \frac{3.00 \times 10^8 \, \text{m/s}}{408 \times 10^{-9} \, \text{m}} = \textbf{7.35} \times \textbf{10}^{\textbf{14}}\textbf{/s}$$

(continued)

(continued)

In 1854, Kirchhoff found that each element has a unique spectrum. Later, Bunsen and Kirchhoff developed the prism spectroscope and used it to confirm their discovery of two new elements, cesium (in 1860) and rubidium (in 1861).

Answer Check Use the same units for wavelength and speed of light; then check that the units cancel after substituting to give the units of frequency. We have used SI units here.

Exercise 7.2 The element cesium was discovered in 1860 by Robert Bunsen and Gustav Kirchhoff, who found two bright blue lines in the spectrum of a substance isolated from a mineral water. One of the spectral lines of cesium has a wavelength of 456 nm. What is its frequency? ◀

■ See Problems 7.37 and 7.38.

The range of frequencies or wavelengths of electromagnetic radiation is called the **electromagnetic spectrum,** shown in Figure 7.5. Visible light extends from the violet end of the spectrum, which has a wavelength of about 400 nm, to the red end, with a wavelength of less than 800 nm. Beyond these extremes, electromagnetic radiation is not visible to the human eye. Infrared radiation has wavelengths greater than 800 nm (greater than the wavelength of red light), and ultraviolet radiation has wavelengths less than 400 nm (less than the wavelength of violet light).

CONCEPT CHECK 7.1

Laser light of a specific frequency falls on a crystal that converts a portion of this light into light with double the original frequency. How is the wavelength of this frequency-doubled light related to the wavelength of the original laser light? Suppose the original laser light was red. In which region of the spectrum would the frequency-doubled light be? (If this is in the visible region, what color is the light?)

Figure 7.5 ▼

The electromagnetic spectrum Divisions between regions are not defined precisely.

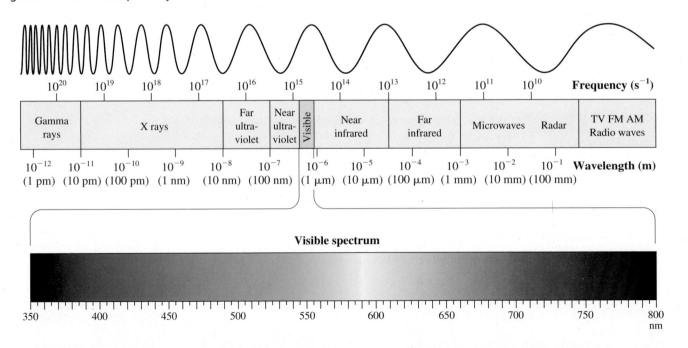

7.2 Quantum Effects and Photons

Isaac Newton, who studied the properties of light in the seventeenth century, believed that light consisted of a beam of particles. In 1801, however, British physicist Thomas Young showed that light, like waves, could be diffracted. *Diffraction* is a property of waves in which the waves spread out when they encounter an obstruction or small hole about the size of the wavelength. You can observe diffraction by viewing a light source through a hole—for example, a streetlight through a mesh curtain. The image of the streetlight is blurred by diffraction.

By the early part of the twentieth century, the wave theory of light appeared to be well entrenched. But in 1905 the German physicist Albert Einstein (1879–1955; emigrated to the United States in 1933) discovered that he could explain a phenomenon known as the *photoelectric effect* by postulating that light had both wave and particle properties. Einstein based this idea on the work of the German physicist Max Planck (1858–1947).

Planck's Quantization of Energy

In 1900 Max Planck found a theoretical formula that exactly describes the intensity of light of various frequencies emitted by a hot solid at different temperatures. ▶ Earlier, others had shown experimentally that the light of maximum intensity from a hot solid varies in a definite way with temperature. A solid glows red at 750°C, then white as the temperature increases to 1200°C. At the lower temperature, chiefly red light is emitted. As the temperature increases, more yellow and blue light become mixed with the red, giving white light. ▶

According to Planck, the atoms of the solid oscillate, or vibrate, with a definite frequency ν, depending on the solid. But in order to reproduce the results of experiments on glowing solids, he found it necessary to accept a strange idea. An atom could have only certain energies of vibration, E, those allowed by the formula

$$E = nh\nu, \qquad n = 1, 2, 3, \ldots$$

where h is a constant, now called **Planck's constant,** *a physical constant relating energy and frequency, having the value* 6.63×10^{-34} *J·s.* The value of n must be 1 or 2 or some other whole number. Thus, the only energies a vibrating atom can have are $h\nu$, $2h\nu$, $3h\nu$, and so forth.

The numbers symbolized by n are called *quantum numbers.* The vibrational energies of the atoms are said to be *quantized;* that is, the possible energies are limited to certain values.

The quantization of energy seems contradicted by everyday experience. Consider the potential energy of an object, such as a tennis ball. Its potential energy depends on its height above the surface of the earth: the greater the upward height, the greater the potential energy. (Recall the discussion of potential energy in Section 6.1.) We have no problem in placing the tennis ball at any height, so it can have any energy. Imagine, however, that you could only place the tennis ball on the steps of a stairway. In that case, you could only put the tennis ball on one of the steps, so the potential energy of the tennis ball could have only certain values; its energy would be quantized. Of course, this restriction of the tennis ball is artificial; in fact, a tennis ball can have a range of energies, not just particular values. As we will see, quantum effects depend on the mass of the object: the smaller the mass, the more likely you will see quantum effects. Atoms, and particularly electrons, have small enough masses to exhibit quantization of energy; tennis balls do not.

Photoelectric Effect

Planck himself was uneasy with the quantization assumption and tried unsuccessfully to eliminate it from his theory. Albert Einstein, on the other hand, boldly extended Planck's work to include the structure of light itself. Einstein reasoned that if a vibrating atom changed energy, say from $3h\nu$ to $2h\nu$, it would decrease in energy by $h\nu$, and this energy would be emitted as a bit (or quantum) of light

Max Planck was professor of physics at the University of Berlin when he did this research. He received the Nobel Prize in physics for it in 1918.

Radiation emitted from the human body and warm objects is mostly infrared, which is detected by burglar alarms, military night-vision scopes, and similar equipment.

Albert Einstein obtained his Ph.D. in 1905. In the same year he published five groundbreaking research papers: one on the photoelectric effect (for which he received the Nobel Prize in physics in 1921); two on special relativity; a paper on the determination of molecular size; and one on Brownian motion (which led to experiments that tested kinetic-molecular theory and ended remaining doubts about the existence of atoms and molecules).

energy. He therefore postulated that light consists of quanta (now called **photons**), or *particles of electromagnetic energy, with energy E proportional to the observed frequency of the light:* ◀

$$E = h\nu$$

In 1905 Einstein used this photon concept to explain the photoelectric effect.

The **photoelectric effect** is *the ejection of electrons from the surface of a metal or from another material when light shines on it* (see Figure 7.6). Electrons are ejected, however, only when the frequency of light exceeds a certain *threshold value* characteristic of the particular metal. For example, although violet light will cause potassium metal to eject electrons, no amount of red light (which has a lower frequency) has any effect.

To explain this dependence of the photoelectric effect on the frequency, Einstein assumed that an electron is ejected from a metal when it is struck by a single photon. Therefore, this photon must have at least enough energy to remove the electron from the attractive forces of the metal. No matter how many photons strike the metal, if no single one has sufficient energy, an electron cannot be ejected. A photon of red light has insufficient energy to remove an electron from potassium. But a photon corresponding to the threshold frequency has just enough energy, and at higher frequencies it has more than enough energy. When the photon hits the metal, its energy $h\nu$ is taken up by the electron. The photon ceases to exist as a particle; it is said to be *absorbed.*

The minimum energy that a photon can have and still eject an electron from a material in the photoelectric effect is characteristic of the material; this energy is called its photoelectric *work function.* The work function is frequently expressed in electron volts, eV. (An electron volt is the energy imparted to an electron when it is accelerated by an electric potential of one volt; 1 eV equals 1.602×10^{-19} J.) For example, the work function of potassium metal is 2.30 eV. If a photon has energy greater than the work function, the excess energy appears as kinetic energy of the ejected electron.

The wave and particle pictures of light should be regarded as complementary views of the same physical entity. This is called the *wave–particle duality* of light. The equation $E = h\nu$ displays this duality; E is the energy of a light particle or photon, and ν is the frequency of the associated wave. Neither the wave nor the particle view alone is a complete description of light.

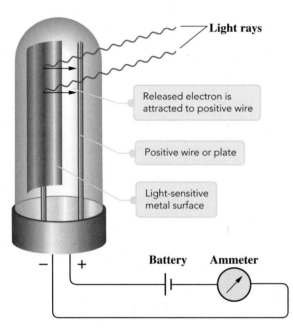

Light rays

Released electron is attracted to positive wire

Positive wire or plate

Light-sensitive metal surface

Battery   **Ammeter**

−   +

Figure 7.6 ▲

The photoelectric effect Light shines on a metal surface, knocking out electrons. The metal surface is contained in an evacuated tube, which allows the ejected electrons to be accelerated to a positively charged plate. As long as light of sufficient frequency shines on the metal, free electrons are produced and a current flows through the tube. (The current is measured by an ammeter.) When the light is turned off, the current stops flowing.

⭘WL **Interactive Example** 7.3   **Calculating the Energy of a Photon**

Gaining Mastery Toolbox

Critical Concept 7.3
The energy of a photon, E, is related to Planck's constant, h, times the frequency of the corresponding light wave, ν.

Solution Essentials:
• Equation relating the photon energy to frequency:
  $E = h\nu$
• Equation relating frequency and wavelength: $c = \nu\lambda$

The red spectral line of lithium occurs at 671 nm (6.71×10^{-7} m). Calculate the energy of one photon of this light.

Problem Strategy Note that the energy of a photon is related to its corresponding frequency. Therefore, you will first need to obtain this frequency from the wavelength of the spectral line.

Solution The frequency of this light is

$$\nu = \frac{c}{\lambda} = \frac{3.00 \times 10^8 \text{ m/s}}{6.71 \times 10^{-7} \text{ m}} = 4.47 \times 10^{14}/\text{s}$$

(continued)

Zapping Hamburger with Gamma Rays

In 1993 in Seattle, Washington, four children died and hundreds of people became sick from food poisoning when they ate undercooked hamburgers containing a dangerous strain of a normally harmless bacterium, *Escherichia coli*. A similar bout of food poisoning from hamburgers containing the dangerous strain of *E. coli* occurred in the summer of 1997; 17 people became ill. This time the tainted hamburger was traced to a meat processor in Nebraska, which immediately recalled 25 million pounds of ground beef. Within months, the U.S. Food and Drug Administration approved the irradiation of red meat by gamma rays to kill harmful bacteria. Gamma rays are a form of electromagnetic radiation similar to x rays, but the photons of gamma rays have higher energy.

The idea of using high-energy radiation to disinfect foods is not new (Figure 7.7). It has been known for about 50 years that high-energy radiation kills bacteria and molds in foods, prolonging their shelf life. Gamma rays kill organisms by breaking up the DNA molecules within their cells. This DNA holds the information for producing vital cell proteins.

Currently, the gamma rays used in food irradiation come from the radioactive decay of cobalt-60. It is perhaps this association with radioactivity that has slowed the acceptance of irradiation as a food disinfectant. However, food is not made radioactive by irradiating it with gamma rays. The two major concerns about

food irradiation—whether a food's nutritional value might be significantly diminished and whether decomposition products might cause cancer—appear to be without foundation. Irradiation of foods with gamma rays may soon join pasteurization as a way to protect our food supply from harmful bacteria.

Figure 7.7 ▲

Gamma irradiation of foods Here, fresh produce has been irradiated to reduce spoilage and to ensure that the produce does not contain harmful bacteria. Now, meat can also be irradiated.

■ See Problems 7.91 and 7.92.

(*continued*)

Hence, the energy of one photon is

$$E = h\nu = 6.63 \times 10^{-34}\,\text{J} \cdot \text{s} \times 4.47 \times 10^{14}/\text{s} = \mathbf{2.96 \times 10^{-19}\,J}$$

Answer Check Be sure to use the same system of units for the wavelength, the speed of light, and Planck's constant. Check that units cancel to give the proper units for energy in the final answer.

Exercise 7.3 The following are representative wavelengths in the infrared, ultraviolet, and x-ray regions of the electromagnetic spectrum, respectively: $1.0 \times 10^{-6}\,\text{m}$, $1.0 \times 10^{-8}\,\text{m}$, and $1.0 \times 10^{-10}\,\text{m}$. What is the energy of a photon of each radiation? Which has the greatest amount of energy per photon? Which has the least?

■ See Problems 7.43, 7.44, 7.45, and 7.46.

© AP Photo/Alan Richard

Figure 7.8 ▲

Niels Bohr (1885–1962) After Bohr developed his quantum theory of the hydrogen atom, he used his ideas to explain the periodic behavior of the elements. Later, when the new quantum mechanics was discovered by Schrödinger and Heisenberg, Bohr spent much of his time developing its philosophical basis. He received the Nobel Prize in physics in 1922.

7.3 The Bohr Theory of the Hydrogen Atom

According to Rutherford's nuclear model, the atom consists of a nucleus with most of the mass of the atom and a positive charge, around which move enough electrons to make the atom electrically neutral. But this model, offered in 1911, posed a dilemma. Using the then-current theory, one could show that an electrically charged particle (such as an electron) that revolves around a center would continuously lose energy as electromagnetic radiation. As an electron in an atom lost energy, it would spiral into the nucleus (in about 10^{-10} s, according to available theory). The stability of the atom could not be explained.

A solution to this theoretical dilemma was found in 1913 by Niels Bohr (1885–1962), a Danish physicist, who at the time was working with Rutherford (see Figure 7.8). Using the work of Planck and Einstein, Bohr applied a new theory to the simplest atom, hydrogen. Before we look at Bohr's theory, we need to consider the line spectra of atoms, which we looked at briefly in the chapter opening.

Atomic Line Spectra

As described in the previous section, a heated solid emits light. A heated tungsten filament in an ordinary lightbulb is a typical example. With a prism we can spread out the light from a bulb to give a **continuous spectrum**—that is, *a spectrum containing light of all wavelengths,* like that of a rainbow (see Figure 7.9). The light emitted by a heated gas, however, yields different results. Rather than a continuous spectrum, with all colors of the rainbow, we obtain a **line spectrum**—*a spectrum showing only certain colors or specific wavelengths of light.* When the light from a hydrogen gas discharge tube (in which an electrical discharge heats hydrogen gas) is separated into its components by a prism, it gives a spectrum of lines, each line corresponding to light of a given wavelength. The light produced in the discharge tube is emitted by hydrogen atoms. Figure 7.2 shows the line spectrum of the hydrogen atom, as well as line spectra of other atoms.

The line spectrum of the hydrogen atom is especially simple. In the visible region, it consists of only four lines (a red, a blue-green, a blue, and a violet), although others appear in the infrared and ultraviolet regions. In 1885 J. J. Balmer showed that the wavelengths λ in the visible spectrum of hydrogen could be reproduced by a simple formula:

$$\frac{1}{\lambda} = 1.097 \times 10^7/\text{m}\left(\frac{1}{2^2} - \frac{1}{n^2}\right)$$

Figure 7.9 ▶

Dispersion of white light by a prism White light, entering at the left, strikes a prism, which disperses the light into a continuous spectrum of wavelengths.

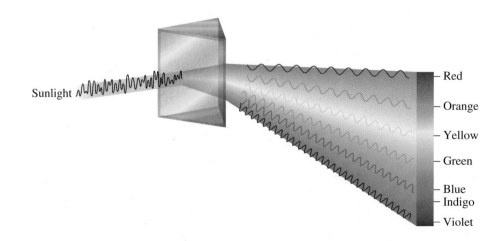

Sunlight

Red
Orange
Yellow
Green
Blue
Indigo
Violet

Here n is some whole number (integer) greater than 2. By substituting $n = 3$, for example, and calculating $1/\lambda$ and then λ, one finds $\lambda = 6.56 \times 10^{-7}$ m, or 656 nm, a wavelength corresponding to red light. The wavelengths of the other lines in the hydrogen atom visible spectrum are obtained by successively substituting $n = 4$, $n = 5$, and $n = 6$.

Bohr's Postulates

Bohr set down the following postulates to account for (1) the stability of the hydrogen atom (that the atom exists and its electron does not continuously radiate energy and spiral into the nucleus) and (2) the line spectrum of the atom.

1. Energy-level Postulate An electron can have only *specific energy values in an atom,* which are called its **energy levels.** Therefore, the atom itself can have only specific total energy values.

Bohr borrowed the idea of quantization of energy from Planck. Bohr, however, devised a rule for this quantization that could be applied to the motion of an electron in an atom. From this he derived the following formula for the energy levels of the electron in the hydrogen atom:

$$E = -\frac{R_H}{n^2} \quad n = 1, 2, 3, \ldots \infty \quad \text{(for H atom)}$$

The energies have negative values because the energy of the separated nucleus and electron is taken to be zero. As the nucleus and electron come together to form a stable state of the atom, energy is released and the energy becomes less than zero, or negative.

where R_H is the *Rydberg constant* (expressed in energy units) with the value 2.179×10^{-18} J. Different values of the possible energies of the electron are obtained by putting in different values of n, which can have only the integral values 1, 2, 3, and so forth (up to infinity, ∞). Here n is called the *principal quantum number.* The diagram in Figure 7.10 shows the energy levels of the electron in the H atom. ▶

2. Transitions Between Energy Levels An electron in an atom can change energy only by going from one energy level to another energy level. By so doing, the electron undergoes a *transition.*

By using only these two postulates, Bohr was able to explain Balmer's formula for the wavelengths in the spectrum of the hydrogen atom. According to Bohr, the emission of light from an atom occurs as follows. An electron in a higher energy level (initial energy level, E_i) undergoes a transition to a lower energy level (final energy level, E_f) (see Figure 7.10). In this process, the electron loses energy, which is emitted as a photon. Balmer's formula follows by this reasoning: The energy lost by the hydrogen atom is $\Delta E = E_f - E_i$. If we write n_i for the principal quantum number of the initial energy level, and n_f for the principal quantum number of the final energy level, then from Postulate 1,

$$E_i = -\frac{R_H}{n_i^2} \quad \text{and} \quad E_f = -\frac{R_H}{n_f^2}$$

so

$$\Delta E = \left(-\frac{R_H}{n_f^2}\right) - \left(-\frac{R_H}{n_i^2}\right) = -R_H\left(\frac{1}{n_f^2} - \frac{1}{n_i^2}\right)$$

For example, if the electron undergoes a transition from $n_i = 4$ to $n_f = 2$,

$$\Delta E = -2.179 \times 10^{-18} \text{ J} \left(\frac{1}{2^2} - \frac{1}{4^2}\right) = -4.086 \times 10^{-19} \text{ J}$$

The sign of the energy change is negative. This means that energy equal to 4.086×10^{-19} J is lost by the atom in the form of a photon.

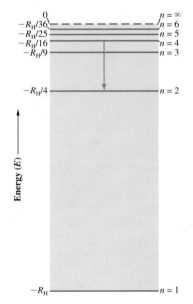

Figure 7.10 ▲

Energy-level diagram for the electron in the hydrogen atom Energy is plotted on the vertical axis (in fractional multiples of R_H). The arrow represents an electron transition (discussed in Postulate 2) from level $n = 4$ to level $n = 2$. Light of wavelength 486 nm (blue-green) is emitted. (See Example 7.4 for the calculation of this wavelength.)

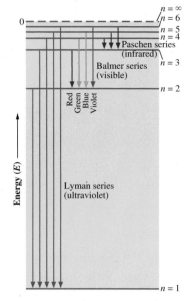

Figure 7.11 ▲

Transitions of the electron in the hydrogen atom The diagram shows the Lyman, Balmer, and Paschen series of transitions that occur for n_f = 1, 2, and 3, respectively.

In general, the energy of the emitted photon, $h\nu$, equals the positive energy lost by the atom ($-\Delta E$):

$$\text{Energy of emitted photon} = h\nu = -\Delta E = -(E_f - E_i)$$

That is,

$$h\nu = R_H\left(\frac{1}{n_f^2} - \frac{1}{n_i^2}\right)$$

Recalling that $\nu = c/\lambda$, you can rewrite this as

$$\frac{1}{\lambda} = \frac{R_H}{hc}\left(\frac{1}{n_f^2} - \frac{1}{n_i^2}\right)$$

By substituting $R_H = 2.179 \times 10^{-18}$ J, $h = 6.626 \times 10^{-34}$ J·s, and $c = 2.998 \times 10^8$ m/s, you find that $R_H/hc = 1.097 \times 10^7$/m, which is the constant given in the Balmer formula. In Balmer's formula, the quantum number n_f is 2. This means that Balmer's formula gives wavelengths that occur when electrons in H atoms undergo transitions from energy levels $n_i > 2$ to level $n_f = 2$. If you change n_f to other integers, you obtain different series of lines (or wavelengths) for the spectrum of the H atom (see Figure 7.11).

Example **7.4**

Determining the Wavelength or Frequency of a Hydrogen Atom Transition

Gaining Mastery Toolbox

Critical Concept 7.4
The energy levels of the hydrogen atom, E, are related to the inverse of the square of the principal quantum number, n. The wavelength or frequency of a transition depends on the difference in energies of the levels involved.

Solution Essentials:
• Energy levels of the H atom: $E = -R_H/n^2$
• Equation relating frequency and wavelength: $c = \nu\lambda$

What is the wavelength of light emitted when the electron in a hydrogen atom undergoes a transition from energy level $n = 4$ to level $n = 2$?

Problem Strategy Remember that the wavelength or frequency of a transition depends on the difference in energies of the levels involved. For an H atom, the energy levels are $E = -R_H/n^2$. You calculate the difference in energy for the two levels *to obtain the energy of the photon.* Then, calculate the frequency and wavelength of the emitted light. (Although you could do this problem by "plugging into" the previously derived equation for $1/\lambda$, the method followed here requires you to remember only a minimum number of key formulas.)

Solution From the formula for the energy levels, you know that

$$E_i = \frac{-R_H}{4^2} = \frac{-R_H}{16} \quad \text{and} \quad E_f = \frac{-R_H}{2^2} = \frac{-R_H}{4}$$

You subtract the lower value from the higher value, to get a positive result. (The energy of the photon is positive. If your result is negative, it means you calculated $E_f - E_i$, rather than $E_i - E_f$. Simply reverse the subtraction to obtain a positive result.) Because this result equals the energy of the photon, you equate it to $h\nu$:

$$\left(\frac{-R_H}{16}\right) - \left(\frac{-R_H}{4}\right) = \frac{-4R_H + 16R_H}{64} = \frac{-R_H + 4R_H}{16} = \frac{3R_H}{16} = h\nu$$

The frequency of the light emitted is

$$\nu = \frac{3R_H}{16h} = \frac{3}{16} \times \frac{2.179 \times 10^{-18} \text{ J}}{6.626 \times 10^{-34} \text{ J·s}} = 6.17 \times 10^{14}/\text{s}$$

(continued)

(continued)

Since $\lambda = c/\nu$,

$$\lambda = \frac{3.00 \times 10^8 \text{ m/s}}{6.17 \times 10^{14}/\text{s}} = 4.86 \times 10^{-7} \text{ m, or } \textbf{486 nm}$$

The color is blue-green (see Figure 7.10).

Answer Check Balmer lines (lines ending with $n = 2$) are in the visible region of the spectrum. Wavelengths of visible light are in the range of 400 nm to about 800 nm. Note that the answer lies in this range.

Exercise 7.4 Calculate the wavelength of light emitted from the hydrogen atom when the electron undergoes a transition from level $n = 3$ to level $n = 1$.

■ See Problems 7.49, 7.50, 7.51, and 7.52.

According to Bohr's theory, the *emission* of light from an atom occurs when an electron undergoes a transition from an upper energy level to a lower one. To complete the explanation, we need to describe how the electron gets into the upper level prior to emission. Normally, the electron in a hydrogen atom exists in its lowest, or $n = 1$, level. To get into a higher energy level, the electron must gain energy, or be *excited*. One way this can happen is through the collision of two hydrogen atoms. During this collision, some of the kinetic energy of one atom can be gained by the electron of another atom, thereby boosting, or exciting, the electron from the $n = 1$ level to a higher energy level. The excitation of atoms and the subsequent emission of light are most likely to occur in a hot gas, where atoms have large kinetic energies.

Bohr's theory explains not only the emission but also the *absorption* of light. When an electron in the hydrogen atom undergoes a transition from $n = 3$ to $n = 2$, a photon of red light (wavelength 656 nm) is emitted. When red light of wavelength 656 nm shines on a hydrogen atom in the $n = 2$ level, a photon can be absorbed. If the photon is absorbed, the energy is gained by the electron, which undergoes a transition to the $n = 3$ level. (This is the reverse of the emission process we just discussed.) Materials that have a color, such as dyed textiles and painted walls, appear colored because of the absorption of light. For example, when white light falls on a substance that absorbs red light, the color components that are not absorbed, the yellow and blue light, are reflected. The substance appears blue-green.

Postulates 1 and 2 hold for atoms other than hydrogen, except that the energy levels cannot be obtained by a simple formula. However, if you know the wavelength of the emitted light, you can relate it to ν and then to the difference in energy levels of the atom. The energy levels of atoms have been experimentally determined in this way.

Exercise 7.5 What is the difference in energy levels of the sodium atom if emitted light has a wavelength of 589 nm?

■ See Problems 7.55 and 7.56.

CONCEPT CHECK 7.2

An atom has a line spectrum consisting of a red line and a blue line. Assume that each line corresponds to a transition between two adjacent energy levels. Sketch an energy-level diagram with three energy levels that might explain this line spectrum, indicating the transitions on this diagram. Consider the transition from the highest energy level on this diagram to the lowest energy level. How would you describe the color or region of the spectrum corresponding to this transition?

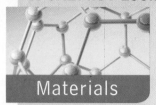
Lasers and CD and DVD Players

Lasers are sources of intense, highly directed beams of *monochromatic* light—light of very narrow wavelength range. The word *laser* is an acronym meaning **l**ight **a**mplification by **s**timulated **e**mission of **r**adiation. Many different kinds of lasers now exist, but the general principle of a laser can be understood by looking at the ruby laser, the first type constructed (in 1960).

Ruby is aluminum oxide containing a small concentration of chromium(III) ions, Cr^{3+}, in place of some aluminum ions. The electron transitions in a ruby laser are those of Cr^{3+} ions. Figure 7.12 shows an energy-level diagram of this ion in ruby. When you shine light of wavelength 545 nm on a ruby crystal, the light is absorbed and Cr^{3+} undergoes a transition from level 1 to level 3. Most of the ions in level 3 then undergo *radiationless transitions* to level 2. (In these transitions, the ions lose energy as heat to the crystal, rather than by emitting photons.) When an intense green light at 545 nm is flashed on a ruby crystal, Cr^{3+} ions of the ruby end up in level 2. Like all excited states, level 2 spontaneously emits photons, going to the ground state. But whereas most excited states quickly (within 10^{-8} s) emit photons, level 2 has a much longer lifetime (a fraction of a millisecond), resulting in a buildup of excited Cr^{3+} ions in this level. If these accumulated excited ions can be triggered to emit all at once, or nearly so, an intense emission of monochromatic light at 694 nm will occur.

The process of *stimulated emission* is ideal for this triggering. When a photon corresponding to 694 nm encounters a Cr^{3+} ion in level 2, it stimulates the ion to undergo the transition from level 2 to level 1. The ion emits a photon corresponding to exactly the same wavelength as the original photon. In place of just one photon, there are now two photons, the original one and the one obtained by stimulated emission. The net effect is to increase the intensity of the light at this wavelength. Thus, a weak light at 694 nm can be amplified by stimulated emission of the excited ruby.

Figure 7.13 shows a sketch of a ruby laser. A flash lamp emits green light (at 545 nm) that is absorbed by the ruby, building up a concentration of Cr^{3+} in level 2. A few of these excited ions spontaneously emit photons corresponding to 694 nm (red light), and these photons then stimulate other ions to emit, which in turn stimulate more ions to emit, and so forth, resulting in a pulse of laser light at 694 nm.

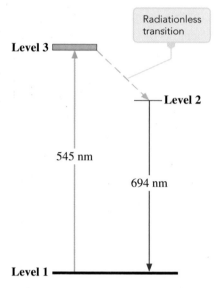

Figure 7.12 ▲

Energy levels of the chromium(III) ion in ruby The energy levels have been numbered by increasing energy from 1 to 3. (Level 3 is broadened in the solid state.) When chromium(III) ions absorb the 545-nm portion of the light from a flash lamp, electrons in the ions first undergo transitions to level 3 and then radiationless transitions to level 2. (The energy is lost as heat.) Electrons accumulate in level 2 until more exist in level 2 than in level 1. Then a photon from a spontaneous emission from level 2 to level 1 can stimulate the emission from other atoms in level 2. These photons move back and forth between reflective surfaces, stimulating additional emissions and forming a laser pulse at 694 nm.

Laser light is *coherent*. This means that the waves forming the beam are all in phase; that is, the waves have their maxima and minima at the same points in space and time. The property of coherence of a laser beam is used in CDs (compact discs) and DVDs. Music, video, or other information is encoded on the discs in the form of pits, or indentations, on a spiral track, starting at the center of the disc. Figure 7.14 shows how a CD works; DVDs work similarly but with shorter wavelengths of laser light.

When the disc is played, a small laser beam scans the track and is reflected back to a detector. Light reflected from an indentation is out of phase with light from the laser and interferes with it. Because of this interference, the reflected beam is diminished in intensity and gives a diminished detector signal. Fluctuations in the signal are then converted to sound, video, or other information. The difference between CDs and DVDs is in the wavelength of laser light used. A shorter wavelength allows the indentations on a disc to be smaller so that more information can be placed on a disc of the same size. A CD disc uses laser light at 780 nm (in the near infrared), whereas a DVD disc uses light at 650 nm (red). A Blu-ray disc uses laser light at 405 nm, which is in the violet (although commonly called a "blue" laser).

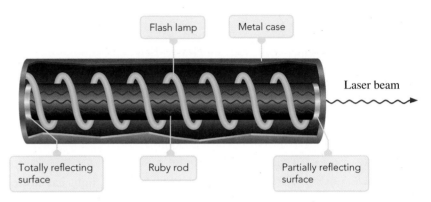

Figure 7.13 ▲

A ruby laser A flash lamp encircles a ruby rod. Light of 545 nm (green) from the flash "pumps" electrons from level 1 to level 3, then level 2, from which stimulated emission forms a laser pulse at 694 nm (red). The stimulated emission bounces back and forth between reflective surfaces at the ends of the ruby rod, building up a coherent laser beam. One end has a partially reflective surface to allow the laser beam to exit from the ruby.

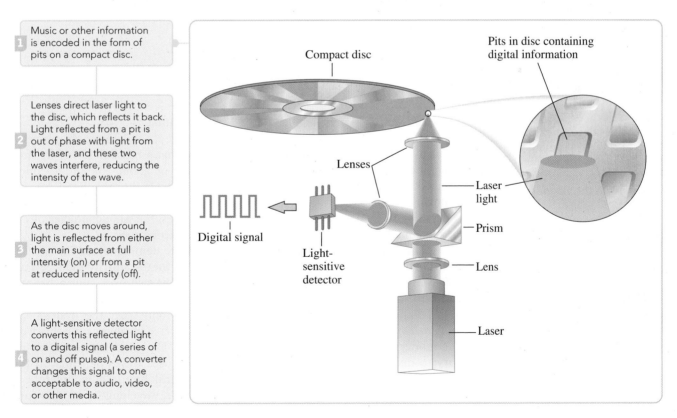

1. Music or other information is encoded in the form of pits on a compact disc.

2. Lenses direct laser light to the disc, which reflects it back. Light reflected from a pit is out of phase with light from the laser, and these two waves interfere, reducing the intensity of the wave.

3. As the disc moves around, light is reflected from either the main surface at full intensity (on) or from a pit at reduced intensity (off).

4. A light-sensitive detector converts this reflected light to a digital signal (a series of on and off pulses). A converter changes this signal to one acceptable to audio, video, or other media.

Figure 7.14 ▲

How a compact disc player works

■ See Problems 7.93 and 7.94.

Quantum Mechanics and Quantum Numbers

Bohr's theory firmly established the concept of atomic energy levels. It was unsuccessful, however, in accounting for the details of atomic structure and in predicting energy levels for atoms other than hydrogen. Further understanding of atomic structure required other theoretical developments.

7.4 Quantum Mechanics

Current ideas about atomic structure depend on the principles of *quantum mechanics,* a theory that applies to submicroscopic (that is, extremely small) particles of matter, such as electrons. The development of this theory was stimulated by the discovery of the de Broglie relation.

de Broglie Relation

A photon has a rest mass of zero but a relativistic mass m as a result of its motion. Einstein's equation $E = mc^2$ relates this relativistic mass to the energy of the photon, which also equals hν. Therefore, $mc^2 = h\nu$, or $mc = h\nu/c = h/\lambda$.

According to Einstein, light has not only wave properties, which we characterize by frequency and wavelength, but also particle properties. For example, a particle of light, the photon, has a definite energy $E = h\nu$. One can also show that the photon has momentum. (The momentum of a particle is the product of its mass and speed.) This momentum, mc, is related to the wavelength of the light: $mc = h/\lambda$ or $\lambda = h/mc$. ◄

In 1923 the French physicist Louis de Broglie reasoned that if light (considered as a wave) exhibits particle aspects, then perhaps particles of matter show characteristics of waves under the proper circumstances. He therefore postulated that a particle of matter of mass m and speed v has an associated wavelength, by analogy with light:

$$\lambda = \frac{h}{mv}$$

Figure 7.15 ▲

Scanning electron microscope This microscope can resolve details down to 3 nm in a sample. The operator places the sample inside the chamber at the left and views the image on the video screens.

Courtesy of JEOL USA

The equation $\lambda = h/mv$ is called the **de Broglie relation.**

If matter has wave properties, why are they not commonly observed? Calculation using the de Broglie relation shows that a baseball (0.145 kg) moving at about 60 mi/hr (27 m/s) has a wavelength of about 10^{-34} m, a value so incredibly small that such waves cannot be detected. On the other hand, electrons moving at only moderate speeds have wavelengths of a few hundred picometers (1 pm = 10^{-12} m). Under the proper circumstances, the wave character of electrons and even molecules (such as C_{60}) have been observed.

The wave property of electrons was first demonstrated in 1927 by C. Davisson and L. H. Germer in the United States and by George Paget Thomson (son of J. J. Thomson) in Britain. They showed that a beam of electrons, just like x rays, can be diffracted by a crystal. The German physicist Ernst Ruska used this wave property to construct the first *electron microscope* in 1933; he shared the 1986 Nobel Prize in physics for this work. A modern instrument is shown in Figure 7.15. The *resolving power,* or ability to distinguish detail, of a microscope that uses waves depends on their wavelength. To resolve detail the size of several hundred picometers, we need a wavelength on that order. X rays have wavelengths in this range but are difficult to focus. Electrons, on the other hand, are readily focused with electric and magnetic fields. Figure 7.16 shows a photograph taken with an electron microscope, displaying the detail that is possible with this instrument.

Figure 7.16 ▲

Scanning electron microscope image This image is of a wasp's head. Color has been added by computer for contrast in discerning different parts of the image.

© Biophoto Associates/Photo Researchers, Inc.

Example 7.5 **Calculate the Wavelength of a Moving Particle**

Gaining Mastery Toolbox

Critical Concept 7.5
A particle has an associated wavelength, λ, that depends inversely on the mass, m, and speed of the particle, v.

Solution Essentials:
- DeBroglie relation: $\lambda = h/mv$

a Calculate the wavelength (in meters) of the wave associated with a 1.00-kg mass moving at 1.00 km/hr. b What is the wavelength (in picometers) associated with an electron, whose mass is 9.11×10^{-31} kg, traveling at a speed of 4.19×10^6 m/s? (This speed can be attained by an electron accelerated between two charged plates differing by 50.0 volts; voltages in the kilovolt range are used in electron microscopes.)

Problem Strategy The questions ask for the wavelength associated with a moving mass, which is given by de Broglie's relation, $\lambda = h/mv$. Be careful to use consistent units in substituting into this equation. If you use SI units, then in part a you will need to convert km/hr to m/s. In part b, you will need to convert the answer in meters to picometers.

Solution

a A speed v of 1.00 km/hr equals

$$1.00 \frac{\text{km}}{\text{hr}} \times \frac{1\text{hr}}{3600\text{ s}} \times \frac{10^3\text{ m}}{1\text{ km}} = 0.278\text{ m/s}$$

Substituting quantities (all expressed in SI units for consistency), you get

$$\lambda = \frac{h}{mv} = \frac{6.63 \times 10^{-34}\text{ J}\cdot\text{s}}{1.00\text{ kg} \times 0.278\text{ m/s}} = \mathbf{2.38 \times 10^{-33}\ m}$$

b $$\lambda = \frac{h}{mv} = \frac{6.63 \times 10^{-34}\text{ J}\cdot\text{s}}{9.11 \times 10^{-31}\text{ kg} \times 4.19 \times 10^{6}\text{ m/s}} = 1.74 \times 10^{-10}\text{ m} = \mathbf{174\ pm}$$

Answer Check Make sure that you are using the correct equation for the de Broglie relation; units should cancel to give units of length (for wavelength). Check that the final units are those asked for in the problem statement.

Exercise 7.6 Calculate the wavelength (in picometers) associated with an electron traveling at a speed of 2.19×10^6 m/s.

See Problems 7.57 and 7.58.

CONCEPT CHECK 7.3

A proton is approximately 2000 times heavier than an electron. How would the speeds of these particles compare if their corresponding wavelengths were about equal?

Wave Functions

De Broglie's relation applies quantitatively only to particles in a force-free environment. It cannot be applied directly to an electron in an atom, where the electron is subject to the attractive force of the nucleus. But in 1926 Erwin Schrödinger, guided by de Broglie's work, devised a theory that could be used to find the wave properties of electrons in atoms and molecules. *The branch of physics that mathematically describes the wave properties of submicroscopic particles* is called **quantum mechanics** or **wave mechanics.** ▶

Schrödinger received the Nobel Prize in physics in 1933 for his wave formulation of quantum mechanics. Werner Heisenberg won the Nobel Prize the previous year for his matrix-algebra formulation of quantum mechanics. The two formulations yield identical results.

Without going into the mathematics of quantum mechanics here, we will discuss some of the most important conclusions of the theory. In particular, quantum mechanics alters the way we think about the motion of particles.

Our usual concept of motion comes from what we see in the everyday world. We might, for instance, visually follow a ball that has been thrown. The path of the ball is given by its position and velocity (or momentum) at various times. We are therefore conditioned to think in terms of a continuous path for moving objects. In Bohr's theory, the electron was thought of as moving about, or orbiting, the nucleus in the way the earth orbits the sun. Quantum mechanics vastly changes this view of motion. We can no longer think of an electron as having a precise orbit in an atom. To describe such an orbit, we would have to know the exact position of the electron at various times and exactly how long it would take it to travel to a nearby position in the orbit. That is, at any moment we would have to know not only the precise position but also the precise momentum (mass times speed) of the electron.

In 1927 Werner Heisenberg showed from quantum mechanics that it is impossible to know simultaneously, with absolute precision, both the position and the momentum of a particle such as an electron. Heisenberg's **uncertainty principle** is *a relation that states that the product of the uncertainty in position and the uncertainty in momentum of a particle can be no smaller than Planck's constant divided by 4π.* Thus, letting Δx be the uncertainty in the x coordinate of the particle and letting Δp_x be the uncertainty in the momentum in the x direction, we have

$$(\Delta x)(\Delta p_x) \geq \frac{h}{4\pi}$$

There are similar relations in the y and z directions. The uncertainty principle says that the more precisely you know the position (the smaller Δx, Δy, and Δz), the less well you know the momentum of the particle (the larger Δp_x, Δp_y, and Δp_z). In other words, if you know very well where a particle is, you cannot know where it is going!

The uncertainty principle is significant only for particles of very small mass such as electrons. You can see this by noting that the momentum equals mass times velocity, so $p_x = m v_x$. The preceding relation becomes

$$(\Delta x)(\Delta v_x) \geq \frac{h}{4\pi m}$$

For dust particles and baseballs, where m is relatively large, the term on the right becomes nearly zero, and the uncertainties of position and velocity are quite small. The path of a baseball has meaning. For electrons, however, the uncertainties in position and momentum are normally quite large relative to atomic scale dimensions. We cannot describe the electron in an atom as moving in a definite orbit.

Although quantum mechanics does not allow us to describe the electron in the hydrogen atom as moving in an orbit, it does allow us to make *statistical* statements about where we would find the electron if we were to look for it. For example, we can obtain the *probability* of finding an electron at a certain point in a hydrogen atom. Although we cannot say that an electron will definitely be at a particular position at a given time, we can say that the electron is likely (or not likely) to be at this position.

Information about a particle in a given energy level (such as an electron in an atom) is contained in a mathematical expression called a *wave function*, denoted by the Greek letter psi, ψ. The wave function is obtained by solving an equation of quantum mechanics (Schrödinger's equation). Its square, ψ^2, gives the probability of finding the particle within a region of space.

The wave function and its square, ψ^2, have values for all locations about a nucleus. Figure 7.17 shows values of ψ^2 for the electron in the lowest energy level of the hydrogen atom along a line starting from the nucleus. Note that ψ^2 is large near the nucleus ($r = 0$), indicating that the electron is most likely to be found in this region. The value of ψ^2 decreases rapidly as the distance from the nucleus increases, but ψ^2 never goes to exactly zero, although the probability does become extremely small at large distances from the nucleus. This means that an atom does not have a definite boundary, unlike in the Bohr model of the atom.

Figure 7.18 shows another view of this electron probability. The graph plots the probability of finding the electron in different spherical shells at particular

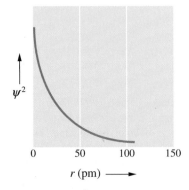

Figure 7.17 ▲

Plot of ψ^2 for the lowest energy level of the hydrogen atom The square of the wave function is plotted versus the distance, r, from the nucleus.

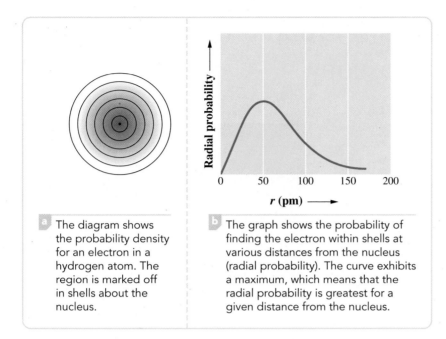

a The diagram shows the probability density for an electron in a hydrogen atom. The region is marked off in shells about the nucleus.

b The graph shows the probability of finding the electron within shells at various distances from the nucleus (radial probability). The curve exhibits a maximum, which means that the radial probability is greatest for a given distance from the nucleus.

distances from the nucleus, rather than the probability at a point. Even though the probability of finding the electron at a point near the nucleus is high, the volume of any shell there is small. Therefore, the probability of finding the electron *within a shell* is greatest at some distance from the nucleus. This distance just happens to equal the radius that Bohr calculated for an electron orbit in his model.

7.5 Quantum Numbers and Atomic Orbitals

According to quantum mechanics, each electron in an atom is described by four different quantum numbers, three of which (n, l, and m_l) specify the wave function that gives the probability of finding the electron at various points in space. ▶ *A wave function for an electron in an atom* is called an **atomic orbital.** An atomic orbital is pictured qualitatively by describing the region of space where there is high probability of finding the electrons. The atomic orbital so pictured has a definite shape. A fourth quantum number (m_s) refers to a magnetic property of electrons called *spin*. We first look at quantum numbers, then at atomic orbitals.

Three different quantum numbers are needed because there are three dimensions to space.

Quantum Numbers

The allowed values and general meaning of each of the four quantum numbers of an electron in an atom are as follows:

1. Principal Quantum Number (n) *This quantum number is the one on which the energy of an electron in an atom principally depends; it can have any positive value: 1, 2, 3, and so on.* The energy of an electron in an atom depends *principally* on n. The smaller n is, the lower the energy. In the case of the hydrogen atom or single-electron atomic ions, such as Li^{2+} and He^+, n is the only quantum number determining the energy (which is given by Bohr's formula, discussed in Section 7.3). For other atoms, the energy also depends to a slight extent on the l quantum number.

The *size* of an orbital also depends on n. The larger the value of n is, the larger the orbital. Orbitals of the same quantum state n are said to belong to the same *shell*. Shells are sometimes designated by the following letters:

Letter	K	L	M	N ...
n	1	2	3	4 ...

2. Angular Momentum Quantum Number (*l*) (Also Called *Azimuthal Quantum Number*)
This quantum number distinguishes orbitals of given n having different shapes; it can have any integer value from 0 to n − 1. Within each shell of quantum number n, there

Scanning Tunneling Microscopy

The 1986 Nobel Prize in physics went to three physicists for their work in developing microscopes for viewing extremely small objects. Half of the prize went to Ernst Ruska for the development of the electron microscope, described earlier. The other half was awarded to Gerd Binnig and Heinrich Rohrer, at IBM's research laboratory in Zurich, Switzerland, for their invention of the *scanning tunneling microscope* in 1981. This instrument makes possible the viewing of atoms and molecules on a solid surface (Figure 7.19).

The scanning tunneling microscope is based on the concept of quantum mechanical *tunneling,* which in turn depends on the probability interpretation of quantum mechanics. Consider a hydrogen atom, which consists of an electron about a proton (call it A), and imagine another proton (called B) some distance from the first. In classical terms, if we were to move the electron from the region of proton A to the region of proton B, energy would have to be supplied to remove the electron from the attractive field of proton A. Quantum mechanics, however, gives us a different picture. The probability of the electron in the hydrogen atom being at a location far from proton A, say near proton B, is very small but not zero. In effect, this means that the electron, which nominally belongs to proton A, may find itself near proton B without extra energy having been supplied. The electron is said to have *tunneled* from one atom to another.

The scanning tunneling microscope consists of a tungsten metal needle with an extremely fine point (the probe) placed close to the sample to be viewed (Figure 7.20). If the probe is close enough to the sample, electrons can tunnel from the probe to the sample. The probability for this can be increased by having a small voltage applied between the probe and sample. Electrons tunneling from the probe to the sample give rise to a measurable electric current. The magnitude of the current depends on the

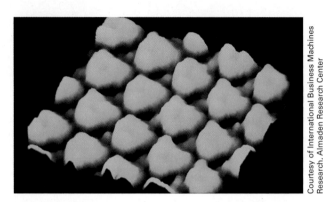

Courtesy of International Business Machines Research, Almaden Research Center

Figure 7.19 ▲

Scanning tunneling microscope image of benzene molecules on a metal surface Benzene molecules, C_6H_6, are arranged in a regular array on a rhodium metal surface.

distance between the probe and the sample (as well as on the wave function of the atom in the sample). By adjusting this distance, the current can be maintained at a fixed value. As the probe scans the sample, it moves toward or away from the sample to maintain a fixed current; in effect, the probe follows the contours of the sample. (The probe is attached to the end of a *piezoelectric* rod, which undergoes very small changes in length when small voltages are applied to it.)

Researchers routinely use the scanning tunneling microscope to study the arrangement of atoms and molecules on surfaces. Lately, scientists have also used the microscope probe to move atoms about a surface in an effort to construct miniature devices.

are *n* different kinds of orbitals, each with a distinctive shape denoted by an *l* quantum number. For example, if an electron has a principal quantum number of 3, the possible values for *l* are 0, 1, and 2. Thus, within the *M* shell (*n* = 3), there are three kinds of orbitals, each having a different shape for the region where the electron is most likely to be found. These orbital shapes will be discussed later in this section.

Although the energy of an orbital is principally determined by the *n* quantum number, the energy also depends somewhat on the *l* quantum number (except for the H atom). For a given *n*, the energy of an orbital increases with *l*.

Orbitals of the same *n* but different *l* are said to belong to different *subshells* of a given shell. The different subshells are usually denoted by letters as follows:

Letter	s	p	d	f	g . . .
l	0	1	2	3	4 . . .

The rather odd choice of letter symbols for l quantum numbers survives from old spectroscopic terminology (describing the lines in a spectrum as sharp, principal, diffuse, and fundamental).

To denote a subshell within a particular shell, we write the value of the *n* quantum number for the shell, followed by the letter designation for the subshell. For example, 2*p* denotes a subshell with quantum numbers *n* = 2 and *l* = 1. ◄

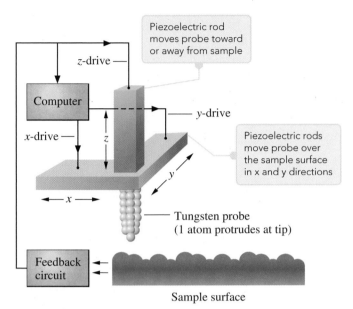

Figure 7.20 ▲

The scanning tunneling microscope A tunneling current flows between the probe and the sample when there is a small voltage between them. A feedback circuit, which provides this voltage, senses the current and varies the voltage on a piezoelectric rod (*z*-drive) in order to keep the distance constant between the probe and sample. A computer has been programmed to provide voltages to the *x*-drive and the *y*-drive to move the probe over the surface of the sample.

Figure 7.21 shows 48 iron atoms that were arranged in a circle on a copper surface. Each of the iron atoms looks like a sharp mountain peak rising from a plain. Inside this "quantum corral" (as it has come to be called), you can see the wavelike distribution of electrons trapped within. Here is graphic proof of the wavelike nature of electrons!

Figure 7.21 ▲

Quantum corral IBM scientists used a scanning tunneling microscope probe to arrange 48 iron atoms in a circle on a copper metal surface. The iron atoms appear as the tall peaks in the diagram. Note the wavelike ripple of electrons trapped within the circle of atoms (called a *quantum corral*). [IBM Research Division, Almaden Research Center; research done by Donald M. Eigler and coworkers.]

■ See Problems 7.95 and 7.96.

3. Magnetic Quantum Number (m_l) *This quantum number distinguishes orbitals of given n and l—that is, of given energy and shape but having a different orientation in space; the allowed values are the integers from −l to +l.* For $l = 0$ (*s* subshell), the allowed m_l quantum number is 0 only; there is only one orbital in the *s* subshell. For $l = 1$ (*p* subshell), $m_l = -1$, 0, and $+1$; there are three different orbitals in the *p* subshell. The orbitals have the same shape but different orientations in space. In addition, all orbitals of a given subshell have the same energy. Note that there are $2l + 1$ orbitals in each subshell of quantum number *l*.

4. Spin Quantum Number (m_s) *This quantum number refers to the two possible orientations of the spin axis of an electron; possible values are $+\frac{1}{2}$ and $-\frac{1}{2}$.* An electron acts as though it were spinning on its axis like the earth. Such an electron spin would give rise to a circulating electric charge that would generate a magnetic field. In this way, an electron behaves like a small bar magnet, with a north and a south pole. ▶

Table 7.1 lists the permissible quantum numbers for all orbitals through the $n = 4$ shell. These values follow from the rules just given. Energies for the orbitals are shown in Figure 7.22 for the hydrogen atom. Note that all orbitals with the same

Electron spin will be discussed further in Section 8.1.

287

Table 7.1		Permissible Values of Quantum Numbers for Atomic Orbitals			
n	l	m_l*	Subshell Notation	Number of Orbitals in the Subshell	
1	0	0	$1s$	1	
2	0	0	$2s$	1	
2	1	$-1, 0, +1$	$2p$	3	
3	0	0	$3s$	1	
3	1	$-1, 0, +1$	$3p$	3	
3	2	$-2, -1, 0, +1, +2$	$3d$	5	
4	0	0	$4s$	1	
4	1	$-1, 0, +1$	$4p$	3	
4	2	$-2, -1, 0, +1, +2$	$4d$	5	
4	3	$-3, -2, -1, 0, +1, +2, +3$	$4f$	7	

*Any one of the m_l quantum numbers may be associated with the n and l quantum numbers on the same line.

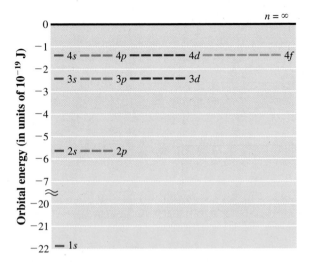

Figure 7.22 ▲

Orbital energies of the hydrogen atom The lines for each subshell indicate the number of different orbitals of that subshell. (Note break in the energy scale.)

principal quantum number n have the same energy. For atoms with more than one electron, however, only orbitals in the same subshell (denoted by a given n and l) have the same energy. We will have more to say about orbital energies in Chapter 8.

Example 7.6

Applying the Rules for Quantum Numbers

State whether each of the following sets of quantum numbers is permissible for an electron in an atom. If a set is not permissible, explain why.

a $n = 1, l = 1, m_l = 0, m_s = +\frac{1}{2}$

b $n = 3, l = 1, m_l = -2, m_s = -\frac{1}{2}$

c $n = 2, l = 1, m_l = 0, m_s = +\frac{1}{2}$

d $n = 2, l = 0, m_l = 0, m_s = 1$

Problem Strategy Apply the rules for quantum numbers in order, first to n, then to l and m_l, and finally to m_s. A set of quantum numbers is impermissible if it disobeys any rule.

Solution

a **Not permissible.** The l quantum number is equal to n; it must be less than n.

b **Not permissible.** The magnitude of the m_l quantum number (that is, the m_l value, ignoring its sign) must not be greater than l.

c **Permissible.**

d **Not permissible.** The m_s quantum number can be only $+\frac{1}{2}$ or $-\frac{1}{2}$.

Answer Check Check that n is a positive integer (it cannot be zero). Also, check that l is a positive integer (but zero is allowed) and that m_l is an integer whose magnitude (its value except for sign) is equal to or less than l. The m_s quantum number can be only $+\frac{1}{2}$ or $-\frac{1}{2}$.

Exercise 7.7 Explain why each of the following sets of quantum numbers is not permissible for an orbital.

a $n = 0, l = 1, m_l = 0, m_s = +\frac{1}{2}$

b $n = 2, l = 3, m_l = 0, m_s = -\frac{1}{2}$

c $n = 3, l = 2, m_l = +3, m_s = +\frac{1}{2}$

d $n = 3, l = 2, m_l = +2, m_s = 0$

■ See Problems 7.69 and 7.70.

Atomic Orbital Shapes

An *s* orbital has a spherical shape, though specific details of the probability distribution depend on the value of *n*. Figure 7.23 shows cross-sectional representations of the probability distributions of a 1*s* and a 2*s* orbital. The color shading is darker where the electron is more likely to be found. In the case of a 1*s* orbital, the electron is most likely to be found near the nucleus. The shading becomes lighter as the distance from the nucleus increases, indicating that the electron is less likely to be found there.

The orbital does not abruptly end at some particular distance from the nucleus. An atom, therefore, has an indefinite extension, or "size." We can gauge the "size" of the orbital by means of the *99% contour*. The electron has a 99% probability of being found within the space of the 99% contour (the sphere indicated by the dashed line in the diagram).

A 2*s* orbital differs in detail from a 1*s* orbital. The electron in a 2*s* orbital is likely to be found in two regions, one near the nucleus and the other in a spherical shell about the nucleus. (The electron is most likely to be here.) The 99% contour shows that the 2*s* orbital is larger than the 1*s* orbital.

A cross-sectional diagram cannot portray the three-dimensional aspect of the 1*s* and 2*s* atomic orbitals. Figure 7.24 shows cutaway diagrams, which better illustrate this three-dimensionality.

There are three *p* orbitals in each *p* subshell. All *p* orbitals have the same basic shape (two lobes arranged along a straight line with the nucleus between the lobes) but differ in their orientations in space. Because the three orbitals are set at right angles to each other, we can show each one as oriented along a different coordinate axis (Figure 7.25). We denote these orbitals as $2p_x$, $2p_y$, and $2p_z$. A $2p_x$ orbital has its greatest electron probability along the *x*-axis, a $2p_y$ orbital along the *y*-axis, and a $2p_z$ orbital along the *z*-axis. Other *p* orbitals, such as 3*p*, have this same general shape, with differences in detail depending on *n*. We will discuss *s* and *p* orbital shapes again in Chapter 10 in reference to chemical bonding.

There are five *d* orbitals, which have more complicated shapes than do *s* and *p* orbitals. These are represented in Figure 7.26.

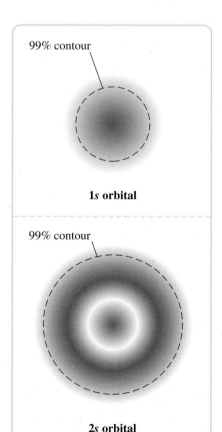

1s orbital

2s orbital

Figure 7.23 ◀

Cross-sectional representations of the probability distributions of *s* orbitals
In a 1*s* orbital, the probability distribution is largest near the nucleus. In a 2*s* orbital, it is greatest in a spherical shell about the nucleus. Note the relative "size" of the orbitals, indicated by the 99% contours.

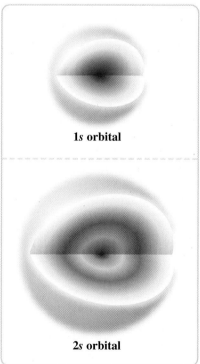

1s orbital

2s orbital

Figure 7.24 ▲

Cutaway diagrams showing the spherical shape of *s* orbitals
In both diagrams, a segment of each orbital is cut away to reveal the electron distribution of the orbital.

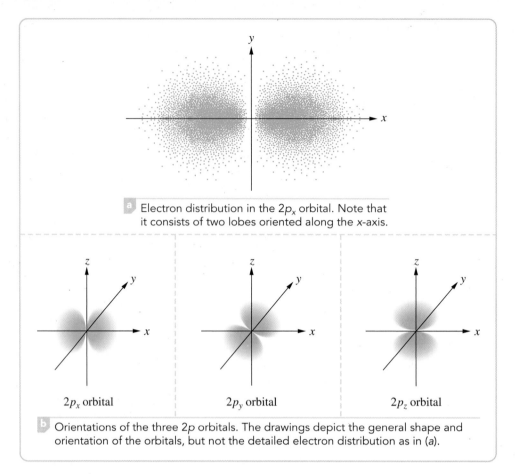

a Electron distribution in the $2p_x$ orbital. Note that it consists of two lobes oriented along the x-axis.

$2p_x$ orbital $2p_y$ orbital $2p_z$ orbital

b Orientations of the three $2p$ orbitals. The drawings depict the general shape and orientation of the orbitals, but not the detailed electron distribution as in (a).

Figure 7.25 ▲

The 2p orbitals

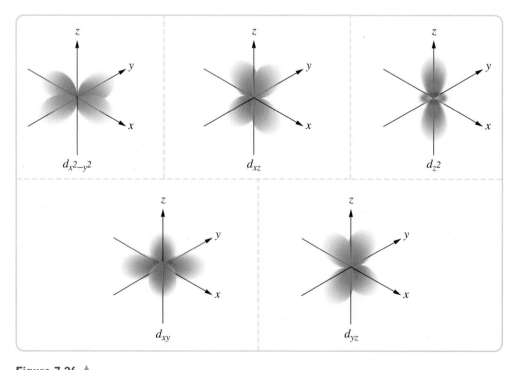

$d_{x^2-y^2}$ d_{xz} d_{z^2}

d_{xy} d_{yz}

Figure 7.26 ▲

The five 3d orbitals These are labeled by subscripts, as in d_{xy}, that describe their mathematical characteristics.

A Checklist for Review

◌OWL and GoChemistry Sign in at **www.cengage.com/owl** to:
- View tutorials and simulations, develop problem-solving skills, and complete online homework assigned by your professor.
- For quick review and exam prep, download Go Chemistry mini lecture modules from OWL or purchase them at **www.cengagebrain.com**.

Summary of Facts and Concepts

One way to study the electronic structure of the atom is to analyze the electromagnetic radiation that is emitted from an atom. Electromagnetic radiation is characterized by its *wavelength* λ and frequency ν, and these quantities are related to the speed of light c ($c = \nu\lambda$).

Einstein showed that light consists of particles (*photons*), each of energy $E = h\nu$, where h is Planck's constant. According to Bohr, electrons in an atom have *energy levels,* and when an electron in a higher energy level drops (or undergoes a transition) to a lower energy level, a photon is emitted. The energy of the photon equals the difference in energy between the two levels.

Electrons and other particles of matter have both particle and wave properties. For a particle of mass m and speed v, the wavelength is related to momentum mv by the *de Broglie relation:* $\lambda = h/mv$. The wave properties of a particle are described by a *wave function,* from which we can get the probability of finding the particle in different regions of space.

Each electron in an atom is characterized by four different quantum numbers. The distribution of an electron in space—its *atomic orbital*—is characterized by three of these quantum numbers: the principal quantum number, the angular momentum quantum number, and the magnetic quantum number. The fourth quantum number (spin quantum number) describes the magnetism of the electron.

Learning Objectives	Important Terms
7.1 The Wave Nature of Light ■ Define the *wavelength* and *frequency* of a wave. ■ Relate the wavelength, frequency, and speed of light. Examples 7.1 and 7.2 ■ Describe the different regions of the electromagnetic spectrum.	**wavelength (λ)** **frequency (ν)** **electromagnetic spectrum**
7.2 Quantum Effects and Photons ■ State Planck's quantization of vibrational energy. ■ Define *Planck's constant* and *photon.* ■ Describe the photoelectric effect. ■ Calculate the energy of a photon from its frequency or wavelength. Example 7.3	**Planck's constant** **photons** **photoelectric effect**
7.3 The Bohr Theory of the Hydrogen Atom ■ State the postulates of Bohr's theory of the hydrogen atom. ■ Relate the energy of a photon to the associated energy levels of an atom. ■ Determine the wavelength or frequency of a hydrogen atom transition. Example 7.4 ■ Describe the difference between *emission* and *absorption* of light by an atom.	**continuous spectrum** **line spectrum** **energy levels**
7.4 Quantum Mechanics ■ State the de Broglie relation. ■ Calculate the wavelength of a moving particle. Example 7.5 ■ Define *quantum mechanics.* ■ State Heisenberg's uncertainty principle. ■ Relate the wave function for an electron to the probability of finding it at a location in space.	**de Broglie relation** **quantum (wave) mechanics** **uncertainty principle**

7.5 Quantum Numbers and Atomic Orbitals

- Define *atomic orbital*.
- Define each of the quantum numbers for an atomic orbital.
- State the rules for the allowed values for each quantum number.
- Apply the rules for quantum numbers. Example 7.6
- Describe the shapes of *s*, *p*, and *d* orbitals.

atomic orbital
principal quantum number (*n*)
angular momentum quantum number (*l*)
magnetic quantum number (*m_l*)
spin quantum number (*m_s*)

Key Equations

$c = \nu\lambda$

$E = h\nu$

$E = -\dfrac{R_H}{n^2} \quad n = 1, 2, 3, \ldots \infty \quad$ (for H atom)

Energy of emitted photon $= h\nu = -(E_f - E_i)$

$\lambda = \dfrac{h}{m\upsilon}$

Questions and Problems

⏾WL Interactive versions of these problems may be assigned in OWL.

Self-Assessment and Review Questions

Key: These questions test your understanding of the ideas you worked with in the chapter. These problems vary in difficulty and often can be used for the basis of discussion.

7.1 Give a brief wave description of light. What are two characteristics of light waves?

7.2 Briefly describe the portions of the electromagnetic spectrum, starting with shortest wavelengths and going to longer wavelengths.

7.3 What is the mathematical relationship among the different characteristics of light waves? State the meaning of each of the terms in the equation.

7.4 Planck originated the idea that energies can be quantized. What does the term *quantized* mean? What was Planck trying to explain when he was led to the concept of quantization of energy? Give the formula he arrived at and explain each of the terms in the formula.

7.5 In your own words, explain the photoelectric effect. How does the photon concept explain this effect?

7.6 Describe the wave–particle picture of light.

7.7 Give the equation that relates particle properties of light. Explain the meaning of each symbol in the equation.

7.8 Physical theory at the time Rutherford proposed his nuclear model of the atom was not able to explain how this model could give a stable atom. Explain the nature of this difficulty.

7.9 Explain the main features of Bohr's theory. Do these features solve the difficulty alluded to in Question 7.8?

7.10 Explain the process of absorption of light by an atom.

7.11 Explain the process of emission of light by an atom.

7.12 What kind of information does a wave function give about an electron in an atom?

7.13 What is the evidence for electron waves? Give a practical application.

7.14 The atom is sometimes said to be similar to a miniature planetary system, with electrons orbiting the nucleus. What does the uncertainty principle have to say about this view of the atom?

7.15 Bohr described the hydrogen atom as an electron orbiting a hydrogen nucleus. Although certain aspects of his theory are still valid, his theory agreed quantitatively with experiment only in the case of the hydrogen atom. In what way does quantum mechanics change Bohr's original picture of the hydrogen atom?

7.16 Give the possible values of ⓐ the principal quantum number, ⓑ the angular momentum quantum number, ⓒ the magnetic quantum number, and ⓓ the spin quantum number.

7.17 What is the general shape of an *s* orbital? of a *p* orbital?

7.18 What is the notation for the subshell in which $n = 4$ and $l = 3$? How many orbitals are in this subshell?

7.19 Which of the following statements about a hydrogen atom is *false*?

ⓐ The wavelength of light emitted when the electron goes from the $n = 3$ level to the $n = 1$ level is the same as the wavelength of light absorbed when the electron goes from the $n = 1$ level to $n = 3$ level.

ⓑ An electron in the $n = 1$ level of the hydrogen atom is in its ground state.

ⓒ An electron in the $n = 1$ level is higher in energy than an electron in the $n = 4$ level.

ⓓ On average, an electron in the $n = 3$ level is farther from the nucleus than an electron in the $n = 2$ state.

e Light of greater frequency is required for a transition from the $n = 1$ level to $n = 3$ level than is required for a transition from the $n = 2$ level to $n = 3$ level.

7.20 Which of the following statements is (are) true?
   I. The product of wavelength and frequency of light is a constant.
   II. As the energy of electromagnetic radiation increases, its frequency decreases.
   III. As the wavelength of light increases, its frequency increases.
   a I only
   b II only
   c III only
   d I and III only
   e II and III only

7.21 Of the following possible transitions of an electron in a hydrogen atom, which emits light of the highest energy?
   a Transition from the $n = 2$ to the $n = 1$ level
   b Transition from the $n = 1$ to the $n = 2$ level
   c Transition from the $n = 5$ to the $n = 4$ level
   d Transition from the $n = 3$ to the $n = 1$ level
   e Transition from the $n = 1$ to the $n = 3$ level

7.22 What wavelength of electromagnetic radiation corresponds to a frequency of $3.46 \times 10^{13} \text{ s}^{-1}$?
   a 8.66×10^{-6} m
   b 1.15×10^{5} m
   c 7.65×10^{-29} m
   d 9.10×10^{-6} m
   e 8.99×10^{-6} m

Concept Explorations

Key: Concept explorations are comprehensive problems that provide a framework that will enable you to explore and learn many of the critical concepts and ideas in each chapter. If you master the concepts associated with these explorations, you will have a better understanding of many important chemistry ideas and will be more successful in solving all types of chemistry problems. These problems are well suited for group work and for use as in-class activities.

7.23 Light, Energy, and the Hydrogen Atom
   a Which has the greater wavelength, blue light or red light?
   b How do the frequencies of blue light and red light compare?
   c How does the energy of blue light compare with that of red light?
   d Does blue light have a greater speed than red light?
   e How does the energy of three photons from a blue light source compare with the energy of one photon of blue light from the same source? How does the energy of two photons corresponding to a wavelength of 451 nm (blue light) compare with the energy of three photons corresponding to a wavelength of 704 nm (red light)?
   f A hydrogen atom with an electron in its ground state interacts with a photon of light with a wavelength of 1.22×10^{-6} m. Could the electron make a transition from the ground state to a higher energy level? If it does make a transition, indicate which one. If no transition can occur, explain.
   g If you have one mole of hydrogen atoms with their electrons in the $n = 1$ level, what is the minimum number of photons you would need to interact with these atoms in order to have all of their electrons promoted to the $n = 3$ level? What wavelength of light would you need to perform this experiment?

7.24 Investigating Energy Levels
Consider the hypothetical atom X that has one electron like the H atom but has different energy levels. The energies of an electron in an X atom are described by the equation

$$E = -\frac{R_H}{n^3}$$

where R_H is the same as for hydrogen (2.179×10^{-18} J). Answer the following questions, *without calculating energy values.*

   a How would the ground-state energy levels of X and H compare?
   b Would the energy of an electron in the $n = 2$ level of H be higher or lower than that of an electron in the $n = 2$ level of X? Explain your answer.
   c How do the spacings of the energy levels of X and H compare?
   d Which would involve the emission of a higher frequency of light, the transition of an electron in an H atom from the $n = 5$ to the $n = 3$ level or a similar transition in an X atom?
   e Which atom, X or H, would require more energy to completely remove its electron?
   f A photon corresponding to a particular frequency of blue light produces a transition from the $n = 2$ to the $n = 5$ level of a hydrogen atom. Could this photon produce the same transition ($n = 2$ to $n = 5$) in an atom of X? Explain.

Conceptual Problems

Key: These problems are designed to check your understanding of the concepts associated with some of the main topics presented in each chapter. A strong conceptual understanding of chemistry is the foundation for both applying chemical knowledge and solving chemical problems. These problems vary in level of difficulty and often can be used as a basis for group discussion.

7.25 Some infrared radiation has a wavelength that is 1000 times larger than that of a certain visible light. This visible light has a frequency that is 1000 times smaller than that of some X radiation. How many times more energy is there in a photon of this X radiation than there is in a photon of the infrared radiation?

7.26 Consider two beams of the same yellow light. Imagine that one beam has its wavelength doubled; the other has its frequency doubled. Which of these two beams is then in the ultraviolet region?

7.27 An atom in its ground state absorbs a photon (photon 1), then quickly emits another photon (photon 2). One of these photons corresponds to ultraviolet radiation, whereas the other one corresponds to red light. Explain what is happening. Which electromagnetic radiation, ultraviolet or red light, is associated with the emitted photon (photon 2)?

7.28 One photon of green light has less than twice the energy of two photons of red light. Consider two hypothetical experiments. In one experiment, potassium metal is exposed to one photon of green light; in another experiment, potassium metal is exposed to two photons of red light. In one of these experiments, no electrons are ejected by the photoelectric effect (no matter how many times this experiment is repeated). In the other experiment, at least one electron was observed to be ejected. What is the maximum number of electrons that could be ejected during this other experiment, one or two?

7.29 Three emission lines involving three energy levels in an atom occur at wavelengths x, $1.5x$, and $3.0x$ nanometers. Which wavelength corresponds to the transition from the highest to the lowest of the three energy levels?

7.30 An atom emits yellow light when an electron makes the transition from the $n = 5$ to the $n = 1$ level. In separate experiments, suppose you bombarded the $n = 1$ level of this atom with red light, yellow light (obtained from the previous emission), and blue light. In which experiment or experiments would the electron be promoted to the $n = 5$ level?

7.31 Which of the following particles has the longest wavelength?

　a　a proton traveling at $2x$ meters per second
　b　an electron traveling at x meters per second
　c　a proton traveling at x meters per second

7.32 Imagine a world in which the rule for the l quantum number is that values start with 1 and go up to n. The rules for the n and m_l quantum numbers are unchanged from those of our world. Write the quantum numbers for the first two shells (i.e., $n = 1$ and $n = 2$).

7.33 Given the following energy level diagram for an atom that contains an electron in the $n = 3$ level, answer the following questions.

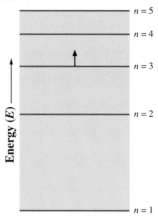

　a　Which transition of the electron will emit light of the lowest frequency?
　b　Using only those levels depicted in the diagram, which transition of the electron would require the highest-frequency light?
　c　If the transition from the $n = 3$ level to the $n = 1$ level emits green light, what color light is absorbed when an electron makes the transition from the $n = 1$ to $n = 3$ level?

7.34 The following shapes each represent an orbital of an atom in a hypothetical universe. The small circle is the location of the nucleus in each orbital.

　a　If you placed an electron in each orbital, which one would be higher in energy?
　b　When an electron makes a transition from the orbital represented on the right to the orbital on the left, would you expect energy to be absorbed or released?
　c　Draw a sketch of an orbital of the same type that would be higher in energy than either of the two pictured orbitals.

Practice Problems

Key: These problems are for practice in applying problem-solving skills. They are divided by topic, and some are keyed to exercises (see the ends of the exercises). The problems are arranged in matching pairs; the odd-numbered problem of each pair is listed first, and its answer is given in the back of the book.

Electromagnetic Waves

7.35 Microwaves have frequencies in the range 10^9 to 10^{12}/s (cycles per second), equivalent to between 1 gigahertz and 1 terahertz. What is the wavelength of microwave radiation whose frequency is 1.395×10^{10}/s?

7.36 Radio waves in the AM region have frequencies in the range 530 to 1700 kilocycles per second (530 to 1700 kHz). Calculate the wavelength corresponding to a radio wave of frequency 1.365×10^6/s (that is, 1365 kHz).

7.37 Calculate the frequency associated with light of wavelength 434 nm. (This corresponds to one of the wavelengths of light emitted by the hydrogen atom.)

7.38 Light with a wavelength of 478 nm lies in the blue region of the visible spectrum. Calculate the frequency of this light.

7.39 At its closest approach, Mars is 56 million km from Earth. How long would it take to send a radio message from a space probe of Mars to Earth when the planets are at this closest distance?

7.40 The space probe *Pioneer 11* was launched April 5, 1973, and reached Jupiter in December 1974, traveling a distance of 998 million km. How long did it take an electromagnetic signal to travel to Earth from *Pioneer 11* when it was near Jupiter?

7.41 The second is defined as the time it takes for 9,192,631,770 wavelengths of a certain transition of the cesium-133 atom to pass a fixed point. What is the frequency of this electromagnetic radiation? What is the wavelength?

7.42 The meter was defined in 1963 as the length equal to 1,650,763.73 wavelengths of the orange-red radiation emitted by the krypton-86 atom (the meter has since been redefined). What is the wavelength of this transition? What is the frequency?

Photons

7.43 What is the energy of a photon corresponding to radio waves of frequency 1.365×10^6/s?

7.44 What is the energy of a photon corresponding to microwave radiation of frequency 1.258×10^{10}/s?

7.45 The green line in the atomic spectrum of thallium has a wavelength of 535 nm. Calculate the energy of a photon of this light.

7.46 Molybdenum compounds give a yellowish-green flame test. The atomic emission responsible for this color has a wavelength of 551 nm. Obtain the energy of a single photon of this wavelength.

7.47 Selenium atoms have a particular transition that emits light of frequency 1.53×10^{15} Hz. (Hz is the abbreviation for *hertz*, which is equivalent to the unit/s, or s^{-1}.) Is this light in the visible spectrum? If so, what is the color of the light? (See Figure 7.5.)

7.48 A particular transition of the rubidium atom emits light whose frequency is 3.84×10^{14} Hz. (Hz is the abbreviation for *hertz*, which is equivalent to the unit/s, or s^{-1}.) Is this light in the visible spectrum? If so, what is the color of the light? (See Figure 7.5.)

Bohr Theory

7.49 An electron in a hydrogen atom in the level $n = 5$ undergoes a transition to level $n = 3$. What is the frequency of the emitted radiation?

7.50 Calculate the frequency of electromagnetic radiation emitted by the hydrogen atom in the electron transition from $n = 6$ to $n = 3$.

7.51 The first line of the Lyman series of the hydrogen atom emission results from a transition from the $n = 2$ level to the $n = 1$ level. What is the wavelength of the emitted photon? Using Figure 7.5, describe the region of the electromagnetic spectrum in which this emission lies.

7.52 What is the wavelength of the electromagnetic radiation emitted from a hydrogen atom when the electron undergoes the transition $n = 4$ to $n = 1$? In what region of the spectrum does this line occur? (See Figure 7.5.)

7.53 Calculate the shortest wavelength of the electromagnetic radiation emitted by the hydrogen atom in undergoing a transition from the $n = 6$ level.

7.54 Calculate the longest wavelength of the electromagnetic radiation emitted by the hydrogen atom in undergoing a transition from the $n = 6$ level.

7.55 What is the difference in energy between the two levels responsible for the violet emission line of the calcium atom at 422.7 nm?

7.56 What is the difference in energy between the two levels responsible for a red emission line of the strontium atom at 640.8 nm?

de Broglie Waves

Note: Masses of the electron, proton, and neutron are listed on the inside back cover of this book.

7.57 What is the wavelength of a neutron traveling at a speed of 4.15 km/s? (Neutrons of these speeds are obtained from a nuclear pile.)

7.58 What is the wavelength of a proton traveling at a speed of 6.21 km/s? What would be the region of the spectrum for electromagnetic radiation of this wavelength?

7.59 At what speed must a neutron travel to have a wavelength of 10.6 pm?

7.60 At what speed must an electron travel to have a wavelength of 10.0 pm?

7.61 What is the de Broglie wavelength of a 145-g baseball traveling at 30.0 m/s (67.1 mph)? Is the wavelength much smaller or much larger than the diameter of an atom (on the order of 100 pm)?

7.62 What is the de Broglie wavelength of an oxygen molecule, O_2, traveling at 535 m/s? Is the wavelength much smaller or much larger than the diameter of an atom (on the order of 100 pm)?

Atomic Orbitals

7.63 If the n quantum number of an atomic orbital is 4, what are the possible values of l? If the l quantum number is 3, what are the possible values of m_l?

7.64 The n quantum number of an atomic orbital is 5. What are the possible values of l? What are the possible values of m_l if the l quantum number is 4?

7.65 How many subshells are there in the M shell? How many orbitals are there in the d subshell?

7.66 How many subshells are there in the N shell? How many orbitals are there in the f subshell?

7.67 Give the notation (using letter designations for l) for the subshells denoted by the following quantum numbers.
- a. $n = 6, l = 2$
- b. $n = 5, l = 4$
- c. $n = 4, l = 3$
- d. $n = 6, l = 1$

7.68 Give the notation (using letter designations for l) for the subshells denoted by the following quantum numbers.
- a. $n = 5, l = 3$
- b. $n = 3, l = 1$
- c. $n = 4, l = 2$
- d. $n = 4, l = 0$

7.69 Explain why each of the following sets of quantum numbers would not be permissible for an electron, according to the rules for quantum numbers.

a $n = 0, l = 1, m_l = 0, m_s = +\frac{1}{2}$

b $n = 3, l = 2, m_l = +3, m_s = -\frac{1}{2}$

c $n = 2, l = 1, m_l = -1, m_s = +\frac{3}{2}$

d $n = 1, l = 3, m_l = +3, m_s = +\frac{1}{2}$

e $n = 1, l = 0, m_l = 0, m_s = +1$

7.70 State which of the following sets of quantum numbers would be possible and which impossible for an electron in an atom.

a $n = 2, l = 0, m_l = 0, m_s = +\frac{1}{2}$

b $n = 1, l = 1, m_l = 0, m_s = +\frac{1}{2}$

c $n = 2, l = 1, m_l = -1, m_s = +\frac{1}{2}$

d $n = 2, l = 1, m_l = -2, m_s = +\frac{1}{2}$

e $n = 0, l = 0, m_l = 0, m_s = -\frac{1}{2}$

General Problems

Key: These problems provide more practice but are not divided by topic or keyed to exercises. Each section ends with essay questions, each of which is color coded to refer to the *A Chemist Looks at Daily Life* (orange), *A Chemist Looks at Materials* (blue), and *Instrumental Methods* (brown) chapter essay on which it is based. Odd-numbered problems and the even-numbered problems that follow are similar; answers to all odd-numbered problems except the essay questions are given in the back of the book.

7.71 The barium atom has an emission with wavelength 554 nm (green). Calculate the frequency of this light and the energy of a photon of this light.

7.72 The blue line of the strontium atom emission has a wavelength of 461 nm. What is the frequency of this light? What is the energy of a photon of this light?

7.73 The energy of a photon is 4.10×10^{-19} J. What is the wavelength of the corresponding light? What is the color of this light?

7.74 The energy of a photon is 3.05×10^{-19} J. What is the wavelength of the corresponding light? What is the color of this light?

7.75 The photoelectric work function of a metal is the minimum energy needed to eject an electron by irradiating the metal with light. For calcium, this work function equals 4.34×10^{-19} J. What is the minimum frequency of light for the photoelectric effect in calcium?

7.76 The photoelectric work function for magnesium is 2.95×10^{-19} J. Calculate the minimum frequency of light required to eject electrons from magnesium.

7.77 Light of wavelength 345 nm shines on a piece of calcium metal. What is the speed of the ejected electron? (Light energy greater than that of the work function of calcium ends up as kinetic energy of the ejected electron. See Problem 7.75 for the definition of work function and its value for calcium.)

7.78 Light of wavelength 285 nm shines on a piece of magnesium metal. What is the speed of the ejected electron? (Light energy greater than that of the work function of magnesium ends up as kinetic energy of the ejected electron. See Problem 7.76 for the definition of work function and its value for magnesium.)

7.79 Calculate the wavelength of the Balmer line of the hydrogen spectrum in which the initial n quantum number is 5 and the final n quantum number is 2.

7.80 Calculate the wavelength of the Balmer line of the hydrogen spectrum in which the initial n quantum number is 3 and the final n quantum number is 2.

7.81 A line of the Lyman series of the hydrogen atom spectrum has the wavelength 9.50×10^{-8} m. It results from a transition from an upper energy level to $n = 1$. What is the principal quantum number of the upper level?

7.82 One of the lines in the Balmer series of the hydrogen atom emission spectrum is at 397 nm. It results from a transition from an upper energy level to $n = 2$. What is the principal quantum number of the upper level?

7.83 A hydrogen-like ion has a nucleus of charge $+Ze$ and a single electron outside this nucleus. The energy levels of these ions are $-Z^2 R_H/n^2$ (where Z = atomic number). Calculate the wavelength of the transition from $n = 3$ to $n = 2$ for He^+, a hydrogen-like ion. In what region of the spectrum does this emission occur?

7.84 What is the wavelength of the transition from $n = 4$ to $n = 2$ for Li^{2+}? In what region of the spectrum does this emission occur? Li^{2+} is a hydrogen-like ion. Such an ion has a nucleus of charge $+Ze$ and a single electron outside this nucleus. The energy levels of the ion are $-Z^2 R_H/n^2$, where Z is the atomic number.

7.85 An electron microscope employs a beam of electrons to obtain an image of an object. What energy must be imparted to each electron of the beam to obtain a wavelength of 10.0 pm? Obtain the energy in electron volts (eV) (1 eV = 1.602×10^{-19} J).

7.86 Neutrons are used to obtain images of the hydrogen atoms in molecules. What energy must be imparted to each neutron in a neutron beam to obtain a wavelength of 10.0 pm? Obtain the energy in electron volts (eV) (1 eV = 1.602×10^{-19} J).

7.87 What is the number of different orbitals in each of the following subshells?

a $5s$ b $3d$ c $4f$ d $4p$

7.88 What is the number of different orbitals in each of the following subshells?

a $5f$ b $5g$ c $6s$ d $5p$

7.89 List the possible subshells for the $n = 8$ shell.

7.90 List the possible subshells for the $n = 6$ shell.

■ **7.91** What are gamma rays? How does the gamma radiation of foods improve their shelf life?

■ **7.92** How can gamma rays that are used in food irradiation be produced? Does such irradiated food show any radioactivity?

■ **7.93** The word *laser* is an acronym meaning *l*ight *a*mplification by *s*timulated *e*mission of *r*adiation. What is the stimulated emission of radiation?

■ **7.94** Explain how lasers are used to "read" a compact disc.

■ **7.95** Explain how the probe in a scanning tunneling microscope scans a sample on the surface of a metal.

■ **7.96** Explain the concept of quantum mechanical tunneling.

Strategy Problems

Key: As noted earlier, all of the practice and general problems are matched pairs. This section is a selection of problems that are not in matched-pair format. These challenging problems require that you employ many of the concepts and strategies that were developed in the chapter. In some cases, you will have to integrate several concepts and operational skills in order to solve the problem successfully.

Note: See back cover for physical constants and conversion factors.

7.97 What wavelength of electromagnetic radiation corresponds to a frequency of 7.76×10^9 s^{-1}? Note that Planck's constant is 6.63×10^{-34} J·s, and the speed of light is 3.00×10^8 m/s.

7.98 AM radio stations broadcast at frequencies between 530 kHz and 1700 kHz. (1 kHz = 10^3 s^{-1}.) For a station broadcasting at 1.69×10^3 kHz, what is the energy of this radio wave? Note that Planck's constant is 6.63×10^{-34} J·s, and the speed of light is 3.00×10^8 m/s.

7.99 A ruby laser puts out a pulse of red light at a wavelength of 694 nm. (The ruby laser is described in the essay on p. 280.) If a pulse delivers 1.05×10^5 watts of power for 280 μs, how many photons are there in this pulse? (1 watt = 1 J/s.)

7.100 A particular microwave oven delivers 800 watts. (A watt is a unit of power, which is the joules of energy delivered, or used, per second.) If the oven uses microwave radiation of wavelength 12.2 cm, how many photons of this radiation are required to heat 1.00 g of water 1.00°C, assuming that all of the photons are absorbed?

7.101 The retina of the eye contains two types of light-sensitive cells: rods (responsible for night vision) and cones (responsible for color vision). Rod cells are about a hundred times more sensitive to light than cone cells and are able to detect a single photon. Suppose a group of rod cells are radiated with a pulse of light having an energy equal to 1.60×10^{-16} J. If the wavelength of this light was 498 nm (the wavelength at which rod cells are most sensitive), how many photons are in this light pulse?

7.102 Ozone in the stratosphere absorbs ultraviolet light of wavelengths shorter than 320 nm, thus filtering out the most energetic radiation from sunlight. During this absorption, an ozone molecule absorbs a photon, which breaks an oxygen-oxygen bond, yielding an oxygen molecule and an oxygen atom:

$$O_3(g) + h\nu \longrightarrow O_2(g) + O(g)$$

(Here, $h\nu$ denotes a photon.) Suppose a flask of ozone is irradiated with a pulse of UV light of wavelength 275 nm. Assuming that each photon of this pulse that is absorbed breaks up one ozone molecule, calculate the energy absorbed per mole of O_2 produced, giving the answer in kJ/mol.

7.103 The photoelectric *work function* of a metal is the minimum energy required to eject an electron by shining light on the metal. The work function of calcium is 4.60×10^{-19} J. What is the longest wavelength of light (in nanometers) that can cause an electron to be ejected from calcium metal.

7.104 The surface of a metal was illuminated with light whose wavelength is 389 nm. Upon illumination, the metal ejected an electron with a speed of 3.34×10^5 m/s. What is the photoelectric work function of the metal (in eV)? Assuming that the metal is an element, use values given in the *CRC Handbook of Chemistry and Physics* to discover the element.

7.105 The photoelectric work function of potassium is 2.29 eV. A photon of energy greater than this ejects the electron with the excess as kinetic energy. Suppose light of wavelength 455 nm ejects an electron from the surface of potassium. What is the speed of the ejected electron?

7.106 Calculate the shortest wavelength of *visible* light (in nanometers) seen in the spectrum of the hydrogen atom. What are the principal quantum numbers for the levels in this transition? Does Figure 7.11 include all visible lines?

7.107 Light of wavelength 1.03×10^{-7} m is emitted when an electron in an excited level of a hydrogen atom undergoes a transition to the $n = 1$ level. What is the region of the spectrum of this light? What is the principal quantum number of the excited level?

7.108 A hydrogen atom in the ground state absorbs a photon whose wavelength is 95.0 nm. The resulting excited atom then emits a photon of 1282 nm. What are the regions of the electromagnetic spectrum for the radiations involved in these transitions? What is the principal quantum number of the final state resulting from the emission from the excited atom?

7.109 Hydrogen-like ions are atomic ions consisting of just one electron about a nucleus. The energy levels of such an ion are given by a formula similar to that of the hydrogen atom except that the Rydberg constant for a hydrogen-like ion of atomic number Z has a value of $2.179 \times 10^{-18} Z^2$. What is the shortest wavelength transition in the Li^{2+} ion?

7.110 The root-mean-square speed of an oxygen molecule, O_2, at 21°C is 479 m/s. Calculate the de Broglie wavelength for an O_2 molecule traveling at this speed. How does this wavelength compare with the approximate length of this molecule, which is about 242 pm? (For this comparison, state the wavelength as a percentage of the molecular length.)

7.111 It requires 799 kJ of energy to break one mole of carbon–oxygen double bonds in carbon dioxide. What wavelength of light does this correspond to per bond? Is there any transition in the hydrogen atom that has at least this quantity of energy in one photon?

7.112 An electron is accelerated through a potential difference of 15.6 kilovolts (giving the electron a kinetic energy of 15.6 keV). What is the associated wavelength of the electron in angstroms?

7.113 In X-ray fluorescence spectroscopy, a material can be analyzed for its constituent elements by radiating the material with short-wavelength X rays, which induce the atoms to emit longer-wavelength X rays characteristic of those atoms. Tungsten, for example, emits characteristic X rays of wavelength 0.1476 nm. If an electron has an equivalent wavelength, what is its kinetic energy?

7.114 For each of the following combinations of quantum numbers, make changes that produce an allowed combination. Count 3 for each change of n, 2 for each change of l, and 1 for each change of m_l. What is the lowest possible count that you can obtain?

a $n = 5, l = 6, m_l = 3$
b $n = 3, l = 3, m_l = -3$
c $n = 5, l = 5, m_l = 4$
d $n = 3, l = 0, m_l = -2$

7.115 The term *degeneracy* means the number of different quantum states of an atom or molecule having the same energy. For example, the degeneracy of the $n = 2$ level of the hydrogen atom is 4 (a $2s$ quantum state, and three different $2p$ states). What is the degeneracy of the $n = 5$ level?

7.116 In a hypothetical universe, the quantum numbers for an atomic orbital follow these rules:

n = any positive integer value from 2 to ∞
l = any positive integer value from 2 to $n + 1$
m_l = any integer from $-(l - 1)$ to $(l - 1)$

How many orbitals are possible for the $n = 2$ shell?

Cumulative-Skills Problems

Key: The problems under this heading combine skills introduced in previous chapters with those given in the current one.

7.117 The energy required to dissociate the H_2 molecule to H atoms is 432 kJ/mol H_2. If the dissociation of an H_2 molecule were accomplished by the absorption of a single photon whose energy was exactly the quantity required, what would be its wavelength (in meters)?

7.118 The energy required to dissociate the Cl_2 molecule to Cl atoms is 239 kJ/mol Cl_2. If the dissociation of a Cl_2 molecule were accomplished by the absorption of a single photon whose energy was exactly the quantity required, what would be its wavelength (in meters)?

7.119 A microwave oven heats by radiating food with microwave radiation, which is absorbed by the food and converted to heat. Suppose an oven's radiation wavelength is 12.5 cm. A container with 0.250 L of water was placed in the oven, and the temperature of the water rose from 20.0°C to 100.0°C. How many photons of this microwave radiation were required? Assume that all the energy from the radiation was used to raise the temperature of the water.

7.120 Warm objects emit electromagnetic radiation in the infrared region. Heat lamps employ this principle to generate infrared radiation. Water absorbs infrared radiation with wavelengths near 2.80 μm. Suppose this radiation is absorbed by the water and converted to heat. A 1.00-L sample of water absorbs infrared radiation, and its temperature increases from 20.0°C to 30.0°C. How many photons of this radiation are used to heat the water?

7.121 Light with a wavelength of 425 nm fell on a potassium surface, and electrons were ejected at a speed of 4.88×10^5 m/s. What energy was expended in removing an electron from the metal? Express the answer in joules (per electron) and in kilojoules per mole (of electrons).

7.122 Light with a wavelength of 405 nm fell on a strontium surface, and electrons were ejected. If the speed of an ejected electron is 3.36×10^5 m/s, what energy was expended in removing the electron from the metal? Express the answer in joules (per electron) and in kilojoules per mole (of electrons).

7.123 When an electron is accelerated by a voltage difference, the kinetic energy acquired by the electron equals the voltage times the charge on the electron. Thus, one volt imparts a kinetic energy of 1.602×10^{-19} volt-coulombs, or 1.602×10^{-19} J. What is the wavelength for electrons accelerated by 1.00×10^4 volts?

7.124 When an electron is accelerated by a voltage difference, the kinetic energy acquired by the electron equals the voltage times the charge on the electron. Thus, one volt imparts a kinetic energy of 1.602×10^{-19} volt-coulombs, which equals 1.602×10^{-19} J. What is the wavelength associated with electrons accelerated by 4.00×10^3 volts?

Electron Configurations and Periodicity

Sodium metal reacts vigorously with water to produce hydrogen gas (which catches fire); other Group IA metals similarly react with water. The other product is sodium hydroxide, which is used in products such as oven cleaner and soap.

Contents and Concepts

Electronic Structure of Atoms

In the previous chapter, you learned that we characterize an atomic orbital by four quantum numbers: n, l, m_l, and m_s. In the first section, we look further at electron spin; then we discuss how electrons are distributed among the possible orbitals of an atom.

8.1 Electron Spin and the Pauli Exclusion Principle

8.2 Building-Up Principle and the Periodic Table

8.3 Writing Electron Configurations Using the Periodic Table

8.4 Orbital Diagrams of Atoms; Hund's Rule

Periodicity of the Elements

You learned how the periodic table can be explained by the periodicity of the ground-state configurations of the elements. Now we will look at various aspects of the periodicity of the elements.

8.5 Mendeleev's Predictions from the Periodic Table

8.6 Some Periodic Properties

8.7 Periodicity in the Main-Group Elements

OWL Sign in to OWL at **www.cengage.com/owl** to view tutorials and simulations, develop problem-solving skills, and complete online homework assigned by your professor.

Chemistry Download mini lecture videos for key concept review and exam prep from OWL or purchase them from **www.cengagebrain.com.**

Figure 8.1 ▲

Marie Sklodowska Curie (1867–1934), with Pierre Curie Marie Sklodowska Curie, born in Warsaw, Poland, began her doctoral work with Henri Becquerel soon after he discovered the spontaneous radiation emitted by uranium salts. She found this radiation to be an atomic property and coined the word *radioactivity* for it. In 1903 the Curies and Becquerel were awarded the Nobel Prize in physics for their discovery of radioactivity. Three years later, Pierre Curie was killed in a carriage accident. Marie Curie continued their work on radium and in 1911 was awarded the Nobel Prize in chemistry for the discovery of polonium and radium and the isolation of pure radium metal. This was the first time a scientist had received two Nobel awards. (Since then two others have been so honored.)

Figure 8.2 ▶

The Stern–Gerlach experiment The diagram shows the experiment using hydrogen atoms (simpler to interpret theoretically), although the original experiment employed silver atoms. A beam of hydrogen atoms (shown in blue) is split into two by a nonuniform magnetic field. One beam consists of atoms each with an electron having $m_s = +\frac{1}{2}$; the other beam consists of atoms each having an electron with $m_s = -\frac{1}{2}$.

arie Curie, a Polish-born French chemist, and her husband, Pierre, announced the discovery of radium in 1898 (Figure 8.1). They had separated a very radioactive mixture from pitchblende, an ore of uranium. This mixture was primarily a compound of barium. When the mixture was heated in a flame, however, it gave a new atomic line spectrum, in addition to the spectrum for barium. The Curies based their discovery of a new element on this finding. It took them four more years to obtain a pure compound of radium. Radium, like uranium, is a radioactive element. But in most of its chemical and physical properties, radium is similar to the nonradioactive element barium. It was this similarity that made the final separation of the new element so difficult.

Chemists had long known that groups of elements have similar properties. In 1869 Dmitri Mendeleev found that when the elements were arranged in a particular way, they fell into columns, with elements in the same column displaying similar properties. Thus, Mendeleev placed beryllium, calcium, strontium, and barium in one column. Now, with the Curies' discovery, radium was added to this column.

Mendeleev's arrangement of the elements, the *periodic table,* was originally based on the observed chemical and physical properties of the elements and their compounds. We now explain this arrangement in terms of the electronic structure of atoms. In this chapter we will look at this electronic structure and its relationship to the periodic table of elements.

Electronic Structure of Atoms

In Chapter 7 we found that an electron in an atom has four quantum numbers—n, l, m_l, and m_s—associated with it. The first three quantum numbers characterize the orbital that describes the region of space where an electron is most likely to be found; we say that the electron "occupies" this orbital. The spin quantum number, m_s, describes the spin orientation of an electron. In the first section, we will look further at electron spin; then we will discuss how electrons are distributed among the possible orbitals of an atom.

8.1 Electron Spin and the Pauli Exclusion Principle

Otto Stern and Walther Gerlach first observed electron spin magnetism in 1921. They directed a beam of silver atoms into the field of a specially designed magnet. The same experiment can be done with hydrogen atoms. The beam of hydrogen atoms is split into two by the magnetic field; half of the atoms are bent in one direction and half in the other (see Figure 8.2). The fact that the atoms are affected by the laboratory magnet shows that they themselves act as magnets.

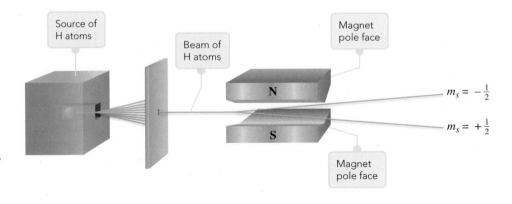

Source of H atoms

Beam of H atoms

Magnet pole face

N

S

$m_s = -\frac{1}{2}$

$m_s = +\frac{1}{2}$

Magnet pole face

The beam of hydrogen atoms is split into two because the electron in each atom behaves as a tiny magnet with only two possible orientations. In effect, the electron acts as though it were a ball of spinning charge (Figure 8.3), and, like a circulating electric charge, the electron would create a magnetic field. Electron spin, however, is subject to a quantum restriction on the possible directions of the spin axis. The resulting directions of spin magnetism correspond to spin quantum numbers $m_s = +\frac{1}{2}$ and $m_s = -\frac{1}{2}$. ▶

Protons and many nuclei also have spin. See the Instrumental Methods essay in this section.

Electron Configurations and Orbital Diagrams

We describe the electrons in an atom by the atom's electron configuration. An **electron configuration** of an atom is *a particular distribution of electrons among the available subshells.* A subshell, as we noted in the previous chapter, consists of a group of orbitals having the same n and l quantum numbers but different m_l values. Recall that we denote each subshell by its principal quantum number, n, followed by a letter standing for its l quantum number (*s, p, d, f,* etc.).

The notation for a configuration lists the subshell symbols one after the other, with a superscript giving the number of electrons in that subshell. For example, a configuration of the lithium atom (atomic number 3) with two electrons in the $1s$ subshell and one electron in the $2s$ subshell is written $1s^2 2s^1$.

Whereas the configuration gives the number of electrons in each subshell, we use *a diagram to show how the orbitals of a subshell are occupied by electrons.* It is called an **orbital diagram**. An orbital is represented by a circle. Each group of orbitals in a subshell is labeled by its subshell notation. An electron in an orbital is shown by an arrow; the arrow points up when $m_s = +\frac{1}{2}$ and down when $m_s = -\frac{1}{2}$. The orbital diagram

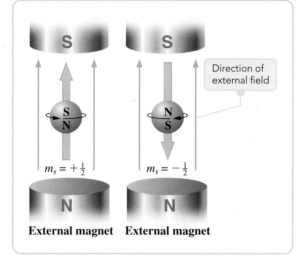

Figure 8.3 ▲

A representation of electron spin The two possible spin orientations are indicated by the models. By convention, the spin direction is given as shown by the large arrow on the spin axis. Electrons behave as tiny bar magnets, as shown in the figure.

$1s$ $2s$ $2p$

shows the electronic structure of an atom in which there are two electrons in the $1s$ subshell, or orbital (one electron with $m_s = +\frac{1}{2}$, the other with $m_s = -\frac{1}{2}$); two electrons in the $2s$ subshell ($m_s = +\frac{1}{2}$, $m_s = -\frac{1}{2}$); and one electron in the $2p$ subshell ($m_s = +\frac{1}{2}$). The electron configuration is $1s^2 2s^2 2p^1$.

Pauli Exclusion Principle

Not all of the conceivable arrangements of electrons among the orbitals of an atom are physically possible. The **Pauli exclusion principle**, which summarizes experimental observations, states that *no two electrons in an atom can have the same four quantum numbers.* If one electron in an atom has the quantum numbers $n = 1$, $l = 0$, $m_l = 0$, and $m_s = +\frac{1}{2}$, no other electron can have these same quantum numbers. In other words, you cannot place two electrons with the same value of m_s in a $1s$ orbital. The orbital diagram

$1s$

is not a possible arrangement of electrons.

Because there are only two possible values of m_s, an orbital can hold no more than two electrons—and then only if the two electrons have different spin quantum numbers. In an orbital diagram, an orbital with two electrons must be written

Nuclear Magnetic Resonance (NMR)

You have just seen that electrons have a spin and as a result behave like tiny magnets. Protons and neutrons similarly have spins. Therefore, depending on the arrangement of protons and neutrons, a nucleus could have spin. A nucleus with spin will act like a bar magnet, similar to the electron although many times smaller in magnitude. Examples of nuclei with spin are hydrogen-1 (proton), carbon-13 (but carbon-12, the most abundant nuclide of carbon, has no spin), and fluorine-19.

Although nuclear magnetism is much smaller than that of electrons, with the correct equipment it is easily seen and in fact forms the basis of *nuclear magnetic resonance* (NMR) spectroscopy, one of the most important methods for determining molecular structure. Nuclear magnetism also forms the basis of the medical diagnostic tool *magnetic resonance imaging,* or MRI (Figure 8.4).

The essential features of NMR can be seen if you consider the proton. Like the electron, the proton has two spin states. In the absence of a magnetic field, these spin states have the same energy, but in the field of a strong magnet (external field), they have different energies. The state in which the proton magnetism is aligned with the external field, so the south pole of the proton magnet faces the north pole of the external magnet, will have lower energy. The state in which the proton magnet is turned 180°, with its south pole facing the south pole of the external magnet, will have higher energy. Now if a proton in the lower spin state is irradiated with electromagnetic waves of the proper frequency (in which the photon has energy equal to the difference in energy of the spin states), the proton can change to the higher spin state. The frequency absorbed by the proton depends on the magnitude of the magnetic field. For the magnets used in these instruments, the radiation lies in the radio-frequency range. Frequencies commonly used are 300 MHz and 900 MHz.

Figure 8.5 shows a diagram of an NMR spectrometer. It consists of a sample in the field of a variable electromagnet and near two coils, one a radio wave transmitter and the other a receiver coil

perpendicular to the transmitter coil (so that the receiver will not pick up the signal from the transmitter). Suppose the transmitter radiates waves of 300 MHz. If the sample absorbs these radio waves, protons will undergo transitions from the lower to the higher spin state. Once protons are in the higher energy state, they tend to lose energy, going back to the lower spin state and radiating 300-MHz radio waves. Thus, the sample acts like a transmitter, but one with coils in various directions, so the signal can be detected by the receiver coil.

In general, the sample will not absorb at the chosen frequency. But you can change the energy difference between spin states, and therefore the frequency that is absorbed by the sample, by increasing or decreasing the magnitude of the external magnetic field using small coils on the magnet pole faces. In so doing you can, in effect, "tune" the sample, or bring it into "resonance" with the transmitter frequency. (Alternatively, many modern instruments vary the transmitter frequency to obtain resonance.)

Because each proton in a substance is surrounded by electrons that have their own magnetic fields, the magnetic environment of a proton depends to some extent on the bonding, or chemical, environment in which the proton is involved. So, the external magnetic field needed to bring a proton into resonance with the 300-MHz radiation, say, varies with the chemical (bonding) environment. A spectrum is produced by recording the magnetic field (or frequency) at which the protons in a molecule produce a resonance signal. Figure 8.6 shows a high-resolution NMR spectrum of ethanol.

Ethanol, whose molecular structure is

$$\begin{array}{ccc} & H & H \\ & | & | \\ H-O-&C-&C-H \\ & | & | \\ & H & H \end{array}$$

has protons in three different chemical environments: a proton bonded to an oxygen atom (H—O—), the protons in a —CH$_2$— group, and the protons in a —CH$_3$ group. Each of these three

Figure 8.4 ▶

Magnetic resonance imaging A patient's head is placed in a large magnet and subjected to a radio pulse. Proton spin transitions give rise to a radio wave emission that can be analyzed electronically and converted by computer to a two-dimensional image of a plane portion of the brain. Paul C. Lauterbur, a chemist at the University of Illinois (Urbana), and Sir Peter Mansfield, University of Nottingham (UK), won the Nobel Prize in Physiology or Medicine in 2003 for their work in the early 1970s in developing MRI.

Figure 8.5 ◄

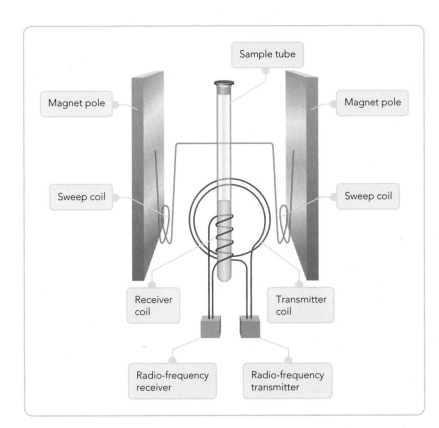

Nuclear magnetic resonance spectrometer
A sample in the tube is in a magnetic field, which can be varied by changing the electric current in the *sweep* coils (shown in yellow). A radio-frequency transmitter radiates radio waves from a coil (shown in red) to the sample. The sample absorbs these waves if the radio frequency corresponds to the difference in energy of the sample's nuclear spin states. When the nuclear spins go back to the ground state, the sample emits radio waves, which are detected by the receiver coil (in black).

types of protons shows up as a peak or, as we will see, a group of peaks in the NMR spectrum. The position of the center of a group of peaks for a given proton relative to some standard is referred to as the *chemical shift* for that proton, because it depends on the chemical environment of the proton. By measuring the area under the peaks for a given type of proton (usually done electronically by the spectrometer), a researcher can discover the number of protons of that type, effectively giving the chemist structural information about the molecule.

As we noted earlier, a given proton may give rise to several peaks in a high-resolution spectrum. (At low resolution, these separate peaks may appear as only one for each proton.) For example, the protons for the —CH₂— group of ethanol give four closely spaced peaks. These peaks arise because of the interaction of the proton spins on this group with those protons on the adjacent —CH₃ group. The number of peaks (four) for the —CH₂— group is one more than the number of protons on the adjacent group (three).

NMR spectroscopy is one of the most important tools a chemist has in determining the identity of a substance and its molecular structure. The method is quick and can yield such information as the chemical bonding

environment of the protons and the number of protons having a given environment. It can also give information about other nuclei, such as carbon-13.

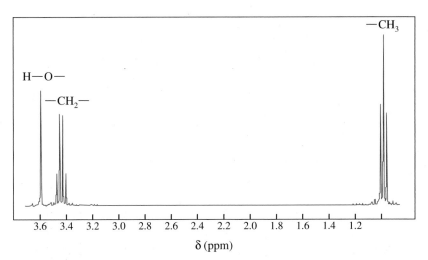

Figure 8.6 ▲

High-resolution NMR spectrum of ethanol, CH₃CH₂OH The protons of a given type occur as a peak or closely spaced group of peaks. The position of the center of such a group of peaks relative to some standard is referred to as the *chemical shift*. The peaks within a group arise from the interaction of the spins between protons of one type with those of another type.

■ See Problems 8.85 and 8.86.

with arrows pointing in opposite directions. The two electrons are said to have opposite spins. We can restate the Pauli exclusion principle:

> **Pauli exclusion principle:** An orbital can hold at most two electrons, and then only if the electrons have opposite spins.

Each subshell holds a maximum of twice as many electrons as the number of orbitals in the subshell. Thus, a $2p$ subshell, which has three orbitals (with $m_l = -1$, 0, and $+1$), can hold a maximum of six electrons. The maximum number of electrons in various subshells is given in the following table.

Subshell	Number of Orbitals	Maximum Number of Electrons
s ($l = 0$)	1	2
p ($l = 1$)	3	6
d ($l = 2$)	5	10
f ($l = 3$)	7	14

Example 8.1

Applying the Pauli Exclusion Principle

Gaining Mastery Toolbox

Critical Concept 8.1
The Pauli exclusion principle states that an orbital can hold at most two electrons and then only if the electrons have opposite spin.

Solution Essentials:
• Pauli exclusion principle
• The number of orbitals in a subshell of given l is $2l + 1$

Which of the following orbital diagrams or electron configurations are possible and which are impossible, according to the Pauli exclusion principle? Explain.

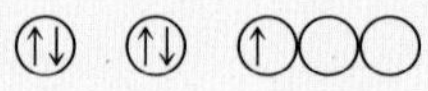

e $1s^2 2s^1 2p^7$ f $1s^2 2s^2 2p^6 3s^2 3p^6 3d^8 4s^2$

Problem Strategy In any orbital diagram, make sure that each orbital contains no more than two electrons, which have opposite spins. In any electron configuration, make sure that a subshell of given l contains no more than $2(2l + 1)$ electrons (there are $2l + 1$ different orbitals of given l, and each of these orbitals can hold two electrons).

Solution a **Possible** orbital diagram. b **Impossible** orbital diagram; there are three electrons in the $2s$ orbital. c **Impossible** orbital diagram; there are two electrons in a $2p$ orbital with the same spin. d **Impossible** electron configuration; there are three electrons in the $1s$ subshell (one orbital). e **Impossible** electron configuration; there are seven electrons in the $2p$ subshell (which can hold only six electrons). f **Possible**. Note that the $3d$ subshell can hold as many as ten electrons.

Answer Check The overarching idea here is that each orbital in an atom can hold a maximum of only two electrons, and then only if the two electrons have opposite spins. For example, a p subshell has three orbitals and therefore holds a maximum of six electrons.

(continued)

(continued)

Exercise 8.1 Look at the following orbital diagrams and electron configurations. Which are possible and which are not, according to the Pauli exclusion principle? Explain.

a. $\uparrow$ $\uparrow$ $\bigcirc\bigcirc\bigcirc$
   1s   2s   2p

b. $\uparrow$ $\uparrow$ $\uparrow\downarrow\uparrow\downarrow\uparrow\downarrow$
   1s   2s   2p

c. $\uparrow\downarrow$ $\uparrow\downarrow$ $\uparrow\downarrow\uparrow\downarrow\uparrow\downarrow$
   1s   2s   2p

d. $1s^2 2s^2 2p^4$

e. $1s^2 2s^4 2p^2$

f. $1s^2 2s^2 2p^6 3s^2 3p^{10} 3d^{10}$

■ See Problems 8.41, 8.42, 8.43, and 8.44.

8.2 Building-Up Principle and the Periodic Table

Every atom has an infinite number of possible electron configurations. The configuration associated with the lowest energy level of the atom corresponds to a quantum-mechanical state called the *ground state*. Other configurations correspond to *excited states*, associated with energy levels other than the lowest. For example, the ground state of the sodium atom is known from experiment to have the electron configuration $1s^2 2s^2 2p^6 3s^1$. The electron configuration $1s^2 2s^2 2p^6 3p^1$ represents an excited state of the sodium atom. ▶

The chemical properties of an atom are related primarily to the electron configuration of its ground state. Table 8.1 lists the experimentally determined

The transition of the sodium atom from the excited state $1s^2 2s^2 2p^6 3p^1$ to the ground state $1s^2 2s^2 2p^6 3s^1$ is accompanied by the emission of yellow light at 589 nm. Excited states of an atom are needed to describe its spectrum.

Table 8.1 Ground-State Electron Configurations of Atoms Z = 1 to 36*

Z	Element	Configuration	Z	Element	Configuration
1	H	$1s^1$	19	K	$1s^2 2s^2 2p^6 3s^2 3p^6 4s^1$
2	He	$1s^2$	20	Ca	$1s^2 2s^2 2p^6 3s^2 3p^6 4s^2$
3	Li	$1s^2 2s^1$	21	Sc	$1s^2 2s^2 2p^6 3s^2 3p^6 3d^1 4s^2$
4	Be	$1s^2 2s^2$	22	Ti	$1s^2 2s^2 2p^6 3s^2 3p^6 3d^2 4s^2$
5	B	$1s^2 2s^2 2p^1$	23	V	$1s^2 2s^2 2p^6 3s^2 3p^6 3d^3 4s^2$
6	C	$1s^2 2s^2 2p^2$	24	Cr	$1s^2 2s^2 2p^6 3s^2 3p^6 3d^5 4s^1$
7	N	$1s^2 2s^2 2p^3$	25	Mn	$1s^2 2s^2 2p^6 3s^2 3p^6 3d^5 4s^2$
8	O	$1s^2 2s^2 2p^4$	26	Fe	$1s^2 2s^2 2p^6 3s^2 3p^6 3d^6 4s^2$
9	F	$1s^2 2s^2 2p^5$	27	Co	$1s^2 2s^2 2p^6 3s^2 3p^6 3d^7 4s^2$
10	Ne	$1s^2 2s^2 2p^6$	28	Ni	$1s^2 2s^2 2p^6 3s^2 3p^6 3d^8 4s^2$
11	Na	$1s^2 2s^2 2p^6 3s^1$	29	Cu	$1s^2 2s^2 2p^6 3s^2 3p^6 3d^{10} 4s^1$
12	Mg	$1s^2 2s^2 2p^6 3s^2$	30	Zn	$1s^2 2s^2 2p^6 3s^2 3p^6 3d^{10} 4s^2$
13	Al	$1s^2 2s^2 2p^6 3s^2 3p^1$	31	Ga	$1s^2 2s^2 2p^6 3s^2 3p^6 3d^{10} 4s^2 4p^1$
14	Si	$1s^2 2s^2 2p^6 3s^2 3p^2$	32	Ge	$1s^2 2s^2 2p^6 3s^2 3p^6 3d^{10} 4s^2 4p^2$
15	P	$1s^2 2s^2 2p^6 3s^2 3p^3$	33	As	$1s^2 2s^2 2p^6 3s^2 3p^6 3d^{10} 4s^2 4p^3$
16	S	$1s^2 2s^2 2p^6 3s^2 3p^4$	34	Se	$1s^2 2s^2 2p^6 3s^2 3p^6 3d^{10} 4s^2 4p^4$
17	Cl	$1s^2 2s^2 2p^6 3s^2 3p^5$	35	Br	$1s^2 2s^2 2p^6 3s^2 3p^6 3d^{10} 4s^2 4p^5$
18	Ar	$1s^2 2s^2 2p^6 3s^2 3p^6$	36	Kr	$1s^2 2s^2 2p^6 3s^2 3p^6 3d^{10} 4s^2 4p^6$

*A complete table appears in Appendix D.

The quantum numbers and characteristics of orbitals were discussed in Section 7.5.

ground-state electron configurations of atoms $Z = 1$ to $Z = 36$. (A complete table appears in Appendix D.)

Building-Up Principle (Aufbau Principle)

Most of the configurations in Table 8.1 can be explained in terms of the **building-up principle** (or **Aufbau principle**), *a scheme used to reproduce the electron configurations of the ground states of atoms by successively filling subshells with electrons in a specific order (the building-up order).* Following this principle, you obtain the electron configuration of an atom by successively filling subshells in the following order: 1s, 2s, 2p, 3s, 3p, 4s, 3d, 4p, 5s, 4d, 5p, 6s, 4f, 5d, 6p, 7s, 5f. This order reproduces the experimentally determined electron configurations (with some exceptions, which we will discuss later). You need not memorize this order. As you will see, you can very easily obtain it from the periodic table. (You can also reproduce this order from the mnemonic diagram shown in Figure 8.7.)

The building-up order corresponds for the most part to increasing energy of the subshells. You might expect this. By filling orbitals of lowest energy first, you usually get the lowest total energy (ground state) of the atom. Recall that the energy of an orbital depends only on the quantum numbers n and l. ◄ (The energy of the H atom, however, depends only on n.) Orbitals with the same n and l but different m_l—that is, different orbitals of the same subshell—have the same energy. The energy depends primarily on n, increasing with its value. For example, a 3s orbital has greater energy than a 2s orbital. Except for the H atom, the energies of orbitals with the same n increase with the l quantum number. A 3p orbital has slightly greater energy than a 3s orbital. The orbital of lowest energy is 1s; next higher are 2s and 2p, then 3s and 3p. The order of these subshells by energy, from 1s to 3p, follows the building-up order as listed earlier.

When subshells have nearly the same energy, however, the building-up order is not strictly determined by the order of their energies. The ground-state configurations, which we are trying to predict by the building-up order, are determined by the *total* energies of the atoms. The total energy of an atom depends not only on the energies of the subshells but also on the energies of interaction among the different subshells. It so happens that for all elements with $Z = 21$ or greater, the energy of the 3d subshell is lower than the energy of the 4s subshell (Figure 8.8), which is opposite to the building-up order. You need the building-up order to predict the electron configurations of the ground states of atoms.

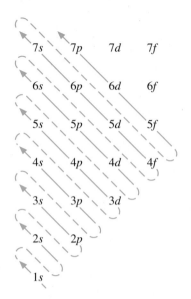

Figure 8.7 ▲

A mnemonic diagram for the building-up order (diagonal rule) You obtain this diagram by writing the subshell(s) in rows, each row having subshell(s) of given *n*. Within each row, you arrange the subshell(s) by increasing *l*. (You can stop after writing the *nf* subshell(s), since no known elements contain *g* or higher subshell(s).) Now, starting with the 1s subshell(s), draw a series of diagonals, as shown. The building-up order is the order in which these diagonals strike the subshell(s).

Figure 8.8 ▶

Orbital energies for the scandium atom (Z = 21) Note that in the scandium atom, unlike the hydrogen atom, the subshells for each *n* are spread apart in energy. Thus, the 2p energy is above the 2s. Similarly, the $n = 3$ subshells are spread to give the order 3s < 3p < 3d. The 3d subshell energy is now just below the 4s. (Values for this figure were calculated from theory by Charlotte F. Fischer, Vanderbilt University.)

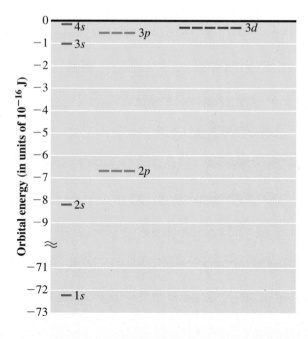

Now you can see how to reproduce the electron configurations of Table 8.1 using the building-up principle. Remember that the number of electrons in a neutral atom equals the atomic number Z. (The nuclear charge is $+Z$.) In the case of the simplest atom, hydrogen ($Z = 1$), you obtain the ground state by placing the single electron into the $1s$ orbital, giving the configuration $1s^1$ (this is read as "one-ess-one"). Now you go to helium ($Z = 2$). The first electron goes into the $1s$ orbital, as in hydrogen, followed by the second electron, because any orbital can hold two electrons. The configuration is $1s^2$. Filling the $n = 1$ shell creates a very stable configuration, and as a result, helium is chemically unreactive.

You continue this way through the elements, each time increasing Z by 1 and adding another electron. You obtain the configuration of an atom from that of the preceding element by adding an electron into the next available orbital, following the building-up order. In lithium ($Z = 3$), the first two electrons give the configuration $1s^2$, like helium, but the third electron goes into the next higher orbital in the building-up order, because the $1s$ orbital is now filled. This gives the configuration $1s^2 2s^1$. In beryllium ($Z = 4$), the fourth electron fills the $2s$ orbital, giving the configuration $1s^2 2s^2$.

Using the abbreviation [He] for $1s^2$, the configurations are

$Z = 3$	lithium	$1s^2 2s^1$	or	[He]$2s^1$
$Z = 4$	beryllium	$1s^2 2s^2$	or	[He]$2s^2$

With boron ($Z = 5$), the electrons begin filling the $2p$ subshell. You get

$Z = 5$	boron	$1s^2 2s^2 2p^1$	or	[He]$2s^2 2p^1$
$Z = 6$	carbon	$1s^2 2s^2 2p^2$	or	[He]$2s^2 2p^2$
$Z = 10$	neon	$1s^2 2s^2 2p^6$	or	[He]$2s^2 2p^6$

Having filled the $2p$ subshell, you again find a particularly stable configuration. Neon is chemically unreactive as a result.

With sodium ($Z = 11$), the $3s$ orbital begins to fill. Using the abbreviation [Ne] for $1s^2 2s^2 2p^6$, you have

$Z = 11$	sodium	$1s^2 2s^2 2p^6 3s^1$	or	[Ne]$3s^1$
$Z = 12$	magnesium	$1s^2 2s^2 2p^6 3s^2$	or	[Ne]$3s^2$

Then the $3p$ subshell begins to fill.

$Z = 13$	aluminum	$1s^2 2s^2 2p^6 3s^2 3p^1$	or	[Ne]$3s^2 3p^1$
$Z = 18$	argon	$1s^2 2s^2 2p^6 3s^2 3p^6$	or	[Ne]$3s^2 3p^6$

With the $3p$ subshell filled, a stable configuration has been attained; argon is an unreactive element.

Now the $4s$ orbital begins to fill. You get [Ar]$4s^1$ for potassium ($Z = 19$) and [Ar]$4s^2$ for calcium ($Z = 20$) ([Ar] $= 1s^2 2s^2 2p^6 3s^2 3p^6$). At this point the $3d$ subshell begins to fill. You get [Ar]$3d^1 4s^2$ for scandium ($Z = 21$), [Ar]$3d^2 4s^2$ for titanium ($Z = 22$), and [Ar]$3d^3 4s^2$ for vanadium ($Z = 23$). *Note that we have written the configurations with subshells arranged in order by shells.* This generally places the subshells in order by energy and puts the subshells involved in chemical reactions at the far right. ▶

Let us skip to zinc ($Z = 30$). The $3d$ subshell has filled; the configuration is [Ar]$3d^{10} 4s^2$. Now the $4p$ subshell begins to fill, starting with gallium ($Z = 31$), configuration [Ar]$3d^{10} 4s^2 4p^1$, and ending with krypton ($Z = 36$), configuration [Ar]$3d^{10} 4s^2 4p^6$.

Although the building-up order reproduces the ground-state electron configurations, it has no other significance. The order by n (and then by l within a given n), however, generally places the most easily ionized orbitals at the far right. For example, the electron configuration of Fe (in order by shells) is $1s^2 2s^2 2p^6 3s^2 3p^6 3d^6 4s^2$. The 4s electrons ionize first.

Electron Configurations and the Periodic Table

By this time you can see a pattern develop among the ground-state electron configurations of the atoms. This pattern explains the periodic table, which was briefly described in Section 2.5. Consider helium, neon, argon, and krypton, elements in Group VIIIA of the periodic table. Neon, argon, and krypton have configurations in which a p subshell has just filled. (Helium has a filled $1s$ subshell; no $1p$ subshell is possible.)

helium	$1s^2$
neon	$1s^2 2s^2 2p^6$
argon	$1s^2 2s^2 2p^6 3s^2 3p^6$
krypton	$1s^2 2s^2 2p^6 3s^2 3p^6 3d^{10} 4s^2 4p^6$

These elements are the first members of the group called *noble gases* because of their relative unreactivity.

Look now at the configurations of beryllium, magnesium, and calcium, members of the group of *alkaline earth metals* (Group IIA), which are similar, moderately reactive elements.

beryllium	$1s^2 2s^2$	or	$[\text{He}]2s^2$
magnesium	$1s^2 2s^2 2p^6 3s^2$	or	$[\text{Ne}]3s^2$
calcium	$1s^2 2s^2 2p^6 3s^2 3p^6 4s^2$	or	$[\text{Ar}]4s^2$

Each of these configurations consists of a **noble-gas core**, that is, *an inner-shell configuration corresponding to one of the noble gases,* plus two outer electrons with an ns^2 configuration.

The elements boron, aluminum, and gallium (Group IIIA) also have similarities. Their configurations are

boron	$1s^2 2s^2 2p^1$	or	$[\text{He}]2s^2 2p^1$
aluminium	$1s^2 2s^2 2p^6 3s^2 3p^1$	or	$[\text{Ne}]3s^2 3p^1$
gallium	$1s^2 2s^2 2p^6 3s^2 3p^6 3d^{10} 4s^2 4p^1$	or	$[\text{Ar}]3d^{10} 4s^2 4p^1$

Boron and aluminum have noble-gas cores plus three electrons with the configuration $ns^2 np^1$. Gallium has an additional filled $3d$ subshell. *The noble-gas core together with $(n - 1)d^{10}$ electrons* is often referred to as a **pseudo-noble-gas core**, because these electrons usually are not involved in chemical reactions.

An electron in an atom outside the noble-gas or pseudo-noble-gas core is called a **valence electron**. Such electrons are primarily involved in chemical reactions, and similarities among the configurations of valence electrons (the *valence-shell configurations*) account for similarities of the chemical properties among groups of elements.

Figure 8.9 shows a periodic table with the valence-shell configurations included. Note the similarity in electron configuration within any group (column) of elements. This similarity explains what chemists since Mendeleev have known—the properties of elements in any group are similar.

The *main-group* (or *representative*) *elements* all have valence-shell configurations $ns^a np^b$, with some choice of a and b. (b could be equal to 0.) In other words, the outer s or p subshell is being filled. Similarly, in the *d-block transition elements* (often called simply *transition elements* or *transition metals*), a d subshell is being filled. In the *f-block transition elements* (or *inner transition elements*), an f subshell is being filled. (See Figure 8.9 or Appendix D for the configurations of these elements.)

Exceptions to the Building-Up Principle

As we have said, the building-up principle reproduces most of the ground-state configurations correctly. There are some exceptions, however, and chromium ($Z = 24$) is the first we encounter. The building-up principle predicts the configuration

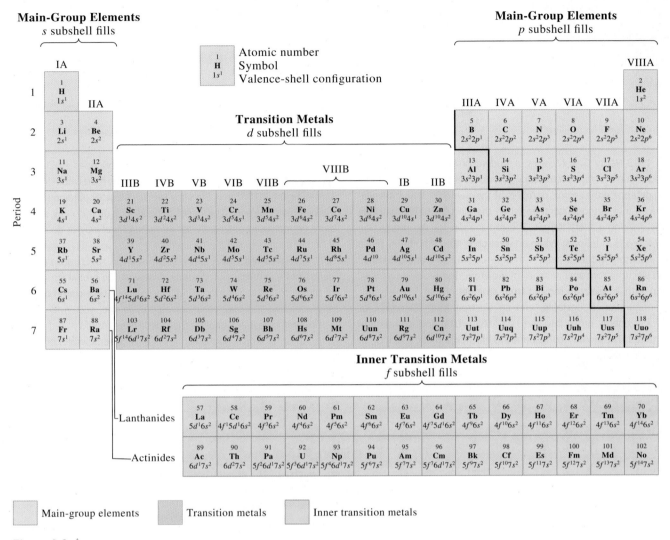

Figure 8.9 ▲

A periodic table This table shows the valence-shell configurations of the elements.

$[Ar]3d^44s^2$, though the correct one is found experimentally to be $[Ar]3d^54s^1$. These two configurations are actually very close in total energy because of the closeness in energies of the $3d$ and $4s$ orbitals (Figure 8.8). For that reason, small effects can influence which of the two configurations is actually lower in energy. Copper ($Z = 29$) is another exception to the building-up principle, which predicts the configuration $[Ar]3d^94s^2$, although experiment shows the ground-state configuration to be $[Ar]3d^{10}4s^1$.

We need not dwell on these exceptions beyond noting that they occur. The point to remember is that the configuration predicted by the building-up principle is very close in energy to the ground-state configuration (if it is not the ground state). Most of the qualitative conclusions regarding the chemistry of an element are not materially affected by arguing from the configuration given by the building-up principle. ▶

More exceptions occur among the heavier transition elements, where the outer subshells are very close together. We must concede that simplicity was not the uppermost concern in the construction of the universe!

CONCEPT CHECK 8.1

Imagine a world in which the Pauli principle is "No more than one electron can occupy an atomic orbital, irrespective of its spin." How many elements would there be in the second row of the periodic table, assuming that nothing else is different about this world?

X Rays, Atomic Numbers, and Orbital Structure (Photoelectron Spectroscopy)

In 1913 Henry G. J. Moseley, a student of Rutherford, used the technique of *x-ray spectroscopy* (just discovered by Max von Laue) to determine the atomic numbers of the elements. X rays are produced in a cathode-ray tube when the electron beam (cathode ray) falls on a metal target. The explanation for the production of x rays is as follows: When an electron in the cathode ray hits a metal atom in the target, it can (if it has sufficient energy) knock an electron from an inner shell of the atom. This produces a metal ion with an electron missing from an inner orbital. The electron configuration is unstable, and an electron from an orbital of higher energy drops into the half-filled orbital and a photon is emitted. The photon corresponds to electromagnetic radiation in the x-ray region.

The energies of the inner orbitals of an atom and the energy changes between them depend on the nuclear charge, $+Z$.

Therefore, the photon energies $h\nu$ and the frequencies ν of emitted x rays depend on the atomic number Z of the metal atom in the target. Figure 8.10 shows the x-ray spectra Moseley obtained with various metal targets. The direct dependence of the x-ray spectrum on atomic number provides an unequivocal way of deciding whether a substance is a pure element or not.

A related technique, *x-ray photoelectron spectroscopy,* experimentally confirms our theoretical view of the orbital structure of the atom. Instead of irradiating a sample with an electron beam and analyzing the frequencies of emitted x rays, you irradiate a sample with x rays and analyze the kinetic energies of ejected electrons. In other words, you observe the *photoelectric effect* on the sample (see Section 7.2).

As an example of photoelectron spectroscopy, consider a sample of neon gas (Ne atoms). Suppose the sample is irradiated

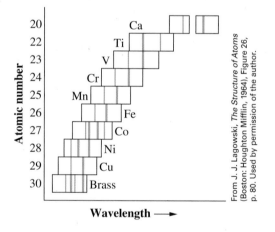

From J. J. Lagowski, *The Structure of Atoms* (Boston: Houghton Mifflin, 1964), Figure 26, p. 80. Used by permission of the author.

Figure 8.10 ◄

X-ray spectra of the elements calcium to zinc, obtained by Moseley Each line results from an emission of given wavelength. Because of the volatility of zinc, Moseley used brass (a copper–zinc alloy) to observe the spectrum of zinc. Note the copper lines in brass. Also note how the lines progress to the right (indicating increasing wavelength or decreasing energy difference) with decreasing atomic number.

8.3 Writing Electron Configurations Using the Periodic Table

To discuss bonding and the chemistry of the elements coherently, you must be able to reproduce the electron configurations with ease, following the building-up principle. All you need is some facility in recall of the building-up order of subshells.

One approach is to recall the structure of the periodic table. Because that structure is basic, it offers a sound way to remember the building-up order. There is a definite pattern to the order of filling of the subshells as you go through the with x rays of a specific frequency great enough to remove a 1s electron from the neon atom. Part of the energy of the x-ray photon, $h\nu$, is used to remove the electron from the atom. (This is the *ionization energy, I.E.,* for that electron.) The remaining energy appears as kinetic energy, E_k, of the ejected electron. From the law of conservation of energy, you can write

$$E_k = h\nu - I.E.$$

Because $h\nu$ is fixed, E_k will depend linearly on *I.E.*, the ionization energy.

If you look at the electrons ejected from neon, you find that they have kinetic energies related to the ionization energies from all possible orbitals (1s, 2s, and 2p) in the atom. When you scan the various kinetic energies of ejected electrons, you see a spectrum with peaks corresponding to the different occupied orbitals (see Figure 8.11a). These ionization energies are approximately equal to the positive values of the orbital energies (Figure 8.11B), so this spectrum provides direct experimental verification of the discrete energy levels associated with the electrons of the atom.

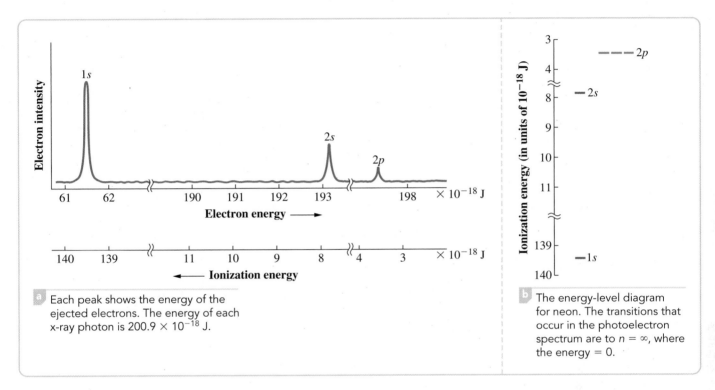

a Each peak shows the energy of the ejected electrons. The energy of each x-ray photon is 200.9×10^{-18} J.

b The energy-level diagram for neon. The transitions that occur in the photoelectron spectrum are to $n = \infty$, where the energy = 0.

Figure 8.11 ▲

X-ray photoelectron spectrum of neon

■ See Problems 8.87 and 8.88.

elements in the periodic table, and from this you can write the building-up order. Figure 8.12 shows a periodic table stressing this pattern. For example, in the violet-colored area, an *ns* subshell is being filled. In the blue-colored area, an *np* subshell is being filled. The value of *n* is obtained from the period (row) number. In the red area, an $(n - 1)d$ subshell is being filled.

You read the building-up order by starting with the first period, in which the 1s subshell is being filled. In the second period, you have 2s (violet area); then, staying in the same period but jumping across, you have 2p (blue area). In the

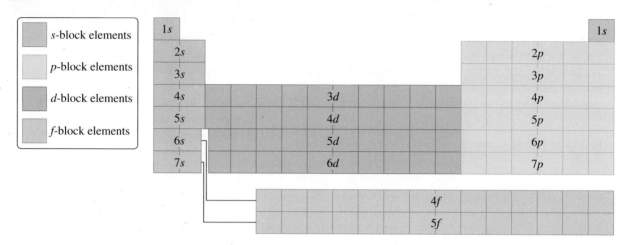

Figure 8.12 ▲

A periodic table illustrating the building-up order The colored areas of elements show the different subshells that are filling with those elements.

third period, you have $3s$ and $3p$; in the fourth period, $4s$ (violet area), $3d$ (red area), and $4p$ (blue area). This pattern should become clear enough to visualize with a periodic table that is not labeled with the subshells, such as the one on the inside front cover of this book. The detailed method is illustrated in the next example. (An alternative way to reconstruct the building-up order of subshells is to write the diagram in Figure 8.7.)

Note on writing electron configurations of atoms: If you write the configuration of an atom with its subshells arranged by shells, the orbitals will usually be in order by energy, with the subshells used to describe chemical reactivity at the far right. For example, the configuration of Br would be written $[Ar]3d^{10}4s^{2}4p^{5}$. However, you will also see the orbitals in a configuration written in the building-up order, which is an empirical order that has been found to reproduce most of the atomic configurations. The configuration of Br would be written $[Ar]4s^{2}3d^{10}4p^{5}$. Ask your instructor whether he or she has a preference.

⬥WL **Interactive Example** 8.2   **Determining the Configuration of an Atom Using the Building-Up Principle**

Gaining Mastery Toolbox

Critical Concept 8.2
To write the electron configuration for the ground state of a neutral atom of atomic number Z, write the subshells in building-up order and then distribute Z electrons into these subshells, starting with the $1s$ subshell.

Solution Essentials:
- Building-up order of subshells
- Pauli exclusion principle
- The number of orbitals in a subshell of given l is $2l + 1$.

Use the building-up principle to obtain the configuration for the ground state of the gallium atom ($Z = 31$). Give the configuration in complete form (do not abbreviate for the core). What is the valence-shell configuration?

Problem Strategy Write the subshells in their building-up order. You can use any periodic table to do this. All you have to remember is the pattern in which subshells are filled as you progress through the table (Figure 8.12). Go through the periods, starting with

hydrogen, writing the subshells that are being filled, and stopping with the element whose configuration you want (gallium in this example). Then you distribute the electrons (equal to the atomic number of the element) to the subshells. The valence-shell configuration includes only the subshells outside the noble-gas or pseudo-noble-gas core.

Solution From a periodic table, you get the following building-up order:

	$1s$	$2s\ 2p$	$3s\ 3p$	$4s\ 3d\ 4p$
Period:	first	second	third	fourth

Now you fill the subshells with electrons, remembering that you have a total of 31 electrons to distribute. You get

$$1s^{2}2s^{2}2p^{6}3s^{2}3p^{6}4s^{2}3d^{10}4p^{1}$$

Or, if you rearrange the subshells by shells, you write

$$\mathbf{1s^{2}2s^{2}2p^{6}3s^{2}3p^{6}3d^{10}4s^{2}4p^{1}}$$

The valence-shell configuration is $\mathbf{4s^{2}4p^{1}}$.

(continued)

(continued)

Answer Check The principal quantum number of the valence shell must equal the period of the element. If your answer is not consistent with this, examine your work for error. Here, gallium is in period 4, and the principal quantum number of the valence shell is 4.

Exercise 8.2   Use the building-up principle to obtain the electron configuration for the ground state of the manganese atom ($Z = 25$).

See Problems 8.47, 8.48, 8.49, and 8.50.

In many cases, you need only the configuration of the outer electrons. You can determine this from the position of the element in the periodic table. Recall that the valence-shell configuration of a main-group element is $ns^a np^b$, where n, the principal quantum number of the outer shell, also equals the period number for the element. The total number of valence electrons, which equals $a + b$, can be obtained from the group number. For example, gallium is in Period 4, so $n = 4$. It is in Group IIIA, so the number of valence electrons is 3. This gives the valence-shell configuration $4s^2 4p^1$. The configuration of outer shells of a transition element is obtained in a similar fashion. The next example gives the details.

Example 8.3   **Determining the Configuration of an Atom Using the Period and Group Numbers**

Gaining Mastery Toolbox

Critical Concept 8.3
To write the valence-shell (or outer-shell) configuration of an element using the periodic table, determine the period and group number of the element, noting whether it is a main-group or transition element. The valence-shell configuration of a main-group element is $ns^a np^b$, where n equals the period of the element and $a + b$ the group number. (Distribute the electrons to the ns subshell, then to the np subshell.) The outer-shell configuration of a transition element is $(n - 1)d^{a-2} ns^2$, where a is the group number.

Solution Essentials:
• Determine the period and group number of an element
• Pauli exclusion principle
• The number of orbitals in a subshell of given l is $2l + 1$.

What are the configurations for the outer electrons of **a** tellurium, $Z = 52$, and **b** nickel, $Z = 28$?

Problem Strategy Find the period and the group number (the Roman numeral) of the element for the atom. Note whether it is a main-group or a transition element.

If the atom is that of a main-group element, the valence configuration is $ns^a np^b$, where n equals the period of the element and $a + b$ equals the group number (which equals the number of valence electrons). Distribute these electrons to the ns orbital; then distribute any remaining electrons to the np orbitals.

If the atom is that of a transition element, the outer-shell configuration is $(n - 1)d^{a-2} ns^2$, where n equals the period and a is the group number—except for Group VIIIB, which has three columns. For Group VIIIB, count 8, 9, then 10 for successive columns, which then equals a.

Solution

a You locate tellurium in a periodic table and find it to be in Period 5, Group VIA. Thus, it is a main-group element, and the outer subshells are $5s$ and $5p$. These subshells contain six electrons, because the group is VIA. The valence-shell configuration is $\mathbf{5s^2 5p^4}$. **b** Nickel is a Period 4 transition element, in which the general form of the outer-shell configuration is $3d^{a-2} 4s^2$. To determine a, you note that it equals the Roman numeral group number up to iron (8). After that you count Co as 9 and Ni as 10. Hence, the outer-shell configuration is $\mathbf{3d^8 4s^2}$.

Answer Check Note whether the configuration does follow the general form, either for a main-group element or a transition element. Check that the principal quantum number, n, equals the period of the element. For a main-group element, the total number of valence electrons should equal the group number. For a transition element, the number of electrons in the outer shell equals a.

Exercise 8.3   Using the periodic table on the inside front cover, write the valence-shell configuration of arsenic (As).

See Problems 8.51, 8.52, 8.53, and 8.54.

Exercise 8.4   The lead atom has the ground-state configuration $[Xe]4f^{14} 5d^{10} 6s^2 6p^2$. Find the period and group for this element. From its position in the periodic table, would you classify lead as a main-group element, a transition element, or an inner transition element?

See Problems 8.55 and 8.56.

Two elements in Period 3 are adjacent to one another in the periodic table. The ground-state atom of one element has only *s* electrons in its valence shell; the other one has at least one *p* electron in its valence shell. Identify the elements.

8.4 Orbital Diagrams of Atoms; Hund's Rule

In discussing the ground states of atoms, we have not yet described how the electrons are arranged within each subshell. There may be several different ways of arranging electrons in a particular configuration. Consider the carbon atom ($Z = 6$) with the ground-state configuration $1s^2 2s^2 2p^2$. Three possible arrangements are given in the following orbital diagrams.

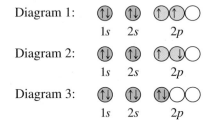

Diagram 1:

       1*s* 2*s*   2*p*

Diagram 2:

       1*s* 2*s*   2*p*

Diagram 3:

       1*s* 2*s*   2*p*

These orbital diagrams show different states of the carbon atom. Each state has a different energy and, as you will see, different magnetic characteristics.

Hund's Rule

In about 1927, Friedrich Hund discovered an empirical rule determining the lowest-energy arrangement of electrons in a subshell. **Hund's rule** states that *the lowest-energy arrangement of electrons in a subshell is obtained by putting electrons into separate orbitals of the subshell with the same spin before pairing electrons.* Let us see how this would apply to the carbon atom, whose ground-state configuration is $1s^2 2s^2 2p^2$. The first four electrons go into the $1s$ and $2s$ orbitals.

       1*s* 2*s*   2*p*

The next two electrons go into separate $2p$ orbitals, with both electrons having the same spin, following Hund's rule.

       1*s* 2*s*   2*p*

We see that the orbital diagram corresponding to the lowest energy is the one we previously labeled Diagram 1.

To apply Hund's rule to the oxygen atom, whose ground-state configuration is $1s^2 2s^2 2p^4$, we place the first seven electrons as follows:

       1*s* 2*s*   2*p*

The last electron is paired with one of the $2p$ electrons to give a doubly occupied orbital. The orbital diagram for the ground state of the oxygen atom is

       1*s* 2*s*   2*p*

In the following example, Hund's rule is used to determine the orbital diagram for the ground state of a more complicated atom. Table 8.2 gives orbital diagrams for the ground states of the first ten elements.

Example 8.4   Applying Hund's Rule

Critical Concept 8.4
Hund's rule states that the lowest-energy arrangement of electrons in a subshell is obtained by putting electrons with the same spin into separate orbitals of the subshell before pairing electrons.

Solution Essentials:
• Hund's rule
• Building-up order of subshells
• Pauli exclusion principle
• The number of orbitals in a subshell of given l is $2l + 1$.

Write an orbital diagram for the ground state of the iron atom.

Problem Strategy First obtain the electron configuration, as described in Example 8.2. Then draw circles for the orbitals of each subshell. A filled subshell should have doubly occupied orbitals (two electrons with opposite spins). For a partially filled subshell, apply Hund's rule, putting electrons into separate orbitals with the same spin (either all up or all down) before pairing electrons.

Solution The electron configuration of the iron atom is $1s^2 2s^2 2p^6 3s^2 3p^6 3d^6 4s^2$. All the subshells except the $3d$

are filled. In placing the six electrons in the $3d$ subshell, you note that the first five go into separate $3d$ orbitals with their spin arrows in the same direction. The sixth electron must doubly occupy a $3d$ orbital. The orbital diagram is

You can write this diagram in abbreviated form using [Ar] for the argon-like core of the iron atom:

Answer Check Check that you have started with the correct configuration for the atom. Then note that you have followed Hund's rule for a partially filled subshell.

Exercise 8.5 Write an orbital diagram for the ground state of the phosphorus atom ($Z = 15$). Write all orbitals.

■ See Problems 8.57 and 8.58.

Table 8.2 Orbital Diagrams for the Ground States of Atoms from $Z = 1$ to $Z = 10$

Atom	Z	Configuration	Orbital Diagram 1s	2s	2p
Hydrogen	1	$1s^1$	↑	○	○○○
Helium	2	$1s^2$	↑↓	○	○○○
Lithium	3	$1s^2 2s^1$	↑↓	↑	○○○
Beryllium	4	$1s^2 2s^2$	↑↓	↑↓	○○○
Boron	5	$1s^2 2s^2 2p^1$	↑↓	↑↓	↑○○
Carbon	6	$1s^2 2s^2 2p^2$	↑↓	↑↓	↑↑○
Nitrogen	7	$1s^2 2s^2 2p^3$	↑↓	↑↓	↑↑↑
Oxygen	8	$1s^2 2s^2 2p^4$	↑↓	↑↓	↑↓↑↑
Fluorine	9	$1s^2 2s^2 2p^5$	↑↓	↑↓	↑↓↑↓↑
Neon	10	$1s^2 2s^2 2p^6$	↑↓	↑↓	↑↓↑↓↑↓

Magnetic Properties of Atoms

The magnetic properties of a substance can reveal certain information about the arrangement of electrons in an atom (or molecule). Although an electron in an atom behaves like a small magnet, the magnetic attractions from two electrons that are opposite in spin cancel each other. As a result, an atom that has only doubly occupied orbitals has no net spin magnetism. However, an atom with *unpaired* electrons—that is, with an excess of one kind of spin—does exhibit a net magnetism.

Levitating Frogs and People

Have you ever seen a magic act in which a person rises above the stage in apparent defiance of gravity? Magicians have often included such acts of "levitation" as part of their performances. Of course, magical levitation uses tricks to deceive you. But levitation can be done, without tricks, using the known laws of electromagnetism.

Recently, researchers wanted to demonstrate that you can levitate almost anything if you have the proper magnetic field. They placed a frog within the magnetic field of a powerful electromagnet (which consists of an electric current flowing through a coil of wire). The frog floated in midair above the coil, at the point where the upward repulsion of the frog from the magnetic field just balanced the downward force of gravity (Figure 8.13).

We don't usually think of things such as frogs as being magnetic. In fact, most materials, including most substances in a biological organism, are diamagnetic. They contain pairs of electrons, in which the magnetism of each electron in a pair offsets the equal and opposite magnetism of the other electron. You might expect diamagnetic materials to be nonmagnetic because of this balance of opposite electron spins. But when you place a diamagnetic material in an external magnetic field, its electrons move so as to induce, or generate, a smaller magnetic field that is opposite in direction to the external field. This results in a repulsive force between the diamagnetic material and the external field. The repulsive force, though generally small, can be easily observed if the external magnet is large enough. Although, with a sufficiently large magnet, it would be possible to levitate a person as well as a frog, the researchers don't plan to do that.

Courtesy of A. Geim, University of Nijmegen

Figure 8.13 ▲

Levitation of a frog A frog placed in a powerful magnetic field, generated by a current flowing through a water-cooled coil of wire, appears to float in midair in defiance of gravity. The frog contains materials that are diamagnetic (composed of electron pairs), which are repelled by the magnetic field.

■ See Problems 8.89 and 8.90.

The strong, permanent magnetism seen in iron objects is called ferromagnetism and is due to the cooperative alignment of electron spins in many iron atoms. Paramagnetism is a much weaker effect. Nevertheless, paramagnetic substances can be attracted to a strong magnet. Liquid oxygen is composed of paramagnetic O_2 molecules. When poured over a magnet, the liquid clings to the poles. (See Figure 10.29.)

The magnetic properties of an atom can be observed. The most direct way is to determine whether the atomic substance is attracted to the field of a strong magnet. A **paramagnetic substance** is *a substance that is weakly attracted by a magnetic field, and this attraction is generally the result of unpaired electrons.* ◄ For example, sodium vapor has been found experimentally to be paramagnetic. The explanation is that the vapor consists primarily of sodium atoms, each containing an unpaired electron. (The configuration is $[Ne]3s^1$.) A **diamagnetic substance** is *a substance that is not attracted by a magnetic field or is very slightly repelled by such a field. This property generally means that the substance has only paired electrons.* Mercury vapor is found experimentally to be diamagnetic. The explanation is that mercury vapor consists of mercury atoms, with the electron configuration $[Xe]4f^{14}5d^{10}6s^2$, which has only paired electrons.

We expect the different orbital diagrams presented at the beginning of this section for the $1s^2 2s^2 2p^2$ configuration of the carbon atom to have different magnetic properties. Diagram 1, predicted by Hund's rule to be the ground state, would give a magnetic atom, whereas the other diagrams would not. If we could prepare a vapor of carbon atoms, it should be attracted to a magnet. (It should be paramagnetic.) It is difficult to prepare a vapor of free carbon atoms in sufficient concentration to observe a result. However, the visible–ultraviolet spectrum of carbon atoms can be obtained easily from dilute vapor. From an analysis of this spectrum, it is possible to show that the ground-state atom is magnetic, which is consistent with the prediction of Hund's rule.

 ## Periodicity of the Elements

You have seen that the periodic table that Mendeleev discovered in 1869 can be explained by the periodicity of the ground-state electron configurations of the atoms. Now we will look at various aspects of the periodicity of the elements.

8.5 Mendeleev's Predictions from the Periodic Table

One of Mendeleev's periodic tables is reproduced in Figure 8.14. Though somewhat different from modern tables, it shows essentially the same arrangement. In this early form of the periodic table, within each column some elements were placed toward the left side and some toward the right. With some exceptions, the elements on a given side have similar properties. Mendeleev left spaces in his periodic table for what he felt were undiscovered elements. There are blank spaces in his row 5—for example, one directly under aluminum and another under silicon (looking at just the elements on the right side of the column). By writing the known elements in this row with their atomic weights, he could determine approximate values (between the known ones) for the missing elements (values in parentheses).

Cu	Zn	—	—	As	Se	Br
63 amu	65 amu	(68 amu)	(72 amu)	75 amu	78 amu	80 amu

The Group III element directly under aluminum Mendeleev called eka-aluminum, with the symbol Ea. (*Eka* is the Sanskrit word meaning "first"; thus eka-aluminum is the first element under aluminum.) The known Group III elements have oxides of the form R_2O_3, so Mendeleev predicted that eka-aluminum would have an oxide with the formula Ea_2O_3.

The physical properties of this undiscovered element could be predicted by comparing values for the neighboring known elements. For eka-aluminum Mendeleev predicted a density of 5.9 g/cm^3, a low *melting point* (the temperature at which a substance melts), and a high *boiling point* (the temperature at which a substance boils).

In 1874 the French chemist Paul-Émile Lecoq de Boisbaudran found two previously unidentified lines in the atomic spectrum of a sample of sphalerite (a zinc sulfide, ZnS, mineral). Realizing he was on the verge of a discovery, Lecoq de Boisbaudran quickly prepared a large batch of the zinc mineral, from which he isolated a gram of a new element. He called this new element gallium. The properties of gallium were remarkably close to those Mendeleev predicted for eka-aluminum.

Figure 8.14 ◀

Mendeleev's periodic table This one was published in 1872.

Reihen	Gruppe I. — R²O	Gruppe II. — RO	Gruppe III. — R²O³	Gruppe IV. RH⁴ RO²	Gruppe V. RH³ R²O⁵	Gruppe VI. RH² RO³	Gruppe VII. RH R²O⁷	Gruppe VIII. — RO⁴
1	H=1							
2	Li=7	Be=9,4	B=11	C=12	N=14	O=16	F=19	
3	Na=23	Mg=24	Al=27,3	Si=28	P=31	S=32	Cl=35,5	
4	K=39	Ca=40	—=44	Ti=48	V=51	Cr=52	Mn=55	Fe=56, Co=59, Ni=59, Cu=63.
5	(Cu=63)	Zn=65	—=68	—=72	As=75	Se=78	Br=80	
6	Rb=85	Sr=87	?Yt=88	Zr=90	Nb=94	Mo=96	—=100	Ru=104, Rh=104, Pd=106, Ag=108.
7	(Ag=108)	Cd=112	In=113	Sn=118	Sb=122	Te=125	J=127	
8	Cs=133	Ba=137	?Di=138	?Ce=140	—	—	—	—
9	(—)	—	—	—	—	—	—	
10	—	—	?Er=178	?La=180	Ta=182	W=184	—	Os=195, Ir=197, Pt=198, Au=199.
11	(Au=199)	Hg=200	Tl=204	Pb=207	Bi=208		—	—
12	—	—	—	Th=231	—	U=240	—	— — — —

Tabelle II.

Property	*Predicted for Eka-Aluminum*	*Found for Gallium*
Atomic weight	68 amu	69.7 amu
Formula of oxide	Ea_2O_3	Ga_2O_3
Density of the element	5.9 g/cm^3	5.91 g/cm^3
Melting point of the element	Low	30.1°C
Boiling point of the element	High	1983°C

The predictive power of Mendeleev's periodic table was demonstrated again when scandium (eka-boron) was discovered in 1879 and germanium (eka-silicon) in 1886. Both elements had properties remarkably like those predicted by Mendeleev. These early successes won acceptance for the organizational and predictive power of the periodic table.

8.6 Some Periodic Properties

The electron configurations of the atoms display a periodic variation with increasing atomic number (nuclear charge). As a result, the elements show periodic variations of physical and chemical behavior. The **periodic law** states that *when the elements are arranged by atomic number, their physical and chemical properties vary periodically*. In this section, we will look at three physical properties of an atom: atomic radius, ionization energy, and electron affinity. These three quantities, especially ionization energy and electron affinity, are important in discussions of chemical bonding (the subject of Chapter 9).

Atomic Radius

The definition of atomic radius is somewhat arbitrary; in fact, different definitions have been given. Because the statistical distribution of electrons does not abruptly end but merely decreases to smaller and smaller values as the distance from the nucleus increases, an atom has no definite boundary. You can see this in the plot of the electron distribution for the argon atom, shown in Figure 8.15.

This distribution has been calculated from an approximate wavefunction obtained for the isolated atom. Note that the distribution shows a maximum at each shell of the atom and falls steadily to a negligible value after the maximum for the outer shell, which occurs at 71 pm.

It seems reasonable, as one possibility, to define the **atomic radius** (calculated from theory for the isolated atom) as *the maximum in the radial distribution function of the outer shell of the atom*. From this definition, we would say that argon has a radius of 71 pm. Similar calculations have been done on most of the atoms in the periodic table. In Figure 8.16, we have graphically plotted the atomic radii from a particular set of calculations versus atomic number to display their periodic variations. Figure 8.17 shows the same data visually.

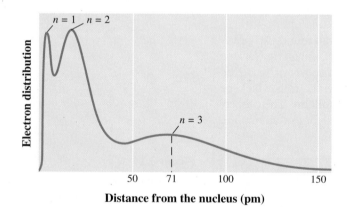

Figure 8.15 ▲

Electron distribution for the argon atom This is a radial distribution, showing the probability of finding an electron at a given distance from the nucleus. The distribution shows three maxima, for the $n = 1$, $n = 2$, and $n = 3$ shells. The outermost maximum occurs at 71 pm; then the distribution falls steadily, becoming negligibly small after several hundred picometers.

There are other definitions of atomic radii. As we will discuss in the next chapter on chemical bonding, the *covalent bond* is one type of chemical bond, and atomic radii have been defined in a way that reproduces the empirically determined covalent bond distances (the distances between atomic nuclei) in chemical compounds. By this definition, the **atomic radius** (in this case, often called the **covalent radius**) is *the value for that atom in a set of covalent radii assigned to atoms in such a way that the sum of the covalent radii of atoms A and B predicts the approximate A—B bond length*. Note that these atomic radii are from empirical data rather than calculated from wavefunctions, and as we noted, they refer to atoms in compounds and not to isolated atoms. We will discuss these covalent radii again in Section 9.10.

Although the definition of atomic radius is somewhat arbitrary, as long as we refer to a set of values obtained in a consistent manner, they can be useful.

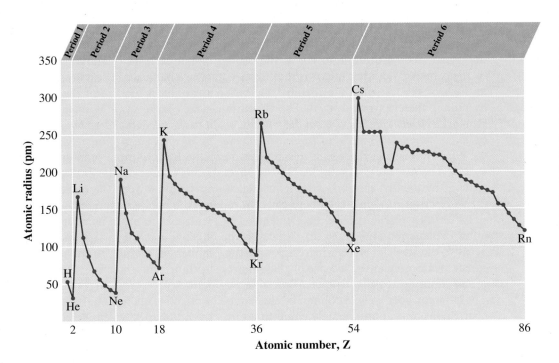

Figure 8.16 ◄

Atomic radius (isolated atom) versus atomic number These radii are calculated values for isolated atoms. Note that the curve is periodic (tends to repeat). Each period of elements begins with the Group IA atom, and the atomic radius tends to decrease until the Group VIIIA atom.

In particular, atomic radii in a given set follow general periodic trends, which can be seen in Figures 8.16 and 8.17. These trends are:

1. **Within each period (horizontal row)**, the atomic radius tends to decrease with increasing atomic number (nuclear charge). Thus, the largest atom in a period tends to be the Group IA atom and the smallest the noble-gas atom.

2. **Within each group (vertical column)**, the atomic radius tends to increase with the period number.

Note that the atomic radius increases greatly going from any noble-gas atom to the following Group IA atom, giving the curve in Figure 8.16 a saw-toothed appearance. We should stress that these are general trends, and you may see occasional exceptions from these trends in some lists of atomic radii.

Figure 8.17 ▼

Representation of atomic radii (isolated atoms) The atomic radii are in picometers. Note the trends within each period and each group.

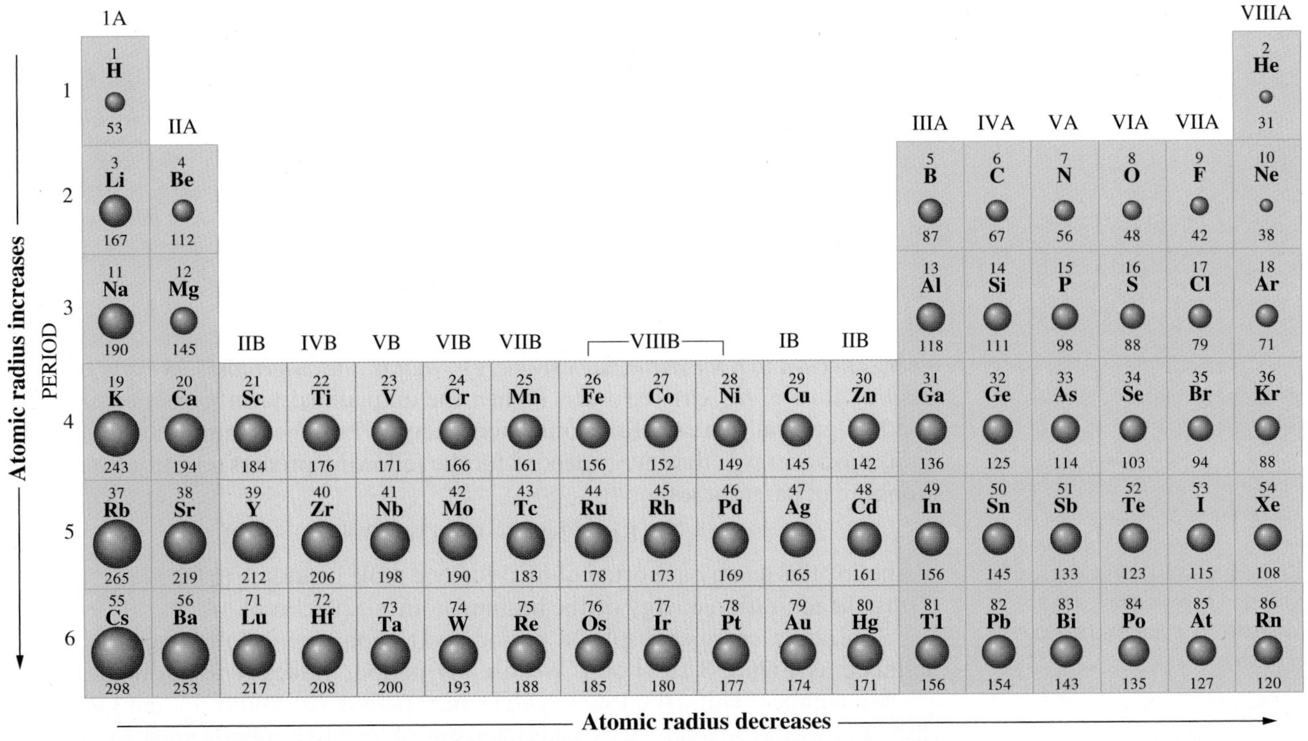

Atomic radius increases →

PERIOD

	1A																	VIIIA
1	1 H 53	IIA											IIIA	IVA	VA	VIA	VIIA	2 He 31
2	3 Li 167	4 Be 112											5 B 87	6 C 67	7 N 56	8 O 48	9 F 42	10 Ne 38
3	11 Na 190	12 Mg 145	IIB	IVB	VB	VIB	VIIB	—VIIIB—			IB	IIB	13 Al 118	14 Si 111	15 P 98	16 S 88	17 Cl 79	18 Ar 71
4	19 K 243	20 Ca 194	21 Sc 184	22 Ti 176	23 V 171	24 Cr 166	25 Mn 161	26 Fe 156	27 Co 152	28 Ni 149	29 Cu 145	30 Zn 142	31 Ga 136	32 Ge 125	33 As 114	34 Se 103	35 Br 94	36 Kr 88
5	37 Rb 265	38 Sr 219	39 Y 212	40 Zr 206	41 Nb 198	42 Mo 190	43 Tc 183	44 Ru 178	45 Rh 173	46 Pd 169	47 Ag 165	48 Cd 161	49 In 156	50 Sn 145	51 Sb 133	52 Te 123	53 I 115	54 Xe 108
6	55 Cs 298	56 Ba 253	71 Lu 217	72 Hf 208	73 Ta 200	74 W 193	75 Re 188	76 Os 185	77 Ir 180	78 Pt 177	79 Au 174	80 Hg 171	81 Tl 156	82 Pb 154	83 Bi 143	84 Po 135	85 At 127	86 Rn 120

→ **Atomic radius decreases** →

These general trends in atomic radius can be explained if you look at the two factors that primarily determine the size of the outermost orbital. One factor is the principal quantum number n of the orbital; the larger n is, the larger the size of the orbital. The other factor is the effective nuclear charge acting on an electron in the orbital; increasing the effective nuclear charge reduces the size of the orbital by pulling the electrons inward. The **effective nuclear charge** is *the positive charge that an electron experiences from the nucleus, equal to the nuclear charge but reduced by any shielding or screening from any intervening electron distribution.* Consider the effective nuclear charge on the 2s electron in the lithium atom (configuration $1s^2 2s^1$). The nuclear charge is $3e$, but the effect of this charge on the 2s electron is reduced by the distribution of the two 1s electrons lying between the nucleus and the 2s electron (roughly, each core electron reduces the nuclear charge by $1e$).

Consider a given period of elements. The principal quantum number of the outer orbitals remains constant. However, the effective nuclear charge increases, because the nuclear charge increases and the number of core electrons remains constant. Consequently, the size of the outermost orbital and, therefore, the radius of the atom decrease with increasing Z in any period.

Now consider a given column of elements. The effective nuclear charge remains nearly constant (approximately equal to e times the number of valence electrons), but n gets larger. You observe that the atomic radius increases.

☹WL **Interactive Example** 8.5   **Determining Relative Atomic Sizes from Periodic Trends**

Gaining Mastery Toolbox

Critical Concept 8.5
The general trends in atomic radii are that they decrease left to right in a period and increase from top to bottom in a group.

Solution Essentials:
• General trends of atomic radii

Refer to a periodic table and use the trends noted for size of atomic radii to arrange the following in order of increasing atomic radius: Al, C, Si.

Problem Strategy The general trends in atomic radius are that it decreases left to right in a period and that it increases from top to bottom of a group. Compare any two elements in the same period, or in the same group, to decide which of the two is smaller. From this, decide the order of the elements.

Solution Note that C is above Si in Group IVA. Therefore, the radius of C is smaller than that of Si (the atomic radius increases going down a group of elements). Note that Al and Si are in the same period. Therefore, the radius of Si is smaller than that of Al (radius decreases with Z in a period). Hence the order of elements by increasing radius is **C, Si, Al**.

Answer Check After you have decided on the order of the atoms, check that the atomic radius does indeed increase, left to right.

Exercise 8.6   Using a periodic table, arrange the following in order of increasing atomic radius: Na, Be, Mg.

■ See Problems 8.61 and 8.62.

Ionization Energy

The **first ionization energy** (or **first ionization potential**) of an atom is *the minimum energy needed to remove the highest-energy (that is, the outermost) electron from the neutral atom in the gaseous state.* (When the unqualified term *ionization energy* is used, it generally means first ionization energy.) For the lithium atom, the first ionization energy is the energy needed for the following process (electron configurations are in parentheses):

$$\text{Li}(1s^2 2s^1) \longrightarrow \text{Li}^+(1s^2) + e^-$$

Values of this energy are often quoted for one mole of atoms (6.02×10^{23} atoms). Thus, the ionization energy of the lithium atom is 520 kJ/mol. ◀

Ionization energies display a periodic variation when plotted against atomic number, as Figure 8.18 shows. Within any period, values tend to increase with atomic number. Thus, the lowest values in a period are found for the Group IA elements *(alkali metals)*. It is characteristic of reactive metals such as these to

Ionization energies are also given in electron volts (eV). This is the amount of energy imparted to an electron when it is accelerated through an electrical potential of one volt. One electron volt is equivalent to 96.5 kJ/mol. (This is an easy conversion to remember; it is approximately one hundred kilojoules per mole.)

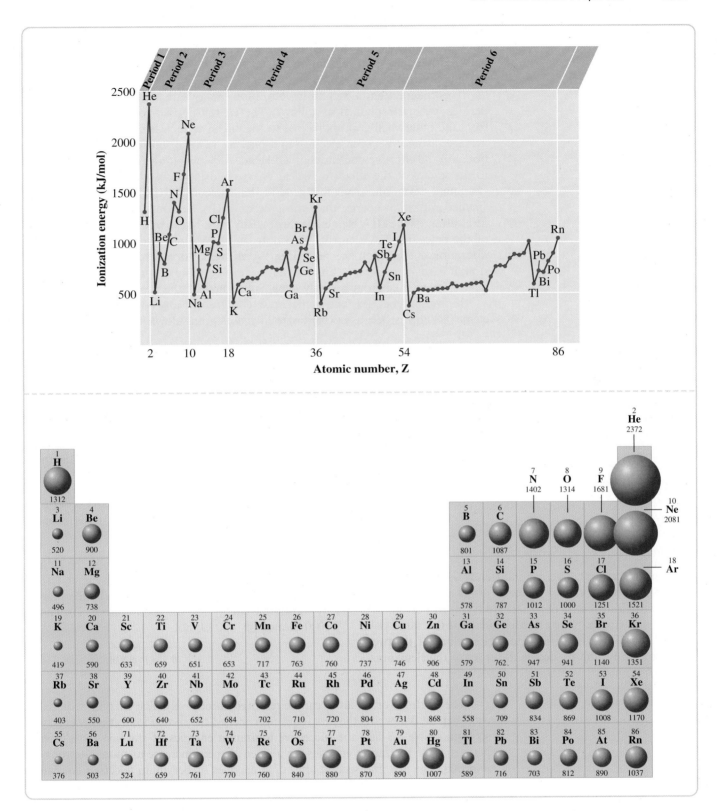

Figure 8.18 ▲

Ionization energy versus atomic number *Top:* Note that the values tend to increase within each period, except for small drops in ionization energy at the Group IIIA and VIA elements. Large drops occur when a new period begins. *Bottom:* Note the trends within each period. The size of the sphere for each atom indicates the relative magnitude of ionization energy. The values given are in kilojoules/mole.

lose electrons easily. The largest ionization energies in any period occur for the noble-gas elements. In other words, a noble-gas atom loses electrons with difficulty, which is partly responsible for the stability of the noble-gas configurations and the unreactivity of the noble gases.

This general trend—increasing ionization energy with atomic number in a given period—can be explained as follows: The energy needed to remove an electron from the outer shell is proportional to the effective nuclear charge divided by the average distance between electron and nucleus. (This distance is inversely proportional to the effective nuclear charge.) Hence, the ionization energy is proportional to the square of the effective nuclear charge and increases going across a period.

Small deviations from this general trend occur. A IIIA element (ns^2np^1) has smaller ionization energy than the preceding IIA element (ns^2). Apparently, the np electron of the IIIA element is more easily removed than one of the ns electrons of the preceding IIA element. Also note that a VIA element (ns^2np^4) has smaller ionization energy than the preceding VA element. As a result of electron repulsion, it is easier to remove an electron from the doubly occupied np orbital of the VIA element than from a singly occupied orbital of the preceding VA element.

Ionization energies tend to decrease going down any column of main-group elements. This is because atomic size increases going down the column.

Example 8.6   Determining Relative Ionization Energies from Periodic Trends

Gaining Mastery Toolbox

Critical Concept 8.6
The general trends in first ionization energy are that it increases left to right in a period and decreases from top to bottom in a group.

Solution Essentials:
• General trends of first ionization energy

Using a periodic table only, arrange the following elements in order of increasing ionization energy: Ar, Se, S.

Problem Strategy The general trends in (first) ionization energy are that it increases from left to right in a period and that it decreases from top to bottom of a group. Compare any two elements in the same period, or in the same group, to decide which of the two is smaller. From this, decide the order of the elements.

Solution Note that Se is below S in Group VIA. Therefore, the ionization energy of Se should be less than that of S. Also, S and Ar are in the same period, with Z increasing from S to Ar. Therefore, the ionization energy of S should be less than that of Ar. Hence the order is **Se, S, Ar**.

Answer Check After you have decided on the order of the atoms, check that the ionization energy does indeed increase, left to right.

Exercise 8.7 The first ionization energy of the chlorine atom is 1251 kJ/mol. Without looking at Figure 8.18, state which of the following values would be the more likely ionization energy for the iodine atom. Explain.
a 1000 kJ/mol. b 1400 kJ/mol.

■ See Problems 8.63 and 8.64.

The electrons of an atom can be removed successively. The energies required at each step are known as the *first* ionization energy, the *second* ionization energy, and so forth. Table 8.3 lists the successive ionization energies of the first ten elements. Note that the ionization energies for a given element increase as more electrons are removed. The first and second ionization energies of beryllium (electron configuration $1s^22s^2$) are 900 kJ/mol and 1757 kJ/mol, respectively. The first ionization energy is the energy needed to remove a $2s$ electron from the Be atom. The second ionization energy is the energy needed to remove a $2s$ electron from the positive ion Be^+. Its value is greater than that of the first ionization energy because the electron is being removed from a positive ion, which strongly attracts the electron.

Note that there is a large jump in value from the second ionization energy of Be (1757 kJ/mol) to the third ionization energy (14,848 kJ/mol). The second ionization energy corresponds to removing a valence electron (from the $2s$ orbital), which is relatively easy. The third ionization energy corresponds to removing an electron from the core of the atom (from the $1s$ orbital)—that is, from a noble-gas configuration ($1s^2$). A vertical line in Table 8.3 separates the energies needed to

Table 8.3	Successive Ionization Energies of the First Ten Elements (kJ/mol)*						
Element	First	Second	Third	Fourth	Fifth	Sixth	Seventh
H	1312						
He	2372	5250					
Li	520	7298	11,815				
Be	900	1757	14,848	21,006			
B	801	2427	3660	25,026	32,827		
C	1086	2353	4620	6223	37,831	47,277	
N	1402	2856	4578	7475	9445	53,267	64,360
O	1314	3388	5300	7469	10,990	13,326	71,330
F	1681	3374	6050	8408	11,023	15,164	17,868
Ne	2081	3952	6122	9371	12,177	15,238	19,999

*Ionization energies to the right of a vertical line correspond to removal of electrons from the core of the atom.

remove valence electrons from those needed to remove core electrons. For each element, a large increase in ionization energy occurs when this line is crossed. The large increase results from the fact that once the valence electrons are removed, a stable noble-gas configuration is obtained. Further ionizations become much more difficult. We will see in Chapter 9 that metal atoms often form compounds by losing valence electrons; core electrons are not significantly involved in the formation of compounds.

Electron Affinity

When a neutral atom in the gaseous state picks up an electron to form a stable negative ion, energy is released. For example, a chlorine atom can pick up an electron to give a chloride ion, Cl^-, and 349 kJ/mol of energy is released. We can write this process, including the electron configurations, as follows:

$$Cl([Ne]3s^23p^5) + e^- \longrightarrow Cl^-([Ne]3s^23p^6); \Delta E = -349 \text{ kJ/mol}$$

What is often measured is the energy of the *electron detachment* from the negative ion, which is essentially the ionization energy of the negative ion.

$$Cl^-([Ne]3s^23p^6) \longrightarrow Cl([Ne]3s^23p^5) + e^-; \Delta E = 349 \text{ kJ/mol}$$

The **electron affinity** of an atom is defined as *the energy required to remove an electron from the atom's negative ion (in its ground state)*. Or, if you wish, the electron affinity is the negative of the energy change obtained when the neutral atom picks up an electron. The quantity is positive if a stable negative ion forms. A large positive value means that the neutral atom has a strong affinity for an electron (picks up an electron easily). Table 8.4 displays the electron affinities of the main-group elements. ▶

It should be noted that electron affinities are also defined with the opposite sign. In this convention, the electron affinity of the chlorine atom is −349 kJ/mol. This is essentially the energy change (energy is released) in adding an electron to the chlorine atom, Cl.

Table 8.4	Electron Affinities of the Main-Group Elements (kJ/mol)							
Period	IA	IIA	IIIA	IVA	VA	VIA	VIIA	VIIIA
1	H 73							He ≤0
2	Li 60	Be ≤0	B 27	C 122	N ≤0	O 141	F 328	Ne ≤0
3	Na 53	Mg ≤0	Al 44	Si 134	P 72	S 200	Cl 349	Ar ≤0
4	K 48	Ca 2	Ga 41	Ge 119	As 78	Se 195	Br 325	Kr ≤0
5	Rb 47	Sr 5	In 37	Sn 107	Sb 101	Te 190	I 295	Xe ≤0
6	Cs 46	Ba 14	Tl 36	Pb 35	Bi 91	Po 180	At 270	Rn ≤0

Figure 8.19 ▶

Electron affinity versus atomic number in the main-group elements Electron affinities in the main-group elements show a periodic variation when plotted against atomic number, although this variation is somewhat more complicated than that displayed by ionization energies. In a given period, the electron affinity rises from the Group IA element to the Group VIIA element but with sharp drops in the Group IIA and Group VA elements.

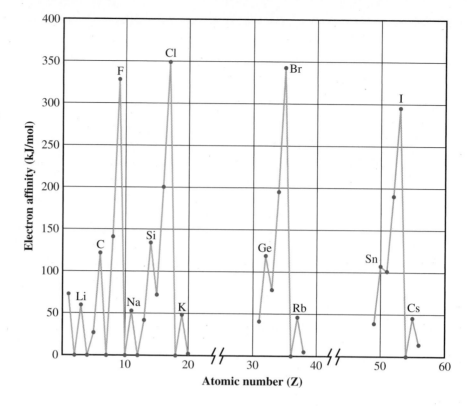

If we look at these main-group elements, we see that the electron affinities vary in going across any period in a fairly predictable fashion. See the graph in Figure 8.19. Broadly speaking, the general trend in any period of the main-group elements is an increase in electron affinities from left to right, although the IIA element and the VA element tend to have smaller electron affinities than the preceding element, and the Group VIIIA elements (noble gases) have zero or small negative values (indicating unstable negative ions). Note especially that the Group VIA and Group VIIA elements have the largest electron affinities of any of the other main-group elements. The stability of the negative ions of these elements explains why they have compounds with monatomic anions (such as F⁻). While the trend in going across a period is rather clear, this is not the case in going down a column of elements; there is no simple trend.

Let's look more closely at these main-group elements. Consider those in Group IA. All of the Group IA elements have moderately positive electron affinities, as you might expect since the added electron just completes the outer ns subshell. When you add an electron to a potassium atom, for example, it goes into the $4s$ orbital to form a moderately stable negative ion, releasing energy and yielding a positive electron affinity (EA):

$$K([Ar]4s^1) + e^- \longrightarrow K^-([Ar]4s^2); \Delta E = -48 \text{ kJ/mol (EA} = 48 \text{ kJ/mol)}$$

The Group IIA elements have a filled outer ns subshell, so the added electron has to go into the next higher subshell, the np subshell. Thus, we expect the electron affinities of these elements to be smaller than those for the IA elements. Beryllium and magnesium appear to have values close to zero (no stable negative ions found), whereas the elements from calcium on have small positive values (weakly stable negative ions). For the calcium atom, we have

$$Ca([Ar]4s^2) + e^- \longrightarrow Ca^-([Ar]4s^24p^1); \Delta E = -2 \text{ kJ/mol (EA} = 2 \text{ kJ/mol)}$$

From the Group IIIA elements to the Group VIIA elements, the added electron goes into the outer np subshell, which already has one or more electrons in it. With the exception of the Group VA element, the electron affinity in any period tends to rise from the Group IIA element to the Group VIIA element. For example, in the Period 4 elements from gallium to bromine, we observe:

$$Ga([Ar]4s^24p^1) + e^- \longrightarrow Ga^-([Ar]4s^24p^2); \Delta E = -41 \text{ kJ/mol (EA = 41 kJ/mol)}$$

$$Ge([Ar]4s^24p^2) + e^- \longrightarrow Ge^-([Ar]4s^24p^3); \Delta E = -119 \text{ kJ/mol (EA = 119 kJ/mol)}$$

$$As([Ar]4s^24p^3) + e^- \longrightarrow As^-([Ar]4s^24p^4); \Delta E = -78 \text{ kJ/mol (EA = 78 kJ/mol)}$$

$$Se([Ar]4s^24p^4) + e^- \longrightarrow Se^-([Ar]4s^24p^5); \Delta E = -195 \text{ kJ/mol (EA = 195 kJ/mol)}$$

$$Br([Ar]4s^24p^5) + e^- \longrightarrow Br^-([Ar]4s^24p^6); \Delta E = -325 \text{ kJ/mol (EA = 325 kJ/mol)}$$

The drop in electron affinity of the Group VA element from the preceding element is easily explained. In the Group VA element (arsenic, As, in the above list), the added electron must pair up with one of the *np* electrons since all of the *np* orbitals have one electron, whereas in the preceding element the extra electron goes into an empty *np* orbital. This pairing of electrons in an orbital requires some energy, resulting in a smaller electron affinity for the Group VA element compared with the preceding IVA element.

No stable negative ions of the Group VIIIA elements (noble-gas elements) have been found. We might expect this since these atoms have filled outer shells so that the addition of an electron would have to go into the next higher shell. The electron affinities are therefore zero (or possibly negative), as we noted earlier.

> **Exercise 8.8** Without looking at Table 8.4 but using the general comments in this section, decide which has the larger electron affinity, C or F.
>
> ■ See Problems 8.65 and 8.66.

CONCEPT CHECK 8.3

Given the following information for element E, identify the element's group in the periodic table: The electron affinity of E is negative (that is, it does not form a stable negative ion). The first ionization energy of E is less than the second ionization energy, which in turn is very much less than its third ionization energy.

8.7 Periodicity in the Main-Group Elements

The chemical and physical properties of the main-group elements clearly display periodic character. For instance, as pointed out in Section 2.5, the metallic elements lie to the left of the "staircase" line in the periodic table (see inside front cover), nonmetals lie to the right, and the metalloids (with intermediate characteristics) lie along this line. So, as you move left to right in any row of the periodic table, the metallic character of the elements decreases. As you progress down a column, however, the elements tend to increase in metallic character.

These variations of metallic–nonmetallic character can be attributed in part to variations in the ionization energies of the corresponding atoms. Elements with low ionization energy tend to be metallic, whereas those with high ionization energy tend to be nonmetallic. As you saw in the previous section, ionization energy is a periodic property, so it is not surprising that the metallic–nonmetallic character of an element is similarly periodic.

The basic–acidic behavior of the oxides of the elements is a good indicator of the metallic–nonmetallic character of the elements. Oxides are classified as basic or acidic depending on their reactions with acids and bases. A **basic oxide** is *an oxide that reacts with acids.* Most metal oxides are basic. An **acidic oxide** is *an oxide that reacts with bases.* Most nonmetal oxides are acidic oxides. An **amphoteric oxide** is *an oxide that has both basic and acidic properties.*

In the following brief descriptions of the main-group elements, we will note the metallic–nonmetallic behavior of the elements, as well as the basic–acidic character of the oxides. Although elements in a given group are expected to be similar, the degree of similarity does vary among the groups. The alkali metals (Group IA) show marked similarities, as do the halogens (Group VIIA). On the other hand, the Group IVA elements range from a nonmetal (carbon) at the top of the column to a metal (lead) at the bottom. In either case, however, the changes from one element in a column to the next lower one are systematic, and the periodic table helps us to correlate these systematic changes.

Hydrogen ($1s^1$)

At very high pressures, however, hydrogen is believed to have metallic properties.

Although the electron configuration of hydrogen would seem to place the element in Group IA, its properties are quite different, and it seems best to consider this element as belonging in a group by itself. The element is a colorless gas composed of H_2 molecules. ◀

Group IA Elements, the Alkali Metals (ns^1)

The *alkali metals* are soft and reactive, with the reactivities increasing as you move down the column of elements. All of the metals react with water to produce hydrogen.

$$2Li(s) + 2H_2O(l) \longrightarrow 2LiOH(aq) + H_2(g)$$

The vigor of the reaction increases from lithium (moderate) to rubidium (violent). All of the alkali metals form basic oxides with the general formula R_2O.

Group IIA Elements, the Alkaline Earth Metals (ns^2)

The *alkaline earth metals* are also chemically reactive but much less so than the alkali metals. Reactivities increase going down the group. The alkaline earth metals form basic oxides with the general formula RO.

Group IIIA Elements (ns^2np^1)

Figure 8.20 ▲

Gallium This metal melts from the heat of a hand.

Groups IA and IIA exhibit only slight increases in metallic character down a column, but with Group IIIA we see a significant increase. The first Group IIIA element, boron, is a metalloid. Other elements in this group—aluminum, gallium, indium, and thallium—are metals. (Gallium is a curious metal; it melts readily in the palm of the hand—see Figure 8.20.)

The oxides in this group have the general formula R_2O_3. Boron oxide, B_2O_3, is an acidic oxide; aluminum oxide, Al_2O_3, and gallium oxide, Ga_2O_3, are amphoteric oxides. The change in the oxides from acidic to amphoteric to basic is indicative of an increase in metallic character of the elements.

Bronze, one of the first alloys (metallic mixtures) used by humans, contains about 90% copper and 10% tin. Bronze melts at a lower temperature than copper but is much harder.

Group IVA Elements (ns^2np^2)

This group shows the most distinct change in metallic character. It begins with the nonmetal carbon, C, followed by the metalloids silicon, Si, and germanium, Ge, and then the metals tin, Sn, and lead, Pb. Both tin and lead were known to the ancients. ◀

All the elements in this group form oxides with the general formula RO_2, which progress from acidic to amphoteric. Carbon dioxide, CO_2, an acidic oxide, is a gas. (Carbon also forms the monoxide, CO.) Silicon dioxide, SiO_2, an acidic oxide, exists as quartz and white sand (particles of quartz). Germanium dioxide, GeO_2, is acidic, though less so than silicon dioxide. Tin dioxide, SnO_2, an amphoteric oxide, is found as the mineral cassiterite, the principal ore of tin. Lead(IV) oxide, PbO_2, is amphoteric. Lead has a more stable monoxide, PbO. Figure 8.21 shows oxides of some Group IVA elements.

Figure 8.21 ▲

Oxides of some Group IVA elements Powdered lead(II) oxide (yellow), lead(IV) oxide (dark brown), tin dioxide (white), and crystalline silicon dioxide (clear quartz).

Group VA Elements (ns^2np^3)

The Group VA elements also show the distinct transition from nonmetal (nitrogen, N, and phosphorus, P) to metalloid (arsenic, As, and antimony, Sb) to metal (bismuth, Bi). Nitrogen occurs as a colorless, odorless, relatively unreactive gas with N_2 molecules; white phosphorus is a white, waxy solid with P_4 molecules. Gray arsenic is a brittle solid with metallic luster; antimony is a brittle solid with a silvery, metallic luster. Bismuth is a hard, lustrous metal with a pinkish tinge.

The Group VA elements form oxides with empirical formulas R_2O_3 and R_2O_5. In some cases, the molecular formulas are twice these formulas—that is, R_4O_6 and R_4O_{10}. Nitrogen has the acidic oxides N_2O_3 and N_2O_5, although it also has other, better known oxides, such as NO. Phosphorus has the acidic oxides P_4O_6 and P_4O_{10}. Arsenic has the acidic oxides As_2O_3 and As_2O_5; antimony has the amphoteric oxides Sb_2O_3 and Sb_2O_5; and bismuth has the basic oxide Bi_2O_3.

Group VIA Elements, the Chalcogens (ns^2np^4)

These elements, the *chalcogens* (pronounced kal'-ke-jens), show the transition from nonmetal (oxygen, O, sulfur, S, and selenium, Se) to metalloid (tellurium, Te) to metal (polonium, Po). Oxygen occurs as a colorless, odorless gas with O_2 molecules. It also has an allotrope, ozone, with molecular formula O_3. Sulfur is a brittle, yellow solid with molecular formula S_8. Tellurium is a shiny gray, brittle solid; polonium is a silvery metal.

Sulfur, selenium, and tellurium form oxides with the formulas RO_2 and RO_3. (Sulfur burns in air to form sulfur dioxide; see Figure 8.22.) These oxides, except for TeO_2, are acidic; TeO_2 is amphoteric. Polonium has an oxide PoO_2, which is amphoteric, though more basic in character than TeO_2.

Group VIIA Elements, the Halogens (ns^2np^5)

The *halogens* are reactive nonmetals with the general molecular formula X_2, where X symbolizes a halogen. Fluorine, F_2, is a pale yellow gas; chlorine, Cl_2, a pale greenish yellow gas; bromine, Br_2, a reddish brown liquid; and iodine, I_2, a bluish black solid that has a violet vapor (see Figure 8.23). Little is known about the chemistry of astatine, At, because all isotopes are radioactive with very short half-lives. (The half-life of a radioactive isotope is the time it takes for half of the isotope to decay, or break down, to another element.) It might be expected to be a metalloid.

Each halogen forms several compounds with oxygen; these are generally unstable, acidic oxides.

Group VIIIA Elements, the Noble Gases (ns^2np^6)

The Group VIIIA elements exist as gases consisting of uncombined atoms. For a long time these elements were thought to be chemically inert, because no compounds were known. Then, in the early 1960s, several compounds of xenon were prepared. Now compounds are also known for argon, krypton, and radon. These elements are known as the *noble gases* because of their relative unreactivity.

CONCEPT CHECK 8.4

A certain element is a metalloid that forms an acidic oxide with the formula R_2O_5. Identify the element.

Figure 8.22 ▲

Burning sulfur The combustion of sulfur in air gives a blue flame. The product is primarily sulfur dioxide, detectable by its acrid odor.

Figure 8.23 ▲

The halogens From left to right, the flasks contain chlorine, bromine, and iodine. (Bromine and iodine were warmed to produce the vapors; bromine is normally a reddish brown liquid and iodine a bluish black solid.)

A Checklist for Review

OWL and **Go Chemistry** Sign in at **www.cengage.com/owl** to:

- View tutorials and simulations, develop problem-solving skills, and complete online homework assigned by your professor.
- For quick review and exam prep, download Go Chemistry mini lecture modules from OWL or purchase them at **www.cengagebrain.com**.

Summary of Facts and Concepts

To understand the similarities that exist among the members of a group of elements, it is necessary to know the *electron configurations* for the ground states of atoms. Only those arrangements of electrons allowed by the *Pauli exclusion principle* are possible. The ground-state configuration of an atom represents the electron arrangement that has the lowest total energy. This arrangement can be reproduced by the *building-up principle* (Aufbau principle), where electrons fill the subshells in particular order (the building-up order) consistent with the Pauli exclusion principle. The arrangement of electrons in partially filled subshells is governed by *Hund's rule*.

Elements in the same group of the periodic table have similar valence-shell configurations. As a result, chemical and physical properties of the elements show periodic behavior. *Atomic radii,* for example, tend to decrease across any period (left to right) and increase down any group (top to bottom). *First ionization energies* tend to increase across a period and decrease down a group. *Electron affinities* of the Group VIA and Group VIIA elements have large values.

Learning Objectives	Important Terms
8.1 Electron Spin and the Pauli Exclusion Principle	
▪ Define *electron configuration* and *orbital diagram.* ▪ State the Pauli exclusion principle. ▪ Apply the Pauli exclusion principle. Example 8.1	**electron configuration** **orbital diagram** **Pauli exclusion principle**
8.2 Building-Up Principle and the Periodic Table	
▪ Define *building-up principle.* ▪ Define *noble-gas core, pseudo-noble-gas core,* and *valence electron.* ▪ Define *main-group element* and *(d-block* and *f-block) transition element.*	**building-up (Aufbau) principle** **noble-gas core** **pseudo-noble-gas core** **valence electron**
8.3 Writing Electron Configurations Using the Periodic Table	
▪ Determine the configuration of an atom using the building-up principle. Example 8.2 ▪ Determine the configuration of an atom using the period and group numbers. Example 8.3	
8.4 Orbital Diagrams of Atoms; Hund's Rule	
▪ State Hund's rule. ▪ Apply Hund's rule. Example 8.4 ▪ Define *paramagnetic substance* and *diamagnetic substance.*	**Hund's rule** **paramagnetic substance** **diamagnetic substance**
8.5 Mendeleev's Predictions from the Periodic Table	
▪ Describe how Mendeleev predicted the properties of undiscovered elements.	
8.6 Some Periodic Properties	
▪ State the periodic law. ▪ State the general periodic trends in size of atomic radii. ▪ Define *effective nuclear charge.* ▪ Determine relative atomic sizes from periodic trends. Example 8.5 ▪ State the general periodic trends in ionization energy. ▪ Define *first ionization energy.* ▪ Determine relative ionization energies from periodic trends. Example 8.6 ▪ Define *electron affinity.* ▪ State the broad general trend in electron affinity across any period.	**periodic law** **atomic radius** **effective nuclear charge** **first ionization energy (first ionization potential)** **electron affinity**
8.7 Periodicity in the Main-Group Elements	
▪ Define *basic oxide, acidic oxide,* and *amphoteric oxide.* ▪ State the main group corresponding to an *alkali metal,* an *alkaline earth metal,* a *chalcogen,* a *halogen,* and a *noble gas.* ▪ Describe the change in metallic/nonmetallic character (or reactivities) in going through any main group of elements.	**basic oxide** **acidic oxide** **amphoteric oxide**

Questions and Problems

Self-Assessment and Review Questions

Key: These questions test your understanding of the ideas you worked with in the chapter. These problems vary in difficulty and often can be used for the basis of discussion.

8.1 Describe the experiment of Stern and Gerlach. How are the results for the hydrogen atom explained?

8.2 Describe the model of electron spin given in the text. What are the restrictions on electron spin?

8.3 How does the Pauli exclusion principle limit the possible electron configurations of an atom?

8.4 List the orbitals in order of increasing orbital energy up to and including $3p$ orbitals.

8.5 What is the maximum number of electrons that can occupy a g subshell ($l = 4$)?

8.6 Define each of the following: noble-gas core, pseudo-noble-gas core, valence electron.

8.7 Define the terms *diamagnetic substance* and *paramagnetic substance.* Does the ground-state oxygen atom give a diamagnetic or a paramagnetic substance? Explain.

8.8 Give two different possible orbital diagrams for the $1s^2 2s^2 2p^4$ configuration of the oxygen atom, one of which should correspond to the ground state. Label the diagram for the ground state.

8.9 What kind of subshell is being filled in Groups IA and IIA? in Groups IIIA to VIIIA? in the transition elements? in the lanthanides and actinides?

8.10 How was Mendeleev able to predict the properties of gallium before it was discovered?

8.11 Describe the major trends that emerge when atomic radii are plotted against atomic number. Describe the trends observed when first ionization energies are plotted against atomic number.

8.12 What atom has the smallest radius among the alkaline earth elements?

8.13 What main group in the periodic table has elements with the largest electron affinities for each period? What group of elements has only unstable negative ions?

8.14 The ions Na^+ and Mg^{2+} occur in chemical compounds, but the ions Na^{2+} and Mg^{3+} do not. Explain.

8.15 Describe the major trends in metallic character observed in the periodic table of the elements.

8.16 Distinguish between an acidic and a basic oxide. Give examples of each.

8.17 What would you predict for the atomic number of the halogen in Period 7?

8.18 What is the name of the alkali metal atom with valence-shell configuration $5s^1$?

8.19 List the elements in Groups IIIA to VIA in the same order as in the periodic table. Label each element as a metal, a metalloid, or a nonmetal. Does each column of elements display the expected trend of increasing metallic characteristics?

8.20 For the list of elements you made for Question 8.19, note whether the oxides of each element are acidic, basic, or amphoteric.

8.21 From what is said in Section 8.7 about Group IIA elements, list some properties of barium.

8.22 Write an equation for the reaction of potassium metal with water.

8.23 Give the names and formulas of two oxides of carbon.

8.24 Match each description in the left column with the appropriate element in the right column.

a	A soft, light metal that reacts vigorously with water	Sulfur
		Sodium
b	A reddish brown liquid	White phosphorus
c	A yellow solid that burns in air	Bromine
d	A waxy, white solid, normally stored under water	

8.25 Phosphorus has _____ unpaired electrons.

   a 1   b 2   c 3   d 4   e 5

8.26 Which of the following atoms, designated by their electron configurations, has the *highest* ionization energy?

  a $[Kr]4d^{10}5s^2 5p^3$
  b $[Xe]4f^{14}5d^{10}6s^2 6p^3$
  c $[Ne]3s^2 3p^2$
  d $[Ne]3s^2 3p^3$
  e $[Ar]3d^{10}4s^2 4p^3$

8.27 When trying to remove electrons from Be, which of the following sets of ionization energy makes the most sense going from first to third ionization energy? Explain your answer.

  a First IE 900 KJ/mol, second IE 1750 kJ/mol, third IE 15,000 kJ/mol
  b First IE 1750 KJ/mol, second IE 900 kJ/mol, third IE 15,000 kJ/mol
  c First IE 15,000 KJ/mol, second IE 1750 kJ/mol, third IE 900 kJ/mol
  d First IE 900 KJ/mol, second IE 15,000 kJ/mol, third IE 22,000 kJ/mol
  e First IE 900 KJ/mol, second IE 1750 kJ/mol, third IE 1850 kJ/mol

8.28 Consider the following orderings.

  I. Al < Si < P < S
  II. Be < Mg < Ca < Sr
  III. I < Br < Cl < F

Which of these give(s) a correct trend in atomic size?

  a I only
  b II only
  c III only
  d I and II only
  e II and III only

Concept Explorations

Key: Concept explorations are comprehensive problems that provide a framework that will enable you to explore and learn many of the critical concepts and ideas in each chapter. If you master the concepts associated with these explorations, you will have a better understanding of many important chemistry ideas and will be more successful in solving all types of chemistry problems. These problems are well suited for group work and for use as in-class activities.

8.29 Periodic Properties I

A hypothetical element, X, has the following ionization energy values:

First ionization energy: 900 kJ/mol

Second ionization energy: 1750 kJ/mol

Third ionization energy: 14,900 kJ/mol

Fourth ionization energy: 21,000 kJ/mol

Another element, Y, has the following ionization energy values:

First ionization energy: 1200 kJ/mol

Second ionization energy: 2500 kJ/mol

Third ionization energy: 19,900 kJ/mol

Fourth ionization energy: 26,000 kJ/mol

a To what family of the periodic table would element X be most likely to belong? Explain?

b What charge would you expect element X to have when it forms an ion?

c If you were to place elements X and Y into the periodic table, would element Y be in the same period as element X? If not in the same period, where might they be relative to each other in the periodic table?

d Would an atom of Y be smaller or larger than an atom of X? Explain your reasoning.

8.30 Periodic Properties II

Consider two hypothetical elements, X and Z. Element X has an electron affinity of 300 kJ/mol, and element Z has an electron affinity of 75 kJ/mol.

a If you have an X^- ion and Z^- ion, from which ion would it require more energy to remove an electron? Explain your answer.

b If elements X and Z are in the same period of the periodic table, which atom would you expect to have the greater atomic radius? Why?

c Assuming that the elements are in the same period, which element would you expect to have the smaller first ionization energy?

d Do the valence electrons in element Z feel a greater effective nuclear charge than those in element X? Explain how you arrived at your answer.

Conceptual Problems

Key: These problems are designed to check your understanding of the concepts associated with some of the main topics presented in each chapter. A strong conceptual understanding of chemistry is the foundation for both applying chemical knowledge and solving chemical problems. These problems vary in level of difficulty and often can be used as a basis for group discussion.

8.31 Imagine a world in which all quantum numbers, except the l quantum number, are as they are in the real world. In this imaginary world, l begins with 1 and goes up to n (the value of the principal quantum number). Assume that the orbitals fill in the order by n, then l; that is, the first orbital to fill is for $n = 1, l = 1$; the next orbital to fill is for $n = 2, l = 1$, and so forth. How many elements would there be in the first period of the periodic table?

8.32 Suppose that the Pauli exclusion principle were "No more than two electrons can have the same four quantum numbers." What would be the electron configurations of the ground states for the first six elements of the periodic table, assuming that, except for the Pauli principle, the usual building-up principle held?

8.33 Two elements in Period 5 are adjacent to one another in the periodic table. The ground-state atom of one element has only s electrons in its valence shell; the other has at least one d electron in an unfilled shell. Identify the elements.

8.34 You travel to an alternate universe where the atomic orbitals are different from those on earth, but all other aspects of the atoms are the same. In this universe, you find that the first (lowest energy) orbital is filled with three electrons and the second orbital can hold a maximum of nine electrons. You discover an element Z that has five electrons in its atom. Would you expect Z to be more likely to form a cation or an anion? Indicate a possible charge on this ion.

8.35 Two elements are in the same column of the periodic table, one above the other. The ground-state atom of one element has two s electrons in its outer shell, and no d electrons anywhere in its configuration. The other element has d electrons in its configuration. Identify the elements.

8.36 Two elements are in the same group, one following the other. One is a metalloid; the other is a metal. Both form oxides of the formula RO_2. The first is acidic; the next is amphoteric. Identify the two elements.

8.37 Would you expect to find an element having both a very large (positive) first ionization energy and an electron affinity that is also a large positive number? Explain.

8.38 A metalloid has an acidic oxide of the formula R_2O_3. The element has no oxide of the formula R_2O_5. What is the name of the element?

8.39 Given the following information, identify the group from the periodic table that contains elements that behave like main-group element "E."

i. The electron affinity of E is small.

ii. The ionization energy (IE) trend for element E is: first ionization energy < second ionization energy <<< third ionization energy.

iii. Samples of E are lustrous and are good electrical conductors.

8.40 A hypothetical element A has the following properties:

First ionization energy: 850 kJ/mol

Second ionization energy: 1700 kJ/mol

Third ionization energy: 13,999 kJ/mol

a If you were to react element A with oxygen, what would be the chemical formula of the resulting compound?

b Write the balanced chemical reaction of A reacting with oxygen to give the product from part a.

c Would you expect the product of the chemical reaction of A with oxygen to be a basic, an acidic, or an amphoteric oxide?

Practice Problems

Key: These problems are for practice in applying problem-solving skills. They are divided by topic, and some are keyed to exercises (see the ends of the exercises). The problems are arranged in matching pairs; the odd-numbered problem of each pair is listed first, and its answer is given in the back of the book.

Pauli Exclusion Principle

8.41 Which of the following orbital diagrams are allowed and which are not allowed by the Pauli exclusion principle? Explain. For those that are allowed, write the electron configuration.

8.42 Which of the following orbital diagrams are allowed by the Pauli exclusion principle? Explain how you arrived at this decision. Give the electron configuration for the allowed ones.

8.43 Which of the following electron configurations are possible? Explain why the others are not.

  a $1s^22s^22p^63s^33d^7$ b $1s^22s^22p^5$
  c $1s^22s^22p^63s^33d^8$ d $1s^12s^22p^7$

8.44 Choose the electron configurations that are possible from among the following. Explain why the others are impossible.

  a $1s^22s^12p^6$ b $1s^22s^22p^8$
  c $1s^22s^22p^63s^23p^63d^7$ d $1s^22s^32p^63s^13d^9$

8.45 Write all of the possible orbital diagrams for the electron configuration $1s^22p^1$. (There are six different diagrams.)

8.46 Give the different orbital diagrams for the configuration $1s^12p^1$. (There are twelve different diagrams.)

Building-Up Principle and Hund's Rule

8.47 Use the building-up principle to obtain the ground-state configuration of arsenic.

8.48 Give the electron configuration of the ground state of iodine, using the building-up principle.

8.49 Use the building-up principle to obtain the electron configuration of the ground state of manganese.

8.50 Give the electron configuration of the ground state of nickel, using the building-up principle.

8.51 Bismuth is a Group VA element in Period 6. Write the valence-shell configuration of bismuth.

8.52 Bromine is a Group VIIA element in Period 4. Deduce the valence-shell configuration of bromine.

8.53 Zirconium is a Group IVB element in Period 5. What would you expect for the valence-shell configuration of zirconium?

8.54 Manganese is a Group VIIB element in Period 4. What would you expect for the configuration of outer electrons of manganese?

8.55 The configuration for the ground state of iridium is $[Xe]4f^{14}5d^76s^2$. What are the group and period for this element? Is it a main-group, a *d*-transition, or an *f*-transition element?

8.56 Thallium has the ground-state configuration $[Xe]4f^{14}5d^{10}6s^26p^1$. Give the group and period for this element. Classify it as a main-group, a *d*-transition, or an *f*-transition element.

8.57 Write the orbital diagram for the ground state of cobalt. The electron configuration is $[Ar]3d^74s^2$.

8.58 Write the orbital diagram for the ground state of samarium. The electron configuration is $[Xe]4f^66s^2$.

8.59 Write an orbital diagram for the ground state of the potassium atom. Is the atomic substance diamagnetic or paramagnetic?

8.60 Write an orbital diagram for the ground state of the zinc atom. Is the atomic substance diamagnetic or paramagnetic?

Periodic Trends

8.61 Order the following elements by increasing atomic radius according to what you expect from periodic trends: Se, As, S.

8.62 Using periodic trends, arrange the following elements in order of increasing atomic radius: F, S, Cl.

8.63 Using periodic trends, arrange the following elements by increasing ionization energy: Ar, Cl, Na, Al.

8.64 Arrange the following elements in order of increasing ionization energy: Mg, Ca, S. Do not look at Figure 8.18.

8.65 From what you know in a general way about electron affinities, state which member of each of the following pairs has the greater value: a F, Li b As, Br.

8.66 From what you know in a general way about electron affinities, state which member of each of the following pairs has the greater value: a Cl, S b Se, K.

8.67 If potassium chlorate has the formula $KClO_3$, what formula would you expect for lithium bromate?

8.68 Write the simplest formulas expected for two oxides of selenium.

General Problems

Key: These problems provide more practice but are not divided by topic or keyed to exercises. Each section ends with essay questions, each of which is color coded to refer to the *Instrumental Methods* (brown) and *A Chemist Looks at Frontiers* (purple) chapter essay on which it is based. Odd-numbered problems and the even-numbered problems that follow are similar; answers to all odd-numbered problems except the essay questions are given in the back of the book.

8.69 Write the complete ground-state electron configuration of the strontium atom, Sr, using the building-up principle.

8.70 Write the complete ground-state electron configuration of the tellurium atom, Te, using the building-up principle.

8.71 Obtain the valence-shell configuration of the polonium atom, Po, using the position of this atom in the periodic table.

8.72 Obtain the valence-shell configuration of the lead atom, Pb, using the position of this atom in the periodic table.

8.73 Write the orbital diagram for the ground state of the arsenic atom. Give all orbitals.

8.74 Write the orbital diagram for the ground state of the selenium atom. Give all orbitals.

8.75 For eka-bismuth, predict the electron configuration, whether the element is a metal or nonmetal, and the formula of an oxide.

8.76 For eka-lead, predict the electron configuration, whether the element is a metal or nonmetal, and the formula of an oxide.

8.77 From Figure 8.18, predict the first ionization energy of francium ($Z = 87$).

8.78 From Figure 8.18, predict the first ionization energy of ununseptium ($Z = 117$).

8.79 Write the orbital diagram for the ground state of rhodium. The configuration is $[Kr]4d^85s^1$.

8.80 Write the orbital diagram corresponding to the ground state of Nb, whose configuration is $[Kr]4d^45s^1$.

8.81 Match each set of characteristics on the left with an element in the column at the right.

a reactive nonmetal; the atom has a large electron affinity

Sodium (Na)
Antimony (Sb)
Argon (Ar)
Chlorine (Cl_2)

b A soft metal; the atom has a low ionization energy
c A metalloid that forms an oxide of formula R_2O_3
d A chemically unreactive gas

8.82 Match each element on the right with a set of characteristics on the left.

a A metal that forms an oxide of formula R_2O_3

Oxygen (O_2)
Gallium (Ga)
Barium (Ba)
Fluorine (F_2)

b A reactive, pale yellow gas; the atom has a large electron affinity
c A soft metal that reacts with water to produce hydrogen
d A colorless gas; the atom has a moderately large negative electron affinity

8.83 Find the electron configuration of the element with $Z = 23$. From this, give its group and period in the periodic table. Classify the element as a main-group, a d-block transition, or an f-block transition element.

8.84 Find the electron configuration of the element with $Z = 33$. From this, give its group and period in the periodic table. Is this a main-group, a d-block transition, or an f-block transition element?

■ **8.85** Explain the origin of the "chemical shift" seen for a proton in a molecule.

■ **8.86** What property of an atom does nuclear magnetic resonance depend on? What frequency range of the electromagnetic spectrum does NMR use?

■ **8.87** Explain how x rays are produced when an electron beam falls on a metal target.

■ **8.88** Describe how x-ray photoelectron spectroscopy is done. What kind of information is obtained from photoelectron spectroscopy?

■ **8.89** Researchers have shown that it is possible to levitate diamagnetic materials. What animal did they use in their experiments to show this?

■ **8.90** How is it possible for a diamagnetic material to be levitated by a magnetic field?

Strategy Problems

Key: As noted earlier, all of the practice and general problems are matched pairs. This section is a selection of problems that are not in matched-pair format. These challenging problems require that you employ many of the concepts and strategies that were developed in the chapter. In some cases, you will have to integrate several concepts and operational skills in order to solve the problem successfully.

8.91 The following are orbital diagrams for presumed ground-state atoms. Several, though, violate Pauli's principle or Hund's rule. Which of the these follow both Pauli's principle and Hund's rule, and which violate one or the other (state whether Pauli's principle or Hund's rule is violated)?

8.92 As we noted at the beginning of Section 8.2, an atom has an infinite number of electron configurations, including not just the ground state but excited states as well. The following are excited states of neutral atoms. Identify the element and write the ground-state electron configuration for that atom.

a $1s^22s^22p^63s^23p^63d^1$
b $1s^22s^22p^53s^23p^1$
c $1s^22s^22p^63s^23p^63d^{10}4s^14p^1$
d $1s^22s^22p^63s^23p^63d^94s^24p^5$
e $1s^12s^22p^63s^23p^6$

8.93 A metallic element, M, reacts vigorously with water to form a solution of MOH. If M is in Period 5, what is the ground-state electron configuration of the atom?

8.94 A nonmetallic element, R, burns brightly in air to give the oxide R_4O_{10}. If R is in Period 3, what is the ground-state valence-shell configuration of the atom?

8.95 An atom of an element has the following ground-state configuration: $[Xe]4f^{14}5d^96s^1$. Using this configuration, give the symbol of the element. Explain your reasoning.

8.96 The ground-state electron configuration of an atom is $1s^22s^22p^63s^23p^4$. What is the valence-shell configuration of the atom in the same group, but in Period 5?

8.97 An atom of an element has the following ground-state configuration: $[Kr]4d^{10}5s^25p^b$. If the element is in Group VIA, what is the symbol of the element? What is b?

8.98 A main-group atom in Period 5 has the following orbital diagram for its valence shell (ground state): (↑↓) (↑)(↑)(↑). What is the symbol for the element? Write the formula of the oxide containing the most oxygen.

8.99 A neutral atom has the electron configuration $1s^2 2s^2 2p^6 3s^1 3p^2$. Is this the ground-state configuration? What is the period and column of the element? What is the name of the element? What is the formula of the oxide of this element?

8.100 The following are electron configurations of neutral atoms, which may or may not be in ground states: $1s^2 2s^2 2p^5 3s^2 3p^2$ and $1s^2 2s^2 2p^6 3s^2 3p^5$. Write the formula of a compound of these two elements.

8.101 A moderately reactive metallic element (whose melting point is 850°C) burns in a container of a greenish gas (also an element) to produce a white solid. The white solid dissolves in water to form a clear solution. When carbon dioxide gas, CO_2, is bubbled into this solution, it produces a white precipitate. Identify, by names and formulas, the substances referred to as the white solid and the white precipitate. Use a handbook of chemistry to obtain any data you may need.

8.102 A metallic element reacts vigorously with water, evolving hydrogen gas. An excited atom of this element has its outer electron in the $3p$ orbital. When this electron drops to its ground state in the $3s$ orbital, light is emitted of wavelength 589 nm. What is the identity of the element? Explain how you arrived at your answer. What is the color of the emitted light?

8.103 The densities (in g/cm³) of some elements in Period 5 that span indium, In, are as follows: Ag, 10.49; Cd, 8.65; Sn, 7.31; and Sb, 6.68. Plot the density of each element against its atomic number. Using the graph, estimate the density of indium. Compare your estimate to the value that you find in a handbook of chemistry or online.

8.104 All elements up to atomic number 118 have been discovered. What electron configuration would you expect for the yet-to-be discovered element 120? Explain your answer.

8.105 Using the periodic table, decide which of the following atoms you expect to have the largest radius: H, Sc, As, Br, V, K. Why?

8.106 Write the orbital diagram for the ground-state valence electrons of the main-group atom in Period 5 that has the smallest radius. Explain how you got this answer.

8.107 An atom easily loses two electrons to form the ion R^{2+}. The element, which is in Period 6, forms the oxides RO and RO_2. Give the orbital diagram for the ground-state valence shell of the element that immediately follows R in the periodic table.

8.108 The energy of dissociation of the Cl_2 molecule is 240 kJ/mol. Using the appropriate data from this chapter, calculate the energy obtained (or released) in adding electrons to the Cl_2 molecule to produce Cl^- ions. Is this process endothermic or exothermic?

8.109 List three elements in order of increasing ionization energy. Two of the elements should be halogens (but not F), and the other should be Se.

8.110 The electron affinity of the lutetium atom (element 71) was measured using the technique of photoelectron spectroscopy with an infrared laser (the essay on p. 310 describes this instrumental method, using x rays). In this experiment, a beam of lutetium negative ions, Lu^-, was prepared and irradiated with a laser beam having a wavelength at 1064 nm. The energy supplied by a photon in this laser beam removes an electron from a negative ion, leaving the neutral atom. The energy needed to remove the electron from the negative ion to give the neutral atom (both in their ground states) is the electron affinity of lutetium. Any excess energy of the photon shows up as kinetic energy of the emitted electron. If the emitted electron in this experiment has a kinetic energy of 0.825 eV, what is the electron affinity of lutetium?

Cumulative-Skills Problems

Key: The problems under this heading combine skills introduced in previous chapters with those given in the current one.

8.111 A sample of cesium metal reacted completely with water, evolving 48.1 mL of dry H_2 at 19°C and 768 mmHg. What is the equation for the reaction? What was the mass of cesium in the sample?

8.112 A 2.50-g sample of barium reacted completely with water. What is the equation for the reaction? How many milliliters of dry H_2 evolved at 21°C and 748 mmHg?

8.113 What is the formula of radium oxide? What is the percentage of radium in this oxide?

8.114 What is the formula of hydrogen telluride? What is the percentage of tellurium in this compound?

8.115 How much energy is evolved when 2.65 mg of Cl(*g*) atoms adds electrons to give $Cl^-(g)$ ions?

8.116 How much energy would be required to ionize 5.00 mg of Na(*g*) atoms to $Na^+(g)$ ions? The first ionization energy of Na atoms is 496 kJ/mol.

8.117 Using the Bohr formula for the energy levels, calculate the energy required to raise the electron in a hydrogen atom from $n = 1$ to $n = \infty$. Express the result for 1 mol H atoms. Because the $n = \infty$ level corresponds to removal of the electron from the atom, this energy equals the ionization energy of the H atom.

8.118 Calculate the ionization energy of the He^+ ion in kJ/mol (this would be the second ionization energy of He). See Problem 8.117. The Bohr formula for the energy levels of an ion consisting of a nucleus of charge Z and a single electron is $-R_H Z^2 / n^2$.

8.119 The lattice energy of an ionic solid such as NaCl is the enthalpy change $\Delta H°$ for the process in which the solid changes to ions. For example,

$$NaCl(s) \longrightarrow Na^+(g) + Cl^-(g) \quad \Delta H = 786 \text{ kJ/mol}$$

Assume that the ionization energy and electron affinity are ΔH values for the processes defined by those terms. The ionization energy of Na is 496 kJ/mol. Use this, the electron affinity from Table 8.4, and the lattice energy of NaCl to calculate ΔH for the following process:

$$Na(g) + Cl(g) \longrightarrow NaCl(s)$$

8.120 Calculate ΔH for the following process:

$$K(g) + Br(g) \longrightarrow KBr(s)$$

The lattice energy of KBr is 689 kJ/mol, and the ionization energy of K is 419 kJ/mol. The electron affinity of Br is given in Table 8.4. See Problem 8.119.

9

Ionic and Covalent Bonding

The shape of snowflakes results from bonding (and intermolecular) forces in H_2O.

Contents and Concepts

Ionic Bonds

Molten salts and aqueous solutions of salts are electrically conducting. This conductivity results from the motion of ions in the liquids. This suggests the possibility that ions exist in certain solids, held together by the attraction of ions of opposite charge.

9.1 Describing Ionic Bonds

9.2 Electron Configurations of Ions

9.3 Ionic Radii

Covalent Bonds

Not all bonds can be ionic. Hydrogen, H_2, is a clear example, in which there is a strong bond between two like atoms. The bonding in the hydrogen molecule is covalent. A *covalent bond* forms between atoms by the sharing of a pair of electrons.

9.4 Describing Covalent Bonds

9.5 Polar Covalent Bonds; Electronegativity

9.6 Writing Lewis Electron-Dot Formulas

9.7 Delocalized Bonding: Resonance

9.8 Exceptions to the Octet Rule

9.9 Formal Charge and Lewis Formulas

9.10 Bond Length and Bond Order

9.11 Bond Enthalpy

The properties of a substance, such as sodium chloride (Figure 9.1), are determined in part by the chemical bonds that hold the atoms together. A *chemical bond* is a strong attractive force that exists between certain atoms in a substance. In Chapter 2 we described how sodium (a silvery metal) reacts with chlorine (a pale greenish yellow gas) to produce sodium chloride (table salt, a white solid). The substances in this reaction are quite different, as are their chemical bonds. Sodium chloride, NaCl, consists of Na^+ and Cl^- ions held in a regular arrangement, or crystal, by *ionic bonds.* Ionic bonding results from the attractive force of oppositely charged ions.

A second kind of chemical bond is a *covalent bond.* In a covalent bond, two atoms share valence electrons (outer-shell electrons), which are attracted to the positively charged cores of both atoms, thus linking them. For example, chlorine gas consists of Cl_2 molecules. You would not expect the two Cl atoms in each Cl_2 molecule to acquire the opposite charges required for ionic bonding. Rather, a covalent bond holds the two atoms together. This is consistent with the equal sharing of electrons that you would expect between identical atoms. In most molecules, the atoms are linked by covalent bonds.

Metallic bonding, seen in sodium and other metals, represents another important type of bonding. A crystal of sodium metal consists of a regular arrangement of sodium atoms. The valence electrons of these atoms move throughout the crystal, attracted to the positive cores of all Na^+ ions. This attraction holds the crystal together.

What determines the type of bonding in each substance? How do you describe the bonding in various substances? In this chapter we will look at some simple, useful concepts of bonding that can help us answer these questions. We will be concerned with ionic and covalent bonds in particular.

Figure 9.1 ▲

Sodium chloride crystals Natural crystals of sodium chloride mineral (halite).

Ionic Bonds

The first explanation of chemical bonding was suggested by the properties of *salts,* substances now known to be ionic. Salts are generally crystalline solids that melt at high temperatures. Sodium chloride, for example, melts at 801°C. A molten salt (the liquid after melting) conducts an electric current. A salt dissolved in water gives a solution that also conducts an electric current. The electrical conductivity of the molten salt and the salt solution results from the motion of ions in the liquids. This suggests the possibility that ions exist in certain solids, held together by the attraction of opposite charges.

9.1 Describing Ionic Bonds

An **ionic bond** is *a chemical bond formed by the electrostatic attraction between positive and negative ions.* The bond forms between two atoms when one or more electrons are transferred from the valence shell of one atom to the valence shell of the other. The atom that loses electrons becomes a *cation* (positive ion), and the atom that gains electrons becomes an *anion* (negative ion). Any given ion tends to attract as many neighboring ions of opposite charge as possible. When large numbers of ions gather together, they form an ionic solid. The solid normally has a regular, crystalline structure that allows for the maximum attraction of ions, given their particular sizes.

To understand why ionic bonding occurs, consider the transfer of a valence electron from a sodium atom (electron configuration $[Ne]3s^1$) to the valence shell of a chlorine atom ($[Ne]3s^23p^5$). You can represent the electron transfer by the following equation:

$$Na([Ne]3s^1) + Cl([Ne]3s^23p^5) \longrightarrow Na^+([Ne]) + Cl^-([Ne]3s^23p^6)$$

As a result of the electron transfer, ions are formed, each of which has a noble-gas configuration. The sodium atom has lost its $3s$ electron and has taken on the neon

Table 9.1   **Lewis Electron-Dot Symbols for Atoms of the Second and Third Periods**

Period	IA ns^1	IIA ns^2	IIIA ns^2np^1	IVA ns^2np^2	VA ns^2np^3	VIA ns^2np^4	VIIA ns^2np^5	VIIIA ns^2np^6
Second	Li·	·Be·	·B·	·C·	:N·	:O·	:F·	:Ne:
Third	Na·	·Mg·	·Al·	·Si·	:P·	:S·	:Cl·	:Ar:

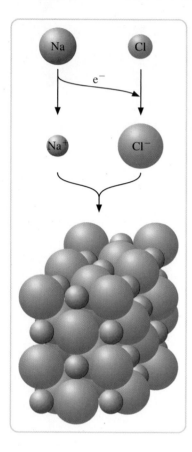

configuration, [Ne]. The chlorine atom has accepted the electron into its $3p$ subshell and has taken on the argon configuration, $[Ne]3s^23p^6$. Such noble-gas configurations and the corresponding ions are particularly stable. This stability of the ions accounts in part for the formation of the ionic solid NaCl. Once a cation or anion forms, it attracts ions of opposite charge. Within the sodium chloride crystal, NaCl, every Na^+ ion is surrounded by six Cl^- ions, and every Cl^- ion by six Na^+ ions. (Figure 2.19 shows the arrangement of ions in the NaCl crystal.)

Lewis Electron-Dot Symbols

You can simplify the preceding equation for the electron transfer between Na and Cl by writing Lewis electron-dot symbols for the atoms and monatomic ions. A **Lewis electron-dot symbol** is *a symbol in which the electrons in the valence shell of an atom or ion are represented by dots placed around the letter symbol of the element.* Table 9.1 lists Lewis symbols and corresponding valence-shell electron configurations for the atoms of the second and third periods. Note that dots are placed one to each side of a letter symbol until all four sides are occupied. Then the dots are written two to a side until all valence electrons are accounted for. The exact placement of the single dots is immaterial. For example, the single dot in the Lewis symbol for chlorine can be written on any one of the four sides. [This pairing of dots does not always correspond to the pairing of electrons in the ground state. Thus, you write ·Ḃ· for boron, rather than :B·, which more closely corresponds to the ground-state configuration $[He]2s^22p^1$. The first symbol better reflects boron's chemistry, in which each single electron (single dot) tends to be involved in bond formation.]

The equation representing the transfer of an electron from the sodium atom to the chlorine atom is

$$Na· \ + \ ·\ddot{\underset{..}{Cl}}: \ \longrightarrow \ Na^+ \ + \ [:\ddot{\underset{..}{Cl}}:]^-$$

The noble-gas configurations of the ions are apparent from the symbols. No dots are shown for the cation. (All valence electrons have been removed, leaving the noble-gas core.) Eight dots are shown in brackets for the anion (noble-gas configuration ns^2np^6).

Example 9.1   Using Lewis Symbols to Represent Ionic Bond Formation

Gaining Mastery Toolbox

Critical Concept 9.1
To represent the formation of the ions of an ionic compound from the atoms using Lewis symbols, you show dots being transferred from the metal to nonmetal atoms to give the ions with noble-gas configurations.

Solution Essentials:
• Lewis electron-dot symbols
• Noble-gas configuration
• Formula of a binary ionic compound

Use Lewis electron-dot symbols to represent the transfer of electrons from magnesium to fluorine atoms to form ions with noble-gas configurations.

Problem Strategy Write down the Lewis symbols for the two atoms. Note how many electrons the metal atom should lose to assume a noble-gas configuration (all of the dots, representing valence electrons) and how many electrons the nonmetal atom should gain to assume such a configuration (enough to give eight dots about the atomic symbol). Represent the transfer of electrons between ions diagrammatically in the form of an equation.

(continued)

(*continued*)

Solution The Lewis symbols for the atoms are $\ddot{\text{:}}\overset{..}{\text{F}}\cdot$ and $\cdot\text{Mg}\cdot$ (see Table 9.1). The magnesium atom loses two electrons to assume a noble-gas configuration. But because a fluorine atom can accept only one electron to fill its valence shell, two fluorine atoms must take part in the electron transfer. We can represent this electron transfer as follows:

$$:\overset{..}{\underset{..}{\text{F}}}\cdot \; + \; \cdot\text{Mg}\cdot \; + \; \cdot\overset{..}{\underset{..}{\text{F}}}: \longrightarrow \; [:\overset{..}{\underset{..}{\text{F}}}:]^- \; + \; \text{Mg}^{2+} \; + \; [:\overset{..}{\underset{..}{\text{F}}}:]^-$$

Answer Check Check that you have the correct symbols for the atoms and that the symbols for the ions do have noble-gas configurations.

Exercise 9.1 Represent the transfer of electrons from magnesium to oxygen atoms to assume noble-gas configurations. Use Lewis electron-dot symbols.

■ See Problems 9.37 and 9.38.

Energy Involved in Ionic Bonding

You have seen in a qualitative way why a sodium atom and a chlorine atom might be expected to form an ionic bond. It is instructive, however, to look at the energy changes involved in ionic bond formation. From this analysis, you can gain further understanding of why certain atoms bond ionically and others do not.

If atoms come together and bond, there should be a net decrease in energy, because the bonded state should be more stable and therefore at a lower energy level. Consider again the formation of an ionic bond between a sodium atom and a chlorine atom. You can think of this as occurring in two steps: (1) An electron is transferred between the two separate atoms to give ions. (2) The ions then attract one another to form an ionic bond. In reality, the transfer of the electron and the formation of an ionic bond occur simultaneously, rather than in discrete steps, as the atoms approach one another. But the *net* quantity of energy involved is the same whether the steps occur one after the other or at the same time.

The first step requires removal of the $3s$ electron from the sodium atom and the addition of this electron to the valence shell of the chlorine atom. Removing the electron from the sodium atom requires energy (the first ionization energy of the sodium atom, which equals 496 kJ/mol). Adding the electron to the chlorine atom releases energy (equal to -349 kJ/mol, which is the negative of the electron affinity of the chlorine atom). ▶ The overall energy of this step is $(496 - 349)$ kJ/mol, or 147 kJ/mol (Figure 9.2, Step 1). So, the process requires more energy to remove an electron from the sodium atom than is gained when the electron is added to the chlorine atom. The formation of ions from the atoms is not in itself energetically favorable.

When positive and negative ions bond, however, more than enough energy is released to make the overall process favorable. What principally determines the

Ionization energies and electron affinities of atoms were discussed in Section 8.6.

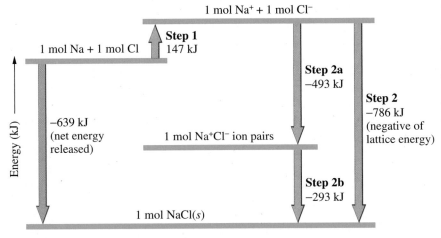

Figure 9.2 ◀

Energetics of ionic bonding The transfer of an electron from a Na atom to a Cl atom is not in itself energetically favorable; it requires 147 kJ/mol of energy (Step 1). However, 493 kJ of energy is released when these oppositely charged ions come together to form ion pairs (Step 2a). Moreover, additional energy (293 kJ) is released when these ion pairs form the solid NaCl crystal (Step 2b). The lattice energy released when one mole each of Na⁺ and Cl⁻ ions react to produce NaCl(*s*) is 786 kJ/mol, and the overall process of NaCl formation is energetically favorable, releasing 639 kJ/mol if one starts with gaseous Na and Cl atoms.

energy released when ions bond is the attraction of oppositely charged ions. To see this, let's look first at the energy obtained when a Na$^+$ ion and a Cl$^-$ ion come together to form an ion-pair molecule. We will estimate this energy by making the simplifying assumption that the ions are spheres, just touching, with the distance between the nuclei of the ions equal to this distance in the NaCl crystal. From experiment, this distance is known to be 282 pm, or 2.82×10^{-10} m. To calculate the energy obtained when the ion spheres come together, we use Coulomb's law.

Coulomb's law states that *the potential energy obtained in bringing two charges Q$_1$ and Q$_2$, initially far apart, up to a distance r apart is directly proportional to the product of the charges and inversely proportional to the distance between them:*

$$E = \frac{kQ_1Q_2}{r}$$

Here k is a physical constant, equal to 8.99×10^9 J·m/C^2 (C is the symbol for coulomb). The charge on Na$^+$ is $+e$ and that on Cl$^-$ is $-e$, where e equals 1.602×10^{-19} C. Thus, our estimate of the energy of attraction of Na$^+$ and Cl$^-$ ions to form an ion pair is

$$E = \frac{-(8.99 \times 10^9 \text{ J·m/C}^2) \times (1.602 \times 10^{-19} \text{ C})^2}{2.82 \times 10^{-10} \text{ m}} = -8.18 \times 10^{-19} \text{ J}$$

The minus sign means energy is released. This energy is for the formation of one ion pair. To express this for one mole of Na$^+$Cl$^-$ ion pairs, we multiply by Avogadro's number, 6.02×10^{23}. We obtain -493 kJ/mol for the energy obtained when one mole of Na$^+$ and one mole of Cl$^-$ come together to form Na$^+$Cl$^-$ ion pairs (Figure 9.2, Step 2a). ◄

The energy value −493 kJ/mol is approximate, because of the simplifying assumption we made (that the ions are spheres, just touching).

What we see is that the formation of ion pairs from sodium and chlorine atoms is energetically favorable. The attraction of oppositely charged ions, however, does not stop with the bonding of pairs of ions. The maximum attraction of ions of opposite charge with the minimum repulsion of ions of the same charge is obtained with the formation of the crystalline solid. Then additional energy is released. The energy of this step (Figure 9.2, Step 2b) is most easily obtained as the difference between Figure 9.2, Step 2a, which we just calculated, and Figure 9.2, Step 2, the energy released when a crystalline solid forms from the ions. This is the negative of the *lattice energy* of NaCl. The additional energy (in going from ion pairs to the crystalline solid), Figure 9.2, Step 2b, equals -293 kJ/mol.

The **lattice energy** is *the change in energy that occurs when an ionic solid is separated into isolated ions in the gas phase.* For sodium chloride, the process is

$$\text{NaCl}(s) \longrightarrow \text{Na}^+(g) + \text{Cl}^-(g)$$

The distances between ions in the crystal are continuously enlarged until the ions are very far apart. You can obtain an experimental value for this process from thermodynamic data (see the Born-Haber cycle on the next page). The lattice energy for NaCl is 786 kJ/mol so that for the reverse process, when the ions come together to bond, the energy is -786 kJ/mol (Figure 9.2, Step 2). Consequently, the net energy obtained when gaseous Na and Cl atoms form solid NaCl is $(-786 + 147)$ kJ/mol $= -639$ kJ/mol. The negative sign shows that there has been a net decrease in energy, which you expect when stable bonding has occurred.

From this energy analysis, you can see that two elements bond ionically if the ionization energy of one is sufficiently small and the electron affinity of the other is sufficiently large. This situation exists between a reactive metal (which has low ionization energy) and a reactive nonmetal (which has large electron affinity). In general, bonding between a metal and a nonmetal is ionic. This energy analysis also explains why ionic bonding normally results in a solid rather than in ion-pair molecules.

Lattice Energies from the Born–Haber Cycle

The preceding energy analysis requires us to know the lattice energy of solid sodium chloride. Direct experimental determination of the lattice energy of an ionic solid is difficult. However, this quantity can be indirectly determined from experiment by means of a thermochemical "cycle" originated by Max Born and Fritz Haber in 1919 and now called the *Born–Haber cycle*. The reasoning is based on Hess's law.

To obtain the lattice energy of NaCl, you think of solid sodium chloride being formed from the elements by two different routes, as shown in Figure 9.3. In one route, NaCl(*s*) is formed directly from the elements, Na(*s*) and $\frac{1}{2}Cl_2(g)$. The enthalpy change for this is $\Delta H_f°$, which is given in Table 6.2 as −411 kJ per mole of NaCl. The second route consists of the following five steps, along with the enthalpy change for each. (To be precise, the ionization energy and electron affinity are energy changes, ΔE, and we should add small corrections to give the enthalpy changes, ΔH.)

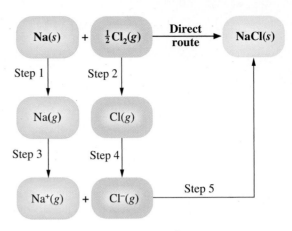

Figure 9.3 ▲

Born–Haber cycle for NaCl The formation of NaCl(s) from the elements is accomplished by two different routes. The direct route is the formation reaction (shown in boldface), and the enthalpy change is $\Delta H_f°$. The indirect route occurs in five steps.

1. *Sublimation of sodium.* Metallic sodium is vaporized to a gas of sodium atoms. (*Sublimation* is the transformation of a solid to a gas.) The enthalpy change for this process, measured experimentally, is 108 kJ per mole of sodium.

2. *Dissociation of chlorine.* Chlorine molecules are dissociated to atoms. The enthalpy change for this equals the Cl—Cl bond dissociation energy, which is 240 kJ per mole of bonds, or 120 kJ per mole of Cl atoms.

3. *Ionization of sodium.* Sodium atoms are ionized to Na^+ ions. The enthalpy change is essentially the ionization energy of atomic sodium, which equals 496 kJ per mole of Na.

4. *Formation of chloride ion.* The electrons from the ionization of sodium atoms are transferred to chlorine atoms. The enthalpy change for this is the negative of the electron affinity of atomic chlorine, equal to −349 kJ per mole of Cl atoms.

5. *Formation of NaCl(s) from ions.* The ions Na^+ and Cl^- formed in Steps 3 and 4 combine to give solid sodium chloride. Because this process is just the reverse of the one corresponding to the lattice energy (breaking the solid into ions), the enthalpy change is the negative of the lattice energy. If we let U be the lattice energy, the enthalpy change for Step 5 is $-U$.

Let us write these five steps and add them. We also add the corresponding enthalpy changes, following Hess's law.

$$
\begin{array}{lll}
Na(s) & \longrightarrow Na(g) & \Delta H_1 = \quad 108 \text{ kJ} \\
\tfrac{1}{2}Cl_2(g) & \longrightarrow Cl(g) & \Delta H_2 = \quad 120 \text{ kJ} \\
Na(g) & \longrightarrow Na^+(g) + e^-(g) & \Delta H_3 = \quad 496 \text{ kJ} \\
Cl(g) + e^-(g) & \longrightarrow Cl^-(g) & \Delta H_4 = -349 \text{ kJ} \\
Na^+(g) + Cl^-(g) & \longrightarrow NaCl(s) & \Delta H_5 = -U \\
\hline
Na(s) + \tfrac{1}{2}Cl_2(g) & \longrightarrow NaCl(s) & \Delta H_f° = \quad 375 \text{ kJ} - U
\end{array}
$$

In summing the equations, we have canceled terms that appear on both the left and right sides of the arrows. The final equation is simply the formation reaction for NaCl(*s*). Adding the enthalpy changes, we find that the enthalpy change for this formation reaction is 375 kJ − U. But the enthalpy of formation has been determined calorimetrically and equals −411 kJ. Equating these two values, we get

$$375 \text{ kJ} - U = -411 \text{ kJ}$$

Solving for U yields the lattice energy of NaCl:

$$U = (375 + 411) \text{ kJ} = 786 \text{ kJ}$$

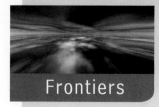
Ionic Liquids and Green Chemistry

New substances with peculiar or strange behavior have always intrigued chemists. The German chemist Hennig Brand in 1669 discovered a white, waxy solid that he called "cold fire," because it glowed in the dark—a strange property then and still captivating today. He tried to keep the recipe (lengthy boiling of putrid urine) secret, but people's fascination with the waxy solid, which was white phosphorus, proved irresistible.

Modern chemists continue to expand the boundaries of known materials. They have discovered plastics that conduct electricity like metals and materials that look solid but are so porous they are almost as light as air. Now chemists have produced room-temperature ionic liquids (Figure 9.4). Most of these are clear, well-behaved substances that look and pour much like water, and like water their strength is as solvents (liquids that dissolve other substances). In fact, they promise to be "super solvents." Given a material—an organic substance, a plastic, or even a rock—researchers believe you will be able to find an ionic liquid that is capable of dissolving it!

Compare room-temperature ionic liquids with their more prosaic cousins. Sodium chloride, a typical example, is solid at room

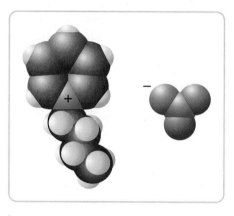

Figure 9.5 ▲

The ions composing an ionic liquid Shown here are space-filling models of the ions composing *N*-butylpyridinium nitrate. Note the bulky cation and small nitrate anion.

Figure 9.4 ▲

Room-temperature ionic liquids A series of tubes containing luminescent ionic liquids, spelling out QUILL (Queen's University Ionic Liquid Laboratories Research Centre, Belfast. Ireland).

© Martyn Earle 2007/Quill Research Center

temperature and doesn't melt until exposed to temperatures over 800°C. The molten liquid, though clear, is very corrosive (chemically reactive). The high melting point of sodium chloride is easily explained. It consists of small, spherical ions that pack closely together. Thus, the ions interact strongly, giving a solid with a high melting point. Room-temperature ionic liquids, by contrast, consist of large, nonspherical cations with various anions (Figure 9.5). It is the large, bulky cations that keep the ions from packing closely; the large distances between ions result in weak interactions, yielding a substance whose melting point is often well below room temperature.

The demand for *green chemistry,* the commercial production of chemicals using environmentally sound methods, has spurred much of the research into ionic liquids. Many chemical processes use volatile organic solvents. These solvents are liquids that evaporate easily into the surrounding air, where they can contribute to air pollution. Organic solvents are often flammable, too. Ionic liquids are neither volatile nor flammable. In addition to these environmental rewards, however, ionic liquids appear to offer another bonus: The proper choice of ionic liquid may improve the yield and lower the costs of a chemical process.

■ See Problems 9.111 and 9.112.

Properties of Ionic Substances

Typically, ionic substances are high-melting solids. Sodium chloride, NaCl, ordinary salt, melts at 801°C, and magnesium oxide, MgO, a ceramic, melts at 2800°C. The explanation for the high melting points of these substances is simple.

Small, spherical cations and anions interact by strong bonds that essentially depend on the electrical force of attraction described by Coulomb's law. Large groups of such cations and anions attract one another with strong ionic bonds linking all of the ions together, forming a crystalline solid. When you heat any solid, the atoms or ions of that solid begin to vibrate; and, as you raise the temperature, these atoms or ions vibrate through larger and larger distances. At a high enough temperature, the atoms or ions of the solid may move sufficiently far apart that the solid melts to a liquid (or the solid may simply decompose to different substances). The temperature at which melting occurs depends on the strength of the interaction between the atoms or ions. Typical ionic solids require high temperatures for this process to occur because of the strong interactions between the ions.

Coulomb's law also explains why magnesium oxide, which is composed of ions having double charges (Mg^{2+} and O^{2-}), has such a high melting point compared with sodium chloride (Na^+ and Cl^-). Coulombic interactions depend on the product of the ion charges. For NaCl, this is 1×1, whereas for MgO, this is 2×2, or 4 times greater. (Coulomb's law also depends on the distance between charges, or ions, which is about the same in NaCl and MgO.) The much larger Coulombic interaction in MgO requires a much greater temperature to initiate melting.

The liquid melt from an ionic solid consists of ions, and so the liquid conducts an electric current. If the ionic solid dissolves in a molecular liquid, such as water, the resulting solution consists of ions dispersed among molecules; the solution also conducts an electric current.

Recently, as the accompanying essay on ionic liquids describes, chemists have prepared ionic substances that behave atypically. Their unusual properties are the result of large, nonspherical cations that lead to especially weak ionic bonding. The melting points of these ionic substances are unusually low; often they are liquids at room temperature (that is, their melting points are below room temperature).

9.2 Electron Configurations of Ions

In the previous section, we described the formation of ions from atoms. Often you can understand what monatomic ions form by looking at the electron configurations of the atoms and deciding what configurations you would expect for the ions.

Ions of the Main-Group Elements

In Chapter 2, we listed the common monatomic ions of the main-group elements (Table 2.3). Most of the cations are obtained by removing all the valence electrons from the atoms of metallic elements. Once these atoms have lost their valence electrons, they have stable noble-gas or pseudo-noble-gas configurations. The stability of these configurations can be seen by looking at the successive ionization energies of some atoms. Table 9.2 lists the first through the fourth

Table 9.2   Ionization Energies of Na, Mg, and Al (in kJ/mol)*

	Successive Ionization Energies			
Element	First	Second	Third	Fourth
Na	496	4,562	6,910	9,543
Mg	738	1,451	7,733	10,542
Al	578	1,817	2,745	11,577

*Energies for the ionization of valence electrons lie to the left of the colored line.

ionization energies of Na, Mg, and Al. The energy needed to remove the first electron from the Na atom is only 496 kJ/mol (first ionization energy). But the energy required to remove another electron (the second ionization energy) is nearly ten times greater (4562 kJ/mol). The electron in this case must be taken from Na$^+$, which has a neon configuration. Magnesium and aluminum atoms are similar. Their valence electrons are easily removed, but the energy needed to take an electron from either of the ions that result (Mg^{2+} and Al^{3+}) is extremely high. That is why no compounds are found with ions having charges greater than the group number.

The loss of successive electrons from an atom requires increasingly more energy. Consequently, Group IIIA elements show less tendency to form ionic compounds than do Group IA and IIA elements, which primarily form ionic compounds. Boron, in fact, forms no compounds with B^{3+} ions. The bonding is normally covalent, a topic discussed later in this chapter. However, the tendency to form ions becomes greater going down any column of the periodic table because of decreasing ionization energy. The remaining elements of Group IIIA do form compounds containing 3+ ions.

There is also a tendency for Group IIIA to VA elements of higher periods, particularly Period 6, to form compounds with ions having a positive charge of two less than the group number. Thallium in IIIA, Period 6, has compounds with 1+ ions and compounds with 3+ ions. Ions with charge equal to the group number minus two are obtained when the np electrons of an atom are lost but the ns^2 electrons are retained. For example,

$$Tl([Xe]4f^{14}5d^{10}6s^26p^1) \longrightarrow Tl^+([Xe]4f^{14}5d^{10}6s^2) + e^-$$

Few compounds of 4+ ions are known because the energy required to form 4+ ions is so great. The first three elements of Group IVA—C, Si, and Ge—are nonmetals (or metalloids) and usually form covalent rather than ionic bonds. Tin and lead, however, the fourth and fifth elements of Group IVA, commonly form compounds with 2+ ions (ionic charge equal to the group number minus two). For example, tin forms tin(II) chloride, SnCl$_2$, which is an ionic compound. It also forms tin(IV) chloride, SnCl$_4$, but this is a covalent, not an ionic, compound. Bismuth, in Group VA, is a metallic element that forms compounds of Bi^{3+} (ionic charge equal to the group number minus two), where only the $6p$ electrons have been lost.

Group VIA and Group VIIA elements, whose atoms have the largest electron affinities, would be expected to form monatomic ions by gaining electrons to give noble-gas or pseudo-noble-gas configurations. An atom of a Group VIIA element (valence-shell configuration ns^2np^5) picks up one electron to give a 1− anion (ns^2np^6); examples are F$^-$ and Cl$^-$. (Hydrogen also forms compounds of the 1− ion, H$^-$. The hydride ion, H$^-$, has a $1s^2$ configuration, like the noble-gas atom helium.) An atom of a Group VIA element (valence-shell configuration ns^2np^4) picks up two electrons to give a 2− anion (ns^2np^6); examples are O^{2-} and S^{2-}. Although the electron affinity of nitrogen ($2s^22p^3$) is essentially zero, the N^{3-} ion ($2s^22p^6$) is stable in the presence of certain positive ions, including Li$^+$ and those of the alkaline earth elements. ◄

To summarize, the common monatomic ions found in compounds of the main-group elements fall into three categories (see Table 2.3):

1. Cations of Groups IA to IIIA having noble-gas or pseudo-noble-gas configurations. The ion charges equal the group numbers.

2. Cations of Groups IIIA to VA having ns^2 electron configurations. The ion charges equal the group numbers minus two. Examples are Tl$^+$, Sn^{2+}, Pb^{2+}, and Bi^{3+}.

3. Anions of Groups VA to VIIA having noble-gas or pseudo-noble-gas configurations. The ion charges equal the group numbers minus eight.

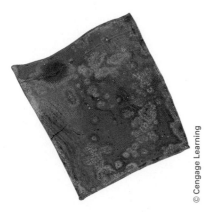

© Cengage Learning

Interestingly, lithium metal reacts with nitrogen at room temperature to form a layer of lithium nitride, Li$_3$N, on the metal surface (see photo above).

Example 9.2   Writing the Electron Configuration and Lewis Symbol for a Main-Group Ion

Gaining Mastery Toolbox

Critical Concept 9.2
The common monatomic ions in the main-group elements fall into three categories: (1) cations of Groups IA to IIIA usually have ion charges equal to group numbers, (2) cations of Groups IIIA to VA may have ion charges equal to group numbers minus two, and (3) anions of Groups VA to VIIA have charges equal to the group numbers minus 8.

Solution Essentials:
- Categories of main-group monatomic ions
- Number of valence electrons in a main-group atom
- Electron configuration of an atom

Write the electron configuration and the Lewis symbol for N^{3-}.

Problem Strategy Recall the three categories of main-group monatomic ions: (1) cations of Groups IA to IIIA usually have ion charges equal to group numbers; (2) cations of Groups IIIA to VA may have ion charges equal to group numbers minus two—an example is Tl^+; and (3) anions of Groups VA to VIIA have charges equal to the group numbers minus eight. Write the electron configuration of the atom, and then subtract or add electrons to give the ion. Similarly, write the Lewis symbol of the atom, and subtract or add dots to give the ion.

Solution The electron configuration of the N atom is $[He]2s^2 2p^3$. By gaining three electrons, the atom assumes a 3− charge and the neon configuration **$[He]2s^2 2p^6$**. The Lewis symbol is

$$[:\ddot{N}:]^{3-}$$

Answer Check Check that the ion has a noble-gas or pseudo-noble-gas configuration and that the Lewis symbol has eight dots, or none, around the atomic symbol.

Exercise 9.2 Write the electron configuration and the Lewis symbol for Ca^{2+} and for S^{2-}.

■ See Problems 9.39 and 9.40.

Exercise 9.3 Write the electron configurations of Pb and Pb^{2+}.

■ See Problems 9.41 and 9.42.

Polyatomic Ions

Many ions, particularly anions, are polyatomic. Some common polyatomic ions are listed in Table 2.5. The atoms in these ions are held together by covalent bonds, which we will discuss in Section 9.4.

Transition-Metal Ions

Most transition elements form several cations of different charges. For example, iron has the cations Fe^{2+} and Fe^{3+}. Neither has a noble-gas configuration; that would require the energetically impossible loss of eight electrons from the neutral atom.

In forming ions in compounds, the atoms of transition elements generally lose the ns electrons first; then they may lose one or more $(n-1)d$ electrons. The 2+ ions are common for the transition elements and are obtained by the loss of the highest-energy s electrons from the atom. Many transition elements also form 3+ ions by losing one $(n-1)d$ electron in addition to the two ns electrons. Table 2.4 lists some common transition-metal ions. Many compounds of transition-metal ions are colored because of transitions involving d electrons, whereas the compounds of the main-group elements are usually colorless (Figure 9.6).

Cr^{3+} Fe^{2+} Co^{2+} Cu^{2+}

Mn^{2+} Fe^{3+} Ni^{2+} Zn^{2+}

© James Scherer/Cengage Learning

Figure 9.6 ◀

Common transition-metal cations in aqueous solution *Left to right:* Cr^{3+} (red-violet), Mn^{2+} (pale pink), Fe^{2+} (pale green), Fe^{3+} (pale yellow), Co^{2+} (pink), Ni^{2+} (green), Cu^{2+} (blue), and Zn^{2+} (colorless).

⊙WL Interactive Example 9.3   Writing Electron Configurations of Transition-Metal Ions

Gaining Mastery Toolbox

Critical Concept 9.3
The atoms of transition elements generally lose the ns electrons first; then they may lose one or more $(n - 1)d$ electrons.

Solution Essentials:
• Electron configuration of a transition-metal atom
• Building-up principle

Write the electron configurations of Fe^{2+} and Fe^{3+}.

Problem Strategy After obtaining the electron configuration of the atom (using the building-up principle as in Example 8.2 or Example 8.3), remove ns electrons, then $(n - 1)d$ electrons, until you have the correct positive charge on the ion.

Solution The electron configuration of the Fe atom $(Z = 26)$ is $[Ar]3d^6 4s^2$. To obtain the configuration of Fe^{2+}, remove the $4s^2$ electrons. To obtain the configuration of Fe^{3+}, also remove a $3d$ electron. **The configuration of Fe^{2+} is $[Ar]3d^6$, and that of Fe^{3+} is $[Ar]3d^5$.**

Answer Check Check that the number of outer electrons in an ion is correct. The number of outer electrons equals the group number (but count the columns of Group VIIIB as 8, 9, and 10) minus the ion charge. For example, the configuration of Fe^{2+} is given as $[Ar]3d^6$. Iron is in the first column of Group VIIIB (count 8); the ion has two fewer electrons ($8 - 2 = 6$). This checks with the configuration as written.

Exercise 9.4   Write the electron configuration of Mn^{2+}.

■ See Problems 9.43 and 9.44.

CONCEPT CHECK 9.1

The following are electron configurations for some ions. Which ones would you expect to see in chemical compounds? State the concept or rule you used to decide for or against any ion.

a Fe^{2+} $[Ar]3d^4 4s^2$ b N^{2-} $[He]2s^2 2p^5$ c Zn^{2+} $[Ar]3d^{10}$

d Na^{2+} $[He]2s^2 2p^5$ e Ca^{2+} $[Ne]3s^2 3p^6$

9.3 Ionic Radii

A monatomic ion, like an atom, is a nucleus surrounded by a distribution of electrons. The **ionic radius** is *a measure of the size of the spherical region around the nucleus of an ion within which the electrons are most likely to be found.* As for an atomic radius, defining an ionic radius is somewhat arbitrary, because an electron distribution never abruptly ends. However, if we imagine ions to be spheres of definite size, we can obtain their radii from known distances between nuclei in crystals. (These distances can be determined accurately by observing how crystals diffract x rays.) ◄

The study of crystal structure by x-ray diffraction is discussed in Chapter 11.

To understand how you compute ionic radii, consider the determination of the radius of an I^- ion in a lithium iodide (LiI) crystal. Figure 9.7 depicts a layer of ions in LiI. The distance between adjacent iodine nuclei equals twice the I^- radius. From x-ray diffraction experiments, the iodine–iodine distance is found to be 426 pm (1 pm = 1×10^{-12} m). Therefore, the I^- radius in LiI is $\frac{1}{2} \times 426$ pm = 213 pm. Other crystals give approximately the same radius for the I^- ion. Table 9.3, which lists average values of ionic radii obtained from many compounds of the main-group elements, gives 216 pm for the I^- radius.

That you can find values of ionic radii that agree with the known structures of many crystals is strong evidence for the existence of ions in the solid state. Moreover, these values of ionic radii compare with atomic radii in ways that you might expect. For example, you expect a cation to be smaller and an anion to be larger than the corresponding atom (see Figure 9.8).

A cation formed when an atom loses all its valence electrons is smaller than the atom because it has one less shell of electrons. But even when only some of the valence electrons are lost from an atom, the ion is smaller. With fewer electrons in the valence orbitals, the electron–electron repulsion is initially less, so these

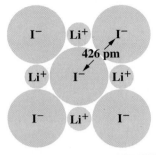

Figure 9.7 ◄

Determining the iodide ion radius in the lithium iodide (LiI) crystal

a A three-dimensional view of the crystal.

b Cross section through a layer of ions. Iodide ions are assumed to be spheres in contact with one another. The distance between iodine nuclei (426 pm) is determined experimentally. One-half of this distance (213 pm) equals the iodide ion radius.

Table 9.3	Ionic Radii (in pm) of Some Main-Group Elements				
Period	**IA**	**IIA**	**IIIA**	**VIA**	**VIIA**
2	Li^+	Be^{2+}		O^{2-}	F^-
	60	31		140	136
3	Na^+	Mg^{2+}	Al^{3+}	S^{2-}	Cl^-
	95	65	50	184	181
4	K^+	Ca^{2+}	Ga^{3+}	Se^{2-}	Br^-
	133	99	62	198	195
5	Rb^+	Sr^{2+}	In^{3+}	Te^{2-}	I^-
	148	113	81	221	216
6	Cs^+	Ba^{2+}	Tl^{3+}		
	169	135	95		

Na [He] $2s^2 2p^6 3s^1$

Na^+ [He] $2s^2 2p^6$

Cl [Ne] $3s^2 3p^5$

Cl^- [Ne] $3s^2 3p^6$

Figure 9.8 ▲

Comparison of atomic and ionic radii Note that the sodium atom loses its outer shell in forming the Na^+ ion. Thus, the cation is smaller than the atom. The Cl^- ion is larger than the Cl atom, because the same nuclear charge holds a greater number of electrons less strongly.

orbitals can shrink to increase the attraction of the electrons for the nucleus. Similarly, because an anion has more electrons than the atom, the electron–electron repulsion is greater, so the valence orbitals expand. Thus, the anion radius is larger than the atomic radius.

Exercise 9.5 Which has the larger radius, S or S^{2-}? Explain.

■ See Problems 9.45 and 9.46.

The ionic radii of the main-group elements shown in Table 9.3 follow a regular pattern, just as atomic radii do. *Ionic radii increase down any column because of the addition of electron shells.*

Exercise 9.6 Without looking at Table 9.3, arrange the following ions in order of increasing ionic radius: Sr^{2+}, Mg^{2+}, Ca^{2+}. (You may use a periodic table.)

■ See Problems 9.47 and 9.48.

The pattern across a period becomes clear if you look first at the cations and then at the anions. For example, in the third period we have

Cation	Na^+	Mg^{2+}	Al^{3+}
Radius (pm)	95	65	50

All of these ions have the neon configuration $1s^2 2s^2 2p^6$ but different nuclear charges; that is, they are isoelectronic. **Isoelectronic** *refers to different species having the same number and configuration of electrons.* To understand the decrease in radius from Na^+ to Al^{3+}, imagine the nuclear charge (atomic number) of Na^+ to increase. With each increase of charge, the orbitals contract due to the greater attractive force of the nucleus. Thus, in any isoelectronic sequence of atomic ions, the ionic radius decreases with increasing atomic number (just as it does for the atoms).

If you look at the anions in the third-period elements, you notice that they are much larger than the cations in the same period. This abrupt increase in ionic radius is due to the fact that the anions S^{2-} and Cl^- have configurations with one more shell of electrons than the cations. And because these anions also constitute an isoelectronic sequence (with argon configuration), the ionic radius decreases with atomic number (as with the atoms):

Anion	S^{2-}	Cl^-
Radius (pm)	184	181

Thus, *in general, across a period the cations decrease in radius. When you reach the anions, there is an abrupt increase in radius, and then the radius again decreases.*

⊙WL Interactive Example 9.4 **Using Periodic Trends to Obtain Relative Ionic Radii**

Gaining Mastery Toolbox

Critical Concept 9.4
In any series of isoelectronic ions, the ionic radius decreases with increasing atomic number. Thus, across a period, the cations decrease in radius; however, when you reach the anions in that period, you find an abrupt increase in radius, followed by a further decrease in radius.

Solution Essentials:
• Obtain valence-shell configuration using periodic table
• Building-up principle

Without looking at Table 9.3, arrange the following ions in order of decreasing ionic radius: F^-, Mg^{2+}, O^{2-}. (You may use a periodic table.)

Problem Strategy Note that in a series of isoelectronic ions, the ion radius decreases with an increase in nuclear charge (or atomic number).

Solution Note that F^-, Mg^{2+}, and O^{2-} are isoelectronic. If you arrange them by increasing nuclear charge, they will be in order of decreasing ionic radius. **The order is O^{2-}, F^-, Mg^{2+}.**

Answer Check Check that the ions are isoelectronic by writing their electron configurations.

Exercise 9.7 Without looking at Table 9.3, arrange the following ions in order of increasing ionic radius: Cl^-, Ca^{2+}, P^{3-}. (You may use a periodic table.)

■ See Problems 9.49 and 9.50.

Covalent Bonds

In the preceding sections we looked at ionic substances, which are typically high-melting solids. Many substances, however, are molecular—gases, liquids, or low-melting solids consisting of molecules (Figure 9.9). A molecule is a group of atoms, frequently nonmetal atoms, strongly linked by chemical bonds. Often the forces that hold atoms together in a molecular substance cannot be understood on the basis of the attraction of oppositely charged ions (the ionic model). An obvious example is the molecule H_2, in which the two H atoms are held together tightly and no ions are present. In 1916 Gilbert Newton Lewis proposed that the strong attractive force between two atoms in a molecule results from a **covalent bond,** *a chemical*

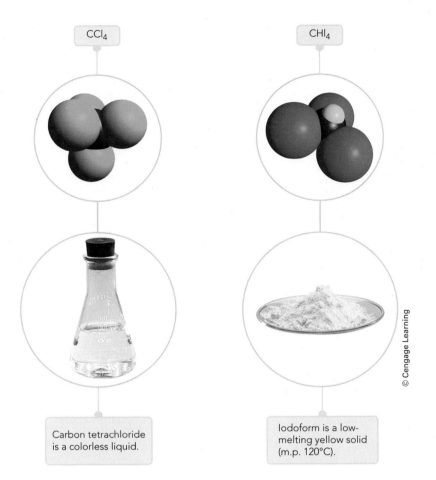

CCl₄

CHI₄

© Cengage Learning

Carbon tetrachloride is a colorless liquid.

Iodoform is a low-melting yellow solid (m.p. 120°C).

Figure 9.9 ◀

Two molecular substances

bond formed by the sharing of a pair of electrons between atoms. ▶ In 1926 Walter Heitler and Fritz London showed that the covalent bond in H_2 could be quantitatively explained by the newly discovered theory of quantum mechanics. We will discuss the descriptive aspects of covalent bonding in the following sections.

G. N. Lewis (1875–1946) was professor of chemistry at the University of California at Berkeley. Well known for work on chemical bonding, he was also noted for his research in molecular spectroscopy and thermodynamics (the study of heat involved in chemical and physical processes).

9.4 Describing Covalent Bonds

Consider the formation of a covalent bond between two H atoms to give the H_2 molecule. As the atoms approach one another, their $1s$ orbitals begin to overlap. Each electron can then occupy the space around both atoms. In other words, the two electrons can be shared by the atoms (Figure 9.10). The electrons are attracted simultaneously by the positive charges of the two hydrogen nuclei. This attraction that bonds the electrons to both nuclei is the force holding the atoms together. Although ions do not exist in H_2, the force that holds the atoms together can still be regarded as arising from the attraction of oppositely charged particles: nuclei and electrons.

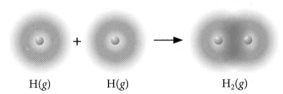

H(g)   +   H(g)   →   H₂(g)

Figure 9.10 ▲

The electron probability distribution for the H_2 molecule The electron density (shown in red) occupies the space around both atoms.

It is interesting to see how the potential energy of the atoms changes as they approach and then bond. Figure 9.11 shows the potential energy of the atoms for various distances between nuclei. The potential energy of the atoms when they are some distance apart is indicated by a position on the potential-energy curve at the far right. As the atoms approach (moving from right to left on the potential-energy curve), the potential energy gets lower and lower. The decrease in energy is a reflection of the bonding of the atoms. Eventually, as the atoms get close enough, the repulsion of the positive charges on the nuclei becomes larger than the attraction of electrons for nuclei. In other words, the potential energy reaches a minimum value and then increases. The distance between nuclei at this minimum energy is called the *bond length* of H_2. It is the normal distance between nuclei in the molecule.

Figure 9.11 ▶

Potential-energy curve for H₂ The stable molecule occurs at the bond distance corresponding to the minimum in the potential-energy curve.

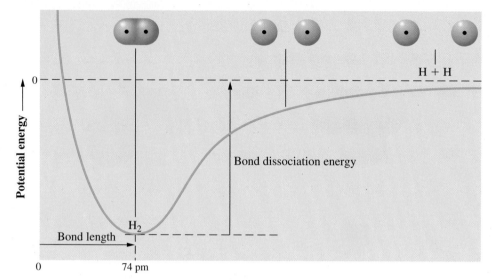

Now imagine the reverse process. You start with the H₂ molecule, the atoms at their normal bond length apart (at the minimum of the potential-energy curve). To separate the atoms in the molecule, energy must be added (you move along the curve to the flat portion at the right). The energy that must be added is called the *bond dissociation energy*. The larger the bond dissociation energy, the stronger the bond.

Lewis Formulas

You can represent the formation of the covalent bond in H₂ from atoms as follows:

$$\text{H} \cdot + \cdot \text{H} \longrightarrow \text{H} : \text{H}$$

This uses the Lewis electron-dot symbol for the hydrogen atoms and represents the covalent bond by a pair of dots. Recall that the two electrons from the covalent bond spend part of the time in the region of each atom. In this sense, each atom in H₂ has a helium configuration. We can draw a circle about each atom to emphasize this.

$$\text{H} (\!:\!) \text{H}$$

The formation of a bond between H and Cl to give an HCl molecule can be represented in a similar way.

$$\text{H} \cdot \ + \ \cdot \overset{..}{\underset{..}{\text{Cl}}} : \ \longrightarrow \ \left(\text{H} (\!:\!) \overset{..}{\underset{..}{\text{Cl}}} : \right)$$

As the two atoms approach each other, unpaired electrons on each atom pair up to form a covalent bond. The pair of electrons is shared by the two atoms. Each atom then acquires a noble-gas configuration of electrons, the H atom having two electrons about it (as in He), and the Cl atom having eight valence electrons about it (as in Ar).

A formula using dots to represent valence electrons is called a **Lewis electron-dot formula.** An electron pair represented by a pair of dots in such a formula is either a **bonding pair** *(an electron pair shared between two atoms)* or a **lone,** or **nonbonding, pair** *(an electron pair that remains on one atom and is not shared).* For example,

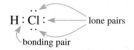

Bonding pairs are often represented by dashes rather than by pairs of dots.

Frequently, the number of covalent bonds formed by an atom equals the number of unpaired electrons shown in its Lewis symbol. Consider the formation of NH_3.

$$3H\cdot \; + \; \cdot\ddot{\underset{\cdot\cdot}{N}}: \; \longrightarrow \; H:\underset{\cdot\cdot}{\overset{\overset{H}{\cdot\cdot}}{N}}:$$
$$\underset{\cdot}{H}$$

Each bond is formed between an unpaired electron on one atom and an unpaired electron on another atom. In many instances, the number of bonds formed by an atom in Groups IVA to VIIA equals the number of unpaired electrons, which is eight minus the group number. For example, a nitrogen atom (Group VA) forms $8 - 5 = 3$ covalent bonds. ▶

Coordinate Covalent Bonds

When bonds form between atoms that both donate an electron, you have

$$A\cdot + \cdot B \longrightarrow A:B$$

However, it is possible for both electrons to come from the same atom. A **coordinate covalent bond** is *a bond formed when both electrons of the bond are donated by one atom*:

$$A + :B \longrightarrow A:B$$

A coordinate covalent bond is not essentially different from other covalent bonds; it involves the sharing of a pair of electrons between two atoms. An example is the formation of the ammonium ion, in which an electron pair on the N atom of NH_3 forms a bond with H^+.

$$H^+ + :NH_3 \longrightarrow \left[H:\underset{\underset{H}{\overset{\overset{H}{\cdot\cdot}}{N}}}{\overset{\cdot\cdot}{N}}:H \right]^+$$

The new N—H bond is clearly identical to the other N—H bonds.

Octet Rule

In each of the molecules we have described so far, the atoms have obtained noble-gas configurations through the sharing of electrons. Atoms other than hydrogen have obtained eight electrons in their valence shells; hydrogen atoms have obtained two. *The tendency of atoms in molecules to have eight electrons in their valence shells (two for hydrogen atoms)* is known as the **octet rule.** Many of the molecules we will discuss follow the octet rule. Some do not.

Multiple Bonds

In the molecules we have described so far, each of the bonds has been a **single bond**—that is, *a covalent bond in which a single pair of electrons is shared by two atoms*. But it is possible for atoms to share two or more electron pairs. A **double bond** is *a covalent bond in which two pairs of electrons are shared by two atoms*. A **triple bond** is *a covalent bond in which three pairs of electrons are shared by two atoms*. As examples, consider ethylene, C_2H_4, and acetylene, C_2H_2. Their Lewis formulas are

$$\underset{H}{\overset{H}{}}C::C\overset{H}{\underset{H}{}} \quad \text{or} \quad \underset{H}{\overset{H}{}}C{=}C\overset{H}{\underset{H}{}}$$
Ethylene

$$H:C:::C:H \quad \text{or} \quad H{-}C{\equiv}C{-}H$$
Acetylene

Note the octet of electrons on each C atom. Double bonds form primarily with C, N, O, and S atoms. Triple bonds form mostly to C and N atoms.

The numbers of unpaired electrons in the Lewis symbols for the atoms of elements in Groups IA and IIIA equal the group numbers. But except for the first elements of these groups, the atoms usually form ionic bonds.

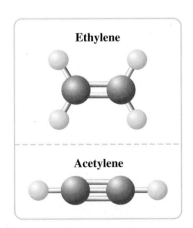

Ethylene

Acetylene

Chemical Bonds in Nitroglycerin

Nitroglycerin gained a nasty reputation soon after its discovery in 1846. Unless kept cold, the pale yellow, oily liquid detonates from even the mildest vibration. In the French film *The Wages of Fear,* four men agree to drive two trucks loaded with nitroglycerin over mountainous roads from a remote village in South America. They are to drive the nitroglycerin to an out-of-control oil-well fire, where the explosive will be used to close off the well. On the way, they find the road obstructed by a huge boulder that has rolled across it, and they decide to blast the boulder out of their way using some of their explosive cargo. You see one of the men gingerly pouring nitroglycerin from a thermos bottle down a stick and into a hole in the rock. Beads of sweat form on his contorted face, while he tries desperately to suspend his breathing and any extra movement that might set off the nitroglycerin.

Here is the structure of nitroglycerin (also see the molecular model in Figure 9.12):

With just a little jostling, nitroglycerin, $C_3H_5(ONO_2)_3$, can rearrange its atoms to give stable products:

$$4C_3H_5(ONO_2)_3(l) \longrightarrow$$
$$6N_2(g) + 12CO_2(g) + 10H_2O(g) + O_2(g)$$

The stability of the products results from their strong bonds, which are much stronger than those in nitroglycerin. Nitrogen, for example, has a strong nitrogen–nitrogen triple bond, and carbon dioxide has two strong carbon–oxygen double bonds. The explosive force of the reaction results from both the rapid reaction and from the large volume increase on forming gaseous products.

In 1867, the Swedish chemist Alfred Nobel discovered that nitroglycerin behaved better when absorbed on diatomaceous earth, a crumbly rock, giving an explosive mixture that Nobel called *dynamite.* Part of the wealth he made from his explosive factories was left in trust to establish the Nobel Prizes. Although explosives are often associated in people's minds with war, they do have many peacetime uses, including road building, mining, and demolition (Figure 9.13).

Figure 9.13 ▲

Demolition of a building Explosives are placed at predetermined positions so that when they are detonated, the building collapses on itself.

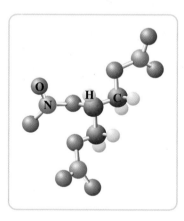

Figure 9.12 ▲

Molecular model of nitroglycerin In this ball-and-stick model, carbon atoms are gray, hydrogen atoms are light blue, oxygen atoms are red, and nitrogen atoms are dark blue.

■ See Problems 9.113 and 9.114.

9.5 Polar Covalent Bonds; Electronegativity

A covalent bond involves the sharing of at least one pair of electrons between two atoms. When the atoms are alike, as in the case of the H—H bond of H_2, the bonding electrons are shared equally. That is, the electrons spend the same amount of time in the vicinity of each atom. But when the two atoms are of different elements, the bonding electrons need not be shared equally. A **polar covalent bond** (or simply *polar bond*) is *a covalent bond in which the bonding electrons spend more time near one atom than the other.* For example, in the case of the HCl molecule (Figure 9.14), the bonding electrons spend more time near the chlorine atom than the hydrogen atom.

Figure 9.14 ▲

The HCl molecule A molecular model.

You can consider the polar covalent bond as intermediate between a *nonpolar* covalent bond, as in H_2, and an ionic bond, as in NaCl. From this point of view, an ionic bond is simply an extreme example of a polar covalent bond. To illustrate, we can represent the bonding in H_2, HCl, and NaCl with electron-dot formulas as follows:

$$H : H \qquad H : \overset{\cdot\cdot}{\underset{\cdot\cdot}{Cl}} : \qquad Na^+ \quad : \overset{\cdot\cdot}{\underset{\cdot\cdot}{Cl}} :^-$$

Nonpolar covalent   Polar covalent   Ionic

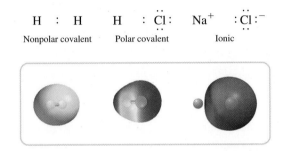

The bonding pairs of electrons are equally shared in H_2, unequally shared in HCl, and essentially not shared in NaCl. Thus, it is possible to arrange different bonds to form a gradual transition from nonpolar covalent to ionic.

Note that a polar bond results when the bonding pair is drawn more toward one atom than the other. The concept of electronegativity is useful in judging whether a bond will be polar or not. **Electronegativity** is *a measure of the ability of an atom in a molecule to draw bonding electrons to itself.* Several electronegativity scales have been proposed. In 1934 Robert S. Mulliken suggested on theoretical grounds that the electronegativity (X) of an atom be given as half its ionization energy (*I.E.*) plus electron affinity (*E.A.*). ▶

$$X = \frac{I.E. + E.A.}{2}$$

An atom such as fluorine that tends to pick up electrons easily (large *E.A.*) and hold on to them strongly (large *I.E.*) has a large electronegativity. On the other hand, an atom such as lithium or cesium that loses electrons readily (small *I.E.*) and has little tendency to gain electrons (small *E.A.*) has a small electronegativity. Until recently, only a few electron affinities had been measured. For this reason, Mulliken's scale has had limited utility. A more widely used scale was derived earlier by Linus Pauling from bond enthalpies, which are discussed later in this chapter. Pauling's electronegativity values are given in Figure 9.15. Because electronegativities depend somewhat on the bonds formed, these values are actually average ones. ▶

Fluorine, the most electronegative element, was assigned a value of 4.0 on Pauling's scale. Lithium, at the left end of the same period, has a value of 1.0. Cesium, in the same column but below lithium, has a value of 0.7. *In general, electronegativity increases from left to right and decreases from top to bottom in the periodic table.* Metals are the least electronegative elements (they are *electropositive*) and nonmetals the most electronegative.

The absolute value of the difference in electronegativity of two bonded atoms gives a rough measure of the polarity to be expected in a bond. When this difference

Robert S. Mulliken received the Nobel Prize in chemistry in 1966 for his work on molecular orbital theory (discussed in Chapter 10).

Linus Pauling received the Nobel Prize in chemistry in 1954 for his work on the nature of the chemical bond. In 1962 he received the Nobel Peace Prize.

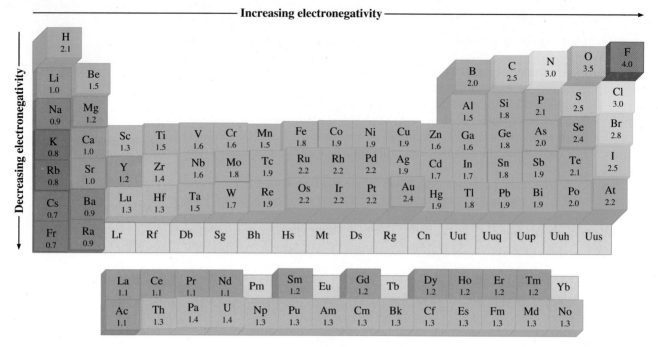

Figure 9.15 ▲

Electronegativities of the elements
The values given are those of Pauling (those in gray are not known).

The electronegativity difference in metal–metal bonding would also be small. This bonding frequently involves delocalized metallic bonding, briefly described in Section 9.7.

is small, the bond is nonpolar. When it is large, the bond is polar (or, if the difference is very large, perhaps ionic). The electronegativity differences for the bonds H—H, H—Cl, and Na—Cl are 0.0, 0.9, and 2.1, respectively, following the expected order. Differences in electronegativity explain why ionic bonds usually form between a metal atom and a nonmetal atom; the electronegativity difference would be largest between these elements. On the other hand, covalent bonds form primarily between two nonmetals because the electronegativity differences are small. ◀

Example 9.5   **Using Electronegativities to Obtain Relative Bond Polarities**

Gaining Mastery Toolbox

Critical Concept 9.5
The absolute value of the difference in electronegativity of two bonded atoms is a rough measure of the polarity of the bond. (A useful rule to remember is that electronegativity increases left to right and decreases top to bottom in the periodic table.)

Solution Essentials:
• Electronegativity

Use electronegativity values (Figure 9.15) to arrange the following bonds in order of increasing polarity: P—H, H—O, C—Cl.

Problem Strategy Order the bonds by the increasing positive value of the difference of electronegativities of the atoms forming the bond. The bonds should then be roughly in order of increasing polarity.

Solution The absolute values of the electronegativity differences are P—H, 0.0; H—O, 1.4; C—Cl, 0.5. Hence, the order is **P—H, C—Cl, H—O**.

Answer Check Make sure you have the correct electronegativities and differences.

Exercise 9.8 Using electronegativities, decide which of the following bonds is most polar: C—O, C—S, H—Br.

■ See Problems 9.57 and 9.58.

You can use an electronegativity scale to predict the direction in which the electrons shift during bond formation; the electrons are pulled toward the more electronegative atom. For example, consider the H—Cl bond. Because the Cl atom ($X = 3.0$) is more electronegative than the H atom ($X = 2.1$), the bonding electrons in H—Cl are pulled toward Cl. Because the bonding electrons spend most of their time around the Cl atom, that end of the bond acquires a partial negative charge (indicated $\delta-$). The H-atom end of the bond has a partial positive charge ($\delta+$). You can show this as follows:

$$\overset{\delta+}{\text{H}}-\overset{\delta-}{\text{Cl}}$$

The HCl molecule is said to be a *polar molecule*. We will say more about polar molecules when we look at molecular structures in Chapter 10.

9.6 Writing Lewis Electron-Dot Formulas

The Lewis electron-dot formula of a molecule is similar to the structural formula in that it shows how atoms are bonded. Bonding electron pairs are indicated either by two dots or by a dash. In addition to the bonding electrons, however, an electron-dot formula shows the positions of lone pairs of electrons, whereas the structural formula does not. Thus, the electron-dot formula is a simple two-dimensional representation of the positions of electrons in a molecule. In the next chapter we will see how to predict the three-dimensional shape of a molecule from the two-dimensional electron-dot formula. In this section we will discuss the steps for writing the electron-dot formula for a molecule made from atoms of the main-group elements. ▶

Before you can write the Lewis formula of a molecule (or a polyatomic ion), you must know the *skeleton structure* of the molecule. The skeleton structure tells you which atoms are bonded to one another (without regard to whether the bonds are single or not).

Normally the skeleton structure must be found by experiment. For simple molecules, you can often predict the skeleton structure. For instance, many small molecules or polyatomic ions consist of a central atom around which are bonded atoms of greater electronegativity, such as F, Cl, and O. Note the structural formula (and molecular model) of phosphorus oxychloride, $POCl_3$ (Figure 9.16). The phosphorus atom is surrounded by more electronegative atoms. (In some cases, H atoms surround a more electronegative atom as in H_2O and NH_3, but H cannot be a central atom because it normally forms only one bond.) Similarly, *oxoacids* are substances in which O atoms (and possibly other electronegative atoms) are bonded to a central atom, with one or more H atoms usually bonded to O atoms. An example is chlorosulfonic acid, HSO_3Cl (Figure 9.17).

Lewis formulas do not directly convey information about molecular shape. For example, the Lewis formula of methane, CH_4, is written as the flat (two-dimensional) formula

$$\text{H}:\overset{\displaystyle \text{H}}{\underset{\displaystyle \text{H}}{\text{C}}}:\text{H}$$

The actual methane molecule, however, is not flat; it has a three-dimensional structure, as explained in Chapter 10.

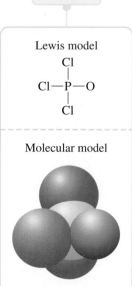

Phosphorus oxychloride $POCl_3$

Figure 9.16 ◀

Lewis formula and molecular model of phosphorus oxychloride molecule The central atom, P, is surrounded by O and Cl atoms.

Lewis model

$$\text{Cl}-\overset{\displaystyle \text{Cl}}{\underset{\displaystyle \text{Cl}}{\text{P}}}-\text{O}$$

Molecular model

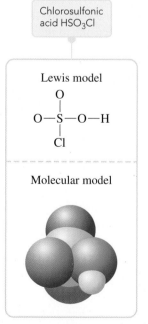

Chlorosulfonic acid HSO_3Cl

Lewis model

$$\text{O}-\overset{\displaystyle \text{O}}{\underset{\displaystyle \text{Cl}}{\text{S}}}-\text{O}-\text{H}$$

Molecular model

Figure 9.17 ◀

Lewis formula and molecular model of chlorosulfonic acid molecule This oxoacid has the central atom, S, surrounded by O atoms and a Cl atom.

Another useful idea for predicting skeleton structures is that molecules or polyatomic ions with symmetrical formulas often have symmetrical structures. For example, disulfur dichloride, S_2Cl_2, is symmetrical, with the more electronegative Cl atoms around the S atoms: Cl—S—S—Cl.

Once you know the skeleton structure of a molecule, you can find the Lewis formula by applying the following four steps. These steps allow you to write electron-dot formulas even when the central atom does not follow the octet rule.

Step 1: *Calculate the total number of valence electrons* for the molecule by summing the number of valence electrons (= group number) for each atom. If you are writing the Lewis formula of a polyatomic anion, you *add* the number of negative charges to this total. (For CO_3^{2-} you add 2 because the -2 charge indicates that there are two more electrons than are provided by the neutral atoms.) For a polyatomic cation, you *subtract* the number of positive charges from the total. (For NH_4^+ you subtract 1.)

Step 2: *Write the skeleton structure of the molecule or ion,* connecting every bonded pair of atoms by a pair of dots (or a dash).

Step 3: *Distribute electrons to the atoms surrounding the central atom (or atoms)* to satisfy the octet rule for these surrounding atoms.

Step 4: *Distribute the remaining electrons as pairs to the central atom (or atoms),* after subtracting the number of electrons already distributed from the total found in Step 1. If there are fewer than eight electrons on the central atom, this suggests that a multiple bond is present. (Two electrons fewer than an octet suggests a double bond; four fewer suggests a triple bond or two double bonds.) To obtain a multiple bond, move one or two electron pairs (depending on whether the bond is to be double or triple) from a surrounding atom to the bond connecting the central atom. Atoms that often form multiple bonds are C, N, O, and S.

The next several examples illustrate how to write the Lewis electron-dot formula for a small molecule, given the molecular formula.

Example 9.6

Gaining Mastery Toolbox

Critical Concept 9.6
To write the Lewis electron-dot formula for a molecule, first obtain the total number of valence electrons for the molecule; second, write the skeleton structure, allowing two electrons for each bond between atoms; third, distribute electrons to the outer atoms in order to satisfy the octet rule for each atom; and finally distribute the remaining electrons to the central atom.

Solution Essentials:
• Skeleton structure of a molecule
• Octet rule
• Number of valence electrons in a main-group atom

Writing Lewis Formulas (Single Bonds Only)

Sulfur dichloride, SCl_2, is a red, fuming liquid used in the manufacture of insecticides. Write the Lewis formula for the molecule.

Problem Strategy You follow the four steps: (1) Calculate the total number of valence electrons. (2) Write the skeleton structure with two electrons to each bond between atoms. (3) Distribute electrons to the outer atoms to satisfy the octet rule. (4) Distribute the remaining electrons to the central atom.

Solution The number of valence electrons from an atom equals the group number: 6 for S, 7 for each Cl, for a total of 20 electrons. You expect the skeleton structure to have S as the central atom, with the more electronegative Cl atoms bonded to it. After connecting atoms by electron pairs and distributing electrons to the outer atoms, you have

$$:\!\ddot{Cl}:S:\ddot{Cl}\!: \quad \text{or} \quad :\!\ddot{Cl}\!-\!S\!-\!\ddot{Cl}\!:$$

This accounts for 8 electron pairs, or 16 electrons. Subtracting this from the total number of electrons (20) gives 4 electrons, or 2 electron pairs. You place these on the central atom (S). The final Lewis formula is

$$:\!\ddot{Cl}:\ddot{S}:\ddot{Cl}\!: \quad \text{or} \quad :\!\ddot{Cl}\!-\!\ddot{S}\!-\!\ddot{Cl}\!:$$

SCl_2

(continued)

(*continued*)

Answer Check In general, atoms surrounding the central atom have four electron pairs (an octet) about them. Note that our answer follows this "octet" rule. Even the central atom often follows this octet rule, as it does here. (We will see some exceptions to this octet rule for the central atom, however.)

Exercise 9.9 Dichlorodifluoromethane, CCl_2F_2, is a gas used as a refrigerant and aerosol propellant. Write the Lewis formula for CCl_2F_2.

■ See Problems 9.61 and 9.62.

Example 9.7

Gaining Mastery Toolbox

Critical Concept 9.7
If after following the rules for writing an electron-dot formula, you find that the central atom does not have enough electrons for an octet, transfer electron pairs from outer atoms to form multiple bonds in order to give an octet to the central atom.

Solution Essentials:
• Skeleton structure of a molecule
• Lewis representation of multiple bonds
• Octet rule
• Number of valence electrons in a main-group atom

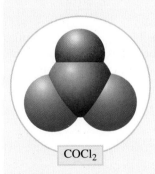

$COCl_2$

Writing Lewis Formulas (Including Multiple Bonds)

Carbonyl chloride, or phosgene, $COCl_2$, is a highly toxic gas used as a starting material for the preparation of polyurethane plastics. What is the electron-dot formula of $COCl_2$?

Problem Strategy You follow the four steps as in the previous example, but after distributing the remaining electrons to the central atom, you find that it does not have enough electrons to give an octet. To rectify this, you transfer an electron pair from an outer atom to form a double bond (part of Step 4). Do you move the electron pair from the Cl or the O atom? It helps to remember that C, N, O, and S atoms often form double bonds. In this case, you move the electron pair from the O atom.

Solution The total number of valence electrons is $4 + 6 + (2 \times 7) = 24$. You expect the more electropositive atom, C, to be central, with the O and Cl atoms bonded to it. After distributing electron pairs to these surrounding atoms, you have

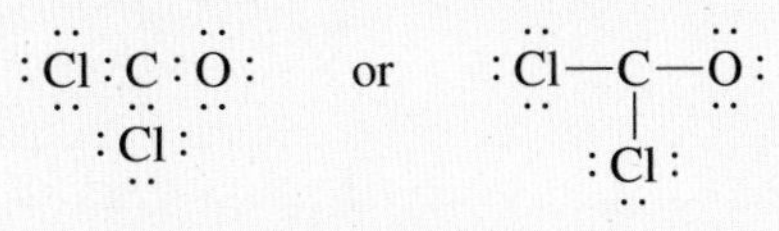

$$:\ddot{C}l:C:\ddot{O}: \quad \text{or} \quad :\ddot{C}l—C—\ddot{O}: \\ \quad\quad :\ddot{C}l: \quad\quad\quad\quad\quad\quad :\ddot{C}l:$$

This accounts for all 24 valence electrons, leaving only 6 electrons on C. This is two fewer than an octet, so you move a pair of electrons on the O atom to give a carbon–oxygen double bond. The electron-dot formula of $COCl_2$ is

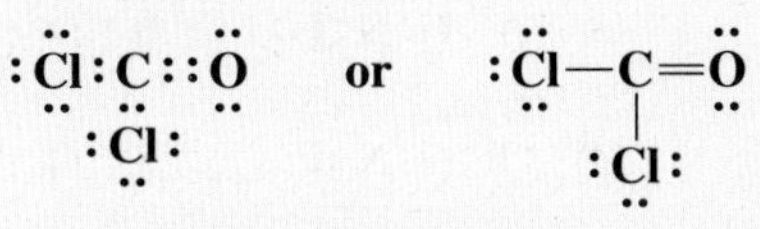

$$:\ddot{C}l:C::\ddot{O} \quad \text{or} \quad :\ddot{C}l—C=\ddot{O} \\ \quad :\ddot{C}l: \quad\quad\quad\quad\quad\quad :\ddot{C}l:$$

Answer Check Note that the atoms surrounding the central atom have four electron pairs about them; that is, they follow the octet rule. The central atom also follows the octet rule, although exceptions occur.

Exercise 9.10 Write the electron-dot formula of carbon dioxide, CO_2.

■ See Problems 9.63 and 9.64.

Example 9.8

Writing Lewis Formulas (Ionic Species)

Gaining Mastery Toolbox

Critical Concept 9.8
In obtaining the Lewis formula for an ion, note that the total number of valence electrons equals that for the neutral molecule less the charge on the ion (that is, subtract if a positive ion and add if a negative ion).

Solution Essentials:
• Skeleton structure of a molecule
• Lewis representation of multiple bonds
• Octet rule
• Number of valence electrons in a main-group atom

Obtain the electron-dot formula (Lewis formula) of the BF_4^- ion.

Problem Strategy After calculating the total number of valence electrons for the neutral atoms, you add or subtract electrons to account for the ion charge. If the ion is negative, add electrons to give the negative charge. If the ion is positive, subtract electrons to give the positive charge. Once you have the total number of valence electrons in the ion, proceed with the rest of the steps to write the electron-dot formula.

Solution The total valence electrons provided by the boron and four fluorine atoms is $3 + (4 \times 7) = 31$. Because the anion has a charge of -1, it has one more electron than is provided by the neutral atoms. Thus, the total number of valence electrons is 32 (or 16 electron pairs). You assume that the skeleton structure has boron as the central atom, with the more electronegative F atoms bonded to it. After connecting the B and F atoms by bonds and placing electron pairs around the F atoms to satisfy the octet rule, you obtain

$$
\begin{array}{ccc}
& :\!\ddot{F}\!: & \\
& \vdots & \\
:\!\ddot{F}\!:\!B\!:\!\ddot{F}\!: & \text{or} & :\!\ddot{F}\!-\!B\!-\!\ddot{F}\!: \\
& :\!\ddot{F}\!: &
\end{array}
$$

This uses up all 32 electrons. The charge on the entire ion is indicated by the minus sign written as a superscript to square brackets enclosing the electron-dot formula.

$$
\left[\begin{array}{c} :\!\ddot{F}\!: \\ :\!\ddot{F}\!:\!B\!:\!\ddot{F}\!: \\ :\!\ddot{F}\!: \end{array} \right]^- \quad \text{or} \quad \left[\begin{array}{c} :\!\ddot{F}\!: \\ :\!\ddot{F}\!-\!B\!-\!\ddot{F}\!: \\ :\!\ddot{F}\!: \end{array} \right]^-
$$

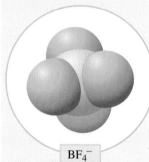

BF_4^-

Answer Check Note that the atoms surrounding the central atom have four electron pairs about them; that is, they follow the octet rule. The central atom also follows the octet rule, although exceptions occur.

Exercise 9.11 Write the electron-dot formula of a the hydronium ion, H_3O^+; b the chlorite ion, ClO_2^-.

■ See Problems 9.65 and 9.66.

CONCEPT CHECK 9.2

Each of the following may seem, at first glance, to be plausible electron-dot formulas for the molecule N_2F_2. Most, however, are incorrect for some reason. What concepts or rules apply to each, either to cast it aside or to keep it as the correct formula?

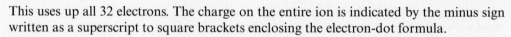

a $:\!\ddot{F}\!:\!N\!:\!N\!:\!\ddot{F}\!:$ b $:\!\ddot{F}\!:\!N\!::\!N\!:\!\ddot{F}\!:$ c $:\!\ddot{F}\!::\!N\!:\!N\!:\!\ddot{F}\!:$

d $:\!\ddot{F}\!:\!N\!:\!N\!:\!\ddot{F}\!:$ e $:\!\ddot{F}\!:\!N\!::\!\ddot{F}\!:\!N\!:$ f $:\!\ddot{F}\!:\!N\ N\!:\!\ddot{F}\!:$

9.7 Delocalized Bonding: Resonance

We have assumed up to now that the bonding electrons are localized in the region between two atoms. In some cases, however, this assumption does not fit the experimental data. Suppose, for example, that you try to write an electron-dot formula for ozone, O_3. You find that you can write two formulas:

$$
\underset{A}{:\!\ddot{O}\!:\!\overset{\ddot{O}}{}\!:\!\ddot{O}\!:} \quad \text{and} \quad \underset{B}{:\!\ddot{O}\!:\!\overset{\ddot{O}}{}\!:\!\ddot{O}\!:}
$$

In formula A, the oxygen–oxygen bond on the left is a double bond and the oxygen–oxygen bond on the right is a single bond. In formula B, the situation is just the opposite. Experiment shows, however, that both bonds in O_3 are identical. Therefore, neither formula can be correct. ▶

According to theory, one of the bonding pairs in ozone is spread over the region of all three atoms rather than associated with a particular oxygen–oxygen bond. This is called **delocalized bonding,** *a type of bonding in which a bonding pair of electrons is spread over a number of atoms rather than localized between two.* We might symbolically describe the delocalized bonding in ozone as follows:

The lengths of the two oxygen–oxygen bonds (that is, the distances between the atomic nuclei) are both 128 pm.

(For clarity, only the bonding pairs are given.) The broken line indicates a bonding pair of electrons that spans three nuclei rather than only two. In effect, the oxygen–oxygen bond is neither a single bond nor a double bond but an intermediate type.

A single electron-dot formula cannot properly describe delocalized bonding. Instead, a resonance description is often used. According to the **resonance description,** *you describe the electron structure of a molecule having delocalized bonding by writing all possible electron-dot formulas.* These are called the *resonance formulas* of the molecule. The actual electron distribution of the molecule is a composite of these resonance formulas.

The electron structure of ozone can be described in terms of the two resonance formulas presented at the start of this section. By convention, we usually write all of the resonance formulas and connect them by double-headed arrows. For ozone we would write

Unfortunately, this notation can be misinterpreted. It does not mean that the ozone molecule flips back and forth between two forms. There is only one ozone molecule. The double-headed arrow means that you should form a mental picture of the molecule by fusing the various resonance formulas. The left oxygen–oxygen bond is double in formula A and the right one is double in formula B, so you must picture an electron pair that actually encompasses both bonds.

Attempting to write electron-dot formulas leads you to recognize that delocalized bonding exists in many molecules. Whenever you can write several plausible electron-dot formulas—which often differ merely in the allocation of single and double bonds to the same kinds of atoms (as in ozone)—you can expect delocalized bonding.

Example 9.9   Writing Resonance Formulas

Gaining Mastery Toolbox

Critical Concept 9.9
Having written one electron-dot formula containing single and double bonds, note whether it is possible to write other electron-dot formulas that differ in the placement of these single and double bonds.

Solution Essentials:
• Resonance
• Skeleton structure of a molecule
• Lewis representation of multiple bonds
• Octet rule
• Number of valence electrons in a main-group atom

Describe the electron structure of the carbonate ion, CO_3^{2-}, in terms of electron-dot formulas.

Problem Strategy Write at least one electron-dot formula for a molecule or ion. If the formula has both single and multiple bonds, note whether it is possible to write other electron-dot formulas differing only in the placement of single and double bonds.

Solution One possible electron-dot formula for the carbonate ion is

(continued)

(continued)

Because you expect all carbon–oxygen bonds to be equivalent, you must describe the electron structure in resonance terms.

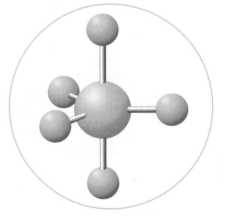

You expect one electron pair to be delocalized over the region of all three carbon–oxygen bonds.

CO_3^{2-}

Answer Check Note that the skeleton structure is the same for all resonance formulas. This must be so, because they represent the same molecule. Also, note that the atoms about the central atom follow the octet rule. (The central atom in this case also follows the octet rule.)

Exercise 9.12 Describe the bonding in NO_3^- using resonance formulas.

■ See Problems 9.67, 9.68, 9.69, and 9.70.

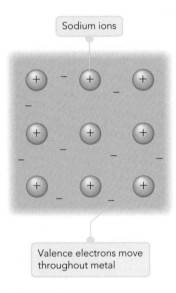

Sodium ions

Valence electrons move throughout metal

Figure 9.18 ▲

Delocalized bonding in sodium metal The metal consists of positive sodium ions in a "sea" of valence electrons. Valence (bonding) electrons are free to move throughout the metal crystal (beige area).

Metals are extreme examples of delocalized bonding. A sodium metal crystal, for example, can be regarded as an array of Na^+ ions surrounded by a "sea" of electrons (Figure 9.18). The valence, or bonding, electrons are delocalized over the entire metal crystal. The freedom of these electrons to move throughout the crystal is responsible for the electrical conductivity of a metal.

9.8 Exceptions to the Octet Rule

Many molecules composed of atoms of the main-group elements have electronic structures that satisfy the octet rule, but a number of them do not. A few molecules, such as NO, have an odd number of electrons and so cannot satisfy the octet rule. Other exceptions to the octet rule fall into two groups—one a group of molecules with an atom having fewer than eight valence electrons around it and the other a group of molecules with an atom having more than eight valence electrons around it.

The exceptions in which the central atom has more than eight valence electrons around it are fairly numerous. Phosphorus pentafluoride is a simple example. This is a colorless gas of PF_5 molecules. Each molecule consists of a phosphorus atom surrounded by the more electronegative fluorine atoms (Figure 9.19). Following the steps outlined in Section 9.6, you arrive at the following electron-dot formula, in which the phosphorus atom has ten valence electrons around it:

Figure 9.19 ▲

Phosphorus pentafluoride, PF₅ A ball-and-stick molecular model.

The octet rule stems from the fact that the main-group elements in most cases employ only an *ns* and three *np* valence-shell orbitals in bonding, and these orbitals hold eight electrons. Elements of the second period are restricted to these orbitals, but from the third period on, the elements also have unfilled *nd* orbitals, which may be used in bonding. For example, the valence-shell configuration of phosphorus is $3s^2 3p^3$. Using just these $3s$ and $3p$ orbitals, the phosphorus atom can accept only three additional electrons, forming three covalent bonds (as in PF_3). However, more bonds can be formed if the empty $3d$ orbitals of the atom are used. If each of the five electrons of the phosphorus atom is paired with unpaired electrons of fluorine atoms, PF_5 can be formed. In this way, phosphorus forms both the trifluoride and the pentafluoride. By contrast, nitrogen (which has no available d orbitals in its valence shell) forms only the trifluoride, NF_3.

Example 9.10 Writing Lewis Formulas (Exceptions to the Octet Rule)

Gaining Mastery Toolbox

Critical Concept 9.10
In the final step in writing an electron-dot formula (in which you distribute the remaining electrons to the central atom), you may discover that this atom now has more than an octet of electrons.

Solution Essentials:
- Skeleton structure of a molecule
- Lewis representation of multiple bonds
- Octet rule
- Number of valence electrons in a main-group atom

Xenon, a noble gas, forms a number of compounds. One of these is xenon tetrafluoride, XeF_4, a white, crystalline solid first prepared in 1962. What is the electron-dot formula of the XeF_4 molecule?

Problem Strategy You follow the four steps outlined in Section 9.6. In Step 4, when you distribute the remaining electrons to the central atom, you find that it has more than an octet.

Solution There are 8 valence electrons from the Xe atom and 7 from each F atom, for a total of 36 valence electrons. For the skeleton structure, you draw the Xe atom surrounded by the electronegative F atoms. After placing electron pairs on the F atoms to satisfy the octet rule for them, you have:

This accounts for 16 pairs, or 32 electrons. A total of 36 electrons is available, so you put an additional $36 - 32 = 4$ electrons (2 pairs) on the Xe atom. The Lewis formula is

Answer Check Note that the atoms surrounding the central atom follow the octet rule, although the central atom itself does not. The central atom is from a period greater than 2, so it has a valence shell that can accommodate more than eight electrons.

Exercise 9.13 Sulfur tetrafluoride, SF_4, is a colorless gas. Write the electron-dot formula of the SF_4 molecule.

■ See Problems 9.71 and 9.72.

The other group of exceptions to the octet rule consists mostly of molecules containing Group IIA or IIIA atoms. Consider boron trifluoride, BF_3. The molecule consists of boron surrounded by the much more electronegative fluorine atoms. The total number of valence electrons is $3 + (3 \times 7) = 24$. If you connect boron and fluorine atoms by electron pairs and fill out the fluorine atoms with octets of electrons, you obtain

$$
\begin{array}{c}
\ddot{\text{F}}: \\
| \\
:\ddot{\text{F}}\text{—B} \\
| \\
:\ddot{\text{F}}:
\end{array}
$$

If you now follow with Step 4 of the procedure outlined in Section 9.6, you note that the boron atom has only six electrons on it. You write the following resonance formulas, each having a double bond:

$$
\begin{array}{ccccc}
:\ddot{\text{F}}: & & :\ddot{\text{F}}: & & :\ddot{\text{F}}: \\
\| & & | & & | \\
:\ddot{\text{F}}\text{—B} & \longleftrightarrow & \ddot{\text{F}}\text{=B} & \longleftrightarrow & :\ddot{\text{F}}\text{—B} \\
| & & | & & \| \\
:\ddot{\text{F}}: & & :\ddot{\text{F}}: & & :\ddot{\text{F}}:
\end{array}
$$

In fact, there is evidence to suggest that the first formula written—the one with all single bonds and in which boron has only six electrons around it—describes the chemistry of boron trifluoride very well. For example, boron trifluoride reacts with molecules having a lone pair, such as with ammonia, NH_3, to give the compound BF_3NH_3. The reaction is easy to describe in terms of the formula in which boron has only six electrons around it.

$$:\overset{..}{\underset{..}{F}}-\overset{\overset{\displaystyle :\overset{..}{F}:}{|}}{\underset{\underset{\displaystyle :\overset{..}{F}:}{|}}{B}} \;+\; :\overset{\overset{\displaystyle H}{|}}{\underset{\underset{\displaystyle H}{|}}{N}}-H \;\longrightarrow\; :\overset{..}{\underset{..}{F}}-\overset{\overset{\displaystyle :\overset{..}{F}:}{|}}{\underset{\underset{\displaystyle :\overset{..}{F}:}{|}}{B}}\;:\;\overset{\overset{\displaystyle H}{|}}{\underset{\underset{\displaystyle H}{|}}{N}}-H$$

coordinate covalent bond

In this reaction, a coordinate covalent bond forms between the boron and nitrogen atoms, and the boron achieves an octet of electrons.

The chemistry of BF_3 thus appears to support an electron structure with boron having only six electrons around it. No doubt, resonance involving all four of the Lewis formulas that we have drawn best describes the actual electron structure of boron trifluoride, but the relative importance of the different resonance formulas is not settled. ◄

Other examples of molecules with Group IIIA atoms (such as Al) or Group IIA atoms (such as Be) display electron structures similar to that of boron trifluoride. For example, the BeF_2 molecule (found in the vapor over the heated solid BeF_2) has the Lewis formula

$$:\overset{..}{\underset{..}{F}}:Be:\overset{..}{\underset{..}{F}}:$$

Aluminum chloride, $AlCl_3$, offers an interesting study in bonding. At room temperature, the substance is a white, crystalline solid and an ionic compound, as might be expected for a binary compound of a metal and a nonmetal. However, the substance has a relatively low melting point (192°C) for an ionic compound. Apparently this is due to the fact that instead of melting to a liquid of ions, as happens with most ionic solids, the compound forms Al_2Cl_6 molecules (Figure 9.20), with Lewis formula

$$\overset{\displaystyle \overset{..}{Cl}\quad\;\overset{..}{Cl}\quad\;\overset{..}{Cl}}{\underset{\displaystyle \overset{..}{\underset{..}{Cl}}\quad\overset{..}{\underset{..}{Cl}}\quad\overset{..}{\underset{..}{Cl}}}{\overset{\diagdown\quad\diagup\quad\diagdown\quad\diagup}{Al\qquad Al}}}$$

Each atom has an octet of electrons around it. Note that two of the Cl atoms are in *bridge* positions, with each Cl atom having two covalent bonds. When this liquid is heated, it vaporizes as Al_2Cl_6 molecules. As the vapor is further heated, these molecules break up into $AlCl_3$ molecules. These molecules have an electron structure similar to that of BF_3.

The B–F bond length (130 pm) is shorter than expected for a single bond (152 pm), which indicates partial double-bond character. See the discussion in Section 9.10 on bond length and bond order.

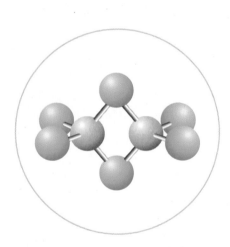

Figure 9.20 ▲

The Al_2Cl_6 molecule A ball-and-stick molecular model.

Exercise 9.14 Beryllium chloride, $BeCl_2$, is a solid substance consisting of long (essentially infinite) chains of atoms with Cl atoms in bridge positions.

$$\overset{\displaystyle \overset{..}{Cl}\qquad\overset{..}{Cl}\qquad\overset{..}{Cl}}{\underset{\displaystyle \overset{..}{\underset{..}{Cl}}\qquad\overset{..}{\underset{..}{Cl}}\qquad\overset{..}{\underset{..}{Cl}}}{\overset{\diagup\;\diagdown\quad\diagup\;\diagdown\quad\diagup\;\diagdown}{\quad Be\qquad Be\qquad Be}}}$$

However, if the solid is heated, it forms a vapor of $BeCl_2$ molecules. Write the electron-dot formula of the $BeCl_2$ molecule.

■ See Problems 9.73 and 9.74.

9.9 Formal Charge and Lewis Formulas

Earlier we looked at writing the Lewis formula for carbonyl chloride, $COCl_2$ (Example 9.7). In the solution to that example, we wrote the answer as follows:

$$:\overset{..}{\underset{}{Cl}}-C=\overset{..}{\underset{}{O}}$$
$$|$$
$$:\overset{..}{\underset{..}{Cl}}:$$

You may recall that we had to decide whether to draw the double bond between C and O or between C and Cl. We decided in favor of a carbon-oxygen double bond by applying the simple idea that multiple bonds most likely involve C, N, O, and S atoms. In this section, we will describe how you can use the concept of formal charge to determine the best Lewis formula. Formal charge is also a help in writing the skeleton structure of a molecule.

The **formal charge** of an atom in a Lewis formula is *the hypothetical charge you obtain by assuming that bonding electrons are equally shared between bonded atoms and that the electrons of each lone pair belong completely to one atom.* ▶ When you use the Lewis formula that most clearly describes the bonding, the formal charges give you an approximate distribution of electrons in the molecule. Let us see how you would obtain the formal charges, given several possible Lewis formulas of a molecule, and how you would use these formal charges to determine the best Lewis formula.

You begin by writing the possible Lewis formulas. For $COCl_2$, you can write

$$:\overset{..}{\underset{}{Cl}}-C=\overset{..}{\underset{}{O}} \qquad \overset{..}{\underset{}{Cl}}=C-\overset{..}{\underset{}{O}}: \qquad :\overset{..}{\underset{}{Cl}}-C-\overset{..}{\underset{}{O}}:$$
$$|\qquad\qquad\quad |\qquad\qquad\quad \|$$
$$:\overset{}{\underset{..}{Cl}}:\qquad :\overset{}{\underset{..}{Cl}}:\qquad :\overset{}{\underset{..}{Cl}}:$$

The rules for computing oxidation number are similar, except that both bonding electrons are assigned to the more electronegative atom.

For each formula, you apply the following *rules for formal charge* to assign the valence electrons to individual atoms:

1. Half of the electrons of a bond are assigned to each atom in the bond (counting each dash as two electrons).

2. Both electrons of a lone pair are assigned to the atom to which the lone pair belongs.

You calculate the formal charge on an atom by taking the number of valence electrons on the free atom (equal to the group number) and subtracting the number of electrons you assigned the atom by the rules of formal charge; that is,

Formal charge = valence electrons on free atom $-\frac{1}{2}$ (number of electrons in a bond) $-$ (number of lone-pair electrons)

As a check on your work, you should note that the sum of the formal charges equals the charge on the molecular species (zero for a neutral molecule).

Consider a Cl atom in the first Lewis formula we wrote for $COCl_2$. You begin by counting the electrons assigned to the Cl atom by the rules of formal charge. You have 1 electron from the single bond and 6 from the lone-pair electrons, for a total of 7. The formal charge of Cl equals the number of valence electrons (7) less the number of assigned electrons (7). This gives a formal charge of 0 for the Cl atom. Similar calculations of formal charge give 0 for each of the other atoms in the Lewis formula.

Now look at the second Lewis formula we wrote for $COCl_2$ and assign the electrons by the rules for formal charge. For the Cl atom having the double bond, you count 2 electrons from the double bond and 4 from the lone pairs, for a total of 6. The formal charge of the Cl atom is $7 - 6 = +1$. For the O atom, you count 1 electron from the single bond and 6 from the lone-pair electrons, for a total of 7. The number of valence electrons on the free O atom is 6, so the formal charge of the O atom is $6 - 7 = -1$. The formal charge of each of the other atoms is 0.

We indicate the formal charges in a Lewis formula by inserting circled numbers near the atoms (writing + for +1 and − for −1). For the formulas given earlier for $COCl_2$, the formal charges are shown as follows:

$$:\ddot{C}l-C=\ddot{O} \quad \oplus \ddot{C}l=C-\ddot{O}:\ominus \quad :\ddot{C}l-C-\ddot{O}:\ominus$$
$$\quad\quad\quad\;|\quad\quad\quad\quad\quad\quad\;|\quad\quad\quad\quad\quad\quad\;\|$$
$$\quad\quad\quad:\ddot{C}l:\quad\quad\quad\quad\quad\quad:\ddot{C}l:\quad\quad\quad\quad\quad:\ddot{C}l:$$
$$\quad\quad\quad\quad\quad\quad\quad\quad\quad\quad\quad\quad\quad\quad\quad\quad\quad\quad\oplus$$

Now we can discuss three rules that are useful in deciding which of several resonance formulas best approximates the electron distribution of a molecule or ion.

> **RULE A** Whenever you can write several Lewis formulas for a molecule, choose the one having the lowest magnitudes of formal charges.
>
> **RULE B** When two proposed Lewis formulas for a molecule have the same magnitudes of formal charges, choose the one having the negative formal charge on the more electronegative atom.
>
> **RULE C** When possible, choose Lewis formulas that do not have like charges on adjacent atoms.

The second and third formulas for $COCl_2$ have formal charges on the Cl and O atoms. The first formula, however, has zero formal charges for all atoms. So by Rule A, this formula most closely approximates the actual electron distribution. You would write it as the best description by a single Lewis formula. (To be more precise, you could consider a resonance description involving all three Lewis formulas. However, the Lewis formulas would not participate equally in the resonance description; the first one would predominate.)

You can also use formal charges to help you choose the most likely skeleton structure from several possibilities. Consider thionyl chloride, $SOCl_2$. Which of the following Lewis formulas is most plausible? Note that these Lewis formulas have quite different atomic arrangements, so they should not be considered as simply different resonance formulas of the same molecule.

$$\begin{array}{ccc} :\ddot{C}l: & :\ddot{C}l: & \overset{\ominus}{:\ddot{S}:} \\ | & | & | \\ :\ddot{C}l-\ddot{S}-\ddot{O}: & :\ddot{C}l-\ddot{O}-\ddot{S}: & :\ddot{C}l-\ddot{C}l-\ddot{O}: \\ \quad\oplus\;\ominus & \quad\oplus\;\ominus & \overset{+2}{}\;\ominus \end{array}$$

You can verify these formal charges using the method we have outlined here. The last formula has a +2 formal charge on the Cl atom, which is larger in magnitude than the formal charges on any atoms in the other two formulas. Therefore, using Rule A, you would not consider the last formula a plausible structure. To choose between the first and second structures, you apply Rule B. According to this rule, you would choose the first structure over the second, because the first one associates a negative formal charge with the more electronegative atom (O), whereas the second structure associates a positive charge with the more electronegative atom. Note that this application of formal charge gives the same result that you would obtain by assuming that the least electronegative atom is the central atom.

Now that you have one Lewis formula with the appropriate skeleton structure for $SOCl_2$, you should explore the possibility of other Lewis formulas with this skeleton structure. You can write a Lewis formula that has a sulfur–oxygen double bond.

$$:\ddot{C}l:$$
$$\quad|$$
$$:\ddot{C}l-\ddot{S}=\ddot{O}$$

$SOCl_2$

This formula has formal charges of zero for each atom; so according to Rule A, it should be closer to the actual electron distribution than the previous formula. (You could write a resonance description involving both formulas.) Note that the sulfur atom has 10 electrons around it. As long as the atom has available d orbitals for bonding (as any element in the third and higher periods will have), the atom need not obey the octet rule. The next example further illustrates the application of formal charges in choosing the appropriate Lewis formula.

Example 9.11

Using Formal Charges to Determine the Best Lewis Formula

Gaining Mastery Toolbox

Critical Concept 9.11
Given several resonance formulas for a molecule, decide the most important contributor by applying one or more of the following rules: (a) choose the formula with the lowest magnitudes of formal charges, (b) choose the formula with a negative formal charge on the more electronegative atom, and (c) when possible, choose the formula not having like charges on adjacent atoms.

Solution Essentials:
- Formal charge
- Resonance
- Skeleton structure of a molecule
- Lewis representation of multiple bonds
- Octet rule
- Number of valence electrons in a main-group atom

H_2SO_4

Write the Lewis formula that best describes the charge distribution in the sulfuric acid molecule, H_2SO_4, according to the rules of formal charge.

Problem Strategy Draw the skeleton structure of the molecule or ion, and then follow the steps for writing electron-dot formulas. Note that you can move electron pairs into bonding regions to obtain additional formulas (having multiple bonds). The formal charge of an atom in a formula equals the group number of the atom minus the number of electrons assigned to the atom by the rules of formal charge (half of the electrons of connecting bonds plus all of the lone-pair electrons on the atom). Apply Rules A, B, and C given before this example to decide which Lewis formula is best.

Solution Assume a skeleton structure in which the S atom is surrounded by the more electronegative O atoms; the H atoms are then attached to two of the O atoms. The following is a possible Lewis formula having single bonds:

$$
\begin{array}{c}
\ddot{\text{:}}\ddot{\text{O}}\text{:} \\
| \\
\text{H}-\ddot{\text{O}}-\text{S}-\ddot{\text{O}}-\text{H} \\
| \\
\text{:}\ddot{\text{O}}\text{:}
\end{array}
$$

You calculate the formal charge of each O atom bonded to an H atom as follows: You assign the O atom 2 electrons from the two single bonds and 4 electrons from the two lone pairs, for a total of 6 electrons. Because the number of valence electrons on the O atom is 6, the formal charge is $6 - 6 = 0$. For each of the other two O atoms, you assign 1 electron from the one single bond and 6 electrons from the three lone pairs, for a total of 7. The formal charge is $6 - 7 = -1$. For the S atom, you assign 4 electrons from the four single bonds, for a total of 4. Because the number of valence electrons is 6, the formal charge is $6 - 4 = +2$. The formal charge of each of the H atoms is $1 - 1 = 0$. The Lewis formula with formal charges is

$$
\begin{array}{c}
\ddot{\text{:}}\ddot{\text{O}}\text{:}^{\ominus} \\
| ^{(+2)} \\
\text{H}-\ddot{\text{O}}-\text{S}-\ddot{\text{O}}-\text{H} \\
| \\
\text{:}\ddot{\text{O}}\text{:}^{\ominus}
\end{array}
$$

You can also write a Lewis formula that has zero formal charges for atoms, if you form sulfur–oxygen double bonds:

$$
\begin{array}{c}
\ddot{\text{:}}\text{O}\text{:} \\
\| \\
\text{H}-\ddot{\text{O}}-\text{S}-\ddot{\text{O}}-\text{H} \\
\| \\
\text{:}\text{O}\text{:}
\end{array}
$$

To the top O atom, you assign 2 electrons from the double bond and 4 electrons from the two lone pairs, for a total of 6 electrons. The formal charge is $6 - 6 = 0$. To the S atom, you assign a total of 6 electrons from bonds. The formal charge is $6 - 6 = 0$. The formal

(*continued*)

(*continued*)

charges of the other atoms are also 0. Thus, this Lewis formula should be a better representation of the electron distribution in the H_2SO_4 molecule.

Answer Check The sum of the formal charges in any formula should equal the charge on the species, which is zero for a neutral molecule.

Exercise 9.15 Write the Lewis formula that best describes the phosphoric acid molecule, H_3PO_4.

■ See Problems 9.77 and 9.78.

CONCEPT CHECK 9.3

Which of the models shown below most accurately represents the hydrogen cyanide molecule, HCN? Write the electron-dot formula that most closely agrees with this model. State any concept or rule you used in arriving at your answer.

9.10 Bond Length and Bond Order

Bond length (or **bond distance**) is *the distance between the nuclei in a bond.* Bond lengths are determined experimentally using x-ray diffraction or the analysis of molecular spectra. Knowing the bond length in a particular molecule can sometimes provide a clue to the type of bonding present.

Bond lengths for a given bonding situation can often be predicted within a few picometers from a set of covalent radii. The **covalent radius** of an atom is *the value for that atom in a set of covalent radii assigned to atoms in such a way that the sum of the covalent radii of atoms A and B predicts the approximate A—B bond length.* A number of different sets of covalent radii have been derived for various bonding situations. Those given in Table 9.4 were obtained by a statistical analysis of known *single bond* lengths. To illustrate how we might use this table, let's suppose we would like to have an approximate C—Cl single bond length. Looking at Table 9.4, we see that the C and Cl covalent radii are 76 pm and 102 pm, respectively. The sum of these two radii (76 + 102 pm = 178 pm) gives us an approximate value for the C—Cl single bond length. Here are actual C—Cl bond lengths in several compounds:

chloromethane, CH_3Cl, 178.4 pm;

tetrachloromethane, CCl_4, 176.6 pm;

hexachloroethane, CCl_3CCl_3, 174 pm.

The value predicted from Table 9.4 compares favorably with these known bond lengths.

Covalent radii are one type of atomic radii (recall the discussion in Section 8.6), and as such, the values for any given set of radii should follow the two major periodic trends for atomic radii:

1. Within a period, the covalent radius tends to decrease with increasing atomic number.

2. Within a group, the covalent radius tends to increase with period number.

Note that the radii in Table 9.4 do follow these trends rather well.

Table 9.4   **Single-Bond Covalent Radii**

Atomic Number	Symbol	Name	Covalent Radius (pm)	Atomic Number	Symbol	Name	Covalent Radius (pm)
1	H	Hydrogen	31	44	Ru	Ruthenium	146
2	He	Helium	28	45	Rh	Rhodium	142
3	Li	Lithium	128	46	Pd	Palladium	139
4	Be	Beryllium	96	47	Ag	Silver	145
5	B	Boron	84	48	Cd	Cadmium	144
6	C	Carbon	76	49	In	Indium	142
7	N	Nitrogen	71	50	Sn	Tin	139
8	O	Oxygen	66	51	Sb	Antimony	139
9	F	Fluorine	57	52	Te	Tellurium	138
10	Ne	Neon	58	53	I	Iodine	139
11	Na	Sodium	166	54	Xe	Xenon	140
12	Mg	Magnesium	141	55	Cs	Cesium	244
13	Al	Aluminum	121	56	Ba	Barium	215
14	Si	Silicon	111	57	La	Lanthanum	207
15	P	Phosphorus	107	58	Ce	Cerium	204
16	S	Sulfur	105	59	Pr	Praseodymium	203
17	Cl	Chlorine	102	60	Nd	Neodymium	201
18	Ar	Argon	106	61	Pm	Promethium	199
19	K	Potassium	203	62	Sm	Samarium	198
20	Ca	Calcium	176	63	Eu	Europium	198
21	Sc	Scandium	170	64	Gd	Gadolinium	196
22	Ti	Titanium	160	65	Tb	Terbium	194
23	V	Vanadium	153	66	Dy	Dysprosium	192
24	Cr	Chromium	139	67	Ho	Holmium	192
25	Mn	Manganese	139	68	Er	Erbium	189
26	Fe	Iron	132	69	Tm	Thulium	190
27	Co	Cobalt	126	70	Yb	Ytterbium	187
28	Ni	Nickel	124	71	Lu	Lutetium	187
29	Cu	Copper	132	72	Hf	Hafnium	175
30	Zn	Zinc	122	73	Ta	Tantalum	170
31	Ga	Gallium	122	74	W	Tungsten	162
32	Ge	Germanium	120	75	Re	Rhenium	151
33	As	Arsenic	119	76	Os	Osmium	144
34	Se	Selenium	120	77	Ir	Iridium	141
35	Br	Bromine	120	78	Pt	Platinum	136
36	Kr	Krypton	116	79	Au	Gold	136
37	Rb	Rubidium	220	80	Hg	Mercury	132
38	Sr	Strontium	195	81	Tl	Thallium	145
39	Y	Yttrium	190	82	Pb	Lead	146
40	Zr	Zirconium	175	83	Bi	Bismuth	148
41	Nb	Niobium	164	84	Po	Polonium	140
42	Mo	Molybdenum	154	85	At	Astatine	150
43	Tc	Technetium	147	86	Rn	Radon	150

Exercise 9.16 Estimate the O—H bond length in H_2O from the covalent radii listed in Table 9.4.

■ See Problems 9.79, 9.80, 9.81, and 9.82.

Although a double bond is stronger than a single bond, it is not necessarily less reactive. Ethylene, $CH_2{=}CH_2$, for example, is more reactive than ethane, $CH_3{-}CH_3$, where carbon atoms are linked through a single bond.

The **bond order,** defined in terms of the Lewis formula, is *the number of pairs of electrons in a bond.* For example, in C : C the bond order is 1 (single bond); in C : : C the bond order is 2 (double bond). Bond length depends on bond order. As the bond order increases, the bond strength increases and the nuclei are pulled inward, decreasing the bond length. Look at carbon–carbon bonds. The average C—C bond length is 154 pm, whereas C=C is 134 pm long and C≡C is 120 pm long. ◀

Example 9.12   Relating Bond Order and Bond Length

Gaining Mastery Toolbox

Critical Concept 9.12
As the bond order between two atoms increases, the bond length decreases.

Solution Essentials:
• Bond order
• Bond length

Consider the molecules N_2H_4, N_2, and N_2F_2. Which molecule has the shortest nitrogen–nitrogen bond? Which has the longest nitrogen–nitrogen bond?

Problem Strategy First, write the electron-dot formulas for the molecules. Then note that the length of a bond between atoms decreases with bond order.

Solution First write the Lewis formulas:

$$H{-}\overset{..}{N}{-}\overset{..}{N}{-}H \qquad :N{\equiv}N: \qquad :\overset{..}{F}{-}\overset{..}{N}{=}\overset{..}{N}{-}\overset{..}{F}:$$
$$\overset{|}{H}\ \ \overset{|}{H}$$
$$N_2H_4 \qquad\qquad N_2 \qquad\qquad N_2F_2$$

The **nitrogen–nitrogen bond should be shortest in N_2, where it is a triple bond, and longest in N_2H_4, where it is a single bond.** (Experimental values for the nitrogen–nitrogen bond lengths are 109 pm for N_2, 122 pm for N_2F_2, and 147 pm for N_2H_4.)

Answer Check Check that you have written the electron-dot formulas correctly.

Exercise 9.17 Formic acid, isolated in 1670, is the irritant in ant bites. The structure of formic acid is

$$H{-}C{=}O$$
$$\overset{|}{O}$$
$$\overset{|}{H}$$

One of the carbon–oxygen bonds has a length of 136 pm; the other is 123 pm long. What is the length of the C=O bond in formic acid?

■ See Problems 9.83 and 9.84.

9.11 Bond Enthalpy

In Section 9.4, when we described the formation of a covalent bond, we introduced the concept of bond dissociation energy, the energy required to break a particular bond in a molecule (see Figure 9.11). This bond dissociation energy is one measure of the strength of a bond. In this section, we want to look at a related idea, the *bond enthalpy,* which we can use as a measure of the average strength of a bond in its compounds. Bond enthalpies are often obtained from enthalpies of reaction, ΔH, so the term "bond enthalpies" is appropriate, but you will often see the term *bond energies* used for the same values. As a practical matter, the approximations made in arriving at average bond values tend to obscure the small differences between enthalpy changes and energy changes for bonds (which are small). So, for that reason, the terms "bond enthalpy" and "bond energy" are often used interchangeably. We will, however, continue to use the term bond enthalpy.

Let's explore the possibility of assigning a value for the enthalpy change involved in breaking a particular type of bond, whatever the compound. Consider the experimentally determined enthalpy changes for the breaking, or dissociation, of a C—H bond in methane, CH_4, and in ethane, C_2H_6, in the gas phase:

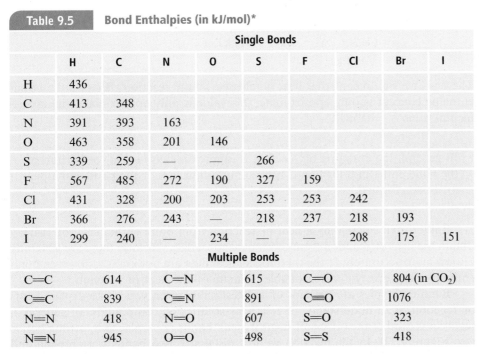

$$\text{H—C—H}(g) \longrightarrow \text{H—C}(g) + \text{H}(g) \qquad \Delta H = 435 \text{ kJ}$$

$$\text{H—C—C—H}(g) \longrightarrow \text{H—C—C}(g) + \text{H}(g) \qquad \Delta H = 410 \text{ kJ}$$

Note that the ΔH values are approximately the same in the two cases. This suggests that the enthalpy change for the dissociation of a C—H bond may be about the same in other molecules. Comparisons of this sort lead to the conclusion that we can obtain approximate values of the enthalpy changes for given bond types.

We define the A—B **bond enthalpy** (denoted BE) as *the average enthalpy change for the breaking of an A—B bond in a molecule in the gas phase.* For example, to calculate a value for the C—H bond enthalpy, or BE(C—H), we might look at the experimentally determined enthalpy change for the breaking of all the C—H bonds in methane:

$$CH_4(g) \longrightarrow C(g) + 4H(g); \Delta H = 1662 \text{ kJ}$$

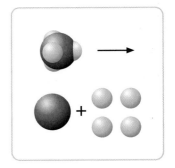

Because four C—H bonds are broken, we obtain an average value for breaking one C—H bond by dividing the enthalpy change for the reaction by 4:

$$BE(\text{C—H}) = \tfrac{1}{4} \times 1662 \text{ kJ} = 416 \text{ kJ}$$

Similar calculations with other molecules, such as ethane, would yield approximately the same values for the C—H bond enthalpy.

Table 9.5 lists values of some bond enthalpies, which are based on an approximate fit to known thermodynamic data. Note that the value given in this table for the C—H bond is 413 kJ/mol, which is in fair agreement with the value we just obtained (416 kJ/mol). Because it takes energy to break a bond, bond enthalpies are always positive numbers. When a bond is formed, the enthalpy change is equal to the negative of the bond enthalpy (heat is released).

Table 9.5	Bond Enthalpies (in kJ/mol)*								
	Single Bonds								
	H	**C**	**N**	**O**	**S**	**F**	**Cl**	**Br**	**I**
H	436								
C	413	348							
N	391	393	163						
O	463	358	201	146					
S	339	259	—	—	266				
F	567	485	272	190	327	159			
Cl	431	328	200	203	253	253	242		
Br	366	276	243	—	218	237	218	193	
I	299	240	—	234	—	—	208	175	151
	Multiple Bonds								
C=C	614		C=N	615		C=O		804 (in CO_2)	
C≡C	839		C≡N	891		C≡O		1076	
N=N	418		N=O	607		S=O		323	
N≡N	945		O=O	498		S=S		418	

*Data are taken from http://wiki.chemeddl.org/index.php/15.4_Bond_Enthalpies.

Figure 9.21 ▶

Reaction of methane with chlorine Molecular models illustrate the reaction $CH_4 + Cl_2 \longrightarrow CH_3Cl + HCl$.

Bond enthalpy is a measure of the strength of a bond: the larger the bond enthalpy, the stronger the chemical bond. Note from Table 9.5 that the bonds C—C, C=C, and C≡C have bond enthalpies of 348, 614, and 839 kJ/mol, respectively. These numbers indicate that the triple bond is stronger than the double bond, which in turn is stronger than the single bond.

You can use this table of bond enthalpies to estimate heats of reaction, or enthalpy changes, ΔH, for gaseous reactions. To illustrate this, let's find ΔH for the following reaction (Figure 9.21):

$$CH_4(g) + Cl_2(g) \longrightarrow CH_3Cl(g) + HCl(g)$$

You can *imagine* that the reaction takes place in steps involving the breaking and forming of bonds. Starting with the reactants, you suppose that one C—H bond and the Cl—Cl bond break.

$$
\begin{array}{c}
H \\
| \\
H-C-H + Cl-Cl \longrightarrow H-C + H + Cl + Cl \\
| | \\
H H
\end{array}
$$

The enthalpy change is $BE(C-H) + BE(Cl-Cl)$. Now you reassemble the fragments to give the products.

$$
\begin{array}{c}
H \\
| \\
H-C + H + Cl + Cl \longrightarrow H-C-Cl + H-Cl \\
| | \\
H H
\end{array}
$$

In this case, C—Cl and H—Cl bonds are formed, and the enthalpy change equals the negative of the bond enthalpies: $-BE(C-Cl) - BE(H-Cl)$. Substituting bond-enthalpy values from Table 9.5, you get the enthalpy of reaction.

$$
\begin{aligned}
\Delta H &\simeq BE(C-H) + BE(Cl-Cl) - BE(C-Cl) - BE(H-Cl) \\
&= (413 + 242 - 328 - 431) \text{ kJ} \\
&= -104 \text{ kJ}
\end{aligned}
$$

The negative sign means that heat is released by the reaction. Because the bond-enthalpy concept is only approximate, this value is only approximate (the experimental value is -101 kJ).

> In general, the enthalpy of reaction is (approximately) equal to the sum of the bond enthalpies for bonds broken minus the sum of the bond enthalpies for bonds formed.

Rather than calculating a heat of reaction from bond enthalpies, you usually obtain the heat of reaction from thermochemical data, because these are generally known more accurately. If the thermochemical data are not known, however, bond enthalpies can give you a reasonable estimate of the heat of reaction.

Bond enthalpies are perhaps of greatest value when you try to explain heats of reaction or to understand the relative stabilities of compounds. In general, a reaction is exothermic (gives off heat) if weak bonds are replaced by strong bonds (see Figure 9.22). In the reaction we just discussed, two bonds were broken and replaced by two new, stronger bonds. In the following example, a strong bond

Figure 9.22 ▲

Explosion of nitrogen triiodide–ammonia complex The red-brown complex of nitrogen triiodide and ammonia is so sensitive to explosion that it can be detonated with a feather. Nitrogen–iodine single bonds are replaced by very stable nitrogen–nitrogen triple bonds (N_2) and iodine–iodine single bonds (I_2).

(C=C) is replaced by two weaker ones (two C—C bonds). Although the single bond is weaker than the double bond, as you would expect, two C—C bonds *together* create an energetically more stable situation than one C=C bond.

Example 9.13

Gaining Mastery Toolbox

Critical Concept 9.13
The enthalpy of a reaction equals approximately the sum of the bond enthalpies for bonds broken minus the sum of bond enthalpies formed.

Solution Essentials:
• Bond enthalpy
• Enthalpy of reaction

Estimating ΔH from Bond Enthalpies

Polyethylene is formed by linking many ethylene molecules into long chains. Estimate the enthalpy change per mole of ethylene for this reaction (shown below), using bond enthalpies.

Problem Strategy Break bonds in the reactants, and then form new bonds to give the products. The approximate enthalpy change equals the sum of the bond enthalpies for the bonds broken minus the sum of the bond enthalpies for the bonds formed.

Solution Imagine the reaction to involve the breaking of the carbon–carbon double bonds and the formation of carbon–carbon single bonds. For a very long chain, the net result is that for every C=C bond broken, two C—C bonds are formed.

$$\Delta H \simeq 614 - (2 \times 348) = -82\ \text{kJ}$$

Answer Check Note whether the sign of ΔH agrees with what you would expect. Perhaps weak bonds have been replaced by strong bonds, in which case ΔH should be negative. Here, every double bond is replaced by two single bonds, so ΔH should be negative, as it is.

Exercise 9.18 Use bond enthalpies to estimate the enthalpy change for the combustion of ethylene, C_2H_4, according to the equation

$$C_2H_4(g) + 3O_2(g) \longrightarrow 2CO_2(g) + 2H_2O(g)$$

■ See Problems 9.85 and 9.86.

Infrared Spectroscopy and Vibrations of Chemical Bonds

A chemical bond acts like a stiff spring connecting nuclei. As a result, the nuclei in a molecule vibrate, rather than maintaining fixed positions relative to each other. Nuclear vibration is depicted in Figure 9.23, which shows a spring model of HCl.

This vibration of molecules is revealed in their absorption of infrared radiation. (An instrument for observing the absorption of infrared radiation is shown in Figure 9.24.) The frequency of radiation absorbed equals the frequencies of nuclear vibrations. For example, the H—Cl bond vibrates at a frequency of 8.652×10^{13} vibrations per second. If radiation of this frequency falls on the molecule, it absorbs the radiation, which is in the infrared region, and begins vibrating more strongly.

The infrared absorption spectrum of a molecule of even moderate size can have a rather complicated appearance. Figure 9.25 shows the infrared (IR) spectrum of ethyl butyrate, a compound present in pineapple flavor. The complicated appearance of the IR spectrum is actually an advantage. Two different compounds are unlikely to have exactly the same IR spectrum. Therefore, the IR spectrum can act as a compound's "fingerprint."

Figure 9.24 ▲

A Fourier transform infrared (FTIR) spectrometer A Nicolet 560 E.S.P.FT-IR spectrometer.

Courtesy of Nicolet

The IR spectrum of a compound can also yield structural information. Suppose you would like to obtain the structural formula of ethyl butyrate. The molecular formula, determined from combustion analysis, is $C_6H_{12}O_2$. Important information about this structure can be obtained from Figure 9.25.

You first need to be able to read such a spectrum. Instead of plotting an IR spectrum in frequency units (since the frequencies are very large), you usually would give the frequencies in *wavenumbers*, which are proportional to frequency. To get the wavenumber, you divide the frequency by the speed of light expressed in centimeters per second. For example, HCl absorbs at $(8.652 \times 10^{13}\ \text{s}^{-1})/(2.998 \times 10^{10}\ \text{cm/s}) = 2886\ \text{cm}^{-1}$ (wavenumbers).

Figure 9.23 ▲

Vibration of the HCl molecule The vibrating molecule is represented here by a spring model. The atoms in the molecule vibrate; that is, they move constantly back and forth.

A Checklist for Review

Summary of Facts and Concepts

An *ionic bond* is a strong attractive force holding ions together. An ionic bond can form between two atoms by the transfer of electrons from the valence shell of one atom to the valence shell of the other. Many similar ions attract one another to form a crystalline solid, in which positive ions are surrounded by negative ions and negative ions are surrounded by positive ions. As a result, ionic solids are typically high-melting solids. Monatomic cations of the main-group elements have charges equal to the group number (or in some cases, the group number minus two). Monatomic anions of the main-group elements have charges equal to the group number minus eight.

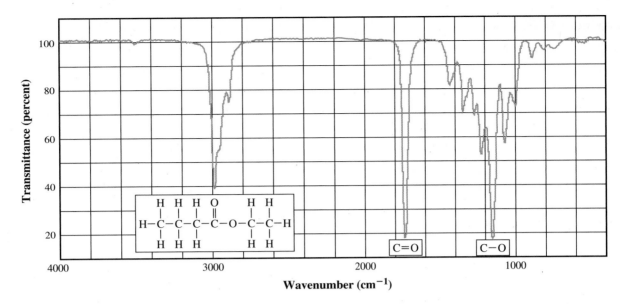

Figure 9.25 ▲

Infrared spectrum of ethyl butyrate Note the peaks corresponding to vibrations of C=O and C—O bonds. The molecular structure is shown at the bottom left. (From NIST Mass Spec Data Center, S.E. Stein, director, "IR and Mass Spectra" in *NIST Chemistry WebBook, NIST Standard Reference Database Number 69,* Eds. W.G. Mallard and P.J. Linstrom, February 2000, National Institute of Standards and Technology, Gaithersburg, MD, 20899 [http://webbook.nist.gov]. © 1991, 1994, 1996, 1997, 1998, 1999, 2000. Copyright by the U.S. Secretary of Commerce on behalf of the United States of America. All rights reserved.)

Wavenumber, or sometimes wavelength, is plotted along the horizontal axis.

Percent transmittance—that is, the percent of radiation that passes through a sample—is plotted on the vertical axis. When a molecule absorbs radiation of a given frequency or wavenumber, this is seen in the spectrum as an inverted spike (peak) at that wavenumber. Certain structural features of molecules appear as absorption peaks in definite regions of the infrared spectrum. For example, the absorption peak at 1730 cm^{-1} is characteristic of the C=O bond. With some knowledge of where various bonds absorb, one can identify other peaks, including that of C—O at 1180 cm^{-1}. (Generally, the IR peak for an A—B bond occurs at lower wavenumber than for an A=B bond.) The IR spectrum does not reveal the complete structure, but it provides important clues. Data from other instruments, such as the mass spectrometer (page 100), give additional clues.

■ See Problems 9.115 and 9.116.

A *covalent bond* is a strong attractive force that holds two atoms together by their sharing of electrons. These bonding electrons are attracted simultaneously to both atomic nuclei, and they spend part of the time near one atom and part of the time near the other. If an electron pair is not equally shared, the bond is *polar.* This polarity results from the difference in *electronegativities* of the atoms—that is, from the unequal abilities of the atoms to draw bonding electrons to themselves.

Lewis electron-dot formulas are simple representations of the valence-shell electrons of atoms in molecules and ions. You can apply simple rules to draw these formulas. In molecules with *delocalized bonding,* it is not possible to describe accurately the electron distribution with a single Lewis formula. For these molecules, you must use *resonance.* Although the atoms in Lewis formulas often satisfy the octet rule, *exceptions to the octet rule* are not uncommon. You can obtain the Lewis formulas for these exceptions by following the rules for writing Lewis formulas. The concept of *formal charge* will often help you decide which of several Lewis formulas gives the best description of a molecule or ion.

Bond lengths can be estimated from the *covalent radii* of atoms. Bond length depends on *bond order;* as the bond order increases, the bond length decreases. The A—B *bond enthalpy* is the average enthalpy change when an A—B bond is broken. You can use bond enthalpies to estimate ΔH for gas-phase reactions.

Learning Objectives	Important Terms
9.1 Describing Ionic Bonds	
■ Define *ionic bond*. ■ Explain the *Lewis electron-dot symbol* of an atom. ■ Use Lewis symbols to represent ionic bond formation. Example 9.1 ■ Describe the energetics of ionic bonding. ■ Define *lattice energy*. ■ Describe the Born–Haber cycle to obtain a lattice energy from thermodynamic data. ■ Describe some general properties of ionic substances.	**ionic bond** **Lewis electron-dot symbol** **Coulomb's law** **lattice energy**
9.2 Electron Configurations of Ions	
■ State the three categories of monatomic ions of the main-group elements. ■ Write the electron configuration and Lewis symbol for a main-group ion. Example 9.2 ■ Note the polyatomic ions given earlier in Table 2.5. ■ Note the formation of 2+ and 3+ transition-metal ions. ■ Write electron configurations of transition-metal ions. Example 9.3	
9.3 Ionic Radii	
■ Define *ionic radius*. ■ Define *isoelectronic ions*. ■ Use periodic trends to obtain relative ionic radii. Example 9.4	**ionic radius** **isoelectronic**
9.4 Describing Covalent Bonds	
■ Describe the formation of a covalent bond between two atoms. ■ Define *Lewis electron-dot formula*. ■ Define *bonding pair* and *lone (nonbonding) pair* of electrons. ■ Define *coordinate covalent bond*. ■ State the octet rule. ■ Define *single bond, double bond,* and *triple bond*.	**covalent bond** **Lewis electron-dot formula** **bonding pair** **lone (nonbonding) pair** **coordinate covalent bond** **octet rule** **single bond** **double bond** **triple bond**
9.5 Polar Covalent Bonds; Electronegativity	
■ Define *polar covalent bond*. ■ Define *electronegativity*. ■ State the general periodic trends in electronegativities. ■ Use electronegativity to obtain relative bond polarity. Example 9.5	**polar covalent bond** **electronegativity**
9.6 Writing Lewis Electron-Dot Formulas	
■ Write Lewis formulas with single bonds only. Example 9.6 ■ Write Lewis formulas having multiple bonds. Example 9.7 ■ Write Lewis formulas for ionic species. Example 9.8	
9.7 Delocalized Bonding; Resonance	
■ Define *delocalized bonding*. ■ Define *resonance description*. ■ Write resonance formulas. Example 9.9	**delocalized bonding** **resonance description**

9.8 Exceptions to the Octet Rule

- Write Lewis formulas (exceptions to the octet rule). Example 9.10
- Note exceptions to the octet rule in Group IIA and Group IIIA elements.

9.9 Formal Charge and Lewis Formulas

- Define *formal charge*.
- State the rules for obtaining formal charge.
- State two rules useful in writing Lewis formulas.
- Use formal charges to determine the best Lewis formula. Example 9.11

formal charge

9.10 Bond Length and Bond Order

- Define *bond length* (*bond distance*).
- Define *covalent radii*.
- Define *bond order*.
- Explain how bond order and bond length are related. Example 9.12

bond length (bond distance)
covalent radii
bond order

9.11 Bond Enthalpy

- Define *bond enthalpy*.
- Estimate ΔH from bond enthalpies. Example 9.13

bond enthalpy

Questions and Problems

WL Interactive versions of these problems may be assigned in OWL.

Self-Assessment and Review Questions

Key: These questions test your understanding of the ideas you worked with in the chapter. These problems vary in difficulty and often can be used for the basis of discussion.

9.1 Describe the formation of a sodium chloride crystal from atoms.

9.2 Why does sodium chloride normally exist as a crystal rather than as a molecule composed of one cation and one anion?

9.3 Explain what energy terms are involved in the formation of an ionic solid from atoms. In what way should these terms change (become larger or smaller) to give the lowest energy possible for the solid?

9.4 Define lattice energy for potassium bromide.

9.5 Why do most monatomic cations of the main-group elements have a charge equal to the group number? Why do most monatomic anions of these elements have a charge equal to the group number minus eight?

9.6 The 2+ ions of transition elements are common. Explain why this might be expected.

9.7 Describe the trends shown by the radii of the monatomic ions for the main-group elements both across a period and down a column.

9.8 Explain how ionic radii are obtained from known distances between nuclei in crystals.

9.9 Describe the formation of a covalent bond in H_2 from atoms. What does it mean to say that the bonding electrons are shared by the two atoms?

9.10 Give an example of a molecule that has a coordinate covalent bond.

9.11 Draw a potential-energy diagram for a molecule such as Cl_2. Indicate the bond length (194 pm) and the bond dissociation energy (240 kJ/mol).

9.12 The octet rule correctly predicts the Lewis formula of many molecules involving main-group elements. Explain why this is so.

9.13 Describe the kinds of exceptions to the octet rule that we encounter in compounds of the main-group elements. Give examples.

9.14 Describe the general trends in electronegativities of the elements in the periodic table both across a period and down a column.

9.15 What is the qualitative relationship between bond polarity and electronegativity difference?

9.16 What is a resonance description of a molecule? Why is this concept required if we wish to retain Lewis formulas as a description of the electron structure of molecules?

9.17 What is the relationship between bond order and bond length? Use an example to illustrate it.

9.18 Define bond enthalpy. Explain how one can use bond enthalpies to estimate the heat of reaction.

9.19 Which of the following contains *both* ionic and covalent bonds in the same compound?

a $MgCl_2$
b BaO
c SO_4^{2-}
d $BaCO_3$
e H_2S

9.20 The radii of the species S, S$^+$, and S$^-$ *decrease* in the following order:

a S$^-$ > S > S$^+$ b S > S$^+$ > S$^-$

c S > S$^-$ > S$^+$ d S$^+$ > S > S$^-$

e S$^+$ > S > S$^-$

9.21 Which of the following is the ground-state electron configuration of a C^{3-} ion?

a $1s^22s^22p^5$ b $1s^22s^22p^63s^1$

c $1s^22s^22p^4$ d $[He]2s^22p^6$

e $[He]2s^1$

9.22 The element X below could be

$$\cdot \overset{\cdot\cdot}{\underset{\cdot\cdot}{X}} \cdot$$

a aluminum
b phosphorus
c silicon
d sulfur
e iodine

Concept Explorations

Key: Concept explorations are comprehensive problems that provide a framework that will enable you to explore and learn many of the critical concepts and ideas in each chapter. If you master the concepts associated with these explorations, you will have a better understanding of many important chemistry ideas and will be more successful in solving all types of chemistry problems. These problems are well suited for group work and for use as in-class activities.

9.23 Forming Ionic Compounds

a Consider a metal atom, which we will give the symbol M. Metal M can readily form the M$^+$ cation. If the sphere on the left below represents the metal atom M, which of the other three spheres would probably represent the M$^+$ cation? Explain your choice.

b Consider a nonmetal atom, which we will give the symbol X. The element X is a gas at room temperature and can easily form the X$^-$ anion. If atom X is represented by the sphere on the left below, which of the other three spheres would probably represent the X$^-$ anion. Explain your choice.

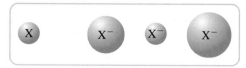

c On the basis of the information given previously, write a balanced chemical equation for the reaction that occurs when metal M reacts with nonmetal X. Include state symbols for the reactants and products.

d Is the product of the reaction of M and X a molecule or an ionic compound? Provide justification for your answer.

e Which of the representations below is the most appropriate to describe the compound of M and X, according to your answer for part d? Explain how you arrived at your choice.

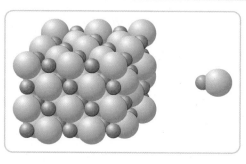

9.24 Bond Enthalpy

When atoms of the hypothetical element X are placed together, they rapidly undergo reaction to form the X$_2$ molecule:

$$X(g) + X(g) \longrightarrow X_2(g)$$

a Would you predict that this reaction is exothermic or endothermic? Explain.

b Is the bond enthalpy of X$_2$ a positive or a negative quantity? Why?

c Suppose ΔH for the reaction is -500 kJ/mol. Estimate the bond enthalpy of the X$_2$ molecule.

d Another hypothetical molecular compound, Y$_2(g)$, has a bond enthalpy of 750 kJ/mol, and the molecular compound XY(g) has a bond enthalpy of 1500 kJ/mol. Using bond enthalpy information, calculate ΔH for the following reaction.

$$X_2(g) + Y_2(g) \longrightarrow 2XY(g)$$

e Given the following information, as well as the information previously presented, predict whether or not the hypothetical ionic compound AX is likely to form. In this compound, A forms the A$^+$ cation, and X forms the X$^-$ anion. Be sure to justify your answer.

Reaction: A(g) + $\frac{1}{2}$X$_2(g)$ $\longrightarrow$ AX(s)

The first ionization energy of A(g) is 400 kJ/mol.

The electron affinity of X(g) is 525 kJ/mol.

The lattice energy of AX(s) is 100 kJ/mol.

f If you predicted that no ionic compound would form from the reaction in part e, what minimum amount of AX(s) lattice energy might lead to compound formation?

Conceptual Problems

Key: These problems are designed to check your understanding of the concepts associated with some of the main topics presented in each chapter. A strong conceptual understanding of chemistry is the foundation for both applying chemical knowledge and solving chemical problems. These problems vary in level of difficulty and often can be used as a basis for group discussion.

9.25 You land on a distant planet in another universe and find that the $n = 1$ level can hold a maximum of 4 electrons, the $n = 2$ level can hold a maximum of 5 electrons, and the $n = 3$ level can hold a maximum of 3 electrons. Like our universe, protons have a charge of $+1$, electrons have a charge of -1, and opposite charges attract. Also, a filled shell results in greater stability of an atom, so the atom tends to gain or lose electrons to give a filled shell. Predict the formula of a compound that results from the reaction of a neutral metal atom X, which has 7 electrons, and a neutral nonmetal atom Y, which has 3 electrons.

9.26 Below on the left side are models of two atoms, one from a metal, the other from a nonmetal. On the right side are corresponding monatomic ions of those atoms. Decide which of these ions is the cation and which is the anion. Label atoms as "metal" or "nonmetal," as appropriate.

9.27 Which of the following represent configurations of thallium ions in compounds? Explain your decision in each case.

- **a** Tl^{2+} $[Xe]4f^{14}5d^{10}6p^1$
- **b** Tl^{3+} $[Xe]4f^{14}5d^{10}$
- **c** Tl^{4+} $[Xe]4f^{14}5d^9$
- **d** Tl^+ $[Xe]4f^{14}5d^{10}6s^2$

9.28 Predict a possible monatomic ion for Element 117, Uus. Given the spherical models below, which should be labeled "atom" and which should be labeled "ion"?

9.29 Examine each of the following electron-dot formulas and decide whether the formula is correct, or whether you could write a formula that better approximates the electron structure of the molecule. State which concepts or rules you use in each case to arrive at your conclusion.

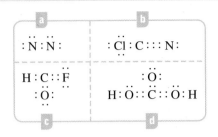

9.30 For each of the following molecular models, write an appropriate Lewis formula.

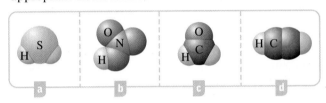

9.31 For each of the following molecular formulas, draw the most reasonable skeleton structure.

- **a** CH_2Cl_2
- **b** HNO_2
- **c** NOF
- **d** N_2O_4

What rule or concept did you use to obtain each structure?

9.32 Below are three resonance formulas for N_2O (nitrous oxide). Rank these in terms of how closely you think each one represents the true electron structure of the molecule. State the rules and concepts you use to do this ranking.

9.33 Sodium, Na, reacts with element X to form an ionic compound with the formula Na_3X.

- **a** What is the formula of the compound you expect to form when calcium, Ca, reacts with element X?
- **b** Would you expect this compound to be ionic or molecular?

9.34 The enthalpy change for each of the following reactions was calculated using bond enthalpies. The bond enthalpies of X—O, Y—O, and Z—O are all equal.

$$X-X + O=O \longrightarrow X-O-O-X; \Delta H = -275 \text{ kJ}$$
$$Y-Y + O=O \longrightarrow Y-O-O-Y; \Delta H = +275 \text{ kJ}$$
$$Z-Z + O=O \longrightarrow Z-O-O-Z; \Delta H = -100 \text{ kJ}$$

- **a** Rank the bonds X—X, Y—Y, and Z—Z from strongest to weakest.
- **b** Compare the enthalpies required to completely dissociate each of the products to atoms.
- **c** If O_2 molecules were O—O instead of O=O, how would this change ΔH for each reaction?

Practice Problems

Key: These problems are for practice in applying problem-solving skills. They are divided by topic, and some are keyed to exercises (see the ends of the exercises). The problems are arranged in matching pairs; the odd-numbered problem of each pair is listed first, and its answer is given in the back of the book.

Ionic Bonding

9.35 Write Lewis symbols for the following:
a O   b O^{2-} c Ca   d Ca^{2+}

9.36 Write Lewis symbols for the following:
a P   b P^{3-} c Ga   d Ga^{3+}

9.37 Use Lewis symbols to represent the transfer of electrons between the following atoms to form ions with noble-gas configurations:
a K and I   b Ca and Br

9.38 Use Lewis symbols to represent the electron transfer between the following atoms to give ions with noble-gas configurations:
a Sr and O   b Ba and I

9.39 For each of the following, write the electron configuration and Lewis symbol:
a As   b As^{3+} c Se   d Se^{2-}

9.40 For each of the following, write the electron configuration and Lewis symbol:
a Ge   b Ge^{2+} c K^+ d I^-

9.41 Write the electron configurations of Sn and Sn^{2+}.

9.42 Write the electron configurations of Bi and Bi^{3+}.

9.43 Give the electron configurations of Ni^{2+} and Ni^{3+}.

9.44 Give the electron configurations of Cu^+ and Cu^{2+}.

Ionic Radii

9.45 Arrange the members of each of the following pairs in order of increasing radius and explain the order:
a Br, Br^- b Sr, Sr^{2+}

9.46 Arrange the members of each of the following pairs in order of increasing radius and explain the order:
a Al, Al^{3+} b Te, Te^{2-}

9.47 Without looking at Table 9.3, arrange the following in order of increasing ionic radius: Se^{2-}, Te^{2-}, S^{2-}. Explain how you arrived at this order. (You may use a periodic table.)

9.48 Which has the larger radius, N^{3-} or P^{3-}? Explain. (You may use a periodic table.)

9.49 Arrange the following in order of increasing ionic radius: F^-, Na^+, and N^{3-}. Explain this order. (You may use a periodic table.)

9.50 Arrange the following in order of increasing ionic radius: I^-, Cs^+, and Te^{2+}. Explain this order. (You may use a periodic table.)

Covalent Bonding

9.51 Use Lewis symbols to show the reaction of atoms to form arsine, AsH_3. Indicate which electron pairs in the Lewis formula of AsH_3 are bonding and which are lone pairs.

9.52 Use Lewis symbols to show the reaction of atoms to form hydrogen selenide, H_2Se. Indicate bonding pairs and lone pairs in the electron-dot formula of this compound.

9.53 Assuming that the atoms form the normal number of covalent bonds, give the molecular formula of the simplest compound of arsenic and bromine atoms.

9.54 Assuming that the atoms form the normal number of covalent bonds, give the molecular formula of the simplest compound of germanium and fluorine atoms.

Polar Covalent Bonds; Electronegativity

9.55 With the aid of a periodic table (not Figure 9.15), arrange the following in order of increasing electronegativity:
a Sr, Cs, Ba   b Ca, Ge, Ga   c P, As, S

9.56 Using a periodic table (not Figure 9.15), arrange the following in order of increasing electronegativity:
a P, O, N   b Na, Al, Mg   c C, Al, Si

9.57 Arrange the following bonds in order of increasing polarity using electronegativities of atoms: P—O, C—Cl, As—Br.

9.58 Decide which of the following bonds is least polar on the basis of electronegativities of atoms: H—S, Si—Cl, N—Cl.

9.59 Indicate the partial charges for the bonds given in Problem 9.57, using the symbols δ^+ and δ^-.

9.60 Indicate the partial charges for the bonds given in Problem 9.58, using the symbols δ^+ and δ^-.

Writing Lewis Formulas

9.61 Write Lewis formulas for the following molecules:
a H_2S b NF_3 c Br_2

9.62 Write Lewis formulas for the following molecules:
a BrF   b PBr_3 c NOF

9.63 Write Lewis formulas for the following molecules:
a HNO_2 b $COBr_2$ c P_2

9.64 Write Lewis formulas for the following molecules:
a CO   b BrCN   c N_2F_2

9.65 Write Lewis formulas for the following ions:
a ClO^- b $SnCl_3^-$ c S_2^{2-}

9.66 Write Lewis formulas for the following ions:
a CN^- b IBr_2^+ c ClF_2^+

Resonance

9.67 Write resonance descriptions for the following:
a SO_3 b HNO_3

9.68 Write resonance descriptions for the following:
　a　$ClNO_2$　　　b　NO_2^-

9.69 Give the resonance description of the formate ion. The skeleton structure is

$$\left[\begin{array}{c} O \\ | \\ H-C-O \end{array}\right]^-$$

9.70 Use resonance to describe the electron structure of nitromethane, CH_3NO_2. The skeleton structure is

$$\begin{array}{ccc} & H & O \\ & | & | \\ H- & C-N-O \\ & | \\ & H \end{array}$$

Exceptions to the Octet Rule

9.71 Write Lewis formulas for the following:
　a　XeF_2　　b　SeF_4　　c　TeF_6　　d　XeF_5^+

9.72 Write Lewis formulas for the following:
　a　ClF_3　　b　IF_4^-　　c　BrF_5　　d　I_3^-

9.73 Write Lewis formulas for the following:
　a　BCl_3　　b　$TlCl_2^+$　　c　$BeBr_2$

9.74 Write Lewis formulas for the following:
　a　BeF_3^-　　b　$AlBr_3$　　c　BeF_2

Formal Charge and Lewis Formulas

9.75 Write a Lewis formula for each of the following, assuming that the octet rule holds for the atoms. Then obtain the formal charges of the atoms.
　a　O_3　　　b　CO　　　c　HNO_3

9.76 Write a Lewis formula for each of the following, assuming that the octet rule holds for the atoms. Then obtain the formal charges of the atoms.
　a　$ClNO$　　b　$POCl_3$　　c　N_2O (NNO)

9.77 For each of the following, use formal charges to choose the Lewis formula that gives the best description of the electron distribution:
　a　ClO_2F
　b　SO_2
　c　ClO_3^-

9.78 For each of the following, use formal charges to choose the Lewis formula that gives the best description of the electron distribution:
　a　SOF_2
　b　H_2SO_3
　c　$HClO_2$

Bond Length, Bond Order, and Bond Enthalpy

9.79 Use covalent radii (Table 9.4) to estimate the length of the P—F bond in phosphorus trifluoride, PF_3.

9.80 What do you expect for the B—Cl bond length in boron trichloride, BCl_3, on the basis of covalent radii (Table 9.4)?

9.81 Calculate the bond length for each of the following single bonds, using covalent radii (Table 9.4):
　a　Si—O
　b　Br—Cl
　c　S—Cl
　d　C—H

9.82 Calculate the C—H and C—Cl bond lengths in chloroform, $CHCl_3$, using values for the covalent radii from Table 9.4. How do these values compare with the experimental values: C—H, 107 pm; C—Cl, 177 pm?

9.83 Which of the following two compounds has the shorter carbon–oxygen bond?

$$\begin{array}{cc} \begin{array}{c} H \\ | \\ H-C-O-H \\ | \\ H \end{array} & \begin{array}{c} :O: \\ \| \\ H-C-H \end{array} \\ \text{Methanol} & \text{Formaldehyde} \end{array}$$

9.84 One of the following compounds has a carbon–nitrogen bond length of 116 pm; the other has a carbon–nitrogen bond length of 147 pm. Match a bond length with each compound.

$$\begin{array}{cc} \begin{array}{c} H \\ | \\ H-C-N-H \\ | \quad | \\ H \quad H \end{array} & \begin{array}{c} H \\ | \\ H-C-C\equiv N: \\ | \\ H \end{array} \\ \text{Methylamine} & \text{Acetonitrile} \end{array}$$

9.85 Use bond enthalpies (Table 9.5) to estimate ΔH for the following gas-phase reaction.

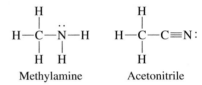

This is called an "addition" reaction, because a compound (HBr) is added across the double bond.

9.86 A commercial process for preparing ethanol (ethyl alcohol), C_2H_5OH, consists of passing ethylene gas, C_2H_4, and steam over an acid catalyst (to speed up the reaction). The gas-phase reaction is

$$\begin{array}{c} H \quad H \\ | \quad | \\ C=C + H-O-H \longrightarrow \\ | \quad | \\ H \quad H \end{array} \begin{array}{c} H \quad H \\ | \quad | \\ H-C-C-O-H \\ | \quad | \\ H \quad H \end{array}$$

Estimate ΔH for this reaction, using bond enthalpies (Table 9.5).

General Problems

Key: These problems provide more practice but are not divided by topic or keyed to exercises. Each section ends with essay questions, each of which is color coded to refer to the *A Chemist Looks at Frontiers* (purple), *A Chemist Looks at Daily Life* (orange), and *Instrumental Methods* (brown) chapter essay on which it is based. Odd-numbered problems and the even-numbered problems that follow are similar; answers to all odd-numbered problems except the essay questions are given in the back of the book.

9.87 For each of the following pairs of elements, state whether the binary compound formed is likely to be ionic or covalent. Give the formula and name of the compound.
  a K, Se   **b** Al, F   **c** Ba, Br   **d** Si, Cl

9.88 For each of the following pairs of elements, state whether the binary compound formed is likely to be ionic or covalent. Give the formula and name of the compound.
  a Sr, O   **b** C, Br   **c** Ga, F   **d** N, Br

9.89 Give the Lewis formula for the selenite ion, SeO_3^{2-}. Write the formula of aluminum selenite.

9.90 Give the Lewis formula for the arsenate ion, AsO_4^{3-}. Write the formula of lead(II) arsenate.

9.91 Selenic acid, H_2SeO_4, is a crystalline substance and a strong acid. Write the electron-dot formula of selenic acid.

9.92 Iodic acid, HIO_3, is a colorless, crystalline compound. What is the electron-dot formula of iodic acid?

9.93 Sodium amide, known commercially as sodamide, is used in preparing indigo, the dye used to color blue jeans. It is an ionic compound with the formula $NaNH_2$. What is the electron-dot formula of the amide anion, NH_2^-?

9.94 Lithium aluminum hydride, $LiAlH_4$, is an important reducing agent (an element or compound that generally has a strong tendency to give up electrons in its chemical reactions). Write the electron-dot formula of the AlH_4^- ion.

9.95 Nitronium perchlorate, NO_2ClO_4, is a reactive salt of the nitronium ion NO_2^+. Write the electron-dot formula of NO_2^+.

9.96 Solid phosphorus pentabromide, PBr_5, has been shown to have the ionic structure $[PBr_4^+][Br^-]$. Write the electron-dot formula of the PBr_4^+ cation.

9.97 Write electron-dot formulas for the following:
  a C_2^{2-} **b** $GaCl_4^-$ **c** CSe_2 **d** $SeOCl_2$

9.98 Write electron-dot formulas for the following:
  a NO^+ **b** IF_2^+ **c** Si_2H_6 **d** $POBr_3$

9.99 Write Lewis formulas for the following:
  a ICN   **b** $SbCl_3$ **c** ICl_3 **d** IF_5

9.100 Write Lewis formulas for the following:
  a $AlCl_4^-$ **b** AlF_6^{3-} **c** BrF_3 **d** IF_6^+

9.101 Give resonance descriptions for the following:
  a SeO_2 **b** N_2O_4

9.102 Give resonance descriptions for the following:
  a CH_3NO_2 **b** $C_2O_4^{2-}$

9.103 Acetic acid has the structure $CH_3CO(OH)$, in which the OH group is bonded to a C atom. The two carbon–oxygen bonds have different lengths. When an acetic acid molecule loses the H from the OH group to form the acetate ion, the two carbon–oxygen bonds become equal in length. Explain.

9.104 The compound S_2N_2 has a cyclic structure with alternating sulfur and nitrogen atoms. Draw all resonance formulas in which the atoms obey the octet rule. Of these, select those in which the formal charges of all atoms are closest to zero.

9.105 The atoms in N_2O_5 are connected as follows:

No attempt has been made here to indicate whether a bond is single or double or whether there is resonance. Obtain the Lewis formula (or formulas). The N—O bond lengths are 118 pm and 136 pm. Indicate the lengths of the bonds in the compound.

9.106 Methyl nitrite has the structure

No attempt has been made here to indicate whether a bond is single or double or whether there is resonance. Obtain the Lewis formula (or formulas). The N—O bond lengths are 122 pm and 137 pm. Indicate the lengths of the N—O bonds in the compound.

9.107 Use bond enthalpies to estimate ΔH for the reaction
$$H_2(g) + O_2(g) \longrightarrow H_2O_2(g)$$

9.108 Use bond enthalpies to estimate ΔH for the reaction
$$2F_2(g) + N_2(g) \longrightarrow N_2F_4(g)$$

9.109 Use bond enthalpies to estimate ΔH for the reaction
$$HCN(g) + 2H_2(g) \longrightarrow CH_3NH_2(g)$$

9.110 Use bond enthalpies to estimate ΔH for the reaction
$$N_2F_2(g) + F_2(g) \longrightarrow N_2F_4(g)$$

■ **9.111** What advantages does using an ionic liquid as a solvent have over using an organic solvent?

■ **9.112** Compare the properties of an ionic material such as sodium chloride with a room-temperature ionic liquid. Explain this difference.

■ **9.113** Explain the decomposition of nitroglycerin in terms of relative bond enthalpies.

■ **9.114** How did the Swedish chemist Alfred Nobel manage to tame the decomposition of nitroglycerin?

■ **9.115** What property of a chemical bond gives rise to the infrared spectrum of a compound?

■ **9.116** What kind of information can be obtained about a compound from its infrared spectrum?

Strategy Problems

Key: As noted earlier, all of the practice and general problems are matched pairs. This section is a selection of problems that are not in matched-pair format. These challenging problems require that you employ many of the concepts and strategies that were developed in the chapter. In some cases, you will have to integrate several concepts and operational skills in order to solve the problem successfully.

9.117 Which of the following Lewis symbols is (or are) *not* correct?

   a $[:\ddot{O}:]^{2-}$ b $[\text{Al}\cdot]^{3+}$ c $[:\ddot{F}:]^-$ d $\cdot\dot{\text{Si}}\cdot$ e $\cdot\text{Mg}$

9.118 Calculate the lattice energy of potassium fluoride, KF, using the Born-Haber cycle. Use thermodynamic data from Appendix C to obtain the enthalpy changes for each step. (Note: You will obtain a slightly different answer if you use values given in Chapter 8 for the ionization energy and electron affinity, which are energy values at 0 K rather than the enthalpy changes at 298 K.)

9.119 Draw a figure similar to Figure 9.2 but for the energetics of ionic bonding in KF. Use the ionization energy and electron affinity from values given in Tables 8.3 and 8.4 in Chapter 8. Calculate the sublimation energy from thermodynamic data (Appendix C). For the lattice energy, see Problem 9.118.

9.120 Which of the following are isoelectronic with the potassium ion, K^+? What are the electron configurations of these species?

   a Ar b Cl^+ c Kr d Cl^- e Ca^+

9.121 Consider the following ions: Al^{3+}, Mg^{2+}, Na^+, S^{2-}, Cl^-. Write down all possible formulas of ionic compounds between each of the oppositely charged ions. Of these compounds, note the ones with a metal-to-nonmetal ratio of one-to-one. Which compound do you expect has the highest melting point? Why?

9.122 An ion M^{2+} has the configuration $[Ar]3d^2$, and an atom has the configuration $[Ar]4s^2$. Identify the ion and the atom.

9.123 Give the symbol of an atomic ion for each of the following electron configurations:

   a $1s^22s^22p^63s^23p^63d^{10}4s^2$
   b $1s^22s^22p^63s^23p^63d^{10}4s^24p^6$
   c $1s^22s^22p^63s^23p^6$
   d $1s^22s^22p^63s^23p^64s^1$
   e $1s^22s^22p^63s^23p^63d^8$

9.124 Calculate the difference in electronegativities between the atoms in SrF_2 and between the atoms in SnF_2. Which substance would you expect to be more ionic in character? One of these compounds melts at 213°C, the other one at about 1400°C. What is the melting point of tin(II) fluoride?

9.125 Using the ionic radii given in Table 9.3, estimate the energy to form a mole of Na^+F^- ion pairs from the corresponding atomic ions.

9.126 Which of the following molecules possesses a double bond? If none do, so state. Which species has an atom with a nonzero formal charge?

   a I_2 b SF_4 c $COCl_2$ d C_2H_4 e CN^-

9.127 Which of the following molecules contains *only* double bonds? If none do, so state.

   a NCCN b CO_2 c C_2H_4 d O_3 e N_2

9.128 Two fourth-period atoms, one of a transition metal, M, and the other of a main-group nonmetal, X, form a compound with the formula M_2X_3. What is the electron configuration of atom X if M is Fe? What is the configuration of X if M is Co?

9.129 Draw resonance formulas for the azide ion, N_3^-, and for the nitronium ion, NO_2^+. Decide which resonance formula is the best description of each ion.

9.130 In Section 9.8, several resonance formulas of BF_3 were discussed. Which of these would be favored based on the concept of formal charge? Explain how you arrived at this answer. Does this answer agree with the chemistry described in the text?

9.131 Consider all A—B bonds that can be formed between any two of the four elements germanium (Ge) to bromine (Br) in Period 4. Which bond should be the most polar? What is the covalent single bond length for this bond? What is the likely molecular formula of a binary compound of these two elements? Draw an electron-dot formula for the molecule.

9.132 Draw resonance formulas of the phosphoric acid molecule, $(HO)_3PO$. Obtain formal charges for the atoms in these resonance formulas. From this result, which resonance formula would you expect to most closely approximate the actual electron distribution?

9.133 Consider hypothetical elements X and Y. Suppose the enthalpy of formation of the compound XY is -336 kJ/mol, the bond enthalpy for X_2 is 414 kJ/mol, and the bond enthalpy for Y_2 is 159 kJ/mol. Estimate the XY bond enthalpy in units of kJ/mol.

9.134 Nitrous oxide, N_2O, has a linear structure NNO. Write resonance formulas for this molecule and from them estimate the NN bond length in the molecule. Use the data in Example 9.12.

9.135 Using bond enthalpies, estimate the heat obtained in burning 10.0 g of methane gas, CH_4, in oxygen gas, O_2, to obtain carbon dioxide gas and water vapor.

9.136 Write the electron-dot formula for the nitrous acid molecule, H—O—N—O (bond lines are drawn only to show how the atoms are bonded, not the multiplicity of the bonds). The experimental values for the nitrogen-oxygen bonds in this molecule are 120 pm and 146 pm. Assign these values to the N—O bonds in the nitrous acid molecule. Explain how you arrived at your answer.

Cumulative-Skills Problems

Key: The problems under this heading combine skills introduced in previous chapters with those given in the current one.

9.137 Phosphorous acid, H_3PO_3, has the structure $(HO)_2PHO$, in which one H atom is bonded to the P atom, and two H atoms are bonded to O atoms. For each bond to an H atom, decide whether it is polar or nonpolar. Assume that only polar-bonded H atoms are acidic. Write the balanced equation for the complete neutralization of phosphorous acid with sodium hydroxide. A 200.0-mL sample of H_3PO_3 requires 22.50 mL of 0.1250 *M* NaOH for complete neutralization. What is the molarity of the H_3PO_3 solution?

9.138 Hypophosphorous acid, H_3PO_2, has the structure $(HO)PH_2O$, in which two H atoms are bonded to the P atom, and one H atom is bonded to an O atom. For each bond to an H atom, decide whether it is polar or nonpolar. Assume that only polar-bonded H atoms are acidic. Write the balanced equation for the complete neutralization of hypophosphorous acid with sodium hydroxide. A 200.0-mL sample of H_3PO_2 requires 22.50 mL of 0.1250 *M* NaOH for complete neutralization. What is the molarity of the H_3PO_2 solution?

9.139 An ionic compound has the following composition (by mass): Ca, 30.3%; N, 21.2%; O, 48.5%. What are the formula and name of the compound? Write the Lewis formulas for the ions.

9.140 An ionic compound has the following composition (by mass): Mg, 10.9%; Cl, 31.8%; O, 57.3%. What are the formula and name of the compound? Write the Lewis formulas for the ions.

9.141 A gaseous compound has the following composition by mass: C, 25.0%; H, 2.1%; F, 39.6%; O, 33.3%. Its molecular mass is 48.0 amu. Write the Lewis formula for the molecule.

9.142 A liquid compound used in dry cleaning contains 14.5% C and 85.5% Cl by mass and has a molecular mass of 166 amu. Write the Lewis formula for the molecule.

9.143 A compound of arsenic and fluorine is a gas. A sample weighing 0.100 g occupies 14.2 mL at 23°C and 765 mmHg. What is the molecular mass of the compound? Write the Lewis formula for the molecule.

9.144 A compound of tin and chlorine is a colorless liquid. The vapor has a density of 7.49 g/L at 151°C and 1.00 atm. What is the molecular mass of the compound? Why do you think the compound is molecular and not ionic? Write the Lewis formula for the molecule.

9.145 Calculate the enthalpy of reaction for

$$HCN(g) \longrightarrow H(g) + C(g) + N(g)$$

from enthalpies of formation (see Appendix C). Given that the C—H bond enthalpy is 411 kJ/mol, obtain a value for the C≡N bond enthalpy. Compare your result with the value given in Table 9.5.

9.146 Assume the values of the C—H and C—C bond enthalpies given in Table 9.5. Then, using data given in Appendix C, calculate the C=O bond enthalpy in acetaldehyde,

$$\begin{array}{ccc} & H & O \\ & | & \| \\ H- & C-C & -H \\ & | & \\ & H & \end{array}$$

Compare your result with the value given in Table 9.5.

9.147 According to Pauling, the A—B bond enthalpy is equal to the average of the A—A and B—B bond enthalpies plus an enthalpy contribution from the polar character of the bond:

$$BE(A-B) = \tfrac{1}{2}[BE(A-A) + BE(B-B)] + k(X_A - X_B)^2]$$

Here X_A and X_B are the electronegativities of atoms A and B, and k is a constant equal to 98.6 kJ. Assume that the electronegativity of H is 2.1. Use the formula to calculate the electronegativity of oxygen.

9.148 Because known compounds with N—I bonds tend to be unstable, there are no thermodynamic data available with which to calculate the N—I bond enthalpy. However, we can estimate a value from Pauling's formula relating electronegativities and bond enthalpies (see Problem 9.147). Using Pauling's electronegativities and the bond enthalpies given in Table 9.5, calculate the N—I bond enthalpy.

9.149 Using Mulliken's formula, calculate a value for the electronegativity of chlorine. Use values of the ionization energy from Figure 8.18 and values of the electron affinity from Table 8.4. Divide this result (in kJ/mol) by 230 to get a value comparable to Pauling's scale.

9.150 Using Mulliken's formula, calculate a value for the electronegativity of oxygen. Convert the result to a value on Pauling's scale. See Problem 9.149.

10

Molecular Geometry and Chemical Bonding Theory

© Michael Rosenfeld/Getty Images; Cengage Learning

Vitamin C

Humans are one of the few animals unable to produce vitamin C (ascorbic acid) naturally and must obtain it through the diet (such as citrus fruits) or from supplements. The space-filling model shows the three-dimensional shape of the ascorbic acid molecule.

OWL Sign in to OWL at **www.cengage.com/owl** to view tutorials and simulations, develop problem-solving skills, and complete online homework assigned by your professor.

GO Chemistry Download mini lecture videos for key concept review and exam prep from OWL or purchase them from **www.cengagebrain.com**.

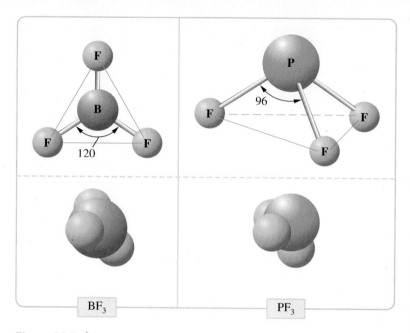

BF₃

PF₃

Figure 10.1 ▲

Molecular models of BF₃ and PF₃ Although both molecules have the general formula AX₃, boron trifluoride is planar (flat), whereas phosphorus trifluoride is pyramidal (pyramid shaped).

W e know from various experiments that molecules have definite structures; that is, the atoms of a molecule have definite positions relative to one another in three-dimensional space. Consider two molecules: boron trifluoride, BF₃, and phosphorus trifluoride, PF₃. Both have the same general formula, AX₃, but their molecular structures are very different (Figure 10.1). Boron trifluoride is a planar, or flat, molecule. The angle between any two B—F bonds is 120°. The geometry or general shape of the BF₃ molecule is said to be *trigonal planar* (trigonal because the figure described by imaginary lines connecting the F atoms has three sides). Phosphorus trifluoride is nonplanar, and the angle between any two P—F bonds is 96°. If you imagine lines connecting the fluorine atoms, these lines and those of the P—F bonds describe a pyramid with the phosphorus atom at its apex. The geometry of the PF₃ molecule is said to be *trigonal pyramidal*. How do you explain such different molecular geometries?

Subtle differences in structure are also possible. Consider the following molecular structures, which differ only in the arrangement of the atoms about the C=C bond; such molecules are referred to as *isomers*.

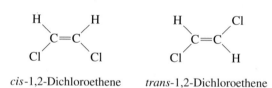

cis-1,2-Dichloroethene   *trans*-1,2-Dichloroethene

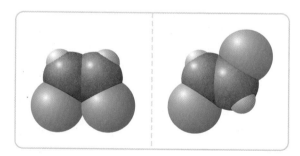

In the *cis* compound, both H atoms are on the same side of the C=C bond; in the *trans* compound, they are on opposite sides. These structural formulas represent entirely different compounds, as their boiling points clearly demonstrate. *Cis*-1,2-dichloroethene boils at 60°C; *trans*-1,2-dichloroethene boils at 48°C. The differences between such *cis* and *trans* compounds can be quite important. The central molecular event in the detection of light by the human eye involves the transformation of a *cis* compound to its corresponding *trans* compound after the absorption of a photon of light. How do you explain the existence of *cis* and *trans* compounds?

In this chapter we discuss how to explain the geometries of molecules in terms of their electronic structures. We also explore two theories of chemical bonding: valence bond theory and molecular orbital theory.

Molecular Geometry and Directional Bonding

Molecular geometry is *the general shape of a molecule, as determined by the relative positions of the atomic nuclei.* There is a simple model that allows you to predict molecular geometries, or shapes, from Lewis formulas. This valence-shell electron-pair model usually predicts the correct general shape of a molecule. It does not explain chemical bonding, however. For this you must look at a theory, such as valence bond theory, that is based on quantum mechanics. Valence bond theory provides further insight into why bonds form and, at the same time, reveals that bonds have definite directions in space, resulting in particular molecular geometries.

10.1 The Valence-Shell Electron-Pair Repulsion (VSEPR) Model

The **valence-shell electron-pair repulsion (VSEPR) model** *predicts the shapes of molecules and ions by assuming that the valence-shell electron pairs are arranged about each atom so that electron pairs are kept as far away from one another as possible, thus minimizing electron-pair repulsions.* ▶ The possible arrangements assumed by different numbers of electron pairs about an atom are shown in Figure 10.2. (If you tie various numbers of similar balloons together, they assume these same arrangements; see Figure 10.3.)

For example, if there are only two electron pairs in the valence shell of an atom, these pairs tend to be at opposite sides of the nucleus, so that repulsion is minimized. This gives a *linear* arrangement of electron pairs; that is, the electron pairs mainly occupy regions of space at an angle of 180° to one another (Figure 10.2).

If three electron pairs are in the valence shell of an atom, they tend to be arranged in a plane directed toward the corners of a triangle of equal sides (equilateral triangle). This arrangement is *trigonal planar,* in which the regions of space occupied by electron pairs are directed at 120° angles to one another (Figure 10.2).

The acronym VSEPR is pronounced "vesper."

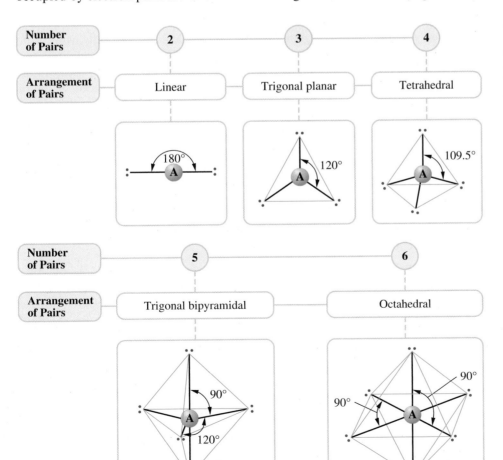

Figure 10.2 ◀

Arrangement of electron pairs about an atom Black lines give the directions of electron pairs about atom A. The blue lines merely help depict the geometric arrangement of electron pairs.

Number of Pairs	2	3	4
Arrangement of Pairs	Linear	Trigonal planar	Tetrahedral

180° 120° 109.5°

Number of Pairs	5	6
Arrangement of Pairs	Trigonal bipyramidal	Octahedral

90° 120° 90° 90°

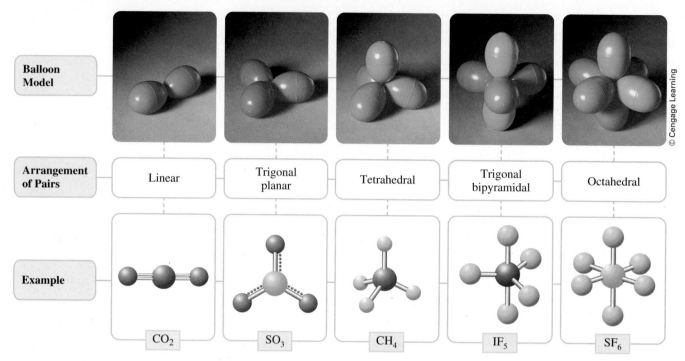

Balloon Model				
Arrangement of Pairs	Linear	Trigonal planar	Tetrahedral	Trigonal bipyramidal	Octahedral
Example	CO_2	SO_3	CH_4	IF_5	SF_6

© Cengage Learning

Figure 10.3 ▲

Balloon models of electron-pair arrangements When you tie similar balloons together, they assume the arrangements shown. These arrangements are the same as those assumed by valence electron pairs about an atom, as in the examples.

Four electron pairs in the valence shell of an atom tend to have a *tetrahedral* arrangement. That is, if you imagine the atom at the center of a regular tetrahedron, each region of space in which an electron pair mainly lies extends toward a corner, or vertex (Figure 10.2). (A regular tetrahedron is a geometrical shape with four faces, each an equilateral triangle. Thus, it has the form of a triangular pyramid.) The regions of space mainly occupied by electron pairs are directed at approximately 109.5° angles to one another.

When you determine the geometry of a molecule experimentally, you locate the positions of the atoms, not the electron pairs. To predict the relative positions of atoms around a given atom using the VSEPR model, you first note the arrangement of valence-shell electron pairs around that central atom. Some of these electron pairs are bonding pairs and some are lone pairs. *The direction in space of the bonding pairs gives you the molecular geometry.* To consider examples of the method, we will first look at molecules AX_n in which the central atom A is surrounded by two, three, or four valence-shell electron pairs. Later we will look at molecules in which A is surrounded by five or six valence-shell electron pairs.

Central Atom with Two, Three, or Four Valence-Shell Electron Pairs

Examples of geometries in which two, three, or four valence-shell electron pairs surround a central atom are shown in Figure 10.4.

Two Electron Pairs (Linear Arrangement) To find the geometry of the molecule AX_n, you first determine the number of valence-shell electron pairs around atom A. You can get this information from the electron-dot formula. For example, following the rules given in Section 9.6, you can find the electron-dot formula for the BeF_2 molecule. ◀

$$\ddot{\overset{..}{F}} : Be : \ddot{\overset{..}{F}} :$$

There are two electron pairs in the valence shell for beryllium, and the VSEPR model predicts that they will have a linear arrangement (see Figure 10.2). Fluorine atoms are bonded in the same direction as the electron pairs. Hence, the geometry of the BeF_2 molecule is *linear*—that is, the atoms are arranged in a straight line (see Figure 10.4).

Beryllium fluoride, BeF_2, is normally a solid. The BeF_2 molecule exists in the vapor phase at high temperature.

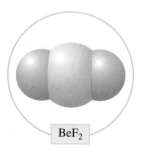

BeF_2

Electron Pairs			Arrangement of Pairs	Molecular Geometry		Example	
Total	Bonding	Lone					
2	2	0	Linear	Linear AX$_2$		BeF$_2$	F — Be — F
3	3	0	Trigonal planar	Trigonal planar AX$_3$		BF$_3$	
	2	1		Bent (or angular) AX$_2$	—Lone pair	SO$_2$	
4	4	0	Tetrahedral	Tetrahedral AX$_4$		CH$_4$	
	3	1		Trigonal pyramidal AX$_3$		NH$_3$	
	2	2		Bent (or angular) AX$_2$		H$_2$O	

Figure 10.4 ▲

Molecular geometries Central atom with two, three, or four valence-shell electron pairs. The notation in the column at the right provides a convenient way to represent the shape of a molecule drawn on a flat surface. A straight line represents a bond or electron pair in the plane of the page. A dashed line represents a bond or electron pair extending behind the page. A wedge represents a bond or electron pair extending above (in front of) the page.

The VSEPR model can also be applied to molecules with multiple bonds. In this case, each multiple bond is treated as though it were a single electron pair. (All pairs of a multiple bond are required to be in approximately the same region.) To predict the geometry of carbon dioxide, for example, you first write the electron-dot formula.

$$:\overset{..}{O}::C::\overset{..}{O}:$$

You have two double bonds, or two electron groups, about the carbon atom, and these are treated as though there were two pairs on carbon. Thus, according to the VSEPR model, the bonds are arranged linearly and the geometry of carbon dioxide is linear.

Three Electron Pairs (Trigonal Planar Arrangement) Let us now predict the geometry of the boron trifluoride molecule, BF$_3$, introduced in the chapter opening. The electron-dot formula is

$$\overset{\displaystyle :\overset{..}{F}:}{:\overset{..}{F}:B:\overset{..}{F}:}$$

The three electron pairs on boron have a trigonal planar arrangement, and the molecular geometry, which is determined by the directions of the bonding pairs, is also *trigonal planar* (Figures 10.1 and 10.4).

BF$_3$

Sulfur dioxide, SO_2, provides an example of a molecule with three electron groups about the S atom; one group is a lone pair. This molecule requires a resonance description, so it gives us a chance to see how the VSEPR model can be applied in such cases. The resonance description can be written

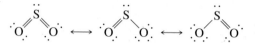

Whichever formula you consider, sulfur has three groups of electrons about it. These groups have a trigonal planar arrangement, so sulfur dioxide is a *bent,* or *angular,* molecule (Figure 10.4). In general, whenever you consider a resonance description, you can use any of the resonance formulas to predict the molecular geometry with the VSEPR model.

Note the difference between the *arrangement of electron pairs* and the *molecular geometry,* or the arrangement of nuclei. What you usually "see" by means of x-ray diffraction and similar methods are the nuclear positions—that is, the molecular geometry. Unseen, but nevertheless important, are the lone pairs, which occupy definite positions about an atom according to the VSEPR model.

Four Electron Pairs (Tetrahedral Arrangement) The common and most important case of four electron pairs about the central atom (the octet rule) leads to three different molecular geometries, depending on the number of bonds formed. Note the following examples.

H	H	H
H:C:H	:N:H	:O:H
H	H	

Molecular geometry: tetrahedral   trigonal pyramidal   bent

CH_4 NH_3 H_2O

In each of these examples, the electron pairs are arranged tetrahedrally, and two or more atoms are bonded in these tetrahedral directions to give the different geometries (Figure 10.4).

When all electron pairs are bonding, as in methane, CH_4, the molecular geometry is *tetrahedral* (Figure 10.5). When three of the pairs are bonding and one pair is nonbonding, as in ammonia, NH_3, the molecular geometry is *trigonal pyramidal.* Note that the nitrogen atom is at the apex of the pyramid, and the three hydrogen atoms extend downward to form the triangular base of the pyramid.

Now you can answer the question raised in the chapter opening: Why is the BF_3 molecule trigonal planar and the PF_3 molecule trigonal pyramidal? The difference occurs because in BF_3 there are three pairs of electrons around boron, and in PF_3 there are four pairs of electrons around phosphorus. As in NH_3, there are three bonding pairs and one lone pair in PF_3, which give rise to the trigonal pyramidal geometry.

Steps in the Prediction of Geometry by the VSEPR Model Let us summarize the steps to follow in order to predict the geometry of an AX_n molecule or ion by the VSEPR method. (All X atoms of AX_n need not be identical.)

Step 1: Write the electron-dot formula from the molecular formula.

Step 2: Determine from the electron-dot formula the number of electron pairs around the central atom, including bonding and nonbonding pairs. (Count a multiple bond as one pair. If resonance occurs, use one resonance formula to determine this number.)

Step 3: Determine the arrangement of these electron pairs about the central atom (see Figure 10.2).

Step 4: Obtain the molecular geometry from the directions of the bonding pairs for this arrangement (see Figure 10.4).

CH_4

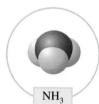

NH_3

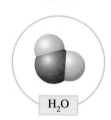

H_2O

Figure 10.5 ▲

The methane molecule A ball-and-stick model shows the tetrahedral arrangement of bonds. This model is within an electrostatic-potential map, showing the electron density about the molecule by gradations of color. Thus, the red color means high density, which occurs in the C—H bonding regions.

CONCEPT CHECK 10.1

An atom in a molecule is surrounded by four pairs of electrons: one lone pair and three bonding pairs. Describe how the four electron pairs are arranged about the atom. How are any three of these pairs arranged in space? What is the geometry about this central atom, taking into account just the bonded atoms?

⚫WL Interactive Example | **10.1** | **Predicting Molecular Geometries (Two, Three, or Four Electron Pairs)**

Gaining Mastery Toolbox

Critical Concept 10.1
The arrangement of electron groups about a central atom depends on the number of electron groups: two electron groups have a linear arrangement, three electron groups a trigonal planar arrangement, and four electron groups a tetrahedral arrangement.

Solution Essentials:
• Molecular geometries for a given arrangement of electrons
• Arrangement of electron groups about an atom
• Electron-dot formulas

BeCl₂

NO₂⁻

SiCl₄

Predict the geometry of the following molecules or ions, using the VSEPR method: **a** $BeCl_2$; **b** NO_2^-; **c** $SiCl_4$.

Problem Strategy After drawing the Lewis formula for a molecule, determine the number of electron groups about the central atom. This number determines the arrangement of the electron groups (Figure 10.2). The geometry is determined by the arrangement of bonds to the central atom (Figure 10.4).

Solution

a Following the steps outlined in Sections 9.6 and 9.8 for writing Lewis formulas, you distribute the valence electrons to the skeleton structure of $BeCl_2$ as follows:

$$: \ddot{Cl} : Be : \ddot{Cl} :$$

The two pairs on Be have a linear arrangement, indicating a **linear** molecular geometry for $BeCl_2$. (See Figure 10.4; 2 electron pairs, 0 lone pairs.)

b The nitrite ion, NO_2^-, has the following resonance description:

$$\left[:\ddot{O}\diagup^{\!\!N}\!\diagdown\ddot{O}: \right]^- \longleftrightarrow \left[:\ddot{O}\diagdown_{\!\!N}\!\diagup\ddot{O}: \right]^-$$

The N atom has three electron groups (one single bond, one double bond, and one lone pair) about it. Therefore the molecular geometry of the NO_2^- ion is **bent.** (See Figure 10.4; 2 bonding pairs, 1 lone pair.)

c The electron-dot formula of $SiCl_4$ is

$$\begin{array}{c} : \ddot{Cl} : \\ : \ddot{Cl} : Si : \ddot{Cl} : \\ : \ddot{Cl} : \end{array}$$

The molecular geometry is **tetrahedral.** (See Figure 10.4; 4 bonding pairs, 0 lone pairs.)

Answer Check Be careful not to confuse the name for the arrangement of valence electron pairs with the name for the molecular geometry. The geometry refers only to the arrangement of bonds. In this example, the nitrite ion, NO_2^-, has a trigonal planar arrangement of valence electron pairs and a bent geometry (arrangement of the bonds).

Exercise 10.1 Use the VSEPR method to predict the geometry of the following ion and molecules: **a** ClO_3^-; **b** OF_2; **c** SiF_4.

■ See Problems 10.33, 10.34, 10.35, and 10.36.

Bond Angles and the Effect of Lone Pairs

The VSEPR model allows you to predict the approximate angles between bonds in molecules. For example, it tells you that CH_4 should have a tetrahedral geometry and that the H—C—H bond angles should be 109.5° (see Figure 10.2). Because all of the

Figure 10.6 ▶

H—A—H bond angles in some molecules
Experimentally determined bond angles are shown for CH_4, CH_3Cl, NH_3, and H_2O, represented here by models.

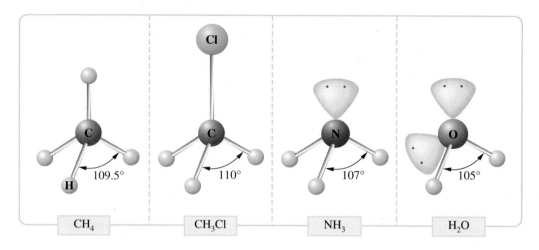

valence-shell electron pairs about the carbon atom are bonding and all of the bonds are alike, you expect the CH_4 molecule to have an exact tetrahedral geometry. However, if one or more of the electron pairs is nonbonding or if there are dissimilar bonds, all four valence-shell electron pairs will not be alike. Then you expect the bond angles to deviate from the 109.5° predicted by an exact tetrahedral geometry. This is indeed what you see. Experimentally determined H—A—H bond angles (A is the central atom) in CH_4, CH_3Cl, NH_3, and H_2O are shown in Figure 10.6.

The increase or decrease of bond angles from the ideal values is often predictable. *A lone pair tends to require more space than a corresponding bonding pair.* You can explain this as follows: A lone pair of electrons is attracted to only one atomic core, whereas a bonding pair is attracted to two. As a result, the lone pair is spatially more diffuse, while the bonding pair is drawn more tightly to the nuclei. Consider the trigonal pyramidal molecule NH_3. The lone pair on the nitrogen atom requires more space than the bonding pairs. Therefore, the N—H bonds are effectively pushed away from the lone pair, and the H—N—H bond angles become smaller than the tetrahedral value of 109.5°. How much smaller, the VSEPR model cannot tell you. The experimental value of an H—N—H bond angle is a few degrees smaller (107°). But the trigonal pyramidal molecule PF_3, mentioned in the chapter opening, has an F—P—F bond angle of 96°, which is significantly smaller than the exact tetrahedral value.

Multiple bonds require more space than single bonds because of the greater number of electrons. You therefore expect the C=O bond in the formaldehyde molecule, CH_2O, to require more space than the C—H bonds. You predict that the H—C—H bond angle will be smaller than the 120° seen in an exact trigonal planar geometry, such as the F—B—F angle in BF_3. Similarly, you expect the H—C—H bond angles in the ethylene molecule, CH_2CH_2, to be smaller than the trigonal planar value. Experimental values are shown in Figure 10.7.

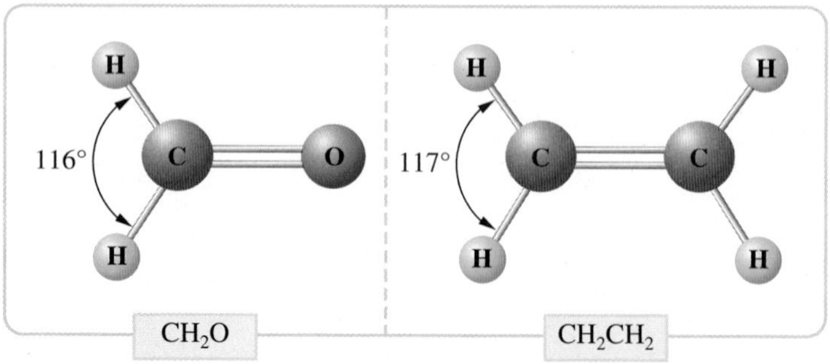

Central Atom with Five or Six Valence-Shell Electron Pairs

Five electron pairs tend to have a *trigonal bipyramidal arrangement.* The electron pairs tend to be directed to the corners of a trigonal bipyramid, a figure formed by placing the face of one tetrahedron onto the face of another tetrahedron (see Figure 10.2).

Six electron pairs tend to have an *octahedral arrangement.* The electron pairs tend to be directed to the corners of a regular octahedron, a figure that has eight triangular faces and six vertexes, or corners (see Figure 10.2).

Figure 10.7 ▲

H—C—H bond angles in molecules with a carbon double bond Bond angles are shown for formaldehyde, CH_2O, and ethylene, CH_2CH_2, represented here by models.

Five Electron Pairs (Trigonal Bipyramidal Arrangement)   Large atoms like phosphorus can accommodate more than eight valence electrons. The phosphorus atom in phosphorus pentachloride, PCl_5, has five electron pairs in its valence shell. With five electron pairs around phosphorus, all bonding, PCl_5 should have a *trigonal bipyramidal geometry* (Figure 10.8). Note, however, that the vertexes of the trigonal bipyramid are not all equivalent (that is, the angles between the electron pairs are not all the same). Thus, the directions to which the electron pairs point are

Figure 10.8 ▼

Molecular geometries
Central atom with five or six valence-shell electron pairs.

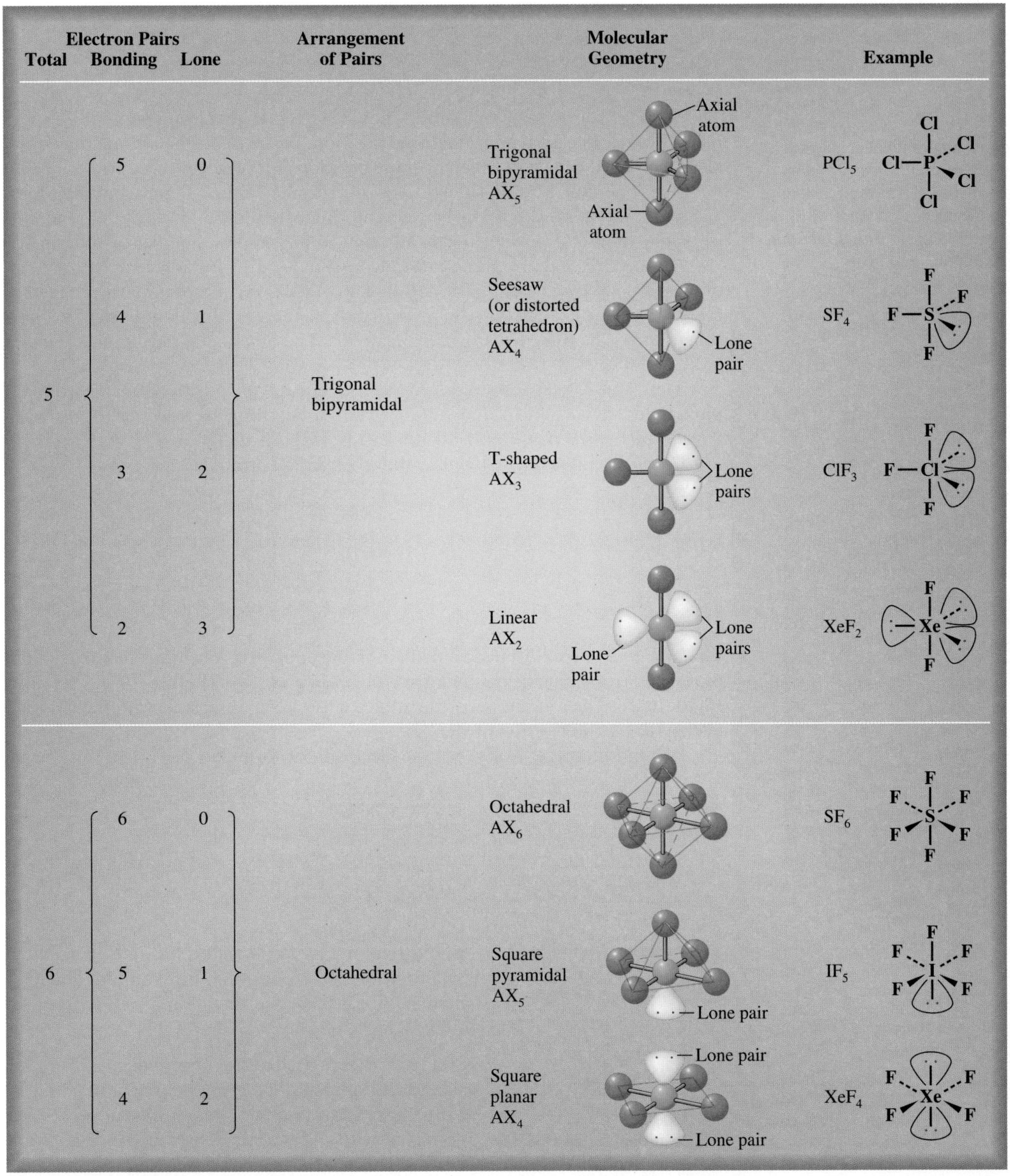

Electron Pairs			Arrangement of Pairs	Molecular Geometry		Example
Total	**Bonding**	**Lone**				
5	5	0	Trigonal bipyramidal	Trigonal bipyramidal AX$_5$		PCl_5
	4	1		Seesaw (or distorted tetrahedron) AX$_4$		SF_4
	3	2		T-shaped AX$_3$		ClF_3
	2	3		Linear AX$_2$		XeF_2
6	6	0	Octahedral	Octahedral AX$_6$		SF_6
	5	1		Square pyramidal AX$_5$		IF_5
	4	2		Square planar AX$_4$		XeF_4

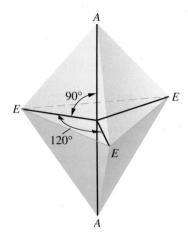

Figure 10.9 ▲

Trigonal bipyramidal arrangement of electron pairs Electron pairs are directed along the black lines to the vertices of a trigonal bipyramid. Equatorial directions are labeled *E*, and axial directions are labeled *A*.

Figure 10.10 ▲

The sulfur tetrafluoride molecule Molecular model.

Figure 10.11 ▲

The chlorine trifluoride molecule Molecular model.

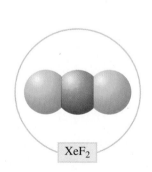

Figure 10.12 ▲

The xenon difluoride molecule Molecular model.

not equivalent. Two of the directions, called *axial directions,* form an axis through the central atom (see Figure 10.9). They are 180° apart. The other three directions are called *equatorial directions.* These point toward the corners of the equilateral triangle that lies on a plane through the central atom, perpendicular (at 90°) to the axial directions. The equatorial directions are 120° from each other.

Molecular geometries other than trigonal bipyramidal are possible when one or more of the five electron pairs are lone pairs. Consider the sulfur tetrafluoride molecule, SF_4. The Lewis electron-dot formula is

The five electron pairs about sulfur should have a trigonal bipyramidal arrangement. Because the axial and equatorial positions of the electron pairs are not equivalent, you must decide in which of these positions the lone pair appears.

The lone pair acts as though it is somewhat "larger" than a bonding pair. Therefore, the total repulsion between all pairs should be lower if the lone pair is in a position that puts it directly adjacent to the smallest number of other pairs.

An electron pair in an equatorial position is directly adjacent to only *two* other pairs—the two axial pairs at 90°. The other two equatorial pairs are farther away at 120°. An electron pair in an axial position is directly adjacent to *three* other pairs—the three equatorial pairs at 90°. The other axial pair is much farther away at 180°.

Thus you expect lone pairs to occupy equatorial positions in the trigonal bipyramidal arrangement. For sulfur tetrafluoride, this gives a *seesaw* (or *distorted tetrahedral*) *geometry* (Figure 10.10; see also Figure 10.8).

Chlorine trifluoride, ClF_3, has the electron-dot formula

The lone pairs on chlorine occupy two of the equatorial positions of the trigonal bipyramidal arrangement, giving a *T-shaped geometry* (Figure 10.11; see also Figure 10.8). The four atoms of the molecule all lie in one plane, with the chlorine nucleus at the intersection of the "T."

Xenon difluoride, XeF_2, has the electron-dot formula

The three lone pairs on xenon occupy the equatorial positions of the trigonal bipyramidal arrangement, giving a *linear geometry* (Figure 10.12; see also Figure 10.8).

Six Electron Pairs (Octahedral Arrangement) There are six electron pairs (all bonding pairs) about sulfur in sulfur hexafluoride, SF_6. Thus, it has an *octahedral geometry,* with sulfur at the center of the octahedron and fluorine atoms at the vertexes (Figure 10.8).

Iodine pentafluoride, IF_5, has the electron-dot formula

The lone pair on iodine occupies one of the six equivalent positions in the octahedral arrangement, giving a *square pyramidal geometry* (Figure 10.13; see also Figure 10.8). The name derives from the shape formed by drawing lines between atoms.

Xenon tetrafluoride, XeF_4, has the electron-dot formula

Figure 10.13 ▲

The iodine pentafluoride molecule Molecular model.

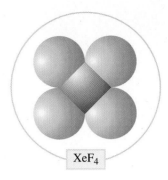

Figure 10.14 ▲

The xenon tetrafluoride molecule Molecular model.

The two lone pairs on xenon occupy opposing positions in the octahedral arrangement to minimize their repulsion. The result is a *square planar geometry* (Figure 10.14; see also Figure 10.8).

⊙WL Interactive Example | 10.2 | **Predicting Molecular Geometries (Five or Six Electron Pairs)**

Gaining Mastery Toolbox

Critical Concept 10.2
The arrangement of electron groups about a central atom depends on the number of electron groups: five electron groups have a trigonal bipyramidal arrangement and six electron groups an octahedral arrangement.

Solution Essentials:
• Molecular geometries for a given arrangement of electrons
• Arrangement of electron groups about an atom
• Electron-dot formulas

What do you expect for the geometry of tellurium tetrachloride, $TeCl_4$?

Problem Strategy After drawing the Lewis formula for a molecule, determine the number of electron groups about the central atom. This number determines the arrangement of the electron groups (Figure 10.2). The geometry is determined by the arrangement of bonds to the central atom (Figure 10.8).

Solution First you distribute the valence electrons to the Cl atoms to satisfy the octet rule. Then you allocate the remaining valence electrons to the central atom,

Te (following the steps outlined in Section 9.6). The electron-dot formula is

$$\ddot{\underset{\ddot{}}{Cl}} \; \overset{\ddot{}}{\underset{\ddot{}}{Cl}} \\ \qquad Te \\ \ddot{\underset{\ddot{}}{Cl}} \; \overset{}{\underset{\ddot{}}{Cl}}$$

There are five electron pairs in the valence shell of Te in $TeCl_4$. Of these, four are bonding pairs and one is a lone pair. The arrangement of electron pairs is trigonal bipyramidal. You expect the lone pair to occupy an equatorial position, so $TeCl_4$ has a **seesaw** molecular geometry. (See Figure 10.8; 4 bonding pairs, 1 lone pair.)

Answer Check Again, be careful to distinguish the name for the arrangement of valence electron pairs from the name for the molecular geometry.

Exercise 10.2 According to the VSEPR model, what molecular geometry would you predict for iodine trichloride, ICl_3?

■ See Problems 10.39, 10.40, 10.41, and 10.42.

Applying the VSEPR Model to Larger Molecules

In the preceding discussion, we looked at the geometry of molecules having a single central atom with other atoms or groups attached to that central atom. We can also apply what we have learned there to more complicated molecules. What we need to

A CHEMIST Looks at...

Daily Life

Left-Handed and Right-Handed Molecules

This morning, half asleep, I (D. E.) tried to put my left glove on my right hand. Silly mistake, of course. A person's two hands are similar but not identical. If I look at my hands, say, with both palms toward me, the thumbs point in opposite directions. But, if I now hold the palm of my right hand toward a mirror and compare that image with my actual left hand, I see that the thumbs point in the same direction. A person's left hand looks like the mirror image of his or her right hand, and vice versa. You can mentally superimpose one hand onto the mirror image of the other hand. Yet the two hands are themselves nonsuperimposable. Many molecules have this "handedness" property too, and this possibility gives rise to a subtle kind of isomerism.

The presence of handedness in molecules depends on the fact that the atoms in the molecules occupy specific places in three-dimensional space. In most organic molecules, this property depends on the tetrahedral bonding of the carbon atom. Consider a carbon atom bonded to four different kinds of groups, say, —H, —OH, —CH₃, and —COOH. (The resulting molecule is called lactic acid.) Two isomers are possible in which one isomer is the mirror image of the other, but the two isomers are themselves not identical (they are nonsuperimposable; see Figure 10.15). However, all four groups bonded to this carbon atom must be different to have this isomerism. If you replace the —OH group of lactic acid by, say, an H atom, the molecule and its mirror image become identical. There are no isomers; there is just one kind of molecule.

The two lactic acid isomers shown in Figure 10.15 are labeled D-lactic acid and L-lactic acid (D is for *dextro*, meaning right; L is for *levo*, meaning left). They might be expected to have quite similar properties, and they do. They have identical melting points, boiling points, and solubilities, for instance. And if we try to prepare lactic acid in the laboratory, we almost always get an equal molar mixture of the D and L isomers. But the biological origins of the molecules are different. Only D-lactic acid occurs in sour milk, whereas only L-lactic acid forms in muscle tissue after exercising.

Biological molecules frequently have the handedness character. The chemical substance responsible for the flavor of spearmint is L-carvone. The substance responsible for the flavor of caraway seeds, which you see in some rye bread, is D-carvone (see Figure 10.16). Here are two substances whose molecules differ only in being mirror images of one another. Yet, they have strikingly different flavors: one is minty; the other is pungent aromatic.

do is to imagine a large molecule broken into pieces to which we can apply the VSEPR model. As an example, let's consider the lactic acid molecule.

Lactic acid has the following electron-dot formula:

$$\begin{array}{c} \text{H} \quad\; \text{H} \quad\; :\text{O}: \\ | \qquad | \qquad || \\ \text{H}-\text{C}-\text{C}-\text{C}-\overset{..}{\underset{..}{\text{O}}}-\text{H} \\ | \qquad | \\ \text{H} \quad\; :\text{O}: \\ | \\ \text{H} \end{array}$$

We have divided this formula into several "central atoms," each surrounded by a number of bonding or lone pairs of electrons. Note that the carbon atom at the far left is surrounded by bonds to three hydrogen atoms plus a bond to another carbon atom. These four bonds would be expected to have a tetrahedral arrangement about this carbon atom. Similarly, the next carbon atom is surrounded by four bonds and would also be expected to have a tetrahedral arrangement of bonds about it. The carbon atom on the right, though, has a single bond to the carbon atom on its left and a single bond to the oxygen atom on its right plus a double bond to the oxygen atom above it. We would expect these three groups of electrons to have a trigonal

Figure 10.15 ▶

Isomers of lactic acid Note that the two molecules labeled D-lactic acid and L-lactic acid are mirror images and cannot be superimposed on one another. See Figure 10.4 for an explanation of the notation used to represent the tetrahedral geometry.

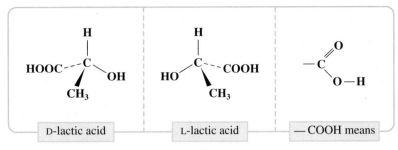

D-lactic acid   L-lactic acid   —COOH means

D-Carvone

caraway seeds

spearmint

L-Carvone

© Cengage Learning

Figure 10.16 ▲

Spearmint and caraway
Spearmint leaves contain L-carvone; caraway seeds contain D-carvone.

■ See Problems 10.83 and 10.84.

planar arrangement about that carbon atom. Then, there are two —O—H groups, with each oxygen atom having a tetrahedral arrangement of electron pairs and a bent or angular geometry of bonds. Here is a molecular model of lactic acid.

You might find it instructive to use a molecular model kit to build this structure. The essay on the preceding page describes a subtle type of isomerism that arises in lactic acid.

10.2 Dipole Moment and Molecular Geometry

The VSEPR model provides a simple procedure for predicting the geometry of a molecule. However, predictions must be verified by experiment. Information about the geometry of a molecule can sometimes be obtained from an experimental quantity called the dipole moment, which is related to the polarity of the bonds in a molecule.

Figure 10.17 ▶

Alignment of polar molecules by an electric field

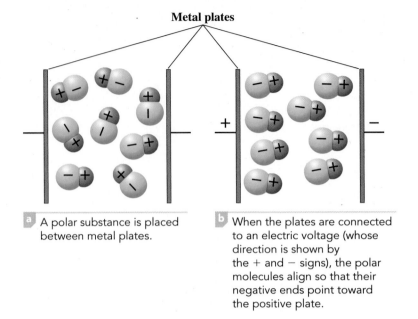

Metal plates

a A polar substance is placed between metal plates.

b When the plates are connected to an electric voltage (whose direction is shown by the + and − signs), the polar molecules align so that their negative ends point toward the positive plate.

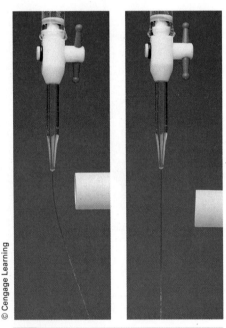

© Cengage Learning

left Water is a polar liquid; it is attracted to the electrically charged rod. The capacitance of the polar liquid (due to the alignment of polar molecules) stabilizes a charge separation induced in the column of liquid surface near the rod, resulting in a net attraction.

right Carbon tetrachloride, CCl_4, is a nonpolar liquid; it is not attracted to the glass rod.

Figure 10.18 ▲

Attraction of a polar liquid to an electrified rod

The **dipole moment** is *a quantitative measure of the degree of charge separation in a molecule and is therefore an indicator of the polarity of the molecule.* In general, a positive charge q and a negative charge $-q$, separated by a distance d has a dipole moment, μ, equal to

$$\mu = q \times d$$

Consider the bond in the H—Cl molecule. Because the Cl atom is more electronegative than the H atom, the electrons of the bond are pulled toward it, giving this atom a partial negative charge, which we indicate as δ^-. The H atom then has an equal but positive charge, δ^+:

$$\overset{\delta^+ \quad \delta^-}{\text{H—Cl}}$$

So, the dipole moment in HCl is δd. Dipole moments are usually measured in units of *debyes* (D); the dipole moment of HCl is 1.08 D. In SI units, dipole moments are measured in coulomb-meters (C·m), and $1\,D = 3.34 \times 10^{-30}\,C\cdot m$.

Measurements of dipole moments are based on the fact that polar molecules (molecules having a dipole moment) can be oriented by an electric field. Figure 10.17 shows an electric field generated by charged plates. Note that the polar molecules tend to align themselves so that the negative ends of the molecules point toward the positive plate, and the positive ends point toward the negative plate. This orientation of the molecules affects the *capacitance* of the charged plates (the capacity of the plates to hold a charge). Consequently, measurements of the capacitance of plates separated by different substances can be used to obtain the dipole moments of those substances. Figure 10.18 shows a simple demonstration that distinguishes between a polar liquid (one whose molecules have a dipole moment) and a nonpolar liquid. The polar liquid, but not the nonpolar one, is attracted to a charged rod.

You can sometimes relate the presence or absence of a dipole moment in a molecule to its molecular geometry. For example, consider the carbon dioxide molecule. Each carbon–oxygen bond has a polarity in which the more electronegative oxygen atom has a partial negative charge.

$$\overset{\delta^- \quad 2\delta^+ \quad \delta^-}{\text{O═C═O}}$$

We will denote the dipole-moment contribution from each bond (the bond dipole) by an arrow with a positive sign at one end (↦). The dipole-moment arrow points from the positive partial charge toward the negative partial charge. Thus, you can rewrite the formula for carbon dioxide as

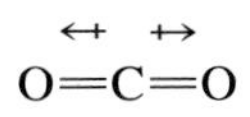

$$\overset{\leftharpoondown}{O}=C=\overset{\rightharpoonup}{O}$$

Each bond dipole, like a force, is a *vector* quantity; that is, it has both magnitude and direction. Like forces, two bond dipoles of equal magnitude but opposite direction cancel each other. (Think of two groups of people in a tug of war. As long as each group pulls on the rope with the same force but in the opposite direction, there is no movement—the net force is zero.) Because the two carbon–oxygen bond dipoles in CO_2 are equal but point in opposite directions, they give a net dipole moment of zero for the molecule.

For comparison, consider the water molecule. The bond dipoles point from the hydrogen atoms toward the more electronegative oxygen.

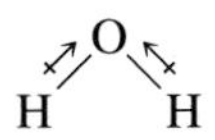

Here, the two bond dipoles do not point directly toward or away from each other. As a result, they add together to give a nonzero dipole moment for the molecule.

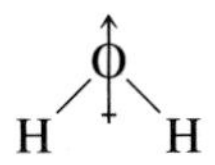

The dipole moment of H_2O has been observed to be 1.94 D.

The fact that the water molecule has a dipole moment is excellent experimental evidence for a bent geometry. If the H_2O molecule were linear, the dipole moment would be zero.

The analysis we have just made for two different geometries of AX_2 molecules can be extended to other AX_n molecules (in which all X atoms are identical). Table 10.1 summarizes the relationship between molecular geometry and dipole moment. Those geometries in which A—X bonds are directed symmetrically about the central atom (for example, linear, trigonal planar, and tetrahedral) give molecules of zero dipole moment; that is, the molecules are *nonpolar*. Those geometries in which the X atoms tend to be on one side of the molecule (for example, bent and trigonal pyramidal) can have nonzero dipole moments; that is, they can give *polar* molecules.

Table 10.1	Relationship Between Molecular Geometry and Dipole Moment	
Formula	**Molecular Geometry**	**Dipole Moment***
AX	Linear	Can be nonzero
AX_2	Linear	Zero
	Bent	Can be nonzero
AX_3	Trigonal planar	Zero
	Trigonal pyramidal	Can be nonzero
	T-shaped	Can be nonzero
AX_4	Tetrahedral	Zero
	Square planar	Zero
	Seesaw	Can be nonzero
AX_5	Trigonal bipyramidal	Zero
	Square pyramidal	Can be nonzero
AX_6	Octahedral	Zero

*All X atoms are assumed to be identical.

Example 10.3 Relating Dipole Moment and Molecular Geometry

Each of the following molecules has a nonzero dipole moment. Select the molecular geometry that is consistent with this information. Explain your reasoning.

a SO_2 linear, bent

b PH_3 trigonal planar, trigonal pyramidal

Problem Strategy Because the dipole moment is a *vector* quantity, it can happen that nonzero bond dipoles cancel one another to give a molecule with zero dipole moment. Look for bonds that are symmetrically arranged to oppose one another (think of a tug of war between bonds). Table 10.1 summarizes the relationship between geometry and dipole moment.

Solution **a** In the linear geometry, the S—O bond contributions to the dipole moment would cancel, giving a zero dipole moment. That would not happen in the **bent** geometry; hence, this must be the geometry for the SO_2 molecule. **b** In the trigonal planar geometry, the bond contributions to the dipole moment would cancel, giving a zero dipole moment. That would not occur in the **trigonal pyramidal** geometry; hence, this is a possible molecular geometry for PH_3.

Answer Check Make sure that you have the correct geometry for the molecule, and note how the bonds either oppose one another, to give zero dipole moment, or else do not oppose and so give a dipole moment.

Exercise 10.3 Bromine trifluoride, BrF_3, has a nonzero dipole moment. Indicate which of the following geometries are consistent with this information: **a** trigonal planar; **b** trigonal pyramidal; **c** T-shaped.

■ See Problems 10.43 and 10.44.

Exercise 10.4 Which of the following would be expected to have a dipole moment of zero on the basis of symmetry? Explain. **a** $SOCl_2$; **b** SiF_4; **c** OF_2.

■ See Problems 10.45 and 10.46.

In Example 10.3, the fact that certain molecular geometries necessarily imply a zero dipole moment allowed us to eliminate these geometries when considering a molecule with a nonzero dipole moment. The reverse argument does not follow, however. For example, suppose a molecule of the type AX_3 is found to have no measurable dipole moment. It may be that the geometry is trigonal planar. Another possibility, however, is that the geometry is trigonal pyramidal or T-shaped, but the lone pairs on the central atom offset the polarity of the bonds.

You can see this effect of lone pairs on the dipole moment of nitrogen trifluoride, NF_3. Judging by the electronegativity difference, you would expect each N—F bond to be quite polar. Yet the dipole moment of nitrogen trifluoride is only 0.2 D. (By contrast, ammonia has a dipole moment of 1.47 D.) The explanation for the small dipole moment in NF_3 appears in Figure 10.19. The lone pair on nitrogen has a dipole-moment contribution that is directed outward from the nucleus, because the electrons are offset from the nuclear center. This dipole-moment contribution thus opposes the N—F bond moments. ◀

Can you explain why NH_3 has such a large dipole moment compared with NF_3?

Figure 10.19 ▶

Explanation of the small dipole moment of NF_3

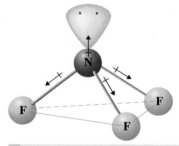

a The lone-pair and bond contributions to the dipole moment of NF_3 tend to offset one another.

b The electrostatic-potential model of NF_3 shows the electron density, which increases as the color changes from blue to yellow to red.

Effect of Polarity on Molecular Properties

Whether a molecule has a dipole moment (that is, whether it is polar or nonpolar) can affect its properties. We will explore this issue in detail in the next chapter (Section 11.5). Briefly, though, we can see why *cis*- and *trans*-1,2-dichloroethene, mentioned in the chapter opening, have different boiling points. The C—Cl bonds are quite polar, because of the relatively large electronegativity difference between C and Cl atoms (0.9); the H—C bonds have moderate polarity (electronegativity difference of 0.4). The bond dipoles (light arrows) add together in *cis*-1,2-dichloroethene to give a polar molecule (the total dipole moment is indicated by the heavy arrow). (Note that the electronegative Cl atoms are on the same side in the *cis* molecule.) In the *trans* molecule, the bond dipoles subtract to give a nonpolar molecule.

cis-1,2-Dichloroethene   *trans*-1,2-Dichloroethene

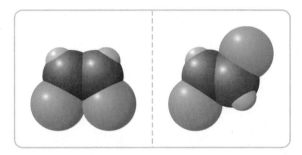

In a liquid, polar molecules tend to orient so that positive ends of molecules are attracted to negative ends. This results in an attractive force between molecules (an *intermolecular* force), which increases the energy required for the liquid to boil, and therefore increases the boiling point. Recall that the *trans* compound boils at 48°C, whereas the *cis* compound boils at a higher temperature, 60°C.

CONCEPT CHECK 10.2

Two molecules, each with the general formula AX_3, have different dipole moments. Molecule Y has a dipole moment of zero, whereas molecule Z has a nonzero dipole moment. From this information, what can you say about the geometries of Y and Z?

10.3 Valence Bond Theory

The VSEPR model is usually a satisfactory method for predicting molecular geometries. To understand bonding and electronic structure, however, you must look to quantum mechanics. We will consider two theories stemming from quantum mechanics: valence bond theory and molecular orbital theory. Both use the methods of quantum mechanics but make different simplifying assumptions. In this section, we will look in a qualitative way at the basic ideas involved in **valence bond theory,** *an approximate theory to explain the electron pair or covalent bond by quantum mechanics.*

Basic Theory

According to valence bond theory, a bond forms between two atoms when the following conditions are met:

1. An orbital on one atom comes to occupy a portion of the same region of space as an orbital on the other atom. The two orbitals are said to *overlap*.

2. The total number of electrons in both orbitals is no more than two.

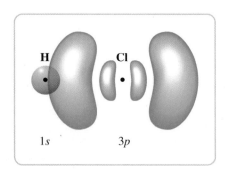

Figure 10.20 ▲

Formation of H₂ The H—H bond forms when the 1s orbitals, one from each atom, overlap.

Figure 10.21 ▲

Bonding in HCl The bond forms by the overlap of a hydrogen 1s orbital (blue) along the axis of a chlorine 3p orbital (green).

As the orbital of one atom overlaps the orbital of another, the electrons in the orbitals begin to move about both atoms. Because the electrons are attracted to both nuclei at once, they pull the atoms together. Strength of bonding depends on the amount of overlap; the greater the overlap, the greater the bond strength. The two orbitals cannot contain more than two electrons because a given region of space can hold only two electrons with opposite spin.

For example, consider the formation of the H_2 molecule from atoms. Each atom has the electron configuration $1s^1$. As the H atoms approach each other, their 1s orbitals begin to overlap and a covalent bond forms (Figure 10.20). Valence bond theory also explains why two He atoms (each with electron configuration $1s^2$) do not bond. Suppose two He atoms approach one another and their 1s orbitals begin to overlap. Each orbital is doubly occupied, so the sharing of electrons between atoms would place the four valence electrons from the two atoms in the same region. This, of course, could not happen. As the orbitals begin to overlap, each electron pair strongly repels the other. The atoms come together, then fly apart.

Because the strength of bonding depends on orbital overlap, orbitals other than s orbitals bond only in given directions. *Orbitals bond in the directions in which they protrude or point, to obtain maximum overlap.* Consider the bonding between a hydrogen atom and a chlorine atom to give the HCl molecule. A chlorine atom has the electron configuration $[Ne]3s^2 3p^5$. Of the orbitals in the valence shell of the chlorine atom, three are doubly occupied by electrons and one (a 3p orbital) is singly occupied. The bonding of the hydrogen atom has to occur with the singly occupied 3p orbital of chlorine. For the strongest bonding to occur, the maximum overlap of orbitals is required. The 1s orbital of hydrogen must overlap along the axis of the singly occupied 3p orbital of chlorine (see Figure 10.21).

Hybrid Orbitals

From what has been said, you might expect the number of bonds formed by a given atom to equal the number of unpaired electrons in its valence shell. Chlorine, whose orbital diagram is

has one unpaired electron and forms one bond. Oxygen, whose orbital diagram is

has two unpaired electrons and forms two bonds, as in H_2O.

However, consider the carbon atom, whose orbital diagram is

You might expect this atom to bond to two hydrogen atoms to form the CH_2 molecule. Although this molecule is known to be present momentarily during some reactions, it is very reactive and cannot be isolated. But methane, CH_4, in which the carbon atom bonds to four hydrogen atoms, is well known. In fact, a carbon atom usually forms four bonds.

You might explain this as follows: Four unpaired electrons are formed when an electron from the 2s orbital of the carbon atom is *promoted* (excited) to the vacant 2p orbital.

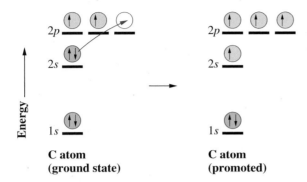

It would require energy to promote the carbon atom this way, but more than enough energy would be obtained from the formation of two additional covalent bonds. One bond would form from the overlap of the carbon 2s orbital with a hydrogen 1s orbital. Each of the other three bonds would form from a carbon 2p orbital and a hydrogen 1s orbital.

Experiment shows, however, that the four C—H bonds in methane are identical. ▶ This implies that the carbon orbitals involved in bonding are also equivalent. For this reason, valence bond theory assumes that the four valence orbitals of the carbon atom combine during the bonding process to form four new, but equivalent, hybrid orbitals. **Hybrid orbitals** are *orbitals used to describe bonding that are obtained by taking combinations of atomic orbitals of the isolated atoms.* In this case, a set of hybrid orbitals is constructed from one *s* orbital and three *p* orbitals, so they are called sp^3 hybrid orbitals. Calculations from theory show that each sp^3 hybrid orbital has a large lobe pointing in one direction and a small lobe pointing in the opposite direction. The four sp^3 hybrid orbitals point in tetrahedral directions. Figure 10.22a shows the shape of a single sp^3 orbital; Figure 10.22b shows a stylized set of four orbitals pointing in tetrahedral directions.

Nuclear magnetic resonance (page 302) and infrared spectroscopy (page 370) both show that CH_4 has four equivalent C—H bonds.

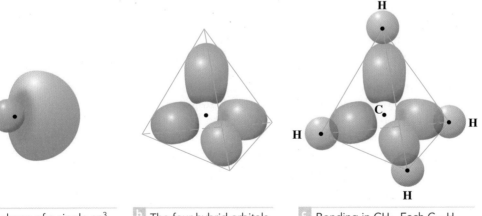

Figure 10.22 ◀

Spatial arrangement of sp^3 hybrid orbitals

a The shape of a single sp^3 hybrid orbital. Each orbital consists of two lobes. One lobe is small, but dense, and concentrated near the nucleus. The other lobe is large, but diffuse. Bonding occurs with the large lobe, since it extends farther from the nucleus.

b The four hybrid orbitals are arranged tetrahedrally in space. (Small lobes are omitted here for clarity, and large lobes are stylized and greatly narrowed for ease in depicting the directional bonding)

c Bonding in CH_4. Each C—H bond is formed by the overlap of a 1s orbital from a hydrogen atom and an sp^3 hybrid orbital of the carbon atom.

The C—H bonds in methane, CH_4, are described by valence bond theory as the overlapping of each sp^3 hybrid orbital of the carbon atom with $1s$ orbitals of hydrogen atoms (see Figure 10.22C). Thus, the bonds are arranged tetrahedrally, which is predicted by the VSEPR model. You can represent the hybridization of carbon and the bonding of hydrogen to the carbon atom in methane as follows:

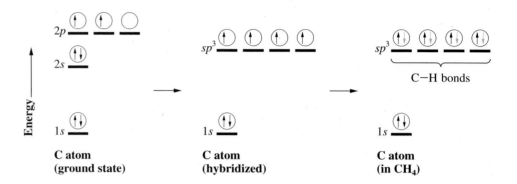

Here the blue arrows represent electrons originally belonging to hydrogen atoms.

Hybrid orbitals can be formed from various numbers of atomic orbitals. *The number of hybrid orbitals formed always equals the number of atomic orbitals used.* For example, if you combine an s orbital and two p orbitals to get a set of equivalent orbitals, you get three hybrid orbitals (called sp^2 hybrid orbitals). A set of hybrid orbitals always has definite directional characteristics. Here, all three sp^2 hybrid orbitals lie in a plane and are directed at $120°$ angles to one another; that is, they have a trigonal planar arrangement. Some possible hybrid orbitals and their geometric arrangements are listed in Table 10.2 and shown in Figure 10.23. Note that the geometric arrangements of hybrid orbitals are the same as those for electron pairs in the VSEPR model.

Only two of the three p orbitals are used to form sp^2 hybrid orbitals. The unhybridized p orbital is perpendicular to the plane of the sp^2 hybrid orbitals. Similarly, only one of the three p orbitals is used to form sp hybrid orbitals. The two unhybridized p orbitals are perpendicular to the axis of the sp hybrid orbitals and perpendicular to each other. We will use these facts when we discuss multiple bonding in the next section.

Now that you know something about hybrid orbitals, let us develop a general scheme for describing the bonding about any atom (we will call this the central atom). First, notice from Table 10.2 that there is a relationship between the type of hybrid orbitals on an atom and the geometric arrangement of those orbitals. If you know one, you can infer the other. Thus, if you know the geometric arrangement, you know what hybrid orbitals to use in the bond description of the central atom. Your first task in describing the bonding, then, is to obtain the geometric arrangement about the central atom. In lieu of experimental information about the geometry, you can use the VSEPR model, since it proves to be a fairly good

Table 10.2	Kinds of Hybrid Orbitals		
Hybrid Orbitals	**Geometric Arrangement**	**Number of Orbitals**	**Example**
sp	Linear	2	Be in BeF_2
sp^2	Trigonal planar	3	B in BF_3
sp^3	Tetrahedral	4	C in CH_4
sp^3d	Trigonal bipyramidal	5	P in PCl_5
sp^3d^2	Octahedral	6	S in SF_6

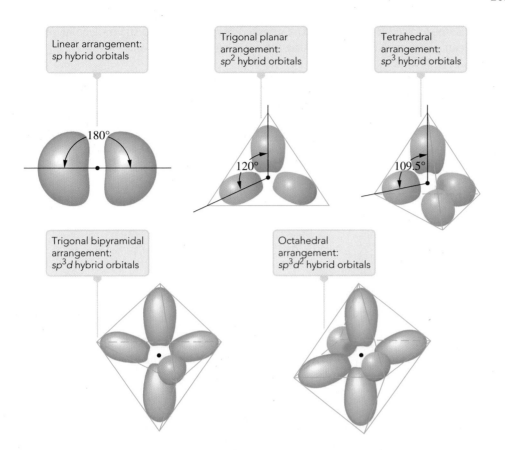

Linear arrangement:
sp hybrid orbitals

180°

Trigonal planar
arrangement:
sp² hybrid orbitals

120°

Tetrahedral
arrangement:
sp³ hybrid orbitals

109.5°

Trigonal bipyramidal
arrangement:
sp³d hybrid orbitals

Octahedral
arrangement:
sp³d² hybrid orbitals

Figure 10.23 ◀

Diagrams of hybrid orbitals showing their spatial arrangements
Each lobe shown is one hybrid orbital (small lobes are omitted for clarity).

predictor of molecular geometry. To obtain the bonding description about any atom in a molecule, you proceed as follows:

Step 1: Write the Lewis electron-dot formula of the molecule.

Step 2: From the Lewis formula, use the VSEPR model to obtain the arrangement of electron pairs about this atom.

Step 3: From the geometric arrangement of the electron pairs, deduce the type of hybrid orbitals on this atom required for the bonding description (see Table 10.2).

Step 4: Assign valence electrons to the hybrid orbitals of this atom one at a time, pairing them only when necessary.

Step 5: Form bonds to this atom by overlapping singly occupied orbitals of other atoms with the singly occupied hybrid orbitals of this atom.

As an application of this scheme, let us look at the BF_3 molecule and obtain the bond description of the boron atom. Following Step 1, you write the Lewis formula of BF_3:

$$\ddot{:}\underset{\displaystyle ..}{\overset{\displaystyle ..}{F}}\ddot{:}$$
$$|$$
$$B - \ddot{F}\ddot{:}$$
$$|$$
$$\ddot{:}\underset{\displaystyle ..}{F}\ddot{:}$$

Now you apply the VSEPR model to the boron atom (Step 2). There are three electron pairs about the boron atom, so they are expected to have a planar trigonal arrangement. Looking at Table 10.2 (Step 3), you note that three sp^2 hybrid orbitals have a trigonal planar arrangement. In Step 4, you assign the valence electrons of the boron atom to the hybrid orbitals. Finally, in Step 5, you imagine three fluorine atoms approaching the boron atom. The singly occupied $2p$ orbital on a fluorine atom overlaps one of the sp^2 hybrid orbitals on boron, forming a covalent bond. Three such B—F bonds form. Note that one of the $2p$ orbitals of boron remains unhybridized and is unoccupied by electrons. It is oriented perpendicular to the molecular plane. You can summarize these steps using orbital diagrams as follows:

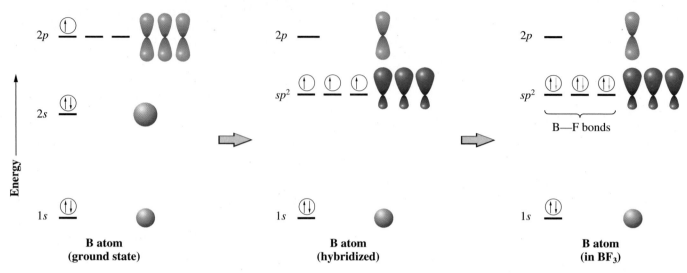

The next example further illustrates how to describe the bonding about an atom in a molecule. To solve problems of this type, you need to know the geometric arrangements of the hybrid orbitals listed in Table 10.2.

Example 10.4

Applying Valence Bond Theory (Two, Three, or Four Electron Pairs)

Gaining Mastery Toolbox

Critical Concept 10.4
Using the VSEPR model, determine the geometry about the central atom and then the hybrid orbitals of the atom. Now form bonds by overlapping these hybrid orbitals with orbitals from bonding atoms; use nonbonded hybrid orbitals for the lone pairs.

Solution Essentials:
- Bond formation by orbital overlap
- Kinds of hybrid orbitals
- VSEPR model
- Electron-dot formulas

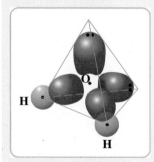

Figure 10.24 ▲

Bonding in H$_2$O Orbitals on oxygen are sp^3 hybridized; bonding is tetrahedral.

Describe the bonding in H$_2$O according to valence bond theory. Assume that the molecular geometry is the same as given by the VSEPR model.

Problem Strategy From the Lewis formula for a molecule, determine its geometry about the central atom using the VSEPR model. From this geometry, determine the hybrid orbitals on this atom, assigning its valence electrons to these orbitals one at a time. Finally, form bonds by overlapping these hybrid orbitals with orbitals of the other atoms.

Solution The Lewis formula for H$_2$O is

$$H - \ddot{O}:$$
$$\quad\;\; |$$
$$\quad\;\; H$$

Note that there are four pairs of electrons about the oxygen atom. According to the VSEPR model, these are directed tetrahedrally, and from Table 10.2 you see that you should use sp^3 hybrid orbitals. Each O—H bond is formed by the overlap of a $1s$ orbital of a hydrogen atom with one of the singly occupied sp^3 hybrid orbitals of the oxygen atom. You can represent the bonding to the oxygen atom in H$_2$O as follows (see also Figure 10.24):

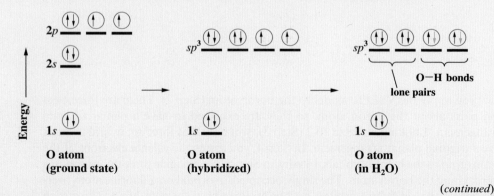

(*continued*)

(continued)

Answer Check Check that you have the correct Lewis formula for the molecule and that you have the correct geometry about the central atom and the correct hybrid orbitals (Table 10.2).

Exercise 10.5 Using hybrid orbitals, describe the bonding in NH_3 according to valence bond theory.

■ See Problems 10.47, 10.48, 10.49, and 10.50.

The actual angle between O—H bonds in H_2O has been experimentally determined to be 104.5°. Because this is close to the tetrahedral angle (109.5°), the description of bonding given in Example 10.4 is essentially correct. To be more precise, you should note that the hybrid orbitals for bonds and lone pairs are not exactly equivalent. The lone pairs are somewhat larger than bonding pairs. Because they take up more space, the lone pairs push the bonding pairs closer together than they are in the exact tetrahedral case.

As the next example illustrates, hybrid orbitals are also useful for describing bonding when the central atom of a molecule is surrounded by more than eight valence electrons.

Example 10.5   **Applying Valence Bond Theory (Five or Six Electron Pairs)**

Describe the bonding in XeF_4 using hybrid orbitals.

Gaining Mastery Toolbox

Critical Concept 10.5
Using the VSEPR model, determine the geometry about the central atom and then the hybrid orbitals of the atom. Now form bonds by overlapping these hybrid orbitals with orbitals from bonding atoms; use nonbonded hybrid orbitals for the lone pairs.

Solution Essentials:
• Bond formation by orbital overlap
• Kinds of hybrid orbitals
• VSEPR model
• Electron-dot formula

Problem Strategy From the Lewis formula for a molecule, determine its geometry about the central atom using the VSEPR model. From this geometry, determine the hybrid orbitals on this atom, assigning its valence electrons to these orbitals one at a time. Finally, form bonds by overlapping these hybrid orbitals with orbitals of the other atoms.

Solution The Lewis formula of XeF_4 is

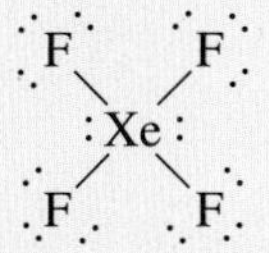

The xenon atom has four single bonds and two lone pairs. It will require six orbitals to describe the bonding. This suggests that you use sp^3d^2 hybrid orbitals on xenon (according to Table 10.2). Each fluorine atom (valence-shell configuration $2s^2 2p^5$) has one singly occupied orbital, so you assume that this orbital is used in bonding. Each Xe—F bond is formed by the overlap of a xenon sp^3d^2 hybrid orbital with a singly occupied fluorine $2p$ orbital. You can summarize this as follows:

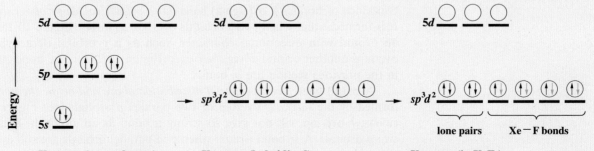

(continued)

(continued)

Answer Check Check that you have the correct Lewis formula for the molecule and that you have the correct geometry about the central atom and the correct hybrid orbitals (Table 10.2).

Exercise 10.6 Describe the bonding in PCl_5 using hybrid orbitals.

■ See Problems 10.53, 10.54, 10.55, and 10.56.

10.4 Description of Multiple Bonding

In the previous section, we described bonding as the overlap of *one* orbital from each of the bonding atoms. Now we consider the possibility that *more than one* orbital from each bonding atom might overlap, resulting in a multiple bond.

As an example, consider the ethylene molecule.

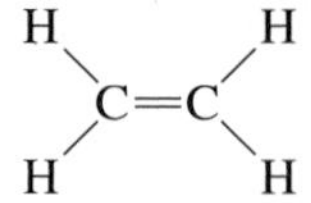

$$
\begin{array}{cc}
H & H \\
\diagdown & \diagup \\
C{=}C \\
\diagup & \diagdown \\
H & H
\end{array}
$$

One hybrid orbital is needed for each bond (whether a single or a multiple bond) and for each lone pair. Because each carbon atom is bonded to three other atoms and there are no lone pairs, three hybrid orbitals are needed. This suggests the use of sp^2 hybrid orbitals on each carbon atom (there are three sp^2 hybrid orbitals; see Table 10.2). Thus, during bonding, the $2s$ orbital and two of the $2p$ orbitals of each carbon atom form three hybrid orbitals having trigonal planar orientation. A third $2p$ orbital on each carbon atom remains unhybridized and is perpendicular to the plane of the three sp^2 hybrid orbitals.

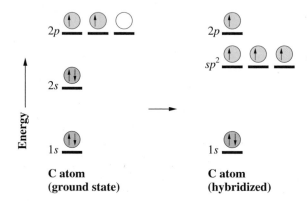

C atom (ground state)   **C atom (hybridized)**

To describe the multiple bonding in ethylene, we must distinguish between two kinds of bonds. A **σ (sigma) bond** *has a cylindrical shape about the bond axis.* It is formed either when two *s* orbitals overlap, as in H_2 (Figure 10.25A), or when an orbital with directional character, such as a *p* orbital or a hybrid orbital, overlaps another orbital *along their axis* (Figure 10.25B). The bonds we discussed in the previous section are σ bonds.

A **π (pi) bond** *has an electron distribution above and below the bond axis.* It is formed by the *sideways* overlap of two parallel *p* orbitals (see Figure 10.25C). A sideways overlap will not give so strong a bond as an along-the-axis overlap of two *p* orbitals. A π bond occurs when two parallel orbitals are still available after strong σ bonds have formed.

Now imagine that the separate atoms of ethylene move into their normal molecular positions. Each sp^2 hybrid carbon orbital overlaps a $1s$ orbital of a

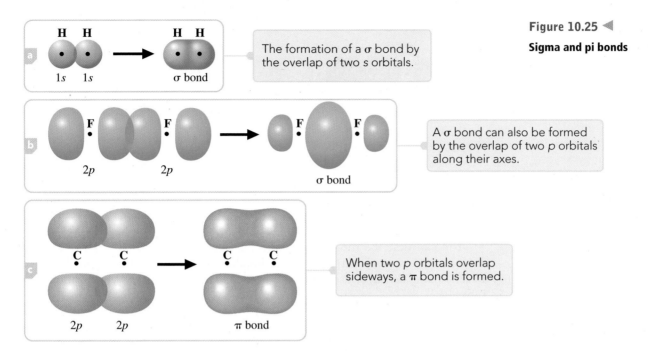

Figure 10.25 ◀

Sigma and pi bonds

a
H H   H H
1s 1s   σ bond

The formation of a σ bond by the overlap of two s orbitals.

b
F   F   F   F
2p   2p   σ bond

A σ bond can also be formed by the overlap of two p orbitals along their axes.

c
C   C   C   C
2p   2p   π bond

When two p orbitals overlap sideways, a π bond is formed.

hydrogen atom or an sp^2 hybrid orbital of another carbon atom to form a σ bond (Figure 10.26A). Together, the σ bonds give the molecular framework of ethylene.

As you see from the orbital diagram for the hybridized C atom, a single $2p$ orbital still remains on each carbon atom. These orbitals are perpendicular to the plane of the hybrid orbitals; that is, they are perpendicular to the —CH₂ plane. Note that the two —CH₂ planes can rotate about the carbon–carbon axis without affecting the overlap of the hybrid orbitals. As these planes rotate, the $2p$ orbitals also rotate. When the —CH₂ planes rotate so that the $2p$ orbitals become parallel, the orbitals overlap to give a π bond (Figure 10.26B).

You therefore describe the carbon–carbon double bond as one σ bond and one π bond. Note that when the two $2p$ orbitals are parallel, the two —CH₂ ends of the molecule lie in the same plane. Thus, the formation of a π bond "locks" the two ends into a flat, rigid molecule.

Figure 10.26 ▼

Bonding in ethylene

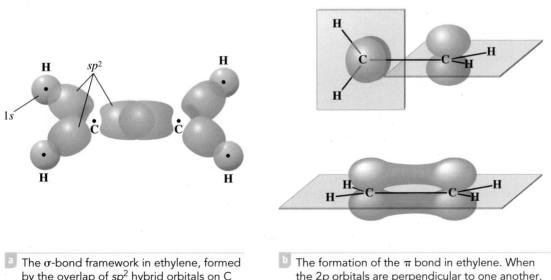

a The σ-bond framework in ethylene, formed by the overlap of sp^2 hybrid orbitals on C atoms and 1s orbitals on H atoms.

b The formation of the π bond in ethylene. When the $2p$ orbitals are perpendicular to one another, there is no overlap and no bond formation. when the two —CH₂ groups rotate so that the $2p$ orbitals are parallel, a π bond forms.

Figure 10.27 ▶

Bonding in acetylene

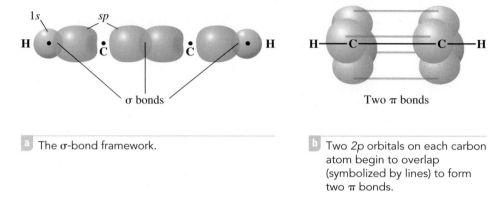

a The σ-bond framework.

b Two 2p orbitals on each carbon atom begin to overlap (symbolized by lines) to form two π bonds.

You can describe the triple bonding in acetylene, H—C≡C—H, in similar fashion. Because each carbon atom is bonded to two other atoms and there are no lone pairs, two hybrid orbitals are needed. This suggests *sp* hybridization (see Table 10.2).

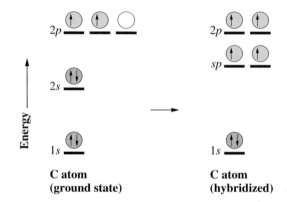

These *sp* hybrid orbitals have a linear arrangement, so the H—C—C—H geometry is linear. Bonds formed by the overlap of these hybrid orbitals are σ bonds. The two 2p orbitals not used to construct hybrid orbitals are perpendicular to the bond axis and to each other. They are used to form two π bonds. Thus, the carbon–carbon triple bond consists of one σ bond and two π bonds (see Figure 10.27).

Example **10.6**

Gaining Mastery Toolbox

Critical Concept 10.6
A double bond is described as a π bond plus a σ bond. To describe the bonding, first determine the number of hybrid orbitals on the central atom (equal to the number of electron groups about the atom). Use these orbitals to form sigma bonds and lone pairs. Finally, use unhybridized *p* orbitals to form π bonds.

Solution Essentials:
• σ- and π-bond formation by orbital overlap
• Kinds of hybrid orbitals
• VSEPR model
• Electron-dot formulas

Applying Valence Bond Theory (Multiple Bonding)

Describe the bonding on a given N atom in dinitrogen difluoride, N_2F_2, using valence bond theory.

Problem Strategy From the Lewis formula, determine the hybrid orbitals on the double-bonded atoms. Do this by counting the number of electron groups about each atom, and from this determine the kind of hybrid orbitals used. Form a σ bond between the double-bonded atoms by overlapping a hybrid orbital of one atom with a hybrid orbital of the other atom. Also, form σ bonds from these atoms to other atoms. Use any unhybridized *p* orbitals on the central atoms to form π bonds between them.

Solution The electron-dot formula of N_2F_2 is

$$:\!\overset{..}{\underset{..}{F}}\!\!-\!\!\overset{..}{N}\!\!=\!\!\overset{..}{N}\!\!-\!\!\overset{..}{\underset{..}{F}}\!:$$

π
σ

(*continued*)

(continued)

Note that a double bond is described as a π bond plus a σ bond. Hybrid orbitals are needed to describe each σ bond and each lone pair (a total of three hybrid orbitals for each N atom). This suggests sp^2 hybridization (see Table 10.2). According to this description, one of the sp^2 hybrid orbitals is used to form the N—F bond, another to form the σ bond of N=N, and the third to hold the lone pair on the N atom. The $2p$ orbitals on each N atom overlap to form the π bond of N=N. Hybridization and bonding of the N atoms are shown as follows:

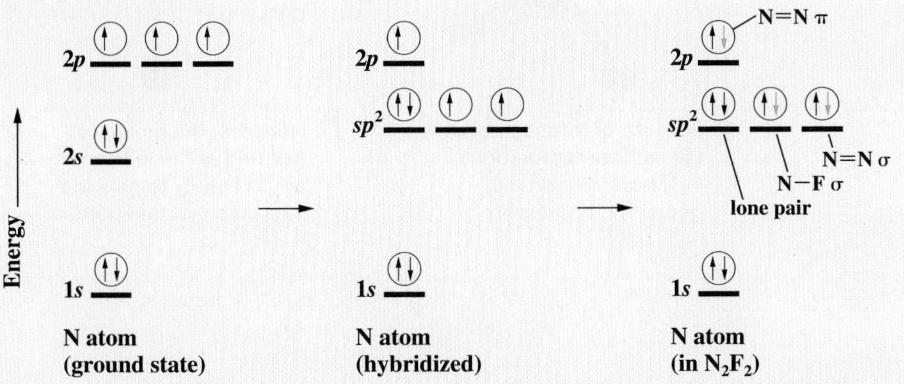

Answer Check Check that the number of hybrid orbitals on an atom of a multiple bond equals the number of sigma bonds plus lone pairs on that atom.

Exercise 10.7 Describe the bonding on the carbon atom in carbon dioxide, CO_2, using valence bond theory.

■ See Problems 10.57 and 10.58.

The π-bond description of the double bond agrees well with experiment. The *geometric*, or *cis–trans*, *isomers* of the compound 1,2-dichloroethene (described in the chapter opening) illustrate this. *Isomers* are compounds of the same molecular formula but with different arrangements of the atoms. (The numbers in the name 1,2-dichloroethene refer to the positions of the chlorine atoms; one chlorine atom is attached to carbon atom 1 and the other to carbon atom 2.) The structures of these isomers of 1,2-dichloroethene are

cis-1,2-Dichloroethene   *trans*-1,2-Dichloroethene

To transform one isomer into the other, one end of the molecule must be rotated as the other remains fixed. For this to happen, the π bond must be broken. Breaking the π bond requires considerable energy, so the *cis* and *trans* compounds are not easily interconverted. Contrast this with 1,2-dichloroethane (Figure 10.28), in which the two ends of the molecule can rotate without breaking any bonds. Here isomers corresponding to different spatial orientations of the two chlorine atoms cannot be prepared, because the two ends rotate freely with respect to one another. There is only one compound.

As we noted in the chapter opening, *cis* and *trans* isomers have different properties. The *cis* isomer of 1,2-dichloroethene boils at 60°C, the *trans* compound at 48°C.

Figure 10.28 ▶

Lack of geometric isomers in 1,2-dichloroethane

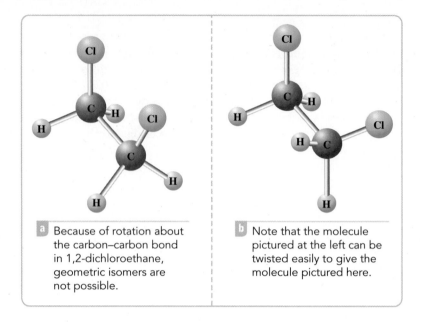

a   Because of rotation about the carbon–carbon bond in 1,2-dichloroethane, geometric isomers are not possible.

b   Note that the molecule pictured at the left can be twisted easily to give the molecule pictured here.

These isomers can also be differentiated on the basis of dipole moment. The *trans* compound has no dipole moment because it is symmetrical (the polar C—Cl bonds point in opposite directions and so cancel). However, the *cis* compound has a dipole moment of 1.85 D.

It is possible to convert one isomer to another if sufficient energy is supplied— say by chemical reaction. The role of the conversion of a *cis* isomer to its *trans* isomer in human vision is discussed in the essay at the end of Section 10.7.

Exercise 10.8   Dinitrogen difluoride (see Example 10.6) exists as *cis* and *trans* isomers. Write structural formulas for these isomers and explain (in terms of the valence bond theory of the double bond) why they exist.

■ See Problems 10.59 and 10.60.

CONCEPT CHECK 10.3

An atom in a molecule has one single bond and one triple bond to other atoms. What hybrid orbitals do you expect for this atom? Describe how you arrive at your answer.

Molecular Orbital Theory

In the preceding sections, we looked at a simple version of valence bond theory. Although simple valence bond theory satisfactorily describes most of the molecules we encounter, it does not apply to all molecules. For example, according to this theory any molecule with an even number of electrons should be diamagnetic (not attracted to a magnet), because we assume the electrons to be paired and to have opposite spins. In fact, a few molecules with an even number of electrons are paramagnetic (attracted to a magnet), indicating that some of the electrons are not paired. The best-known example of such a paramagnetic molecule is O_2. Because of the paramagnetism of O_2, liquid oxygen sticks to a magnet when poured over it

(Figure 10.29). Although valence bond theory can be extended to explain the electronic structure of O_2, molecular orbital theory (an alternative bonding theory) provides a straightforward explanation of the paramagnetism of O_2.

Molecular orbital theory is *a theory of the electronic structure of molecules in terms of molecular orbitals, which may spread over several atoms or the entire molecule.* This theory views the electronic structure of molecules to be much like the electronic structure of atoms. Each molecular orbital has a definite energy. To obtain the ground state of a molecule, electrons are put into orbitals of lowest energy, consistent with the Pauli exclusion principle, just as in atoms.

10.5 Principles of Molecular Orbital Theory

You can think of a molecular orbital as being formed from a combination of atomic orbitals. As atoms approach each other and their atomic orbitals overlap, molecular orbitals are formed.

Bonding and Antibonding Orbitals

Consider the H_2 molecule. As the atoms approach to form the molecule, their $1s$ orbitals overlap. One molecular orbital is obtained by adding the two $1s$ orbitals (see Figure 10.30). Note that where the atomic orbitals overlap, their values sum to give a larger result. This means that in this molecular orbital, electrons are often found in the region between the two nuclei where the electrons can hold the nuclei together. *Molecular orbitals that are concentrated in regions between nuclei* are called **bonding orbitals.** The bonding orbital in H_2, which we have just described, is denoted σ_{1s}. The σ (sigma) means that the molecular orbital has a cylindrical shape about the bond axis. The subscript $1s$ tells us that the molecular orbital was obtained from $1s$ atomic orbitals.

Another molecular orbital is obtained by subtracting the $1s$ orbital on one atom from the $1s$ orbital on the other (Figure 10.30). When the orbitals are subtracted, the resulting values in the region of the overlap are close to zero. This means that in this molecular orbital, the electrons spend little time between the nuclei. *Molecular orbitals having zero values in the region between two nuclei and therefore concentrated in other regions* are called **antibonding orbitals.** The antibonding orbital in H_2, which we have just described, is denoted σ_{1s}^*. The asterisk, which you read as "star," tells us that the molecular orbital is antibonding.

Figure 10.31 shows the energies of the molecular orbitals σ_{1s} and σ_{1s}^* relative to the atomic orbitals. Energies of the separate atomic orbitals are represented

Figure 10.29 ▲

Paramagnetism of oxygen, O_2 Liquid oxygen is poured between the poles of a strong magnet. Oxygen adheres to the poles, showing that it is paramagnetic.

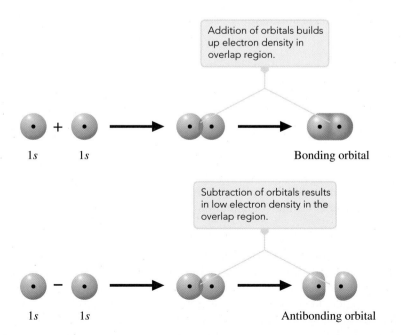

Addition of orbitals builds up electron density in overlap region.

$1s$ $1s$ Bonding orbital

Subtraction of orbitals results in low electron density in the overlap region.

$1s$ $1s$ Antibonding orbital

Figure 10.30 ◄

Formation of bonding and antibonding orbitals from 1s orbitals of hydrogen atoms When the two $1s$ orbitals overlap, they can either add to give a bonding molecular orbital or subtract to give an antibonding molecular orbital.

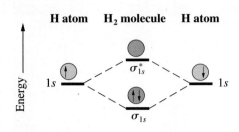

Figure 10.31 ▲

Relative energies of the 1s orbital of the H atom and the σ_{1s} and σ_{1s}^* molecular orbitals of H₂ Arrows denote occupation of the σ_{1s} orbital by electrons in the ground state of H₂.

by heavy black lines at the far left and right. The energies of the molecular orbitals are shown by heavy black lines in the center. (These are connected by dashed lines to show which atomic orbitals were used to obtain the molecular orbitals.) Note that the energy of a bonding orbital is less than that of the separate atomic orbitals, whereas the energy of an antibonding orbital is higher.

You obtain the electron configuration for the ground state of H₂ by placing the two electrons (one from each atom) into the lower-energy orbital (see Figure 10.31). The orbital diagram is

$$\sigma_{1s} \quad \sigma_{1s}^*$$

and the electron configuration is $(\sigma_{1s})^2$. Because the energy of the two electrons is lower than their energies in the isolated atoms, the H₂ molecule is stable.

Configurations involving the σ_{1s}^* orbital describe excited states of the molecule. As an example of an excited state, you can write the orbital diagram

$$\sigma_{1s} \quad \sigma_{1s}^*$$

The corresponding electron configuration is $(\sigma_{1s})^1(\sigma_{1s}^*)^1$.

You obtain a similar set of orbitals when you consider the approach of two helium atoms. To obtain the ground state of He₂, you note that there are four electrons, two from each atom, that would fill the molecular orbitals. Two electrons go into the σ_{1s} orbital and two go into the σ_{1s}^* orbital. The orbital diagram is

$$\sigma_{1s} \quad \sigma_{1s}^*$$

and the configuration is $(\sigma_{1s})^2(\sigma_{1s}^*)^2$. The energy decrease from the bonding electrons is offset by the energy increase from the antibonding electrons. Hence, He₂ is not a stable molecule. Molecular orbital theory therefore explains why the element helium exists as a monatomic gas, whereas hydrogen is diatomic.

Bond Order

The term *bond order* refers to the number of bonds that exist between two atoms. In molecular orbital theory, the bond order of a diatomic molecule is defined as one-half the difference between the number of electrons in bonding orbitals, n_b, and the number of electrons in antibonding orbitals, n_a. ◄

$$\text{Bond order} = \tfrac{1}{2}(n_b - n_a)$$

For H₂, which has two bonding electrons,

$$\text{Bond order} = \tfrac{1}{2}(2 - 0) = 1$$

That is, H₂ has a single bond. For He₂, which has two bonding and two antibonding electrons,

$$\text{Bond order} = \tfrac{1}{2}(2 - 2) = 0$$

For a Lewis formula, or electron-dot formula, the bond order equals the number of electron pairs shared between two atoms (see Section 9.10).

Bond orders need not be whole numbers; half-integral bond orders of $\tfrac{1}{2}, \tfrac{3}{2}$, and so forth are also possible. For example, the H₂⁺ molecular ion, which is formed in mass spectrometers, has the configuration $(\sigma_{1s})^1$ and a bond order of $\tfrac{1}{2}(1 - 0) = \tfrac{1}{2}$.

Factors That Determine Orbital Interaction

The H₂ and He₂ molecules are relatively simple. A 1s orbital on one atom interacts with the 1s orbital on the other atom. But now let us consider the Li₂ molecule. Each Li atom has 1s and 2s orbitals. Which orbitals interact to form molecular orbitals? To find out, you need to understand the factors that determine orbital interaction.

The strength of the interaction between two atomic orbitals to form molecular orbitals is determined by two factors: (1) the energy difference between the interacting orbitals and (2) the magnitude of their overlap. *For the interaction to be strong, the energies of the two orbitals must be approximately equal and the overlap must be large.*

From this last statement, you see that when two Li atoms approach one another to form Li_2, only like orbitals on the two atoms interact appreciably. The 2s orbital of one lithium atom interacts with the 2s orbital of the other atom, but the 2s orbital from one atom does not interact with the 1s orbital of the other atom, because their energies are quite different. Also, because the 2s orbitals are outer orbitals, they are able to overlap and interact strongly when the atoms approach. As in H_2, these atomic orbitals interact to give a bonding orbital (denoted σ_{2s}) and an antibonding orbital (denoted σ_{2s}^*). However, even though the 1s orbitals of the two atoms have the same energy, they do not overlap appreciably and so interact weakly. (The difference in energy between σ_{1s} and σ_{1s}^* is very small.) Figure 10.32 gives the relative energies of the orbitals.

You obtain the ground-state configuration of Li_2 by putting six electrons (three from each atom) into the molecular orbitals of lowest energy. The configuration of the diatomic molecule Li_2 is

$$Li_2 \quad (\sigma_{1s})^2(\sigma_{1s}^*)^2(\sigma_{2s})^2$$

The $(\sigma_{1s})^2(\sigma_{1s}^*)^2$ part of the configuration is often abbreviated KK (which denotes the K shells, or inner shells, of the two atoms). These electrons do not have a significant effect on bonding. ▶

$$Li_2 \quad KK(\sigma_{2s})^2$$

In calculating bond order, we can ignore KK (it includes two bonding and two antibonding electrons). Thus, the bond order is 1, as in H_2.

In Be_2, the energy diagram is similar to that in Figure 10.32. You have eight electrons to distribute (four from each Be atom). The ground-state configuration of Be_2 is

$$Be_2 \quad KK(\sigma_{2s})^2(\sigma_{2s}^*)^2$$

Note that the configuration has two bonding and two antibonding electrons outside the K shells. Thus, the bond order is $\frac{1}{2}(2 - 2) = 0$. No bond is formed, so the Be_2 molecule, like the He_2 molecule, is unstable.

10.6 Electron Configurations of Diatomic Molecules of the Second-Period Elements

The previous section looked at the electron configurations of some simple molecules: H_2, He_2, Li_2, and Be_2. These are **homonuclear diatomic molecules**—that is, *molecules composed of two like nuclei.* (**Heteronuclear diatomic molecules** are *molecules composed of two different nuclei*—for example, CO and NO.) To find the electron configurations of other homonuclear diatomic molecules, we need to have additional molecular orbitals.

We have already looked at the formation of molecular orbitals from s atomic orbitals. Now we need to consider the formation of molecular orbitals from p atomic orbitals. There are two different ways in which 2p atomic orbitals can interact. One set of 2p orbitals can overlap along their axes to give one bonding and one antibonding σ orbital (σ_{2p} and σ_{2p}^*). The other two sets of 2p orbitals then overlap sideways to give two bonding and two antibonding π orbitals (π_{2p} and π_{2p}^*). (See Figure 10.33.)

Figure 10.34 shows the relative energies of the molecular orbitals obtained from 2s and 2p atomic orbitals. This order of molecular orbitals reproduces the known electron configurations of homonuclear diatomic molecules composed of elements in the second row of the periodic table. ▶ The order of filling is

$$\sigma_{2s} \quad \sigma_{2s}^* \quad \pi_{2p} \quad \sigma_{2p} \quad \pi_{2p}^* \quad \sigma_{2p}^*$$

Li atom Li_2 molecule   Li atom

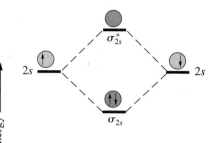

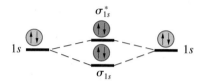

Figure 10.32 ▲

The energies of the molecular orbitals of Li_2 Arrows indicate the occupation of these orbitals by electrons in the ground state.

However, small shifts in the energy of the core electrons due to bonding can be measured by x-ray photoelectron spectroscopy (page 310).

This order gives the correct number of electrons in subshells, even though in O_2 and F_2 the energy of σ_{2p} is below π_{2p}.

Figure 10.33 ▶

The different ways in which 2p orbitals can interact Let the z axis be the bond axis between two atoms. When a $2p_z$ orbital (lying along the bond axis) on one atom overlaps with a similar $2p_z$ orbital on the other atom, they interact to form σ_{2p} (bonding sigma orbital) and σ_{2p}^* (anti-bonding sigma orbital). See at the top of the figure. When a $2p_x$ orbital (which is perpendicular to the bond axis) overlaps a $2p_x$ on the other atom, they interact to form π_{2px} (bonding pi orbital) and π_{2px}^* (antibonding pi orbital). These are shown at the bottom of the figure. Similar pi orbitals (π_{2py} and π_{2py}^*) are formed by overlap of the $2p_y$ orbitals on the two atoms. (The $2p_y$ orbital is perpendicular to the bond axis and perpendicular to the $2p_x$ orbital.)

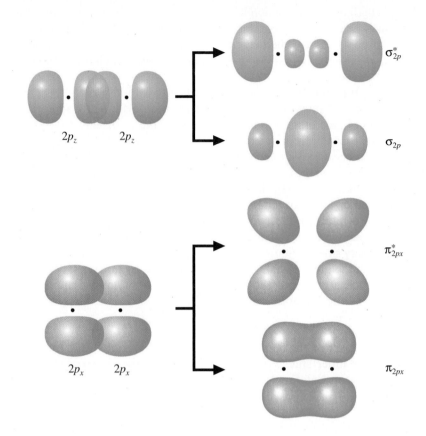

Note that there are two orbitals in the π subshell and two orbitals in the π^* subshell. Because each orbital can hold two electrons, a π or π^* subshell can hold four electrons. The next example shows how you can use the order of filling to obtain the orbital diagram, magnetic character, electron configuration, and bond order of a homonuclear diatomic molecule.

Figure 10.34 ▶

Relative energies of molecular orbitals of homonuclear diatomic molecules (excluding K shells) The arrows show the occupation of molecular orbitals by the valence electrons of N_2.

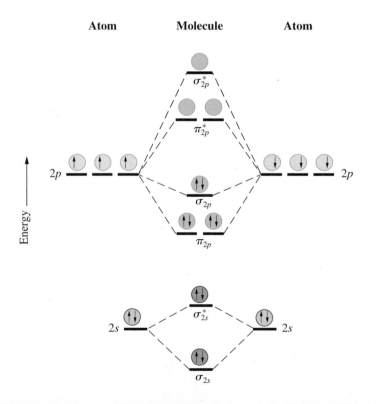

Example 10.7   **Describing Molecular Orbital Configurations (Homonuclear Diatomic Molecules)**

Gaining Mastery Toolbox

Critical Concept 10.7
Place the electrons of a homonuclear diatomic molecule into the molecular orbitals. As in an atom, observe the order of filling of the orbitals, the Pauli exclusion principle, and Hund's rule.

Solution Essentials:
- Order of filling of molecular orbitals
- Hund's rule
- Pauli exclusion principle

Give the orbital diagram of the O_2 molecule. Is the molecular substance diamagnetic or paramagnetic? What is the electron configuration? What is the bond order of O_2?

Problem Strategy Note the order of filling of molecular orbitals for the second-row homonuclear diatomic molecules (given just before this example). Now assign the valence electrons to these orbitals in this order. Pay attention to both the Pauli exclusion principle (no more than two electrons to an orbital, and then only with opposed spins) and Hund's rule (fill all orbitals of a subshell with one electron each—and with the same spins—before pairing electrons in any orbital). Decide whether the substance is diamagnetic (all electrons paired) or paramagnetic (unpaired electrons). Then calculate the bond order equal to $\frac{1}{2}$(number of bonding electrons − number of antibonding electrons).

Solution There are 12 valence electrons in O_2 (six from each atom), which occupy the molecular orbitals as shown in the following orbital diagram:

$$\text{KK} \;\uparrow\downarrow\;\; \uparrow\downarrow\;\; \uparrow\downarrow\;\uparrow\downarrow\;\; \uparrow\downarrow\;\; \uparrow\;\uparrow\;\; \bigcirc$$
$$\quad\quad \sigma_{2s} \;\; \sigma_{2s}^* \;\; \pi_{2p} \;\;\; \sigma_{2p} \;\; \pi_{2p}^* \;\; \sigma_{2p}^*$$

Note that the two electrons in the π_{2p}^* subshell must go into different orbitals with their spins in the same direction (Hund's rule). Because there are two unpaired electrons, the molecular substance is **paramagnetic.** The electron configuration is

$$\text{KK}(\sigma_{2s})^2(\sigma_{2s}^*)^2(\pi_{2p})^4(\sigma_{2p})^2(\pi_{2p}^*)^2$$

There are eight bonding electrons and four antibonding electrons. Therefore,

$$\text{Bond order} = \tfrac{1}{2}(8 - 4) = \mathbf{2}$$

Answer Check Check that the number of electrons assigned to orbitals equals the number of valence electrons of the two atoms. Make sure that the Pauli exclusion principle and Hund's rule have been properly applied.

Exercise 10.9 The C_2 molecule exists in the vapor phase over carbon at high temperature. Describe the molecular orbital structure of this molecule; that is, give the orbital diagram and electron configuration. Would you expect the molecular substance to be diamagnetic or paramagnetic? What is the bond order for C_2?

■ See Problems 10.61 and 10.62.

Table 10.3 compares experimentally determined bond lengths, bond dissociation energies, and magnetic character of the second-period homonuclear diatomic molecules with the bond order calculated from molecular orbital theory, as in the preceding example. Note that as the bond order increases, bond length tends to decrease and bond dissociation energy tends to increase. You should be able to verify that

Table 10.3   **Theoretical Bond Orders and Experimental Data for the Second-Period Homonuclear Diatomic Molecules**

Molecule	Bond Order	Bond Length (pm)	Bond Dissociation Energy (kJ/mol)	Magnetic Character
Li_2	1	267	110	Diamagnetic
Be_2	0	*	*	*
B_2	1	159	290	Paramagnetic
C_2	2	124	602	Diamagnetic
N_2	3	110	942	Diamagnetic
O_2	2	121	494	Paramagnetic
F_2	1	142	155	Diamagnetic
Ne_2	0	*	*	*

The symbol * means that no stable molecule has been observed.

the experimentally determined magnetic character of the molecule, given in the last column of the table, is correctly predicted by molecular orbital theory.

When the atoms in a heteronuclear diatomic molecule are close to one another in a row of the periodic table, the molecular orbitals have the same relative order of energies as those for homonuclear diatomic molecules. In this case you can obtain the electron configurations in the same way, as the next example illustrates.

Example **10.8** **Describing Molecular Orbital Configurations (Heteronuclear Diatomic Molecules)**

Gaining Mastery Toolbox

Critical Concept 10.8
If the two different atoms of a heteronuclear diatomic molecule are close to one another in the periodic table, you can use the same order of the filling of orbitals that you used for homonuclear diatomic molecules.

Solution Essentials:
- Order of filling of molecular orbitals
- Hund's rule
- Pauli exclusion principle

Write the molecular orbital diagram for nitrogen monoxide (nitric oxide), NO. What is the bond order of NO?

Problem Strategy If the two different atoms of the molecule are close to one another in the periodic table, then you can assume that the order of filling is the same as that for homonuclear diatomic molecules. After that, the procedure is similar to that for a homonuclear molecule (see Example 10.7).

Solution You assume that the order of filling of orbitals is the same as for homonuclear diatomic molecules.

There are 11 valence electrons in NO. Thus, the orbital diagram is

$$\text{KK}\ (\uparrow\downarrow)\ (\uparrow\downarrow)\ \underbrace{(\uparrow\downarrow)\ (\uparrow\downarrow)}\ (\uparrow\downarrow)\ \underbrace{(\uparrow)\ (\ )}\ (\ )$$
$$\quad\quad\ \sigma_{2s}\ \ \sigma_{2s}^{*}\quad\ \pi_{2p}\quad\ \ \sigma_{2p}\ \ \ \pi_{2p}^{*}\ \ \ \sigma_{2p}^{*}$$

Because there are eight bonding and three antibonding electrons,

$$\text{Bond order} = \tfrac{1}{2}(8 - 3) = \tfrac{5}{2}$$

Answer Check Check that the number of electrons assigned to orbitals equals the number of valence electrons of the two atoms. Make sure that the Pauli exclusion principle and Hund's rule have been properly applied.

Exercise 10.10 Give the orbital diagram and electron configuration for the carbon monoxide molecule, CO. What is the bond order of CO? Is the molecule diamagnetic or paramagnetic?

■ See Problems 10.63 and 10.64.

When the two atoms in a heteronuclear diatomic molecule differ appreciably, you can no longer use the scheme appropriate for homonuclear diatomic molecules. The HF molecule is an example. Figure 10.35 shows the relative energies of the molecular orbitals that form. Sigma bonding and antibonding orbitals are formed by combining the 1s orbital on the H atom with the 2p orbital on the F atom that lies along the bond axis. The 2s orbital and the other 2p orbitals on fluorine remain as *nonbonding* orbitals (neither bonding nor antibonding). Note that the energy of the 2p subshell in fluorine is lower than the energy of the 1s orbital in hydrogen, because the fluorine electrons are more tightly held. As a result, the bonding orbital is made up of a greater percentage of the 2p fluorine orbital than of the 1s hydrogen orbital. This means that electrons in the bonding molecular orbital spend more time in the vicinity of the fluorine atom; that is, the H—F bond is polar, with the F atom having a small negative charge.

10.7 Molecular Orbitals and Delocalized Bonding

One of the advantages of using molecular orbital theory is the simple way in which it describes molecules with delocalized bonding. Whereas valence bond theory requires two or more resonance formulas, molecular orbital theory describes the bonding in terms of a single electron configuration.

Consider the ozone molecule as an example. In the previous chapter, we described this molecule in terms of resonance as follows:

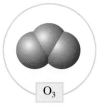
O_3

$$\ddot{\underset{..}{O}}\!\!:\!\!-\!\!\overset{\displaystyle \overset{..}{O}}{}\!\!-\!\!:\!\!\ddot{\underset{..}{O}}\!\!: \longleftrightarrow :\!\!\ddot{\underset{..}{O}}\!\!-\!\!\overset{\displaystyle \overset{..}{O}}{}\!\!-\!\!:\!\!\ddot{\underset{..}{O}}\!\!:$$

Some of these electrons are localized either on one atom or between an oxygen–oxygen bond (these are printed in black). The other electrons are delocalized (these are printed in blue). They are the electron pairs that change position in going from one resonance formula to the other.

Each O atom has three localized electron pairs about it, which suggests that each atom uses sp^2 hybrid orbitals (see Table 10.2). The overlap of one hybrid orbital on the center O atom with a hybrid orbital on the end O atom at the left forms the left O—O bond (Figure 10.36). The overlap of another hybrid orbital on the center O atom with a hybrid orbital on the end O atom at the right forms the other O—O bond. This leaves one hybrid orbital on the center O atom and two on each of the end O atoms for lone pairs.

After the hybrid orbitals are formed, one unhybridized p orbital remains on each O atom. These three p orbitals are perpendicular to the plane of the molecule and parallel to one another. They can overlap sideways to give three π molecular orbitals. (Three interacting atomic orbitals produce three molecular orbitals.) All three will span the entire molecule (Figure 10.37). One of these orbitals turns out to be antibonding; one is nonbonding (that is, it has an energy equal to that of the isolated atoms); and the third is bonding.

The bonding and nonbonding π orbitals are both doubly occupied. This agrees with the resonance description, in which one delocalized pair is bonding and the other is a lone (nonbonding) pair on the end oxygen atoms.

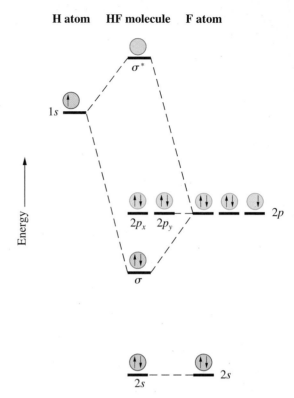

Figure 10.35 ▲

Relative energies of molecular orbitals in the valence shell of HF Arrows show the occupation of orbitals by electrons.

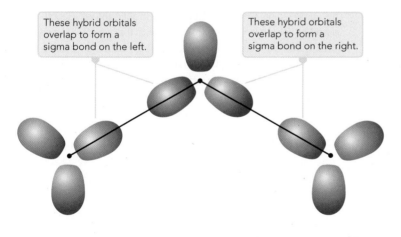

Figure 10.36 ◄

Hybrid orbitals on oxygen atoms of ozone, O_3 Each oxygen atom has sp^2 hybrid orbitals. One hybrid orbital on the center atom and one hybrid orbital on the left atom overlap to form a single bond at the left. Similar orbitals at the right form another single bond. Each atom has a π orbital perpendicular to the plane of these hybrid orbitals.

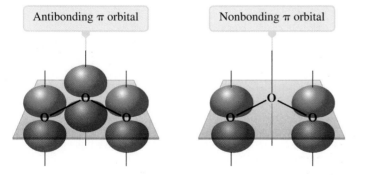
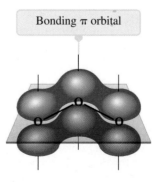

Figure 10.37 ▲

The electron distributions of the π molecular orbitals of O_3 Note that one molecular orbital is nonbonding. It describes a lone pair of electrons that can be on either end-oxygen atom.

Human Vision

Human vision, as well as the detection of light by other organisms, is known to involve a compound called 11-*cis*-retinal. The structural formula is shown on the left side of Figure 10.38. This compound, which is derived from vitamin A, is contained in two kinds of cells—rod cells and cone cells—located in the retina of the eye. Rod cells are responsible for vision in dim light. They give the sensation of light or dark but no sensation of color. Cone cells are responsible for color vision, but they require bright light. Color vision is possible because three types of cone cells exist: one type absorbs light in the red region of the spectrum, another absorbs in the green region, and a third absorbs in the blue region. In each of these different cells, 11-*cis*-retinal is attached to a different protein molecule, which affects the region of light that is absorbed by retinal.

When a photon of light is absorbed by 11-*cis*-retinal, it is transformed to 11-*trans*-retinal (see Figure 10.38). This small change at one carbon–carbon double bond results in a large movement of one end of the molecule with respect to the other. In other words, the absorption of a single photon results in a significant change in molecular geometry. This change in shape of retinal also affects the shape of the protein molecule to which it is attached, and *this* change results in a sequence of events, not completely understood, in which an electrical signal is generated.

To understand the *cis–trans* conversion when a photon is absorbed, consider the similar, but simpler, molecule *cis*-1,2-dichloroethene (see structural formula on page 382). The double bond consists of σ bond and a π bond. The π bond results from two electrons in a π-type molecular orbital. A corresponding antibonding

orbital, π^*, has higher energy than the π orbital and is unoccupied in the ground state. When a photon of light in the ultraviolet region (near 180 nm) is absorbed by *cis*-1,2-dichloroethene, an electron in the π orbital undergoes a transition to the π^* orbital. Whereas there were previously two electrons in the bonding π orbital (contributing one bond), there is now one electron in the bonding orbital and one in the antibonding orbital (with zero net bonding). As a result of the absorption of a photon, the double bond becomes a single bond and rotation about the bond is now possible. The *cis* isomer tends to rotate to the *trans* isomer, which is more stable. Then energy is lost as heat, and the electron in the π^* orbital undergoes a transition back to the π orbital; the single bond becomes a double bond again. The net result is the conversion of the *cis* isomer to the *trans* isomer.

Retinal differs from dichloroethene in having six double bonds alternating with single bonds instead of only one double bond. The alternating double and single bonds give a rigid structure, and when the middle double bond undergoes the *cis*-to-*trans* change, a large movement of atoms occurs. Another result of a long chain of alternating double and single bonds is a change in the wavelength of light absorbed in the π-to-π^* transition. Dichloroethene absorbs in the ultraviolet region, but retinal attached to its protein absorbs in the visible region. (The attached protein also alters the wavelength of absorbed light.) Many biological compounds that are colored consist of long chains of alternating double and single bonds. β-Carotene, a yellow pigment in carrots, has this structure. It is broken down in the body to give vitamin A, from which retinal is derived.

11-*cis*-Retinal   11-*trans*-Retinal

Figure 10.38 ▲

The *cis–trans* conversion of retinal on absorption of light Note the large movement of atoms caused by the rotation about the double bond between carbon atoms 11 and 12.

■ See Problems 10.85 and 10.86.

A CHEMIST Looks at...

Environment

Stratospheric Ozone (An Absorber of Ultraviolet Rays)

Ozone (also known as trioxygen, O_3) is a reactive form of oxygen having a bent molecular geometry, as you would predict from its resonance formulas:

Ozone forms in the atmosphere whenever O_2 is irradiated by ultraviolet light or subjected to electrical discharges. On bright sunny days, you might notice the fresh-air smell of dilute ozone (trioxygen, O_3). Around electrical equipment, however, ozone often occurs in higher concentrations and has a disagreeable odor.

Although ozone occurs at ground level and is an ingredient of *photochemical smog,* it is an essential component of the *stratosphere,* which occurs at about 10 to 15 km above ground level. Ozone in the stratosphere absorbs ultraviolet radiation between 200 and 300 nm, which is vitally important to life on earth. Radiation from the sun contains ultraviolet rays that are harmful to the DNA of biological organisms. Oxygen, O_2, absorbs the most energetic of these ultraviolet rays in the earth's upper atmosphere, but only ozone in the stratosphere absorbs the remaining ultraviolet radiation that is destructive to life on earth (Figure 10.39).

In 1974, Mario J. Molina and F. Sherwood Rowland expressed concern that chlorofluorocarbons (CFCs), such as $CClF_3$ and CCl_2F_2, would be a source of chlorine atoms that could catalyze the decomposition of ozone in the stratosphere, so that the ozone would be destroyed faster than it could be produced. The chlorofluorocarbons are relatively inert compounds that were used as refrigerants, spray-can propellants, and blowing agents (substances used to produce plastic foams). As a result of their inertness, however, CFCs concentrate in the atmosphere, where they steadily rise into the stratosphere. Once they are in the stratosphere, ultraviolet light decomposes them to form chlorine atoms, which react with ozone to form ClO and O_2. The ClO molecules react with oxygen atoms in the stratosphere to regenerate Cl atoms.

$$Cl(g) + O_3(g) \longrightarrow ClO(g) + O_2(g)$$
$$ClO(g) + O(g) \longrightarrow Cl(g) + O_2(g)$$
$$\overline{O_3(g) + O(g) \longrightarrow 2O_2(g)}$$

The net result is the decomposition of ozone to dioxygen. Chlorine atoms are consumed in the first step but regenerated in the second. Thus, chlorine atoms are not used up, so they function as a catalyst.

In 1985, British researchers reported that the ozone over the Antarctic had been declining precipitously each spring for several years, although it returned the following winter. Then in August 1987, an expedition was assembled to begin flights through the "ozone hole" with instruments (Figure 10.40). Researchers found that wherever ozone is depleted in the stratosphere, chlorine monoxide (ClO) appears, as expected if chlorine atoms are catalyzing the depletion (see the previous reactions). The chlorofluorocarbons are the suspected source of these chlorine atoms, and by international treaty they are being phased out of production. Refrigerators and

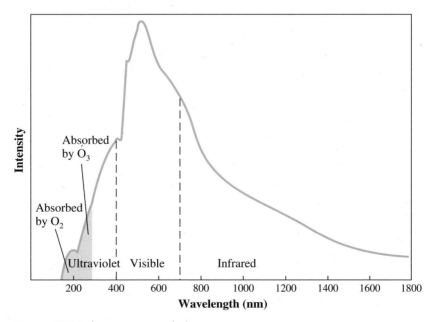

Figure 10.39 ▲

The solar spectrum Radiation below 200 nm is absorbed in the upper atmosphere by O_2. Ultraviolet radiation between 200 and 300 nm is absorbed by O_3 in the stratosphere.

(continued)

air conditioners, for example, now use hydrofluorocarbons (HFCs), such as R410A, which is a mixture of CH_2F_2 and CF_3CHF_2.

In October 1995, Sherwood Rowland (University of California, Irvine), Mario Molina (Massachusetts Institute of Technology), and

Paul Crutzen (Max Planck Institute, Mainz, Germany) were awarded the Nobel Prize in chemistry for their work on stratospheric ozone depletion. Crutzen studied the effect of nitrogen oxides in catalyzing the decomposition of ozone.

Figure 10.40 ▶

Total stratospheric ozone over the Southern Hemisphere This computer image shows the total quantity of ozone over the Southern Hemisphere in Dobson units (equivalent to the thickness, in units of 10^{-2} mm, of total ozone, assuming it to be compressed to 1 atm at 0°C). The ozone depletion is shown near the South Pole (pink color; black color indicates no data). The data are from the SBUV/2 instrument on board the National Oceanic and Atmospheric Administration's polar orbiting satellite.

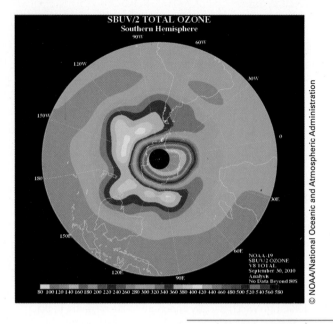

© NOAA/National Oceanic and Atmospheric Administration

■ See Problems 10.87 and 10.88.

A Checklist for Review

Summary of Facts and Concepts

Molecular geometry refers to the spatial arrangement of atoms in a molecule. The *valence-shell electron-pair repulsion (VSEPR) model* is a simple model for predicting molecular geometries. It is based on the idea that the valence-shell electron pairs are arranged symmetrically about an atom to minimize electron-pair repulsion. The geometry about an atom is then determined by the directions of the bonding pairs. Information about the geometry of a molecule can sometimes be obtained from the experimentally determined presence or absence of a *dipole moment*.

The bonding and geometry in a molecule can be described in terms of *valence bond theory*. In this theory, a bond is formed by the overlap of orbitals from two atoms. *Hybrid orbitals*, a set of equivalent orbitals formed by combining atomic orbitals, are often needed to describe

this bond. Multiple bonds occur via the overlap of atomic orbitals to give σ bonds and π bonds. *Cis–trans* isomers result from the molecular rigidity imposed by a π bond.

Molecular orbital theory can also be used to explain bonding in molecules. According to this theory, electrons in a molecule occupy orbitals that may spread over the entire molecule. You can think of these molecular orbitals as constructed from atomic orbitals. Thus, when two atoms approach to form a diatomic molecule, the atomic orbitals interact to form *bonding* and *antibonding* molecular orbitals. The electron configuration of a diatomic molecule such as O_2 can be predicted from the order of filling of the molecular orbitals. From this configuration, you can predict the bond order and whether the molecular substance is diamagnetic or paramagnetic.

Learning Objectives	Important Terms
10.1 The Valence-Shell Electron-Pair Repulsion (VSEPR) Model	
■ Define *molecular geometry*. ■ Define *valence-shell electron-pair repulsion model*.	**molecular geometry** **valence-shell electron-pair repulsion (VSEPR) model**

- Note the difference between the arrangement of electron pairs about a central atom and molecular geometry.
- Note the four steps in the prediction of geometry by the VSEPR model.
- Predict the molecular geometry (two, three, or four electron pairs). Example 10.1
- Note that a lone pair tends to require more space than a corresponding bonding pair and that a multiple bond requires more space than a single bond.
- Predict the molecular geometry (five or six electron pairs). Example 10.2

10.2 Dipole Moment and Molecular Geometry

- Define *dipole moment*.
- Explain the relationship between dipole moment and molecular geometry. Example 10.3
- Note that the polarity of a molecule can affect certain properties, such as a boiling point.

dipole moment

10.3 Valence Bond Theory

- Define *valence bond theory*.
- State the two conditions needed for bond formation, according to valence bond theory.
- Define *hybrid orbitals*.
- State the five steps in describing bonding, following the valence bond theory.
- Apply valence bond theory (two, three, or four electron pairs). Example 10.4
- Apply valence bond theory (five or six electron pairs). Example 10.5

valence bond theory
hybrid orbitals

10.4 Description of Multiple Bonding

- Define σ (*sigma*) *bond*.
- Define π (*pi*) *bond*.
- Apply valence bond theory (multiple bonding). Example 10.6
- Explain geometric, or *cis–trans*, isomers in terms of the π-bond description of a double bond.

σ **(sigma) bond**
π **(pi) bond**

10.5 Principles of Molecular Orbital Theory

- Define *molecular orbital theory*.
- Define *bonding orbitals* and *antibonding orbitals*.
- Define *bond order*.
- State the two factors that determine the strength of interaction between two atomic orbitals.
- Describe the electron configurations of H_2, He_2, Li_2, and Be_2.

molecular orbital theory
bonding orbitals
antibonding orbitals

10.6 Electron Configurations of Diatomic Molecules of the Second-Period Elements

- Define *homonuclear diatomic molecules* and *heteronuclear diatomic molecules*.
- Describe molecular orbital configurations (homonuclear diatomic molecules). Example 10.7
- Describe molecular orbital configurations (heteronuclear diatomic molecules). Example 10.8

homonuclear diatomic molecules
heteronuclear diatomic molecules

10.7 Molecular Orbitals and Delocalized Bonding

- Describe the delocalized bonding in molecules such as O_3.

Questions and Problems

OWL Interactive versions of these problems may be assigned in OWL.

Self-Assessment and Review Questions

Key: These questions test your understanding of the ideas you worked with in the chapter. These problems vary in difficulty and often can be used for the basis of discussion.

10.1 Describe the main features of the VSEPR model.

10.2 According to the VSEPR model, what are the arrangements of two, three, four, five, and six valence-shell electron pairs about an atom?

10.3 Why is a lone pair expected to occupy an equatorial position instead of an axial position in the trigonal bipyramidal arrangement?

10.4 Why is it possible for a molecule to have polar bonds, yet have a dipole moment of zero?

10.5 Explain why nitrogen trifluoride has a small dipole moment even though it has polar bonds in a trigonal pyramidal arrangement.

10.6 Explain in terms of valence bond theory why the orbitals used on an atom give rise to a particular geometry about that atom.

10.7 What is the difference between a sigma bond and a pi bond?

10.8 What is the angle between two sp^3 hybrid orbitals?

10.9 Describe the bonding in ethylene, C_2H_4, in terms of valence bond theory.

10.10 What are the differences between a bonding and an antibonding molecular orbital of a diatomic molecule?

10.11 How does the valence bond description of a carbon–carbon double bond account for *cis–trans* isomers?

10.12 What factors determine the strength of interaction between two atomic orbitals to form a molecular orbital?

10.13 Describe the formation of bonding and antibonding molecular orbitals resulting from the interaction of two $2s$ orbitals.

10.14 How does molecular orbital theory describe the bonding in the HF molecule?

10.15 Describe the formation of molecular orbitals resulting from the interaction of two $2p$ orbitals.

10.16 Describe the bonding in O_3, using molecular orbital theory. Compare this with the resonance description.

10.17 According to the VSEPR model, the H—P—H bond angle in PH_3 is
- a $109.5°$
- b $90°$
- c $120°$
- d a little less than $109.5°$
- e a little less than $120°$

10.18 Which of the following molecular geometries does the XeF_5^+ cation exhibit?
- a tetradral
- b T-shaped
- c octahedral
- d square pyramidal
- e trigonal bipyramidal

10.19 Which of the following would be a polar molecule?
- a BeF_2
- b SF_6
- c H_2S
- d CO_2
- e CH_4

10.20 What is the bond order of NO?
- a $\frac{1}{2}$
- b $\frac{3}{2}$
- c $\frac{5}{2}$
- d 3
- e 1

Concept Explorations

Key: Concept explorations are comprehensive problems that provide a framework that will enable you to explore and learn many of the critical concepts and ideas in each chapter. If you master the concepts associated with these explorations, you will have a better understanding of many important chemistry ideas and will be more successful in solving all types of chemistry problems. These problems are well suited for group work and for use as in-class activities.

10.21 Best Lewis Formula and Molecular Geometry

A student writes the Lewis electron-dot formula for the carbonate anion, CO_3^{2-}, as

$$\left[\begin{array}{c} :\ddot{O}: \\ \| \\ C \\ :\ddot{O} \quad \ddot{O}: \end{array} \right]^{2-}$$

- a Does this Lewis formula obey the octet rule? Explain. What are the formal charges on the atoms? Try describing the bonding for this formula in valence bond terms. Do you have any difficulty doing this?

- b Does this Lewis formula give a reasonable description of the electron structure, or is there a better one? If there is a better Lewis formula, write it down and explain why it is better.

- c The same student writes the following resonance description for CO_2:

$$\ddot{O}—\ddot{C}=\ddot{O} \longleftrightarrow :\ddot{O}—C=\ddot{O}$$

Is there something wrong with this description? (What would you predict as the geometries of these formulas?)

- d Is one or the other formula a better description? Could a value for the dipole moment help you decide?

- e Can you write a Lewis formula that gives an even better description of CO_2? Explain your answer.

10.22 Molecular Geometry and Bonding

- a Shown here are several potential Lewis electron-dot formulas for the SeO_2 molecule.

Are there any electron-dot formulas that you expect to give the best description? How would you describe the bonding in terms of electron-dot formulas?

- b Describe the bonding in SeO_2 in valence bond terms. If there is delocalized bonding, note this in your description and explain how molecular orbital theory can be useful here.

c Determine the arrangement of electron pairs about Se in the SeO_2 molecule. What would you expect for the molecular geometry of the SeO_2 molecule?

d Would you expect the O—Se—O angle in the SeO_2 molecule to be greater than, equal to, or less than 120°. Explain your answer.

e Draw an electron-dot formula for the H_2Se molecule and determine its molecular geometry.

f Compare the H—Se—H bond angle to the O—Se—O bond angle. Is it larger or smaller? Explain.

g Determine whether the H_2Se and SeO_2 molecules have dipole moments. Describe how you arrived at your answer.

Conceptual Problems

Key: These problems are designed to check your understanding of the concepts associated with some of the main topics presented in each chapter. A strong conceptual understanding of chemistry is the foundation for both applying chemical knowledge and solving chemical problems. These problems vary in level of difficulty and often can be used as a basis for group discussion.

10.23 Match the following molecular substances a–d with one of the molecular models (i) to (iv) that correctly depicts the geometry of the corresponding molecule.

a SeO_2 b $BeCl_2$ c PBr_3 d BCl_3

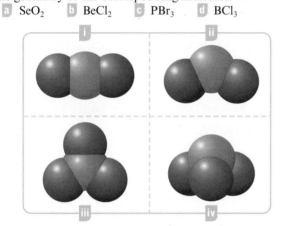

10.24 Which of the following molecular models correctly depicts the geometry of ClCN?

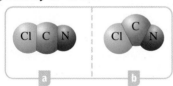

10.25 Suppose that a BF_3 molecule approaches the lone pair on the N atom of an $:NH_3$ molecule, and that a bond forms between the B atom and the N atom. Consider the arrangement of electron pairs about the B atom at the moment of this bond formation, and describe the repulsions among the electron pairs and how they might be expected to change the geometry about the B atom.

10.26 Suppose that an ethane molecule, CH_3CH_3, is broken into two CH_3 molecules in such a way that one $:CH_3$ molecule retains the electron pair that was originally the one making up the C—C bond. The other CH_3 molecule has two fewer electrons. Imagine that momentarily this CH_3 molecule has the geometry it had in the ethane molecule. Describe the electron repulsions present in this molecule, and indicate how they would be expected to rearrange its geometry.

10.27 Indicate what hybrid orbital depicted below is expected for the central atom of each of the following:

a SeF_4 b RnF_4 c BeF_2 d SiF_4

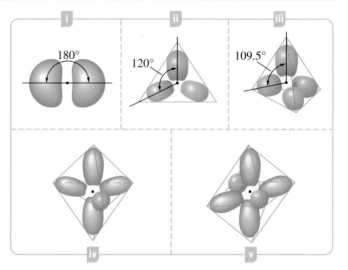

10.28 An atom in a molecule has two bonds to other atoms and one lone pair. What kind of hybrid orbitals do you expect for this atom? Describe how you arrived at your answer.

10.29 A neutral molecule is identified as a tetrafluoride, XF_4, where X is an unknown atom. If the molecule has a dipole moment of 0.63 D, can you give some possibilities for the identity of X?

10.30 Two compounds have the same molecular formula, $C_2H_2Br_2$. One has a dipole moment; the other does not. Both compounds react with bromine, Br_2, to produce the same compound. This reaction is a generally accepted test for double bonds, and each bromine atom of Br_2 adds to a different atom of the double bond. What is the identity of the original compounds? Describe the argument you use.

10.31 Acetic acid, the sour constituent of vinegar, has the following structure:

$$\underset{\text{a} \quad \text{b} \quad \text{c}}{CH_3\overset{\displaystyle \overset{O}{\|}}{C}\!\!-\!\!OH}$$

Indicate what geometry i–iv is expected to be found about each of the atoms labeled a, b, and c.

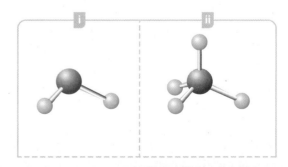

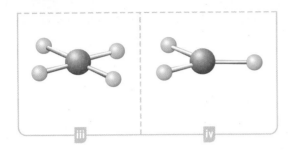

10.32 What are the bond angles predicted by the VSEPR model about the carbon atom in the formate ion, HCO_2^-? Considering that the bonds to this atom are not identical, would you expect the experimental values to agree precisely with the VSEPR values? How might they differ?

Practice Problems

Key: These problems are for practice in applying problem-solving skills. They are divided by topic, and some are keyed to exercises (see the ends of the exercises). The problems are arranged in matching pairs; the odd-numbered problem of each pair is listed first, and its answer is given in the back of the book.

The VSEPR Model

10.33 Predict the shape or geometry of the following molecules, using the VSEPR model.
   a PCl_3 b COF_2 c SF_2 d SiF_4

10.34 Use the electron-pair repulsion model to predict the geometry of the following molecules:
   a $GeCl_2$ b $AsCl_3$ c SO_3 d XeO_4

10.35 Predict the geometry of the following ions, using the electron-pair repulsion model.
   a ClO_3^- b SCN^- c PO_4^{3-} d H_3O^+

10.36 Use the VSEPR model to predict the geometry of the following ions:
   a NO_2^- b SO_3^{2-} c N_3^- d BH_4^-

10.37 For each of the following molecules, state the bond angle (or bond angles, as appropriate) that you would expect to see on the central atom based on the simple VSEPR model. Would you expect the actual bond angles to be greater or less than this?
   a CCl_4 b SCl_2 c $COCl_2$ d AsH_3

10.38 For each of the following molecules, state the bond angle (or bond angles, as appropriate) that you would expect to see on the central atom based on the VSEPR model. Would you expect the actual bond angles to be greater or less than this?
   a OF_2 b NCl_3 c CF_2CF_2 d GeF_4

10.39 From the electron-pair repulsion model, predict the geometry of the following molecules:
   a TeF_6 b ClF_5 c SeF_4 d SbF_5

10.40 What geometry is expected for the following molecules, according to the VSEPR model?
   a PF_5 b BrF_3 c BrF_5 d SCl_4

10.41 Predict the geometries of the following ions, using the VSEPR model.
   a $SnCl_5^-$ b PF_6^- c ClF_2^- d IF_4^-

10.42 Name the geometries expected for the following ions, according to the electron-pair repulsion model.
   a BrF_6^+ b IF_2^- c ICl_4 d IF_4^+

Dipole Moment and Molecular Geometry

10.43 a The molecule BrF_3 has a dipole moment of 1.19 D. Which of the following geometries are possible: trigonal planar, trigonal pyramidal, or T-shaped? b The molecule $TeCl_4$ has a dipole moment of 2.54 D. Is the geometry tetrahedral, seesaw, or square planar?

10.44 a The molecule AsF_3 has a dipole moment of 2.59 D. Which of the following geometries are possible: trigonal planar, trigonal pyramidal, or T-shaped? b The molecule H_2S has a dipole moment of 0.97 D. Is the geometry linear or bent?

10.45 Which of the following molecules would be expected to have a dipole moment of zero because of symmetry?
   a $BeBr_2$ b H_2Se c AsF_3 d SeF_6

10.46 Which of the following molecules would be expected to have zero dipole moment on the basis of their geometry?
   a CS_2 b TeF_2 c $SeCl_4$ d XeF_4

Valence Bond Theory

10.47 What hybrid orbitals would be expected for the central atom in each of the following molecules or ions?

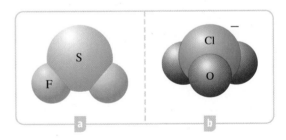

10.48 What hybrid orbitals would be expected for the central atom in each of the following molecules or ions?

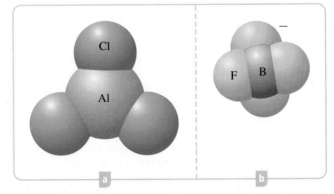

10.49 What hybrid orbitals would be expected for the central atom in each of the following molecules or ions?
  a COF_2 b $SeCl_2$ c CO_2 d NO_2^-

10.50 What hybrid orbitals would be expected for the central atom in each of the following molecules or ions?
  a PCl_3 b $SiCl_4$ c BeF_2 d SO_2

10.51 a Mercury(II) chloride dissolves in water to give poorly conducting solutions, indicating that the compound is largely nonionized in solution—it dissolves as $HgCl_2$ molecules. Describe the bonding of the $HgCl_2$ molecule, using valence bond theory. b Phosphorus trichloride, PCl_3, is a colorless liquid with a highly irritating vapor. Describe the bonding in the PCl_3 molecule, using valence bond theory. Use hybrid orbitals.

10.52 What hybrid orbitals would be expected for the central atom in each of the following?
  a XeF_2 b BrF_5 c PCl_5 d ClF_4^+

10.53 a Nitrogen trifluoride, NF_3, is a relatively unreactive, colorless gas. How would you describe the bonding in the NF_3 molecule in terms of valence bond theory? Use hybrid orbitals. b Silicon tetrafluoride, SiF_4, is a colorless gas formed when hydrofluoric acid attacks silica (SiO_2) or glass. Describe the bonding in the SiF_4 molecule, using valence bond theory.

10.54 What hybrid orbitals would be expected for the central atom in each of the following?
  a SF_4 b $AsCl_5$ c XeF_4 d IF_4^-

10.55 Phosphorus pentachloride is normally a white solid. It exists in this state as the ionic compound $[PCl_4^+]$ $[PCl_6^-]$. Describe the electron structure of the PCl_6^- ion in terms of valence bond theory.

10.56 Iodine, I_2, dissolves in an aqueous solution of iodide ion, I^-, to give the triiodide ion, I_3^-. This ion consists of one iodine atom bonded to two others. Describe the bonding of I_3^- in terms of valence bond theory.

10.57 a Formaldehyde, H_2CO, is a colorless, pungent gas used to make plastics. Give the valence bond description of the formaldehyde molecule. (Both hydrogen atoms are attached to the carbon atom.) b Nitrogen, N_2, makes up about 80% of the earth's atmosphere. Give the valence bond description of this molecule.

10.58 a The molecule HN=NH exists as a transient species in certain reactions. Give the valence bond description

of this species. b Hydrogen cyanide, HCN, is a very poisonous gas or liquid with the odor of bitter almonds. Give the valence bond description of HCN. (Carbon is the central atom.)

10.59 The hyponitrite ion, $^-O-N=N-O^-$, exists in solid compounds as the *trans* isomer. Using valence bond theory, explain why *cis–trans* isomers might be expected for this ion. Draw structural formulas of the *cis–trans* isomers.

10.60 Fumaric acid, $C_4H_4O_4$, occurs in the metabolism of glucose in the cells of plants and animals. It is used commercially in beverages. The structural formula of fumaric acid is

Maleic acid is the *cis* isomer of fumaric acid. Using valence bond theory, explain why these isomers are possible.

Molecular Orbital Theory

10.61 Describe the electronic structure of each of the following, using molecular orbital theory. Calculate the bond order of each and decide whether it should be stable. For each, state whether the substance is diamagnetic or paramagnetic.
  a O_2^- b B_2^+ c B_2

10.62 Use molecular orbital theory to describe the bonding in the following. For each one, find the bond order and decide whether it is stable. Is the substance diamagnetic or paramagnetic?
  a Be_2^+ b Ne_2 c C_2^+

10.63 Assume that the cyanide ion, CN^-, has molecular orbitals similar to those of a homonuclear diatomic molecule. Write the configuration and bond order of CN^-. Is a substance of the ion diamagnetic or paramagnetic?

10.64 Write the molecular orbital configuration of the diatomic molecule BN. What is the bond order of BN? Is the substance diamagnetic or paramagnetic? Use the order of energies that was given for homonuclear diatomic molecules.

General Problems

Key: These problems provide more practice but are not divided by topic or keyed to exercises. Each section ends with essay questions, each of which is color coded to refer to the *A Chemist Looks at Daily Life* (orange), *A Chemist Looks at Life Science* (pink), and *A Chemist Looks at Environment* (green) chapter essay on which it is based. Odd-numbered problems and the even-numbered problems that follow are similar; answers to all odd-numbered problems except the essay questions are given in the back of the book.

10.65 Predict the molecular geometry of the following:
  a NH_4^+ b HOF c ClF_4^+ d PF_5

10.66 Predict the molecular geometry of the following:
  a $SnCl_2$ b $COBr_2$ c ICl_2^- d PCl_6^-

10.67 Which of the following molecules or ions are linear?
  a NH_4^+ b HOF c ClF_4^+ d PF_5

10.68 Which of the following molecules or ions are trigonal planar?
  a ClF_4^- b TeF_5^- c AsH_3 d BBr_3

10.69 Describe the hybrid orbitals used by each nitrogen atom in the following molecules:

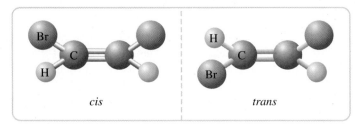

10.70 Describe the hybrid orbitals used by each carbon atom in the following molecules:

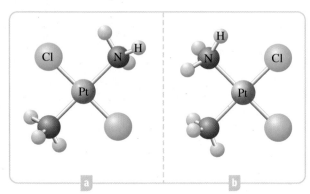

10.71 Explain how the dipole moment could be used to distinguish between the *cis* and *trans* isomers of 1,2-dibromoethene:

cis   *trans*

10.72 Two compounds have the formula $Pt(NH_3)_2Cl_2$. (Compound B is *cisplatin,* mentioned in the opening to Chapter 1.) They have square planar structures. One is expected to have a dipole moment; the other is not. Which one would have a dipole moment?

10.73 Explain in terms of bonding theory why all four hydrogen atoms of allene, $H_2C=C=CH_2$, cannot lie in the same plane.

10.74 Explain in terms of bonding theory why all atoms of $H_2C=C=C=CH_2$ must lie in the same plane.

10.75 What is the molecular orbital configuration of He_2^+? Do you expect the ion to be stable?

10.76 What is the molecular orbital configuration of HeH^+? Do you expect the ion to be stable?

10.77 Calcium carbide, CaC_2, consists of Ca^{2+} and C_2^{2-} (acetylide) ions. Write the molecular orbital configuration and bond order of the acetylide ion, C_2^{2-}.

10.78 Sodium peroxide, Na_2O_2, consists of Na^+ and O_2^{2-} (peroxide) ions. Write the molecular orbital configuration and bond order of the peroxide ion, O_2^{2-}.

10.79 The nitrogen–nitrogen bond distance in N_2 is 109 pm. On the basis of bond orders, would you expect the bond distance in N_2^+ to be less than or greater than 109 pm? Answer the same question for N_2^-.

10.80 The oxygen–oxygen bond in O_2^+ is 112 pm and in O_2 is 121 pm. Explain why the bond length in O_2^+ is shorter than in O_2. Would you expect the bond length in O_2^- to be longer or shorter than that in O_2? Why?

10.81 Using molecular orbital theory, determine the electronic structure of the first excited electronic state of N_2. What differences are expected in the properties of the excited state of N_2 compared with the same properties of the ground state?

10.82 The ionization energy of O_2 is smaller than the ionization energy of atomic O; the opposite is true for the ionization energies of N_2 and atomic N. Explain this behavior in terms of the molecular orbital energy diagrams of O_2 and N_2.

■ **10.83** Describe the property of "handedness." What is it about the bonding of atoms in molecules that can give rise to handedness in molecules?

■ **10.84** Biological molecules frequently display handedness. Describe the difference between the molecule responsible for the flavor of spearmint and the molecule responsible for the flavor of caraway seeds. In which way are the molecules similar and in which way do they differ?

■ **10.85** Describe the *cis–trans* conversion that occurs in the ultraviolet absorption of *cis*-1,2-dichloroethene.

■ **10.86** Color vision results from the absorption of light by the cone cells of the retina. Briefly describe how different types of cone cells give rise to color vision. What is the chemical substance primarily responsible for human vision? How can its color absorption be changed?

■ **10.87** In 1995, Sherwood Rowland and Mario Molina received the Nobel Prize in chemistry. Explain in some detail the work they did for which they were awarded this prize.

■ **10.88** What is the biological importance of stratospheric ozone? Explain.

Strategy Problems

Key: As noted earlier, all of the practice and general problems are matched pairs. This section is a selection of problems that are not in matched-pair format. These challenging problems require that you employ many of the concepts and strategies that were developed in the chapter. In some cases, you will have to integrate several concepts and operational skills in order to solve the problem successfully.

10.89 The bond length in C_2 is 131 pm. Compare this with the bond lengths in C_2H_2 (120 pm), C_2H_4 (134 pm), and C_2H_6 (153 pm). What bond order would you predict for C_2 from its bond length? Does this agree with the molecular orbital configuration you would predict for C_2?

10.90 Each of the following compounds has a nitrogen–nitrogen bond: N_2, N_2H_4, N_2F_2. Match each compound with one of the following bond lengths: 110 pm, 122 pm, 145 pm. Describe the geometry about one of the N atoms in each compound. What hybrid orbitals are needed to describe the bonding in valence bond theory?

10.91 Calcium carbide, CaC_2, has an ionic structure with ions Ca^{2+} and C_2^{2-}. Give the valence bond description of the bonding in the C_2^{2-} ion. Now write the MO configuration of this ion. What is the bond order according to this configuration? Compare the valence bond and molecular orbital descriptions.

10.92 Write Lewis formulas for the BF molecule (two with a single B—F bond, two with a double B—F bond, and one with a triple B—F bond) in which the octet rule is satisfied for at least one of the atoms. Obtain the formal charges of the atoms. Based on the information you have, is there one formula that you think best describes the molecule? Explain. For this best Lewis formula, give the valence bond description of the bonding.

10.93 Boron trifluoride, BF_3, reacts with ammonia, NH_3, to form an addition compound, BF_3NH_3. Describe the geometries about the B and the N atoms in this compound. Describe the hybridization on these two atoms. Now describe the bonding between the B and N atoms, using valence bond theory. Compare the geometries and hybridization of these atoms in this addition compound with that in the reactant molecules BF_3 and NH_3.

10.94 Acetylsalicylic acid (aspirin) has the structure shown by the following molecular model:

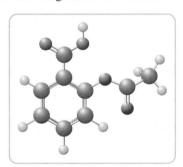

Describe the geometry and bonding about the carbon atom shown at the top of the model, in which the carbon atom (in black) is bonded to two oxygen atoms (red) and to another carbon atom. One of the oxygen atoms is bonded to a hydrogen atom (blue).

10.95 Allene (1,2-propadiene), a gas, has the following structure:

$$\underset{H}{\overset{H}{\diagdown}}C=C=C\underset{H}{\overset{H}{\diagup}}$$

What is the hybridization at each carbon atom? What is the geometry about each carbon atom? Using valence-bond theory, describe the bonding about the center carbon

atom. 1,3-dichloroallene is a derivative of allene with the following structure:

$$\underset{H}{\overset{Cl}{\diagdown}}C=C=C\underset{Cl}{\overset{H}{\diagup}}$$

Is this a polar molecule? (If you have difficulty visualizing the structure of this molecule, you might build a molecular model of it.)

10.96 The triiodide ion, I_3^-, and the azide ion, N_3^-, have similar skeleton formulas:

$$I—I—I \qquad N—N—N$$

Otherwise, how similar are they? To answer, draw electron-dot formulas for each ion, including resonance formulas, if there are any. What is the geometry about the center atom in each molecule? Describe the hybridization about this atom in each molecule.

10.97 When allene (described in the previous problem) is heated with sodium metal, it rearranges to give the isomeric molecule methyl acetylene, which has the following structure:

$$\underset{H}{\overset{H}{\underset{|}{\overset{|}{H—C}}}}—C\equiv C—H$$

What is the hybridization at each carbon atom? What is the geometry about each carbon atom? Using valence-bond theory, describe the bonding about the center carbon atom.

10.98 Hydrogen azide, HN_3, is a covalent molecule in which the hydrogen atom is bonded to a nitrogen atom at one end of the azide ion (see previous problem). Draw an electron-dot formula of this molecule. Describe the hybridization about the nitrogen atom to which the hydrogen atom is bonded. Use valence-bond theory to describe the bonding about this nitrogen atom. What do you expect for the approximate H—N—N angle?

10.99 There are a number of known pentahalides of phosphorus. One of them is the trifluorodibromophosphorus compound, PF_3Br_2, an unstable liquid. Draw the Lewis electron-dot formula of PF_3Br_2. What is the hybridization of the phosphorus atom expected in this molecule? What is the geometry about this atom? Describe the placement of fluorine and bromine atoms about the phosphorus atom in this geometry. Explain how you come to this conclusion.

10.100 A molecule XF_6 (having no lone pairs) has a dipole moment of zero. (X denotes an unidentified element.) When two atoms of fluorine have been taken away, you get the molecule XF_4, which has dipole moment of 0.632 D.

$$XF_6 \longrightarrow XF_4 + 2F$$

Describe the molecular geometry and bonding for each molecule. What are some possibilities for X?

10.101 Describe the molecular orbital configurations of C_2^+, C_2, and C_2^{2-}. What are the bond orders of these species? Arrange the three species by increasing bond length. Arrange the species by increasing bond enthalpy. Explain these arrangements of bond length and bond enthalpy.

10.102 If the HCl molecule were 100% ionic, the molecule would consist of a positive charge e and a negative charge $-e$ separated by a distance d equal to the bond length. The experimental value of the bond length is 127 pm. Calculate the dipole moment for such a completely ionic molecule. The actual dipole moment, μ, of HCl is 1.08 D. Linus Pauling, Nobel laureate, used the ratio of the actual dipole moment to that of the 100% ionic molecule as an estimate of the fraction of ionic character in the HCl molecule. What is the fraction of ionic character in this molecule according to this estimate?

10.103 Three different compounds have the same molecular formula, $C_2H_2F_2$; call them A, B, and C. Compound A has a dipole moment of 2.42 D, compound B has a dipole moment of 1.38 D, and compound C has a dipole moment of 0 D. All three compounds react with hydrogen, H_2:

$$C_2H_2F_2 + H_2 \longrightarrow C_2H_4F_2$$

Compound A gives product D (molecular formula $C_2H_4F_2$), compound B gives product E (same molecular formula, $C_2H_4F_2$, but different from product D), and compound C gives product D. Identify the structural formulas of compounds A, B, and C.

10.104 Obtain a Lewis formula of H_2N_2. This molecule exhibits isomerism (it has two isomers). Sketch the electron-dot formulas of these isomers. Describe the bonding in terms of valence bond theory, and explain this isomerism in terms of this bonding description.

10.105 Dioxygen, O_2, has sometimes been represented by the electron-dot formula

$$\ddot{O}=\ddot{O}$$

Note, however, that this formula implies that all electrons are paired. Give the molecular orbital description

of the state of the O_2 molecule of lowest energy *in which all electrons are paired in this way*. Compare this configuration with that of the ground state of O_2. In what way are they similar, and in what way are they different?

10.106 Solid sulfur normally consists of crystals of S_8 molecules, but when heated strongly, the solid vaporizes to give S_2 molecules (among other molecular species). Describe the bonding in S_2 in molecular orbital terms, assuming the orbitals are analogous to those of the preceding period. What would you expect to happen to the sulfur–sulfur bond length if two electrons were added to give the S_2^{2-} ion? What would you expect to happen to the bond length if, instead, two electrons were taken away to give S_2^{2+}?

10.107 Consider the bonding in nitrate ion, NO_3^-. First draw resonance formulas of this ion. Now describe the bonding of this ion in terms of molecular orbitals. (Refer to the delocalized bonding of the ozone molecule described in the text.) Suppose each atom uses sp^2 hybrid orbitals. How many π molecular orbitals can you form from the $2p$ orbitals that remain on these atoms? How many of these π orbitals will be occupied?

10.108 Formaldehyde has the formula H_2CO. Describe the bonding between the C and O atoms in VB terms. Formaldehyde absorbs UV light at about 270 nm. The absorption is described as a π to π^* transition. Explain in molecular orbital terms what is involved.

Cumulative-Skills Problems

Key: The problems under this heading combine skills introduced in previous chapters with those given in the current one.

10.109 A molecular compound is composed of 58.8% Xe, 7.2% O, and 34.0% F, by mass. If the molecular weight is 223 amu, what is the molecular formula? What is the Lewis formula? Predict the molecular geometry using the VSEPR model. Describe the bonding, using valence bond theory.

10.110 A molecular compound is composed of 60.4% Xe, 22.1% O, and 17.5% F, by mass. If the molecular weight is 217.3 amu, what is the molecular formula? What is the Lewis formula? Predict the molecular geometry using the VSEPR model. Describe the bonding, using valence bond theory.

10.111 Excess fluorine, $F_2(g)$, reacts at 150°C with bromine, $Br_2(g)$, to give a compound BrF_n. If 423 mL $Br_2(g)$ at 150°C and 748 mmHg produced 4.20 g BrF_n, what is n? Describe the bonding in the molecule, using valence bond theory.

10.112 A compound of chlorine and fluorine, ClF_n, reacts at about 75°C with uranium metal to produce uranium hexafluoride, UF_6, and chlorine monofluoride, $ClF(g)$. A quantity of uranium produced 3.53 g UF_6 and 343 mL ClF at 75°C and 2.50 atm. What is the formula (n) of the compound? Describe the bonding in the molecule, using valence bond theory.

10.113 Draw resonance formulas of the nitric acid molecule, HNO_3. What is the geometry about the N atom? What is the hybridization on N? Use bond enthalpies and one Lewis formula for HNO_3 to estimate $\Delta H°_f$ for $HNO_3(g)$. The actual value of $\Delta H°_f$ for $HNO_3(g)$ is -135 kJ/mol, which is lower than the estimated value because of stabilization of HNO_3 by resonance. The *resonance energy* is defined as $\Delta H°_f$ (estimated) $-\Delta H°_f$ (actual). What is the resonance energy of HNO_3?

10.114 One resonance formula of benzene, C_6H_6, is

What is the other resonance formula? What is the geometry about a carbon atom? What hybridization would be used in valence bond theory to describe the bonding? The $\Delta H°_f$ for $C_6H_6(g)$ is -83 kJ/mol; $\Delta H°_f$ for C(g) is 715 kJ/mol. Obtain the resonance energy of benzene. (See Problem 10.113.)

11

States of Matter; Liquids and Solids

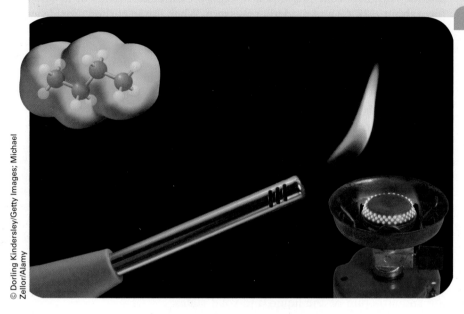

© Dorling Kindersley/Getty Images; Michael Zellor/Alamy

Liquid butane, C_4H_{10}, in the lighter changes to the gaseous state, which burns when ignited. Butane stoves also employ this change of state.

OWL Sign in to OWL at **www.cengage.com/owl** to view tutorials and simulations, develop problem-solving skills, and complete online homework assigned by your professor.

Go Chemistry Download mini lecture videos for key concept review and exam prep from OWL or purchase them from **www.cengagebrain.com**.

© Cengage Learning

Figure 11.1 ▲

Dry ice Dry ice is solid carbon dioxide, CO_2, whose temperature is about −78°C. It passes directly from the solid to the gaseous state. The white plumes are composed of ice crystals (H_2O, not CO_2) formed by condensation from the cold air.

Solid carbon dioxide is shown in Figure 11.1. It has an interesting property: at normal pressures it passes directly to the gaseous state without first melting to the liquid. This property, together with the fact that this change occurs at −78°C, makes solid carbon dioxide useful for keeping materials very cold. Because solid CO_2 cools other objects and does not leave a liquid residue, it is called *dry ice.*

Liquid carbon dioxide is obtained by putting carbon dioxide gas under pressure; it can exist only at pressures greater than 5 atm. In most cases, carbon dioxide is transported as a liquid in high-pressure steel tanks or at low temperature (about −18°C) and moderate pressure. When liquid carbon dioxide evaporates (changes to vapor), it absorbs large quantities of heat, cooling as low as −57°C. Because of this property, it is used as a refrigerant. Liquid carbon dioxide is often used to freeze foods for grocery stores and fast-food restaurants. If the compressed gas from the evaporating liquid is allowed to expand through a valve, the rapidly cooled vapor forms solid carbon dioxide "snow." (Most gases cool when they expand in this manner; for example, air cools when it escapes from the valve of an inflated tire.) This carbon dioxide snow is compacted into blocks and is the source of dry ice.

Carbon dioxide, like water and most other pure substances, exists in solid, liquid, and gaseous states and can undergo changes from one state to another. Substances change state under various temperature and pressure conditions. Can we obtain some understanding of these conditions for a change of state? Why, for example, is carbon dioxide normally a gas, whereas water is normally a liquid? What are the conditions under which carbon dioxide gas changes to a liquid or to a solid?

These are some of the questions we will examine in this chapter. They concern certain physical properties that, as you will see, can often be related to the bonding and structure of the liquid and solid states.

11.1 Comparison of Gases, Liquids, and Solids

The different states of matter were defined in Section 1.4, and the gas laws and the kinetic theory of gases were defined in Chapter 5. Here we want to recall the salient features of those discussions and then compare gases, liquids, and solids. We especially want to compare how these states are viewed in terms of kinetic-molecular theory.

Gases are compressible fluids. According to kinetic-molecular theory, gases are composed of molecules or single atoms that are in constant random motion throughout mostly empty space (unless the gas is highly compressed). A gas is easily compressed because the molecules can be pushed into a smaller space. A gas is fluid because individual molecules can move easily relative to one another.

Liquids are relatively incompressible fluids. According to kinetic-molecular theory, the molecules of a liquid are in constant random motion (as in a gas) but are more tightly packed, so there is much less free space. Because the molecules can move relative to one another as in a gas, a liquid can flow (it is fluid). But the lack of empty space explains why a liquid, unlike a gas, is nearly incompressible.

Solids are nearly incompressible and are rigid, not fluid. Kinetic-molecular theory explains this by saying that the particles making up a solid (which may be atoms, molecules, or ions) exist in close contact and (unlike those in a gas or liquid) do not move about but oscillate or vibrate about fixed sites. This explains the rigidity of a solid. And, of course, the compact structure explains its incompressibility. Figure 11.2 compares the gaseous, liquid, and solid states in kinetic-molecular terms.

Recall that gases normally follow closely the ideal gas law, $PV = nRT$. The simplicity of this equation is a result of the nearly negligible forces of interaction between molecules and the nearly negligible molecular size compared with the

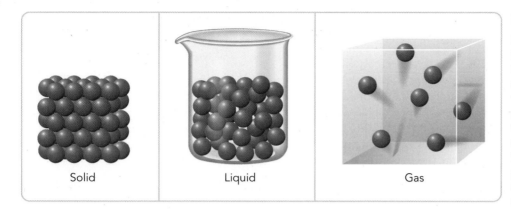

Figure 11.2 ◄

Representation of the states of matter In a solid, the basic units (atoms, ions, or molecules) are closely packed and vibrate about fixed sites. In a liquid, molecules are rather closely packed, similar to those in a molecular solid, but can slip by one another and are in constant random motion. In a gas, molecules are widely spaced and, similar to those in a liquid, are in constant random motion.

total volume of gas. Still, to be accurate we should account for both of these quantities, and the van der Waals equation, which we discussed in Section 5.8, is one attempt to do this.

$$(P + n^2a/V^2)(V - nb) = nRT$$

In this equation, the constants a and b depend on the nature of the gas. The constant a depends on the magnitude of the intermolecular forces; the constant b depends on the relative size of the molecules.

No such simple equations exist for the liquid and solid states. The size of the particles making up the liquid and solid states cannot be neglected relative to the total volume or accounted for in a simple way. Similarly, the forces of interaction between particles in the liquid and the solid states cannot be neglected; indeed, these forces of interaction are crucial when you describe the properties of the liquid and solid states.

Some of the most important properties of liquids and solids are concerned with changes of physical state. Therefore, we begin with this topic. In later sections, we will look specifically at the liquid and solid states and see how their properties depend on the forces of attraction between particles.

Changes of State

In the chapter opener, we discussed the change of solid carbon dioxide (dry ice) directly to a gas. Such *a change of a substance from one state to another* is called a **change of state** or **phase transition.** In the next sections, we will look at the different kinds of phase transitions and the conditions under which they occur. (A *phase* is a homogeneous portion of a system—that is, a given state of either a substance or a solution.)

11.2 Phase Transitions

In general, each of the three states of a substance can change into either of the other states by undergoing a phase transition.

Melting is *the change of a solid to the liquid state* (melting is also referred to as *fusion*). For example,

$$H_2O(s) \longrightarrow H_2O(l) \qquad \text{(melting, fusion)}$$
$$\text{ice, snow} \qquad \text{liquid water}$$

Freezing is *the change of a liquid to the solid state.* The freezing of liquid water to ice is a common example.

$$H_2O(l) \longrightarrow H_2O(s) \qquad \text{(freezing)}$$
$$\text{liquid water} \qquad \text{ice}$$

The casting of a metal object involves the melting of the metal, then its freezing in a mold.

Vaporization is *the change of a solid or a liquid to the vapor.* For example,

$$H_2O(l) \longrightarrow H_2O(g) \qquad \text{(vaporization)}$$

liquid water   water vapor

The change of a solid directly to the vapor is specifically referred to as **sublimation.** Although the vapor pressure of many solids is quite low, some (usually molecular solids) have appreciable vapor pressure. Ice, for instance, has a vapor pressure of 4.7 mmHg at 0°C. For this reason, a pile of snow slowly disappears in winter even though the temperature is too low for it to melt. The snow is being changed directly to water vapor.

$$H_2O(s) \longrightarrow H_2O(g) \qquad \text{(sublimation)}$$

ice, snow   water vapor

Sublimation can be used to purify solids that readily vaporize. Figure 11.3 shows a simple way to purify iodine by sublimation. Impure iodine is heated in a beaker so that it vaporizes, leaving nonvolatile impurities behind. The vapor crystallizes on the bottom surface of a dish containing ice that rests on top of the beaker. Freeze-drying of foods is a commercial application of sublimation. Brewed coffee, for example, is frozen and placed in a vacuum to remove water vapor. The ice continues to sublime until it is all gone, leaving freeze-dried coffee. Most freeze-dried foods are easily reconstituted by adding water.

Condensation is *the change of a gas to either the liquid or the solid state* (the change of vapor to solid is sometimes called *deposition*). Dew is liquid water formed by condensation of water vapor from the atmosphere. Frost is solid water formed by direct condensation of water vapor from the atmosphere without first forming liquid water. Snow is formed by a similar process in the upper atmosphere.

$$H_2O(g) \longrightarrow H_2O(l) \qquad \text{(condensation)}$$

water vapor   dew

$$H_2O(g) \longrightarrow H_2O(s) \qquad \text{(condensation, deposition)}$$

water vapor   frost, snow

Figure 11.3 ▼

Sublimation of iodine

When a substance that is normally a gas, such as carbon dioxide, changes to the liquid state, the phase transition is often referred to as *liquefaction.* The following diagram summarizes these phase transitions.

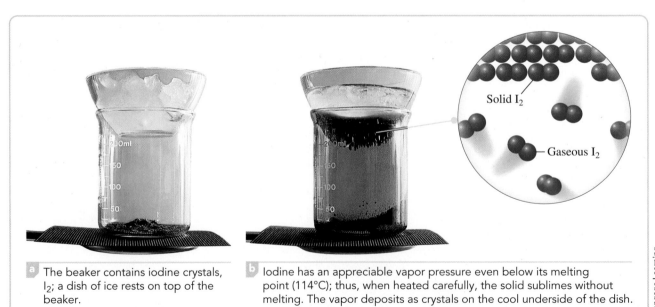

a The beaker contains iodine crystals, I_2; a dish of ice rests on top of the beaker.

b Iodine has an appreciable vapor pressure even below its melting point (114°C); thus, when heated carefully, the solid sublimes without melting. The vapor deposits as crystals on the cool underside of the dish.

Solid I_2

Gaseous I_2

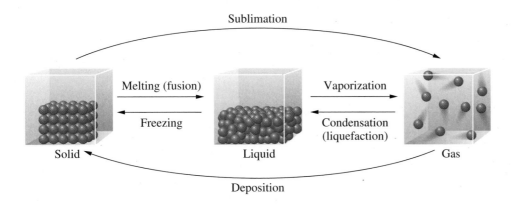

Sublimation

Melting (fusion)

Freezing

Vaporization

Condensation
(liquefaction)

Solid   Liquid   Gas

Deposition

Vapor Pressure

Liquids, and even some solids, are continuously vaporizing. If a liquid is in a closed vessel with space above it, a partial pressure of the vapor state builds up in this space. The **vapor pressure** of a liquid is *the partial pressure of the vapor over the liquid, measured at equilibrium at a given temperature.* To understand what we mean by equilibrium, let us look at a simple method for measuring vapor pressure.

You introduce a few drops of water from a medicine dropper into the mercury column of a barometer (Figure 11.4*a*). Being less dense than mercury, the water rises in the tube to the top of the mercury, where it evaporates, or vaporizes.

To understand the process of vaporization, it is necessary to realize that the molecules in a liquid have a distribution of kinetic energies (Figure 11.5). Those molecules moving away from the surface and toward the vapor phase will escape only if their kinetic energies are greater than a certain minimum value equal to the potential energy from the attraction of molecules in the body of the liquid.

Figure 11.4 ◄

Measurement of the vapor pressure of water

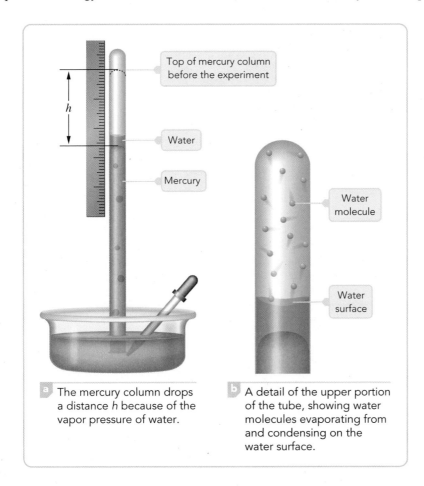

h

Top of mercury column before the experiment

Water

Mercury

Water molecule

Water surface

a The mercury column drops a distance *h* because of the vapor pressure of water.

b A detail of the upper portion of the tube, showing water molecules evaporating from and condensing on the water surface.

Figure 11.5 ▶

Distribution of kinetic energies of molecules in a liquid Distributions of kinetic energies are shown for two different temperatures. The fraction of molecules having kinetic energies greater than the minimum necessary for escape is given by the colored areas (orange for the lower temperature, red area plus orange area at the higher temperature). Note that the fraction of molecules having kinetic energies greater than the minimum value for vaporization increases with temperature.

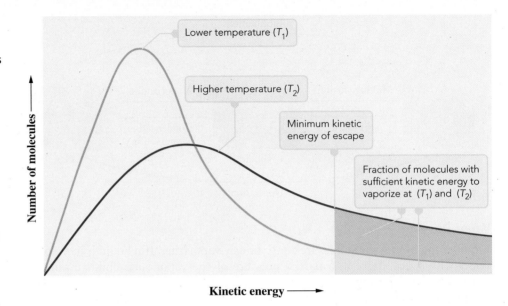

As molecules at the surface gain sufficient kinetic energy (through collisions with neighboring molecules), they escape the liquid and go into the space at the top of the barometer column (see Figure 11.4*B*). More and more water molecules begin to fill this space. As the number of molecules in the vapor state increases, more and more gaseous molecules collide with the liquid water surface, exerting pressure on it. So, as vaporization of water proceeds, the mercury column moves downward.

Some of the molecules in the vapor collide with the liquid surface and stick; that is, the vapor condenses to liquid. The rate of condensation steadily increases as the number of molecules in the volume of vapor increases, until the rate at which molecules are condensing on the liquid equals the rate at which molecules are vaporizing (Figure 11.6).

Figure 11.6 ▶

Rates of vaporization and condensation of a liquid over time As vaporization continues at a constant rate, the rate of condensation increases. Eventually, the two rates become equal. The partial pressure of vapor reaches a steady value at a given temperature, which is the vapor pressure of the liquid. The process is said to reach equilibrium. The flasks, A, B, and C, present the molecular view of the process at positions A, B, and C on the curve.

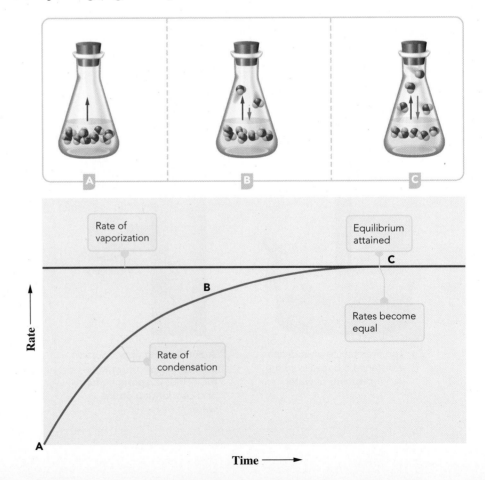

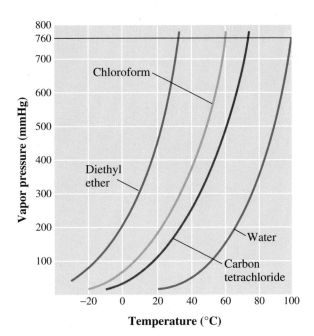

Figure 11.7 ◀

Variation of vapor pressure with temperature Several common liquids are shown. Note that the vapor pressure increases rapidly with temperature.

When the rates of vaporization and condensation have become equal, the number of molecules in the vapor at the top of the column in Figure 11.4 stops increasing and remains unchanged. The partial pressure of the vapor (as measured by the change in height of the mercury column) reaches a steady value, which is its vapor pressure. The liquid and vapor are in *equilibrium*. Although the partial pressure of the vapor is unchanging, molecules are still leaving the liquid and coming back to the liquid. For this reason, you speak of this as a *dynamic equilibrium*—one in which the molecular processes (in this case, vaporization and condensation) are continuously occurring. You represent this dynamic equilibrium for the vaporization and condensation of water by the equation with a double arrow.

$$H_2O(l) \rightleftharpoons H_2O(g)$$

As Figure 11.7 demonstrates, the vapor pressure of a substance depends on the temperature. (Appendix B gives a table of the vapor pressures of water at various temperatures.) As the temperature increases, the kinetic energy of molecular motion becomes greater, and the vapor pressure increases. Liquids and solids with relatively high vapor pressure at normal temperatures are said to be *volatile*. Chloroform and carbon tetrachloride are volatile liquids. Naphthalene, $C_{10}H_8$, and *para*-dichlorobenzene, $C_6H_4Cl_2$, are volatile solids: they have appreciable vapor pressures at room temperature. Both are used to make mothballs.

Boiling Point and Melting Point

The temperature at which the vapor pressure of a liquid equals the pressure exerted on the liquid (atmospheric pressure, unless the vessel containing the liquid is closed) is called the **boiling point** of the liquid. As the temperature of a liquid is raised, the vapor pressure increases. When the vapor pressure equals the atmospheric pressure, stable bubbles of vapor form within the body of the liquid (Figure 11.8). This process is called boiling. Once boiling starts, the temperature of the liquid remains at the boiling point (as long as sufficient heat is supplied).

Because the pressure exerted on a liquid can vary, the boiling point of a liquid can vary. For instance, at 1.00 atm (the average atmospheric pressure at sea level), the boiling point of water is 100°C. But at 0.83 atm (the average atmospheric pressure in Denver, Colorado, at 1609 m above sea level), the boiling point of water is 95°C. The *normal boiling point* of a liquid is the boiling point at 1 atm. ▶

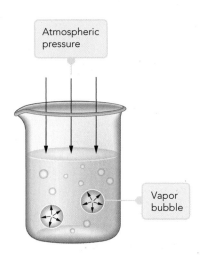

Figure 11.8 ▲

Boiling of a liquid A liquid boils when the vapor pressure inside a bubble of vapor equals the external pressure. The temperature at which this occurs is called the boiling point.

If you want to cook a stew in less time than is possible at atmospheric pressure, you can use a pressure cooker. This is especially valuable at high altitudes, where water boils at a lower temperature. The steam pressure inside the cooker is allowed to build up to a given value before being released by a valve. The stew then boils at a higher temperature and is cooked more quickly.

Table 11.1	Melting Points and Boiling Points (at 1 atm) of Several Substances		
Name	Type of Solid*	Melting Point, °C	Boiling Point, °C
Neon, Ne	Molecular	−249	−246
Hydrogen sulfide, H_2S	Molecular	−86	−61
Chloroform, $CHCl_3$	Molecular	−64	62
Water, H_2O	Molecular	0	100
Acetic acid, $HC_2H_3O_2$	Molecular	17	118
Mercury, Hg	Metallic	−39	357
Sodium, Na	Metallic	98	883
Tungsten, W	Metallic	3410	5660
Cesium chloride, CsCl	Ionic	645	1290
Sodium chloride, NaCl	Ionic	801	1413
Magnesium oxide, MgO	Ionic	2800	3600
Quartz, SiO_2	Covalent network	1610	2230
Diamond, C	Covalent network	3550	4827

*Types of solids are discussed in Section 11.6.

The temperature at which a pure liquid changes to a crystalline solid, or freezes, is called the **freezing point.** *The temperature at which a crystalline solid changes to a liquid, or melts,* is called the **melting point;** it is identical to the freezing point. The melting or freezing occurs at the temperature where the liquid and solid are in dynamic equilibrium.

$$\text{Solid} \rightleftharpoons \text{liquid}$$

Unlike boiling points, melting points are affected significantly only by large pressure changes.

Both the melting point and the boiling point are characteristic physical properties of a substance and can be used to help identify it. Table 11.1 gives melting points and boiling points of several substances.

Heat of Phase Transition

Any change of state involves the addition or removal of energy as heat to or from the substance. A simple experiment shows that this is the case. Suppose you add heat at a constant rate to a beaker containing ice at −20°C. In Figure 11.9, we have plotted the temperature of the different phases of water as heat is added. The temperature of the ice begins to rise from −20°C, as you would expect; the addition of heat normally raises the temperature of a substance. At 0°C, the ice begins to melt, so that you get a beaker of ice in water. Note the flat region in the curve, labeled ice and water. Why is this region flat? It means that heat is being added to the system without a change in temperature; the temperature remains at 0°C. This temperature, of course, is the melting point of ice. The heat being added is energy required to melt ice to water at the same temperature. The intermolecular forces binding water molecules to specific sites in the solid phase must be partially broken to allow water molecules the ability to slide over one another easily, as happens in the liquid state. Note the flat regions for each of the phase transitions. Because heat is being added at a constant rate, the length of each flat region is proportional to the *heat of phase transition.*

The heat needed for the melting of a solid is called the **heat of fusion** (or *enthalpy of fusion*) and is denoted ΔH_{fus}. For ice, the heat of fusion is 6.01 kJ per mole.

$$H_2O(s) \longrightarrow H_2O(l); \Delta H_{fus} = 6.01 \text{ kJ/mol}$$

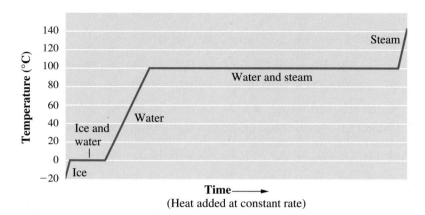

Figure 11.9 ◄

Heating curve for water Heat is being added at a constant rate to a system containing water. Note the flat regions of the curve. When heat is added during a phase transition, the temperature does not change.

The heat needed for the vaporization of a liquid is called the **heat of vaporization** (or *enthalpy of vaporization*) and is denoted ΔH_{vap}. At 100°C, the heat of vaporization of water is 40.7 kJ per mole. ▶

$$H_2O(l) \longrightarrow H_2O(g); \Delta H_{vap} = 40.7 \text{ kJ/mol}$$

Note that much more heat is required for vaporization than for melting. Melting needs only enough energy for the molecules to escape from their sites in the solid. For vaporization, enough energy must be supplied to break most of the intermolecular attractions.

A refrigerator relies on the cooling effect accompanying vaporization. The mechanism contains an enclosed gas that can be liquefied under pressure, such as ammonia or 1,1,1,2-tetrafluoroethane (NFC-134a), CH_2FCF_3. As the liquid is allowed to evaporate, it absorbs heat and thus cools its surroundings (the interior space of the refrigerator). Gas from the evaporation is recycled to a compressor and then to a condenser, where it is liquefied again. Heat leaves the condenser, going into the surrounding air.

The fact that heat is required for vaporization can be demonstrated by evaporating a dish of water inside a vessel attached to a vacuum pump. With the pressure kept low, water evaporates quickly enough to freeze the water remaining in the dish.

⌾WL Interactive Example | **11.1** | **Calculating the Heat Required for a Phase Change of a Given Mass of Substance**

Gaining Mastery Toolbox

Critical Concept 11.1
Substances undergoing a phase transition do not change temperature. The energy associated with phase transitions is linked to structural changes in the substance and not to changes in the kinetic energy of the molecules (temperature). For example, when ice melts at 0°C and 1 atm of pressure by absorbing energy, the molecular structure of water undergoes a change without any change in temperature.

Solution Essentials:
• Heat of fusion
• Heat of vaporization
• Freezing point
• Phase transition

A particular refrigerator cools by evaporating liquefied dichlorodifluoromethane, CCl_2F_2. How many kilograms of this liquid must be evaporated to freeze a tray of water at 0°C to ice at 0°C? The mass of the water is 525 g, the heat of fusion of ice is 6.01 kJ/mol, and the heat of vaporization of dichlorodifluoromethane is 17.4 kJ/mol.

Problem Strategy The cooling effect as CCl_2F_2 liquid evaporates is used to freeze water at 0°C to ice. You first need to obtain the quantity of heat removed from the water to freeze it (using the heat of fusion of ice as a conversion factor, converting moles of H_2O to the heat required). Then you use this quantity of heat to determine the amount of CCl_2F_2 to evaporate (now using the heat of vaporization of CCl_2F_2 to convert this quantity of heat to the amount of CCl_2F_2 that must be vaporized).

Solution The heat that must be removed to freeze 525 g of water at 0°C is

$$525 \text{ g } H_2O \times \frac{1 \text{ mol } H_2O}{18.0 \text{ g } H_2O} \times \frac{-6.01 \text{ kJ}}{1 \text{ mol } H_2O} = -175 \text{ kJ}$$

The minus sign indicates that heat energy is taken away from the water. Consequently, the vaporization of dichlorodifluoromethane absorbs 175 kJ of heat. The mass of CCl_2F_2 that must be vaporized to absorb this quantity of heat is

$$175 \text{ kJ} \times \frac{1 \text{ mol } CCl_2F_2}{17.4 \text{ kJ}} \times \frac{121 \text{ g } CCl_2F_2}{1 \text{ mol } CCl_2F_2} = 1.22 \times 10^3 \text{ g } CCl_2F_2$$

1.22 kg of dichlorodifluoromethane must be evaporated.

(continued)

(continued)

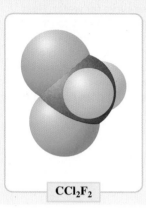

CCl$_2$F$_2$

Answer Check The calculations are similar to the stoichiometry calculations we did in Chapter 3. As in those calculations, make sure that the units properly cancel to convert those of the known quantity to the units of the desired quantity.

Exercise 11.1 The heat of vaporization of ammonia is 23.4 kJ/mol. How much heat is required to vaporize 1.00 kg of ammonia? How many grams of water at 0°C could be frozen to ice at 0°C by the evaporation of this amount of ammonia?

■ See Problems 11.43, 11.44, 11.45, and 11.46.

CONCEPT CHECK 11.1

Shown here is a representation of a closed container in which you have just placed 10 L of H$_2$O. In our experiment, we are going to call this starting point in time $t = 0$ and assume that all of the H$_2$O is in the liquid phase. We have represented a few of the H$_2$O molecules in the water as dots.

$t = 0$

a Consider a time $t = 1$, at which some time has passed but the system has not reached equilibrium.

  i. How will the level of the liquid H$_2$O compare to that at $t = 0$?
  ii. How will the vapor pressure in the flask compare to that at $t = 0$?
  iii. How will the number of H$_2$O molecules in the vapor state compare to that at $t = 0$?
  iv. How does the rate of evaporation in this system compare to the rate of condensation?
  v. Draw a picture of the system at $t = 1$.

b Consider a time $t = 2$, at which enough time has passed for the system to reach equilibrium.

  i. How will the level of the liquid H$_2$O compare to that at $t = 1$?
  ii. How will the vapor pressure in the flask compare to that at $t = 1$?
  iii. How will the number of H$_2$O molecules in the vapor state compare to that at $t = 1$?
  iv. How does the rate of evaporation in this system compare to the rate of condensation?
  v. Draw a picture of the system at $t = 2$.

Clausius–Clapeyron Equation: Relating Vapor Pressure and Liquid Temperature

We noted earlier that the vapor pressure of a substance depends on temperature. The variation of vapor pressure with temperature of some liquids was given in Figure 11.7. It can be shown that the logarithm of the vapor pressure of a liquid or solid varies with the absolute temperature according to the following approximate relation:

$$\ln P = -\frac{A}{T} + B$$

Here ln P is the natural logarithm of the vapor pressure, and A and B are positive constants. You can confirm this relation for the liquids shown in Figure 11.7 by replotting the data. If you put $y = \ln P$ and $x = 1/T$, the previous relation becomes

$$y = -Ax + B$$

This means that if you plot $\ln P$ versus $1/T$, you should get a straight line with slope $-A$. The data of Figure 11.7 are replotted this way in Figure 11.10. ▶

The previous equation has been derived from thermodynamics, by assuming the vapor behaves like an ideal gas. The result, known as the *Clausius–Clapeyron equation,* shows that the constant A is proportional to the heat of vaporization of the liquid, ΔH_{vap}.

$$\ln P = \frac{-\Delta H_{vap}}{RT} + B$$

A two-point form of the Clausius–Clapeyron equation is very useful for calculations. Let us write the previous equation for two different temperatures:

$$\ln P_2 = \frac{-\Delta H_{vap}}{RT_2} + B$$

$$\ln P_1 = \frac{-\Delta H_{vap}}{RT_1} + B$$

Here P_1 is the vapor pressure at absolute temperature T_1, and P_2 is the vapor pressure at absolute temperature T_2. If you subtract the second equation from the first, you get

$$\ln P_2 - \ln P_1 = \frac{-\Delta H_{vap}}{RT_2} - \frac{-\Delta H_{vap}}{RT_1} + B - B$$

which you can write as

$$\ln \frac{P_2}{P_1} = \frac{\Delta H_{vap}}{R}\left(\frac{1}{T_1} - \frac{1}{T_2}\right)$$

The next two examples illustrate the use of this two-point form of the Clausius– Clapeyron equation.

See Appendix A for the graphing of a straight line.

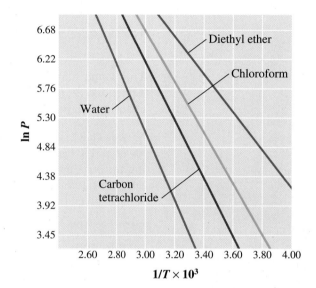

Figure 11.10 ▲

Plot of the logarithm of vapor pressure versus 1/T The liquids shown in Figure 11.7 are replotted in this graph. Note the straight-line relationship.

Example 11.2 **Calculating the Vapor Pressure at a Given Temperature**

Critical Concept 11.2
The vapor pressure of a liquid varies with temperature in a predictable manner. The variation of vapor pressure with temperature of a liquid can be expressed as the Clausius-Clapeyron equation.

Solution Essentials:
• Clausius-Clapeyron equation
$\left(\ln \frac{P_2}{P_1} = \frac{\Delta H_{vap}}{R}\left(\frac{1}{T_1} - \frac{1}{T_2}\right)\right)$
• Heat of vaporization
• Vapor pressure
• Normal boiling point

Estimate the vapor pressure of water at 85°C. Note that the normal boiling point of water is 100°C and that its heat of vaporization is 40.7 kJ/mol.

Problem Strategy The two-point form of the Clausius–Clapeyron equation relates five quantities: P_1, T_1, P_2, T_2, and ΔH_{vap}. The normal boiling point is the temperature at which the vapor pressure equals 760 mmHg. Thus, you can let P_1 equal 760 mmHg and T_1 equal 373 K (100°C). Then P_2 would be the vapor pressure at T_2, or 358 K (85°C). The heat of vaporization, ΔH_{vap}, is 40.7×10^3 J/mol. Note that we have expressed this quantity in joules per mole to agree with the units of the molar gas constant, R, which equals 8.31 J/(K · mol).

Solution You can now substitute into the two-point equation (the equation shaded in color just before this example).

$$\ln \frac{P_2}{760 \text{ mmHg}} = \frac{40.7 \times 10^3 \text{ J/mol}}{8.31 \text{ J/(K· mol)}}\left(\frac{1}{373 \text{ K}} - \frac{1}{358 \text{ K}}\right)$$

$$= (4898 \text{ K}) \times (-1.123 \times 10^{-4}/\text{K}) = -0.550$$

(continued)

(continued)

You now want to solve this equation for P_2, which is the vapor pressure of water at 85°C. To do so, you take the antilogarithm (antiln) of both sides of the equation. This gives

$$\frac{P_2}{760 \text{ mmHg}} = \text{antiln}(-0.550)$$

The antiln (-0.550) is $e^{-0.550} = 0.577$. (Appendix A describes how to obtain the antilogarithm.) Thus

$$\frac{P_2}{760 \text{ mmHg}} = 0.577$$

Therefore,

$$P_2 = 0.577 \times 760 \text{ mmHg} = \textbf{439 mmHg}$$

This value may be compared with the experimental value of 434 mmHg, given in Appendix B. Differences from the experimental value stem from the fact that the Clausius–Clapeyron equation is an approximate relation.

Answer Check The result should be less than 760 mmHg, because the vapor pressure is lower at a lower temperature. If you inadvertently dropped the minus sign before taking the antiln, you would have gotten a result larger than 760 mmHg. Also, if you had entered ΔH_{vap} in kilojoules, rather than joules, the result would be close to zero, which is not reasonable given that the water is at 85°C.

Exercise 11.2 Carbon disulfide, CS_2, has a normal boiling point of 46°C and a heat of vaporization of 26.8 kJ/mol. What is the vapor pressure of carbon disulfide at 35°C?

■ See Problems 11.49 and 11.50.

Example 11.3

Calculating the Heat of Vaporization from Vapor Pressures

Gaining Mastery Toolbox

Critical Concept 11.3
The heat of vaporization of a liquid, ΔH_{vap}, does not depend upon temperature. The two-point form of the Clausius-Clapeyron equation can be used to calculate the ΔH_{vap} of a liquid.

Solution Essentials:
• Clausius-Clapeyron equation
$\left(\ln \dfrac{P_2}{P_1} = \dfrac{\Delta H_{vap}}{R}\left(\dfrac{1}{T_1} - \dfrac{1}{T_2}\right) \right)$
• Heat of vaporization
• Vapor pressure

Calculate the heat of vaporization of diethyl ether (often called simply "ether"), $C_4H_{10}O$, from the following vapor pressures: 400 mmHg at 18°C and 760 mmHg at 35°C. (Each pressure value has three significant figures.)

Problem Strategy If you let P_1 be 400 mmHg, then T_1 is 291 K (18°C), P_2 is 760 mmHg, and T_2 is 308 K (35°C). You substitute these values into the Clausius–Clapeyron equation.

Solution

$$\ln \frac{760 \text{ mmHg}}{400 \text{ mmHg}} = \frac{\Delta H_{vap}}{8.31 \text{ J/(K·mol)}} \times \left(\frac{1}{291 \text{ K}} - \frac{1}{308 \text{ K}}\right)$$

$$0.642 = (2.28 \times 10^{-5}) \times \Delta H_{vap}/(\text{J/mol})$$

Therefore,

$$\Delta H_{vap} = 2.82 \times 10^4 \text{ J/mol} \quad \textbf{(28.2 kJ/mol)}$$

Answer Check Whenever you are dealing with calculated quantities of heat, always make sure that the sign of your answer makes sense. In this case, because vaporization is a process that requires energy, the answer should always be a positive quantity as it is here. Also, enthalpies of vaporization generally fall in the range of tens of kilojoules per mole, so anything much outside that range should be suspect.

Exercise 11.3 Selenium tetrafluoride, SeF_4, is a colorless liquid. It has a vapor pressure of 757 mmHg at 105°C and 522 mmHg at 95°C. What is the heat of vaporization of selenium tetrafluoride?

■ See Problems 11.51 and 11.52.

11.3 Phase Diagrams

As we mentioned in the chapter opening, the solid, liquid, and gaseous states of carbon dioxide exist under different temperature and pressure conditions. A **phase diagram** is *a graphical way to summarize the conditions under which the different states of a substance are stable.*

Melting-Point Curve

Figure 11.11 is a phase diagram for water. It consists of three curves that divide the diagram into regions labeled "solid," "liquid," and "gas." In each region, the indicated state is stable. Every point on each of the curves indicates experimentally determined temperatures and pressures at which two states are in equilibrium. Thus, the curve *AB*, dividing the solid region from the liquid region, represents the conditions under which the solid and liquid are in equilibrium.

$$\text{Solid} \rightleftharpoons \text{liquid}$$

This curve gives the melting points of the solid at various pressures.

Usually, the melting point is only slightly affected by pressure. For this reason, the melting-point curve *AB* is nearly vertical. If a liquid is more dense than the solid, as is the case for water, the melting point decreases with pressure. ▶ The melting-point curve in such cases leans slightly to the left. In the case of ice, the decrease is indeed slight—only 1°C for a pressure increase of 133 atm. For most substances, the liquid state is less dense than the solid. In that case, the melting-point curve leans slightly to the right.

Vapor-Pressure Curves for the Liquid and the Solid

The curve *AC* that divides the liquid region from the gaseous region gives the vapor pressures of the liquid at various temperatures. It also gives the boiling points of the liquid for various pressures. The boiling point of water at 1 atm is shown on the phase diagram (Figure 11.11).

The curve *AD* that divides the solid region from the gaseous region gives the vapor pressures of the solid at various temperatures. This curve intersects the other curves at point *A*, the **triple point,** which is *the point on a phase diagram representing the temperature and pressure at which three phases of a substance coexist in equilibrium.* For water, the triple point occurs at 0.01°C, 0.00603 atm (4.58 mmHg), and the solid, liquid, and vapor phases coexist. ▶

Suppose a solid is warmed at a pressure below the pressure at the triple point. In a phase diagram, this corresponds to moving along a horizontal line below the triple point. You can see from Figure 11.11 that such a line will intersect curve *AD*, the vapor-pressure curve for the solid. Thus, the solid will pass directly into the gas; that is, the solid will sublime. Freeze-drying of a food (or brewed coffee) is accomplished by placing the frozen food in a vacuum (below 0.00603 atm) so

The following is a dramatic demonstration of the effect of pressure on the melting of ice: Suspend a block of ice between two chairs. Then loop a length of wire over the block and place weights on the ends of the wire. The pressure of the wire will melt the ice under it. Liquid water will flow over the top of the wire and freeze again, because it is no longer under the pressure of the wire. As a result, the wire will cut through the ice and fall to the floor, but the block of ice will remain as one piece.

Because the triple point for water occurs at a definite temperature, it is used to define the Kelvin temperature scale. The temperature of water at its triple point is defined to be 273.16 K (0.01°C).

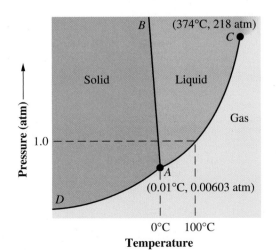

Figure 11.11 ◀

Phase diagram for water (not to scale) The curves *AB*, *AC*, and *AD* divide the diagram into regions that give combinations of temperature and pressure for which only one state is stable. Along any curve, the two states from the adjoining regions are in equilibrium.

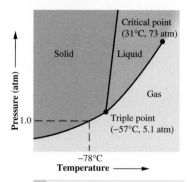

a Carbon dioxide. At normal atmospheric pressure (1 atm), the solid sublimes when warmed.

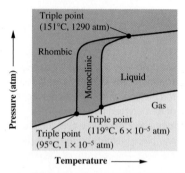

b Sulfur. This substance has a complicated phase diagram with three triple points.

Figure 11.12 ▲

Phase diagrams for carbon dioxide and sulfur (not to scale)

Figure 11.13 ▶

Observing the critical phenomenon

that the ice in it sublimes. Because the food can be dried at a lower temperature than if heat-dried, it retains more flavor and can often be reconstituted by simply adding water.

The triple point of carbon dioxide is at −57°C and 5.1 atm (Figure 11.12*a*). Therefore, the solid sublimes if warmed at any pressure below 5.1 atm. This is why solid carbon dioxide sublimes at normal atmospheric pressure (1 atm). Above 5.1 atm, however, the solid melts if warmed. Sulfur has a more complicated phase diagram (Figure 11.12*b*). It displays three triple points, one of them involving two different solid forms of sulfur (called rhombic sulfur and monoclinic sulfur), as well as the vapor.

Critical Temperature and Pressure

Imagine an experiment in which liquid and gaseous carbon dioxide are sealed in a thick-walled glass vessel at 20°C. At this temperature, the liquid is in equilibrium with its vapor at a pressure of 57 atm. You observe that the liquid and vapor are separated by a well-defined boundary, or meniscus. Now suppose the temperature is raised. The vapor pressure increases, and at 30°C it is 71 atm. Then, as the temperature approaches 31°C, a curious thing happens. The meniscus becomes fuzzy and less well defined. At 31°C, the meniscus disappears altogether. Above this temperature and above the critical pressure, there is only one fluid state, called a *supercritical fluid* (see Figure 11.13).

The temperature above which the liquid state of a substance no longer exists regardless of the pressure is called the **critical temperature.** For carbon dioxide it is 31°C. *The vapor pressure at the critical temperature* is called the **critical pressure** (73 atm for carbon dioxide). It is the minimum pressure that must be applied to a gas at the critical temperature to liquefy it.

On a phase diagram, the preceding experiment corresponds to following the vapor-pressure curve where the liquid and vapor are in equilibrium. Note that this curve in Figure 11.12*a* ends at a point at which the temperature and pressure have their critical values. This is the *critical point.* If you look at the phase diagram

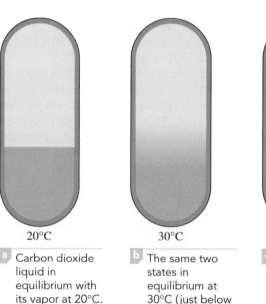

20°C	30°C	31°C

a Carbon dioxide liquid in equilibrium with its vapor at 20°C.

b The same two states in equilibrium at 30°C (just below the critical point); the densities of the liquid and vapor are becoming equal.

c Carbon dioxide at 31°C (the critical temperature); the liquid and vapor now have the same densities—in fact, the distinction between liquid and vapor has disappeared, resulting in what is called a supercritical fluid.

for water, you will see that the vapor-pressure curve for the liquid similarly ends, at point *C*, which is the critical point for water. In this case, the critical temperature is 374°C and the critical pressure is 218 atm.

Many important gases cannot be liquefied at room temperature. Nitrogen, for example, has a critical temperature of −147°C. This means the gas cannot be liquefied until the temperature is below −147°C.

Example 11.4   **Relating the Conditions for the Liquefaction of a Gas to Its Critical Temperature**

Gaining Mastery Toolbox

Critical Concept 11.4
No substance can be liquefied if its temperature is at or above the critical temperature. There are substances that cannot be liquefied at normal (room) temperatures no matter what the pressure.

Solution Essentials:
• Critical temperature
• Phase transition

The critical temperatures of ammonia and nitrogen are 132°C and −147°C, respectively. Explain why ammonia can be liquefied at room temperature by merely compressing the gas to a high enough pressure, whereas the compression of nitrogen requires lower temperatures for liquefaction to occur.

Problem Strategy Note that above the critical temperature, a gas cannot be liquefied (that is, transformed so that the gas coexists in equilibrium with the liquid). This can occur only below the critical temperature.

Solution **The critical temperature of ammonia is well above room temperature.** If ammonia gas at room temperature is compressed, it will liquefy. **Nitrogen, however, has a critical temperature well below room temperature.** It cannot be liquefied by compression unless the temperature is below the critical temperature.

Answer Check Make sure that you understand the critical temperature correctly; liquefaction can occur only *below* the critical temperature if sufficient pressure has been applied to the gas.

Exercise 11.4 Describe how you could liquefy the following gases: a methyl chloride, CH_3Cl (critical point, 144°C, 66 atm); b oxygen, O_2 (critical point, −119°C, 50 atm).

■ See Problems 11.57 and 11.58.

CONCEPT CHECK 11.2

When camping at high altitude, you need to pay particular attention to changes in cooking times for foods that are boiled in water. If you like eggs that are boiled for 10 minutes near sea level, would you have to cook them for a longer or a shorter time at 3200 m to get the egg you like? Be sure to explain your answer.

Liquid State

Now we will look at some physical properties of liquids. After that, we will explain the experimental values of these properties in terms of intermolecular forces.

11.4 Properties of Liquids; Surface Tension and Viscosity

You have seen that molecules tend to escape the liquid state and form a vapor. The vapor pressure is the equilibrium partial pressure of this vapor over the liquid; it increases with temperature. The boiling point is the temperature at which the vapor pressure equals the pressure applied to the liquid. Both vapor pressure and boiling point are important properties of a liquid. Values of the vapor pressure for some liquids at 20°C are listed in Table 11.2 (column 3). Two additional properties given in Table 11.2 are surface tension and viscosity. All of these properties, as you will see in the next section, depend on intermolecular forces, which are related to molecular structure.

Removing Caffeine from Coffee

Legend has it that a goat herder named Kaldi, in the ancient country of Abyssinia (now Ethiopia, in northeast Africa), discovered the pleasant effect of coffee beans when he noted that his goats danced wildly about after eating the shiny leaves and red berries of a small tree. Later he tried eating the whole berries himself, and he soon found that he was dancing with his goats. Whether the legend is true or not, it does effectively describe the stimulant effect of this berry from the coffee tree. As we now know, coffee, which today is obtained by roasting the berries' green pits, or beans as they are called, contains the stimulant caffeine, a bitter-tasting white substance with the formula $C_8H_{10}N_4O_2$. (Figure 11.14 shows the structural formula of caffeine.)

For those who like the taste of roasted coffee but don't want the caffeine, decaffeinated coffee is available. A German chemist, Ludwig Roselius, first made "decaf" coffee about 1900 by extracting the caffeine from green coffee beans with the solvent chloroform, $CHCl_3$. Later, commercial processes replaced chloroform with the safer organic solvent methylene chloride, CH_2Cl_2. Today, though, most of the commercial decaffeinated coffee produced (Figure 11.15) uses *supercritical* carbon dioxide as the extracting fluid.

In a tank of carbon dioxide under pressure (for example, in a CO_2 fire extinguisher), the substance normally exists as the liquid in equilibrium with its gas phase. But we know from the previous text discussion that above 31°C (88°F), the two phases, gas and liquid, are replaced by a single fluid phase. So, on a hot summer day (above 31°C, or 88°F), the carbon dioxide in such a tank is above its critical temperature and pressure and exists as the supercritical fluid.

Supercritical carbon dioxide is a near-ideal solvent. Under normal conditions, carbon dioxide is not a very good solvent for organic substances, but supercritical carbon dioxide readily dissolves many of these substances, including caffeine. It is nontoxic and nonflammable. It also has no effect on the stratospheric ozone layer, whereas methylene chloride does. (See the essay on stratospheric ozone at the end of Chapter 10.) Carbon dioxide does contribute to the greenhouse effect (discussed in the essay at the end of Chapter 5), but the gas once used can be recirculated for solvent use and not vented to the atmosphere.

Supercritical fluids have gained much attention recently because of the possibility of replacing toxic and environmentally less desirable solvents. For example, at the moment, the usual solvent used to dry-clean clothes is perchloroethylene, CCl_2CCl_2. Although nonflammable and less toxic than carbon tetrachloride, which was the solvent previously used, perchloroethylene is regulated as an air pollutant under the Clean Air Act. Some scientists have shown that you can dry-clean with supercritical carbon dioxide if you use a special detergent.

Substances other than carbon dioxide have also shown intriguing solvent properties. For example, whereas water under normal conditions dissolves ionic and polar substances, above its critical point (374°C, 217 atm) it becomes an excellent solvent for nonpolar substances. Supercritical water and carbon dioxide promise to replace many toxic or environmentally unfriendly organic solvents.

Space-filling model of caffeine, $C_8H_{10}N_4O_2$.

Figure 11.14 ▲

The caffeine molecule Space-filling model of caffeine, $C_8H_{10}N_4O_2$.

© Cengage Learning

Figure 11.15 ▲

Decaf coffee Coffee that was "naturally decaffeinated" using supercritical carbon dioxide.

■ See Problems 11.123 and 11.124.

Table 11.2	Properties of Some Liquids at 20°C			
Substance	Molecular Weight (amu)	Vapor Pressure (mmHg)	Surface Tension (J/m^2)	Viscosity (kg/m·s)
Water, H_2O	18	1.8×10^1	7.3×10^{-2}	1.0×10^{-3}
Carbon dioxide, CO_2	44	4.3×10^4	1.2×10^{-3}	7.1×10^{-5}
Pentane, C_5H_{12}	72	4.4×10^2	1.6×10^{-2}	2.4×10^{-4}
Glycerol, $C_3H_8O_3$	92	1.6×10^{-4}	6.3×10^{-2}	1.5×10^0
Chloroform, $CHCl_3$	119	1.7×10^2	2.7×10^{-2}	5.8×10^{-4}
Carbon tetrachloride, CCl_4	154	8.7×10^1	2.7×10^{-2}	9.7×10^{-4}
Bromoform, $CHBr_3$	253	3.9×10^0	4.2×10^{-2}	2.0×10^{-3}

Surface Tension

As Figure 11.16 illustrates, a molecule within the body of a liquid tends to be attracted equally in all directions, so that it experiences no net force. On the other hand, a molecule at the surface of a liquid experiences a net attraction by other molecules toward the interior of the liquid. As a result, there is a tendency for the surface area of a liquid to be reduced as much as possible. This explains why falling raindrops are nearly spherical. (The sphere has the smallest surface area for a given volume of any geometrical shape.)

Energy is needed to reverse the tendency toward reduction of surface area in liquids. **Surface tension** is *the energy required to increase the surface area of a liquid by a unit amount.* Column 4 in Table 11.2 lists values of the surface tension of some liquids at 20°C.

The surface tension of a liquid can be affected by dissolved substances. Soaps and detergents, in particular, drastically decrease the surface tension of water. An interesting but simple experiment shows the effect of soap on surface tension. Because of surface tension, a liquid behaves as though it had a skin. Water bugs (striders) seem to skitter across this skin as if ice-skating (Figure 11.17). You can actually float a pin on water, if you carefully lay it across the surface (Figure 11.18). If you then put a drop of soap solution onto the water, the soap spreads across the surface and the pin sinks. Water bugs also sink in soapy water.

Capillary rise is a phenomenon related to surface tension. When a small-diameter glass tube, or capillary, is placed upright in water, a column of the liquid rises in the tube (Figure 11.19*a*). This capillary rise can be explained in the following way: water molecules happen to be attracted to glass. Because of this attraction, a thin film of water starts to move up the inside of the glass

Figure 11.16 ▲

Explaining surface tension Note that a molecule at the surface experiences a net force toward the interior of the liquid, whereas a molecule in the interior experiences no net force.

Figure 11.17 ◄

Water strider on a water surface The water strider seems to skate on the surface "skin" of the water, which is actually a result of surface tension.

Figure 11.18 ▲

Demonstration of surface tension of water A steel pin will float on the surface of water because of surface tension (*top*). The water surface is depressed and slightly stretched by the pin (*bottom*), but even more surface would have to be created for the pin to submerge, which takes more energy. The easiest way to float a pin is to place a square of tissue paper, with the pin on top of it, on the prongs of a fork. When the fork is lowered into the water, the tissue and pin float on the surface. The paper soon becomes water-logged and sinks, leaving the pin floating.

capillary. But in order to reduce the surface area of this film, the water level begins to rise also. The final water level is a balance between the surface tension and the potential energy required to lift the water against gravity. Note that the *meniscus,* or liquid surface within the tube, has its edges curved upward; that is, it has a concave shape.

In the case of mercury, the liquid level in the capillary is lower than for the liquid outside (Figure 11.19*b*). Also, the meniscus has its edges curved downward; that is, it has a convex shape. The difference between Figures 11.19*a* and 11.19*b* results from the fact that the attraction between mercury atoms is greater than the attraction of mercury atoms for glass, in contrast to the situation for water and glass.

Viscosity

Viscosity is *the resistance to flow that is exhibited by all liquids and gases.* The viscosity of a liquid can be obtained by measuring the time it takes for a given quantity to flow through a capillary tube or, alternatively, by the time it takes for a steel ball of given radius to fall through a column of liquid (Figure 11.20). Even without such quantitative measurements, you know from experience that the viscosity of syrup is greater than that of water; syrup is a *viscous* liquid. Viscosity is an important characteristic of motor oils. In the United States, the Society of Automotive Engineers (SAE) has established numbers to indicate the viscosity of motor oils at a given temperature (0°F, or −18°C); higher numbers indicate greater viscosities at that temperature. The viscosity of a simple oil increases as the temperature decreases, so an SAE 40 oil that is appropriate for an auto engine in the summer would be too viscous in very cold weather. An oil designated as SAE 10W/40 is an oil mixture that behaves like an SAE 10 oil in the winter (W), but like an SAE 40 oil in the summer. Column 5 in Table 11.2 gives the viscosities (in SI units) of some liquids at 20°C.

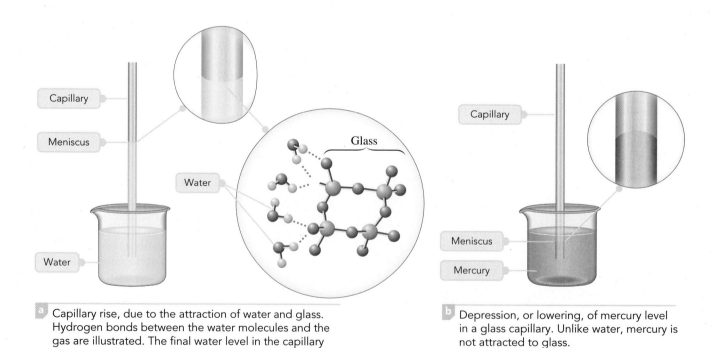

a Capillary rise, due to the attraction of water and glass. Hydrogen bonds between the water molecules and the gas are illustrated. The final water level in the capillary is a balance between the force of gravity and the surface tension of water.

b Depression, or lowering, of mercury level in a glass capillary. Unlike water, mercury is not attracted to glass.

Figure 11.19 ▲

Liquid levels in capillaries

11.5 Intermolecular Forces; Explaining Liquid Properties

In the preceding section, we described a number of properties of liquids. Now we want to explain these properties. Many of the physical properties of liquids (and certain solids too) can be explained in terms of **intermolecular forces,** *the forces of interaction between molecules.* These forces are normally weakly attractive.

One of the most direct indications of the attraction between molecules is the heat of vaporization of liquids. Consider a substance like neon, which consists of molecules that are single atoms. (There is no tendency for these atoms to bond chemically.) Neon is normally a gas, but it liquefies when the temperature is lowered to −246°C at 1.00 atm pressure. The heat of vaporization of the liquid at this temperature is 1.77 kJ/mol. Some of this energy (0.23 kJ/mol) is needed to push back the atmosphere when the vapor forms. The remaining energy (1.54 kJ/mol) must be supplied to overcome intermolecular attractions. Because a molecule in a liquid is surrounded by several neighbor molecules, this remaining energy is some multiple of a single molecule–molecule interaction. Typically, this multiple is about 5, so we expect the neon–neon interaction energy to be about 0.3 kJ/mol. Other experiments yield a similar value. By comparison, the energy of attraction between two hydrogen atoms in the hydrogen molecule is 432 kJ/mol. Thus, the energy of attraction between neon atoms is about a thousand times smaller than that between atoms in a chemical bond.

Attractive intermolecular forces can be larger than those in neon. For example, chlorine, Cl_2, and bromine, Br_2, have intermolecular attractive energies of 3.0 kJ/mol and 4.3 kJ/mol, respectively. But even these values are much smaller than bond energies.

Three types of attractive forces are known to exist between neutral molecules: dipole–dipole forces, London (or dispersion) forces, and hydrogen bonding forces. Each of these will be explained in this section. The term **van der Waals forces** is *a general term for those intermolecular forces that include dipole–dipole and London forces.* ▶ Van der Waals forces are the weak attractive forces in a large number of substances, including Ne, Cl_2, and Br_2, which we just discussed. Hydrogen bonding occurs in substances containing hydrogen atoms bonded to certain very electronegative atoms. Approximate energies of intermolecular attractions are compared with those of chemical bonds in Table 11.3.

Dipole–Dipole Forces

Polar molecules can attract one another through dipole–dipole forces. The **dipole–dipole force** is *an attractive intermolecular force resulting from the tendency of polar molecules to align themselves such that the positive end of one molecule is near the negative end of another.* Recall that a polar molecule has a dipole moment as a result of the electronic structure of the molecule. ▶ For example, hydrogen chloride, HCl, is a polar molecule because of the difference in electronegativities of the H and Cl atoms.

$$\overset{\delta+}{H} - \overset{\delta-}{Cl}$$

Figure 11.21 shows the alignment of polar molecules in the case of HCl. Note that this alignment creates a net attraction between molecules. This attractive force is partly responsible for the fact that hydrogen chloride, a gas at room temperature, becomes a liquid when cooled to −85°C. At this temperature, HCl molecules have slowed enough (that is, have a low enough kinetic energy) for the intermolecular forces to hold the molecules in the liquid state.

London (Dispersion) Forces

In nonpolar molecules, there can be no dipole–dipole force. Yet such substances liquefy, so there must be another type of attractive intermolecular force. Even in

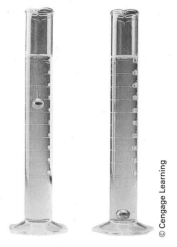

Figure 11.20 ▲

Comparison of the viscosities of two liquids Similar steel balls were dropped simultaneously into two graduated cylinders, the right one containing water and the left one glycerol. A steel ball takes considerably longer to fall through a column of glycerol than through a similar column of water, because glycerol has a greater viscosity than water.

The Dutch physicist Johannes van der Waals (1837–1923) was the first to suggest the importance of intermolecular forces and used the concept to derive his equation for gases. (See van der Waals equation, Section 5.8.)

Dipole moments were discussed in Section 10.2.

Table 11.3	Types of Intermolecular and Chemical Bonding Interactions	
Type of Interaction		**Approximate Energy (kJ/mol)**
Intermolecular		
Van der Waals		
(dipole–dipole, London)		0.1 to 10
Hydrogen bonding		10 to 40
Chemical bonding		
Ionic		100 to 1000
Covalent		100 to 1000

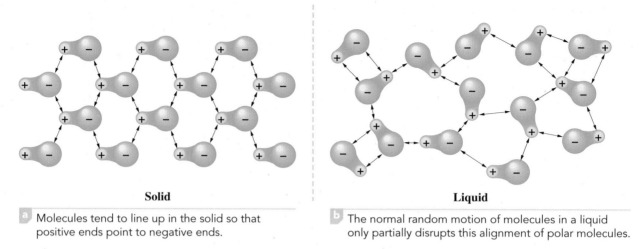

Solid

a Molecules tend to line up in the solid so that positive ends point to negative ends.

Liquid

b The normal random motion of molecules in a liquid only partially disrupts this alignment of polar molecules.

Figure 11.21 ▲

Alignment of polar molecules of HCl

hydrogen chloride, calculations show that the dipole–dipole force accounts for less than 20% of the total energy of attraction. What is the nature of the additional attractive force?

In 1930 Fritz London found he could account for a weak attraction between any two molecules. Let us use neon to illustrate his argument. The electrons of this atom move about the nucleus so that over a period of time they have a spherical distribution. The electrons spend as much time on one side of the nucleus as on the other. Experimentally, you find that the atom has no dipole moment. However, at any instant, more electrons may be on one side of the nucleus than on the other. The atom becomes a small, instantaneous dipole, with one side having a partial negative charge, δ^-, and the other side having a partial positive charge, δ^+ (see Figure 11.22a).

Now imagine that another neon atom is close to this first atom—say, next to the partial negative charge that appeared at that instant. This partial negative charge repels the electrons in the second atom, so at this instant it too becomes a small dipole (see Figure 11.22b). Note that these two dipoles are oriented with the partial negative charge of one atom next to the partial positive charge of the other atom. Therefore, an attractive force exists between the two atoms.

Electrons are in constant motion, but the motion of electrons on one atom affects the motion of electrons on the other atom. As a result, the *instantaneous dipoles* of the atoms tend to change together, maintaining a net attractive force (compare *b* and *c* of Figure 11.22). Such instantaneous changes of electron distributions occur in all molecules. For this reason, an attractive force always exists between any two molecules.

Figure 11.22 ▼

Origin of the London force

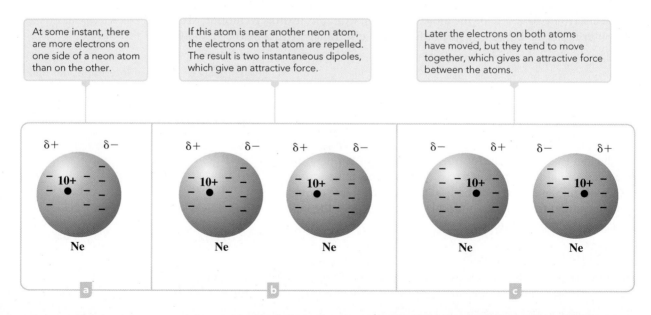

At some instant, there are more electrons on one side of a neon atom than on the other.

If this atom is near another neon atom, the electrons on that atom are repelled. The result is two instantaneous dipoles, which give an attractive force.

Later the electrons on both atoms have moved, but they tend to move together, which gives an attractive force between the atoms.

Thus, you see that **London forces** (also called **dispersion forces**) are *the weak attractive forces between molecules resulting from the small, instantaneous dipoles that occur because of the varying positions of the electrons during their motion about nuclei.*

London forces tend to increase with molecular mass. This is because molecules with larger molecular mass usually have more electrons, and London forces increase in strength with the number of electrons. Also, larger molecular mass often means larger atoms, which are more *polarizable* (more easily distorted to give instantaneous dipoles because the electrons are farther from the nuclei). The relationship of London forces to molecular mass is only approximate, however. For molecules of about the same molecular mass, the more compact one is less polarizable, so the London forces are smaller. Consider pentane, 2-methylbutane, and 2,2-dimethylpropane.

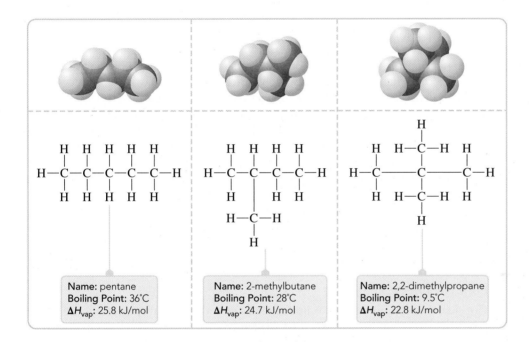

Name: pentane
Boiling Point: 36°C
ΔH_{vap}: 25.8 kJ/mol

Name: 2-methylbutane
Boiling Point: 28°C
ΔH_{vap}: 24.7 kJ/mol

Name: 2,2-dimethylpropane
Boiling Point: 9.5°C
ΔH_{vap}: 22.8 kJ/mol

All have the same molecular formula, C_5H_{12}, and thus the same molecular mass, but they differ in the arrangement of the atoms. Pentane is a long chain of carbon atoms to which hydrogen atoms are attached. However, 2-methylbutane and neopentane have increasingly more compact arrangements of atoms. As a result, you expect London forces to decrease from pentane to 2-methylbutane and from 2-methylbutane to neopentane. This is confirmed in the heats of vaporization and boiling points listed above.

Intermolecular Forces and the Properties of Liquids

The vapor pressure of a liquid depends on intermolecular forces, because the ease or difficulty with which a molecule leaves the liquid depends on the strength of its attraction to other molecules. When the intermolecular forces in the liquid are strong, you expect the vapor pressure to be low.

Let us look again at Table 11.2, which lists vapor pressures as well as surface tensions and viscosities of some liquids at 20°C. The intermolecular attractions in these liquids (except for water and glycerol) are due entirely to van der Waals forces. Of the van der Waals forces, the London force is always present and usually dominant. The dipole–dipole force is usually appreciable only in small polar molecules or in large molecules having very large dipole moments. As noted earlier, the dipole–dipole attraction in the small, polar molecule HCl accounts for less than 20% of its interaction energy.

The liquids in Table 11.2 are listed in order of increasing molecular mass. Recall that London forces tend to increase with molecular mass. Therefore, London forces would be expected to increase for these liquids from the top to the bottom of the table. If London forces are the dominant attractive forces in these liquids, the vapor pressures should decrease from top to bottom in the table. And this is what you see, except for water and glycerol. (Their vapor pressures are relatively low, and an additional force, hydrogen bonding, is needed to explain them.)

The normal boiling point of a liquid must depend on intermolecular forces, because it is related to the vapor pressure. Normal boiling points are approximately proportional to the energy of intermolecular attraction. Thus, they are lowest for liquids with the weakest intermolecular forces.

Surface tension also depends on intermolecular forces. Surface tension is the energy needed to increase the surface area of a liquid. To increase the surface area, it is necessary to pull molecules apart against the intermolecular forces of attraction. Thus, surface tension would be expected to increase with the strength of attractive forces. If London forces are the dominant attractive forces in the liquid, surface tension should increase with molecular mass—that is, it should increase from top to bottom in Table 11.2. This is what you see, except for water and glycerol, which have relatively high surface tensions.

The viscosity of a liquid depends in part on intermolecular forces, because increasing the attractive forces between molecules increases the resistance to flow. If London forces are dominant, you should expect the viscosity of the liquids in Table 11.2 to increase from the top of the column to the bottom, which it does (again with the exception of water and glycerol; these liquids have relatively high viscosities). Note, however, that the viscosity also depends on other factors, such as the possibility of molecules tangling together. Liquids with long molecules that tangle together are expected to have high viscosities.

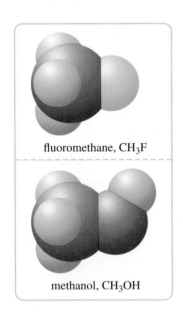

fluoromethane, CH₃F

methanol, CH₃OH

Figure 11.23 ▲

Fluoromethane and methanol Space-filling molecular models.

CONCEPT CHECK 11.3

Consider two liquids, labeled A and B, that are both pure substances. Liquid A has relatively strong intermolecular forces when compared to liquid B. The molecular structures of A and B are similar. Which one of the statements regarding liquids A and B is false?

a Liquid A would have a higher boiling point than liquid B.

b Liquid B would be more viscous than liquid A.

c Liquid B would have a lower freezing point than liquid A.

d Liquid A would have a lower vapor pressure than Liquid B.

e Liquid A and B would melt at different temperatures.

Hydrogen Bonding

It is interesting to compare fluoromethane, CH_3F, and methanol, CH_3OH (Figure 11.23). They have about the same molecular mass (34 for CH_3F and 32 for CH_3OH) and about the same dipole moment (1.81 D for CH_3F and 1.70 D for CH_3OH). You might expect these substances to have about the same intermolecular attractive forces and therefore about the same boiling points. In fact, the boiling points are quite different. Fluoromethane boils at −78°C and is a gas under normal conditions. Methanol boils at 65°C and is normally a liquid. We must conclude that there is an intermolecular attraction in methanol that is not present in fluoromethane.

We have already seen that the properties of water and glycerol cannot be explained in terms of van der Waals forces alone. What water, glycerol, and methanol have in common is one or more —OH groups.

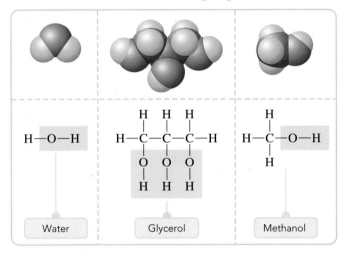

Molecules that have the —OH group are subject to an additional attractive force called hydrogen bonding. **Hydrogen bonding** is *a weak to moderate attractive force that exists between a hydrogen atom covalently bonded to a very electronegative atom, X, and a lone pair of electrons on another small, electronegative atom, Y.* Hydrogen bonding is represented in structural drawings by a series of dots.

$$—X—H\cdots Y—$$

Usually, hydrogen bonding is seen in cases where X and Y are the atoms F, O, or N.

Figure 11.24*a* shows a plot of boiling point versus molecular mass for the hydrides (binary compounds with hydrogen) of the Group VIA elements. If

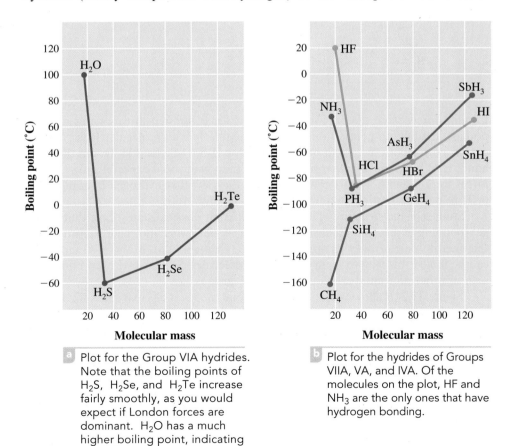

a Plot for the Group VIA hydrides. Note that the boiling points of H_2S, H_2Se, and H_2Te increase fairly smoothly, as you would expect if London forces are dominant. H_2O has a much higher boiling point, indicating an additional intermolecular attraction.

b Plot for the hydrides of Groups VIIA, VA, and IVA. Of the molecules on the plot, HF and NH_3 are the only ones that have hydrogen bonding.

Figure 11.24 ◀

Boiling point versus molecular mass for hydrides

Figure 11.25 ▶

Hydrogen bonding in water

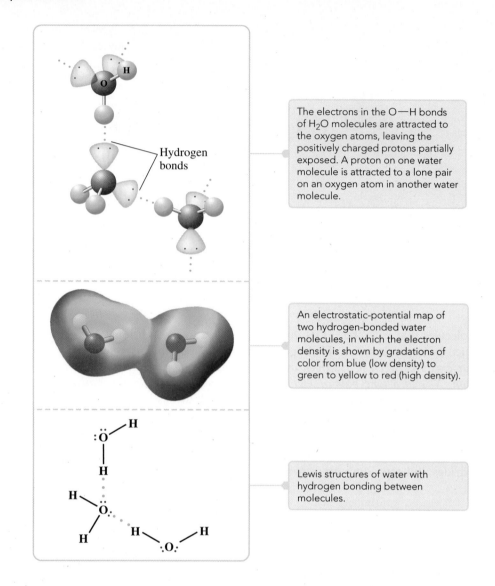

The electrons in the O—H bonds of H_2O molecules are attracted to the oxygen atoms, leaving the positively charged protons partially exposed. A proton on one water molecule is attracted to a lone pair on an oxygen atom in another water molecule.

Hydrogen bonds

An electrostatic-potential map of two hydrogen-bonded water molecules, in which the electron density is shown by gradations of color from blue (low density) to green to yellow to red (high density).

Lewis structures of water with hydrogen bonding between molecules.

London forces were the only intermolecular forces present in this series of compounds, the boiling points should increase regularly with molecular mass. You do see such a regular increase in boiling point for H_2S, H_2Se, and H_2Te. On the other hand, H_2O has a much higher boiling point than you would expect if only London forces were present. This is consistent with the view that hydrogen bonding exists in H_2O but is essentially absent in H_2S, H_2Se, and H_2Te.

Studies of the structure of H_2O in its different physical states show that hydrogen bonding is present. The hydrogen atom of one water molecule is attracted to the electron pair of another water molecule (Figure 11.25). Many H_2O molecules can be linked this way to form clusters of molecules. Similar hydrogen bonding is seen in other molecules containing —O—H groups.

Figure 11.24*B* shows plots of boiling point versus molecular mass for Groups VIIA, VA, and IVA hydrides. Hydrogen fluoride, HF, and ammonia, NH_3, have particularly high boiling points compared with other hydrides of their periodic group. Definite structural evidence has established hydrogen bonding in HF. In solid hydrogen fluoride, the structure shows a zigzag arrangement involving hydrogen bonding.

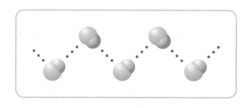

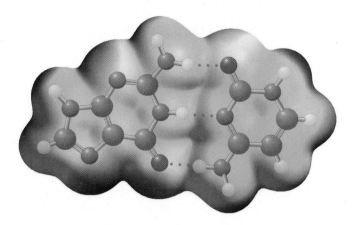

Figure 11.26 ◀

Hydrogen bonding between two biologically important molecules These *electrostatic-potential maps* show the electron densities (relative to the nuclei) of two molecules (guanine and cytosine). The electron densities increase as the color changes from blue to green to yellow to red. Note the three H bonds between the two molecules. Each H bond is between an H atom with low electron density (exposing the positive charge on the atom) and the lone electron pair (high density) of either an O or N atom. (The molecules are guanine and cytosine, which are two of the four bases used as codes in DNA; this is explained in Chapter 24.)

The hydrogen bonding attraction between two water molecules may be explained in part on the basis of the dipole moment of the —O—H bond.

$$\overset{\delta-}{-}\text{O}\overset{\delta+}{-}\text{H}$$

The partial positive charge on the hydrogen atom of one water molecule is attracted to the partial negative charge of a lone pair on another water molecule. In addition, however, a hydrogen atom covalently bonded to an electronegative atom appears to be special. The electrons in the —O—H bond are drawn to the O atom, leaving the concentrated positive charge of the proton partially exposed. This concentrated positive charge is strongly attracted to a lone pair of electrons on another O atom. Figure 11.26 shows the hydrogen bonding between two biologically important molecules.

⚘WL Interactive Example 11.5 Identifying Intermolecular Forces in a Substance

Gaining Mastery Toolbox

Critical Concept 11.5
There are three types of intermolecular forces possible in molecular substances: London, dipole-dipole, and hydrogen bonding. All molecular substances have London forces. Molecular structure and chemical composition determine if dipole-dipole forces are expected. Hydrogen bonding is an important special case of very strong dipole-dipole forces.

Solution Essentials:
• Hydrogen bonding
• London forces
• Dipole-dipole forces
• Intermolecular forces

What kinds of intermolecular forces (London, dipole–dipole, hydrogen bonding) are expected in the following substances?
a methane, CH_4
b (trichloromethane see first model at left)
c butanol (butyl alcohol), $CH_3CH_2CH_2CH_2OH$

Problem Strategy Note that London forces are always present. Then ask yourself whether the molecule is polar—in which case, dipole–dipole forces exist. (Refer to Section 10.2 on dipole moment and molecular geometry.) Now note whether hydrogen bonding is present. (Is a hydrogen atom bonded to F, O, or N?)

Solution

a Methane is a nonpolar molecule. Hence, the only intermolecular attractions are **London forces.** b Trichloromethane is an unsymmetrical molecule with polar bonds. Thus, we expect **dipole–dipole forces,** in addition to **London forces.** c Butanol has a hydrogen atom attached to an oxygen atom. Therefore, you expect **hydrogen bonding.** Because the molecule is polar (from the O—H bond), you also expect **dipole–dipole forces. London forces** exist too, because such forces exist between all molecules.

Answer Check Remember that London forces are always present, even when other types of intermolecular forces exist.

Exercise 11.5 List the different intermolecular forces you would expect for each of the following compounds:
a propanol, $CH_3CH_2CH_2OH$
b (carbon dioxide see second model at left)
c sulfur dioxide, SO_2

■ See Problems 11.63 and 11.64.

Example 11.6   Determining Relative Vapor Pressures on the Basis of Intermolecular Attractions

Gaining Mastery Toolbox

Critical Concept 11.6
The strength of the intermolecular forces in a liquid determines the properties of the liquid: vapor pressure, freezing point, and boiling point. The stronger the intermolecular forces in a substance the higher the vapor pressure, freezing point, and boiling point.

Solution Essentials:
- Hydrogen bonding
- London forces
- Dipole-dipole forces
- Intermolecular forces
- Vapor pressure
- Freezing point
- Boiling point

For each of the following pairs, choose the substance you expect to have the lower vapor pressure at a given temperature: a carbon dioxide (CO_2) or sulfur dioxide (SO_2); b dimethyl ether (CH_3OCH_3) or ethanol (CH_3CH_2OH).

Problem Strategy There are a couple of basic ideas to keep in mind. First, London forces tend to increase with molecular mass. Second, hydrogen bonding tends to dominate other types of forces if it is present.

Solution

a The molecular weights for SO_2 and CO_2 are 64 amu and 44 amu, respectively. Therefore, the London forces between SO_2 molecules should be greater than those between CO_2 molecules. Moreover, because SO_2 is polar but CO_2 is not, there are dipole–dipole forces between SO_2 molecules but not between CO_2 molecules. **You conclude that sulfur dioxide has the lower vapor pressure.** Experimental values of the vapor pressure at 20°C are: CO_2, 56.3 atm; SO_2, 3.3 atm. b Dimethyl ether and

ethanol have the same molecular formulas but different structural formulas. The structural formulas are

Dimethyl ether   Ethanol

The molecular masses are equal, and therefore the London forces are approximately the same. However, there will be strong hydrogen bonding in ethanol but not in dimethyl ether. **You expect ethanol to have the lower vapor pressure.** Experimental values of vapor pressure at 20°C are: CH_3OCH_3, 4.88 atm; CH_3CH_2OH, 0.056 atm.

Answer Check Make sure you have the correct molecular masses for the substances and that you have the correct molecular structures so that you will see hydrogen bonding if it is present.

Exercise 11.6 Arrange the following hydrocarbons in order of increasing vapor pressure: ethane, C_2H_6; propane, C_3H_8; and butane, C_4H_{10}. Explain your answer.

■ See Problems 11.67 and 11.68.

Exercise 11.7 At the same temperature, methyl chloride, CH_3Cl, has a vapor pressure of 1490 mmHg, and ethanol has a vapor pressure of 42 mmHg. Explain why you might expect methyl chloride to have a higher vapor pressure than ethanol, even though methyl chloride has a somewhat larger molecular weight.

■ See Problems 11.69 and 11.70.

CONCEPT CHECK 11.4

A common misconception is that the following chemical reaction occurs when water is boiled:

$$2H_2O(l) \longrightarrow 2H_2(g) + O_2(g)$$

instead of

$$H_2O(l) \longrightarrow H_2O(g)$$

a What evidence do you have that the second reaction is correct?

b How would the enthalpy of the wrong reaction compare with that of the correct reaction?

c How could you calculate the enthalpy change for the wrong reaction (Chapter 6)?

Solid State

A solid is a nearly incompressible state of matter with a well-defined shape, because the units (atoms, molecules, or ions) making up the solid are in close contact and in fixed positions or sites. In the next section, we will look at the kinds of forces

Gecko Toes, Sticky But Not Tacky

A gecko (Figure 11.27) can effortlessly climb a wall or walk across a glass ceiling. And like a superacrobat, it can catch itself by a toe while falling. How does it do this? Does the gecko have a gluey substance, or something like sticky tape, on its toes? But if so, how is it possible for the gecko to plant its feet and remove them 15 times a second, which it does when it scurries up a tree? Having a toe stick to a leaf is one thing; being able to remove it easily is another. Recently, the biologist Kellar Autumn at Lewis and Clark College, with other scientists, discovered that the gecko uses van der Waals forces to attach itself to surfaces and employs a special technique to disengage from that surface.

Van der Waals forces exist between any two surfaces, but they are extremely weak unless relatively large areas of the two surfaces come quite close together. The toe of a gecko is covered with fine hairs, each hair having over a thousand split ends. As the gecko walks across a surface, it presses these stalks of hairs against the surface. The intimate contact of a billion or so split ends of hairs with the surface results in a large, attractive force that holds the gecko fast. Just as easily, a gecko's foot comes cleanly away. As the gecko walks, its foot naturally bends so the hairs at the back edge of its toes disengage, row after row, until the toe is free. It is the mechanics of the gecko's walk that allows it to connect to and disengage easily from a surface.

It is interesting to compare gecko toes to adhesive tapes. These tapes are covered with a soft, tacky material that flows when pressure is applied. The tacky adhesive and a surface can then come in close contact, where intermolecular forces provide effective attraction. The problem with a tacky adhesive is that it sticks to not only the surface but dirt as well. You can pull a Post-it

Figure 11.27 ▲

A Tokay gecko This is a common gecko, a small lizard of the type studied by Kellar Autumn.

note off one surface and stick it to another surface several times. Eventually, the adhesive gets dirty, and the note loses its adhesive quality and refuses to stick. Gecko toes, though, are sticky but not tacky. The fine hairs do not continue to stick to dirt and can stick and unstick from surfaces indefinitely.

Materials scientists are busy trying to imitate the gecko by developing a "gecko tape," a plastic tape with the property that it can stick and unstick from surfaces many times. The plastic tape is covered in many hairs with split ends, like those on the gecko's toes. Other scientists are trying to design robots that mimic the way a gecko walks. Using gecko tape on the robot's feet might allow it to climb walls. Would we have tried to design such devices if no one had discovered how the gecko does it?

■ See Problems 11.125 and 11.126.

holding the units together in different types of solids. In later sections, we will look at crystalline solids and their structure.

11.6 Classification of Solids by Type of Attraction of Units

A solid consists of structural units—atoms, molecules, or ions—that are attracted to one another strongly enough to give a rigid substance. One way to classify solids is by the type of force holding the structural units together. In some cases, these forces are intermolecular and hold molecules together. In other cases, these forces are chemical bonds (metallic, ionic, or covalent) and hold atoms together. From this point of view, then, there are four different types of solids: molecular, metallic, ionic, and covalent. It should be noted, however, that a given ionic bond may have considerable covalent character, and vice versa, so the distinction between ionic and covalent solids is not a precise one.

Types of Solids

A **molecular solid** is *a solid that consists of atoms or molecules held together by intermolecular forces.* Many solids are of this type. Examples include solid neon, solid water (ice), and solid carbon dioxide (dry ice).

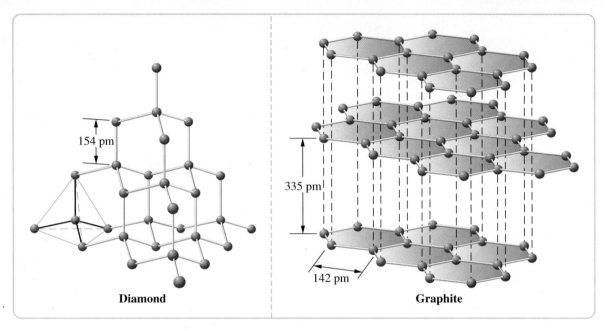

Diamond   Graphite

Figure 11.28 ▲

Structures of diamond and graphite Diamond is a three-dimensional network solid; every carbon atom is covalently bonded to four others. Graphite consists of sheets of carbon atoms covalently bonded to form a hexagonal pattern of atoms; the sheets are held together by van der Waals forces.

A **metallic solid** is *a solid that consists of positive cores of atoms held together by a surrounding "sea" of electrons (metallic bonding).* In this kind of bonding, positively charged atomic cores are surrounded by delocalized electrons. Examples include iron, copper, and silver.

An **ionic solid** is *a solid that consists of cations and anions held together by the electrical attraction of opposite charges (ionic bonds).* Examples are cesium chloride, sodium chloride, and zinc sulfide (but ZnS has considerable covalent character).

A **covalent network solid** is *a solid that consists of atoms held together in large networks or chains by covalent bonds.* Diamond is an example of a three-dimensional network solid. Every carbon atom in diamond is covalently bonded to four others, so an entire crystal might be considered an immense molecule. It is possible to have similar two-dimensional "sheet" and one-dimensional "chain" molecules, with atoms held together by covalent bonds. Examples are diamond (three-dimensional network), graphite (sheets), and asbestos (chains). Figure 11.28 shows the structures of diamond and graphite. Table 11.4 summarizes these four types of solids.

Table 11.4	Types of Solids		
Type of Solid	**Structural Units**	**Attractive Forces Between Structural Units**	**Examples**
Molecular	Atoms or molecules	Intermolecular forces	Ne, H_2O, CO_2
Metallic	Atoms (positive cores surrounded by electron "sea")	Metallic bonding (extreme delocalized bonding)	Fe, Cu, Ag
Ionic	Ions	Ionic bonding	CsCl, NaCl, ZnS
Covalent network	Atoms	Covalent bonding	Diamond, graphite, asbestos

Example 11.7   Identifying Types of Solids

Gaining Mastery Toolbox

Critical Concept 11.7
The type of solid (molecular, metallic, ionic, or covalent network) is determined by the bonding between the structural units. Molecular solids have molecules as the structural units, metals have metal atoms as the structural units, ionic solids have ions (cations and anions) as the structural units, and covalent network solids have covalently bonded atoms as the structural units.

Solution Essentials:
• Covalent network solid
• Ionic solid
• Metallic solid
• Molecular solid
• Ionic bond
• Molecule
• Ionic compound

Which of the four basic types of solids would you expect the following substances to be? a solid ammonia, NH_3; b cesium, Cs; c cesium iodide, CsI; d silicon, Si.

Problem Strategy Some general notions of bonding can be helpful, from which the type of solid formed follows. The metallic elements, of course, have metallic bonding. The non-metallic elements in Groups VA to VIIA, which are rather reactive, are usually molecular, whereas those in Group VIIIA (the noble gases) are atomic. The nonmetals in Groups IIIA (B) and IVA (C and Si) often occur with network bonding. A compound of two non-metals is likely to be molecular, although in some cases network bonding occurs (involving Groups IIIA and IVA elements especially, such as BN, SiO_2, and SiC, with strong single bonds). A compound of a metal and a nonmetal is likely to have ionic bonds.

Solution

a Ammonia has a molecular structure; therefore, it freezes as a **molecular solid.** b Cesium is a metal; it is a **metallic solid.** c Cesium iodide is an ionic substance; it exists as an **ionic solid.** d Silicon atoms might be expected to form covalent bonds to other silicon atoms, as carbon does in diamond. A **covalent network solid** would result.

Answer Check Make sure you correctly understand the bonding involved in each case.

Exercise 11.8   Classify each of the following solids according to the forces of attraction that exist between the structural units: a zinc, Zn; b sodium iodide, NaI; c silicon carbide, SiC; d methane, CH_4.

■ See Problems 11.73, 11.74, 11.75, and 11.76.

Physical Properties

Many physical properties of a solid can be directly related to its structure. Let us look at several of these properties.

Melting Point and Structure Table 11.1 lists the melting points for various solids and also gives the type of solid. For a solid to melt, the forces holding the structural units in their sites must be overcome, at least partially. In a molecular solid, these forces are weak intermolecular attractions. Thus, molecular solids tend to have low melting points (usually below 300°C). At room temperature, many molecular substances are either liquids (such as water and ethanol) or gases (such as carbon dioxide and ammonia). Note also that the melting points of molecular solids reflect the strengths of intermolecular attractions, just as boiling points do. This can be seen in the melting points of the molecular solids listed in Table 11.1.

For ionic solids and covalent network solids to melt, chemical bonds must be broken. For that reason, the melting points of these types of solids are relatively high. The melting point of the ionic solid sodium chloride is 801°C; that of magnesium oxide is 2800°C. Melting points of covalent network solids are generally quite high. Quartz, for example, melts at 1610°C; diamond, at 3550°C. See Table 11.1. ▶

Both molecular and covalent network solids have covalent bonds, but none of the covalent bonds are broken during the melting of a molecular solid.

The difference between the melting points of sodium chloride (801°C) and magnesium oxide (2800°C) can be explained in terms of the charges on the ions. You expect the melting point to rise with lattice energy. ▶ Lattice energies, however, depend on the product of the ionic charges (and inversely on the distance between charges). For sodium chloride (Na^+Cl^-) this ion-charge product is $(+1) \times (-1) = -1$, whereas for magnesium oxide ($Mg^{2+}O^{2-}$) it is $(+2) \times (-2) = -4$. Thus, the lattice energy is greater for magnesium oxide, and this is reflected in its much higher melting point.

Lattice energy is the energy needed to separate a crystal into isolated ions in the gaseous state. It represents the strength of attraction of ions in the solid.

Metals often have high melting points, but there is considerable variability. Mercury, a liquid at room temperature, melts at −39°C. Tungsten melts at 3410°C, the highest melting point of any metallic element. In general, melting points are low for the Group IA and IIA elements but increase as you move right in the periodic table to the transition metals. The elements in the center of the transition series (such as tungsten) have high melting points. Then, as you move farther to the right, the melting points decrease and are again low for the Group IIB elements.

Example 11.8

Determining Relative Melting Points Based on Types of Solids

Gaining Mastery Toolbox

Critical Concept 11.8
The strength of the forces between structural units in a solid determines the melting point of the substance. Ionic solids and covalent network solids generally have the strongest forces between structural units. Molecular substances have relatively weak forces between structural units when compared to ionic and covalent network solids.

Solution Essentials:
• Covalent network solid
• Ionic solid
• Metallic solid
• Molecular solid
• Intermolecular forces
• Ionic bond

Arrange the following elements in order of increasing melting point: silicon, hydrogen, lithium. Explain your reasoning.

Problem Strategy You first need to decide the type of solid expected for each substance. Of these types, molecular substances tend to have relatively low melting points. In comparing the melting points of two molecular solids, it is helpful to understand the intermolecular forces involved and their relative strengths. (Review Examples 11.5 and 11.6.) Stronger intermolecular forces result in higher melting points. Ionic solids and covalent network solids usually have relatively high melting points. Metals have variable melting points, but roughly these increase in moving right through the periodic table.

Solution Hydrogen is a molecular substance (H_2) with a very low molecular mass (2 amu). You expect it to have a very low melting point. Lithium is a Group IA element and is expected to be a relatively low-melting metal. Except for mercury, however, all metals are solids below 25°C. So the melting point of lithium is well above that of hydrogen. Silicon might be expected to have a covalent network structure like carbon and to have a high melting point. Therefore, the list of these elements in order of increasing melting point is **hydrogen, lithium, silicon.**

Answer Check Make sure you understand the bonding and type of solid involved in each substance. Note that solving this problem draws on many skills, including your understanding of periodic behavior, bonding, polarity, and intermolecular forces and your ability to distinguish among types of solids.

Exercise 11.9 Decide what type of solid is formed for each of the following substances: C_2H_5OH, CH_4, CH_3Cl, $MgSO_4$. On the basis of the type of solid and the expected magnitude of intermolecular forces (for molecular crystals), arrange these substances in order of increasing melting point. Explain your reasoning.

■ See Problems 11.77 and 11.78.

Figure 11.29 ▲

Behavior of crystals when struck The lead in the fishing weight is malleable; rock salt, NaCl, cracks along the crystal planes.

© Cengage Learning

Hardness and Structure Hardness depends on how easily the structural units of a solid can be moved relative to one another and therefore on the strength of attractive forces between the units. Thus, molecular crystals, with weak intermolecular forces, are rather soft compared with ionic crystals, in which the attractive forces are much stronger. A three-dimensional covalent network solid is usually quite hard because of the rigidity given to the structure by strong covalent bonds throughout. Diamond and silicon carbide (SiC), three-dimensional covalent networks, are among the hardest substances known.

We should add that molecular and ionic crystals are generally brittle because they fracture easily along crystal planes. Metallic crystals, by contrast, are *malleable;* that is, they can be shaped by hammering (Figure 11.29).

Electrical Conductivity and Structure One of the characteristic properties of metals is good electrical conductivity. The delocalized valence electrons are easily moved by an electrical field and are responsible for carrying the electric current. By contrast, most covalent and ionic solids are nonconductors, because the electrons are localized to particular atoms or bonds. Ionic substances do become conducting in the liquid state, however, because the ions can move. In an ionic liquid, it is the ions that carry the electric current.

Table 11.5	Properties of the Different Types of Solids		
Type of Solid	**Melting Point**	**Hardness and Brittleness**	**Electrical Conductivity**
Molecular	Low	Soft and brittle	Nonconducting
Metallic	Variable	Variable hardness; malleable	Conducting
Ionic	High to very high	Hard and brittle	Nonconducting solid (conducting liquid)
Covalent network	Very high	Very hard	Usually nonconducting

Table 11.5 summarizes the properties we discussed for the different types of solids.

11.7 Crystalline Solids; Crystal Lattices and Unit Cells

Solids can be crystalline or amorphous. A **crystalline solid** *is composed of one or more crystals; each crystal has a well-defined, ordered structure in three dimensions.* Sodium chloride (table salt) and sucrose (table sugar) are examples of crystalline substances. Metals are usually compact masses of crystals. An **amorphous solid** *has a disordered structure; it lacks the well-defined arrangement of basic units (atoms, molecules, or ions) found in a crystal.* A glass is an amorphous solid obtained by cooling a liquid rapidly enough that its basic units are "frozen" in random positions before they can assume an ordered crystalline arrangement. Common window glass is an example, as is obsidian, a natural glass formed when molten rock from a volcanic eruption cools quickly. ▶

Crystal Lattices

A crystal is a three-dimensional ordered arrangement of basic units. (The basic unit may be an atom, molecule, or ion, depending on the type of crystal.) The ordered structure of a crystal is conveniently described in terms of a *crystal lattice.* The idea of a crystal lattice is most easily grasped by looking at two-dimensional repeating patterns such as those seen on many fabrics or wallpapers. For simplicity, consider the pattern of A's shown in Figure 11.30a. This pattern consists of a repetition of basic units A. A lattice of points is obtained by choosing the same point in each basic unit. (The choice of this point is immaterial as long as the same point is chosen in each basic unit.) In the pattern of Figure 11.30, we have chosen the points to be midway on the crossbar of each A. The collection of these points constitutes the lattice of the pattern (Figure 11.30b). This lattice shows the essential arrangement of the basic units (A's) in the pattern. We could have chosen a different location for the point in each A, but we would have obtained the same lattice.

When quartz crystals (SiO_2) are melted and then cooled rapidly, they form a glass called silica glass. Though ideal for some purposes, silica glass is difficult to work with because of its high temperature of softening. This temperature can be lowered by adding various oxides, such as Na_2O and CaO. In practice, ordinary glass is formed by melting sand (quartz crystals) with sodium carbonate and calcium carbonate. The carbonates decompose to the oxides plus carbon dioxide.

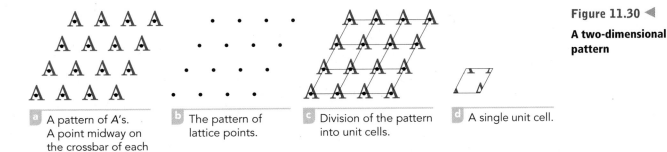

Figure 11.30 ◀

A two-dimensional pattern

a A pattern of A's. A point midway on the crossbar of each A has been selected for a lattice point.

b The pattern of lattice points.

c Division of the pattern into unit cells.

d A single unit cell.

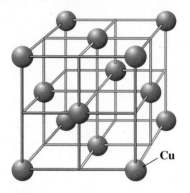

Figure 11.31 ▲

Crystal structure and crystal lattice of copper The copper atoms have been shrunk so that the crystal structure is more visible. The crystal lattice is the geometric arrangement of lattice points, which you can take to be the centers of the atoms. Lines have been drawn to emphasize the geometry of the lattice.

Figure 11.32 ▼

Unit-cell shapes of the different crystal systems The unit cells of the seven different crystal systems are displayed, showing the relationships between edge lengths and angles (summarized in Table 11.6).

Similarly, we define a **crystal lattice** as *the geometric arrangement of lattice points of a crystal, in which we choose one lattice point at the same location within each of the basic units of the crystal.* Figure 11.31 shows the crystal structure of copper metal. If we place a point at the same location in each copper atom (say at the center of the atom), we obtain the crystal lattice for copper. The crystal lattice and crystal structure are not the same. We can imagine building the crystal structure for copper from its crystal lattice by placing a copper atom at each lattice point. Many different crystals have the same crystal lattice. For instance, the crystal lattice for solid methane, CH_4, is the same (except for the distances between points) as that for copper. We obtain the methane crystal structure by placing CH_4 molecules at the lattice points. The crystal lattice shows only the arrangement of basic units of the crystal.

You can imagine dividing a crystal into many equivalent cells, or *unit cells,* so that these cells become the "bricks" from which you mentally construct the crystal. Let us look at the two-dimensional pattern in Figure 11.30c, in which we have connected lattice points by straight lines. The parallelograms formed by the straight lines divide the pattern into small cells. These are the smallest possible cells from which you could imagine constructing the pattern by stacking the individual cells—they are the unit cells of the pattern (Figure 11.30d).

In the same way, you can divide any crystal lattice into boxlike cells, or unit cells. The **unit cell** of a crystal is *the smallest boxlike unit (each box having faces that are parallelograms) from which you can imagine constructing a crystal by stacking the units in three dimensions.*

There are seven basic shapes possible for unit cells, which give rise to the seven *crystal systems* used to classify crystals. A crystal belonging to a given crystal system has a unit cell with one of the seven shapes shown in Figure 11.32. Each unit-cell shape can be characterized by the angles between the edges of the unit cell and the relative lengths of those edges. A crystal belonging to the cubic system, for example, has 90° angles between the edges of the unit cell, and all edges are of equal length. Table 11.6 summarizes the relationships between the angles and edge lengths for the unit cell of each crystal system.

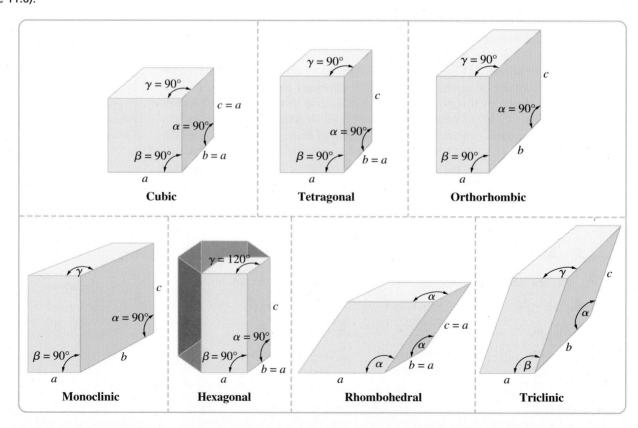

Table 11.6   **The Seven Crystal Systems**

Crystal System	Edge Length	Angles	Examples
Cubic	$a = b = c$	$\alpha = \beta = \gamma = 90°$	NaCl, Cu
Tetragonal	$a = b \neq c$	$\alpha = \beta = \gamma = 90°$	TiO_2 (rutile), Sn (white tin)
Orthorhombic	$a \neq b \neq c$	$\alpha = \beta = \gamma = 90°$	$CaCO_3$ (aragonite), $BaSO_4$
Monoclinic	$a \neq b \neq c$	$\alpha = \beta = 90°, \gamma \neq 90°$	$PbCrO_4$
Hexagonal	$a = b \neq c$	$\alpha = \beta = 90°, \gamma = 120°$	C (graphite), ZnO
Rhombohedral	$a = b = c$	$\alpha = \beta = \gamma \neq 90°$	$CaCO_3$ (calcite), HgS (cinnabar)
Triclinic	$a \neq b \neq c$	$\alpha \neq \beta \neq \gamma \neq 90°$	$K_2Cr_2O_7$, $CuSO_4 \cdot 5H_2O$

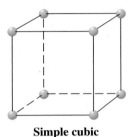

Simple cubic

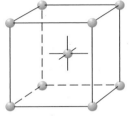

Body-centered cubic

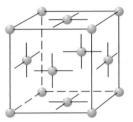

Face-centered cubic

Figure 11.33 ◀

Cubic unit cells The simple cubic unit cell has lattice points only at the corners; the body-centered cubic unit cell also has a lattice point at the center of the cell; and the face-centered cubic unit cell has lattice points at the centers of each face in addition to those at the corners.

Most of the crystal systems have more than one possible crystal lattice. A *simple* (or *primitive*) lattice has a unit cell in which there are lattice points only at the corners of the unit cell. Other lattices in the same crystal system have additional lattice points either within the *body* of the unit cell or on *faces* of the unit cell. As an example, we will describe the cubic crystal system in some detail.

Cubic Unit Cells

The cubic crystal system has three possible cubic unit cells: simple (or primitive) cubic, body-centered cubic, and face-centered cubic. These cells are illustrated in Figure 11.33. A **simple cubic unit cell** is *a cubic unit cell in which lattice points are situated only at the corners.* A **body-centered cubic unit cell** is *a cubic unit cell in which there is a lattice point at the center of the cubic cell in addition to those at the corners.* A **face-centered cubic unit cell** is *a cubic unit cell in which there are lattice points at the centers of each face of the unit cell in addition to those at the corners.*

The simplest crystal structures are those in which there is only a single atom at each lattice point. Most metals are examples. Copper metal has a face-centered cubic unit cell with one copper atom at each lattice point (see Figure 11.31). The unit cells for this and other cubic cells of simple atomic crystals are shown in Figure 11.34. Note that only the portions of the atoms actually within the unit cell are shown. For certain applications, you will need to know the number of atoms in a unit cell of such an atomic crystal. The next example shows how you count atoms in a unit cell.

Simple cubic

Body-centered cubic

Face-centered cubic

Figure 11.34 ◀

Space-filling representation of cubic unit cells Only that portion of each atom belonging to a unit cell is shown. Note that a corner atom is shared with eight unit cells and a face atom is shared with two.

Example 11.9   Determining the Number of Atoms in a Unit Cell

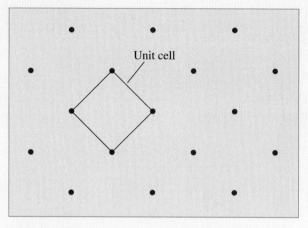

Figure 11.35 ▲

A two-dimensional lattice

How many atoms are there in the face-centered cubic unit cell of an atomic crystal having one atom at each lattice point?

Problem Strategy Remember that atoms at the corners and faces of a unit cell are shared with adjoining unit cells (see Figure 11.34). Each corner atom is shared with eight unit cells (so there is only one-eighth of any corner atom in the unit cell), while each face is shared between two unit cells (giving one-half of a face atom per unit cell). Now add up the atoms for corners and faces, and then add the number of atoms totally within the unit cell (if any) to obtain the total number of atoms per unit cell.

Solution There are six faces on a cube (each face giving one-half atom), yielding a total of three atoms from faces. Also, there are eight corners (each corner giving one-eighth atom), yielding one atom from corners. There are no atoms totally within the unit cell. Thus,

there are a total of four atoms in a face-centered cubic unit cell. To summarize:

$$6 \text{ faces} \times \frac{\frac{1}{2} \text{ atom}}{\text{face}} = 3 \text{ atoms}$$

$$8 \text{ corners} \times \frac{\frac{1}{8} \text{ atom}}{\text{corner}} = 1 \text{ atom}$$

$$\text{Total} \quad \textbf{4 atoms}$$

Answer Check Again, look at Figure 11.34, and see that your calculation agrees with it.

Exercise 11.10   Figure 11.35 shows solid dots ("atoms") forming a two-dimensional lattice. A unit cell is marked off by the black line. How many "atoms" are there in this unit cell in such a two-dimensional lattice?

■ See Problems 11.83 and 11.84.

Crystal Defects

So far we have assumed that crystals have perfect order. In fact, real crystals have various defects or imperfections. These are principally of two kinds: chemical impurities and defects in the formation of the lattice.

Ruby is an example of a crystal with a chemical impurity. The crystal is mainly colorless aluminum oxide, Al_2O_3, but occasional aluminum ions, Al^{3+}, are replaced by chromium(III) ions, Cr^{3+}, which give a red color.

Various lattice defects occur during crystallization. Crystal planes may be misaligned, or sites in the crystal lattice may remain vacant. For example, there might be an equal number of sodium ion vacancies and chloride ion vacancies in a sodium chloride crystal. It is also possible to have an unequal number of cation and anion vacancies in an ionic crystal. For example, iron(II) oxide, FeO, usually crystallizes with some iron(II) sites left unoccupied. Enough of the remaining iron atoms have 3+ charges to give an electrically balanced crystal. As a result, there are more oxygen atoms than iron atoms in the crystal. Moreover, the exact composition of the crystal can vary, so the formula FeO is only approximate. Such a compound whose composition varies slightly from its idealized formula is

Liquid-Crystal Displays

Liquid-crystal displays (LCDs) are now commonly used in cell phones, portable computers, and flat-panel televisions. The liquid crystals that are the basis of these displays consist of rodlike, polar molecules in a phase that is intermediate in order between that of a liquid and that of a crystalline solid. The molecules tend to align or orient themselves in the same direction along their long axes, like matches in a matchbox, but they occupy random positions within the substance. This particular type of liquid crystal, having orientational order but no positional order, is referred to as *nematic* (Figure 11.36). The molecular order in a nematic liquid crystal, which results from weak intermolecular forces, is easily disrupted. And because these molecules are polar, they can be reoriented by the application of an electric field. A liquid-crystal display uses this ease of molecular reorientation by an electric field to change small areas, or *pixels,* of the display from light to dark.

An LCD display consists of many pixel cells arranged in rows and columns, from which an image is formed. Each pixel consists of liquid-crystal molecules contained between glass plates whose

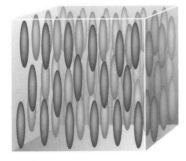

Figure 11.36 ▲

Nematic liquid crystal The rodlike molecules are aligned in the same direction, although the positions of the molecules are random.

interior surfaces are covered with a transparent electrode material (Figure 11.37, *top*). Outside each glass plate is a *polarizer.* The polarizer is similar to a Polaroid lens in sunglasses. Only those light waves that are vibrating in a particular direction are allowed to pass—say, up and down at the left side. (We say that the light is *polarized.*) On the right, the polarizer is oriented so that only light vibrating at a 90° angle to that on the left can pass. Imagine light polarized in the up-and-down direction entering the glass plate at the left. Without liquid crystals between the glass plates, the polarized light passes to the right side, where it is blocked by the polarizer on that side. You can demonstrate this blocking effect (if you are willing to sacrifice a pair of sunglasses) by placing one Polaroid lens in front of the other, rotating one lens while holding the other fixed. The image you see varies from light to dark as you rotate.

Liquid-crystal molecules in an LCD cell can redirect polarized light so that it enters at the left vibrating up and down, but exits at the right vibrating at a 90° angle to that at the left. This is accomplished by treating the inner electrode surfaces so that the rodlike liquid-crystal molecules align themselves against the surface in a definite direction. At the left, these molecules are aligned up and down; at the right, they are aligned at a 90° angle to those on the left. From left to right, the molecules tend to line up one against the other, but they twist their orientations from up and down at the left to 90° at the right. Now, polarized light entering at the left is redirected by each molecule until it is vibrating so that it can exit the polarizer on the right. The light is no longer blocked.

Suppose we apply an electric charge to the electrodes. The rodlike, polar molecules of the liquid crystals now align themselves perpendicularly to the glass plates (Figure 11.37, *bottom*). The molecules can no longer reorient the polarized light; the light is blocked and the pixel is dark. By turning the electrodes on or off, it is possible to make the pixels either dark or bright. An image is formed by turning various screen pixels on and others off by means of software.

(continued)

said to be *nonstoichiometric.* Other examples of nonstoichiometric compounds are Cu_2O and Cu_2S. Each usually has less copper than expected from the formula. Some ceramic materials having superconducting properties at relatively high temperatures (page 876) are nonstoichiometric compounds. An example is yttrium barium copper oxide, $YBa_2Cu_3O_{7-x}$, where x is approximately 0.1. It is a compound with oxygen atom vacancies.

11.8 Structures of Some Crystalline Solids

We have described the structure of crystals in a general way. Now we want to look in detail at the structure of several crystalline solids that represent the different types: molecular, metallic, ionic, and covalent network.

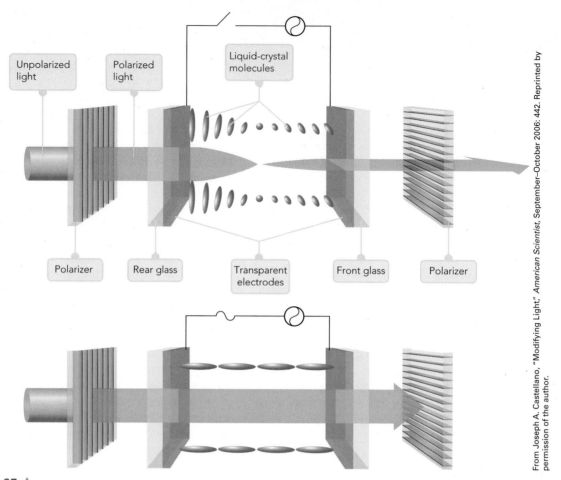

From Joseph A. Castellano, "Modifying Light," *American Scientist*, September–October 2006: 442. Reprinted by permission of the author.

Figure 11.37 ▲

Liquid-crystal display *Top:* Liquid crystals are sandwiched between glass plates. The rodlike molecules are aligned up and down at the left plate but gradually twist until, at the right plate, they are oriented at a 90° angle to the original direction. The polarizer at the left allows light to pass that is vibrating up and down. The liquid-crystal molecules turn the plane of polarization of the light so that it can exit at the right. The pixel is on (bright). *Bottom:* When an electric charge is placed on the electrodes, the molecules orient perpendicularly to the plates and can no longer twist the polarized light so that it can exit. The pixel is now off (dark).

A color display has each pixel subdivided into three subpixels. Each subpixel has a color filter in front of it, a filter that is red, green, or blue. The color image is obtained by turning appropriate subpixels on or off.

■ See Problems 11.127 and 11.128.

Molecular Solids; Closest Packing

The simplest molecular solids are the frozen noble gases—for example, solid neon. In this case, the molecules are single atoms and the intermolecular interactions are London forces. These forces are nondirectional (in contrast to covalent bonding, which is directional), and the maximal attraction is obtained when each atom is surrounded by the largest possible number of other atoms. The problem, then, is simply to find how identical spheres can be packed as tightly as possible into a given space.

To form a layer of close-packed spheres, you place each row of spheres in the crevices of the adjoining rows. This is illustrated in Figure 11.38a. (You may find it helpful to stack identical coins in the manner described here for spheres.) Place the next layer of spheres in the hollows in the first layer. Only half of these hollows can be filled with spheres. Once you have placed a sphere in a given hollow, it partially covers three neighboring hollows (Figure 11.38b) and completely determines the pattern of the layer (Figure 11.38c).

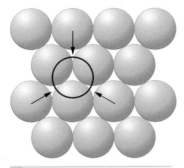

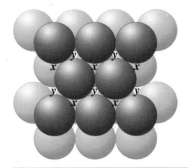

a Each layer is formed by placing spheres in the crevices of the adjoining row.

b When a sphere is placed in a hollow on top of a layer, three other hollows are partially covered.

c There are two types of hollows on top of the second layer, labeled x and y. The x sites are directly above the spheres in the first layer; the y sites are not. Spheres of a third layer may occupy either all of the x sites or all of the y sites.

Figure 11.38 ▲

Closest packing of spheres

If you pack into a crate as many perfectly round oranges of the same size as possible, they will be close packed. However, they need not be either hexagonal or cubic close packed, because the A-, B-, and C-type layers can be placed at random (except that similar layers cannot be adjacent). A possible arrangement might be ABCABABABC · · · It is the forces between structural units in a close-packed crystal that determine whether one of the ordered structures, either cubic or hexagonal, will occur.

When you come to fill the third layer, you find that you have a choice of sites for spheres, labeled x and y in Figure 11.38c. Note that each of the x sites has a sphere in the first layer directly beneath it, but the y sites do not. When the spheres in the third layer are placed in the x sites, so that the third layer repeats the first layer, you label this stacking *ABA*. When successive layers are placed so that the spheres of each layer are directly over a layer that is one layer away, you get a stacking that you label *ABABABA* . . . . (This notation indicates that the *A* layers are directly over other *A* layers, whereas the *B* layers are directly over other *B* layers.) The result is a **hexagonal close-packed structure** (hcp), *a crystal structure composed of close-packed atoms (or other units) with the stacking* ABABABA . . .; *the structure has a hexagonal unit cell.*

When the spheres in the third layer are placed in y sites, so that the third layer is over neither the first nor the second layer, you get a stacking that you label *ABC*. The fourth layer must be over either the *A* or the *B* layer. If subsequent layers are stacked so that they are over the layer two layers below, you get the stacking *ABCABCABCA* . . ., which results in a **cubic close-packed structure** (ccp), *a crystal structure composed of close-packed atoms (or other units) with the stacking ABCABCABCA* . . . . ▶

The cubic close-packed lattice is identical to the lattice having a face-centered cubic unit cell. To see this, you take portions of four layers from the cubic close-packed array (Figure 11.39, *left*). When these are placed together, they form a cube, as shown in Figure 11.39, *right*.

In any close-packed arrangement, each interior atom is surrounded by 12 nearest-neighbor atoms. *The number of nearest-neighbor atoms of an atom* is called its **coordination number.** Thinking of atoms as hard spheres, one can calculate that the spheres occupy 74% of the space of the crystal. There is no way of packing identical spheres so that an atom has a coordination number greater than 12 or the spheres occupy more than 74% of the space of the crystal. All the noble-gas solids have cubic close-packed crystals except helium, which has a hexagonal close-packed crystal.

A substance composed of polyatomic molecules with approximately spherical shapes might be expected to crystallize in an approximately close-packed structure. Methane, CH_4, is an example. Solid methane has a face-centered cubic lattice with one CH_4 molecule at each lattice point.

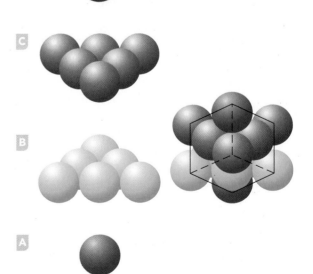

Figure 11.39 ▲

The cubic close-packed structure has a face-centered cubic lattice *Left:* An "exploded" view of portions of four layers of the cubic close-packed structure. *Right:* These layers form a face-centered cubic unit cell.

Figure 11.40 ▶

Crystal structures of metals
Most metals have one of the close-packed structures (hcp = hexagonal close-packed; ccp = cubic close-packed) or a body-centered cubic structure (bcc; manganese has a complicated bcc lattice, with several atoms at each lattice point). Other structures are body-centered tetragonal (bct), orthorhombic (or), and rhombohedral (rh).

IA	IIA	IIIB	IVB	VB	VIB	VIIB		VIIIB		IB	IIB	IIIA	IVA	VA
Li bcc	**Be** hcp													
Na bcc	**Mg** hcp											**Al** ccp		
K bcc	**Ca** ccp	**Sc** hcp	**Ti** hcp	**V** bcc	**Cr** bcc	**Mn** bcc	**Fe** bcc	**Co** hcp	**Ni** ccp	**Cu** ccp	**Zn** hcp	**Ga** or		
Rb bcc	**Sr** ccp	**Y** hcp	**Zr** hcp	**Nb** bcc	**Mo** bcc	**Tc** hcp	**Ru** hcp	**Rh** ccp	**Pd** ccp	**Ag** ccp	**Cd** hcp	**In** bct	**Sn** bct	
Cs bcc	**Ba** bcc	**La** hcp	**Hf** hcp	**Ta** bcc	**W** bcc	**Re** hcp	**Os** hcp	**Ir** ccp	**Pt** ccp	**Au** ccp	**Hg** rh	**Tl** hcp	**Pb** ccp	**Bi** rh

Metallic Solids

If you were to assume that metallic bonding is completely nondirectional, you would expect metals to crystallize in one of the close-packed structures, as the noble-gas elements do. Indeed, many of the metallic elements do have either cubic or hexagonal close-packed crystals. Copper and silver, for example, are cubic close packed (that is, face-centered cubic with one atom at each lattice point of a face-centered cubic lattice). However, a body-centered cubic arrangement of spheres (having one sphere at each lattice point of a body-centered cubic lattice) occupies almost as great a percentage of space as the close-packed ones. In a body-centered arrangement, 68% of the space is occupied by spheres, compared with the maximal value of 74% in the close-packed arrangement. Each atom in a body-centered lattice has a coordination number of 8, rather than the 12 of each atom in a close-packed lattice.

Figure 11.40 lists the structures of the metallic elements. With few exceptions, they have either a close-packed or a body-centered structure.

Ionic Solids

The description of ionic crystals is complicated by the fact that you must give the positions in the crystal structure of both the cations and the anions. We will look at the structures of three different cubic crystals with the general formula MX (where M is the metal and X is the nonmetal): cesium chloride (CsCl), sodium chloride (NaCl), and cubic zinc sulfide or zinc blende (ZnS). Many ionic substances of the general formula MX have crystal structures that are one of these types. Once you have learned these structures, you can relate the structures of many other compounds to them.

Cesium chloride consists of positive ions and negative ions of about equal size arranged in a cubic array. Figure 11.41 shows the unit cell of cesium chloride. It shows both a space-filling model (ions are shown as spheres of approximately the correct relative sizes) and a model in which the ions are shrunk in size in order to display more clearly their relative positions. Note that there are chloride ions at each corner of the cube and a cesium ion at the center. Alternatively, the unit cell may be taken to have cesium ions at the corners of a cube and a chloride ion at the center. In either case, the unit cell contains one Cs^+ ion and one Cl^-

Figure 11.41 ▶

Cesium chloride unit cell
Cesium chloride consists of interpenetrating simple cubic lattices of Cs^+ and Cl^- ions. The figure shows a unit cell with Cl^- ions at the corners of the unit cell and a Cs^+ ion at the center. (A space-filling model is on the left; a model with ions shrunk in size to emphasize the structure is on the right.) An alternative unit cell would have Cs^+ ions at the corners with a Cl^- ion at the center.

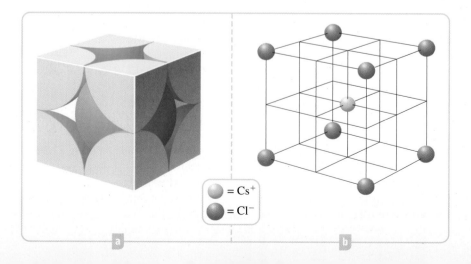

= Cs^+
= Cl^-

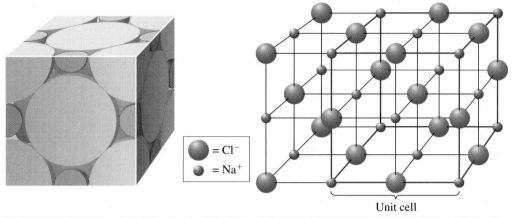

a The unit cell shown here (as a space-filling model) has Cl⁻ ions at the corners and at the centers of the faces of the cube. The Cl⁻ ions are much larger than Na⁺ ions, so the Cl⁻ ions are nearly touching and have approximately a cubic close-packed structure; the Na⁺ ions are in cavities in this structure.

b The alternative unit cell shown here has Na⁺ ions at the corners and the centers of the faces of the unit cell.

$= Cl^-$

$= Na^+$

Unit cell

ion. (Figure 11.41 has $\frac{1}{8}$ Cl⁻ ion at each of eight corners plus one Cs⁺ ion at the center of the unit cell.) Ammonium chloride, NH_4Cl, and thallium chloride, $TlCl$, are examples of other ionic compounds having this "cesium chloride" structure.

Figure 11.42 shows the unit cell of sodium chloride. It has Cl⁻ ions at the corners as well as at the center of each cube face. Because the chloride ions are so much larger than the sodium ions, the Cl⁻ ions are nearly touching and form an approximate cubic close-packed structure of Cl⁻ ions. The Na⁺ ions are arranged in cavities of this close-packed structure. Some other compounds that crystallize in this structure are potassium chloride, KCl; calcium oxide, CaO; and silver chloride, $AgCl$.

Zinc sulfide, ZnS, crystallizes in either of two structures. The zinc sulfide mineral having cubic structure is called zinc blende or sphalerite. Zinc sulfide also exists as hexagonal crystals. (The mineral is called wurtzite.) A solid substance that can occur in more than one crystal structure is said to be *polymorphic*. We will discuss only the zinc blende structure (cubic ZnS).

Figure 11.43 shows the unit cell of zinc blende. Sulfide ions are at the corners and at the center of each face of the unit cell. The positions of the zinc ions can be described if you imagine the unit cell divided into eight cubic parts, or subcubes. Zinc ions are at the centers of alternate subcubes. The number of S^{2-} ions per unit cell is four (8 corner ions $\times \frac{1}{8}$ + 6 face ions $\times \frac{1}{2}$), and the number of Zn^{2+} ions is four (all are completely within the unit cell). Thus, there are equal numbers of Zn^{2+} and S^{2-} ions, as expected by the formula ZnS.

The unit cell of zinc blende can also be taken to have zinc ions at the corners and face centers, with sulfide ions in alternate subcubes. Zinc oxide, ZnO, and beryllium oxide, BeO, also have the "zinc blende" structure.

Covalent Network Solids

The structures of covalent network solids are determined primarily by the directions of covalent bonds. Diamond is a simple example. It is one allotropic form of the element carbon, in which each carbon atom is covalently bonded to four other carbon atoms in tetrahedral directions to give a three-dimensional covalent network (see Figure 11.28). You can describe the bonding by assuming that every carbon atom is sp^3 hybridized. Each C—C bond forms by the overlap of one sp^3 hybrid orbital on one carbon atom with an sp^3 hybrid orbital on the other atom. The unit cell of diamond is similar to that of zinc blende (Figure 11.43); the Zn^{2+} and S^{2-} ions are replaced by carbon atoms, as in Figure 11.44. The crystal lattice, which is face-centered cubic, can be obtained by placing lattice points at the centers of alternate carbon atoms (either those shown in dark gray or those shown in light gray). Silicon, germanium, and gray tin (a nonmetallic allotrope) have "diamond" structures.

Figure 11.42 ▲

Sodium chloride unit cell Sodium chloride has a face-centered cubic lattice.

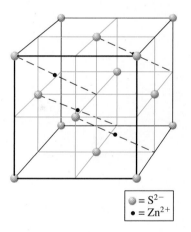

$= S^{2-}$

$= Zn^{2+}$

Figure 11.43 ▲

Zinc blende (cubic ZnS) unit cell Zinc sulfide, ZnS, crystallizes in two different forms, or polymorphs: a hexagonal form (wurtzite) and a face-centered cubic form (zinc blende, or sphalerite), which is shown here. Sulfide ions, S^{2-}, are at the corners and at the centers of the faces of the unit cell; Zn^{2+} ions are in alternate subcubes of the unit cell. The unit cell can also be described as having Zn^{2+} ions at each corner and at the center of each face, with S^{2-} ions in alternate subcubes.

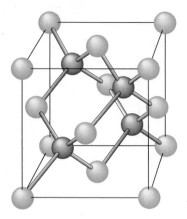

Figure 11.44 ▲

Diamond unit cell The unit cell of diamond can be obtained from the unit cell of zinc blende by replacing Zn^{2+} ions with carbon atoms (shown here as dark spheres) and S^{2-} ions by carbon atoms (shown here as light spheres). Note that each carbon atom is bonded to four others.

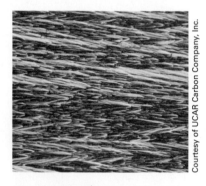

Figure 11.45 ▲

Electron micrograph of graphite Note the layer structure.

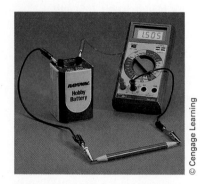

Figure 11.46 ▲

Demonstration of the electrical conductivity of graphite The nonzero reading on the meter shows that the pencil lead (graphite) conducts electricity.

Graphite is another allotrope of carbon. It consists of large, flat sheets of carbon atoms covalently bonded to form hexagonal arrays, and these sheets are stacked one on top of the other (see Figure 11.27). You can describe each sheet in terms of resonance formulas such as the following:

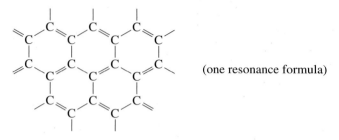

(one resonance formula)

This is just one of many possible resonance formulas, because others can be drawn by moving the double bonds around. This means that the π electrons are delocalized over a plane of carbon atoms. You can also describe this delocalization in molecular orbital terms. You imagine that each carbon atom is sp^2 hybridized, so each atom has three sp^2 hybrid orbitals in the molecular plane and an additional $2p$ orbital perpendicular to this plane. A σ bond forms between two carbon atoms when an sp^2 hybrid orbital on one atom overlaps an sp^2 hybrid orbital on another. The result is a planar σ-bond framework. The $2p$ orbitals of all the atoms overlap to form π orbitals encompassing the entire plane of atoms.

Each carbon–carbon distance within a graphite sheet is 142 pm, which is almost midway between the length of a C—C bond (154 pm) and a C=C bond (134 pm) and which indicates an intermediate bond order (see Figure 11.28). The two-dimensional sheets, or layers, are stacked one on top of the other, and the attraction between sheets results from London forces. The distance between layers is 335 pm, which is greater than the carbon–carbon bond length, because London forces are weaker than covalent bonding forces.

You can explain several properties of graphite on the basis of its structure. The layer structure, which you see in electron micrographs (Figure 11.45), tends to separate easily, so that sheets of graphite slide over one another. The "lead" in a pencil is graphite. As you write with a pencil, layers of graphite rub off to form the pencil mark. Similarly, the use of graphite to lubricate locks and other closely fitting metal surfaces depends on the ability of graphite particles to move over one another as the layers slide. Another property of graphite is its electrical conductivity. Electrical conductivity is unusual in covalent network solids, because normally the electrons in such bonds are localized. Graphite is a moderately good conductor, however, because delocalization leads to mobile electrons within each carbon atom layer. Figure 11.46 shows a demonstration of the electrical conductivity of graphite.

CONCEPT CHECK 11.5

Shown here is a representation of a unit cell for a crystal. The orange balls are atom A, and the gray balls are atom B.

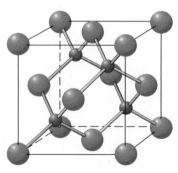

a What is the chemical formula of the compound that has this unit cell ($A_?B_?$)?

b Consider the configuration of the A atoms. Is this a cubic unit cell? If so, which type?

11.9 Calculations Involving Unit-Cell Dimensions

You can determine the structure and dimensions of a unit cell by diffraction methods, which we will describe briefly in the next section. Once you know the unit-cell dimensions and the structure of a crystal, however, some interesting calculations are possible. For instance, suppose you find that a metallic solid has a face-centered cubic unit cell with all atoms at lattice points, and you determine the edge length of the unit cell. From this unit-cell dimension, you can calculate the volume of the unit cell. Then, knowing the density of the metal, you can calculate the mass of the atoms in the unit cell. Because you know that the unit cell is face-centered cubic with all atoms at lattice points and that such a unit cell has four atoms, you can obtain the mass of an individual atom. This determination of the mass of a single atom gave one of the first accurate determinations of Avogadro's number. The calculations are shown in the next example.

Example `11.10`

Gaining Mastery Toolbox

Critical Concept 11.10
The face-centered cubic (FCC) unit cell contains four atoms. Each corner of the unit cell contains $\frac{1}{8}$ of an atom and each face of the unit cell contains $\frac{1}{2}$ of an atom.

Solution Essentials:
• Face-centered unit cell
• Unit cell

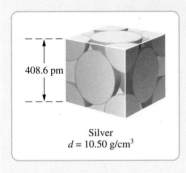

408.6 pm

Silver
$d = 10.50 \text{ g/cm}^3$

Calculating Atomic Mass from Unit-Cell Dimension and Density

X-ray diffraction from crystals provides one of the most accurate ways of determining Avogadro's number. Silver crystallizes in a face-centered cubic lattice with all atoms at the lattice points. The length of an edge of the unit cell was determined by x-ray diffraction to be 408.6 pm (4.086 Å). The density of silver is 10.50 g/cm^3. Calculate the mass of a silver atom. Then, using the known value of the atomic mass, calculate Avogadro's number.

Problem Strategy Knowing the edge length of a unit cell, you can calculate the unit-cell volume. Then, from the density, you can find the mass of the unit cell and hence the mass of a silver atom.

Solution You obtain the volume, V, of the unit cell by cubing the length of an edge.

$$V = (4.086 \times 10^{-10}\,\text{m})^3 = 6.822 \times 10^{-29}\,\text{m}^3$$

The density, d, of silver in grams per cubic meter is

$$d = 10.50\,\frac{\text{g}}{\text{cm}^3} \times \left(\frac{1\,\text{cm}}{10^{-2}\,\text{m}}\right)^3 = 1.050 \times 10^7\,\text{g/m}^3$$

Density is mass per volume; hence, the mass of a unit cell equals the density times the volume of the unit cell.

$$
\begin{aligned}
m &= dV \\
&= 1.050 \times 10^7\,\text{g/m}^3 \times 6.822 \times 10^{-29}\,\text{m}^3 \\
&= 7.163 \times 10^{-22}\,g
\end{aligned}
$$

Because there are four atoms in a face-centered unit cell having one atom at each lattice point (see Example 11.9), the mass of a silver atom is

$$\text{Mass of 1 Ag atom} = \tfrac{1}{4} \times 7.163 \times 10^{-22}\,\text{g} = \mathbf{1.791 \times 10^{-22}\,g}$$

The known atomic mass of silver is 107.87 amu. Thus, the molar mass (Avogadro's number of atoms) is 107.87 g/mol, and Avogadro's number, N_A, is

$$N_A = \frac{107.87\,\text{g/mol}}{1.791 \times 10^{-22}\,\text{g}} = \mathbf{6.023 \times 10^{23}\text{/mol}}$$

(continued)

(continued)

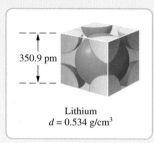

Lithium
$d = 0.534 \text{ g/cm}^3$

Answer Check In any problem of this type where you have to calculate the mass of an atom, you can always check the validity of your answer by calculating the mass of a mole of atoms (the molar mass). For example, because the calculated mass of the silver atom in this problem is 1.791×10^{-22} g, the molar mass is $(6.02 \times 10^{23}) \times (1.791 \times 10^{-22}) = 108$ g/mol, which is the expected molar mass of silver to three significant figures.

Exercise 11.11 Lithium metal has a body-centered cubic structure with all atoms at the lattice points and a unit-cell length of 350.9 pm (see art at left). Calculate Avogadro's number. The density of lithium is 0.534 g/cm³.

■ See Problems 11.85 and 11.86.

If you know or assume the structure of an atomic crystal, you can calculate the length of the unit-cell edge from the density of the substance. This is illustrated in the next example. Agreement of this value with that obtained from x-ray diffraction confirms that your view of the structure of the crystal is correct.

Example 11.11   **Calculating Unit-Cell Dimensions from Unit-Cell Type and Density**

Gaining Mastery Toolbox

Critical Concept 11.11
The edges of any cubic unit cell are of equal length. If you know the volume of a cubic unit cell, the length of an edge is the cube root of the unit-cell volume ($l = \sqrt[3]{V}$).

Solution Essentials:
- Face-centered unit cell
- Unit cell

Platinum crystallizes in a face-centered cubic lattice with all atoms at the lattice points. It has a density of 21.45 g/cm³ and an atomic mass of 195.08 amu. From these data, calculate the length of a unit-cell edge. Compare this with the value of 392.4 pm obtained from x-ray diffraction.

Problem Strategy The mass of an atom, and hence the mass of a unit cell, can be calculated from the atomic mass. Knowing the density and the mass of the unit cell, we can calculate the volume and the edge length of a unit cell.

Solution You can use Avogadro's number (6.022×10^{23}/mol) to convert the molar mass of platinum (195.08 g/mol) to the mass per atom.

$$\frac{195.08 \text{ g Pt}}{1 \text{ mol Pt}} \times \frac{1 \text{ mol Pt}}{6.022 \times 10^{23} \text{ Pt atoms}} = \frac{3.239 \times 10^{-22} \text{ g Pt}}{1 \text{ Pt atom}}$$

Because there are four atoms per unit cell, the mass per unit cell can be calculated as follows:

$$\frac{3.239 \times 10^{-22} \text{ g Pt}}{1 \text{ Pt atom}} \times \frac{4 \text{ Pt atoms}}{1 \text{ unit cell}} = \frac{1.296 \times 10^{-21} \text{ g}}{1 \text{ unit cell}}$$

The volume of the unit cell is

$$V = \frac{m}{d} = \frac{1.296 \times 10^{-21} \text{ g}}{21.45 \text{ g/cm}^3} = 6.042 \times 10^{-23} \text{ cm}^3$$

392.4 pm

Platinum
$d = 21.45 \text{ g/cm}^3$
atomic mass = 195.08 amu

(continued)

(*continued*)

If the edge length of the unit cell is denoted as *l*, the volume is $V = l^3$. Hence, the edge length is

$$l = \sqrt[3]{V}$$
$$= \sqrt[3]{6.042 \times 10^{-23} \text{ cm}^3}$$
$$= 3.924 \times 10^{-8} \text{ cm (or } 3.924 \times 10^{-10} \text{ m)}$$

Thus, the edge length is 392.4 pm, which is in excellent agreement with the x-ray diffraction value.

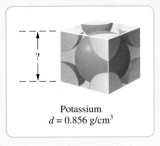

Potassium
$d = 0.856 \text{ g/cm}^3$

Answer Check Note that the edge length for the unit cell will be greater by a small factor than the values you see for atomic diameters, so you should obtain something on the order of a few hundred picometers. You should suspect an error if you obtain anything much smaller or larger than this.

Exercise 11.12 Potassium metal has a body-centered cubic structure with all atoms at the lattice points (see art at left). The density of the metal is 0.856 g/cm³. Calculate the edge length of a unit cell.

■ See Problems 11.87 and 11.88.

11.10 Determining Crystal Structure by X-Ray Diffraction

The determination of crystal structure by x-ray diffraction is one of the most important ways of determining the structures of molecules. Because of its ordered structure, a crystal consists of repeating planes of the same kind of atom. These planes can act as reflecting surfaces for x rays. When x rays are reflected from these planes, they show a *diffraction pattern,* which can be recorded on a photographic plate as a series of spots (see Figure 11.47). By analyzing the diffraction pattern, you can determine the positions of the atoms in the unit cell of the crystal. Once you have determined the positions of each atom in the unit cell of a molecular solid, you have also found the positions of the atoms in the molecule. ▶

To understand how such a diffraction pattern occurs, let us look at the diffraction of two waves. Suppose two waves of the same wavelength come together in phase. By *in phase* we mean that the waves come together so that their peaks (maxima) and troughs (minima) match (see Figure 11.48*a*). Two waves that come together in phase form a wave with higher peaks and lower troughs than either wave. We say that the waves reinforce each other, or undergo *constructive interference.* The height of an x-ray wave (its *amplitude*) is related to the intensity of the x ray, so constructive interference gives a resultant wave having increased intensity.

Alternatively, suppose two waves of the same wavelength come together out of phase. By *out of phase* we mean that where one wave has its peaks, the other has its troughs (see Figure 11.48*b*). When two waves come together out of phase, each peak of one wave combines with a trough of the other so that the resultant wave amplitude is smaller than that of either wave (zero, if the two waves have the same amplitudes). We say that the waves undergo *destructive interference.* The resultant wave has decreased intensity.

Now consider the scattering, or reflection, of x rays from a crystal. Figure 11.49 shows two rays: one ray reflected from one plane, the other from another plane. The two rays start out in phase, but because one ray travels farther, they may end up out of phase after reflection. Only at certain angles of reflection do the rays remain in phase. At some angles the rays constructively interfere (giving dark areas on a negative, corresponding to large x-ray intensity). At other angles the

X-ray diffraction has been used to obtain the structure of proteins, which are large molecules essential to life. John Kendrew and Max Perutz received the Nobel Prize in chemistry in 1962 for their x-ray work in determining the structure of myoglobin (MM = 17,600 amu) and hemoglobin (MM = 66,000 amu). Hemoglobin is the oxygen-carrying protein of the red blood cells, and myoglobin is the oxygen-carrying protein in muscle tissue.

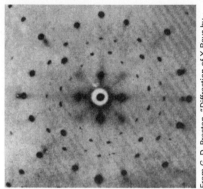

Figure 11.47 ▲

A crystal diffraction pattern This diffraction pattern was obtained by diffracting x rays from a sodium chloride crystal and collecting the image on photographic film. The resulting negative shows dark spots on a lighter background.

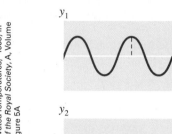

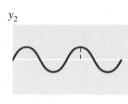

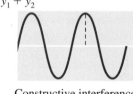

Constructive interference

Destructive interference

a Two waves constructively interfere when they are in phase (when their peaks and troughs match). The two waves add to give a resultant wave with higher peaks and deeper troughs than either wave.

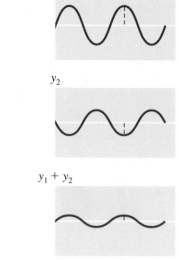

b Two waves destructively interfere when they are out of phase (when the peaks of one wave match the troughs of the other wave). The two waves add to give a resultant wave with lower peaks and more shallow troughs than either wave.

Figure 11.48 ▲

Wave interference

rays destructively interfere (giving light areas on a photographic plate, corresponding to low x-ray intensity). Note the dark and light areas on the photographic negative of the diffraction pattern shown in Figure 11.47.

You can relate the resulting pattern produced by the diffraction to the structure of the crystal. You can determine the type of unit cell and its size; if the solid is composed of molecules, you can determine the position of each atom in the molecule.

Figure 11.49 ▶

Diffraction of x rays from crystal planes

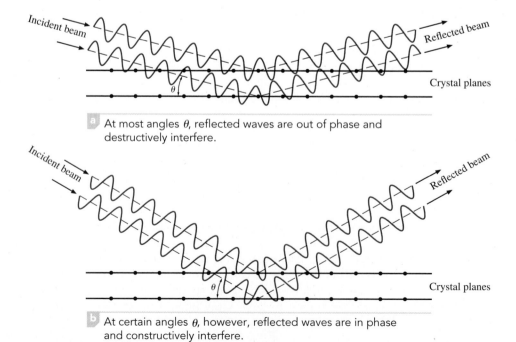

a At most angles θ, reflected waves are out of phase and destructively interfere.

b At certain angles θ, however, reflected waves are in phase and constructively interfere.

Automated X-Ray Diffractometry

Max von Laue, a German physicist, was the first to suggest the use of x rays for the determination of crystal structure. Soon afterward, in 1913, the British physicists William Bragg and his son Lawrence developed the method on which modern crystal-structure determination is based. They realized that the atoms in a crystal form reflecting planes for x rays, and from this idea they derived the fundamental equation of crystal-structure determination.

$$n\lambda = 2d \sin \theta, \; n = 1, 2, 3, \ldots$$

The *Bragg equation* relates the wavelength of x rays, λ, to the distance between atomic planes, d, and the angle of reflection, θ. Note that reflections occur at several angles, corresponding to different integer values of n.

A molecular crystal has many different atomic planes, so that it reflects an x-ray beam in many different directions. By analyzing

Courtesy of Bruker Analytical X-Ray Systems, Inc., Madison, Wisconsin, USA

the intensities and angular directions of the reflected beams, you can determine the exact positions of all the atoms in the unit cell of the crystal and therefore obtain the structure of the molecule. The problem of obtaining the x-ray data (intensities and angular directions of the reflections) and then analyzing them, however, is not trivial. Originally, the reflected x rays were recorded on photographic plates. After taking many pictures, the scientist would pore over the negatives, measuring the densities of the spots and their positions on the plates. Then he or she would work through lengthy and laborious calculations to analyze the data. Even with early computers, the determination of a molecular structure required a year or more.

With the development of electronic x-ray detectors and mini-computers, x-ray diffraction has become automated, so that the time and effort of determining the structure of a molecule have been substantially reduced. Now frequently the most difficult task is preparing a suitable crystal. The crystal should be several tenths of a millimeter in each dimension and without significant defects. Such crystals of protein molecules, for example, can be especially difficult to prepare.

Once a suitable crystal has been obtained, the structure of a molecule of moderate size can often be determined in a day or so. The crystal is mounted on a glass fiber (or in a glass capillary containing an inert gas, if the substance reacts with air) and placed on a pin or spindle within the circular assembly of the x-ray diffractometer (Figure 11.50). The crystal and x-ray detector (placed on the opposite side of the crystal from the x-ray tube) rotate under computer control, while the computer records the intensities and angles of thousands of x-ray reflection spots. After computer analysis of the data, the molecular structure is printed out.

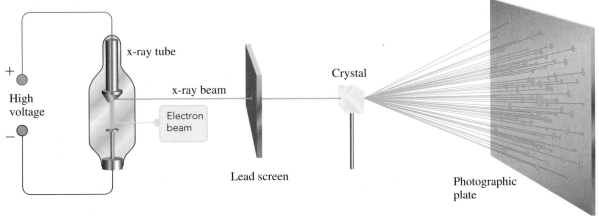

Figure 11.50 ▲

Automated x-ray diffractometer *(Top)* The single-crystal specimen is mounted on a glass fiber, which is placed on a spindle within the circular assembly of the diffractometer. A new data collection system (left of specimen) reduces data collection time from several days to several hours. *(Bottom)* The schematic diagram shows the diffracted rays being detected by a photograph. In a modern diffractometer, the final data collection is done with a fixed electronic detector, and the crystal is rotated. The data are collected and analyzed by a computer interfaced with the diffractometer.

■ See Problems 11.129 and 11.130.

Water (A Special Substance for Planet Earth)

Water is the only liquid substance (other than petroleum) to be found on earth in significant amounts. This liquid is also readily convertible under conditions on earth to the solid and gaseous forms. Water has several unusual properties that set it apart from other substances. For example, its solid phase, ice, is less dense than liquid water, whereas for most substances the solid phase is more dense than the liquid. In addition, water has an unusually large heat capacity. These are some of the properties that are important in determining conditions favorable to life. In fact, it is difficult to think of life without water.

The unusual properties of water are largely linked to its ability to form hydrogen bonds. For example, ice is less dense than liquid water because ice has an open, hydrogen-bonded structure (see Figure 11.51). Each oxygen atom in the structure of ice is surrounded tetrahedrally by four hydrogen atoms: two that are close and covalently bonded to give the H_2O molecule, and two others that are farther away and held by hydrogen bonds. The tetrahedral angles give rise to a three-dimensional structure that contains open space. When ice melts, hydrogen bonds break, the ice structure partially disintegrates, and the molecules become more compactly arranged, leading to a more dense liquid.

Even liquid water has significant hydrogen bonding. Only about 15% of the hydrogen bonds are broken when ice melts. You might view the liquid as composed of icelike clusters in which hydrogen bonds are continually breaking and forming, so clusters disappear and new ones appear. This is sometimes referred to as the "flickering cluster" model of liquid water.

As the temperature rises from 0°C, the clusters tend to break down further, giving an even more compact liquid. Thus, the density rises (see Figure 11.52). However, as is normal in any liquid when the temperature rises, the molecules begin to move and vibrate faster, and the space occupied by the average molecule increases. For most other liquids, this results in a continuous decrease in density with temperature increase. In water, this normal effect is countered by the density increase due to breaking of hydrogen bonds, but at 4°C the normal effect begins to predominate. Water shows a maximum density at 4°C and becomes less dense at higher temperatures.

Figure 11.51 ▲

The structure of ice Oxygen atoms are represented by large red spheres; hydrogen atoms, by small blue spheres. Each oxygen atom is tetrahedrally surrounded by four hydrogen atoms. Two are close, giving the H_2O molecule. Two are farther away, held by hydrogen bonding (represented by three dots). The distribution of hydrogen atoms in these two types of positions is random.

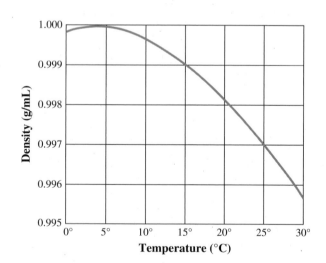

Figure 11.52 ▲

Density of water versus temperature Water has a maximum density at 4°C.

The unusually large heat of vaporization of water has an important effect on the earth's weather. Evaporation of the surface waters absorbs over 30% of the solar energy reaching the earth's surface. This energy is released when the water vapor condenses, and thunderstorms and hurricanes may result. In the process, the waters of the earth are circulated and the freshwater sources are replenished. This natural cycle of water from the oceans to freshwater sources and its return to the ocean is called the *hydrologic cycle* (Figure 11.53).

Although evaporation and condensation of water play a dominant role in our weather, other properties are important as well. The exceptionally large heat capacity of bodies of water has an important moderating effect on the surrounding temperature by warming cold air masses in winter and cooling warm air masses in summer. Worldwide, the oceans are most important, but even inland lakes have a pronounced effect. For example, the Great Lakes give Detroit a more moderate winter than cities somewhat farther south but with no lakes nearby.

The fact that ice is less dense than water means that it forms on top of the liquid when freezing occurs. This has far-reaching effects, both for weather and for aquatic animals. When ice forms on a body of water, it insulates the underlying water from the cold air and limits further freezing. Fish depend on this for winter survival. Consider what would happen to a lake if ice were more dense than water. The ice would freeze from the bottom of the lake upward. Without the insulating effect at the surface, the lake could well freeze solid, killing the fish. Spring thaw would be prolonged, because the insulating effect of the surface water would make it take much longer for the ice at the bottom of a lake to melt.

The solvent properties of water are also unusual. Water is both a polar substance and a hydrogen-bonding molecule. As a result, water dissolves many substances, including ionic and polar compounds. These properties make water the premier solvent, biologically and industrially.

As a result of the solvent properties of water, the naturally occurring liquid always contains dissolved materials, particularly ionic substances. *Hard water* contains certain metal ions, such as Ca^{2+} and Mg^{2+}. These ions react with soaps, which are sodium salts of stearic acid and similar organic acids, to produce a curdy precipitate of calcium and magnesium salts. This precipitate adheres to clothing and bathtubs (as bathtub ring). Removing Ca^{2+} and Mg^{2+} ions from hard water is referred to as *water softening*.

Water is often softened by *ion exchange*. Ion exchange is a process whereby a water solution is passed through a column of a material that replaces one kind of ion in solution with another kind. Home and commercial water softeners contain cation-exchange resins, which are insoluble macromolecular substances (substances consisting of giant molecules) to which negatively charged groups are chemically bonded. The negative charges are counterbalanced by ions such as Na^+. When hard water containing the Ca^{2+} ion passes through a column of this resin, the Na^+ ion in the resin is replaced by Ca^{2+} ion.

$$2NaR(s) + Ca^{2+}(aq) \longrightarrow CaR_2(s) + 2Na^+(aq)$$

$$(R^- = \text{anion of exchange resin})$$

The water passing through the column now contains Na^+ in place of Ca^{2+} and has been softened. Once the resin has been completely converted to the calcium salt, it can be regenerated by flushing the column with a concentrated solution of NaCl to reverse the previous reaction.

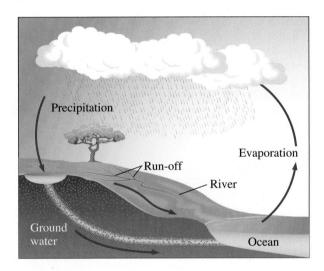

Figure 11.53 ▲

The hydrologic cycle Ocean water evaporates to form clouds. Then, precipitation (rain and snow) from these clouds replenishes freshwater sources. Water from these sources eventually returns to the oceans via rivers, run-off, or groundwater.

■ See Problems 11.131 and 11.132.

A Checklist for Review

OWL and **Go Chemistry** Sign in at **www.cengage.com/owl** to:

• View tutorials and simulations, develop problem-solving skills, and complete online homework assigned by your professor.

• Download Go Chemistry mini lecture modules for quick review and exam prep from OWL or purchase them at **www.cengagebrain.com**.

Summary of Facts and Concepts

Gases are composed of molecules in constant random motion throughout mostly empty space. This explains why gases are compressible fluids. Liquids are also composed of molecules in constant random motion, but the molecules are more tightly packed. Thus, liquids are incompressible fluids. Solids are composed of atoms, molecules, or ions that are in close contact and oscillate about fixed sites. Thus, solids are incompressible and rigid rather than fluid.

Any state of matter may change to another state. Melting and freezing are examples of such changes of state, or *phase transitions*. Vapor pressure, boiling point, and melting point are properties of substances that involve phase transitions. Vapor pressure increases with temperature; according to the *Clausius–Clapeyron equation*, the logarithm of the vapor pressure varies inversely with the absolute temperature. The conditions under which a given state of a substance exists are graphically summarized in a *phase diagram*.

Like vapor pressure and boiling point, *surface tension* and *viscosity* are important properties of liquids. These properties can be explained in terms of intermolecular forces. The three kinds of attractive intermolecular forces

are *dipole–dipole forces, London forces,* and *hydrogen bonding*. London forces are weak attractive forces present in all molecules; London forces tend to increase with increasing molecular weight. Vapor pressure tends to decrease with increasing molecular weight, whereas boiling point, surface tension, and viscosity tend to increase (unless hydrogen bonding is present). Hydrogen bonding occurs in substances containing H atoms bonded to F, O, or N atoms. When hydrogen bonding is present, vapor pressure tends to be lower than otherwise expected, whereas boiling point, surface tension, and viscosity tend to be higher.

Solids can be classified in terms of the type of force between the structural units; there are molecular, metallic, ionic, and covalent network solids. Melting point, hardness, and electrical conductivity are properties that can be related to the structure of the solid.

Solids can be crystalline or amorphous. A crystalline solid has an ordered arrangement of structural units placed at *crystal lattice* points. We may think of a crystal as constructed from *unit cells*. Cubic unit cells are of three kinds: simple cubic, body-centered cubic, and face-centered cubic. One of the most important ways of determining the structure of a crystalline solid is by *x-ray diffraction*.

Learning Objectives	Important Terms
11.1 Comparison of Gases, Liquids, and Solids ■ Recall the definitions of *gas, liquid,* and *solid* given in Section 1.4. ■ Compare a gas, a liquid, and a solid using a kinetic-molecular theory description. ■ Recall the ideal gas law and the van der Waals equation for gases (there are no similar simple equations for liquids and solids).	
11.2 Phase Transitions ■ Define *change of state (phase transition)*. ■ Define *melting, freezing, vaporization, sublimation,* and *condensation*. ■ Define *vapor pressure*. ■ Describe the process of reaching a dynamic equilibrium that involves the vaporization of a liquid and condensation of its vapor. ■ Define *boiling point*. ■ Describe the process of boiling. ■ Define *freezing point* and *melting point*. ■ Define *heat (enthalpy) of fusion* and *heat (enthalpy) of vaporization*. ■ Calculate the heat required for a phase change of a given mass of substance. Example 11.1 ■ Describe the general dependence of the vapor pressure (ln *P*) on the temperature (*T*). ■ State the Clausius–Clapeyron equation (the two-point form).	**change of state (phase transition)** **melting** **freezing** **vaporization** **sublimation** **condensation** **vapor pressure** **boiling point** **freezing point** **melting point** **heat of fusion** **heat of vaporization**

- Calculate the vapor pressure at a given temperature.
  Example 11.2
- Calculate the heat of vaporization from vapor
  pressure. Example 11.3

11.3 Phase Diagrams

- Define *phase diagram.*
- Describe the melting-point curve and the vapor-
  pressure curves (for the liquid and the solid) in a
  phase diagram.
- Define *triple point.*
- Define *critical temperature* and *critical pressure.*
- Relate the conditions for the liquefaction of a gas to
  its critical temperature. Example 11.4

phase diagram
triple point
critical temperature
critical pressure

11.4 Properties of Liquids; Surface Tension and Viscosity

- Define *surface tension.*
- Describe the phenomenon of capillary rise.
- Define *viscosity.*

surface tension
viscosity

11.5 Intermolecular Forces; Explaining Liquid Properties

- Define *intermolecular forces.*
- Define *dipole–dipole force.*
- Describe the alignment of polar molecules in a sub-
  stance.
- Define *London (dispersion) forces.*
- Note that London forces tend to increase with molecu-
  lar mass.
- Relate the properties of liquids to the intermolecular
  forces involved.
- Define *hydrogen bonding.*
- Identify the intermolecular forces in a substance.
  Example 11.5
- Determine relative vapor pressures on the basis of
  intermolecular attractions. Example 11.6

intermolecular forces
van der Waals forces
dipole–dipole force
London (dispersion) forces
hydrogen bonding

11.6 Classification of Solids by Type of Attraction of Units

- Define *molecular solid, metallic solid, ionic solid,* and
  covalent network solid.
- Identify types of solids. Example 11.7
- Relate the melting point of a solid to its structure.
- Determine relative melting points based on types of
  solids. Example 11.8
- Relate the hardness and electrical conductivity of a
  solid to its structure.

molecular solid
metallic solid
ionic solid
covalent network solid

11.7 Crystalline Solids; Crystal Lattices and Unit Cells

- Define *crystalline solid* and *amorphous solid.*
- Define *crystal lattice* and *unit cell* of a crystal lattice.
- Define *simple cubic unit cell, body-centered cubic unit
  cell,* and *face-centered cubic unit cell.*
- Determine the number of atoms in a unit cell.
  Example 11.9
- Describe the two kinds of crystal defects.

crystalline solid
amorphous solid
crystal lattice
unit cell
simple cubic unit cell
body-centered cubic unit cell
face-centered cubic unit cell

11.8 Structures of Some Crystalline Solids

- Define *hexagonal close-packed structure* and *cubic
  close-packed structure.*
- Define *coordination number.*

hexagonal close-packed structure
cubic close-packed structure
coordination number

- Note the common structures (face-centered cubic and body-centered cubic) of metallic solids.
- Describe the three types of cubic structures of ionic solids.
- Describe the covalent network structure of diamond and graphite.

11.9 Calculations Involving Unit-Cell Dimensions

- Calculate atomic mass from unit-cell dimension and density. Example 11.10
- Calculate unit-cell dimension from unit-cell type and density. Example 11.11

11.10 Determining Crystal Structure by X-Ray Diffraction

- Describe how constructive and destructive interference give rise to a diffraction pattern.
- Note that diffraction of x rays from a crystal gives information about the positions of atoms in the crystal.

Key Equation

$$\ln \frac{P_2}{P_1} = \frac{\Delta H_{vap}}{R}\left(\frac{1}{T_1} - \frac{1}{T_2}\right)$$

Questions and Problems

OWL Interactive versions of these problems may be assigned in OWL.

Self-Assessment and Review Questions

Key: These questions test your understanding of the ideas you worked with in the chapter. These problems vary in difficulty and often can be used for the basis of discussion.

11.1 List the different phase transitions that are possible and give examples of each.

11.2 Describe how you could purify iodine by sublimation.

11.3 Describe vapor pressure in molecular terms. What do we mean by saying it involves a dynamic equilibrium?

11.4 Explain why 15 g of steam at 100°C melts more ice than 15 g of liquid water at 100°C.

11.5 Why is the heat of fusion of a substance smaller than its heat of vaporization?

11.6 Explain why evaporation leads to cooling of the liquid.

11.7 Describe the behavior of a liquid and its vapor in a closed vessel as the temperature increases.

11.8 The pressure in a cylinder of nitrogen continuously decreases as gas is released from it. On the other hand, a cylinder of propane maintains a constant pressure as propane is released. Explain this difference in behavior.

11.9 Gases that cannot be liquefied at room temperature merely by compression are called "permanent" gases. How could you liquefy such a gas?

11.10 Why does the vapor pressure of a liquid depend on the intermolecular forces?

11.11 Explain the surface tension of a liquid in molecular terms. How does the surface tension make a liquid act as though it had a "skin"?

11.12 Explain the origin of the London force that exists between two molecules.

11.13 Explain what is meant by hydrogen bonding. Describe the hydrogen bonding between two H_2O molecules.

11.14 Why do molecular substances have relatively low melting points?

11.15 Describe the distinguishing characteristics of a crystalline solid and an amorphous solid.

11.16 Describe the face-centered cubic unit cell.

11.17 Describe the structure of thallium(I) iodide, which has the same structure as cesium chloride.

11.18 Explain in words how Avogadro's number could be obtained from the unit-cell edge length of a cubic crystal. What other data are required?

11.19 What is the coordination number of Cs^+ in CsCl? of Na^+ in NaCl? of Zn^{2+} in ZnS?

11.20 Explain the production of an x-ray diffraction pattern by a crystal in terms of the interference of waves.

11.21 Which of the following would release the most heat? Assume the same mass of H_2O in each case.
- a Cooling the H_2O sample from 123°C to 101°C
- b Cooling the H_2O sample from 20°C to −1.0°C
- c Cooling the H_2O sample from 105°C to 84°C
- d Heating the H_2O sample from −15°C to 60°C
- e Cooling the H_2O sample from 95°C to 74°C

11.22 The triple point of a solid is at 5.2 atm and −57°C. Under typical laboratory conditions of $P = 0.98$ atm and $T = 23$°C, this solid will
- a remain solid indefinitely
- b boil
- c melt
- d sublime
- e condense

11.23 Under the right conditions, hydrogen gas, H_2, can be liquefied. Which is the most important intermolecular force that is responsible for allowing hydrogen molecules to be liquefied?

- a. hydrogen bonding
- b. dipole–dipole interactions
- c. London (dispersion) forces
- d. covalent bonds
- e. ion–dipole forces

11.24 An element crystallizes with a simple cubic lattice with atoms at all the lattice points. If the radius of the atom is 200. pm, what is the volume of the unit cell?

- a. 8.00×10^6 pm^3
- b. 6.00×10^7 pm^3
- c. 4.00×10^4 pm^3
- d. 6.40×10^7 pm^3
- e. 1.60×10^5 pm^3

Concept Explorations

Key: Concept explorations are comprehensive problems that provide a framework that will enable you to explore and learn many of the critical concepts and ideas in each chapter. If you master the concepts associated with these explorations, you will have a better understanding of many important chemistry ideas and will be more successful in solving all types of chemistry problems. These problems are well suited for group work and for use as in-class activities.

11.25 Intermolecular Forces

The following picture represents atoms of hypothetical, nonmetallic, monatomic elements A, B, and C in a container at a temperature of 4 K (the piston maintains the pressure at 1 atm). None of these elements reacts with the others.

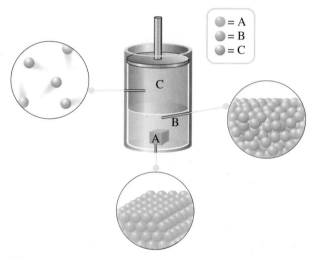

- a. What is the state (solid, liquid, or gas) of each of the elements represented in the container?
- b. Rank the elements in the container from greatest to least, in terms of intermolecular interactions. Explain your answer.
- c. What type(s) of intermolecular attractions are present in each of these elements?
- d. Explain which element has the greatest atomic mass.
- e. One of the elements in the container has a normal boiling point of 2 K. Which element would that be (A, B, or C)? How do you know?
- f. One of the elements has a melting point of 50 K. Which element would that be (A, B, or C)? Why?
- g. The remaining element (the one you have yet to choose) has a normal boiling point of 25 K. Identify the element. Could this element have a freezing point of 7 K? Explain.

- h. If you started heating the sample to 20 K, explain what you would observe with regard to the container and its contents during the heating.
- i. Describe the container and its contents at 20 K. Describe (include a drawing) how the container and its contents look at 20 K.
- j. Now you increase the temperature of the container to 30 K. Describe (include a drawing) how the container and its contents look at 30 K. Be sure to note any changes in going from 20 K to 30 K.
- k. Finally, you heat the container to 60 K. Describe (include a drawing) how the container and its contents look at this temperature. Be sure to note any changes in going from 30 K to 60 K.

11.26 Heat and Molecular Behavior
Part 1:

- a. Is it possible to add heat to a pure substance and not observe a temperature change? If so, provide examples.
- b. Describe, on a molecular level, what happens to the heat being added to a substance just before and during melting. Do any of these molecular changes cause a change in temperature?

Part 2: Consider two pure substances with equal molar masses: substance A, having very strong intermolecular attractions, and substance B, having relatively weak intermolecular attractions. Draw two separate heating curves for 0.25-mol samples of substance A and substance B in going from the solid to the vapor state. You decide on the freezing point and boiling point for each substance, keeping in mind the information provided in this problem. Here is some additional information for constructing the curves. In both cases, the rate at which you add heat is the same. Prior to heating, both substances are at −50°C, which is below their freezing points. The heat capacities of A and B are very similar in all states.

- a. As you were heating substances A and B, did they melt after equal quantities of heat were added to each substance? Explain how your heating curves support your answer.
- b. What were the boiling points you assigned to the substances? Are the boiling points the same? If not, explain how you decided to display them on your curves.
- c. According to your heating curves, which substance reached the boiling point first? Justify your answer.
- d. Is the quantity of heat added to melt substance A at its melting point the same as the quantity of heat required to convert all of substance A to a gas at its boiling point? Should these quantities be equal? Explain.

Conceptual Problems

Key: These problems are designed to check your understanding of the concepts associated with some of the main topics presented in each chapter. A strong conceptual understanding of chemistry is the foundation for both applying chemical knowledge and solving chemical problems. These problems vary in level of difficulty and often can be used as a basis for group discussion.

11.27 Shown here is a curve of the distribution of kinetic energies of the molecules in a liquid at an arbitrary temperature *T*.

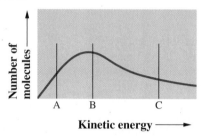

The lines marked A, B, and C represent the point where each of the molecules for three different liquids (liquid A, liquid B, and liquid C) has the minimum kinetic energy to escape into the gas phase (see Figure 11.5 for more information). Write a brief explanation for each of your answers to the following questions.

- a Which of the molecules—A, B, or C—would have the majority of the molecules in the gas phase at temperature *T*?
- b Which of the molecules—A, B, or C—has the strongest intermolecular attractions?
- c Which of the molecules would have the lowest vapor pressure at temperature *T*?

11.28 Consider a substance X with a $\Delta H_{vap} = 20.3$ kJ/mol and $\Delta H_{fus} = 9.0$ kJ/mol. The melting point, freezing point, and heat capacities of both the solid and liquid X are identical to those of water.

- a If you place one beaker containing 50 g of X at $-10°C$ and another beaker with 50 g of H_2O at $-10°C$ on a hot plate and start heating them, which material will reach the boiling point first?
- b Which of the materials from part a, X or H_2O, would completely boil away first?
- c On a piece of graph paper, draw the heating curve for H_2O and X. How do the heating curves reflect your answers from parts a and b?

11.29 Using the information presented in this chapter, explain why farmers spray water above and on their fruit trees on still nights when they know the temperature is going to drop below 0°C. (*Hint:* Totally frozen fruit is what the farmers are trying to avoid.)

11.30 You are presented with three bottles, each containing a different liquid: bottle A, bottle B, and bottle C. Bottle A's label states that it is an ionic compound with a boiling point of 35°C. Bottle B's label states that it is a molecular compound with a boiling point of 29.2°C. Bottle C's label states that it is a molecular compound with a boiling point of 67.1°C.

- a Which of the compounds is most likely to be incorrectly identified?
- b If Bottle A were a molecular compound, which of the compounds has the strongest intermolecular attractions?
- c If Bottle A were a molecular compound, which of the compounds would have the highest vapor pressure?

11.31 Shown here is a representation of a unit cell for a crystal. The orange balls are atom A, and the gray balls are atom B.

- a What is the chemical formula of the compound that has this unit cell ($A_?B_?$)?
- b Consider the configuration of the A atoms. Is this a cubic unit cell? If so, which type?

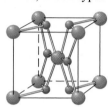

11.32 As a demonstration, an instructor had a block of ice suspended between two chairs. She then hung a wire over the block of ice, with weights attached to the ends of the wire:

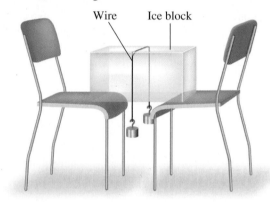

Wire   Ice block

Later, at the end of the lecture, she pointed out that the wire was lying on the floor, apparently having cut its way through the block of ice, although the ice block remained intact. Explain, by referring to the phase diagram for water (Figure 11.11), what has happened.

11.33 The heats of vaporization for water and carbon disulfide are 40.7 kJ/mol and 26.8 kJ/mol, respectively. A vapor (steam) burn occurs when the concentrated vapor of a substance condenses on your skin. Which of these substances, water or carbon disulfide, will result in the most severe burn if identical quantities of each vapor at a temperature just above their boiling point came in contact with your skin?

11.34 If you place water at room temperature in a well-insulated cup and allow some of the water to evaporate, the temperature of the water in the cup will drop lower than room temperature. Come up with an explanation for this observation.

11.35 When hypothetical element X forms a solid, it can crystallize in three ways: with unit cells being either simple cubic, face-centered cubic, or body-centered cubic.

- a Which crystalline form of solid X has the highest density?
- b Which crystalline form of solid X has the most empty space?

11.36 Consider two flasks that contain different pure liquids at 20°C.

The liquid in one flask, substance A, has a molar mass of 100 g/mol and has hydrogen bonding. The liquid in the other flask, substance B, has a molar mass of 105 g/mol and has dipole–dipole interactions.

a If the molecular structures of the compounds are very similar, which flask probably contains substance A?

b If you were to increase the temperature of each of the flasks by 15°C, how would the pictures change (assume that you stay below the boiling points of the liquids)?

Practice Problems

Key: These problems are for practice in applying problem-solving skills. They are divided by topic, and some are keyed to exercises (see the ends of the exercises). The problems are arranged in matching pairs; the odd-numbered problem of each pair is listed first, and its answer is given in the back of the book.

Phase Transitions

11.37 Identify the phase transition occurring in each of the following.

a Molten lava from a volcano cools and turns to solid rock.

b When carbon dioxide gas under pressure exits from a small orifice, it turns to a white "snow."

c Chlorine gas is passed into a very cold test tube where it turns to a yellow liquid.

d A mixture of scrambled eggs placed in a cold vacuum chamber slowly turns to a powdery solid.

e The water level in an aquarium tank falls continuously (the tank has no leak).

11.38 Identify the phase transition occurring in each of the following.

a Mothballs slowly become smaller and eventually disappear.

b Rubbing alcohol spilled on the palm of the hand feels cool as the volume of liquid decreases.

c A black deposit of tungsten metal collects on the inside of a lightbulb whose filament is tungsten metal.

d Raindrops hit a cold metal surface, which becomes covered with ice.

e Candle wax turns to liquid under the heat of the candle flame.

11.39 Use Figure 11.7 to estimate the boiling point of carbon tetrachloride, CCl_4, under an external pressure of 250 mmHg.

11.40 Use Figure 11.7 to estimate the boiling point of diethyl ether, $(C_2H_5)_2O$, under an external pressure of 350 mmHg.

11.41 An electric heater coil provided heat to a 15.5-g sample of iodine, I_2, at the rate of 3.48 J/s. It took 4.54 min from the time the iodine began to melt until the iodine was completely melted. What is the heat of fusion per mole of iodine?

11.42 A 35.8-g sample of cadmium metal was melted by an electric heater providing 4.66 J/s of heat. If it took 6.92 min from the time the metal began to melt until it was completely melted, what is the heat of fusion per mole of cadmium?

11.43 Isopropyl alcohol, $CH_3CHOHCH_3$, is used in rubbing alcohol mixtures. Alcohol on the skin cools by evaporation. How much heat is absorbed by the alcohol if 4.50 g evaporates? The heat of vaporization of isopropyl alcohol is 42.1 kJ/mol.

11.44 Liquid butane, C_4H_{10}, is stored in cylinders to be used as a fuel. Suppose 35.5 g of butane gas is removed from a cylinder. How much heat must be provided to vaporize this much gas? The heat of vaporization of butane is 21.3 kJ/mol.

11.45 A quantity of ice at 0.0°C was added to 33.6 g of water at 21.0°C to give water at 0.0°C. How much ice was added? The heat of fusion of water is 6.01 kJ/mol and the specific heat is 4.18 J/(g·°C).

11.46 Water at 0°C was placed in a dish inside a vessel maintained at low pressure by a vacuum pump. After a quantity of water had evaporated, the remainder froze. If 9.31 g of ice at 0°C was obtained, how much liquid water must have evaporated? The heat of fusion of water is 6.01 kJ/mol and its heat of vaporization is 44.9 kJ/mol at 0°C.

11.47 A quantity of ice at 0°C is added to 64.3 g of water in a glass at 55°C. After the ice melted, the temperature of the water in the glass was 15°C. How much ice was added? The heat of fusion of water is 6.01 kJ/mol and the specific heat is 4.18 J/(g·°C).

11.48 Steam at 100°C was passed into a flask containing 275 g of water at 21°C, where the steam condensed. How many grams of steam must have condensed if the temperature of the water in the flask was raised to 83°C? The heat of vaporization of water at 100°C is 40.7 kJ/mol and the specific heat is 4.18 J/(g·°C).

11.49 Methanol, CH_3OH, a colorless, volatile liquid, was formerly known as wood alcohol. It boils at 65.0°C and has a heat of vaporization of 37.4 kJ/mol. What is its vapor pressure at 22.0°C?

11.50 Chloroform, $CHCl_3$, a volatile liquid, was once used as an anesthetic but has been replaced by safer compounds. Chloroform boils at 61.7°C and has a heat of vaporization of 31.4 kJ/mol. What is its vapor pressure at 36.2°C?

11.51 Carbon disulfide, CS_2, is a volatile, flammable liquid. It has a vapor pressure of 400.0 mmHg at 28.0°C and 760.0 mmHg at 46.5°C. What is the heat of vaporization of this substance?

11.52 White phosphorus, P_4, is normally a white, waxy solid melting at 44°C to a colorless liquid. The liquid has a vapor pressure of 400.0 mmHg at 251.0°C and 760.0 mmHg at 280.0°C. What is the heat of vaporization of this substance?

Phase Diagrams

11.53 Shown here is the phase diagram for compound X. The triple point of X is −25.1°C at 0.50 atm and the critical point is 22°C and 21.3 atm.

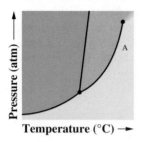

a What is the state of X at position A?

b If we decrease the temperature from the compound at position A to −28.2°C while holding the pressure constant, what is the state of X?

c If we take the compound starting under the conditions of part b and increase the temperature to 15.3°C and decrease the pressure to 0.002 atm, what is the state of X?

d Would it be possible to make the compound a solid by starting with the conditions of part c and increasing just the pressure?

11.54 Shown here is the phase diagram for compound Z. The triple point of Z is −5.1°C at 3.3 atm and the critical point is 51°C and 99.1 atm.

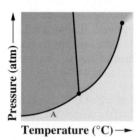

a What is the state of Z at position A?

b If we increase the temperature of the compound at position A to 60°C while holding the pressure constant, what is the state of Z?

c If we take the compound starting under the conditions of part b and reduce the temperature to 20°C and increase the pressure to 65 atm, what is the state of Z?

d Would it be possible to make the compound a solid by starting with the conditions of part c and increasing just the pressure?

11.55 Use graph paper and sketch the phase diagram of argon, Ar, from the following information: normal melting point, −187°C; normal boiling point, −186°C; triple point, −189°C, 0.68 atm; critical point, −122°C, 48 atm. Label each phase region on the diagram.

11.56 Use graph paper and sketch the phase diagram of oxygen, O_2, from the following information: normal melting point, −218°C; normal boiling point, −183°C; triple point, −219°C, 1.10 mmHg; critical point, −118°C, 50.1 atm. Label each phase region on the diagram.

11.57 Which of the following substances can be liquefied by applying pressure at 25°C? For those that cannot, describe the conditions under which they *can* be liquefied.

Substance	Critical Temperature	Critical Pressure
Sulfur dioxide, SO_2	158°C	78 atm
Acetylene, C_2H_2	36°C	62 atm
Methane, CH_4	−82°C	46 atm
Carbon monoxide, CO	−140°C	35 atm

11.58 A tank of gas at 21°C has a pressure of 1.0 atm. Using the data in the table, answer the following questions. Explain your answers.

a If the tank contains butane, C_4H_{10}, is the liquid state also present?

b If the tank contains carbon tetrafluoride, CF_4, is the liquid state also present?

Substance	Boiling Point at 1 atm	Critical Temperature	Critical Pressure
C_4H_{10}	−0.5°C	152°C	38 atm
CF_4	−128°C	−46°C	41 atm

11.59 Bromine, Br_2, has a triple point at −7.3°C and 44 mmHg and a critical point at 315°C and 102 atm. The density of the solid is 3.4 g/cm³, and the density of the liquid is 3.1 g/cm³. Sketch a rough phase diagram of bromine, labeling all important features. Circle the correct word in each of the following sentences (and explain your answers).

a Bromine vapor at 40 mmHg condenses to the (liquid, solid) when cooled sufficiently.

b Bromine vapor at 400 mmHg condenses to the (liquid, solid) when cooled sufficiently.

11.60 Krypton, Kr, has a triple point at −169°C and 133 mmHg and a critical point at −63°C and 54 atm. The density of the solid is 2.8 g/cm³, and the density of the liquid is 2.4 g/cm³. Sketch a rough phase diagram of krypton. Circle the correct word in each of the following sentences (and explain your answers).

a Solid krypton at 130 mmHg (melts, sublimes without melting) when the temperature is raised.

b Solid krypton at 760 mmHg (melts, sublimes without melting) when the temperature is raised.

Intermolecular Forces and Properties of Liquids

11.61 The heats of vaporization of liquid O_2, liquid Ne, and liquid methanol, CH_3OH, are 6.8 kJ/mol, 1.8 kJ/mol, and 34.5 kJ/mol, respectively. Are the relative values as you would expect? Explain.

11.62 The heats of vaporization of liquid Cl_2, liquid H_2, and liquid N_2 are 20.4 kJ/mol, 0.9 kJ/mol, and 5.6 kJ/mol, respectively. Are the relative values as you would expect? Explain.

11.63 For each of the following substances, list the kinds of intermolecular forces expected.

a
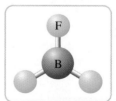

b isopropyl alcohol, $CH_3CHOHCH_3$

c hydrogen iodide, HI
d krypton, Kr

11.64 Which of the following compounds would you expect to exhibit *only* London forces?
a silicon tetrafluoride, SiF_4
b

c potassium chloride, KCl
d phosphorus pentachloride, PCl_5

11.65 Arrange the following substances in order of increasing magnitude of the London forces: $SiCl_4$, CCl_4, $GeCl_4$.

11.66 Arrange the following substances in order of increasing magnitude of the London forces: Cl_2, F_2, Br_2.

11.67 The halogens form a series of compounds with each other, which are called interhalogens. Examples are bromine chloride (BrCl), iodine bromide (IBr), bromine fluoride (BrF), and chlorine fluoride (ClF). Which compound is expected to have the highest boiling point at any given pressure? Explain.

11.68 Methane, CH_4, reacts with chlorine, Cl_2, to produce a series of chlorinated hydrocarbons: methyl chloride (CH_3Cl), methylene chloride (CH_2Cl_2), chloroform ($CHCl_3$), and carbon tetrachloride (CCl_4). Which compound has the lowest vapor pressure at room temperature? Explain.

11.69 Predict the order of increasing vapor pressure at a given temperature for the following compounds:
a $HOCH_2CH_2OH$ b FCH_2CH_2F c FCH_2CH_2OH
Explain why you chose this order.

11.70 Predict the order of increasing vapor pressure at a given temperature for the following compounds:
a $CH_3CH_2CH_2CH_2OH$
b $CH_3CH_2OCH_2CH_3$
c $HOCH_2CH_2CH_2OH$
Explain why you chose this order.

11.71 List the following substances in order of increasing boiling point.

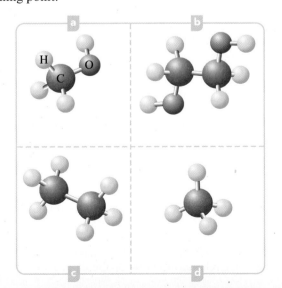

11.72 Arrange the following compounds in order of increasing boiling point.

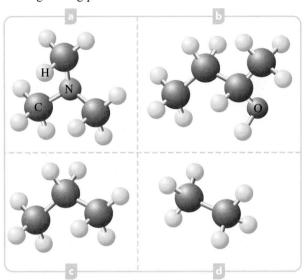

Types of Solids

11.73 Classify each of the following by the type of solid it forms: a F_2;   b $BaCl_2$;   c BCl_3;   d CCl_4;   e CaO.

11.74 Classify each of the following by the type of solid it forms: a Na;   b Fe;   c B;   d H_2O;   e KF.

11.75 Classify each of the following solid elements as molecular, metallic, ionic, or covalent network.
a sulfur, S_8 b iodine, I_2
c tin, Sn   d germanium, Ge

11.76 Which of the following do you expect to be molecular solids?
a bromine chloride, BrCl
b sodium fluoride, NaF
c lithium bromide, LiBr
d silicon tetrachloride, $SiCl_4$

11.77 Arrange the following compounds in order of increasing melting point.

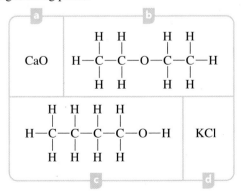

11.78 Arrange the following substances in order of increasing melting point.

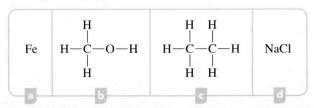

11.79 Associate each type of solid in the left-hand column with two of the properties in the right-hand column. Each property may be used more than once.

- a ionic solid high-melting
- b molecular solid low-melting
- c covalent network malleable
   solid hard
    electrically conducting
- d metallic solid brittle

11.80 On the basis of the description given, classify each of the following solids as molecular, metallic, ionic, or covalent network. Explain your answers.

- a a lustrous, yellow solid that conducts electricity
- b a hard, black solid melting at 2350°C to give a non-conducting liquid
- c a nonconducting, pink solid melting at 650°C to give an electrically conducting liquid
- d red crystals having a characteristic odor and melting at 171°C

11.81 Associate each of the solids Co, LiCl, SiC, and CHI$_3$ with one of the following sets of properties.

- a A very hard, blackish solid subliming at 2700°C.
- b A gray, lustrous solid melting at 1495°C; both the solid and liquid are electrical conductors.
- c A white solid melting at 613°C; the liquid is electrically conducting, although the solid is not.
- d A yellow solid with a characteristic odor having a melting point of 120°C.

11.82 Associate each of the solids BN, P$_4$S$_3$, Pb, and CaCl$_2$ with one of the following sets of properties.

- a A bluish white, lustrous solid melting at 327°C; the solid is soft and malleable.
- b A white solid melting at 772°C; the solid is an electrical nonconductor but dissolves in water to give a conducting solution.
- c A yellowish green solid melting at 172°C.
- d A very hard, colorless substance melting at about 3000°C.

Crystal Structure

11.83 How many atoms are there in a body-centered cubic unit cell of an atomic crystal in which all atoms are at lattice points?

11.84 How many atoms are there in a simple cubic unit cell of an atomic crystal in which all atoms are at lattice points?

11.85 Nickel has a face-centered unit cell with all atoms at lattice points and an edge length of 352.4 pm. The density of metallic nickel is 8.91 g/cm^3. What is the mass of a nickel atom? From the atomic mass, calculate Avogadro's number.

11.86 Metallic iron has a body-centered cubic lattice with all atoms at lattice points and a unit cell whose edge length is 286.6 pm. The density of iron is 7.87 g/cm^3. What is the mass of an iron atom? Compare this value with the value you obtain from the molar mass.

11.87 Copper metal has a face-centered cubic structure with all atoms at lattice points and a density of 8.93 g/cm^3. Its atomic mass is 63.5 amu. Calculate the edge length of the unit cell.

11.88 Barium metal has a body-centered cubic lattice with all atoms at lattice points; its density is 3.51 g/cm^3. From these data and the atomic weight, calculate the edge length of a unit cell.

11.89 Gold has cubic crystals whose unit cell has an edge length of 407.9 pm. The density of the metal is 19.3 g/cm^3. From these data and the atomic mass, calculate the number of gold atoms in a unit cell, assuming all atoms are at lattice points. What type of cubic lattice does gold have?

11.90 Chromium forms cubic crystals whose unit cell has an edge length of 288.5 pm. The density of the metal is 7.20 g/cm^3. Use these data and the atomic mass to calculate the number of atoms in a unit cell, assuming all atoms are at lattice points. What type of cubic lattice does chromium have?

11.91 Lead has a face-centered cubic lattice with all atoms at lattice points and a unit-cell edge length of 495.0 pm. Its atomic mass is 207.2 amu. What is the density of lead?

11.92 Tungsten has a body-centered cubic lattice with all atoms at the lattice points. The edge length of the unit cell is 316.5 pm. The atomic mass of tungsten is 183.8 amu. Calculate its density.

11.93 Metallic magnesium has a hexagonal close-packed structure and a density of 1.74 g/cm^3. Assume magnesium atoms to be spheres of radius r. Because magnesium has a close-packed structure, 74.1% of the space is occupied by atoms. Calculate the volume of each atom; then find the atomic radius, r. The volume of a sphere is equal to $4\pi r^3/3$.

11.94 Metallic barium has a body-centered cubic structure (all atoms at the lattice points) and a density of 3.51 g/cm^3. Assume barium atoms to be spheres. The spheres in a body-centered array occupy 68.0% of the total space. Find the atomic radius of barium. (See Problem 11.93.)

General Problems

Key: These problems provide more practice but are not divided by topic or keyed to exercises. Each section ends with essay questions, each of which is color coded to refer to the *A Chemist Looks at Daily Life* (orange), *A Chemist Looks at Frontiers* (purple), *A Chemist Looks at Environment* (green), and *Instrumental Methods* (brown) chapter essay on which it is based. Odd-numbered problems and the even-numbered problems that follow are similar; answers to all odd-numbered problems except the essay questions are given in the back of the book.

11.95 Snow forms in the upper atmosphere in a cold air mass that is supersaturated with water vapor. When the snow later falls through a lower, warm air mass, rain forms.

When this rain falls on a sunny spot, the drops evaporate. Describe all of the phase changes that have occurred.

11.96 If you leave your car parked outdoors in the winter, you may find frost on the windows in the morning. If you then start the car and let the heater warm the windows, after some minutes the windows will be dry. Describe all of the phase changes that have occurred.

11.97 The percent relative humidity of a sample of air is found as follows: (partial pressure of water vapor/vapor pressure of water) × 100. A sample of air at 21°C was cooled to 15°C, where moisture began to condense as dew. What was the relative humidity of the air at 21°C?

11.98 A sample of air at 21°C has a relative humidity of 55%. At what temperature will water begin to condense as dew? (See Problem 11.97.)

11.99 The vapor pressure of water is 17.5 mmHg at 20.0°C and 355.1 mmHg at 80.0°C. Calculate the boiling point of water at 755.0 mmHg.

11.100 The vapor pressure of benzene is 100.0 mmHg at 26.1°C and 400.0 mmHg at 60.6°C. What is the boiling point of benzene at 760.0 mmHg?

11.101 Describe the behavior of carbon dioxide gas when compressed at the following temperatures:
   a −70°C   b 40°C   c 20°C

The triple point of carbon dioxide is −57°C and 5.1 atm, and the critical point is 31°C and 73 atm.

11.102 Describe the behavior of iodine vapor when cooled at the following pressures:
   a 120 atm   b 1 atm   c 50 mmHg

The triple point of iodine is 114°C and 90.1 mmHg, and the critical point is 512°C and 116 atm.

11.103 Describe the formation of hydrogen bonds in hydrogen peroxide, H_2O_2. Represent possible hydrogen bonding structures in hydrogen peroxide by using structural formulas and the conventional notation for a hydrogen bond.

11.104 Describe the formation of hydrogen bonds in propanol, $CH_3CH_2CH_2OH$. Represent possible hydrogen bonding structures in propanol by using structural formulas and the conventional notation for a hydrogen bond.

11.105 Ethylene glycol (CH_2OHCH_2OH) is a slightly viscous liquid that boils at 198°C. Pentane (C_5H_{12}), which has approximately the same molecular weight as ethylene glycol, is a nonviscous liquid that boils at 36°C. Explain the differences in physical characteristics of these two compounds.

11.106 Pentylamine, $CH_3CH_2CH_2CH_2CH_2NH_2$, is a liquid that boils at 104°C and has a viscosity of 10×10^{-4} kg/(m·s). Triethylamine, $(CH_3CH_2)_3N$, is a liquid that boils at 89°C and has a viscosity of about 4×10^{-4} kg/(m·s). Explain the differences in properties of these two compounds.

11.107 The elements in Problem 11.107 form the fluorides AlF_3, SiF_4, PF_3, and SF_4. In each case, identify the type of solid formed by the fluoride.

11.108 Consider the elements Al, Si, P, and S from the third row of the periodic table. In each case, identify the type of solid the element would form.

11.109 Decide which substance in each of the following pairs has the lower melting point. Explain how you made each choice.
   a potassium chloride, KCl; or calcium oxide, CaO
   b carbon tetrachloride,

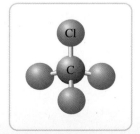

or hexachloroethane,

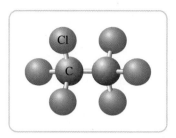

   c zinc, Zn; or chromium, Cr
   d acetic acid, CH_3COOH; or ethyl chloride, C_2H_5Cl

11.110 Decide which substance in each of the following pairs has the lower melting point. Explain how you made each choice.
   a magnesium oxide, MgO; or hexane, C_6H_{14}
   b 1-propanol,

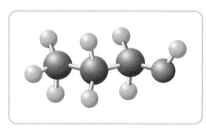

or ethylene glycol,

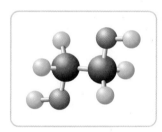

   c silicon, Si; or sodium, Na
   d methane, CH_4; or silane, SiH_4

11.111 The edge length of the unit cell of tantalum metal, Ta, is 330.6 pm; the unit cell is body-centered cubic (one atom at each lattice point). Tantalum has a density of 16.69 g/cm³. What is the mass of a tantalum atom? Use Avogadro's number to calculate the atomic mass of tantalum.

11.112 Iridium metal, Ir, crystallizes in a face-centered cubic (close-packed) structure. The edge length of the unit cell was found by x-ray diffraction to be 383.9 pm. The density of iridium is 22.42 g/cm³. Calculate the mass of an iridium atom. Use Avogadro's number to calculate the atomic mass of iridium.

11.113 Use your answer to Problem 11.87 to calculate the radius of the copper atom. Assume that copper atoms are spheres. Then note that the spheres on any face of a unit cell touch along the diagonal.

11.114 Rubidium metal has a body-centered cubic structure (with one atom at each lattice point). The density of the metal is 1.532 g/cm³. From this information and the atomic mass, calculate the edge length of the unit cell. Now assume that rubidium atoms are spheres. Each corner sphere of the unit cell touches the body-centered sphere. Calculate the radius of a rubidium atom.

11.115 Calculate the percent of volume that is actually occupied by spheres in a face-centered cubic lattice of identical spheres. You can do this by first relating the radius of a sphere, r, to the length of an edge of a unit cell, l. (Note that the spheres do not touch along an edge but do touch along the diagonal of a face.) Then calculate the volume of a unit cell in terms of r. The volume occupied by spheres equals the number of spheres per unit cell times the volume of a sphere ($4\pi r^3/3$).

11.116 Calculate the percent of volume that is actually occupied by spheres in a body-centered cubic lattice of identical spheres. You can do this by first relating the radius of a sphere, r, to the length of an edge of a unit cell, l. (Note that the spheres do not touch along an edge but do touch along a diagonal passing through the body-centered sphere.) Then calculate the volume of a unit cell in terms of r. The volume occupied by spheres equals the number of spheres per unit cell times the volume of a sphere ($4\pi r^3/3$).

11.117 For the hydrogen halides and the noble gases, we have the following boiling points:

Halogen Family, °C	Noble Gases, °C
HF, 19	Ne, −246
HCl, −115	Ar, −186
HBr, −67	Kr, −152
HI, −35	Xe, −108

Account for the following:

a The general trend in the boiling points of the hydrides and the noble gases.

b The unusual boiling point of hydrogen fluoride.

c The observation that the hydrogen halides have boiling points that are significantly higher than the noble gases.

11.118 For the carbon and nitrogen family hydrides, we have the following boiling points:

Carbon Family, °C	Nitrogen Family, °C
CH$_4$, −164	NH$_3$, −33
SiH$_4$, −112	PH$_3$, −88
GeH$_4$, −88	AsH$_3$, −55
SnH$_4$, −52	SbH$_3$, −17

Account for the following:

a The general trend in the boiling points of the binary hydrides.

b The unusual boiling point of ammonia.

c The observation that the nitrogen family hydrides have boiling points that are notably higher than those of the carbon family.

11.119 Account for the following observations:

a Both diamond and silicon carbide are very hard, whereas graphite is both soft and slippery.

b Carbon dioxide is a gas, whereas silicon dioxide is a high-melting solid.

11.120 Greater variation exists between the properties of the first and second members of a family in the periodic table than between other members. Discuss this observation for the oxygen family using the following data.

Element	Boiling Point, °C	Compound	Boiling Point, °C
O$_2$	−183	H$_2$O	100
S$_8$	445	H$_2$S	−61
Se$_8$	685	H$_2$Se	−42

11.121 Use chemical principles to discuss the following observations:

a CO$_2$ sublimes at −78°C, whereas SiO$_2$ boils at 2200°C.

b CF$_4$ boils at −128°C, whereas SiF$_4$ boils at −86°C.

c HF boils at 19°C, whereas HCl boils at −85°C.

11.122 Draw Lewis structures of each of the following compounds: LiH, NH$_3$, CH$_4$, CO$_2$. Rank these compounds from lowest to highest boiling point. Explain your reasoning.

■ **11.123** Describe the contents of a carbon dioxide fire extinguisher at 20°C. Then describe it at 35°C. Explain the difference.

■ **11.124** Discuss why supercritical carbon dioxide is a nearly ideal solvent.

■ **11.125** A gecko's toes have been shown to stick to walls through van der Waals forces. Van der Waals forces also exist between your finger and a wall. Why, then, doesn't your finger stick to the wall in the same way as the gecko's toes?

■ **11.126** Although a gecko's toes stick easily to a wall, their toes lift off a surface just as easily. Explain.

■ **11.127** Describe the structure of a nematic liquid crystal. How is it similar to a liquid? How is it similar to a crystalline solid?

■ **11.128** What properties of nematic liquid crystals are employed in LCD displays?

■ **11.129** Briefly describe what it is that the Bragg equation relates?

■ **11.130** How is it possible to obtain the structure of a molecule using x-ray diffraction from the molecular crystal?

■ **11.131** What properties of water are unusual? How does hydrogen bonding explain some of these unusual properties?

■ **11.132** The fact that solid ice is less dense than the liquid water is important to weather and aquatic life. Explain.

Strategy Problems

Key: As noted earlier, all of the practice and general problems are matched pairs. This section is a selection of problems that are not in matched-pair format. These challenging problems require that you employ many of the concepts and strategies that were developed in the chapter. In some cases, you will have to integrate several concepts and operational skills in order to solve the problem successfully.

11.133 In an experiment, 20.00 L of dry nitrogen gas, N$_2$, at 20.0°C and 750.0 mmHg is slowly bubbled into water in a flask to determine its vapor pressure (see the figure below).

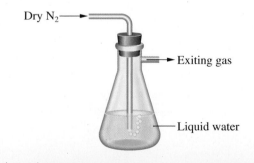

The liquid water is weighed before and after the experiment, from which the experimenter determines that it loses 353.6 mg in mass. Answer the following questions.

a. How many moles of nitrogen were bubbled into the water?

b. The liquid water diminished by how many moles? What happened to the liquid?

c. How many moles of gas exit the flask during the experiment? What is the partial pressure of nitrogen gas exiting? The total gas pressure is 750.0 mmHg.

d. From these data, calculate the vapor pressure of water at 20°C. Does this agree with what you expect?

11.134 Look at the phase diagram for sulfur in Figure 11.12*b*. What states of sulfur would you expect to see at each of the triple points? What solid phase would you expect to freeze out if you cooled liquid sulfur? Is this the phase that you usually see at room temperature? Explain your answers.

11.135 On a particular summer day, the temperature is 30.0°C and the relative humidity is 80.0% (which means that the partial pressure of water vapor in the ambient air is 80.0% of the equilibrium vapor pressure of water). A sample of this air is placed in a 1.00-L flask, which is then closed and cooled to 5.0°C. What is the mass (in grams) of water that still exists as vapor in the flask? How much liquid water (in grams) condenses out?

11.136 Carbon dioxide is often sold in steel cylinders. Use the data given in Figure 11.12*a*, to answer the following questions.

a. Describe what you would expect to see in the cylinder if it were transparent and at room temperature. Explain.

b. Suppose you have attached a pressure gauge to the cylinder to register the pressure within the cylinder. How would you expect the gauge to behave as you continue to use CO_2 from the cylinder until you can no longer obtain any CO_2 gas from it?

11.137 You may have seen the statement made that the liquid state is the stable state of water below 100°C (but above 0°C), whereas the vapor state is the stable state above 100°C. Yet you also know that a pan of water set out on a table at 20°C will probably evaporate completely in a few days, in which case, liquid water has changed to the vapor state. Explain what is happening here. What is wrong with the simple statement given at the beginning of this problem? Give a better statement.

11.138 Consider the following two compounds:

$CH_3CH_2CH_2CH_2CH_2OH$ 1-pentanol

$CH_3CH_2CH_2CH_2CH_2CH_3$ hexane

a. What are the different types of intermolecular forces that exist in each compound?

b. One of these compounds has a normal boiling point of 69°C, and the other has a normal boiling point of 138°C. What is the normal boiling point of hexane? Explain.

c. One of these compounds has a viscosity of 0.313 g/(cm·s), and the other has a viscosity of 2.987 g/(cm·s). Assign viscosities to 1-pentanol and to hexane.

d. Consider the compound $CH_3CH_2CH_2CH_2CH_2Cl$. Where do you think the boiling point of this compound might lie: above both 1-pentanol and hexane, intermediate between these two compounds, or below both of these two compounds?

11.139 Consider the following three compounds:

CH_3CHO, $CH_3CH_2CH_3$, CH_3CH_2OH

a. Describe the types of intermolecular forces that you expect to see in each. Explain how you arrived at these types.

b. The heats of vaporization of these compounds are (in no particular order): 25.8 kJ/mol, 38.6 kJ/mol, and 19.0 kJ/mol. What is the heat of vaporization of CH_3CHO?

c. If the normal boiling point of CH_3CHO is 21°C, what is its vapor pressure at 15°C?

11.140 Here is some information about a compound. It has a melting point of 33°C, a normal boiling point of 81°C, a heat of fusion of 11 kJ/mol, and a heat of vaporization of 32 kJ/mol. What is its vapor pressure at 15°C?

11.141 Rhenium forms a series of solid oxides: Re_2O_7 (yellow), ReO_3 (red), Re_2O_5 (blue), and ReO_2 (brown). One of them has a crystal structure with the following unit cell:

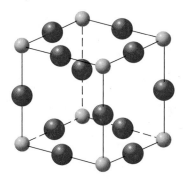

a. How many rhenium atoms (gray spheres) are there in a unit cell? How many oxygen atoms are there in a unit cell?

b. What is the color of the compound? Explain.

11.142 Shown below is the cubic unit cell of an ionic compound. Answer the questions by referring to this structure. Be careful to note that some atoms are hidden by those in front.

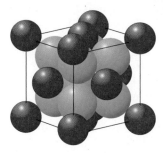

a. One of the spheres (red or green) represents a monatomic, metallic ion. The other color sphere represents a monatomic, nonmetal ion. Which spheres probably represent the metal ion? Explain.

b. How many red spheres are there in the unit cell? How many green ones?

c. From the information you have, deduce the general formula of the compound, using M for the metallic element and X for the nonmetallic element. What are the formulas of the ions?

d. Give an example of a compound that might have this structure. Explain why you think this compound might have this structure. Which ion of this compound would be represented by the red spheres?

11.143 Two pure solid substances, substance A and substance B, are heated until a small amount of liquid begins to appear. After the appearance of the liquid, the two samples are heated at a rate of 5 kJ/min. After 10 minutes of applying heat, 10 mL of liquid A and 5 mL of liquid B have formed. In both samples there is still solid present.

a Which substance has stronger intermolecular attractions?

b Which substance has a smaller heat of fusion?

c Which substance would you expect to boil at a higher temperature?

11.144 Strontium crystallizes as a face-centered cubic cell. What is the mass of a single unit cell of Sr?

11.145 A puddle in the street has a volume of 11.5 L. Over the course of 5 days, the puddle freezes at night and then thaws during the day. The puddle reaches a low temperature of −4.5°C during each night and a high temperature of 3.4°C during each day.

a When does energy transfer from the puddle to the street occur (day or night)? Justify your answer.

b What is the total amount of energy that is transferred from the puddle to the surroundings during a 3-day/night period beginning the second evening at sunset (assume that the puddle is at its highest temperature at sunset)?

c What is the total amount of energy that is transferred to the puddle during a 3-day/night period beginning the second morning at sunrise?

11.146 Describe how you would adjust the conditions (temperature and pressure) if you wanted to increase the solubility of a gas in a liquid.

11.147 Compare the quantities of heat required for each of the following pairings. Justify your answers without performing calculations.

a Evaporating a 1.00-g sample of liquid water at 100°C or subliming a 1.00 g sample of water at −10°C?

b Condensing a 0.5-g sample of water vapor at 60°C or melting a 0.5-g sample of water at 0°C.

c Vaporizing a 50-g sample of liquid water placed into an oven set at 200°C or vaporizing a 100-g sample of liquid water placed into an oven set at 100°C. Both water samples are at 100°C when they are placed in the ovens.

11.148 A cubic unit cell is found to have calcium ions at the corners, a titanium atom in the center of the cell, and oxide ions at the center of each face. What is the empirical formula of this compound?

Cumulative-Skills Problems

Key: The problems under this heading combine skills introduced in previous chapters with those given in the current one.

11.149 The vapor pressure of a volatile liquid can be determined by slowly bubbling a known volume of gas through the liquid at a given temperature and pressure. In an experiment, a 5.40-L sample of nitrogen gas, N_2, at 20.0°C and 745 mmHg is bubbled through liquid isopropyl alcohol, C_3H_8O, at 20.0°C. Nitrogen containing the vapor of C_3H_8O at its vapor pressure leaves the vessel at 20.0°C and 745 mmHg. It is found that 0.6149 g C_3H_8O has evaporated. How many moles of N_2 are in the gas leaving the liquid? How many moles of alcohol are in this gaseous mixture? What is the mole fraction of alcohol vapor in the gaseous mixture? What is the partial pressure of the alcohol in the gaseous mixture? What is the vapor pressure of C_3H_8O at 20.0°C?

11.150 In an experiment, a sample of 6.35 L of nitrogen at 25.0°C and 768 mmHg is bubbled through liquid acetone, C_3H_6O. The gas plus vapor at its equilibrium partial pressure leaves the liquid at the same temperature and pressure. If 6.550 g of acetone has evaporated, what is the vapor pressure of acetone at 25.0°C? See Problem 11.149.

11.151 How much heat is needed to vaporize 25.0 mL of liquid methanol, CH_3OH, at 25.0°C? The density of the liquid is 0.787 g/mL. Use standard heats of formation, which are given in Appendix C.

11.152 How much heat is needed to vaporize 10.0 mL of liquid hydrogen cyanide, HCN, at 25.0°C? The density of

the liquid is 0.687 g/mL. Use standard heats of formation, which are given in Appendix C.

11.153 How much heat must be added to 50.0 g of solid sodium, Na, at 25.0°C to give the liquid at its melting point, 97.8°C? The heat capacity of solid sodium is 28.2 J/(K·mol), and its heat of fusion is 2.60 kJ/mol.

11.154 How much heat must be added to 12.5 g of solid white phosphorus, P_4, at 25.0°C to give the liquid at its melting point, 44.1°C? The heat capacity of solid white phosphorus is 95.4 J/(K·mol); its heat of fusion is 2.63 kJ/mol.

11.155 Acetic acid, CH_3COOH, forms stable pairs of molecules held together by two hydrogen bonds.

Such molecules—themselves formed by the association of two simpler molecules—are called dimers. The vapor over liquid acetic acid consists of a mixture of monomers (single acetic acid molecules) and dimers. At 100.6°C the total pressure of vapor over liquid acetic acid is 436 mmHg. If the vapor consists of 0.630 mole fraction of the dimer, what are the masses of monomer and dimer in 1.000 L of the vapor? What is the density of the vapor?

11.156 The total pressure of vapor over liquid acetic acid at 71.38°C is 146 mmHg. If the density of the vapor is 0.702 g/L, what is the mole fraction of dimer in the vapor? See Problem 11.155.

12 Solutions

Sugarcane produces a juice that is a solution containing sucrose (sugar). Soft drinks are flavored solutions of dissolved sucrose.

Contents and Concepts

▶ Solution Formation

Here you will have the opportunity to develop a more thorough understanding of the solution process and of what constitutes a solution.

▶ Colligative Properties

We will explore how properties of solutions are affected by changes in concentration. You will also be introduced to several ways of expressing concentration.

▶ Colloid Formation

When relatively large particles of a substance disperse throughout another substance, the result is called a colloid.

⚙WL Sign in to OWL at **www.cengage.com/owl** to view tutorials and simulations, develop problem-solving skills, and complete online homework assigned by your professor.

go Chemistry Download mini lecture videos for key concept review and exam prep from OWL or purchase them from **www.cengagebrain.com**.

© Radius Images/Alamy; SambaPhoto/Edu Lyra/Getty Images

There are various practical reasons for preparing solutions. For instance, most chemical reactions are run in solution. Also, solutions have particular properties that are useful. When gold is used for jewelry, it is mixed, or alloyed, with a small amount of silver. Gold–silver alloys are not only harder than pure gold, but they also melt at lower temperatures and are therefore easier to cast.

Solubility, the amount of one substance that dissolves in another, varies with temperature and, in some cases, with pressure. The variation of solubility with pressure can be a useful property. Acetylene gas, C_2H_2, for example, is used as a fuel in welding torches. It can be transported safely under pressure in cylinders as a solution in acetone, CH_3COCH_3. Acetone is a liquid, and at 1 atm pressure 1 L of the liquid dissolves 27 g of acetylene. But at 12 atm, the pressure in a full cylinder, the same quantity of acetone dissolves 320 g of acetylene, so more can be transported. When the valve on a cylinder is opened, reducing the pressure, acetylene gas comes out of solution.

Another useful property of solutions is their lower melting, or freezing, points compared with those of the major component. We have already mentioned the lowering of the melting point of gold when a small amount of silver is added. The use of ethylene glycol, CH_2OHCH_2OH, as an automobile antifreeze depends on the same property. Water containing ethylene glycol freezes at temperatures below the freezing point of pure water.

What determines how much the freezing point of a solution is lowered? How does the solubility of a substance change when conditions such as pressure and temperature change? These are some of the questions we will address in this chapter.

Solution Formation

When sodium chloride dissolves in water, the resulting uniform dispersion of ions in water is called a solution. In general, a *solution* is a homogeneous mixture of two or more substances, consisting of ions or molecules. A *colloid* is similar in that it appears to be homogeneous like a solution. In fact, it consists of comparatively large particles of one substance dispersed throughout another substance or solution.

From the examples of solutions mentioned in the chapter opening, you see that they may be quite varied in their characteristics. To begin our discussion, let us look at the different types of solutions we might encounter. We will discuss colloids at the end of the chapter.

12.1 Types of Solutions

Solutions may exist in any of the three states of matter; that is, they may be gases, liquids, or solids. Some examples are listed in Table 12.1. The terms *solute* and *solvent* refer to the components of a solution. The **solute,** *in the case of a solution of a gas or solid dissolved in a liquid, is the gas or solid; in other cases, the solute is the component in smaller amount.* The **solvent,** *in a solution of a gas or solid dissolved in a liquid, is the liquid; in other cases, the solvent is the component in greater amount.* Thus, when sodium chloride is dissolved in water, we have created a solution in which sodium chloride is the solute and water the solvent.

Gaseous Solutions

In general, nonreactive gases or vapors can mix in all proportions to give a gaseous mixture. *Fluids that mix with or dissolve in each other in all proportions* are said to be **miscible fluids.** Gases are thus miscible. (If two fluids do not mix but, rather, form two layers, they are said to be *immiscible*.) Air, which is a mixture of oxygen, nitrogen, and smaller amounts of other gases, is an example of a gaseous solution.

Table 12.1 Examples of Solutions

Solution	State of Matter	Description
Air	Gas	Homogeneous mixture of gases (O_2, N_2, others)
Soda water	Liquid	Gas (CO_2) dissolved in a liquid (H_2O)
Ethanol in water	Liquid	Liquid solution of two completely miscible liquids
Brine	Liquid	Solid (NaCl) dissolved in a liquid (H_2O)
Potassium–sodium alloy	Liquid	Solution of two solids (K + Na)
Dental-filling alloy	Solid	Solution of a liquid (Hg) in a solid (Ag plus other metals)
Gold–silver alloy	Solid	Solution of two solids (Au + Ag)

Liquid Solutions

Most liquid solutions are obtained by dissolving a gas, liquid, or solid in some liquid. Soda water, for example, consists of a solution of carbon dioxide gas in water. Acetone, C_3H_6O, in water is an example of a liquid–liquid solution. (Immiscible and miscible liquids are shown in Figure 12.1.) Brine is water with sodium chloride (a solid) dissolved in it. Seawater contains both dissolved gases (from air) and solids (mostly sodium chloride).

It is also possible to make a liquid solution by mixing two solids together. Consider a potassium–sodium alloy. Both potassium and sodium are solid at room temperature, but a liquid solution results when the mixture contains 10% to 50% sodium. ▶

Sodium metal melts at 98°C and potassium metal melts at 63°C. When the two metals are mixed to give a solution that is 20% sodium, for example, the melting point is lowered to −10°C. This is another example of the lowering of the freezing, or melting, point of solutions, which we will discuss later in the chapter. Potassium–sodium alloy is used as a heat-transfer medium in nuclear reactors.

Solid Solutions

Solid solutions are also possible. In the chapter opening, we mentioned gold–silver alloys. Dental-filling alloy is a solution of mercury (a liquid) in silver (a solid), with small amounts of other metals.

Figure 12.1 ▼

Immiscible and miscible liquids

Acetone and water are miscible liquids; that is, the two substances dissolve in each other in all proportions.

Water and methylene chloride are immiscible and form two layers.

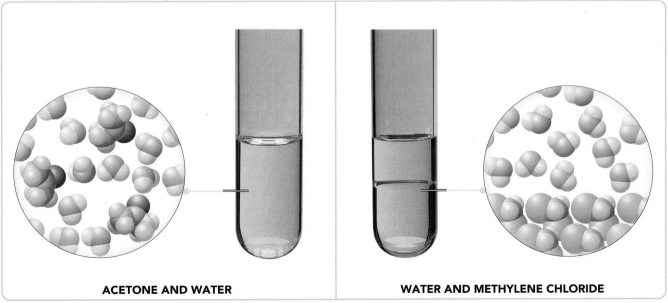

ACETONE AND WATER   **WATER AND METHYLENE CHLORIDE**

© Cengage Learning

Exercise 12.1 Give an example of a solid solution prepared from a liquid and a solid.

■ See Problems 12.37 and 12.38.

CONCEPT CHECK 12.1

Identify the solute(s) and solvent(s) in the following solutions.

a 80 g of Cr and 5 g of Mo

b 5 g of $MgCl_2$ dissolved in 1000 g of H_2O

c 39% N_2, 41% Ar, and the rest O_2

12.2 Solubility and the Solution Process

The amount of substance that will dissolve in a solvent depends on both the substance and the solvent. We describe the amount that dissolves in terms of solubility.

Solubility; Saturated Solutions

To understand the concept of solubility, consider the process of dissolving sodium chloride in water. Sodium chloride is an ionic substance, and it dissolves in water as Na^+ and Cl^- ions. If you could view the dissolving of sodium chloride at the level of ions, you would see a dynamic process. Suppose you stir 40.0 g of sodium chloride crystals into 100 mL of water at 20°C. Sodium ions and chloride ions leave the surface of the crystals and enter the solution. The ions move about at random in the solution and may by chance collide with a crystal and stick, thus returning to the crystalline state. As the sodium chloride continues to dissolve, more ions enter the solution, and the rate at which they return to the crystalline state increases (the more ions in solution, the more likely ions are to collide with the crystals and stick). Eventually, a *dynamic equilibrium* is reached in which the rate at which ions leave the crystals equals the rate at which ions return to the crystals (see Figure 12.2). You write the dynamic equilibrium this way:

$$NaCl(s) \xrightleftharpoons{H_2O} Na^+(aq) + Cl^-(aq)$$

At equilibrium, no more sodium chloride appears to dissolve; 36.0 g has gone into solution, leaving 4.0 g of crystals at the bottom of the vessel. You have a **saturated solution**—that is, *a solution that is in equilibrium with respect to a given dissolved substance*. The solution is saturated with respect to NaCl, and no more NaCl can dissolve. The **solubility** of sodium chloride in water *(the amount that dissolves in a given quantity of water at a given temperature to give a saturated solution)* is 36.0 g/100 mL at 20°C. Note that if you had mixed 30.0 g of sodium chloride with 100 mL of water, all of the crystals would have dissolved. You would have an **unsaturated solution,** *a solution not in equilibrium with respect to a given dissolved substance and in which more of the substance can dissolve.* (Saturated and unsaturated solutions are compared in Figure 12.3.)

Sometimes it is possible to obtain a **supersaturated solution,** *a solution that contains more dissolved substance than a saturated solution does.* For example, the solubility of sodium thiosulfate, $Na_2S_2O_3$, in water at 100°C is 231 g/100 mL. But at room temperature, the solubility is much less—about 50 g/100 mL. Suppose you prepare a solution saturated with sodium thiosulfate at 100°C. You might expect that as the water solution was cooled, sodium thiosulfate would crystallize out. In fact, if the solution is slowly cooled to room temperature, this does not occur. Instead the result is a solution in which 231 g of sodium thiosulfate is dissolved in 100 mL of cold water, compared with the 50 g you would normally expect to find dissolved.

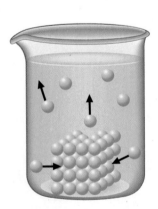

Figure 12.2 ▲

Solubility equilibrium The solid crystalline phase is in dynamic equilibrium with species (ions or molecules) in a saturated solution. The rate at which species leave the crystals equals the rate at which species return to the crystals.

Figure 12.3 ◄

Comparison of unsaturated and saturated solutions

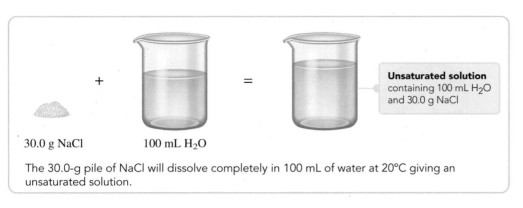

+   30.0 g NaCl   100 mL H₂O   =

Unsaturated solution
containing 100 mL H₂O
and 30.0 g NaCl

The 30.0-g pile of NaCl will dissolve completely in 100 mL of water at 20°C giving an unsaturated solution.

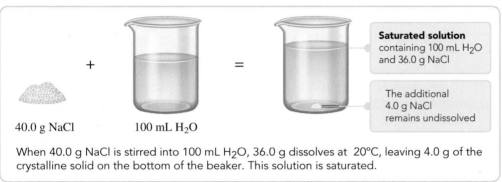

+   40.0 g NaCl   100 mL H₂O   =

Saturated solution
containing 100 mL H₂O
and 36.0 g NaCl

The additional
4.0 g NaCl
remains undissolved

When 40.0 g NaCl is stirred into 100 mL H₂O, 36.0 g dissolves at 20°C, leaving 4.0 g of the crystalline solid on the bottom of the beaker. This solution is saturated.

Supersaturated solutions are not in equilibrium with the solid substance. If a small crystal of sodium thiosulfate is added to a supersaturated solution, the excess immediately crystallizes out. Crystallization from a supersaturated solution is usually quite fast and dramatic (see Figure 12.4).

Factors in Explaining Solubility

The solubilities of substances in one another vary widely. You might find a substance miscible in one solvent but nearly insoluble in another. As a general rule, "like dissolves like." That is, similar substances dissolve one another. Oil is miscible in gasoline. Both are mixtures of hydrocarbon substances (compounds of hydrogen and carbon only). On the other hand, oil does not mix with water. Water is a polar substance, whereas hydrocarbons are not. Why do similar substances dissolve in one another to greater extents than do dissimilar substances? What factors are involved in solubility?

Figure 12.4 ◄

Crystallization from a supersaturated solution of sodium acetate

Crystallization begins immediately when a small crystal of sodium acetate is added.

Within seconds, crystal growth spreads from the original crystal throughout the solution.

Crystal growth will rapidly continue until the entire solution contains solid sodium acetate.

© Cengage Learning

Figure 12.5 ▶

The mixing of gas molecules

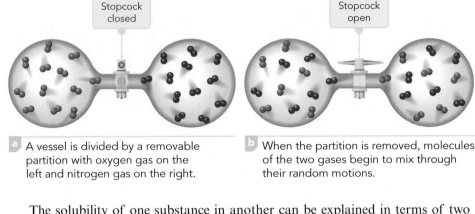

a A vessel is divided by a removable partition with oxygen gas on the left and nitrogen gas on the right.

b When the partition is removed, molecules of the two gases begin to mix through their random motions.

The solubility of one substance in another can be explained in terms of two factors. One is the natural tendency of substances to mix. This is sometimes also referred to as the natural tendency toward disorder. Figure 12.5a shows a vessel divided into two parts, with oxygen gas on the left and nitrogen gas on the right. If you remove the partition, the molecules of the two gases begin to mix. Ultimately, the molecules become thoroughly mixed through their random motions (Figure 12.5b). You might expect a similar mixing of molecules or ions in other types of solutions.

If the process of dissolving one substance in another involved nothing more than simple mixing, you would expect substances to be completely soluble in one another; that is, you would expect substances to be miscible. You know that this is only sometimes the case. Usually, substances have limited solubility in one another. A factor that can limit solubility is the relative forces of attraction between species (molecules or ions). Suppose there are strong attractions between solute species and strong attractions between solvent species, but weak attractions between solute and solvent species. In that case, the strongest attractions are maintained so long as the solute and solvent species do not mix. The lowest energy of the solute–solvent system is obtained then also.

The solubility of a solute in a solvent (that is, the extent of the mixing of the solute and solvent species) depends on a balance between the natural tendency for the solute and solvent species to mix and the tendency for a system to have the lowest energy possible.

Molecular Solutions

The simplest example of a molecular solution is one gas dissolved in another gas. Air, essentially a solution of oxygen and nitrogen, is an example. The intermolecular forces in gases are weak. The only solubility factor of importance is the tendency for molecules to mix (this will be discussed in Chapter 18). Gases are therefore miscible.

Substances may be miscible even when the intermolecular forces are not negligible. Consider the solution of the two similar liquid hydrocarbons heptane, C_7H_{16}, and octane, C_8H_{18}, which are components of gasoline. The intermolecular attractions are due to London forces, and those between heptane and octane molecules are nearly equal to those between octane and octane molecules and heptane and heptane molecules. The different intermolecular attractions are about the same strength, so there are no favored attractions. Octane and heptane molecules tend to move freely through one another. Therefore, the tendency of molecules to mix results in miscibility of the substances.

As a counterexample, consider the mixing of octane with water. There are strong hydrogen bonding forces between water molecules. For octane to mix with water, hydrogen bonds must be broken and replaced by much weaker London forces between water and octane. In this case, the maximum forces of attraction among molecules (and therefore the lower energy) result if the octane and water remain unmixed. Therefore, octane and water are nearly immiscible. (See Figure 12.6.)

The statement "like dissolves like" succinctly expresses these observations. That is, substances with similar intermolecular attractions are usually soluble in

= A molecule

= B molecule

Figure 12.6 ▲

The immiscibility of liquids Suppose an A molecule moves from liquid A into liquid B. If the intermolecular attraction between two A molecules is much stronger than the intermolecular attraction between an A molecule and a B molecule, the net force of attraction tends to pull the A molecule back into liquid A. Thus, liquid A will be immiscible with liquid B.

Table 12.2	Solubilities of Alcohols in Water	
Name	Formula	Solubility in H_2O (g/100 g H_2O at 20°C)
Methanol	CH_3OH	Miscible
Ethanol	CH_3CH_2OH	Miscible
1-Propanol	$CH_3CH_2CH_2OH$	Miscible
1-Butanol	$CH_3CH_2CH_2CH_2OH$	7.9
1-Pentanol	$CH_3CH_2CH_2CH_2CH_2OH$	2.7
1-Hexanol	$CH_3CH_2CH_2CH_2CH_2CH_2OH$	0.6

one another. The two similar hydrocarbons heptane and octane are completely miscible, whereas octane and water (with dissimilar intermolecular attractions) are immiscible.

For the series of alcohols (organic, or carbon-containing, compounds with an —OH group) listed in Table 12.2, the solubility in water decreases from miscible to slightly soluble. Water and alcohols are alike in having —OH groups through which strong hydrogen bonding attractions arise (see Figure 12.7).

The attraction between a methanol molecule, CH_3OH, and a water molecule is nearly as strong as that between two methanol molecules or between two water molecules. Methanol and water molecules tend to mix freely. Methanol and water are miscible, as are ethanol and water and 1-propanol and water. However, as the hydrocarbon end, R—, of the alcohol becomes the more prominent portion of the molecule, the alcohol becomes less like water. Now the forces of attraction between alcohol and water molecules are weaker than those between two alcohol molecules or between two water molecules. Therefore, the solubilities of alcohols decrease with increasing length of R.

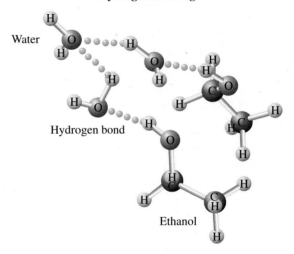

Ethanol–Water Hydrogen Bonding

Figure 12.7 ▲

Hydrogen bonding between water and ethanol molecules The dots depict the hydrogen bonding between the oxygen and hydrogen atoms on adjacent molecules.

CONCEPT CHECK 12.2

You have two molecular compounds, X and Y. Compound X has stronger intermolecular forces than compound Y. X has no dipole-dipole bonding of any kind, and Y exhibits hydrogen bonding. Predict the relative solubility of compounds X and Y in water and in a nonpolar solvent. Explain your reasoning.

Exercise 12.2 Which of the following compounds is likely to be more soluble in water: C_4H_9OH or C_4H_9SH? Explain.

■ See Problems 12.39, 12.40, 12.41, and 12.42.

Ionic Solutions

Ionic substances differ markedly in their solubilities in water. For example, sodium chloride, NaCl, has a solubility of 36 g per 100 mL of water at room temperature, whereas calcium phosphate, $Ca_3(PO_4)_2$, has a solubility of only 0.002 g per 100 mL of water. In most cases, these differences in solubility can be explained in terms of the different energies of attraction between ions in the crystal and between ions and water.

The energy of attraction between an ion and a water molecule is due to an *ion–dipole force*. Water molecules are polar, so they tend to orient with respect to nearby ions. In the case of a positive ion (Li^+, for example), water molecules orient with their oxygen atoms (the negative ends of the molecular dipoles) toward the ion. In the case of a negative ion (for instance, F^-), water molecules orient with

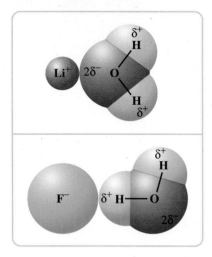

Figure 12.8 ▲

Attraction of water molecules to ions because of the ion–dipole force The O end of H_2O orients toward the cation, whereas an H atom of H_2O orients toward the anion.

Hydration of ions also occurs in crystalline solids. For example, copper(II) sulfate pentahydrate, $CuSO_4 \cdot 5H_2O$, contains five water molecules for each $CuSO_4$ formula unit in the crystal. Substances like this, which are called hydrates, were discussed in Chapter 2.

According to Coulomb's law, the energy of attraction of two ions is proportional to the product of the ion charges and inversely proportional to the distance between the centers of the ions.

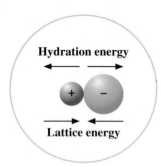

Figure 12.10 ▲

The effects of lattice energy and hydration energy on the solution process of ionic solids Hydration energy pulls ions apart, whereas the lattice energy keeps the ions together. If the lattice energy is large relative to the hydration energy, the ions are likely to remain together, resulting in an insoluble compound.

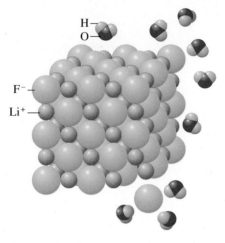

Figure 12.9 ▲

The dissolving of lithium fluoride in water Ions on the surface of the crystal can hydrate—that is, associate with water molecules; ions at the corners are especially easy to remove because they are held by fewer lattice forces. Ions are completely hydrated in the aqueous phase and move off into the body of the liquid.

their hydrogen atoms (the positive ends of the molecular dipoles) toward the ion (see Figure 12.8).

The attraction of ions for water molecules is called **hydration.** Hydration of ions favors the dissolving of an ionic solid in water. Figure 12.9 illustrates the hydration of ions in the dissolving of a lithium fluoride crystal. As ions on the surface become partially hydrated, bonds to the crystal weaken to the point the ions break free and subsequently become completely hydrated in solution.

If the hydration of ions were the only factor in the solution process, you would expect all ionic solids to be soluble in water. The ions in a crystal, however, are very strongly attracted to one another. Therefore, the solubility of an ionic solid depends not only on the energy of hydration of ions (energy associated with the attraction between ions and water molecules) but also on *lattice energy,* the energy holding ions together in the crystal lattice. Lattice energy works against the solution process, so an ionic solid with relatively large lattice energy is usually insoluble (Figure 12.10). ◀

Lattice energies depend on the charges on the ions (as well as on the distance between the centers of neighboring positive and negative ions). The greater the magnitude of ion charge, the greater is the lattice energy. ◀ For this reason, you might expect substances with singly charged ions to be comparatively soluble and those with multiply charged ions to be less soluble. This is borne out by the fact that compounds of the alkali metal ions (such as Na^+ and K^+) and ammonium ions (NH_4^+) are generally soluble, whereas those of phosphate ions (PO_4^{3-}) combined with multiply-charged cations, for example, are generally insoluble.

Lattice energy is also inversely proportional to the distance between neighboring ions, and this distance depends on the sum of the radii of the ions. For example, the lattice energy of magnesium hydroxide, $Mg(OH)_2$, is inversely proportional to the sum of the radii of Mg^{2+} and OH^-. In the series of alkaline earth hydroxides— $Mg(OH)_2$, $Ca(OH)_2$, $Sr(OH)_2$, and $Ba(OH)_2$—the lattice energy decreases as the radius of the alkaline earth ion increases (from Mg^{2+} to Ba^{2+}). If the lattice energy alone determines the trend in solubilities, you should expect the solubility to increase from magnesium hydroxide to barium hydroxide: the expected solubility ranking being $Mg(OH)_2 < Ca(OH)_2 < Sr(OH)_2 < Ba(OH)_2$. In fact, this is what you find. Magnesium hydroxide is insoluble in water, and barium hydroxide is soluble. But this is not the whole story. The energy of hydration also depends on ionic radius. A small ion has a concentrated electric charge and a strong electric field that attracts water molecules. Therefore, the energy of hydration is greatest for a small ion such as Mg^{2+} and least for a large ion such as Ba^{2+}. If the energy of hydration of ions alone determined the trend in solubilities, you would expect the solubilities to decrease from magnesium hydroxide to barium hydroxide, rather than to increase. (We should also add that the energy of hydration increases with the charge on the ion; energy of hydration is greater for Mg^{2+} than for Na^+. Magnesium ion has a larger charge, and also is a smaller ion than Na^+.)

The explanation for the observed solubility trend in the alkaline earth hydroxides is that the lattice energy decreases more rapidly in the series $Mg(OH)_2$, $Ca(OH)_2$, $Sr(OH)_2$, and $Ba(OH)_2$ than does the energy of hydration in the series of ions Mg^{2+}, Ca^{2+}, Sr^{2+}, and Ba^{2+}. For this reason, the lattice-energy factor dominates this solubility trend.

Hemoglobin Solubility and Sickle-Cell Anemia

Sickle-cell anemia was the first inherited disease shown to have a specific molecular basis. In people with the disease, the red blood cells tend to become elongated (sickle-shaped) when the concentration of oxygen (O_2) is low, as it is, for example, in the venous blood supply (see Figure 12.11). Once the red blood cells have sickled, they can no longer function in their normal capacity as oxygen carriers, and they often break apart. Moreover, the sickled cells clog capillaries, interfering with the blood supply to vital organs.

In 1949 Linus Pauling showed that people with sickle-cell anemia have abnormal hemoglobin. Hemoglobin is the substance in red blood cells that carries oxygen. Hemoglobin is normally present in solution within the red blood cells. But in people with sickle-cell anemia, the unoxygenated hemoglobin readily comes out of solution. It produces a fibrous precipitate that deforms the cell, giving it the characteristic sickle shape.

Hemoglobins are large molecules (molecular weight about 64,000 amu) consisting of four protein chains, two of one kind (α chains) and two of another kind (β chains). Normal and sickle-cell hemoglobins are almost exactly alike, except that each

β chain of the hemoglobin responsible for sickle-cell anemia differs from normal hemoglobin in one place. In this place, the normal hemoglobin has the group

$$
\begin{array}{c}
| \\
CH_2 \\
| \\
CH_2 \\
| \\
HO{-}C{=}O
\end{array}
$$

which helps confer water solubility on the molecule because of the polarity of the group and its ability to form hydrogen bonds. The abnormal hemoglobin has the following hydrocarbon group:

$$
\begin{array}{c}
| \\
H{-}C{-}CH_3 \\
| \\
CH_3
\end{array}
$$

Hydrocarbon groups are nonpolar. This small change makes the molecule less water-soluble.

Figure 12.11 ▲

Red blood cells Scanning electron micrographs of normal (*left*) and sickled (*right*) red blood cells. The color is added by computer.

■ See Problems 12.111 and 12.112.

You see the opposite solubility trend when the energy of hydration decreases more rapidly so that it dominates the trend. Consider the alkaline earth sulfates. Here the lattice energy depends on the sum of the radius of the cation and the radius of the sulfate ion. Because the sulfate ion, SO_4^{2-}, is much larger than the hydroxide ion, OH^-, the percent change in lattice energy in going through the series of sulfates from $MgSO_4$ to $BaSO_4$ is smaller than in the hydroxides. The lattice energy changes less, and the energy of hydration of the cation decreases by a greater amount. Now the energy of hydration dominates the solubility trend, and the solubility decreases from magnesium sulfate to barium sulfate. Magnesium sulfate is soluble in water, and barium sulfate is insoluble.

Exercise 12.3 Which ion has the larger hydration energy, Na^+ or K^+?

■ See Problems 12.43 and 12.44.

CONCEPT CHECK 12.3

The hypothetical ionic compound AB_2 is very soluble in water. Another hypothetical ionic compound, CB_2, is only slightly soluble in water. The lattice energies for these compounds are about the same. Provide an explanation for the solubility difference between these compounds.

12.3 Effects of Temperature and Pressure on Solubility

In general, the solubility of a substance depends on temperature. For example, the solubility of ammonium nitrate in 100 mL of water is 118 g at 0°C and 811 g at 100°C. Pressure may also have an effect on solubility, as you will see.

Temperature Change

Most gases become less soluble in water at higher temperatures. The first bubbles that appear when tap water is heated are bubbles of air released as the increasing temperature reduces the solubility of air in water. In contrast, most ionic solids become more soluble in water with rising temperature.

The variations of the solubilities of the salts KNO_3, $CuSO_4$, $NaCl$, and $Ce_2(SeO_4)_3$ are shown in Figure 12.12. Three of the salts show the usual behavior; their solubilities increase with rising temperature. For example, potassium nitrate, KNO_3, changes solubility dramatically from 14 g/100 g H_2O at 0°C to 245 g/100 g H_2O at 100°C. Copper(II) sulfate, $CuSO_4$, shows a moderate increase in solubility over this temperature interval. Sodium chloride, $NaCl$, increases only slightly in solubility with temperature.

A number of ionic compounds decrease in solubility with increasing temperature. Calcium sulfate, $CaSO_4$, and calcium hydroxide, $Ca(OH)_2$, are common

Figure 12.12 ▶

Solubility of some ionic salts at different temperatures The solubilities of the salts NaCl, KNO_3, and $CuSO_4$ rise with increasing temperature, as is the case with most ionic solids. The solubility of $Ce_2(SeO_4)_3$, however, falls with increasing temperature.

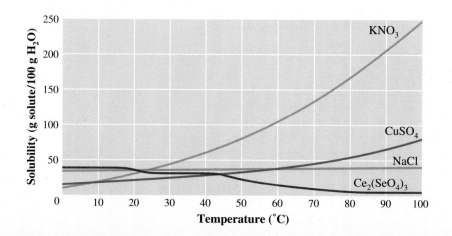

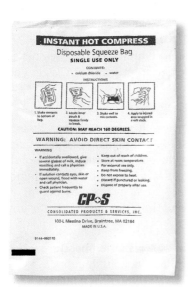

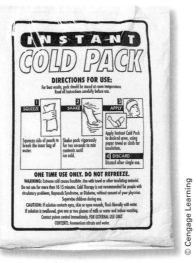

Figure 12.13 ◄

Instant hot and cold compress packs

© Cengage Learning

a The inner bag of the instant hot compress contains calcium chloride, $CaCl_2$. When the inner bag is broken, the $CaCl_2$ dissolves exothermically in water (heat is released).

b The instant cold compress contains an inner bag of ammonium nitrate, NH_4NO_3, which when broken apart allows the NH_4NO_3 to dissolve in the water in the outer bag. The solution process is endothermic (heat is absorbed), so the bag feels cold.

examples. They are slightly soluble compounds that become even less soluble at higher temperatures. Cerium selenate, $Ce_2(SeO_4)_3$, shown in Figure 12.12, is very soluble at 0°C but much less soluble at 100°C.

Heat can be released or absorbed when ionic substances are dissolved in water. In some cases, this *heat of solution* is quite noticeable. When sodium hydroxide is dissolved in water, the solution becomes hot (the solution process is exothermic). On the other hand, when ammonium nitrate is dissolved in water, the solution becomes very cold (the solution process is endothermic). This cooling effect from the dissolving of ammonium nitrate in water is exploited in instant cold packs used in hospitals and elsewhere. An instant cold pack consists of a bag of NH_4NO_3 crystals inside a bag of water (Figure 12.13). When the inner bag is broken, NH_4NO_3 dissolves in the water. Heat is absorbed, so the bag feels cold. Hot packs, by contrast, containing either $CaCl_2$ or $MgSO_4$, dissolve in water with the evolution of heat.

Pressure Change

In general, pressure change has little effect on the solubility of a liquid or solid in water, but the solubility of a gas is very much affected by pressure. The qualitative effect of a change in pressure on the solubility of a gas can be predicted from Le Châtelier's principle. **Le Châtelier's principle** *states that when a system in equilibrium is disturbed by a change of temperature, pressure, or concentration variable, the system shifts in equilibrium composition in a way that tends to counteract this change of variable.* Let us see how Le Châtelier's principle can predict the effect of a change in pressure on gas solubility.

Imagine a cylindrical vessel that is fitted with a movable piston and contains carbon dioxide gas over its saturated water solution (Figure 12.14). The equilibrium is

$$CO_2(g) \rightleftharpoons CO_2(aq)$$

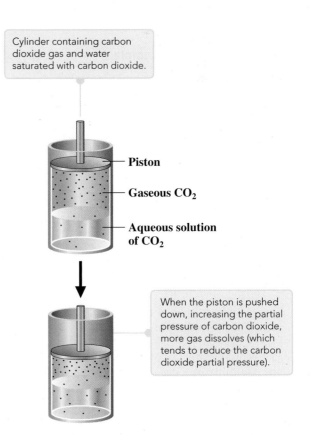

Cylinder containing carbon dioxide gas and water saturated with carbon dioxide.

— Piston

— Gaseous CO_2

— Aqueous solution of CO_2

When the piston is pushed down, increasing the partial pressure of carbon dioxide, more gas dissolves (which tends to reduce the carbon dioxide partial pressure).

Figure 12.14 ▲

Effect of pressure on gas solubility

Figure 12.15 ▲

Sudden release of pressure from a carbonated beverage
A carbonated beverage is produced by dissolving carbon dioxide in beverage solution under pressure. More carbon dioxide dissolves at the higher pressure than otherwise. When this pressure is suddenly released, carbon dioxide is less soluble, and the excess bubbles from the solution.

Suppose you increase the partial pressure of CO_2 gas by pushing the piston down. This change of partial pressure, according to Le Châtelier's principle, shifts the equilibrium composition in a way that tends to counteract the pressure increase. From the preceding equation, you see that the partial pressure of CO_2 gas decreases if more CO_2 dissolves.

$$CO_2(g) \longrightarrow CO_2(aq)$$

The system comes to a new equilibrium, in which more CO_2 has dissolved. So you can predict that carbon dioxide is more soluble at higher pressures. Conversely, when the partial pressure of carbon dioxide gas is reduced, its solubility is decreased. A bottle of carbonated beverage fizzes when the cap is removed: as the partial pressure of carbon dioxide is reduced, gas comes out of solution (Figure 12.15). The argument given here for carbon dioxide holds for any gas:

> All gases become more soluble in a liquid at a given temperature when the partial pressure of the gas over the solution is increased.

Henry's Law: Relating Pressure to the Solubility of a Gas in a Liquid

The effect of pressure on the solubility of a gas in a liquid can be predicted quantitatively. According to **Henry's law,** *the solubility of a gas is directly proportional to the partial pressure of the gas above the solution.* Expressed mathematically, the law is

$$S = k_H P$$

where S is the solubility of the gas (expressed as mass of solute per unit volume of solvent), k_H is Henry's law constant for the gas for a particular liquid at a given temperature, and P is the partial pressure of the gas. The next example shows how this formula is used.

⊙WL Interactive Example   **12.1**   **Applying Henry's Law**

In the chapter opening, we noted that 27 g of acetylene, C_2H_2, dissolves in 1 L of acetone at 1.0 atm pressure. If the partial pressure of acetylene is increased to 12 atm, what is its solubility in acetone?

Problem Strategy Because we are trying to determine the solubility of a gas in a liquid, this problem is an application of the Henry's law equation: $S = k_H P$. Let S_1 be the solubility of the gas at partial pressure P_1, and let S_2 be the solubility at partial pressure P_2. Then you can write Henry's law for both pressures.

$$S_1 = k_H P_1$$
$$S_2 = k_H P_2$$

Dividing the second equation by the first, you get

$$\frac{S_2}{S_1} = \frac{k_H P_2}{k_H P_1} \quad \text{or} \quad \frac{S_2}{S_1} = \frac{P_2}{P_1}$$

You can use this relation to find the solubility at one pressure given the solubility at another.

Solution At 1.0 atm partial pressure of acetylene (P_1), the solubility S_1 is 27 g C_2H_2 per liter of acetone. For a partial pressure of 12 atm (P_2),

$$\frac{S_2}{27 \text{ g } C_2H_2/\text{L acetone}} = \frac{12 \text{ atm}}{1.0 \text{ atm}}$$

Then

$$S_2 = \frac{27 \text{ g } C_2H_2}{\text{L acetone}} \times \frac{12}{1.0} = \frac{3.2 \times 10^2 \text{ g } C_2H_2}{\text{L acetone}}$$

At 12 atm partial pressure of acetylene, 1 L of acetone dissolves **3.2×10^2 g of acetylene.**

(continued)

(*continued*)

Answer Check Because the solubility of a gas is directly proportional to the pressure of the gas above the solution, check your answer to make certain that the gas at higher pressure has the greater solubility.

Exercise 12.4 A liter of water at 25°C dissolves 0.0404 g O_2 when the partial pressure of the oxygen is 1.00 atm. What is the solubility of oxygen from air, in which the partial pressure of O_2 is 159 mmHg?

■ See Problems 12.47 and 12.48.

CONCEPT CHECK 12.4

Most fish have a very difficult time surviving at elevations much above 3500 m. How could Henry's law be used to account for this fact?

Colligative Properties

In the chapter opening, we mentioned that the addition of ethylene glycol, CH_2OHCH_2OH, to water lowers the freezing point of water below 0°C. For example, if 0.010 mol of ethylene glycol is added to 1 kg of water, the freezing point is lowered to −0.019°C. The magnitude of freezing-point lowering is directly proportional to the number of ethylene glycol molecules added to a quantity of water. Thus, if you add 0.020 (=0.010 × 2) mol of ethylene glycol to 1 kg of water, the freezing point is lowered to −0.038°C (=−0.019°C × 2). The same lowering is observed with the addition of other nonelectrolyte substances. For example, an aqueous solution of 0.020 mol of urea, $(NH_2)_2CO$, in 1 kg of water also freezes at −0.038°C.

Freezing-point lowering is a colligative property. **Colligative properties** of solutions are *properties that depend on the concentration of solute molecules or ions in solution but not on the chemical identity of the solute* (whether it is ethylene glycol or urea, for instance). In the next sections, we will discuss several colligative properties, which are expressed quantitatively in terms of various concentration units. We will first look into ways of expressing the concentration of a solution.

12.4 Ways of Expressing Concentration

The *concentration* of a solute is the amount of solute dissolved in a given quantity of solvent or solution. The quantity of solvent or solution can be expressed in terms of volume or in terms of mass or molar amount. Thus, there are several ways of expressing the concentration of a solution.

Recall that the *molarity* of a solution is the moles of solute in a liter of solution. ▶

Molarity was discussed in Section 4.7.

$$\text{Molarity} = \frac{\text{moles of solute}}{\text{liters of solution}}$$

For example, 0.20 mol of ethylene glycol dissolved in enough water to give 2.0 L of solution has a molarity of

$$\frac{0.20 \text{ mol ethylene glycol}}{2.0 \text{ L solution}} = 0.10 \, M \text{ ethylene glycol}$$

The solution is 0.10 molar (denoted M). This unit is especially useful when you wish to dispense a given amount of solute, because the unit directly relates the amount of solute to the volume of solution.

Some other concentration units are defined in terms of the mass or molar amount of solvent or solution. The most important of these are mass percentage of solute, molality, and mole fraction.

Mass Percentage of Solute

Solution concentration is sometimes expressed in terms of the **mass percentage of solute**—that is, *the percentage by mass of solute contained in a solution.*

$$\text{Mass percentage of solute} = \frac{\text{mass of solute}}{\text{mass of solution}} \times 100\%$$

For example, an aqueous solution that is 3.5% sodium chloride by mass contains 3.5 g of NaCl in 100.0 g of solution. It could be prepared by dissolving 3.5 g of NaCl in 96.5 g of water (100.0 − 3.5 = 96.5).

Example 12.2

Critical Concept 12.2
Solution concentration can be expressed as a mass percentage of solute. When expressing solution concentration as a mass percent, the units of mass for the solute and solvent must be the same, for example, *grams* of solute and *grams* of solution. The mass of solution is the sum of the mass(es) of the solute(s) and the solvent.

Solution Essentials:
- Mass percentage of solute
  (% mass $= \dfrac{\text{mass of solute}}{\text{mass of solution}} \times 100\%$)
- Solute
- Solution
- Concentration

Calculating with Mass Percentage of Solute

How would you prepare 425 g of an aqueous solution containing 2.40% by mass of sodium acetate, $NaC_2H_3O_2$?

Problem Strategy In order to know how to prepare this solution, we need to know the mass of solute ($NaC_2H_3O_2$) and the mass of solution. The mass percent given in the problem (2.40%) will enable us to determine the mass of solute present in the 425-g sample. Keeping in mind that the total mass of solution equals the sum of the mass of solute and the mass of solvent (mass solution = mass solute + mass solvent), we will be able to determine the mass of H_2O needed to prepare the solution.

Solution The mass of sodium acetate (solute) in 425 g of solution is

$$\text{Mass of } NaC_2H_3O_2 = 425 \text{ g} \times 0.0240 = 10.2 \text{ g}$$

We have the relationship

$$\text{Mass of solution} = \text{mass } NaC_2H_3O_2 + \text{mass } H_2O$$

Therefore, the quantity of water in the solution is

$$\text{Mass of } H_2O = \text{mass of solution} - \text{mass of } NaC_2H_3O_2 = 425 \text{ g} - 10.2 \text{ g} = 415 \text{ g}$$

You would prepare the solution by dissolving 10.2 g of sodium acetate in 415 g of water.

Answer Check A quick way to check your work is to substitute the values you determined into the formula for mass percentage of solute; make sure that the calculated value agrees with that given in the problem.

Exercise 12.5 An experiment calls for 35.0 g of hydrochloric acid that is 20.2% HCl by mass. How many grams of HCl is this? How many grams of water?

■ See Problems 12.49, 12.50, 12.51, and 12.52.

Molality

The **molality** of a solution is *the moles of solute per kilogram of solvent.*

$$\text{Molality} = \frac{\text{moles of solute}}{\text{kilograms of solvent}}$$

For example, 0.20 mol of ethylene glycol dissolved in 2.0×10^3 g (=2.0 kg) of water has a molality of

$$\frac{0.20 \text{ mol ethylene glycol}}{2.0 \text{ kg solvent}} = 0.10 \; m \text{ ethylene glycol}$$

That is, the solution is 0.10 molal (denoted m). The units of molality and molarity are sometimes confused. Note that molality is defined in terms of *mass of solvent,* and molarity is defined in terms of *volume of solution.*

⏻WL Interactive Example 12.3   Calculating the Molality of Solute

Gaining Mastery Toolbox

Critical Concept 12.3
Solution concentration can be expressed as the ratio of the moles of solute to the mass of solvent: molality (m). When calculating solution molality, the mass of solute must be expressed in kg.

Solution Essentials:
• Molality
$$\left(m = \frac{\text{moles of solute}}{\text{kg of solvent}} \right)$$
• Solute
• Solvent
• Concentration

Glucose, $C_6H_{12}O_6$, is a sugar that occurs in fruits. It is also known as "blood sugar" because it is found in blood and is the body's main source of energy (see Figure 12.16). What is the molality of a solution containing 5.67 g of glucose dissolved in 25.2 g of water?

Problem Strategy Recall that the definition of *molality* is moles of solute per kilogram of solvent. Therefore, we need to determine these two quantities to solve this problem. Start by converting mass of solute (glucose) to moles, and next determine the mass of solvent (water) in kilograms. Then use these quantities to calculate the solution molality.

Solution The moles of glucose (MM = 180.2 amu) in 5.67 g are found as follows:

$$5.67 \text{ g } C_6H_{12}O_6 \times \frac{1 \text{ mol } C_6H_{12}O_6}{180.2 \text{ g } C_6H_{12}O_6} = 0.0315 \text{ mol } C_6H_{12}O_6$$

The mass of water is 25.2 g, or 25.2×10^{-3} kg.

$$\text{Molality} = \frac{0.0315 \text{ mol } C_6H_{12}O_6}{25.2 \times 10^{-3} \text{ kg solvent}} = \textbf{1.25 } \boldsymbol{m} \textbf{ } C_6H_{12}O_6$$

Answer Check Any time you calculate a molality that is greater than about 15 m, unless it is a very concentrated solution or a pure solvent, you probably made an error.

Glucose

Figure 12.16 ▲

Intravenous feeding of glucose Glucose, $C_6H_{12}O_6$, is the principal energy substance of the body and is transported to various cells by the blood. Here a glucose solution is being prepared for administration to a patient.

Exercise 12.6 Toluene, $C_6H_5CH_3$, is a liquid compound similar to benzene, C_6H_6. It is the starting material for other substances, including trinitrotoluene (TNT). Find the molality of toluene in a solution that contains 35.6 g of toluene and 125 g of benzene.

■ See Problems 12.53, 12.54, 12.55, and 12.56.

© ER Productions/Corbis

Mole Fraction

The **mole fraction** of a component substance A (X_A) in a solution is defined as *the moles of component substance divided by the total moles of solution* (that is, moles of solute plus solvent).

$$X_A = \frac{\text{moles of substance } A}{\text{total moles of solution}}$$

For example, if a solution is made up of 1 mol of ethylene glycol and 9 mol of water, the total moles of solution are 1 mol + 9 mol = 10 mol. The mole fraction of ethylene glycol is 1/10 = 0.1, and the mole fraction of water is 9/10 = 0.9. Multiplying mole fractions by 100 gives *mole percent*. Hence, this solution is 10 mole percent ethylene glycol and 90 mole percent water. You can also say that 10% of the molecules in the solution are ethylene glycol and 90% are water. The sum of the mole fractions of all the components of a solution equals 1.

Example 12.4   Calculating the Mole Fractions of Components

Gaining Mastery Toolbox

Critical Concept 12.4
Solution concentration can be expressed as the ratio of the moles of one of the substances in the solution to the total number of moles of all substances that comprise the solution, that is, the mole fraction. Both solute and solvent concentration can be expressed as a mole fraction. The sum of the mole fractions of all the components that make up a solution must be equal to 1.

Solution Essentials:
• Mole fraction $\left(X_A = \dfrac{\text{moles of substance } A}{\text{total moles of solution}} \right)$
• Solution

What are the mole fractions of glucose and water in a solution containing 5.67 g of glucose, $C_6H_{12}O_6$, dissolved in 25.2 g of water?

Problem Strategy The mole fraction of a solution is the moles of the substance of interest divided by the total number of moles of solution. For this problem, the total number of moles of solution is the sum of the moles of glucose and the moles of water. The moles of glucose and water can be calculated directly from their masses. Add these mole quantities together to obtain the total moles of the solution. Divide each substance by the total moles to obtain the mole fractions.

Solution This is the glucose solution described in Example 12.3. There you found that 5.67 g of glucose equals 0.0315 mol of glucose. The moles of water in the solution are

$$25.2 \text{ g } H_2O \times \frac{1 \text{ mol } H_2O}{18.0 \text{ g } H_2O} = 1.40 \text{ mol } H_2O$$

Hence, the total moles of solution are

$$1.40 \text{ mol} + 0.0315 \text{ mol} = 1.432 \text{ mol}$$

(You retain an extra figure in the answer for further computation.) Finally, you get

$$\textbf{Mole fraction glucose} = \frac{0.0315 \text{ mol}}{1.432 \text{ mol}} = \textbf{0.0220}$$

$$\textbf{Mole fraction water} = \frac{1.40 \text{ mol}}{1.432 \text{ mol}} = \textbf{0.978}$$

Answer Check Note that the sum of the mole fractions is 1.00. Whenever you calculate mole fractions, you can check your answer by making sure that they sum to 1.

Exercise 12.7 Calculate the mole fractions of toluene and benzene in the solution described in Exercise 12.6.

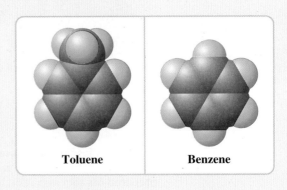

Toluene   Benzene

■ See Problems 12.57 and 12.58.

Conversion of Concentration Units

It is relatively easy to interconvert concentration units when they are expressed in terms of mass or moles of solute and solvent, as the following examples show.

Example 12.5   Converting Molality to Mole Fractions

Critical Concept 12.5
The concentration of a solution can be expressed in many different ways. Given the concentration of a solution in one set of units, it is often possible to convert to a different set of units. It is critical to keep track of all of the components that define a solution (solute, solvent, and solution) because the concentration units are defined using these quantities. In order to solve the problem in Example 12.5, we must first determine the moles of solute (glucose), moles of solvent (water), and moles of solution (glucose and water).

Solution Essentials:
- Molality $\left(m = \dfrac{\text{moles of solute}}{\text{kg of solvent}} \right)$
- Mole fraction $\left(X_A = \dfrac{\text{moles of substance } A}{\text{total moles of solution}} \right)$
- Solute
- Solvent
- Solution

An aqueous solution is 0.120 m glucose, $C_6H_{12}O_6$. What are the mole fractions of each component in the solution?

Problem Strategy Because molality is the moles of solute per kilogram of solvent, a 0.120 m glucose solution contains 0.120 mol of glucose in 1.00 kg of water. After converting 1.00 kg H_2O to moles, you can calculate the mole fraction using its definition.

Solution The moles of H_2O in 1.00 kg of water are

$$1.00 \times 10^3 \text{ g } H_2O \times \frac{1 \text{ mol } H_2O}{18.0 \text{ g } H_2O} = 55.6 \text{ mol } H_2O$$

Hence,

$$\text{Mole fraction glucose} = \frac{0.120 \text{ mol}}{(0.120 + 55.6) \text{ mol}} = \mathbf{0.00215}$$

$$\text{Mole fraction water} = \frac{55.6 \text{ mol}}{(0.120 + 55.6) \text{ mol}} = \mathbf{0.998}$$

Answer Check Check to see that the mole fractions of each substance sum to 1.

Exercise 12.8 A solution is 0.120 m methanol dissolved in ethanol. Calculate the mole fractions of methanol, CH_3OH, and ethanol, C_2H_5OH, in the solution.

■ See Problems 12.59 and 12.60.

Example 12.6   Converting Mole Fractions to Molality

Critical Concept 12.6
The concentration of a solution can be expressed in many different ways. Given the concentration of a solution in one set of units, it is often possible to convert to a different set of units. It is critical to keep track of all of the components that define a solution (solute, solvent, and solution) because the concentration units are defined using these quantities. The solution to the problem in Example 12.6 requires that we first determine the moles of solute (glucose) and the mass of solvent (water).

Solution Essentials:
- Molality $\left(m = \dfrac{\text{moles of solute}}{\text{kg of solvent}} \right)$
- Mole fraction $\left(X_A = \dfrac{\text{moles of substance } A}{\text{total moles of solution}} \right)$
- Solute
- Solvent
- Solution

A solution is 0.150 mole fraction glucose, $C_6H_{12}O_6$, and 0.850 mole fraction water. What is the molality of glucose in the solution?

Problem Strategy The molality of the solution is the moles of solute (glucose) per kilogram of solvent (H_2O). Given the mole fraction information in the problem, we know that each mole of solution contains 0.150 mol of glucose and 0.850 mol of water. If we convert the moles of water (0.850 mol H_2O) to mass of water, in kilograms, we can then calculate the molality.

Solution The mass of this amount of water is

Glucose

$$0.850 \text{ mol } H_2O \times \frac{18.0 \text{ g } H_2O}{1 \text{ mol } H_2O} = 15.3 \text{ g } H_2O \ (=0.0153 \text{ kg } H_2O)$$

The molality of glucose, $C_6H_{12}O_6$, in the solution is

$$\text{Molality of } C_6H_{12}O_6 = \frac{0.150 \text{ mol } C_6H_{12}O_6}{0.0153 \text{ kg solvent}} = \mathbf{9.80 \ \textit{m} \ C_6H_{12}O_6}$$

Answer Check When calculating molality of solutions, always make certain that the mass of solvent is expressed in kilograms.

Exercise 12.9 A solution is 0.250 mole fraction methanol, CH_3OH, and 0.750 mole fraction ethanol, C_2H_5OH. What is the molality of methanol in the solution?

■ See Problems 12.61 and 12.62.

To convert molality to molarity, and vice versa, you must know the density of the solution. The calculations are described in the next two examples. As the first example shows, molality and molarity are approximately equal in dilute aqueous solutions.

Example 12.7

Converting Molality to Molarity

An aqueous solution is 0.273 m KCl. What is the molar concentration of potassium chloride, KCl? The density of the solution is 1.011×10^3 g/L.

Problem Strategy You are asked to determine the molarity of the solution, which is the moles of solute (KCl) per liter of solution (volume of H_2O and KCl). The molality tells you that the solution contains 0.273 mol KCl in 1 kg of water. If we determine the volume of solution that contains 0.273 mol KCl, we should be able to calculate the molarity. Because the solution is made up of only KCl and water, the solution mass is the mass of 0.273 mol KCl plus the mass of 1.000 kg of water. Knowing the total mass of solution, we then can use the solution density to convert from the mass of solution to the volume of solution. The molarity is obtained by dividing the moles of solute by the volume of solution in liters.

Solution The first step in determining the volume of solution is to find the mass of potassium chloride in this quantity of solution:

$$0.273 \; \text{mol KCl} \times \frac{74.6 \text{ g KCl}}{1 \text{ mol KCl}} = 20.4 \text{ g KCl}$$

The total mass of the solution equals the mass of water plus the mass of potassium chloride.

$$1.000 \times 10^3 \text{ g} + 20.4 \text{ g} = 1.020 \times 10^3 \text{ g}$$

The volume of the solution can be calculated using the inverse of the density of the solution.

$$\text{Volume of solution} = 1.020 \times 10^3 \text{ g} \times \frac{1 \text{ L}}{1.011 \times 10^3 \text{ g}} = 1.009 \text{ L}$$

There is 0.273 mol KCl in every 1.009 L of solution. Hence, the molarity of the solution is

$$\frac{0.273 \text{ mol KCl}}{1.009 \text{ L solution}} = \textbf{0.271 } \boldsymbol{M} \textbf{ KCl}$$

Note that the molarity and the molality of this solution are approximately equal. This happens in cases where the solutions are dilute and the density is about 1 g/mL.

Answer Check If you are converting between molarity and molality of dilute aqueous solutions, the concentrations expressed either way generally do not differ greatly. You can see that is the case here: 0.271 M KCl versus 0.273 m KCl. Use this fact to check your answers for this type of problem.

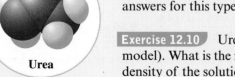

Urea

Exercise 12.10 Urea, $(NH_2)_2CO$, is used as a fertilizer (see the accompanying molecular model). What is the molar concentration of an aqueous solution that is 3.42 m urea? The density of the solution is 1.045 g/mL.

■ See Problems 12.63 and 12.64.

Example 12.8   Converting Molarity to Molality

Critical Concept 12.8
The concentration of a solution can be expressed in many different ways. Given the concentration of a solution in one set of units, it is often possible to convert to a different set of units. It is critical to keep track of all of the components that define a solution (solute, solvent, and solution) because the concentration units are defined using these quantities. When solving the problem in Example 12.8, it helps to assume that you have a 1-L sample of the solution.

Solution Essentials:
• Molality
$$\left(m = \frac{\text{moles of solute}}{\text{kg of solvent}} \right)$$
• Molarity
$$\left(M = \frac{\text{moles of solute}}{\text{L of solution}} \right)$$
• Solute
• Solvent
• Solution

An aqueous solution is 0.907 M Pb(NO$_3$)$_2$. What is the molality of lead(II) nitrate, Pb(NO$_3$)$_2$, in this solution? The density of the solution is 1.252 g/mL.

Problem Strategy You are asked to calculate the molality of the solution, which is the moles of solute (Pb(NO$_3$)$_2$) per kilogram of solvent (H$_2$O). From the molarity, we know that there are 0.907 mol Pb(NO$_3$)$_2$ per liter of solution. If we consider a 1-L sample of this solution, we can use the density to determine the mass of solution. Because the mass of solution is the mass of lead(II) nitrate plus the mass of water, you can calculate the mass of lead(II) nitrate by difference (mass H$_2$O = mass of solution − mass of Pb(NO$_3$)$_2$). Now you can calculate the molality.

Solution Assuming 1.000 L (1.000 × 10^3 mL) of solution,

$$\text{Mass of solution} = \text{density} \times \text{volume}$$
$$= 1.252 \text{ g/mL} \times 1.000 \times 10^3 \text{ mL}$$
$$= 1.252 \times 10^3 \text{ g}$$

The mass of lead nitrate is

$$0.907 \text{ mol Pb(NO}_3)_2 \times \frac{331.2 \text{ g Pb(NO}_3)_2}{1 \text{ mol Pb(NO}_3)_2} = 3.00 \times 10^2 \text{ g Pb(NO}_3)_2$$

The mass of water in 1.000 L of solution is

$$\text{Mass of H}_2\text{O} = \text{mass of solution} - \text{mass of Pb(NO}_3)_2$$
$$= 1.252 \times 10^3 \text{ g} - 3.00 \times 10^2 \text{ g}$$
$$= 9.52 \times 10^2 \text{ g} \, (=0.952 \text{ kg H}_2\text{O})$$

Hence, the molality of lead(II) nitrate in this solution is

$$\frac{0.907 \text{ mol Pb(NO}_3)_2}{0.952 \text{ kg solvent}} = \textbf{0.953 } \textit{m} \textbf{ Pb(NO}_3)_2$$

Answer Check A very common error that you want to avoid when doing this kind of problem is to forget that the solution density is typically the mass of solute per volume of *solution,* not the mass of solute per volume of solvent.

Exercise 12.11 An aqueous solution is 2.00 M urea. The density of the solution is 1.029 g/mL. What is the molal concentration of urea in the solution?

■ See Problems 12.65 and 12.66.

12.5 Vapor Pressure of a Solution

During the nineteenth century, chemists observed that the vapor pressure of a volatile solvent was lowered by addition of a nonvolatile solute. **Vapor-pressure lowering** of a solvent is *a colligative property equal to the vapor pressure of the pure solvent minus the vapor pressure of the solution.* For example, water at 20°C has a vapor pressure of 17.54 mmHg. Ethylene glycol, CH$_2$OHCH$_2$OH, is a liquid whose vapor pressure at 20°C is relatively low; it can be considered to be nonvolatile compared with water. An aqueous solution containing 0.0100 mole fraction of ethylene glycol has a vapor pressure of 17.36 mmHg. Thus, the vapor-pressure lowering, ΔP, of water is

$$\Delta P = 17.54 \text{ mmHg} - 17.36 \text{ mmHg} = 0.18 \text{ mmHg}$$

Figure 12.17 shows a demonstration of vapor-pressure lowering.

In about 1886, the French chemist François Marie Raoult observed that the partial vapor pressure of solvent over a solution of a nonelectrolyte solute depends on the mole fraction of solvent in the solution. Consider a solution of volatile

Figure 12.17 ▶

Demonstration of Raoult's law of partial vapor pressures

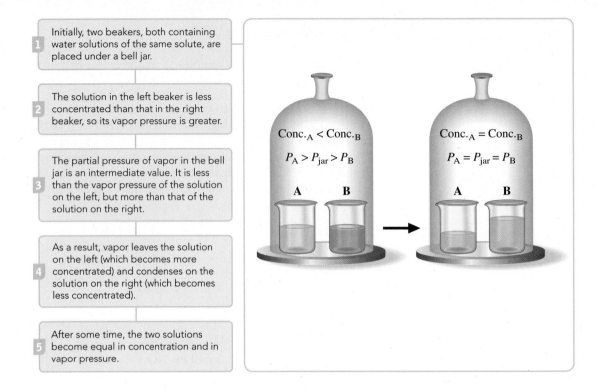

1 Initially, two beakers, both containing water solutions of the same solute, are placed under a bell jar.

2 The solution in the left beaker is less concentrated than that in the right beaker, so its vapor pressure is greater.

3 The partial pressure of vapor in the bell jar is an intermediate value. It is less than the vapor pressure of the solution on the left, but more than that of the solution on the right.

4 As a result, vapor leaves the solution on the left (which becomes more concentrated) and condenses on the solution on the right (which becomes less concentrated).

5 After some time, the two solutions become equal in concentration and in vapor pressure.

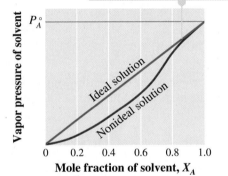

Portion of curve where nonideal solution follows Raoult's law for dilute solution

Figure 12.18 ▲

Plot of vapor pressures of solutions showing Raoult's law Vapor pressures of solvent A for two solutions have been plotted against mole fraction of solvent. In one case (labeled "ideal solution"), the vapor pressure is proportional to the mole fraction of solvent for all mole fractions; it follows Raoult's law for all concentrations of solute. For the "nonideal solution," Raoult's law is followed for low solute concentrations (mole fraction of solvent near 1), but the vapor pressure deviates at other concentrations.

solvent, A, and nonelectrolyte solute, B, which may be volatile or non-volatile. According to **Raoult's law,** *the partial pressure of solvent, P_A, over a solution equals the vapor pressure of the pure solvent, P_A°, times the mole fraction of solvent, X_A, in the solution.*

$$P_A = P_A^\circ X_A$$

If the solute is nonvolatile, P_A is the total vapor pressure of the solution. Because the mole fraction of solvent in a solution is always less than 1, the vapor pressure of the solution of a nonvolatile solute is less than that for the pure solvent; the vapor pressure is lowered. In general, Raoult's law is observed to hold for dilute solutions—that is, solutions in which X_A is close to 1. If the solvent and solute are chemically similar, Raoult's law may hold for all mole fractions. Raoult's law is displayed graphically for two solutions in Figure 12.18.

You can obtain an explicit expression for the vapor-pressure lowering of a solvent in a solution, assuming that Raoult's law holds and that the solute is a nonvolatile nonelectrolyte. The vapor-pressure lowering, ΔP, is

$$\Delta P = P_A^\circ - P_A$$

Substituting Raoult's law gives

$$\Delta P = P_A^\circ - P_A^\circ X_A = P_A^\circ(1 - X_A)$$

But the sum of the mole fractions of the components of a solution must equal 1; that is, $X_A + X_B = 1$. So $X_B = 1 - X_A$. Therefore,

$$\Delta P = P_A^\circ X_B$$

From this equation, you can see that the vapor-pressure lowering is a colligative property—one that depends on the concentration, but not on the nature, of the solute. Thus, if the mole fraction of ethylene glycol, X_B, in an aqueous solution is doubled from 0.010 to 0.020, the vapor-pressure lowering is doubled from 0.18 mmHg to 0.36 mmHg. Also, because the previous equation does not depend on the characteristics of the solute (other than its being nonvolatile and a nonelectrolyte), a solution that is 0.010 mole fraction urea, $(NH_2)_2CO$, has the same vapor-pressure lowering as one that is 0.010 mole fraction ethylene glycol. The next example illustrates the use of the previous equation.

Example 12.9   Calculating Vapor-Pressure Lowering

Critical Concept 12.9
The vapor pressure of a solution is lower than the vapor pressure of the pure solvent. Raoult's law ($P_A = P_A^\circ X_A$) can be used to calculate the vapor pressure of dilute solutions. Raoult's law requires that the solution concentration be expressed as a mole fraction. The vapor pressure lowering of a solution can be expressed as a function of the mole fraction of solute: $\Delta P = P_A^\circ X_B$.

Solution Essentials:
- Raoult's law ($P_A = P_A^\circ X_A$)
- Equation for vapor pressure lowering ($\Delta P = P_A^\circ X_B$)
- Mole fraction
- Solute
- Partial pressure
- Vapor pressure

Calculate the vapor-pressure lowering of water when 5.67 g of glucose, $C_6H_{12}O_6$, is dissolved in 25.2 g of water at 25°C. The vapor pressure of water at 25°C is 23.8 mmHg. What is the vapor pressure of the solution?

Problem Strategy In order to calculate the vapor-pressure lowering of the solution described in this problem, we need to determine the amount that the vapor pressure of the pure solvent is changed (ΔP of water) with the addition of a solute (glucose). We can use Raoult's law to calculate this quantity, noting that we need to know the mole fraction of the solute (glucose). The vapor pressure of the solution is found by subtracting the vapor-pressure change of the pure solvent from the vapor pressure of the pure solvent.

Solution This is the glucose solution described in Example 12.4. According to the calculations performed there, the solution is 0.0220 mole fraction glucose. Therefore, the vapor-pressure lowering is

$$\Delta P = P_A^\circ X_B = 23.8 \text{ mmHg} \times 0.0220 = \textbf{0.524 mmHg}$$

The vapor pressure of the solution is

$$P_A = P_A^\circ - \Delta P = (23.8 - 0.524) \text{ mmHg} = \textbf{23.3 mmHg}$$

Answer Check In this type of calculation, a correct answer will always have the vapor pressure of the solution lower than the vapor pressure of the pure solvent.

Exercise 12.12   Naphthalene, $C_{10}H_8$, is used to make mothballs. Suppose a solution is made by dissolving 0.515 g of naphthalene in 60.8 g of chloroform, $CHCl_3$. Calculate the vapor-pressure lowering of chloroform at 20°C from the naphthalene. The vapor pressure of chloroform at 20°C is 156 mmHg. Naphthalene can be assumed to be nonvolatile compared with chloroform. What is the vapor pressure of the solution?

■ See Problems 12.67 and 12.68.

An *ideal solution* of substances A and B is one in which both substances follow Raoult's law for all values of mole fractions. Such solutions occur when the substances are chemically similar so that the intermolecular forces between A and B molecules are similar to those between two A molecules or between two B molecules. In this case, we are not restricted to the solute being nonvolatile: both the solute and solvent have significant vapor pressure. Therefore, the total vapor pressure over an ideal solution equals the sum of the partial vapor pressures, each of which is given by Raoult's law:

$$P = P_A^\circ X_A + P_B^\circ X_B$$

Solutions of benzene, C_6H_6, and toluene, $C_6H_5CH_3$, are ideal. Note the similarity in their structures (see Figure 12.19).

Suppose a solution is 0.70 mole fraction benzene and 0.30 mole fraction toluene. The vapor pressures of pure benzene and pure toluene are 75 mmHg and 22 mmHg, respectively. Hence, the total vapor pressure is

$$P = (75 \text{ mmHg} \times 0.70) + (22 \text{ mmHg} \times 0.30) = 59 \text{ mmHg}$$

You can easily see that the vapor over this solution is richer in the more volatile component (benzene). The partial vapor pressure of benzene over the solution is

$$75 \text{ mmHg} \times 0.70 = 53 \text{ mmHg}$$

Because the total vapor pressure is 59 mmHg, the mole fraction of benzene in the vapor is

$$\frac{53 \text{ mmHg}}{59 \text{ mmHg}} = 0.90$$

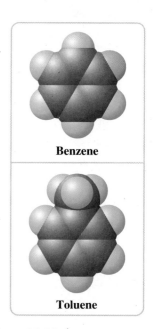

Benzene

Toluene

Figure 12.19 ▲

Molecular models of benzene and toluene *Top:* Benzene. *Bottom:* Toluene, which differs from benzene, where one hydrogen is replaced by a CH_3 group.

Figure 12.20 ▶

Fractional distillation This is a typical laboratory apparatus for the fractional distillation of a liquid solution to obtain pure liquid components. The fractionating column contains a packing material, such as glass beads, on which vapor condenses and liquid redistills. The stopper on the receiver is loosely fitted to allow ventilation to the atmosphere.

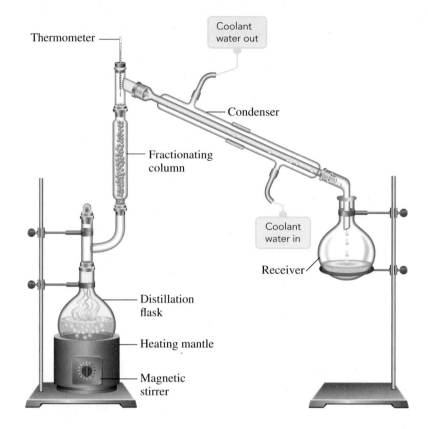

The vapor is 0.90 mole fraction benzene, whereas the liquid solution is 0.70 mole fraction benzene. This is a general principle: the vapor over a solution is richer in the more volatile component.

Use is made of this principle to separate a more volatile component from a liquid mixture. If you distill a mixture of benzene and toluene, the vapor and the resulting liquid that distills over (the distillate) will be richer in benzene, the more volatile component. If you then take this distillate and distill *it,* the vapor and the resulting liquid that comes over will be even richer in benzene. After many such distillations, you can obtain nearly pure benzene. ◀

Simple distillation was discussed in Section 1.4.

In practice, instead of actually performing a series of simple distillations to separate the volatile components of a mixture, you perform a single *fractional distillation* using a fractionating column, such as the one shown in Figure 12.20. Vapor from the liquid in the flask condenses on cooler packing material (which might be glass beads) in the lower part of the column. In turn, this liquid (which is slightly richer in benzene) vaporizes and condenses farther up the column, where it is even cooler. The liquid that results is even richer in benzene. The distillation process continues up the column, so the composition of the liquid varies continuously from a mixture of benzene and toluene at the bottom of the column to pure benzene at the top. The temperature also varies continuously up the column, being cooler at the top than at the bottom. The liquid that distills over at first is pure benzene. As the liquid in the distillation flask becomes richer in the less volatile component (toluene), the boiling point of the liquid in the flask increases. Fractional distillation is a common procedure in the laboratory and in industrial processing. For example, it is used to separate the components of petroleum into gasoline, diesel fuel, and so forth. ◀

Table 23.4 lists the various fractions obtained from petroleum, along with their boiling-point ranges.

CONCEPT CHECK 12.5

Suppose you need to boil a water-based solution at a temperature lower than 100°C. What kind of liquid could you add to the water to make this happen?

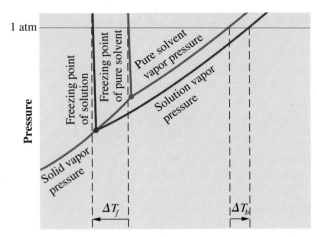

Figure 12.21 ◄

Phase diagram showing the effect of a nonvolatile solute on freezing point and boiling point Note that the freezing point (T_f) is lowered and the boiling point (T_b) is elevated.

Temperature (°C)

12.6 Boiling-Point Elevation and Freezing-Point Depression

The *normal* boiling point of a liquid is the temperature at which its vapor pressure equals 1 atm. Because the addition of a nonvolatile solute to a liquid reduces its vapor pressure, the temperature must be increased to a value greater than the normal boiling point to achieve a vapor pressure of 1 atm. Figure 12.21 shows the vapor-pressure curve of a solution. The curve is below the vapor-pressure curve of the pure liquid solvent. The **boiling-point elevation,** ΔT_b, is *a colligative property of a solution equal to the boiling point of the solution minus the boiling point of the pure solvent.*

The boiling-point elevation, ΔT_b, is found to be proportional to the molal concentration, c_m, of the solution (for dilute solutions).

$$\Delta T_b = K_b c_m$$

The constant of proportionality, K_b (called the *boiling-point-elevation constant*), depends only on the solvent. Table 12.3 lists values of K_b, as well as boiling points, for some solvents. Benzene, for example, has a boiling-point-elevation constant of 2.61°C/m. This means that a 0.100 *m* solution of a nonvolatile, undissociated solute in benzene boils at 0.261°C above the boiling point of pure benzene. Pure benzene boils at 80.2°C, so a 0.100 *m* solution boils at 80.2°C + 0.261°C = 80.5°C.

Figure 12.21 also shows the effect of a dissolved solute on the freezing point of a solution. Usually it is the pure solvent that freezes out of solution. Sea ice, for example, is almost pure water. For that reason, the vapor-pressure curve for the solid is unchanged. Therefore, the freezing point shifts to a lower temperature.

Table 12.3   Boiling-Point-Elevation Constants (K_b) and Freezing-Point-Depression Constants (K_f)

Solvent	Formula	Boiling Point (°C)	Freezing Point (°C)	K_b (°C/m)	K_f (°C/m)
Acetic acid	$HC_2H_3O_2$	118.5	16.60	3.08	3.59
Benzene	C_6H_6	80.2	5.455	2.61	5.065
Camphor	$C_{10}H_{16}O$	—	179.5	—	40
Carbon disulfide	CS_2	46.3	—	2.40	—
Cyclohexane	C_6H_{12}	80.74	6.55	2.79	20.0
Ethanol	C_2H_5OH	78.3	—	1.07	—
Water	H_2O	100.000	0.000	0.512	1.858

Data are taken from Landolt-Börnstein, 6th ed., *Zahlenwerte und Functionen aus Physik, Chemie, Astronomie, Geophysik, und Technik,* Vol. II, Part IIa (Heidelberg: © Springer-Verlag, 1960), pp. 844–849 and pp. 918–919.

The **freezing-point depression**, ΔT_f, is *a colligative property of a solution equal to the freezing point of the pure solvent minus the freezing point of the solution.* (ΔT_f is shown in Figure 12.21.)

Freezing-point depression, ΔT_f, like boiling-point elevation, is proportional to the molal concentration, c_m (for dilute solutions).

$$\Delta T_f = K_f c_m$$

Here K_f is the *freezing-point-depression constant* and depends only on the solvent. Table 12.3 gives values of K_f for some solvents. The freezing-point-depression constant for benzene is 5.07°C/*m*. Thus a 0.100 *m* solution freezes at 0.507°C below the freezing point of pure benzene. Pure benzene freezes at 5.46°C; the freezing point of the solution is 5.46°C − 0.507°C = 4.95°C.

Example 12.10   Calculating Boiling-Point Elevation and Freezing-Point Depression

Gaining Mastery Toolbox

Critical Concept 12.10
Solutions have higher boiling points and lower freezing points than the pure solvent. The magnitude of the changes in boiling and freezing points of a particular solvent depends on the concentration of the solution: the more concentrated the solution, the more the boiling and freezing points change from their normal values.

Solution Essentials:
- Boiling-point elevation ($\Delta T_b = K_b c_m$)
- Freezing-point depression ($\Delta T_f = K_f c_m$)
- Boiling-point-elevation constant (K_b)
- Freezing-point-depression constant (K_f)
- Molality $\left(m = \dfrac{\text{moles of solute}}{\text{kg of solvent}} \right)$
- Solute
- Solvent

An aqueous solution is 0.0222 *m* glucose. What are the boiling point and the freezing point of this solution?

Problem Strategy In this problem we will need to use the equations for both boiling-point elevation and freezing-point depression. To use these equations, we need solution molality and the appropriate constants. The constants are found in Table 12.3.

Solution Table 12.3 gives K_b and K_f for water as 0.512°C/*m* and 1.86°C/*m*, respectively. Therefore,

$$\Delta T_b = K_b c_m = 0.512°C/m \times 0.0222\ m = 0.0114°C$$
$$\Delta T_f = K_f c_m = 1.86°C/m \times 0.0222\ m = 0.0413°C$$

The boiling point of the solution is 100.000°C + 0.0114°C = **100.011°C**, and the freezing point is 0.000°C − 0.0413°C = **−0.041°C**. Note that ΔT_b is added and ΔT_f is subtracted.

Answer Check When reporting the final answer for either freezing-point depression or boiling-point elevation, always make sure that the final answer makes sense. For example, a solvent that contains a nonvolatile solute will always boil at a higher than normal boiling-point temperature and freeze at a lower than normal freezing-point temperature, as is the case here. If you find a solution like the one in this problem that has water as the solvent boiling at less than 100°C or freezing above 0°C, you will know that you have made an error.

Exercise 12.13 How many grams of ethylene glycol, CH_2OHCH_2OH, must be added to 37.8 g of water to give a freezing point of −0.150°C?

■ See Problems 12.69 and 12.70.

The boiling-point elevation and the freezing-point depression of solutions have a number of practical applications. We mentioned in the chapter opening that ethylene glycol is used in automobile radiators as an antifreeze because it lowers the freezing point of the coolant (Figure 12.22). The same substance also helps prevent the radiator coolant from boiling away by elevating the boiling point. Sodium chloride and calcium chloride are spread on icy roads in the winter to lower the melting point of ice and snow below the temperature of the surrounding air. Salt–ice mixtures are used as freezing mixtures in home ice cream makers.

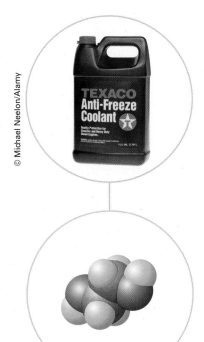

Figure 12.22 ◀

Automobile antifreeze mixtures The main ingredient of automobile antifreeze mixtures is ethylene glycol, CH_2OHCH_2OH. This substance has low vapor pressure, so it does not easily vaporize away. It is relatively cheap, being manufactured from ethylene, C_2H_4, which is obtained from petroleum. Ethylene glycol is also used to produce polyester fibers and plastics. The substance has a sweet taste (the name glycol derives from the Greek word *glukus,* meaning "sweet") but is poisonous.

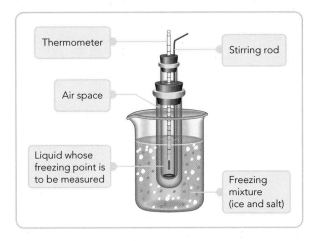

Figure 12.23 ▲

Determination of the freezing point of a liquid The liquid is cooled by means of a freezing mixture. To control the rate of temperature decrease, the liquid is separated from the freezing mixture by an air space.

Melting ice cools the ice cream mixture to the freezing point of the solution, which is well below the freezing point of pure water.

Colligative properties are also used to obtain molecular masses. Although the mass spectrometer is now often used for routine determinations of the molecular mass of pure substances, colligative properties are still employed to obtain information about the species in solution. Freezing-point depression is often used because it is simple to determine a melting point or a freezing point (Figure 12.23). From the freezing-point lowering, you can calculate the molal concentration, and from the molality, you can obtain the molecular mass. The next two examples illustrate these calculations.

Example 12.11

Critical Concept 12.11
Molality is the ratio of moles of solute per kg of solvent. For a given solution, this ratio will be constant regardless of the temperature of the solution.

Solution Essentials:

• Molality
$$\left(m = \frac{\text{moles of solute}}{\text{kg of solvent}} \right)$$

• Molecular mass

Calculating the Molecular Mass of a Solute from Molality

A solution is prepared by dissolving 0.131 g of a substance in 25.4 g of water. The molality of the solution is determined by freezing-point depression to be 0.056 m. What is the molecular mass of the substance?

Problem Strategy The key to this problem is to determine the moles of substance in the sample. Recall that if you know the molality of a substance (given in the problem), you can determine the moles of substance present in 1 kg of solvent (0.056 mol substance/1.0 kg H_2O). You then can use this ratio to determine the moles of substance present in the mass of water used to prepare the sample solution (moles of substance/25.4 × 10^{-3} kg H_2O). Now you know the moles of substance, its molar mass equals mass of substance (0.131 g) divided by moles of substance. The molecular mass (in amu) has the same numerical value as molar mass in g/mol.

Solution The molality of an aqueous solution is

$$\text{Molality} = \frac{\text{moles of substance}}{\text{kg } H_2O}$$

(continued)

(*continued*)

Hence, for the given solution, we can use the following proportional relationship.

$$\frac{0.056 \text{ mol}}{1 \text{ kg H}_2\text{O}} = \frac{\text{moles of substance}}{25.4 \times 10^{-3} \text{ kg H}_2\text{O}}$$

Rearranging this equation, you get

$$\text{Moles of substance} = \frac{0.056 \text{ mol}}{\text{kg H}_2\text{O}} \times 25.4 \times 10^{-3} \text{ kg H}_2\text{O} = 1.42 \times 10^{-3} \text{ mol}$$

(An extra digit is retained for further computation.) The molar mass of the substance equals

$$\text{Molar mass} = \frac{0.131 \text{ g}}{1.42 \times 10^{-3} \text{ mol}} = 92.3 \text{ g/mol}$$

You round the answer to two significant figures. The molecular mass is **92 amu.**

Answer Check Mistakes often give unreasonable values. For example, if you calculated the molality by dividing moles of substance by g H_2O (instead of kg H_2O), you would have obtained a molecular mass of 0.092 amu. Because the atomic mass of H (the lightest element) is 1.01 amu, a value of 0.092 amu is not possible. Very large molecular masses, like 92,000 amu, should also set off alarm bells (unless you have reason to believe the material is polymeric or biological). Typical molecular substances have molecular masses that range from tens up to hundreds of amu.

Exercise 12.14 A 0.930-g sample of ascorbic acid (vitamin C) was dissolved in 95.0 g of water. The concentration of ascorbic acid, as determined by freezing-point depression, was 0.0555 *m*. What is the molecular mass of ascorbic acid?

■ See Problems 12.73 and 12.74.

Example 12.12

Calculating the Molecular Mass from Freezing-Point Depression

Critical Concept 12.12
The change in freezing point of a solution is directly proportional to the molality of the solution. For a given solution, if the change in freezing point is known, then the molality of the solution can be calculated.

Solution Essentials:
- Freezing-point depression $(\Delta T_f = K_f c_m)$
- Freezing-point-depression constant (K_f)
- Molality $\left(m = \dfrac{\text{moles of solute}}{\text{kg of solvent}}\right)$
- Molecular mass

Camphor is a white solid that melts at 179.5°C. It has been used to determine the molecular masses of organic compounds because of its unusually large freezing-point-depression constant (40°C/*m*), which allows ordinary thermometers to be used. The organic substance is dissolved in melted camphor, and then the melting point of the solution is determined. **a** A 1.07-mg sample of a compound was dissolved in 78.1 mg of camphor. The solution melted at 176.0°C. What is the molecular mass of the compound? **b** If the empirical formula of the compound is CH, what is the molecular formula?

Problem Strategy

a The key to this problem is to determine the molality of the solution. Given the information in the problem, the equation for freezing-point depression provides a path for determining the solution molality. From the melting point of the solution, calculate the freezing-point depression, ΔT_f; then solve the equation $\Delta T_f = K_f c_m$ for the molality, *m*. Use the method of the preceding example to obtain the molecular mass, given the mass of the sample and the molality of the solution. **b** See Example 3.12 for the relation between number of empirical formula units *n*, empirical formula mass, and molecular mass.

Solution

a The freezing-point lowering is

$$\Delta T_f = (179.5 - 176.0)°\text{C} = 3.5°\text{C}$$

so the molality of the solution is

$$\frac{\Delta T_f}{K_f} = \frac{3.5°\text{C}}{40°\text{C}/m} = 0.088 \ m$$

(*continued*)

(*continued*)

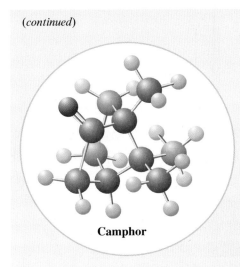

Camphor

From this you compute the moles of the compound that are dissolved in 78.1 mg of camphor (see Example 12.11).

$$0.088 \text{ mol/kg} \times 78.1 \times 10^{-6} \text{ kg} = 6.9 \times 10^{-6} \text{ mol}$$

The molar mass of the compound is

$$M_m = \frac{1.07 \times 10^{-3} \text{ g}}{6.9 \times 10^{-6} \text{ mol}} = 1.6 \times 10^2 \text{ g/mol}$$

The molecular mass is **160 amu** (two significant figures).

b The empirical formula mass of CH is 13 amu. Therefore, the number of CH units in the molecule is

$$n = \frac{\text{molecular mass}}{\text{empirical formula mass}} = \frac{1.6 \times 10^2 \text{ amu}}{13 \text{ amu}} = 12$$

The molecular formula is **$C_{12}H_{12}$.**

Answer Check In calculations like this, which focus on molecular mass determinations, the concentrations (molality) of the solutions are typically on the order of that in this problem. If you calculate a concentration of solute that is very large, check for errors.

Exercise 12.15 A 0.205-g sample of white phosphorus was dissolved in 25.0 g of carbon disulfide, CS_2. The boiling-point elevation of the carbon disulfide solution was found to be 0.159°C. What is the molecular mass of the phosphorus in solution? What is the formula of molecular phosphorus?

■ See Problems 12.75 and 12.76.

12.7 Osmosis

Certain membranes allow smaller solvent molecules to pass through them but not larger, solvated solute particles, particularly not those of large molecular mass. Such a membrane is called *semipermeable* and might be an animal bladder, a vegetable tissue, or a piece of cellophane. Figure 12.24 depicts the operation of a semipermeable membrane. **Osmosis** is *the phenomenon of solvent flow through a semipermeable membrane to equalize the solute concentrations on both sides of the membrane.* When two solutions of the same solvent are separated by a semipermeable membrane, solvent molecules migrate through the membrane in both directions. However, the solvent migration is faster from the solution of low solute concentration to the solution of high solute concentration.

Figure 12.25 shows an experiment that demonstrates osmosis. A concentrated glucose solution is placed in an inverted funnel whose mouth is sealed with a semipermeable membrane. The funnel containing the glucose solution is then placed in a beaker of pure water. As water flows from the beaker through the membrane into the funnel, the liquid level rises in the stem of the funnel.

The glucose solution continues to rise up the funnel stem until the downward pressure exerted by the solution above the membrane eventually stops the upward flow of solvent (water). In general, **osmotic pressure** is *a colligative property of a solution equal to the pressure that, when applied to the solution, just stops osmosis.*

The osmotic pressure, π, of a solution is related to the molar concentration of solute, M:

$$\pi = MRT$$

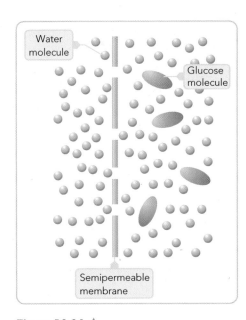

Figure 12.24 ▲

A semipermeable membrane separating water and an aqueous solution of glucose
Here the membrane is depicted as being similar to a sieve—the pores are too small to allow the glucose molecules to pass. The actual mechanism in particular cases may be more complicated. For example, water may be absorbed on one side of the membrane and diffuse through it to the other side.

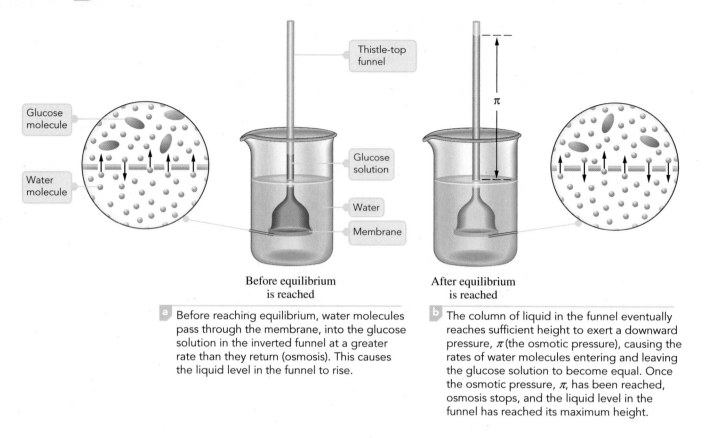

Thistle-top funnel

π

Glucose molecule

Water molecule

Glucose solution

Water

Membrane

Before equilibrium is reached

After equilibrium is reached

a Before reaching equilibrium, water molecules pass through the membrane, into the glucose solution in the inverted funnel at a greater rate than they return (osmosis). This causes the liquid level in the funnel to rise.

b The column of liquid in the funnel eventually reaches sufficient height to exert a downward pressure, π (the osmotic pressure), causing the rates of water molecules entering and leaving the glucose solution to become equal. Once the osmotic pressure, π, has been reached, osmosis stops, and the liquid level in the funnel has reached its maximum height.

Figure 12.25 ▲

An experiment in osmosis

Here R is the gas constant and T is the absolute temperature. There is a formal similarity between this equation for osmotic pressure and the equation for an ideal gas:

$$PV = nRT$$

The molar concentration M of a gas equals n/V; therefore, $P = (n/V)RT = MRT$.

To see the order of magnitude of the osmotic pressure of a solution, consider the aqueous solution described in Example 12.10. There you found that a solution that is about 0.02 m has a freezing-point depression of about 0.04°C. The molarity and molality of a dilute aqueous solution are approximately equal (see Example 12.7). Therefore, this solution will also be about 0.02 M. Hence, the osmotic pressure at 25°C (298 K) is

$$\pi = MRT$$

$$= 0.02 \; \text{mol/L} \times 0.082 \; \text{L·atm/(K·mol)} \times 298 \; \text{K}$$

$$= 0.5 \; \text{atm}$$

If this pressure were to be exerted by a column of a water solution, as in Figure 12.25, the column would have to be more than 4.5 m high.

In most osmotic-pressure experiments, much more dilute solutions are employed. Often osmosis is used to determine the molecular masses of macromolecular or polymeric substances. Polymers are very large molecules generally made up from a simple, repeating unit. A typical molecule of polyethylene, for instance, might have the formula $CH_3(CH_2CH_2)_{2000}CH_3$. Although polymer solutions can be fairly concentrated in terms of grams per liter, on a mole-per-liter basis they may be quite dilute. The freezing-point depression is usually too small to measure, although the osmotic pressure may be appreciable.

Example 12.13   **Calculating Osmotic Pressure**

Gaining Mastery Toolbox

Critical Concept 12.13
At a given temperature, the osmotic pressure of a solution is directly proportional to the molarity of the solution. A concentrated solution can have a great osmotic pressure.

Solution Essentials:
• Osmotic pressure ($\pi = MRT$)
• Molarity
• Ideal gas constant
• Kelvin temperature

The formula for low-molecular-mass starch is $(C_6H_{10}O_5)_n$, where n averages 2.00×10^2. When 0.798 g of starch is dissolved in 100.0 mL of water solution, what is the osmotic pressure, in mmHg, at 25°C?

Problem Strategy Calculations of osmotic pressure involve the formula $\pi = MRT$. In order to employ this equation here, we need to determine the molarity (M) of the solution. If we obtain the molecular mass of the starch, we can determine the number of moles of starch in the 0.798-g sample used to prepare the solution. Then, knowing the moles of starch and the volume of the solution (100.0 mL), we can calculate the molarity. After using the formula to determine osmotic pressure, convert the pressure from units of atm to mmHg.

Solution The molecular mass of $(C_6H_{10}O_5)_{200}$ is 32,400 amu. The number of moles in 0.798 g of starch is

$$0.798 \text{ g starch} \times \frac{1 \text{ mol starch}}{32,400 \text{ g starch}} = 2.46 \times 10^{-5} \text{ mol starch}$$

The molarity of the solution is

$$\frac{2.46 \times 10^{-5} \text{ mol}}{0.1000 \text{ L solution}} = 2.46 \times 10^{-4} \text{ mol/L solution}$$

and the osmotic pressure at 25°C is

$$\pi = MRT$$
$$= 2.46 \times 10^{-4} \text{ mol/L} \times 0.0821 \text{ L·atm/(K·mol)} \times 298 \text{ K}$$
$$= 6.02 \times 10^{-3} \text{ atm} = 6.02 \times 10^{-3} \text{ atm} \times \frac{760 \text{ mmHg}}{1 \text{ atm}}$$
$$= \textbf{4.58 mmHg}$$

For comparison, you can calculate the freezing-point depression. Assume that the molality is equal to the molarity (this is in effect the case for very dilute aqueous solutions). Then,

$$\Delta T_f = 1.86°C/m \times 2.46 \times 10^{-4} m = 4.58 \times 10^{-4} °C$$

which is barely detectable with generally available equipment.

Answer Check When performing calculations that involve the formula for osmotic pressure, always make certain that your temperature is in kelvins and that you use the correct value of the ideal gas constant, R.

Exercise 12.16   Calculate the osmotic pressure at 20°C of an aqueous solution containing 5.0 g of sucrose, $C_{12}H_{22}O_{11}$, in 100.0 mL of solution.

■ See Problems 12.77 and 12.78.

Osmosis is important in many biological processes. ▶ A cell might be thought of (simplistically) as an aqueous solution enclosed by a semipermeable membrane. The solution surrounding the cell must have an osmotic pressure equal to that within the cell. Otherwise, water would either leave the cell, dehydrating it, or enter the cell and possibly burst the membrane. For intravenous feeding (adding a nutrient solution to the venous blood supply of a patient), it is necessary that the nutrient solution have exactly the same osmotic pressure as blood plasma. If it does not, the blood cells may collapse or burst as a result of osmosis (Figure 12.26).

The body has portions of organs that are at different osmotic pressures, and an active pumping mechanism is required to offset osmosis. For example, cells of the

Osmotic pressure appears to be important for the rising of sap in a tree. During the day, water evaporates from the leaves of the tree, so the aqueous solution in the leaves becomes more concentrated and the osmotic pressure increases. Sap flows upward to dilute the water solution in the leaves.

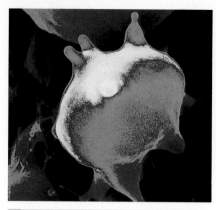

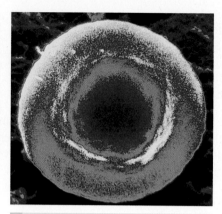

© David Phillips/Photo Researchers, Inc.

a The osmotic pressure of the solution is greater than that of the cell; the cell has collapsed.

b The osmotic pressure of the solution is equal to that of the cell; the cell has its normal round shape with depressed center.

c The osmotic pressure of the solution is less than that of the cell; the cell has a bloated shape.

Figure 12.26 ▲

The importance of osmotic pressure to cells Shown here are color-enhanced electron micrographs of red blood cells in solutions of various osmotic pressures.

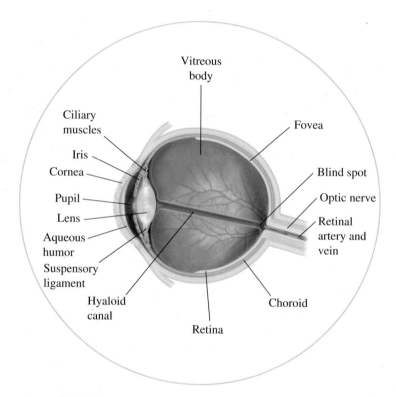

Figure 12.27 ▲

Parts of the eye Certain cells in the cornea act as pumps to prevent osmosis of water from the aqueous humor.

transparent tissue of the exterior eye, the cornea, have a more concentrated optical fluid than does the aqueous humor, a solution just behind the cornea (Figure 12.27). To prevent the cornea from taking up additional water from the aqueous humor, cells that pump water are located in the corneal tissue adjacent to the aqueous humor. Corneas that are to be stored and used for transplants must be removed from the globe of the eye soon after the donor's death. This prevents the clouding that occurs when the pumping mechanism fails (as its energy supply is depleted at death).

The process of *reverse osmosis* has been applied to the problem of purifying water. In particular, the method has been used to *desalinate* ocean water (that is, remove salts from seawater to make drinkable or industrially usable water). In normal osmosis, the solvent flows from a dilute solution through a membrane to a more concentrated solution. By applying a pressure equal to the osmotic pressure to the more concentrated solution, it is possible to stop the process of osmosis. By application of an even greater pressure, the osmotic process can be reversed. Then, solvent flows from the concentrated solution (which could be ocean water) through a membrane to the more dilute solution (which could be more or less pure water).

CONCEPT CHECK 12.6

Explain why pickles are stored in a brine (salt) solution. What would the pickles look like if they were stored in water?

12.8 Colligative Properties of Ionic Solutions

To explain the colligative properties of ionic solutions, you must realize that it is the total concentration of ions, rather than the concentration of ionic substance, that is important. For example, the freezing-point depression of 0.100 *m* sodium chloride solution is nearly twice that of 0.100 m glucose solution. You can explain this by saying that sodium chloride dissolves in water to form the ions Na^+ and Cl^-. Each formula unit of NaCl gives two particles. You can write the freezing-point lowering more generally as

$$\Delta T_f = iK_f c_m$$

where i is the number of ions resulting from each formula unit and c_m is the molality computed on the basis of the formula of the ionic compound. With ionic solutions, the equations for the other colligative properties must be modified in a similar fashion to include the factor i.

Vapor-pressure lowering: $\Delta P = iP_A^\circ X_B$

Boiling-point elevation: $\Delta T_b = iK_b c_m$

Osmotic pressure: $\pi = iMRT$

The actual values for the colligative properties of ionic solutions agree with the previous equations only when the solutions are quite dilute. For example, the i values calculated from the freezing-point depression are usually smaller than the number of ions obtained from the formula unit. For example, a 0.029 m aqueous solution of potassium sulfate, K_2SO_4, has a freezing point of $-0.14°C$. Hence,

$$i = \frac{\Delta T_f}{K_f c_m} = \frac{0.14°C}{1.86°C/m \times 0.029\ m} = 2.6$$

You might have expected a value of 3, on the basis of the fact that K_2SO_4 ionizes to give three ions. ▶ At first, this was taken as evidence that salts were not completely ionized in solution. In 1923, however, Peter Debye and Erich Hückel were able to show that the colligative properties of salt solutions could be explained by assuming that the salt is completely ionized in solution but that the *activities,* or effective concentrations, of the ions are less than their actual concentrations as a result of the electrical interactions of the ions in solution. The *Debye–Hückel theory* allows us to calculate these activities. When this is done, excellent agreement is obtained for dilute solutions.

The value of i equal to $\Delta T_f/K_f c_m$ is often called the van 't Hoff factor. *Thus, the van 't Hoff factor for 0.029 m K_2SO_4 is 2.6.*

Example 12.14   Determining Colligative Properties of Ionic Solutions

Gaining Mastery Toolbox

Critical Concept 12.14
Colligative properties depend on the total number of particles in solution. For dilute ionic solutions, the chemical formula of the soluble ionic compound is used to determine the number of ions (particles) in solution resulting from each formula unit (i). For example, a solution of $MgCl_2$ would have $i = 3$ because one mol of $MgCl_2$ produces three moles of ions in solution.

Solution Essentials:
- Number of ions from each formula unit (i)
- Freezing-point depression of ionic solution ($\Delta T_f = iK_f c_m$)
- Freezing-point-depression constant (K_f)
- Molality $\left(m = \dfrac{\text{moles of solute}}{\text{kg of solvent}} \right)$

Estimate the freezing point of a 0.010 m aqueous solution of aluminum sulfate, $Al_2(SO_4)_3$. Assume the value of i based on the formula of the compound.

Problem Strategy Because aluminum sulfate is a soluble ionic compound, we need to use a modified form of the freezing-point depression formula, $\Delta T_f = iK_f c_m$. To obtain i, we need to determine the number of ions produced when $Al_2(SO_4)_3$ dissolves in water.

Solution When aluminum sulfate, $Al_2(SO_4)_3$, dissolves in water, it dissociates into five ions.

$$Al_2(SO_4)_3(s) \xrightarrow{\text{H}_2\text{O}} 2Al^{3+}(aq) + 3SO_4^{2-}(aq)$$

Therefore, you assume $i = 5$. The freezing-point depression is

$$\Delta T_f = iK_f c_m = 5 \times 1.86°C/m \times 0.010\ m = 0.093°C$$

The estimated freezing point of the solution is $0.000°C - 0.093°C = \mathbf{-0.093°C}.$

Answer Check Whenever you have to solve a problem involving the colligative properties of aqueous solutions, always make your first step a determination of whether the solute is an ionic or a molecular compound. This information will point you to the correct form of the equation.

Exercise 12.17   Estimate the boiling point of a 0.050 m aqueous $MgCl_2$ solution. Assume a value of i based on the formula.

■ See Problems 12.79 and 12.80.

CONCEPT CHECK 12.7

Each of the following substances is dissolved in a separate 10.0-L container of water: 1.5 mol NaCl, 1.3 mol Na_2SO_4, 2.0 mol $MgCl_2$, and 2.0 mol KBr. Without doing extensive calculations, rank the boiling points of the solutions from highest to lowest.

Colloid Formation

As we noted at the beginning of the chapter, colloids appear homogeneous like solutions, but they consist of comparatively large particles of one substance dispersed throughout another substance. We will look at colloids in the following section.

12.9 Colloids

A **colloid** is *a dispersion of particles of one substance (the dispersed phase) throughout another substance or solution (the continuous phase).* Fog is an example of a colloid: it consists of very small water droplets (dispersed phase) in air (continuous phase). A colloid differs from a true solution in that the dispersed particles are larger than normal molecules, though they are too small to be seen with a microscope. The particles range from about 1×10^3 pm to about 2×10^5 pm in size.

Tyndall Effect

Although a colloid appears to be homogeneous because the dispersed particles are quite small, it can be distinguished from a true solution by its ability to scatter light. *The scattering of light by colloidal-size particles* is known as the **Tyndall effect.** ◀ For example, the atmosphere appears to be a clear gas, but a ray of sunshine against a dark background shows up many fine dust particles by light scattering. Similarly, when a beam of light is directed through clear gelatin (a colloid, not a true solution), the beam becomes visible by the scattering of light from colloidal gelatin particles. The beam appears as a ray passing through the solution (Figure 12.28). When the same experiment is performed with a true solution, such as an aqueous solution of sodium chloride, the beam of light is not visible.

Although all gases and liquids scatter light, the scattering from a pure substance or true solution is quite small and usually not detectable. However, because of the considerable depth of the atmosphere, the scattering of light by air molecules can be seen. The blue color of the sky is due to the fact that blue light is scattered more easily than red light.

© Cengage Learning

Figure 12.28 ▲

A demonstration of the Tyndall effect by a colloid
A light beam is visible perpendicular to its path only if light is scattered toward the viewer. The vessel on the left contains a colloid, which scatters light. The vessel on the right contains a true solution, which scatters negligible light.

Types of Colloids

Colloids are characterized according to the state (solid, liquid, or gas) of the dispersed phase and of the continuous phase. Table 12.4 lists various types of colloids and some examples of each. Fog and smoke are **aerosols,** which are *liquid droplets or solid particles dispersed throughout a gas.* An **emulsion** consists of *liquid droplets dispersed throughout another liquid* (as particles

Table 12.4 **Types of Colloids**

Continuous Phase	Dispersed Phase	Name	Example
Gas	Liquid	Aerosol	Fog, mist
Gas	Solid	Aerosol	Smoke
Liquid	Gas	Foam	Whipped cream
Liquid	Liquid	Emulsion	Mayonnaise (oil dispersed in water)
Liquid	Solid	Sol	AgCl(s) dispersed in H_2O
Solid	Gas	Foam	Pumice, plastic foams
Solid	Liquid	Gel	Jelly, opal (mineral with liquid inclusions)
Solid	Solid	Solid sol	Ruby glass (glass with dispersed metal)

of butterfat are dispersed through homogenized milk). A **sol** consists of *solid particles dispersed in a liquid.*

Hydrophilic and Hydrophobic Colloids

Colloids in which the continuous phase is water are divided into two major classes: hydrophilic colloids and hydrophobic colloids. A **hydrophilic colloid** is *a colloid in which there is a strong attraction between the dispersed phase and the continuous phase (water).* Many such colloids consist of macromolecules (very large molecules) dispersed in water. Except for the large size of the dispersed molecules, these colloids are like normal solutions. Protein solutions, such as gelatin in water, are hydrophilic colloids. Gelatin molecules are attracted to water molecules by London forces and hydrogen bonding.

A **hydrophobic colloid** is *a colloid in which there is a lack of attraction between the dispersed phase and the continuous phase (water).* Hydrophobic colloids are basically unstable. Given sufficient time, the dispersed phase aggregates into larger particles. In this behavior, they are quite unlike true solutions and hydrophilic colloids. The time taken to separate may be extremely long, however. A colloid of gold particles in water prepared by Michael Faraday in 1857 is still preserved in the British Museum in London. This colloid is hydrophobic as well as a sol (solid particles dispersed in water).

Hydrophobic sols are often formed when a solid crystallizes rapidly from a chemical reaction or a supersaturated solution. When crystallization occurs rapidly, many centers of crystallization (called *nuclei*) form at the same time. Ions are attracted to these nuclei, and very small crystals form. These small crystals are prevented from settling out by the random thermal motion of the solvent molecules, which continue to buffet them.

You might expect these very small crystals to aggregate into larger crystals because the aggregation would bring ions of opposite charge into contact. However, sol formation appears to happen when, for some reason, each of the small crystals gets a preponderance of one kind of charge on its surface. For example, iron(III) hydroxide forms a colloid because an excess of iron(III) ion (Fe^{3+}) is present on the surface, giving each crystal an excess of positive charge. These positively charged crystals repel one another, so aggregation to larger particles is prevented.

Coagulation

An iron(III) hydroxide sol can be made to aggregate by the addition of an ionic solution, particularly if the solution contains anions with multiple charges (such as phosphate ions, PO_4^{3-}). **Coagulation** is *the process by which the dispersed phase of a colloid is made to aggregate and thereby separate from the continuous phase.* You can picture what happens in the following way: A positively charged colloidal particle of iron(III) hydroxide gathers a layer of anions around it. The thickness of this layer is determined by the charge on the anions. The greater the magnitude of the negative charge, the more compact is the layer of charge. Phosphate ions, for example, gather more closely to the positively charged colloidal particles than do chloride ions (see Figure 12.29). If the ion layer is gathered close to the colloidal particle, the overall charge is effectively neutralized. In that case, two colloidal particles can approach close enough to aggregate.

The curdling of milk when it sours is another example of coagulation. Milk is a colloidal suspension in which the particles are prevented from aggregating because they have electric charges of the same sign. The ions responsible for the coagulation (curdling) are formed when lactose (milk sugar) ferments to lactic acid. A third example is the coagulation of a colloidal suspension of soil in river water when the water meets the concentrated ionic solution of an ocean. The Mississippi Delta was formed in this way.

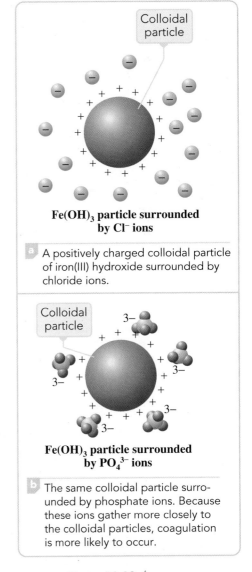

$Fe(OH)_3$ particle surrounded by Cl^- ions

a A positively charged colloidal particle of iron(III) hydroxide surrounded by chloride ions.

$Fe(OH)_3$ particle surrounded by PO_4^{3-} ions

b The same colloidal particle surrounded by phosphate ions. Because these ions gather more closely to the colloidal particles, coagulation is more likely to occur.

Figure 12.29 ▲

Layers of ions surrounding charged colloidal particles

Exercise 12.18   Colloidal sulfur particles are negatively charged with thiosulfate ions, $S_2O_3^{2-}$, and other ions on the surface of the sulfur. Indicate which of the following would be most effective in coagulating colloidal sulfur: NaCl, MgCl$_2$, or AlCl$_3$.

■ See Problems 12.85 and 12.86.

Association Colloids

When molecules or ions that have both a hydrophobic and a hydrophilic end are dispersed in water, they associate, or aggregate, to form colloidal-sized particles, or micelles. A **micelle** is *a colloidal-sized particle formed in water by the association of molecules or ions that each have a hydrophobic end and a hydrophilic end.* The hydrophobic ends point inward toward one another, and the hydrophilic ends are on the outside of the micelle facing the water. *A colloid in which the dispersed phase consists of micelles* is called an **association colloid.**

Ordinary soap in water provides an example of an association colloid. Soap consists of compounds such as sodium stearate, C$_{17}$H$_{35}$COONa. The stearate ion has a long hydrocarbon end that is hydrophobic (because it is nonpolar) and a carboxyl group (COO$^-$) at the other end that is hydrophilic (because it is ionic).

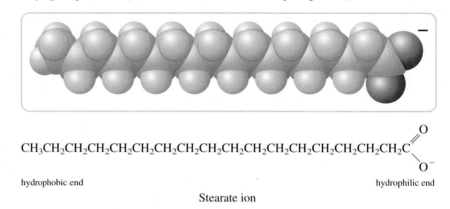

$$CH_3CH_2CH_2CH_2CH_2CH_2CH_2CH_2CH_2CH_2CH_2CH_2CH_2CH_2CH_2CH_2CH_2C \overset{O}{\underset{O^-}{\big<}}$$

hydrophobic end   hydrophilic end

Stearate ion

In water solution, the stearate ions associate into micelles in which the hydrocarbon ends point inward toward one another and away from the water, and ionic carboxyl groups are on the outside of the micelle facing the water (Figure 12.30).

Figure 12.30 ▶

A stearate micelle in a water solution Stearate ions associate in groups (micelles), with their hydrocarbon ends pointing inward. The ionic ends, on the outside of the micelle, point into the water solution.

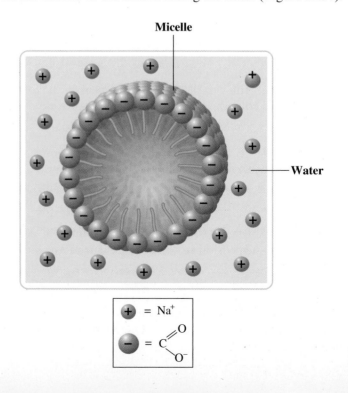

Micelle

Water

$\oplus$ = Na$^+$

$\ominus$ = C $\overset{O}{\underset{O^-}{\big<}}$

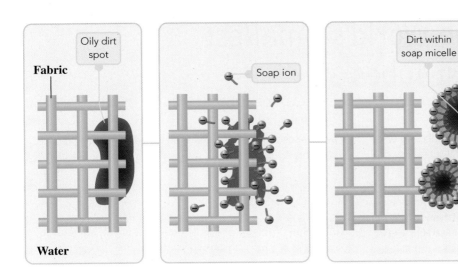

Figure 12.31 ◀

The cleansing action of soap The hydrocarbon ends of soap ions gather around an oil spot, forming a micelle that can be washed away in the water.

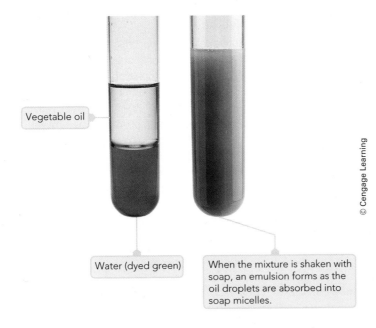

Figure 12.32 ◀

Formation of an association colloid of vegetable oil with soap

© Cengage Learning

The cleansing action of soap occurs because oil and grease can be absorbed into the hydrophobic centers of soap micelles and washed away (Figures 12.31 and 12.32). Synthetic detergents also form association colloids. Sodium lauryl sulfate is a synthetic detergent present in toothpastes and shampoos.

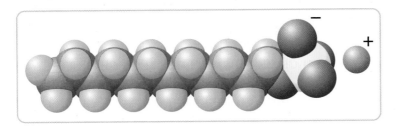

$$CH_3CH_2CH_2CH_2CH_2CH_2CH_2CH_2CH_2CH_2CH_2CH_2OSO_3^- \, Na^+$$

Sodium lauryl sulfate

It has a hydrophilic sulfate group ($-OSO_3^-$) and a hydrophobic dodecyl group ($C_{12}H_{25}-$, the hydrocarbon end).

The detergent molecules we have discussed so far are classified in the trade as "anionics," because they have a negative charge at the hydrophilic end. Other

The World's Smallest Test Tubes

Phospholipids are the main constituents of the membranes of living cells. These membranes enclose the chemical substances of a cell and allow these substances to carry on the reactions needed for life processes unimpeded by random events external to the membrane. Recently, chemists have prepared cell-like structures—*lipid vesicles*—from phospholipids and have used them to carry out chemical reactions by combining the contents of two lipid vesicles, with diameters from about 50 nanometers (billionths of a meter) to several micrometers (millionths of a meter). Richard Zare, at Stanford University, who was one of those who described these experiments, has called these lipid vesicles "the world's smallest test tubes."

Like the soap ions, phospholipids have the intriguing property of self-assembly into groups. Soap ions in water naturally group themselves into micelles. Phospholipids, however, have a chemical structure that, while similar to that of soap ions, precludes their assembly into micelles. Rather, they form layer structures. And under the right circumstances, a layer may fold back on itself like the skin of a basketball to enclose a space, forming a vesicle.

Lecithins are typical phospholipids. Lipids are biological substances like fats and oils that are soluble in organic solvents; phospholipids contain phosphate groups. (If you are a label reader, you may have seen soy lecithin listed as an ingredient in food products such as bakery goods.) Here is the general structure of a lecithin:

$$
\begin{array}{l}
\overset{\displaystyle O}{\overset{\displaystyle \|}{R-C}}-O-CH_2 \\
R'-\overset{\displaystyle }{\underset{\displaystyle \|}{C}}-O-\overset{\displaystyle }{CH}\qquad\ \ \overset{\displaystyle O}{} \\
O\quad H_2C-O-\overset{\displaystyle \|}{\underset{\displaystyle |}{P}}-O-CH_2-CH_2-\overset{+}{N}(CH_3)_3 \\
O^-
\end{array}
$$

R and R' denote long hydrocarbon groups, with perhaps 14 to 18 carbon atoms.

Note that, like a soap ion, a phospholipid has a hydrophobic end (the two hydrocarbon groups) and a hydrophilic end (an ionic portion containing a phosphate group, $-PO_4^--$). Like soap ions, phospholipid molecules in water tend to associate so that their hydrophilic (or ionic) "heads" dip into the water, with their hydrophobic (or hydrocarbon) "tails" pointing away. However, the hydrocarbon tails of a phospholipid molecule are too bulky to associate into micelles and instead form a *bilayer*, a layer two molecules thick. Figure 12.34 shows a model of a cell membrane that displays this bilayer structure, with two phospholipid layers, their hydrocarbon

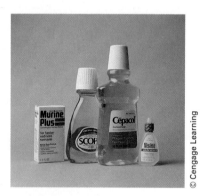

Figure 12.33 ▲

Commercial preparations containing cationic detergents Many cationic detergents exhibit germicidal properties. For that reason, they are used in certain hospital antiseptics, mouthwashes, and eye-wetting solutions.

detergent molecules are "cationics," because they have a positive charge at the hydrophilic end. An example is

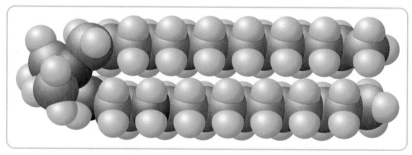

$$
\begin{array}{l}
H_3C\quad CH_2CH_2CH_2CH_2CH_2CH_2CH_2CH_2CH_2CH_2CH_2CH_2CH_2CH_2CH_3 \\
\overset{+}{N} \\
H_3C\quad CH_2CH_2CH_2CH_2CH_2CH_2CH_2CH_2CH_2CH_2CH_2CH_2CH_2CH_2CH_3
\end{array}
$$

hydrophilic end   hydrophobic end

Many cationic detergents also have germicidal properties and are used in hospital disinfectants and in mouthwashes (Figure 12.33).

CONCEPT CHECK 12.8

If electrodes that are connected to a direct current (DC) source are dipped into a beaker of colloidal iron(III) hydroxide, a precipitate collects at the negative electrode. Explain why this happens.

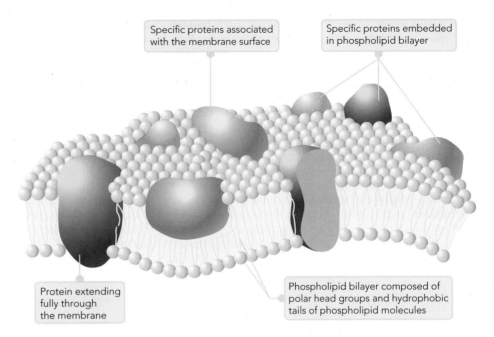

Figure 12.34 ▶

A model of a cell membrane
The membrane consists of a phospholipid bilayer with protein molecules inserted in it.

Specific proteins associated with the membrane surface

Specific proteins embedded in phospholipid bilayer

Protein extending fully through the membrane

Phospholipid bilayer composed of polar head groups and hydrophobic tails of phospholipid molecules

tails pointing inward toward each other. The hydrophilic heads are rendered in blue; the hydrophobic tails are in yellow. Note that the interior of a membrane (shown in pale green) is "oily" from the hydrocarbon groups and therefore repels molecules or ions dissolved in the water solutions outside and inside a cell. (A cell, though, must interact with its environment. Protein molecules embedded in the cell membrane form channels or active pumps to transport specific molecules and ions across the cell membrane.)

In one of their experiments, Zare and his colleagues brought two phospholipid vesicles, each about a micrometer in diameter and containing different reactant substances, together under a

special microscope. When the vesicles just touched, the researchers delivered an electrical pulse to them that opened a small pore in each vesicle. The two vesicles coalesced into one, and their contents reacted. The scientists detected the product molecules from the orange fluorescence they gave off when radiated with ultraviolet light. The techniques these researchers are developing should allow scientists to mimic the conditions in biological cells. And lipid vesicles might one day be designed to deliver drugs to specific cells, say cancer cells.

■ See Problems 12.113 and 12.114.

A Checklist for Review

⚙OWL and 🔲Chemistry Sign in at **www.cengage.com/owl** to:
• View tutorials and simulations, develop problem-solving skills, and complete online homework assigned by your professor.
• Download Go Chemistry mini lecture modules for quick review and exam prep from OWL or purchase them at **www.cengagebrain.com**.

Summary of Facts and Concepts

Solutions are homogeneous mixtures and can be gases, liquids, or solids. Two gases, for example, will mix in all proportions to give a gaseous solution, because gases are *miscible* in one another. Often one substance will dissolve in another only to a limited extent. The maximum amount that dissolves at equilibrium is the *solubility* of that substance. Solubility is explained in terms of the natural tendency toward disorder (say, by the mixing of two substances) and by the tendency to result in the strongest forces of attraction between species. The dissolving of one molecular substance in another is limited when intermolecular forces strongly favor the unmixed substances. When two substances have similar types of intermolecular forces, they tend to be miscible ("like dissolves like"). Ionic substances differ greatly in their solubilities in water, because solubility depends on the relative effects of *lattice energy* and *hydration energy.*

Solubilities usually vary with temperature. At higher temperatures, most gases will become less soluble in water, whereas most ionic solids will become more soluble. Pressure has a significant effect on the solubility only of a gas. A gas is more soluble in a liquid if the partial pressure of the gas is increased, which is in agreement with *Le Châtelier's principle.* According to *Henry's law,* the solubility of a gas is directly proportional to the partial pressure of the gas.

Colligative properties of solutions depend only on the concentration of solute particles. The concentration may be defined by the *molarity, mass percentage of solute, molality,* or *mole fraction.* One example of a colligative property is the *vapor-pressure lowering* of a volatile solvent by the addition of a nonvolatile solute. According to *Raoult's law,* the vapor pressure of this solution depends on the mole fraction of solvent. Because adding a nonvolatile solute lowers the vapor pressure, the boiling point must be raised to bring the vapor pressure back to 1 atm *(boiling-point elevation)*. Such a solution also exhibits a *freezing-point depression.* Boiling-point elevation and freezing-point depression are colligative properties. *Osmosis* is another colligative property. In osmosis, there is a flow of solvent through a semipermeable membrane in order to equalize the concentrations of solutions on the two sides of the membrane. Colligative properties may be used to measure solute concentration. In this way, one can determine the molecular mass of the solute. The colligative properties of an ionic solution depend on the concentration of ions, which you can obtain from the concentration of ionic compound and its formula. Interactions of ions in solution do affect the properties significantly, however.

A *colloid* is a dispersion of particles of one substance (about 1×10^3 pm to 2×10^5 pm in size) throughout another. Colloids can be detected by the *Tyndall effect* (the scattering of light by colloidal-size particles). Colloids are characterized by the state of the dispersed phase and the state of the continuous phase. An aerosol, for example, consists of liquid droplets or solid particles dispersed in a gas. Colloids in water are hydrophilic or hydrophobic. A *hydrophilic colloid* consists of a dispersed phase that is attracted to the water. Many of these colloids have macromolecules dissolved in water. In *hydrophobic colloids,* there is little attraction between the dispersed phase and the water. The colloidal particles may be electrically charged and repel one another, preventing aggregation to larger particles. An ionic solution can neutralize this charge so that the colloid will *coagulate*—that is, aggregate. An *association colloid* consists of molecules with a hydrophobic end and a hydrophilic end dispersed in water. These molecules associate into colloidal-size groups, or *micelles.*

Learning Objectives	Important Terms
12.1 Types of Solutions	
■ Define *solute* and *solvent.* ■ Define *miscible fluid.* ■ Provide examples of *gaseous solutions, liquid solutions,* and *solid solutions.*	**solute** **solvent** **miscible fluids**
12.2 Solubility and the Solution Process	
■ List the conditions that must be present to have a *saturated solution,* to have an *unsaturated solution,* and to have a *supersaturated solution.* ■ Describe the factors that make one substance soluble in another. ■ Determine when a *molecular solution* will form when substances are mixed. ■ Learn what conditions must be met in order to create an *ionic solution.*	**saturated solution** **solubility** **unsaturated solution** **supersaturated solution** **hydration (of ions)**
12.3 Effects of Temperature and Pressure on Solubility	
■ State the general trends of the solubility of gases and solids with temperature. ■ Explain how the solubility of a gas changes with temperature. ■ Apply Henry's law. Example 12.1	**Le Châtelier's principle** **Henry's law**
12.4 Ways of Expressing Concentration	
■ Define *colligative property.* ■ Define *molarity.* ■ Define *mass percentage of solute.* ■ Calculate with mass percentage of solute. Example 12.2 ■ Define *molality.* ■ Calculate the molality of solute. Example 12.3 ■ Define *mole fraction.* ■ Calculate the mole fraction of components. Example 12.4 ■ Convert molality to mole fractions. Example 12.5 ■ Convert mole fractions to molality. Example 12.6 ■ Convert molality to molarity. Example 12.7 ■ Convert molarity to molality. Example 12.8	**colligative properties** **mass percentage of solute** **molality** **mole fraction**

12.5 Vapor Pressure of a Solution

- Explain *vapor-pressure lowering* of a solvent.
- State *Raoult's law.*
- Calculate vapor-pressure lowering. Example 12.9
- Describe an *ideal solution.*

vapor-pressure lowering
Raoult's law

12.6 Boiling-Point Elevation and Freezing-Point Depression

- Define *boiling-point elevation* and *freezing-point depression.*
- Calculate boiling-point elevation and freezing-point depression. Example 12.10
- Calculate the molecular mass of a solute from molality. Example 12.11
- Calculate the molecular mass from freezing-point depression. Example 12.12

boiling-point elevation
freezing-point depression

12.7 Osmosis

- Describe a system where *osmosis* will take place.
- Calculate osmotic pressure. Example 12.13

osmosis
osmotic pressure

12.8 Colligative Properties of Ionic Solutions

- Determine the colligative properties of ionic solutions. Example 12.14

12.9 Colloids

- Define *colloid.*
- Explain the *Tyndall effect.*
- Give examples of *hydrophilic colloids* and *hydrophobic colloids.*
- Describe *coagulation.*
- Explain how *micelles* can form an *association colloid.*

colloid
Tyndall effect
aerosols
emulsion
sol
hydrophilic colloid
hydrophobic colloid
coagulation
micelle
association colloid

Key Equations

$S = k_H P$

$$\text{Mass percentage of solute} = \frac{\text{mass of solute}}{\text{mass of solution}} \times 100\%$$

$$\text{Molality} = \frac{\text{moles of solute}}{\text{kilograms of solvent}}$$

$$X_A = \frac{\text{moles of substance } A}{\text{total moles of solution}}$$

$P_A = P_A^\circ X_A$

$\Delta P = i P_A^\circ X_B$

$\Delta T_b = i K_b c_m$

$\Delta T_f = i K_f c_m$

$\pi = iMRT$

Questions and Problems

OWL Interactive versions of these problems may be assigned in OWL.

Self-Assessment and Review Questions

Key: These questions test your understanding of the ideas you worked with in the chapter. These problems vary in difficulty and often can be used for the basis of discussion.

12.1 Give one example of each: a gaseous solution, a liquid solution, a solid solution.

12.2 What are the two factors needed to explain the differences in solubilities of substances?

12.3 Explain in terms of intermolecular attractions why octane is immiscible in water.

12.4 Explain why ionic substances show a wide range of solubilities in water.

12.5 Using the concept of hydration, describe the process of dissolving a sodium chloride crystal in water.

12.6 What is the usual solubility behavior of an ionic compound in water when the temperature is raised? Give an example of an exception to this behavior.

12.7 Give one example of each: a salt whose heat of solution is exothermic and a salt whose heat of solution is endothermic.

12.8 What do you expect to happen to a concentration of dissolved gas in a solution as the solution is heated?

12.9 Pressure has an effect on the solubility of oxygen in water but a negligible effect on the solubility of sugar in water. Why?

12.10 Explain why a carbonated beverage must be stored in a closed container.

12.11 Four ways were discussed to express the concentration of a solute in solution. Identify them and define each concentration unit.

12.12 When two beakers containing different concentrations of a solute in water are placed in a closed cabinet for a time, one beaker gains solvent and the other loses it, so that the concentrations of solute in the two beakers become equal. Explain what is happening.

12.13 Explain the process of fractional distillation to separate a solution of two liquids into pure components.

12.14 Explain why the boiling point of a solution containing a nonvolatile solute is higher than the boiling point of a pure solvent.

12.15 List two applications of freezing-point depression.

12.16 One can often see "sunbeams" passing through the less dense portions of clouds. What is the explanation for this?

12.17 Explain the process of reverse osmosis to produce drinkable water from ocean water.

12.18 Give an example of an aerosol, a foam, an emulsion, a sol, and a gel.

12.19 Explain how soap removes oil from a fabric.

12.20 Explain on the basis that "like dissolves like" why glycerol, $CH_2OHCHOHCH_2OH$, is miscible in water but benzene, C_6H_6, has very limited solubility in water.

12.21 Calculate the number of moles of barium chloride in 427 g of a 3.17% by mass barium chloride solution?
- a 6.50×10^{-2} mol
- b 7.83×10^{-2} mol
- c 4.31×10^{-2} mol
- d 7.81×10^{-2} mol
- e 4.27×10^{-2} mol

12.22 The mole fraction of copper(II) nitrate in an aqueous copper(II) nitrate solution is 0.135. What is the molality of the copper(II) nitrate solution?
- a $7.49\ m$
- b $8.01\ m$
- c $8.66\ m$
- d $9.34\ m$
- e $8.09\ m$

12.23 A 5.1-g sample of $CaCl_2$ is dissolved in a beaker of water. Which of the following statements is true of this solution?
- a The solution will freeze at a lower temperature than pure water.
- b The solution has a higher vapor pressure than pure water.
- c The solution will boil at a lower temperature than pure water.
- d Water is the solute in this solution.
- e None of the other statements (a–d) are true.

12.24 If 291g of a compound is added to 1.02 kg of water to increase the boiling point by 5.77°C, what is the molar mass of the added compound? (Assume a van 't Hoff factor of 1.)
- a 31.1 g/mol
- b 30.3 g/mol
- c 28.5 g/mol
- d 18.3 g/mol
- e 25.3 g/mol

Concept Explorations

Key: Concept explorations are comprehensive problems that provide a framework that will enable you to explore and learn many of the critical concepts and ideas in each chapter. If you master the concepts associated with these explorations, you will have a better understanding of many important chemistry ideas and will be more successful in solving all types of chemistry problems. These problems are well suited for group work and for use as in-class activities.

Colligative Properties I

12.25 Consider two hypothetical pure substances, AB(*s*) and XY(*s*). When equal molar amounts of these substances are

placed in separate 500-mL samples of water, they undergo the following reactions:

$$AB(s) \longrightarrow A^+(aq) + B^-(aq)$$
$$XY(s) \longrightarrow XY(aq)$$

- a Which solution would you expect to have the lower boiling point? Why?
- b Would you expect the vapor pressures of the two solutions to be equal? If not, which one would you expect to have the higher vapor pressure?
- c Describe a procedure that would make the two solutions have the same boiling point.

d If you took 250 mL of the AB(*aq*) solution prepared above, would it have the same boiling point as the original solution? Be sure to explain your answer.

e The container of XY(*aq*) is left out on the bench top for several days, which allows some of the water to evaporate from the solution. How would the melting point of this solution compare to the melting point of the original solution?

Colligative Properties II

12.26 Equal numbers of moles of two soluble, substances, substance A and substance B, are placed into separate 1.0-L samples of water.

a The water samples are cooled. Sample A freezes at −0.50°C, and Sample B freezes at −1.00°C. Explain how the solutions can have different freezing points.

b You pour 500 mL of the solution containing substance B into a different beaker. How would the freezing point of this 500-mL portion of solution B compare to the freezing point of the 1.0-L sample of solution A?

c Calculate the molality of the solutions of A and B. Assume that $i = 1$ for substance A.

d If you were to add an additional 1.0 kg of water to solution B, what would be the new freezing point of the solution? Try to write an answer to this question without using a mathematical formula.

e What concentration (molality) of substances A and B would result in both solutions having a freezing point of −0.25°C?

f Compare the boiling points, vapor pressure, and osmotic pressure of the original solutions of A and B. Don't perform the calculations; just state which is the greater in each case.

Conceptual Problems

Key: These problems are designed to check your understanding of the concepts associated with some of the main topics presented in each chapter. A strong conceptual understanding of chemistry is the foundation for both applying chemical knowledge and solving chemical problems. These problems vary in level of difficulty and often can be used as a basis for group discussion.

12.27 You want to purchase a salt to melt snow and ice on your sidewalk. Which one of the following salts would best accomplish your task using the least amount: KCl, $CaCl_2$, PbS_2, $MgSO_4$, or AgCl?

12.28 Even though the oxygen demands of trout and bass are different, they can exist in the same body of water. However, if the temperature of the water in the summer gets above about 23°C, the trout begin to die, but not the bass. Why is this the case?

12.29 Ten grams of the hypothetical ionic compounds XZ and YZ are each placed in a separate 2.0-L beaker of water. XZ completely dissolves, whereas YZ is insoluble. The energy of hydration of the Y^+ ion is greater than the X^+ ion. Explain this difference in solubility.

12.30 Small amounts of a nonvolatile, nonelectrolyte solute and a volatile solute are each dissolved in separate beakers containing 1 kg of water. If the number of moles of each solute is equal:

a Which solution will boil at a higher temperature?

b Which solution will have the higher vapor pressure?

12.31 A Cottrell precipitator consists of a column containing electrodes that are connected to a high-voltage direct current (DC) source. The Cottrell precipitator is placed in smokestacks to remove smoke particles from the gas discharged from an industrial plant. Explain how you think this works.

12.32 Consider the following dilute NaCl(*aq*) solutions.

a Which one will freeze at a lower temperature?

b Which one will boil at a higher temperature?

c If the solutions were separated by a semipermeable membrane that allowed only water to pass, which solution would you expect to show an increase in the concentration of NaCl?

12.33 A green leafy salad wilts if left too long in a salad dressing containing vinegar and salt. Explain what happens.

12.34 People have proposed towing icebergs to arid parts of the earth as a way to deliver freshwater. Explain why icebergs do not contain salts although they are formed by the freezing of ocean water (i.e., saltwater).

12.35 If 1-mol samples of urea, a nonelectrolyte, sodium chloride, and calcium chloride are each dissolved in equal volumes of water in separate containers:

a Which solution would have the highest freezing point?

b Which solution would have the highest boiling point?

12.36 Consider the following three beakers that contain water and a non-volatile solute. The solute is represented by the orange spheres.

a Which solution would have the lowest boiling point?

b Which solution would have the highest vapor pressure?

c What could you do in the laboratory to make each solution have the same freezing point?

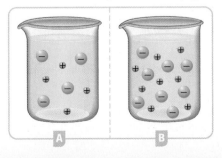

Practice Problems

Key: These problems are for practice in applying problem-solving skills. They are divided by topic, and some are keyed to exercises (see the ends of the exercises). The problems are arranged in matching pairs; the odd-numbered problem of each pair is listed first, and its answer is given in the back of the book.

Types of Solutions

12.37 Give an example of a solid solution prepared from two solids.

12.38 Give an example of a liquid solution prepared by dissolving a gas in a liquid.

Solubility

12.39 Would boric acid, $B(OH)_3$, be more soluble in ethanol, C_2H_5OH, or in benzene, C_6H_6?

12.40 Would naphthalene, $C_{10}H_8$, be more soluble in ethanol, C_2H_5OH, or in benzene, C_6H_6?

12.41 Indicate which of the following is more soluble in ethanol, C_2H_5OH: acetic acid, CH_3COOH, or stearic acid, $C_{17}H_{35}COOH$.

12.42 Arrange the following substances in order of increasing solubility in hexane, C_6H_{14}: CH_2OHCH_2OH, $C_{10}H_{22}$, H_2O.

12.43 Which of the following ions would be expected to have the greater energy of hydration, Mg^{2+} or Al^{3+}?

12.44 Which of the following ions would be expected to have the greater energy of hydration, F^- or Cl^-?

12.45 Arrange the following alkaline-earth-metal iodates in order of increasing solubility in water and explain your reasoning: $Ba(IO_3)_2$, $Ca(IO_3)_2$, $Sr(IO_3)_2$, $Mg(IO_3)_2$. Note that IO_3^- is a large anion.

12.46 Explain the trends in solubility (grams per 100 mL of water) of the alkali-metal fluorides and permanganates.

	Li^+	Na^+	K^+	Rb^+	Cs^+
F^-	0.27	4.2	92	131	367
MnO_4^-	71	Very soluble	6.4	0.5	0.097

12.47 The solubility of carbon dioxide in water is 0.161 g CO_2 in 100 mL of water at 20°C and 1.00 atm. A soft drink is carbonated with carbon dioxide gas at 5.50 atm pressure. What is the solubility of carbon dioxide in water at this pressure?

12.48 Nitrogen, N_2, is soluble in blood and can cause intoxication at sufficient concentration. For this reason, the U.S. Navy advises divers using compressed air not to go below 125 feet. The total pressure at this depth is 4.79 atm. If the solubility of nitrogen at 1.00 atm is 1.75×10^{-3} g/100 mL of water, and the mole percent of nitrogen in air is 78.1, what is the solubility of nitrogen in water from air at 4.79 atm?

Solution Concentration

12.49 How would you prepare 72.5 g of an aqueous solution that is 5.00% potassium iodide, KI, by mass?

12.50 How would you prepare 455 g of an aqueous solution that is 7.00% sodium sulfate, Na_2SO_4, by mass?

12.51 What mass of solution containing 5.00% potassium iodide, KI, by mass contains 516 mg KI?

12.52 What mass of solution containing 6.50% sodium sulfate, Na_2SO_4, by mass contains 1.75 g Na_2SO_4?

12.53 Lauryl alcohol, $C_{12}H_{25}OH$, is prepared from coconut oil; it is used to make sodium lauryl sulfate, a synthetic detergent. What is the molality of lauryl alcohol in a solution of 17.1 g lauryl alcohol dissolved in 185 g ethanol, C_2H_5OH?

12.54 Vanillin, $C_8H_8O_3$, occurs naturally in vanilla extract and is used as a flavoring agent. A 39.1-mg sample of vanillin was dissolved in 168.5 mg of diphenyl ether, $(C_6H_5)_2O$. What is the molality of vanillin in the solution?

12.55 Fructose, $C_6H_{12}O_6$, is a sugar occurring in honey and fruits. The sweetest sugar, it is nearly twice as sweet as sucrose (cane or beet sugar). How much water should be added to 1.75 g of fructose to give a 0.125 m solution?

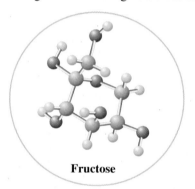

Fructose

12.56 Caffeine, $C_8H_{10}N_4O_2$, is a stimulant found in tea and coffee. A sample of the substance was dissolved in 45.0 g of chloroform, $CHCl_3$, to give a 0.0946 m solution. How many grams of caffeine were in the sample?

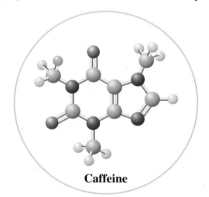

Caffeine

12.57 An automobile antifreeze solution contains 2.50 kg of ethylene glycol, CH_2OHCH_2OH, and 2.25 kg of water. Find the mole fraction of ethylene glycol in this solution. What is the mole fraction of water?

12.58 A 100.0-g sample of a brand of rubbing alcohol contains 65.0 g of isopropyl alcohol, C_3H_7OH, and 35.0 g of water. What is the mole fraction of isopropyl alcohol in the solution? What is the mole fraction of water?

12.59 A bleaching solution contains sodium hypochlorite, NaClO, dissolved in water. The solution is 0.650 m NaClO. What is the mole fraction of sodium hypochlorite?

12.60 An antiseptic solution contains hydrogen peroxide, H_2O_2, in water. The solution is 0.610 m H_2O_2. What is the mole fraction of hydrogen peroxide?

12.61 Concentrated aqueous ammonia contains 1.00 mol NH_3 dissolved in 2.50 mol H_2O. What is the mole fraction of NH_3 in concentrated aqueous ammonia? What is the molal concentration of NH_3?

12.62 Concentrated hydrochloric acid contains 1.00 mol HCl dissolved in 3.31 mol H_2O. What is the mole fraction of HCl in concentrated hydrochloric acid? What is the molal concentration of HCl?

12.63 Oxalic acid, $H_2C_2O_4$, occurs as the potassium or calcium salt in many plants, including rhubarb and spinach. An aqueous solution of oxalic acid is 0.580 m $H_2C_2O_4$. The density of the solution is 1.022 g/mL. What is the molar concentration?

12.64 Citric acid, $H_3C_6H_5O_7$, occurs in plants. Lemons contain 5% to 8% citric acid by mass. The acid is added to beverages and candy. An aqueous solution is 0.688 m citric acid. The density is 1.049 g/mL. What is the molar concentration?

12.65 A beverage contains tartaric acid, $H_2C_4H_4O_6$, a substance obtained from grapes during wine making. If the beverage is 0.265 M tartaric acid, what is the molal concentration? The density of the solution is 1.016 g/mL.

12.66 A solution of vinegar is 0.763 M acetic acid, $HC_2H_3O_2$. The density of the vinegar is 1.004 g/mL. What is the molal concentration of acetic acid?

Colligative Properties

12.67 Calculate the vapor pressure at 35°C of a solution made by dissolving 20.2 g of sucrose, $C_{12}H_{22}O_{11}$, in 70.1 g of water. The vapor pressure of pure water at 35°C is 42.2 mmHg. What is the vapor-pressure lowering of the solution? (Sucrose is nonvolatile.)

12.68 What is the vapor pressure at 23°C of a solution of 1.20 g of naphthalene, $C_{10}H_8$, in 25.6 g of benzene, C_6H_6? The vapor pressure of pure benzene at 23°C is 86.0 mmHg; the vapor pressure of naphthalene can be neglected. Calculate the vapor-pressure lowering of the solution.

12.69 A solution was prepared by dissolving 0.800 g of sulfur, S_8, in 100.0 g of acetic acid, $HC_2H_3O_2$. Calculate the freezing point and boiling point of the solution.

12.70 What is the boiling point of a solution of 0.150 g of glycerol, $C_3H_8O_3$, in 20.0 g of water? What is the freezing point?

12.71 An aqueous solution of a molecular compound freezes at −0.086°C. What is the molality of the solution?

12.72 Urea, $(NH_2)_2CO$, is dissolved in 250.0 g of water. The solution freezes at −0.085°C. How many grams of urea were dissolved to make this solution?

12.73 A 0.0182-g sample of an unknown substance was dissolved in 2.135 g of benzene. The molality of this solution, determined by freezing-point depression, was 0.0698 m. What is the molecular mass of the unknown substance?

12.74 A solution contains 0.0653 g of a compound in 8.31 g of ethanol. The molality of the solution is 0.0368 m. Calculate the molecular mass of the compound.

12.75 Safrole is contained in oil of sassafras and was once used to flavor root beer. A 2.39-mg sample of safrole was dissolved in 103.0 mg of diphenyl ether. The solution had a melting point of 25.70°C. Calculate the molecular mass of safrole. The freezing point of pure diphenyl ether is 26.84°C, and the freezing-point-depression constant, K_f, is 8.00°C/m.

12.76 Butylated hydroxytoluene (BHT) is used as an antioxidant in processed foods. (It prevents fats and oils from becoming rancid.) A solution of 2.500 g of BHT in 100.0 g of benzene had a freezing point of 4.880°C. What is the molecular mass of BHT?

12.77 Arginine vasopressin is a pituitary hormone. It helps regulate the amount of water in the blood by reducing the flow of urine from the kidneys. An aqueous solution containing 21.6 mg of vasopressin in 100.0 mL of solution has an osmotic pressure at 25°C of 3.70 mmHg. What is the molecular mass of the hormone?

12.78 Dextran is a polymeric carbohydrate produced by certain bacteria. It is used as a blood plasma substitute. An aqueous solution contains 0.582 g of dextran in 106 mL of solution at 21°C. It has an osmotic pressure of 1.47 mmHg. What is the average molecular mass of the dextran?

12.79 What is the freezing point of 0.0075 m aqueous calcium chloride, $CaCl_2$? Use the formula of the salt to obtain i.

12.80 What is the freezing point of 0.0088 m aqueous sodium phosphate, Na_3PO_4? Use the formula of the salt to obtain i.

12.81 A 0.0140-g sample of an ionic compound with the formula $Cr(NH_3)_5Cl_3$ was dissolved in water to give 25.0 mL of solution at 25°C. The osmotic pressure was determined to be 119 mmHg. How many ions are obtained from each formula unit when the compound is dissolved in water?

12.82 In a mountainous location, the boiling point of pure water is found to be 95°C. How many grams of sodium chloride must be added to 1 kg of water to bring the boiling point back to 100°C? Assume that $i = 2$.

Colloids

12.83 Give the type of colloid (aerosol, foam, emulsion, sol, or gel) that each of the following represents.
- a milk of magnesia
- b silt in water
- c rain cloud
- d soapsuds

12.84 Give the type of colloid (aerosol, foam, emulsion, sol, or gel) that each of the following represents.
- a salad dressing
- b beaten egg white
- c dust cloud
- d ocean spray

12.85 Arsenic(III) sulfide forms a sol with a negative charge. Which of the following ionic substances should be most effective in coagulating the sol?
- a KCl
- b $MgCl_2$
- c $Al_2(SO_4)_3$
- d Na_3PO_4

12.86 Aluminum hydroxide forms a positively charged sol. Which of the following ionic substances should be most effective in coagulating the sol?
- a NaCl
- b $CaCl_2$
- c $Fe_2(SO_4)_3$
- d K_3PO_4

General Problems

Key: These problems provide more practice but are not divided by topic or keyed to exercises. Each section ends with essay questions, each of which is color coded to refer to the *A Chemist Looks at Life Science* (pink) and *A Chemist Looks at Frontiers* (purple) chapter essay on which it is based. Odd-numbered problems and the even-numbered problems that follow are similar; answers to all odd-numbered problems except the essay questions are given in the back of the book.

12.87 A natural gas mixture consists of 88.0 mole percent CH_4 (methane) and 12.0 mole percent C_2H_6 (ethane). Suppose water is saturated with the gas mixture at 20°C and 1.00 atm total pressure, and the gas is then expelled from the water by heating. What is the composition in mole fractions of the gas mixture that is expelled? The solubilities of CH_4 and C_2H_6 at 20°C and 1.00 atm are 0.023 g/L H_2O and 0.059 g/L H_2O, respectively.

12.88 A gaseous mixture consists of 80.0 mole percent N_2 and 20.0 mole percent O_2 (the approximate composition of air). Suppose water is saturated with the gas mixture at 25°C and 1.00 atm total pressure, and then the gas is expelled from the water by heating. What is the composition in mole fractions of the gas mixture that is expelled? The solubilities of N_2 and O_2 at 25°C and 1.00 atm are 0.0175 g/L H_2O and 0.0393 g/L H_2O, respectively.

12.89 An aqueous solution is 8.50% ammonium chloride, NH_4Cl, by mass. The density of the solution is 1.024 g/mL. What are the molality, mole fraction, and molarity of NH_4Cl in the solution?

12.90 An aqueous solution is 27.0% lithium chloride, LiCl, by mass. The density of the solution is 1.127 g/mL. What are the molality, mole fraction, and molarity of LiCl in the solution?

12.91 The diving atmosphere used by the U.S. Navy in its undersea Sea-Lab experiments consisted of 0.036 mole fraction O_2 and 0.056 mole fraction N_2, with helium (He) making up the remainder. What are the masses of nitrogen, oxygen, and helium in a 8.32-g sample of this atmosphere?

12.92 A 55-g sample of a gaseous fuel mixture contains 0.51 mole fraction propane, C_3H_8; the remainder of the mixture is butane, C_4H_{10}. What are the masses of propane and butane in the sample?

12.93 A liquid solution consists of 0.30 mole fraction ethylene dibromide, $C_2H_4Br_2$, and 0.70 mole fraction propylene dibromide, $C_3H_6Br_2$. Both ethylene dibromide and propylene dibromide are volatile liquids; their vapor pressures at 85°C are 173 mmHg and 127 mmHg, respectively. Assume that each compound follows Raoult's law in the solution. Calculate the total vapor pressure of the solution.

12.94 What is the total vapor pressure at 20°C of a liquid solution containing 0.30 mole fraction benzene, C_6H_6, and 0.70 mole fraction toluene, $C_6H_5CH_3$? Assume that Raoult's law holds for each component of the solution. The vapor pressure of pure benzene at 20°C is 75 mmHg; that of toluene at 20°C is 22 mmHg.

12.95 A sample of potassium aluminum sulfate 12-hydrate, $KAl(SO_4)_2 \cdot 12H_2O$, containing 101.5 mg is dissolved in 1.000 L of solution. Calculate the following for the solution:
- **a** The molarity of $KAl(SO_4)_2$.
- **b** The molarity of SO_4^{2-}.
- **c** The molality of $KAl(SO_4)_2$, assuming that the density of the solution is 1.00 g/mL.

12.96 A sample of aluminum sulfate 18-hydrate, $Al_2(SO_4)_3 \cdot 18H_2O$, containing 125.0 mg is dissolved in 1.000 L of solution. Calculate the following for the solution:
- **a** The molarity of $Al_2(SO_4)_3$.
- **b** The molarity of SO_4^{2-}.
- **c** The molality of $Al_2(SO_4)_3$, assuming that the density of the solution is 1.00 g/mL.

12.97 Urea, $(NH_2)_2CO$, has been used to melt ice from sidewalks, because the use of salt is harmful to plants. If the saturated aqueous solution contains 41% urea by mass, what is the freezing point? (The answer will be approximate, because the equation in the text applies accurately only to dilute solutions.)

12.98 Calcium chloride, $CaCl_2$, has been used to melt ice from roadways. Given that the saturated solution is 32% $CaCl_2$ by mass, estimate the freezing point.

12.99 The osmotic pressure of blood at 37°C is 7.7 atm. A solution that is given intravenously must have the same osmotic pressure as the blood. What should be the molarity of a glucose solution to give an osmotic pressure of 7.7 atm at 37°C?

12.100 Maltose, $C_{12}H_{22}O_{11}$, is a sugar produced by malting (sprouting) grain. A solution of maltose at 25°C has an osmotic pressure of 11.0 atm. What is the molar concentration of maltose?

12.101 Which aqueous solution has the lower freezing point, 0.10 m $CaCl_2$ or 0.10 m glucose?

12.102 Which aqueous solution has the lower boiling point, 0.10 m NaBr or 0.10 m $CaCl_2$?

12.103 Commercially, sulfuric acid is obtained as a 98% solution. If this solution is 18 M, what is its density? What is its molality?

12.104 Phosphoric acid is usually obtained as an 85% phosphoric acid solution. If it is 15 M, what is the density of this solution? What is its molality?

12.105 A compound of cobalt, carbon, and oxygen contains 28.10% C and 34.47% Co. When 0.147 g of this compound is dissolved in 6.72 g of cyclohexane, the solution freezes at 5.23°C. What is the molecular formula of this compound?

12.106 A compound of manganese, carbon, and oxygen contains 28.17% Mn and 30.80% C. When 0.125 g of this compound is dissolved in 5.38 g of cyclohexane, the solution freezes at 5.28°C. What is the molecular formula of this compound?

12.107 The carbohydrate digitoxose contains 48.64% carbon and 8.16% hydrogen. The addition of 18.0 g of this compound to 100. g of water gives a solution that has a freezing point of $-2.2°C$.

a What is the molecular formula of the compound?

b What is the molar mass of this compound to the nearest tenth of a gram?

12.108 Analysis of a compound gave 39.50% C, 2.21% H, and 58.30% Cl. When 0.855 g of this solid was dissolved in 7.50 g of naphthalene, the solution had a freezing point of 78.0°C. The pure solvent freezes at 80.0°C; its molal freezing point constant is 6.8°C/m.

a What is the molecular formula of the compound?

b What is its molecular mass to the nearest 0.1 g?

12.109 A salt solution has an osmotic pressure of 16 atmospheres at 22°C. What is the freezing point of this solution? What assumptions must be made to solve this problem?

12.110 Fish blood has an osmotic pressure equal to that of seawater. If seawater freezes at $-2.3°C$, what is the osmotic pressure of the blood at 25°C? To solve this problem, what assumptions must be made?

■ **12.111** What is the difference between normal hemoglobin and the hemoglobin associated with sickle-cell anemia? How does this affect the solubility of the hemoglobin?

■ **12.112** What tends to happen to the hemoglobin in a blood cell of a person who has sickle-cell anemia? How does this affect the shape and oxygen-carrying capacity of the cell?

■ **12.113** How are phospholipids similar in structure to a soap molecule? What structural difference accounts for the fact that phospholipids form lipid bilayers rather than the micelles that soaps form?

■ **12.114** How have chemists used these phospholipids to form "the world's smallest test tubes"? Describe how these phospholipid vesicles might be used.

Strategy Problems

Key: As noted earlier, all of the practice and general problems are matched pairs. This section is a selection of problems that are not in matched-pair format. These challenging problems require that you employ many of the concepts and strategies that were developed in the chapter. In some cases, you will have to integrate several concepts and operational skills in order to solve the problem successfully.

12.115 Two samples of sodium chloride solutions are brought to a boil on a stove. One of the solutions boils at 100.10°C and the other at 100.15°C.

a Which of the solutions is more concentrated?

b Which of the solutions would have a lower freezing point?

c If you split the solution that boils at 100.1°C into two portions, how would the boiling points of the samples compare?

Which of the following statements do you agree with regarding the determination of your answer for part c?

i The question cannot be answered with certainty without knowing the volumes of each portion.

ii Making the necessary assumption that the two samples have equal volumes, I was able to correctly answer the question.

iii The volumes that the sample was split into are irrelevant when determining the correct answer.

12.116 A sample of an ionic solid is dissolved in 1.00 kg of water. The freezing point of the water is $-0.01°C$. If three times the mass of ionic solid is dissolved in 1.00 kg of water and the resulting freezing point of the solution is $-0.09°C$, which of the following would be possible formulas for the solid: MX, MX_2, or MX_3, where M represents a metal cation and X an anion with a 1-charge? Explain your reasoning.

12.117 You have an aqueous, dilute solution of a nonvolatile nonelectrolyte. Assuming that the solution is ideal, choose the following statement(s) that are true.

a If the concentration of the solute is increased, then the osmotic pressure of the solution will increase.

b The addition of more solute will cause an increase in the boiling point of the solution.

c The vapor pressure of this solution would decrease if some of the water were allowed to evaporate.

d The solution will freeze at a temperature below 0°C at 1.0 atm.

e The addition of pure water to the solution will cause the boiling point to decrease.

12.118 Consider the three ionic compounds, NaCl, KCl, and LiCl.

a Which compound would you expect to have the greatest lattice energy?

b Which compound would have the greatest energy of hydration?

c Are the three ionic compounds soluble in water?

d What is the relationship between the lattice energy and the energy of hydration that will make an ionic compound soluble in water?

12.119 The concentrations of pollutants are often reported as parts per million (ppm). Parts per million of a solution is defined as:

$$ppm = \frac{mass\ of\ solute}{mass\ of\ solution} \times 10^6$$

The U.S. Environmental Protection Agency (EPA) limit for barium in municipal drinking water is 2 ppm.

a Calculate the maximum mass percentage of barium allowed in drinking water.

b Calculate the molarity of a solution of barium that is 2 ppm. Assume that the density of the solution is 1.00 g/mL.

c Express the concentration of barium as mg per liter.

12.120 When 79.3 g of a particular compound is dissolved in 0.878 kg of water at 1 atm pressure, the solution freezes at $-1.34°C$. If the compound does not undergo ionization in solution and is nonvolatile, determine the molecular mass of the compound.

12.121 Which of the following ways of expressing concentration would be most likely to change with a change in temperature: molarity, molality, or mass percentage? Explain your reasoning.

12.122 What is the boiling point of a solution made by adding 6.69 g of magnesium chloride to 250.0 g of water? Use the formula of the salt to obtain i.

12.123 At 15°C and a partial pressure of 2.14 atm, the solubility of an unknown gas in a liquid is 15.6 g/L. What is the solubility of the gas in the liquid when the partial pressure of the gas is 24.6 atm? What would happen to the solubility of the gas if the temperature of the liquid were increased?

12.124 Methanol, CH_3OH, is a compound that is used in car windshield washer fluid to lower the freezing point of the solution. If an aqueous solution of windshield washer fluid has a mole fraction of methanol of 0.525, what is the mass percentage of water in this solution?

12.125 How many grams of potassium chloride must be added to 372 g of water in order to prepare a 0.220 m potassium chloride solution?

12.126 Although somewhat toxic, barium chloride has been used as a cardiac stimulant. Calculate the number of moles of barium chloride in 427 g of a 3.25% by mass barium chloride solution?

12.127 Ethanol, CH_3CH_2OH, is known as grain alcohol and is the alcohol found in alcoholic beverages. Calculate the mole fraction of ethanol in a solution that contains 4.76 g of ethanol dissolved in 50.0 g of water.

12.128 An aqueous solution is 0.797 M magnesium chloride. Given that the density of the solution is 1.108 g/mL, what is the molality of magnesium chloride in this solution?

12.129 A starch has a molar mass of 3.20×10^4 g/mol. If 0.759 g of this starch is dissolved in 112 mL of solution, what is the osmotic pressure, *in torr*, at 25.00°C?

12.130 A $CaCl_2$ solution at 25°C has an osmotic pressure of 16 atm and a density of 1.108 g/mL. What is the freezing point of this solution?

Cumulative-Skills Problems

Key: The problems under this heading combine skills introduced in previous chapters with those given in the current one.

12.131 An experimenter makes up a solution of 0.325 mol Na_3PO_4, 0.100 mol $Ca(NO_3)_2$, and 0.150 mol $AgNO_3$ in 4.000 L of water solution. Note any precipitations that occur, writing a balanced equation for each. Then, calculate the molarities of each ion in the solution.

12.132 An experimenter makes up a solution of 0.375 mol Na_2CO_3, 0.125 mol $Ca(NO_3)_2$, and 0.200 mol $AgNO_3$ in 2.000 L of water solution. Note any precipitations that occur, writing a balanced equation for each. Then, calculate the molarities of each ion in the solution.

12.133 The lattice enthalpy of sodium chloride, $\Delta H°$ for

$$NaCl(s) \longrightarrow Na^+(g) + Cl^-(g)$$

is 787 kJ/mol; the heat of solution in making up 1 M $NaCl(aq)$ is +4.0 kJ/mol. From these data, obtain the sum of the heats of hydration of Na^+ and Cl^-. That is, obtain the sum of $\Delta H°$ values for

$$Na^+(g) \longrightarrow Na^+(aq)$$
$$Cl^-(g) \longrightarrow Cl^-(aq)$$

If the heat of hydration of Cl^- is -338 kJ/mol, what is the heat of hydration of Na^+?

12.134 The lattice enthalpy of potassium chloride is 717 kJ/mol; the heat of solution in making up 1 M $KCl(aq)$ is +18.0 kJ/mol. Using the value for the heat of hydration of Cl^- given in Problem 12.133, obtain the heat of hydration of K^+. Compare this with the value you obtained for Na^+ in Problem 12.133. Explain the relative values of Na^+ and K^+.

12.135 A solution is made up by dissolving 15.0 g $MgSO_4 \cdot 7H_2O$ in 100.0 g of water. What is the molality of $MgSO_4$ in this solution?

12.136 A solution is made up by dissolving 18.0 g $Na_2CO_3 \cdot 10H_2O$ in 100.0 g of water. What is the molality of Na_2CO_3 in this solution?

12.137 An aqueous solution is 15.0% by mass of copper(II) sulfate pentahydrate, $CuSO_4 \cdot 5H_2O$. What is the molarity of $CuSO_4$ in this solution at 20°C? The density of this solution at 20°C is 1.167 g/mL.

12.138 An aqueous solution is 20.0% by mass of sodium thiosulfate pentahydrate, $Na_2S_2O_3 \cdot 5H_2O$. What is the molarity of $Na_2S_2O_3$ in this solution at 20°C? The density of this solution at 20°C is 1.174 g/mL.

12.139 The freezing point of 0.0830 m aqueous acetic acid is -0.159°C. Acetic acid, $HC_2H_3O_2$, is partially dissociated according to the equation

$$HC_2H_3O_2(aq) \rightleftharpoons H^+(aq) + C_2H_3O_2^-(aq)$$

Calculate the percentage of $HC_2H_3O_2$ molecules that are dissociated, assuming the equation for the freezing-point depression holds for the total concentration of molecules and ions in the solution.

12.140 The freezing point of 0.109 m aqueous formic acid is -0.210°C. Formic acid, $HCHO_2$, is partially dissociated according to the equation

$$HCHO_2(aq) \rightleftharpoons H^+(aq) + CHO_2^-(aq)$$

Calculate the percentage of $HCHO_2$ molecules that are dissociated, assuming the equation for the freezing-point depression holds for the total concentration of molecules and ions in the solution.

12.141 A compound of carbon, hydrogen, and oxygen was burned in oxygen, and 1.000 g of the compound produced 1.418 g CO_2 and 0.871 g H_2O. In another experiment, 0.1103 g of the compound was dissolved in 45.0 g of water. This solution had a freezing point of -0.0734°C. What is the molecular formula of the compound?

12.142 A compound of carbon, hydrogen, and oxygen was burned in oxygen, and 1.000 g of the compound produced 1.434 g CO_2 and 0.783 g H_2O. In another experiment, 0.1107 g of the compound was dissolved in 25.0 g of water. This solution had a freezing point of -0.0894°C. What is the molecular formula of the compound?

13 Rates of Reaction

©Thomas Eisner and Daniel Aneshansley, Cornell University/Visuals Unlimited; AP Photo/David Duprey

The bombardier beetle and the rocket belt use hydrogen peroxide (H_2O_2) to fuel energetic reactions. In both reactions, a catalyst rapidly breaks down the H_2O_2, releasing oxygen, heat, and water vapor. The pressure from these hot gas mixtures being released is enough to expel a caustic chemical defense or propel a person into the air.

OWL Sign in to OWL at **www.cengage.com/owl** to view tutorials and simulations, develop problem-solving skills, and complete online homework assigned by your professor.

goChemistry Download mini lecture videos for key concept review and exam prep from OWL or purchase them from **www.cengagebrain.com**.

Figure 13.1 ▲

Reaction time A chemical reaction takes time. Some reactions are complete in a fraction of a second; others take years. The reaction shown here is complete in less than a minute. The reaction is the formation of a product of formaldehyde with hydrogen sulfite ion: $H_2C{=}O(aq)$ + $HSO_3^-(aq) \longrightarrow H_2C(OH)SO_3^-(aq)$. As the hydrogen sulfite ion is used up, the solution becomes less acidic and then changes to basic. Bromthymol-blue indicator marks the change from acidic to basic by changing from yellow to blue.

hemical reactions require varying lengths of time for completion, depending on the characteristics of the reactants and products and the conditions under which the reaction is run (Figure 13.1). Many reactions are over in a fraction of a second, whereas others can take much longer. If you add barium ion to an aqueous solution of sulfate ion, a precipitate of barium sulfate forms almost immediately. On the other hand, the reactions that occur in a cement mixture as it hardens to concrete require years for completion.

The study of the rate, or speed, of a reaction has important applications. In the manufacture of ammonia from nitrogen and hydrogen, you may wish to know what conditions will help the reaction proceed in a commercially feasible length of time. You may also wish to know whether the nitrogen monoxide, NO, in the exhaust gases of high-altitude aircraft will destroy the ozone in the stratosphere faster than the ozone is produced. Answering these questions requires knowledge of the rates of reactions.

Another reason for studying reaction rates is to understand how chemical reactions occur. By noting how the rate of a reaction is affected by changing conditions, you can sometimes learn the details of what is happening at the molecular level.

A reaction whose rate has been extensively studied under various conditions is the decomposition of dinitrogen pentoxide, N_2O_5. When this substance is heated in the gas phase, it decomposes to nitrogen dioxide and oxygen:

$$2N_2O_5(g) \longrightarrow 4NO_2(g) + O_2(g)$$

We will look at this reaction in some detail. The questions we will pose include: How is the rate of a reaction like this measured? What conditions affect the rate of a reaction? How do you express the relationship of rate of a reaction to the variables that affect rate? What happens at the molecular level when N_2O_5 decomposes to NO_2 and O_2?

Reaction Rates

Chemical kinetics is the study of reaction rates, how reaction rates change under varying conditions, and what molecular events occur during the overall reaction. In the first part of this chapter, we will look at reaction rates and the variables that affect them.

What variables affect reaction rates? As we noted in the chapter opening, the rate depends on the characteristics of the reactants in a particular reaction. Some reactions are fast and others are slow, but the rate of any given reaction may be affected by the following factors:

1. *Concentrations of reactants.* Often the rate of reaction increases when the concentration of a reactant is increased. A piece of steel wool burns with some difficulty in air (20% O_2) but bursts into a dazzling white flame in pure oxygen. The rate of burning increases with the concentration of O_2. In some reactions, however, the rate is unaffected by the concentration of a particular reactant, as long as it is present at some concentration.

2. *Concentration of catalyst.* A **catalyst** is *a substance that increases the rate of reaction without being consumed in the overall reaction.* Because the catalyst is not consumed by the reaction, it does not appear in the balanced chemical equation (although its presence may be indicated by writing its formula over the arrow). A solution of pure hydrogen peroxide, H_2O_2, is stable, but when hydrobromic acid, HBr(*aq*), is added, H_2O_2 decomposes rapidly into H_2O and O_2 (Figure 13.2).

© Cengage Learning

Figure 13.2 ◀

Catalytic decomposition of hydrogen peroxide The hydrogen peroxide decomposes rapidly when hydrobromic acid is added to an aqueous solution. One of the products is oxygen gas, which bubbles vigorously from the solution. In addition, some HBr is oxidized to Br_2, as can be seen from the red color of the liquid and vapor.

$$2H_2O_2(aq) \xrightarrow{\text{HBr}(aq)} 2H_2O(l) + O_2(g)$$

Here HBr acts as a catalyst to speed decomposition.

3. *Temperature at which the reaction occurs.* Usually reactions speed up when the temperature increases. It takes less time to boil an egg at sea level than on a mountaintop, where water boils at a lower temperature. Reactions during cooking go faster at higher temperature.

4. *Surface area of a solid reactant or catalyst.* If a reaction involves a solid with a gas or liquid, the surface area of the solid affects the reaction rate. Because the reaction occurs at the surface of the solid, the rate increases with increasing surface area. A wood fire burns faster if the logs are chopped into smaller pieces. Similarly, the surface area of a solid catalyst is important to the rate of reaction. The greater the surface area per unit volume, the faster the reaction (Figure 13.3).

In the first sections of this chapter, we will look primarily at reactions in the gas phase and in liquid solution where factors 1 to 3 are important. Thus, we will look at the effect of *concentrations* and *temperature* on reaction rates. Before we can explore these factors, however, we need a precise definition of reaction rate.

Figure 13.3 ▲

Effect of large surface area on the rate of reaction Lycopodium powder (from the tiny spores of a club moss) ignites easily to produce a yellow flame. The powder has a large surface area per volume and burns rapidly in air.

13.1 Definition of Reaction Rate

The rate of a reaction is the amount of product formed or the amount of reactant used up per unit of time. So that a rate calculation does not depend on the total quantity of reaction mixture used, you express the rate for a unit volume of the mixture. Therefore, the **reaction rate** is *the increase in molar concentration of product of a reaction per unit time or the decrease in molar concentration of reactant per unit time.* The usual unit of reaction rate is moles per liter per second, mol/(L·s).

Consider the gas-phase reaction discussed in the chapter opening:

$$2N_2O_5(g) \longrightarrow 4NO_2(g) + O_2(g)$$

The rate for this reaction could be found by observing the increase in molar concentration of O_2 produced. You denote the molar concentration of a substance by enclosing its formula in square brackets. Thus, $[O_2]$ is the molar concentration of O_2. In a given time interval Δt, the molar concentration of oxygen, $[O_2]$, in the reaction vessel increases by the amount $\Delta[O_2]$. The symbol Δ (capital Greek delta) means "change in"; you obtain the change by subtracting the initial value from the final value. The rate of the reaction is given by

$$\text{Rate of formation of oxygen} = \frac{\Delta[O_2]}{\Delta t}$$

This equation gives the *average* rate over the time interval Δt. If the time interval is very short, the equation gives the *instantaneous* rate—that is, the rate at a particular

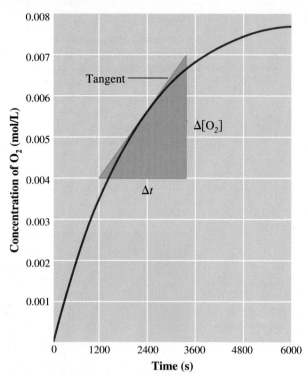

Figure 13.4 ▲

The instantaneous rate of reaction
In the reaction $2N_2O_5(g) \longrightarrow$ $4NO_2(g) + O_2(g)$, the concentration of O_2 increases over time. You obtain the instantaneous rate at a given time from the slope of the tangent at the point on the curve corresponding to that time. In this diagram, the slope equals $\Delta[O_2]/\Delta t$ obtained from the tangent.

In calculus, the rate at a given moment (the instantaneous rate) is given by the derivative $d[O_2]/dt$.

Figure 13.5 ▶

Calculation of the average rate The average rate of formation of O_2 during the decomposition of N_2O_5 was calculated during two different time intervals. When the time changes from 600 s to 1200 s, the average rate is 2.5×10^{-6} mol/(L·s). Later, when the time changes from 4200 s to 4800 s, the average rate has slowed to 5×10^{-7} mol/(L·s). Thus, the rate of the reaction decreases as the reaction proceeds.

instant of time. The instantaneous rate is also the value of $\Delta[O_2]/\Delta t$ for the tangent at a given instant (the straight line that just touches the curve of concentration versus time at a given point). See Figure 13.4. ◀

To understand the difference between average rate and instantaneous rate, it may help to think of the speed of an automobile. Speed can be defined as the rate of change of position, x; that is, speed equals $\Delta x/\Delta t$, where Δx is the distance traveled. If an automobile travels 84 miles in 2.0 hours, the average speed over this time interval is 84 mi/2.0 hr = 42 mi/hr. However, at any instant during this interval, the speedometer, which registers instantaneous speed, may read more or less than 42 mi/hr. At some moment on the highway, it may read 55 mi/hr, whereas on a congested city street it may read only 20 mi/hr. The quantity $\Delta x/\Delta t$ becomes more nearly an instantaneous speed as the time interval Δt is made smaller.

Figure 13.5 shows the increase in concentration of O_2 during the decomposition of N_2O_5. It shows the calculation of average rates at two positions on the curve. For example, when the time changes from 600 s to 1200 s ($\Delta t = 600$ s), the O_2 concentration increases by 0.0015 mol/L (= $\Delta[O_2]$). Therefore, the average rate = $\Delta[O_2]/\Delta t$ = (0.0015 mol/L)/ 600 s = 2.5×10^{-6} mol/(L·s). Later, during the time interval from 4200 s to 4800 s, the average rate is 5×10^{-7} mol/(L·s). Note that the rate decreases as the reaction proceeds.

Because the amounts of products and reactants are related by stoichiometry, any substance in the reaction can be used to express the rate of reaction. In the case of the decomposition of N_2O_5 to NO_2 and O_2, we gave the rate in terms of

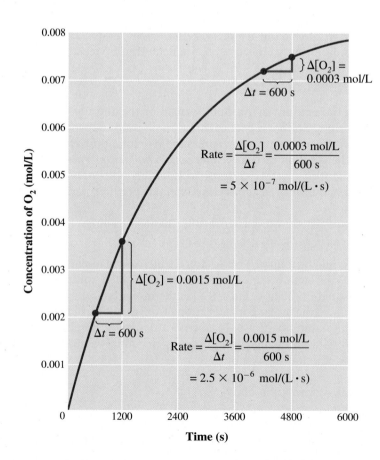

Time	[O₂]
0	0.0000
600	0.0021
1200	0.0036
1800	0.0048
2400	0.0057
3000	0.0063
3600	0.0068
4200	0.0072
4800	0.0075
5400	0.0077
6000	0.0078

the rate of formation of oxygen, $\Delta[O_2]/\Delta t$. However, you can also express it in terms of the rate of decomposition of N_2O_5.

$$\text{Rate of decomposition of } N_2O_5 = -\frac{\Delta[N_2O_5]}{\Delta t}$$

Note the negative sign. It always occurs in a rate expression for a reactant in order to indicate a decrease in concentration and to give a positive value for the rate. Thus, because $[N_2O_5]$ decreases, $\Delta[N_2O_5]$ is negative and $-\Delta[N_2O_5]/\Delta t$ is positive.

The rate of decomposition of N_2O_5 and the rate of formation of oxygen are easily related. Two moles of N_2O_5 decompose for each mole of oxygen formed, so the rate of decomposition of N_2O_5 is twice the rate of formation of oxygen. To equate the rates, you must divide the rate of decomposition of N_2O_5 by 2 (its coefficient in the balanced chemical equation).

$$\text{Rate of formation of } O_2 = \tfrac{1}{2}(\text{rate of decomposition of } N_2O_5)$$

or

$$\frac{\Delta[O_2]}{\Delta t} = -\tfrac{1}{2}\frac{\Delta[N_2O_5]}{\Delta t}$$

⏻WL Interactive Example 13.1   Relating the Different Ways of Expressing Reaction Rates

Consider the reaction of nitrogen dioxide with fluorine to give nitryl fluoride, NO_2F.

$$2NO_2(g) + F_2(g) \longrightarrow 2NO_2F(g)$$

How is the rate of formation of NO_2F related to the rate of reaction of fluorine?

Problem Strategy We need to express the rate of this reaction in terms of concentration changes with time of the product, NO_2F, and reactant, F_2, and then relate these two rates. The rate of disappearance of reactants is expressed as a negative quantity of concentration change per some time interval. The rate of formation of products is expressed as a positive quantity of concentration change per some time interval. In order to equate rate expressions, we need to divide each by the coefficient of the corresponding substance in the chemical equation.

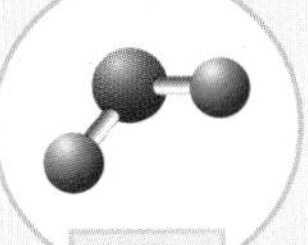

Solution You write

$$\text{Rate of formation of } NO_2F = \frac{\Delta[NO_2F]}{\Delta t}$$

and

$$\text{Rate of reaction of } F_2 = -\frac{\Delta[F_2]}{\Delta t}$$

You divide each rate by the corresponding coefficient and then equate them:

$$\frac{1}{2}\frac{\Delta[NO_2F]}{\Delta t} = -\frac{\Delta[F_2]}{\Delta t}$$

Answer Check When solving problems such as this, always make sure that you have expressed the rates with the correct sign: negative for reactants, positive for products.

NO₂

Exercise 13.1   For the reaction given in Example 13.1, how is the rate of formation of NO_2F related to the rate of reaction of NO_2?

■ See Problems 13.39 and 13.40.

Example 13.2

Gaining Mastery Toolbox

Critical Concept 13.2
The average rate of a chemical reaction is the change in concentration of a reactant or product over a defined time interval. The average rate of a chemical reaction decreases with time as the concentration of the reactants decrease.

Solution Essentials:
• Average reaction rate
• Reaction rate
• Concentration (molarity)

Calculating the Average Reaction Rate

Calculate the average rate of decomposition of N_2O_5, $-\Delta[N_2O_5]/\Delta t$, by the reaction

$$2N_2O_5(g) \longrightarrow 4NO_2(g) + O_2(g)$$

during the time interval from $t = 600$ s to $t = 1200$ s (regard all time figures as significant). Use the following data:

Time	$[N_2O_5]$
600 s	1.24×10^{-2} M
1200 s	0.93×10^{-2} M

Problem Strategy An average reaction rate is the change in concentration of a reactant or product over a time interval; in this case it's the rate of decomposition of the reactant N_2O_5 ($-\Delta[N_2O_5]/\Delta t$). The $\Delta[N_2O_5]$ in the equation is the change in concentration of N_2O_5 (final value minus initial value). The Δt is the time interval (final minus initial) over which the concentration change occurred.

Solution Average rate of decomposition of $N_2O_5 = -\dfrac{\Delta[N_2O_5]}{\Delta t} =$

$$-\frac{(0.93 - 1.24) \times 10^{-2}\ M}{(1200 - 600)\ s} = -\frac{-0.31 \times 10^{-2}\ M}{600\ s} = \mathbf{5.2 \times 10^{-6}\ M/s}$$

Note that this rate is twice the rate of formation of O_2 in the same time interval (within the experimental error of the value given in the preceding text discussion).

Answer Check Always make sure that your reaction rates are expressed as a positive number. If you end up with a negative rate, it is likely that you did not treat Δ quantities as a final value minus an initial value.

Exercise 13.2   Iodide ion is oxidized by hypochlorite ion in basic solution.

$$I^-(aq) + ClO^-(aq) \longrightarrow Cl^-(aq) + IO^-(aq)$$

In 1.00 M NaOH at 25°C, the iodide-ion concentration (equal to the ClO^- concentration) at different times was as follows:

Time	$[I^-]$
2.00 s	0.00169 M
8.00 s	0.00101 M

Calculate the average rate of reaction of I^- during this time interval.

■ See Problems 13.46 and 13.47.

CONCEPT CHECK 13.1

Shown here is a plot of the concentration of a reactant D versus time.

a How do the instantaneous rates at points A and B compare?

b Is the rate for this reaction constant at all points in time?

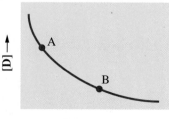

$[D] \longrightarrow$

Time (s) $\longrightarrow$

13.2 Experimental Determination of Rate

To obtain the rate of a reaction, you must determine the concentration of a reactant or product during the course of the reaction. One way to do this for a slow reaction is to withdraw samples from the reaction vessel at various times and analyze them. The rate of the reaction of ethyl acetate with water in acidic solution was one of the first to be determined this way. ▶

$$CH_3CH_2O\overset{\overset{\displaystyle O}{\|}}{C}CH_3 + H_2O \xrightarrow{\ H^+\ } CH_3CH_2OH + HO\overset{\overset{\displaystyle O}{\|}}{C}CH_3$$

Ethyl acetate   Ethanol   Ecetic acid

Ethyl acetate is a liquid with a fruity odor that belongs to a class of organic (carbon-containing) compounds called esters.

This reaction is slow, so the amount of acetic acid produced is easily obtained by titration before any significant further reaction occurs.

More convenient are techniques that can continuously follow the progress of a reaction by observing the change in some physical property of the system. These physical methods are often adaptable to fast reactions as well as slow ones. For example, if a gas reaction involves a change in the number of molecules, the pressure of the system changes when the volume and temperature are held constant. By following the pressure change as the reaction proceeds, you can obtain the reaction rate. The decomposition of dinitrogen pentoxide in the gas phase, which we mentioned earlier, has been studied this way. Dinitrogen pentoxide crystals are sealed in a vessel equipped with a manometer (a pressure-measuring device; see Figure 13.6). The vessel is then plunged into a water bath at 45°C, at which temperature the solid vaporizes and the gas decomposes.

$$2N_2O_5(g) \longrightarrow 4NO_2(g) + O_2(g)$$

Manometer readings are taken at regular time intervals, and the pressure values are converted to partial pressures or concentrations of N_2O_5. The rates of reaction during various time intervals can be calculated as described in the previous section.

Another physical property used to follow the progress of a reaction is color, or the absorption of light by some species. Consider the reaction

$$ClO^-(aq) + I^-(aq) \longrightarrow IO^-(aq) + Cl^-(aq)$$

The hypoiodite ion, IO^-, absorbs at the blue end of the spectrum near 400 nm. The intensity of this absorption is proportional to $[IO^-]$, and you can use the absorption to determine the reaction rate. You can also follow the decomposition of N_2O_5 from the intensity of the red-brown color of the product NO_2.

Figure 13.6 ◀

An experiment to follow the concentration of N_2O_5 as the decomposition proceeds
The total pressure is measured during the reaction at 45°C. Pressure values can be related to the concentrations of N_2O_5, NO_2, and O_2 in the flask.

Opening for adding $N_2O_5(s)$

Flask containing $N_2O_5(g)$, $NO_2(g)$, and $O_2(g)$

Insulated water bath

45°C

Digital manometer

In these experiments, the intensity of absorption at a particular wavelength is measured by a spectrometer appropriate for the visible region of the spectrum. You are not limited to the use of visible-region spectrometers for the determination of reaction rates, however. Depending on the reaction, other types of instruments are used, including infrared (IR) and nuclear magnetic resonance (NMR) spectrometers, which are described briefly in Instrumental Methods essays in earlier chapters.

13.3 Dependence of Rate on Concentration

Experimentally, it has been found that a reaction rate depends on the concentrations of certain reactants as well as the concentration of catalyst, if there is one (Figure 13.7). Consider the reaction of nitrogen dioxide with fluorine to give nitryl fluoride, NO_2F.

$$2NO_2(g) + F_2(g) \longrightarrow 2NO_2F(g)$$

The rate of this reaction is observed to be proportional to the concentration of nitrogen dioxide. When the concentration of nitrogen dioxide is doubled, the rate doubles. The rate is also proportional to the concentration of fluorine; doubling the concentration of fluorine also doubles the rate.

A **rate law** is *an equation that relates the rate of a reaction to the concentrations of reactants (and catalyst) raised to various powers.* The following equation is the rate law for the foregoing reaction:

$$\text{Rate} = k[NO_2][F_2]$$

Note that in this rate law both reactant concentrations have an exponent of 1. Here k, called the **rate constant,** is *a proportionality constant in the relationship between rate and concentrations.* It has a fixed value at any given temperature, but it varies with temperature. Whereas the units of rate are usually given as mol/(L·s), the units of k depend on the form of the rate law. For the previous rate law,

$$k = \frac{\text{rate}}{[NO_2][F_2]}$$

Figure 13.7 ▼

Effect of reactant concentrations on rate of reaction

Time = 0 seconds

Time = 20 seconds

Time = 40 seconds

© Cengage Learning

a Both beakers contain the same amounts of reactants: sodium thiosulfate, $Na_2S_2O_3$, in acidic solution and sodium arsenite, Na_3AsO_3. However, the beaker on the right contains more water and thus lower concentrations of reactants.

b $Na_2S_2O_3$ decomposes slowly in acidic solution to yield H_2S, which reacts quickly with Na_3AsO_3 to give a bright yellow precipitate of As_2S_3. The time it takes the precipitate to form depends on the decomposition rate of $Na_2S_2O_3$.

c Note that the precipitate forms more slowly in the solution of lower concentrations (the beaker on the right).

from which you get the following unit for k:

$$\frac{mol/(L \cdot s)}{(mol/L)^2} = L/(mol \cdot s)$$

As a more general example, consider the reaction of substances A and B to give D and E, according to the balanced equation

$$aA + bB \xrightarrow{C} dD + eE \quad C = catalyst$$

You could write the rate law in the form

$$Rate = k[A]^m[B]^n[C]^p$$

The exponents m, n, and p are frequently, but not always, integers. *They must be determined experimentally* and they cannot be obtained simply by looking at the balanced equation. For example, note that the exponents in the equation Rate $= k[NO_2]$ $[F_2]$ have no relationship to the coefficients in the balanced equation $2NO_2 + F_2 \longrightarrow 2NO_2F$.

Once you know the rate law for a reaction and have found the value of the rate constant, you can calculate the rate of a reaction for any values of reactant concentrations. As you will see later, knowledge of the rate law is also useful in understanding how the reaction occurs at the molecular level.

Reaction Order

You can classify a reaction by its orders. The **reaction order** with respect to a given reactant species equals *the exponent of the concentration of that species in the rate law, as determined experimentally.* For the reaction of NO_2 with F_2 to give NO_2F, the reaction is first order with respect to the NO_2 because the exponent of $[NO_2]$ in the rate law is 1. Similarly, the reaction is first order with respect to F_2.

The **overall order of a reaction** equals *the sum of the orders of the reactant species in the rate law.* In this example, the overall order is 2; that is, the reaction is second order overall.

Reactions display a variety of reaction orders. Some examples follow.

1. Cyclopropane, C_3H_6, has the molecular structure

$$\begin{array}{c} CH_2 \\ \diagup \quad \diagdown \\ H_2C \!-\!\!\!-\!\!\!- CH_2 \end{array}$$

When heated, the carbon ring (the triangle) opens up to give propylene, $CH_2\!=\!CHCH_3$. Because the compounds are isomers (different compounds with the same molecular formula), the reaction is called an *isomerization*.

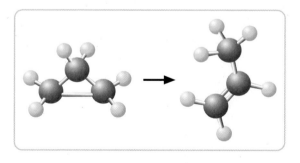

$$C_3H_6(g) \longrightarrow CH_2\!=\!CHCH_3(g)$$
Cyclopropane   Propylene

The reaction has the experimentally determined rate law

$$Rate = k[C_3H_6]$$

The reaction is first order in cyclopropane and first order overall.

2. Nitrogen monoxide, NO, reacts with hydrogen according to the equation

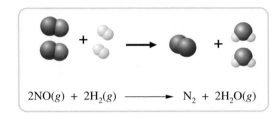

$$2NO(g) + 2H_2(g) \longrightarrow N_2 + 2H_2O(g)$$

The experimentally determined rate law is

$$\text{Rate} = k[NO]^2[H_2]$$

Thus, the reaction is second order in NO, first order in H_2, and third order overall.

3. Acetone, CH_3COCH_3, reacts with iodine in acidic solution.

$$CH_3COCH_3(aq) + I_2(aq) \xrightarrow{H^+} CH_3COCH_2I(aq) + HI(aq)$$

The experimentally determined rate law is

$$\text{Rate} = k[CH_3COCH_3][H^+]$$

The reaction is first order in acetone. It is zero order in iodine; that is, the rate law contains the factor $[I_2]^0 = 1$. Therefore, the rate does not depend on the concentration of I_2, as long as some concentration of I_2 is present. Note that the reaction is first order in the catalyst, H^+. Thus, the overall order is 2.

Although reaction orders frequently have whole-number values (particularly 1 or 2), they can be fractional. Zero and negative orders are also possible.

OWL Interactive Example 13.3 Determining the Order of Reaction from the Rate Law

Gaining Mastery Toolbox

Critical Concept 13.3
For a chemical reaction, the rate law specifies the reaction order of each reactant and allows the determination of the overall order. The reactant order for each reactant is the exponent in the rate law. The overall order of the reaction is the sum of the exponents (orders) of each reactant in the rate law. Rate laws must be determined experimentally.

Solution Essentials:
- Overall reaction order
- Reaction order
- Rate law

Bromide ion is oxidized by bromate ion in acidic solution.

$$5Br^-(aq) + BrO_3^-(aq) + 6H^+(aq) \longrightarrow 3Br_2(aq) + 3H_2O(l)$$

The experimentally determined rate law is

$$\text{Rate} = k[Br^-][BrO_3^-][H^+]^2$$

What is the order of reaction with respect to each reactant species? What is the overall order of the reaction?

Problem Strategy The exponent of each of the reactants given in the experimentally determined rate law is the reaction order of that reactant. The overall order of the reaction is the sum of the orders of the reactant species in the rate law.

Solution The reaction is **first order with respect to Br^-** and **first order with respect to BrO_3^-; it is second order with respect to H^+. The reaction is fourth order overall** $(=1 + 1 + 2)$.

Answer Check Whenever you are determining the order of reactions or of reactant species, never rely on the stoichiometry of the reaction. You must use the experimentally determined rate law.

Exercise 13.3 What are the reaction orders with respect to each reactant species for the following reaction?

$$NO_2(g) + CO(g) \longrightarrow NO(g) + CO_2(g)$$

Assume the rate law is

$$\text{Rate} = k[NO_2]^2$$

What is the overall order?

See Problems 13.47, 13.48, 13.49, and 13.50.

CONCEPT CHECK 13.2

Consider the reaction $Q + R \longrightarrow S + T$ and the rate law for the reaction:

$$\text{Rate} = k[Q]^0[R]^2$$

a You run the reaction three times, each time starting with $[R] = 2.0\ M$. For each run you change the starting concentration of $[Q]$: run 1, $[Q] = 0.0\ M$; run 2, $[Q] = 1.0\ M$; run 3, $[Q] = 2.0\ M$. Rank the rate of the three reactions using each of these concentrations.

b The way the rate law is written in this problem is not typical for expressions containing reactants that are zero order in the rate law. Write the rate law in the more typical fashion.

Determining the Rate Law

The experimental determination of the rate law for a reaction requires that you find the order of the reaction with respect to each reactant and any catalyst. The *initial-rate method* is a simple way to obtain reaction orders. It consists of doing a series of experiments in which the initial, or starting, concentrations of reactants are varied. Then the initial rates are compared, from which the reaction orders can be deduced.

To see how this method works, again consider the reaction

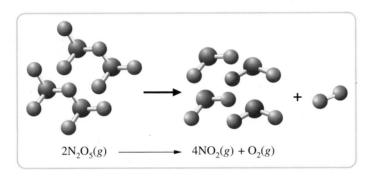

$$2N_2O_5(g) \longrightarrow 4NO_2(g) + O_2(g)$$

You observe this reaction in two experiments. In Experiment 2, the initial concentration of N_2O_5 is twice that in Experiment 1. You then note the initial rate of disappearance of N_2O_5 in each case. The initial concentrations and corresponding initial rates for the two experiments are given in the following table: ▶

In Figure 13.4, the slope of the tangent to the curve at t = 0 equals the initial rate of appearance of O_2, which equals one-half the initial rate of disappearance of N_2O_5.

	Initial N_2O_5 Concentration	Initial Rate of Disappearance of N_2O_5
Experiment 1	1.0×10^{-2} mol/L	4.8×10^{-6} mol/(L·s)
Experiment 2	2.0×10^{-2} mol/L	9.6×10^{-6} mol/(L·s)

The rate law for this reaction will have the concentration of reactant raised to a power m.

$$\text{Rate} = k[N_2O_5]^m$$

The value of m (the reaction order) must be determined from the experimental data. Note that when the N_2O_5 concentration is doubled, you get a new rate, Rate′, given by the following equation:

$$\text{Rate}' = k(2[N_2O_5])^m = 2^m k[N_2O_5]^m$$

This rate is 2^m times the original rate.

You can now see how the rate is affected when the concentration is doubled for various choices of m. Suppose $m = 2$. You get $2^m = 2^2 = 4$. That is, when the initial concentration is doubled, the rate is multiplied by 4. We summarize the results for various choices of m in Table 13.1.

Table 13.1	Effect on Rate of Doubling the Initial Concentration of Reactant
m	**Rate Is Multiplied by:**
-1	$\frac{1}{2}$
0	1
1	2
2	4

Suppose you divide the initial rate of reaction of N_2O_5 from Experiment 2 by the initial rate from Experiment 1.

$$\frac{9.6 \times 10^{-6}\ \text{mol/(L}\cdot\text{s)}}{4.8 \times 10^{-6}\ \text{mol/(L}\cdot\text{s)}} = 2.0$$

You see that when the N_2O_5 concentration is doubled, the rate is doubled. This corresponds to the case $m = 1$. The rate law must have the form

$$\text{Rate} = k[N_2O_5]$$

You can determine the value of the rate constant k by substituting values of the rate and N_2O_5 concentrations from any of the experiments into the rate law. Using values from Experiment 2, you get

$$9.6 \times 10^{-6}\ \text{mol/(L}\cdot\text{s)} = k \times 2.0 \times 10^{-2}\ \text{mol/L}$$

Hence,

$$k = \frac{9.6 \times 10^{-6}/\text{s}}{2.0 \times 10^{-2}} = 4.8 \times 10^{-4}/\text{s}$$

Example 13.4 Determining the Rate Law from Initial Rates

Gaining Mastery Toolbox

Critical Concept 13.4
The rate law is a function of the concentrations of the reactants of a chemical reaction and the rate constant. The exponents of each reactant (order) in the rate law must be experimentally determined. The initial rate method is an experimental approach that allows for the determination of the rate law for a chemical reaction: the order of each reactant and the rate constant, *k*.

Solution Essentials:
- Initial rate method
- Reaction order
- Rate constant
- Rate law

H_2O_2

Iodide ion is oxidized in acidic solution to triiodide ion, I_3^-, by hydrogen peroxide.

$$H_2O_2(aq) + 3I^-(aq) + 2H^+(aq) \longrightarrow I_3^-(aq) + 2H_2O(l)$$

A series of four experiments was run at different concentrations, and the initial rates of I_3^- formation were determined (see table). a From these data, obtain the reaction orders with respect to H_2O_2, I^-, and H^+. b Then find the rate constant.

	Initial Concentrations (mol/L)			*Initial Rate*
	H_2O_2	I^-	H^+	*[mol/(L·s)]*
Exp. 1	0.010	0.010	0.00050	1.15×10^{-6}
Exp. 2	0.020	0.010	0.00050	2.30×10^{-6}
Exp. 3	0.010	0.020	0.00050	2.30×10^{-6}
Exp. 4	0.010	0.010	0.00100	1.15×10^{-6}

Problem Strategy

a In order to know the reaction orders of the reactants, you need to know the rate law. Using the reactants in the chemical equation, you assume the rate law has the following form:

$$\text{Rate} = k[H_2O_2]^m[I^-]^n[H^+]^p$$

You can determine each of the reaction orders (exponents m, n, and p) by choosing experiments in which all concentrations of reactants except one are held constant. Solving for each exponent will require comparing different experiments. Once you know the reaction orders, you will know the rate law. b To determine the rate constant, you can take the data from one of the experiments and substitute them into the rate law; the only unknown in the equation will be the rate constant, k.

Solution

a Comparing Experiment 1 and Experiment 2, you see that when the H_2O_2 concentration is doubled (with other concentrations constant), the rate is doubled. From Table 13.1, you see that $m = 1$. The reaction is first order in H_2O_2.

To solve a problem such as this in a general way, you approach it algebraically. You write the rate law for two experiments. (The subscripts denote the experiments.)

$$(\text{Rate})_1 = k[H_2O_2]_1^m[I^-]_1^n[H^+]_1^p$$

$$(\text{Rate})_2 = k[H_2O_2]_2^m[I^-]_2^n[H^+]_2^p$$

(continued)

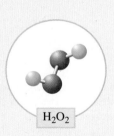

(*continued*)

Now you divide the second equation by the first.

$$\frac{(\text{Rate})_2}{(\text{Rate})_1} = \frac{k[H_2O_2]_2^m[I^-]_2^n[H^+]_2^p}{k[H_2O_2]_1^m[I^-]_1^n[H^+]_1^p}$$

The rate constant cancels. Grouping the terms, you obtain

$$\frac{(\text{Rate})_2}{(\text{Rate})_1} = \left(\frac{[H_2O_2]_2}{[H_2O_2]_1}\right)^m\left(\frac{[I^-]_2}{[I^-]_1}\right)^n\left(\frac{[H^+]_2}{[H^+]_1}\right)^p$$

Now you substitute values from Experiment 1 and Experiment 2. (In a ratio the units cancel and can be omitted.)

$$\frac{2.30 \times 10^{-6}}{1.15 \times 10^{-6}} = \left[\frac{0.020}{0.010}\right]^m\left[\frac{0.010}{0.010}\right]^n\left[\frac{0.00050}{0.00050}\right]^p$$

This gives $2 = 2^m$, from which you obtain $m = 1$. That is, doubling the H_2O_2 concentration doubles the rate.

The exponents n and p are obtained in the same way. Comparing Experiment 1 and Experiment 3, you see that doubling the I^- concentration (with the other concentrations constant) doubles the rate. Therefore, $n = 1$ (the reaction is first order in I^-). Finally, comparing Experiment 1 and Experiment 4, you see that doubling the H^+ concentration (holding other concentrations constant) has no effect on the rate. Therefore, $p = 0$ (the reaction is zero order in H^+). Because $[H^+]^0 = 1$, the rate law is

Rate $= k[H_2O_2][I^-]$

The reaction orders with respect to H_2O_2, I^-, and H^+, are **1, 1, and 0,** respectively. Note that the orders are *not* related to the coefficients of the overall equation.

b You calculate the rate constant by substituting values from any of the experiments into the rate law. Using Experiment 1, you obtain

$$1.15 \times 10^{-6}\,\frac{\text{mol}}{\text{L}\cdot\text{s}} = k \times 0.010\,\frac{\text{mol}}{\text{L}} \times 0.010\,\frac{\text{mol}}{\text{L}}$$

$$k = \frac{1.15 \times 10^{-6}/\text{s}}{0.010 \times 0.010 \times \text{mol}/\text{L}} = \mathbf{1.2 \times 10^{-2}\ L/(mol \cdot s)}$$

Answer Check Although reaction orders of greater than 2 are certainly possible, they are not nearly as common as those less than or equal to 2. If you keep this fact in mind when solving problems of this type, a reaction order greater than 2 should be looked at carefully to make sure that a calculation error wasn't made.

Exercise 13.4 The initial-rate method was applied to the decomposition of nitrogen dioxide.

$$2NO_2(g) \longrightarrow 2NO(g) + O_2(g)$$

It yielded the following results:

	Initial NO₂ Concentration	Initial Rate of Formation of O₂
Exp.1	0.010 mol/L	7.1×10^{-5} mol/(L·s)
Exp.2	0.020 mol/L	28×10^{-5} mol/(L·s)

Find the rate law and the value of the rate constant with respect to O_2 formation.

■ See Problems 13.51, 13.52, 13.53, 13.54, 13.55, and 13.56.

CONCEPT CHECK 13.3

Rate laws are not restricted to chemical systems; they are used to help describe many "everyday" events. For example, a rate law for tree growth might look something like this:

$$\text{Rate of growth} = (\text{soil type})^w(\text{temperature})^x(\text{light})^y(\text{fertilizer})^z$$

In this equation, like chemical rate equations, the exponents need to be determined by experiment. (Can you think of some other factors?)

a Say you are a famous physician trying to determine the factors that influence the rate of aging in humans. Develop a rate law, like the one above, that would take into account at least four factors that affect the rate of aging.

b Explain what you would need to do in order to determine the exponents in your rate law.

c Consider smoking to be one of the factors in your rate law. You conduct an experiment and find that a person smoking two packs of cigarettes a day quadruples ($4\times$) the rate of aging over that of a one-pack-a-day smoker. Assuming that you could hold all other factors in your rate law constant, what would be the exponent of the smoking term in your rate law?

13.4 Change of Concentration with Time

A rate law tells you how the rate of a reaction depends on reactant concentrations at a particular moment. But often you would like to have a mathematical relationship showing how a reactant concentration changes over a period of time. Such an equation would be directly comparable to the experimental data, which are usually obtained as concentrations at various times. In addition to summarizing the experimental data, this equation would predict concentrations for all times. Using it, you could answer questions such as: How long does it take for this reaction to be 50% complete? to be 90% complete?

Moreover, as you will see, knowing exactly how the concentrations change with time for different rate laws suggests ways of plotting the experimental data on a graph. Graphical plotting provides an alternative to the initial-rate method for determining the rate law.

Using calculus, we can transform a rate law into a *mathematical relationship between concentration and time* called an **integrated rate law.** Because we will work only with the final equations, we need not go into the derivations here. We will look in some depth at first-order reactions and briefly at second-order and zero-order reactions.

Integrated Rate Laws (Concentration—Time Equations)

First-Order Rate Law Let us first look at first-order rate laws. The decomposition of dinitrogen pentoxide,

$$2N_2O_5(g) \longrightarrow 4NO_2(g) + O_2(g)$$

has the following rate law:

$$\text{Rate} = -\frac{\Delta[N_2O_5]}{\Delta t} = k[N_2O_5]$$

Using calculus, one can show that such a first-order rate law leads to the following relationship between N_2O_5 concentration and time: ◀

$$\ln\frac{[N_2O_5]_t}{[N_2O_5]_0} = -kt$$

Here $[N_2O_5]_t$ is the concentration at time t, and $[N_2O_5]_0$ is the initial concentration of N_2O_5 (that is, the concentration at $t = 0$). The symbol "ln" denotes the natural logarithm (base $e = 2.718 \ldots$).

The general derivation using calculus is as follows: Substituting [A] for [N_2O_5], the rate law becomes

$$\frac{-d[A]}{dt} = k[A]$$

You rearrange this to give

$$\frac{-d[A]}{[A]} = k\,dt$$

Integrating from time = 0 to time = t,

$$-\int_{[A]_0}^{[A]_t} \frac{d[A]}{[A]} = k\int_0^t dt$$

gives

$$-\{\ln[A]_t - \ln[A]_0\} = k(t - 0)$$

This can be rearranged to give the equation in the text.

This equation enables you to calculate the concentration of N_2O_5 at any time, once you are given the initial concentration and the rate constant. Also, you can find the time it takes for the N_2O_5 concentration to decrease to a particular value.

More generally, let A be a substance that reacts to give products according to the equation

$$aA \longrightarrow products$$

where a is the stoichiometric coefficient of reactant A. Suppose that this reaction has a first-order rate law

$$Rate = -\frac{\Delta[A]}{\Delta t} = k[A]$$

Using calculus, you get the following integrated rate law equation:

$$\ln\frac{[A]_t}{[A]_0} = -kt \quad \text{(first-order integrated rate law)}$$

Here $[A]_t$ is the concentration of reactant A at time t, and $[A]_0$ is the initial concentration. The ratio $[A]_t/[A]_0$ is the fraction of reactant remaining at time t.

Second-Order Rate Law Consider the reaction

$$aA \longrightarrow products$$

and suppose it has the second-order rate law

$$Rate = -\frac{\Delta[A]}{\Delta t} = k[A]^2$$

An example is the decomposition of nitrogen dioxide at moderately high temperatures (300°C to 400°C).

$$2NO_2(g) \longrightarrow 2NO(g) + O_2(g)$$

Using calculus, you can obtain the following relationship between the concentration of A and the time.

$$\frac{1}{[A]_t} = kt + \frac{1}{[A]_0} \quad \text{(second-order integrated rate law)}$$

Using this equation, you can calculate the concentration of NO_2 at any time during its decomposition if you know the rate constant and the initial concentration. At 330°C, the rate constant for the decomposition of NO_2 is 0.775 L/(mol·s). Suppose the initial concentration is 0.0030 mol/L. What is the concentration of NO_2 after 645 s? By substituting into the previous equation, you get

$$\frac{1}{[NO_2]_t} = 0.775 \text{L/(mol·s)} \times 645 \text{ s} + \frac{1}{0.0030 \text{ mol/L}} = 8.3 \times 10^2 \text{ L/mol}$$

If you invert both sides of the equation, you find that $[NO_2]_t = 0.0012$ mol/L. Thus, after 645 s, the concentration of NO_2 decreased from 0.0030 mol/L to 0.0012 mol/L.

Zero-Order Reactions

There are instances where reactions are zero-order. An example includes the decomposition of ethyl alcohol in the liver in the presence of the enzyme liver alcohol dehydrogenase. Once again, consider the reaction with the general form

$$aA \longrightarrow products$$

which has a zero-order rate law of

$$Rate = k[A]^0$$

Keeping in mind that any real number raised to the zero power is 1, the expression for zero-order rate law is usually written as

$$Rate = k$$

This rate law indicates that the rate of a zero-order reaction does not change with concentration. For example, if you have a zero-order reaction

$$A \longrightarrow B + C$$

the rate of this reaction has no dependence upon the concentration of A; however, there must be some A present for the reaction to occur. Many zero-order reactions also require some minimum reactant concentration for the reaction to behave as zero-order. This is true in the case of enzyme reactions like that mentioned above as well as in reactions that are catalyzed by a metal surface, such as the decomposition reaction of N_2O in the presence of platinum metal.

$$2N_2O(g) \xrightarrow{Pt} 2N_2(g) + O_2(g)$$

As long as this reaction is run at a high enough pressure of N_2O to saturate the metal surface, it will behave as a zero-order reaction.

The relationship between concentration and time for a zero-order reaction is

$$[A]_t = -kt + [A]_0 \quad \text{(zero-order integrated rate law)}$$

Below is an example of a calculation that requires an integrated rate law.

Example 13.5

Using an Integrated Rate Law

Gaining Mastery Toolbox

Critical Concept 13.5
An integrated rate law describes mathematically how the concentration of a reactant changes with time. An integrated rate law allows one to calculate reactant concentrations at any point in time during a chemical reaction.

Solution Essentials:
- Integrated rate law for a first-order reaction
$$\left(\ln \frac{[A]_t}{[A]_0} = -kt \right)$$
- Integrated rate law
- Rate law
- Reaction order

N_2O_5

The decomposition of N_2O_5 to NO_2 and O_2 is first order, with a rate constant of 4.80×10^{-4}/s at 45°C. **a** If the initial concentration is 1.65×10^{-2} mol/L, what is the concentration after 825 s? **b** How long would it take for the concentration of N_2O_5 to decrease to 1.00×10^{-2} mol/L from its initial value, given in a?

Problem Strategy Part a of this problem wants us to determine the concentration of a reactant (N_2O_5) after some period of time has elapsed. Part b of the problem asks us to determine the amount of time that has elapsed for a certain change in concentration of the reactant. An integrated rate law (first-order here), which relates concentration changes to time, will enable us to complete both of these problems.

Solution

a In this case, you need to use the equation relating concentration to time for a first-order reaction, which is

$$\ln \frac{[A]_t}{[A]_0} = -kt$$

Substituting the appropriate values, you get

$$\ln \frac{[N_2O_5]_t}{1.65 \times 10^{-2} \text{ mol/L}} = -4.80 \times 10^{-4}\text{/s} \times 825 \text{ s} = -0.396$$

To solve for $[N_2O_5]_t$, you take the antilogarithm (antiln) of both sides. This removes the ln from the left and gives antiln(−0.396), or $e^{-0.396}$, on the right, which equals 0.673.

$$\frac{[N_2O_5]_t}{1.65 \times 10^{-2} \text{ mol/L}} = 0.673$$

Hence,

$$[N_2O_5]_t = 1.65 \times 10^{-2}\,\text{mol/L} \times 0.673 = \textbf{0.0111 mol/L}$$

b You substitute into the same first-order equation relating concentration to time.

$$\ln \frac{1.00 \times 10^{-2} \text{ mol/L}}{1.65 \times 10^{-2} \text{ mol/L}} = -4.80 \times 10^{-4}\text{/s} \times t$$

The left side equals −0.501; the right side equals -4.80×10^{-4}/s $\times t$. Hence,

$$0.501 = 4.80 \times 10^{-4}\text{/s} \times t$$

(continued)

(continued)

or

$$t = \frac{0.501}{4.80 \times 10^{-4}/s} = \mathbf{1.04 \times 10^3\,s} \quad \textbf{(17.4 min)}$$

Answer Check It is easy to make calculation errors when working with exponents and logarithms. Because of this, it is a good idea to check whether your final answer yields expected results when placed into a starting equation. For example, the first step of part a of this calculation involved the equation

$$\ln \frac{[N_2O_5]_t}{1.65 \times 10^{-2}\,\text{mol/L}} = -0.396$$

If you plug the answer that you obtained for $[N_2O_5]_t$ into the left-hand side of this equation, it should yield -0.396.

Exercise 13.5 ⓐ What would be the concentration of dinitrogen pentoxide in the experiment described in Example 13.5 after 6.00×10^2 s? ⓑ How long would it take for the concentration of N_2O_5 to decrease to 10.0% of its initial value?

■ See Problems 13.57, 13.58, 13.59, 13.60, 13.61, and 13.62.

Half-Life of a Reaction

As a reaction proceeds, the concentration of a reactant decreases, because it is being consumed. The **half-life**, $t_{1/2}$, of a reaction is *the time it takes for the reactant concentration to decrease to one-half of its initial value.* ▶

The concept of half-life is also used to characterize a radioactive nucleus, whose radioactive decay is a first-order process. This is discussed in Chapter 20.

For a first-order reaction, such as the decomposition of dinitrogen pentoxide, the half-life is independent of the initial concentration. To see this, substitute into the equation

$$\ln \frac{[N_2O_5]_t}{[N_2O_5]_0} = -kt$$

In one half-life, the N_2O_5 concentration decreases by one-half, from its initial value, $[N_2O_5]_0$, to the value $[N_2O_5]_t = \frac{1}{2}[N_2O_5]_0$. After substituting, the equation becomes

$$\ln \frac{\frac{1}{2}[N_2O_5]_0}{[N_2O_5]_0} = -kt_{1/2}$$

The expression on the left equals $\ln \frac{1}{2} = -0.693$. Hence,

$$0.693 = kt_{1/2}$$

Solving for the half-life, $t_{1/2}$, you get

$$t_{1/2} = \frac{0.693}{k}$$

Because the rate constant for the decomposition of N_2O_5 at 45°C is 4.80×10^{-4}/s, the half-life is

$$t_{1/2} = \frac{0.693}{4.80 \times 10^{-4}/s} = 1.44 \times 10^3\,s$$

Thus, the half-life is 1.44×10^3 s, or 24.0 min.

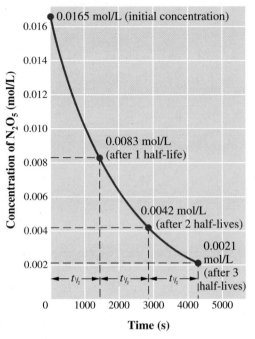

Figure 13.8 ▲

A graph illustrating that the half-life of a first-order reaction is independent of initial concentration In one half-life (1440 s, or 24.0 min), the concentration decreases by one-half, from 0.0165 mol/L to $\frac{1}{2}$ × 0.0165 mol/L = 0.0083 mol/L. After each succeeding half-life (24.0 min), the concentration decreases by one-half again, from 0.0083 mol/L to $\frac{1}{2}$ × 0.0083 mol/L = 0.0042 mol/L, then to $\frac{1}{2}$ × 0.0042 mol/L = 0.0021 mol/L, and so forth.

You see that the half-life of N_2O_5 does not depend on the initial concentration of N_2O_5. This means that the half-life is the same at any time during the reaction. If the initial concentration is 0.0165 mol/L, after one half-life (24.0 min) the concentration decreases to $\frac{1}{2}$ × 0.0165 mol/L = 0.0083 mol/L. After another half-life (another 24.0 min), the N_2O_5 concentration decreases to $\frac{1}{2}$ × 0.0083 mol/L = 0.0042 mol/L. Every time one half-life passes, the N_2O_5 concentration decreases by one-half again (see Figure 13.8).

The foregoing result for the half-life for the first-order decomposition of N_2O_5 is perfectly general. That is, for the general first-order rate law

$$\text{Rate} = -\frac{\Delta[A]}{\Delta t} = k[A]$$

the half-life is related to the rate constant but is independent of concentration of A.

$$t_{1/2} = \frac{0.693}{k} \quad \text{(first-order)}$$

Example 13.6   Relating the Half-Life of a Reaction to the Rate Constant

Gaining Mastery Toolbox

Critical Concept 13.6
The half-life of a chemical reaction is mathematically related to the rate constant. For a first-order reaction, the rate constant and half-life are independent of the reactant concentration.

Solution Essentials:
• Half-life
• First-order reaction half-life expression
$\left(t_{1/2} = \dfrac{0.693}{k} \right)$
• Rate constant

Sulfuryl chloride, SO_2Cl_2, is a colorless, corrosive liquid whose vapor decomposes in a first-order reaction to sulfur dioxide and chlorine.

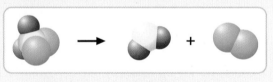

$$SO_2Cl_2(g) \longrightarrow SO_2(g) + Cl_2(g)$$

At 320°C, the rate constant is 2.20×10^{-5}/s. **a** What is the half-life of SO_2Cl_2 vapor at this temperature? **b** How long (in hours) would it take for 50.0% of the SO_2Cl_2 to decompose? How long would it take for 75.0% of the SO_2Cl_2 to decompose?

Problem Strategy In part a of this problem we are asked about the half-life of a first-order reaction, so a logical starting point is to use the appropriate half-life equation. For part b, consider that each half-life reduces the amount of material by 50.0%; thus two half-lives will reduce a starting amount of material by 75%, three half lives by 87.5%, and so on.

Solution

a $t_{1/2} = \dfrac{0.693}{k} = \dfrac{0.693}{2.20 \times 10^{-5}/s} = \mathbf{3.15 \times 10^4 \ s}$

b The half-life is the time required for one-half (50.0%) of the SO_2Cl_2 to decompose. This is 3.15×10^4 s, or **8.75 hr.** After another half-life, one-half of the remaining SO_2Cl_2 decomposes. The total decomposed is $\frac{1}{2} + (\frac{1}{2} \times \frac{1}{2}) = \frac{3}{4}$, or 75.0%. The time required is two half-lives, or 2 × 8.75 hr = **17.5 hr.**

(continued)

(*continued*)

An alternative approach is to use the integrated rate law for a first-order reaction:

$$\ln\frac{[SO_2Cl_2]_t}{[SO_2Cl_2]_0} = -kt$$

Given the data in the problem statement, we can let $[SO_2Cl_2]_0 = 1.00$ and $[SO_2Cl_2]_t = 0.250$ (the 0.250 represents how much would be remaining at time t after 75% of the SO_2Cl_2 had decomposed). Using these values and the given rate constant, solve for t.

Answer Check In a problem such as this, where the initial concentration of reactant is not given, the alternative approach works *only* for first-order reactions.

Exercise 13.6 The isomerization of cyclopropane, C_3H_6, to propylene, $CH_2\!=\!CHCH_3$, is first order in cyclopropane and first order overall. At 1000°C, the rate constant is 9.2/s. What is the half-life of cyclopropane at 1000°C? How long would it take for the concentration of cyclopropane to decrease to 50% of its initial value? to 25% of its initial value?

■ See Problems 13.63 and 13.64.

It can be shown by reasoning similar to that given previously that the half-life of a second-order rate law, Rate $= k[A]^2$, is

$$t_{1/2} = \frac{1}{k[A]_0} \quad \text{(second-order)}$$

In this case, the half-life depends on initial concentration and each subsequent half-life becomes larger as time goes on. Consider the decomposition of NO_2 at 330°C. It takes 430 s for the concentration to decrease by one-half from 0.0030 mol/L to 0.0015 mol/L. However, it takes 860 s (twice as long) for the concentration to decrease by one-half again. The fact that the half-life increases or changes with time is evidence that the reaction is not first order.

For zero-order reactions, the half-life is dependent upon the initial concentration of the reactant. However, in contrast to the second-order reaction, as a zero-order reaction proceeds, each half-life gets shorter.

$$t_{1/2} = \frac{[A]_0}{2k} \quad \text{(zero-order)}$$

Graphing of Kinetic Data

Earlier you saw that the order of a reaction can be determined by comparing initial rates for several experiments in which different initial concentrations are used (initial-rate method). It is also possible to determine the order of a reaction by graphical plotting of the data for a particular experiment. The experimental data are plotted in several different ways, first assuming a first-order reaction, then a second-order reaction, and so forth. The order of the reaction is determined by which graph gives the best fit to the experimental data. To illustrate, we will look at how the plotting is done for first-order and second-order reactions and then compare graphs for a specific reaction.

You have seen that the first-order rate law, $-\Delta[A]/\Delta t = k[A]$, gives the following relationship between concentration of A and time:

$$\ln\frac{[A]_t}{[A]_0} = -kt$$

This equation can be rewritten in a slightly different form, which you can identify with the equation of a straight line. Using the property of logarithms that $\ln(A/B) = \ln A - \ln B$, you get

$$\ln[A]_t = -kt + \ln[A]_0$$

A straight line has the mathematical form $y = mx + b$, when y is plotted on the vertical axis against x on the horizontal axis. ◄ Let us now make the following identifications:

See Appendix A for a discussion of the mathematics of a straight line.

$$\underset{y}{\underline{\ln[A]_t}} = \underset{mx}{\underline{-k \quad t}} + \underset{b}{\underline{\ln[A]_0}}$$

This means that if you plot $\ln[A]_t$ on the vertical axis against the time t on the horizontal axis, you will get a straight line for a first-order reaction. Figure 13.9 shows a plot of $\ln[N_2O_5]$ at various times during the decomposition reaction. The fact that the points lie on a straight line is confirmation that the rate law is first order.

You can obtain the rate constant for the reaction from the slope, m, of the straight line.

$$-k = m \quad \text{or} \quad k = -m$$

You calculate the slope of this curve in the same way you obtained the average rate of reaction from kinetic data (Example 13.2). You select two points far enough apart that when you subtract to obtain Δx and Δy for the slope, you do not lose significant figures. Using the first and last points in Figure 13.9, you get

$$m = \frac{\Delta y}{\Delta x} = \frac{(-5.843) - (-4.104)}{(3600 - 0)\ \text{s}} = \frac{-1.739}{3600\ \text{s}} = -4.83 \times 10^{-4}/\text{s}$$

Therefore, $k = 4.84 \times 10^{-4}/\text{s}$. (Two points were selected directly from the experimental data. In precise work, you would first draw the straight line that best fits the experimental data points and then calculate the slope of this line.)

The second-order rate law, $-\Delta[A]/\Delta t = k[A]^2$, gives the following relationship between concentration of A and time:

$$\underset{y}{\underline{\frac{1}{[A]_t}}} = \underset{mx}{\underline{kt}} + \underset{b}{\underline{\frac{1}{[A]_0}}}$$

In this case, you get a straight line if you plot $1/[A]_t$ on the vertical axis against the time t on the horizontal axis for a second-order reaction.

Figure 13.9 ▶

Plot of ln[N₂O₅] versus time A straight line can be drawn through the experimental points (colored dots). The fact that the straight line fits the experimental data so well confirms that the rate law is first order.

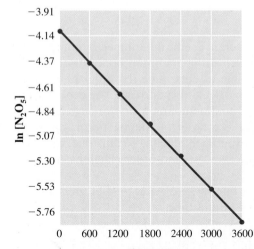

Time	[N₂O₅]	ln[N₂O₅]
0	0.0165	−4.104
600	0.0124	−4.390
1200	0.0093	−4.678
1800	0.0071	−4.948
2400	0.0053	−5.240
3000	0.0039	−5.547
3600	0.0029	−5.843

As an illustration of the determination of reaction order by graphical plotting, consider the following data for the decomposition of NO_2 at 330°C.

$$2NO_2(g) \longrightarrow 2NO(g) + O_2(g)$$

The concentrations of NO_2 for various times are given in the table of data in Figure 13.10. In Figure 13.10a we have plotted $\ln[NO_2]$ against t, and in Figure 13.10b we have plotted $1/[NO_2]$ against t. Only in Figure 13.10b do the points closely follow a straight line, indicating that the rate law is second order. That is,

$$\text{Rate of disappearance of } NO_2 = -\frac{\Delta[NO_2]}{\Delta t} = k[NO_2]^2$$

You can obtain the rate constant, k, for a second-order reaction from the slope of the line, similar to the way you did for a first-order reaction. In this case, however, the slope equals k, as you can see from the preceding equation. Choosing the first and last points in Figure 13.10b, you get

$$k = \frac{\Delta y}{\Delta x} = \frac{(379 - 100) \text{ L/mol}}{(360 - 0) \text{ s}} = 0.775 \text{ L/(mol·s)}$$

Graphical methods can also be used to determine the rate constant for zero-order reactions. Table 13.2 summarizes this and the other relationships discussed in this section for zero-order, first-order, and second-order reactions.

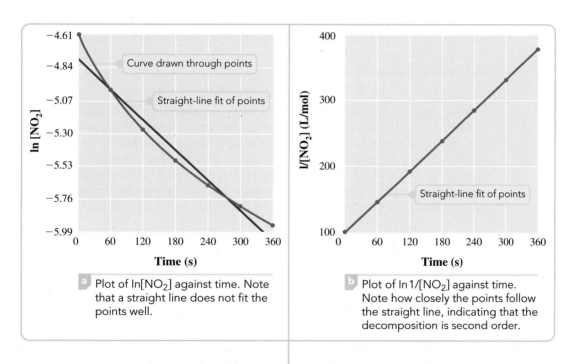

Figure 13.10 ◀

Plotting the data for the decomposition of nitrogen dioxide at 330°C

a Plot of ln[NO₂] against time. Note that a straight line does not fit the points well.

b Plot of ln 1/[NO₂] against time. Note how closely the points follow the straight line, indicating that the decomposition is second order.

Time (s)	[NO₂] (mol/L)	(A) ln [NO₂]	(B) 1/[NO₂]
0	1.00×10^{-2}	−4.605	100
60	0.683×10^{-2}	−4.984	146
120	0.518×10^{-2}	−5.263	193
180	0.418×10^{-2}	−5.476	239
240	0.350×10^{-2}	−5.656	286
300	0.301×10^{-2}	−5.805	332
360	0.264×10^{-2}	−5.938	379

Order	Rate Law	Integrated Rate Law	Half-Life	Straight-Line Plot

Table 13.2 Relationships for Zero-Order, First-Order, and Second-Order Reactions

Order	Rate Law	Integrated Rate Law	Half-Life	Straight-Line Plot
0	Rate $= k$	$[A]_t = -kt + [A]_0$	$\dfrac{[A]_0}{2k}$	$[A]$ vs t
1	Rate $= k[A]$	$\ln\dfrac{[A]_t}{[A]_0} = -kt$	$0.693/k$	$\ln[A]$ vs t
2	Rate $= k[A]^2$	$\dfrac{1}{[A]_t} = kt + \dfrac{1}{[A]_0}$	$1/(k[A]_0)$	$\dfrac{1}{[A]}$ vs t

CONCEPT CHECK 13.4

A reaction believed to be either first or second order has a half-life of 20 s at the beginning of the reaction but a half-life of 40 s sometime later. What is the order of the reaction?

13.5 Temperature and Rate; Collision and Transition-State Theories

As we noted earlier, the rate of reaction depends on temperature. This shows up in the rate law through the rate constant, which is found to vary with temperature. In most cases, the rate increases with temperature (Figure 13.11). Consider the reaction of nitrogen monoxide, NO, with chlorine, Cl_2, to give nitrosyl chloride, NOCl, and chlorine atoms.

$$NO(g) + Cl_2(g) \longrightarrow NOCl(g) + Cl(g)$$

The rate constant k for this reaction is 4.9×10^{-6} L/(mol·s) at 25°C and 1.5×10^{-5} L/(mol·s) at 35°C. In this case the rate constant and therefore the rate are more than tripled for a 10°C rise in temperature. The change in rate constant with temperature varies considerably from one reaction to another. In many cases, the rate of reaction approximately doubles for a 10°C rise, and this is often given as an approximate rule. How do you explain this strong dependence of reaction rate on temperature? To understand it, you need to look at a simple theoretical explanation of reaction rates.

Figure 13.11 ▶

Effect of temperature on reaction rate

© Cengage Learning

a Each test tube contains potassium permanganate, $KMnO_4$, and oxalic acid, $H_2C_2O_4$, at the same concentrations. Permanganate ion oxidizes oxalic acid to CO_2 and H_2O. One test tube was placed in a beaker of warm water (40°C); the other was kept at room temperature (20°C).

b After 10 minutes, the test tube at 40°C (in the beaker over the burner) showed a noticeable reaction, whereas the other did not.

Collision Theory

Why the rate constant depends on temperature can be explained by collision theory. **Collision theory** of reaction rates is *a theory that assumes that, for reaction to occur, reactant molecules must collide with an energy greater than some minimum value and with the proper orientation. The minimum energy of collision required for two molecules to react* is called the **activation energy, E_a.** The value of E_a depends on the particular reaction.

In collision theory, the rate constant for a reaction is given as a product of three factors: (1) Z, the collision frequency, (2) f, the fraction of collisions having energy greater than the activation energy, and (3) p, the fraction of collisions that occur with the reactant molecules properly oriented. Thus,

$$k = Zfp$$

We will discuss each of these factors in turn.

To have a specific reaction to relate the concepts to, consider the gas-phase reaction of NO with Cl_2, mentioned previously. This reaction is believed to occur in a single step. An NO molecule collides with a Cl_2 molecule. If the collision has sufficient energy and if the molecules are properly oriented, they react to produce NOCl and Cl.

Collision frequency Z, the frequency with which the reactant molecules collide, depends on temperature. As you will see, however, this dependence of collision frequency on temperature does not explain why reaction rates usually change greatly with small temperature increases. You can easily explain why the collision frequency depends on temperature. As the temperature rises, the gas molecules move faster and therefore collide more frequently. Thus, collision frequency is proportional to the root-mean-square (rms) molecular speed, which in turn is proportional to the square root of the absolute temperature, according to the kinetic theory of gases. ▶ From kinetic theory, one can show that at 25°C, a 10°C rise in temperature increases the collision frequency by about 2%. If you were to assume that each collision of reactant molecules resulted in reaction, you would conclude that the rate would increase with temperature at the same rate as the collision frequency—that is, by 2% for a 10°C rise in temperature. This clearly does not explain the tripling of the rate (a 300% increase) that you see in the reaction of NO with Cl_2 when the temperature is raised from 25°C to 35°C.

According to the kinetic theory of gases, the rms molecular speed equals $\sqrt{3RT/M_m}$ (see Section 5.6).

You see that the collision frequency varies only slowly with temperature. However, f, the fraction of molecular collisions having energy greater than the activation energy, changes rapidly in most reactions with even small temperature changes. It can be shown that f is related to the activation energy, E_a, this way:

$$f = e^{-E_a/RT}$$

Here $e = 2.718\ldots$, and R is the gas constant, which equals 8.31 J/(mol·K). For the reaction of NO with Cl_2, the activation energy is 8.5×10^4 J/mol. At 25°C (298 K), the fraction of collisions with sufficient energy for reaction is 1.2×10^{-15}. Thus, the number of collisions of reactant molecules that actually result in reaction is extremely small. But the frequency of collisions is very large, so the reaction rate, which depends on the product of these quantities, is not small. If the temperature is raised by 10°C to 35°C, the fraction of collisions of NO and Cl_2 molecules with sufficient energy for reaction is 3.8×10^{-15}, over three times larger than the value at 25°C! The tripling of the reaction rate is explained by the temperature dependence of f.

From the previous equation relating f to E_a, you see that f decreases with increasing values of E_a. Because the rate constant depends on f, this means that reactions with large activation energies have small rate constants and that reactions with small activation energies have large rate constants.

The reaction rate also depends on p, the proper orientation of the reactant molecules when they collide. This factor is independent of temperature changes. You can see why it is important that the reactant molecules be properly oriented by looking in some detail at the reaction of NO with Cl_2. Figure 13.12 shows two possible molecular collisions. In Figure 13.12a, the NO and Cl_2 molecules collide

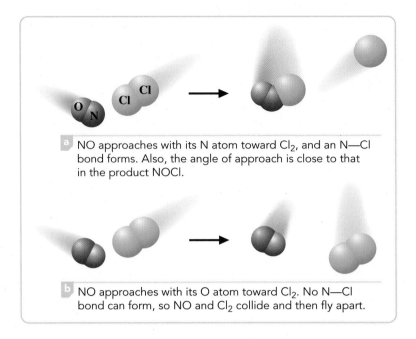

a NO approaches with its N atom toward Cl$_2$, and an N—Cl bond forms. Also, the angle of approach is close to that in the product NOCl.

b NO approaches with its O atom toward Cl$_2$. No N—Cl bond can form, so NO and Cl$_2$ collide and then fly apart.

properly oriented for reaction. The NO molecule approaches with its N atom toward the Cl$_2$ molecule. In addition, the angle of approach is about that expected for the formation of bonds in the product molecule NOCl. In Figure 13.12*b*, an NO molecule approaches with its O atom toward the Cl$_2$ molecule. Because this orientation does not allow the formation of a bond between the N atom and a Cl atom, it is ineffective for reaction. The NO and Cl$_2$ molecules come together and then fly apart. All orientations except those close to that shown in (*a*) are ineffective.

Transition-State Theory

Collision theory explains some important features of a reaction, but it is limited in that it does not explain the role of activation energy. **Transition-state theory** *explains the reaction resulting from the collision of two molecules in terms of an activated complex.* An **activated complex** (transition state) is *an unstable grouping of atoms that can break up to form products.* We can represent the formation of the activated complex this way:

$$O{=}N + Cl{-}Cl \longrightarrow [O{=}N{\cdots}Cl{\cdots}Cl]$$

When the molecules come together with proper orientation, an N—Cl bond begins to form. At the same time, the kinetic energy of the collision is absorbed by the activated complex as a vibrational motion of the atoms. This energy becomes concentrated in the bonds denoted by the dashed lines and can flow between them. If, at some moment, sufficient energy becomes concentrated in one of the bonds of the activated complex, that bond breaks or falls apart. Depending on whether the N----Cl or Cl----Cl bond breaks, the activated complex either reverts to the reactants or yields the products.

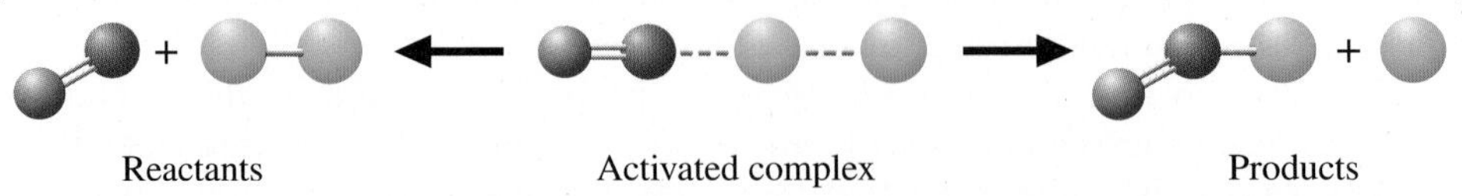

Reactants   Activated complex   Products

$$O{=}N + Cl_2 \longleftarrow [O{=}N{\cdots}Cl{\cdots}Cl] \longrightarrow O{=}N{-}Cl + Cl$$

Potential-Energy Diagrams for Reactions

It is instructive to consider a potential-energy diagram for the reaction of NO with Cl_2. We can represent this reaction by the equation

$$NO + Cl_2 \longrightarrow NOCl_2^{\ddagger} \longrightarrow NOCl + Cl$$

Here $NOCl_2^{\ddagger}$ denotes the activated complex. Figure 13.13 shows the change in potential energy (indicated by the solid curve) that occurs during the progress of the reaction. The potential-energy curve starts at the left with the potential energy of the reactants, $NO + Cl_2$. Moving along the curve toward the right, the potential energy increases to a maximum, corresponding to the activated complex. Farther to the right, the potential energy decreases to that of the products, $NOCl + Cl$.

At the start, the NO and Cl_2 molecules have a certain quantity of kinetic energy. The total energy of these molecules equals their potential energy plus their kinetic energy, and it remains constant throughout the reaction (according to the law of conservation of energy). As the reaction progresses (going from left to right in the diagram), the reactants come together. The potential energy increases because the outer electrons of the two molecules repel. The kinetic energy decreases, and the molecules slow down. Only if the reactant molecules have sufficient kinetic energy is it possible for the potential energy to increase to the value for the activated complex. This kinetic energy must be equal to or greater than the difference in energy between the activated complex and the reactant molecules (85 kJ/mol). The energy difference is the activation energy for the forward reaction.

At the maximum in the potential-energy curve, the reactant molecules have come together as the activated complex. When the activated complex breaks up into products, the products go to lower potential energy (by 2 kJ/mol) and gain in kinetic energy. Note that the energy of the products is higher than the energy of the reactants. The difference in energy equals the heat of reaction, ΔH. Because the energy increases, ΔH is positive and the reaction is endothermic.

Now look at the reverse reaction:

$$NOCl + Cl \longrightarrow NO + Cl_2$$

The activation energy is 2 kJ/mol, the difference in energy of the initial species, $NOCl + Cl$, and the activated complex. This is a smaller quantity than that for the forward reaction, so the rate constant for the reverse reaction is larger.

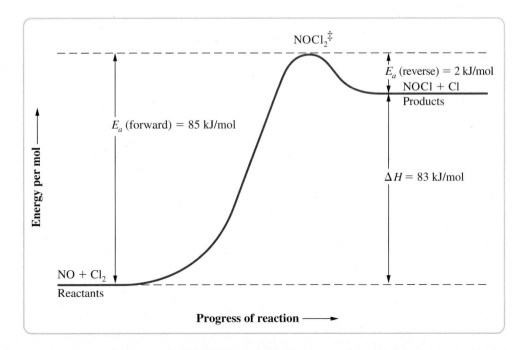

Figure 13.13 ◀

Potential-energy curve (not to scale) for the endothermic reaction NO + Cl$_2$ ⟶ NOCl + Cl For NO and Cl_2 to react, at least 85 kJ/mol of energy must be supplied by the collision of reactant molecules. Once the activated complex forms, it may break up to products, releasing 2 kJ/mol of energy. The difference, (85 − 2) kJ/mol = 83 kJ/mol, is the heat energy absorbed, ΔH.

Figure 13.14 ▶

Potential-energy curve for an exothermic reaction The energy of the reactants is higher than that of the products, so heat energy is released when the reaction goes in the forward direction.

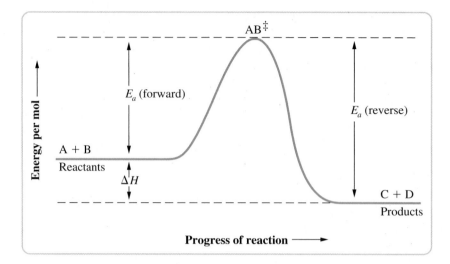

Figure 13.14 shows the potential-energy curve for an exothermic reaction. In this case, the energy of the reactants is higher than that of the products, so heat energy is released when the reaction goes in the forward direction.

CONCEPT CHECK 13.5

Consider the following potential-energy curves for two different reactions:

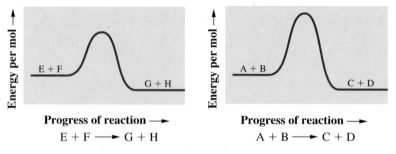

a Which reaction has a higher activation energy for the forward reaction?

b If both reactions were run at the same temperature and have the same orientation requirements to react, which one would have the larger rate constant?

c Are these reactions exothermic or endothermic?

Arrhenius published this equation in 1889 and suggested that molecules must be given enough energy to become "activated" before they could react. Collision and transition-state theories, which enlarged on this concept, were developed later (1920s and 1930s, respectively).

13.6 Arrhenius Equation

Rate constants for most chemical reactions closely follow an equation of the form

$$k = Ae^{-E_a/RT}$$

The mathematical equation $k = Ae^{-E_a/RT}$, *which expresses the dependence of the rate constant on temperature,* is called the **Arrhenius equation,** after its formulator, the Swedish chemist Svante Arrhenius. ◀ Here e is the base of natural logarithms, $2.718 \ldots$; E_a is the activation energy; R is the gas constant, 8.31 J/(K·mol); and T is the absolute temperature. *The symbol A in the Arrhenius equation, which is assumed to be a constant,* is called the **frequency factor.** The frequency factor is related to the frequency of collisions with proper orientation (pZ). (The frequency factor does have a slight dependence on temperature, as you see from collision theory, but usually this can be ignored.)

It is useful to recast Arrhenius's equation in logarithmic form. Taking the natural logarithm of both sides of the Arrhenius equation gives

$$\ln k = \ln A - \frac{E_a}{RT}$$

Let us make the following identification of symbols:

$$\underbrace{\ln k}_{y} = \underbrace{\ln A}_{b} + \underbrace{\left(\frac{-E_a}{R}\right)}_{m}\underbrace{\left(\frac{1}{T}\right)}_{x}$$

This shows that if you plot ln k against $1/T$, you should get a straight line. The slope of this line is $-E_a/R$, from which you can obtain the activation energy E_a. The intercept is ln A. Figure 13.15 shows a plot of ln k versus $1/T$ for the data given in Table 13.3. It demonstrates that the points do lie on a straight line.

You can put the previous equation into a form that is useful for computation. First you write the equation for two different absolute temperatures T_1 and T_2. You write k_1 for the rate constant at temperature T_1 and k_2 for the rate constant at temperature T_2.

$$\ln k_2 = \ln A - \frac{E_a}{RT_2}$$

$$\ln k_1 = \ln A - \frac{E_a}{RT_1}$$

You eliminate ln A by subtracting the equations.

$$\ln k_2 - \ln k_1 = -\frac{E_a}{RT_2} + \frac{E_a}{RT_1}$$

or

$$\ln\frac{k_2}{k_1} = \frac{E_a}{R}\left(\frac{1}{T_1} - \frac{1}{T_2}\right)$$

This form of the Arrhenius equation is called the two-point form. The next example illustrates the use of the two-point Arrhenius equation.

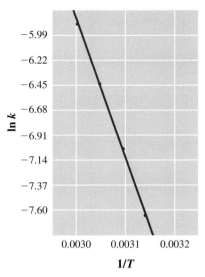

Figure 13.15 ▲

Plot of ln k versus 1/T The logarithm of the rate constant for the decomposition of N_2O_5 (data from Table 13.3) is plotted versus 1/T. A straight line is then fitted to the points; the slope equals $-E_a/R$.

Table 13.3   Rate Constant for the Decomposition of N_2O_5 at Various Temperatures

Temperature (°C)	k (/s)
45.0	4.8×10^{-4}
50.0	8.8×10^{-4}
55.0	1.6×10^{-3}
60.0	2.8×10^{-3}

♥WL Interactive Example 13.7   Using the Two-point Arrhenius Equation

Gaining Mastery Toolbox

Critical Concept 13.7
As reaction temperature changes, the rate constant changes while the activation energy remains constant. The two-point form of the Arrhenius equation can be used to calculate changes in rate constants with temperature or the activation energy of a reaction.

Solution Essentials:
- Two-point form of the Arrhenius equation:
$\ln\frac{k_2}{k_1} = \frac{E_a}{R}\left(\frac{1}{T_1} - \frac{1}{T_2}\right)$
- Activation energy
- Rate constant

The rate constant for the formation of hydrogen iodide from the elements

$$H_2(g) + I_2(g) \longrightarrow 2HI(g)$$

is 2.7×10^{-4} L/(mol·s) at 600 K and 3.5×10^{-3} L/(mol·s) at 650 K. **a** Find the activation energy E_a. **b** Then calculate the rate constant at 700 K.

Problem Strategy a Substitute the data given in the problem statement into the equation noted just before this example, then solve for E_a. **b** Use the same equation, but substitute for k_1, T_1, T_2, and E_a obtained in a, and solve for k_2.

Solution

a
$$\ln\frac{3.5 \times 10^{-3}}{2.7 \times 10^{-4}} = \frac{E_a}{8.31 \text{ J/(mol·K)}}\left(\frac{1}{600 \text{ K}} - \frac{1}{650 \text{ K}}\right)$$

$$\ln 1.30 \times 10^1 = 2.56 = \frac{E_a}{8.31 \text{ J/mol}} \times (1.28 \times 10^{-4})$$

(continued)

(continued)

Hence,

$$E_a = \frac{2.56 \times 8.31 \text{ J/mol}}{1.28 \times 10^{-4}} = 1.66 \times 10^5 \text{ J/mol}$$

b Substitute $E_a = 1.66 \times 10^5$ J/mol and

$$k_1 = 2.7 \times 10^{-4} \text{ L/(mol·s)} \qquad (T_1 = 600 \text{ K})$$
$$k_2 = \text{unknown as yet} \qquad (T_2 = 700 \text{ K})$$

You get

$$\ln \frac{k_2}{2.7 \times 10^{-4} \text{ L/(mol·s)}} = \frac{1.66 \times 10^5 \text{ J/mol}}{8.31 \text{ J/(mol·K)}} \times \left(\frac{1}{600 \text{ K}} - \frac{1}{700 \text{ K}} \right) = 4.77$$

Taking antilogarithms,

$$\frac{k_2}{2.7 \times 10^{-4} \text{ L/(mol·s)}} = e^{4.77} = 1.2 \times 10^2$$

Hence,

$$k_2 = (1.2 \times 10^2) \times (2.7 \times 10^{-4}) \text{ L/(mol·s)} = 3.2 \times 10^{-2} \text{ L/(mol·s)}$$

Answer Check Activation energies are typically in the range of 10–100 kJ/mol, as is true here, where the value is 166 kJ. Anything outside that range should be suspect.

Exercise 13.7 Acetaldehyde, CH_3CHO, decomposes when heated.

$$CH_3CHO(g) \longrightarrow CH_4(g) + CO(g)$$

The rate constant for the decomposition is $1.05 \times 10^{-3}/(M^{1/2}·s)$ at 759 K and $2.14 \times 10^{-2}/(M^{1/2}·s)$ at 836 K. What is the activation energy for this decomposition? What is the rate constant at 865 K?

■ See Problems 13.77, 13.78, 13.79, and 13.80.

Reaction Mechanisms

A balanced chemical equation is a description of the overall result of a chemical reaction. However, what actually happens at the molecular level may be more involved than is represented by this single equation. The reaction may take place in several steps. In the next sections, we will examine some reactions and see how the rate law can give us information about these steps, or *elementary reactions*.

13.7 Elementary Reactions

Consider the reaction of nitrogen dioxide with carbon monoxide.

$$NO_2(g) + CO(g) \longrightarrow NO(g) + CO_2(g) \qquad \text{(net chemical equation)}$$

At temperatures below 500 K, this gas-phase reaction is believed to take place in two steps.

$$NO_2 + NO_2 \longrightarrow NO_3 + NO \qquad \text{(elementary reaction)}$$
$$NO_3 + CO \longrightarrow NO_2 + CO_2 \qquad \text{(elementary reaction)}$$

Each step, called an **elementary reaction,** is *a single molecular event, such as a collision of molecules, resulting in a reaction. The set of elementary reactions whose overall effect is given by the net chemical equation* is called the **reaction mechanism.**

According to the reaction mechanism just given, two NO_2 molecules collide and react to give the product molecule NO and the reaction intermediate NO_3. A **reaction intermediate** is *a species produced during a reaction that does not appear in the net equation because it reacts in a subsequent step in the mechanism.* Often the reaction intermediate has a fleeting existence and cannot be isolated from the reaction mixture. The NO_3 molecule is known only from its visible light spectrum. It reacts quickly with CO to give the product molecules NO_2 and CO_2.

The overall chemical equation, which represents the net result of these two elementary reactions in the mechanism, is obtained by adding the steps and canceling species that occur on both sides.

$$NO_2 + NO_2 \longrightarrow NO_3 + NO \qquad \text{(elementary reaction)}$$
$$NO_3 + CO \longrightarrow NO_2 + CO_2 \qquad \text{(elementary reaction)}$$
$$\overline{NO_2 + \cancel{NO_2} + \cancel{NO_3} + CO \longrightarrow \cancel{NO_3} + NO + \cancel{NO_2} + CO_2} \quad \text{(overall equation)}$$

Example 13.8

Writing the Overall Chemical Equation from a Mechanism

Gaining Mastery Toolbox

Critical Concept 13.8
When added together, the elementary reactions of a mechanism yield the balanced, overall chemical equation. Each elementary reaction represents a single molecular event. The overall chemical equation does not contain reaction intermediates.

Solution Essentials:
• Elementary reaction
• Reaction mechanism
• Reaction intermediate

Carbon tetrachloride, CCl_4, is obtained by chlorinating methane or an incompletely chlorinated methane such as chloroform, $CHCl_3$. The mechanism for the gas-phase chlorination of $CHCl_3$ is

$$Cl_2 \rightleftharpoons 2Cl \qquad \text{(elementary reaction)}$$
$$Cl + CHCl_3 \longrightarrow HCl + CCl_3 \qquad \text{(elementary reaction)}$$
$$Cl + CCl_3 \longrightarrow CCl_4 \qquad \text{(elementary reaction)}$$

Obtain the net, or overall, chemical equation from this mechanism.

Problem Strategy To obtain the overall chemical equation, we need to add up the steps of the mechanism and cancel the species that occur on both sides.

Solution The first step produces two Cl atoms (a reaction intermediate). One Cl atom is used in the second step and another is used in the third step. Thus all Cl atoms cancel. Similarly, the intermediate CCl_3, produced in the second step, is used up in the third step. You can cancel all Cl and CCl_3 species.

$$Cl_2 \rightleftharpoons 2\cancel{Cl}$$
$$\cancel{Cl} + CHCl_3 \longrightarrow HCl + \cancel{CCl_3}$$
$$\cancel{Cl} + \cancel{CCl_3} \longrightarrow CCl_4$$
$$\overline{Cl_2 + CHCl_3 \longrightarrow HCl + CCl_4} \qquad \text{(overall equation)}$$

Answer Check As a final check of the overall equation, make sure that it is balanced. If it is not balanced, you probably missed a reaction intermediate and/or canceled a species that was not a reaction intermediate.

(continued)

(continued)

Exercise 13.8 The iodide ion catalyzes the decomposition of aqueous hydrogen peroxide, H_2O_2. This decomposition is believed to occur in two steps.

$$H_2O_2 + I^- \longrightarrow H_2O + IO^- \qquad \text{(elementary reaction)}$$
$$H_2O_2 + IO^- \longrightarrow H_2O + O_2 + I^- \qquad \text{(elementary reaction)}$$

What is the overall equation representing this decomposition? Note that IO^- is a reaction intermediate. The iodide ion is not an intermediate; it was added to the reaction mixture.

■ See Problems 13.83 and 13.84.

Molecularity

Elementary reactions are classified according to their molecularity. The **molecularity** is *the number of molecules on the reactant side of an elementary reaction. A* **unimolecular reaction** is *an elementary reaction that involves one reactant molecule;* a **bimolecular reaction** is *an elementary reaction that involves two reactant molecules.* Bimolecular reactions are the most common. Unimolecular reactions are best illustrated by decomposition of some previously excited species. Some gas-phase reactions are thought to occur in a **termolecular reaction,** *an elementary reaction that involves three reactant molecules.* Higher molecularities are not encountered, presumably because the chance of the correct four molecules coming together at once is extremely small.

As an example of a unimolecular reaction, consider the elementary process in which an energetically excited ozone molecule (symbolized by O_3^*) spontaneously decomposes.

$$O_3^* \longrightarrow O_2 + O$$

Normally, a molecule in a sample of ozone gas is in a lower energy level. But such a molecule may be excited to a higher level if it collides with another molecule or absorbs a photon. The energy of the excited molecule is distributed among its three nuclei as vibrational energy. After a period of time, this energy becomes redistributed. If by chance most of the energy finds its way to one end oxygen atom, that atom will fly off. In other words, the excited ozone molecule decomposes into an oxygen molecule and an oxygen atom.

All of the steps in the mechanism of the reaction of NO_2 with CO, given earlier in this section, are bimolecular. Consider the first step, which involves the reaction of two NO_2 molecules.

$$NO_2 + NO_2 \longrightarrow NO_3 + NO$$

When these two NO_2 molecules come together, they form an activated complex of the six atoms, $(NO_2)_2$, which immediately breaks into two new molecules, NO_3 and NO.

The overall reaction of two atoms—say, two Br atoms—to form a diatomic molecule (Br_2) is normally a termolecular process. When two bromine atoms collide, they form an excited bromine molecule, Br_2^*. This excited molecule immediately flies apart, re-forming the atoms, unless another atom or molecule is present just at the moment of molecule formation to take away the excess energy. Suppose an argon atom and the two bromine atoms all collide at the same moment.

$$Br + Br + Ar \longrightarrow Br_2 + Ar^*$$

Energy that would have been left with the bromine molecule is now picked up by the argon atom (giving the energized atom Ar^*). The bromine molecule is stabilized by being left in a lower energy level.

Example 13.9   Determining the Molecularity of an Elementary Reaction

Critical Concept 13.9
Counting the number of molecules that are reactants in an elementary reaction defines the molecularity of the reaction. One reactant is unimolecular, two reactants bimolecular, and three reactants termolecular.

Solution Essentials:
- Molecularity
- Unimolecular reaction
- Bimolecular reaction
- Termolecular reaction
- Elementary reaction
- Reaction mechanism

What is the molecularity of each step in the mechanism described in Example 13.8?

$$Cl_2 \rightleftharpoons 2Cl$$
$$Cl + CHCl_3 \longrightarrow HCl + CCl_3$$
$$Cl + CCl_3 \longrightarrow CCl_4$$

Problem Strategy The molecularity of any elementary reaction equals the number of reactant molecules.

Solution Counting the reactant molecules, the forward part of the first step is **unimolecular;** the reverse of the first step, the second step, and the third step are each **bimolecular.**

Answer Check If you end up with a molecularity of a reaction greater than 3, you have made a mistake.

Exercise 13.9 The following is an elementary reaction that occurs in the decomposition of ozone in the stratosphere by nitrogen monoxide.

$$O_3 + NO \longrightarrow O_2 + NO_2$$

What is the molecularity of this reaction? That is, is the reaction unimolecular, bimolecular, or termolecular?

■ See Problems 13.85 and 13.86.

Rate Equation for an Elementary Reaction

There is no necessarily simple relationship between the overall reaction and the rate law that you observe for it. As we stressed before, the rate law must be obtained experimentally. However, when you are dealing with an elementary reaction, the rate does have a simple, predictable form.

> For an elementary reaction, the rate is proportional to the product of the concentration of each reactant molecule.

To understand this, let us look at the different possibilities. Consider the unimolecular elementary reaction

$$A \longrightarrow B + C$$

For each A molecule there is a definite probability, or chance, that it will decompose into B and C molecules. The more A molecules there are in a given volume, the more A molecules that can decompose in that volume per unit time. In other words, the rate of reaction is proportional to the concentration of A.

$$\text{Rate} = k[A]$$

Now consider a bimolecular elementary reaction, such as

$$A + B \longrightarrow C + D$$

For the reaction to occur, the reactant molecules A and B must collide. Reaction does not occur with every collision. Nevertheless, the rate of formation of product is proportional to the frequency of molecular collisions, because a definite fraction of the collisions produces reaction. Within a given volume, the frequency of collisions is proportional to the number of A molecules, n_A, times the number of B molecules, n_B. Furthermore, the concentration of A is proportional to n_A, and the concentration of B is proportional to n_B. Therefore, the rate of this elementary reaction is proportional to [A][B].

$$\text{Rate} = k[A][B]$$

A termolecular elementary reaction has a rate equation that is obtained by similar reasoning. For the elementary reaction

$$A + B + C \longrightarrow D + E$$

the rate is proportional to the concentrations of A, B, and C.

$$Rate = k[A][B][C]$$

Any reaction you observe is likely to consist of several elementary steps, and the rate law that you find is the combined result of these steps. This is why you cannot predict the rate law by looking at the overall equation.

Example 13.10 **Writing the Rate Equation for an Elementary Reaction**

Critical Concept 13.10
If the overall reaction is an elementary reaction, then the rate law can be written directly from the overall reaction. This is a special case; it is the *only* instance in which the rate law can be written from the overall reaction. Otherwise, one can never assume that the rate law can be written from the overall chemical equation.

Solution Essentials:
• Elementary reaction
• Reaction mechanism
• Rate law

Write rate equations for each of the following elementary reactions.

a Ozone is converted to O_2 by NO in a single step.

$$O_3 + NO \longrightarrow O_2 + NO_2$$

b The recombination of iodine atoms occurs as follows:

$$I + I + M \longrightarrow I_2 + M^*$$

where M is some atom or molecule that absorbs energy from the reaction.

c An H_2O molecule absorbs energy; some time later enough of this energy flows into one O—H bond to break it.

$$H_2O \longrightarrow H + O—H$$

Problem Strategy The rate equation can be written directly from the elementary reaction (but *only* for an elementary reaction).

Solution
a Rate = $k[O_3][NO]$
b Rate = $k[I]^2[M]$
c Rate = $k[H_2O]$

Answer Check Keep in mind that writing the rate equation of the reaction works *only* for an elementary reaction.

Exercise 13.10 Write the rate equation, showing the dependence of rate on concentrations, for the elementary reaction

$$NO_2 + NO_2 \longrightarrow N_2O_4$$

■ See Problems 13.87 and 13.88.

You see the scientific method in operation here, which recalls the discussion in Chapter 1. Experiments have been made from which you determine the rate law. Then a mechanism is devised to explain the rate law. This mechanism in turn suggests more experiments. These may confirm the explanation or may disagree with it. If they disagree, a new mechanism must be devised that explains all of the experimental evidence.

In the next section, we look at the relationship between a reaction mechanism and the observed rate law.

13.8 The Rate Law and the Mechanism

The mechanism of a reaction cannot be observed directly. A mechanism is devised to explain the experimental observations. It is like the explanation provided by a detective to explain a crime in terms of the clues found. Other explanations may be possible, and further clues may make one of these other explanations seem more plausible than the currently accepted one. So it is with reaction mechanisms. They are accepted provisionally, with the understanding that further experiments may lead you to accept another mechanism as the more probable explanation. ◄

An important clue in understanding the mechanism of a reaction is the rate law. The reason for its importance is that once you assume a mechanism, you can predict the rate law. If this prediction does not agree with the experimental rate law, the assumed mechanism must be wrong. Take, for example, the overall equation

$$2NO_2(g) + F_2(g) \longrightarrow 2NO_2F(g)$$

If you follow the rate of disappearance of F_2, you observe that it is directly proportional to the concentration of NO_2 and F_2.

$$Rate = k[NO_2][F_2] \qquad \text{(experimental rate law)}$$

This rate law is a summary of the experimental data. Assume that the reaction occurs in a single elementary reaction.

$$NO_2 + NO_2 + F_2 \longrightarrow NO_2F + NO_2F \quad \text{(elementary reaction)}$$

This, then, is your assumed mechanism. Because this is an elementary reaction, you can immediately write the rate law predicted by it.

$$\text{Rate} = k[NO_2]^2[F_2] \quad \text{(predicted rate law)}$$

However, this does not agree with experiment, and your assumed mechanism must be discarded. You conclude that the reaction occurs in more than one step.

Rate-Determining Step

The reaction of NO_2 with F_2 is believed to occur in the following steps (elementary reactions).

$$NO_2 + F_2 \xrightarrow{k_1} NO_2F + F \quad \text{(slow step)}$$
$$F + NO_2 \xrightarrow{k_2} NO_2F \quad \text{(fast step)}$$

The rate constant for each step has been written over the arrow. The net result of the mechanism must be equivalent to the net result of the overall equation. By adding the two steps together, you can see that this is the case.

$$NO_2 + F_2 \longrightarrow NO_2F + F$$
$$\underline{F + NO_2 \longrightarrow NO_2F}$$
$$2NO_2 + F_2 \longrightarrow 2NO_2F$$

The F atom is a reaction intermediate.

The mechanism must also be in agreement with the experimental rate law. Let us look at that. Note that the second step is assumed to be much faster than the first, so that as soon as NO_2 and F_2 react, the F atom that is formed reacts with an NO_2 molecule to give another NO_2F molecule. Therefore, the rate of disappearance of F_2 (and therefore the rate law) is determined completely by the slow step, or rate-determining step. The **rate-determining step** is *the slowest step in the reaction mechanism.*

To understand better the significance of the rate-determining step, suppose you and a friend want to start a study group. You and your friend decide to send invitation cards to some students in your class. You write a lengthy note on each card explaining the study group, taking an average of 2.0 minutes per card. Your friend puts the card in an envelope, affixes a computer-printed address label, seals the envelope, and stamps it, taking 0.5 minute per card. How long does it take to do 100 cards? What is the average time taken per card (*rate* of completing the cards)?

Because you take longer than your friend to do each card, your friend can complete the task for one card while you are working on another. It takes you a total of (100 cards × 2.0 min/card) to do your step in the task. When you have finished the last card, your friend still has to place the card in the envelope, and so forth. Therefore, you have to add an additional 0.5 min to the total time.

$$\text{Total time for 100 cards} = (100 \text{ cards} \times 2.\underline{0} \text{ min/card}) \times 0.\underline{5} \text{ min}$$
$$= 2\underline{0}0.5 \text{ min}$$

We have underlined the last significant figure.

The rate of completing the invitation cards is

$$\text{Rate} = 2\underline{0}0.5 \text{ min}/100 \text{ cards} = 2.0 \text{ min/card}$$

The rate is essentially the time it takes you to compose a note. The time for your friend to finish the task is insignificant compared with the total time. The rate for the task equals the rate for the slower step (the rate-determining step).

The rate-determining step in the reaction of NO_2 and F_2 is the first step in the mechanism, in which NO_2 reacts with F_2 to produce NO_2F and an F atom. The rate equation for this rate-determining step of the mechanism is

$$\text{Rate} = k_1[NO_2][F_2]$$

This should equal the experimental rate law (otherwise the mechanism cannot be correct), which it does if you equate k_1 to k (the experimental rate constant). This agreement is not absolute evidence that the mechanism is correct. However, one can perform experiments to see whether fluorine atoms react very quickly with nitrogen dioxide. Such experiments show that they do.

Example 13.11

Gaining Mastery Toolbox

Critical Concept 13.11
If a reaction mechanism consists of a slow first step and a fast second step, the rate law is written based only on the slow first step. In this instance, the first step in the mechanism is the rate-determining step.

Solution Essentials:
- Rate-determining step
- Reaction mechanism
- Rate law

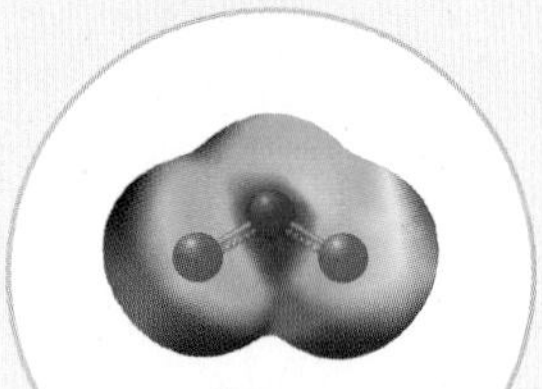

Ozone

O_3

Determining the Rate Law from a Mechanism with an Initial Slow Step

Ozone reacts with nitrogen dioxide to produce oxygen and dinitrogen pentoxide.

$$O_3(g) + 2NO_2(g) \longrightarrow O_2(g) + N_2O_5(g)$$

The proposed mechanism is

$$O_3 + NO_2 \longrightarrow NO_3 + O_2 \quad \text{(slow)}$$
$$NO_3 + NO_2 \longrightarrow N_2O_5 \quad \text{(fast)}$$

What is the rate law predicted by this mechanism?

Problem Strategy Consider the mechanism of the reaction and look to see whether the first step is slow compared to the other steps. If this is the case, then the rate law can be written on the basis of this first step.

Solution The rate law from the first step is

$$\textbf{Rate} = k[O_3][NO_2]$$

Answer Check Keep in mind that writing the rate law as we have done in this problem works when the first step of the mechanism is much slower than the other steps. Reaction mechanisms that are not structured like this one will require different approaches to determine the mechanism.

Exercise 13.11 The iodide-ion-catalyzed decomposition of hydrogen peroxide, H_2O_2, is believed to follow the mechanism

$$H_2O_2 + I^- \xrightarrow{k_1} H_2O + IO^- \quad \text{(slow)}$$
$$H_2O_2 + IO^- \xrightarrow{k_2} H_2O + O_2 + I^- \quad \text{(fast)}$$

What rate law is predicted by this mechanism? Explain.

■ See Problems 13.89 and 13.90.

Mechanisms with an Initial Fast Step

A somewhat more complicated situation occurs when the rate-determining step follows an initial fast, equilibrium step. The decomposition of dinitrogen pentoxide,

$$2N_2O_5(g) \longrightarrow 4NO_2(g) + O_2(g) \quad \text{(overall equation)}$$

which we discussed in the chapter opening section, is believed to follow this type of mechanism.

$$N_2O_5 \underset{k_{-1}}{\overset{k_1}{\rightleftharpoons}} NO_2 + NO_3 \quad \text{(fast, equilibrium)}$$
$$NO_2 + NO_3 \xrightarrow{k_2} NO + NO_2 + O_2 \quad \text{(slow)}$$
$$NO_3 + NO \xrightarrow{k_3} 2NO_2 \quad \text{(fast)}$$

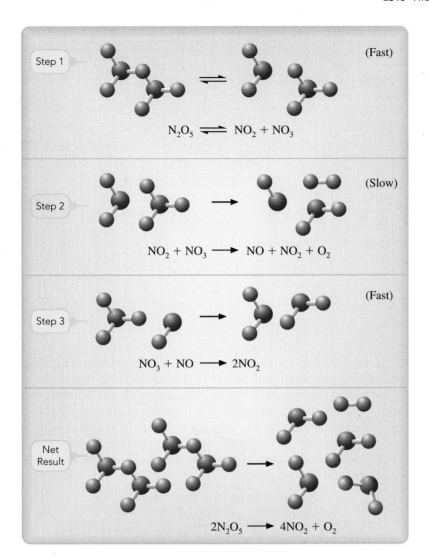

Figure 13.16 ◀

Representation of the mechanism of decomposition of N_2O_5, using molecular models Note that little atomic rearrangement occurs during each step.

Step 1   (Fast)

$$N_2O_5 \rightleftharpoons NO_2 + NO_3$$

Step 2   (Slow)

$$NO_2 + NO_3 \longrightarrow NO + NO_2 + O_2$$

Step 3   (Fast)

$$NO_3 + NO \longrightarrow 2NO_2$$

Net Result

$$2N_2O_5 \longrightarrow 4NO_2 + O_2$$

To give the overall stoichiometry, you need to multiply the first step by two. Note that there are two reaction intermediates, NO_3 and NO. Figure 13.16 represents the mechanism by means of molecular models. Let us show that this mechanism is consistent with the experimentally determined rate law,

$$\text{Rate} = k[N_2O_5]$$

The second step in the mechanism is assumed to be much slower than the other steps and is therefore rate-determining. Hence, the rate law predicted from this mechanism is

$$\text{Rate} = k_2[NO_2][NO_3]$$

However, this equation cannot be compared directly with experiment because it is written in terms of the reaction intermediate, NO_3. The experimental rate law will be written in terms of substances that occur in the chemical equation, not of reaction intermediates. For purposes of comparison, it is necessary to re-express the rate equation, eliminating $[NO_3]$. To do this, you must look at the first step in the mechanism.

This step is fast and reversible. That is, N_2O_5 dissociates rapidly into NO_2 and NO_3, and these products in turn react to re-form N_2O_5. The rate of the forward reaction (dissociation of N_2O_5) is

$$\text{Forward rate} = k_1[N_2O_5]$$

and the rate of the reverse reaction (formation of N_2O_5 from NO_2 and NO_3) is

$$\text{Reverse rate} = k_{-1}[NO_2][NO_3]$$

When the reaction first begins, there are no NO_2 or NO_3 molecules, and the reverse rate is zero. But as N_2O_5 dissociates, the concentration of N_2O_5 decreases and the concentrations of NO_2 and NO_3 increase. Therefore, the forward rate decreases and the reverse rate increases. Soon the two rates become equal, such that N_2O_5 molecules form as often as other N_2O_5 molecules dissociate. The first step has reached *dynamic equilibrium.* Because these elementary reactions are much faster than the second step, this equilibrium is reached before any significant reaction by the second step occurs. Moreover, this equilibrium is maintained throughout the reaction. ◀

At equilibrium, the forward and reverse rates are equal, so you can write

$$k_1[N_2O_5] = k_{-1}[NO_2][NO_3]$$

$$[NO_3] = \frac{k_1}{k_{-1}} \frac{[N_2O_5]}{[NO_2]}$$

Substituting into the rate law, you get

$$\text{Rate} = k_2[NO_2][NO_3] = k_2[\cancel{NO_2}] \times \frac{k_1}{k_{-1}} \frac{[N_2O_5]}{[\cancel{NO_2}]}$$

or

$$\text{Rate} = k_2 \frac{k_1}{k_{-1}}[N_2O_5]$$

Thus, if you identify $k_1 k_2/k_{-1}$ as k, you reproduce the experimental rate law.

In addition to correctly predicting the experimental rate law, the mechanism must also be in agreement with the overall equation for the reaction. Although the first step is essentially in equilibrium, the products of this step (NO_2 and NO_3) are being continuously used in the subsequent steps. Note that each of these subsequent steps uses up a molecule of NO_3. For these steps to proceed, the first step must effectively produce two molecules of NO_3. Thus, the net result of the mechanism is as follows:

$$2(N_2O_5 \longrightarrow NO_2 + \cancel{NO_3})$$
$$\cancel{NO_2} + \cancel{NO_3} \longrightarrow \cancel{NO} + \cancel{NO_2} + O_2$$
$$\underline{\cancel{NO_3} + \cancel{NO} \longrightarrow 2NO_2}$$
$$2N_2O_5 \longrightarrow 4NO_2 + O_2$$

The net result of the mechanism is equivalent to the overall equation for the reaction, as it should be.

Chemical equilibrium is discussed in detail in the next chapter.

Example 13.12   Determining the Rate Law from a Mechanism with an Initial Fast, Equilibrium Step

Gaining Mastery Toolbox

Critical Concept 13.12
If a reaction mechanism consists of a fast, equilibrium first step and a slow second step, the rate law will be based on the second slow step. The final rate law, derived from the rate-determining (slow) step, must be expressed in terms of only the concentrations of reactants present in the overall equation.

Solution Essentials:
• Fast, equilibrium step
• Rate-determining step
• Reaction mechanism
• Rate law

Nitrogen monoxide can be reduced with hydrogen gas to give nitrogen and water vapor.

$$2NO(g) + 2H_2(g) \longrightarrow N_2(g) + 2H_2O(g)$$
$$\text{(overall equation)}$$

A proposed mechanism is

$$2NO \underset{k_{-1}}{\overset{k_1}{\rightleftharpoons}} N_2O_2 \qquad \text{(fast, equilibrium)}$$

$$N_2O_2 + H_2 \overset{k_2}{\longrightarrow} N_2O + H_2O \qquad \text{(slow)}$$

$$N_2O + H_2 \overset{k_3}{\longrightarrow} N_2 + H_2O \qquad \text{(fast)}$$

What rate law is predicted by this mechanism?

Problem Strategy Examining the reaction mechanism, you can see that it has an initial fast, equilibrium step, followed by a slow step. You write the rate equation for the rate-determining (slow) step, just as in the previous example. In this case, however, the equation contains a species, N_2O_2, that does not appear in the overall equation for the reaction. You need to eliminate it from the

(continued)

(continued)

final form of the rate law. Note that the first step is fast and reaches equilibrium. You use this fact to write an expression for [N_2O_2]. Then you substitute this expression for [N_2O_2] into the rate equation.

Solution According to the rate-determining step,

$$\text{Rate} = k_2[N_2O_2][H_2]$$

You try to eliminate N_2O_2 from the rate law by looking at the first step, which is fast and reaches equilibrium. At equilibrium, the forward rate and the reverse rate are equal.

$$k_1[NO]^2 = k_{-1}[N_2O_2]$$

Therefore, [N_2O_2] = $(k_1/k_{-1})[NO]^2$, so

$$\text{Rate} = \frac{k_2 k_1}{k_{-1}}[NO]^2[H_2]$$

Experimentally, you should observe the rate law

$$\textbf{Rate} = \textbf{\textit{k}[NO]}^2\textbf{[H}_2\textbf{]}$$

where we have replaced the constants $k_2 k_1/k_{-1}$ with k, which represents the experimentally observed rate constant.

Answer Check Always examine your answer to make certain that the rate law is expressed in terms of the concentrations of the reactants in the overall equation. However, keep in mind that the exponents do not necessarily reflect the stoichiometry of the overall equation.

Exercise 13.12 Nitrogen monoxide, NO, reacts with oxygen to produce nitrogen dioxide.

$$2NO(g) + O_2(g) \longrightarrow 2NO_2(g) \quad \text{(overall equation)}$$

If the mechanism is

$$NO + O_2 \underset{k_{-1}}{\overset{k_1}{\rightleftharpoons}} NO_3 \quad \text{(fast, equilibrium)}$$

$$NO_3 + NO \overset{k_2}{\longrightarrow} NO_2 + NO_2 \quad \text{(slow)}$$

what is the predicted rate law? Remember to express this in terms of substances in the chemical equation.

■ See Problems 13.91 and 13.92.

CONCEPT CHECK 13.6

You are a chemist in charge of a research laboratory that is trying to increase the reaction rate for the balanced chemical reaction

$$X + 2Y \longrightarrow Z$$

a One of your researchers comes into your office and states that she has found a material that significantly lowers the activation energy of the reaction. Explain the effect this will have on the rate of the reaction.

b Another researcher states that, after doing some experiments, he has determined that the rate law is

$$\text{Rate} = k[X][Y]$$

Is this possible?

c Yet another person in the lab reports that the mechanism for the reaction is:

$$2Y \longrightarrow I \qquad \text{(slow)}$$
$$X + I \longrightarrow Z \qquad \text{(fast)}$$

Is the rate law from part *b* consistent with this mechanism? If not, what should the rate law be?

13.9 Catalysis

A catalyst is a substance that has the seemingly miraculous power of speeding up a reaction without being consumed by it. In theory, you could add a catalyst to a reaction mixture and, after the reaction, separate that catalyst and use it over and over again. In practice, there is often some loss of catalyst through other reactions that can occur at the same time. **Catalysis** is *the increase in rate of a reaction that results from the addition of a catalyst.*

Catalysts are of enormous importance to the chemical industry, because they allow a reaction to occur with a reasonable rate at a much lower temperature than otherwise; lower temperatures translate into lower energy costs. Moreover, catalysts are often quite specific—they increase the rate of certain reactions, but not others. For instance, an industrial chemist can start with a mixture of carbon monoxide and hydrogen and produce methane gas using one catalyst or produce gasoline using another catalyst. The most remarkable catalysts are enzymes. *Enzymes* are the marvelously selective catalysts employed by biological organisms. A biological cell contains thousands of different enzymes that in effect direct all of the chemical processes that occur in the cell.

How can we explain how a catalyst can influence a reaction without being consumed by it? Briefly, the catalyst must participate in at least one step of a reaction and be regenerated in a later step. Consider the commercial preparation of sulfuric acid, H_2SO_4, from sulfur dioxide, SO_2. The first step involves the reaction of SO_2 with O_2 to produce sulfur trioxide, SO_3. For this reaction to occur at an economical rate, it requires a catalyst. An early industrial process employed nitrogen monoxide, NO, as the catalyst.

$$2SO_2(g) + O_2(g) \xrightarrow{\text{NO}} 2SO_3(g)$$

Nitrogen monoxide does not appear in the overall equation but must participate in the reaction mechanism. Here is a proposed mechanism:

$$2NO + O_2 \longrightarrow 2NO_2$$
$$NO_2 + SO_2 \longrightarrow NO + SO_3$$

To obtain the overall reaction, the last step must occur twice each time the first step occurs once. As you can see from the mechanism, two molecules of NO are used up in the first step and are regenerated in the second step.

Note that the catalyst is an active participant in the reaction. But how does this participation explain the increase in speed of the catalyzed reaction over the uncatalyzed reaction? The Arrhenius equation (given at the beginning of Section 13.6) provides an answer. The catalyzed reaction mechanism makes available a reaction path having an increased overall rate of reaction. It increases this rate either by increasing the frequency factor A or, more commonly, by decreasing the activation energy E_a. The most dramatic effect comes from decreasing the activation energy, because it occurs as an exponent in the Arrhenius equation.

The depletion of ozone in the stratosphere by Cl atoms provides an example of the lowering of activation energy by a catalyst. Ozone is normally present in the stratosphere and provides protection against biologically destructive, short-wavelength ultraviolet radiation from the sun. Some recent ozone depletion in the stratosphere is believed to result from the Cl-catalyzed decomposition of O_3. Cl atoms in the stratosphere originate from the decomposition of chlorofluorocarbons (CFCs), which are compounds manufactured as refrigerants, aerosol propellants, and so forth. These Cl atoms react with ozone to form ClO and O_2, and the ClO reacts with O atoms (normally in the stratosphere) to produce Cl and O_2.

$$Cl(g) + O_3(g) \longrightarrow ClO(g) + O_2(g)$$
$$\underline{ClO(g) + O(g) \longrightarrow Cl(g) + O_2(g)}$$
$$O_3(g) + O(g) \longrightarrow 2O_2(g)$$

The net result is the decomposition of ozone with O atoms to produce O_2. Figure 13.17 shows the potential-energy curves for the uncatalyzed and the catalyzed reactions. The uncatalyzed reaction has such a large activation energy that its rate is extremely low. The addition of chlorine atoms provides an alternative pathway to the same overall reaction, but has a much lower activation energy and therefore an increased rate. ◀

Ozone depletion in the stratosphere was discussed in an essay at the end of Chapter 10.

Homogeneous Catalysis

The oxidation of sulfur dioxide using nitrogen monoxide as a catalyst is an example of **homogeneous catalysis,** which is *the use of a catalyst in the same phase as the reacting species.* The catalyst NO and the reacting species SO_2 and O_2 are all in the gas phase. Another example occurs in the oxidation of thallium(I) to thallium(III) by cerium(IV) in aqueous solution.

$$2Ce^{4+}(aq) + Tl^+(aq) \longrightarrow 2Ce^{3+}(aq) + Tl^{3+}(aq)$$

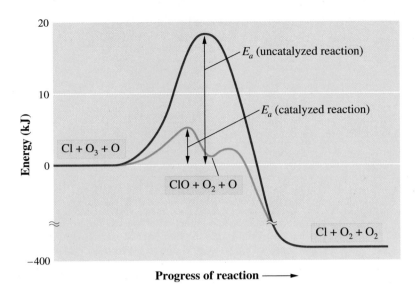

Figure 13.17 ◄

Comparison of activation energies in the uncatalyzed and catalyzed decompositions of ozone
The uncatalyzed reaction is O_3 + $O \longrightarrow O_2$ + O_2. Catalysis by Cl atoms provides an alternative pathway with lower activation energy, and therefore a faster reaction.

The uncatalyzed reaction is very slow; presumably it involves the collision of three positive ions. The reaction can be catalyzed by manganese(II) ion, however. The mechanism is thought to be

$$Ce^{4+} + Mn^{2+} \longrightarrow Ce^{3+} + Mn^{3+}$$
$$Ce^{4+} + Mn^{3+} \longrightarrow Ce^{3+} + Mn^{4+}$$
$$Mn^{4+} + Tl^{+} \longrightarrow Tl^{3+} + Mn^{2+}$$

Each step is bimolecular.

A striking example of a homogeneous catalyst increasing the rate of a reaction is the decomposition of aqueous hydrogen peroxide to form water and oxygen.

$$2H_2O_2(aq) \longrightarrow O_2(g) + 2H_2O(l)$$

Under standard conditions without a catalyst, this reaction is very slow, so slow in fact that when you observe a solution of hydrogen peroxide, you cannot visibly detect the reaction. However, when a solution containing aqueous potassium iodide is added to hydrogen peroxide, the rate of the reaction increases dramatically (see Figure 13.18). In this reaction, the aqueous iodide anion is the homogeneous catalyst. The proposed mechanism for the most important reaction in the solution is

Step 1: $H_2O_2(aq) + I^-(aq) \longrightarrow IO^-(aq) + H_2O$	(slow)
Step 2: $IO^-(aq) + H_2O_2(aq) \longrightarrow H_2O(l) + O_2(g) + I^-(aq)$	(fast)

Overall: $2H_2O_2(aq) \longrightarrow O_2(g) + 2H_2O(l)$

Heterogeneous Catalysis

Some of the most important industrial reactions involve **heterogeneous catalysis**— that is, *the use of a catalyst that exists in a different phase from the reacting species, usually a solid catalyst in contact with a gaseous or liquid solution of reactants.* Such surface, or heterogeneous, catalysis is thought to occur by chemical adsorption of the reactants onto the surface of the catalyst. *Adsorption* is the attraction of molecules to a surface. In *physical adsorption,* the attraction is provided by weak intermolecular forces. **Chemisorption,** by contrast, is *the binding of a species to a surface by chemical bonding forces.* It may happen that bonds in the species are broken during chemisorption, and this may provide the basis of catalytic action in certain cases. ▶

An example of heterogeneous catalysis involving chemisorption is provided by *catalytic hydrogenation.* This is the addition of H_2 to a compound, such as one with a carbon–carbon double bond, using a catalyst of platinum or nickel metal. Vegetable oils, which contain carbon–carbon double bonds, are changed

Understanding the chemical processes occurring at surfaces is being advanced by new techniques, including x-ray photoelectron spectroscopy and scanning tunneling microscopy.

a The beakers contain a 30% solution of $H_2O_2(aq)$ and a saturated solution of KI(aq).

b The catalyzed decomposition reaction of hydrogen peroxide forming water and oxygen gas. The $I^-(aq)$ is the homogeneous catalyst for this reaction. The brown color of the solution is due to the formation of iodine (I_2) in the reaction mixture from reactions of $I^-(aq)$ with hydrogen peroxide that are not part of the mechanism described in the text.

Figure 13.18 ▲

Catalyzed decomposition reaction of H_2O_2

to solid fats (shortening) when the bonds are catalytically hydrogenated. In the case of ethylene, C_2H_4, the equation is

$$\underset{\text{Ethylene}}{\overset{H}{\underset{H}{>}}C=C\overset{H}{\underset{H}{<}}} + H_2 \xrightarrow{Pt} \underset{\text{Ethane}}{H-\overset{H}{\underset{H}{C}}-\overset{H}{\underset{H}{C}}-H}$$

A mechanism for this reaction is represented by the four steps shown in Figure 13.19. In (a), ethylene and hydrogen molecules diffuse to the catalyst surface, where, as shown in (b), they undergo chemisorption. The pi electrons of ethylene form bonds to the metal, and hydrogen molecules break into H atoms that bond to the metal. In (c), two H atoms migrate to an ethylene molecule bonded to the catalyst, where they react to form ethane. Then, in (d), because ethane does not bond to the metal, it diffuses away. Note that the catalyst surface that was used in (b) is regenerated in (d).

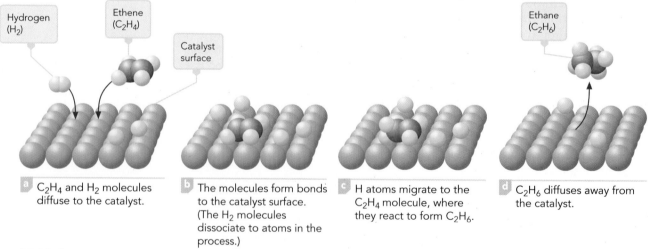

Hydrogen (H_2) Ethene (C_2H_4) Catalyst surface Ethane (C_2H_6)

a C_2H_4 and H_2 molecules diffuse to the catalyst.

b The molecules form bonds to the catalyst surface. (The H_2 molecules dissociate to atoms in the process.)

c H atoms migrate to the C_2H_4 molecule, where they react to form C_2H_6.

d C_2H_6 diffuses away from the catalyst.

Figure 13.19 ▲

Proposed mechanism of catalytic hydrogenation of C_2H_4

At the beginning of this section, we described the catalytic oxidation of SO_2 to SO_3, the anhydride of sulfuric acid. In an early process, the homogeneous catalyst NO was used. Today, in the *contact process*, a heterogeneous catalyst, Pt or V_2O_5, is used. Surface catalysts are used in the catalytic converters of automobiles to convert substances that would be atmospheric pollutants, such as CO and NO, into harmless substances, such as CO_2 and N_2 (Figure 13.20).

Enzyme Catalysis

Almost all enzymes, the catalysts of biological organisms, are protein molecules with molecular masses ranging to over a million amu. An enzyme has enormous

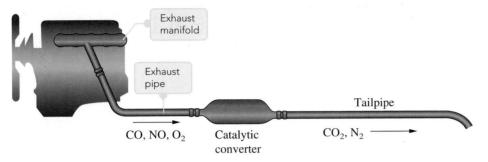

© Cengage Learning

Figure 13.20 ▲

Automobile catalytic converter *Left:* Exhaust gases from the automobile engine pass to the exhaust manifold and then to the catalytic converter, where pollutants CO and NO are converted to CO_2 and N_2. *Right:* Cross-sectional views of some automobile catalytic converters, showing the packing material with catalyst on its surface.

catalytic activity, converting a thousand or so reactant molecules to products in a second. Enzymes are also highly specific, each enzyme acting only on a specific substance, or a specific type of substance, catalyzing it to undergo a particular reaction. For example, the enzyme sucrase, which is present in the digestive fluid of the small intestine, catalyzes the reaction of sucrose (table sugar) with water to form the simpler sugars glucose and fructose. *The substance whose reaction the enzyme catalyzes* is called the **substrate.** Thus, the enzyme sucrase acts on the substrate sucrose.

Figure 13.21 shows schematically how an enzyme acts. The enzyme molecule is a protein chain that tends to fold into a roughly spherical form with an *active site* at which the substrate molecule binds and the catalysis takes place. The substrate molecule, S, fits into the active site on the enzyme molecule, E, somewhat in the way a key fits into a lock, forming an enzyme–substrate complex, ES. (The lock-and-key model is only a rough approximation, because the active site on an enzyme deforms somewhat to fit the substrate molecule.) In effect, the active site "recognizes" the substrate and gives the enzyme its specificity. On binding to the enzyme, the substrate may have bonds weakened or new bonds formed that help yield the products, P.

$$E + S \longrightarrow ES \longrightarrow E + P$$

The formation of the enzyme–substrate complex provides a new pathway to products with a lower activation energy. Figure 13.22 compares the potential-energy curves of the uncatalyzed and catalyzed reactions. Note the lower activation energy of the catalyzed reaction.

Figure 13.21 ▼

Enzyme action (lock-and-key model) The enzyme has an active site to which the substrate binds to form an enzyme–substrate complex. The active site of the enzyme acts like a lock into which the substrate (key) fits. While bound to the enzyme, the substrate may have bonds weakened or new bonds formed to yield the products, which leave the enzyme.

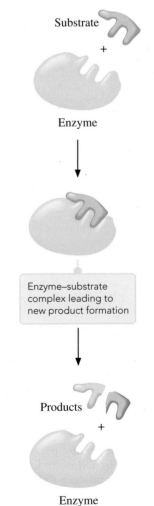

Substrate

+

Enzyme

Enzyme–substrate complex leading to new product formation

Products

+

Enzyme

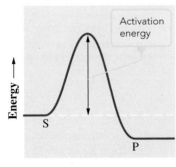

Progress of reaction ⟶

a The uncatalyzed reaction of substrate to product.

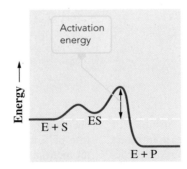

Progress of reaction ⟶

b The enzyme-catalyzed reaction of substrate to product. Note that the catalyzed reaction provides a pathway with lower activation energy than does the uncatalyzed reaction.

Figure 13.22 ◀

Potential-energy curves for the reaction of substrate, S, to products, P

Seeing Molecules React

Exactly what happens when molecules react? Transition-state theory and, more recently, calculations from quantum-mechanical theory have given us insight into how molecules behave in elementary reactions. But can we actually observe *through experiment* how molecules react? Until fairly recently, this seemed an impossible dream. Molecules are so small, the distances that atoms in molecules move to break bonds and form new ones are so minuscule, and the time that it takes a reaction to occur is so short that to be able to observe molecules while they react seemed well beyond human ability.

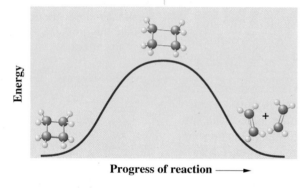

The cyclobutane molecule goes to a transition state in which two carbon-carbon bonds are stretched. When the transition state goes over to products, these stretched bonds break.

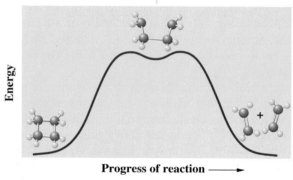

The cyclobutane molecule goes to an intermediate molecule in which one carbon-carbon bond has broken. When this intermediate goes over to products, the other carbon-carbon bond breaks. Experiments show that this is what actually happens.

Reprinted with the permission of The Nobel e-Museum from wysiwyg://26/http://www.nobel.se/chemistry/laureates/1999/press.html

Figure 13.23 ▲

Two possible potential-energy curves for the decomposition of cyclobutane to ethylene

A Checklist for Review

OWL and Go Chemistry Sign in at **www.cengage.com/owl** to:

- View tutorials and simulations, develop problem-solving skills, and complete online homework assigned by your professor.
- For quick review and exam prep, download Go Chemistry mini lecture modules from OWL or purchase them at **www.cengagebrain.com**.

Summary of Facts and Concepts

The *reaction rate* is defined as the increase in moles of product per liter per second (or as the decrease in moles of reactant per liter per second). Rates of reaction are determined by following the change of concentration of a reactant or product, either by chemical analysis or by observing a physical property. It is found that reaction rates are proportional to concentrations of reactants raised to various powers (usually 1 or 2, but they can be fractional or negative). The *rate law* mathematically expresses the relationship between rates and concentrations for a chemical reaction. Although the rate law tells you how the rate of a reaction depends on concentrations at a given moment, it is possible to transform a rate law to show how concentrations change with time, producing the *integrated rate law*. The *half-life* of a reaction is the time it takes for the reaction concentration to decrease to one-half of its original concentration.

Reaction rates can often double or triple with a 10°C rise in temperature. The effect of temperature on the rate can be explained by *collision theory*. According to this

Within the last couple of decades, however, chemistry has progressed enormously in its quest to deal with substances at the molecular level. In 1987, Ahmed H. Zewail, professor of chemistry at California Institute of Technology, managed to devise an experimental technique that can follow molecules while they are reacting. In effect, Zewail developed "the world's smallest camera." In the process, he founded an area of research that has come to be called "femtochemistry," because reactions between molecules happen on the scale of femtoseconds. A femtosecond (fs) is 10^{-15} s, which is a million times smaller than a billionth of a second!

The experiment involves using a laser, first to send a *pump pulse* to excite a reactant molecule to a higher energy state. (See the essay on lasers at the end of Section 7.3.) Then, a weaker laser pulse or *probe pulse* of properly chosen wavelength is used to detect the molecule as it transforms to products. By varying the time between the pump pulse and the probe pulse, the experimenter can follow a molecule throughout the course of its reaction. The change of wavelength of the probe pulse tells the experimenter something about the change in character of the molecule as it reacts.

As an example of a reaction, Zewail looked at the decomposition of cyclobutane to ethylene:

$$
\begin{array}{cc}
\underset{\text{Cyclobutane}}{\left. \begin{array}{c} \text{H}\quad\text{H} \\ \text{H}-\text{C}-\text{C}-\text{H} \\ \text{H}-\text{C}-\text{C}-\text{H} \\ \text{H}\quad\text{H} \end{array} \right.} \longrightarrow & \underset{\text{Ethylene}}{2\ \begin{array}{c} \text{H}\qquad\text{H} \\ \text{C}=\text{C} \\ \text{H}\qquad\text{H} \end{array}}
\end{array}
$$

Figure 13.23 shows two possible potential-energy diagrams for this reaction. The left diagram shows the cyclobutane molecule passing through a transition state (at the peak of the potential curve) in which two carbon–carbon bonds have simultaneously stretched. When the transition state goes over to products, these stretched bonds break. The right diagram shows the cyclobutane

Figure 13.24 ▲

Ahmed H. Zewail The chemist is shown here with his laser apparatus.

molecule going first to an intermediate molecule (between the two peaks). In this intermediate molecule, one carbon–carbon bond is broken. When the intermediate molecule goes over to products, another bond breaks, giving two ethylene molecules. Zewail and his coworkers were able to show that cyclobutane decomposed to ethylene by first going to an intermediate molecule (that is, the reaction actually goes by the potential curve on the right in Figure 13.23). The total time of the reaction was 700 fs! In 1999, Zewail (Figure 13.24) won the Nobel Prize in chemistry for his pioneering work in femtochemistry.

■ See Problems 13.129 and 13.130.

theory, two molecules react after colliding only when the energy of collision is greater than the *activation energy* and when the molecules are properly oriented. It is the rapid increase in the fraction of collisions having energy greater than the activation energy that explains the large temperature dependence of reaction rates. *Transition-state theory* explains reaction rates in terms of the formation of an *activated complex* of the colliding molecules. The *Arrhenius equation* is a mathematical relationship showing the dependence of a rate constant on temperature.

A chemical equation describes the overall result of a chemical reaction that may take place in one or more steps. The steps are called *elementary reactions,* and the set of elementary reactions that describes what is happening at the molecular level in an overall reaction is called the *reaction mechanism.* In some cases, the overall reaction involves a *reaction intermediate*—a species produced in one step but used up in a subsequent one. The rate of the overall reaction depends on the rate of the slowest step (the *rate-determining step*). This rate is proportional to the product of the concentrations of the reactant molecules in that step. If this step involves a reaction intermediate, its concentration can be eliminated by using the relationship of the concentrations in the preceding fast, *equilibrium* step.

A *catalyst* is a substance that speeds up a chemical reaction without being consumed by it. The catalyst is used up in one step of the reaction mechanism but regenerated in a later step. Catalytic activity operates by providing a reaction mechanism that has lower activation energy. Catalysis is classified as *homogeneous catalysis* if the substances react within one phase and as *heterogeneous catalysis* if the substances in a gas or liquid phase react at the surface of a solid catalyst. Many industrial reactions involve heterogeneous catalysis. Enzymes are biological catalysts (usually proteins). *Enzyme catalysis* is highly specific, so each enzyme acts on a particular kind of *substrate.*

Learning Objectives	Important Terms
13.1 Definition of a Reaction Rate ■ Define *reaction rate*. ■ Explain *instantaneous* rate and *average* rate of a reaction. ■ Explain how the different ways of expressing reaction rates are related. Example 13.1 ■ Calculate average reaction rate. Example 13.2	catalyst reaction rate
13.2 Experimental Determination of Rate ■ Describe how reaction rates may be experimentally determined.	
13.3 Dependence of Rate on Concentration ■ Define and provide examples of a *rate law*, *rate constant*, and *reaction order*. ■ Determine the order of reaction from the rate law. Example 13.3 ■ Determine the rate law from initial rates. Example 13.4	rate law rate constant reaction order overall order of a reaction
13.4 Change of Concentration with Time ■ Learn the *integrated rate laws* for *first-order*, *second-order*, and *zero-order* reactions. ■ Use an integrated rate law. Example 13.5 ■ Define *half-life* of a reaction. ■ Learn the half-life equations for first-order, second-order, and zero-order reactions. ■ Relate the half-life of a reaction to the rate constant. Example 13.6 ■ Plot kinetic data to determine the order of a reaction.	integrated rate law half-life
13.5 Temperature and Rate; Collision and Transition-State Theories ■ State the postulates of *collision theory*. ■ Explain *activation energy* (E_a). ■ Describe how temperature, activation energy, and molecular orientation influence reaction rates. ■ State the *transition-state theory*. ■ Define *activated complex*. ■ Describe and interpret *potential-energy curves* for endothermic and exothermic reactions.	collision theory activation energy transition-state theory activated complex
13.6 Arrhenius Equation ■ Use the *Arrhenius equation*. Example 13.7	Arrhenius equation frequency factor
13.7 Elementary Reactions ■ Define *elementary reaction, reaction mechanism*, and *reaction intermediate*. ■ Write the overall chemical equation from a mechanism. Example 13.8 ■ Define *molecularity*. ■ Give examples of *unimolecular, bimolecular,* and *termolecular* reactions. ■ Determine the molecularity of an elementary reaction. Example 13.9 ■ Write the rate equation for an elementary reaction. Example 13.10	elementary reaction reaction mechanism reaction intermediate molecularity unimolecular reaction bimolecular reaction termolecular reaction

13.8 The Rate Law and the Mechanism

- Explain the *rate-determining* step of a mechanism.
- Determine the rate law from a mechanism with an initial slow step. Example 13.11
- Determine the rate law from a mechanism with an initial fast, equilibrium step. Example 13.12

rate-determining step

13.9 Catalysis

- Describe how a *catalyst* influences the rate of a reaction.
- Indicate how a catalyst changes the potential-energy curve of a reaction.
- Define *homogeneous catalysis* and *heterogeneous catalysis*.
- Explain *enzyme catalysis*.

catalysis
homogeneous catalysis
heterogeneous catalysis
chemisorption
substrate

Key Equations

$$\ln \frac{[A]_t}{[A]_0} = -kt \quad \text{(first-order)}$$

$$t_{1/2} = \frac{0.693}{k} \quad \text{(first-order)}$$

$$\frac{1}{[A]_t} = kt + \frac{1}{[A]_0} \quad \text{(second-order)}$$

$$t_{1/2} = \frac{1}{k[A]_0} \quad \text{(second-order)}$$

$$[A]_t = -kt + [A]_0 \quad \text{(zero-order)}$$

$$t_{1/2} = \frac{[A]_0}{2k} \quad \text{(zero-order)}$$

$$\ln \frac{k_2}{k_1} = \frac{E_a}{R} \left(\frac{1}{T_1} - \frac{1}{T_2} \right)$$

Questions and Problems

⊌WL Interactive versions of these problems may be assigned in OWL.

Self-Assessment and Review Questions

Key: These questions test your understanding of the ideas you worked with in the chapter. These problems vary in difficulty and often can be used for the basis of discussion.

13.1 List the four variables or factors that can affect the rate of reaction.

13.2 Give at least two physical properties that might be used to determine the rate of a reaction.

13.3 Define the rate of reaction of HBr in the following reaction. How is this related to the rate of formation of Br_2?

$$4HBr(g) + O_2(g) \longrightarrow 2Br_2(g) + 2H_2O(g)$$

13.4 A rate of reaction depends on four variables (Question 13.1). Explain by means of an example how the rate law deals with each of these variables.

13.5 The exponents in a rate law have no relationship to the coefficients in the overall balanced equation for the reaction. Give an example of a balanced equation and the rate law for a reaction that clearly demonstrates this.

13.6 The reaction
$$3I^-(aq) + H_3AsO_4(aq) + 2H^+(aq) \longrightarrow$$
$$I_3^-(aq) + H_3AsO_3(aq) + H_2O(l)$$

is found to be first order with respect to each of the reactants. Write the rate law. What is the overall order?

13.7 A rate law is one-half order with respect to a reactant. What is the effect on the rate when the concentration of this reactant is doubled?

13.8 The rate of a reaction is quadrupled when the concentration of one reactant is doubled. What is the order of the reaction with respect to this reactant?

13.9 The reaction $A(g) \longrightarrow B(g) + C(g)$ is known to be first order in $A(g)$. It takes 25 s for the concentration of $A(g)$ to decrease by one-half of its initial value. How long does it take for the concentration of $A(g)$ to decrease to one-fourth of its initial value? to one-eighth of its initial value?

13.10 Compare the half-life equations for a first-order and a second-order reaction. For which reaction order is the value of the half-life independent of the reactant concentration?

13.11 Sketch a potential-energy diagram for the exothermic, elementary reaction

$$A + B \longrightarrow C + D$$

and on it denote the activation energies for the forward and reverse reactions. Also indicate the reactants, products, and activated complex.

13.12 What two factors determine whether a collision between two reactant molecules will result in reaction?

13.13 Rate constants for reactions often follow the Arrhenius equation. Write this equation and then identify each term in it with the corresponding factor or factors from collision theory. Give a physical interpretation of each of those factors.

13.14 Draw a structural formula for the activated complex in the following reaction:

$$NO_2 + NO_3 \longrightarrow NO + NO_2 + O_2$$

Refer to Figure 13.16, Step 2, for details of this reaction. Use dashed lines for bonds about to form or break. Use a single line for all other bonds.

13.15 By means of an example, explain what is meant by the term *reaction intermediate*.

13.16 Why is it generally impossible to predict the rate law for a reaction on the basis of the chemical equation only?

13.17 The rate law for the reaction

$$2NO_2Cl(g) \longrightarrow 2NO_2(g) + Cl_2(g) \quad \text{(overall equation)}$$

is first order in nitryl chloride, NO_2Cl.

$$\text{Rate} = k[NO_2Cl]$$

Explain why the mechanism for this reaction cannot be the single elementary reaction

$$2NO_2Cl \longrightarrow 2NO_2 + Cl_2 \quad \text{(elementary reaction)}$$

13.18 There is often one step in a reaction mechanism that is rate-determining. What characteristic of such a step makes it rate-determining? Explain.

13.19 The dissociation of N_2O_4 into NO_2,

$$N_2O_4(g) \rightleftharpoons 2NO_2(g)$$

is believed to occur in one step. Obtain the concentration of N_2O_4 in terms of the concentration of NO_2 and the rate constants for the forward and reverse reactions, when the reactions have come to equilibrium.

13.20 How does a catalyst speed up a reaction? How can a catalyst be involved in a reaction without being consumed by it?

13.21 Compare physical adsorption and chemisorption (chemical adsorption).

13.22 Describe the steps in the catalytic hydrogenation of ethylene.

13.23 You are running the reaction $2A + B \longrightarrow C + 3D$. Your lab partner has conducted the first two experiments to determine the rate law for the reaction by the method of initial rates.

Experiment Number	Concentration of A (M)	Concentration of B (M)
1	0.0250	0.0330
2	0.0500	0.0330
3		

Presuming that you can measure the initial rate of each experiment, which of the following concentrations for Experiment 3 would help you to determine the rate law easily?

a $[A] = 0.0330\ M,\ [B] = 0.0330\ M$
b $[A] = 0.0125\ M,\ [B] = 0.0500\ M$
c $[A] = 0.0250\ M,\ [B] = 0.0400\ M$
d $[A] = 0.0250\ M,\ [B] = 0.0330\ M$
e $[A] = 0.0500\ M,\ [B] = 0.0330\ M$

13.24 At a constant temperature, which of the following would be expected to affect the rate of a given chemical reaction, $A + B \longrightarrow C + D$?

I. The reaction temperature
II. The concentration of the products
III. A catalyst

a I only   b II only
c III only   d I and II only
e I and III only

13.25 Consider the reaction $E + F \longrightarrow G + H$, which has the following reaction coordinate diagram.

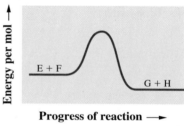

$$E + F \longrightarrow G + H$$

Which one of the following statements is true?

a The activation energy is greatest for the forward reaction.
b The reactants are at a lower energy than the products.
c The reaction is endothermic.
d The rate of the forward reaction would be slowed by an increase in temperature.
e If the rate law for the reaction is Rate = $k[E]^2[F]$, then doubling the concentration of E will cause the rate to increase by a factor of 4 (that is, the reaction will proceed four times faster).

13.26 The hypothetical reaction

$$A + B + C \longrightarrow D + E$$

has the rate law Rate = $k[A]^2[B]$. Which of the following changes would have the *least* effect on increasing the rate of this reaction?

a Doubling $[A]_0$ while keeping $[B]_0$ and $[C]_0$ constant.
b Doubling $[B]_0$ while keeping $[A]_0$ and $[C]_0$ constant.
c Doubling $[C]_0$ while keeping $[A]_0$ and $[B]_0$ constant.
d Increasing the rate constant by running the reaction at a higher temperature.
e Adding a catalyst to the reaction mixture.

Concept Explorations

Key: Concept explorations are comprehensive problems that provide a framework that will enable you to explore and learn many of the critical concepts and ideas in each chapter. If you master the concepts associated with these explorations, you will have a better understanding of many important chemistry ideas and will be more successful in solving all types of chemistry problems. These problems are well suited for group work and for use as in-class activities.

13.27 Kinetics I

Consider the hypothetical reaction $A(g) + 2B(g) \longrightarrow C(g)$. The four containers below represent this reaction being run with different initial amounts of A and B. Assume that the volume of each container is 1.0 L. The reaction is second order with respect to A and first order with respect to B.

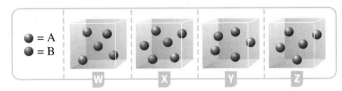

a Based on the information presented in the problem, write the rate law for the reaction.

b Which of the containers, W, X, Y, or Z, would have the greatest reaction rate? Justify your answer.

c Which of the containers would have the lowest reaction rate? Explain.

d If the volume of the container X were increased to 2.0 L, how would the rate of the reaction in this larger container compare to the rate of reaction run in the 1.0-L container X? (Assume that the number of A and B atoms is the same in each case.)

e If the temperature in container W were increased, what impact would this probably have on the rate of reaction? Why?

f If you want to double the rate of reaction in container X, what are some things that you could do to the concentration(s) of A and B?

g In which container would you observe the slowest rate of formation of C?

h Assuming that A and B are not in great excess, which would have the greater impact on the rate of reaction in container W: removing a unit of B or removing a unit of A? Explain.

i Describe how the rate of consumption of A compares to the rate of consumption of B. If you cannot answer this question, what additional information do you need to provide an answer?

j If the product C were removed from the container as it formed, what effect would this have on the rate of the reaction?

13.28 Kinetics II

You and a friend are working together in order to obtain as much kinetic information as possible about the reaction $A(g) \longrightarrow B(g) + C(g)$. One thing you know before performing the experiments is that the reaction is zero order, first order, or second order with respect to A. Your friend goes off, runs the experiment, and brings back the following graph.

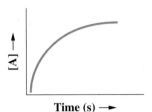

a After studying the curve of the graph, she declares that the reaction is second order, with a corresponding rate law of Rate = $k[A]^2$. Judging solely on the basis of the information presented in this plot, is she correct in her statement that the reaction must be second order?

Here are some data collected from her experiment:

Time (s)	[A]
0.0	1.0
1.0	0.14
3.0	2.5×10^{-3}
5.0	4.5×10^{-5}
7.0	8.3×10^{-7}

b The half-life of the reaction is 0.35 s. Do these data support the reaction being second order, or is it something else? Try to reach a conclusive answer without graphing the data.

c What is the rate constant for the reaction?

d The mechanism for this reaction is found to be a two-step process, with intermediates X and Y. The first step of the reaction is the rate-determining step. Write a possible mechanism for the reaction.

e You perform additional experiments and find that the rate constant doubles in value when you increase the temperature by 10°C. Your lab partner doesn't understand why the rate constant changes in this manner. What could you say to your partner to help her understand? Feel free to use figures and pictures as part of your explanation.

Conceptual Problems

Key: These problems are designed to check your understanding of the concepts associated with some of the main topics presented in each chapter. A strong conceptual understanding of chemistry is the foundation for both applying chemical knowledge and solving chemical problems. These problems vary in level of difficulty and often can be used as a basis for group discussion.

13.29 Consider the reaction $3A \longrightarrow 2B + C$.

a One rate expression for the reaction is

$$\text{Rate of formation of C} = +\frac{\Delta[C]}{\Delta t}$$

Write two other rate expressions for this reaction in this form.

b Using your two rate expressions, if you calculated the average rate of the reaction over the same time interval, would the rates be equal?

c If your answer to part b was no, write two rate expressions that would give an equal rate when calculated over the same time interval.

13.30 Given the reaction $2A + B \longrightarrow C + 3D$, can you write the rate law for this reaction? If so, write the rate law; if not, why?

13.31 The reaction $2A(g) \longrightarrow A_2(g)$ is being run in each of the following containers. The reaction is found to be second-order with respect to A.

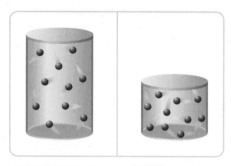

a Write the rate law for the reaction.
b Which reaction container will have a greater reaction rate?
c In which container will the reaction have a shorter half-life?
d What are the relative rates of the reactions in each container?
e After a set amount of time has elapsed, which container will contain fewer A atoms?

13.32 When viewed from a molecular perspective, a particular reaction is written as

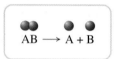

$$AB \longrightarrow A + B$$

a If the reaction is first-order with a half-life of 10 seconds, complete the following pictures after 10 and 20 seconds have elapsed.

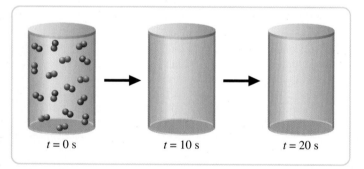

$t = 0$ s $t = 10$ s $t = 20$ s

b How would the pictures from part a change if the reaction were second-order with the same half-life?
c For the first-order case, what are the relative reaction rates at the start of the reaction and after 10 seconds have elapsed?
d For the second-order case, what are the relative reaction rates at the start of the reaction and after 10 seconds have elapsed?

13.33 You perform some experiments for the reaction $A \longrightarrow B + C$ and determine the rate law has the form

$$\text{Rate} = k[A]^x$$

Calculate the value of exponent x for each of the following.
a [A] is doubled and the rate doubles.
b [A] is tripled and you observe no rate change.
c [A] is tripled and the rate goes up by a factor of 27.

13.34 A friend of yours runs a reaction and generates the following plot. She explains that in following the reaction, she measured the concentration of a compound that she calls "E."

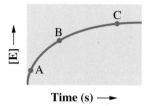

a Is the average rate faster between points A and B or B and C? Why?
b Your friend tells you that E is either a reactant or a product. Which is it and why?

13.35 Given the hypothetical plot shown here for the concentration of compound Y versus time, answer the following questions.

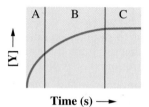

a In which region of the curve is the rate the fastest (A, B, or C)?
b In which region of the curve does the rate have a constant value (A, B, or C)?

13.36 You carry out the following reaction by introducing N_2O_4 into an evacuated flask and observing the concentration change of the product over time.

$$N_2O_4(g) \longrightarrow 2NO_2(g)$$

Which one of the curves shown here reflects the data collected for this reaction?

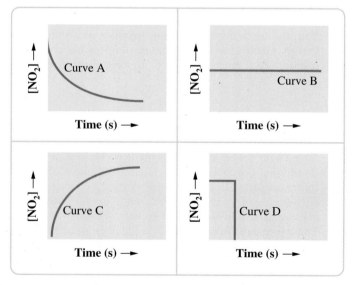

13.37 You are running the reaction $2A + B \longrightarrow C + 3D$. Your lab partner has conducted the first two experiments to determine the rate law for the reaction. He has recorded the initial rates for these experiments in another data table.

Come up with some reactant concentrations for Experiment 3 that will allow you to determine the rate law by measuring the initial rate.

Experiment Number	Concentration of A (M)	Concentration of B (M)
1	1.0	1.0
2	2.0	1.0
3		

13.38 The chemical reaction A $\longrightarrow$ B + C has a rate constant that obeys the Arrhenius equation. Predict what happens to both the rate constant k and the rate of the reaction if the following were to occur.

a a decrease in temperature

b an increase in both activation energy and temperature

c an increase in the activation energy of the forward and reverse reactions

Practice Problems

Key: These problems are for practice in applying problem-solving skills. They are divided by topic, and some are keyed to exercises (see the ends of the exercises). The problems are arranged in matching pairs; the odd-numbered problem of each pair is listed first, and its answer is given in the back of the book.

Reaction Rates

13.39 For the reaction of hydrogen with iodine

$$H_2(g) + I_2(g) \longrightarrow 2HI(g)$$

relate the rate of disappearance of hydrogen gas to the rate of formation of hydrogen iodide.

13.40 Relate the rate of decomposition of NO_2 to the rate of formation of O_2 for the following reaction:

$$2NO_2(g) \longrightarrow 2NO(g) + O_2(g)$$

13.41 To obtain the rate of the reaction

$$5Br^-(aq) + BrO_3^-(aq) + 6H^+(aq) \longrightarrow 3Br_2(aq) + 3H_2O(l)$$

you might follow the Br^- concentration or the BrO_3^- concentration. How are the rates in terms of these species related?

13.42 To obtain the rate of the reaction

$$3I^-(aq) + H_3AsO_4(aq) + 2H^+(aq) \longrightarrow$$
$$I_3^-(aq) + H_3AsO_3(aq) + H_2O(l)$$

you might follow the H^+ concentration or the I_3^- concentration. How are the rates in terms of these species related?

13.43 Ammonium nitrite, NH_4NO_2, decomposes in solution, as shown here.

$$NH_4NO_2(aq) \longrightarrow N_2(g) + 2H_2O(l)$$

The concentration of NH_4^+ ion at the beginning of an experiment was 0.500 M. After 3.00 hours, it was 0.432 M. What is the average rate of decomposition of NH_4NO_2 in this time interval?

13.44 Iron(III) chloride is reduced by tin(II) chloride.

$$2FeCl_3(aq) + SnCl_2(aq) \longrightarrow 2FeCl_2(aq) + SnCl_4(aq)$$

The concentration of Fe^{3+} ion at the beginning of an experiment was 0.03586 M. After 4.00 min, it was 0.02715 M. What is the average rate of reaction of $FeCl_3$ in this time interval?

13.45 Azomethane, CH_3NNCH_3, decomposes according to the following equation:

$$CH_3NNCH_3(g) \longrightarrow C_2H_6(g) + N_2(g)$$

The initial concentration of azomethane was 1.50 $\times$ 10^{-2} mol/L. After 7.00 min, this concentration decreased to 1.01 $\times$ 10^{-2} mol/L. Obtain the average rate of reaction during this time interval. Express the answer in units of mol/(L·s).

13.46 Nitrogen dioxide, NO_2, decomposes upon heating to form nitric oxide and oxygen according to the following equation:

$$2NO_2(g) \longrightarrow 2NO(g) + O_2(g)$$

At the beginning of an experiment, the concentration of nitrogen dioxide in a reaction vessel was 0.1103 mol/L. After 65.0 s, the concentration decreased to 0.1076 mol/L. What is the average rate of decomposition of NO_2 during this time interval, in mol/(L·s)?

Rate Laws

13.47 For the reaction of nitrogen monoxide, NO, with chlorine, Cl_2,

$$2NO(g) + Cl_2(g) \longrightarrow 2NOCl(g)$$

the observed rate law is

$$Rate = k[NO]^2[Cl_2]$$

What is the reaction order with respect to nitrogen monoxide and with respect to Cl_2? What is the overall order?

13.48 Hydrogen sulfide is oxidized by chlorine in aqueous solution.

$$H_2S(aq) + Cl_2(aq) \longrightarrow S(s) + 2HCl(aq)$$

The experimental rate law is

$$Rate = k[H_2S][Cl_2]$$

What is the reaction order with respect to H_2S and with respect to Cl_2? What is the overall order?

13.49 Oxalic acid, $H_2C_2O_4$, is oxidized by permanganate ion to CO_2 and H_2O.

$$2MnO_4^-(aq) + 5H_2C_2O_4(aq) + 6H^+(aq) \longrightarrow$$
$$2Mn^{2+}(aq) + 10CO_2(g) + 8H_2O(l)$$

The rate law is

$$Rate = k[MnO_4^-][H_2C_2O_4]$$

What is the order with respect to each reactant? What is the overall order?

13.50 Iron(II) ion is oxidized by hydrogen peroxide in acidic solution.

$$H_2O_2(aq) + 2Fe^{2+}(aq) + 2H^+(aq) \longrightarrow$$
$$2Fe^{3+}(aq) + 2H_2O(l)$$

The rate law is

$$Rate = k[H_2O_2][Fe^{2+}]$$

What is the order with respect to each reactant? What is the overall order?

13.51 In experiments on the decomposition of azomethane,

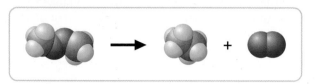

$$CH_3NNCH_3(g) \longrightarrow C_2H_6(g) + N_2(g)$$

the following data were obtained:

	Initial Concentration of Azomethane	Initial Rate
Exp. 1	1.13×10^{-2} M	2.8×10^{-6} M/s
Exp. 2	2.26×10^{-2} M	5.6×10^{-6} M/s

What is the rate law? What is the value of the rate constant?

13.52 Ethylene oxide, C_2H_4O, decomposes when heated to give methane and carbon monoxide.

$$C_2H_4O(g) \longrightarrow CH_4(g) + CO(g)$$

The following kinetic data were observed for the reaction at 688 K:

	Initial Concentration of Ethylene Oxide	Initial Rate
Exp. 1	0.00272 M	5.57×10^{-7} M/s
Exp. 2	0.00544 M	1.11×10^{-6} M/s

Find the rate law and the value of the rate constant for this reaction.

13.53 In a kinetic study of the reaction

$$2NO(g) + O_2(g) \longrightarrow 2NO_2(g)$$

the following data were obtained for the initial rates of disappearance of NO:

	Initial Concentrations		Initial Rate of Reaction of NO
	NO	O_2	
Exp. 1	0.0125 M	0.0253 M	0.0281 M/s
Exp. 2	0.0250 M	0.0253 M	0.112 M/s
Exp. 3	0.0125 M	0.0506 M	0.0561 M/s

Obtain the rate law. What is the value of the rate constant?

13.54 Nitrogen monoxide, NO, reacts with hydrogen to give nitrous oxide, N_2O, and water.

$$2NO(g) + H_2(g) \longrightarrow N_2O(g) + H_2O(g)$$

In a series of experiments, the following initial rates of disappearance of NO were obtained:

	Initial Concentration		Initial Rate of Reaction of NO
	NO	H_2	
Exp. 1	6.4×10^{-3} M	2.2×10^{-3} M	2.6×10^{-5} M/s
Exp. 2	12.8×10^{-3} M	2.2×10^{-3} M	1.0×10^{-4} M/s
Exp. 3	6.4×10^{-3} M	4.5×10^{-3} M	5.1×10^{-5} M/s

Find the rate law and the value of the rate constant for the reaction of NO.

13.55 Chlorine dioxide, ClO_2, is a reddish-yellow gas that is soluble in water. In basic solution it gives ClO_3^- and ClO_2^- ions.

$$2ClO_2(aq) + 2OH^-(aq) \longrightarrow \\ ClO_3^-(aq) + ClO_2^-(aq) + H_2O$$

To obtain the rate law for this reaction, the following experiments were run and, for each, the initial rate of reaction of ClO_2 was determined. Obtain the rate law and the value of the rate constant.

	Initial Concentrations (mol/L)		Initial Rate (mol/(L·s))
	ClO_2	OH^-	
Exp. 1	0.060	0.030	0.0248
Exp. 2	0.020	0.030	0.00276
Exp. 3	0.020	0.090	0.00828

13.56 Iodide ion is oxidized to hypoiodite ion, IO^-, by hypochlorite ion, ClO^-, in basic solution. The equation is

$$I^-(aq) + ClO^-(aq) \xrightarrow{OH^-} IO^-(aq) + Cl^-(aq)$$

The following initial-rate experiments were run and, for each, the initial rate of formation of IO^- was determined. Find the rate law and the value of the rate constant.

	Initial Concentrations (mol/L)			Initial Rate (mol/(L·s))
	I^-	ClO^-	OH^-	
Exp. 1	0.010	0.020	0.010	12.2×10^{-2}
Exp. 2	0.020	0.010	0.010	12.2×10^{-2}
Exp. 3	0.010	0.010	0.010	6.1×10^{-2}
Exp. 4	0.010	0.010	0.020	3.0×10^{-2}

Integrated Rate Laws (Change of Concentration with Time); Half-Life

13.57 Cyclopropane, C_3H_6, is converted to its isomer propylene, $CH_2\!=\!CHCH_3$, when heated. The rate law is first order in cyclopropane, and the rate constant is 6.0×10^{-4}/s at 500°C. If the initial concentration of cyclopropane is 0.0226 mol/L, what is the concentration after 525 s?

13.58 Sulfuryl chloride, SO_2Cl_2, decomposes when heated.

$$SO_2Cl_2(g) \longrightarrow SO_2(g) + Cl_2(g)$$

In an experiment, the initial concentration of SO_2Cl_2 was 0.0248 mol/L. If the rate constant is 2.2×10^{-5}/s, what is the concentration of SO_2Cl_2 after 2.0 hr? The reaction is first order.

13.59 A reaction of the form $aA \longrightarrow$ Products is second order with a rate constant of 0.169 L/(mol·s). If the initial concentration of A is 0.159 mol/L, how many seconds would it take for the concentration of A to decrease to 5.50×10^{-3} mol/L?

13.60 A reaction of the form $aA \longrightarrow$ Products is second-order with a rate constant of 0.225 L/(mol·s). If the initial concentration of A is 0.293 mol/L, what is the molar concentration of A after 35.4 s?

13.61 Ethyl chloride, CH_3CH_2Cl, used to produce tetraethyl-lead gasoline additive, decomposes, when heated, to give ethylene and hydrogen chloride.

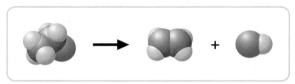

$$CH_3CH_2Cl(g) \longrightarrow C_2H_4(g) + HCl(g)$$

The reaction is first order. In an experiment, the initial concentration of ethyl chloride was 0.00100 M. After heating at 500°C for 155 s, this was reduced to 0.00067 M. What was the concentration of ethyl chloride after a total of 256 s?

13.62 Cyclobutane, C_4H_8, consisting of molecules in which four carbon atoms form a ring, decomposes, when heated, to give ethylene.

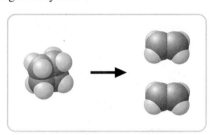

$$C_4H_8(g) \longrightarrow 2C_2H_4(g)$$

The reaction is first order. In an experiment, the initial concentration of cyclobutane was 0.00150 M. After heating at 450°C for 455 s, this was reduced to 0.00119 M. What was the concentration of cyclobutane after a total of 875 s?

13.63 Dinitrogen pentoxide, N_2O_5, decomposes when heated in carbon tetrachloride solvent.

$$N_2O_5 \longrightarrow 2NO_2 + \tfrac{1}{2}O_2(g)$$

If the rate constant for the decomposition of N_2O_5 is 6.2×10^{-4}/min, what is the half-life? (The rate law is first order in N_2O_5.) How long would it take for the concentration of N_2O_5 to decrease to 25% of its initial value? to 6.25% of its initial value?

13.64 Methyl isocyanide, CH_3NC, isomerizes, when heated, to give acetonitrile (methyl cyanide), CH_3CN.

$$CH_3NC(g) \longrightarrow CH_3CN(g)$$

The reaction is first order. At 230°C, the rate constant for the isomerization is 6.3×10^{-4}/s. What is the half-life? How long would it take for the concentration of CH_3NC to decrease to 50.0% of its initial value? to 25.0% of its initial value?

13.65 In the presence of excess thiocyanate ion, SCN^-, the following reaction is first order in chromium(III) ion, Cr^{3+}; the rate constant is 2.0×10^{-6}/s.

$$Cr^{3+}(aq) + SCN^-(aq) \longrightarrow Cr(SCN)^{2+}(aq)$$

What is the half-life in hours? How many hours would be required for the initial concentration of Cr^{3+} to decrease to each of the following values: 25.0% left, 12.5% left, 6.25% left, 3.125% left?

13.66 In the presence of excess thiocyanate ion, SCN^-, the following reaction is first order in iron(III) ion, Fe^{3+}; the rate constant is 1.27/s.

$$Fe^{3+}(aq) + SCN^-(aq) \longrightarrow Fe(SCN)^{2+}(aq)$$

What is the half-life in seconds? How many seconds would be required for the initial concentration of Fe^{3+} to decrease to each of the following values: 25.0% left, 12.5% left, 6.25% left, 3.125% left? What is the relationship between these times and the half-life?

13.67 A reaction of the form $aA \longrightarrow$ Products is second order with a half-life of 485 s. What is the rate constant of the reaction if the initial concentration of A is 5.99×10^{-3} mol/L?

13.68 A reaction of the form $aA \longrightarrow$ Products is second order with a rate constant of 0.413 L/(mol·s). What is the half-life, in seconds, of the reaction if the initial concentration of A is 5.25×10^{-3} mol/L?

13.69 In the presence of excess thiocyanate ion, SCN^-, the following reaction is first order in chromium(III) ion, Cr^{3+}; the rate constant is 2.0×10^{-6}/s.

$$Cr^{3+}(aq) + SCN^-(aq) \longrightarrow Cr(SCN)^{2+}(aq)$$

If 85.0% reaction is required to obtain a noticeable color from the formation of the $Cr(SCN)^{2+}$ ion, how many hours are required?

13.70 In the presence of excess thiocyanate ion, SCN^-, the following reaction is first order in iron(III) ion, Fe^{3+}; the rate constant is 1.27/s.

$$Fe^{3+}(aq) + SCN^-(aq) \longrightarrow Fe(SCN)^{2+}(aq)$$

If 90.0% reaction is required to obtain a noticeable color from the formation of the $Fe(SCN)^{2+}$ ion, how many seconds are required?

13.71 The reaction $A \longrightarrow B + C$ is found to be zero order. If it takes 5.2×10^2 seconds for an initial concentration of A to go from 0.50 M to 0.25 M, what is the rate constant for the reaction?

13.72 It is found that a gas undergoes a zero-order decomposition reaction in the presence of a nickel catalyst. If the rate constant for this reaction is 8.1×10^{-2} mol/(L·s), how long will it take for the concentration of the gas to change from an initial concentration of 0.10 M to 1.0×10^{-2} M?

13.73 Chlorine dioxide oxidizes iodide ion in aqueous solution to iodine; chlorine dioxide is reduced to chlorite ion.

$$2ClO_2(aq) + 2I^-(aq) \longrightarrow 2ClO_2^-(aq) + I_2(aq)$$

The order of the reaction with respect to ClO_2 was determined by starting with a large excess of I^-, so that its concentration was essentially constant. Then

$$\text{Rate} = k[ClO_2]^m[I^-]^n = k'[ClO_2]^m$$

where $k' = k[I^-]^n$. Determine the order with respect to ClO_2 and the rate constant k' by plotting the following data assuming first- and then second-order kinetics. [Data from H. Fukutomi and G. Gordon, *J. Am. Chem. Soc., 89,* 1362 (1967).]

Time (s)	$[ClO_2]$ (mol/L)
0.00	4.77×10^{-4}
1.00	4.31×10^{-4}
2.00	3.91×10^{-4}
3.00	3.53×10^{-4}
5.00	2.89×10^{-4}
10.00	1.76×10^{-4}
30.00	2.4×10^{-5}
50.00	3.2×10^{-6}

13.74 Methyl acetate, CH_3COOCH_3, reacts in basic solution to give acetate ion, CH_3COO^-, and methanol, CH_3OH.

$$CH_3COOCH_3(aq) + OH^-(aq) \longrightarrow$$
$$CH_3COO^-(aq) + CH_3OH(aq)$$

The overall order of the reaction was determined by starting with methyl acetate and hydroxide ion at the same concentrations, so $[CH_3COOCH_3] = [OH^-] = x$. Then

$$\text{Rate} = k[CH_3COOCH_3]^m[OH^-]^n = kx^{m+n}$$

Determine the overall order and the value of the rate constant by plotting the following data assuming first- and then second-order kinetics.

Time (min)	$[CH_3COOCH_3]$ (mol/L)
0.00	0.01000
3.00	0.00740
4.00	0.00683
5.00	0.00634
10.00	0.00463
20.00	0.00304
30.00	0.00224

Rate and Temperature

13.75 Sketch a potential-energy diagram for the decomposition of nitrous oxide.

$$N_2O(g) \longrightarrow N_2(g) + O(g)$$

The activation energy for the forward reaction is 251 kJ; the $\Delta H°$ is $+167$ kJ. What is the activation energy for the reverse reaction? Label your diagram appropriately.

13.76 Sketch a potential-energy diagram for the reaction of nitric oxide with ozone.

$$NO(g) + O_3(g) \longrightarrow NO_2(g) + O_2(g)$$

The activation energy for the forward reaction is 10 kJ; the $\Delta H°$ is -200 kJ. What is the activation energy for the reverse reaction? Label your diagram appropriately.

13.77 In a series of experiments on the decomposition of dinitrogen pentoxide, N_2O_5, rate constants were determined at two different temperatures. At 35°C, the rate constant was 1.4×10^{-4}/s; at 45°C, the rate constant was 5.0×10^{-4}/s. What is the activation energy for this reaction? What is the value of the rate constant at 55°C?

13.78 The reaction

$$2NOCl(g) \longrightarrow 2NO(g) + Cl_2(g)$$

has rate-constant values for the reaction of NOCl of 9.3×10^{-6}/s at 350 K and 6.9×10^{-4}/s at 400 K. Calculate the activation energy for the reaction. What is the rate constant at 435 K?

13.79 The rate of a particular reaction triples when the temperature is increased from 25°C to 35°C. Calculate the activation energy for this reaction.

13.80 The rate of a particular reaction quadruples when the temperature is increased from 25°C to 42°C. Calculate the activation energy for this reaction.

13.81 The following values of the rate constant were obtained for the decomposition of nitrogen dioxide at various temperatures. Plot the natural logarithm of k versus $1/T$ and from the graph obtain the energy of activation.

Temperature (°C)	k (L/mol·s)
320	0.527
330	0.776
340	1.121
350	1.607

13.82 The following values of the rate constant were obtained for the decomposition of hydrogen iodide at various temperatures. Plot the natural logarithm of k versus $1/T$ and from the graph obtain the energy of activation.

Temperature (°C)	k (L/mol·s)
440	2.69×10^{-3}
460	6.21×10^{-3}
480	1.40×10^{-2}
500	3.93×10^{-2}

Reaction Mechanisms; Catalysis

13.83 The decomposition of ozone is believed to occur in two steps:

$$O_3 \rightleftharpoons O_2 + O \quad \text{(elementary reaction)}$$
$$O_3 + O \longrightarrow 2O_2 \quad \text{(elementary reaction)}$$

Identify any reaction intermediate. What is the overall reaction?

13.84 Nitrogen monoxide, NO, is believed to react with chlorine according to the following mechanism:

$$NO + Cl_2 \rightleftharpoons NOCl_2 \quad \text{(elementary reaction)}$$
$$NOCl_2 + NO \longrightarrow 2NOCl \quad \text{(elementary reaction)}$$

Identify any reaction intermediate. What is the overall equation?

13.85 Identify the molecularity of each of the following elementary reactions.

a. $O_3 \longrightarrow O_2 + O$
b. $H + H + N_2 \longrightarrow H_2 + N_2^*$
c. $NO + O_3 \longrightarrow NO_2 + O_2$
d. $NOCl_2 + NO \longrightarrow 2NOCl$

13.86 What is the molecularity of each of the following elementary reactions?

a. $CS_2 \longrightarrow CS + S$
b. $NO_2Cl + Cl \longrightarrow NO_2 + Cl_2$

c $O + O_2 + N_2 \longrightarrow O_3 + N_2^*$

d $Cl + H_2 \longrightarrow HCl + H$

13.87 Write a rate equation, showing the dependence of rate on reactant concentrations, for each of the following elementary reactions.

a $NOCl_2 + NO \longrightarrow 2NOCl$

b $O_3 \longrightarrow O_2 + O$

13.88 Write a rate equation, showing the dependence of rate on reactant concentrations, for each of the following elementary reactions.

a $CH_3Br + OH^- \longrightarrow CH_3OH + Br^-$

b $CS_2 \longrightarrow CS + S$

13.89 The isomerization of cyclopropane, C_3H_6, is believed to occur by the mechanism shown in the following equations:

$$C_3H_6 + C_3H_6 \xrightarrow{k_1} C_3H_6 + C_3H_6^* \qquad \text{(Step 1)}$$
$$C_3H_6^* \xrightarrow{k_2} CH_2{=}CHCH_3 \qquad \text{(Step 2)}$$

Here $C_3H_6^*$ is an excited cyclopropane molecule. At low pressure, Step 1 is much slower than Step 2. Derive the rate law for this mechanism at low pressure. Explain.

13.90 The thermal decomposition of nitryl chloride, NO_2Cl,

$$2NO_2Cl(g) \longrightarrow 2NO_2(g) + Cl_2(g)$$

is thought to occur by the mechanism shown in the following equations:

$$NO_2Cl \xrightarrow{k_1} NO_2 + Cl \qquad \text{(slow step)}$$
$$NO_2Cl + Cl \xrightarrow{k_2} NO_2 + Cl_2 \qquad \text{(fast step)}$$

What rate law is predicted by this mechanism?

13.91 The reaction

$$H_2(g) + I_2(g) \longrightarrow 2HI(g)$$

may occur by the following mechanism:

$$I_2 \underset{k_{-1}}{\overset{k_1}{\rightleftharpoons}} 2I \qquad \text{(fast, equilibrium)}$$
$$I + I + H_2 \xrightarrow{k_2} 2HI \qquad \text{(slow)}$$

What rate law is predicted by this mechanism?

13.92 Ozone decomposes to oxygen gas.

$$2O_3(g) \longrightarrow 3O_2(g)$$

A proposed mechanism for this decomposition is

$$O_3 \underset{k_{-1}}{\overset{k_1}{\rightleftharpoons}} O_2 + O \qquad \text{(fast, equilibrium)}$$
$$O_3 + O \xrightarrow{k_2} 2O_2 \qquad \text{(slow)}$$

What is the rate law derived from this mechanism?

13.93 Consider the following mechanism for a reaction in aqueous solution and indicate the species acting as a catalyst:

$$NH_2NO_2 + OH^- \rightleftharpoons H_2O + NHNO_2^-$$
$$NHNO_2^- \longrightarrow N_2O(g) + OH^-$$

Explain why you believe this species is a catalyst. What is the overall reaction? What substance might be added to the reaction mixture to give the catalytic activity?

13.94 The following is a possible mechanism for a reaction involving hydrogen peroxide in aqueous solution; only a small amount of sodium bromide was added to the reaction mixture.

$$H_2O_2 + Br^- \longrightarrow BrO^- + H_2O$$
$$H_2O_2 + BrO^- \longrightarrow Br^- + H_2O + O_2$$

What is the overall reaction? What species is acting as a catalyst? Are there any reaction intermediates?

General Problems

Key: These problems provide more practice but are not divided by topic or keyed to exercises. Each section ends with essay questions, each of which is color coded to refer to the *A Chemist Looks at Frontiers* (purple) chapter essay on which it is based. Odd-numbered problems and the even-numbered problems that follow are similar; answers to all odd-numbered problems except the essay questions are given in the back of the book.

13.95 A study of the decomposition of azomethane,

$$CH_3NNCH_3(g) \longrightarrow C_2H_6(g) + N_2(g)$$

gave the following concentrations of azomethane at various times:

Time	[CH₃NNCH₃]
0 min	$1.50 \times 10^{-2} M$
10 min	$1.29 \times 10^{-2} M$
20 min	$1.10 \times 10^{-2} M$
30 min	$0.95 \times 10^{-2} M$

Obtain the average rate of decomposition in units of M/s for each time interval.

13.96 Nitrogen dioxide decomposes when heated.

$$2NO_2(g) \longrightarrow 2NO(g) + O_2(g)$$

During an experiment, the concentration of NO_2 varied with time in the following way:

Time	[NO₂]
0.0 min	0.1103 M
1.0 min	0.1076 M
2.0 min	0.1050 M
3.0 min	0.1026 M

Obtain the average rate of decomposition of NO_2 in units of M/s for each time interval.

13.97 You can write the rate law for the decomposition of azomethane as

$$\text{Rate} = k[CH_3NNCH_3]^n \quad \text{or} \quad k = \frac{\text{rate}}{[CH_3NNCH_3]^n}$$

when n is the order of the reaction. Note that when you divide the rates at various times by the concentrations raised to the correct power n, you should get the same number (the rate constant k). Verify that the decomposition of azomethane is first order by dividing each average rate in a time interval (obtained in Problem 13.95) by the average concentration in that interval. Note that each calculation gives nearly the same value. Take the average of these values to obtain the rate constant.

13.98 Use the technique described in Problem 13.97 to verify that the decomposition of nitrogen dioxide is second order. That is, divide each average rate in a time interval (obtained in Problem 13.96) by the square of the average concentration in that interval. Note that each calculation gives nearly the same value. Take the average of these calculated values to obtain the rate constant.

13.99 Methyl acetate reacts in acidic solution.

$$CH_3COOCH_3 + H_2O \xrightarrow{H^+} CH_3OH + CH_3COOH$$
$$\text{methyl acetate} \qquad\qquad\qquad \text{methanol} \quad \text{acetic acid}$$

The rate law is first order in methyl acetate in acidic solution, and the rate constant at 25°C is 1.26×10^{-4}/s. How long will it take for 65% of the methyl acetate to react?

13.100 Benzene diazonium chloride, C_6H_5NNCl, decomposes by a first-order rate law.

$$C_6H_5NNCl \longrightarrow C_6H_5Cl + N_2(g)$$

If the rate constant at 20°C is 4.3×10^{-5}/s, how long will it take for 65% of the compound to decompose?

13.101 What is the half-life of methyl acetate hydrolysis at 25°C in the acidic solution described in Problem 13.99?

13.102 What is the half-life of benzene diazonium chloride decomposition at 20°C? See Problem 13.100 for data.

13.103 A compound decomposes by a first-order reaction. The concentration of compound decreases from 0.1180 M to 0.0950 M in 5.2 min. What fraction of the compound remains after 7.1 min?

13.104 A compound decomposes by a first-order reaction. If the concentration of the compound is 0.025 M after 65 s when the initial concentration was 0.0350 M, what is the concentration of the compound after 75 s?

13.105 Butadiene can undergo the following reaction to form a dimer (two butadiene molecules hooked together).

$$2C_4H_8(g) \longrightarrow C_8H_{12}(g)$$

The half-life for the reaction at a given temperature is 5.92×10^{-2} s. The reaction is second order.

a If the initial concentration of C_4H_8 is 0.50 M, what is the rate constant for the reaction?

b If the initial concentration of C_4H_8 is 0.010 M, what will be the concentration of C_4H_8 after 3.6×10^2 s?

13.106 A plot of 1/[A] versus time for the hypothetical reaction A $\longrightarrow$ B + C yields a straight line.

a What is the order of the reaction?

b If it took 66 s for the concentration to drop 40 percent from its initial value of 0.50 M, what is the rate constant for the reaction?

13.107 At 330°C, the rate constant for the decomposition of NO_2 is 0.775 L/(mol·s). If the reaction is second order, what is the concentration of NO_2 after 2.5×10^2 seconds if the starting concentration was 0.050 M? What is the half-life of this reaction under these conditions?

13.108 A second-order decomposition reaction run at 550°C has a rate constant of 3.1×10^{-2} L/(mol·s). If the initial concentration of the reactant is 0.10 M, what is the concentration of this reactant after 1.5×10^2 s? What is the half-life of this reaction under these conditions?

13.109 Plot the data given in Problem 13.95 to verify that the decomposition of azomethane is first order. Determine the rate constant from the slope of the straight-line plot of log [CH_3NNCH_3] versus time.

13.110 The decomposition of aqueous hydrogen peroxide in a given concentration of catalyst yielded the following data:

Time	0.0 min	5.0 min	10.0 min	15.0 min
[H_2O_2]	0.1000 M	0.0804 M	0.0648 M	0.0519 M

Verify that the reaction is first order. Determine the rate constant for the decomposition of H_2O_2 (in units of /s) from the slope of the straight-line plot of ln [H_2O_2] versus time.

13.111 The reaction 2B $\longrightarrow$ C + 2D is found to be zero-order when run at 990°C. If it takes 4.4×10^2 s for an initial concentration of B to go from 0.50 M to 0.20 M, what is the rate constant for the reaction? What is the half-life of the reaction under these conditions?

13.112 In the presence of a tungsten catalyst at high temperatures, the decomposition of ammonia to nitrogen and hydrogen is a zero-order process. If the rate constant at a particular temperature is 3.7×10^{-6} mol/(L·s), how long will it take for the ammonia concentration to drop from an initial concentration of 5.0×10^{-4} M to 5.0×10^{-5} M? What is the half-life of the reaction under these conditions?

13.113 The decomposition of nitrogen dioxide,

$$2NO_2(g) \longrightarrow 2NO(g) + O_2(g)$$

has a rate constant of 0.498 M/s at 319°C and a rate constant of 1.81 M/s at 354°C. What are the values of the activation energy and the frequency factor for this reaction? What is the rate constant at 420°C?

13.114 A second-order reaction has a rate constant of 8.7×10^{-4}/(M·s) at 30°C. At 40°C, the rate constant is 1.8×10^{-3}/(M·s). What are the activation energy and frequency factor for this reaction? Predict the value of the rate constant at 45°C.

13.115 Methyl chloride, CH_3Cl, reacts in basic solution to give methanol.

$$CH_3Cl + OH^- \longrightarrow CH_3OH + Cl^-$$

This reaction is believed to occur in a single step. If so, what should be the rate law?

13.116 At high temperature, the reaction

$$NO_2(g) + CO(g) \longrightarrow NO(g) + CO_2(g)$$

is thought to occur in a single step. What should be the rate law in that case?

13.117 Nitryl bromide, NO_2Br, decomposes into nitrogen dioxide and bromine.

$$2NO_2Br(g) \longrightarrow 2NO_2(g) + Br_2(g)$$

A proposed mechanism is

$$NO_2Br \longrightarrow NO_2 + Br \qquad\qquad \text{(slow)}$$
$$NO_2Br + Br \longrightarrow NO_2 + Br_2 \qquad \text{(fast)}$$

Write the rate law predicted by this mechanism.

13.118 Tertiary butyl chloride reacts in basic solution according to the equation

$$(CH_3)_3CCl + OH^- \longrightarrow (CH_3)_3COH + Cl^-$$

The accepted mechanism for this reaction is

$$(CH_3)_3CCl \longrightarrow (CH_3)_3C^+ + Cl^- \quad \text{(slow)}$$
$$(CH_3)_3C^+ + OH^- \longrightarrow (CH_3)_3COH \quad \text{(fast)}$$

What should be the rate law for this reaction?

13.119 Acetone reacts with iodine in acidic aqueous solution to give monoiodoacetone.

$$\underset{\text{acetone}}{CH_3COCH_3} + I_2 \xrightarrow{H^+} \underset{\text{monoiodoacetone}}{CH_3COCH_2I} + HI$$

A possible mechanism for this reaction is

$$CH_3\overset{O}{\overset{\|}{C}}CH_3 + H_3O^+ \underset{k_{-1}}{\overset{k_1}{\rightleftharpoons}} CH_3\overset{\overset{\displaystyle H}{|}O^+}{\overset{\|}{C}}CH_3 + H_2O$$
$$\text{(fast, equilibrium)}$$

$$CH_3\overset{\overset{\displaystyle H}{|}O^+}{\overset{\|}{C}}CH_3 + H_2O \xrightarrow{k_2} CH_3\overset{\overset{\displaystyle H}{|}O}{C}=CH_2 + H_3O^+ \quad \text{(slow)}$$

$$CH_3\overset{\overset{\displaystyle H}{|}O}{C}=CH_2 + I_2 \xrightarrow{k_3} CH_3\overset{O}{\overset{\|}{C}}CH_2I + HI \quad \text{(fast)}$$

Write the rate law that you derive from this mechanism.

13.120 Urea, $(NH_2)_2CO$, can be prepared by heating ammonium cyanate, NH_4OCN.

$$NH_4OCN \longrightarrow (NH_2)_2CO$$

This reaction may occur by the following mechanism:

$$NH_4^+ + OCN^- \underset{k_{-1}}{\overset{k_1}{\rightleftharpoons}} NH_3 + HOCN \quad \text{(fast, equilibrium)}$$
$$NH_3 + HOCN \xrightarrow{k_2} (NH_2)_2CO \quad \text{(slow)}$$

What is the rate law predicted by this mechanism?

13.121 A study of the gas-phase oxidation of nitrogen monoxide at 25°C and 1.00 atm pressure gave the following results:

$$2NO(g) + O_2(g) \longrightarrow NO_2(g)$$

	Conc. NO, mol/L	Conc. O_2, mol/L	Initial Rate
Exp. 1	4.5×10^{-2}	2.2×10^{-2}	0.80×10^{-2} mol/(L·s)
Exp. 2	4.5×10^{-2}	4.5×10^{-2}	1.60×10^{-2} mol/(L·s)
Exp. 3	9.0×10^{-2}	9.0×10^{-2}	1.28×10^{-1} mol/(L·s)
Exp. 4	3.8×10^{-1}	4.6×10^{-3}	?

a What is the experimental rate law for the reaction above?
b What is the initial rate of the reaction in Experiment 4?

13.122 The reaction of water with CH_3Cl in acetone as a solvent is represented by the equation

$$CH_3Cl + H_2O \longrightarrow CH_3OH + HCl$$

The rate of the reaction doubles when the concentration of CH_3Cl is doubled and it quadruples when the concentration of H_2O is doubled.

a What is the unit for k?
b Calculate k if CH_3OH is formed at a rate of 1.50 M/s when $[CH_3Cl] = [H_2O] = 0.40 \ M$.

13.123 The reaction of thioacetamide with water is shown by the equation below:

$$CH_3C(S)NH_2(aq) + H_2O \longrightarrow H_2S(aq) + CH_3C(O)NH_2(aq)$$

The rate of reaction is given by the rate law:

$$\text{Rate} = k[H_3O^+][CH_3C(S)NH_2]$$

Consider one liter of solution that is 0.20 M in $CH_3C(S)NH_2$ and 0.15 M in HCl at 25°C.

a For each of the changes listed below, state whether the rate of reaction increases, decreases, or remains the same. Why?
   i. A 4-g sample of NaOH is added to the solution.
   ii. 500 mL of water is added to the solution.
b For each of the changes listed below, state whether the value of k will increase, decrease, or remain the same. Why?
   i. A catalyst is added to the solution.
   ii. The reaction is carried out at 15°C instead of 25°C.

13.124 The reaction of methylacetate with water is shown by the equation below:

$$CH_3COOCH_3(aq) + H_2O \longrightarrow CH_3COOH(aq) + CH_3OH(aq)$$

The rate of the reaction is given by the rate law:

$$\text{Rate} = k[H_3O^+][CH_3COOCH_3]$$

Consider one liter of solution that is 0.15 M in CH_3COOCH_3 and 0.015 M in H_3O^+ at 25°C.

a For each of the changes listed below, state whether the rate of reaction increases, decreases, or remains the same. Why?
   i. Some concentrated sulfuric acid is added to the solution.
   ii. Water is added to the solution.
b For each of the changes listed below, state whether the value of k will increase, decrease, or remain the same. Why?
   i. Some concentrated sulfuric acid is added to the solution.
   ii. The reaction is carried out at 35°C instead of 25°C.

13.125 Draw and label the potential-energy curve for the reaction

$$N_2O_4(g) \rightleftharpoons 2NO_2(g); \Delta H = 57 \text{ kJ}$$

The activation energy for the reverse reaction is 23 kJ. Note ΔH and E_a on the diagram. What is the activation energy for the forward reaction? For which reaction (forward or reverse) will the reaction rate be most sensitive to a temperature increase? Explain.

13.126 Draw a potential-energy diagram for an uncatalyzed exothermic reaction. On the same diagram, indicate the change that results on the addition of a catalyst. Discuss the role of a catalyst in changing the rate of reaction.

13.127 What is meant by the term *rate of a chemical reaction*? Why does the rate of a reaction normally change with time? When does the rate of a chemical reaction equal the rate constant?

13.128 Briefly discuss the factors that affect the rates of chemical reactions. Which of the factors affect the magnitude of the rate constant? Which factor(s) do not affect the magnitude of the rate constant? Why?

■ **13.129** Describe in general terms how chemists can follow the character of a reacting molecule using lasers.

■ **13.130** Describe the potential-energy curve for the decomposition of cyclobutane to ethylene.

Strategy Problems

Key: As noted earlier, all of the practice and general problems are matched pairs. This section is a selection of problems that are not in matched-pair format. These challenging problems require that you employ many of the concepts and strategies that were developed in the chapter. In some cases, you will have to integrate several concepts and operational skills in order to solve the problem successfully.

13.131 The rate constant for a certain reaction is $1.4 \times 10^{-5} \, M^{-1} \, min^{-1}$ at 483 K. The activation energy for the reaction is 2.11×10^3 J/mol. What is the rate constant for the reaction at 611 K?

13.132 The decomposition of hydrogen peroxide is a first-order reaction:

$$H_2O_2(aq) \longrightarrow H_2O(l) + \tfrac{1}{2}O_2(g)$$

The half-life of the reaction is 17.0 minutes.

a What is the rate constant of the reaction?

b If you started the reaction with $[H_2O_2] = 0.100 \, M$, what would be the hydrogen peroxide concentration after 15.0 minutes?

c If you had a bottle of H_2O_2, how long would it take for 86.0% to decompose?

13.133 A reaction of the form A $\longrightarrow$ Products is second order with a half-life of 574 s. What is the rate constant of the reaction if the initial concentration of A is 1.87×10^{-2} mol/L?

13.134 What is the rate law for the following gas-phase elementary reaction?

$$2I + H_2 \longrightarrow 2HI$$

13.135 A possible mechanism for a gas-phase reaction is given below. What is the rate law predicted by this mechanism?

$$NO + Cl_2 \underset{k_r}{\overset{k_f}{\rightleftharpoons}} NOCl_2 \qquad \text{(fast equilibrium)}$$

$$NOCl_2 + NO \overset{k_2}{\longrightarrow} 2NOCl \qquad \text{(slow)}$$

13.136 The hypothetical reaction A + 2B $\longrightarrow$ Products has the rate law Rate = $k[A]^2[B]^3$. If the reaction is run two separate times, holding the concentration of A constant while doubling the concentration of B from one run to the next, how would the rate of the second run compare to the rate of the first run?

13.137 Say you run the following elementary, termolecular reaction:

$$2A + B \longrightarrow D$$

Using the starting concentrations $[A] = 2.0 \, M$ and $[B] = 2.0 \, M$, you measure the rate to be $16 \, M^{-2} \cdot s^{-1}$. What is the value of the rate constant?

13.138 For the decomposition of one mole of nitrosyl chloride, $\Delta H = 38$ kJ.

$$NOCl(g) \longrightarrow NO(g) + \tfrac{1}{2}Cl_2(g)$$

The activation energy for this reaction is 100 kJ.

a Is this reaction exothermic or endothermic?

b What is the activation energy for the reverse reaction?

c If a catalyst were added to the reaction, how would this affect the activation energy?

13.139 The following data were collected for the reaction A(g) + B(g) $\longrightarrow$ Products.

Experiment	$[A]_0$ *(M)*	$[B]_0$ *(M)*	Rate *(M/s)*
1	0.0100	0.100	1.0×10^{-3}
2	0.0300	0.100	3.0×10^{-3}
3	0.0300	0.300	2.7×10^{-2}

a Determine the rate law for this reaction.

b Calculate the rate constant.

c Calculate the rate when $[A] = 0.200 \, M$ and $[B] = 0.200 \, M$.

13.140 Given the following mechanism for a chemical reaction:

$$H_2O_2 + I^- \longrightarrow H_2O + IO^-$$

$$H_2O_2 + IO^- \longrightarrow H_2O + O_2 + I^-$$

a Write the overall reaction.

b Identify the catalyst and the reaction intermediate.

c With the information given in this problem, can you write the rate law? Explain.

13.141 A hypothetical reaction has the two-step mechanism

$$AB + C \longrightarrow ABC$$

$$ABC + C \longrightarrow AC + BC$$

The potential energy curve for the reaction is

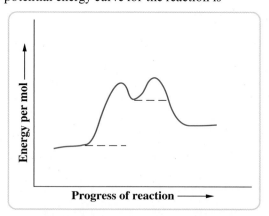

a Write the chemical formulas of the reactants, products, and the reaction intermediate on the potential energy curve.

b From the mechanism, what is the overall reaction?

c What is the rate-limiting step for the reaction?

d Propose a rate law based on the rate-limiting step.

e Is the reaction exothermic or endothermic?

13.142 The rate law for reaction $O_3(g) + NO(g) \longrightarrow O_2(g) + NO_2(g)$ is first order in O_3 and first order in NO. The reaction is exothermic, and the temperature dependence of the rate constant can be modeled by the Arrhenius equation. Which of the following statements regarding this reaction are true as the temperature of the reaction is increased (choose all that apply)?

- a. The reaction rate would increase.
- b. The activation energy would increase.
- c. A greater percentage of the molecular collisions between O_3 and NO in a reaction mixture would exceed the activation energy.
- d. The reaction would become more exothermic.
- e. The magnitude of the rate constant would increase.

13.143 A series of kinetics experiments are conducted on the hypothetical reaction $A + 2B \longrightarrow AB_2$. Prior to conducting experiments, there are four proposed rate laws

$$\text{rate} = k \, [A]^2[B]$$
$$\text{rate} = k \, [A][B]$$
$$\text{rate} = k \, [A][B]^2$$
$$\text{rate} = k \, [A][B]^4$$

Experimental data indicates that when the concentration of A is doubled while the concentration of B is held constant, the rate of the reaction doubles. The data also shows that when the concentration of A is held constant and the concentration of B is doubled, the rate of the reaction increases by a factor of four.

- a. Which of the proposed rate laws is consistent with the experimental data?
- b. Using your answer from part a, predict the change in the initial rate for a reaction in which the concentration of A is reduced by 50% and the concentration of B is increased by 100%.
- c. If you were to run the reaction with $[A]_0 = 0.10 \, M$ and $[B]_0 = 0.10 \, M$, sketch a concentration versus time graph depicting the changes in concentration of both A and B.

13.144 For many reactions, the reaction rate approximately doubles with every $10°C$ increase in temperature. Using this approximation, if a reaction were run at $35°C$ and then at $75°C$, how would the reaction rates compare at these two temperatures?

13.145 A substance is found to have a reaction with the form $aA \longrightarrow$ products. A $1/[A]$ versus time plot of experimental data at $25°C$ produced a straight line with a slope of 2.8×10^{-2} L/(mol·sec).

- a. What is the rate constant for the reaction?
- b. If the initial concentration of A is $1.9 \times 10^{-4} \, M$, what is the half-life of the reaction?
- c. If the initial concentration of A is $2.1 \times 10^{-3} \, M$, what is the concentration of A after 275 seconds?

13.146 Dinitrogen pentoxide decomposes according to the reaction

$$2N_2O_5(g) \longrightarrow 4NO_2(g) + O_2(g)$$

At $45°C$ the rate constant is $6.2 \times 10^{-4} \, s^{-1}$. The reaction is first order.

- a. Write the rate law for the reaction.
- b. Calculate the rate of the reaction when $[N_2O_5]$ is $0.20 \, M$.
- c. Calculate the half-life of the reaction.
- d. If the initial concentration of N_2O_5 is $0.50 \, M$, what is the concentration of $[N_2O_5]$ after 1.5 hours?

Cumulative-Skills Problems

Key: The problems under this heading combine skills introduced in previous chapters with those given in the current one.

13.147 Hydrogen peroxide undergoes a first-order decomposition to water and O_2 in aqueous solution. The rate constant at $25°C$ is 7.40×10^{-4}/s. Calculate the volume of O_2 obtained from the decomposition reaction of 1.00 mol H_2O_2 at $25°C$ and 740 mmHg after 30.0 min.

13.148 Dinitrogen pentoxide, N_2O_5, undergoes first-order decomposition in chloroform solvent to yield NO_2 and O_2. The rate constant at $45°C$ is 6.2×10^{-4}/min. Calculate the volume of O_2 obtained from the reaction of 1.00 mol N_2O_5 at $45°C$ and 770 mmHg after 20.0 hr.

13.149 Hydrogen peroxide in aqueous solution decomposes by a first-order reaction to water and oxygen. The rate constant for this decomposition is 7.40×10^{-4}/s. What quantity of heat energy is initially liberated per second from 2.00 L of solution that is $1.50 \, M \, H_2O_2$? See Appendix C for data.

13.150 Nitrogen dioxide reacts with carbon monoxide by the overall equation

$$NO_2(g) + CO(g) \longrightarrow NO(g) + CO_2(g)$$

At a particular temperature, the reaction is second order in NO_2 and zero order in CO. The rate constant is 0.515 L/(mol·s). How much heat energy evolves per second initially from 3.50 L of reaction mixture containing $0.0275 \, M \, NO_2$? See Appendix C for data. Assume the enthalpy change is constant with temperature.

13.151 Nitrogen monoxide reacts with hydrogen as follows:

$$2NO(g) + H_2(g) \longrightarrow N_2O(g) + H_2O(g)$$

The rate law is $-\Delta[H_2]/\Delta t = k[NO]^2[H_2]$, where k is $1.10 \times 10^{-7} \, L^2/(mol^2 \cdot s)$ at $826°C$. A vessel contains NO and H_2 at $826°C$. The partial pressures of NO and H_2 are 144 mmHg and 315 mmHg, respectively. What is the rate of decrease of partial pressure of NO? See Problem 13.151.

13.152 Nitrogen monoxide reacts with oxygen to give nitrogen dioxide.

$$2NO(g) + O_2(g) \longrightarrow 2NO_2(g)$$

The rate law is $-\Delta[NO]/\Delta t = k[NO]^2[O_2]$, where the rate constant is $1.16 \times 10^{-5} \, L^2/(mol^2 \cdot s)$ at $339°C$. A vessel contains NO and O_2 at $339°C$. The initial partial pressures of NO and O_2 are 155 mmHg and 345 mmHg, respectively. What is the rate of decrease of partial pressure of NO (in mmHg per second)? (*Hint:* From the ideal gas law, obtain an expression for the molar concentration of a particular gas in terms of its partial pressure.)

14 Chemical Equilibrium

Nitrogen dioxide, NO_2 (a reddish brown gas), is in equilibrium with dinitrogen tetroxide, N_2O_4 (a colorless gas). Nitrogen dioxide is present in many urban smogs, giving them a characteristic reddish brown color.

OWL Sign in to OWL at **www.cengage.com/owl** to view tutorials and simulations, develop problem-solving skills, and complete online homework assigned by your professor.

Go Chemistry Download mini lecture videos for key concept review and exam prep from OWL or purchase them from **www.cengagebrain.com**.

© Novastock/Photo Researchers, Inc.; Cengage Learning

Chemical reactions often seem to stop before they are complete. Such reactions are *reversible*. That is, the original reactants form products, but then the products react with themselves to give back the original reactants. Actually two reactions are occurring, and the eventual result is a mixture of reactants and products, rather than simply a mixture of products.

Consider the gaseous reaction in which carbon monoxide and hydrogen react to produce methane and steam:

$$CO(g) + 3H_2(g) \longrightarrow \underset{\text{methane}}{CH_4(g)} + H_2O(g)$$

This reaction, which requires a catalyst to occur at a reasonable rate, is called *catalytic methanation*. It is a potentially useful reaction because it can be used to produce methane from coal or organic wastes from which carbon monoxide can be obtained (Figure 14.1). Methane, in turn, can be used as a fuel and as a starting material for the production of organic chemicals. Although, at the moment, natural gas (which is mostly methane) is plentiful, analysts predict that oil and natural gas production will peak sometime in the early part of this century. Then the demand for other sources of fuels and organic chemicals, such as coal and organic wastes, will rise.

Catalytic methanation is a reversible reaction, and depending on the reaction conditions, the final reaction mixture will have varying amounts of the products methane and steam, as well as the starting substances carbon monoxide and hydrogen. It is also possible to start with methane and steam and, under the right conditions, form a mixture that is predominantly carbon monoxide and hydrogen. The process is called *steam reforming*.

$$CH_4(g) + H_2O(g) \longrightarrow CO(g) + 3H_2(g)$$

The product mixture of CO and H_2 (synthesis gas) is used to prepare a number of industrial chemicals.

The processes of catalytic methanation and steam reforming illustrate the reversibility of chemical reactions. Starting with CO and H_2 and using the right conditions, you can form predominantly CH_4 and H_2O. Starting with CH_4 and H_2O and using different conditions, you can obtain a reaction mixture that is predominantly CO and H_2. An important question is, What conditions favor the production of CH_4 and H_2O, and what conditions favor the production of CO and H_2?

As noted earlier, certain reactions (such as catalytic methanation) appear to stop before they are complete. The reaction mixture ceases to change in any of its properties and consists of both reactants and products in definite concentrations. Such a reaction mixture is said to have reached *chemical equilibrium*. In earlier chapters, we discussed other types of equilibria, including the equilibrium between a liquid and its vapor and the equilibrium between a solid and its saturated solution. In this chapter, we will see how to determine the composition of a reaction mixture at equilibrium and how to alter this composition by changing the conditions for the reaction.

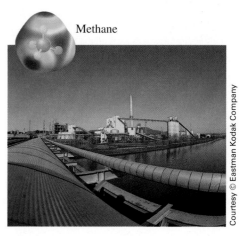

Figure 14.1 ▲

Coal gasification plant Gaseous fuels can be made from coal in a number of ways. One way (described in the text) is to react coal with steam to produce a mixture of CO and H_2, which then reacts by catalytic methanation to yield CH_4, the major component of natural gas.

Describing Chemical Equilibrium

Many chemical reactions are like the catalytic methanation reaction. Such reactions can be made to go predominantly in one direction or the other, depending on the conditions. Let us look more closely at this reversibility and see how to characterize it quantitatively.

14.1 Chemical Equilibrium—A Dynamic Equilibrium

When substances react, they eventually form a mixture of reactants and products in *dynamic equilibrium*. ▶ This dynamic equilibrium consists of a forward reaction, in which substances react to give products, and a reverse reaction, in which products

The concept of dynamic equilibrium in chemical reactions was briefly mentioned in Section 13.8. Discussions on vapor pressure (Section 11.2) and solubility (Section 12.2) give detailed explanations of the role of dynamic equilibria.

react to give the original reactants. Both forward and reverse reactions occur at the same rate, or speed.

Consider the catalytic methanation reaction discussed in the chapter opening. It consists of forward and reverse reactions, as represented by the chemical equation

$$CO(g) + 3H_2(g) \rightleftharpoons CH_4(g) + H_2O(g)$$

Suppose you put 1.000 mol CO and 3.000 mol H_2 into a 10.00-L vessel at 1200 K (927°C). The rate of the reaction of CO and H_2 depends on the concentrations of CO and H_2. At first these concentrations are large, but as the substances react their concentrations decrease (Figure 14.2). The rate of the forward reaction, which depends on reactant concentrations, is large at first but steadily decreases. On the other hand, the concentrations of CH_4 and H_2O, which are zero at first, increase with time. The rate of the reverse reaction starts at zero and steadily increases. The forward rate decreases and the reverse rate increases until eventually the rates become equal. When that happens, CO and H_2 molecules are formed as fast as they react. The concentrations of reactants and products no longer change, and the reaction mixture has reached *equilibrium*. Figure 14.3 shows how the amounts of substances in the reaction mixture become constant when equilibrium is reached.

Chemical equilibrium is *the state reached by a reaction mixture when the rates of forward and reverse reactions have become equal.* If you observe the reaction mixture, you see no net change, although the forward and reverse reactions are continuing. The continuing forward and reverse reactions make the equilibrium a *dynamic* process.

Suppose you place known amounts of reactants in a vessel and let the mixture come to equilibrium. To obtain the composition of the equilibrium mixture, you need to determine the amount of only one of the substances. The amounts of the others can be calculated from the amounts originally placed in the vessel and the equation that represents the reaction. For example, suppose you place 1.000 mol CO and 3.000 mol H_2 in a reaction vessel at 1200 K. After equilibrium is reached, you chill the reaction mixture quickly to condense the

Figure 14.2 ▶

Change of rates as reaction proceeds

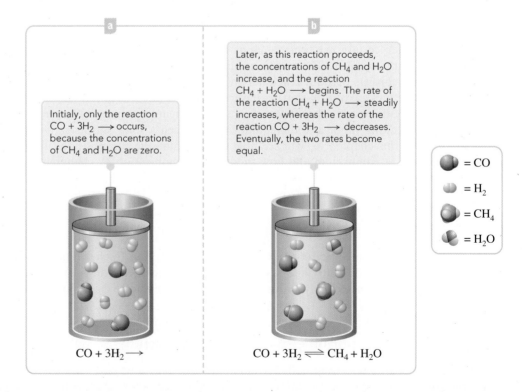

Initialy, only the reaction CO + 3H₂ ⟶ occurs, because the concentrations of CH₄ and H₂O are zero.

Later, as this reaction proceeds, the concentrations of CH₄ and H₂O increase, and the reaction CH₄ + H₂O ⟶ begins. The rate of the reaction CH₄ + H₂O ⟶ steadily increases, whereas the rate of the reaction CO + 3H₂ ⟶ decreases. Eventually, the two rates become equal.

= CO
= H₂
= CH₄
= H₂O

CO + 3H₂ ⟶

CO + 3H₂ ⇌ CH₄ + H₂O

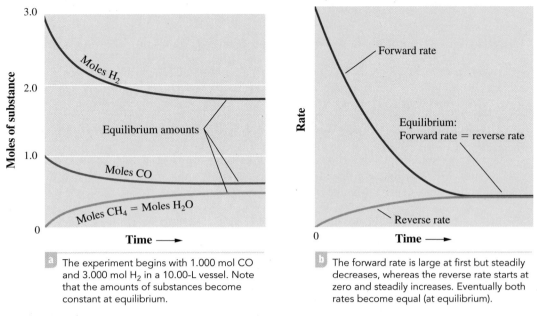

a) The experiment begins with 1.000 mol CO and 3.000 mol H_2 in a 10.00-L vessel. Note that the amounts of substances become constant at equilibrium.

b) The forward rate is large at first but steadily decreases, whereas the reverse rate starts at zero and steadily increases. Eventually both rates become equal (at equilibrium).

Figure 14.3 ▲

Catalytic methanation reaction approaches equilibrium

water vapor to liquid, which you then weigh and find to be 0.387 mol. You calculate the other substances from the stoichiometry of the reaction, as shown in the following example.

⧫WL Interactive Example 14.1   **Applying Stoichiometry to an Equilibrium Mixture**

Gaining Mastery Toolbox

Critical Concept 14.1
This is a problem in stoichiometry; we set it up in tabular form with the chemical equation forming the heading, a useful procedure for the equilibrium calculations in this chapter.

Solution Essentials:
• Table for change amounts in a reaction
• Stoichiometry

Carbon monoxide and hydrogen react according to the following equation:

$$CO(g) + 3H_2(g) \rightleftharpoons CH_4(g) + H_2O(g)$$

When 1.000 mol CO and 3.000 mol H_2 are placed in a 10.00-L vessel at 927°C (1200 K) and allowed to come to equilibrium, the mixture is found to contain 0.387 mol H_2O. What is the molar composition of the equilibrium mixture? That is, how many moles of each substance are present?

Problem Strategy The problem is essentially one of stoichiometry. It involves initial, or *starting,* amounts of reactants. These amounts *change* as the reaction comes to equilibrium. At *equilibrium,* you analyze the reaction mixture for the amount of one of the reactants or products. It is convenient to solve this problem by first setting up a table in which you write the starting, change, and equilibrium values of each substance under the balanced equation. This way you can easily see what you have to calculate.

Amount (mol)	$CO(g)$ +	$3H_2(g)$ ⇌	$CH_4(g)$ +	$H_2O(g)$
Starting				
Change				
Equilibrium				

You fill in the starting amounts from the values given in the problem statement. You are not given explicit values for the changes that occur in going to equilibrium, so you let x be the molar change. That is, each product increases by x moles multiplied by the coefficients of the substances in the balanced equation. Reactants decrease by x moles multiplied by the corresponding coefficients. The decrease is indicated by a negative sign. Equilibrium values are equal to starting values plus the changes.

(continued)

(continued)

Solution Using the information given in the problem, you set up the following table:

Amount (mol)	CO(g) +	3H$_2$(g) $\rightleftharpoons$	CH$_4$(g) +	H$_2$O(g)
Starting	1.000	3.000	0	0
Change	−x	−3x	+x	+x
Equilibrium	1.000 − x	3.000 − 3x	x	x = 0.387

The problem statement gives the equilibrium amount for H$_2$O. This tells you that $x = 0.387$ mol. You calculate equilibrium amounts for other substances from the expressions given in the table, using this value of x.

$$\text{Equilibrium amount CO} = (1.000 - x) \text{ mol}$$
$$= (1.000 - 0.387) \text{ mol}$$
$$= 0.613 \text{ mol}$$

$$\text{Equilibrium amount H}_2 = (3.000 - 3x) \text{ mol}$$
$$= (3.000 - 3 \times 0.387) \text{ mol}$$
$$= 1.839 \text{ mol}$$

$$\text{Equilibrium amount CH}_4 = x \text{ mol} = 0.387 \text{ mol}$$

Therefore, the amounts of substances in the equilibrium mixture are **0.613 mol CO, 1.839 mol H$_2$, 0.387 mol CH$_4$, and 0.387 mol H$_2$O.**

Answer Check You start with 4.000 mol reactants (1.000 mol CO + 3.000 mol H$_2$). If the reaction were to go to completion, you would obtain 2.000 mol products (1.000 mol CH$_4$ + 1.000 mol H$_2$O). The equilibrium quantities will be intermediate. Check that this is the case.

Exercise 14.1 Synthesis gas (a mixture of CO and H$_2$) is increased in the concentration of hydrogen by passing it with steam over a catalyst. This is the so-called water–gas shift reaction. Some of the CO is converted to CO$_2$, which can be removed:

$$CO(g) + H_2O(g) \rightleftharpoons CO_2(g) + H_2(g)$$

Suppose you start with a gaseous mixture containing 1.00 mol CO and 1.00 mol H$_2$O. When equilibrium is reached at 1000°C, the mixture contains 0.43 mol H$_2$. What is the molar composition of the equilibrium mixture?

■ See Problems 14.29, 14.30, 14.31, 14.32, 14.33, and 14.34.

CONCEPT CHECK 14.1

Two substances A and B react to produce substance C. When reactant A decreases by molar amount x, product C increases by molar amount x. When reactant B decreases by molar amount x, product C increases by molar amount $2x$. Write the chemical equation for the reaction.

14.2 The Equilibrium Constant

In the preceding example, we found that when 1.000 mol CO and 3.000 mol H$_2$ react in a 10.00-L vessel by catalytic methanation at 1200 K, they give an equilibrium mixture containing 0.613 mol CO, 1.839 mol H$_2$, 0.387 mol CH$_4$, and 0.387 mol H$_2$O.

Let us call this Experiment 1. Now consider a similar experiment, Experiment 2, in which you start with an additional mole of carbon monoxide. That is, you place 2.000 mol CO and 3.000 mol H_2 in a 10.00-L vessel at 1200 K. At equilibrium, you find that the vessel contains 1.522 mol CO, 1.566 mol H_2, 0.478 mol CH_4, and 0.478 mol H_2O. What you observe from the results of Experiments 1 and 2 is that the equilibrium composition depends on the amounts of starting substances. Nevertheless, you will see that all of the equilibrium compositions for a reaction at a given temperature are related by a quantity called the *equilibrium constant.*

Definition of the Equilibrium Constant K_c

Consider the reaction

$$a\text{A} + b\text{B} \rightleftharpoons c\text{C} + d\text{D}$$

where A, B, C, and D denote reactants and products, and a, b, c, and d are coefficients in the balanced chemical equation. The **equilibrium-constant expression** for a reaction is *an expression obtained by multiplying the concentrations of products, dividing by the concentrations of reactants, and raising each concentration term to a power equal to the coefficient in the chemical equation.* The **equilibrium constant K_c** is *the value obtained for the equilibrium-constant expression when equilibrium concentrations are substituted.* ▶ For the previous reaction, you have

$$K_c = \frac{[\text{C}]^c[\text{D}]^d}{[\text{A}]^a[\text{B}]^b}$$

Here you denote the molar concentration of a substance by writing its formula in square brackets. The subscript c on the equilibrium constant means that it is defined in terms of molar concentrations. The **law of mass action** is *a relation that states that the values of the equilibrium-constant expression K_c are constant for a particular reaction at a given temperature, whatever equilibrium concentrations are substituted.* ▶

As the following example illustrates, the equilibrium-constant expression is defined in terms of the balanced chemical equation. If the equation is rewritten with different coefficients, the equilibrium-constant expression will be changed.

In Section 14.3, you will see that you ignore concentrations of pure liquids and solids when you write the equilibrium-constant expression. The concentrations of these substances are constant and are incorporated into the value of K_c.

"Active mass" was an early term for concentration—hence the term mass action.

✪WL Interactive Example 14.2 Writing Equilibrium-Constant Expressions

Gaining Mastery Toolbox

Critical Concept 14.2
You can write the equilibrium-constant expression given any balanced chemical equation.

Solution Essentials:
- Equilibrium-constant expression, K_c
- Balanced equation

a Write the equilibrium-constant expression K_c for catalytic methanation.

$$CO(g) + 3H_2(g) \rightleftharpoons CH_4(g) + H_2O(g)$$

b Write the equilibrium-constant expression K_c for the reverse of the previous reaction.

$$CH_4(g) + H_2O(g) \rightleftharpoons CO(g) + 3H_2(g)$$

c Write the equilibrium-constant expression K_c for the synthesis of ammonia.

$$N_2(g) + 3H_2(g) \rightleftharpoons 2NH_3(g)$$

d Write the equilibrium-constant expression K_c when the equation for the previous reaction is written

$$\tfrac{1}{2}N_2(g) + \tfrac{3}{2}H_2(g) \rightleftharpoons NH_3(g)$$

Problem Strategy Write the concentrations of products in the top (numerator) of the equilibrium-constant expression, and write the concentrations of reactants in the bottom (denominator). Raise each concentration term to the power equal to the coefficient of the substance in the chemical equation.

Solution

a The expression for the equilibrium constant is

$$K_c = \frac{[CH_4][H_2O]}{[CO][H_2]^3}$$

Note that concentrations of products are on the top and concentrations of reactants are on the bottom. Also note that each concentration term is raised to a power equal to the coefficient of the substance in the chemical equation.

(continued)

(continued)

b When the equation is written in reverse order, the expression for K_c is inverted:

$$K_c = \frac{[CO][H_2]^3}{[CH_4][H_2O]}$$

c The equilibrium constant for $N_2 + 3H_2 \rightleftharpoons 2NH_3$ is

$$K_c = \frac{[NH_3]^2}{[N_2][H_2]^3}$$

d When the coefficients in the equation in part c are multiplied by $\frac{1}{2}$ to give

$$\tfrac{1}{2}N_2 + \tfrac{3}{2}H_2 \rightleftharpoons NH_3$$

the equilibrium-constant expression becomes

$$K_c = \frac{[NH_3]}{[N_2]^{1/2}[H_2]^{3/2}}$$

which is the square root of the previous expression.

Answer Check Check your answers. Make sure that the products are on top and the reactants are on the bottom of each equilibrium-constant expression. Then check that each reactant and product term is raised to the correct power (equal to the stoichiometric coefficient in the chemical equation).

Exercise 14.2 **a** Write the equilibrium-constant expression K_c for the equation

$$2NO_2(g) + 7H_2(g) \rightleftharpoons 2NH_3(g) + 4H_2O(g)$$

b Write the equilibrium-constant expression K_c when this reaction is written

$$NO_2(g) + \tfrac{7}{2}H_2(g) \rightleftharpoons NH_3(g) + 2H_2O(g)$$

■ See Problems 14.35 and 14.36.

Equilibrium: A Kinetics Argument

The law of mass action was first stated by the Norwegian chemists Cato Guldberg and Peter Waage in 1867. They were led to this law by a kinetics argument. To understand the argument, consider the decomposition of dinitrogen tetroxide, N_2O_4.

Dinitrogen tetroxide is a colorless substance that decomposes, when warmed, to give nitrogen dioxide, NO_2, a brown substance (see Figure 14.4). Suppose you start with dinitrogen tetroxide gas and heat it above room temperature. At this higher temperature, N_2O_4 decomposes to NO_2, which can be noted by the increasing intensity of the brown color of the reaction mixture.

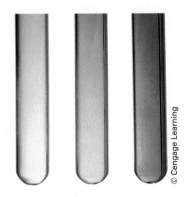

© Cengage Learning

Figure 14.4 ▲

Temperature effect on NO₂–N₂O₄ equilibrium The temperature in the test tubes increases from left to right. N_2O_4 (colorless) predominates at cooler temperatures, whereas NO_2 (reddish brown) predominates at higher temperatures.

$$N_2O_4(g) \longrightarrow 2NO_2(g)$$

Once some NO_2 is produced, it can react to re-form N_2O_4. Call the decomposition of N_2O_4 the forward reaction and the formation of N_2O_4 the reverse reaction. These are elementary reactions, and you can immediately write the rate equations. The rate of the forward reaction is $k_f[N_2O_4]$, where k_f is the rate constant for the forward reaction. The rate of the reverse reaction is $k_r[NO_2]^2$, where k_r is the rate constant for the reverse reaction.

At first, the reaction vessel contains mostly N_2O_4, so the concentration of NO_2 is very small. Therefore, the rate of the forward reaction is relatively large, and the rate of the reverse reaction is relatively small. However, as the decomposition proceeds, the concentration of N_2O_4 decreases, so the forward rate decreases. At the same time, the concentration of NO_2 increases, so the reverse rate increases. Eventually the two rates become equal. After that, on average, every time an N_2O_4 molecule decomposes, two NO_2 molecules recombine. Overall, the concentrations of N_2O_4 and NO_2 no longer change, although forward and reverse reactions are still occurring. The reaction has reached a dynamic equilibrium.

At equilibrium, you can write: Rate of forward reaction = rate of reverse reaction, or

$$k_f[N_2O_4] = k_r[NO_2]^2$$

Rearranging gives

$$\frac{k_f}{k_r} = \frac{[NO_2]^2}{[N_2O_4]}$$

The right side is the equilibrium-constant expression for the decomposition of dinitrogen tetroxide. The left side is the ratio of rate constants. Thus, the equilibrium-constant expression is constant for a given temperature, and you can identify the equilibrium constant K_c as the ratio of rate constants for the forward and reverse reactions.

$$K_c = \frac{k_f}{k_r}$$

If the overall reaction occurs by a multistep mechanism, one can show that the equilibrium constant equals a product of ratios of rate constants—one ratio for each step in the mechanism.

Obtaining Equilibrium Constants for Reactions

At the beginning of this section, we gave data from the results of two experiments, Experiments 1 and 2, involving catalytic methanation. By substituting the molar concentrations from these two experiments into the equilibrium-constant expression for the reaction, you can show that you get the same value for both experiments, as you expect from the law of mass action. This value equals K_c for methanation at 1200 K. Thus, besides verifying the validity of the law of mass action in this case, you can see how an equilibrium constant can be obtained from concentration data. ▶

It is also possible to determine equilibrium constants from thermochemical data, as described in Section 18.6.

Experiment 1 The equilibrium composition is 0.613 mol CO, 1.839 mol H_2, 0.387 mol CH_4, and 0.387 mol H_2O. The volume of the reaction vessel is 10.00 L, so the equilibrium concentration of CO is

$$[CO] = \frac{0.613 \text{ mol}}{10.00 \text{ L}} = 0.0613 \text{ } M$$

Similarly, the other equilibrium concentrations are $[H_2] = 0.1839$ M, $[CH_4] = 0.0387$ M, and $[H_2O] = 0.0387$ M.

Substitute these values into the equilibrium expression for catalytic methanation. (The expression was obtained in Example 14.2.) Although until now we have consistently carried units along with numbers in calculations, it is the usual practice to write equilibrium constants without units. We will follow that practice here. ▶ Substitution of concentrations into the equilibrium-constant expression gives

$$K_c = \frac{[CH_4][H_2O]}{[CO][H_2]^3} = \frac{(0.0387)(0.0387)}{(0.0613)(0.1839)^3} = 3.93$$

Experiment 2 The equilibrium composition is 1.522 mol CO, 1.566 mol H_2, 0.478 mol CH_4, and 0.478 mol H_2O. Therefore, the concentrations, which are obtained by dividing by 10.00 L, are $[CO] = 0.1522$ M, $[H_2] = 0.1566$ M, $[CH_4] = 0.0478$ M, and $[H_2O] = 0.0478$ M. Substituting into the equilibrium-constant expression gives

In thermodynamics, the equilibrium constant is defined in terms of activities, rather than concentrations. For an ideal mixture, the activity of a substance is the ratio of its concentration (or partial pressure if a gas) to a standard concentration of 1 M (or partial pressure of 1 atm), so that units cancel. Thus, activities have numerical values but no units.

$$K_c = \frac{[CH_4][H_2O]}{[CO][H_2]^3} = \frac{(0.0478)(0.0478)}{(0.1522)(0.1566)^3} = 3.91$$

Within the precision of the data, these values (3.93 and 3.91) of the equilibrium expression for different starting mixtures of these gases at 1200 K are the same. Moreover, experiment shows that when you start with CH_4 and H_2O, instead of CO and H_2, an equilibrium mixture is again reached that yields the same value of K_c (see Experiment 3, Figure 14.5). You can take the equilibrium constant for catalytic methanation at 1200 K to be 3.92, the average of these values.

Figure 14.5 ▶

Some equilibrium compositions for the methanation reaction
Different starting concentrations have been used in each experiment (all at 1200 K). Experiments 1 and 2 start with different concentrations of reactants CO and H_2. Experiment 3 starts with the products CH_4 and H_2O. All three experiments yield essentially the same value of K_c.

	Starting Concentrations	Equilibrium Concentrations	Calculated Value of K_c
Experiment 1	0.1000 M CO 0.3000 M H_2	0.0613 M CO 0.1839 M H_2 0.0387 M CH_4 0.0387 M H_2O	$K_c = 3.93$
Experiment 2	0.2000 M CO 0.3000 M H_2	0.1522 M CO 0.1566 M H_2 0.0478 M CH_4 0.0478 M H_2O	$K_c = 3.91$
Experiment 3	0.1000 M CH_4 0.1000 M H_2O	0.0613 M CO 0.1839 M H_2 0.0387 M CH_4 0.0387 M H_2O	$K_c = 3.93$

The reaction described here is used industrially to adjust the ratio of H_2 to CO in synthesis gas (mixture of CO and H_2). In this process, CO reacts with H_2O to give H_2, so the ratio of H_2 to CO is increased.

The following exercise provides additional practice in evaluating an equilibrium constant from equilibrium compositions that are experimentally determined. Example 14.3 gives the detailed solution of a more complicated problem of this type.

Exercise 14.3 When 1.00 mol each of carbon monoxide and water reach equilibrium at 1000°C in a 10.0-L vessel, the equilibrium mixture contains 0.57 mol CO, 0.57 mol H_2O, 0.43 mol CO_2, and 0.43 mol H_2. Write the chemical equation for the equilibrium. What is the value of K_c? ◄

■ See Problems 14.43 and 14.44.

◑WL Interactive Example 14.3 **Obtaining an Equilibrium Constant from Reaction Composition**

Gaining Mastery Toolbox

Critical Concept 14.3
From the equilibrium amounts of substances in a reaction, calculate the equilibrium constant.

Solution Essentials:
• Equilibrium constant, K_c
• Table for change amount in a reaction
• Stoichiometry

Hydrogen iodide, HI, decomposes at moderate temperatures according to the equation

$$2HI(g) \rightleftharpoons H_2(g) + I_2(g)$$

The amount of I_2 in the reaction mixture can be determined from the intensity of the violet color of I_2; the more intense the color, the more I_2 in the reaction vessel. When 4.00 mol HI was placed in a 5.00-L vessel at 458°C, the equilibrium mixture was found to contain 0.442 mol I_2. What is the value of K_c for the decomposition of HI at this temperature?

Problem Strategy You will first need to calculate the molar concentrations of substances, because the equilibrium constant is stated in terms of these quantities. The problem statement gives the starting amount of HI and the equilibrium amount of I_2. As in Example 14.1, it is convenient to set up a table of starting, change, and equilibrium concentrations. You use the equilibrium values to calculate the equilibrium constant K_c.

Solution To obtain the concentrations of HI and I_2, you divide the molar amounts by the volume of the reaction vessel (5.00 L).

$$\text{Starting concentration of HI} = \frac{4.00 \text{ mol}}{5.00 \text{ L}} = 0.800 \, M$$

$$\text{Equilibrium concentration of } I_2 = \frac{0.442 \text{ mol}}{5.00 \text{ L}} = 0.0884 \, M$$

(continued)

(continued)

From these values, you set up the following table:

Concentration (M)	2HI(g)	⇌	H₂(g)	+	I₂(g)
Starting	0.800		0		0
Change	−2x		+x		+x
Equilibrium	0.800 − 2x		x		x = 0.0884

The equilibrium concentrations of substances can be evaluated from the expressions given in the last line of this table. You know that the equilibrium concentration of I_2 is 0.0884 *M* and that this equals *x*. Therefore,

$$[HI] = (0.800 - 2x)\ M = (0.800 - 2 \times 0.0884)\ M = 0.623\ M$$
$$[H_2] = x = 0.0884\ M$$

Now you substitute into the equilibrium-constant expression for the reaction. From the chemical equation, you write

$$K_c = \frac{[H_2][I_2]}{[HI]^2}$$

Substituting yields

$$K_c = \frac{(0.0884)(0.0884)}{(0.623)^2} = \mathbf{0.0201}$$

Answer Check Arithmetic mistakes are frequent sources of error. If you can obtain even a rough estimate of the answer, you can often spot such errors. For example, suppose the decomposition of HI could go completely to products; then, 4 mol HI would give 2 mol I_2. The reaction actually gives only a little over 0.4 mol I_2, which suggests that K_c is small, probably much less than 1. (Remember, K_c has products over reactants.) If your calculation gives you something greater than 1, you probably made an arithmetic error. Recheck your calculations.

Exercise 14.4 Hydrogen sulfide, a colorless gas with a foul odor, dissociates on heating:

$$2H_2S(g) \rightleftharpoons 2H_2(g) + S_2(g)$$

When 0.100 mol H_2S was put into a 10.0-L vessel and heated to 1132°C, it gave an equilibrium mixture containing 0.0285 mol H_2. What is the value of K_c at this temperature?

■ See Problems 14.45, 14.46, 14.47, and 14.48.

The Equilibrium Constant K_p

In discussing gas-phase equilibria, it is often convenient to write the equilibrium constant in terms of partial pressures of gases rather than concentrations. Note that the concentration of a gas is proportional to its partial pressure at a fixed temperature (Figure 14.6). You can see this by looking at the ideal gas law, $PV = nRT$, and

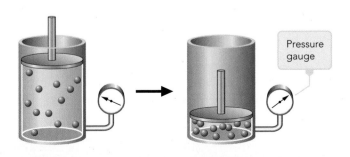

Figure 14.6 ◀

The concentration of a gas at a given temperature is proportional to the pressure When the gas is compressed, the pressure increases (as noted from the pressure gauge) and the concentration of the gas (molecules per unit volume) increases.

solving for n/V, which is the molar concentration of the gas. You get $n/V = P/RT$. In other words, the molar concentration of a gas equals its partial pressure divided by RT, which is constant at a given temperature.

When you express *an equilibrium constant for a gaseous reaction in terms of partial pressures,* you call it the **equilibrium constant K_p.** For catalytic methanation,

$$CO(g) + 3H_2(g) \rightleftharpoons CH_4(g) + H_2O(g)$$

the equilibrium expression in terms of partial pressures becomes

$$K_p = \frac{P_{CH_4}P_{H_2O}}{P_{CO}P_{H_2}^{3}}$$

In general, the value of K_p is different from that of K_c. From the relationship for molar concentration $n/V = P/RT$, one can show that

$$K_p = K_c(RT)^{\Delta n}$$

where Δn is the sum of the coefficients of *gaseous* products in the chemical equation minus the sum of the coefficients of *gaseous* reactants. For the methanation reaction ($K_c = 3.92$), in which 2 mol of gaseous products ($CH_4 + H_2O$) are obtained from 4 mol of gaseous reactants ($CO + 3H_2$), Δn equals $2 - 4 = -2$. The usual unit of partial pressures in K_p is atmospheres; therefore, the value of R is 0.0821 L·atm/(K·mol). Hence,

$$K_p = 3.92 \times (0.0821 \times 1200)^{-2} = 4.04 \times 10^{-4}$$

Exercise 14.5 Phosphorus pentachloride dissociates on heating:

$$PCl_5(g) \rightleftharpoons PCl_3(g) + Cl_2(g)$$

If K_c equals 3.26×10^{-2} at 191°C, what is K_p at this temperature?

■ See Problems 14.51, 14.52, 14.53, and 14.54.

Equilibrium Constant for the Sum of Reactions

It is possible to determine the equilibrium constants for various chemical reactions and then use them to obtain the equilibrium constants of other reactions. A useful rule is as follows: *If a given chemical equation can be obtained by taking the sum of other equations, the equilibrium constant for the given equation equals the product of the equilibrium constants of the added equations.*

As an application of this rule, consider the following reactions at 1200 K:

$$CO(g) + 3H_2(g) \rightleftharpoons CH_4(g) + H_2O(g); K_1 = 3.92 \qquad \text{(Reaction 1)}$$
$$CH_4(g) + 2H_2S(g) \rightleftharpoons CS_2(g) + 4H_2(g); K_2 = 3.3 \times 10^4 \qquad \text{(Reaction 2)}$$

When you take the sum of these two equations, you get

$$CO(g) + 2H_2S(g) \rightleftharpoons CS_2(g) + H_2O(g) + H_2(g|) \qquad \text{(Reaction 3)}$$

According to the rule, the equilibrium constant for this reaction, K_3, is $K_3 = K_1K_2$. This result is easy to verify. Substituting the expressions for K_1 and K_2 in the product K_1K_2 gives

You may also use equilibrium expressions in terms of pressures in this calculation.

$$K_1K_2 = \frac{[\cancel{CH_4}][H_2O]}{[CO][\cancel{H_2}]^3} \times \frac{[CS_2][H_2]^4}{[\cancel{CH_4}][H_2S]^2} = \frac{[CS_2][H_2O][H_2]}{[CO][H_2S]^2}$$

which is the equilibrium-constant expression for Reaction 3. The value of the equilibrium constant is $K_1K_2 = 3.92 \times (3.3 \times 10^4) = 1.3 \times 10^5$. ◄

Consider the following hypothetical reactions. The equilibrium constants K given for each reaction are defined in terms of a concentration unit of molecules per liter.

$$A(g) \rightleftharpoons B(g) \quad K = 2$$
$$X(g) \rightleftharpoons 2Y(g) \quad K = 6$$
$$2C(g) \rightleftharpoons D(g) \quad K = 1$$

Assume that the reactions have reached equilibrium. Match each of these reactions with one of the containers **I** to **IV** (each of which has a volume of 1 L). Identify the "color" of each molecule (that is, is A red or blue?).

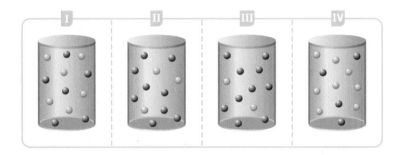

14.3 Heterogeneous Equilibria; Solvents in Homogeneous Equilibria

A **homogeneous equilibrium** is *an equilibrium that involves reactants and products in a single phase.* Catalytic methanation is an example of a homogeneous equilibrium; it involves only gaseous reactants and products. On the other hand, a **heterogeneous equilibrium** is *an equilibrium involving reactants and products in more than one phase.* For example, the reaction of iron metal filings with steam to produce iron oxide, Fe_3O_4, and hydrogen involves solid phases, Fe and Fe_3O_4, in addition to a gaseous phase.

$$3Fe(s) + 4H_2O(g) \rightleftharpoons Fe_3O_4(s) + 4H_2(g)$$

In writing the equilibrium-constant expression for a heterogeneous equilibrium, you omit concentration terms for pure solids and liquids. For the reaction of iron with steam, you would write

$$K_c = \frac{[H_2]^4}{[H_2O]^4}$$

Concentrations of Fe and Fe_3O_4 are omitted, because whereas the concentration of a gas can have various values, the concentration of a pure solid or a pure liquid is a constant at a given temperature.

To understand why such constant concentrations are omitted in writing K_c, write the equilibrium-constant expression for the reaction of iron with steam, but include [Fe] and $[Fe_3O_4]$. Call this K_c'.

$$K_c' = \frac{[Fe_3O_4][H_2]^4}{[Fe]^3[H_2O]^4}$$

You can rearrange the above equation, putting all of the constant factors on the left side.

$$\underbrace{\frac{[Fe]^3}{[Fe_3O_4]}}_{\substack{\text{constant} \\ \text{factors}}} \times K_c' = \underbrace{\frac{[H_2]^4}{[H_2O]^4}}_{\substack{\text{variable} \\ \text{factors}}}$$

Slime Molds and Leopards' Spots

A bright yellow splotch highlights a patch of wood chips in a garden. It looks as if a prankster has poured yellow paint over the chips. Next day the yellow splotch has moved a few inches, and now appears to be climbing up the base of a wall by the garden! A biologist would tell you the yellow splotch is a slime mold (Figure 14.7).

Less noticeable are the cellular slime molds. Much of the time they consist of a group of single-celled amoebas searching for bacteria to eat. But when their food supply is gone, the amoebas release a chemical that signals them to clump together, forming a multicelled slug that moves to a dry, sunlit patch where it forms spores that can be blown to better feeding grounds. During the initial stages of aggregation, the amoebas form a distinctive pattern reminiscent of that seen by chemists in an *oscillating reaction,* which occurs as substances diffuse through a gel (Figure 14.8).

The Russian chemist Boris P. Belousov reported one of the first oscillating reactions in 1958, in which the reaction cycled back and forth over time from one color to another. At first, his work was ignored. You might expect a reaction to form products of a given color, with the reaction gradually slowing as it comes to equilibrium, but how could a chemical reaction oscillate, say, from red to blue and back, over a period of hours? Another Russian chemist, Anatol Zhabotinsky, improved the experiment, leaving

no doubt that some chemical reactions do oscillate. And when the chemicals are placed in a gel in a flat dish, the colors appear as a pattern of waves in space rather than as oscillations in time. Chemists have since shown that the Belousov–Zhabotinsky reaction occurs by two different mechanisms, first by one, then by the other. These mechanisms are repeated in space or time, depending on the concentrations of intermediate substances. During the reaction, an indicator changes color depending on which mechanism is active. Although the complete set of elementary steps of the Belousov–Zhabotinsky reaction is complicated, the overall reaction occurs just as you would expect. The initial reactants continue to decrease over time and the final products increase as the substances come to equilibrium.

An oscillating mechanism also appears to be at work in the aggregation phase of cellular slime molds. And these are not the only biological examples of chemical oscillations. One theory of the formation of fur patterns in animals, such as a leopard's spots, assumes that an oscillating reaction occurs in the skin of the embryo, which eventually generates the fur pattern you see in the adult. A similar theory has been used to explain striped patterns on seashells. The rhythmic beat of your heart is an example where the concentrations of substances oscillate in time. Who would have thought that slime molds, leopards' spots, and cardiac rhythm had anything in common?

Figure 14.7 ▲

A slime mold The fruiting body (yellow) of the slime mold *Fuligo septica* appears on a patch of leaves.

© Jeffrey Lepore/Photo Researchers, Inc.

© Cengage Learning

Figure 14.8 ▲

An oscillating reaction This is a plate showing the wave patterns of the Belousov–Zhabotinsky reaction (in a gel).

■ See Problems 14.113 and 14.114.

Note that all terms on the left are constant but that those on the right (the concentrations of H_2 and H_2O) are variable. The left side equals K_c. Thus, in effect, concentrations of pure solids or pure liquids are incorporated in the value of K_c. Following similar reasoning, you also omit the concentration of solvent from K_c for a homogeneous reaction, as long as the concentration of solvent remains essentially constant during reaction. ▶

The fact that the concentrations of both Fe and Fe_3O_4 do not appear in the equilibrium-constant expression means that the equilibrium is not affected by the amounts of these substances, *as long as some of each is present.*

> *If you use the thermodynamic equilibrium constant, which is defined in terms of activities, you find that the activity of a pure solid or liquid is 1. Therefore, when you write the equilibrium constant, the activities of pure solids and liquids need not be given explicitly.*

✪WL Interactive Example 14.4   Writing K_c for a Reaction with Pure Solids or Liquids

Gaining Mastery Toolbox

Critical Concept 14.4
Pure solids and pure liquids do not appear in the equilibrium-constant expression.

Solution Essentials:
- Equilibrium-constant expression, K_c or K_p
- Balanced equation

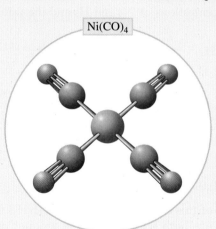

Ni(CO)₄

a Quicklime (calcium oxide, CaO) is prepared by heating a source of calcium carbonate, $CaCO_3$, such as limestone or seashells.

$$CaCO_3(s) \rightleftharpoons CaO(s) + CO_2(g)$$

Write the expression for K_c.

b You can write the equilibrium-constant expression for a physical equilibrium, such as vaporization, as well as for a chemical equilibrium. Write the expression for K_c for the vaporization of water.

$$H_2O(l) \rightleftharpoons H_2O(g)$$

Problem Strategy Write the equilibrium-constant expression excluding terms for pure solids and liquids.

Solution In writing the equilibrium expressions, you ignore pure liquid and solid phases. **a** $K_c = [CO_2]$; **b** $K_c = [H_2O(g)]$.

Answer Check The equilibrium-constant expressions should include all reactants and products, except those of pure solids and liquids. Note that the equilibrium between liquid water and water vapor depends only on the vapor (the vapor pressure).

Exercise 14.6 The Mond process for purifying nickel involves the formation of nickel tetracarbonyl, $Ni(CO)_4$, a volatile liquid, from nickel metal and carbon monoxide. Carbon monoxide is passed over impure nickel to form nickel carbonyl vapor, which, when heated, decomposes and deposits pure nickel.

$$Ni(s) + 4CO(g) \rightleftharpoons Ni(CO)_4(g)$$

Write the expression for K_c for this reaction.

■ See Problems 14.55 and 14.56.

Using the Equilibrium Constant

In the preceding sections, we described how a chemical reaction reaches equilibrium and how you can characterize this equilibrium by the equilibrium constant. Now we want to see the ways in which an equilibrium constant can be used to answer important questions. We will look at the following uses:

1. *Qualitatively interpreting the equilibrium constant.* By merely looking at the magnitude of K_c, you can tell whether a particular equilibrium favors products or reactants.

2. *Predicting the direction of reaction.* Consider a reaction mixture that is not at equilibrium. By substituting the concentrations of substances that exist in a reaction mixture into an expression similar to the equilibrium constant and comparing with K_c, you can predict whether the reaction will proceed toward products or toward reactants (as defined by the way you write the chemical equation).

3. *Calculating equilibrium concentrations.* Once you know the value of K_c for a reaction, you can determine the composition at equilibrium for any set of starting concentrations.

Let us see what meaning we can attach to the value of K_c.

14.4 Qualitatively Interpreting the Equilibrium Constant

If the value of the equilibrium constant is large, you know immediately that the products are favored at equilibrium. Consider the synthesis of ammonia from its elements.

$$N_2(g) + 3H_2(g) \rightleftharpoons 2NH_3(g)$$

At 25°C the equilibrium constant K_c equals 4.1×10^8. This means that the numerator (related to product concentrations) is 4.1×10^8 times larger than the denominator (related to reactant concentrations). In other words, at this temperature the reaction favors the formation of ammonia at equilibrium.

You can verify this by calculating one possible equilibrium composition for this reaction. Suppose the equilibrium mixture is 0.010 M in N_2 and 0.010 M in H_2. From these concentrations you can calculate the concentration of ammonia necessary to give equilibrium. Substitute the concentrations of N_2 and H_2 and the value of K_c into the equilibrium expression

$$K_c = \frac{[NH_3]^2}{[N_2][H_2]^3}$$

You get

$$\frac{[NH_3]^2}{(0.010)(0.010)^3} = 4.1 \times 10^8$$

Now you can solve for the concentration of ammonia.

$$[NH_3]^2 = 4.1 \times 10^8 \times (0.010)(0.010)^3 = 4.1$$

After you take the square root of both sides of this equation, you find that $[NH_3] = 2.0$ M. The concentrations of N_2 and H_2 are both 0.010 M, so the amount of ammonia formed at equilibrium is 200 times that of any one reactant. Figure 14.9 shows the burning of methane, another reaction with an enormously large equilibrium constant.

If the value of the equilibrium constant is small, the reactants are favored at equilibrium. As an example, consider the reaction of nitrogen and oxygen to give nitrogen monoxide, NO:

$$N_2(g) + O_2(g) \rightleftharpoons 2NO(g)$$

The equilibrium constant K_c equals 4.6×10^{-31} at 25°C. If you assume that the concentrations of N_2 and O_2 are 1.0 M, you find that the concentration of NO is 6.8×10^{-16} M. In this case the equilibrium constant is very small, and the concentration of product is not detectable. Reaction occurs to only a very limited extent. ◀

When the equilibrium constant is neither large nor small (around 1), neither reactants nor products are strongly favored. The equilibrium mixture contains significant amounts of all substances in the reaction. For example, in the case of the methanation reaction, the equilibrium constant K_c equals 3.92 at 1200 K.

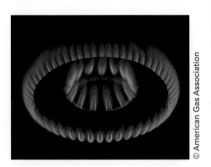

© American Gas Association

Figure 14.9 ▲

Methane (natural gas) reacts with oxygen The equilibrium mixture is almost entirely carbon dioxide and water. The equilibrium constant K_c for the reaction $CH_4 + 2O_2 \rightleftharpoons CO_2 + 2H_2O$ is 10^{140} at 25°C.

The equilibrium constant does become large enough at higher temperatures (about 2000°C) to give appreciable amounts of nitrogen monoxide (nitric oxide). This is why air ($N_2 + O_2$) in flames and in auto engines becomes a source of the pollutant NO.

You found that if you start with 1.000 mol CO and 3.000 mol H_2 in a 10.00-L vessel, the equilibrium composition is 0.613 mol CO, 1.839 mol H_2, 0.387 mol CH_4, and 0.387 mol H_2O. Neither the reactants nor the products are predominant.

If K_c for a reaction,

$$\underbrace{a\text{A} + b\text{B}}_{\text{reactants}} \rightleftharpoons \underbrace{c\text{C} + d\text{D}}_{\text{products}}$$

is large, the equilibrium mixture is mostly products. If K_c is small, the equilibrium mixture is mostly reactants. When K_c is around 1, the equilibrium mixture contains appreciable amounts of both reactants and products.

Exercise 14.7   The equilibrium constant K_c for the reaction

$$2NO(g) + O_2(g) \rightleftharpoons 2NO_2(g)$$

equals 4.0×10^{13} at 25°C. Does the equilibrium mixture contain predominantly reactants or products? If [NO] = [O_2] = 2.0×10^{-6} M at equilibrium, what is the equilibrium concentration of NO_2?

■ See Problems 14.57, 14.58, 14.59, and 14.60.

14.5 Predicting the Direction of Reaction

Suppose a gaseous mixture from an industrial plant has the following composition at 1200 K: 0.0200 M CO, 0.0200 M H_2, 0.00100 M CH_4, and 0.00100 M H_2O. If the mixture is passed over a catalyst at 1200 K, would the reaction

$$CO(g) + 3H_2(g) \rightleftharpoons CH_4(g) + H_2O(g)$$

go toward the right or the left? That is, would the mixture form more CH_4 and H_2O in going toward equilibrium, or more CO and H_2?

To answer this question, you substitute the concentrations of substances into the *reaction quotient* and compare its value to K_c. The **reaction quotient, Q_c,** is *an expression that has the same form as the equilibrium-constant expression but whose concentration values are not necessarily those at equilibrium.* For catalytic methanation, the reaction quotient is

$$Q_c = \frac{[CH_4]_i[H_2O]_i}{[CO]_i[H_2]_i^3}$$

where the subscript i indicates concentrations at a particular instant i. When you substitute the concentrations of the gaseous mixture given above, you get

$$Q_c = \frac{(0.00100)(0.00100)}{(0.0200)(0.0200)^3} = 6.25$$

Recall that the equilibrium constant K_c for catalytic methanation is 3.92 at 1200 K. For the reaction mixture to go to equilibrium, the value of Q_c must decrease from 6.25 to 3.92. This will happen if the reaction goes to the left. In that case, the numerator of Q_c ([CH_4]$_i$[H_2O]$_i$) will decrease, and the denominator ([CO]$_i$[H_2]$_i^3$) will increase. Thus, the gaseous mixture will give more CO and H_2.

Consider the problem more generally. You are given a reaction mixture that is not yet at equilibrium. You would like to know in what direction the reaction will go as it approaches equilibrium. To answer this, you substitute the concentrations of substances from the mixture into the reaction quotient Q_c. Then, you compare Q_c to the equilibrium constant K_c.

Figure 14.10 ▶

Direction of reaction The figure shows the relative sizes of Q_c and K_c. Reaction proceeds in the direction of the arrows. Thus, on the left, $Q_c < K_c$, so the reaction goes to the right, from reactants to products.

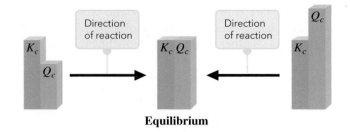

Equilibrium

For the general reaction $aA + bB \rightleftharpoons cC + dD$,

$$Q_c = \frac{[C]_i^c [D]_i^d}{[A]_i^a [B]_i^b}$$

Then

If $Q_c > K_c$, the reaction will go to the left.
If $Q_c < K_c$, the reaction will go to the right.
If $Q_c = K_c$, the reaction mixture is at equilibrium.

(See Figure 14.10.)

⬛WL Interactive Example 14.5 **Using the Reaction Quotient**

Gaining Mastery Toolbox

Critical Concept 14.5
The reaction quotient is an expression with the same form as the equilibrium constant, but whose concentrations are those of the reaction, which may not be at equilibrium.

Solution Essentials:
• Reaction quotient
• Equilibrium-constant expression, K_c or K_p
• Balanced equations

A 50.0-L reaction vessel contains 1.00 mol N_2, 3.00 mol H_2, and 0.500 mol NH_3. Will more ammonia, NH_3, be formed or will it decompose when the mixture goes to equilibrium at 400°C? The equation is

$$N_2(g) + 3H_2(g) \rightleftharpoons 2NH_3(g)$$

K_c is 0.500 at 400°C.

Problem Strategy First, write the reaction quotient, Q_c, which has the same form as the equilibrium-constant expression, but whose concentrations are those for a reaction mixture. Then, compare the value of Q_c with that of K_c. If Q_c is less than K_c, the reaction goes in the direction written; otherwise it goes in the opposite direction (unless $Q_c = K_c$, in which case the reaction is at equilibrium).

Solution The composition of the gas has been given in terms of moles. You convert these to molar concentrations by dividing by the volume (50.0 L). This gives 0.0200 M N_2, 0.0600 M H_2, and 0.0100 M NH_3. Substituting these concentrations into the reaction quotient gives

$$Q_c = \frac{[NH_3]_i^2}{[N_2]_i [H_2]_i^3} = \frac{(0.0100)^2}{(0.0200)(0.0600)^3} = 23.1$$

Because $Q_c = 23.1$ is greater than $K_c = 0.500$, the reaction will go to the left as it approaches equilibrium. Therefore, **ammonia will decompose**.

Answer Check Here is a check on your result. A reaction tends to go toward equilibrium. This means that Q goes toward K_c. If Q is larger than K_c, it must become smaller; the top values in Q (products) need to get smaller, and the bottom values (reactants) need to get larger. Thus, the reaction goes from right to left.

Exercise 14.8 A 10.0-L vessel contains 0.0015 mol CO_2 and 0.10 mol CO. If a small amount of carbon is added to this vessel and the temperature raised to 1000°C, will more CO form? The reaction is

$$CO_2(g) + C(s) \rightleftharpoons 2CO(g)$$

The value of K_c is 1.17 at 1000°C. Assume that the volume of gas in the vessel is 10.0 L.

■ See Problems 14.61 and 14.62.

CONCEPT CHECK 14.3

Carbon monoxide and hydrogen react in the presence of a catalyst to form methanol, CH_3OH:

$$CO(g) + 2H_2(g) \rightleftharpoons CH_3OH(g)$$

An equilibrium mixture of these three substances is suddenly compressed so that the concentrations of all substances initially double. In what direction does the reaction go as a new equilibrium is attained?

14.6 Calculating Equilibrium Concentrations

Once you have determined the equilibrium constant for a reaction, you can use it to calculate the concentrations of substances in an equilibrium mixture. The next example illustrates a simple type of equilibrium problem.

⬤WL Interactive Example | **14.6** | **Obtaining One Equilibrium Concentration Given the Others**

Gaining Mastery Toolbox

Critical Concept 14.6
If you have all but one of the concentrations of substances in a chemical equilibrium, you can use the equilibrium constant to calculate this one concentration.

Solution Essentials:
• Equilibrium-constant expression, K_c or K_p
• Balanced equation

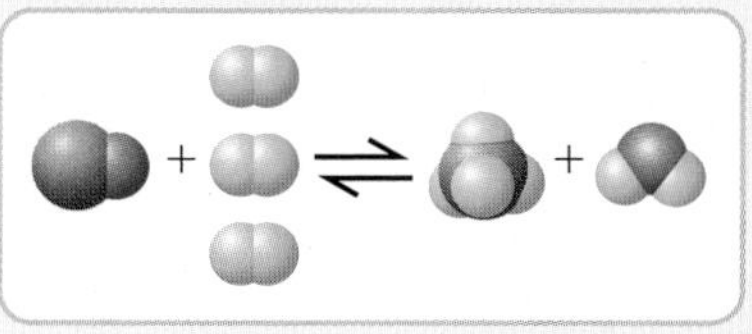

A gaseous mixture contains 0.30 mol CO, 0.10 mol H_2, and 0.020 mol H_2O, plus an unknown amount of CH_4, in each liter. This mixture is at equilibrium at 1200 K.

$$CO(g) + 3H_2(g) \rightleftharpoons CH_4(g) + H_2O(g)$$

What is the concentration of CH_4 in this mixture? The equilibrium constant K_c equals 3.92.

Problem Strategy Calculate concentrations from moles of substances. Substitute these concentrations into the equilibrium-constant equation and solve for the unknown concentration.

Solution The equilibrium-constant equation is

$$K_c = \frac{[CH_4][H_2O]}{[CO][H_2]^3}$$

Substituting the known concentrations and the value of K_c gives

$$3.92 = \frac{[CH_4](0.020)}{(0.30)(0.10)^3}$$

You can now solve for $[CH_4]$.

$$[CH_4] = \frac{(3.92)(0.30)(0.10)^3}{(0.020)} = 0.059$$

The concentration of CH_4 in the mixture is **0.059 mol/L.**

Answer Check A simple check is to put the equilibrium concentrations you have calculated into the equilibrium-constant expression. You should obtain the correct value of K_c (within rounding error).

Exercise 14.9 Phosphorus pentachloride gives an equilibrium mixture of PCl_5, PCl_3, and Cl_2 when heated.

$$PCl_5(g) \rightleftharpoons PCl_3(g) + Cl_2(g)$$

A 1.00-L vessel contains an unknown amount of PCl_5 and 0.020 mol each of PCl_3 and Cl_2 at equilibrium at 250°C. How many moles of PCl_5 are in the vessel if K_c for this reaction is 0.0415 at 250°C?

■ See Problems 14.65 and 14.66.

Usually you begin a reaction with known starting quantities of substances and want to calculate the quantities at equilibrium. The next example illustrates the steps used to solve this type of problem.

Example 14.7

Gaining Mastery Toolbox

Critical Concept 14.7
You use the table of change amounts (involving x as the change) for an equilibrium mixture to set up the equilibrium constant as an equation in x.

Solution Essentials:
• Solving a linear equation in x
• Equilibrium-constant expression, K_c or K_p
• Table for change amounts in a reaction (expressing the changes in terms of x)

See the "Equilibrium Calculator (Equilib)," by Robert Allendorfer, JCE: Software, Madison, WI (requires MS Windows).

Solving an Equilibrium Problem (Involving a Linear Equation in x)

The reaction

$$CO(g) + H_2O(g) \rightleftharpoons CO_2(g) + H_2(g)$$

is used to increase the ratio of hydrogen in synthesis gas (mixtures of CO and H_2). Suppose you start with 1.00 mol each of carbon monoxide and water in a 50.0-L vessel. How many moles of each substance are in the equilibrium mixture at 1000°C? The equilibrium constant K_c at this temperature is 0.58.

Problem Strategy The solution of an equilibrium problem involves three steps. In Step 1, you use the given information to set up a table of starting, change, and equilibrium concentrations. In this problem, you are given starting amounts of reactants. You use these to obtain the starting concentrations. Express the change concentrations in terms of x, as we did in Example 14.1. In Step 2, you substitute the equilibrium concentrations in the table into the equilibrium-constant expression and equate this to the value of the equilibrium constant. In Step 3, you solve this equilibrium-constant equation. This example gives an equation that is a perfect square; taking the square root of it gives a linear equation, which is easy to solve. (More generally, an equilibrium equation involves solving an nth-order equation for its roots. For $n = 2$, you can use the quadratic formula, as illustrated in Example 14.8. For $n > 2$, it is time to think about using a computer.) ◄

Solution

Step 1: The *starting concentrations* of CO and H_2O are

$$[CO] = [H_2O] = \frac{1.00 \text{ mol}}{50.0 \text{ L}} = 0.0200 \text{ mol/L}$$

The starting concentrations of the products, CO_2 and H_2, are 0. The *changes in concentrations* when the mixture goes to equilibrium are not given. However, you can write them all in terms of a single unknown. If you let x be the moles of CO_2 formed per liter, then the moles of H_2 formed per liter is also x. Similarly, x moles each of CO and H_2O are consumed. You write the changes for CO and H_2O as $-x$. You obtain the *equilibrium concentrations* by adding the change in concentrations to the starting concentrations, as shown in the following table:

Concentration (M)	CO(g) +	$H_2O(g)$ $\rightleftharpoons$	$CO_2(g)$ +	$H_2(g)$
Starting	0.0200	0.0200	0	0
Change	$-x$	$-x$	$+x$	$+x$
Equilibrium	$0.0200 - x$	$0.0200 - x$	x	x

Step 2: You then substitute the values for the equilibrium concentrations into the equilibrium-constant equation,

$$K_c = \frac{[CO_2][H_2]}{[CO][H_2O]}$$

and you get

$$0.58 = \frac{(x)(x)}{(0.0200 - x)(0.0200 - x)}$$

or

$$0.58 = \frac{x^2}{(0.0200 - x)^2}$$

(continued)

(*continued*)

Step 3: You now solve this equilibrium equation for the value of x. Note that the right-hand side is a perfect square. If you take the square root of both sides, you get

$$\pm\, 0.76 = \frac{x}{0.0200 - x}$$

We have written $\pm$ to indicate that you should consider both positive and negative values, because both are mathematically possible. Rearranging the equation gives

$$x = \frac{0.0200 \times 0.76}{1.76} = 0.0086 \quad \text{and} \quad x = \frac{-0.0200 \times 0.76}{0.24} = -0.063$$

You can dismiss the negative value as physically impossible (x can be only positive; it represents the concentration of CO_2 formed). If you substitute for x in the last line of the table, the equilibrium concentrations are 0.0114 M CO, 0.0114 M H_2O, 0.0086 M CO_2, and 0.0086 M H_2. To find the moles of each substance in the 50.0-L vessel, you multiply the concentrations by the volume of the vessel. For example, the amount of CO is

$$0.0114 \text{ mol/L} \times 50.0 \text{ L} = 0.570 \text{ mol}$$

The equilibrium composition of the reaction mixture is **0.570 mol CO, 0.570 mol H_2O, 0.43 mol CO_2, and 0.43 mol H_2.**

Answer Check A rough estimate of the answer can sometimes save you from giving the wrong answer. If this reaction could go completely to products, 1 mol CO could give only 1 mol CO_2. (Is the molar amount that you obtained for CO_2 less than 1?) The magnitude of K_c suggests that the reaction will give a mixture of approximately similar molar amounts of reactants and products. (Does your answer agree with this?)

Exercise 14.10 What is the equilibrium composition of a reaction mixture if you start with 0.500 mol each of H_2 and I_2 in a 1.0-L vessel? The reaction is

$$H_2(g) + I_2(g) \rightleftharpoons 2HI(g) \qquad K_c = 49.7 \text{ at } 458°C$$

■ See Problems 14.67 and 14.68.

The preceding example illustrates the three steps used to solve for equilibrium concentrations, as summarized in the following list:

1. Set up a table of concentrations (starting, change, and equilibrium expressions in x).
2. Substitute the expressions in x for equilibrium concentrations into the equilibrium-constant equation.
3. Solve the equilibrium-constant equation for the values of the equilibrium concentrations.

In the previous example, if you had not started with the same number of moles of reactants, you would not have gotten an equation with a perfect square. In that case you would have had to solve a quadratic equation. The next example illustrates how to solve such an equation. ▶

A quadratic equation of the form

$$ax^2 + bx + c = 0$$

has the solutions

$$x = \frac{-b \pm \sqrt{b^2 - 4ac}}{2a}$$

This equation for x is called the quadratic formula. (See Appendix A.)

Example 14.8

Solving an Equilibrium Problem (Involving a Quadratic Equation in *x*)

Hydrogen and iodine react according to the equation

$$H_2(g) + I_2(g) \rightleftharpoons 2HI(g)$$

Suppose 1.00 mol H_2 and 2.00 mol I_2 are placed in a 1.00-L vessel. How many moles of each substance are in the gaseous mixture when it comes to equilibrium at 458°C? The equilibrium constant K_c at this temperature is 49.7.

Problem Strategy You follow the three-step approach outlined just before this example. The equilibrium equation will be a quadratic equation, which you can solve by the quadratic formula. (See Appendix A for a discussion of this equation and formula.)

Solution

Step 1: The concentrations of substances are as follows:

Concentration (M)	$H_2(g)$	+	$I_2(g)$	$\rightleftharpoons$	2HI(g)
Starting	1.00		2.00		0
Change	−x		−x		+2x
Equilibrium	1.00 − x		2.00 − x		2x

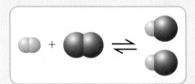

Note that the changes in concentrations equal *x* multiplied by the coefficient of that substance in the balanced chemical equation. The change is negative for a reactant and positive for a product. Equilibrium concentrations equal starting concentrations plus the changes in concentrations.

Step 2: Substituting into the equilibrium-constant equation,

$$K_c = \frac{[HI]^2}{[H_2][I_2]}$$

you get

$$49.7 = \frac{(2x)^2}{(1.00 - x)(2.00 - x)}$$

Step 3: Because the right-hand side is not a perfect square, you must use the quadratic formula to solve for *x*. The equation rearranges to give

$$(1.00 - x)(2.00 - x) = (2x)^2/49.7 = 0.0805x^2$$

or

$$0.920x^2 - 3.00x + 2.00 = 0$$

Hence,

$$x = \frac{3.00 \pm \sqrt{9.00 - 7.36}}{1.84} = 1.63 \pm 0.70$$

A quadratic equation has two mathematical solutions. You obtain one solution by taking the upper (positive) sign in ± and the other by taking the lower (negative) sign. You get

$$x = 2.33 \quad \text{and} \quad x = 0.93$$

However, x = 2.33 gives a negative value to 1.00 − x (the equilibrium concentration of H_2), which is physically impossible. Only x = 0.93 remains. You substitute this value of *x* into the last line of the table in Step 1 to get the equilibrium concentrations and then multiply these by the volume of the vessel (1.00 L) to get the amounts of substances. The last line of the table is rewritten

Concentration (M)	$H_2(g)$	+	$I_2(g)$	$\rightleftharpoons$	2HI(g)
Equilibrium	1.00 − x = 0.07		2.00 − x = 1.07		2x = 1.86

The equilibrium composition is **0.07 mol H_2, 1.07 mol I_2, and 1.86 mol HI.**

(continued)

(continued)

Answer Check If you substitute the equilibrium concentrations of reactants and product into the equilibrium-constant expression, you should get the value of K_c. Check that you do.

Exercise 14.11 Phosphorus pentachloride, PCl_5, decomposes when heated.

$$PCl_5(g) \rightleftharpoons PCl_3(g) + Cl_2(g)$$

If the initial concentration of PCl_5 is 1.00 mol/L, what is the equilibrium composition of the gaseous mixture at 160°C? The equilibrium constant K_c at 160°C is 0.0211.

■ See Problems 14.69 and 14.70.

CONCEPT CHECK 14.4

A and B react to produce C according to the following chemical equation:

$$A + B \longrightarrow C$$

Amounts of A and B are added to an equilibrium reaction mixture of A, B, and C such that when equilibrium is again attained, the amounts of A and B are doubled in the same volume. How is the amount of C changed?

Changing the Reaction Conditions; Le Châtelier's Principle

Obtaining the maximum amount of product from a reaction depends on the proper selection of reaction conditions. By changing these conditions, you can increase or decrease the yield of product. There are three ways to alter the equilibrium composition of a gaseous reaction mixture and possibly increase the yield of product:

1. Changing the concentrations by removing products or adding reactants to the reaction vessel.
2. Changing the partial pressure of gaseous reactants and products by changing the volume.
3. Changing the temperature.

Note that a catalyst cannot alter equilibrium composition, although it can change the rate at which a product is formed and affect the result.

14.7 Removing Products or Adding Reactants

One way to increase the yield of a desired product is to change concentrations in a reaction mixture by removing a product or adding a reactant. Consider the methanation reaction,

$$CO(g) + 3H_2(g) \rightleftharpoons CH_4(g) + H_2O(g)$$

If you place 1.000 mol CO and 3.000 mol H_2 in a 10.00-L reaction vessel, the equilibrium composition at 1200 K is 0.613 mol CO, 1.839 mol H_2, 0.387 mol CH_4, and 0.387 mol H_2O. Can you alter this composition by removing or adding one of the substances to improve the yield of methane?

To answer this question, you can apply **Le Châtelier's principle,** which states that *when a system in chemical equilibrium is disturbed by a change of temperature, pressure, or a concentration, the system shifts in equilibrium composition in a way*

*Le Châtelier's principle was intro-
duced in Section 12.3, where it was
used to determine the effect of pressure on solubility of gases in liquids.*

that tends to counteract this change of variable. ◀ Suppose you remove a substance from or add a substance to the equilibrium mixture in order to alter the concentration of the substance. Chemical reaction then occurs to partially restore the initial concentration of the removed or added substance. (However, if the concentration of substance cannot be changed, as in the case of a pure solid or liquid reactant or product, changes in amount will have no effect on the equilibrium composition.)

For example, suppose that water vapor is removed from the reaction vessel containing the equilibrium mixture for methanation. Le Châtelier's principle predicts that net chemical change will occur to partially reinstate the original concentration of water vapor. This means that the methanation reaction momentarily goes in the forward direction,

$$CO(g) + 3H_2(g) \longrightarrow CH_4(g) + H_2O(g)$$

until equilibrium is re-established. Going in the forward direction, the concentrations of both water vapor and methane increase.

A practical way to remove water vapor in this reaction would be to cool the reaction mixture quickly to condense the water. Liquid water could be removed and the gases reheated until equilibrium is again established. The concentration of water vapor would build up as the concentration of methane increases. Table 14.1 lists the amounts of each substance at each stage of this process (carried out in a 10.00-L reaction vessel). Note how the yield of methane has been improved.

It is often useful to add an excess of a cheap reactant in order to force the reaction toward more products. In this way, the more expensive reactant is made to react to a greater extent than it would otherwise.

Consider the ammonia synthesis

$$N_2(g) + 3H_2(g) \rightleftharpoons 2NH_3(g)$$

If you wished to convert as much hydrogen to ammonia as possible, you might increase the concentration of nitrogen. To understand the effect of this, first suppose that a mixture of nitrogen, hydrogen, and ammonia is at equilibrium. If nitrogen is now added to this mixture, the equilibrium is disturbed. According to Le Châtelier's principle, the reaction will now go in the direction that will use up some of the added nitrogen.

$$N_2(g) + 3H_2(g) \longrightarrow 2NH_3(g)$$

Consequently, adding more nitrogen has the effect of converting a greater quantity of hydrogen to ammonia. Figure 14.11 illustrates the effect of adding or removing a reactant on another chemical equilibrium.

You can look at these situations in terms of the reaction quotient, Q_c. Consider the methanation reaction, in which

$$Q_c = \frac{[CH_4]_i[H_2O]_i}{[CO]_i[H_2]_i^3}$$

If the reaction mixture is at equilibrium, $Q_c = K_c$. Suppose you remove some H_2O from the equilibrium mixture. Now $Q_c < K_c$, and from what we said in Section 14.5, you know the reaction will proceed in the forward direction to restore equilibrium.

Table 14.1 **Effect of Removing Water Vapor from a Methanation Mixture (in a 10.00-L Vessel)**

Stage of Process	Mol CO	Mol H_2	Mol CH_4	Mol H_2O
Original reaction mixure	0.613	1.839	0.387	0.387
After removing water (before equilibrium)	0.613	1.839	0.387	0
When equilibrium is reestablished	0.491	1.473	0.509	0.122

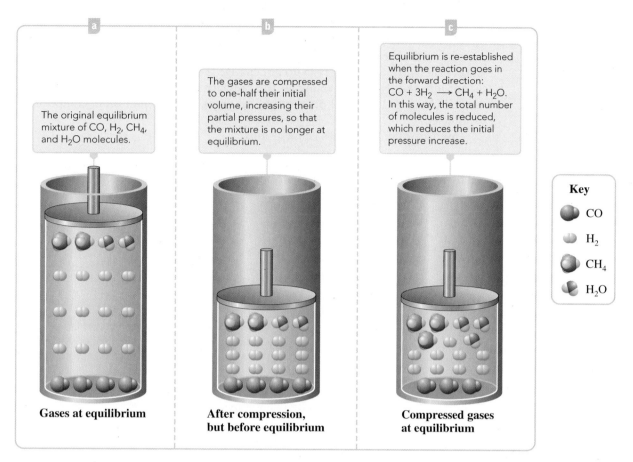

Figure 14.12 ▲

Effect on chemical equilibrium of changing the pressure The effect on the methanation reaction is shown; the approximate composition is represented by the proportion of different "molecules" (see key at right). The actual arrangement of molecules in a gas is random.

You can see the quantitative effect of the pressure change by solving the equilibrium problem. Recall that if you put 1.000 mol CO and 3.000 mol H_2 into a 10.00-L vessel, the equilibrium composition at 1200 K, where K_c equals 3.92, is 0.613 mol CO, 1.839 mol H_2, 0.387 mol CH_4, and 0.387 mol H_2O. Suppose the volume of the reaction gases is halved so that the initial concentrations are doubled. The temperature remains at 1200 K, so K_c is still 3.92. But if you solve for the new equilibrium composition (following the method of Example 14.8), you obtain 0.495 mol CO, 1.485 mol H_2, 0.505 mol CH_4, and 0.505 mol H_2O. Note that the amount of CH_4 has increased from 0.387 mol to 0.505 mol. You conclude that high pressure of reaction gases favors high yields of methane.

To decide the direction of reaction when the pressure of the reaction mixture is increased (say, by decreasing the volume), you ignore liquids and solids. They are not much affected by pressure changes, because they are nearly incompressible. Consider the reaction

$$C(s) + CO_2(g) \rightleftharpoons 2CO(g)$$

The moles *of gas* increase when the reaction goes in the forward direction (1 mol CO_2 goes to 2 mol CO). Therefore, when you increase the pressure of the reaction mixture by decreasing its volume, the reaction goes in the reverse direction. The moles of gas decrease, and the initial increase of pressure is partially reduced, as Le Châtelier's principle leads you to expect.

It is important to note that an increase or a decrease in pressure of a gaseous reaction must result in changes of partial pressures of substances in the chemical equation if it is to have an effect on the equilibrium composition. Only changes in partial pressures, or concentrations, of these substances can affect the reaction quotient. Consider the effect of increasing the pressure in the methanation reaction by adding helium gas. Although the total pressure increases, the partial pressures of CO, H_2, CH_4, and H_2O do not change. Thus, the equilibrium composition is not affected. However, changing the pressure by changing the volume of this system changes the partial pressures of all gases, so the equilibrium composition is affected.

We can summarize these conclusions:

> If the pressure is increased by decreasing the volume of a reaction mixture, the reaction shifts in the direction of fewer moles of gas.

Example 14.10 Applying Le Châtelier's Principle When the Pressure Is Altered

Gaining Mastery Toolbox

Critical Concept 14.10
Increasing the gas pressure by decreasing the volume of a reaction mixture shifts the reaction in the direction of fewer moles of gas.

Solution Essentials:
- Note whether moles of gases increase or decrease during reaction.
- Le Chatelier's principle
- Balanced equation

Look at each of the following equations and decide whether an increase of pressure obtained by decreasing the volume will increase, decrease, or have no effect on the amounts of products.

a $CO(g) + Cl_2(g) \rightleftharpoons COCl_2(g)$

b $2H_2S(g) \rightleftharpoons 2H_2(g) + S_2(g)$

c $C(graphite) + S_2(g) \rightleftharpoons CS_2(g)$

Problem Strategy According to Le Châtelier's principle, if you apply pressure to an equilibrium mixture, the reaction goes in the direction to partially offset this applied pressure. Ask whether the molecules of *gas* increase or decrease during reaction from left to right. If they decrease, this should partially offset any increase in applied pressure; the reaction should go from left to right. Otherwise, the reaction should go from right to left.

Solution

a Reaction decreases the number of molecules of gas (from two to one). According to Le Châtelier's principle, an increase of pressure **increases** the amount of product.

b Reaction increases the number of molecules of gas (from two to three); hence, an increase of pressure **decreases** the amounts of products.

c Reaction does not change the number of molecules of gas. (You ignore the change in volume due to consumption of solid carbon, because the change in volume of a solid is insignificant. Look only at gas volumes when deciding the effect of pressure change on equilibrium composition.) Pressure change has **no effect.**

Answer Check You look only at the gaseous components when deciding the direction of reaction from an applied pressure. Make sure you have done this correctly.

Exercise 14.13 Can you increase the amount of product in each of the following reactions by increasing the pressure? Explain.

a $CO_2(g) + H_2(g) \rightleftharpoons CO(g) + H_2O(g)$

b $4CuO(s) \rightleftharpoons 2Cu_2O(s) + O_2(s)$

c $2SO_2(g) + O_2(g) \rightleftharpoons 2SO_3(g)$

■ See Problems 14.75 and 14.76.

Effect of Temperature Change

Temperature has a profound effect on most reactions. In the first place, reaction rates usually increase with an increase in temperature, meaning that equilibrium is reached sooner. Many gaseous reactions are sluggish or have imperceptible rates at room temperature but speed up enough at higher temperature to become commercially feasible processes.

Second, equilibrium constants vary with temperature. Table 14.2 gives values of K_c for methanation at various temperatures. Note that K_c equals 4.9×10^{27} at 298 K. Therefore, an equilibrium mixture at room temperature is mostly methane and water.

Whether you should raise or lower the temperature of a reaction mixture to increase the equilibrium amount of product can be shown by Le Châtelier's principle. Again, consider the methanation reaction,

$$CO(g) + 3H_2(g) \rightleftharpoons CH_4(g) + H_2O(g); \Delta H° = -206.2 \text{ kJ}$$

The value of $\Delta H°$ shows this reaction to be quite exothermic. As products are formed, considerable heat is released. According to Le Châtelier's principle, as the temperature is raised, the reaction shifts to form more reactants, thereby absorbing heat and "attempting" to counter the increase in temperature.

$$CO(g) + 3H_2(g) \longleftarrow CH_4(g) + H_2O(g) + \text{heat}$$

You predict that the equilibrium constant will be smaller for higher temperatures, in agreement with the values of K_c given in Table 14.2. Figure 14.13 illustrates the effect of temperature on another chemical equilibrium. ▶

The conclusions from Le Châtelier's principle regarding temperature effects on an equilibrium can be summarized this way:

Table 14.2	Equilibrium Constant for Methanation at Different Temperatures
Temperature (K)	K_c
298	4.9×10^{27}
800	1.38×10^5
1000	2.54×10^2
1200	3.92

The quantitative effect of temperature on the equilibrium constant is discussed in Chapter 18.

For an endothermic reaction ($\Delta H°$ positive), the amounts of products are increased at equilibrium by an increase in temperature (K_c is larger at higher T). For an exothermic reaction ($\Delta H°$ negative), the amounts of products are increased at equilibrium by a decrease in temperature (K_c is larger at lower T).

✪WL Interactive Example 14.11   Applying Le Châtelier's Principle When the Temperature Is Altered

Gaining Mastery Toolbox

Critical Concept 14.11
If a reaction evolves heat, that reaction is shifted to the left to absorb heat when the temperature is raised. If the reaction absorbs heat, it shifts to the right when the temperature is raised.

Solution Essentials:
• Raising the temperature shifts reaction away from side where heat appears.
• Le Chatelier's principle
• Enthalpy of reaction
• Balanced equation

Carbon monoxide is formed when carbon dioxide reacts with solid carbon (graphite).

$$CO_2(g) + C(\text{graphite}) \rightleftharpoons 2CO(g); \Delta H° = 172.5 \text{ kJ}$$

Is a high or low temperature more favorable to the formation of carbon monoxide?

Problem Strategy Decide in which direction you want the reaction to go. (If you want more product, then you want to shift the reaction to the right.) Is this endothermic or exothermic? According to Le Châtelier's principle, if you raise the temperature of an equilibrium mixture, reaction goes in the direction that absorbs heat (that is, it goes in the endothermic direction), to partially offset the heat added for the temperature increase. (Similarly, if you lower the temperature, the reaction goes in the exothermic direction.) Now you can determine whether you must increase or decrease temperature.

Solution You want to increase the yield of CO, so you need to shift the equilibrium toward the right, which is endothermic (absorbs heat). You need to raise the temperature (add heat).

Alternatively, you can look at the problem this way. Write the chemical equation with heat on the left or right, depending on whether the heat is absorbed (consumed) or given off. The reaction to produce CO absorbs heat, so you write the equation with heat on the left.

$$\text{Heat} + CO_2(g) + C(\text{graphite}) \longrightarrow 2CO(g)$$

If you add heat (increasing the temperature), you push the reaction to the right, using up some of that heat.

Thus, **high temperature** is more favorable to the formation of carbon monoxide. This explains why combustion of carbon, or of materials containing carbon, can produce significant amounts of carbon monoxide, in addition to carbon dioxide.

Answer Check Check that your answer is consistent with Le Châtelier's principle. If you increase the temperature of the reaction (add heat), the reaction shifts in a direction to use up some of that added heat.

Exercise 14.14 Consider the possibility of converting carbon dioxide to carbon monoxide by the endothermic reaction

$$CO_2(g) + H_2(g) \rightleftharpoons CO(g) + H_2O(g)$$

Is a high or a low temperature more favorable to the production of carbon monoxide? Explain.

■ See Problems 14.77 and 14.78.

Figure 14.13 ▲

Effect on chemical equilibrium of changing the temperature The reaction shown is endothermic: $Co(H_2O)_6^{2+}(aq) + 4Cl^-(aq) \rightleftharpoons$ $CoCl_4^{2-}(aq) + 6H_2O(l)$. At room temperature, the equilibrium mixture is blue, from $CoCl_4^{2-}$. When cooled by the ice bath, the equilibrium mixture turns pink, from $Co(H_2O)_6^{2+}$.

© Cengage Learning

CONCEPT CHECK 14.5

Given the hypothetical exothermic reaction $A_2(g) + 2B(g) \longrightarrow 2AB(g)$ at equilibrium, decide which of the following containers represents the reaction mixture at the higher temperature? (The other container represents the reaction at a lower temperature.)

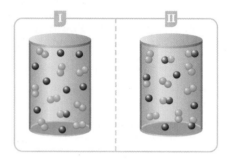

Choosing the Optimum Conditions for Reaction

You are now in a position to understand the optimum conditions for the methanation reaction. Because the reaction is exothermic, low temperatures should favor high yields of methane; that is, the equilibrium constant is large for low temperature. But gaseous reactions are often very slow at room temperature. In practice, the methanation reaction is run at moderately elevated temperatures (230°C–450°C) in the presence of a nickel catalyst. Under these conditions, the rate of reaction is sufficiently fast, but the equilibrium constant is not too small. Because the methanation reaction involves a decrease in moles of gas, the yield of methane should increase as the pressure increases. However, the equilibrium constant is large at the usual operating temperatures, so very high pressures are not needed to obtain economical yields of methane. Pressures of 1 atm to 100 atm are usual for this reaction.

As another example, consider the Haber process for the synthesis of ammonia:

$$N_2(g) + 3H_2(g) \xrightleftharpoons[\text{}]{\text{Fe catalyst}} 2NH_3(g); \Delta H° = -91.8 \text{ kJ}$$

Because the reaction is exothermic, the equilibrium constant is larger for lower temperatures. But the reaction proceeds too slowly at room temperature to be practical, even in the presence of the best available catalysts. ◄ The optimum choice of temperature, found experimentally to be about 450°C, is a compromise between an increased rate of reaction at higher temperature and an increased yield of ammonia at lower temperature. Because the formation of ammonia decreases the moles of gases, the yield of product is improved by high pressures. The equilibrium constant K_c is only 0.159 at 450°C, so higher pressures (about 200 atm) are required for an economical yield of ammonia. Ammonia from the Haber reactor is removed from the reaction mixture by cooling the compressed gases until NH_3 liquefies. Unreacted N_2 and H_2 circulate back to the reactor. See Figure 14.14.

This technological problem has stimulated a great deal of basic research into understanding how certain bacteria "fix" nitrogen at atmospheric pressure to make NH_3. An enzyme called nitrogenase is responsible for N_2 fixation in these bacteria. This enzyme contains Fe and Mo, which may play a catalytic role.

Exercise 14.15 Consider the reaction

$$2CO_2(g) \rightleftharpoons 2CO(g) + O_2(g); \Delta H° = 566 \text{ kJ}$$

Discuss the temperature and pressure conditions that would give the best yield of carbon monoxide.

■ See Problems 14.81 and 14.82.

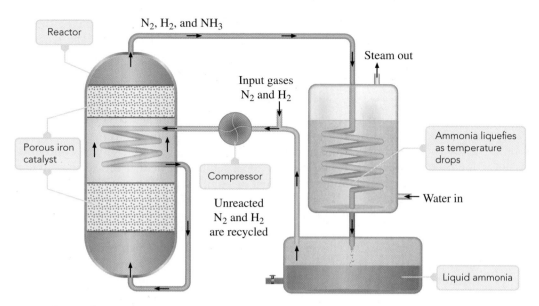

Figure 14.14 ▲

Industrial production of ammonia Shown is a schematic diagram of the production of ammonia from nitrogen (obtained from air) and hydrogen (from natural gas). Input gases (N_2 and H_2) are compressed to about 200 atm and enter the reactor (the vessel on the left) where they are heated by the exothermic reaction mixture to about 450°C. The hot reaction gas then passes through the porous iron catalyst to give an equilibrium mixture of N_2, H_2, and NH_3. This mixture exits the reactor and goes to a vessel where it is cooled by water. In the process, steam is produced, which is used to power the compressor and other parts of the ammonia plant. When the reaction gases are sufficiently cooled (at high pressure), the ammonia liquefies. The unreacted N_2 and H_2 are recycled back to the reactor.

14.9 Effect of a Catalyst

A *catalyst* is a substance that increases the rate of a reaction but is not consumed by it. The significance of a catalyst can be seen in the reaction of sulfur dioxide with oxygen to give sulfur trioxide.

$$2SO_2(g) + O_2(g) \rightleftharpoons 2SO_3(g)$$

The equilibrium constant K_c for this reaction is 1.7×10^{26}, which indicates that for all practical purposes the reaction should go almost completely to products. Yet when sulfur is burned in air or oxygen, it forms predominantly SO_2 and very little SO_3. Oxidation of SO_2 to SO_3 is simply too slow to give a significant amount of product. However, the rate of the reaction is appreciable in the presence of a platinum or vanadium(V) oxide catalyst. The oxidation of SO_2 in the presence of a catalyst is the main step in the *contact process* for the industrial production of sulfuric acid, H_2SO_4. Sulfur trioxide reacts with water to form sulfuric acid. (In the industrial process, SO_3 is dissolved in concentrated H_2SO_4, which is then diluted.) ▶

It is important to understand that *a catalyst has no effect on the equilibrium composition of a reaction mixture. A catalyst merely speeds up the attainment of equilibrium.* For example, suppose you mix 2.00 mol SO_2 and 1.00 mol O_2 in a 100.0-L vessel. In the absence of a catalyst, these substances appear unreactive. Much later, if you analyze the mixture, you find essentially the same amounts of SO_2 and O_2. But when a catalyst is added, the rates of both forward and reverse reactions are very much increased. As a result, the reaction mixture comes to equilibrium in a short time. The amounts of SO_2, O_2, and SO_3 can be calculated from the equilibrium constant. You find that the mixture is mostly SO_3 (2.00 mol), with only 1.7×10^{-8} mol SO_2 and 8.4×10^{-9} mol O_2.

Sulfur dioxide from the combustion of coal and from other sources appears to be a major cause of the marked increase in acidity of rain in the eastern United States in the past few decades. This acid rain has been shown to contain sulfuric and nitric acids. The SO_2 is oxidized in moist, polluted air to H_2SO_4. Acid rain is discussed in the essay on acid rain at the end of Section 16.2.

A catalyst is useful for a reaction, such as $2SO_2 + O_2 \rightleftharpoons 2SO_3$, that is normally slow but has a large equilibrium constant. However, if the reaction has an exceedingly small equilibrium constant, a catalyst is of little help. The reaction

$$N_2(g) + O_2(g) \rightleftharpoons 2NO(g)$$

has been considered for the industrial production of nitric acid. (NO reacts with O_2 and H_2O to give nitric acid.) At 25°C, however, the equilibrium constant K_c equals 4.6×10^{-31}. An equilibrium mixture would contain an extremely small concentration of NO. A catalyst would not provide a significant yield at this temperature; a catalyst merely speeds up the attainment of equilibrium. The equilibrium constant increases as the temperature is raised, so that at 2000°C, air (a mixture of N_2 and O_2) forms about 0.4% NO at equilibrium. An industrial plant was set up in Norway in 1905 to prepare nitrate fertilizers using this reaction. The plant was eventually made obsolete by the *Ostwald process* for making nitric acid, in which NO is prepared by the oxidation of ammonia. This latter reaction is more economical than the direct reaction of N_2 and O_2, in part because the equilibrium constant is larger at moderate temperatures.

Although a catalyst cannot affect the composition at true equilibrium, in some cases it can affect the product in a reaction because it affects the rate of one reaction out of several possible reactions. The importance of rate for determining the product in a reaction is illustrated by a simple example in Figure 14.15. Mercury(II) ion reacts with iodide ion to form a precipitate of mercury(II) iodide.

$$Hg^{2+}(aq) + 2I^-(aq) \longrightarrow HgI_2(s)$$

In concentrated solutions, orange tetragonal crystals of HgI_2 form. In dilute solution, however, yellow rhombic crystals form faster and are the initial product. Later, when true equilibrium is attained, the orange crystals appear.

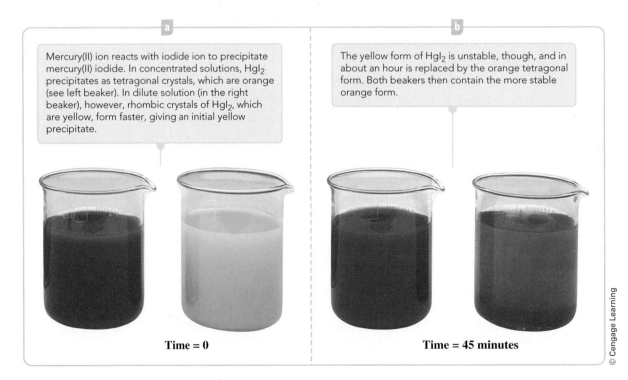

a	b
Mercury(II) ion reacts with iodide ion to precipitate mercury(II) iodide. In concentrated solutions, HgI₂ precipitates as tetragonal crystals, which are orange (see left beaker). In dilute solution (in the right beaker), however, rhombic crystals of HgI₂, which are yellow, form faster, giving an initial yellow precipitate.	The yellow form of HgI₂ is unstable, though, and in about an hour is replaced by the orange tetragonal form. Both beakers then contain the more stable orange form.
Time = 0	Time = 45 minutes

Figure 14.15 ▲

An example of how rates of reaction can affect the kind of product

The Ostwald process presents an interesting example of the effect of a catalyst in determining a product when several possibilities exist. Two reactions of ammonia with oxygen are possible. The reaction used in the Ostwald process is

$$4NH_3(g) + 5O_2(g) \rightleftharpoons 4NO(g) + 6H_2O(g)$$

However, nitrogen monoxide decomposes to its elements,

$$2NO(g) \rightleftharpoons N_2(g) + O_2(g)$$

and the equilibrium constant for this is quite large (2.2×10^{30} at 25°C), so at true equilibrium the products of the reaction of ammonia with oxygen might be expected to be nitrogen and water.

$$4NH_3(g) + 3O_2(g) \rightleftharpoons 2N_2(g) + 6H_2O(g)$$

Ammonia can be made to burn in oxygen, and N_2 and H_2O are the products. (The reaction occurs most readily in the presence of a copper catalyst; see Figure 14.16.) However, what Ostwald discovered was that the first reaction, to form NO from NH_3 and O_2, is catalyzed by platinum. Therefore, by using this catalyst at moderate temperatures, nitric oxide can be selectively formed. The dissociation of nitric oxide to its elements is normally too slow at moderate temperatures to be significant. Many other examples could be cited in which the products of given reactants depend on the catalyst.

© Cengage Learning

Figure 14.16 ▲

Oxidation of ammonia using a copper catalyst The products (formed on the glowing copper coil) are N_2 and H_2O, whereas a platinum catalyst results in NO and H_2O as products.

A Checklist for Review

⏻WL and [Chemistry] Sign in at **www.cengage.com/owl** to:

• View tutorials and simulations, develop problem-solving skills, and complete online homework assigned by your professor.

• For quick review and exam prep, download Go Chemistry mini lecture modules from OWL or purchase them at **www.cengagebrain.com.**

Summary of Facts and Concepts

Chemical equilibrium can be characterized by the *equilibrium constant K_c*. In the expression for K_c, the concentration of products is in the numerator and the concentration of reactants is in the denominator. Pure liquids and solids are ignored in writing the equilibrium-constant expression. When K_c is very large, the equilibrium mixture is mostly products, and when K_c is very small, the equilibrium mixture is mostly reactants. The reaction quotient Q_c takes the form of the equilibrium-constant expression. If you substitute the concentrations of substances in a

reaction mixture into Q_c, you can predict the direction the reaction must go to attain equilibrium. You can use K_c to calculate the composition of the reaction mixture at equilibrium, starting from various initial compositions.

The *choice of conditions*, including *catalysts*, can be very important to the success of a reaction. Removing a product from the reaction mixture, for example, shifts the equilibrium composition to give more product. Changing the pressure and temperature can also affect the product yield. *Le Châtelier's principle* is useful in predicting the effect of such changes.

Learning Objectives

14.1 Chemical Equilibrium—A Dynamic Equilibrium

▪ Define *dynamic equilibrium* and *chemical equilibrium.*
▪ Apply stoichiometry to an equilibrium mixture. Example 14.1

14.2 The Equilibrium Constant

▪ Define *equilibrium-constant expression* and *equilibrium constant.*
▪ State the *law of mass action.*
▪ Write equilibrium-constant expressions. Example 14.2
▪ Describe the kinetics argument for the approach to chemical equilibrium.
▪ Obtain an equilibrium constant from reaction composition. Example 14.3

Important Terms

chemical equilibrium

equilibrium-constant expression
equilibrium constant K_c
equilibrium constant K_p
law of mass action

- Describe the equilibrium constant K_p; indicate how K_p and K_c are related.
- Obtain K_c for a reaction that can be written as a sum of other reactions of known K_c values.

14.3 Heterogeneous Equilibria; Solvents in Homogeneous Equilibria	
- Define *homogeneous equilibrium* and *heterogeneous equilibrium*. - Write K_c for a reaction with pure solids or liquids. Example 14.4	homogeneous equilibrium heterogeneous equilibrium
14.4 Qualitatively Interpreting the Equilibrium Constant	
- Give a qualitative interpretation of the equilibrium constant based on its value.	
14.5 Predicting the Direction of Reaction	
- Define *reaction quotient, Q.* - Describe the direction of reaction after comparing Q with K_c. - Use the reaction quotient. Example 14.5	reaction quotient
14.6 Calculating Equilibrium Concentrations	
- Obtain one equilibrium concentration given the others. Example 14.6 - Solve an equilibrium problem (involving a linear equation in x). Example 14.7 - Solve an equilibrium problem (involving a quadratic equation in x). Example 14.8	
14.7 Removing Products or Adding Reactants	
- State *Le Châtelier's principle.* - State what happens to an equilibrium when a reactant or product is added or removed. - Apply Le Châtelier's principle when a concentration is altered. Example 14.9	Le Châtelier's principle
14.8 Changing the Pressure and Temperature	
- Describe the effect of a pressure change on chemical equilibrium. - Apply Le Châtelier's principle when the pressure is altered. Example 14.10 - Describe the effect of a temperature change on chemical equilibrium. - Apply Le Châtelier's principle when the temperature is altered. Example 14.11 - Describe how the optimum conditions for a reaction are chosen.	
14.9 Effect of a Catalyst	
- Define *catalyst.* - Compare the effect of a catalyst on rate of reaction with its effect on equilibrium. - Describe how a catalyst can affect the product formed.	

Key Equations

$$K_c = \frac{[C]^c[D]^d}{[A]^a[B]^b} \qquad\qquad Q_c = \frac{[C]_i^c[D]_i^d}{[A]_i^a[B]_i^b}$$

Questions and Problems

⊍WL Interactive versions of these problems may be assigned in OWL.

Self-Assessment and Review Questions

Key: These questions test your understanding of the ideas you worked with in the chapter. These problems vary in difficulty and often can be used for the basis of discussion.

14.1 Consider the reaction $N_2O_4(g) \rightleftharpoons 2NO_2(g)$. Draw a graph illustrating the changes of concentrations of N_2O_4 and NO_2 as equilibrium is approached. Describe how the rates of the forward and reverse reactions change as the mixture approaches dynamic equilibrium. Why is this called a *dynamic* equilibrium?

14.2 When 1.0 mol each of $H_2(g)$ and $I_2(g)$ are mixed at a certain high temperature, they react to give a final mixture consisting of 0.5 mol each of $H_2(g)$ and $I_2(g)$ and 1.0 mol $HI(g)$. Why do you obtain the same final mixture when you bring 2.0 mol $HI(g)$ to the same temperature?

14.3 Explain why the equilibrium constant for a gaseous reaction can be written in terms of partial pressures instead of concentrations.

14.4 Obtain the equilibrium constant for the reaction

$$HCN(aq) \rightleftharpoons H^+(aq) + CN^-(aq)$$

from the following

$$HCN(aq) + OH^-(aq) \rightleftharpoons CN^-(aq) + H_2O(l);$$
$$K_1 = 4.9 \times 10^4$$
$$H_2O(l) \rightleftharpoons H^+(aq) + OH^-(aq); K_2 = 1.0 \times 10^{-14}$$

14.5 Which of the following reactions involve homogeneous equilibria and which involve heterogeneous equilibria? Explain the difference.

a $2Cu(NO_3)_2(s) \rightleftharpoons 2CuO(s) + 4NO_2(g) + O_2(g)$
b $2N_2O(g) \rightleftharpoons 2N_2(g) + O_2(g)$
c $2NO(g) + O_2(g) \rightleftharpoons 2NO_2(g)$
d $2NH_3(g) + 3CuO(s) \rightleftharpoons$
$$3H_2O(g) + N_2(g) + 3Cu(s)$$

14.6 Explain why pure liquids and solids can be ignored when writing the equilibrium-constant expression.

14.7 What qualitative information can you get from the magnitude of the equilibrium constant?

14.8 List the possible ways in which you can alter the equilibrium composition of a reaction mixture.

14.9 What is the reaction quotient? How is it useful?

14.10 Two moles of H_2 are mixed with 1 mol O_2 at 25°C. No observable reaction takes place, although K_c for the reaction to form water is very large at this temperature. When a piece of platinum is added, however, the gases react rapidly. Explain the role of platinum in this reaction. How does it affect the equilibrium composition of the reaction mixture?

14.11 List four ways in which the yield of ammonia in the reaction

$$N_2(g) + 3H_2(g) \rightleftharpoons 2NH_3(g); \Delta H° < 0$$

can be improved for a given amount of H_2. Explain the principle behind each way.

14.12 How is it possible for a catalyst to give products from a reaction mixture that are different from those obtained when no catalyst or a different catalyst is used? Give an example.

14.13 A chemist put 1.18 mol of substance A and 2.85 mol of substance B into a 10.0-L flask, which she then closed. A and B react by the following equation:

$$A(g) + 2B(g) \rightleftharpoons 3C(g) + D(g)$$

She found that the equilibrium mixture at 25°C contained 0.376 mol of D. How many moles of B are in the flask at equilibrium at 25°C?

a 2.52 mol
b 2.10 mol
c 2.41 mol
d 3.60 mol
e 2.47 mol

14.14 The reaction $3A(g) + B(s) \rightleftharpoons 2C(aq) + D(aq)$ occurs at 25°C in a flask, which has 1.87 L available for gas. After the reaction attains equilibrium, the amounts (mol) *or* concentrations (*M*) of substances are as follows:

2.48 mol A	1.13 *M* C
2.41 mol B	2.27 *M* D

What is the equilibrium constant K_c for this reaction at 25°C?

a 1.24
b 0.190
c 0.516
d 1.15
e 0.939

14.15 A graduate student places 0.272 mol of $PCl_3(g)$ and 8.56×10^{-4} mol of $PCl_5(g)$ into a 0.718-L flask at a certain temperature. $PCl_5(g)$ is known to decompose as follows:

$$PCl_5(g) \rightleftharpoons PCl_3(g) + Cl_2(g)$$

After the reaction attains equilibrium, the student finds that the flask contains 2.51×10^{-4} mol of Cl_2. Calculate the equilibrium constant K_c for the reaction at this temperature.

a 8.51×10^{-2}
b 0.157
c 0.114
d 2.40×10^4
e 8.88×10^4

14.16 An experimenter places the following concentrations of gases in a closed container: [NOBr] = 7.13×10^{-2} *M*, [NO] = 1.58×10^{-2} *M*, [Br$_2$] = 1.29×10^{-2} *M*. These gases then react:

$$2NOBr(g) \rightleftharpoons 2NO(g) + Br_2(g)$$

At the temperature of the reaction, the equilibrium constant K_c is 3.07×10^{-4}. Calculate the reaction quotient, Q_c, from the initial concentrations and determine whether the concentration of NOBr increases or decreases as the reaction approaches equilibrium.

a $Q_c = 6.33 \times 10^{-4}$; the concentration of NOBr decreases
b $Q_c = 6.33 \times 10^{-4}$; the concentration of NOBr increases
c $Q_c = 1.58 \times 10^4$; the concentration of NOBr increases
d $Q_c = 4.65 \times 10^{-4}$; the concentration of NOBr decreases
e $Q_c = 4.65 \times 10^{-4}$; the concentration of NOBr increases

Concept Explorations

Key: Concept explorations are comprehensive problems that provide a framework that will enable you to explore and learn many of the critical concepts and ideas in each chapter. If you master the concepts associated with these explorations, you will have a better understanding of many important chemistry ideas and will be more successful in solving all types of chemistry problems. These problems are well suited for group work and for use as in-class activities.

14.17 Chemical Equilibrium I

Part 1: You run the chemical reaction $C(aq) + D(aq) \rightleftharpoons 2E(aq)$ at 25°C. The equilibrium constant K_c for the reaction at this temperature is 2.0.

a Write the equilibrium-constant expression for the reaction.

b Can you come up with some possible concentrations of C, D, and E that you might observe when the reaction has reached equilibrium at 25°C? What are these values?

c A student says that only a very limited number of concentrations for C, D, and E are possible at equilibrium. Is this true? State why you think this is true or is not true.

d If you start with 1.0 M concentrations of both C and D and allow the reaction to come to equilibrium, would you expect the concentration of C to have decreased to zero? If not, what would you expect for the concentration of C? (An approximate value is fine.)

Part 2: Consider the reaction $A(aq) + B(aq) \rightleftharpoons F(aq) + G(aq)$, whose equilibrium constant is 1.0×10^{-5} at 20°C. For each of the situations described below, indicate whether any reaction occurs. If reaction does occur, then indicate the direction of that reaction and describe how the concentrations of A, B, F, and G change during this reaction.

a $A(aq)$ and $B(aq)$ are mixed together in a container.

b $F(aq)$ and $G(aq)$ are mixed together in a container.

c $A(aq)$ and $F(aq)$ are mixed together in a container.

d $B(aq)$ and $G(aq)$ are mixed together in a container.

e Just $B(aq)$ is placed into a container.

f Just $G(aq)$ is placed into a container.

Consider any one of these situations in which a reaction does occur. At equilibrium, does the reaction mixture have appreciably more products than reactants? If not, how would you describe the equilibrium composition of the reaction mixture? How did you arrive at this answer?

14.18 Chemical Equilibrium II

Magnesium hydroxide, $Mg(OH)_2$, is a white, partially soluble solid that is used in many antacids. The chemical equation for the dissolving of $Mg(OH)_2(s)$ in water is

$$Mg(OH)_2(s) \rightleftharpoons Mg^{2+}(aq) + 2OH^-(aq)$$

a Describe a simple experimental procedure that you could use to study this solubility equilibrium. In your experiment, how would you determine when the solution process has attained equilibrium?

b Write the equilibrium-constant expression for this dissolving of magnesium hydroxide.

c Suppose equilibrium has been established in a container of magnesium hydroxide in water, and you decide to add more solid $Mg(OH)_2$. What would you expect to observe? What effect will this addition of $Mg(OH)_2$ have on the concentrations of $Mg^{2+}(aq)$ and $OH^-(aq)$?

d Say you have prepared an equilibrium solution of $Mg(OH)_2$ by adding pure solid $Mg(OH)_2$ to water. If you know the concentration of $OH^-(aq)$, can you determine the concentration of $Mg^{2+}(aq)$? If not, what information do you need that will allow you to determine the answer?

e You slowly add OH^- from another source (say, NaOH) to an equilibrium mixture of $Mg(OH)_2$ and water. How do you expect the concentration of the $Mg^{2+}(aq)$ to change? What might you be able to observe happening to the $Mg(OH)_2(s)$ as you add the OH^-?

f Next you remove some, but not all, of the $Mg(OH)_2(s)$ from the mixture. How will this affect the concentrations of the $Mg^{2+}(aq)$ and $OH^-(aq)$?

g If someone hands you a container of $Mg(OH)_2(aq)$ and there is no solid $Mg(OH)_2$ present, is this solution at equilibrium? If it is not at equilibrium, what could you add to or remove from the container that would give an equilibrium system?

h Consider an individual $OH^-(aq)$ ion in an $Mg(OH)_2$ solution at equilibrium. If you could follow this ion over a long period of time, would you expect it always to remain as an $OH^-(aq)$ ion, or could it change in some way?

Conceptual Problems

Key: These problems are designed to check your understanding of the concepts associated with some of the main topics presented in each chapter. A strong conceptual understanding of chemistry is the foundation for both applying chemical knowledge and solving chemical problems. These problems vary in level of difficulty and often can be used as a basis for group discussion.

14.19 During an experiment with the Haber process, a researcher put 1 mol N_2 and 1 mol H_2 into a reaction vessel to observe the equilibrium formation of ammonia, NH_3.

$$N_2(g) + 3H_2(g) \rightleftharpoons 2NH_3(g)$$

When these reactants come to equilibrium, assume that x mol H_2 react. How many moles of ammonia form?

14.20 Suppose liquid water and water vapor exist in equilibrium in a closed container. If you add a small amount of liquid water to the container, how does this affect the amount of water vapor in the container? If, instead, you add a small amount of water vapor to the container, how does this affect the amount of liquid water in the container?

14.21 A mixture initially consisting of 2 mol CO and 2 mol H_2 comes to equilibrium with methanol, CH_3OH, as the product:

$$CO(g) + 2H_2(g) \rightleftharpoons CH_3OH(g)$$

At equilibrium, the mixture will contain which of the following?

a less than 1 mol CH_3OH

b 1 mol CH_3OH

c more than 1 mol CH₃OH but less than 2 mol
d 2 mol CH₃OH
e more than 2 mol CH₃OH

14.22 When a continuous stream of hydrogen gas, H_2, passes over hot magnetic iron oxide, Fe_3O_4, metallic iron and water vapor form. When a continuous stream of water vapor passes over hot metallic iron, the oxide Fe_3O_4 and H_2 form. Explain why the reaction goes in one direction in one case but in the reverse direction in the other.

14.23 For the reaction $2HI(g) \rightleftharpoons H_2(g) + I_2(g)$ carried out at some fixed temperature, the equilibrium constant is 2.0.

a Which of the following pictures correctly depicts the reaction mixture at equilibrium?

b For the pictures that represent nonequilibrium situations, describe which way the reaction will shift to attain equilibrium.

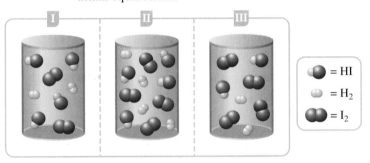

14.24 An experimenter introduces 4.0 mol of gas A into a 1.0-L container at 200 K to form product B according to the reaction

$$2A(g) \rightleftharpoons B(g)$$

Using the experimenter's data (one curve is for A, the other is for B), calculate the equilibrium constant at 200 K.

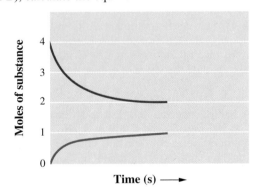

14.25 The following reaction is carried out at 500 K in a container equipped with a movable piston.

$$A(g) + B(g) \rightleftharpoons C(g); \quad K_c = 10 \text{ (at 500 K)}$$

After the reaction has reached equilibrium, the container has the composition depicted here.

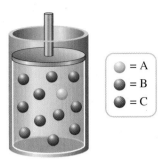

Suppose the container volume is doubled.

a How does the equilibrium composition shift?

b How does the concentration of each of the reactants and the product change? (That is, does the concentration increase, decrease, or stay the same?)

14.26 During the commercial preparation of sulfuric acid, sulfur dioxide reacts with oxygen in an exothermic reaction to produce sulfur trioxide. In this step, sulfur dioxide mixed with oxygen-enriched air passes into a reaction tower at about 420°C, where reaction occurs on a vanadium(V) oxide catalyst. Discuss the conditions used in this reaction in terms of its effect on the yield of sulfur trioxide. Are there any other conditions that you might explore in order to increase the yield of sulfur trioxide?

14.27 For the endothermic reaction $AB(g) \rightleftharpoons A(g) + B(g)$ the following represents a reaction container at two different temperatures. Which one (I or II) is at the lower temperature?

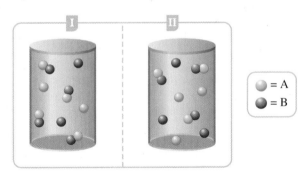

14.28 At some temperature, a 100-L reaction vessel contains a mixture that is initially 1.00 mol CO and 2.00 mol H_2. The vessel also contains a catalyst so that the following equilibrium is attained:

$$CO(g) + 2H_2(g) \rightleftharpoons CH_3OH(g)$$

At equilibrium, the mixture contains 0.100 mol CH₃OH (methanol). In a later experiment in the same vessel, you start with 1.00 mol CH₃OH. How much methanol is there at equilibrium? Explain.

Practice Problems

Key: These problems are for practice in applying problem-solving skills. They are divided by topic, and some are keyed to exercises (see the ends of the exercises). The problems are arranged in matching pairs; the odd-numbered problem of each pair is listed first, and its answer is given in the back of the book.

Reaction Stoichiometry

14.29 A 2.500-mol sample of phosphorus pentachloride, PCl_5, decomposes at 160°C and 1.00 atm to give 0.338 mol of phosphorus trichloride, PCl_3, at equilibrium.

$$PCl_5(g) \rightleftharpoons PCl_3(g) + Cl_2(g)$$

What is the composition of the final reaction mixture?

14.30 You place 4.00 mol of dinitrogen trioxide, N_2O_3, into a flask, where it decomposes at 25.0°C and 1.00 atm:

$$N_2O_3(g) \rightleftharpoons NO_2(g) + NO(g)$$

What is the composition of the reaction mixture at equilibrium if it contains 1.20 mol of nitrogen dioxide, NO_2?

14.31 Nitrogen monoxide, NO, reacts with bromine, Br_2, to give nitrosyl bromide, NOBr.

$$2NO(g) + Br_2(g) \rightleftharpoons 2NOBr(g)$$

A sample of 0.0524 mol NO with 0.0262 mol Br_2 gives an equilibrium mixture containing 0.0311 mol NOBr. What is the composition of the equilibrium mixture?

14.32 You place 0.600 mol of nitrogen, N_2, and 1.800 mol of hydrogen, H_2, into a reaction vessel at 450°C and 10.0 atm. The reaction is

$$N_2(g) + 3H_2(g) \rightleftharpoons 2NH_3(g)$$

What is the composition of the equilibrium mixture if you obtain 0.048 mol of ammonia, NH_3, from it?

14.33 In the contact process, sulfuric acid is manufactured by first oxidizing SO_2 to SO_3, which is then reacted with water. The reaction of SO_2 with O_2 is

$$2SO_2(g) + O_2(g) \rightleftharpoons 2SO_3(g)$$

A 2.000-L flask was filled with 0.0400 mol SO_2 and 0.0200 mol O_2. At equilibrium at 900 K, the flask contained 0.0296 mol SO_3. How many moles of each substance were in the flask at equilibrium?

14.34 Methanol, CH_3OH, formerly known as wood alcohol, is manufactured commercially by the following reaction:

$$CO(g) + 2H_2(g) \rightleftharpoons CH_3OH(g)$$

A 1.500-L vessel was filled with 0.1500 mol CO and 0.3000 mol H_2. When this mixture came to equilibrium at 500 K, the vessel contained 0.1187 mol CO. How many moles of each substance were in the vessel?

The Equilibrium Constant and Its Evaluation

14.35 Write equilibrium-constant expressions K_c for each of the following reactions.

a $PCl_3(g) + 3NH_3(g) \rightleftharpoons P(NH_2)_3(g) + 3HCl(g)$
b $N_2O_3(g) \rightleftharpoons NO_2(g) + NO(g)$
c $2H_2S(g) \rightleftharpoons 2H_2(g) + S_2(g)$
d $2NO(g) + O_2(g) \rightleftharpoons 2NO_2(g)$

14.36 Write equilibrium-constant expressions K_c for each of the following reactions.

a $2NO(g) + Cl_2(g) \rightleftharpoons 2NOCl(g)$
b $2HCl(g) + \frac{1}{2}O_2(g) \rightleftharpoons Cl_2(g) + H_2O(g)$
c $2NO(g) + 2H_2(g) \rightleftharpoons N_2(g) + 2H_2O(g)$
d $N_2H_4(g) \rightleftharpoons N_2(g) + 2H_2(g)$

14.37 The equilibrium-constant expression for a gas reaction is

$$K_c = \frac{[H_2O]^2[SO_2]^2}{[H_2S]^2[O_2]^3}$$

Write the balanced chemical equation corresponding to this expression.

14.38 The equilibrium-constant expression for a gas reaction is

$$K_c = \frac{[CO_2]^3[H_2O]^4}{[C_3H_8][O_2]^5}$$

Write the balanced chemical equation corresponding to this expression.

14.39 The equilibrium-constant expression for a reaction is

$$K_c = \frac{[NH_3]^4[O_2]^5}{[NO]^4[H_2O]^6}$$

What is the equilibrium-constant expression when the equation for this reaction is halved and then reversed?

14.40 The equilibrium-constant expression for a reaction is

$$K_c = \frac{[NO_2]^4[O_2]}{[N_2O_5]^2}$$

What is the equilibrium-constant expression when the equation for this reaction is halved and then reversed?

14.41 The equilibrium constant K_c for the equation

$$2HI(g) \rightleftharpoons H_2(g) + I_2(g)$$

at 425°C is 1.84. What is the value of K_c for the following equation?

$$H_2(g) + I_2(g) \rightleftharpoons 2HI(g)$$

14.42 The equilibrium constant K_c for the equation

$$CS_2(g) + 4H_2(g) \rightleftharpoons CH_4(g) + 2H_2S(g)$$

at 900°C is 27.8. What is the value of K_c for the following equation?

$$\tfrac{1}{2}CH_4(g) + H_2S(g) \rightleftharpoons \tfrac{1}{2}CS_2(g) + 2H_2(g)$$

14.43 A 4.00-L vessel contained 0.0148 mol of phosphorus trichloride, 0.0126 mol of phosphorus pentachloride, and 0.0870 mol of chlorine at 230°C in an equilibrium mixture. Calculate the value of K_c for the reaction

$$PCl_3(g) + Cl_2(g) \rightleftharpoons PCl_5(g)$$

14.44 A 6.00-L reaction vessel at 491°C contained 0.488 mol H_2, 0.206 mol I_2, and 2.250 mol HI. Assuming that the substances are at equilibrium, find the value of K_c at 491°C for the reaction of hydrogen and iodine to give hydrogen iodide. The equation is

$$H_2(g) + I_2(g) \rightleftharpoons 2HI(g)$$

14.45 Obtain the value of K_c for the following reaction at 900 K:

$$2SO_2(g) + O_2(g) \rightleftharpoons 2SO_3(g)$$

Use the data given in Problem 14.33.

14.46 Obtain the value of K_c for the following reaction at 500 K:

$$CO(g) + 2H_2(g) \rightleftharpoons CH_3OH(g)$$

Use the data given in Problem 14.34.

14.47 At 77°C, 2.00 mol of nitrosyl bromide, NOBr, placed in a 1.00-L flask dissociates to the extent of 9.4%; that is, for each mole of NOBr before reaction, (1.000 − 0.094) mol

NOBr remains after dissociation. Calculate the value of K_c for the dissociation reaction

$$2NOBr(g) \rightleftharpoons 2NO(g) + Br_2(g)$$

14.48 A 2.00-mol sample of nitrogen dioxide was placed in an 80.0-L vessel. At 200°C, the nitrogen dioxide was 6.0% decomposed according to the equation

$$2NO_2(g) \rightleftharpoons 2NO(g) + O_2(g)$$

Calculate the value of K_c for this reaction at 200°C. (See Problem 14.47.)

14.49 Write equilibrium-constant expressions K_p for each of the following reactions:

a $CS_2(g) + 4H_2(g) \rightleftharpoons CH_4(g) + 2H_2S(g)$
b $4HCl(g) + O_2(g) \rightleftharpoons 2H_2O(g) + 2Cl_2(g)$
c $CO(g) + 2H_2(g) \rightleftharpoons CH_3OH(g)$
d $H_2(g) + Br_2(g) \rightleftharpoons 2HBr(g)$

14.50 Write equilibrium-constant expressions K_p for each of the following reactions:

a $2NOBr(g) \rightleftharpoons 2NO(g) + Br_2(g)$
b $2SO_3(g) \rightleftharpoons 2SO_2(g) + O_2(g)$
c $N_2O_4(g) \rightleftharpoons 2NO_2(g)$
d $4NH_3(g) + 5O_2(g) \rightleftharpoons 4NO(g) + 6H_2O(g)$

14.51 The value of K_c for the following reaction at 900°C is 0.28.

$$CS_2(g) + 4H_2(g) \rightleftharpoons CH_4(g) + 2H_2S(g)$$

What is K_p at this temperature?

14.52 The equilibrium constant K_c equals 0.0952 for the following reaction at 227°C.

$$CH_3OH(g) \rightleftharpoons CO(g) + 2H_2(g)$$

What is the value of K_p at this temperature?

14.53 The reaction

$$SO_2(g) + \tfrac{1}{2}O_2(g) \rightleftharpoons SO_3(g)$$

has K_p equal to 6.55 at 627°C. What is the value of K_c at this temperature?

14.54 Fluorine, F_2, dissociates into atoms on heating.

$$\tfrac{1}{2}F_2(g) \rightleftharpoons F(g)$$

The value of K_p at 842°C is 7.55×10^{-2}. What is the value of K_c at this temperature?

14.55 Write the expression for the equilibrium constant K_c for each of the following equations:

a $FeO(s) + CO(g) \rightleftharpoons Fe(s) + CO_2(g)$
b $Na_2CO_3(s) + SO_2(g) + \tfrac{1}{2}O_2(g) \rightleftharpoons Na_2SO_4(s) + CO_2(g)$
c $PbI_2(s) \rightleftharpoons Pb^{2+}(aq) + 2I^-(aq)$
d $C(s) + CO_2(g) \rightleftharpoons 2CO(g)$

14.56 For each of the following equations, give the expression for the equilibrium constant K_c:

a $Na_2CO_3(s) + H_2O(l) + CO_2(g) \rightleftharpoons 2NaHCO_3(s)$
b $Fe^{3+}(aq) + 3OH^-(aq) \rightleftharpoons Fe(OH)_3(s)$
c $C(s) + 2N_2O(g) \rightleftharpoons CO_2(g) + 2N_2(g)$
d $NH_4Cl(s) \rightleftharpoons NH_3(g) + HCl(g)$

Using the Equilibrium Constant

14.57 On the basis of the value of K_c, decide whether or not you expect nearly complete reaction at equilibrium for each of the following:

a $C_2H_4(g) + H_2(g) \rightleftharpoons C_2H_6(g)$; $K_c = 1.3 \times 10^{21}$
b $N_2(g) + O_2(g) \rightleftharpoons 2NO(g)$; $K_c = 4.6 \times 10^{-31}$

14.58 Would either of the following reactions go almost completely to product at equilibrium?

a $COCl_2(g) \rightleftharpoons CO(g) + Cl_2(g)$; $K_c = 36 \times 10^{-16}$
b $2NO(g) + 2H_2(g) \rightleftharpoons N_2(g) + 2H_2O(g)$; $K_c = 6.5 \times 10^{113}$

14.59 Hydrogen fluoride decomposes according to the following equation:

$$2HF(g) \rightleftharpoons H_2(g) + F_2(g)$$

The value of K_c at room temperature is 1.0×10^{-95}. From the magnitude of K_c, do you think the decomposition occurs to any great extent at room temperature? If an equilibrium mixture in a 1.0-L vessel contains 1.0 mol HF, what is the amount of H_2 formed? Does this result agree with what you expect from the magnitude of K_c?

14.60 Suppose sulfur dioxide reacts with oxygen at 25°C.

$$2SO_2(g) + O_2(g) \rightleftharpoons 2SO_3(g)$$

The equilibrium constant K_c equals 8.0×10^{35} at this temperature. From the magnitude of K_c, do you think this reaction occurs to any great extent when equilibrium is reached at room temperature? If an equilibrium mixture is 1.0 M SO_3 and has equal concentrations of SO_2 and O_2, what is the concentration of SO_2 in the mixture? Does this result agree with what you expect from the magnitude of K_c?

14.61 The following reaction has an equilibrium constant K_c equal to 3.07×10^{-4} at 24°C.

$$2NOBr(g) \rightleftharpoons 2NO(g) + Br_2(g)$$

For each of the following compositions, decide whether the reaction mixture is at equilibrium. If it is not, decide which direction the reaction should go.

a $[NOBr] = 0.121\ M$, $[NO] = 0.0159\ M$, $[Br_2] = 0.0139\ M$
b $[NOBr] = 0.103\ M$, $[NO] = 0.0134\ M$, $[Br_2] = 0.0181\ M$
c $[NOBr] = 0.0472\ M$, $[NO] = 0.0121\ M$, $[Br_2] = 0.0105\ M$
d $[NOBr] = 0.0720\ M$, $[NO] = 0.0162\ M$, $[Br_2] = 0.0123\ M$

14.62 The following reaction has an equilibrium constant K_c equal to 3.59 at 900°C.

$$CH_4(g) + 2H_2S(g) \rightleftharpoons CS_2(g) + 4H_2(g)$$

For each of the following compositions, decide whether the reaction mixture is at equilibrium. If it is not, decide which direction the reaction should go.

a $[CH_4] = 1.26\ M$, $[H_2S] = 1.32\ M$, $[CS_2] = 1.43\ M$, $[H_2] = 1.12\ M$
b $[CH_4] = 1.25\ M$, $[H_2S] = 1.52\ M$, $[CS_2] = 1.15\ M$, $[H_2] = 1.73\ M$
c $[CH_4] = 1.30\ M$, $[H_2S] = 1.41\ M$, $[CS_2] = 1.10\ M$, $[H_2] = 1.20\ M$
d $[CH_4] = 1.56\ M$, $[H_2S] = 1.43\ M$, $[CS_2] = 1.23\ M$, $[H_2] = 1.91\ M$

14.63 Methanol, CH_3OH, is manufactured industrially by the reaction

$$CO(g) + 2H_2(g) \rightleftharpoons CH_3OH(g)$$

A gaseous mixture at 500 K is 0.020 M CH_3OH, 0.10 M CO, and 0.10 M H_2. What will be the direction of reaction if this mixture goes to equilibrium? The equilibrium constant K_c equals 10.5 at 500 K.

14.64 Sulfur trioxide, used to manufacture sulfuric acid, is obtained commercially from sulfur dioxide.

$$2SO_2(g) + O_2(g) \rightleftharpoons 2SO_3(g)$$

The equilibrium constant K_c for this reaction is 4.17×10^{-2} at 727°C. What is the direction of reaction when a mixture that is 0.80 M SO_2, 0.60 M O_2, and 0.10 M SO_3 approaches equilibrium at 727°C?

14.65 Nitrogen monoxide, NO, is formed in automobile exhaust by the reaction of N_2 and O_2 (from air).

$$N_2(g) + O_2(g) \rightleftharpoons 2NO(g)$$

The equilibrium constant K_c is 0.0025 at 2127°C. If an equilibrium mixture at this temperature contains 0.023 mol N_2 and 0.031 mol O_2 per liter, what is the concentration of NO?

14.66 Phosgene, $COCl_2$, used in the manufacture of polyurethane plastics, is prepared from CO and Cl_2.

$$CO(g) + Cl_2(g) \rightleftharpoons COCl_2(g)$$

An equilibrium mixture at 395°C contains 0.012 mol CO and 0.025 mol Cl_2 per liter, as well as $COCl_2$. If K_c at 395°C is 1.23×10^3, what is the concentration of $COCl_2$?

14.67 Iodine and bromine react to give iodine monobromide, IBr.

$$I_2(g) + Br_2(g) \rightleftharpoons 2IBr(g)$$

What is the equilibrium composition of a mixture at 150°C that initially contained 0.0015 mol each of iodine and bromine in a 5.0-L vessel? The equilibrium constant K_c for this reaction at 150°C is 1.2×10^2.

14.68 Initially a mixture contains 0.850 mol each of N_2 and O_2 in an 8.00-L vessel. Find the composition of the mixture when equilibrium is reached at 3900°C. The reaction is

$$N_2(g) + O_2(g) \rightleftharpoons 2NO(g)$$

and $K_c = 0.0123$ at 3900°C.

14.69 The equilibrium constant K_c for the reaction

$$PCl_3(g) + Cl_2(g) \rightleftharpoons PCl_5(g)$$

equals 49 at 230°C. If 0.400 mol each of phosphorus trichloride and chlorine are added to a 4.0-L reaction vessel, what is the equilibrium composition of the mixture at 230°C?

14.70 Calculate the composition of the gaseous mixture obtained when 1.25 mol of carbon dioxide is exposed to hot carbon at 800°C in a 1.25-L vessel. The equilibrium constant K_c at 800°C is 14.0 for the reaction

$$CO_2(g) + C(s) \rightleftharpoons 2CO(g)$$

14.71 Suppose 1.000 mol CO and 3.000 mol H_2 are put in a 10.00-L vessel at 1200 K. The equilibrium constant K_c for

$$CO(g) + 3H_2(g) \rightleftharpoons CH_4(g) + H_2O(g)$$

equals 3.92. Find the equilibrium composition of the reaction mixture.

14.72 The equilibrium constant K_c for the reaction

$$N_2(g) + 3H_2(g) \rightleftharpoons 2NH_3(g)$$

at 450°C is 0.159. Calculate the equilibrium composition when 1.00 mol N_2 is mixed with 3.00 mol H_2 in a 2.00-L vessel.

Le Châtelier's Principle

14.73 Consider the equilibrium

$$FeO(s) + CO(g) \rightleftharpoons Fe(s) + CO_2(g)$$

When carbon dioxide is removed from the equilibrium mixture (say, by passing the gases through water to absorb CO_2), what is the direction of net reaction as the new equilibrium is achieved?

14.74 a Predict the direction of reaction when chlorine gas is added to an equilibrium mixture of PCl_3, PCl_5, and Cl_2. The reaction is

$$PCl_3(g) + Cl_2(g) \rightleftharpoons PCl_5(g)$$

b What is the direction of reaction when chlorine gas is removed from an equilibrium mixture of these gases?

14.75 What would you expect to be the effect of an increase of pressure on each of the following reactions? Would the pressure change cause the reaction to go to the right or left?
a $H_2(g) + Br_2(g) \rightleftharpoons 2HBr(g)$
b $CO_2(g) + C(s) \rightleftharpoons 2CO(g)$
c $CH_4(g) + 2S_2(g) \rightleftharpoons CS_2(g) + 2H_2S(g)$

14.76 Indicate whether either an increase or a decrease of pressure obtained by changing the volume would increase the amount of product in each of the following reactions.
a $CO(g) + 2H_2(g) \rightleftharpoons CH_3OH(g)$
b $2SO_2(g) + O_2(g) \rightleftharpoons 2SO_3(g)$
c $2NO_2(g) \rightleftharpoons N_2O_4(g)$

14.77 One way of preparing hydrogen is by the decomposition of water.

$$2H_2O(g) \rightleftharpoons 2H_2(g) + O_2(g); \Delta H° = 484 \text{ kJ}$$

Would you expect the decomposition to be favorable at high or low temperature? Explain.

14.78 Methanol is prepared industrially from synthesis gas (CO and H_2).

$$CO(g) + 2H_2(g) \rightleftharpoons CH_3OH(g); \Delta H° = -21.7 \text{ kcal}$$

Would the fraction of methanol obtained at equilibrium be increased by raising the temperature? Explain.

14.79 Use thermochemical data (Appendix C) to decide whether the equilibrium constant for the following reaction will increase or decrease with temperature.

$$2NO_2(g) + 7H_2(g) \rightleftharpoons 2NH_3(g) + 4H_2O(g)$$

14.80 Use thermochemical data (Appendix C) to decide whether the equilibrium constant for the following reaction will increase or decrease with temperature.

$$CH_4(g) + 2H_2S(g) \rightleftharpoons CS_2(g) + 4H_2(g)$$

14.81 What would you expect to be the general temperature and pressure conditions for an optimum yield of nitrogen monoxide, NO, by the oxidation of ammonia?

$$4NH_3(g) + 5O_2(g) \rightleftharpoons 4NO(g) + 6H_2O(g); \Delta H° < 0$$

14.82 Predict the general temperature and pressure conditions for the optimum conversion of ethylene (C_2H_4) to ethane (C_2H_6).

$$C_2H_4(g) + H_2(g) \rightleftharpoons C_2H_6(g); \Delta H° < 0$$

General Problems

Key: These problems provide more practice but are not divided by topic or keyed to exercises. Each section ends with essay questions, each of which is color coded to refer to the *A Chemist Looks at Daily Life* (orange) chapter essay on which it is based. Odd-numbered problems and the even-numbered problems that follow are similar; answers to all odd-numbered problems except the essay questions are given in the back of the book.

14.83 A mixture of carbon monoxide, hydrogen, and methanol, CH_3OH, is at equilibrium according to the equation

$$CO(g) + 2H_2(g) \rightleftharpoons CH_3OH(g)$$

At 250°C, the mixture is 0.096 M CO, 0.191 M H_2, and 0.030 M CH_3OH. What is K_c for this reaction at 250°C?

14.84 An equilibrium mixture of SO_3, SO_2, and O_2 at 727°C is 0.0160 M SO_3, 0.0056 M SO_2, and 0.0021 M O_2. What is the value of K_c for the following reaction?

$$SO_2(g) + \tfrac{1}{2}O_2(g) \rightleftharpoons SO_3(g)$$

14.85 An equilibrium mixture of dinitrogen tetroxide, N_2O_4, and nitrogen dioxide, NO_2, is 65.8% NO_2 by mass at 1.00 atm pressure and 25°C. Calculate K_c at 25°C for the reaction

$$N_2O_4(g) \rightleftharpoons 2NO_2(g)$$

14.86 At 850°C and 1.000 atm pressure, a gaseous mixture of carbon monoxide and carbon dioxide in equilibrium with solid carbon is 90.55% CO by mass.

$$C(s) + CO_2(g) \rightleftharpoons 2CO(g)$$

Calculate K_c for this reaction at 850°C.

14.87 A 2.00-L vessel contains 1.00 mol N_2, 1.00 mol H_2, and 2.00 mol NH_3. What is the direction of reaction (forward or reverse) needed to attain equilibrium at 400°C? The equilibrium constant K_c for the reaction

$$N_2(g) + 3H_2(g) \rightleftharpoons 2NH_3(g)$$

is 0.51 at 400°C.

14.88 A vessel originally contained 0.0200 mol iodine monobromide (IBr), 0.050 mol I_2, and 0.050 mol Br_2. The equilibrium constant K_c for the reaction

$$I_2(g) + Br_2(g) \rightleftharpoons 2IBr(g)$$

is 1.2×10^2 at 150°C. What is the direction (forward or reverse) needed to attain equilibrium at 150°C?

14.89 A 2.0-L reaction flask initially contains 0.010 mol CO, 0.80 mol H_2, and 0.50 mol CH_3OH (methanol). If this mixture is brought in contact with a zinc oxide–chromium(III) oxide catalyst, the equilibrium

$$CO(g) + 2H_2(g) \rightleftharpoons CH_3OH(g)$$

is attained. The equilibrium constant K_c for this reaction at 300°C is 1.1×10^{-2}. What is the direction of reaction (forward or reverse) as the mixture attains equilibrium?

14.90 A gaseous mixture containing 1.00 mol each of CO, H_2O, CO_2, and H_2 is exposed to a zinc oxide–copper oxide catalyst at 1000°C. The reaction is

$$CO(g) + H_2O(g) \rightleftharpoons CO_2(g) + H_2(g)$$

and the equilibrium constant K_c is 0.58 at 1000°C. What is the direction of reaction (forward or reverse) as the mixture attains equilibrium?

14.91 Hydrogen bromide decomposes when heated according to the equation

$$2HBr(g) \rightleftharpoons H_2(g) + Br_2(g)$$

The equilibrium constant K_c equals 1.6×10^{-2} at 200°C. What are the moles of substances in the equilibrium mixture at 200°C if we start with 0.010 mol HBr in a 1.0-L vessel?

14.92 Iodine monobromide, IBr, occurs as brownish-black crystals that vaporize with decomposition:

$$2IBr(g) \rightleftharpoons I_2(g) + Br_2(g)$$

The equilibrium constant K_c at 100°C is 0.026. If 0.010 mol IBr is placed in a 1.0-L vessel at 100°C, what are the moles of substances at equilibrium in the vapor?

14.93 Dinitrogen tetroxide, N_2O_4, is a colorless gas (boiling point, 21°C), which dissociates to give nitrogen dioxide, NO_2, a reddish brown gas.

$$N_2O_4(g) \rightleftharpoons 2NO_2(g)$$

The equilibrium constant K_c at 25°C is 0.125. What percentage of dinitrogen tetroxide is dissociated when 0.0400 mol N_2O_4 is placed in a 1.00-L flask at 25°C?

14.94 Phosgene, $COCl_2$, is a toxic gas used in the manufacture of urethane plastics. The gas dissociates at high temperature.

$$COCl_2(g) \rightleftharpoons CO(g) + Cl_2(g)$$

At 400°C, the equilibrium constant K_c is 8.05×10^{-4}. Find the percentage of phosgene that dissociates at this temperature when 1.00 mol of phosgene is placed in a 25.0-L vessel.

14.95 Suppose you start with a mixture of 1.00 mol CO and 4.00 mol H_2 in a 10.00-L vessel. Find the moles of substances present at equilibrium at 1200 K for the reaction

$$CO(g) + 3H_2(g) \rightleftharpoons CH_4(g) + H_2O(g); \quad K_c = 3.92$$

You will get an equation of the form

$$f(x) = 3.92$$

where $f(x)$ is an expression in the unknown x (the amount of CH_4). Solve this equation by guessing values of x, then computing values of $f(x)$. Find values of x such that the values of $f(x)$ bracket 3.92. Then choose values of x to get a smaller bracket around 3.92. Obtain x to two significant figures.

14.96 What are the moles of substances present at equilibrium at 450°C if 1.00 mol N_2 and 4.00 mol H_2 in a 10.0-L vessel react according to the following equation?

$$N_2(g) + 3H_2(g) \rightleftharpoons 2NH_3(g)$$

The equilibrium constant K_c is 0.153 at 450°C. Use the numerical procedure described in Problem 14.95.

14.97 The equilibrium constant K_c for the synthesis of methanol, CH_3OH,

$$CO(g) + 2H_2(g) \rightleftharpoons CH_3OH(g)$$

is 4.3 at 250°C and 1.8 at 275°C. Is this reaction endothermic or exothermic?

14.98 The amount of nitrogen dioxide formed by dissociation of dinitrogen tetroxide,

$$N_2O_4(g) \rightleftharpoons 2NO_2(g)$$

increases as the temperature rises. Is the dissociation of N_2O_4 endothermic or exothermic?

14.99 For the reaction

$$N_2(g) + 3H_2(g) \rightleftharpoons 2NH_3(g)$$

show that $K_c = K_p(RT)^2$

Do not use the formula $K_p = K_c(RT)^{\Delta n}$ given in the text. Start from the fact that $P_i = [i]RT$, where P_i is the partial pressure of substance i and $[i]$ is its molar concentration. Substitute into K_c.

14.100 For the reaction

$$COCl_2(g) \rightleftharpoons CO(g) + Cl_2(g)$$

show that $K_c = K_p/(RT)$

Do not use the formula $K_p = K_c(RT)^{\Delta n}$ given in the text. See Problem 14.99.

14.101 At high temperatures, a dynamic equilibrium exists between carbon monoxide, carbon dioxide, and solid carbon.

$$C(s) + CO_2(g) \rightleftharpoons 2CO(g); \Delta H° = 172.5 \text{ kJ}$$

At 900°C, K_c is 0.238.

a   What is the value of K_p?
b   Some carbon dioxide is added to the hot carbon and the system is brought to equilibrium at 900°C. If the total pressure in the system is 6.80 atm, what are the partial pressures of CO and CO_2?
c   How will the equilibrium system respond if additional carbon at 900°C is added to the reaction system? Why?

14.102 At high temperatures, a dynamic equilibrium exists between carbon monoxide, carbon dioxide, and solid carbon.

$$C(s) + CO_2(g) \rightleftharpoons 2CO(g); \Delta H° = 172.5 \text{ kJ}$$

At 850°C, K_c is 0.153.

a   What is the value of K_p?
b   If the original reaction system consisted of just carbon and 1.50 atm of CO_2, what are the pressures of CO_2 and CO when equilibrium has been established?
c   How will the equilibrium pressure of CO change if the temperature is decreased?

14.103 The equilibrium constant K_c for the reaction

$$PCl_3(g) + Cl_2(g) \rightleftharpoons PCl_5(g)$$

equals 4.1 at 300°C.

a   A sample of 35.8 g of PCl_5 is placed in a 5.0-L reaction vessel and heated to 300°C. What are the equilibrium concentrations of all of the species?
b   What fraction of PCl_5 has decomposed?
c   If 35.8 g of PCl_5 were placed in a 1.0-L vessel, what qualitative effect would this have on the fraction of PCl_5 that has decomposed (give a qualitative answer only; do not do the calculation)? Why?

14.104 At 25°C in a closed system, ammonium hydrogen sulfide exists as the following equilibrium:

$$NH_4HS(s) \rightleftharpoons NH_3(g) + H_2S(g)$$

a   When a sample of pure $NH_4HS(s)$ is placed in an evacuated reaction vessel and allowed to come to equilibrium at 25°C, the total pressure is 0.660 atm. What is the value of K_p?
b   To this system, sufficient $H_2S(g)$ is injected until the pressure of H_2S is three times that of the ammonia at equilibrium. What are the partial pressures of NH_3 and H_2S?

c   In a different experiment, 0.750 atm of NH_3 and 0.500 atm of H_2S are introduced into a 1.00-L vessel at 25°C. How many moles of NH_4HS are present when equilibrium is established?

14.105 At moderately high temperatures, $SbCl_5$ decomposes into $SbCl_3$ and Cl_2 as follows:

$$SbCl_5(g) \rightleftharpoons SbCl_3(g) + Cl_2(g)$$

A 65.4-g sample of $SbCl_5$ is placed in an evacuated 5.00-L vessel. It is raised to 195°C, and the system comes to equilibrium. If at this temperature 35.8% of the $SbCl_5$ is decomposed, what is the value of K_p?

14.106 The following reaction is important in the manufacture of sulfuric acid.

$$SO_2(g) + \tfrac{1}{2}O_2(g) \rightleftharpoons SO_3(g)$$

At 900 K, 0.0216 mol of SO_2 and 0.0148 mol of O_2 are sealed in a 1.00-L reaction vessel. When equilibrium is reached, the concentration of SO_3 is determined to be 0.0175 M. Calculate K_c for this reaction.

14.107 Phosgene was used as a poisonous gas in World War I. At high temperatures it decomposes as follows:

$$COCl_2(g) \rightleftharpoons CO(g) + Cl_2(g)$$

with $K_c = 4.6 \times 10^{-3}$ at 800 K.

a   A sample of 7.80 g of $COCl_2$ is placed in a 1.00-L reaction vessel and heated to 800 K. What are the equilibrium concentrations of all of the species?
b   What fraction of $COCl_2$ has decomposed?
c   If 3 g of carbon monoxide is inserted into the reaction vessel, what qualitative effect would this have on the fraction of $COCl_2$ that has decomposed?

14.108 Sulfuryl chloride is used in organic chemistry as a chlorinating agent. At moderately high temperatures it decomposes as follows:

$$SO_2Cl_2(g) \rightleftharpoons SO_2(g) + Cl_2(g)$$

with $K_c = 0.045$ at 650 K.

a   A sample of 8.25 g of SO_2Cl_2 is placed in a 1.00-L reaction vessel and heated to 650 K. What are the equilibrium concentrations of all of the species?
b   What fraction of SO_2Cl_2 has decomposed?
c   If 5 g of chlorine is inserted into the reaction vessel, what qualitative effect would this have on the fraction of SO_2Cl_2 that has decomposed?

14.109 Gaseous acetic acid molecules have a certain tendency to form dimers. (A *dimer* is a molecule formed by the association of two identical, simpler molecules.) The equilibrium constant K_c at 25°C for this reaction is 3.2×10^4.

a   If the initial concentration of CH_3COOH monomer (the simpler molecule) is 4.0×10^{-4} M, what are the concentrations of monomer and dimer when the system comes to equilibrium? (The simpler quadratic equation is obtained by assuming that all of the acid molecules have dimerized and then some of it dissociates to monomer.)
b   Why do acetic acid molecules dimerize? What type of structure would you draw for the dimer?
c   As the temperature increases, would you expect the percentage of dimer to increase or decrease? Why?

14.110 Gaseous acetic acid molecules have a certain tendency to form dimers. (A *dimer* is a molecule formed by the association of two identical, simpler molecules.) The equilibrium constant K_p at 25°C for this reaction is 1.3×10^3.

 a If the initial pressure of CH_3COOH monomer (the simpler molecule) is 7.5×10^{-3} atm, what are the pressures of monomer and dimer when the system comes to equilibrium? (The simpler quadratic equation is obtained by assuming that all of the acid molecules have dimerized and then some of it dissociates to monomer.)

 b Why do acetic acid molecules dimerize? What type of structure would you draw for the dimer?

 c As the temperature decreases, would you expect the percentage of dimer to increase or decrease? Why?

14.111 When 0.112 mol of NO and 18.22 g of bromine are placed in a 1.00-L reaction vessel and sealed, the mixture is heated to 350 K and the following equilibrium is established:

$$2NO(g) + Br_2(g) \rightleftharpoons 2NOBr(g)$$

If the equilibrium concentration of nitrosyl bromide is 0.0824 M, what is K_c?

14.112 When 0.0322 mol of NO and 1.52 g of bromine are placed in a 1.00-L reaction vessel and sealed, the mixture reacts and the following equilibrium is established:

$$2NO(g) + Br_2(g) \rightleftharpoons 2NOBr(g)$$

At 25°C the equilibrium pressure of nitrosyl bromide is 0.438 atm. What is K_p?

■ **14.113** Oscillating reactions were first reported in 1958. How is it possible for a reaction to oscillate from one color to another?

■ **14.114** In what way are slime molds and leopards' spots related?

Strategy Problems

Key: As noted earlier, all of the practice and general problems are matched pairs. This section is a selection of problems that are not in matched-pair format. These challenging problems require that you employ many of the concepts and strategies that were developed in the chapter. In some cases, you will have to integrate several concepts and operational skills in order to solve the problem successfully.

14.115 A chemist placed a mixture of $CO_2(g)$ and $CF_4(g)$ into a flask at a particular temperature. These gases react:

$$CO_2(g) + CF_4(g) \rightleftharpoons 2COF_2(g)$$

After this mixture came to equilibrium, she found that it contained 0.40 mole fraction of CF_4 and 0.20 mole fraction of COF_2. Calculate K_p at this temperature for this reaction.

14.116 You mix equal moles of N_2, O_2, and NO and place them into a container fitted with a movable piston. Then you heat the mixture to 2127°C, compressing the mixture with the piston to 10.0 atm, where the following reaction may occur:

$$N_2(g) + O_2(g) \rightleftharpoons 2NO(g)$$

K_p for this reaction at 2127°C is 0.0025. Is the mixture of gases initially at equilibrium at this temperature and pressure, or does it undergo reaction to the left or to the right? Do you have enough information to answer this question? Explain your answer.

14.117 The equilibrium constant K_c for the equation

$$2HI(g) \rightleftharpoons H_2(g) + I_2(g)$$

is 1.84 at 425°C. The equilibrium constant K_c for the equation

$$H_2(g) + I_2(g) \rightleftharpoons 2HI(g)$$

is 49.7 at 458°C. Is this second equation endothermic or exothermic? How do you know? Describe the effect on this second equation of an increase in temperature. What would be the effect of an increase in pressure?

14.118 Consider the mixture described in the previous problem, but suppose it is heated to 3900°C and compressed to 15.0 atm. K_p at this temperature is 0.0123. What is the equilibrium composition of the mixture? Would it be best to describe this composition in moles or in mole fractions?

14.119 Consider the reaction $N_2O_4(g) \rightleftharpoons 2NO_2(g)$. Would you expect this reaction to be endothermic or exothermic? Why? N_2O_4 is a colorless gas; NO_2 is red-brown. Would you expect a mixture of these gases to become more or less red-brown as you raise the temperature? Explain.

14.120 A researcher put 0.400 mol PCl_3 and 0.600 mol Cl_2 into a 5.00-L vessel at a given temperature to produce phosphorus pentachloride, PCl_5:

$$PCl_3(g) + Cl_2(g) \rightleftharpoons PCl_5(g)$$

What will be the composition of this gaseous mixture at equilibrium? $K_c = 25.6$ at the temperature of this experiment.

14.121 Ammonium hydrogen sulfide, NH_4HS, is unstable at room temperature and decomposes:

$$NH_4HS(s) \rightleftharpoons NH_3(g) + H_2S(g)$$

You have placed some solid ammonium hydrogen sulfide in a closed flask. Which of the following would produce less hydrogen sulfide, H_2S, which is a poisonous gas?

 a Removing some NH_3 from the flask
 b Adding some NH_3 to the flask
 c Removing some of the NH_4HS
 d Increasing the pressure in the flask by adding helium gas
Explain each of your answers.

14.122 Iodine, I_2, is a blue-black solid, but it easily vaporizes to give a violet vapor. At high temperatures, this molecular substance dissociates to atoms:

$$I_2(g) \rightleftharpoons 2I(g)$$

An absent-minded professor measured equilibrium constants for this dissociation at two temperatures, 700°C and 800°C. He obtained the K_p values 0.01106 and 0.001745, but he couldn't remember which value went with what temperature. Can you please help him assign temperatures to the equilibrium constants? State your argument carefully.

14.123 A chemist wants to prepare phosgene, $COCl_2$, by the following reaction:

$$CO(g) + Cl_2(g) \rightleftharpoons COCl_2(g)$$

He places 4.00 g of chlorine, Cl_2, and an equal molar amount of carbon monoxide, CO, into a 10.00-L reaction vessel at 395°C. After the reaction comes to equilibrium, he adds another 4.00 g of chlorine to the vessel in order to push the reaction to the right to get more product. What is the partial pressure of phosgene when the reaction again comes to equilibrium? $K_c = 1.23 \times 10^3$.

14.124 Nitrogen dioxide, NO_2, is a red-brown gas that reacts with itself to produce N_2O_4, a colorless gas:

$$2NO_2(g) \rightleftharpoons N_2O_4(g)$$

A flask of "nitrogen dioxide" gas is actually a mixture of NO_2 and N_2O_4. It is possible to obtain the equilibrium constant K_p for this reaction from the density of an equilibrium mixture at a given temperature and total pressure. To see how to do that, answer the following questions. (You might find it helpful to refer to the derivation of the equation relating molar mass and density at the end of Section 5.3, on ideal gases.)

 a Use the ideal gas law to obtain an equation at a given T and P for the mass of a gas, m, in terms of its molar mass, M_m.

 b Suppose you have a mixture of two gases, A and B. Write an equation for the mass of gas A (m_A) in terms of the molar mass of A, M_{mA}, and its partial pressure, P_A. Write a similar equation for B. Now write an equation for the total mass of gas, $m = m_A + m_B$. Finally, obtain an equation for the density of the gas mixture, d, in terms of the partial pressures and molar masses of the components A and B.

 c Let A be NO_2 and B be N_2O_4. Express the equation you just obtained for the density, d, in terms of the molar mass, M_{mA}, eliminating M_{mB}. Also, eliminate P_B by expressing it in terms of P_A and P (total pressure). Now solve this to obtain an equation for P_A.

 d In an experiment, a chemist discovered that the density of nitrogen dioxide in equilibrium with dinitrogen tetroxide at 1.00 atm and 35.0°C is 2.86 g/L. What is the equilibrium constant K_p for the reaction as written at the beginning of this problem?

14.125 A container with a volume of 1.500 L was evacuated and then filled at low temperature with 0.0500 mol liquid dinitrogen tetroxide, N_2O_4. As the temperature was raised, this substance formed nitrogen dioxide, NO_2:

$$N_2O_4(g) \rightleftharpoons 2NO_2(g)$$

If the pressure of this reaction mixture at equilibrium at 25.0°C was 0.971 atm, what was the total moles of gas in this mixture? What are the moles of NO_2 and moles of N_2O_4? What are the partial pressures of NO_2 and N_2O_4? What is K_p at 25.0°C?

14.126 An evacuated 1.000-L flask was filled at room temperature with 0.01308 mol NO (nitrogen monoxide) and 0.00559 mol Br_2 (bromine). The temperature was then raised to 50.5°C, and these substances reacted to give nitrosyl bromide, NOBr:

$$2NO(g) + Br_2(g) \rightleftharpoons 2NOBr(g)$$

When equilibrium was achieved, the pressure of the gases in the flask was found to be 0.4027 atm. What was the total moles of gases in the flask at equilibrium? How many moles of nitrosyl bromide were formed? What was the moles of each substance in the flask at equilibrium? What are the partial pressures of the different gases in this equilibrium mixture? What is K_p at this temperature?

14.127 Nitrosyl chloride gas, NOCl, was introduced into a 2.000-L flask. The gas was then heated to 240°C, where it decomposed according to the equation

$$2NOCl(g) \rightleftharpoons 2NO(g) + Cl_2(g)$$

The pressure was maintained at 1.000 atm by a piston mechanism attached to the flask. When the gas reached equilibrium, it was analyzed and found to contain 0.404 g chlorine gas, Cl_2, in the 2.00-L flask. What is the equilibrium constant K_p for the reaction?

14.128 The major constituents of the atmosphere are nitrogen, N_2, and oxygen, O_2. Dry atmospheric air at a pressure of 1.000 atm has a partial pressure of N_2 of 0.781 atm and a partial pressure of O_2 of 0.209 atm. At high temperatures, N_2 and O_2 react to produce nitrogen monoxide (nitric oxide), NO.

$$N_2(g) + O_2(g) \rightleftharpoons 2NO(g)$$

Suppose a sample of dry air is raised to 2127°C in a hot flame and then rapidly cooled to fix the amount of NO formed. If K_p for this reaction at this temperature is 0.0025, how many grams of NO would be produced from 100.0 g of dry air?

14.129 Iodine, I_2, and bromine, Br_2, react to produce iodine monobromide, IBr.

$$I_2(g) + Br_2(g) \rightleftharpoons 2IBr(g)$$

A starting mixture of 0.5000 mol I_2 and 0.5000 mol Br_2 reacts at 150°C to produce 0.4221 mol IBr at equilibrium. What would be the equilibrium composition (in moles) of a mixture that starts with 1.000 mol I_2 and 2.000 mol Br_2?

14.130 Sulfur dioxide reacts with oxygen to produce sulfur trioxide in the presence of a catalyst:

$$2SO_2(g) + O_2(g) \rightleftharpoons 2SO_3(g)$$

In an experiment carried out at 727°C, the partial pressures of SO_2 and O_2 at equilibrium were found to be 0.721 atm and 0.581 atm, respectively, with a total pressure of 2.325 atm. What is the value of K_p for this reaction at this temperature? What would be the effect of an increase in pressure on the amount of SO_3 produced? What would be the effect of an increase in temperature on the amount of SO_3 produced? (Use relevant data from Appendix C.) Explain your answers.

14.131 Consider the production of ammonia from its elements:

$$N_2(g) + 3H_2(g) \rightleftharpoons 2NH_3(g)$$

In a pilot experiment, a mixture of 1.00 part nitrogen gas and 3.00 parts hydrogen gas at a total pressure of 10.00 atm was passed through a hot tube over a catalyst at 450.0°C. What are the total moles of gas in a 1.000-L volume of this equilibrium mixture? If the equilibrium mixture contains a partial pressure of 0.204 atm of ammonia, how many moles of ammonia are in this 1.000-L volume? How many moles of N_2 are in this same volume? What is the partial pressure of N_2 in this equilibrium mixture? What is the equilibrium constant K_p for this reaction?

14.132 Molecular bromine, Br_2, dissociates at elevated temperatures into bromine atoms, Br.

$$Br_2(g) \rightleftharpoons 2Br(g)$$

A 3.000-L flask initially contains pure molecular bromine. The temperature is then raised to 1600 K. If the total pressure of this equilibrium mixture at this elevated temperature is 1.000 atm, what are the total moles of gas in the container? A spectroscopic analysis of this mixture showed that it contained 1.395 g of Br atoms. What is the partial pressure of Br? What is K_p for the dissociation of molecular bromine to bromine atoms?

14.133 A mixture of 0.0565 mol phosphorus pentachloride, PCl_5, and 0.0800 mol helium gas, He, was placed in a 1.000-L flask and heated to 250.0°C. The phosphorus pentachloride decomposes at this temperature to give phosphorus trichloride, PCl_3, and chlorine gas, Cl_2. The helium gas is inert.

$$PCl_5(g) \rightleftharpoons PCl_3(g) + Cl_2(g)$$

What is the partial pressure of helium in this equilibrium mixture at 250.0°C? At equilibrium, the total pressure is found to be 6.505 atm. What is K_c for the dissociation of PCl_5?

14.134 Calcium carbonate, $CaCO_3$, decomposes when heated to give calcium oxide, CaO, and carbon dioxide, CO_2.

$$CaCO_3(s) \rightleftharpoons CaO(s) + CO_2(g)$$

K_p for this reaction at 900°C is 1.040. What would be the yield of carbon dioxide (in grams) when 1.000 g of $CaCO_3$ and 1.000 g CaO are heated to 900°C in a 1.000-L vessel. (Ignore the volume occupied by the solids.) What would be the effect of adding a similar quantity of carbon dioxide to this equilibrium mixture? What would happen if the quantity of calcium carbonate were doubled?

Cumulative-Skills Problems

Key: The problems under this heading combine skills introduced in previous chapters with those given in the current one.

14.135 The following equilibrium was studied by analyzing the equilibrium mixture for the amount of H_2S produced.

$$Sb_2S_3(s) + 3H_2(g) \rightleftharpoons 2Sb(s) + 3H_2S(g)$$

A vessel whose volume was 2.50 L was filled with 0.0100 mol of antimony(III) sulfide, Sb_2S_3, and 0.0100 mol H_2. After the mixture came to equilibrium in the closed vessel at 440°C, the gaseous mixture was removed, and the hydrogen sulfide was dissolved in water. Sufficient lead(II) ion was added to react completely with the H_2S to precipitate lead(II) sulfide, PbS. If 1.029 g PbS was obtained, what is the value of K_c at 440°C?

14.136 The following equilibrium was studied by analyzing the equilibrium mixture for the amount of HCl produced.

$$LaCl_3(s) + H_2O(g) \rightleftharpoons LaOCl(s) + 2HCl(g)$$

A vessel whose volume was 1.25 L was filled with 0.0125 mol of lanthanum(III) chloride, $LaCl_3$, and 0.0250 mol H_2O. After the mixture came to equilibrium in the closed vessel at 619°C, the gaseous mixture was removed and dissolved in more water. Sufficient silver ion was added to precipitate the chloride ion completely as silver chloride. If 3.59 g AgCl was obtained, what is the value of K_c at 619°C?

14.137 Antimony(V) chloride, $SbCl_5$, decomposes on heating to give antimony(III) chloride, $SbCl_3$, and chlorine.

$$SbCl_5(g) \rightleftharpoons SbCl_3(g) + Cl_2(g)$$

A closed 3.50-L vessel initially contains 0.0125 mol $SbCl_5$. What is the total pressure at 248°C when equilibrium is achieved? The value of K_c at 248°C is 2.50×10^{-2}.

14.138 Phosphorus pentachloride, PCl_5, decomposes on heating to give phosphorus trichloride, PCl_3, and chlorine.

$$PCl_5(g) \rightleftharpoons PCl_3(g) + Cl_2(g)$$

A closed 2.00-L vessel initially contains 0.0100 mol PCl_5. What is the total pressure at 250°C when equilibrium is achieved? The value of K_c at 250°C is 4.15×10^{-2}.

15 Acids and Bases

Red cabbage juice is green in basic solution (left beaker) but turns red when dry ice, solid CO_2, is added (center) to give an acidic solution (right). The colors of red cabbage and many fruits, such as blueberries, result from compounds called *anthocyanins* (see model of the one present in red cabbage), which are important in human nutrition.

© Royalty-Free/Corbis; Cengage Learning

Acids and bases were first recognized by simple properties, such as taste. Acids have a sour taste; bases are bitter. Also, acids and bases change the color of certain dyes called indicators, such as litmus and phenolphthalein. Acids change litmus from blue to red and basic phenolphthalein from red to colorless. Bases change litmus from red to blue and phenolphthalein from colorless to pink. As you can see from these color changes, acids and bases neutralize, or reverse, the action of one another. During neutralization, acids and bases react with each other to produce ionic substances called salts. Acids react with active metals, such as magnesium and zinc, to release hydrogen.

The Swedish chemist Svante Arrhenius framed the first successful concept of acids and bases. He defined acids and bases in terms of the effect these substances have on water. According to Arrhenius, acids are substances that increase the concentration of H^+ ion in aqueous solution, and bases increase the concentration of OH^- ion in aqueous solution. But many reactions that have characteristics of acid-base reactions in aqueous solution occur in other solvents or without a solvent. For example, hydrochloric acid reacts with aqueous ammonia, which in the Arrhenius view is a base because it increases the concentration of OH^- ion in aqueous solution. The reaction can be written

$$HCl(aq) + NH_3(aq) \longrightarrow NH_4Cl(aq)$$

The product is a solution of NH_4Cl—that is, a solution of NH_4^+ and Cl^- ions. A very similar reaction occurs between hydrogen chloride and ammonia dissolved in benzene, C_6H_6. The product is again NH_4Cl, which in this case precipitates from the solution.

$$HCl(benzene) + NH_3(benzene) \longrightarrow NH_4Cl(s)$$

Hydrogen chloride and ammonia react even in the gas phase. If watch glasses (shallow glass dishes) of concentrated hydrochloric acid and concentrated ammonia are placed next to each other, dense, white fumes of NH_4Cl form where HCl gas and NH_3 gas come into contact (Figure 15.1).

$$HCl(g) + NH_3(g) \longrightarrow NH_4Cl(s)$$

These reactions of HCl and NH_3, in benzene and in the gas phase, are similar to the reaction in aqueous solution but cannot be explained by the Arrhenius concept. Broader acid–base concepts are needed.

In the first part of this chapter, we will discuss the Arrhenius, the Brønsted–Lowry, and the Lewis concepts of acids and bases. The Brønsted–Lowry and Lewis concepts apply to nonaqueous as well as aqueous solutions and also expand on the Arrhenius concept in other ways. This chapter expands on what you learned in Chapter 4 about acids and bases.

Figure 15.1 ▲

Reaction of HCl(g) and NH₃(g) to form NH₄Cl(s) Gases from the concentrated solutions diffuse from their watch glasses (shallow dishes) and react to give a smoke of ammonium chloride.

© Cengage Learning

Acid–Base Concepts

Antoine Lavoisier was one of the first chemists to try to explain what makes a substance acidic. In 1777 he proposed that oxygen was an essential element in acids. (*Oxygen,* which he named, means "acid-former" in Greek.) But in 1808 Humphry Davy showed that hydrogen chloride, which dissolves in water to give hydrochloric acid, contains only hydrogen and chlorine. Although some chemists argued that chlorine was a compound of oxygen, chlorine was eventually proved to be an element. Chemists then noted that hydrogen, not oxygen, must be the essential constituent of acids. The cause of acidity and basicity was first explained in 1884 by Svante Arrhenius.

15.1 Arrhenius Concept of Acids and Bases

We can state the Arrhenius concept of an acid as follows: *An acid is a substance that, when dissolved in water, increases the concentration of hydronium ion, $H_3O^+(aq)$.* For simplicity, chemists frequently use the notation $H^+(aq)$ for the $H_3O^+(aq)$ ion and call

This section summarizes the discussion in section 4.4.

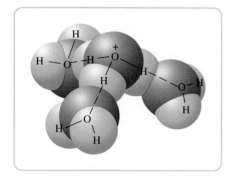

Figure 15.2 ▲

The hydronium ion, H_3O^+
The species is shown here hydrogen-bonded to three water molecules. The positive charge shown is actually distributed over the ion.

We introduced strong and weak acids and bases in Section 4.4.

Table 15.1	Common Strong Acids and Bases
Strong Acids	**Strong Bases***
$HClO_4$	LiOH
H_2SO_4	NaOH
HI	KOH
HBr	$Ca(OH)_2$
HCl	$Sr(OH)_2$
HNO_3	$Ba(OH)_2$

*In general, the Group IA and Group IIA hydroxides (except beryllium hydroxide) are strong bases.

it the hydrogen ion. Remember, however, that the aqueous hydrogen ion is not a bare proton in water, but a proton chemically bonded to water—that is, $H_3O^+(aq)$. The $H_3O^+(aq)$ ion is itself associated through hydrogen bonding with a variable number of water molecules. One such species is shown in Figure 15.2. *A base, in the Arrhenius concept, is a substance that, when dissolved in water, increases the concentration of hydroxide ion, $OH^-(aq)$.* ◄

The special role of the hydronium ion (or hydrogen ion) and the hydroxide ion in aqueous solutions arises from the following reaction:

$$H_2O(l) + H_2O(l) \rightleftharpoons H_3O^+(aq) + OH^-(aq)$$

The addition of acids and bases alters the concentrations of these ions in water.

In Arrhenius's theory, a *strong acid* is a substance that completely ionizes in aqueous solution to give $H_3O^+(aq)$ and an anion. An example is perchloric acid, $HClO_4$.

$$HClO_4(aq) + H_2O(l) \longrightarrow H_3O^+(aq) + ClO_4^-(aq)$$

Other examples of strong acids are H_2SO_4, HI, HBr, HCl, and HNO_3. A *strong base* completely ionizes in aqueous solution to give OH^- and a cation. Sodium hydroxide is an example of a strong base.

$$NaOH(s) \xrightarrow{H_2O} Na^+(aq) + OH^-(aq)$$

The principal strong bases are the hydroxides of Group IA elements and Group IIA elements (except Be). See Table 15.1. ◄

Most of the other acids and bases you encounter are *weak*. They are not completely ionized in solution and exist in reversible reaction with the corresponding ions. Consider acetic acid, $HC_2H_3O_2$. The reaction with water is

$$HC_2H_3O_2(aq) + H_2O(l) \rightleftharpoons H_3O^+(aq) + C_2H_3O_2^-(aq)$$

Evidence for the Arrhenius theory comes from the heat of reaction, $\Delta H°$, for the neutralization of a strong acid by a strong base. This neutralization is essentially the reaction of $H_3O^+(aq)$ with $OH^-(aq)$ and should therefore always give the same $\Delta H°$ per mole of water formed. For example, if you write the neutralization of $HClO_4$ with NaOH in ionic form, you have

$$H_3O^+(aq) + \cancel{ClO_4^-(aq)} + \cancel{Na^+(aq)} + OH^-(aq) \longrightarrow$$
$$\cancel{Na^+(aq)} + \cancel{ClO_4^-(aq)} + 2H_2O(l)$$

After canceling, you get the net ionic equation

$$H_3O^+(aq) + OH^-(aq) \longrightarrow 2H_2O(l)$$

Experimentally, it is found that all neutralizations involving strong acids and bases have the same $\Delta H°$: -55.90 kJ per mole of H^+. This indicates that the same reaction occurs in each neutralization, as Arrhenius's theory predicts.

Despite its successes, the Arrhenius concept is limited. In addition to looking at acid–base reactions only in aqueous solutions, it singles out the OH^- ion as the source of base character, when other species can play a similar role. Broader definitions of acids and bases are described in the next sections.

15.2 Brønsted–Lowry Concept of Acids and Bases

In 1923 the Danish chemist Johannes N. Brønsted (1879–1947) and, independently, the British chemist Thomas M. Lowry (1874–1936) pointed out that acid–base reactions can be seen as proton-transfer reactions and that acids and bases can be defined in terms of this proton (H^+) transfer. According to the Brønsted–Lowry concept, an **acid** is *the species donating a proton in a proton-transfer reaction*. A **base** is *the species accepting the proton in a proton-transfer reaction*. ◄

Consider, for example, the reaction of hydrochloric acid with ammonia, which was mentioned in the chapter opening. Writing it as an ionic equation, you have

$$H_3O^+(aq) + \cancel{Cl^-(aq)} + NH_3(aq) \longrightarrow H_2O(l) + NH_4^+(aq) + \cancel{Cl^-(aq)}$$

To be precise, we should say hydrogen nucleus instead of proton (because natural hydrogen contains some ^2H as well as ^1H). The term proton is conventional in this context, however.

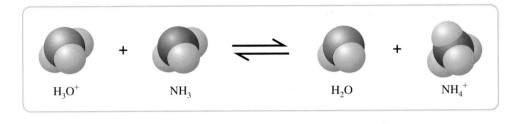

Figure 15.3 ◄

Representation of the reaction $H_3O^+ + NH_3 \rightleftharpoons H_2O + NH_4^+$
Note the transfer of a proton, H^+, from H_3O^+ to NH_3. The charges indicated for ions are *overall* charges. They are not to be associated with specific locations on the ions.

After canceling Cl^-, you obtain the net ionic equation.

$$H_3O^+(aq) + NH_3(aq) \longrightarrow H_2O(l) + NH_4^+(aq)$$

In this reaction in aqueous solution, a proton, H^+, is transferred from the H_3O^+ ion to the NH_3 molecule, giving H_2O and NH_4^+ (Figure 15.3). Here H_3O^+ is the proton donor, or acid, and NH_3 is the proton acceptor, or base. Note that in the Brønsted–Lowry concept, acids (and bases) can be ions as well as molecular substances.

You can also apply the Brønsted–Lowry concept to the reaction of HCl and NH_3 dissolved in benzene, C_6H_6, which was mentioned in the chapter opening. In benzene, HCl and NH_3 are not ionized. The equation is

$$\underset{\text{acid}}{HCl(benzene)} + \underset{\text{base}}{NH_3(benzene)} \longrightarrow NH_4Cl(s)$$

Here the HCl molecule is the proton donor, or acid, and the NH_3 molecule is the proton acceptor, or base.

In any reversible acid–base reaction, both forward and reverse reactions involve proton transfers. Consider the reaction of NH_3 with H_2O.

$$\underset{\text{base}}{NH_3(aq)} + \underset{\text{acid}}{H_2O(l)} \rightleftharpoons \underset{\text{acid}}{NH_4^+(aq)} + \underset{\text{base}}{OH^-(aq)}$$

In the forward reaction, NH_3 accepts a proton from H_2O. Thus, NH_3 is a base and H_2O is an acid. In the reverse reaction, NH_4^+ donates a proton to OH^-. NH_4^+ ion is the acid and OH^- is the base.

Note that NH_3 and NH_4^+ differ by a proton. That is, NH_3 becomes the NH_4^+ ion by gaining a proton, whereas the NH_4^+ ion becomes the NH_3 molecule by losing a proton. The species NH_4^+ and NH_3 are a conjugate acid–base pair. A **conjugate acid–base pair** consists of *two species in an acid–base reaction, one acid and one base, that differ by the loss or gain of a proton.* The acid in such a pair is called the *conjugate acid* of the base, whereas the base is the *conjugate base* of the acid. Here NH_4^+ is the conjugate acid of NH_3, and NH_3 is the conjugate base of NH_4^+.

Example 15.1

Gaining Mastery Toolbox

Critical Concept 15.1
A Brønsted–Lowry acid is a proton donor and a Brønsted–Lowry base is a proton acceptor.

Solution Essentials:
• Conjugate acid–base pairs
• Acid–base reactions

Identifying Acid and Base Species

In the following equations, label each species as an acid or a base. Show the conjugate acid–base pairs.

a $HCO_3^-(aq) + HF(aq) \rightleftharpoons H_2CO_3(aq) + F^-(aq)$

b $HCO_3^-(aq) + OH^-(aq) \rightleftharpoons CO_3^{2-}(aq) + H_2O(l)$

Problem Strategy Recall that a Brønsted–Lowry acid is a proton donor, and the base is a proton acceptor. Examine each equation to find the proton donor on each side. Then label the acids and bases.

(continued)

(continued)

Solution

a On the left, HF is the proton donor; on the right, H_2CO_3 is the proton donor. The proton acceptors are HCO_3^- and F^-. Once the proton donors and acceptors are identified, the acids and bases can be labeled.

$$\text{HCO}_3^-(aq) + \text{(H)F}(aq) \rightleftharpoons \text{(H)}_2\text{CO}_3(aq) + \text{F}^-(aq)$$
$$\qquad\text{base}\qquad\qquad\text{acid}\qquad\qquad\text{acid}\qquad\qquad\text{base}$$

In this reaction, H_2CO_3 and HCO_3^- are a conjugate acid–base pair, as are HF and F^-.

b You have

$$\text{(H)CO}_3^-(aq) + \text{OH}^-(aq) \rightleftharpoons \text{CO}_3^{2-}(aq) + \text{(H)}_2\text{O}(l)$$
$$\qquad\text{acid}\qquad\qquad\text{base}\qquad\qquad\text{base}\qquad\qquad\text{acid}$$

Here HCO_3^- and CO_3^{2-} are a conjugate acid–base pair, as are H_2O and OH^-. Note that although HCO_3^- functions as an acid in this reaction, it functions as a base in part a.

Answer Check Check that each conjugate acid–base pair consists of one species from the left side of the equation and one species from the right side, and that the two species differ by only a proton.

Exercise 15.1 For the reaction

$$\text{H}_2\text{CO}_3(aq) + \text{CN}^-(aq) \rightleftharpoons \text{HCN}(aq) + \text{HCO}_3^-(aq)$$

label each species as an acid or a base. For the base on the left, what is the conjugate acid?

■ See Problems 15.35 and 15.36

The Brønsted–Lowry concept defines a species as an acid or a base according to its function in the acid–base, or proton-transfer, reaction. As you saw in Example 15.1, some species can act as either an acid or a base. An **amphiprotic species** is *a species that can act as either an acid or a base (it can lose or gain a proton)*, depending on the other reactant. ◄ For example, HCO_3^- acts as an acid in the presence of OH^- but as a base in the presence of HF. Anions with ionizable hydrogens, such as HCO_3^-, and certain solvents, such as water, are amphiprotic.

The amphiprotic characteristic of water is important in the acid–base properties of aqueous solutions. Consider, for example, the reactions of water with the base NH_3 and with the acid $HC_2H_3O_2$ (acetic acid).

$$\text{NH}_3(aq) + \text{(H)}_2\text{O}(l) \rightleftharpoons \text{NH}_4^+(aq) + \text{OH}^-(aq)$$
$$\qquad\text{base}\qquad\qquad\text{acid}\qquad\qquad\text{acid}\qquad\qquad\text{base}$$

$$\text{(H)C}_2\text{H}_3\text{O}_2(aq) + \text{H}_2\text{O}(aq) \rightleftharpoons \text{C}_2\text{H}_3\text{O}_2^-(aq) + \text{H}_3\text{O}^+(aq)$$
$$\qquad\qquad\text{acid}\qquad\qquad\text{base}\qquad\qquad\text{base}\qquad\qquad\text{acid}$$

In the first case, water reacts as an acid with the base NH_3. In the second case, water reacts as a base with the acid $HC_2H_3O_2$.

You have now seen several ways in which the Brønsted–Lowry concept of acids and bases has greater scope than the Arrhenius concept. In the Brønsted–Lowry concept:

1. A base is a species that accepts protons; OH^- is only one example of a base.
2. Acids and bases can be ions as well as molecular substances.

Amphoteric is a general term referring to a species that can act as an acid or a base. The species need not be amphiprotic, however. For example, aluminum oxide is an amphoteric oxide, because it reacts with acids and bases. It is not amphiprotic, because it has no protons.

3. Acid–base reactions are not restricted to aqueous solution.

4. Some species can act as either acids or bases, depending on what the other reactant is.

CONCEPT CHECK 15.1

Chemists in the seventeenth century discovered that the substance that gives red ants their irritating bite is an acid with the formula $HCHO_2$. They called this substance formic acid after the ant, whose Latin name is *formica rufus*. Formic acid has the following structural formula and molecular model:

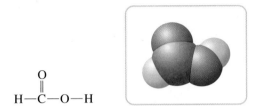

$$\overset{\displaystyle O}{\underset{\displaystyle }{\overset{\displaystyle \|}{H-C-O-H}}}$$

Write the acid–base equilibria connecting all components in the aqueous solution of formic acid. Now list all of the species present.

15.3 Lewis Concept of Acids and Bases

Certain reactions have the characteristics of acid–base reactions but do not fit the Brønsted–Lowry concept. An example is the reaction of the basic oxide Na_2O with the acidic oxide SO_3 to give the salt Na_2SO_4. ▶

$$Na_2O(s) + SO_3(g) \longrightarrow Na_2SO_4(s)$$

Acidic and basic oxides were discussed in Section 8.7.

G. N. Lewis, who proposed the electron-pair theory of covalent bonding, realized that the concept of acids and bases could be generalized to include reactions of acidic and basic oxides and many other reactions, as well as proton-transfer reactions. According to this concept, a **Lewis acid** is *a species that can form a covalent bond by accepting an electron pair from another species;* a **Lewis base** is *a species that can form a covalent bond by donating an electron pair to another species.* The Lewis and the Brønsted–Lowry concepts are simply different ways of looking at certain chemical reactions. Such different views are often helpful in devising new reactions.

Consider again the neutralization of NH_3 by HCl in aqueous solution, mentioned earlier. It consists of the reaction of a proton from H_3O^+ with $:NH_3$.

$$H^+ \quad + \quad \overset{\displaystyle H}{\underset{\displaystyle H}{:N:H}} \quad \longrightarrow \quad \left[\overset{\displaystyle H}{\underset{\displaystyle H}{H:N:H}}\right]^+$$

electron-pair   electron-pair
acceptor   donor

Here the red arrow shows the proton accepting an electron pair from NH_3 and an H—N bond being formed. The proton is an electron-pair acceptor, so it is a Lewis acid. Ammonia, NH_3, which has a lone pair of electrons, is an electron-pair donor and therefore a Lewis base.

Now let us look at the reaction of Na_2O with SO_3. It involves the reaction of the oxide ion, O^{2-}, from the ionic solid, Na_2O, with SO_3.

$$:\!\overset{..}{\underset{..}{O}}\!:^{2-} \; + \; \overset{\displaystyle :\overset{..}{O}:}{\underset{\displaystyle :\overset{..}{O}:}{S::\overset{..}{O}}} \; \longrightarrow \; \left[\overset{\displaystyle :\overset{..}{O}:}{\underset{\displaystyle :\overset{..}{O}:}{:\overset{..}{O}:S:\overset{..}{O}:}}\right]^{2-}$$

Lewis   Lewis
base   acid

Here SO_3 accepts the electron pair from the O^{2-} ion. At the same time, an electron pair from the S=O bond moves to the O atom. Thus, O^{2-} is the Lewis base and SO_3 is the Lewis acid.

The Lewis concept embraces many reactions that we might not think of as acid–base reactions. The reaction of boron trifluoride with ammonia is an example.

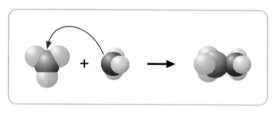

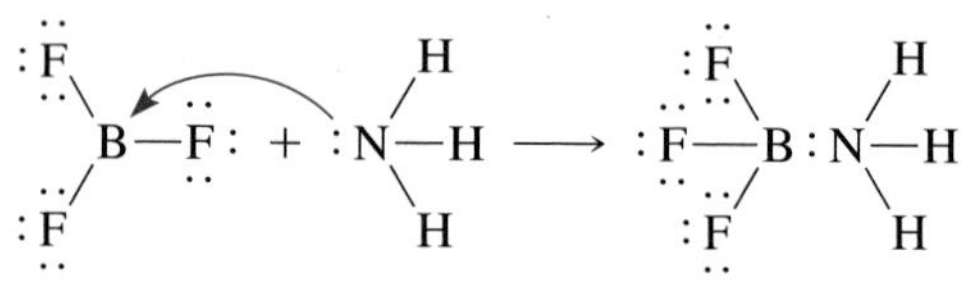

In this reaction, the NH_3 molecule donates the lone pair of electrons on the nitrogen atom to the boron atom of BF_3. Boron trifluoride accepts the electron pair and so is a Lewis acid. Ammonia donates the electron pair and so is a Lewis base. The electron pair originally on the nitrogen atom is now shared between the nitrogen and boron atoms, forming a B–N bond.

The formation of *complex ions* can also be looked at as Lewis acid–base reactions. Complex ions are formed when a metal ion bonds to electron pairs from molecules such as H_2O or NH_3 or from anions such as :C≡N:⁻. An example of a complex ion is $Al(H_2O)_6^{3+}$. Hydrated ions like $Al(H_2O)_6^{3+}$ are present in compounds (hydrates) and in aqueous solution. The formation of a hydrated metal ion, such as $Al(H_2O)_6^{3+}$, involves a Lewis acid–base reaction.

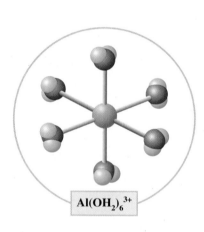

$Al(OH_2)_6^{3+}$

$$Al^{3+} + 6(\overset{\cdot\cdot}{:}\overset{\cdot\cdot}{O}-H) \longrightarrow Al(\overset{\cdot\cdot}{:}\overset{\cdot\cdot}{O}-H)_6^{3+}$$

Lewis acid   Lewis base

Example 15.2   Identifying Lewis Acid and Base Species

Critical Concept 15.2
A Lewis acid is an electron-pair acceptor and a Lewis base is an electron-pair donor.

Solution Essentials:
• Complex ions
• Lewis electron-dot formulas

In the following reactions, identify the Lewis acid and the Lewis base.

a $Ag^+ + 2NH_3 \rightleftharpoons Ag(NH_3)_2^+$
b $B(OH)_3 + H_2O \rightleftharpoons B(OH)_4^- + H^+$

Problem Strategy Write the equations using Lewis electron-dot formulas. Then identify the electron-pair acceptor, or Lewis acid, and the electron-pair donor, or Lewis base.

Solution

a The silver ion, Ag^+, forms a complex ion with two NH_3 molecules.

$$Ag^+ + 2:NH_3 \rightleftharpoons Ag(:NH_3)_2^+$$

Lewis acid   Lewis base

b The reaction is

$$H-\overset{\overset{\displaystyle H}{|}}{\underset{\underset{\displaystyle H}{|}}{\overset{\cdot\cdot}{O}:}}B\overset{\overset{\displaystyle :\overset{\cdot\cdot}{O}:}{|}}{\underset{\underset{\displaystyle :\overset{\cdot\cdot}{O}:}{|}}{}} + :\overset{\cdot\cdot}{O}:H \rightleftharpoons \left[H-\overset{\overset{\displaystyle H}{|}}{\underset{\underset{\displaystyle H}{|}}{\overset{\cdot\cdot}{O}:}}B:\overset{\cdot\cdot}{O}-H \right]^- + H^+$$

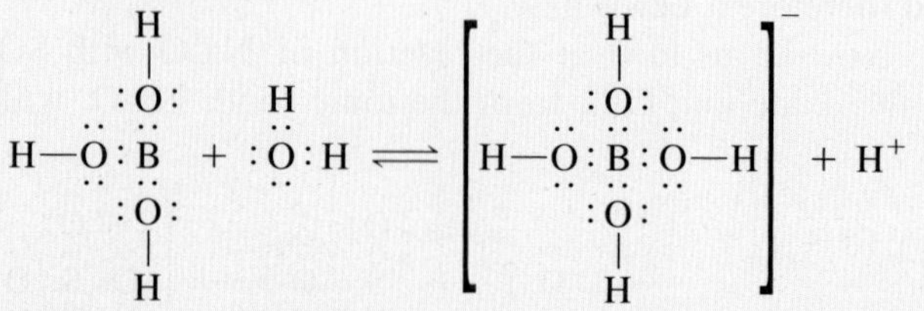

Lewis acid   Lewis base

(*continued*)

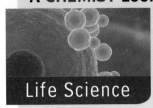
Taking Your Medicine

The expressions "take your medicine" and "a bitter pill to swallow" are metaphors that refer to doing whatever is necessary to solve a difficult personal problem. Taking medicine might be unpleasant if the medicine is bitter—and medicines frequently taste bitter. Why?

A bitter taste appears to be a common feature of a base. It is a fact that many medicinal substances are nitrogen bases, substances that organic chemists call *amines*. Such substances are considered derivatives of ammonia, in which one or more hydrogen atoms have been substituted by carbon-containing groups. Here, for example, are the structures of some amines:

$$\begin{array}{ccc} CH_3 & CH_3 & CH_3 \\ | & | & | \\ H\!-\!\underset{..}{N}\!-\!H & H_3C\!-\!\underset{..}{N}\!-\!H & H_3C\!-\!\underset{..}{N}\!-\!CH_3 \end{array}$$

The nitrogen atom in an amine has a lone pair of electrons that can be donated to form a covalent bond, so an amine is a Lewis base.

But like ammonia, an amine accepts a hydrogen ion from an acid to form an amine ion, so it is also a Brønsted–Lowry base:

$$\begin{array}{c} CH_3 \\ | \\ H\!-\!\underset{..}{N}\!-\!H + H_3O^+ \end{array} \rightleftharpoons \left[\begin{array}{c} CH_3 \\ | \\ H\!-\!N\!-\!H \\ | \\ H \end{array}\right]^+ + H_2O$$

Some of our most important medicinal drugs have originated from plants. Lewis and Clark took Peruvian bark, or cinchona bark, with them as a medicine on their 1804 expedition from the eastern United States to the Pacific Coast and back. The bitter essence of cinchona bark is quinine, an amine drug that has been used to combat malaria. Quinine is responsible for the bitter taste of tonic water, a carbonated beverage (Figure 15.4). Some other amines that come from plants are caffeine (from coffee, a stimulant), atropine (from the deadly nightshade, used to dilate the pupil of the eye for eye exams), and codeine (from the opium poppy, used as a painkiller).

Figure 15.4 ▶

Tonic water This carbonated beverage contains quinine, which accounts for its bitter taste. Note the N atoms (in blue) in the structural formula of quinine, which give the substance its basic character.

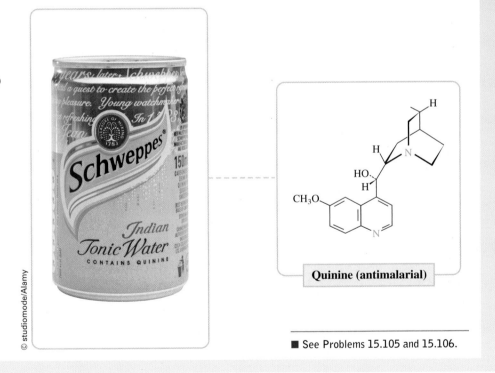

© studiomode/Alamy

Quinine (antimalarial)

■ See Problems 15.105 and 15.106.

(*continued*)

Answer Check Check the electron-dot formulas of the species in each acid–base reaction. The Lewis base donates an electron pair to the Lewis acid, which accepts that pair.

Exercise 15.2 Identify the Lewis acid and the Lewis base in each of the following reactions. Write the chemical equations using electron-dot formulas.

a $BF_3 + CH_3OH \longrightarrow F_3B:\overset{H}{\underset{}{\underset{..}{O}}}CH_3$

b $O^{2-} + CO_2 \longrightarrow CO_3^{2-}$

■ See Problems 15.39, 15.40, 15.41, and 15.42.

Acid and Base Strengths

The Brønsted–Lowry concept considers an acid–base reaction as a proton-transfer reaction. It is useful to consider such acid–base reactions as a competition between species for protons. From this point of view, you can order acids by their relative strengths as proton donors. The stronger acids are those that lose their protons more easily than other acids. Similarly, the stronger bases are those that hold on to protons more strongly than other bases.

15.4 Relative Strengths of Acids and Bases

By comparing various acid–base reactions, you can construct a table of relative strengths of acids and bases (see Table 15.2). To see how you might do this, look again at what we have called a strong acid. Recall that an acid is strong if it completely ionizes in water. In the reaction of hydrogen chloride with water, for example, water acts as a base, accepting the proton from HCl.

$$HCl(aq) + H_2O(l) \longrightarrow Cl^-(aq) + H_3O^+(aq)$$

$$\underset{\text{acid}}{HCl(aq)} + \underset{\text{base}}{H_2O(l)} \longrightarrow \underset{\text{base}}{Cl^-(aq)} + \underset{\text{acid}}{H_3O^+(aq)}$$

Table 15.2 Relative Strengths of Acids and Bases

	Acid	Base	
Strongest acids	$HClO_4$	ClO_4^-	Weakest bases
	H_2SO_4	HSO_4^-	
	HI	I^-	
	HBr	Br^-	
	HCl	Cl^-	
	HNO_3	NO_3^-	
	H_3O^+	H_2O	
	HSO_4^-	SO_4^{2-}	
	H_2SO_3	HSO_3^-	
	H_3PO_4	$H_2PO_4^-$	
	HNO_2	NO_2^-	
	HF	F^-	
	$HC_2H_3O_2$	$C_2H_3O_2^-$	
	$Al(H_2O)_6^{3+}$	$Al(H_2O)_5OH^{2+}$	
	H_2CO_3	HCO_3^-	
	H_2S	HS^-	
	HClO	ClO^-	
	HBrO	BrO^-	
	NH_4^+	NH_3	
	HCN	CN^-	
	HCO_3^-	CO_3^{2-}	
	H_2O_2	HO_2^-	
	HS^-	S^{2-}	
Weakest acids	H_2O	OH^-	Strongest bases

The reverse reaction occurs only to an extremely small extent. Because the reaction goes almost completely to the right, you say that HCl is a strong acid. However, even though the reaction goes almost completely to products, you can consider the reverse reaction. In it, the Cl^- ion acts as the base, accepting a proton from the acid H_3O^+.

Look at this reaction in terms of the relative strengths of the two acids, HCl and H_3O^+. Because HCl is a strong acid, it must lose its proton readily, more readily than H_3O^+ does. You would say that HCl is a stronger acid than H_3O^+, or that, of the two, H_3O^+ is the weaker acid.

$$HCl(aq) + H_2O(l) \rightleftharpoons Cl^-(aq) + H_3O^+(aq)$$
$$\text{stronger} \qquad\qquad\qquad \text{weaker}$$
$$\text{acid} \qquad\qquad\qquad\qquad \text{acid}$$

It is important to understand that the terms *stronger* and *weaker* are used here only in a comparative sense. The H_3O^+ ion is a relatively strong acid.

An acid–base reaction normally goes in the direction of the weaker acid. You can use this fact to compare the relative strengths of any two acids, as we did in comparing the relative strengths of HCl and H_3O^+. As another example, look at the ionization of acetic acid, $HC_2H_3O_2$, in water.

$$HC_2H_3O_2(aq) + H_2O(l) \rightleftharpoons C_2H_3O_2^-(aq) + H_3O^+(aq)$$

Experiment reveals that in a 0.1 M acetic acid solution, only about 1% of the acetic acid molecules have ionized by this reaction. This implies that $HC_2H_3O_2$ is a weaker acid than H_3O^+. If you look at the similar situation for 0.1 M HF, you find that about 3% of the HF molecules have dissociated. Thus, HF is a weaker acid than H_3O^+ but a stronger acid than $HC_2H_3O_2$. You have already seen that HCl is stronger than H_3O^+. Thus, you have determined that the acid strengths for these four acids are in the order $HCl > H_3O^+ > HF > HC_2H_3O_2$.

This procedure of determining the relative order of acid strengths by comparing their relative ionizations in water cannot be used to obtain the relative strengths of two strong acids such as HCl and HI. When these acids are dissolved in water, they are essentially 100% ionized. However, if you look at solutions of equal concentrations of these acids in another solvent that is less basic than water—say, pure acetic acid—you do see a difference. Neither acid is completely ionized, but a greater fraction of HI molecules is found to be ionized. Thus, HI is a stronger acid than HCl. In water, the acid strengths of the strong acids appear to be the same; that is, they are "leveled out." We say that water exhibits a *leveling effect* on the strengths of the strong acids.

The first column of Table 15.2 lists acids by their strength; the strongest is at the top of the table. Note that the arrow down the left side of the table points toward the weaker acid and is in the direction the reaction goes. For example, in the reaction involving HCl and H_3O^+, the arrow points from HCl toward H_3O^+ (the weaker acid) and is in the direction in which the reaction occurs.

You can also view this same reaction in terms of the bases, H_2O and Cl^-. A stronger base picks up a proton more readily than does a weaker one. Water has greater base strength than the Cl^- ion; that is, H_2O picks up protons more readily than Cl^- does. In fact, Cl^- has little attraction for protons. (If it had more, HCl would not lose its proton so readily.) Because H_2O molecules compete more successfully for protons than Cl^- ions do, the reaction goes almost completely to the right. (Note that the arrow on the right in Table 15.2 points from H_2O to Cl^-.) That is, at the completion of the reaction, concentrations of product species (Cl^- and H_3O^+) are much greater than the concentration of reactant HCl.

$$HCl(aq) + H_2O(l) \longrightarrow Cl^-(aq) + H_3O^+(aq)$$
$$\text{stronger} \qquad\qquad \text{weaker}$$
$$\text{base} \qquad\qquad\qquad \text{base}$$

The reaction goes in the direction of the weaker base.

By comparing reactions between different pairs of bases, you can arrive at a relative order for base strengths, just as for acids. A definite relationship exists between acid and base strengths. When you say an acid loses its proton readily, you can also say that its conjugate base does not hold the proton very tightly. *The strongest acids have the weakest conjugate bases, and the strongest bases have the weakest conjugate acids.* This means that a list of conjugate bases of the acids in Table 15.2 will be in order of increasing base strength. The weakest bases will be at the top of the table and the strongest at the bottom.

You can use Table 15.2 to predict the direction of an acid–base reaction. The direction for an acid–base reaction always favors the weaker acid and weaker base; that is, the normal direction of reaction is from the stronger acid and base to the weaker acid and base. For the reaction we have been discussing, you have

$$HCl(aq) + H_2O(l) \longrightarrow Cl^-(aq) + H_3O^+(aq)$$

| stronger | stronger | weaker | weaker |
| acid | base | base | acid |

The reaction follows the direction of the arrows at the left and right of Table 15.2.

⚙WL Interactive Example 15.3   Deciding Whether Reactants or Products Are Favored in an Acid–Base Reaction

Gaining Mastery Toolbox

Critical Concept 15.3
An acid–base reaction goes in the direction of the weaker acid (and weaker base).

Solution Essentials:
• Table 15.2 showing relative strengths of acids and bases

For the following reaction, decide which species (reactants or products) are favored at the completion of the reaction.

$$SO_4^{2-}(aq) + HCN(aq) \rightleftharpoons HSO_4^-(aq) + CN^-(aq)$$

Problem Strategy Use Table 15.2 to compare the relative strengths of acids and bases.

Solution If you compare the relative strengths of the two acids HCN and HSO_4^-, you see that HCN is weaker. Or, comparing the bases SO_4^{2-} and CN^-, you see that SO_4^{2-} is weaker. Hence, the reaction would normally go from right to left.

$$SO_4^{2-}(aq) + HCN(aq) \longleftarrow HSO_4^-(aq) + CN^-(aq)$$

| weaker | weaker | stronger | stronger |
| base | acid | acid | base |

The reactants are favored.

Answer Check Check that the stronger acid has a weaker conjugate base, and vice versa. The reaction goes from stronger acid and base to weaker acid and base.

Exercise 15.3   Determine the direction of the following reaction from the relative strengths of acids and bases.

$$H_2S(aq) + C_2H_3O_2^-(aq) \rightleftharpoons HC_2H_3O_2(aq) + HS^-(aq)$$

■ See Problems 15.45, 15.46, 15.47, and 15.48.

CONCEPT CHECK 15.2

Formic acid, $HCHO_2$, is a stronger acid than acetic acid, $HC_2H_3O_2$. Which is the stronger base, formate ion, CHO_2^-, or acetate ion, $C_2H_3O_2^-$?

15.5 Molecular Structure and Acid Strength

The strength of an acid depends on how easily the proton, H^+, is lost or removed from an H—X bond in the acid species. By understanding the factors that determine the ease of proton loss, you will be able to predict the relative strengths of similar acids.

Two factors are important in determining relative acid strengths. One is the polarity of the bond to which the H atom is attached. The H atom should have a positive partial charge:

$$\overset{\delta+}{H}—\overset{\delta-}{X}$$

The more polarized the bond is in this direction, the more easily the proton is removed and the greater the acid strength. Figure 15.5 dramatically illustrates this idea.

Figure 15.5 illustrates this idea. It shows *electrostatic-potential maps* of acetic acid, $HC_2H_3O_2$ or CH_3COOH, and sulfuric acid, H_2SO_4, whose molecular formulas and molecular models are

Acetic acid   Sulfuric acid

CH₃COOH   **H₂SO₄**

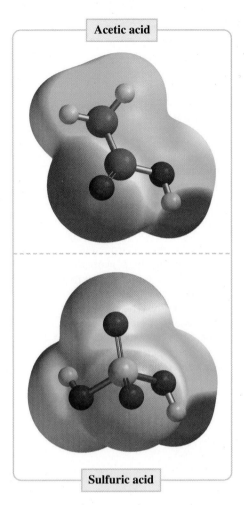

Acetic acid

Sulfuric acid

An electrostatic-potential map shows the electrostatic potential on a surface of fixed electron density. A partial negative potential or polarity ($\delta-$) is shown in red, and a partial positive polarity ($\delta+$) in blue, with intermediate values shown from orange to green. Acetic acid has a hydrogen atom attached to an oxygen atom, giving a polar bond, which Figure 15.5 shows in dark blue, as expected, and this hydrogen atom is acidic. Three hydrogen atoms are attached to a carbon atom (two pointing at the reader, the other below and not visible) and are more negative and are not acidic. The hydrogen atoms in sulfuric acid are dark blue and acidic, as expected.

The second factor determining acid strength is the strength of the bond—that is, how tightly the proton is held. This, in turn, depends on the size of atom X. The larger atom X, the weaker the bond and the greater the acid strength.

Consider a series of binary acids, HX, formed from a given column of elements of the periodic table. The acids would be compounds of these elements with hydrogen, such as the binary acids of the Group VIIA elements: HF, HCl, HBr, and HI. As you go down the column of elements, each time adding a shell of electrons to the atom, the radius increases markedly. For this reason, the size of atom X is the dominant factor in determining the acid strength. *In going down a column of elements of the periodic table, the size of atom X increases, the H—X bond strength decreases, and the strength of the binary acid increases.* You predict the following order of acid strength:

$$HF < HCl < HBr < HI$$

This is the same order shown in Table 15.2.

As you go across a row of elements of the periodic table, the atomic radius decreases slowly. For this reason, the relative strengths of the binary acids of these elements are less dependent on the size of atom X. Now the polarity of the H—X bond becomes the dominant factor in determining acid strength. *Going across a row of elements of the periodic table, the electronegativity increases, the H—X bond polarity increases, and the acid strength increases.* For example, the binary acids of the last two elements of the second period are H_2O and HF. The acid strengths are

$$H_2O < HF$$

Figure 15.5 ▲

Electrostatic-potential maps of acetic acid and sulfuric acid molecules Colors near red represent large negative values of the electrostatic potential, while colors near blue represent large positive values. (Orange, yellow, and green colors represent intermediate values.) Note the hydrogen atoms at the right are colored blue, indicating positive potential values, which are indicative of relative acidity of these atoms.

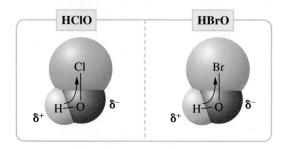

Figure 15.6 ▲

Effect of atom Y on the acid strengths of acids H—O—Y Electronegative atoms tend to pull electrons toward themselves. Thus, the O atom in the H—O bond of an oxoacid pulls the bonding electrons toward itself, giving the H atom a positive polarity (δ^+) compared with that of the O atom (δ^-). Other electronegative atoms attached to this O atom increase its effective electronegativity. (The arrow indicates the resulting movement of electrons in each molecule.) Thus, in a series of acids H—O—Y, the acidity increases with the electronegativity of Y. Here H—O—Cl is more acidic than H—O—Br.

The formulas of these acids may be written HXO or HOX, depending on the convention used. Formulas of oxoacids are generally written with the acidic H atoms first, followed by the characteristic central element (X), then O atoms. However, the formulas of molecules composed of three atoms are also often written in the order in which the atoms are bonded, which in this case is HOX.

This again is the order shown in Table 15.2. Hydrogen fluoride, HF, is a weak acid, and H_2O is a very weak acid.

Now consider the oxoacids. An oxoacid has the structure

$$H—O—Y—$$

The acidic H atom is always attached to an O atom, which, in turn, is attached to an atom Y. Other groups, such as O atoms or O—H groups, may also be attached to Y. Bond polarity appears to be the dominant factor determining relative strengths of the oxoacids. This, in turn, depends on the electronegativity of atom Y (see Figure 15.6). If the electronegativity of atom Y is large, the H—O bond is relatively polar and the acid strength large. *For a series of oxoacids of the same structure, differing only in the atom Y, the acid strength increases with the electronegativity of Y.* Consider, for example, the acids HClO, HBrO, and HIO. ◀ The structures are

$$H—\overset{..}{\underset{..}{O}}—\overset{..}{\underset{..}{Cl}}: \qquad H—\overset{..}{\underset{..}{O}}—\overset{..}{\underset{..}{Br}}: \qquad H—\overset{..}{\underset{..}{O}}—\overset{..}{\underset{..}{I}}:$$

The electronegativity of Group VIIA elements decreases going down the column of elements, so the electronegativity decreases from Cl to Br to I. The order of acid strengths is

$$HIO < HBrO < HClO$$

For a series of oxoacids, $(HO)_mYO_n$, the acid strength increases with n, the number of O atoms bonded to Y (excluding O atoms in OH groups). The oxoacids of chlorine provide an example:

$$H—\overset{..}{\underset{..}{O}}—\overset{..}{\underset{..}{Cl}}: \qquad H—\overset{..}{\underset{..}{O}}—\overset{..}{\underset{..}{Cl}}—\overset{..}{\underset{..}{O}}: \qquad H—\overset{..}{\underset{..}{O}}—\overset{\overset{\overset{..}{O}:}{|}}{\underset{..}{Cl}}—\overset{..}{\underset{..}{O}}: \qquad H—\overset{..}{\underset{..}{O}}—\overset{\overset{\overset{..}{O}:}{|}}{\underset{\underset{:\overset{..}{O}:}{|}}{Cl}}—\overset{..}{\underset{..}{O}}:$$

With each additional O atom, the Cl atom becomes effectively more electronegative (see Figure 15.7). As a result, the H atom becomes more acidic. The acid strengths increase in the following order:

$$HClO < HClO_2 < HClO_3 < HClO_4$$

Before we leave the subject of molecular structure and acid strength, let us look at the relative acid strengths of a polyprotic acid and its corresponding acid anions. For example, H_2SO_4 ionizes by losing a proton to give HSO_4^-, which, in turn, ionizes to give SO_4^{2-}. HSO_4^- can lose a proton, so it is acidic. However, because of the negative charge of the ion, which tends to attract protons, its acid strength is reduced from that of the uncharged species. That is, the acid strengths are in the order

$$HSO_4^- < H_2SO_4$$

We conclude that *the acid strength of a polyprotic acid and its anions decreases with increasing negative charge* (see Table 15.2).

Figure 15.7 ▶

Acid strengths of a series of oxoacids $(HO)_mYO_n$ Consider the acids where Y is Cl. The Cl atom becomes effectively more electronegative as more O atoms are attached to it. As a result, the O atom bonded to the H atom becomes effectively more electronegative and so attracts electrons more easily; the H atom then becomes more acidic. Thus, $HClO_4$ has the greatest acid strength of this series of oxoacids.

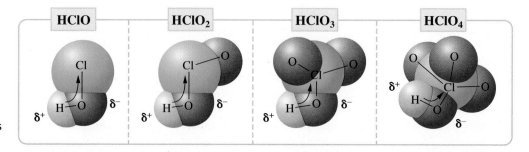

Exercise 15.4   Which member of each of the following pairs is the stronger acid?
a NH_3, PH_3; b HI, H_2Te; c HSO_3^-, H_2SO_3; d H_3AsO_4, H_3AsO_3; e HSO_4^-, $HSeO_4^-$

■ See Problems 15.51 and 15.52.

Autoionization of Water and pH

In aqueous solutions, two ions have dominant roles. These ions, the hydronium ion H_3O^+(or hydrogen ion, H^+) and the hydroxide ion OH^-, are available in any aqueous solution as a result of the *autoionization* of water, a reaction of water with itself, which we will describe in the next section. This will also give us some background to acid–base equilibrium calculations, which we will discuss in Chapter 16.

15.6 Autoionization of Water

Although pure water is often considered a nonelectrolyte (nonconductor of electricity), precise measurements do show a very small conduction. This conduction results from **autoionization** (or **self-ionization**), *a reaction in which two like molecules react to give ions.* In the case of water, a proton from one H_2O molecule is transferred to another H_2O molecule, leaving behind an OH^- ion and forming a hydronium ion, $H_3O^+(aq)$.

$$H_2O(l) + H_2O(l) \rightleftharpoons H_3O^+(aq) + OH^-(aq)$$

You can see the slight extent to which the autoionization of water occurs by noting the small value of its equilibrium constant, which we indicate here as K_w.

$$K_w = [H_3O^+][OH^-]$$

The concentration of H_2O is excluded here because its concentration remains essentially constant (as an essentially pure liquid, H_2O is excluded from the equilibrium constant; see Section 14.3). We call K_w, *the equilibrium value of the ion product* $[H_3O^+][OH^-]$, the **ion product constant for water**. At 25°C, the value of K_w is 1.0×10^{-14}. Like any equilibrium constant, K_w varies with temperature. At body temperature (37°C), K_w equals 2.5×10^{-14}.

$$K_w = [H_3O^+][OH^-] = 1.0 \times 10^{-14} \text{ at } 25°C$$

Because we often write $H^+(aq)$ for $H_3O^+(aq)$, the ion-product constant for water can be written $K_w = [H^+][OH^-]$.

Using K_w, you can calculate the concentrations of H_3O^+ and OH^- ions in pure water. These ions are produced in equal numbers in pure water, so their concentrations are equal. Let $x = [H_3O^+] = [OH^-]$. Then, substituting into the equation for the ion-product constant,

$$K_w = [H_3O^+][OH^-]$$

you get, at 25°C,

$$1.0 \times 10^{-14} = x^2$$

Hence, x equals 1.0×10^{-7}. Thus, the concentrations of H_3O^+ and OH^- are both 1.0×10^{-7} *M* in pure water at 25°C.

If you add an acid or a base to water, the concentrations of H_3O^+ and OH^- will no longer be equal. The equilibrium-constant equation $K_w = [H_3O^+][OH^-]$ will still hold, however, as we will discuss in the next section.

15.7 Solutions of a Strong Acid or Base

Consider an aqueous solution of a strong acid or base. Suppose you dissolve 0.10 mol HCl in 1.0 L of aqueous solution, giving 0.10 M HCl. You would like to know the concentration of H_3O^+ ion in this solution. In addition to the autoionzation of water, you have the reaction of HCl with water, which also produces H_3O^+ ion. A strong acid such as hydrochloric acid, HCl(aq), essentially reacts completely with water.

$$HCl(aq) + H_2O(l) \longrightarrow H_3O^+(aq) + Cl^-(aq)$$

Because you started with 0.10 mol HCl in 1.0 L of solution, the reaction will produce 0.10 mol H_3O^+, so the concentration of H_3O^+ ion from HCl is 0.10 M.

Now consider the concentration of H_3O^+ ion produced by the autoionization of water. In pure water, the concentration of H_3O^+ produced is 1.0×10^{-7} M; in an acid solution, the contribution of H_3O^+ from water will be even smaller. You can see this by applying Le Châtelier's principle to the autoionization reaction. When you increase the concentration of H_3O^+ in water by adding an acid, the autoionization of water reverses until a new equilibrium is obtained.

$$H_2O(l) + H_2O(l) \longleftarrow H_3O^+(aq) + OH^-(aq)$$

Consequently, the concentration of H_3O^+ produced by the autoionization of water ($< 1 \times 10^{-7}$ M) is negligible in comparison with that produced from HCl (0.10 M). So, 0.10 M HCl has a concentration of H_3O^+ ion equal to 0.10 M.

In a solution of a strong acid, you can normally ignore the autoionization of water as a source of H_3O^+. The H_3O^+ concentration is usually determined by the strong acid concentration. (This is not true when the acid solution is extremely dilute, however. In a solution that is 1.0×10^{-7} M HCl, the autoionization of water produces an amount of H_3O^+ comparable with that produced by HCl.)

Although you normally ignore the autoionization of water in calculating the H_3O^+ concentration in a solution of a strong acid, the autoionization equilibrium still exists and is responsible for a small concentration of OH^- ion. You can use the ion-product constant for water to calculate this concentration. As an example, calculate the concentration of OH^- ion in 0.10 M HCl. You substitute $[H_3O^+] = 0.10$ M into the equilibrium equation for K_w (for 25°C).

$$K_w = [H_3O^+][OH^-]$$
$$1.0 \times 10^{-14} = 0.10 \times [OH^-]$$

Solving for $[OH^-]$,

$$[OH^-] = \frac{1.0 \times 10^{-14}}{0.10} = 1.0 \times 10^{-13}$$

The OH^- concentration is 1.0×10^{-13} M.

Now consider a solution of a strong base, such as 0.010 M NaOH. What are the OH^- and H_3O^+ concentrations in this solution? Because NaOH is a strong base, all of the NaOH is present in the solution as ions. One mole of NaOH dissolves in water as one mole of Na^+ and one mole of OH^-. Therefore, the concentration of OH^- obtained from NaOH in 0.010 M NaOH solution is 0.010 M. The concentration of OH^- produced from the autoionization of water in this solution ($< 1 \times 10^{-7}$ M) is negligible and can be ignored. Therefore, the concentration of OH^- ion in the solution is 0.010 M. Hydronium ion, H_3O^+, is produced by the autoionization of water. To obtain its concentration, you substitute into the equilibrium equation for K_w (for 25°C).

$$K_w = [H_3O^+][OH^-]$$
$$1.0 \times 10^{-14} = [H_3O^+] \times 0.010$$

Solving for H_3O^+ concentration,

$$[H_3O^+] = \frac{1.0 \times 10^{-14}}{0.010} = 1.0 \times 10^{-12}$$

The H_3O^+ concentration is 1.0×10^{-12} *M*.

The following example further illustrates the calculation of H_3O^+ and OH^- concentrations in solutions of a strong acid or base.

☯WL Interactive Example `15.4` **Calculating Concentrations of H_3O^+ and OH^- in Solutions of a Strong Acid or Base**

Gaining Mastery Toolbox

Critical Concept 15.4
At equilibrium in water solutions, the ion product $[H_3O^+][OH^-] = 1.0 \times 10^{-14}$ at 25°C.

Solution Essentials:
• Equilibrium constant involving a pure liquid
• Chemical equilibrium

Calculate the concentrations of hydronium ion and hydroxide ion at 25°C in: a 0.15 *M* HNO_3, b 0.010 *M* $Ca(OH)_2$.

Problem Strategy A strong acid (or base) is fully ionized in solution, so you can use its formula and molar concentration to determine the H_3O^+ (or OH^-) ion concentration. (If not very dilute, any contribution to this ion concentration from the autoionization of water is negligible.) K_w relates the H_3O^+ and OH^- ion concentrations; therefore, you use this relation to obtain the OH^- concentration, given the H_3O^+ concentration, or vice versa.

Solution

a Every mole of HNO_3 contributes one mole of H_3O^+ ion, so the H_3O^+ concentration is **0.15 *M***. The OH^- concentration is obtained from the equation for K_w.

$$K_w = [H_3O^+][OH^-]$$
$$1.0 \times 10^{-14} = 0.15 \times [OH^-]$$

so

$$[OH^-] = \frac{1.0 \times 10^{-14}}{0.15} = 6.7 \times 10^{-14}$$

The concentration of OH^- in 0.15 *M* HNO_3 is **6.7 × 10^{-14} *M***.

b $Ca(OH)_2(s) \xrightarrow{H_2O} Ca^{2+}(aq) + 2OH^-(aq)$

Every mole of $Ca(OH)_2$ that dissolves yields two moles of OH^-. Therefore, 0.010 *M* $Ca(OH)_2$ contains 2 × 0.010 *M* $OH^- =$ **0.020 *M* OH^-**. The H_3O^+ concentration is obtained from

$$K_w = [H_3O^+][OH^-]$$
$$1.0 \times 10^{-14} = [H_3O^+] \times 0.020$$

so

$$[H^+] = \frac{1.0 \times 10^{-14}}{0.020} = 5.0 \times 10^{-13}$$

The hydronium-ion concentration is **5.0 × 10^{-13} *M***.

Answer Check The hydronium-ion concentration of a solution of an acid should be greater than that in pure water (1.0×10^{-7} *M*); the hydroxide-ion concentration of a solution of an acid should be less than that in pure water (1.0×10^{-7} *M*). The hydroxide-ion concentration of a solution of a base should be greater than that in pure water; the hydronium-ion concentration of a solution of a base should be less than that in pure water. Always check your calculations to see that this is true.

Exercise 15.5 A solution of barium hydroxide at 25°C is 0.125 *M* $Ba(OH)_2$. What are the concentrations of hydronium ion and hydroxide ion in the solution?

■ See Problems 15.53 and 15.54.

By dissolving substances in water, you can alter the concentrations of H_3O^+ and OH^- ions. In a *neutral* solution, the concentrations of H_3O^+ and OH^- remain equal, as they are in pure water. In an *acidic* solution, the concentration of H_3O^+ is greater than that of OH^-. In a *basic* solution, the concentration of OH^- is greater than that of H_3O^+. At 25°C, you observe the following conditions:

In an acidic solution, $[H_3O^+] > 1.0 \times 10^{-7}$ *M*.
In a neutral solution, $[H_3O^+] = 1.0 \times 10^{-7}$ *M*.
In a basic solution, $[H_3O^+] < 1.0 \times 10^{-7}$ *M*

Exercise 15.6   A solution has a hydroxide-ion concentration of 1.0×10^{-5} *M* at 25°C. Is the solution acidic, neutral, or basic?

■ See Problems 15.59 and 15.60.

CONCEPT CHECK 15.3

Rank the following solutions from most acidic to most basic (water molecules have been omitted for clarity).

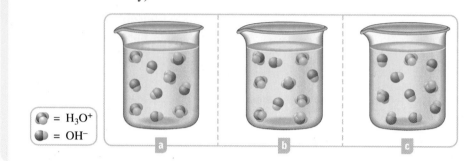

= H_3O^+

= OH^-

15.8 The pH of a Solution

You see that whether an aqueous solution is acidic, neutral, or basic depends on the hydronium-ion concentration. You can quantitatively describe the acidity by giving the hydronium-ion concentration. But because these concentration values may be very small, it is often more convenient to give the acidity in terms of **pH**, which is defined as *the negative of the base 10 logarithm of the molar hydronium-ion concentration:* ◄

The Danish biochemist S. P. L. Sørensen devised the pH scale while working on the brewing of beer.

$$pH = -\log [H_3O^+]$$

(You can also write $pH = -\log [H^+]$.) For a solution in which the hydronium-ion concentration is 1.0×10^{-3} *M*, the pH is

$$pH = -\log(1.0 \times 10^{-3}) = 3.00$$

See Appendix A for a discussion of logarithms.

Note that the number of places after the decimal point in the pH equals the number of significant figures reported in the hydronium-ion concentration. ◄

A neutral solution, whose hydronium-ion concentration at 25°C is 1.0×10^{-7} *M*, has a pH of 7.00. For an acidic solution, the hydronium-ion concentration is greater than 1.0×10^{-7} *M*, so the pH is less than 7.00. Similarly, a basic solution has a pH greater than 7.00. Figure 15.8 shows a diagram of the pH scale and lists pH values of some common solutions.

Figure 15.8 ▼

The pH scale Solutions having pH less than 7 are acidic; those having pH more than 7 are basic.

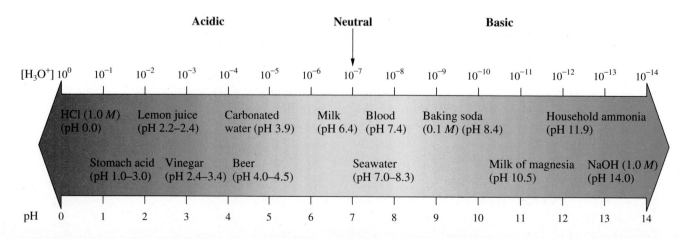

Example 15.5   Calculating the pH from the Hydronium-Ion Concentration

Critical Concept 15.5
The pH of a solution equals –log [H⁺].

Solution Essentials:
• $K_w = [H_3O^+][OH^-] = 1.0 \times 10^{-14}$ at 25°C
• Molar concentration

A sample of orange juice has a hydronium-ion concentration of 2.9×10^{-4} *M*. What is the pH? Is the solution acidic?

Problem Strategy pH is the negative base 10 logarithm of the molar hydronium-ion concentration. Determine the log using an electronic calculator; instructions vary, so refer to those for your calculator. On many calculators, you enter the H_3O^+ concentration, press *log*, and change the sign of the result. You should write the pH with the same number of decimal places as there are significant figures in $[H_3O^+]$.

Solution

$$pH = -\log[H_3O^+] = -\log(2.9 \times 10^{-4}) = \textbf{3.54}$$

The pH is less than 7.00, so the solution is **acidic** (as you expect for orange juice).

Answer Check Having obtained the pH, you should check the result by calculating 10^{-pH}, which should give you back the hydronium-ion concentration of the solution. Remember that a pH greater than 7 is basic and a pH less than 7 is acidic.

Exercise 15.7   What is the pH of a sample of gastric juice (digestive juice in the stomach) whose hydronium-ion concentration is 0.045 *M*?

■ See Problems 15.67 and 15.68.

You can find the pH of a solution of known hydroxide-ion concentration by first solving for the hydronium-ion concentration, as shown in Example 15.4 b. However, you can also find the pH simply from the pOH, a measure of hydroxide-ion concentration similar to the pH:

$$pOH = -\log[OH^-]$$

Then, because $K_w = [H_3O^+][OH^-] = 1.0 \times 10^{-14}$ at 25°C, you can show that ▶

$$pH + pOH = 14.00$$

Taking the logarithm of both sides of the equation
$[H_3O^+][OH^-] = 1.0 \times 10^{-14}$
you get
$\log[H_3O^+] + \log[OH^-] = -14.00$
or
$(-\log[H_3O^+]) + (-\log[OH^-]) = 14.00$
Hence, pH + pOH = 14.00.

You can rearrange this equation to obtain the pH from the value of pOH. For example, suppose you wish to find the pH of an ammonia solution whose hydroxide-ion concentration is 1.9×10^{-3} *M*. You first calculate the pOH.

$$pOH = -\log(1.9 \times 10^{-3}) = 2.72$$

Then the pH is

$$pH = 14.00 - pOH$$
$$= 14.00 - 2.72 = 11.28$$

Exercise 15.8   A saturated solution of calcium hydroxide has a hydroxide-ion concentration of 0.025 *M*. What is the pH of the solution?

■ See Problems 15.73 and 15.74.

The usefulness of the pH scale depends on the ease with which you can interconvert pH and hydronium-ion concentrations. In the next example, you will find the hydronium-ion concentration, given the pH value.

Example 15.6   Calculating the Hydronium-Ion Concentration from the pH

Critical Concept 15.6
You can obtain the hydronium-ion concentration from pH: $[H_3O^+] = 10^{-pH}$

Solution Essentials:
- $K_w = [H_3O^+][OH^-] = 1.0 \times 10^{-14}$ at 25°C
- Molar concentration

The pH of human arterial blood is 7.40. What is the hydronium-ion concentration?

Problem Strategy The hydronium-ion concentration is
$$[H_3O^+] = \text{antilog}(-pH) = 10^{-pH}$$
On some calculators, you obtain the antilogarithm by taking the *inverse* of the logarithm. On others, you use a 10^x key.

Solution The result is
$$[H_3O^+] = \text{antilog}(-7.40) = 10^{-7.40} = \mathbf{4.0 \times 10^{-8} \, M}$$
Because the pH was given to two decimal places, you write the hydronium-ion concentration with two significant figures.

Answer Check Check your result by using your answer (the hydronium-ion concentration) to obtain the pH ($= -\log [H_3O^+]$).

Exercise 15.9   A brand of carbonated beverage has a pH of 3.16. What is the hydronium-ion concentration of the beverage?

■ See Problems 15.75 and 15.76.

Exercise 15.10   A 0.010 *M* solution of ammonia, NH_3, has a pH of 10.6 at 25°C. What is the concentration of hydroxide ion? (One way to solve this problem is to find the pOH first and then calculate the hydroxide-ion concentration.)

■ See Problems 15.77 and 15.78.

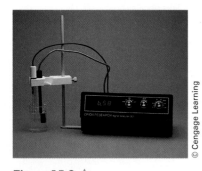

Figure 15.9 ▲

A digital pH meter The experimenter places a probe containing electrodes into the solution and reads the pH on the meter.

Litmus contains several substances and changes color over a broader range, from about pH 5 to pH 8.

The pH of a solution can be accurately measured by a pH meter (Figure 15.9). This instrument consists of specially designed electrodes that are dipped into the solution. A voltage, which depends on the pH, is generated between the electrodes and is read on a meter calibrated directly in pH. (The principle behind the operation of a pH meter is discussed in Chapter 19.)

Although less precise, acid–base indicators are often used to measure pH, because they usually change color within a small pH range. The color change of an indicator involves establishment of an equilibrium between an acid form and a base form that have different colors. If we denote the acid form as HIn, then the base form is In⁻ and the equilibrium is

$$\text{HIn}(aq) + H_2O(l) \rightleftharpoons H_3O^+(aq) + \text{In}^-(aq)$$

Consider the case of phenolphthalein. The acid form is colorless and the base form is pink. When a base is added to an acidic solution of phenolphthalein, OH⁻ ion from the base reacts with H_3O^+ in the solution. According to Le Châtelier's principle, the equilibrium is shifted to the right (to replenish H_3O^+). Thus, the colorless acid form of the indicator, HIn, is converted to the pink base form, In⁻. A solution of phenolphthalein begins to turn pink at about pH 8.0. By pH 9.7, the color change is essentially complete. ◄ Figure 15.10 shows the color changes for various acid–base indicators. (See also Figure 15.11.) Paper strips impregnated with several

© Cengage Learning

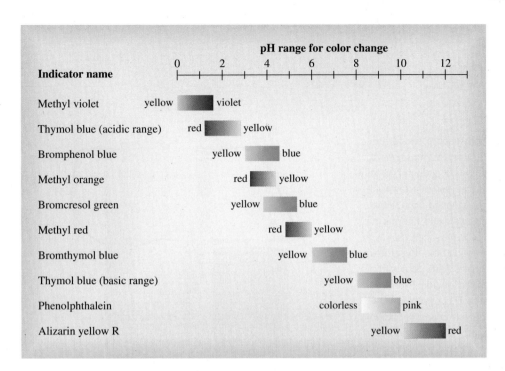

Figure 15.10 ◄

Color changes of some acid–base indicators Acid–base indicators are dyes whose acid form has one color and whose base form has another color. Thymol blue is a diprotic acid and has two different color changes (listed here in the acidic range and the basic range).

indicators are often used to measure pH values. Such "pH paper" gives a definite color for different pH ranges and can give the pH to the nearest integer value or better.

CONCEPT CHECK 15.4

You have solutions of NH_3, HCl, $NaOH$, and $HC_2H_3O_2$ (acetic acid), all with the same solute concentrations. Rank these solutions in order of pH, from the highest to the lowest.

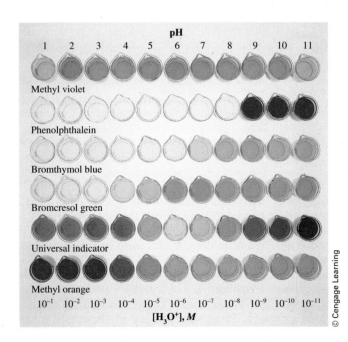

Figure 15.11 ◄

Some acid–base indicators in solutions of different H_3O^+ concentrations.

© Cengage Learning

Unclogging the Sink and Other Chores

When the sink drain is clogged, you might need a strong arm, but a strong base might work just as well. Drain cleaner solutions, available in grocery stores, are simply solutions of sodium hydroxide, a strong base. Such solutions work by chemically reacting with fat and with hair, the usual ingredients of a stopped drain. When fat reacts with a strong base such as sodium hydroxide, it forms the salt of a fatty acid, a product otherwise known as soap. Hair is a protein material, and in the presence of a strong base, the protein breaks up into the salts of its constituent amino acids. So a wad of greasy hair packed in a drain, in the presence of sodium hydroxide, becomes a soapy solution of amino acid salts. That solution washes easily down the drain. Oven cleaner is a paste made of sodium hydroxide. It cleans by reacting with the fatty deposits and other residues in the oven.

Sodium hydroxide is produced commercially by sending a direct current through an aqueous solution of sodium chloride, a process referred to as electrolysis (Figure 15.12). Hydrogen gas is released at the negative pole and chlorine gas is released at the positive pole. At the end of the electrolysis, the solution contains sodium hydroxide. The overall reaction is

$$2NaCl(aq) + 2H_2O(l) \xrightarrow{\text{electrolysis}} H_2(g) + Cl_2(g) + 2NaOH(aq)$$

Chlorine is prepared commercially in this same process.

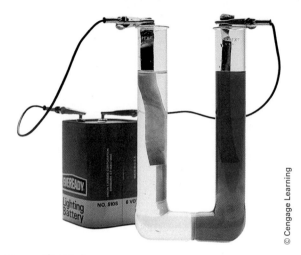

Figure 15.12 ▲

Preparation of sodium hydroxide by hydrolysis A solution of sodium chloride is electrolyzed. Note the pink color at the negative electrode. The phenolphthalein indicator turns pink from the hydroxide ion formed at the electrode. Later, if we evaporate the solution, sodium hydroxide crystallizes out.

■ See Problems 15.107 and 15.108.

A Checklist for Review

OWL and **Go Chemistry** Sign in at **www.cengage.com/owl** to:

• View tutorials and simulations, develop problem-solving skills, and complete online homework assigned by your professor.

• For quick review and exam prep, download Go Chemistry mini lecture modules from OWL or purchase them at **www.cengagebrain.com**.

Summary of Facts and Concepts

The *Arrhenius concept* was the first successful theory of acids and bases. Then, in 1923, Brønsted and Lowry characterized acid–base reactions as proton-transfer reactions. According to the *Brønsted–Lowry concept*, an acid is a proton donor and a base is a proton acceptor. The *Lewis concept* is even more general than the Brønsted–Lowry concept. A Lewis acid is an electron-pair acceptor and a Lewis base is an electron-pair donor. Reactions of acidic and basic oxides and the formation of complex ions, as well as proton-transfer reactions, can be described in terms of the Lewis concept.

Acid–base reactions can be viewed as a competition for protons. Using this idea, you can construct a table of *relative strengths of acids and bases*. Reaction favors the weaker acid and the weaker base. Acid strength depends on the polarity and the strength of the bond involving the acidic hydrogen atom. Therefore, you can *relate molecular structure and acid strength*. The text gives a number of rules that allow you to predict the relative strengths of similar acids based on molecular structure.

Water ionizes to give hydronium ion and hydroxide ion. The concentrations of these ions in aqueous solution are related by the *ion-product constant for water* (K_w). Thus, you can describe the acidity or basicity of a solution by the hydronium-ion concentration. You often use the *pH*, which is the negative base 10 logarithm of the hydronium-ion concentration, as a measure of acidity.

Learning Objectives	Important Terms
15.1 Arrhenius Concept of Acids and Bases ■ Define *acid* and *base* according to the Arrhenius concept.	
15.2 Brønsted–Lowry Concept of Acids and Bases ■ Define *acid* and *base* according to the Brønsted–Lowry concept. ■ Define the term *conjugate acid–base pair.* ■ Identify acid and base species. Example 15.1 ■ Define *amphiprotic species.*	**acid (Brønsted–Lowry)** **base (Brønsted–Lowry)** **conjugate acid–base pair** **amphiprotic species**
15.3 Lewis Concept of Acids and Bases ■ Define *Lewis acid* and *Lewis base.* ■ Identify Lewis acid and Lewis base species. Example 15.2	**Lewis acid** **Lewis base**
15.4 Relative Strengths of Acids and Bases ■ Understand the relationship between the strength of an acid and that of its conjugate base. ■ Decide whether reactants or products are favored in an acid–base reaction. Example 15.3	
15.5 Molecular Structure and Acid Strength ■ Note the two factors that determine relative acid strengths. ■ Understand the periodic trends in the strengths of the binary acids HX. ■ Understand the rules for determining the relative strengths of oxoacids. ■ Understand the relative acid strengths of a polyprotic acid and its anions.	
15.6 Autionization of Water ■ Define *autoionization* (or *self-ionization*). ■ Define the *ion-product constant for water.*	**autoionization (self-ionization)** **ion-product constant for water (K_w)**
15.7 Solutions of a Strong Acid or Base ■ Calculate the concentrations of H_3O^+ and OH^- in solutions of a strong acid or base. Example 15.4	
15.8 The pH of a Solution ■ Define *pH.* ■ Calculate the pH from the hydronium-ion concentration. Example 15.5 ■ Calculate the hydronium-ion concentration from the pH. Example 15.6 ■ Describe the determination of pH by a pH meter and by acid–base indicators.	**pH**

Key Equations

$K_w = [H_3O^+][OH^-]$

  $= 1.0 \times 10^{-14}$ at 25°C

$pH = -\log[H_3O^+]$

$pH + pOH = 14.00$

Questions and Problems

🐱WL Interactive versions of these problems may be assigned in OWL.

Self-Assessment and Review Questions

Key: These questions test your understanding of the ideas you worked with in the chapter. These problems vary in difficulty and often can be used for the basis of discussion.

15.1 Which of the following are strong acids? Which are weak acids? a HNO_3; b HNO_2; c HCN; d $HC_2H_3O_2$; e $HClO$; f HCl;

15.2 Describe any thermochemical (heat of reaction) evidence for the Arrhenius concept.

15.3 Define an acid and a base according to the Brønsted–Lowry concept. Give an acid–base equation and identify each species as an acid or a base.

15.4 What is meant by the conjugate acid of a base?

15.5 Write an equation in which $H_2PO_3^-$ acts as an acid and another in which it acts as a base.

15.6 Describe four ways in which the Brønsted–Lowry concept expands on the Arrhenius concept.

15.7 Define an acid and a base according to the Lewis concept. Give a chemical equation to illustrate.

15.8 Give two important factors that determine the strength of an acid. How does an increase in each factor affect the acid strength?

15.9 Explain why an acid–base reaction favors the weaker acid.

15.10 What is meant by the autoionization of water? Write the expression for K_w. What is its value at 25°C?

15.11 What is meant by the pH of a solution? Describe two ways of measuring pH.

15.12 Which is more acidic, a solution having a pH of 4 or one having a pH of 5?

15.13 What is the pH of a neutral solution at 37°C, where K_w equals 2.5×10^{-14}?

15.14 What is the sum of the pH and the pOH for a solution at 37°C, where K_w equals 2.5×10^{-14}?

15.15 The conjugate base of hydrofluoric acid dissolved in water is:

a F^- b OH^-
c H_3O^+ d HF
e F_2

15.16 In the following reaction, identify the Lewis acid.

$$Fe^{3+}(aq) + 6CN^-(aq) \rightleftharpoons Fe(CN)_6^{3-}(aq)$$

a CN^- b $Fe(CN)_6^{3-}$
c H^+ d Fe^{3+}
e OH^-

15.17 Calculate the hydronium-ion concentration at 25°C in a 1.3×10^{-2} M $Ba(OH)_2$ solution.

a 1.3×10^{-2} M b 7.7×10^{-13} M
c 2.6×10^{-2} M d 3.8×10^{-13} M
e 1.5×10^2 M

15.18 The pH of a solution is 9.55 at 25°C. What is the hydroxide-ion concentration in the solution?

a 2.8×10^{-10} M b 3.5×10^{-5} M
c 3.1×10^{-10} M d 3.2×10^{-5} M
e 3.8×10^{-5} M

Concept Explorations

Key: Concept explorations are comprehensive problems that provide a framework that will enable you to explore and learn many of the critical concepts and ideas in each chapter. If you master the concepts associated with these explorations, you will have a better understanding of many important chemistry ideas and will be more successful in solving all types of chemistry problems. These problems are well suited for group work and for use as in-class activities.

15.19 Acids

You make a solution by dissolving 0.0010 mol of HCl in enough water to make 1.0 L of solution.

a Write the chemical equation for the reaction of HCl(*aq*) and water.

b Without performing calculations, give a rough estimate of the pH of the HCl solution. Justify your answer.

c Calculate the H_3O^+ concentration and the pH of the solution.

d Is there any concentration of the base OH^- present in this solution of HCl(*aq*)? If so, where did it come from?

e If you increase the OH^- concentration of the solution by adding NaOH, does the H_3O^+ concentration change? If you think it does, explain why this change occurs and whether the H_3O^+ concentration increases or decreases.

f If you were to measure the pH of 10 drops of the original HCl solution, would you expect it to be different from the pH of the entire sample? Explain.

g Explain how two different volumes of your original HCl solution can have the same pH yet contain different moles of H_3O^+.

h If 1.0 L of pure water were added to the HCl solution, would this have any impact on the pH? Explain.

15.20 Strong Acids, Weak Acids, and pH

Two 0.10-mol samples of the hypothetical monoprotic acids HA(*aq*) and HB(*aq*) are used to prepare 1.0-L stock solutions of each acid.

a Write the chemical reactions for these acids in water. What are the concentrations of the two acid solutions?

b One of these acids is a strong acid, and one is weak. What could you measure that would tell you which acid was strong and which was weak?

c Say that the HA(*aq*) solution has a pH of 3.7. Is this the stronger of the two acids? How did you arrive at your answer?

d What is the concentration of $A^-(aq)$ in the HA solution described in part c?

e If HB(*aq*) is a strong acid, what is the hydronium-ion concentration?

f In the solution of HB(*aq*), which of the following would you expect to be in the greatest concentration: $H_3O^+(aq)$, $B^-(aq)$, HB(*aq*), or $OH^-(aq)$? How did you decide?

g In the solution of HA(*aq*), which of the following would you expect to be in the greatest concentration: $H_3O^+(aq)$, $A^-(aq)$, HA(*aq*), or $OH^-(aq)$? How did you decide?

h Say you add 1.0 L of pure water to a solution of HB. Would this water addition make the solution more acidic, make it less acidic, or not change the acidity of the original solution? Be sure to fully justify your answer.

i You prepare a 1.0-L solution of HA. You then take a 200-mL sample of this solution and place it into a separate container. Would this 200 mL sample be more acidic, be less acidic, or have the same acidity as the original 1.0-L solution of HA (*aq*)? Be sure to support your answer.

Conceptual Problems

Key: These problems are designed to check your understanding of the concepts associated with some of the main topics presented in each chapter. A strong conceptual understanding of chemistry is the foundation for both applying chemical knowledge and solving chemical problems. These problems vary in level of difficulty and often can be used as a basis for group discussion.

15.21 Aqueous solutions of ammonia, NH_3, were once thought to be solutions of an ionic compound (ammonium hydroxide, NH_4OH) in order to explain how the solutions could contain hydroxide ion. Using the Brønsted–Lowry concept, show how NH_3 yields hydroxide ion in aqueous solution without involving the species NH_4OH.

15.22 Self-contained environments, such as that of a space station, require that the carbon dioxide exhaled by people be continuously removed. This can be done by passing the air over solid alkali hydroxide, in which carbon dioxide reacts with hydroxide ion. What ion is produced by the addition of OH^- ion to CO_2? Use the Lewis concept to explain this.

15.23 Blood contains several substances that minimize changes in its acidity by reacting with either an acid or a base. One of these is the hydrogen phosphate ion, HPO_4^{2-}. Write one equation showing this species acting as a Brønsted–Lowry acid and another in which the species acts as a Brønsted–Lowry base.

15.24 The value of the ion-product constant for water, K_w, increases with temperature. What will be the effect of lowering the temperature on the pH of pure water?

15.25 Compare the structures of HNO_2 and H_2CO_3. Which would you expect to be the stronger acid? Explain your choice.

15.26 You make solutions of ammonia and sodium hydroxide by adding the same moles of each solute to equal volumes of water. Which solution would you expect to have the higher pH?

15.27 A strong monoprotic acid, with the molecular structure ⬤, is dissolved in a beaker of water. Which of the following pictures best represents the acid solution (water molecules have been omitted for clarity)?

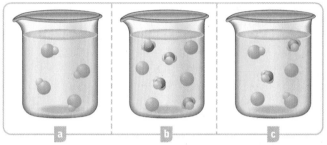

15.28 A weak acid, HA, is dissolved in water. Which one of the following beakers represents the resulting solution? (Water molecules have been omitted for clarity.)

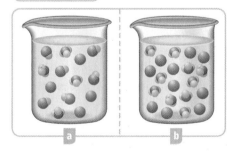

Practice Problems

Key: These problems are for practice in applying problem-solving skills. They are divided by topic, and some are keyed to exercises (see the ends of the exercises). The problems are arranged in matching pairs; the odd-numbered problem of each pair is listed first, and its answer is given in the back of the book.

Brønsted-Lowry Concept

15.29 Write the balanced reaction of hypobromous acid, HOBr, with water to form hydronium ion and the conjugate base of HOBr. Identify each species as either an acid or a base.

15.30 Write the balanced reaction of hydroxide ion with hydrofluoric acid, HF, to form fluoride ion and water. Identify each species as either an acid or a base.

15.31 Give the conjugate base to each of the following species regarded as acids.

a H_2S
b $H_2AsO_4^-$
c HPO_4^{2-}
d HNO_2

15.32 Give the conjugate base to each of the following species regarded as acids:

a HSO_4^- b PH_4^+

c HSe^- d $HOBr$

15.33 Give the conjugate acid to each of the following species regarded as bases.

a ClO^- b TeO_3^{2-}

c AsH_3 d $H_2PO_4^-$

15.34 Give the conjugate acid to each of the following species regarded as bases.

a HS^- b NH_2^-

c BrO_2^- d N_2H_4

15.35 For the following reactions, label each species as an acid or a base. Indicate the species that are conjugates of one another.

a $SO_3^{2-} + NH_4^+ \rightleftharpoons HSO_3^- + NH_3$

b $HSO_4^- + NH_3 \rightleftharpoons SO_4^{2-} + NH_4^+$

c $HPO_4^{2-} + NH_4^+ \rightleftharpoons H_2PO_4^- + NH_3$

d $Al(H_2O)_6^{3+} + H_2O \rightleftharpoons Al(H_2O)_5OH^{2+} + H_3O^+$

15.36 For the following reactions, label each species as an acid or a base. Indicate the species that are conjugates of one another.

a $H_2PO_4^- + HCO_3^- \rightleftharpoons HPO_4^{2-} + H_2CO_3$

b $F^- + HSO_4^- \rightleftharpoons HF + SO_4^{2-}$

c $HSO_4^- + H_2O \rightleftharpoons SO_4^{2-} + H_3O^+$

d $H_2S + CN^- \rightleftharpoons HS^- + HCN$

Lewis Acid–Base Concept

15.37 The following shows ball-and-stick models of the reactants in a Lewis acid–base reaction.

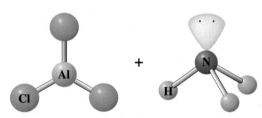

Write the complete equation for the reaction, including the product. Identify each reactant as a Lewis acid or a Lewis base.

15.38 The following shows ball-and-stick models of the reactants in a Lewis acid–base reaction.

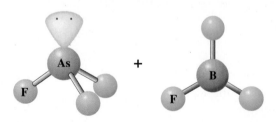

Write the complete equation for the reaction, including the product. Identify each reactant as a Lewis acid or a Lewis base.

15.39 Complete each of the following equations. Then write the Lewis formulas of the reactants and products and identify each reactant as a Lewis acid or a Lewis base.

a $I^- + I_2 \longrightarrow$

b $AlCl_3 + Cl^- \longrightarrow$

15.40 Complete each of the following equations. Then write the Lewis formulas of the reactants and products and identify each reactant as a Lewis acid or a Lewis base.

a $BBr_3 + Br^- \longrightarrow$

b $InCl_3 + Cl^- \longrightarrow$

15.41 In the following reactions, identify each reactant as a Lewis acid or a Lewis base.

a $Cu^{2+} + 6H_2O \longrightarrow Cu(H_2O)_6^{2+}$

b $BBr_3 + :AsH_3 \longrightarrow Br_3B:AsH_3$

15.42 In the following reactions, label each reactant as a Lewis acid or a Lewis base.

a $Ag^+ + 2CN^- \longrightarrow Ag(CN)_2^-$

b $BeF_2 + 2F^- \longrightarrow BeF_4^{2-}$

15.43 Natural gas frequently contains hydrogen sulfide, H_2S. H_2S is removed from natural gas by passing it through aqueous ethanolamine, $HOCH_2CH_2NH_2$ (an ammonia derivative), which reacts with the hydrogen sulfide. Write the equation for the reaction. Identify each reactant as either a Lewis acid or a Lewis base. Explain how you arrived at your answer.

15.44 Coal and other fossil fuels usually contain sulfur compounds that produce sulfur dioxide, SO_2, when burned. One possible way to remove the sulfur dioxide is to pass the combustion gases into a tower packed with calcium oxide, CaO. Write the equation for the reaction. Identify each reactant as either a Lewis acid or a Lewis base. Explain how you arrived at your answer.

Acid and Base Strengths

15.45 Complete the following equation. Using Table 15.2, predict whether you would expect the reaction to occur to any significant extent or whether the reaction is more likely to occur in the opposite direction.

$$HCN(aq) + SO_4^{2-}(aq) \longrightarrow$$

15.46 Complete the following equation. Using Table 15.2, predict whether you would expect the reaction to occur to any significant extent or whether the reaction is more likely to occur in the opposite direction.

$$HSO_4^-(aq) + ClO^-(aq) \longrightarrow$$

15.47 Use Table 15.2 to decide whether the species on the left or those on the right are favored by the reaction.

a $HCO_3^- + OH^- \rightleftharpoons CO_3^{2-} + H_2O$

b $NH_4^+ + H_2PO_4^- \rightleftharpoons NH_3 + H_3PO_4$

c $HCN + HS^- \rightleftharpoons CN^- + H_2S$

d $Al(H_2O)_6^{3+} + OH^- \rightleftharpoons Al(H_2O)_5OH^{2+} + H_2O$

15.48 Use Table 15.2 to decide whether the species on the left or those on the right are favored by the reaction.

a $CN^- + H_2CO_3 \rightleftharpoons HCN + HCO_3^-$

b $CN^- + H_2O \rightleftharpoons HCN + OH^-$

c $HCO_3^- + H_2S \rightleftharpoons H_2CO_3 + HS^-$

d $NH_4^+ + CO_3^{2-} \rightleftharpoons NH_3 + HCO_3^-$

15.49 In the following reaction of trichloroacetic acid, $HC_2Cl_3O_2$, with formate ion, CHO_2^-, the formation of trichloroacetate ion, $C_2Cl_3O_2^-$, and formic acid, $HCHO_2$ is favored.

$$HC_2Cl_3O_2 + CHO_2^- \longrightarrow C_2Cl_3O_2^- + HCHO_2$$

Which is the stronger acid, trichloroacetic acid or formic acid? Explain.

15.50 In the following reaction of tetrafluoroboric acid, HBF_4, with the acetate ion, $C_2H_3O_2^-$, the formation of tetrafluoroborate ion, BF_4^-, and acetic acid, $HC_2H_3O_2$ is favored.

$$HBF_4 + C_2H_3O_2^- \longrightarrow BF_4^- + HC_2H_3O_2$$

Which is the weaker base, BF_4^- or acetate ion?

15.51 For each of the following pairs, give the stronger acid. Explain your answer.
- a H_2S, H_2O
- b H_2S, HS^-
- c HIO_4, HIO_3
- d HBr, H_2Se
- e H_2SO_3, H_2SeO_3

15.52 Order each of the following pairs by acid strength, giving the weaker acid first. Explain your answer.
- a HNO_3, HNO_2
- b HCO_3^-, H_2CO_3
- c H_2S, H_2Te
- d HCl, H_2S
- e H_3PO_4, H_3AsO_4

Solutions of a Strong Acid or Base

15.53 What are the concentrations of H_3O^+ and OH^- in each of the following?
- a $1.55\ M$ NaOH
- b $0.15\ M$ $Sr(OH)_2$
- c $0.056\ M$ $HClO_4$
- d $0.47\ M$ HCl

15.54 What are the concentrations of H_3O^+ and OH^- in each of the following?
- a $1.2\ M$ HBr
- b $0.32\ M$ KOH
- c $0.085\ M$ $Ca(OH)_2$
- d $0.38\ M$ HNO_3

15.55 A solution of hydrochloric acid is $0.050\ M$ HCl. What is the hydronium-ion concentration at 25°C? What is the hydroxide-ion concentration at 25°C?

15.56 A solution is $0.030\ M$ HNO_3 (nitric acid). What is the hydronium-ion concentration at 25°C? What is the hydroxide-ion concentration at 25°C?

15.57 A saturated solution of magnesium hydroxide is $3.2 \times 10^{-4}\ M$ $Mg(OH)_2$. What are the hydronium-ion and hydroxide-ion concentrations in the solution at 25°C?

15.58 What are the hydronium-ion and the hydroxide-ion concentrations of a solution at 25°C that is $0.0085\ M$ barium hydroxide, $Ba(OH)_2$?

15.59 The following are solution concentrations. Indicate whether each solution is acidic, basic, or neutral.
- a $1 \times 10^{-7}\ M$ OH^-
- b $5 \times 10^{-6}\ M$ H_3O^+
- c $2 \times 10^{-9}\ M$ H_3O^+
- d $5 \times 10^{-9}\ M$ OH^-

15.60 The following are solution concentrations. Indicate whether each solution is acidic, basic, or neutral.
- a $2 \times 10^{-11}\ M$ OH^-
- b $2 \times 10^{-6}\ M$ H_3O^+
- c $6 \times 10^{-10}\ M$ OH^-
- d $6 \times 10^{-3}\ M$ H_3O^+

15.61 An antiseptic solution at 25°C has a hydroxide-ion concentration of $8.4 \times 10^{-5}\ M$. Is the solution acidic, neutral, or basic?

15.62 A shampoo solution at 25°C has a hydroxide-ion concentration of $1.5 \times 10^{-9}\ M$. Is the solution acidic, neutral, or basic?

Calculations Involving pH

15.63 Which of the following pH values indicate an acidic solution at 25°C? Which are basic and which are neutral?
- a 10.5
- b 1.6
- c 7.0
- d 4.6

15.64 Which of the following pH values indicate an acidic solution at 25°C? Which are basic and which are neutral?
- a 1.6
- b 12.9
- c 8.9
- d 5.1

15.65 For each of the following, state whether the solution at 25°C is acidic, neutral, or basic: a A $0.1\ M$ solution of trisodium phosphate, Na_3PO_4, has a pH of 12.0. b A $0.1\ M$ solution of calcium chloride, $CaCl_2$, has a pH of 7.0. c A $0.2\ M$ solution of copper(II) sulfate, $CuSO_4$, has a pH of 4.0. d A sample of rainwater has a pH of 5.7.

15.66 For each of the following, state whether the solution at 25°C is acidic, neutral, or basic: a A beverage solution has a pH of 3.5. b A $0.50\ M$ solution of potassium bromide, KBr, has a pH of 7.0. c A $0.050\ M$ solution of pyridine, C_5H_5N, has a pH of 9.0. d A solution of iron(III) chloride has a pH of 5.5.

15.67 Obtain the pH corresponding to the following hydronium-ion concentrations.
- a $1.0 \times 10^{-4}\ M$
- b $3.2 \times 10^{-10}\ M$
- c $2.3 \times 10^{-5}\ M$
- d $2.91 \times 10^{-11}\ M$

15.68 Obtain the pH corresponding to the following hydronium-ion concentrations.
- a $1.0 \times 10^{-8}\ M$
- b $5.0 \times 10^{-12}\ M$
- c $7.5 \times 10^{-3}\ M$
- d $6.35 \times 10^{-9}\ M$

15.69 A sample of vinegar has a hydronium-ion concentration of $7.5 \times 10^{-3}\ M$. What is the pH of the vinegar?

15.70 Some lemon juice has a hydronium-ion concentration of $2.5 \times 10^{-3}\ M$. What is the pH of the lemon juice?

15.71 Obtain the pH corresponding to the following hydroxide-ion concentrations.
- a $5.25 \times 10^{-9}\ M$
- b $8.3 \times 10^{-3}\ M$
- c $3.6 \times 10^{-12}\ M$
- d $2.1 \times 10^{-8}\ M$

15.72 Obtain the pH corresponding to the following hydroxide-ion concentrations.
- a $4.83 \times 10^{-11}\ M$
- b $3.2 \times 10^{-5}\ M$
- c $2.7 \times 10^{-10}\ M$
- d $5.0 \times 10^{-4}\ M$

15.73 A solution of lye (sodium hydroxide, NaOH) has a hydroxide-ion concentration of $0.050\ M$. What is the pH at 25°C?

15.74 A solution of washing soda (sodium carbonate, Na_2CO_3) has a hydroxide-ion concentration of $0.0040\ M$. What is the pH at 25°C?

15.75 The pH of a cup of coffee (at 25°C) was found to be 5.12. What is the hydronium-ion concentration?

15.76 A wine was tested for acidity, and its pH was found to be 3.85 at 25°C. What is the hydronium-ion concentration?

15.77 A detergent solution has a pH of 11.63 at 25°C. What is the hydroxide-ion concentration?

15.78 Morphine is a narcotic that is used to relieve pain. A solution of morphine has a pH of 9.61 at 25°C. What is the hydroxide-ion concentration?

15.79 A 1.00-L aqueous solution contained 5.80 g of sodium hydroxide, NaOH. What was the pH of the solution at 25°C?

15.80 A 1.00-L aqueous solution contained 6.78 g of barium hydroxide, $Ba(OH)_2$. What was the pH of the solution at 25°C?

15.81 A certain sample of rainwater gives a yellow color with methyl red and a yellow color with bromthymol blue. What is the approximate pH of the water? Is the rainwater acidic, neutral, or basic? (See Figure 15.10.)

15.82 A drop of thymol blue gave a yellow color with a solution of aspirin. A sample of the same aspirin solution gave a yellow color with bromphenol blue. What was the pH of the solution? Was the solution acidic, neutral, or basic? (See Figure 15.10.)

General Problems

Key: These problems provide more practice but are not divided by topic or keyed to exercises. Each section ends with essay questions, each of which is color coded to refer to the *A Chemist Looks at Life Science* (pink) and *A Chemist Looks at Daily Life* (orange) chapter essay on which it is based. Odd-numbered problems and the even-numbered problems that follow are similar; answers to all odd-numbered problems except the essay questions are given in the back of the book.

15.83 Which of the following substances are acids in terms of the Arrhenius concept? Which are bases? Show the acid or base character by using chemical equations.
a P_4O_{10} b Na_2O c N_2H_4 d H_2Te

15.84 Identify each of the following as an acid or a base in terms of the Arrhenius concept. Give the chemical equation for the reaction of the substance with water, showing the origin of the acidity or basicity.
a BaO b H_2S c CH_3NH_2 d SO_2

15.85 Write a reaction for each of the following in which the species acts as a Brønsted acid. The equilibrium should favor the product side.
a H_2O_2 b HCO_3^- c NH_4^+ d $H_2PO_4^-$

15.86 Write a reaction for each of the following in which the species acts as a Brønsted base. The equilibrium should favor the product side.
a H_2O b $H_2PO_4^-$ c NH_3 d HCO_3^-

15.87 For each of the following, write the complete chemical equation for the acid–base reaction that occurs. Describe each using Brønsted language (if appropriate) and then using Lewis language (show electron-dot formulas).
a The ClO^- ion reacts with water.
b The reaction of NH_4^+ and NH_2^- in liquid ammonia to produce NH_3.

15.88 For each of the following, write the complete chemical equation for the acid–base reaction that occurs. Describe each using Brønsted language (if appropriate) and then using Lewis language (show electron-dot formulas).
a The HS^- ion reacts in water to produce H_2S.
b Cyanide ion, CN^-, reacts with Fe^{3+}.

15.89 Complete the following reaction. Decide on the basis of relative acid strengths whether the reaction is more likely to go in the direction written or in the opposite direction.

$$H_2S + CN^- \rightleftharpoons$$

15.90 Complete the following proton-transfer reaction. Decide on the basis of relative acid strengths whether the reaction is more likely to go in the direction written or in the opposite direction.

$$HNO_2 + F^- \rightleftharpoons$$

15.91 List the following compounds in order of increasing acid strength: HBr, H_2Se, H_2S.

15.92 List the following compounds in order of increasing acid strength: $HBrO_2$, $HBrO$, $HClO_2$.

15.93 A solution is 0.25 M KOH. What are the concentrations of H_3O^+ and OH^- in this solution?

15.94 A solution is 0.25 M Sr(OH)$_2$. What are the concentrations of H_3O^+ and OH^- in this solution?

15.95 A wine has a hydronium-ion concentration equal to 1.5×10^{-3} M. What is the pH of this wine?

15.96 A sample of lemon juice has a hydronium-ion concentration equal to 2.6×10^{-2} M. What is the pH of this sample?

15.97 A sample of grape juice has a pH of 4.15. What is the hydroxide-ion concentration of this solution?

15.98 A sample of apple cider has a pH of 3.15. What is the hydroxide-ion concentration of this solution?

15.99 A 2.500-g sample of a mixture of sodium hydrogen carbonate and potassium chloride is dissolved in 25.00 mL of 0.437 M H$_2$SO$_4$. Some acid remains after treatment of the sample.
a Write both the net ionic and the molecular equations for the complete reaction of sodium hydrogen carbonate with sulfuric acid.
b If 35.4-mL of 0.108 M NaOH were required to titrate the excess sulfuric acid, how many moles of sodium hydrogen carbonate were present in the original sample?
c What is the percent composition of the original sample?

15.100 A 2.500-g sample of a mixture of sodium carbonate and sodium chloride is dissolved in 25.00 mL of 0.798 M HCl. Some acid remains after the treatment of the sample.
a Write the net ionic equation for the complete reaction of sodium carbonate with hydrochloric acid.
b If 28.7 mL of 0.108 M NaOH were required to titrate the excess hydrochloric acid, how many moles of sodium carbonate were present in the original sample?
c What is the percent composition of the original sample?

15.101 The dihydrogen phosphate ion has the ability to act as an acid in the presence of a base and as a base in the presence of an acid. What is this property called? Illustrate this behavior with water by writing Brønsted–Lowry acid–base reactions. Also illustrate this property by selecting a common acid and base to react with the dihydrogen phosphate ion.

15.102 The bicarbonate ion has the ability to act as an acid in the presence of a base or as a base in the presence of an acid, so it is said to be amphiprotic. Illustrate this behavior with water by writing Brønsted–Lowry acid–base reactions. Also illustrate this property by selecting a common strong acid and base to react with the bicarbonate ion.

15.103 The hydride ion does not exist in water because it has a greater attraction for the hydronium ion than the hydroxide ion does. Write the equation for the reaction that occurs when calcium hydride is added to water. Which is the stronger base, the hydride ion or the hydroxide ion? What is the meaning of the statement that the strongest base that can exist in water is the hydroxide ion?

15.104 The nitride ion and the amide ion, NH_2^-, have greater attractions for the hydronium ion than the hydroxide ion does. Write the equations for the reactions that occur when calcium nitride and sodium amide are added to water (each gives NH_3). Which is the stronger base, the

nitride ion or the amide ion? Why? What is the meaning of the statement that the hydroxide ion is the strongest base that can exist in water?

■ **15.105** Why do medicines frequently taste bitter?

■ **15.106** Show how methylamine, CH_3NH_2, is a Bronsted–Lowry base. Is methylamine also a Lewis base? Explain.

■ **15.107** How is sodium hydroxide prepared commercially?

■ **15.108** Explain how sodium hydroxide dissolves grease and hair from a stopped drain.

Strategy Problems

Key: As noted earlier, all of the practice and general problems are matched pairs. This section is a selection of problems that are not in matched-pair format. These challenging problems require that you employ many of the concepts and strategies that were developed in the chapter. In some cases, you will have to integrate several concepts and operational skills in order to solve the problem successfully.

15.109 ⓐ Consider the hydrated aluminum ion $Al(H_2O)_6^{3+}$ as a Brønsted–Lowry acid. Write the chemical equation in which this ion loses a proton in a reaction with ammonia, NH_3. Identify the conjugate acids and bases in this reaction. ⓑ Ethanethiol, CH_3CH_2SH, is a malodorous compound present in petroleum. It is removed from petroleum by reaction with sulfuric acid. Write a chemical equation for this reaction, identifying acid and base conjugates.

15.110 Saccharin, $HC_7H_4NO_3S$, is a well-known sweetener. It is a weak acid, and as the sweetener, it is generally supplied as the sodium salt. Here is the structural formula for the saccharin molecule. From the structural formula, decide which H atom is acidic. What is the basis for your decision.

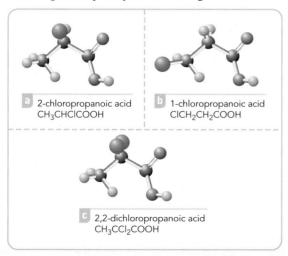

15.111 Consider the acids listed and decide on the basis of their structures how you would order them, from weakest to strongest. Explain your reasoning.

ⓐ 2-chloropropanoic acid
$CH_3CHClCOOH$

ⓑ 1-chloropropanoic acid
$ClCH_2CH_2COOH$

ⓒ 2,2-dichloropropanoic acid
CH_3CCl_2COOH

15.112 Consider the substances arsine, AsH_3, and phosphine, PH_3, that are analogous to ammonia, NH_3. Write chemical equations illustrating the basic character of arsine and phosphine. Write formulas for the ions formed by NH_3, AsH_3, and PH_3 with H^+. Order these ions by their acidic character. Explain how you arrived at this order. Now compare the basicities of the substances AsH_3, PH_3, and NH_3, ordering them by their basic character, from least to most basic. Explain how you arrived at this order.

15.113 Hydrazine, N_2H_4 (having the structure H_2NNH_2), and its derivatives have been used as rocket fuels. Draw the Lewis electron-dot formula for the hydrazine molecule. Describe the geometries expected about the nitrogen atoms in this molecule. Why would you expect hydrazine to be basic? Which substance, NH_3 or N_2H_4, would you expect to be more basic? Why? Write the chemical equation in which hydrazine reacts with hydrochloric acids to form the salt N_2H_5Cl. Consider the positive ion of this salt. How does its basic character compare with that of NH_3 and N_2H_4? Explain.

15.114 Liquid ammonia undergoes autoionization similar to that of water:

$$NH_3(l) + NH_3(l) \rightleftharpoons NH_4^+(l) + NH_2^-(l)$$

How would you define an acid and a base similar to the way these terms are defined by the Brønsted-Lowry concept for aqueous solutions? Write the expression for the ion-product constant, K_{am}, for this autoionization. The value of K_{am} is 5.1×10^{-27}. What is the concentration of NH_4^+ in a neutral solution of liquid ammonia? Suppose you dissolve NH_4I in liquid ammonia to give NH_4^+ and I^- ions. Similarly, you dissolve KNH_2 (potassium amide) in liquid ammonia to give a solution of K^+ and NH_2^- ions. Which of these two solutions is acidic and which basic according to your definitions? Now, suppose you add the two solutions together. Write an equation for the neutralization reaction.

15.115 Ethanol (ethyl alcohol), CH_3CH_2OH, can act as a Brønsted–Lowry acid. Write the chemical equation for the reaction of ethanol as an acid with hydroxide ion, OH^-. Ethanol can also react as a Brønsted–Lowry base. Write the chemical equation for the reaction of ethanol as a base with hydronium ion, H_3O^+. Explain how you arrived at these chemical equations. Both of these reactions can also be considered Lewis acid–base reactions. Explain this.

15.116 Ethanol, CH_3CH_2OH, can undergo auto-ionization. Write the chemical equation for this auto-ionization. Explain how you arrived at this equation. At room temperature, the ion product for this self-ionization is 1.0×10^{-20}. What is the concentration of each ion at this temperature? Show how you arrived at these concentrations.

15.117 Aluminum chloride, $AlCl_3$, reacts with trimethyl-amine, $N(CH_3)_3$. What would you guess to be the product of this reaction? Explain why you think so. Describe the reaction in terms of one of the acid–base concepts. Write an appropriate equation to go with this description. Which substance is the acid according to this acid–base concept? Explain.

15.118 Consider each of the following pairs of compounds, and indicate which one of each pair is the stronger Lewis acid. Explain how you arrived at your answers.

  a BF_3, BCl_3 b Fe^{2+}, Fe^{3+}

15.119 Arrange each of the following in order of increasing acidity. You may need to use a couple of rules to decide the order for a given series. Explain the reasoning you use in each case.

  a $HBrO_2$, $HBrO_3$, $HBrO$
  b H_2TeO_3, H_2SO_3, H_2SeO_3
  c HI, SbH_3, H_2Te
  d H_2S, HBr, H_2Se
  e $HClO_2$, $HClO_3$, $HBrO_2$

15.120 The following is a table of some of the oxoacids of the halogens. Which acid in the table is the strongest? Which is the weakest? Explain how you arrived at your answers.

Chlorine	Bromine	Iodine
HOCl	HOBr	HOI
HOClO	HOBrO	HOIO
HOClO_2	HOBrO_2	HOIO_2

15.121 How many grams of water are in dissociated form (as H^+ and OH^- ions) in 1.00 L of pure water? How many hydrogen ions, $H^+(aq)$, are there in 1.00 L of pure water?

15.122 Suppose you have 557 mL of 0.0300 M HCl, and you want to make up a solution of HCl that has a pH of 1.831. What is the maximum volume (in liters) that you can make of this solution?

15.123 You want to make up 3.00 L of aqueous hydrochloric acid, $HCl(aq)$, that has a pH of 2.00. How many grams of

concentrated hydrochloric acid will you need? Concentrated hydrochloric acid contains 37.2 mass percent of HCl.

15.124 You can obtain the pH of a 0.100 M HCl solution by assuming that all of the H_3O^+ ion comes from the HCl, in which case the pH equals $-\log 0.100 = 1.00$. But if you want the pH of a solution that is 1.00×10^{-7} M HCl, you also need to account for H_3O^+ ion coming from water. (Why?) Note that the auto-ionization of water is the only equilibrium you need to account for. What is the pH of 1.00×10^{-7} M HCl?

15.125 Pure liquid hydrogen fluoride ionizes in a way similar to that of water.

  a Write the equilibrium reaction for the auto-ionization of liquid hydrogen fluoride.
  b Will sodium fluoride be an acid or a base in liquid hydrogen fluoride? Why?
  c Perchloric acid is a strong acid in liquid hydrogen fluoride. Write the chemical equation for the ionization reaction. What is the conjugate acid in this medium?

15.126 Pure liquid ammonia ionizes in a way similar to that of water.

  a Write the equilibrium reaction of liquid ammonia.
  b Will sodium amide, $NaNH_2$, be an acid or a base in liquid ammonia? Why?
  c Ammonium chloride can be used to titrate sodium amide in liquid ammonia. Write the chemical equation for the titration of sodium amide with ammonium chloride in liquid ammonia.

15.127 A solution contains 0.675 g of ethylamine, $C_2H_5NH_2$, per 100.0 mL of solution. Electrical conductivity measurements at 20°C show that 0.98% of the ethylamine has reacted with water. Write the equation for this reaction. Calculate the pH of the solution.

15.128 A solution contains 4.25 g of ammonia per 250.0 mL of solution. Electrical conductivity measurements at 25°C show that 0.42% of the ammonia has reacted with water. Write the equation for this reaction and calculate the pH of the solution.

Cumulative-Skills Problems

Key: The problems under this heading combine skills introduced in previous chapters with those given in the current one.

15.129 Phosphorous acid, H_3PO_3, and phosphoric acid, H_3PO_4, have approximately the same acid strengths. From this information, and noting the possibility that one or more hydrogen atoms may be directly bonded to the phosphorus atom, draw the structural formula of phosphorous acid. How many grams of sodium hydroxide would be required to completely neutralize 1.00 g of this acid?

15.130 Hypophosphorous acid, H_3PO_2, and phosphoric acid, H_3PO_4, have approximately the same acid strengths. From this information, and noting the possibility that one or more hydrogen atoms may be directly bonded to the phosphorus atom, draw the structural formula of hypophosphorous acid. How many grams of sodium hydroxide would be required to completely neutralize 1.00 g of this acid?

15.131 Boron trifluoride, BF_3, and ammonia, NH_3, react to produce $BF_3 : NH_3$. A coordinate covalent bond is formed between the boron atom on BF_3 and the nitrogen atom on NH_3. Write the equation for this reaction, using Lewis electron-dot formulas. Label the Lewis acid and the Lewis base. Determine how many grams of $BF_3 : NH_3$ are formed when 10.0 g BF_3 and 10.0 g NH_3 are placed in a reaction vessel, assuming that the reaction goes to completion.

15.132 Boron trifluoride, BF_3, and diethyl ether, $(C_2H_5)_2O$, react to produce a compound with the formula $BF_3 : (C_2H_5)_2O$. A coordinate covalent bond is formed between the boron atom on BF_3 and the oxygen atom on $(C_2H_5)_2O$. Write the equation for this reaction, using Lewis electron-dot formulas. Label the Lewis acid and the Lewis base. Determine how many grams of $BF_3 : (C_2H_5)_2O$ are formed when 10.0 g BF_3 and 20.0 g $(C_2H_5)_2O$ are placed in a reaction vessel, assuming that the reaction goes to completion.

Acid–Base Equilibria

Limestone, which is mainly $CaCO_3$, is slightly soluble in acidic solution, so these limestone gargoyles are easily damaged by acid rain. The basic property of limestone makes the powdered form useful as a soil conditioner, to adjust soil pH. Sulfuric acid is a component of acid rain; a major commercial use of this acid is to produce fertilizers, such as ammonium phosphate, from phosphate rock.

OWL Sign in to OWL at **www.cengage.com/owl** to view tutorials and simulations, develop problem-solving skills, and complete online homework assigned by your professor.

goChemistry Download mini lecture videos for key concept review and exam prep from OWL or purchase them from **www.cengagebrain.com**.

Many well-known substances are weak acids or bases. The following are weak acids: aspirin (acetylsalicylic acid, a headache remedy), phenobarbital (a sedative), saccharin (a sweetener), and niacin (nicotinic acid, a B vitamin). That these are *weak* acids means that their reactions with water do not go to completion. To discuss such acid–base reactions, you need to look at the equilibria involved and be able to calculate the concentrations of species in a reaction mixture.

Consider, for example, how you could answer the following questions: What is the hydronium-ion concentration of 0.10 *M* niacin (nicotinic acid)? What is the hydronium-ion concentration of the solution obtained by dissolving one 5.00-grain tablet of aspirin (acetylsalicylic acid) in 0.500 L of water? If these were solutions of strong acids, the calculations would be simple; 0.10 *M* monoprotic acid would yield 0.10 *M* H_3O^+ ion. However, because niacin is a weak monoprotic acid, the H_3O^+ concentration is less than 0.10 *M*. To find the concentration, you need the equilibrium constant for the reaction involved, and you need to solve an equilibrium problem.

A similar process is involved in finding the hydronium-ion concentration of 0.10 *M* sodium nicotinate (sodium salt of niacin). In this case, you need to look at the acid–base equilibrium of the nicotinate ion. Let us see how you answer questions of this sort.

Solutions of a Weak Acid or Base

The simplest acid–base equilibria are those in which a single acid or base solute reacts with water. We will look first at solutions of weak acids, next at solutions of weak bases. Then we will consider solutions of salts, which can have acidic or basic properties as a result of reactions of their ions with water.

16.1 Acid-Ionization Equilibria

An acid reacts with water to produce hydronium ion (hydrogen ion) and the conjugate base ion. The process is called *acid ionization* or *acid dissociation*. Consider an aqueous solution of the weak acid acetic acid, $HC_2H_3O_2$, the sour constituent of vinegar (Figure 16.1). When the acid is added to water, it reacts according to the equation

$$HC_2H_3O_2(aq) + H_2O(l) \rightleftharpoons H_3O^+(aq) + C_2H_3O_2^-(aq)$$

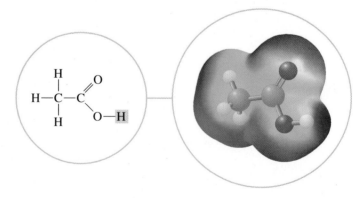

Figure 16.1 ▲

The structure of acetic acid *Left:* Structural formula. Only the hydrogen atom (in color) attached to the oxygen atom is acidic. The hydrogen atoms attached to carbon are not acidic. The acidic group

(often written —COOH and called the carboxylic acid group) occurs in many organic acids. *Right:* Electrostatic potential map of acetic acid.

Because acetic acid is a weak electrolyte, the acid *ionizes* to a small extent in water (about 5% or less, depending on the concentration of acid).

For a strong acid, which ionizes completely in solution, the concentrations of ions are determined by the stoichiometry of the reaction from the initial concentration of acid. However, for a weak acid such as acetic acid, the concentrations of ions in solution are determined from the **acid-ionization constant** (also called the acid-dissociation constant), which is *the equilibrium constant for the ionization of a weak acid.*

To find the acid-ionization constant, write HA for the general formula of a weak, monoprotic acid. The acid-ionization equilibrium in aqueous solution is

$$HA(aq) + H_2O(l) \rightleftharpoons H_3O^+(aq) + A^-(aq)$$

The corresponding equilibrium constant is

$$K_c = \frac{[H_3O^+][A^-]}{[HA][H_2O]}$$

Assuming that this is a dilute solution and that the reaction occurs to only a small extent, the concentration of water will be nearly constant. Rearranging this equation gives

$$K_a = [H_2O]K_c = \frac{[H_3O^+][A^-]}{[HA]}$$

Thus, K_a, the acid-ionization constant, equals the constant $[H_2O]K_c$. ▶

$$K_a = \frac{[H_3O^+][A^-]}{[HA]}$$

Experimental Determination of K_a

The ionization constant for a weak acid is usually determined experimentally by one of two methods. In one method, the electrical conductivity or some colligative property of a solution of the acid is measured to obtain its degree of ionization. The **degree of ionization** of a weak electrolyte is *the fraction of molecules that react with water to give ions.* This may also be expressed as a percentage, giving the *percent ionization.* In the other method, the pH of a solution of the weak acid is determined. From the pH one finds the concentration of H_3O^+ ion and then the concentrations of other ions. The following example shows how to calculate K_a from the pH of a solution of nicotinic acid (Figure 16.2). It also shows how to calculate the degree of ionization and the percent ionization.

In terms of the thermodynamic equilibrium constant, the activity of H_2O is nearly constant and essentially 1, so it does not appear explicitly in the equilibrium constant.

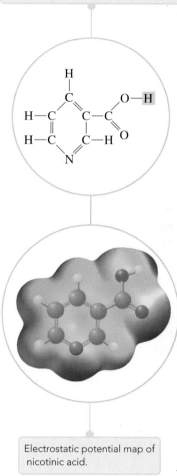

The structural formula of nicotinic acid (niacin); note the acidic hydrogen atom (in color). Nicotinic acid was first prepared by oxidizing nicotine, a basic substance present in tobacco.

Electrostatic potential map of nicotinic acid.

Figure 16.2 ▲

The structure of nicotinic acid

Example 16.1   Determining K_a from the Solution pH

Gaining Mastery Toolbox

Critical Concept 16.1
Weak acids are partially ionized in solution. Therefore, calculating the pH or [H_3O^+] of a weak acid solution requires performing an equilibrium calculation. Equilibrium calculations were introduced in Chapter 14.

Solution Essentials:
- Acid ionization constant (K_a) • Acid
- Degree of ionization   • Equilibrium-constant expression
- pH   • Concentration (molarity)
- Weak acid

Nicotinic acid (niacin) is a monoprotic acid with the formula $HC_6H_4NO_2$. A solution that is 0.012 M in nicotinic acid has a pH of 3.39 at 25°C. What is the acid-ionization constant, K_a, for this acid at 25°C? What is the degree of ionization of nicotinic acid in this solution?

Problem Strategy It is important to realize that when we say the solution is 0.012 M, this refers to *how the solution is prepared.* The solution is made up by adding 0.012 mol of substance to enough water to give a liter

(continued)

(continued)

of solution. Once the solution is prepared, some molecules ionize, so the actual concentration is somewhat less than 0.012 M. To solve for K_a, you follow the three steps outlined in Chapter 14 for solving equilibrium problems.

Solution

Step 1: Abbreviate the formula for nicotinic acid as HNic. Then 1 L of solution contains 0.012 mol HNic and 0 mol Nic⁻, the acid anion, before ionization. The H_3O^+ concentration at the start is that from the self-ionization of water. It is usually much smaller than that obtained from the acid (unless the solution is extremely dilute or K_a is quite small), so you can write $[H_3O^+] = \sim0$ (meaning approximately zero). If x mol HNic ionizes, x mol each of H_3O^+ and Nic⁻ is formed, leaving $(0.012 - x)$ mol HNic in solution. You can summarize the situation as follows:

Concentration (M)	HNic(aq) + H₂O(l) ⇌ H₃O⁺(aq) + Nic⁻(aq)		
Starting	0.012	~0	0
Change	−x	+x	+x
Equilibrium	0.012 − x	x	x

Thus, the molar concentrations of HNic, H_3O^+, and Nic⁻ at equilibrium are $(0.012 - x)$, x, and x, respectively.

Step 2: The equilibrium-constant equation is

$$K_a = \frac{[H_3O^+][Nic^-]}{[HNic]}$$

When you substitute the expressions for the equilibrium concentrations, you get

$$K_a = \frac{x^2}{(0.012 - x)}$$

Step 3: The value of x equals the numerical value of the molar hydronium-ion concentration and can be obtained from the pH of the solution.

$$x = [H_3O^+] = \text{antilog}(-\text{pH}) = \text{antilog}(-3.39) = 4.1 \times 10^{-4} = 0.00041$$

You can substitute this value of x into the equation obtained in Step 2. Note first, however, that

$$0.012 - x = 0.012 - 0.00041 = 0.01159 \simeq 0.012 \quad \text{(to two significant figures)}$$

This means that the concentration of undissolved acid is equal to the original concentration of the acid within the precision of the data. (We will make use of this type of observation in later problem solving.) Therefore,

$$K_a = \frac{x^2}{(0.012 - x)} \simeq \frac{x^2}{0.012} \simeq \frac{(0.00041)^2}{0.012} = \mathbf{1.4 \times 10^{-5}}$$

To obtain the degree of ionization, note that x mol out of 0.012 mol of nicotinic acid ionizes. Hence,

$$\text{Degree of ionization} = \frac{x}{0.012} = \frac{0.00041}{0.012} = \mathbf{0.034}$$

The percent ionization is obtained by multiplying this by 100, which gives 3.4%.

Lactic acid has the structural formula

and the electrostatic potential map

Answer Check In order to be a weak acid, the K_a value must be very small, as it is here. Therefore, if you calculate a K_a for a weak acid that is greater than 10^{-1}, you should check for errors in your work.

Exercise 16.1 Lactic acid, $HC_3H_5O_3$, is found in sour milk, where it is produced by the action of lactobacilli on lactose, or milk sugar. A 0.025 M solution of lactic acid has a pH of 2.75. What is the ionization constant K_a for this acid? What is the degree of ionization? ◀

■ See Problems 16.35 and 16.36.

Table 16.1	Acid-Ionization Constants at 25°C*	
Substance	**Formula**	**K_a**
Acetic acid	$HC_2H_3O_2$	1.7×10^{-5}
Benzoic acid	$HC_7H_5O_2$	6.3×10^{-5}
Boric acid	H_3BO_3	5.9×10^{-10}
Carbonic acid	H_2CO_3	4.3×10^{-7}
	HCO_3^-	4.8×10^{-11}
Cyanic acid	$HOCN$	3.5×10^{-4}
Formic acid	$HCHO_2$	1.7×10^{-4}
Hydrocyanic acid	HCN	4.9×10^{-10}
Hydrofluoric acid	HF	6.8×10^{-4}
Hydrogen sulfate ion	HSO_4^-	1.1×10^{-2}
Hydrogen sulfide	H_2S	8.9×10^{-8}
	HS^-	$1.2 \times 10^{-13\dagger}$
Hypochlorous acid	$HClO$	3.5×10^{-8}
Nitrous acid	HNO_2	4.5×10^{-4}
Oxalic acid	$H_2C_2O_4$	5.6×10^{-2}
	$HC_2O_4^-$	5.1×10^{-5}
Phosphoric acid	H_3PO_4	6.9×10^{-3}
	$H_2PO_4^-$	6.2×10^{-8}
	HPO_4^{2-}	4.8×10^{-13}
Phosphorous acid	H_2PHO_3	1.6×10^{-2}
	$HPHO_3^-$	7×10^{-7}
Propionic acid	$HC_3H_5O_2$	1.3×10^{-5}
Pyruvic acid	$HC_3H_3O_3$	1.4×10^{-4}
Sulfurous acid	H_2SO_3	1.3×10^{-2}
	HSO_3^-	6.3×10^{-8}

*The ionization constants for polyprotic acids are for successive ionizations. For example, for H_3PO_4, the equilibrium is $H_3PO_4 + H_2O \rightleftharpoons H_3O^+ + H_2PO_4^-$. For $H_2PO_4^-$, the equilibrium is $H_2PO_4^- + H_2O \rightleftharpoons H_3O^+ + HPO_4^{2-}$.

†This value is in doubt. Some evidence suggests that it is about 10^{-19}. See R. J. Myers, *J. Chem. Educ.*, **63**, 687 (1986).

Table 16.1 lists acid-ionization constants for various weak acids. The weakest acids have the smallest values of K_a.

Calculations with K_a

Once you know the value of K_a for an acid HA, you can calculate the equilibrium concentrations of species HA, A^-, and H_3O^+ for solutions of different molarities. The general method for doing this was discussed in Chapter 14. Here we illustrate the use of a simplifying approximation that can often be used for weak acids. In the next example, we will look at an amplification of the first question posed in the chapter opening.

Example 16.2

Calculating Concentrations of Species in a Weak Acid Solution Using K_a (Approximation Method)

Gaining Mastery Toolbox

Critical Concept 16.2
In some cases, assumptions can be made that simplify equilibrium calculations. When making an assumption, it is essential that you perform a final check to see if the assumption is valid. For our purposes, the assumption is generally valid if it introduces an error of 5% or less.

Solution Essentials:
- Acid ionization constant (K_a)
- Degree of ionization
- pH
- Weak acid
- Acid
- Equilibrium-constant expression

What are the concentrations of nicotinic acid, hydrogen ion, and nicotinate ion in a solution of 0.10 *M* nicotinic acid, $HC_6H_4NO_2$, at 25°C? What is the pH of the solution? What is the degree of ionization of nicotinic acid? The acid-ionization constant, K_a, was determined in the previous example to be 1.4×10^{-5}.

Problem Strategy You follow the three steps for solving equilibrium problems that were introduced in Example 14.7. In the last step, you solve the equilibrium-constant equation for the equilibrium concentrations. The resulting equation is quadratic, but because the equilibrium concentration of a weak acid is usually negligibly different from its starting value, the equation simplifies so that it involves only the square of the unknown, which is easily solved by taking the square root. (You will need to check that this assumption is valid.)

Solution

Step 1: At the start (before ionization), the concentration of nicotinic acid, HNic, is 0.10 *M* and that of its conjugate base, Nic^-, is 0. The concentration of H_3O^+ is essentially zero (~0), assuming that the contribution from the self-ionization of water can be neglected. In 1 L of solution, the nicotinic acid ionizes to give x mol H_3O^+ and x mol Nic^-, leaving $(0.10 - x)$ mol of nicotinic acid. These data are summarized in the following table:

Concentration (M)	$HNic(aq) + H_2O(l) \rightleftharpoons$	$H_3O^+(aq) +$	$Nic^-(aq)$
Starting	0.10	~0	0
Change	$-x$	$+x$	$+x$
Equilibrium	$0.10 - x$	x	x

The equilibrium concentrations of HNic, H_3O^+, and Nic^- are $(0.10 - x)$, x, and x, respectively.

Step 2: Now substitute these concentrations and the value of K_a into the equilibrium-constant equation for acid ionization:

$$\frac{[H_3O^+][Nic^-]}{[HNic]} = K_a$$

You get

$$\frac{x^2}{(0.10 - x)} = 1.4 \times 10^{-5}$$

Step 3: Now solve this equation for x. This is actually a quadratic equation, but it can be simplified so that the value of x is easily found. Because the acid-ionization constant is small, the value of x is small. Assume that x is much smaller than 0.10, so that

$$0.10 - x \simeq 0.10$$

You will need to check that this assumption is valid after you obtain a value for x. The equilibrium-constant equation becomes

$$\frac{x^2}{0.10} \simeq 1.4 \times 10^{-5}$$

or

$$x^2 \simeq 1.4 \times 10^{-5} \times 0.10 = 1.4 \times 10^{-6}$$

Hence,

$$x \simeq 1.2 \times 10^{-3} = 0.0012$$

(continued)

(continued)

At this point, you should check to make sure that the assumption $0.10 - x \simeq 0.10$ is valid. You substitute the value obtained for x into $0.10 - x$.

$$0.10 - x = 0.10 - 0.0012 = 0.10 \qquad \text{(to two significant figures)}$$

The assumption is indeed valid.

Now you can substitute the value of x into the last line of the table written in Step 1 to find the concentrations of species. The concentrations of nicotinic acid, hydronium ion, and nicotinate ion are **0.10 M, 0.0012 M, and 0.0012 M**, respectively.

The pH of the solution is

$$\text{pH} = -\log{[H_3O^+]} = -\log{(0.0012)} = \mathbf{2.92}$$

The degree of ionization equals the amount per liter of nicotinic acid that ionizes ($x = 0.0012$) divided by the total amount per liter of nicotinic acid initially present (0.10). Thus, the degree of ionization is 0.0012/0.10 = **0.012**.

Answer Check As a quick check on your work, substitute the concentrations that you calculated for the acid (nicotinic), its conjugate base (nicotinate), and hydronium ion into the equilibrium expression for K_a. You should get the value of K_a given in the problem.

$$K_a = \frac{[H_3O^+][\text{Nic}^-]}{[\text{HNic}]} = \frac{0.0012 \times 0.0012}{0.10} = 1.4 \times 10^{-5}$$

The pH, of course, should be in the acid range, below 7. Check that this is the case.

Exercise 16.2 What are the concentrations of hydrogen ion and acetate ion in a solution of 0.10 M acetic acid, $HC_2H_3O_2$? What is the pH of the solution? What is the degree of ionization? See Table 16.1 for the value of K_a.

■ See Problems 16.37 and 16.38.

In Example 16.2, the degree of ionization of nicotinic acid (0.012) is relatively small in a 0.10 M solution. That is, only 1.2% of the molecules ionize. It is the small value of the degree of ionization that allows you to neglect x in the term $0.10 - x$ and thereby simplify the calculation.

The degree of ionization of a weak acid depends on both K_a and the concentration of the acid solution. For a given concentration, the larger the K_a, the greater is the degree of ionization. For a given value of K_a, however, the more dilute the solution, the greater is the degree of ionization. Figure 16.3 shows how the percent ionization (degree of ionization $\times$ 100) varies with the concentration of solution.

How do you know when you can use the simplifying assumption of Example 16.2, where you neglected x in the denominator of the equilibrium equation? *It can be shown that this simplifying assumption gives an error of less than 5% if the concentration of acid, C_a, divided by K_a equals 100 or more.* For example, for an acid concentration of 10^{-2} M and K_a of 10^{-5}, $C_a/K_a = 10^{-2}/10^{-5} = 10^3$, so the assumption is valid. It would not be valid, however, if the concentration of the same acid were 10^{-4} M.

If the simplifying assumption of Example 16.2 is not valid, you can solve the equilibrium equation exactly by using the *quadratic formula*. We illustrate this in the next example with a solution of aspirin, $HC_9H_7O_4$, a common headache remedy. (Figure 16.4 shows the structural formula.)

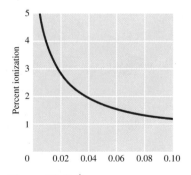

Figure 16.3 ▲

Variation of percent ionization of a weak acid with concentration
On this curve for nicotinic acid, note that the percent ionization is greatest for the most dilute solutions.

Example 16.3

Calculating Concentrations of Species in a Weak Acid Solution Using K_a (Quadratic Formula)

Gaining Mastery Toolbox

Critical Concept 16.3
Some equilibrium problems require an exact solution (no assumptions). When the concentration of acid C_a divided by K_a is less than 100 ($C_a/K_a < 100$) no assumptions should be made when solving the problem. This approach of prechecking an assumption limits errors to 5% or less.

Solution Essentials:
• Acid ionization constant (K_a)
• pH
• Relative strengths of acids and bases
• Equilibrium-constant expression
• Quadratic formula

What is the pH at 25°C of the solution obtained by dissolving a 5.00-grain tablet of aspirin (acetylsalicylic acid) in 0.500 L of water? The tablet contains 5.00 grains, or 0.325 g, of acetylsalicylic acid, $HC_9H_7O_4$. The acid is monoprotic, and K_a equals 3.3×10^{-4} at 25°C.

Problem Strategy You follow the three steps for equilibrium problems. However, the approximation used in the previous example to solve the equilibrium-constant equation is not valid here. Verify this by calculating C_a/K_a; it is not greater than 100. You solve the equation exactly using the quadratic formula.

Solution The molar mass of $HC_9H_7O_4$ is 180.2 g. From this you find that an aspirin tablet contains 0.00180 mol of the acid. Hence, the concentration of the aspirin solution is 0.00180 mol/0.500 L, or 0.0036 M. (Retain two significant figures, the number of significant figures in K_a.) $C_a/K_a = 0.0036/3.3 \times 10^{-4} = 11$, which is less than 100, so we must solve the equilibrium equation exactly.

Step 1: Abbreviate the formula for acetylsalicylic acid as HAcs and let x be the amount of H_3O^+ ion formed in 1 L of solution. The amount of acetylsalicylate ion formed is also x mol; the amount of nonionized acetylsalicylic acid is $(0.0036 - x)$ mol. These data are summarized in the following table:

Concentration (M)	$HAcs(aq)$ + $H_2O(l)$	$\rightleftharpoons$	$H_3O^+(aq)$ +	$Acs^-(aq)$
Starting	0.0036		~0	0
Change	$-x$		$+x$	$+x$
Equilibrium	$0.0036 - x$		x	x

Step 2: If you substitute the equilibrium concentrations and the value of K_a into the equilibrium-constant equation, you get

$$\frac{[H_3O^+][Acs^-]}{[HAcs]} = K_a$$

$$\frac{x^2}{0.0036 - x} = 3.3 \times 10^{-4}$$

Step 3: You can solve the equation exactly using the quadratic formula. Rearrange the preceding equation to put it in the form $ax^2 + bx + c = 0$. You get

$$x^2 = (0.0036 - x) \times 3.3 \times 10^{-4} = 1.2 \times 10^{-6} - 3.3 \times 10^{-4} x$$

or

$$x^2 + 3.3 \times 10^{-4} x - 1.2 \times 10^{-6} = 0$$

Now substitute into the quadratic formula.

$$x = \frac{-b \pm \sqrt{b^2 - 4ac}}{2a}$$

$$= \frac{-3.3 \times 10^{-4} \pm \sqrt{(3.3 \times 10^{-4})^2 + 4(1.2 \times 10^{-6})}}{2}$$

$$= \frac{-3.3 \times 10^{-4} \pm 2.2 \times 10^{-3}}{2}$$

Structural formula of aspirin (acetylsalicylic acid), a well-known headache remedy. Note the acidic hydrogen atom (in color).

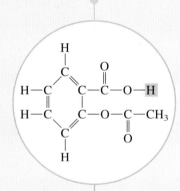

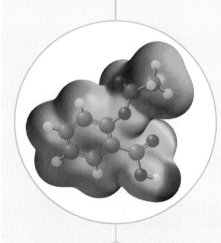

Electrostatic potential map of aspirin.

Figure 16.4 ▲

The structure of aspirin

(continued)

(continued)

If you take the lower sign in $\pm$, you will get a negative value for x. But x equals the concentration of H_3O^+ ion, so it must be positive. Therefore, you ignore the negative solution. Taking the upper sign, you get

$$x = [H_3O^+] = 9.4 \times 10^{-4}$$

Now you can calculate the pH.

$$pH = -\log[H_3O^+] = -\log(9.4 \times 10^{-4}) = \textbf{3.03}$$

Answer Check As in the previous example, you can do a quick check of your work by essentially reversing the problem, calculating K_a from the equilibrium concentrations of acid, conjugate base, and hydronium ion. You can obtain this from the last line of the equilibrium table, now that you have the value of x.

$$[HAcs] = 0.0036 - x = 0.0036 - 9.4 \times 10^{-4} = 0.0027$$
$$[Acs^-] = [H_3O^+] = x = 9.4 \times 10^{-4}$$

Substituting these values into the expression for K_a gives the value of the equilibrium constant.

$$K_a = \frac{[H_3O^+][Acs^-]}{[HAcs]} = \frac{(9.4 \times 10^{-4})^2}{0.0027} = 3.3 \times 10^{-4}$$

Exercise 16.3 What is the pH of an aqueous solution that is 0.0030 M pyruvic acid, $HC_3H_3O_3$? (Pyruvic acid forms during the breakdown of glucose in a cell.) ▶

Exercise 16.3 mentions that pyruvic acid is formed in biological cells. In muscle tissue, glucose is broken down to pyruvic acid, which is eventually oxidized to carbon dioxide and water. The structural formula of pyruvic acid is

and the electrostatic potential map is

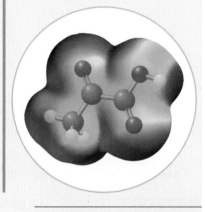

■ See Problems 16.43 and 16.44.

CONCEPT CHECK 16.1

You have prepared dilute solutions of equal molar concentrations of $HC_2H_3O_2$ (acetic acid), HNO_2, HF, and HCN. Rank the solutions from the highest pH to the lowest pH. (Refer to Table 16.1.)

16.2 Polyprotic Acids

In the preceding section, we dealt only with acids releasing one H_3O^+ ion or proton. Some acids, however, have two or more such protons; these acids are called *polyprotic acids*. Sulfuric acid, for example, can lose two protons in aqueous solution. One proton is lost completely to form H_3O^+ (sulfuric acid is a strong acid).

$$H_2SO_4(aq) + H_2O(l) \longrightarrow H_3O^+(aq) + HSO_4^-(aq)$$

The hydrogen sulfate ion, HSO_4^-, may then lose the other proton. In this case, an equilibrium exists.

$$HSO_4^-(aq) + H_2O(l) \rightleftharpoons H_3O^+(aq) + SO_4^{2-}(aq)$$

For a weak diprotic acid like carbonic acid, H_2CO_3, there are two simultaneous equilibria to consider.

$$H_2CO_3(aq) + H_2O(l) \rightleftharpoons H_3O^+(aq) + HCO_3^-(aq)$$
$$HCO_3^-(aq) + H_2O(l) \rightleftharpoons H_3O^+(aq) + CO_3^{2-}(aq)$$

Carbonic acid is in equilibrium with the hydrogen carbonate ion, HCO_3^-, which is in turn equilibrated with the carbonate ion, CO_3^{2-}. Each equilibrium has an associated acid-ionization constant. For the loss of the first proton,

$$K_{a1} = \frac{[H_3O^+][HCO_3^-]}{[H_2CO_3]} = 4.3 \times 10^{-7}$$

and for the loss of the second proton,

$$K_{a2} = \frac{[H_3O^+][CO_3^{2-}]}{[HCO_3^-]} = 4.8 \times 10^{-11}$$

Note that K_{a1} for carbonic acid is much larger than K_{a2} (by a factor of about 1×10^4). This indicates that carbonic acid loses the first proton more easily than the second one, because the first proton separates from an ion of single negative charge, whereas the second proton separates from an ion of double negative charge. This double negative charge strongly attracts the proton back to it.

> In general, the second ionization constant, K_{a2}, of a polyprotic acid is much smaller than the first ionization constant, K_{a1}. In the case of a triprotic acid, the third ionization constant, K_{a3}, is much smaller than the second one, K_{a2}.

See the values for phosphoric acid, H_3PO_4, in Table 16.1.

Calculating the concentrations of various species in a solution of a polyprotic acid might appear complicated, because several equilibria occur at once. However, reasonable assumptions can be made that simplify the calculation, as we show in the next example.

Example 16.4

Critical Concept 16.4
For many polyprotic acids, the first ionization constant, K_{a1}, can be used to calculate the pH of the solution. Typically, the first ionization constant of a polyprotic acid is much greater than the subsequent ionization constant(s). Therefore, it is the first ionization step of the acid that creates the bulk of the $H_3O^+(aq)$ in solution.

Solution Essentials:
• Polyprotic acid
• Acid ionization constant (K_a)
• pH
• Relative strengths of acids and bases
• Equilibrium-constant expression

Calculating Concentrations of Species in a Solution of a Diprotic Acid

Ascorbic acid (vitamin C) is a diprotic acid, $H_2C_6H_6O_6$ (Figure 16.5). What is the pH of a 0.10 M solution? What is the concentration of ascorbate ion, $C_6H_6O_6^{2-}$? The acid ionization constants are $K_{a1} = 7.9 \times 10^{-5}$ and $K_{a2} = 1.6 \times 10^{-12}$.

Problem Strategy Diprotic acids have two ionization constants, one for the loss of each proton. To be exact, you should account for both reactions. However, K_{a2} is so much smaller than K_{a1} that the smaller amount of hydronium ion produced in the second reaction can be neglected. Follow the three steps for solving equilibrium problems.

Considering only the first ionization, set up a table of concentrations (starting, change, and equilibrium). Substitute these values into the equilibrium equation for K_{a1}, solve for $x = [H_3O^+]$, and then solve for pH (as in Example 16.2). Ascorbate ion is produced in the second ionization; its concentration, as we show, equals K_{a2}.

Solution CALCULATION OF pH Abbreviate the formula for ascorbic acid as H_2Asc. Hydronium ions are produced by two successive acid ionizations.

$$H_2Asc(aq) + H_2O(l) \rightleftharpoons H_3O^+(aq) + HAsc^-(aq); \quad K_{a1} = 7.9 \times 10^{-5}$$

$$HAsc^-(aq) + H_2O(l) \rightleftharpoons H_3O^+(aq) + Asc^{2-}(aq); \quad K_{a2} = 1.6 \times 10^{-12}$$

If you let x be the amount of H_3O^+ formed in the first ionization, you get the following results:

Concentration (M)	H_2Asc(aq) + H_2O(l) $\rightleftharpoons$ H_3O^+(aq) +		$HAsc^-$(aq)
Starting	0.10	~0	0
Change	$-x$	$+x$	$+x$
Equilibrium	$0.10 - x$	x	x

(continued)

(continued)

Now substitute into the equilibrium-constant equation for the first ionization.

$$\frac{[H_3O^+][HAsc^-]}{[H_2Asc]} = K_{a1}$$

$$\frac{x^2}{0.10 - x} = 7.9 \times 10^{-5}$$

Assuming x to be much smaller than 0.10, you get

$$\frac{x^2}{0.10} \simeq 7.9 \times 10^{-5}$$

or

$$x^2 \simeq 7.9 \times 10^{-5} \times 0.10$$

$$x \simeq 2.8 \times 10^{-3} = 0.0028$$

(Note that $0.10 - x = 0.10 - 0.0028 = 0.10$, correct to two significant figures. Therefore, the assumption that $0.10 - x \simeq 0.10$ is correct.) The hydronium-ion concentration is 0.0028 M, so

$$pH = -\log[H_3O^+] = -\log(0.0028) = \textbf{2.55}$$

ASCORBATE-ION CONCENTRATION Ascorbate ion, Asc^{2-}, which we will call y, is produced only in the second reaction. Assume the starting concentrations of H_3O^+ and $HAsc^-$ for this reaction to be those from the first equilibrium. The amounts of the species in 1 L of solution are as follows:

Concentration (M)	$HAsc^-(aq)$	$+ H_2O(l)$	$\rightleftharpoons$	$H_3O^+(aq)$	$+ Asc^{2-}(aq)$
Starting	0.0028			0.0028	0
Change	$-y$			$+y$	$+y$
Equilibrium	$0.0028 - y$			$0.0028 + y$	y

Now substitute into the equilibrium-constant equation for the second ionization.

$$\frac{[H_3O^+][Asc^{2-}]}{[HAsc^-]} = K_{a2}$$

$$\frac{(0.0028 + y)y}{0.0028 - y} = 1.6 \times 10^{-12}$$

This equation can be simplified if you assume that y is much smaller than 0.0028. (Again, this assumes that the reaction occurs to only a small extent, as you expect from the magnitude of the equilibrium constant.) That is,

$$0.0028 + y \simeq 0.0028$$

$$0.0028 - y \simeq 0.0028$$

Then the equilibrium equation reads

$$\frac{(0.0028)y}{0.0028} \simeq 1.6 \times 10^{-12}$$

Hence,

$$y \simeq 1.6 \times 10^{-12}$$

(Note that 1.6×10^{-12} is indeed much smaller than 0.0028, as you assumed.) The concentration of ascorbate ion equals K_{a2}, or **1.6×10^{-12} M**.

Answer Check Because the second ionization constant of an acid, K_{a2}, is always smaller than the first ionization constant, K_{a1}, you should expect any species produced during the second ionization to be in lower concentration than those from the first ionization.

Structural formula of ascorbic acid (vitamin C); note the acidic hydrogen atoms (in color).

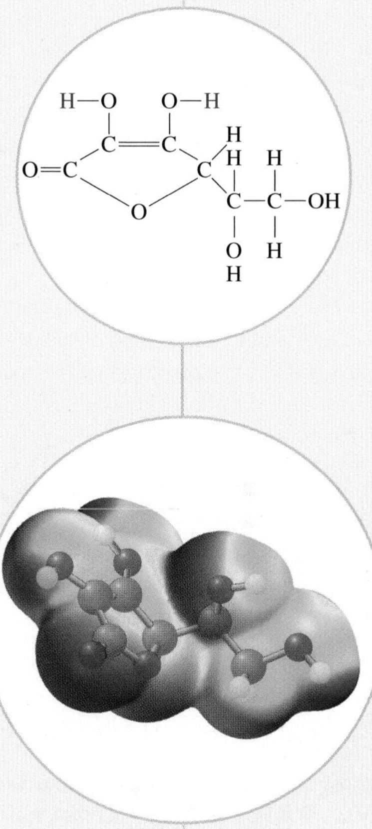

Electrostatic potential map of ascorbic acid.

Figure 16.5 ▲
The structure of ascorbic acid

(continued)

Acid Rain

Acid rain means rain having a pH lower than that of natural rain (pH 5.6). Natural rain dissolves carbon dioxide from the atmosphere to give a slightly acidic solution. The pH of rain in eastern North America and western Europe, however, is approximately 4 and sometimes lower. This acidity is primarily the result of the dissolving in rainwater of sulfur oxides and nitrogen oxides from human activities. In the northeastern United States, the strong acid components in acid rain are about 62% sulfuric acid, 32% nitric acid, and 6% hydrochloric acid.

The sulfuric acid in acid rain has been traced to the burning of fossil fuels and to the burning of sulfide ores in the production of metals, such as zinc and copper. Coal, for example, contains some sulfur mainly as pyrite, or iron(II) disulfide, FeS_2. When this burns in air, it produces sulfur dioxide.

$$4FeS_2(s) + 11O_2(g) \longrightarrow 2Fe_2O_3(s) + 8SO_2(g)$$

In the presence of dust particles and other substances in polluted air, the sulfur dioxide oxidizes further to give sulfur trioxide, which reacts with water to form sulfuric acid.

$$2SO_2(g) + O_2(g) \longrightarrow 2SO_3(g)$$

$$SO_3(g) + H_2O(l) \longrightarrow H_2SO_4(aq)$$

Almost all of the sulfuric acid dissociates into the ions H_3O^+, HSO_4^-, and SO_4^{2-}. The first ionization of H_2SO_4 is complete; the second ionization (that of HSO_4^-) is partial ($K_{a2} = 1.1 \times 10^{-2}$).

Acid rain can be harmful to some plants, to fish (by changing the pH of lake water), and to structural materials and monuments (Figure 16.6). Marble, for example, is composed of calcium carbonate, $CaCO_3$, which dissolves in water of low pH.

$$H_3O^+(aq) + CaCO_3(s) \longrightarrow Ca^{2+}(aq) + HCO_3^-(aq) + H_2O(l)$$

It is instructive to calculate the pH of natural rain—that is, water saturated with carbon dioxide from the atmosphere. For this, you need to solve two equilibrium problems. First, you need the equilibrium for the solution of CO_2 in water.

$$CO_2(g) + H_2O(l) \rightleftharpoons H_2CO_3(aq)$$

$$K = \frac{[H_2CO_3]}{P_{CO_2}} = 3.5 \times 10^{-2}$$

Here $[H_2CO_3]$ represents the concentration of the molecular species formed when CO_2 dissolves in water. The partial pressure of CO_2 in the atmosphere is 0.00033 atm. If you substitute this for P_{CO_2} and solve for $[H_2CO_3]$, you obtain 1.2×10^{-5} M. (Note that the equilibrium-constant equation involving P_{CO_2} is essentially Henry's law, discussed in Section 12.3.) Now you need to account for the acid ionization of H_2CO_3. You look only at the first ionization, which produces almost all of the H_3O^+.

$$H_2CO_3(aq) + H_2O(l) \rightleftharpoons H_3O^+(aq) + HCO_3^-(aq)$$

$$K_{a1} = \frac{[H_3O^+][HCO_3^-]}{[H_2CO_3]} = 4.3 \times 10^{-7}$$

(*continued*)

Exercise 16.4 Sulfurous acid, H_2SO_3, is a diprotic acid with $K_{a1} = 1.3 \times 10^{-2}$ and $K_{a2} = 6.3 \times 10^{-8}$. The acid forms when sulfur dioxide (a gas with a suffocating odor) dissolves in water. What is the pH of a 0.25 M solution of sulfurous acid? What is the concentration of sulfite ion, SO_3^{2-}, in the solution? Note that K_{a1} is relatively large.

■ See Problems 16.47 and 16.48.

Example 16.4 shows that $[H_3O^+]$ and $[HA^-]$ in a solution of a diprotic acid H_2A can be calculated from the first ionization constant, K_{a1}.

The concentration of the ion A^{2-} equals the second ionization constant, K_{a2}.

16.3 Base-Ionization Equilibria

Equilibria involving weak bases are treated similarly to those for weak acids. Ammonia, for example, ionizes in water as follows:

$$NH_3(aq) + H_2O(l) \rightleftharpoons NH_4^+(aq) + OH^-(aq)$$

Denoting $[H_3O^+] = [HCO_3^-]$ by x, you can write the equilibrium equation

$$\frac{x^2}{1.2 \times 10^{-5} - x} = 4.3 \times 10^{-7}$$

Making the usual assumption about neglecting x in the denominator, you obtain $x = [H_3O^+] = 2.3 \times 10^{-6}$. This assumption and the neglect of the self-ionization of water affect the answer in the second significant figure. Retaining one significant figure, pH = $-\log(2 \times 10^{-6}) = 5.7$, which is essentially what we gave earlier for the pH of natural rain.

Figure 16.6 ▲

Effect of acid rain on marble These photographs of a marble statue of George Washington, located in New York, were taken 60 years apart. The more recent photograph (*right*) shows that much of the detail of the statue has been dissolved by acid rain. Marble ($CaCO_3$) reacts with H_3O^+ ion to form soluble $Ca(HCO_3)_2$.

■ See Problems 16.133 and 16.134.

The corresponding equilibrium constant is

$$K_c = \frac{[NH_4^+][OH^-]}{[NH_3][H_2O]}$$

In terms of the thermodynamic equilibrium constant, the activity of H_2O is 1, so it does not appear explicitly in the equilibrium constant.

Because the concentration of pure H_2O is constant, you can rearrange this equation as you did for acid ionization. ▶

$$K_b = [H_2O]K_c = \frac{[NH_4^+][OH^-]}{[NH_3]}$$

In general, a weak base B with the *base ionization*

$$B(aq) + H_2O(l) \rightleftharpoons HB^+(aq) + OH^-(aq)$$

has a **base-ionization constant**, K_b (*the equilibrium constant for the ionization of a weak base*), equal to

$$K_b = \frac{[HB^+][OH^-]}{[B]}$$

Table 16.2 lists ionization constants for some weak bases. All of these bases are similar to ammonia in that they have a N atom with a lone pair of electrons that accepts an H^+ ion from water. For example, the base ionization of methylamine, CH_3NH_2, is

Table 16.2 Base-Ionization Constants at 25°C

Substance	Formula	K_b
Ammonia	NH_3	1.8×10^{-5}
Aniline	$C_6H_5NH_2$	4.2×10^{-10}
Dimethylamine	$(CH_3)_2NH$	5.1×10^{-4}
Ethylamine	$C_2H_5NH_2$	4.7×10^{-4}
Hydrazine	N_2H_4	1.7×10^{-6}
Hydroxylamine	NH_2OH	1.1×10^{-8}
Methylamine	CH_3NH_2	4.4×10^{-4}
Pyridine	C_5H_5N	1.4×10^{-9}
Urea	NH_2CONH_2	1.5×10^{-14}

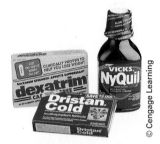

Figure 16.7 ▲

Many over-the-counter drugs are salts of bases These salts are often chlorides (frequently called hydrochlorides) and are analogous to ammonium chloride. You might find it instructive to read the ingredients from the labels for these products.

$$H-\underset{\underset{H}{|}}{\overset{\overset{H}{|}}{C}}-\underset{\underset{H}{|}}{\overset{\overset{H}{|}}{N}}: + H_2O \rightleftharpoons H-\underset{\underset{H}{|}}{\overset{\overset{H}{|}}{C}}-\overset{+}{\underset{\underset{H}{|}}{\overset{\overset{H}{|}}{N}}}-H + OH^-$$

Many drugs and plant substances such as caffeine are bases. Often a prescription or over-the-counter drug is a salt of a base, because the salt is more soluble in water (Figure 16.7). These salts are analogous to the salts of ammonia, such as NH_4Cl. Other common bases are anions of weak acids. Examples are F^- and $C_2H_3O_2^-$ (acetate ion). We will discuss these in the next section.

In the following exercise, you will determine the K_b of a molecular base from the pH of a solution. For these calculations you can check to see if a simplifying assumption is valid by making sure that the concentration of base C_b divided by the K_b equals 100 or more. Note that this is essentially the problem you looked at in Example 16.1.

Some other well-known alkaloids and their plant sources are nicotine (tobacco), cocaine (coca bush), mescaline (species of cactus), and caffeine (coffee and tea).

Exercise 16.5 Quinine is an alkaloid, or naturally occurring base, used to treat malaria. A 0.0015 M solution of quinine has a pH of 9.84. The basicity of alkaloids is due to a nitrogen atom that picks up protons from water in the same manner as ammonia does. ◀ What is K_b?

■ See Problems 16.51 and 16.52.

Example 16.5   **Calculating Concentrations of Species in a Weak Base Solution Using K_b**

Gaining Mastery Toolbox

Critical Concept 16.5
Calculation of the species in a solution of a weak base are performed in the same manner as for a weak acid. For a weak base, the ionization constant is K_b, and the chemical reaction is written as the base ionizing in water.

Solution Essentials:
• Base ionization constant (K_b)
• pH
• Self-ionization of water ($K_w = [H_3O^+][OH^-]$)
• Relative strengths of acids and bases
• Equilibrium-constant expression

Morphine, $C_{17}H_{19}NO_3$, is administered medically to relieve pain. It is a naturally occurring base, or alkaloid. What is the pH of a 0.0075 M solution of morphine at 25°C? The base-ionization constant, K_b, is 1.6×10^{-6} at 25°C.

Problem Strategy First of all, you need to be aware that the compound $C_{17}H_{19}NO_3$ is a weak base. The calculation for the ionization of a weak base parallels that used with weak acids in Examples 16.2 and 16.3: you write

(continued)

(continued)

the equation, make a table of concentrations (Step 1), set up the equilibrium-constant equation for K_b (Step 2), and solve for $x = [OH^-]$ (Step 3). Because $C_b(0.0075)$ divided by K_b (1.5×10^{-6}) is greater than 100, the self-ionization of water can be neglected. You obtain $[H_3O^+]$, and then the pH, by solving $K_w = [H_3O^+][OH^-]$.

Solution Morphine, which we abbreviate as Mor, ionizes by picking up a proton from water (as does ammonia).

$$Mor(aq) + H_2O \rightleftharpoons HMor^+(aq) + OH^-(aq)$$

Step 1: Summarize the concentrations in a table. For the *change* values, note that the morphine in 1 L of solution ionizes to give x mol $HMor^+$ and x mol OH^-.

Concentration (M)	Mor(aq) + H₂O(l) ⇌	HMor⁺(aq) +	OH⁻(aq)
Starting	0.0075	0	~0
Change	$-x$	$+x$	$+x$
Equilibrium	$0.0075 - x$	x	x

Step 2: Substituting into the equilibrium equation

$$\frac{[HMor^+][OH^-]}{[Mor]} = K_b$$

gives

$$\frac{x^2}{0.0075 - x} = 1.6 \times 10^{-6}$$

Step 3: If you assume that x is small enough to neglect compared with 0.0075, you have

$$0.0075 - x \simeq 0.0075$$

and you can write the equilibrium-constant equation as

$$\frac{x^2}{0.0075} \simeq 1.6 \times 10^{-6}$$

Hence,

$$x^2 \simeq 1.6 \times 10^{-6} \times 0.0075 = 1.2 \times 10^{-8}$$

$$x = [OH^-] \simeq 1.1 \times 10^{-4}$$

You can now calculate the hydronium-ion concentration. The product of the hydronium-ion concentration and the hydroxide-ion concentration equals 1.0×10^{-14} at 25°C:

$$[H_3O^+][OH^-] = 1.0 \times 10^{-14}$$

Substituting into this equation gives

$$[H_3O^+] \times 1.1 \times 10^{-4} = 1.0 \times 10^{-14}$$

Hence,

$$[H_3O^+] = \frac{1.0 \times 10^{-14}}{1.1 \times 10^{-4}} = 9.1 \times 10^{-11}$$

The pH of the solution is

$$pH = -\log[H_3O^+] = -\log(9.1 \times 10^{-11}) = \textbf{10.04}$$

Note that you could also calculate the pH using the formula pH + pOH = 14.00. You would first calculate the pOH; then pH = 14.00 − pOH.

Answer Check A reasonable answer for the pH of a weak base in aqueous solution is greater than 7.

Exercise 16.6 What is the hydronium-ion concentration of a 0.20 *M* solution of ammonia in water? See Table 16.2 for K_b.

■ See Problems 16.53 and 16.54.

A base B is placed into a beaker of water with the result depicted below (water molecules have been omitted for clarity). Write the ionization reaction for this base and classify the base as being strong or weak.

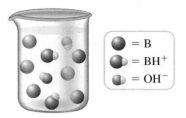

16.4 Acid–Base Properties of Salt Solutions

A salt may be regarded as an ionic compound obtained by a neutralization reaction in aqueous solution. The resulting salt solution may be neutral, but often it is acidic or basic (Figure 16.8). One of the successes of the Brønsted–Lowry concept of acids and bases was in pointing out that some ions can act as acids or bases. The acidity or basicity of a salt solution is explained in terms of the acidity or basicity of individual ions in the solution.

Consider a solution of sodium cyanide, NaCN. A 0.1 M solution has a pH of 11.1 and is therefore fairly basic. Sodium cyanide dissolves in water to give Na^+ and CN^- ions.

$$NaCN(s) \xrightarrow{H_2O} Na^+(aq) + CN^-(aq)$$

Sodium ion, Na^+, is unreactive with water, but the cyanide ion, CN^-, reacts to produce HCN and OH^-.

$$CN^-(aq) + H_2O(l) \rightleftharpoons HCN(aq) + OH^-(aq)$$

From the Brønsted–Lowry point of view, the CN^- ion acts as a base, because it accepts a proton from H_2O. You can also see, however, that OH^- ion is a product, so you expect the solution to have a basic pH. This explains why solutions of NaCN are basic.

The reaction of the CN^- ion with water is referred to as the *hydrolysis* of CN^-. The **hydrolysis** of an ion is *the reaction of an ion with water to produce the conjugate acid and hydroxide ion or the conjugate base and hydronium ion.* The CN^- ion hydrolyzes to give the conjugate acid and OH^- ion. As another example, consider the ammonium ion, NH_4^+, which hydrolyzes as follows:

$$NH_4^+(aq) + H_2O(l) \rightleftharpoons NH_3(aq) + H_3O^+(aq)$$

The ammonium ion acts as an acid, donating a proton to H_2O and forming NH_3. Note that this equation has the form of an acid ionization, so you could write the K_a expression for it. Similarly, the hydrolysis reaction for the CN^- ion has the form of a base ionization, so you could write the K_b expression for it.

Two questions become apparent. First, how can you predict whether a particular salt solution will be acidic, basic, or neutral? Then, how can you calculate the concentrations of H_3O^+ ion and OH^- ion in the salt solution (or, equivalently, how do you predict the pH of the salt solution)? We will now look at the first of these questions.

Prediction of Whether a Salt Solution Is Acidic, Basic, or Neutral

When the CN^- ion hydrolyzes, it produces the conjugate acid, HCN. Hydrogen cyanide, HCN, is a weak acid. That means that it tends to hold on to the proton strongly (does not ionize readily). In other words, the cyanide ion, CN^-, tends to pick up a proton easily, so it acts as a base. This argument can be generalized: *The anions of weak acids are basic.* On the other hand, *the anions of strong acids have hardly any basic character; that is, these ions do not hydrolyze.* For example, the Cl^- ion, which is conjugate to the strong acid HCl, shows no appreciable reaction with water.

$$Cl^-(aq) + H_2O(l) \longrightarrow \text{no reaction}$$

NH₄Cl

NaCl

Na₂CO₃

Figure 16.8 ▲

Salt solutions are acidic, neutral, or basic A pH meter measures the pH of solutions of the salts NH₄Cl, NaCl, and Na₂CO₃.

© Cengage Learning

Now consider a cation conjugate to a weak base. The simplest example is the NH_4^+ ion, which we just discussed. It behaves like an acid. *The cations of weak bases are acidic. On the other hand, the cations of strong bases (metal ions of Groups IA and IIA elements—except Be) have hardly any acidic character; that is, these ions do not hydrolyze.* For example,

$$Na^+(aq) + H_2O(l) \longrightarrow \text{no reaction}$$

Aqueous metal ions, other than the cations of the strong bases, usually hydrolyze by acting as acids. These ions usually form hydrated metal ions. For example, the aluminum ion Al^{3+} forms the hydrated ion $Al(H_2O)_6^{3+}$. The bare Al^{3+} ion acts as a Lewis acid, forming bonds to the electron pairs on O atoms of H_2O molecules. Because the electrons are drawn away from the O atoms by the Al atom, the O atoms in turn tend to draw electrons from the O—H bonds, making them highly polar. As a result, the H_2O molecules in $Al(H_2O)_6^{3+}$ are acidic. The $Al(H_2O)_6^{3+}$ ion hydrolyzes in aqueous solution by donating a proton (from one of the acidic H_2O molecules on the ion) to water molecules in the solvent. After the proton leaves the hydrated metal ion, the ion charge is reduced by 1 and one H_2O molecule on the ion becomes a hydroxide ion bonded to the metal. Thus, the formula of the ion becomes $Al(H_2O)_5(OH)^{2+}$.

$$Al(H_2O)_6^{3+}(aq) + H_2O(l) \rightleftharpoons Al(H_2O)_5(OH)^{2+}(aq) + H_3O^+(aq)$$

To predict the acidity or basicity of a salt solution, you need to examine the acidity or basicity of the ions composing the salt. Consider potassium acetate, $KC_2H_3O_2$. The ions are K^+ and $C_2H_3O_2^-$ (acetate ion). Potassium is a Group IA element, so K^+ does not hydrolyze. However, the acetate ion is the conjugate of acetic acid, a weak acid. Therefore, the acetate ion is basic.

$$K^+(aq) + H_2O(l) \longrightarrow \text{no reaction}$$
$$C_2H_3O_2^-(aq) + H_2O(l) \rightleftharpoons HC_2H_3O_2(aq) + OH^-(aq)$$

A solution of potassium acetate is predicted to be basic.

From this discussion, we can derive a set of rules for deciding whether a salt solution will be neutral, acidic, or basic. These rules apply to normal salts (those in which the anion has no acidic hydrogen atoms).

1. **A salt of a strong base and a strong acid.** The salt has no hydrolyzable ions and so gives a neutral aqueous solution. An example is NaCl.

$$Na^+(aq) + H_2O(l) \longrightarrow \text{No reaction}$$
$$Cl^-(aq) + H_2O(l) \longrightarrow \text{No reaction}$$

2. **A salt of a strong base and a weak acid.** The anion of the salt is the conjugate of the weak acid. It hydrolyzes to give a basic solution. An example is NaCN.

$$Na^+(aq) + H_2O(l) \longrightarrow \text{No reaction}$$
$$CN^-(aq) + H_2O(l) \longrightarrow HCN(aq) + \textbf{OH}^-(aq)$$

3. **A salt of a weak base and a strong acid.** The cation of the salt is the conjugate of the weak base. It hydrolyzes to give an acidic solution. An example is NH_4Cl.

$$NH_4^+(aq) + H_2O(l) \longrightarrow NH_3(aq) + \textbf{H}_3\textbf{O}^+(aq)$$
$$Cl^-(aq) + H_2O(l) \longrightarrow \text{No reaction}$$

4. **A salt of a weak base and a weak acid.** Both ions hydrolyze. Whether the solution is acidic or basic depends on the relative acid–base strengths of the two ions. To determine this, you need to compare the K_a of the cation with the K_b of the anion. If the K_a is larger, the solution is acidic. If the K_b is larger, the solution is basic. Consider solutions of ammonium formate, NH_4CHO_2. These solutions are slightly acidic, because the K_a for NH_4^+ ($= 5.6 \times 10^{-10}$) is somewhat larger than the K_b for formate ion, CHO_2^- ($= 5.9 \times 10^{-11}$).

$$NH_4^+(aq) + H_2O(l) \longrightarrow NH_3(aq) + \textbf{H}_3\textbf{O}^+(aq)$$
$$CHO_2^-(aq) + H_2O(l) \longrightarrow HCHO_2(aq) + \textbf{OH}^-(aq)$$

The next example illustrates the application of these rules.

Example 16.6   **Predicting Whether a Salt Solution Is Acidic, Basic, or Neutral**

Decide whether aqueous solutions of the following salts are acidic, basic, or neutral: a KCl; b NaF; c $Zn(NO_3)_2$; d NH_4CN.

Problem Strategy Start by determining what type of acid–base reaction occurred to make the salt. Once you know this, you can apply the rules above to predict whether a solution will be acidic, basic, or neutral.

Solution

a KCl produces $K^+(aq)$ and $Cl^-(aq)$ ions in solution and is a salt of a strong base (KOH) and a strong acid (HCl), so none of the ions hydrolyze, and a solution of KCl is **neutral**.

b NaF produces $Na^+(aq)$ and $F^-(aq)$ in solution and is a salt of a strong base (NaOH) and a weak acid (HF), so a solution of NaF is **basic** because of the hydrolysis of F^-. (You can assume HF is a weak acid, because it is not one of the common strong acids listed in Table 15.1.)

c $Zn(NO_3)_2$ produces $Zn^{2+}(aq)$ and $NO_3^-(aq)$ in solution and is the salt of a weak base ($Zn(OH)_2$) and a

strong acid (HNO_3), so a solution of $Zn(NO_3)_2$ is **acidic**. Note also that Zn^{2+} is not an ion of a Group IA or Group IIA element, so you can expect it to form a metal hydrate ion that hydrolyzes to give an acidic solution.

d NH_4CN produces $NH_4^+(aq)$ and $CN^-(aq)$ in solution and is a salt of a weak base (NH_3) and a weak acid (HCN). Therefore, to answer the question, you would need the K_a for NH_4^+ and the K_b for CN^-. Although these are listed in some tables, more frequently they are not. However, as you will see later in this section, values are easily calculated from the molecular substances that are conjugates (NH_3 and HCN). The K_a for NH_4^+ is 5.6×10^{-10}, and the K_b for CN^- is 2.0×10^{-5}. From these values, you conclude that a solution of NH_4CN is **basic**.

Answer Check You can check your answers by writing hydrolysis reactions for each of the ions in solution. If either or both of these reactions produce a weak acid or base, the solution will not be neutral. You then can analyze the products of these reactions for OH^- and/or H_3O^+ to check your final answer (keep in mind you may have to look at the K_a and K_b information as well). For example, using NaF from part b, the hydrolysis reactions for the ions that make up this salt are

$$Na^+(aq) + H_2O(l) \longrightarrow \text{no reaction}$$
$$F^-(aq) + H_2O(l) \longrightarrow HF(aq) + OH^-(aq)$$

Because there was no reaction for the Na^+, and OH^- was produced by the hydrolysis of F^-, the solution containing the NaF salt is basic.

Exercise 16.7   Consider solutions of the following salts: a NH_4NO_3; b KNO_3; c $Al(NO_3)_3$. For each, state whether the solution is acidic, basic, or neutral.

■ See Problems 16.59 and 16.60.

The pH of a Salt Solution

Suppose you wish to calculate the pH of 0.10 M NaCN. As you have seen, the solution is basic because of the hydrolysis of the CN^- ion. The calculation is essentially the same as that for any weak base (Example 16.5). You would require the base-ionization constant of the cyanide ion. However, frequently the ionization constants for ions are not listed directly in tables, because the values are easily related to those for the conjugate molecular species. Thus, the K_b for CN^- is related to the K_a for HCN.

To see this relationship between K_a and K_b for conjugate acid–base pairs, consider the acid ionization of HCN and the base ionization of CN^-. When these two reactions are added, you get the ionization of water.

$$HCN(aq) + H_2O(l) \rightleftharpoons H_3O^+(aq) + CN^-(aq) \quad K_a$$
$$\underline{CN^-(aq) + H_2O(l) \rightleftharpoons HCN(aq) + OH^-(aq) \quad K_b}$$
$$2H_2O(l) \rightleftharpoons H_3O^+(aq) + OH^-(aq) \quad K_w$$

The equilibrium constants for the reactions are shown symbolically at the right. In Section 14.2, it was shown that when two reactions are added, their equilibrium constants are multiplied. Therefore, ▶

$$K_a K_b = K_w$$

This relationship is general and shows that the product of acid- and base-ionization constants in aqueous solution for conjugate acid–base pairs equals the ion-product constant for water, K_w.

Another way to obtain this result is by multiplying the expression for K_b by $[H_3O^+]/[H_3O^+]$ and rearranging.

$$K_b = \frac{[HCN][OH^-]}{[CN^-]} \times \frac{[H_3O^+]}{[H_3O^+]}$$

$$= [H_3O^+][OH^-] \Big/ \frac{[H_3O^+][CN^-]}{[HCN]}$$

This is equivalent to

$$K_b = K_w/K_a \quad or \quad K_a K_b = K_w$$

Example 16.7 Obtaining K_a from K_b or K_b from K_a

Gaining Mastery Toolbox

Critical Concept 16.7
Acid- and base-ionization constants are related. The equation that relates K_a and K_b is $K_a K_b = K_w$.

Solution Essentials:
- Mathematical relationship between K_a and K_b ($K_a K_b = K_w$)
- Base-ionization constant (K_b)
- Acid-ionization constant (K_a)
- Ion-product constant for water ($K_w = 1.0 \times 10^{-14}$)
- Conjugate acid–base pair
- Equilibrium-constant expression

Use Tables 16.1 and 16.2 to obtain the following at 25°C: **a** K_b for CN^-; **b** K_a for NH_4^+.

Problem Strategy Keeping in mind that K_w is known, we can apply the formula $K_a K_b = K_w$ to find the unknown ionization constant (K_a or K_b).

Solution

a The conjugate acid of CN^- is HCN, whose K_a is 4.9×10^{-10} (Table 16.1). Hence,

$$K_b = \frac{K_w}{K_a} = \frac{1.0 \times 10^{-14}}{4.9 \times 10^{-10}} = \textbf{2.0} \times \textbf{10}^{-5}$$

Note that K_b is approximately equal to K_b for ammonia (1.8×10^{-5}). This means that the base strength of CN^- is comparable to that of NH_3.

b The conjugate base of NH_4^+ is NH_3, whose K_b is 1.8×10^{-5} (Table 16.2). Hence,

$$K_a = \frac{K_w}{K_b} = \frac{1.0 \times 10^{-14}}{1.8 \times 10^{-5}} = \textbf{5.6} \times \textbf{10}^{-10}$$

NH_4^+ is a relatively weak acid. Acetic acid, by way of comparison, has K_a equal to 1.7×10^{-5}.

Answer Check Be aware that the values of K_a, K_b, and K_w are all temperature dependent, so make sure that you are using the appropriate values when performing calculations.

Exercise 16.8 Calculate the following at 25°C, using Tables 16.1 and 16.2: **a** K_b for F^-; **b** K_a for $C_6H_5NH_3^+$ (conjugate acid of aniline, $C_6H_5NH_2$).

■ See Problems 16.63 and 16.64.

For a solution of a salt in which only one of the ions hydrolyzes, the calculation of the concentrations of species present follows that for solutions of weak acids or bases. The only difference is that you must first obtain K_a or K_b for the ion that hydrolyzes. The next example illustrates the reasoning and calculations involved.

Example 16.8 Calculating Concentrations of Species in a Salt Solution

Gaining Mastery Toolbox

Critical Concept 16.8
When salts dissolve in solution, the aqueous anions may hydrolyze with water. Recognize the conjugates of weak acids and bases and be able to write the chemical equations for the hydrolysis reactions of these conjugates.

Solution Essentials:
- Rules for hydrolysis reactions of salts in solution
- Hydrolysis reaction
- Mathematical relationship between K_a and K_b ($K_a K_b = K_w$)
- Base-ionization constant (K_b)
- Acid-ionization constant (K_a)
- pH
- Ion-product constant for water ($K_w = 1.0 \times 10^{-14}$)
- Relative strengths of acids and bases
- Conjugate acid–base pair
- Equilibrium-constant expression

What is the pH of 0.10 M sodium nicotinate at 25°C (a problem posed in the chapter opening)? The K_a for nicotinic acid was determined in Example 16.1 to be 1.4×10^{-5} at 25°C.

Problem Strategy We first need to determine the relevant reactions of sodium nicotinate in water. If there is a reaction, we need to write it down and determine whether the reaction is that of an acid or of a base. If the reaction is that of an acid, we will need to use K_a in our calculations; for a base, we will need K_b. If necessary,

(continued)

(continued)

we can use the equation $K_a K_b = K_w$ to find the needed ionization constant. We then can perform the equilibrium calculation in the usual manner.

Solution Sodium nicotinate gives Na^+ and nicotinate ions in solution. Only the nicotinate ion hydrolyzes. Write HNic for nicotinic acid and Nic^- for the nicotinate ion. The hydrolysis of nicotinate ion is

$$Nic^-(aq) + H_2O(l) \rightleftharpoons HNic(aq) + OH^-(aq)$$

Nicotinate ion acts as a base, and you can calculate the concentration of species in solution as in Example 16.5. First, however, you need K_b for the nicotinate ion. This is related to K_a for nicotinic acid by the equation $K_a K_b = K_w$. Substituting, you get

$$K_b = \frac{K_w}{K_a} = \frac{1.0 \times 10^{-14}}{1.4 \times 10^{-5}} = 7.1 \times 10^{-10}$$

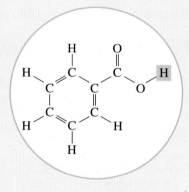

Now you can proceed with the equilibrium calculation. We will only sketch this calculation, because it is similar to that in Example 16.5. You let $x = [HNic] = [OH^-]$ and then substitute into the equilibrium-constant equation.

$$\frac{[HNic][OH^-]}{[Nic^-]} = K_b$$

This gives

$$\frac{x^2}{0.10 - x} = 7.1 \times 10^{-10}$$

Solving this equation, you find that $x = [OH^-] = 8.4 \times 10^{-6}$. Hence,

$$pH = 14.00 - pOH = 14.00 + \log(8.4 \times 10^{-6}) = \textbf{8.92}$$

Answer Check Use the rules for predicting whether a salt is acidic, basic, or neutral to assess your answer. In this case, the solution was prepared by dissolving the salt of a strong base and a weak acid, so we should expect the correct answer to have a pH greater than 7.

Exercise 16.9 Benzoic acid, $HC_7H_5O_2$, and its salts are used as food preservatives. What is the concentration of benzoic acid in an aqueous solution of 0.015 M sodium benzoate? What is the pH of the solution? K_a for benzoic acid is 6.3×10^{-5}.

■ See Problems 16.65, 16.66, 16.67, and 16.68.

CONCEPT CHECK 16.3

Which of the following aqueous solutions has the highest pH and which has the lowest? **a** 0.1 M NH_3; **b** 0.1 M NH_4Br; **c** 0.1 M NaF; **d** 0.1 M NaCl.

Solutions of a Weak Acid or Base with Another Solute

In the preceding sections, we looked at solutions that contained a weak acid, a weak base, or a salt of a weak acid or base. In the remaining sections of this chapter, we will look at the effect of adding another solute to a solution of a weak acid or base. The solutes we will look at are those that significantly affect acid or base ionization—that is, strong acids and bases, and salts that contain an ion that is produced in the acid or base ionization. These solutes affect the equilibrium through the *common-ion effect*.

16.5 Common-Ion Effect

The **common-ion effect** is *the shift in an ionic equilibrium caused by the addition of a solute that provides an ion that takes part in the equilibrium.* Consider a solution of acetic acid, $HC_2H_3O_2$, in which you have the following acid-ionization equilibrium:

$$HC_2H_3O_2(aq) + H_2O(l) \rightleftharpoons C_2H_3O_2^-(aq) + H_3O^+(aq)$$

Suppose you add $HCl(aq)$ to this solution. What is the effect on the acid-ionization equilibrium? Because $HCl(aq)$ is a strong acid, it provides H_3O^+ ion, which is present on the right side of the equation for acetic acid ionization. According to Le Châtelier's principle, the equilibrium composition should shift to the left. ▶

$$HC_2H_3O_2(aq) + H_2O(l) \longleftarrow C_2H_3O_2^-(aq) + \underset{\text{added}}{H_3O^+(aq)}$$

The degree of ionization of acetic acid is decreased by the addition of a strong acid. This repression of the ionization of acetic acid by $HCl(aq)$ is an example of the common-ion effect.

> *Le Châtelier's principle was applied in Section 14.7 to the problem of adding substances to an equilibrium mixture.*

CONCEPT CHECK 16.4

The chemical equation for the hydrolysis of nitrous acid is

$$HNO_2(aq) + H_2O(l) \rightleftharpoons H_3O^+(aq) + NO_2^-(aq)$$

Which of the following, when added to a nitrous acid solution, would reduce the degree of ionization of the acid: $NaNO_3(aq)$, $KNO_2(aq)$, $HNO_3(aq)$, or $NaCl(aq)$?

Example 16.9

Calculating the Common-Ion Effect on Acid Ionization (Effect of a Strong Acid)

Gaining Mastery Toolbox

Critical Concept 16.9
Chemical equilibrium will be shifted by the addition of an aqueous ion that is present in a chemical reaction (Le Châtelier's principle). The concentrations of all the aqueous reactants and products change when the equilibrium is shifted.

Solution Essentials:
• Common-ion effect
• Acid-ionization constant (K_a)
• Degree of ionization
• Relative strengths of acids and bases
• Le Châtelier's principle
• Equilibrium-constant expression

The degree of ionization of acetic acid, $HC_2H_3O_2$, in a 0.10 M aqueous solution at 25°C is 0.013. K_a at this temperature is 1.7×10^{-5}. Calculate the degree of ionization of $HC_2H_3O_2$ in a 0.10 M solution at 25°C to which sufficient HCl is added to make it 0.010 M HCl. How is the degree of ionization affected?

Problem Strategy This is an acid-ionization problem, but it differs from the simple ionization illustrated by Example 16.2. Here you have a starting concentration of H_3O^+ (= 0.010 M) from the addition of a strong acid (HCl). This gives a different type of equation in Step 2, but you solve it in Step 3 by using a similar approximation method. You assume that x is small compared with starting concentrations of acid and H_3O^+, so that the resulting equation is linear, rather than quadratic. The calculation of degree of ionization follows that in Example 16.2. Note that the degree of ionization given in the problem (0.013) refers to the 0.10 M $HC_2H_3O_2$ solution prior to the addition of the HCl. Therefore, this information is not used in solving the problem but is used for comparison with the calculated answer.

Solution

Step 1: Starting concentrations are $[HC_2H_3O_2] = 0.10\ M$, $[H_3O^+] = 0.010\ M$ (from HCl), and $[C_2H_3O_2^-] = 0$. The acetic acid ionizes to give an additional x mol/L of H_3O^+ and x mol/L of $C_2H_3O_2^-$. The table is

Concentration (M)	$HC_2H_3O_2$(aq) + H_2O(l) $\rightleftharpoons$ H_3O^+(aq) +		$C_2H_3O_2^-$(aq)
Starting	0.10	0.010	0
Change	$-x$	$+x$	$+x$
Equilibrium	$0.10 - x$	$0.010 + x$	x

Step 2: You substitute into the equilibrium-constant equation

$$\frac{[H_3O^+][C_2H_3O_2^-]}{[HC_2H_3O_2]} = K_a$$

(continued)

(*continued*)

obtaining

$$\frac{(0.010 + x)x}{0.10 - x} = 1.7 \times 10^{-5}$$

Step 3: To solve this equation, assume that x is small compared with 0.010. Then

$$0.010 + x \approx 0.010$$
$$0.10 - x \approx 0.10$$

The equation becomes

$$\frac{0.010x}{0.10} \approx 1.7 \times 10^{-5}$$

Solving for x, you get

$$x = 1.7 \times 10^{-5} \times \frac{0.10}{0.010} = 1.7 \times 10^{-4}$$

The degree of ionization of $HC_2H_3O_2$ is $x/0.10 = \mathbf{0.0017}$. This is much smaller than the value for 0.10 M $HC_2H_3O_2$ (0.013), because **the addition of HCl suppresses the ionization of $HC_2H_3O_2$.**

Answer Check As with all problems of this type, check to see that all approximations made are valid. In this case, make certain that x can be neglected in the $0.010 + x$ terms.

Exercise 16.10 What is the concentration of formate ion, CHO_2^-, in a solution at 25°C that is 0.10 M $HCHO_2$ and 0.20 M HCl? What is the degree of ionization of formic acid, $HCHO_2$?

■ See Problems 16.69 and 16.70.

As you will see in the following section, solutions that contain a weak acid or a weak base and a corresponding salt are especially important. Therefore, you will need to be able to calculate the concentrations of species present in such solutions. A solution of acetic acid and sodium acetate is an example. From Le Châtelier's principle, you can see that the addition of acetate ion (from sodium acetate) suppresses the ionization of the acetic acid by the common-ion effect:

$$HC_2H_3O_2(aq) + H_2O(l) \longleftarrow \underset{\text{added}}{C_2H_3O_2^-(aq)} + H_3O^+(aq)$$

The next example illustrates the calculation of pH, which is similar to that of the previous example.

Example 16.10 Calculating the Common-Ion Effect on Acid Ionization (Effect of a Conjugate Base)

Gaining Mastery Toolbox

Critical Concept 16.10
Before performing calculations, the relevant hydrolysis reaction for the acid or base must be determined. When you add a soluble salt to a solution containing an acid or base, you must determine if there is an ion (a common ion) that will affect the chemical reaction.

Solution Essentials:
• Common-ion effect
• Acid ionization constant (K_a)
• pH
• Le Châtelier's principle
• Equilibrium-constant expression

A solution is prepared to be 0.10 M acetic acid, $HC_2H_3O_2$, and 0.20 M sodium acetate, $NaC_2H_3O_2$. What is the pH of this solution at 25°C? K_a for acetic acid at 25°C is 1.7×10^{-5}.

Problem Strategy First you need to identify the relevant reaction for the acid or base in solution. In this case,

$$HC_2H_3O_2(aq) + H_2O(l) \rightleftharpoons C_2H_3O_2^-(aq) + H_3O^+(aq)$$

(*continued*)

(continued)

Once you have the reaction, you need to see whether there are any ions in solution from the added salt ($NaC_2H_3O_2$) that will affect the chemical reaction (a common ion). If there is a common ion, as there is here ($C_2H_3O_2^-$), you need to take into account its concentration when creating your table and performing calculations.

Solution

Step 1: Consider the equilibrium

$$HC_2H_3O_2(aq) + H_2O(l) \rightleftharpoons H_3O^+(aq) + C_2H_3O_2^-(aq)$$

Initially, 1 L of solution contains 0.10 mol of acetic acid. Sodium acetate is a strong electrolyte, so 1 L of solution contains 0.20 mol of acetate ion (a common ion). When the acetic acid ionizes, it gives x mol of hydronium ion and x mol of acetate ion. This is summarized in the following table:

Concentration (M)	$HC_2H_3O_2(aq)$ + $H_2O(l)$ $\rightleftharpoons$	$H_3O^+(aq)$	+ $C_2H_3O_2^-(aq)$
Starting	0.10	~0	0.20
Change	−x	+x	+x
Equilibrium	0.10 − x	x	0.20 + x

Step 2: The equilibrium-constant equation is

$$\frac{[H_3O^+][C_2H_3O_2^-]}{[HC_2H_3O_2]} = K_a$$

Substituting into this equation gives

$$\frac{x(0.20 + x)}{0.10 - x} = 1.7 \times 10^{-5}$$

Step 3: To solve the equation, assume that x is small compared with 0.10 and 0.20. Then

$$0.20 + x \simeq 0.20$$
$$0.10 - x \simeq 0.10$$

The equilibrium equation becomes

$$\frac{x(0.20)}{0.10} \simeq 1.7 \times 10^{-5}$$

Hence,

$$x \simeq 1.7 \times 10^{-5} \times \frac{0.10}{0.20} = 8.5 \times 10^{-6}$$

(Note that x is indeed much smaller than 0.10 or 0.20.) The hydronium-ion concentration is 8.5×10^{-6} M, and

$$pH = -\log [H_3O^+] = -\log(8.5 \times 10^{-6}) = \textbf{5.07}$$

For comparison, the pH of 0.10 M acetic acid is 2.88.

Answer Check Always make sure that the ions in solution will have an effect on the equilibrium. For example, if the solution were 0.20 M NaCl instead of 0.20 M $HC_2H_3O_2$, there would have been no common ion, and therefore the calculated pH would be the same as that of a solution of only 0.20 M $HC_2H_3O_2$.

Exercise 16.11 One liter of solution was prepared by dissolving 0.025 mol of formic acid, $HCHO_2$, and 0.018 mol of sodium formate, $NaCHO_2$, in water. What was the pH of the solution? K_a for formic acid is 1.7×10^{-4}.

■ See Problems 16.71 and 16.72.

The previous examples considered the common-ion effect in solutions of weak acids. You also encounter the common-ion effect in solutions of weak bases. For example, the base ionization of NH_3 is

$$NH_3(aq) + H_2O(l) \rightleftharpoons NH_4^+(aq) + OH^-(aq)$$

This suggests that you could suppress, or decrease the effect of, the base ionization by adding an ammonium salt (providing NH_4^+ ion) or a strong base (providing OH^- ion). The calculations are similar to those for acid ionization, except that you use the appropriate base ionization.

16.6 Buffers

A **buffer** is *a solution characterized by the ability to resist changes in pH when limited amounts of acid or base are added to it.* If 0.01 mol of hydrochloric acid is added to 1 L of pure water, the pH changes from 7.0 to 2.0—a pH change of 5.0 units. By contrast, the addition of this amount of hydrochloric acid to 1 L of buffered solution might change the pH by only 0.1 unit. Biological fluids, such as blood, are usually buffer solutions; the control of pH is vital to proper functioning of these fluids. The oxygen-carrying function of blood depends on its being maintained very near a pH of 7.4. If the pH were to change by a tenth of a unit, the capacity of the blood to carry oxygen would be lost.

Buffers contain either a weak acid and its conjugate base or a weak base and its conjugate acid. Thus, a buffer solution contains both an acid species and a base species in equilibrium. To understand the action of a buffer, consider one that contains approximately equal molar amounts of a weak acid HA and its conjugate base A^-. When a strong acid is added to the buffer, it supplies hydronium ions that react with the base A^-.

$$H_3O^+(aq) + A^-(aq) \longrightarrow HA(aq) + H_2O(l)$$

On the other hand, when a strong base is added to the buffer, it supplies hydroxide ions. These react with the acid HA.

$$OH^-(aq) + HA(aq) \longrightarrow H_2O(l) + A^-(aq)$$

A buffer solution resists changes in pH through its ability to combine with both H_3O^+ and OH^- ions. Figure 16.9 depicts this buffer action.

Blood, as a buffer solution, contains H_2CO_3 and HCO_3^-, as well as other conjugate acid–base pairs. A buffer frequently used in the laboratory contains varying proportions of the conjugate acid–base pair $H_2PO_4^-$ and HPO_4^{2-} (Figure 16.10).

Buffers also have commercial applications. For example, the label on a package of artificial fruit juice mix says that it contains "citric acid to provide tartness (acidity) and sodium citrate to regulate tartness." A solution of citric acid and its conjugate base, citrate ion (provided by sodium citrate), functions as an acid–base

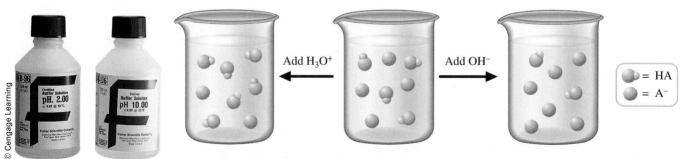

Figure 16.9 ▲

Buffer action A buffer is a solution of about equal molar amounts of a weak acid, HA, and its conjugate base, A^-, in equilibrium (center beaker). When an acid, such as HCl, is added to the buffer, H_3O^+ from this acid reacts with A^- in the buffer to produce HA (left beaker). However, when a base, such as NaOH, is added to the buffer, OH^- from this base reacts with HA in the buffer to produce A^- (right beaker). The buffer pH changes only slightly in both cases.

Figure 16.10 ▲

Some laboratory buffers These commercially prepared buffer solutions have different proportions of the same conjugate acid and base.

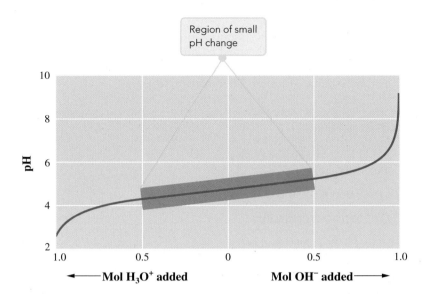

Figure 16.11 ◄

Effect of added acid or base on a buffer solution The buffer contains 1.0 mol of acetic acid and 1.0 mol of acetate ion in 1.00 L of solution. Note that the addition of 0.5 mol or less of strong acid or base gives only a small change of pH.

buffer, which is what "to regulate tartness" means. The pH of the buffer is in the acid range.

Two important characteristics of a buffer are its pH and its *buffer capacity,* which is the amount of acid or base the buffer can react with before giving a significant pH change. Buffer capacity depends on the amount of acid and conjugate base in the solution. Figure 16.11 illustrates the change in pH of a buffer solution containing 1.0 mol of acetic acid and 1.0 mol of acetate ion to which varying amounts of H_3O^+ and OH^- are added. This buffer changes less than 0.5 pH unit as long as no more than 0.5 mol H_3O^+ or OH^- ion is added. Note that this is one-half (or less than one-half) the amounts of acid and conjugate base in the solution. The ratio of amount of acid to amount of conjugate base is also important. Unless this ratio is close to 1 (between 1:10 and 10:1), the buffer capacity will be too low to be useful.

The other important characteristic of a buffer is its pH. Let us now look at how to calculate the pH of a buffer.

The pH of a Buffer

The solution described in Example 16.10 is a buffer, because it is a solution of a weak acid (0.10 *M* acetic acid) and its conjugate base (0.20 *M* acetate ion from sodium acetate). The example described how to calculate the pH of such a buffer (we obtained a pH of 5.07). The next example is similar, except now you are given a recipe for making up a buffer from volumes of solutions and are asked to calculate the pH. In this example, the buffer consists of a molecular base (NH_3) and its conjugate acid (NH_4^+).

⛵WL Interactive Example 16.11   Calculating the pH of a Buffer from Given Volumes of Solution

Gaining Mastery Toolbox

Critical Concept 16.11
Buffer solutions can be prepared by mixing an acid and the conjugate base of the acid or a base and the conjugate acid of the base. A buffer solution, regardless of the way it is prepared, will resist pH changes when limited amounts of acid or base are added to the solution because it contains a conjugate acid–base pair.

Solution Essentials:
- Buffer
- Base-ionization constant (K_b)
- Acid-ionization constant (K_a)
- pH
- Conjugate acid–base pair
- Relative strengths of acids and bases
- Equilibrium-constant expression

Instructions for making up a buffer say to mix 60. mL (0.060 L) of 0.100 *M* NH_3 with 40. mL (0.040 L) of 0.100 *M* NH_4Cl. What is the pH of this buffer?

Problem Strategy The buffer contains a base and its conjugate acid in equilibrium. The equation is

$$NH_3(aq) + H_2O(l) \rightleftharpoons NH_4^+(aq) + OH^-(aq)$$

The problem is to obtain the concentration of H_3O^+ or OH^- in this equilibrium mixture. To do the equilibrium

(continued)

(continued)

calculation, you need the starting concentrations in the solution obtained by mixing the NH_3 and NH_4Cl solutions. For this, you calculate the moles of NH_3 and moles of NH_4^+ added to the buffer solution, and divide by the total volume of buffer. Then you are ready to do the equilibrium calculation.

Solution How many moles of NH_3 are added? Recall that

$$\text{Molarity } NH_3 = \frac{\text{moles } NH_3}{\text{liters } NH_3 \text{ solution}}$$

Note that the instructions say to add 60. mL (or 0.060 L) of 0.100 M NH_3. So

$$\text{Moles } NH_3 = \text{molarity } NH_3 \times \text{liters } NH_3 \text{ solution}$$
$$= 0.100 \frac{\text{mol}}{\text{L}} \times 0.060 \text{ L} = 0.0060 \text{ mol } NH_3$$

In the same way, you find that you have added 0.0040 mol NH_4^+ (from NH_4Cl). Assume that the total volume of buffer equals the sum of the volumes of the two solutions.

$$\text{Total volume buffer} = 60. \text{ mL} + 40. \text{ mL} = 100. \text{ mL} (0.100 \text{ L})$$

Therefore, the concentrations of base and conjugate acid are

$$[NH_3] = \frac{0.0060 \text{ mol}}{0.100 \text{ L}} = 0.060 \; M$$

$$[NH_4^+] = \frac{0.0040 \text{ mol}}{0.100 \text{ L}} = 0.040 \; M$$

Step 1: Fill in the concentration table for the acid–base equilibrium (base ionization of NH_3).

Concentration (M)	$NH_3(aq)$ + $H_2O(l)$ $\rightleftharpoons$	$NH_4^+(aq)$ +	$OH^-(aq)$
Starting	0.060	0.040	0
Change	$-x$	$+x$	$+x$
Equilibrium	$0.060 - x$	$0.040 + x$	x

Step 2: You substitute the equilibrium concentrations into the equilibrium-constant equation.

$$\frac{[NH_4^+][OH^-]}{[NH_3]} = K_b$$

$$\frac{(0.040 + x)x}{0.060 - x} = 1.8 \times 10^{-5}$$

Step 3: To solve this equation, you assume that x is small compared with 0.040 and 0.060. So this equation becomes

$$\frac{0.040x}{0.060} = 1.8 \times 10^{-5}$$

Therefore, $x = (0.060/0.040) \times (1.8 \times 10^{-5}) = 2.7 \times 10^{-5}$. (Check that x can be neglected in $0.040 + x$ and $0.060 - x$.) Thus, the hydroxide-ion concentration is $2.7 \times 10^{-5} \; M$. The pH of the buffer is

$$pH = 14.00 - pOH = 14.00 + \log(2.7 \times 10^{-5}) = \mathbf{9.43}$$

(If you have 2.7×10^{-5} in your calculator from the previous calculation, you obtain the pH by pressing the *log* button, then adding 14.)

Answer Check To avoid a very common mistake, prior to setting up the equilibrium problem, always make sure that you first take into account the total volume of solution and then calculate the concentrations.

(continued)

(continued)

Exercise 16.12 What is the pH of a buffer prepared by adding 30.0 mL of 0.15 M $HC_2H_3O_2$ (acetic acid) to 70.0 mL of 0.20 M $NaC_2H_3O_2$ (sodium acetate)?

See Problems 16.75 and 16.76.

Adding an Acid or a Base to a Buffer

We noted earlier that a buffer resists changes in pH, and we also described qualitatively how this occurs. What we want to do here is to show quantitatively that a buffer does indeed tend to resist pH change when a small quantity of an acid or a base is added to it.

All of the previous acid–base examples dealt with situations in which an acid and its conjugate base were close to equilibrium. In the following example, we consider the addition of a strong acid to an acetic acid/acetate ion buffer. Hydronium ion reacts with the conjugate base (acetate ion) in the buffer, so that the concentrations of species in solution change markedly from their initial value. (If you add a strong base to this buffer, hydroxide ion reacts with acetic acid in the buffer.)

Example 16.12

Calculating the pH of a Buffer When a Strong Acid or Strong Base Is Added

Gaining Mastery Toolbox

Critical Concept 16.12
Strong acids react completely with weak bases, and strong bases react completely with weak acids. When either a strong acid or strong base is added to a buffer, a stoichiometric calculation is performed prior to the equilibrium calculation. The result of the stoichiometric calculation is the starting point for the equilibrium calculation.

Solution Essentials:
• Buffer
• Base-ionization constant (K_b)
• Acid-ionization constant (K_a)
• pH
• Relative strengths of acids and bases
• Equilibrium-constant expression
• Volumetric analysis

Calculate the pH of 75 mL of the buffer solution described in Example 16.10 (0.10 M $HC_2H_3O_2$ and 0.20 M $NaC_2H_3O_2$) to which 9.5 mL of 0.10 M hydrochloric acid is added. Compare the pH change with what would occur if this amount of acid were added to pure water.

Problem Strategy Do the problem in two parts. To get started, you assume that the H_3O^+ ion from the strong acid and the conjugate base from the buffer react completely. This is a stoichiometric calculation. Next, you need to refine the assumption from the stoichiometric calculation that you started with, because in reality the H_3O^+ ion and the base from the buffer reach equilibrium just before complete reaction. So you now solve the equilibrium problem using concentrations from the stoichiometric calculation. Because these concentrations are not far from equilibrium, you can use the usual simplifying assumption about x.

Solution When hydronium ion (from hydrochloric acid) is added to the buffer, it reacts with acetate ion.

$$H_3O^+(aq) + C_2H_3O_2^-(aq) \longrightarrow HC_2H_3O_2(aq) + H_2O(l)$$

Because acetic acid is a weak acid, you can assume as a first approximation that the reaction goes to completion. This part of the problem is simply a stoichiometric calculation. Then you assume that the acetic acid ionizes slightly. This part of the problem involves an acid-ionization equilibrium.

STOICHIOMETRIC CALCULATION You must first calculate the amounts of hydrogen ion, acetate ion, and acetic acid present in the solution before reaction. The molar amount of hydrogen ion will equal the molar amount of hydrochloric acid added, which you obtain by converting the volume of hydrochloric acid, HCl, to moles of HCl. To do this, you note for 0.10 M HCl that 1 L HCl is equivalent to 0.10 mol HCl. Hence, to convert 9.5 mL ($= 9.5 \times 10^{-3}$ L) of hydrochloric acid, you have

$$9.5 \times 10^{-3} \text{ L HCl} \times \frac{0.10 \text{ mol HCl}}{1 \text{ L HCl}} = 0.00095 \text{ mol HCl}$$

Hydrochloric acid is a strong acid, so it exists in solution as the ions H_3O^+ and Cl^-. Therefore, the amount of H_3O^+ added is 0.00095 mol.

(continued)

(*continued*)

The amounts of acetate ion and acetic acid in the 75-mL sample are found in a similar way. The buffer in Example 16.10 contains 0.20 mol of acetate ion and 0.10 mol of acetic acid in 1 L of solution. The amounts in 75 mL (= 0.075 L) of solution are obtained by converting to moles.

$$0.075 \text{ L soln} \times \frac{0.20 \text{ mol } C_2H_3O_2^-}{1 \text{ L soln}} = 0.015 \text{ mol } C_2H_3O_2^-$$

$$0.075 \text{ L soln} \times \frac{0.10 \text{ mol } HC_2H_3O_2}{1 \text{ L soln}} = 0.0075 \text{ mol } HC_2H_3O_2$$

You now assume that all of the hydronium ion added (0.00095 mol) reacts with acetate ion. Therefore, 0.00095 mol of acetic acid is produced and 0.00095 mol of acetate ion is used up. Hence, after reaction you have

Moles of acetate ion = $(0.015 - 0.00095)$ mol $C_2H_3O_2^-$ = 0.014 mol $C_2H_3O_2^-$

Moles of acetic acid = $(0.0075 + 0.00095)$ mol $HC_2H_3O_2$ = 0.0085 mol $HC_2H_3O_2$

EQUILIBRIUM CALCULATION You first calculate the concentrations of $HC_2H_3O_2$ and $C_2H_3O_2^-$ present in the solution before you consider the acid-ionization equilibrium. Note that the total volume of solution (buffer plus hydrochloric acid) is 75 mL + 9.5 mL, or 85 mL (0.085 L). Hence, the starting concentrations are

$$[HC_2H_3O_2] = \frac{0.0085 \text{ mol}}{0.085 \text{ L}} = 0.10 \, M$$

$$[C_2H_3O_2^-] = \frac{0.014 \text{ mol}}{0.085 \text{ L}} = 0.16 \, M$$

From this, you construct the following table:

Concentration (M)	$HC_2H_3O_2(aq)$	+ $H_2O(l)$	$\rightleftharpoons$	$H_3O^+(aq)$	+ $C_2H_3O_2^-(aq)$
Starting	0.10			~0	0.16
Change	$-x$			$+x$	$+x$
Equilibrium	$0.10 - x$			x	$0.16 + x$

The equilibrium-constant equation is

$$\frac{[H_3O^+][C_2H_3O_2^-]}{[HC_2H_3O_2]} = K_a$$

Substituting, you get

$$\frac{x(0.16 + x)}{0.10 - x} = 1.7 \times 10^{-5}$$

If you assume that x is small enough that $0.16 + x \approx 0.16$ and $0.10 - x \approx 0.10$, this equation becomes

$$\frac{x(0.16)}{0.10} = 1.7 \times 10^{-5}$$

or

$$x = 1.7 \times 10^{-5} \times \frac{0.10}{0.16} = 1.1 \times 10^{-5}$$

Note that x is indeed small, so the assumptions you made earlier are correct. The H_3O^+ concentration is $1.1 \times 10^{-5} \, M$. The pH is

$$pH = -\log[H_3O^+] = -\log(1.1 \times 10^{-5}) = \textbf{4.96}$$

Because the pH of the buffer was 5.07 (see Example 16.10), the pH has changed by $5.07 - 4.96 = 0.11$ unit.

(*continued*)

(*continued*)

ADDING HCl TO PURE WATER If 9.5 mL of 0.10 M hydrochloric acid were added to 75 mL of pure water, the hydronium-ion concentration would change to

$$[H_3O^+] = \frac{\text{amount of } H_3O^+ \text{ added}}{\text{total volume of solution}}$$

$$[H_3O^+] = \frac{0.00095 \text{ mol } H_3O^+}{0.085 \text{ L solution}} = 0.011 \; M$$

(The total volume is 75 mL of water plus 9.5 mL HCl, assuming no change of volume on mixing.) The pH is

$$pH = -\log [H_3O^+] = -\log(0.011) = 1.96$$

The pH of pure water is 7.00, so the change in pH is 7.00 − 1.96 = 5.04 units, compared with 0.11 unit for the buffered solution.

Answer Check Generally speaking, buffer solutions will not undergo large changes in pH by the addition of small amounts of a strong acid or strong base, so this is something to look for in your answers. However, if the buffer capacity of the solution is exceeded, you will see significant changes in pH.

Exercise 16.13 Suppose you add 50.0 mL of 0.10 M sodium hydroxide to 1 L of the solution described in Exercise 16.11. What would be the pH of the result?

■ See Problems 16.77 and 16.78.

Henderson–Hasselbalch Equation

How do you prepare a buffer of given pH? We can show that the buffer must be prepared from a conjugate acid–base pair in which the acid-ionization constant is approximately equal to the desired H_3O^+ concentration. To illustrate, consider a buffer made up of a weak acid HA and its conjugate base A^-. The acid-ionization equilibrium is

$$HA(aq) + H_2O(l) \rightleftharpoons H_3O^+(aq) + A^-(aq)$$

and the acid-ionization constant is

$$K_a = \frac{[H_3O^+][A^-]}{[HA]}$$

By rearranging, you get an equation for the H_3O^+ concentration.

$$[H_3O^+] = K_a \times \frac{[HA]}{[A^-]}$$

This equation expresses the H_3O^+ concentration in terms of the K_a for the acid and the ratio of concentrations of HA and A^-. This equation was derived from the equilibrium constant, so the concentrations of HA and A^- should be equilibrium values. But because the presence of A^- suppresses the ionization of HA, these concentrations do not differ significantly from the values used to prepare the buffer. If [HA] and $[A^-]$ are approximately equal, the hydronium-ion concentration of the buffer is approximately equal to K_a.

You can use the preceding equation to derive an equation for the pH of a buffer. Take the negative logarithm of both sides of the equation. That is,

$$-\log[H_3O^+] = -\log\left(K_a \times \frac{[HA]}{[A^-]}\right) = -\log K_a - \log\frac{[HA]}{[A^-]}$$

The left side equals the pH. You can also simplify the right side. The pK_a of a weak acid is defined in a manner similar to pH and pOH. ▶

$$pK_a = -\log K_a$$

Acid- and base-ionization constants are often listed as pK_as and pK_bs (pK_b = −log K_b).

The previous equation can be written

$$pH = pK_a - \log\frac{[HA]}{[A^-]} = pK_a + \log\frac{[A^-]}{[HA]}$$

More generally, you can write

$$pH = pK_a + \log\frac{[base]}{[acid]}$$

This is *an equation relating the pH of a buffer for different concentrations of conjugate acid and base;* it is known as the **Henderson–Hasselbalch equation**. By substituting the value of pK_a for the conjugate acid and the ratio [base]/[acid], you obtain the pH of the buffer.

The question we asked earlier was how to prepare a buffer of a given pH—for example, pH 4.90. You can see that you need to find a conjugate acid–base pair in which the pK_a of the acid is close to the desired pH. K_a for acetic acid is 1.7×10^{-5}, and its pK_a is $-\log(1.7 \times 10^{-5}) = 4.77$. You can get a pH somewhat higher by increasing the ratio [base]/[acid].

Consider the calculation of the pH of a buffer containing 0.10 M NH_3 and 0.20 M NH_4Cl. The conjugate acid is NH_4^+, whose K_a you can calculate from K_b for NH_3 ($= 1.8 \times 10^{-5}$). K_a for NH_4^+ is 5.6×10^{-10}, and the pK_a is $-\log(5.6 \times 10^{-10}) = 9.25$. Hence,

$$pH = 9.25 + \log\frac{0.10}{0.20} = 8.95$$

The solution is basic, as you could have guessed, because the buffer contains the weak base NH_3 and the very weak acid NH_4^+.

CONCEPT CHECK 16.5

You add 1.5 mL of 1 M HCl to each of the following solutions. Which one will show the least change of pH?

a 15 mL of 0.1 M NaOH

b 15 mL of 0.1 M $HC_2H_3O_2$

c 30 mL of 0.1 M NaOH and 30 mL of 0.1 M $HC_2H_3O_2$

d 30 mL of 0.1 M NaOH and 60 mL of 0.1 M $HC_2H_3O_2$

CONCEPT CHECK 16.6

The beaker on the left below represents a buffer solution of a weak acid HA and its conjugate base, A^-. (Water molecules and spectator ions have been omitted for clarity.)

a Which beaker on the right, X or Y, depicts the solution after the addition of two formula units of NaOH?

b How many HCl molecules can be added to the original buffer solution before the buffer capacity is exceeded?

c Draw a picture of the solution where the buffer capacity has been exceeded by the addition of HCl.

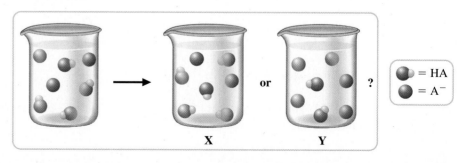

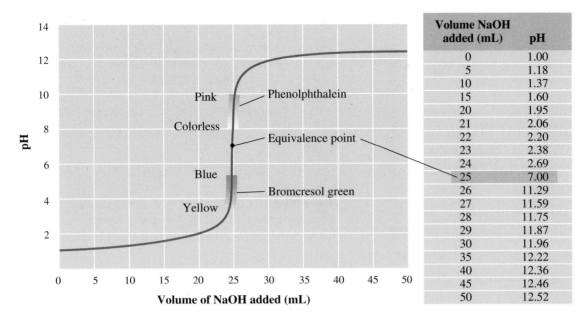

Volume NaOH added (mL)	pH
0	1.00
5	1.18
10	1.37
15	1.60
20	1.95
21	2.06
22	2.20
23	2.38
24	2.69
25	7.00
26	11.29
27	11.59
28	11.75
29	11.87
30	11.96
35	12.22
40	12.36
45	12.46
50	12.52

16.7 Acid–Base Titration Curves

An acid–base titration is a procedure for determining the amount of acid (or base) in a solution by determining the volume of base (or acid) of known concentration that will completely react with it. ▶ An **acid–base titration curve** is *a plot of the pH of a solution of acid (or base) against the volume of added base (or acid)*. Such curves are used to gain insight into the titration process. You can use the titration curve to choose an indicator that will show when the titration is complete.

Titration of a Strong Acid by a Strong Base

Figure 16.12 shows a curve for the titration of 25.0 mL of 0.100 *M* HCl by 0.100 *M* NaOH. Note that the pH changes slowly at first until the molar amount of base added nearly equals that of the acid—that is, until the titration is near the equivalence point. The **equivalence point** is *the point in a titration when a stoichiometric amount of reactant has been added.* At the equivalence point, the pH of this solution of NaOH and HCl is 7.0, because it contains a salt, NaCl, that does not hydrolyze. However, the pH changes rapidly near the equivalence point, from a pH of about 3 to a pH of about 11. To detect the equivalence point, you add an indicator that changes color within the pH range 3–11. Phenolphthalein can be used, because it changes from colorless to pink in the pH range 8.2–10.0. (Figure 15.10 showed the pH ranges for the color changes of indicators.) Even though this color change occurs on the basic side, only a fraction of a drop of base is required to change the pH several units when the titration is near the equivalence point. The indicator bromcresol green, whose color changes in the pH range 3.8–5.4, would also work. Because the pH change is so large, many other indicators could be used.

The following example shows how to calculate a point on the titration curve of a strong acid and a strong base.

Figure 16.12 ▲

Curve for the titration of a strong acid by a strong base Here 25.0 mL of 0.100 *M* HCl is titrated by 0.100 *M* NaOH. The portions of the curve where indicators bromcresol green and phenolphthalein change color are shown. Note that both indicators change color where the pH changes rapidly (the nearly vertical part of the curve).

The technique of titration was discussed in Section 4.10.

◉WL Interactive Example | 16.13 | **Calculating the pH of a Solution of a Strong Acid and a Strong Base**

Gaining Mastery Toolbox

Critical Concept 16.13
Strong acids react completely with strong bases. Calculations involving reactions of strong acids and strong bases are stoichiometric (non-equilibrium).

Solution Essentials:
• pH
• Relative strengths of acids and bases
• Volumetric analysis

Calculate the pH of a solution in which 10.0 mL of 0.100 *M* NaOH is added to 25.0 mL of 0.100 *M* HCl.

Problem Strategy Whenever we are performing calculations that involve titrations, we need to write down the balanced chemical equation and determine the extent of the reaction. In this case, the reaction is between a

(continued)

(continued)

strong acid and strong base, so we assume that the reaction is complete. Therefore, we will be performing a stoichiometric calculation.

Solution The equation is

$$H_3O^+(aq) + OH^-(aq) \longrightarrow H_2O(l) + H_2O(l)$$

You get the amounts of reactants by multiplying the volume (in liters) of each solution by its molar concentration.

$$Mol\ H_3O^+ = 0.0250\ L \times 0.100\ mol/L = 0.00250\ mol$$
$$Mol\ OH^- = 0.0100\ L \times 0.100\ mol/L = 0.00100\ mol$$

All of the OH^- reacts, leaving an excess of H_3O^+.

$$Excess\ H_3O^+ = (0.00250 - 0.00100)\ mol = 0.00150\ mol\ H_3O^+$$

You obtain the concentration of H_3O^+ by dividing this amount of H_3O^+ by the total volume of solution ($= 0.0250\ L + 0.0100\ L = 0.0350\ L$).

$$[H_3O^+] = \frac{0.00150\ mol}{0.0350\ L} = 0.0429$$

Hence,

$$pH = -\log[H_3O^+] = -\log(0.0429) = \mathbf{1.368}$$

Answer Check A quick check of the concentrations and volumes of the HCl and NaOH being reacted indicates that we would expect the answer to have a pH less than 7.

Exercise 16.14 What is the pH of a solution in which 15 mL of 0.10 *M* NaOH has been added to 25 mL of 0.10 *M* HCl?

■ See Problems 16.85 and 16.86.

Titration of a Weak Acid by a Strong Base

The titration of a weak acid by a strong base gives a somewhat different curve. Figure 16.13 shows the curve for the titration of 25.0 mL of 0.100 *M* nicotinic

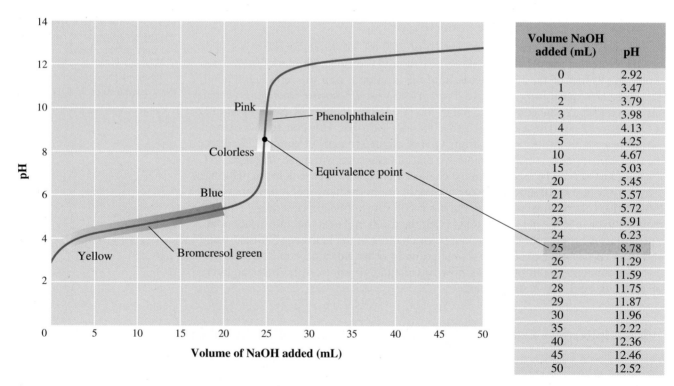

Volume NaOH added (mL)	pH
0	2.92
1	3.47
2	3.79
3	3.98
4	4.13
5	4.25
10	4.67
15	5.03
20	5.45
21	5.57
22	5.72
23	5.91
24	6.23
25	8.78
26	11.29
27	11.59
28	11.75
29	11.87
30	11.96
35	12.22
40	12.36
45	12.46
50	12.52

Figure 16.13 ▲

Curve for the titration of a weak acid by a strong base Here 25.0 mL of 0.100 *M* nicotinic acid, a weak acid, is titrated by 0.100 *M* NaOH. Note that bromcresol green changes color during the early part of the titration, well before the equivalence point. Phenolphthalein changes color where the pH changes rapidly (near the equivalence point). Thus, phenolphthalein could be used as an indicator for the titration, whereas bromcresol green could not.

acid, $HC_6H_4NO_2$, by 0.100 M NaOH. The titration starts at a higher pH than the titration of HCl, because nicotinic acid is a weak acid. As before, the pH changes slowly at first, then rapidly near the equivalence point. The pH range in which the rapid change is seen occurs from about pH 7 to pH 11. Note that the pH range is shorter than that for the titration of a strong acid by a strong base. This means that the choice of an indicator is more critical. Phenolphthalein would work; it changes color in the range 8.2–10.0. Bromcresol green would not work because it changes color in the range 3.8–5.4, which occurs before the titration curve rises steeply.

Note also that the equivalence point for the titration curve of nicotinic acid occurs on the basic side. This happens because at the equivalence point the solution is that of the salt, sodium nicotinate, which is basic from the hydrolysis of the nicotinate ion. The optimum choice of indicator would be one that changes color over a range that includes the pH of the equivalence point.

The following example shows how to calculate the pH at the equivalence point in the titration of a weak acid and a strong base.

⬛WL Interactive Example ▮16.14▮   **Calculating the pH at the Equivalence Point in the Titration of a Weak Acid by a Strong Base**

Gaining Mastery Toolbox

Critical Concept 16.14
A strong base will react completely with a weak acid. When a strong base is added to a solution containing a weak acid, a stoichiometric calculation is performed prior to the equilibrium calculation.

Solution Essentials:
- Acid–base titration
- Base-ionization constant (K_b)
- pH
- Relative strengths of acids and bases

Calculate the pH of the solution at the equivalence point when 25 mL of 0.10 M nicotinic acid is titrated by 0.10 M sodium hydroxide. K_a for nicotinic acid equals 1.4×10^{-5}.

Problem Strategy Because the titration is at the equivalence point, we know that equal molar amounts of nicotinic acid and sodium hydroxide have been reacted. From the chemical reaction we can determine that 1 mol of nicotinic acid reacts with 1 mol of sodium hydroxide to produce 1 mol of sodium nicotinate and 1 mol of water. We now can perform the stoichiometric calculation to determine the moles of nicotinate ion. Using this quantity and the total volume of solution, we can calculate the nicontinate ion concentration. We then can find the pH of this solution by treating it as a hydrolysis problem.

Solution **CONCENTRATION OF NICOTINATE ION:** Because the reaction is at the equivalence point, we assume that the reaction of the base with the acid is complete. The chemical equation for the titration is

$$HNic(aq) + NaOH(aq) \longrightarrow H_2O(aq) + NaNic(aq)$$

Therefore, the calculation of the moles of nicotinate ion is

$$2.5 \times 10^{-3} \, \text{L HNic} \times \frac{0.10 \, \text{mol HNic}}{1 \, \text{L HNic}} \times \frac{1 \, \text{mol NaNic}}{1 \, \text{mol HNic}} \times \frac{1 \, \text{mol Nic}^-}{1 \, \text{mol NaNic}} =$$
$$2.5 \times 10^{-4} \, \text{mol Nic}^-$$

From the reaction stoichiometry, we know that 25 mL of 0.10 M sodium hydroxide is needed to completely react with 25 mL of 0.10 M nicotinic acid, so the total volume of solution is 50 mL (assuming there is no volume change on mixing). Dividing the molar amount of nicotinate ion by the volume of solution in liters gives the molar concentration of nicotinate ion.

$$\text{Molar concentration} = \frac{2.5 \times 10^{-3} \, \text{mol}}{50 \times 10^{-3} \, \text{L}} = 0.050 \, M$$

HYDROLYSIS OF NICOTINATE ION This portion of the calculation follows the method given in Example 16.8. You find that K_b for nicotinate ion is 7.1×10^{-10} and that the concentration of hydroxide ion is $6.0 \times 10^{-6} \, M$. The pH is **8.78**.

(continued)

(continued)

Answer Check One bad assumption that is often made concerning acid–base titrations is that the pH is always 7 at the equivalence point. Although this is true when strong acids and bases are used, it is not the case here. Note that you can predict whether the solution will be acidic or basic at the equivalence point by examining the products of the chemical reaction to see if an acid or base was produced. Here, the weak base Nic^- was produced; therefore, we would expect a correct answer to be a pH greater than 7.

Exercise 16.15 What is the pH at the equivalence point when 25 mL of 0.10 M HF is titrated by 0.15 M NaOH?

■ See Problems 16.87 and 16.88.

Titration of a Weak Base by a Strong Acid

When you titrate a weak base by a strong acid, you get a titration curve similar to that obtained when a weak acid is titrated by a strong base. Figure 16.14 shows the pH changes during the titration of 25.0 mL of 0.100 M NH$_3$ by 0.100 M HCl. In this case, the pH declines slowly at first, then falls abruptly from about pH 7 to pH 3. Methyl red, which changes color from yellow at pH 6.0 to red at pH 4.8, is a possible indicator for this titration. Note that phenolphthalein could not be used to find the equivalence point.

This last section covering acid–base titrations is an ideal place to illustrate how titration problems can be configured to encompass many of the ideas and problem-solving approaches encountered in this chapter. Although Example 16.15 is focused on the titration of a weak base by a strong acid, the concepts and operational skills apply to the other types of acid–base titrations.

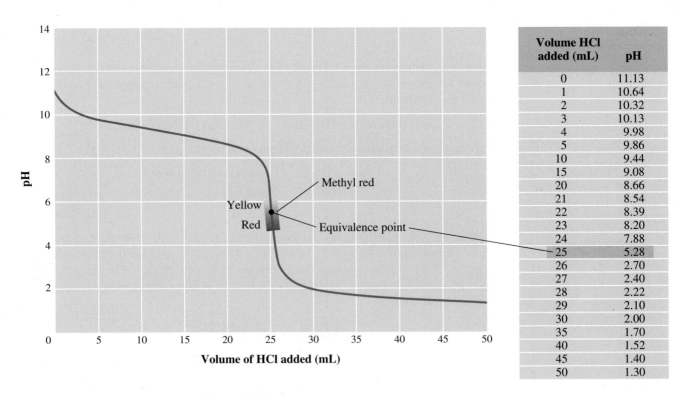

Volume HCl added (mL)	pH
0	11.13
1	10.64
2	10.32
3	10.13
4	9.98
5	9.86
10	9.44
15	9.08
20	8.66
21	8.54
22	8.39
23	8.20
24	7.88
25	5.28
26	2.70
27	2.40
28	2.22
29	2.10
30	2.00
35	1.70
40	1.52
45	1.40
50	1.30

Figure 16.14 ▲

Curve for the titration of a weak base by a strong acid Here 25.0 mL of 0.100 M NH$_3$ is titrated by 0.100 M HCl. Methyl red can be used as an indicator for the titration.

Example 16.15

Calculation of the pH of a Solution at Several Points of a Titration of a Weak Base by a Strong Acid

Gaining Mastery Toolbox

Critical Concept 16.15
A weak base will react completely with a strong acid. When a strong acid is added to a solution containing a weak base, a stoichiometric calculation is performed prior to the equilibrium calculation.

Solution Essentials:
• Acid–base titration
• Buffer
• Acid-ionization constant (K_a)
• Base-ionization constant (K_b)
• pH
• Relative strengths of acids and bases
• Equilibrium-constant expression

Figure 16.14 is the curve for the titration of a 25-mL sample of 0.100 M NH$_3$ by 0.100 M HCl. Calculate the pH of the NH$_3$ solution at the following points during the titration: **a** Prior to the addition of any HCl. **b** After the addition of 12 mL of 0.100 M HCl. **c** At the equivalence point. **d** After the addition of 31 mL of 0.100 M HCl.

Problem Strategy The key to solving this problem is to relate each part to a problem that you have already encountered. Part a involves a weak base (NH$_3$) in water, which was done in Example 16.5. For part b, you need to recognize that all of the H$_3$O$^+$ from the strong acid (HCl) will completely react with the NH$_3$, creating a buffer solution. This is similar to Example 16.12. Part c is at the equivalence point, so you know that the moles of HCl added to the solution are equal to the moles of NH$_3$. As in Example 16.14, you will need to perform a stoichiometry problem followed by an equilibrium calculation, only this time it is for a weak base with a strong acid. Finally, in part d, the titration is past the equivalence point. Because you are adding a strong acid beyond the equivalence point, the pH of the solution can be calculated based *only* on the moles of unreacted HCl (H$_3$O$^+$) that remain in the solution.

Solution

a NH$_3$, a weak base, undergoes the hydrolysis reaction

$$NH_3(aq) + H_2O(aq) \longrightarrow NH_4^+(aq) + OH^-(aq) \qquad K_b = 1.8 \times 10^{-5}$$

Starting with the initial concentration of NH$_3$ of 0.100 M, you perform an equilibrium calculation by setting up a table and solving for the concentration of OH$^-$. Then calculate the pH.

The pH prior to the start of the titration is **11.13**.

b When HCl is added to the NH$_3$ solution, it undergoes the reaction

$$H_3O^+(aq) + NH_3(aq) \longrightarrow NH_4^+(aq) + H_2O(l)$$

During the reaction all of the added HCl is consumed. (When looking at the reaction, keep in mind that because HCl is a strong acid, we represent it as H$_3$O$^+$ in aqueous solution.) Solving this stoichiometry problem, you can determine the moles of NH$_3$ that remain and the number of moles of NH$_4^+$ that were produced. Accounting for the volume of the solution, you calculate the following concentrations:

$$[NH_3] = 3.5 \times 10^{-2} M$$
$$[NH_4^+] = 3.2 \times 10^{-2} M$$

Recognizing that this is a buffer solution, you can use the Henderson–Hasselbalch equation (K_a for NH$_4^+$ = 5.6 × 10^{-10}) to solve for the pH, which is **9.29**. If you don't recognize that this is a buffer solution, you can solve this as an equilibrium problem based on the hydrolysis reaction of NH$_3$.

c This part is solved in nearly the same manner as part b, only now the number of moles of H$_3$O$^+$ is equal to the number of moles of NH$_3$. Using the reaction stoichiometry and keeping track of the total solution volume (50 mL), you determine

$$[NH_4^+] = 5.0 \times 10^{-2} M$$
$$[NH_3] = 0 M$$

This is not a buffer solution because the [NH$_3$] = 0 M; therefore, we need to solve an equilibrium problem of the hydrolysis of NH$_4^+$, which indicates pH = **5.28**. Note how this agrees with Figure 16.14.

d Finally, you need to calculate the concentration of H$_3$O$^+$ that is due to the excess HCl added beyond the equivalence point. Remembering the titration reaction

$$H_3O^+(aq) + NH_3(aq) \longrightarrow NH_4^+(aq) + H_2O(l)$$

(*continued*)

(continued)

You perform the stoichiometry calculation by first determing the moles of H_3O^+ that reacted with the NH_3. Then you use this quantity to determine the amount of unreacted H_3O^+ that remains in solution.

$$\text{Mol } H_3O^+ \text{ reacted} = 2.5 \times 10^{-3} \text{ mol}$$

$$\text{Mol } H_3O^+ \text{ remaining} = (3.1 \times 10^{-3} \text{ mol} - 2.5 \times 10^{-3} \text{ mol}) = 6.0 \times 10^{-4} \text{ mol}$$

In order to calculate the pH, you first need to know the concentration of the H_3O^+. Taking into account the total volume of solution (56 mL), you determine the concentration of H_3O^+.

$$[H_3O^+] = \frac{6.0 \times 10^{-4} \text{ mol}}{0.056 \text{ L}} = 1.1 \times 10^{-2} \text{ M}$$

Therefore, the pH = **1.96**.

Answer Check If you know the general attributes of the different titration curves, you can make sure that your numerical answers fall on the curve. Compare the answers in this problem with the curve in Figure 16.14.

Exercise 12.16 An 80.0-mL sample of 0.200 *M* ammonia is titrated with 0.100 *M* hydrochloric acid. K_b for ammonia is 1.8×10^{-5}. Calculate the pH of the solution at each of the following points of the titration: **a** before the addition of any HCl; **b** halfway to the equivalence point; **c** at the equivalence point; **d** after the addition of 175 mL of 0.100 *M* HCl.

■ See Problems 16.91 and 16.92.

A Checklist for Review

OWL and **GO Chemistry** Sign in at **www.cengage.com/owl** to:
• View tutorials and simulations, develop problem-solving skills, and complete online homework assigned by your professor.
• Download Go Chemistry mini lecture modules for quick review and exam prep from OWL or purchase them at **www.cengagebrain.com**.

Summary of Facts and Concepts

When a weak acid dissolves in water, it ionizes to give hydronium ions and the conjugate base ions. The equilibrium constant for this *acid ionization* is

$$K_a = [H_3O^+][A^-]/[HA],$$

where HA is the general formula for the acid. The constant K_a can be determined from the pH of an acid solution of known concentration. Once obtained, the acid-ionization constant can be used to find the concentrations of species in any solution of the acid. In the case of a *diprotic* acid, H_2A, the concentration of H_3O^+ and HA^- are calculated from K_{a1}, and the concentration of A^{2-} equals K_{a2}. Similar considerations apply to *base ionizations*.

Solutions of salts may be acidic or basic because of *hydrolysis* of the ions. The equilibrium constant for hydrolysis equals K_a for a cation or K_b for an anion. Calculation of the pH of a solution of a salt in which one ion hydrolyzes is fundamentally the same as calculation of the pH of a solution of an acid or base. However, K_a or K_b for an ion is usually obtained from the conjugate base or acid by applying the equation $K_a K_b = K_w$.

A *buffer* is a solution that can resist changes in pH when small amounts of acid or base are added to it. A buffer contains either a weak acid and its conjugate base or a weak base and its conjugate acid. The concentrations of acid and base conjugates are approximately equal. Different pH ranges are possible for acid–base conjugates, depending on their ionization constants.

An *acid–base titration curve* is a plot of the pH of the solution against the volume of reactant added. During the titration of a strong acid by a strong base, the pH changes slowly at first. Then, as the amount of base nears the stoichiometric value—that is, nears the *equivalence point*—the pH rises abruptly, changing by several units. The pH at the equivalence point is 7.0. A similar curve is obtained when a weak acid is titrated by a strong base. However, the pH changes less at the equivalence point. Moreover, the pH at the equivalence point is greater than 7.0 because of hydrolysis of the salt produced. An indicator must be chosen that changes color within a pH range near the equivalence point, where the pH changes rapidly.

Learning Objectives	Important Terms
16.1 Acid-Ionization Equilibria ■ Write the chemical equation for a weak acid undergoing *acid ionization* in aqueous solution. ■ Define *acid-ionization constant* and *degree of ionization*. ■ Determine K_a from the solution pH. Example 16.1 ■ Calculate concentrations of species in a weak acid solution using K_a (approximation method). Example 16.2 ■ State the assumption that allows for using approximations when solving problems. ■ Calculate concentrations of species in a weak acid solution using K_a (quadratic formula). Example 16.3	**acid-ionization constant** **degree of ionization**
16.2 Polyprotic Acids ■ State the general trend in the ionization constants of a polyprotic acid. ■ Calculate concentrations of species in a solution of a diprotic acid. Example 16.4	
16.3 Base-Ionization Equilibria ■ Write the chemical equation for a weak base undergoing ionization in aqueous solution. ■ Define *base-ionization constant*. ■ Calculate concentrations of species in a weak base solution using K_b. Example 16.5	**base-ionization constant**
16.4 Acid–Base Properties of Salt Solutions ■ Write the *hydrolysis* reaction of an ion to form an acidic solution. ■ Write the hydrolysis reaction of an ion to form a basic solution. ■ Predict whether a salt solution is acidic, basic, or neutral. Example 16.6 ■ Obtain K_a from K_b or K_b from K_a. Example 16.7 ■ Calculating concentrations of species in a salt solution. Example 16.8	**hydrolysis**
16.5 Common-Ion Effect ■ Explain the common-ion effect. ■ Calculate the common-ion effect on acid ionization (effect of a strong acid). Example 16.9 ■ Calculate the common-ion effect on acid ionization (effect of a conjugate base). Example 16.10	**common-ion effect**
16.6 Buffers ■ Define *buffer* and *buffer capacity*. ■ Describe the pH change of a buffer solution with the addition of acid or base. ■ Calculate the pH of a buffer from given volumes of solution. Example 16.11 ■ Calculate the pH of a buffer when a strong acid or a strong base is added. Example 16.12 ■ Learn the *Henderson–Hasselbalch* equation. ■ State when the Henderson–Hasselbalch equation can be applied.	**buffer** **Henderson–Hasselbalch equation**

16.7 Acid–Base Titration Curves

- Define *equivalence point.*
- Describe the curve for the titration of a strong acid by a strong base.
- Calculate the pH of a solution of a strong acid and a strong base. Example 16.13
- Describe the curve for the titration of a weak acid by a strong base.
- Calculate the pH at the equivalence point in the titration of a weak acid by a strong base. Example 16.14
- Describe the curve for the titration of a weak base by a strong acid.
- Calculate the pH of a solution at several points of a titration of weak base by a strong acid. Example 16.15

acid–base titration curve
equivalence point

Key Equations

$$K_a = \frac{[H_3O^+][A^-]}{[HA]}$$

$$K_b = \frac{[HB^+][OH^-]}{[B]}$$

$$K_a K_b = K_w$$

$$pH = pK_a + \log\frac{[base]}{[acid]}$$

Questions and Problems

⊙WL Interactive versions of these problems may be assigned in OWL.

Self-Assessment and Review Questions

Key: These questions test your understanding of the ideas you worked with in the chapter. These problems vary in difficulty and often can be used for the basis of discussion.

16.1 Write an equation for the ionization of hydrogen cyanide, HCN, in aqueous solution. What is the equilibrium expression K_a for this acid ionization?

16.2 Which of the following is the weakest acid: $HClO_4$, HCN, or $HC_2H_3O_2$? See Table 15.1 and Table 16.1.

16.3 Describe how the degree of ionization of a weak acid changes as the concentration increases.

16.4 Briefly describe two methods for determining K_a for a weak acid.

16.5 Consider a solution of 0.0010 M HF ($K_a = 6.8 \times 10^{-4}$). In solving for the concentrations of species in this solution, could you use the simplifying assumption in which you neglect x in the denominator of the equilibrium equation? Explain.

16.6 Phosphorous acid, H_2PHO_3, is a diprotic acid. Write equations for the acid ionizations. Write the expressions for K_{a1} and K_{a2}.

16.7 What is the concentration of oxalate ion, $C_2O_4^{2-}$, in 0.10 M oxalic acid, $H_2C_2O_4$? K_{a1} is 5.6×10^{-2}, and K_{a2} is 5.1×10^{-5}.

16.8 Write the equation for the ionization of aniline, $C_6H_5NH_2$, in aqueous solution. Write the expression for K_b.

16.9 Which of the following is the strongest base: NH_3, $C_6H_5NH_2$, or CH_3NH_2? See Table 16.2.

16.10 Do you expect a solution of anilinium chloride (aniline hydrochloride), $C_6H_5NH_3Cl$, to be acidic or basic? (Anilinium chloride is the salt of aniline and hydrochloric acid.) Write the equation for the reaction involved. What is the equilibrium expression? How would you obtain the value for the equilibrium constant from Table 16.2?

16.11 What is meant by the common-ion effect? Give an example.

16.12 The pH of 0.10 M CH_3NH_2 (methylamine) is 11.8. When the chloride salt of methylamine, CH_3NH_3Cl, is added to this solution, does the pH increase or decrease? Explain, using Le Châtelier's principle and the common-ion effect.

16.13 Define the term *buffer.* Give an example.

16.14 What is meant by the capacity of a buffer? Describe a buffer with low capacity and the same buffer with greater capacity.

16.15 Describe the pH changes that occur during the titration of a weak base by a strong acid. What is meant by the term *equivalence point*?

16.16 If the pH is 8.0 at the equivalence point for the titration of a certain weak acid with sodium hydroxide, what indicator might you use? (See Figure 15.10.) Explain your choice.

16.17 Which of the following salts would produce the most basic aqueous solution?
a KBr b NaF c $LiNO_3$
d NH_4Cl e $MgCl_2$

16.18 Hydrogen sulfide, H_2S, is a very weak diprotic acid. In a 0.10 M solution of this acid, which of the following would you expect to find in the *highest* concentration?
a H_2S b H_3O^+ c HS^- d S^{2-} e OH^-

16.19 If you mix 0.10 mol of NH_3 and 0.10 mol of HCl in a bucket of water, you would expect the resulting solution to be
a very basic b slightly basic c neutral
d slightly acidic e very acidic

16.20 If 20.0 mL of a 0.10 M NaOH solution is added to a 30.0-mL sample of a 0.10 M weak acid, HA, what is the pH of the resulting solution? ($K_a = 1.8 \times 10^{-5}$ for HA)
a 8.73 b 4.74 c 2.74 d 2.87 e 5.05

Concept Explorations

Key: Concept explorations are comprehensive problems that provide a framework that will enable you to explore and learn many of the critical concepts and ideas in each chapter. If you master the concepts associated with these explorations, you will have a better understanding of many important chemistry ideas and will be more successful in solving all types of chemistry problems. These problems are well suited for group work and for use as in-class activities.

16.21 Aqueous Solutions of Acids, Bases, and Salts

a For each of the following salts, write the reaction that occurs when it dissociates in water: NaCl(*s*), NaCN(*s*), KClO$_2$(*s*), NH$_4$NO$_3$(*s*), KBr(*aq*), and NaF(*s*).

b Consider each of the reactions that you wrote above, and identify the aqueous ions that *could* be proton donors (acids) or proton acceptors (bases). Briefly explain how you decided which ions to choose.

c For each of the acids and bases that you identified in part b, write the chemical reaction it can undergo in aqueous solution (its reaction with water).

d Are there any reactions that you have written above that you anticipate will occur to such an extent that the pH of the solution will be affected? As part of your answer, be sure to explain how you decided.

e Assume that in each case above, 0.01 mol of the salt was dissolved in enough water at 25°C to make 1.0 L of solution. In each case, what additional information would you need in order to calculate the pH? If there are cases where no additional information is required, be sure to state that as well.

f Say you take 0.01 mol of NH$_4$CN and dissolve it in enough water at 25°C to make 1.0 L of solution. Using chemical reactions and words, explain how you would go about determining what effect this salt will have on the pH of the solution. Be sure to list any additional information you would need to arrive at an answer.

16.22 The pH of Mixtures of Acid, Base, and Salt Solutions

a When 0.10 mol of the ionic solid NaX, where X is an unknown anion, is dissolved in enough water to make 1.0 L of solution, the pH of the solution is 9.12. When 0.10 mol of the ionic solid ACl, where A is an unknown cation, is dissolved in enough water to make 1.0 L of solution, the pH of the solution is 7.00. What would be the pH of 1.0 L of solution that contained 0.10 mol of AX? Be sure to document how you arrived at your answer.

b In the AX solution prepared above, is there any OH$^-$ present? If so, compare the [OH$^-$] in the solution to the [H$_3$O$^+$].

c From the information presented in part a, calculate K_b for the X$^-$(*aq*) anion and K_a for the conjugate acid of X$^-$(*aq*).

d To 1.0 L of solution that contains 0.10 mol of AX, you add 0.025 mol of HCl. How will the pH of this solution compare to that of the solution that contained only NaX? Use chemical reactions as part of your explanation; you do not need to solve for a numerical answer.

e Another 1.0 L sample of solution is prepared by mixing 0.10 mol of AX and 0.10 mol of HCl. The pH of the resulting solution is found to be 3.12. Explain why the pH of this solution is 3.12.

f Finally, consider a different 1.0-L sample of solution that contains 0.10 mol of AX and 0.1 mol of NaOH. The pH of this solution is found to be 13.00. Explain why the pH of this solution is 13.00.

g Some students mistakenly think that a solution that contains 0.10 mol of AX and 0.10 mol of HCl should have a pH of 1.00. Can you come up with a reason why students have this misconception? Write an approach that you would use to help these students understand what they are doing wrong.

Conceptual Problems

Key: These problems are designed to check your understanding of the concepts associated with some of the main topics presented in each chapter. A strong conceptual understanding of chemistry is the foundation for both applying chemical knowledge and solving chemical problems. These problems vary in level of difficulty and often can be used as a basis for group discussion.

16.23 Which of the following beakers best represents a container of a weak acid, HA, in water? (Water molecules have been omitted for clarity.)

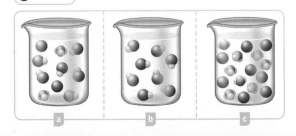

16.24 You have 0.10-mol samples of three acids identified simply as HX, HY, and HZ. For each acid, you make up 0.10 *M* solutions by adding sufficient water to each of the acid samples. When you measure the pH of these samples, you find that the pH of HX is greater than the pH of HY, which in turn is greater than the pH of HZ.

a Which of the acids is the least ionized in its solution?
b Which acid has the largest K_a?

16.25 What reaction occurs when each of the following is dissolved in water?

a C$_6$H$_5$NH$_3$Cl   b C$_6$H$_5$NH$_2$
c NaF   d HF

16.26 You have the following solutions, all of the same molar concentration: KBr, HBr, CH$_3$NH$_2$, and NH$_4$Cl. Rank them from the lowest to the highest hydroxide-ion concentration.

16.27 Rantidine is a nitrogen base that is used to control stomach acidity by suppressing the stomach's production of hydrochloric acid. The compound is present in Zantac® as the chloride salt (rantidinium chloride; also called rantidine hydrochloride). Do you expect a solution of rantidine hydrochloride to be acidic, basic, or neutral? Explain by means of a general chemical equation.

16.28 A chemist prepares dilute solutions of equal molar concentrations of NH_3, NH_4Br, NaF, and $NaCl$. Rank these solutions from highest pH to lowest pH.

16.29 A friend of yours has performed three titrations: strong acid with a strong base, weak acid with a strong base, and weak base with a strong acid. He hands you the three titration curves, saying he has forgotten which is which. What attributes of the curves would you look at to identify each curve correctly?

16.30 You want to prepare a buffer solution that has a pH equal to the pK_a of the acid component of the buffer. If you have 100 mL of a 0.10 M solution of the acid HA, what volume and concentration of NaA solution could you use in order to prepare the buffer?

16.31 You are given the following acid–base titration data, where each point on the graph represents the pH after adding a given volume of titrant (the substance being added during the titration).

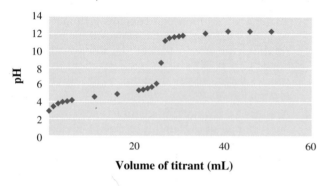

a What substance is being titrated, a strong acid, strong base, weak acid, or weak base?
b What is the pH at the equivalence point of the titration?
c What indicator might you use to perform this titration? Explain.

16.32 The three flasks shown below depict the titration of an aqueous NaOH solution with HCl at different points. One represents the titration prior to the equivalence point, another represents the titration at the equivalence point, and the other represents the titration past the equivalence point. (Sodium ions and solvent water molecules have been omitted for clarity.)

a Write the balanced chemical equation for the titration.
b Label each of the beakers shown to indicate which point in the titration they represent.
c For each solution, indicate whether you expect it to be acidic, basic, or neutral.

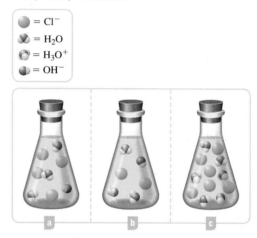

Practice Problems

Key: These problems are for practice in applying problem-solving skills. They are divided by topic, and some are keyed to exercises (see the ends of the exercises). The problems are arranged in matching pairs; the odd-numbered problem of each pair is listed first, and its answer is given in the back of the book.

Note: For values of K_a and K_b that are not given in the following problems, see Tables 16.1 and 16.2.

Acid Ionization

16.33 Write chemical equations for the acid ionizations of each of the following weak acids (express these in terms of H_3O^+).
a $HClO_2$ (chlorous acid)   b HBrO (hypobromous acid)
c HCN (hydrocyanic acid)   d HNO_2 (nitrous acid)

16.34 Write chemical equations for the acid ionizations of each of the following weak acids (express these in terms of H_3O^+).
a HCO_2H (formic acid)   b HF (hydrofluoric acid)
c HN_3 (hydrazoic acid)   d HOCN (cyanic acid)

16.35 Heavy metal azides, which are salts of hydrazoic acid, HN_3, are used as explosive detonators. A solution of 0.20 M hydrazoic acid has a pH of 3.21. What is the K_a for hydrazoic acid?

16.36 Acrylic acid, whose formula is $HC_3H_3O_2$ or $HO_2CCH=CH_2$, is used in the manufacture of plastics. A 0.10 M aqueous solution of acrylic acid has a pH of 2.63. What is K_a for acrylic acid?

16.37 Boric acid, $B(OH)_3$, is used as a mild antiseptic. What is the pH of a 0.015 M aqueous solution of boric acid? What is the degree of ionization of boric acid in this solution? The hydronium ion arises principally from the reaction

$$B(OH)_3(aq) + 2H_2O(l) \rightleftharpoons B(OH)_4^-(aq) + H_3O^+(aq)$$

The equilibrium constant for this reaction is 5.9×10^{-10}.

16.38 Formic acid, $HCHO_2$, is used to make methyl formate (a fumigant for dried fruit) and ethyl formate (an artificial rum flavor). What is the pH of a 0.12 M solution of formic acid? What is the degree of ionization of $HCHO_2$ in this solution? See Table 16.1 for K_a.

16.39 $C_6H_4NH_2COOH$, *para*-aminobenzoic acid (PABA), is used in some sunscreen agents. Calculate the concentrations of hydronium ion and *para*-aminobenzoate ion, $C_6H_4NH_2COO^-$, in a 0.055 M solution of the acid. The value of K_a is 2.2×10^{-5}.

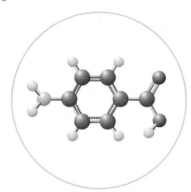

16.40 Barbituric acid, $HC_4H_3N_2O_3$, is used to prepare various barbiturate drugs (used as sedatives). Calculate the concentrations of hydronium ion and barbiturate ion in a 0.10 M solution of the acid. The value of K_a is 9.8×10^{-5}.

16.41 A chemist wanted to determine the concentration of a solution of lactic acid, $HC_3H_5O_3$. She found that the pH of the solution was 2.60. What was the concentration of the solution? The K_a of lactic acid is 1.4×10^{-4}.

16.42 A solution of acetic acid, $HC_2H_3O_2$, on a laboratory shelf was of undetermined concentration. If the pH of the solution was found to be 2.68, what was the concentration of the acetic acid?

16.43 Chloroacetic acid, $HC_2H_2ClO_2$, has a greater acid strength than acetic acid, because the electronegative chlorine atom pulls electrons away from the O—H bond and thus weakens it. Calculate the hydronium-ion concentration and the pH of a 0.0020 M solution of chloroacetic acid. K_a is 1.3×10^{-3}.

16.44 Hydrofluoric acid, HF, unlike hydrochloric acid, is a weak electrolyte. What is the hydronium-ion concentration and the pH of a 0.040 M aqueous solution of HF?

16.45 What is the hydronium-ion concentration of a 2.00 M solution of 2,6-dinitrobenzoic acid, $(NO_2)_2C_6H_3COOH$, for which $K_a = 7.94 \times 10^{-2}$?

16.46 What is the hydronium-ion concentration of a 3.00×10^{-4} M solution of p-bromobenzoic acid, BrC_6H_4COOH, for which $K_a = 1.00 \times 10^{-4}$?

16.47 Phthalic acid, $H_2C_8H_4O_4$, is a diprotic acid used in the synthesis of phenolphthalein indicator. $K_{a1} = 1.2 \times 10^{-3}$, and $K_{a2} = 3.9 \times 10^{-6}$. **a** Calculate the hydronium-ion

concentration of a 0.015 M solution. **b** What is the concentration of the $C_8H_4O_4^{2-}$ ion in the solution?

16.48 Carbonic acid, H_2CO_3, can be found in a wide variety of body fluids (from dissolved CO_2). **a** Calculate the hydronium-ion concentration of a 6.00×10^{-4} M H_2CO_3 solution. **b** What is the concentration of CO_3^{2-}?

Base Ionization

16.49 Write the chemical equation for the base ionization of aniline, $C_6H_5NH_2$. Write the K_b expression for aniline.

16.50 Write the chemical equation for the base ionization of methylamine, CH_3NH_2. Write the K_b expression for methylamine.

16.51 Ethanolamine, $HOC_2H_4NH_2$, is a viscous liquid with an ammonialike odor; it is used to remove hydrogen sulfide from natural gas. A 0.15 M aqueous solution of ethanolamine has a pH of 11.34. What is K_b for ethanolamine?

16.52 Trimethylamine, $(CH_3)_3N$, is a gas with a fishy, ammonialike odor. An aqueous solution that is 0.25 M trimethylamine has a pH of 11.63. What is K_b for trimethylamine?

16.53 What is the concentration of hydroxide ion in a 0.21 M aqueous solution of hydroxylamine, NH_2OH? What is the pH?

16.54 What is the concentration of hydroxide ion in a 0.060 M aqueous solution of methylamine, CH_3NH_2? What is the pH?

Acid–Base Properties of Salt Solutions; Hydrolysis

16.55 Note whether hydrolysis occurs for each of the following ions. If hydrolysis does occur, write the chemical equation for it. Then write the equilibrium expression for the acid or base ionization (whichever occurs).
a $NH_2NH_3^+$ **b** Br^-
c NO_3^- **d** OCl^-

16.56 Note whether hydrolysis occurs for each of the following ions. If hydrolysis does occur, write the chemical equation for it. Then write the equilibrium expression for the acid or base ionization (whichever occurs).
a $CH_3NH_3^+$ **b** Cl^-
c ClO_2^- **d** CO_3^{2-}

16.57 Write the equation for the acid ionization of the $Cu(H_2O)_6^{2+}$ ion.

16.58 Write the equation for the acid ionization of the $Zn(H_2O)_6^{2+}$ ion.

16.59 For each of the following salts, indicate whether the aqueous solution will be acidic, basic, or neutral.
a $Fe(NO_3)_3$ **b** Na_2CO_3
c $Ca(CN)_2$ **d** NH_4ClO_4

16.60 Note whether the aqueous solution of each of the following salts will be acidic, basic, or neutral.
a Na_2S **b** $Cu(NO_2)_2$
c KNO_3 **d** CH_3NH_3Cl

16.61 Decide whether solutions of the following salts are acidic, neutral, or basic.
a anilinium acetate   **b** ammonium acetate

16.62 Decide whether solutions of the following salts are acidic, neutral, or basic.
a anilinium cyanate b ammonium cyanate

16.63 Obtain a the K_b value for NO_2^-; b the K_a value for $C_5H_5NH^+$ (pyridinium ion).

16.64 Obtain a the K_a value for NH_3OH^+ (hydroxylammonium ion); b the K_b value for $C_3H_3O_2^-$.

16.65 What is the pH of a 0.025 M aqueous solution of sodium propionate, $NaC_3H_5O_2$? What is the concentration of propionic acid in the solution?

16.66 Calculate the OH^- concentration and pH of a 0.0025 M aqueous solution of sodium cyanide, NaCN. Finally, obtain the hydronium-ion, CN^-, and HCN concentrations.

16.67 What is the pH of a 0.30 M solution of methylammonium chloride, CH_3NH_3Cl? What is the concentration of methylamine in the solution?

16.68 Calculate the concentration of pyridine, C_5H_5N, in a solution that is 0.15 M pyridinium bromide, C_5H_5NHBr. What is the pH of the solution?

Common-Ion Effect

16.69 Calculate the degree of ionization of a 0.22 M $HCHO_2$ (formic acid); b the same solution that is also 0.15 M HCl.

16.70 Calculate the degree of ionization of a 0.75 M HF (hydrofluoric acid); b the same solution that is also 0.12 M HCl.

16.71 What is the pH of a solution that is 0.10 M KNO_2 and 0.15 M HNO_2 (nitrous acid)?

16.72 What is the pH of a solution that is 0.20 M KOCN and 0.10 M HOCN (cyanic acid)?

16.73 What is the pH of a solution that is 0.15 M $C_2H_5NH_2$ (ethylamine) and 0.10 M $C_2H_5NH_3Br$ (ethylammonium bromide)?

16.74 What is the pH of a solution that is 0.10 M CH_3NH_2 (methylamine) and 0.15 M CH_3NH_3Cl (methylammonium chloride)?

Buffers

16.75 A buffer is prepared by adding 115 mL of 0.30 M NH_3 to 135 mL of 0.15 M NH_4NO_3. What is the pH of the final solution?

16.76 A buffer is prepared by adding 45.0 mL of 0.15 M NaF to 35.0 mL of 0.10 M HF. What is the pH of the final solution?

16.77 What is the pH of a buffer solution that is 0.10 M NH_3 and 0.10 M NH_4^+? What is the pH if 12 mL of 0.20 M hydrochloric acid is added to 125 mL of buffer?

16.78 A buffer is prepared by mixing 525 mL of 0.50 M formic acid, $HCHO_2$, and 475 mL of 0.50 M sodium formate, $NaCHO_2$. Calculate the pH. What would be the pH of 85 mL of the buffer to which 8.6 mL of 0.15 M hydrochloric acid had been added?

16.79 What is the pH of a buffer solution that is 0.15 M chloroacetic acid and 0.10 M sodium chloroacetate? $K_a = 1.3 \times 10^{-3}$.

16.80 What is the pH of a buffer solution that is 0.10 M propionic acid and 0.20 M sodium propionate?

16.81 What is the pH of a buffer solution that is 0.15 M methylamine and 0.20 M methylammonium chloride?

16.82 What is the pH of a buffer solution that is 0.15 M pyridine and 0.10 M pyridinium bromide?

16.83 How many moles of sodium acetate must be added to 2.0 L of 0.10 M acetic acid to give a solution that has a pH equal to 5.00? Ignore the volume change due to the addition of sodium acetate.

16.84 How many moles of hydrofluoric acid, HF, must be added to 250 mL of 0.25 M sodium fluoride to give a buffer of pH 3.50? Ignore the volume change due to the addition of hydrofluoric acid.

Titration Curves

16.85 What is the pH of a solution in which 35 mL of 0.10 M NaOH is added to 25 mL of 0.10 M HCl?

16.86 What is the pH of a solution in which 15 mL of 0.10 M NaOH is added to 25 mL of 0.10 M HCl?

16.87 A 1.24-g sample of benzoic acid was dissolved in water to give 50.0 mL of solution. This solution was titrated with 0.180 M NaOH. What was the pH of the solution when the equivalence point was reached?

16.88 A 0.400-g sample of propionic acid was dissolved in water to give 50.0 mL of solution. This solution was titrated with 0.150 M NaOH. What was the pH of the solution when the equivalence point was reached?

16.89 Find the pH of the solution obtained when 32 mL of 0.087 M ethylamine is titrated to the equivalence point with 0.15 M HCl.

16.90 What is the pH at the equivalence point when 22 mL of 0.20 M hydroxylamine is titrated with 0.10 M HCl?

16.91 A 50.0-mL sample of a 0.100 M solution of NaCN is titrated by 0.200 M HCl. K_b for CN^- is 2.0×10^{-5}. Calculate the pH of the solution: a prior to the start of the titration; b after the addition of 15.0 mL of 0.100 M HCl; c at the equivalence point; d after the addition of 30.0 mL of 0.200 M HCl.

16.92 Sodium benzoate, $NaC_7H_5O_2$, is used as a preservative in foods. Consider a 50.0-mL sample of 0.250 M $NaC_7H_5O_2$ being titrated by 0.200 M HBr. Calculate the pH of the solution: a when no HBr has been added; b after the addition of 50.0 mL of the HBr solution; c at the equivalence point; d after the addition of 75.00 mL of the HBr solution. The K_b value for the benzoate ion is 1.6×10^{-10}.

16.93 Calculate the pH of a solution obtained by mixing 500.0 mL of 0.10 M NH_3 with 200.0 mL of 0.15 M HCl.

16.94 Calculate the pH of a solution obtained by mixing 35.0 mL of 0.15 M acetic acid with 27.0 mL of 0.10 M sodium acetate.

General Problems

Key: These problems provide more practice but are not divided by topic or keyed to exercises. Each section ends with essay questions, each of which is color coded to refer to the *A Chemist Looks at Environment* (green) chapter essay on which it is based. Odd-numbered problems and the even-numbered problems that follow are similar; answers to all odd-numbered problems except the essay questions are given in the back of the book.

16.95 Cyanoacetic acid, $CH_2CNCOOH$, is used in the manufacture of barbiturate drugs. An aqueous solution containing 5.0 g in a liter of solution has a pH of 1.89. What is the value of K_a?

16.96 Salicylic acid, $C_6H_4OHCOOH$, is used in the manufacture of acetylsalicylic acid (aspirin) and methyl salicylate (wintergreen flavor). A saturated solution of salicylic acid contains 2.2 g of the acid per liter of solution and has a pH of 2.43. What is the value of K_a?

16.97 A 0.050 *M* aqueous solution of sodium hydrogen sulfate, $NaHSO_4$, has a pH of 1.73. Calculate K_{a2} for sulfuric acid. Sulfuric acid is a strong acid, so you can ignore hydrolysis of the HSO_4^- ion.

16.98 A 0.10 *M* aqueous solution of sodium dihydrogen phosphate, NaH_2PO_4, has a pH of 4.10. Calculate K_{a2} for phosphoric acid. You can ignore hydrolysis of the $H_2PO_4^-$ ion.

16.99 Calculate the base-ionization constants for CN^- and CO_3^{2-}. Which ion is the stronger base?

16.100 Calculate the base-ionization constants for PO_4^{3-} and SO_4^{2-}. Which ion is the stronger base?

16.101 Calculate the pH of a 0.15 *M* aqueous solution of aluminum chloride, $AlCl_3$. The acid ionization of hydrated aluminum ion is

$$Al(H_2O)_6^{3+}(aq) + H_2O(l) \rightleftharpoons Al(H_2O)_5OH^{2+}(aq) + H_3O^+(aq)$$

and K_a is 1.4×10^{-5}.

16.102 Calculate the pH of a 0.10 *M* aqueous solution of zinc chloride, $ZnCl_2$. The acid ionization of hydrated zinc ion is

$$Zn(H_2O)_6^{2+}(aq) + H_2O(l) \rightleftharpoons Zn(H_2O)_5OH^+(aq) + H_3O^+(aq)$$

and K_a is 2.5×10^{-10}.

16.103 A buffer is made by dissolving 12.5 g of sodium dihydrogen phosphate, NaH_2PO_4, and 15.0 g of disodium hydrogen phosphate, Na_2HPO_4, in a liter of solution. What is the pH of the buffer?

16.104 An artificial fruit beverage contains 11.0 g of tartaric acid, $H_2C_4H_4O_6$, and 20.0 g of its salt, potassium hydrogen tartrate, per liter. What is the pH of the beverage? $K_{a1} = 1.0 \times 10^{-3}$.

16.105 Blood contains several acid–base systems that tend to keep its pH constant at about 7.4. One of the most important buffer systems involves carbonic acid and hydrogen carbonate ion. What must be the ratio of $[HCO_3^-]$ to $[H_2CO_3]$ in the blood if the pH is 7.40?

16.106 Codeine, $C_{18}H_{21}NO_3$, is an alkaloid ($K_b = 6.2 \times 10^{-9}$) used as a painkiller and cough suppressant. A solution of codeine is acidified with hydrochloric acid to pH 4.50. What is the ratio of the concentration of the conjugate acid of codeine to that of the base codeine?

16.107 Calculate the pH of a solution obtained by mixing 456 mL of 0.10 *M* hydrochloric acid with 285 mL of 0.15 *M* sodium hydroxide. Assume the combined volume is the sum of the two original volumes.

16.108 Calculate the pH of a solution made up from 2.0 g of potassium hydroxide dissolved in 115 mL of 0.19 *M* perchloric acid. Assume the change in volume due to adding potassium hydroxide is negligible.

16.109 What is the pH of the solution obtained by titrating 1.30 g of sodium hydrogen sulfate, $NaHSO_4$, dissolved in 50.0 mL of water with 0.175 *M* sodium hydroxide until the equivalence point is reached? Assume that any volume change due to adding the sodium hydrogen sulfate or to mixing the solutions is negligible.

16.110 Find the pH of the solution obtained when 25 mL of 0.065 *M* benzylamine, $C_7H_7NH_2$, is titrated to the equivalence point with 0.050 *M* hydrochloric acid. K_b for benzylamine is 4.7×10^{-10}.

16.111 Ionization of the first proton from H_2SO_4 is complete (H_2SO_4 is a strong acid); the acid-ionization constant for the second proton is 1.1×10^{-2}. **a** What would be the approximate hydronium-ion concentration in 0.100 *M* H_2SO_4 if ionization of the second proton were ignored? **b** The ionization of the second proton must be considered for a more exact answer, however. Calculate the hydronium-ion concentration in 0.100 *M* H_2SO_4, accounting for the ionization of both protons.

16.112 Ionization of the first proton from H_2SeO_4 is complete (H_2SeO_4 is a strong acid); the acid-ionization constant for the second proton is 1.2×10^{-2}. **a** What would be the approximate hydronium-ion concentration in 0.150 *M* H_2SeO_4 if ionization of the second proton were ignored? **b** The ionization of the second proton must be considered for a more exact answer, however. Calculate the hydronium-ion concentration in 0.150 *M* H_2SeO_4, accounting for the ionization of both protons.

16.113 Methylammonium chloride is a salt of methylamine, CH_3NH_2. A 0.10 *M* solution of this salt has a pH of 5.82.
a Calculate the value for the equilibrium constant for the reaction

$$CH_3NH_3^+ + H_2O \rightleftharpoons CH_3NH_2 + H_3O^+$$

b What is the K_b value for methylamine?
c What is the pH of a solution in which 0.450 mol of solid methylammonium chloride is added to 1.00 L of a 0.250 *M* solution of methylamine? Assume no volume change.

16.114 Sodium benzoate is a salt of benzoic acid, C_6H_5COOH. A 0.15 *M* solution of this salt has a pOH of 5.31 at room temperature.

a Calculate the value for the equilibrium constant for the reaction

$$C_6H_5COO^- + H_2O \rightleftharpoons C_6H_5COOH + OH^-$$

b What is the K_a value for benzoic acid?

c Benzoic acid has a low solubility in water. What is its molar solubility if a saturated solution has a pH of 2.83 at room temperature?

16.115 Each of the following statements concerns a 0.010 M solution of a weak acid, HA. Briefly describe why each statement is either true or false.

a [HA] is much greater than [A$^-$].

b [OH$^-$] is approximately equal to [H$_3$O$^+$].

c The pH is 2.

d [HA] is approximately equal to 0.010 M.

e [H$_3$O$^+$] is approximately equal to [A$^-$].

f The H$_3$O$^+$ concentration is 0.010 M.

16.116 Each of the following statements concerns a 0.10 M solution of a weak organic base, B. Briefly describe why each statement is either true or false.

a [B] is approximately equal to 0.10 M.

b [B] is much greater than [HB$^+$].

c [H$_3$O$^+$] is greater than [HB$^+$].

d The pH is 13.

e [HB$^+$] is approximately equal to [OH$^-$].

f [OH$^-$] equals 0.10 M.

16.117 A 0.239-g sample of unknown organic base is dissolved in water and titrated with a 0.135 M hydrochloric acid solution. After the addition of 18.35 mL of acid, a pH of 10.73 is recorded. The equivalence point is reached when a total of 39.24 mL of HCl is added. The base and acid combine in a 1:1 ratio.

a What is the molar mass of the organic base?

b What is the K_b value for the base? The K_b value could have been determined very easily if a pH measurement had been made after the addition of 19.62 mL of HCl. Why?

16.118 A 0.288-g sample of an unknown monoprotic organic acid is dissolved in water and titrated with a 0.115 M sodium hydroxide solution. After the addition of 17.54 mL of base, a pH of 4.92 is recorded. The equivalence point is reached when a total of 33.83 mL of NaOH is added.

a What is the molar mass of the organic acid?

b What is the K_a value for the acid? The K_a value could have been determined very easily if a pH measurement had been made after the addition of 16.92 mL of NaOH. Why?

16.119 **a** Draw a pH titration curve that represents the titration of 50.0 mL of 0.10 M NH$_3$ by the addition of 0.10 M HCl from a buret. Label the axes and put a scale on each axis. Show where the equivalence point and the buffer region are on the titration curve. You should do calculations for the 0%, 30%, 50%, and 100% titration points. **b** Is the solution neutral, acidic, or basic at the equivalence point? Why?

16.120 **a** Draw a pH titration curve that represents the titration of 25.0 mL of 0.15 M propionic acid, CH$_3$CH$_2$COOH,

by the addition of 0.15 M KOH from a buret. Label the axes and put a scale on each axis. Show where the equivalence point and the buffer region are on the titration curve. You should do calculations for the 0%, 50%, 60%, and 100% titration points. **b** Is the solution neutral, acidic, or basic at the equivalence point? Why?

16.121 The equilibrium equations and K_a values for three reaction systems are given below.

$$NH_4^+(aq) + H_2O \rightleftharpoons H_3O^+ + NH_3(aq); K_a = 5.6 \times 10^{-10}$$

$$H_2CO_3(aq) + H_2O \rightleftharpoons$$
$$H_3O^+(aq) + HCO_3^-(aq); K_a = 4.3 \times 10^{-7}$$

$$H_2PO_4^-(aq) + H_2O \rightleftharpoons$$
$$H_3O^+(aq) + HPO_4^{2-}(aq); K_a = 6.2 \times 10^{-8}$$

a Which conjugate pair would be best for preparing a buffer with a pH of 6.96? Why?

b How would you prepare 100 mL of a buffer with a pH of 6.96 assuming that you had available 0.10 M solutions of each pair?

16.122 The equilibrium equations and K_a values for three reaction systems are given below.

$$H_2C_2O_4(aq) + H_2O \rightleftharpoons$$
$$H_3O^+(aq) + HC_2O_4^-(aq); K_a = 5.6 \times 10^{-2}$$

$$H_3PO_4(aq) + H_2O \rightleftharpoons$$
$$H_3O^+(aq) + H_2PO_4^-(aq); K_a = 6.9 \times 10^{-3}$$

$$HCOOH(aq) + H_2O \rightleftharpoons$$
$$H_3O^+(aq) + HCOO^-(aq); K_a = 1.7 \times 10^{-4}$$

a Which conjugate pair would be best for preparing a buffer with a pH of 2.88?

b How would you prepare 50 mL of a buffer with a pH of 2.88 assuming that you had available 0.10 M solutions of each pair?

16.123 A 25.0-mL sample of hydroxylamine is titrated to the equivalence point with 35.8 mL of 0.150 M HCl.

a What was the concentration of the original hydroxylamine solution?

b What is the pH at the equivalence point?

c Which indicators, bromphenol blue, methyl red, or phenolphthalein, should be used to detect the end point of the titration? Why?

16.124 A 25.00-mL sample contains 0.562 g of NaHCO$_3$. This sample is used to standardize an NaOH solution. At the equivalence point, 40.95 mL of NaOH has been added.

a What was the concentration of the NaOH?

b What is the pH at the equivalence point?

c Which indicator, bromthymol blue, methyl violet, or alizarin yellow R, should be used in the titration? Why?

16.125 A solution made up of 1.0 M NH$_3$ and 0.50 M (NH$_4$)$_2$SO$_4$ has a pH of 9.26.

a Write the net ionic equation that represents the reaction of this solution with a strong acid.

b Write the net ionic equation that represents the reaction of this solution with a strong base.

c To 100. mL of this solution, 10.0 mL of 1.00 M HCl is added. How many moles of NH$_3$ and NH$_4^+$ are present in the reaction system before and after the addition of the HCl? What is the pH of the resulting solution?

d Why did the pH change only slightly upon the addition of HCl?

16.126 A solution is prepared from 0.150 mol of formic acid and enough water to make 0.425 L of solution.

a Determine the concentrations of H_3O^+ and $HCOO^-$ in this solution.

b Determine the H_3O^+ concentration that would be necessary to decrease the $HCOO^-$ concentration above by a factor of 10. How many milliliters of 2.00 M HCl would be required to produce this solution? Consider that the solution was made by combining the HCl, the HCOOH, and enough water to make 0.425 L of solution.

c Qualitatively, how can you account for the differences in the percentage dissociation of formic acid in parts a and b of this problem?

16.127 An important component of blood is the buffer combination of dihydrogen phosphate ion and the hydrogen phosphate ion. Consider blood with a pH of 7.44.

a What is the ratio of $[H_2PO_4^-]$ to $[HPO_4^{2-}]$?

b What does the pH become if 25% of the hydrogen phosphate ions are converted to dihydrogen phosphate ion?

c What does the pH become if 15% of the dihydrogen phosphate ions are converted to hydrogen phosphate ions?

16.128 An important component of blood is the buffer combination of bicarbonate ion and carbonic acid. Consider blood with a pH of 7.42.

a What is the ratio of $[H_2CO_3]$ to $[HCO_3^-]$?

b What does the pH become if 15% of the bicarbonate ions are converted to carbonic acid?

c What does the pH become if 25% of the carbonic acid molecules are converted to bicarbonate ions?

16.129 Tartaric acid is a weak diprotic fruit acid with $K_{a1} = 1.0 \times 10^{-3}$ and $K_{a2} = 4.6 \times 10^{-5}$.

a Letting the symbol H_2A represent tartaric acid, write the chemical equations that represent K_{a1} and K_{a2}. Write the chemical equation that represents $K_{a1} \times K_{a2}$.

b Qualitatively describe the relative concentrations of H_2A, HA^-, A^{2-}, and H_3O^+ in a solution that is about 0.5 M in tartaric acid.

c Calculate the pH of a 0.0250 M tartaric acid solution and the equilibrium concentration of $[H_2A]$.

d What is the A^{2-} concentration in solutions b and c?

16.130 Malic acid is a weak diprotic organic acid with $K_{a1} = 4.0 \times 10^{-4}$ and $K_{a2} = 9.0 \times 10^{-6}$.

a Letting the symbol H_2A represent malic acid, write the chemical equations that represent K_{a1} and K_{a2}. Write the chemical equation that represents $K_{a1} \times K_{a2}$.

b Qualitatively describe the relative concentrations of H_2A, HA^-, A^{2-}, and H_3O^+ in a solution that is about one molar in malic acid.

c Calculate the pH of a 0.0175 M malic acid solution and the equilibrium concentration of $[H_2A]$.

d What is the A^{2-} concentrationin in solutions b and c?

16.131 A quantity of 0.15 M hydrochloric acid is added to a solution containing 0.10 mol of sodium acetate. Some of the sodium acetate is converted to acetic acid, resulting in a final volume of 650 mL of solution. The pH of the final solution is 4.56.

a What is the molar concentration of the acetic acid?

b How many milliliters of hydrochloric acid were added to the original solution?

c What was the original concentration of the sodium acetate?

16.132 A quantity of 0.25 M sodium hydroxide is added to a solution containing 0.15 mol of acetic acid. The final volume of the solution is 375 mL and the pH of this solution is 4.45.

a What is the molar concentration of the sodium acetate?

b How many milliliters of sodium hydroxide were added to the original solution?

c What was the original concentration of the acetic acid?

■ **16.133** How is acid rain defined? What is the source of the low pH of acid rain?

■ **16.134** How can you account for the fact that normal rain is slightly acidic?

Strategy Problems

Key: As noted earlier, all of the practice and general problems are matched pairs. This section is a selection of problems that are not in matched-pair format. These challenging problems require that you employ many of the concepts and strategies that were developed in the chapter. In some cases, you will have to integrate several concepts and operational skills in order to solve the problem successfully.

16.135 A 30.0-mL sample of 0.05 M HClO is titrated by a 0.0250 M KOH solution. K_a for HClO is 3.5×10^{-8}. Calculate a the pH when no base has been added; b the pH when 30.00 mL of the base has been added; c the pH at the equivalence point; d the pH when an additional 4.00 mL of the KOH solution has been added beyond the equivalence point.

16.136 A 0.108 M sample of a weak acid is 3.95% ionized in solution. What is the hydroxide concentration of this solution?

16.137 A generic base, B^-, is added to 2.25 L of water. The pH of the solution is found to be 10.10. What is the concentration of the base B^- in this solution? K_a for the acid HB at 25°C is 1.99×10^{-9}.

16.138 Calculate the pH of a solution made by mixing 0.62 L of 0.10 M NH_4Cl with 0.50 L of 0.10 M NaOH. K_b for NH_3 is 1.8×10^{-5}.

16.139 Cyanic acid, HOCN, is a weak acid with a K_a value of 3.5×10^{-4} at 25°C. In a 0.293 M solution of the acid, the degree of ionization is 3.5×10^{-2}. Calculate the degree of ionization in a 0.293 M solution to which sufficient HCl is added to make it 4.19×10^{-2} M HCl in the given volume.

16.140 The K_b for NH_3 is 1.8×10^{-5} at 25°C. Calculate the pH of a buffer solution made by mixing 65.1 mL of 0.142 M NH_3 with 38.0 mL of 0.172 M NH_4Cl at 25°C. Assume that the volumes of the solutions are additive.

16.141 K_a for formic acid is 1.7×10^{-4} at 25°C. A buffer is made by mixing 529 mL of 0.465 M formic acid, $HCHO_2$, and 494 mL of 0.524 M sodium formate, $NaCHO_2$. Calculate the pH of this solution at 25°C after 110. mL of 0.152 M HCl has been added to this buffer.

16.142 Calculate the pH of a solution made by mixing 7.52 mL of 4.9×10^{-2} M $Ca(OH)_2$ with 22.5 mL of 0.11 M HCl.

16.143 K_a for acetic acid is 1.7×10^{-5} at 25°C. A buffer solution is made by mixing 52.1 mL of 0.122 M acetic acid with 46.1 mL of 0.182 M sodium acetate. Calculate the pH of this solution at 25°C after the addition of 5.82 mL of 0.125 M NaOH.

16.144 A 0.150 M solution of NaClO is prepared by dissolving NaClO in water. A 50.0-mL sample of this solution is titrated with 0.100 M HCl. Calculate the pH of the solution at each of the following points of the titration: a prior to the addition of any HCl; b halfway to the equivalence point; c at the equivalence point; d after 5.00 mL of HCl has been added beyond the equivalence point. K_a for HClO is 3.5×10^{-8}.

16.145 A solution is prepared by dissolving ammonium nitrite in water. Predict whether the solution would be acidic or basic. If you want to make the solution have a neutral pH, which of the following could be added to achieve this result: HCl, NaCl, or KOH? Justify your answer.

16.146 Two samples of 1.00 M HCl of equivalent volumes are prepared. One sample is titrated to the equivalence point with a 1.00 M solution of sodium hydroxide, while the other sample is titrated to the equivalence point with a 1.00 M solution of calcium hydroxide.

a Compare the volumes of sodium hydroxide and calcium hydroxide required to reach the equivalence point for each titration.

b Determine the pH of each solution halfway to the equivalence point.

c Determine the pH of each solution at the equivalence point.

16.147 You have the following solutions at your disposal to prepare a buffer solution with a pH greater than 7.0:

50.0 mL of 0.10 M NH_3 20.0 mL of 0.10 M NaCl
55.0 mL of 0.10 M NaOH   20.0 mL of 0.10 M HNO_3
50.0 mL of 0.10 M HCl   20.0 mL of 0.10 M KOH.

a Assuming that you are going to mix the entire quantity of the listed solutions to prepare the buffer, which two solutions would you use?

b Calculate the pH of the buffer solution that you prepared in part a.

16.148 A sample of NH_4Cl is prepared for titration by dissolving the salt in water, making 100.0 mL of solution. A second sample of NH_4Cl of identical mass is used to prepare 200.0 mL of solution. A solution of 0.10 M NaOH or 0.10 M HCl may be used to perform the titration.

a Which solution, the 0.10 M NaOH or 0.10 M HCl, should be used to perform the titration? Justify your answer?

b Compare the volume of titrant selected in part a that you would need to reach the equivalence point of the titrations for each of the samples.

c Compare the pH of the two solutions at the equivalence point of the titrations (assume additive volumes).

16.149 A solution of weak base is titrated to the equivalence point with a strong acid. Which one of the following statements is most likely to be correct?

a The pH of the solution at the equivalence point is 7.0.

b The pH of the solution is greater than 13.0.

c The pH of the solution is less than 2.0.

d The pH of the solution is between 2.0 and 7.0.

e The pH of the solution is between 7.0 and 13.0.

The reason that best supports my choosing the answer above is

a Whenever a solution is titrated with a strong acid, the solution will be very acidic.

b Because the solution contains a weak base and the acid (titrant) is used up at the equivalence point, the solution will be basic.

c Because the solution contains the conjugate acid of the weak base at the equivalence point, the solution will be acidic.

16.150 A buffer solution is prepared by mixing equal volumes of 0.10 M $NaNO_2$ and 0.10 M HNO_2 solutions.

a Calculate the pH of this solution.

b How would the buffer capacity of 500 mL of this buffer solution compare to the buffer capacity of 1000 mL of the solution? Explain your reasoning.

c If you had 2.0 L of this buffer solution, what is the minimum volume of 0.10 M $Ca(OH)_2$ required to exceed the buffer capacity?

Cumulative-Skills Problems

Key: The problems under this heading combine skills introduced in previous chapters with those given in the current one.

16.151 The pH of a household cleaning solution is 11.50. This cleanser is an aqueous solution of ammonia with a density of 1.00 g/mL. What is the mass percentage of ammonia in the solution?

16.152 The pH of a white vinegar solution is 2.45. This vinegar is an aqueous solution of acetic acid with a density of 1.09 g/mL. What is the mass percentage of acetic acid in the solution?

16.153 What is the freezing point of 0.87 M aqueous ammonia? The density of this solution is 0.992 g/cm^3.

16.154 What is the freezing point of 0.92 M aqueous acetic acid? The density of this solution is 1.008 g/cm^3.

16.155 A chemist needs a buffer with pH 4.35. How many milliliters of pure acetic acid (density = 1.049 g/mL) must be added to 465 mL of 0.0941 M NaOH solution to obtain such a buffer?

16.156 A chemist needs a buffer with pH 3.50. How many milliliters of pure formic acid (density = 1.220 g/mL) must be added to 375 mL of 0.0857 M NaOH solution to obtain such a buffer?

Example 17.1   Writing Solubility Product Expressions

Critical Concept 17.1
When an excess of a slightly soluble ionic solid exists in a water solution, an equilibrium occurs between the solid and its ions in solution.

Solution Essentials:
- Solubility product constant, K_{sp}
- Equilibrium constant, K_c
- Molar concentration
- Solubility rules

Write the solubility product expressions for the following salts: a AgCl; b Hg_2Cl_2; c $Pb_3(AsO_4)_2$.

Problem Strategy The equilibrium constant for a slightly soluble ionic compound is simply the product of the concentrations of the ions, each raised to a power equal to the number of such ions in the formula of the compound.

Solution The equilibria and solubility product expressions are

a $AgCl(s) \rightleftharpoons Ag^+(aq) + Cl^-(aq); K_{sp} = [Ag^+][Cl^-]$

b $Hg_2Cl_2(s) \rightleftharpoons Hg_2^{2+}(aq) + 2Cl^-(aq); K_{sp} = [Hg_2^{2+}][Cl^-]^2$

Note that the mercury(I) ion is Hg_2^{2+}. (Mercury also has salts with the mercury(II) ion, Hg^{2+}.)

c $Pb_3(AsO_4)_2(s) \rightleftharpoons 3Pb^{2+}(aq) + 2AsO_4^{3-}(aq); K_{sp} = [Pb^{2+}]^3[AsO_4^{3-}]^2$

Answer Check Make sure that you have the correct formula of the compound and that you understand the ions that are present. For example, Hg_2Cl_2 is made up of the ions Hg_2^{2+} and Cl^- (not Hg^+ and Cl^-).

Exercise 17.1   Give solubility product expressions for the following: a barium sulfate; b iron(III) hydroxide; c calcium phosphate.

■ See Problems 17.27 and 17.28.

The solubility product constant, K_{sp}, of a slightly soluble ionic compound is expressed in terms of the molar concentrations of ions in the saturated solution. These ion concentrations are in turn related to the *molar solubility* of the ionic compound, which is the moles of compound that dissolve to give a liter of saturated solution. The next two examples show how to determine the solubility product constant from the solubility of a slightly soluble ionic compound.

⏻WL Interactive Example 17.2   Calculating K_{sp} from the Solubility (Simple Example)

Critical Concept 17.2
You obtain the solubility product constant, K_{sp}, of a slightly soluble compound from the molar concentrations of the compound's ions in solution; these ion concentrations can be related to the molar solubility of the compound.

Solution Essentials:
- Solubility product constant, K_{sp}
- Molar solubility
- Equilibrium constant, K_c
- Molar concentration

A liter of a solution saturated at 25°C with calcium oxalate, CaC_2O_4, is evaporated to dryness, giving a 0.0061-g residue of CaC_2O_4. Calculate the solubility product constant for this salt at 25°C.

Problem Strategy The grams of residue from a liter of saturated solution equals the solubility in grams per liter. Convert this to molar solubility (solubility in moles per liter). Then follow the steps similar to those for an equilibrium calculation: set up a table of ion concentrations (starting, change, and equilibrium) and calculate K_{sp} from the concentrations. The starting concentrations are those before any solid has dissolved ($= 0$), and the change values are obtained from the molar solubility.

Solution The solubility of calcium oxalate is 0.0061 g/L of *solution*, and the formula mass of CaC_2O_4 is 128 amu. Convert grams per liter to moles per liter:

$$\text{Molar solubility of } CaC_2O_4 = 0.0061 \text{ g } \cancel{CaC_2O_4}/L \times \frac{1 \text{ mol } CaC_2O_4}{128 \text{ g } \cancel{CaC_2O_4}}$$

$$= 4.8 \times 10^{-5} \text{ mol } CaC_2O_4/L$$

Now look at the equilibrium problem.

(continued)

(continued)

Step 1: Suppose you mix solid CaC_2O_4 in a liter of solution. Of this solid, 4.8×10^{-5} mol will dissolve to form 4.8×10^{-5} mol of each ion. The results are summarized in the following table. (Because the concentration of the solid does not appear in K_{sp}, we do not include it in the table.)

Concentration (M)	$CaC_2O_4(s) \rightleftharpoons$	Ca^{2+} (aq)	+	$C_2O_4^{2-}$ (aq)
Starting		0		0
Change		$+4.8 \times 10^{-5}$		$+4.8 \times 10^{-5}$
Equilibrium		4.8×10^{-5}		4.8×10^{-5}

Step 2: You now substitute into the equilibrium-constant equation:

$$K_{sp} = [Ca^{2+}][C_2O_4^{2-}] = (4.8 \times 10^{-5})(4.8 \times 10^{-5}) = \mathbf{2.3 \times 10^{-9}}$$

Answer Check Often, the most effective way to check your work is to invert the problem, using your answer to calculate the information given in the problem statement; the actual steps often involve different operations. Here, you would first calculate the ion concentrations by taking the square root of the K_{sp} that you obtained. This ion concentration is, in this case, equal to the molar solubility. Multiplying the molar solubility by the formula mass gives the solubility of the compound, which should match the value given in the problem.

Exercise 17.2 Silver ion may be recovered from used photographic fixing solution by precipitating it as silver chloride. The solubility of silver chloride is 1.9×10^{-3} g/L. Calculate K_{sp}.

■ See Problems 17.29 and 17.30.

♥WL Interactive Example **17.3** **Calculating K_{sp} from the Solubility (More Complicated Example)**

Gaining Mastery Toolbox

Critical Concept 17.3
The number of moles of ion formed when one mole of slightly soluble compound dissolves affects the calculation of K_{sp} from molar solubility in two places: the molar concentration of ion (which is some multiple of molar solubility) and its exponent in the expression for the solubility product.

Solution Essentials:
• Solubility product constant, K_{sp}
• Molar solubility
• Equilibrium constant, K_c
• Molar concentration

By experiment, it is found that 1.2×10^{-3} mol of lead(II) iodide, PbI_2, dissolves in 1 L of aqueous solution at 25°C. What is the solubility product constant at this temperature?

Problem Strategy This is similar to the previous example, except that every mole of the compound (PbI_2) dissolves to form *two* moles of the anion (I^-). This difference affects the calculation in two places: it affects the molar concentration of the ion (the molar concentration of the anion is *twice* that of the molar solubility), and it affects the equilibrium-constant expression (which contains the anion concentration *squared*).

Solution

Step 1: Suppose the solid lead(II) iodide is mixed into 1 L of solution. You find that 1.2×10^{-3} mol dissolves to form 1.2×10^{-3} mol Pb^{2+} and $2 \times (1.2 \times 10^{-3})$ mol I^-, as summarized in the following table:

Concentration (M)	$PbI_2(s) \rightleftharpoons$	Pb^{2+} (aq)	+	$2I^-$ (aq)
Starting		0		0
Change		$+1.2 \times 10^{-3}$		$+2 \times (1.2 \times 10^{-3})$
Equilibrium		1.2×10^{-3}		$2 \times (1.2 \times 10^{-3})$

Step 2: You substitute into the equilibrium-constant equation:

$$K_{sp} = [Pb^{2+}][I^-]^2 = (1.2 \times 10^{-3}) \times (2 \times 1.2 \times 10^{-3})^2 = \mathbf{6.9 \times 10^{-9}}$$

(continued)

(continued)

Lead(II) iodide

Answer Check As in the previous example, an effective check is simply to reverse the calculation to see whether you do get the value given in the problem statement. Suppose you let the molar solubility be x (which is the value given in the problem statement). Then x is the concentration of Pb^{2+}, and $2x$ is the concentration of I^- (because there are twice as many I^- ions as Pb^{2+} ions). Thus, $K_{sp} = [Pb^{2+}][I^-]^2 = x \times (2x)^2 = 4x^3$, and

$$x = \sqrt[3]{K_{sp}/4} = \sqrt[3]{6.9 \times 10^{-9}/4} = 1.2 \times 10^{-3}$$

Exercise 17.3   Lead(II) arsenate, $Pb_3(AsO_4)_2$, has been used as an insecticide. It is only slightly soluble in water. If the solubility is 3.0×10^{-5} g/L, what is the solubility product constant? Assume that the solubility equilibrium is the only important one.

■ See Problems 17.31 and 17.32.

Table 17.1 lists the solubility product constants for various ionic compounds. If the solubility product constant is known, the solubility of the compound can be calculated.

Table 17.1 Solubility Product Constants, K_{sp}, at 25°C

Substance	Formula	K_{sp}	Substance	Formula	K_{sp}
Aluminum hydroxide	$Al(OH)_3$	4.6×10^{-33}	Lead(II) sulfide	PbS	2.5×10^{-27}
Barium chromate	$BaCrO_4$	1.2×10^{-10}	Magnesium arsenate	$Mg_3(AsO_4)_2$	2×10^{-20}
Barium fluoride	BaF_2	1.0×10^{-6}	Magnesium carbonate	$MgCO_3$	1.0×10^{-5}
Barium sulfate	$BaSO_4$	1.1×10^{-10}	Magnesium hydroxide	$Mg(OH)_2$	1.8×10^{-11}
Cadmium oxalate	CdC_2O_4	1.5×10^{-8}	Magnesium oxalate	MgC_2O_4	8.5×10^{-5}
Cadmium sulfide	CdS	8×10^{-27}	Manganese(II) sulfide	MnS	2.5×10^{-10}
Calcium carbonate	$CaCO_3$	3.8×10^{-9}	Mercury(I) chloride	Hg_2Cl_2	1.3×10^{-18}
Calcium fluoride	CaF_2	3.4×10^{-11}	Mercury(II) sulfide	HgS	1.6×10^{-52}
Calcium oxalate	CaC_2O_4	2.3×10^{-9}	Nickel(II) hydroxide	$Ni(OH)_2$	2.0×10^{-15}
Calcium phosphate	$Ca_3(PO_4)_2$	1×10^{-26}	Nickel(II) sulfide	NiS	3×10^{-19}
Calcium sulfate	$CaSO_4$	2.4×10^{-5}	Silver acetate	$AgC_2H_3O_2$	2.0×10^{-3}
Cobalt(II) sulfide	CoS	4×10^{-21}	Silver bromide	AgBr	5.0×10^{-13}
Copper(II) hydroxide	$Cu(OH)_2$	2.6×10^{-19}	Silver chloride	AgCl	1.8×10^{-10}
Copper(II) sulfide	CuS	6×10^{-36}	Silver chromate	Ag_2CrO_4	1.1×10^{-12}
Iron(II) hydroxide	$Fe(OH)_2$	8×10^{-16}	Silver iodide	AgI	8.3×10^{-17}
Iron(II) sulfide	FeS	6×10^{-18}	Silver sulfide	Ag_2S	6×10^{-50}
Iron(III) hydroxide	$Fe(OH)_3$	2.5×10^{-39}	Strontium carbonate	$SrCO_3$	9.3×10^{-10}
Lead(II) arsenate	$Pb_3(AsO_4)_2$	4×10^{-36}	Strontium chromate	$SrCrO_4$	3.5×10^{-5}
Lead(II) chloride	$PbCl_2$	1.6×10^{-5}	Strontium sulfate	$SrSO_4$	2.5×10^{-7}
Lead(II) chromate	$PbCrO_4$	1.8×10^{-14}	Zinc hydroxide	$Zn(OH)_2$	2.1×10^{-16}
Lead(II) iodide	PbI_2	6.5×10^{-9}	Zinc sulfide	ZnS	1.1×10^{-21}
Lead(II) sulfate	$PbSO_4$	1.7×10^{-8}			

Example 17.4

Calculating the Solubility from K_{sp}

The mineral fluorite is calcium fluoride, CaF_2. Calculate the solubility (in grams per liter) of calcium fluoride in water from the solubility product constant (3.4×10^{-11}).

Problem Strategy This problem is the reverse of the preceding ones; instead of finding K_{sp} from the solubility, here you calculate solubility from the K_{sp}. You follow the three steps for equilibrium problems, but since the molar solubility is not immediately known, you assign it the value x. For Step 1, you obtain the concentration of each ion by multiplying x by the coefficient of the ion in the chemical equation. In Step 2, you obtain K_{sp} as a cubic in x. In Step 3, you solve the equilibrium-constant equation for x, which will involve taking a cube root.

Solution

Step 1: Let x be the molar solubility of CaF_2. When solid CaF_2 is mixed into a liter of solution, x mol dissolves, forming x mol Ca^{2+} and $2x$ mol F^-.

Concentration (M)	$CaF_2(s) \rightleftharpoons Ca^{2+}(aq)$	$+ 2F^-(aq)$
Starting	0	0
Change	$+x$	$+2x$
Equilibrium	x	$2x$

The mineral fluorite (CaF_2)

Step 2: You substitute into the equilibrium-constant equation.

$$[Ca^{2+}][F^-]^2 = K_{sp}$$
$$(x) \times (2x)^2 = 3.4 \times 10^{-11}$$
$$4x^3 = 3.4 \times 10^{-11}$$

Step 3: You now solve for x.

$$x = \sqrt[3]{\frac{3.4 \times 10^{-11}}{4}} = 2.0 \times 10^{-4}$$

The molar solubility is 2.0×10^{-4} mol CaF_2 per liter. To get the solubility in grams per liter, you convert, using the molar mass of CaF_2 (78.1 g/mol).

$$\text{Solubility} = 2.0 \times 10^{-4} \text{ mol } CaF_2/L \times \frac{78.1 \text{ g } CaF_2}{1 \text{ mol } CaF_2} = \mathbf{1.6 \times 10^{-2} \text{ g } CaF_2/L}$$

Answer Check As a quick check of your work, calculate K_{sp} for CaF_2 from the values you obtained for the molar concentrations of ions:

$$[Ca^{2+}] = x = 2.0 \times 10^{-4}$$
$$[F^-] = 2x = 4.0 \times 10^{-4}$$
$$K_{sp} = [Ca^{2+}][F^-]^2 = (2.0 \times 10^{-4})(4.0 \times 10^{-4})^2 = 3.2 \times 10^{-11}$$

Note the agreement (within computational error) with the value given in the problem (3.4×10^{-11}).

Exercise 17.4 Anhydrite is a calcium sulfate mineral deposited when seawater evaporates. What is the solubility of calcium sulfate, in grams per liter? Table 17.1 gives the solubility product constant for calcium sulfate.

■ See Problems 17.35, 17.36, 17.37, and 17.38.

These examples illustrate the relationship between the solubility of a slightly soluble ionic compound in pure water and its solubility product constant. In the next section, you will see how the solubility product constant can be used to calculate the solubility in the presence of other ions. K_{sp} is also useful in deciding whether to expect precipitation under given conditions.

Before leaving this section, we need to add a note of caution. We have calculated the solubility product of calcium oxalate, CaC_2O_4, and the solubility of calcium fluoride, CaF_2, assuming that only the solubility equilibrium is important. You need to be aware, however, that hydrolysis of the anion can be important, too, particularly if the anion is quite basic, which will be the case if the conjugate acid is rather weak. In the two examples given in this section, hydrolysis is relatively unimportant. (You may want to look at Problem 17.117, which examines this issue.)

CONCEPT CHECK 17.1

Lead compounds have been used as paint pigments, but because the lead(II) ion is toxic, the use of lead paints in homes is now prohibited. Which of the following lead(II) compounds would yield the greatest number of lead(II) ions when added to the same quantity of water (assuming that some undissolved solid always remains): $PbCrO_4$, $PbSO_4$, or PbS?

17.2 Solubility and the Common-Ion Effect

The importance of the solubility product constant becomes apparent when you consider the solubility of one salt in the solution of another salt having the same cation or anion. For example, suppose you wish to know the solubility of calcium oxalate in a solution of calcium chloride. Each salt contributes the same cation (Ca^{2+}). The effect of the calcium ion provided by the calcium chloride is to make calcium oxalate less soluble than it would be in pure water.

You can explain this decrease in solubility in terms of Le Châtelier's principle. Suppose you first mix crystals of calcium oxalate in a quantity of pure water to establish the equilibrium:

$$CaC_2O_4(s) \rightleftharpoons Ca^{2+}(aq) + C_2O_4^{2-}(aq)$$

Now imagine that you add some calcium chloride. Calcium chloride is a soluble salt, so it dissolves to give an increase in calcium-ion concentration. You can regard this increase as a stress on the original equilibrium. According to Le Châtelier's principle, the ions will react to remove some of the added calcium ion.

$$CaC_2O_4(s) \longleftarrow Ca^{2+}(aq) + C_2O_4^{2-}(aq)$$

In other words, some calcium oxalate precipitates from the solution. The solution now contains less calcium oxalate. You conclude that calcium oxalate is less soluble in a solution of calcium chloride than in pure water.

A solution of soluble calcium chloride has an ion *in common* with the slightly soluble calcium oxalate. The decrease in solubility of calcium oxalate in a solution of calcium chloride is an example of the *common-ion effect* (first discussed in Section 16.5). In general, any ionic equilibrium is affected by a substance producing an ion involved in the equilibrium, as you would predict from Le Châtelier's principle. Figure 17.2 illustrates the principle of the common-ion effect for lead(II) chromate, $PbCrO_4$, which is only slightly soluble in water at 25°C. When the very soluble $Pb(NO_3)_2$ is added to a saturated solution of $PbCrO_4$, the concentration of the common ion, Pb^{2+}, increases and $PbCrO_4$ precipitates. The equilibrium is

$$PbCrO_4(s) \rightleftharpoons Pb^{2+}(aq) + CrO_4^{2-}(aq)$$

The next example shows how you can calculate the solubility of a slightly soluble salt in a solution containing a substance with a common ion.

Figure 17.2 ▲

Demonstration of the common-ion effect When the experimenter adds lead(II) nitrate solution (colorless) from the dropper to the saturated solution of lead(II) chromate (pale yellow), a yellow precipitate of lead chromate forms.

© Cengage Learning

Example 17.5

Calculating the Solubility of a Slightly Soluble Salt in a Solution of a Common Ion

Gaining Mastery Toolbox

Critical Concept 17.5
The solubility of an ionic compound is less in a solution that contains an ion also present in the compound than it is in pure water. Note that the equilibrium concentration of this ion equals the initial concentration of the ion plus whatever concentration is contributed by the dissolution of the compound.

Solution Essentials:
• Solubility product constant, K_{sp}
• Molar solubility
• Common-ion effect
• Le Châtelier's principle
• Equilibrium constant, K_c
• Molar concentration

What is the molar solubility of calcium oxalate in 0.15 M calcium chloride? Compare this molar solubility with that found earlier (Example 17.2) for CaC_2O_4 in pure water (4.8 × 10^{-5} M). The solubility product constant for calcium oxalate is 2.3 × 10^{-9}.

Problem Strategy The two salts have a common ion (Ca^{2+}): one salt ($CaCl_2$) is soluble and provides Ca^{2+} ion that suppresses the solubility of the slightly soluble salt (CaC_2O_4). You solve the solubility equilibrium for the slightly soluble salt, noting that when you set up the concentration table you have a starting concentration of Ca^{2+} ion from $CaCl_2$. The equilibrium-constant equation you obtain is a quadratic, which you can reduce to a linear equation using the same approximation used in Chapter 16; you assume x is small compared with the number you add it to. (That is, the starting Ca^{2+} concentration is not much affected by the dissolution of the slightly soluble salt.) The value of x equals the molar solubility of the slightly soluble salt in the presence of the common ion.

Solution

Step 1: Suppose solid CaC_2O_4 is mixed into 1 L of 0.15 M $CaCl_2$ and that the molar solubility of the CaC_2O_4 is x M. At the start (before CaC_2O_4 dissolves), there is 0.15 mol Ca^{2+} in the solution. Of the solid CaC_2O_4, x mol dissolves to give x mol of additional Ca^{2+} and x mol $C_2O_4^{2-}$. The following table summarizes these results:

Concentration (M)	$CaC_2O_4(s) \rightleftharpoons Ca^{2+}(aq) +$	$C_2O_4^{2-}(aq)$
Starting	0.15	0
Change	+x	+x
Equilibrium	0.15 + x	x

Step 2: You substitute into the equilibrium-constant equation:

$$[Ca^{2+}][C_2O_4^{2-}] = K_{sp}$$
$$(0.15 + x)x = 2.3 \times 10^{-9}$$

Step 3: Now rearrange this equation to give

$$x = \frac{2.3 \times 10^{-9}}{0.15 + x}$$

Because calcium oxalate is only slightly soluble, you might expect x to be negligible compared with 0.15. In that case,

$$0.15 + x \simeq 0.15$$

and the previous equation becomes

$$x \simeq \frac{2.3 \times 10^{-9}}{0.15} = 1.5 \times 10^{-8}$$

Note that x is indeed much smaller than 0.15, so your assumption was correct. Therefore, the molar solubility of calcium oxalate in 0.15 M $CaCl_2$ is **1.5 × 10^{-8} M.** In pure water, the molar solubility is 4.8 × 10^{-5} M, which is over **3000 times greater.**

Answer Check Calculate K_{sp} from the calculated molar solubility of calcium oxalate (1.5 × 10^{-8} M). The concentration of Ca^{2+} ion is 0.15 M from $CaCl_2$ (plus 1.5 × 10^{-8} M from dissolved calcium oxalate, which is negligible), and the concentration of oxalate ion is 1.5 × 10^{-8} M. Therefore,

$$K_{sp} = [Ca^{2+}][C_2O_4^{2-}] = (0.15)(1.5 \times 10^{-8}) = 2.3 \times 10^{-9}$$

which agrees with the value given in the problem statement.

(continued)

(continued)

Exercise 17.5 **a** Calculate the molar solubility of barium fluoride, BaF_2, in water at 25°C. The solubility product constant for BaF_2 at this temperature is 1.0×10^{-6}.

b What is the molar solubility of barium fluoride in 0.15 M NaF at 25°C? Compare the solubility in this case with that of BaF_2 in pure water.

■ See Problems 17.39, 17.40, 17.41, and 17.42.

CONCEPT CHECK 17.2

Suppose you have equal volumes of saturated solutions of $NaNO_3$, Na_2SO_4, and PbS. Which solution would dissolve the most lead(II) sulfate, $PbSO_4$?

CONCEPT CHECK 17.3

Consider the beaker below, which represents a saturated solution of AgCl(*aq*). Draw a picture of the solution in the beaker after the addition of NaCl(*aq*).

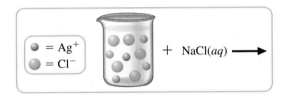

17.3 Precipitation Calculations

In the chapter opening, we mentioned that calcium oxalate precipitates to form kidney stones. The same salt would precipitate in the body if oxalic acid (a poison) were accidentally ingested, because Ca^{2+} ion is present in the blood. (See Figure 17.3 for another example of a precipitation.) To understand processes such as these, you must understand the conditions under which precipitation occurs. Precipitation is merely another way of looking at a solubility equilibrium. Rather than ask how much of a substance will dissolve in a solution, you ask: Will precipitation occur for given starting ion concentrations?

Criterion for Precipitation

The question just asked can be stated more generally: Given the concentrations of substances in a reaction mixture, will the reaction go in the forward or the reverse direction? To answer this, you evaluate the *reaction quotient* Q_c and compare it with the equilibrium constant K_c. ▶ The reaction quotient has the same form as the equilibrium-constant expression, but the concentrations of substances are not necessarily equilibrium values. Rather, they are concentrations at the start of a reaction. To predict the direction of reaction, you compare Q_c with K_c.

If $Q_c < K_c$, the reaction should go in the forward direction.
If $Q_c = K_c$, the reaction mixture is at equilibrium.
If $Q_c > K_c$, the reaction should go in the reverse direction.

Suppose you add lead(II) nitrate, $Pb(NO_3)_2$, and sodium chloride, NaCl, to water to give a solution that is 0.050 M Pb^{2+} and 0.10 M Cl^-. Will lead(II) chloride, $PbCl_2$, precipitate? To answer this, you first write the solubility equilibrium.

$$PbCl_2(s) \xrightarrow{H_2O} Pb^{2+}(aq) + 2Cl^-(aq)$$

© Cengage Learning

Figure 17.3 ▲

Formation of a precipitate Silver chromate, Ag_2CrO_4, precipitates when a demonstrator adds aqueous potassium chromate, K_2CrO_4(aq), to aqueous silver nitrate, $AgNO_3$(aq).

| The reaction quotient was discussed in Section 14.5.

The reaction quotient has the form of the equilibrium-constant expression, which in this case is the K_{sp} expression, but the concentrations of the products are starting values, denoted by the subscript i.

$$Q_c = [\text{Pb}^{2+}]_i[\text{Cl}^-]_i^{\,2}$$

Here Q_c for a solubility reaction is often called the **ion product** (rather than reaction quotient), because it is *the product of ion concentrations in a solution, each concentration raised to a power equal to the number of ions in the formula of the ionic compound.*

To evaluate the ion product Q_c, you substitute the concentrations of Pb^{2+} and Cl^- ions in the solution at the start of the reaction. You have 0.050 M Pb^{2+} and 0.10 M Cl^-. Substituting, you find that

$$Q_c = (0.050)(0.10)^2 = 5.0 \times 10^{-4}$$

K_{sp} for PbCl_2 is 1.6×10^{-5} (Table 17.1), so Q_c is greater than K_{sp}. Therefore, the reaction goes in the reverse direction. That is,

$$\text{PbCl}_2(s) \longleftarrow \text{Pb}^{2+}(aq) + 2\text{Cl}^-(aq)$$

In other words, Pb^{2+} and Cl^- react to precipitate PbCl_2. As precipitation occurs, the ion concentrations, and hence the ion product, decrease. Precipitation ceases when the ion product equals K_{sp}. The reaction mixture is then at equilibrium.

We can summarize our conclusions in terms of the following criterion for precipitation.

Precipitation may not occur even though the ion product has been exceeded. In such a case, the solution is supersaturated. Usually a small crystal forms after a time, and then precipitation occurs rapidly.

> Precipitation is expected to occur if the ion product for a solubility reaction is greater than K_{sp}. If the ion product is less than K_{sp}, precipitation will not occur (the solution is unsaturated with respect to the ionic compound). If the ion product equals K_{sp}, the reaction is at equilibrium (the solution is saturated with the ionic compound). ◄

⚘WL Interactive Example   **17.6**   **Predicting Whether Precipitation Will Occur (Given Ion Concentrations)**

Gaining Mastery Toolbox

Critical Concept 17.6
Precipitation is expected to occur when you mix two solutions and the ion product for the precipitation reaction exceeds the solubility product.

Solution Essentials:
- Solubility product constant, K_{sp}
- Molar solubility
- Reaction quotient
- Equilibrium constant, K_c
- Molar concentration

The concentration of calcium ion in blood plasma is 0.0025 M. If the concentration of oxalate ion is 1.0×10^{-7} M, do you expect calcium oxalate to precipitate? K_{sp} for calcium oxalate is 2.3×10^{-9}.

Problem Strategy Precipitation is simply another way of looking at solubility equilibrium.

$$\text{CaC}_2\text{O}_4(s) \rightleftharpoons \text{Ca}^{2+}(aq) + \text{C}_2\text{O}_4^{2-}(aq)$$

Going to the right, calcium oxalate dissolves, but going to the left, Ca^{2+} and $\text{C}_2\text{O}_4^{2-}$ react to precipitate calcium oxalate. The reaction quotient, Q_c, for this equation as written is the ion product $[\text{Ca}^{2+}]_i[\text{C}_2\text{O}_4^{2-}]_i$. This ion product decreases as precipitation occurs (and the ion concentrations decrease). As the reaction approaches equilibrium, it approaches the equilibrium constant (the solubility product) K_{sp}. This means that Q_c should be greater than K_{sp} for precipitation to occur; otherwise, precipitation cannot occur. Thus, you calculate Q_c and compare it with K_{sp}.

Solution The ion product for calcium oxalate is

$$\text{Ion product} = [\text{Ca}^{2+}]_i[\text{C}_2\text{O}_4^{2-}]_i = (0.0025) \times (1.0 \times 10^{-7}) = 2.5 \times 10^{-10}$$

This value is smaller than the solubility product constant, so **you do not expect precipitation to occur.**

(continued)

(*continued*)

Answer Check Be sure that you have the correct expression for the reaction quotient and that you correctly compare the reaction quotient to the solubility product.

Exercise 17.6 Anhydrite is a mineral composed of $CaSO_4$ (calcium sulfate). An inland lake has Ca^{2+} and SO_4^{2-} concentrations of 0.0052 *M* and 0.0041 *M*, respectively. If these concentrations were doubled by evaporation, would you expect calcium sulfate to precipitate?

■ See Problems 17.47 and 17.48.

Precipitation is an important industrial and laboratory process. Figure 17.4 (next page) shows the vats used to precipitate magnesium ion from seawater, which is a source of magnesium metal. The following example is a typical problem in precipitation. You are given the volumes and concentrations of two solutions and asked whether a precipitate will form when the solutions are mixed.

Example 17.7 **Predicting Whether Precipitation Will Occur (Given Solution Volumes and Concentrations)**

Gaining Mastery Toolbox

Critical Concept 17.7
To see if precipitation is expected, you need the ion product, which is given in terms of concentrations of ions. You obtain these concentrations from the moles of ions added and the total volume of the solution.

Solution Essentials:
- Solubility product constant, K_{sp}
- Molar solubility
- Reaction quotient
- Equilibrium constant, K_c
- Molar concentration

Sulfate ion, SO_4^{2-}, in solution is often determined quantitatively by precipitating it as barium sulfate, $BaSO_4$. The sulfate ion may have been formed from a sulfur compound. Analysis for the amount of sulfate ion then indicates the percentage of sulfur in the compound. Is a precipitate expected to form when 50.0 mL of 0.0010 *M* $BaCl_2$ is added to 50.0 mL of 0.00010 *M* Na_2SO_4? The solubility product constant for barium sulfate is 1.1×10^{-10}. Assume that the total volume of solution, after mixing, equals the sum of the volumes of the separate solutions.

Problem Strategy You need to calculate Q_c, the ion product, to compare it with K_{sp}, the solubility product. For Q_c, you need to obtain the concentrations of ions in the solution just after mixing, but before precipitation, if any occurs. First, calculate the moles of each ion added. Then, using the total volume of solution (assumed to equal the sum of the two added solutions),

calculate the molar concentration of each ion present. Finally, calculate Q_c and compare it with K_{sp}.

Solution The molar amount of Ba^{2+} present in 50.0 mL (0.0500 L) of 0.0010 *M* $BaCl_2$ is

$$\text{Amount of } Ba^{2+} = \frac{0.0010 \text{ mol } Ba^{2+}}{1 \text{ L soln}} \times 0.050 \text{ L soln} =$$
$$5.0 \times 10^{-5} \text{ mol } Ba^{2+}$$

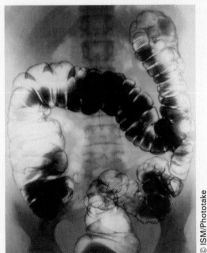

An x-ray photograph of a patient's colon (large intestine)
The colon shows up after the patient has ingested an aqueous suspension of barium sulfate, which is opaque to x rays. Even though free barium ion is toxic, barium sulfate is harmless because it is so insoluble. (Color enhanced by computer for contrast.)

© ISM/Phototake

(*continued*)

(continued)

The molar concentration of Ba^{2+} in the total solution equals the molar amount of Ba^{2+} divided by the total volume (0.0500 L $BaCl_2$ + 0.0500 L Na_2SO_4 = 0.1000 L).

$$[Ba^{2+}] = \frac{5.0 \times 10^{-5}\ mol}{0.1000\ L\ soln} = 5.0 \times 10^{-4} M$$

Similarly, you find $[SO_4^{2-}] = 5.0 \times 10^{-5}\ M$

(Try the calculations for SO_4^{2-}.) The ion product is

$$Q_c = [Ba^{2+}]_i[SO_4^{2-}]_i = (5.0 \times 10^{-4}) \times (5.0 \times 10^{-5})$$
$$= 2.5 \times 10^{-8}$$

Because the ion product is greater than the solubility product constant (1.1×10^{-10}), **you expect barium sulfate to precipitate.**

Answer Check Note that you calculate the *amount* of barium ion added using the volume of barium chloride solution added (0.050 L). Similarly, you calculate the amount of sulfate ion present from the *initial* volume of sodium sulfate (0.050 L). But to calculate the *concentrations* of these ions when both solutions are present together, you use the total volume of the two solutions (0.1000 L).

Exercise 17.7   A solution of 0.00016 M lead(II) nitrate, $Pb(NO_3)_2$, was poured into 456 mL of 0.00023 M sodium sulfate, Na_2SO_4. Would a precipitate of lead(II) sulfate, $PbSO_4$, be expected to form if 255 mL of the lead nitrate solution were added?

■ See Problems 17.49 and 17.50.

Figure 17.4 ▲

Vats for the precipitation of magnesium hydroxide from seawater Seawater contains magnesium ion (in addition to other ions). When base is added, magnesium hydroxide precipitates. This is the source of magnesium metal.

© Keystone/Getty Images

Fractional Precipitation

Fractional precipitation is *the technique of separating two or more ions from a solution by adding a reactant that precipitates first one ion, then another, and so forth.* For example, suppose a solution is 0.10 M Ba^{2+} and 0.10 M Sr^{2+}. As we will show in the following paragraphs, when you slowly add a concentrated solution of potassium chromate, K_2CrO_4, to the solution of Ba^{2+} and Sr^{2+} ions, barium chromate precipitates first. After most of the Ba^{2+} ion has precipitated, strontium chromate begins to come out of solution. It is therefore possible to separate Ba^{2+} and Sr^{2+} ions from a solution by fractional precipitation using K_2CrO_4.

To understand why Ba^{2+} and Sr^{2+} ions can be separated in this way, calculate (1) the concentration of CrO_4^{2-} necessary to just begin the precipitation of $BaCrO_4$, and (2) the concentration of CrO_4^{2-} necessary to just begin the precipitation of $SrCrO_4$. Ignore any volume change in the solution of Ba^{2+} and Sr^{2+} ions resulting from the addition of the concentrated K_2CrO_4 solution. To calculate the CrO_4^{2-} concentration when $BaCrO_4$ begins to precipitate, you substitute the initial Ba^{2+} concentration into the solubility product equation. K_{sp} for $BaCrO_4$ is 1.2×10^{-10}.

$$[Ba^{2+}][CrO_4^{2-}] = K_{sp} \quad (\text{for } BaCrO_4)$$
$$(0.10)[CrO_4^{2-}] = 1.2 \times 10^{-10}$$
$$[CrO_4^{2-}] = \frac{1.2 \times 10^{-10}}{0.10} = 1.2 \times 10^{-9} M$$

In the same way, you can calculate the CrO_4^{2-} concentration when $SrCrO_4$ begins to precipitate. K_{sp} for $SrCrO_4$ is 3.5×10^{-5}.

$$[Sr^{2+}][CrO_4^{2-}] = K_{sp} \quad (\text{for } SrCrO_4)$$
$$(0.10)[CrO_4^{2-}] = 3.5 \times 10^{-5}$$
$$[CrO_4^{2-}] = \frac{3.5 \times 10^{-5}}{0.10} = 3.5 \times 10^{-4} M$$

Note that $BaCrO_4$ precipitates first, because the CrO_4^{2-} concentration necessary to form the $BaCrO_4$ precipitate is smaller.

These results reveal that as the solution of K_2CrO_4 is slowly added to the solution of Ba^{2+} and Sr^{2+}, barium chromate begins to precipitate when the CrO_4^{2-} concentration reaches 1.2×10^{-9} M. Barium chromate continues to precipitate

© Cengage Learning

Figure 17.5 ◀

Titration of chloride ion by silver nitrate using potassium chromate as an indicator

A small amount of K_2CrO_4 (yellow) has been added to a solution containing an unknown amount of Cl^- ion.

The solution is titrated by $AgNO_3$ solution, giving a white precipitate of AgCl.

When nearly all of the Cl^- ion has precipitated as AgCl, silver chromate begins to precipitate. Silver chromate, Ag_2CrO_4, has a red-brown color, and the appearance of this color signals the end of the titration. An excess of Ag^+ was added to show the color of Ag_2CrO_4 more clearly in this photo.

as K_2CrO_4 is added. When the concentration of CrO_4^{2-} reaches 3.5×10^{-4} M, strontium chromate begins to precipitate.

What is the percentage of Ba^{2+} ion remaining just as $SrCrO_4$ begins to precipitate? First calculate the concentration of Ba^{2+} at this point. You write the solubility product equation and substitute $[CrO_4^{2-}] = 3.5 \times 10^{-4}$, the concentration of chromate ion when $SrCrO_4$ begins to precipitate.

$$[Ba^{2+}][CrO_4^{2-}] = K_{sp} \text{ (for BaCrO}_4)$$

$$[Ba^{2+}](3.5 \times 10^{-4}) = 1.2 \times 10^{-10}$$

$$[Ba^{2+}] = \frac{1.2 \times 10^{-10}}{3.5 \times 10^{-4}} = 3.4 \times 10^{-7} M$$

To calculate the percentage of Ba^{2+} ion remaining, you divide this concentration of Ba^{2+} by the initial concentration (0.10 M) and multiply by 100%.

$$\frac{3.4 \times 10^{-7}}{0.10} \times 100\% = 0.00034\%$$

The percentage of Ba^{2+} ion remaining in solution is quite low, so most of the Ba^{2+} ion has precipitated by the time $SrCrO_4$ begins to precipitate. You conclude that Ba^{2+} and Sr^{2+} can indeed be separated by fractional precipitation. (Another application of fractional precipitation is shown in Figure 17.5.)

17.4 Effect of pH on Solubility

In discussing solubility, we have assumed that the only equilibrium of interest was the one between the solid ionic compound and its ions in solution. Sometimes, however, it is necessary to account for other reactions the ions might undergo. For example, if the anion is the conjugate base of a weak acid, the anion reacts with H_3O^+ ion. You should expect the solubility to be affected by pH. We will look into this possibility now.

Qualitative Effect of pH

Consider the equilibrium between solid calcium oxalate, CaC_2O_4, and its ions in aqueous solution:

$$CaC_2O_4(s) \rightleftharpoons Ca^{2+}(aq) + C_2O_4^{2-}(aq)$$

Because the oxalate ion is conjugate to a weak acid (hydrogen oxalate ion, $HC_2O_4^-$), you would expect it to react with any H_3O^+ ion that is added—say, from a strong acid:

$$C_2O_4^{2-}(aq) + H_3O^+(aq) \rightleftharpoons HC_2O_4^-(aq) + H_2O(l)$$

According to Le Châtelier's principle, as $C_2O_4^{2-}$ ion is removed by reaction with H_3O^+ ion, more calcium oxalate dissolves to replenish some of the $C_2O_4^{2-}$ ion.

$$CaC_2O_4(s) \longrightarrow Ca^{2+}(aq) + C_2O_4^{2-}(aq)$$

Therefore, you expect calcium oxalate to be more soluble in acidic solution (low pH) than in pure water.

In general, salts of weak acids should be expected to be more soluble in acidic solutions. As an example, consider the process of tooth decay. Bacteria on the teeth produce an acidic medium as a result of the metabolism of sugar. Teeth are normally composed of a calcium phosphate mineral hydroxyapatite, which you can denote as either $Ca_5(PO_4)_3OH$ or $3Ca_3(PO_4)_2 \cdot Ca(OH)_2$. This mineral salt of the weak acid H_2O (conjugate to OH^-) dissolves in the presence of the acid medium, producing cavities in the teeth. Fluoride toothpastes provide F^- ion, which gradually replaces the OH^- ion in the teeth to produce fluorapatite, $Ca_5(PO_4)_3F$ or $3Ca_3(PO_4)_2 \cdot CaF_2$, which is much less soluble than hydroxyapatite.

Example   17.8

Determining the Qualitative Effect of pH on Solubility

Gaining Mastery Toolbox

Critical Concept 17.8
Salts of weak acids are more soluble in acidic solution. In comparing two salts, the solubility of the one whose anion corresponds to the weaker acid is most affected by the addition of a strong acid.

Solution Essentials:
- Solubility equilibrium
- Weak acid
- Acid-ionization constant, K_a
- Le Châtelier's principle

Consider two slightly soluble salts, calcium carbonate and calcium sulfate. Which of these would have its solubility more affected by the addition of HCl, a strong acid? Would the solubility of the salt increase or decrease?

Problem Strategy A strong acid affects the solubility of a salt of a weak acid by providing hydronium ion that reacts with the anion of the salt in the solution. As the anion is removed by this reaction, more salt dissolves to replenish the anion concentration. You need to determine which salt has the anion corresponding to the weaker acid.

Solution Calcium carbonate gives the solubility equilibrium

$$CaCO_3(s) \rightleftharpoons Ca^{2+}(aq) + CO_3^{2-}(aq)$$

When a strong acid is added, the hydrogen ion reacts with carbonate ion, because it is conjugate to a weak acid (HCO_3^-).

$$H_3O^+(aq) + CO_3^{2-}(aq) \rightleftharpoons H_2O(l) + HCO_3^-(aq)$$

As carbonate ion is removed, calcium carbonate dissolves. Moreover, the hydrogen carbonate ion itself is removed in further reaction.

$$H_3O^+(aq) + HCO_3^-(aq) \longrightarrow H_2O(l) + H_2CO_3(aq) \longrightarrow 2H_2O(l) + CO_2(g)$$

Bubbles of carbon dioxide gas appear as more calcium carbonate dissolves.

For calcium sulfate, the corresponding equilibria are

$$CaSO_4(s) \rightleftharpoons Ca^{2+}(aq) + SO_4^{2-}(aq)$$

$$H_3O^+(aq) + SO_4^{2-}(aq) \rightleftharpoons H_2O(l) + HSO_4^-(aq)$$

Again, the anion of the insoluble salt is removed by reaction with hydronium ion. You would expect calcium sulfate to become more soluble in strong acid. However, HSO_4^- is a much stronger acid than HCO_3^-, as you can see by comparing acid-ionization constants. (The values K_{a2} for H_2CO_3 and K_a for HSO_4^- in Table 16.1 are 4.8×10^{-11} and 1.1×10^{-2}, respectively.) **Thus, calcium carbonate is much more soluble in acidic solution, whereas the solubility of calcium sulfate is only slightly affected.**

(continued)

Limestone Caves

On the side of a hill, covered by tall grass and shrubs, the cave entrance might easily be missed. It is barely large enough for a person to enter, even if hunched over, but the space soon expands into a chamber that one can easily stand in. Rock columns rise to an arched ceiling, and there are caverns leading mysteriously away on both sides of the chamber. Limestone caves, like the one described here, which is in Tennessee, dot the landscapes of the eastern and southwestern United States and of Europe and other places as well. Though many limestone caves are inaccessible, a few, like Mammouth Cave in Kentucky and Carlsbad Caverns in New Mexico, are spectacular tourist attractions with multicolored rock draperies and icicles hanging from the ceilings of large underground rooms (Figure 17.6). How do such caves form?

Limestone itself formed millions of years ago. Seashells accumulated on ocean floors, and then layers of these shells were subsequently compacted by great pressure from overlying sediment. Later movements of the earth's crust pushed the limestone layers out of the sea, often leaving them well inland.

Seashells and limestone are primarily calcium carbonate. Calcium carbonate is fairly insoluble, as you would expect from its solubility product constant (3.8×10^{-9}). But it dissolves readily in acidic solution. Water that has filtered through decomposing vegetation contains carbonic acid, as well as other acids. When such an acidic solution comes into contact with limestone, it carves out caverns. The water is now a solution of calcium hydrogen carbonate.

$$CaCO_3(s) + H_2O(l) + CO_2(aq) \longrightarrow Ca^{2+}(aq) + 2HCO_3^-(aq)$$

This process is reversible, which you would predict from Le Châtelier's principle. As the calcium hydrogen carbonate solution

Figure 17.6 ▲

A limestone cave (Luray Caverns, Virginia) Such caves are formed by the action of acidic groundwater on limestone deposits. The icicle-like formations (stalagmites and stalactites) in the caves are caused by the reprecipitation of calcium carbonate as carbon dioxide in the solution equilibrates with the surrounding air.

percolates through holes in the limestone and drips into caverns, water evaporates and carbon dioxide gas escapes; calcium carbonate precipitates onto icicle-shaped *stalactites*.

$$Ca^{2+}(aq) + 2HCO_3^-(aq) \longrightarrow CaCO_3(s) + H_2O(l) + CO_2(g)$$

Excess solution drips onto the cavern floor, where calcium carbonate precipitates on upward-growing *stalagmites*. Iron compounds in the solution add yellow and brown colors to the columns of rock. Stalactites and stalagmites grow at the rate of about a centimeter every 300 years.

■ See Problems 17.107 and 17.108.

(continued)

Answer Check Make sure you compare the correct acid-ionization constants. For each salt, you look at the acid corresponding to the anion plus a hydrogen ion. For $CaCO_3$, you look at CO_3^{2-} plus H^+ (HCO_3^-, not H_2CO_3). For $CaSO_4$, you look at SO_4^{2-} plus H^+ (HSO_4^-, not H_2SO_4).

Exercise 17.8 Which salt would have its solubility more affected by changes in pH, silver chloride or silver cyanide?

■ See Problems 17.59 and 17.60.

CONCEPT CHECK 17.4

If you add a dilute acidic solution to a mixture containing magnesium oxalate and calcium oxalate, which of the two compounds is more likely to dissolve?

Separation of Metal Ions by Sulfide Precipitation

Many metal sulfides are insoluble in water but dissolve in acidic solution. The sulfide scheme discussed later, in Section 17.7, uses this change in solubility of the metal sulfides with pH, or hydronium-ion concentration, to separate a mixture of metal ions. Here is how the separation scheme works.

Consider a solution that is 0.10 M in zinc ion and in lead(II) ion. You add hydrogen sulfide, H_2S, to this solution. Hydrogen sulfide ionizes in water as a diprotic acid:

$$H_2S(aq) + H_2O(l) \rightleftharpoons H_3O^+(aq) + HS^-(aq)$$
$$HS^-(aq) + H_2O(l) \rightleftharpoons H_3O^+(aq) + S^{2-}(aq)$$

The ionization forms sulfide ion, S^{2-}, which can combine with the metal ions to precipitate the sulfides. Note that by increasing the hydronium-ion concentration, you can reverse both of the previous equilibria and in that way reduce the sulfide-ion concentration. This allows you to control the sulfide-ion concentration of a hydrogen sulfide solution by varying its pH; the lower the pH, the lower the sulfide-ion concentration.

Whether lead(II) sulfide and zinc sulfide precipitate from the hydrogen sulfide solution depends on the solubility product constants of the sulfides and on the sulfide-ion concentration, which in turn depends on the pH of the solution. Here are the solubility equilibria:

$$ZnS(s) \rightleftharpoons Zn^{2+}(aq) + S^{2-}(aq); \ K_{sp} = 1.1 \times 10^{-21}$$
$$PbS(s) \rightleftharpoons Pb^{2+}(aq) + S^{2-}(aq); \ K_{sp} = 2.5 \times 10^{-27}$$

By adjusting the pH of the solution, you adjust the sulfide-ion concentration in order to precipitate the least soluble metal sulfide while maintaining the other metal ion in solution. The solubility product constant of lead(II) sulfide is much smaller than that of zinc sulfide, so it is the least soluble sulfide. When a solution that is 0.10 M in each metal ion and 0.30 M in hydronium ion is saturated with hydrogen sulfide, lead(II) sulfide precipitates, but zinc ion remains in solution. You can now filter off the precipitate of lead(II) sulfide, leaving a solution containing the zinc ion.

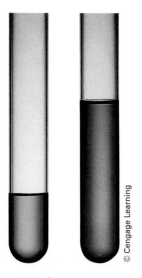

© Cengage Learning

Figure 17.7 ▲

Complex ions of nickel(II) The test tube on the left contains a solution of $Ni(H_2O)_6{}^{2+}$ ion (green). Adding ammonia to a similar test tube containing this ion, on the right, yields the ammonia complex ion $Ni(NH_3)_6{}^{2+}$ (purple).

The Lewis concept of acids and bases was discussed in Section 15.3.

Complex-Ion Equilibria

Many metal ions, especially those of the transition elements, form coordinate covalent bonds with molecules or anions having lone pairs of electrons. For example, the silver ion, Ag^+, can react with NH_3 to form the $Ag(NH_3)_2{}^+$ ion. The lone pair of electrons on the N atom of NH_3 forms a coordinate covalent bond to the silver ion to form what is called a complex ion. (See Figure 17.7.)

$$Ag^+ + 2(:NH_3) \longrightarrow (H_3N:Ag:NH_3)^+$$

Note that the metal ion acts as a Lewis acid, and the molecule or anion having the lone pair of electrons acts as a Lewis base. ◀

A **complex ion** is an *ion formed from a metal ion with a Lewis base attached to it by a coordinate covalent bond.* A *complex* is a compound containing complex ions. A **ligand** is *a Lewis base that bonds to a metal ion to form a complex ion.* For example, $Ag(NH_3)_2{}^+$ is a complex ion formed from the Ag^+ ion and two NH_3 molecules. The NH_3 molecules are the ligands.

17.5 Complex-Ion Formation

The aqueous silver ion forms a complex ion with ammonia by reacting with NH_3 in steps.

$$Ag^+(aq) + NH_3(aq) \rightleftharpoons Ag(NH_3)^+(aq)$$
$$Ag(NH_3)^+(aq) + NH_3(aq) \rightleftharpoons Ag(NH_3)_2{}^+(aq)$$

When you add these equations, you get the overall equation for the formation of the complex ion $Ag(NH_3)_2^+$.

$$Ag^+(aq) + 2NH_3(aq) \rightleftharpoons Ag(NH_3)_2^+(aq)$$

The **formation constant**, or **stability constant, K_f,** of a complex ion is *the equilibrium constant for the formation of the complex ion from the aqueous metal ion and the ligands.* The formation constant of $Ag(NH_3)_2^+$ is

$$K_f = \frac{[Ag(NH_3)_2^+]}{[Ag^+][NH_3]^2}$$

The value of K_f for $Ag(NH_3)_2^+$ is 1.7×10^7. Its large value means that the complex ion is quite stable. When a large amount of NH_3 is added to a solution of Ag^+, you expect most of the Ag^+ ion to react to form the complex ion. Table 17.2 to the right lists values for the formation constants of some complex ions.

The **dissociation constant (K_d)** for a complex ion is *the reciprocal, or inverse, value of K_f.* The equation for the dissociation of $Ag(NH_3)_2^+$ is

$$Ag(NH_3)_2^+(aq) \rightleftharpoons Ag^+(aq) + 2NH_3(aq)$$

and its equilibrium constant is

$$K_d = \frac{1}{K_f} = \frac{[Ag^+][NH_3]^2}{[Ag(NH_3)_2^+]}$$

Table 17.2	Formation Constants of Complex Ions at 25°C
Complex Ion	**K_f**
$Ag(CN)_2^-$	5.6×10^{18}
$Ag(NH_3)_2^+$	1.7×10^7
$Ag(S_2O_3)_2^{3-}$	2.9×10^{13}
$Cd(NH_3)_4^{2+}$	1.0×10^7
$Cu(CN)_2^-$	1.0×10^{16}
$Cu(NH_3)_4^{2+}$	4.8×10^{12}
$Fe(CN)_6^{4-}$	1.0×10^{35}
$Fe(CN)_6^{3-}$	9.1×10^{41}
$Ni(CN)_4^{2-}$	1.0×10^{31}
$Ni(NH_3)_6^{2+}$	5.6×10^8
$Zn(NH_3)_4^{2+}$	2.9×10^9
$Zn(OH)_4^{2-}$	2.8×10^{15}

Equilibrium Calculations with K_f

The next example shows how to calculate the concentration of an aqueous metal ion in equilibrium with a complex ion.

Example 17.9	Calculating the Concentration of a Metal Ion in Equilibrium with a Complex Ion

What is the concentration of $Ag^+(aq)$ ion in 0.010 M $AgNO_3$ that is also 1.00 M NH_3? K_f for $Ag(NH_3)_2^+$ ion is 1.7×10^7.

Problem Strategy The formation constant for $Ag(NH_3)_2^+$ is large, so silver exists primarily as this ion. This suggests that you do this problem in two parts. First you do the *stoichiometry calculation,* in which you assume that $Ag^+(aq)$ reacts completely to form $Ag(NH_3)_2^+(aq)$. Then you do the *equilibrium calculation,* in which this complex ion dissociates to give a small amount of $Ag^+(aq)$.

Solution STOICHIOMETRY CALCULATION In 1 L of solution, you initially have 0.010 mol $Ag^+(aq)$ from $AgNO_3$. This reacts to give 0.010 mol $Ag(NH_3)_2^+$, leaving $(1.00 - 2 \times 0.010)$ mol NH_3, which equals 0.98 mol NH_3 to two significant figures. You now look at the equilibrium for the dissociation of $Ag(NH_3)_2^+$.

EQUILIBRIUM CALCULATION

Step 1: One liter of the solution contains 0.010 mol $Ag(NH_3)_2^+$ and 0.98 mol NH_3. The complex ion dissociates slightly, so that 1 L of solution contains x mol Ag^+. These data are summarized in the following table:

Concentration (M)	$Ag(NH_3)_2^+(aq)$ ⇌	$Ag^+(aq)$ +	$2NH_3(aq)$
Starting	0.010	0	0.98
Change	$-x$	$+x$	$+2x$
Equilibrium	$0.010 - x$	x	$0.98 + 2x$

(continued)

(continued)

Formation of copper(II)–ammonia complex ion, Cu(NH₃)₄²⁺ (dark blue) Ammonia is being added to copper sulfate solution (light blue), giving a precipitate of copper(II) hydroxide, which dissolves as the copper(II) ion complexes with more ammonia.

Step 2: The formation constant is

$$K_f = \frac{[\text{Ag(NH}_3)_2{}^+]}{[\text{Ag}^+][\text{NH}_3]^2} = 1.7 \times 10^7$$

From this you can write the dissociation constant, which corresponds to the reaction as shown in the table.

$$\frac{[\text{Ag}^+][\text{NH}_3]^2}{[\text{Ag(NH}_3)_2{}^+]} = K_d = \frac{1}{K_f}$$

Substituting into this equation gives

$$\frac{x(0.98 + 2x)^2}{(0.010 - x)} = \frac{1}{1.7 \times 10^7}$$

Step 3: The right-hand side of the equation equals 5.9×10^{-8}. If you assume x to be small compared with 0.010,

$$\frac{x(0.98)^2}{0.010} \approx 5.9 \times 10^{-8}$$

and

$$x \approx 5.9 \times 10^{-8} \times 0.010/(0.98)^2 = 6.1 \times 10^{-10}$$

The silver-ion concentration is $\mathbf{6.1 \times 10^{-10}}$ **M** (over 10 million times smaller than its value in 0.010 M AgNO₃ that does not contain ammonia).

Answer Check Calculate the value of K_f from the equilibrium concentrations of species, using the value of x (6.1×10^{-10}) that you just obtained.

$$K_f = \frac{[\text{Ag(NH}_3)^{2+}]}{[\text{Ag}^+][\text{NH}_3]^2} = \frac{(0.010 - x)}{x(0.98 + 2x)^2} = \frac{0.010}{6.1 \times 10^{-10}(0.98)^2} = 1.7 \times 10^7$$

Exercise 17.9 What is the concentration of $\text{Cu}^{2+}(aq)$ in a solution that was originally 0.015 M Cu(NO₃)₂ and 0.100 M NH₃? The Cu^{2+} ion forms the complex ion $\text{Cu(NH}_3)_4{}^{2+}$. Its formation constant is given in Table 17.2.

■ See Problems 17.64 and 17.65.

Amphoteric Hydroxides

An **amphoteric hydroxide** is *a metal hydroxide that reacts with both bases and acids.* Zinc is an example of a metal that forms such a hydroxide. Zinc hydroxide, Zn(OH)₂, is an insoluble hydroxide, as are most of the metal hydroxides except those of Groups IA and IIA elements. Zinc hydroxide reacts with a strong acid, as you would expect of a base, and the metal hydroxide dissolves.

$$\text{Zn(OH)}_2(s) + 2\text{H}_3\text{O}^+(aq) \longrightarrow \text{Zn}^{2+}(aq) + 4\text{H}_2\text{O}(l)$$

With a base, however, Zn(OH)₂ reacts to form the complex ion $\text{Zn(OH)}_4{}^{2-}$.

$$\text{Zn(OH)}_2(s) + 2\text{OH}^-(aq) \longrightarrow \text{Zn(OH)}_4{}^{2-}(aq)$$

In this way the hydroxide also dissolves in a strong base.

When a strong base is slowly added to a solution of ZnCl₂, a white precipitate of Zn(OH)₂ first forms.

$$\text{Zn}^{2+}(aq) + 2\text{OH}^-(aq) \longrightarrow \text{Zn(OH)}_2(s)$$

But as more base is added, the white precipitate dissolves, forming the complex ion $\text{Zn(OH)}_4{}^{2-}$. (See Figure 17.8.)

| The beaker on the left contains ZnCl₂; the one on the right contains NaOH. | As some of the solution of NaOH is added to the solution of ZnCl₂, a white precipitate of Zn(OH)₂ forms. | After more NaOH is added, the precipitate dissolves by forming the hydroxo complex Zn(OH)₄²⁻. |

© Cengage Learning

Other common amphoteric hydroxides are those of aluminum, chromium(III), lead(II), tin(II), and tin(IV). The amphoterism of $Al(OH)_3$ is commercially used to separate aluminum oxide from the aluminum ore bauxite. Bauxite contains hydrated Al_2O_3 plus impurities, such as silica sand (SiO_2) and iron oxide (Fe_2O_3). The aluminum oxide dissolves in NaOH solution as the $Al(OH)_4^-$ ion. After the insoluble impurities are filtered off, the solution of $Al(OH)_4^-$ can be slightly acidified to precipitate pure $Al(OH)_3$.

$$Al(OH)_4^-(aq) + H_3O^+(aq) \longrightarrow Al(OH)_3(s) + 2H_2O(l)$$

Figure 17.8 ▲

Demonstration of the amphoteric behavior of zinc hydroxide

17.6 Complex Ions and Solubility

Example 17.9 demonstrates that the formation of the complex ion $Ag(NH_3)_2^+$ reduces the concentration of silver ion, $Ag^+(aq)$, in solution. The solution was initially 0.010 M $AgNO_3$, or 0.010 M $Ag^+(aq)$. When this solution is made 1.00 M in NH_3, the $Ag^+(aq)$ ion concentration decreases to 6.1×10^{-10} M. You can see that the ion product for a slightly soluble silver salt could be decreased to below K_{sp}. Then the slightly soluble salt might not precipitate in a solution containing ammonia, whereas it otherwise would.

Example | **17.10** | **Predicting Whether a Precipitate Will Form in the Presence of the Complex Ion**

Gaining Mastery Toolbox

Critical Concept 17.10
The formation of a complex ion can reduce the concentration of a metal ion to the extent that it affects whether a slightly soluble salt of this metal ion precipitates.

Solution Essentials:
• Solubility product constant, K_{sp}
• Molar solubility
• Complex ion
• Formation constant, K_f, of a complex ion
• Reaction quotient
• Equilibrium constant, K_c
• Molar concentration

a Will silver chloride precipitate from a solution that is 0.010 M $AgNO_3$ and 0.010 M NaCl? **b** Will silver chloride precipitate from this solution if it is also 1.00 M NH_3?

Problem Strategy Part a is a simple precipitation problem in which you compare Q_c with K_{sp}. See Example 17.6. Part b differs only in that you need to know the Ag^+ concentration in equilibrium with a complex ion.

Solution **a** To determine whether a precipitate should form, you calculate the ion product and compare it with K_{sp} for AgCl (1.8×10^{-10}).

$$\text{Ion product} = [Ag^+]_i[Cl^-]_i = (0.010)(0.010) = 1.0 \times 10^{-4}$$

This is greater than $K_{sp} = 1.8 \times 10^{-10}$, so **a precipitate should form.**

b You first need to calculate the concentration of $Ag^+(aq)$ in a solution containing 1.00 M NH_3. We did this in Example 17.9 and found that $[Ag^+]$ equals 6.1×10^{-10} M. Hence,

$$\text{Ion product} = [Ag^+]_i[Cl^-]_i = (6.1 \times 10^{-10})(0.010) = 6.1 \times 10^{-12}$$

Because the ion product is smaller than $K_{sp} = 1.8 \times 10^{-10}$, **no precipitate should form.**

(continued)

(continued)

Answer Check Check your arithmetic in calculating the ion product. Then note that when a precipitate forms, the ion product decreases as it approaches the solubility product. Therefore, the ion product, Q_c, should be larger than K_{sp} if a precipitate is to form.

Exercise 17.10 Will silver iodide precipitate from a solution that is 0.0045 M AgNO$_3$, 0.15 M NaI, and 0.20 M KCN?

■ See Problems 17.65 and 17.66.

In Example 17.10, we determined whether a precipitate of AgCl is expected to form from a solution made up with given concentrations of Ag$^+$, Cl$^-$, and NH$_3$. The problem involved the Ag(NH$_3$)$_2$$^+$ complex in equilibrium. From it, we determined the concentration of free Ag$^+$(aq) and then calculated the ion product of AgCl.

Suppose you want to find the solubility of AgCl in a solution of aqueous ammonia of given concentration. This problem involves the solubility equilibrium for AgCl, in addition to the complex-ion equilibrium. As silver chloride dissolves to give ions, the Ag$^+$ ion reacts with NH$_3$ to give the complex ion Ag(NH$_3$)$_2$$^+$. The equilibria are

$$AgCl(s) \rightleftharpoons Ag^+(aq) + Cl^-(aq)$$
$$Ag^+(aq) + 2NH_3(aq) \rightleftharpoons Ag(NH_3)_2^+(aq)$$

When Ag$^+$(aq) reacts to give the complex ion, more AgCl dissolves to partially replenish the Ag$^+$(aq) ion, according to Le Châtelier's principle. Therefore, silver chloride is more soluble in aqueous ammonia than in pure water. The next example shows how you can calculate this solubility.

Example 17.11

Calculating the Solubility of a Slightly Soluble Ionic Compound in a Solution of the Complex Ion

Gaining Mastery Toolbox

Critical Concept 17.11
The formation of a complex ion reduces the concentration of a metal ion in solution and consequently increases the solubility of a slightly soluble salt of the metal ion.

Solution Essentials:
- Solubility product constant, K_{sp}
- Molar solubility
- Complex ion
- Formation constant, K_f, of a complex ion
- Reaction quotient
- Equilibrium constant, K_c
- Molar concentration

Calculate the molar solubility of AgCl in 1.0 M NH$_3$.

Problem Strategy If you add the solubility equilibrium and the complex-ion equilibrium (the two equations just before this example), you obtain the overall equilibrium for the dissolving of AgCl in NH$_3$. The equilibrium constant for this reaction equals the product of the equilibrium constants for solubility and complex-ion formation. (We will show that in the solution to this example; however, we discussed the general principle at the end of Section 14.2.) Once you have the equilibrium constant, you can solve the equilibrium problem.

Solution You obtain the overall equation for the process by adding the solubility and complex-ion equilibria.

$$AgCl(s) \rightleftharpoons \cancel{Ag^+(aq)} + Cl^-(aq)$$
$$\underline{\cancel{Ag^+(aq)} + 2NH_3(aq) \rightleftharpoons Ag(NH_3)_2^+(aq)}$$
$$AgCl(s) + 2NH_3(aq) \rightleftharpoons Ag(NH_3)_2^+(aq) + Cl^-(aq)$$

The equilibrium constant for the overall equation is

$$K_c = \frac{[Ag(NH_3)_2^+][Cl^-]}{[NH_3]^2}$$

(continued)

(*continued*)

Because the overall equation is the sum of the solubility and complex-ion equilibria, the equilibrium constant K_c equals the product of the equilibrium constants for the solubility and complex-ion equilibria. You can easily verify this.

$$K_c = K_{sp}K_f = [Ag^+][Cl^-]\frac{[Ag(NH_3)_2{}^+]}{[Ag^+][NH_3]^2}$$

The value of $K_c = K_{sp}K_f$ can be obtained from K_{sp} for AgCl in Table 17.1 and K_f for $Ag(NH_3)_2{}^+$ in Table 17.2.

$$K_{sp}K_f = (1.8 \times 10^{-10})(1.7 \times 10^7) = 3.1 \times 10^{-3}$$

You now solve the equilibrium problem. The concentration table follows:

Concentration (M)	AgCl(s) +	2NH₃(aq)	⇌	Ag(NH₃)₂⁺(aq) +	Cl⁻(aq)
Starting		1.0		0	0
Change		−2x		+x	+x
Equilibrium		1.0 − 2x		x	x

Substituting into the equilibrium-constant equation

$$\frac{[Ag(NH_3)_2{}^+][Cl^-]}{[NH_3]^2} = K_{sp}K_f$$

gives

$$\frac{x^2}{(1.0 - 2x)^2} = 3.1 \times 10^{-3}$$

You solve this equation by taking the square root of both sides.

$$\frac{x}{1.0 - 2x} = 0.056$$

Rearranging yields

$$x = 0.056(1.0 - 2x) = 0.056 - 0.11x$$

Hence,

$$x = \frac{0.056}{1.11} = 0.050$$

Note that the molar solubility of AgCl equals the molar concentration of silver in the solution. Because of the stability of $Ag(NH_3)_2{}^+$, most of the silver in solution will be in the form of this complex ion. Therefore, because x equals the concentration of $Ag(NH_3)_2{}^+$, the molar solubility of AgCl equals **0.050 M**.

Answer Check Check that you have the correct overall equilibrium for the problem and that you have then done the arithmetic correctly for the equilibrium constant. Once you have the molar solubility of the compound, x, you can put x back into the equilibrium-constant expression and calculate the value of K_c as a final check.

Exercise 17.11 What is the molar solubility of AgBr in 1.0 M $Na_2S_2O_3$ (sodium thiosulfate)? Silver ion forms the complex ion $Ag(S_2O_3)_2{}^{3-}$. See Tables 17.1 and 17.2 for data.

■ See Problems 17.67 and 17.68.

An Application of Solubility Equilibria

Qualitative analysis involves *the determination of the identity of the substances present in a mixture.* One scheme for the qualitative analysis of a solution of metal ions is based on the relative solubilities of the metal sulfides. This scheme is no longer widely used in practical analysis, having been largely replaced by modern instrumental methods. But it is still taught in general chemistry laboratory as a way of imparting some knowledge of inorganic chemistry while illustrating the concepts of solubility, buffers, and so forth. We will look at the main outline of the sulfide scheme.

17.7 Qualitative Analysis of Metal Ions

In the qualitative analysis scheme for metal ions, a cation is usually detected by the presence of a characteristic precipitate. For example, silver ion gives a white precipitate with chloride ion. But other ions also give a white precipitate with chloride ion. Therefore, you must first subject a mixture to a procedure that separates individual ions before applying a precipitation test to any particular ion.

Figure 17.9 shows a flowchart illustrating how the metal ions in an aqueous solution are first separated into five analytical groups. The Analytical Group I ions, Ag^+, Hg_2^{2+}, and Pb^{2+}, are separated from a solution of other ions by adding dilute hydrochloric acid. The ions are precipitated as the chlorides AgCl, Hg_2Cl_2, and $PbCl_2$, which are removed by filtering.

The other metal ions remain in the filtrate, the solution that passes through the filter. Many of these ions can be separated by precipitating them as metal

Figure 17.9 ▶

Flowchart of the qualitative analysis scheme for the separation of metal ions The first group of ions consists of those precipitated by HCl. The separation of the next two groups is based on the differences in solubilities of the metal sulfides. The fourth group consists of ions of the alkaline earth elements, which are precipitated as carbonates or phosphates. The final group of ions is in the filtrate after these precipitations.

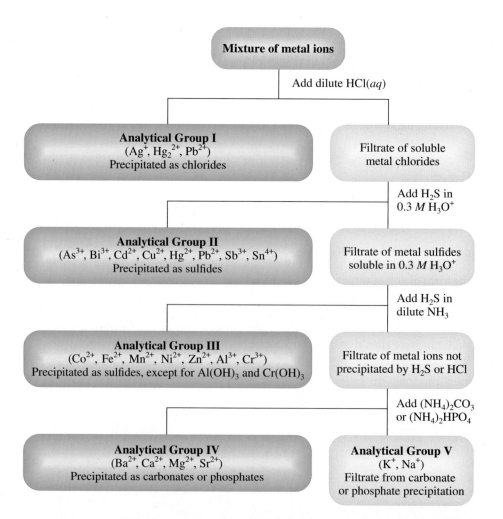

Mixture of metal ions

Add dilute HCl(*aq*)

Analytical Group I
(Ag^+, Hg_2^{2+}, Pb^{2+})
Precipitated as chlorides

Filtrate of soluble metal chlorides

Add H_2S in 0.3 *M* H_3O^+

Analytical Group II
(As^{3+}, Bi^{3+}, Cd^{2+}, Cu^{2+}, Hg^{2+}, Pb^{2+}, Sb^{3+}, Sn^{4+})
Precipitated as sulfides

Filtrate of metal sulfides soluble in 0.3 *M* H_3O^+

Add H_2S in dilute NH_3

Analytical Group III
(Co^{2+}, Fe^{2+}, Mn^{2+}, Ni^{2+}, Zn^{2+}, Al^{3+}, Cr^{3+})
Precipitated as sulfides, except for $Al(OH)_3$ and $Cr(OH)_3$

Filtrate of metal ions not precipitated by H_2S or HCl

Add $(NH_4)_2CO_3$ or $(NH_4)_2HPO_4$

Analytical Group IV
(Ba^{2+}, Ca^{2+}, Mg^{2+}, Sr^{2+})
Precipitated as carbonates or phosphates

Analytical Group V
(K^+, Na^+)
Filtrate from carbonate or phosphate precipitation

sulfides with H_2S. This separation takes advantage of the fact that only the least soluble metal sulfides precipitate in acidic solution. After these are removed and the solution is made basic, other metal sulfides precipitate. ▶

Analytical Group II consists of metal ions precipitated as metal sulfides from a solution that is 0.3 M H_3O^+ and is saturated with H_2S. ▶ The ions are As^{3+}, Bi^{3+}, Cd^{2+}, Cu^{2+}, Hg^{2+}, Pb^{2+}, Sb^{3+}, and Sn^{4+}. Lead(II) ion can appear in Analytical Group II as well as in Group I, because $PbCl_2$ is somewhat soluble and therefore Pb^{2+} may not be completely precipitated by HCl.

Analytical Group III consists of the metal ions in the filtrate from Group II that are precipitated in weakly basic solution with H_2S. The ions Co^{2+}, Fe^{2+}, Mn^{2+}, Ni^{2+}, and Zn^{2+} are precipitated as the sulfides. The ions Al^{3+} and Cr^{3+} are precipitated as hydroxides.

The filtrate obtained after the Group III metal ions have precipitated contains the alkali metal ions and the alkaline earth ions. The Analytical Group IV ions, Ba^{2+}, Ca^{2+}, Mg^{2+}, and Sr^{2+}, are precipitated as carbonates or phosphates by adding $(NH_4)_2CO_3$ or $(NH_4)_2HPO_4$. The filtrate from this separation contains the Analytical Group V ions, K^+ and Na^+.

To illustrate the separation scheme, suppose you have a solution containing the metal cations Ag^+, Cu^{2+}, Zn^{2+}, Ca^{2+}, and Na^+. When you add dilute HCl(aq) to this solution, Ag^+ precipitates as AgCl.

$$Ag^+(aq) + Cl^-(aq) \longrightarrow \underset{\text{white}}{AgCl(s)}$$

You filter off the precipitate of AgCl. The filtrate contains the remaining ions.

When you make the filtrate 0.3 M H_3O^+ and then saturate it with H_2S, the Cu^{2+} ion precipitates as CuS.

$$Cu^{2+}(aq) + S^{2-}(aq) \longrightarrow \underset{\text{black}}{CuS(s)}$$

Zinc(II) ion also forms a sulfide, but it is soluble in acidic solution. After removal of the precipitate of CuS, the solution is made weakly basic with $NH_3(aq)$. This increases the S^{2-} concentration sufficiently to precipitate ZnS.

$$Zn^{2+}(aq) + S^{2-}(aq) \longrightarrow \underset{\text{white}}{ZnS(s)}$$

The filtrate from this separation contains Ca^{2+} and Na^+. Calcium ion can be precipitated as the phosphate or the carbonate. For example, by adding $(NH_4)_2CO_3$, you get the reaction

$$Ca^{2+}(aq) + CO_3^{2-}(aq) \longrightarrow \underset{\text{white}}{CaCO_3(s)}$$

The filtrate from this solution contains Na^+ ion. Now the five ions have been separated.

In the complete analysis, once the ions are separated into analytical groups, these are further separated into the individual ions. We can illustrate this with Analytical Group I, which is precipitated as a mixture of the chlorides AgCl, Hg_2Cl_2, and $PbCl_2$. Of these, only $PbCl_2$ is significantly soluble in hot water. Thus, if the Analytical Group I precipitate is mixed with hot water, lead chloride dissolves and the silver and mercury(I) chlorides can be filtered off. Any lead(II) ion is in the filtrate and can be revealed by adding potassium chromate. Lead(II) ion gives a bright yellow precipitate of lead(II) chromate, $PbCrO_4$.

$$Pb^{2+}(aq) + CrO_4^{2-}(aq) \longrightarrow \underset{\text{yellow}}{PbCrO_4(s)}$$

If there is precipitate remaining after the extraction with hot water, it consists of AgCl, Hg_2Cl_2, or both. However, only silver ion forms a stable ammonia complex ion. Therefore, silver chloride dissolves in ammonia.

$$AgCl(s) + 2NH_3(aq) \longrightarrow Ag(NH_3)_2^+(aq) + Cl^-(aq)$$

The solubility of the metal sulfides at different pH was discussed in Section 17.4.

Hydrogen sulfide is often prepared in the solution by adding thioacetamide, CH_3CSNH_2, and warming.

$$CH_3CSNH_2 + H_2O \longrightarrow$$
$$CH_3CONH_2 + H_2S$$

Mercury(I) chloride is simultaneously oxidized and reduced in ammonia solution, giving a precipitate of mercury(II) amido chloride, $HgNH_2Cl$, and mercury metal, which appears black because of its finely divided state.

$$Hg_2Cl_2(s) + 2NH_3(aq) \longrightarrow \underbrace{HgNH_2Cl(s) + Hg(l)}_{\text{black or gray}} + NH_4^+(aq) + Cl^-(aq)$$

The presence of silver ion in the filtrate is revealed by adding hydrochloric acid, which combines with the NH_3, releasing $Ag^+(aq)$ ion. The $Ag^+(aq)$ ion reacts with $Cl^-(aq)$ to give a white precipitate of AgCl.

$$Ag(NH_3)_2^+(aq) + 2H_3O^+(aq) + Cl^-(aq) \longrightarrow AgCl(s) + 2NH_4^+(aq) + 2H_2O(l)$$

The other analytical groups are separated in similar fashion. Once a given ion has been separated, its presence is usually confirmed by a particular reactant—perhaps one that gives a distinctive precipitate.

A Checklist for Review

Summary of Facts and Concepts

The equilibrium constant for the equilibrium between a slightly soluble ionic solid and its ions in solution is called the *solubility product constant, K_{sp}*. Its value can be determined from the solubility of the solid. Conversely, when the solubility product constant is known, the solubility of the solid can be calculated. The solubility is decreased by the addition of a soluble salt that supplies a common ion. Qualitatively, this can be seen to follow from Le Châtelier's principle. Quantitatively, the *common-ion effect* on solubility can be evaluated from the solubility product constant.

Rather than look at the solubility process as the dissolving of a solid in a solution, you can look at it as the *precipitation* of the solid from the solution. You can decide whether precipitation will occur by computing the *ion product*. Precipitation occurs when the ion product is greater than K_{sp}.

Solubility is affected by the pH if the compound supplies an anion conjugate to a weak acid. As the pH decreases (H_3O^+ ion concentration increases), the anion concentration decreases because the anion forms the weak acid. Therefore, the ion product decreases and the solubility increases.

The concentration of a metal ion in solution is decreased by *complex-ion formation*. The equilibrium constant for the formation of the complex ion from the aqueous metal ion and the ligands is called the *formation constant* (or *stability constant*), K_f. Because complex-ion formation reduces the concentration of aqueous metal ion, an ionic compound of the metal is more soluble in a solution of the ligand.

The sulfide scheme of *qualitative analysis* separates the metal ions in a mixture by using precipitation reactions. Variation of solubility with pH and with complex-ion formation is used to aid in the separation.

Learning Objectives

17.1 The Solubility Product Constant

- Define the *solubility product constant, K_{sp}*.
- Write solubility product expressions. Example 17.1
- Define *molar solubility*.
- Calculate K_{sp} from the solubility (simple example). Example 17.2
- Calculate K_{sp} from the solubility (more complicated example). Example 17.3
- Calculate the solubility from K_{sp}. Example 17.4

17.2 Solubility and the Common-Ion Effect

- Explain how the solubility of a salt is affected by another salt that has the same cation or anion (*common ion*).
- Calculate the solubility of a slightly soluble salt in a solution of a common ion. Example 17.5

Important Terms

solubility product constant (K_{sp})

17.3 Precipitation Calculations

- Define *ion product.*
- State the criterion for precipitation.
- Predict whether precipitation will occur, given ion concentrations. Example 17.6
- Predict whether precipitation will occur, given solution volumes and concentrations. Example 17.7
- Define *fractional precipitation.*
- Explain how two ions can be separated using fractional precipitation.

ion product
fractional precipitation

17.4 Effect of pH on Solubility

- Explain the qualitative effect of pH on solubility of a slightly soluble salt.
- Determine the qualitative effect of pH on solubility. Example 17.8
- Explain the basis for the sulfide scheme to separate a mixture of metal ions.

17.5 Complex-Ion Formation

- Define *complex ion* and *ligand.*
- Define *formation constant* or *stability constant, K_f,* and *dissociation constant, K_d.*
- Calculate the concentration of a metal ion in equilibrium with a complex ion. Example 17.9
- Define *amphoteric hydroxide.*

complex ion
ligand
formation (stability) constant (K_f)
dissociation constant (of a complex ion) (K_d)
amphoteric hydroxide

17.6 Complex Ions and Solubility

- Predict whether a precipitate will form in the presence of the complex ion. Example 17.10
- Calculate the solubility of a slightly soluble ionic compound in a solution of the complex ion. Example 17.11

17.7 Qualitative Analysis of Metal Ions

- Define *qualitative analysis.*
- Describe the main outline of the sulfide scheme for qualitative analysis.

qualitative analysis

Questions and Problems

⚈WL Interactive versions of these problems may be assigned in OWL.

Self-Assessment and Review Questions

Key: These questions test your understanding of the ideas you worked with in the chapter. These problems vary in difficulty and often can be used for the basis of discussion.

17.1 Suppose the molar solubility of nickel hydroxide, $Ni(OH)_2$, is x M. Show that K_{sp} for nickel hydroxide equals $4x^3$.
17.2 Explain why calcium sulfate is less soluble in sodium sulfate solution than in pure water.
17.3 What must be the concentration of silver ion in a solution that is in equilibrium with solid silver chloride and that is 0.10 M in Cl^-?
17.4 Discuss briefly how you could predict whether a precipitate will form when solutions of lead nitrate and potassium iodide are mixed. What information do you need to have?

17.5 Explain how metal ions such as Pb^{2+} and Zn^{2+} are separated by precipitation with hydrogen sulfide.
17.6 Explain why barium fluoride dissolves in dilute hydrochloric acid but is insoluble in water.
17.7 A precipitate forms when a small amount of sodium hydroxide is added to a solution of aluminum sulfate. This precipitate dissolves when more sodium hydroxide is added. Explain what is happening.
17.8 Lead chloride at first precipitates when sodium chloride is added to a solution of lead nitrate. Later, when the solution is made more concentrated in chloride ion, the precipitate dissolves. Explain what is happening. What equilibria are involved? Note that lead ion forms the complex ion $PbCl_4^{2-}$.

17.9 A solution containing calcium ion and magnesium ion is buffered with ammonia–ammonium chloride. When carbonate ion is added to the solution, calcium carbonate precipitates but magnesium carbonate does not. Explain.

17.10 Describe how you would separate the metal ions in a solution containing silver ion, copper(II) ion, and nickel(II) ion, using the sulfide scheme of qualitative analysis.

17.11 The solubility of the hypothetical salt A_3B_2 is 6.1×10^{-9} mol/L at a certain temperature. What is K_{sp} for the salt? A_3B_2 dissolves according to the equation

$$A_3B_2(s) \rightleftharpoons 3A^{2+}(aq) + 2B^{3-}(aq)$$

a. 3.7×10^{-17}
b. 9.1×10^{-25}
c. 9.1×10^{-40}
d. 3.7×10^{-32}
e. 1.4×10^{-33}

17.12 Write the solubility product expression for the salt Ag_3PO_4.

a. $K_{sp} = [Ag^+]^3[P^{3-}][O^{2-}]^4$
b. $K_{sp} = [Ag^{3+}][PO_4^-]^3$
c. $K_{sp} = 3[Ag^+] \times [PO_4^{3-}]$
d. $K_{sp} = [Ag^+]^3[PO_4^{3-}]$
e. $K_{sp} = [Ag^+][PO_4^{3-}]^3$

17.13 What is the molar solubility of calcium oxalate, CaC_2O_4, in a 0.203 M calcium chloride solution? The solubility product constant for calcium oxalate is 2.3×10^{-9}.

a. $4.8 \times 10^{-5}\ M$
b. $6.6 \times 10^{-6}\ M$
c. $8.0 \times 10^{-6}\ M$
d. $1.1 \times 10^{-8}\ M$
e. $1.2 \times 10^{-5}\ M$

17.14 Suppose the K_a values for the hypothetical acids HX, HY, and HZ are 9.5×10^{-5}, 7.6×10^{-4}, and 1.2×10^{-2}, respectively. Assuming that "A" is a cation that produces a slightly soluble salt with all three acids, which of the following salts would have its solubility most affected (compared with its solubility in pure water) when it is dissolved in a 0.10 M solution of HCl?

a. AX
b. AY
c. AZ
d. The solubility of all the salts would be affected by the same amount.
e. There is not enough information given in the problem to determine the answer.

Concept Explorations

Key: Concept explorations are comprehensive problems that provide a framework that will enable you to explore and learn many of the critical concepts and ideas in each chapter. If you master the concepts associated with these explorations, you will have a better understanding of many important chemistry ideas and will be more successful in solving all types of chemistry problems. These problems are well suited for group work and for use as in-class activities.

17.15 Solubility and Solubility Product

You put 0.10-mol samples of KNO_3, $(NH_4)_2S$, K_2S, MnS, AgCl, and $BaSO_4$ into separate flasks and add 1.0 L of water to each one. Then you stir the solutions for 5 minutes at room temperature. Assume that you have 1.0 L of solution in each case.

a. Are there any beakers where you would observe solid still present? How do you know?

b. Can you calculate the potassium ion concentration, K^+, for the solutions of KNO_3 and K_2S? If so, do the calculations, and then compare these K^+ concentrations.

c. For the solutions of $(NH_4)_2S$, K_2S, and MnS, how do the concentrations of sulfide ion, S^{2-}, compare? (You don't need to calculate an answer at this point; just provide a rough comparison.) Be sure to justify your answer.

d. Are there any cases where you need more information to calculate the sulfide-ion concentration for the solutions of $(NH_4)_2S$, K_2S, and MnS from part c? If so, what additional information do you need?

e. Consider all of the solutions listed at the beginning of this problem. For which ones do you need more information than is given in the question to determine the concentrations of the ions present? Where can you find this information?

f. How is the solubility of an ionic compound related to the concentrations of the ions of the dissolved compound in solution?

17.16 Solubility Equilibria

Consider three hypothetical ionic solids: AX, AX_2, and AX_3 (each X forms X^-). Each of these solids has the same K_{sp} value, 5.5×10^{-7}. You place 0.25 mol of each compound in a separate container and add enough water to bring the volume to 1.0 L in each case.

a. Write the chemical equation for each of the solids dissolving in water.

b. Would you expect the concentration of each solution to be 0.25 M in the compound? Explain, in some detail, why or why not.

c. Would you expect the concentrations of the A cations (A^+, A^{2+}, and A^{3+}) in the three solutions to be the same? Does just knowing the stoichiometry of each reaction help you determine the answer, or do you need something else? Explain your answer in detail, but without doing any arithmetic calculations.

d. Of the three solids, which one would you expect to have the greatest molar solubility? Explain in detail, but without doing any arithmetic calculations.

e. Calculate the molar solubility of each compound.

Conceptual Problems

Key: These problems are designed to check your understanding of the concepts associated with some of the main topics presented in each chapter. A strong conceptual understanding of chemistry is the foundation for both applying chemical knowledge and solving chemical problems. These problems vary in level of difficulty and often can be used as a basis for group discussion.

17.17 Which compound in each of the following pairs of compounds is the more soluble one?

- a magnesium hydroxide or copper(II) hydroxide
- b silver chloride or silver iodide

17.18 You are given two mineral samples: halite, which is NaCl, and fluorite, which is CaF_2. Describe a simple test you could use to discover which mineral is fluorite.

17.19 You are given a saturated solution of lead(II) chloride. Which one of the following solutions would be most effective in yielding a precipitate when added to the lead(II) chloride solution?

- a saturated PbS(*aq*)
- b 0.1 *M* Na_2SO_4(*aq*)
- c 0.1 *M* NaCl(*aq*)

17.20 Which of the following pictures best represents a solution made by adding 10 g of silver chloride, AgCl, to a liter of water? In these pictures, the gray spheres represent Ag^+ ions and the green spheres represent chloride ions. For clarity, water molecules are not shown.

17.21 Which of the following pictures best represents an unsaturated solution of sodium chloride, NaCl? In these pictures, the dark gray spheres represent Na^+ ions and the green spheres represent chloride ions. For clarity, water molecules are not shown.

17.22 When ammonia is first added to a solution of copper(II) nitrate, a pale blue precipitate of copper(II) hydroxide forms. As more ammonia is added, however, this precipitate dissolves. Describe what is happening.

17.23 You add dilute hydrochloric acid to a solution containing a metal ion. No precipitate forms. After the acidity is adjusted to 0.3 *M* hydronium ion, you bubble hydrogen sulfide into the solution. Again no precipitate forms. Is it possible that the original solution contained silver ion? Could it have contained copper(II) ion?

17.24 You are given a solution of the ions Mg^{2+}, Ca^{2+}, and Ba^{2+}. Devise a scheme to separate these ions using sodium sulfate. Note that magnesium sulfate is soluble.

Practice Problems

Key: These problems are for practice in applying problem-solving skills. They are divided by topic, and some are keyed to exercises (see the ends of the exercises). The problems are arranged in matching pairs; the odd-numbered problem of each pair is listed first, and its answer is given in the back of the book.

Note: For values of K_{sp} and K_f that are not given in the following problems, see Tables 17.1 and 17.2.

Solubility and K_{sp}

17.25 Use the solubility rules (Table 4.1) to decide which of the following compounds are expected to be soluble and which insoluble.

- a NaBr
- b PbI_2
- c $BaCO_3$
- d $(NH_4)_2SO_4$

17.26 Use the solubility rules (Table 4.1) to decide which of the following compounds are expected to be soluble and which insoluble.

- a $Ca(NO_3)_2$
- b MgI_2
- c $PbSO_4$
- d AgBr

17.27 Write solubility product expressions for the following compounds.

- a $Ca_3(AsO_4)_2$
- b $Fe(OH)_3$
- c $Mg(OH)_2$
- d $SrCO_3$

17.28 Write solubility product expressions for the following compounds.

- a $Ba_3(PO_4)_2$
- b $FePO_4$
- c Ag_2S
- d PbI_2

17.29 The solubility of magnesium oxalate, MgC_2O_4, in water is 0.0093 mol/L. Calculate K_{sp}.

17.30 The solubility of silver bromate, $AgBrO_3$, in water is 0.0072 g/L. Calculate K_{sp}.

17.31 The solubility of copper(II) iodate in water is 0.13 g/100 mL. Calculate the solubility product constant for copper(II) iodate, $Cu(IO_3)_2$.

17.32 The solubility of nickel(II) carbonate, $NiCO_3$, in water is 0.0448 g/L. Calculate K_{sp}.

17.33 The pH of a saturated solution of magnesium hydroxide (milk of magnesia) was found to be 10.52. From this, find K_{sp} for magnesium hydroxide.

17.34 A solution saturated in calcium hydroxide (limewater) has a pH of 12.35. What is K_{sp} for calcium hydroxide?

17.35 Magnesite (magnesium carbonate, $MgCO_3$) is a common magnesium mineral. From the solubility product constant, find the solubility of magnesium carbonate in grams per liter of water.

17.36 Strontianite (strontium carbonate) is an important mineral of strontium. Calculate the solubility of strontium carbonate, $SrCO_3$, from the solubility product constant.

17.37 What is the solubility of strontium iodate, $Sr(IO_3)_2$, in water? The K_{sp} for $Sr(IO_3)_2$ is 1.14×10^{-7}.

17.38 What is the solubility of PbF_2 in water? The K_{sp} for PbF_2 is 2.7×10^{-8}.

Common-Ion Effect

17.39 What is the solubility (in grams per liter) of strontium sulfate, $SrSO_4$, in 0.23 M sodium sulfate, Na_2SO_4?

17.40 What is the solubility (in grams per liter) of lead(II) chromate, $PbCrO_4$, in 0.13 M potassium chromate, K_2CrO_4?

17.41 The solubility of magnesium fluoride, MgF_2, in water is 0.016 g/L. What is the solubility (in grams per liter) of magnesium fluoride in 0.020 M sodium fluoride, NaF?

17.42 The solubility of silver sulfate, Ag_2SO_4, in water has been determined to be 8.0 g/L. What is the solubility in 0.45 M sodium sulfate, Na_2SO_4?

17.43 What is the solubility (in grams per liter) of magnesium oxalate, MgC_2O_4, in 0.020 M sodium oxalate, $Na_2C_2O_4$? Solve the equation exactly.

17.44 Calculate the molar solubility of strontium sulfate, $SrSO_4$, in 0.0012 M sodium sulfate, Na_2SO_4. Solve the equation exactly.

Precipitation

17.45 From each of the following ion concentrations in a solution, predict whether a precipitate will form in the solution.

a $[Pb^{2+}] = 0.035\ M$, $[Cl^-] = 0.15\ M$
b $[Ba^{2+}] = 0.020\ M$, $[F^-] = 0.015\ M$

17.46 From each of the following ion concentrations in a solution, predict whether a precipitate will form in the solution.

a $[Pb^{2+}] = 0.0048\ M$, $[Cl^-] = 0.082\ M$
b $[Sr^{2+}] = 0.012\ M$, $[CO_3^{2-}] = 0.0015\ M$

17.47 Lead(II) chromate, $PbCrO_4$, was used as a yellow paint pigment ("chrome yellow"). When a solution is prepared that is $5.0 \times 10^{-4}\ M$ in lead ion, Pb^{2+}, and $5.0 \times 10^{-5}\ M$ in chromate ion, CrO_4^{2-}, would you expect some of the lead(II) chromate to precipitate?

17.48 Lead sulfate, $PbSO_4$, was used as a white paint pigment. When a solution is prepared that is $5.0 \times 10^{-3}\ M$ in lead ion, Pb^{2+}, and $1.0 \times 10^{-4}\ M$ in sulfate ion, SO_4^{2-}, would you expect some of the lead sulfate to precipitate?

17.49 The following solutions are mixed: 1.0 L of 0.00010 M NaOH and 1.0 L of 0.0020 M $MgSO_4$. Is a precipitate expected? Explain.

17.50 A 45-mL sample of 0.015 M calcium chloride, $CaCl_2$, is added to 55 mL of 0.010 M sodium sulfate, Na_2SO_4. Is a precipitate expected? Explain.

17.51 A 65.0-mL sample of 0.010 M $Pb(NO_3)_2$ was added to a beaker containing 40.0 mL of 0.035 M KCl. Will a precipitate form?

17.52 A 45.0-mL sample of 0.0015 M $BaCl_2$ was added to a beaker containing 75.0 mL of 0.0025 M KF. Will a precipitate form?

17.53 How many moles of calcium chloride, $CaCl_2$, can be added to 1.5 L of 0.020 M potassium sulfate, K_2SO_4, before a precipitate is expected? Assume that the volume of the solution is not changed significantly by the addition of calcium chloride.

17.54 Magnesium sulfate, $MgSO_4$, is added to 456 mL of 0.040 M sodium hydroxide, NaOH, until a precipitate just forms. How many grams of magnesium sulfate were added? Assume that the volume of the solution is not changed significantly by the addition of magnesium sulfate.

17.55 What is the I^- concentration just as AgCl begins to precipitate when 1.0 M $AgNO_3$ is slowly added to a solution containing 0.015 M Cl^- and 0.015 M I^-?

17.56 What is the Cl^- concentration just as Ag_2CrO_4 begins to precipitate when 0.90 M $AgNO_3$ is slowly added to a solution containing 0.015 M Cl^- and 0.015 M CrO_4^{2-}?

Effect of pH on Solubility

17.57 Write the net ionic equation in which the slightly soluble salt magnesium carbonate, $MgCO_3$, dissolves in dilute hydrochloric acid.

17.58 Write the net ionic equation in which the slightly soluble salt barium fluoride, BaF_2, dissolves in dilute hydrochloric acid.

17.59 Which salt would you expect to dissolve more readily in acidic solution, barium sulfate or barium fluoride? Explain.

17.60 Which salt would you expect to dissolve more readily in acidic solution, strontium phosphate, $Sr_3(PO_4)_2$, or strontium sulfate, $SrSO_4$? Explain.

Complex Ions

17.61 Write the chemical equation for the formation of the $Fe(CN)_6^{3-}$ ion. Write the K_f expression.

17.62 Write the chemical equation for the formation of the $Cu(CN)_2^-$ ion. Write the K_f expression.

17.63 Sufficient sodium cyanide, NaCN, was added to 0.015 M silver nitrate, $AgNO_3$, to give a solution that was initially 0.100 M in cyanide ion, CN^-. What is the concentration of silver ion, Ag^+, in this solution after $Ag(CN)_2^-$ forms? The formation constant K_f for the complex ion $Ag(CN)_2^-$ is 5.6×10^{18}.

17.64 The formation constant K_f for the complex ion $Zn(OH)_4^{2-}$ is 2.8×10^{15}. What is the concentration of zinc ion, Zn^{2+}, in a solution that is initially 0.20 M in $Zn(OH)_4^{2-}$?

17.65 Predict whether nickel(II) hydroxide, $Ni(OH)_2$, will precipitate from a solution that is 0.0020 M $NiSO_4$, 0.010 M NaOH, and 0.10 M NH_3. Note that nickel(II) ion forms the $Ni(NH_3)_6^{2+}$ complex ion.

17.66 Predict whether cadmium oxalate, CdC_2O_4, will precipitate from a solution that is 0.0020 M $Cd(NO_3)_2$, 0.010 M $Na_2C_2O_4$, and 0.10 M NH_3. Note that cadmium ion forms the $Cd(NH_3)_4^{2+}$ complex ion.

17.67 What is the molar solubility of CdC_2O_4 in 0.10 M NH_3?

17.68 What is the molar solubility of ZnS in 0.10 M NH_3?

Qualitative Analysis

17.69 Describe how you could separate the following mixture of metal ions: Cd^{2+}, Sr^{2+}, and Pb^{2+}.

17.70 Describe how you could separate the following mixture of metal ions: Na^+, Hg^{2+}, and Ca^{2+}.

17.71 A student dissolved a compound in water and added hydrochloric acid. No precipitate formed. Next she bubbled H_2S into this solution, but again no precipitate formed. However, when she made the solution basic with ammonia and bubbled in H_2S, a precipitate formed. Which of the following are possible as the cation in the compound?

a Ca^{2+} b Mn^{2+} c Ag^+ d Cd^{2+}

17.72 A student was asked to identify a compound. In an effort to do so, he first dissolved the compound in water. He found that no precipitate formed when hydrochloric acid was added, but when H_2S was bubbled into this acidic solution, a precipitate formed. Which one of the following could be the precipitate?

a Ag_2S b CdS   c MnS   d PbS

General Problems

Key: These problems provide more practice but are not divided by topic or keyed to exercises. Each section ends with essay questions, each of which is color coded to refer to the *A Chemist Looks at Environment* (green) chapter essay on which it is based. Odd-numbered problems and the even-numbered problems that follow are similar; answers to all odd-numbered problems except the essay questions are given in the back of the book.

17.73 Mercury(II) ion is often precipitated as mercury(II) sulfide in qualitative analysis. Using the solubility product constant, calculate the molar solubility of mercury(II) sulfide in water, assuming that no other reactions occur.

17.74 Lead(II) sulfate is often used as a test for lead(II) ion in qualitative analysis. Using the solubility product constant, calculate the molar solubility of lead(II) sulfate in water.

17.75 Mercury(I) chloride, Hg_2Cl_2, is an unusual salt in that it dissolves to form Hg_2^{2+} and $2Cl^-$. Use the solubility product constant to calculate the following: a the molar solubility of Hg_2Cl_2; b the solubility of Hg_2Cl_2 in grams per liter.

17.76 Magnesium ammonium phosphate is an unusual salt in that it dissolves to form Mg^{2+}, NH_4^+, and PO_4^{3-} ions. K_{sp} for magnesium ammonium phosphate equals 2.5×10^{-13}. Calculate: a its molar solubility; b its solubility in grams per liter.

17.77 For cerium(III) hydroxide, $Ce(OH)_3$, K_{sp} equals 2.0×10^{-20}. Calculate: a its molar solubility (recall that taking the square root twice gives the fourth root); b the pOH of the saturated solution.

17.78 Copper(II) ferrocyanide, $Cu_2Fe(CN)_6$, dissolves to give Cu^{2+} and $[Fe(CN)_6]^{4-}$ ions; K_{sp} for $Cu_2Fe(CN)_6$ equals 1.3×10^{-16}. Calculate: a the molar solubility, and b the solubility in grams per liter of copper(II) ferrocyanide.

17.79 What is the solubility of magnesium hydroxide in a solution buffered at pH 8.80?

17.80 What is the solubility of silver oxide, Ag_2O, in a solution buffered at pH 10.78? The equilibrium is $Ag_2O(s) + H_2O(l) \rightleftharpoons 2Ag^+(aq) + 2OH^-(aq)$; $K_c = 2.0 \times 10^{-8}$.

17.81 What is the molar solubility of $Al(OH)_3$ in a solution containing 1.5×10^{-3} M NaOH?

17.82 What is the molar solubility of $Mg(OH)_2$ in a solution containing 1.0×10^{-1} M NaOH?

17.83 What must be the concentration of sulfate ion in order to precipitate calcium sulfate, $CaSO_4$, from a solution that is 0.0030 M Ca^{2+}?

17.84 What must be the concentration of chromate ion in order to precipitate strontium chromate, $SrCrO_4$, from a solution that is 0.0034 M Sr^{2+}?

17.85 A 3.20-L solution of 1.25×10^{-3} M $Pb(NO_3)_2$ is mixed with a 0.80-L solution of 5.0×10^{-1} M NaCl. Calculate Q_c for the dissolution of $PbCl_2$. No precipitate has formed. Is the solution supersaturated, saturated, or unsaturated?

17.86 A 0.150-L solution of 2.4×10^{-5} M $MgCl_2$ is mixed with 0.050 L of 4.0×10^{-3} M NaOH. Calculate Q_c for the dissolution of $Mg(OH)_2$. No precipitate has formed. Is the solution supersaturated, saturated, or unsaturated?

17.87 How many grams of sodium sulfate can be added to 446 mL of 0.0032 M barium chloride before a precipitate forms?

17.88 How many grams of sodium chloride can be added to 785 mL of 0.0015 M silver nitrate before a precipitate forms?

17.89 Solid KSCN was added to a 2.00 M Co^{2+} solution so that it was also initially 2.00 M SCN^-. These ions then reacted to give the complex ion $Co(SCN)^+$, whose formation constant is 1.0×10^2. What is the concentration of $Co^{2+}(aq)$ at equilibrium? Be sure to check any simplifying assumption you make.

17.90 Solid KSCN was added to a 2.00 M Fe^{3+} solution so that it was also initially 2.00 M SCN^-. These ions then reacted to give the complex ion $Fe(SCN)^{2+}$, whose formation constant is 9.0×10^2. What is the concentration of $Fe^{3+}(aq)$ at equilibrium? Be sure to check any simplifying assumption you make.

17.91 Calculate the molar solubility of silver iodide, AgI, in 2.2 M NH_3.

17.92 Calculate the molar solubility of silver bromide, AgBr, in 5.0 M NH_3.

17.93 The solubility of zinc oxalate, ZnC_2O_4, in 0.0150 M ammonia is 3.6×10^{-4} mol/L. What is the oxalate-ion concentration in the saturated solution? If the solubility product constant for zinc oxalate is 1.5×10^{-9}, what must be the zinc-ion concentration in the solution? Now calculate the formation constant for the complex ion $Zn(NH_3)_4^{2+}$.

17.94 The solubility of cadmium oxalate, CdC_2O_4, in 0.150 M ammonia is 6.1×10^{-3} mol/L. What is the oxalate-ion concentration in the saturated solution? If the solubility product constant for cadmium oxalate is 1.5×10^{-8}, what must be the cadmium-ion concentration in the solution? Now calculate the formation constant for the complex ion $Cd(NH_3)_4^{2+}$.

17.95 Hydrazine, N_2H_4, is a base that ionizes to give $N_2H_5^+$ and OH^- ($K_b = 1.7 \times 10^{-6}$). You add magnesium sulfate to a hydrazine solution. Calculate the concentration of Mg^{2+} ion when magnesium hydroxide, $Mg(OH)_2$, just begins to precipitate from 0.20 M N_2H_4. K_{sp} for $Mg(OH)_2$ is 1.8×10^{-11}.

17.96 Ammonia, NH_3, is a base that ionizes to give NH_4^+ and OH^- ($K_b = 1.8 \times 10^{-5}$). You add magnesium sulfate to an ammonia solution. Calculate the concentration of Mg^{2+} ion when magnesium hydroxide, $Mg(OH)_2$, just begins to precipitate from 0.10 M NH_3. K_{sp} for $Mg(OH)_2$ is 1.8×10^{-11}.

17.97 A saturated solution of copper(II) iodate in pure water has a copper-ion concentration of 2.7×10^{-3} M.
- **a** What is the molar solubility of copper iodate in a 0.35 M potassium iodate solution?
- **b** What is the molar solubility of copper iodate in a 0.35 M copper nitrate solution?
- **c** Should there be a difference in the answers to parts a and b? Why?

17.98 A saturated solution of lead iodate in pure water has an iodate-ion concentration of 8.0×10^{-5} M.
- **a** What is the molar solubility of lead iodate in a 0.15 M lead nitrate solution at the same temperature?
- **b** Should the molar solubility of lead iodate in part a be the same as, greater than, or less than that of lead iodate in pure water? Why?

17.99 A solution contains 0.0150 M lead(II) ion. A concentrated sodium iodide solution is added dropwise to precipitate lead iodide (assume no volume change).
- **a** At what concentration of I^- does precipitate start to form?
- **b** When $[I^-] = 2.0 \times 10^{-3}$ M, what is the lead-ion concentration? What percentage of the lead(II) originally present remains in solution?

17.100 A solution contains 0.00740 M calcium ion. A concentrated sodium fluoride solution is added dropwise to precipitate calcium fluoride (assume no volume change).
- **a** At what concentration of F^- does precipitate start to form?
- **b** When $[F^-] = 9.5 \times 10^{-4}$ M, what is the calcium-ion concentration? What percentage of the calcium ion has precipitated?

17.101 a If the molar solubility of beryllium(II) hydroxide is 8.6×10^{-7} M in pure water, what is its K_{sp} value?
- **b** What is the molar solubility of beryllium(II) hydroxide in a solution that is 1.50 M in NH_3 and 0.25 M in NH_4Cl?
- **c** Account for the differences in molar solubility in parts a and b.

17.102 a If the molar solubility of cobalt(II) hydroxide is 5.4×10^{-6} mol/L in pure water, what is its K_{sp} value?
- **b** What is the molar solubility of $Co(OH)_2$ in a buffered solution that has a pH of 10.43?
- **c** Account for the differences in molar solubility in parts a and b.

17.103 Although silver chloride is insoluble in water, it readily dissolves upon the addition of ammonia.

$$AgCl(s) + 2NH_3(aq) \rightleftharpoons Ag(NH_3)_2^+(aq) + Cl^-(aq)$$

- **a** What is the equilibrium constant for this dissolving process?
- **b** Ammonia is added to a solution containing excess AgCl(s). The final volume is 1.00 L and the resulting equilibrium concentration of NH_3 is 0.80 M. Calculate the number of moles of AgCl dissolved, the molar concentration of $Ag(NH_3)_2^+$, and the number of moles of NH_3 added to the original solution.

17.104 Crystals of AgBr can be removed from black-and-white photographic film by reacting the AgBr with sodium thiosulfate.

$$AgBr(s) + 2S_2O_3^{2-}(aq) \rightleftharpoons [Ag(S_2O_3)_2]^{3-}(aq) + Br^-(aq)$$

- **a** What is the equilibrium constant for this dissolving process?
- **b** In order to dissolve 2.5 g of AgBr in 1.0 L of solution how many moles of $Na_2S_2O_3$ must be added?

17.105 A 1.0-L solution that is 4.2 M in ammonia is mixed with 26.7 g of ammonium chloride.
- **a** What is the hydroxide-ion concentration of this solution?
- **b** 0.075 mol of $MgCl_2$ is added to the above solution. Assume that there is no volume change. After $Mg(OH)_2$ has precipitated, what is the molar concentration of magnesium ion? What percent of the Mg^{2+} is removed from solution?

17.106 A 1.0-L solution that is 1.6 M in ammonia is mixed with 75.8 g of ammonium sulfate.

  a What is the hydroxide-ion concentration of this solution?

  b 0.058 mol of $MnCl_2$ is added to the above solution. Assume that there is no volume change. After $Mn(OH)_2$ has precipitated, what is the molar concentration of manganese ion? What percent of the Mn^{2+} is removed from solution (K_{sp} of $Mn(OH)_2$ is 4.6×10^{-14})?

■ **17.107** What chemical reaction is responsible for the dissolving out of caverns in limestone caves?

■ **17.108** What chemical reaction is responsible for the formation of stalactites and stalagmites in limestone caves? How is this reaction related to the one involved in the dissolving out of caverns?

Strategy Problems

Key: As noted earlier, all of the practice and general problems are matched pairs. This section is a selection of problems that are not in matched-pair format. These challenging problems require that you employ many of the concepts and strategies that were developed in the chapter. In some cases, you will have to integrate several concepts and operational skills in order to solve the problem successfully.

17.109 You add 50.0 mL of 0.100 M HCl to 50.0 mL of 0.100 M $AgNO_3$. What are the final concentrations of H_3O^+ and Cl^- in the solution?

17.110 A researcher found the solubility of lead(II) bromide in 0.100 M NaBr to be 1.31 g/L. What is the solubility product of lead(II) bromide?

17.111 A chemist mixes 1.00 L each of 0.100 M Na_2CO_3 and 0.200 M $CaCl_2$ in a beaker. What is the concentration of carbonate ion, CO_3^{2-}, in the final solution?

17.112 Gout is a painful inflammation of the joints caused by an excess of uric acid in the blood and its precipitation as sodium urate (the sodium salt of uric acid), $NaC_5H_3N_4O_3$, in tissues of the joints. The solubility of sodium urate in aqueous solution is 7.0 mg per 100 mL at body temperature, 37°C. The normal level of sodium ion in blood plasma is 3.2 g per liter. What would be the concentration of urate ion in blood plasma if it were saturated with sodium urate?

Uric acid

17.113 An analytical chemist has a solution containing chloride ion, Cl^-. She decides to determine the amount of chloride ion in the solution by titrating 50.0 mL of this solution by 0.100 M $AgNO_3$. As a way to indicate the endpoint of the titration, she added 1.00 g of potassium chromate, K_2CrO_4 (see Figure 17.5). As she slowly added the silver nitrate to the solution, a white precipitate formed. She continued the titration, with more white precipitate forming. Finally, the solution turned red, from another precipitate. The volume of the solution at this point was 60.3 mL. How many moles of chloride ion were there in

the original solution? How many moles of chloride ion were there in the final solution? You may make any reasonable approximations.

17.114 A scientist was interested in how soluble rust is in acidic soils, so she set up an idealized problem to get an initial feel for the situation. A fairly acidic soil has a pH of 4.50. Also, rust is essentially $Fe(OH)_3$. Therefore, she considered the following problem: Suppose a 1.00-g sample of iron(III) hydroxide is exposed to 1.00 L of a buffer with a pH of 4.50. She then calculated the nanograms of Fe^{3+} that dissolve in a liter of this buffer. Show how you would do this problem. Explain your work.

17.115 How would the solubility of calcium fluoride be affected by the presence of fluoride ion from another source? What is the solubility of calcium fluoride in a saturated solution of barium fluoride? How does this compare with the value of the solubility of calcium fluoride found in Example 17.4? Is this what you expect?

17.116 A 1.00-g sample of solid barium sulfate and a similar 1.00-g sample of solid calcium sulfate were added to water at 25°C to give 1.00 L of solution. Calculate the concentrations of Ba^{2+}, Ca^{2+}, and SO_4^{2-} ions present in the solution.

17.117 Calcium fluoride, CaF_2, is a very slightly soluble salt. In Example 17.4, we calculated its solubility from the solubility constant, assuming that the solubility equilibrium represented by K_{sp} was the only important one. However, a solution of the salt does contain the species HF. Where does this come from? Would you expect this to increase or decrease the solubility as we calculated it in the example? Using the fluoride-ion concentration that we calculated in the example, obtain the concentration of HF in the solution. From this, decide how important the formation of fluoride ion is in determining the solubility of calcium fluoride. Without doing any calculations, decide how important you think a similar effect would be in determining the solubility of calcium carbonate. Explain.

17.118 You slowly add ammonia, NH_3, to 1.00 L of water containing 5.00 g of solid nickel(II) iodate, $Ni(IO_3)_2$. How much ammonia (in moles) would you have to add to just dissolve the nickel(II) iodate? What would be the molar concentration of ammonia in the solution at that point? K_{sp} for nickel iodate is 1.4×10^{-8}.

17.119 A chemist mixes equal volumes of 0.10 M $CaCl_2$ and 0.15 M NaF in a beaker. What are the concentrations of Ca^{2+}, Na^+, Cl^-, and F^- in the solution in the beaker at 25°C?

17.120 Suppose you add 35.6 mL of 0.578 M H_2SO_4 to 55.6 mL of 0.491 M $Ba(OH)_2$. What would be the concentrations of Ba^{2+} and SO_4^{2-} in the final solution at 25°C? What would be the pH of this solution?

17.121 Suppose you add 35.6 mL of 0.578 M $Ba(OH)_2$ to 55.6 mL of 0.491 M H_2SO_4. What would be the concentrations of Ba^{2+} and SO_4^{2-} in the final solution at 25°C? What would be the pH of this solution?

17.122 What volume of water would be required to dissolve 10.0 g of silver chloride, AgCl, at 25°C? What volume of 0.15 M NH_3 would be required to dissolve the same quantity of silver chloride?

17.123 Natural water is said to be "hard" when it contains metal ions such as Ca^{2+} and Mg^{2+} that interfere with the use of detergents. These ions are often present in natural water as the soluble hydrogen carbonates, $Ca(HCO_3)_2$ and $Mg(HCO_3)_2$. One way to remove these ions from natural water is to add lime, $Ca(OH)_2$, to it. This addition of a calcium compound may seem paradoxical. What happens to a solution of calcium hydrogen carbonate when the pH is raised? Does this solve the paradox? Explain.

17.124 Suppose you have a water solution that is 0.0010 M Ca^{2+} and 0.0010 M Mg^{2+}. Describe the precipitation reactions as you slowly add a concentrated solution of sodium carbonate, Na_2CO_3, to this solution. What would be the concentration of carbonate ion, CO_3^{2-}, when at least 90.0% of Ca^{2+} and Mg^{2+} have been precipitated.

17.125 Magnesium metal is produced from seawater that contains magnesium ion, Mg^{2+}; seawater contains 1.35 mg of magnesium ion per liter. Magnesium ion is precipitated from seawater by adding calcium hydroxide, $Ca(OH)_2$, to it. (Calcium hydroxide can be obtained from seashells, which are mostly calcium carbonate.) What would be the pH of the seawater after 90.0% of the magnesium ion has been precipitated by calcium hydroxide? How many grams of $Ca(OH)_2$ would have to be added per liter of seawater to precipitate this percentage of magnesium ion?

17.126 Solid magnesium hydroxide, $Mg(OH)_2$, is added to a solution that is 1.00 M NH_4Cl. What is the concentration of Mg^{2+} in this solution at equilibrium?

17.127 Suppose that an aqueous solution is in equilibrium with solid calcium sulfate and solid barium sulfate. What are the concentrations of calcium ion and barium ion in the solution?

17.128 The text describes zinc hydroxide as an amphoteric hydroxide, so $Zn(OH)_2$ is soluble in basic solution. What is the molar solubility of $Zn(OH)_2$ in 1.00 M NaOH? What is the pH of the equilibrium solution?

Cumulative-Skills Problems

Key: The problems under this heading combine skills introduced in previous chapters with those given in the current one.

17.129 A solution is 1.8×10^{-4} M Co^{2+} and 0.20 M HSO_4^-. The solution also contains Na_2SO_4. What should be the minimum molarity of Na_2SO_4 to prevent the precipitation of cobalt(II) sulfide when the solution is saturated with hydrogen sulfide (0.10 M H_2S)?

17.130 A solution is 1.5×10^{-4} M Zn^{2+} and 0.20 M HSO_4^-. The solution also contains Na_2SO_4. What should be the minimum molarity of Na_2SO_4 to prevent the precipitation of zinc sulfide when the solution is saturated with hydrogen sulfide (0.10 M H_2S)?

17.131 What is the solubility of calcium fluoride in a buffer solution containing 0.45 M $HCHO_2$ (formic acid) and 0.20 M $NaCHO_2$? (*Hint:* Consider the equation $CaF_2(s) + 2H^+(aq) \rightleftharpoons Ca^{2+}(aq) + 2HF(aq)$, and solve the equilibrium problem.)

17.132 What is the solubility of magnesium fluoride in a buffer solution containing 0.45 M $HC_2H_3O_2$ (acetic acid) and 0.20 M $NaC_2H_3O_2$? The K_{sp} for magnesium fluoride is 6.5×10^{-9}. (See the hint for Problem 17.131.)

17.133 A 50.0-mL sample of 0.0150 M Ag_2SO_4 is added to 25.0 mL of 0.0100 M $PbCl_2$. What is the net ionic equation for the reaction that occurs? What are the concentrations of ions in the mixture at equilibrium?

17.134 A 67.0-mL sample of 0.350 M $MgSO_4$ is added to 45.0 mL of 0.250 M $Ba(OH)_2$. What is the net ionic equation for the reaction that occurs? What are the concentrations of ions in the mixture at equilibrium?

18 Thermodynamics and Equilibrium

© Paul Silverman/Fundamental Photographs, NYC; Cengage Learning

The spontaneous reaction of iron with oxygen in air produces a coating of rust, a form of iron(III) oxide. Hematite, a mineral form of iron(III) oxide, is a commercial source of iron metal (produced from a spontanous reaction with carbon and oxygen).

Contents and Concepts

⛯WL Sign in to OWL at **www.cengage.com/owl** to view tutorials and simulations, develop problem-solving skills, and complete online homework assigned by your professor.

🞂Chemistry Download mini lecture videos for key concept review and exam prep from OWL or purchase them from **www.cengagebrain.com**.

Urea is used as a plant fertilizer because it slowly decomposes in the soil to provide ammonia.

A molecular model of urea.

Figure 18.1 ▲

Urea

Urea, NH_2CONH_2, is an important industrial chemical. It is used to make synthetic resins for adhesives and melamine plastics (Figure 18.1). Its major use, however, is as a nitrogen fertilizer for plants. Urea is produced by reacting ammonia with carbon dioxide. We can write the overall reaction as:

$$2NH_3(g) + CO_2(g) \longrightarrow NH_2CONH_2(aq) + H_2O(l)$$

Suppose you want to determine whether this or some other reaction could be useful for the industrial preparation of urea. Some of your immediate questions might be the following: Does the reaction naturally go in the direction it is written? Will the reaction mixture contain a sufficient amount of the product at equilibrium? We addressed these questions in the preceding chapters by looking at the equilibrium constant. Now we want to discuss these questions from the point of view of thermodynamics.

Thermodynamics is the study of the relationship between heat and other forms of energy involved in a chemical or physical process. With only heat measurements of substances, you can answer the questions just posed. You can predict the natural direction of a chemical reaction, and you can also determine the composition of a reaction mixture at equilibrium. Consider, then, the possibility of reacting NH_3 and CO_2 to give urea. Just how do you apply thermodynamics to such a reaction?

18.1 First Law of Thermodynamics: A Review

Thermodynamics is described in terms of three laws. *The first law of thermodynamics is essentially the law of conservation of energy applied to thermodynamic systems.* The law of conservation of energy, you may recall, says that the total energy remains constant.

Mathematical Statement of the First Law

In a thermodynamic system, the sum of the kinetic and potential energies of the particles making up the system is referred to as its *internal energy, U.* The internal energy can change as the result of transfers of energy (or energy flows) into or out of the system, and the first law simply says that the change of internal energy equals the sum of these energy transfers.

These transfers, or flows of energy, are of two kinds: heat, q, and work, w. If a chemical reaction occurs in the system, *heat of reaction* may evolve or perhaps be absorbed. Energy as heat leaves or enters the system. The system may also do work on the surroundings or have work done on it by the surroundings. *Pressure-volume work* is a common form of work.

$$\text{Pressure-volume work} = -P\Delta V$$

Suppose the volume, V, of the system increases so that the system pushes against the external pressure of the atmosphere, P. Effectively the system lifts the atmosphere against gravity and does work on it. In doing this work, energy leaves the system (flows out of it) and goes into the surroundings. Or perhaps the volume of the system decreases so that the external pressure pushes against the system. In this case, work is done on the system; energy flows from the surroundings to the system.

According to the first law of thermodynamics, whenever a thermodynamic system undergoes a physical or chemical change, the change of internal energy, ΔU, of the system equals the sum of the heat and work done in that physical or chemical change.

$$\Delta U = q + w$$

Enthalpy of Reaction

When a physical or chemical change occurs against a constant external pressure, as would happen when the change occurs in a flask open to the atmosphere, it is useful to define the thermodynamic quantity called *enthalpy, H.* The enthalpy of a system is defined as the system's internal energy plus pressure times volume; it is a state function since U, P, and V are state functions.

$$H = U + PV$$

With this definition, one can show that the heat involved in a physical or chemical change for a given fixed pressure and temperature simply equals the change in enthalpy for the system.

$$q = \Delta H$$

It is useful to tabulate *standard enthalpies of formation*, ΔH_f°, for substances. Appendix C lists values at 25°C and 1 atm pressure for selected substances. Using these values, you can calculate the standard enthalpy change for a reaction, which equals the heat of reaction.

$$\Delta H^\circ = \Sigma\, n\Delta H_f^\circ(\text{products}) - \Sigma\, m\Delta H_f^\circ(\text{reactants})$$

Consider the reaction between NH_3 and CO_2 to produce urea and water, which is the reaction given in the chapter opening.

$$2NH_3(g) + CO_2(g) \longrightarrow NH_2CONH_2(aq) + H_2O(l)$$

The standard enthalpies of formation at 25°C for these substances are (in kJ/mol): $NH_3(g)$, −45.9; $CO_2(g)$, −393.5; $NH_2CONH_2(aq)$, −319.2; and $H_2O(l)$, −285.8. Substituting these values into the equation for the standard enthalpy change yields

$$\Delta H^\circ = [(-319.2 - 285.8) - (-2 \times 45.9 - 393.5)]\text{ kJ} = -119.7 \text{ kJ}$$

From the minus sign of ΔH°, you can conclude that heat is evolved. The reaction is exothermic.

Spontaneous Processes and Entropy

Why does a chemical reaction go naturally in a particular direction? To answer this question, we need to look at spontaneous processes. A **spontaneous process** is *a physical or chemical change that occurs by itself.* It requires no continuing outside agency to make it happen. A rock at the top of a hill rolls down (Figure 18.2, *top*). Heat flows from a hot object to a cold one. An iron object rusts in moist air (Figure 18.3). These processes occur spontaneously, or naturally, without requiring an outside force or agency. They continue until equilibrium is reached. If these processes were to go in the opposite direction, they would be *nonspontaneous.* The rolling of a rock uphill by itself is not a natural process; it is nonspontaneous (see Figure 18.2, *bottom*). The rock could be moved to the top of the hill, but work would have to be expended. Heat can be made to flow from a cold to a hot object, but a heat pump or refrigerator is needed. Rust can be converted to iron, but the process requires chemical reactions used in the manufacture of iron from its ore (iron oxide).

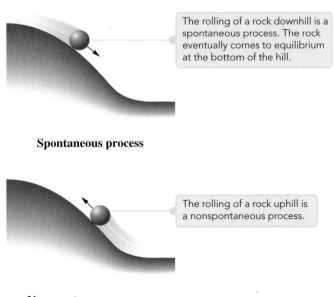

The rolling of a rock downhill is a spontaneous process. The rock eventually comes to equilibrium at the bottom of the hill.

Spontaneous process

The rolling of a rock uphill is a nonspontaneous process.

Nonspontaneous process

Figure 18.2 ▲

Examples of a spontaneous and a nonspontaneous process

© Nelson Morris/Photo Researchers, Inc.

Figure 18.3 ▲

The rusting of iron in moist air is a spontaneous reaction This sculpture by Picasso in Chicago (Daley Plaza) was allowed to rust to give a pleasing effect.

18.2 Entropy and the Second Law of Thermodynamics

When you ask whether a chemical reaction goes in the direction in which it is written, as we did in the chapter opening, you are asking whether the reaction is spontaneous in this direction. The first law of thermodynamics cannot help you answer such a question. It does help you keep track of the various forms of energy in a chemical change, using the constancy of total energy (conservation of energy). But although at one time it was thought that spontaneous reactions must be exothermic ($\Delta H < 0$), many spontaneous reactions are now known to be endothermic ($\Delta H > 0$); recall the reaction shown in Figure 6.1.

The second law of thermodynamics, which we will discuss in this section, provides a way to answer questions about the spontaneity of a reaction. The second law is expressed in terms of a quantity called *entropy*.

Entropy

Entropy, S, is *a thermodynamic quantity that is a measure of how dispersed the energy of a system is among the different possible ways that system can contain energy.* When the energy of a thermodynamic system is concentrated in a relatively few energy states, the entropy of the system is low. When that same energy, however, is dispersed or spread out over a great many energy states, the entropy of the system is high. What you will see is that the entropy (energy dispersal) of a system plus its surroundings increases in a spontaneous process.

Let's look at some simple spontaneous processes. Suppose you place a hot cup of coffee on the table. Heat energy from the hot coffee flows slowly to the table and to the air surrounding the cup. In this process, energy spreads out or disperses. The entropy of the system (coffee cup) and its surroundings (table and surrounding air) increases in this spontaneous process.

As a slightly more complicated example, consider a rock at the top of a hill. The rock has potential energy relative to what it would have at the bottom of the hill. Imagine the rock rolling downhill. As it does so, its potential energy changes to kinetic energy, and during its descent atoms of the rock collide with those of the hillside and surrounding air, dispersing energy from the rock to its surroundings. The entropy of the system (rock) plus its surroundings (hillside and surrounding air) increases in this spontaneous process.

As a final example, imagine a flask containing a gas connected to an evacuated flask by a valve or stopcock (Figure 18.4). When the valve is opened, gas in the flask flows into the space of the evacuated flask. In this case, the kinetic energy of the gas molecules spreads out or disperses over the volumes of both flasks. The entropy of this system (both flasks) increases in this spontaneous process. In this case, we assume that the surroundings do not participate in the overall change (there is no entropy change in the surroundings); the entropy change occurs entirely within the system.

In each of these spontaneous processes, energy has been dispersed, or spread out. The entropy of the system plus surroundings has increased.

As we will discuss later, the SI unit of entropy is joules per kelvin (J/K). Entropy, like enthalpy, is a state function. That is, the quantity of entropy in a given amount of substance depends only on variables, such as temperature and pressure, that determine the state of the substance. If these variables are fixed,

Figure 18.4 ▶

Expansion of a gas into a vacuum Initially, the flask on the left contains a gas, whereas the flask on the right is evacuated. Then the valve is opened and gas spontaneously flows into the evacuated flask (the vacuum).

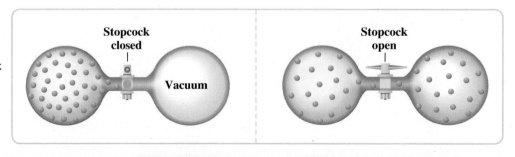

the quantity of entropy is fixed. For example, 1 mol of ice at 0°C and 1 atm pressure has an entropy that has been determined experimentally to be 41 J/K. One mole of liquid water at 0°C and 1 atm has an entropy of 63 J/K. (In the next section, we will describe how you measure these values.) Why is the entropy of liquid water higher than that of solid water (ice)? In an ice crystal, water molecules occupy regular fixed positions in the crystal lattice. Molecules can vibrate or oscillate about these fixed positions, but otherwise are restricted in their motions, and therefore the energies available to them. In the liquid state, water molecules can rotate as well as vibrate internally and can move around somewhat (though not as freely as molecules can move in the gaseous state). Thus, the entropy of liquid water is expected to be higher than that of ice—the energy is dispersed over more available energy states in the liquid.

You calculate the entropy change, ΔS, for a process similarly to the way you calculate ΔH. If S_i is the initial entropy and S_f is the final entropy, the change in entropy is

$$\Delta S = S_f - S_i$$

For the melting of ice to liquid water,

$$H_2O(s) \longrightarrow H_2O(l)$$

$$\Delta S = (63 - 41) \text{ J/K} = 22 \text{ J/K}$$

When 1 mol of ice melts at 0°C, the water increases in entropy by 22 J/K. The entropy increases, as you expect, because the energy of water becomes more dispersed when it melts.

CONCEPT CHECK 18.1

You have a sample of 1.0 mg of solid iodine at room temperature. Later, you notice that the iodine has sublimed (passed into the vapor state). What can you say about the change of entropy of the iodine?

Second Law of Thermodynamics

A process occurs naturally as a result of the dispersal of energy in the system plus surroundings. We can state this precisely in terms of the **second law of thermodynamics**, which states that *the total entropy of a system and its surroundings always increases for a spontaneous process.* Note that entropy (energy dispersal) is quite different from energy itself. Energy can be neither created nor destroyed during a spontaneous, or natural, process—its total amount remains fixed. But energy is dispersed in a spontaneous process, which means that entropy is produced (or created) during such a process.

For a spontaneous process carried out at a given temperature, the second law can be restated in a form that refers only to the system (and not to the system plus surroundings, as in the previous statement of the second law). We will find this new statement particularly useful in analyzing chemical problems.

Suppose a spontaneous process occurs within a system that is in thermal contact with its surroundings at a given temperature T—say, a chemical reaction in a flask. As the chemical reaction occurs, entropy is produced (or created) within the system (the flask) as a result of this spontaneous process. At the same time, heat might flow into or out of the system as a result of the thermal contact. (In other words, the reaction may be endothermic or exothermic.) *Heat flow is also a flow of entropy, because it is a dispersal of energy,* either into the flask or outside of it. In general, the entropy change associated with a flow of heat q at an absolute temperature T can be shown to equal q/T. The *net change* of entropy in the system (in the flask) is the entropy created during the spontaneous chemical reaction that occurs plus the entropy change that is associated with the heat flow (entropy flow).

$$\Delta S = \text{entropy created} + \frac{q}{T}$$

It might be helpful to consider an analogy. A baker makes (creates) cookies at her store (the system). She also buys cookies from a wholesaler and sells cookies at her store. The change in number of cookies in her store in any given interval of time equals the number made (number created) plus the number bought minus the number sold (flow of cookies into or out of her store). We obtain an equation similar to the one we wrote for the entropy change, ΔS.

Change of cookies in store = cookies created + flow of cookies into or out of store

However, our analogy breaks down if someone eats cookies in the store. Whereas entropy can only be created, cookies can be created and destroyed (eaten).

The quantity of entropy created during a spontaneous process is a positive quantity—the entropy increases as it is created. If we delete "entropy created" from the right side of the equation for ΔS, we know that the left side is then greater than the right side:

$$\Delta S > \frac{q}{T}$$

We can now restate the second law as follows:

Second Law of Thermodynamics: *For a spontaneous process at a given temperature T, the change in entropy of the system is greater than the heat divided by the absolute temperature, q/T.*

Entropy and Molecular Disorder

Fundamentally, entropy is related to energy dispersal. The energy of a molecular system may at first be concentrated in a few energy states and then later dispersed among many more energy states. In such a case, the entropy of the system increases. You might also say that the molecular system has increased in "disorder," in the sense that the energy is strewn or distributed over many energy states, rather than concentrated in a few energy states.

Consider the pendulum depicted in Figure 18.5. Suppose you start the pendulum by moving it upward and releasing it. The pendulum begins to swing to and fro. On each swing, it moves through a smaller and smaller arc, eventually coming to rest. While the pendulum is moving, its atoms have a concerted motion, which is the kinetic energy of the pendulum (Figure 18.5a). All of the atoms of the pendulum are moving in the same direction, though with small oscillational motions (or thermal motions) superimposed on this concerted motion. As these atoms collide with molecules of the surrounding air, however, some of this concerted motion is transferred to the random motion of air molecules (Figure 18.5b). Air molecules buffeted by the pendulum begin to move with it, but soon, their motions become random as they collide with many other air molecules. The molecular motion becomes more disordered. Eventually, the pendulum comes to rest with its energy dispersed, partly to the surrounding air (Figure 18.5c). Entropy has increased.

Imagine that you had captured this motion on film or video, and later you show this film in reverse. The pendulum, initially at rest, begins to swing through larger and larger arcs. This motion is clearly not what you would expect to see. A pendulum at rest does not spontaneously begin to move to and fro. Let's look at this process in molecular detail. At first the pendulum is at rest with surrounding molecules of air in random motion (Figure 18.5d). Then, by chance a large number of air molecules gather together to buffet the stationary pendulum (Figure 18.5e). Molecules of air, at first moving randomly, begin to move together, and start the pendulum moving upward (Figure 18.5f). What does thermodynamics have to say about this process? Note that this process does not violate the first law; energy remains constant; it merely exchanges from one molecule to another. This process does violate the second law, however. Energy dispersed over many energy states (molecules moving randomly) becomes more concentrated in those states in which the molecules move together. A disordered state becomes ordered, and entropy decreases. This process is nonspontaneous.

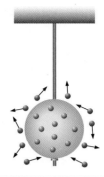

a A pendulum is placed in motion. Molecules of the pendulum (red spheres) move in the same direction (but with thermal motions superimposed on this concerted motion).

b As a result of collisions of the pendulum with air molecules (blue spheres), some of the concerted motion is degraded to random motion of surrounding air molecules.

c Eventually, the pendulum comes to rest with its energy dispersed.

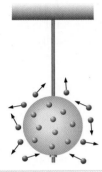

d The pendulum at rest is surrounded by air molecules in random motion.

e By chance, a group of air molecules gather together to buffet the pendulum on one side.

f The pendulum starts moving. Although E to F is not impossible, it is highly unlikely.

Let's look at a few other examples. Earlier we considered the cooling of a cup of coffee. Heat moves from the hot coffee to its surroundings. Molecular motion becomes more disordered, in the sense that it is dispersed over a larger volume. Similarly, when a gas flows from a flask into a connected evacuated flask, molecules become more disordered in the sense that they are strewn over a larger volume. When ice melts, water molecules leave an ordered crystal and enter a more disordered liquid state. As a final example, consider the chemical reaction in which dinitrogen tetroxide, N_2O_4, dissociates into two nitrogen dioxide molecules, NO_2:

$$N_2O_4(g) \longrightarrow 2NO_2(g)$$

The reaction, in breaking one molecule into two smaller ones, creates more disorder in the sense that the two NO_2 molecules can now move independently. In each of these changes, molecular disorder increases, as does entropy.

Unfortunately, the picture of entropy as a measure of disorder has been misapplied to situations that are not molecular. For example, imagine your desk with books and papers placed carefully to give an "orderly" arrangement. Later, as you work, your books and papers become strewn about—your desk becomes disordered. This has sometimes been described as a spontaneous process in which entropy increases. Of course, the books and papers do not move spontaneously— you move them. In fact, there is essentially no difference between the thermodynamic entropies of the ordered and disordered desks. All that has happened is that normal-sized objects (books and papers) have been moved by you from human-defined order to human-defined disorder.

Again, fundamentally entropy is related to the dispersal of energy. The entropy of a system can be defined rigorously by the methods of *statistical thermodynamics*

Figure 18.5 ▲

Motion of a pendulum *A* to *C* depicts the spontaneous motion of the pendulum. *D* to *F* depicts the nonspontaneous motion of a pendulum.

in terms of a sum over the energy levels available to the system. The picture of thermodynamic entropy as disorder may sometimes be useful, but it should be applied carefully to molecular systems.

Entropy Change for a Phase Transition

Certain processes occur at equilibrium or, more precisely, very close to equilibrium. For example, ice at 0°C is in equilibrium with liquid water at 0°C. If heat is slowly absorbed by the system, it remains very near equilibrium, but the ice melts. Under these essentially equilibrium conditions, no significant amount of entropy is created. The entropy change results entirely from the absorption of heat. Therefore,

$$\Delta S = \frac{q}{T} \quad \text{(equilibrium process)}$$

Other phase changes, such as the vaporization of a liquid, also occur under equilibrium conditions. ◀

When a system is at equilibrium, a small change in a condition can make the process go in one direction or the other. The process is said to be reversible. Think of a two-pan balance in which the weight on each pan has been adjusted to make the balance beam level—that is, so the beam is at equilibrium. A small weight on either pan will tip the scale one way or the other.

You can use the previous equation to obtain the entropy change for a phase change. Consider the melting of ice. The heat absorbed is the heat of fusion, ΔH_{fus}, which is known from experiment to be 6.0 kJ for 1 mol of ice. You get the entropy change for melting by dividing ΔH_{fus} by the absolute temperature of the phase transition, 273 K (0°C). Because entropy changes are usually expressed in joules per kelvin, you convert ΔH_{fus} to 6.0×10^3 J.

$$\Delta S = \frac{\Delta H_{fus}}{T} = \frac{6.0 \times 10^3 \text{ J}}{273 \text{ K}} = 22 \text{ J/K}$$

Note that this is the ΔS value we obtained earlier for the conversion of ice to water.

Example 18.1   Calculating the Entropy Change for a Phase Transition

The heat of vaporization, ΔH_{vap}, of carbon tetrachloride, CCl_4, at 25°C is 39.4 kJ/mol.

$$CCl_4(l) \longrightarrow CCl_4(g); \Delta H_{vap} = 39.4 \text{ kJ/mol}$$

If 1 mol of liquid carbon tetrachloride at 25°C has an entropy of 216 J/K, what is the entropy of 1 mol of the vapor in equilibrium with the liquid at this temperature?

Problem Strategy The entropy change for this equilibrium vaporization is $\Delta H_{vap}/T$. The entropy of the vapor equals the entropy of the liquid plus the entropy change.

Solution When liquid CCl_4 evaporates, it absorbs heat: $\Delta H_{vap} = 39.4$ kJ/mol (39.4×10^3 J/mol) at 25°C, or 298 K. The entropy change, ΔS, is

$$\Delta S = \frac{\Delta H_{vap}}{T} = \frac{39.4 \times 10^3 \text{ J/mol}}{298 \text{ K}} = 132 \text{ J/(mol·K)}$$

In other words, 1 mol of carbon tetrachloride increases in entropy by 132 J/K when it vaporizes. The entropy of 1 mol of the vapor equals the entropy of 1 mol of liquid (216 J/K) plus 132 J/K.

Entropy of vapor $= (216 + 132)$ J/(mol·K) $=$

348 J/(mol·K)

Note: This is the entropy at the equilibrium vapor pressure, which differs from $S°$ (entropy at 1 atm).

Answer Check A simple check is to reverse the problem. At equilibrium, the enthalpy change, ΔH, equals $T\Delta S$.

$$\Delta H = 298 \text{ K} \times 132 \text{ J/(mol·K)} = 39.3 \times 10^3 \text{ J/mol}$$

This is the enthalpy of vaporization given in the problem statement (within rounding error).

Exercise 18.1 Liquid ethanol, $C_2H_5OH(l)$, at 25°C has an entropy of 161 J/(mol·K). If the heat of vaporization, ΔH_{vap}, at 25°C is 42.6 kJ/mol, what is the entropy of the vapor in equilibrium with the liquid at 25°C?

■ See Problems 18.35 and 18.36.

Criterion for a Spontaneous Reaction

Now you can see how thermodynamics is applied to the question of whether a reaction is spontaneous. Suppose that you propose a chemical reaction for the preparation of a substance. You also assume that the reaction occurs at constant temperature and pressure. For urea, you propose the reaction

$$2NH_3(g) + CO_2(g) \longrightarrow NH_2CONH_2(aq) + H_2O(l)$$

Is this a spontaneous reaction? That is, does it go left to right as written? You can use the second law in the form $\Delta S > q/T$ to answer this question if you know both ΔH and ΔS for the reaction.

Recall that the heat of reaction at constant pressure, q_p, equals the enthalpy change ΔH. The second law for a spontaneous reaction at constant temperature and pressure becomes

$$\Delta S > \frac{q_p}{T} = \frac{\Delta H}{T} \quad \text{(spontaneous reaction, constant } T \text{ and } P)$$

ΔS is greater than $\Delta H/T$ for a spontaneous reaction at constant temperature and pressure. Therefore, if you subtract ΔS from $\Delta H/T$, you get a negative quantity. That is,

$$\frac{\Delta H}{T} - \Delta S < 0 \quad \text{(spontaneous reaction, constant } T \text{ and } P)$$

Multiplying each term of this inequality by the positive quantity T, you get

$$\Delta H - T\Delta S < 0 \quad \text{(spontaneous reaction, constant } T \text{ and } P)$$

This inequality is important. If you have a table of entropies of substances, you can calculate ΔS for the proposed reaction. From Table 6.2, you can also calculate ΔH. If $\Delta H - T\Delta S$ is negative for the reaction, you would predict that it is spontaneous left to right, as written. However, if $\Delta H - T\Delta S$ is positive, you would predict that the reaction is nonspontaneous in the direction written but spontaneous in the opposite direction. If $\Delta H - T\Delta S$ is zero, the reaction is at equilibrium.

18.3 Standard Entropies and the Third Law of Thermodynamics

To determine experimentally the entropy of a substance, you first measure the heat absorbed by the substance by warming it at various temperatures. That is, you find the heat capacity at different temperatures. ▶ You then calculate the entropy as we will describe. This determination of the entropy is based on the third law of thermodynamics.

Heat capacity was discussed in Section 6.6.

Third Law of Thermodynamics

The **third law of thermodynamics** states that *a substance that is perfectly crystalline at 0 K has an entropy of zero.* This seems reasonable. A perfectly crystalline substance at 0 K should have perfect order. When the temperature is raised, however, the substance increases in entropy as it absorbs heat and energy disperses through it.

You can determine the entropy of a substance at a temperature other than 0 K—say, at 298 K (25°C)—by slowly heating the substance from near 0 K to 298 K. Recall that the entropy change ΔS that occurs when heat is absorbed at a temperature T is q/T. Suppose you heat the substance from near 0.0 K to 2.0 K, and the heat absorbed is 0.19 J. You find the entropy change by dividing the heat absorbed by the average absolute temperature $[\frac{1}{2}(0.0 + 2.0)K = 1.0 \text{ K}]$. Therefore, ΔS equals 0.19 J/1.0 K = 0.19 J/K. This gives you the entropy of the substance at 2.0 K. Now you heat the substance from 2.0 K to 4.0 K, and this time 0.88 J of heat is absorbed. The average temperature is $\frac{1}{2}(2.0 + 4.0)K = 3.0 \text{ K}$, and the entropy change is 0.88 J/3.0 K = 0.29 J/K. The entropy of the substance at 4.0 K is (0.19 + 0.29) J/K = 0.48 J/K. Proceeding this way, you can eventually get the entropy at 298 K. ▶

The process just described is essentially the numerical evaluation of an integral, which you can obtain as follows: The heat absorbed for temperature change dT is $C_p(T)dT$, where C_p is the heat capacity at constant pressure, and the entropy change is $C_p(T)dT/T$. The standard entropy at temperature T is

$$\int_0^T \frac{C_p(T)dT}{T}$$

Figure 18.6 ▶

Standard entropy of bromine, Br₂, at various temperatures The entropy rises gradually as the temperature increases, but it jumps sharply at each phase transition.

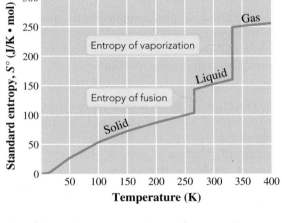

Figure 18.6 shows how the entropy of a substance changes with temperature. Note that the entropy increases gradually as the temperature increases. But when there is a phase change (for example, from solid to liquid), the entropy increases sharply. The entropy change for the phase transition is calculated from the enthalpy of the phase transition, as described earlier (see Example 18.1). The **standard entropy** of a substance or ion (Table 18.1), also called its **absolute entropy, $S°$**, is *the entropy value for the standard state of the species* (indicated by the superscript degree sign). For a substance, the standard state is the pure substance at 1 atm pressure. For a species in solution, the standard state is the 1 M solution. Table 18.1 gives standard entropies of various substances at 25°C and 1 atm. Note that the elements have nonzero values, unlike standard enthalpies of formation, $\Delta H_f°$, which by convention are zero. The symbol $S°$, rather than $\Delta S°$, is chosen for standard entropies to emphasize that they originate from the third law. ◀

Entropies of substances must be positive. Those for ions, however, can be negative because they are derived in a different way by arbitrarily setting $S°$ for $H_3O^+(aq)$ equal to zero.

Entropy Change for a Reaction

Once you determine the standard entropies of all substances in a reaction, you can calculate the change of entropy, $\Delta S°$, for the reaction. A sample calculation is described in Example 18.3. Even without knowing values for the entropies of substances, you can sometimes predict the sign of $\Delta S°$ for a reaction. The entropy usually increases in the following situations:

1. A reaction in which a molecule is broken into two or more smaller molecules.
2. A reaction in which there is an increase in moles of gas. (This may result from a molecule breaking up, in which case Rules 1 and 2 are related.)
3. A process in which a solid changes to liquid or gas or a liquid changes to a gas.

◔WL Interactive Example 18.2 **Predicting the Sign of the Entropy Change of a Reaction**

Gaining Mastery Toolbox

Critical Concept 18.2
The entropy of a system usually increases during a change that involves one or more of the following processes: a molecule breaks into smaller parts, moles of gas increase, a solid changes to a liquid or a gas, or a liquid changes to a gas. The entropy usually decreases if the change involves a reverse of any of these processes.

Solution Essentials:
• Second law of thermodynamics
• Entropy, S

a. The following equation represents the essential change that takes place during the fermentation of glucose (grape sugar) to ethanol (ethyl alcohol).

$$\underset{\text{glucose}}{C_6H_{12}O_6(s)} \longrightarrow \underset{\text{ethanol}}{2C_2H_5OH(l)} + 2CO_2(g)$$

Is $\Delta S°$ positive or negative? Explain.

b. Do you expect the entropy to increase or decrease in the preparation of urea from NH_3 and CO_2?

$$2NH_3(g) + CO_2(g) \longrightarrow NH_2CONH_2(aq) + H_2O(l)$$

Explain.

(continued)

Table 18.1   Standard Entropies (at 25°C)*

Substance or Ion	$S°$ J/(mol·K)	Substance or Ion	$S°$ J/(mol·K)	Substance or Ion	$S°$ J/(mol·K)
$e^-(g)$	20.87	*Aldehydes (continued)*		**Nitrogen** *(continued)*	
Bromine		$CH_3CHO(g)$	246.4	$NH_3(g)$	192.7
$Br(g)$	174.9	$CH_3CHO(l)$	160.4	$NH_4^+(aq)$	113.4
$Br^-(aq)$	82.4	**Chlorine**		$NO(g)$	210.6
$Br^-(g)$	163.4	$Cl(g)$	165.1	$NO_2(g)$	239.9
$Br_2(g)$	245.3	$Cl^-(aq)$	56.5	$HNO_3(aq)$	146.4
$Br_2(l)$	152.2	$Cl^-(g)$	153.2	**Oxygen**	
$HBr(g)$	198.6	$Cl_2(g)$	223.0	$O(g)$	160.9
Calcium		$HCl(g)$	186.8	$O_2(g)$	205.0
$Ca(s)$	41.59	**Fluorine**		$O_3(g)$	238.8
$Ca^{2+}(aq)$	−53.1	$F(g)$	158.6	**Silicon**	
$CaCO_3(s, calcite)$	92.9	$F^-(g)$	145.5	$Si(s)$	18.82
$CaO(s)$	38.21	$F^-(aq)$	−13.8	$SiCl_4(l)$	239.7
Carbon		$F_2(g)$	202.7	$SiF_4(g)$	282.7
$C(g)$	158.0	$HF(g)$	173.7	$SiO_2(s, quartz)$	41.46
$C(s, diamond)$	2.377	**Hydrogen**		**Silver**	
$C(s, graphite)$	5.740	$H(g)$	114.6	$Ag(s)$	42.55
$CCl_4(g)$	309.7	$H^+(aq)$	0	$Ag^+(aq)$	72.68
$CCl_4(l)$	216.4	$H^+(g)$	108.8	$AgBr(s)$	107.1
$CO(g)$	197.5	$H_2(g)$	130.6	$AgCl(s)$	96.2
$CO_2(g)$	213.7	$H_2O(g)$	188.7	$AgF(s)$	83.7
$CO_3^{2-}(aq)$	−56.9	$H_2O(l)$	69.95	$AgI(s)$	115.5
$CS_2(g)$	237.9	$OH^-(aq)$	−10.75	**Sodium**	
$CS_2(l)$	151.3	**Iodine**		$Na(g)$	153.6
$HCN(g)$	201.7	$I(g)$	180.7	$Na(s)$	51.46
$HCN(l)$	112.8	$I^-(aq)$	109.6	$Na^+(aq)$	59.1
$HCO_3^-(aq)$	91.2	$I^-(g)$	169.2	$Na^+(g)$	147.8
Hydrocarbons		$I_2(s)$	116.1	$Na_2CO_3(s)$	138.8
$CH_4(g)$	186.1	$HI(g)$	206.5	$NaCl(s)$	72.12
$C_2H_4(g)$	219.2	**Lead**		$NaHCO_3(s)$	101.7
$C_2H_6(g)$	229.5	$Pb(s)$	64.78	**Sulfur**	
$C_6H_6(l)$	173.4	$Pb^{2+}(aq)$	10.5	$S(g)$	167.7
Alcohols		$PbO(s)$	66.32	$S(s, monoclinic)$	33.03
$CH_3OH(l)$	126.8	$PbS(s)$	91.34	$S(s, rhombic)$	32.06
$C_2H_5OH(l)$	160.7	**Nitrogen**		$S_2(g)$	228.1
Aldehydes		$N(g)$	153.2	$SO_2(g)$	248.1
$HCHO(g)$	219.0	$N_2(g)$	191.6	$H_2S(g)$	205.6

*See Appendix C for additional values.

(continued)

c What is the sign of $\Delta S°$ for the following reaction?

$$CO(g) + H_2O(g) \longrightarrow CO_2(g) + H_2(g)$$

Problem Strategy Compare the reactants and products in each reaction. Look for the following: molecules breaking into smaller parts or aggregrating into larger parts, moles of gas increasing or decreasing, and changes of state. Then apply the three rules preceding this example.

Solution

a A molecule (glucose) breaks into smaller molecules (C_2H_5OH and CO_2). Moreover, this results in a gas being released. You predict that $\Delta S°$ for this reaction is **positive**. That is, the entropy increases.

b In this reaction, the moles of gas decrease (by 3 mol), which would decrease the entropy. You predict that the entropy should **decrease**. That is, $\Delta S°$ is negative.

c Because there is no change in the number of moles of gas, **you cannot predict the sign of $\Delta S°$ from the rules given**.

Answer Check Review the situations listed before this example, and note that the entropy increases whenever a molecule breaks up, moles of gas increase, or a phase change occurs from solid to liquid or from liquid to gas.

Exercise 18.2 Predict the sign of $\Delta S°$ for each of the following reactions.

a $CaCO_3(s) \longrightarrow CaO(s) + CO_2(g)$

b $CS_2(l) \longrightarrow CS_2(g)$

c $2Hg(l) + O_2(g) \longrightarrow 2HgO(s)$

d $2Na_2O_2(s) + 2H_2O(l) \longrightarrow 4NaOH(aq) + O_2(g)$

■ See Problems 18.39 and 18.40.

It is useful to be able to predict the sign of $\Delta S°$. You gain some understanding of the reaction, and you can use the prediction for qualitative work. For quantitative work, however, you need to find the value of $\Delta S°$. You can find the standard change of entropy, $\Delta S°$, for a reaction by subtracting the standard entropies of reactants from the standard entropies of products, similar to the way you obtained $\Delta H°$.

$$\Delta S° = \Sigma nS°(\text{products}) - \Sigma mS°(\text{reactants})$$

The next example illustrates the calculations.

ⓌWL Interactive Example 18.3   Calculating $\Delta S°$ for a Reaction

Gaining Mastery Toolbox

Critical Concept 18.3
You calculate the standard entropy change for a physical or chemical change from the standard entropies of reactants and products similar to the way in which you calculate the standard enthalpy of reaction from standard enthalpies of formation. $\Delta S° = \Sigma nΔS°(\text{products}) - \Sigma mΔS°(\text{reactants})$

Solution Essentials:
• Standard entropy, $S°$
• Entropy, S

Calculate the change of entropy, $\Delta S°$, at 25°C for the reaction in which urea is formed from NH_3 and CO_2.

$$2NH_3(g) + CO_2(g) \longrightarrow NH_2CONH_2(aq) + H_2O(l)$$

The standard entropy of $NH_2CONH_2(aq)$ is 174 J/(mol·K). See Table 18.1 for other values.

Problem Strategy The calculation is similar to that used to obtain $\Delta H°$ from $\Delta H_f°$ values.

Solution It is convenient to put the standard entropy values multiplied by stoichiometric coefficients below the formulas in the balanced equation.

$$2NH_3(g) + CO_2(g) \longrightarrow NH_2CONH_2(aq) + H_2O(l)$$
$$S°: 2 \times 193 \quad 214 \quad\quad 174 \quad\quad\quad 70$$

You calculate the entropy change by subtracting the entropy of the reactants from the entropy of the products.

$$\Delta S° = \Sigma nS°(\text{products}) - \Sigma mS°(\text{reactants})$$
$$= [(174 + 70) - (2 \times 193 + 214)] \text{ J/K} = -356 \text{ J/K}$$

Answer Check Carefully review your arithmetic. Also, note that moles of gas decrease, so we expect that the entropy decreases (and, thus, the sign should be negative; see Example 18.2b).

Exercise 18.3 Calculate the change of entropy, $\Delta S°$, for the reaction given in Example 18.2a. The standard entropy of glucose, $C_6H_{12}O_6(s)$, is 212 J/(mol·K). See Table 18.1 for other values.

■ See Problems 18.41 and 18.42.

Free-Energy Concept

At the end of Section 18.2, you saw that the quantity $\Delta H - T\Delta S$ can serve as a criterion for spontaneity of a reaction at constant temperature and pressure. If the value of this quantity is negative, the reaction is spontaneous. If it is positive, the reaction is nonspontaneous. If it equals zero, the reaction is at equilibrium.

As an application of this criterion, consider the reaction described in the chapter opening, in which urea is prepared from NH_3 and CO_2. The heat of reaction $\Delta H°$ was calculated from enthalpies of formation in Section 18.1, where we obtained -119.7 kJ. Then, in Example 18.3, we calculated the entropy of reaction $\Delta S°$ and found a value of -356 J/K, or -0.356 kJ/K. Substitute these values and $T = 298$ K (25°C) into the expression $\Delta H° - T\Delta S°$.

$$\Delta H° - T\Delta S° = (-119.7 \text{ kJ}) - (298 \text{ K})(-0.356 \text{ kJ/K}) = -13.6 \text{ kJ}$$

We see that $\Delta H° - T\Delta S°$ is a negative quantity, from which we conclude that the reaction is spontaneous under standard conditions.

18.4 Free Energy and Spontaneity

It is very convenient to define a new thermodynamic quantity in terms of H and S that will be directly useful as a criterion of spontaneity. For this purpose, the American physicist J. Willard Gibbs (1839–1903) introduced the concept of **free energy, G,** which is *a thermodynamic quantity defined by the equation $G = H - TS$*. This quantity gives a direct criterion for spontaneity of reaction. ▶

The quantity G is called Gibbs energy, Gibbs free energy, or the Gibbs function—hence the symbol G.

As a reaction proceeds at a given temperature and pressure, reactants form products and the enthalpy H and entropy S change. These changes in H and S, denoted ΔH and ΔS, result in a change in free energy, ΔG, given by the equation

$$\Delta G = \Delta H - T\Delta S$$

Note that the change in free energy, ΔG, equals the quantity $\Delta H - T\Delta S$ that you just saw serves as a criterion for spontaneity of a reaction. Thus, if you can show that ΔG for a reaction at a given temperature and pressure is negative, you can predict that the reaction will be spontaneous.

Standard Free-Energy Change

Recall that for purposes of tabulating thermodynamic data, certain *standard states* are chosen, which are indicated by a superscript degree sign on the symbol of the quantity. The standard states are as follows: for pure liquids and solids, 1 atm pressure; for gases, 1 atm partial pressure; for solutions, 1 *M* concentration. The temperature is the temperature of interest, usually 25°C (298 K).

The standard free-energy change, $\Delta G°$, is the free-energy change that occurs when reactants in their standard states are converted to products in their standard states. Example 18.4 illustrates the calculation of the standard free-energy change, $\Delta G°$, from $\Delta H°$ and $\Delta S°$.

$$\Delta G° = \Delta H° - T\Delta S°$$

⚙WL Interactive Example 18.4 **Calculating $\Delta G°$ from $\Delta H°$ and $\Delta S°$**

Gaining Mastery Toolbox

Critical Concept 18.4
The standard free-energy change for a physical or chemical change equals the standard enthalpy change minus the absolute temperature times the standard entropy change, $\Delta G° = \Delta H° - T\Delta S°$.

Solution Essentials:
- Standard free-energy change, $\Delta G°$
- Free-energy change, ΔG
- Standard entropy change, $\Delta S°$
- Standard enthalpy of formation, $\Delta H_f°$

What is the standard free-energy change, $\Delta G°$, for the following reaction at 25°C?

$$N_2(g) + 3H_2(g) \longrightarrow 2NH_3(g)$$

Use values of $\Delta H_f°$ and $S°$ from Tables 6.2 and 18.1.

Problem Strategy Calculate $\Delta H°$ and $\Delta S°$, then substitute these values into $\Delta H° - T\Delta S°$ to obtain $\Delta G°$.

(continued)

(continued)

Solution Write the balanced equation and place below each formula the values of ΔH_f° and S° multiplied by stoichiometric coefficients.

$$N_2(g) + 3H_2(g) \longrightarrow 2NH_3(g)$$

ΔH_f°:   0   0 $2 \times (-45.9)$ kJ

S°: 191.6 3×130.6 2×192.7 J/K

You calculate ΔH° and ΔS° by taking values for products and subtracting values for reactants.

$$\Delta H^{\circ} = \Sigma\, n\Delta H_f^{\circ}(\text{products}) - \Sigma\, m\Delta H_f^{\circ}(\text{reactants})$$

$$= [2 \times (-45.9) - 0]\,\text{kJ} = -91.8\,\text{kJ}$$

$$\Delta S^{\circ} = \Sigma\, nS^{\circ}(\text{products}) - \Sigma\, mS^{\circ}(\text{reactants})$$

$$= [2 \times 192.7 - (191.6 + 3 \times 130.6)]\,\text{J/K} = -198.0\,\text{J/K}$$

You now substitute into the equation for ΔG° in terms of ΔH° and ΔS°. Note that you substitute ΔS° in units of kJ/K.

$$\Delta G^{\circ} = \Delta H^{\circ} - T\Delta S^{\circ} = -91.8\,\text{kJ} - (298\,\text{K})(-0.1980\,\text{kJ/K}) = \mathbf{-32.8\,kJ}$$

Answer Check Check your arithmetic. You need to be careful about the signs, especially when calculating ΔH. Individual enthalpies of formation may be positive or negative. Also, you will be subtracting reactant data from product data.

Exercise 18.4 Calculate ΔG° for the following reaction at 25°C. Use data given in Tables 6.2 and 18.1.

$$CH_4(g) + 2O_2(g) \longrightarrow CO_2(g) + 2H_2O(g)$$

■ See Problems 18.45 and 18.46.

Standard Free Energies of Formation

The **standard free energy of formation, ΔG_f°**, of a substance is defined similarly to the standard enthalpy of formation. That is, ΔG_f° is *the free-energy change that occurs when 1 mol of substance is formed from its elements in their reference forms (usually the stablest states) at 1 atm and at a specified temperature* (usually 25°C). (For a species in solution, the standard state is the 1 *M* solution.) For example, the standard free energy of formation of $NH_3(g)$ is the free-energy change for the reaction

$$\tfrac{1}{2}N_2(g) + \tfrac{3}{2}H_2(g) \longrightarrow NH_3(g)$$

The reactants, N_2 and H_2, each at 1 atm, are converted to the product, NH_3, at 1 atm pressure. In Example 18.4, you found ΔG° for the formation of 2 mol NH_3 from its elements to be -32.8 kJ. Hence, $\Delta G_f^{\circ}(NH_3) = -32.8$ kJ/2 mol $= -16.4$ kJ/mol.

As in the case of standard enthalpies of formation, the standard free energies of formation of elements in their stablest states are assigned the value zero. By tabulating ΔG_f° for substances, as in Table 18.2, you can easily calculate ΔG° for any reaction involving those substances. You simply subtract the standard free energies of reactants from the standard free energies of products:

$$\Delta G^{\circ} = \Sigma\, n\Delta G_f^{\circ}(\text{products}) - \Sigma\, m\Delta G_f^{\circ}(\text{reactants})$$

The next example illustrates the calculations.

Table 18.2 Standard Free Energies of Formation (at 25°C)*

Substance or Ion	ΔG_f° (kJ/mol)	Substance or Ion	ΔG_f° (kJ/mol)	Substance or Ion	ΔG_f° (kJ/mol)
$e^-(g)$	0	*Aldehydes (continued)*		**Nitrogen** *(continued)*	
Bromine		$CH_3CHO(g)$	−133.4	$NH_3(g)$	−16.40
$Br(g)$	82.40	$CH_3CHO(l)$	−128.3	$NH_4^+(aq)$	−79.37
$Br^-(aq)$	−104.0	**Chlorine**		$NO(g)$	86.60
$Br^-(g)$	−238.8	$Cl(g)$	105.3	$NO_2(g)$	51.24
$Br_2(g)$	3.159	$Cl^-(aq)$	−131.3	$HNO_3(aq)$	−111.3
$Br_2(l)$	0	$Cl^-(g)$	−240.2	**Oxygen**	
$HBr(g)$	−53.50	$Cl_2(g)$	0	$O(g)$	231.8
Calcium		$HCl(g)$	−95.30	$O_2(g)$	0
$Ca(s)$	0	**Fluorine**		$O_3(g)$	163.2
$Ca^{2+}(aq)$	−553.5	$F(g)$	62.31	**Silicon**	
$CaCO_3(s, \text{calcite})$	−1128.8	$F^-(g)$	−262.0	$Si(s)$	0
$CaO(s)$	−603.5	$F^-(aq)$	−278.8	$SiCl_4(l)$	−619.9
Carbon		$F_2(g)$	0	$SiF_4(g)$	−1572.7
$C(g)$	671.3	$HF(g)$	−274.6	$SiO_2(s, \text{quartz})$	−856.4
$C(s, \text{diamond})$	2.900	**Hydrogen**		**Silver**	
$C(s, \text{graphite})$	0	$H(g)$	203.3	$Ag(s)$	0
$CCl_4(g)$	−53.65	$H^+(aq)$	0	$Ag^+(aq)$	77.12
$CCl_4(l)$	−65.27	$H^+(g)$	1517.0	$AgBr(s)$	−96.90
$CO(g)$	−137.2	$H_2(g)$	0	$AgCl(s)$	−109.8
$CO_2(g)$	−394.4	$H_2O(g)$	−228.6	$AgF(s)$	—
$CO_3^{2-}(aq)$	−527.9	$H_2O(l)$	−237.1	$AgI(s)$	−66.19
$CS_2(g)$	66.85	$OH^-(aq)$	−157.3	**Sodium**	
$CS_2(l)$	65.27	**Iodine**		$Na(g)$	76.86
$HCN(g)$	124.7	$I(g)$	70.21	$Na(s)$	0
$HCN(l)$	124.9	$I^-(aq)$	−51.59	$Na^+(aq)$	−261.9
$HCO_3^-(aq)$	−586.8	$I^-(g)$	−221.5	$Na^+(g)$	574.4
Hydrocarbons		$I_2(s)$	0	$Na_2CO_3(s)$	−1048.0
$CH_4(g)$	−50.80	$HI(g)$	1.576	$NaCl(s)$	−384.0
$C_2H_4(g)$	68.39	**Lead**		$NaHCO_3(s)$	−851.0
$C_2H_6(g)$	−32.89	$Pb(s)$	0	**Sulfur**	
$C_6H_6(l)$	124.4	$Pb^{2+}(aq)$	−24.39	$S(g)$	236.5
Alcohols		$PbO(s)$	−189.3	$S(s, \text{monoclinic})$	0.070
$CH_3OH(l)$	−166.4	$PbS(s)$	−96.68	$S(s, \text{rhombic})$	0
$C_2H_5OH(l)$	−174.9	**Nitrogen**		$S_2(g)$	79.7
Aldehydes		$N(g)$	455.6	$SO_2(g)$	−300.1
$HCHO(g)$	−113	$N_2(g)$	0	$H_2S(g)$	−33.33

*See Appendix C for additional values.

OWL Interactive Example 18.5 Calculating $\Delta G°$ from Standard Free Energies of Formation

Gaining Mastery Toolbox

Critical Concept 18.5
You calculate the standard free-energy change for a physical or chemical change from the standard free energies of formation of reactants and products similar to the way in which you calculate the standard enthalpy of reaction from standard enthalpies of formation.
$\Delta G° = \Sigma\, n\Delta G_f°(\text{products}) - \Sigma\, m\Delta G_f°(\text{reactants})$

Solution Essentials:
- Standard free-energy of formation, $\Delta G_f°$
- Standard free-energy change, $\Delta G°$
- Free-energy change, ΔG

Calculate $\Delta G°$ for the combustion of 1 mol of ethanol, C_2H_5OH, at 25°C.

$$C_2H_5OH(l) + 3O_2(g) \longrightarrow 2CO_2(g) + 3H_2O(g)$$

Use the standard free energies of formation given in Table 18.2.

Problem Strategy Calculate $\Delta G°$ from $\Delta G_f°$ values, similar to the way you calculate $\Delta H°$ from $\Delta H_f°$.

Solution Write the balanced equation with values of $\Delta G_f°$ multiplied by stoichiometric coefficients below each formula.

	$C_2H_5OH(l)$	$+\ 3O_2(g)$	$\longrightarrow\ 2CO_2(g)$	$+\ 3H_2O(g)$
$\Delta G_f°$:	-174.9	0	$2(-394.4)$	$3(-228.6)$ kJ

The calculation is

$$\begin{aligned}\Delta G° &= \Sigma\, n\Delta G_f°(\text{products}) - \Sigma\, m\Delta G_f°(\text{reactants})\\ &= [2(-394.4) + 3(-228.6) - (-174.9)]\text{ kJ}\\ &= \mathbf{-1299.7\ kJ}\end{aligned}$$

Answer Check As in calculating ΔH values, you need to be careful of signs. Also, note that you can obtain standard free energies from Appendix C. Just be careful to use the correct column for $\Delta G°$ values.

Exercise 18.5 Calculate $\Delta G°$ at 25°C for the following reaction, using values of $\Delta G_f°$.

$$CaCO_3(s) \longrightarrow CaO(s) + CO_2(g)$$

See Problems 18.49 and 18.50.

$\Delta G°$ as a Criterion for Spontaneity

You have already seen that the quantity $\Delta G = \Delta H - T\Delta S$ can be used as a criterion for the spontaneity of a reaction. The change of free energy, ΔG, should be calculated for the conditions at which the reaction occurs. If the reactants are at standard conditions and give products at standard conditions, the free-energy change you need to look at is $\Delta G°$. The calculation is simple, as shown in Example 18.5. For other conditions, you should look at the appropriate ΔG value. This would be a more complicated calculation. Nevertheless, the standard free-energy change $\Delta G°$ is still a useful guide to the spontaneity of reaction in these cases. The following rules are useful in judging the spontaneity of a reaction:

1. When $\Delta G°$ is a large negative number (more negative than about -10 kJ), the reaction is spontaneous as written, and reactants transform almost entirely to products when equilibrium is reached.
2. When $\Delta G°$ is a large positive number (larger than about 10 kJ), the reaction is nonspontaneous as written, and reactants do not give significant amounts of products at equilibrium.
3. When $\Delta G°$ has a small negative or positive value (less than about 10 kJ), the reaction gives an equilibrium mixture with significant amounts of both reactants and products.

☯WL Interactive Example 18.6   **Interpreting the Sign of $\Delta G°$**

Calculate $\Delta H°$ and $\Delta G°$ for the reaction

$$2KClO_3(s) \longrightarrow 2KCl(s) + 3O_2(g)$$

Interpret the signs obtained for $\Delta H°$ and $\Delta G°$. Values of $\Delta H_f°$ (in kJ/mol) are as follows: $KClO_3(s)$, -397.7; $KCl(s)$, -436.7. Similarly, values of $\Delta G_f°$ (in kJ/mol) are as follows: $KClO_3(s)$, -296.3; $KCl(s)$, -408.8. Note that $O_2(g)$ is the reference form of the element, so $\Delta H_f° = \Delta G_f° = 0$ for it.

Problem Strategy Calculate $\Delta H°$ and $\Delta G°$; then interpret the signs of these quantities. The interpretation of the sign of ΔH was discussed in Chapter 6. As long as $\Delta G°$ is not close to zero, you can interpret its sign as follows: a negative $\Delta G°$ means the reactants tend to go mostly to products; a positive $\Delta G°$ means that the reaction is nonspontaneous as written. When the value of $\Delta G°$ is close to zero, the reaction gives an equilibrium mixture.

Solution The problem is set up as follows:

$$2KClO_3(s) \longrightarrow 2KCl(s) + 3O_2(g)$$

$\Delta H_f°$: $2 \times (-397.7)$ $2 \times (-436.7)$ 0 kJ

$\Delta G_f°$: $2 \times (-296.3)$ $2 \times (-408.8)$ 0 kJ

Then,

$$\Delta H° = [2 \times (-436.7) - 2 \times (-397.7)] \text{ kJ} = \mathbf{-78.0 \text{ kJ}}$$

$$\Delta G° = [2 \times (-408.8) - 2 \times (-296.3)] \text{ kJ} = \mathbf{-225.0 \text{ kJ}}$$

The reaction is exothermic, liberating 78.0 kJ of heat. The large negative value for $\Delta G°$ indicates that the equilibrium composition is mostly potassium chloride and oxygen.

Answer Check Check your arithmetic. Then remember that a negative sign for $\Delta H°$ means that heat was lost from the system; a large negative sign for $\Delta G°$ means that products predominate.

Exercise 18.6   Which of the following reactions are spontaneous in the direction written? See Table 18.2 for data.

a $C(graphite) + 2H_2(g) \longrightarrow CH_4(g)$
b $2H_2(g) + O_2(g) \longrightarrow 2H_2O(l)$
c $4HCN(g) + 5O_2(g) \longrightarrow 2H_2O(l) + 4CO_2(g) + 2N_2(g)$
d $Ag^+(aq) + I^-(aq) \longrightarrow AgI(s)$

■ See Problems 18.53 and 18.54.

CONCEPT CHECK 18.2

Consider the reaction of nitrogen, N_2, and oxygen, O_2, to form nitrogen monoxide, NO:

$$N_2(g) + O_2(g) \longrightarrow 2NO(g)$$

From the standard free energy of formation of NO, what can you say about this reaction?

18.5 Interpretation of Free Energy

You have seen that the free-energy change serves as a criterion for spontaneity of a chemical reaction. This gives you some idea of what free energy is and how you interpret it. In this section, we will look more closely at the meaning of free energy.

Maximum Work

Theoretically, spontaneous reactions can be used to obtain useful work. By useful work, we mean energy that can be used directly to move objects of normal size. We use the combustion of gasoline to move an automobile, and we use a reaction in a battery to generate electricity to drive a motor. Similarly, biochemical reactions in muscle tissue occur in such a way as to contract muscle fibers and lift a weight.

Coupling of Reactions

Many important processes are thermodynamically unfavorable, or *nonspontaneous*. Assembling many small amino acid molecules into a large protein molecule is an example of a nonspontaneous process. Yet biological organisms do assemble proteins from amino acids. Biological organisms accomplish many similar nonspontaneous processes. How do they do this?

Let's look at a simpler example. We would like to obtain iron metal from iron ore. Here is the equation for the direct decomposition of iron(III) oxide, the iron-containing substance in the iron ore hematite:

$$2Fe_2O_3(s) \longrightarrow 4Fe(s) + 3O_2(g); \Delta G° = +1487 \text{ kJ}$$

This reaction is clearly nonspontaneous, because $\Delta G°$ is a large positive quantity. The nonspontaneity of this reaction is in agreement with common knowledge. Iron tends to rust (forms iron oxide) in air, by combining with oxygen. You do not expect a rusty wrench to turn spontaneously into shiny iron and oxygen. But this does not mean you cannot change iron(III) oxide to iron metal. It merely means that you have to do work on the iron(III) oxide to reduce it to the metal. You must, in effect, find a way to couple this nonspontaneous reaction to one that is sufficiently spontaneous; that is, you must couple this reaction to one having a more negative $\Delta G°$.

Consider the spontaneous reaction of carbon monoxide with oxygen (the burning of CO):

$$2CO(g) + O_2(g) \longrightarrow 2CO_2(g); \Delta G° = -514.4 \text{ kJ}$$

For 3 mol O_2, $\Delta G°$ is -1543 kJ, which is more negative than for the direct decomposition of 2 mol Fe_2O_3 to its elements. Let us add the two reactions:

$$2Fe_2O_3(s) \longrightarrow 4Fe(s) + 3O_2(g); \Delta G° = 1487 \text{ kJ}$$
$$6CO(g) + 3O_2(g) \longrightarrow 6CO_2(g); \Delta G° = -1543 \text{ kJ}$$

$$\overline{2Fe_2O_3(s) + 6CO(g) \longrightarrow 4Fe(s) + 6CO_2(g); \Delta G° = -56 \text{ kJ}}$$

The net result is to reduce iron(III) oxide to the metal by reacting the oxide with carbon monoxide. This is the reaction that occurs in a blast furnace, where iron ore is commercially reduced to iron.

The concept of the coupling of two chemical reactions, a nonspontaneous reaction with a spontaneous one, to give an overall spontaneous change is a very useful one, especially in biochemistry. Adenosine triphosphate, or ATP, is a molecule containing a triphosphate group. It reacts with water, in the presence of an enzyme (biochemical catalyst), to give adenosine diphosphate, ADP, and a phosphate ion. (Figure 18.7 shows the structure of ATP and ADP.) The reaction has a negative $\Delta G°$ (and is a spontaneous reaction).

$$ATP + H_2O \longrightarrow ADP + \text{phosphate ion}; \Delta G° = -31 \text{ kJ}$$

ATP is first synthesized in a living organism; the energy for this synthesis is obtained from food. The spontaneous reaction of ATP to give ADP is then coupled to various nonspontaneous reactions in the organism to accomplish necessary life processes (see Figure 18. 8).

Often reactions are not carried out in a way that does useful work. The reactants are simply poured together in a reaction vessel, and products are separated from the mixture. As the reaction occurs, the free energy of the system decreases and entropy is produced, but no useful work is obtained.

In principle, if a reaction is carried out to obtain the maximum useful work, no entropy is produced. It can be shown that the maximum useful work, w_{max}, for a spontaneous reaction is ΔG.

$$w_{max} = \Delta G$$

The term *free energy* comes from this result. *The free-energy change is the maximum energy available, or free, to do useful work.* As a reaction occurs in such a way as to give the maximum useful work, the free energy decreases and a corresponding quantity of useful work is obtained. ◄

The concept of maximum work from a chemical reaction is an idealization. In any real situation, less than this maximum work is obtained and some entropy

The sign of w (work) is defined so that a negative value means work is obtained from the system (that is, energy is subtracted from the system). You can obtain work from a reaction if its ΔG is negative.

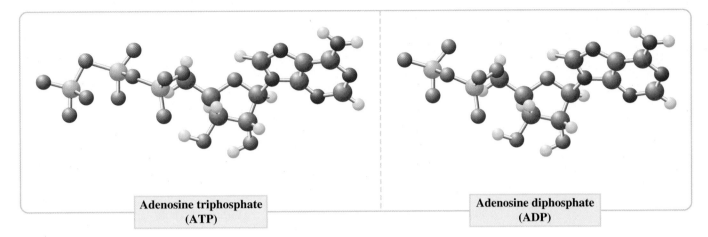

Adenosine triphosphate
(ATP)

Adenosine diphosphate
(ADP)

Figure 18.7 ▲

Molecular models of ATP and ADP The phosphate groups are on the left, with the phosphorus atoms shown in orange, oxygen atoms in red, nitrogen atoms in blue, carbon atoms in gray, and hydrogen atoms in light blue.

Figure 18.8 ▶

Coupling of reactions in a biological cell The mitochondria are parts of a cell that obtain energy in the form of ATP from glucose (from food), $C_6H_{12}O_6$. ATP molecules are used in various cell processes, such as protein synthesis from amino acids, which occurs in the ribosomes.

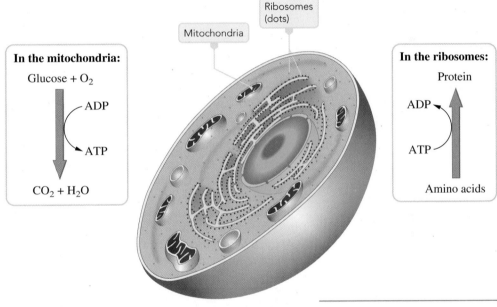

In the mitochondria:

Glucose + O_2

ADP

ATP

$CO_2 + H_2O$

Ribosomes (dots)

Mitochondria

In the ribosomes:

Protein

ADP

ATP

Amino acids

■ See Problems 18.101 and 18.102.

is created. When this work is eventually expended, it appears in the environment as additional entropy.

Free-Energy Change During Reaction

You have seen that the free-energy change is related to the work done during a chemical reaction. Consider the combustion of gasoline in O_2. The reaction is spontaneous, so the free-energy change is negative. That is, the free energy of the system changes to a lower value as the reactants are converted to products. Figure 18.9 shows the free-energy change that occurs during this reaction. When the gasoline is burned in a gasoline stove, the decrease in free energy shows up as an increase in entropy of the system and the surroundings. However, when the gasoline is burned in an automobile engine, some of the decrease in free energy shows up as work done. Theoretically, all of the free-energy decrease can be used to do work. This gives the maximum work. In practice, less work is obtained, and the difference appears as an increase of entropy. Ultimately, the work is itself used up—that is, changed to entropy. ▶

What happens in the gasoline engine is complicated, but you need look at only the initial and final states. Reactants at the surrounding temperature and pressure go into the engine, which produces work and heat, and the products leave the engine, eventually reaching the initial temperature and pressure of the reactants.

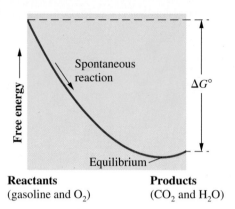

Figure 18.9 ▲

Free-energy change during a spontaneous reaction (combustion of gasoline) The free energy decreases as the reaction proceeds. At equilibrium, the free energy becomes a minimum, and the equilibrium mixture is mostly products.

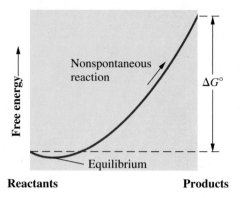

Figure 18.10 ▲

Free-energy change during a nonspontaneous reaction The free energy decreases until equilibrium, at which the minimum value is reached. There is very little reaction, however, because the equilibrium mixture is mostly reactants. To change reactants to mostly products is to undergo a nonspontaneous reaction.

Look at Figure 18.9 again. At the start of the reaction, the system contains only reactants: gasoline and O_2. The free energy has a relatively high value. This decreases as the reaction proceeds. The decrease in free energy appears either as an increase in entropy or as work done. Eventually, the free energy of the system reaches its minimum value. Then the net reaction stops; it comes to equilibrium.

Before we leave this section, consider a reaction in which $\Delta G°$ is positive. You would predict that the reaction is nonspontaneous. Figure 18.10 shows the free-energy change as the reaction proceeds. Note that there is a small decrease in free energy as the system goes to equilibrium. Some reaction occurs in order to give the equilibrium mixture. But this mixture consists primarily of reactants, because the reaction does not go very far before coming to equilibrium. To change reactants completely to products is a nonspontaneous reaction (shown by the arrow along the curve in Figure 18.10).

Free Energy and Equilibrium Constants

In the previous section, you saw that for any spontaneous reaction, the total free energy of the substances in the reaction mixture decreases as the reaction proceeds (that is, the free-energy change is negative). Earlier, you saw that the standard free-energy change, $\Delta G°$, can be used as a criterion for the spontaneity of a reaction. When $\Delta G°$ is negative, the reaction is spontaneous. We will now describe how the standard free-energy change is related to the equilibrium constant.

18.6 Relating $\Delta G°$ to the Equilibrium Constant

One of the most important results of chemical thermodynamics is an equation relating the standard free-energy change for a reaction to the equilibrium constant. Before we look at this equation, we must discuss the thermodynamic form of the equilibrium constant that occurs in that equation.

The **thermodynamic equilibrium constant, K,** is *the equilibrium constant in which the concentrations of gases are expressed in partial pressures in atmospheres, whereas the concentrations of solutes in liquid solutions are expressed in molarities.* For a reaction involving only solutes in liquid solution, K is identical to K_c; for reactions involving only gases, K equals K_p. ◄

To be precise, K is defined in terms of activities, which are dimensionless quantities numerically equal to "effective" concentrations and pressures.

⭘WL Interactive Example 18.7   Writing the Expression for a Thermodynamic Equilibrium Constant

Critical Concept 18.7
You write the expression for the thermodynamic equilibrium constant with gases expressed as partial pressures and concentrations of solutes as molarities.

Solution Essentials:
- Equilibrium-constant expression for gases K_p
- Equilibrium-constant expression K_c

Write expressions for the thermodynamic equilibrium constants for each of the following reactions:

a The reaction given in the chapter opening,

$$2NH_3(g) + CO_2(g) \rightleftharpoons NH_2CONH_2(aq) + H_2O(l)$$

b The solubility process

$$AgCl(s) \rightleftharpoons Ag^+(aq) + Cl^-(aq)$$

Problem Strategy The thermodynamic equilibrium constant, K, has characteristics of K_p and K_c. All gases appear in K as partial pressures; all solutes appear as concentrations. Pure solids and liquids do not appear in the equilibrium constant.

Solution

a Note that H_2O is a solvent, so it does not appear explicitly in K. The gases appear in K as partial pressures, and the solute appears as a molar concentration.

$$K = \frac{[NH_2CONH_2]}{P^2_{NH_3}P_{CO_2}}$$

b Note that AgCl is a solid, so it does not appear explicitly in K. The solutes appear as molar concentrations.

$$K = [Ag^+][Cl^-]$$

This is identical to K_{sp}.

Answer Check Make sure that the products are on top and the reactants are on the bottom of each equilibrium-constant expression. Then check that each reactant and product term is raised to the correct power (equal to the stoichiometric coefficient in the chemical equation).

Exercise 18.7   Give the expression for K for each of the following reactions.

a $CaCO_3(s) \rightleftharpoons CaO(s) + CO_2(g)$
b $PbI_2(s) \rightleftharpoons Pb^{2+}(aq) + 2I^-(aq)$
c $H^+(aq) + HCO_3^-(aq) \rightleftharpoons H_2O(l) + CO_2(g)$

■ See Problems 18.59 and 18.60.

The standard free-energy change, $\Delta G°$, for a reaction can be calculated using data from thermodynamic tables. If you want the free-energy change when reactants under nonstandard conditions are changed to products under nonstandard conditions (ΔG), you can obtain it from the standard free-energy change $\Delta G°$ using the following equation:

$$\Delta G = \Delta G° + RT \ln Q$$

Here Q is the thermodynamic form of the reaction quotient. It has the same general appearance as the thermodynamic equilibrium constant K, but the concentrations and partial pressures are those for a mixture at some instant, perhaps at the beginning of a reaction. You obtain ΔG from $\Delta G°$ by adding $RT \ln Q$, where $\ln Q$ is the natural logarithm of Q ($\ln Q = 2.303 \log Q$). Note that ΔG is the change of free energy that occurs when the reaction mixture has the composition expressed by Q. ▶

You can derive the equation relating $\Delta G°$ to the equilibrium constant K from the preceding equation. In the previous section, you saw that as a chemical reaction approaches equilibrium, the free energy decreases and continues to decrease until equilibrium is reached. At equilibrium, the free energy ceases to change; then $\Delta G = 0$. Also, the reaction quotient Q becomes equal to the equilibrium constant K. If you substitute $\Delta G = 0$ and $Q = K$ into the preceding equation, you obtain

$$0 = \Delta G° + RT \ln K$$

This result is easily rearranged to give the basic equation relating the standard free-energy change to the equilibrium constant.

$$\Delta G° = -RT \ln K$$

Strictly speaking, because the composition changes with reaction, ΔG is for an infinitesimal change in G per infinitesimal change in reaction amounts; that is, ΔG equals the slope of the reaction curve (like the curves in Figures 18.9 and 18.10).

The next example illustrates the calculation of a thermodynamic equilibrium constant from the standard free-energy change, $\Delta G°$.

⏱WL Interactive Example 18.8 Calculating *K* from the Standard Free-Energy Change (Molecular Equation)

Gaining Mastery Toolbox

Critical Concept 18.8
The standard free-energy change and the thermodynamic equilibrium constant for a reaction at a given temperature are related: $\Delta G° = -RT \ln K$.

Solution Essentials:
• Thermodynamic equilibrium constant, K
• Standard free-energy change, $\Delta G°$

Find the value of the equilibrium constant K at 25°C (298 K) for the reaction

$$2NH_3(g) + CO_2(g) \rightleftharpoons NH_2CONH_2(aq) + H_2O(l)$$

The standard free-energy change, $\Delta G°$, at 25°C equals −13.6 kJ. (We calculated this value of $\Delta G° = \Delta H° - T\Delta S°$ just before Section 18.4.)

Problem Strategy Rearrange the equation $\Delta G° = -RT \ln K$ to give

$$\ln K = \frac{\Delta G°}{-RT}$$

$\Delta G°$ and R must be in compatible units. You normally express $\Delta G°$ in joules and set R equal to 8.31 J/(K·mol).

Solution Substituting numerical values into this equation,

$$\ln K = \frac{-13.6 \times 10^3}{-8.31 \times 298} = 5.49$$

Hence,

$$K = e^{5.49} = \mathbf{2.4 \times 10^2}$$

Note that although the value of K indicates that products predominate at equilibrium, K is only moderately large. You would expect that the composition could be easily shifted toward reactants if you could remove either NH_3 or CO_2 (according to Le Châtelier's principle). This is what happens when urea is used as a fertilizer. As NH_3 is used up, more NH_3 is produced by the decomposition of urea.

Answer Check Make sure that you have expressed $\Delta G°$ and R in similar units. If $\Delta G°$ is given in kJ/mol, whereas R is given in J/(K·mol), express both in joules or both in kilojoules. (If you always carry units in your calculations, you will not make a mistake in units.) Also, note that ln is the natural logarithm, and its inverse or antilogarithm is the exponential function. Thus, the inverse of ln x is e^x.

Exercise 18.8 Use the data from Table 18.2 to obtain the equilibrium constant K_p at 25°C for the reaction

$$CaCO_3(s) \rightleftharpoons CaO(s) + CO_2(g)$$

Note that values of $\Delta G_f°$ are needed for $CaCO_3$ and CaO, even though the substances do not appear in $K_p = P_{CO_2}$.

━━━━━━━━━━━━━━━━━━━━━━━━━━━━

■ See Problems 18.61, 18.62, 18.63, and 18.64.

The method of calculating K from $\Delta G°$ also works for net ionic equations. In Example 18.9, you obtain a solubility product constant from thermodynamic data.

Example 18.9 Calculating *K* from the Standard Free-Energy Change (Net Ionic Equation)

Gaining Mastery Toolbox

Critical Concept 18.9
The standard free-energy change and the thermodynamic equilibrium constant for a reaction at a given temperature are related: $\Delta G° = -RT \ln K$.

Solution Essentials:
• Thermodynamic equilibrium constant, K
• Standard free-energy change, $\Delta G°$

Calculate the equilibrium constant K_{sp} at 25°C for the reaction

$$AgCl(s) \rightleftharpoons Ag^+(aq) + Cl^-(aq)$$

using standard free energies of formation.

Problem Strategy This is similar to the preceding example, except that you must first calculate $\Delta G_f°$ for the ionic equation. Note that K_{sp} equals K, the thermodynamic equilibrium constant.

Solution You first calculate $\Delta G°$. Writing $\Delta G_f°$ values below the formulas in the equation gives

$$AgCl(s) \rightleftharpoons Ag^+(aq) + Cl^-(aq)$$

$\Delta G_f°$: −109.8　　　77.1　　−131.3 kJ

Hence,

$$\Delta G° = [(77.1 - 131.3) - (-109.8)]\,kJ$$
$$= 55.6\,kJ\,(or\,55.6 \times 10^3\,J)$$

(continued)

(continued)

You now substitute numerical values into the equation relating ln K and $\Delta G°$.

$$\ln K = \frac{\Delta G°}{-RT} = \frac{55.6 \times 10^3}{-8.31 \times 298} = -22.45$$

Therefore,

$$K = e^{-22.45} = 1.8 \times 10^{-10}$$

Answer Check When you calculate $\Delta G°$, be careful of signs. In this example, note that the last term in calculating $\Delta G°$ has a minus sign before -109.8, which gives $+109.8$.

Exercise 18.9 Calculate the solubility product constant for $Mg(OH)_2$ at 25°C. The $\Delta G_f°$ values (in kJ/mol) are as follows: $Mg^{2+}(aq)$, -454.8; $OH^-(aq)$, -157.3; $Mg(OH)_2(s)$, -833.7.

■ See Problems 18.65 and 18.66.

You can understand the use of $\Delta G°$ as a criterion for spontaneity by looking at its relationship to the equilibrium constant, $\Delta G° = -RT \ln K$. When the equilibrium constant is greater than 1, ln K is positive and $\Delta G°$ is negative. Similarly, when the equilibrium constant is less than 1, ln K is negative and $\Delta G°$ is positive. This agrees with the first two rules listed at the end of Section 18.4. You can get the third rule by substituting $\Delta G° = \pm 10 \times 10^3$ J into $\Delta G° = -RT \ln K$ and solving for K. You find that K is between 0.018 and 57. In this range, the equilibrium mixture contains significant amounts of reactants as well as products.

CONCEPT CHECK 18.3

The following reaction is spontaneous in the direction given.

$$A(g) + B(g) \longrightarrow C(g) + D(g)$$

Suppose you are given a vessel containing an equilibrium mixture of A, B, C, and D, and you increase the concentration of C by increasing its partial pressure.

a　How is the value of $\Delta G°$ affected by the addition of C to the vessel?

b　How is the value of ΔG affected by the addition of C to the vessel?

18.7 Change of Free Energy with Temperature

In the previous sections, you obtained the free-energy change and equilibrium constant for a reaction at 25°C, the temperature at which the thermodynamic data were given. How do you find $\Delta G°$ or K at another temperature? Precise calculations are possible but are rather involved. Instead, we will look at a simple method that gives approximate results.

In this method, you assume that $\Delta H°$ and $\Delta S°$ are essentially constant with respect to temperature. (This is only approximately true.) You get the value of $\Delta G_T°$ at any temperature T by substituting values of $\Delta H°$ and $\Delta S°$ at 25°C (obtained from appropriate tables) into the following equation. ▶

$$\Delta G_T° = \Delta H° - T\Delta S° \quad \text{(approximation for } \Delta G_T°)$$

This approximation is most accurate for temperatures not too different from the temperature for which the $\Delta H°$ and $\Delta S°$ values are obtained. Much different temperatures give greater error.

Remember that the superscript degree sign (°) refers to substances in standard states, that is, at 1 atm and at the *specified temperature*. Although until now this was 25°C (298 K), we now consider other temperatures. Note that, in general, $\Delta G°$ depends strongly on temperature.

Spontaneity and Temperature Change

All of the four possible choices of signs for $\Delta H°$ and $\Delta S°$, listed in Table 18.3, give different temperature behaviors for $\Delta G°$. Consider the case in which $\Delta H°$ is negative and $\Delta S°$ is positive. An example is the reaction

$$C_6H_{12}O_6(s) \longrightarrow 2C_2H_5OH(l) + 2CO_2(g)$$
$$\text{glucose} \qquad \qquad \text{ethanol}$$

This represents the overall change of glucose (grape sugar) to ethanol (ethyl alcohol). The signs of $\Delta H°$ and $\Delta S°$ are easily explained. The formation of more stable bonds, such as occur in CO_2, releases energy as heat. The reaction is exothermic and $\Delta H°$ is negative. As explained in Example 18.2, the breaking up of a molecule ($C_6H_{12}O_6$) into smaller ones and the formation of a gas are expected to increase the entropy, so $\Delta S°$ is positive.

When $\Delta H°$ is negative and $\Delta S°$ is positive, both terms in $\Delta G°$ (that is, $\Delta H°$ and $-T\Delta S°$) are negative. Therefore, $\Delta G°$ is always negative and the reaction is spontaneous at all temperatures.

If the signs of $\Delta H°$ and $\Delta S°$ are reversed (that is, if $\Delta H°$ is positive, or endothermic, and $\Delta S°$ is negative), $\Delta G°$ is always positive. Thus, the reaction is nonspontaneous at all temperatures. An example is the reaction in which oxygen gas is converted to ozone.

$$3O_2(g) \longrightarrow 2O_3(g)$$

To accomplish this conversion, oxygen is passed through a tube in which an electrical discharge occurs. The electrical discharge supplies the necessary free energy for this otherwise nonspontaneous change.

In the reaction described in the chapter opening,

$$2NH_3(g) + CO_2(g) \longrightarrow NH_2CONH_2(aq) + H_2O(l)$$

both $\Delta H°$ and $\Delta S°$ are negative. In this case, the sign of $\Delta G°$ depends on the relative magnitudes of the terms $\Delta H°$ and $-T\Delta S°$, which have opposite signs. At some temperature, these terms just cancel and $\Delta G°$ equals zero. Below this temperature, $\Delta G°$ is negative. Above it, $\Delta G°$ is positive. Therefore, this reaction is spontaneous at low temperatures but nonspontaneous at sufficiently high temperatures. This particular reaction is spontaneous at 25°C but nonspontaneous at about 60°C.

A reaction in which both $\Delta H°$ and $\Delta S°$ are positive is the one described in the opening to Chapter 6. ◀

| The reaction mixture spontaneously cools enough to freeze water. See Figure 6.1.

$$Ba(OH)_2 \cdot 8H_2O(s) + 2NH_4NO_3(s) \longrightarrow Ba(NO_3)_2(aq) + 2NH_3(g) + 10H_2O(l)$$

The reaction is endothermic, and because crystalline solids change to a solution and a gas, the entropy increases. Again, the sign of $\Delta G°$ depends on the relative magnitudes of the terms $\Delta H°$ and $-T\Delta S°$. The reaction is spontaneous at room temperature but would be nonspontaneous at a sufficiently low temperature. Table 18.3 summarizes this discussion.

CONCEPT CHECK 18.4

Consider the decomposition of dinitrogen tetroxide, N_2O_4, to nitrogen dioxide, NO_2:

$$N_2O_4(g) \longrightarrow 2NO_2(g)$$

How would you expect the spontaneity of the reaction to behave with temperature change?

Table 18.3   Effect of Temperature on the Spontaneity of Reactions

$\Delta H°$	$\Delta S°$	$\Delta G°$	Description*	Example
−	+	−	Spontaneous at all T	$C_6H_{12}O_6(s) \longrightarrow 2C_2H_5OH(l) + 2CO_2(g)$
+	−	+	Nonspontaneous at all T	$3O_2(g) \longrightarrow 2O_3(g)$
−	−	+ or −	Spontaneous at low T; nonspontaneous at high T	$2NH_3(g) + CO_2(g) \longrightarrow NH_2CONH_2(aq) + H_2O(l)$
+	+	+ or −	Nonspontaneous at low T; spontaneous at high T	$Ba(OH)_2 \cdot 8H_2O(s) + 2NH_4NO_3(s) \longrightarrow$ $Ba(NO_3)_2(aq) + 2NH_3(g) + 10H_2O(l)$

*The terms *low temperature* and *high temperature* are relative. For a particular reaction, *high temperature* could mean room temperature.

Calculation of $\Delta G°$ at Various Temperatures

As an application of the method of calculating $\Delta G°$ at various temperatures, assuming $\Delta H°$ and $\Delta S°$ are constant, look at the following reaction:

$$CaCO_3(s) \longrightarrow CaO(s) + CO_2(g)$$

At 25°C, $\Delta G°$ equals $+130.9$ kJ, and the equilibrium partial pressure of CO_2 calculated from this is 1.1×10^{-23} atm. The very small value of this partial pressure shows that $CaCO_3$ is quite stable at room temperature. In the next example, you will see how $\Delta G°$ and K_p for this reaction change at higher temperature.

♥WL Interactive Example | 18.10 | **Calculating $\Delta G°$ and K at Various Temperatures**

Gaining Mastery Toolbox

Critical Concept 18.10
You can estimate the free-energy change for a reaction at a temperature T from the standard enthalpy and standard entropy changes at 25°C: $\Delta G_T° = \Delta H° - T\Delta S°$.

Solution Essentials:
- Relation between standard free-energy change and equilibrium constant, $\Delta G° = -RT \ln K$
- Standard free-energy change, $\Delta G°$
- Free-energy change, ΔG
- Standard entropy change, $\Delta S°$
- Standard enthalpy of reaction, $\Delta H°$

a What is $\Delta G°$ at 1000°C for the following reaction?

$$CaCO_3(s) \rightleftharpoons CaO(s) + CO_2(g)$$

Is this reaction spontaneous at 1000°C and 1 atm?
b What is the value of K_p at 1000°C for this reaction? What is the partial pressure of CO_2?

Problem Strategy

a Calculate $\Delta H°$ and $\Delta S°$ at 25°C using standard enthalpies of formation and standard entropies. Then substitute into the equation for $\Delta G_T°$.
b Use the value of $\Delta G_T°$ to find $K (= K_p)$, as in Example 18.8.

Solution

a From Tables 6.2 and 18.1, you have

$$CaCO_3(s) \rightleftharpoons CaO(s) + CO_2(g)$$

$\Delta H_f°$:	-1206.9	-635.1	-393.5 kJ
$S°$:	92.9	38.2	213.7 J/K

You calculate $\Delta H°$ and $\Delta S°$ from these values.

$$\Delta H° = [(-635.1 - 393.5) - (-1206.9)]\, kJ = 178.3\, kJ$$

$$\Delta S° = [(38.2 + 213.7) - (92.9)]\, J/K = 159.0\, J/K$$

Now you substitute $\Delta H°$, $\Delta S°$ ($= 0.1590$ kJ/K), and T ($= 1273$ K) into the equation for $\Delta G_T°$.

$$\Delta G_T° = \Delta H° - T\Delta S° = 178.3\, kJ - (1273\, K)(0.1590\, kJ/K) = \mathbf{-24.1\, kJ}$$

Because $\Delta G_T°$ is negative, the reaction should be **spontaneous** at 1000°C and 1 atm
b Substitute the value of $\Delta G°$ at 1273 K, which equals -24.1×10^3 J, into the equation relating $\ln K$ and $\Delta G°$.

$$\ln K = \frac{\Delta G°}{-RT} = \frac{-24.1 \times 10^3}{-8.31 \times 1273} = 2.278$$

$$K = K_p = e^{2.278} = \mathbf{9.8}$$

$K_p = P_{CO_2}$, so the partial pressure of CO_2 is **9.8 atm.**

Answer Check The things to be careful of have been noted before: watch the signs in calculating $\Delta H°$ and $\Delta S°$, and always use compatible units. Usually, $\Delta H°$ is given in kJ, whereas $\Delta S°$ is given in J/K. Here, we converted the units for $\Delta S°$ to kJ/K. As long as you carry units in your calculations, you will detect any error in choice of units that you might make.

Exercise 18.10 The thermodynamic equilibrium constant for the vaporization of water,

$$H_2O(l) \rightleftharpoons H_2O(g)$$

is $K_p = P_{H_2O}$. Use thermodynamic data to calculate the vapor pressure of water at 45°C. Compare your answer with the value given in Appendix B.

■ See Problems 18.67 and 18.68.

You can use the ideas described in this section to find the temperature at which a reaction such as the decomposition of $CaCO_3$ changes from nonspontaneous to spontaneous under standard conditions (1 atm for reactants and products). At this temperature, $\Delta G°$ equals zero.

$$\Delta G° = 0 = \Delta H° - T\Delta S°$$

Solving for T gives

$$T = \frac{\Delta H°}{\Delta S°}$$

For the decomposition of $CaCO_3$, using values obtained in Example 18.10, you get

$$T = \frac{178.3\,kJ}{0.1590\,kJ/K} = 1121\,K\,(848°C)$$

Thus, $CaCO_3$ should be stable to thermal decomposition to CaO and CO_2 at 1 atm until heated to 848°C. This is only approximate, of course, because you assumed that $\Delta H°$ and $\Delta S°$ are constant with temperature.

Exercise 18.11 To what temperature must magnesium carbonate be heated to decompose it to MgO and CO_2 at 1 atm? Is this higher or lower than the temperature required to decompose $CaCO_3$? Values of $\Delta H_f°$ (in kJ/mol) are as follows: MgO(s), −601.2; $MgCO_3$(s), −1111.7. Values of $S°$ (in J/K) are as follows: MgO(s), 26.9; $MgCO_3$(s), 65.9. Data for CO_2 are given in Tables 6.2 and 18.1.

■ See Problems 18.69 and 18.70.

A Checklist for Review

Summary of Facts and Concepts

Entropy, S, is a thermodynamic measure of energy dispersal in a system. According to the *second law of thermodynamics,* the total entropy of a system and its surroundings increases for a *spontaneous process,* one that occurs of its own accord. For a spontaneous reaction at constant T and P, you can show that the entropy change ΔS is greater than $\Delta H/T$. For an equilibrium process, such as a phase transition, $\Delta S = \Delta H/T$. The standard entropy $S°$ of a substance is determined by measuring the heat absorbed in warming it at increasing temperatures at 1atm. The method depends on the *third law of thermodynamics,* which states that perfectly crystalline substances at 0 K have zero entropy.

Free energy, G, is defined as $H - TS$. The change in free energy, ΔG, for a reaction at constant T and P equals $\Delta H - T\Delta S$, which is negative for a spontaneous change. *Standard free-energy changes,* $\Delta G°$, can be calculated from $\Delta H_f°$ and $S°$ values or from *standard free energies of formation,* $\Delta G_f°$. $\Delta G_f°$ for a substance is the standard free-energy change for the formation of the substance from the elements in their stablest states. The standard free-energy change, $\Delta G°$, serves as a *criterion for spontaneity* of a reaction. A negative value means the reaction is spontaneous. A value of zero means the reaction is at equilibrium. A positive value means the reaction is nonspontaneous.

The free-energy change, ΔG, is related to the *maximum work* that can be done by a reaction. This maximum work equals ΔG. A nonspontaneous reaction can be made to go by coupling it with a reaction that has a negative ΔG, so that ΔG for the overall result is negative. The reaction with the negative ΔG does work on the nonspontaneous reaction.

The *thermodynamic equilibrium constant, K,* can be calculated from the standard free-energy change by the relationship $\Delta G° = -RT \ln K$. $\Delta G°$ at various temperatures can be obtained by using the approximation that $\Delta H°$ and $\Delta S°$ are temperature independent. Therefore, $\Delta G°$, which equals $\Delta H° - T\Delta S°$, can be easily calculated for any T.

Learning Objectives	Important Terms
18.1 First Law of Thermodynamics: A Review ■ Recall the statement of the first law of thermodynamics. ■ Recall the definition of enthalpy. ■ Recall that the heat of reaction at constant pressure equals the enthalpy change. ■ Recall how to calculate the enthalpy of reaction from standard enthalpies of formation.	
18.2 Entropy and the Second Law of Thermodynamics ■ Define *spontaneous process.* ■ Define *entropy.* ■ Relate entropy to disorder in a molecular system (energy dispersal). ■ State the *second law of thermodynamics* in terms of system plus surroundings. ■ State the second law of thermodynamics in terms of the system only. ■ Calculate the entropy change for a phase transition. Example 18.1 ■ Describe how $\Delta H - T\Delta S$ functions as a criterion of a spontaneous reaction.	**spontaneous process** **entropy, S** **second law of thermodynamics**
18.3 Standard Entropies and the Third Law of Thermodynamics ■ State the *third law of thermodynamics.* ■ Define *standard entropy (absolute entropy).* ■ State the situations in which the entropy usually increases. ■ Predict the sign of the entropy change of a reaction. Example 18.2 ■ Express the standard change of entropy of a reaction in terms of standard entropies of products and reactants. ■ Calculate $\Delta S°$ for a reaction. Example 18.3	**third law of thermodynamics** **standard (absolute) entropy**
18.4 Free Energy and Spontaneity ■ Define *free energy, G.* ■ Define the *standard free-energy change.* ■ Calculate $\Delta G°$ from $\Delta H°$ and $\Delta S°$. Example 18.4 ■ Define the *standard free energy of formation, $\Delta G°$.* ■ Calculate $\Delta G°$ from standard free energies of formation. Example 18.5 ■ State the rules for using $\Delta G°$ as a criterion for spontaneity. ■ Interpret the sign of $\Delta G°$. Example 18.6	**free energy, G** **standard free energy of formation, $\Delta G°_f$**
18.5 Interpretation of Free Energy ■ Relate the free-energy change to maximum useful work. ■ Describe how the free energy changes during a chemical reaction.	
18.6 Relating $\Delta G°$ to the Equilibrium Constant ■ Define the *thermodynamic equilibrium constant, K.* ■ Write the expression for a thermodynamic equilibrium constant. Example 18.7	**thermodynamic equilibrium constant, K**

- Indicate how the free-energy change of a reaction and the reaction quotient are related.
- Relate the standard free-energy change to the thermodynamic equilibrium constant.
- Calculate K from the standard free-energy change (molecular equation). Example 18.8
- Calculate K from the standard free-energy change (net ionic equation). Example 18.9

18.7 Change of Free Energy with Temperature

- Describe how $\Delta G°$ at a given temperature ($\Delta G_T°$) is approximately related to $\Delta H°$ and $\Delta S°$ at 25°C.
- Describe how the spontaneity or nonspontaneity of a reaction is related to each of the four possible combinations of signs of $\Delta H°$ and $\Delta S°$.
- Calculate $\Delta G°$ and K at various temperatures. Example 18.10

Key Equations

$$\Delta S > \frac{q}{T} \quad \text{(spontaneous process)}$$

$$\Delta S = \frac{q}{T} \quad \text{(equilibrium process)}$$

$$\Delta S° = \Sigma\, nS°(\text{products}) - \Sigma\, mS°(\text{reactants})$$

$$\Delta G° = \Delta H° - T\Delta S°$$

$$\Delta G° = \Sigma\, n\Delta G_f°(\text{products}) - \Sigma\, m\Delta G_f°(\text{reactants})$$

$$\Delta G = \Delta G° + RT \ln Q$$

$$\Delta G° = -RT \ln K$$

$$\Delta G_T° = \Delta H° - T\Delta S° \quad \text{(approximation for } \Delta G_T°)$$

Questions and Problems

OWL Interactive versions of these problems may be assigned in OWL.

Self-Assessment and Review Questions

Key: These questions test your understanding of the ideas you worked with in the chapter. These problems vary in difficulty and often can be used for the basis of discussion.

18.1 What is a spontaneous process? Give three examples of spontaneous processes. Give three examples of nonspontaneous processes.

18.2 Which contains greater entropy, a quantity of frozen benzene or the same quantity of liquid benzene at the same temperature? Explain in terms of the dispersal of energy in the substance.

18.3 State the second law of thermodynamics.

18.4 The entropy change ΔS for a phase transition equals $\Delta H/T$, where ΔH is the enthalpy change. Why is it that the entropy change for a system in which a chemical reaction occurs spontaneously does not equal $\Delta H/T$?

18.5 Describe what you would look for in a reaction involving gases in order to predict the sign of $\Delta S°$. Explain.

18.6 Describe how the standard entropy of hydrogen gas at 25°C can be obtained from heat measurements.

18.7 Define the free energy G. How is ΔG related to ΔH and ΔS?

18.8 What is meant by the standard free-energy change $\Delta G°$ for a reaction? What is meant by the standard free energy of formation $\Delta G_f°$ of a substance?

18.9 Explain how $\Delta G°$ can be used to decide whether a chemical equation is spontaneous in the direction written.

18.10 What is the useful work obtained in the ideal situation in which a chemical reaction with free-energy change ΔG is run so that it produces no entropy?

18.11 Give an example of a chemical reaction used to obtain useful work.

18.12 How is the concept of coupling of reactions useful in explaining how a nonspontaneous change could be made to occur?

18.13 Explain how the free energy changes as a spontaneous reaction occurs. Show by means of a diagram how G changes with the extent of reaction.

18.14 Explain how an equilibrium constant can be obtained from thermal data alone (that is, from measurements of heat only).

18.15 Discuss the different sign combinations of $\Delta H°$ and $\Delta S°$ that are possible for a process carried out at constant temperature and pressure. For each combination, state whether the process must be spontaneous or not, or whether both situations are possible. Explain.

18.16 Consider a reaction in which $\Delta H°$ and $\Delta S°$ are positive. Suppose the reaction is nonspontaneous at room temperature. How would you estimate the temperature at which the reaction becomes spontaneous?

18.17 Which of the following are true about the process of water making the transition from the liquid to the gaseous state at 110°C?

a $\Delta G > 0$, $\Delta H < 0$, and $\Delta S = 0$
b $\Delta G < 0$, $\Delta H > 0$, and $\Delta S < 0$
c $\Delta G > 0$, $\Delta H > 0$, and $\Delta S > 0$
d $\Delta G < 0$, $\Delta H < 0$, and $\Delta S > 0$
e $\Delta G < 0$, $\Delta H > 0$, and $\Delta S > 0$

18.18 You run a reaction that has a negative entropy change and is exothermic. Assuming that the entropy and enthalpy do not change with temperature, you could predict that as you increase the temperature:

I. the equilibrium shifts to favor the reaction products.
II. the reaction becomes more spontaneous.
III. ΔG for the reaction increases (becomes more positive).

a I only
b II only
c III only
d I and III only
e II and III only

18.19 A reaction has a $\Delta G = -10.0$ kJ and a $\Delta H = -20.0$ kJ. If $\Delta S = -1.82 \times 10^3$ J/K, what was the temperature at which the reaction occurred?

a 5.49×10^{-3} K
b 5.49 K
c 8.55 K
d 6.04×10^{-3} K
e 6.04 K

18.20 Given the following information at 25°C, calculate $\Delta G°$ at 25°C for the reaction

$$2A(g) + B(g) \longrightarrow 3C(g)$$

Substance	$\Delta H_f°$ *(kJ/mol)*	S°*(J/mol·K)*
A(g)	191	244
B(g)	70.8	300
C(g)	-197	164

a -346 kJ
b 346 kJ
c -956 kJ
d -1.03×10^3 kJ
e 956 kJ

Concept Explorations

Key: Concept explorations are comprehensive problems that provide a framework that will enable you to explore and learn many of the critical concepts and ideas in each chapter. If you master the concepts associated with these explorations, you will have a better understanding of many important chemistry ideas and will be more successful in solving all types of chemistry problems. These problems are well suited for group work and for use as in-class activities.

18.21 Thermodynamics and Spontaneous Processes

Consider a sample of water at 25°C in a beaker in a room at 50°C.

a What change do you expect to observe in the water sample? Would this be a spontaneous process or not?
b What are the enthalpy and entropy changes for this change in the water sample? (Just indicate the sign of the changes.) Explain your answers.
c Does the entropy of the water increase or decrease during the change? How do you know?
d Is there a change in free energy for the water sample? If so, indicate the sign of the free-energy change and explain how you arrived at your answer.

Consider the same sample of water, but starting at 75°C in a room at 50°C.

e What change would you observe in the water sample? Would this change be a spontaneous process or not?
f What are the enthalpy and entropy changes for the water sample? (Just indicate the sign of the changes.) Explain your answers.
g Does the entropy of the water increase or decrease during the change? How do you know?
h Is there a change in free energy for the water sample? If so, indicate the sign of the free-energy change and explain how you arrived at your answer.

Finally, consider the same sample of water, starting at 50°C in a room at 50°C.

i What would you observe in the water sample? Is this a spontaneous process?

j What are the enthalpy and entropy changes for the water sample? (Just indicate the sign of the changes.) Be sure to justify your answer.
k Did the entropy of the water increase or decrease during the change? How do you know?
l Can you determine the exact free-energy change of the water in this case? If you can make this determination, what is the significance of this value?

18.22 Free Energy and the Equilibrium Constant

You place the substance A(g) in a container. Consider the following reaction under standard conditions to produce the substance B(g):

$$A(g) \rightleftharpoons B(g)$$

For this reaction as written, the equilibrium constant is a very large, positive number.

a When A(g) reacts to give B(g), does the standard free energy ($G°$) of the reaction change as the reaction proceeds or does it remain constant? Explain.
b When A(g) reacts to give B(g), does the free energy (G) of the reaction change as the reaction proceeds, or does it remain constant? Explain.
c Is this reaction spontaneous? How do you know?
d When the reaction reaches equilibrium, is the following statement true: $\Delta G° = \Delta G = 0$? If not, what can you say about the values of $\Delta G°$ and ΔG when equilibrium has been reached?
e When the reaction has reached equilibrium, what can you say about the composition of the reaction mixture? Is it mostly A(g), is it mostly B(g), or is it something close to equal amounts of A(g) and B(g)?
f Now consider running the reaction in reverse: B(g) $\longrightarrow$ A(g). For the reaction as written, what can you say about $\Delta G°$, ΔG, the equilibrium constant, and the composition of the reaction mixture at equilibrium? Also, is the reaction spontaneous in this direction?

Conceptual Problems

Key: These problems are designed to check your understanding of the concepts associated with some of the main topics presented in each chapter. A strong conceptual understanding of chemistry is the foundation for both applying chemical knowledge and solving chemical problems. These problems vary in level of difficulty and often can be used as a basis for group discussion.

18.23 For each of the following statements, indicate whether it is true or false.

a The entropy of a system always increases for a spontaneous change.

b The entropy of a system and its surroundings always increases for a spontaneous change.

c The free energy of a system always increases for a spontaneous change.

d A spontaneous reaction is always a fast reaction.

e A spontaneous reaction always releases heat.

18.24 Which of the following are spontaneous processes?

a A rusty crowbar turns shiny.

b Butane from a lighter burns in air.

c A cube of sugar dissolves in a cup of hot tea.

d Hydrogen and oxygen gases bubble out from a glass of pure water.

e A clock pendulum, initially stopped, begins swinging.

18.25 For each of the following series of pairs, indicate which one of the pair has the greater quantity of entropy.

a 1.0 mol of carbon dioxide gas at 20°C, 1 atm, or 2.0 mol of carbon dioxide gas at 20°C, 1 atm

b 1.0 mol of butane liquid at 20°C, 10 atm, or 1.0 mol of butane gas at 20°C, 10 atm

c 1.0 mol of solid carbon dioxide at −80°C, 1 atm, or 1.0 mol of solid carbon dioxide at −90°C, 1 atm

d 25 g of solid bromine at −7°C, 1 atm, or 25 g of bromine vapor at −7°C, 1 atm

18.26 Predict the sign of the entropy change for each of the following processes.

a Rainwater on the pavement evaporates.

b A lake freezes over in the winter.

c A tree leafs out in the spring.

d A drop of food coloring diffuses throughout a glass of water.

e Flowers wilt and stems decompose in the fall.

18.27 Hypothetical elements A(g) and B(g) are introduced into a container and allowed to react according to the reaction A(g) + 2B(g) $\longrightarrow$ AB$_2$(g). The container depicts the reaction mixture after equilibrium has been attained.

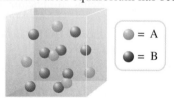

a Is the value of ΔS for the reaction positive, negative, or zero?

b Is the value of ΔH for the reaction positive, negative, or zero?

c Prior to equilibrium, is the value of ΔG for the reaction positive, negative, or zero?

d At equilibrium, is the value of ΔG for the reaction positive, negative, or zero?

18.28 Here is a simple experiment. Take a rubber band and stretch it. (Is this a spontaneous process? How does the Gibbs free energy change?) Place the rubber band against your lips; note how warm the rubber band has become. (How does the enthalpy change?) According to polymer chemists, the rubber band consists of long, coiled molecules. On stretching the rubber band, these long molecules uncoil and align themselves in a more ordered state. Show how the experiment given here is in accord with this molecular view of the rubber band.

18.29 Hypothetical elements X(g) and Y(g) are introduced into a container and allowed to react according to the reaction: X(g) + Y(g) $\longrightarrow$ XY(g). The container depicts the reaction mixture after equilibrium has been attained.

Next, the volume of the container is doubled by rapidly moving the plunger upward.

a How will this volume change affect the equilibrium constant of the reaction?

b Immediately after the volume change, how does Q compare to the equilibrium constant and which way will the reaction proceed?

c Immediately after the volume change, are the values of ΔG and ΔS positive, negative, or zero for the forward reaction?

18.30 Describe how you would expect the spontaneity ($\Delta G°$) for each of the following reactions to behave with a change of temperature.

a Chlorine is added to ethylene to produce dichloroethane, a solvent:

$$Cl_2(g) + C_2H_4(g) \longrightarrow C_2H_4Cl_2(l)$$

b Phosgene, COCl$_2$, the starting material for the preparation of polyurethane plastics, decomposes as follows:

$$COCl_2(g) \longrightarrow CO(g) + Cl_2(g)$$

Practice Problems

Key: These problems are for practice in applying problem-solving skills. They are divided by topic, and some are keyed to exercises (see the ends of the exercises). The problems are arranged in matching pairs; the odd-numbered problem of each pair is listed first, and its answer is given in the back of the book.

Entropy Changes

18.31 Chloroform, CHCl$_3$, is a solvent and has been used as an anesthetic. The heat of vaporization of chloroform at its boiling point (61.2°C) is 29.6 kJ/mol. What is the entropy change when 1.20 mol CHCl$_3$ vaporizes at its boiling point?

18.32 Diethyl ether (known simply as ether), $(C_2H_5)_2O$, is a solvent and anesthetic. The heat of vaporization of diethyl ether at its boiling point (35.6°C) is 26.7 kJ/mol. What is the entropy change when 1.34 mol $(C_2H_5)_2O$ vaporizes at its boiling point?

18.33 The heat of vaporization of carbon disulfide, CS_2, at 25°C is 27.2 kJ/mol. What is the entropy change when 1.00 mol of vapor in equilibrium with liquid condenses to liquid at 25°C? The entropy of this vapor at 25°C is 243 J/(mol·K). What is the entropy of the liquid at this temperature?

18.34 The enthalpy change when liquid methanol, CH_3OH, vaporizes at 25°C is 38.0 kJ/mol. What is the entropy change when 1.00 mol of vapor in equilibrium with liquid condenses to liquid at 25°C? The entropy of this vapor at 25°C is 255 J/(mol·K). What is the entropy of the liquid at this temperature?

18.35 Predict the sign of $\Delta S°$, if possible, for each of the following reactions. If you cannot predict the sign for any reaction, state why.

- a $2C(s) + O_2(g) \longrightarrow 2CO(g)$
- b $N_2(g) + O_2(g) \longrightarrow 2NO(g)$
- c $C_2H_2(g) + 2H_2(g) \longrightarrow C_2H_6(g)$
- d $2C_2H_2(g) + 3O_2(g) \longrightarrow 4CO(g) + 2H_2O(g)$

18.36 Predict the sign of $\Delta S°$, if possible, for each of the following reactions. If you cannot predict the sign for any reaction, state why.

- a $2SO_3(g) \longrightarrow 2SO_2(g) + O_2(g)$
- b $C_2H_5OH(l) + 3O_2(g) \longrightarrow 2CO_2(g) + 3H_2O(l)$
- c $P_4(g) \longrightarrow P_4(s)$
- d $2NaHCO_3(s) \longrightarrow Na_2CO_3(s) + H_2O(g) + CO_2(g)$

18.37 Calculate $\Delta S°$ for the following reactions, using standard entropy values.

- a $CS_2(l) + 3O_2(g) \longrightarrow CO_2(g) + 2SO_2(g)$
- b $2CH_3OH(l) + 3O_2(g) \longrightarrow 2CO_2(g) + 4H_2O(g)$
- c $Ag(s) + \frac{1}{2}Cl_2(g) \longrightarrow AgCl(s)$
- d $2Na(s) + Cl_2(g) \longrightarrow 2NaCl(s)$

18.38 Calculate $\Delta S°$ for the following reactions, using standard entropy values.

- a $2Ca(s) + O_2(g) \longrightarrow 2CaO(s)$
- b $Pb(s) + \frac{1}{2}O_2(g) \longrightarrow PbO(s)$
- c $CS_2(g) + 4H_2(g) \longrightarrow CH_4(g) + 2H_2S(g)$
- d $C_2H_4(g) + 3O_2(g) \longrightarrow 2CO_2(g) + 2H_2O(g)$

18.39 Calculate $\Delta S°$ for the reaction

$$C_2H_4(g) + 3O_2(g) \longrightarrow 2CO_2(g) + 2H_2O(l)$$

See Table 18.1 for values of standard entropies. Does the entropy of the chemical system increase or decrease as you expect? Explain.

18.40 What is the change in entropy, $\Delta S°$, for the reaction

$$CaCO_3(s) + 2H^+(aq) \longrightarrow Ca^{2+}(aq) + H_2O(l) + CO_2(g)$$

See Table 18.1 for values of standard entropies. Does the entropy of the chemical system increase or decrease as you expect? Explain.

Free-Energy Change and Spontaneity

18.41 Using enthalpies of formation (Appendix C), calculate $\Delta H°$ for the following reaction at 25°C. Also calculate $\Delta S°$ for this reaction from standard entropies at 25°C. Use these values to calculate $\Delta G°$ for the reaction at this temperature.

$$2CH_3OH(l) + 3O_2(g) \longrightarrow 2CO_2(g) + 4H_2O(l)$$

18.42 Using enthalpies of formation (Appendix C), calculate $\Delta H°$ for the following reaction at 25°C. Also calculate $\Delta S°$ for this reaction from standard entropies at 25°C. Use these values to calculate $\Delta G°$ for the reaction at this temperature.

$$4HCN(l) + 5O_2(g) \longrightarrow 2H_2O(g) + 4CO(g) + 2N_2(g)$$

18.43 The free energy of formation of one mole of compound refers to a particular chemical equation. For each of the following, write that equation.

- a $MgO(s)$
- b $COCl_2(g)$
- c $CF_4(g)$
- d $PCl_5(g)$

18.44 The free energy of formation of one mole of compound refers to a particular chemical equation. For each of the following, write that equation.

- a $KBr(s)$
- b $CH_3Cl(l)$
- c $H_2S(g)$
- d $AsH_3(g)$

18.45 Calculate the standard free energy of the following reactions at 25°C, using standard free energies of formation.

- a $CaCO_3(s) + 2H^+(aq) \longrightarrow Ca^{2+}(aq) + H_2O(l) + CO_2(g)$
- b $C_2H_4(g) + 3O_2(g) \longrightarrow 2CO_2(g) + 2H_2O(l)$

18.46 Calculate the standard free energy of the following reactions at 25°C, using standard free energies of formation.

- a $Na_2CO_3(s) + H^+(aq) \longrightarrow 2Na^+(aq) + HCO_3^-(aq)$
- b $C_2H_4(g) + 3O_2(g) \longrightarrow 2CO_2(g) + 2H_2O(g)$

18.47 On the basis of $\Delta G°$ for each of the following reactions, decide whether the reaction is spontaneous or nonspontaneous as written. Or, if you expect an equilibrium mixture with significant amounts of both reactants and products, say so.

- a $SO_2(g) + 2H_2S(g) \longrightarrow 3S(s) + 2H_2O(g)$; $\Delta G° = -91$ kJ
- b $2H_2O_2(aq) \longrightarrow O_2(g) + 2H_2O(l)$; $\Delta G° = -211$ kJ
- c $HCOOH(l) \longrightarrow CO_2(g) + H_2(g)$; $\Delta G° = 119$ kJ
- d $I_2(s) + Br_2(l) \longrightarrow 2IBr(g)$; $\Delta G° = 7.5$ kJ
- e $NH_4Cl(s) \longrightarrow NH_3(g) + HCl(g)$; $\Delta G° = 92$ kJ

18.48 For each of the following reactions, state whether the reaction is spontaneous or nonspontaneous as written or is easily reversible (that is, is a mixture with significant amounts of reactants and products).

- a $2NO(g) + 3H_2O(g) \longrightarrow 2NH_3(g) + \frac{5}{2}O_2(g)$; $\Delta G° = 479$ kJ
- b $2HBr(g) \longrightarrow H_2(g) + Br_2(g)$; $\Delta G° = 110$ kJ
- c $HCN(g) + 2H_2(g) \longrightarrow CH_3NH_2(g)$; $\Delta G° = -92$ kJ
- d $2NO(g) \longrightarrow N_2(g) + O_2(g)$; $\Delta G° = -173$ kJ
- e $H_2(g) + I_2(s) \longrightarrow 2HI(g)$; $\Delta G° = 2.6$ kJ

18.49 Calculate $\Delta H°$ and $\Delta G°$ for the following reactions at 25°C, using thermodynamic data from Appendix C; interpret the signs of $\Delta H°$ and $\Delta G°$.

- a $Al_2O_3(s) + 2Fe(s) \longrightarrow Fe_2O_3(s) + 2Al(s)$
- b $COCl_2(g) + H_2O(l) \longrightarrow CO_2(g) + 2HCl(g)$

18.50 Calculate $\Delta H°$ and $\Delta G°$ for the following reactions at 25°C, using thermodynamic data from Appendix C; interpret the signs of $\Delta H°$ and $\Delta G°$.

a $CS_2(l) + 2H_2O(l) \longrightarrow CO_2(g) + 2H_2S(g)$
b $2PbO(s) + N_2(g) \longrightarrow 2Pb(s) + 2NO(g)$

Maximum Work

18.51 Consider the reaction of 2 mol $H_2(g)$ at 25°C and 1 atm with 1 mol $O_2(g)$ at the same temperature and pressure to produce liquid water at these conditions. If this reaction is run in a controlled way to generate work, what is the maximum useful work that can be obtained? How much entropy is produced in this case?

18.52 Consider the reaction of 1 mol $H_2(g)$ at 25°C and 1 atm with 1 mol $Br_2(l)$ at the same temperature and pressure to produce gaseous HBr at these conditions. If this reaction is run in a controlled way to generate work, what is the maximum useful work that can be obtained? How much entropy is produced in this case?

18.53 What is the maximum work that could be obtained from 4.85 g of zinc metal in the following reaction at 25°C?

$$Zn(s) + Cu^{2+}(aq) \longrightarrow Zn^{2+}(aq) + Cu(s)$$

18.54 What is the maximum work that could be obtained from 3.65 g of zinc metal in the following reaction at 25°C?

$$Zn(s) + 2H^+(aq) \longrightarrow Zn^{2+}(aq) + H_2(g)$$

Calculation of Equilibrium Constants

18.55 Give the expression for the thermodynamic equilibrium constant for each of the following reactions.

a $2Li(s) + 2H_2O(l) \rightleftharpoons 2Li^+(aq) + 2OH^-(aq) + H_2(g)$
b $CO(g) + H_2O(g) \rightleftharpoons CO_2(g) + H_2(g)$
c $Mg(OH)_2(s) \rightleftharpoons Mg^{2+}(aq) + 2OH^-(aq)$

18.56 Write the expression for the thermodynamic equilibrium constant for each of the following reactions.

a $CO(g) + 2H_2(g) \rightleftharpoons CH_3OH(g)$
b $2Ag^+(aq) + CrO_4^{2-}(aq) \rightleftharpoons Ag_2CrO_4(s)$
c $CaCO_3(s) + 2H^+(aq) \rightleftharpoons Ca^{2+}(aq) + H_2O(l) + CO_2(g)$

18.57 What is the standard free-energy change $\Delta G°$ at 25°C for the following reaction? Obtain necessary information from Appendix C.

$$H_2(g) + Br_2(l) \longrightarrow 2HBr(g)$$

What is the value of the thermodynamic equilibrium constant K?

18.58 What is the standard free-energy change $\Delta G°$ at 25°C for the following reaction? See Table 18.2 for data.

$$C(graphite) + 2Cl_2(g) \longrightarrow CCl_4(l)$$

Calculate the value of the thermodynamic equilibrium constant K.

18.59 Calculate the standard free-energy change and the equilibrium constant K_p for the following reaction at 25°C. See Appendix C for data.

$$CO(g) + 2H_2(g) \rightleftharpoons CH_3OH(g)$$

18.60 Calculate the standard free-energy change and the equilibrium constant K_p for the following reaction at 25°C. See Table 18.2 for data.

$$CO(g) + 3H_2(g) \rightleftharpoons CH_4(g) + H_2O(g)$$

18.61 Obtain the equilibrium constant K_c at 25°C from the free-energy change for the reaction

$$Fe(s) + Cu^{2+}(aq) \rightleftharpoons Fe^{2+}(aq) + Cu(s)$$

See Appendix C for data.

18.62 Calculate the equilibrium constant K_c at 25°C from the free-energy change for the following reaction:

$$Zn(s) + 2Ag^+(aq) \rightleftharpoons Zn^{2+}(aq) + Ag(s)$$

See Appendix C for data.

Free Energy and Temperature Change

18.63 Use data given in Tables 6.2 and 18.1 to obtain the value of K_p at 1000°C for the reaction

$$C(graphite) + CO_2(g) \rightleftharpoons 2CO(g)$$

Carbon monoxide is known to form during combustion of carbon at high temperatures. Do the data agree with this? Explain.

18.64 Use data given in Tables 6.2 and 18.1 to obtain the value of K_p at 2000°C for the reaction

$$N_2(g) + O_2(g) \rightleftharpoons 2NO(g)$$

Nitric oxide is known to form in hot flames in air, which is a mixture of N_2 and O_2. It is present in auto exhaust from this reaction. Are the data in agreement with this result? Explain.

18.65 Oxygen was first prepared by heating mercury(II) oxide, HgO.

$$2HgO(s) \longrightarrow 2Hg(g) + O_2(g)$$

Estimate the temperature at which HgO decomposes to O_2 at 1 atm. See Appendix C for data.

18.66 Sodium carbonate, Na_2CO_3, can be prepared by heating sodium hydrogen carbonate, $NaHCO_3$.

$$2NaHCO_3(s) \longrightarrow Na_2CO_3(s) + H_2O(g) + CO_2(g)$$

Estimate the temperature at which $NaHCO_3$ decomposes to products at 1 atm. See Appendix C for data.

General Problems

Key: These problems provide more practice but are not divided by topic or keyed to exercises. Each section ends with essay questions, each of which is color coded to refer to the *A Chemist Looks at Life Science* (pink) chapter essay on which it is based. Odd-numbered problems and the even-numbered problems that follow are similar; answers to all odd-numbered problems except the essay questions are given in the back of the book.

18.67 Find the sign of $\Delta S°$ for the reaction

$$2N_2O_5(s) \longrightarrow 4NO_2(g) + O_2(g)$$

The reaction is endothermic and spontaneous at 25°C. Explain the spontaneity of the reaction in terms of enthalpy and entropy changes.

18.68 The combustion of acetylene, C_2H_2, is a spontaneous reaction given by the equation

$$2C_2H_2(g) + 5O_2(g) \longrightarrow 4CO_2(g) + 2H_2O(l)$$

As expected for a combustion, the reaction is exothermic. What is the sign of $\Delta H°$? What do you expect for the sign of $\Delta S°$? Explain the spontaneity of the reaction in terms of the enthalpy and entropy changes.

18.69 Estimate the value of $\Delta H°$ for the following reaction from bond energies (Table 9.5).

$$H_2(g) + Cl_2(g) \longrightarrow 2HCl(g)$$

Is the reaction exothermic or endothermic? Note that the reaction involves the breaking of symmetrical molecules (H_2 and Cl_2) and the formation of a less symmetrical product (HCl). From this, would you expect $\Delta S°$ to be positive or negative? Comment on the spontaneity of the reaction in terms of the changes in enthalpy and entropy.

18.70 Compare the energies of the bonds broken and those formed (see Table 9.5) for the reaction

$$HCN(g) + 2H_2(g) \longrightarrow CH_3NH_2(g)$$

From this, conclude whether the reaction is exothermic or endothermic. What is the sign of $\Delta S°$? Explain. The reaction is spontaneous at 25°C. Explain this in terms of the enthalpy and entropy changes.

18.71 Acetic acid, CH_3COOH, freezes at 16.6°C. The heat of fusion, ΔH_{fus}, is 69.0 J/g. What is the change of entropy, ΔS, when 1 mol of liquid acetic acid freezes to the solid?

18.72 Acetone, CH_3COCH_3, boils at 56°C. The heat of vaporization of acetone at this temperature is 29.1 kJ/mol. What is the entropy change when 1 mol of liquid acetone vaporizes at 56°C?

18.73 Without doing any calculations, decide what the sign of $\Delta S°$ will be for each of the following reactions.
- a $O_2(g) + 2F_2(g) \longrightarrow 2OF_2(g)$
- b $2N_2O_5(g) \longrightarrow 4NO_2(g) + O_2(g)$
- c $2LiOH(aq) + CO_2(g) \longrightarrow Li_2CO_3(aq) + H_2O(l)$
- d $(NH_4)_2Cr_2O_7(s) \longrightarrow N_2(g) + 4H_2O(g) + Cr_2O_3(s)$

18.74 For each of the following reactions, decide whether there is an increase or a decrease in entropy. Why do you think so? (No calculations are needed.)
- a $N_2(g) + 3H_2(g) \longrightarrow 2NH_3(g)$
- b $NH_4Cl(s) \longrightarrow NH_3(g) + HCl(g)$
- c $CO(g) + 2H_2(g) \longrightarrow CH_3OH(l)$
- d $Li_3N(s) + 3H_2O(l) \longrightarrow 3LiOH(aq) + NH_3(g)$

18.75 The following equation shows how nitrogen dioxide reacts with water to produce nitric acid:

$$3NO_2(g) + H_2O(l) \longrightarrow 2HNO_3(l) + NO(g)$$

Predict the sign of $\Delta S°$ for this reaction.

18.76 Ethanol burns in air or oxygen according to the equation

$$C_2H_5OH(l) + 3O_2(g) \longrightarrow 2CO_2(g) + 3H_2O(g)$$

Predict the sign of $\Delta S°$ for this reaction.

18.77 Acetic acid in vinegar results from the bacterial oxidation of ethanol.

$$C_2H_5OH(l) + O_2(g) \longrightarrow CH_3COOH(l) + H_2O(l)$$

What is $\Delta S°$ for this reaction? Use standard entropy values. (See Appendix C for data.)

18.78 Methanol is produced commercially from carbon monoxide and hydrogen.

$$CO(g) + 2H_2(g) \longrightarrow CH_3OH(l)$$

What is $\Delta S°$ for this reaction? Use standard entropy values.

18.79 Is the following reaction spontaneous as written? Explain. Do whatever calculation is needed to answer the question.

$$CH_4(g) + N_2(g) \longrightarrow HCN(g) + NH_3(g)$$

18.80 Is the following reaction spontaneous as written? Explain. Do whatever calculation is needed to answer the question.

$$SO_2(g) + H_2(g) \longrightarrow H_2S(g) + O_2(g)$$

18.81 The reaction

$$CO_2(g) + H_2(g) \longrightarrow CO(g) + H_2O(g)$$

is nonspontaneous at room temperature but becomes spontaneous at a much higher temperature. What can you conclude from this about the signs of $\Delta H°$ and $\Delta S°$, assuming that the enthalpy and entropy changes are not greatly affected by the temperature change? Explain your reasoning.

18.82 The reaction

$$N_2(g) + 3H_2(g) \longrightarrow 2NH_3(g)$$

is spontaneous at room temperature but becomes nonspontaneous at a much higher temperature. From this fact alone, obtain the signs of $\Delta H°$, and $\Delta S°$, assuming that $\Delta H°$ and $\Delta S°$ do not change much with temperature. Explain your reasoning.

18.83 Calculate $\Delta G°$ at 25°C for the reaction

$$BaSO_4(s) \rightleftharpoons Ba^{2+}(aq) + SO_4^{2-}(aq)$$

See Appendix C for values of $\Delta G°_f$. What is the value of the solubility product constant, K_{sp}, for this reaction at 25°C?

18.84 Calculate $\Delta G°$ at 25°C for the reaction

$$CaF_2(s) \rightleftharpoons Ca^{2+}(aq) + 2F^-(aq)$$

See Appendix C for values of $\Delta G°_f$. What is the value of the solubility product constant, K_{sp}, for this reaction at 25°C?

18.85 Consider the decomposition of phosgene, $COCl_2$.

$$COCl_2(g) \longrightarrow CO(g) + Cl_2(g)$$

Calculate $\Delta H°$ and $\Delta S°$ at 25°C for this reaction. See Appendix C for data. What is $\Delta G°$ at 25°C? Assume that $\Delta H°$ and $\Delta S°$ are constant with respect to a change of temperature. Now calculate $\Delta G°$ at 800°C. Compare the two values of $\Delta G°$. Briefly discuss the spontaneity of the reaction at 25°C and at 800°C.

18.86 Consider the reaction

$$CS_2(g) + 4H_2(g) \rightleftharpoons CH_4(g) + 2H_2S(g)$$

Calculate $\Delta H°$, $\Delta S°$, and $\Delta G°$ at 25°C for this reaction. Assume $\Delta H°$ and $\Delta S°$ are constant with respect to a change of temperature. Now calculate $\Delta G°$ at 650°C. Compare the two values of $\Delta G°$. Briefly discuss the spontaneity of the reaction at 25°C and at 650°C.

18.87 a From a consideration of the following reactions, calculate $\Delta H_f°$ for ethane, $C_2H_6(g)$.

$$CH_3CHO(g) + 2H_2(g) \longrightarrow C_2H_6(g) + H_2O(l);$$
$$\Delta H° = -204 \text{ kJ}$$

$$2H_2(g) + O_2(g) \longrightarrow 2H_2O(g); \Delta H° = -484 \text{ kJ}$$

$$2C_2H_5OH(l) + O_2(g) \longrightarrow 2CH_3CHO(g) + 2H_2O(l);$$
$$\Delta H° = -348 \text{ kJ}$$

$$H_2O(l) \longrightarrow H_2O(g); \Delta H°(kJ) = 44$$

$$2C_2H_5OH(l) \longrightarrow 4C(s) + 6H_2(g) + O_2(g); \Delta H°(kJ) = 555$$

b From appropriate entropy data, calculate $\Delta S°$ for the formation of ethane from the elements.

c Using the information from above, calculate $\Delta G_f°$ for ethane at 25°C.

18.88 a From a consideration of the following reactions, calculate $\Delta H_f°$ for methane, $CH_4(g)$.

$$CO_2(g) + 2H_2(g) \longrightarrow HCHO(g) + H_2O(g); \Delta H° = 35 \text{ kJ}$$

$$CO_2(g) \longrightarrow C(s) + O_2(g); \Delta H° = 393 \text{ kJ}$$

$$HCHO(g) + 2H_2(g) \longrightarrow CH_4(g) + H_2O(g); \Delta H° = -201 \text{ kJ}$$

$$2H_2(g) + O_2(g) \longrightarrow 2H_2O(g); \Delta H° = -484 \text{ kJ}$$

b From appropriate entropy data, calculate $\Delta S°$ for the formation of methane from the elements.

c Using the information from above, calculate $\Delta G_f°$ for methane at 25°C.

18.89 For the reaction

$$CH_3OH(l) + \tfrac{3}{2}O_2(g) \longrightarrow 2H_2O(l) + CO_2(g)$$

the value of $\Delta G°$ is -702.2 kJ at 25°C. Other data are as follows:

	$\Delta H_f°$ (kJ/mol) at 25°C	$S°$ (J/mol·K) at 25°C
$CH_3OH(l)$	-238.7	126.8
$H_2O(l)$	-285.8	70.0
$CO_2(g)$	-393.5	213.7

Calculate the standard entropy, $S°$, per mole of $O_2(g)$.

18.90 For the reaction

$$HCHO(g) + \tfrac{2}{3}O_3(g) \longrightarrow CO_2(g) + H_2O(g)$$

the value of $\Delta G°$ is -618.8 kJ/mol at 25°C. Other data are as follows:

	$\Delta H_f°$ (kJ/mol) at 25°C	$S°$ (J/mol·K) at 25°C
HCHO	-117.0	219.0
$H_2O(g)$	-241.8	188.7
$CO_2(g)$	-393.5	213.7
$O_3(g)$	142.7	?

Calculate the entropy, $S°$, per mole of $O_3(g)$.

18.91 Tungsten is usually produced by the reduction of WO_3 with hydrogen:

$$WO_3(s) + 3H_2(g) \longrightarrow W(s) + 3H_2O(g)$$

Consider the following data:

	$WO_3(s)$	$H_2O(g)$
$\Delta H_f°$ (kJ/mol)	-839.9	-241.8
$\Delta G_f°$ (kJ/mol)	-763.1	-228.6

a Is $K > 1$ or < 1 at 25°C? Explain your answer.

b What is the value of $\Delta S°$ at 25°C?

c What is the temperature at which $\Delta G°$ equals zero for this reaction at 1 atm pressure?

d What is the driving force of this reaction?

18.92 Tin(IV) oxide can be reacted with either hydrogen or carbon to form tin and water vapor or carbon dioxide, respectively. See Appendix C for data.

a Calculate $\Delta H°$ and $\Delta S°$ at 25°C for the reaction of SnO_2 with H_2 and for the reaction of SnO_2 with C.

b At what temperature will each of these processes become spontaneous?

c Industrially, which process is preferred? Why?

18.93 For the decomposition of formic acid,

$$HCOOH(l) \longrightarrow H_2O(l) + CO(g)$$

$\Delta H° = +29$ kJ/mol at 25°C.

a Does the tendency of this reaction to proceed to a state of minimum energy favor the formation of water and carbon monoxide or formic acid? Explain.

b Does the tendency of this reaction to proceed to a state of maximum entropy favor the formation of products or reactants? Explain.

18.94 a Why are some reactions exothermic and others endothermic?

b Discuss the driving force in a spontaneous reaction that is highly exothermic and in one that is endothermic.

18.95 For the reaction

$$2Cu(s) + S(s) \longrightarrow Cu_2S(s)$$

$\Delta H°$ and $\Delta G°$ are negative and $\Delta S°$ is positive.

a At equilibrium, will reactants or products predominate? Why?

b Why must the reaction system be heated in order to produce copper(I) sulfide?

18.96 For the reaction

$$C_6H_6(l) + Br_2(l) \longrightarrow C_6H_5Br(l) + HBr(g)$$

the products have a lower free energy than the reactants.

a Predict the sign of the entropy change for this reaction. Explain.

b Predict whether reactants or products would predominate in an equilibrium mixture. Why?

c How can you account for the fact that a mixture of the reactants must be heated in the presence of a catalyst in order to produce products?

18.97 When 1.000 g of ethylene glycol, $C_2H_6O_2$, is burned at 25°C and 1.00 atmosphere pressure, $H_2O(l)$ and $CO_2(g)$ are formed with the evolution of 19.18 kJ of heat.

a Calculate the molar enthalpy of formation of ethylene glycol. (It will be necessary to use data from Appendix C.)

b ΔG_f° of ethylene glycol is -322.5 kJ/mol. What is ΔG° for the combustion of 1 mol ethylene glycol?

c What is ΔS° for the combustion of 1 mol ethylene glycol?

18.98 When 1.000 g of gaseous butane, C_4H_{10}, is burned at 25°C and 1.00 atm pressure, $H_2O(l)$ and $CO_2(g)$ are formed with the evolution of 49.50 kJ of heat.

a Calculate the molar enthalpy of formation of butane. (Use enthalpy of formation data for H_2O and CO_2.)

b ΔG_f° of butane is -17.2 kJ/mol. What is ΔG° for the combustion of 1 mol butane?

c From a and b, calculate ΔS° for the combustion of 1 mol butane.

18.99 a Calculate K_1 at 25°C for sulfurous acid:

$$H_2SO_3(aq) \rightleftharpoons H^+(aq) + HSO_3^-(aq)$$

	H_2SO_3(aq)	H^+(aq)	HSO_3^-(aq)
ΔH_f° (kJ/mol)	-608.8	0	-626.2
S° (J/mol·K)	232.2	0	139.8

b Which thermodynamic factor is the most significant in accounting for the fact that sulfurous acid is a weak acid? Why?

18.100 a Calculate K_1 at 25°C for phosphoric acid:

$$H_3PO_4(aq) \rightleftharpoons H^+(aq) + H_2PO_4^-(aq)$$

	H_3PO_4(aq)	H^+(aq)	$H_2PO_4^-$(aq)
ΔH_f° (kJ/mol)	-1288.3	0	-1285
S° (J/mol·K)	158.2	0	89

b Which thermodynamic factor is the most significant in accounting for the fact that phosphoric acid is a weak acid? Why?

■ **18.101** The direct reaction of iron(III) oxide, Fe_2O_3, to give iron and oxygen gas is a nonspontaneous reaction; normally, iron combines with oxygen to give rust (the oxide). Yet we do change iron(III) oxide, as iron ore, into iron metal. How is this possible? Explain.

■ **18.102** Adenosine triphosphate (ATP) is often referred to as a biological "energy" source. What does this mean? Explain how ATP is used in biological organisms to produce proteins from amino acids.

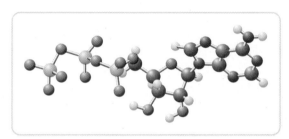

Strategy Problems

Key: As noted earlier, all of the practice and general problems are matched pairs. This section is a selection of problems that are not in matched-pair format. These challenging problems require that you employ many of the concepts and strategies that were developed in the chapter. In some cases, you will have to integrate several concepts and operational skills in order to solve the problem successfully.

18.103 Consider the reaction in which hydrazine vapor decomposes to its elements:

$$N_2H_4(g) \rightleftharpoons N_2(g) + 2H_2(g)$$

Use data from Appendix C to answer the following questions.

a What is the standard enthalpy change for this reaction at 25°C?

b Calculate the standard entropy change at 25°C. Does the sign of the entropy change agree with what you would expect? Explain.

c Describe the spontaneity or nonspontaneity of the reaction at various temperatures.

d What happens to ΔG° as you lower the temperature? Would this reaction give greater product at higher or at lower temperature? Explain.

18.104 Consider the reaction for the formation of $H_2O(g)$ from its elements. Use the data in Appendix C to answer the following questions.

a What is the standard enthalpy of formation of $H_2O(g)$ at 25°C?

b What is the standard entropy of formation of $H_2O(g)$ at 25°C?

c Use these values to calculate the standard free energy of formation of $H_2O(g)$ at 25°C. Check this value against the one given in Appendix C.

d A student obtained ΔG_f° for $H_2O(g)$ by calculating $\Delta H_f^\circ - TS^\circ$, using ΔH_f° and S° for $H_2O(g)$ from Appendix C. Is this correct? Explain.

18.105 Acetone, CH_3COCH_3, is a fragrant liquid that is used as a solvent for lacquers, paint removers, and nail polish remover. It burns in oxygen to give carbon dioxide and water:

$$CH_3COCH_3(l) + 4O_2(g) \longrightarrow 3CO_2(g) + 3H_2O(l)$$

If the standard free-energy change for this reaction is -1739 kJ/mol, what is the standard free energy of formation of acetone? Use data from Appendix C.

18.106 Cobalt(II) chloride hexahydrate, $CoCl_2\cdot6H_2O$, is a bright pink compound, but in the presence of very dry air it loses water vapor to the air to produce the light blue anhydrous salt $CoCl_2$. Calculate the standard free-energy change for the reaction at 25°C:

$$CoCl_2\cdot6H_2O(s) \rightleftharpoons CoCl_2(s) + 6H_2O(g)$$

Here are some thermodynamic data at 25°C:

	ΔH_f° *(kJ/mol)*	ΔG_f° *(kJ/mol)*	S° *(J/mol·K)*
$CoCl_2(s)$	$-313.$	$-270.$	109.
$CoCl_2\cdot6H_2O(s)$	$-2115.$	$-1725.$	343.

What is the partial pressure of water vapor in equilibrium with the anhydrous salt and the hexahydrate at 25°C? (Give the value in mmHg.) What is the relative humidity of air that has this partial pressure of water? The relative humidity of a sample of air is

$$\text{Relative humidity} = \frac{\text{partial pressure of } H_2O(g) \text{ in air}}{\text{vapor pressure of water}} \times 100\%$$

What do you expect to happen to the equilibrium partial pressure over the hexahydrate as the temperature is raised? Explain.

18.107 An experimenter placed 1.000 mol of H_2 and 1.000 mol of I_2 in a 1.000-L flask. The substances reacted to produce hydrogen iodide:

$$H_2(g) + I_2(g) \rightleftharpoons 2HI(g)$$

When the contents of the flask came to equilibrium at 458°C, the experimenter found that it contained 0.225 mol H_2, 0.225 mol I_2, and 1.550 mol HI.

a Calculate the standard free-energy change for the reaction at 458°C.

b Assume that the standard enthalpy change for the reaction at 458°C is the same as it is at 298°C. Calculate the standard entropy change for the reaction at 458°C.

18.108 Hydrogen gas and iodine vapor react to produce hydrogen iodide gas:

$$H_2(g) + I_2(g) \rightleftharpoons 2HI(g)$$

Calculate the free-energy change ΔG for the following two conditions, at 25°C. Which one is closer to equilibrium? Explain.

a The partial pressures of $H_2(g)$, $I_2(g)$, and $HI(g)$ are each $\frac{1}{3}$ atm.

b The partial pressures of $H_2(g)$ and $I_2(g)$ are each $\frac{1}{3}$ atm; the partial pressure of $HI(g)$ is $\frac{1}{90}$ atm.

18.109 Sulfur is produced in volcanic gases by the reaction of hydrogen sulfide with sulfur dioxide:

$$2H_2S(g) + SO_2(g) \rightleftharpoons 3S(s, \text{rhombic}) + 2H_2O(g)$$

A particular sample of volcanic gas at 1.00 atm has a partial pressure of sulfur dioxide of 89 mmHg. If this gas at 1170°C also contains hydrogen sulfide and water vapor in equilibrium with rhombic sulfur, what is the ratio of the partial pressure of water vapor to the partial pressure of hydrogen sulfide? Use any reasonable approximations.

18.110 Silver carbonate, Ag_2CO_3, is a light yellow compound that decomposes when heated to give silver oxide and carbon dioxide:

$$Ag_2CO_3(s) \rightleftharpoons Ag_2O(s) + CO_2(g)$$

A researcher measured the partial pressure of carbon dioxide over a sample of silver carbonate at 220°C and found that it was 1.37 atm. Calculate the partial pressure of carbon dioxide at 25°C. The standard enthalpies of formation of silver carbonate and silver oxide at 25°C are −505.9 kJ/mol and −31.05 kJ/mol, respectively. Make any reasonable assumptions in your calculations. State the assumptions that you make, and note why you think they are reasonable.

18.111 Adenosine triphosphate, ATP, is used as a free-energy source by biological cells. (See the essay on page 758.) ATP hydrolyzes in the presence of enzymes to give ADP:

$$ATP(aq) + H_2O(l) \longrightarrow ADP(aq) + H_2PO_4^-(aq);$$
$$\Delta G° = -30.5 \text{ kJ/mol at } 25°C$$

Consider a hypothetical biochemical reaction of molecule A to give molecule B:

$$A(aq) \longrightarrow B(aq); \Delta G° = +15.0 \text{ kJ/mol at } 25°C$$

a Calculate the ratio [B]/[A] at 25°C at equilibrium.

b Now consider this reaction "coupled" to the reaction for the hydrolysis of ATP:

$$A(aq) + ATP(aq) + H_2O(l) \longrightarrow B(aq) + ADP(aq) + \\ H_2PO_4^-(aq)$$

If a cell maintains a high ratio of ATP to ADP and $H_2PO_4^-$ by continuously making ATP, the conversion of A to B can be made highly spontaneous. A characteristic value of this ratio is

$$\frac{[ATP]}{[ADP][H_2PO_4^-]} = 500$$

Calculate the ratio [B][A] in this case and compare it with the uncoupled reaction. Compared with the uncoupled reaction, how much larger is this ratio when coupled to the hydrolysis of ATP?

18.112 A car has been fitted with an internal combustion engine that uses propane as a fuel.

$$C_3H_8(g) + 5O_2(g) \longrightarrow 3CO_2(g) + 4H_2O(g)$$

What is the maximum work that you can obtain from 1.00 kg of propane, C_3H_8? Suppose the car is driving uphill. What is the maximum height, in miles, that the car could go to, assuming that the car's mass is 4000. lb? The work done in moving a mass, m, up to a height h is mgh, where g is the constant acceleration of gravity (equal to 9.81 m/s²).

18.113 A quantity of strong acid is added to pure water to give a solution that is 0.100 M H^+. Calculate Q for the following reaction just as the H^+ ion is added and before equilibrium is established.

$$H^+(aq) + OH^-(aq) \longrightarrow H_2O(l)$$

What is the value of the free-energy change, ΔG, for this reaction starting from these nonstandard conditions?

18.114 Given the following information, calculate the standard free energy of formation of $OH^-(aq)$ at 25°C and compare the value you calculate with that given in Table 18.2. At 25°C, the pH of pure water is 7.00 and the free energy of formation of $H_2O(aq)$ is −231.71 kJ/mol. Note that $\Delta G_f°$ for $H^+(aq)$ equals zero by convention.

18.115 Sodium acetate crystallizes from a supersaturated solution (see Figure 12.4). What can you say about the sign of $\Delta G°$? What would you expect for the sign of $\Delta S°$? What about the sign of $\Delta H°$? Is the crystallization exothermic or endothermic? Explain your answers.

18.116 According to a source, lithium peroxide (Li_2O_2) decomposes to lithium oxide (Li_2O) and oxygen gas at about 195°C. If the standard enthalpy change for this decomposition is 33.9 kJ/mol, what would you give as an estimate for the standard entropy change for this reaction? Explain.

18.117 Tetrachloromethane (carbon tetrachloride), CCl_4, has a normal boiling point of 76.7°C and an enthalpy of vaporization, $\Delta H_{vap}°$, of 29.82 kJ/mol. Estimate the entropy of vaporization, $\Delta S_{vap}°$. Estimate the free energy of vaporization, $\Delta G_{vap}°$, at 25°C.

18.118 A truck weighing 1700 kg uses natural gas as fuel. Natural gas is mainly methane, CH_4, and though it is normally a gas, it is stored in a tank as compressed liquid. Assume that the fuel is methane gas and that it burns completely in the engine to carbon dioxide gas and water vapor. What is the maximum work that 1.00 mol $CH_4(g)$ can do if it burns at 25°C? What would be the maximum height that this truck could travel up a hill on 1.00 m^3 of liquid methane fuel? The density of liquid methane is 416 kg/m^3. (Note that the potential energy of a mass m raised to a height h is mgh, where g is the acceleration of gravity and equals 9.81 m/s^2.)

18.119 Ammonia, NH_3, is produced from its elements in the industrial Haber process. Calculate $\Delta H°$, $\Delta G°$, and $\Delta S°$ at 25°C. See Appendix C for data. The industrial process is run at temperatures around 450°C. Would the equilibrium yield be more or less at 25°C? Explain. Estimate the equilibrium constant K_p at 450°C.

18.120 Consider the reaction of methane, $CH_4(g)$, with oxygen, $O_2(g)$, to produce carbon dioxide, $CO_2(g)$, and water, $H_2O(l)$, taking place in a closed container. What is the entropy change at 25°C due to the reaction of 1.00 mol CH_4 in the container? When this reaction occurs, heat is produced. Suppose this heat leaves the container at 25°C. What is the entropy that flows out of the container as a result of this heat flow? What is the net change of entropy of the container and its surroundings?

18.121 Hydrogen gas is important industrially and has been used as a fuel. Water is an obvious source of hydrogen, but it requires free energy to break water into its elements. What is the standard free-energy change for the reaction at 25°C:

$$2H_2O(g) \longrightarrow 2H_2(g) + O_2(g)$$

Is this reaction spontaneous or not? One way to obtain the free energy needed for the production of hydrogen is to effectively couple this reaction to one that is quite spontaneous. Consider the reaction in which carbon reacts with oxygen to form carbon monoxide:

$$2C(s) + O_2(g) \longrightarrow 2CO(g)$$

What is the standard free-energy change for this reaction at 25°C? Is the reaction spontaneous or not? See if you can effectively couple these two reactions to get one that is spontaneous and would be useful as a way to manufacture hydrogen. What is the standard free-energy change for this reaction?

18.122 Coal is used as a fuel in some electric-generating plants. Coal is a complex material, but for simplicity we may consider it to be a form of carbon. The energy that can be derived from a fuel is sometimes compared with the enthalpy of the combustion reaction:

$$C(s) + O_2(g) \longrightarrow CO_2(g)$$

Calculate the standard enthalpy change for this reaction at 25°C. Actually, only a fraction of the heat from this reaction is available to produce electric energy. In electric generating plants, this reaction is used to generate heat for a steam engine, which turns the generator. Basically the steam engine is a type of heat engine in which steam enters the engine at high temperature (T_h), work is done, and the steam then exits at a lower temperature (T_l). The maximum fraction, f, of heat available to produce useful energy depends on the difference between these temperatures (expressed in kelvins), $f = (T_h - T_l)/T_h$. What is the maximum heat energy available for useful work from the combustion of 1.00 mol of $C(s)$ to $CO_2(g)$? (Assume the value of $\Delta H°$ calculated at 25°C for the heat obtained in the generator.) It is possible to consider more efficient ways to obtain useful energy from a fuel. For example, methane can be burned in a fuel cell to generate electricity directly. The maximum useful energy obtained in these cases is the maximum work, which equals the free-energy change. Calculate the standard free-energy change for the combustion of 1.00 mol of $C(s)$ to $CO_2(g)$. Compare this value with the maximum obtained with the heat engine described here.

Cumulative-Skills Problems

Key: The problems under this heading combine skills introduced in previous chapters with those given in the current one.

18.123 Hydrogen gas and iodine gas react to form hydrogen iodide. If 0.500 mol H_2 and 1.00 mol I_2 are placed in a closed 10.0-L vessel, what is the mole fraction of HI in the mixture when equilibrium is reached at 205°C? Use data from Appendix C and any reasonable approximations to obtain K.

18.124 Hydrogen bromide dissociates into its gaseous elements, H_2 and Br_2, at elevated temperatures. Calculate the percent dissociation at 375°C and 1.00 atm. What would be the percent dissociation at 375°C and 10.0 atm? Use data from Appendix C and any reasonable approximation to obtain K.

18.125 A 20.0-L vessel is filled with 1.00 mol of ammonia, NH_3. What percent of ammonia dissociates to the elements if equilibrium is reached at 345°C? Use data from Appendix C and make any reasonable approximations to obtain K.

18.126 A 25.0-L vessel is filled with 0.0100 mol CO and 0.0300 mol H_2. How many moles of CH_4 and how many moles of H_2O are produced when equilibrium is reached at 785°C? Use data from Appendix C and make any reasonable approximations to obtain K.

18.127 K_{sp} for silver chloride at 25.0°C is 1.782×10^{-10}. At 35.0°C, K_{sp} is 4.159×10^{-10}. What are $\Delta H°$ and $\Delta S°$ for the reaction?

18.128 K_a for acetic acid at 25.0°C is 1.754×10^{-5}. At 50.0°C, K_a is 1.633×10^{-5}. What are $\Delta H°$ and $\Delta S°$ for the ionization of acetic acid?

19 Electrochemistry

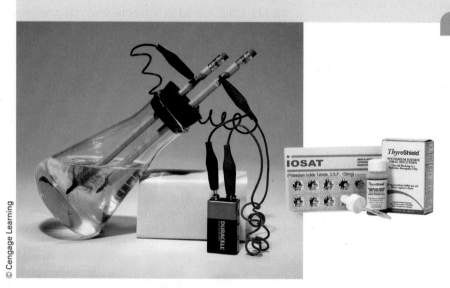

Electrolysis of a potassium iodide solution producing iodine and hydrogen gas. Potassium iodide pills can be used to protect against thyroid cancer that results from heavy exposure to nuclear fallout by blocking the absorption of radioactive iodine by the thyroid gland.

◙WL Sign in to OWL at www.cengage.com/owl to view tutorials and simulations, develop problem-solving skills, and complete online homework assigned by your professor.

GoChemistry Download mini lecture videos for key concept review and exam prep from OWL or purchase them from www.cengagebrain.com.

The first battery was invented by Alessandro Volta about 1800. He assembled a pile consisting of pairs of zinc and silver disks separated by paper disks soaked in salt water. With a tall pile, he could detect a weak electric shock when he touched the two ends of the pile. Later Volta showed that any two different metals could be used to make such a voltaic pile (Figure 19.1).

A battery cell that became popular during the nineteenth century was constructed in 1836 by the English chemist John Frederick Daniell. This cell used zinc and copper. The basic principle was that of Volta's battery pile, but each metal was surrounded by a solution of the metal ion, and the solutions were kept separate by a porous ceramic barrier. Each metal with its solution was a half-cell; a zinc half-cell and a copper half-cell made up one voltaic cell. This construction became the standard form of such cells, which exploit the spontaneous chemical reaction

Figure 19.1 ▲

Lemon battery Zinc and copper strips in a lemon generate a voltage.

$$Zn(s) + Cu^{2+}(aq) \longrightarrow Zn^{2+}(aq) + Cu(s)$$

to generate electrical energy.

Today many different kinds of batteries are in use. They include miniature button batteries for pocket calculators and watches, as well as larger batteries to power electric cars or to store the energy from the solar collectors of communications satellites. But all such batteries operate on the same general principles of Volta's pile or Daniell's cell. In this chapter, we will look at the general principles involved in setting up a chemical reaction as a battery. We will answer such questions as the following: What voltage can you expect from a particular battery? How can you relate the battery voltage to the equilibrium constant for the reaction?

Half-Reactions

A voltaic cell employs a spontaneous oxidation–reduction reaction as a source of energy. It is constructed to separate the reaction physically into two half-reactions, one involving oxidation and the other reduction, with electrons moving from the oxidation half-reaction through the external circuit to the reduction half-reaction. ▶

You may wish to review Sections 4.5 and 4.6.

19.1 Balancing Oxidation–Reduction Reactions in Acidic and Basic Solutions

In Chapter 4 (Section 4.6) we introduced the *half-reaction method* of balancing simple oxidation–reduction reactions. We now extend this method to reactions that occur in acidic or basic solutions. The steps used to balance these equations successfully are built upon those presented in Chapter 4. Keep in mind that oxidation–reduction reactions involve a transfer of electrons from one species to another. For example, in the reaction described in the chapter opener, zinc metal becomes zinc(II) ion: each zinc atom loses two electrons, and each copper(II) ion becomes copper metal (copper ion gains two electrons).

In this chapter, we will represent the hydronium ion, $H_3O^+(aq)$, by its simpler notation $H^+(aq)$, which chemists call the hydrogen ion. In the chapters that focused on acid–base reactions, we stressed proton transfer in solution, so we adopted the hydronium-ion notation. Here, we shift our focus to the electron-transfer aspect of certain reactions, so we adopt the simpler hydrogen-ion notation. Remember, however, that this is simply a notation change. It does not change the chemistry.

Skeleton Oxidation–Reduction Equations

To tackle more complex oxidation–reduction reactions in acidic and basic solutions, we need to review and discuss the essential information required to describe an oxidation–reduction reaction, which is called a *skeleton equation*. To set up the skeleton equation and then balance it, you need answers to the following questions:

1. What species is being oxidized (or, what is the reducing agent)? What species is being reduced (or, what is the oxidizing agent)?
2. What species result from the oxidation and reduction?
3. Does the reaction occur in acidic or basic solution?

For example, iron(II) ion may be oxidized easily by a number of oxidizing agents to yield the iron(III) ion. Permanganate ion in acidic, aqueous solution is a strong oxidizing agent and can oxidize Fe^{2+} to Fe^{3+}. In the process of oxidizing the iron(II) ion, the permanganate ion (MnO_4^-) is reduced to manganese(II) ion (Mn^{2+}). We can express these facts in the following skeleton equation:

$$\overset{+2}{Fe^{2+}}(aq) + \overset{+7}{MnO_4^-}(aq) \longrightarrow \overset{+3}{Fe^{3+}}(aq) + \overset{+2}{Mn^{2+}}(aq) \quad \text{(acidic solution)}$$

We have written oxidation numbers over the appropriate atoms. Note that the skeleton equation is not balanced. Nor is it complete; in an acidic, aqueous solution, $H^+(aq)$ and $H_2O(l)$ may be possible reactants or products in the equation. (If this were a basic solution, OH^- ion would replace the H^+ ion.)

Half-Reaction Method in Acidic and Basic Solutions

Although the preceding skeleton equation is not complete, it does give the essential information about the oxidation–reduction reaction. Moreover, given the skeleton equation, you can complete and balance the equation using the half-reaction method. Let us see how to do that. We first look at balancing oxidation–reduction equations in acidic solution. To balance such equations in basic solution requires additional steps.

Steps in Balancing Oxidation–Reduction Equations in Acidic Solution

Step 1: Assign oxidation numbers to each atom so you know what is oxidized and what is reduced.

Step 2: Split the skeleton equation into two half-reactions, proceeding as follows. Note the species containing the element that increases in oxidation number, and write those species to give the oxidation half-reaction. Similarly, note the species containing the element that decreases in oxidation number, and write the reduction half-reaction.

Step 3: Complete and balance each half-reaction.

a. Balance all atoms except O and H.

b. Balance O atoms by adding H_2O's to one side of the equation.

c. Balance H atoms by adding H^+ ions to one side of the equation.

d. Balance electric charge by adding electrons (e^-) to the more positive side.

Step 4: Combine the two half-reactions to obtain the final balanced oxidation–reduction equation.

a. Multiply each half-reaction by a factor such that when the half-reactions are added, the electrons cancel. (Electrons cannot appear in the final equation.)

b. Simplify the balanced equation by canceling species that occur on both sides, and reduce the coefficients to the smallest whole numbers. Check that the equation is indeed balanced.

Let us see how we would apply this method to the skeleton equation we wrote for the oxidation of iron(II) ion by permanganate ion in acidic solution. We have already determined the elements that change oxidation number (Step 1). We now write the incomplete equation representing the oxidation half-reaction. Iron(II) is oxidized to iron(III) ion. Therefore, following Step 2, we write

$$\overset{+2}{Fe^{2+}}(aq) \longrightarrow \overset{+3}{Fe^{3+}}(aq)$$

Note that the equation is balanced in Fe as it stands (Step 3a). Steps 3b and 3c require that the half-reaction be balanced in O and H. However, since these atoms do not appear in the half-reaction, we can ignore those steps. Then (Step 3d) you need to balance the half-reaction in electric charge. You do that by adding one electron to the more positive side.

$$Fe^{2+}(aq) \longrightarrow Fe^{3+}(aq) + e^- \quad \text{(oxidation half-reaction)}$$

Check that the half-reaction is balanced in Fe and in electric charge. Note that electrons are lost, as you expect for oxidation.

Now let us look at the reduction half-reaction. From the skeleton equation, you write the species containing the atom for which the oxidation number decreases (Step 2):

$$\overset{+7}{Mn}O_4^-(aq) \longrightarrow \overset{+2}{Mn}^{2+}(aq)$$

The equation is balanced in Mn as it stands (Step 3a). For Step 3b, you add four H_2O's to the right side of the equation to balance the O atoms. (Water is present, since this is an aqueous solution.)

$$MnO_4^-(aq) \longrightarrow Mn^{2+}(aq) + 4H_2O(l)$$

Step 3c says to balance the H atoms by adding H^+ ions. (Hydrogen ions are present, since this is an acidic solution.) You add eight H^+ ions to the left side of the equation.

$$MnO_4^-(aq) + 8H^+(aq) \longrightarrow Mn^{2+}(aq) + 4H_2O(l)$$

Finally, you balance the half-reaction in electric charge by adding electrons to the more positive side of the equation (Step 3d). Note that the left side has a charge of $(-1 + 8) = +7$, whereas the right side has a charge of $+2$. So you add five electrons to the left side:

$$MnO_4^-(aq) + 8H^+(aq) + 5e^- \longrightarrow Mn^{2+}(aq) + 4H_2O(l) \quad \text{(reduction half-reaction)}$$

Check that this half-reaction is balanced in the different kinds of atoms and in electric charge. Note that electrons are gained, as you expect for reduction.

In Step 4a, you need to multiply each half-reaction by a factor such that when the two half-reactions are added, the electrons cancel. You multiply the oxidation half-reaction by 5 (the number of electrons in the reduction half-reaction) and multiply the reduction half-reaction by 1 (the number of electrons in the oxidation half-reaction). Then you add the two half-reactions:

$$5 \times (Fe^{2+} \longrightarrow Fe^{3+} + e^-)$$
$$\underline{1 \times (MnO_4^- + 8H^+ + 5e^- \longrightarrow Mn^{2+} + 4H_2O)}$$
$$5Fe^{2+} + MnO_4^- + 8H^+ + \cancel{5e^-} \longrightarrow 5Fe^{3+} + Mn^{2+} + 4H_2O + \cancel{5e^-}$$

After canceling electrons, the equation is in simplest form (Step 4b). The final balanced equation is

$$5Fe^{2+}(aq) + MnO_4^-(aq) + 8H^+(aq) \longrightarrow 5Fe^{3+}(aq) + Mn^{2+}(aq) + 4H_2O(l)$$

You can write this equation in terms of the hydronium-ion notation, $H_3O^+(aq)$, if you wish, by changing $8H^+(aq)$ to $8H_3O^+(aq)$ and adding $8H_2O(l)$ to the other side of the equation to give a total of $12H_2O(l)$.

The following example further illustrates the half-reaction method.

◔WL Interactive Example | 19.1 | **Balancing Equations by the Half-Reaction Method (Acidic Solution)**

Gaining Mastery Toolbox

Critical Concept 19.1
When an oxidation–reduction reaction occurs, there will be elements that change oxidation state. Oxidation numbers, introduced in Chapter 4, are used to track the oxidation states of elements involved in chemical reactions. For acidic solutions, there are excess protons (H^+) available to balance the chemical equation.

Solution Essentials:
• Half-reaction-balancing method for acidic solutions
• Skeleton oxidation–reduction equations
• Balanced oxidation–reduction reaction
• Half-reaction
• Oxidation numbers
• Oxidation–reduction reaction

Zinc metal reacts with nitric acid, HNO_3, to produce a number of products, depending on how dilute the acid solution is. In a concentrated solution, zinc reduces nitrate ion to ammonium ion; zinc is oxidized to zinc ion, Zn^{2+}. Write the net ionic equation for this reaction.

Problem Strategy From the word statement of the problem, you write the skeleton equation representing the essential features of the oxidation–reduction reaction. Then, assign oxidation numbers to the atoms (Step 1) and separate the skeleton equation into two incomplete half-reactions (Step 2). Now complete and

(continued)

(continued)

balance each half-reaction (Step 3). (Balance the half-reaction in all elements except O and H; then balance O atoms, then H atoms, and finally electric charge.) In Step 4, you combine the half-reactions to eliminate electrons and simplify the result.

Solution Note that nitric acid is a strong acid, so it exists in solution as H^+ and NO_3^- ions. For the skeleton equation, write just the NO_3^- ion, because you will be adding H^+ later as needed to balance the equation. The skeleton equation, with oxidation numbers for those atoms that change values (following Step 1), is

$$\overset{0}{Zn}(s) + \overset{+5}{NO_3^-}(aq) \longrightarrow \overset{+2}{Zn^{2+}}(aq) + \overset{-3}{NH_4^+}(aq)$$

Now you separate this equation into two incomplete half-reactions (Step 2). Note that zinc is oxidized (increases in oxidation number), and nitrogen is reduced (decreases in oxidation number). The two incomplete half-reactions are

$$Zn(s) \longrightarrow Zn^{2+}(aq)$$
$$NO_3^-(aq) \longrightarrow NH_4^+(aq)$$

The oxidation half-reaction is balanced in Zn, as it stands (Step 3a). There are no O or H atoms, so you can ignore Steps 3b and 3c. Following (Step 3d, you add electrons to the more positive side to balance the charge.

$$Zn(s) \longrightarrow Zn^{2+}(aq) + 2e^- \qquad \text{(oxidation half-reaction)}$$

The reduction half-reaction is balanced in N (Step 3a) but is not complete in either O or H. Therefore, you add three H_2O's to the right side to balance O atoms (Step 3b) and add ten H^+ ions to the left side to balance the four H atoms in the NH_4^+ ion and the six H atoms in the H_2O's you just added (Step 3c).

$$10H^+(aq) + NO_3^-(aq) \longrightarrow NH_4^+(aq) + 3H_2O(l)$$

Now, you add electrons to the more positive side to balance the charge.

$$10H^+(aq) + NO_3^-(aq) + 8e^- \longrightarrow NH_4^+(aq) + 3H_2O(l) \qquad \text{(reduction half-reaction)}$$

Check that the half-reaction is balanced in charge, as well as in the different kinds of atoms.

Finally (Step 4a), you multiply the two half-reactions by factors so that when added, the electrons cancel. You multiply the oxidation half-reaction by 4 and the reduction half-reaction by 1:

$$4 \times (Zn \longrightarrow Zn^{2+} + 2e^-)$$
$$\underline{10H^+ + NO_3^- + 8e^- \longrightarrow NH_4^+ + 3H_2O}$$
$$4Zn + 10H^+ + NO_3^- + \cancel{8e^-} \longrightarrow 4Zn^{2+} + NH_4^+ + 3H_2O + \cancel{8e^-}$$

The equation requires no further simplification beyond canceling electrons (Step 4b). The net ionic equation is

$$\mathbf{4Zn(s) + 10H^+(aq) + NO_3^-(aq) \longrightarrow 4Zn^{2+}(aq) + NH_4^+(aq) + 3H_2O(l)}$$

Answer Check After you have completed the process for balancing this type of equation, you can perform a final check of your answer by making sure that you have both mass and charge balance.

Exercise 19.1 Iodic acid, HIO_3, can be prepared by reacting iodine, I_2, with concentrated nitric acid. The skeleton equation is

$$I_2(s) + NO_3^-(aq) \longrightarrow IO_3^-(aq) + NO_2(g)$$

Balance this equation.

■ See Problems 19.35 and 19.36.

To balance an oxidation–reduction equation in basic solution, you begin by balancing the equation *as if it were a reaction in acidic solution*. Then, you add the following steps. ▶

Additional Steps for Balancing Oxidation–Reduction Equations in Basic Solution

Step 5: Note the number of H^+ ions in the equation. Add this number of OH^- ions to both sides of the equation.

Step 6: Simplify the equation by noting that H^+ reacts with OH^- to give H_2O. Cancel any H_2O's that occur on both sides of the equation and reduce the equation to simplest terms.

You can apply Steps 5 and 6 to the individual half-reactions, rather than to the overall equation. Although this is more work, it does yield the complete half-reactions, which can be useful in discussing batteries (later in this chapter).

The following example illustrates how to apply these rules.

✿WL Interactive Example 19.2   Balancing Equations by the Half-Reaction Method (Basic Solution)

Gaining Mastery Toolbox

Critical Concept 19.2
When an oxidation–reduction reaction occurs, there will be elements that change oxidation state. Oxidation numbers, introduced in Chapter 4, are used to track the oxidation states of elements involved in chemical reactions. For basic solutions, start the balancing process as if the solution is acidic. In the final balancing step, use excess hydroxide ions (OH^-) to balance the chemical equation.

Solution Essentials:
- Half-reaction-balancing method for basic solutions
- Half-reaction-balancing method for acidic solutions
- Skeleton oxidation–reduction equations
- Balanced oxidation–reduction reaction
- Half-reaction
- Oxidation numbers
- Oxidation–reduction reaction

Permanganate ion oxidizes sulfite ion in basic solution according to the following skeleton equation:

$$MnO_4^-(aq) + SO_3^{2-}(aq) \longrightarrow MnO_2(s) + SO_4^{2-}(aq)$$

Use the half-reaction method to complete and balance this equation.

Problem Strategy First complete and balance this equation as if it were in acid solution, as in the previous example. Then, following Step 5, add as many OH^- ions to both sides of the equation as there are H^+ ions. After noting that each H^+ ion reacts with an OH^- ion to give H_2O, simplify the equation (Step 6).

Solution After balancing the equation as if it were in acid solution, you obtain the following:

$$2MnO_4^- + 3SO_3^{2-} + 2H^+ \longrightarrow 2MnO_2 + 3SO_4^{2-} + H_2O$$

Following Step 5, you add $2OH^-$ to both sides of the equation.

$$2MnO_4^- + 3SO_3^{2-} + \underbrace{2H^+ + 2OH^-}_{2H_2O} \longrightarrow 2MnO_2 + 3SO_4^{2-} + H_2O + 2OH^-$$

You replace $2H^+ + 2OH^-$ by $2H_2O$ on the left side, then cancel one of these with the H_2O on the right side. The balanced equation for the reaction in basic solution is

$$\mathbf{2MnO_4^-(aq) + 3SO_3^{2-}(aq) + H_2O(l) \longrightarrow 2MnO_2(s) + 3SO_4^{2-}(aq) + 2OH^-(aq)}$$

Answer Check Just as in balancing oxidation–reduction reactions in acidic solution, as a final check, be certain that the masses and charges balance in the equation.

Exercise 19.2   Balance the following equation using the half-reaction method.

$$H_2O_2 + ClO_2 \longrightarrow ClO_2^- + O_2 \qquad \text{(basic solution)}$$

■ See Problems 19.36 and 19.37.

Voltaic Cells

The next several sections describe battery cells, or voltaic cells (also called galvanic cells). These are a kind of electrochemical cell. An **electrochemical cell** is *a system consisting of electrodes that dip into an electrolyte and in which a chemical reaction either uses or generates an electric current.* A **voltaic,** or **galvanic, cell** is *an electrochemical cell in which a spontaneous reaction generates an electric current.* An **electrolytic cell** is *an electrochemical cell in which an external energy source drives an otherwise nonspontaneous reaction.* In the next sections, we will discuss the basic principles behind voltaic cells and then explore some of their commercial uses.

19.2 Construction of Voltaic Cells

A voltaic cell consists of two *half-cells* that are electrically connected. Each **half-cell** is *the portion of an electrochemical cell in which a half-reaction takes place*. A simple half-cell can be made from a metal strip that dips into a solution of its metal ion. An example is the zinc–zinc ion half-cell (often called simply a zinc electrode), which consists of a zinc metal strip dipping into a solution of a zinc salt. Another simple half-cell consists of a copper metal strip dipping into a solution of a copper salt (copper electrode).

In a voltaic cell, two half-cells are connected in such a way that electrons flow from one metal electrode to another through an external circuit, and ions flow from one half-cell to another through an internal cell connection. Figure 19.2 illustrates an atomic view of a voltaic cell consisting of a zinc electrode and a copper electrode. As long as there is an external circuit, electrons can flow through it from one electrode to another. Because zinc tends to lose electrons more readily than copper, zinc atoms in the zinc electrode lose electrons to produce zinc ions. These electrons flow through the external circuit to the copper electrode, where they react with the copper ions in that half-cell to deposit copper metal atoms. The net result is that zinc metal reacts with copper ions to produce zinc ions and copper metal, and an electric current flows through the external circuit.

The two half-cells must be connected internally to allow ions to flow between them. As zinc ions continue to be produced, the zinc ion solution begins to build up a positive charge. Similarly, as copper ions plate out as copper, the solution builds up a negative charge. The half-cell reactions will stop unless positive ions can move from the zinc half-cell to the copper half-cell, and negative ions from the copper half-cell can move to the zinc half-cell. It is necessary that these ion flows occur without mixing of the zinc ion and copper ion solutions. If copper ion were to come in contact with the zinc metal, for example, direct reaction would occur without an electric current being generated. The voltage would drop, and the battery would run down quickly.

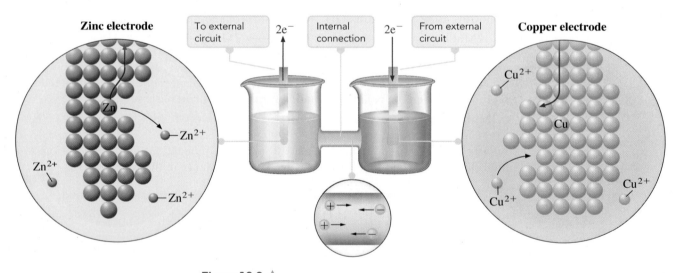

Zinc electrode   To external circuit $2e^-$ Internal connection $2e^-$ From external circuit   **Copper electrode**

Zn Zn^{2+} Zn^{2+} Zn^{2+}

Cu^{2+} Cu Cu^{2+} Cu^{2+} Cu^{2+}

Figure 19.2 ▲

Atomic view of a voltaic cell In the half-cell on the left, a zinc metal atom loses two electrons. These flow through the zinc electrode to the external circuit, then to the copper electrode in the half-cell on the right. Negative charge on the copper electrode (from the electrons) attracts a copper ion, which reacts with the electrons on the electrode to form a copper metal atom. The internal connection, discussed later in the text, allows ions to flow between the two half-cells; it is required to maintain charge balance.

Figure 19.3*a* shows the two half-cells of a voltaic cell connected by a salt bridge. A **salt bridge** is *a tube of an electrolyte in a gel that is connected to the two half-cells of a voltaic cell; the salt bridge allows the flow of ions but prevents the mixing of the different solutions that would allow direct reaction of the cell reactants.* In Figure 19.3*b*, the half-cells are connected externally so that an electric current flows. Figure 19.3*c* shows an actual setup of the zinc–copper cell.

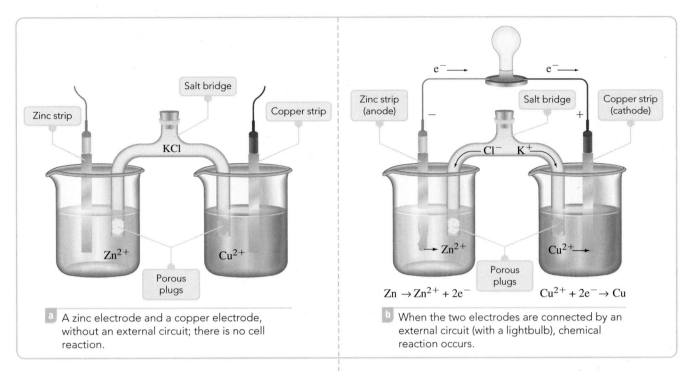

a A zinc electrode and a copper electrode, without an external circuit; there is no cell reaction.

b When the two electrodes are connected by an external circuit (with a lightbulb), chemical reaction occurs.

$$Zn \rightarrow Zn^{2+} + 2e^-$$ $$Cu^{2+} + 2e^- \rightarrow Cu$$

Figure 19.3 ◀

A zinc–copper voltaic cell

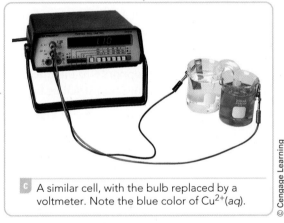

c A similar cell, with the bulb replaced by a voltmeter. Note the blue color of $Cu^{2+}(aq)$.

© Cengage Learning

The two half-cell reactions, as noted earlier, are

$$Zn(s) \longrightarrow Zn^{2+}(aq) + 2e^-$$ (oxidation half-reaction)

$$Cu^{2+}(aq) + 2e^- \longrightarrow Cu(s)$$ (reduction half-reaction)

The first half-reaction, in which a species loses electrons, is the oxidation half-reaction. *The electrode at which oxidation occurs* is called the **anode.** The second half-reaction, in which a species gains electrons, is the reduction half-reaction. *The electrode at which reduction occurs* is called the **cathode.** These definitions of anode and cathode hold for all electrochemical cells, including electrolytic cells.

Note that the sum of the two half-reactions

$$Zn(s) + Cu^{2+}(aq) \longrightarrow Zn^{2+}(aq) + Cu(s)$$

is *the net reaction that occurs in the voltaic cell;* it is called the **cell reaction.**

Once you know which electrode is the anode and which is the cathode, you can determine the direction of electron flow in the external portion of the circuit. Electrons are given up by the anode (from the oxidation half-reaction) and thus flow from it, whereas electrons are used up by the cathode (by the reduction half-reaction) and so flow into this electrode. The anode in a voltaic cell has a negative sign, because electrons flow from it. The cathode in a voltaic cell has a positive sign. Look again at Figure 19.3*b* and note the labeling of the electrodes as anode and cathode; also note the direction of electron flow in the external circuit and the signs of the electrodes. Note too the migration of ions in the solutions. The following example further illustrates these points about a voltaic cell.

Example 19.3   Sketching and Labeling a Voltaic Cell

Gaining Mastery Toolbox

Critical Concept 19.3
An electrochemical cell is composed of two half-cells. Each half-reaction is contained in a separate half-cell. Electrons flow from the anode (oxidation) to the cathode (reduction).

Solution Essentials:
- Anode and cathode
- Salt bridge
- Half-cell
- Voltaic cell
- Electrochemical cell
- Balanced oxidation–reduction reaction
- Half-reaction
- Oxidation–reduction reaction

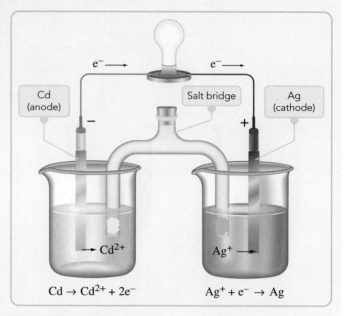

Figure 19.4 ▲

Cell described in the example A voltaic cell consisting of cadmium and silver electrodes.

A voltaic cell is constructed from a half-cell in which a cadmium rod dips into a solution of cadmium nitrate, $Cd(NO_3)_2$, and another half-cell in which a silver rod dips into a solution of silver nitrate, $AgNO_3$. The two half-cells are connected by a salt bridge. Silver ion is reduced during operation of the voltaic cell. Draw a sketch of the cell. Label the anode and cathode, showing the corresponding half-reactions at these electrodes. Indicate the electron flow in the external circuit (with a lightbulb), the signs of the electrodes, and the direction of cation migration in the half-cells.

Problem Strategy The problem states that silver ion is reduced during the cell operation, so we know that silver is the cathode. If silver is the cathode, then cadmium must be the anode. With this information we can then write the half-reactions. The half-reactions provide all of the information we need to draw a complete sketch of the cell.

Solution Because silver ion is reduced at the silver electrode, the silver electrode is the cathode. The half-reaction is

$$Ag^+(aq) + e^- \longrightarrow Ag(s)$$

The cadmium electrode must be the anode (the electrode at which oxidation occurs); the half-reaction is

$$Cd(s) \longrightarrow Cd^{2+}(aq) + 2e^-$$

Now that you have labeled the electrodes, you can see that the electron flow in the external circuit is from the cadmium electrode (anode) to the silver electrode (cathode).

Positive ions will flow in the solution portion of the circuit opposite to the direction of the electrons. **The sketch for this cell, including the labeling, is given in Figure 19.4.**

Answer Check Make sure that the direction of electron flow is consistent with the half-reactions; the electrons always flow *from* the material undergoing oxidation.

Exercise 19.3   A voltaic cell consists of a silver–silver ion half-cell and a nickel–nickel(II) ion half-cell. Silver ion is reduced during operation of the cell. Sketch the cell, labeling the anode and cathode and indicating the corresponding electrode reactions. Show the direction of electron flow in the external circuit and the direction of cation movement in the half-cells.

■ See Problems 19.43 and 19.44.

CONCEPT CHECK 19.1

If you were to construct a wet cell and decided to replace the salt bridge with a piece of copper wire, would the cell produce a sustainable current? Explain your answer.

19.3 Notation for Voltaic Cells

It is convenient to have a shorthand way of designating particular voltaic cells. The cell described earlier, consisting of a zinc metal–zinc ion half-cell and a copper metal–copper ion half-cell, is written

$$Zn(s)\,|\,Zn^{2+}(aq)\,\|\,Cu^{2+}(aq)\,|\,Cu(s)$$

In this notation, the anode, or oxidation half-cell, is always written on the left; the cathode, or reduction half-cell, is written on the right. The two electrodes are electrically connected by means of a salt bridge, denoted by two vertical bars.

$$\underset{\text{anode}}{Zn(s)\,|\,Zn^{2+}(aq)} \quad \underset{\text{salt bridge}}{\|} \quad \underset{\text{cathode}}{Cu^{2+}(aq)\,|\,Cu(s)}$$

The cell terminals are at the extreme ends in this cell notation, and a single vertical bar indicates a phase boundary—say, between a solid terminal and the electrode solution. For the anode of the same cell, you have

$$\underset{\text{anode terminal}}{Zn(s)} \quad \underset{\text{phase boundary}}{|} \quad \underset{\text{anode electrolyte}}{Zn^{2+}(aq)}$$

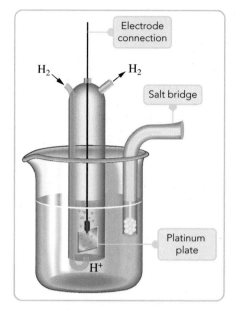

When the half-reaction involves a gas, an inert material such as platinum serves as a terminal and as an electrode surface on which the half-reaction occurs. The platinum catalyzes the half-reaction but otherwise is not involved in it. Figure 19.5 shows a hydrogen electrode; hydrogen bubbles over a platinum plate that is immersed in an acidic solution. The cathode half-reaction is

$$2H^+(aq) + 2e^- \rightleftharpoons H_2(g)$$

The notation for the hydrogen electrode, written as a cathode, is

$$H^+(aq)|H_2(g)|Pt$$

To write such an electrode as an anode, you simply reverse the notation:

$$Pt|H_2(g)|H^+(aq)$$

Here are several additional examples of this notation for electrodes (written as cathodes). A comma separates ions present in the same solution. We will write the cathode with oxidized species before reduced species, in the same order as in the half-reaction.

Cathode	*Cathode Reaction*
$Cl_2(g)|Cl^-(aq)|Pt$	$Cl_2(g) + 2e^- \rightleftharpoons 2Cl^-(aq)$
$Fe^{3+}(aq), Fe^{2+}(aq)|Pt$	$Fe^{3+}(aq) + e^- \rightleftharpoons Fe^{2+}(aq)$
$Cd^{2+}(aq)|Cd(s)$	$Cd^{2+}(aq) + 2e^- \rightleftharpoons Cd(s)$

Figure 19.5 ▲

A hydrogen electrode Hydrogen gas bubbles over a platinum surface, where the half-reaction $2H^+(aq) + 2e^- \rightleftharpoons H_2(g)$ occurs.

Exercise 19.4 Write the notation for a cell in which the electrode reactions are

$$2H^+(aq) + 2e^- \longrightarrow H_2(g)$$

$$Zn(s) \longrightarrow Zn^{2+}(aq) + 2e^-$$

■ See Problems 19.49 and 19.50.

You can write the overall cell reaction from the cell notation by first writing the appropriate half-cell reactions, multiplying as necessary, and then summing so that the electrons cancel. Example 19.4 shows how this is done.

◉WL Interactive Example 19.4   **Writing the Cell Reaction from the Cell Notation**

Gaining Mastery Toolbox

Critical Concept 19.4
Cell notation fully describes the half-reactions of a voltaic cell. This notation can include the phases of the substances, ion concentrations, and gas pressures. The anode half-cell reaction is always represented on the left.

Solution Essentials:
- Cell notation
- Anode and cathode
- Salt bridge
- Half-cell
- Voltaic cell
- Electrochemical cell
- Balanced oxidation–reduction reaction
- Half-reaction
- Oxidation–reduction reaction

ⓐ Write the cell reaction for the voltaic cell

$$Tl(s)|Tl^+(aq)\|Sn^{2+}(aq)|Sn(s)$$

ⓑ Write the cell reaction for the voltaic cell

$$Zn(s)|Zn^{2+}(aq)\|Fe^{3+}(aq), Fe^{2+}(aq)|Pt$$

Problem Strategy The cell notation gives the species involved in each half-reaction. Complete and balance each half-reaction, and then multiply by factors so that when you add the half-reactions, the electrons cancel. The result is the cell reaction.

Solution

ⓐ The half-cell reactions are

$$Tl(s) \longrightarrow Tl^+(aq) + e^-$$

$$Sn^{2+}(aq) + 2e^- \longrightarrow Sn(s)$$

(continued)

(continued)

Multiplying the anode reaction by 2 and then summing the half-cell reactions gives

$$2Tl(s) + Sn^{2+}(aq) \longrightarrow 2Tl^+(aq) + Sn(s)$$

b The half-cell reactions are

$$Zn(s) \longrightarrow Zn^{2+}(aq) + 2e^-$$
$$Fe^{3+}(aq) + e^- \longrightarrow Fe^{2+}(aq)$$

and the cell reaction is

$$Zn(s) + 2Fe^{3+}(aq) \longrightarrow Zn^{2+}(aq) + 2Fe^{2+}(aq)$$

Answer Check Always make sure that you multiplied the half-reactions by the correct coefficients to ensure that the electrons cancel.

Exercise 19.5 Give the overall cell reaction for the voltaic cell

$$Cd(s)|Cd^{2+}(aq)\|H^+(aq)|H_2(g)|Pt$$

■ See Problems 19.53 and 19.54.

To fully specify a voltaic cell, it is necessary to give the concentrations of solutions or ions and the pressure of gases. In the cell notation, these are written within parentheses for each species. For example,

$$Zn(s)|Zn^{2+}(1.0\ M)\|H^+(1.0\ M)|H_2(1.0\ atm)|Pt$$

19.4 Cell Potential

Work is needed to move electrons in a wire or to move ions through a solution to an electrode. The situation is analogous to pumping water from one point to another. Work must be expended to pump the water. The analogy may be extended. Water moves from a point at high pressure to a point at low pressure. Thus, a pressure difference is required. The work expended in moving water through a pipe depends on the volume of water and the pressure difference. The situation with electricity is similar. An electric charge moves from a point at high electric potential (high electrical pressure) to a point at low electric potential (low electrical pressure). The work needed to move an electric charge through a conductor depends on the total charge moved and the potential difference. **Potential difference** is *the difference in electric potential (electrical pressure) between two points.* You measure this quantity in volts. The **volt, V,** is *the SI unit of potential difference.* The electrical work expended in moving a charge through a conductor is

Electrical work = charge × potential difference

Corresponding SI units for the terms in this equation are

Joules = coulombs × volts

The **Faraday constant, F,** is *the magnitude of charge on one mole of electrons; it equals* 9.6485×10^4 *C per mole of electrons* (96,485 C/mol e⁻). The *faraday* is a unit of charge equal to 9.6485×10^4 C. In moving this quantity of charge (1 faraday of charge) from one electrode to another, the numerical value of the work done by a voltaic cell is the product of the faraday constant F times the potential difference between the electrodes. The work w is the negative of this, because the voltaic cell loses energy as it does work on the surroundings.

$$w = -F \times \text{potential difference}$$

In the normal operation of a voltaic cell, the potential difference (voltage) across the electrodes is less than the maximum possible voltage of the cell. One reason for this is that it takes energy or work to drive a current through the cell itself. The decrease in cell voltage as current is drawn reflects this energy expenditure within the cell; and the greater the current, the lower the voltage. Thus, the cell voltage has its maximum value only when no current flows. The situation is analogous to measuring the difference between the pressure of water in a faucet and that of the outside atmosphere. You expect this difference to be higher if you measure it by a device that does not require you to run water from the faucet. When you do run water, the pressure difference drops.

The maximum potential difference between the electrodes of a voltaic cell is referred to as the **cell potential** or **electromotive force (emf)** of the cell, or E_{cell}. It can be measured by an electronic digital voltmeter (Figure 19.6).

You can also determine which electrode of a cell is the anode and which is the cathode from this measurement. The digital voltmeter gives the reading with a sign attached, from which you can deduce the sign, or polarity, of each electrode. The anode of a voltaic cell has negative polarity, and the cathode has positive polarity.

We can now write an expression for the maximum work obtainable from a voltaic cell. Let n be the number of moles of electrons transferred in the overall cell equation. The maximum electrical work of a voltaic cell for molar amounts of reactants (according to the cell equation *as written*) is

$$w_{max} = -nFE_{cell}$$

Here E_{cell} is the cell potential, and F is the Faraday constant, 9.6485×10^4 C/mol e$^-$.

Figure 19.6 ▲

A digital voltmeter A digital voltmeter draws negligible current, so it can be used to measure cell potential.

© Cengage Learning

◉WL Interactive Example 19.5   Calculating the Quantity of Work from a Given Amount of Cell Reactant

Gaining Mastery Toolbox

Critical Concept 19.5
The equation for the maximum electrical work ($w_{max} = -nFE_{cell}$) is only for the molar amounts of reactants in the balanced chemical equation. Balancing the electrochemical equation will provide the values of n and E_{cell} for the equation and the subsequent calculation of w_{max}. If the amount of reactants is *not* the same as that in the balanced electrochemical chemical equation, then the maximum work must be calculated using the quantities of reactant consumed.

Solution Essentials:
- Maximum electrical work of voltaic cell ($w_{max} = -nFE_{cell}$)
- Cell potential
- Balanced oxidation–reduction reaction
- Half-reaction

The cell potential of a particular voltaic cell with the cell reaction

$$Hg_2^{2+}(aq) + H_2(g) \rightleftharpoons 2Hg(l) + 2H^+(aq)$$

is 0.650 V. Calculate the maximum electrical work of this cell when 0.500 g H_2 is consumed.

Problem Strategy We can determine the maximum work of a cell by using the formula $w_{max} = -nFE_{cell}$, where E_{cell} is given in the problem, F is the Faraday constant, and n is the number of moles of electrons transferred in the overall cell equation. To get n, you split the cell reaction into half-reactions, multiplying each by a factor so that when added together, they give the cell reaction. Then n equals the number of moles of electrons in either half-reaction.

Solution The half-reactions are

$$Hg_2^{2+}(aq) + 2e^- \rightleftharpoons 2Hg(l)$$
$$H_2(g) \rightleftharpoons 2H^+(aq) + 2e^-$$

n equals 2, and the maximum work for the reaction as written is

$$w_{max} = -nFE_{cell} = -2 \text{ mol e}^- \times 96,485 \text{ C/mol e}^- \times 0.650 \text{ V}$$
$$= -1.25 \times 10^5 \text{ V·C}$$
$$= -1.25 \times 10^5 \text{ J}$$

(Remember that a joule is equal to a volt-coulomb. Also, note that the negative sign means work is done by the cell—that is, energy is lost by the cell.) For 0.500 g H_2, the maximum work is

$$0.500 \text{ g } H_2 \times \frac{1 \text{ mol } H_2}{2.02 \text{ g } H_2} \times \frac{-1.25 \times 10^5 \text{ J}}{1 \text{ mol } H_2} = -3.09 \times 10^4 \text{ J}$$

Note that the conversion factor -1.25×10^5 J/1 mol H_2 is determined by the chemical equation as written. In the equation, 1 mol H_2 reacts, and the maximum work produced is -1.25×10^5 J.

(continued)

(continued)

Answer Check The maximum work per mole should be in the range of values expected for ΔG, which is on the order of tens to hundreds of kilojoules per mole, as it is here. You should be suspicious of any values that you obtain that lie outside this range.

Exercise 19.6 What is the maximum electrical work that can be obtained from 6.54 g of zinc metal that reacts in a Daniell cell, described in the chapter opening, whose cell potential is 1.10 V? The overall cell reaction is

$$Zn(s) + Cu^{2+}(aq) \longrightarrow Zn^{2+}(aq) + Cu(s)$$

■ See Problems 19.57, 19.58, 19.59, and 19.60.

19.5 Standard Cell Potentials and Standard Electrode Potentials

A cell potential is a measure of the driving force of the cell reaction. This reaction occurs in the cell as separate half-reactions: an oxidation half-reaction and a reduction half-reaction. The general forms of these half-reactions are

$$\text{Reduced species} \longrightarrow \text{oxidized species} + ne^- \quad \text{(oxidation/anode)}$$

$$\text{Oxidized species} + ne^- \longrightarrow \text{reduced species} \quad \text{(reduction/cathode)}$$

You can imagine that the cell potential is composed of a contribution from the anode (whose value depends on the ability of the oxidation half-reaction to lose electrons) and a contribution from the cathode (whose value depends on the ability of the reduction half-reaction to gain electrons). We call these contributions the *oxidation potential* and the *reduction potential,* respectively. Then

$$E_{cell} = \text{oxidation potential} + \text{reduction potential}$$

A reduction potential is a measure of the tendency for an oxidized species to gain electrons in the reduction half-reaction (in other words, a measure of the ability of the species to act as an oxidizing agent). If you can construct a table of reduction potentials, you will have a list of strengths of oxidizing agents, in addition to having a way of calculating cell potentials.

You can look at an oxidation half-reaction as the reverse of a corresponding reduction half-reaction. The oxidation potential for an oxidation half-reaction equals the negative of the reduction potential for the reverse half-reaction (which is a reduction).

Oxidation potential for a half-reaction = −reduction potential for the reverse half-reaction

This means that in practice you need to tabulate only oxidation potentials or reduction potentials. The choice, by convention, is to tabulate reduction potentials. We call these *electrode potentials,* and we denote them by the symbol E (without the subscript cell, as in E_{cell}).

Consider the zinc–copper cell described earlier.

$$Zn(s)|Zn^{2+}(aq) \| Cu^{2+}(aq)|Cu(s)$$

The half-reactions are

$$Zn(s) \longrightarrow Zn^{2+}(aq) + 2e^-$$

$$Cu^{2+}(aq) + 2e^- \longrightarrow Cu(s)$$

The first half-reaction is an oxidation. If you write E_{Zn} for the electrode potential corresponding to the reduction half-reaction $Zn^{2+}(aq) + 2e^- \longrightarrow Zn(s)$, then $-E_{Zn}$ is the potential for the oxidation half-reaction $Zn(s) \longrightarrow Zn^{2+}(aq) + 2e^-$. The copper half-reaction is a reduction. Write E_{Cu} for the electrode potential.

The cell potential is the sum of the potentials for the reduction and oxidation half-reactions. For the cell we have been describing, the cell potential is the sum of the reduction potential (electrode potential) for the copper half-cell and the oxidation potential (negative of the electrode potential) for the zinc half-cell.

$$E_{cell} = E_{Cu} + (-E_{Zn}) = E_{Cu} - E_{Zn}$$

Note that the cell potential equals the difference between the two electrode potentials. You can think of the electrode potential as the electric potential on the electrode, and you obtain the cell potential as a potential difference in which you subtract the anode potential from the cathode potential.

$$E_{cell} = E_{cathode} - E_{anode}$$

The electrode potential is an *intensive property*. This means that its value is independent of the amount of species in the reaction. Thus, the electrode potential for the half-reaction

$$2Cu^{2+}(aq) + 4e^- \longrightarrow 2Cu(s)$$

is the same as for

$$Cu^{2+}(aq) + 2e^- \longrightarrow Cu(s)$$

Tabulating Standard Electrode Potentials

The cell potential of a voltaic cell depends on the concentrations of substances and the temperature of the cell. For purposes of tabulating electrochemical data, it is usual to choose thermodynamic standard-state conditions for voltaic cells. The **standard cell potential,** $E°_{cell}$, is *the emf of a voltaic cell operating under standard-state conditions (solute concentrations are each 1 M, gas pressures are each 1 atm, and the temperature has a specified value—usually 25°C).* Note the superscript degree sign (°), which signifies standard-state conditions. ▶

If you can derive a table of electrode potentials, you can calculate cell potentials from them. This provides a great advantage over tabulating cell potentials. From a small table of electrode potentials, you can obtain all the cell potentials you could construct from pairs of electrodes. For instance, a table of 40 electrode potentials would give the cell potentials of nearly 800 voltaic cells.

However, it is not possible to measure the potential of a single electrode; only the cell potentials of cells can be measured. What you can do is measure the cell potentials of cells constructed from various electrodes connected in turn to one particular electrode, which you choose as a reference. You arbitrarily assign this reference electrode a potential equal to zero and obtain the potentials for the other electrodes by measuring the cell potentials. By convention, the reference chosen for comparing electrode potentials is the *standard hydrogen electrode,* and it is assigned a potential of 0.00 V. This is an electrode like that pictured earlier in Figure 19.5 operating under standard-state conditions.

The **standard electrode potential,** $E°$, is *the electrode potential when the concentrations of solutes are 1 M, the gas pressures are 1 atm, and the temperature has a specified value (usually 25°C).* The superscript degree sign (°) signifies standard-state conditions.

To understand how standard electrode potentials are obtained, look at how you would find the standard electrode potential, $E°$, for the zinc electrode. You connect a standard zinc electrode to a standard hydrogen electrode. When you measure the cell potential with a voltmeter, you obtain 0.76 V, with the zinc electrode acting as the anode.

Now write the cell potential in terms of the electrode potentials. The cell is

$$Zn(s)|Zn^{2+}(aq)\|H^+(aq)|H_2(g)|Pt$$

and the half-reactions with corresponding half-cell potentials (oxidation or reduction potentials) are

$$Zn(s) \longrightarrow Zn^{2+}(aq) + 2e^-; \; -E°_{Zn}$$
$$2H^+(aq) + 2e^- \longrightarrow H_2(g); \; E°_{H_2}$$

The cell potential is the sum of the half-cell potentials.

$$E_{cell} = E°_{H_2} + (-E°_{Zn})$$

Substitute 0.76 V for the cell potential and 0.00 V for the standard hydrogen electrode potential. This gives $E°_{Zn} = -0.76$ V.

Proceeding in this way, you can obtain the electrode potential for a series of half-cell reactions. Table 19.1 lists standard electrode potentials for selected half-cells at 25°C.

To be precise, the standard-state conditions are those in which the activities (ideal concentrations) of solutes and gases equal 1. This is equivalent to 1 M only for ideal solutions. Because of the strong attractions between ions, substantial deviations from ideal conditions exist in electrolyte solutions, except when very dilute. We will ignore these deviations here.

Strengths of Oxidizing and Reducing Agents

Standard electrode potentials are useful in determining the strengths of oxidizing and reducing agents under standard-state conditions. Because electrode potentials are reduction potentials, those reduction half-reactions in Table 19.1 with the larger (that is, more positive) electrode potentials have the greater tendency to go left to right as written. A reduction half-reaction has the general form

$$\text{Oxidized species} + ne^- \longrightarrow \text{reduced species}$$

Table 19.1 Standard Electrode (Reduction) Potentials in Aqueous Solution at 25°C*	
Cathode (Reduction) Half-Reaction	**Standard Potential, $E°$(V)**
$Li^+(aq) + e^- \rightleftharpoons Li(s)$	-3.04
$Na^+(aq) + e^- \rightleftharpoons Na(s)$	-2.71
$Mg^{2+}(aq) + 2e^- \rightleftharpoons Mg(s)$	-2.38
$Al^{3+}(aq) + 3e^- \rightleftharpoons Al(s)$	-1.66
$2H_2O(l) + 2e^- \rightleftharpoons H_2(g) + 2OH^-(aq)$	-0.83
$Zn^{2+}(aq) + 2e^- \rightleftharpoons Zn(s)$	-0.76
$Cr^{3+}(aq) + 3e^- \rightleftharpoons Cr(s)$	-0.74
$Fe^{2+}(aq) + 2e^- \rightleftharpoons Fe(s)$	-0.41
$Cd^{2+}(aq) + 2e^- \rightleftharpoons Cd(s)$	-0.40
$Ni^{2+}(aq) + 2e^- \rightleftharpoons Ni(s)$	-0.23
$Sn^{2+}(aq) + 2e^- \rightleftharpoons Sn(s)$	-0.14
$Pb^{2+}(aq) + 2e^- \rightleftharpoons Pb(s)$	-0.13
$Fe^{3+}(aq) + 3e^- \rightleftharpoons Fe(s)$	-0.04
$2H^+(aq) + 2e^- \rightleftharpoons H_2(g)$	0.00
$Sn^{4+}(aq) + 2e^- \rightleftharpoons Sn^{2+}(aq)$	0.15
$Cu^{2+}(aq) + e^- \rightleftharpoons Cu^+(aq)$	0.16
$Cu^{2+}(aq) + 2e^- \rightleftharpoons Cu(s)$	0.34
$IO^-(aq) + H_2O(l) + 2e^- \rightleftharpoons I^-(aq) + 2OH^-(aq)$	0.49
$Cu^+(aq) + e^- \rightleftharpoons Cu(s)$	0.52
$I_2(s) + 2e^- \rightleftharpoons 2I^-(aq)$	0.54
$Fe^{3+}(aq) + e^- \rightleftharpoons Fe^{2+}(aq)$	0.77
$Hg_2^{2+}(aq) + 2e^- \rightleftharpoons 2Hg(l)$	0.80
$Ag^+(aq) + e^- \rightleftharpoons Ag(s)$	0.80
$Hg^{2+}(aq) + 2e^- \rightleftharpoons Hg(l)$	0.85
$ClO^-(aq) + H_2O(l) + 2e^- \rightleftharpoons Cl^-(aq) + 2OH^-(aq)$	0.90
$2Hg^{2+}(aq) + 2e^- \rightleftharpoons Hg_2^{2+}(aq)$	0.90
$NO_3^-(aq) + 4H^+(aq) + 3e^- \rightleftharpoons NO(g) + 2H_2O(l)$	0.96
$Br_2(l) + 2e^- \rightleftharpoons 2Br^-(aq)$	1.07
$O_2(g) + 4H^+(aq) + 4e^- \rightleftharpoons 2H_2O(l)$	1.23
$Cr_2O_7^{2-}(aq) + 14H^+(aq) + 6e^- \rightleftharpoons 2Cr^{3+}(aq) + 7H_2O(l)$	1.33
$Cl_2(g) + 2e^- \rightleftharpoons 2Cl^-(aq)$	1.36
$MnO_4^-(aq) + 8H^+(aq) + 5e^- \rightleftharpoons Mn^{2+}(aq) + 4H_2O(l)$	1.49
$H_2O_2(aq) + 2H^+(aq) + 2e^- \rightleftharpoons 2H_2O(l)$	1.78
$S_2O_8^{2-}(aq) + 2e^- \rightleftharpoons 2SO_4^{2-}(aq)$	2.01
$F_2(g) + 2e^- \rightleftharpoons 2F^-(aq)$	2.87

*See Appendix I for a more extensive table.

The oxidized species acts as an oxidizing agent. Consequently, *the strongest oxidizing agents in a table of standard electrode potentials are the oxidized species corresponding to half-reactions with the largest (most positive) $E°$ values.*

Those reduction half-reactions with lower (that is, more negative) electrode potentials have a greater tendency to go right to left. That is,

$$\text{Reduced species} \longrightarrow \text{oxidized species} + ne^-$$

The reduced species acts as a reducing agent. Consequently, *the strongest reducing agents in a table of standard electrode potentials are the reduced species corresponding to half-reactions with the smallest (most negative) $E°$ values.*

The first two and the last two entries in Table 19.1 are as follows:

$$Li^+(aq) + e^- \longrightarrow Li(s)$$
$$Na^+(aq) + e^- \longrightarrow Na(s)$$
$$\cdots$$
$$\cdots$$
$$S_2O_8^{2-}(aq) + 2e^- \longrightarrow 2SO_4^{2-}(aq)$$
$$F_2(g) + 2e^- \longrightarrow 2F^-(aq)$$

The strongest oxidizing agents are the species on the left side in the last two entries in the table (given here in red). The strongest reducing agents are the species on the right side in the first two entries in the table (given here in blue).

Example 19.6   Determining the Relative Strengths of Oxidizing and Reducing Agents

Gaining Mastery Toolbox

Critical Concept 19.6
The strongest oxidizing agents are *reactants* at the bottom of the table of standard electrode potentials (Table 19.1), while the strongest reducing agents are *products* at the top of the table. A strong oxidizing agent is favored to gain electrons during an electrochemical reaction. A strong reducing agent is favored to lose electrons during an electrochemical reaction.

Solution Essentials:
- Table of Standard Electrode (Reduction) Potentials (Table 19.1)
- Standard electrode potential
- Half-reaction
- Oxidation–reduction reaction
- Oxidizing agent
- Reducing agent

a Order the following oxidizing agents by increasing strength under standard-state conditions: $Cl_2(g)$, $H_2O_2(aq)$, $Fe^{3+}(aq)$.

b Order the following reducing agents by increasing strength under standard-state conditions: $H_2(g)$, $Al(s)$, $Cu(s)$.

Problem Strategy

a Read down the table of electrode potentials, picking out the reduction half-reactions in which the species of interest occur as reactants. The order from *top to bottom* in which these species occur in the table is the order of increasing oxidizing power.

b Read down the table of electrode potentials, picking out the reduction half-reactions in which the species of interest occur as products. The order from *bottom*

to top in which these species occur in the table is the order of increasing reducing strength.

Solution

a The half-reactions and corresponding electrode potentials are as follows:

$Fe^{3+}(aq) + e^- \longrightarrow Fe^{2+}(aq)$	0.77 V
$Cl_2(g) + 2e^- \longrightarrow 2Cl^-(aq)$	1.36 V
$H_2O_2(aq) + 2H^+(aq) + 2e^- \longrightarrow 2H_2O(l)$	1.78 V

The order by increasing oxidizing strength is **$Fe^{3+}(aq)$, $Cl_2(g)$, $H_2O_2(aq)$.**

b The half-reactions and corresponding electrode potentials are

$Al^{3+}(aq) + 3e^- \longrightarrow Al(s)$	-1.66 V
$2H^+(aq) + 2e^- \longrightarrow H_2(g)$	0.00 V
$Cu^{2+}(aq) + 2e^- \longrightarrow Cu(s)$	0.34 V

The order by increasing reducing strength is **$Cu(s)$, $H_2(g)$, $Al(s)$.**

Answer Check To consistently get this type of problem right, make certain that you understand that the strongest oxidizing agents are those species that are most likely to undergo reduction, and the strongest reducing agents are the least likely to undergo reduction.

Exercise 19.7   Which is the stronger oxidizing agent, $NO_3^-(aq)$ in acidic solution (to NO) or $Ag^+(aq)$?

■ See Problems 19.61, 19.62, 19.63, and 19.64.

You can use a table of electrode potentials to predict the direction of spontaneity of an oxidation–reduction reaction. You need note only the relative strengths of the oxidizing agents on the left and right sides of the equation. The stronger oxidizing agent will be on the reactant side when the equation is written as a spontaneous reaction. (Alternatively, you can look at the reducing agents; the stronger reducing agent will be on the reactant side of the spontaneous reaction.)

Example 19.7   Determining the Direction of Spontaneity from Electrode Potentials

Gaining Mastery Toolbox

Critical Concept 19.7
Spontaneous reactions have the species undergoing reduction with the more positive standard electron potential. Only when the reaction is written in such a way that the oxidizing agent (the species undergoing reduction) is strong enough to remove electrons from the reducing agent (the species undergoing oxidation) will a spontaneous reaction occur.

Solution Essentials:
- Table of Standard Electrode (Reduction) Potentials (Table 19.1)
- Standard electrode potential
- Spontaneous reaction
- Half-reaction
- Oxidation–reduction reaction
- Oxidizing agent
- Reducing Agent

Consider the reaction

$$Zn^{2+}(aq) + 2Fe^{2+}(aq) \longrightarrow Zn(s) + 2Fe^{3+}(aq)$$

Does the reaction go spontaneously in the direction indicated, under standard conditions?

Problem Strategy Find the oxidizing agents in the equation; one is on the left side and the other on the right side. Locate these oxidizing agents in a table of electrode potentials. (The oxidizing agent is on the left side of the reduction half-reaction.) The stronger oxidizing agent is the one involved in the half-reaction with the more positive standard electrode potential.

Solution In this reaction, Zn^{2+} is the oxidizing agent on the left side; Fe^{3+} is the oxidizing agent on the right side. The corresponding standard electrode potentials are

$$Zn^{2+}(aq) + 2e^- \longrightarrow Zn(s); E° = -0.76 \text{ V}$$
$$Fe^{3+}(aq) + e^- \longrightarrow Fe^{2+}(aq); E° = 0.77 \text{ V}$$

The stronger oxidizing agent is the one involved in the half-reaction with the more positive standard electrode potential, so Fe^{3+} is the stronger oxidizing agent. The reaction is **nonspontaneous** as written.

Answer Check When writing the half-reactions for comparison, make sure that you have noted the correct sign of $E°$ for each reaction.

Exercise 19.8 Does the following reaction occur spontaneously in the direction indicated, under standard conditions?

$$Cu^{2+}(aq) + 2I^-(aq) \longrightarrow Cu(s) + I_2(s)$$

■ See Problems 19.65 and 19.66.

Calculating Cell Potentials

When calculating cell potential values, chemists employ two approaches, both of which rely on using standard reduction potential values. One approach requires that you combine half-reactions and their standard potential values (reduction and oxidation); the other involves using an equation to calculate the difference between standard reduction potentials. The advantage of the first approach is that while determining a cell potential value, you will develop a complete understanding of the half-reactions that are taking place in the cell, identification of the species that are undergoing oxidation and reduction, as well as the number of electrons involved in the balanced oxidation–reduction reaction. The second approach involves determination of the anode and cathode in the reaction and using an equation with the appropriate reduction potentials. The advantage of this approach is that you can quickly calculate the cell potential from tabulated standard reduction potential values and that you are calculating the cell potential by difference.

Calculating Cell Potentials by Addition of Half-Reactions and Standard Potentials Consider a voltaic cell constructed from a cadmium electrode and a silver electrode having the following reduction half-reactions and corresponding standard electrode potentials (reduction potentials):

$$Cd^{2+}(aq) + 2e^- \longrightarrow Cd(s); E°_{Cd} = -0.40 \text{ V}$$
$$Ag^+(aq) + e^- \longrightarrow Ag(s); E°_{Ag} = 0.80 \text{ V}$$

You will need to reverse one of these half-reactions to obtain the oxidation part of the cell reaction. The cell reaction is spontaneous, with the stronger reducing agent on the left. This will be Cd, because it is the reactant in the half-reaction with the more negative electrode potential. Therefore, you reverse the first half-reaction, as well as the sign of the half-cell potential (to obtain the oxidation potential, which equals $-E^\circ_{Cd}$).

$$Cd(s) \longrightarrow Cd^{2+}(aq) + 2e^-; \quad -E^\circ_{Cd} = 0.40 \text{ V}$$
$$Ag^+(aq) + e^- \longrightarrow Ag(s); \quad E^\circ_{Ag} = 0.80 \text{ V}$$

To obtain the cell reaction, you must multiply the half-reactions by factors so that when the half-reactions are added together, the electrons cancel. This does not affect the half-cell potentials, which do not depend on amount of substance. You multiply the second half-reaction by 2 and then add.

$$Cd(s) \longrightarrow Cd^{2+}(aq) + 2e^- \qquad\qquad -E^\circ_{Cd} = 0.40 \text{ V}$$
$$\underline{2Ag^+(aq) + 2e^- \longrightarrow 2Ag(s)} \qquad\qquad E^\circ_{Ag} = 0.80 \text{ V}$$
$$Cd(s) + 2Ag^+(aq) \longrightarrow Cd^{2+}(aq) + 2Ag(s) \qquad E^\circ_{Ag} - E^\circ_{Cd} = 1.20 \text{ V}$$

The cell potential, E°_{cell}, is 1.20 V. The cell diagram is

$$Cd(s)\,|\,Cd^{2+}(aq)\,\|\,Ag^+(aq)\,|\,Ag(s)$$

Suppose that at the beginning of this problem, you had mistakenly reversed the silver half-reaction, instead of correctly reversing the cadmium half-reaction. Then your work would have looked like the following:

$$Cd^{2+}(aq) + 2e^- \longrightarrow Cd(s) \qquad\qquad E^\circ_{Cd} = -0.40 \text{ V}$$
$$\underline{2Ag(s) \longrightarrow 2Ag^+(aq) + 2e^-} \qquad\qquad -E^\circ_{Ag} = -0.80 \text{ V}$$
$$Cd^{2+}(aq) + 2Ag(s) \longrightarrow Cd(s) + 2Ag^+(aq) \qquad E^\circ_{Cd} - E^\circ_{Ag} = -1.20 \text{ V}$$

The corresponding cell notation would have been

$$Ag(s)\,|\,Ag^+(aq)\,\|\,Cd^{2+}(aq)\,|\,Cd(s)$$

Everything is simply the reverse of the correct result. At this point, you realize your mistake, because the cell potential is found to be negative. A negative potential merely indicates that the cell reaction is nonspontaneous as written. To obtain the spontaneous reaction and a positive cell potential, you simply reverse both half-reactions and corresponding half-cell potentials. (You would also reverse the cell notation.) This changes the sign of the cell potential.

Calculation of Cell Potentials Using Standard Reduction Potentials Note that in the case of the cadmium–silver cell above, the cell potential equals the standard electrode potential of the cathode minus the standard electrode potential of the anode. Thus, alternatively, you can calculate any cell potential using the equation

$$E^\circ_{cell} = E^\circ_{cathode} - E^\circ_{anode}$$

Applying this equation to our example, you obtain

$$E^\circ_{cell} = E^\circ_{Ag} - E^\circ_{Cd} = 0.80 \text{ V} - (-0.40 \text{ V}) = 1.20 \text{ V}$$

Example 19.8   **Calculating the Cell Potential from Standard Potentials**

Gaining Mastery Toolbox

Critical Concept 19.8
A voltaic cell always has a positive standard cell potential. When employing the addition of half-reaction approach, the half-reactions must be combined in such a way that the sum of the half-cell potentials add up to a voltage greater than zero. Direct calculation of the standard cell potential can be accomplished by using the equation $E^\circ_{cell} = E^\circ_{cathode} - E^\circ_{anode}$

Solution Essentials:
- Standard cell potential (E°_{cell})
- Direct calculation of standard cell potential ($E^\circ_{cell} = E^\circ_{cathode} - E^\circ_{anode}$)
- Table of Standard Electrode (Reduction) Potentials (Table 19.1)
- Cell notation
- Voltaic cell
- Anode and cathode
- Balanced oxidation–reduction reaction
- Half-reaction

Calculate the standard cell potential of the following voltaic cell at 25°C using standard electrode potentials.

$$Al(s)\,|\,Al^{3+}(aq)\,\|\,Fe^{2+}(aq)\,|\,Fe(s)$$

What is the cell reaction?

Problem Strategy From a table of electrode potentials, write the two reduction half-reactions and standard electrode potentials for the cell. The cell notation assumes that the anode (the oxidation half-cell) is on the left. Change the direction of this half-reaction and the sign of its electrode potential. (Assuming that

(*continued*)

(continued)

the cell notation was written correctly, you change the direction of the half-reaction corresponding to the smaller, or more negative, electrode potential.) Multiply the half-reactions (but not the electrode potentials) by factors so that when the half-reactions are added, the electrons cancel. The sum of the half-reactions is the cell reaction. Add the electrode potentials to get the cell potential.

An alternative strategy is to apply the equation

$$E^\circ_{cell} = E^\circ_{cathode} - E^\circ_{anode}$$

Solution (Both Approaches) The reduction half-reactions and standard electrode potentials are

$$Al^{3+}(aq) + 3e^- \longrightarrow Al(s); E^\circ_{Al} = -1.66 \text{ V}$$
$$Fe^{2+}(aq) + 2e^- \longrightarrow Fe(s); E^\circ_{Fe} = -0.41 \text{ V}$$

You reverse the first half-reaction and its half-cell potential to obtain

$$Al(s) \longrightarrow Al^{3+}(aq) + 3e^-; -E^\circ_{Al} = 1.66 \text{ V}$$
$$Fe^{2+}(aq) + 2e^- \longrightarrow Fe(s); E^\circ_{Fe} = -0.41 \text{ V}$$

You obtain the cell potential by adding the half-cell potentials. Because you also want the cell reaction, you multiply the first half-reaction by 2 and the second half-reaction by 3, so that when the half-reactions are added, the electrons cancel. The addition of half-reactions is displayed below. The cell potential is **1.25 V.**

$$2Al(s) \longrightarrow 2Al^{3+}(aq) + 6e^- \qquad\qquad -E^\circ_{Al} = 1.66 \text{ V}$$
$$\underline{3Fe^{2+}(aq) + 6e^- \longrightarrow 3Fe(s) \qquad\qquad\qquad E^\circ_{Fe} = -0.41 \text{ V}}$$
$$\mathbf{2Al(s) + 3Fe^{2+}(aq) \longrightarrow 2Al^{3+}(aq) + 3Fe(s)} \quad E^\circ_{Fe} - E^\circ_{Al} = 1.25 \text{ V}$$

Alternatively, you can calculate the cell potential from the formula

$$E^\circ_{cell} = E^\circ_{cathode} - E^\circ_{anode}$$

to obtain

$$E^\circ_{cell} = E^\circ_{Fe} - E^\circ_{Al}$$
$$= -0.41 \text{ V} - (-1.66 \text{ V})$$
$$= 1.25 \text{ V}$$

Answer Check If the problem states that the cell is voltaic, the cell potential should always be positive as it is here. If you calculate a negative cell potential for a voltaic cell, it is likely that you reversed the half-reactions or mislabeled the anode and cathode.

Exercise 19.9 Using standard electrode potentials, calculate E°_{cell} at 25°C for the following cell.

$$Zn(s)|Zn^{2+}(aq)\|Cu^{2+}(aq)|Cu(s)$$

■ See Problems 19.69 and 19.70.

CONCEPT CHECK 19.2

Let us define the reduction of I_2 to I^- ions, $I_2(s) + 2e^- \longrightarrow 2I^-(aq)$, as the standard reduction reaction with $E^\circ = 0.00$ V. We then construct a new standard reduction table based on this definition.

a What would be the new standard reduction potential of H^+?

b Would using a new standard reduction table change the *measured* value of a freshly prepared voltaic cell made from Cu and Zn? (Assume you have the appropriate solutions and equipment to construct the cell.)

c Would the calculated voltage for the cell in part b be different if you were using the values presented in Table 19.1? Do the calculations to justify your answer.

19.6 Equilibrium Constants from Cell Potentials

Some of the most important results of electrochemistry are the relationships among cell potential, free-energy change, and equilibrium constant. Recall that the free-energy change ΔG for a reaction equals the maximum useful work of the reaction (Section 18.5).

$$\Delta G = w_{max}$$

For a voltaic cell, this work is the electrical work, $-nFE_{cell}$ (where n is the number of moles of electrons transferred in a reaction), so when the reactants and products are in their standard states, you have

$$\Delta G^\circ = -nFE^\circ_{cell}$$

With this equation, cell potential measurements become an important source of thermodynamic information. Alternatively, thermodynamic data can be used to calculate cell potentials. These calculations are shown in the following examples.

Example 19.9

Calculating the Free-Energy Change from Electrode Potentials

Gaining Mastery Toolbox

Critical Concept 19.9
A spontaneous chemical reaction will have a standard cell potential (E°_{cell}) that is a positive number (voltaic cell). The standard cell potential, E°_{cell}, is related to the standard free-energy change, ΔG°, of a chemical reaction by the equation $\Delta G^\circ = -nFE^\circ_{cell}$.

Solution Essentials:
• Free-energy change of electrochemical reaction ($\Delta G^\circ = -nFE^\circ_{cell}$)
• Direct calculation of standard cell potential ($E^\circ_{cell} = E^\circ_{cathode} - E^\circ_{anode}$)
• Table of Standard Electrode (Reduction) Potentials (Table 19.1)
• Voltaic cell
• Standard free energy (ΔG°)
• Balanced oxidation–reduction reaction
• Half-reaction

Using standard electrode potentials, calculate the standard free-energy change at 25°C for the reaction

$$Zn(s) + 2Ag^+(aq) \longrightarrow Zn^{2+}(aq) + 2Ag(s)$$

Problem Strategy Substitute into $\Delta G^\circ = -nFE^\circ_{cell}$. Use a table of standard electrode potentials to obtain E°_{cell}. The cell reaction equals the sum of the half-reactions after they have been multiplied by factors so that the electrons cancel in the summation. Note that n is the number of moles of electrons involved in each half-reaction.

Solution The half-reactions, corresponding half-cell potentials, and their sums are displayed below.

$$
\begin{array}{ll}
Zn(s) \longrightarrow Zn^{2+}(aq) + 2e^- & -E^\circ = 0.76 \text{ V} \\
2Ag^+(aq) + 2e^- \longrightarrow 2Ag(s) & E^\circ = 0.80 \text{ V} \\
\hline
Zn(s) + 2Ag^+(aq) \longrightarrow Zn^{2+}(aq) + 2Ag(s) & E^\circ_{cell} = 1.56 \text{ V}
\end{array}
$$

Note that each half-reaction involves the transfer of two moles of electrons; hence, $n = 2$. Also, $E^\circ_{cell} = 1.56$ V, and the faraday constant, F, is 9.6485×10^4 C/mol e$^-$. Therefore,

$$\Delta G^\circ = -nFE^\circ_{cell} = -2 \text{ mol e}^- \times 96{,}485 \text{ C/mol e}^- \times 1.56 \text{ V} = -3.01 \times 10^5 \text{ J}$$

Recall that (coulombs) × (volts) = joules. Thus, the standard free-energy change is **−301 kJ.**

Answer Check Free energies of reaction are generally on the order of tens to hundreds of kilojoules, as shown in this answer. Values outside of this range would indicate that you probably made a calculation error.

Exercise 19.10 What is ΔG° at 25°C for the reaction

$$Sn^{2+}(aq) + 2Hg^{2+}(aq) \longrightarrow Sn^{4+}(aq) + Hg_2^{2+}(aq)$$

For data, see Table 19.1.

■ See Problems 19.73 and 19.74.

Example 19.10

Critical Concept 19.10
The standard cell potential, $E°_{cell}$, is related to the standard free energy change, $\Delta G°$, of a chemical reaction by the equation $\Delta G° = -nFE°_{cell}$ For a spontaneous reaction, $\Delta G° < 0$, and $E°_{cell} > 0$.

Solution Essentials:
• Free-energy change of electrochemical reaction: ($\Delta G° = -nFE°_{cell}$)
• Direct calculation of standard cell potential ($E°_{cell} = E°_{cathode} - E°_{anode}$)
• Table of Standard Electrode (Reduction) Potentials (Table 19.1)
• Voltaic cell
• Standard free energy ($\Delta G°$)
• Balanced oxidation–reduction reaction
• Half-reaction

Calculating the Cell Potential from Free-Energy Change

Suppose the reaction of zinc metal and chlorine gas is utilized in a voltaic cell in which zinc ions and chloride ions are formed in aqueous solution.

$$Zn(s) + Cl_2(g) \xrightarrow{H_2O} Zn^{2+}(aq) + 2Cl^-(aq)$$

Calculate the standard potential for this cell at 25°C from standard free energies of formation (see Appendix C).

Problem Strategy Calculate $\Delta G°$ and substitute it and the value of n into the equation $\Delta G° = -nFE°_{cell}$. Solve for $E°_{cell}$.

Solution Write the equation with $\Delta G°_f$'s beneath.

$$Zn(s) + Cl_2(g) \longrightarrow Zn^{2+}(aq) + 2Cl^-(aq)$$
$$\Delta G°_f: \quad 0 \qquad 0 \qquad -147.0 \quad 2 \times (-131.3) \text{ kJ}$$

Hence,

$$\Delta G° = \Sigma n\Delta G°_f(\text{products}) - \Sigma m\Delta G°_f(\text{reactants})$$
$$= [-147.0 + 2 \times (-131.3)] \text{ kJ}$$
$$= -410 \text{ kJ} = -4.10 \times 10^5 \text{ J}$$

You obtain n by splitting the reaction into half-reactions.

$$Zn(s) \longrightarrow Zn^{2+}(aq) + 2e^-$$
$$Cl_2(g) + 2e^- \longrightarrow 2Cl^-(aq)$$

Each half-reaction involves the transfer of two moles of electrons, so $n = 2$. Now you substitute into

$$\Delta G° = -nFE°_{cell}$$
$$-4.10 \times 10^5 \text{ J} = -2 \text{ mol } e^- \times 96,485 \text{ C/mol } e^- \times E°_{cell}$$

Solving for $E°_{cell}$, you get $E°_{cell} = \textbf{2.12 V}$.

Answer Check Because this problem describes a voltaic cell, we should expect the value of $E°_{cell}$ to be positive.

Exercise 19.11 Use standard free energies of formation (Appendix C) to obtain the standard cell potential of a cell at 25°C with the reaction

$$Mg(s) + Cu^{2+}(aq) \longrightarrow Mg^{2+}(aq) + Cu(s)$$

■ See Problems 19.77 and 19.78.

The measurement of cell potentials gives you yet another way to obtain equilibrium constants. Combining the previous equation, $\Delta G° = -nFE°_{cell}$, with the equation $\Delta G° = -RT \ln K$ from Section 18.6, you get

$$nFE°_{cell} = RT \ln K$$

or

$$E°_{cell} = \frac{RT}{nF} \ln K = \frac{2.303 \, RT}{nF} \log K$$

Substituting values for the constants R and F at 25°C gives the equation

$$E°_{cell} = \frac{0.0592}{n} \log K \quad \text{(values in volts at 25°C)}$$

This formula is typically written using the common logarithm (base 10) function as it is shown here. The factor 2.303 allows for conversion between the common logarithm and base e logarithm ($2.303 \log x = \ln x$).

Figure 19.7 summarizes the various relationships among K, $\Delta G°$, and $E°_{cell}$. ◄

The next example illustrates calculation of the thermodynamic equilibrium constant from the cell potential.

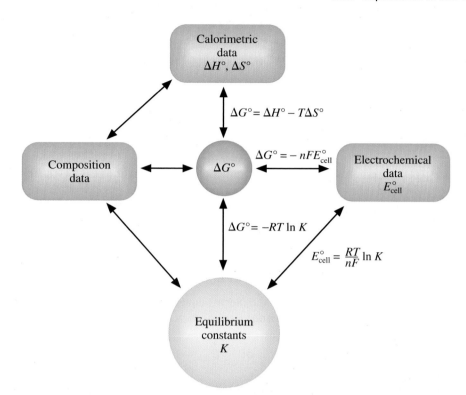

Figure 19.7 ◄

The relationships among K, $\Delta G°$, and $E°_{cell}$ The diagram shows how composition data, calorimetric data, and electrochemical data are related.

Example **19.11**   **Calculating the Equilibrium Constant from Cell Potential**

Gaining Mastery Toolbox

Critical Concept 19.11
For a spontaneous chemical reaction, the equilibrium constant will have a value greater than 1. Under standard conditions, the standard cell potential is related to the equilibrium constant by the equation $E°_{cell} = \frac{0.0592}{n} \log K$.

Solution Essentials:
• Relationship between the standard cell potential and the equilibrium constant ($E°_{cell} = \frac{0.0592}{n} \log K$)
• Standard cell potential ($E°_{cell}$)
• Cell notation
• Equilibrium constant

The standard cell potential for the following voltaic cell is 1.10 V:

$$Zn(s)|Zn^{2+}(aq)\|Cu^{2+}(aq)|Cu(s)$$

Calculate the equilibrium constant K_c for the reaction

$$Zn(s) + Cu^{2+}(aq) \rightleftharpoons Zn^{2+}(aq) + Cu(s)$$

Problem Strategy Substitute the standard cell potential into the equation relating this quantity to the thermodynamic equilibrium constant, K. Solve for K. Note that $K = K_c$.

Solution The reaction corresponds to the one for the voltaic cell. Note that $n = 2$. Substituting into the equation relating $E°_{cell}$ and K gives

$$1.10 = \frac{0.0592}{2} \log K_c$$

Solving for $\log K_c$, you find

$$\log K_c = 37.2$$

Now take the antilog of both sides:

$$K_c = \text{antilog } (37.2) = 1.6 \times 10^{37}$$

The number of significant figures in the answer equals the number of decimal places in 37.2 (one). Thus

$$\boxed{K_c = 2 \times 10^{37}}$$

Answer Check Whenever the cell potential is a positive number (voltaic cell), you should calculate an equilibrium constant greater than 1.

Exercise 19.12 Calculate the equilibrium constant K_c for the following reaction from standard electrode potentials.

$$Fe(s) + Sn^{4+}(aq) \rightleftharpoons Fe^{2+}(aq) + Sn^{2+}(aq)$$

■ See Problems 19.81 and 19.82.

19.7 Dependence of Cell Potential on Concentration

The cell potential of a cell depends on the concentrations of ions and on gas pressures. For that reason, cell potentials provide a way to measure ion concentrations. The pH meter, for example, depends on the variation of cell potential with hydrogen-ion concentration. You can relate cell potentials for various concentrations of ions and various gas pressures to standard electrode potentials by means of an equation first derived by the German chemist Walther Nernst (1864–1941). ◀

Nernst also formulated the third law of thermodynamics, for which he received the Nobel Prize in 1920.

Nernst Equation

Recall that the free-energy change, ΔG, is related to the standard free-energy change, $\Delta G°$, by the following equation (Section 18.6):

$$\Delta G = \Delta G° + RT \ln Q$$

Here Q is the thermodynamic reaction quotient. The reaction quotient has the form of the equilibrium constant, except that the concentrations and gas pressures are those that exist in a reaction mixture at a given instant. You can apply this equation to a voltaic cell. In that case, the concentrations and gas pressures are those that exist in the cell at a particular instant. If you substitute $\Delta G = -nFE_{cell}$ and $\Delta G° = -nFE°_{cell}$ into this equation, you obtain

$$-nFE_{cell} = -nFE°_{cell} + RT \ln Q$$

This result rearranges to give the **Nernst equation,** *an equation relating the cell potential to its standard cell potential and the reaction quotient.*

$$E_{cell} = E°_{cell} - \frac{RT}{nF} \ln Q \quad \text{or} \quad E_{cell} = E°_{cell} - \frac{2.303\ RT}{nF} \log Q$$

If you substitute 298 K (25°C) for the temperature in the Nernst equation and put in values for R and F, you get (using common logarithms)

$$E_{cell} = E°_{cell} - \frac{0.0592}{n} \log Q \quad \text{(values in volts at 25°C)}$$

You can show from the Nernst equation that the cell potential, E_{cell}, decreases as the cell reaction proceeds. As the reaction occurs in the voltaic cell, the concentrations of products increase and the concentrations of reactants decrease. Therefore, Q and $\log Q$ increase. The second term in the Nernst equation, $(0.0592/n) \log Q$, increases, so that the difference $E°_{cell} - (0.0592/n) \log Q$ decreases. Thus, the cell potential, E_{cell}, becomes smaller. Eventually the cell potential goes to zero, and the cell reaction comes to equilibrium.

As an example of the computation of the reaction quotient, consider the following voltaic cell:

$$Cd(s)|Cd^{2+}(0.0100M)\|H^+(1.00M)|H_2(1.00\ atm)|Pt$$

The cell reaction is

$$Cd(s) + 2H^+(aq) \rightleftharpoons Cd^{2+}(aq) + H_2(g)$$

and the expression for the equilibrium constant is

$$K = \frac{[Cd^{2+}]P_{H_2}}{[H^+]^2}$$

Note that the hydrogen-gas concentration is given here in terms of the pressure (in atmospheres). The expression for the reaction quotient has the same form as K, except that the values for the ion concentrations and hydrogen-gas pressures are those that exist in the cell. ◀ Hence,

We have omitted subscript i's on the concentrations and pressures in Q to simplify the notation.

$$Q = \frac{[Cd^{2+}]P_{H_2}}{[H^+]^2} = \frac{0.0100 \times 1.00}{(1.00)^2} = 0.0100$$

The next example illustrates a complete calculation of cell potential from the ion concentrations in a voltaic cell.

OWL Interactive Example 19.12 Calculating the Cell Potential for Nonstandard Conditions

Gaining Mastery Toolbox

Critical Concept 19.12
The cell potential (E_{cell}) is the cell potential when the reaction is run under nonstandard conditions. Nonstandard conditions are when the temperature, concentration(s), or gas pressures differ from 298 K, 1.0 *M*, and 1.00 atm, respectively. Le Châtelier's principle can be used to estimate the effect that concentration and gas pressure changes will have on the cell potential.

Solution Essentials:
- Nernst Equation at 298 K (25°C) ($E_{cell} = E^\circ_{cell} - \frac{0.0592}{n} \log Q$)
- Standard cell potential (E°_{cell})
- Cell potential (E_{cell})
- Reaction quotient (Q)
- Balanced oxidation–reduction reaction

What is the cell potential of the following voltaic cell at 25°C?

$$Zn(s)|Zn^{2+}(1.00 \times 10^{-5} M)||Cu^{2+}(0.100\ M)|Cu(s)$$

The standard cell potential of this cell is 1.10 V.

Problem Strategy From the cell reaction, deduce the value of n; then calculate Q. Now substitute n, Q, and E°_{cell} into the Nernst equation to obtain E_{cell}.

Solution The cell reaction is

$$Zn(s) + Cu^{2+}(aq) \rightleftharpoons Zn^{2+}(aq) + Cu(s)$$

The number of moles of electrons transferred is two; hence, $n = 2$. The reaction quotient is

$$Q = \frac{[Zn^{2+}]}{[Cu^{2+}]} = \frac{1.00 \times 10^{-5}}{0.100} = 1.00 \times 10^{-4}$$

The standard cell potential is 1.10 V, so for this reaction running at 25°C, the Nernst equation becomes

$$\begin{aligned} E_{cell} &= E^\circ_{cell} - \frac{0.0592}{n} \log Q \\ &= 1.10 - \frac{0.0592}{2} \log (1.0 \times 10^{-4}) \\ &= 1.10 - (-0.12) = 1.22 \text{ V} \end{aligned}$$

The cell potential is **1.22 V.**

Answer Check Whenever you perform a calculation involving nonstandard cell concentrations, you can use Le Châtelier's principle to see if your answer is reasonable. Recall that standard cell concentrations are equivalent and are 1.00 *M*. At standard concentrations the cell potential is 1.10 V. In this problem, the concentration of the product (Zn^{2+}) is 1.0×10^{-5} *M*, whereas the concentration of the reactant (Cu^{2+}) is 0.100 *M*. With these concentrations we would expect this spontaneous reaction to be shifted toward products, thereby resulting in a value of E_{cell} greater than the standard value, which is what we observe.

Exercise 19.13 What is the cell potential of the following voltaic cell at 25°C?

$$Zn(s)|Zn^{2+}(0.200\ M)||Ag^{+}(0.00200\ M)|Ag(s)$$

See Problems 19.85 and 19.86.

Determination of pH

The pH of a solution can be obtained very accurately from cell potential measurements, using the Nernst equation to relate cell potential to pH. To see how this is done, suppose you have a test solution whose pH you would like to determine. You set up a voltaic cell as follows: You use the test solution as the electrolyte for a hydrogen electrode and bubble in hydrogen gas at 1 atm. (The hydrogen electrode is shown in Figure 19.5.) Now connect this hydrogen electrode to a standard zinc electrode to give the following cell:

$$Zn|Zn^{2+}(1\ M)||H^{+}(\text{test solution})|H_2(1\text{ atm})|Pt$$

The cell reaction is

$$Zn(s) + 2H^{+}(\text{test solution}) \longrightarrow Zn^{2+}(1\ M) + H_2(1\text{ atm})$$

The cell potential depends on the hydrogen-ion concentration of the test solution, according to the Nernst equation. The standard cell potential of the cell equals 0.76 V, and

$$Q = \frac{[Zn^{2+}]P_{H_2}}{[H^{+}]^2} = \frac{1}{[H^{+}]^2}$$

Substituting into the Nernst equation, you obtain

$$E_{cell} = 0.76 - \frac{0.0592}{2} \log \frac{1}{[H^+]^2} = 0.76 + 0.0592 \log [H^+]$$

where $[H^+]$ is the hydrogen-ion concentration of the test solution.

To obtain the relationship between the cell potential (E_{cell}) and pH, you substitute the following into the preceding equation: ◀

$$pH = -\log [H^+]$$

The result is

$$E_{cell} = 0.76 - 0.0592 \, pH$$

Which you can rearrange to give the pH directly in terms of the cell potential:

$$pH = \frac{0.76 - E_{cell}}{0.0592}$$

In this way, measurement of the cell potential gives you the pH of the solution.

The hydrogen electrode is seldom employed in routine laboratory work, because it is awkward to use. It is often replaced by a *glass electrode*. This compact electrode (see Figure 19.8) consists of a silver wire coated with silver chloride immersed in a solution of dilute hydrochloric acid. The electrode solution is separated from the test solution by a thin glass membrane, which develops a potential across it depending on the hydrogen-ion concentrations on its inner and outer surfaces. A mercury–mercury(I) chloride (calomel) electrode is often used as the other electrode. The cell potential depends linearly on the pH. In a common arrangement, the cell potential is measured with a voltmeter that reads pH directly (see Figure 15.7).

The glass electrode is an example of an *ion-selective electrode*. Many electrodes have been developed that are sensitive to a particular ion, such as K^+, NH_4^+, Ca^{2+}, or Mg^{2+}. They can be used to monitor solutions of that ion. It is even possible to measure the concentration of a nonelectrolyte. To measure urea, NH_2CONH_2, in solution, you can use an electrode selective to NH_4^+ that is coated with a gel containing the enzyme urease. (An enzyme is a biochemical catalyst.) The gel is held in place around the electrode by means of a nylon net. Urease catalyzes the decomposition of urea to ammonium ion, whose concentration is measured.

$$NH_2CONH_2(aq) + 2H_2O(l) + H^+(aq) \xrightarrow{urease} 2NH_4^+(aq) + HCO_3^-(aq)$$

Keeping in mind that H^+ is a shorthand representation of H_3O^+, the equation for pH used here is the same as presented in Chapter 15 ($pH = -\log [H_3O^+]$).

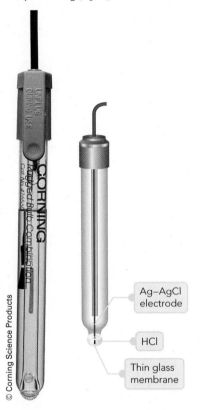

Ag–AgCl electrode

HCl

Thin glass membrane

Figure 19.8 ▲

A glass electrode *Left:* A small, commercial glass electrode. *Right:* A sketch showing the construction of a glass electrode for measuring hydrogen-ion concentrations.

Exercise 19.14 What is the nickel(II)-ion concentration in the voltaic cell

$$Zn(s)|Zn^{2+}(1.00M)\|Ni^{2+}(aq)|Ni(s)$$

if the cell potential is 0.34 V at 25°C?

■ See Problems 19.89 and 19.90.

CONCEPT CHECK 19.3

Consider a voltaic cell, $Fe(s)\,|\,Fe^{2+}(aq)\,\|\,Cu^{2+}(aq)\,|\,Cu(s)$, being run under standard conditions.

a Is $\Delta G°$ positive or negative for this process?

b Change the concentrations from their standard values in such a way that E_{cell} is reduced. Write your answer using the shorthand notation of Section 19.3.

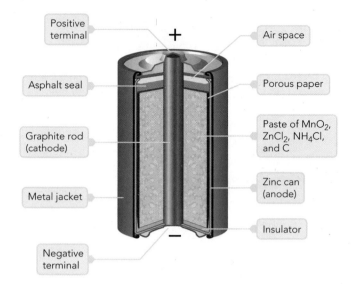

Figure 19.9 ▲

Leclanché dry cell The cell has a zinc anode and a graphite rod with a paste of MnO_2, $ZnCl_2$, NH_4Cl, and C as the cathode.

19.8 Some Commercial Voltaic Cells

We commonly use voltaic cells as convenient, portable sources of energy. Flashlights and radios are examples of devices that are often powered by the **zinc–carbon,** or **Leclanché, dry cell** (Figure 19.9). *This voltaic cell has a zinc can as the anode; a graphite rod in the center, surrounded by a paste of manganese dioxide, ammonium and zinc chlorides, and carbon black, is the cathode.* The electrode reactions are complicated but are approximately these:

$$Zn(s) \longrightarrow Zn^{2+}(aq) + 2e^- \qquad \text{(anode)}$$

$$2NH_4^+(aq) + 2MnO_2(s) + 2e^- \longrightarrow Mn_2O_3(s) + H_2O(l) + 2NH_3(aq) \quad \text{(cathode)}$$

The voltage of this dry cell is initially about 1.5 V, but it decreases as current is drawn off. The voltage also deteriorates rapidly in cold weather.

An **alkaline dry cell** (Figure 19.10) is *similar to the Leclanché cell, but it has potassium hydroxide in place of ammonium chloride.* This cell performs better under current drain and in cold weather. The half-reactions are

$$Zn(s) + 2OH^-(aq) \longrightarrow Zn(OH)_2(s) + 2e^- \qquad \text{(anode)}$$

$$2MnO_2(s) + H_2O(l) + 2e^- \longrightarrow Mn_2O_3(s) + 2OH^-(aq) \qquad \text{(cathode)}$$

A dry cell is not truly "dry," because the electrolyte is an aqueous paste. Solid-state batteries have been developed, however. One of these is a **lithium–iodine battery,** *a voltaic cell in which the anode is lithium metal and the cathode is an I_2 complex.* These solid-state electrodes are separated by a thin crystalline layer of lithium iodide (Figure 19.11). Current is carried through the crystal by diffusion of Li^+ ions. Although the cell has high resistance and therefore low current, the battery is very reliable and is used to power heart pacemakers. The battery is implanted within the patient's chest and lasts about ten years before it has to be replaced.

Once a dry cell is completely discharged (has come to equilibrium), the cell is not easily reversed, or recharged, and is normally discarded. Some types of cells are rechargeable after use, however. An important example is the **lead storage cell.** *This voltaic cell consists of electrodes of lead alloy grids; one electrode is packed with a spongy lead to form the anode, and the other electrode is packed with*

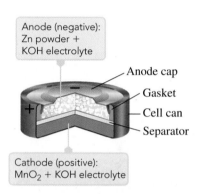

Figure 19.10 ▲

A small alkaline dry cell The anode is zinc powder and the cathode is MnO_2, as in the Leclanché cell; however, the electrolyte is KOH.

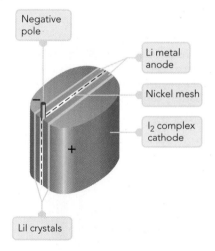

Figure 19.11 ▲

A solid-state lithium–iodine battery The anode is lithium metal and the cathode is a complex of iodine, I_2; the electrodes are separated by a thin crystalline layer of lithium iodide. The battery consists of two cells enclosed in a titanium shell.

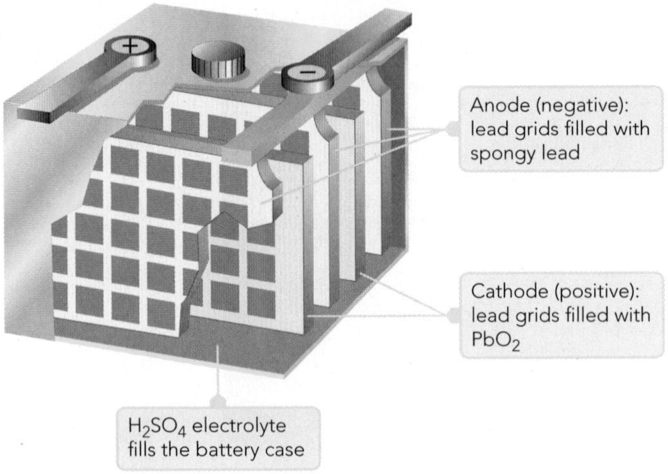

Figure 19.12 ▲

A lead storage battery Each cell delivers about 2 V, and a battery consisting of six cells in series gives about 12 V.

lead(IV) oxide to form the cathode (see Figure 19.12). Both are bathed in an aqueous solution of sulfuric acid, H_2SO_4. The half-cell reactions during discharge are

$$Pb(s) + HSO_4^-(aq) \longrightarrow PbSO_4(s) + H^+(aq) + 2e^- \qquad \text{(anode)}$$
$$PbO_2(s) + 3H^+(aq) + HSO_4^-(aq) + 2e^- \longrightarrow PbSO_4(s) + 2H_2O(l) \quad \text{(cathode)}$$

White lead(II) sulfate coats each electrode during discharge, and sulfuric acid is consumed. Each cell delivers about 2 V, and a battery consisting of six cells in series gives about 12 V.

After the lead storage battery is discharged, it is recharged from an external electric current. The previous half-reactions are reversed. Some water is decomposed into hydrogen and oxygen gas during this recharging, so more water may have to be added at intervals. However, newer batteries use lead electrodes containing some calcium metal; the calcium–lead alloy resists the decomposition of water. These *maintenance-free* batteries are sealed (Figure 19.13).

The **nickel–cadmium cell** (nicad cell) is a common storage battery. It is *a voltaic cell consisting of an anode of cadmium and a cathode of hydrated nickel oxide (approximately NiOOH) on nickel; the electrolyte is potassium hydroxide.* Nicad batteries are used in calculators, portable power tools, shavers, and toothbrushes. The half-cell reactions during discharge are

$$Cd(s) + 2OH^-(aq) \longrightarrow Cd(OH)_2(s) + 2e^- \qquad \text{(anode)}$$
$$NiOOH(s) + H_2O(l) + e^- \longrightarrow Ni(OH)_2(s) + OH^-(aq) \qquad \text{(cathode)}$$

These half-reactions are reversed when the cell is recharged. Nicad batteries can be recharged and discharged many times (see Figure 19.14). Recently, *nickel metal hydride* (NiMH) and *lithium-ion* batteries have been replacing nicad batteries in many applications that require rechargeable batteries. The NiMH batteries use a metal hydride for the anode instead of the more toxic cadmium and will continue to function with significantly more discharge-recharge cycles than the nicad batteries. NiMH batteries are currently used to supply electric power for hybrid-electric vehicles. Lithium-ion batteries typically use a carbon anode and lithium cobalt dioxide or a lithium manganese compound as the cathode. Lithium-ion batteries are commonly used to power consumer electronics because they are relatively lightweight for the amount of power that they can supply and they can be cycled through the discharged-recharge cycle many times before failure.

A **fuel cell** is *essentially a battery, but it differs in operating with a continuous supply of energetic reactants, or fuel.* Figure 19.15 shows a proton-exchange membrane (PEM) fuel cell that uses hydrogen and oxygen. On one side of the cell, the

© Cengage Learning

Figure 19.13 ▲

A maintenance-free battery This lead storage battery contains calcium–lead alloy grids for electrodes. This alloy resists the decompostion of water, so the battery can be sealed.

© Cengage Learning

Figure 19.14 ▲

Nicad storage batteries These rechargeable cells have a cadmium anode and a hydrated nickel oxide cathode.

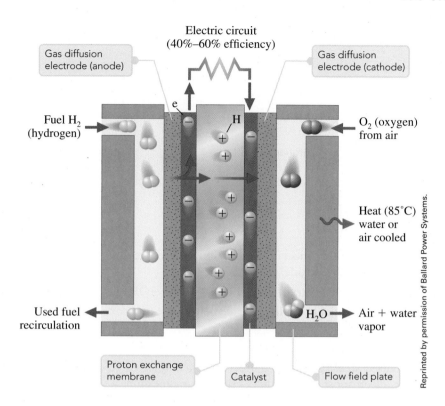

Electric circuit
(40%–60% efficiency)

Gas diffusion
electrode (anode)

Gas diffusion
electrode (cathode)

e

H

Fuel H₂
(hydrogen)

O₂ (oxygen)
from air

Heat (85°C)
water or
air cooled

Used fuel
recirculation

H₂O

Air + water
vapor

Proton exchange
membrane

Catalyst

Flow field plate

Reprinted by permission of Ballard Power Systems.

Figure 19.15 ◀

A hydrogen fuel cell At the anode, hydrogen gas enters a chamber where a porous platinum catalyst oxidizes the hydrogen to H⁺(*aq*) ions at the anode. The H⁺ ions then migrate through a special proton-exchange membrane (PEM) and move to the cathode; the proton-exchange membrane allows only the very small H⁺(*aq*) ions to pass from the anode to the cathode. At the cathode, oxygen gas enters the cell, where it comes into contact with a platinum catalyst that facilitates the reduction of O₂(*g*) in the presence of the H⁺(*aq*) and the electrons from the oxidation reaction to produce water. The net chemical change is the reaction of hydrogen with oxygen to produce water. This fuel cell has a measured potential of 0.7 V and, unlike many other types of fuel cells, can run at temperatures under 100°C.

anode, hydrogen passes through a porous material containing a platinum catalyst, allowing the following reaction to occur:

$$H_2(g) \longrightarrow 2H^+(aq) + 2e^- \qquad \text{(anode)}$$

The $H^+(aq)$ ions then migrate through a proton-exchange membrane to the other side of the cell to participate in the cathode reaction with $O_2(g)$:

$$O_2(g) + 4H^+(aq) + 4e^- \longrightarrow 2H_2O(l) \qquad \text{(cathode)}$$

The sum of the half-reactions (note that the anode reaction must be multiplied by 2 prior to adding) is

$$2H_2(g) + O_2(g) \longrightarrow 2H_2O(l)$$

which is the net reaction in the fuel cell. The first applications of PEM fuel cells were in space, but more recently, they have provided power for lighting, emergency power generators, communications equipment, automobiles, and buses. Other types of cells using other materials and fuels such as hydrocarbons or methanol are either in commercial production or under development.

Another use of voltaic cells is to control the corrosion of underground pipelines and tanks. Such pipelines and tanks are usually made of steel, an alloy of iron, and their corrosion or rusting is an electrochemical process.

Consider the rusting that occurs when a drop of water is in contact with iron. The edge of the water drop exposed to the air becomes one pole of a voltaic cell (see Figure 19.16). At this edge, molecular oxygen from air is reduced to hydroxide ion in solution.

$$O_2(g) + 2H_2O(l) + 4e^- \longrightarrow 4OH^-(aq)$$

The electrons for this reduction are supplied by the oxidation of metallic iron at the center of the drop, which acts as the other pole of the voltaic cell.

$$Fe(s) \longrightarrow Fe^{2+}(aq) + 2e^-$$

These electrons flow from the center of the drop through the metallic iron to the edge of the drop. The metallic iron functions as the external circuit between the cell poles.

Ions move within the water drop, completing the electric circuit. Iron(II) ions move outward from the center of the drop, and hydroxide ions move inward from

Figure 19.16 ▶

The electrochemical process involved in the rusting of iron Here a single drop of water containing ions forms a voltaic cell in which iron is oxidized to iron(II) ion at the center of the drop (this is the anode). Oxygen gas from air is reduced to hydroxide ion at the periphery of the drop (the cathode). Hydroxide ions and iron(II) ions migrate together and react to form iron(II) hydroxide. This is oxidized to iron(III) hydroxide by more O_2 that dissolves at the surface of the drop. Iron(III) hydroxide precipitates, and this settles to form rust on the surface of the iron.

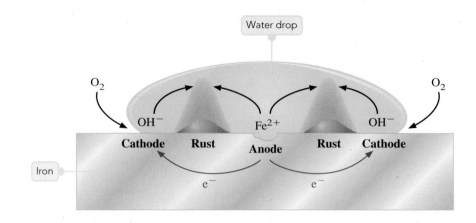

the edge. The two ions meet in a doughnut-shaped region, where they react to precipitate iron(II) hydroxide.

$$Fe^{2+}(aq) + 2OH^-(aq) \longrightarrow Fe(OH)_2(s)$$

This precipitate is quickly oxidized by oxygen to rust (approximated by the formula $Fe_2O_3 \cdot H_2O$).

$$4Fe(OH)_2(s) + O_2(g) \longrightarrow 2Fe_2O_3 \cdot H_2O(s) + 2H_2O(l)$$

If a buried steel pipeline (Figure 19.17) is connected to an active metal (that is, a highly electropositive substance) such as magnesium, a voltaic cell is formed; the active metal is the anode and iron becomes the cathode. Wet soil forms the electrolyte, and the electrode reactions are

$$Mg(s) \longrightarrow Mg^{2+}(aq) + 2e^- \qquad \text{(anode)}$$
$$O_2(g) + 2H_2O(l) + 4e^- \longrightarrow 4OH^-(aq) \qquad \text{(cathode)}$$

As the cathode, the iron-containing steel pipe is protected from oxidation. Of course, the magnesium rod is eventually consumed and must be replaced, but this is cheaper than digging up the pipeline. This use of an active metal to protect iron from corrosion is called *cathodic protection*. See Figure 19.18 for a laboratory demonstration of cathodic protection.

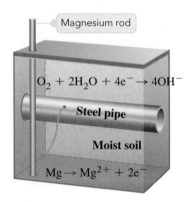

Figure 19.17 ▲

Cathodic protection of a buried steel pipe Iron in the steel becomes the cathode in an iron–magnesium voltaic cell. Magnesium is then oxidized in preference to iron.

CONCEPT CHECK 19.4

Keeping in mind that seawater contains a number of ions, explain why seawater corrodes iron much faster than freshwater.

Figure 19.18 ▶

A demonstration of cathodic protection The nails are in a gel containing phenolphthalein indicator and potassium ferricyanide. Iron corrosion yields Fe^{2+}, which reacts with ferricyanide ion to give a dark blue precipitate. Where OH^- forms, phenolphthalein appears pink.

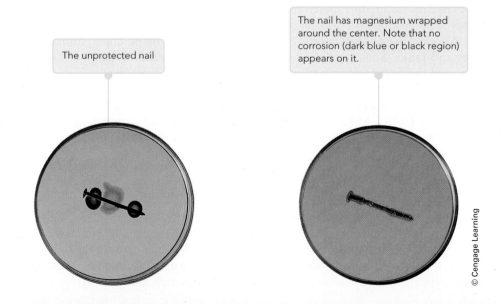

The unprotected nail

The nail has magnesium wrapped around the center. Note that no corrosion (dark blue or black region) appears on it.

Electrolytic Cells

An *electrolytic cell,* you may recall, is an electrochemical cell in which an external energy source drives an otherwise nonspontaneous reaction. *The process of producing a chemical change in an electrolytic cell* is called **electrolysis.** Many important substances, including aluminum and chlorine, are produced commercially by electrolysis. We will begin by looking at the electrolysis of molten salts.

19.9 Electrolysis of Molten Salts

Figure 19.19 shows a simple electrolytic cell. Wires from a battery are connected to electrodes that dip into molten sodium chloride. (NaCl melts at 801°C.) At the electrode connected to the negative pole of the battery, globules of sodium metal form; chlorine gas evolves from the other electrode. The half-reactions are

$$Na^+(l) + e^- \longrightarrow Na(l)$$

$$Cl^-(l) \longrightarrow \tfrac{1}{2}Cl_2(g) + e^-$$

As noted earlier (Section 19.2), the *anode* is the electrode at which oxidation occurs, and the *cathode* is the electrode at which reduction occurs (these definitions hold for electrolytic cells as well as for voltaic cells). Thus, during the electrolysis of molten NaCl, the reduction of Na^+ to Na occurs at the cathode, and the oxidation of Cl^- to Cl_2 occurs at the anode (note the labeling of electrodes in Figure 19.19).

This electrolysis of molten NaCl is used commercially to obtain sodium metal from sodium chloride. A **Downs cell** is *a commercial electrochemical cell used to obtain sodium metal by the electrolysis of molten sodium chloride* (Figure 19.20). The cell is constructed to keep the products of the electrolysis separate, because they would otherwise react. Calcium chloride is added to the sodium chloride to lower the melting point from 801°C for NaCl to about 580°C for the mixture. (Remember that the melting point, or freezing point, of a substance is lowered by the addition of a solute.) You obtain the cell reaction by adding the half-reactions.

$$Na^+(l) + e^- \longrightarrow Na(l)$$
$$\underline{Cl^-(l) \longrightarrow \tfrac{1}{2}Cl_2(g) + e^-}$$
$$Na^+(l) + Cl^-(l) \longrightarrow Na(l) + \tfrac{1}{2}Cl_2(g)$$

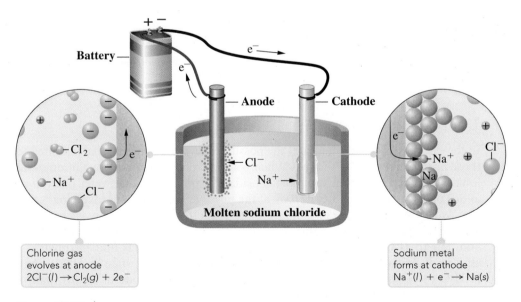

Chlorine gas evolves at anode
$2Cl^-(l) \rightarrow Cl_2(g) + 2e^-$

Sodium metal forms at cathode
$Na^+(l) + e^- \rightarrow Na(s)$

Figure 19.19 ▲

Electrolysis of molten sodium chloride Sodium metal forms at the cathode from the reduction of Na^+ ion; chlorine gas forms at the anode from the oxidation of Cl^- ion. Sodium metal is produced commercially this way, although the commercial cell must be designed to collect the products and to keep them away from one another.

Figure 19.20 ▶

A Downs cell for the preparation of sodium metal In this commercial cell, sodium is produced by the electrolysis of molten sodium chloride. The salt contains calcium chloride, which is added to lower the melting point of the mixture. Liquid sodium forms at the cathode, where it rises to the top of the molten salt and collects in a tank. Chlorine gas is a by-product.

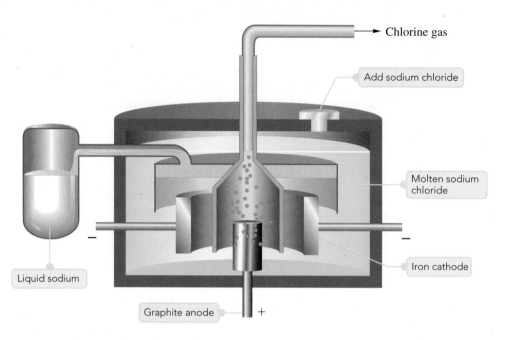

Chlorine gas

Add sodium chloride

Molten sodium chloride

Iron cathode

Liquid sodium

Graphite anode

A number of other reactive metals are obtained by the electrolysis of a molten salt or ionic compound. Lithium, magnesium, and calcium metals are all obtained by the electrolysis of the chlorides. The first commercial preparation of sodium metal adapted the method used by Humphry Davy when he discovered the element in 1807. Davy electrolyzed molten sodium hydroxide, NaOH, whose melting point (318°C) is relatively low for an ionic compound. The half-reactions are

$$Na^+(l) + e^- \longrightarrow Na(l) \qquad \text{(cathode)}$$
$$4OH^-(l) \longrightarrow O_2(g) + 2H_2O(g) + 4e^- \qquad \text{(anode)}$$

Many of the commercial uses of electrolysis involve aqueous solutions. We will look at electrolysis in aqueous solution in the next section.

Exercise 19.15 Write the half-reactions for the electrolysis of the following molten compounds: **a** KCl; **b** KOH.

■ See Problems 19.91 and 19.92.

19.10 Aqueous Electrolysis

In the electrolysis of a molten salt, the possible half-reactions are usually limited to those involving ions from the salt. When you electrolyze an aqueous solution of an ionic compound, however, you must consider the possibility that water is involved at one or both electrodes. Let us look at the possible half-reactions involving water.

Water can be reduced or oxidized in half-reactions, and you can easily obtain these half-reactions. To do this, first note the species likely to be involved. In addition to H_2O, these are H_2, O_2, H^+, and OH^-. Only H_2 and O_2 involve a change of oxidation state. Hydrogen has a lower oxidation state in H_2 (0) than in H_2O (+1), whereas the oxygen has a higher oxidation state in O_2 (0) than in H_2O (−2). Therefore, you can reduce water to H_2 or oxidize it to O_2.

Consider the reduction half-reaction. It must involve the reduction of H_2O to H_2. You need to balance the half-reaction by putting an oxygen-containing species on the right. The only species in which there are no changes in oxidation states from those in H_2O is OH^-. The balanced half-reaction is

$$2H_2O(l) + 2e^- \longrightarrow H_2(g) + 2OH^-(aq)$$

You can obtain the oxidation half-reaction for water in a similar way. It involves the oxidation of H_2O to O_2. You need to put a hydrogen-containing species on the right side of the equation to balance it. The only species in which there is no change in oxidation state is H^+. The balanced half-reaction is

$$2H_2O(l) \longrightarrow O_2(g) + 4H^+(aq) + 4e^-$$

Now let us consider the electrolysis of different aqueous solutions and try to decide what the half-reactions might be. Once you have the half-reactions, you can obtain the overall chemical change due to the electrolysis.

Electrolysis of Sulfuric Acid Solutions

To decide what is likely to happen during the electrolysis of a solution of sulfuric acid, H_2SO_4, you must consider the half-reactions involving ionic species from H_2SO_4, in addition to those involving water. Sulfuric acid is a strong acid and ionizes completely into H^+ and HSO_4^-. The HSO_4^- ion is relatively strong and for the most part ionizes into H^+ and SO_4^{2-}. Therefore, the species you have to look at are H^+, SO_4^{2-}, and H_2O.

At the cathode, the possible reduction half-reactions are

$$2H^+(aq) + 2e^- \longrightarrow H_2(g)$$
$$2H_2O(l) + 2e^- \longrightarrow H_2(g) + 2OH^-(aq)$$

At the anode, the possible oxidation half-reactions are

$$2SO_4^{2-}(aq) \longrightarrow S_2O_8^{2-}(aq) + 2e^-$$
$$2H_2O(l) \longrightarrow O_2(g) + 4H^+(aq) + 4e^-$$

In the first half-reaction, the sulfate ion, SO_4^{2-}, is oxidized to the peroxydisulfate ion, $S_2O_8^{2-}$. The $S_2O_8^{2-}$ ion contains the peroxy group —O—O— and therefore has O in the -1 oxidation state. Oxygen in this half-reaction is oxidized from oxidation state -2 to -1.

Consider the cathode reaction. As you can easily show, the two reduction half-reactions (reduction of H^+ and reduction of H_2O) in acidic solution are essentially equivalent, although this is not immediately obvious. Consider the reduction of water to give H_2 and OH^-. In acidic solution, OH^- reacts with H^+ to give H_2O. The net result is obtained by adding the two reactions.

$$2H_2O(l) + 2e^- \longrightarrow H_2(g) + 2OH^-(aq)$$
$$\underline{2H^+(aq) + 2OH^-(aq) \longrightarrow 2H_2O(l)}$$
$$2H^+(aq) + 2e^- \longrightarrow H_2(g)$$

You can consider the cathode reaction to be the reduction of H^+.

For the anode, consider the oxidation half-reactions and their oxidation potentials (which equal the electrode potentials with signs reversed).

$$2SO_4^{2-}(aq) \longrightarrow S_2O_8^{2-}(aq) + 2e^-; -E° = -2.01 \text{ V}$$
$$2H_2O(l) \longrightarrow O_2(g) + 4H^+(aq) + 4e^-; -E° = -1.23 \text{ V}$$

The species whose oxidation half-reaction has the larger (less negative) oxidation potential is the more easily oxidized. Therefore, under standard conditions, you expect H_2O to be oxidized in preference to SO_4^{2-}.

A note of caution must be added here. Electrode potentials are measured under conditions in which the half-reactions are at or very near equilibrium. In electrolysis, in which the half-reactions can be far from equilibrium, you may have to supply a larger voltage than predicted from electrode potentials. This additional voltage, called *overvoltage,* can be fairly large (several tenths of a volt), particularly for forming a gas. This means that in trying to predict which of two half-reactions is the one that actually occurs at an electrode, you must be especially careful if the electrode potentials are within several tenths of a volt of one another. For example, the standard oxidation potential for O_2 is -1.23 V, but because of overvoltage, the effective oxidation potential may be several tenths of a volt lower (more negative). The sulfate ion is so difficult to oxidize ($-E° = -2.01$ V), however, that the previous conclusion about the anode half-reaction in electrolyzing aqueous sulfuric acid is unaltered.

You obtain the cell reaction by adding the half-reactions that occur at the electrodes.

$$2 \times [2H^+(aq) + 2e^- \longrightarrow H_2(g)] \qquad \text{(cathode)}$$
$$\underline{2H_2O(l) \longrightarrow O_2(g) + 4H^+(aq) + 4e^-} \qquad \text{(anode)}$$
$$2H_2O(l) \longrightarrow 2H_2(g) + O_2(g)$$

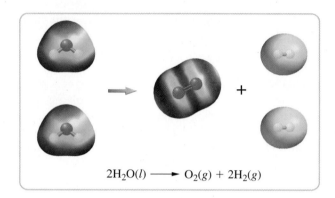

$$2H_2O(l) \longrightarrow O_2(g) + 2H_2(g)$$

The net cell reaction is simply the *electrolysis of water*. Where electricity is cheap, hydrogen can be prepared commercially by the electrolysis of water. Here H_2SO_4 was used as an electrolyte; however, various electrolytes can be used, including NaCl.

Electrolysis of Sodium Chloride Solutions

When you electrolyze an aqueous solution of sodium chloride, NaCl, the possible species involved in half-reactions are Na^+, Cl^-, and H_2O. The possible cathode half-reactions are

$$Na^+(aq) + e^- \longrightarrow Na(s); \; E° = -2.71 \text{ V}$$
$$2H_2O(l) + 2e^- \longrightarrow H_2(g) + 2OH^-(aq); \; E° = -0.83 \text{ V}$$

Under standard conditions, you expect H_2O to be reduced in preference to Na^+, which agrees with what you observe. Hydrogen gas evolves at the cathode. ◄

The possible anode half-reactions are

$$2Cl^-(aq) \longrightarrow Cl_2(g) + 2e^-; \; -E° = -1.36 \text{ V}$$
$$2H_2O(l) \longrightarrow O_2(g) + 4H^+(aq) + 4e^-; \; -E° = -1.23 \text{ V}$$

The OH^- concentration is actually 1×10^{-7} M. From Nernst's equation, E for the reduction of H_2O is -0.41 V, so you still conclude that H_2O is reduced in preference to Na^+.

Under standard-state conditions, you might expect H_2O to be oxidized in preference to Cl^-. However, the potentials are close and overvoltages at the electrodes could alter this conclusion.

It is possible, nevertheless, to give a general statement about the product expected at the anode. Electrode potentials, as you have seen, depend on concentrations. It turns out that when the solution is concentrated enough in Cl^-, Cl_2 is the product, but in dilute solution, O_2 is the product. To see this, you would simply apply the Nernst equation to the $Cl^-|Cl_2$ half-reaction.

$$2Cl^-(aq) \longrightarrow Cl_2(g) + 2e^-$$

Starting with very dilute NaCl solutions, you would find that the oxidation potential of Cl^- is very negative, so H_2O is reduced in preference to Cl^-. But as you increased the NaCl concentration, you would find that the oxidation potential of Cl^- increases until eventually Cl^- is oxidized in preference to H_2O. The product changes from O_2 to Cl_2.

The half-reactions and cell reaction for the electrolysis of aqueous sodium chloride to chlorine and hydroxide ion are as follows:

$$2H_2O(l) + 2e^- \longrightarrow H_2(g) + 2OH^-(aq) \qquad \text{(cathode)}$$
$$\underline{2Cl^-(aq) \longrightarrow Cl_2(g) + 2e^-} \qquad \text{(anode)}$$
$$2H_2O(l) + 2Cl^-(aq) \longrightarrow H_2(g) + Cl_2(g) + 2OH^-(aq)$$

Because the electrolysis started with sodium chloride, the cation in the electrolyte solution is Na^+. When you evaporate the electrolyte solution at the cathode, you obtain sodium hydroxide, NaOH. Figure 19.21 shows the similar electrolysis of aqueous potassium iodide, KI.

The electrolysis of aqueous sodium chloride is the basis of the *chlor-alkali industry,* the major commercial source of chlorine and sodium hydroxide. Commercial cells are of several types, but in each the main problem is to keep the products separate, because chlorine reacts with aqueous sodium hydroxide.

Figure 19.22 shows a modern **chlor-alkali membrane cell,** *a cell for the electrolysis of aqueous sodium chloride in which the anode and cathode compartments are separated by a special plastic membrane that allows only cations to pass through it.* Sodium chloride solution is added to the anode compartment, where chloride ion is oxidized to chlorine. The sodium ions carry the current from the anode to the cathode by passing through the membrane. Water is added at the top of the cathode compartment, where it is reduced to hydroxide ion, and sodium hydroxide solution is removed at the bottom of the cathode compartment.

The older **chlor-alkali mercury cell** is *a cell for the electrolysis of aqueous sodium chloride in which mercury metal is used as the cathode* (Figure 19.23).

Figure 19.21 ▲

Electrolysis of aqueous potassium iodide Iodine forms at the anode by the oxidation of I⁻ ion. Iodine reacts with iodide ion to form I_3^- ion, which is red-brown. Hydrogen bubbles and hydroxide ion form at the cathode by the reduction of water.

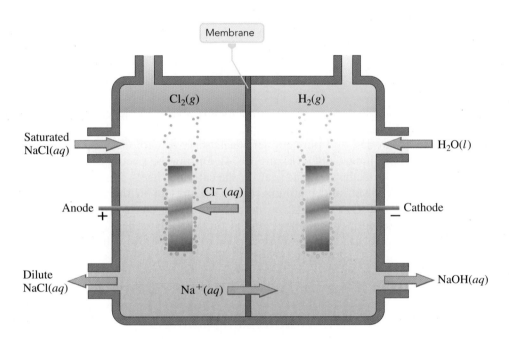

Figure 19.22 ◄

A chlor-alkali membrane cell Sodium chloride solution enters the anode compartment, where Cl⁻ ion is oxidized to Cl_2. Sodium ion migrates from the anode compartment through the membrane to the cathode compartment. Here water is reduced to hydrogen and OH⁻ ion. Sodium hydroxide solution is removed at the bottom of the cathode compartment.

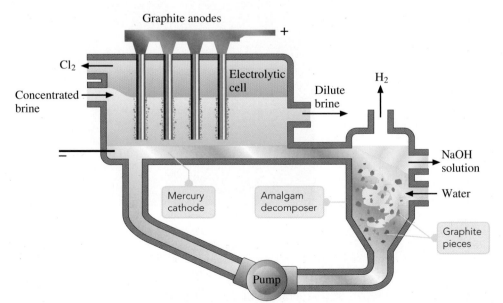

Figure 19.23 ◄

A chlor-alkali mercury cell Chloride ion oxidizes to chlorine gas at the graphite anodes (+ electrodes), and sodium deposits at the mercury cathode. Sodium amalgam (a liquid sodium–mercury alloy) circulates to the amalgam decomposer, where sodium reacts with water to form NaOH(aq) and H_2.

At the mercury cathode, sodium ion is reduced in preference to water. Sodium ion is reduced to sodium to form a liquid sodium–mercury alloy called sodium amalgam. (An *amalgam* is an alloy of mercury with any of various other metals.)

$$Na^+(aq) + e^- \longrightarrow Na(amalgam)$$

Sodium amalgam circulates from the electrolytic cell to the amalgam decomposer. The amalgam and graphite particles in the decomposer form the electrodes of many voltaic cells, in which sodium reacts with water to give sodium hydroxide solution and hydrogen gas. Historically, loss of mercury from these cells has been a source of mercury pollution of waterways.

Electroplating of Metals

Many metals are protected from corrosion by plating them with other metals. Zinc coatings are often used to protect steel, because the coat protects the steel by cathodic protection even when the zinc coat is scratched (see end of Section 19.8). A thin zinc coating can be applied to steel by *electrogalvanizing,* or zinc electroplating. (Galvanized steel has a thick zinc coating obtained by dipping the object in molten zinc.) The steel object is placed in a bath of zinc salts and made the cathode in an electrolytic cell. The cathode half-reaction is

$$Zn^{2+}(aq) + 2e^- \longrightarrow Zn(s)$$

Electrolysis is also used to purify some metals. For example, copper for electrical use, which must be very pure, is purified by electrolysis. Slabs of impure copper serve as anodes, and pure copper sheets serve as cathodes; the electrolyte bath is copper(II) sulfate, $CuSO_4$ (Figure 19.24). During the electrolysis, copper(II) ions leave the anode slabs and plate out on the cathode sheets. Less reactive metals, such as gold, silver, and platinum, that were present in the impure copper form a valuable mud that collects on the bottom of the electrolytic cell. Metals more reactive than copper remain as ions in the electrolytic bath. After about a month in the electrolytic cell, the pure copper cathodes are much enlarged and are removed from the cell bath.

Figure 19.24 ▼

Purification of copper by electrolysis

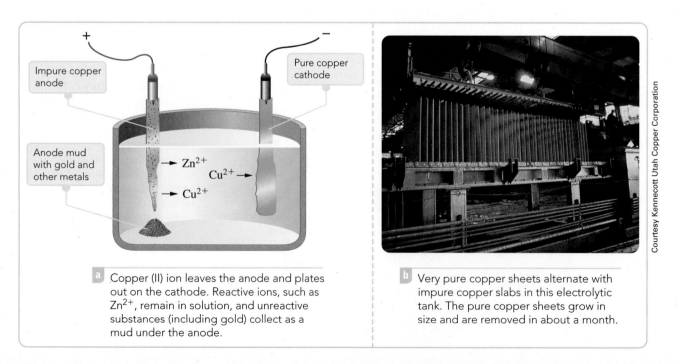

a Copper (II) ion leaves the anode and plates out on the cathode. Reactive ions, such as Zn^{2+}, remain in solution, and unreactive substances (including gold) collect as a mud under the anode.

b Very pure copper sheets alternate with impure copper slabs in this electrolytic tank. The pure copper sheets grow in size and are removed in about a month.

Courtesy Kennecott Utah Copper Corporation

Example 19.13   Predicting the Half-Reactions in an Aqueous Electrolysis

Critical Concept 19.13
With aqueous electrolysis, one must consider the possibility of the oxidation or reduction of water. For the reduction of water, $E° = -0.83$ V, and for the oxidation of water, $-E° = -1.23$ V. If the reduction half-reaction is less (more negative) than -0.83 V, water will be reduced during electrolysis. If the oxidation half-reaction is more negative than -1.23 V, water will be oxidized by electrolysis.

Solution Essentials:
• Water-reduction half-reaction: $2H_2O(l) + 2e^- \longrightarrow H_2(g) + 2OH^-(aq)$;
$$E° = -0.83\,V$$
• Water-oxidation half-reaction: $2H_2O(l) \longrightarrow O_2(g) + 4H^+(aq) + 4e^-$;
$$-E° = -1.23\,V$$
• Aqueous electrolysis
• Electrolysis
• Standard electrode potential
• Half-reaction

What do you expect to be the half-reactions in the electrolysis of aqueous copper(II) sulfate?

Problem Strategy From the problem, determine which species might be involved in the electrolysis. Write the half-reactions for these species, identifying those that could undergo reduction (cathode) and those that could undergo oxidation (anode). Using the electrode potentials of the half-reactions, determine which are most likely to occur.

Solution The species you should consider for half-reactions are $Cu^{2+}(aq)$, $SO_4^{2-}(aq)$, and H_2O.

Possible cathode half-reactions are

$$Cu^{2+}(aq) + 2e^- \longrightarrow Cu(s); E° = 0.34\,V$$
$$2H_2O(l) + 2e^- \longrightarrow H_2(g) + 2OH^-(aq); E° = -0.83\,V$$

Because the copper electrode potential is much larger than the reduction potential of water, you expect Cu^{2+} to be reduced.

Possible anode half-reactions, with the signs reversed for the electrode potentials, are

$$2SO_4^{2-}(aq) \longrightarrow S_2O_8^{2-}(aq) + 2e^-; -E° = -2.01\,V$$
$$2H_2O(l) \longrightarrow O_2(g) + 4H^+(aq) + 4e^-; -E° = -1.23\,V$$

You expect H_2O to be oxidized.
The expected half-reactions are

$$\mathbf{Cu^{2+}(aq) + 2e^- \longrightarrow Cu(s)}$$
$$\mathbf{2H_2O(l) \longrightarrow O_2(g) + 4H^+(aq) + 4e^-}$$

Answer Check For aqueous electrolysis, always make sure that you consider the possibility of water undergoing oxidation at the anode or reduction at the cathode.

Exercise 19.16 Give the half-reactions that occur when aqueous silver nitrate is electrolyzed. Nitrate ion is not oxidized during the electrolysis.

■ See Problems 19.93 and 19.94.

19.11 Stoichiometry of Electrolysis

In 1831 and 1832, the British chemist and physicist Michael Faraday showed that the amounts of substances released at the electrodes during electrolysis are related to the total charge that has flowed in the electric circuit. ▶ If you look at the electrode reactions, you see that the relationship is stoichiometric.

When molten sodium chloride is electrolyzed, sodium ions migrate to the cathode, where they react with electrons.

$$Na^+ + e^- \longrightarrow Na(l)$$

Similarly, chloride ions migrate to the anode and release electrons.

$$Cl^- \longrightarrow \tfrac{1}{2}Cl_2(g) + e^-$$

Therefore, when one mole of electrons reacts with sodium ions, one faraday of charge (the magnitude of charge on one mole of electrons) passes through the circuit. One mole of sodium metal is deposited at one electrode, and one-half mole of chlorine gas evolves at the other electrode.

What is new in this type of stoichiometric problem is the measurement of numbers of electrons. You do not weigh them as you do substances. Rather, you measure the quantity of electric charge that has passed through the circuit. You use the following fact:

One faraday $(9.6485 \times 10^4\,C)$ is equivalent to the charge on one mole of electrons.

If you know the current in a circuit and the length of time it has been flowing, you can calculate the electric charge.

$$\text{Electric charge} = \text{electric current} \times \text{time lapse}$$

Faraday's results were summarized in two laws: (1) The quantity of a substance liberated at an electrode is directly proportional to the quantity of electric charge that has flowed in the circuit. (2) For a given quantity of electric charge, the amount of any metal deposited is proportional to its equivalent mass (atomic mass divided by the charge on the metal ion). These laws follow directly from the stoichiometry of electrolysis.

Corresponding units are

$$\text{Coulombs} = \text{amperes} \times \text{seconds}$$

The **ampere (A)** is *the base unit of current in the International System (SI).* The coulomb (C), the SI unit of electric charge, is equivalent to an ampere-second. Thus, a current of 0.50 amperes flowing for 84 seconds gives a charge of 0.50 A $\times$ 84 s = 42 A·s, or 42 C.

If you are given the amount of substance produced at an electrode and the time of electrolysis, you can determine the current. If you are given the current and the time of electrolysis, you can calculate the amount of substance produced at an electrode. The next two examples illustrate these calculations.

♥WL Interactive Example 19.14   Calculating the Amount of Charge from the Amount of Product in an Electrolysis

Gaining Mastery Toolbox

Critical Concept 19.14
The faraday (9.6485×10^4 C) is a conversion factor that relates a mole of electrons to its charge. The SI unit of charge is the coulomb (C), and an ampere (A) is the SI unit of current.

Solution Essentials:
- Charge, current, and time relationship (C = A × sec)
- Ampere (A)
- Coulomb (C)
- Faraday (9.6485×10^4 C), the charge on one mole of electrons
- Aqueous electrolysis
- Electrolysis
- Half-reaction

When an aqueous solution of copper(II) sulfate, $CuSO_4$, is electrolyzed, copper metal is deposited.

$$Cu^{2+}(aq) + 2e^- \longrightarrow Cu(s)$$

(The other electrode reaction gives oxygen: $2H_2O \longrightarrow O_2 + 4H^+ + 4e^-$.) If a constant current was passed for 5.00 h and 404 mg of copper metal was deposited, what was the current?

Problem Strategy The electrode equation for copper says that 1 mol Cu is equivalent to 2 mol e^-. You can use this in the conversion of grams of Cu to moles of electrons needed to deposit this amount of copper. The Faraday constant (which says that one mole of electrons is equivalent to 9.6485×10^4 C) converts moles of electrons to coulombs.

Diagramming the solution clearly shows how three conversion factors (3 ➤) are used to solve the problem.

Starting units   Goal units

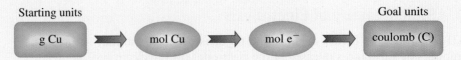

g Cu ➤ mol Cu ➤ mol e^- ➤ coulomb (C)

The current in amperes (A) equals the charge in coulombs divided by the time in seconds.

Solution The conversion of grams of Cu to coulombs required to deposit 404 mg of copper is

$$0.404 \text{ g Cu} \times \frac{1 \text{ mol Cu}}{63.6 \text{ g Cu}} \times \frac{2 \text{ mol } e^-}{1 \text{ mol Cu}} \times \frac{9.6485 \times 10^4 \text{ C}}{1 \text{ mol } e^-} = 1.\underline{2}26 \times 10^3 \text{ C}$$

The time lapse, 5.00 h, equals 1.80×10^4 s. Thus,

$$\text{Current} = \frac{\text{charge}}{\text{time}} = \frac{1.226 \times 10^3 \text{ C}}{1.80 \times 10^4 \text{ s}} = \mathbf{6.81 \times 10^{-2} A}$$

Answer Check A common mistake that you want to avoid when you work this type of problem is thinking that the coulomb (C) is the current. Therefore, always make sure that when a problem asks for the current, you convert from coulombs to amps (A).

Exercise 19.17 A constant electric current deposits 365 mg of silver in 216 min from an aqueous silver nitrate solution. What is the current?

■ See Problems 19.95 and 19.96.

⚙WL Interactive Example `19.15` **Calculating the Amount of Product from the Amount of Charge in an Electrolysis**

Gaining Mastery Toolbox

Critical Concept 19.15
The faraday (9.6485×10^4 C) is a conversion factor that relates a mole of electrons to its charge. The SI unit of charge is the coulomb (C), and an ampere (A) is the SI unit of current.

Solution Essentials:
- Charge, current, and time relationship (C = A × sec)
- Ampere (A)
- Coulomb (C)
- Faraday (9.6485×10^4 C), the charge on one mole of electrons
- Electrolysis
- Half-reaction

When an aqueous solution of potassium iodide is electrolyzed using platinum electrodes, the half-reactions are

$$2I^-(aq) \longrightarrow I_2(aq) + 2e^-$$
$$2H_2O(l) + 2e^- \longrightarrow H_2(g) + 2OH^-(aq)$$

How many grams of iodine are produced when a current of 8.52 mA flows through the cell for 10.0 min?

Problem Strategy The current in amperes multiplied by the time in seconds equals the charge in coulombs. Convert the charge in coulombs to moles of electrons. Then, noting that each mole of iodine produced requires 2 mol e^-, convert mol e^- to mol I_2. Finally, convert mol I_2 to g I_2.

Diagramming the solution clearly shows how three conversion factors (3 ➡) are used to solve the problem.

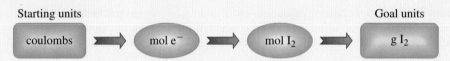

Starting units ... Goal units

Note that two moles of electrons are equivalent to one mole of I_2. Hence,

Solution When the current flows for 6.00×10^2 s (10.0 min), the amount of charge is

$$8.52 \times 10^{-3}\,A \times 6.00 \times 10^2\,s = 5.11\,C$$

$$5.11\,C \times \frac{1\,\text{mol }e^-}{9.6485 \times 10^4\,C} \times \frac{1\,\text{mol }I_2}{2\,\text{mol }e^-} \times \frac{254\,g\,I_2}{1\,\text{mol }I_2} = \mathbf{6.73 \times 10^{-3}\,g\,I_2}$$

Answer Check It is wise to see whether the calculated amount of material deposited during electrolysis is consistent with the amount and duration of the current. In this problem, where a small amount of current (8.52 mA) flows for a short period of time (10.0 min), we should not expect a large mass of product (I_2) to be deposited.

Exercise 19.18 How many grams of oxygen are liberated by the electrolysis of water with a current of 0.0565 A after 1.85×10^4 s?

■ See Problems 19.97 and 19.98.

A Checklist for Review

⚙**WL** and 〔Go Chemistry〕 Sign in at **www.cengage.com/owl** to:

- View tutorials and simulations, develop problem-solving skills, and complete online homework assigned by your professor.
- For quick review and exam prep, download Go Chemistry mini lecture modules from OWL or purchase them at **www.cengagebrain.com**.

Summary of Facts and Concepts

Oxidation–reduction reactions involve a transfer of electrons from one species to another. The *half-reaction method* can be applied to balancing oxidation–reduction reactions in acidic and basic solutions. Many of the principles required for balancing these reactions were presented in Chapter 4.

Electrochemical cells are of two types: voltaic and electrolytic. A *voltaic cell* uses a spontaneous chemical reaction to generate an electric current. It does this by physically separating the reaction into its oxidation and reduction half-reactions. These half-reactions take place

in *half-cells*. The half-cell in which reduction occurs is called the *cathode;* the half-cell in which oxidation occurs is called the *anode*. Electrons flow in the external circuit from the anode to the cathode.

The *cell potential* is the maximum voltage of a voltaic cell. It can be directly related to the maximum work that can be done by the cell. A *standard electrode potential*, or reduction potential, refers to the potential of an electrode in which molar concentrations and gas pressures (in atmospheres) have unit values. A table of standard electrode potentials is useful for establishing the direction of spontaneity of an oxidation–reduction reaction and for calculating the *standard cell potential* of a cell.

The standard free-energy change, standard cell potential, and equilibrium constant are all related. Knowing one, you can calculate the others. Electrochemical measurements can therefore provide equilibrium or thermodynamic information.

An electrode potential depends on concentrations of the electrode substances, according to the *Nernst equation.* Because of this relationship, cell potentials can be used to measure ion concentrations. This is the basic principle of a pH meter, a device that measures the hydrogen-ion concentration.

Voltaic cells are used commercially as portable energy sources (batteries). In addition, the basic principle of the voltaic cell is employed in the *cathodic protection* of buried pipelines and tanks.

Electrolytic cells represent another type of electrochemical cell. They use an external voltage source to push a reaction in a nonspontaneous direction. The electrolysis of an aqueous solution often involves the oxidation or reduction of water at the electrodes. Electrolysis of concentrated sodium chloride solution, for example, gives hydrogen at the cathode. The amounts of substances released at an electrode are related to the amount of charge passed through the cell. This relationship is stoichiometric and follows from the electrode reactions.

Learning Objectives	Important Terms
19.1 Balancing Oxidation–Reduction Reactions in Acidic and Basic Solutions	
■ Learn the steps for balancing oxidation–reduction reactions in acidic solution using the half-reaction method. ■ Balance equations by the half-reaction method (acidic solution). Example 19.1 ■ Learn the additional steps for balancing oxidation–reduction reactions in basic solution using the half-reaction method. ■ Balance equations by the half-reaction method (basic solution). Example 19.2	electrochemical cell voltaic (galvanic) cell electrolytic cell
19.2 Construction of Voltaic Cells	
■ Define *electrochemical cell, voltaic (galvanic) cell, electrolytic cell,* and *half-cell.* ■ Describe the function of the *salt bridge* in a voltaic cell. ■ State the reactions that occur at the *anode* and the *cathode* in an electrochemical cell. ■ Define *cell reaction.* ■ Sketch and label a voltaic cell. Example 19.3	half-cell salt bridge anode cathode cell reaction
19.3 Notation for Voltaic Cells	
■ Write the cell reaction from the cell notation. Example 19.4	
19.4 Cell Potential	
■ Define *cell potential* and *volt.* ■ Calculate the quantity of work from a given amount of cell reactant. Example 19.5	potential difference volt (V) Faraday constant (F) cell potential or electromotive force (emf)
19.5 Standard Cell Potentials and Standard Electrode Potentials	
■ Explain how the electrode potential of a cell is an *intensive property.* ■ Define *standard cell potential* and *standard electrode potential.* ■ Interpret the table of standard reduction potentials. ■ Determine the relative strengths of oxidizing and reducing agents. Example 19.6 ■ Determine the direction of spontaneity from electrode potentials. Example 19.7 ■ Calculate cell potential from standard potentials. Example 19.8	standard cell potential standard electrode potential

19.6 Equilibrium Constants from Cell Potentials

- Calculate the free-energy change from electrode potentials. Example 19.9
- Calculate the cell potential from free-energy change. Example 19.10
- Calculate the equilibrium constant from cell potential. Example 19.11

19.7 Dependence of Cell Potential on Concentration

- Calculate the cell potential for nonstandard conditions. Example 19.12
- Describe how pH can be determined using a glass electrode.

Nernst equation

19.8 Some Commercial Voltaic Cells

- Describe the construction and reactions of a *zinc–carbon dry cell*, a *lithium–iodine battery*, a *lead storage cell*, and a *nickel-cadmium cell*.
- Explain the operation of a proton-exchange membrane *fuel cell*.
- Explain the electrochemical process of the rusting of iron.
- Define *cathodic protection*.

zinc–carbon (Leclanché) dry cell
alkaline dry cell
lithium–iodine battery
lead storage cell
nickel–cadmium cell
fuel cell

19.9 Electrolysis of Molten Salts

- Define *electrolysis*.

electrolysis
Downs cell

19.10 Aqueous Electrolysis

- Learn the half-reactions for water undergoing oxidation and reduction.
- Predict the half-reactions in an aqueous electrolysis. Example 19.13

chlor-alkali membrane cell
chlor-alkali mercury cell

19.11 Stoichiometry of Electrolysis

- Calculate the amount of charge from the amount of product in an electrolysis. Example 19.14
- Calculate the amount of product from the amount of charge in an electrolysis. Example 19.15

ampere (A)

Key Equations

$$w_{max} = -nFE_{cell}$$

$$E°_{cell} = E°_{cathode} - E°_{anode}$$

$$\Delta G° = -nFE°_{cell}$$

$$E°_{cell} = \frac{0.0592}{n} \log K \quad \text{(values in volts at 25°C)}$$

$$E_{cell} = E°_{cell} - \frac{0.0592}{n} \log Q \quad \text{(values in volts at 25°C)}$$

Questions and Problems

⚫WL Interactive versions of these problems may be assigned in OWL.

Self-Assessment and Review Questions

Key: These questions test your understanding of the ideas you worked with in the chapter. These problems vary in difficulty and often can be used for the basis of discussion.

19.1 Describe the difference between a voltaic cell and an electrolytic cell.

19.2 Define *cathode* and *anode*.

19.3 What is the SI unit of electrical potential?

19.4 Define the *faraday*.

19.5 Why is it necessary to measure the voltage of a voltaic cell when no current is flowing to obtain the cell potential?

19.6 How are standard electrode potentials defined?

19.7 Express the SI unit of energy as the product of two electrical units.

19.8 Give the mathematical relationships between the members of each possible pair of the three quantities $\Delta G°$, $E°_{cell}$, and K.

19.9 Using the Nernst equation, explain how the corrosion of iron would be affected by the pH of a water drop.

19.10 Describe the zinc–carbon, or Leclanché, dry cell and the lead storage battery.

19.11 What is a fuel cell? Describe an example.

19.12 Explain the electrochemistry of rusting.

19.13 Iron may be protected by coating with tin (tin cans) or with zinc (galvanized iron). Galvanized iron does not corrode as long as zinc is present. By contrast, when a tin can is scratched, the exposed iron underneath corrodes rapidly. Explain the difference between zinc and tin as protective coatings against iron corrosion.

19.14 The electrolysis of water is often done by passing a current through a dilute solution of sulfuric acid. What is the function of the sulfuric acid?

19.15 Describe a method for the preparation of sodium metal from sodium chloride.

19.16 Potassium was discovered by the British chemist Humphry Davy when he electrolyzed molten potassium hydroxide. What would be the anode reaction?

19.17 Briefly explain why different products are obtained from the electrolysis of molten NaCl and the electrolysis of a dilute aqueous solution of NaCl.

19.18 Write the Nernst equation for the electrode reaction $2Cl^-(aq) \longrightarrow Cl_2(g) + 2e^-$. With this equation, explain why the electrolysis of concentrated sodium chloride solution might be expected to release chlorine gas rather than oxygen gas at the anode.

19.19 The following oxidation–reduction reaction occurs in acidic solution. When the equation is balanced using the smallest set of whole-number stoichiometric coefficients possible, what is the stoichiometric coefficient for water? (*Hint:* In addition to the species shown in the original equation, $H^+(aq)$ and $H_2O(l)$ can also appear in the balanced equation.)

$$MnO_4^-(aq) + CH_3OH(aq) \longrightarrow Mn^{2+}(aq) + HCO_2H(aq)$$

a 13   b 19   c 11   d 7   e 4

19.20 What half-reaction would be expected to occur at the cathode in the electrolysis of aqueous sodium fluoride?

a $2F^-(aq) \longrightarrow F_2(g) + 2e^-$

b $2H_2O(l) \longrightarrow O_2(g) + 4H^+(aq) + 4e^-$

c $2H_2O(l) + 2e^- \longrightarrow H_2(g) + 2OH^-(aq)$

d $Na^+(aq) + e^- \longrightarrow Na(s)$

e $NaF(aq) \longrightarrow NaF(s)$

19.21 Electrolysis of an aqueous solution of sodium chloride produces chlorine, $Cl_2(g)$. The reaction that occurs at the anode is

$$2Cl^-(aq) \longrightarrow Cl_2(g) + 2e^-$$

A constant electric current produces 152 g Cl_2 in 56.1 min from an aqueous solution of sodium chloride. What is the current?

a 123 A   b 61.5 A   c 30.7 A

d 6.18×10^5 A   e 4.36×10^3 A

19.22 The voltaic cell is represented as

$$Zn(s)|Zn^{2+}(1.0M)\|Cu^{2+}(1.0M)|Cu(s)$$

Which of the following statements is *not* true of this cell?

a The mass of the zinc electrode, $Zn(s)$, decreases as the cell runs.

b The copper electrode is the anode.

c Electrons flow through the external circuit from the zinc electrode to the copper electrode.

d Reduction occurs at the copper electrode as the cell runs.

e The concentration of Cu^{2+} decreases as the cell runs.

Concept Explorations

Key: Concept explorations are comprehensive problems that provide a framework that will enable you to explore and learn many of the critical concepts and ideas in each chapter. If you master the concepts associated with these explorations, you will have a better understanding of many important chemistry ideas and will be more successful in solving all types of chemistry problems. These problems are well suited for group work and for use as in-class activities.

19.23 Electrochemical Cells I

You have the following setup to construct a cell under standard conditions. The anode and cathode are iron and silver rods. Using cell notation, the cell is

$$Fe(s)|Fe^{2+}(aq)\|Ag^+(aq)|Ag(s).$$

a Complete the figure of the cell by labeling the anode and cathode and showing the corresponding reactions at the electrodes. Indicate the electron flow in the external circuit, the signs of the electrodes, and the direction of cation migration in the half-cells.

b What is the cell potential, $E°_{cell}$? Is this a voltaic or an electrolytic cell? How do you know?

c You send your lab partner to find the solutions for the beakers. What would you tell him to look for? Provide specific examples of solutions as part of your answer.

d When you are constructing the cell, what impact does the volume of solution in each of the beakers have on the cell potential? (Assume that there is enough volume to cover at least one-third of the anode and cathode rods.)

e If you were to add deionized (distilled) water to the solution that contains the $Ag^+(aq)$, would this have any effect on the cell voltage? If so, explain how the cell voltage would change.

f When constructing the cell, you decide to swap the silver rod that is used as an anode for a copper rod, $Cu(s)$. Will this change have any effect on the cell potential or the operation of the cell? Explain.

g Starting with the original cell, this time you swap the iron rod used as the cathode for a copper rod, Cu(s). How will this change affect the cell potential and the operation of the cell?

19.24 Electrochemical Cells II

Consider this cell running under standard conditions:

$$Ni(s)|Ni^{2+}(aq)\|Cu^+(aq)|Cu(s)$$

a Is this cell a voltaic or an electrolytic cell? How do you know?

b Does current flow in this cell spontaneously?

c What is the maximum cell potential for this cell?

d Say the cell is connected to a voltmeter. Describe what you might see for an initial voltage and what voltage changes, if any, you would observe as time went by.

e What is the free energy of this cell when it is first constructed?

f Does the free energy of the cell change over time as the cell runs? If so, how does it change?

Conceptual Problems

Key: These problems are designed to check your understanding of the concepts associated with some of the main topics presented in each chapter. A strong conceptual understanding of chemistry is the foundation for both applying chemical knowledge and solving chemical problems. These problems vary in level of difficulty and often can be used as a basis for group discussion.

19.25 Keeping in mind that aqueous Cu^{2+} is blue and aqueous Zn^{2+} is colorless, predict what you would observe over a several-day period if you performed the following experiments.

a A strip of Zn is placed into a beaker containing aqueous Zn^{2+}.

b A strip of Cu is placed into a beaker containing aqueous Cu^{2+}.

c A strip of Zn is placed into a beaker containing aqueous Cu^{2+}.

d A strip of Cu is placed into a beaker containing aqueous Zn^{2+}.

19.26 You are working at a plant that manufactures batteries. A client comes to you and asks for a 6.0-V battery that is made from silver and cadmium. Assuming that you are running the battery under standard conditions, how should it be constructed?

19.27 The composition of the hull of a submarine is mostly iron. Pieces of zinc, called "zincs," are placed in contact with the hull throughout the inside of the submarine. Why is this done?

19.28 You place a battery in a flashlight in which all of the electrochemical reactions have reached equilibrium. What do you expect to observe when you turn on the flashlight? Explain your answer.

19.29 The difference between a "heavy-duty" and a regular zinc–carbon battery is that the zinc can in the heavy-duty battery has thicker walls. What makes this battery heavy-duty in terms of output?

19.30 From an electrochemical standpoint, what metal, other than zinc, would be a reasonable candidate to coat a piece of iron to prevent corrosion (oxidation)?

19.31 Pick a combination of two metals from the Standard Reduction Potential table (Table 19.1 or Appendix I) that would result in a cell with a potential of about +0.90 V. For your answer, write both the half-reactions, write the overall balanced reaction, and calculate the cell potential for your choice.

19.32 You have 1.0 *M* solutions of $Al(NO_3)_3$ and $AgNO_3$ along with Al and Ag electrodes to construct a voltaic cell. The salt bridge contains a saturated solution of KCl. Complete the picture associated with this problem by

a writing the symbols of the elements and ions in the appropriate areas (both solutions and electrodes).

b identifying the anode and cathode.

c indicating the direction of electron flow through the external circuit.

d indicating the cell potential (assume standard conditions, with no current flowing).

e writing the appropriate half-reaction under each of the containers.

f indicating the direction of ion flow in the salt bridge.

g identifying the species undergoing oxidation and reduction.

h writing the balanced overall reaction for the cell.

19.33 The zinc–copper voltaic cell shown with this problem is currently running under standard conditions. How would the intensity of light from the bulb change if you were to

a dissolve some additional $CuSO_4(s)$ in the $CuSO_4$ solution?

b dissolve some additional $Zn(NO_3)_2(s)$ in the $Zn(NO_3)_2$ solution?

c add H_2O to the $CuSO_4$ solution?

d remove the salt bridge?

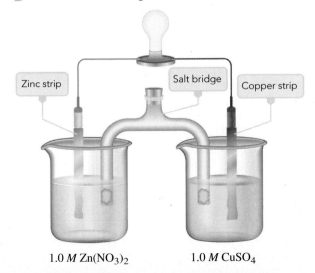

Zinc strip   Salt bridge   Copper strip

1.0 *M* $Zn(NO_3)_2$ 1.0 *M* $CuSO_4$

19.34 The development of lightweight batteries is an ongoing research effort combining many of the physical sciences. You are a member of an engineering team trying to develop a lightweight battery that will effectively react with $O_2(g)$ from the atmosphere as an oxidizing agent.

A reducing agent must be chosen for this battery that will be lightweight, have nontoxic products, and react spontaneously with oxygen. Using data from Appendix I, suggest a likely reducing agent, being sure these conditions are met. Are there any drawbacks to your selection?

Practice Problems

Key: These problems are for practice in applying problem-solving skills. They are divided by topic, and some are keyed to exercises (see the ends of the exercises). The problems are arranged in matching pairs; the odd-numbered problem of each pair is listed first, and its answer is given in the back of the book.

Balancing Oxidation–Reduction Equations

19.35 Balance the following oxidation–reduction equations. The reactions occur in acidic solution.

a $Cr_2O_7^{2-} + C_2O_4^{2-} \longrightarrow Cr^{3+} + CO_2$
b $Cu + NO_3^- \longrightarrow Cu^{2+} + NO$
c $MnO_2 + HNO_2 \longrightarrow Mn^{2+} + NO_3^-$
d $PbO_2 + Mn^{2+} + SO_4^{2-} \longrightarrow PbSO_4 + MnO_4^-$
e $HNO_2 + Cr_2O_7^{2-} \longrightarrow Cr^{3+} + NO_3^-$

19.36 Balance the following oxidation–reduction equations. The reactions occur in acidic solution.

a $Mn^{2+} + BiO_3^- \longrightarrow MnO_4^- + Bi^{3+}$
b $Cr_2O_7^{2-} + I^- \longrightarrow Cr^{3+} + IO_3^-$
c $MnO_4^- + H_2SO_3 \longrightarrow Mn^{2+} + SO_4^{2-}$
d $Cr_2O_7^{2-} + Fe^{2+} \longrightarrow Cr^{3+} + Fe^{3+}$
e $As + ClO_3^- \longrightarrow H_3AsO_3 + HClO$

19.37 Balance the following oxidation–reduction equations. The reactions occur in basic solution.

a $Mn^{2+} + H_2O_2 \longrightarrow MnO_2 + H_2O$
b $MnO_4^- + NO_2^- \longrightarrow MnO_2 + NO_3^-$
c $Mn^{2+} + ClO_3^- \longrightarrow MnO_2 + ClO_2$
d $MnO_4^- + NO_2 \longrightarrow MnO_2 + NO_3^-$
e $Cl_2 \longrightarrow Cl^- + ClO_3^-$

19.38 Balance the following oxidation–reduction equations. The reactions occur in basic solution.

a $Cr(OH)_4^- + H_2O_2 \longrightarrow CrO_4^{2-} + H_2O$
b $MnO_4^- + Br^- \longrightarrow MnO_2 + BrO_3^-$
c $Co^{2+} + H_2O_2 \longrightarrow Co(OH)_3 + H_2O$
d $Pb(OH)_4^{2-} + ClO^- \longrightarrow PbO_2 + Cl^-$
e $Zn + NO_3^- \longrightarrow NH_3 + Zn(OH)_4^{2-}$

19.39 Balance the following oxidation–reduction equations. The reactions occur in acidic or basic aqueous solution, as indicated.

a $H_2S + NO_3^- \longrightarrow NO_2 + S_8$ (acidic)
b $NO_3^- + Cu \longrightarrow NO + Cu^{2+}$ (acidic)
c $MnO_4^- + SO_2 \longrightarrow SO_4^{2-} + Mn^{2+}$ (acidic)
d $Bi(OH)_3 + Sn(OH)_3^- \longrightarrow Sn(OH)_6^{2-} + Bi$ (basic)

19.40 Balance the following oxidation–reduction equations. The reactions occur in acidic or basic aqueous solution, as indicated.

a $Hg_2^{2+} + H_2S \longrightarrow Hg + S_8$ (acidic)
b $S^{2-} + I_2 \longrightarrow SO_4^{2-} + I^-$ (basic)
c $Al + NO_3^- \longrightarrow Al(OH)_4^- + NH_3$ (basic)
d $MnO_4^- + C_2O_4^{2-} \longrightarrow MnO_2 + CO_2$ (basic)

19.41 Balance the following oxidation–reduction equations. The reactions occur in acidic or basic aqueous solution, as indicated.

a $MnO_4^- + I^- \longrightarrow MnO_2 + IO_3^-$ (basic)
b $Cr_2O_7^{2-} + Cl^- \longrightarrow Cr^{3+} + Cl_2$ (acidic)
c $S_8 + NO_3^- \longrightarrow SO_2 + NO$ (acidic)
d $H_2O_2 + MnO_4^- \longrightarrow O_2 + MnO_2$ (basic)
e $Zn + NO_3^- \longrightarrow Zn^{2+} + N_2$ (acidic)

19.42 Balance the following oxidation–reduction equations. The reactions occur in acidic or basic aqueous solution, as indicated.

a $Cr_2O_7^{2-} + H_2O_2 \longrightarrow Cr^{3+} + O_2$ (acidic)
b $CN^- + MnO_4^- \longrightarrow CNO^- + MnO_2$ (basic)
c $Cr(OH)_4^- + OCl^- \longrightarrow CrO_4^{2-} + Cl^-$ (basic)
d $Br_2 + SO_2 \longrightarrow Br^- + SO_4^{2-}$ (acidic)
e $CuS + NO_3^- \longrightarrow Cu^{2+} + NO + S_8$ (acidic)

Electrochemical Cells

19.43 A voltaic cell is constructed from the following half-cells: a magnesium electrode in magnesium sulfate solution and a nickel electrode in nickel sulfate solution. The half-reactions are

$$Mg(s) \longrightarrow Mg^{2+}(aq) + 2e^-$$
$$Ni^{2+}(aq) + 2e^- \longrightarrow Ni(s)$$

Sketch the cell, labeling the anode and cathode (and the electrode reactions), and show the direction of electron flow and the movement of cations.

19.44 Half-cells were made from a nickel rod dipping in a nickel sulfate solution and a silver rod dipping in a silver nitrate solution. The half-reactions in a voltaic cell using these half-cells were

$$Ag^+(aq) + e^- \longrightarrow Ag(s)$$
$$Ni(s) \longrightarrow Ni^{2+}(aq) + 2e^-$$

Sketch the cell and label the anode and cathode, showing the corresponding electrode reactions. Give the direction of electron flow and the movement of cations.

19.45 Zinc reacts spontaneously with silver ion.

$$Zn(s) + 2Ag^+(aq) \longrightarrow Zn^{2+}(aq) + 2Ag(s)$$

Describe a voltaic cell using this reaction. What are the half-reactions?

19.46 Iron reacts spontaneously with copper(II) ion.

$$Fe(s) + Cu^{2+}(aq) \longrightarrow Fe^{2+}(aq) + Cu(s)$$

Obtain half-reactions for this, and then describe a voltaic cell using these half-reactions.

19.47 A silver oxide–zinc cell maintains a fairly constant voltage during discharge (1.60 V). The button form of this cell is used in watches, hearing aids, and other electronic devices. The half-reactions are

$$Zn(s) + 2OH^-(aq) \longrightarrow Zn(OH)_2(s) + 2e^-$$
$$Ag_2O(s) + H_2O(l) + 2e^- \longrightarrow 2Ag(s) + 2OH^-(aq)$$

Identify the anode and the cathode reactions. What is the overall reaction in the voltaic cell?

19.48 A mercury battery, used for hearing aids and electric watches, delivers a constant voltage (1.35 V) for long periods. The half-reactions are

$$HgO(s) + H_2O(l) + 2e^- \longrightarrow Hg(l) + 2OH^-(aq)$$
$$Zn(s) + 2OH^-(aq) \longrightarrow Zn(OH)_2(s) + 2e^-$$

Which half-reaction occurs at the anode and which occurs at the cathode? What is the overall cell reaction?

Voltaic Cell Notation

19.49 Write the cell notation for a voltaic cell with the following half-reactions.

$$Ni(s) \longrightarrow Ni^{2+}(aq) + 2e^-$$
$$Pb^{2+}(aq) + 2e^- \longrightarrow Pb(s)$$

19.50 Write the cell notation for a voltaic cell with the following half-reactions.

$$Al(s) \longrightarrow Al^{3+}(aq) + 3e^-$$
$$2H^+(aq) + 2e^- \longrightarrow H_2(g)$$

19.51 Give the notation for a voltaic cell constructed from a hydrogen electrode (cathode) in 1.0 M HCl and a nickel electrode (anode) in 1.0 M NiSO$_4$ solution. The electrodes are connected by a salt bridge.

19.52 A voltaic cell has an iron rod in 0.30 M iron(III) chloride solution for the cathode and a zinc rod in 0.20 M zinc sulfate solution for the anode. The half-cells are connected by a salt bridge. Write the notation for this cell.

19.53 Write the overall cell reaction for the following voltaic cell.

$$Fe(s)|Fe^{2+}(aq)\|Ag^+(aq)|Ag(s)$$

19.54 Write the overall cell reaction for the following voltaic cell.

$$Pt|H_2(g)|H^+(aq)\|Br_2(l)|Br^-(aq)|Pt$$

19.55 Consider the voltaic cell

$$Cd(s)|Cd^{2+}(aq)\|Ni^{2+}(aq)|Ni(s)$$

Write the half-cell reactions and the overall cell reaction. Make a sketch of this cell and label it. Include labels showing the anode, cathode, and direction of electron flow.

19.56 Consider the voltaic cell

$$Zn(s)|Zn^{2+}(aq)\|Cr^{3+}(aq)|Cr(s)$$

Write the half-cell reactions and the overall cell reaction. Make a sketch of this cell and label it. Include labels showing the anode, cathode, and direction of electron flow.

Electrode Potentials and Cell Potentials

19.57 A voltaic cell whose cell reaction is

$$2Fe^{3+}(aq) + Zn(s) \longrightarrow 2Fe^{2+}(aq) + Zn^{2+}(aq)$$

has a cell potential of 0.72 V. What is the maximum electrical work that can be obtained from this cell per mole of iron(III) ion?

19.58 A particular voltaic cell operates on the reaction

$$Zn(s) + Cl_2(g) \longrightarrow Zn^{2+}(aq) + 2Cl^-(aq)$$

giving a cell potential of 0.853 V. Calculate the maximum electrical work generated when 15.0 g of zinc metal is consumed.

19.59 What is the maximum work you can obtain from 30.0 g of nickel in the following cell when the cell potential is 0.97 V?

$$Ni(s)|Ni^{2+}(aq)\|Ag^+(aq)|Ag(s)$$

19.60 Calculate the maximum work available from 50.0 g of aluminum in the following cell when the cell potential is 1.15 V.

$$Al(s)|Al^{3+}(aq)\|H^+(aq)|O_2(g)|Pt$$

Note that O$_2$ is reduced to H$_2$O.

19.61 Order the following oxidizing agents by increasing strength under standard-state conditions: O$_2(g)$; MnO$_4^-(aq)$; NO$_3^-(aq)$ (in acidic solution).

19.62 Order the following oxidizing agents by increasing strength under standard-state conditions: Ag$^+(aq)$; Cd$^{2+}(aq)$; MnO$_4^-(aq)$ (in acidic solution).

19.63 Consider the reducing agents Cu$^+(aq)$, Zn(s), and Fe(s). Which is strongest? Which is weakest?

19.64 Consider the reducing agents Sn$^{2+}(aq)$, Cl$_2(g)$, and I$^-(aq)$. Which is strongest? Which is weakest?

19.65 Consider the following reactions. Are they spontaneous in the direction written, under standard conditions at 25°C?

a $Sn^{4+}(aq) + 2Fe^{2+}(aq) \longrightarrow Sn^{2+}(aq) + 2Fe^{3+}(aq)$
b $4MnO_4^-(aq) + 12H^+(aq) \longrightarrow$
$$4Mn^{2+}(aq) + 5O_2(g) + 6H_2O(l)$$

19.66 Answer the following questions by referring to standard electrode potentials at 25°C.

a Will oxygen, O$_2$, oxidize iron(II) ion in solution under standard conditions?
b Will copper metal reduce 1.0 M Ni$^{2+}(aq)$ to metallic nickel?

19.67 What would you expect to happen when chlorine gas, Cl$_2$, at 1 atm pressure is bubbled into a solution containing 1.0 M F$^-$ and 1.0 M Br$^-$ at 25°C? Write a balanced equation for the reaction that occurs.

19.68 Dichromate ion, Cr$_2$O$_7^{2-}$, is added to an acidic solution containing Br$^-$ and Mn^{2+}. Write a balanced equation for any reaction that occurs. Assume standard conditions at 25°C.

19.69 Calculate the standard cell potential of the following cell at 25°C.

$$Cr(s)|Cr^{3+}(aq)\|Hg_2^{2+}(aq)|Hg(l)$$

19.70 Calculate the standard cell potential of the following cell at 25°C.

$$Sn(s)|Sn^{2+}(aq)\|I_2(aq)|I^-(aq)$$

19.71 What is the standard cell potential you would obtain from a cell at 25°C using an electrode in which I$^-(aq)$ is in contact with I$_2(s)$ and an electrode in which a chromium strip dips into a solution of Cr$^{3+}(aq)$?

19.72 What is the standard cell potential you would obtain from a cell at 25°C using an electrode in which $Hg_2^{2+}(aq)$ is in contact with mercury metal and an electrode in which an aluminum strip dips into a solution of $Al^{3+}(aq)$?

Relationships Among $E°_{cell}$, $\Delta G°$, and K

19.73 Calculate the standard free-energy change at 25°C for the following reaction.

$$3Cu(s) + 2NO_3^-(aq) + 8H^+(aq) \longrightarrow \\ 3Cu^{2+}(aq) + 2NO(g) + 4H_2O(l)$$

Use standard electrode potentials.

19.74 Calculate the standard free-energy change at 25°C for the following reaction.

$$4Al(s) + 3O_2(g) + 12H^+(aq) \longrightarrow 4Al^{3+}(aq) + 6H_2O(l)$$

Use standard electrode potentials.

19.75 What is $\Delta G°$ for the following reaction?

$$2I^-(aq) + Cl_2(g) \longrightarrow I_2(s) + 2Cl^-(aq)$$

Use data given in Table 19.1.

19.76 Using electrode potentials, calculate the standard free-energy change for the reaction

$$Na(s) + \tfrac{1}{2}Br_2(g) \xrightarrow{H_2O} Na^+(aq) + Br^-(aq)$$

19.77 Calculate the standard cell potential at 25°C for the following cell reaction from standard free energies of formation (Appendix C).

$$Mg(s) + 2Ag^+(aq) \longrightarrow Mg^{2+}(aq) + 2Ag(s)$$

19.78 Calculate the standard cell potential at 25°C for the following cell reaction from standard free energies of formation (Appendix C).

$$2Al(s) + 3Cu^{2+}(aq) \longrightarrow 2Al^{3+}(aq) + 3Cu(s)$$

19.79 Calculate the standard cell potential of the lead storage cell whose overall reaction is

$$PbO_2(s) + 2HSO_4^-(aq) + 2H^+(aq) + Pb(s) \longrightarrow \\ 2PbSO_4(s) + 2H_2O(l)$$

See Appendix C for free energies of formation.

19.80 Calculate the standard cell potential of the cell corresponding to the oxidation of oxalic acid, $H_2C_2O_4$, by permanganate ion, MnO_4^-.

$$5H_2C_2O_4(aq) + 2MnO_4^-(aq) + 6H^+(aq) \longrightarrow \\ 10CO_2(g) + 2Mn^{2+}(aq) + 8H_2O(l)$$

See Appendix C for free energies of formation; $\Delta G_f°$ for $H_2C_2O_4(aq)$ is -698 kJ.

19.81 Calculate the equilibrium constant K for the following reaction at 25°C from standard electrode potentials.

$$Fe^{3+}(aq) + Cu(s) \longrightarrow Fe^{2+}(aq) + Cu^{2+}(aq)$$

The equation is not balanced.

19.82 Calculate the equilibrium constant K for the following reaction at 25°C from standard electrode potentials.

$$Sn^{4+}(aq) + 2Hg(l) \longrightarrow Sn^{2+}(aq) + Hg_2^{2+}(aq)$$

The equation is not balanced.

19.83 Copper(I) ion can act as both an oxidizing agent and a reducing agent. Hence, it can react with itself.

$$2Cu^+(aq) \longrightarrow Cu(s) + Cu^{2+}(aq)$$

Calculate the equilibrium constant at 25°C for this reaction, using appropriate values of electrode potentials.

19.84 Use electrode potentials to calculate the equilibrium constant at 25°C for the reaction

$$ClO_4^-(aq) + ClO_2^-(aq) \rightleftharpoons 2ClO_3^-(aq)$$

See Appendix I for data.

Nernst Equation

19.85 Calculate the cell potential of the following cell at 25°C.

$$Cr(s)|Cr^{3+}(1.0 \times 10^{-3}\ M)\|Ni^{2+}(1.5\ M)|Ni(s)$$

19.86 What is the cell potential of the following cell at 25°C?

$$Ni(s)|Ni^{2+}(1.0\ M)\|Sn^{2+}(1.5 \times 10^{-4}\ M)|Sn(s)$$

19.87 Calculate the cell potential of a cell operating with the following reaction at 25°C, in which $[MnO_4^-] = 0.010\ M$, $[Br^-] = 0.010\ M$, $[Mn^{2+}] = 0.15\ M$, and $[H^+] = 1.0\ M$.

$$2MnO_4^-(aq) + 10Br^-(aq) + 16H^+(aq) \longrightarrow \\ 2Mn^{2+}(aq) + 5Br_2(l) + 8H_2O(l)$$

19.88 Calculate the cell potential of a cell operating with the following reaction at 25°C, in which $[Cr_2O_7^{2-}] = 0.020\ M$, $[I^-] = 0.015\ M$, $[Cr^{3+}] = 0.40\ M$, and $[H^+] = 0.60\ M$.

$$Cr_2O_7^{2-}(aq) + 6I^-(aq) + 14H^+(aq) \longrightarrow \\ 2Cr^{3+}(aq) + 3I_2(s) + 7H_2O(l)$$

19.89 The voltaic cell

$$Cd(s)|Cd^{2+}(aq)\|Ni^{2+}(1.0\ M)|Ni(s)$$

has a cell potential of 0.240 V at 25°C. What is the concentration of cadmium ion? ($E°_{cell} = 0.170$ V.)

19.90 The cell potential of the following cell at 25°C is 0.480 V.

$$Zn|Zn^{2+}(1\ M)\|H^+(\text{test solution})|H_2(1\ \text{atm})|Pt$$

What is the pH of the test solution?

Electrolysis

19.91 What are the half-reactions in the electrolysis of a $CaS(l)$; b $CsOH(l)$?

19.92 What are the half-reactions in the electrolysis of a $MgBr_2(l)$; b $Ca(OH)_2(l)$?

19.93 Describe what you expect to happen when the following solutions are electrolyzed: a aqueous Na_2SO_4; b aqueous KBr. That is, what are the electrode reactions? What is the overall reaction?

19.94 Describe what you expect to happen when the following solutions are electrolyzed: a aqueous $CuCl_2$; b aqueous $Cu(NO_3)_2$. That is, what are the electrode reactions? What is the overall reaction? Nitrate ion is not oxidized.

19.95 In the commercial preparation of aluminum, aluminum oxide, Al_2O_3, is electrolyzed at 1000°C. (The mineral cryolite is added as a solvent.) Assume that the cathode reaction is

$$Al^{3+} + 3e^- \longrightarrow Al$$

How many coulombs of electricity are required to give 3.61 kg of aluminum?

19.96 Chlorine, Cl_2, is produced commercially by the electrolysis of aqueous sodium chloride. The anode reaction is

$$2Cl^-(aq) \longrightarrow Cl_2(g) + 2e^-$$

How long will it take to produce 2.00 kg of chlorine if the current is 5.00×10^2 A?

19.97 When molten lithium chloride, LiCl, is electrolyzed, lithium metal is liberated at the cathode. How many grams of lithium are liberated when 5.00×10^3 C of charge passes through the cell?

19.98 How many grams of cadmium are deposited from an aqueous solution of cadmium sulfate, $CdSO_4$, when an electric current of 1.75 A flows through the solution for 325 min?

General Problems

Key: These problems provide more practice but are not divided by topic or keyed to exercises. Each section ends with essay questions. Odd-numbered problems and the even-numbered problems that follow are similar; answers to all odd-numbered problems except the essay questions are given in the back of the book.

19.99 Some metals, such as iron, can be oxidized to more than one oxidation state. Obtain the balanced net ionic equations for the following oxidation–reduction reactions, in which nitric acid is reduced to nitric oxide, NO.
  a Oxidation of iron metal to iron(II) ion by nitric acid.
  b Oxidation of iron(II) ion to iron(III) ion by nitric acid.
  c Oxidation of iron metal to iron(III) by nitric acid. [Consider adding the a and b equations.]

19.100 Some metals, such as thallium, can be oxidized to more than one oxidation state. Obtain the balanced net ionic equations for the following oxidation–reduction reactions, in which nitric acid is reduced to nitric oxide, NO.
  a Oxidation of thallium metal to thallium(I) ion by nitric acid.
  b Oxidation of thallium(I) ion to thallium(III) ion by nitric acid.
  c Oxidation of thallium metal to thallium(III) by nitric acid. [Consider adding the a and b equations.]

19.101 Balance the following skeleton equations. The reactions occur in acidic or basic aqueous solution, as indicated.
  a $MnO_4^- + S^{2-} \longrightarrow MnO_2 + S_8$ (basic)
  b $IO_3^- + HSO_3^- \longrightarrow I^- + SO_4^{2-}$ (acidic)
  c $Fe(OH)_2 + CrO_4^{2-} \longrightarrow$
    $Fe(OH)_3 + Cr(OH)_4^-$ (basic)
  d $Cl_2 \longrightarrow Cl^- + ClO^-$ (basic)

19.102 Balance the following skeleton equations. The reactions occur in acidic or basic aqueous solution, as indicated.
  a $MnO_4^- + H_2S \longrightarrow Mn^{2+} + S_8$ (acidic)
  b $Zn + NO_3^- \longrightarrow Zn^{2+} + N_2O$ (acidic)
  c $MnO_4^{2-} \longrightarrow MnO_4^- + MnO_2$ (basic)
  d $Br_2 \longrightarrow Br^- + BrO_3^-$ (basic)

19.103 Iron(II) hydroxide is a greenish precipitate that is formed from iron(II) ion by the addition of a base.

This precipitate gradually turns to the yellowish-brown iron(III) hydroxide from oxidation by O_2 in the air. Write a balanced equation for this oxidation by O_2.

19.104 A sensitive test for bismuth(III) ion consists of shaking a solution suspected of containing the ion with a basic solution of sodium stannite, Na_2SnO_2. A positive test consists of the formation of a black precipitate of bismuth metal. Stannite ion is oxidized by bismuth(III) ion to stannate ion, SnO_3^{2-}. Write a balanced equation for the reaction.

19.105 Give the notation for a voltaic cell that uses the reaction

$$Ca(s) + Cl_2(g) \longrightarrow Ca^{2+}(aq) + 2Cl^-(aq)$$

What is the half-cell reaction for the anode? for the cathode? What is the standard cell potential of the cell?

19.106 Give the notation for a voltaic cell whose overall cell reaction is

$$Mg(s) + 2Ag^+(aq) \longrightarrow Mg^{2+}(aq) + 2Ag(s)$$

What are the half-cell reactions? Label them as anode or cathode reactions. What is the standard cell potential of this cell?

19.107 Use electrode potentials to answer the following questions. a Is the oxidation of nickel by iron(III) ion a spontaneous reaction under standard conditions?

$$Ni(s) + 2Fe^{3+}(aq) \longrightarrow Ni^{2+}(aq) + 2Fe^{2+}(aq)$$

b Will iron(III) ion oxidize tin(II) ion to tin(IV) ion under standard conditions?

$$2Fe^{3+}(aq) + Sn^{2+}(aq) \longrightarrow 2Fe^{2+}(aq) + Sn^{4+}(aq)$$

19.108 Use electrode potentials to answer the following questions, assuming standard conditions. a Do you expect permanganate ion (MnO_4^-) to oxidize chloride ion to chlorine gas in acidic solution? b Will dichromate ion ($Cr_2O_7^{2-}$) oxidize chloride ion to chlorine gas in acidic solution?

19.109 Determine the cell potential of the following cell.

$$Pb|PbSO_4(s), SO_4^{2-}(1.0\ M)||H^+(1.0\ M)|H_2(1.0\ atm)|Pt$$

The anode is essentially a lead electrode, $Pb|Pb^{2+}(aq)$. However, the anode solution is saturated with lead sulfate, so that the lead(II)-ion concentration is determined by the solubility product of $PbSO_4$ ($K_{sp} = 1.7 \times 10^{-8}$).

19.110 Determine the cell potential of the following cell.

$$Pt|H_2(1.0 \text{ atm})|H^+(1.0 \text{ } M)\|Cl^-(1.0 \text{ } M), AgCl(s)|Ag$$

The cathode is essentially a silver electrode, $Ag^+(aq)|Ag$. However, the cathode solution is saturated with silver chloride, so that the silver-ion concentration is determined by the solubility product of AgCl ($K_{sp} = 1.8 \times 10^{-10}$).

19.111 a Calculate the equilibrium constant for the following reaction at 25°C.

$$Sn(s) + Pb^{2+}(aq) \rightleftharpoons Sn^{2+}(aq) + Pb(s)$$

The standard cell potential of the corresponding voltaic cell is 0.010 V. b If an excess of tin metal is added to 1.0 M Pb^{2+}, what is the concentration of Pb^{2+} at equilibrium?

19.112 a Calculate the equilibrium constant for the following reaction at 25°C.

$$Ag^+(aq) + Fe^{2+}(aq) \rightleftharpoons Ag(s) + Fe^{3+}(aq)$$

The standard cell potential of the corresponding voltaic cell is 0.030 V. b When equal volumes of 0.75 M solutions of Ag^+ and Fe^{2+} are mixed, what is the equilibrium concentration of Fe^{2+}?

19.113 How many faradays are required for each of the following processes? How many coulombs are required?

a Reduction of 1.0 mol Na^+ to Na
b Reduction of 1.0 mol Cu^{2+} to Cu
c Oxidation of 1.0 g H_2O to O_2
d Oxidation of 1.0 g Cl^- to Cl_2

19.114 How many faradays are required for each of the following processes? How many coulombs are required?

a Reduction of 2.0 mol Fe^{3+} to Fe^{2+}
b Reduction of 2.0 mol Fe^{3+} to Fe
c Oxidation of 2.5 g Sn^{2+} to Sn^{4+}
d Reduction of 2.5 g Au^{3+} to Au

19.115 In an analytical determination of arsenic, a solution containing arsenious acid, H_3AsO_3, potassium iodide, and a small amount of starch is electrolyzed. The electrolysis produces free iodine from iodide ion, and the iodine immediately oxidizes the arsenious acid to hydrogen arsenate ion, $HAsO_4^{2-}$.

$$I_2(aq) + H_3AsO_3(aq) + H_2O(l) \longrightarrow$$
$$2I^-(aq) + HAsO_4^{2-}(aq) + 4H^+(aq)$$

When the oxidation of arsenic is complete, the free iodine combines with the starch to give a deep blue color. If, during a particular run, it takes 65.4 s for a current of 10.5 mA to give an endpoint (indicated by the blue color), how many grams of arsenic are present in the solution?

19.116 The amount of lactic acid, $HC_3H_5O_3$, produced in a sample of muscle tissue was analyzed by reaction with hydroxide ion. Hydroxide ion was produced in the sample mixture by electrolysis. The cathode reaction was

$$2H_2O(l) + 2e^- \longrightarrow H_2(g) + 2OH^-(aq)$$

Hydroxide ion reacts with lactic acid as soon as it is produced. The endpoint of the reaction is detected with an acid–base indicator. It required 105 s for a current of 15.6 mA to reach the endpoint. How many grams of lactic acid (a monoprotic acid) were present in the sample?

19.117 A standard electrochemical cell is made by dipping a silver electrode into a 1.0 M Ag^+ solution and a cadmium electrode into a 1.0 M Cd^{2+} solution.

a What is the spontaneous chemical reaction, and what is the maximum potential produced by this cell?
b What would be the effect on the potential of this cell if sodium sulfide were added to the Cd^{2+} half-cell and CdS were precipitated? Why?
c What would be the effect on the potential of the cell if the size of the silver electrode were doubled?

19.118 A standard electrochemical cell is made by dipping an iron electrode into a 1.0 M Fe^{2+} solution and a copper electrode into a 1.0 M Cu^{2+} solution.

a What is the spontaneous chemical reaction, and what is the maximum potential produced by this cell?
b What would be the effect on the potential of this cell if sodium carbonate were added to the Cu^{2+} half-cell and $CuCO_3$ were precipitated? Why?
c What would be the effect on the potential of the cell if the size of the iron electrode were halved?

19.119 A solution of copper(II) sulfate is electrolyzed by passing a current through the solution using inert electrodes. Consequently, there is a decrease in the Cu^{2+} concentration and an increase in the hydronium ion concentration. Also, one electrode increases in mass and a gas evolves at the other electrode. Write half-reactions that occur at the anode and at the cathode.

19.120 A potassium chloride solution is electrolyzed by passing a current through the solution using inert electrodes. A gas evolves at each electrode, and there is a large increase in pH of the solution. Write the half-reactions that occur at the anode and at the cathode.

19.121 A constant current of 1.50 amp is passed through an electrolytic cell containing a 0.10 M solution of $AgNO_3$ and a silver anode and a platinum cathode until 2.48 g of silver is deposited.

a How long does the current flow to obtain this deposit?
b What mass of chromium would be deposited in a similar cell containing 0.10 M Cr^{3+} if the same amount of current were used?

19.122 A constant current of 1.25 amp is passed through an electrolytic cell containing a 0.050 M solution of $CuSO_4$ and a copper anode and a platinum cathode until 3.00 g of copper is deposited.

a How long does the current flow to obtain this deposit?
b What mass of silver would be deposited in a similar cell containing 0.15 M Ag^+ if the same amount of current were used?

19.123 An aqueous solution of an unknown salt of gold is electrolyzed by a current of 2.75 amps for 3.50 hours. The electroplating is carried out with an efficiency of 90.0%, resulting in a deposit of 21.221 g of gold.

a How many faradays are required to deposit the gold?
b What is the charge on the gold ions (based on your calculations)?

19.124 An aqueous solution of an unknown salt of vanadium is electrolyzed by a current of 2.50 amps for 1.90 hours. The electroplating is carried out with an efficiency of 95.0%, resulting in a deposit of 2.850 g of vanadium.

a How many faradays are required to deposit the vanadium?

b What is the charge on the vanadium ions (based on your calculations)?

19.125 An electrochemical cell is made by placing a zinc electrode in 1.00 L of 0.200 M $ZnSO_4$ solution and a copper electrode in 1.00 L of 0.0100 M $CuCl_2$ solution.

a What is the initial voltage of this cell when it is properly constructed?

b Calculate the final concentration of Cu^{2+} in this cell if it is allowed to produce an average current of 1.00 amp for 225 s.

19.126 An electrochemical cell is made by placing an iron electrode in 1.00 L of 0.15 M $FeSO_4$ solution and a copper electrode in 1.00 L of 0.040 M $CuSO_4$ solution.

a What is the initial voltage of this cell when it is properly constructed?

b Calculate the final concentration of Cu^{2+} in this cell if it is allowed to produce an average current of 1.25 amp for 375 s.

19.127 **a** Calculate $\Delta G°$ for the following cell reaction:

$$Cd(s)|Cd^{2+}(aq)\|Co^{2+}(aq)|Co(s)$$

The $\Delta G_f°$ for $Cd^{2+}(aq)$ is -77.6 kJ/mol.

b From $\Delta G°$, calculate the standard cell potential for the cell reaction and from this, determine the standard potential for $Co^{2+}(aq) + 2e^- \longrightarrow Co(s)$.

19.128 **a** Calculate $\Delta G°$ for the following cell reaction:

$$Tl(s)|Tl^+(aq)\|Pb^{2+}(aq)|Pb(s)$$

The $\Delta G_f°$ for $Tl^+(aq)$ is -32.4 kJ/mol.

b From $\Delta G°$, calculate the standard cell potential for the cell reaction and from this, determine the standard potential for $Tl^+(aq) + e^- \longrightarrow Tl(s)$.

Strategy Problems

Key: As noted earlier, all of the practice and general problems are matched pairs. This section is a selection of problems that are not in matched-pair format. These challenging problems require that you employ many of the concepts and strategies that were developed in the chapter. In some cases, you will have to integrate several concepts and operational skills in order to solve the problem successfully.

19.129 A voltaic cell is constructed from a half-cell in which a chromium rod dips into a solution of chromium(III) nitrate, $Cr(NO_3)_3$, and another half-cell in which a lead rod dips into a solution of lead(II) nitrate, $Pb(NO_3)_2$. The two half-cells are connected by a salt bridge. Chromium metal is oxidized during operation of the voltaic cell. What is the half-reaction that occurs at the cathode?

19.130 What are the overall cell reaction and the cell potential for this voltaic cell?

$$Mg(s)|Mg^{2+}(aq)\|Fe^{3+}(aq)|Fe(s)$$

19.131 The cell potential of a particular voltaic cell with the cell reaction

$$2Cr(s) + 3Cu^{2+}(aq) \longrightarrow 2Cr^{3+}(aq) + 3\,Cu(s)$$

is 1.14 V. What is the maximum electrical work, per mole, that can be obtained from 6.61 g of chromium metal?

19.132 Order the following oxidizing agents by increasing strength under standard-state conditions: $Mg^{2+}(aq)$, $Hg^{2+}(aq)$, $Pb^{2+}(aq)$.

19.133 What is the cell potential (E_{cell}) of a spontaneous cell that is run at 25°C and contains $[Cr^{3+}] = 0.10$ M and $[Ag^+] = 1.0 \times 10^{-4}$ M?

19.134 A strip of aluminum metal is placed into a beaker containing 1.0 M $CuSO_4$. **a** Will a spontaneous reaction occur? **b** If a spontaneous reaction does occur, write the half-reactions and describe what you would observe.

19.135 Which of the following reactions occur spontaneously *as written*, with the production of a measurable electric current?

a $I_2 + NaBr \longrightarrow Br_2 + 2NaI$

b $Li + NaCl \longrightarrow LiCl + Na$

c $Li^+ + Na^+ \longrightarrow NaLi$

d $Ag + Li \longrightarrow Ag^+ + Li^+$

e $2Li^+ + F_2 \longrightarrow 2LiF$

19.136 How many grams of hydrogen can be produced by the electrolysis of water with a current of 0.433 A running for 5.00 min?

19.137 The following two half-reactions are involved in a voltaic cell. At standard conditions, what species is produced at each electrode?

$$Ag^+ + e^- \longrightarrow Ag \qquad E° = 0.80\ V$$
$$Ni^{2+} + 2e^- \longrightarrow Ni \qquad E° = -0.25\ V$$

19.138 Consider the following reaction run at standard conditions:

$$Al(s) + Fe^{2+}(aq) \longrightarrow Fe(s) + Al^{3+}(aq)$$

a Calculate the standard cell potential for this cell from standard free energies of formation (see Appendix C).

b Calculate the equilibrium constant for the reaction.

c How would the value of the equilibrium constant be affected if the concentration of $Fe^{2+}(aq)$ were changed to 1.1 M?

19.139 A beaker is found to contain $I^-(aq)$, $Cu^{2+}(aq)$, and $Mg^{2+}(aq)$ ions with $Ag(s)$.

a Would there be any spontaneous electrochemical reactions occurring in the mixture?

b A strip of aluminum metal is added to the beaker. What would you observe?

c Write the half-reactions for any of the ions that would spontaneously react with the $Al(s)$ placed in the beaker.

d If you were to add a solution that contains $Al^{3+}(aq)$ to the mixture, would you observe any reaction? Justify your answer.

19.140 A 1.0-L sample of 1.0 M HCl solution has a 10.0 A current applied for 45 minutes. What is the pH of the solution after the electricity has been turned off?

19.141 Consider the following cell running under standard conditions:

$$Fe(s)|Fe^{2+}(aq)\ \|Al^{3+}(aq)|Al(s)$$

a Is this a voltaic cell?

b Which species is being reduced during the chemical reaction?

c Which species is the oxidizing agent?

d What happens to the concentration of $Fe^{2+}(aq)$ as the reaction proceeds?

e How does the mass of $Al(s)$ change as the reaction proceeds?

19.142 A cell is constructed from some unknown elements designated as X and Y. Under standard conditions, a cell is created: $X(s)|X^{3+}(aq)||Y^+(aq)|Y(s)$. The maximum free-energy change for the reaction is -350 kJ.

a Is this an electrolytic cell? Justify your answer.

b If you were to decrease the concentration of $Y^{3+}(aq)$, how would this affect the cell potential of the reaction?

c When the reaction reaches equilibrium, what is the value of $\Delta G°$.

d What is the value of the equilibrium constant for the reaction?

19.143 Two identical pieces of the same unknown metal are placed into 1.0 M solutions of $LiNO_3(aq)$ and $Fe(NO_3)_2(aq)$. No reaction is observed in the $LiNO_3(aq)$; however, the metal appears to dissolve in the $Fe(NO_3)_2(aq)$. If the possible metals are nickel and zinc, which metal would be expected to react? Calculate the cell potential under standard conditions for the metal that did react.

19.144 A solution is prepared by dissolving 145 g of silver nitrate in enough water to make a total volume of 1.0 L. A different solution is prepared by dissolving 105 grams of zinc nitrate in enough water to make a total volume of 1.1 L. A 375-mL portion of each solution is used to prepare an electrochemical cell using a salt bridge, a 20.0-g strip of zinc metal in the zinc nitrate solution, and a 15.0-g strip of silver metal in the silver nitrate solution.

a Calculate the cell potential at 25°C.

b Calculate $\Delta G°$ for this cell.

c How would the cell potential change using an additional 25.0 mL of the silver nitrate solution?

d If the mass of the zinc metal strip were 40.0 grams, how would this affect the cell potential?

e If 0.5 g of sodium chloride were added to the silver nitrate solution in the half-cell, what affect would this have on the cell potential (increase or decrease)?

Cumulative-Skills Problems

Key: The problems under this heading combine skills introduced in previous chapters with those given in the current one.

19.145 Consider the following cell reaction at 25°C.

$$2Cr^{3+}(aq) + 3Zn(s) \longrightarrow 3Zn^{2+}(aq) + 2Cr(s)$$

Calculate the standard cell potential of this cell from the standard electrode potentials, and from this obtain $\Delta G°$ for the cell reaction. Use data in Appendix C to calculate $\Delta H°$; note that $Cr(H_2O)_6^{3+}(aq)$ equals $Cr^{3+}(aq)$. Use these values of $\Delta H°$ and $\Delta G°$ to obtain $\Delta S°$ for the cell reaction.

19.146 Consider the following cell reaction at 25°C.

$$2Cr(s) + 3Fe^{2+}(aq) \longrightarrow 2Cr^{3+}(aq) + 3Fe(s)$$

Calculate the standard cell potential of this cell from the standard electrode potentials, and from this obtain $\Delta G°$ for the cell reaction. Use data in Appendix C to calculate $\Delta H°$; note that $Cr(H_2O)_6^{3+}(aq)$ equals $Cr^{3+}(aq)$. Use these values of $\Delta H°$ and $\Delta G°$ to obtain $\Delta S°$ for the cell reaction.

19.147 Under standard conditions for all concentrations, the following reaction is spontaneous at 25°C.

$$O_2(g) + 4H^+(aq) + 4Br^-(aq) \longrightarrow 2H_2O(l) + 2Br_2(l)$$

If $[H^+]$ is decreased so that the pH = 3.60, what value will E_{cell} have, and will the reaction be spontaneous at this $[H^+]$?

19.148 Under standard conditions for all concentrations, the following reaction is spontaneous at 25°C.

$$O_3(g) + 2H^+(aq) + 2Co^{2+}(aq) \longrightarrow$$
$$O_2(g) + H_2O(l) + 2Co^{3+}(l)$$

If $[H^+]$ is decreased so that the pH = 9.00, what value will E_{cell} have, and will the reaction be spontaneous at this $[H^+]$?

19.149 Under standard conditions for all concentrations, the following reaction is spontaneous at 25°C.

$$O_2(g) + 4H^+(aq) + 4Br^-(aq) \longrightarrow 2H_2O(l) + 2Br_2(l)$$

If $[H^+]$ is adjusted by adding a buffer of 0.10 M NaOCN and 0.10 M HOCN ($K_a = 3.5 \times 10^{-4}$), what value will E_{cell} have, and will the reaction be spontaneous at this $[H^+]$?

19.150 Under standard conditions for all concentrations, the following reaction is spontaneous at 25°C.

$$O_3(g) + 2H^+(aq) + 2Co^{2+}(aq) \longrightarrow$$
$$O_2(g) + H_2O(l) + 2Co^{3+}(aq)$$

If $[H^+]$ is adjusted by adding a buffer of 0.10 M NaClO and 0.10 M HClO ($K_a = 3.5 \times 10^{-8}$), what value will E_{cell} have, and will the reaction be spontaneous at this $[H^+]$?

19.151 An electrode is prepared by dipping a silver strip into a solution saturated with silver thiocyanate, AgSCN, and containing 0.10 M SCN$^-$. The cell potential of the voltaic cell constructed by connecting this electrode as the cathode to the standard hydrogen half-cell as the anode is 0.45 V. What is the solubility product of silver thiocyanate?

19.152 An electrode is prepared from liquid mercury in contact with a saturated solution of mercury(I) chloride, Hg_2Cl_2, containing 1.00 M Cl$^-$. The cell potential of the voltaic cell constructed by connecting this electrode as the cathode to the standard hydrogen half-cell as the anode is 0.268 V. What is the solubility product of mercury(I) chloride?

20 Nuclear Chemistry

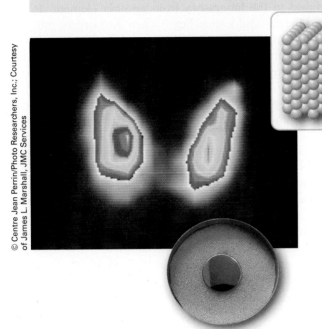

Technetium-99m, Tc-99m, is the most widely used radioisotope for medical diagnostics. Shown here is a scintigram (two-dimensional scan) of a patient's thyroid gland. It was produced by injecting the patient intravenously with a small amount of technetium-99m in the form of a solution of sodium pertechnetate, NaTcO$_4$. Gamma rays emitted by the technetium were detected to produce the scintigram. Below this scintigram is a disk with a thin coating of technetium. The close-packed crystal structure of technetium is shown above right.

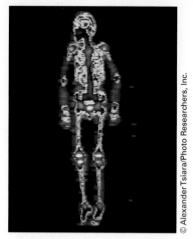

Figure 20.1 ▲

Image of a person's skeleton obtained using an excited form of technetium-99 A technetium compound was injected into the body, where it concentrated in bone tissue. Gamma rays (similar to x rays) emitted by technetium were detected by special equipment to produce this image.

Technetium is an unusual element. Although a *d*-transition element (under manganese in Group VIIB) with a small atomic number (Z = 43), it has no stable isotopes. The nucleus of every technetium isotope is radioactive and decays, or disintegrates, to give an isotope of another element. Many of the technetium isotopes decay by emitting an electron from the nucleus.

Because of its nuclear instability, technetium is not found naturally on earth. Nevertheless, it is produced commercially in kilogram quantities from other elements by nuclear reactions, processes in which nuclei are transformed into different nuclei. Technetium (from the Greek *tekhnetos,* meaning "artificial") was the first new element produced in the laboratory from another element. It was discovered in 1937 by Carlo Perrier and Emilio Segrè when the element molybdenum was bombarded with deuterons (nuclei of hydrogen, each having one proton and one neutron). Later, technetium was found to be a product of the fission, or splitting, of uranium nuclei. Technetium is produced in nuclear fission reactors used to generate electricity.

Technetium is one of the principal isotopes used in medical diagnostics based on radioactivity. A compound of technetium is injected into a vein, where it concentrates in certain body organs. The energy emitted by technetium nuclei is detected by special equipment and gives an image of these body organs. Figure 20.1 shows the image of a person's skeleton obtained from technetium administered in this manner. The technetium is eliminated by the body after several hours.

In this chapter, we will look at nuclear processes such as those we have described for technetium. We will answer such questions as the following: How do you describe the radioactive decay of technetium? How do you describe the transformation of a molybdenum nucleus into technetium? How is technetium produced from uranium by nuclear fission, or splitting? What are some practical applications of nuclear processes?

Radioactivity and Nuclear Bombardment Reactions

In chemical reactions, only the outer electrons of the atoms are disturbed. The nuclei of the atoms are not affected. In nuclear reactions, however, the nuclear changes that occur are independent of the chemical environment of the atom. For example, the nuclear changes in a radioactive ^3_1H atom are the same if the atom is part of an H_2 molecule or incorporated into H_2O.

We will look at two types of nuclear reactions. One type is **radioactive decay**, *the process in which a nucleus spontaneously disintegrates, giving off radiation.* The radiation consists of one or more of the following, depending on the nucleus: electrons, nuclear particles (such as neutrons), smaller nuclei (usually helium-4 nuclei), and electromagnetic radiation.

The second type of nuclear reaction is a **nuclear bombardment reaction**, *a nuclear reaction in which a nucleus is bombarded, or struck, by another nucleus or by a nuclear particle.* If there is sufficient energy in this collision, the nuclear particles of the reactants rearrange to give a product nucleus or nuclei. First, we will look at radioactive decay.

20.1 Radioactivity

The phenomenon of radioactivity was discovered by Antoine Henri Becquerel in 1896. He discovered that photographic plates develop bright spots when exposed to uranium minerals, and he concluded that the minerals give off some sort of radiation.

The radiation from uranium minerals was later shown to be separable by electric (and magnetic) fields into three types, alpha (α), beta (β), and gamma (γ) rays (Figure 20.2). *Alpha rays* bend away from a positive plate and toward a negative plate, indicating that they have a positive charge; they are now known to consist of helium-4 nuclei (nuclei with two protons and two neutrons). *Beta rays*

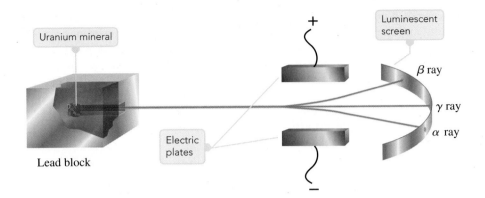

Figure 20.2 ◄

Separation of the radiation from a radioactive material (uranium mineral) The radiation separates into alpha (α), beta (β), and gamma (γ) rays when it passes through an electric field.

bend in the opposite direction, indicating that they have a negative charge; they are now known to consist of high-speed electrons. *Gamma rays* are unaffected by electric and magnetic fields: they have been shown to be a form of electromagnetic radiation that is similar to x rays, except they are higher in energy with shorter wavelengths (about 1 pm, or 1×10^{-12} m). Uranium minerals contain a number of radioactive elements, each emitting one or more of these radiations. Uranium-238, the main uranium isotope in uranium minerals, emits alpha rays and thereby decays, or disintegrates, to thorium-234 nuclei.

A sample of uranium-238 decays, or disintegrates, spontaneously over a period of billions of years. After about 30 billion years, the sample would be nearly gone. Strontium-90, formed by nuclear reactions that occur in nuclear weapons testing and nuclear power reactors, decays more rapidly. A sample of strontium-90 would be nearly gone after several hundred years. In either case, it is impossible to know when a particular nucleus will decay, although, as you will see in Section 20.4, precise information can be given about the rate of decay of any radioactive sample.

Nuclear Equations

You can write an equation for the nuclear reaction corresponding to the decay of uranium-238 much as you would write an equation for a chemical reaction. You represent the uranium-238 nucleus by the *nuclide symbol* $^{238}_{92}\text{U}$. ▶ The radioactive decay of $^{238}_{92}\text{U}$ by alpha-particle emission (loss of a $^{4}_{2}\text{He}$ nucleus) is written

Nuclide symbols were introduced in Section 2.3. For uranium-238, you write

$$\text{mass number} \longrightarrow \, ^{238}_{92}\text{U} \longleftarrow \text{atomic number}$$

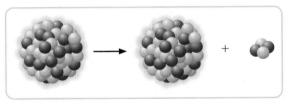

$$^{238}_{92}\text{U} \longrightarrow \, ^{234}_{90}\text{Th} + \, ^{4}_{2}\text{He}$$

The product, in addition to being a helium-4 alpha particle, is thorium-234. This is an example of a **nuclear equation**, which is *a symbolic representation of a nuclear reaction*. Normally, only the nuclei are represented. It is not necessary to indicate the chemical compound or the electron charges for any ions involved, because the chemical environment has no effect on nuclear processes.

Reactant and product nuclei are represented in nuclear equations by their nuclide symbols. Other particles are given the following symbols, in which the subscript equals the charge and the superscript equals the total number of protons and neutrons in the particle (mass number):

Proton	$^{1}_{1}\text{H}$	or	$^{1}_{1}\text{P}$
Neutron	$^{1}_{0}\text{n}$		
Electron	$^{0}_{-1}\text{e}$	or	$^{0}_{-1}\beta$
Positron	$^{0}_{1}\text{e}$	or	$^{0}_{1}\beta$
Gamma photon	$^{0}_{0}\gamma$		

The decay of a nucleus with the emission of an electron, $_{-1}^{0}e$, is usually called beta emission, and the emitted electron is sometimes labeled $_{-1}^{0}\beta$. A **positron** is *a particle similar to an electron, having the same mass but a positive charge.* A **gamma photon** is *a particle of electromagnetic radiation of short wavelength (about 1 pm, or 10^{-12} m) and high energy.*

ⓌL Interactive Example 20.1 Writing a Nuclear Equation

Gaining Mastery Toolbox

Critical Concept 20.1
There are six particles that can be involved in nuclear reactions: proton, neutron, beta (electron), positron, gamma photon, and alpha. When writing the particle symbol, the subscript equals the number of protons, and the superscript equals the sum of the protons and neutrons. In order to complete and balance nuclear reactions, you must know the name and symbol for each of these particles.

Solution Essentials:
- Nuclear particles: proton ($_{1}^{1}H$), neutron ($_{0}^{1}n$), beta ($_{-1}^{0}e$), positron ($_{1}^{0}e$), gamma photon ($_{0}^{0}\gamma$), and alpha ($_{2}^{4}He$)
- Nuclear equation
- Nuclide symbol

Write the nuclear equation for the radioactive decay of radium-226 by alpha decay to give radon-222. A radium-226 nucleus emits one alpha particle, leaving behind a radon-222 nucleus.

Problem Strategy Use the list of elements on the inside back cover to obtain the atomic numbers of radium and radon, so you can write their nuclide symbols for the nuclear equation. The nuclide symbol of the alpha particle is $_{2}^{4}He$.

Solution The nuclear equation is

$$_{88}^{226}Ra \longrightarrow \,_{86}^{222}Rn + \,_{2}^{4}He$$

Answer Check Make sure that you have written the nuclide symbols with the correct superscripts and subscripts.

Exercise 20.1 Potassium-40 is a naturally occurring radioactive isotope. It decays to calcium-40 by beta emission. When a potassium-40 nucleus decays by beta emission, it emits one beta particle and gives a calcium-40 nucleus. Write the nuclear equation for this decay.

■ See Problems 20.33, 20.34, 20.35, and 20.36.

The total charge is conserved, or remains constant, during a nuclear reaction. This means that the sum of the subscripts (number of protons, or positive charges, in the nuclei) for the products must equal the sum of the subscripts for the reactants. For the equation in Example 20.1, the subscript for the reactant $_{88}^{226}Ra$ is 88. For the products, the sum of the subscripts is $86 + 2 = 88$.

Similarly, the total number of *nucleons* (protons and neutrons) is conserved, or remains constant, during a nuclear reaction. This means that the sum of the superscripts (the mass numbers) for the reactants equals the sum of the superscripts for the products. For the equation in Example 20.1, the superscript for the reactant nucleus is 226. For the products, the sum of the superscripts is $222 + 4 = 226$.

Note that if all reactants and products but one are known in a nuclear equation, the identity of that one nucleus or particle can be easily obtained. This is illustrated in the next example.

Example 20.2 Deducing a Product or Reactant in a Nuclear Equation

Gaining Mastery Toolbox

Critical Concept 20.2
The total charge (sum of all charges) and total number of nucleons (protons and neutrons) is conserved during a nuclear reaction. In a balanced nuclear reaction, the sum of all reactant superscripts must equal the sum of all product superscripts, and the sum of all reactant subscripts must equal the sum of all product subscripts.

Solution Essentials:
- Nucleon
- Nuclear particles: proton ($_{1}^{1}H$), neutron ($_{0}^{1}n$), beta ($_{-1}^{0}e$), positron ($_{1}^{0}e$), gamma photon ($_{0}^{0}\gamma$), and alpha ($_{2}^{4}He$)
- Nuclear equation
- Nuclide symbol

Technetium-99 is a long-lived radioactive isotope of technetium. Each nucleus decays by emitting one beta particle. What is the product nucleus?

Problem Strategy First, write the nuclear equation using the symbol $_{Z}^{A}X$ for the unknown nuclide or particle. Then, solve for A and Z using the following relations: (1) the sum of the subscripts for the reactants equals the sum of the subscripts for the products, and (2) the sum of the superscripts for the reactants equals the sum of the superscripts for the products.

(continued)

(continued)

Solution Technetium-99 has the nuclide symbol $^{99}_{43}\text{Tc}$. A beta particle is an electron; the symbol is $^{0}_{-1}\text{e}$. The nuclear equation is

$$^{99}_{43}\text{Tc} \longrightarrow {}^{A}_{Z}\text{X} + {}^{0}_{-1}\text{e}$$

From the superscripts, you can write

$$99 = A + 0, \text{ or } A = 99$$

Similarly, from the subscripts, you get

$$43 = Z - 1, \text{ or } Z = 43 + 1 = 44$$

Hence $A = 99$ and $Z = 44$, so the product is $^{99}_{44}\text{X}$. Because element 44 is ruthenium, symbol Ru, you write the product nucleus as $^{99}_{44}\textbf{Ru}$.

Answer Check Although problems of this type are relatively simple, a very common error is not to account properly for the -1 subscript of the beta particle.

Exercise 20.2 Plutonium-239 decays by alpha emission, with each nucleus emitting one alpha particle. What is the other product of this decay?

■ See Problems 20.37, 20.38, 20.39, and 20.40.

Nuclear Stability

At first glance, the existence of several protons in the small space of a nucleus is puzzling. Why wouldn't the protons be strongly repelled by their like electric charges? The existence of stable nuclei with more than one proton is due to the nuclear force. The **nuclear force** is *a strong force of attraction between nucleons that acts only at very short distances (about 10^{-15} m)*. Beyond nuclear distances, these nuclear forces become negligible. Therefore, two protons that are much farther apart than 10^{-15} m repel one another by their like electric charges. Inside the nucleus, however, two protons are close enough together for the nuclear force between them to be effective. This force in a nucleus can more than compensate for the repulsion of electric charges and thereby give a stable nucleus.

The protons and neutrons in a nucleus appear to have energy levels much as the electrons in an atom have energy levels. The **shell model of the nucleus** is *a nuclear model in which protons and neutrons exist in levels, or shells, analogous to the shell structure that exists for electrons in an atom*. Recall that in an atom, filled shells of electrons are associated with the special stability of the noble gases. The total numbers of electrons for these stable atoms are 2 (for He), 10 (for Ne), 18 (for Ar), and so forth. Experimentally, note that nuclei with certain numbers of protons or neutrons appear to be very stable. These numbers, called *magic numbers* and associated with specially stable nuclei, were later explained by the shell model. According to this theory, a **magic number** is *the number of nuclear particles in a completed shell of protons or neutrons*. Because nuclear forces differ from electrical forces, these numbers are not the same as those for electrons in atoms. For protons, the magic numbers are 2, 8, 20, 28, 50, and 82. Neutrons have these same magic numbers, as well as the magic number 126. For protons, calculations show that 114 should also be a magic number.

Some of the evidence for these magic numbers, and therefore for the shell model of the nucleus, is as follows. Many radioactive nuclei decay by emitting alpha particles, or $^{4}_{2}\text{He}$ nuclei. There appears to be special stability in the $^{4}_{2}\text{He}$ nucleus. It contains two protons and two neutrons; that is, it contains a magic number of protons (2) and a magic number of neutrons (also 2).

Another piece of evidence is seen in the final products obtained in natural radioactive decay. For example, uranium-238 decays to thorium-234, which in turn decays to protactinium-234, and so forth. Each product is radioactive and decays to another nucleus until the final product, $^{206}_{82}\text{Pb}$, is reached. This nucleus is stable. Note that it contains 82 protons, a magic number. Other radioactive decay series end at $^{207}_{82}\text{Pb}$ and $^{208}_{82}\text{Pb}$, each of which has a magic number of protons. Note that $^{208}_{82}\text{Pb}$ also has a magic number of neutrons ($208 - 82 = 126$).

Evidence also points to the special stability of pairs of protons and pairs of neutrons, analogous to the stability of pairs of electrons in molecules. Table 20.1 lists the number of stable isotopes that have an even number of protons and an

Table 20.1	Number of Stable Isotopes with Even and Odd Numbers of Protons and Neutrons			
	Number of Stable Isotopes			
	157	52	50	5
Number of protons	Even	Even	Odd	Odd
Number of neutrons	Even	Odd	Even	Odd

even number of neutrons (157). By comparison, only 5 stable isotopes have an odd number of protons and an odd number of neutrons.

When you plot each stable nuclide on a graph with the number of protons (Z) on the horizontal axis and the number of neutrons (N) on the vertical axis, these stable nuclides fall in a certain region, or band, of the graph. The **band of stability** is *the region in which stable nuclides lie in a plot of number of protons against number of neutrons.* Figure 20.3 shows the band of stability; the rest of the figure is explained later in this section. For nuclides up to $Z = 20$, the ratio of neutrons to protons is about 1.0 to 1.1. As Z increases, however, the neutron-to-proton ratio increases to about 1.5. This increase in neutron-to-proton ratio with increasing Z is believed to result from the increasing repulsions of protons from their electric charges. More neutrons are required to give attractive nuclear forces to offset these repulsions.

It appears that when the number of protons becomes very large, the proton–proton repulsions become so great that stable nuclides are impossible. No stable nuclides are known with atomic numbers greater than 83. On the other hand, all elements with Z equal to 83 or less have one or more stable nuclides, with the exception of technetium ($Z = 43$), as noted in the chapter opening, and promethium ($Z = 61$).

Figure 20.3 ▶

Band of stability The stable nuclides, indicated by black dots, cluster in a band. Nuclides to the left of the band of stability usually decay by beta emission, whereas those to the right usually decay by positron emission or electron capture. Nuclides of $Z > 83$ often decay by alpha emission.

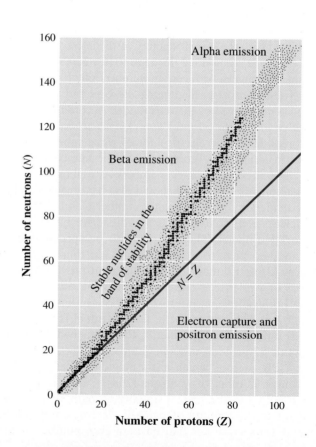

Example 20.3   Predicting the Relative Stabilities of Nuclides

Gaining Mastery Toolbox

Critical Concept 20.3
There are several factors that contribute to nuclear stability: a full nuclear shell (magic number), an even number of protons and neutrons, and a proton-to-neutron ratio that lies close to the band of stability. When an element has an atomic number greater than 83, it will not be stable.

Solution Essentials:
• Band of stability
• Proton to neutron ratio
• Number of protons and neutrons in nuclide
• Magic number and the shell model of the nucleus
• Radioactive decay

One of the nuclides in each of the following pairs is radioactive; the other is stable. Which is radioactive and which is stable? Explain.

a $^{208}_{84}$Po, $^{209}_{83}$Bi   b $^{39}_{19}$K, $^{40}_{19}$K   c $^{71}_{31}$Ga, $^{76}_{31}$Ga

Problem Strategy You must decide which nuclide of each pair is more likely to be stable, based on the general principles stated in the preceding text.

Solution a Polonium has an atomic number greater than 83, so $^{208}_{84}$**Po is radioactive**. Bismuth-209 has

126 neutrons (a magic number), so $^{209}_{83}$**Bi is expected to be stable**.

b Of these two isotopes, $^{39}_{19}$K has a magic number of neutrons (20), so $^{39}_{19}$**K is expected to be stable**. The isotope $^{40}_{19}$K has an odd number of protons (19) and an odd number of neutrons (21). Because stable odd–odd nuclei are rare, **you might expect $^{40}_{19}$K to be radioactive**.

c Of the two isotopes, $^{76}_{31}$Ga lies farther from the center of the band of stability, so it is more likely to be radioactive. For this reason, **you expect $^{76}_{31}$Ga to be radioactive and $^{71}_{31}$Ga, which has an even number of neutrons and is close to the atomic mass of gallium, to be stable**.

Answer Check When trying to determine whether a nuclide is radioactive, always first perform the simple check to see if the atomic number is greater than 83.

Exercise 20.3 Of the following nuclides, two are radioactive. Which are radioactive and which is stable? Explain. a $^{118}_{50}$Sn; b $^{76}_{33}$As; c $^{227}_{89}$Ac.

■ See Problems 20.41 and 20.42.

Types of Radioactive Decay

There are six common types of radioactive decay; the first five are listed in Table 20.2. Note how the last column in Table 20.2 indicates the conditions where one would expect to observe a particular type of emission.

1. **Alpha emission** (abbreviated α): *emission of a $^{4}_{2}$He nucleus, or alpha particle, from an unstable nucleus.* An example is the radioactive decay of radium-226.

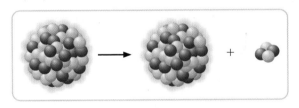

$$^{226}_{88}\text{Ra} \longrightarrow ^{222}_{86}\text{Rn} + ^{4}_{2}\text{He}$$

Table 20.2   **Types of Radioactive Decay**

Type of Decay	Radiation	Equivalent Process	Resulting Nuclear Change		Usual Nuclear Condition
			Atomic Number	Mass Number	
Alpha emission (α)	$^{4}_{2}$He	—	−2	−4	$Z > 83$
Beta emission (β)	$^{0}_{-1}$e	$^{1}_{0}$n $\longrightarrow$ $^{1}_{1}$p + $^{0}_{-1}$e	+1	0	N/Z too large
Positron emission (β^{+})	$^{0}_{1}$e	$^{1}_{1}$p $\longrightarrow$ $^{1}_{0}$n + $^{0}_{1}$e	−1	0	N/Z too small
Electron capture (EC)	x rays	$^{1}_{1}$p + $^{0}_{-1}$e $\longrightarrow$ $^{1}_{0}$n	−1	0	N/Z too small
Gamma emission (γ)	$^{0}_{0}\gamma$	—	0	0	Excited

A CHEMIST Looks at...

Magic Numbers

Maria Goeppert-Mayer (Figure 20.4) (1906–1972) received the 1963 Nobel Prize in Physics for her discovery of the shell structure of the nucleus (independently discovered by the German physicist J. Hans D. Jensen). Notably, she was the second woman to receive the Nobel Prize (Marie Curie was the first). In 1948, Goeppert-Mayer had noticed that the most abundant nuclei were those with special numbers of protons and neutrons. (These special numbers came to be called *magic numbers*.) Although these numbers had been noted earlier, her work was especially detailed and convincing, and it strongly pointed to a shell structure of the nucleus.

Figure 20.4 ▲

Maria Goeppert-Mayer

According to the shell model of the nucleus, protons and neutrons occupy independent sets of orbitals, similar to those of electrons in atoms. Nucleon orbitals have similar quantum numbers too: n (principal quantum number) and l (orbital angular momentum quantum number). Each nucleon also has a spin of 1/2. Unlike electrons, though, the l quantum number of a nucleon is not restricted by n. It can have any positive integer value, whatever the value of n, beginning with $l = 0$. For example, the orbitals corresponding to $n = 1$ are $1s$, $1p$, $1d$, $1f$, . . . (where s stands for $l = 0$, p for $l = 1$, and so forth).

A simple scheme based on nucleons oscillating within the nucleus gives energy levels proportional to $2(n - 1) + l$, which leads to the following increasing order of energy levels:

$1s$, $1p$, $1d$ and $2s$ (with about the same energies), $1f$, . . .

Similar to electrons, an s orbital holds 2 nucleons, a p orbital holds 6 nucleons, a d orbital holds 10 nucleons, and an f orbital holds 14 nucleons. Thus it takes 2 nucleons to fill a $1s$ level, 8 nucleons to fill the $1s$ and $1p$ levels, 20 nucleons to fill the $1s$, $1p$, $1d$, and $2s$ levels, and 34 nucleons to fill the $1s$, $1p$, $1d$, $2s$ and $1f$ levels. The numbers 2, 8, and 20 are the first three magic numbers. After that, however, this simple scheme fails.

Goeppert-Mayer was able to solve this problem by noting that in nuclei, the orbital angular momentum of each nucleon couples to its spin (equal to 1/2) to give a total angular momentum of either $l + 1/2$ or $l - 1/2$. Each level of given l thus splits by this *spin-orbit coupling* into two sublevels, with the sublevel of angular momentum $l + 1/2$ being lower in energy. For example, a $1f$ level is split to give a lower level that Goeppert-Mayer showed can hold 8 nucleons and an upper level that can hold 6 nucleons for a total of 14 for an f shell. Therefore, the magic number after 20 is 20 + 8 = 28 (not 20 + 14 = 34). She was able to show that the shell model reproduced the other magic numbers, as well.

■ See Problems 20.103 and 20.104.

The product nucleus has an atomic number that is two less, and a mass number that is four less, than that of the original nucleus.

2. **Beta emission** (abbreviated β or β^-): *emission of a high-speed electron from an unstable nucleus.* Beta emission is equivalent to the conversion of a neutron to a proton.

$$^1_0n \longrightarrow \, ^1_1p \, + \, ^{\ 0}_{-1}e$$

$$^1_0n \longrightarrow \, ^1_1p \, + \, ^{\ 0}_{-1}e$$

An example of beta emission is the radioactive decay of carbon-14.

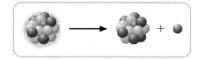

$$^{14}_{6}\text{C} \longrightarrow {}^{14}_{7}\text{N} + {}^{0}_{-1}\text{e}$$

The product nucleus has an atomic number that is one more than that of the original nucleus. The mass number remains the same.

3. **Positron emission** (abbreviated β^+): *emission of a positron from an unstable nucleus.* A positron, denoted in nuclear equations as ${}^{0}_{1}\text{e}$, is a particle identical to an electron in mass but having a positive instead of a negative charge. Positron emission is equivalent to the conversion of a proton to a neutron. ▶

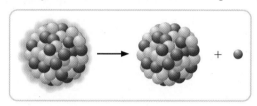

$$^{1}_{1}\text{p} \longrightarrow {}^{1}_{0}\text{n} + {}^{0}_{1}\text{e}$$

The radioactive decay of technetium-95 is an example of positron emission.

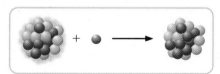

$$^{95}_{43}\text{Tc} \longrightarrow {}^{95}_{42}\text{Mo} + {}^{0}_{1}\text{e}$$

The product nucleus has an atomic number that is one less than that of the original nucleus. The mass number remains the same.

4. **Electron capture** (abbreviated EC): *the decay of an unstable nucleus by capturing, or picking up, an electron from an inner orbital of an atom.* In effect, a proton is changed to a neutron, as in positron emission.

$$^{1}_{1}\text{p} + {}^{0}_{-1}\text{e} \longrightarrow {}^{1}_{0}\text{n}$$

An example is given by potassium-40, which has a natural abundance of 0.012%. ▶ Potassium-40 can decay by electron capture, as well as by beta and positron emissions. The equation for electron capture is

$$^{40}_{19}\text{K} + {}^{0}_{-1}\text{e} \longrightarrow {}^{40}_{18}\text{Ar}$$

The product nucleus has an atomic number that is one less than that of the original nucleus. The mass number remains the same. When another orbital electron fills the vacancy in the inner-shell orbital created by electron capture, an x-ray photon is emitted.

5. **Gamma emission** (abbreviated γ): *emission from an excited nucleus of a gamma photon, corresponding to radiation with a wavelength of about 10^{-12} m.* In many cases, radioactive decay results in a product nucleus that is in an excited state. As in the case of atoms, the excited state is unstable and goes to a lower-energy state with the emission of electromagnetic radiation. For nuclei, this radiation is in the gamma-ray region of the spectrum.

Positrons are annihilated as soon as they encounter electrons. When a positron and an electron collide, both particles vanish with the emission of two gamma photons that carry away the energy.

$$^{0}_{1}\text{e} + {}^{0}_{-1}\text{e} \longrightarrow 2{}^{0}_{0}\gamma$$

Most of the argon in the atmosphere is believed to have resulted from the radioactive decay of ${}^{40}_{19}K$.

Often gamma emission occurs very quickly after radioactive decay. In some cases, however, an excited state has significant lifetime before it emits a gamma photon. A **metastable nucleus** is *a nucleus in an excited state with a lifetime of at least one nanosecond (10^{-9} s)*. In time, the metastable nucleus decays by gamma emission. An example is metastable technetium-99, denoted $^{99m}_{43}\text{Tc}$, which is used in medical diagnosis, as discussed in Section 20.5.

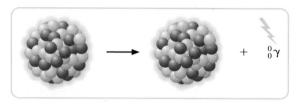

$$^{99m}_{43}\text{Tc} \longrightarrow \, ^{99}_{43}\text{Tc} + \, ^{0}_{0}\gamma$$

The product nucleus is simply a lower-energy state of the original nucleus, so there is no change of atomic number or mass number.

6. **Spontaneous fission**: *the spontaneous decay of an unstable nucleus in which a heavy nucleus of mass number greater than 89 splits into lighter nuclei and energy is released.* For example, a uranium-236 atom can spontaneously undergo the following nuclear reaction:

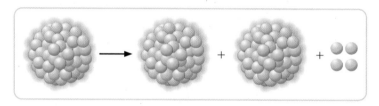

$$^{236}_{92}\text{U} \longrightarrow \, ^{96}_{39}\text{Y} + \, ^{136}_{53}\text{I} + 4\,^{1}_{0}\text{n}$$

This process will be discussed further in Section 20.6.

Nuclides outside the band of stability (Figure 20.3) are generally radioactive. Nuclides to the left of the band of stability have a neutron-to-proton ratio (N/Z) larger than that needed for stability. These nuclides tend to decay by beta emission. Beta emission reduces the neutron-to-proton ratio, because in this process a neutron is changed to a proton. The product is a stabler nuclide. In contrast, nuclides to the right of the band of stability have a neutron-to-proton ratio smaller than that needed for stability. These nuclides tend to decay by either positron emission or electron capture. Both processes convert a proton to a neutron, increasing the neutron-to-proton ratio and giving a stabler product nuclide. The types of radioactive decay expected of unstable nuclides are noted in Figure 20.3.

Consider a series of isotopes of a given element, such as carbon. Carbon-12 and carbon-13 are stable isotopes, whereas the other isotopes of carbon are radioactive. The isotopes of mass number smaller than 12 decay by positron emission. For example, carbon-11 decays by positron emission to boron-11.

$$^{11}_{6}\text{C} \longrightarrow \, ^{11}_{5}\text{B} + \, ^{0}_{1}\text{e}$$

Carbon-11 has a neutron-to-proton ratio of 5/6 (= 0.8), which increases in the product boron-11 to 6/5 (1.2). The isotopes of carbon with mass number greater than 13 decay by beta emission. Carbon-14 decays by beta emission to produce nitrogen-14.

$$^{14}_{6}\text{C} \longrightarrow \, ^{14}_{7}\text{N} + \, ^{0}_{-1}\text{e}$$

Carbon-14 has a neutron-to-proton ratio of 8/6 (1.3), which decreases in the product nitrogen-14 to 7/7 (1.0).

Now consider the radioactive isotope phosphorus-30. You can predict the expected type of radioactive decay of this isotope by noting whether the mass number is less than or greater than the mass number of stable isotopes. Generally, the mass numbers of stable isotopes will be close to the numerical value of the atomic mass of the element. The atomic mass of phosphorus is 31.0 amu, so you might expect phosphorus-31 to be a stable isotope (which it is). Phosphorus-30 has a mass number less than that of the stable isotope phosphorus-31. Therefore, you expect that phosphorus-30 will decay by either positron emission or electron capture. Positron emission is actually observed.

Positron emission and electron capture are competing radioactive decay processes, and what is observed depends on the relative rates of the two processes. The rate of electron capture increases with atomic number of the decaying nuclide and therefore becomes important in heavier elements. Positron emission is generally seen in lighter elements (recall that phosphorus-30 decays by positron emission). However, in the very heavy elements, especially those with Z greater than 83, radioactive decay is often by alpha emission (noted in Figure 20.3 at the top of the figure). $^{238}_{92}U$, $^{226}_{88}Ra$, and $^{232}_{90}Th$ are examples of alpha emitters.

Example 20.4   Predicting the Type of Radioactive Decay

Gaining Mastery Toolbox

Critical Concept 20.4
The total number of protons or the proton-to-neutron ration (N/Z) can be used to predict the type of nuclear decay of a nuclide. The last column of Table 20.2 provides guidance for linking the type of nuclear decay to these quantities.

Solution Essentials:
- Alpha emission, beta emission, positron emission, electron capture, and gamma emission
- Nuclear conditions that cause a particular type of decay (Table 20.2)
- Nuclear particles: proton (^1_1H), neutron (^1_0n), beta ($^{\,0}_{-1}e$), positron (^0_1e), gamma photon ($^0_0\gamma$), and alpha (^4_2He)
- Nuclear equation
- Nuclide symbol

Predict the expected type of radioactive decay for each of the following radioactive nuclides: **a** $^{47}_{20}Ca$; **b** $^{25}_{13}Al$.

Problem Strategy Compare each nuclide with the stable nuclides of the same element. A nuclide with an N/Z ratio greater than that of the stable nuclides is expected to exhibit beta emission. A nuclide with an N/Z ratio less than that of the stable nuclides is expected to exhibit positron emission or electron capture; electron

capture is important with heavier elements. Since you are comparing nuclides of the same Z, you can compare mass numbers ($= N + Z$), rather than N/Z.

Solution

a The atomic mass of calcium is 40.1 amu, so you expect calcium-40 to be a stable isotope. Calcium-47 has a mass number greater than that of the stable isotope, so you expect it to decay by **beta emission**. (This is the observed behavior of calcium-47.)

b The atomic mass of aluminum is 27.0 amu, so you expect aluminum-27 to be a stable isotope. The mass number of aluminum-25 is less than 27, so you expect aluminum-25 to decay by either **positron emission or electron capture**. Positron emission is actually observed.

Answer Check To answer these types of questions correctly, you must understand how the N/Z ratio of nuclides determines the type of emission from radioactive nuclides.

Exercise 20.4 Predict the type of decay expected for each of the following radioactive nuclides: **a** $^{13}_7N$; **b** $^{26}_{11}Na$.

■ See Problems 20.43 and 20.44.

Radioactive Decay Series

All nuclides with atomic number greater than $Z = 83$ are radioactive, as we have noted. Many of these nuclides decay by alpha emission. Alpha particles, or ^4_2He nuclei, are especially stable and are formed in the radioactive nucleus at the moment of decay. By emitting an alpha particle, the nucleus reduces its atomic number, becoming more stable. However, if the nucleus has a very large atomic number, the product nucleus is also radioactive. Natural radioactive elements, such as uranium-238, give a **radioactive decay series**, *a sequence in which one radioactive nucleus decays to a second, which then decays to a third, and so forth.* Eventually, a stable nucleus, which is an isotope of lead, is reached.

Figure 20.5 ▶

Uranium-238 radioactive decay series Each nuclide occupies a position on the graph determined by its atomic number and mass number. Alpha decay is shown by a red diagonal line. Beta decay is shown by a short blue horizontal line.

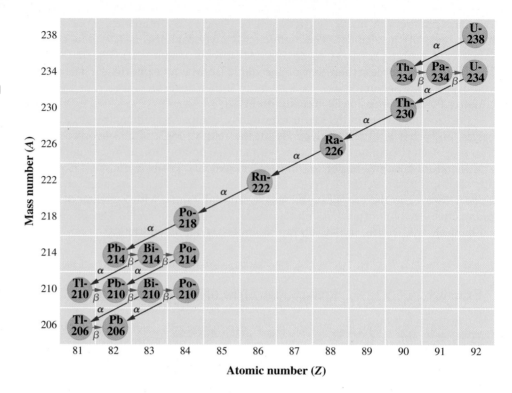

Three radioactive decay series are found naturally. One of these series begins with uranium-238. Figure 20.5 shows the sequence of nuclear decay processes. In the first step, uranium-238 decays by alpha emission to thorium-234.

$$^{238}_{92}\text{U} \longrightarrow \, ^{234}_{90}\text{Th} + \, ^{4}_{2}\text{He}$$

This step is represented in Figure 20.5 by a red arrow labeled α. Each alpha decay reduces the atomic number by 2 and the mass number by 4. The atomic number and mass number of each nuclide are given by its horizontal and vertical position on the graph. Thorium-234 in turn decays by beta emission to protactinium-234, which decays by beta emission to uranium-234.

$$^{234}_{90}\text{Th} \longrightarrow \, ^{234}_{91}\text{Pa} + \, ^{0}_{-1}\text{e}$$

$$^{234}_{91}\text{Pa} \longrightarrow \, ^{234}_{92}\text{U} + \, ^{0}_{-1}\text{e}$$

Beta emission is represented in Figure 20.5 by a blue arrow labeled β. Each beta decay increases the atomic number by one but has no effect on the mass number. After the decay of protactinium-234 and the formation of uranium-234, there are a number of alpha-decay steps. The final product of the series is lead-206.

Natural uranium is 99.28% $^{238}_{92}\text{U}$, which decays as we have described. However, the natural element also contains 0.72% $^{235}_{92}\text{U}$. This isotope starts a second radioactive decay series, which consists of a sequence of alpha and beta decays, ending with lead-207. The third naturally occurring radioactive decay series begins with thorium-232 and ends with lead-208. All three radioactive decay series found naturally end with an isotope of lead.

CONCEPT CHECK 20.1

You have two samples of water, each made up of different isotopes of hydrogen: one contains $^{1}_{1}\text{H}_2\text{O}$ and the other, $^{3}_{1}\text{H}_2\text{O}$.

a Would you expect these two water samples to be chemically similar?

b Would you expect these two water samples to be physically the same?

c Which one of these water samples would you expect to be radioactive?

20.2 Nuclear Bombardment Reactions

The nuclear reactions discussed in the previous section are radioactive decay reactions, in which a nucleus spontaneously decays to another nucleus and emits a particle, such as an alpha or beta particle. In 1919, Ernest Rutherford discovered that it is possible to change the nucleus of one element into the nucleus of another element by processes that can be controlled in the laboratory. **Transmutation** is *the change of one element to another by bombarding the nucleus of the element with nuclear particles or nuclei.*

Transmutation

Rutherford used a radioactive element as a source of alpha particles and allowed these particles to collide with nitrogen nuclei. He discovered that protons are ejected in the process. The equation for the nuclear reaction is

$$^{14}_{7}\text{N} + ^{4}_{2}\text{He} \longrightarrow ^{17}_{8}\text{O} + ^{1}_{1}\text{H}$$

The experiments were repeated on other light nuclei, most of which were transmuted to other elements with the ejection of a proton. These experiments yielded two significant results. First, they strengthened the view that all nuclei contain protons. Second, they showed for the first time that it is possible to change one element into another under laboratory control.

When beryllium is bombarded with alpha particles, a penetrating radiation is given off that is not deflected by electric or magnetic fields. Therefore, the radiation does not consist of charged particles. The British physicist James Chadwick (1891–1974) suggested in 1932 that the radiation from beryllium consists of neutral particles, each with a mass approximately that of a proton. The particles are called neutrons. The reaction that resulted in the discovery of the neutron is

$$^{9}_{4}\text{Be} + ^{4}_{2}\text{He} \longrightarrow ^{12}_{6}\text{C} + ^{1}_{0}\text{n}$$

In 1933, a nuclear bombardment reaction was used to produce the first artificial radioactive isotope. Irène and Frédéric Joliot-Curie found that aluminum bombarded with alpha particles produces phosphorus-30, which decays by emitting positrons. The reactions are

$$^{27}_{13}\text{Al} + ^{4}_{2}\text{He} \longrightarrow ^{30}_{15}\text{P} + ^{1}_{0}\text{n}$$

$$^{30}_{15}\text{P} \longrightarrow ^{30}_{14}\text{Si} + ^{0}_{1}\text{e}$$

Phosphorus-30 was the first radioactive nucleus produced in the laboratory. Since then over a thousand radioactive isotopes have been made. ▶

Some of the uses of radioactive isotopes are discussed in Section 20.5.

Nuclear bombardment reactions are often referred to by an abbreviated notation. For example, the reaction

$$^{14}_{7}\text{N} + ^{4}_{2}\text{He} \longrightarrow ^{17}_{8}\text{O} + ^{1}_{1}\text{H}$$

is abbreviated $^{14}_{7}\text{N}(\alpha, \text{p})^{17}_{8}\text{O}$. In this notation, you first write the nuclide symbol for the original nucleus (target). Then, in parentheses, you write the symbol for the projectile particle (incoming particle), followed by a comma and the symbol for the ejected particle. After the last parenthesis, you write the nuclide symbol for the product nucleus. The following symbols are used for particles:

Neutron	n
Proton	p
Deuteron, $^{2}_{1}\text{H}$	d
Alpha, $^{4}_{2}\text{He}$	α

Example 20.5 Using the Notation for a Bombardment Reaction

Gaining Mastery Toolbox

Critical Concept 20.5
The abbreviated notation for nuclear bombardment reactions requires that symbols be used to denote the nuclear particles. The symbols are: n for neutron, p for proton, d for deuteron ($^2_1$H), and α for alpha ($^4_2$He).

Solution Essentials:
• Particle symbols for nuclear bombardment reactions
• Abbreviated nuclear bombardment notation
• Nuclear equation

a Write the abbreviated notation for the following bombardment reaction, in which neutrons were first discovered.

$$^9_4\text{Be} + ^4_2\text{He} \longrightarrow ^{12}_6\text{C} + ^1_0\text{n}$$

b Write the nuclear equation for the bombardment reaction denoted $^{27}_{13}\text{Al(p, d)}^{26}_{13}\text{Al}$.

Problem Strategy Understand and use the abbreviated notation and particle symbols for nuclear reactions.

Solution a The notation is $^9_4\text{Be}(\alpha, \text{n})^{12}_6\text{C}$.
b The nuclear equation is

$$^{27}_{13}\text{Al} + ^1_1\text{H} \longrightarrow ^{26}_{13}\text{Al} + ^2_1\text{H}$$

Answer Check You can check your answers by working backward to see whether they lead to the representations given in the problem.

Exercise 20.5 a Write the abbreviated notation for the reaction

$$^{40}_{20}\text{Ca} + ^2_1\text{H} \longrightarrow ^{41}_{20}\text{Ca} + ^1_1\text{H} .$$

b Write the nuclear equation for the bombardment reaction $^{12}_6\text{C(d, p)}^{13}_6\text{C}$.

■ See Problems 20.47, 20.48, 20.49, and 20.50.

Elements of large atomic number merely scatter, or deflect, alpha particles from natural sources, rather than giving a transmutation reaction. These elements have nuclei of large positive charge, and the alpha particle must be traveling very fast in order to penetrate the nucleus and react. Alpha particles from natural sources do not have sufficient kinetic energy. To shoot charged particles into heavy nuclei, it is necessary to accelerate the charged particles.

A **particle accelerator** is *a device used to accelerate electrons, protons, alpha particles, and other ions to very high speeds.* The basis of a particle accelerator is the fact that a charged particle will accelerate toward a plate having a charge opposite in sign to that of the particle. It is customary to measure the kinetic energies of these particles in units of electron volts. An **electron volt (eV)** is *the quantity of energy that would have to be imparted to an electron (whose charge is 1.602×10^{-19} C) to accelerate it by one volt potential difference.*

$$1 \text{ eV} = (1.602 \times 10^{-19} \text{ C}) \times (1 \text{ V}) = 1.602 \times 10^{-19} \text{ J}$$

Typically, particle accelerators give charged particles energies of millions of electron volts (MeV). To keep the accelerated particles from colliding with molecules of gas, the apparatus is enclosed and evacuated to low pressures, about 10^{-6} mmHg or less.

A charged particle can be accelerated in stages to very high kinetic energy. Figure 20.6 shows a diagram of a **cyclotron**, *a type of particle accelerator consisting of two hollow, semicircular metal electrodes called dees (because the shape resembles the letter D), in which charged particles are accelerated by stages to higher and higher kinetic energies.* Ions introduced at the center of the cyclotron are accelerated in the space between the two dees. Magnet poles (not shown in the figure) above and below the dees keep the ions moving in an enlarging spiral path. The dees are connected to a high-frequency electric current that changes their polarity so that each time an ion moves into the space between the dees, it is accelerated. Thus, the ion is continually accelerated until it finally leaves the cyclotron at high speed. Outside the cyclotron, the ions are directed toward a target element so that investigators may study nuclear reactions or prepare isotopes.

Technetium was first prepared by directing **deuterons**, or *nuclei of hydrogen-2 atoms,* from a cyclotron to a molybdenum target. ◀ The nuclear reaction is $^{96}_{42}\text{Mo(d, n)}^{97}_{43}\text{Tc}$, or

$$^{96}_{42}\text{Mo} + ^2_1\text{H} \longrightarrow ^{97}_{43}\text{Tc} + ^1_0\text{n}$$

The hydrogen-2 atom is often called deuterium and given the symbol D. It is a stable isotope with a natural abundance of 0.015%. Deuterium was discovered in 1931 by Harold Urey and coworkers and was first prepared in pure form by G. N. Lewis.

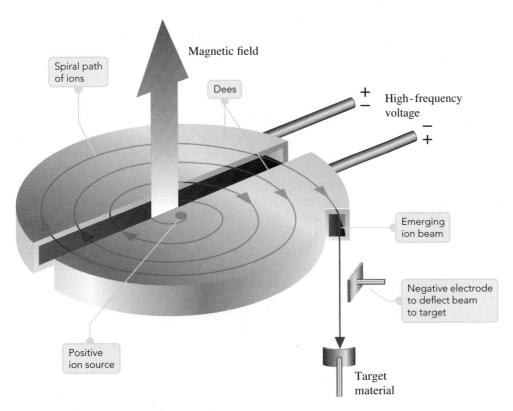

Spiral path of ions

Magnetic field

Dees

High-frequency voltage

Emerging ion beam

Negative electrode to deflect beam to target

Positive ion source

Target material

© Fermilab Visual Media Services

Figure 20.6 ◄

A cyclotron Positive ions are introduced at the center of the cyclotron. A fraction of the ions will be crossing the gap between the dees when the electric polarities are just right to accelerate them. (The positive ions are accelerated toward the negative dee and away from the positive dee.) The dees alternate in polarities, so that these ions continue to be accelerated each time they pass between the gap in the dees. Magnet poles above and below the dees produce a magnetic field that keeps the ions moving in a spiral path within the dees. At the end of their path, the ions encounter a negative electrode that deflects them to a target material.

The energies of particles obtained from a cyclotron are limited (to about 20 MeV for deuterons, for example). For higher energies, more sophisticated accelerators are required. Figure 20.7 shows the accelerator at Fermilab in Batavia, Illinois. Larger and larger particle accelerators are being built to study the basic constituents of matter.

Figure 20.7 ▲

The Fermilab accelerator in Batavia, Illinois The tunnel of the main accelerator is shown here. Protons are accelerated in the upper ring of conventional magnets (red and blue) to 400 billion electron volts. These protons are then injected into the lower ring of superconducting magnets (yellow), where they are accelerated to almost a trillion electron volts.

Transuranium Elements

The **transuranium elements** are *elements with atomic numbers greater than that of uranium (Z = 92), the naturally occurring element of greatest Z.* In 1940, E. M. McMillan and P. H. Abelson, at the University of California at Berkeley, discovered the first transuranium element. They produced an isotope of element 93, which they named neptunium, by bombarding uranium-238 with neutrons. This gave uranium-239, by the capture of a neutron, and this nucleus decayed in a few days by beta emission to neptunium-239.

$$^{238}_{92}\text{U} + ^{1}_{0}\text{n} \longrightarrow ^{239}_{92}\text{U}$$

$$^{239}_{92}\text{U} \longrightarrow ^{239}_{93}\text{Np} + ^{0}_{-1}\text{e}$$

The next transuranium element to be discovered was plutonium (Z = 94). Deuterons, the positively charged nuclei of hydrogen-2, were accelerated by a cyclotron and directed at a uranium target to give neptunium-238, which decayed to plutonium-238.

$$^{238}_{92}\text{U} + ^{2}_{1}\text{H} \longrightarrow ^{238}_{93}\text{Np} + 2^{1}_{0}\text{n}$$

$$^{238}_{93}\text{Np} \longrightarrow ^{238}_{94}\text{Pu} + ^{0}_{-1}\text{e}$$

Another isotope of plutonium, plutonium-239, is now produced in large quantity in nuclear reactors, as described in Section 20.7. Plutonium-239 is used for nuclear weapons.

The discovery of the next two transuranium elements, americium (Z = 95) and curium (Z = 96), depended on an understanding of the correct positions in

Figure 20.8 ▲

The vertical wheel, an apparatus used in the discovery of elements 104, 105, and 106 The apparatus was designed by Albert Ghiorso and colleagues. They synthesized element 106 by bombarding californium-249 with oxygen-18. The products were whisked by gas jets to the vertical wheel, where they were deposited (at the top). Detectors around the periphery of the wheel monitored the alpha activity to identify the products. Binary-coded holes label the positions around the wheel for computer collection of data.

the periodic table of the elements beyond actinium ($Z = 89$). It had been thought that these elements should be placed after actinium under the *d*-transition elements, so uranium was placed in Group VIB under tungsten. However, Glenn T. Seaborg, then at the University of California, Berkeley, postulated a second series of elements to be placed at the bottom of the periodic table, under the lanthanides, as shown in modern tables (see inside front cover). These elements, the actinides, would be expected to have chemical properties similar to those of the lanthanides. Once they understood this, Seaborg and others were able to use the predicted chemical behaviors of the actinides to separate americium and curium.

Figure 20.8 shows the apparatus used by Albert Ghiorso at the Lawrence Berkeley Laboratory in Berkeley, California, for the discovery of elements 104, 105, and 106. A team of scientists led by Peter Armbruster in Darmstadt, Germany, created elements 107 to 109 in the early 1980s by bombarding other elements with heavy-metal ions. During the 1994–1996 period, they reported the discovery of elements 110, 111, and 112. Recent work has yielded other elements, including the heaviest to date, 118.

The transuranium elements have a number of commercial uses. Plutonium-238 emits only alpha radiation, which is easily stopped by shielding. The isotope has been used as a power source for space satellites, navigation buoys, and heart pacemakers. Americium-241 is both an alpha-ray and a gamma-ray emitter. The gamma rays are used in devices that measure the thickness of materials such as metal sheets. Americium-241 is also used in home smoke detectors, in which the alpha radiation ionizes the air in a chamber within the detector and renders it electrically conducting. Smoke reduces the conductivity of the air, and this reduced conductivity is detected by an alarm circuit.

Figure 20.9 shows the High Flux Isotope Reactor at Oak Ridge National Laboratory, where technicians produce transuranium elements by bombarding plutonium-239 with neutrons. Neutrons are produced by nuclear fission, a process discussed in Section 20.7.

Example 20.6   Determining the Product Nucleus in a Nuclear Bombardment Reaction

Gaining Mastery Toolbox

Critical Concept 20.6
The total charge (sum of all charges) and total number of nucleons (protons and neutrons) is conserved during a nuclear reaction. In a balanced nuclear reaction, the sum of all reactant superscripts must equal the sum of all product superscripts, and the sum of all reactant subscripts must equal the sum of all product subscripts. Coefficients are used to indicate the number of nuclides.

Solution Essentials:
- Nucleon
- Nuclear particles: proton (^1_1H), neutron (^1_0n), beta ($^{0}_{-1}e$), positron (^0_1e), gamma photon ($^0_0\gamma$), and alpha (^4_2He)
- Nuclear equation
- Nuclide symbol

Plutonium-239 was bombarded by alpha particles. Each $^{239}_{94}Pu$ nucleus was struck by one alpha particle and emitted one neutron. What was the product nucleus?

Problem Strategy Apply a strategy similar to that used to solve Example 20.2.

Solution You can write the nuclear equation as follows:

$$^{239}_{94}Pu + {}^4_2He \longrightarrow {}^A_ZX + {}^1_0n$$

To balance this equation, write

$$239 + 4 = A + 1 \quad \text{(from superscripts)}$$
$$94 + 2 = Z + 0 \quad \text{(from subscripts)}$$

Hence,

$$A = 239 + 4 - 1 = 242$$
$$Z = 94 + 2 = 96$$

The product is $^{242}_{96}Cm$.

Answer Check Be careful in problems like this where the nuclides have coefficients; in those cases, make sure that you count correctly.

Exercise 20.6 Carbon-14 is produced in the upper atmosphere when a particular nucleus is bombarded with neutrons. A proton is ejected for each nucleus that reacts. What is the identity of the nucleus that produces carbon-14 by this reaction?

■ See Problems 20.53, 20.54, 20.55, and 20.56.

20.3 Radiations and Matter: Detection and Biological Effects

Radiations from nuclear processes affect matter in part by dissipating energy in it. An alpha, beta, or gamma particle traveling through matter dissipates energy by ionizing atoms or molecules, producing positive ions and electrons. In some cases, these radiations may also excite electrons in matter. When these electrons undergo transitions back to their ground states, light is emitted. The ions, free electrons, and light produced in matter can be used to detect nuclear radiations. Because nuclear radiations can ionize molecules and break chemical bonds, they adversely affect biological organisms. We will first look at the detection of nuclear radiations and then briefly discuss biological effects and radiation dosage in humans.

Radiation Counters

Two types of devices—*ionization counters* and *scintillation counters*—are used to count particles emitted from radioactive nuclei and other nuclear processes. Ionization counters detect the production of ions in matter. Scintillation counters detect the production of scintillations, or flashes of light.

A **Geiger counter** (Figure 20.10), *a kind of ionization counter used to count particles emitted by radioactive nuclei, consists of a metal tube filled with gas, such as argon.* The tube is fitted with a thin glass or plastic window through which radiation enters. A wire runs down the tube's center, from which the wire is insulated. The tube and wire are connected to a high-voltage source so that the tube becomes the negative electrode and the wire the positive electrode. Normally the gas in the tube is an insulator and no current flows through it. However, when radiation, such as an alpha particle, $_2^4He^{2+}$, passes through the window of the tube and into the gas, atoms are ionized. Free electrons are quickly accelerated to the wire. As they are accelerated to the wire, additional atoms may be ionized from collisions with these electrons and more electrons set free. An avalanche of electrons is created, and this gives a pulse of current that is detected by electronic equipment. The amplified pulse activates a digital counter or gives an audible "click."

Alpha and beta particles can be detected directly by a Geiger counter. ▶ To detect neutrons, boron trifluoride is added to the gas in the tube. Neutrons react with boron-10 nuclei to produce alpha particles, which can then be detected.

$$_0^1n + \,_5^{10}B \longrightarrow \,_3^7Li + \,_2^4He$$

A **scintillation counter** (Figure 20.11) is *a device that detects nuclear radiation from flashes of light generated in a material by the radiation.* A *phosphor* is a substance that emits flashes of light when struck by radiation. Rutherford used zinc sulfide as a phosphor to detect alpha particles. A sodium iodide crystal containing thallium(I) iodide is used as a phosphor for gamma radiation. (Excited technetium-99 emits gamma rays and is used for medical diagnostics. The gamma rays are detected by a scintillation counter.)

Figure 20.9 ▲

The High Flux Isotope Reactor at Oak Ridge National Laboratory The reactor in which transuranium isotopes are produced lies beneath a protecting pool of water. Visible light is emitted by high-energy particles moving through the water.

Gamma and low-energy alpha particles are better detected by scintillation counters. These counters are filled with solids or liquids, which are more likely to stop gamma rays. Also, low-energy alpha particles may be absorbed by the window of the Geiger counter and go undetected.

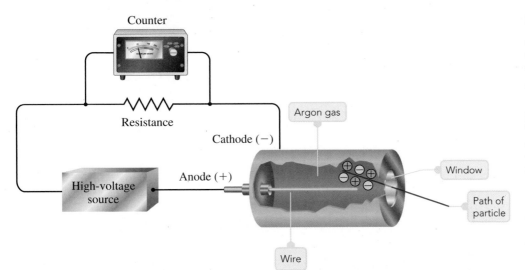

Counter

Resistance

Cathode (−)

Anode (+)

High-voltage source

Argon gas

Window

Path of particle

Wire

Figure 20.10 ◀

A Geiger counter A particle of radiation enters the thin window and passes into the gas. Energy from the particle ionizes gas molecules, giving positive ions and electrons, which are accelerated to the electrodes. The electrons, which move faster, strike the wire anode and create a pulse of current. Pulses from radiation particles are counted.

Figure 20.11 ▶

A scintillation counter probe

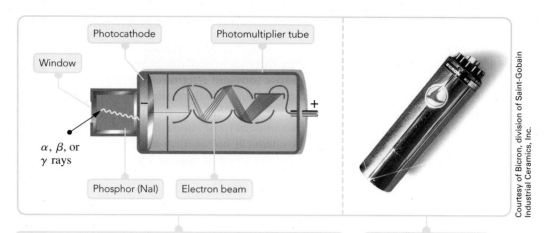

Photocathode

Photomultiplier tube

Window

α, β, or γ rays

Phosphor (NaI)   Electron beam

Courtesy of Bicron, division of Saint-Gobain Industrial Ceramics, Inc.

Radiation passes through the window into a phosphor (here, NaI). Flashes of light from the phosphor fall on the photocathode, which ejects electrons by the photoelectric effect. The electrons are accelerated to electrodes of increasing positive voltage, each electrode ejecting more electrons, so that the original signal is magnified.

Photograph of a scintillation probe (the window is on the left).

The flashes of light from the phosphor are detected by a *photomultiplier*. A photon of light from the phosphor hits a photoelectric-sensitive surface (the photocathode). ◀ This emits an electron, which is accelerated by a positive voltage to another electrode, from which several electrons are emitted. These electrons are accelerated by a higher voltage to the next electrode, from which more electrons are emitted, and so forth. The result is that a single electron may produce a million electrons and therefore a detectable pulse of electric current.

The photoelectric effect was discussed in Section 7.2.

A radiation counter can be used to measure the rate of nuclear disintegrations in a radioactive material. The **activity of a radioactive source** is *the number of nuclear disintegrations per unit time occurring in a radioactive material.* A **curie (Ci)** is *a unit of activity equal to 3.700×10^{10} disintegrations per second.* ◀ For example, a sample of technetium having an activity of 1.0×10^{-2} Ci is decaying at the rate of $(1.0 \times 10^{-2}) \times (3.7 \times 10^{10}) = 3.7 \times 10^{8}$ nuclei per second.

The curie was originally defined as the number of disintegrations per second from 1.0 g of radium-226.

Biological Effects and Radiation Dosage

Although the quantity of energy dissipated in a biological organism from a radiation dosage might be small, the effects can be quite damaging because important chemical bonds may be broken. DNA in the chromosomes of the cell is especially affected, which interferes with cell division (Figure 20.12). Cells that divide the fastest, such as those in the blood-forming tissue in bone marrow, are most affected by nuclear radiations.

To monitor the effect of nuclear radiations on biological tissue, it is necessary to have a measure of radiation dosage. The **rad** (from *r*adiation *a*bsorbed *d*ose) is *the dosage of radiation that deposits 1×10^{-2} J of energy per kilogram of tissue.* However, the biological effect of radiation depends not only on the energy deposited in the tissue but also on the type of radiation. For example, neutrons are more destructive than gamma rays of the same radiation dosage measured in rads. A **rem** is *a unit of radiation dosage used to relate various kinds of radiation in terms*

Figure 20.12 ▶

Examples of damage to DNA from nuclear radiation Damaged DNA can prevent cells from properly functioning and increase the likelihood of tumors.

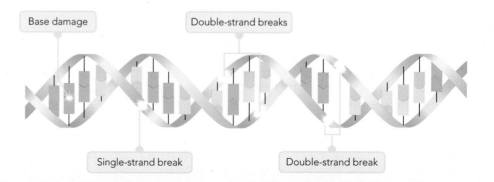

Base damage

Double-strand breaks

Single-strand break

Double-strand break

of biological destruction. It equals the rad times a factor for the type of radiation, called the relative biological effectiveness (RBE).

$$\text{rems} = \text{rads} \times \text{RBE}$$

Beta and gamma radiations have an RBE of about 1, whereas neutron radiation has an RBE of about 5 and alpha radiation an RBE of about 10. ▶

The effects of radiation on a person depend not only on the dosage but also on the length of time in which the dose was received. A series of small doses has less overall effect than these dosages given all at once. A single dose of about 500 rems is fatal to most people, and survival from a much smaller dose can be uncertain or leave the person chronically ill. Detectable effects are seen with doses as low as 30 rems. Continuous exposure to such low levels of radiation may result in cancer or leukemia. At even lower levels, whether the radiation dose is safe depends on the possible genetic effects of the radiation. Because radiation can cause chromosome damage, heritable defects are possible.

Safe limits for radiations are much debated. Although there may not be a strictly safe limit, it is important to have in mind the magnitude of the radiations humans may be subjected to. The background radiation we all receive results from cosmic rays (radiations from space) and natural radioactivity. This averages about 0.1 rem per year but varies considerably with location. Uranium and its decay products in the soil are an important source of background radiation. ▶ Another source is potassium-40, a radioactive isotope with a natural abundance of 0.012%. In addition to natural background radiation, we may receive radiation from other fairly common sources. The most important of these are x rays used in medical diagnosis. The average person receives a radiation dose from this source that is about equal to that of the natural background. Very small radiation sources include consumer items such as television sets and luminous watches.

The background radiation to which we are all subjected has increased slightly since the advent of nuclear technology. Fallout from atmospheric testing of nuclear weapons increased this background by several percent, but it has decreased since atmospheric testing was banned. The radiation contributed by nuclear power plants is only a fraction of a percent of the natural background.

Sources of alpha radiation outside the body are relatively harmless, because the radiation is absorbed by the skin. Internal sources, however, are very destructive.

Soils contain varying amounts of uranium-238, which decays in several steps to radium-226, then to radon-222, a gas (see Figure 20.5). Some homes situated in areas of high uranium content have been found to accumulate radon gas. Radon has a half-life of 3.8 days, decaying by alpha emission to radioactive lead, bismuth, and polonium. These decay products can remain in the lungs and may lead to lung cancer.

CONCEPT CHECK 20.2

If you are internally exposed to 10 rads of α, β, and γ radiation, which form of radiation will cause the greatest biological damage?

20.4 Rate of Radioactive Decay

Although technetium-99 is radioactive and decays by emitting beta particles (electrons), it is impossible to say when a particular nucleus will disintegrate. A sample of technetium-99 continues to give off beta rays for millions of years. Thus, a particular nucleus might disintegrate the next instant or several million years later. The rate of this radioactive decay cannot be changed by varying the temperature, the pressure, or the chemical environment of the technetium nucleus. Radioactivity, whether from technetium or some other nucleus, is not affected by the variables that affect the rate of a chemical reaction. In this section, we will look at how to express quantitatively the rate of radioactive decay.

Rate of Radioactive Decay and Half-Life

The rate of radioactive decay—that is, the number of nuclei disintegrating per unit time—is found to be proportional to the number of radioactive nuclei in the sample. You can express this rate mathematically as

$$\text{Rate} = kN_t$$

Here N_t is the number of radioactive nuclei at time t, and k is the **radioactive decay constant**, or *rate constant for radioactive decay*. This rate constant is a characteristic of the radioactive nuclide, each nuclide having a different value. ▶

The rate equation for radioactive decay has the same form as the rate law for a first-order chemical reaction. Indeed, radioactive decay is a first-order rate process, and the mathematical relationships used in Chapter 13 for first-order reactions apply here also.

You can obtain the decay constant for a radioactive nucleus by counting the nuclear disintegrations over a period of time. The original definition of the curie (now 3.7×10^{10} disintegrations per second) was the activity or decay rate of 1.0 g of radium-226. You can use this with the rate equation to obtain the decay constant of radium-226. Radium-226 has a molar mass of 226 g. A 1.0-g sample of radium-226 contains the following number of nuclei:

$$1.0 \text{ g Ra-226} \times \frac{1 \text{ mol Ra-226}}{226 \text{ g Ra-226}} \times \frac{6.02 \times 10^{23} \text{ Ra-226 nuclei}}{1 \text{ mol Ra-226}} =$$

$$2.7 \times 10^{21} \text{ Ra-226 nuclei}$$

This equals the value of N_t. When you solve Rate $= kN_t$ for k, you get

$$k = \frac{\text{rate}}{N_t}$$

Substituting into this gives

Recall from Chapter 7 that /s is equivalent to the unit s^{-1}.

$$k = \frac{3.7 \times 10^{10} \text{ nuclei/s}}{2.7 \times 10^{21} \text{ nuclei}} = 1.4 \times 10^{-11}\text{/s} \blacktriangleleft$$

Example 20.7

Calculating the Decay Constant from the Activity

Gaining Mastery Toolbox

Critical Concept 20.7
The radioactive decay constant (k) is independent of the amount of radioactive material and depends only on the identity of the nuclide. Different nuclides have different decay constants. The rate of radioactive decay depends upon the decay constant and the *number* of radioactive nuclei at a particular time (N_t). The rate of nuclear decay is proportional to the number of radioactive nuclei in a sample, expressed mathematically as Rate $= kN_t$.

Solution Essentials:
• Curie (Ci)
• Activity
• Radioactive decay constant
• Radioactive decay equation (Rate $= kN_t$)
• Nuclear equation
• Nuclide symbol

A 1.0-mg sample of technetium-99 has an activity of 1.7×10^{-5} Ci (Ci = curies), decaying by beta emission. What is the decay constant for $^{99}_{43}$Tc?

Problem Strategy Solve the equation Rate $= kN_t$ for k; obtain Rate and N_t from values in the problem statement. The rate of decay (Rate) of Tc-99 equals the activity in curies expressed as nuclei decaying per second, noting that 1 Ci $= 3.7 \times 10^{10}$ nuclei/s. Obtain the number of Tc-99 nuclei, N_t, by converting 1.0×10^{-3} g of Tc-99 as follows:

$$\text{g Tc-99} \longrightarrow \text{mol Tc-99} \longrightarrow \text{number Tc-99 nuclei}$$

For the first conversion, note that the numerical value of the molar mass of Tc-99 is approximately equal to the mass number of Tc(99).

Solution The rate of decay in this sample is

$$\text{Rate} = 1.7 \times 10^{-5} \text{ Ci} \times \frac{3.7 \times 10^{10} \text{ nuclei/s}}{1.0 \text{ Ci}} = 6.3 \times 10^{5} \text{ nuclei/s}$$

The number of nuclei in this sample of 1.0×10^{-3} g $^{99}_{43}$Tc is

$$1.0 \times 10^{-3} \text{ g Tc-99} \times \frac{1 \text{ mol Tc-99}}{99 \text{ g Tc-99}} \times \frac{6.02 \times 10^{23} \text{ Tc-99 nuclei}}{1 \text{ mol Tc-99}}$$

$$= 6.1 \times 10^{18} \text{ Tc-99 nuclei}$$

The decay constant is

$$k = \frac{\text{rate}}{N_i} = \frac{6.3 \times 10^{5} \text{ nuclei/s}}{6.1 \times 10^{18} \text{ nuclei}} = \mathbf{1.0 \times 10^{-13}\text{/s}}$$

Answer Check You should typically expect decay constants to range from 10^{-16}/s to 10^{-3}/s, so anything outside this range should be carefully checked.

Exercise 20.7 The nucleus $^{99m}_{43}$Tc is a metastable nucleus of technetium-99; it is used in medical diagnostic work. Technetium-99m decays by emitting gamma rays. A 2.5-μg (microgram) sample has an activity of 13 Ci. What is the decay constant (in units of /s)?

■ See Problems 20.57, 20.58, 20.59, and 20.60.

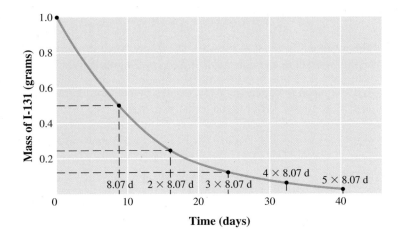

Figure 20.13 ◀

Radioactive decay of a 1.000-g sample of iodine-131 The sample decays by one-half in each half-life of 8.07 days.

We define the **half-life** of a radioactive nucleus as *the time it takes for one-half of the nuclei in a sample to decay.* ▶ The half-life is independent of the amount of sample. In some cases, you can find the half-life of a radioactive sample by directly observing how long it takes for one-half of the sample to decay. For example, you find that 1.000 g of iodine-131, an isotope used in treating thyroid cancer, decays to 0.500 g in 8.07 days. ▶ Thus, its half-life is 8.07 days. In another 8.07 days, this sample would decay to one-half of 0.500 g, or 0.250 g, and so forth. Figure 20.13 shows this decay pattern.

Although you might be able to obtain the half-life of a radioactive nucleus by direct observation in some cases, this is impossible for many nuclei because they decay too quickly or too slowly. Uranium-238, for example, has a half-life of 4.51 billion years, much too long to be directly observed! The usual method of determining the half-life is by measuring decay rates and relating them to half-lives.

You relate the half-life for radioactive decay, $t_{1/2}$, to the decay constant k by the equation

$$t_{1/2} = \frac{0.693}{k}$$

The next example illustrates the use of this relation.

The half-life of a reaction was discussed in Section 13.4.

Iodine is attracted to the thyroid gland, which incorporates the element into the growth hormone thyroxine. Radiation from iodine-131 kills cancer cells in the thyroid gland.

Example 20.8 — Calculating the Half-Life from the Decay Constant

Gaining Mastery Toolbox

Critical Concept 20.8
The half-life of a nuclide undergoing nuclear decay is the time it takes for one-half of the nuclei in the sample to decay. The half-life of a sample is independent of the amount of radioactive material and depends only upon the decay constant of the nuclide in the sample.

Solution Essentials:
• Half-life equation for nuclear decay ($t_{1/2} = \frac{0.693}{k}$)
• Half-life
• Radioactive decay constant

The decay constant for the beta decay of $^{99}_{43}\text{Tc}$ was obtained in Example 20.7. We found that k equals 1.0×10^{-13}/s. What is the half-life of this isotope in years?

Problem Strategy Because the decay constant is given in the problem, the preceding equation can be used to calculate the half-life ($t_{1/2}$) from k.

Solution Substitute the value of k into the preceding equation.

$$t_{1/2} = \frac{0.693}{k} = \frac{0.693}{1.0 \times 10^{-13}/\text{s}} = 6.9 \times 10^{12}\ \text{s}$$

Then you convert this half-life in seconds to years.

$$6.9 \times 10^{12}\ \text{s} \times \frac{1\ \text{min}}{60\ \text{s}} \times \frac{1\ \text{h}}{60\ \text{min}} \times \frac{1\ \text{d}}{24\ \text{h}} \times \frac{1\ \text{y}}{365\ \text{d}} = \textbf{2.2} \times \textbf{10}^{\textbf{5}}\ \textbf{y}$$

(continued)

(*continued*)

Answer Check Note the inverse relationship between the decay constant and the half-life; those compounds with very small decay constants have very long half-lives.

> **Exercise 20.8** Cobalt-60, used in cancer therapy, decays by beta and gamma emission. The decay constant is 4.18×10^{-9}/s. What is the half-life in years?

■ See Problems 20.61 and 20.62.

Tables of radioactive nuclei often list the half-life. When you want the decay constant or the activity of a sample, you can calculate them from the half-life. This calculation is illustrated in the next example.

Example 20.9

Calculating the Decay Constant and Activity from the Half-Life

Gaining Mastery Toolbox

Critical Concept 20.9
The half-life of a nuclide undergoing nuclear decay is the time it takes for one-half of the nuclei in the sample to decay. The half-life of a sample is independent of the amount of radioactive material and depends only upon the decay constant of the nuclide in the sample.

Solution Essentials:
- Half-life equation for nuclear decay ($t_{1/2} = \frac{0.693}{k}$)
- Half-life
- Curie (Ci)
- Radioactive decay constant
- Radioactive decay equation (Rate = kN_t)
- Nuclide symbol

Tritium, $_1^3H$, is a radioactive nucleus of hydrogen. It is used in luminous watch dials. Tritium decays by beta emission with a half-life of 12.3 years. What is the decay constant (in /s)? What is the activity (in Ci) of a sample containing 2.5 μg of tritium? The atomic mass of $_1^3H$ is 3.02 amu.

Problem Strategy First convert the half-life to its value in seconds and then calculate k. Then you can use the rate equation to find the rate of decay of the sample (in nuclei/s) and finally the activity.

Solution The conversion of the half-life to seconds gives

$$12.3 \text{ y} \times \frac{365 \text{ d}}{1 \text{ y}} \times \frac{24 \text{ h}}{1 \text{ d}} \times \frac{60 \text{ min}}{1 \text{ h}} \times \frac{60 \text{ s}}{1 \text{ min}} = 3.88 \times 10^8 \text{ s}$$

Because $t_{1/2} = 0.693/k$, you solve this for k and substitute the half-life in seconds.

$$k = \frac{0.693}{t_{1/2}} = \frac{0.693}{3.88 \times 10^8 \text{ s}} = \mathbf{1.79 \times 10^{-9}\text{/s}}$$

Before substituting into the rate equation, you need to know the number of tritium nuclei in a sample containing 2.5×10^{-6} g of tritium. You get

$$2.5 \times 10^{-6} \text{ g H-3} \times \frac{1 \text{ mol H-3}}{3.02 \text{ g H-3}} \times \frac{6.02 \times 10^{23} \text{ H-3 nuclei}}{1 \text{ mol H-3}} = 5.0 \times 10^{17} \text{ H-3 nuclei}$$

Now you substitute into the rate equation.

$$\text{Rate} = kN_t = 1.79 \times 10^{-9}\text{/s} \times 5.0 \times 10^{17} \text{ nuclei} = 9.0 \times 10^8 \text{ nuclei/s}$$

You obtain the activity of the sample by converting the rate in disintegrations of nuclei per second to curies.

$$\text{Activity} = 9.0 \times 10^8 \text{ nuclei/s} \times \frac{1.0 \text{ Ci}}{3.7 \times 10^{10} \text{ nuclei/s}} = \mathbf{0.024 \text{ Ci}}$$

Answer Check Note that the answer is a relatively small activity (0.024 Ci). If we consider the very small size of the sample of tritium (2.5 μg), we should expect an answer of this magnitude.

(*continued*)

(continued)

Exercise 20.9 Strontium-90, $^{90}_{38}Sr$, is a radioactive decay product of nuclear fallout from nuclear weapons testing. Because of its chemical similarity to calcium, it is incorporated into the bones if present in food. The half-life of strontium-90 is 28.1 y. What is the decay constant of this isotope? What is the activity of a sample containing 5.2 ng (5.2 × 10^{-9} g) of strontium-90?

■ See Problems 20.63, 20.64, 20.65, and 20.66.

Once you know the decay constant for a radioactive isotope, you can calculate the fraction of the radioactive nuclei that remains after a given period of time by the following equation. ▶

$$\ln\frac{N_t}{N_0} = -kt$$

Here N_0 is the number of nuclei in the original sample ($t = 0$). After a period of time t, the number of nuclei decreases by decay to the number N_t. The fraction of nuclei remaining after time t is N_t/N_0. The next example illustrates the use of this equation.

A similar equation is given in Chapter 13 for the reactant concentration at time t, $[A]_t$

$$\ln\frac{[A]_t}{[A]_0} = -kt$$

where $[A]_0$ is the concentration of A at t = 0.

⚙WL Interactive Example 20.10   Determining the Fraction of Nuclei Remaining after a Specified Time

Gaining Mastery Toolbox

Critical Concept 20.10
The decay constant can be used to determine the amount of radioactive material that remains after a period of time has elapsed. The fraction of material that remains after a period of time has elapsed (t) is the ratio of the number of radioactive nuclei (N_t) to the number of radioactive nuclei in the original sample when the time measurement started (N_0). This fraction is often expressed as a percent.

Solution Essentials:
• Equation relating fraction of nuclei after time has elapsed to the decay constant $\left(\ln\frac{N_t}{N_0} = -kt\right)$
• Half-life equation for nuclear decay ($t_{1/2} = \frac{0.693}{k}$)
• Half-life
• Radioactive decay constant
• Nuclide symbol

Phosphorus-32 is a radioactive isotope with a half-life of 14.3 d (d = days). A biochemist has a vial containing a compound of phosphorus-32. If the compound is used in an experiment 5.5 d after the compound was prepared, what fraction of the radioactive isotope originally present remains? Suppose the sample in the vial originally contained 0.28 g of phosphorus-32. How many grams remain after 5.5 d?

Problem Strategy In the first part of the problem, we are asked to determine what fractional amount of radioactive P-32 in a compound remains after 5.5 d. For this, we can use the preceding equation. Once we know the fractional amount of P-32 that remains, we can multiply this quantity by the original amount of compound (0.28 g) to find the mass of compound that remains after 5.5 d.

Solution If N_0 is the original number of P-32 nuclei in the vial and N_t is the number after 5.5 d, the fraction remaining is N_t/N_0. You can obtain this fraction from the equation

$$\ln\frac{N_t}{N_0} = -kt$$

You substitute $k = 0.693/t_{1/2}$.

$$\ln\frac{N_t}{N_0} = \frac{-0.693t}{t_{1/2}}$$

Because $t = 5.5$ d and $t_{1/2} = 14.3$ d, you obtain

$$\ln\frac{N_t}{N_0} = \frac{-0.693 \times 5.5 \text{ d}}{14.3 \text{ d}} = -0.267$$

(continued)

(*continued*)

(An additional digit has been retained for further calculation.) Hence,

$$\text{Fraction nuclei remaining} = \frac{N_t}{N_0} = e^{-0.267} = \mathbf{0.77}$$

(Note that 0.766 is rounded to two significant figures.) Thus, 77% of the nuclei remain. The mass of $^{32}_{15}\text{P}$ in the vial after 5.5 d is

$$0.28 \text{ g} \times 0.766 = \mathbf{0.21 \text{ g}}$$

Answer Check After one half-life of 14.3 days, only one-half of the sample would be left. Since only 5.5 days have elapsed, which is less than the half-life, you should expect less than half of the 0.28 g sample to have decayed. This means that there should be more than 0.14 g remaining, which is the case. For any problems of this type, this approach is an effective method for quickly checking your answers.

Exercise 20.10 A nuclear power plant emits into the atmosphere a very small amount of krypton-85, a radioactive isotope with a half-life of 10.76 y. What fraction of this krypton-85 remains after 25.0 y?

■ See Problems 20.69 and 20.70.

Radioactive Dating

Fixing the dates of relics and stone implements or pieces of charcoal from ancient campsites is an application based on radioactive decay rates. Because the rate of radioactive decay of a nuclide is constant, this rate can serve as a clock for dating very old rocks and human implements. Dating wood and similar carbon-containing objects that are several thousand to fifty thousand years old can be done with radioactive carbon, carbon-14, which has a half-life of 5730 y.

Carbon-14 is present in the atmosphere as a result of cosmic-ray bombardment of earth. Cosmic rays are radiations from space that consist of protons and alpha particles, as well as heavier ions. These radiations produce other kinds of particles, including neutrons, as they bombard the upper atmosphere. The collision of a neutron with a nitrogen-14 nucleus (the most abundant nitrogen nuclide) can produce a carbon-14 nucleus.

$$^{14}_{7}\text{N} + ^{1}_{0}\text{n} \longrightarrow ^{14}_{6}\text{C} + ^{1}_{1}\text{H}$$

Carbon dioxide containing carbon-14 mixes with the lower atmosphere. Because of the constant production of $^{14}_{6}\text{C}$ and its radioactive decay, a small, constant fractional abundance of carbon-14 is maintained in the atmosphere.

Living plants, which continuously use atmospheric carbon dioxide, also maintain a constant abundance of carbon-14. Similarly, living animals, by feeding on plants, have a constant fractional abundance of carbon-14. But once an organism dies, it is no longer in chemical equilibrium with atmospheric CO_2. The ratio of carbon-14 to carbon-12 begins to decrease by radioactive decay of carbon-14. In this way, this ratio of carbon isotopes becomes a clock measuring the time since the death of the organism.

If you assume that the ratio of carbon isotopes in the lower atmosphere has remained at the present level for the last 50,000 years (presently 1 out of 10^{12} carbon atoms is carbon-14), you can deduce the age of any dead organic object by measuring the level of beta emissions that arise from the radioactive decay of carbon-14. ◀ This is the decay reaction:

$$^{14}_{6}\text{C} \longrightarrow ^{14}_{7}\text{N} + ^{0}_{-1}\text{e}$$

In this way, bits of campfire charcoal, parchment, jawbones, and the like have been dated.

Analyses of tree rings have shown that this assumption is not quite valid. Before 1000 B.C., the levels of carbon-14 were somewhat higher than they are today. Moreover, recent human activities (burning of fossil fuels and atmospheric nuclear testing) have changed the fraction of carbon-14 in atmospheric CO_2.

Example 20.11   **Applying the Carbon-14 Dating Method**

Gaining Mastery Toolbox

Critical Concept 20.11
Radioactive dating requires that we know the starting amount of the radioactive isotope (N_0) when the material was created, the current amount of radioactive isotope (N_t), and the half-life of the radioactive isotope. For carbon-14 dating, we assume that the amount of carbon-14 in a living organism has remained constant during the past 50,000 years; therefore, we know the starting amount of carbon-14 (N_t) of any plants that were alive during the last 50,000 years.

Solution Essentials:
• Radioactive dating
• Equation relating fraction of nuclei after time has elapsed to the decay constant ($\ln\dfrac{N_t}{N_0} = -kt$)
• Half-life equation for nuclear decay ($t_{1/2} = \frac{0.693}{k}$)
• Half-life
• Radioactive decay constant
• Nuclide symbol

A piece of charcoal from a tree killed by the volcanic eruption that formed the crater in Crater Lake (in Oregon) gave 7.0 disintegrations of carbon-14 nuclei per minute per gram of total carbon. Present-day carbon (in living matter) gives 15.3 disintegrations per minute per gram of total carbon. Determine the date of the volcanic eruption. Recall that the half-life of carbon-14 is 5730 y.

Problem Strategy The solution to the problem is determining the amount of time (t) that it takes for the amount of carbon-14 in the living sample (N_0) to decrease to the amount present today (N_t). The equation $\ln\dfrac{N_t}{N_0} = -kt$ relates these quantities; however, we do not have the decay constant (k), and thus, you cannot directly use this equation to determine the elapsed time. However, the half-life of carbon-14 is given in the problem so that we can employ the relationship between the decay constant and the half-life of carbon-14 to solve first for the decay constant, $k = 0.693/t_{1/2}$, and then substitute the value for the decay constant into the equation $\ln\dfrac{N_t}{N_0} = -kt$.

Recognizing that the process described above is a combination of equations, we can do some algebra to simplify our solving the problem.
Start with

$$\ln\frac{N_t}{N_0} = -kt$$

substitute $k = 0.693/t_{1/2}$ into the proceeding equation to yield

$$\ln\frac{N_t}{N_0} = 0.693t/t_{1/2}$$

Rearranging the equation to solve for t yields

$$t = t_{1/2}/0.693 \ln\frac{N_t}{N_0}$$

To get N_0/N_t, you assume that the ratio of $^{14}_6C$ to $^{12}_6C$ in the atmosphere has remained constant. Then you can say that 1.00 g of total carbon from the living tree gave 15.3 disintegrations per minute. The ratio of the number of $^{14}_6C$ nuclei originally present to the number that existed at the time of dating equals the ratio of rates of disintegration.

Solution The ratio of rates of disintegration is

$$\frac{N_0}{N_t} = \frac{15.3}{7.0} = 2.2$$

Substituting this value of N_0/N_t and $t_{1/2} = 5730$ y into the equation developed in the Problem Strategy gives

$$t = \frac{t_{1/2}}{0.693}\ln\frac{N_0}{N_t} = \frac{5730\text{ y}}{0.693}\ln 2.2 = 6.5\times10^3\text{ y}$$

Thus, the date of the eruption was about **4500 B.C.**, 6500 years ago.

Answer Check You can obtain a quick check by comparing the disintegration rate expected after various half-lives. For example, the original rate is 15.3 disintegrations per minute, whereas the rate after half the sample decays is one-half this, or 7.6 disintegrations per minute. (After two half-lives, the disintegration rate is one-half of 7.6, or 3.8 disintegrations per minute.) Thus, a rate of 7.0 disintegrations per minute is expected to occur at a time somewhat greater than the half-life of carbon-14 (that is, somewhat greater than 5730 y).

Exercise 20.11 A jawbone from the archaeological site at Folsom, New Mexico, was dated by analysis of its radioactive carbon. The activity of the carbon from the jawbone was 4.5 disintegrations per minute per gram of total carbon. What was the age of the jawbone? Carbon from living material gives 15.3 disintegrations per minute per gram of carbon.

■ See Problems 20.75 and 20.76.

For the age of rocks and meteorites, other, similar methods of dating have been used. One method depends on the radioactivity of naturally occurring potassium-40, which decays by positron emission and electron capture (as well as by beta emission).

$$^{40}_{19}K \longrightarrow {}^{40}_{18}Ar + {}^0_1e$$

$$^{40}_{19}K + {}^0_{-1}e \longrightarrow {}^{40}_{18}Ar$$

Potassium occurs in many rocks. Once such a rock forms by solidification of molten material, the argon from the decay of potassium-40 is trapped. To obtain the age of a rock, you first determine the number of $^{40}_{19}K$ atoms and the number of $^{40}_{18}Ar$ atoms in a sample. The number of $^{40}_{19}K$ atoms equals N_t. The number originally present, N_0, equals N_t plus the number of argon atoms, because each argon atom resulted from the decay of a $^{40}_{19}K$ nucleus. You then calculate the age of the rock from the ratio N_t/N_0.

The oldest rocks on earth have been dated at 3.8×10^9 y. The rocks at the earth's surface have been subjected to extensive weathering, so even older rocks may have existed. This age, 3.8×10^9 y, therefore represents the minimum possible age of the earth—the time since the solid crust first formed. Ages of meteorites, assumed to have solidified at the same time as other solid objects in the solar system, including earth, have been determined to be 4.4×10^9 y to 4.6×10^9 y. It is now believed from this and other evidence that the age of the earth is 4.6×10^9 y.

CONCEPT CHECK 20.3

Why do you think that carbon-14 dating is limited to less than 50,000 years?

20.5 Applications of Radioactive Isotopes

We have already described two applications of nuclear chemistry. One was the preparation of elements not available naturally. We noted that the discovery of the transuranium elements clarified the position of the heavy elements in the periodic table. In the section just completed, we discussed the use of radioactivity in dating objects. We will discuss practical uses of nuclear energy later in the chapter. Here we will look at the applications of radioactive isotopes to chemical analysis and to medicine.

Chemical Analysis

A **radioactive tracer** is *a very small amount of radioactive isotope added to a chemical, biological, or physical system to study the system.* The advantage of a radioactive tracer is that it behaves chemically just as a nonradioactive isotope does, but it can be detected in exceedingly small amounts by measuring the radiations emitted.

As an illustration of the use of radioactive tracers, consider the problem of establishing that chemical equilibrium is a dynamic process. Let us look at the equilibrium of solid lead(II) iodide and its saturated solution, containing $Pb^{2+}(aq)$ and $I^-(aq)$. The equilibrium is

$$PbI_2(s) \underset{}{\overset{H_2O}{\rightleftharpoons}} Pb^{2+}(aq) + 2I^-(aq)$$

In two separate beakers, you prepare saturated solutions of PbI_2 in contact with the solid. One beaker contains only natural iodine atoms with nonradioactive isotopes. The other beaker contains radioactive iodide ion, $^{131}I^-$. Some of the solution, but no solid, containing the radioactive iodide ion is now added to the beaker containing nonradioactive iodide ion. Both solutions are saturated, so the amount of solid in this beaker remains constant. Yet after a time the solid lead iodide, which was originally nonradioactive, becomes radioactive. This is evidence for a dynamic equilibrium, in which radioactive iodide ions in the solution substitute for nonradioactive iodide ions in the solid.

With only naturally occurring iodine available, it would have been impossible to detect the dynamic equilibrium. By using $^{131}I^-$ as a radioactive tracer, you can easily follow the substitution of radioactive iodine into the solid by measuring its increase in radioactivity.

A series of experiments using tracers was carried out in the late 1950s by Melvin Calvin at the University of California, Berkeley, to discover the mechanism of photosynthesis in plants. ◀ The overall process of photosynthesis involves the reaction of CO_2 and H_2O to give glucose, $C_6H_{12}O_6$, and O_2. Energy for photosynthesis comes from the sun.

Melvin Calvin received the Nobel Prize in Chemistry in 1961 for his work on photosynthesis.

$$6CO_2(g) + 6H_2O(l) \xrightarrow{\text{sunlight}} C_6H_{12}O_6(aq) + 6O_2(g)$$

This equation represents only the net result of photosynthesis. As Calvin was able to show, the actual process consists of many separate steps. In several experiments, algae (single-celled plants) were exposed to carbon dioxide containing much more radioactive carbon-14 than occurs naturally. Then the algae were extracted with a solution of alcohol and water. The various compounds in this solution were separated by chromatography and identified. ▶ Compounds that contained radioactive carbon were produced in the different steps of photosynthesis. Eventually, Calvin was able to use tracers to show the main steps in photosynthesis.

Chromatography was discussed in the essay at the end of Section 1.4.

Another example of the use of radioactive tracers in chemistry is **isotope dilution**, *a technique to determine the quantity of a substance in a mixture or of the total volume of solution by adding a known amount of an isotope to it.* After removal of a portion of the mixture, the fraction by which the isotope has been diluted provides a way of determining the quantity of substance or volume of solution.

As an example, suppose you wish to obtain the volume of water in a tank but are unable to drain the tank. You add 100 mL of water containing a radioactive isotope. After allowing this to mix completely with the water in the tank, you withdraw 100 mL of solution from the tank. You find that the activity of this solution in curies is 1/1000 that of the original radioactive solution. The isotope has been diluted by a factor of 1000, so the volume of the tank is 1000×100 mL $= 100,000$ mL (100 L).

A typical chemical example of isotope dilution is the determination of the amount of vitamin B_{12}, a cobalt-containing substance, in a sample of food. Although part of the vitamin in food can be obtained in pure form, not all of the vitamin can be separated. Therefore, you cannot precisely determine the quantity of vitamin B_{12} in a sample of food by separating the pure vitamin and weighing it. But you can determine the amount of vitamin B_{12} by isotope dilution. Suppose you add 2.0×10^{-7} g of vitamin B_{12} containing radioactive cobalt-60 to 125 g of food and mix well. You then separate 5.4×10^{-7} g of pure vitamin B_{12} from the food and find that the activity in curies of this quantity of the vitamin contains 5.6% of the activity added from the radioactive cobalt. This indicates that you have recovered 5.6% of the total amount of vitamin B_{12} in the sample. Therefore, the mass of vitamin B_{12} in the food, including the amount added (2.0×10^{-7} g), is

$$5.4 \times 10^{-7} \text{ g} \times \frac{100}{5.6} = 9.6 \times 10^{-6} \text{ g}$$

Subtracting the amount added in the analysis gives

$$9.6 \times 10^{-6} \text{ g} - 2.0 \times 10^{-7} \text{ g} = 9.4 \times 10^{-6} \text{ g}$$

Neutron activation analysis is *an analysis of elements in a sample based on the conversion of stable isotopes to radioactive isotopes by bombarding a sample with neutrons.* Human hair contains trace amounts of many elements. By determining the exact amounts and the position of the elements in the hair shaft, you can identify whom the hair comes from (assuming you have a sample known to be that person's hair for comparison). Consider the analysis of human hair for arsenic, for example. When the natural isotope $^{75}_{33}$As is bombarded with neutrons, a metastable nucleus $^{76m}_{33}$As is obtained.

$$^{75}_{33}\text{As} + {}^{1}_{0}\text{n} \longrightarrow {}^{76m}_{33}\text{As}$$

A metastable nucleus is in an excited state. It decays, or undergoes a transition, to a lower state by emitting gamma rays. The frequencies, or energies, of the gamma rays emitted are characteristic of the element and serve to identify it. Also, the intensities of the gamma rays emitted are proportional to the amount of the element present. The method is very sensitive; it can identify as little as 10^{-9} g of arsenic. ▶

Neutron activation analysis has been used to authenticate oil paintings by giving an exact analysis of pigments used. Pigment compositions have changed, so it is possible to detect fraudulent paintings done with more modern pigments. The analysis can be done without affecting the painting.

Medical Therapy and Diagnosis

The use of radioactive isotopes has had a profound effect on the practice of medicine. Radioisotopes were first used in medicine in the treatment of cancer. This treatment is based on the fact that rapidly dividing cells, such as those in cancer, are more adversely affected by radiation from radioactive substances than are cells that divide more slowly. Radium-226 and its decay product radon-222 were used for cancer therapy a few years after the discovery of radioactivity. Today gamma radiation from cobalt-60 is more commonly used.

Cancer therapy is only one of the ways in which radioactive isotopes are used in medicine. The greatest advances in the use of radioactive isotopes have been in the diagnosis of disease. Radioactive isotopes are used for diagnosis in two ways. They are used to develop images of internal body organs so that their functioning can be examined. And they are used as tracers in the analysis of minute amounts of substances, such as a growth hormone in blood, to deduce possible disease conditions.

Technetium-99m is the radioactive isotope used most often to develop images (pictures) of internal body organs. It has a half-life of 6.02 h, decaying by gamma emission to technetium-99 in its nuclear ground state. The image is prepared by scanning part of the body for gamma rays with a scintillation detector. (Figure 20.1 shows the image of a person's skeleton obtained with technetium-99m.)

The technetium isotope is produced in a special container, or generator, shown in Figure 20.14. The generator contains radioactive molybdate ion, MoO_4^{2-}, adsorbed on alumina granules. Radioactive molybdenum-99 is produced from uranium-235 nuclear fission products (see Section 20.7). This radioactive molybdenum, adsorbed on alumina, is placed in the generator and sent to the hospital. Pertechnetate ion is obtained when the molybdenum-99 nucleus in MoO_4^{2-} decays. The nuclear equation is

$$^{99}_{42}\text{Mo} \longrightarrow \,^{99m}_{43}\text{Tc} + \,^{0}_{-1}\text{e}$$

Each day pertechnetate ion, TcO_4^-, is leached from the generator with a salt solution whose osmotic pressure is the same as that of blood. One use of this pertechnetate ion is in assessing the performance of a patient's heart. The physician injects tin(II) ion into a vein and, a few minutes later, administers a similar injection of the pertechnetate ion. In the presence of tin(II) ion, the pertechnetate ion binds to the red blood cells. The heart then becomes visible in gamma-ray imaging equipment,

Figure 20.14 ▶

A technetium-99m generator

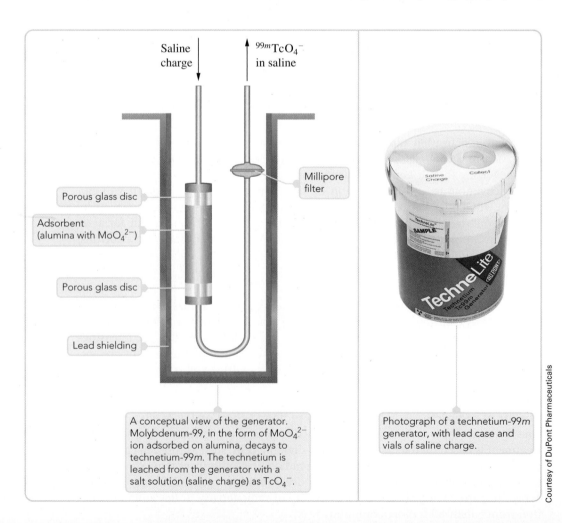

A conceptual view of the generator. Molybdenum-99, in the form of MoO_4^{2-} ion adsorbed on alumina, decays to technetium-99m. The technetium is leached from the generator with a salt solution (saline charge) as TcO_4^-.

Photograph of a technetium-99m generator, with lead case and vials of saline charge.

Courtesy of DuPont Pharmaceuticals

because of the quantity of blood in the heart. Technetium-99*m* pyrophosphate ($Tc_2P_2O_7$) is another technetium species useful for gamma-ray imaging. This compound binds especially strongly to recently damaged heart muscle, so these gamma-ray images can be used to assess the extent of damage from a heart attack.

Thallium-201 is a radioisotope used to determine whether a person has heart disease (caused by narrowing of the arteries to the heart). This isotope decays by electron capture and emits x rays and gamma rays, which can be used to obtain images similar to those obtained from technetium-99*m* (Figure 20.15). Thallium-201 injected into the blood binds particularly strongly to heart muscle. Diagnosis of heart disease depends on the fact that only tissue that receives sufficient blood flow binds thallium-201. When someone exercises strenuously, some part of the person's heart tissue may not receive sufficient blood because of narrowed arteries. These areas do not bind thallium-201 and show up on an image as dark spots.

More than a hundred different radioactive isotopes have been used in medicine. Besides technetium-99*m* and thallium-201, other examples include iodine-131, used to measure thyroid activity; phosphorus-32, used to locate tumors; and iron-59, used to measure the rate of formation of red blood cells.

Radioimmunoassay is a technique for analyzing blood and other body fluids for very small quantities of biologically active substances. The technique depends on the reversible binding of the substance to an antibody. Antibodies are produced in animals as protection against foreign substances. They protect by binding to the substance and countering its biological activity. Consider, for example, the analysis for insulin in a sample of blood from a patient. Before the analysis, a solution of insulin-binding antibodies has been prepared from laboratory animals. This solution is combined with insulin containing a radioactive isotope, in which the antibodies bind with radioactive insulin. Now the blood sample containing an unknown amount of insulin is added to the antibody–radioactive insulin mixture. The nonradioactive insulin replaces some of the radioactive insulin bound to the antibody. As a result, the antibody loses some of its radioactivity. The loss in radioactivity can be related to the amount of insulin in the blood sample.

Figure 20.15 ▼

Using thallium-201 to diagnose heart disease

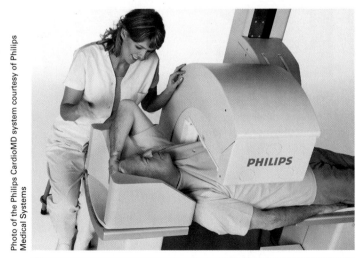

Photo of the Philips CardioMD system courtesy of Philips Medical Systems

a A nuclear camera system used to detect the gamma emissions from thallium-201 or technetium-99*m* for cardiac studies.

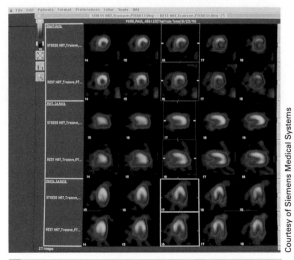

Courtesy of Siemens Medical Systems

b A series of cross-sectional images of a patient's heart after exercise and then some time afterward. By comparing the stress and rest images, a physician can see if there is impaired blood flow to an area of the heart (the area is dark in the stress image but light in the rest image) or if the heart muscle has been damaged through a heart attack (the area is dark in both stress and rest images).

Positron Emission Tomography (PET)

Positron emission tomography (PET) is a technique for following biochemical processes within the organs (brain, heart, and so forth) of the human body. Like magnetic resonance imaging (see essay at the end of Section 8.1), a PET scan produces an image of a two-dimensional slice through a body organ of a patient. The image shows the distribution of some positron-emitting isotope present in a compound that was administered earlier by injection. By comparing the PET scan of the patient with that of a healthy subject, a physician can diagnose the presence or absence of disease (Figure 20.16). The PET scan of the brain of an Alzheimer's patient differs markedly from that of a healthy subject. Similarly, the PET scan of a patient with a heart damaged by a coronary attack clearly shows the damaged area. Moreover, the PET scan may help the physician assess the likelihood of success of bypass surgery.

Some isotopes used in PET scans are carbon-11, nitrogen-13, oxygen-15, and fluorine-18. All have short half-lives, so the radiation dosage to the patient is minimal. However, because of the short half-life, a chemist must prepare the diagnostic compound containing the radioactive nucleus shortly before the physician administers it. The preparation of this compound requires a cyclotron, whose cost (several million dollars) is a major deterrent to the general use of PET scans.

Figure 20.17 shows a patient undergoing a PET scan of the brain. The instrument actually detects gamma radiation. When a nucleus emits a positron within the body, the positron travels only a few millimeters before it reacts with an electron. This reaction is an example of the annihilation of matter (an electron) by antimatter (a positron). Both the electron and the positron disappear and produce two gamma photons. The gamma photons easily pass through human tissue, so they can be recorded by scintillation detectors placed around the body. You can see the circular bank of detectors in Figure 20.17. The detectors record the distribution of gamma radiation, and from this information a computer constructs images that can be used by the physician.

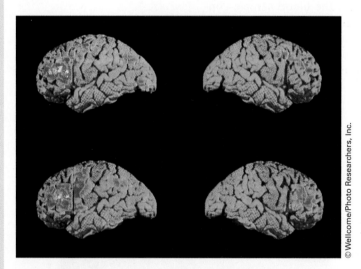

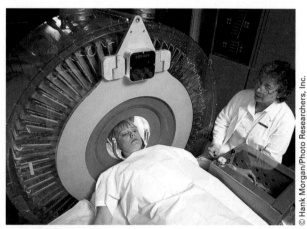

© Wellcome/Photo Researchers, Inc.

© Hank Morgan/Photo Researchers, Inc.

Figure 20.17 ▲

A patient undergoing a PET scan of the brain Note the circular bank of gamma-ray detectors.

Figure 20.16 ▲

PET scans of normal and schizophrenic patients Colored positron emission tomography (PET) brain scans of a schizophrenic (bottom) versus normal patient (top), speaking. The red/yellow highlighted areas are active when patients generate words. In the schizophrenic brain, there is additional brain activity not seen in normal subjects. Schizophrenic patients can suffer from delusions and depression.

■ See Problems 20.105 and 20.106.

Energy of Nuclear Reactions

Nuclear reactions, like chemical reactions, involve changes of energy. However, the changes of energy in nuclear reactions are enormous by comparison with those in chemical reactions. The energy released by certain nuclear reactions is used in nuclear power reactors and to provide the energy for nuclear weapons. In the case of a power plant, the energy from the reactions is released in controlled, small amounts by the fission of uranium-235. With a nuclear weapon, the objective is to release the energy as rapidly as possible and produce a nuclear explosion. Although the source of energy for nuclear power plants and weapons can be the same, a typical nuclear power plant does not contain enough fissionable material in high enough concentration to produce a nuclear explosion.

20.6 Mass–Energy Calculations

When nuclei decay, they form products of lower energy. The change of energy is related to changes of mass, according to the mass–energy equivalence relation derived by Albert Einstein in 1905. Using this relation, you can obtain the energies of nuclear reactions from mass changes.

Mass–Energy Equivalence

One of the conclusions from Einstein's theory of special relativity is that the mass of a particle changes with its speed: the greater the speed, the greater the mass. Or, because kinetic energy depends on speed, the greater the kinetic energy of a particle, the greater its mass. This result, according to Einstein, is even more general. Energy and mass are equivalent and are related by the equation

$$E = mc^2$$

Here c is the speed of light, 3.00×10^8 m/s.

The meaning of this equation is that for any mass there is an associated energy, or for any energy there is an associated mass. If a system loses energy, it must also lose mass. For example, when carbon burns in oxygen, it releases heat energy.

$$C(graphite) + O_2(g) \longrightarrow CO_2(g); \Delta H = -393.5 \text{ kJ}$$

Because this chemical system loses energy, it should also lose mass. In principle, you could obtain ΔH for the reaction by measuring the change in mass and relating this by Einstein's equation to the change in energy. ▶ However, weight measurements are of no practical value in determining heats of reaction—the changes in mass are simply too small to detect.

The change in mass gives the change in internal energy, ΔU. The enthalpy change, ΔH, equals $\Delta U + P\Delta V$. For the reaction given in the text, $P\Delta V$ is essentially zero.

Calculation of the mass change for a typical chemical reaction, the burning of carbon in oxygen, will show just how small this quantity is. When the energy changes by an amount ΔE, the mass changes by an amount Δm. You can write Einstein's equation in the form

$$\Delta E = (\Delta m)c^2$$

The change in energy when 1 mol of carbon reacts is -3.935×10^5 J, or -3.935×10^5 kg·m^2/s^2. Hence,

$$\Delta m = \frac{\Delta E}{c^2} = \frac{-3.935 \times 10^5 \text{ kg·m}^2/s^2}{(3.00 \times 10^8 \text{ m/s})^2} = -4.37 \times 10^{-12} \text{ kg}$$

For comparison, a good analytical balance can detect a mass as small as 1×10^{-7} kg, but this is ten thousand times greater than the mass change caused by the release of heat during the combustion of 1 mol of carbon.

By contrast, the mass changes in nuclear reactions are approximately a million times larger per mole of reactant than those in chemical reactions. Consider the alpha decay of uranium-238 to thorium-234. The nuclear equation is

$$^{238}_{92}\text{U} \longrightarrow ^{234}_{90}\text{Th} + ^{4}_{2}\text{He}$$
$$238.0003 \qquad 233.9942 \qquad 4.00150 \text{ amu}$$

Table 20.3 Masses of Some Elements and Other Particles

Symbol	Z	A	Mass (amu)	Symbol	Z	A	Mass (amu)
e^-	−1	0	0.000549	Co	27	59	58.93320
n	0	1	1.008665	Ni	28	58	57.93534
p	1	1	1.00728		28	60	59.93079
H	1	1	1.00783	Pb	82	206	205.97444
	1	2	2.01400		82	207	206.97587
	1	3	3.01605		82	208	207.97663
He	2	3	3.01603	Po	84	208	207.98122
	2	4	4.00260		84	210	209.98285
Li	3	6	6.01512	Rn	86	222	222.01757
	3	7	7.01600	Ra	88	226	226.02540
Be	4	9	9.01218	Th	90	230	230.03313
B	5	10	10.01294		90	234	234.03660
	5	11	11.00931	Pa	91	234	234.04330
C	6	12	12.00000	U	92	233	233.03963
	6	13	13.00336		92	234	234.04095
O	8	16	15.99492		92	235	235.04392
Cr	24	52	51.94051		92	238	238.05078
Fe	26	56	55.93494	Pu	94	239	239.05216

Here we have written the nuclear mass (in amu) beneath each nuclide symbol. (Table 20.3 lists masses of some elements and other particles.) The change in mass for this nuclear reaction, starting with molar amounts, is

$$\Delta m = (233.9942 + 4.00150 - 238.0003) \text{ g} = -0.0046 \text{ g}$$

As in calculating ΔH and similar quantities, you subtract the value for the reactant from the sum of the values for the products. The minus sign indicates a loss of mass. This loss of mass is clearly large enough to detect.

From a table of atomic masses, such as Table 20.3, you can use Einstein's equation to calculate the energy change for a nuclear reaction. This is illustrated in the next example. Recall from Section 20.2 that 1 eV = 1.602×10^{-19} J. Therefore, 1 MeV equals 1.602×10^{-13} J.

Example 20.12 Calculating the Energy Change for a Nuclear Reaction

Gaining Mastery Toolbox

Critical Concept 20.12
All types of reactions, both chemical and nuclear, undergo a mass change (Δm) when going from reactants to products. The change in mass is related to a change in energy by the equation $\Delta E = (\Delta m)c^2$; a reduction in mass leads to the production of energy. Nuclear reactions undergo a relatively significant reduction in mass when going from reactants to products, thereby producing great amounts of energy. Ordinary chemical reactions do not undergo significant mass changes so that they produce far less energy than nuclear reactions.

Solution Essentials:
- Einstein's equation ($\Delta E = (\Delta m)c^2$)
- Nuclear mass
- Electron volt
- Metric prefixes

a Calculate the energy change in joules (four significant figures) for the following nuclear reaction per mole of ^2_1H:

$$^2_1\text{H} + ^3_2\text{He} \longrightarrow ^4_2\text{He} + ^1_1\text{H}$$

Atomic and particle masses are given in Table 20.3; the speed of light is given on the inside back cover of this book.

b What is the energy change in MeV for one ^2_1H nucleus?

Problem Strategy In order to calculate the energy changes for both parts of this problem, we will need to use Einstein's equation in the form $\Delta E = (\Delta m)c^2$.

(continued)

(*continued*)

When using this equation for this type of calculation, we will use the nuclear mass of the nuclides. For part a, we will need to determine the mass change (Δm) of the given nuclear reaction per mole of $^{2}_{1}\text{H}$. For part b, we calculate Δm for one $^{2}_{1}\text{H}$ nucleus. Finally, we convert our answer from joules to MeV.

Solution

a You need to write the *nuclear* masses below each nuclide symbol and then calculate Δm. Once you have Δm, you can obtain ΔE. The nuclear mass of a nuclide is the mass of the atom minus the mass of the electrons. For example, using mass information from Table 20.3, you calculate the nuclear mass of $^{4}_{1}\text{He}$ by starting with the atomic mass of $^{4}_{1}\text{He}$ and subtracting the mass of two electrons.

$$\text{Nuclear mass of } ^{4}_{1}\text{He} = 4.00260 \text{ amu} - (2 \times 0.000549 \text{ amu}) = 4.001502 \text{ amu}$$

$$^{2}_{1}\text{H} \;+\; ^{3}_{2}\text{He} \longrightarrow \; ^{4}_{2}\text{He} \;+\; \; ^{1}_{1}\text{H}$$
$$\;\;\;2.01345 \quad\;\; 3.01493 \qquad\;\; 4.00150 \quad 1.00728 \text{ amu}$$

Hence,

$$\Delta m = (4.00150 + 1.00728 - 2.01345 - 3.01493) \text{ amu}$$
$$= -0.01960 \text{ amu}$$

To obtain the energy change for molar amounts, note that the molar mass of an atom in grams is numerically equal to the mass of a single atom in amu. Therefore, the mass change for molar amounts in this nuclear reaction is -0.01960 g, or -1.960×10^{-5} kg. The energy change is

$$\Delta E = (\Delta m)c^2 = (-1.960 \times 10^{-5} \text{ kg})(2.998 \times 10^{8} \text{ m/s})^2$$
$$= -1.762 \times 10^{12} \text{ kg·m}^2/\text{s}^2 \qquad \text{or} \qquad \mathbf{-1.762 \times 10^{12} \text{ J}}$$

b The mass change for the reaction of one $^{2}_{1}\text{H}$ atom is -0.01960 amu. First change this to grams. Recall that 1 amu equals 1/12 the mass of a $^{12}_{6}\text{C}$ atom, whose mass is 12 g/6.022×10^{23}. Thus, 1 amu = 1 g/6.022×10^{23}. Hence, the mass change in grams is

$$\Delta m = -0.01960 \text{ amu} \times \frac{1 \text{ g}}{1 \text{ amu} \times 6.022 \times 10^{23}}$$
$$= -3.255 \times 10^{-26} \text{ g} \qquad (\text{or} -3.255 \times 10^{-29} \text{ kg})$$

Then,

$$\Delta E = (\Delta m)c^2 = (-3.255 \times 10^{-29} \text{ kg})(2.998 \times 10^{8} \text{ m/s})^2$$
$$= -2.926 \times 10^{-12} \text{ J}$$

Now convert this to MeV:

$$\Delta E = -2.926 \times 10^{-12} \text{ J} \times \frac{1 \text{ MeV}}{1.602 \times 10^{-13} \text{ J}} = \mathbf{-18.26 \text{ MeV}}$$

Answer Check When calculating mass changes for nuclear reactions, you should expect answers in the small magnitude shown in this example (10^{-26} g). Any mass change significantly larger than this value should be suspect.

Exercise 20.12 a Calculate the energy change in joules when 1.00 g $^{234}_{90}\text{Th}$ decays to $^{234}_{91}\text{Pa}$ by beta emission.

b What is the energy change in MeV when one $^{234}_{90}\text{Th}$ nucleus decays? Use Table 20.3 for these calculations.

■ See Problems 20.81 and 20.82.

Nuclear Binding Energy

The equivalence of mass and energy explains the otherwise puzzling fact that the mass of an atom is always less than the sum of the masses of its constituent particles. For example, the helium-4 atom consists of two protons, two neutrons, and two electrons, giving the following sum:

$$\text{Mass of 2 electrons} = 2 \times 0.000549 \text{ amu} = 0.00110 \text{ amu}$$
$$\text{Mass of 2 protons} = 2 \times 1.00728 \text{ amu} = 2.01456 \text{ amu}$$
$$\text{Mass of 2 neutrons} = 2 \times 1.00867 \text{ amu} = \underline{2.01734 \text{ amu}}$$
$$\text{Total mass of particles} = 4.03300 \text{ amu}$$

The mass of the helium-4 atom is 4.00260 amu (see Table 20.3), so the mass difference is

$$\Delta m = (4.00260 - 4.03300) \text{ amu} = -0.03040 \text{ amu}$$

This mass difference is explained as follows. When the nucleons come together to form a nucleus, energy is released. (The nucleus has lower energy and is therefore more stable than the separate nucleons.) According to Einstein's equation, there must be an equivalent decrease in mass.

The **binding energy** of a nucleus is *the energy needed to break a nucleus into its individual protons and neutrons.* Thus, the binding energy of the helium-4 nucleus is the energy change for the reaction

$$^{4}_{2}\text{He}^{2+} \longrightarrow 2\,^{1}_{1}\text{p} + 2\,^{1}_{0}\text{n}$$

The **mass defect** of a nucleus is *the nucleon mass minus the atomic mass.* In the case of helium-4, the mass defect is 4.03300 amu − 4.00260 amu = 0.03040 amu. (This is the positive value of the mass difference we calculated earlier.) Both the binding energy and the corresponding mass defect are reflections of the stability of the nucleus.

To compare the stabilities of various nuclei, it is useful to compare binding energies per nucleon. Figure 20.18 shows values of this quantity (in MeV per nucleon) plotted against the mass number for various nuclides. Most of the points lie near the smooth curve drawn on the graph.

Note that nuclides near mass number 50 have the largest binding energies per nucleon. This means that a group of nucleons would tend to form those nuclides, because they would form nuclei of the lowest energy. For this reason, heavy nuclei might be expected to split to give lighter nuclei, and light nuclei might be expected to react, or combine, to form heavier nuclei.

Figure 20.18 ▶

Plot of binding energy per nucleon versus mass number The binding energy of each nuclide is divided by the number of nucleons (total of protons and neutrons), then plotted against the mass number of the nuclide. Note that nuclides near mass number 50 have the largest binding energies per nucleon. Thus, heavy nuclei are expected to undergo fission to approach this mass number, whereas light nuclei are expected to undergo fusion.

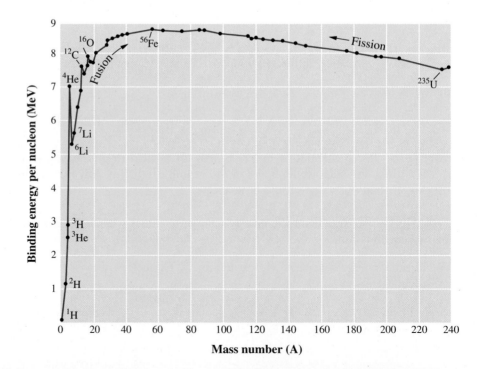

Nuclear fission is *a nuclear reaction in which a heavy nucleus splits into lighter nuclei and energy is released.* For example, californium-252 decays both by alpha emission (97%) and by spontaneous fission (3%). There are many possible ways in which the nucleus can split. One way is represented by the following equation:

$$^{252}_{98}\text{Cf} \longrightarrow {}^{142}_{56}\text{Ba} + {}^{106}_{42}\text{Mo} + 4{}^{1}_{0}\text{n}$$

In some cases, a nucleus can be induced to undergo fission by bombarding it with neutrons. An example is the nuclear fission of uranium-235. When a neutron strikes the $^{235}_{92}\text{U}$ nucleus, the nucleus splits into roughly equal parts, giving off several neutrons. Three possible splittings are shown in the following equations:

$$^{1}_{0}\text{n} + {}^{235}_{92}\text{U} \begin{cases} \longrightarrow {}^{142}_{54}\text{Xe} + {}^{90}_{38}\text{Sr} + 4{}^{1}_{0}\text{n} \\ \longrightarrow {}^{139}_{56}\text{Ba} + {}^{94}_{36}\text{Kr} + 3{}^{1}_{0}\text{n} \\ \longrightarrow {}^{144}_{55}\text{Cs} + {}^{90}_{37}\text{Rb} + 2{}^{1}_{0}\text{n} \end{cases}$$

Nuclear fission is discussed further in Section 20.7.

Nuclear fusion is *a nuclear reaction in which light nuclei combine to give a stabler, heavier nucleus plus possibly several neutrons, and energy is released.* An example of nuclear fusion is

$$^{2}_{1}\text{H} + {}^{3}_{1}\text{H} \longrightarrow {}^{4}_{2}\text{He} + {}^{1}_{0}\text{n}$$

Even though a nuclear reaction is energetically favorable, the reaction may be imperceptibly slow unless the proper conditions are present (see Section 20.7).

20.7 Nuclear Fission and Nuclear Fusion

We have seen that the stablest nuclei are those of intermediate size (with mass numbers around 50). Nuclear fission and nuclear fusion are reactions in which nuclei attain sizes closer to this intermediate range. In doing so, these reactions release tremendous amounts of energy. Nuclear fission of uranium-235 is employed in nuclear power plants to generate electricity. Nuclear fusion may supply us with energy in the future.

Nuclear Fission; Nuclear Reactors

Nuclear fission was discovered as a result of experiments to produce transuranium elements. Soon after the neutron was discovered in 1932, experimenters realized that this particle, being electrically neutral, should easily penetrate heavy nuclei. They began using neutrons in bombardment reactions, hoping to produce isotopes that would decay to new elements. In 1938, Otto Hahn, Lise Meitner, and Fritz Strassmann in Berlin identified barium in uranium samples that had been bombarded with neutrons. Soon afterward, the presence of barium was explained as a result of fission of the uranium-235 nucleus. When this nucleus is struck by a neutron, it splits into two nuclei. Fissions of uranium nuclei produce approximately 30 different elements of intermediate mass, including barium.

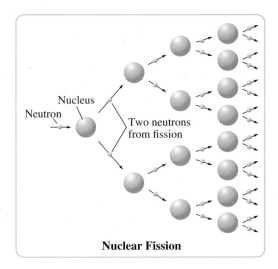

Nuclear Fission

When the uranium-235 nucleus splits, approximately two or three neutrons are released. If the neutrons from each nuclear fission are absorbed by other uranium-235 nuclei, these nuclei split and release even more neutrons. In this way, a chain reaction can occur. A nuclear **chain reaction** is *a self-sustaining series of nuclear fissions caused by the absorption of neutrons released from previous nuclear fissions.* The numbers of nuclei that split quickly multiply as a result of the absorption of neutrons released from previous nuclear fissions. Figure 20.19 shows how such a chain reaction occurs. The chain reaction of nuclear fissions is the basis of nuclear power and nuclear weapons.

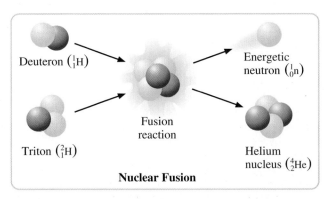

Deuteron ($^{1}_{1}\text{H}$)

Triton ($^{2}_{1}\text{H}$)

Fusion reaction

Energetic neutron ($^{1}_{0}\text{n}$)

Helium nucleus ($^{4}_{2}\text{He}$)

Nuclear Fusion

Figure 20.19 ▶

Representation of a chain reaction of nuclear fissions Each nuclear fission produces two or more neutrons, which can in turn cause more nuclear fissions. At each stage, a greater number of neutrons are produced, so the number of nuclear fissions multiplies quickly. Such a chain reaction is the basis of nuclear power and nuclear weapons. The original nucleus splits into pieces of varying mass number.

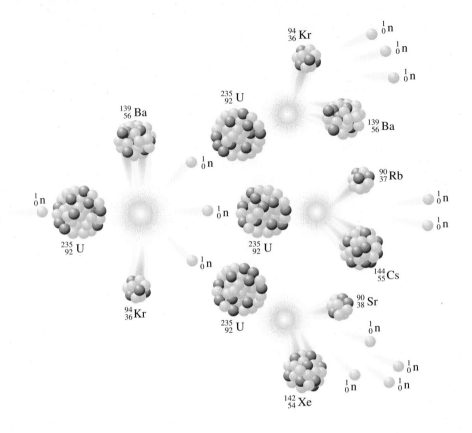

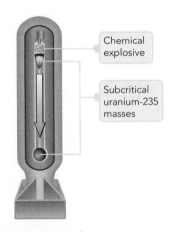

Figure 20.20 ▲

An atomic bomb In the gun-type bomb, one piece of uranium-235 of subcritical mass is hurled into another piece by a chemical explosive. The two pieces together give a supercritical mass, and a nuclear explosion results. Bombs using plutonium-239 require an implosion technique, in which wedges of plutonium arranged on a spherical surface are pushed into the center of the sphere by a chemical explosive, where a supercritical mass of plutonium results in a nuclear explosion.

To sustain a chain reaction in a sample of fissionable material, a nucleus that splits must give an average of one neutron that results in the fission of another nucleus, and so on. If the sample is too small, many of the neutrons leave the sample before they have a chance to be absorbed. There is thus a **critical mass** for a particular fissionable material, which is *the smallest mass of fissionable material in which a chain reaction can be sustained.* If the mass is much larger than this (a *supercritical* mass), the numbers of nuclei that split multiply rapidly. An atomic bomb is detonated with a small amount of chemical explosive that pushes together two or more masses of fissionable material to get a supercritical mass (Figure 20.20). A rapid chain reaction results in the splitting of most of the fissionable nuclei—and in the explosive release of an enormous amount of energy.

A **nuclear fission reactor** is *a device that permits a controlled chain reaction of nuclear fissions.* In power plants, a nuclear reactor is used to produce heat, which in turn is used to produce steam to drive an electric generator. A nuclear reactor consists of fuel rods alternating with control rods contained within a vessel. The **fuel rods** are *the cylinders that contain fissionable material.* In the light-water (ordinary water) reactors commonly used in the United States (see Figure 20.21), these fuel rods contain uranium dioxide pellets in a zirconium alloy tube. Natural uranium contains only 0.72% uranium-235, which is the isotope that undergoes fission. The uranium used for fuel in these reactors is "enriched" so that it contains about 3% of the uranium-235 isotope. **Control rods** are *cylinders composed of substances that absorb neutrons, such as boron and cadmium, and can therefore slow the chain reaction.* By varying the depth of the control rods within the fuel-rod assembly (reactor core), one can increase or decrease the absorption of neutrons. If necessary, these rods can be dropped all the way into the fuel-rod assembly to stop the chain reaction.

A **moderator**, *a substance that slows down neutrons,* is required if uranium-235 is the fuel and this isotope is present as a small fraction of the total fuel. The neutrons that are released by the splitting of uranium-235 nuclei are absorbed more readily by uranium-238 than by other uranium-235 nuclei. However, when the neutrons are slowed down by a moderator, they are more readily absorbed by

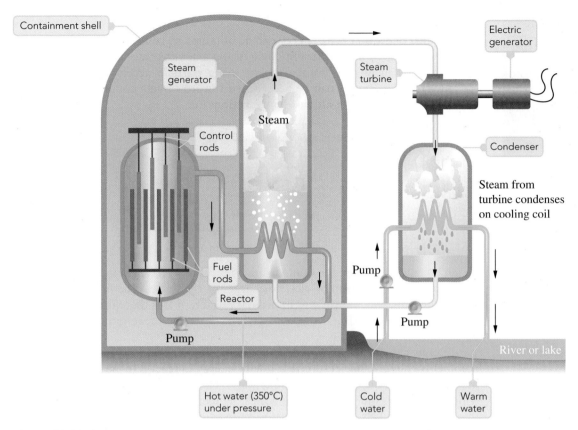

Figure 20.21 ▲

Light-water nuclear reactor The nuclear reactor consists of fuel rods with interspersed control rods. By raising or lowering the control rods, an operator can increase or decrease the rate of energy release from the fuel rods. Heat from the fuel rods raises the temperature of the liquid water in the reactor. A pump circulates the hot water to a steam generator, and the resulting steam passes through a turbine that operates an electric generator. The steam leaves the turbine and goes into the condenser, where it liquefies on the cooling coil. A nearby river or lake provides the cold water for the condenser.

uranium-235, so it is possible to sustain a chain reaction with low fractional abundance of this isotope. Commonly used moderators are heavy water ($_1^2H_2O$), light water, and graphite.

In the light-water reactor, ordinary water acts as both a moderator and a coolant. Figure 20.21 shows a pressurized-water design of this type of reactor. Water in the reactor is maintained at about 350°C under high pressure (150 atm) so it does not boil. The hot water is circulated to a heat exchanger, where the heat is used to produce steam to run a turbine and generate electricity.

After a period of time, fission products that absorb neutrons accumulate in the fuel rods. This interferes with the chain reaction, so eventually the fuel rods must be replaced. Originally, the intention was to send these fuel rods to *reprocessing plants,* where fuel material could be chemically separated from the radioactive wastes. Opposition to constructing these plants has been intense, however. Plutonium-239 would be one of the fuel materials separated from the spent fuel rods. This isotope is produced during the operation of the reactor when uranium-238 is bombarded with neutrons. It is fissionable and can be used to construct atomic bombs. For this reason, many people believe that the availability of this element in large quantities would increase the chance that many countries and terrorist groups could divert enough plutonium to produce atomic bombs.

Whether or not the spent fuel rods are reprocessed, a pressing problem facing the nuclear power industry is how to dispose of radioactive wastes safely. One of many proposals is to encase the waste in a ceramic material and store the solids deep in the earth, perhaps in salt mines.

© Princeton Plasma Physics Laboratory

Figure 20.22 ▲

Torus Fusion Reactor The National Spherical Torus Experiment at the Princeton Plasma Physics Laboratory uses this reactor to test an advanced magnetic field configuration to confine a compact, spherically shaped plasma.

Natural hydrogen contains 0.015% deuterium.

Even though the plasma is nearly 100 million °C above the melting point of any material, the total quantity of heat that could be transferred from the plasma is very small, because its concentration is extremely low. Therefore, if the plasma were to touch the walls of the reactor, the plasma would cool but the walls would not melt.

Nuclear Fusion

As we noted in Section 20.6, energy can be obtained by combining light nuclei into a heavier nucleus by nuclear fusion. Such fusion reactions have been observed in the laboratory by means of bombardment using particle accelerators. Deuterons ($_1^2$H nuclei), for example, can be accelerated toward targets containing deuterium ($_1^2$H atoms) or tritium ($_1^3$H atoms). The reactions are

$$_1^2\text{H} + _1^2\text{H} \longrightarrow _2^3\text{He} + _0^1\text{n}$$

$$_1^2\text{H} + _1^3\text{H} \longrightarrow _2^4\text{He} + _0^1\text{n}$$

To get the nuclei to react, the bombarding nucleus must have enough kinetic energy to overcome the repulsion of electric charges of the nuclei. The first reaction uses only deuterium, which is present in ordinary water. ◀ It is therefore very attractive as a source of energy. But, as we will discuss, the second reaction is more likely to be used first.

Energy cannot be obtained in a practical way using particle accelerators. Another way to give nuclei sufficient kinetic energy to react is by heating the nuclear materials to a sufficiently high temperature. The reaction of deuterium, $_1^2$H, and tritium, $_1^3$H, turns out to require the lowest temperature of any fusion reaction. For this reason it is likely to be the first fusion reaction developed as an energy source. For practical purposes, the temperature will have to be about 100 million °C. At this temperature, all atoms have been stripped of their electrons, so a plasma results. A **plasma** is *an electrically neutral gas of ions and electrons.* At 100 million °C, the plasma is essentially separate nuclei and electrons. Thus, the development of nuclear fusion requires study of the properties of plasmas at high temperatures.

It is now believed that the energy of stars, including our sun, where extremely high temperatures exist, derives from nuclear fusion. The hydrogen bomb also employs nuclear fusion for its destructive power. High temperature is first attained by a fission bomb. This then ignites fusion reactions in surrounding material of deuterium and tritium.

The main problem in developing controlled nuclear fusion is how to heat a plasma to high temperature and maintain those temperatures. When a plasma touches any material whatsoever, heat is quickly conducted away, and the plasma temperature quickly falls. ◀ A *magnetic fusion reactor* uses a magnetic field to hold the plasma away from any material (see Figure 20.22). A *laser fusion reactor* employs a bank of lasers aimed at a single point. Pellets containing deuterium and tritium would drop into the reactor, where they would be heated to 100 million °C by bursts of laser light.

A Checklist for Review

⊙WL and **GO Chemistry** Sign in at **www.cengage.com/owl** to:

- View tutorials and simulations, develop problem-solving skills, and complete online homework assigned by your professor.
- For quick review and exam prep, download Go Chemistry mini lecture modules from OWL or purchase them at **www.cengagebrain.com**.

Summary of Facts and Concepts

Nuclear reactions are of two types, *radioactive decay* and *nuclear bombardment*. Such reactions are represented by nuclear equations, each nucleus being denoted by a nuclide symbol. The equations must be balanced in charge (subscripts) and in nucleons (superscripts).

According to the *nuclear shell model*, the nucleons are arranged in shells. *Magic numbers* are the numbers of nucleons in a completed shell of protons or neutrons. Nuclei with magic numbers of protons or neutrons are especially stable. Pairs of protons and pairs of neutrons are also especially stable. When placed on a plot of N versus Z, stable nuclei fall in a *band of stability*. Those radioactive nuclides that fall to the left of the band of stability in this plot usually decay by beta emission. Those radioactive nuclides that fall to the right of the band of stability usually decay by positron emission or electron capture. However, nuclides with $Z > 83$ often decay by alpha emission. Uranium-238 forms a *radioactive decay series*. In such series, one element decays to another, which decays to another, and so forth, until a stable isotope is reached (lead-206, in the case of the uranium-238 series).

Transmutation of elements has been carried out in the laboratory by bombarding nuclei with various atomic particles. Alpha particles from natural sources can be used as reactants with light nuclei. For heavier nuclei, positive ions such as alpha particles must first be accelerated in a *particle accelerator*. Many of the *transuranium elements* have been obtained by bombardment of elements with accelerated particles. For example, plutonium was first made by bombarding uranium-238 with deuterons (^2_1H nuclei) from a *cyclotron*, a type of particle accelerator.

Particles of radiation from nuclear processes can be counted by Geiger counters or scintillation counters. In a *Geiger counter*, the particle ionizes a gas, which then conducts a pulse of electricity between two electrodes. In a *scintillation counter*, the particle hits a phosphor, and this emits a flash of light that is detected by a photomultiplier tube. The activity of a radioactive source, or the number of nuclear disintegrations per unit time, is measured in units of *curies* (3.700×10^{10} disintegrations per second).

Radiation affects biological organisms by breaking chemical bonds. The *rad* is the measure of radiation dosage that deposits 1×10^{-2} J of energy per kilogram of tissue. A *rem* equals the number of rads times a factor to account for the relative biological effectiveness of the radiation.

Radioactive decay is a first-order rate process. The rate is characterized by the *decay constant, k,* or by the *half-life, $t_{1/2}$*. The quantities k and $t_{1/2}$ are related. Knowing one or the other, you can calculate how long it will take for a given radioactive sample to decay by a certain fraction. Methods of *radioactive dating* depend on determining the fraction of a radioactive isotope that has decayed and, from this, the time that has elapsed.

Radioactive isotopes are used as *radioactive tracers* in chemical analysis and medicine. *Isotope dilution* is one application of radioactive tracers in which the dilution of the tracer can be related to the original quantity of nonradioactive isotope. *Neutron activation analysis* is a method of analysis that depends on the conversion of elements to radioactive isotopes by neutron bombardment.

According to Einstein's *mass–energy equivalence*, mass is related to energy by the equation $E = mc^2$. A nucleus has less mass than the sum of the masses of the separate nucleons. The positive value of this mass difference is called the *mass defect;* it is equivalent to the *binding energy* of the nucleus. Nuclides having mass numbers near 50 have the largest binding energies per nucleon. It follows that heavy nuclei should tend to split, a process called *nuclear fission,* and that light nuclei should tend to combine, a process called *nuclear fusion*. Tremendous amounts of energy are released in both processes. Nuclear fission is used in conventional nuclear power reactors. Nuclear fusion reactors are in the experimental stage.

Learning Objectives

20.1 Radioactivity

- Define *radioactive decay* and *nuclear bombardment reaction.*
- Learn the nuclear symbols for *positron, gamma photon, electron, neutron,* and *proton.*
- Write a nuclear equation. Example 20.1
- Deduce a product or reactant in a nuclear equation. Example 20.2
- Describe the *shell model of the nucleus.*
- Explain the *band of stability.*
- Predict the relative stabilities of nuclides. Example 20.3
- List the six types of radioactive decay.
- Predict the type of radioactive decay. Example 20.4
- Define *radioactive decay series.*

20.2 Nuclear Bombardment Reactions

- Define *transmutation.*
- Use the notation for a bombardment reaction. Example 20.5
- Locate the *transuranium elements* on the periodic table.
- Determine the product nucleus in a nuclear bombardment reaction. Example 20.6

Important Terms

radioactive decay
nuclear bombardment reaction
nuclear equation
positron
gamma photon
nuclear force
shell model of the nucleus
magic number
band of stability
alpha emission
beta emission
positron emission
electron capture
gamma emission
metastable nucleus
spontaneous fission
radioactive decay series

transmutation
particle accelerator
electron volt (eV)
cyclotron
deuterons
transuranium elements

20.3 Radiations and Matter: Detection and Biological Effects

- State the purposes of a *Geiger counter* and a *scintillation counter*.
- Define *activity of a radioactive source* and *curie (Ci)*.
- State the relationship between a *rad* and a *rem*.

Geiger counter
scintillation counter
activity of a radioactive source
curie (Ci)
rad
rem

20.4 Rate of Radioactive Decay

- Define *radioactive decay constant.*
- Calculate the decay constant from activity. Example 20.7
- Define *half-life.*
- Draw a typical half-life decay curve of a radioactive element.
- Calculate the half-life from the decay constant. Example 20.8
- Calculate the decay constant and activity from half-life. Example 20.9
- Determine the fraction of nuclei remaining after a specified time. Example 20.10
- Apply the carbon-14 dating method. Example 20.11

radioactive decay constant
half-life

20.5 Applications of Radioactive Isotopes

- State the ways in which radioactive isotopes are used for chemical analysis.
- Describe how isotopes are used for medical therapy and diagnosis.

radioactive tracer
isotope dilution
neutron activation analysis

20.6 Mass–Energy Calculations

- Calculate the energy change for a nuclear reaction. Example 20.12
- Define *nuclear binding energy* and *mass defect.*
- Compare and contrast *nuclear fission* and *nuclear fusion.*

binding energy
mass defect
nuclear fission
nuclear fusion

20.7 Nuclear Fission and Nuclear Fusion

- Explain how a controlled *chain reaction* is applied in a *nuclear fission reactor* using a *critical mass* of fissionable material.
- Write the reaction of the nuclear fusion of deuterium and tritium.

chain reaction
critical mass
nuclear fission reactor
fuel rods
control rods
moderator
plasma

Key Equations

$$\text{Rate} = kN_t$$

$$\ln\frac{N_t}{N_0} = -kt$$

$$t_{1/2} = \frac{0.693}{k}$$

$$\Delta E = (\Delta m)c^2$$

Questions and Problems

OWL Interactive versions of these problems may be assigned in OWL.

Self-Assessment and Review Questions

Key: These questions test your understanding of the ideas you worked with in the chapter. These problems vary in difficulty and often can be used for the basis of discussion.

20.1 What are the two types of nuclear reactions? Give an example of a nuclear equation for each type.

20.2 What are *magic numbers*? Give several examples of nuclei with magic numbers of protons.

20.3 List characteristics to look for in a nucleus to predict whether it is stable.

20.4 What are the six common types of radioactive decay? What condition usually leads to each type of decay?

20.5 What are the isotopes that begin each of the naturally occurring radioactive decay series?

20.6 Give equations for **a** the first transmutation of an element obtained in the laboratory by nuclear bombardment, and for **b** the reaction that produced the first artificial radioactive isotope.

20.7 What is a particle accelerator, and how does one operate? Why are they required for certain nuclear reactions?

20.8 In what major way has the discovery of transuranium elements affected the form of modern periodic tables?

20.9 Describe how a Geiger counter works. How does a scintillation counter work?

20.10 Define the units *curie, rad,* and *rem.*

20.11 The half-life of cesium-137 is 30.2 y. How long will it take for a sample of cesium-137 to decay to 1/8 of its original mass?

20.12 What is the age of a rock that contains equal numbers of $^{40}_{19}K$ and $^{40}_{18}Ar$ nuclei? The half-life of $^{40}_{19}K$ is 1.28×10^9 y.

20.13 What is a radioactive tracer? Give an example of the use of such a tracer in chemistry.

20.14 Isotope dilution has been used to obtain the volume of blood supply in a living animal. Explain how this could be done.

20.15 Briefly describe neutron activation analysis.

20.16 The deuteron, ^2_1H, has a mass that is smaller than the sum of the masses of its constituents, the proton plus the neutron. Explain why this is so.

20.17 Certain stars obtain their energy from nuclear reactions such as

$$^{12}_6C + {}^{12}_6C \longrightarrow {}^{23}_{11}Na + {}^1_1H$$

Explain in a sentence or two why this reaction might be expected to release energy.

20.18 Briefly describe how a nuclear fission reactor operates.

20.19 Polonium-216 decays to lead-212 by emission of an alpha particle. Which of the following is the nuclear equation for this radioactive decay?

a $^{216}_{84}Po \longrightarrow {}^{212}_{82}Pb + {}^0_{-1}e$

b $^{216}_{84}Po \longrightarrow {}^{212}_{82}Pb + {}^4_2He$

c $^{216}_{84}Po + {}^4_2He \longrightarrow {}^{212}_{82}Pb$

d $^{216}_{84}Po \longrightarrow {}^{212}_{82}Pb + {}^0_1e$

e $^{216}_{84}Po + 2{}^0_{-1}e \longrightarrow {}^{212}_{82}Pb + 4{}^1_0n$

20.20 The half-life for the decay of carbon-14 is 5730 years. Present-day carbon (in living matter) gives 15.3 disintegrations of ^{14}C per minute per gram of total carbon. An archaeological sample containing carbon was dated by analysis of its radioactive carbon. The activity of the carbon from the sample was 5.4 disintegrations of ^{14}C per minute per gram of total carbon. What was the age of the sample?

a 2.4×10^{-5} y

b 1.2×10^4 y

c 3.7×10^3 y

d 7.2×10^3 y

e 8.6×10^3 y

20.21 A radioactive isotope has a half-life of 56.6 days. What fraction of the isotope remains after 449 days?

a 3.2×10^{-6}

b 1.00

c 0.92

d 4.1×10^{-3}

e 0.83

20.22 Neutron activation analysis is an analytical technique in which a sample of material is bombarded with neutrons from a fission reactor. When a 35.0-g aluminum can is irradiated, it has an initial activity of 40.0 curies (Ci). The safety office won't let you touch anything having an activity in excess of 0.100 Ci. Assuming all the radioactivity is from ^{28}Al, which has a half-life of 2.28 min, how many minutes do you have to wait after bombardment before you can handle the can safely?

a 17.0 min

b 1.32×10^3 min

c 19.7 min

d 478 min

e 476 min

Conceptual Problems

Key: These problems are designed to check your understanding of the concepts associated with some of the main topics presented in each chapter. A strong conceptual understanding of chemistry is the foundation for both applying chemical knowledge and solving chemical problems. These problems vary in level of difficulty and often can be used as a basis for group discussion.

20.23 When considering the lifetime of a radioactive species, a general rule of thumb is that after 10 half-lives have passed, the amount of radioactive material left in the sample is negligible. The disposal of some radioactive materials is based on this rule.

a What percentage of the original material is left after 10 half-lives?

b When would it be a bad idea to apply this rule?

20.24 Use drawings to complete the following nuclear reactions (yellow circles represent neutrons and blue circles represent protons). Once you have completed the drawings, write the nuclide symbols under each reaction.

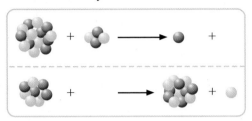

20.25 Sodium has only one naturally occurring isotope, sodium-23. Using the data presented in Table 20.3, explain how the molecular mass of sodium is 22.98976 amu and not the sum of the masses of the protons, neutrons, and electrons.

20.26 Identify each of the following reactions as fission, fusion, a transmutation, or radioactive decay.

a $4^1_1H \longrightarrow {}^4_2He + 2^0_1e$
b ${}^{14}_6C \longrightarrow {}^{14}_7N + {}^0_{-1}e$
c ${}^1_0n + {}^{235}_{92}U \longrightarrow {}^{140}_{56}Ba + {}^{93}_{36}Kr + 3^1_0e$
d ${}^{14}_7N + {}^4_2He \longrightarrow {}^{17}_8O + {}^1_1H$

20.27 A radioactive sample with a half-life of 10 minutes is placed in a container.

a Complete the pictures below depicting the amount of this sample at the beginning of the experiment ($t = 0$ min) and 30 minutes into the experiment ($t = 30$ min). Each sphere represents one radioactive nuclide of the sample.

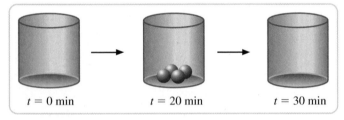

b A friend working with 1000 atoms of the same nuclide represents the amount of nuclide at three time intervals in the following manner.

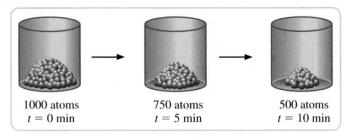

Is his drawing correct? If not, explain where and why it is incorrect.

20.28 You have a mixture that contains 10 g of Pu-239 with a half-life of 2.4×10^4 years and 10 g of Np-239 with a half-life of 2.4 days. Estimate how much time must elapse before the quantity of radioactive material is reduced by 50%.

20.29 Come up with an explanation as to why α radiation is easily blocked by materials such as a piece of wood, whereas γ radiation easily passes through those same materials.

20.30 You have an acquaintance who tells you that he is going to reduce his radiation exposure to zero. What examples could you present that would illustrate this to be an impossible goal?

20.31 In Chapter 7 (A Chemist Looks At: Zapping Hamburger with Gamma Rays) there is a discussion of how gamma radiation is used to kill bacteria in food. As indicated in the feature, there is concern on the part of some people that the irradiated food is radioactive. Why is this not the case? If you wanted to make the food radioactive, what would you have to do?

20.32 You have a pile of I-131 atoms with a half-life of 8 days. A portion of the solid I-131 is represented below. Can you predict how many half-lives will occur before the green I-131 atom undergoes decay?

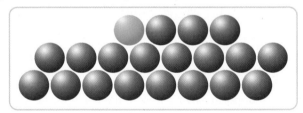

Practice Problems

Key: These problems are for practice in applying problem-solving skills. They are divided by topic, and some are keyed to exercises (see the ends of the exercises). The problems are arranged in matching pairs; the odd-numbered problem of each pair is listed first, and its answer is given in the back of the book.

Radioactivity

20.33 Rubidium-87, which forms about 28% of natural rubidium, is radioactive, decaying by the emission of a single beta particle to strontium-87. Write the nuclear equation for this decay of rubidium-87.

20.34 Write the nuclear equation for the decay of phosphorus-32 to sulfur-32 by beta emission. A phosphorus-32 nucleus emits a beta particle and gives a sulfur-32 nucleus.

20.35 Thorium is a naturally occurring radioactive element. Thorium-232 decays by emitting a single alpha particle to produce radium-228. Write the nuclear equation for this decay of thorium-232.

20.36 Radon is a radioactive noble gas formed in soil containing radium. Radium-226 decays by emitting a single alpha particle to produce radon-222. Write the nuclear equation for this decay of radium-226.

20.37 Fluorine-18 is an artificially produced radioactive isotope. It decays by emitting a single positron. Write the nuclear equation for this decay.

20.38 Scandium-41 is an artificially produced radioactive isotope. It decays by emitting a single positron. Write the nuclear equation for this decay.

20.39 Polonium was discovered in uranium ores by Marie and Pierre Curie. Polonium-210 decays by emitting a single alpha particle. Write the nuclear equation for this decay.

20.40 Actinium was discovered in uranium ore residues by André-Louis Debierne. Actinium-227 decays by emitting a single alpha particle. Write the nuclear equation for this decay.

20.41 From each of the following pairs, choose the nuclide that is radioactive. (One is known to be radioactive, the other stable.) Explain your choice.
[a] $^{122}_{51}Sb$, $^{136}_{54}Xe$ [b] $^{204}_{82}Pb$, $^{204}_{85}At$ [c] $^{87}_{37}Rb$, $^{80}_{37}Rb$

20.42 From each of the following pairs, choose the nuclide that is radioactive. (One is known to be radioactive, the other stable.) Explain your choice.
[a] $^{102}_{47}Ag$, $^{109}_{47}Ag$ [b] $^{25}_{12}Mg$, $^{24}_{10}Ne$ [c] $^{203}_{81}Tl$, $^{223}_{90}Th$

20.43 Predict the type of radioactive decay process that is likely for each of the following nuclides.
[a] $^{228}_{92}U$ [b] $^{8}_{5}B$ [c] $^{68}_{29}Cu$

20.44 Predict the type of radioactive decay process that is likely for each of the following nuclides.
[a] $^{60}_{30}Zn$ [b] $^{10}_{6}C$ [c] $^{241}_{93}Np$

20.45 Four radioactive decay series are known—three naturally occurring, and one beginning with the synthetic isotope $^{241}_{94}Pu$. To which of these decay series does the isotope $^{219}_{86}Rn$ belong? To which series does $^{220}_{86}Rn$ belong? Each isotope in these series decays by either alpha emission or beta emission. How do these decay processes affect the mass number?

20.46 Four radioactive decay series are known—three naturally occurring, and one beginning with the synthetic isotope $^{241}_{94}Pu$. To which of these decay series does the isotope $^{227}_{89}Ac$ belong? To which series does $^{225}_{89}Ac$ belong? Each isotope in these series decays by either alpha emission or beta emission. How do these decay processes affect the mass number?

Nuclear Bombardment Reactions

20.47 Write the abbreviated notations for the following bombardment reactions.
[a] $^{26}_{12}Mg + ^{2}_{1}H \longrightarrow ^{24}_{11}Na + ^{4}_{2}He$
[b] $^{16}_{8}O + ^{1}_{0}n \longrightarrow ^{16}_{7}N + ^{1}_{1}p$

20.48 Write abbreviated notations for the following bombardment reactions.
[a] $^{14}_{7}N + ^{1}_{0}n \longrightarrow ^{14}_{6}C + ^{1}_{1}H$
[b] $^{63}_{29}Cu + ^{4}_{2}He \longrightarrow ^{66}_{31}Ga + ^{1}_{0}n$

20.49 Write the nuclear equations for the following bombardment reactions.
[a] $^{27}_{13}Al(d, \alpha)^{25}_{12}Mg$ [b] $^{10}_{5}B(\alpha, p)^{13}_{6}C$

20.50 Write the nuclear equations for the following bombardment reactions.
[a] $^{45}_{21}Sc(n, \alpha)^{42}_{19}K$ [b] $^{63}_{29}Cu(p, n)^{63}_{30}Zn$

20.51 A proton is accelerated to 12.6 MeV per particle. What is this energy in kJ/mol?

20.52 An alpha particle is accelerated to 23.1 MeV per particle. What is this energy in kJ/mol?

20.53 Fill in the missing parts of the following reactions.
[a] $^{6}_{3}Li + ^{1}_{0}n \longrightarrow ? + ^{3}_{1}H$
[b] $^{232}_{90}Th(?, n)^{235}_{92}U$

20.54 Fill in the missing parts of the following reactions.
[a] $^{27}_{13}Al + ^{3}_{1}H \longrightarrow ^{27}_{12}Mg + ?$
[b] $^{12}_{6}C(^{3}_{1}H, ?)^{14}_{6}C$

20.55 Curium was first synthesized by bombarding an element with alpha particles. The products were curium-242 and a neutron. What was the target element?

20.56 Californium was first synthesized by bombarding an element with alpha particles. The products were californium-245 and a neutron. What was the target element?

Rate of Radioactive Decay

20.57 Tritium, or hydrogen-3, is prepared by bombarding lithium-6 with neutrons. A 0.250-mg sample of tritium decays at the rate of 8.94×10^{10} disintegrations per second. What is the decay constant (in /s) of tritium, whose atomic mass is 3.02 amu?

20.58 The first isotope of plutonium discovered was plutonium-238. It is used to power batteries for heart pacemakers. A sample of plutonium-238 weighing 2.8×10^{-6} g decays at the rate of 1.8×10^{6} disintegrations per second. What is the decay constant of plutonium-238 in reciprocal seconds (/s)?

20.59 Sulfur-35 is a radioactive isotope used in chemical and medical research. A 0.48-mg sample of sulfur-35 has an activity of 20.4 Ci. What is the decay constant of sulfur-35 (in /s)?

20.60 Sodium-24 is used in medicine to study the circulatory system. A sample weighing 5.2×10^{-6} g has an activity of 45.3 Ci. What is the decay constant of sodium-24 (in /s)?

20.61 Tellurium-123 is a radioactive isotope occurring in natural tellurium. The decay constant is 1.7×10^{-21}/s. What is the half-life in years?

20.62 Neptunium-237 was the first isotope of a transuranium element to be discovered. The decay constant is 1.03×10^{-14}/s. What is the half-life in years?

20.63 Carbon-14 has been used to study the mechanisms of reactions that involve organic compounds. The half-life of carbon-14 is 5.73×10^{3} y. What is the decay constant (in /s)?

20.64 Promethium-147 has been used in luminous paint for dials. The half-life of this isotope is 2.5 y. What is the decay constant (in /s)?

20.65 Gold-198 has a half-life of 2.69 d. What is the activity (in curies) of a 0.86-mg sample?

20.66 Cesium-134 has a half-life of 2.05 y. What is the activity (in curies) of a 0.50-mg sample?

20.67 A sample of a phosphorus compound contains phosphorus-32. This sample of radioactive isotope is decaying at the rate of 6.3×10^{12} disintegrations per second. How many grams of ^{32}P are in the sample? The half-life of ^{32}P is 14.3 d.

20.68 A sample of sodium thiosulfate, $Na_2S_2O_3$, contains sulfur-35. Determine the mass of ^{35}S in the sample from the decay rate, which was determined to be 7.3×10^{11} disintegrations per second. The half-life of ^{35}S is 88 d.

20.69 A sample of sodium-24 was administered to a patient to test for faulty blood circulation by comparing the radioactivity reaching various parts of the body. What fraction of the sodium-24 nuclei would remain undecayed after 12.0 h? The half-life is 15.0 h. If a sample contains 6.0 μg of ^{24}Na, how many micrograms remain after 12.0 h?

20.70 A solution of sodium iodide containing iodine-131 was given to a patient to test for malfunctioning of the thyroid gland. What fraction of the iodine-131 nuclei would remain undecayed after 5.0 d? If a sample contains 2.5 μg of ^{131}I, how many micrograms remain after 5.0 d? The half-life of I-131 is 8.07 d.

20.71 If 28.0% of a sample of nitrogen-17 decays in 1.97 s, what is the half-life of this isotope (in seconds)?

20.72 If 20.0% of a sample of zinc-65 decays in 69.9 d, what is the half-life of this isotope (in days)?

20.73 A sample of iron-59 initially registers 125 counts per second on a radiation counter. After 10.0 d, the sample registers 107 counts per second. What is the half-life (in days) of iron-59?

20.74 A sample of silver-112 gives a reading of 875 counts per hour on a radiation counter. After 0.25 h, the sample gives a reading of 400 counts per hour. What is the half-life (in hours) of silver-112?

20.75 Carbon from a cypress beam obtained from the tomb of Sneferu, a king of ancient Egypt, gave 8.1 disintegrations of ^{14}C per minute per gram of carbon. How old is the cypress beam? Carbon from living material gives 15.3 disintegrations of ^{14}C per minute per gram of carbon.

20.76 Carbon from the Dead Sea Scrolls, very old manuscripts found in Israel, gave 12.1 disintegrations of ^{14}C per minute per gram of carbon. How old are the manuscripts? Carbon from living material gives 15.3 disintegrations of ^{14}C per minute per gram of carbon.

20.77 Several hundred pairs of sandals found in a cave in Oregon were found by carbon-14 dating to be 9.0×10^3 years old. What must have been the activity of the carbon-14 in the sandals in disintegrations per minute per gram? Assume the original activity was 15.3 disintegrations per minute per gram.

20.78 Some mammoth bones found in Arizona were found by carbon-14 dating to be 1.13×10^4 years old. What must have been the activity of the carbon-14 in the bones in disintegrations per minute per gram? Assume the original activity was 15.3 disintegrations per minute per gram.

Mass–Energy Equivalence

20.79 Find the change of mass (in grams) resulting from the release of heat when 1 mol C reacts with 1 mol O_2.

$$C(s) + O_2(g) \longrightarrow CO_2(g); \Delta H = -393.5 \text{ kJ}$$

20.80 Find the change of mass (in grams) resulting from the release of heat when 1 mol SO_2 is formed from the elements.

$$S(s) + O_2(g) \longrightarrow SO_2(g); \Delta H = -297 \text{ kJ}$$

20.81 Calculate the energy change for the following nuclear reaction (in joules per mole of ^2_1H).

$$^2_1H + ^3_1H \longrightarrow ^4_2He + ^1_0n$$

Give the energy change in MeV per ^2_1H nucleus. See Table 20.3.

20.82 Calculate the change in energy, in joules per mole of ^1_1H, for the following nuclear reaction.

$$^1_1H + ^1_1H \longrightarrow ^2_1H + ^0_1e$$

Give the energy change in MeV per ^1_1H nucleus. See Table 20.3.

20.83 Obtain the mass defect (in amu) and binding energy (in MeV) for the ^6_3Li nucleus. What is the binding energy (in MeV) per nucleon? See Table 20.3.

20.84 Obtain the mass defect (in amu) and binding energy (in MeV) for the $^{58}_{28}Ni$ nucleus. What is the binding energy (in MeV) per nucleon? See Table 20.3.

General Problems

Key: These problems provide more practice but are not divided by topic or keyed to exercises. Each section ends with essay questions, each of which is color added to refer to the *A Chemist Looks at Frontiers* (purple) or *A Chemist Looks at Life Science* (pink) chapter essay on which it is based. Odd-numbered problems and the even-numbered problems that follow are similar; answers to all odd-numbered problems except the essay questions are given in the back of the book.

20.85 Sodium-23 is the only stable isotope of sodium. Predict how sodium-20 will decay and how sodium-26 will decay.

20.86 Aluminum-27 is the only stable isotope of aluminum. Predict how aluminum-24 will decay and how aluminum-30 will decay.

20.87 A uranium-235 nucleus decays by a series of alpha and beta emissions until it reaches lead-207. How many alpha emissions and how many beta emissions occur in this series of decays?

20.88 A thorium-232 nucleus decays by a series of alpha and beta emissions until it reaches lead-208. How many

alpha emissions and how many beta emissions occur in this series of decays?

20.89 A bismuth-209 nucleus reacts with an alpha particle to produce an astatine nucleus and two neutrons. Write the complete nuclear equation for this reaction.

20.90 A bismuth-209 nucleus reacts with a deuteron to produce a polonium nucleus and a neutron. Write the complete nuclear equation for this reaction.

20.91 Complete the following equation by filling in the blank.

$$^{238}_{92}U + ^{12}_{6}C \longrightarrow \underline{\quad} + 4^1_0n$$

20.92 Complete the following equation by filling in the blank.

$$^{246}_{96}Cm + ^{12}_{6}C \longrightarrow \underline{\quad} + 4^1_0n$$

20.93 Tritium, or hydrogen-3, is formed in the upper atmosphere by cosmic rays, similar to the formation of carbon-14. Tritium has been used to determine the age of wines. A certain wine that has been aged in a bottle has a tritium content only 70% of that in a similar wine of the same mass that has just been bottled. How long has the aged wine been in the bottle? The half-life of tritium is 12.3 y.

20.94 The naturally occurring isotope rubidium-87 decays by beta emission to strontium-87. This decay is the basis of a method for determining the ages of rocks. A sample of rock contains 102.1 μg ^{87}Rb and 5.0 μg ^{87}Sr. What is the age of the rock? The half-life of rubidium-87 is 4.8×10^{10} y.

20.95 When a positron and an electron collide, they are annihilated and two gamma photons of equal energy are emitted. Calculate the wavelength corresponding to this gamma emission.

20.96 When technetium-99m decays to technetium-99, a gamma photon corresponding to an energy of 0.143 MeV is emitted. What is the wavelength of this gamma emission? What is the difference in mass between Tc-99m and Tc-99?

20.97 Calculate the energy released when 5.00 kg of uranium-235 undergoes the following fission process.

$$^{1}_{0}n + {}^{235}_{92}U \longrightarrow {}^{136}_{53}I + {}^{96}_{39}Y + 4{}^{1}_{0}n$$

The masses of $^{136}_{53}$I and $^{96}_{39}$Y nuclei are 135.8401 amu and 95.8629 amu, respectively. Other masses are given in Table 20.3. Compare this energy with the heat released when 5.00 kg C(graphite) burns to $CO_2(g)$.

20.98 Calculate the energy released when 1.00 kg of hydrogen-1 undergoes fusion to helium-4, according to the following reaction.

$$4{}^{1}_{1}H \longrightarrow {}^{4}_{2}He + 2{}^{0}_{1}e$$

This reaction is one of the principal sources of energy from the sun. Compare the energy released by 1.00 kg of $^{1}_{1}$H in this reaction to the heat released when 1.00 kg of C(graphite) burns to $CO_2(g)$. See Table 20.3 for data.

20.99 The half-life of calcium-47 is 4.536 days and it decays by the emission of a beta particle.

  a Write a balanced equation for the decay of Ca-47.

  b If 10.0 μg of Ca-47 is needed for an experiment, what mass of ^{47}CaSO$_4$ must be ordered if it takes 48 h for it to arrive from the supplier?

20.100 The radioactive isotope phosphorus-32 is often used in biochemical research. Its half-life is 14.28 days and it decays by beta emission.

  a Write a balanced equation for the decomposition of P-32.

  b If the original sample were 275 mg of K$_3$^{32}PO$_4$, what amount remains after 37.0 d? What percent of the sample has undergone decay?

20.101 The half-life of $^{82}_{35}$Br is 1.471 days. This isotope decays by the emission of a beta particle.

  a Gaseous HBr is made with Br-82. When the bromine isotope decays, the HBr produces H$_2$ and the bromine decay product. Write a balanced equation for the decay of Br-82. Now, write a balanced equation for the decomposition of H^{82}Br.

  b If a pure sample of 0.0150 mol of HBr made entirely with Br-82 is placed in an evacuated 1.00-L flask, how much HBr remains after 10.0 h? If the temperature is 22°C, what is the pressure in the flask?

20.102 The half-life of $^{132}_{52}$Te is 3.26 days. This isotope decays by beta emission to a highly unstable intermediate that decays rapidly to a stable product by beta emission.

  a What is the ultimate product obtained from Te-132? Write a balanced equation for this reaction.

  b Gaseous H$_2$Te is made with Te-132. When the tellurium isotope decays, the H$_2$Te produces H$_2$ and the ultimate decay product of Te-132. Write a balanced equation for the formation of stable products from H$_2$^{132}Te.

  c If a pure sample of 0.0125 mol of H$_2$Te made entirely of Te-132 is placed in an evacuated 1.50-L flask, how much H$_2$Te remains after 93.0 h? If the temperature is 25°C, what is the pressure in the flask?

■ **20.103** How would you describe the structure of the alpha particle in terms of the shell model? Note that the protons and neutrons have their own shell structures. Why would you expect the alpha particle to be especially stable?

■ **20.104** What are the next three energy levels following the 1f level (before spin-orbit coupling is accounted for)?

■ **20.105** What is the purpose of a PET scan? What type of substance is administered to a person undergoing a PET scan?

■ **20.106** The instrument for PET scans actually detects gamma rays. However, the radioactive substance emits another particle. Describe what is happening.

Strategy Problems

Key: As noted earlier, all of the practice and general problems are matched pairs. This section is a selection of problems that are not in matched-pair format. These challenging problems require that you employ many of the concepts and strategies that were developed in the chapter. In some cases, you will have to integrate several concepts and operational skills in order to solve the problem successfully.

20.107 Complete the following nuclear reactions.

  a $^{31}_{14}$Si $\longrightarrow$ $^{31}_{15}$P + ?

  b $^{44}_{22}$Ti + ? $\longrightarrow$ $^{44}_{21}$Sc

  c $^{252}_{98}$Cf $\longrightarrow$ $^{142}_{56}$Ba + ? + 4$^{1}_{0}$n

20.108 What nuclide is formed when americium-241 undergoes alpha decay?

20.109 Radioisotope thermoelectric generators can be used by satellites to obtain power from radioactive decay of various isotopes, plutonium-238 being the preferred fuel. Plutonium-238 decays via alpha emission and has a half-life of 87.7 years.

  a Write the nuclear equation for the alpha decay of plutonium-238.

  b If you were to start with 250.0 g of Pu-238 to power a satellite, what mass would remain after 45 years?

20.110 Calculate the energy released, in joules per mole, when uranium-238 undergoes alpha decay. See Table 20.3.

20.111 Consider Rn-222.

  a Predict the type of radioactive decay that Rn-222 is most likely to undergo.

  b Write the equation for the decay process.

  c The half-life of Rn-222 is 3.82 days. If you start with a 150.0-g sample of Rn-222, how much remains after 15.0 days?

20.112 The gold-198 isotope is used in the treatment of brain, prostate, and ovarian cancer. Au-198 has a half-life of 2.69 d. If a hospital needs to have 25 mg of Au-198 on hand for treatments on a particular day, and shipping takes 72 h, what mass of Au-198 needs to be ordered?

20.113 The bromine isotope Br-75 is used for imaging in positron emission tomography. Br-75 has a half-life of 57 h. How much time must elapse for a given dose of Br-75 to drop to 25% of its initial amount?

20.114 The decay of Rb-87 ($t_{1/2} = 4.8 \times 10^{10}$ y) to Sr-87 has been used to determine the age of ancient rocks and minerals.

- a. Write the balanced nuclear equation for this decay.
- b. If a sample of rock is found to be 0.100% by mass Rb-87 and 0.00250% by mass Sr-87, what is the age of the rock? Assume that there was no Sr-87 present when the rock formed.

20.115 You have two piles of different unknown radioactive substances: pile A with a mass of 200 g, and pile B with a mass of 100 g. Would it be possible for these two piles to have the same rate of radioactive decay? Explain.

20.116 Cobalt-60 has a half-life of 5.26 y. Gamma radiation from this isotope is used to treat malignant tumors. What mass of cobalt-60 is required to generate a 500-Ci gamma-ray source?

20.117 Consider a 1.0-g sample of carbon.

- a. Calculate the amount of energy released (kJ) if the entire mass of the carbon sample could be converted to energy.
- b. Using data from Appendix C, calculate the amount of energy released when the sample of graphite is combusted.
- c. If you had a sample of coal that was pure graphite, what mass of coal would need to be combusted in order to produce an equivalent energy to that calculated in part a?

20.118 Samarium-153 is used in a drug to treat bone disorders. Sm has an activity of 8.6 Ci per mg.

- a. What is the rate constant of Sm-153?

- b. What is the half-life of Sm-153.
- c. In the United States, pharmaceutical companies can purchase Sm-153 from the National Isotope Development Center at Oak Ridge National Laboratory. A company needs a 500.0-mg sample of Sm-153 to manufacture a bone treatment drug. However, if production is expected to be delayed by 8 weeks after the company has obtained the Sm-153, what mass of Sm-153 will it need to purchase?

20.119 Radon-222 gas can be found seeping from granite that contains uranium-238. Radon-222 is a nuclide in the radioactive decay series of uranium-238. Radon is an element with a half-life of 3.82 days.

- a. Predict the most likely particle emitted when radon-222 undergoes nuclear decay and write the nuclear equation for the decay.
- b. What is the decay constant (k) for radon-222.
- c. The U.S. Environmental Protection Agency recommends that the concentration of radon gas in dwellings not exceed 4 piCi per liter of air. What mass of Rn-222 would be in a 1-L sample of air that had a decay rate of 4 piCi?

20.120 Plutonium-239 is a fissionable material used in the construction of nuclear weapons and has a half-life of 24,100 years. Approximately 2×10^4 kg of plutonium is produced per year. How long would it take for 99.99% of the plutonium produced in one year to decay?

20.121 Californium-251 was synthesized at the University of California in 1950. What type of decay would you predict for Ca-251? Write the balanced nuclear reaction for this decay.

20.122 Sodium has only one naturally occurring isotope that is not radioactive.

- a. What is the mass defect of sodium?
- b. What is the nuclear binding energy (MeV) of a sodium nucleon?

Cumulative-Skills Problems

Key: The problems under this heading combine skills introduced in previous chapters with those given in the current one.

20.123 A sample of sodium phosphate, Na_3PO_4, weighing 54.5 mg contains radioactive phosphorus-32 (with mass 32.0 amu). If 15.6% of the phosphorus atoms in the compound is phosphorus-32 (the remainder is naturally occurring phosphorus), how many disintegrations of this nucleus occur per second in this sample? Phosphorus-32 has a half-life of 14.3 d.

20.124 A sample of sodium thiosulfate, $Na_2S_2O_3$, weighing 43.6 mg contains radioactive sulfur-35 (with mass 35.0 amu). If 22.3% of the sulfur atoms in the compound is sulfur-35 (the remainder is naturally occurring sulfur), how many disintegrations of this nucleus occur per second in this sample? Sulfur-35 has a half-life of 87.9 d.

20.125 Polonium-210 has a half-life of 138.4 days, decaying by alpha emission. Suppose the helium gas originating from the alpha particles in this decay were collected. What volume of helium at 25°C and 735 mmHg could be

obtained from 1.0000 g of polonium dioxide, PoO_2, in a period of 48.0 h?

20.126 Radium-226 decays by alpha emission to radon-222, a noble gas. What volume of pure radon-222 at 23°C and 755 mmHg could be obtained from 543.0 mg of radium bromide, $RaBr_2$, in a period of 37.5 y? The half-life of radium-226 is 1602 years.

20.127 What is the energy (in joules) evolved when 1 mol of helium-4 nuclei is produced from protons and neutrons? How many liters of ethane, $C_2H_6(g)$, at 25°C and 725 mmHg are needed to evolve the same quantity of energy when the ethane is burned in oxygen to $CO_2(g)$ and $H_2O(g)$? See Table 20.3 and Appendix C for data.

20.128 Plutonium-239 has been used as a power source for heart pacemakers. What is the energy obtained from the following decay of 215 mg of plutonium-239?

$$^{239}_{94}Pu \longrightarrow {}^{4}_{2}He + {}^{235}_{92}U$$

Suppose the electric energy produced from this amount of plutonium-239 is 25.0% of this value. What is the minimum grams of zinc that would be needed for the standard voltaic cell $Zn|Zn^{2+}||Cu^{2+}|Cu$ to obtain the same electric energy?

21

Chemistry of the Main-Group Elements

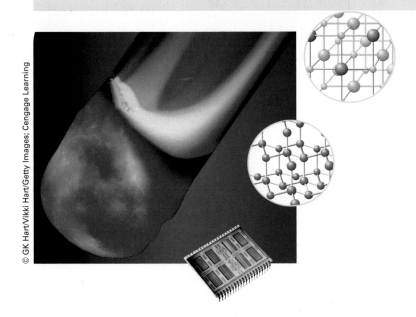

The exothermic reaction of silicon dioxide (silica sand) and magnesium forms silicon and magnesium oxide. (The models depict these products.) A similar reaction of silicon dioxide with carbon occurs in the production of silicon-based semiconductor chips for use in computers and cell phones.

Contents and Concepts

We begin our study with some general observations about the main-group elements.

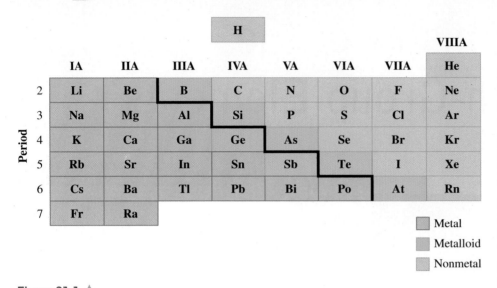

Figure 21.1 ▲

An abridged periodic table showing the main-group elements Elements to the left of the heavy staircase line are largely metallic in character; those to the right are largely nonmetallic.

Suppose you were asked to name some important elements. Let's see. There is oxygen in air, which you need to breath to support the oxidation reactions in the mitochondria of your cells that supply your body with energy. Then there is hydrogen, which combines with oxygen to form water. Water is the indispensable medium for most of the life processes that occur in your cells and those of other living things. Then there is carbon, which is central to biological compounds. Carbohydrates (including sugars, starch, and cellulose) and fats and oils contain carbon with hydrogen and oxygen. Proteins are the major structural and functional compounds of living organisms. They are made principally of carbon, hydrogen, oxygen, and nitrogen. These four elements—along with a few others, such as aluminum and silicon—are also among the most abundant elements on earth. They are all *main-group elements.*

Figure 21.1 is an abridged periodic table showing just these elements. Hydrogen is shown here, as it often is, in a group by itself at the top of the table, although it is sometimes placed in Groups IA and VIIA. Carbon is in Group IVA, nitrogen is in Group VA, and oxygen is in Group VIA. The main-group elements are definitely important to life, but they also play central roles in many everyday applications, as we will see.

A dominant feature of these elements is the pronounced trends in the properties as you go from one group to the other in the periodic table. We will begin this chapter by looking at these and other *general observations* about the main-group elements. Then we will look at the *metallic elements,* describing how the metals are produced from their sources and how the metal atoms chemically bond. And, of course, we will want to look at some individual elements, exploring their physical and chemical properties, as well as their commercial applications. In the final sections of this chapter, we will explore the *nonmetallic elements.*

21.1 General Observations About the Main-Group Elements

In this section, we want to consider some periodic trends that you will find useful as you study the chemistry of specific elements. The elements on the left side of the periodic table in Figure 21.1 are largely metallic in character; those on the right side are largely nonmetallic. Table 21.1 compares the characteristics of metallic and nonmetallic elements, and the following discussion describes some of these characteristics in greater detail.

The metallic elements generally have low ionization energies and low electronegativities compared with the nonmetallic elements. As a result, the metals tend to lose their valence electrons to form cations (Na^+, Ca^{2+}, Al^{3+}) in compounds or in aqueous solution. Nonmetals, on the other hand, form monatomic anions (O^{2-}, Cl^-) and oxoanions (NO_3^-, SO_4^{2-}).

Also, as noted earlier (Section 8.7), the oxides of the metals are usually basic. The oxides of the most reactive metals react with water to give basic solutions. For example,

$$CaO(s) + H_2O(l) \longrightarrow Ca^{2+}(aq) + 2OH^-(aq)$$

Table 21.1	Comparison of Metallic and Nonmetallic Elements
Metals	**Nonmetals**
Lustrous	Nonlustrous
Solids at 20°C (except Hg, which is a liquid)	Solids or gases at 20°C (except Br_2, which is a liquid)
Solids are malleable and ductile	Solids are usually hard and brittle
Conductors of heat and electricity	Nonconductors of heat and electricity (except graphite, an allotrope of C)
Low ionization energies	Moderate to high ionization energies
Low electronegativities	Moderate to high electronegativities
Form cations	Form monatomic anions or oxoanions
Oxides are basic (unless metal is in a high oxidation state)	Oxides are acidic

[Oxides of metals in high oxidation states, which you encounter in some transition metals, can be acidic; chromium(VI) oxide, CrO_3, is an acidic oxide.] The oxides of the nonmetals are acidic. Sulfur dioxide, an oxide of a nonmetal, dissolves in water to form an acidic solution.

$$SO_2(g) + 2H_2O(l) \rightleftharpoons H_3O^+(aq) + HSO_3^-(aq)$$

Silicon dioxide (silica sand) does not dissolve in water, but it does react when melted with basic oxides such as calcium oxide to form salts.

$$SiO_2(l) + CaO(s) \longrightarrow CaSiO_3(l)$$

Table 21.2 shows the oxidation states displayed by compounds of the main-group elements. The principal conclusions are as follows: the metallic elements generally have oxidation states equal to the group number (the Roman numeral for a column), representing a loss of the valence electrons in forming compounds. Some of the metallic elements in the fifth and sixth periods also have oxidation states equal to the group number minus two (for example, Pb^{2+}). However, the nonmetals (except for the most electronegative elements fluorine and oxygen) have a variety of oxidation states extending from the group number (the most positive value) to the group number minus eight (the most negative value). For example, chlorine (a Group VIIA element) has the following oxidation states in compounds: +7, +5, +3, +1, –1.

The metallic–nonmetallic characteristics of the elements change in definite ways as you move from left to right across a period or down a column of the periodic table. *The metallic characteristics of the main-group elements in the periodic table generally decrease in going across a period from left to right.* Figure 21.1 illustrates this in a broad way. In any period, the elements at the far left are metals and those at the far right are nonmetals. The trend of decreasing metallic character can be seen clearly in the third-period elements. Sodium, magnesium, and aluminum are metals. Silicon is a metalloid, whereas phosphorus, sulfur, and chlorine are nonmetals. The oxides of these elements show the expected trend from basic to acidic. Sodium oxide and magnesium oxide are basic. Aluminum oxide is amphoteric (it has basic and acidic character). The oxides of the elements silicon to chlorine are acidic.

The metallic characteristics of the main-group elements in the periodic table become more important going down any column (group). This trend is more pronounced in the middle groups of the periodic table (Groups IIIA to VA). For example, in Group IVA, carbon is a nonmetal, silicon and germanium are metalloids, and tin and lead are metals. The metallic elements tend to become more reactive as you progress down a column. You can see this most clearly in the Group IIA elements; beryllium is much less reactive than strontium and barium.

A second-period element is often rather different from the remaining elements in its group. The second-period element generally has a small atom that tends to hold electrons strongly, giving rise to a relatively high electronegativity (electron-withdrawing power). For example, nitrogen has an electronegativity of 3.0, but

Table 21.2   **Oxidation States in Compounds of the Main-Group Elements***

Period	Group						
	IA	IIA	IIIA	IVA	VA	VIA	VIIA
2	Li	Be	B	C	N	O	F
	+1	+2	+3	+4	+5	−1	−1
				+2	+4	−2	
				−4	+3		
					+2		
					+1		
					−3		
3	Na	Mg	Al	Si	P	S	Cl
	+1	+2	+3	+4	+5	+6	+7
				−4	+3	+4	+5
					−3	+2	+3
						−2	+1
							−1
4	K	Ca	Ga	Ge	As	Se	Br
	+1	+2	+3	+4	+5	+6	+7
				+2	+3	+4	+5
					−3	−2	+1
							−1
5	Rb	Sr	In	Sn	Sb	Te	I
	+1	+2	+3	+4	+5	+6	+7
			+1	+2	+3	+4	+5
					−3	−2	+1
							−1
6	Cs	Ba	Tl	Pb	Bi	Po	At
	+1	+2	+3	+4	+5	+4	+5
			+1	+2	+3	+2	−1

*The most common oxidation state is shown in color. Some uncommon oxidation states are not shown.

the other Group VA elements have electronegativities between 1.9 and 2.1. Another reason for the difference in behavior between a second-period element and the other elements of the same group has to do with the fact that bonding in the second-period elements involves only *s* and *p* orbitals, whereas the other elements may use *d* orbitals. This places a limit on the types of compounds formed by the second-period elements. For example, although nitrogen forms only the trihalides (such as NCl_3), phosphorus has both trihalides (PCl_3) and pentahalides (PCl_5), which it forms using 3*d* orbitals.

Chemistry of the Main-Group Metals

In the first half of this chapter, we will focus on the metals in Groups IA, IIA, IIIA, and IVA. We will discuss the preparation of several of these metals in Section 21.2 and the bonding of metals in Section 21.3. Then, in Sections 21.4 to 21.6, we will address these questions: What are the physical and chemical properties of some of the main-group metallic elements and their compounds? What are some of the commercial uses of these substances?

21.2 Metals: Characteristics and Production

One of the obvious characteristics of a metal is its luster, or shine. But metals have other interesting properties, such as their relatively high electrical and heat conductivities. Metals are also more or less malleable and ductile. By *malleable,* we mean you can pound the material into flat sheets; by *ductile,* we mean you can draw it into a wire. Thus we can define a **metal** as *a material that is lustrous (shiny), has high electrical and heat conductivities, and is malleable and ductile.*

A metal need not be a pure element; it can be a compound or mixture. An **alloy** is *a material with metallic properties that is either a compound or a mixture.* If the alloy is a mixture, it may be homogeneous (a solution) or heterogeneous. Most commercial metals are alloys consisting of one metal with small quantities of other metals to add desirable characteristics. For example, aluminum is often alloyed with magnesium and copper for strength.

Sources of Metals

Metals and their compounds come from various minerals. A **mineral** is *a naturally occurring inorganic solid substance or solid solution with a definite crystalline structure.* Thus a mineral might be a definite chemical substance, or it might be a homogeneous solid mixture. For example, bauxite, from which we obtain aluminum, is a *rock* (a naturally occurring solid material composed of one or more minerals) that contains several aluminum minerals, including gibbsite, a mineral form of aluminum hydroxide. Corundum is an oxide mineral of aluminum, Al_2O_3. Some of the aluminum atoms in corundum may be replaced by chromium atoms, forming a mineral that is a red solution of Al_2O_3 and Cr_2O_3. When this mineral is of gemstone quality, it is known as ruby (Figure 21.2). An **ore** is *a rock or mineral from which a metal or nonmetal can be economically produced.* Thus bauxite is the principal ore of aluminum.

Metallurgy

No doubt nuggets of gleaming metal in sand and rock initially fascinated early humans simply because of their appearance. Eventually, however, someone discovered that metals could be fashioned into tools. The first metals used were probably those that occurred naturally in the free state, such as gold. Later, it was found that heating certain rocks in a hot charcoal fire yielded metals. We realize now that these rocks contain metal compounds that can be reduced to the metallic state by reaction with charcoal (carbon) or the carbon monoxide obtained from the partial burning of charcoal. Copper was probably the first metal produced this way. Later, it was discovered that rocks containing copper and tin compounds yielded bronze, the first manufactured alloy. It was from these beginnings that metallurgy arose. **Metallurgy** is *the scientific study of the production of metals from their ores and the making of alloys having various useful properties.* In this section, we will look at the basic steps in the production of a metal from its ore:

1. *Preliminary treatment.* Usually, an ore must first be treated in some way to concentrate its metal-containing portion. Ores are usually mixtures of a mineral containing the metal along with economically worthless rock material that must be discarded. During the preliminary treatment, the metal-containing mineral is separated from these less desirable parts of the ore. It may also be necessary to transform the metal-containing mineral by chemical reaction to a metal compound that is more easily reduced to the free metal.

2. *Reduction.* Unless the metal occurs free, the metal compound obtained from preliminary treatment has to be reduced. Electrolysis or chemical reduction may be used, depending on the metal.

3. *Refining.* Once the free metal is obtained, it may have to be purified before it can be used. This purification process is referred to as metal refining.

Preliminary Treatment

A metal ore contains varying quantities of economically worthless material along with the mineral containing the metallic element. To separate the useful mineral from an ore, both physical and chemical methods are used. Panning for gold nuggets is a

Figure 21.2 ▲

Ruby Ruby is aluminum oxide, Al_2O_3, containing a small percentage of chromium(III) oxide, Cr_2O_3.

simple physical separation method based on differences in densities of gold and other minerals. Flushing water over a pan of gold-bearing earth easily washes sand and dirt aside, leaving the more dense grains of gold at the bottom of the pan.

The Bayer process is an example of a chemical method of concentrating an ore in its metal-bearing fraction. The **Bayer process** is *a chemical procedure in which purified aluminum oxide, Al₂O₃, is separated from the aluminum ore bauxite.* It depends on the fact that aluminum hydroxide is amphoteric, meaning that it behaves both as a base (reacting with acid to form the metal ion), which is expected of a metal hydroxide, and as an acid (reacting with base to form a hydroxo metal ion). Bauxite, as we noted earlier, contains aluminum hydroxide, $Al(OH)_3$, and aluminum oxide hydroxide, $AlO(OH)$, as well as relatively worthless constituents such as silicate minerals with some iron oxides. When bauxite is mixed with hot, aqueous sodium hydroxide solution, the amphoteric aluminum minerals, along with some silicates, dissolve in the strong base. The aluminum minerals give the aluminate ion, $Al(OH)_4^-$.

$$Al(OH)_3(s) + OH^-(aq) \longrightarrow Al(OH)_4^-(aq)$$

$$AlO(OH)(s) + OH^-(aq) + H_2O(l) \longrightarrow Al(OH)_4^-(aq)$$

Other substances, including the iron oxides, remain undissolved and are filtered off. As the hot solution of sodium aluminate cools, aluminum hydroxide precipitates, leaving the soluble silicates in solution. In practice, the solution is seeded with aluminum hydroxide to start the precipitation.

$$Al(OH)_4^-(aq) \longrightarrow Al(OH)_3(s) + OH^-(aq)$$

Aluminum hydroxide can also be precipitated from the basic solution by acidifying it with carbon dioxide gas (Figure 21.3). The aluminum hydroxide is finally *calcined* (heated strongly in a furnace) to produce aluminum oxide.

$$2Al(OH)_3(s) \xrightarrow{\Delta} Al_2O_3(s) + 3H_2O(g)$$

Figure 21.3 ▶

Precipitation of aluminum hydroxide by acidifying an aluminate ion solution

Carbon dioxide gas is bubbled into a solution of aluminate ion, Al(OH)₄−, containing aluminon dye.

Aluminum hydroxide, Al(OH)₃, precipitates from the acidic solution. The precipitate is normally white but adsorbs the dye to form a pink-colored precipitate called a *lake*. A standard test for aluminum ion is its precipitation as the hydroxide in the presence of aluminon dye to form this lake.

© Cengage Learning

Bauxite is also a source of gallium, the element under aluminum in the periodic table. Gallium is important in preparing gallium arsenide for solid-state devices, including lasers for compact disc players and light-emitting diodes, or LEDs (Figure 21.4). Sodium gallate, $Na[Ga(OH)_4]$, is more soluble than the corresponding aluminum compound, and it remains in the filtrate from the Bayer process after aluminum hydroxide has been filtered off.

Once an ore is concentrated, it may be necessary to convert the mineral to a compound more suitable for reduction. **Roasting** is *the process of heating a mineral in air to obtain the oxide.* Sulfide minerals, such as lead ore (containing the mineral galena, PbS), are usually roasted before reducing them to the metal.

$$2PbS(s) + 3O_2(g) \longrightarrow 2PbO(s) + 2SO_2(g); \Delta H° = -836 \text{ kJ}$$

The roasting process is exothermic, because it is essentially the burning of the sulfide ore. Once the ore has been heated to initiate roasting, additional heating is unnecessary.

Reduction

A metal, if it is not already free, must be obtained from one of its compounds by reduction, using either electrolysis or chemical reduction.

Electrolysis The reactive main-group metals, such as lithium, are obtained by *electrolysis.* Electrolysis uses an electric current to reduce a metal compound to the metal. Lithium, for example, is obtained commercially by electrolysis of molten lithium chloride, LiCl. (The industrial process uses a mixture of LiCl and KCl melting at 450°C, which is lower than the melting point of pure LiCl at 613°C.) Molten lithium chloride is a liquid consisting of lithium ions, Li^+, and chloride ions, Cl^-. Electrons from the negative pole of a battery flow to an attached electrode and react with lithium ions in the melt, forming lithium metal:

$$Li^+ + e^- \longrightarrow Li(l)$$

At the other electrode, chloride ions yield chlorine gas, and electrons flow back to the battery.

$$Cl^- \longrightarrow \tfrac{1}{2}Cl_2(g) + e^-$$

The net result of these electrode reactions is the production of lithium metal and chlorine from lithium chloride. The electric current supplies the energy for the reaction.

$$LiCl(l) \longrightarrow Li(l) + \tfrac{1}{2}Cl_2(g)$$

Magnesium is another reactive metal obtained by electrolysis of the chloride. Magnesium ion, Mg^{2+}, is the third most abundant dissolved ion in the oceans, after Cl^- and Na^+. The oceans, therefore, are an essentially inexhaustible supply of magnesium ion, from which the metal can be obtained. The **Dow process,** *a commercial method for isolating magnesium from seawater,* depends on the fact that magnesium ion can be precipitated from aqueous solution by adding a base. Figure 21.5 shows a flowchart for the process. Seashells provide the source of the base (calcium hydroxide). Seashells are principally calcium carbonate, $CaCO_3$, which when heated decomposes to calcium oxide, CaO, and carbon dioxide, CO_2. Calcium oxide reacts with water to produce calcium hydroxide. The base, calcium hydroxide, provides hydroxide ion that reacts with magnesium ion in seawater to precipitate magnesium hydroxide:

$$Mg^{2+}(aq) + 2OH^-(aq) \longrightarrow Mg(OH)_2(s)$$

The magnesium hydroxide is allowed to settle out in ponds; here it gives a suspension, or slurry, of $Mg(OH)_2$. The slurry is filtered to recover the magnesium hydroxide precipitate, which is then treated with hydrochloric acid to yield magnesium chloride. The dry magnesium chloride salt is melted and electrolyzed at 700°C to yield magnesium metal and chlorine gas.

$$MgCl_2(l) \xrightarrow{\text{electrolysis}} Mg(l) + Cl_2(g)$$

Aluminum is manufactured from the aluminum oxide obtained in the Bayer process using electrolysis. The method for doing this was discovered in 1886 by Charles Martin Hall (then a student at Oberlin College) in the United States and, independently, by Paul Héroult in France.

Figure 21.4 ▲

Preparation of gallium arsenide, GaAs, semiconductor material
The bars in the tube are gallium; the small pieces in the tube near the furnace are arsenic. Gallium arsenide is used to make semiconductor diode lasers for use in fiber optic communications, compact disc players, and laser printers. Also, computer chips are being developed using gallium arsenide, because it conducts an electrical signal five times faster than presently used silicon.

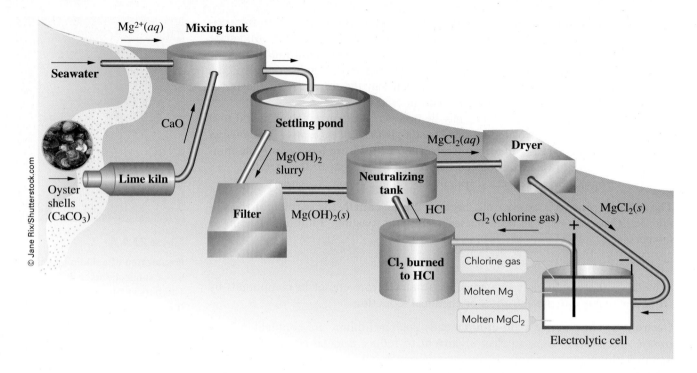

Figure 21.5 ▲

Dow process for producing magnesium from seawater Oyster shells or other seashells are calcined in a kiln to produce calcium oxide, which when added to seawater precipitates magnesium ion as magnesium hydroxide. This is neutralized with hydrochloric acid to give magnesium chloride. Electrolysis of molten $MgCl_2$ yields magnesium metal and chlorine. Hydrochloric acid can be recovered by burning the chlorine with natural gas.

The **Hall–Héroult process** is *the commercial method for producing aluminum by the electrolysis of a molten mixture of aluminum oxide in cryolite, Na_3AlF_6.* Originally, the process used natural cryolite as the molten solvent. Today, the Hall–Héroult process uses synthetic cryolite, or sodium hexafluoroaluminate, produced by reacting aluminum hydroxide from the Bayer process with sodium hydroxide and hydrofluoric acid.

$$3NaOH(aq) + Al(OH)_3(s) + 6HF(aq) \longrightarrow Na_3AlF_6(aq) + 6H_2O(l)$$

A Hall–Héroult electrolytic cell consists of a rectangular steel shell lined first with an insulating material, then with carbon (from baked petroleum coke) to form the negative electrode (Figure 21.6). The other (positive) electrode is also made from carbon. The electrolyte consists of molten cryolite, at about 1000°C, into which some aluminum oxide is dissolved. The overall electrolysis reaction is

$$2Al_2O_3(s) + 3C(s) \xrightarrow{\text{electrolysis}} 4Al(l) + 3CO_2(g)$$

The positive carbon electrodes are consumed in the electrolysis and must be replaced periodically; aluminum oxide is continually added to the electrolyte bath.

Chemical Reduction The cheapest chemical reducing agent is some form of carbon, such as hard coal or coke. Coke is the solid residue left from coal after its volatile constituents have been driven off by heating in the absence of air. Some important metals, including iron and zinc, are obtained by reduction of their compounds with carbon (or carbon monoxide, which is a product of partial oxidation of carbon).

Figure 21.6 ▶

Hall–Héroult cell for the production of aluminum Aluminum oxide is electrolyzed in molten cryolite (the electrolyte). Molten aluminum forms at the negative electrode (tank lining), where it is periodically withdrawn.

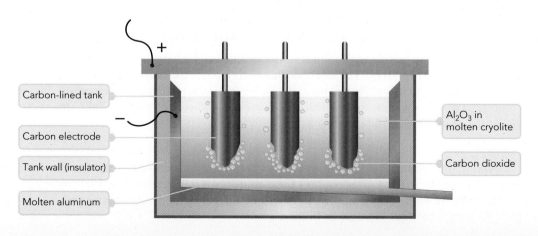

Earlier, we described how lead is obtained from its sulfide ore by roasting it to lead(II) oxide. When the oxide is heated strongly with carbon in a *blast furnace,* it is reduced to the metal. Lead(II) oxide and coke are added at the top of the furnace; a blast of air enters at the bottom. Some of the coke burns in the air and heats the material to a high temperature. The lead(II) oxide reacts with carbon:

$$PbO(s) + C(s) \longrightarrow Pb(l) + CO(g)$$

Liquid lead collects at the bottom of the furnace, where it is drawn off.

Refining

Often the metal obtained from the reduction process contains impurities and must be purified, or *refined.* Various refining techniques are used. For example, zinc is easily vaporized (it boils at 908°C), so it is purified by distillation. Lead is purified by blowing air through the liquid metal to oxidize impurities, including carbon and sulfur. Copper is purified by electrolysis, in which the impure copper forms the positive pole and pure copper the negative. Copper ions flow from the impure copper and collect on the pure copper. See Figure 19.24.

21.3 Bonding in Metals

The special properties of a metal result from its delocalized bonding, in which bonding electrons are spread over a number of atoms. In this section, we will look first at the "electron-sea" model of a metal and then at the molecular orbital theory of bonding in metals.

Electron-Sea Model of Metals

A very simple picture of a metal depicts an array of positive ions surrounded by a "sea" of valence electrons free to move over the entire metal crystal. When the metal is connected to a source of electric current, the electrons easily move away from the negative side of the electric source and toward the positive side, forming an electric current in the metal. In other words, the metal is a conductor of electric current because of the mobility of the valence electrons. A metal is also a good heat conductor because the mobile electrons can carry additional kinetic energy across the metal. ▶

The electron-sea model of metallic bonding is described briefly in Section 9.7; see Figure 9.18.

Molecular Orbital Theory of Metals

The electron-sea model of metals is a simplified view that accounts in only a qualitative way for properties of a metal such as electrical conductivity. Molecular orbital theory gives a more detailed picture of the bonding in a metal and other solids as well. ▶

Molecular orbital theory is discussed in Sections 10.5 to 10.7.

Recall that molecular orbitals form between two atoms when atomic orbitals on the atoms overlap. In some cases, the atomic orbitals on three or more atoms overlap to form molecular orbitals that encompass all of the atoms. These molecular orbitals are said to be *delocalized.* The number of molecular orbitals that form by the overlap of atomic orbitals always equals the number of atomic orbitals. In a metal, the outer orbitals of an enormous number of metal atoms overlap to form an enormous number of molecular orbitals that are delocalized over the metal. As a result, a large number of energy levels are crowded together into "bands." Because of these energy bands, the molecular orbital theory of metals is often referred to as *band theory.*

Sodium metal provides a simple illustration of band theory. Imagine that you build a crystal of sodium by bringing sodium atoms together one at a time, and during this process you follow the formation of molecular orbitals and associated energy levels. Each isolated sodium atom has the electron configuration $1s^2 2s^2 2p^6 3s^1$, or $[Ne]3s^1$. When two sodium atoms approach each other, their $3s$ orbitals overlap to form two molecular orbitals (a bonding molecular orbital and an antibonding molecular orbital). The inner-core electrons, represented as [Ne] in the electron configuration, remain essentially nonbonding. Now imagine that a third atom is brought up to this diatomic molecule. The three $3s$ orbitals overlap to form three molecular orbitals, each orbital encompassing the entire Na_3 molecule (that is, they are delocalized over the molecule). When a large number N (on the order of Avogadro's number)

Figure 21.7 ▶

Formation of an energy band in sodium metal Note that the number of energy levels grows until the levels merge into a continuous band of energies. The lower half of the band is occupied by electrons.

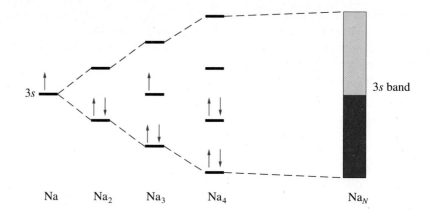

of sodium atoms have been brought together to form a crystal, the atoms will have formed N molecular orbitals encompassing the entire crystal.

Figure 21.7 shows that at each stage the number of energy levels grows. Eventually, the energy levels effectively merge into a *band* of continuous energies. We call this the $3s$ band of the sodium metal. A $3s$ band formed from N atoms will have N orbitals that hold a maximum of $2N$ electrons. Because each sodium atom has one valence electron, N atoms will supply N $3s$ electrons and will therefore half-fill the $3s$ band.

We can now explain the electrical conductivity of sodium metal. Electrons become free to move throughout a crystal when they are excited to unoccupied orbitals of a band. In sodium metal, this requires very little energy; because of the half-filled $3s$ band, unoccupied orbitals lie just above the occupied orbitals of highest energy. When a voltage is applied to a metal crystal, electrons are excited to the unoccupied orbitals and move toward the positive pole of the voltage source. A current flows in the metal.

The explanation of the electrical conductivity of magnesium metal is somewhat more complicated than in the case of sodium. A magnesium atom has the configuration $[Ne]3s^2$. As in sodium, the $3s$ orbitals of magnesium metal overlap to form a $3s$ band. If this were the whole story, you would expect the $2N$ valence electrons of the atoms to completely fill the $3s$ band. Therefore, if a small voltage were applied, the electrons would have no place to go. You would expect magnesium to be an insulator (a nonconducting material), but in fact it is a conductor.

Both sodium and magnesium have unoccupied $3p$ bands formed from unoccupied $3p$ orbitals of the atoms. The presence of this band in sodium has no effect on its electrical conductivity. In magnesium, however, the situation is different, as Figure 21.8 shows. As the orbitals of the individual atoms interact in forming the metal, the energy levels spread so that the bottom of the $3p$ band merges with the

Figure 21.8 ▶

Formation of 3s and 3p bands in magnesium metal Note that the $3s$ and $3p$ bands merge. As a result, the bands are only partially filled.

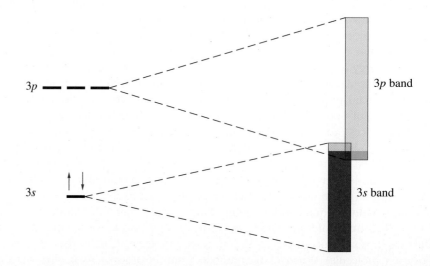

Superconductivity

A *superconductor* is a material that abruptly loses its resistance to an electric current when cooled to a definite characteristic temperature. This means that an electric current will flow in a superconductor without heat loss, unlike the current in a typical conductor. Once a current has been started in a superconducting circuit, it continues to flow indefinitely. Another intriguing property of a superconductor is its perfect diamagnetism. This means that the superconductor completely repels magnetic field lines. Figure 21.9 shows a magnet suspended in midair over one of the newly discovered ceramic superconductors. The magnet seems to levitate—it appears to hover in air in defiance of gravity. In fact, the repulsion of magnetic field lines by the superconductor holds the magnet aloft.

Superconductivity was discovered in 1911 by the Dutch physicist Heike Kamerlingh Onnes soon after he found a way to liquefy helium. By evaporating liquid helium, he could obtain temperatures near absolute zero. Kamerlingh Onnes found that mercury metal suddenly loses all resistance to an electric current when cooled to 4 K. Superconductors first became useful when a niobium metal alloy was found to become superconducting at about 23 K and to remain superconducting even when large currents flow through it. (Many superconducting materials lose their superconductivity when even moderate currents flow through them.) It became possible to construct superconducting magnets with high magnetic fields by starting an electric current in a superconducting circuit. Such magnets are being used in medical magnetic resonance imaging. (See Instrumental Methods: Nuclear Magnetic Resonance in Chapter 8.) Even though expensive liquid helium (more than $20 per gallon) is needed to operate these magnets, they are still cheaper to use than the usual electromagnets because of their much lower power requirements.

In 1986, Johannes Georg Bednorz and Karl Alexander Müller of IBM discovered that certain copper oxide ceramic materials became superconducting at 30 K, and within months researchers had found similar materials that become superconducting at 125 K. (They won the 1987 Nobel Prize in physics for their discovery.) This means that superconductors can be operated using a cheap refrigerant, such as liquid nitrogen, at a few cents per gallon. Perhaps materials can be found that are superconducting even at room temperature. What remains is to determine how to fabricate such superconducting materials into wires and similar objects with the ability to carry large currents.

In 2001, Japanese investigators discovered that an alloy of titanium, magnesium, and boron becomes superconducting below 40 K. Later, they found that it was magnesium diboride, MgB_2, in this alloy that was responsible for the superconductivity. Although its temperature of superconductivity is lower than that of the copper oxide ceramics, magnesium diboride behaves more like the well-known metallic superconductors and might serve as a cheap, efficient replacement for them. In any case, it offers a new avenue of research into this promising arena.

© Argonne National Laboratory

Figure 21.9 ▲

Levitation of a magnet by a superconductor The magnet (samarium–cobalt alloy) is supported above the ceramic superconductor (approximate formula $YBa_2Cu_3O_7$). The ceramic becomes superconducting when it is cooled by liquid nitrogen.

■ See Problems 21.191 and 21.192.

top of the 3*s* band. Imagine electrons filling the 3*s* band. When the electrons reach the energy where the two bands have merged, electrons begin to fill orbitals in both bands. As a result, the 3*s* and 3*p* bands of magnesium metal are only partially filled by the time you have accounted for all 2*N* valence electrons. Therefore, when a voltage is applied to the metal, the highest-energy electrons are easily excited into the unoccupied orbitals, giving an electrical conductor.

In general, a solid that has a partially filled band will be an electrical conductor. A solid that has only completely filled bands (without a nearby unfilled band) will be a nonconductor of electricity, or electrical insulator.

21.4 Group IA: The Alkali Metals

The alkali metals, the elements in the first column of the periodic table except hydrogen, are all soft, silvery metals. Figure 21.10 shows sodium metal being cut with a knife. The alkali metals are the most reactive of all metals, reacting readily with air and water. Because the valence configuration of these elements is ns^1, the alkali metals usually react by losing this electron to form the +1 ions, such as Li^+ and Na^+. Most of the compounds of these ions are soluble in water.

Because of their chemical reactivity, the Group IA elements never occur as free metals in nature. They do occur extensively in silicate minerals, which weather to form soluble compounds of the elements (particularly of sodium and potassium). These soluble compounds eventually find their way to landlocked lakes and oceans, where they concentrate. Enormous underground beds of sodium and potassium compounds formed when lakes and seas became isolated by geologic events; the water eventually evaporated, leaving solid deposits of alkali metal compounds. Commercially, sodium and potassium compounds are common, and both sodium and lithium metals are available in quantity.

Lithium

In recent years, the commercial importance of lithium metal has risen markedly. Among the alkali metals, only sodium metal is more important.

As you might expect for an alkali metal, lithium is chemically reactive. Like the other alkali metals, it is a relatively soft metal, although the hardest of the Group IA elements. Lithium does exhibit properties that are somewhat different from those of the lower members of the Group IA elements. Many of its ionic compounds are much less soluble than are similar compounds of the other alkali metals. Lithium carbonate, Li_2CO_3, for example, is only slightly soluble in water at room temperature, whereas sodium carbonate, Na_2CO_3, is soluble.

Lithium Metallurgy The commercial source of lithium metal is the ore spodumene, which is a lithium aluminum silicate mineral, $LiAl(SiO_3)_2$. The ore is heated and then washed with sulfuric acid to obtain a solution of lithium ion, which is precipitated as lithium carbonate. Lithium carbonate from lithium ore is the primary starting substance for the production of lithium metal and lithium compounds. Lithium metal is obtained by electrolysis of the chloride.

The use of lithium metal has greatly expanded in recent years (Figure 21.11). A major use is in the production of low-density aluminum alloy for aircraft construction. Batteries with lithium metal anodes have also become common (Figure 21.12). The anode reaction is the oxidation of lithium to the ion:

$$Li(s) \longrightarrow Li^+ + e^-$$

The cathode material varies, but the cathodes in batteries used in cameras, calculators, and similar devices consist of manganese(II) oxide, the same substance used in common dry cells. The electrolyte in this Li/MnO_2 cell is a solution of lithium perchlorate, $LiClO_4$, dissolved in an organic solvent.

Reactions of Lithium Metal Lithium, like the other alkali metals, reacts readily with water and with moisture in the air to produce lithium hydroxide and hydrogen gas:

$$2Li(s) + 2H_2O(l) \longrightarrow 2LiOH(aq) + H_2(g)$$

When a pellet of lithium is placed in water, the pellet spins around on the surface of the water, evolving bubbles of hydrogen gas. (Lithium floats because it is less dense than water.) The reaction is not as vigorous, however, as the similar reaction of sodium and water.

Lithium burns in air to produce lithium oxide, Li_2O, a white powder.

$$4Li(s) + O_2(g) \longrightarrow 2Li_2O(s)$$

When heated with nitrogen gas, lithium reacts to form lithium nitride, Li_3N, the only stable alkali-metal nitride.

$$6Li(s) + N_2(g) \longrightarrow 2Li_3N(s)$$

Lithium nitride is an ionic compound of the nitride ion, N^{3-}.

Figure 21.10 ▲

Cutting of sodium metal The metal is easily cut with a sharp knife.

Figure 21.11 ▲

A roll of lithium metal for batteries The metal must be handled in humidity-free rooms to prevent corrosion.

Lithium Compounds Lithium carbonate, which is obtained from lithium ore, is the primary source of other lithium compounds. Significant quantities of lithium carbonate are also used in the preparation of porcelain enamels, glazes, and special glass. When purified, the carbonate is used as a source of lithium ion for the treatment of bipolar disorder. The physiological action of lithium ion in this treatment is not completely understood, although lithium ion is believed to inhibit certain biochemical reactions involved in the action of neurotransmitters in the brain.

Lithium hydroxide, LiOH, is a strong base. It is produced by reaction of lithium carbonate with calcium hydroxide in a precipitation reaction. Calcium carbonate is much less soluble than lithium carbonate:

$$Li_2CO_3(aq) + Ca(OH)_2(aq) \longrightarrow 2LiOH(aq) + CaCO_3(s)$$

The calcium carbonate is filtered off, and lithium hydroxide is recovered from the filtrate. Lithium hydroxide is used in the production of lithium soaps, which are used in making lubricating greases from oil. (Lithium soap thickens the oil.) Soaps are salts of fatty acids; they are prepared by heating a fat with a strong base.

Lithium hydroxide is used to remove carbon dioxide from the air in spacecraft and submarines. Humans exhale carbon dioxide that is produced by the normal biochemical reactions of the body. For people to live in a closed space, it is necessary to remove this carbon dioxide. Although carbon dioxide is not normally toxic, increasing concentrations do have a physiological effect by interfering with the blood's capacity to carry oxygen, and high concentrations of carbon dioxide are hazardous. Like the other alkali-metal hydroxides, lithium hydroxide absorbs carbon dioxide from air by forming the carbonate and hydrogen carbonate:

$$2LiOH(s) + CO_2(g) \longrightarrow Li_2CO_3(s) + H_2O(l)$$

$$Li_2CO_3(s) + H_2O(l) + CO_2(g) \longrightarrow 2LiHCO_3(s)$$

Sodium hydroxide could be used to remove carbon dioxide from air, since it undergoes the same reactions. However, the formula weight of LiOH (24 amu) is much lower than that of NaOH (40 amu). Where light weight is important, lithium hydroxide is preferred, even though it is more expensive than sodium hydroxide.

Table 21.3 lists some uses of lithium and other alkali metal compounds.

Courtesy Medtronic, Inc.

Figure 21.12 ▲

Lithium batteries Lithium batteries are used wherever a reliable current is required for a lengthy period.

Sodium and Potassium

Sodium compounds are of enormous economic importance. Common table salt, which is sodium chloride, has been an important article of commerce since earliest recorded history. Salt was of such importance in Roman times that a specific allowance of salt was part of the soldiers' pay. The word *salary* derives from the Latin word for this salt allowance (*salarium,* which comes from the Latin word *sal,* for "salt"). Sodium metal and sodium compounds are produced from sodium chloride.

The economic importance of potassium stems in large part from its role as a plant nutrient. Enormous quantities of potassium chloride from underground deposits are used as fertilizer to increase the world's food supply. In fact, plants were an early source of potassium compounds. Wood and other plant materials were burned to give ashes that were leached in pots to give potash, which consists primarily of potassium carbonate. The name potassium has its origin in the word *potash* from this earlier source of the element.

Sodium metal is more reactive than lithium metal, and potassium metal is still more reactive. In general, as you move down the column of alkali metals in Group IA of the periodic table, you find that the metals become more chemically reactive (forming the +1 ions). This is partly a result of the decrease in ionization energy of the atom (from 520 kJ/mol for the first ionization of lithium to 376 kJ/mol for that of cesium). As the atomic size increases going from the top of the column to the bottom, the valence electrons are held less strongly and so are lost more easily.

The Group IA metals also increase in softness going from the top of the column to the bottom. Lithium is a moderately firm metal, whereas potassium is soft enough to be cut with a butter knife. This increasing softness of the metals is in part a result of increasing atomic size, which results in decreasing strength of bonding by the valence-shell *s* electrons.

Table 21.3	Uses of Alkali Metal Compounds
Compound	**Use**
Li_2CO_3	Preparation of porcelain, glazes, special glasses
	Preparation of LiOH
	Treatment of bipolar disorder
LiOH	Manufacture of lithium soaps for lubricating greases
	In air-regeneration systems
LiH	Reducing agent in organic syntheses
$LiNH_2$	Preparation of antihistamines and other pharmaceuticals
NaCl	Source of sodium and sodium compounds
	Condiment and food preservative
	Soap manufacture (precipitates soap from reaction mixture)
NaOH	Pulp and paper industry
	Extraction of aluminum oxide from ore
	Manufacture of viscose rayon
	Petroleum refining
	Manufacture of soap
Na_2CO_3	Manufacture of glass
	In detergents and water softeners
Na_2O_2	Textile bleach
$NaNH_2$	Preparation of indigo dye for denim (blue jeans)
KCl	Fertilizer
	Source of other potassium compounds
KOH	Manufacture of soft soap
	Manufacture of other potassium compounds
K_2CO_3	Manufacture of glass
KNO_3	Fertilizers
	Explosives and fireworks

Sodium Metallurgy Sodium metal is obtained by the electrolysis of molten, or fused, sodium chloride. Sodium chloride is mined from huge underground deposits. The other source of sodium chloride is seawater, which is a solution of many dissolved substances, but sodium chloride is the principal one. Sodium chloride melts at 801°C, but commercial electrolysis employs a mixture of NaCl and $CaCl_2$, which melts at 580°C. We discussed the electrolysis in some detail in Section 19.9. ◄

Figure 19.20 depicts the commercial Downs cell for the preparation of sodium from molten sodium chloride.

Sodium metal is a strong reducing agent, and this accounts for many of its major uses. For example, it is used to obtain metals such as titanium and zirconium by reduction of their compounds. Titanium is a strong metal used in airplane and spacecraft manufacture. It is produced by reduction of titanium tetrachloride, $TiCl_4$, obtained by chemical processing of titanium ores. The overall process can be written as follows:

$$TiCl_4(g) + 4Na(l) \longrightarrow Ti(s) + 4NaCl(s)$$

Sodium is also employed as a reducing agent in the production of a number of organic compounds, including dyes and pharmaceutical drugs.

Reactions of Sodium Metal Like lithium, sodium metal reacts with water, but with even greater vigor. The reaction is sufficiently exothermic to ignite the hydrogen gas produced in the reaction:

$$2Na(s) + 2H_2O(l) \longrightarrow 2NaOH(aq) + H_2(g)$$

Sodium burns in air, producing some sodium oxide, Na_2O, but mainly sodium peroxide, Na_2O_2, a compound containing the peroxide ion O_2^{2-} (Figure 21.13).

$$2Na(s) + O_2(g) \longrightarrow Na_2O_2(s)$$

The peroxide ion acts as an oxidizing agent (Figure 21.14).

Sodium Compounds Sodium hydroxide, NaOH, is among the top ten industrial chemicals. It is produced by the electrolysis of aqueous sodium chloride. The overall electrolysis, which was described in Section 19.10, can be written as

$$2NaCl(aq) + 2H_2O(l) \xrightarrow{\text{electrolysis}} H_2(g) + Cl_2(g) + 2NaOH(aq)$$

Both chlorine gas and sodium hydroxide are major products of this electrolysis. ▶

Sodium hydroxide is a strong base, and this property is useful in many applications. It is used in large quantities in aluminum production (which depends on its reaction with the amphoteric aluminum hydroxide) and in the refining of petroleum (where it chemically reacts with acid constituents during the refining process). Large quantities of sodium hydroxide are also used to produce sodium compounds, which occur in a wide variety of commercial products.

For example, soap is made by heating a fat with sodium hydroxide solution. The product is a mixture of sodium salts of fatty acids, obtained from the fat. Fatty acids are chemically similar to acetic acid, but they consist of long chains of carbon atoms. Compare the structure of the acetate ion with that of the stearate ion. (Stearic acid is a typical fatty acid.) ▶

Figures 19.22 and 19.23 show two different commercial cells for the electrolysis of aqueous sodium chloride.

The action of soap in dispersing oil in water is described in Section 12.9.

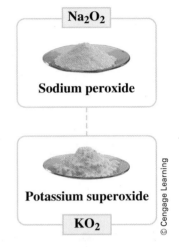

Na_2O_2

Sodium peroxide

Potassium superoxide

KO_2

© Cengage Learning

Figure 21.13 ▲

Oxygen compounds of alkali metals Sodium peroxide, Na_2O_2 (yellowish white, *top*), and potassium superoxide, KO_2 (orange-yellow, *bottom*). Potassium superoxide is the principal product formed when potassium metal burns in air.

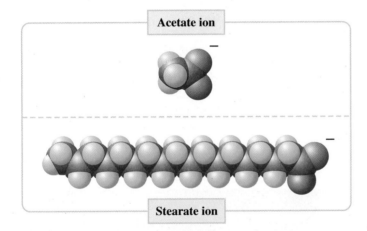

Acetate ion

Stearate ion

A couple of drops of water are added to a mixture of sodium peroxide and sulfur to initiate a reaction.

During the violent reaction, sulfur is oxidized to sulfur dioxide.

© Cengage Learning

Figure 21.14 ◀

Sodium peroxide as an oxidizing agent

Soda ash decomposes to sodium oxide when heated. Sodium and calcium oxides, which are basic oxides, react with fused silicon dioxide, SiO_2, an acidic oxide, to produce silicate glass.

Sodium hydroxide is commonly known as lye or caustic soda. You can buy it in a grocery store as a drain opener and as an oven cleaner (Figure 21.15). Both usages depend on the reaction of sodium hydroxide with organic materials, such as fats and hair, to produce soluble materials. Because of these reactions, sodium hydroxide and its solutions require careful handling.

Sodium carbonate is another important compound of sodium. The anhydrous compound, Na_2CO_3, is known commercially as *soda ash*. Large quantities of soda ash are consumed with sand and lime in making glass. ◄ The decahydrate, $Na_2CO_3 \cdot 10H_2O$, is added to many detergent preparations and is sold commercially as washing soda. In the United States, most sodium carbonate is produced from the mineral trona, whose chemical formula is approximately $Na_2CO_3 \cdot NaHCO_3 \cdot 2H_2O$. There are large deposits of trona in southwestern Wyoming.

Worldwide, a large fraction of sodium carbonate is still produced by the Solvay process. The **Solvay process** is *an industrial method for obtaining sodium carbonate from sodium chloride and limestone.* In the main step of this process, ammonia is first dissolved in a saturated solution of sodium chloride. Then carbon dioxide is bubbled in, and sodium hydrogen carbonate (baking soda) precipitates. Figure 21.16 shows a laboratory demonstration of this reaction.

$$NH_3(g) + \underbrace{H_2O(l) + CO_2(g)}_{H_2CO_3(aq)} + NaCl(aq) \longrightarrow NaHCO_3(s) + NH_4Cl(aq)$$

You can think of this as an exchange reaction of NH_4HCO_3 (from $NH_3 + H_2CO_3$) with NaCl, to give the products $NaHCO_3$ and NH_4Cl. The sodium hydrogen carbonate is filtered from the solution. When heated to 175°C, it decomposes to sodium carbonate.

$$2NaHCO_3(s) \xrightarrow{\Delta} Na_2CO_3(s) + CO_2(g) + H_2O(g)$$

The reaction to produce sodium hydrogen carbonate uses relatively expensive ammonia. The industrial Solvay process illustrates how you can affect the economics of a process by the ingenious use of raw materials and by recycling intermediate products. Carbon dioxide for the industrial process is obtained by heating limestone ($CaCO_3$):

$$CaCO_3(s) \xrightarrow{\Delta} CaO(s) + CO_2(g)$$

Salt

Baking soda

Washing soda

Drain opener

Oven cleaner

Figure 21.15 ▲

Sodium compounds available at the grocery store

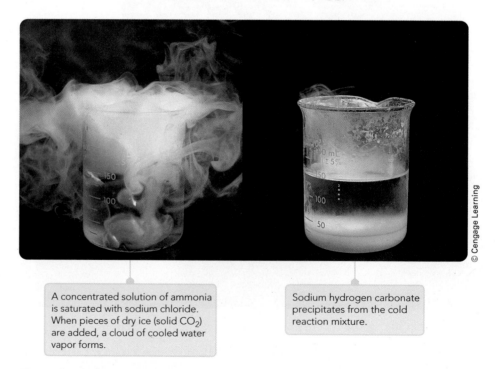

A concentrated solution of ammonia is saturated with sodium chloride. When pieces of dry ice (solid CO_2) are added, a cloud of cooled water vapor forms.

Sodium hydrogen carbonate precipitates from the cold reaction mixture.

Figure 21.16 ▲

Laboratory demonstration of the Solvay process

The calcium oxide is used to recover ammonia from the ammonium chloride obtained from the preparation of sodium hydrogen carbonate.

$$CaO(s) + 2NH_4Cl(aq) \xrightarrow{\Delta} 2NH_3(g) + CaCl_2(aq) + H_2O(l)$$

The carbon dioxide obtained when sodium hydrogen carbonate is heated to form sodium carbonate is also recycled to prepare more sodium hydrogen carbonate. The overall result of these steps in the industrial Solvay process is the reaction of limestone ($CaCO_3$) and salt (NaCl) to produce sodium carbonate and calcium chloride:

$$CaCO_3 + 2NaCl \longrightarrow Na_2CO_3 + CaCl_2$$

The raw materials (limestone and salt) are cheap, and for many years the Solvay process was the principal source of sodium carbonate. Unfortunately, there is not sufficient demand for the by-product calcium chloride. Some is used for deicing of roads, but much of it has to be disposed of as waste. The environmental cost of this waste disposal has made the Solvay process increasingly less attractive.

Potassium and Potassium Compounds The principal source of potassium and potassium compounds is potassium chloride, KCl, which is obtained from underground deposits. Small amounts of this salt are used to prepare the metal. Potassium is prepared by the chemical reduction of potassium chloride rather than by electrolysis of the molten chloride, as in the preparation of sodium. In the commercial process, potassium chloride is melted with sodium metal by heating to 870°C.

$$Na(l) + KCl(l) \longrightarrow NaCl(l) + K(g)$$

At this temperature, potassium forms as a vapor. The reaction goes in the direction written because potassium vapor leaves the reaction chamber and is condensed.

Almost all of the potassium metal produced is used in the preparation of potassium superoxide, KO_2, for self-contained breathing apparatus used in situations, such as firefighting, where toxic fumes may be present. When potassium burns in air, it produces potassium superoxide:

$$K(s) + O_2(g) \longrightarrow KO_2(s)$$

(The superoxide ion is O_2^-, and the oxidation state of oxygen in this ion is $-\frac{1}{2}$.) In a self-contained breathing apparatus, the potassium superoxide is contained in a canister through which one's breath passes. Moisture in the breath attacks the superoxide, releasing oxygen.

$$4KO_2(s) + 2H_2O(l) \longrightarrow 4KOH(s) + 3O_2(g)$$

The potassium hydroxide produced in this reaction removes the carbon dioxide from the exhaled air. The net effect of the self-contained breathing apparatus is to remove moisture and carbon dioxide from exhaled air and provide oxygen.

More than 90% of the potassium chloride that is mined is used directly as a plant fertilizer. The rest is used in the preparation of potassium compounds. Potassium hydroxide, used in the manufacture of liquid soaps, is obtained by the electrolysis of aqueous potassium chloride. Potassium nitrate is prepared by the reaction of potassium hydroxide and nitric acid. This compound is used in fertilizers and for explosives and fireworks. (Table 21.3 summarizes the major uses of potassium compounds, as well as those of lithium and sodium.)

21.5 Group IIA: The Alkaline Earth Metals

The Group IIA elements are also known as the alkaline earth metals. These metals exhibit the expected periodic trends. If you compare an alkaline earth metal with the alkali metal in the same period, you see that the alkaline earth metal is less reactive and a harder metal. For example, lithium reacts readily with water, but beryllium reacts hardly at all even with steam. Lithium is a soft metal, whereas beryllium is hard enough to scratch glass. You also see the expected trend within the column of alkaline earth metals; the elements at the bottom of the column are more reactive and are softer metals than those at the top of the column (Figure 21.17). Barium, in the sixth period, is a very reactive metal and, when placed in water, reacts much like an alkali metal. It is also a soft metal, much like the alkali metals.

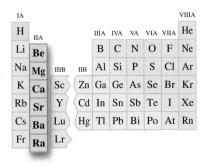

Figure 21.17 ▲

Magnesium and barium metals
Magnesium metal turnings are in the beaker (*left*). Barium metal (*right*) is much more reactive than magnesium and must be stored in a bottle of kerosene to exclude moisture and oxygen, with which barium reacts.

Magnesium, in the third period, is much less reactive and a harder metal (comparable to aluminum).

As we noted in Section 21.1, a second-period element is often considerably different from the other elements in its column. Those differences become more pronounced as you progress to the right in the periodic table. We briefly noted differences in properties of lithium from those of the other alkali metals, though those differences are not great. Beryllium, however, shows rather marked differences from the other alkaline earth elements. We have already noted that beryllium differs in its lack of reactivity compared with the other Group IIA metals. Another notable difference is in the properties of the hydroxides. Whereas those of the elements magnesium to barium are basic, beryllium hydroxide is amphoteric, reacting with both acids and bases.

Like the alkali metals, the Group IIA elements occur in nature as silicate rocks. They also occur as carbonates and sulfates, and many of these are commercial sources of alkaline earth metals and compounds.

Calcium and magnesium are the most common and commercially useful of the alkaline earth elements, and we will consider them in some detail in the next two subsections. Calcium is the fifth, and magnesium the eighth, most abundant element in the earth's crust.

Magnesium

The name *magnesium* comes from the name of the mineral magnesite, which in turn is believed to stem from the name *Magnesia,* a site in northern Greece where various minerals, including those of magnesium, have been mined since ancient times. The British chemist Humphry Davy discovered the pure element magnesium in 1808. Davy had already discovered the alkali metals potassium and sodium in late 1807. Several months later, he managed to isolate in quick succession the alkaline earth metals barium, strontium, calcium, and finally magnesium. He obtained magnesium by a procedure similar to the one he used for the other alkaline earths. He electrolyzed a moist mixture of magnesium oxide and mercury(II) oxide, from which he obtained magnesium amalgam (an alloy of magnesium dissolved in mercury). To obtain pure magnesium, Davy distilled the mercury from the amalgam.

Magnesium metal is isolated from seawater by the Dow process, which is described in Section 21.2.

Figure 21.18 ▲

Magnesium in motorcycles
Magnesium alloys are used in many motorcycle parts. Shown here is a closeup of the engine area of a Harley-Davidson motorcycle. The V engine (upper right), engine block (lower right), and transmission case (lower left) are made of an aluminum–magnesium alloy.

Magnesium Metallurgy Magnesium has become increasingly important as a structural metal. Its great advantages are its low density (1.74 g/cm^3, which compares with 7.87 g/cm^3 for iron and 2.70 g/cm^3 for aluminum, the other important structural metals) and the relative strength of its alloys.

Some important commercial sources of magnesium are the minerals dolomite, $CaCO_3 \cdot MgCO_3$, and magnesite, $MgCO_3$. However, magnesium ion, Mg^{2+}, is the third most abundant dissolved ion in the oceans, after Cl^- and Na^+. The oceans, therefore, are an essentially inexhaustible supply of the ion, from which the metal can be obtained. ◀

Pure magnesium metal is a relatively reactive element. Its alloys, however, contain aluminum and small quantities of other metals to impart both strength and corrosion resistance. Increasing quantities of magnesium alloy are used to make automobile and aircraft parts, as well as consumer materials such as power-tool and lawn-mower housings (Figure 21.18). Most commercial aluminum metal contains some percentage of magnesium, which improves the hardness and corrosion resistance of the aluminum.

Magnesium metal is also used as a reducing agent in the manufacture of titanium and zirconium from their tetrachlorides. (Sodium is also used as the reducing agent in the production of titanium, as we mentioned earlier.)

$$ZrCl_4(g) + 2Mg(l) \longrightarrow Zr(s) + 2MgCl_2(s)$$

Zirconium metal is used to make containment vessels for uranium-235 fuel rods for nuclear power reactors, because of its low absorption of the neutrons needed in the nuclear fission of uranium-235.

Reactions of Magnesium Metal Once magnesium metal has been ignited, it burns vigorously in air to form the oxide.

$$2Mg(s) + O_2(g) \longrightarrow 2MgO(s)$$

The metal powder and fine wire burn readily. Burning magnesium metal gives off an intense white light; this is the white light you see in the burning of some fireworks and flares.

Magnesium also burns in carbon dioxide, producing magnesium oxide and soot, or carbon (Figure 21.19).

$$2Mg(s) + CO_2(g) \longrightarrow 2MgO(s) + C(s)$$

Many fire extinguishers contain carbon dioxide, which normally smothers a fire by preventing oxygen from air getting to combustible materials. Carbon dioxide, however, cannot be used to extinguish magnesium fires; burning magnesium must be smothered with sand.

The pure metal reacts slowly with water. (Magnesium alloys are even less reactive.) Magnesium does react readily with steam, however.

$$Mg(s) + H_2O(g) \longrightarrow MgO(s) + H_2(g)$$

Magnesium Compounds When magnesite, a magnesium carbonate mineral, is heated above 350°C, it decomposes to the oxide, a white solid:

$$MgCO_3(s) \xrightarrow{\Delta} MgO(s) + CO_2(g)$$

Careful heating at low temperature results in a powdery form of the oxide that is relatively reactive. It reacts slowly with water to produce magnesium hydroxide but reacts readily with acids to yield the corresponding salts:

$$MgO(s) + H_2O(l) \longrightarrow Mg(OH)_2(s)$$

$$MgO(s) + 2HCl(aq) \longrightarrow MgCl_2(aq) + H_2O(l)$$

Large quantities of magnesium oxide are used in animal feed supplements, because magnesium ion is an important nutrient for animals. (The ion is also important for human nutrition.) Firebricks are produced by strongly heating magnesite (to about 1400°C) to give a hard, relatively inert form of magnesium oxide. Magnesium oxide is quite stable at high temperatures and is a good thermal insulator.

Magnesium hydroxide is only slightly soluble in water. (The solubility product, K_{sp}, is 1.8×10^{-11}.) A suspension of the white compound in water is called *milk of magnesia* (Figure 21.20). This suspension has a pH of about 10, which means the solution is mildly basic. Milk of magnesia is sold as an antacid, because the magnesium hydroxide reacts to neutralize excess hydrochloric acid in the stomach. Table 21.4 gives a summary list of the uses of compounds of magnesium and other Group IIA elements.

Calcium

Calcium is a common element. It is present in the earth's crust as silicates, which weather to give free calcium ion, Ca^{2+}. The ion is about as abundant in seawater as magnesium ion. Calcium ion is an important nutrient for living organisms. As noted in the preceding section, seashells are principally calcium carbonate. Corals are marine organisms that grow in colonies; their calcium carbonate skeletons eventually form enormous coral reefs in warm waters. The Bahamas and the Florida Keys originated as coral reefs. Deposits of limestone (mostly $CaCO_3$) formed in earlier times as sediments of seashells and coral and by the direct precipitation of calcium carbonate from seawater. Gypsum, $CaSO_4·2H_2O$, is another important mineral of calcium; deposits originated from evaporation of inland lakes and seas.

Calcium Metallurgy Most calcium metal is obtained by the reduction of calcium oxide with aluminum. Calcium metal is used mainly in some alloys. For example, the addition of a small quantity of calcium to the lead used for the electrodes of lead storage batteries substantially reduces the decomposition of water into its

Figure 21.19 ▲

Reaction of magnesium and carbon dioxide When a glowing ribbon of magnesium metal is thrust into a beaker of carbon dioxide (from the sublimation of the dry ice at the bottom of the beaker), the metal bursts into a bright flame, producing a smoke of magnesium oxide and carbon.

Figure 21.20 ▲

Milk of magnesia Milk of magnesia is a suspension of $Mg(OH)_2$ in water. It is used as an antacid and a laxative.

Table 21.4	Uses of Alkaline Earth Compounds
Compound	Use
MgO	Refractory bricks (for furnaces)
	Animal feeds
$Mg(OH)_2$	Source of magnesium for the metal and compounds
	Milk of magnesia (antacid and laxative)
$MgSO_4 \cdot 7H_2O$	Fertilizer
	Medicinal uses (laxative and analgesic)
	Mordant (used in dyeing fabrics)
CaO and $Ca(OH)_2$	Manufacture of steel
	Neutralizer for chemical processing
	Water treatment
	Mortar
	Stack-gas scrubber (to remove H_2S and SO_2)
$CaCO_3$	Paper coating and filter
	Antacids, dentifrices
$CaSO_4$	Plaster, wallboard
	Portland cement
$Ca(H_2PO_4)_2$	Soluble phosphate fertilizer
$BaSO_4$	Oil-well drilling mud
	Gastrointestinal x-ray photography
	Paint pigment

elements during the recharging of the battery. Since gases are not produced during recharging, the battery can be sealed so it requires less maintenance.

Reactions of Calcium Metal Calcium is a soft, reactive metal. It reacts with water, like the alkali metals do, to produce the metal hydroxide and hydrogen; the reaction, however, is much less vigorous.

$$Ca(s) + 2H_2O(l) \longrightarrow Ca(OH)_2(aq) + H_2(g)$$

Calcium burns in air to produce the oxide, CaO, and with chlorine to produce calcium chloride. Calcium also reacts directly with hydrogen to give the hydride. Calcium hydride, CaH_2, is a convenient source of hydrogen, because it reacts easily with water to yield H_2.

$$CaH_2(s) + 2H_2O(l) \longrightarrow Ca(OH)_2(aq) + 2H_2(g)$$

Calcium Compounds Calcium compounds are extremely important commercially (see Table 21.4). Enormous quantities of them are used in metallurgy, in building materials, in the making of glass and paper, and in other products. Limestone ($CaCO_3$), dolomite ($CaCO_3 \cdot MgCO_3$), anhydrite ($CaSO_4$), and gypsum ($CaSO_4 \cdot 2H_2O$) are all important sources of calcium compounds. An interesting use of ground limestone is shown in Figure 21.21; this application depends on the basic character of the carbonate ion. The mineral gypsum, $CaSO_4 \cdot 2H_2O$, forms plaster of Paris, $CaSO_4 \cdot \frac{1}{2}H_2O$, when heated to 100°C. Plaster of Paris is used to make wallboard.

Calcium oxide, CaO, is among the top ten industrial chemicals. It is prepared by calcining calcium carbonate. Limestone and seashells are all possible starting materials.

$$CaCO_3(s) \xrightarrow[900°C]{\Delta} CaO(s) + CO_2(g)$$

Calcium oxide is known commercially as *quicklime* or simply *lime*. Much of the calcium oxide produced is used in the manufacture of iron from its ores. Iron is obtained by reducing iron ore (containing iron oxides with silicate impurities) in a blast furnace. A charge of iron ore, carbon, and calcium oxide is added at the top

Figure 21.21 ◄

Raising the pH of a lake with ground limestone A specially designed barge applies a slurry of finely ground limestone ($CaCO_3$) to a lake to neutralize its acidity, caused by acid-rain pollution. Acid rain results from the burning of sulfur-containing materials (coal and sulfide ores) and from nitrogen oxide pollution from automobiles.

of the furnace, and a stream of compressed air flows in from the bottom. The purpose of calcium oxide is to combine with silicon dioxide (sand) and silicate impurities in the iron ore to produce a glassy material (called *slag*). The slag is molten at the temperatures of the blast furnace, and it flows to the bottom of the furnace and is withdrawn. The essential reaction is that between a basic oxide (calcium oxide) and an acidic oxide (silicon dioxide).

$$CaO(s) + SiO_2(s) \longrightarrow CaSiO_3(l)$$

Basic oxide   Acidic oxide   Calcium silicate slag

Calcium oxide reacts exothermically with water to produce calcium hydroxide. The considerable heat released is apparent when you add a few drops of water to a pile of calcium oxide powder; there is a hiss, and puffs of steam rise from the mixture.

$$CaO(s) + H_2O(l) \longrightarrow Ca(OH)_2(s); \Delta H° = -65.2 \text{ kJ}$$

Commercially, calcium hydroxide is referred to as *slaked lime*. Calcium hydroxide is a strong base, and many of its uses depend on this fact. For example, hydrochloric and sulfuric acids are used to remove rust from steel products. After the steel has been cleaned this way, calcium hydroxide is used to neutralize the excess acid.

Calcium hydroxide solutions react with gaseous carbon dioxide (the acid oxide of carbonic acid) to give a white, milky precipitate of calcium carbonate.

$$Ca(OH)_2(aq) + CO_2(g) \longrightarrow CaCO_3(s) + H_2O(l)$$

This reaction, with its formation of a milky precipitate, is the basis of a simple test for carbon dioxide. Commercially, the reaction is important as a method of preparing precipitated calcium carbonate, a pure form of the finely divided compound. An important use of precipitated calcium carbonate is as a filler for paper. (The purpose of a filler is to improve the paper's characteristics, such as brightness and inking ability.) Precipitated calcium carbonate is also used in toothpowders, antacids, and nutritional supplements (Figure 21.22).

Bricklaying mortar is made by mixing slaked lime, $Ca(OH)_2$, with sand and water. The mortar hardens as the mixture dries and calcium hydroxide crystallizes. Over time, however, the mortar sets to a harder solid as calcium hydroxide reacts with carbon dioxide in the air to form calcium carbonate crystals that intertwine with the sand particles. ►

$$Ca(OH)_2(s) + CO_2(g) \longrightarrow CaCO_3(s) + H_2O(l)$$

Large amounts of quicklime, CaO, and slaked lime, $Ca(OH)_2$, are used to *soften* municipal water supplies. This may seem paradoxical at first, because the process of "softening" water refers to the removal of certain metal ions, particularly calcium ions, from the water. Water that is "hard" contains these metal ions. When soap, which is a sodium salt of a fatty acid, is added to hard water, a curdy precipitate of the calcium salt of the fatty acid forms. (When this precipitate

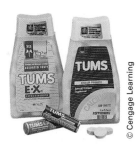

Figure 21.22 ▲

Some antacids contain calcium carbonate These antacid preparations consist of purified calcium carbonate with flavorings and a binder. Such antacids are sometimes prescribed as a nutrient calcium supplement.

Mortar was used by the ancient Romans for buildings and roads.

Figure 21.23 ▶

Reaction of carbon dioxide with calcium hydroxide solution

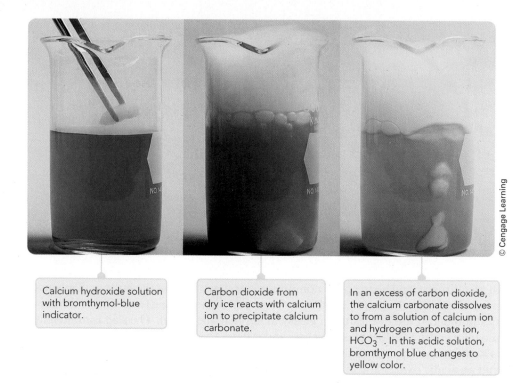

Calcium hydroxide solution with bromthymol-blue indicator.

Carbon dioxide from dry ice reacts with calcium ion to precipitate calcium carbonate.

In an excess of carbon dioxide, the calcium carbonate dissolves to from a solution of calcium ion and hydrogen carbonate ion, HCO_3^-. In this acidic solution, bromthymol blue changes to yellow color.

© Cengage Learning

adheres to the tub, it is called bathtub ring.) The calcium ion in hard water results when water containing carbon dioxide from air (and from rotting leaves and other organic matter) passes through limestone ($CaCO_3$). The calcium carbonate dissolves in this water to produce soluble calcium hydrogen carbonate. ◀

A brief discussion of hard water and water softening appears in the essay on water at the end of Chapter 11.

$$CaCO_3(s) + H_2O(l) + CO_2(g) \longrightarrow Ca^{2+}(aq) + 2HCO_3^-(aq)$$

Limestone   Calcium hydrogen carbonate

Figure 21.23 illustrates the dissolution of calcium carbonate by carbon dioxide. The apparent paradox is resolved when we observe that the addition of a stoichiometric amount of calcium hydroxide to hard water containing calcium hydrogen carbonate can precipitate all of the calcium ions as calcium carbonate.

$$Ca^{2+}(aq) + 2HCO_3^-(aq) + Ca(OH)_2(aq) \longrightarrow 2CaCO_3(s) + 2H_2O(l)$$

21.6 Group IIIA and Group IVA Metals

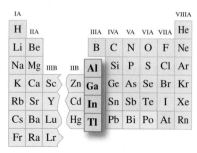

The Group IIIA elements clearly show the trend of increasing metallic character in going down any column of elements in the periodic table. Boron, at the top of Group IIIA, is a metalloid, and its chemistry is typical of a nonmetal. The compound $B(OH)_3$ is actually acidic (boric acid). The rest of the elements (aluminum, gallium, indium, and thallium) are metals, but their hydroxides change from amphoteric (acidic and basic) for aluminum and gallium to basic for indium and thallium.

Bonding to boron is covalent; the atom shares its $2s^2 2p^1$ valence electrons to give the +3 oxidation state. Although aluminum has many covalent compounds, it also has definitely ionic ones, such as AlF_3. The cation $Al^{3+}(aq)$ is present in aqueous solutions of aluminum salts. Gallium, indium, and thallium also have ionic compounds and give cations in aqueous solution. The +3 oxidation state is the only important one for aluminum, but some +1 compounds are known for gallium and indium, and in the case of thallium, both +1 and +3 states are important.

The Group IVA elements also show the trend of greater metallic character in going down the column of elements. The last two elements of the column, tin and lead, are well-known metals, and we will discuss them in this section. Tin and lead, in their compounds, exist in the +2 and +4 oxidation states. In tin, both oxidation states are common. In most lead compounds, lead is in the +2 oxidation state. Lead(IV) compounds, such as PbO_2, are strong oxidizing agents, indicating their tendency to revert to lead(II).

Aluminum

Aluminum is the third most abundant element in the earth's crust (after oxygen and silicon). It occurs primarily in aluminosilicate minerals. The weathering of these rocks results in aluminum-containing clays, which are an essential part of most soils. Further weathering of clay yields bauxite, the principal ore of aluminum, which contains aluminum hydroxide, $Al(OH)_3$, and aluminum oxide hydroxide, $AlO(OH)$. Deposits of bauxite occur throughout the world but are particularly common in tropical and subtropical regions. Corundum is a hard mineral of aluminum oxide, Al_2O_3. The pure oxide is colorless, but the presence of impurities can give various colors to it. Sapphire (usually blue) and ruby (deep red) are gem-quality corundum (Figure 21.24).

Figure 21.24 ▲

Gem-quality corundum This is the Logan sapphire, on exhibit at the Smithsonian Institution in Washington, D.C. It is the largest sapphire on public display in the United States.

Aluminum Metallurgy We discussed the details of aluminum metallurgy in Section 21.2. Aluminum is the most important commercial metal after iron. Despite the fact that pure aluminum is soft and chemically reactive, the addition of a small quantity of other metals, such as copper and magnesium, yields hard, corrosion-resistant alloys. The corrosion resistance of these alloys, together with their relatively low densities, allows their use in structural applications and for containers and packaging (Figure 21.25). Aluminum is also a very good conductor of electricity. Because of this property and its low density, aluminum is used to make electrical transmission wire.

Reactions of Aluminum Metal Aluminum is a chemically reactive metal, although much less reactive than the alkali metals and alkaline earth metals. It does not react at an appreciable rate with water at room temperature. In air, aluminum metal reacts readily with oxygen, but the aluminum oxide that forms gives an adherent coat, which protects the underlying metal from further reaction. (The metal will burn vigorously once started, however.) Thus, unlike iron, which is chemically less reactive than aluminum but corrodes quickly in a moist environment, aluminum is corrosion resistant. The demonstration depicted in Figure 21.26 shows that aluminum does corrode quickly in air in the absence of an oxide coating. That aluminum metal does not normally corrode, or rust, makes the metal extremely useful in many practical applications. However, this property is also the source of an environmental problem. Tin cans (nowadays mostly steel) disintegrate quickly in the environment through rusting, but aluminum cans remain intact for decades. The solution to this problem is to recycle aluminum cans, which also saves on the energy otherwise required in the electrolytic production of the metal.

Aluminum metal is also used to produce other metals. The **Goldschmidt process** is *a method of preparing a metal by reduction of its oxide with powdered aluminum.* Chromium metal is obtained this way; the reaction is highly exothermic, because of the large negative heat of formation of Al_2O_3:

$$Cr_2O_3(s) + 2Al(s) \longrightarrow Al_2O_3(l) + 2Cr(l); \Delta H^\circ = -536 \text{ kJ}$$

A similar reaction with a mixture of iron(III) oxide and aluminum powder (called *thermite*) produces iron for certain kinds of welding. Once the thermite powder is ignited, the reaction is self-sustaining and gives a spectacular incandescent shower. The reaction is also the basis of certain kinds of incendiary bombs.

Aluminum Compounds The most important compound of aluminum is aluminum oxide, Al_2O_3, or alumina. It is prepared by heating aluminum hydroxide, obtained from bauxite, at low temperature (550°C). Alumina is a white powder or porous solid. Although most alumina is used to make aluminum metal, large quantities are used for other purposes. For example, alumina is used as a carrier, or support, for many of the heterogeneous catalysts required in chemical processes, including those used in the production of gasoline (Figure 21.27).

When aluminum oxide is fused (melted) at high temperature (2045°C), it forms corundum, one of the hardest materials known. Corundum is used as an abrasive for grinding tools. When aluminum oxide is fused with small quantities of other

Figure 21.25 ▲

Aluminum products Common items made from aluminum metal.

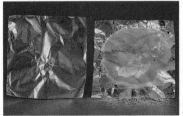

Figure 21.26 ▲

Protective oxide coating on aluminum In air, aluminum metal forms an adherent coating of aluminum oxide that protects the metal from further oxidation. Note the shiny surface of the foil (*left*). When a similar sheet of aluminum (*right*) is coated with mercury, however, needle-like crystals of aluminum oxide form at the aluminum–mercury surface. Instead of adhering to the metal, the aluminum oxide flakes away from the metal, and the aluminum continuously oxidizes in air.

The ruby laser is described in the essay at the end of Section 7.3.

Courtesy of Engelhard Corporation

Figure 21.27 ▲

Heterogeneous catalysts Many heterogeneous catalysts use aluminum oxide as a carrier, or support.

metal oxides, synthetic sapphires and rubies are obtained. Synthetic ruby, for example, contains about 2.5% chromium oxide, Cr_2O_3. Ruby is used in fine instrument bearings ("jewel" bearings) and to make lasers. ◄

Aluminum oxide is used in the manufacture of industrial ceramics. Industrial ceramics are materials made by high-temperature firing (heating) of minerals or inorganic substances. (The term *ceramics* derives from the Greek word *kerimikos*, which means "of pottery," referring to objects made by firing clay.) Ceramics made from aluminum oxide are used to line metallurgical furnaces, and the white ceramic material in automobile spark plugs is made from aluminum oxide. Ceramic fibers composed of aluminum oxide with other metal oxides have been developed for special applications, including ceramic-fiber-reinforced aluminum.

Aluminum sulfate octadecahydrate, $Al_2(SO_4)_3 \cdot 18H_2O$, is the most common soluble salt of aluminum. It is prepared by dissolving bauxite in sulfuric acid. The salt is acidic in aqueous solution. In water, aluminum ion forms a strong hydration complex, $Al(H_2O)_6^{3+}$, and this ion in turn hydrolyzes.

$$Al(H_2O)_6^{3+}(aq) + H_2O(l) \rightleftharpoons Al(H_2O)_5OH^{2+}(aq) + H_3O^+(aq)$$

Until recently, large quantities of aluminum sulfate were used to make paper. Printing paper requires the addition of materials to improve the capacity of the paper to hold ink without the ink spreading. A mixture of rosin (a tree resin) and aluminum sulfate was traditionally used for this purpose. The aluminum ion, however, makes the paper fairly acidic, and the paper fibers deteriorate over time; see Figure 21.28. Acid-free paper is now being produced that is less prone to deterioration.

Aluminum sulfate is also used to treat the wastewater obtained from the process of making paper pulp, a water slurry of fibers obtained from wood. Aluminum sulfate and a base, such as calcium hydroxide, are added to the wastewater, and a gelatinous precipitate of aluminum hydroxide forms.

$$Al^{3+}(aq) + 3OH^-(aq) \longrightarrow Al(OH)_3(s)$$

Colloidal particles of clay and other substances adhere to the precipitate, which is then filtered from the water. The same procedure is one of the steps in the purification of municipal water supplies. The process removes colloidal particles of clay and some bacteria.

Aluminum hydroxide is amphoteric. With acids, aluminum hydroxide acts as a base, as most metal hydroxides do. The reaction is simply a neutralization.

$$Al(OH)_3(s) + 3H_3O^+(aq) \longrightarrow Al^{3+}(aq) + 6H_2O(l)$$

With bases, aluminum hydroxide forms a hydroxo ion (tetrahydroxoaluminate ion, often simply called the aluminate ion). The acidic behavior of the hydroxide is that expected of a nonmetal hydroxide.

$$Al(OH)_3(s) + OH^-(aq) \longrightarrow Al(OH)_4^-(aq)$$

The amphoteric character of aluminum hydroxide is a reflection of the partial nonmetallic character of aluminum. Table 21.5 lists the major uses of aluminum compounds.

Figure 21.28 ▶

Deterioration of paper in books Aluminum sulfate and similar acidic compounds mixed with materials such as clay and rosin were added to paper to improve its printing characteristics. However, the acids decompose the cellulose fibers in paper, causing the paper to deteriorate over time.

© Istockphoto.com/gaffera

Table 21.5   Uses of Aluminum Compounds

Compound	Use
Al_2O_3	Source of aluminum and its compounds
	Abrasive
	Refractory bricks and furnace linings
	Synthetic sapphires and rubies
$Al_2(SO_4)_3 \cdot 18H_2O$	Making of paper
	Water purification
$AlCl_3$	Catalyst in organic reactions
$AlCl_3 \cdot 6H_2O$	Antiperspirant

Tin and Lead

Tin and lead were both known in ancient times. The Egyptians used lead coins and made lead sculptures perhaps as early as 5000 B.C. The use of tin in the form of bronze, an alloy of copper and tin, dates from about 3500 B.C.

Tin is a relatively rare element (ranking 50th or so in abundance in the earth's crust). However, the element occurs in localized deposits of the tin ore cassiterite (SnO_2), so the metal is much more common than you might expect from its abundance in the earth's crust. Lead is more abundant than tin. Its most important ore is galena, a lead(II) sulfide mineral (PbS). Galena is a common mineral and often occurs in association with other important metallic elements, such as silver.

As noted earlier, metallic character increases in moving down a group of elements. Tin, which is between the metalloid germanium and the metal lead in Group IVA, illustrates this periodic trend in a very interesting way: it has two different forms, or allotropes; one is a metal and the other is a nonmetal.

The nonmetallic form of tin, called *gray tin,* is a brittle, gray powder. The metallic form of tin is called *white tin.* White tin is stable above 13°C, but at lower temperatures white tin slowly undergoes a transition to gray tin.

$$\text{White tin} \underset{}{\overset{13°C}{\rightleftharpoons}} \text{gray tin}$$

The transition from white tin to gray tin becomes more rapid if the temperature is much lower than 13°C. During a cold winter in the 1850s, the tin pipes of some church organs in Russia and other parts of Europe began crumbling from what was described then as "tin disease." Tin disease, as we now know, is simply the transition from white to gray tin (Figure 21.29).

Metallurgy of Tin and Lead   Tin metal is obtained from cassiterite. Purified tin(IV) oxide, SnO_2, from tin ore is reduced to the metal by heating with carbon in a furnace.

$$SnO_2(s) + 2C(s) \xrightarrow{\Delta} Sn(l) + 2CO(g)$$

Lead metal is obtained from ores containing galena. The ore is first concentrated in the lead(II) sulfide mineral using physical separation techniques. The concentrated ore is then roasted; that is, the sulfide ore is burned in air to yield lead(II) oxide.

$$2PbS(s) + 3O_2(g) \longrightarrow 2PbO(s) + 2SO_2(g)$$

The fused mass from the roasting is broken up, mixed with coke (carbon), and fed into a blast furnace. Here the lead(II) oxide is reduced with carbon monoxide produced in the blast furnace by partial oxidation of the carbon.

$$PbO(s) + CO(g) \longrightarrow Pb(l) + CO_2(g)$$

Tin is used to make tinplate, which is steel (iron alloy) with a thin coat of tin. The tin coating protects the iron from corrosion from air or acidic materials. Tin cans were originally made from tinplate, although these cans are now mostly aluminum or plastic-coated steel with perhaps a tinplate rim. Tin is also used to make a number of alloys. Solder is a low-melting alloy of tin and lead; bronze is an alloy of copper and tin (Figure 21.30).

Figure 21.29 ▲

Allotropes of tin A bar of metallic tin (white tin) was cooled to −45°C in a solution of tin(IV) ion catalyst. After several hours, an area of gray tin formed on the bar and then grew rapidly until the bar broke apart.

Figure 21.30 ▲

Tin alloys Shown here are solder (and a soldering iron) and bronze.

Lead storage batteries are discussed in Section 19.8.

More than half of the lead produced is used to make electrodes for lead storage batteries. ◄ The manufacture of military and sporting ammunition also consumes a significant fraction of the lead produced. Because lead resists attack from many corrosive substances, the metal is also used to make chemical plant equipment.

Reactions of Tin and Lead Metals Tin and lead are much less reactive than the metals of Groups IA, IIA, and IIIA. Whereas aluminum reacts vigorously with dilute hydrochloric and sulfuric acids, tin reacts only slowly with these acids. Tin reacts more rapidly with the concentrated acids. The products are tin(II) ion and hydrogen:

$$Sn(s) + 2HCl(aq) \longrightarrow SnCl_2(aq) + H_2(g)$$
$$Sn(s) + H_2SO_4(aq) \longrightarrow SnSO_4(aq) + H_2(g)$$

Lead metal reacts with these acids, but the products $PbCl_2$ and $PbSO_4$ are insoluble and adhere to the metal. As a result, the reaction soon stops.

$$Pb(s) + 2HCl(aq) \longrightarrow PbCl_2(s) + H_2(g)$$
$$Pb(s) + H_2SO_4(aq) \longrightarrow PbSO_4(s) + H_2(g)$$

Tin and Lead Compounds Tin(II) chloride, $SnCl_2$, is used as a reducing agent in the preparation of dyes and other organic compounds. In reactions where tin(II) ion acts as a reducing agent, tin(II) ion oxidizes to tin(IV) species. Tin(II) compounds are commonly referred to as *stannous* compounds, using an older naming system. Thus, tin(II) chloride is commonly called stannous chloride. Tin also has a number of tin(IV), or *stannic,* compounds. Tin(IV) oxide, SnO_2, is, as we have seen, the chemical substance in the mineral cassiterite. Tin(IV) chloride, $SnCl_4$, is a liquid; it freezes at $-33°C$. (These properties indicate that the substance is molecular, rather than ionic.)

Lead also exists in compounds in the $+2$ and $+4$ oxidation states, although lead(II) compounds are the more common. The starting compound for preparing most lead compounds is lead(II) oxide, PbO. This is a reddish yellow solid, commercially called *litharge.* Lead(II) oxide is prepared by exposing molten lead to air. Lead(IV) oxide, PbO_2, is a dark brown or black powder; it forms the cathode of lead storage batteries. The cathode is made by packing a paste of lead(II) oxide (litharge) into a lead metal grid. When the battery is charged, the lead(II) oxide is oxidized to lead(IV) oxide. Table 21.6 gives a summary list of the uses of tin and lead compounds.

Table 21.6   **Uses of Tin and Lead Compounds**

Compound	Use
SnO_2	Manufacture of tin compounds
	Glazes and enamels
$SnCl_2$	Reducing agent in preparing organic compounds
PbO	Lead storage batteries
	Lead glass
PbO_2	Cathode in lead storage batteries
Pb_3O_4	Pigment for painting structural steel

Chemistry of the Nonmetals

The nonmetals are elements that do not exhibit the characteristics of a metal. Nearly half of them are colorless gases; and until the eighteenth century, gases were referred to as "airs" and were not well differentiated. In contrast, the special characteristics of metals held special fascination for early humans—only fire was more fascinating. And both of the two nonmetals known to the ancients, carbon and sulfur, are associated with fire.

Carbon was known in the form of charcoal and lampblack, or soot, which are products of fire. Charcoal may have been so common that it was hardly noticed at first, until its role in the reduction of metal ores was discovered. Lampblack was used by the ancient Egyptians to produce ink for writing on papyrus.

Free sulfur was less widely available than carbon, although it was probably well known because of its ready occurrence in volcanic areas. No doubt its yellow color made it stand out among other rocks. That sulfur also burned with a beautiful blue flame made it especially distinctive. The old English name of sulfur was *brimstone,* which means "a stone that burns." This term survives today in the expression "fire and brimstone."

These two nonmetals have rather distinctive physical characteristics, as do some of the nonmetals discovered later. For example, white phosphorus is a waxy, white solid; bromine is a reddish brown liquid; chlorine is a greenish yellow gas. Some nonmetallic elements are shown in Figure 21.31.

In the remainder of this chapter, we look at the chemistry of the nonmetals. Some questions we will address are these: What are the chemical and physical properties of the more important nonmetallic elements? What are some of the commercial uses of these substances and their compounds?

Figure 21.31 ▲

Some nonmetals *Left to right:* Sulfur, bromine, white phosphorus, and carbon.

21.7 Hydrogen

Hydrogen is the most abundant element in the universe, comprising nearly 90% of all atoms, and is the third most abundant element on the surface of the earth. (Oxygen and silicon are the most abundant.) Most stars, including our sun, consist primarily of hydrogen. The hydrogen in our sun is the fuel for the nuclear fusion reactions that produce the life-sustaining energy that reaches our planet. On earth, the majority of hydrogen is found in oceans combined with oxygen as water. Hydrogen occurs in a variety of compounds that you have had and will have the opportunity to study as a part of your general chemistry course. These include the organic compounds (Chapter 23), biologically important compounds (Chapter 24), and acids and bases (Chapters 4 and 15).

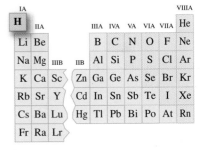

Properties and Preparation of Hydrogen

Hydrogen was first isolated and identified by Henry Cavendish in 1766. His experiments consisted of reacting iron, zinc, and tin with several different binary acids. For example, $H_2(g)$ can easily be produced on a small scale according to the reaction

$$2HCl(g) + Zn(s) \longrightarrow ZnCl_2(aq) + H_2(g)$$

Hydrogen is a colorless, odorless gas that is less dense than air. Even though hydrogen is often placed in Group IA of the periodic table, aside from its valence electron configuration, it has little in common with the alkali metals that make up the rest of Group IA. It is much less likely to form a cation than any of the alkali metals, with a first ionization energy of 1312 kJ/mol versus 520 kJ/mol for lithium. Because of this, hydrogen generally combines with nonmetallic elements to form covalent compounds such as CH_4, H_2S, and PH_3.

There are three isotopes of hydrogen: *protium,* ${}^1_1H$ or H, which is the most abundant; *deuterium,* ${}^2_1H$ or D; and *tritium,* ${}^3_1H$ or T. All three isotopes are naturally occurring; however, only 0.0156% is D and a trace is T. Because an atom of D has about twice the mass of protium, compounds that contain deuterium often have different properties than those that contain only protium. For example, the normal boiling point of D_2O is 101.42°C versus 100.00°C for H_2O. Tritium is produced naturally in the upper atmosphere by nuclear reactions that are induced by cosmic rays or in a nuclear reactor by bombarding lithium-6 with neutrons:

$$^6_3Li + {}^1_0n \longrightarrow {}^3_1H + {}^4_2He$$

Tritium is radioactive with a half-life of 12.3 years, hence very little of what is naturally produced in the upper atmosphere reaches the surface of the earth.

The isotopes of hydrogen find many applications. They are used as markers or labels that can be followed during chemical reactions. For example, a chemist who was interested in determining whether hydrogen atoms move between water molecules could make a solution that contains DOD and HOH. If hydrogen atom transfer takes place (which is the case in this example), then the solution, after a period of time, would be expected to contain DOD, HOH, and the new compound DOH. Information about the exchange rate of H with D can also be obtained from the experiment by measuring the rate at which the DOH forms. Because so many compounds contain hydrogen, the hydrogen isotopes are widely used in this manner.

The elemental form of hydrogen is a diatomic molecule having a bond dissociation energy of 432 kJ/mol. This is a large value when compared with chlorine at 240 kJ/mol. This relatively high bond dissociation energy indicates why hydrogen is less reactive than its halogen counterparts. However, with the addition of heat or light, or in the presence of a suitable catalyst, hydrogen can be induced to react.

Hydrogen is currently produced on a massive industrial scale, with an annual U.S. production on the order of 10^{10} m^3. Approximately 40% of this production is used to manufacture ammonia by the Haber process:

$$N_2(g) + 3H_2(g) \longrightarrow 2NH_3(g)$$

The hydrogen for this reaction is generally prepared using the **steam-reforming process** *where steam and hydrocarbons from natural gas or petroleum react at high temperature and pressure in the presence of a catalyst to form carbon monoxide and hydrogen.* For example:

$$C_3H_8(g) + 3H_2O(g) \xrightarrow[\Delta]{Ni} 3CO(g) + 7H_2(g)$$

Another route for hydrogen production is the *water–gas reaction,* which is no longer used commercially but may become important in the future as natural gas and petroleum become more expensive and scarce. In this reaction, steam is passed over red-hot coke or coal.

$$C(s) + H_2O(g) \longrightarrow CO(g) + H_2(g)$$

Both of these reactions produce a mixture of hydrogen and carbon monoxide. Such mixtures are used to produce various organic compounds, but to obtain pure hydrogen the carbon monoxide must be removed. First the carbon monoxide is reacted with steam in the presence of a catalyst to give carbon dioxide and more hydrogen.

$$CO(g) + H_2O(g) \xrightarrow[\Delta]{catalyst} CO_2(g) + H_2(g)$$

The carbon dioxide is then removed by passing the mixture of gases through a basic aqueous solution.

Hydrogen can also be produced via the electrolysis of water. The net reaction is

$$2H_2O(l) \longrightarrow 2H_2(g) + O_2(g)$$

Due to the cost of electricity, electrolysis is often not economical. However, hydrogen is produced economically as a by-product of the electrolysis of aqueous NaCl solutions during the production of NaOH and chlorine. ◄

See Section 19.10 for a discussion of the electrolysis of sodium chloride solutions.

Hydrogen Reactions and Compounds

In addition to the preparation of ammonia, the other major use of hydrogen is in the petrochemical industry. In many cases the reaction is one where hydrogen is added to hydrocarbon compounds containing carbon–carbon double bonds to produce compounds that contain carbon–carbon single bonds. For example, 1-butene can be reacted with hydrogen using a platinum or palladium catalyst to produce butane.

$$CH_3CH_2CH{=}CH_2 + H_2 \xrightarrow{Pt} CH_3CH_2CH_2CH_3$$

This process, called *hydrogenation,* is used in the food processing industry where oils (liquids) that contain many carbon–carbon double bonds are converted to fats (solids) that contain few or no carbon–carbon double bonds. Another important process that requires hydrogen is the cobalt-catalyzed *synthesis gas* reaction with carbon monoxide to produce methanol (vapor).

$$CO(g) + 2H_2(g) \xrightarrow{\text{cobalt catalyst}} CH_3OH(g)$$

Hydrogen is also used to reduce metal oxides to extract pure metals. For example, tungsten(VI) oxide can be reduced at high temperatures via the reaction

$$WO_3(s) + 3H_2(g) \longrightarrow W(s) + 3H_2O(g)$$

When hydrogen combines with another element it forms a **binary hydride**—that is, *a compound that contains hydrogen and one other element*. There are three categories of binary hydrides: ionic hydrides, covalent hydrides, and metallic hydrides.

Ionic hydrides, which contain the hydride ion, H^-, can be directly formed via the reaction of an alkali metal or the larger Group IIA metals (Ca, Sr, and Ba) with hydrogen gas near 400°C.

$$2Li(s) + H_2(g) \longrightarrow 2LiH(s)$$

$$Ba(s) + H_2(g) \longrightarrow BaH_2(s)$$

These hydrides are white crystalline compounds in which the H atoms have an oxidation state of -1. Ionic hydrides can undergo an oxidation–reduction reaction with water to produce hydrogen and a basic solution. For example:

$$LiH(s) + H_2O(l) \longrightarrow H_2(g) + LiOH(aq)$$

Because of this, hydrides can be used as a source of hydrogen gas where transportation of $H_2(g)$ is impractical, such as for inflating weather balloons. Ionic hydrides are also used as reducing agents (a source of electrons) during chemical reactions.

Covalent hydrides are molecular compounds in which hydrogen is covalently bonded to another element. Examples of these compounds are NH_3, H_2O, H_2O_2, and HF. Some of these compounds often can be formed from the direct reaction of the elements. If the nonmetal reacting with hydrogen is reactive, the reaction will readily occur without the need for elevated temperatures or a catalyst:

$$F_2(g) + H_2(g) \longrightarrow 2HF(g)$$

The reaction of hydrogen with oxygen to form water is an example of a reaction that requires the input of energy to get started; however, once it does, the reaction is rapid and exothermic.

$$2H_2(g) + O_2(g) \longrightarrow 2H_2O(g) \qquad \Delta H = -484 \text{ kJ}$$

Because it is such an exothermic reaction and the product is a gas, it is an ideal rocket fuel (Figure 21.32).

The combustion of hydrogen produces more heat per gram than any other fuel (120 kJ/g). Unlike hydrocarbons, it is a "clean fuel" because the product (water) is environmentally benign. Because of these features, hydrogen may become the favorite fuel of the twenty-first century. (An energy source will be needed to produce the hydrogen, however.)

Metallic hydrides are compounds containing a transition metal and hydrogen. Generally, the formula of these compounds is MH_x, where x is often not an integer. These compounds contain hydrogen atoms that are spread throughout a metal crystal occupying the holes in the crystal lattice. Often, hydrogen atoms enter the holes in nonstoichiometric amounts. The result is that the composition of the metallic hydride is variable. For example, under one pressure of H_2 the composition of the metallic hydride MH_x might be $MH_{0.4}$, whereas at a higher pressure of H_2 it might be $MH_{0.5}$. Examples of these compounds are $TiH_{1.7}$ and $ZrH_{1.9}$.

21.8 Group IVA: The Carbon Family

The elements of Group IVA show in more striking fashion than the previous groups the normal periodic trend of greater metallic character going down a column. Carbon, the first element in the group, is distinctly nonmetallic. Silicon and germanium are metalloids, or semimetals, although their chemical properties are primarily those of nonmetals. Tin and lead, the last two elements in the group, are metals.

© Robert Sullivan/AFP/Getty Images

Figure 21.32 ▲

A liquid-hydrogen storage tank
Liquid hydrogen is used as a rocket fuel.

In the following subsections, we will look at carbon and silicon. Both elements have many important compounds that exhibit tetrahedral (sp^3) bonding. Carbon, however, also has many compounds that contain multiple bonds in which sp and sp^2 bonding occurs; silicon has very few such compounds. Another difference in the bonding characteristics of these elements is silicon's ability to use sp^3d^2 hybrid orbitals in octahedral bonding, which is not possible with carbon. The hexafluorosilicate ion, SiF_6^{2-}, is an example of such bonding.

Carbon

One of the most important features of the carbon atom is its ability to bond to other carbon atoms to form chains and rings of enormous variety. *The covalent bonding of two or more atoms of the same element to one another* is referred to as **catenation**. Although other elements display catenation, none show it to the same degree as carbon. Millions of carbon compounds are known, most classified as *organic*. These can be thought of as derivatives of hydrocarbons, each consisting of a chain (or more complicated arrangement) of carbon atoms to which hydrogen atoms are bonded. In this section, we will confine ourselves to the element and its oxides and carbonates.

Allotropes of Carbon Until recently, carbon was thought to occur in only two principal allotropic forms: diamond and graphite. Both allotropes are covalent network solids, whose structures we discussed in detail in Section 11.8. In diamond, each carbon atom is tetrahedrally (sp^3) bonded to four other carbon atoms. To move one plane of atoms in the diamond crystal relative to another requires the breaking of many strong carbon–carbon bonds. Because of this, diamond is one of the hardest substances known. As a pure substance, diamond is colorless, although natural diamond may be colored by impurities.

Graphite is a black substance having a layer structure. Each layer consists of carbon atoms bonded to three other carbon atoms to give a hexagonal pattern of carbon atoms arranged in a plane. The bonding involves sp^2 hybridization of the carbon atoms with delocalized π electrons. You can also describe the bonding in terms of resonance formulas with alternating single and double bonds. One layer of carbon atoms in graphite is held to another layer only by van der Waals forces. Because of the relative weakness of these forces, the layers in graphite easily slide over one another, resulting in a substance that is soft and slippery. Graphite, unlike diamond, is a good electrical conductor, because of the delocalized bonding within layers.

In 1985 a third allotropic form of carbon, known as buckminsterfullerene (C_{60}), was discovered. This and similar forms of carbon have been studied extensively. The buckminsterfullerene molecule has a stable "soccer-ball" structure, described in the essay at the end of this section.

Carbon Black The form of carbon known as *carbon black* is composed of extremely small crystals of carbon having an amorphous, or imperfect, graphite structure. It is produced in large quantities by burning natural gas (CH_4) or petroleum hydrocarbons in a limited supply of air so that heat "cracks" (or breaks bonds in) the hydrocarbon. Lampblack is a form of carbon black.

$$CH_4(g) \xrightarrow{\Delta} \underset{\text{Carbon black}}{C(s)} + 2H_2(g)$$

Carbon black is used in the manufacture of rubber tires (to increase wear) and as a pigment in black printing inks. Coke is an amorphous carbon obtained by heating coal in the absence of air; it is used in large quantities in the making of iron.

Oxides of Carbon Carbon has two principal oxides: carbon monoxide, CO, and carbon dioxide, CO_2. Carbon and organic compounds burn in an excess of oxygen to give carbon dioxide, CO_2. However, an equilibrium exists among carbon, carbon monoxide, and carbon dioxide that favors carbon monoxide above 700°C.

$$CO_2(g) + C(s) \rightleftharpoons 2CO(g)$$

For this reason, carbon monoxide is almost always one of the products of combustion of carbon and organic compounds, unless an excess of oxygen is present, in which case the carbon monoxide burns to carbon dioxide.

These two oxides of carbon are quite different in their chemical and physiological properties. Carbon monoxide is a colorless, odorless gas that burns in air with a blue flame. It is a toxic gas, which poisons by attaching strongly to iron atoms in the hemoglobin of red blood cells, preventing them from carrying out their normal oxygen-carrying function. As a result, the cells of the body become starved for oxygen.

Carbon monoxide is manufactured industrially from natural gas (CH_4) and petroleum hydrocarbons, either by reaction with steam or by partial oxidation. For example,

$$CH_4(g) + H_2O(g) \xrightarrow{Ni} CO(g) + 3H_2(g)$$
$$2CH_4(g) + O_2(g) \longrightarrow 2CO(g) + 4H_2(g)$$

The product in these reactions is a mixture of carbon monoxide and hydrogen, which is called *synthesis gas*. Synthesis gas can yield any of a number of organic products depending on the reaction conditions and catalyst. Methanol, CH_3OH, for example, is produced in large quantities from synthesis gas.

$$CO(g) + 2H_2(g) \xrightarrow{catalyst} CH_3OH(l)$$

Carbon dioxide is a colorless, odorless gas with a faint acid taste. Under normal circumstances, the gas is nontoxic, although at high concentrations it interferes with respiration. Carbon dioxide does not support most combustions, which makes CO_2 useful as a fire extinguisher.

Carbon dioxide is produced whenever carbon or organic materials are burned. For example,

$$CH_4(g) + 2O_2(g) \longrightarrow CO_2(g) + 2H_2O(g)$$
$$CS_2(l) + 3O_2(g) \longrightarrow CO_2(g) + 2SO_2(g)$$
$$C_2H_5OH(l) + 3O_2(g) \longrightarrow 2CO_2(g) + 3H_2O(g)$$

Carbon dioxide is obtained commercially as a by-product in the production of ammonia (see Section 21.9) and in the calcining (strong heating) of limestone to give calcium oxide. Liquid carbon dioxide and solid carbon dioxide (dry ice) are used in large quantities as refrigerants. Carbonated beverages are made by dissolving carbon dioxide gas under pressure in an aqueous solution of sugar and flavorings. Carbonated water is acidic as the result of the formation of carbonic acid, although carbonated beverages frequently also contain fruit acids and phosphoric acid.

Carbonates Carbon dioxide dissolves in water to form an aqueous solution of carbonic acid.

$$CO_2(g) + H_2O(l) \rightleftharpoons H_2CO_3(aq)$$

The acid is diprotic (has two acidic H atoms per molecule) and dissociates to form hydrogen carbonate ion and carbonate ion.

$$H_2CO_3(aq) + H_2O(l) \rightleftharpoons H_3O^+(aq) + HCO_3^-(aq)$$
$$HCO_3^-(aq) + H_2O(l) \rightleftharpoons H_3O^+(aq) + CO_3^{2-}(aq)$$

Carbonic acid has never been isolated from solution, but its salts, hydrogen carbonates and carbonates, are well known. When you bubble carbon dioxide gas into an aqueous solution of calcium hydroxide, a milky white precipitate of calcium carbonate forms (Figure 21.33).

$$CO_2(g) + Ca(OH)_2(aq) \longrightarrow CaCO_3(s) + H_2O(l)$$

This is a standard test for carbon dioxide. The reaction is also used to manufacture a pure calcium carbonate for antacids and other products.

Carbonate minerals are very common and many are of commercial importance. Limestone contains the mineral calcite, which is calcium carbonate, $CaCO_3$. Much of this was formed by marine organisms, although some limestone was also formed by direct precipitation from water solution.

© Cengage Learning

Figure 21.33 ▲

Test for carbon dioxide When carbon dioxide is bubbled into a solution of calcium hydroxide (limewater), a milky white precipitate of calcium carbonate forms. This is the basis of a test for carbon dioxide. The reaction is also used to manufacture pure calcium carbonate. (The needle at the right is to provide an escape for excess gas.)

Table 21.7 Uses of Some Compounds of Carbon and of Silicon

Compound	Use
CO	Fuel; reducing agent
	Synthesis of methanol, CH_3OH
CO_2	Refrigerant
	Carbonation of beverages
SiO_2	Source of silicon and its compounds
	Abrasives
	Glass
$(CH_3)_2SiCl_2$	Manufacture of silicones (used as lubricants, hydraulic fluids, caulking compounds, and medical implants)

Table 21.7 lists uses of some compounds of carbon (and of silicon, discussed in the next subsection).

Silicon

If we were to list the materials of technology, silicon would certainly be prominent on that list. Silicon is the basic material in the semiconductor devices that make up CD players, computers, and other electronics gear. Oxygen compounds of silicon also are important to technology. Quartz crystals, a form of silica (also called silicon dioxide, SiO_2), for example, are used to control frequencies in radio transmitters and watches. And many practical materials such as cement and bricks are silicate materials. Silicates are compounds of silicon and oxygen with one or more metallic elements.

Silicon occurs in the earth's crust, the outermost solid layer of the planet, as compounds with oxygen. About 95% of the earth's crust is silica and silicate rocks and minerals. Elemental silicon is obtained by reducing quartz sand (SiO_2) with coke (C) in an electric furnace at 3000°C.

$$SiO_2(l) + 2C(s) \longrightarrow Si(l) + 2CO(g)$$

Silicon has a diamond-like structure (in which silicon atoms bond tetrahedrally to four other silicon atoms). It is a hard, lustrous gray solid and is used to make alloys and solid-state electronic devices.

For the manufacture of solid-state devices, it is necessary to start with extremely pure silicon (no more than 10^{-8}% impurities). See Figure 21.34. You first convert the impure element to silicon tetrachloride, $SiCl_4$, which is a low-boiling liquid (b.p. 58°C) that can be purified by distillation.

$$Si(s) + 2Cl_2(g) \longrightarrow SiCl_4(l)$$

You then reduce the purified silicon tetrachloride by passing the vapor with hydrogen through a hot tube, where pure silicon crystallizes on the surface of a pure silicon rod.

$$SiCl_4(g) + 2H_2(g) \longrightarrow Si(s) + 4HCl(g)$$

Silica (Silicon Dioxide) Silica (whose formal chemical name is silicon dioxide, SiO_2) has several different crystalline forms. The most important of these is quartz, which is a constituent of many rocks. It is the weathering of these rocks that releases quartz particles, a major component of most kinds of sands. Amethyst is a gem form of quartz; it contains a small quantity of Fe_2O_3, which is believed to be responsible for its purple color. **Silica**, SiO_2, is *a covalent network solid in which each silicon atom is covalently bonded in tetrahedral directions to four oxygen atoms; each oxygen atom is in turn bonded to another silicon atom* (Figure 21.35).

Quartz crystals have a very interesting and useful property: they exhibit the *piezoelectric effect.* In a piezoelectric crystal, such as quartz, compression of the crystal in a particular direction causes an electric voltage to develop across it.

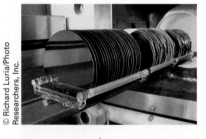

Figure 21.34 ▲

High-purity silicon rod and wafers cut from it Silicon wafers form the base material for integrated circuit chips used in solid-state electronic devices.

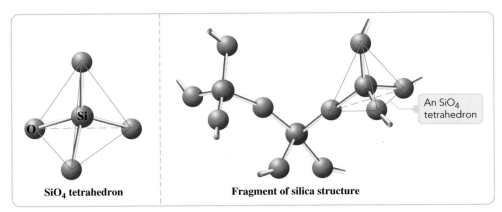

SiO4 tetrahedron   Fragment of silica structure

An SiO4 tetrahedron

Figure 21.35 ◀

Structure of silica (SiO₂)
Left: A silicon atom is bonded tetrahedrally to four oxygen atoms giving an SiO₄ tetrahedron. Each of the oxygen atoms on this tetrahedron bonds to silicon atoms on other tetrahedra, giving a three-dimensional structure. *Right:* Shown is a fragment of silica structure with three SiO₄ tetrahedra bonded together.

Such crystals are used in phonograph pickups and microphones to convert sound vibrations to alternating electric currents. The opposite effect is also possible: an alternating electric current applied to a piezoelectric crystal can make it vibrate. When cut to precise dimensions, the crystal responds most strongly to a certain vibrational frequency. Such crystals are used to control the frequency of an alternating electric current. When the alternating current frequency deviates from the natural frequency of the crystal, a feedback mechanism adjusts the alternating current frequency. Quartz crystals are used to control radio and television frequencies, as well as clocks.

Amethyst

Silicates A **silicate** is *a compound of silicon and oxygen (with one or more metals) that may be formally regarded as a derivative of silicic acid, H_4SiO_4 or $Si(OH)_4$.* Silicic acid has never been isolated, although solutions containing anions of the acid are well known. When silica is melted with sodium carbonate, it forms a soluble material referred to commercially as *water glass.* These solutions contain various silicate ions, such as

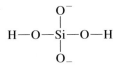

The solution also contains ions with two or more silicon atoms, which form by condensation reactions. A **condensation reaction** is *a reaction in which two molecules or ions are chemically joined by the elimination of a small molecule such as H_2O.* For example, the silicic acid anion can react in a condensation reaction with another such anion to form a disilicate anion (anion of two silicon atoms):

$$
H{-}O{-}\underset{\underset{O_-}{|}}{\overset{\overset{O^-}{|}}{Si}}{-}O{-}H \;+\; H{-}O{-}\underset{\underset{O_-}{|}}{\overset{\overset{O^-}{|}}{Si}}{-}O{-}H \;\longrightarrow\; H{-}O{-}\underset{\underset{O_-}{|}}{\overset{\overset{O^-}{|}}{Si}}{-}O{-}\underset{\underset{O_-}{|}}{\overset{\overset{O^-}{|}}{Si}}{-}O{-}H \;+\; H_2O
$$

Silicate anions containing long Si—O—S— . . . chains can result from such condensation reactions. Materials exhibiting chains and networks of this sort of silicon–oxygen bonding are common. The silicate minerals of the earth all have this type of structure. (See Figure 21.36.)

Silicones A **silicone** is *a polymer containing chains of silicon–oxygen bonds, with hydrocarbon groups (such as $CH_3{-}$) attached to silicon atoms.* (Polymers are very large molecules made up of smaller molecules repeatedly linked together. See Section 2.6.) The preparation of many silicones begins with the reaction of silicon with methyl chloride at elevated temperature in the presence of a copper catalyst.

$$
Si(s) + 2CH_3Cl(g) \xrightarrow[\text{300°C}]{\text{catalyst}} (CH_3)_2SiCl_2(g)
$$

© Karuka/Shutterstock.com

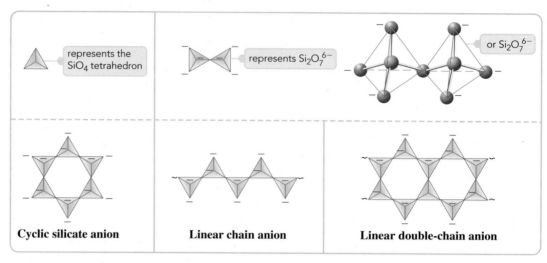

Cyclic silicate anion   **Linear chain anion**   **Linear double-chain anion**

Figure 21.36 ▲

Structures of some silicate minerals A number of silicate minerals consist of finite silicate anions, such as SiO_4^{4-} and $Si_2O_7^{6-}$. Beryl contains the cyclic anion $Si_6O_{18}^{12-}$. Other silicate minerals have anions with very long chains or double chains.

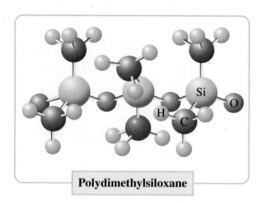

Polydimethylsiloxane

Figure 21.37 ▲

Molecular model of a silicone This model shows a fragment of the polydimethylsiloxane molecule. Note how the chain structure is created by linking $(CH_3)_2SiO_2$ tetrahedra through Si—O bonds.

Figure 21.38 ▲

Products that contain silicone

If a long chain of Si—O bonds is desired, the product of this reaction is reacted with water.

$$(CH_3)_2SiCl_2(l) + 2H_2O(l) \longrightarrow (CH_3)_2Si(OH)_2(l) + 2HCl(g)$$

The silicon-containing product of this reaction then undergoes a condensation reaction to form a silicone oil (polydimethylsiloxane) and water. (See Figure 21.37.)

$$n(CH_3)Si(OH)_2 \longrightarrow \left[\begin{array}{c} CH_3 \\ | \\ Si-O- \\ | \\ CH_3 \end{array} \right]_n + nH_2O$$

Depending on the application for the silicone, the chain length can be varied to make oils of different weights, or the chains can be linked together to form *elastomers* (rubbers). The methyl group, CH_3, can also be replaced by a different organic group, which can lead to the formation of hard materials called resins. Silicones have a wide variety of applications (Figure 21.38).

21.9 Group VA: Nitrogen and the Phosphorus Family

Like the carbon family of elements, the Group VA elements show the distinct trend of increasing metallic character as you go from top to bottom of the column. The first members, nitrogen and phosphorus, are nonmetallic; arsenic and antimony are metalloids; bismuth is a metal.

As expected for a second-period element, nitrogen is in many respects different from the other elements in its group. You can see this in the formulas of the elements and compounds. Elementary nitrogen is N_2, while white phosphorus is P_4. Similarly, the common +5 oxoacid of nitrogen is HNO_3, that of phosphorus is H_3PO_4, and that of arsenic is H_3AsO_4. (Nitrogen exists in many molecular compounds in a variety of oxidation states. Phosphorus exists in many compounds in the +5 and +3 oxidation states. Many phosphorus compounds have P—O—P bonding.)

Nitrogen

The element nitrogen is crucial to life: it is a component of all proteins, which are involved in almost every biochemical process that occurs in living organisms. Most of the available nitrogen on earth, however, is present as nitrogen gas (dinitrogen, N_2)

Buckminsterfullerene— A Third Form of Carbon

Until recently, carbon was thought to occur in only two principal forms: diamond and graphite, both network solids. In 1985, Harold W. Kroto (from the University of Sussex in Brighton, England) approached Richard E. Smalley and Robert F. Curl (at Rice University, in Houston, Texas) to do some experiments to simulate the conditions in certain stars to see what sorts of carbon-containing molecules might be produced. The research group at Rice had previously constructed an instrument in which they used an intense laser beam to vaporize solids. The hot vapor produced in this way could then be directed as a molecular beam into a mass spectrometer, where the molecular masses of the species in the vapor could be measured.

The experiments on the vaporization of graphite produced surprising results. Molecular clusters of 2 to 30 carbon atoms were found, as expected; but in addition, the mass spectrum of the vapor consistently showed the presence of a particularly abundant molecule, C_{60}. Why was this molecule so stable? Kroto, Smalley, and Curl wrestled with this problem and eventually came to the conclusion that the molecule must be like a piece of graphite sheet that somehow closed back on itself to form a closed-dome structure. Kroto recalled how he once built a cardboard dome with the night sky printed on it for his children; he thought it contained not only hexagons, as in graphite, but also pentagons. Smalley set about trying to construct a model of C_{60} by gluing paper polygons together. He discovered that he could obtain a very stable, closed polygon with 60 vertices by starting with a pentagon and attaching hexagons to each of its five edges. To this bowl-shaped structure he attached more pentagons and hexagons, producing a paper soccer ball stable enough to bounce on the floor. The molecular structure is shown in Figure 21.39. Kroto and Smalley named the molecule buckminsterfullerene, after R. Buckminster Fuller, who studied closed-dome architectural structures constructed from polygons.

In 1990, buckminsterfullerene was prepared in gram quantities. Once the reddish brown substance was available in sufficient quantity, researchers were able to verify the soccer-ball structure of the C_{60} molecule. Since the discovery of buckminsterfullerene, many similar molecular forms of carbon, called fullerenes, have been found and studied.

In 1996, the Nobel Prize in chemistry was awarded to professors Curl, Kroto, and Smalley for their discovery.

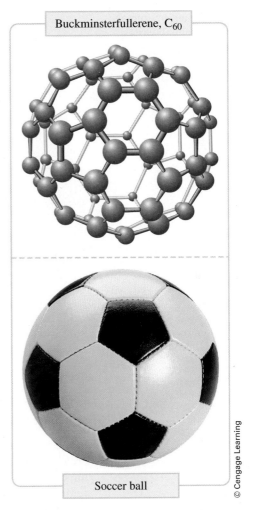

Buckminsterfullerene, C_{60}

Soccer ball

© Cengage Learning

Figure 21.39 ▲

Structure of buckminsterfullerene *Top:* A molecular model of C_{60}. Carbon atoms are at the corners of each polygon. Each bond is intermediate between a single and a double bond, similar to the bonding in graphite. *Bottom:* The buckminsterfullerene molecule is often called a "buckyball," because of its soccer-ball appearance.

■ See Problems 21.193 and 21.194.

in the atmosphere, which consists of 78.1% N_2, by mass. Dinitrogen, also simply called "nitrogen," has collected in the atmosphere because of its relative chemical unreactivity. Most organisms cannot use dinitrogen from the atmosphere as their source of the element. However, certain soil bacteria, as well as bacteria that live in nodules on the roots of beans, clover, and similar plants, can "fix" nitrogen; that is, they convert dinitrogen to ammonium and nitrate compounds. Plants use these simple nitrogen compounds to make proteins and other complex nitrogen compounds. Animals eat these plants, and other animals eat those animals. Finally,

Figure 21.40 ▶

The nitrogen cycle Nitrogen, N_2, is fixed (converted to compounds) by bacteria, by lightning, and by the industrial synthesis of ammonia. Fixed nitrogen is used by plants and enters the food chain of animals. Later, plant and animal wastes decompose. Denitrifying bacteria complete the cycle by producing free nitrogen again.

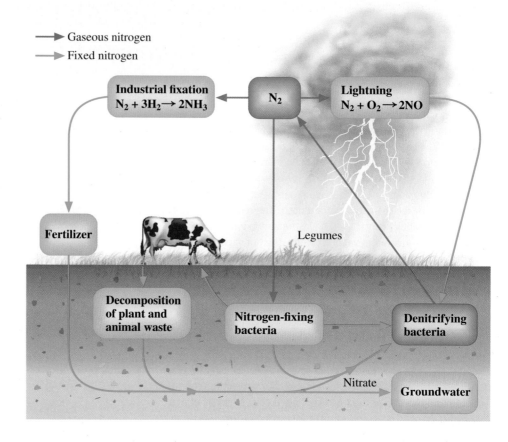

bacteria in decaying organic matter convert the nitrogen compounds back to dinitrogen. In this way, nitrogen in our environment continually cycles from dinitrogen to living organisms and back. Figure 21.40 depicts this *nitrogen cycle.*

Properties and Uses of Nitrogen Daniel Rutherford, a chemist and physician, discovered nitrogen, N_2, in air in 1772. In his experiments, he removed the oxygen from air by burning a substance in it. When burning carbon-containing substances, he removed the carbon dioxide that formed by reacting it with aqueous potassium hydroxide. He then showed that the residual gas would no longer support either combustion or living organisms. Although this residual gas is mostly N_2, it does contain small amounts of noble gases.

Nitrogen, N_2, is a relatively unreactive element because of the stability of the nitrogen–nitrogen triple bond. (The N≡N bond energy is 945 kJ/mol, compared with 163 kJ/mol for the N—N bond energy.) When substances burn in air, they generally react with oxygen, leaving the nitrogen unreacted. Some very reactive metals do react directly with nitrogen, however. For example, when magnesium metal burns in air, it forms the nitride, as well as the oxide.

$$3Mg(s) + N_2(g) \longrightarrow Mg_3N_2(s)$$

Because the nitride ion, N^{3-}, is a very strong base, ionic nitrides react with water, producing ammonia.

$$N^{3-}(aq) + 3H_2O(l) \longrightarrow NH_3(g) + 3OH^-(aq)$$

Magnesium nitride reacts with water to give magnesium hydroxide and ammonia.

$$Mg_3N_2(s) + 6H_2O(l) \longrightarrow 3Mg(OH)_2(s) + 2NH_3(g)$$

Air is the major commercial source of nitrogen. The components of air are separated by liquefaction, followed by distillation. Nitrogen is the most volatile component in liquid air, so it is the first to distill off, leaving behind a liquid that is primarily oxygen with a small amount of noble gases (mostly argon).

Liquid nitrogen is used as a refrigerant to freeze foods, to freeze soft or rubbery materials prior to grinding them, and to freeze biological materials (Figure 21.41). Large quantities of nitrogen are also used as a *blanketing gas,* whose purpose is to protect a material from oxygen during processing or storage. Thus, electronic components are often made under a nitrogen atmosphere. The other principal use of nitrogen is to prepare nitrogen compounds.

Nitrogen Compounds Ammonia, NH_3, is the most important commercial compound of nitrogen. A colorless gas with a characteristic irritating or pungent odor, it is prepared commercially from N_2 and H_2 by the *Haber process*. Figure 21.42 shows a flowchart of the industrial preparation of ammonia from natural gas, steam, and air. Small amounts can be prepared in the laboratory by heating a solution of an ammonium salt with a strong base, such as NaOH or $Ca(OH)_2$ (Figure 21.43).

$$NH_4Cl(aq) + NaOH(aq) \xrightarrow{\Delta} NH_3(g) + H_2O(l) + NaCl(aq)$$

Ammonia is easily liquefied, and the liquid is used as a nitrogen fertilizer. Ammonium salts, such as the sulfate and nitrate, are also sold as fertilizers. Large quantities of ammonia are converted to urea, NH_2CONH_2, which is used as a fertilizer, as a livestock feed supplement, and in the manufacture of urea–formaldehyde plastics. Ammonia is also the starting compound for the preparation of most other nitrogen compounds.

Dinitrogen monoxide, commonly known as nitrous oxide, N_2O, is a colorless gas with a sweet odor. It can be prepared by careful heating of molten ammonium nitrate. (If heated strongly, it explodes.)

$$NH_4NO_3(s) \xrightarrow{\Delta} N_2O(g) + 2H_2O(g)$$

Nitrous oxide, or laughing gas, is used as a dental anesthetic. It is also useful as a propellant in whipped-cream dispensers. The gas dissolves in cream under pressure. When the cream is dispensed, the gas bubbles out, forming a foam.

Nitrogen monoxide, commonly known as nitric oxide, NO, is a colorless gas that is of great industrial and biological importance. ▶ Although it can be prepared by the direct combination of the elements at elevated temperatures, large amounts are prepared from ammonia as the first step in the commercial preparation of nitric acid. The ammonia is oxidized in the presence of a platinum catalyst.

$$4NH_3(g) + 5O_2(g) \xrightarrow{Pt} 4NO(g) + 6H_2O(g)$$

© wunkley/Alamy

Figure 21.41 ▲

Liquid nitrogen When liquid nitrogen is poured on a table (which, although at room temperature, is about 220°C above the boiling point of nitrogen), the liquid sizzles and boils away violently.

The biological importance of NO was discussed in an essay in Chapter 5.

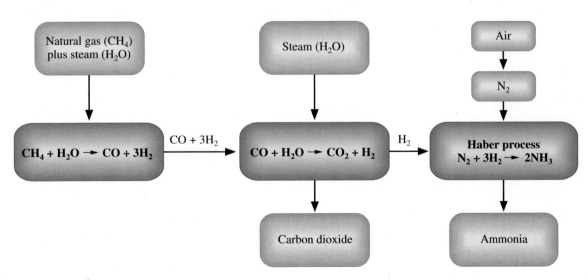

Figure 21.42 ▲

Industrial preparation of ammonia (flowchart) The raw materials are natural gas, water, and air. Hydrogen for the Haber process is obtained by reacting natural gas with steam to give carbon monoxide and hydrogen. In the next step, carbon monoxide is reacted with steam to give carbon dioxide and additional hydrogen. The carbon dioxide is removed by dissolving it in water solution.

Figure 21.43 ▲

Preparation of ammonia from an ammonium salt Sodium hydroxide solution was added to ammonium chloride. Ammonia gas, formed in the reaction, turns colorless phenolphthalein in the moist paper to a bright pink.

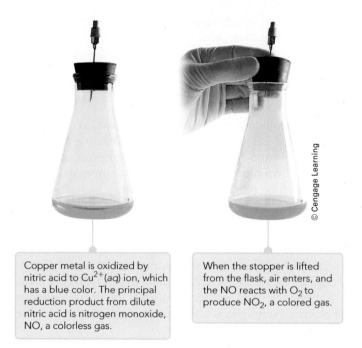

Copper metal is oxidized by nitric acid to $Cu^{2+}(aq)$ ion, which has a blue color. The principal reduction product from dilute nitric acid is nitrogen monoxide, NO, a colorless gas.

When the stopper is lifted from the flask, air enters, and the NO reacts with O_2 to produce NO_2, a colored gas.

Figure 21.44 ▲

Reaction of copper metal with dilute nitric acid

Nitric acid, HNO_3, is the third most important industrial acid (after sulfuric and phosphoric acids). It is used to prepare explosives, nylon, and polyurethane plastics. Nitric acid is produced commercially by the **Ostwald process**, which is *an industrial preparation of nitric acid starting from the catalytic oxidation of ammonia to nitrogen monoxide.* In this process, ammonia is burned in the presence of a platinum catalyst to give NO, which is then reacted with oxygen to form NO_2. The NO_2 is dissolved in water, where it reacts to form nitric acid and nitrogen monoxide.

$$4NH_3(g) + 5O_2(g) \xrightarrow{\text{Pt}} 4NO(g) + 6H_2O(g)$$
$$2NO(g) + O_2(g) \longrightarrow 2NO_2(g)$$
$$3NO_2(g) + H_2O(l) \longrightarrow 2HNO_3(aq) + NO(g)$$

The nitrogen monoxide produced in the last step is recycled for use in the second step.

Nitric acid is a strong oxidizing agent. Although copper metal is unreactive to most acids, it is oxidized by the nitrate ion in acid solution. In dilute acid, nitrogen monoxide is the principal reduction product (Figure 21.44).

$$3Cu(s) + 8H_3O^+(aq) + 2NO_3^-(aq) \longrightarrow 3Cu^{2+}(aq) + 2NO(g) + 12H_2O(l)$$

With concentrated nitric acid, nitrogen dioxide is obtained (Figure 21.45).

$$Cu(s) + 4H_3O^+(aq) + 2NO_3^-(aq) \longrightarrow Cu^{2+}(aq) + 2NO_2(g) + 6H_2O(l)$$

When certain nitrates, such as sodium nitrate, are heated, they decompose to the nitrites.

Figure 21.45 ▲

Reaction of copper metal with concentrated nitric acid The principal reduction product from concentrated nitric acid is nitrogen dioxide, NO_2 (reddish brown gas).

$$2NaNO_3(s) \xrightarrow{\Delta} 2NaNO_2(s) + O_2(g)$$

The corresponding acid, nitrous acid, is unstable and is usually prepared when needed as an aqueous solution. When the stoichiometric amount of sulfuric acid is added to an aqueous solution of barium nitrite, $Ba(NO_2)_2$, barium sulfate precipitates, leaving a solution of nitrous acid.

$$Ba(NO_2)_2(aq) + H_2SO_4(aq) \longrightarrow BaSO_4(s) + 2HNO_2(aq)$$

Table 21.8 lists uses of some compounds of nitrogen (and of phosphorus, discussed in the next section).

Table 21.8   Uses of Some Compounds of Nitrogen and of Phosphorus

Compound	Use
NH_3	Nitrogen fertilizer
	Manufacture of nitrogen compounds
N_2H_4	Blowing agent for foamed plastics
	Water treatment
HNO_3	Explosives
	Polyurethane plastics
$Ca(H_2PO_4)_2 \cdot H_2O$	Phosphate fertilizer
	Baking powder
$CaHPO_4 \cdot 2H_2O$	Animal feed additive
	Toothpowder
H_3PO_4	Manufacture of phosphate fertilizers
	Soft drinks
PCl_3	Manufacture of $POCl_3$
	Manufacture of pesticides
$POCl_3$	Manufacture of plasticizers (substances that keep plastics pliable)
	Manufacture of flame retardants
$Na_5P_3O_{10}$	Detergent additive

CONCEPT CHECK 21.1

Considering the fact that N_2 makes up about 80% of the atmosphere, why don't animals use the abundant N_2 instead of O_2 for biological reactions?

Phosphorus

Phosphorus, the most abundant of the Group VA elements, occurs in phosphate minerals, such as fluorapatite, whose formula is written either $Ca_5(PO_4)_3F$ or $3Ca_3(PO_4)_2 \cdot CaF_2$. Unlike nitrogen, which exists in important compounds with oxidation states between -3 and $+5$, the most important oxidation states of phosphorus are $+3$ and $+5$. Like nitrogen, however, phosphorus is an important element in living organisms. DNA (deoxyribonucleic acid), a chainlike biological molecule in which information about inheritable traits resides, contains phosphate groups along the length of its chain. Similarly, ATP (adenosine triphosphate), the energy-containing molecule of living organisms, contains phosphate groups.

Allotropes of Phosphorus   Phosphorus has two common allotropes: white phosphorus and red phosphorus (Figure 21.46). White phosphorus, a waxy, white solid,

Figure 21.46 ◄

Allotropes of phosphorus
Left: White phosphorus.
Right: Red phosphorus.

© Cengage Learning

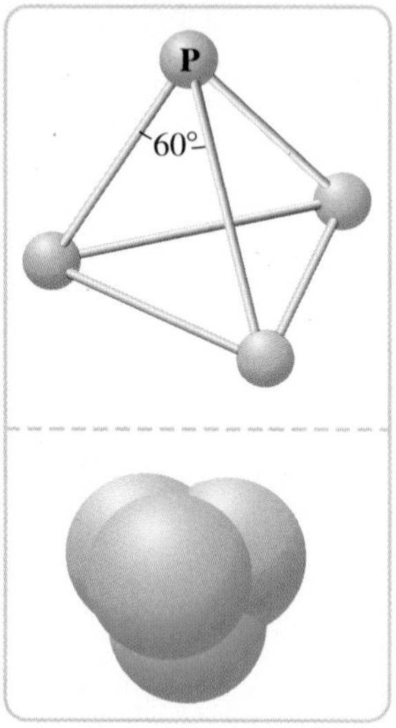

Figure 21.47 ▲

Structure of the P$_4$ molecule
Top: The reactivity of white phosphorus results from the small P—P—P angle (60°).
Bottom: Space-filling molecular model.

is very poisonous and very reactive. If white phosphorus is left exposed to air, it bursts spontaneously into flame. Because of its reactivity with oxygen, white phosphorus is stored under water, in which it is insoluble. As you might expect from its low melting point (44°C), white phosphorus is a molecular solid, with the formula P$_4$.

The phosphorus atoms in the P$_4$ molecule are arranged at the corners of a regular tetrahedron such that each atom is single-bonded to the other three (Figure 21.47). The experimentally determined P—P—P bond angle is 60°. Because this is considerably smaller than the normal bond angle, the bonding in P$_4$ is strained and therefore weaker. This accounts for the reactivity seen in this phosphorus allotrope; chemical reactions of P$_4$ replace its weak bonds by stronger ones.

White phosphorus is a major industrial chemical and is prepared by heating phosphate rock with coke (C) and quartz sand (SiO$_2$) in an electric furnace. The overall reaction is

$$2Ca_3(PO_4)_2(s) + 6SiO_2(s) + 10C(s) \xrightarrow{1500°C} 6CaSiO_3(l) + 10CO(g) + P_4(g)$$
Calcium silicate

The gases from the furnace are cooled by water to condense the phosphorus vapor to the liquid; the carbon monoxide gas is used as a fuel. The other product, calcium silicate glass (called *slag*), is drained periodically from the bottom of the furnace.

Most of the white phosphorus produced is used to manufacture phosphoric acid, H$_3$PO$_4$. Some white phosphorus is converted to red phosphorus, an amorphous material that can be crystallized to give an allotrope with a polymeric chain structure (Figure 21.48). Red phosphorus is much less reactive than white phosphorus and can be stored in the presence of air. This phosphorus allotrope is relatively nontoxic and is used in the striking surface for safety matches. Red phosphorus is made by heating white phosphorus at about 400°C for several hours.

Phosphorus Oxides Phosphorus has two common oxides, tetraphosphorus hexoxide, P$_4$O$_6$, and tetraphosphorus decoxide, P$_4$O$_{10}$. Their common names are phosphorus trioxide and phosphorus pentoxide, respectively. The names are at odds with present rules of nomenclature but stem from their empirical formulas, P$_2$O$_3$ and P$_2$O$_5$.

These oxides have interesting structures (Figure 21.49). Tetraphosphorus hexoxide has a tetrahedron of phosphorus atoms, as in P$_4$, but with oxygen atoms between each pair of phosphorus atoms. Tetraphosphorus decoxide is similar, except that each phosphorus atom has an additional oxygen atom bonded to it. These phosphorus–oxygen bonds are much shorter than the ones in the P—O—P bridges (139 pm versus 162 pm); hence, they are best represented as double bonds.

Tetraphosphorus hexoxide is a low-melting solid (m.p. 23°C) and the anhydride of phosphorous acid, H$_3$PO$_3$. Tetraphosphorus decoxide is the most important oxide; it is a white solid that sublimes at 360°C. This oxide is the anhydride of phosphoric acid, H$_3$PO$_4$. The reaction with water is quite vigorous, making tetraphosphorus decoxide useful in the laboratory as a drying agent. It is prepared by burning white phosphorus in air.

$$P_4(s) + 5O_2(g) \longrightarrow P_4O_{10}(s)$$

Tetraphosphorus decoxide is used to manufacture phosphoric acid, an oxoacid.

Figure 21.48 ▶

Chain structure of red phosphorus
The structure is obtained by linking P$_4$ tetrahedra together after breaking a bond in each tetrahedron.

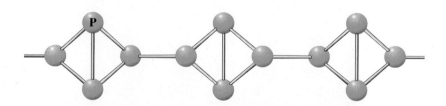

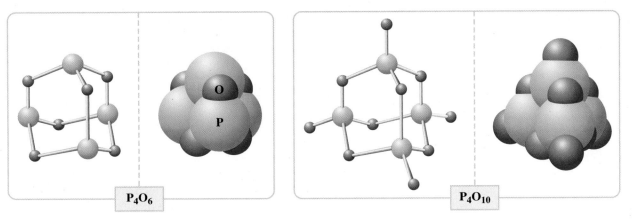

P_4O_6 P_4O_{10}

Figure 21.49 ▲

Structures of the phosphorus oxides *Left:* The phosphorus atoms in P_4O_6 have tetrahedral positions, as in P_4; however, the phosphorus atoms are bonded to oxygen atoms, forming P—O—P bridges between each pair of phosphorus atoms. *Right:* The P_4O_{10} molecule is similar, except that an additional oxygen atom is bonded to each phosphorus atom. Both ball-and-stick and space-filling models are shown.

Oxoacids of Phosphorus Phosphorus has many oxoacids, but the most important of these can be thought of as derivatives of orthophosphoric acid (often called simply phosphoric acid), H_3PO_4. Orthophosphoric acid is a colorless solid, melting at 42°C when pure. It is usually sold as an aqueous solution, however. Orthophosphoric acid is triprotic (three acidic H atoms per molecule); possible sodium salts are sodium dihydrogen phosphate (NaH_2PO_4), disodium hydrogen phosphate (Na_2HPO_4), and trisodium phosphate (Na_3PO_4). The acid has the following electron-dot formula:

$$H-\overset{..}{\underset{..}{O}}-\overset{\overset{\displaystyle :O:}{\|}}{\underset{\underset{\displaystyle H}{|}}{\underset{\displaystyle :O:}{P}}}-\overset{..}{\underset{..}{O}}-H$$

Orthophosphoric acid is produced in enormous quantity either directly from phosphate rock or from tetraphosphorus decoxide, which in turn is obtained by burning white phosphorus. In the direct process, phosphate rock is treated with sulfuric acid, from which a solution of phosphoric acid is obtained by filtering off the calcium sulfate and other solid materials. The product is an impure phosphoric acid, which is used in the manufacture of phosphate fertilizers. ▶

$$Ca_3(PO_4)_2(s) + 3H_2SO_4(aq) \longrightarrow 3CaSO_4(s) + 2H_3PO_4(aq)$$

A purer grade of orthophosphoric acid is obtained by reacting tetraphosphorus decoxide with water. It is used in soft drinks for tartness and also in making phosphates for detergent formulations.

Orthophosphoric acid, H_3PO_4, undergoes condensation reactions to form other phosphoric acids. For example, two orthophosphoric acid molecules condense to form diphosphoric acid (also called pyrophosphoric acid).

Because phosphate rock contains CaF_2, hydrofluoric acid is a by-product in the preparation of phosphoric acid. HF is used in aluminum production.

$$H-O-\overset{\overset{\displaystyle O}{\|}}{\underset{\underset{\displaystyle H}{|}}{\underset{\displaystyle O}{P}}}-O-H \ + \ H-O-\overset{\overset{\displaystyle O}{\|}}{\underset{\underset{\displaystyle H}{|}}{\underset{\displaystyle O}{P}}}-O-H \ \longrightarrow \ H-O-\overset{\overset{\displaystyle O}{\|}}{\underset{\underset{\displaystyle H}{|}}{\underset{\displaystyle O}{P}}}-O-\overset{\overset{\displaystyle O}{\|}}{\underset{\underset{\displaystyle H}{|}}{\underset{\displaystyle O}{P}}}-O-H \ + \ H_2O$$

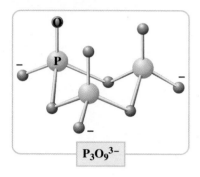

Figure 21.50 ▲

The structure of trimetaphosphate ion, $P_3O_9^{3-}$ A cyclic metaphosphate ion; the general formula of the metaphosphate ions is $(PO_3^-)_n$.

Two series of these phosphoric acids exist (all having phosphorus in the +5 oxidation state). One series consists of the linear **polyphosphoric acids**, which are *acids with the general formula $H_{n+2}P_nO_{3n+1}$ formed from linear chains of P—O bonds.*

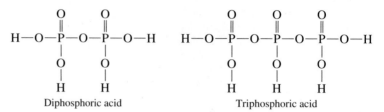

Diphosphoric acid   Triphosphoric acid

The other series consists of the **metaphosphoric acids**, which are *acids with the general formula $(HPO_3)_n$*. Figure 21.50 shows the structure of a cyclic, or ring, metaphosphate anion. When a linear polyphosphoric acid chain is very long, the formula becomes $(HPO_3)_n$, with n very large, and the acid is called *polymetaphosphoric acid.*

The polyphosphates and metaphosphates are used in detergents, where they act as water softeners by complexing with metal ions in the water. Sodium triphosphate, $Na_5P_3O_{10}$, one of the most commonly used polyphosphates, is manufactured by adding sufficient sodium carbonate to orthophosphoric acid to give a solution of the salts NaH_2PO_4 and Na_2HPO_4. When this solution is sprayed into a hot kiln, the orthophosphate ions condense to give sodium triphosphate.

The use of phosphates in detergents has been criticized for contributing to the overfertilization of plants and algae in lakes (a process referred to as *eutrophication*). Such lakes become oxygen-deficient from decomposing plants and algae, and the fish die. Detergents with phosphates have been banned in some states of the United States, and in response, manufacturers have introduced phosphate-free products.

Table 21.8 lists some uses of compounds of phosphorus.

21.10 Group VIA: Oxygen and the Sulfur Family

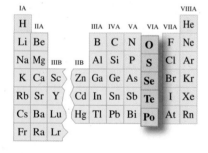

Group VIA, like the preceding groups, shows the trend from nonmetallic to metallic as you proceed from top to bottom of the column of elements. Oxygen and sulfur are strictly nonmetallic. Although the chemical properties of selenium and tellurium are predominantly those of nonmetals, they do have semiconducting allotropes as expected of metalloids. Polonium is a radioactive metal.

Oxygen, a second-period element, has rather different properties from those of the other members of Group VIA. It is very electronegative, and bonding involves only s and p orbitals. For the other members of Group VIA, d orbitals become a factor in bonding. Oxygen exists in compounds mainly in the −2 oxidation state. The other Group VIA elements exist in compounds in this state also, but the +4 and +6 states are common. (Oxygen is the most abundant element on earth, making up 48% by mass of the earth's crust. Sulfur, although not as abundant as oxygen, is a common element. Both oxygen and sulfur are important to living organisms.)

Oxygen

Oxygen in the form of dioxygen, O_2, makes up 20.9 mole percent of the atmosphere. Although this might seem to constitute a considerable quantity of the element, most of the oxygen on earth is present as oxide and oxoanion minerals (in silicates, carbonates, sulfates, and so forth). Indeed, oxygen combines with almost every element—only some of the noble gases have no known oxygen compounds. The chemistry of oxygen is therefore very important, though we frequently discuss it in the context of other elements. In fact, much of the chemistry of the elements discussed in this chapter concerns either their reactions with oxygen or the properties of their oxides and oxoacids.

Properties and Preparation of Oxygen The common form of the element oxygen is dioxygen, O_2. (The element also exists as the allotrope ozone, O_3.) Dioxygen, usually called simply oxygen, is a colorless, odorless gas under standard conditions. The critical temperature is $-118°C$. Therefore, you can liquefy oxygen if you first cool the gas below this temperature and then compress it. Both liquid and solid O_2 have a pale blue color. The melting point of the solid is $-218°C$, and the boiling point at 1 atm is $-183°C$. ▶

Oxygen is produced in enormous quantity from air. As described in the discussion of nitrogen, air is first liquefied, then distilled. Nitrogen and argon are more volatile components of air and distill off, leaving liquid oxygen behind.

Oxygen can be prepared in small quantities by decomposing certain oxygen-containing compounds. Both the Swedish chemist Karl Wilhelm Scheele and the British chemist Joseph Priestley are credited with the discovery of oxygen. Priestley obtained the gas in 1774 by heating mercury(II) oxide.

$$2HgO(s) \xrightarrow{\Delta} 2Hg(l) + O_2(g)$$

In one laboratory preparation, potassium chlorate, $KClO_3$, is heated with pure manganese(IV) oxide, MnO_2, as a catalyst.

$$2KClO_3(s) \xrightarrow[MnO_2]{\Delta} 2KCl(s) + 3O_2(g)$$

Reactions of Oxygen Molecular oxygen is a very reactive gas and combines directly with many substances. The products are usually oxides. An **oxide** is *a binary compound with oxygen in the −2 oxidation state.*

Most metals react readily with oxygen to form oxides, especially if the metal is in a form that exposes sufficient surface area. For example, magnesium wire and iron wool burn brightly in air to yield the oxides.

$$2Mg(s) + O_2(g) \longrightarrow 2MgO(s)$$

$$3Fe(s) + 2O_2(g) \longrightarrow Fe_3O_4(s)$$

The resulting oxides MgO and Fe_3O_4 are basic oxides, as is true of most metal oxides. If the metal is in a high oxidation state, however, the oxide may be acidic. For example, chromium(III) oxide, Cr_2O_3, is a basic oxide, but chromium(VI) oxide, CrO_3, is an acidic oxide (Figure 21.51).

The alkali metals form an interesting series of binary compounds with oxygen. When an alkali metal burns in air, the principal product with oxygen depends on the metal. With lithium, the product is the basic oxide, Li_2O. With the other alkali metals, the product is predominantly the peroxide and superoxide. A **peroxide** is *a compound with oxygen in the −1 oxidation state.* (Peroxides contain either the O_2^{2-} ion or the covalently bonded group —O—O—.) A **superoxide** is *a binary compound with oxygen in the $-\frac{1}{2}$ oxidation state;* superoxides contain the superoxide ion, O_2^-. Sodium metal burns in air to give mainly the peroxide.

$$2Na(s) + O_2(g) \longrightarrow Na_2O_2(s)$$

Potassium and the other alkali metals form mainly the superoxides.

$$K(s) + O_2(g) \longrightarrow KO_2(s)$$

Nonmetals react with oxygen to form covalent oxides, most of which are acidic. For example, carbon burns in an excess of oxygen to give carbon dioxide, which is the acid anhydride of carbonic acid (that is, carbon dioxide produces carbonic acid when it reacts with water). Sulfur, S_8, burns in oxygen to give sulfur dioxide, SO_2, the acid anhydride of sulfurous acid.

$$S_8(s) + 8O_2(g) \longrightarrow 8SO_2(g)$$

Sulfur forms another oxide, sulfur trioxide, SO_3, but only small amounts are obtained during the burning of sulfur in air. Sulfur trioxide is the acid anhydride of sulfuric acid.

Compounds in which at least one element is in a reduced state are oxidized by oxygen, giving compounds that would be expected to form when the individual elements are burned in oxygen. For example, a hydrocarbon such as octane, C_8H_{18}, burns to give carbon dioxide and water.

$$2C_8H_{18}(l) + 25O_2(g) \longrightarrow 16CO_2(g) + 18H_2O(g)$$

Ozone and its presence in the atmosphere are discussed in an essay at the end of Chapter 10.

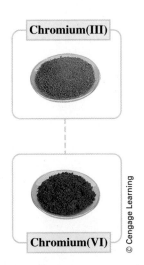

Chromium(III)

Chromium(VI)

© Cengage Learning

Figure 21.51 ▲

Chromium oxides *Top:* Chromium(III) oxide, a basic oxide. *Bottom:* Chromium(VI) oxide, an acidic oxide.

Some other examples are given in the following equations:

$$2H_2S(g) + 3O_2(g) \longrightarrow 2H_2O(g) + 2SO_2(g)$$
$$CS_2(l) + 3O_2(g) \longrightarrow CO_2(g) + 2SO_2(g)$$
$$2ZnS(s) + 3O_2(g) \longrightarrow 2ZnO(s) + 2SO_2(g)$$

Note the products that are formed; sulfur compounds usually form SO_2.

CONCEPT CHECK 21.2

Why do we need such low temperatures to liquefy gases such as nitrogen, oxygen, and helium?

© Farrell Grehan/Photo Researchers, Inc.

Figure 21.52 ▲

Sulfur obtained from underground deposits Sulfur is obtained from these deposits by pumping in superheated water; molten sulfur is then pumped to the earth's surface. (Details of this *Frasch process* are described later.) Here, molten sulfur from a well is being directed to an area where it can cool and solidify.

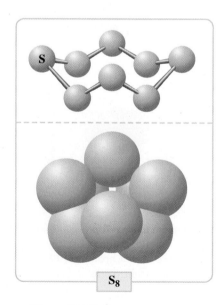

Figure 21.53 ▲

Structure of the S_8 molecule *Top:* Each molecule consists of eight S atoms arranged in a ring (in the shape of a crown). *Bottom:* Space-filling molecular model.

Sulfur

Sulfur occurs in sulfate minerals, such as gypsum ($CaSO_4 \cdot 2H_2O$), and in sulfide and disulfide minerals, many of which are important metal ores. Pyrite is iron(II) disulfide, FeS_2; it consists of Fe^{2+} and S_2^{2-} ions. This mineral is sometimes called "fool's gold," because its golden color often fooled novice miners into thinking they had found gold. Sulfur is also present in coal and petroleum as organic sulfur compounds and in natural gas as hydrogen sulfide, H_2S. Free sulfur occurs in some volcanic areas, perhaps formed by the reaction of hydrogen sulfide and sulfur dioxide, which are present in volcanic gases.

$$16H_2S(g) + 8SO_2(g) \longrightarrow 16H_2O(l) + 3S_8(s)$$

Commercial deposits of free sulfur also occur in the rock at the top of salt domes, which are massive columns of salt embedded in rock a hundred meters or more below the earth's surface. These deposits are believed to have formed by bacterial action involving calcium sulfate minerals. Figure 21.52 shows free sulfur obtained from such deposits that occur in the United States along the Gulf of Mexico.

Sulfur also occurs in several amino acids, which are the building blocks of the proteins in living organisms. Plants are able to use sulfate ion as a source of sulfur for the synthesis of amino acids. Animals and decay bacteria derive most of their nutritional sulfur from organic sources.

Allotropes of Sulfur Sulfur has a fascinating array of allotropes, including two common crystal forms, rhombic sulfur and monoclinic sulfur (see Figure 6.18). *Rhombic sulfur* is the stablest form of the element under normal conditions; natural sulfur is rhombic sulfur. It is a yellow, crystalline solid with a lattice consisting of crown-shaped S_8 molecules (Figure 21.53). The relative stability of this molecule results in part from the ability of sulfur atoms to undergo catenation—that is, to form stable bonds to other sulfur atoms. Rhombic sulfur melts at 113°C to give an orange-colored liquid.

When this liquid is cooled, it crystallizes to give *monoclinic sulfur*. This allotrope also consists of S_8 molecules; it differs from rhombic sulfur only in the way the molecules are packed to form crystals. Monoclinic sulfur melts at 119°C. It is unstable below 96°C, and in a few weeks at room temperature it reverts to rhombic sulfur.

If, instead of cooling the liquid sulfur, you raise its temperature above 160°C, the sulfur begins to darken and at somewhat higher temperatures changes to a dark reddish brown, viscous liquid. The original melt consists of S_8 molecules, but these rings of sulfur atoms open up, and the fragments join to give long spiral chains of sulfur atoms. The spiral chains have unpaired electrons at their ends, and these unpaired electrons are responsible for the color. The viscosity of the liquid increases as compact S_8 molecules are replaced by long spiral chains that can intertwine. At temperatures greater than 200°C, the chains begin to break apart and the viscosity decreases. Figure 21.54 shows the appearance of sulfur at various temperatures.

When molten sulfur above 160°C (but below 200°C) is poured into water, the liquid changes to a rubbery mass, called *plastic sulfur*. Plastic sulfur is an amorphous mixture of sulfur chains. The rubberiness of this sulfur results from the

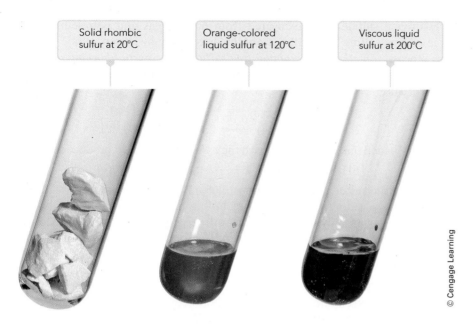

Figure 21.54 ◄

Sulfur at various temperatures

Solid rhombic sulfur at 20°C

Orange-colored liquid sulfur at 120°C

Viscous liquid sulfur at 200°C

© Cengage Learning

ability of the spiral chains of sulfur atoms to stretch and then relax to their original length. Plastic sulfur reverts to rhombic sulfur after a period of time.

Sulfur boils at 445°C, giving a vapor of S_8, S_6, S_4, and S_2 molecules. Measurements of vapor density depend on the temperature, as a result of these different species of molecular sulfur in the vapor. (The relationship between gas density and molecular mass is described in Section 5.3.)

Production of Sulfur Free sulfur that occurs in deep underground deposits is mined by the **Frasch process**, *a mining procedure in which underground deposits of solid sulfur are melted in place with superheated water, and the molten sulfur is forced upward as a froth using air under pressure* (see Figure 21.55). A sulfur well is similar to an oil well but consists of three concentric pipes. Superheated water flows down the outer pipe, and compressed air flows down the inner pipe. The superheated water melts the sulfur, which is then pushed up the middle pipe by the compressed air. Molten sulfur spews from the well and solidifies in large storage areas. Sulfur obtained this way is 99.6% pure.

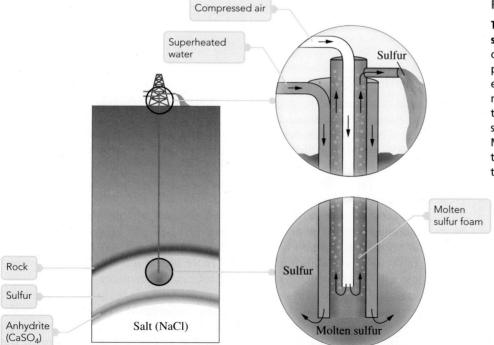

Compressed air

Superheated water

Sulfur

Rock

Sulfur

Anhydrite ($CaSO_4$)

Salt (NaCl)

Sulfur

Molten sulfur foam

Molten sulfur

Figure 21.55 ◄

The Frasch process for mining sulfur The well consists of concentric pipes. Superheated water passing down the outer pipe exits into the sulfur deposit, melting it. Compressed air from the inner pipe pushes the molten sulfur up the middle pipe. Molten sulfur flows from the top of the well onto the ground to cool.

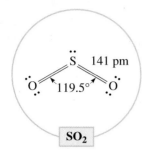

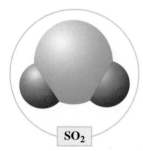

Figure 21.56 ▲

Structure of the SO$_2$ molecule The molecule has a bent geometry.

Hydrogen sulfide, H$_2$S, recovered from natural gas and petroleum is also a source of free sulfur. The sulfur is obtained from the hydrogen sulfide gas by the **Claus process**, *a method of obtaining free sulfur by the partial burning of hydrogen sulfide.* The partial burning of hydrogen sulfide produces some sulfur, as well as sulfur dioxide.

$$8H_2S(g) + 4O_2(g) \longrightarrow S_8(s) + 8H_2O(g)$$
$$2H_2S(g) + 3O_2(g) \longrightarrow 2SO_2(g) + 2H_2O(g)$$

The sulfur dioxide that forms reacts with the hydrogen sulfide to produce more sulfur.

$$16H_2S(g) + 8SO_2(g) \longrightarrow 3S_8(s) + 16H_2O(g)$$

Most of the sulfur produced (almost 90%) is used to make sulfuric acid. The remainder has a wide variety of uses, including the vulcanization of rubber (sulfur converts the initially tacky material into useful rubber), the production of carbon disulfide (to make cellophane), and the preparation of sulfur dioxide for bleaching.

Sulfur Oxides and Oxoacids Sulfur dioxide, SO$_2$, is a colorless, toxic gas with a characteristic suffocating odor. Its presence in polluted air (from the burning of fossil fuels) is known to cause respiratory ailments. The SO$_2$ molecule has a bent geometry with a bond angle of 119.5°, very close to that predicted by the VSEPR model (Figure 21.56).

Sulfur dioxide gas is very soluble in water, producing acidic solutions. Although these solutions are often referred to as solutions of *sulfurous acid,* they appear to be composed primarily of hydrated species of SO$_2$; the acid H$_2$SO$_3$ has never been isolated. An aqueous solution of sulfur dioxide, SO$_2(aq)$, does apparently contain small amounts of the ions HSO$_3^-$ and SO$_3^{2-}$, which would be expected to be produced by the ionization of H$_2$SO$_3$.

$$H_2O(l) + SO_2(g) \rightleftharpoons H_2SO_3(aq)$$
$$H_2SO_3(aq) + H_2O(l) \rightleftharpoons H_3O^+(aq) + HSO_3^-(aq)$$
$$HSO_3^-(aq) + H_2O(l) \rightleftharpoons H_3O^+(aq) + SO_3^{2-}(aq)$$

When an appropriate amount of base is added to an aqueous solution of sulfur dioxide, the corresponding hydrogen sulfite salt or sulfite salt is obtained. Sodium hydrogen sulfite (also called sodium bisulfite) and sodium sulfite are produced this way commercially using sodium carbonate as the base:

$$Na_2CO_3(aq) + 2SO_2(aq) + H_2O(l) \longrightarrow 2NaHSO_3(aq) + CO_2(g)$$
$$Na_2CO_3(aq) + SO_2(aq) \longrightarrow Na_2SO_3(aq) + CO_2(g)$$

Sulfites and hydrogen sulfites decompose when treated with acid to give sulfur dioxide. For example,

$$NaHSO_3(aq) + HCl(aq) \longrightarrow NaCl(aq) + H_2O(l) + SO_2(g)$$

This reaction can be used to prepare small amounts of sulfur dioxide in the laboratory.

Sulfur dioxide is produced on a large scale by burning sulfur, S$_8$. It is also obtained as a by-product of the roasting of sulfide ores (such as FeS$_2$, CuS, ZnS, and PbS). Most of this sulfur dioxide is used to prepare sulfuric acid. Some is used as a bleach for wood pulp and textiles (Figure 21.57) and as a disinfectant and food preservative (for example, in wine and dried fruit). Its use as a food preservative depends on the fact that sulfur dioxide is especially toxic to yeasts, molds, and certain bacteria. Because some people are allergic to sulfur dioxide, foods containing it must be properly labeled.

Sulfur trioxide is a liquid at room temperature. The liquid actually consists of S$_3$O$_9$ molecules in equilibrium with SO$_3$ molecules (Figure 21.58). The vapor-phase molecule is SO$_3$, which has a planar triangular geometry.

Sulfur trioxide is formed in small amounts when sulfur is burned in air, although the principal product is sulfur dioxide. Thermodynamically, sulfur trioxide is the preferred product of sulfur and oxygen. Sulfur dioxide does react slowly with oxygen in air to produce sulfur trioxide, but the reaction is much faster in the

© Cengage Learning

Figure 21.57 ▲

Bleaching of a rose by sulfur dioxide The dye in the rose is reduced by sulfur dioxide (contained in the beaker) to a colorless substance.

presence of a catalyst, such as platinum. Sulfur trioxide is produced commercially by the oxidation of sulfur dioxide in the presence of vanadium(V) oxide catalyst.

$$2SO_2(g) + O_2(g) \xrightarrow{\text{V}_2\text{O}_5} 2SO_3(g)$$

Sulfur trioxide reacts vigorously and exothermically with water to produce sulfuric acid.

$$SO_3(g) + H_2O(l) \longrightarrow H_2SO_4(aq)$$

The **contact process** is *an industrial method for the manufacture of sulfuric acid. It consists of the reaction of sulfur dioxide with oxygen to form sulfur trioxide using a catalyst of vanadium(V) oxide, followed by the reaction of sulfur trioxide with water.* Because the direct reaction of sulfur trioxide with water produces mists that are unmanageable, the sulfur trioxide is actually dissolved in concentrated sulfuric acid, which is then diluted with water. ▶

Sulfuric acid is a component of acid rain and forms in air from sulfur dioxide, following reactions that are similar to those involved in the contact process. Atmospheric sulfur dioxide has both natural and human origins. Natural sources include plant and animal decomposition and volcanic emissions. However, the burning of coal, oil, and natural gas has been identified as a major source of acid rain pollution. After persisting in the atmosphere for some time, sulfur dioxide is oxidized to sulfur trioxide, which dissolves in rain to give $H_2SO_4(aq)$.

Concentrated sulfuric acid is a viscous liquid and a powerful dehydrating agent. The concentrated acid is also an oxidizing agent. Copper is not dissolved by most acids. $E°$ for $Cu^{2+}|Cu$ is positive, so Cu is not oxidized by H^+ (H_3O^+). It is, however, dissolved by hot, concentrated sulfuric acid. In this reaction, sulfate ion in acid solution is reduced to sulfur dioxide:

$$\overset{0}{Cu}(s) + 2H_2\overset{+6}{S}O_4(l) \longrightarrow \overset{+2}{Cu}SO_4(aq) + 2H_2O(l) + \overset{+4}{S}O_2(g)$$

More sulfuric acid is made than any other industrial chemical. Most of this acid is used to make soluble phosphate and ammonium sulfate fertilizers. Sulfuric acid is also used in petroleum refining and in the manufacture of many chemicals. Table 21.9 lists uses of some sulfur compounds.

Sulfuric acid and acid rain are discussed in an essay in Section 16.2.

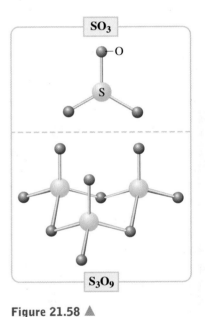

Figure 21.58 ▲

Structures of SO₃ and S₃O₉ molecules These molecules are in equilibrium in liquid sulfur trioxide. The vapor consists mostly of SO_3.

Table 21.9	Uses of Some Sulfur Compounds
Compound	**Use**
CS_2	Manufacture of rayon and cellophane
	Manufacture of CCl_4
SO_2	Manufacture of H_2SO_4
	Food preservative
	Textile bleach
H_2SO_4	Manufacture of phosphate fertilizers
	Petroleum refining
	Manufacture of various chemicals
$Na_2S_2O_3$	Photographic fixer

21.11 Group VIIA: The Halogens

The Group VIIA elements, called the halogens, have very similar properties, or at least they have properties that change smoothly in progressing down the column. All are reactive nonmetals, except perhaps for astatine, whose chemistry is not well known.

As a second-period element, fluorine does exhibit some differences from the other elements of Group VIIA, although these are not so pronounced as those of the second-period elements in Groups IIIA to VIA. The solubilities of the fluorides in water, for example, are often quite different from those of the chlorides, bromides, and iodides. Calcium chloride, bromide, and iodide are very soluble in water.

Calcium fluoride, however, is insoluble. Silver chloride, bromide, and iodide are insoluble, but silver fluoride is soluble.

All of the halogens form stable compounds in which the element is in the -1 oxidation state. In fluorine compounds, this is the only oxidation state. Chlorine, bromine, and iodine also have compounds in which the halogen is in one of the positive oxidation states $+1$, $+3$, $+5$, or $+7$. The higher positive oxidation states ($> +1$) are due to the involvement of d orbitals in bonding.

Chlorine

Chlorine gas, Cl_2, was discovered in 1774 by the Swedish chemist Karl Wilhelm Scheele by heating hydrochloric acid with manganese dioxide.

$$4HCl(aq) + MnO_2(s) \longrightarrow MnCl_2(aq) + Cl_2(g) + 2H_2O(l)$$

He immediately noted the suffocating odor of the gas. Scheele also discovered that chlorine solutions could bleach cotton cloth. Within a few decades, chlorine-based bleaches became major items of commerce. Currently, chlorine is the most commercially important halogen.

Properties of Chlorine Chlorine gas has a pale greenish yellow color. It is a very reactive oxidizing agent and supports the oxidation of many substances in a manner similar to oxygen. We discussed the reaction of chlorine with sodium in Chapter 2. ◀

See the introduction to chapter 2.

All of the halogens are oxidizing agents, though the oxidizing power decreases from fluorine to iodine. Thus, chlorine is a stronger oxidizing agent than either bromine or iodine. When chlorine gas is bubbled into a bromide solution, free bromine is obtained. Similarly, chlorine oxidizes iodide ion to iodine.

$$Cl_2(g) + 2KBr(aq) \longrightarrow 2KCl(aq) + Br_2(aq)$$
$$Cl_2(g) + 2KI(aq) \longrightarrow 2KCl(aq) + I_2(aq)$$

These reactions can be used as a test for bromide and iodide ions. Suppose an aqueous solution of chlorine is added to a test tube containing either bromide or iodide ion. The corresponding free halogen is formed in the water solution. It is readily identified by adding the organic solvent methylene chloride, CH_2Cl_2, which dissolves the halogen, forming a colored layer at the bottom of the test tube. Bromide ion gives an orange layer; iodide ion gives a violet layer. Of course, neither bromine nor iodine is strong enough to oxidize chloride ion.

Chlorine reacts with water by being both oxidized and reduced:

$$\overset{0}{Cl_2}(g) + H_2O(l) \rightleftharpoons \overset{+1}{HClO}(aq) + \overset{-1}{HCl}(aq)$$

In an aqueous solution of chlorine at 25°C, about two-thirds of the chlorine is present as $Cl_2(aq)$; the rest is HClO and HCl.

Preparation and Uses of Chlorine Chlorine is a major industrial chemical. It is prepared commercially by the electrolysis of aqueous sodium chloride. ◀ Chlorine can be prepared in small amounts for laboratory use by the reaction of chloride ion with a strong oxidizing agent, such as potassium dichromate or manganese dioxide. However, chlorine is readily available in steel cylinders for laboratory work.

The electrolysis of aqueous sodium chloride is discussed in Section 19.10.

The principal use of chlorine is in the preparation of chlorinated hydrocarbons, such as vinyl chloride, $CH_2{=}CHCl$ (for polyvinyl chloride plastics), and methyl chloride, CH_3Cl (for the manufacture of silicones, polymers with Si—O bonds and organic groups). Various insecticides are also chlorinated hydrocarbons; many of these (such as DDT) are now restricted in their use because of possible environmental damage.

Other major uses of chlorine are as a bleaching agent for textiles and paper pulp and as a disinfectant. Not long after the discovery of chlorine, chlorine bleaches became available commercially. Chlorine solutions were used as disinfectants early in the nineteenth century. Today, chlorine gas is commonly used for disinfecting municipal water supplies.

Hydrogen Chloride Hydrogen chloride, HCl, is a colorless gas with a sharp, penetrating odor. The gas is very soluble in water (Figure 21.59), and the water solution is commonly referred to as hydrochloric acid. The molecular species HCl ionizes nearly completely in aqueous solution:

$$HCl(g) + H_2O(l) \longrightarrow H_3O^+(aq) + Cl^-(aq)$$

Hydrogen chloride can be produced by heating sodium chloride with concentrated sulfuric acid.

$$NaCl(s) + H_2SO_4(l) \xrightarrow{\Delta} NaHSO_4(s) + HCl(g)$$

On stronger heating, the sodium hydrogen sulfate reacts with sodium chloride to produce additional hydrogen chloride.

$$NaCl(s) + NaHSO_4(s) \xrightarrow{\Delta} Na_2SO_4(s) + HCl(g)$$

Hydrogen bromide and hydrogen iodide can also be produced from their salts by a similar replacement reaction, but in these cases phosphoric acid is used instead of sulfuric acid, which tends to oxidize the bromide and iodide ions to the respective elements (Figure 21.60).

Most of the hydrogen chloride available commercially is obtained as a by-product in the manufacture of chlorinated hydrocarbons. In these reactions, hydrogen bonded to carbon is replaced by chlorine, forming the chlorinated compound and HCl. An example is the preparation of methyl chloride, CH_3Cl, from methane.

$$CH_4(g) + Cl_2(g) \longrightarrow CH_3Cl(g) + HCl(g)$$

Hydrochloric acid is the fourth most important industrial acid (after sulfuric, phosphoric, and nitric acids). It is used to clean metal surfaces of oxides (a process called *pickling*) and to extract certain metal ores, such as those of tungsten.

Oxoacids of Chlorine The halogens form a variety of oxoacids (Table 21.10). Figure 21.61 shows the structures of the chlorine oxoacids. The acidic character of these acids increases with the number of oxygen atoms bonded to the halogen atom—that is, in the order HClO, $HClO_2$, $HClO_3$, $HClO_4$. (See Section 15.5 for a discussion of molecular structure and acid strength.) Perchloric acid, $HClO_4$, is the strongest of the common acids. Of the chlorine oxoacids, only perchloric acid is stable; the other oxoacids have never been isolated and are known only in aqueous solution.

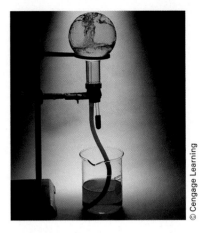

Figure 21.59 ▲

The hydrogen chloride fountain The flask contains hydrogen chloride gas. When water is added to the flask from the dropper, the hydrogen chloride dissolves in it, reducing the pressure in the flask. Atmospheric pressure pushes the water in the beaker into the flask, forming a stream or fountain. Water in the beaker has bromthymol-blue indicator in it. The indicator solution in the flask is yellow because it is acidic.

Figure 21.60 ◄

The action of concentrated sulfuric acid on halide salts Concentrated sulfuric acid was added to watch glasses containing, from left to right, NaCl, NaBr, and NaI. Note the formation of some Br_2 in the center watch glass (brown) and the formation of I_2 vapor (purple) over the watch glass on the right. These halogens form when concentrated sulfuric acid oxidizes the corresponding halide ions. Chloride ion is not oxidized by H_2SO_4.

Table 21.10 Halogen Oxoacids

Oxidation State	Fluorine Oxoacids	Chlorine Oxoacids	Bromine Oxoacids	Iodine Oxoacids	General Name
+1	HFO*	HClO†	HBrO†	HIO†	Hypohalous acid
+3	—	$HClO_2$†	$HBrO_2$†	—	Halous acid
+5	—	$HClO_3$†	$HBrO_3$†	HIO_3	Halic acid
+7	—	$HClO_4$	$HBrO_4$†	HIO_4 H_5IO_6	Perhalic acid

*The oxidation state of F in HFO is −1.
†These acids are known only in aqueous solution.

| HClO | HClO$_2$ | HClO$_3$ | HClO$_4$ |

Figure 21.61 ▲

Structures of the chlorine oxoacids
The models of the oxoacids also
include lone pairs on the Cl atom.

© Cengage Learning

Figure 21.62 ▲

**Solution of sodium hypochlorite
bleach** The solution is manufactured by allowing chlorine and
sodium hydroxide solution (from
the electrolysis of aqueous
sodium chloride) to react.

Hypochlorous acid, HClO, is produced when chlorine disproportionates (is
oxidized and reduced) in water. In basic solution, the equilibrium is very far
toward the acid anions:

$$Cl_2(g) + 2OH^-(aq) \longrightarrow Cl^-(aq) + ClO^-(aq) + H_2O(l)$$

Solutions of sodium hypochlorite are manufactured by allowing the chlorine gas
released by the electrolysis of aqueous sodium chloride to mix with the cold solution of sodium hydroxide that is also obtained in the electrolysis. These solutions
are sold as a bleach (Figure 21.62).

Hypochlorite ion itself is unstable, disproportionating into chlorate ion,
ClO_3^-, and chloride ion.

$$3ClO^-(aq) \longrightarrow ClO_3^-(aq) + 2Cl^-(aq)$$

At room temperature the reaction is slow; but in hot solution in the presence of
base, the reaction is fast. Therefore, when chlorine reacts with hot sodium hydroxide solution, sodium chlorate is the product instead of sodium hypochlorite.

$$3Cl_2(g) + 6NaOH(aq) \longrightarrow NaClO_3(aq) + 5NaCl(aq) + 3H_2O(l)$$

Sodium chlorate can be crystallized from the solution. Solutions of chloric acid,
$HClO_3$, can be prepared, although the pure acid cannot be isolated.

Sodium perchlorate and potassium perchlorate are produced commercially by
the electrolysis of a saturated solution of the corresponding chlorate. The anode
reaction is

$$ClO_3^-(aq) + 3H_2O(l) \longrightarrow ClO_4^-(aq) + 2H_3O^+(aq) + 2e^-$$

Hydrogen evolves at the cathode. Perchloric acid can be prepared by treating a perchlorate salt with sulfuric acid.

$$KClO_4(s) + H_2SO_4(l) \longrightarrow KHSO_4(s) + HClO_4(l)$$

The perchloric acid is distilled from the mixture at reduced pressure (to keep the
temperature below 92°C, where perchloric acid decomposes explosively).

Table 21.11 lists uses of some halogen compounds.

Table 21.11	Uses of Some Halogen Compounds
Compound	**Use**
AgBr, AgI	Photographic film
CCl$_4$	Manufacture of fluorocarbons
CH$_3$Br	Pesticide
C$_2$H$_4$Cl$_2$	Manufacture of vinyl chloride (plastics)
HCl	Metal treating
	Food processing
NaClO	Household bleach
	Manufacture of hydrazine for rocket fuel
NaClO$_3$	Paper pulp bleaching (with ClO$_2$)
KI	Human nutritional and animal feed supplement

21.12 Group VIIIA: The Noble Gases

In our discussions of bonding, we pointed out the relative stability of the electron configurations of the Group VIIIA noble gases. For many years, it was thought that because the atoms of these elements had completed octets, the noble gases would be completely unreactive. Consequently, these elements were known as *inert gases*. Compounds of argon, krypton, xenon, and radon have since been prepared, however, so the term is not quite appropriate.

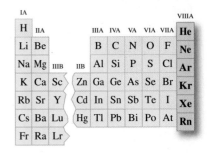

Helium and the Other Noble Gases

The noble gases were not known until 1894. A couple of years earlier, the English physicist Lord Rayleigh discovered that the density of nitrogen gas obtained from air (1.2561 g/L at STP) was noticeably greater than the density of nitrogen obtained by decomposition of nitrogen compounds (1.2498 g/L at STP). He concluded that one of these two nitrogen sources was contaminated with another substance.

Soon after this, Rayleigh began collaborating with the Scottish chemist William Ramsay. Ramsay passed atmospheric nitrogen over hot magnesium to remove the nitrogen as magnesium nitride, Mg_3N_2, and obtained a nonreactive residual gas. He placed this gas in a sealed glass tube and passed a high-voltage electrical discharge through it to observe its emission spectrum. The spectrum showed a series of red and green lines and was unlike that of any known element. Ramsay and Rayleigh concluded that they had discovered a new element, which they called argon (from the Greek word *argos,* meaning "lazy"—referring to argon's lack of chemical reactivity). They also surmised that argon was a member of a new column of elements in the periodic table.

Before the discovery of argon, some lines in the spectrum of the sun were ascribed to an element not yet known on earth. This element was called helium (from the Greek *helios,* meaning "sun"). In 1895, Ramsay and the Swedish chemist Per Theodor Cleve (working independently) announced the discovery of helium gas in the mineral cleveite (a uranium ore) by identifying the spectrum of helium. Several years later, Ramsay discovered neon, krypton, and xenon by fractional distillation of liquid air. Radon was discovered in 1900 as a gaseous decay product of radium.

Preparation and Uses of the Noble Gases Commercially, all the noble gases except helium and radon are obtained by the distillation of liquid air. The principal sources of helium are certain natural-gas wells. Helium has the lowest boiling point (−268.9°C) of any substance and is very important in low-temperature research. The major use of argon is as a blanketing gas (inert gas) in metallurgical processes. It is also used as a mixture with nitrogen to fill incandescent lightbulbs. In these bulbs the gas mixture conducts heat away from the hot tungsten filament, without reacting with it. All the noble gases are used in gas discharge tubes (containing gas through which a high-voltage electric current can be discharged, giving light from atomic emission). Neon gives a highly visible red-orange emission and has long been used in advertising signs. The noble gases are also used in a number of lasers. The helium–neon laser was the first continuously operating gas laser. It emits red light at a wavelength of 632.8 nm.

Compounds of the Noble Gases Neil Bartlett, working at the University of British Columbia, prepared the first noble-gas compound after he discovered that molecular oxygen reacts with platinum hexafluoride, PtF_6, to form the ionic solid $[O_2^+]$ $[PtF_6^-]$. Because the ionization energy of xenon (1.17×10^3 kJ/mol) is close to that of molecular oxygen (1.21×10^3 kJ/mol), Bartlett reasoned that xenon should also react with platinum hexafluoride. In 1962 he reported the synthesis of an orange-yellow compound with the approximate formula $XePtF_6$. ▶ Later in the same year, chemists from Argonne National Laboratory near Chicago reported that xenon

The product actually has variable composition and can be represented by the formula $Xe(PtF_6)_n$, where n is between 1 and 2.

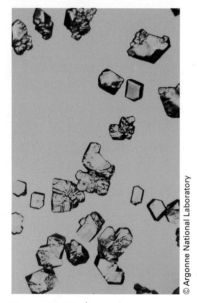

© Argonne National Laboratory

Figure 21.63 ▲

Crystals of xenon tetrafluoride This photomicrograph shows crystals obtained in the experiment that first produced a binary compound of xenon.

reacts directly with fluorine at 400°C to give the tetrafluoride. The elements react even at room temperature when exposed to sunlight.

$$Xe(g) + 2F_2(g) \longrightarrow XeF_4(s)$$

The product is a volatile, colorless solid (Figure 21.63). Since then, a number of noble-gas compounds have been prepared that typically involve bonds to the highly electronegative elements fluorine and oxygen. Most of these are compounds of xenon (see Table 21.12), but a few are compounds of krypton and radon. Recently, an argon compound, HArF, was synthesized.

Table 21.12   Some Compounds of Xenon

Compound	Formula	Description
Xenon difluoride	XeF_2	Colorless crystals
Xenon tetrafluoride	XeF_4	Colorless crystals
Xenon hexafluoride	XeF_6	Colorless crystals
Xenon trioxide	XeO_3	Colorless crystals, explosive
Xenon tetroxide	XeO_4	Colorless gas, explosive

A Checklist for Review

Summary of Facts and Concepts

Several general observations can be made about the main-group elements. First, the metallic characteristics of these elements generally decrease across a period from left to right in the periodic table. Second, metallic characteristics of the main-group elements become more pronounced going down any column (group). Finally, a second-period element is usually rather different from the other elements in its group.

Metals may be pure elements, or they may be *alloys,* which can be either compounds or mixtures. Metallic elements and their compounds are obtained principally from the earth's crust, most of which is composed of metal silicates. The chief sources of metals, however, are not silicates but oxide, carbonate, and sulfide *minerals,* which exist in *ore* deposits widely scattered over the earth.

An important part of *metallurgy* is the production of metals from their ores. This involves three basic steps: *preliminary treatment, reduction,* and *refining.* In preliminary treatment, the ore is concentrated in its metal-containing mineral. The concentrated ore may also require a process such as *roasting,* in which the metal compound is transformed to one that is more easily reduced. In reduction, the metal compound is reduced to the metal, by either electrolysis or chemical reduction. In refining, the metal is purified, or freed of contaminants.

The *electron-sea model* is a simple depiction of a metal as an array of positive ions surrounded by delocalized valence electrons. Molecular orbital theory gives a more detailed picture of the bonding in metals. Because the energy levels in a metal crowd together into bands, this picture of metal bonding is often referred to as *band theory.* According to band theory, the electrons in a crystal become free to move when they are excited to the unoccupied orbitals of a band. In a metal, this requires very little energy, because unoccupied orbitals lie just above the occupied orbitals of highest energy.

The Group IA metals (alkali metals) are soft, chemically reactive elements. Lithium, sodium, and potassium are important alkali metals. In recent years, the commercial uses of lithium have grown markedly. The metal is obtained by the electrolysis of molten lithium chloride and is used in the production of low-density alloys and as a battery anode. Lithium hydroxide is used to make lithium soap for lubricating greases; it is produced by the reaction of lithium carbonate and calcium hydroxide.

Sodium metal is prepared in large quantities. It is used as a reducing agent in the preparation of other metals, such as titanium and zirconium, and in the preparation of dyes and pharmaceuticals. Sodium compounds are of enormous economic importance. Sodium chloride is the source of sodium and most of its compounds. Sodium hydroxide is prepared by the electrolysis of aqueous sodium chloride; as a strong base, it has many useful commercial applications, including aluminum production and petroleum refining. Sodium carbonate is obtained from the mineral trona, which contains sodium carbonate and sodium hydrogen carbonate, and by the *Solvay process* from salt ($NaCl$) and limestone ($CaCO_3$). Sodium carbonate is used to make glass. Potassium metal is produced in relatively small quantities, but potassium compounds are important. Large quantities of potassium chloride are used as a plant fertilizer.

Magnesium and calcium are the most important of the Group IIA (alkaline earth) metals. Magnesium and its alloys are important structural metals. Calcium is important primarily as its compounds, which are prepared from natural carbonates, such as limestone, and the sulfates, such as gypsum. When limestone is heated strongly, it decomposes to calcium oxide (lime). Enormous quantities of lime are used in the production of iron from its ores.

Of the Group IIIA and Group IVA metals, aluminum, tin, and lead are especially important. Aluminum is the third most abundant element in the earth's crust. It is obtained commercially from bauxite; through chemical processing bauxite yields pure aluminum oxide. Most of this aluminum oxide is used in the production of aluminum by electrolysis. Some aluminum oxide is used as a carrier for heterogeneous catalysts and in manufacturing industrial ceramic materials.

Tin is normally a metal (called white tin) but does undergo a low-temperature conversion to a nonmetallic form (called gray tin). Tin is obtained by reduction of cassiterite, a mineral form of SnO_2. Tin is used to make tin plate, bronze, and solder. Lead is obtained from galena, which is a sulfide ore, PbS. More than half of the lead produced is used to make electrodes for lead storage batteries. Litharge, or lead(II) oxide, is an important lead compound from which other lead compounds are prepared.

Carbon and tin are the least metallic of the Group IVA elements. *Catenation* is an important feature of carbon chemistry and is responsible for the enormous number of organic compounds. Carbon has several allotropes, the principal ones being diamond and graphite, which are covalent-network solids, and buckminsterfullerene, which is molecular (C_{60}). The element has important industrial uses, including carbon black for rubber tires. The principal oxides of carbon are CO and CO_2. Mixtures of carbon monoxide and hydrogen (synthesis gas) are used to prepare various organic compounds. Liquid and solid carbon dioxide are used as refrigerants, and the gas is used to make carbonated beverages.

Hydrogen is the most abundant element in the universe and is the third most abundant element on the surface of the earth. Most of the hydrogen on earth is found in water. Hydrogen has three isotopes: *protium, deuterium,* and *tritium.* Protium is the most abundant, with less than 0.02% being deuterium and only a trace being radioactive tritium. Deuterium and tritium isotopes can be substituted for protium in chemical compounds in order to provide markers that can be followed during a chemical reaction or to change the chemical and physical properties of the compound. Elemental hydrogen is produced on an industrial scale by the *steam-reforming process* in which a hydrocarbon is reacted with water in the presence of a catalyst at high temperature. The bulk of the hydrogen produced in this manner is used to make organic compounds including methanol. Hydrogen forms three classes of binary compounds called binary hydrides: ionic hydrides, covalent hydrides, and metallic hydrides. The *ionic hydrides* are reactive solids formed either by the reaction of hydrogen with an alkali metal to form compounds with the general formula MH, or with larger alkaline earth metals to form MH_2. The *covalent hydrides* are compounds in which hydrogen is covalently bonded to another element. The *metallic hydrides* contain a transition metal element and hydrogen. In these compounds, the lattice of metal atoms forms a porous structure that allows hydrogen atoms to enter and bond. Metallic hydrides are often nonstoichiometric, meaning that the ratio of hydrogen atoms to metal atoms is not a whole number.

Silicon is a prominent material of technology. The element is the basic material used in semiconductor devices for solid-state electronics. These devices require silicon of extreme purity. The impure element is converted to silicon tetrachloride, which is purified by distillation, after which the silicon tetrachloride is reduced to the pure silicon. *Silica* is chemically known as silicon dioxide; quartz is a common form of silica. It exhibits the *piezoelectric effect,* which is used to control radio, television, and clock frequencies. *Silicates* are formed by *condensation reactions,* and an enormous variety of silicate minerals exists. The *silicones* are materials that contain chains of Si—O bonds, with hydrocarbon groups attached to Si atoms. They have wide applications, including their use as elastomers and lubricants.

Of the Group VA elements, nitrogen and phosphorus are particularly important. Nitrogen, N_2, is obtained from liquid air by fractional distillation; liquid nitrogen is used as a refrigerant. Ammonia, NH_3, is the most important compound of nitrogen. It is prepared from the elements and is used as a fertilizer. Ammonia is also the starting compound for the manufacture of other nitrogen compounds. For example, in the *Ostwald process* for the preparation of nitric acid, ammonia is burned in the presence of a catalyst to nitrogen monoxide (nitric oxide). NO reacts with oxygen to give nitrogen dioxide, which reacts with water to give nitric acid.

Phosphorus has two common allotropes, white phosphorus (P_4) and red phosphorus (chain structure). White phosphorus is obtained by heating a phosphate mineral with sand and coke in an electric furnace. When phosphorus burns in air, it forms tetraphosphorus decoxide, P_4O_{10}. This oxide reacts with water to give orthophosphoric acid, H_3PO_4. Phosphorus has many oxoacids; most are obtained by condensation reactions with orthophosphoric acid. One series is called the *polyphosphoric acids;* they have the general formula $H_{n+2}P_nO_{3n+1}$. Triphosphoric acid is an example; sodium triphosphate, $Na_5P_3O_{10}$, is used in detergents. The *metaphosphoric acids* have the general formula $(HPO_3)_n$.

Oxygen, a Group VIA element, occurs in the atmosphere (as O_2), but mostly it is present on earth as oxide and oxoanion minerals. Oxygen has two allotropes: dioxygen, O_2, and ozone, O_3. Dioxygen, usually called simply oxygen, is obtained commercially from liquid air. Oxygen reacts with almost all elements to give *oxides* or, in some cases, *peroxides* or *superoxides.*

Sulfur, another Group VIA element, occurs in sulfate and sulfide minerals. Free sulfur, S_8, occurring in deep underground deposits is mined by the *Frasch process*. Sulfur is also produced by the *Claus process,* in which hydrogen sulfide (obtained from natural gas and petroleum) is partially burned. Most of the sulfur is used to prepare sulfuric acid by the *contact process*. In this process, sulfur is burned to sulfur dioxide, SO_2, which in the presence of a catalyst and oxygen forms sulfur trioxide, SO_3. This oxide dissolves in concentrated sulfuric acid, which when diluted with water gives additional sulfuric acid. Sulfuric acid is the most important compound of sulfur.

The Group VIIA elements, or halogens, are reactive. Chlorine (Cl_2), a pale greenish yellow gas, is prepared commercially by the electrolysis of aqueous sodium chloride. Its principal uses are in the preparation of chlorinated hydrocarbons and as a bleaching agent and disinfectant. Hydrogen chloride, HCl, is one of the most important compounds of chlorine; aqueous solutions of HCl are known as hydrochloric acid.

The Group VIIIA elements, the noble gases, were discovered at the end of the nineteenth century. Although the noble gases were at first thought to be unreactive, compounds of these gases are now known.

Learning Objectives	Important Terms
21.1 General Observations About the Main-Group Elements	
■ Note the low ionization energies and electronegativities of the metals. ■ Give the principal oxidation states of the main-group elements. ■ State the periodic trends in metallic characteristics.	
21.2 Metals: Characteristics and Production	
■ Define *metal, alloy, mineral,* and *ore.* ■ Define *metallurgy.* ■ State the basic steps in the production of a metal. ■ Define the *Bayer process*. ■ Describe the roasting of lead sulfide ore. ■ Describe the electrolysis of molten lithium chloride. ■ Define the *Dow process.* ■ Define the *Hall–Héroult process.* ■ Describe the chemical reduction of lead(II) oxide to lead metal. ■ Give some methods for refining metals.	metal alloy mineral ore metallurgy Bayer process roasting Dow process Hall–Héroult process
21.3 Bonding in Metals	
■ Describe the electron-sea model of metals. ■ Describe the molecular orbital theory of sodium and magnesium metals.	
21.4 Group IA: The Alkali Metals	
■ Note the reactivity of the alkali metals. ■ Describe the metallurgy, reactions, and compounds of lithium. ■ Describe the metallurgy, reactions, and compounds of sodium. ■ Define the *Solvay process.* ■ Describe some compounds of potassium.	Solvay process
21.5 Group IIA: The Alkaline Earth Metals	
■ Describe the metallurgy, reactions, and compounds of magnesium. ■ Describe the metallurgy, reactions, and compounds of calcium.	
21.6 Group IIIA and Group IVA Metals	
■ Describe the metallurgy, reactions, and compounds of aluminum.	Goldschmidt process

- Define the *Goldschmidt process.*
- Describe the metallurgy, reactions, and compounds of tin and lead.

21.7 Hydrogen

- Describe some properties of hydrogen.
- Describe the commercial preparation of hydrogen.
- Define the *steam-reforming process.*
- Describe some reactions and compounds of hydrogen.
- Define a *binary hydride.*

steam-reforming process
binary hydride

21.8 Group IVA: The Carbon Family

- Define *catenation.*
- Describe some allotropes of carbon.
- Describe the chemical properties of the oxides of carbon.
- Describe the chemical properties of the carbonates.
- Describe the preparation of extremely pure silicon.
- Define *silica.*
- Describe the uses of quartz.
- Define *silicate, condensation reaction,* and *silicone.*

catenation
silica
silicate
condensation reaction
silicone

21.9 Group VA: Nitrogen and the Phosphorus Family

- Describe the properties and uses of nitrogen.
- Describe some nitrogen compounds.
- Define the *Ostwald process.*
- Describe the allotropes of phosphorus.
- Describe the phosphorus oxides and the oxoacids of phosphorus.
- Define *polyphosphoric acids* and *metaphosphoric acids.*

Ostwald process
polyphosphoric acids
metaphosphoric acids

21.10 Group VIA: Oxygen and the Sulfur Family

- Describe the properties and preparation of oxygen.
- Describe some reactions of oxygen.
- Define *oxide, peroxide,* and *superoxide.*
- Describe the allotropes of sulfur.
- Describe the production of sulfur.
- Define the *Frasch process* and the *Claus process.*
- Describe the sulfur oxides and oxoacids.
- Define the *contact process.*

oxide
peroxide
superoxide
Frasch process
Claus process
contact process

21.11 Group VIIA: The Halogens

- Describe chlorine and its properties, preparation, and uses.
- Describe the preparation of hydrogen chloride and its uses.
- Describe the preparation and uses of the oxoacids of chlorine.

21.12 Group VIIIA: The Noble Gases

- Describe the discovery, preparation, and uses of the noble gases.
- Describe some compounds of the noble gases.

Questions and Problems

Self-Assessment and Review Questions

Key: These questions test your understanding of the ideas you worked with in the chapter. These problems vary in difficulty and often can be used for the basis of discussion.

21.1 Give four characteristics of a metal.

21.2 What is an alloy? Give an example.

21.3 What are the basic steps in the production of a pure metal from a natural source? Illustrate each step with the preparation of aluminum.

21.4 Define *mineral, rock,* and *ore.* Is bauxite a mineral or a rock? Explain.

21.5 What is the purpose of roasting lead(II) ore?

21.6 Briefly describe three different methods for refining a metal.

21.7 Write equations for the preparation of magnesium metal from seawater, limestone, and hydrochloric acid.

21.8 Draw a flowchart for the preparation of aluminum from its ore. Label each step.

21.9 Explain metallic conduction in terms of molecular orbital theory (band theory).

21.10 Give the commercial source of each of the following metals: Li, Na, Mg, Ca, Al, Sn, Pb.

21.11 Give equations for the reactions of lithium and sodium metals with water and with oxygen in air.

21.12 Which do you expect to be more reactive, lithium or potassium? Explain.

21.13 Write the equation for the reaction of lithium carbonate with barium hydroxide.

21.14 Ethanol, C_2H_5OH, reacts with sodium metal because the hydrogen atom attached to the oxygen atom is slightly acidic. Write a balanced equation for the reaction of sodium with ethanol to give the salt sodium ethoxide, $NaOC_2H_5$.

21.15 How is sodium hydroxide manufactured? What is another product in this process?

21.16 a Write electrode half-reactions for the electrolysis of fused sodium chloride. b Do the same for fused sodium hydroxide. The hydroxide ion is oxidized to oxygen and water.

21.17 Describe common uses of the following sodium compounds: sodium chloride, sodium hydroxide, and sodium carbonate.

21.18 Describe the main step in the Solvay process. Give the balanced equation for this reaction.

21.19 Because magnesium reacts with oxygen, steam, and carbon dioxide, magnesium fires are extinguished only by smothering the fire with sand. Write balanced equations for the reactions of magnesium with O_2, $H_2O(g)$, and CO_2.

21.20 a Calcium oxide is prepared industrially from what calcium compound? Write the chemical equation for the reaction. b Write the equation for the preparation of calcium hydroxide.

21.21 Purified calcium carbonate is prepared by precipitation. Write the equation for the commercial process using carbon dioxide. You can also prepare calcium carbonate by precipitation using sodium carbonate. Write the equation.

21.22 Calcium carbonate is used in some antacid preparations to neutralize the hydrochloric acid in the stomach. Write the equation for this neutralization.

21.23 Write the chemical equation of the thermite reaction, in which iron(III) oxide is reduced by aluminum.

21.24 What are some major uses of aluminum oxide?

21.25 How is aluminum sulfate used to purify municipal water supplies?

21.26 Lead(IV) oxide forms the cathode of lead storage batteries. How is this substance produced for these batteries?

21.27 The yellow paint pigment chrome yellow [lead(II) chromate] is produced by a precipitation reaction. Write an equation in which chrome yellow is produced from lead(II) nitrate.

21.28 Why are lead pigments no longer used in house paints?

21.29 Describe the reactions that are used in the steam-reforming process for the production of hydrogen.

21.30 Give the names and symbols for the three isotopes of hydrogen. Which isotope is radioactive?

21.31 Explain why hydrogen has the potential to be widely used as a fuel.

21.32 Give an example of a compound of each of the binary hydrides.

21.33 What is meant by the term *catenation*? Give an example of a carbon compound that displays catenation.

21.34 Carbon monoxide is a poisonous gas. What is the mechanism of this poisoning?

21.35 Write equations for the equilibria involving carbon dioxide in water.

21.36 Describe the steps in preparing ultrapure silicon from quartz sand.

21.37 Use structural formulas to illustrate the hypothetical condensation reaction of two silicic acid molecules, $Si(OH)_4$.

21.38 Describe the natural cycle of nitrogen from the atmosphere to biological organisms and back to the atmosphere.

21.39 Describe Rutherford's preparation of nitrogen from air. Was the gas he obtained pure nitrogen? Explain.

21.40 List the different nitrogen oxides. What is the oxidation number of nitrogen in each?

21.41 In your own words, describe the manufacture of ammonia from natural gas, steam, and air.

21.42 Describe the steps in the Ostwald process for the manufacture of nitric acid from ammonia.

21.43 Describe the structure of white phosphorus. How does the structure account for the reactivity of this substance?

21.44 Give chemical equations for the reaction of white phosphorus in an excess of air and for the reaction of the product with water.

21.45 Describe two different methods used to manufacture phosphoric acid, H_3PO_4, starting from $Ca_3(PO_4)_2$.

21.46 What is the purpose of adding polyphosphates to a detergent?

21.47 By means of an equation, show how triphosphoric acid could be formed from orthophosphoric acid and diphosphoric acid.

21.48 What reaction was used by Priestley in preparing pure oxygen?

21.49 What is the most important commercial means of producing oxygen?

21.50 Define *oxide, peroxide,* and *superoxide.* Give an example of each.

21.51 Give an example of an acidic oxide and a basic oxide.

21.52 List three natural sources of sulfur or sulfur compounds.

21.53 Describe the structure of the rhombic sulfur molecule.

21.54 Discuss the various allotropes of sulfur and describe how they can be prepared from rhombic sulfur.

21.55 Describe the Frasch process for mining sulfur.

21.56 With the aid of chemical equations, describe the Claus process for the production of sulfur from hydrogen sulfide.

21.57 Give equations for preparing each of the following:
   a SO_2 b H_2S

21.58 Write equations for each of the following.
   a reaction of H_2S with SO_2
   b oxidation of SO_2 with $Cr_2O_7^{2-}(aq)$ to SO_4^{2-}
   c reaction of hot, concentrated H_2SO_4 with Cu
   d reaction of sulfur with Na_2SO_3

21.59 Give equations for the steps in the contact process for the manufacture of sulfuric acid from sulfur.

21.60 Give the equation for the reaction used by the Swedish chemist Scheele to prepare chlorine.

21.61 Complete and balance the following equations. Write *NR* if no reaction occurs.
   a $Cl_2(aq) + Br^-(aq) \longrightarrow$
   b $Br_2(aq) + I^-(aq) \longrightarrow$
   c $Br_2(aq) + Cl^-(aq) \longrightarrow$
   d $I_2(aq) + Cl^-(aq) \longrightarrow$

21.62 A test tube contains a solution of one of the following salts: NaCl, NaBr, NaI. Describe a single test that can distinguish among these possibilities.

21.63 What are some major commercial uses of chlorine?

21.64 How is sodium hypochlorite prepared? Give the balanced chemical equation.

21.65 Describe the preparation of perchloric acid from sodium chloride.

21.66 What was the argument used by Bartlett that led him to the first synthesis of a noble-gas compound?

21.67 Which of the following are used in the production of $H_2(g)$?
  I. steam-reforming process
  II. water–gas reaction
 III. Haber process
   a I only
   b II only
   c III only
   d I and II only
   e II and III only

21.68 The major use of $Ca_3(PO_4)_2$ is in
   a the tempering of metals
   b phosphorus for fireworks
   c acid neutralization
   d the production of phosphoric acid
   e the vulcanization of rubber

21.69 In which of the following is nitrogen or nitrogen-containing compounds *not* used?
   a fertilizer   b explosives   c dental anesthetic
   d blanketing gas   e jet fuel

21.70 Sodium hypochlorite is used in laundry bleach. The formula of sodium hypochlorite is
   a $NaClO_4$ b $NaClO_2$ c NaClO
   d NaCl   e $NaClO_3$

Conceptual Problems

Key: These problems are designed to check your understanding of the concepts associated with some of the main topics presented in each chapter. A strong conceptual understanding of chemistry is the foundation for both applying chemical knowledge and solving chemical problems. These problems vary in level of difficulty and often can be used as a basis for group discussion.

21.71 When producing coke, why is the coal heated in the absence of air? Write the chemical reaction for what would happen if it were to be heated in air.

21.72 Even though hydrogen isn't a metal, why is it in Group IA of most periodic tables?

21.73 What happens to the metallic character of the main-group elements as you move left to right across any row of the periodic table? What happens to the metallic character of the main-group elements as you move down a column (group)?

21.74 Aluminum hydroxide is an amphoteric substance. What does this mean? Write equations to illustrate.

21.75 Lithium hydroxide, like sodium hydroxide, becomes contaminated when exposed to air. What is the source of this contamination? What reactions take place?

21.76 Tin metal would not make a very good structural metal in cold climates. Why?

21.77 Oxygen, like other second-period elements, is somewhat different from the other elements in its group. List some of these differences.

21.78 Hydrogen chloride can be prepared by heating NaCl with concentrated sulfuric acid. Why is substituting NaBr for NaCl in this reaction not a satisfactory way to prepare HBr?

21.79 Given that the reaction $Cl_2(g) + 2KBr(aq) \longrightarrow 2KCl(aq) + Br_2(aq)$ readily occurs, would you expect the reaction $I_2(s) + 2KCl(aq) \longrightarrow 2KI(aq) + Cl_2(aq)$ to occur?

21.80 Do you expect an aqueous solution of sodium hypochlorite to be acidic, neutral, or basic? What about an aqueous solution of sodium perchlorate?

Practice Problems

Key: These problems are for practice in applying problem-solving skills. They are divided by topic, and some are keyed to exercises (see the ends of the exercises). The problems are arranged in matching pairs; the odd-numbered problem of each pair is listed first, and its answer is given in the back of the book.

Metallurgy

21.81 Pure iron is prepared for special purposes by precipitating iron(III) oxide and reducing the dry oxide with hydrogen gas. Write the balanced equation for this reduction of iron(III) oxide.

21.82 Manganese is available from its ore pyrolusite, which is manganese(IV) oxide, MnO_2. Some manganese metal is prepared from MnO_2 by reduction with aluminum metal. Write the balanced equation for this reduction.

21.83 How many kilograms of iron can be produced from 2.00 kg of hydrogen, H_2, when you reduce iron(III) oxide? (See Problem 21.81.)

21.84 How many kilograms of manganese can be produced from 1.00 kg of aluminum, Al, when you reduce manganese(IV) oxide? (See Problem 21.82.)

21.85 Using thermodynamic data given in Appendix C, obtain $\Delta H°$ for the roasting of galena, PbS:

$$PbS(s) + \tfrac{3}{2}O_2(g) \longrightarrow PbO(s) + SO_2(g)$$

Is the reaction endothermic or exothermic?

21.86 Using thermodynamic data given in Appendix C, obtain $\Delta H°$ for the reduction of pyrolusite, MnO_2, by aluminum (see Problem 21.82). Is this reaction endothermic or exothermic?

Bonding in Metals

21.87 Sketch a diagram showing the formation of energy levels from the valence orbitals for Ca, Ca_2, Ca_3, and Ca_n. On the diagram, place arrows indicating how the electrons fill these energy levels.

21.88 Sketch a diagram showing the formation of energy levels from the valence orbitals for K, K_2, K_3, and K_n. On the diagram, place arrows indicating how the electrons fill these energy levels.

21.89 How many energy levels are there in the $3p$ band of a single crystal of magnesium with a mass of 1.00 mg?

21.90 How many energy levels are there in the valence band of a single crystal of sodium with a mass of 1.00 mg?

Group IA: The Alkali Metals

21.91 Caustic soda, NaOH, can be manufactured from sodium carbonate in a manner similar to the preparation of lithium hydroxide. Write balanced equations (in three steps) for the preparation of NaOH from slaked lime, $Ca(OH)_2$, salt (NaCl), carbon dioxide, ammonia, and water.

21.92 Sodium phosphate, Na_3PO_4, is produced by the neutralization reaction. Phosphoric acid, H_3PO_4, is obtained by burning phosphorus to P_4O_{10}, then reacting the oxide with water to give the acid. Write balanced equations (in four steps) for the preparation of Na_3PO_4 from P_4, H_2O, air, and NaCl.

21.93 Complete and balance the following equations.
a $NaOH(s) + CO_2(g) \longrightarrow$
b $K_2SO_4(aq) + Pb(NO_3)_2(aq) \longrightarrow$
c $K(s) + Br_2(l) \longrightarrow$
d $K(s) + H_2O(l) \longrightarrow$
e $Li_2CO_3(aq) + HNO_3(aq) \longrightarrow$

21.94 Complete and balance the following equations.
a $LiHCO_3(s) \overset{\Delta}{\longrightarrow}$
b $Na_2SO_4(aq) + BaCl_2(aq) \longrightarrow$
c $K_2CO_3(aq) + Ca(OH)_2(aq) \longrightarrow$
d $Li(s) + HCl(aq) \longrightarrow$
e $Na(s) + ZrCl_4(g) \longrightarrow$

21.95 Francium was discovered as a minor decay product of actinium-227. Write the nuclear equation for the decay of actinium-227 by alpha emission.

21.96 Francium-223 is a radioactive alkali metal that decays by beta emission. Write the nuclear equation for this decay process.

Group IIA: The Alkaline Earth Metals

21.97 Potassium hydroxide and barium hydroxide are strong bases. What simple chemical test could you use to distinguish between solutions of these two bases?

21.98 Sodium hydroxide and calcium hydroxide are strong bases. What simple chemical test could you use to distinguish between solutions of these two bases?

21.99 Devise a chemical method for separating a solution containing $MgCl_2$ and $BaCl_2$ to give two solutions or compounds each containing only one of the metal ions.

21.100 Devise a chemical method for separating a solution containing NaCl and $MgCl_2$ to give two solutions or compounds each containing only one of the metal ions.

21.101 A short-lived isotope of radium decays by alpha emission to radon-219. Write the nuclear equation for this decay process.

21.102 Thorium-230, which occurs in uranium minerals, decays by alpha emission to radium. Write the nuclear equation for this decay process.

21.103 Complete and balance the following equations.
a $Ba(s) + H_2O(l) \longrightarrow$
b $Mg(s) + NiCl_2(aq) \longrightarrow$
c $NaOH(aq) + MgSO_4(aq) \longrightarrow$
d $Mg(OH)_2(s) + HNO_3(aq) \longrightarrow$
e $BaCO_3(s) \overset{\Delta}{\longrightarrow}$

21.104 Complete and balance the following equations.
a. $KOH(aq) + MgCl_2(aq) \longrightarrow$
b. $Mg(s) + CuSO_4(aq) \longrightarrow$
c. $Sr(s) + H_2O(l) \longrightarrow$
d. $SrCO_3(s) + HCl(aq) \longrightarrow$
e. $Ba(OH)_2(aq) + CO_2(g) \longrightarrow$

21.105 You have a 0.060 M solution of calcium hydrogen carbonate. How many grams of calcium hydroxide must be added to 25.0 mL of the solution to precipitate all of the calcium?

21.106 You have a 0.21 M solution of magnesium ion. How many grams of calcium hydroxide must be added to 50.0 mL of the solution to precipitate all of the magnesium ion?

Group IIIA and Group IVA Metals

21.107 Baking powders contain sodium (or potassium) hydrogen carbonate and an acidic substance. When water is added to a baking powder, carbon dioxide is released. One kind of baking powder contains $NaHCO_3$ and sodium aluminum sulfate, $NaAl(SO_4)_2$. Write the net ionic equation for the reaction that occurs in water solution.

21.108 When aluminum sulfate is dissolved in water, it produces an acidic solution. Suppose the pH of this solution is raised by the dropwise addition of aqueous sodium hydroxide. a. Describe what you would observe as the pH continues to rise. b. Write balanced equations for any reactions that occur.

21.109 Unlabeled test tubes contain solid $AlCl_3 \cdot 6H_2O$ in one, $Ba(OH)_2 \cdot 8H_2O$ in another, and $MgSO_4 \cdot 7H_2O$ in the other. How could you find out what is in each test tube, using chemical tests that involve only these compounds plus water?

21.110 The following solid substances are in separate but unlabeled test tubes: $Al_2(SO_4)_3 \cdot 18H_2O$, $BaCl_2 \cdot 2H_2O$, KOH. Describe how you could identify the compounds by chemical tests using only these substances and water.

21.111 Lead(II) nitrate, one of the few soluble lead salts, gives a solution with a pH of about 3 to 4. Write a chemical equation to explain why the solution is acidic.

21.112 The $Sn^{2+}(aq)$ ion can be written in more detail as $Sn(H_2O)_6^{2+}$. This ion is acidic by hydrolysis. Write a possible equation for this hydrolysis.

21.113 Lead(IV) oxide is a strong oxidizing agent. For example, lead(IV) oxide will oxidize hydrochloric acid to chlorine, Cl_2. Write the balanced equation for this reaction.

21.114 Lead(IV) oxide can be prepared by oxidizing plumbite ion, $Pb(OH)_3^-$, which exists in a basic solution of Pb^{2+}. Write the balanced equation for this oxidation by OCl^- in basic solution.

21.115 Complete and balance the following equations.
a. $Ga(OH)_3(s) \xrightarrow{\Delta}$
b. $Pb(NO_3)_2(aq) + NaI(aq) \longrightarrow$
c. $Al(s) + Mn_3O_4(s) \longrightarrow$
d. $Al(s) + AgNO_3(aq) \longrightarrow$
e. $Al_2O_3(s) + H_2SO_4(aq) \longrightarrow$

21.116 Complete and balance the following equations.
a. $Pb(NO_3)_2(aq) + Al(s) \longrightarrow$
b. $Pb(NO_3)_2(aq) + Na_2CrO_4(aq) \longrightarrow$
c. $Al_2(SO_4)_3(aq) + $ dilute $LiOH(aq) \longrightarrow$
d. $Al(s) + HCl(aq) \longrightarrow$
e. $Sn(s) + HBr(aq) \longrightarrow$

Hydrogen

21.117 How much heat will be evolved when 20.0 grams of the binary covalent hydride HF is produced via the following reaction?

$$F_2(g) + H_2(g) \longrightarrow 2HF(g) \quad \Delta H = -545 \text{ kJ}$$

21.118 Calculate the amount of heat evolved when 3.5×10^4 kg of hydrogen is combusted.

$$2H_2(g) + O_2(g) \longrightarrow 2H_2O(g) \quad \Delta H = -484 \text{ kJ}$$

21.119 Indicate the oxidation state for the element noted in each of the following:
a. H in CaH_2
b. H in H_2O
c. C in CH_4
d. S in H_2SO_4

21.120 Indicate the oxidation state for the element noted in each of the following:
a. H in H_2
b. H in C_2H_4
c. Si in SiH_4
d. N in HNO_3

Group IVA: The Carbon Family

21.121 Describe the bonding (using valence bond theory) of the Group IVA atoms in each of the following:
a. CCl_4
b. HCN
c. SiF_4
d. CH_3COOH (acetic acid)

21.122 Describe the bonding (using valence bond theory) of the Group IVA atoms in each of the following:
a. C_2H_6
b. SiF_6^{2-}
c. $CH_3CH=CH_2$
d. SiH_4

21.123 Calculate the standard enthalpy change for the following "cracking" reactions.
a. $CH_4(g) \longrightarrow C(\text{graphite}) + 2H_2(g)$
b. $C_2H_6(g) \longrightarrow C_2H_4(g) + H_2(g)$

21.124 Calculate the standard enthalpy change for the following reactions involving synthesis gas.
a. $CO(g) + 3H_2(g) \longrightarrow CH_4(g) + H_2O(g)$
b. $CO(g) + 2H_2(g) \longrightarrow CH_3OH(g)$

21.125 Complete and balance the following equations.
a. $CO_2(g) + Ba(OH)_2(aq) \longrightarrow$
b. $MgCO_3(s) + HBr(aq) \longrightarrow$

21.126 Complete and balance the following equations.
a. $Ca(HCO_3)_2(aq) + Ca(OH)_2(aq) \longrightarrow$
b. $NaHCO_3(aq) + HC_2H_3O_2(aq) \longrightarrow$

21.127 Use balanced equations to show how you could prepare Na_2CO_3 from carbon, NaOH, air, and H_2O.

21.128 Use balanced equations to show how you could prepare methanol, CH_3OH, from ethane, C_2H_6, and water.

21.129 Silicon has a diamond-like structure, with each Si atom bonded to four other Si atoms. Describe the bonding in silicon in terms of hybrid orbitals.

21.130 Silicon carbide, SiC, has a structure in which each Si atom is bonded to four C atoms, and each C atom is bonded to four Si atoms. Describe the bonding in terms of hybrid orbitals.

21.131 Silicon is purified by first converting it to silicon tetrachloride, which is then distilled. How many kilograms of silicon are required to prepare 5.00 kg of $SiCl_4$?

21.132 Silicon is prepared by reducing silicon dioxide with carbon at high temperature. How many kilograms of silicon dioxide are required to prepare 5.00 kg of silicon?

21.133 Draw a portion of the structure of the mineral spodumene, $LiAl(SiO_3)_2$, which is a mineral consisting of a long chain of SiO_4 tetrahedra. Use this drawing to verify the empirical formula given.

21.134 Draw a portion of the structure of the mineral beryl, $Be_3Al_2(SiO_3)_6$, which is a mineral consisting of cyclic silicate anions. Use this drawing to verify the empirical formula given.

Group VA: Nitrogen and the Phosphorus Family

21.135 Magnesium nitride, Mg_3N_2, reacts with water to produce magnesium hydroxide and ammonia. How many grams of ammonia can you obtain from 7.50 g of magnesium nitride?

21.136 Ammonia reacts with oxygen in the presence of a platinum catalyst to give nitric oxide, NO. How many grams of oxygen are required in this reaction to give 1.50 g NO?

21.137 You have the following substances: NH_3, O_2, Pt, and H_2O. Write equations for the preparation of N_2O from these substances.

21.138 Give equations for the preparation of N_2O. You can use NaOH, $NaNO_3$, H_2SO_4, and $(NH_4)_2SO_4$ (plus H_2O). Several steps are required.

21.139 Silver metal reacts with nitric acid to give silver ion and nitric oxide. Write the balanced equation for this reaction.

21.140 Zinc metal reacts with concentrated nitric acid to give zinc ion and ammonium ion. Write the balanced equation for this reaction.

21.141 Although phosphorus pentabromide exists in the vapor phase as PBr_5 molecules, in the solid phase the substance is ionic and has the structure $[PBr_4^+]Br^-$. What is the expected geometry of PBr_4^+? Describe the bonding to phosphorus in PBr_4^+.

21.142 Although phosphorus pentachloride exists in the vapor phase as PCl_5 molecules, in the solid phase the substance is ionic and has the structure $[PCl_4^+][PCl_6^-]$. What is the expected geometry of PCl_6^-? Describe the bonding to phosphorus in PCl_6^-.

21.143 Phosphorous acid, H_3PO_3, is oxidized to phosphoric acid, H_3PO_4, by hot, concentrated sulfuric acid, which is reduced to SO_2. Write the balanced equation for this reaction.

21.144 Phosphorous acid, H_3PO_3, is oxidized to phosphoric acid, H_3PO_4, by nitric acid, which is reduced to nitrogen monoxide, NO. Write the balanced equation for this reaction.

21.145 According to an analysis, a sample of phosphate rock contains 74.6% $Ca_3(PO_4)_2$, by mass. How many grams of phosphoric acid, H_3PO_4, can be obtained from 30.0 g of the phosphate rock according to the following reaction?

$$Ca_3(PO_4)_2(s) + 3H_2SO_4(aq) \longrightarrow 3CaSO_4(s) + 2H_3PO_4(aq)$$

21.146 According to an analysis, a sample of phosphate rock contains 71.2% $Ca_3(PO_4)_2$, by mass. How many grams of calcium dihydrogen phosphate, $Ca(H_2PO_4)_2$, can be obtained from 10.0 g of phosphate rock from the following reaction?

$$Ca_3(PO_4)_2(s) + 4H_3PO_4(aq) \longrightarrow 3Ca(H_2PO_4)_2(aq)$$

Group VIA: Oxygen and the Sulfur Family

21.147 Write an equation for each of the following.
- a burning of calcium metal in oxygen
- b burning of phosphine, PH_3, in excess oxygen
- c burning of ethanolamine, $HOCH_2CH_2NH_2$, in excess oxygen (N ends up as N_2)

21.148 Write an equation for each of the following.
- a burning of lithium metal in oxygen
- b burning of methylamine, CH_3NH_2, in excess oxygen (N ends up as N_2)
- c burning of diethyl sulfide, $(C_2H_5)_2S$, in excess oxygen

21.149 What are the oxidation numbers of sulfur in each of the following?
- a SF_6
- b SO_3
- c H_2S
- d $CaSO_3$

21.150 What are the oxidation numbers of sulfur in each of the following?
- a CaS
- b SCl_4
- c S_8
- d $CaSO_4$

21.151 Selenous acid, H_2SeO_3, is reduced by H_2S to sulfur, S_8, and selenium, Se_8. Write the balanced equation for this reaction.

21.152 Concentrated sulfuric acid oxidizes iodide ion to iodine, I_2. Write the balanced equation for this reaction.

21.153 Sodium hydrogen sulfite is prepared from sodium carbonate and sulfur dioxide:

$$Na_2CO_3(s) + 2SO_2(g) + H_2O(l) \longrightarrow$$
$$2NaHSO_3(aq) + CO_2(g)$$

How many grams of $NaHSO_3$ can be obtained from 25.0 g of Na_2CO_3?

21.154 Sodium thiosulfate, $Na_2S_2O_3$, is prepared from sodium sulfite and sulfur:

$$8Na_2SO_3(aq) + S_8(s) \longrightarrow 8Na_2S_2O_3(aq)$$

How many grams of $Na_2S_2O_3$ can be obtained from 10.0 g of sulfur?

Group VIIA: The Halogens

21.155 A solution of chloric acid may be prepared by reacting a solution of barium chlorate with sulfuric acid.

Barium sulfate precipitates. Write the balanced equation for the reaction.

21.156 A solution of chlorous acid may be prepared by reacting a solution of barium chlorite with sulfuric acid. Barium sulfate precipitates. Write the balanced equation for the reaction.

21.157 Iodic acid, HIO_3, can be prepared by oxidizing elemental iodine with concentrated nitric acid, which is reduced to nitrogen dioxide, $NO_2(g)$. Write the balanced equation for the reaction.

21.158 Chlorine can be prepared by oxidizing chloride ion (from hydrochloric acid) with potassium dichromate, $K_2Cr_2O_7$, which is reduced to Cr^{3+}. Write the balanced equation for the reaction.

21.159 Discuss the bonding in each of the following molecules or ions. What is the expected molecular geometry?
   a BrO_3^- b BrF_3 c Cl_2O

21.160 Discuss the bonding in each of the following molecules or ions. What is the expected molecular geometry?
   a $HClO$ b ClO_4^- c ClF_5

21.161 Write balanced equations for each of the following.
   a Hydrogen bromide gas forms when sodium bromide is heated with phosphoric acid.
   b Bromine reacts with aqueous sodium hydroxide to give hypobromite and bromide ions.

21.162 Write balanced equations for each of the following.
   a Solid calcium fluoride is heated with sulfuric acid to give hydrogen fluoride vapor.

   b Solid potassium chlorate is carefully heated to yield potassium chloride and potassium perchlorate.

21.163 Using standard electrode potentials, decide whether aqueous sodium hypochlorite solution will oxidize Br^- to Br_2 in basic solution under standard conditions. See Appendix I for data.

21.164 By calculating the standard emf, decide whether aqueous sodium hypochlorite solution will oxidize $Fe^{2+}(aq)$ to $Fe^{3+}(aq)$ in basic solution under standard conditions. See Appendix I for data.

Group VIIIA: The Noble Gases

21.165 Xenon tetrafluoride, XeF_4, is a colorless solid. Give the Lewis formula for the XeF_4 molecule. What is the hybridization of the xenon atom in this compound? What geometry is predicted by the VSEPR model for this molecule?

21.166 Xenon tetroxide, XeO_4, is a colorless, unstable gas. Give the Lewis formula for the XeO_4 molecule. What is the hybridization of the xenon atom in this compound? What geometry would you expect for this molecule?

21.167 Xenon trioxide, XeO_3, is reduced to xenon in acidic solution by iodide ion. Iodide ion is oxidized to iodine, I_2. Write a balanced chemical equation for the reaction.

21.168 Xenon difluoride, XeF_2, is hydrolyzed (broken up by water) in basic solution to give xenon, fluoride ion, and O_2 as products. Write a balanced equation for the reaction.

General Problems

Key: These problems provide more practice but are not divided by topic or keyed to exercises. Each section ends with essay questions, each of which is color coded to refer to the *A Chemist Looks at Materials* (blue) chapter essay on which it is based. Odd-numbered problems and the even-numbered problems that follow are similar; answers to all odd-numbered problems except the essay questions are given in the back of the book.

21.169 A sample of lead ore contains 78.0% galena, PbS. How many metric tons of Pb can be obtained from 1.00 metric ton of the lead ore?

21.170 A sample of zinc ore contains 87.0% sphalerite, ZnS. How many metric tons of Zn can be obtained from 1.00 metric ton of the zinc ore?

21.171 An antacid tablet consists of calcium carbonate with other ingredients. The calcium carbonate in a 0.9863-g sample of the antacid was dissolved in 50.00 mL of 0.5068 *M* HCl, then titrated with 41.23 mL of 0.2590 *M* NaOH. What was the mass percentage of $CaCO_3$ in the antacid?

21.172 A 50.00-mL volume of 0.4987 *M* HCl was added to a 5.436-g sample of milk of magnesia. This solution was then titrated with 0.2456 *M* NaOH. If it required 39.42 mL of NaOH to reach the endpoint, what was the mass percentage of $Mg(OH)_2$ in the milk of magnesia?

21.173 Calculate the standard enthalpy change, $\Delta H°$, for the following reaction at 25°C.

$$Fe_2O_3(s) + 2Al(s) \longrightarrow 2Fe(s) + Al_2O_3(s)$$

What is the enthalpy change per mole of iron?

21.174 Calculate the standard enthalpy change, $\Delta H°$, for the following reaction at 25°C.

$$3CaO(s) + 2Al(s) \longrightarrow 3Ca(s) + Al_2O_3(s)$$

What is the enthalpy change per mole of calcium?

21.175 A sample of limestone was dissolved in hydrochloric acid, and the carbon dioxide gas that evolved was collected. If a 0.1662-g sample of limestone gave 34.56 mL of dry carbon dioxide gas at 745 mmHg and 21°C, what was the mass percentage of $CaCO_3$ in the limestone?

21.176 A sample of rock containing magnesite, $MgCO_3$, was dissolved in hydrochloric acid, and the carbon dioxide gas that evolved was collected. If a 0.1504-g sample of the rock gave 37.71 mL of dry carbon dioxide gas at 758 mmHg and 22°C, what was the mass percentage of $MgCO_3$ in the rock?

21.177 How many grams of sodium chloride are required to produce 20.00 g of $NaHCO_3$ by the Solvay process?

21.178 How many grams of aluminum are required to react with 15.00 g of chromium(III) oxide by the Goldschmidt process for the production of chromium metal?

21.179 Estimate the temperature at which barium carbonate decomposes to barium oxide and CO_2 at 1 atm.

$$BaCO_3(s) \longrightarrow BaO(s) + CO_2(g)$$

Use thermodynamic data in Appendix C.

21.180 Estimate the temperature at which strontium carbonate begins to decompose to strontium oxide and CO_2 at 1 atm.

$$SrCO_3(s) \longrightarrow SrO(s) + CO_2(g)$$

Use thermodynamic data in Appendix C.

21.181 Calculate $E°$ for the disproportionation of $In^+(aq)$.

$$3In^+(aq) \rightleftharpoons 2In(s) + In^{3+}(aq)$$

(*Disproportionation* is a reaction in which a species undergoes both oxidation and reduction.) Use the following standard potentials:

$$In^+(aq) + e^- \rightleftharpoons In(s); \; E° = -0.21 \text{ V}$$
$$In^{3+}(aq) + 2e^- \rightleftharpoons In^+(aq); \; E° = -0.40 \text{ V}$$

From $E°$, calculate $\Delta G°$ for the disproportionation (in kilojoules). Does this reaction occur spontaneously?

21.182 Calculate $E°$ for the disproportionation of $Tl^+(aq)$.

$$3Tl^+(aq) \rightleftharpoons 2Tl(s) + Tl^{3+}(aq)$$

(*Disproportionation* is a reaction in which a species undergoes both oxidation and reduction.) Use the following standard potentials:

$$Tl^+(aq) + e^- \rightleftharpoons Tl(s); \; E° = -0.34 \text{ V}$$
$$Tl^{3+}(aq) + 2e^- \rightleftharpoons Tl^+(aq); \; E° = 1.25 \text{ V}$$

From $E°$, calculate $\Delta G°$ for the disproportionation (in kilojoules). Does this reaction occur spontaneously?

21.183 Potassium chlorate, $KClO_3$, is used in fireworks and explosives. It can be prepared by bubbling chlorine into hot aqueous potassium hydroxide; $KCl(aq)$ and H_2O are the other products in the reaction. How many grams of $KClO_3$ can be obtained from 138 L of Cl_2 whose pressure is 784 mmHg at 25°C?

21.184 Lithium hydroxide has been used in spaceships to absorb carbon dioxide exhaled by astronauts. Assuming that the product is lithium carbonate, determine what mass of lithium hydroxide is needed to absorb the carbon dioxide from 1.00 L of air containing 30.0 mmHg partial pressure of CO_2 at 25°C.

21.185 The main ingredient in many phosphate fertilizers is $Ca(H_2PO_4)_2 \cdot H_2O$. If a fertilizer is 17.1% P (by mass), and all of this phosphorus is present as $Ca(H_2PO_4)_2 \cdot H_2O$, what is the mass percentage of this salt in the fertilizer?

21.186 A fertilizer contains phosphorus in two compounds, $Ca(H_2PO_4)_2 \cdot H_2O$ and $CaHPO_4$. The fertilizer contains 30.0% $Ca(H_2PO_4)_2 \cdot H_2O$ and 10.0% $CaHPO_4$ (by mass). What is the mass percentage of phosphorus in the fertilizer?

21.187 NaClO solution is made by electrolysis of $NaCl(aq)$ by allowing the products NaOH and Cl_2 to mix. How long must a cell operate to produce 1.00×10^3 L of 5.25% NaClO solution (density = 1.00 g/mL) if the cell current is 3.00×10^3 A?

21.188 Sodium perchlorate, $NaClO_4$, is produced by electrolysis of sodium chlorate, $NaClO_3$. If a current of 2.50×10^3 A passes through an electrolytic cell, how many kilograms of sodium perchlorate are produced per hour?

21.189 The amount of sodium hypochlorite in a bleach solution can be determined by using a given volume of bleach to oxidize excess iodide ion to iodine, because the reaction goes to completion. The amount of iodine produced is then determined by titration with sodium thiosulfate, $Na_2S_2O_3$, which is oxidized to sodium tetrathionate, $Na_2S_4O_6$. Potassium iodide was added in excess to 5.00 mL of bleach (density = 1.00 g/mL). This solution, containing the iodine released in the reaction, was titrated with 0.100 M $Na_2S_2O_3$. If 34.6 mL of sodium thiosulfate was required to reach the endpoint (detected by disappearance of the blue color of the starch–iodine complex), what was the mass percentage of NaClO in the bleach?

21.190 Ascorbic acid, $C_6H_8O_6$ (vitamin C), is a reducing agent. The amount of this acid in solution can be determined quantitatively by a titration procedure involving iodine, I_2, in which ascorbic acid is oxidized to dehydroascorbic acid, $C_6H_6O_6$.

$$C_6H_8O_6(aq) + I_2(aq) \longrightarrow C_6H_6O_6(aq) + 2HI(aq)$$

A 30.0-g sample of an orange-flavored beverage mix was placed in a flask to which 10.00 mL of 0.0500 M KIO_3 and excess KI were added. The IO_3^- and I^- ions react in acid solution to give I_2, which then reacts with ascorbic acid. Excess iodine is titrated with sodium thiosulfate (see Problem 21.189). If 31.2 mL of 0.0300 M $Na_2S_2O_3$ is required to titrate the excess I_2, how many grams of ascorbic acid are there in 100.0 g of beverage mix?

■ **21.191** What is a superconductor? What is one use of superconductors?

■ **21.192** In 1986, Bernorz and Müller received the Nobel Prize in physics for what discovery? Why was this discovery important?

■ **21.193** How was the molecule C_{60} first discovered? Describe the experiment in some detail.

■ **21.194** What makes C_{60} such a stable molecule? What shape does the molecule have?

Strategy Problems

Key: As noted earlier, all of the practice and general problems are matched pairs. This section is a selection of problems that are not in matched-pair format. These challenging problems require that you employ many of the concepts and strategies that were developed in the chapter. In some cases, you will have to integrate several concepts and operational skills in order to solve the problem successfully.

21.195 A 0.325-g sample of sulfur was burned in an excess of oxygen in the presence of a platinum catalyst. The final product of this combustion was then dissolved in water to give 1.000 L of solution. To this solution was added 75.6 mL of 0.1028 M aqueous $BaCl_2$, resulting in a white precipitate. This precipitate was filtered off, dried, and weighed. What was the product of the combustion of sulfur in excess oxygen? What was the solution obtained by dissolving this product in water? What was the white precipitate? What was the mass of this dried precipitate?

21.196

[a] Nitrogen dioxide can be prepared by heating lead nitrate to about 400°C. The products, in addition to nitrogen dioxide, are lead(II) oxide and oxygen. Write the equation for this decomposition.

[b] Nitrogen dioxide gas is always in equilibrium with dinitrogen tetroxide gas. Write the chemical equation for this equilibrium.

[c] A sample of an equilibrium mixture of these gases at 738 mmHg and 25°C weighing 1.427 g had a volume of 456 mL. According to these data, what is the value of K_p for the equilibrium of nitrogen dioxide going to dinitrogen tetroxide?

21.197 Use thermodynamic data to calculate the standard enthalpy change for the decomposition of water vapor to atoms in the gas phase. From this, obtain a value for the O—H bond enthalpy. Similarly, obtain the standard enthalpy change for the decomposition of hydrogen peroxide vapor to atoms in the gas phase. Then, using the value you found for the O—H bond enthalpy, obtain a value for the O—O bond enthalpy. Compare these values with those listed in Table 9.5.

21.198 Sulfur is formed in volcanic gases when sulfur dioxide reacts with hydrogen sulfide. The same reaction has been proposed as a method of removing sulfur dioxide from the gases emitted from coal-fired electricity generation plants. The emitted gases, containing sulfur dioxide, are bubbled through an aqueous solution of hydrogen sulfide.

[a] Write the balanced chemical equation for the reaction of sulfur dioxide gas with hydrogen sulfide gas to produce solid sulfur and water vapor.

[b] Suppose 5.00 L of sulfur dioxide at 748 mmHg and 22°C reacts with 15.1 g of hydrogen sulfide. How many grams of sulfur would be produced?

21.199 A sample of a metal sulfide weighing 6.125 g was roasted in air to obtain 5.714 g of the oxide. The oxide was then dissolved in dilute sulfuric acid, which yielded a white precipitate. The mass of the dry precipitate was 7.763 g.

[a] Show that these data are consistent with the precipitate being a sulfate.

[b] From this information, decide what metal or metals are consistent with the information given. Justify your answer.

21.200 A mixture contained aluminum sulfate and sodium sulfate. A sample of the mixture weighing 3.458 g was dissolved in water and treated with sodium hydroxide solution to yield a precipitate. The precipitate was incinerated to yield 0.474 g aluminum oxide. Write chemical equations for the reactions that occur. Calculate the percentage of aluminum sulfate in the mixture.

21.201 A quantity of white phosphorus was burned in an excess of oxygen. The product of this combustion was dissolved in water to give 1.258 L of solution. This solution was found to have a pH of 2.091.

[a] Write the equation for the combustion.

[b] What was the mass of the white phosphorus?

[c] The solution was treated with a solution of calcium nitrate, yielding a white precipitate. Write the chemical equation for this precipitation reaction.

[d] Assuming that sufficient calcium nitrate was added to give complete reaction, what was the mass of precipitate obtained?

21.202 A white crystalline material contained sodium acetate and sodium chloride. To determine the composition of this mixture, a researcher dissolved 0.613 g of this mixture in water to give 25.0 mL of solution. She then determined that the pH of this solution was 9.03. What was the percentage of sodium acetate in the mixture, assuming that it contained only sodium chloride and sodium acetate?

21.203 Write the chemical equation for the reaction of iron metal with hydrochloric acid. Similarly, write the chemical equation for the reaction of copper metal with nitric acid. Discuss the difference between copper and iron metals in reacting with dilute hydrochloric acid and with dilute nitric acid. Use electrode potentials in your discussion.

21.204 When carbon burns in air, carbon dioxide is a product. Carbon monoxide can also be present in this combustion because it is the product of the reaction of carbon dioxide with hot carbon. Write the chemical equation for this production of carbon monoxide. Using thermodynamic data and any reasonable approximations, obtain the equilibrium constant for this reaction at 2000°C. Assume the carbon is in the form of graphite.

The Transition Elements and Coordination Compounds

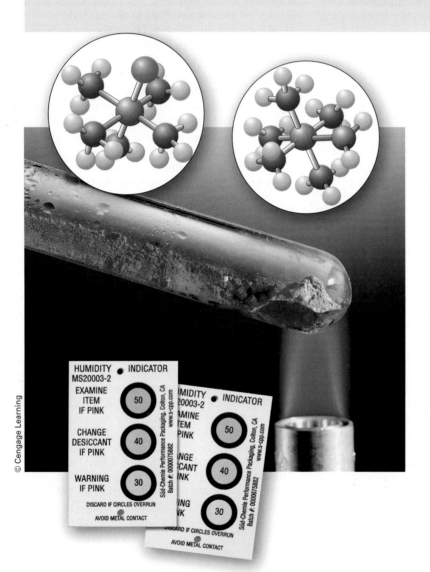

© Cengage Learning

Heating the red cobalt(III) complex $[Co(NH_3)_5H_2O]Cl_3(s)$ drives off the water molecule, forming the purple complex $[Co(NH_3)_5Cl]Cl_2(s)$. Similarly, complexes of cobalt(II) undergo color changes with hydration or dehydration. The complex $[CoCl_2(H_2O)_2]$ is used in humidity indicator cards. It is blue, but picks up water molecules in a humid atmosphere to form the pink complex $[Co(H_2O)_6]Cl_2$.

OWL Sign in to OWL at **www.cengage.com/owl** to view tutorials and simulations, develop problem-solving skills, and complete online homework assigned by your professor.

GO Chemistry Download mini lecture videos for key concept review and exam prep from OWL or purchase them from **www.cengagebrain.com**.

In the previous chapter, we studied the main-group elements—the A groups in the periodic table. Between columns IIA and IIIA are ten columns of the transition elements—the B groups. Among these elements are metals with familiar commercial applications: iron tools, copper wire, silver jewelry, and coins. Many catalysts for important industrial reactions involve transition elements. Examples of these are found in petroleum refining and in the synthesis of ammonia from N_2 and H_2.

In addition to their commercial usefulness, many transition elements have biological importance. Iron compounds, for example, are found throughout the plant and animal kingdoms. Iron is present in hemoglobin, the molecule in red blood cells that is responsible for the transport of oxygen, O_2, from the lungs to other body tissue. Myoglobin, found in muscle tissue, is a similar molecule containing iron. Myoglobin takes oxygen from hemoglobin, holding it until it is required by the muscle cells. Cytochromes, iron-containing compounds within each cell, are involved in the oxidation of food molecules. In these cases, the transition element is central to the structure and function of the biological molecule. Hemoglobin and myoglobin are examples of *metal complexes,* or *coordination compounds,* in which the metal atom is surrounded by other atoms bonded to it by the electron pairs these atoms donate. In hemoglobin and myoglobin, the O_2 molecule bonds to the iron atom.

We will look at the general properties of transition metals in the first section of the chapter; in the second section we will investigate the chemistry of two transition elements, chromium and copper. The remaining sections of the chapter will cover the structure, naming, and bonding of complex ions and coordination compounds.

Properties of the Transition Elements

Often, all of the B elements of the periodic table are classified as transition elements, though strictly speaking, the **transition elements** are defined as *those metallic elements that have an incompletely filled d subshell or easily give rise to common ions that have incompletely filled d subshells.* Iron, whose electron configuration is $[Ar]3d^64s^2$, is an example of such an element; iron has an incompletely filled $3d$ subshell. Another example is copper. Although the free element has the configuration $[Ar]3d^{10}4s^1$, in which the $3d$ subshell is complete, copper readily forms the copper(II) ion, whose configuration $[Ar]3d^9$ has an incomplete $3d$ subshell. The Group IIB elements zinc, cadmium, and mercury have filled d subshells in the element and in the common ions, so in the strict sense these are not transition elements, although they are often included with them.

Sometimes, too, chemists include the two rows of elements at the bottom of the periodic table with the transition elements. These two rows, often referred to as the *inner transition elements*, have partially filled f subshells in common oxidation states. The elements in the first row are called the *lanthanides,* or *rare earths,* and the elements in the second row are called the *actinides.* Figure 22.1 shows the divisions of the transition elements. The B columns of transition elements, as well as the inner transition elements, frequently form complex ions and coordination compounds.

22.1 Periodic Trends in the Transition Elements

The transition elements have a number of characteristics that set them apart from the main-group elements:

1. All of the transition elements are metals and, except for the IIB elements, have high melting points and high boiling points and are hard solids. For instance, of the fourth-period elements from scandium to copper, the lowest-melting

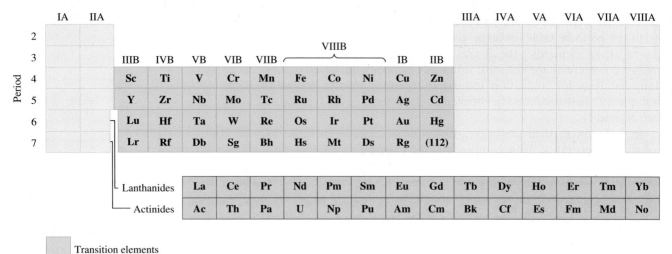

	IA	IIA							VIIIB			IB	IIB	IIIA	IVA	VA	VIA	VIIA	VIIIA
2																			
3			IIIB	IVB	VB	VIB	VIIB				IB	IIB							
4			Sc	Ti	V	Cr	Mn	Fe	Co	Ni	Cu	Zn							
5			Y	Zr	Nb	Mo	Tc	Ru	Rh	Pd	Ag	Cd							
6			Lu	Hf	Ta	W	Re	Os	Ir	Pt	Au	Hg							
7			Lr	Rf	Db	Sg	Bh	Hs	Mt	Ds	Rg	(112)							

Lanthanides

La	Ce	Pr	Nd	Pm	Sm	Eu	Gd	Tb	Dy	Ho	Er	Tm	Yb
Ac	Th	Pa	U	Np	Pu	Am	Cm	Bk	Cf	Es	Fm	Md	No

Actinides

☐ Transition elements

☐ Inner transition elements
(lanthanides and actinides)

Figure 22.1 ▲

Classification of the transition elements The classification into the B groups of transition elements and the inner transition elements.

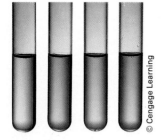

© Cengage Learning

Figure 22.2 ▲

Oxidation states of vanadium The test tubes contain, left to right: VO_3^- (aq), pale yellow; VO^{2+}(aq), bright blue; V^{3+}, gray-blue; V^{2+}, violet. The oxidation states of vanadium are, left to right, +5, +4, +3, and +2.

metal is copper (1083°C) and the highest-melting metal is vanadium (1890°C). Of the main-group metals, only beryllium melts above 1000°C; the rest melt at appreciably lower temperatures.

2. With the exception of the IIIB and IIB elements, every transition element has several oxidation states. Vanadium, for example, exists as aqueous ions in all oxidation states from +2 to +5 (Figure 22.2). Of the main-group metals, only the heavier ones display several oxidation states. Because of their multiplicity of oxidation states, the transition elements are often involved in oxidation–reduction reactions.

3. Transition-metal compounds are often colored, and many are paramagnetic. Most compounds of the main-group metals are colorless and diamagnetic.

We will look at points 1 and 2 in some detail in this section. In particular, we will examine trends in melting points, boiling points, hardness, and oxidation states. We will also look at trends in covalent radii and ionization energies, which we can relate to chemical properties. We will discuss the color and paramagnetism of transition-metal complexes later in the chapter.

Electron Configurations

Electronic structure is central to any discussion of the transition elements. Table 22.1 lists the ground-state electron configurations of the fourth-period transition elements. For the most part, these configurations are predicted by the building-up principle. Following the building-up principle, the 3d subshell begins to fill after calcium (configuration $[Ar]4s^2$). Thus scandium has the configuration $[Ar]3d^1 4s^2$, and as you go across the period, additional electrons go into the 3d subshell. You get the configuration $[Ar]3d^2 4s^2$ for titanium and $[Ar]3d^3 4s^2$ for vanadium. Then in chromium the configuration predicted by the building-up principle is $[Ar]3d^4 4s^2$, but the actual configuration is $[Ar]3d^5 4s^1$. The configurations for the rest of the elements are those predicted by the building-up principle, until you get to copper. Here, the predicted configuration is $[Ar]3d^9 4s^2$, but the actual configuration is $[Ar]3d^{10} 4s^1$.

Melting Points and Boiling Points

Table 22.1 reveals that the melting points of the transition metals increase from 1541°C for scandium to 1890°C for vanadium and 1857°C for chromium, then decrease from 1535°C for iron to 1083°C for copper and 420°C for zinc. The same pattern is observed in the fifth-period and sixth-period elements. As you read across

Table 22.1	Properties of the Fourth-Period Transition Elements				
Property	Scandium	Titanium	Vanadium	Chromium	Manganese
Electron configuration	$[Ar]3d^14s^2$	$[Ar]3d^24s^2$	$[Ar]3d^34s^2$	$[Ar]3d^54s^1$	$[Ar]3d^54s^2$
Melting point, °C	1541	1660	1890	1857	1244
Boiling point, °C	2831	3287	3380	2672	1962
Density, g/cm³	3.0	4.5	6.0	7.2	7.2
Electronegativity (Pauling scale)	1.3	1.5	1.6	1.6	1.5
Covalent radius, pm	144	132	122	118	117
Ionic radius (for M^{2+}), pm	—	100	93	87	81
Property	Iron	Cobalt	Nickel	Copper	Zinc
Electron configuration	$[Ar]3d^64s^2$	$[Ar]3d^74s^2$	$[Ar]3d^84s^2$	$[Ar]3d^{10}4s^1$	$[Ar]3d^{10}4s^2$
Melting point, °C	1535	1495	1453	1083	420
Boiling point, °C	2750	2870	2732	2567	907
Density, g/cm³	7.9	8.9	8.9	8.9	7.1
Electronegativity (Pauling scale)	1.8	1.8	1.8	1.9	1.6
Covalent radius, pm	117	116	115	117	125
Ionic radius (for M^{2+}), pm	75	79	83	87	88

a row of transition elements, the melting points increase, reaching a maximum at the Group VB or VIB elements, after which the melting points decrease. For instance, tungsten in Group VIB has the highest melting point (3410°C) of all metallic elements, and mercury in Group IIB has the lowest (-39°C). Similar trends can be seen in the boiling points of the metals. These properties depend on the strength of metal bonding, which in turn depends roughly on the number of unpaired electrons in the metal atoms. At the beginning of a period of transition elements, there is one unpaired d electron. The number of unpaired d electrons increases across a period until Group VIB, after which the electrons begin to pair.

Atomic Radii

Trends in atomic radii are of concern because chemical properties are determined in part by atomic size. Looking at the fourth-period covalent radii (one measure of atomic size) in Table 22.1, you see that they decrease quickly from scandium (144 pm) to titanium (132 pm) and vanadium (122 pm). This decrease in atomic size across a row is also observed in the main-group elements. It is due to an increase in *effective nuclear charge* that acts on the outer electrons and pulls them in more strongly. The effective nuclear charge is the positive charge "felt" by an electron; it equals the nuclear charge minus the shielding or screening of the positive charge by intervening electrons. After vanadium, the covalent radii decrease slowly from 118 pm for chromium to 115 pm for nickel. Then the covalent radii increase slightly to 117 pm for copper and 125 pm for zinc. ▶ The relative constancy in covalent radii for the later elements is partly responsible for the similarity in properties of the Group VIIIB elements iron, cobalt, and nickel.

The small increase in covalent radius for copper and zinc has no simple explanation.

Figure 22.3 compares the covalent radii of the transition elements. The atomic radii increase in going from a fourth-period to a fifth-period element within any column. For example, reading down the Group IIIB elements, you find that the covalent radius of scandium is 144 pm and of yttrium is 162 pm. You would expect an increase of radius from scandium to yttrium because of the addition

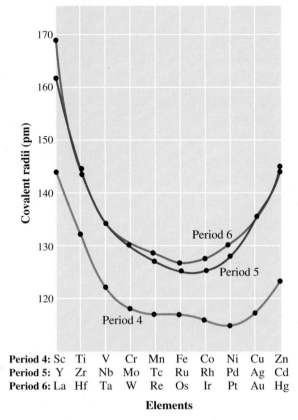

Period 4: Sc Ti V Cr Mn Fe Co Ni Cu Zn
Period 5: Y Zr Nb Mo Tc Ru Rh Pd Ag Cd
Period 6: La Hf Ta W Re Os Ir Pt Au Hg

Elements

Figure 22.3 ▲

Covalent radii for transition elements of Periods 4 to 6 Note that elements of the fifth and sixth periods of the same column have approximately the same radii.

The atomic number of an atom can be obtained from its x-ray spectrum, which therefore gives unequivocal identification of an element. See the essay at the end of Section 8.2.

of a shell of electrons. Continuing down the column of IIIB elements, you find a small increase of covalent radius to 169 pm in lanthanum. But all of the remaining elements of the sixth period have nearly the same covalent radii as the corresponding elements in the fifth period.

This similarity of radii in the fifth- and sixth-period transition elements can be explained in terms of the *lanthanide contraction.* Between lanthanum and hafnium in the sixth period are 14 lanthanide elements (cerium to lutetium), in which the 4*f* subshell fills. The covalent radius decreases slowly from cerium (165 pm) to lutetium (156 pm), but the total decrease is substantial. By the time the 4*f* subshell is complete, the covalent radii of the transition elements from hafnium on are similar to those of the elements in the preceding row of the periodic table. Therefore, the covalent radius of hafnium (146 pm) is approximately the same as that of zirconium (145 pm).

The chemical properties of the transition elements parallel the pattern seen in covalent radii. That is, the fourth-period elements have substantially different properties from the elements in the same group in the fifth and sixth periods. However, elements in the same group of the fifth and sixth periods are very much alike. Hafnium, for example, is so much like zirconium that it remained undiscovered until it was identified in 1923 in zirconium ores from its x-ray spectrum. ◀ Zirconium (mixed with hafnium as an impurity) was discovered more than a hundred years earlier, in 1787.

Ionization Energies

Looking at the first ionization energies of the fourth-period transition elements (Table 22.2), you see that although they vary somewhat irregularly, they tend to increase from left to right. The other rows of transition elements behave similarly. Most noteworthy, however, is that all of the sixth-period elements after lanthanum have ionization energies higher than those of the fourth-period and fifth-period transition elements in the same group. This behavior is opposite to that of the main-group elements, where the ionization energies decrease regularly down a column. The high ionization energies of the sixth-period elements from osmium to mercury are no doubt one determinant of the relative unreactivity of these elements.

Oxidation States

Table 22.3 gives oxidation states in compounds of the fourth-period transition elements. Scandium occurs as the 3+ ion and zinc as the 2+ ion. The other elements exhibit several oxidation states. This multiplicity of oxidation states is due to the varying involvement of the *d* electrons in bonding.

Table 22.2	First Ionization Energies of the Transition Elements (kJ/mol)									
Period	**IIIB**	**IVB**	**VB**	**VIB**	**VIIB**		**VIIIB**		**IB**	**IIB**
Fourth	Sc	Ti	V	Cr	Mn	Fe	Co	Ni	Cu	Zn
	631	658	650	652	717	759	758	737	745	906
Fifth	Y	Zr	Nb	Mo	Tc	Ru	Rh	Pd	Ag	Cd
	616	660	664	685	702	711	720	805	731	868
Sixth	La	Hf	Ta	W	Re	Os	Ir	Pt	Au	Hg
	538	680	761	770	760	840	880	870	890	1007

Table 22.3   Oxidation States of the Fourth-Period Transition Elements

IIIB	IVB	VB	VIB	VIIB	VIIIB			IB	IIB
Sc	Ti	V	Cr	Mn	Fe	Co	Ni	Cu	Zn
							+1	+1	
	+2	+2	+2	+2	+2	+2	+2	+2	+2
+3	+3	+3	+3	+3	+3	+3	+3	+3	
	+4	+4	+4	+4	+4	+4	+4		
		+5	+5	+5	+5				
			+6	+6	+6				
				+7					

Key: Common oxidation states are in boldface. Additional oxidation states, particularly zero and negative values, may be observed in complexes with CO and with organic compounds.

Most of the transition elements have a doubly filled *ns* orbital. Because the *ns* electrons ionize before the $(n - 1)d$ electrons, you might expect the +2 oxidation state to be common. This oxidation state is in fact seen in all of the fourth-period elements except scandium, where the 3+ ion with an Ar configuration is especially stable.

The maximum oxidation state possible equals the number of *s* and *d* electrons in the valence shell (which equals the group number for the elements up to iron). Titanium in Group IVB has a maximum oxidation state of +4. Similarly, vanadium (Group VB), chromium (Group VIB), and manganese (Group VIIB) exhibit maximum oxidation states of +5, +6, and +7, respectively. ▶

The total number of oxidation states actually observed increases from scandium (+3) to manganese (all states from +2 to +7). From iron on, however, the maximum oxidation state is not attained, so the number of observed states decreases. The highest oxidation state seen in iron is +6. ▶ Thereafter, the highest observed oxidation number decreases until, for zinc compounds, you find only the +2 oxidation state. The Zn^{2+} ion has a stable $[Ar]3d^{10}$ configuration.

The maximum oxidation state is generally found in compounds of the transition elements with very electronegative elements, such as F and O (in oxides and oxoanions).

Presumably, Fe(VIII) would oxidize any element to which it would bond.

22.2 The Chemistry of Two Transition Elements

In this section, we look at the chemical properties of two transition elements, chromium and copper. Both elements are well known and commercially important; in addition, they provide many examples of colorful compounds.

Chromium

A fascinating feature of chromium chemistry is the many colorful compounds of this element. In fact, this feature is the origin of the name *chromium,* which derives from the Greek word *khroma,* meaning "color." The common oxidation states of chromium in compounds are +2, +3, and +6.

The chromium(II) ion, $[Cr(H_2O)_6]^{2+}(aq)$, has a bright blue color. The ion is obtained when an acid, such as $HCl(aq)$, reacts with chromium metal (see Figure 22.4, *left*).

$$Cr(s) + 2HCl(aq) + 6H_2O(l) \longrightarrow Cr(H_2O)_6^{2+}(aq) + 2Cl^-(aq) + H_2(g)$$

The chromium(II) ion is easily oxidized to chromium(III) ion by O_2 (Figure 22.4, *right*). Therefore, to prepare chromium(II) salts, we must carry out this reaction in the absence of air.

The most common oxidation state of chromium is +3. Chromium metal burns in air to give chromium(III) oxide, Cr_2O_3

Chromium metal reacts with hydrochloric acid to produce $[Cr(H_2O)_6]^{2+}$, which is blue.

$[Cr(H_2O)_6]^{2+}$ oxidizes with air to give a chromium(III) species, which is green.

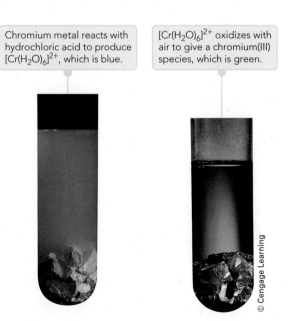

Figure 22.4 ▲

Aqueous chromium ion

© Cengage Learning

(also called chromic oxide). This oxide is a dark green solid, which has been used as a paint pigment (chrome green). Chromium(III) oxide dissolves in acid solution to form the violet-colored ion $[Cr(H_2O)_6]^{3+}(aq)$. (Hydrochloric acid solution, however, yields green-colored complex ions of Cr^{3+} with Cl^-.) Potassium chromium(III) sulfate (or chrome alum), $KCr(SO_4)_2 \cdot 12H_2O$, is a common chromium(III) salt. It is a deep purple compound used for leather tanning.

The $+6$ oxidation state is represented by the chromate ion, CrO_4^{2-}, and the dichromate ion, $Cr_2O_7^{2-}$. Sodium chromate, Na_2CrO_4, which is yellow, and sodium dichromate, $Na_2Cr_2O_7$, which is orange, are the primary sources of compounds of chromium. Sodium chromate, Na_2CrO_4, is prepared from chromite, $FeCr_2O_4(s)$, the principal chromium ore, by strongly heating it with sodium carbonate in air. Chromite is an iron(II) chromium(III) oxide, and both iron and chromium are oxidized in this reaction.

$$4FeCr_2O_4(s) + 8Na_2CO_3(s) + 7O_2(g) \xrightarrow{1100°C} 8Na_2CrO_4(s) + 2Fe_2O_3(s) + 8CO_2(g)$$

Sodium chromate is soluble and can be leached from the mixture with water.

Sodium chromate can be easily converted to sodium dichromate by treating it with an acid. When the yellow solution of a chromate salt is acidified, it turns orange from the formation of dichromate ion, $Cr_2O_7^{2-}(aq)$ (Figure 22.5).

$$\underset{\text{yellow}}{2CrO_4^{2-}(aq)} + 2H^+(aq) \rightleftharpoons \underset{\text{orange}}{Cr_2O_7^{2-}(aq)} + H_2O(l)$$

The chromate and dichromate ions are in equilibrium, which is sensitive to pH changes; lower pH favors dichromate ion. (Increasing the H^+ concentration pushes the equilibrium to the right, according to Le Châtelier's principle.)

The dichromate ion is a strong oxidizing agent in acid solution.

$$Cr_2O_7^{2-}(aq) + 14H^+(aq) + 6e^- \longrightarrow 2Cr^{3+}(aq) + 7H_2O(l); \quad E° = 1.33 \text{ V}$$

The color of the chromium(III) species depends on anions in solution that can form complexes with Cr^{3+}. Frequently the ion is green, so there is a definite color change from orange to green during the reaction.

Chromium(VI) oxide (also called chromium trioxide, CrO_3) is a red, crystalline compound. It precipitates when concentrated sulfuric acid is added to concentrated solutions of a dichromate salt. ◄

Chromium trioxide is a strong oxidizing agent. Vapors of ethanol, C_2H_5OH, spontaneously ignite in the presence of the solid oxide.

$$K_2Cr_2O_7(aq) + 2H_2SO_4(l) \longrightarrow 2KHSO_4(aq) + 2CrO_3(s) + H_2O(l)$$

Chromium trioxide is an acidic oxide, and its aqueous solutions are referred to as chromic acid. Chromic acid is used in chromium plating.

Pure chromium metal is prepared by the exothermic reaction of chromium(III) oxide, obtained from chromite ore, with aluminum (the *Goldschmidt process*). Once the reaction mixture is ignited, the large heat of reaction produces molten chromium.

$$Cr_2O_3(s) + 2Al(s) \longrightarrow 2Cr(l) + Al_2O_3(s); \quad \Delta H° = -488 \text{ kJ}$$

Figure 22.5 ▶

Chromate–dichromate equilibrium The beaker on the left contains CrO_4^{2-} (yellow). When the experimenter adds sulfuric acid to a similar solution in the beaker on the right, CrO_4^{2-} is converted to $Cr_2O_7^{2-}$ (orange).

CrO_4^{2-} $Cr_2O_7^{2-}$

© Cengage Learning

One of the chief uses of chromium is in steel making. For this purpose, chromium is usually prepared directly as an alloy of iron and chromium, called ferrochromium, by reducing chromite ore with carbon in an electric furnace.

$$FeCr_2O_4(s) + 4C(s) \xrightarrow{\Delta} Fe(l) + 2Cr(l) + 4CO(g)$$

Copper

Copper and the other Group IB elements, silver and gold, can be found in nature as the free metals. This is a reflection of the stability of their 0 oxidation states. For example, copper is not attacked by most acids.

Copper metal does react with hot, concentrated sulfuric acid and with nitric acid. In these cases, the anion of the acid acts as the oxidizing agent (rather than H^+, the usual oxidizing agent in acids). Sulfuric acid is reduced to sulfur dioxide:

$$Cu(s) + 2H_2SO_4(l) \longrightarrow Cu^{2+}(aq) + SO_4^{2-}(aq) + SO_2(g) + 2H_2O(l)$$

Dilute nitric acid is reduced to NO, concentrated nitric acid to NO_2:

$$3Cu(s) + 2NO_3^-(aq) + 8H^+(aq) \longrightarrow 3Cu^{2+}(aq) + 2NO(g) + 4H_2O(l)$$
$$Cu(s) + 2NO_3^-(aq) + 4H^+(aq) \longrightarrow Cu^{2+}(aq) + 2NO_2(g) + 2H_2O(l)$$

The copper(II) ion, or cupric ion, given in these equations as $Cu^{2+}(aq)$, is written more precisely as $[Cu(H_2O)_6]^{2+}$. It has a bright blue color. Hydrated copper(II) salts, such as copper(II) sulfate pentahydrate, $CuSO_4 \cdot 5H_2O$, also have a blue color. Four of the water molecules of copper(II) sulfate pentahydrate are associated with Cu^{2+}, and the fifth is hydrogen-bonded to the sulfate ion as well as to the water molecules on the copper ion. You can write the formula as $[Cu(H_2O)_4]SO_4 \cdot H_2O$ to better represent its structure. When this salt is heated, it loses its water of hydration in stages. The hydrate $CuSO_4 \cdot H_2O$ and the anhydrous salt $CuSO_4$ are nearly colorless substances; the complexing of water molecules to the copper ion is responsible for the blue color of the pentahydrate.

Although most of the aqueous chemistry of copper involves the +2 oxidation state, there are a number of important compounds of copper(I). When copper is heated in oxygen below 1000°C, it forms the black copper(II) oxide, CuO. But above this temperature, it forms the brick-red copper(I) oxide, Cu_2O. This oxide is found naturally as the mineral cuprite.

The formation of copper(I) oxide by the reduction of copper(II) ion in basic solution forms the basis of a common test for glucose in urine.

$$\underset{\text{blue}}{RCHO(aq) + 2Cu^{2+}(aq)} + 5OH^-(aq) \longrightarrow RCOO^-(aq) + \underset{\text{brick-red}}{Cu_2O(s)} + 3H_2O(l)$$

When the copper(II) ion is heated in basic solution with a reducing sugar, such as glucose (represented here by the general formula RCHO), the ion is reduced to copper(I) oxide. The blue-green solution of Cu^{2+} gives a brick-red precipitate of Cu_2O (Figure 22.6).

The principal commercial use of copper metal is as an electrical conductor. Common ores of copper are native copper (the free metal), copper oxides, and copper sulfides. Most copper is presently obtained by open-pit mining of low-grade rock containing only a small percentage of copper as copper sulfides. The ore is concentrated in copper by flotation. In this process, a slurry of the crushed ore is agitated with air, and the copper sulfides are carried away in the froth. This concentrated ore is then treated in several steps that result in the production of molten copper(I) sulfide, Cu_2S. This molten material, called *matte,* is reduced to copper by blowing air through it.

$$Cu_2S(l) + O_2(g) \longrightarrow 2Cu(l) + SO_2(g)$$

The metal produced is called *blister copper* and is about 99% pure. For electrical use, the copper must be further purified or refined by electrolysis (see Figure 19.24).

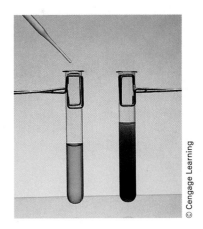

Figure 22.6 ▲

Benedict's test for glucose The experimenter adds glucose solution to a basic solution of copper(II) ion (blue-green solution in left test tube); the solution also contains citrate ion to complex copper(II) ion, so that it will not precipitate as copper(II) hydroxide. When this solution is heated, the glucose reduces copper(II) ion to copper(I) oxide (brick-red precipitate in right test tube).

© Cengage Learning

Complex Ions and Coordination Compounds

As shown in the previous section, ions of the transition elements exist in aqueous solution as *complex ions.* Iron(II) ion, for example, exists in water as $[Fe(H_2O)_6]^{2+}$. The water molecules in this ion are arranged about the iron atom with their oxygen atoms bonded to the metal by donating electron pairs to it. Replacing the H_2O molecules by six CN^- ions gives the $[Fe(CN)_6]^{4-}$ ion. Some of the transition elements have biological activity, and their role in human nutrition (Table 22.4) depends in most cases on the formation of *complexes,* or *coordination compounds,* which exhibit the type of bonding that occurs in $[Fe(H_2O)_6]^{2+}$ and $[Fe(CN)_6]^{4-}$. In the chapter opening, we saw that hemoglobin, a complex of iron, is vital to the transport of oxygen by the red blood cells.

22.3 Formation and Structure of Complexes

Lewis acid–base reactions are discussed in Section 15.3.

A metal atom, particularly a transition-metal atom, often functions in chemical reactions as a Lewis acid, accepting electron pairs from molecules or ions. ◀ For example, Fe^{2+} and H_2O can bond to one another in a Lewis acid–base reaction.

coordinate covalent bond

$$Fe^{2+} + :\ddot{O}-H \longrightarrow \left[Fe:\ddot{O}-H\right]^{2+}$$

Lewis acid   Lewis base

A pair of electrons on the oxygen atom of H_2O forms a coordinate covalent bond to Fe^{2+}. In water, the Fe^{2+} ion ultimately bonds to six H_2O molecules, forming the $[Fe(H_2O)_6]^{2+}$ ion.

The Fe^{2+} ion also undergoes a similar Lewis acid–base reaction with cyanide ions. In this case, the Fe^{2+} ion bonds to the electron pair on the carbon atom of CN^-.

$$Fe^{2+} + (:C{\equiv}N:)^- \longrightarrow (Fe:C{\equiv}N:)^+$$

Although the ion $[Fe(CN)_6]^{4-}$ is a complex of Fe^{2+} and CN^- ions, a solution of $[Fe(CN)_6]^{4-}$ contains negligible concentration of CN^-. Therefore, a substance such as $K_4[Fe(CN)_6]$ is relatively nontoxic, even though the free cyanide ion is a potent poison.

Finally, a very stable ion, $[Fe(CN)_6]^{4-}$, is obtained, which consists of one Fe^{2+} bonded to six cyanide ions. ◀ Note that the charge on the $[Fe(CN)_6]^{4-}$ ion equals the sum of the charges on the ions from which it is formed: $+2 + 6(-1) = -4$.

In some cases, a neutral species is produced from a metal ion and anions. Cisplatin, the anticancer drug discussed in the opening of Chapter 1, has the structure

$$H_3N:Pt:\ddot{Cl}:$$
with $:\ddot{Cl}:$ above and NH_3 below

Table 22.4   Transition Elements Essential to Human Nutrition

Element	Some Biochemical Substances	Function
Chromium	Glucose tolerance factor	Utilization of glucose
Manganese	Isocitrate dehydrogenase	Cell energetics
Iron	Hemoglobin and myoglobin	Transport and storage of oxygen
	Cytochrome *c*	Cell energetics
	Catalase	Decomposition of hydrogen peroxide
Cobalt	Cobalamin (vitamin B_{12})	Development of red blood cells
Copper	Ceruloplasmin	Synthesis of hemoglobin
	Cytochrome oxidase	Cell energetics
Zinc	Carbonic anhydrase	Elimination of carbon dioxide
	Carboxypeptidase A (pancreatic juice)	Protein digestion
	Alcohol dehydrogenase	Oxidation of ethanol

It consists of Pt^{2+} with two NH_3 molecules (neutral) and two Cl^- ions, giving a neutral species. Iron pentacarbonyl, $[Fe(CO)_5]$, is an example of a neutral species formed from a neutral iron atom and CO molecules.

Basic Definitions

A **complex ion** is *a metal ion with Lewis bases attached to it through coordinate covalent bonds.* A **complex** (or **coordination compound**) is *a compound consisting either of complex ions and other ions of opposite charge* (for example, the compound $K_4[Fe(CN)_6]$ of the complex ion $[Fe(CN)_6]^{4-}$ and four K^+ ions) *or of a neutral complex species* (such as cisplatin).

Ligands are *the Lewis bases attached to the metal atom in a complex.* They are electron-pair donors, so ligands may be neutral molecules (such as H_2O or NH_3) or anions (such as CN^- or Cl^-) that have at least one atom with a lone pair of electrons. Cations only rarely function as ligands. You might expect this, because an electron pair on a cation is held securely by the positive charge, so it would not be involved in coordinate bonding. ▶

The **coordination number** of a metal atom in a complex is *the total number of bonds the metal atom forms with ligands.* In $[Fe(H_2O)_6]^{2+}$, the iron atom bonds to each oxygen atom in the six water molecules. Therefore, the coordination number of iron in this ion is 6, by far the most common coordination number. Coordination number 4 is also well known, and many examples of number 5 have been discovered. Table 22.5 gives some examples of complexes for the coordination numbers 2 to 8. The coordination number for an atom depends on several factors, but size of the metal atom is important. For example, coordination numbers 7 and 8 are seen primarily in fifth- and sixth-period elements, whose atoms are relatively large. ▶

Polydentate Ligands

The ligands we have discussed so far bond to the metal atom through one atom of the ligand. For instance, ammonia bonds through the nitrogen atom. This type of bonding indicates a **monodentate ligand** (meaning "one-toothed" ligand)—that is, *a ligand that bonds to a metal atom through one atom of the ligand.* A **bidentate ligand** ("two-toothed" ligand) is *a ligand that bonds to a metal atom through two atoms of the ligand.* Ethylenediamine is an example.

A cation in which the positive charge is far removed from an electron pair that could be donated can function as a ligand. An example is the pyrazinium ion:

Very high coordination numbers (9 to 12) are known for some complex ions of the lanthanide elements.

Table 22.5	Examples of Complexes of Various Coordination Numbers
Complex	**Coordination Number**
$[Ag(NH_3)_2]^+$	2
$[HgI_3]^-$	3
$[PtCl_4]^{2-}$, $[Ni(CO)_4]$	4
$[Fe(CO)_5]$, $[Co(CN)_5]^{3-}$	5
$[Co(NH_3)_6]^{3+}$, $[W(CO)_6]$	6
$[Mo(CN)_7]^{3-}$	7
$[W(CN)_8]^{4-}$	8

Ethylenediamine

Nitrogen atoms at the ends of the molecule have lone pairs of electrons that can form coordinate covalent bonds. In forming a complex, the ethylenediamine molecule bends around so that both nitrogen atoms coordinate to the metal atom, M.

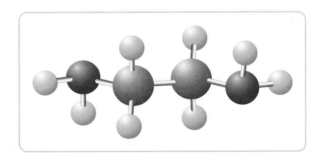

Figure 22.7 ▶

Structure of tris(ethylenediamine) cobalt(III) ion, [Co(en)₃]³⁺

Ball-and-stick model.

Shorthand notation. Note that N⌒N represents ethylenediamine.

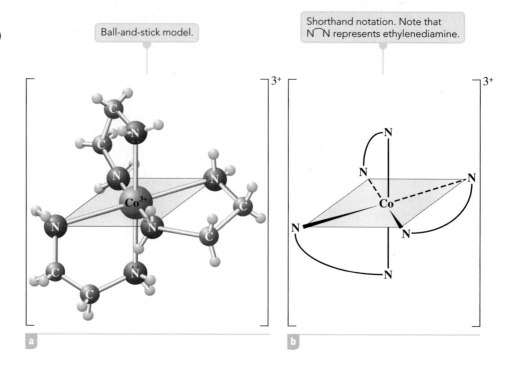

Because ethylenediamine is a common bidentate ligand, it is frequently abbreviated in formulas as "en." Figure 22.7 shows the structure of the stable ion $[Co(en)_3]^{3+}$. The oxalate ion, $C_2O_4^{2-}$, is another common bidentate ligand.

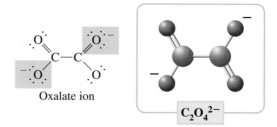

Oxalate ion

$C_2O_4^{2-}$

The hemoglobin molecule in red blood cells is an example of a complex with a *quadridentate ligand*—one that bonds to the metal atom through four ligand atoms. Hemoglobin consists of the protein *globin* chemically bonded to *heme*, whose structure is shown in Figure 22.8 (*left*). Heme is a planar molecule consisting of iron(II) to which a quadridentate ligand is bonded through its four nitrogen atoms.

Figure 22.8 ▶

Complexes with ligands that bond with more than one atom to the metal atom (polydentate ligands) *Left:* Structure of heme. *Right:* Complex of Fe²⁺ and ethylenediaminetetraacetate ion (EDTA). Note how the EDTA ion envelops the metal ion.

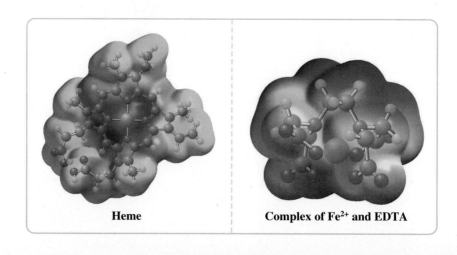

Heme　　　　　　**Complex of Fe²⁺ and EDTA**

Ethylenediaminetetraacetate ion (EDTA) is a ligand that bonds through six of its atoms.

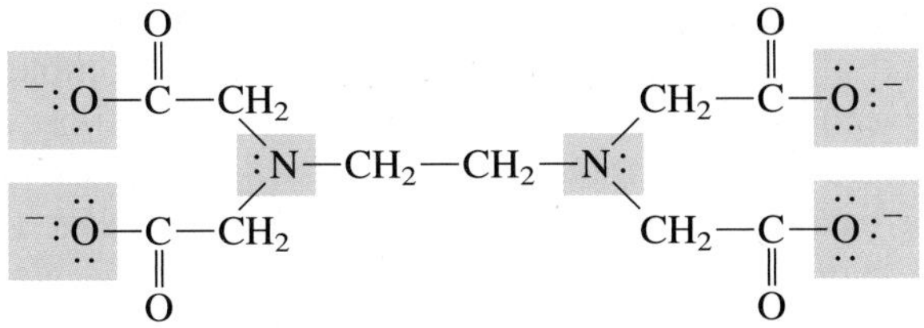

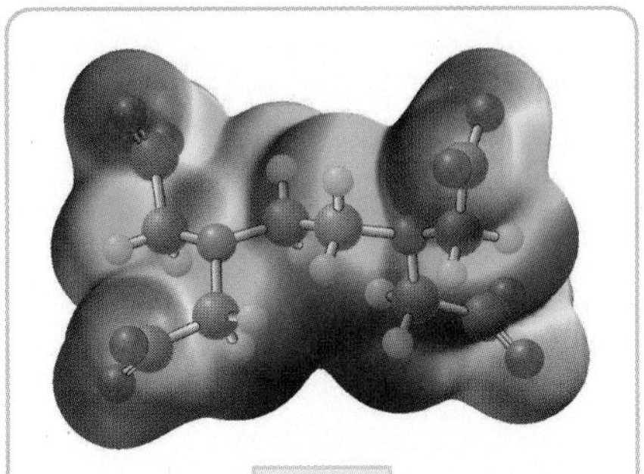

EDTA

It can completely envelop a metal atom, simultaneously occupying all six positions in an octahedral geometry (Figure 22.8, *right*).

A **polydentate ligand** ("having many teeth") is *a ligand that can bond with two or more atoms to a metal atom. A complex formed by polydentate ligands is fre-quently quite stable and is called a* **chelate** *(pronounced "key-late")*. ▶ Because of the stability of chelates, polydentate ligands (also called *chelating agents*) are often used to remove metal ions from a chemical system. EDTA, for example, is added to certain canned foods to remove transition-metal ions that can catalyze the deterioration of the food. The same chelating agent has been used to treat lead poisoning because it binds Pb^{2+} ions as the chelate, forming a substance that can then be excreted by the kidneys.

The term chelate *is derived from the Greek* chele *for "claw," because a polydentate ligand appears to attach itself to the metal atom like crab claws to some object.*

Discovery of Complexes; Formula of a Complex

In 1798 B. M. Tassaert found that a solution of cobalt(II) chloride in aqueous ammonia exposed to air (an oxidizing agent) deposits orange-yellow crystals. He assigned the formula $CoCl_3 \cdot 6NH_3$ to these crystals. Later, a similar compound of platinum was assigned the formula $PtCl_4 \cdot 6NH_3$. Because these formulas suggested that the substances were somehow composed of two stable compounds—platinum(IV) chloride and ammonia in the latter case—they were called complex compounds, or simply complexes.

The basic explanation for the structure of these complexes was given by the Swiss chemist Alfred Werner in 1893. According to Werner, a metal atom exhib-its two kinds of valences, a primary valence and a secondary valence. The primary valence is what we now call the oxidation number of the metal. The secondary valence corresponds to what we now call the coordination number, which is often 6. In Werner's view, the substance previously represented by the formula $PtCl_4 \cdot 6NH_3$ is composed of the ion $[Pt(NH_3)_6]^{4+}$, with six ammonia molecules directly attached to the platinum atom. The charge of this ion is balanced by four Cl^- ions, giving a neutral compound with the structural formula $[Pt(NH_3)_6]Cl_4$.

Table 22.6 lists a series of platinum(IV) complexes studied by Werner and both the old and Werner's modern formulas. (Note that square brackets are used in modern formulas to group the metal and its associated ligands.) According to Werner's theory, these complexes should dissolve to give different numbers of ions per formula unit. For example, $[Pt(NH_3)_4Cl_2]Cl_2$ dissolves to give three ions: $[Pt(NH_3)_4Cl_2]^{2+}$ and two Cl^- ions. Werner was able to show that the electrical conductances of solutions of these complexes were equal to what was expected

Table 22.6 Some Platinum(IV) Complexes Studied by Werner

Old Formula	Modern Formula	Number of Ions	Number of Free Cl^- Ions
$PtCl_4 \cdot 6NH_3$	$[Pt(NH_3)_6]Cl_4$	5	4
$PtCl_4 \cdot 4NH_3$	$[Pt(NH_3)_4Cl_2]Cl_2$	3	2
$PtCl_4 \cdot 3NH_3$	$[Pt(NH_3)_3Cl_3]Cl$	2	1
$PtCl_4 \cdot 2NH_3$	$[Pt(NH_3)_2Cl_4]$	0	0

for the number of ions predicted by his formulas. He also demonstrated that the chloride ions in the platinum complexes were of two kinds: those that could be precipitated from solution as AgCl using silver nitrate and those that could not. He explained that the chloride ions within the platinum complex ion are securely attached to the metal atom and that only those outside the complex ion can be precipitated with silver nitrate. The number of free Cl^- ions (those not attached to platinum) determined this way agreed exactly with his formulas. Over a period of 20 years, Werner carried out many experiments on complexes, the results of all of them in basic agreement with his theory. He received the Nobel Prize for his work in 1913.

CONCEPT CHECK 22.1

Another complex studied by Werner had a composition corresponding to the formula $PtCl_4 \cdot 2KCl$. From electrical-conductance measurements, he determined that each formula unit contained three ions. He also found that silver nitrate did not give a precipitate of AgCl with this complex. Write a formula for this complex that agrees with this information.

22.4 Naming Coordination Compounds

Thousands of coordination compounds are now known. A systematic method of naming such compounds, or *nomenclature,* needs to provide basic information about the structure of a coordination compound. What is the metal in the complex? Does the metal atom occur in the cation or the anion? What is the oxidation state of the metal? What are the ligands? You can answer these questions by following the rules of nomenclature agreed upon by the International Union of Pure and Applied Chemistry (IUPAC). These rules are essentially an extension of those originally given by Werner.

1. In naming a salt, the name of the cation precedes the name of the anion. (This is a familiar rule.)

$$K_4[Fe(CN)_6] \text{ is named } \underbrace{\text{potassium}}_{\text{cation}} \underbrace{\text{hexacyanoferrate(II)}}_{\text{anion}}$$

$$[Co(NH_3)_6]Cl_3 \text{ is named } \underbrace{\text{hexaamminecobalt(III)}}_{\text{cation}} \underbrace{\text{chloride}}_{\text{anion}}$$

2. The name of the complex—whether anion, cation, or neutral species—consists of two parts written together as one word. Ligands are named first, and the metal atom is named second.

$$[Fe(CN)_6]^{4-} \text{ is named } \underbrace{\text{hexacyanoferrate(II)}}_{\substack{\text{ligand} \\ \text{name}}} \underbrace{\text{ferrate(II)}}_{\substack{\text{metal} \\ \text{name}}} \text{ ion}$$

$$[Co(NH_3)_6]^{3+} \text{ is named } \underbrace{\text{hexaamminecobalt(III)}}_{\substack{\text{ligand} \\ \text{name}}} \underbrace{\text{cobalt(III)}}_{\substack{\text{metal} \\ \text{name}}} \text{ ion}$$

Salad Dressing and Chelate Stability

The list of ingredients for a particular mayonnaise reads: vegetable oil, eggs, vinegar, and calcium disodium EDTA (ethylenediaminetetraacetate). EDTA? Yes, commercial mayonnaise and salad dressings use EDTA to remove traces of metal ions. Metal ions can catalyze undesirable reactions or else provide nutrient for bacteria, resulting in off-flavors and spoilage of the product. EDTA is a polydentate ligand that forms particularly stable chelates with many metal ions, effectively removing those ions from the product. Many commercial products contain chelating agents such as EDTA (Figure 22.9).

What accounts for the special stability of chelates? Their stability stems from the additional entropy obtained when they are formed. This leads to a large negative $\Delta G°$, which is equivalent to a large equilibrium constant for the formation of the chelate. To understand how this happens, consider the formation of the chelate $[Co(en)_3]^{3+}$ from the complex ion $[Co(NH_3)_6]^{3+}$, with monodentate ligands NH_3.

$$[Co(NH_3)_6]^{3+} + 3en \rightleftharpoons [Co(en)_3]^{3+} + 6NH_3$$

Each en molecule replaces two NH_3 molecules. Therefore, the number of particles in the reaction mixture is increased when the reaction goes to the right. In most cases, an increase in number of particles increases the possibilities for randomness or disorder. Therefore, when the reaction goes to the right, there is an increase in entropy; that is, $\Delta S°$ is positive.

At the same time, the reaction involves very little change of internal energy ($\Delta U°$) or enthalpy ($\Delta H° = \Delta U° + P\Delta V$), because the bonds are similar; all consist of a nitrogen atom coordinated to a cobalt atom. Six nitrogen-metal bonds in $Co(NH_3)_6{}^{3+}$ are broken, and six new nitrogen-metal bonds of about the same energy are formed in $Co(en)_3{}^{3+}$. Therefore, $\Delta H° \simeq 0$.

The spontaneity of a reaction depends on $\Delta G°$, which equals $\Delta H° - T\Delta S°$. But because $\Delta H° \simeq 0$,

$$\Delta G° = \Delta H° - T\Delta S°$$
$$= -T\Delta S°$$

The entropy change is positive, so $\Delta G°$ is negative, and the reaction is spontaneous from left to right.

The fact that the equilibrium for the reaction favors the chelate $[Co(en)_3]^{3+}$ means that the chelate has thermodynamic stability. A similar argument could be made for any equilibrium involving the replacement of monodentate ligands by polydentate ligands. Reaction tends to favor the chelate.

Figure 22.9 ▲

EDTA *Left:* Mayonnaise and salad dressings contain EDTA to reduce the concentrations of certain metal ions. Some plant fertilizers contain EDTA chelates of copper and other metals, providing the metal in a soluble form that plants can utilize. *Right:* An electrostatic potential map of disodium EDTA.

■ See Problems 22.73 and 22.74.

3. The complete ligand name consists of a Greek prefix denoting the number of ligands, followed by the specific name of the ligand. When there are two or more ligands, the ligands are written in alphabetical order (disregarding Greek prefixes).
   a. Anionic ligands end in *-o*. Some examples:

Anion Name	*Ligand Name*
Bromide, Br^-	Bromo
Carbonate, CO_3^{2-}	Carbonato
Chloride, Cl^-	Chloro
Cyanide, CN^-	Cyano
Fluoride, F^-	Fluoro
Hydroxide, OH^-	Hydroxo
Oxalate, $C_2O_4^{2-}$	Oxalato
Oxide, O^{2-}	Oxo
Sulfate, SO_4^{2-}	Sulfato

   b. Neutral ligands are usually given the name of the molecule. There are, however, several important exceptions:

Molecule	*Ligand Name*
Ammonia, NH_3	Ammine
Carbon monoxide, CO	Carbonyl
Water, H_2O	Aqua

   c. The prefixes used to denote the number of ligands are *mono-* (1, usually omitted); *di-* (2); *tri-* (3); *tetra-* (4); *penta-* (5); *hexa-* (6); and so forth. To see how the ligand name is formed, consider the complex ions

   $$[Fe(CN)_6]^{4-} \text{ or hexacyanoferrate(II) ion}$$
   6   CN^- ligands

   $$[Co(NH_3)_6]^{3+} \text{ or hexaamminecobalt(III) ion}$$
   6   NH_3 ligands

   d. When the name of the ligand also has a number prefix, the number of ligands is denoted with *bis* (2), *tris* (3), *tetrakis* (4), and so forth. The name of the ligand follows in parentheses. For example, the complex $[Co(en)_3]Cl_3$ is named as follows:

   tris(ethylenediamine)cobalt(III) chloride
   3   ligand name

4. The complete metal name consists of the name of the metal, followed by *-ate* if the complex is an anion, followed by the oxidation number of the metal as a Roman numeral in parentheses. (An oxidation state of zero is indicated by 0 in parentheses.) When there is a Latin name for the metal, it is usually used to name the anion.

English Name	*Latin Name*	*Anion Name*
Copper	Cuprum	Cuprate
Gold	Aurum	Aurate
Iron	Ferrum	Ferrate
Lead	Plumbum	Plumbate
Silver	Argentum	Argentate
Tin	Stannum	Stannate

   Examples are

   hexacyanoferrate(II)   hexaamminecobalt(III)
   ferrum | oxidation   metal oxidation
   = iron | number 2   name   number 3
   indicates
   an anion

⚙WL Interactive Example 22.1   **Writing the IUPAC Name Given the Structural Formula of a Coordination Compound**

Gaining Mastery Toolbox

Critical Concept 22.1
There are naming rules that apply to coordination compounds. The correct name of a coordination compound conveys important information about the compound, including the name of the metal, the metal's oxidation state, and the name and number of ligands attached to the metal.

Solution Essentials:
- Rules for naming coordination compounds
- Ligand
- Coordination compound (complex)
- Complex ion
- Oxidation state

Give the IUPAC name of each of the following coordination compounds:

a $[Pt(NH_3)_4Cl_2]Cl_2$;   b $[Pt(NH_3)_2Cl_2]$;   c $K_2[PtCl_6]$.

Problem Strategy Using the naming rules from this section, determine the name of the coordination compounds.

Solution a The cation is listed first in the formula.

$$\underbrace{[Pt(NH_3)_4Cl_2]}_{\text{Cation}}\underbrace{Cl_2}_{\text{Anions}}$$

There are two Cl^- anions, so the charge on the cation is $+2$: $[Pt(NH_3)_4Cl_2]^{2+}$. The oxidation number of platinum plus the sum of the charges on the ligands (-2) equals the cation charge $+2$. Therefore, the oxidation number of Pt is $+4$. Hence, the name of the compound is **tetraamminedichloroplatinum(IV) chloride**. Note that the ligands are listed in alphabetical order (that is, ammine before chloro).

b This is a neutral complex species. The oxidation number of platinum must balance that of the two chloride ions. The name of the compound is **diamminedichloroplatinum(II)**.

c The complex anion is $[PtCl_6]^{2-}$. The oxidation number of platinum is 4, and the name of the compound is **potassium hexachloroplatinate(IV)**.

Answer Check As a final check to problems of this type, take the name that you have come up with for an answer and see if it leads you to writing the correct formula.

Exercise 22.1 Give the IUPAC names of
a $[Co(NH_3)_5Cl]Cl_2$;   b $K_2[Co(H_2O)(CN)_5]$;
c $[Fe(H_2O)_5(OH)]^{2+}$.

■ See Problems 22.47, 22.48, 22.49, and 22.50.

Example 22.2   **Writing the Structural Formula Given the IUPAC Name of a Coordination Compound**

Gaining Mastery Toolbox

Critical Concept 22.2
The name of a coordination compound provides information about the structure of the compound as well as the oxidation state of the metal. In order to write the structural formula of a coordination compound, it is critical to know how to apply the naming rules for coordination compounds.

Solution Essentials:
- Rules for naming coordination compounds
- Ligand
- Coordination compound (complex)
- Complex ion
- Oxidation state
- Structural formula

Write the structural formula corresponding to each of the following IUPAC names: a hexaaquairon(II) chloride; b potassium tetrafluoroargentate(III); c pentachlorotitanate(II) ion.

Problem Strategy Examine the complex ion and coordination compound names and work "backward" to find the chemical formula using information from this section.

Solution a The complex cation hexaaquairon(II) is Fe^{2+} with six H_2O ligands: $[Fe(H_2O)_6]^{2+}$. The formula of the compound is $[Fe(H_2O)_6]Cl_2$. (Remember to enclose the formula of the complex ion in square brackets.)

b The compound contains the complex anion tetrafluoroargentate(III)—that is, Ag^{3+} with four F^- ligands. The formula of the ion is $[AgF_4]^-$. Hence, the formula of the compound is $K[AgF_4]$.

c The ion contains Ti^{2+} and five Cl^- ligands. The charge on the ion is $+2 + 5(-1) = -3$. The formula of the complex ion is $[TiCl_5]^{3-}$.

Answer Check As a check, once you have written the structural formula, try naming the compound to see if you come up with the IUPAC name.

Exercise 22.2 Write structural formulas for each of the following: a potassium hexacyanoferrate(II), b tetraamminedichlorocobalt(III) chloride, c tetrachloroplatinate(II) ion.

■ See Problems 22.51 and 22.52.

22.5 Structure and Isomerism in Coordination Compounds

Although we described the formation of a complex as a Lewis acid–base reaction (Section 22.3), we did not go into any details of structure. We did not look at the geometry of complex ions or inquire about the precise nature of the bonding. Three properties of complexes have proved pivotal in determining these details.

1. *Isomerism* Isomers are compounds with the same molecular formula (or the same simplest formula, in the case of ionic compounds) but with different arrangements of atoms. Because their atoms are differently arranged, isomers have different properties. There are many possibilities for isomerism in coordination compounds. The study of isomerism can lead to information about atomic arrangement in coordination compounds. Werner's research on the isomerism of coordination compounds finally convinced others that his views were essentially correct.

2. *Paramagnetism* Paramagnetic substances are attracted to a strong magnetic field. Paramagnetism is due to unpaired electrons in a substance. (*Ferromagnetism* in solid iron is also due to unpaired electrons, but in this case the magnetism of many iron atoms is aligned, giving a magnetic effect perhaps a million times stronger than that seen in paramagnetic substances.) Many complex compounds are paramagnetic. The magnitude of this paramagnetism can be measured with a *Gouy balance,* in which the force of magnetic attraction is balanced with weights (Figure 22.10). Because paramagnetism is related to the electron configuration of the complex, these measurements can give information about the bonding.

3. *Color* A substance is colored because it absorbs light in the visible region of the spectrum. The absorption of visible light is due to a transition between closely spaced electronic energy levels. As you have seen, many coordination compounds are highly colored. The color is related to the electronic structure of the compounds. ◀

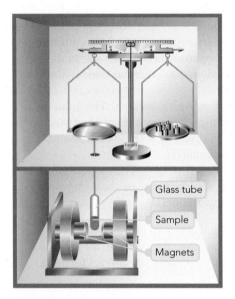

Figure 22.10 ▲

Gouy balance for measuring the paramagnetism of a substance If the sample is attracted into the field of the magnet, the left pan of the balance will be subjected to a downward force. This force is balanced by weights added to the right pan.

Recall that the wavelength of light absorbed by a substance is related to the difference between two energy levels in the substance. See Section 7.3.

In this section, we will investigate the relationship between structure and isomerism. In the following sections, we will look into explanations of the paramagnetism and color of coordination compounds.

There are two major kinds of isomers. **Constitutional isomers** are *isomers that differ in how the atoms are joined together*—that is, in the order in which the atoms are bonded to each other. (For example, H—N=C=O and N≡C—O—H are constitutional isomers because the H atom is bonded to N in one case and to O in the other.) **Stereoisomers**, on the other hand, are *isomers in which the atoms are bonded to each other in the same order but differ in the precise arrangement of the atoms in space.*

Constitutional Isomerism

Coordination compounds provide many special types of constitutional isomers. Here, for example, are two cobalt isomers:

$$[Co(NH_3)_5(SO_4)]Br \qquad \text{a red compound}$$
$$[Co(NH_3)_5Br]SO_4 \qquad \text{a violet compound}$$

This type of constitutional isomerism is called coordination isomerism.

In the first isomer, the sulfate ion is attached to the cobalt atom, and the bromide ion is in the crystal lattice as the anion countering the metal complex cation. In the second isomer, the sulfate ion is in the crystal lattice, with the bromide ion attached to the cobalt atom. ◀

Both the cation and the anion can be metal complex ions, yielding another type of isomer. The following copper–platinum compounds provide an example:

$$[Cu(NH_3)_4][PtCl_4] \quad \text{and} \quad [Pt(NH_3)_4][CuCl_4]$$

In the first compound, the NH_3 ligands are associated with a copper atom and the Cl^- ligands with a platinum atom. In the second compound, the ligands are transposed.

A more subtle type of constitutional isomer is displayed by the following compounds:

$$[Co(NH_3)_5(ONO)]Cl_2$$
$$[Co(NH_3)_5(NO_2)]Cl_2$$

In the first compound, the nitrite ligand, NO_2^-, bonds to the cobalt atom through an electron pair on an oxygen atom. This is a red substance that slowly changes to the second compound. In this yellow-brown compound, the nitrite ligand bonds to the cobalt atom through an electron pair on the nitrogen atom. ▶

This type of constitutional isomerism is called linkage isomerism.

CONCEPT CHECK 22.2

A complex has the composition $Co(NH_3)_4(H_2O)BrCl_2$. Conductance measurements show that there are three ions per formula unit, and precipitation of AgCl with silver nitrate shows that there are two Cl^- ions not coordinated to cobalt. What is the structural formula of the compound?

Stereoisomerism

The existence of structural isomers is strong evidence for the view that complexes consist of groups directly bonded to a central metal atom. The existence of stereoisomers not only strengthens this view but also helps explain how the groups are arranged about the central atom.

Geometric isomers are *isomers in which the atoms are joined to one another in the same way but differ because some atoms occupy different relative positions in space.* Consider the complex diamminedichloroplatinum(II), $[Pt(NH_3)_2Cl_2]$. Two compounds of this composition are known. One is an orange-yellow compound with solubility at 25°C of 0.252 g per 100 g of water. The other compound is pale yellow and much less soluble (0.037 g per 100 g of water at 25°C). ▶ How do you explain the occurrence of these two isomers?

The isomers of [Pt(NH₃)₂Cl₂] also differ in their biological properties, as noted in Chapter 1. The orange-yellow compound acts as an anticancer drug; the other isomer does not.

There are two symmetrical geometries of complexes of four ligands: tetrahedral and square planar. Let us write MA_2B_2 for a complex with ligands A and B about the metal atom M. As Figure 22.11 shows, a tetrahedral geometry for MA_2B_2 allows only one arrangement of ligands. The square planar geometry for MA_2B_2, however, gives two possible arrangements. (The two arrangements for $[Pt(NH_3)_2Cl_2]$ are shown in Figure 22.12.) One, labeled *cis,* has the two A ligands on one side of the square and the two B ligands on the other side. The other arrangement, labeled *trans,* has A and B ligands across the square from one another. The *cis* and *trans* arrangements of MA_2B_2 are examples of geometric isomers.

That there are two isomers of $[Pt(NH_3)_2Cl_2]$ is evidence for the square planar geometry in this complex. But how do you identify the substances and their properties with the *cis* and *trans* arrangements? You can distinguish between them by predicting their polarity. The *trans* arrangement, being completely symmetrical, is nonpolar. The *cis* arrangement, with electronegative Cl atoms on one side of the platinum atom, is polar. This difference between *cis* and *trans* isomers should be reflected in their solubilities in water, because polar substances are more soluble in water (itself a polar substance) than are nonpolar substances. Thus, you would expect the more soluble isomer to have the *cis* arrangement.

Although the difference in solubility is revealing, the most direct evidence of polarity in a molecule comes from measurements of dipole moment. (The dipole moment is a measure of charge separation in a molecule.) These values show that the less soluble platinum(II) complex has no dipole moment and must be *trans* and that the other isomer does have a dipole moment and must be *cis.* Figure 22.12 shows the two isomers.

Six-coordinate complexes have only one symmetrical geometry: octahedral (Figure 22.13). Geometric isomers are possible for this geometry also. Consider the complex MA_4B_2 in which two of the A ligands occupy positions just opposite one another. The other four ligands have a square planar arrangement in which *cis–trans* isomers are possible. Tetraamminedichlorocobalt(III) chloride, $[Co(NH_3)_4Cl_2]Cl$, is an example of a complex with such geometric isomers. The *cis* compound is purple; the *trans* compound is green. See Figure 22.14.

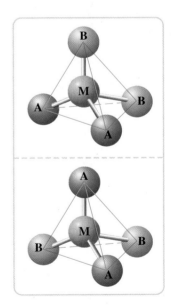

Figure 22.11 ▲

The tetrahedral complex MA_2B_2 has no geometric isomers The molecule on the top can be rotated to look like the molecule on the bottom. A and B are any two ligands.

Figure 22.12 ▶

Geometric isomers of the square planar complex diamminedichloroplatinum(II) The existence of two isomers of $[Pt(NH_3)_2Cl_2]$ is evidence of a square planar geometry. (The tetrahedral geometry would not give isomers.)

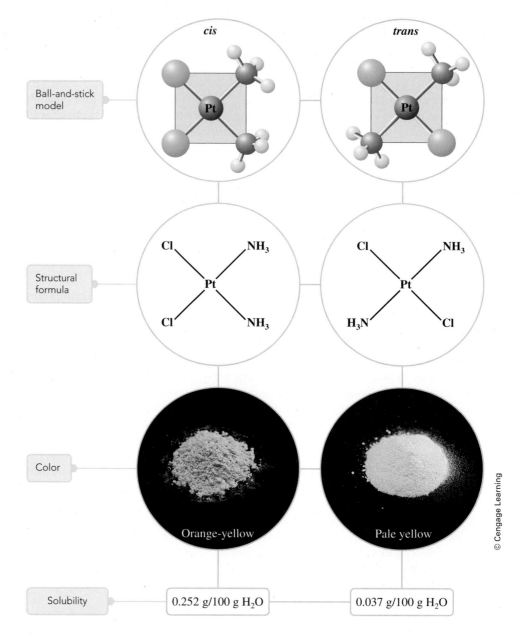

cis

trans

Ball-and-stick model

Structural formula

Color

Orange-yellow

Pale yellow

Solubility

0.252 g/100 g H_2O

0.037 g/100 g H_2O

© Cengage Learning

Figure 22.13 ▶

Octahedral geometry (*a*) All positions in this geometry are equivalent. (*b*) Octahedral geometry is often represented like this.

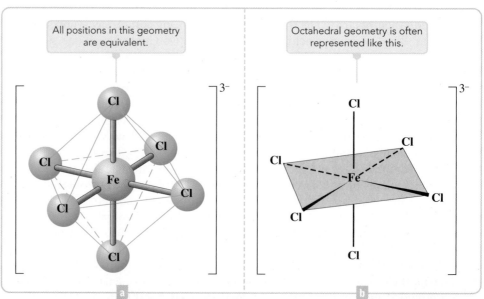

All positions in this geometry are equivalent.

Octahedral geometry is often represented like this.

a

b

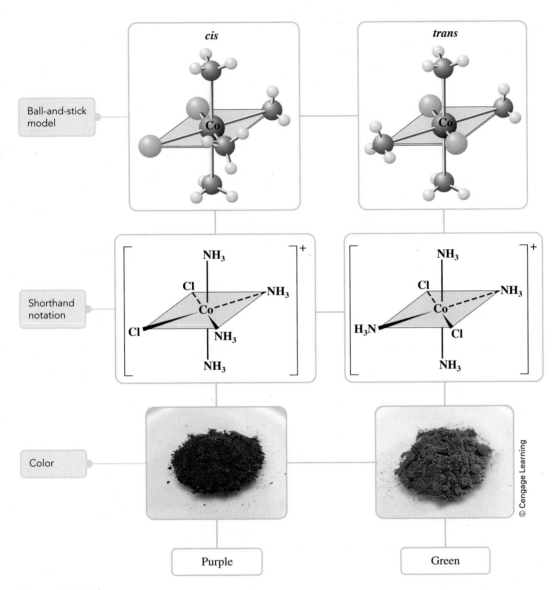

Ball-and-stick model

Shorthand notation

Color

Purple

Green

Figure 22.14 ▲

Geometric isomers of tetraamminedichlorocobalt(III) ion, [Co(NH$_3$)$_4$Cl$_2$]$^+$ The two isomers have different properties. For example, the *cis* compound is purple; the *trans* compound is green.

Example 22.3   **Deciding Whether Geometric Isomers Are Possible**

Gaining Mastery Toolbox

Critical Concept 22.3
Geometric isomers only occur if the complex has either square planar or octahedral geometry. If the complex has either of these two geometries, determine if pairs of ligands can be placed adjacent to each other, which is *cis*, or across from each other, which is *trans*. Geometric isomers are stereoisomers.

Solution Essentials:
- *Cis–trans* isomers
- Geometric isomer
- Stereoisomer
- Complex ion
- Octahedral, tetrahedral, and square planar geometries

Are there any geometric isomers of the complex cation [Co(NH$_3$)$_4$(NO$_2$)$_2$]$^+$? If so, draw them.

Problem Strategy From the formula of the complex ion, determine the geometry. Draw the possible structures, keeping in mind that if the structure is square planar or octahedral, then geometric (*cis–trans*) isomers are possible.

Solution Yes, there are geometric isomers. This is easily seen if you first draw two NH$_3$ ligands opposite one another in an octahedral geometry. Then the NO$_2^-$

(continued)

(continued)

ligands can have *cis* or *trans* arrangements on the plane perpendicular to the axis of the first NH_3 ligands. (See Figure 22.15.)

Figure 22.15 ▶

Cis–trans isomers of [Co(NH₃)₄(NO₂)₂]⁺ Note the arrangements of groups on the gray planes.

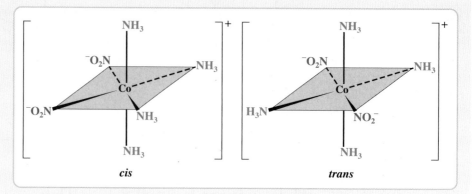

Answer Check When working a problem such as this that requires thinking in three dimensions, it is very helpful to check your work by building molecular models to represent the structure. For example, for this problem you could build an octahedral model with blue balls representing the NH_3 ligands and red balls representing the NO_2^- ligands.

Exercise 22.3 Do any of the following stable octahedral complexes have geometric isomers? If so, draw them.
a $[Co(NH_3)_5Cl]^{2+}$ **b** $[Co(NH_3)_4(H_2O)_2]^{3+}$

■ See Problems 22.53 and 22.54.

Compounds with both bidentate and monodentate ligands increase the potential for isomerism in octahedral complexes. Figure 22.16 shows the isomers of the dichlorobis (ethylenediamine) cobalt (III) ion, $[CoCl_2(en)_2]^+$. Isomer A is the *trans*

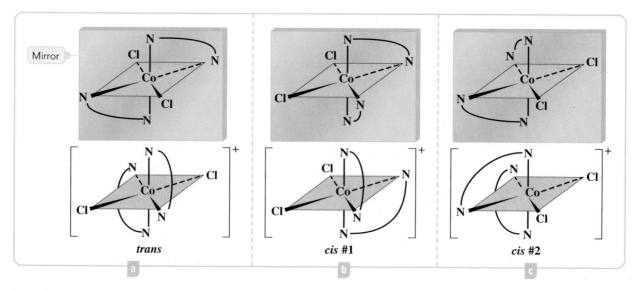

Figure 22.16 ▲

Isomers of dichlorobis (ethylenediamine)cobalt(III) ion, [CoCl₂(en)₂]⁺ Each isomer is shown with its mirror image. N N represents ethylenediamine. (*a*) The *trans* isomer and its mirror image are superimposable, so they represent the same molecule. You can see this by rotating the original structure 90° counterclockwise (about the axis perpendicular to the gray plane). (*b, c*) *Cis* #1 and *cis* #2 are enantiomers (mirror-image, or optical, isomers) of the *cis* form. You cannot rotate one molecule to superimpose it on the other. However, the mirror image of *cis* #1 is identical to *cis* #2. To see this, rotate the mirror image of *cis* #1 by 180° and compare it with *cis* #2.

isomer (it has a green color). Both B and C are *cis* isomers (both have a violet color). Yet isomers B and C are not identical molecules. They are **enantiomers**, or **optical isomers**; that is, they are *isomers that are nonsuperimposable mirror images of one another.*

To better understand the nature of enantiomers, note that the two *cis* isomers have the same relationship to each other as the relationship of your left hand and right hand. The mirror image of a left hand looks like the right hand, and vice versa (Figure 22.17). But neither the right hand nor the mirror image of the left hand can be turned in any way to look exactly like the left hand; that is, the right and left hands cannot be *superimposed* on one another. (Remember that a left-handed glove does not fit a right hand.) Any physical object *possessing the quality of handedness*—whose mirror image is not identical with itself—is said to be **chiral** (from the Greek *cheir,* meaning "hand"). A glove is chiral. However, a pencil, whose mirror image looks identical to the real pencil, is *achiral*. Isomer A in Figure 22.16 is achiral; its mirror image is identical to itself. Isomer B is chiral; its mirror image, isomer C, is not superimposable on B.

Enantiomers (optical isomers) have identical properties in a symmetrical environment. They have identical melting points and the same solubilities and colors. Enantiomers are usually differentiated by the manner in which they affect *plane-polarized* light. Normally, a light beam consists of electromagnetic waves vibrating in all possible planes about the direction of the beam (Figure 22.18). One of these planes may be selected out by passing a beam of light through a *polarizer* (say through a Polaroid lens). When this plane-polarized light is passed through a solution containing an enantiomer, such as one isomer of *cis*-[CoCl₂(en)₂]Cl, the plane of the light wave is twisted. One of the enantiomers twists the plane to the right (as the light comes out toward you); the other isomer twists the plane to the left by the same angle. Because of *the ability to rotate the plane of light waves, either as pure substances or in solution,* enantiomers are said to be **optically active**. (This is also the origin of the term *optical isomer.*) Figure 22.19 shows a sketch of a *polarimeter,* an instrument that determines the angular change in the plane of a light wave made by an optically active compound.

A compound whose solution rotates the plane of polarized light to the right (when looking toward the source of light) is called **dextrorotatory** and is labeled *d. A compound whose solution rotates the plane of polarized light to the left (when looking toward the source of light)* is called **levorotatory** and is labeled *l.* Thus, the dextrorotatory isomer of *cis*-[CoCl₂(en)₂]⁺ is *d-cis*-dichlorobis(ethylenediamine)cobalt(III).

A chemical reaction normally produces *a mixture of equal amounts of optical isomers,* called a **racemic mixture**. Because the two isomers rotate the plane of polarized light in equal but opposite directions, a racemic mixture has no net effect on polarized light.

To show that optical isomers exist, a racemic mixture must be separated into its *d* and *l* isomers; that is, the racemic mixture must be *resolved.* One way to resolve

Figure 22.17 ▲

Nonsuperimposable mirror images The mirror image of the left hand looks like the right hand, but the left hand itself cannot be superimposed on the right hand.

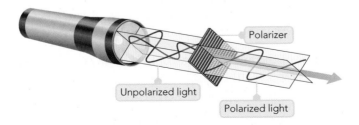

Figure 22.18 ◄

Polarization of light Light from the source consists of waves vibrating in various planes along any axis. The polarizer filters out all waves except those vibrating in a particular plane.

Figure 22.19 ▶

Sketch of a polarimeter Light from a source (at the left) enters a polarizer prism, which splits the light into polarized components. The component that is polarized vertically enters the sample tube containing a solution of an optically active compound, which rotates the plane of the polarized light by an angle that depends on the compound and its concentration. The experimenter can see the light coming through the analyzer prism only if that prism is rotated by an equivalent angle (degree of rotation).

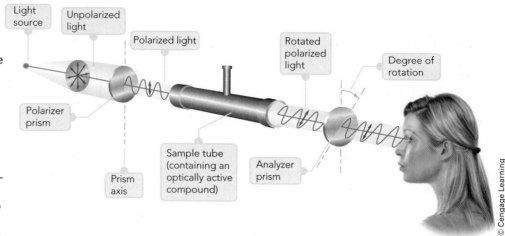

© Cengage Learning

a mixture containing d and l complex ions is to prepare a salt with an optically active ion of opposite charge. For example, the tartaric acid, $H_2C_4H_4O_6$, prepared from the white substance in wine vats is the optically active isomer d-tartaric acid. When the racemic mixture of cis-$[CoCl_2(en)_2]Cl$ is treated with d-tartaric acid, the d-tartrate salts of d- and l-cis-$[CoCl_2(en)_2]^+$ may be crystallized. These salts will no longer be optical isomers of one another and will have different solubilities.

Example 22.4

Deciding Whether Optical Isomers Are Possible

Gaining Mastery Toolbox

Critical Concept 22.4
Structures that are mirror images and cannot be superimposed are optical isomers. Optical isomerism can only be determined by comparing mirror images.

Solution Essentials:
• Optical isomer
• Mirror image
• Geometric isomer
• Stereoisomer
• Complex ion
• Octahedral, tetrahedral, and square planar geometries

Figure 22.20 ▶

Optical isomers of the complex ion Co(en)₃³⁺
Note that the isomers are nonsuperimposable mirror images of one another.

Are there optical isomers of the complex $[Co(en)_3]Cl_3$? If so, draw them.

Problem Strategy Determine the geometry of the complex ion and draw the structures, looking to see whether any structures are nonsuperimposable mirror images.

Solution Yes, there are optical isomers, because the complex $[Co(en)_3]^{3+}$ has nonsuperimposable mirror images. (See Figure 22.20.)

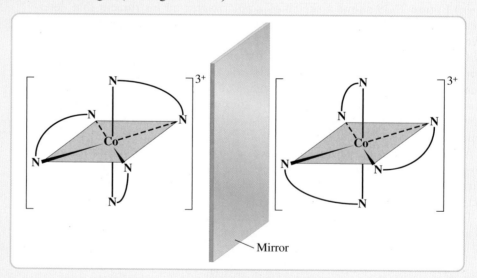

Answer Check As suggested in the Answer Check to Problem 22.3, you should consider building molecular models to check your work. In this case, be sure to build correctly the linkage that makes up the bidentate ethylenediamine ligand.

Exercise 22.4 Does either of the following have optical isomers? If so, draw them.
a $trans$-$[Co(en)_2(NO_2)_2]^+$ b cis-$[Co(en)_2(NO_2)_2]^+$

■ See Problems 22.55 and 22.56.

CONCEPT CHECK 22.3

a Which of the following molecular models of octahedral complexes are mirror images of the complex X? Keep in mind that you can rotate the models when performing comparisons.

b Which complexes are optical isomers of molecule X?

c Identify the distinct geometric isomers of complex X. (Note that some of the models may represent the same molecule.)

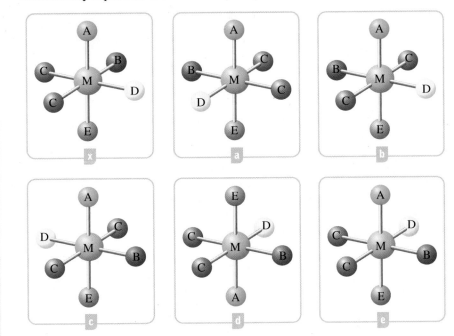

22.6 Valence Bond Theory of Complexes

A complex is formed when electron pairs from ligands are donated to a metal ion. This does not explain how the metal ion can accept electron pairs. Nor does it explain the paramagnetism often observed in complexes. *Valence bond theory* provided the first detailed explanation of the electronic structure of complexes. This explanation is essentially an extension of the view of covalent bonding we described earlier. ▶ According to this view, a covalent bond is formed by the overlap of two orbitals, one from each bonding atom. In the usual covalent bond formation, each orbital originally holds one electron, and after the orbitals overlap, a bond is formed that holds two electrons. In the formation of a coordinate covalent bond in a complex, however, a ligand orbital containing two electrons overlaps an unoccupied orbital on the metal atom. Figure 22.21 diagrams these two bond formations.

See Section 10.3.

As an example of complex formation in a transition metal ion, let's look at the formation of the yellow $[Cr(NH_3)_6]^{3+}$ ion. This ion is known from experiment to be paramagnetic. We can think of this ion as formed by complexing $:NH_3$ ligands with the free chromium(III) ion. First, we need the configuration of the free ion. The chromium atom configuration is $[Ar]3d^54s^1$. In the formation of a transition-metal ion, the outer s electrons (here, $4s^1$) are lost first, then the outer d electrons (here, two of the $3d^5$ electrons). Therefore, Cr^{3+} has the configuration $[Ar]3d^3$. The orbital diagram is

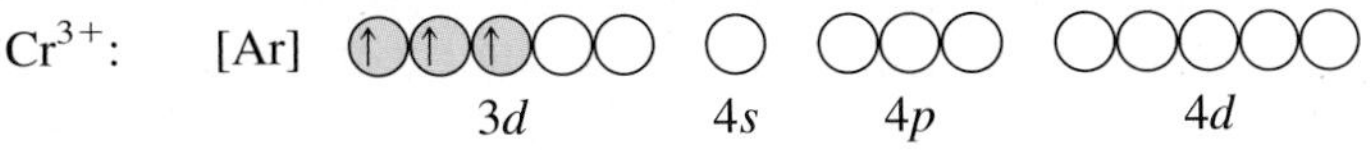

$$Cr^{3+}: \quad [Ar] \; \boxed{\uparrow}\boxed{\uparrow}\boxed{\uparrow}\boxed{\;}\boxed{\;} \quad \boxed{\;} \quad \boxed{\;}\boxed{\;}\boxed{\;} \quad \boxed{\;}\boxed{\;}\boxed{\;}\boxed{\;}\boxed{\;}$$
$$\qquad\qquad\qquad 3d \qquad\quad 4s \quad\; 4p \qquad\qquad 4d$$

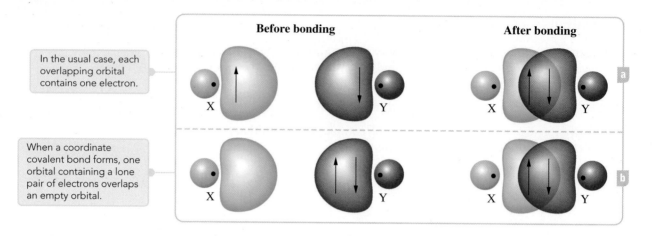

In the usual case, each overlapping orbital contains one electron.

When a coordinate covalent bond forms, one orbital containing a lone pair of electrons overlaps an empty orbital.

Figure 22.21 ▲

Covalent bond formation between atoms X and Y

Note that the $3d^3$ electrons are placed in separate orbitals with their spins in the same direction, following Hund's rule. Two empty $3d$ orbitals, in addition to orbitals of the $n = 4$ shell, can be used for bonding to ligands.

To bond electron pairs from six $:NH_3$ ligands to Cr^{3+}, forming six equivalent bonds, octahedral hybrid orbitals are required. These hybrid orbitals will use two d orbitals, the $4s$ orbital, and the three $4p$ orbitals. The d orbitals could be either $3d$ or $4d$; the two available $3d$ orbitals are used because they have lower energy. We can now write the orbital diagram for the metal atom in the complex:

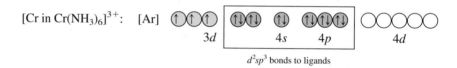

Electron pairs donated from ligands are shown in color. Note that there are three unpaired electrons in $3d$ orbitals on the chromium atom, which explains the paramagnetism of the complex ion. The bonding in other complexes can be explained in a similar fashion.

22.7 Crystal Field Theory

Although valence bond theory explains the bonding and magnetic properties of complexes, it is limited in two important ways. First, the theory cannot easily explain the color of complexes. Second, the theory is difficult to extend quantitatively. Consequently, another theory—crystal field theory—has emerged as the prevailing view of transition-metal complexes.

Crystal field theory is *a model of the electronic structure of transition-metal complexes that considers how the energies of the d orbitals of a metal ion are affected by the electric field of the ligands.* According to this theory, the ligands in a transition-metal complex are treated as point charges. So a ligand anion becomes simply a point of negative charge. A neutral molecule, with its electron pair that it donates to the metal atom, is replaced by a partial negative charge, representing the negative end of the molecular dipole. In the electric field of these negative charges, the five d orbitals of the metal atom no longer have exactly the same energy. The result, as you will see, explains both the paramagnetism and the color observed in certain complexes.

The simplifications used in crystal field theory are drastic. Treating the ligands as point charges is essentially the same as treating the bonding as ionic. However, it turns out that the theory can be extended to include covalent character in the bonding. This simple extension is usually referred to as *ligand field theory,* but after including several levels of refinements, the theory becomes equivalent to molecular orbital theory.

Effect of an Octahedral Field on the *d* Orbitals

All five *d* orbitals of an isolated metal atom have the same energy. But if the atom is brought into the electric field of several point charges, these *d* orbitals may be affected in different ways and therefore may have different energies. To understand how this can happen, you must first see what these *d* orbitals look like. You will then be able to picture what happens to them in the crystal field theory of an octahedral complex.

Figure 22.22 shows the shapes of the five *d* orbitals. The orbital labeled d_{z^2} has a dumbbell shape along the *z*-axis, with a collar in the $x-y$ plane surrounding this dumbbell. Remember that this shape represents the volume most likely to be occupied by an electron in this orbital. The other four *d* orbitals have "cloverleaf" shapes, differing only in the orientation of the orbitals in space. The "cloverleaf" orbital $d_{x^2-y^2}$ has its lobes along the *x*-axis and the *y*-axis. Orbitals d_{xy}, d_{yz}, and d_{xz} have their lobes directed between the two sets of axes designated in the orbital label. Orbital d_{xy}, for example, has its lobes lying between the *x*- and *y*-axes.

A complex ion with six ligands will have the ligands arranged octahedrally about the metal atom to reduce mutual repulsion. Imagine that the ligands are replaced by negative charges. If the ligands are anions, they are replaced by the anion charge. If the ligands are neutral molecules, they are replaced by the partial negative charge from the molecular dipole. The six charges are placed at equal distances from the metal atom, one charge on each of the positive and negative sides of the *x*-, *y*-, and *z*-axes. (See Figure 22.22.)

Fundamentally, the bonding in this model of a complex is due to the attraction of the positive metal ion for the negative charges of the ligands. However, an electron in a *d* orbital of the metal atom is repelled by the negative charge of the ligands. This repulsion alters the energy of the *d* orbital depending on whether

Figure 22.22 ▼

The five *d* orbitals The d_{z^2} orbital has a dumbbell shape with a collar; the other orbitals have cloverleaf shapes. In an isolated atom, these orbitals have the same energy. However, in an octahedral complex, the orbitals split into two sets, with the d_{z^2} and $d_{x^2-y^2}$ orbitals having higher energy than the other three. Note that the lobes of the d_{z^2} and $d_{x^2-y^2}$ orbitals point toward the ligands (represented here by negative charges), whereas the lobes of the other orbitals point between ligands. The repulsion is greater in the case of the d_{z^2} and $d_{x^2-y^2}$ orbitals.

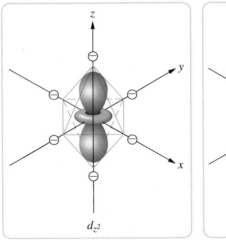

d_{z^2}

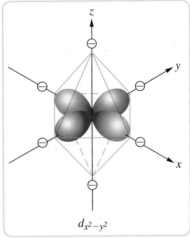

$d_{x^2-y^2}$

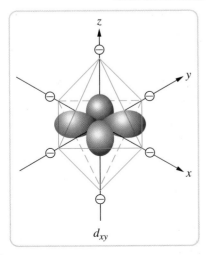

d_{xy}

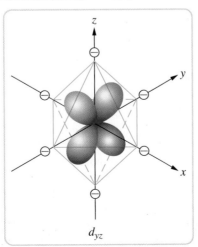

d_{yz}

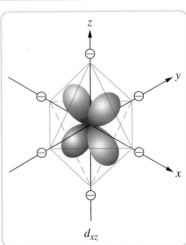

d_{xz}

Figure 22.23 ▶

Energy levels of *d* orbitals in an octahedral field The positive metal ion is attracted to the negative charges (ligands), but electrons in the *d* orbitals are repelled by them. Thus, although there is an overall attraction, the *d* orbitals no longer have the same energy.

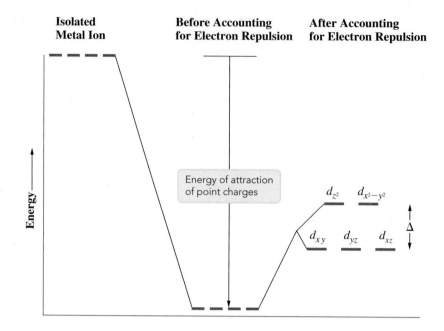

it is directed *toward* or *between* ligands. For example, consider the difference in the repulsive effect of ligands on metal-ion electrons in the d_{z^2} and d_{xy} orbitals. Because the d_{z^2} orbital is directed at the two ligands on the *z*-axis (one on the $-z$ side and the other on the $+z$ side), an electron in the orbital is rather strongly repelled by them. The energy of the d_{z^2} orbital becomes greater. Similarly, an electron in the d_{xy} orbital is repelled by the negative charge of the ligands, but because the orbital is not pointed directly at the ligands, the repulsive effect is smaller. The energy is raised, but less than the energy of the d_{z^2} orbital is raised.

If you look at the five *d* orbitals in an octahedral field (electric field of octahedrally arranged charges), you see that you can divide them into two sets. Orbitals d_{z^2} and $d_{x^2-y^2}$ are directed toward ligands, and orbitals d_{xy}, d_{yz}, and d_{xz} are directed between ligands. The orbitals in the first set (d_{z^2} and $d_{x^2-y^2}$) have higher energy than those in the second set (d_{xy}, d_{yz}, and d_{xz}). Figure 22.23 shows the energy levels of the *d* orbitals in an octahedral field. *The difference in energy between the two sets of d orbitals on a central metal ion that arises from the interaction of the orbitals with the electric field of the ligands* is called the **crystal field splitting, Δ**.

High-Spin and Low-Spin Complexes

Once you have the energy levels for the *d* orbitals in an octahedral complex, you can decide how the *d* electrons of the metal ion are distributed in them. Knowing this distribution, you can predict the magnetic characteristics of the complex.

Consider the complex ion $[Cr(NH_3)_6]^{3+}$. According to crystal field theory, it consists of the Cr^{3+} ion surrounded by NH_3 molecules treated as partial negative charges. The effect of these charges is to split the *d* orbitals of Cr^{3+} into two sets as shown in Figure 22.23. The question now is how the *d* electrons are distributed among the *d* orbitals of the Cr^{3+} ion. Because the electron configuration of Cr^{3+} is $[Ar]3d^3$, there are three *d* electrons to distribute. They are placed in the *d* orbitals of lower energy, following Hund's rule (Figure 22.24). You see that the complex ion $[Cr(NH_3)_6]^{3+}$ has three unpaired electrons and is therefore paramagnetic.

If you examine any transition-metal ion that has configuration d^4, d^5, d^6, or d^7, you will find two bonding possibilities, which yield high-spin complexes in one case and low-spin complexes in the other case. A **high-spin complex** is *a complex in which there is minimum pairing of electrons in the d orbitals of the metal atom.* A **low-spin complex** is *a complex in which there is more pairing of electrons in the d orbitals of the metal atom than in a corresponding high-spin complex.*

Consider the complex ion $[Fe(H_2O)_6]^{2+}$. What are its magnetic characteristics? Remember, you need to look at only the *d* electrons of the metal ion, Fe^{2+}. The

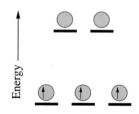

Figure 22.24 ▲

Occupation of the 3*d* orbitals in an octahedral complex of Cr^{3+} The electrons occupy different lower-energy orbitals but have the same spin (Hund's rule).

electron configuration of the ion is $[Ar]3d^6$. You distribute six electrons among the d orbitals of the complex in such a way as to get the lowest total energy. If you place all six electrons into the lower three d orbitals, you get the distribution shown by the energy-level diagram in Figure 22.25a. All of the electrons are paired, so you would predict that this distribution gives a diamagnetic complex. The $[Fe(H_2O)_6]^{2+}$ ion, however, is paramagnetic. Therefore, the distribution given in Figure 22.25a is not correct. What was the mistake?

The mistake was to ignore the **pairing energy, P,** *the energy required to put two electrons into the same orbital.* When an orbital is already occupied by an electron, it requires energy to put another electron into that orbital because of their mutual repulsion. Suppose that this pairing energy is greater than the crystal field splitting; that is, suppose $P > \Delta$. In that case, once the first three electrons have singly occupied the three lower-energy d orbitals, the fourth electron will go into one of the higher-energy d orbitals. It will take less energy to do that than to pair up with an electron in one of the lower-energy orbitals. Similarly, the fifth electron will go into the last empty d orbital. The sixth electron must pair up, so it goes into one of the lower-energy orbitals. Figure 22.25b shows this electron distribution. In this case, there are four unpaired electrons and the complex is paramagnetic.

We see that crystal field theory predicts two possibilities: a *low-spin complex* when $P < \Delta$ and a *high-spin complex* when $P > \Delta$. The value of Δ, as we will explain later, can be obtained from the spectrum of a complex, and the value of P can be calculated theoretically. Even in the absence of these numbers, however, the theory predicts that a paramagnetic octahedral complex of Fe^{2+} should have a magnetism equal to that of four unpaired electrons. This is what you find for the $[Fe(H_2O)_6]^{2+}$ ion.

You would expect low-spin diamagnetic Fe^{2+} complexes to occur for ligands that bond strongly to the metal ion—that is, for those giving large Δ. Ligands that might give a low-spin complex are suggested by a look at the spectrochemical series. The **spectrochemical series** is *an arrangement of ligands according to the relative magnitudes of the crystal field splittings they induce in the d orbitals of a metal ion.* The following is a short version of the spectrochemical series:

weak-binding ligands   strong-bonding ligands

$$I^- < Br^- < Cl^- < F^- < OH^- < H_2O < NH_3 < en < NO_2^- < CN^- < CO$$
increasing $\Delta \longrightarrow$

From this series, you see that the CN^- ion bonds more strongly than H_2O, which explains why $[Fe(CN)_6]^{4-}$ is a low-spin complex ion and $[Fe(H_2O)_6]^{2+}$ is a high-spin complex ion. You can also see why carbon monoxide might be expected to be poisonous. You know that O_2 bonds reversibly to the Fe(II) atom of hemoglobin, so the bonding is only moderately strong. According to the spectrochemical series, however, carbon monoxide, CO, forms a strong bond. The bonding in this case is irreversible (or practically so). It results in a very stable complex of CO and hemoglobin, which cannot function then as a transporter of O_2.

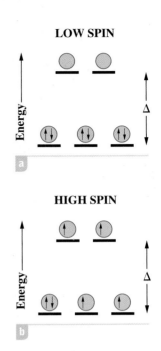

Figure 22.25 ▲

Occupation of the 3d orbitals in complexes of Fe²⁺ (*a*) Low spin. (*b*) High spin.

Example 22.5   **Describing the Bonding in an Octahedral Complex Ion (Crystal Field Theory)**

Gaining Mastery Toolbox

Critical Concept 22.5
In an octahedral complex, the *d*-orbital energy levels of a metal ion are split. This split results in three of the five *d* orbitals being lower energy relative to the other two *d* orbitals. If the energy difference between the lower- and higher-energy group of orbitals is great enough, electrons will pair up in the lower-energy orbitals prior to entering the higher-energy orbitals.

Solution Essentials:
- Pairing energy
- Low-spin complex
- High-spin complex
- Crystal field splitting
- Octahedral field
- Octahedral complex
- *d* orbitals

Describe the distribution of d electrons in the complex ion $[Co(H_2O)_6]^{2+}$, using crystal field theory. The hexaaquacobalt(II) ion is a high-spin complex ion. What would be the distribution of d electrons in an octahedral cobalt(II) complex ion that is low spin? How many unpaired electrons are there in each ion?

Problem Strategy First you need to determine the number of d electrons in the Co^{2+}. Because you are dealing with an octahedral complex, write down the energy levels of

(*continued*)

(continued)

the *d* orbitals in an octahedral field. Then fill these orbitals with the *d* electrons in high-spin (maximum number of unpaired electrons) and low-spin (minimum number of unpaired electrons) configurations, counting the number of unpaired electrons in each case.

Solution The electron configuration of Co^{2+} is $[Ar]3d^7$. The high-spin and low-spin distributions in the *d* orbitals are

High-spin   Low-spin

$[Co(H_2O)_6]^{2+}$, a high-spin complex, has three unpaired electrons. A low-spin complex would have one unpaired electron.

Answer Check Try to avoid the common mistake of assigning an incorrect charge on the metal ion and hence determining the wrong electron configuration.

Exercise 22.5 Describe the distribution of *d* electrons in $[Ni(H_2O)_6]^{2+}$, using crystal field theory. How many unpaired electrons are there in this ion?

■ See Problems 22.57 and 22.58.

Tetrahedral and Square Planar Complexes

When a metal ion bonds with tetrahedrally arranged ligands, the *d* orbitals of the ion split to give two *d* orbitals at lower energy and three *d* orbitals at higher energy (just the opposite of what is found for an octahedral field). (See Figure 22.26*a*.) In the field of ligands in a square planar arrangement, the *d* orbitals split as shown in Figure 22.26*b*.

The observed splittings, Δ, in a tetrahedral field are approximately one-half the size of those in comparable octahedral complexes. Only high-spin complexes are observed, because the pairing energy is always greater than Δ. In the square planar case, only low-spin complexes have been found.

Example 22.6   Describing the Bonding in a Four-Coordinate Complex Ion (Crystal Field Theory)

Gaining Mastery Toolbox

Critical Concept 22.6
In tetrahedral and square planar complexes, the *d*-orbital energy levels of a metal ion are split. For tetrahedral complexes, this split results in two of the five *d* orbitals being lower energy relative to the other three *d* orbitals. For square planar complexes, the splitting is more complex. The energy difference of the splitting in a tetrahedral complex is always less than the pairing energy so that all tetrahedral complexes will have the electrons occupy each orbital prior to pairing up. Square planar complexes always have the electrons pair up prior to occupying a higher-energy orbital.

Solution Essentials:
• Pairing energy
• Low-spin complex
• High-spin complex
• Crystal field splitting
• Tetrahedral complex
• Square planar complex
• *d* orbitals

Describe the *d*-electron distributions of the complexes $[Ni(NH_3)_4]^{2+}$ and $[Ni(CN)_4]^{2-}$, according to crystal field theory. The tetraamminenickel(II) ion is paramagnetic, and the tetracyanonickelate(II) ion is diamagnetic.

Problem Strategy You need to start by recognizing that the complex ions given in the problem are either tetrahedral or square planar. Determine the number of *d* electrons in each complex ion. Consulting

Figure 22.26, write down the energy splittings of the *d* orbitals for tetrahedral and square planar complexes. Then place the electrons in each of your energy splitting diagrams and see which one is consistent with the electron information (paramagnetic or diamagnetic) given in the problem.

Solution You expect the tetrahedral field to give high-spin complexes and the square planar field to give low-spin complexes. Therefore, the geometry of the $[Ni(NH_3)_4]^{2+}$ ion, which is paramagnetic, is probably tetrahedral. The distribution of *d* electrons in the Ni^{2+} ion (configuration d^8) is

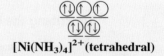

$[Ni(NH_3)_4]^{2+}$ **(tetrahedral)**

The geometry of $[Ni(CN)_4]^{2-}$, which is diamagnetic, is probably square planar; the distribution of *d* electrons is

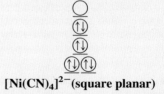

$[Ni(CN)_4]^{2-}$ **(square planar)**

(continued)

(*continued*)

Answer Check Always check the formula of the complex ion or coordination complex to make certain that you are working with the correct *d* orbital energy diagram.

Exercise 22.6 Describe the distribution of *d* electrons in the [CoCl$_4$]$^{2-}$ ion. The ion has a tetrahedral geometry. Assume a high-spin complex.

■ See Problems 22.59 and 22.60.

Visible Spectra of Transition-Metal Complexes

Frequently, substances absorb light only in regions outside the visible spectrum and reflect, or pass on (transmit), all of the visible wavelengths. As a result, these substances appear white or colorless. (White light is a mixture of all visible wavelengths.) However, some substances absorb certain wavelengths in the visible spectrum and transmit the remaining ones; they appear colored. Many transition-metal complexes, as we have noted, are colored substances. The color results from electron jumps, or transitions, between the two closely spaced *d* orbital energy levels that come from the crystal field splitting. ▶

The spectrum of a d^1 configuration complex is particularly simple. Hexaaquatitanium(III) ion, [Ti(H$_2$O)$_6$]$^{3+}$, is an example. Titanium has the configuration [Ar]3$d^2$4s^2, and Ti^{3+} has the configuration [Ar]3d^1. According to crystal field theory, the *d* electron occupies one of the lower-energy *d* orbitals of the octahedral complex. Figure 22.27 shows the visible spectrum of [Ti(H$_2$O)$_6$]$^{3+}$. It results from a transition, or jump, of the *d* electron from a lower-energy *d* orbital to a higher-energy *d* orbital,

The colors of many gemstones are due to transition-metal-ion impurities in the mineral. Ruby has Cr^{3+} in alumina, Al$_2$O$_3$, and emerald has Cr^{3+} in beryl, Be$_3$Al$_2$(SiO$_3$)$_6$.

Figure 22.26 ▼

Energy splittings of the *d* orbitals in a complex with four ligands
(a) A tetrahedral field. The crystal field splitting, Δ, is smaller than in a comparable octahedral complex.
(b) A square planar field.

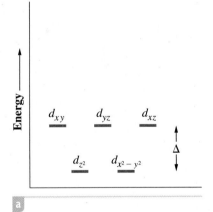

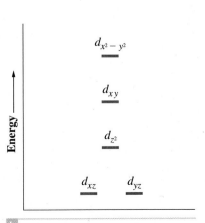

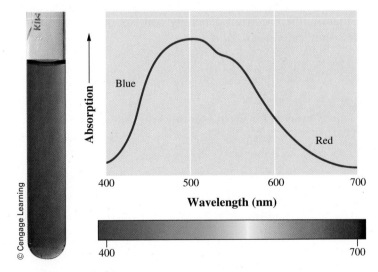

Figure 22.27 ▲

Color and visible spectrum of [Ti(H$_2$O)$_6$]$^{3+}$ *Left:* A test tube containing a solution of [Ti(H$_2$O)$_6$]$^{3+}$. *Right:* Visible spectrum of [Ti(H$_2$O)$_6$]$^{3+}$. Unlike atomic spectra, which show absorption lines, spectra of ions and molecules in solution yield broad bands resulting from changes in nuclear motion that accompany the electronic transitions.

Figure 22.28 ▶

The electronic transition responsible for the visible absorption in [Ti(H₂O)₆]³⁺
An electron undergoes a transition from a lower-energy d orbital to a higher-energy d orbital. The energy change equals the crystal field splitting, Δ.

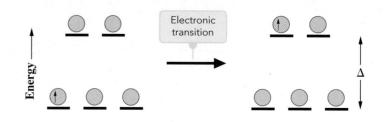

Recall from Section 7.3 that the energy change during a transition equals hν.

as shown in Figure 22.28. Note that the energy change equals the crystal field splitting, Δ. Consequently, the wavelength, λ, of light absorbed is related to Δ. ◀

$$\Delta = h\nu = \frac{hc}{\lambda} \qquad \text{or} \qquad \lambda = \frac{hc}{\Delta}$$

When white light, which contains all visible wavelengths (400 nm to 750 nm), falls on a solution containing $[Ti(H_2O)_6]^{3+}$, blue-green light is absorbed. (The maximum light absorption is observed at 500 nm, which is blue-green light. See Table 22.7.) The other wavelengths of visible light, including red and some blue light, pass through the solution, giving it a red-purple color.

When the ligands in the Ti^{3+} complex are changed, Δ changes and therefore the color of the complex changes. For example, replacing H_2O ligands by weaker F^- ligands should give a smaller crystal field splitting and therefore an absorption at longer wavelengths. The absorption of $[TiF_6]^{3-}$ is at 590 nm, in the yellow rather than the blue-green, and the color observed is violet-blue.

From this discussion, you see that the visible spectrum can be related to the crystal field splitting, and values of Δ can be obtained by spectroscopic analysis. However, when there is more than one d electron, several excited states can be formed. Consequently, the spectrum generally consists of several lines, and the analysis is more complicated than for Ti^{3+}.

Table 22.7	Color Observed for Given Absorption of Light by an Object	
Wavelength Absorbed (nm)	**Color Absorbed**	**Approximate Color Observed***
410	Violet	Green-yellow
430	Violet-blue	Yellow
480	Blue	Orange
500	Blue-green	Red
530	Green	Purple
560	Green-yellow	Violet
580	Yellow	Violet-blue
610	Orange	Blue
680	Red	Blue-green
720	Purple-red	Green

*The exact color depends on the relative intensities of various wavelengths coming from the object and on the response of the eye to those wavelengths.

The Cooperative Release of Oxygen from Oxyhemoglobin

Hemoglobin is an iron-containing substance in red blood cells that is responsible for the transport of O_2 from the lungs to various parts of the body. Myoglobin is a similar substance in muscle tissue, acting as a reservoir for the storage of O_2 and as a transporter of O_2 within muscle cells. The explanation for the different actions of these two substances involves some fascinating transition-metal chemistry.

Myoglobin consists of heme—a complex of Fe(II) bonded to a quadridentate ligand (Figure 22.8)—and globin. Globin, a protein, is attached through a nitrogen atom to one of the octahedral positions of Fe(II). The sixth position is vacant in free myoglobin but is occupied by O_2 in oxymyoglobin. Hemoglobin is essentially a four-unit structure of myoglobinlike units—that is, a *tetramer* of myoglobin (Figure 22.29).

Myoglobin and hemoglobin exist in equilibrium with the oxygenated forms oxymyoglobin and oxyhemoglobin, respectively. For example, hemoglobin (Hb) and O_2 are in equilibrium with oxyhemoglobin (HbO$_2$).

$$\text{Hb} + \text{O}_2 \rightleftharpoons \text{HbO}_2$$

Although hemoglobin is a tetramer of myoglobin, it does not function simply as four independent units of myoglobin. For it to function efficiently as a transporter of O_2 from the lungs and then be able to release that O_2 easily to myoglobin, hemoglobin must be less strongly attached to O_2 in the vicinity of a muscle cell than is myoglobin. In hemoglobin, the release of O_2 from one heme group triggers the release of O_2 from another heme group of the same molecule. In other words, there is a *cooperative release* of O_2 from hemoglobin that makes it possible for it to give up its O_2 to myoglobin.

The mechanism postulated for this cooperative release of O_2 depends on a change of iron(II) from a low-spin to a high-spin form, with a corresponding change in radius of the iron atom. In oxyhemoglobin, iron(II) exists in the low-spin form. When O_2 leaves, the iron atom goes to a high-spin form with two electrons in the higher-energy d orbital. These higher-energy orbitals are somewhat larger than the lower-energy d orbitals.

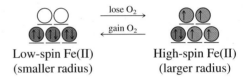

Low-spin Fe(II)　　　High-spin Fe(II)
(smaller radius)　　　(larger radius)

When an O_2 molecule leaves a heme group, the radius of the iron atom increases, and the atom pops out of the heme plane by about 70 pm. In hemoglobin, this change triggers the cooperative release of another O_2 molecule. As the iron atom moves, the attached globin group moves with it. This motion of one globin group causes an adjacent globin group in the tetramer to alter its shape, which in turn makes possible the easy release of an O_2 molecule from its heme unit.

■ See Problems 22.75 and 22.76.

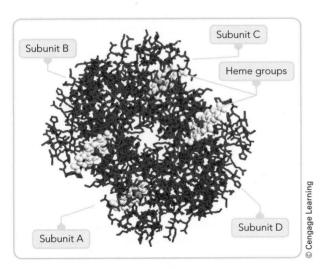

Figure 22.29 ▲

Hemoglobin A molecular model of hemoglobin that depicts each of the ion-containing heme groups contained in the four subunits.

© Cengage Learning

Example **22.7**   **Predicting the Relative Wavelengths of Absorption of Complex Ions**

Critical Concept 22.7
The greater the crystal field splitting in the complex, the higher the energy of light absorbed by the complex. The spectrochemical series can be used to determine the relative amount of ligand field splitting produced by different ligands. Those ligands that bond to the metal most strongly cause the complex to absorb the highest-energy light (shortest wavelength).

Solution Essentials:
• Ligand
• Coordination compound (complex)
• Crystal field splitting
• Spectrochemical series
• Energy of light
• Frequency and wavelength of light

When water ligands in $[Ti(H_2O)_6]^{3+}$ are replaced by CN^- ligands to give $[Ti(CN)_6]^{3-}$, the maximum absorption shifts from 500 nm to 450 nm. Is this shift in the expected direction? Explain. What color do you expect to observe for this ion?

Problem Strategy We are going to need to consult the spectrochemical series in order to do this problem.

Keeping this in mind, those ligands that cause a greater crystal field splitting will absorb higher-energy light.

Solution According to the spectrochemical series, CN^- is a more strongly bonding ligand than H_2O. Consequently, Δ should increase, and the wavelength of the absorption ($\lambda = hc/\Delta$) should decrease. **So the shift of the absorption is in the expected direction.** Because the absorbed light is between blue and violet-blue (see Table 22.7), the observed color is **orange-yellow** (this is the *complementary color* of the color between blue and violet-blue).

Answer Check Check to make certain that you have correctly applied the relationship between the wavelength of light and its energy: the longer the wavelength, the lower the energy.

Exercise 22.7   The $[Fe(H_2O)_6]^{3+}$ ion has a pale purple color, and the $[Fe(CN)_6]^{3-}$ ion has a ruby-red color. What are the approximate wavelengths of the maximum absorption for each ion? Is the shift of wavelength in the expected direction? Explain.

■ See Problems 22.61, 22.62, 22.63, and 22.64.

A Checklist for Review

Summary of Facts and Concepts

The *d-block transition elements* are defined as those elements having a partially filled *d* subshell in any common oxidation state. They have a number of characteristics, including high melting points and a multiplicity of oxidation states. Compounds of transition elements are frequently colored and many are paramagnetic. These properties are due to the participation of *d* orbitals in bonding. We described the chemical properties of two transition elements: Cr and Cu. Chromium metal reacts with acids to give Cr^{2+} ion, which is readily oxidized to Cr^{3+}. The +3 state is the most common oxidation state; chromium(III) oxide is a green pigment. The +6 oxidation state is represented by chromates and dichromates. The dichromate ion, $Cr_2O_7^{2-}$, in acid solution is a strong oxidizing agent. Chromium metal is obtained by reduction of the ore chromite, $FeCr_2O_4$. Copper metal reacts only with acids having strongly oxidizing anions, such as HNO_3; it gives Cu^{2+} ion. Copper(I) oxide, Cu_2O, occurs naturally as a copper mineral. This oxide also forms as a brick-red precipitate when copper(II) ion is reduced in basic solution. The reaction is used as a test for glucose.

Transition-metal atoms often function as Lewis acids, reacting with groups called *ligands* by forming coordinate

covalent bonds to them. The metal atom with its ligands is a *complex ion* or neutral *complex*. Ligands that bond through more than one atom are called *polydentate,* and the complex formed is called a *chelate*. The IUPAC has agreed on a nomenclature of complexes that gives basic structural information about the species. The presence of isomers in coordination compounds is evidence for particular geometries. For example, $[Pt(NH_3)_2Cl_2]$ has two isomers, which is evidence for a square planar geometry having *cis–trans isomers*. Octahedral complexes, often those that have bidentate ligands, may have *optical isomers*—that is, isomers that are mirror images of one another.

Valence bond theory gave the earliest theoretical description of the electronic structure of a complex. According to this theory, a complex forms when doubly occupied ligand orbitals overlap unoccupied orbitals of the metal atom.

Crystal field theory treats the ligands as electric charge points that affect the energy of the *d* orbitals of the metal ion. In an octahedral complex, two of the *d* orbitals have higher energy than the other three. A high-spin complex forms when the *pairing energy* is greater than the *crystal field splitting,* so that electrons would "prefer" to occupy a higher-energy *d* orbital rather than pair up with an electron in a lower-energy orbital. When the pairing energy is smaller than the crystal field splitting, the *d* orbitals are occupied in the normal fashion, giving a low-spin complex. Color in transition-metal complexes is explained as due to a transition of an electron from the lower-energy to the higher-energy *d* orbitals. The crystal field splitting can be obtained experimentally from the visible spectrum of a complex.

Learning Objectives	Important Terms
22.1 Periodic Trends in the Transition Elements ■ Identify the *transition elements* on the periodic table. ■ State the three characteristics that set the transition elements apart from the main-group elements. ■ Write electron configurations of the transition elements. ■ Describe trends in atomic radii of the transition elements. ■ Learn the common oxidation states of the fourth-period transition elements.	**transition elements**
22.2 The Chemistry of Two Transition Elements ■ Learn some of the common chromium compounds and their chemistry. ■ Learn some of the common copper compounds and their chemistry.	
22.3 Formation and Structure of Complexes ■ Define *complex ion, complex (coordination compound), ligand,* and *coordination number.* ■ Give examples of a *monodentate ligand, bidentate ligand,* and *polydentate ligand.*	**complex ion** **complex (coordination compound)** **ligands** **coordination number** **monodentate ligand** **bidentate ligand** **polydentate ligand** **chelate**
22.4 Naming Coordination Compounds ■ Learn the four rules for naming coordination compounds. ■ Write the IUPAC name given the structural formula of a coordination compound. Example 22.1 ■ Write the structural formula given the IUPAC name of a coordination compound. Example 22.2	
22.5 Structure and Isomerism in Coordination Compounds ■ Define *constitutional isomers, stereoisomers,* and *geometric isomers.* ■ Decide whether geometric isomers are possible. Example 22.3 ■ Define *enantiomers (optical isomers).* ■ Explain how structures are used to determine if an isomer is *chiral.* ■ Describe how enantiomers are *optically active.* ■ Define *dextrorotatory, levorotatory,* and *racemic mixture.* ■ Decide whether optical isomers are possible. Example 22.4	**constitutional isomers** **stereoisomers** **geometric isomers** **enantiomers (optical isomers)** **chiral** **optically active** **dextrorotatory** **levorotatory** **racemic mixture**
22.6 Valence Bond Theory of Complexes ■ Write the orbital diagram of a transition metal ion in a complex.	

22.7 Crystal Field Theory

- Define *crystal field splitting, high-spin complex, low-spin complex,* and *pairing energy.*
- Describe the bonding in an octahedral complex ion (crystal field theory). Example 22.5
- Describe the bonding in a four-coordinate complex ion (crystal field theory). Example 22.6
- Predict the relative wavelengths of absorption of complex ions. Example 22.7

crystal field theory
crystal field splitting, Δ
high-spin complex
low-spin complex
pairing energy, *P*
spectrochemical series

Questions and Problems

OWL Interactive versions of these problems may be assigned in OWL.

Self-Assessment and Review Questions

Key: These questions test your understanding of the ideas you worked with in the chapter. These problems vary in difficulty and often can be used for the basis of discussion.

22.1 What characteristics of the transition elements set them apart from the main-group elements?

22.2 According to the building-up principle, what is the electron configuration of the ground state of the technetium atom (atomic number 43)?

22.3 The highest melting point for metals in the fifth period occurs for molybdenum. Explain why this is expected.

22.4 Iron, cobalt, and nickel are similar in properties and are sometimes studied together as the "iron triad." For example, each is a fairly active metal that reacts with acids to give hydrogen and the +2 ions. In addition to the +2 ions, the +3 ions of the metals also figure prominently in the chemistries of the elements. Explain why these elements are similar.

22.5 Write balanced equations for the reactions of Cr and Cu metals with HCl(*aq*). If no reaction occurs, write *NR*.

22.6 Palladium and platinum are very similar to one another. Both are unreactive toward most acids. However, nickel, which is in the same column of the periodic table, is an active metal. Explain why this difference exists.

22.7 Write balanced equations for the reactions of chromium(III) oxide with a strong acid.

22.8 Copper(II) ion in basic solution is reduced by formaldehyde, HCHO, to copper(I) oxide. Formaldehyde is oxidized to formate ion, $HCOO^-$. Write the balanced equation for this reaction.

22.9 Describe the structure of copper(II) sulfate pentahydrate. What color change occurs when the salt is heated? What causes the color change?

22.10 Define the terms *complex ion, ligand,* and *coordination number.* Use an example to illustrate the use of these terms.

22.11 What evidence did Werner obtain to show that the platinum complex $PtCl_4 \cdot 4NH_3$ has the structural formula $[Pt(NH_3)_4Cl_2]Cl_2$?

22.12 Define the term *bidentate ligand.* Give two examples.

22.13 Rust spots on clothes can be removed by dissolving the rust in oxalic acid. The oxalate ion forms a stable complex with Fe^{3+}. Using an electron-dot formula, indicate how an oxalate ion bonds to the metal ion.

22.14 What three properties of coordination compounds have been important in determining the details of their structure and bonding?

22.15 Define each of the following and give an example of each: a linkage isomerism, b coordination isomerism, c hydrate isomerism, d ionization isomerism.

22.16 Define the terms *geometric isomerism* and *optical isomerism* and give an example of each.

22.17 Explain the difference in behavior of *d* and *l* isomers with respect to polarized light.

22.18 What is a racemic mixture? Describe one method of resolving a racemic mixture.

22.19 Describe the formation of a coordinate covalent bond between a metal-ion orbital and a ligand orbital.

22.20 a Describe the steps in the formation of a high-spin octahedral complex of Fe^{2+} in valence bond terms. b Do the same for a low-spin complex.

22.21 Explain why *d* orbitals of a transition-metal atom may have different energies in the octahedral field of six negative charges. Describe how each of the *d* orbitals is affected by the octahedral field.

22.22 a Use crystal field theory to describe a high-spin octahedral complex of Fe^{2+}. b Do the same for a low-spin complex.

22.23 What is meant by the term *crystal field splitting*? How is it determined experimentally?

22.24 What is the spectrochemical series? Use the ligands CN^-, H_2O, Cl^-, and NH_3 to illustrate the term. Then arrange them in order, describing the meaning of this order.

22.25 What is meant by the term *pairing energy*? How do the relative values of pairing energy and crystal field splitting determine whether a complex is low-spin or high-spin?

22.26 A complex absorbs red light from a single electron transition. What color is this complex?

22.27 Which of the following is most likely a tetrahedral complex?
- a $[Zn(NH_3)_2Cl_2]$
- b $[Co(NH_3)_3Cl_3]$
- c $[Co(NH_3)_6]^{2+}$
- d $[Ni(H_2O)_6]SO_4$
- e $[Co(en)_2(H_2O)Cl]^{2+}$

22.28 What is the coordination number of cobalt in the complex $[Co(en)(NH_3)_2Cl_2]ClO_4$?
- a 2
- b 3
- c 4
- d 5
- e 6

22.29 What is the correct name for the coordination compound $[Cr(en)_2(CN)_2]Cl$?

  a bis(ethylenediamine)dicyanochromium(III) dichloride
  b dicyanobis(ethylenediamine)chromium(III) dichloride
  c bis(ethylenediamine)dicyanochromium(III) chloride
  d chromium(III) (diethylenediamine)biscyano chloride
  e dicyanobis(ethylenediamine)chromium(III) chloride

22.30 What is the number of unpaired electrons in the low-spin complex $[Co(NH_3)_6]Cl_3$

  a 0   b 1   c 2   d 3   e 4

Conceptual Problems

Key: These problems are designed to check your understanding of the concepts associated with some of the main topics presented in each chapter. A strong conceptual understanding of chemistry is the foundation for both applying chemical knowledge and solving chemical problems. These problems vary in level of difficulty and often can be used as a basis for group discussion.

22.31 A cobalt complex whose composition corresponded to the formula $Co(NO_2)_2Cl\cdot4NH_3$ gave an electrical conductance equivalent to two ions per formula unit. Excess silver nitrate solution immediately precipitates 1 mol AgCl per formula unit. Write a structural formula consistent with these results.

22.32 For the following coordination compounds, identify the geometric isomer(s) of compound X.

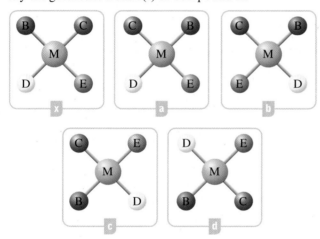

22.33 Describe step by step how the name potassium hexacyanoferrate(II) leads to the structural formula $K_4[Fe(CN)_6]$.

22.34 A complex has a composition corresponding to the formula $CoBr_2Cl\cdot4NH_3$. What is the structural formula if conductance measurements show two ions per formula unit? Silver nitrate solution gives an immediate precipitate of AgCl but no AgBr. Write the structural formula of an isomer.

22.35 Compounds A and B are known to be stereoisomers of one another. Compound A has a violet color; compound B has a green color. Are they geometric or optical isomers?

22.36 For the complexes shown here, which would have the d electron distribution shown in the diagram below: $[MF_6]^{3-}$, $[M(CN)_6]^{3-}$, $[MF_6]^{4-}$, $[M(CN)_6]^{4-}$? Note that the neutral metal atom, M, in each complex is the same and has the ground state electron configuration $[Ar]4s^23d^6$.

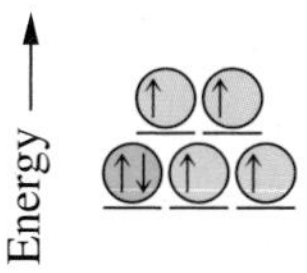

Practice Problems

Key: These problems are for practice in applying problem-solving skills. They are divided by topic, and some are keyed to exercises (see the ends of the exercises). The problems are arranged in matching pairs; the odd-numbered problem of each pair is listed first, and its answer is given in the back of the book.

Properties of the Transition Elements

22.37 Find the oxidation numbers of the transition metal in each of the following compounds:
  a $FeCO_3$ b MnO_2 c $CuCl_2$ d CrO_2Cl_2

22.38 Find the oxidation numbers of the transition metal in each of the following compounds:
  a Ta_2O_5 b $Cu_2(OH)_3Cl$ c $CoSO_4$

22.39 Write the balanced equation for the reaction of iron(II) ion with nitrate ion in acidic solution. Nitrate ion is reduced to NO.

22.40 Write the balanced equation for the reaction of sulfurous acid with dichromate ion.

Structural Formulas and Naming of Complexes

22.41 Give the coordination number of the transition-metal atom in each of the following complexes.
  a $[Cr(en)_2(C_2O_4)]^+$
  b $[Au(en)_2]Cl_3$
  c $[Co(NH_3)_4(H_2O)_2]Cl_3$
  d $[Au(CN)_4]^-$

22.42 Give the coordination number of the transition element in each of the following complexes.

a. $[Ni(NH_3)_6](ClO_3)_2$ b. $[Cu(NH_3)_4]SO_4$

c. $[Cr(en)_3]Cl_3$ d. $K_2[Ni(CN)_4]$

22.43 For each of the following complexes, determine the oxidation state of the transition-metal atom.

a. $[CoCl(en)_2(NO_2)]NO_2$ b. $[PtCl_4]^{2-}$

c. $K_3[Cr(CN)_6]$ d. $[Fe(H_2O)_5(OH)]^{2+}$

22.44 Determine the oxidation number of the transition element in each of the following complexes.

a. $K_2[Ni(CN)_4]$ b. $[Mo(en)_3]^{3+}$

c. $[Cr(C_2O_4)_3]^{3-}$ d. $[Co(NH_3)_5(NO_2)]Cl_2$

22.45 Consider the complex ion $[Mn(NH_3)_2(H_2O)_3(OH)]^{2+}$.

a. What is the oxidation state of the metal atom?

b. Give the formula and name of each ligand in the ion.

c. What is the coordination number of the metal atom?

d. What would be the charge on the complex if all ligands were chloride ions?

22.46 Consider the complex ion $[Cr(NH_3)_2Cl_2(C_2O_4)]^-$.

a. What is the oxidation state of the metal atom?

b. Give the formula and name of each ligand in the ion.

c. What is the coordination number of the metal atom?

d. What would be the charge on the complex if all ligands were chloride ions?

22.47 Write the IUPAC name for each of the following coordination compounds.

a. $K_3[FeF_6]$ b. $[Cu(NH_3)_2(H_2O)_2]^{2+}$

c. $(NH_4)_2[Fe(H_2O)F_5]$ d. $[Ag(CN)_2]^-$

22.48 Name the following complexes, using IUPAC rules.

a. $[CrF_6]^{3-}$ b. $K_2[FeCl_4]$

c. $K_4[Mo(CN)_8]$ d. $[V(C_2O_4)_3]^{2-}$

22.49 Give the IUPAC name for each of the following.

a. $[W(CO)_8]$

b. $[Co(H_2O)_2(en)_2](SO_4)_3$

c. $K[Mo(CN)_8]$

d. $[CrO_4]^{2-}$

22.50 Give the IUPAC name for each of the following.

a. $[Fe(CO)_5]$ b. $[Rh(CN)_2(en)_2]^+$

c. $[Cr(NH_3)_4SO_4]Cl$ d. $[MnO_4]^-$

22.51 Write the structural formula for each of the following compounds.

a. potassium hexacyanomanganate(III)

b. sodium tetracyanozincate(II)

c. tetraamminedichlorocobalt(III) nitrate

d. hexaamminechromium(III) tetrachlorocuprate(II)

22.52 Give the structural formula for each of the following complexes.

a. diaquadicyanocopper(II)

b. potassium hexachloroplatinate(IV)

c. tetraamminenickel(II) perchlorate

d. tetraammineplatinum(II) tetrachlorocuprate(II)

Isomerism

22.53 Draw *cis–trans* structures of any of the following square planar or octahedral complexes that exhibit geometric isomerism. Label the drawings *cis* or *trans*.

a. $[Pd(NH_3)_2Cl_2]$

b. $[Pd(NH_3)_3Cl]^+$

c. $[Pd(NH_3)_4]^{2+}$

d. $[Ru(NH_3)_4Br_2]^+$

22.54 If any of the following octahedral complexes display geometric isomerism, draw the structures and label them *cis* or *trans*.

a. $[Co(NO_2)_4(NH_3)_2]^-$

b. $[Co(NH_3)_5(NO_2)]^{2+}$

c. $[Co(NH_3)_6]^{3+}$

d. $[Cr(NH_3)_5Cl]^{2+}$

22.55 Sketch mirror images of each of the following. From these sketches, determine whether optical isomers exist and note this fact on the drawings.

a. $[Rh(en)_3]^{3+}$

b. *cis*-$[Cr(NH_3)_2(SCN)_4]^-$

22.56 Determine whether there are optical isomers of any of the following. If so, sketch the isomers.

a. *cis*-$[Co(NH_3)_2(en)_2]^{3+}$

b. *trans*-$[IrCl_2(C_2O_4)_2]^{3-}$

Crystal Field Theory

22.57 Using crystal field theory, sketch the energy-level diagram for the *d* orbitals in an octahedral field; then fill in the electrons for the metal ion in each of the following complexes. How many unpaired electrons are there in each case?

a. $[ZrCl_6]^{4-}$

b. $[OsCl_6]^{2-}$ (low-spin)

c. $[MnCl_6]^{4-}$ (high-spin)

22.58 Using crystal field theory, sketch the energy-level diagram for the *d* orbitals in an octahedral field; then fill in the electrons for the metal ion in each of the following complexes. How many unpaired electrons are there in each case?

a. $[V(CN)_6]^{3-}$

b. $[Co(C_2O_4)_3]^{4-}$ (high-spin)

c. $[Mn(CN)_6]^{3-}$ (low-spin)

22.59 Obtain the distribution of *d* electrons in the complex ions listed below, using crystal field theory. Each ion is either tetrahedral or square planar. On the basis of the number of unpaired electrons (given in parentheses), decide the correct geometry.

a. $[Co(NCS)_4]^{2-}$ (3)

b. $[Pt(NH_3)_4]^{2+}$ (0)

c. $[Co(en)_2]^{2+}$ (1)

d. $[FeCl_4]^-$ (5)

22.60 Obtain the distribution of *d* electrons in the complex ions listed below, using crystal field theory. Each ion is either tetrahedral or square planar. On the basis of the number of unpaired electrons (given in parentheses), decide the correct geometry.

- a $[MnCl_4]^{2-}$ (5)
- b $[AuF_4]^-$ (0)
- c $[Pt(NH_3)_2(NO_2)_2]^{2+}$ (2)
- d $[NiCl_4]^{2-}$ (2)

Color

22.61 The $[Co(SCN)_4]^{2-}$ ion has a maximum absorption at 530 nm. What color do you expect for this ion?

22.62 The $[Co(en)_3]^{3+}$ ion has a maximum absorption at 470 nm. What color do you expect for this ion?

22.63 The $[Co(NH_3)_6]^{3+}$ ion has a yellow color, but when one NH_3 ligand is replaced by H_2O to give $[Co(NH_3)_5(H_2O)]^{3+}$, the color shifts to red. Is this shift in the expected direction? Explain.

22.64 The $[Co(en)_3]^{3+}$ ion has a yellow color, but the $[CoF_6]^{3-}$ ion has a blue color. Is the shift from yellow to blue expected when ethylenediamine ligands are replaced by F^- ligands? Explain.

22.65 What is the value of Δ (in kJ/mol) when $\lambda = 680$ nm, corresponding to an electron jump between d-orbital levels in a complex with a d^1 configuration?

22.66 What is the value of Δ (in kJ/mol) when $\lambda = 500$ nm, corresponding to an electron jump between d-orbital levels in a complex with a d^1 configuration?

General Problems

Key: These problems provide more practice but are not divided by topic or keyed to exercises. Each section ends with essay questions, each of which is color coded to refer to the *A Chemist Looks at Daily Life* (orange) or *A Chemist Looks at Life Science* (pink) chapter essay on which it is based. Odd-numbered problems and the even-numbered problems that follow are similar; answers to all odd-numbered problems except the essay questions are given in the back of the book.

22.67 The tetraaquazinc(II) ion, $[Zn(H_2O)_4]^{2+}$, is colorless. Explain why this might be expected.

22.68 The hexaaquascandium(III) ion, $[Sc(H_2O)_6]^{3+}$, is colorless. Explain why this might be expected.

22.69 There are only two geometric isomers of the tetraamminedichlorocobalt(III) ion, $[Co(NH_3)_4Cl_2]^+$. How many geometric isomers would be expected for this ion if it had a regular planar hexagonal geometry? Give drawings for them. Does this rule out a planar hexagonal geometry for $[Co(NH_3)_4Cl_2]^+$? Explain.

22.70 There are only two geometric isomers of the triamminetrichloroplatinum(IV) ion, $[Pt(NH_3)_2Cl_3]^+$. How many geometric isomers would be expected for this ion if it had a regular planar hexagonal geometry? Give drawings for

them. Does this rule out a planar hexagonal geometry for $[Pt(NH_3)_2Cl_4]^+$? Explain.

22.71 Find the concentrations of $Ag^+(aq)$, $NH_3(aq)$, and $[Ag(NH_3)_2]^+(aq)$ at equilibrium when 0.10 mol $Ag^+(aq)$ and 0.10 mol $NH_3(aq)$ are made up to 1.00 L of solution. The dissociation constant, K_d, for the complex $[Ag(NH_3)_2]^+$ is 5.9×10^{-8}.

22.72 Find the concentrations of $Cu^{2+}(aq)$, $NH_3(aq)$, and $[Cu(NH_3)_4]^{2+}(aq)$ at equilibrium when 0.10 mol $Cu^{2+}(aq)$ and 0.40 mol $NH_3(aq)$ are made up to 1.00 L of solution. The dissociation constant, K_d, for the complex $[Cu(NH_3)_4]^{2+}$ is 2.1×10^{-13}.

■ **22.73** Why is EDTA added to commercial mayonnaise and salad dressings? Explain what is happening.

■ **22.74** What accounts for the special stability of chelates? Explain this in terms of an example.

■ **22.75** Describe the functions of hemoglobin and myoglobin in the body. What is similar and what is different about the two functions?

■ **22.76** Describe the mechanism of the cooperative release of oxygen from oxyhemoglobin.

Strategy Problems

Key: As noted earlier, all of the practice and general problems are matched pairs. This section is a selection of problems that are not in matched-pair format. These challenging problems require that you employ many of the concepts and strategies that were developed in the chapter. In some cases, you will have to integrate several concepts and operational skills in order to solve the problem successfully.

22.77 Consider the complex ion $[CoCO_3(NH_3)_4]^+$, where the CO_3^{2-} is a bidentate ligand.

- a Is this complex ion octahedral or square planar?
- b What is the oxidation state of the cobalt?

22.78 Does the complex ion $[Co(en)_2Cl_2]^+$ have *cis–trans* geometric isomers?

22.79 Draw each of the following to determine if they have optical isomers.

- a $Ir(en)_3^{3+}$
- b $[Ir(H_2O)_3Cl_3]$

22.80 Is it possible for a square planar complex to have optical isomers?

22.81 Consider the complex ion $[CoF_6]^{3-}$.

- a What is the geometry?
- b Which is a more likely color for this ion to absorb, red or blue?
- c Would you expect this complex to be high or low spin?

22.82 The complex $[Fe(en)_2Cl_2]Cl$ is low spin.

- a What is the geometry of the complex ion?
- b What is the oxidation state of the Fe?
- c Are there geometric isomers, and is the compound optically active?

22.83 What is the name of $K_2[MoOCl]_4$?

22.84 Write the formula and draw the structure of *cis*-tetraamminedichlorocobalt(III).

22.85 You are studying an octahedral transition metal complex that contains the Co^{2+} ion and discover that it has a strong absorption in the blue region of the visible spectrum. Would you suspect (not conclude) that this complex is high spin or low spin. Explain?

22.86 Consider the low-spin complex ions $[Cr(H_2O)_6]^{3+}$ and $[Mn(CN)_6]^{4-}$.
a Name them.
b Determine the number of unpaired electrons.
c Indicate which complex ion would absorb the highest-frequency light.

22.87 Is it possible to have a paramagnetic d^4-tetrahedral complex ion? Explain.

22.88 Consider the tetracarbonyldibromoiron(II) ion.
a Write the chemical formula of the compound.
b How many geometric isomers are possible for this compound?
c Does the compound have optical isomers?

22.89 Is the complex $[Mn(NH_3)_6]^{2+}$ a low-spin or high-spin complex if it contains five unpaired electrons?

22.90 For the low-spin complex $[Ru(NH_3)_5H_2O]^{2+}$, draw the crystal-field energy diagram. How many unpaired electrons are in the complex ion?

22.91 $[Co(NH_3)_6]^{3+}$ is diamagnetic, and $[Fe(H_2O)_6]^{2+}$ is paramagnetic.
a Name these complex ions.
b What is the geometry of these complex ions?
c Which of these ions has the greater crystal field splitting energy?

22.92 Consult Figure 22.27, which depicts the absorption versus wavelength curve for the $[Ti(H_2O)_6]^{3+}$ ion.
a What is the absorbance maximum (wavelength) of this ion?
b If you were to make the $[Ti(CN)_6]^{3-}$ ion, how would the absorption versus wavelength curve compare to that of the $[Ti(H_2O)_6]^{3+}$ ion?
c How would the energy of the crystal field splitting of the $[Ti(H_2O)_6]^{3+}$ and $[Ti(CN)_6]^{3-}$ compare?

23

Organic Chemistry

© Cengage Learning: Steve Cukrov/Shutterstock.com

Heating fibers made from organic compounds such as acetylnitrile, CH_3CN, yield strong, lightweight carbon fibers, used in composite materials to manufacture the structural components of such things as sailboats, bicycles, and airplanes, as well as golf clubs. The carbon in these fibers has a sheetlike structure similar to graphite, a carbon allotrope.

Contents and Concepts

Organic chemistry is the chemistry of compounds containing carbon. As discussed in Chapter 2, carbon-containing compounds make up the majority of known compounds. Organic chemistry plays a central role in most of the substances that you encounter and use every day: the food you ate this morning, the shampoo that cleans your hair, the fuel used to heat your house and generate electricity, and the list goes on. Also, your life and that of every living organism on earth depends on the chemical reactions of organic molecules.

In this chapter, we introduce this fascinating area of chemistry with a discussion of the structural features of organic molecules, nomenclature, and a few important chemical reactions.

23.1 The Bonding of Carbon

Because carbon is in Group IVA of the periodic table, it has four valence electrons. To fill its octet, it requires four additional electrons, which can be obtained through the formation of four covalent bonds. Carbon forms single, double, and triple bonds to achieve a filled octet. Therefore, the possible bonding combinations for carbon are as follows.

$$-\overset{|}{\underset{|}{C}}- \qquad -\overset{\|}{C}- \qquad =C= \qquad \equiv C-$$

Recall from the VSEPR model (Chapter 10) that the molecular geometry around an atom is dictated by the number of regions of electron density. As we have seen, double and triple bonds count as one area of electron density; therefore, carbon can have a tetrahedral, trigonal planar, or linear geometry.

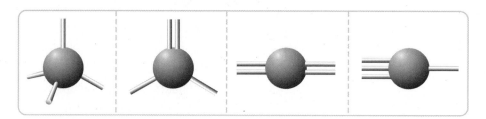

A unique feature of carbon is its ability to bond with other carbon atoms to form chains and rings of various lengths. Several other elements have limited ability to form such chains or rings of like atoms, but only carbon does this with more than a few atoms. Some petroleum-containing products, such as asphalt, consist of molecules with 30 or more carbon atoms bonded together; the molecules that make up polyethylene can have chains with thousands of carbon atoms (Figure 23.1).

Figure 23.1 ▶

Petroleum-based products contain long chains of carbon atoms

a The black, tarry substance in asphalt consists of molecules with 30 or more carbon atoms bonded together.

b Polyethylene contains chains of many thousands of carbon atoms bonded together. Polyethylene is used to create a variety of familiar materials.

© magicinfoto/Shutterstock.com

© Picsfive/Shutterstock.com

Hydrocarbons

The simplest organic compounds are **hydrocarbons**, *compounds containing only carbon and hydrogen.* All other organic compounds—for example, those containing O, N, and the halogen atoms—are classified as being derived from hydrocarbons. At first glance, you might think that the hydrocarbons represent a very limited set of molecules; however, several hundred thousand molecules exist that contain only hydrogen and carbon atoms.

Hydrocarbons can be separated into three main groups:

1. **Saturated hydrocarbons** are *hydrocarbons that contain only single bonds between the carbon atoms.* Saturated hydrocarbon molecules can be *cyclic* or *acyclic.* A cyclic hydrocarbon is one in which a chain of carbon atoms has formed a ring. An acyclic hydrocarbon is one that does not contain a ring of carbon atoms.

2. **Unsaturated hydrocarbons** are *hydrocarbons that contain double or triple bonds between carbon atoms.*

3. **Aromatic hydrocarbons** are *hydrocarbons that contain benzene rings or similar features.*

The saturated and unsaturated hydrocarbons are often referred to as the *aliphatic* hydrocarbons (Figure 23.2).

23.2 Alkanes and Cycloalkanes

The **alkanes** are *acyclic saturated hydrocarbons,* and the **cycloalkanes** are *cyclic saturated* hydrocarbons. The simplest hydrocarbon, an alkane called methane, consists of one carbon atom to which four hydrogen atoms are bonded in a tetrahedral arrangement. You can represent methane by its *molecular formula,* CH_4, which gives the number and kind of atoms in the molecule, or by its *structural formula,* which shows how the atoms are bonded to one another

$$CH_4 \qquad\qquad H-\underset{\underset{\textstyle H}{|}}{\overset{\overset{\textstyle H}{|}}{C}}-H$$

Molecular formula   Structural formula

of methane   of methane

Note that the structural formula does not convey information about the three-dimensional arrangement of the atoms. To do this, you would need to draw the three-dimensional formula depicting the *molecular geometry* (see Figure 23.3). Methane is a very important molecule since it is the principal component of natural gas. In 2010, more than 2.4×10^{11} ft^3 of natural gas were consumed in the United States to supply heating, power generation, transportation, and industrial needs.

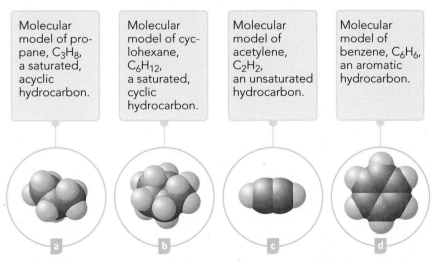

Molecular model of propane, C_3H_8, a saturated, acyclic hydrocarbon.

Molecular model of cyclohexane, C_6H_{12}, a saturated, cyclic hydrocarbon.

Molecular model of acetylene, C_2H_2, an unsaturated hydrocarbon.

Molecular model of benzene, C_6H_6, an aromatic hydrocarbon.

Figure 23.2 ▲

Molecular models for the different hydrocarbons

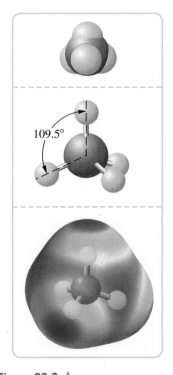

109.5°

Figure 23.3 ▲

Models of methane, CH_4 *Top:* Space-filling model of methane. *Middle:* Ball-and-stick model of methane with bond angle. *Bottom:* Electrostatic potential map.

The term paraffin *comes from the Latin* parum affinus, *meaning "little affinity." The alkanes do not react with many reagents.*

The Alkane Series

The alkanes, also called *paraffins*, ◀ have the general formula C_nH_{2n+2}. For $n = 1$, methane, the formula is CH_4; for $n = 2$, C_2H_6; for $n = 3$, C_3H_8; and so on. Note that the general formula conveys no information about how the atoms are connected. For now, we will assume that the carbon atoms are bonded together in a straight chain with hydrogen atoms completing the four required bonds to each carbon atom; these are called *straight-chain* or *normal* alkanes. The structural formulas for the first four straight chain alkanes are shown.

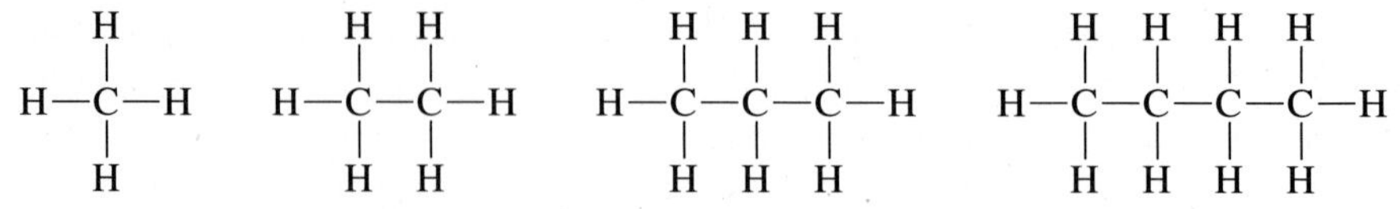

Methane　　　Ethane　　　Propane　　　Butane

Because carbon atoms typically have four bonds, chemists often write the structures of the parts of organic compounds using **condensed structural formulas**, or **condensed formulas**, *where the bonds around each carbon atom in the compound are not explicitly written.* For example,

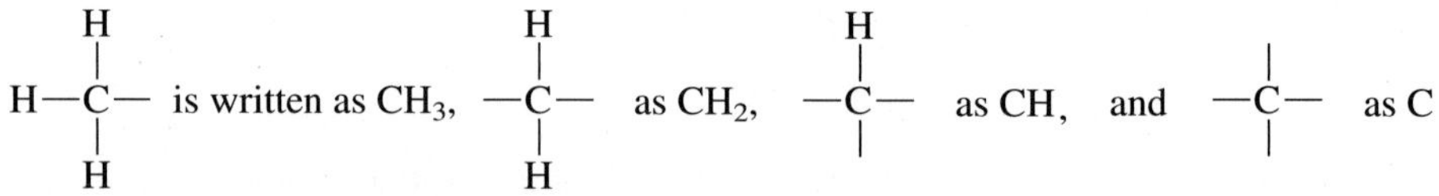

Condensed formulas of the first four alkanes ($n = 1$ to 4) are

$$CH_4 \qquad CH_3CH_3 \qquad CH_3CH_2CH_3 \qquad CH_3CH_2CH_2CH_3$$
Methane　　　Ethane　　　Propane　　　Butane

Note that the condensed formula of an alkane differs from that of the preceding alkane $(n-1)$ by a $—CH_2—$ group. The alkanes constitute a **homologous series**, which is *a series of compounds in which one compound differs from a preceding one by a fixed group of atoms.* Members of a homologous series have similar chemical properties, and their physical properties change throughout the series in a regular way. Table 23.1 lists the melting points and boiling points of the first ten straight-chain alkanes ($n = 1$ to $n = 10$). ◀ Note that the melting points and boiling points generally increase in the series with an increase in the number of carbon atoms in the chain. This is a result of increasing intermolecular forces, which increase with molecular mass.

Formerly, the names of the straight-chain alkanes were distinguished from branched-chain isomers by the prefix n- for normal. This designation is still common (butane is called n-butane) but is not used in the IUPAC name, which we will discuss in Section 23.5.

Table 23.1 Physical Properties of Straight-Chain Alkanes

Name	Number of Carbons	Formula	Melting Point (°C)	Boiling Point (°C)
Methane	1	CH_4	−183	−162
Ethane	2	CH_3CH_3	−172	−89
Propane	3	$CH_3CH_2CH_3$	−187	−42
Butane	4	$CH_3(CH_2)_2CH_3$	−138	0
Pentane	5	$CH_3(CH_2)_3CH_3$	−130	36
Hexane	6	$CH_3(CH_2)_4CH_3$	−95	69
Heptane	7	$CH_3(CH_2)_5CH_3$	−91	98
Octane	8	$CH_3(CH_2)_6CH_3$	−57	126
Nonane	9	$CH_3(CH_2)_7CH_3$	−54	151
Decane	10	$CH_3(CH_2)_8CH_3$	−30	174

Adapted from Robert D. Whitaker et al., *Concepts of General, Organic, and Biological Chemistry,* p. 231. Copyright © 1981 by Houghton Mifflin Company. Used with permission.

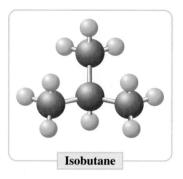 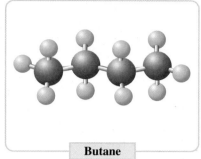

Isobutane   **Butane**

Figure 23.4 ◄

Constitutional isomers of butane
Ball-and-stick models of isobutane
(2-methylpropane) and butane.

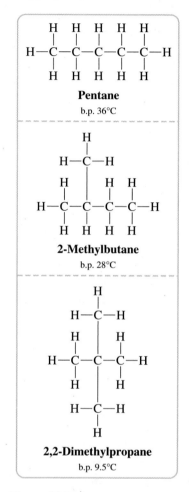

Figure 23.5 ▲

Isomers of pentane Note that
each isomer is a different
compound with a different
boiling point.

Constitutional Isomerism and Branched-Chain Alkanes

In addition to the straight-chain alkanes, *branched-chain* alkanes are possible. For example, isobutane (or 2-methylpropane) has the structure

$$\text{H---C---C---C---H} \quad \text{or} \quad CH_3CHCH_3$$

Isobutane
(2-methylpropane)

Isobutane, C_4H_{10}, has the same molecular formula as butane, the straight-chain hydrocarbon. However, isobutane and butane have different structural formulas and, therefore, different molecular structures. Butane and isobutane are **constitutional** (or **structural**) **isomers**, *compounds with the same molecular formula but different structural formulas.* Figure 23.4 depicts molecular models of isobutane and butane. Because these isomers have different structures, they have different properties. For example, isobutane boils at −12°C, whereas butane boils at 0°C. Here the difference in boiling point can be attributed to the fact that isobutane has a more compact molecular structure than butane, which results in weaker intermolecular interactions between isobutane molecules.

The number of constitutional isomers rapidly increases with the number of carbons in the series. For example, there are three constitutional isomers with the molecular formula C_5H_{12} (pentanes; see Figure 23.5), five of C_6H_{14} (hexanes), and 75 of $C_{10}H_{22}$ (decanes).

Branched alkanes are usually written using condensed structural formulas. The following example and exercise will help you make the transition to writing hydrocarbons in this way.

Example 23.1   Writing Condensed Structural Formulas

Gaining Mastery Toolbox

Critical Concept 23.1
When writing the condensed structural formula of a hydrocarbon, the carbon-hydrogen bonds are not written. Additionally, condensed structural formulas only contain written carbon-carbon bonds if explicitly writing the bonds conveys information about the structure.

Solution Essentials:
• Condensed structural formula
• Structural formula
• Hydrocarbon

Write the condensed structural formula of each of the following alkanes.

a) [structure] b) [structure]

(continued)

(continued)

Problem Strategy A condensed structural formula is one in which as many of the explicit bonds (dashes) as possible are removed from the structural formula. The best way to proceed is to write the carbon backbone and then add in the additional atoms.

Solution

a To write the condensed structural formula, it helps to identify the carbon backbone, or the skeleton of carbon atoms in the molecule. It is good practice to write the longest chain of carbon atoms in the molecule in a straight line. For this case, the skeleton looks like:

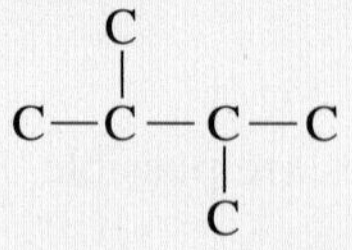

$$
\begin{array}{c}
C \\
| \\
C-C-C-C \\
| \\
C
\end{array}
$$

Next, write the appropriate number of H atoms next to each carbon atom.

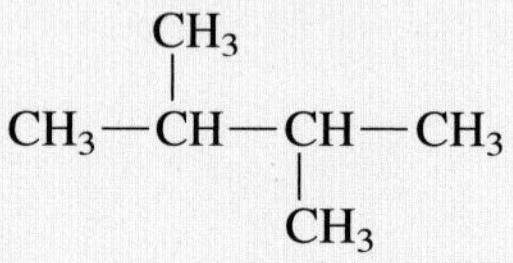

$$
\begin{array}{c}
CH_3 \\
| \\
CH_3-CH-CH-CH_3 \\
| \\
CH_3
\end{array}
$$

Finally, find the straight chain(s) of carbon atoms that contain more than two carbon atoms and remove only the bonds (dashes) that connect those carbon atoms.

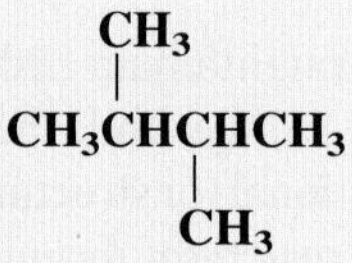

$$
\begin{array}{c}
CH_3 \\
| \\
CH_3CHCHCH_3 \\
| \\
CH_3
\end{array}
$$

After a little practice you will be able to combine these steps.

b Following the steps outlined above, the condensed structural formula for the molecule in part b is

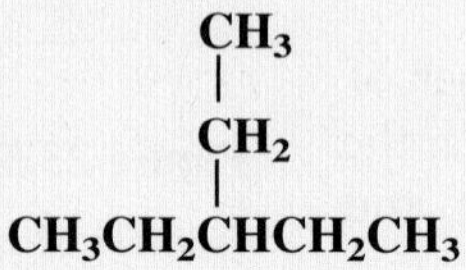

$$
\begin{array}{c}
CH_3 \\
| \\
CH_2 \\
| \\
CH_3CH_2CHCH_2CH_3
\end{array}
\quad \text{or} \quad
\begin{array}{c}
CH_2CH_3 \\
| \\
CH_3CH_2CHCH_2CH_3
\end{array}
$$

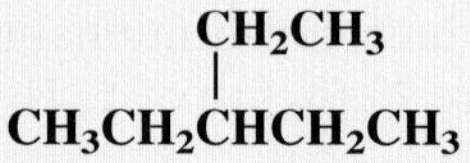

Either of the above structures is correct; which structure you use depends on how much structural information you want to convey. Following the steps in the solution for part a, and recognizing that the $-CH_2-CH_3$ fragment in the structure on the left is a chain of carbons, you should be able to arrive at the structure on the right.

Answer Check Check to see that the number and type of each of the atoms present in the condensed formula agree with the structural formula.

Exercise 23.1 Write the condensed structural formula for the following alkane.

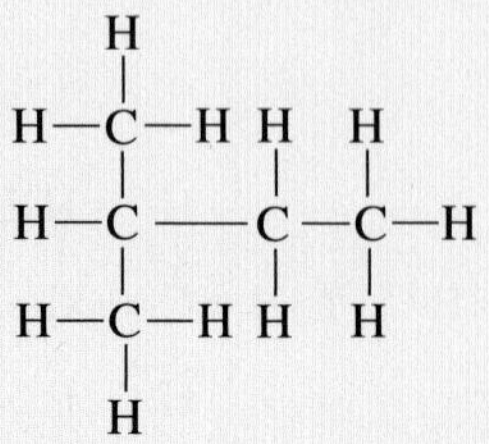

$$
\begin{array}{c}
H \\
| \\
H-C-H\;\;H\;\;\;H \\
||| \\
H-C-\!\!-\!\!-C-C-H \\
||| \\
H-C-H\;\;H\;\;\;H \\
| \\
H
\end{array}
$$

■ See Problems 23.27 and 23.28.

Name	CYCLOPROPANE	CYCLOBUTANE	CYCLOPENTANE	CYCLOHEXANE
Molecular Formula	C_3H_6	C_4H_8	C_5H_{10}	C_6H_{12}
Ball-and-Stick Model				
Full Structural Formula				
Condensed Structural Formula	△	▢	⬠	⬡

Figure 23.6 ▲

First four members of the cycloalkane series These are saturated aliphatic hydrocarbons characterized by carbon-atom rings.

Cycloalkanes

The general formula for cyclic cycloalkanes is C_nH_{2n}. Figure 23.6 gives the names and structural formulas for the first four members of the cycloalkane series. In the condensed structural formulas, a carbon atom and its attached hydrogen atoms are assumed to be at each corner.

CONCEPT CHECK 23.1

In the model shown here, C atoms are black and H atoms are light blue.

a Write the molecular formula.

b Write the condensed structural formula.

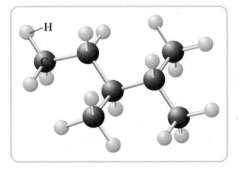

Sources and Uses of Alkanes and Cycloalkanes

Fossil fuels (natural gas, petroleum, and coal) are the principal sources of all types of organic chemicals. Natural gas is a mixture of low-molecular-mass hydrocarbons made up primarily of methane, CH_4, with lesser amounts of ethane, C_2H_6; propane, C_3H_8; and butane, C_4H_{10}. Petroleum, or crude oil, is the raw material extracted from a well. It is a mixture of alkanes and cycloalkanes with small amounts of aromatic hydrocarbons. The composition of petroleum is not consistent; it is dependent on geologic location and the organic matter present during oil

Table 23.2   **Top World Oil Producers**

Country	2008 Production (millions of barrels per day)
Saudi Arabia	11
Russia	10
United States	9
Iran	4
China	4

United States Energy Information Administration (2011), http://www.eia.doe.gov

Table 23.3   **Countries Having the Largest Oil Reserves**

Country	2008 Reserves (billions of barrels)
Saudi Arabia	262
Canada	179
Iran	136
Iraq	115
Kuwait	102

United States Energy Information Administration (2011), http://www.eia.doe.gov

formation. Because of this, a barrel of oil from Saudi Arabia contains a much different mixture of organic materials than a barrel of oil from the North Slope of Canada. The Saudi crude oil might be made up primarily of molecules with 5 to 20 carbon atoms, whereas the Canadian crude might consist mainly of molecules containing 20 to 40 carbon atoms. Crude oils that contain a majority of the lower-molecular-mass hydrocarbons are often more desirable because they are easier to transport and require less refining to convert to high-demand products like gasoline. Tables 23.2 and 23.3 show, respectively, the world's top oil producers and those with the largest reserves. Current world oil consumption is nearly 80 million barrels per day.

Because fossil fuels are extracted from their source as mixtures of hydrocarbons, it is usually necessary to separate these mixtures into various components. Such separations are most easily performed by distilling the mixture into fractions that contain mixtures of compounds of different molecular mass. As you can see from Table 23.4, common names have been given to some of these fractions such as gasoline (C_5 to C_{12}) and kerosene (C_{12} to C_{15}).

Often, there is a need to further separate the fractions listed in Table 23.4 into molecules with the same molecular mass, molecular formula, or structure. Through chemical processes that usually involve catalysts, small molecules can be combined into larger molecules, and large molecules can be broken apart. The processing of petroleum via distillation or chemical reactions is called *petroleum refining*. The type and extent of the petroleum refining that is performed depend on the type of crude oil that is available and on the demand for a particular type of product. One chemical process, called *catalytic cracking,* involves passing hydrocarbon vapor over a heated catalyst of alumina (Al_2O_3) and silica (SiO_2) to break apart, or "crack," high-molecular-mass hydrocarbons to produce hydrocarbons of low molecular mass. For example, this process can be used to convert the fuel oil fraction of petroleum to gasoline.

Petroleum refining also involves the conversion of the relatively abundant alkanes to unsaturated hydrocarbons, aromatic hydrocarbons, and hydrocarbon derivatives. The alkanes serve as the starting point for the majority of organic

Table 23.4   **Fractions from the Distillation of Petroleum**

Boiling Range (°C)	Name	Range of Carbon Atoms per Molecule	Use
Below 20	Gases	C_1 to C_4	Heating, cooking, and petrochemical raw material
20–200	Naphtha; straight-run gasoline	C_5 to C_{12}	Fuel; lighter fractions (such as petroleum ether, b.p. 30°C−60°C) are also used as laboratory solvents
200–300	Kerosene	C_{12} to C_{15}	Fuel
300–400	Fuel oil	C_{15} to C_{18}	Heating homes, diesel fuel
Over 400		Over C_{18}	Lubricating oil, greases, paraffin waxes, asphalt

From Harold Hart, *Organic Chemistry: A Short Course,* Eighth Edition, p. 102. Copyright © 1991 by Houghton Mifflin Company. Used with permission.

compounds, including plastics and pharmaceutical drugs. It is amazing to think about the number of everyday materials we use that started out as an alkane or cycloalkane (Figure 23.7).

Reactions of Alkanes with Oxygen

From your experiences with alkanes, it is probably apparent that they are not particularly reactive molecules at normal temperatures. But imagine if this were not the case; you would have to be extremely careful about how you filled your automobile with fuel, and you couldn't safely use natural gas in your home for heating or cooking. Also, the refining process could not involve the separation of the crude oil components at high temperatures by distillation. However, we do react alkanes every day through combustion with O_2; all hydrocarbons burn (combust) in an excess of O_2 at elevated temperatures to produce carbon dioxide, water, and heat. For example, a propane gas grill uses the reaction

Figure 23.7 ▲

Consumer products derived from petroleum All of these common items contain organic compounds that used petroleum as a starting material.

$$C_3H_8(g) + 5O_2(g) \longrightarrow 3CO_2(g) + 4H_2O(l); \Delta H° = -2220 \text{ kJ/mol}$$

The large negative $\Delta H°$ value for this reaction and all hydrocarbon reactions with oxygen demonstrates why we rely on these molecules to meet our energy needs.

Substitution Reactions of Alkanes

Under the right conditions, alkanes can react with other molecules. An important example is the reaction of alkanes with the halogens F_2, Cl_2, and Br_2. Reaction with Cl_2, for example, requires light (indicated by hv) or heat.

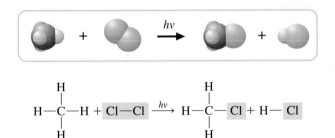

CF_3CH_2F

$$H-\underset{\underset{H}{|}}{\overset{\overset{H}{|}}{C}}-H + Cl-Cl \xrightarrow{hv} H-\underset{\underset{H}{|}}{\overset{\overset{H}{|}}{C}}-Cl + H-Cl$$

This is an example of a substitution reaction. A **substitution reaction** is *a reaction in which a part of the reacting molecule is substituted for an H atom on a hydrocarbon or hydrocarbon group.* All of the H atoms of an alkane may undergo substitution, leading to a mixture of products.

$$CH_3Cl + Cl_2 \xrightarrow{hv} CH_2Cl_2 + HCl$$

$$CH_2Cl_2 + Cl_2 \xrightarrow{hv} CHCl_3 + HCl$$

$$CHCl_3 + Cl_2 \xrightarrow{hv} CCl_4 + HCl$$

CCl_3F

The CCl_4 product can be reacted with HF in the presence of a $SbCl_5$ catalyst to produce trichlorofluoromethane, CCl_3F, also known as CFC-11.

$$CCl_4 + HF \xrightarrow{SbCl_5} CCl_3F + HCl$$

This compound is one of a number of chlorofluorocarbons (CFCs) used for much of the twentieth century as a refrigerant. Data obtained in the 1970s revealed that these compounds survived long enough to travel to the stratospheric region of our atmosphere, where they facilitate the destruction of the ozone layer. (See the essay in Chapter 10 on this topic.) Recent refrigerants that do not as readily contribute to ozone destruction include hydrofluorocarbons (HFCs), such as CF_3CH_2F, which do not contain chlorine atoms. These HFCs are now used to replace CFCs.

CONCEPT CHECK 23.2

For gasoline to function properly in an engine, it should not begin to burn before it is ignited by the spark plug. If it does, it makes the noise we think of as engine "knock." The octane-number scale rates the anti-knock characteristics of a gasoline. This linear scale is based on heptane, given an octane number of 0, and on 2,2,4-trimethylpentane (an octane constitutional isomer), given an octane number of 100. The higher the octane number, the better the anti-knock characteristics. If you had a barrel of heptane and a barrel of 2,2,4-trimethylpentane, how would you blend these compounds to come up with a 90 octane mixture?

23.3 Alkenes and Alkynes

Alkenes and alkynes are unsaturated hydrocarbons. Because they contain carbon–carbon multiple bonds, they are typically much more reactive than alkanes.

Under the proper conditions, molecular hydrogen can be added to an alkane or alkyne to produce a saturated compound in a process called *catalytic hydrogenation.* For example, ethylene adds hydrogen to give ethane.

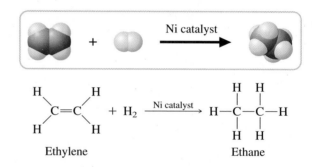

Ethylene   Ethane

Catalytic hydrogenation is also used in the food industry to convert (hydrogenate) carbon–carbon double bonds to carbon–carbon single bonds. For example, margarine can be manufactured by hydrogenating some of the double bonds present in corn oil to change it from oil to a solid (fat).

Alkenes and Geometric Isomerism

Alkenes are *hydrocarbons that have the general formula* C_nH_{2n} *and contain a carbon–carbon double bond.* (These compounds are also called *olefins.*) The simplest alkene, ethylene, has the condensed formula $CH_2{=}CH_2$ and the structural formula

$$
\begin{array}{ccc}
H & & H \\
 \diagdown & & \diagup \\
 & C{=}C & \\
 \diagup & & \diagdown \\
H & & H
\end{array}
$$

Ethylene is a gas with a sweet odor. It is obtained from the refining of petroleum and is an important raw material in the chemical industry. For example, when ethylene molecules are linked they make polyethylene (see Section 23.1), which is commonly used to make soda bottles and milk jugs. Plants also produce small amounts of ethylene. Fruit suppliers have found that exposure of fruit to ethylene speeds ripening.

In ethylene and other alkenes, all of the atoms connected to the two carbon atoms of the double bond lie in a single plane, as Figure 23.8 shows. This is due to the need for maximum overlap of $2p$ orbitals on the carbon atoms to form a pi (π) bond. As a result, rotation about a carbon–carbon double bond cannot occur without breaking the π bond. For rotation to occur, a significant amount of energy must be supplied to break the π bond, so rotation does not normally occur. (This is in contrast to carbon–carbon single bonds, which have

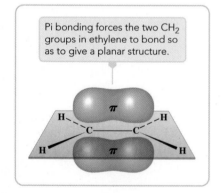

Pi bonding forces the two CH_2 groups in ethylene to bond so as to give a planar structure.

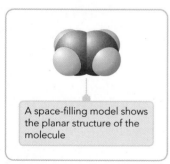

A space-filling model shows the planar structure of the molecule

Figure 23.8 ▲

Representations of the ethylene molecule

very low energy requirements for rotation and so freely rotate under most conditions.) This inability to rotate gives rise to geometric isomers in certain alkenes. **Geometric isomers** are *isomers in which the atoms are joined to one another in the same way but differ because some atoms occupy different relative positions in space.* For example, 2-butene, C_4H_8, has two geometric isomers, called *cis*-2-butene and *trans*-2-butene.

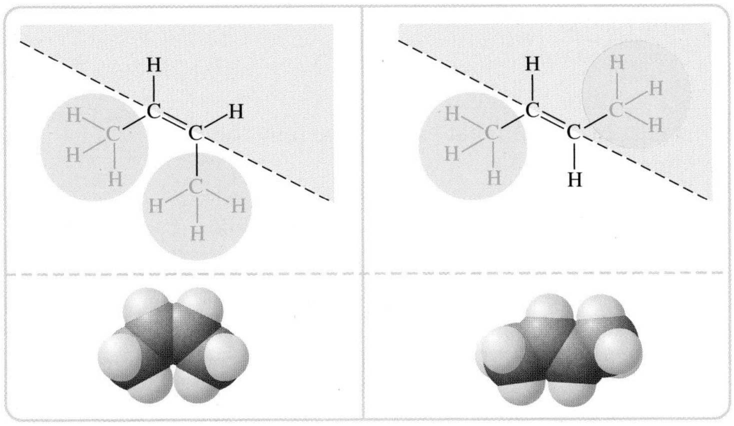

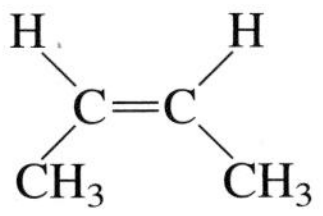

cis-2-Butene
b.p. 3.7°C

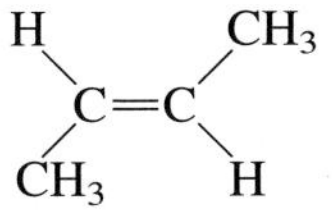

trans-2-Butene
b.p. 0.9°C

The different boiling points confirm that they are different compounds. Note that in the *cis* isomer the two CH_3 groups are attached to the same side of the double bond, whereas in the *trans* isomer the two CH_3 groups are attached on opposite sides of the double bond. An alkene with the general formula

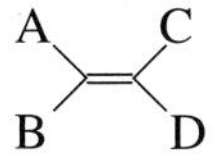

exists as geometric isomers only if groups A and B are different and groups C and D are different. For instance, geometric isomers do not exist for propene, $CH_2\!=\!CHCH_3$.

Example 23.2

Predicting *cis–trans* Isomers

Gaining Mastery Toolbox

Critical Concept 23.2
Cis and *trans* geometric isomers for alkenes can only occur when the groups bonded to the carbon atoms on each side of the double bond are different. This observation means that in every case in which two hydrogen atoms are bonded to a carbon atom on one side of a double bond, the geometry around the double bond in the molecule cannot lead to geometric isomers.

Solution Essentials:
• *Cis* and *trans* isomers
• Geometric isomer
• Alkene

For each of the following alkenes, decide whether *cis–trans* isomers are possible. If so, draw structural formulas of the isomers.

a $CH_3CH_2CH\!=\!C(CH_3)_2$ b $CH_3CH\!=\!CHCH_2CH_3$

Problem Strategy In order to determine whether *cis–trans* isomers of the given compounds are possible, we need to draw the structural formula around the double bond. Then we compare the groups attached to the double bond; if the same groups are attached to the same side of the double bond, geometric isomers are not possible.

Solution a Writing the structure, you have

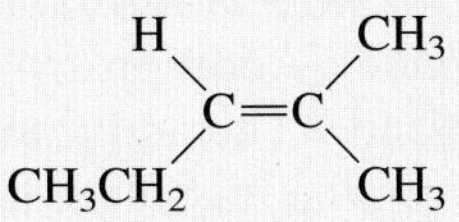

(continued)

(*continued*)

Because two methyl groups are attached to the second carbon atom of the double bond, **geometric isomers are not possible.** ⓑ **Geometric isomers are possible.** They are

$$
\begin{array}{cc}
\underset{\substack{|\\\text{CH}_3}}{\overset{\substack{\text{H}\\|}}{\text{C}}}=\underset{\substack{|\\\text{CH}_2\text{CH}_3}}{\overset{\substack{\text{H}\\|}}{\text{C}}} & \underset{\substack{|\\\text{CH}_3}}{\overset{\substack{\text{H}\\|}}{\text{C}}}=\underset{\substack{|\\\text{H}}}{\overset{\substack{\text{CH}_2\text{CH}_3\\|}}{\text{C}}}
\end{array}
$$

cis-2-Pentene   *trans*-2-Pentene

Answer Check When drawing the structural formula, make sure that you change only the special arrangement of the atoms, not the bonding.

Exercise 23.2 Decide whether *cis–trans* isomers are possible for each of the following compounds. If isomers are possible, draw the structural formulas.

ⓐ $CH_3CH{=}CHCH_2CH_2CH_3$ ⓑ $CH_3CH_2CH{=}CH_2$

■ See Problems 23.29 and 23.30.

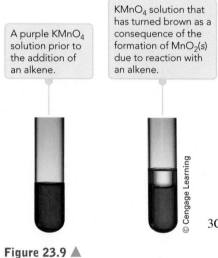

A purple KMnO₄ solution prior to the addition of an alkene.

KMnO₄ solution that has turned brown as a consequence of the formation of MnO₂(s) due to reaction with an alkene.

© Cengage Learning

Figure 23.9 ▲

Test for unsaturation using KMnO₄(*aq*)

Cis and *trans* isomers of dietary fats and oils are suspected to play a role in human health. All unprocessed fats and oils contain only *cis* isomers. During processing, some of the *cis* isomers are converted to *trans* isomers, called *trans* fatty acids. *Trans* fatty acids are suspected of raising the amount of serum cholesterol in the blood stream, which can cause health problems.

Oxidation Reactions of Alkenes

Because alkenes are hydrocarbons, they undergo complete combustion reactions with oxygen at high temperatures to produce carbon dioxide and water. Unsaturated hydrocarbons can also be partially oxidized under relatively mild conditions. For example, when aqueous potassium permanganate, $KMnO_4(aq)$, is added to an alkene (or alkyne), the purple color of $KMnO_4$ fades and a brown precipitate of manganese dioxide forms (Figure 23.9).

$$3C_4H_9CH{=}CH_2 + 2MnO_4^-\,(aq) + 4H_2O \longrightarrow$$
1-Hexene

$$
3C_4H_9\underset{\substack{|\\\text{O}\\|\\\text{H}}}{\overset{\substack{\text{H}\\|}}{\text{C}}}-\underset{\substack{|\\\text{O}\\|\\\text{H}}}{\overset{\substack{\text{H}\\|}}{\text{C}}}-\text{H} + 2MnO_2(s) + 2OH^-\,(aq)
$$

Addition Reactions of Alkenes

Alkenes are more reactive than alkanes because of the presence of the double bond. Many reactants *add* to the double bond. A simple example is the addition of a halogen, such as Br_2, to propene.

$$
CH_3CH{=}CH_2 + Br_2 \longrightarrow CH_3\underset{\substack{|\\\text{Br}}}{\text{CH}}-\underset{\substack{|\\\text{Br}}}{\text{CH}_2}
$$
Propene

An **addition reaction** is *a reaction in which parts of a reactant are added to each carbon atom of a carbon–carbon double bond, which converts to a carbon–carbon single bond.* (Addition to triple bonds is also possible, giving a product with a double bond.) The addition of Br_2 to an alkene is fast. In fact, it occurs so readily that bromine dissolved in carbon tetrachloride, CCl_4, is a useful reagent to test for unsaturation. When a few drops of the solution are added to an alkene, the redbrown color of the Br_2 disappears immediately.

Unsymmetrical reagents, such as HCl and HBr, add to unsymmetrical alkenes to give two products that are constitutional isomers. For example,

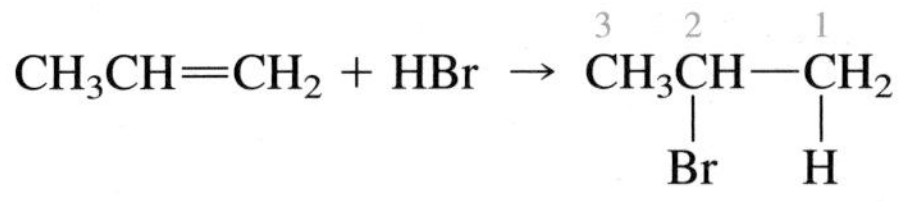

$$CH_3CH{=}CH_2 + HBr \; \rightarrow \; \overset{3}{C}H_3\overset{2}{C}H{-}\overset{1}{C}H_2$$
$$\underset{\text{Br}}{\vert} \quad \underset{\text{H}}{\vert}$$

2-Bromopropane

and

$$CH_3CH{=}CH_2 + HBr \; \rightarrow \; \overset{3}{C}H_3\overset{2}{C}H{-}\overset{1}{C}H_2$$
$$\underset{\text{H}}{\vert} \quad \underset{\text{Br}}{\vert}$$

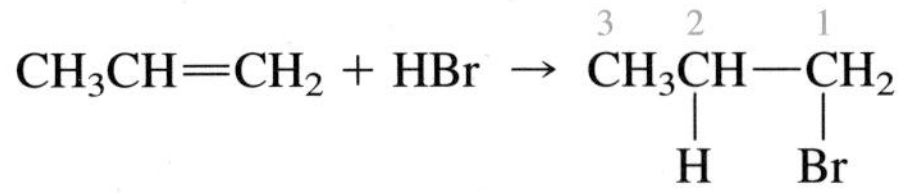

1-Bromopropane

In one case, the hydrogen atom of HBr adds to carbon atom 1, giving 2-bromopropane; in the other case, the hydrogen atom of HBr adds to carbon atom 2, giving 1-bromopropane. (The name 1-bromopropane means that a bromine atom is substituted for a hydrogen atom at carbon atom 1.) However, the two products are not formed in equal amounts; one is more likely to form. **Markownikoff's rule** is *a generalization that states that the major product formed by the addition of an unsymmetrical reagent such as H—Cl, H—Br, or H—OH is the one obtained when the H atom of the reagent adds to the carbon atom of the multiple bond that already has the greater number of hydrogen atoms attached to it.* In the preceding example, the H atom of HBr should add preferentially to carbon atom 1, which has two hydrogen atoms attached to it. The major product then is 2-bromopropane.

Example 23.3

Predicting the Major Product of an Addition Reaction

Gaining Mastery Toolbox

Critical Concept 23.3
When naming hydrocarbons always start the process by determining the longest continuous chain of carbon atoms. In many instances, the longest chain of carbon atoms will consist of some of the carbon atoms on the straight chain as well as those on one of the branches.

Solution Essentials:
- Straight-chain alkanes (Table 23.5)
- Alykyl group names (Table 23.6)
- Rules for naming hydrocarbons
- Alkane
- Hydrocarbon

What is the major product of the following reaction?

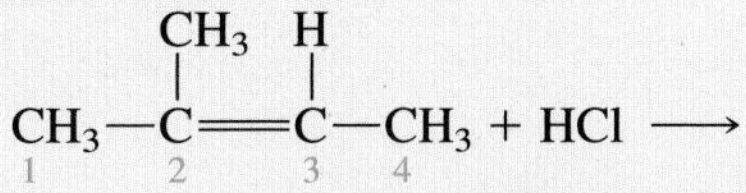

$$\underset{1}{C}H_3{-}\overset{\overset{\displaystyle CH_3}{\vert}}{\underset{2}{C}}{=}\overset{\overset{\displaystyle H}{\vert}}{\underset{3}{C}}{-}\underset{4}{C}H_3 + HCl \longrightarrow$$

Problem Strategy Because this is an addition reaction of an unsymmetrical reagent (HCl), apply Markownikoff's rule to predict the major product.

Solution One H atom is attached to carbon 3 and none to carbon 2. Therefore, the major product is

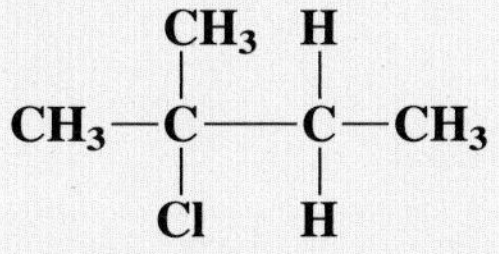

$$CH_3{-}\overset{\overset{\displaystyle CH_3}{\vert}}{\underset{\underset{\displaystyle Cl}{\vert}}{C}}{-}\overset{\overset{\displaystyle H}{\vert}}{\underset{\underset{\displaystyle H}{\vert}}{C}}{-}CH_3$$

Answer Check When writing the structure of compounds that consist of carbon with only single bonds, such as the product here, check to make sure that each carbon has four bonds.

Exercise 23.3 Predict the main product when HBr adds to

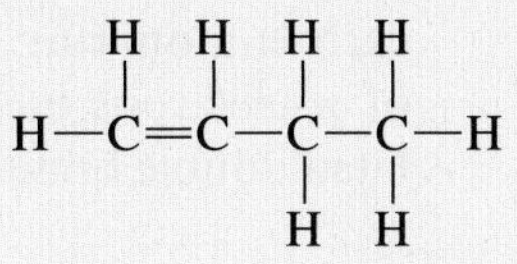

$$H{-}\overset{\overset{\displaystyle H}{\vert}}{C}{=}\overset{\overset{\displaystyle H}{\vert}}{C}{-}\overset{\overset{\displaystyle H}{\vert}}{\underset{\underset{\displaystyle H}{\vert}}{C}}{-}\overset{\overset{\displaystyle H}{\vert}}{\underset{\underset{\displaystyle H}{\vert}}{C}}{-}H$$

■ See Problems 23.35 and 23.36.

Alkynes

Alkynes are *unsaturated hydrocarbons containing a carbon–carbon triple bond.* The general formula is C_nH_{2n-2}. The simplest alkyne is acetylene (ethyne), a linear molecule.

$$H—C≡C—H$$

Acetylene is a very reactive gas that is used to produce a variety of other chemical compounds. It burns with oxygen in the oxyacetylene torch to give a very hot flame (about 3000°C). Acetylene is produced commercially from methane.

$$2CH_4 \xrightarrow{1600°C} CH≡CH + 3H_2$$

Acetylene is also prepared from calcium carbide, CaC_2. Calcium carbide is obtained by heating calcium oxide and coke (carbon) in an electric furnace:

$$CaO(s) + 3C(s) \xrightarrow{2000°C} CaC_2(l) + CO(g)$$

The calcium carbide is cooled until it solidifies. The carbide ion, C_2^{2-}, is strongly basic and reacts with water to produce acetylene (Figure 23.10).

$$CaC_2(s) + 2H_2O(l) \longrightarrow Ca(OH)_2(aq) + C_2H_2(g)$$

The alkynes, like the alkenes, undergo addition reactions, usually adding two molecules of the reagent for each C≡C bond. The major product is the isomer predicted by Markownikoff's rule, as shown in the following reaction.

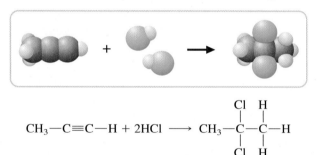

$$CH_3—C≡C—H + 2HCl \longrightarrow CH_3—\overset{\overset{\displaystyle Cl}{|}}{\underset{\underset{\displaystyle Cl}{|}}{C}}—\overset{\overset{\displaystyle H}{|}}{\underset{\underset{\displaystyle H}{|}}{C}}—H$$

Figure 23.10 ▲

Preparation of acetylene gas Here acetylene is prepared by the reaction of water with calcium carbide. The acetylene burns with a sooty flame.

23.4 Aromatic Hydrocarbons

Aromatic hydrocarbons usually contain benzene rings: six-membered rings of carbon atoms with alternating carbon–carbon single and carbon–carbon double bonds. The electronic structure of benzene can be represented by resonance formulas. For benzene,

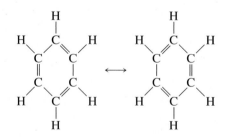

This electronic structure can also be described using molecular orbitals (Figure 23.11). In this description, π molecular orbitals encompass the entire carbon-atom ring, and the π electrons are said to be delocalized. Delocalization of π electrons means that the double bonds in benzene do not behave as isolated double bonds. Two condensed formulas for benzene are

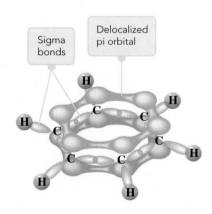

Figure 23.11 ▲

The molecular orbitals of benzene The sigma bond framework (in green) forms by the overlap of sp^2 hybrid orbitals on each carbon atom; H atom 1s orbitals overlap these hybrid orbitals to form C—H bonds. Delocalized pi orbitals form by the overlap of carbon atom 2p orbitals that are perpendicular to the plane of the molecule; they give orbitals with lobes above and below this plane. Only the lowest-energy pi orbital is shown (yellow).

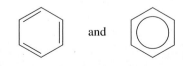

and

where the circle in the formula at right represents the delocalization of the π electrons (and therefore the double bonds). Although you will often encounter benzene represented with alternating double and single bonds as shown on the left, the better representation is the one that indicates the bond delocalization. The space-filling model of benzene is shown in Figure 23.12.

Aromatic compounds are found everywhere. The term *aromatic* implies that compounds that contain a benzene ring have aromas, and this is indeed the case. Flavoring agents that can be synthesized in the laboratory or found in nature include the flavor and aroma of cinnamon, cinnamaldehyde, and the wintergreen flavor of candies and gum, methyl salicylate (Figure 23.13).

Benzene rings also exist in the pain relievers acetylsalicylic acid (aspirin), acetaminophen (Tylenol), and the illegal drug mescaline.

Figure 23.12 ▲

Bond delocalization in benzene
The space-filling model of benzene shows that all carbon–carbon bond distances are identical.

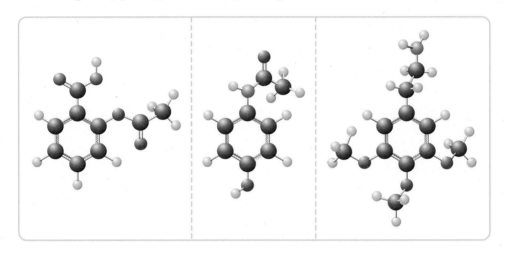

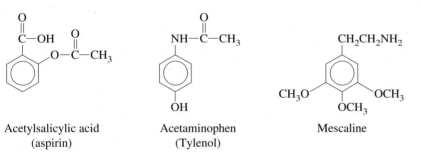

Acetylsalicylic acid
(aspirin)

Acetaminophen
(Tylenol)

Mescaline

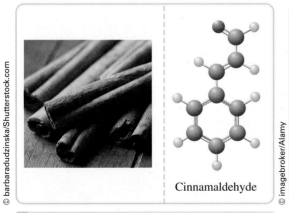

Cinnamaldehyde

a A molecular model of cinnamaldehyde, the molecule responsible for the taste and smell of cinnamon.

Methyl salicylate

b A molecular model of methylsalicylate, the molecule that produces the flavor of wintergreen.

Figure 23.13 ▲

Examples of aromatic compounds

© barbaradudzinska/Shutterstock.com

© imagebroker/Alamy

Other examples include the active ingredients in some sunscreens: *p*-aminobenzoic acid (often abbreviated as PABA on the label) and oxybenzone.

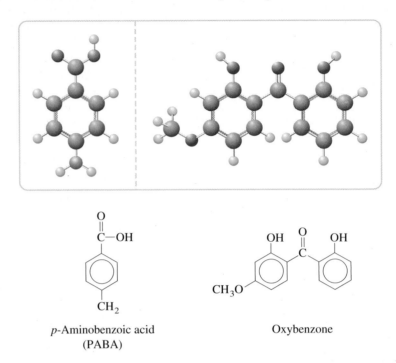

p-Aminobenzoic acid
(PABA)

Oxybenzone

Benzene rings can also fuse together to form polycyclic aromatic hydrocarbons in which two or more rings share carbon atoms. Figure 23.14 gives formulas for some polycyclic aromatic hydrocarbons. Naphthalene is one of the compounds that gives mothballs their characteristic odor.

Substitution Reactions of Aromatic Hydrocarbons

Although benzene, C_6H_6, is an unsaturated hydrocarbon, it does not usually undergo addition reactions because the delocalized π electrons of benzene are more

Figure 23.14 ▶

Some polycyclic aromatic hydrocarbons Naphthalene is the simplest member of this series. It is a white, crystalline substance used in manufacturing plastics and plasticizers (to keep plastics pliable). Small amounts are used for mothballs.

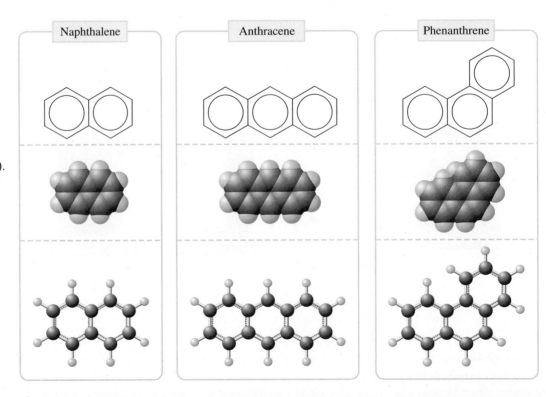

Naphthalene   Anthracene   Phenanthrene

stable than the localized π electrons. Because of this, benzene does not undergo an addition reaction with Br_2 in carbon tetrachloride like alkenes do. The usual reactions of benzene are substitution reactions. For example, in the presence of an iron(III) bromide catalyst, a hydrogen atom on the benzene ring is substituted with a bromine atom.

Benzene Bromobenzene

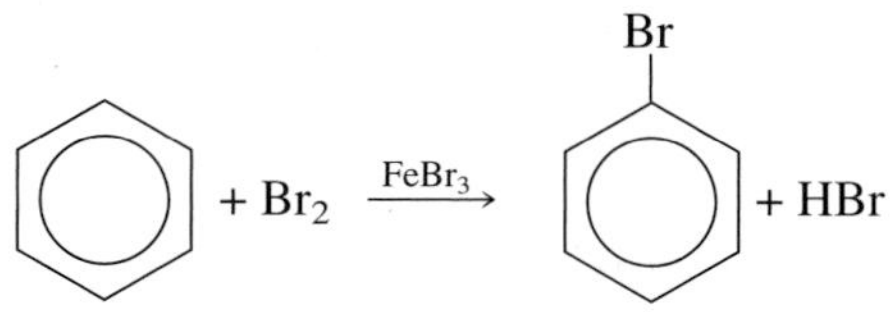

Similarly, benzene undergoes substitution with nitric acid in the presence of sulfuric acid to give nitrobenzene.

Benzene Nitric acid Nitrobenzene

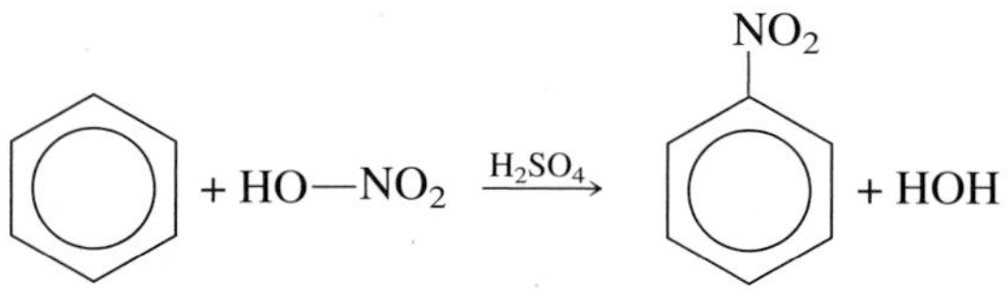

23.5 Naming Hydrocarbons

A nomenclature for organic compounds has developed over the years as a way of understanding and classifying their structures. This nomenclature is now formulated in rules agreed upon by the International Union of Pure and Applied Chemistry (IUPAC).

Nomenclature of Alkanes

The first four straight-chain alkanes (methane, ethane, propane, and butane) have long-established names. Higher members of the series are named from the Greek words indicating the number of carbon atoms in the molecule, plus the suffix *-ane*. For example, the straight-chain alkane C_5H_{12} is named pentane. Table 23.5 gives the formulas, names, and structures of the first ten straight-chain alkanes.

The following four IUPAC rules are applied in naming the branched-chain alkanes:

1. Determine the longest continuous (not necessarily straight) chain of carbon atoms in the molecule. The base name of the branched-chain alkane is that of the straight-chain alkane (Table 23.5) corresponding to the number of carbon atoms in this longest chain. For example, in

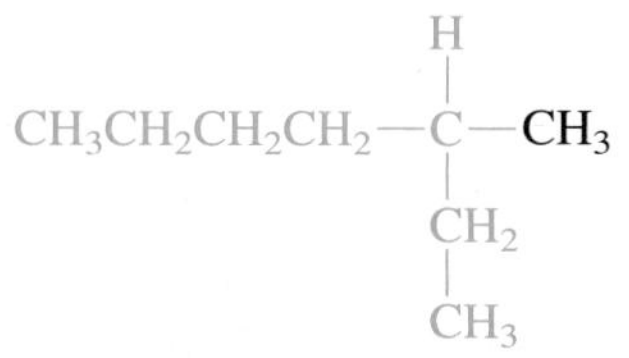

the longest continuous carbon chain, shown in color, has seven carbon atoms, giving the base name heptane. The full name for the alkane includes the names of any branched chains. These names are placed in front of the base name, as described in the remaining rules.

Table 23.5 Formulas, Names, and Structures of the First Ten Straight-Chain Alkanes

Name	Elemental Formula	Condensed Structural Formula	Molecular Model
Methane	CH_4	CH_4	
Ethane	C_2H_6	CH_3CH_3	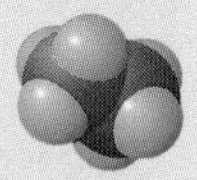
Propane	C_3H_8	$CH_3CH_2CH_3$	
Butane	C_4H_{10}	$CH_3CH_2CH_2CH_3$	
Pentane	C_5H_{12}	$CH_3CH_2CH_2CH_2CH_3$	
Hexane	C_6H_{14}	$CH_3CH_2CH_2CH_2CH_2CH_3$	
Heptane	C_7H_{16}	$CH_3CH_2CH_2CH_2CH_2CH_2CH_3$	
Octane	C_8H_{18}	$CH_3CH_2CH_2CH_2CH_2CH_2CH_2CH_3$	
Nonane	C_9H_{20}	$CH_3CH_2CH_2CH_2CH_2CH_2CH_2CH_2CH_3$	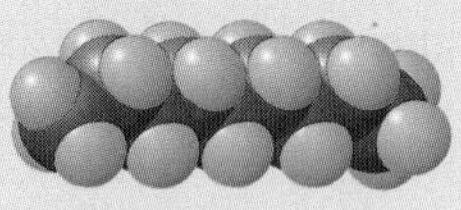
Decane	$C_{10}H_{22}$	$CH_3CH_2CH_2CH_2CH_2CH_2CH_2CH_2CH_2CH_3$	

2. Any chain branching off the longest chain is named as an alkyl group. An alkyl group is an alkane less one hydrogen atom. (Table 23.6 lists some alkyl groups.) When a hydrogen atom is removed from an end carbon atom of a straight-chain alkane, the resulting alkyl group is named by changing the suffix -*ane* of the alkane to -*yl*. For example, removing a hydrogen atom from methane gives the methyl group, $—CH_3$. The structure shown in Rule 1 has a methyl group as a branch on the heptane chain.

3. The complete name of a branch requires a number that locates the branch on the longest chain. For this purpose, you number each carbon atom on the longest

Table 23.6   **Important Alkyl Groups**

Original Alkane	Structure of Alkyl Group	Name of Alkyl Group
Methane, CH_4	CH_3-	Methyl
Ethane, CH_3CH_3	CH_3CH_2-	Ethyl
Propane, $CH_3CH_2CH_3$	$CH_3CH_2CH_2-$	Propyl
Propane, $CH_3CH_2CH_3$	$CH_3\overset{\displaystyle \|}{C}HCH_3$	Isopropyl
Butane, $CH_3CH_2CH_2CH_3$	$CH_3CH_2CH_2CH_2-$	Butyl
Isobutane, $CH_3\underset{\underset{CH_3}{\|}}{C}HCH_3$	$CH_3\underset{\underset{CH_3}{\|}}{\overset{\overset{\|}{}}{C}}CH_3$	*Tertiary*-butyl (*t*-butyl)

chain in the direction that gives the smaller numbers for the locations of all branches. The structural formula in Rule 1 is numbered as shown:

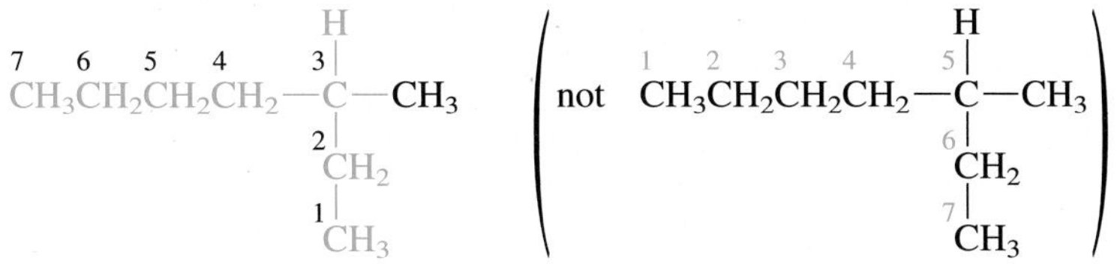

$$
\overset{7\quad 6\quad\ 5\quad\ 4\quad\quad\ \ 3}{CH_3CH_2CH_2CH_2-\underset{\underset{\underset{CH_3}{\overset{1}{|}}}{\underset{CH_2}{\overset{2}{|}}}}{\overset{\overset{H}{\overset{|}{}}}{C}}-CH_3}
\qquad
\left(not \quad \overset{1\quad\ 2\quad\ 3\quad\ 4\quad\quad\ 5}{CH_3CH_2CH_2CH_2-\underset{\underset{\underset{CH_3}{\overset{7}{|}}}{\underset{CH_2}{\overset{6}{|}}}}{\overset{\overset{H}{\overset{|}{}}}{C}}-CH_3} \right)
$$

The methyl branch is located at carbon 3 (not carbon 5) of the heptane chain. The complete name of the branch is 3-methyl, and the compound is named 3-methylheptane. Note that the branch name and the base name are written as a single word, and a hyphen follows the number.

4. When there are more than one alkyl branch of the same kind (say, two methyl groups), this number is indicated by a prefix, such as *di-, tri-,* or *tetra-,* used with the name of the alkyl group. The position of each group on the longest chain is given by numbers. For example,

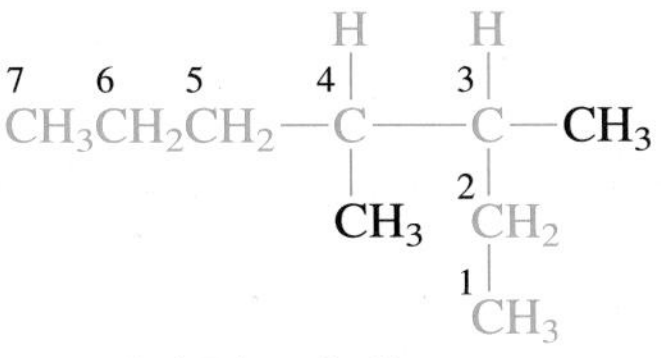

$$
\overset{7\quad\ 6\quad\ 5\quad\ \ 4\quad\quad\ 3}{CH_3CH_2CH_2-\underset{\underset{CH_3}{|}}{\overset{\overset{H}{|}}{C}}-\underset{\underset{\underset{CH_3}{\overset{1}{|}}}{\overset{2}{\underset{CH_2}{|}}}}{\overset{\overset{H}{|}}{C}}-CH_3}
\qquad\qquad
\overset{7\quad\ 6\quad\ 5\quad\ 4\quad\quad\ 3}{CH_3CH_2CH_2CH_2-\underset{\underset{\underset{CH_3}{\overset{1}{|}}}{\overset{2}{\underset{CH_2}{|}}}}{\overset{\overset{CH_3}{|}}{C}}-CH_3}
$$

3,4-Dimethylheptane 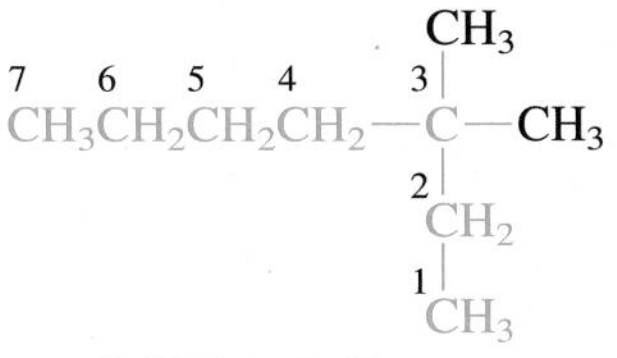   3,3-Dimethylheptane

Note that the position numbers are separated by a comma and are followed by a hyphen.

When there are two or more different alkyl branches, the name of each branch, with its position number, precedes the base name. The branch names are placed in alphabetical order. For example,

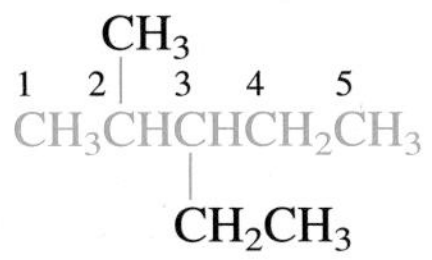

$$
\overset{1\quad\ 2\ \ \ \ 3\quad 4\quad\ 5}{CH_3\underset{\underset{CH_2CH_3}{|}}{\overset{\overset{CH_3}{|}}{C}}HCHCH_2CH_3}
$$

3-Ethyl-2-methylpentane

Note the use of hyphens.

ⓦWL Interactive Example 23.4   Writing the IUPAC Name of an Alkane Given the Structural Formula

Critical Concept 23.4
The name of the alkane contains information about the longest chain of carbons in the molecule. Molecules that have more than four carbon atoms in the longest chain use Greek prefixes (Table 2.7).

Solution Essentials:
- Straight-chain alkanes (Table 23.5)
- Alkyl group names (Table 23.6)
- Rules for naming hydrocarbons
- Alkane
- Hydrocarbon

Give the IUPAC name for each of the following compounds.

a
$$CH_3CH_2CH_2$$
$$CHCH_2CH_2CH_3$$
$$CH_3CH_2CH_2CH_2$$

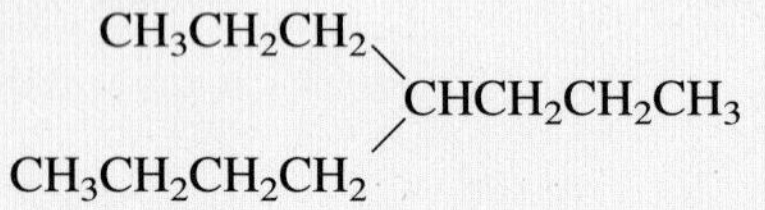

b
$$CH_3$$
$$|$$
$$CH_3-C-CH_3$$
$$|$$
$$CH_2$$
$$|$$
$$CH_2$$
$$|$$
$$CH_2$$
$$|$$
$$CH_3$$

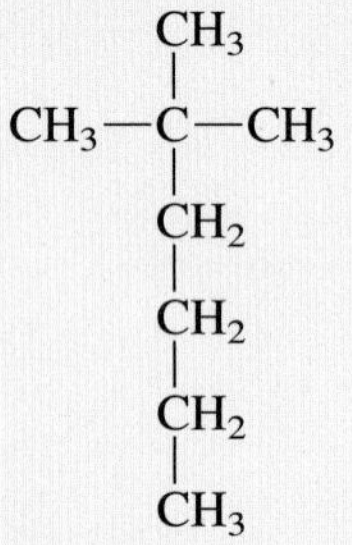

Problem Strategy Given that this molecule consists only of hydrogen and carbon, you need to follow the steps outlined in this section for naming hydrocarbons.

Solution a The longest continuous chain is numbered as follows:

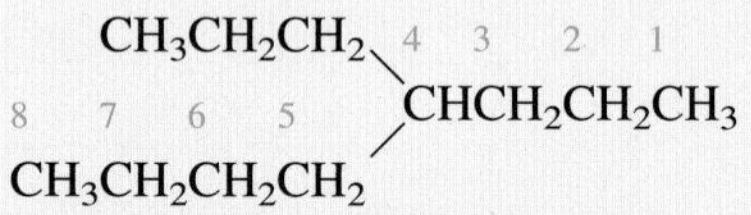

$$CH_3CH_2CH_2 \quad 4 \quad 3 \quad 2 \quad 1$$
$$8 \quad 7 \quad 6 \quad 5 \quad CHCH_2CH_2CH_3$$
$$CH_3CH_2CH_2CH_2$$

The name of the compound is **4-propyloctane**. If the longest chain had been numbered in the opposite direction, you would have the name 5-propyloctane. But because 5 is larger than 4, this name is unacceptable. b The numbering of the longest chain is

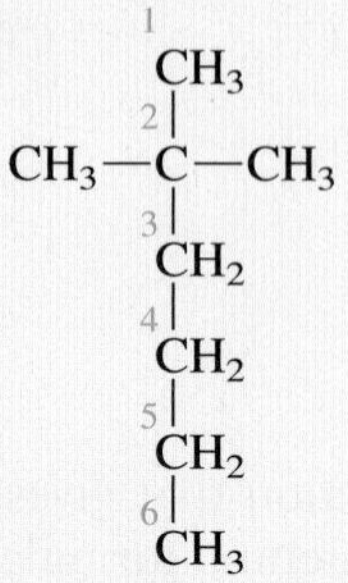

$$^1 CH_3$$
$$^2|$$
$$CH_3-C-CH_3$$
$$^3|$$
$$CH_2$$
$$^4|$$
$$CH_2$$
$$^5|$$
$$CH_2$$
$$^6|$$
$$CH_3$$

Any of the three clustered methyl carbon atoms could be given the number 1, and the other methyl groups branch off carbon atom 2. Hence, the name is **2,2-dimethylhexane**.

Answer Check Once you have come up with the name, check and see whether using this name leads you back to the correct structural formula.

Exercise 23.4 What is the IUPAC name for each of the following hydrocarbons?

a
$$CH_3$$
$$|$$
$$CH_3CHCHCH_3$$
$$|$$
$$CH_3$$

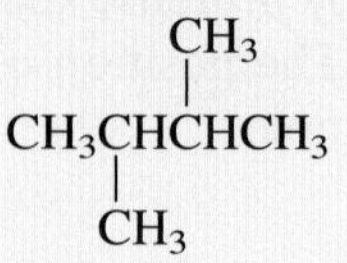

b
$$CH_2CH_2CH_3$$
$$|$$
$$CH_3CHCHCH_2CH_3$$
$$|$$
$$CH_3$$

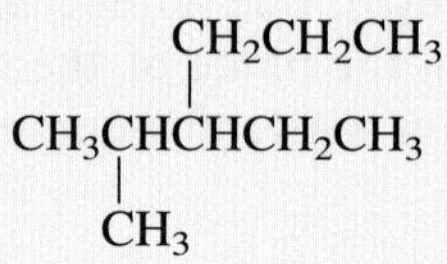

■ See Problems 23.37 and 23.38.

Example 23.5   Writing the Structural Formula of an Alkane Given the IUPAC Name

Critical Concept 23.5
Complex alkane names end with the name of the longest chain hydrocarbon. Be sure to number the hydrocarbon chain prior to attaching the appropriate functional groups.

Solution Essentials:
- Straight-chain alkanes (Table 23.5)
- Alkyl group names (Table 23.6)
- Rules for naming hydrocarbons
- Alkane
- Hydrocarbon

Write the condensed structural formula of 4-ethyl-3-methylheptane.

Problem Strategy In order to write the structural formula of a hydrocarbon, you must first determine the longest carbon chain from the name. Next, number the chain, and then attach the appropriate groups along the chains at their numbered positions. Complete the structure by adding H atoms in such a way that each carbon has four bonds.

Solution First write the carbon skeleton for heptane.

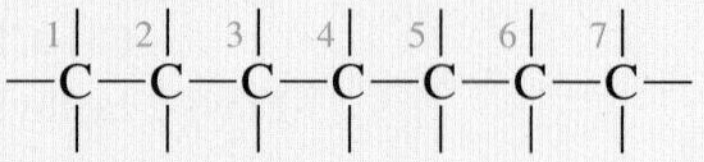

$$
\begin{array}{ccccccc}
1| & 2| & 3| & 4| & 5| & 6| & 7| \\
-C- & C- & C- & C- & C- & C- & C- \\
| & | & | & | & | & | & |
\end{array}
$$

(continued)

(continued)

Then attach the alkyl groups.

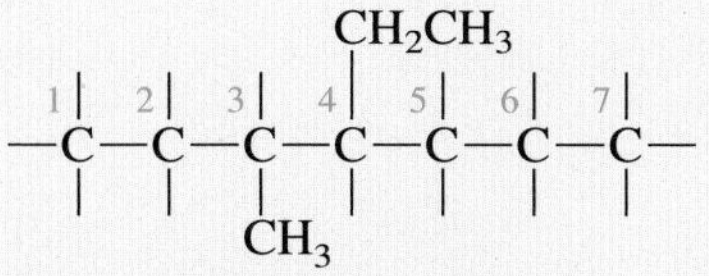

$$\begin{array}{c} \text{CH}_2\text{CH}_3 \\ | \\ \underset{1}{-\text{C}}\underset{2}{-\text{C}}\underset{3}{-\text{C}}\underset{4}{-\text{C}}\underset{5}{-\text{C}}\underset{6}{-\text{C}}\underset{7}{-\text{C}}- \\ | \quad | \quad | \quad | \quad | \quad | \quad | \\ \text{CH}_3 \end{array}$$

After filling out the structure with H atoms, you have

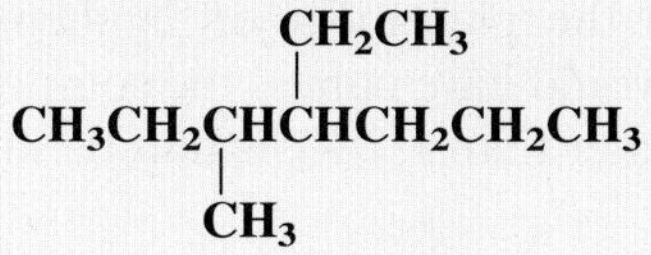

$$\begin{array}{c} \text{CH}_2\text{CH}_3 \\ | \\ \text{CH}_3\text{CH}_2\text{CHCHCH}_2\text{CH}_2\text{CH}_3 \\ | \\ \text{CH}_3 \end{array}$$

Answer Check Once you have written the structural formula of the compound, name it to see if you end up with the compound that you started with.

Exercise 23.5 Write the condensed structural formula of 3,3-dimethyloctane.

■ See Problems 23.39 and 23.40.

Nomenclature of Alkenes and Alkynes

The IUPAC name for an alkene is determined by first identifying the longest chain containing the double bond. As with the alkanes, the longest chain provides the stem name, but the suffix is *-ene* rather than *-ane*. The carbon atoms of the longest chain are then numbered from the end nearer the carbon–carbon double bond, and the position of the double bond is given the number of the first carbon atom of that bond (the smaller number). This number is written in front of the stem name of the alkene. Branched chains are named the same way as the alkanes. The simplest alkene, $\text{CH}_2{=}\text{CH}_2$, is called ethene, although the common name is ethylene.

Example 23.6 **Writing the IUPAC Name of an Alkene Given the Structural Formula**

Gaining Mastery Toolbox

Critical Concept 23.6
Compounds that contain double bonds are named using the suffix *–ene*. Similar to alkanes, the longest carbon chain provides the stem name; however, in this case, the chain must contain the double bond.

Solution Essentials:
• Alkene
• Straight-chain alkanes (Table 23.5)
• Alkyl group names (Table 23.6)
• Rules for naming hydrocarbons
• Alkane
• Hydrocarbon

What is the IUPAC name of each of the following alkenes?

a $\quad \begin{array}{c} \text{CH}_2{=}\text{CHCHCH}_2\text{CH}_3 \\ | \\ \text{CH}_3 \end{array}$

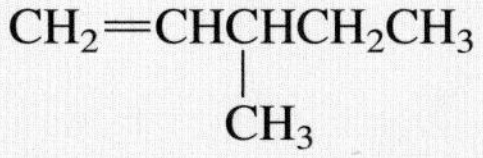

b $\quad \begin{array}{c} \text{CH}_3\text{CH}_2\text{CH}_2\text{CH}_2\text{CHCH}{=}\text{CHCH}_3 \\ | \\ \text{CH}_2 \\ | \\ \text{CH}_2 \\ | \\ \text{CH}_2 \\ | \\ \text{CH}_2 \\ | \\ \text{CH}_3 \end{array}$

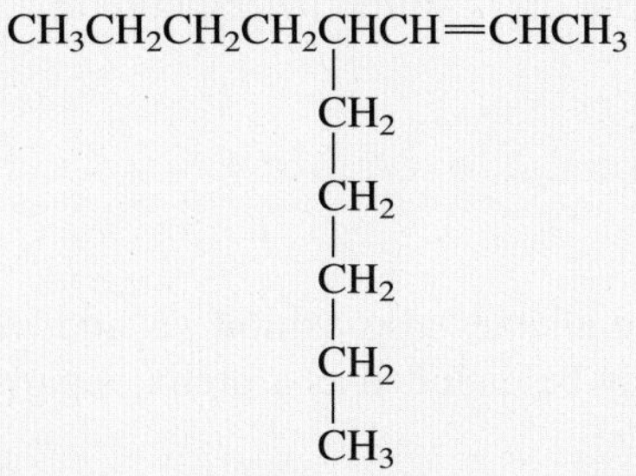

Problem Strategy The compounds in this problem contain double bonds, so they are alkenes. Naming these compounds is done in the same fashion as naming the alkanes, with one exception: the position of the double bond in each structure dictates that we number the longest carbon-atom chain by starting at the end of the chain nearest to the double bond.

(continued)

(continued)

Solution ⓐ The numbering of the carbon chain is

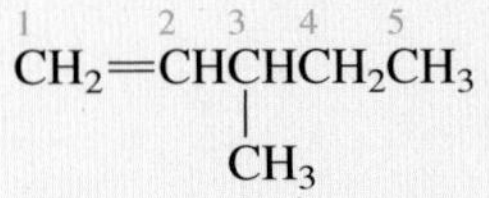

$$\overset{1}{C}H_2 \overset{2}{=} \overset{3}{C}H\overset{4}{C}H\overset{5}{C}H_2CH_3$$
$$|$$
$$CH_3$$

Because the longest chain containing a double bond has five carbon atoms, this is a pentene. It is a 1-pentene, because the double bond is between carbon atoms 1 and 2. The name of the compound is **3-methyl-1-pentene**. Note the placement of hyphens. If the numbering had been in the opposite direction, you would have named the compound as a 4-pentene. But this is unacceptable, because 4 is greater than 1. ⓑ The numbering of the longest chain containing the double bond is

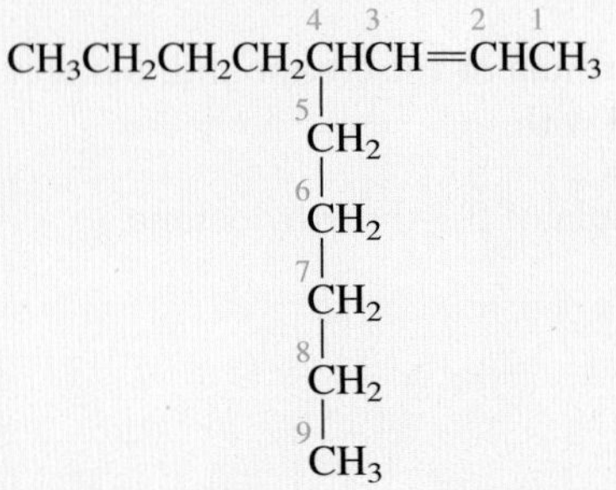

$$CH_3CH_2CH_2\overset{4}{C}H\overset{3}{C}H\overset{2}{=}\overset{1}{C}HCH_3$$
$$\overset{5}{|}$$
$$CH_2$$
$$\overset{6}{|}$$
$$CH_2$$
$$\overset{7}{|}$$
$$CH_2$$
$$\overset{8}{|}$$
$$CH_2$$
$$\overset{9}{|}$$
$$CH_3$$

This gives the name **4-butyl-2-nonene**. There is a longer chain (of ten carbon atoms), but it does not contain the double bond.

Answer Check Here, again, make sure the names correctly lead back to the correct structural formulas.

Exercise 23.6 Give the IUPAC name for each of the following compounds.

ⓐ $CH_3C=CHCHCH_3$
   $\quad\quad|\quad\quad|$
   $\quad\quad CH_3\quad CH_2$
   $\quad\quad\quad\quad\quad|$
   $\quad\quad\quad\quad\quad CH_3$

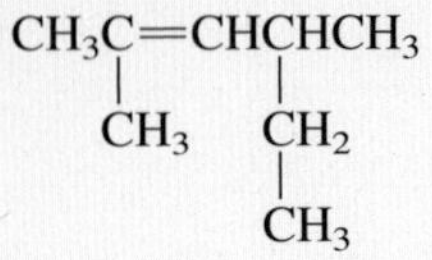

ⓑ $CH_3CH_2CH_2CHCH_2CH_2CH_3$
   $\quad\quad\quad\quad\quad\quad|$
   $\quad\quad\quad\quad\quad\quad CH$
   $\quad\quad\quad\quad\quad\quad\|$
   $\quad\quad\quad\quad\quad\quad CH_2$

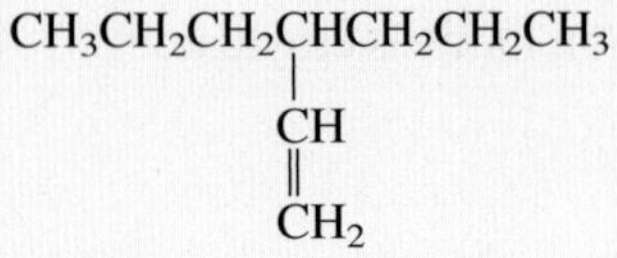

■ See Problems 23.41 and 23.42.

Exercise 23.7 Write the condensed structural formula of 2,5-dimethyl-2-heptene.

■ See Problems 23.43 and 23.44.

Recall that some alkenes also exhibit *cis* and *trans* isomerism. In these cases, either *cis* or *trans* may be included as a prefix, separated by a hyphen, to the name.

Exercise 23.8 Give the IUPAC names, including the *cis* or *trans* label, for each of the geometric isomers of $CH_3CH_2CH=CHCH_2CH_3$.

■ See Problems 23.45 and 23.46.

Alkynes are named using the IUPAC rules in the same way as the alkenes, except that the stem name is determined from the longest chain that contains the carbon–carbon triple bond. The suffix for the stem name is -*yne*.

Exercise 23.9 Give the IUPAC name for each of the following alkynes.

a $CH_3C{\equiv}CH$

b $CH{\equiv}CCHCH_3$
       |
     CH_2CH_3

■ See Problems 23.47 and 23.48.

Nomenclature of Aromatic Hydrocarbons

Simple benzene-containing hydrocarbons that have one group substituted on a benzene ring use the name *benzene* as the suffix and the name of the group as the prefix. For example, the following compound is named ethylbenzene.

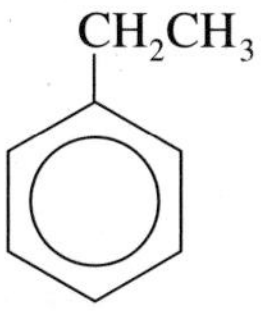

Ethylbenzene

When two groups are on the benzene ring, three isomers are possible. The isomers may be distinguished by using the prefixes *ortho-* (*o*-), *meta-* (*m*-), and *para-* (*p*-). For example,

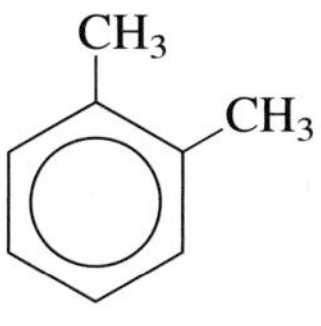

o-Dimethylbenzene   *m*-Dimethylbenzene   *p*-Dimethylbenzene
(*o*-xylene) 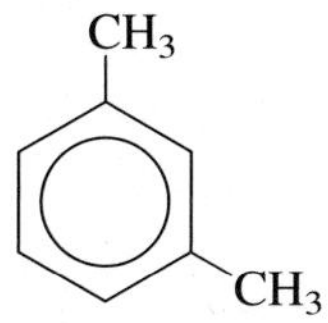   (*m*-xylene)   (*p*-xylene)

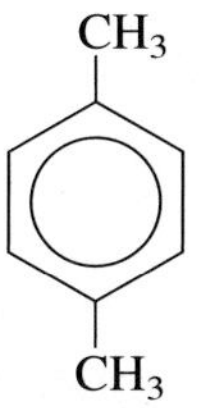

A numbering system is also used to show the positions of two or more groups.

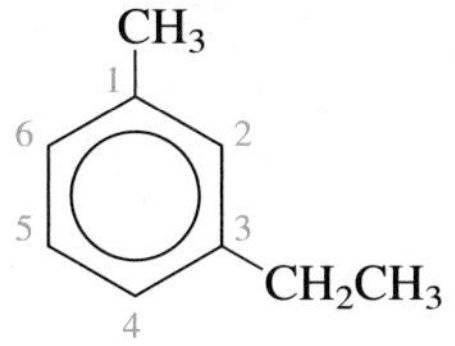

3-Ethyl-1-methylbenzene 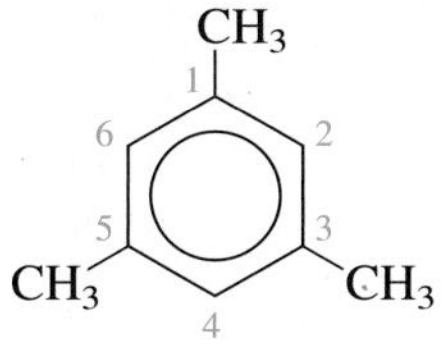   1,3,5-Trimethylbenzene

It is sometimes preferable to name a compound containing a benzene ring by regarding the ring as a group in the same manner as alkyl groups. Pulling a hydrogen atom from benzene leaves the phenyl group, C_6H_5—. For example, you name the following compound diphenylmethane by using methane as the stem name.

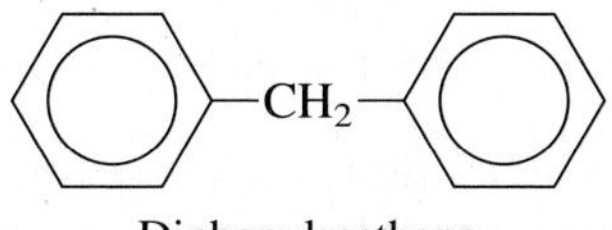

Diphenylmethane

Exercise 23.10 Write the structural formula of a ethylbenzene; b 1,2-diphenylethane.

■ See Problems 23.43 and 23.44.

CONCEPT CHECK 23.3

In the model shown here, C atoms are black and H atoms are light blue.

a Write the molecular formula.

b Write the condensed structural formula.

c Write the IUPAC name.

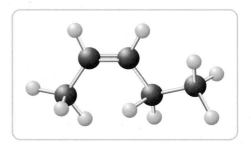

Derivatives of Hydrocarbons

Certain groups of atoms in organic molecules are particularly reactive and have characteristic chemical properties. A **functional group** is *a reactive portion of a molecule that undergoes predictable reactions.* As discussed in Section 23.3, the C=C bond in a compound reacts readily and predictably with the reagents Br_2 and HBr in addition reactions. Thus, the C=C bond is a functional group. Many functional groups contain an atom other than carbon that has lone pair(s) of electrons. These lone pairs of electrons contribute to the reactivity of the functional group. Other functional groups, such as C=O, have multiple bonds that are reactive. Table 23.7 lists some common functional groups.

In the previous sections of the chapter, we discussed the hydrocarbons and their reactions. All other organic compounds can be considered to be derivatives of hydrocarbons. In these compounds, one or more hydrogen atoms of a hydrocarbon have been replaced by atoms other than carbon to give a functional group.

23.6 Organic Compounds Containing Oxygen

Many of the important functional groups in organic compounds contain oxygen. Examples are alcohols, ethers, aldehydes, ketones, carboxylic acids, and esters. In this section, we will look at characteristics of these compounds.

Alcohols and Ethers

Structurally, you may think of an **alcohol** as *a compound obtained by substituting a hydroxyl group (—OH) for an —H atom on a tetrahedral (sp³ hybridized) carbon atom of a hydrocarbon group.* Some examples are

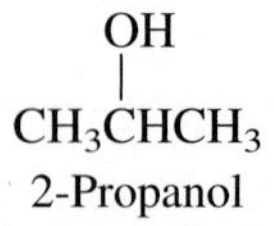

$$CH_3OH \qquad CH_3CH_2OH \qquad CH_3\overset{\overset{\textstyle OH}{|}}{C}HCH_3$$

Methanol   Ethanol   2-Propanol
(methyl alcohol)   (ethyl alcohol)   (isopropyl alcohol)

Alcohols are named by IUPAC rules similar to those for the hydrocarbons, except that the stem name is determined from the longest chain containing the carbon atom to

Table 23.7   Some Organic Functional Groups

Structure of General Compound* (Functional Group in Color)	Name of Functional Group
R—Cl: R—Br:	Organic halide
R—O—H	Alcohol
R—O—R′	Ether
:O: ‖ R—C—H	Aldehyde
:O: ‖ R—C—R′	Ketone
:O: ‖ R—C—O—H	Carboxylic acid
:O: ‖ R—C—O—R′	Ester

R—N—H   R—N—H   R—N—R″
         | |
      H   R′   R′
	Amine

:O: ‖ R—C—N—R′ | H	Amide

*R, R′, and R″ are general hydrocarbon groups.

which the —OH group is attached. The suffix for the stem name is *-ol*. The position of the —OH group is indicated by a number preceding the stem name. (The number is omitted if unnecessary, as in ethanol.)

Alcohols are usually classified by the number of carbon atoms attached to the carbon atom to which the —OH group is bonded. A *primary alcohol* has one such carbon atom, a *secondary alcohol* has two, and a *tertiary alcohol* has three. The following are examples:

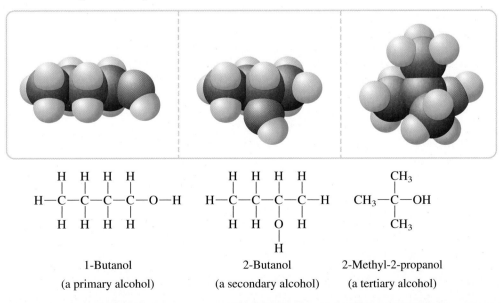

1-Butanol   2-Butanol   2-Methyl-2-propanol

(a primary alcohol)   (a secondary alcohol)   (a tertiary alcohol)

Figure 23.15 ▲

Gasoline containing ethanol
The gasoline in this pump was formulated with ethanol. Ethanol is added to boost octane and oxygenate the fuel.

© Cengage Learning

Methanol, ethanol, ethylene glycol, and glycerol are some common alcohols. Methanol, CH_3OH, was at one time separated from the liquid distilled from saw-dust—hence the common name *wood alcohol*. It is a toxic liquid prepared in large quantities by reacting carbon monoxide with hydrogen at high pressure in the presence of a catalyst. It is used as a solvent and as the starting material for the preparation of formaldehyde. Ethanol is manufactured by the fermentation of glucose (a sugar) or by the addition of water to the double bond of ethylene. The latter reaction is carried out by heating ethylene with water in the presence of sulfuric acid.

$$CH_2\!=\!CH_2 + HOH \xrightarrow{\;H_2SO_4\;} \begin{array}{cc} CH_2\!-\!CH_2 \\ |\quad\;\;\; | \\ H\quad\; OH \end{array}$$

Ethylene   Ethanol

Alcoholic beverages contain ethanol. Ethanol is also a solvent and a starting material for many organic compounds. It is mixed with gasoline and sold as gasohol, an automotive fuel (Figure 23.15). Ethylene glycol (IUPAC name: 1,2-ethanediol) and glycerol (1,2,3-propanetriol) are alcohols containing more than one hydroxyl group.

$$\begin{array}{cc} CH_2\!-\!CH_2 \\ |\quad\;\;\; | \\ OH\quad OH \end{array} \qquad \begin{array}{ccc} CH_2\!-\!CH\!-\!CH_2 \\ |\quad\;\; |\quad\;\; | \\ OH\;\; OH\;\; OH \end{array}$$

Ethylene glycol   Glycerol

Polyesters are discussed in Chapter 24.

Ethylene glycol is a liquid prepared from ethylene and is used as an antifreeze agent. It is also used in the manufacture of polyester plastics and fibers. ◀ Glycerol is a nontoxic, sweet-tasting liquid obtained from fats during the making of soap. It is used in foods and candies to keep them soft and moist.

Just as an alcohol may be thought of as a derivative of water in which one H atom of H_2O has been replaced by a hydrocarbon group, R, an **ether** is *a compound formally obtained by replacing both H atoms of H_2O by the hydrocarbon groups R and R′.*

$$H\!-\!O\!-\!H \qquad R\!-\!O\!-\!H \qquad R\!-\!O\!-\!R'$$

Water   An alcohol   An ether

Common names for ethers are formed from the hydrocarbon groups plus the word *ether*. For example, $CH_3OCH_2CH_2CH_3$ is called methyl propyl ether. By IUPAC rules, the ethers are named as derivatives of the longest hydrocarbon chain. For example, $CH_3OCH_2CH_2CH_3$ is 1-methoxypropane; the methoxy group is $CH_3O\!-\!$. The best-known ether is diethyl ether, $CH_3CH_2OCH_2CH_3$, often called simply ether, a volatile liquid used as a solvent and as an anesthetic.

Exercise 23.11 Give the IUPAC name of the following compound.

$$\begin{array}{c} OH \\ | \\ CH_3CH_2CH_2CCH_2CH_3 \\ | \\ CH_2CH_3 \end{array}$$

■ See Problems 23.53 and 23.54.

Exercise 23.12 Give the common name of each of the following compounds:
a CH_3OCH_3; b $CH_3OCH_2CH_3$.

■ See Problems 23.57 and 23.58.

Aldehydes and Ketones

Aldehydes and ketones are compounds containing a *carbonyl group.* An **aldehyde** is *a compound containing a carbonyl group with at least one H atom attached to it.*

(H)R—$\overset{\displaystyle O}{\overset{\|}{C}}$—H   H—$\overset{\displaystyle O}{\overset{\|}{C}}$—H   CH$_3$—$\overset{\displaystyle O}{\overset{\|}{C}}$—H

An aldehyde   Methanal   Ethanal

(formaldehyde)   (acetaldehyde)

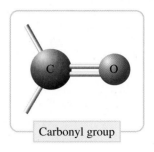

Carbonyl group

Here (H)R indicates a hydrocarbon group or H atom. The aldehyde function is usually abbreviated —CHO, and the structural formula of acetaldehyde is written CH$_3$CHO.

A **ketone** is *a compound containing a carbonyl group with two hydrocarbon groups attached to it.*

R—$\overset{\displaystyle O}{\overset{\|}{C}}$—R′   CH$_3$—$\overset{\displaystyle O}{\overset{\|}{C}}$—CH$_3$ CH$_3$—$\overset{\displaystyle O}{\overset{\|}{C}}$—CH$_2CH_3$

A ketone   Propanone   2-Butanone

(acetone)   (methyl ethyl ketone)

Ethanal

The ketone functional group is abbreviated —CO—, and the structural formula of acetone is written CH$_3$COCH$_3$.

Aldehydes and ketones are named according to IUPAC rules similar to those for naming alcohols. You first locate the longest carbon chain containing the carbonyl group to get the stem hydrocarbon name. Then you change the *-e* ending of the hydrocarbon to *-al* for aldehydes and *-one* for ketones. In the case of aldehydes, the carbon atom of the —CHO group is always the number-1 carbon. In ketones, however, the carbonyl group may occur in various nonequivalent positions on the carbon chain. For ketones, the position of the carbonyl group is indicated by a number before the stem name, just as the position of the hydroxyl group is indicated in alcohols. The carbon chain is numbered to give the smaller number for the position of the carbonyl group.

Propanone

The aldehydes of lower molecular mass have sharp, penetrating odors. Formaldehyde (methanal), HCHO, and acetaldehyde (ethanal), CH$_3$CHO, are examples. With increasing molecular mass, the aldehydes become more fragrant. Some aldehydes of aromatic hydrocarbons have especially pleasant odors (see Figure 23.16). Formaldehyde is a gas produced by the oxidation of methanol. The gas is very soluble in water, and a 37% aqueous solution called Formalin is marketed as a disinfectant and as a preservative of biological specimens. The main use of formaldehyde is in the manufacture of plastics and resins. Acetone, CH$_3$COCH$_3$, is the simplest ketone. It is a liquid with a fragrant odor. The liquid is an important solvent for lacquers, paint removers, and nail polish removers.

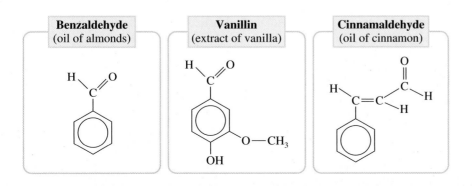

Benzaldehyde (oil of almonds)   **Vanillin** (extract of vanilla)   **Cinnamaldehyde** (oil of cinnamon)

Figure 23.16 ◀

Some aldehydes of aromatic hydrocarbons The name *aromatic* in aromatic hydrocarbons derives from the aromatic odors of many derivatives of these hydrocarbons, including the aldehydes shown here.

Table 23.8   **Common Carboxylic Acids**

Carbon Atoms	Formula	Source	Common Name	IUPAC Name
1	HCOOH	Ants (Latin, *formica*)	Formic acid	Methanoic acid
2	CH_3COOH	Vinegar (Latin, *acetum*)	Acetic acid	Ethanoic acid
3	CH_3CH_2COOH	Milk (Greek, *protos pion,* "first fat")	Propionic acid	Propanoic acid
4	$CH_3(CH_2)_2COOH$	Butter (Latin, *butyrum*)	Butyric acid	Butanoic acid
5	$CH_3(CH_2)_3COOH$	Valerian root (Latin, *valere,* "to be strong")	Valeric acid	Pentanoic acid
6	$CH_3(CH_2)_4COOH$	Goats (Latin, *caper*)	Caproic acid	Hexanoic acid

From Harold Hart, *Organic Chemistry: A Short Course,* Eighth Edition, p. 272. Copyright © 1991 by Houghton Mifflin Company. Used with permission.

Exercise 23.13   Name the following compounds by IUPAC rules:

$$
\text{a } CH_3CH_2CH_2\overset{\overset{\displaystyle O}{\|}}{C}-CH_3
\qquad
\text{b } H\overset{\overset{\displaystyle O}{\|}}{C}-CH_2CH_2CH_3
$$

■ See Problems 23.59 and 23.60.

Carboxylic Acids and Esters

A **carboxylic acid** is *a compound containing the carboxyl group,* —*COOH.*

Ethanoic acid

$$
\underset{\text{Carboxyl group}}{-\overset{\overset{\displaystyle O}{\|}}{C}-O-H}
\qquad
\underset{\text{A carboxylic acid}}{R-\overset{\overset{\displaystyle O}{\|}}{C}-O-H}
\qquad
\underset{\substack{\text{Ethanoic acid}\\ \text{(acetic acid)}}}{CH_3-\overset{\overset{\displaystyle O}{\|}}{C}-O-H}
$$

These compounds are named by IUPAC rules like those for the aldehydes, except that the ending on the stem name is *-oic,* followed by the word *acid.* Many carboxylic acids have been known for a long time and are usually referred to by common names (see Table 23.8). The carboxylic acids are weak acids because of the acidity of the H atom on the carboxyl group. Acid-ionization constants are about 10^{-5}.

An **ester** is *a compound formed from a carboxylic acid, RCOOH, and an alcohol, R′OH.* The general structure is

$$
R\overset{\overset{\displaystyle \colon O\colon}{\|}}{C}-\overset{\cdot\cdot}{\underset{\cdot\cdot}{O}}-R'
$$

23.7 Organic Compounds Containing Nitrogen

Alcohols and ethers, you may recall, can be considered derivatives of H_2O, where one or both H atoms are replaced by hydrocarbon groups, R. Thus, the general formula of an alcohol is ROH, and that of an ether is ROR′. Another important class of organic compounds is obtained by similarly substituting R groups for the H atoms of ammonia, NH_3.

Most organic bases are **amines**, which are *compounds that are structurally derived by replacing one or more hydrogen atoms of ammonia with hydrocarbon groups.*

$$
\underset{\text{A primary amine}}{R-\underset{\underset{\displaystyle H}{|}}{N}-H}
\qquad
\underset{\text{A secondary amine}}{R-\underset{\underset{\displaystyle R'}{|}}{N}-H}
\qquad
\underset{\text{A tertiary amine}}{R-\underset{\underset{\displaystyle R'}{|}}{N}-R''}
$$

Table 23.9   **Some Common Amines**

Name	Formula	Boiling Point °(C)
Methylamine	CH_3-NH_2	−6.5
Dimethylamine	$CH_3-\underset{\underset{H}{\vert}}{N}-CH_3$	7.4
Trimethylamine	$CH_3-\underset{\underset{CH_3}{\vert}}{N}-CH_3$	3.5
Ethylamine	$CH_3CH_2-NH_2$	16.6
Piperidine	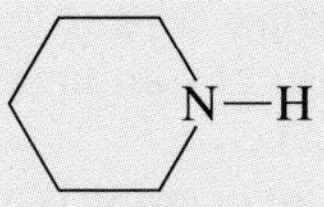 ⬡ N—H	106
Aniline	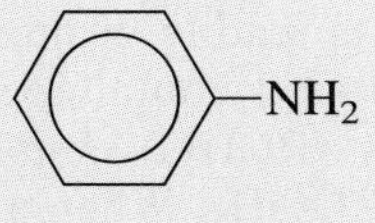 ⬡—NH_2	184

From Robert D. Whitaker et al., *Concepts of General, Organic, and Biological Chemistry*. Copyright © 1981 by Houghton Mifflin Company. Reprinted with permission.

Table 23.9 lists some common amines.

Amines are bases, because the nitrogen atom has an unshared electron pair that can accept a proton to form a substituted ammonium ion. For example,

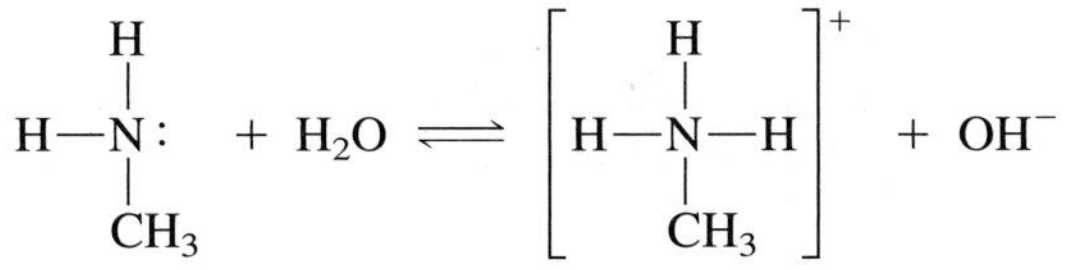

$$H-\underset{\underset{CH_3}{\vert}}{\overset{\overset{H}{\vert}}{N}}: \ + \ H_2O \ \rightleftharpoons \ \left[H-\underset{\underset{CH_3}{\vert}}{\overset{\overset{H}{\vert}}{N}}-H \right]^+ \ + \ OH^-$$

Methylamine   Methylammonium ion

Like ammonia, amines are weak bases. Table 15.2 gives base-ionization constants of some amines.

Amides are *compounds derived from the reaction of ammonia, or of a primary or secondary amine, with a carboxylic acid.* For example, when ammonia is strongly heated with acetic acid, they react to give the amide acetamide. ▶

$$H-\underset{\underset{H}{\vert}}{\overset{\overset{H}{\vert}}{N}}-H \ + \ HO-\underset{\underset{O}{\Vert}}{C}-CH_3 \longrightarrow H-\underset{\underset{O}{\Vert}}{\overset{\overset{H}{\vert}}{N}}-C-CH_3 + H_2O$$

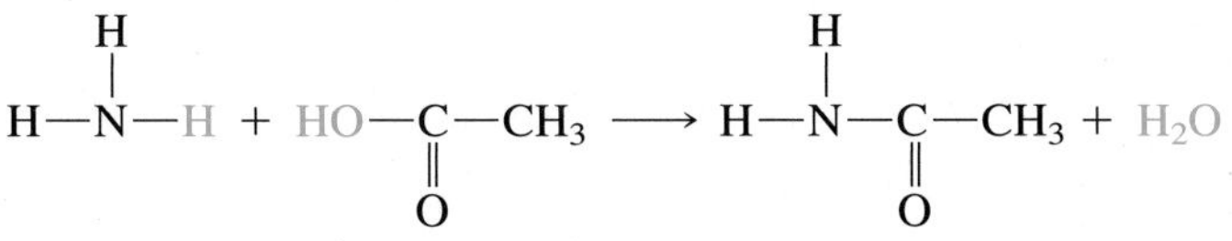

Ammonia   Acetic acid   Acetamide

Methylamine and acetic acid react to give *N*-methylacetamide.

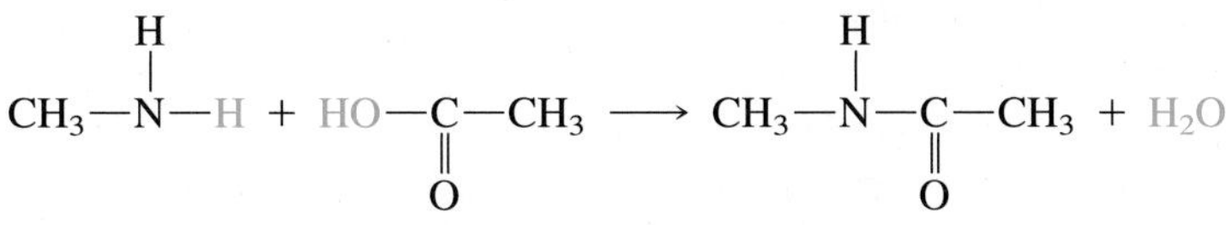

$$CH_3-\underset{\underset{H}{\vert}}{\overset{\overset{H}{\vert}}{N}}-H \ + \ HO-\underset{\underset{O}{\Vert}}{C}-CH_3 \longrightarrow CH_3-\underset{\underset{O}{\Vert}}{\overset{\overset{H}{\vert}}{N}}-C-CH_3 + H_2O$$

Methylamine   Acetic acid   *N*-Methylacetamide

The condensation to give an amide occurs under milder conditions if the ammonia or amine is reacted with the acid chloride, a derivative of the carboxylic acid obtained by replacing

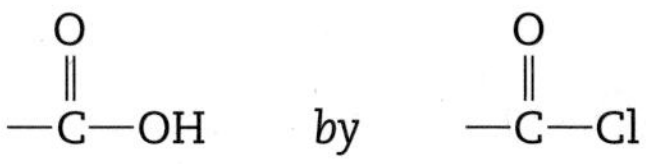

$$-\underset{\underset{}{\Vert}}{\overset{\overset{O}{\Vert}}{C}}-OH \quad by \quad -\underset{\underset{}{\Vert}}{\overset{\overset{O}{\Vert}}{C}}-Cl$$

In this reaction, HCl is a product.

A Checklist for Review

OWL and **Go Chemistry** Sign in at **www.cengage.com/owl** to:

- View tutorials and simulations, develop problem-solving skills, and complete online homework assigned by your professor.
- For quick review and exam prep, download Go Chemistry mini lecture modules from OWL or purchase them at **www.cengagebrain.com**.

Summary of Facts and Concepts

Organic compounds are hydrocarbons or derivatives of hydrocarbons. Hydrocarbons contain only H and C atoms. The three main groups of hydrocarbons are *saturated hydrocarbons,* hydrocarbons with only single bonds between the carbon atoms; *unsaturated hydrocarbons,* hydrocarbons that contain double or triple bonds between carbon atoms; and *aromatic hydrocarbons,* hydrocarbons that contain a benzene ring (a six-membered ring of carbon atoms with alternating single and double carbon–carbon bonds described by resonance formulas).

The *alkanes* are *acyclic,* saturated hydrocarbons that form a *homologous series* of compounds, with the general formula C_nH_{2n+2}. The *cycloalkanes* are *cyclic,* saturated hydrocarbons that form another homologous series with the general formula C_nH_{2n} in which the carbon atoms are joined in a ring. The *alkenes* and *alkynes* are unsaturated hydrocarbons (cyclic or acyclic) that contain carbon–carbon double or triple bonds.

The alkanes and aromatic hydrocarbons usually undergo *substitution reactions.* Alkenes and alkynes undergo *addition reactions. Markownikoff's rule* predicts the major product in the addition of an unsymmetrical reagent to an unsymmetrical alkene.

A *functional group* is a portion of an organic molecule that reacts readily in predictable ways. Important functional groups containing oxygen are *alcohols* (ROH), *aldehydes* (RCHO), *ketones* (RCOR'), and *carboxylic acids* (RCOOH). When a functional group is present in an organic compound, the compound is considered to be a *hydrocarbon derivative. Amines* are organic derivatives of ammonia. They react with carboxylic acids to give *amides.*

Learning Objectives

23.1 The Bonding of Carbon

- Describe the ways in which carbon typically bonds.
- Define *saturated hydrocarbon, unsaturated hydrocarbon,* and *aromatic hydrocarbon.*

23.2 Alkanes and Cycloalkanes

- Define *alkane* and *cycloalkane.*
- Know the general formula of an alkane.
- Give examples of alkanes in a *homologous series.*
- Define *constitutional (structural) isomers.*
- Write the condensed structural formula. **Example 23.1**
- Know the general formula of cyclic *cycloalkanes.*
- State the source of alkanes and cycloalkanes.
- Predict the products of reactions of alkanes with oxygen.
- Predict the products of *substitution reactions* of alkanes with the halogens F_2, Br_2, and Cl_2.

23.3 Alkenes and Alkynes

- Know the general formula of *alkenes* (*olefins*).
- Define *geometric isomers.*
- Predict *cis–trans* isomers. **Example 23.2**
- State Markownikoff's rule.
- Predict the major product of an addition reaction. **Example 23.3**
- Know the general formula of *alkynes.*

23.4 Aromatic Hydrocarbons

- Recognize *aromatic hydrocarbon* molecules.
- Predict the products of substitution reactions of aromatic hydrocarbons.

Important Terms

hydrocarbons
saturated hydrocarbons
unsaturated hydrocarbons
aromatic hydrocarbons

alkanes
cycloalkanes
condensed structural formula
homologous series
constitutional (structural) isomer
substitution reaction

alkenes
geometric isomers
addition reaction
Markownikoff's rule
alkynes

23.5 Naming Hydrocarbons

- Learn the IUPAC rules for naming alkanes.
- Write the IUPAC name of an alkane given the structural formula. **Example 23.4**
- Write the structural formula of an alkane given the IUPAC name. **Example 23.5**
- Learn the additional IUPAC rules for naming alkenes and alkynes.
- Write the IUPAC name of an alkene given the structural formula. **Example 23.6**
- Learn the nomenclature of aromatic hydrocarbons.

23.6 Organic Compounds Containing Oxygen

- Define *functional group*.
- Recognize *alcohols* and *ethers* by functional group.
- Name alcohols using IUPAC rules.
- Use common names for ethers.
- Recognize *aldehydes* and *ketones* by functional group.
- Name aldehydes and ketones using IUPAC rules.
- Recognize *carboxylic acids* and *esters* by functional group.
- Name carboxylic acids and esters using IUPAC rules.

functional group
alcohol
ether
aldehyde
ketone
carboxylic acid
ester

23.7 Organic Compounds Containing Nitrogen

- Recognize *amines* and *amides* by functional group.

amines
amides

Questions and Problems

OWL Interactive versions of these problems may be assigned in OWL.

Self-Assessment and Review Questions

Key: These questions test your understanding of the ideas you worked with in the chapter. These problems vary in difficulty and often can be used for the basis of discussion.

23.1 Give the molecular formula of an alkane with 30 carbon atoms.

23.2 Draw structural formulas of the five isomers of C_6H_{14}.

23.3 Draw structural formulas of an alkane, a cycloalkane, an alkene, and an aromatic hydrocarbon, each with seven carbon atoms.

23.4 Draw structural formulas for the isomers of ethylmethylbenzene.

23.5 Explain why there are two isomers of 2-butene. Draw their structural formulas and name the isomers.

23.6 Fill in the following table.

Hydrocarbon	Source	Use
Methane		
Octane		
Ethylene		
Acetylene		

23.7 Give condensed structural formulas of all possible substitution products of ethane and Cl_2.

23.8 Define the terms *substitution reaction* and *addition reaction*. Give examples of each.

23.9 What would you expect to be the major product when two molecules of HCl add successively to acetylene? Explain.

23.10 Write the structure of propylbenzene. Write the structure of *p*-dichlorobenzene.

23.11 What is a functional group? Give an example and explain how it fits this definition.

23.12 Fill in the following table.

Compound	Source	Use
Methanol		
Ethanol		
Ethylene glycol		
Glycerol		
Formaldehyde		

23.13 An aldehyde contains the carbonyl group. Ketones, carboxylic acids, and esters also contain the carbonyl group. What distinguishes these latter compounds from an aldehyde?

23.14 Identify and name the functional group in each of the following.

a $CH_3OCH_2CH_3$ b CH_3COCH_3
c CH_3CH_2COOH d $CH_3CH=CH_2$
e $CH_3CH_2CH_2OH$ f CH_3CH_2CHO

23.15 How many isomers exist that have the formula C_4H_8?
a two   b three   c four   d five   e six

23.16 The correct formula for butane is
a C_2H_6 b C_8H_{18}
c CH_4 d C_4H_{10}
e C_3H_6

23.17 Which of the following could exist as a *cis–trans* isomer?

a $CH_3CH_2CH_2CH_3$

b $CH_3CHCH_2CH_3$
      |
     CH_3

c CH_3CCH_2
     |
    CH_3

d $CH_3CHCHCH_3$

e none of these

23.18 What is the correct IUPAC name for the following compound?

$$CH_2CH_3$$
$$|$$
$$CH_3CH_2CHCH_3$$
$$|$$
$$OH$$

a 3-methyl-4-pentanol
b 2-ethyl-3-butanol
c 3-ethyl-2-butanol
d 2-ethyl-3-hexanol
e 3-methyl-2-pentanol

Conceptual Problems

Key: These problems are designed to check your understanding of the concepts associated with some of the main topics presented in each chapter. A strong conceptual understanding of chemistry is the foundation for both applying chemical knowledge and solving chemical problems. These problems vary in level of difficulty and often can be used as a basis for group discussion.

23.19 A classmate tells you that the following compound has the name 3-propylhexane.

$$CH_2CH_2CH_3$$
$$|$$
$$CH_3CH_2CHCH_2CH_2CH_3$$

a Is he right? If not, what error did he make and what is the correct name?
b How could you redraw the condensed formula to better illustrate the correct name?

23.20 You are distilling a barrel of oil that contains the hydrocarbons listed in Table 23.4. You heat the contents of the barrel to 200°C.
a What molecules will no longer be present in your sample of oil?
b What molecules will be left in the barrel?
c Provide an explanation for your answers in parts a and b.
d Which molecule would boil off at a lower temperature, hexane or 2,3-dimethylbutane?

23.21 Explain why you wouldn't expect to find a compound with the formula CH_5.

23.22 Catalytic cracking is an industrial process used to convert high-molecular-mass hydrocarbons to low-molecular-mass hydrocarbons. A petroleum company has a huge supply of heating oil stored as straight-chain $C_{17}H_{36}$, and demand has picked up for shorter chain hydrocarbons to be used in formulating gasoline. The company uses catalytic cracking to create the shorter chains necessary for gasoline. If they produce two molecules in the cracking, and 1-octene is one of them, what is the formula of the other molecule produced? As part of your answer, draw the condensed structural formula of the 1-octene.

23.23 In the following four models, C atoms are black, H atoms are light blue, O atoms are red, and N atoms are dark blue:

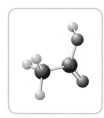

a Write the molecular formula of each molecule.
b Write the condensed structural formula for each molecule.
c Identify the functional group for each molecule.

23.24 In the models shown here, C atoms are black and H atoms are light blue:

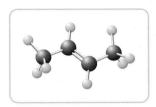

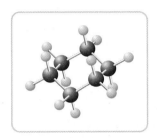

a Write the molecular formula of each molecule.
b Write the condensed structural formula for each molecule.
c Give the IUPAC name of each molecule.

23.25 Why would you expect the melting points of the alkanes to increase in the series methane, ethane, propane, and so on?

23.26 Consider the following formulas of two esters:

$$CH_3CH_2-\overset{\overset{\text{O}}{\|}}{C}-O-CH_3 \qquad CH_3CH_2-O-\overset{\overset{\text{O}}{\|}}{C}-CH_3$$

One of these is ethyl ethanoate (ethyl acetate) and one is methyl propanoate (methyl propionate). Which is which?

Practice Problems

Key: These problems are for practice in applying problem-solving skills. They are divided by topic, and some are keyed to exercises (see the ends of the exercises). The problems are arranged in matching pairs; the odd-numbered problem of each pair is listed first, and its answer is given in the back of the book.

Condensed Structural Formulas

23.27 Write the condensed structural formula of the following alkane.

$$H-\overset{\overset{\displaystyle H}{|}}{\underset{\underset{\displaystyle H}{|}}{C}}-\overset{\overset{\displaystyle H}{|}}{\underset{\underset{\displaystyle H}{|}}{C}}-\overset{\overset{\displaystyle H-\overset{\displaystyle H}{|}}{\underset{\displaystyle |}{C}-H}}{\underset{\underset{\displaystyle H}{|}}{C}}-H$$

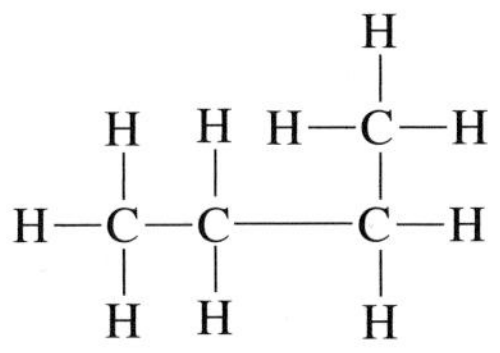

23.28 Write the condensed formula of the following alkane.

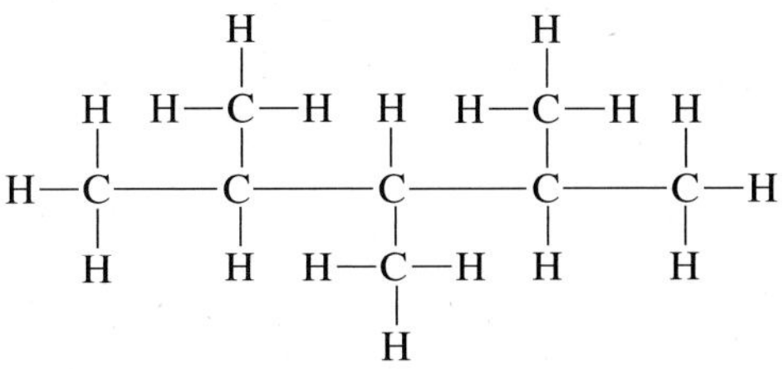

Geometric Isomers

23.29 One or both of the following have geometric isomers. Draw the structures of any geometric isomers.

a $CH_3CHCH=CHCH_3$
        $\overset{|}{CH_3}$

b $CH_3C=CHCH_2CH_3$
        $\overset{|}{CH_3}$

23.30 If there are geometric isomers for the following, draw structural formulas showing the isomers.

a $CH_3CH_2CH=CHCH_2CH_3$
b $CH_3C=CHCH_2CH_3$
        $\overset{|}{CH_2CH_3}$

Reactions of Hydrocarbons

23.31 Complete and balance the following equations. Note any catalyst used.

a $C_2H_4 + O_2 \longrightarrow$
b $CH_2=CH_2 + MnO_4^- + H_2O \longrightarrow$
c $CH_2=CH_2 + Br_2 \longrightarrow$

d

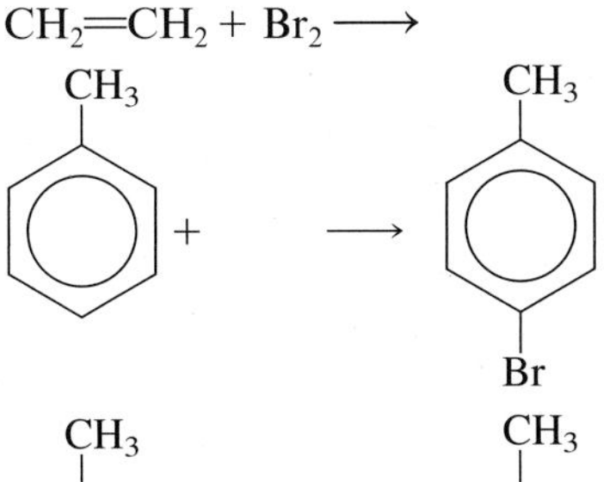

e

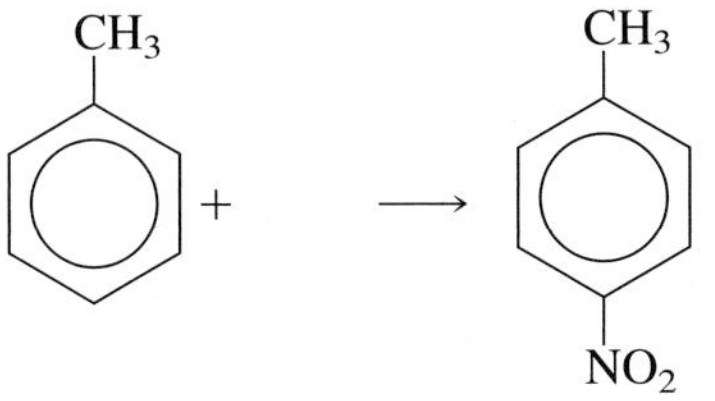

23.32 Complete and balance the following equations. Note any catalyst used.

a $C_6H_6 + O_2 \longrightarrow$

b

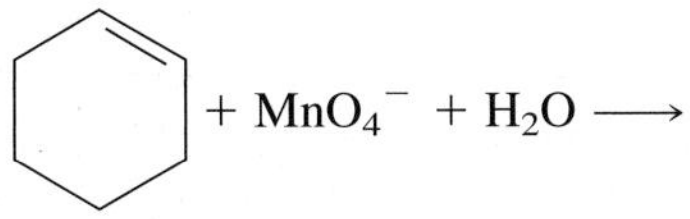

Cyclohexene

$\longrightarrow$ + MnO_4^- + H_2O \longrightarrow$

c $CH_2=CH_2 + HBr \longrightarrow$

23.33 Write an equation for a possible substitution reaction of propane, C_3H_8, with Cl_2.

23.34 Write an equation for a possible substitution reaction of ethane, C_2H_6, with Br_2.

23.35 Complete the following equation, giving only the main product.

$$CH_2\!=\!CHCH_3 + H\!-\!OH \xrightarrow[\text{catalyst}]{H_2SO_4}$$

23.36 What is the major product when HBr is added to methylpropene?

Naming Hydrocarbons

23.37 Give the IUPAC name for each of the following hydrocarbons.

23.38 What is the IUPAC name of each of the following compounds?

23.39 Write the condensed structural formula for each of the following compounds.
- a 2,2,3,3-tetramethylpentane
- b 2-methyl-4-isopropylheptane
- c 2,3-dimethylhexane
- d 3-ethylhexane

23.40 Write the condensed structural formula for each of the following compounds.
- a 2,2-dimethylpentane
- b 2-ethyl-2-methylhexane
- c 3-ethyl-2-methyloctane
- d 3,4,4,5-tetramethylheptane

23.41 Give the IUPAC name of each of the following.
- a $CH_3CH\!=\!CHCH_2CH_3$
- b $CH_3C\!=\!CHCH_2CHCH_3$ with CH_3 groups

23.42 For each of the following, write the IUPAC name.

23.43 Give the condensed structural formula for each of the following compounds.
- a 4-ethyl-2-methyl-2-hexene
- b 3-ethyl-2-pentene

23.44 Write the condensed structural formula for each of the following compounds.
- a 2-methyl-4-propyl-3-decene
- b 2,3-dimethyl-2-hexene

23.45 Give the IUPAC names and include the *cis* or *trans* label for each of the isomers of $CH_3CH\!=\!CHCH_3$.

23.46 Give the IUPAC names and include the *cis* or *trans* label for each of the isomers of $CH_3CH\!=\!CHCH_2CH_3$.

23.47 Give the IUPAC name of each of the following compounds.
- a $CH_3C\!\equiv\!CCH_3$
- b $CH\!\equiv\!CCHCH_2CH_3$ with CH_3

23.48 Write the IUPAC name of each of the following hydrocarbons.
- a $CH_3CHC\!\equiv\!CH$ with CH_3
- b $CH_3C\!\equiv\!CCH_2CH_3$

23.49 Write structural formulas for a *o*-ethylmethylbenzene; b 1,1,1-triphenylethane.

23.50 Write structural formulas for a *p*-diethylbenzene; b 1,2,3-trimethylbenzene.

Naming Oxygen-Containing Organic Compounds

23.51 Circle and name the functional group in each compound.

c HO—C(=O)—CH₂CH₃   **d** H—C(=O)—CH₂CH₃

$$\text{c) } HO-\overset{\displaystyle O}{\overset{\|}{C}}-CH_2CH_3 \qquad \text{d) } H-\overset{\displaystyle O}{\overset{\|}{C}}-CH_2CH_3$$

23.52 Circle and name the functional group in each compound.

a $CH_3-\overset{\displaystyle O}{\overset{\|}{C}}-OH$

b $\overset{\displaystyle O=C-OH}{\underset{|}{CH_3CHCH_3}}$

c $HO-CH_2\overset{|}{\underset{|}{C}}H$ with CH_3 above and CH_3 below
$$HO-CH_2\overset{\displaystyle CH_3}{\underset{\displaystyle CH_3}{\overset{|}{C}H}}$$

d $\overset{\displaystyle O=C-CH_3}{\underset{\displaystyle CH_3CHCH_3}{|}}$

23.53 Give the IUPAC name for each of the following.
a $HOCH_2CH_2CH_2CH_2CH_3$
b $CH_3\overset{|}{\underset{\displaystyle OH}{C}H}CH_2CH_2CH_3$
c $CH_3CH_2CH_2\overset{|}{C}HCH_2CH_2CH_3$
$$H-\overset{|}{\underset{\displaystyle H}{C}}-OH$$
d $CH_3CH_2CH_2\overset{\displaystyle OH}{\overset{|}{C}}HCH_2\overset{\displaystyle CH_3}{\overset{|}{C}}HCH_2CH_3$

23.54 Write the IUPAC name for each of the following.
a $HOCH_2\overset{|}{\underset{\displaystyle CH_2CH_2CH_3}{C}}HCH_2CH_3$
b $HOCH_2CH_2CH_2\overset{|}{\underset{\displaystyle CH_3}{C}}HCH_3$
c $CH_3\overset{|}{C}HCH_2CH_3$
$$H-\overset{|}{\underset{\displaystyle CH_3}{C}}-OH$$
d $HOCH_2\overset{|}{\underset{\displaystyle CH_3}{C}}HCH_2CH_3$

23.55 State whether each of the following alcohols is primary, secondary, or tertiary.
a $CH_3\overset{|}{\underset{\displaystyle OH}{C}}HCH_2CH_3$
b $\overset{\displaystyle CH_2CH_3}{\underset{\displaystyle OH}{H\overset{|}{C}CH_2CH_3}}$
c $HOCH_2\overset{\displaystyle CH_3}{\underset{\displaystyle CH_3}{\overset{|}{C}H}}$
d $HOCH_2\overset{\displaystyle CH_3}{\underset{\displaystyle CH_3}{\overset{|}{C}CH_3}}$

23.56 Classify each of the following as a primary, secondary, or tertiary alcohol.
a $CH_3\overset{\displaystyle CH_3}{\underset{\displaystyle CH_3}{\overset{|}{C}}CH_2\overset{\displaystyle CH_3}{\overset{|}{C}}HOH}$
b $CH_3CH_2CH_2\overset{\displaystyle CH_3}{\underset{\displaystyle CH_3}{\overset{|}{C}}OH}$

c $CH_3CH_2\overset{\displaystyle CH_2OH}{\underset{\displaystyle CH_3}{\overset{|}{C}}CH_3}$
d $CH_3CH_2\overset{\displaystyle CH_2CH_3}{\underset{\displaystyle CH_2CH_3}{\overset{|}{C}}OH}$

23.57 What is the common name of each of the following compounds?
a $CH_3CH_2OCH_2CH_2CH_3$
b $\overset{\displaystyle CH_3}{\underset{\displaystyle CH_3}{H-\overset{|}{C}OCH_3}}$

23.58 What is the common name of each of the following compounds?
a $CH_3O\overset{\displaystyle CH_3}{\underset{\displaystyle CH_3}{\overset{|}{C}CH_3}}$
b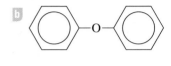

23.59 According to IUPAC rules, what is the name of each of the following compounds?
a $CH_3COCH_2CH_3$
b $CH_3CH_2CH_2CHO$
c $H-\overset{\displaystyle O}{\overset{\|}{C}}CH_2CH_2\overset{\displaystyle CH_3}{\underset{\displaystyle CH_3}{\overset{|}{C}}CH_3}$
d $CH_3\overset{\displaystyle O}{\underset{\displaystyle CH_2CH_3}{\overset{|}{C}}H\overset{\|}{C}CH_3}$

23.60 Write the IUPAC name of each of the following compounds.
a $CH_3\overset{|}{\underset{\displaystyle CHO}{C}}HCH_3$
b $CH_3\overset{|}{\underset{\displaystyle COCH_3}{C}}HCH_3$
c $CH_3CH_2\overset{|}{\underset{\displaystyle O}{C}H_2\overset{\|}{C}-H}$
d $CH_3\overset{|}{C}HCH_3$
$$CH_2-\overset{\displaystyle O}{\overset{\|}{C}}-CH_2CH_3$$

Organic Compounds Containing Nitrogen

23.61 Identify each of the following compounds as a primary, secondary, or tertiary amine, or as an amide.
a $CH_3CH_2NHCH_2CH_3$
b CH_3NH — (attached to benzene ring)

23.62 Identify each of the following compounds as a primary, secondary, or tertiary amine, or as an amide.
a $CH_3CH_2CH_2NH_2$
b $O=C-NH_2$ (attached to benzene ring)

General Problems

Key: These problems provide more practice but are not divided by topic or keyed to exercises. Odd-numbered problems and the even-numbered problems that follow are similar; answers to all odd-numbered problems except the essay questions are given in the back of the book.

23.63 Give the IUPAC name of each of the following compounds.

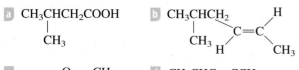

a CH_3CHCH_2COOH
 $\quad\quad\ \ |$
 $\quad\quad\ CH_3$

b CH_3CHCH_2
 $\quad\quad\ \ |$
 $\quad\quad\ CH_3$ ...C=C... H, CH_3, H

c $\quad\ O\quad CH_3$
 $\quad\quad\ \|\quad\ \ |$
 CH_3CHCCH_2CH
 $\quad\ |\quad\quad\ |$
 $\quad CH_2\quad CH_3$
 $\quad\ |$
 $\quad CH_3$

d $CH_3CHC\equiv CCH_3$
 $\quad\quad\ |$
 $\quad\quad CH_3$

23.64 Give the IUPAC name of each of the following compounds

a $\quad\ COOH$
 $\quad\quad\ |$
 $CH_3CCH_2CH_3$
 $\quad\ |$
 $\quad CH_3$

b $\quad\ CH_3$
 $\quad\quad\ |$
 CH_3CCOCH_3
 $\quad\ |$
 $\quad CH_2CH_3$

c $\quad\ CH_3$
 $\quad\quad\ |$
 $CH_3CCH_2CH_3$
 $\quad\ |$
 $\quad CH_3$

d $\quad\ CHO$
 $\quad\quad\ |$
 CH_3CCH_3
 $\quad\ |$
 $\quad CH_2CH_3$

23.65 Write the structural formula for each of the following compounds.
 a isopropyl propionate
 b *t*-butylamine
 c 2,2-dimethylhexanoic acid
 d *cis*-3-hexene

23.66 Write the structural formula for each of the following compounds.
 a 3-ethyl-1-hexene
 b 1,1,2,2-tetraphenylethane
 c 1-phenyl-2-butanone
 d cyclooctanone

23.67 Describe chemical tests that could distinguish between:
 a $CH_2=CH-C\equiv C-CH=CH_2$ and benzene
 b propionaldehyde (propanal) and acetone (propanone)

23.68 Describe chemical tests that could distinguish between:
 a toluene (methylbenzene) and 2-methylcyclohexene
 b acetic acid and acetaldehyde (ethanal)

23.69 Identify each of the following compounds.
 a An alcohol used as a starting material for the manufacture of formaldehyde.
 b A gas with a sweetish odor that promotes the ripening of green fruit.
 c A compound with an ammonia-like odor that acts as a base; its molecular formula is CH_5N.
 d An unsaturated compound of the formula C_7H_8 that gives a negative test with bromine in carbon tetrachloride.

23.70 Identify each of the following compounds.
 a An acidic compound that also has properties of an aldehyde; its molecular formula is CH_2O_2.
 b A compound used as a preservative for biological specimens and as a raw material for plastics.
 c A saturated hydrocarbon boiling at 0°C that is liquefied and sold in cylinders as a fuel.
 d A saturated hydrocarbon that is the main constituent of natural gas.

23.71 A compound with a fragrant odor reacts with dilute acid to give two organic compounds, A and B. Compound A is identified as an alcohol with a molecular mass of 32.0 amu. Compound B is identified as an acid. It can be reduced to give a compound whose composition is 60.0% C, 13.4% H, and 26.6% O and whose molecular mass is 60.1 amu. What is the name of the original compound?

23.72 A compound that is 85.6% C and 14.4% H and has a molecular mass of 56.1 amu reacts with water and sulfuric acid to produce a compound that reacts with acidic potassium dichromate solution to produce a ketone. What is the name of the original hydrocarbon?

Strategy Problems

Key: As noted earlier, all of the practice and general problems are matched pairs. This section is a selection of problems that are not in matched-pair format. These challenging problems require that you employ many of the concepts and strategies that were developed in the chapter. In some cases, you will have to integrate several concepts and operational skills in order to solve the problem successfully.

23.73 For the following molecular model, where C atoms are black, H atoms are light blue, and oxygen is red:
 a Name the functional group.
 b Write the IUPAC name.

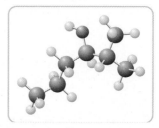

23.74 Name all of the functional groups present in the following molecule.

O   CH
‖ |
HCCH$_2$CHCHCOOH
|
CH$_3$

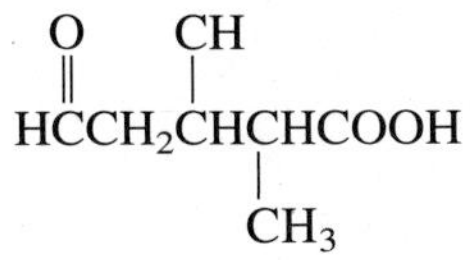

23.75 For each of the following molecules, create the correct condensed structural formulas by adding and/or removing hydrogen atoms (do not add or remove bonds).

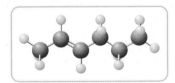

- a CH$_2$CH$_2$CH=CH
- b CH$_3$≡CHCH$_2$CH$_2$
- c CH$_3$CHC=CCH$_2$

23.76 Draw the condensed structural formulas for a primary, a secondary, and a tertiary alcohol, each containing five carbon atoms.

23.77 Write the condensed structural formulas of a ketone and an ester, each containing four carbon atoms.

23.78 For the following molecular model, where C atoms are black and H atoms are light blue:

- a Write the molecular formula.
- b Write the condensed structural formula.
- c Write the IUPAC name.

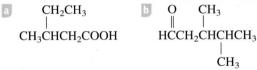

23.79 You encounter a hydrocarbon with the name 2-ethyl-3-methylhexane. Is this a proper IUPAC name? If not, what is the proper IUPAC name? To answer this question, start by writing the condensed structural formula based on this name.

23.80 Consider the molecule CH$_3$CH$_2$CH=CHCH$_3$.

- a Write a structural formula of the *trans* isomer.
- b Name the molecule from part a.
- c Write the condensed structural formula of the major product of the reaction of the molecule with Br$_2$.

23.81 Give the IUPAC name of each of the following.

- a
  CH$_2$CH$_3$
  |
  CH$_3$CHCH$_2$COOH

- b
  O   CH$_3$
  ‖ |
  HCCH$_2$CHCHCH$_3$
  |
  CH$_3$

23.82 Write the condensed structural formula for each of the following.

- a An aldehyde that contains five carbon atoms.
- b An ether that contains four carbon atoms.
- c An aromatic hydrocarbon with methyl groups at positions 1 and 5.
- d A secondary amine that contains five carbon atoms.
- e An alkyne that contains four carbon atoms.

23.83 Draw the structural formula of the following hydrocarbons.

- a chlorobenzene
- b 4-methyl-2-hexene
- c *n*-octane
- d For each answer (a–c), indicate if the molecule is saturated or unsaturated.

23.84 For each of the following, write the condensed structural formula of a molecule that fits the description. Add hydrogen atoms to complete the structure.

- a An aromatic hydrocarbon that contains six carbon atoms and two fluorine atoms.
- b A saturated hydrocarbon that contains seven carbon atoms and one bromine atom at position number three on the carbon chain.
- c An unsaturated alkene that contains five carbon atoms and one double bond.
- d A secondary amine.

23.85 Amino acids are important biological molecules. Using the name *amino acid* as a guide, write the name and structure of the functional groups common to all amino acids.

23.86 Draw the *cis* and *trans* isomers of the compound 3-methyl-3-hexene.

23.87 Indicate the molecular geometry around each carbon atom in the compound

O
‖
H$_3$CCH=CHCH$_2$CCH$_2$COOH

23.88 Complete and balance the reaction for each of the following undergoing a combustion reaction.

- a 2-methylpentane
- b 2,2-dimethylpropane
- c 3-methyl-3-hexene
- d 3-ethyl-3-propanol
- e ethylbenzene

Polymer Materials: Synthetic and Biological

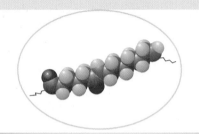

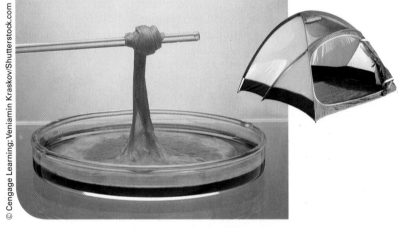

© Cengage Learning; Veniamin Kraskov/Shutterstock.com

Pulling nylon from a reaction mixture (blue dye has been added). Nylon is formed from long, chainlike polymer molecules that line up with one another through intermolecular hydrogen bonding, resulting in a tough, fiberous material with applications ranging from fabrics to molded engineering products (such as gears).

Contents and Concepts

Synthetic Polymers

Polymers are chemical species of high molecular mass made up of many small repeating units. Synthetic polymers are used in many important products, such as plastics, polyurethane coatings, and nylon. Their syntheses have also helped scientists understand the nature of polymers, including many natural materials.

24.1 Synthesis of Organic Polymers

24.2 Electrically Conducting Polymers

Biological Polymers

The biological polymers include carbohydrates such as starch and cellulose, proteins, and nucleic acids (DNA and RNA). We will look at the structure of proteins and nucleic acids in these sections.

24.3 Proteins

24.4 Nucleic Acids

OWL Sign in to OWL at **www.cengage.com/owl** to view tutorials and simulations, develop problem-solving skills, and complete online homework assigned by your professor.

GO Chemistry Download mini lecture videos for key concept review and exam prep from OWL or purchase them from **www.cengagebrain.com.**

C elluloid, discovered around 1850, was the first commercial plastic. Initially it was used as a cheap substitute for ivory in making billiard balls, but it soon became the material of choice for many consumer items such as combs and toothbrushes. Perhaps its best-known application, though, was in producing photographic film for motion pictures. Celluloid is a tough plastic, but it has one nasty property—it catches fire and burns explosively, a property that stems from celluloid's main ingredient, cellulose nitrate. Cellulose nitrate (also known as guncotton) is made by soaking a cellulose material such as cotton in a mixture of nitric and sulfuric acids. Soon after celluloid billiard balls came on the market, there were reports of billiard balls that vanished with a flash and a pop when touched accidentally by a person's cigar! There were also a number of serious theater fires. By the early twentieth century, celluloid was displaced in most of its applications by Bakelite, a hard plastic produced from formaldehyde, HCHO, and phenol, C_6H_5OH.

Plastics, such as celluloid and Bakelite, are *polymers,* which are giant molecules constructed by covalently bonding together many small molecules. The cellulose nitrate of celluloid is a chemically modified form of the natural polymer cellulose, which plants make from the sugar glucose $C_6H_{12}O_6$. Bakelite was the first completely synthetic polymer, produced commercially from nonpolymeric chemicals. Since then, many different plastics with various properties have been developed (Figure 24.1).

In this chapter, we will look at some basic features of synthetic and biological, or natural, polymers. In the part concerning biological polymers, we will concentrate on proteins and how the genetic code, constructed from the polymer DNA, leads to the synthesis of proteins.

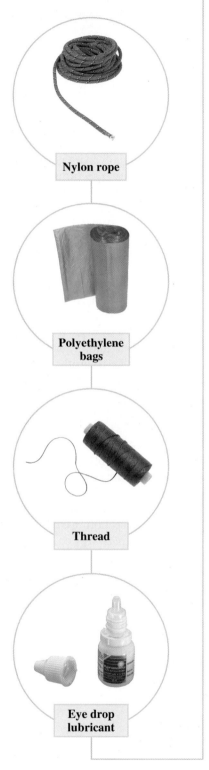

Nylon rope

Polyethylene bags

Thread

Eye drop lubricant

Paper glue

Latex paint

Polyethylene bottle

Figure 24.1 ▶

Products made of synthetic polymers

Synthetic Polymers

A **polymer** is *a chemical species of very high molecular mass made up from many units of low molecular mass covalently linked together. A compound used to make a polymer (and from which the polymer's unit arises) is called a* **monomer**. A simple example is polyethylene, which consists of many ethylene units, —CH_2CH_2—, covalently bonded to one another to form long-chain molecules consisting of thousands of ethylene units. The name *polyethylene* stems from the name of the monomer (ethylene) plus the prefix *poly-*, meaning "many." Shown here is a short piece of a polyethylene molecule.

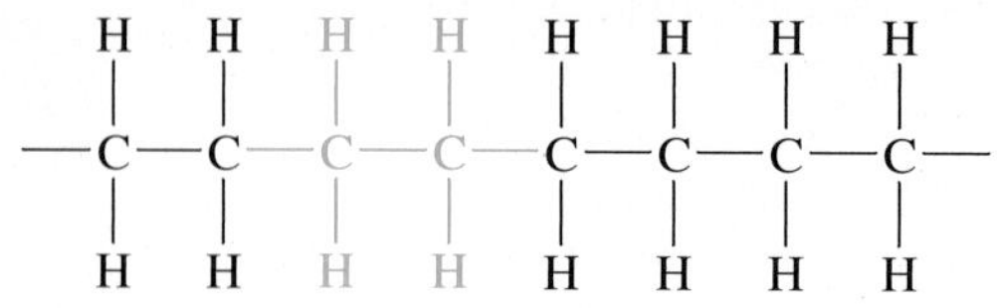

$$
\begin{array}{cccccccc}
\text{H} & \text{H} & \text{H} & \text{H} & \text{H} & \text{H} & \text{H} & \text{H} \\
| & | & | & | & | & | & | & | \\
-\text{C}-\text{C}-\text{C}-\text{C}-\text{C}-\text{C}-\text{C}-\text{C}- \\
| & | & | & | & | & | & | & | \\
\text{H} & \text{H} & \text{H} & \text{H} & \text{H} & \text{H} & \text{H} & \text{H}
\end{array}
$$

One ethylene unit is shown in color.

Polyethylene is an example of a synthetic polymer, one made in the laboratory or commercially from nonpolymeric substances. We will look at synthetic polymers in the next two sections. The first polymers that chemists studied, however, were natural or biological polymers, including rubber, cellulose, starch, and proteins. Proteins and deoxyribonucleic acids (DNA) are polymers that form the basis of life on earth. We will examine these in the last part of the chapter.

24.1 Synthesis of Organic Polymers

The idea that certain natural products such as rubber are composed of giant molecules, or polymers, consisting of many repeating units linked by covalent bonds arose largely from the work of the German chemist Hermann Staudinger (1881–1965) in the early 1920s. His work ushered in an era of systematic study of polymers. Although chemists were initially skeptical of Staudinger's idea of giant molecules, the organic chemist Wallace Carothers proved this idea by using methods of organic chemistry to synthesize polymers (see the essay "The Discovery of Nylon"). Staudinger received the Nobel Prize in 1953 for his contributions to polymer chemistry.

It would be difficult to overemphasize the importance of synthetic polymers to present-day technology. Perhaps you are seated in front of a computer monitor, whose casing is made of hard plastic. The desk on which it sits probably has a coating of a polymer resin, such as polyurethane. Your clothing might well be made of fibers composed of synthetic polymers, such as polyester or nylon. And any elastic bands, such as those on your socks, are made from *elastomers,* an elastic or rubbery type of polymer.

In this section, we describe several synthetic polymers. Polymers are classified as *addition polymers* or *condensation polymers,* depending on the type of reaction used in forming them.

Addition Polymers

An **addition polymer** is *a polymer formed by linking together many molecules by addition reactions.* The monomers must have multiple bonds that will undergo addition reactions. For example, when propylene (IUPAC name, propene) is heated under pressure with a catalyst, it forms polypropylene.

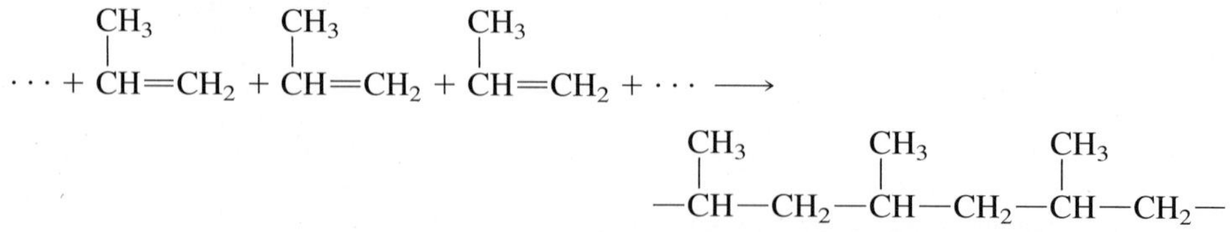

$$
\cdots + \underset{\displaystyle \overset{|}{\text{CH}_3}}{\text{CH}} = \text{CH}_2 + \underset{\displaystyle \overset{|}{\text{CH}_3}}{\text{CH}} = \text{CH}_2 + \underset{\displaystyle \overset{|}{\text{CH}_3}}{\text{CH}} = \text{CH}_2 + \cdots \longrightarrow
$$

$$
-\underset{\displaystyle \overset{|}{\text{CH}_3}}{\text{CH}} - \text{CH}_2 - \underset{\displaystyle \overset{|}{\text{CH}_3}}{\text{CH}} - \text{CH}_2 - \underset{\displaystyle \overset{|}{\text{CH}_3}}{\text{CH}} - \text{CH}_2 -
$$
$$
\text{Polypropylene}
$$

Note that the π electrons in the double bonds of the propylene molecules form new σ bonds that unite the monomers.

Free-Radical Addition The preparation of an addition polymer is often induced by an *initiator,* a compound that produces free radicals (species having an unpaired electron). Organic peroxides (organic compounds with the —O—O— group) are frequently used as initiators. If you write R for an organic group, the general formula of such a peroxide is R—O—O—R. When heated, the —O—O— bond in the organic peroxide breaks, and free radicals form. (The unpaired electron is symbolized by a dot.) Figure 24.2 illustrates the preparation of polystyrene using benzoyl peroxide as an initiator.

$$R-O-O-R \xrightarrow{\Delta} R-O\cdot + \cdot O-R$$

The free radical reacts with the alkene monomer to produce a new free-radical species, which in turn reacts with another monomer molecule. For example, the polymerization of propylene occurs as follows:

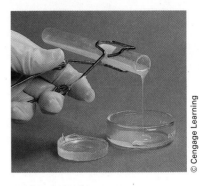

Figure 24.2 ▲

Preparation of polystyrene When styrene, $C_6H_5CH=CH_2$, is heated with a small amount of benzoyl peroxide, ROOR (R = C_6H_5CO-), it yields a viscous liquid. After some time, this liquid sets to a hard plastic (sample shown at left).

At each addition step, the unpaired electron from the free radical pairs with one of the π electrons of the alkene monomer to form a σ bond. One end of the chain retains an unpaired electron and can add another monomer molecule. The reaction terminates when the free electron at the end of the chain reacts with another free radical. Some addition polymers are listed in Table 24.1.

Natural and Synthetic Rubber Natural rubber is an addition polymer of isoprene (formal name, 2-methyl-1,3-butadiene). Rubber has been reproduced in the laboratory by heating isoprene with a catalyst:

Isoprene   Poly-*cis*-isoprene (rubber)

Rubber is obtained from the milky sap, or latex, of the rubber tree (Figure 24.3). After the latex is collected, it is coagulated by the addition of formic or acetic acid.

The elasticity of rubber results from its long, coiled polymer chains. When you stretch a piece of rubber, the long coils unwind. When you let go, the molecules move back to roughly their original positions. Actually, the molecular strands do move relative to one another, which accounts for rubber's pliability and tackiness. Early rubber goods were sticky in the summer and brittle in the winter. In 1839, Charles Goodyear discovered that crude rubber heated with sulfur remained elastic but lacked

Figure 24.3 ▲

Collecting rubber latex The outer bark of the rubber tree (*Hevea brasiliensis*) and the layer just beneath it are cut diagonally so that the latex runs into a collecting cup.

Table 24.1	Some Addition Polymers	
Polymer	**Monomer**	**Uses**
Polyethylene	$CH_2=CH_2$	Bottles, plastic tubing
Polypropylene	$CH_2=\underset{\underset{CH_3}{\shortmid}}{CH}$	Bottles, carpeting, textiles
Polytetrafluoroethylene (Teflon®)	$CF_2=CF_2$	Nonstick surface for frying pans
Poly(vinyl chloride) (PVC)	$CH_2=\underset{\underset{Cl}{\shortmid}}{CH}$	Plastic pipes, floor tile
Polyacrylonitrile (Orlon®, Acrilan®)	$CH_2=\underset{\underset{CN}{\shortmid}}{CH}$	Carpets, textiles
Polystyrene (Styrofoam)	$CH_2=\underset{\underset{C_6H_5}{\shortmid}}{CH}$	Plastic foam insulation, cups

The Discovery of Nylon

The announcement in October 1938 of the discovery of nylon, the first completely synthetic fiber, and the imminent production of nylon stockings was a momentous occasion. The press was fascinated by a fiber described as "strong as steel" and made from nothing more than coal, air, and water. And the public was eager to see stockings made from a fiber that it thought, being as strong as steel, must last forever. From our perspective, the discovery of nylon was important because it was the beginning of the synthetic polymers industry. Its discovery also established the study of polymers as a basic science.

Wallace Hume Carothers (Figure 24.4), the discoverer of nylon, was 31 years old in 1928 when he left Harvard University for the DuPont Company. (He had obtained his Ph.D. four years earlier in organic chemistry at the University of Illinois.) DuPont had decided to establish a group devoted to fundamental research, an idea that was rather novel at the time. The plan was that Carothers would work in an area that, though potentially useful to the company, did not have to have immediate commercial application.

Carothers proposed to work on the structure of materials like rubber, silk, and wool. The German chemist Hermann Staudinger had earlier suggested that these materials were actually macromolecules (that is, huge molecules) consisting of long chains of similar groups of atoms (in other words, polymers). At the time, most chemists thought that these materials were composed of aggregates of many small molecules held together by some unknown force that was different from the normal forces of chemical bonding.

Carothers felt that he could determine the truth of Staudinger's view by using the methods of organic chemistry to synthesize (or build up) macromolecules in such a way as to establish their structure. He had a simple but brilliant idea.

To understand Carothers's idea, consider two functional groups, such as a carboxylic acid group (—COOH) and an alcohol group (HO—), that chemically link together. We can write a reaction between two compounds containing these groups forming an ester (Ch. 23) this way:

$$\underset{\displaystyle \text{—C—OH}}{\overset{\displaystyle O}{\overset{\|}{}}} + \text{HO} \longrightarrow$$

$$\underset{\displaystyle \text{—C—O—}}{\overset{\displaystyle O}{\overset{\|}{}}} + H_2O$$

(where ~~~~ stands for the rest of the molecule)

Even more simply, we can represent the linking of functional groups, such as a carboxyl group and an alcohol (or amine) group, as the linking of a hook and eye (the functional groups).

Hook	Eye	Hook and eye

these undesirable characteristics. Goodyear's process, called *vulcanization,* results in cross-linking of the polymer chains, which anchors them relative to one another.

Natural rubber is a **homopolymer,** *a polymer whose monomer units are all alike.* Most of the plastics we discussed earlier are also homopolymers. On the other hand, some synthetic rubbers (as well as many plastics) are copolymers. A **copolymer** is *a polymer consisting of two or more different monomer units.* Styrene–butadiene rubber (SBR) is an important synthetic rubber; it is a copolymer made from about 25% styrene and 75% 1,3-butadiene. Its approximate structure is

$$-CH_2-CH=CH-CH_2-CH_2-CH=CH-CH_2-CH_2-CH=CH-CH_2-CH_2-\overset{\displaystyle C_6H_5}{\overset{|}{CH}}-$$

| 1,3-Butadiene unit | 1,3-Butadiene unit | 1,3-Butadiene unit | Styrene unit |

Exercise 24.1 An addition polymer is prepared from vinylidene chloride, $CH_2{=}CCl_2$. Write the structure of the addition polymer.

■ See Problems 24.77 and 24.78.

Condensation Polymers

A **condensation polymer** is *a polymer formed by linking together many molecules by condensation reactions.* A condensation reaction is one in which two molecules are

© AP/Wide World Photos

Figure 24.4 ▲

Wallace Carothers Discoverer of nylon.

Now suppose we find a molecule having two of the same functional groups (such as two carboxyl groups) at the ends of the molecule; this would be the equivalent of a molecule with hooks at both ends. Then the reaction of several such molecules would result in a larger molecule.

With many such molecules, the product would be a long-chain molecule—a macromolecule. Carothers felt that if he could

synthesize such a macromolecule with the properties of a textile fiber, he could establish the basis for further study of natural polymers, and he could go on to develop synthetic polymers (such as nylon) with improved properties.

Carothers's first macromolecule of this type was a polyester prepared from two different molecules, a dicarboxylic acid (a molecule with two carboxylic acid groups) and a dialcohol (two alcohol groups). The product, although interesting, tended to decompose in hot water and had a low melting point, so as a fabric it would hardly withstand washing or ironing. For his next experiments he prepared a series of polyamides, each from a dicarboxylic acid and a diamine. One polyamide from this series, called nylon-6,6, had properties that were promising. (The 6,6 refers to the number of carbon atoms in the dicarboxylic acid and the diamine, six in each case.) At this point, other teams at DuPont developed nylon-6,6 as a fiber that could be spun into hosiery.

Carothers might well have been awarded the Nobel Prize for his outstanding contributions to chemistry. In less than ten years, he had established synthetic polymer science and had discovered neoprene rubber and polyesters, as well as nylon. He had numerous interests, including politics, music, and sports, and he also prided himself on his writing skills. Unfortunately, Carothers had been troubled since youth with more and more frequent bouts of depression. After the discovery of nylon, he went into a severe depression, feeling he was a failure as a scientist. On April 29, 1937, he committed suicide by drinking lemonade containing potassium cyanide. His death was a tremendous loss.

■ See Problems 24.67 and 24.68.

joined by the elimination of a small molecule such as water. Wallace Carothers (1896–1937), a chemist at E. I. du Pont de Nemours and Company, realized that polymers could be prepared by condensation reactions. He and his coworkers produced a number of polymers this way, including polyamides (nylon) and polyesters. The first synthetic fibers, produced from nylon, were announced by du Pont in 1938. ▶

See the essay A Chemist Looks At: The Discovery of Nylon.

Polyesters A substance with two alcohol groups reacts with a substance with two carboxylic acid groups to form a *polyester,* a polymer whose repeating units are joined by ester groups. The reactant molecules join as links of a chain to form a very long molecule. The polyester Dacron, used as a textile fiber, is prepared from ethylene glycol and terephthalic acid.

$$\cdots + HOCH_2CH_2OH + HO-\overset{\displaystyle O}{\overset{\displaystyle \|}{C}}-\bigcirc-\overset{\displaystyle O}{\overset{\displaystyle \|}{C}}-OH + HOCH_2CH_2OH + HO-\overset{\displaystyle O}{\overset{\displaystyle \|}{C}}-\bigcirc-\overset{\displaystyle O}{\overset{\displaystyle \|}{C}}-OH + \cdots \longrightarrow$$

Ethylene glycol   Terephthalic acid   Ethylene glycol   Terephthalic acid

$$\sim OCH_2CH_2O-\overset{\displaystyle O}{\overset{\displaystyle \|}{C}}-\bigcirc-\overset{\displaystyle O}{\overset{\displaystyle \|}{C}}-OCH_2CH_2O-\overset{\displaystyle O}{\overset{\displaystyle \|}{C}}-\bigcirc-\overset{\displaystyle O}{\overset{\displaystyle \|}{C}}\sim + nH_2O$$

Dacron (a polyester)

A water molecule is split out during the formation of each ester linkage.

Polyamides When a compound containing two amine groups reacts with a compound containing two carboxylic acid groups, a condensation polymer called a *polyamide* is formed (Figure 24.5). Nylon-6,6 is an example. It is prepared by heating hexamethylene diamine (1,6-diaminohexane) and adipic acid (hexanedioic acid).

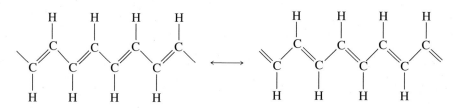

Hexamethylene	Adipic	Hexamethylene	Adipic
diamine	acid	diamine	acid

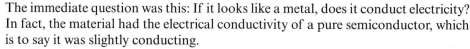

Nylon-6,6 (a polyamide)

24.2 Electrically Conducting Polymers

Sometime during the early 1970s, a graduate student in the laboratory of Professor Hideki Shirakawa in Japan mistakenly added 1000 times too much catalyst in a reaction to polymerize acetylene to polyacetylene. (The basic reaction is similar to the one in which ethylene is polymerized to polyethylene; see the previous section.) To his surprise, the plastic film coating the inside of the vessel, which in previous experiments was black, now had a silvery sheen to it. The silvery film was *trans*-polyacetylene, which has the following Lewis resonance formulas:

Figure 24.5 ▲

Nylon being pulled from a reaction mixture Nylon is formed by reacting hexamethylene diamine (bottom water layer) with adipic acid or, as in this experiment, with adipoyl chloride, $ClCO(CH_2)_4COCl$ (top hexane layer). Nylon, formed at the interface of the two layers, is continuously removed.

The immediate question was this: If it looks like a metal, does it conduct electricity? In fact, the material had the electrical conductivity of a pure semiconductor, which is to say it was slightly conducting.

You can explain the slight conductivity of polyacetylene from its electronic structure. The Lewis resonance formulas of the long-chain molecule have alternating double and single bonds, indicative of delocalized bonding. In a molecular orbital description, the orbitals would encompass the entire length of the molecular chain. Thus, an electron would be able to move easily from one end of the molecule to the other, which means a current could travel the length of the molecule. Pure polyacetylene plastic is only slightly conductive, however, because electric current does not easily pass from one molecule to another.

Sometime later, Shirakawa collaborated with two scientists from the United States, Alan MacDiarmid, a chemist, and Alan Heeger, a physicist. MacDiarmid and Heeger had been doing experiments in which they "doped" (added small amounts of particular substances to) an inorganic polymer to increase its electrical conductivity. In experiments in 1977 with polyacetylene, they found that the electrical conductivity of the material increased dramatically when it was doped with iodine, I_2.

How does doping increase the conductivity of the polymer? Polyacetylene consists of entwined strands of individual polymer molecules. Experiments show that iodine inserts itself between polyacetylene molecules, where it forms triiodide ion, I—I—I⁻. Each ion carries a negative charge that was abstracted from a π orbital of the polymer molecule. The polymer molecule is left with a positive charge, or "hole." The triiodide ion can facilitate the movement of an electron from an uncharged polymer molecule to a positively charged polymer molecule (a hole). In effect, the hole moves

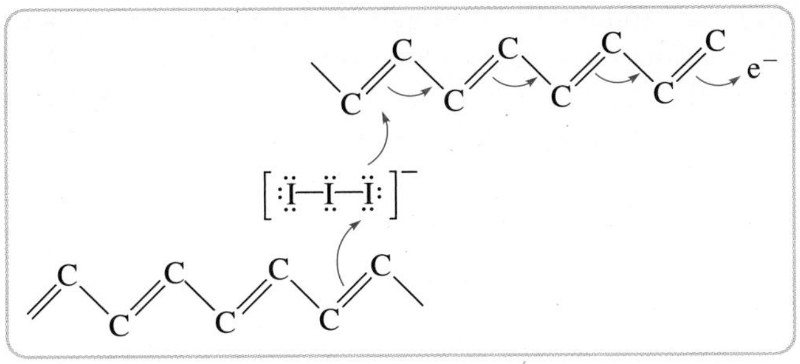

Figure 24.6 ◄

Movement of charge in polyacetylene
The diagram shows the movement of electrons by arrows. When an electron moves away from the polymer chain (perhaps going to an electrode) at the far right, it leaves a hole, or positive charge, in the double bond. An electron from the adjacent double bond can then move into this hole, leaving a hole further to the left. In this way, the hole moves right to left. The triiodide ion facilitates electron transfer between the polymer chains. (Hydrogen atoms in polyacetylene were omitted for clarity.)

(Figure 24.6). The movement of holes through the material constitutes an electric current. Doped polyacetylene was only the first conductive polymer material to be discovered. Since 1977, other conducting polymers have also been developed (Figure 24.7).

During experiments to find alternative ways of doping polymers, a graduate student of Professor MacDiarmid discovered one of the first practical applications of polyacetylene. He placed two strips of polyacetylene into a solution of doping ions and connected the strips to an electric current. The hope was that the ions would penetrate the polymer under the action of the external electric current. But after the student removed the external electric current, the polymer strips retained a charge and acted as a battery. Since then, an all-plastic battery has been developed. Its electrodes are made of conductive polymers, and the electrolyte is a polymer gel.

Scientists have succeeded in making various electronic devices, including transistors and light-emitting diodes (LEDs), from conducting polymers. Commercial applications are just now coming to market. A conductive polymer acting as an LED is being used in a display for a mobile phone. Alan Heeger, Alan MacDiarmid, and Hideki Shirakawa won the Nobel Prize in chemistry for the year 2000 for their discovery and development of conductive polymers.

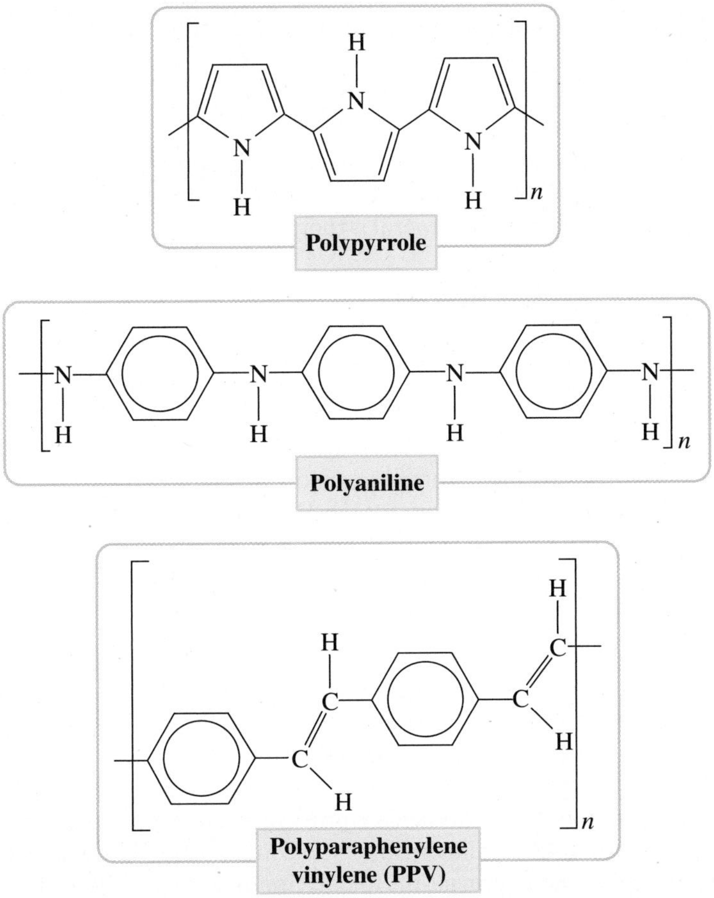

Figure 24.7 ◄

Other conducting polymers
Polypyrrole has been used in the construction of an all-plastic battery. Polyaniline has become a popular conducting polymer.

Polypyrrole

Polyaniline

**Polyparaphenylene
vinylene (PPV)**

Plastics have traditionally been thought of as insulators. In fact, electric wires are generally covered with a coating of polyethylene insulation. Now, some plastics have been found to be as conductive as metals. In chemistry, as in most other contexts, it pays to keep an open mind!

Biological Polymers

The molecules of biological organisms fall into four groups: proteins, carbohydrates, nucleic acids, and lipids (nonpolymeric materials soluble in organic solvents, such as fats and oils). Proteins and nucleic acids, and many carbohydrates, are polymers. In this chapter, we will limit our discussion to proteins and nucleic acids. Proteins are the basic constituents of living organisms; we discuss these substances in the next section. Nucleic acids are important in the biological synthesis of proteins; we discuss these in the final section of the chapter.

24.3 Proteins

Proteins are *biological polymers of small molecules called amino acids*. Their molecular masses range from about 6000 amu to 250,000 amu, so proteins can be very large molecules. Many proteins also contain non-amino-acid components such as metal ions (for example, Fe^{2+}, Zn^{2+}, Cu^+, and Mg^{2+}) or certain complex organic molecules usually derived from vitamins. (Vitamins are organic molecules necessary in very small quantities for normal cell structure and metabolism.)

Proteins are very important molecules in cells and organisms, playing both structural and functional roles. The proteins of bone and connective tissue, for example, are of major structural importance. Some proteins are enzymes, catalyzing specific biochemical reactions; some transport materials in the bloodstream or across biological membranes; and some (hormones) carry chemical messages to coordinate the body's activities. *Insulin* and *glucagon,* for example, are protein hormones made in the pancreas and secreted to regulate the body's blood-sugar level.

Amino Acids

An **amino acid** is *a compound containing an amino group ($—NH_2$) and a carboxyl group ($—COOH$)*. The building blocks of proteins are alpha-amino acids (α-amino acids). An α-amino acid has the general structure

The carbon atom next to the carboxyl carbon is called the α carbon atom. In an α-amino acid the amino group is attached to the α carbon atom. Glycine, NH_2CH_2COOH, is the simplest α-amino acid; R is an H atom.

Glycine

Table 24.2 lists the 20 amino acids from which most proteins are composed. Under the name of each amino acid, its common three-letter abbreviation is shown. For instance, the abbreviation for glycine is gly.

Table 24.2 The Amino Acids Found in Most Proteins

Nonpolar Side Chain		Polar Side Chain	

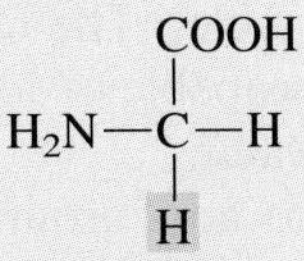

Nonpolar Side Chain

Glycine
gly

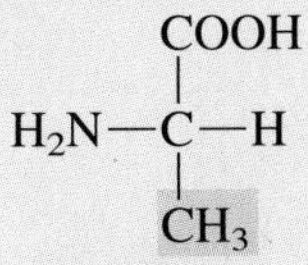

L-Proline
pro

L-Alanine
ala

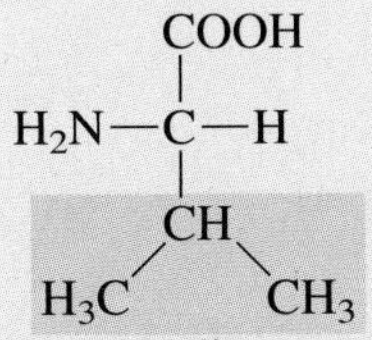

L-Phenylalanine
phe

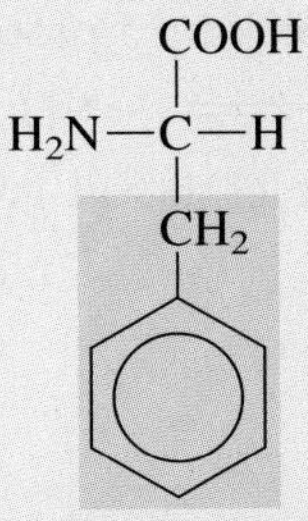

L-Valine
val

L-Tryptophan
trp

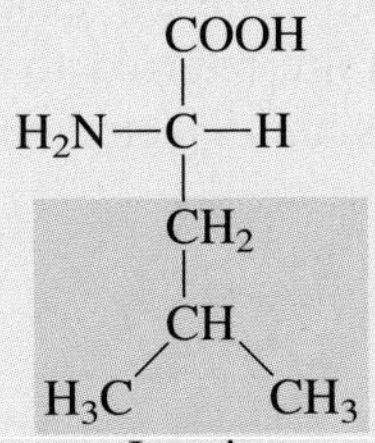

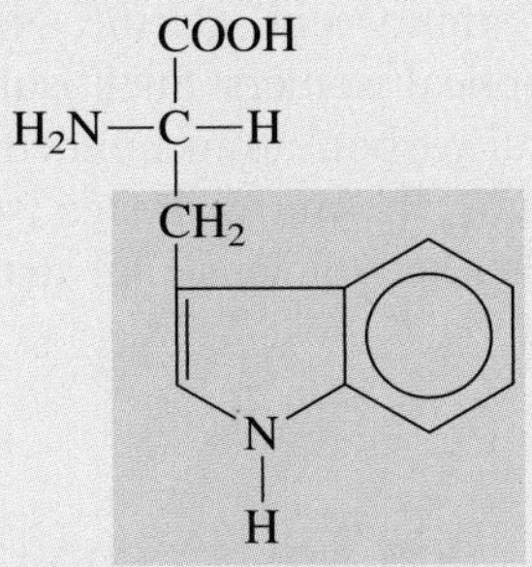

L-Leucine
leu

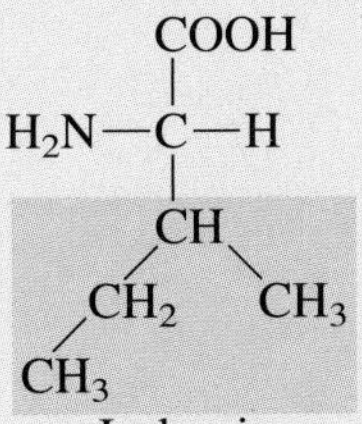

L-Isoleucine
ile

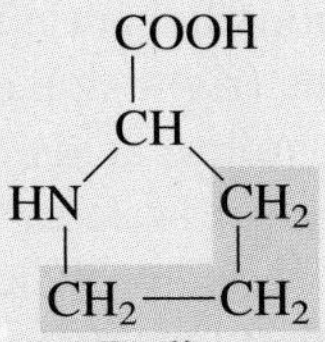

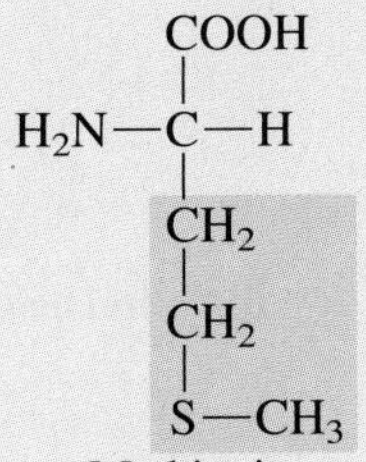

L-Methionine
met

Polar Side Chain

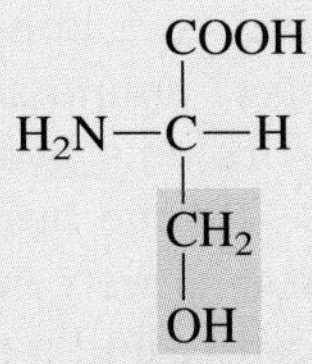

L-Serine
ser

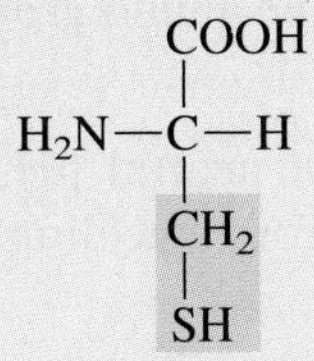

L-Asparagine
asn

L-Cysteine
cys

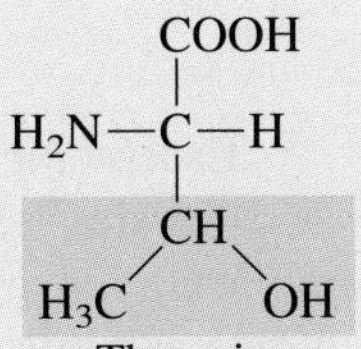

L-Glutamine
gln

L-Threonine
thr

L-Histidine
his

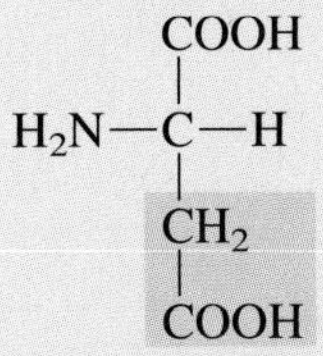

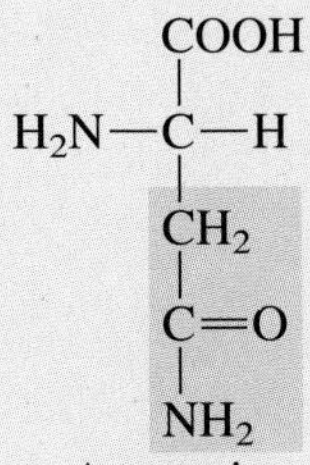

L-Aspartic acid
asp

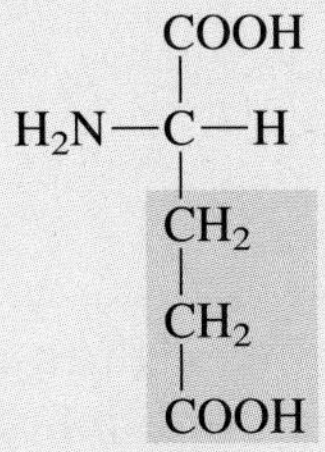

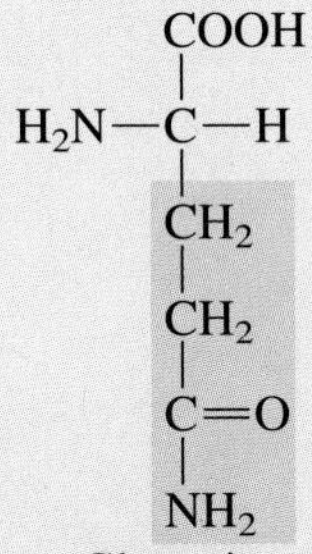

L-Lysine
lys

L-Glutamic acid
glu

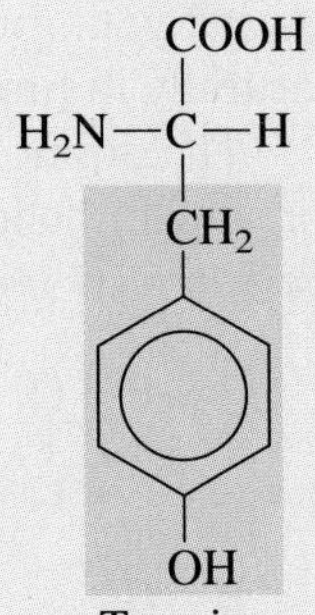

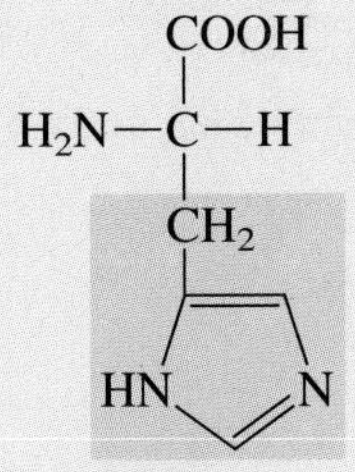

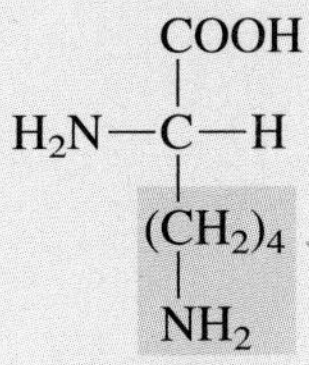

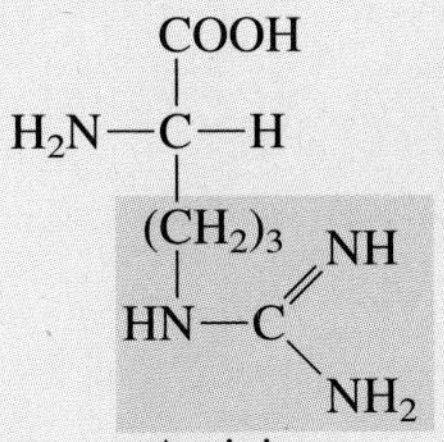

L-Arginine
arg

L-Tyrosine
tyr

Each amino acid has a different R group, or *side chain*. The side chains of the amino acids in a protein determine the protein's properties. In Table 24.2 the side chains of the amino acids are shown within a color screen. Nine of the amino acids have nonpolar, or hydrocarbon, side chains (left side of table). The remaining 11 amino acids have polar side chains (right side of table), capable of ionizing or forming hydrogen bonds with other amino acids or with water.

At the near-neutral pH of a biological system, the amino acid groups in a protein are generally in the form of *zwitterions,* which are species having both a positive and a negative charge. The carboxyl group is acidic, so its proton ionizes:

$$-COOH \rightleftharpoons -COO^- + H^+$$

The amino group is basic, so it tends to pick up a proton:

$$H^+ + -NH_2 \rightleftharpoons -NH_3^+$$

At neutral pH, the net result is that the carboxyl group loses H^+ to the amino group, yielding the zwitterion:

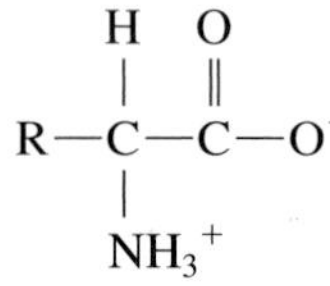

$$
\begin{array}{c}
\quad H \quad O \\
\quad | \quad \parallel \\
R-C-C-O^- \\
\quad | \\
\quad NH_3^+
\end{array}
$$

Except for the simplest α-amino acid, glycine, NH_2CH_2COOH, the α-amino acids exist as *enantiomers,* or optical isomers. Such isomers are mirror images. Any molecule having one tetrahedral carbon atom bonded to four different groups of atoms exhibits optical isomerism. The mirror-image isomers are referred to as the L-isomer and the D-isomer; their three-dimensional structures are depicted here: ◀

Optical isomerism is discussed in Section 22.5.

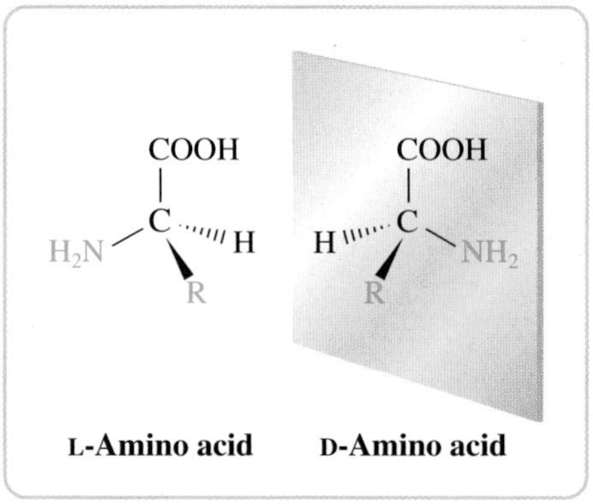

L-Amino acid   **D-Amino acid**

In these three-dimensional perspective formulas, a solid wedge is a bond pointing outward from the page, the hatched wedge is a bond pointing backward, and the straight line is a bond lying in the plane of the page. For instance, in the L-amino acid, the —COOH group and H_2N— groups are in the plane, the R— group points toward you, and the H atom points backward. All of the amino acids that occur naturally in proteins are L-amino acids.

The amino acids in a protein are linked together by peptide bonds. A **peptide** (or **amide**) **bond** is *the C—N bond resulting from a condensation reaction between the carboxyl group of one amino acid and the amino group of a second amino acid:*

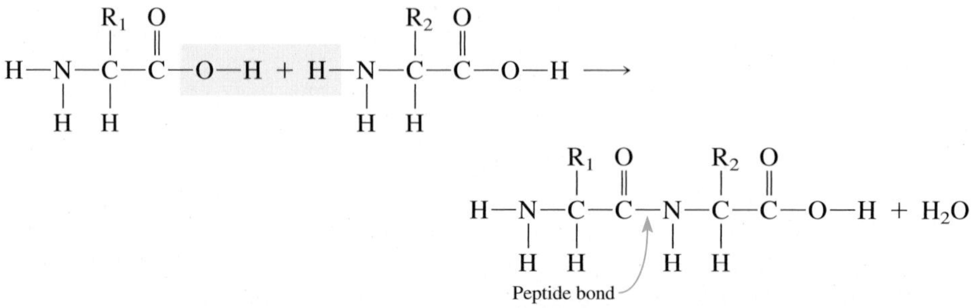

$$
\begin{array}{c}
\quad R_1 \; O \qquad\qquad\qquad R_2 \; O \\
\quad | \quad \parallel \qquad\qquad\qquad | \quad \parallel \\
H-N-C-C-O-H \; + \; H-N-C-C-O-H \longrightarrow \\
\quad | \; | \qquad\qquad\qquad\quad | \; | \\
\quad H \; H \qquad\qquad\qquad\quad H \; H
\end{array}
$$

$$
\begin{array}{c}
\qquad\qquad R_1 \; O \qquad\quad R_2 \; O \\
\qquad\qquad | \quad \parallel \qquad\quad | \quad \parallel \\
\qquad H-N-C-C-N-C-C-O-H \; + \; H_2O \\
\qquad\qquad | \; | \qquad | \; | \\
\qquad\qquad H \; H \qquad H \; H \\
\qquad\qquad\qquad\quad \text{Peptide bond}
\end{array}
$$

The product in this example is a *dipeptide,* a molecule formed by linking together two amino acids. Similarly, a *tripeptide* is formed by linking together three amino acids. A **polypeptide** is *a polymer formed by the linking of many amino acids by peptide bonds.* It may or may not have a biological function. A *protein* is a polypeptide that has a biological function.

CONCEPT CHECK 24.1

Two common amino acids are

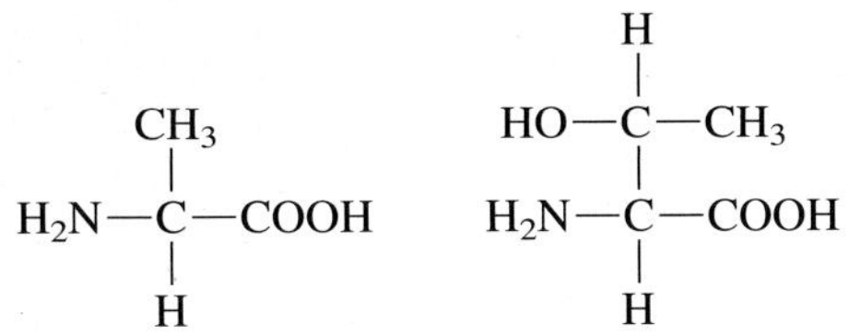

$$
\begin{array}{cc}
\text{CH}_3 & \text{HO}-\overset{\displaystyle H}{\underset{\displaystyle H}{C}}-\text{CH}_3 \\
\text{H}_2\text{N}-\overset{|}{\underset{|}{C}}-\text{COOH} & \text{H}_2\text{N}-\overset{|}{\underset{|}{C}}-\text{COOH} \\
\text{H} & \text{H} \\
\text{Alanine} & \text{Threonine}
\end{array}
$$

Write the structural formulas of all the dipeptides that they could form with each other.

Protein Primary Structure

The **primary structure** of a protein refers to *the order, or sequence, of the amino-acid units in the protein.* This order of amino acids is conveniently shown by denoting the amino acids by their three-letter abbreviations, separated by a dash. In this notation, it is understood that the amino group is on the left and the carboxyl group is on the right. For example, one possible dipeptide formed from alanine and glycine is written ala–gly.

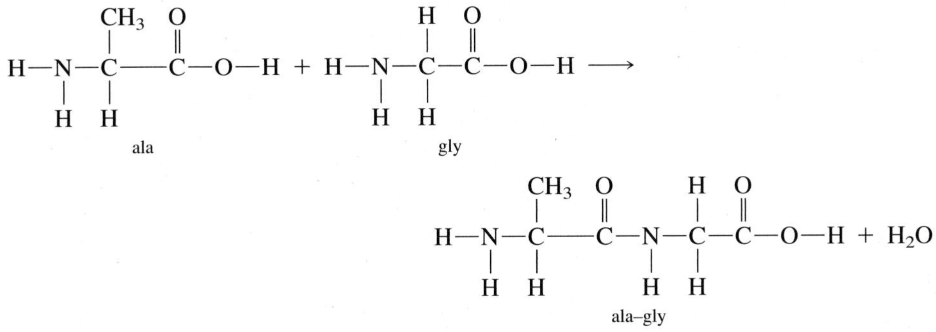

$$
\underset{\text{ala}}{\text{H}-\text{N}-\overset{\text{CH}_3}{\underset{\text{H}}{\text{C}}}-\overset{\text{O}}{\text{C}}-\text{O}-\text{H}} + \underset{\text{gly}}{\text{H}-\text{N}-\overset{\text{H}}{\underset{\text{H}}{\text{C}}}-\overset{\text{O}}{\text{C}}-\text{O}-\text{H}} \longrightarrow
$$

$$
\underset{\text{ala–gly}}{\text{H}-\text{N}-\overset{\text{CH}_3}{\underset{\text{H}}{\text{C}}}-\overset{\text{O}}{\text{C}}-\text{N}-\overset{\text{H}}{\underset{\text{H}}{\text{C}}}-\overset{\text{O}}{\text{C}}-\text{O}-\text{H}} + \text{H}_2\text{O}
$$

Consider a polypeptide containing the following five amino-acid units: glycine, alanine, valine, histidine, and serine. These five units can be ordered in many different ways to make a polypeptide. Three possible primary structures are

gly–ser–ala–val–his

his–ala–ser–val–gly

ser–ala–gly–his–val

Each of the possible sequences would produce a different polypeptide with different properties.

In any protein, which may contain as few as 50 or more than 1000 amino-acid units, the amino acids are arranged in one unique sequence, which constitutes the primary structure of that protein molecule. Figure 24.8 diagrams the primary structure of proinsulin, the precursor to the hormone insulin.

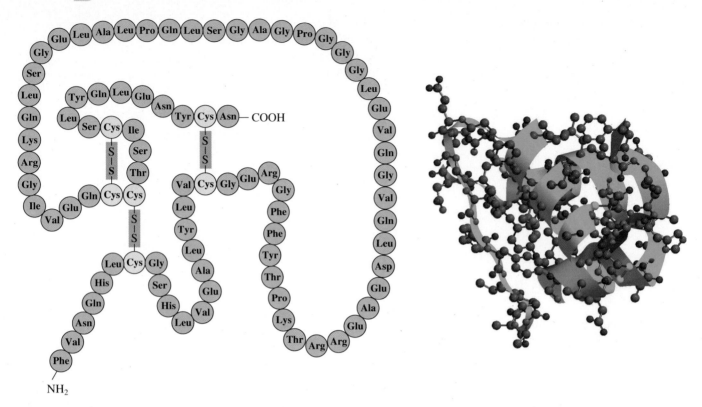

Figure 24.8 ▲

Primary structure of human proinsulin *Left:* The diagram shows the order of the amino acids in the structure of the protein molecule proinsulin. Also shown are the positions of three disulfide linkages (discussed below). *Right:* The actual three-dimensional shape of the molecule is shown (the green ribbon follows the peptide backbone of the protein).

Figure 24.9 ▲

Three-dimensional structure of my-oglobin Myoglobin is present in muscle tissue, and is a reservoir for O₂. In this model, the green ribbons are drawn to follow the backbone of the polypeptide chain. Note the coiled structure, which results from hydrogen bonding. The coiled sections fold so that the nonpolar side chains are inside the globular molecule, and the polar side chains point outward.

Shapes of Proteins

A protein molecule of a unique amino-acid sequence spontaneously folds and coils into a characteristic three-dimensional conformation in aqueous solution (Figure 24.9). The side chains of the amino-acid units, with their different chemical properties (nonpolar or polar), determine the characteristic shape of the protein. For an energetically stable shape, the parts of the chain with non-polar amino-acid side chains are buried within the structure away from water, because nonpolar groups are hydrophobic (not attracted to water). Conversely, most polar groups are stablest on the surface, where they can hydrogen-bond with water or with other polar side chains. Occasionally, side chains form ionic bonds.

One type of covalent linkage, *the disulfide linkage* (—S—S—), is especially important to protein shape. The amino acid cysteine, with its thiol (—SH) side chain, is able to react with a second cysteine in the presence of an oxidiz-ing agent, as follows:

$$\text{cys—SH + HS—cys + (O)} \longrightarrow \text{cys—S—S—cys + H}_2\text{O}$$

When two cysteine side chains in a polypeptide are brought close together by folding of the molecule and then are oxidized, they form a disulfide linkage, which helps anchor the folded chain into position (Figure 24.8). (▶ See first margin note on page 1025.)

Proteins may be classed as fibrous or globular on the basis of their arrange-ments in space. **Fibrous proteins** are *proteins that form long coils or align themselves in parallel to form long, water-insoluble fibers. The relatively simple coiled or parallel arrangement of a protein molecule* is called its **secondary structure** (Figure 24.10). **Globular proteins** are *proteins in which long coils fold into compact, roughly spherical shapes.* In addition to their secondary structure

Figure 24.10 ◄

Protein secondary structures
The alpha helix (coiled structure) and beta sheet, formed by hydrogen bonding between long polypeptide fibers or strands, are common secondary structures.

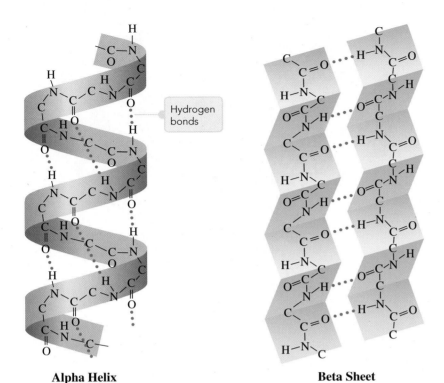

Alpha Helix   **Beta Sheet**

(long coil), globular proteins have a **tertiary structure**, or *structure associated with the way the protein coil is folded.* Most globular proteins are water soluble, because they are relatively small and have hydrophilic ("water-loving") surfaces that bind water molecules.

The characteristic coiling or folding pattern of protein molecules produces unique surface features, such as grooves or indentations, and hydrophobic or charged areas. These surface features are fundamental to protein function. They allow protein molecules to interact with other molecules in specific ways. For example, certain protein molecules characteristically aggregate because of complementary surface features, forming tubules or filaments or other organized structures. Particular surface features of enzyme molecules are responsible for their ability to bind specific molecules and catalyze specific reactions. The surface feature of an enzyme where catalysis occurs is referred to as its *active site* (Figure 24.11). Enzyme catalysis is discussed in Section 13.9.

Hair can be permanently waved by the use of disulfide linkages. Hair proteins have many disulfides that can be chemically reduced and reformed again while the hair is rolled on curlers. The new disulfide links hold the hair shafts in their new, wavy conformation.

24.4 Nucleic Acids

Nucleic acids are vital to the life cycles of cells, because they are the carriers of species inheritance. There are two types of nucleic acids: *deoxyribonucleic acid (DNA)* and *ribonucleic acid (RNA)*. Both are polymers of *nucleotide* building blocks. ▶

Sugar Constituents of Nucleic Acid

The building blocks of nucleic acids are themselves composed of certain organic bases linked to either ribose or deoxyribose, both of which are sugar molecules. In this short subsection, we will briefly discuss the structure of *monosaccharides,* or sugars, before we launch into the structure of nucleic acids.

Monosaccharides are simple sugars, each containing three to nine carbon atoms, generally all but one of which bear a hydroxyl group. The remaining carbon atom is part of a carbonyl group, either an aldehyde or a ketone. The monosaccharides exist as D- and L-isomers, because each molecule has at least one carbon atom with four different groups attached. Only a few of the many known monosaccharides are of major biological importance. All are D-sugars, and most are 5- and 6-carbon sugars (pentoses and hexoses). D-Glucose, D-fructose, D-ribose, and 2-deoxy-D-ribose

Genetic engineering refers to the manipulation of nucleic acids to change the characteristics of organisms (for example, to correct genetic flaws).

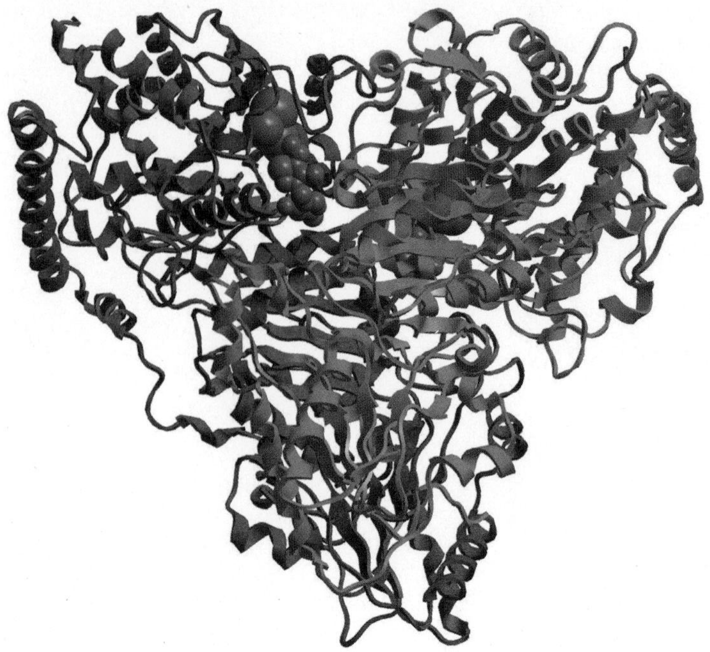

Figure 24.11 ▲

Active site on an enzyme Here a yeast enzyme (represented by the ribbon model showing its polypeptide backbone) holds the thiamine molecule (vitamin B_1, shown here as a space-filling model, in gray) at an active site in the enzyme.

are by far the most important sugars. D-Glucose is a common blood sugar, an important energy source for cell function. D-Fructose is a common sugar in fruits and is involved in cell metabolism (the chemical processes occurring in the cell). D-Ribose and 2-deoxy-D-ribose are parts of nucleic acids, as we noted earlier, where we referred to them simply as ribose and deoxyribose. You could regard 2-deoxy-D-ribose as derived from D-ribose by replacing the hydroxyl group on carbon 2 by a hydrogen atom. (The numbering convention is shown below.)

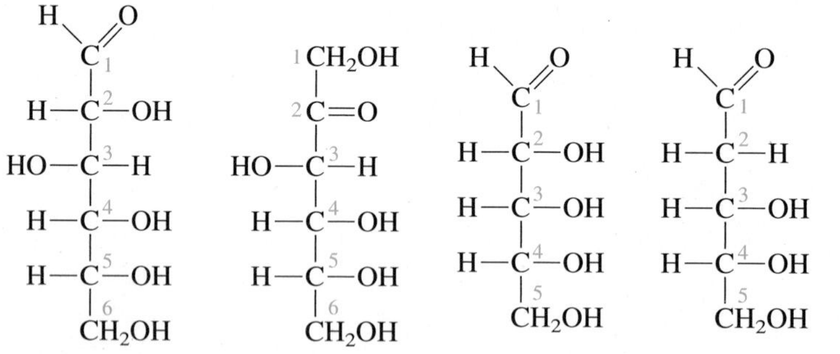

D-Glucose   D-Fructose   D-Ribose   2-Deoxy-D-ribose

Although one often draws the simple sugars as straight-chain molecules, they do not exist predominantly in that form. The chain curls up and the carbonyl group reacts with an alcohol group. With aldehydes, alcohols give *hemiacetals,* or compounds in which an —OH group, an —OR group, and an H atom are attached to the same carbon atom.

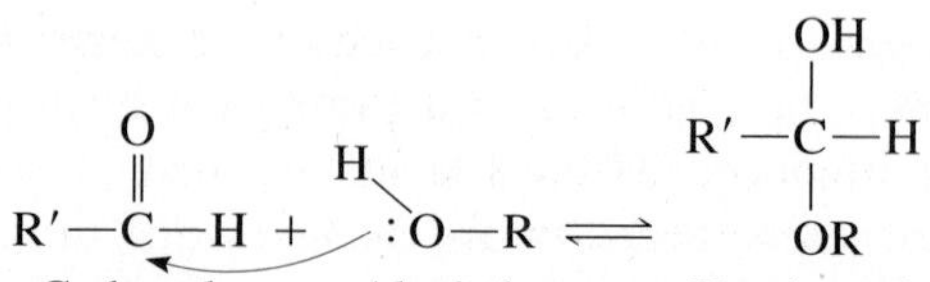

Carbonyl   Alcohol   Hemiacetal

A ketone and an alcohol give a *hemiketal,* which is similar to a hemiacetal except that the H atom is replaced by an R″ group.

The straight-chain form of D-glucose, whose formula we gave earlier, is a flexible molecule that can bend around to give a cyclic hemiacetal.

α-D-Glucose   D-Glucose   β-D-Glucose
(hemiacetal)   (straight-chain form)   (hemiacetal)

The straight-chain form, in the center, is shown bent around so that an —OH group and the —CHO group can react to give a cyclic hemiacetal. A new —OH group is obtained (on the right in each hemiacetal) and can point either down or up. Thus, there are two isomers of the hemiacetal–one labeled α (—OH down) and the other labeled β (—OH up).

Although for simplicity we have drawn the six-membered rings of the hemi-acetals as flat, they are in fact puckered, as shown in these more accurate three-dimensional formulas:

α-D-Glucose   β-D-Glucose

In these three-dimensional formulas, it is understood that there are carbon and hydrogen atoms at each of the corners in the ring, except where there is an oxygen atom.

The structures of the sugars we will need in this section are

β-D-Ribose   2-Deoxy-β-D-ribose

These sugars have five-membered rings. Four of the ring atoms are planar, but either carbon atom 2 or carbon atom 3 (with attached H and OH) points upward or downward from the planar ring atoms.

Nucleotides

Nucleotides are *the building blocks of nucleic acids.* There are two types: *ribonucleotides* and *deoxyribonucleotides.* The general structure of nucleotides consists of an organic base (described below) linked to carbon 1 of a pentose (5-carbon sugar), which is linked in turn at carbon 5 to a phosphate group. The only difference between the two types of nucleotides is in the sugar. Ribonucleotides contain β-D-ribose; deoxyribonucleotides contain 2-deoxy- β-D-ribose. Both sugar groups have five-membered rings.

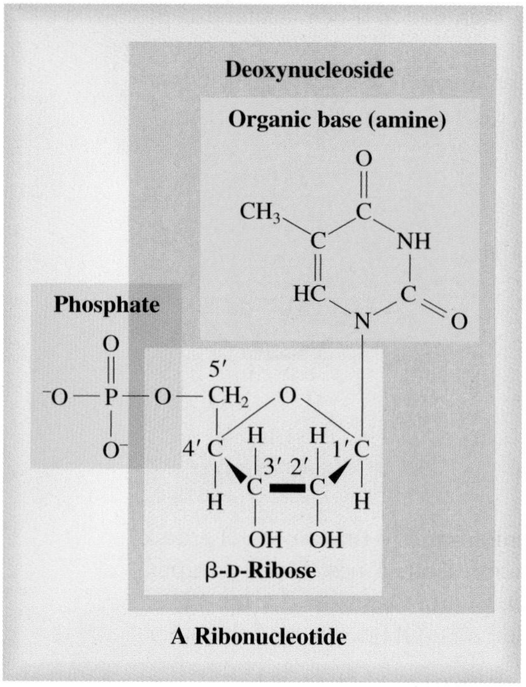

Deoxynucleoside

Organic base (amine)

β-D-Ribose

A Ribonucleotide

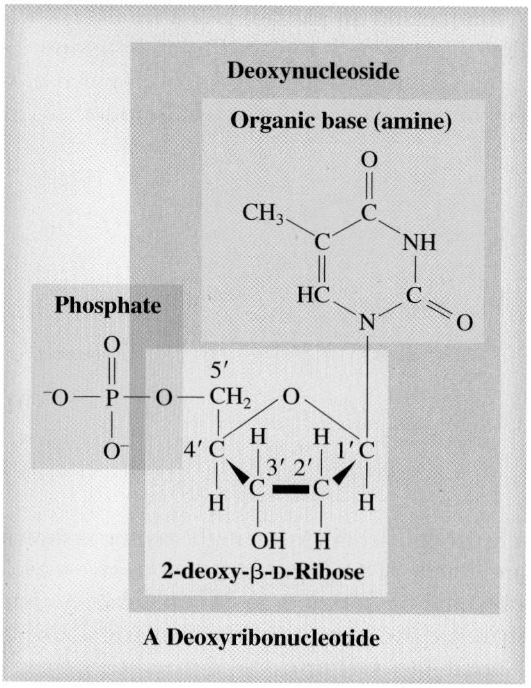

Deoxynucleoside

Organic base (amine)

2-deoxy-β-D-Ribose

A Deoxyribonucleotide

The carbon atoms of ribose are labeled with numbers $1'$, $2'$, . . . , $5'$. (Numbers without primes are used for carbon atoms in the organic base.) Note that the phosphate group is shown in ionized form $-PO_4^{2-}$, as this form predominates near neutral pH in cells.

Five organic bases (amines) are most often found in nucleotides. The bases are *adenine, guanine, cytosine, uracil,* and *thymine.* They link to the sugars through the indicated nitrogen (in color) in each of the following structural formulas:

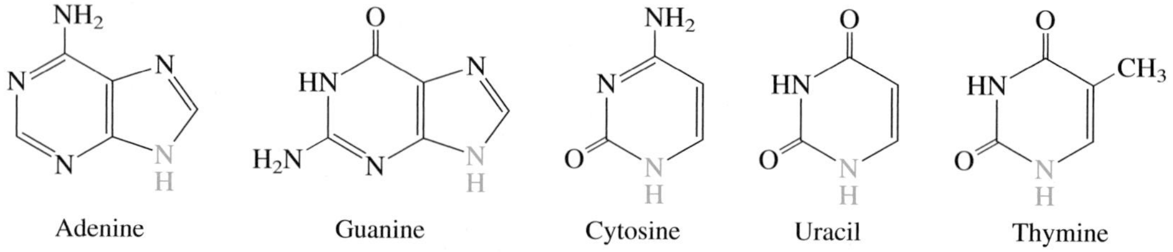

Adenine   Guanine   Cytosine   Uracil   Thymine

The base plus the sugar group (without the phosphate group) is called a *nucleoside.* Nucleosides are named from the bases. For example, the nucleoside composed of adenine with β-D-ribose is called adenosine. The nucleoside composed of adenine with 2-deoxy-β-D-ribose is called deoxyadenosine. A nucleotide is named by adding monophosphate (or diphosphate or triphosphate) after the nucleoside name. A number with a prime indicates the position of the phosphate group on the ribose ring. Thus, adenosine-$5'$-monophosphate is a nucleotide composed of adenine, β-D-ribose, and a phosphate group at the $5'$ position of β-D-ribose.

Polynucleotides and Their Conformations

A **polynucleotide** is *a linear polymer of nucleotide units* linked from the hydroxyl group at the $3'$ carbon of the pentose of one nucleotide to the phosphate group of the other nucleotide. For example,

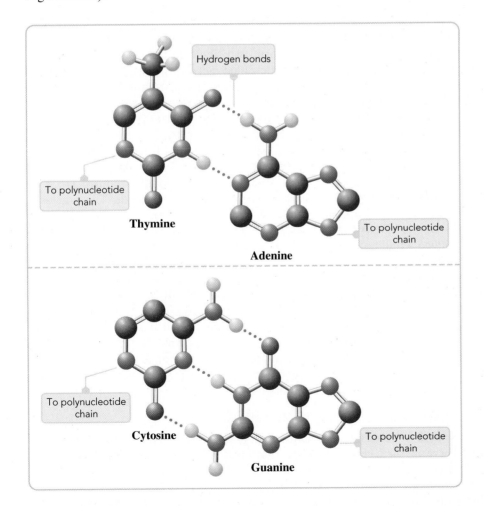

Just as the unique sequence of amino acids in a protein determines the protein's nature, so the sequence of nucleotides determines the particular properties and functions of a polynucleotide. **Nucleic acids** are *polynucleotides folded or coiled into specific three-dimensional shapes.*

Complementary bases are *nucleotide bases that form strong hydrogen bonds with one another.* Adenine and thymine are complementary bases, as are adenine and uracil, and guanine and cytosine. Hydrogen bonding of complementary bases, called *base pairing,* is the key to nucleic acid structure and function (see Figure 24.12).

Figure 24.12 ◀

Hydrogen-bonded complementary bases Each base hydrogen-bonds strongly to only its complementary base. Adenine hydrogen-bonds to thymine or to uracil; guanine hydrogen-bonds to cytosine.

Figure 24.13 ▲

Double helix of the DNA molecule The structure was first deduced by James Watson, an American scientist, and Francis Crick, a British scientist, in the early 1950s.

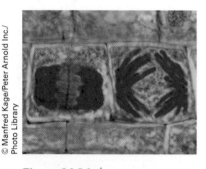

Figure 24.14 ▲

Chromosomes in a dividing cell The chromosomes appear as dense, thick rods.

Deoxyribonucleic acid (DNA) is *the hereditary constituent of cells and consists of two polymer strands of deoxyribonucleotide units.* One of the strands consists of nucleotide bases that are complementary to those of the other strand. The two strands coil about each other in a double helix, with base pairing along the entire lengths of the strands (Figure 24.13). **Ribonucleic acid (RNA)** is *a constituent of cells that is used to manufacture proteins from genetic information. It is a polymer of ribonucleotide units.* Both DNA and RNA have nucleotides with the bases adenine, guanine, and cytosine. In addition, DNA contains thymine but not uracil, and RNA contains uracil but not thymine. It is often convenient in writing the bases in DNA and RNA sequences to use the abbreviations A = adenine, C = cytosine, G = guanine, T = thymine, and U = uracil.

DNA and the Nature of the Genetic Code

Photomicrographs of dividing cells show structures, called chromosomes, as dense, thick rods (Figure 24.14). *Chromosomes are cell structures that contain DNA and proteins;* the DNA contains the genetic inheritance of the cell and organism. Before cell division, the cell synthesizes a new and identical set of chromosomes or, more particularly, a new and identical set of DNA molecules—the genetic information— to be transmitted to the new cell. Thus, the new cell will have all the necessary instructions for normal structure and function.

What is the nature of this genetic information? The genetic information is coded into the linear sequence of nucleotides in the DNA molecule. Each DNA molecule is composed of hundreds of genes. A **gene** is *a sequence of nucleotides in a DNA molecule that codes for a given protein.* The nucleotides in a gene are grouped in sets of three, or triplets. Each triplet codes for one amino acid in a protein.

New DNA must be synthesized during cell division, as we mentioned. This synthesis, or DNA replication, proceeds as follows. The two strands of DNA unravel at one end of the helix to form two single strands. As the two strands unravel, and with the aid of an enzyme called *polymerase,* complementary nucleotides are attached one after the other to these single strands to form two new double strands of DNA. At the completion of the process, two double-stranded DNA chains exist where previously there was one (Figure 24.15).

A similar laboratory technique, based on the *polymerase chain reaction* (PCR), is used to amplify the quantity of DNA in a sample. In the PCR technique, a sample of DNA is heated with polymerase to separate the DNA into single strands. The polymerase then catalyzes the formation of new double-stranded DNA by adding complementary nucleotides to the original single strands. The process is continued, separating double-stranded DNA into single strands and forming new double-stranded DNA from these. Within hours, the original sample of DNA can be amplified a millionfold. Using PCR, researchers can diagnose diseases by identifying the DNA of infectious agents originally present in minute quantities in blood, food, and other samples. The PCR technique also makes it possible to replicate the DNA from Egyptian mummies and ancient bone samples.

Occasionally, an error is made during the synthesis of new DNA. Such a change in the genetic information, or genetic error, is called a *mutation.* One possible consequence could be the synthesis of a faulty or inactive protein. We know of various genetic diseases that result from a single error leading to the change of an amino acid in a protein sequence that has a profound effect on the protein's activity—and so on the life of the organism.

RNA and the Transmission of the Genetic Code

RNA is used to translate the genetic information stored in DNA into protein structure. There are three classes of RNA: ribosomal RNA, messenger RNA, and transfer RNA.

Ribosomes are *tiny cellular particles on which protein synthesis takes place.* They are constructed of numerous proteins plus three or four RNA molecules. *The RNA in a ribosome* is called **ribosomal RNA**. Ribosomes provide a surface on which to organize the process of protein synthesis, and they also contain enzymes that catalyze the process.

A **messenger RNA** (mRNA) molecule is *a relatively small RNA molecule that can diffuse about the cell and attach itself to a ribosome, where it serves as a pattern for protein biosynthesis.* The first step in protein biosynthesis, called *transcription,* is synthesis of a messenger RNA molecule that has a sequence of bases complementary to that of a portion of a DNA strand corresponding to a single protein (Figure 24.16).

A sequence of three bases in a messenger RNA molecule that serves as the code for a particular amino acid is called a **codon.** When the messenger RNA molecule attaches itself to a ribosome, its codons provide the sequence of amino acids in a protein that will be synthesized. *Translation* is the biosynthesis of protein using messenger RNA codons. Table 24.3, the complete messenger RNA code-word dictionary, shows the specific amino acids coded for by messenger RNA codons. There are 64 possible arrangements of the four RNA bases, so there are 64 codons. Three of the codons do not signify amino acids but signify the end of a message and thus are called *termination codons.* The remaining 61 codons signify particular amino acids. Because only 20 different amino acids exist in proteins, there are a number of instances in which 2, 3, 4, or even 6 codons translate to the same amino acid.

A **transfer RNA** (tRNA) molecule is *a small RNA molecule that binds to a particular amino acid and carries it to a ribosome.* At the ribosome, the tRNA molecule attaches itself through base pairing to an mRNA codon. The amino acid is then transferred to a growing protein chain. The process of protein biosynthesis stops when a termination codon is reached on the mRNA.

Figure 24.17 illustrates this process of biosynthesis. At the right, a tRNA molecule carrying an amino acid (labeled aa_8—$tRNA_8$) is about to bind to the mRNA attached to the ribosome. Note that $tRNA_8$ has a sequence of three bases (AGU) complementary to the bases on the mRNA (UCA) to which it is about to bind. The codon UCA is the code for the amino acid serine (see Table 24.3), and this is the amino acid aa_8 carried by $tRNA_8$. On the ribosome, $tRNA_6$ is about to leave as its amino acid aa_6 has been attached to the growing polypeptide chain. (Can you figure out what amino acid this is?) Its place will be taken by $tRNA_7$ as the ribosome moves right, enveloping $tRNA_8$ with its amino acid aa_8. Eventually the completed polypeptide chain will leave the ribosome as a newly synthesized protein molecule.

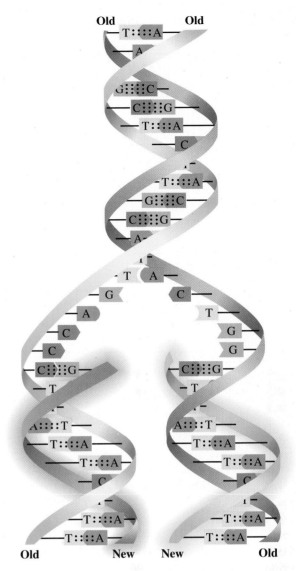

Figure 24.15 ▲

DNA replication The first stage of replication involves the unraveling of the two chains of the DNA double helix. As the chains begin to unravel, the exposed bases on each chain attract the complementary bases. The process is finished when there are two DNA double helixes where there had been only one.

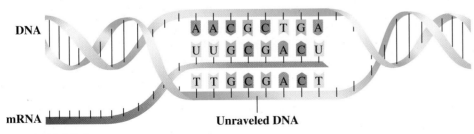

Figure 24.16 ▲

Formation of messenger RNA (transcription) A portion of a DNA double helix corresponding to a single protein unravels, and complementary ribonucleotides line up, attracted by hydrogen bonding to one of the DNA chains. The ribonucleotides are then linked to form messenger RNA. After being released from the DNA, the messenger RNA migrates from the cell nucleus into the cytoplasm to a ribosome, where protein biosynthesis occurs.

Table 24.3 **Genetic Code Dictionary***

First letter	Second letter U		Second letter C		Second letter A		Second letter G		Third letter
U	UUU UUC	Phenyl- alanine	UCU UCC	Serine	UAU UAC	Tyrosine	UGU UGC	Cysteine	U C
	UUA UUG	Leucine	UCA UCG		UAA UAG	Stop codon Stop codon	UGA UGG	Stop codon Tryptophan	A G
C	CUU CUC CUA CUG	Leucine	CCU CCC CCA CCG	Proline	CAU CAC CAA CAG	Histidine Glutamine	CGU CGC CGA CGG	Arginine	U C A G
A	AUU AUC AUA AUG	Isoleucine Methionine, initiation codon	ACU ACC ACA ACG	Threonine	AAU AAC AAA AAG	Asparagine Lysine	AGU AGC AGA AGG	Serine Arginine	U C A G
G	GUU GUC GUA GUG	Valine	GCU GCC GCA GCG	Alanine	GAU GAC GAA GAG	Aspartic acid Glutamic acid	GGU GGC GGA GGG	Glycine	U C A G

*The four nucleotide bases of RNA—U, C, A, and G—are arranged along the left side of the table and along the top. Combining any two of them and then adding a third (again, U, C, A, or G) gives one of the 64 three-nucleotide codons shown in capital letters in the table. Next to each codon appears the name of the amino acid it codes for during protein biosynthesis. In three cases, the phrase *Stop codon* appears because each of these codons serves as a signal to terminate protein biosynthesis.

Figure 24.17 ▶

Synthesis of protein molecules (translation) Messenger RNA (mRNA) provides the code needed to determine the sequence of amino acids to be added to the protein. A transfer RNA molecule (tRNA) is shown here bringing an amino acid (aa$_8$, amino acid 8) to the mRNA bound to a ribosome. The amino-acid-6 (aa$_6$) end of the growing polypeptide chain will be transferred to aa$_7$ as the tRNA with aa$_8$ adds to the mRNA. After transfer of its amino acid, the tRNA molecule leaves the mRNA.

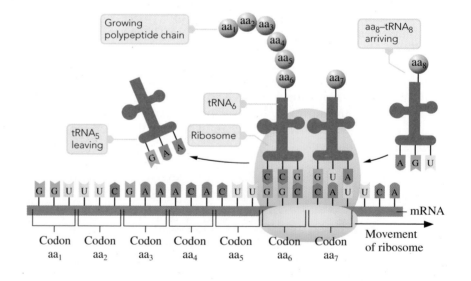

CONCEPT CHECK 24.2

Write the RNA sequence complementary to the following DNA sequence:

ATGCTACGGATTCAA

A CHEMIST Looks at...

Life Science

Tobacco Mosaic Virus and Atomic Force Microscopy

The amount of detail provided by the latest scanning probe microscopes is breathtaking. Figure 24.18 shows what looks like a tumble of logs on a forest floor, each log covered with mossy growth. The image, provided by an atomic force microscope, is of rods of tobacco mosaic virus, a disease agent that infects tobacco and many other crops. Each virus rod is covered by molecules of bovine serum albumin (from the blood serum of cows). Chemists obtained this image as part of a study of the interaction of the albumin protein molecule with the virus. The virus itself consists of a strand of RNA, containing the genetic material of the virus, surrounded by a protective protein coat.

The atomic force microscope (Figure 24.19) is a cousin of the scanning tunneling microscope, which we discussed in an essay at the end of Section 7.4. Both microscopes use a probe to scan a surface; but whereas the scanning tunneling microscope measures an electric current between the probe tip and the sample, the atomic force microscope measures the attractive van der Waals force between the probe tip and the sample. The advantage of the atomic force microscope is that it can be used with almost any surface, whereas a scanning tunneling microscope requires a conductive surface.

By permission of Gang-Yu Liu, University of California at Davis

Figure 24.18 ▲

Tobacco mosaic virus rods The virus rods are covered with adsorbed protein molecules (bovine serum albumin). The protein molecules show up as bright bumps on the dark yellow color of the virus. A computer adds color to the image to create contrast.

Figure 24.19 ▶

Atomic force microscope The sample moves by a scanner under a fixed probe. The probe consists of a silicon nitride tip attached to a cantilever, which can move up and down. A laser beam is reflected from the back of the cantilever to a photodiode divided into four segments (quadrant photodiode). As the cantilever bends up and down (or left and right) from surface forces acting on the probe tip, it deflects the laser beam to different quadrants of the photodiode. A computer coordinates the output from the photodiode with the sample position to create an image that appears on the computer screen.

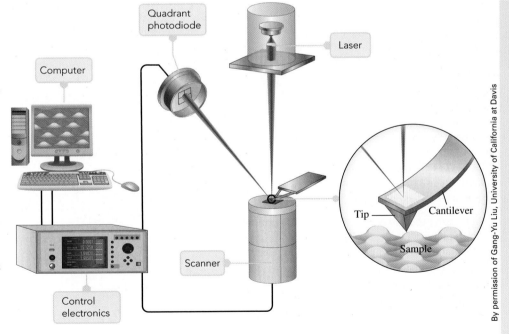

By permission of Gang-Yu Liu, University of California at Davis

See Problems 24.69 and 24.70.

A Checklist for Review

Summary of Facts and Concepts

Polymers are species of very high molecular mass consisting of many units (*monomers*). They are classified as *addition polymers* or *condensation polymers,* depending on the type of reaction used in forming them. Rubber is an addition polymer. Natural rubber is a *homopolymer* of isoprene units; the synthetic rubber SBR (styrene-butadiene rubber) is a *copolymer.* Nylon-6,6 is an example of a condensation polymer; it is produced by condensing hexamethylene diamine with adipic acid.

Certain polymers, such as *trans*-polyacetylene, are semiconductors. By doping them with substances such as iodine, their electrical conductivities can be greatly increased. One of the first applications of a conducting polymer was to produce a battery. These polymers have also been used to produce electronic devices, including transistors and LEDs.

Proteins are polymers of *α-amino acids* linked by *peptide bonds.* The 20 common amino acids of protein have different *side chains,* which may be *polar* or *nonpolar.* The unique *primary structure* of a polypeptide (the amino-acid sequence) is responsible for spontaneous folding into a unique shape maintained by interactions between side chains and disulfide bonds. This folding produces surface features responsible for the protein's function. An enzyme has an *active site* where catalysis takes place.

Nucleic acids are polymers of *nucleotides.* Nucleotides are base–sugar–phosphate compounds. Conformations of nucleic acids are the result of hydrogen bonding, called base pairing, between *complementary bases. DNA* is the genetic material of chromosomes. DNA consists of two complementary polynucleotides coiled into a *double helix;* each has all the information necessary to direct synthesis of a new complementary strand. The genetic code is the relationship between nucleotide sequence and protein amino-acid sequence. Sixty-four *codons* (three-nucleotide sequences) are the basis of the code. *Transcription* is the synthesis of a *messenger RNA* molecule, which represents a copy of the information in a DNA gene. The messenger RNA binds to a *ribosome,* where *translation* of the message into an amino-acid sequence occurs. *Transfer RNA* molecules attach to amino acids and bind to the messenger RNA by base pairing, bringing each amino acid in turn into the position specified by the codon sequence.

Learning Objectives	Important Terms
24.1 Synthesis of Organic Polymers	
▪ Define *polymer* and *monomer.* ▪ Define *addition polymer.* ▪ Describe free-radical addition. ▪ Describe the nature of natural and synthetic rubber. ▪ Define *homopolymer* and *copolymer.* ▪ Define *condensation polymer.* ▪ Describe the synthesis of polyesters and polyamides.	**polymer** **monomer** **addition polymer** **homopolymer** **copolymer** **condensation polymer**
24.2 Electrically Conducting Polymers	
▪ Describe the discovery of electrically conducting polymers. ▪ Explain how these polymers conduct electricity.	
24.3 Proteins	
▪ Define *proteins* and *amino acids.* ▪ Define a *peptide (amide) bond.* ▪ Define a *peptide.* ▪ Define the *primary structure* of a protein. ▪ Describe the shapes of proteins. ▪ Define *fibrous protein* and *globular protein.* ▪ Define the *secondary* and *tertiary structures* of a protein.	**proteins** **amino acid** **peptide (amide) bonds** **polypeptide** **primary structure (of a protein)** **fibrous proteins** **secondary structure (of a protein)** **globular proteins** **tertiary structure (of a protein)**

24.4 Nucleic Acids

- Note the two types of nucleic acids.
- Describe the structure of the monosaccharides.
- Define *nucleotides* and *polynucleotides*.
- Define *nucleic acids, complementary bases, deoxyribonucleic acid (DNA)*, and *ribonucleic acid (RNA)*.
- Describe the nature of the genetic code.
- Describe how RNA translates the genetic code in DNA to protein structure.
- Define *ribosomes, ribosomal RNA, messenger RNA, codon,* and *transfer RNA.*

nucleotides
polynucleotide
nucleic acids
complementary bases
deoxyribonucleic acid (DNA)
ribonucleic acid (RNA)
gene
ribosomes
ribosomal RNA
messenger RNA
codon
transfer RNA

Questions and Problems

OWL Interactive versions of these problems may be assigned in OWL.

Self-Assessment and Review Questions

Key: These questions test your understanding of the ideas you worked with in the chapter. These problems vary in difficulty and often can be used for the basis of discussion.

24.1 What is the difference between an addition polymer and a condensation polymer? Give an example of each, writing the equation for its formation.

24.2 Write the equation for the addition of one isoprene molecule to another.

24.3 Give the structural formula for a portion of the chain polymer obtained by the reaction of ethylene glycol, CH_2OH—CH_2OH, with malonic acid, $HOOC$—CH_2—$COOH$.

24.4 Show how the pi electrons in one resonance formula of polyacetylene would have to shift to obtain the other resonance formula.

24.5 Explain how the doping of polyacetylene increases its electrical conductivity.

24.6 Describe the primary structure of protein. What makes one protein different from another protein of the same size? What is the basis of the unique conformation of a protein?

24.7 Distinguish between secondary and tertiary structures of proteins.

24.8 What is an enzyme?

24.9 What is the difference in structure between D-ribose and 2-deoxy-D-ribose?

24.10 What are the structural forms of D-glucose present in the blood?

24.11 How do ribonucleotides and deoxyribonucleotides differ in structure? Do they form polymers in the same way?

24.12 Name the complementary base pairs. Describe the DNA double helix.

24.13 Explain the nature of the genetic code.

24.14 Define *codon* and *anticodon*. How do they interact?

24.15 Show mathematically why there are 64 possible triplet codons.

24.16 Outline how the genetic message in a gene is translated to produce a polypeptide. Include the roles of the gene, messenger RNA, ribosomes, and transfer RNA.

24.17 An amino acid must contain which of the following functional groups?

- a ketal and amino
- b amino and carboxyl
- c carboxyl and nitro
- d methyl and carboxyl
- e alcohol and amido

24.18 An example of an addition polymer is

- a polyester
- b nylon-6,6
- c rubber
- d Dacron
- e glucose

24.19 The complementary base of thymine is

- a thymine
- b guanine
- c cytosine
- d adenine
- e uracil

24.20 The bonding responsible for linking the two strands that constitute DNA is

- a covalent bonding
- b ionic bonding
- c ion–dipole bonding
- d hydrogen bonding
- e phosphate linkages

Conceptual Problems

Key: These problems are designed to check your understanding of the concepts associated with some of the main topics presented in each chapter. A strong conceptual understanding of chemistry is the foundation for both applying chemical knowledge and solving chemical problems. These problems vary in level of difficulty and often can be used as a basis for group discussion.

24.21 It is your job to manufacture polymers from a series of monomer units. These monomer units are called A, B, and C. In this problem you need to "build " polymers by linking the monomer units. Represent the polymer linkages using dashes. For example, –A–B–C– represents a polymer unit made from linking monomer units A, B, and C.

a Build two different homopolymers from your monomer units.

b Build a copolymer from A and B.

c Build a copolymer that is about 33% C and 66% A.

24.22 Use resonance formulas to explain why polyacetylene has delocalized molecular orbitals extending over the length of the molecule, whereas the following molecule does not.

24.23 Write the structural formulas of all possible tripeptides with the composition of two glycines and one serine. (See the structural formulas in Table 24.2.)

Practice Problems

Key: These problems are for practice in applying problem-solving skills. They are divided by topic. The problems are arranged in matching pairs; the odd-numbered problem of each pair is listed first, and its answer is given in the back of the book.

Synthetic Polymers

24.27 Poly(vinyl chloride) (PVC) is an addition polymer of vinyl chloride, CH_2=$CHCl$. Write the equation for the formation of the polymer.

24.28 Teflon is an addition polymer of 1,1,2,2-tetrafluoroethene. Write the equation for the formation of the polymer.

24.29 Consider the following polymer:

$$—OCH_2CH_2O\overset{O}{\overset{\|}{C}}CH_2CH_2\overset{O}{\overset{\|}{C}}OCH_2CH_2O\overset{O}{\overset{\|}{C}}CH_2CH_2\overset{O}{\overset{\|}{C}}—$$

From which two monomers is the polymer made?

24.30 Consider the following polymer:

$$—\overset{O}{\overset{\|}{C}}(CH_2)_{10}\overset{O}{\overset{\|}{C}}NH(CH_2)_6NH\overset{O}{\overset{\|}{C}}(CH_2)_{10}\overset{O}{\overset{\|}{C}}NH(CH_2)_6NH—$$

From which two monomers is the polymer made?

Conducting Polymers

24.31 Which of the following monomers might you expect would lead to a conducting polymer?

a H_2C=CH—CH_3 b H—C≡C—CH_3

24.32 Which of the following monomers might you expect to lead to a conducting polymer?

a CH_3—C≡C—CH_3 b CH_3—CH=CH—CH_3

24.24 A common amino acid in the body is ornithine. It is involved in the excretion of excess nitrogen into the urine. The structural formula of ornithine is

$$H_2NCH_2CH_2CH_2—\overset{H}{\underset{NH_2}{\overset{|}{\underset{|}{C}}}}—COOH$$

Write the fully ionized form of the molecule. (Note that there are two ionizable amino groups.)

24.25 Give one of the nucleotide sequences that would translate to the peptide lys–pro–ala–phe–trp–glu–his–gly.

24.26 What amino-acid sequence would result if the following messenger RNA sequence were translated from left to right?

AGAGUCCGAGACUUGACGUGA

Amino Acids and Primary Structure

24.33 Valine has the structure

$$CH_3CH—\overset{H}{\underset{NH_2}{\overset{|}{\underset{|}{C}}}}—COOH$$
$$\underset{CH_3}{}$$

Draw the zwitterion that would exist at neutral pH.

24.34 Alanine has the structure

$$CH_3—\overset{H}{\underset{NH_2}{\overset{|}{\underset{|}{C}}}}—COOH$$

Draw the zwitterion that would exist at neutral pH.

24.35 Write the structural formula of a dipeptide formed from the reaction of L-alanine and L-histidine. How many dipeptides are possible?

24.36 Write the structural formulas of two tripeptides formed from the reaction of L-tryptophan, L-glutamic acid, and L-tyrosine. How many tripeptides are possible?

Nucleotides and Polynucleotides

24.37 If adenine, thymine, guanine, and cytosine were each analyzed separately in a sample of DNA, what molar ratios of A : T and G : C would you expect to find?

24.38 If a sample of DNA isolated from a microorganism culture were analyzed and found to contain 1.5 mol of cytosine nucleotides and 0.5 mol of adenosine nucleotides, what would be the amounts of guanine and thymine nucleotides in the sample?

24.39 Write a structural formula for the nucleotide adenosine-5′-monophosphate.

24.40 Write a structural formula for the nucleotide deoxyadenosine-5′-monophosphate.

24.41 How many hydrogen bonds link an adenine–thymine base pair? Would there be any difference in strength between adenine–thymine bonding and adenine–uracil bonding? between adenine–thymine and cytosine–guanine bonding? Explain.

24.42 How many hydrogen bonds link a guanine–cytosine base pair? an adenine–uracil base pair? Would you expect any difference in strength between guanine–cytosine bonding and adenine–uracil bonding? Explain.

DNA, RNA, and Protein Synthesis

24.43 If a codon consists of two nucleotides, how many codons would be possible? Would this be a workable code for the purpose of protein synthesis?

24.44 If a codon consists of four nucleotides, how many codons would be possible? Would this be workable as an amino-acid code?

24.45 Nucleic acids can be denatured by heat, as proteins can. What bonds are broken when a DNA molecule is denatured? Would DNA of greater percentage composition of guanine and cytosine denature more or less readily than DNA of greater percentage composition of adenine and thymine?

24.46 Would RNA of a certain percentage composition of guanine and cytosine denature more or less readily than RNA with a lower percent of guanine and cytosine? Why?

24.47 Write the amino-acid sequence obtained from left-to-right translation of the messenger RNA sequence

AUUGGCGCGAGAUCGAAUGAGCCCAGU

See Table 24.3.

24.48 Consulting Table 24.3, write the amino-acid sequence resulting from left-to-right translation of the messenger RNA sequence

GGAUCCCGCUUUGGGCUGAAAUAG

General Problems

Key: These problems provide more practice but are not divided by topic. Each section ends with essay questions, each of which is color coded to refer to the *A Chemist Looks at Materials* (blue) or *A Chemist Looks at Life Science* (pink) chapter essay on which it is based. Odd-numbered problems and the even-numbered problems that follow are similar; answers to all odd-numbered problems except the essay questions are given in the back of the book.

24.49 List the anticodons to which the following codons would form base pairs:

Codon: UUG CAC ACU GAA

Anticodon:

24.50 List the codons to which the following anticodons would form base pairs:

Anticodon: GAC UGA GGG ACC

Codon:

24.51 Write one nucleotide sequence that would translate to

tyr–ile–pro–his–leu–his–thr–ser–phe–met

24.52 Give one of the nucleotide sequences that would translate to

leu–ala–val–glu–asp–cys–met–trp–lys

24.53 Write the structural formula of the addition polymer made from *trans*-1,2-dichloroethene:

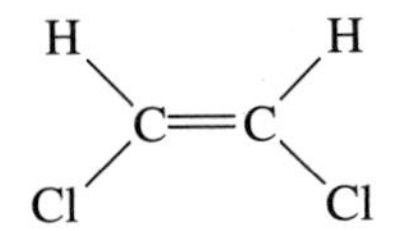

24.54 Neoprene is an addition polymer of chloroprene:

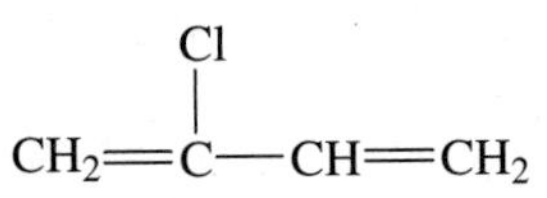

Write the structural formula of neoprene.

24.55 Kevlar® is a polymer fiber used to make bulletproof vests.

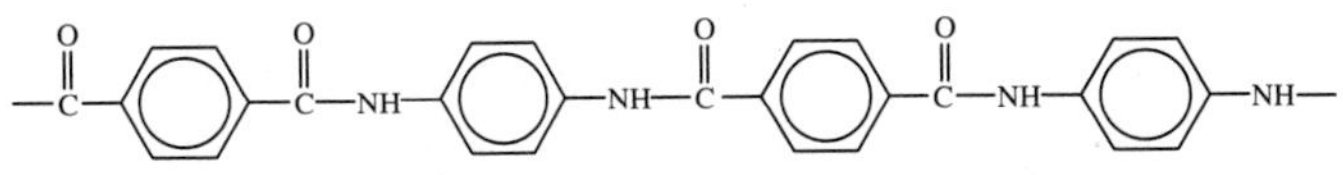

What are the structures of the monomer units in this polymer?

24.56 Nylon-6 has the following structure:

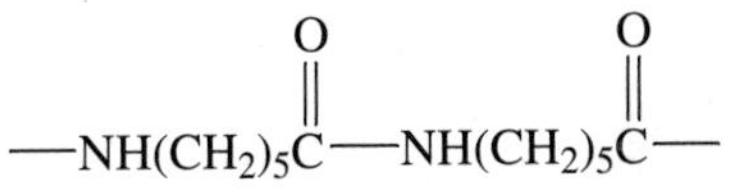

What is the structure of the monomer unit in this polymer? Show the reaction corresponding to the formation of the polymer.

24.57 Which of the following amino acids has a nonpolar side chain?

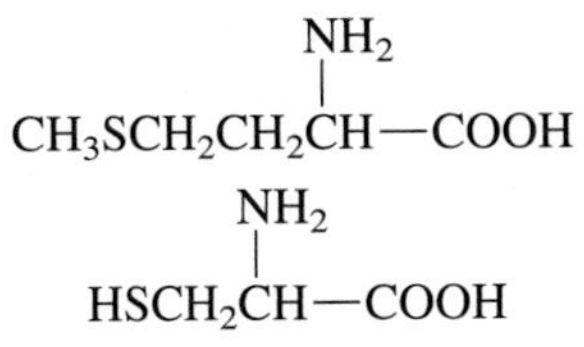

24.58 Which of the following amino acids has a polar side chain?

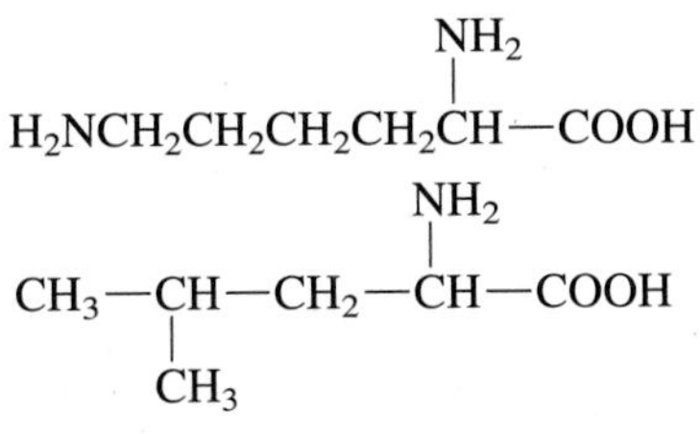

24.59 Draw the zwitterion structure for the amino acid serine.

24.60 Draw the zwitterion structure for the amino acid leucine.

24.61 Write the structural formula of one of the two dipeptides that could be formed from the amino acids in Problem 24.57.

24.62 Write the structural formula of one of the two peptides that could be formed by the amino acids in Problem 24.58.

24.63 A peptide contains six amino acids: L-arginine, L-proline, L-glutamic acid, L-glycine, L-asparagine, and L-glutamine. How many different peptides of this amino-acid composition are possible?

24.64 Using abbreviations for the amino acids, give three possible sequences of the peptide described in Problem 24.63.

24.65 Write the structural formula of deoxyguanosine-5′-monophosphate.

24.66 Name each of the three major constituents of the nucleotide adenosine-5′-monophosphate.

■ **24.67** Wallace Carothers is credited with establishing the science of synthetic polymers. Describe his basic idea that is used in the synthesis of polymers such as nylon and polyesters.

■ **24.68** Prior to Carothers's work, the structures of materials such as rubber, silk, and wool were thought by some to be aggregates of small molecules held together by unknown forces. How did Carothers's work help to elucidate the actual structures?

■ **24.69** Describe the basic principle involved in the atomic force microscope.

■ **24.70** What is the major advantage of the atomic force microscope over the similar scanning tunneling microscope?

Strategy Problems

Key: As noted earlier, all of the practice and general problems are matched pairs. This section is a selection of problems that are not in matched-pair format. These challenging problems require that you employ many of the concepts and strategies that were developed in the chapter. In some cases, you will have to integrate several concepts and operational skills in order to solve the problem successfully.

24.71 Polyesters are made by condensing dicarboxylic acids with glycols. A 1.273-g sample of a dicarboxylic acid was titrated with 0.2183 M NaOH. The sample was completely neutralized with 98.6 mL of the NaOH solution. Draw a possible structural formula for the dicarboxylic acid. Show the condensation reaction of this acid with ethylene glycol to produce the polyester.

24.72 A 10.0-g sample of rubber (poly-*cis*-isoprene) reacts with bromine, Br_2, to produce the polymer in which bromine adds to each of the double bonds. Estimate the quantity of heat that evolves in this process.

24.73 Consider the condensation of the diamine $H_2N(CH_2)_n$ NH_2 with sebacic acid, $(HOOC)(CH_2)_8(COOH)$, to obtain the polyamide. The polyamide has the following composition: 67.2% C, 10.5% H, 10.4% N, and 11.9% O. What is the value of n?

24.74 Aspartame, a low-calorie sweetener, is the methyl ester of a dipeptide. Its structure is

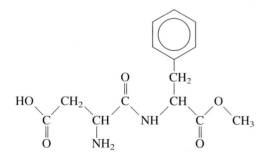

a. Suppose aspartame is hydrolyzed (reacted with water). Give the names of the amino acids obtained from this hydrolysis.

b. Draw the structure of another dipeptide that could be prepared from these amino acids.

c. How many tripeptides could you make from the amino acids valine and alanine?

24.75 The genetic code uses four bases taken in groups of three for each code.

a. How many of these base groups are possible?

b. Is this enough to establish a code for each of the 20 amino acids found in proteins?

c. Imagine a genetic code that uses only two bases but uses groups of four of these bases for the genetic code. How many different groups of this type are possible?

d. Is this enough to establish a code for each of the 20 amino acids found in proteins?

e. Devise a hypothetical code of this type for the 20 amino acids in proteins. Then, using this code, write the base sequence for the tripeptide leu-ala-val.

24.76 A 1.500-g sample of a mixture of ribose and 2-deoxyribose was dissolved in 100.0 g of water. If the freezing point of the solution was −0.0556°C, what is the percentage (by mass) of ribose in the mixture?

24.77 Hemoglobin is the protein in red blood cells that carries oxygen. This protein is composed of two parts: globin and heme, the part that contains all of the iron of hemoglobin. Heme has the following composition: 66.24% C, 5.23% H, 9.06% Fe, 9.09% N, and 10.38% O. Assuming one Fe atom per heme, obtain its molecular mass. The molecular mass of hemoglobin is approximately 64,500 amu. The hemoglobin molecule contains four globin units and four heme units. What is the approximate molecular mass of globin?

24.78 A polymer is hydrolyzed to give an acidic substance and a compound that was found to have properties of an alcohol. When this compound was oxidized, it produced an acid whose molecular mass was found to be 90.0 amu. A 0.145-g sample of this acid is completely neutralized with 30.5 mL of 0.1056 M NaOH to give a salt. The acid produced a white precipitate with calcium chloride. What was the precipitate that formed? Explain what is happening at each step.

24.79 Consider the zwitterion of the amino acid alanine, which is the predominant form of the acid at neutral pH. Draw its structural formula. Imagine that the pH of a solution of alanine is gradually lowered until the zwitterion picks up a proton. Draw the structural formula of the resulting ion. Write the chemical equation for this addition of a proton. Note that the equilibrium constant for the opposite reaction is K_a. The pK_a value for this is 2.3. Use the Henderson–Hasselbalch equation to obtain the ratio of the different ionic forms of alanine in the solution at a pH of 4.0. Which form predominates at this pH?

24.80 Natural rubber is an addition polymer of isoprene. Gutta percha is a similar rubber-like material, although its properties are rather different. (It is a hard material at room temperature but softens when heated.) Its structure has the *trans* form of polyisoprene instead of the *cis* form present in natural rubber.

a Draw the structure of gutta percha.

b If a sample of gutta percha has an average molecular mass of 250,000 amu, what is the average number of isoprene units in the polymer?

c When you stretch a rubber band, it evolves heat. If the stretched rubber band is then spontaneously allowed to contract to its equilibrium shape, it cools. Does the entropy of rubber increase or decrease when it is stretched?

Appendixes

Mathematical Skills

Only a few basic mathematical skills are required for the study of general chemistry. But to concentrate your attention on the concepts of chemistry, you will find it necessary to have a firm grasp of these basic mathematical skills. In this appendix, we will review scientific (or exponential) notation, logarithms, simple algebraic operations, the solution of quadratic equations, and the plotting of straight-line graphs.

A.1 Scientific (Exponential) Notation

In chemistry, you frequently encounter very large and very small numbers. For example, the number of molecules in a liter of air at 20°C and normal barometric pressure is 25,000,000,000,000,000,000,000, and the distance between two hydrogen atoms in a hydrogen molecule is 0.000,000,000,074 meter. In these forms, such numbers are inconvenient to write and difficult to read. For this reason, you normally express them in scientific, or exponential, notation. Scientific calculators also use this notation.

In scientific notation, a number is written in the form $A \times 10^n$. A is a number greater than or equal to 1 and less than 10, and the exponent n (the nth power of 10) is a positive or negative integer. For example, 4853 is written in scientific notation as 4.853×10^3, which is 4.853 multiplied by three factors of 10:

$$4.853 \times 10^3 = 4.853 \times 10 \times 10 \times 10 = 4853$$

The number 0.0568 is written in scientific notation as 5.68×10^{-2}, which is 5.68 divided by two factors of 10:

$$5.68 \times 10^{-2} = \frac{5.68}{10 \times 10} = 0.0568$$

Any number can be conveniently transformed to scientific notation by moving the decimal point to obtain a number, A, greater than or equal to 1 and less than 10. If the decimal point is moved to the left, you multiply A by 10^n, where n equals the number of places moved. If the decimal point is moved to the right, you multiply A by 10^{-n}. Consider the number 0.00731. You must move the decimal point to the right three places. Therefore, 0.00731 equals 7.31×10^{-3}. To transform a number written in scientific notation to the usual form, the process is reversed. If the exponent is positive, the decimal point is shifted right. If the exponent is negative, the decimal point is shifted left.

Example 1   **Expressing Numbers in Scientific Notation**

Express the following numbers in scientific notation:

a. 843.4   b. 0.00421   c. 1.54

Solution Shift the decimal point to get a number between 1 and 10; count the number of positions shifted.

a. $8\,4\,3\,.\,4 = \mathbf{8.434 \times 10^2}$
b. $0\,.\,0\,0\,4\,2\,1 = \mathbf{4.21 \times 10^{-3}}$
c. **leave as is or write 1.54×10^0**

Exercise 1   Express the following numbers in scientific notation:

a. 4.38   b. 4380   c. 0.000483

Example 2   **Converting Numbers in Scientific Notation to Usual Form**

Convert the following numbers in scientific notation to usual form:

a 6.39×10^{-4} b 3.275×10^{2}

Exercise 2   Convert the following numbers in scientific notation to usual form:

a 7.025×10^{3}

b 8.97×10^{-4}

Solution   a $0.0006.39 \times 10^{-4} = \mathbf{0.000639}$

        b $3.27.5 \times 10^{2} = \mathbf{327.5}$

Addition and Subtraction

Before adding or subtracting two numbers written in scientific notation, it is necessary to express both to the same power of 10. After adding or subtracting, it may be necessary to shift the decimal point to express the result in scientific notation.

Example 3   **Adding and Subtracting in Scientific Notation**

Carry out the following arithmetic; give the result in scientific notation.

$$(9.42 \times 10^{-2}) + (7.6 \times 10^{-3})$$

Solution   You can shift the decimal point in either number to obtain both to the same power of 10. For example, to get both numbers to 10^{-2}, you shift the decimal point one place to the left and add 1 to the exponent in the expression 7.6×10^{-3}.

$$7.6 \times 10^{-3} = 0.76 \times 10^{-2}$$

Now you can add the two numbers.

$$(9.42 \times 10^{-2}) + (0.76 \times 10^{-2}) = (9.42 + 0.76) \times 10^{-2} = 10.18 \times 10^{-2}$$

Since 10.18 is not between 1 and 10, you shift the decimal point to express the final result in scientific notation.

$$10.18 \times 10^{-2} = \mathbf{1.018 \times 10^{-1}}$$

Exercise 3   Add the following, and express the sum in scientific notation:

$$(3.142 \times 10^{-4}) + (2.8 \times 10^{-6})$$

Multiplication and Division

To multiply two numbers in scientific notation, you first multiply the two powers of 10 by adding their exponents. Then you multiply the remaining factors. Division is handled similarly. You first move any power of 10 in the denominator to the numerator, changing the sign of the exponent. After multiplying the two powers of 10 by adding their exponents, you carry out the indicated division.

Example 4   **Multiplying and Dividing in Scientific Notation**

Do the following arithmetic, and express the answers in scientific notation.

a $(6.3 \times 10^{2}) \times (2.4 \times 10^{5})$ b $\dfrac{6.4 \times 10^{2}}{2.0 \times 10^{5}}$

(continued)

(*continued*)

Solution a $(6.3 \times 10^2) \times (2.4 \times 10^5) = (6.3 \times 2.4) \times 10^7 = 15.12 \times 10^7 = \mathbf{1.512 \times 10^8}$

b $\dfrac{6.4 \times 10^2}{2.0 \times 10^5} = \dfrac{6.4}{2.0} \times 10^2 \times 10^{-5} = \dfrac{6.4}{2.0} \times 10^{-3} = \mathbf{3.2 \times 10^{-3}}$

Exercise 4 Perform the following operations, expressing the answers in scientific notation:

a $(5.4 \times 10^{-7}) \times (1.8 \times 10^8)$ b $\dfrac{5.4 \times 10^{-7}}{6.0 \times 10^{-5}}$

Powers and Roots

A number $A \times 10^n$ raised to a power p is evaluated by raising A to the power p and multiplying the exponent in the power of 10 by p:

$$(A \times 10^n)^p = A^p \times 10^{n \times p}$$

You extract the rth root of a number $A \times 10^n$ by first moving the decimal point in A so that the exponent in the power of 10 is exactly divisible by r. Suppose this has been done, so that n in the number $A \times 10^n$ is exactly divisible by r. Then

$$\sqrt[r]{A \times 10^n} = \sqrt[r]{A} \times 10^{n/r}$$

Example 5 **Finding Powers and Roots in Scientific Notation**

Evaluate the following expressions:

a $(5.29 \times 10^2)^3$

b $\sqrt{2.31 \times 10^7}$

Solution a $(5.29 \times 10^2)^3 = (5.29)^3 \times 10^6 = 148 \times 10^6 = \mathbf{1.48 \times 10^8}$

b $\sqrt{2.31 \times 10^7} = \sqrt{23.1 \times 10^6} = \sqrt{23.1} \times 10^{6/2} = \mathbf{4.81 \times 10^3}$

Exercise 5 Obtain the values of the following, and express them in scientific notation:

a $(3.56 \times 10^3)^4$ b $\sqrt[3]{4.81 \times 10^2}$

Electronic Calculators

Scientific calculators will perform all of the arithmetic operations we have just described (as well as those discussed in the next section). On most calculators, the basic operations of addition, subtraction, multiplication, and division are similar and straightforward. However, more variation exists in raising a number to a power and extracting a root. If you have the calculator instructions, by all means read them. Otherwise, the following information may help.

Squares and square roots are usually obtained with special keys, perhaps labeled x^2 and $\sqrt{x}$. Thus, to obtain $(5.15)^2$, you enter 5.15 and press x^2. To obtain $\sqrt{5.15}$, you enter 5.15 and press $\sqrt{x}$ (or perhaps INV, for inverse, and x^2). Other powers and roots require a y^x (or a^x) key. The answer to Example 5(a), which is $(5.29 \times 10^2)^3$, would be obtained by a sequence of steps such as the following: enter 5.29×10^2, press the y^x key, enter 3, and press the $=$ key.

The same sequence can be used to extract a root. Suppose you want $\sqrt[5]{2.18 \times 10^6}$. This is equivalent to $(2.18 \times 10^6)^{1/5}$ or $(2.18 \times 10^6)^{0.2}$. If the calculator has a $1/x$ key, the sequence would be as follows: enter 2.18×10^6, press the y^x key, enter 5, press the $1/x$ key, then press the $=$ key. Some calculators have a $\sqrt[x]{y}$ key, so this can be used to extract the xth root of y, using a sequence of steps similar to that for y^x.

A.2 Logarithms

The *logarithm* to the *base a* of a number x, denoted $\log_a x$, is the exponent of the constant a needed to equal the number x. For example, suppose a is 10, and you would like the logarithm of 1000; that is, you would like the value of $\log_{10} 1000$. This is the exponent, y, of 10 such that 10^y equals 1000. The value of y here is 3. Thus, $\log_{10} 1000 = 3$.

Common logarithms are logarithms in which the base is 10. The common logarithm of a number x is often denoted simply as log x. It is easy to see how to obtain the common logarithms of 10, 100, 1000, and so forth. But logarithms are defined for all positive numbers, not just the powers of 10. In general, the exponents, or values, of the logarithm will be decimal numbers. To understand the meaning of a decimal exponent, consider $10^{0.400}$. This is equivalent to $10^{400/1000} = 10^{2/5} = \sqrt[5]{10^2} = 2.51$. Therefore, log $2.51 = 0.400$. Any decimal exponent is essentially a fraction, p/r, so by evaluating the expressions $10^{p/r} = \sqrt[r]{10^p}$ one could construct a table of logarithms. In practice, power series or other methods are used.

The following are fundamental properties of all logarithms:

$$\log_a 1 = 0 \tag{1}$$

$$\log_a(A \times B) = \log_a A + \log_a B \tag{2}$$

$$\log_a \frac{A}{B} = \log_a A - \log_a B \tag{3}$$

$$\log_a A^p = p \log_a A \tag{4}$$

$$\log_a \sqrt[r]{A} = \frac{1}{r}\log_a A \tag{5}$$

These properties are very useful in working with logarithms.

Electronic calculators that evaluate logarithms are now available for about $10 or so. Their simplicity of operation makes them well worth the price. To obtain the logarithm of a number, you enter the number and press the LOG key.

Exercise 6   Find the values of **a** log 0.00582; **b** log 689.

Antilogarithm

The antilogarithm (abbreviated antilog) is the inverse of the common logarithm. Antilog x is simply 10^x. If your electronic calculator has a 10^x key, you obtain the antilogarithm of a number by entering the number and pressing the 10^x key. (It may be necessary to press an inverse key before pressing a 10^x/LOG key.) If your calculator has a y^x (or a^x) key, you enter 10, press y^x, enter x, then press the $=$ key.

Exercise 7   Evaluate **a** antilog 5.728; **b** antilog (-5.728).

Natural Logarithms and Natural Antilogarithms

The mathematical constant $e = 2.71828$, like π, occurs in many scientific and engineering problems. It is frequently seen in the natural exponential function $y = e^x$. The inverse function is called the *natural logarithm*, $x = \ln y$, where $\ln y$ is simplified notation for $\log_e y$.

Finding the natural logarithm with a calculator is analogous to finding the common logarithm, only in this case you enter the number and press the LN key. For example, you may obtain the natural logarithm of 6.2 by entering 6.2 and pressing the LN key ($\ln 6.2 = 1.82$). The natural antilogarithm of a number is the constant e raised to a power equal to the number. Therefore, the expression for the natural antilogarithm of 1.34 is written as

$$\text{natural antilog } 1.34 = e^{1.34}$$

In order to evaluate the expression $e^{1.34}$ on your calculator, you can use the e^x key, or you may have to use the INV key in combination with the LN key ($e^{1.34} = 3.8$).

There are instances where you may want to relate the natural logarithm to the common logarithm; the expression is

$$\ln x = 2.303 \log x$$

Exercise 8 Evaluate a $\ln 9.93$; b $e^{1.10}$.

A.3 Algebraic Operations and Graphing

Often you are given an algebraic formula that you need to rearrange to solve for a particular quantity. As an example, suppose you would like to solve the following equation for V:

$$PV = nRT$$

You can eliminate P from the left-hand side by dividing by P. But to maintain the equality, you must perform the same operation on both sides of the equation:

$$\frac{PV}{P} = \frac{nRT}{P}$$

or

$$V = \frac{nRT}{P}$$

Quadratic Formula

A quadratic equation is one involving only powers of x in which the highest power is 2. The general form of the equation can be written

$$ax^2 + bx + c = 0$$

where a, b, and c are constants. For given values of these constants, only certain values of x are possible. (In general, there will be two values.) These values of x are said to be the solutions of the equation.

These solutions are given by the *quadratic formula:*

$$x = \frac{-b \pm \sqrt{b^2 - 4ac}}{2a}$$

In this formula, the symbol $\pm$ means that there are two possible values of x—one obtained by taking the positive sign, the other by taking the negative sign.

Example 6 — Obtaining the Solutions of a Quadratic Equation

Obtain the solutions of the following quadratic equation:

$$2.00x^2 - 1.72x - 2.86 = 0$$

Solution Using the quadratic formula, you substitute $a = 2.00$, $b = -1.72$, and $c = -2.86$. You get

$$x = \frac{1.72 \pm \sqrt{(-1.72)^2 - 4 \times 2.00 \times (-2.86)}}{2 \times 2.00} = \frac{1.72 \pm 5.08}{4.00} = -0.840 \text{ and } +1.70$$

Mathematically there are two solutions, but in any real problem one solution may not be allowed. For example, if the solution is some physical quantity that can have only positive values, a negative solution must be rejected.

Exercise 9 Find the positive solution (or solutions) to the following equation:

$$1.80x^2 + 0.850x - 9.50 = 0$$

The Straight-Line Graph

A graph is a visual means of representing a mathematical relationship or physical data. Consider the following data table, in which values of y from some experiment are given for four values of x.

x	y
1	-1
2	1
3	3
4	5

By plotting these x, y points on a graph (Figure A.1), you see that they fall on a straight line. This suggests (but does not prove) that other points from this type of

Figure A.1 ▶

A straight-line plot of some data

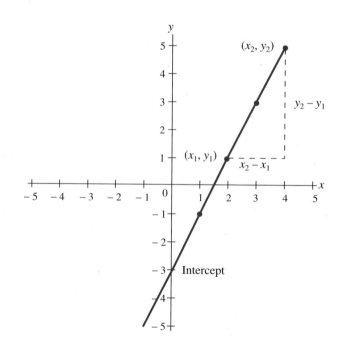

experiment might fall on the same line. It would be useful to have the mathematical equation for this line.

The general equation form of a straight line is

$$y = mx + b$$

The constant m is called the *slope* of the straight line. It is obtained by dividing the vertical distance between any two points on the line by the horizontal distance. If the two points are (x_1, y_1) and (x_2, y_2), the slope is given by the following formula:

$$\text{Slope} = \frac{y_2 - y_1}{x_2 - x_1}$$

Suppose you choose the points (2, 1) and (4, 5) from the data. Then

$$\text{Slope} = \frac{5 - 1}{4 - 2} = 2$$

For the straight line in Figure A.1, $m = 2$.

The constant b is called the *intercept.* It is the value of y at $x = 0$. From Figure A.1, you see that the intercept is -3. Therefore, $b = -3$. Hence, the equation of the straight line is

$$y = 2x - 3$$

Appendix B Vapor Pressure of Water at Various Temperatures

Temperature (°C)	Pressure (mmHg)	Temperature (°C)	Pressure (mmHg)
0	4.6	27	26.7
5	6.5	28	28.3
10	9.2	29	30.0
11	9.8	30	31.8
12	10.5	35	42.2
13	11.2	40	55.3
14	12.0	45	71.9
15	12.8	50	92.5
16	13.6	55	118.0
17	14.5	60	149.4
18	15.5	65	187.5
19	16.5	70	233.7
20	17.5	75	289.1
21	18.7	80	355.1
22	19.8	85	433.6
23	21.1	90	525.8
24	22.4	95	633.9
25	23.8	100	760.0
26	25.2	105	906.1

Appendix C Thermodynamic Quantities for Substances and Ions at 25°C

Substance or Ion	ΔH_f° (kJ/mol)	ΔG_f° (kJ/mol)	S° (J/mol·K)
$e^-(g)$	0	0	20.87
Aluminum			
$Al(s)$	0	0	28.28
$Al^{3+}(aq)$	−531	−485	−321.7
$AlCl_3(s)$	−705.6	−630.0	109.3
$Al_2O_3(s)$	−1675.7	−1582.3	50.95
Barium			
$Ba(g)$	179.1	147.0	170.1
$Ba(s)$	0	0	62.48
$Ba^{2+}(aq)$	−537.6	−560.7	9.6
$BaCO_3(s)$	−1216.3	−1137.6	112.1
$BaCl_2(s)$	−858.6	−810.3	123.7
$Ba(NO_3)_2(aq)$	−952.4	−783.4	302.5
$BaO(s)$	−548.1	−520.4	72.07
$Ba(OH)_2(s)$	−946.3	−859.3	107.1
$Ba(OH)_2 \cdot$ $8H_2O(s)$	−3342.2	−2793	427
$BaSO_4(s)$	−1473.2	−1362.3	132.2
Beryllium			
$Be(s)$	0	0	9.440
$BeO(s)$	−608.4	−579.1	13.77
$Be(OH)_2(s)$	−905.8	−817.9	50.21
Boron			
$B(s)$	0	0	5.834
$BCl_3(l)$	−427.2	−387.4	206
$BF_3(g)$	−1135.6	−1119.0	254.2
$B_2O_3(s)$	−1271.9	−1192.8	53.95
Bromine			
$Br(g)$	111.9	82.40	174.9
$Br^-(aq)$	−121.5	−104.0	82.4
$Br^-(g)$	−219.0	−238.8	163.4
$Br_2(g)$	30.91	3.159	245.3
$Br_2(l)$	0	0	152.2
$HBr(g)$	−36.44	−53.50	198.6
Calcium			
$Ca(g)$	177.8	144.1	154.8
$Ca(s)$	0	0	41.59
$Ca^+(g)$	773.8	732.1	160.5
$Ca^{2+}(aq)$	−542.8	−553.5	−53.1
$Ca^{2+}(g)$	1925.9	—	—
$CaCO_3(s,\ calcite)$	−1206.9	−1128.8	92.9

Substance or Ion	ΔH_f° (kJ/mol)	ΔG_f° (kJ/mol)	S° (J/mol·K)
$CaCl_2(s)$	−795.8	−748.1	104.6
$CaF_2(s)$	−1225.9	−1173.5	68.57
$CaO(s)$	−635.1	−603.5	38.21
$Ca(OH)_2(s)$	−986.1	−898.4	83.39
$Ca_3(PO_4)_2(s)$	−4120.8	−3884.8	236.0
$CaSO_4(s)$	−1434.1	−1321.9	106.7
Carbon			
$C(g)$	716.7	671.3	158.0
$C(s,\ diamond)$	1.897	2.900	2.377
$C(s,\ graphite)$	0	0	5.740
$CCl_4(g)$	−95.98	−53.65	309.7
$CCl_4(l)$	−135.4	−65.27	216.4
$CF_4(g)$	−933.2	−888.5	261.3
$CN^-(aq)$	151	166	118
$CO(g)$	−110.5	−137.2	197.5
$CO_2(g)$	−393.5	−394.4	213.7
$CO_3^{2-}(aq)$	−677.1	−527.9	−56.9
$CS_2(g)$	116.9	66.85	237.9
$CS_2(l)$	89.70	65.27	151.3
$COCl_2(g)$	−220.1	−205.9	283.9
$HCN(aq)$	150.6	172.4	94.1
$HCN(g)$	135.1	124.7	201.7
$HCN(l)$	108.9	124.9	112.8
$HCO_3^-(aq)$	−692.0	−586.8	91.2
Hydrocarbons			
$CH_4(g)$	−74.87	−50.80	186.1
$C_2H_2(g)$	226.7	209.2	200.9
$C_2H_4(g)$	52.47	68.39	219.2
$C_2H_6(g)$	−84.68	−32.89	229.5
$C_3H_8(g)$	−104.7	−23.6	270.2
$C_4H_{10}(g)$	−125.6	−17.2	310.1
$C_6H_6(g)$	82.6	129.7	269.2
$C_6H_6(l)$	49.0	124.4	173.4
Alcohols			
$CH_3OH(g)$	−200.7	−162.0	239.7
$CH_3OH(l)$	−238.7	−166.4	126.8
$C_2H_5OH(g)$	−235.1	−168.6	282.6
$C_2H_5OH(l)$	−277.7	−174.9	160.7
Aldehydes			
$HCHO(g)$	−117	−113	219.0
$CH_3CHO(g)$	−166.1	−133.4	246.4

Substance or Ion	ΔH_f° (kJ/mol)	ΔG_f° (kJ/mol)	S° (J/mol·K)
$CH_3CHO(l)$	−191.8	−128.3	160.4
Carboxylic acids and ions			
$HCOOH(aq)$	−425.6	−351.0	92
$HCOOH(l)$	−424.7	−361.4	129
$HCOO^-(aq)$	−425.6	−351.0	92
$CH_3COOH(aq)$	−486.0	−369.4	86.6
$CH_3COOH(l)$	−484.5	−389.9	159.8
$CH_3COO^-(aq)$	−486.0	−369.4	86.6
Sugars			
$C_6H_{12}O_6$ (s, glucose)	−1273.3	−910.4	212.1
Cesium			
$Cs(g)$	76.50	49.56	175.5
$Cs(l)$	2.087	0.024	92.07
$Cs(s)$	0	0	85.15
$CsBr(s)$	−405.7	−391.4	113.0
$CsCl(s)$	−442.8	−414.4	101.2
$CsF(s)$	−554.7	−525.4	88.28
$CsI(s)$	−346.6	−340.6	123.1
Chlorine			
$Cl(g)$	121.3	105.3	165.1
$Cl^-(aq)$	−167.2	−131.3	56.5
$Cl^-(g)$	−234.0	−240.2	153.2
$Cl_2(g)$	0	0	223.0
$HCl(aq)$	−167.2	−131.3	56.5
$HCl(g)$	−92.31	−95.30	186.8
Chromium			
$Cr(g)$	397.5	352.6	174.2
$Cr(s)$	0	0	23.62
$Cr^{3+}(aq)$	−143.5	—	—
$CrO_4^{2-}(aq)$	−881.2	−727.8	50.21
$Cr_2O_3(s)$	−1134.7	−1053.1	81.15
$Cr_2O_7^{2-}(aq)$	−1490.3	−1301.2	262.9
Cobalt			
$Co(g)$	426.7	382.1	179.4
$Co(s)$	0	0	30.07
$Co^{2+}(aq)$	−58.2	−54.4	113
Copper			
$Cu(g)$	337.6	297.9	166.3
$Cu(s)$	0	0	33.16
$Cu^+(aq)$	71.67	50.00	40.6
$Cu^{2+}(aq)$	64.77	65.52	−99.6
$CuCl_2(s)$	−205.9	−161.7	108.1
$CuO(s)$	−156.1	−128.3	42.59

Substance or Ion	ΔH_f° (kJ/mol)	ΔG_f° (kJ/mol)	S° (J/mol·K)
$Cu_2O(s)$	−170.7	−147.9	92.36
Fluorine			
$F(g)$	79.39	62.31	158.6
$F^-(g)$	−255.1	−262.0	145.5
$F^-(aq)$	−332.6	−278.8	−13.8
$F_2(g)$	0	0	202.7
$HF(aq)$	−332.6	−278.8	−13.8
$HF(g)$	−272.5	−274.6	173.7
Hydrogen			
$H(g)$	218.0	203.3	114.6
$H^+(aq)$	0	0	0
$H^+(g)$	1536.2	1517.0	108.8
$H_2(g)$	0	0	130.6
$H_2O(g)$	−241.8	−228.6	188.7
$H_2O(l)$	−285.8	−237.1	69.95
$H_2O_2(aq)$	−191.2	−134.1	143.9
$H_2O_2(g)$	−136.1	−105.5	232.9
$H_2O_2(l)$	−187.8	−120.4	109.6
OH^-	−230.0	−157.3	−10.75
Iodine			
$I(g)$	106.8	70.21	180.7
$I^-(aq)$	−55.19	−51.59	109.6
$I^-(g)$	−194.6	−221.5	169.2
$I_2(g)$	62.42	19.36	260.6
$I_2(s)$	0	0	116.1
$HI(g)$	26.36	1.576	206.5
Iron			
$Fe(g)$	415.5	369.8	180.4
$Fe(l)$	12.40	10.18	34.76
$Fe(s)$	0	0	27.32
$Fe^{2+}(aq)$	−89.1	−78.87	−137.7
$Fe^{3+}(aq)$	−48.5	−4.6	−315.9
$FeCl_2(s)$	−341.8	−302.3	117.9
$FeCl_3(s)$	−399.4	−333.9	142.3
$FeO(s)$	−272.0	−251.4	60.75
$Fe_2O_3(s)$	−825.5	−743.5	87.40
$Fe_3O_4(s)$	−1120.9	−1017.4	145.3
$FeS_2(s)$	−171.5	−160.1	52.92
Lead			
$Pb(s)$	0	0	64.78
$Pb^{2+}(aq)$	−1.7	−24.39	10.5
$PbBr_2(s)$	−277.4	−260.7	161.1
$PbCO_3(s)$	−699.1	−625.5	131.0
$PbCl_2(s)$	−359.4	−314.1	136.0

(*continued*)

(*continued*)

Substance or Ion	ΔH_f° (kJ/mol)	ΔG_f° (kJ/mol)	S° (J/mol·K)
$PbO(s)$	−219.4	−189.3	66.32
$PbO_2(s)$	−274.5	−215.4	71.80
$PbS(s)$	−98.32	−96.68	91.34
$PbSO_4(s)$	−919.9	−813.2	148.6
Lithium			
$Li(g)$	159.3	126.6	138.7
$Li(s)$	0	0	29.08
$Li^+(aq)$	−278.5	−293.3	13.4
$Li^+(g)$	685.7	648.5	132.9
$LiBr(s)$	−350.9	−341.6	74.06
$LiCl(s)$	−408.3	−384.0	59.30
$LiF(s)$	−616.9	−588.7	35.66
$LiI(s)$	−270.1	−269.7	85.77
Magnesium			
$Mg(g)$	147.1	112.6	148.5
$Mg(s)$	0	0	32.67
$Mg^+(g)$	891.0	848.6	154.3
$Mg^{2+}(aq)$	−466.9	−454.8	−138.1
$Mg^{2+}(g)$	2351	—	—
$MgCO_3(s)$	−1111.7	−1028.1	65.85
$MgCl_2(s)$	−641.6	−592.1	89.63
$Mg_3N_2(s)$	−461.1	−400.9	87.86
$MgO(s)$	−601.2	−568.9	26.92
$Mg(OH)_2(s)$	−924.7	−833.7	63.24
Manganese			
$Mn(g)$	283.3	241.0	173.6
$Mn(s)$	0	0	32.01
$Mn^{2+}(aq)$	−220.7	−228.0	−73.6
$MnO(s)$	−385.2	−362.9	59.71
$MnO_2(s)$	−520.0	−465.2	53.05
$MnO_4^-(aq)$	−541.4	−447.3	191.2
Mercury			
$Hg(g)$	61.38	31.91	174.9
$Hg(l)$	0	0	76.03
$Hg^{2+}(aq)$	171.1	164.4	−32.2
$Hg_2^{2+}(aq)$	172.3	153.6	84.5
$HgCl_2(s)$	−230.1	−184.0	144.5
$Hg_2Cl_2(s)$	−265.2	−210.8	192.4
$HgO(s)$	−90.79	−58.49	70.27
Nickel			
$Ni(g)$	430.1	384.7	182.1
$Ni(s)$	0	0	29.87
$Ni^{2+}(aq)$	−54.0	−45.6	−128.9

Substance or Ion	ΔH_f° (kJ/mol)	ΔG_f° (kJ/mol)	S° (J/mol·K)
$NiCl_2(s)$	−304.9	−258.8	98.16
$NiO(s)$	−239.7	−211.7	37.99
Nitrogen			
$N(g)$	472.7	455.6	153.2
$N_2(g)$	0	0	191.6
$NH_3(aq)$	−80.29	−26.57	111.3
$NH_3(g)$	−45.90	−16.40	192.7
$NH_4^+(aq)$	−132.5	−79.37	113.4
$N_2H_4(g)$	95.35	159.2	238.6
$N_2H_4(l)$	50.63	149.4	121.4
$NH_4Cl(s)$	−314.6	−203.1	94.86
$NH_4NO_3(s)$	−365.6	−184.0	151.1
$NO(g)$	90.29	86.60	210.6
$NO_2(g)$	33.10	51.24	239.9
$NO_3^-(aq)$	−207.4	−111.3	146.4
$N_2O(g)$	82.05	104.2	219.9
$N_2O_4(g)$	9.079	97.72	304.3
$N_2O_4(l)$	−19.56	97.52	209.2
$NOCl(g)$	51.71	66.08	261.6
$HNO_3(aq)$	−207.4	−111.3	146.4
$HNO_3(g)$	−134.3	−73.99	266.3
$HNO_3(l)$	−174.1	−80.79	155.6
Oxygen			
$O(g)$	249.2	231.8	160.9
$O_2(g)$	0	0	205.0
$O_3(g)$	142.7	163.2	238.8
Phosphorus			
$P(g)$	316.4	280.0	163.1
$P(s, red)$	−17.46	−12.03	22.85
$P(s, white)$	0	0	41.08
$P_2(g)$	143.7	103.1	218.0
$P_4(g)$	58.91	24.45	279.9
$PCl_3(g)$	−288.7	−269.6	311.6
$PCl_5(g)$	−360.2	−290.3	364.2
$PF_5(g)$	−1594.4	−1520.7	300.7
$PH_3(g)$	5.439	7.175	210.1
$P_4O_{10}(s)$	−3009.9	−2723.3	228.8
$PO_4^{3-}(aq)$	−1277.4	−1018.8	−222
$POCl_3(g)$	−559.8	−514.3	325.3
$POCl_3(l)$	−597.1	−520.9	222.5
$P_4S_3(s)$	−224.6	−206.9	200.8
$HPO_4^{2-}(aq)$	−1281	−1082	−36
$H_2PO_4^-(aq)$	−1285	−1135	89.1
$H_3PO_4(aq)$	−1288.3	−1142.6	158.2

Substance or Ion	ΔH_f° (kJ/mol)	ΔG_f° (kJ/mol)	S° (J/mol·K)
Potassium			
$K(g)$	89.00	60.51	160.2
$K(s)$	0	0	64.67
$K^+(aq)$	−252.4	−283.3	102.5
$K^+(g)$	514.0	481.0	154.5
$KBr(s)$	−393.8	−380.4	95.94
$KCl(s)$	−436.7	−408.8	82.55
$KF(s)$	−568.6	−538.9	66.55
$KI(s)$	−327.9	−323.0	106.4
$KClO_3(s)$	−397.7	−296.3	143.1
$K_2CO_3(s)$	−1150.2	−1064.5	155.5
$KNO_3(s)$	−494.6	−394.9	133.1
$K_2O(s)$	−363.2	−322.1	94.14
$KO_2(s)$	−284.5	−240.6	122.5
$K_2O_2(s)$	−495.8	−429.8	113.0
$KOH(s)$	−424.7	−378.9	78.91
Rubidium			
$Rb(g)$	80.90	53.11	170.0
$Rb(s)$	0	0	76.78
$RbBr(s)$	−394.6	−381.8	110.0
$RbCl(s)$	−435.5	−407.8	95.90
$RbF(s)$	−549.4	—	75.3
$RbI(s)$	−333.8	−328.9	118.4
Silicon			
$Si(g)$	450.0	405.6	167.9
$Si(s)$	0	0	18.82
$SiC(s)$	−65.3	−62.8	16.61
$SiCl_4(l)$	−687.0	−619.9	239.7
$SiF_4(g)$	−1614.9	−1572.7	282.7
$SiO_2(s, quartz)$	−910.9	−856.4	41.46
Silver			
$Ag(g)$	284.6	245.7	172.9
$Ag(s)$	0	0	42.55
$Ag^+(aq)$	105.6	77.12	72.68
$AgBr(s)$	−100.4	−96.90	107.1
$AgCl(s)$	−127.1	−109.8	96.2
$AgF(s)$	−204.6	—	83.7
$AgI(s)$	−61.84	−66.19	115.5
$Ag_2O(s)$	−31.05	−11.21	121.3
$Ag_2S(s)$	−32.59	−40.67	144.0
$AgNO_3(s)$	−124.4	−33.47	140.9
Sodium			
$Na(g)$	107.3	76.86	153.6
$Na(s)$	0	0	51.46
$Na^+(aq)$	−240.1	−261.9	59.1
$Na^+(g)$	609.3	574.4	147.8

Substance or Ion	ΔH_f° (kJ/mol)	ΔG_f° (kJ/mol)	S° (J/mol·K)
$NaBr(s)$	−361.4	−349.3	86.82
$Na_2CO_3(s)$	−1130.8	−1048.0	138.8
$NaCl(s)$	−411.1	−384.0	72.12
$NaF(s)$	−575.4	−545.1	51.21
$NaHCO_3(s)$	−950.8	−851.0	101.7
$NaI(s)$	−287.9	−284.6	98.50
$NaNO_3(s)$	−467.9	−367.1	116.5
$NaOH(s)$	−425.9	−379.7	64.44
Strontium			
$Sr(g)$	164.0	131.6	164.5
$Sr(s)$	0	0	55.69
$Sr^+(g)$	719.7	679.3	170.3
$Sr^{2+}(aq)$	−545.8	−559.4	−32.6
$Sr^{2+}(g)$	1790.6	—	—
$SrCO_3(s)$	−1220.1	−1140.1	97.1
$SrCl_2(s)$	−828.9	−781.2	114.9
$SrO(s)$	−592.0	−561.4	55.52
$SrSO_4(s)$	−1453.1	−1341.0	117
Sulfur			
$S(g)$	277.0	236.5	167.7
$S(s, monoclinic)$	0.360	0.070	33.03
$S(s, rhombic)$	0	0	32.06
$S^{2-}(aq)$	41.8	83.7	22
$S_2(g)$	128.6	79.7	228.1
$S_8(g)$	100.4	48.61	430.2
$SO_2(g)$	−296.8	−300.1	248.1
$SO_3(g)$	−395.8	−371.0	256.7
$SO_4^{2-}(aq)$	−909.3	−744.6	20.1
$HS^-(aq)$	−17.7	12.6	61.1
$H_2S(aq)$	−39	−27.4	122
$H_2S(g)$	−20.50	−33.33	205.6
$HSO_4^-(aq)$	−887.3	−756.0	131.8
$H_2SO_4(l)$	−814.0	−689.9	156.9
Tin			
$Sn(s, gray)$	−2.09	0.13	44.14
$Sn(s, white)$	0	0	51.55
$SnCl_4(l)$	−511.3	−440.2	258.6
$SnO_2(s)$	−580.7	−519.7	52.3
Zinc			
$Zn(g)$	130.4	94.89	160.9
$Zn(s)$	0	0	41.72
$Zn^{2+}(aq)$	−153.9	−147.0	−112.1
$ZnCl_2(s)$	−415.1	−369.4	111.5
$ZnO(s)$	−348.3	−318.3	43.64
$ZnS(s, sphalerite)$	−206.0	−201.3	−57.7

Appendix D   Electron Configurations of Atoms in the Ground State

Z	Element	Configuration	Z	Element	Configuration
1	H	$1s^1$	44	Ru	$[Kr]4d^75s^1$
2	He	$1s^2$	45	Rh	$[Kr]4d^85s^1$
3	Li	$[He]2s^1$	46	Pd	$[Kr]4d^{10}$
4	Be	$[He]2s^2$	47	Ag	$[Kr]4d^{10}5s^1$
5	B	$[He]2s^22p^1$	48	Cd	$[Kr]4d^{10}5s^2$
6	C	$[He]2s^22p^2$	49	In	$[Kr]4d^{10}5s^25p^1$
7	N	$[He]2s^22p^3$	50	Sn	$[Kr]4d^{10}5s^25p^2$
8	O	$[He]2s^22p^4$	51	Sb	$[Kr]4d^{10}5s^25p^3$
9	F	$[He]2s^22p^5$	52	Te	$[Kr]4d^{10}5s^25p^4$
10	Ne	$[He]2s^22p^6$	53	I	$[Kr]4d^{10}5s^25p^5$
11	Na	$[Ne]3s^1$	54	Xe	$[Kr]4d^{10}5s^25p^6$
12	Mg	$[Ne]3s^2$	55	Cs	$[Xe]6s^1$
13	Al	$[Ne]3s^23p^1$	56	Ba	$[Xe]6s^2$
14	Si	$[Ne]3s^23p^2$	57	La	$[Xe]5d^16s^2$
15	P	$[Ne]3s^23p^3$	58	Ce	$[Xe]4f^15d^16s^2$
16	S	$[Ne]3s^23p^4$	59	Pr	$[Xe]4f^36s^2$
17	Cl	$[Ne]3s^23p^5$	60	Nd	$[Xe]4f^46s^2$
18	Ar	$[Ne]3s^23p^6$	61	Pm	$[Xe]4f^56s^2$
19	K	$[Ar]4s^1$	62	Sm	$[Xe]4f^66s^2$
20	Ca	$[Ar]4s^2$	63	Eu	$[Xe]4f^76s^2$
21	Sc	$[Ar]3d^14s^2$	64	Gd	$[Xe]4f^75d^16s^2$
22	Ti	$[Ar]3d^24s^2$	65	Tb	$[Xe]4f^96s^2$
23	V	$[Ar]3d^34s^2$	66	Dy	$[Xe]4f^{10}6s^2$
24	Cr	$[Ar]3d^54s^1$	67	Ho	$[Xe]4f^{11}6s^2$
25	Mn	$[Ar]3d^54s^2$	68	Er	$[Xe]4f^{12}6s^2$
26	Fe	$[Ar]3d^64s^2$	69	Tm	$[Xe]4f^{13}6s^2$
27	Co	$[Ar]3d^74s^2$	70	Yb	$[Xe]4f^{14}6s^2$
28	Ni	$[Ar]3d^84s^2$	71	Lu	$[Xe]4f^{14}5d^16s^2$
29	Cu	$[Ar]3d^{10}4s^1$	72	Hf	$[Xe]4f^{14}5d^26s^2$
30	Zn	$[Ar]3d^{10}4s^2$	73	Ta	$[Xe]4f^{14}5d^36s^2$
31	Ga	$[Ar]3d^{10}4s^24p^1$	74	W	$[Xe]4f^{14}5d^46s^2$
32	Ge	$[Ar]3d^{10}4s^24p^2$	75	Re	$[Xe]4f^{14}5d^56s^2$
33	As	$[Ar]3d^{10}4s^24p^3$	76	Os	$[Xe]4f^{14}5d^66s^2$
34	Se	$[Ar]3d^{10}4s^24p^4$	77	Ir	$[Xe]4f^{14}5d^76s^2$
35	Br	$[Ar]3d^{10}4s^24p^5$	78	Pt	$[Xe]4f^{14}5d^96s^1$
36	Kr	$[Ar]3d^{10}4s^24p^6$	79	Au	$[Xe]4f^{14}5d^{10}6s^1$
37	Rb	$[Kr]5s^1$	80	Hg	$[Xe]4f^{14}5d^{10}6s^2$
38	Sr	$[Kr]5s^2$	81	Tl	$[Xe]4f^{14}5d^{10}6s^26p^1$
39	Y	$[Kr]4d^15s^2$	82	Pb	$[Xe]4f^{14}5d^{10}6s^26p^2$
40	Zr	$[Kr]4d^25s^2$	83	Bi	$[Xe]4f^{14}5d^{10}6s^26p^3$
41	Nb	$[Kr]4d^45s^1$	84	Po	$[Xe]4f^{14}5d^{10}6s^26p^4$
42	Mo	$[Kr]4d^55s^1$	85	At	$[Xe](4f^{14}5d^{10}6s^26p^5)$
43	Tc	$[Kr]4d^55s^2$	86	Rn	$[Xe]4f^{14}5d^{10}6s^26p^6$

Z	Element	Configuration	Z	Element	Configuration
87	Fr	$[Rn](7s^1)$	103	Lr	$[Rn](5f^{14}6d^17s^2)$
88	Ra	$[Rn]7s^2$	104	Rf	$[Rn](5f^{14}5d^27s^2)$
89	Ac	$[Rn]6d^16s^2$	105	Db	$[Rn](5f^{14}6d^37s^2)$
90	Th	$[Rn]6d^27s^2$	106	Sg	$[Rn](5f^{14}6d^47s^2)$
91	Pa	$[Rn](5f^26d^17s^2)$	107	Bh	$[Rn](5f^{14}6d^57s^2)$
92	U	$[Rn](5f^36d^17s^2)$	108	Hs	$[Rn](5f^{14}6d^67s^2)$
93	Np	$[Rn](5f^46d^17s^2)$	109	Mt	$[Rn](5f^{14}6d^77s^2)$
94	Pu	$[Rn]5f^67s^2$	110	Ds	$[Rn](5f^{14}6d^87s^2)$
95	Am	$[Rn]5f^77s^2$	111	Rg	$[Rn](5f^{14}6d^97s^2)$
96	Cm	$[Rn](5f^76d^17s^2)$	112	Cn	$[Rn](5f^{14}6d^{10}7s^2)$
97	Bk	$[Rn](5f^97s^2)$	113	Uut	$[Rn](5f^{14}6d^{10}7s^27p^1)$
98	Cf	$[Rn](5f^{10}7s^2)$	114	Uuq	$[Rn](5f^{14}6d^{10}7s^27p^2)$
99	Es	$[Rn](5f^{11}7s^2)$	115	Uup	$[Rn](5f^{14}6d^{10}7s^27p^3)$
100	Fm	$[Rn](5f^{12}7s^2)$	116	Uuh	$[Rn](5f^{14}6d^{10}7s^27p^4)$
101	Md	$[Rn](5f^{13}7s^2)$	117	Uus	$[Rn](5f^{14}6d^{10}7s^27p^5)$
102	No	$[Rn](5f^{14}7s^2)$	118	Uuo	$[Rn](5f^{14}6d^{10}7s^27p^6)$

Configurations in this table, except those in parentheses, are taken from Charlotte E. Moore, *National Standard Reference Data Series*, National Bureau of Standards, 34 (September 1970). Those in parentheses were obtained on the basis of their assumed position in the periodic table.

Appendix E Acid-Ionization Constants at 25°C

Substance	Formula	K_a
Acetic acid	$HC_2H_3O_2$	1.7×10^{-5}
Arsenic acid*	H_3AsO_4	6.5×10^{-3}
	$H_2AsO_4^-$	1.2×10^{-7}
	$HAsO_4^{2-}$	3.2×10^{-12}
Ascorbic acid*	$H_2C_6H_6O_6$	6.8×10^{-5}
	$HC_6H_6O_6^-$	2.8×10^{-12}
Benzoic acid	$HC_7H_5O_2$	6.3×10^{-5}
Boric acid	H_3BO_3	5.9×10^{-10}
Carbonic acid*	H_2CO_3	4.3×10^{-7}
	HCO_3^-	4.8×10^{-11}
Chromic acid*	H_2CrO_4	1.5×10^{-1}
	$HCrO_4^-$	3.2×10^{-7}
Cyanic acid	$HOCN$	3.5×10^{-4}
Formic acid	$HCHO_2$	1.7×10^{-4}
Hydrocyanic acid	HCN	4.9×10^{-10}
Hydrofluoric acid	HF	6.8×10^{-4}
Hydrogen peroxide	H_2O_2	1.8×10^{-12}
Hydrogen sulfate ion	HSO_4^-	1.1×10^{-2}
Hydrogen sulfide*	H_2S	8.9×10^{-8}
	HS^-	1.2×10^{-13}
Hypochlorous acid	$HClO$	3.5×10^{-8}

(continued)

(*continued*)

Substance	Formula	K_a
Lactic acid	$HC_3H_5O_3$	1.3×10^{-4}
Nitrous acid	HNO_2	4.5×10^{-4}
Oxalic acid*	$H_2C_2O_4$	5.6×10^{-2}
	$HC_2O_4^-$	5.1×10^{-5}
Phenol	C_6H_5OH	1.1×10^{-10}
Phosphoric acid*	H_3PO_4	6.9×10^{-3}
	$H_2PO_4^-$	6.2×10^{-8}
	HPO_4^{2-}	4.8×10^{-13}
Phosphorous acid*	H_2PHO_3	1.6×10^{-2}
	$HPHO_3^-$	7×10^{-7}
Propionic acid	$HC_3H_5O_2$	1.3×10^{-5}
Pyruvic acid	$HC_3H_3O_3$	1.4×10^{-4}
Sulfurous acid*	H_2SO_3	1.3×10^{-2}
	HSO_3^-	6.3×10^{-8}

*The ionization constants for polyprotic acids are for successive ionizations. For example, for H_3PO_4, the equilibrium is $H_3PO_4 - H^+ + H_2PO_4^-$. For $H_2PO_4^-$, the equilibrium is $H_2PO_4^- - H^+ + HPO_4^{2-}$.

Appendix F   Base-Ionization Constants at 25°C

Substance	Formula	K_b
Ammonia	NH_3	1.8×10^{-5}
Aniline	$C_6H_5NH_2$	4.2×10^{-10}
Dimethylamine	$(CH_3)_2NH$	5.1×10^{-4}
Ethylamine	$C_2H_5NH_2$	4.7×10^{-4}
Ethylenediamine	$NH_2CH_2CH_2NH_2$	5.2×10^{-4}
Hydrazine	N_2H_4	1.7×10^{-6}
Hydroxylamine	NH_2OH	1.1×10^{-8}
Methylamine	CH_3NH_2	4.4×10^{-4}
Pyridine	C_5H_5N	1.4×10^{-9}
Trimethylamine	$(CH_3)_3N$	6.5×10^{-5}
Urea	NH_2CONH_2	1.5×10^{-14}

Appendix G Solubility Product Constants at 25°C

Substance	Formula	K_{sp}
Aluminum hydroxide	$Al(OH)_3$	4.6×10^{-33}
Barium chromate	$BaCrO_4$	1.2×10^{-10}
Barium fluoride	BaF_2	1.0×10^{-6}
Barium sulfate	$BaSO_4$	1.1×10^{-10}
Cadmium oxalate	CdC_2O_4	1.5×10^{-8}
Cadmium sulfide	CdS	8×10^{-27}
Calcium carbonate	$CaCO_3$	3.8×10^{-9}
Calcium fluoride	CaF_2	3.4×10^{-11}
Calcium oxalate	CaC_2O_4	2.3×10^{-9}
Calcium phosphate	$Ca_3(PO_4)_2$	1×10^{-26}
Calcium sulfate	$CaSO_4$	2.4×10^{-5}
Cobalt(II) sulfide	CoS	4×10^{-21}
Copper(II) hydroxide	$Cu(OH)_2$	2.6×10^{-19}
Copper(II) sulfide	CuS	6×10^{-36}
Iron(II) hydroxide	$Fe(OH)_2$	8×10^{-16}
Iron(II) sulfide	FeS	6×10^{-18}
Iron(III) hydroxide	$Fe(OH)_3$	2.5×10^{-39}
Lead(II) arsenate	$Pb_3(AsO_4)_2$	4×10^{-36}
Lead(II) chloride	$PbCl_2$	1.6×10^{-5}
Lead(II) chromate	$PbCrO_4$	1.8×10^{-14}
Lead(II) iodide	PbI_2	6.5×10^{-9}
Lead(II) sulfate	$PbSO_4$	1.7×10^{-8}
Lead(II) sulfide	PbS	2.5×10^{-27}
Magnesium arsenate	$Mg_3(AsO_4)_2$	2×10^{-20}
Magnesium carbonate	$MgCO_3$	1.0×10^{-5}
Magnesium hydroxide	$Mg(OH)_2$	1.8×10^{-11}
Magnesium oxalate	MgC_2O_4	8.5×10^{-5}
Manganese(II) sulfide	MnS	2.5×10^{-10}
Mercury(I) chloride	Hg_2Cl_2	1.3×10^{-18}
Mercury(II) sulfide	HgS	1.6×10^{-52}
Nickel(II) hydroxide	$Ni(OH)_2$	2.0×10^{-15}
Nickel(II) sulfide	NiS	3×10^{-19}
Silver acetate	$AgC_2H_3O_2$	2.0×10^{-3}
Silver bromide	$AgBr$	5.0×10^{-13}
Silver chloride	$AgCl$	1.8×10^{-10}
Silver chromate	Ag_2CrO_4	1.1×10^{-12}
Silver iodide	AgI	8.3×10^{-17}
Silver sulfide	Ag_2S	6×10^{-50}
Strontium carbonate	$SrCO_3$	9.3×10^{-10}
Strontium chromate	$SrCrO_4$	3.5×10^{-5}
Strontium sulfate	$SrSO_4$	2.5×10^{-7}
Zinc hydroxide	$Zn(OH)_2$	2.1×10^{-16}
Zinc sulfide	ZnS	1.1×10^{-21}

Appendix H Formation Constants of Complex Ions at 25°C

Complex Ion	K_f	Complex Ion	K_f
$Ag(CN)_2^-$	5.6×10^{18}	$Fe(CN)_6^{4-}$	1.0×10^{35}
$Ag(NH_3)_2^+$	1.7×10^7	$Fe(CN)_6^{3-}$	9.1×10^{41}
$Ag(S_2O_3)_2^{3-}$	2.9×10^{13}	$Ni(CN)_4^{2-}$	1.0×10^{31}
$Cd(NH_3)_4^{2+}$	1.0×10^7	$Ni(NH_3)_6^{2+}$	5.6×10^8
$Cu(CN)_2^-$	1.0×10^{16}	$Zn(NH_3)_4^{2+}$	2.9×10^9
$Cu(NH_3)_4^{2+}$	4.8×10^{12}	$Zn(OH)_4^{2-}$	2.8×10^{15}

Appendix I Standard Electrode (Reduction) Potentials in Aqueous Solution at 25°C

Cathode (Reduction) Half-Reaction	Standard Potential, $E°$ (V)
$Li^+(aq) + e^- \rightleftharpoons Li(s)$	-3.04
$K^+(aq) + e^- \rightleftharpoons K(s)$	-2.92
$Ca^{2+}(aq) + 2e^- \rightleftharpoons Ca(s)$	-2.76
$Na^+(aq) + e^- \rightleftharpoons Na(s)$	-2.71
$Mg^{2+}(aq) + 2e^- \rightleftharpoons Mg(s)$	-2.38
$Al^{3+}(aq) + 3e^- \rightleftharpoons Al(s)$	-1.66
$2H_2O(l) + 2e^- \rightleftharpoons H_2(g) + 2OH^-(aq)$	-0.83
$Zn^{2+}(aq) + 2e^- \rightleftharpoons Zn(s)$	-0.76
$Cr^{3+}(aq) + 3e^- \rightleftharpoons Cr(s)$	-0.74
$Fe^{2+}(aq) + 2e^- \rightleftharpoons Fe(s)$	-0.41
$Cd^{2+}(aq) + 2e^- \rightleftharpoons Cd(s)$	-0.40
$Ni^{2+}(aq) + 2e^- \rightleftharpoons Ni(s)$	-0.23
$Sn^{2+}(aq) + 2e^- \rightleftharpoons Sn(s)$	-0.14
$Pb^{2+}(aq) + 2e^- \rightleftharpoons Pb(s)$	-0.13
$Fe^{3+}(aq) + 3e^- \rightleftharpoons Fe(s)$	-0.04
$2H^+(aq) + 2e^- \rightleftharpoons H_2(g)$	0.00
$Sn^{4+}(aq) + 2e^- \rightleftharpoons Sn^{2+}(aq)$	0.15
$Cu^{2+}(aq) + e^- \rightleftharpoons Cu^+(aq)$	0.16
$ClO_4^-(aq) + H_2O(l) + 2e^- \rightleftharpoons ClO_3^-(aq) + 2OH^-(aq)$	0.17
$AgCl(s) + e^- \rightleftharpoons Ag(s) + Cl^-(aq)$	0.22
$Cu^{2+}(aq) + 2e^- \rightleftharpoons Cu(s)$	0.34
$ClO_3^-(aq) + H_2O(l) + 2e^- \rightleftharpoons ClO_2^-(aq) + 2OH^-(aq)$	0.35
$IO^-(aq) + H_2O(l) + 2e^- \rightleftharpoons I^-(aq) + 2OH^-(aq)$	0.49
$Cu^+(aq) + e^- \rightleftharpoons Cu(s)$	0.52
$I_2(s) + 2e^- \rightleftharpoons 2I^-(aq)$	0.54
$ClO_2^-(aq) + H_2O(l) + 2e^- \rightleftharpoons ClO^-(aq) + 2OH^-(aq)$	0.59
$Fe^{3+}(aq) + e^- \rightleftharpoons Fe^{2+}(aq)$	0.77
$Hg_2^{2+}(aq) + 2e^- \rightleftharpoons 2Hg(l)$	0.80

Cathode (Reduction) Half-Reaction	Standard Potential, $E°$ (V)
$Ag^+(aq) + e^- \rightleftharpoons Ag(s)$	0.80
$Hg^{2+}(aq) + 2e^- \rightleftharpoons Hg(l)$	0.85
$ClO^-(aq) + H_2O(l) + 2e^- \rightleftharpoons Cl^-(aq) + 2OH^-(aq)$	0.90
$2Hg^{2+}(aq) + 2e^- \rightleftharpoons Hg_2^{2+}(aq)$	0.90
$NO_3^-(aq) + 4H^+(aq) + 3e^- \rightleftharpoons NO(g) + 2H_2O(l)$	0.96
$Br_2(l) + 2e^- \rightleftharpoons 2Br^-(aq)$	1.07
$O_2(g) + 4H^+(aq) + 4e^- \rightleftharpoons 2H_2O(l)$	1.23
$Cr_2O_7^{2-}(aq) + 14H^+(aq) + 6e^- \rightleftharpoons 2Cr^{3+}(aq) + 7H_2O(l)$	1.33
$Cl_2(g) + 2e^- \rightleftharpoons 2Cl^-(aq)$	1.36
$Ce^{4+}(aq) + e^- \rightleftharpoons Ce^{3+}(aq)$	1.44
$MnO_4^-(aq) + 8H^+(aq) + 5e^- \rightleftharpoons Mn^{2+}(aq) + 4H_2O(l)$	1.49
$H_2O_2(aq) + 2H^+(aq) + 2e^- \rightleftharpoons 2H_2O(l)$	1.78
$Co^{3+}(aq) + e^- \rightleftharpoons Co^{2+}(aq)$	1.82
$S_2O_8^{2-}(aq) + 2e^- \rightleftharpoons 2SO_4^{2-}(aq)$	2.01
$O_3(g) + 2H^+(aq) + 2e^- \rightleftharpoons O_2(g) + H_2O(l)$	2.07
$F_2(g) + 2e^- \rightleftharpoons 2F^-(aq)$	2.87

Answers to Exercises

Note: Your answers may differ from those given here in the last significant figure.

CHAPTER 1

1.1 11.02 g   **1.2** Physical properties: soft; silver color; melts at 64°C. Chemical properties: metal, reacts with water; reacts with oxygen; reacts with chlorine.   **1.3** a 4.9   b 2.48   c 0.08   d 3   **1.4** a 1.84 nm   b 5.67 ps   c 7.85 mg   d 9.7 km   e 0.732 ms or 732 μs   f 0.154 nm or 154 pm   **1.5** a 39.2°C   b 195 K   **1.6** 7.87 g/cm^3; the object is made of iron   **1.7** 38.4 cm^3 **1.8** 1.21×10^{-7} mm   **1.9** 6.76×10^{-26} dm^3 **1.10** 3.24 m

CHAPTER 2

2.1 $^{35}_{17}Cl$ **2.2** 35.453 amu   **2.3** a Se: Group VIA, Period 4; nonmetal   b Cs: Group IA, Period 6; metal   c Fe: Group VIIIB, Period 4; metal   d Cu: Group IB, Period 4; metal   e Br: Group VIIA, Period 4; nonmetal   **2.4** K_2CrO_4 **2.5** a calcium oxide   b lead(II) chromate   **2.6** $Tl(NO_3)_3$ **2.7** a dichlorine hex(a)oxide   b phosphorus trichloride   c phosphorus pentachloride   **2.8** a CS_2 b SO_3 **2.9** a boron trifluoride   b hydrogen selenide   **2.10** perbromate ion, BrO_4^- **2.11** sodium carbonate decahydrate   **2.12** $Na_2S_2O_3 \cdot 5H_2O$ **2.13** a $O_2 + 2PCl_3 \longrightarrow 2POCl_3$ b $P_4 + 6N_2O \longrightarrow P_4O_6 + 6N_2$ c $2As_2S_3 + 9O_2 \longrightarrow 2As_2O_3 + 6SO_2$ d $Ca_3(PO_4)_2 + 4H_3PO_4 \longrightarrow 3Ca(H_2PO_4)_2$

CHAPTER 3

3.1 a 46.0 amu   b 180 amu   c 40.0 amu   d 58.3 amu   **3.2** a SO_3, 80.1 amu   b H_2SO_4, 98.1 amu   **3.3** a 6.656×10^{-23} g   b 7.65×10^{-23} g   **3.4** 30.9 g H_2O_2 **3.5** 0.452 mol HNO_3 **3.6** 1.2×10^{21} HCN molecules   **3.7** 35.0% N, 60.0% O, 5.04% H   **3.8** 17.0 g N   **3.9** 40.9% C, 4.57% H, 54.5% O   **3.10** SO_3 **3.11** $C_7H_6O_2$ **3.12** C_2H_4O **3.13**

$$H_2 \quad + \quad Cl_2 \quad \longrightarrow \quad 2HCl$$

1 molecule H_2 + 1 molecule $Cl_2 \longrightarrow$ 2 molecules HCl
1 mol H_2 +   1 mol $Cl_2 \longrightarrow$ 2 mol HCl
2.016 g H_2 +   70.9 g $Cl_2 \longrightarrow 2 \times 36.5$ g HCl

3.14 178 g Na   **3.15** 2.46 kg O_2 **3.16** 81.1 g Hg   **3.17** 0.12 mol $AlCl_3$ **3.18** 11.0 g ZnS   **3.19** 21.4 g; 89.1%

CHAPTER 4

4.1 a soluble   b soluble   c insoluble
4.2 a $2H^+(aq) + 2NO_3^-(aq) + Mg(OH)_2(s) \longrightarrow$
$$2H_2O(l) + Mg^{2+}(aq) + 2NO_3^-(aq)$$
$2H^+(aq) + Mg(OH)_2(s) \longrightarrow 2H_2O(l) + Mg^{2+}(aq)$
b $Pb^{2+}(aq) + 2NO_3^-(aq) + 2Na^+(aq) + SO_4^{2-}(aq) \longrightarrow$
$$PbSO_4(s) + 2Na^+(aq) + 2NO_3^-(aq)$$
$Pb^{2+}(aq) + SO_4^{2-}(aq) \longrightarrow PbSO_4(s)$
4.3 $2NaI(aq) + Pb(C_2H_3O_2)_2(aq) \longrightarrow PbI_2(s) + 2NaC_2H_3O_2(aq)$
$2I^-(aq) + Pb^{2+}(aq) \longrightarrow PbI_2(s)$
4.4 a weak acid   b weak acid   c strong acid   d strong base

4.5 $HCN(aq) + LiOH(aq) \longrightarrow H_2O(l) + LiCN(aq)$
$HCN(aq) + OH^-(aq) \longrightarrow H_2O(l) + CN^-(aq)$
4.6 $H_2SO_4(aq) + KOH(aq) \longrightarrow H_2O(l) + KHSO_4(aq)$
$KHSO_4(aq) + KOH(aq) \longrightarrow H_2O(l) + K_2SO_4(aq)$
$H^+(aq) + OH^-(aq) \longrightarrow H_2O(l)$
$HSO_4^-(aq) + OH^-(aq) \longrightarrow H_2O(l) + SO_4^{2-}(aq)$
4.7 $CaCO_3(s) + 2HNO_3(aq) \longrightarrow$
$$Ca(NO_3)_2(aq) + CO_2(g) + H_2O(l)$$
$CaCO_3(s) + 2H^+(aq) \longrightarrow Ca^{2+}(aq) + CO_2(g) + H_2O(l)$
4.8 a K: +1, Cr: +6, O: −2   b Mn: +7, O: −2
4.9 $Ca + Cl_2 \longrightarrow Ca^{2+} + 2Cl^-$ **4.10** 0.0464 M NaCl
4.11 10.1 mL   **4.12** 0.0075 mol NaCl; 0.44 g NaCl
4.13 12 mL   **4.14** 34.19%   **4.15** 48.4 mL $NiSO_4$
4.16 5.07%

CHAPTER 5

5.1 0.56 atm, 4.3×10^2 mmHg   **5.2** 24.1 L   **5.3** 4.47 dm^3
5.4 5.54 dm^3 **5.5** Use $PV = nRT$. Solve for n: $n = PV/RT = P(V/RT)$. Everything within parentheses is constant; therefore, $n = $ constant $\times P$ (or $\propto P$).   **5.6** 46.0 atm   **5.7** density of He = 0.164 g/L; difference in mass between 1 L of air and 1 L of He = 1.024 g   **5.8** 64.1 amu   **5.9** 2.00 L   **5.10** $P_{O_2} = $ 0.0769 atm; $P_{CO_2} = $ 0.0310 atm; $P = $ 0.1080 atm; $X_{O_2} = $ 0.712   **5.11** 0.01591 mol O_2; $V = $ 0.406 L   **5.12** 219 m/s   **5.13** $T = $ 52.4 K, 728 K   **5.14** 9.96 s   **5.15** 44.0 amu   **5.16** $P_{vdW} = $ 0.992 atm; $P_{ideal} = $ 1.000 atm (larger)

CHAPTER 6

6.1 1.1×10^{-17} J, 2.7×10^{-18} cal   **6.2** 3.89 J   **6.3** exothermic; $q = -1170$ kJ   **6.4** −48.9 kJ; 4.96 kJ; −885.2 kJ
6.5 $2N_2H_4(l) + N_2O_4(l) \longrightarrow 3N_2(g) + 4H_2O(g)$; $\Delta H = -1049$ kJ
6.6 a $N_2H_4(l) + \frac{1}{2}N_2O_4(l) \longrightarrow \frac{3}{2}N_2(g) + 2H_2O(g)$;
$$\Delta H = -5.245 \times 10^2 \text{ kJ}$$
b $4H_2O(g) + 3N_2(g) \longrightarrow N_2O_4(l) + 2N_2H_4(l)$; $\Delta H = 1049$ kJ
6.7 −164 kJ   **6.8** 1.80×10^2 J   **6.9** −54 kJ;
$HCl(aq) + NaOH(aq) \longrightarrow NaCl(aq) + H_2O(l)$; $\Delta H = -54$ kJ
6.10 $\Delta H = -1792$ kJ   **6.11** $\Delta H^\circ_{vap} = 44.0$ kJ
6.12 $\Delta H^\circ = -138.0$ kJ   **6.13** $\Delta H^\circ = 61.6$ kJ

CHAPTER 7

7.1 767 nm   **7.2** 6.58×10^{14}/s   **7.3** 2.0×10^{-19} J, 2.0×10^{-17} J, and 2.0×10^{-15} J; greatest energy: x-ray region; least energy: infrared region   **7.4** 103 nm   **7.5** 3.38×10^{-19} J   **7.6** $\lambda = 332$ pm
7.7 a The value of n must be a positive whole number greater than zero.   b The values of l range from 0 to $n - 1$. Here, l has a value greater than n.   c The values for m_l range from $-l$ to $+l$. Here, m_l has a value greater than that of l.   d The values for m_s are $+\frac{1}{2}$ or $-\frac{1}{2}$, not 0.

CHAPTER 8

8.1 a possible   b possible   c impossible. There are two electrons in a 2p orbital with the same spin.   d possible.

e impossible. Only two electrons are allowed in an *s* subshell. **f** impossible. Only six electrons are allowed in a *p* subshell. **8.2** $1s^2 2s^2 2p^6 3s^2 3p^6 3d^{10} 4s^2$ **8.3** $4s^2 4p^3$ **8.4** Group IVA, Period 6; main-group element

8.5 ⇅ ⇅ ⇅⇅⇅ ⇅ ↑↑↑
1s   2s   2p   3s   3p

8.6 In order of increasing radius: Be, Mg, Na **8.7** It is more likely that 1000 kJ/mol is the ionization energy for iodine because ionization energies tend to decrease as atomic number increases in a column. **8.8** F

CHAPTER 9

9.1 $\cdot \text{Mg} \cdot \ + \ :\overset{\cdot\cdot}{\text{O}}\cdot \ \longrightarrow \ \text{Mg}^{2+} \ + \ \left[:\overset{\cdot\cdot}{\underset{\cdot\cdot}{\text{O}}}: \right]^{2-}$ **9.2** The electron configuration for Ca^{2+} is [Ar] and its Lewis symbol is Ca^{2+}. The electron configuration for S^{2-} is $[\text{Ne}]3s^2 3p^6$ and its Lewis symbol is $\left[:\overset{\cdot\cdot}{\underset{\cdot\cdot}{\text{S}}}: \right]^{2-}$. **9.3** The electron configurations of Pb and Pb^{2+} are $[\text{Xe}]4f^{14}5d^{10}6s^2 6p^2$ and $[\text{Xe}]4f^{14}5d^{10}6s^2$, respectively. **9.4** $[\text{Ar}]3d^5$ **9.5** One would expect S^{2-}, because it has two additional electrons. **9.6** In order of increasing ionic radius: Mg^{2+}, Ca^{2+}, Sr^{2+} **9.7** In order of increasing ionic radius: Ca^{2+}, Cl^-, P^{3-} **9.8** C—O is the most polar bond.

9.9
$$:\overset{\cdot\cdot}{\underset{\cdot\cdot}{\text{Cl}}}:$$
$$:\overset{\cdot\cdot}{\underset{\cdot\cdot}{\text{F}}}-\text{C}-\overset{\cdot\cdot}{\underset{\cdot\cdot}{\text{F}}}:$$
$$:\overset{\cdot\cdot}{\underset{\cdot\cdot}{\text{Cl}}}:$$

9.10 $\overset{\cdot\cdot}{\text{O}}::\text{C}::\overset{\cdot\cdot}{\text{O}}$ or $\overset{\cdot\cdot}{\text{O}}=\text{C}=\overset{\cdot\cdot}{\text{O}}$

9.11 a $\left[\text{H}-\overset{\text{H}}{\underset{}{\overset{|}{\text{O}}}}-\text{H} \right]^{+}$ **b** $\left[:\overset{\cdot\cdot}{\underset{\cdot\cdot}{\text{O}}}-\overset{\cdot\cdot}{\underset{\cdot\cdot}{\text{Cl}}}-\overset{\cdot\cdot}{\underset{\cdot\cdot}{\text{O}}}: \right]^{-}$

9.12 $\left[:\overset{\cdot\cdot}{\underset{\cdot\cdot}{\text{O}}}-\overset{:\text{O}:}{\underset{}{\overset{\|}{\text{N}}}}-\overset{\cdot\cdot}{\underset{\cdot\cdot}{\text{O}}}: \right]^{-} \longleftrightarrow \left[\overset{\cdot\cdot}{\text{O}}=\overset{:\overset{\cdot\cdot}{\underset{\cdot\cdot}{\text{O}}}:}{\underset{}{\text{N}}}-\overset{\cdot\cdot}{\underset{\cdot\cdot}{\text{O}}}: \right]^{-} \longleftrightarrow \left[:\overset{\cdot\cdot}{\underset{\cdot\cdot}{\text{O}}}-\overset{:\overset{\cdot\cdot}{\underset{\cdot\cdot}{\text{O}}}:}{\underset{}{\text{N}}}=\overset{\cdot\cdot}{\text{O}} \right]^{-}$

9.13
$$:\overset{\cdot\cdot}{\text{F}} \qquad \overset{\cdot\cdot}{\text{F}}:$$
$$\underset{:\overset{\cdot\cdot}{\text{F}}\qquad \overset{\cdot\cdot}{\text{F}}:}{\overset{}{\text{S}}}$$

9.14 $:\overset{\cdot\cdot}{\underset{\cdot\cdot}{\text{Cl}}}-\text{Be}-\overset{\cdot\cdot}{\underset{\cdot\cdot}{\text{Cl}}}:$

9.15
$$\text{H}-\overset{\cdot\cdot}{\underset{\cdot\cdot}{\text{O}}}-\overset{:\text{O}:}{\underset{:\underset{|}{\overset{\cdot\cdot}{\text{O}}}:}{\overset{\|}{\text{P}}}}-\overset{\cdot\cdot}{\underset{\cdot\cdot}{\text{O}}}-\text{H}$$
$$\underset{\text{H}}{|}$$

9.16 97 pm **9.17** 123 pm **9.18** −1308 kJ

CHAPTER 10

10.1 a trigonal pyramidal **b** bent **c** tetrahedral **10.2** T-shaped **10.3** **b**, **c** **10.4** **b**; the F atoms are distributed in a tetrahedral arrangement to give a molecule with a dipole moment of zero. **10.5** For the N atom:

N (ground state) = [He] ⇅ ↑↑↑ and
                        2s   2p

N (hybridized) = [He] ⇅↑↑↑
                        sp^3

Each N—H bond is formed by the overlap of a 1*s* orbital of a hydrogen atom with one of the sp^3 hybrid orbitals of the nitrogen atom. **10.6** Each P—Cl bond is formed by the overlap of a phosphorus $sp^3 d$ hybrid orbital with a singly occupied 3*p* chlorine orbital. **10.7** The C atom is *sp* hybridized. A carbon–oxygen bond is double and is composed of a σ bond and a π bond. The σ bond is formed by the overlap of a hybrid C orbital with a 2*p* orbital of O that lies along the axis. The π bond is formed by the sidewise overlap of a 2*p* on C with a 2*p* on O. **10.8** The structural formulas for the isomers are as follows:

$$\overset{\text{F}}{\underset{\cdot\cdot}{\text{N}}}=\overset{\text{F}}{\underset{\cdot\cdot}{\text{N}}} \qquad \qquad \overset{\text{F}}{\underset{\text{F}}{\text{N}}}=\overset{\text{F}}{\underset{\cdot\cdot}{\text{N}}}$$
$$\qquad cis \qquad\qquad\qquad trans$$

These compounds exist as separate isomers. If they were interchangeable, one end of the molecule would have to remain fixed as the other rotated. This would require breaking the π bond and a considerable expenditure of energy.

10.9 KK ⇅ ⇅ ⇅⇅ ○ ○○ ○;
        σ_{2s} σ_{2s}^* π_{2p} σ_{2p} π_{2p}^* σ_{2p}^*
$\text{KK}(\sigma_{2s})^2(\sigma_{2s}^*)^2(\pi_{2p})^4$; diamagnetic; bond order = 2

10.10 KK ⇅ ⇅ ⇅⇅ ⇅ ○○ ○;
         σ_{2s} σ_{2s}^* π_{2p} σ_{2p} π_{2p}^* σ_{2p}^*
$\text{KK}(\sigma_{2s})^2(\sigma_{2s}^*)^2(\pi_{2p})^4(\sigma_{2p})^2$; bond order = 3; diamagnetic

CHAPTER 11

11.1 1.37×10^3 kJ; 4.12×10^3 g **11.2** 5.30×10^2 mmHg **11.3** 43.0 kJ/mol **11.4 a** Methyl chloride can be liquefied by sufficiently increasing the pressure as long as the temperature is below 144°C. **b** Oxygen can be liquefied by compression as long as the temperature is below −119°C. **11.5 a** hydrogen bonding, dipole–dipole forces, and London forces **b** London forces **c** dipole–dipole forces and London forces **11.6** In order of increasing vapor pressure we have: butane (C_4H_{10}), propane (C_3H_8), and ethane (C_2H_6). London forces increase with increasing molecular mass. Therefore, one would expect the lowest vapor pressure to correspond to the highest molecular mass. **11.7** Hydrogen bonding is negligible in methyl chloride but strong in ethanol, which explains the lower vapor pressure of ethanol. **11.8 a** metallic solid **b** ionic solid **c** covalent network solid **d** molecular solid **11.9** C_2H_5OH, molecular; CH_4, molecular; CH_3Cl, molecular; $MgSO_4$, ionic. In order of increasing melting point: CH_4, CH_3Cl, C_2H_5OH, and $MgSO_4$. **11.10** 1 atom **11.11** 6.02×10^{23}/mol **11.12** 533 pm

CHAPTER 12

12.1 A dental filling, made up of liquid mercury and solid silver, is a solid solution. **12.2** C_4H_9OH, because of hydrogen bonding. **12.3** Na^+ **12.4** 8.45×10^{-3} g O_2/L **12.5** 7.07 g HCl; 27.9 g H_2O **12.6** 3.09 *m* $C_6H_5CH_3$ **12.7** mole fraction toluene = 0.194; mole fraction benzene = 0.806 **12.8** mole fraction methanol = 0.00550; mole fraction ethanol = 0.995 **12.9** 7.24 *m* CH_3OH **12.10** 2.96 *M* $(NH_2)_2CO$ **12.11** 2.20 *m* $(NH_2)_2CO$ **12.12** $\Delta P = 1.22$ mmHg; P = 155 mmHg **12.13** 1.89×10^{-1} g CH_2OHCH_2OH **12.14** 176 amu **12.15** 124 amu; P_4 **12.16** $\pi = 3.5$ atm **12.17** 100.077°C **12.18** $AlCl_3$

CHAPTER 13

13.1 $\dfrac{\Delta[NO_2F]}{\Delta t} = -\dfrac{\Delta[NO_2]}{\Delta t}$

13.2 1.1×10^{-4} M/s   **13.3** zero order in [CO]; second order in [NO_2]; second order overall   **13.4** rate $= k[NO_2]^2$; $k = 0.71$ L/(mol·s)   **13.5** a $[N_2O_5]_t = 1.24 \times 10^{-2}$ mol/L   b $t = 4.80 \times 10^3$ s   **13.6** half-life $= 0.075$ s; decreases to 50% of initial concentration in 0.075 s, to 25% in 0.151 s   **13.7** $E_a = 2.06 \times 10^5$ J/mol, $k = 5.80 \times 10^{-2}/(M^{1/2}\cdot s)$ **13.8** $2H_2O_2 \longrightarrow 2H_2O + O_2$ **13.9** bimolecular   **13.10** rate $= k[NO_2]^2$ **13.11** rate $= k_1[H_2O_2][I^-]$ **13.12** rate $= k_2(k_1/k_{-1})[NO]^2[O_2]$

CHAPTER 14

14.1 0.57 mol CO, 0.57 mol H_2O, 0.43 mol CO_2, and 0.43 mol H_2

14.2 a $K_c = \dfrac{[NH_3]^2[H_2O]^4}{[NO_2]^2[H_2]^7}$ b $K_c = \dfrac{[NH_3][H_2O]^2}{[NO_2][H_2]^{7/2}}$

14.3 $CO(g) + H_2O(g) \rightleftharpoons H_2(g) + CO_2(g)$; 0.57
14.4 $K_c = 2.3 \times 10^{-4}$ **14.5** $K_p = 1.24$

14.6 $K_c = \dfrac{[Ni(CO)_4]}{[CO]^4}$ **14.7** products; 0.018 M

14.8 $Q_c = 0.67$; more CO will form   **14.9** 0.0096 moles of PCl_5
14.10 0.11 mol H_2, 0.11 mol I_2, 0.78 mol HI   **14.11** 0.86 M PCl_5, 0.135 M PCl_3, 0.135 M Cl_2 **14.12** a in the reverse direction   b in the reverse direction   **14.13** a no; pressure has no effect.   b no   c yes   **14.14** high temperature   **14.15** High temperatures and low pressure would give the best yield of product.

CHAPTER 15

15.1 $H_2CO_3(aq) + CN^-(aq) \rightleftharpoons HCN(aq) + HCO_3^-(aq)$;
    acid   base   acid   base
HCN is the conjugate acid of CN^-.

15.2 a

BF₃ is the Lewis acid, CH₃OH is the Lewis base.

b

O^{2-} is the Lewis base, CO_2 is the Lewis acid.
15.3 The reactants are favored.   **15.4** a PH_3 b HI   c H_2SO_3 d H_3AsO_4 e HSO_4^- **15.5** 0.250 M OH^-, 4.0×10^{-14} M H^+ **15.6** basic   **15.7** 1.35   **15.8** 12.40
15.9 6.9×10^{-4} M **15.10** 4×10^{-4} M

CHAPTER 16

16.1 1.4×10^{-4}; 0.071   **16.2** $[H^+] = [C_2H_3O_2^-] = 1.3 \times 10^{-3}$ M; pH $= 2.89$; 0.013   **16.3** 3.24   **16.4** pH $= 1.29$; $[SO_3^{2-}] = 6.3 \times 10^{-8}$ M **16.5** 3.3×10^{-6} **16.6** 5.3×10^{-12} M
16.7 a acidic   b neutral   c acidic   **16.8** a 1.5×10^{-11} b 2.4×10^{-5} **16.9** 1.5×10^{-6}; 8.19   **16.10** 8.5×10^{-5} M; 8.5×10^{-4} **16.11** 3.63   **16.12** 5.26   **16.13** 3.83   **16.14** 1.60
16.15 7.97   **16.16** a 11.28   b 9.26   c 5.22   d 2.23

CHAPTER 17

17.1 a $K_{sp} = [Ba^{2+}][SO_4^{2-}]$ b $K_{sp} = [Fe^{3+}][OH^-]^3$ c $K_{sp} = [Ca^{2+}]^3[PO_4^{3-}]^2$ **17.2** 1.8×10^{-10} **17.3** 4.5×10^{-36}
17.4 0.67 g/L   **17.5** a 6.3×10^{-3} mol/L   b 4.4×10^{-5} mol/L
17.6 Precipitation is expected.   **17.7** No precipitate will form.
17.8 silver cyanide   **17.9** 1.2×10^{-9} M **17.10** No precipitate will form.   **17.11** 0.44 mol/L

CHAPTER 18

18.1 304 J/(mol·K)   **18.2** a positive   b positive   c negative   d positive   **18.3** 537 J/K   **18.4** -888.7 kJ
18.5 130.9 kJ   **18.6** All four reactions are spontaneous in the direction written.   **18.7** a $K = P_{CO_2}$ b $K = [Pb^{2+}][I^-]^2$

c $K = \dfrac{P_{CO_2}}{[H^+][HCO_3^-]}$

18.8 1×10^{-23} **18.9** 5×10^{-12} **18.10** 9×10^{-2} atm
18.11 $T = 669.5$ K, which is lower than that for $CaCO_3$.

CHAPTER 19

19.1 $I_2(s) + 10NO_3^-(aq) + 8H^+(aq) \longrightarrow$
$\qquad\qquad\qquad 4H_2O(l) + 2IO_3^-(aq) + 10NO_2(g)$
19.2 $H_2O_2(aq) + 2ClO_2(aq) + 2OH^-(aq) \longrightarrow$
$\qquad\qquad\qquad 2ClO_2^-(aq) + O_2(g) + 2H_2O(l)$
19.3 anode: $Ni(s) \longrightarrow Ni^{2+}(aq) + 2e^-$; cathode: $Ag^+(aq) + e^- \longrightarrow Ag(s)$; electrons flow from anode to cathode; Ag^+ flows to Ag; Ni^{2+} flows away from Ni.
19.4 $Zn(s)|Zn^{2+}(aq)\|H^+(aq)|H_2(g)|Pt$
19.5 $Cd(s) + 2H^+(aq) \longrightarrow Cd^{2+}(aq) + H_2(g)$
19.6 -2.12×10^4 J   **19.7** NO_3^- **19.8** No   **19.9** 1.10 V
19.10 -1.4×10^2 kJ   **19.11** 2.696 V   **19.12** 1×10^{19}
19.13 1.42 V   **19.14** 4×10^{-7} M Ni^{2+}
19.15 a $K^+(l) + e^- \longrightarrow K(l)$
$2Cl^-(l) \longrightarrow Cl_2(g) + 2e^-$ b $K^+(l) + e^- \longrightarrow K(l)$
$4OH^-(l) \longrightarrow O_2(g) + 2H_2O(g) + 4e^-$
19.16 $Ag^+(aq) + e^- \longrightarrow Ag(s)$ (cathode)
$2H_2O(l) \longrightarrow O_2(g) + 4H^+(aq) + 4e^-$ (anode)
19.17 2.52×10^{-2} A
19.18 8.66×10^{-2} g

CHAPTER 20

20.1 $^{40}_{19}K \longrightarrow {}^{40}_{20}Ca + {}^{0}_{-1}e$ **20.2** $^{235}_{92}U$ **20.3** a is stable; b and c are radioactive.   **20.4** a positron emission   b beta emission   **20.5** a $^{40}_{20}Ca(d,p)^{41}_{20}Ca$ b $^{12}_{6}C + {}^{2}_{1}H \longrightarrow {}^{13}_{6}C + {}^{1}_{1}H$ **20.6** $^{14}_{7}N$ **20.7** 3.2×10^{-5}/s   **20.8** 5.26 y
20.9 7.82×10^{-10}/s; 7.4×10^{-7} Ci   **20.10** 0.200
20.11 1.0×10^4 y   **20.12** a 2.58×10^9 J   b 6.25 MeV

CHAPTER 22

22.1 a pentaamminechlorocobalt(III) chloride   b potassium aquapentacyanocobaltate(III)   c pentaaquahydroxoiron(III) ion   **22.2** a $K_4[Fe(CN)_6]$ b $[Co(NH_3)_4Cl_2]Cl$ c $[PtCl_4]^{2-}$
22.3 a no geometric isomers possible   b geometric isomers possible (See Fig. 22.15 for similar example.)

22.4 [a] no optical isomers possible; [b] is similar to Figure 22.16.

22.5

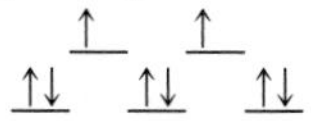

↑   ↑

↑↓   ↑↓   ↑↓

2 unpaired electrons   **22.6**   ↑   ↑   ↑

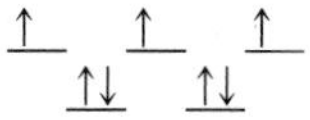

↑↓   ↑↓

22.7 530 nm, 500 nm, yes; CN^- is a more strongly bonding ligand than H_2O.

CHAPTER 23

23.1

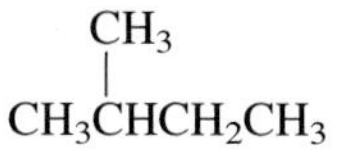

CH_3

$CH_3CHCH_2CH_3$

23.2 [a]

$$\underset{CH_3}{\overset{H}{}}C=C\underset{CH_2CH_2CH_3}{\overset{H}{}} \qquad \underset{CH_3}{\overset{H}{}}C=C\underset{H}{\overset{CH_2CH_2CH_3}{}}$$

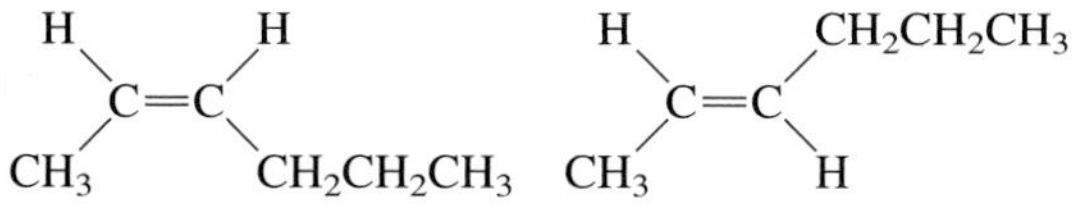

[b] none   **23.3** 2-bromobutane   **23.4** [a] 2,3-dimethylbutane
[b] 3-ethyl-2-methylhexane

CH_3

23.5 $CH_3CH_2-\overset{CH_3}{\underset{CH_3}{C}}-CH_2CH_2CH_2CH_2CH_3$

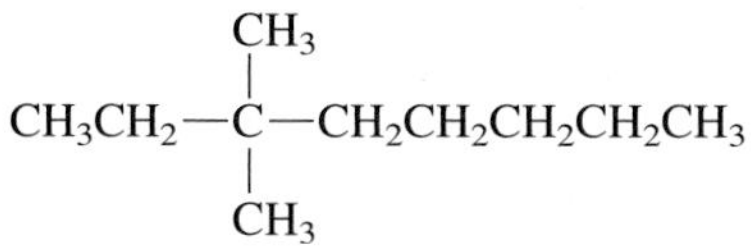

23.6 [a] 2,4-dimethyl-2-hexene   [b] 3-propyl-1-hexene

23.7 $CH_3-\overset{}{\underset{CH_3}{C}}=CH-CH_2-\overset{}{\underset{CH_3}{CH}}-CH_2CH_3$

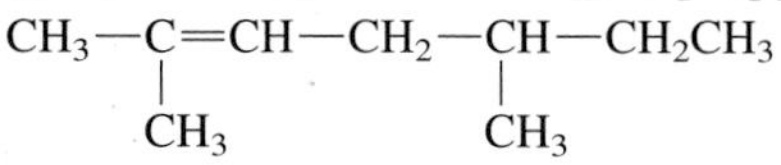

23.8 *cis*-3-hexene, *trans*-3-hexene

23.9 [a] propyne   [b] 3-methyl-1-pentyne

23.10 [a]

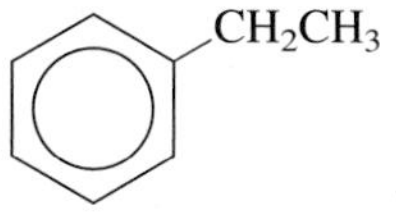

(benzene ring with CH_2CH_3)

[b]

(two benzene rings connected by $H_2C——CH_2$)

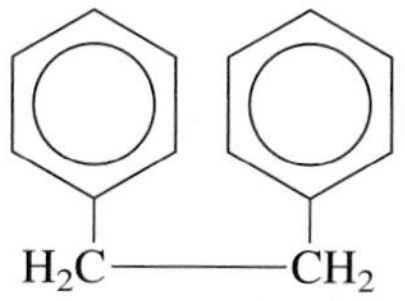

23.11 3-ethyl-3-hexanol   **23.12** [a] dimethyl ether
[b] methyl ethyl ether
23.13 [a] 2-pentanone   [b] butanal

CHAPTER 24

24.1 $-CH_2-CCl_2-CH_2-CCl_2-CH_2-CCl_2-$

Answers to Concept Checks

CHAPTER 1

1.1 a element **b** compound **c** mixture **1.2 a** two if you weigh under 100 lb, three if you weigh 100 lb or more **b** A 165-lb person would be reported as weighing 1.7×10^2 lb. **c** A 74.8-kg or 165-lb male would weigh 50 kg or 200 lb on the truck scale. **1.3 a** Meters, decimeters, or centimeters would be appropriate units. **b** approximately 9 m **c** 39°C would mean that you had a moderate fever. **d** You would probably be comfortable in a 23°C room. **1.4** You need to perform an experiment that will give you the density of the marcasite and then compare that to the density of gold.

CHAPTER 2

2.1 The molecular model on the far right that contains 3 atoms with a ratio of 1:2 **2.2** One would conclude that the atom is made up primarily of a large impenetrable mass with a positive charge. **2.3** They must have similar chemical and/or physical properties. **2.4** Molecular: CH_4, Br_2, CH_4O. Ionic: $CaCl_2$, KNO_3, LiF. Answer **d** is correct. **2.5 a** ether **b** alcohol **c** carboxylic acid **d** hydrocarbon **2.6** Q is likely to be an element in Group VIA.

CHAPTER 3

3.1 a 1.5 mol **b** 4.5 mol **c** 3.0 mol of OH^- since there are 2 mol of OH^- per mol of $Mg(OH)_2$ **3.2 a** The total mass of C and H that you collect will be less than the total mass of material you started with. **b** The calculated percent C and H would be less than the real value. **3.3 a** No, the empirical formula needs to be the smallest whole-number ratio of subscripts. **b** No, the subscripts are not whole numbers. **c** Yes, the empirical and molecular formulas can be the same. **3.4 a** correct **b** incorrect **c** incorrect **d** correct **e** correct **f** incorrect **g** incorrect **h** correct **3.5** Answer **d** is the best answer considering the information provided in the problem. **3.6 a** $X_2(g) + 2Y(g) \longrightarrow 2XY(g)$ **b** container 1 **c** $Y(g)$

CHAPTER 4

4.1 Strong electrolytes: NH_4Cl, $MgBr_2$, and HCl. Answers **b**, **c**, and **e** support the correct choices. **4.2** The left beaker contains the strong electrolyte $LiI(aq)$. The right beaker contains the nonelectrolyte $CH_3OH(aq)$. **4.3 a** $Na_2SO_4(aq) + Sr(C_2H_3O_2)_2(aq) \longrightarrow$
$$SrSO_4(s) + 2NaC_2H_3O_2(aq)$$
b $2Na^+(aq) + SO_4^{2-}(aq) + Sr^{2+}(aq) + 2C_2H_3O_2^-(aq) \longrightarrow$
$$SrSO_4(s) + 2Na^+(aq) + 2C_2H_3O_2^-(aq)$$
c $Sr^{2+}(aq) + SO_4^{2-}(aq) \longrightarrow SrSO_4(s)$ **4.4** $\longrightarrow H_3O^+(aq) + ClO_2^-(aq), \longrightarrow H_3O^+(aq) + Br^-(aq)$. Statement **e** is correct. **4.5 a** $MOH(s)$ is a base. M could be any Group I metal. **b** HA is an acid. A^- is an anion with a −1 charge. **c** H_2A is an acid. A^{2-} is an anion with a −2 charge. **d** $M \longrightarrow Na^+, A \longrightarrow C_2H_3O_2^-, A \longrightarrow SO_3^{2-}$. **4.6** Concentration order: A< C = D < B. Leave A alone; add water to fill beaker B; add water to double the volumes of D and C. **4.7 a** The acid in flask C has three times as many acidic protons as the acid in flask A. The acid in flask B has two times as many acidic protons as the acid in flask A. **b** Yes, if we assume that flask A contains a monoprotic acid.

CHAPTER 5

5.1 Since the height of the column is inversely related to the density of the liquid, the column containing the H_2O would be higher. **5.2 a** Step 1, pressure decrease; Step 2, pressure decrease; total change, final pressure less than starting pressure. **b** Step 1, pressure increase; Step 2, pressure increase; total change, final pressure greater than starting pressure. **5.3 a** All contain same amount of gas. **b** Xe flask **c** He flask **d** They would all still have the same number of moles. **5.4 a** No change in H_2 pressure. **b** They are equal. **c** The total pressure is the sum of the pressures of the H_2 and Ar in the container **5.5** Choice II is correct with answer **c** being the best explanation. **5.6 a** He **b** Raise the temperature of the Ar. **5.7 a** Pressure would be greater for ideal gas. **b** Pressure would be less for ideal gas. **c** Cannot make a determination because the effect of molecular volume and attractions on the pressure of a real gas is opposite.

CHAPTER 6

6.1 Sun's energy is changed to electric energy, to chemical energy (of the battery), and back to electric energy, then to kinetic energy of motor and water, and to potential energy of water. **6.2 d 6.3 c** and **d** are exothermic; **d** is most exothermic **6.4** $\Delta H_{sub} = \Delta H_{fus} + \Delta H_{vap}$

CHAPTER 7

7.1 The wavelength is halved; ultraviolet. **7.2** UV **7.3** The proton speed would be about 2000 times smaller.

CHAPTER 8

8.1 4 **8.2** Mg and Al **8.3** IIA **8.4** As

CHAPTER 9

9.1 a excited state; not expected **b** N^{2-} is not expected. **c** expected configuration **d** Na^{2+} is not expected. **e** expected configuration **9.2 a** has 13 electron pairs; should be 12 **b** correct skeleton structure and distribution of electrons **c** Double bond between F and N atoms is less likely than single bond. **d** no octet on left N atom **e** incorrect skeleton structure **f** no octets on N atoms **9.3 a** most accurate description **b** Electronegative N atom is not expected to be in central position. **c** not enough electron pairs to give octets on C and N atoms (would need 7 pairs) **d** not enough electron pairs to give octets on C and N atoms (would need 6 pairs)

CHAPTER 10

10.1 The four pairs have a tetrahedral arrangement; the molecular geometry is trigonal pyramidal. **10.2** Molecule Y is likely to be trigonal planar, but trigonal pyramidal or T-shaped geometries are possible. Molecule Z cannot have a trigonal planar geometry, but must be either trigonal pyramidal or T-shaped.   **10.3** The single and triple bonds each require a sigma bond orbital, suggesting *sp* hybrids.

CHAPTER 11

11.1 (i) slightly lower   (ii) higher   (iii) more molecules   (iv) Evaporation is greater.
(v)

t = 1

 (i) slightly lower   (ii) higher   (iii) more molecules   (iv) They will be equal.
(v)

t = 2

11.2 Cook it longer because the H_2O temperature would be lower.   **11.3** Answer is an incorrect statement.
11.4 The hydrogen and oxygen would form an explosive mixture. many times greater (more positive) for the wrong reaction Apply Hess's law.   **11.5** AB
 face-centered cubic

CHAPTER 12

12.1 Mo solute, Cr solvent $MgCl_2$ solute, H_2O solvent O_2 and N_2 solutes, Ar solvent   **12.2** Compound X is a large nonpolar molecule. Compound Y is a molecule that exhibits hydrogen bonding. Applying the like dissolves like concept, compound X will be more soluble in the nonpolar solvent and compound Y will be more soluble in a polar solvent.   **12.3** The A^{2+} ion must be smaller so it has a greater energy of hydration.   **12.4** At high elevations, the partial pressure of O_2 is so low that not enough dissolves in the water to sustain fish.   **12.5** A soluble, chemically similar liquid with a higher vapor pressure than water.   **12.6** To prevent water from flowing into the pickle via osmosis. The pickles would be huge and would probably burst.   **12.7** Highest to lowest: $MgCl_2 > KBr > Na_2SO_4 > NaCl$.   **12.8** The iron(III) hydroxide precipitates because the electrons from the negative electrode neutralize the positive charge on the iron(III) hydroxide.

CHAPTER 13

13.1 Point A has a faster instantaneous rate. No, it decreases with time.   **13.2** [Q] = 0.0 *M* slowest (no reaction); the other two are equal in rate because they are zero order with respect to [Q]. Rate = $k[R]^2$ **13.3** Rate of aging = (diet)^w (exercise)^x(sex)^y(occupation)^z Gather a sample of people that have all of the factors the same except one. For example, using the equation given in part , you could determine the effect of diet if you had a sample of people who were the same sex, exercised the same amount, and had the same occupation. You would need to isolate each factor in this fashion to determine the exponent on each

factor. 2   **13.4** Second order   **13.5** the A + B reaction the E + F reaction exothermic   **13.6** It will increase the rate since the activation energy, E_a, is inversely related to the rate constant, *k*. A decrease in E_a results in an increase in the value of *k*. yes No, it should be rate = $k[Y]^2$.

CHAPTER 14

14.1 $2A + B \longrightarrow 2C$ **14.2** IV with A = red, I with X = red or blue, II with C = blue   **14.3** left to right   **14.4** It is quadrupled.   **14.5** I

CHAPTER 15

15.1 $HCHO_2(aq) + H_2O(l) \rightleftharpoons CHO_2^-(aq) + H_3O^+(aq)$ and $H_2O(l) + H_2O(l) \rightleftharpoons H_3O^+(aq) + OH^-(aq)$; $HCHO_2(aq)$, $H_2O(l)$, $CHO_2^-(aq)$, $OH^-(aq)$, and $H_3O^+(aq)$ **15.2** Acetate ion   **15.3** B, A, C   **15.4** NaOH, NH_3, $HC_2H_3O_2$, HCl

CHAPTER 16

16.1 HCN, $HC_2H_3O_2$, HNO_2, HF   **16.2** $B(aq) + H_2O(l) \rightleftharpoons BH^+(aq) + OH^-(aq)$ weak base
16.3 a, b   **16.4** $KNO_2(aq)$ and $HNO_3(aq)$ **16.5**
16.6 Y 3 beaker with 7 HA molecules

CHAPTER 17

17.1 $PbSO_4$ **17.2** $NaNO_3$ **17.3** Picture fewer Ag^+ and more Cl^- ions.   **17.4** Magnesium oxalate

CHAPTER 18

18.1 It has increased.   **18.2** The NO concentration is low at equilibrium.   **18.3** $\Delta G°$ is not changed. ΔG increases.
18.4 The concentration of NO_2 increases with temperature.

CHAPTER 19

19.1 No sustainable current would flow. The wire does not contain mobile positively and negatively charged species, which are necessary to balance the accumulation of charges in each of the half-cells.   **19.2** −0.54 V No They both would be 1.10 V.   **19.3** negative Change the concentrations in a manner to increase Q, where $Q = \dfrac{[Fe^{2+}]}{[Cu^{2+}]}$

For example: $Fe(s)|Fe^{2+} (1.10\ M)\|Cu^{2+} (0.50\ M)|Cu(s)$.
19.4 Many of the ions contained in seawater have very high reduction potentials—higher than Fe(*s*). This means that spontaneous electrochemical reactions will occur with the Fe(*s*), causing the iron to form ions and go into solution, while at the same time, the ions in the sea are reduced and plate out on the surface of the iron.

CHAPTER 20

20.1 yes no, since the $_1^3H_2O$ molecule is more massive than $_1^1H_2O$ $_1^3H_2O$ **20.2** the α particle, since it has the highest RBE   **20.3** After 50,000 years, enough half-lives have passed (about 10) that there would be almost no carbon-14 present to detect and measure.

CHAPTER 21

21.1 Given the high energy demands of animals to move and maintain body temperature, breaking the very strong triple bond of N_2 requires too much energy when compared to the lower-energy double bond of O_2.   **21.2** because the only intermolecular forces in these materials are very weak van der Waals forces

CHAPTER 22

22.1 $K_2[PtCl_6]$ **22.2** $[Co(NH_3)_4(H_2O)Br]Cl_2$
22.3 a Complexes A and E   b A and E   c A and E, B and C

CHAPTER 23

23.1 a C_7H_{16} b

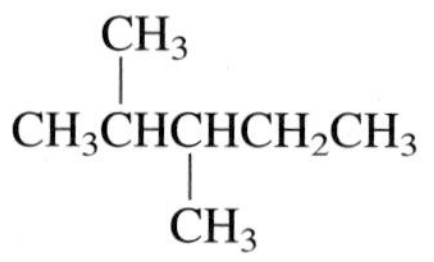

$$CH_3CHCHCH_2CH_3$$

with CH_3 above the second carbon and CH_3 below the third carbon

23.2 Blend of 90% 2,2,4-trimethylpentane and 10% heptane

23.3 a C_5H_{10} b

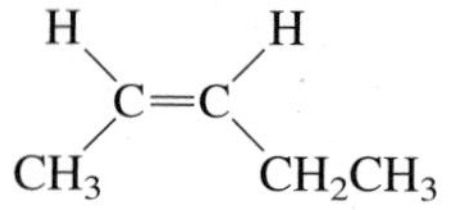

c *cis*-2-pentene

CHAPTER 24

24.1

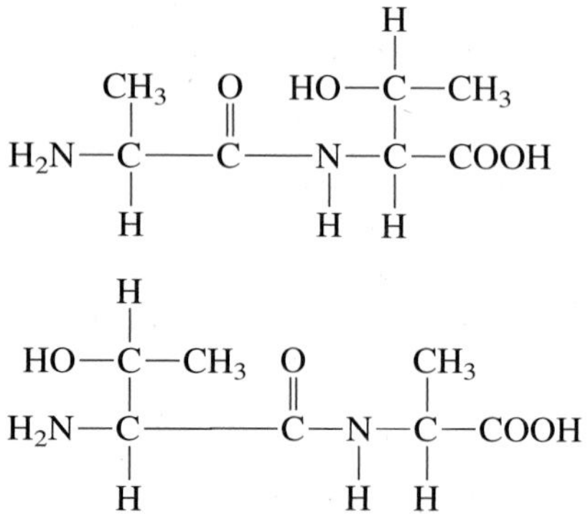

24.2 UACGAUGCCUAAGUU

Answer Section
Selected Odd Problems

CHAPTER 1

1.25 If the material is a pure compound, all samples should have the same melting point, the same color, and the same elemental composition. If it is a mixture, these properties should differ depending on the composition.
1.27 [a] You could first immerse the thermometer in an ice-water bath and mark the level at this point as 0°C. Then, immerse the thermometer in boiling water, and mark the level at this point as 100°C. [b] You could make 19 evenly spaced marks on the thermometer between the two original points, each representing a difference of 5°C.

1.29 [a] $°F = \left(°YS \times \dfrac{120°F}{100°YS}\right) - 100°F$ [b] $-21°F$

1.31 Some physical properties you could measure are density, hardness, color, and conductivity. Chemical properties of sodium would include reaction with air, reaction with water, reaction with chlorine, reaction with acids, bases, etc.
1.33 [a] A paper clip has a mass of about 1 g. [b] Answers will vary depending on your particular sample. Keeping in mind that the SI unit for mass is kg, the approximate weights for the items presented in the problem are as follows: a grain of sand, 1×10^{-5} kg; a paper clip, 1×10^{-3} kg; a nickel, 5×10^{-3} kg; a 5.0-gallon bucket of water, 2.0×10^{1} kg; a brick, 3 kg; a car, 1×10^{3} kg. **1.37** The mass of something (how heavy it is) depends on how much of the item, material, substance, or collection of things you have. The density of something is the mass of a specific amount (volume) of an item, material, substance, or collection of things. You could use 1 kg of feathers and 1 kg of water to illustrate that they have the same mass yet have very different volumes; therefore, they have different densities.
1.39 29.8 g **1.41** [a] Solid [b] Solid [c] Solid [d] Liquid
1.43 [a] Physical change [b] Physical change [c] Chemical change [d] Physical change **1.45** Physical change: Liquid mercury is cooled to solid mercury. Chemical changes: (1) Solid mercury oxide forms liquid mercury metal and gaseous oxygen; (2) glowing wood and oxygen form burning wood (form ash and gaseous products). **1.47** [a] Physical property [b] Chemical property [c] Physical property [d] Physical property [e] Chemical property **1.49** Physical properties: is a solid; has an orange-red color; has a density of 11.1 g/cm³; is insoluble in water. Chemical property: Mercury(II) oxide decomposes when heated to give mercury and oxygen.
1.51 [a] Physical process [b] Chemical reaction [c] Physical process [d] Chemical reaction [e] Physical process
1.53 [a] Homogeneous mixture, if fresh; heterogeneous mixture, if spoiled [b] Substance [c] Solution [d] Substance **1.55** [a] A pure substance with two phases present, liquid and gas. [b] A mixture with two phases present, solid and liquid. [c] A pure substance with two phases present, solid and liquid. [d] A mixture with two phases present, solid and solid. **1.57** [a] 4 [b] 3 [c] 4 [d] 5 [e] 3 [f] 4 **1.59** 4.000×10^{4} **1.61** [a] 8.5 [b] 90.0 [c] 111 [d] 2.3×10^{3} **1.63** 49 cm³ **1.65** [a] 4.851 µg [b] 3.16 cm [c] 2.591 ns [d] 8.93 pg **1.67** [a] 6.15×10^{-12} s [b] 3.781×10^{-6} m [c] 1.546×10^{-10} m [d] 9.7×10^{-3} g **1.69** [a] 20°C [b] $-31°C$ [c] 79°F [d] $-114°F$ **1.71** $-368.9°F$
1.73 7.56 g/cm³ **1.75** cinnabar **1.77** 1.3×10^{2} g **1.79** 25.1 cm³ **1.81** 4.80×10^{5} mg **1.83** 9.6×10^{-8} mm **1.85** 1.3×10^{-15} L **1.87** 3.32×10^{3} J **1.89** 4.435×10^{3} m

1.91 65.5 L **1.93** 0.86 g **1.95** 2.53 g **1.97** 130.0 g **1.99** [a] Chemical [b] Physical [c] Physical [d] Chemical **1.101** yes **1.103** 6.07×10^{4} cm³ **1.105** 8.093×10^{4} gal **1.107** 9.9×10^{1} cm³ **1.109** [a] 7.6×10^{-1} [b] 1.63×10^{1} [c] 1.12×10^{-1} [d] 4.76×10^{2} **1.111** [a] 1.86 cg [b] 77 pm [c] 6.5 nm [d] 0.85 µm **1.113** [a] 1.07×10^{-12} s [b] 5.8×10^{-6} m [c] 3.19×10^{-7} m [d] 1.53×10^{-2} s **1.115** 3051°F **1.117** 1.52×10^{3}°F **1.119** 234.3 K, $-38.0°F$ **1.121** 445°C, 718 K **1.123** 2.70×10^{3} kg/m³ **1.125** 4.88 g/cm³ **1.127** chloroform **1.129** 1.45×10^{3} g **1.131** 33.24 mL **1.133** [a] 9.3×10^{13} nm [b] 3.18×10^{-1} ms [c] 3.71 cm [d] 8.45×10^{9} µg **1.135** [a] 5.91×10^{6} mg [b] 7.53×10^{5} µg [c] 9.01×10^{4} kHz [d] 4.98×10^{-4} kJ **1.137** 1.2230×10^{16} L **1.139** 2.8×10^{4} L **1.141** 55.0 g **1.161** The water will have a lower density than the salt solution. It will not conduct electricity. It will have different chemical properties, such as reaction with silver nitrate solution; and different physical properties, such as the boiling point. Also, you could boil away the water from the salt solution and recover the salt. **1.163** 2.4 L **1.165** 2.4 L **1.167** 3×10^{21} g **1.169** 34%, 79 proof (2 sig. fig.) **1.171** 3.76 g/cm³ **1.173** 2.4 g/cm³

CHAPTER 2

2.27 Once the subscripts of the compounds in the original chemical equation are changed (the molecule N_2 was changed to the atom N), the substances reacting are no longer the same. Your friend may be able to balance the second equation, but it is no longer the same chemical reaction. **2.29** Group them elements by similar physical properties such as density, mass, color, conductivity, etc., or by chemical properties, such as reaction with air, reaction with water, etc. **2.31** XO_4^{2-}, with the greatest number of oxygen atoms in the group, would be perexcate; XO_3^{2-} would be excate; and XO^{2-}, with the fewest oxygen atoms in the group, would be hypoexcite. **2.33** [a] Since the mass of an atom is not due only to the sum of the masses of the protons, neutrons, and electrons, when you change the element in which you are basing the amu, the mass of the amu must change as well. [b] Since the amount of material that makes up a hydrogen atom doesn't change, when the amu gets larger, as in this problem, the hydrogen atom must have a smaller mass in amu. **2.35** [a] $2Li + Cl_2 \longrightarrow 2LiCl$ [b] $Na + S \longrightarrow Na_2S$ [c] $2Al + 3I_2 \longrightarrow 2AlI_3$ [d] $3Ba + N_2 \longrightarrow Ba_3N_2$ [e] $12V + 5P_4 \longrightarrow 4V_3P_5$ **2.37** [a] Argon [b] Zinc [c] Silver [d] Magnesium **2.39** [a] K [b] S [c] Fe [d] Mn **2.41** 3.16×10^{-26} kg **2.43** C; D. **2.45** Cl-35: 17 protons, 18 neutrons, 17 electrons Cl-37: 17 protons, 20 neutrons, 17 electrons **2.47** $^{79}_{34}Se$ **2.49** 13.90 **2.51** 50.94 amu, vanadium **2.53** 39.10 amu, K **2.55** 24.615 amu **2.57** [a] Te [b] Al **2.59** [a] Group IVA, Period 2, nonmetal [b] Group IVA, Period 6, metal [c] Group VIB, Period 4, metal [d] GroupIIA, Period 3, metal **2.61** [a] O (oxygen) [b] F (fluorine) [c] Fe (iron) [d] Ce (cerium) **2.63** They are different in that the solid sulfur consists of S_8 molecules, whereas the hot vapor consists of S_2 molecules. The S_8 molecules are four times as heavy as the S_2 molecules. Hot sulfur is a mixture of S_8 and S_2 molecules, but at high enough temperatures only S_2 molecules are formed. Both hot sulfur and solid sulfur consist of molecules with only sulfur atoms. **2.65** 4.10×10^{22} N atoms; 1.20×10^{24} N atoms **2.67** 7.0×10^{22} C_2H_5OH molecules

2.69 [a] N_2H_4 [b] H_2O_2 [c] C_3H_8O [d] PCl_3 **2.71** [a] PCl_5
[b] NO_2 [c] $C_3H_6O_2$ **2.73** 4 O to 3 N   **2.75** [a] $Fe(CN)_3$
[b] K_2CO_3 [c] Li_3N [d] Ca_3P_2 **2.77** [a] Na_2SO_4: sodium
sulfate   [b] Na_3N: sodium nitride   [c] CuCl: copper(I) chloride
[d] Cr_2O_3: chromium(III) oxide   **2.79** [a] Lead(II) permanganate: $Pb(MnO_4)_2$ [b] Barium hydrogen carbonate: $Ba(HCO_3)_2$
[c] Cesium sulfide: Cs_2S [d] Iron(III) acetate: $Fe(C_2H_3O_2)_3$
2.81 [a] Molecular   [b] Ionic   [c] Molecular   [d] Ionic
2.83 [a] Dinitrogen monoxide   [b] Tetraphosphorus
dec(a)oxide   [c] Arsenic trichloride   [d] Dichlorine
hept(a)oxide   **2.85** [a] P_2O_5 [b] NO_2 [c] N_2F_4 [d] BF_3
2.87 [a] Selenium trioxide   [b] Disulfur dichloride   [c] Carbon
monoxide   **2.89** [a] Sulfurous acid: H_2SO_3 [b] Hyponitrous
acid: $H_2N_2O_2$ [c] Disulfurous acid: $H_2S_2O_5$ [d] Arsenic acid:
H_3AsO_4 **2.91** $Na_2SO_4 \cdot 10H_2O$ is sodium sulfate decahydrate.
2.93 Iron(II) sulfate heptahydrate is $FeSO_4 \cdot 7H_2O$ **2.95** 2 O
atoms, equation not balanced   **2.97** [a] $Sn + 2NaOH \longrightarrow$
$Na_2SnO_2 + H_2$ [b] $8Al + 3Fe_3O_4 \longrightarrow 4Al_2O_3 + 9Fe$
[c] $2CH_3OH + 3O_2 \longrightarrow 2CO_2 + 4H_2O$ [d] $P_4O_{10} + 6H_2O \longrightarrow$
$4H_3PO_4$ [e] $PCl_5 + 4H_2O \longrightarrow H_3PO_4 + 5HCl$
2.99 $Ca_3(PO_4)_2(s) + 3H_2SO_4(aq) \longrightarrow 3CaSO_4(s) + 2H_3PO_4(aq)$

2.101 $PbS(s) + PbSO_4(s) \xrightarrow{\Delta} 2Pb(l) + 2SO_2(g)$

2.103 The ratio of O masses is 3 : 2 (A to B)   **2.105** $-1.6 \times$
10^{-19} C; 2, 4, 6, 7   **2.107** 63, 60   **2.109** $^{74}_{23}V^{5+}$
2.111 Cu-63:0.6916, Cu-65: 0.3084   **2.113** [a] Bromine, Br
[b] Hydrogen, H   [c] Niobium, Nb   [d] Fluorine, F
2.115 [a] Manganese(II) ion   [b] Nickel(II) ion
[c] Cobalt(II) ion   [d] Iron(III) ion   **2.117** Na_2SO_4, NaCl,
$CoSO_4$, and $CoCl_2$ **2.119** [a] Tin(II) phosphate   [b] Ammonium nitrite   [c] Magnesium hydroxide   [d] Nickel(II) sulfite
2.121 [a] Hg_2S [b] $Co_2(SO_3)_3$ [c] $(NH_4)_2Cr_2O_7$
[d] AlF_3 **2.123** [a] Arsenic tribromide   [b] Hydrogen telluride (dihydrogen telluride)   [c] Diphosphorus pent(a)oxide
[d] Silicon dioxide   **2.125** [a] $2NaOH + H_2CO_3 \longrightarrow$
$Na_2CO_3 + 2H_2O$ [b] $SiCl_4 + 2H_2O \longrightarrow SiO_2 + 4HCl$
[c] $Ca_3(PO_4)_2 + 8C \longrightarrow Ca_3P_2 + 8CO$ [d] $2H_2S + 3O_2 \longrightarrow$
$2SO_2 + 2H_2O$ [e] $2N_2O_5 \longrightarrow 4NO_2 + O_2$
2.127 26 electrons   **2.129** 12.507 amu   **2.149** 6.06×10^9 miles
2.151 4.826 g $NiSO_4$ **2.153** 107.9 amu; X is silver

CHAPTER 3

3.19 [a] $3H_2(g) + N_2(g) \longrightarrow 2NH_3(g)$ [b] H_2 [c] six
additional molecules of H_2 **3.21** [a] unreasonable
[b] unreasonable   [c] reasonable   [d] unreasonable
3.23 [a] there is not enough oxygen in the air outside of the bubble for the complete reaction of hydrogen   [b] when 0.5 mole
of O_2 has reacted, all of the H_2 (1 mole) will be consumed, leaving behind 0.5 mole of unreacted O_2.   [c] you have a ratio of
2 moles of H_2 to 1 mole of O_2, which is the correct stoichiometric
amount, so all of the hydrogen and all of the oxygen react
completely.   [d] Oxygen does not combust, and there is no
hydrogen present to burn, so no reaction occurs.
3.25 [a] 1 mol Ca = 40.08 Ca conversion factor correction
[b] 2 mol K^+ ions = 1 mol K_2SO_4 conversion factor correction
[c] 2 mole Na = 1 mol H_2O conversion factor correction
3.27 [a] 32.0 amu   [b] 62.0 amu   [c] 138 amu   [d] 366 amu
3.29 [a] 64.1 amu   [b] 137 amu   **3.31** 97.99 g/mol
3.33 [a] 3.818×10^{-23} g/atom   [b] 2.326×10^{-23} g/atom
[c] 8.383×10^{-23} g/molecule   [d] 5.390×10^{-22} g/unit
3.35 1.231×10^{-22} g/molecule   **3.37** [a] 18.5 g Fe   [b] 40. g F
[c] 1.5 g CO_2 [d] 367 g K_2CrO_4 **3.39** 33.6 g H_3BO_3
3.41 [a] 1.3 mol C_4H_{10} [b] 0.0994 mol Cl_2 [c] 0.238 mol C
[d] 0.112 mol $Al_2(CO_3)_3$ **3.43** 0.03096 mol H_2O; 0.03096 mol
H_2O, 0.006196 mol $CuSO_4$ = 4.99/1, or about 5:1
3.45 [a] 7.12×10^{23} atoms   [b] 2.41×10^{23} atoms   [c] $1.6 \times$
10^{24} molecules   [d] 3.75×10^{23} units   [e] 6.59×10^{22} ions

3.47 2.97×10^{19} molecules   **3.49** 0.661   **3.51** 20.2%
3.53 0.17 g Br   **3.55** 72.3% Al; 27.7% Mg   **3.57** [a] 42.9% C,
57.1% O   [b] 19.2% Na, 1.68% H, 25.8% P, 53.3% O
[c] 32.2% Co, 15.3% N, 52.5% O   [d] 27.3% C, 72.7% O
3.59 91.2% C; 8.75% H   **3.61** sodium sulfite   **3.63** 38.7% C;
9.79% H; 51.5% O   **3.65** OsO_4 **3.67** C_3H_3O **3.69** $C_3H_4O_2$
3.71 [a] CH   [b] C_4H_4; C_6H_6 **3.73** $(BH_3)_2$, or B_2H_6
3.75 $C_2H_2O_4$

3.77
$$C_2H_4 \quad + \quad 3O_2 \quad \longrightarrow$$
1 molecule C_2H_4 +   3 molecules O_2 $\longrightarrow$
1 mole C_2H_4 +   3 moles O_2 $\longrightarrow$
28.052 g C_2H_4 + 3×32.00 g O_2 $\longrightarrow$
$$2CO_2 \quad + \quad 2H_2O$$
2 molecules CO_2 +   2 molecules H_2O
2 moles CO_2 +   2 moles H_2O
2×44.01 g CO_2 + 2×18.016 g H_2O

3.79 2.4 mol CO_2 **3.81** 11.7 mol O_2 **3.83** 150. g $Ca_3(PO_4)_2$
3.85 1.46×10^5 g W   **3.87** 13 g O_2 **3.89** 7.942 g NO_2
3.91 NaOH is the limiting reactant, 0.720 mol of NaClO
3.93 40.5 g CH_3OH; H_2 remains; 5.1 g H_2 **3.95** 9.07 g
3.97 0.461   **3.99** 71.6% C, 6.71% H, 4.91% N, 16.8% O
3.101 $C_6H_4Cl_2$ **3.103** C_4H_4S, C_4H_4S **3.105** 83.7%
3.107 34.0 g $C_6H_5NO_2$, 88.1%   **3.109** 60.3 g   **3.111** 4.94 g HF
3.113 16.2%   **3.133** 0.21 g   **3.135** 616 amu   **3.137** Ag and Cl
3.139 1.97×10^{24} Co atoms

CHAPTER 4

4.21 [a] Any soluble salt that will form a precipitate when reacted
with Ag^+ ions in solution will work, for example: $CaCl_2$, Na_2S,
$(NH_4)_2CO_3$.   [b] No, no precipitate will form.   [c] You would
underestimate the amount of silver present in the solution.
4.23 [a] Sample 1, Sample 2, Sample 3   [b] They all will require
the same volume of 0.1 M NaOH.   **4.25** Probably not, since the
ionic compound that is a nonelectrolyte is not soluble.
4.27 $AlCl_3$ > KCl = NaCl = HCl > $PbCl_2$ > NH_3 = KOH =
HCN.   **4.29** [a] Insoluble   [b] Soluble   [c] Soluble
[d] Soluble   **4.31** [a] Insoluble   [b] Soluble; Li^+ and SO_4^{2-}
[c] Insoluble   [d] Soluble; Na^+ and CO_3^{2-}
4.33 [a] $S^{2-}(aq) + 2H^+(aq) \longrightarrow H_2S(g)$
[b] $H^+(aq) + OH^-(aq) \longrightarrow H_2O(l)$
[c] $OH^-(aq) + NH_4^+(aq) \longrightarrow NH_3(g) + H_2O(l)$
[d] $Ag^+(aq) + Br^-(aq) \longrightarrow AgBr(s)$
4.35 Molecular equation: $Li_2CO_3(aq) + 2HBr(aq) \longrightarrow$
$CO_2(g) + H_2O(l) + 2LiBr(aq)$
Net ionic equation: $CO_3^{2-}(aq) + 2H^+(aq) \longrightarrow CO_2(g) + H_2O(l)$
4.37 [a] $FeSO_4(aq) + NaCl(aq) \longrightarrow$ NR
[b] $Na_2CO_3(aq) + MgBr_2(aq) \longrightarrow MgCO_3(s) + 2NaBr(aq)$
$CO_3^{2-}(aq) + Mg^{2+}(aq) \longrightarrow MgCO_3(s)$
[c] $MgSO_4(aq) + 2NaOH(aq) \longrightarrow Mg(OH)_2(s) + Na_2SO_4(aq)$
$Mg^{2+}(aq) + 2OH^-(aq) \longrightarrow Mg(OH)_2(s)$
[d] $NiCl_2(aq) + NaBr(aq) \longrightarrow$ NR
4.39 [a] $Ba(NO_3)_2(aq) + Li_2SO_4(aq) \longrightarrow$
$$BaSO_4(s) + 2LiNO_3(aq)$$
$Ba^{2+}(aq) + SO_4^{2-}(aq) \longrightarrow BaSO_4(s)$
[b] $Ca(NO_3)_2(aq) + NaBr(aq) \longrightarrow$ NR
[c] $Al_2(SO_4)_3(aq) + 6NaOH(aq) \longrightarrow$
$$2Al(OH)_3(s) + 3Na_2SO_4(aq)$$
$Al^{3+}(aq) + 3OH^-(aq) \longrightarrow Al(OH)_3(s)$
[d] $3CaBr_2(aq) + 2Na_3PO_4(aq) \longrightarrow Ca_3(PO_4)_2(s) + 6NaBr(aq)$
$3Ca^{2+}(aq) + 2PO_4^{3-}(aq) \longrightarrow Ca_3(PO_4)_2(s)$
4.41 [a] Weak base   [b] Weak acid   [c] Strong base
[d] Strong acid
4.43 [a] $NaOH(aq) + HNO_3(aq) \longrightarrow H_2O(l) + NaNO_3(aq)$
$H^+(aq) + OH^-(aq) \longrightarrow H_2O(l)$
[b] $2HCl(aq) + Ba(OH)_2(aq) \longrightarrow 2H_2O(l) + BaCl_2(aq)$
$H^+(aq) + OH^-(aq) \longrightarrow H_2O(l)$

c $2HC_2H_3O_2(aq) + Ca(OH)_2(aq) \longrightarrow$
$$2H_2O(l) + Ca(C_2H_3O_2)_2(aq)$$
$HC_2H_3O_2(aq) + OH^-(aq) \longrightarrow H_2O(l) + C_2H_3O_2^-(aq)$
d $NH_3(aq) + HNO_3(aq) \longrightarrow NH_4NO_3(aq)$
$NH_3(aq) + H^+(aq) \longrightarrow NH_4^+(aq)$

4.45 a $2HBr(aq) + Ca(OH)_2(aq) \longrightarrow 2H_2O(l) + CaBr_2(aq)$
$H^+(aq) + OH^-(aq) \longrightarrow H_2O(l)$
b $3HNO_3(aq) + Al(OH)_3(s) \longrightarrow 3H_2O(l) + Al(NO_3)_3(aq)$
$3H^+(aq) + Al(OH)_3(s) \longrightarrow 3H_2O(l) + Al^{3+}(aq)$
c $2HCN(aq) + Ca(OH)_2(aq) \longrightarrow 2H_2O(l) + Ca(CN)_2(aq)$
$HCN(aq) + OH^-(aq) \longrightarrow H_2O(l) + CN^-(aq)$
d $HCN(aq) + LiOH(aq) \longrightarrow H_2O(l) + LiCN(aq)$
$HCN(aq) + OH^-(aq) \longrightarrow H_2O(l) + CN^-(aq)$

4.47 a $Ca(OH)_2(aq) + 2H_2SO_4(aq) \longrightarrow 2H_2O(l) + Ca(HSO_4)_2(aq)$; $H^+(aq) + OH^-(aq) \longrightarrow H_2O(l)$
b $2H_3PO_4(aq) + Ca(OH)_2(aq) \longrightarrow 2H_2O(l) + Ca(H_2PO_4)_2(aq)$;
$H_3PO_4(aq) + OH^-(aq) \longrightarrow H_2O(l) + H_2PO_4^-(aq)$
c $NaOH(aq) + H_2SO_4(aq) \longrightarrow NaHSO_4(aq) + H_2O(l)$;
$OH^-(aq) + H^+(aq) \longrightarrow H_2O(l)$
d $Sr(OH)_2(aq) + 2H_2CO_3(aq) \longrightarrow Sr(HCO_3)_2(aq) + 2H_2O(l)$;
$OH^-(aq) + H_2CO_3(aq) \longrightarrow HCO_3^-(aq) + H_2O(l)$

4.49 $2H_2SO_3(aq) + Ca(OH)_2(aq) \longrightarrow 2H_2O(l) + Ca(HSO_3)_2(aq)$
$Ca(HSO_3)_2(aq) + Ca(OH)_2(aq) \longrightarrow 2H_2O(l) + 2CaSO_3(s)$
$H_2SO_3(aq) + OH^-(aq) \longrightarrow H_2O(l) + HSO_3^-(aq)$
$Ca^{2+}(aq) + HSO_3^-(aq) + OH^-(aq) \longrightarrow CaSO_3(s) + H_2O(l)$

4.51 a Molecular equation: $BaCO_3(s) + 2HNO_3(aq) \longrightarrow CO_2(g) + H_2O(l) + Ba(NO_3)_2(aq)$
Ionic equation: $BaCO_3(s) + 2H^+(aq) \longrightarrow CO_2(g) + H_2O(l) + Ba^{2+}(aq)$
b Molecular equation: $K_2S(aq) + 2HCl(aq) \longrightarrow H_2S(g) + 2KCl(aq)$
Ionic equation: $S^{2-}(aq) + 2H^+(aq) \longrightarrow H_2S(g)$
c Molecular equation: $CaSO_3(s) + 2HI(aq) \longrightarrow SO_2(g) + H_2O(l) + CaI_2(aq)$
Ionic equation: $CaSO_3(s) + 2H^+(aq) \longrightarrow SO_2(g) + H_2O(l) + Ca^{2+}(aq)$

4.53 $FeS(s) + 2HCl(aq) \longrightarrow H_2S(g) + FeCl_2(aq)$
$FeS(s) + 2H^+(aq) \longrightarrow H_2S(g) + Fe^{2+}(aq)$ **4.55 a** Mn = +6
b Br = +7 **c** Nb = +4 **d** Ga = +3 **4.57 a** −3
b +5 **c** 0 **d** +7 **4.59 a** Hg is +1, O is −2, Cl = +7
b Cr is +3, O is −2, S = +6 **c** Co is +2, O is −2, Se = +6
d Pb is +2, H is +1, O is −2 **4.61 a** Co is reducing agent,
Cl is oxidizing agent **b** P is reducing agent, O is oxidizing
agent **4.63 a** Al is the reducing agent; F_2 is the oxidizing
agent **b** Hg^{2+} is the oxidizing agent; NO^{2-} is the reducing
agent **4.65 a** $3CuCl_2(aq) + 2Al(s) \longrightarrow 2AlCl_3(aq) + 3Cu(s)$
b $2Cr^{3+}(aq) + 3Zn(s) \longrightarrow 2Cr(s) + 3Zn^{2+}(aq)$

4.67 1.02 M **4.69** 0.101 M **4.71** 2.08 L
4.73 35.8 mL **4.75** 2.9×10^{-4} mol **4.77** Weigh out 18 g of
pure Na_2SO_4, and place it in a 250-mL volumetric flask. Add
some water to fill roughly half of the flask to dissolve the solid
with swirling. Then add enough water to fill the flask to the
mark on the neck and mix well by inverting the stoppered flask
several times. **4.79** 7.6 mL **4.81** 0.13 M ClO_3^-, 0.067 M
SO_4^{2-}, 0.27 M Na^+ **4.83** 0.303 g Ba^{2+}; 65.9%
4.85 a 15.25% **b** AuCl **4.87** $FeCl_2$ **4.89** 14.9 mL
4.91 325 mL **4.93** 58.6% **4.95** Molecular equation:
$2Al(s) + 6HClO_4(aq) \longrightarrow 3H_2(g) + 2Al(ClO_4)_3(aq)$
Ionic equation: $2Al(s) + 6H^+(aq) \longrightarrow 3H_2(g) + 2Al^{3+}(aq)$
4.97 Molecular equation: $BaBr_2(aq) + K_2SO_4(aq) \longrightarrow BaSO_4(s) + 2KBr(aq)$
Ionic equation: $Ba^{2+}(aq) + SO_4^{2-}(aq) \longrightarrow BaSO_4(s)$
4.99 a Molecular equation: $LiOH(aq) + HCN(aq) \longrightarrow LiCN(aq) + H_2O(l)$
Ionic equation: $OH^-(aq) + HCN(aq) \longrightarrow CN^-(aq) + H_2O(l)$
b Molecular equation: $Li_2CO_3(aq) + 2HNO_3(aq) \longrightarrow 2LiNO_3(aq) + CO_2(g) + H_2O(l)$

Ionic equation: $CO_3^{2-}(aq) + 2H^+(aq) \longrightarrow CO_2(g) + H_2O(l)$
c Molecular equation: $LiCl(aq) + AgNO_3(aq) \longrightarrow LiNO_3(aq) + AgCl(s)$
Ionic equation: $Cl^-(aq) + Ag^+(aq) \longrightarrow AgCl(s)$ **d** NR
4.101 a $Sr(OH)_2(aq) + 2HC_2H_3O_2(aq) \longrightarrow Sr(C_2H_3O_2)_2(aq) + 2H_2O(l)$
$HC_2H_3O_2(aq) + OH^-(aq) \longrightarrow C_2H_3O_2^-(aq) + H_2O(l)$
b $NH_4I(aq) + CsCl(aq) \longrightarrow$ NR
c $NaNO_3(aq) + CsCl(aq) \longrightarrow$ NR
d $NH_4I(aq) + AgNO_3(aq) \longrightarrow NH_4NO_3(aq) + AgI(s)$
$I^-(aq) + Ag^+(aq) \longrightarrow AgI(s)$
4.103 a add a solution of hydrochloric acid (HCl) to the solid
$MgCO_3$, forming CO_2, H_2O, and aqueous $MgCl_2$. The aqueous $MgCl_2$ can be converted to the solid form by evaporation.
Molecular equation:
$MgCO_3(s) + 2HCl(aq) \longrightarrow MgCl_2(aq) + CO_2(g) + H_2O(l)$
b add a solution of $AgNO_3$ to a solution of the NaCl, precipitating AgCl. The AgCl can be filtered off, leaving aqueous
$NaNO_3$, which can be obtained in solid form by evaporation.
Molecular equation:
$NaCl(aq) + AgNO_3(aq) \longrightarrow AgCl(s) + NaNO_3(aq)$
c add a solution of NaOH to a solution of $Al(NO_3)_3$, precipitating $Al(OH)_3$. The $Al(OH)_3$ can be filtered off and dried to
remove any water adhering to it. Molecular equation:
$Al(NO_3)_3(aq) + 3NaOH(aq) \longrightarrow Al(OH)_3(s) + 3NaNO_3(aq)$
d add a solution of $BaCl_2$ to the solution of H_2SO_4, precipitating $BaSO_4$. The $BaSO_4$ can be filtered off, leaving the desired solution of aqueous HCl. Molecular equation:
$H_2SO_4(aq) + BaCl_2(aq) \longrightarrow BaSO_4(s) + 2HCl(aq)$
4.105 a Decomposition **b** Decomposition **c** Combination
d Displacement **4.107 a** $Pb(NO_3)_2 + H_2SO_4$, $Pb(NO_3)_2 + MgSO_4$, $Pb(NO_3)_2 + Ba(OH)_2$ **b** $Ba(OH)_2 + MgSO_4$
c $Ba(OH)_2 + H_2SO_4$ **4.109** 0.00863 M $Fe_2(SO_4)_3$ molarities:
0.0173, M Fe^{3+} 0.0259, M SO_4^{2-}. **4.111** 329 mL
4.113 0.610 M **4.115** 0.0967 M **4.117** 21.39%
4.119 $CuSO_4 \cdot 2H_2O$ **4.121** Sc = 29.7%, Cl = 70.3%,
$ScScCl_3$ **4.123** 9.66% **4.139** 11.2 g **4.141** Silver nitrate
is $AgNO_3$, and strontium chloride is $SrCl_2$; Molecular equation: $2AgNO_3(aq) + SrCl_2(aq) \longrightarrow 2AgCl(s) + Sr(NO_3)_2(aq)$;
Net ionic equation: $Ag^+(aq) + Cl^-(aq) \longrightarrow AgCl(s)$; AgCl is
silver(I) chloride, and $Sr(NO_3)_2$ is strontium nitrate.
Molecular equation: $AgClO_4(aq) + NaCl(aq) \longrightarrow AgCl(s) + NaClO_4(aq)$
4.143 $CaBr_2(aq) + Cl_2(g) \longrightarrow CaCl_2(aq) + Br_2(l)$
$2Br^-(aq) + Cl_2(g) \longrightarrow 2Cl^-(aq) + Br_2(l)$ 5.67×10^3 g $CaBr_2$
4.145 $Hg(NO_3)_2(aq) + H_2S(g) \longrightarrow HgS(s) + 2HNO_3(aq)$
$Hg^{2+}(aq) + H_2S(g) \longrightarrow HgS(s) + 2H^+(aq)$
The acid formed is nitric acid, a strong acid. The other product is
mercury(II) sulfide. 581.5 g **4.147** 33.3%
4.149 0.0121 L **4.151** 0.930 M **4.153** 87.60 g/mol
(strontium) **4.155** $P_4O_{10} + 6H_2O \longrightarrow 4H_3PO_4$;
55.7 g P_4O_{10} **4.157** 83.57% **4.159** 0.194 L **4.161** 99.00%

CHAPTER 5

5.29 a As the temperature of the water increases, so does its vapor pressure (Table 5.6). **b** If you put an aerosol can in a fire,
you will increase the temperature and thus the pressure. **c** You
would expect the pressure in the tires to decrease, and they would
appear flatter. **d** As you squeeze the balloon, you decrease the
volume, resulting in an increase in pressure.
5.31 a increase by a factor of 2 **b** decrease by a factor of 2
c increase by a factor of 1.5 **d** the pressure would not
change; volume increase by a factor of 2 **5.33 a** The pressure in each of the containers will not change. **b** O_2 **c** ¼
d Both the same **e** H_2 **5.35 a** F_2 flask **b** You could
decrease the temperature or remove some gases from the other

two containers   **5.37** 71 mm Hg   **5.39** 4.24 L   **5.41** 508 L
5.43 3.64×10^{-4} kPa   **5.45** 3.68 mL   **5.47** 1.3 L
5.49 286 K (12.7 or 13°C)   **5.51** 31.6 mL   **5.53** 3 volumes
of H_2 **5.55** $PV = nRT$; $V = nRT$; If the temperature and
number of moles are held constant, then the product nRT is
constant, and volume is inversely proportional to pressure:
$V = $ constant $\times 1/P$.   **5.57** 1.78 atm   **5.59** 22.0 L
5.61 143°C   **5.63** 1.64 g/L   **5.65** 2.24 g/L   **5.67** 47.7 amu
5.69 178 amu   **5.71** For a gas at a given temperature and pressure, the density depends on molecular mass (or, for a mixture,
on average molecular mass). Thus, at the same temperature and
pressure, the density of NH_4Cl gas would be greater than that of
a mixture of NH_3 and HCl, because the average molecular mass
of NH_3 and HCl would be lower than that of NH_4Cl.
5.73 2.41 L   **5.75** 160. L   **5.77** 246 L   **5.79** 4.61 L
5.81 0.421 atm   **5.83** 0.00380 atm O_2; 0.017 atm He;
Total $P = 0.020$ atm   **5.85** P(He) = 5.46 atm; P(N_2) = 1.17 atm;
P(O_2) = 0.28 atm   **5.87** 6.34 g   **5.89** $u_{25} = 5.15 \times 10^2$ m/s;
$u_{125} = 5.95 \times 10^2$ m/s; Graph as in Figure 5.25.
5.91 1.53×10^2 m/s   **5.93** 6.23×10^2°C   **5.95** 6.338 to 1
5.97 3.5 s   **5.99** 146 amu   **5.101** 0.9885 atm, 0.9928 atm
(ideal gas law)   **5.103** At 1.00 atm, $V = 22.4$ L and the van
der Waals equation gives 22.4 L. At 10.0 atm, the van der Waals
equation gives 2.08 L. The ideal gas law gives 2.24 L for 10.0 atm.
5.105 80.7 cm³   **5.107** 181 mL   **5.109** 6.5 dm³   **5.111** $3.01 \times$
10^{20} atoms   **5.113** 28.979 amu   **5.115** 0.0345 g $HC_3H_3O_3$
5.117 20.4 atm O_2 **5.119** 0.440 M   **5.121** HCl
5.123 −27°C   **5.125** 1.0043   **5.127** 28.07 g/mol
5.129 0.167 atm O_2; 0.333 atm CO_2 **5.151** 0.388 g/L
5.153 0.194 g   **5.155** 92.9% $CaCO_3$; 7.1% $MgCO_3$
5.157 8.81 g/L   **5.159** 3.00×10^2 m/s

CHAPTER 6

6.31 [a] Beaker A- exothermic; Beaker B- exothermic;
Beaker C- endothermic   [b] Beaker A- negative; Beaker
B- Negative; Beaker C- Positive   [c] Beaker A- positive;
Beaker B- negative; Beaker C- positive   [d] Beaker A-
decrease; Beaker B- decreace; Beaker C- increase
[e] Beaker A- Hotter; Beaker B- Hotter; Beaker C- Colder
6.33 small car   **6.35** [a] negative   [b] solid   [c] higher than
the final enthalpy   [d] −20°C   **6.37** block of iron
6.39 [a] A   [b] A   **6.41** Equate ΔH for the burning to ΔH_f
(products minus ΔH_f (reactants), then solve for ΔH_f of P_4S_3.
You will need ΔH_f P_4S_{10} and SO_2.   **6.43** −67.16 kcal
6.45 7.1×10^5 J; 1.7×10^5 cal   **6.47** 5.24×10^{-21} J/molecule
6.49 +76 J   **6.51** warm   **6.53** endothermic; +66.2 kJ
6.55 37.56 kJ   **6.57** $2KClO_3(s) \longrightarrow 2KCl(s) + 3O_2(g)$;
$\Delta H = -78.0$ kJ   **6.59** +3010 kJ   **6.61** 227 kJ
6.63 −1.90 kJ/g   **6.65** -6.66×10^2 kJ   **6.67** 7.96 g
6.69 1.15×10^5 J   **6.71** 5.81°C   **6.73** -9.73×10^4 J,
or −97.3 kJ   **6.75** -1.36×10^3 kJ   **6.77** 50.6 kJ
6.79 −906.3 kJ   **6.81** −137 kJ   **6.83** +39.4 kJ
6.85 −1124.2 kJ   **6.87** −23.5 kJ   **6.89** −74.9 kJ
6.91 20.9 kJ   **6.93** m/s²   **6.95** 1.59×10^3 J/g   **6.97** 31.6 m/s
6.99 37.9 kJ   **6.101** −872 kJ/mol   **6.103** 37.4°C
6.105 0.256 J/g C   **6.107** 0.383 J/g°C   **6.109** 27.8 °C
6.111 −23.6 kJ/mol   **6.113** −874.2 kJ/mol   **6.115** −116.3 kJ
6.117 206.2 kJ   **6.119** 178 kJ   **6.121** −143 kJ
6.123 −2225 kJ/mol   **6.125** 90.1 g   **6.149** 37.4°C
6.151 −22 kJ   **6.153** 3.1% CH_4 **6.155** 113 kJ
6.157 [a] 1.88 kJ   [b] 11.9%   **6.159** 84.9 g KO_2 required

CHAPTER 7

7.25 one million times as large   **7.27** red light   **7.29** x
7.31 The electron in part b will have the longest wavelength.
7.33 [a] $n = 3$ to $n = 2$ [b] $n = 3$ to $n = 5$ [c] green

7.35 0.02149 m (2.149 cm)   **7.37** 6.91×10^{14} /s
7.39 1.9×10^2 s   **7.41** 9.192631770×10^9 /s, 0.0326122557 m
7.43 9.044×10^{-28} J   **7.45** 3.71×10^{-19} J   **7.47** not visible
7.49 2.34×10^{14}/s   **7.51** 1.22×10^{-7} m (near UV)   **7.53** 93.8 nm
7.55 4.699×10^{-19} J   **7.57** 9.54 pm   **7.59** 3.73×10^4 m/s
7.61 1.52×10^{-22} pm, much smaller   **7.63** $l = 0, 1, 2,$ or 3;
$ml = -3, -2, -1, 0, +1, +2,$ or $+3$ **7.65** 3, 7   **7.67** [a] 6d
[b] 5g   [c] 4f   [d] 6p   **7.69** [a] Not permissible; m_s can be only
$+1/2$ or $-1/2$.   [b] Not permissible; n cannot be zero.
[c] Not permissible; l can be only as large as $(n - 1)$.
[d] Not permissible; m_s can be only $+1/2$ or $-1/2$.
[e] Not permissible; m_l cannot exceed $+2$ in magnitude.
7.71 5.41×10^{14} /s, 3.59×10^{-19} J   **7.73** 485 nm (blue-green)
7.75 6.55×10^{14}/s   **7.77** 5.58×10^5 m/s   **7.79** 4.34×10^{-7} m
7.81 5   **7.83** 1.641×10^{-7} m (164.1 nm; near UV)
7.85 1.50×10^4 eV   **7.87** [a] Seven   [b] Three   [c] One
[d] Five   **7.89** 8s, 8p, 8d, 8f, 8g, 8h, and 8i.   **7.117** $5.0052 \times$
10^{-7} m   **7.119** 5.27×10^{28} photons   **7.121** 3.59×10^{-19} J;
2.16×10^2 kJ/mol   **7.123** 1.23×10^{-11} m (12.3 pm)

CHAPTER 8

8.31 six   **8.33** Sr, Y   **8.35** calcium (Ca) and strontium (Sr)
8.37 yes, Both electron affinity and ionization energy provide
information about the strength of the attraction between
electrons and a particular nucleus.   **8.39** IIA   **8.41** [a] Not
allowed; the two electrons in the 1s orbital must have opposite spins.   [b] Not allowed; the 2s orbital can hold at most
two electrons, with opposite spins.   [c] Allowed; $1s^22s^22p^4$.
[d] Not allowed; the unpaired electrons in the 2p orbitals
should have parallel spins.   **8.43** [a] Possible state; however,
the 3p and 4s orbitals should be filled before the 3d orbital.
[b] Impossible state; the 2p orbitals can hold no more than six
electrons.   [c] Possible state.   [d] Impossible state; the 3s orbital
can hold no more than two electrons.

8.45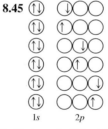

8.47 Arsenic ($Z = 33$): $1s^22s^22p^63s^23p^63d^{10}4s^24p^3$
8.49 $1s^22s^22p^63s^2 3p^63d^54s^2$ **8.51** Bismuth ($Z = 83$): $6s^26p^3$
8.53 $4d^55s^2$ **8.55** Ir is in Group VIIIB, d-transition element
8.57 [Ar] ⇅⇅⇅⇅⇅ ⇅
            3d   4s

8.59 ⇅ ⇅ ⇅⇅⇅ ⇅ ⇅⇅⇅ ↑
       1s 2s   2p   3s   3p   4s
Paramagnetic
8.61 S, Se, As   **8.63** Na, Al, Cl, Ar   **8.65** [a] F   [b] Br
8.67 $LiBrO_3$ **8.69** $1s^22s^22p^63s^23p^63d^{10}4s^24p^65s^2$ **8.71** $6s^26p^4$

8.73 ⇅ ⇅ ⇅⇅⇅ ⇅ ⇅⇅⇅ ⇅⇅⇅⇅⇅
       1s 2s   2p   3s   3p   3d
         ⇅ ↑↑↑
        4s   4p

8.75 [Rn] $5f^{14}6d^{10}7s^27p^3$. It is a metal; the oxide is eka-Bi_2O_3 or
eka-Bi_2O_5 **8.77** 370 kJ/mol
8.81 [a] Cl_2 [b] Na   [c] Sb   [d] Ar   **8.83** $1s^22s^22p^63s^23p^63d^54s^2$,
Group VB, Period 4, d-block transition element
8.111 $2Cs(s) + 2H_2O(l) \longrightarrow 2CsOH(aq) + H_2(g)$, 0.539 g Cs
8.113 RaO; 93.4%   **8.115** -2.61×10^{-2} kJ
8.117 1.312×10^3 kJ/mol H   **8.119** −639 kJ/mol

CHAPTER 9

9.25 XY_3 **9.27** b and d, The 2^+ ion given in configuration a represents a loss of the two outer s electrons, leaving the outer p electrons of higher energy. This configuration would represent an excited state of the 2^+ ion.   **9.29** b
9.31 a Bond Cl and H to C.   b Bond O atoms to N, then bond H to one O.   c Bond O and F to N.   d Place N atoms in center with two O atoms bonded to each N.   **9.33** a Ca_3X_2 b ionic

9.37 a $K\cdot \; +\; \cdot\ddot{\underset{\cdot\cdot}{I}}: \longrightarrow K^+ + \left[:\ddot{\underset{\cdot\cdot}{I}}:\right]^-$

b $:\ddot{\underset{\cdot\cdot}{Br}}\cdot \;+\; \cdot Ca\cdot \;+\; \cdot \ddot{\underset{\cdot\cdot}{Br}}: \longrightarrow$

$\left[:\ddot{\underset{\cdot\cdot}{Br}}:\right]^- + Ca^{2+} + \left[:\ddot{\underset{\cdot\cdot}{Br}}:\right]^-$

9.39 a As: $1s^2 2s^2 2p^6 3s^2 3p^6 3d^{10} 4s^2 4p^3$, $\cdot\ddot{As}\cdot$

b As^{3+}: $1s^2 2s^2 2p^6 3s^2 3p^6 3d^{10} 4s^2$, $\left[\cdot As\cdot\right]^{3+}$

c Se: $1s^2 2s^2 2p^6 3s^2 3p^6 3d^{10} 4s^2 4p^4$, $:\ddot{Se}\cdot$

d Se^{2-}: $1s^2 2s^2 2p^6 3s^2 3p^6 3d^{10} 4s^2 4p^6$, $\left[:\ddot{\underset{\cdot\cdot}{Se}}:\right]^{2-}$

9.41 Sn: $[Kr]4d^{10}5s^25p^2$
Sn^{2+}: The two 5p electrons are lost from the valence shell. $[Kr]4d^{10}5s^2$
9.43 Ni^{2+}: $[Ar]3d^8$ Ni^{3+}: $[Ar]3d^7$ **9.45** a $Br < Br^-$
b $Sr^{2+} < Sr$ **9.47** $S^{2-} < Se^{2-} < Te^{2-}$ **9.49** Smallest Na^+ $(Z = 11)$, F^- $(Z = 9)$, N^{3-} $(Z = 7)$ Largest These ions are isoelectronic. The atomic radius increases with the decreasing nuclear charge (Z).   **9.53** $AsBr_3$ **9.55** a Cs, Ba, Sr   b Ca, Ga, Ge   c As, P, S   **9.57** C—Cl, As—Br, P—O
9.59 a P—O   b C—Cl   c As—Br
$\quad\quad \delta+\;\delta-\quad\quad \delta+\;\delta-\quad\quad\quad \delta+\;\;\delta-$

9.61 a $:\ddot{\underset{:\ddot{F}:}{\overset{:\ddot{F}:}{F}}:\ddot{N}:$ b $:\ddot{Br}-\ddot{Br}:$ c $H-\ddot{S}-H$

9.63 a $H-\ddot{O}-\ddot{N}=\ddot{O}:$ b $:\overset{:\ddot{Br}:}{\underset{\quad}{O}}=C-\ddot{Br}:$ c $:P\equiv P:$

9.65 a $\left[:\ddot{Cl}-\ddot{O}:\right]^-$ b $\left[:\underset{:\ddot{Cl}:}{\overset{:\ddot{Cl}:}{Cl}}-Sn-\ddot{Cl}:\right]^-$ c $\left[:\ddot{S}-\ddot{S}:\right]^{2-}$

9.67 a $:\ddot{O}-\overset{}{S}=\ddot{O} \longleftrightarrow :\ddot{O}-\overset{}{S}-\ddot{O}: \longleftrightarrow \ddot{O}=\overset{}{S}-\ddot{O}:$
$\quad\quad\quad\underset{:\ddot{O}:}{|}\quad\quad\quad\quad\underset{:\ddot{O}:}{||}\quad\quad\quad\quad\underset{:\ddot{O}:}{|}$

b $H-\ddot{O}-\overset{\overset{\ddot{O}}{||}}{N} \longleftrightarrow H-\ddot{O}-\overset{\overset{:\ddot{O}:}{|}}{N}$
with lower oxygens $:\ddot{O}:$ and $\ddot{O}$

9.71 a $:\ddot{F}-\overset{}{Xe}-\ddot{F}:$ b $:\overset{:\ddot{F}\quad\ddot{F}:}{\underset{:\ddot{F}\quad\ddot{F}:}{Te}}$

c $\left[:\overset{:\ddot{F}\quad\ddot{F}}{\underset{:\ddot{F}\quad\ddot{F}}{Xe}}-\ddot{F}:\right]^+$

9.73 a $:\underset{:\ddot{Cl}:}{\overset{:\ddot{Cl}:}{Cl}}-B-\ddot{Cl}:$ b $\left[:\ddot{Cl}-Tl-\ddot{Cl}:\right]^+$

c $:\ddot{Br}-Be-\ddot{Br}:$

9.75 a Starting with the left oxygen, the formal charge of this oxygen is $6 - 1 - 6 = -1$. The formal charge of just the central oxygen is $6 - 3 - 2 = +1$. The formal charge of the right oxygen is $6 - 2 - 4 = 0$.

$\ddot{O}=\overset{\oplus}{O}-\ddot{\underset{\ominus}{O}}:$

b The formal charge of the carbon is $4 - 3 - 2 = -1$. The formal charge of the oxygen is $6 - 3 - 2 = +1$

$:\overset{\ominus}{C}\equiv\overset{\oplus}{O}:$

c The formal charge of the nitrogen is $5 - 4 - 0 = +1$. The formal charge of the hydrogen is $1 - 1 - 0 = 0$. The formal charge of the oxygen bonded to the hydrogen is $6 - 2 - 4 = 0$. The formal charge of the other singly bonded oxygen is $6 - 1 - 6 = -1$. The formal charge of the doubly bonded oxygen is $6 - 2 - 4 = 0$.

$H-\ddot{O}-\overset{\oplus}{\underset{\ddot{\underset{\ominus}{O}}:}{N}}\overset{:\ddot{O}}{}$

9.79 164 pm   **9.81** a 177 pm   b 222 pm   c 207 pm
d 107 pm   **9.83** Formaldehyde   **9.85** -57 kJ
9.87 a ionic, K_2Se, potassium selenide   b ionic, AlF_3, aluminum fluoride   c ionic, $BaBr_2$, barium bromide   d covalent, $SiCl_4$, silicon tetrachloride

9.89 $\left[:\ddot{O}-\overset{:\ddot{O}:}{\underset{:\ddot{O}:}{Se}}-\ddot{O}:\right]^{2-}$ or $\left[:\ddot{O}-\overset{:\ddot{O}:}{\underset{:\ddot{O}:}{Se}}-\ddot{O}:\right]^{2-}$, $Al_2(SeO_3)_3$

9.93 $\left[H-\ddot{N}-H\right]^-$ **9.95** $\left[:\ddot{O}=N=\ddot{O}:\right]^+$

9.97 a $\left[:C\equiv C:\right]^{2-}$ b $\left[:\underset{:\ddot{Cl}:}{\overset{:\ddot{Cl}:}{Cl}}-Ga-\ddot{Cl}:\right]^-$

c $:\overset{:\ddot{O}:}{\underset{}{Se}}=C=\ddot{Se}:$ d $:\ddot{Cl}-\overset{:\ddot{O}:}{\underset{}{Se}}-\ddot{Cl}:$ or $:\ddot{Cl}-\overset{\overset{:\ddot{O}}{||}}{Se}-\ddot{Cl}:$

9.99 a $:\ddot{I}-C\equiv N:$ b $:\underset{:\ddot{Cl}:}{\overset{}{Cl}}-Sb-\ddot{Cl}:$

c $:\ddot{Cl}-\overset{}{I}-\ddot{Cl}:$ with $:\ddot{Cl}:$ d $:\overset{:\ddot{F}\quad\ddot{F}:}{\underset{:\ddot{F}\quad\ddot{F}:}{I}}-\ddot{F}:$

9.101 a $\ddot{O}=\overset{}{Se}-\ddot{O}: \longleftrightarrow :\ddot{O}-\overset{}{Se}=\ddot{O} \longleftrightarrow \ddot{O}=Se=\ddot{O}$

b (resonance structures of N_2O_4-type showing N—N bonds with oxygens)

9.103 when a hydrogen is bonded to the O carrying the formal -1 charge, the acetic acid molecule results. This molecule in fact has one C—O bond and one C=O bond, with the C=O bond being shorter in length. Adding one H to either structure above results in identical structures

9.105

$$:\ddot{O} \quad \ddot{O}: \qquad :\ddot{O} \quad \ddot{O}:$$
$$\backslash N-\ddot{O}-N \diagup \longleftrightarrow \diagup N-\ddot{O}-N \backslash \longleftrightarrow$$
$$:\ddot{O} \diagup \qquad \diagdown \ddot{O}: \qquad :\ddot{O}: \qquad :\ddot{O}:$$

$$:\ddot{O}: \qquad :\ddot{O}: \qquad :\ddot{O}: \qquad :\ddot{O}:$$
$$\diagup N-\ddot{O}-N \diagup \longleftrightarrow \diagup N-\ddot{O}-N \diagup$$
$$:\ddot{O}: \qquad \diagdown \ddot{O}: \qquad :\ddot{O}: \qquad \diagdown \ddot{O}:$$

Because double bonds are shorter, the terminal N—O bonds that resonate between single and double bonds are 118 pm, and the central N—O single bonds are 136 pm.   **9.107** −138 kJ
9.109 −238 kJ   **9.137** P–H nonpolar; O—H 1 polar (acidic)
$H_3PO_3(aq) + 2NaOH(aq) \longrightarrow Na_2HPO_3(aq) + 2H_2O(l)$
0.007031 M H_3PO_3 **9.139** $Ca(NO_2)_2$, calcium nitrite

9.141

$$\begin{array}{c} H \\ \diagdown \\ \diagup C = \ddot{O}: \\ :\ddot{F}: \end{array}$$

9.143 170.02 amu   **9.145** ΔH_{rxn} = 1272.3 kJ/mol; $BE(C \equiv N)$
= 861 kJ/mol (Table 9.5 has 891 kJ/mol)   **9.147** 3.4   **9.149** 3.5

CHAPTER 10

10.23 a ii   b i   c iv   d iii   **10.25** The repulsion between the electron pairs causes the bonding pairs to be pushed away, forcing them closer together.   **10.27** a iv   b v   c i   d iii
10.29 sulfur (S), selenium (Se), or tellurium (Te)
10.31 a ii   b iv   c i   **10.33** a trigonal pyramidal
b trigonal planar   c bent   d tetrahedral
10.35 a trigonal pyramidal   b linear   c tetrahedral
d trigonal pyramidal   **10.37** a 109°, same   b 109°, less
c 120°, the Cl—C—Cl bond to be less than 120°   d 109°, less
10.39 a Octahedral   b Square pyramidal   c Seesaw
d Trigonal bipyramidal   **10.41** a trigonal bipyramidal
b octahedral   c linear   d square planar
10.43 a Trigonal pyramidal and T-shaped   b Seesaw
10.45 CS_2 and XeF_4 **10.47** a uses sp^3 orbitals   b uses sp^3
orbitals   **10.49** a sp^2 b sp c sp d sp^3 **10.51** a The presence of two single bonds and no lone pairs suggests sp hybridization. An Hg—Cl bond is formed by overlapping the Hg sp hybrid orbital with a $3p$ orbital of Cl.   b The presence of three single bonds and one lone pair suggests sp^3 hybridization of the P atom. Three hybrid orbitals each overlap a $3p$ orbital of a Cl atom to form a P—Cl bond. The fourth hybrid orbital contains the lone pair.   **10.53** a sp^3 hybridization of the N atom and 3 hybrid orbitals each overlap a $2p$ orbital of an F atom to form an N—F bond. The fourth hybrid orbital contains the lone pair   b Each hybrid orbital overlaps a 2p orbital of an F atom to form an Si—F bond   **10.55** The P atom in PCl_6^- has 6 single bonds around it and no lone pairs. This suggests sp^3d^2 hybridization. Each bond is a σ bond formed by the overlap of the sp^3d^2 hybrid orbital on the P with a $3p$ orbital on Cl.
10.57 a Because the C is bonded to three other atoms, it is assumed to be sp^2 hybridized. One $2p$ orbital remains unhybridized. The carbon–hydrogen bonds are σ bonds formed by the overlap of an sp^2 hybrid orbital on C with a $1s$ orbital on H. The remaining sp^2 hybrid orbital on C overlaps with a $2p$ orbital on O to form a σ bond. The unhybridized $2p$ orbital on C overlaps with a parallel $2p$ orbital on O to form a π bond. Together, the σ and π bonds constitute a double bond.   b The nitrogen atoms are sp hybridized. A σ bond is formed by the overlap of an sp hybrid orbital from each N. The remaining sp hybrid orbitals contain lone pairs of electrons. The two unhybridized $2p$ orbitals on one N overlap with the parallel unhybridized $2p$ orbitals on the other N to form two π bonds.   **10.59** Each of the N atoms has a lone pair of electrons and is bonded to two atoms. The N atoms are sp^2 hybridized.

The two possible arrangements of the O atoms relative to one another are shown below. Because the π bond between the N atoms must be broken to interconvert these two forms, it is to be expected that the hyponitrite ion will exhibit *cis–trans* isomerism.

$$\begin{bmatrix} :\ddot{N} = \ddot{N}: \\ :\ddot{O}: \quad :\ddot{O}: \end{bmatrix}^{2-} \begin{bmatrix} :\ddot{O}: \\ :\ddot{N} = \ddot{N}: \\ :\ddot{O}: \end{bmatrix}^{2-}$$
$$\quad cis \qquad\qquad trans$$

10.61 a $KK(\sigma_{2s})^2(\sigma_{2s}^*)^2(\pi_{2p})^4(\sigma_{2p})^2(\pi_{2p}^*)^3$, bond order = 3/2, stable, paramagnetic   b $KK(\sigma_{2s})^2(\sigma_{2s}^*)^2(\pi_{2p})^1$, bond order = 1/2, stable, paramagnetic   c $KK(\sigma_{2s})^2(\sigma_{2s}^*)^2(\pi_{2p})^2$, bond order = 1, stable, paramagnetic   **10.63** $KK(\sigma_{2s})^2(\sigma_{2s}^*)^2(\pi_{2p})^4(\sigma_{2p})^2$, bond order = 3, diamagnetic   **10.65** a Tetrahedral,   b Bent, or angular,   c Seesaw,   d Trigonal bypyramidal
10.67 none   **10.69** a sp^3 hybridized,   b sp hybridized
10.71 The *trans* isomer is expected to have a zero dipole moment, whereas the *cis* isomer is not.   **10.73** All four hydrogen atoms of $H_2C=C=CH_2$ cannot lie in the same plane because the second C=C bond forms perpendicular to the plane of the first C=C bond. By looking at Figure 10.26, you can see how this is so. The second C=C bond forms in the plane of the C—H bonds. Thus, the plane of the C—H bonds on the right side will be perpendicular to the plane of the C—H bonds on the left side. This is shown below using the π orbitals of the two C=C double bonds.   **10.75** $(\sigma_{1s})^2(\sigma_{1s}^*)^1He_2^+$ ion is expected to be stable
10.77 $KK(\sigma_{2s})^2(\sigma_{2s}^*)^2(\pi_{2p})^4(\sigma_{2p})^2$, bond order = 3
10.79 both are greater   **10.81** The N_2 ground state is $KK(\sigma_{2s})^2(\sigma_{2s}^*)^2(\pi_{2p})^4(\sigma_{2p})^2(\pi_{2p}^*)^0$, the first excited state is $KK(\sigma_{2s})^2(\sigma_{2s}^*)^2(\pi_{2p})^4(\sigma_{2p})^2(\pi_{2p}^*)^1$. The transition decreases the bond order from 3 to 2, and the ground state is diamagnetic, whereas the excited state is paramagnetic.
10.109 $XeOF_4$ sp^3d^2 hybridization   **10.111** 5.00, Because bromine has six electron pairs, it will require six orbitals to describe the bonding. This suggests sp^3d^2 hybridization.

10.113

$$\begin{array}{cc} H-\ddot{O}: & H-\ddot{O}: \\ | & | \\ N & N \\ \diagup \backslash & \diagup \backslash \\ :\ddot{O}. \quad :\ddot{O}. & :\ddot{O}. \quad :\ddot{O}. \end{array} \longleftrightarrow$$

trigonal planar; therefore, the hybridization is sp^2; $\Delta H_f^\circ = -34.5$ kJ/mol; Resonance energy = 101 kJ/mol

CHAPTER 11

11.27 a A   b C   c C   **11.29** The heat released when the liquid-to-solid phase change occurs prevents the fruit from freezing.   **11.31** a AB_2 b yes, body-centered cell
11.33 water   **11.35** a face-centered cubic   b simple cubic
11.37 a Freezing   b Gas-solid condensation, deposition
c Condensation   d Freezing of eggs and sublimation of ice
e Vaporization   **11.39** 40°C   **11.41** 15.5 kJ/mol
11.43 3.15 kJ   **11.45** 8.84 g   **11.47** 27 g   **11.49** 109 mmHg
11.51 27.8 kJ/mol   **11.53** a gas,   b solid,   c gas,   d No
11.55

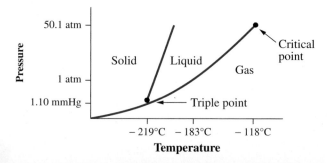

11.57 Liquefied at 25°C: SO_2 and C_2H_2. To liquefy CH_4, lower its temperature below $-82°C$, and then compress it. To liquefy CO, lower its temperature below $-140°C$, and then compress it. **11.59** a solid b liquid **11.61** Yes, The heats of vaporization of 1.8 kJ/mol for Ne and 6.8 kJ/mol for O_2 increase in the order of the respective molecular masses of 20.1 and 32.0 because only London forces are involved. The heat of vaporization of 34.5 kJ/mol for methanol (molecular mass = 32.0) is higher than that of oxygen because of strong hydrogen bonding between H and O. **11.63** a London forces b London and dipole-dipole forces, hydrogen bonding c London and dipole-dipole forces d London forces **11.65** $CCl_4 < SiCl_4 < GeCl_4$ **11.67** Ibr, the largest molecular mass and the largest intermolecular forces and the lowest vapor pressure, so it should have the highest boiling point **11.69** $HOCH_2CH_2OH$, FCH_2CH_2OH, FCH_2CH_2F; hydrogen bonding decreases in magnitude from left to right **11.71** $CH_4 < C_2H_6 < CH_3OH < HOCH_2CH_2OH$ **11.73** a Molecular b Ionic c Molecular d Molecular e Ionic **11.75** a Molecular b Molecular c Metallic d Covalent network **11.77** $(C_2H_5)_2O < C_4H_9OH < KCl < CaO$ **11.79** a High-melting, hard, and brittle b Low-melting and brittle c Hard and high-melting d Malleable and electrically conducting **11.81** a CHI_3 b LiCl c Co d SiC **11.83** two **11.85** 9.75×10^{-23} g/Ni atom **11.87** 361 pm **11.89** 4, fcc **11.91** 19.25 g/cm^3 **11.93** 1.72×10^{-23} cm^3, 160 pm **11.95** Water vapor condenses directly to solid water, the snow melts to liquid water, rain is vaporized to water vapor. **11.97** 68.4% **11.99** 372.3 K (99.1 °C) **11.101** a remains a gas b condenses into a liquid (and to solid at high pressure) c condenses to solid **11.103** hydrogen bonding exists between any hydrogen and the lone pair of electrons of oxygen of an adjacent hydrogen peroxide. For two adjacent hydrogen peroxide molecules, the hydrogen bond may be represented as follows: HO—O—H$\cdots$O(H)—OH **11.105** Ethylene glycol molecules are capable of hydrogen bonding to each other leading to higher boiling point and greater viscosity. **11.107** AlF_3 forms an ionic solid. SiF_4, PF_3, and SF_4 form molecular solids. **11.109** a Lower: KCl, smaller charge on the ions b Lower: CCl_4, lower molecular mass and weaker London forces c Lower: Zn, melting points for Group IIB metals are lower d Lower: C_2H_5Cl, no hydrogen-bond, but acetic acid **11.111** 3.0153×10^{-22} g, 181.6 amu **11.113** 128 pm **11.115** 74.00% **11.117** a The boiling point increases as the size of the molecule increases. b Hydrogen bonding occurs between the H—F molecules. c The hydrogen halides have dipole-dipole interactions. **11.119** a Diamond and silicon carbide are covalent network solids with strong covalent bonds; graphite is a layered structure, with weak forces between the layers. b Silicon dioxide is a covalent network solid. Carbon dioxide is a discrete, small molecule. **11.121** a CO_2 consists of discrete nonpolar molecules; SiO_2 is a covalent network solid b SiF_4 is a larger molecule c HF has hydrogen bonding **11.149** 0.220 mol N_2, 0.01023 mol C_3H_8O, 0.0444, 33.1 mmHg, 33.1 mmHg **11.151** 6.66 kJ **11.153** 10.12 kJ **11.155** 1.83 g/L

CHAPTER 12

12.27 $CaCl_2$ **12.29** The lattice energy must be less for the XZ compound. **12.31** The charged smoke particles are neutralized by the current, which then allows them to aggregate into large particles. These large particles are too big to be carried out of the stack. **12.33** Because the salt concentration is higher outside the lettuce leaf than inside, water will pass out of the lettuce leaf into the dressing via osmosis. **12.35** a urea b $CaCl_2$ **12.37** any alloy, such as 18-kt gold **12.39** Ethanol **12.41** Acetic acid **12.43** Al^{3+} **12.45** $Ba(IO_3)_2 < Sr(IO_3)_2 < Ca(IO_3)_2 < Mg(IO_3)_2$; The iodate ion is fairly large, so the lattice energy for all these iodates should change to a smaller degree than the hydration energy of the cations. Therefore, solubility should increase with decreasing cation radius. **12.47** 0.886 g/100 mL **12.49** Dissolve 3.63 g KI in 68.9 g of water. **12.51** 10.3 g **12.53** 0.496 m LA **12.55** 77.7 g **12.57** 0.244 **12.59** 0.0116 **12.61** 0.286, 22.2 m **12.63** 0.563 M **12.65** 0.271 m **12.67** 41.6 mmHg, 0.631 mmHg **12.69** 118.6°C, 16.49°C **12.71** 4.6×10^{-2} m **12.73** 122 amu **12.75** 163 amu **12.77** 1.09×10^3 amu **12.79** $-0.042°C$ **12.81** $\cong 3$ **12.83** a foam b aerosol c sol d sol **12.85** $Al_2(SO_4)_3$ **12.87** 0.84, 0.16 **12.89** 1.74 m, 0.0303, 1.63 M **12.91** 1.5 g O_2, 2.1 g N_2, 4.8 g He **12.93** 141 mmHg **12.95** a 0.0002140 M $KAl(SO_4)_2$ b 0.0004279 M c 2.14×10^{-4} M **12.97** $-21°C$ **12.99** 0.30 M **12.101** $CaCl_2$ **12.103** 1.8 g/mL, 5×10^2 m **12.105** $Co_2C_8O_8$ **12.107** a $C_6H_{12}O_4$ b 148.2 g/mol **12.109** $-1.2°C$, assume the molality and molarity of the solution are approximately equal and that the electrolyte (salt) is behaving ideally **12.125** 6.10 g **12.131** $3Ca^{2+}(aq) + 2PO_4^{3-}(aq) \longrightarrow Ca_3(PO_4)_2(s)$
$3Ag^+(aq) + PO_4^{3-}(aq) \longrightarrow Ag_3PO_4(s)$
0.0521 M PO_4^{3},
0.0875 M NO_3^-,
0.244 M Na^+
12.133 -783 kJ/mol, -445 kJ/mol **12.135** 0.565 m **12.137** 0.701 mol/L **12.139** 3.1% **12.141** $C_2H_6O_2$

CHAPTER 13

13.29 a Rate of depletion of A $= -\dfrac{\Delta[A]}{\Delta t}$ or Rate of formation of B $= +\dfrac{\Delta[B]}{\Delta t}$.
b No, the rate of depletion of A would be faster than the rate of formation of B. c $-\dfrac{\Delta[A]}{3\Delta t} = \dfrac{\Delta[B]}{2\Delta t}$
13.31 a Rate=$k[A]^2$ b right c right d Left container rate is four times slower. e right **13.33** a $x = 1$ b $x = 0$ c $x = 3$ **13.35** a Region c b Region A **13.37** A number of answers will work as long as you match one of the existing concentrations of A. For example: [A] = 2.0 M with [B] = 2.0 M, or [A] = 1.0 M with [B] = 2.0 M.

13.39 $-\Delta[H_2]/\Delta t = \frac{1}{2}\Delta[HI]/\Delta t$ **13.41** $\dfrac{1}{5}\dfrac{\Delta[Br^-]}{\Delta t} = \dfrac{\Delta[BrO_3^-]}{\Delta t}$

13.43 2.3×10^{-2} M/hr **13.45** 1.2×10^{-5} M/hr **13.47** 2, 1, The overall order is 2 + 1 = 3, third-order **13.49** MnO_4^- is 1, $H_2C_2O_4$ is 1, H^+ is 0; overall order is 0 **13.51** Rate = $k[CH_3NNCH_3]$, $k = 2.5 \times 10^{-4}$/s **13.53** Rate = $k[NO]^2[O_2]$, $7.11 \times 10^3/M^2s$ **13.55** Rate = $k[ClO_2]^2[OH^-]$, $k = 2.3 \times 10^2$ (M^2/s) **13.57** 0.016 M **13.59** 1.04×10^3 s **13.61** 5.2×10^{-4} M **13.63** 2.2×10^3 min, 4.5×10^3 min **13.65** 96 hr, 1.9×10^2 hr, 2.9×10^2 hr, 3.9×10^2 hr, 4.8×10^2 hr **13.67** 0.344 L/(mol·s) **13.69** 2.6×10^2 hr **13.71** 4.8×10^{-4} M/s **13.73** first-order, $k = 0.100$/s **13.75** $+84$ kJ **13.77** 1.1×10^5 J/mol, 1.7×10^{-3}/s **13.79** 8.4×10^4 J/mol **13.81** 1.1×10^2 kJ/mol **13.83** O atom is a reaction intermediate, $2O_3 \longrightarrow 3O_2$ **13.85** a Unimolecular b Termolecular c Bimolecular d Bimolecular **13.87** a Rate = $k[NOCl_2]$ [NO] b Rate = $k[O_3]$ **13.89** Rate = $k[C_3H_6]^2$ **13.91** Rate = $k_2(k_1/k_{-1})[I_2][H_2] = k[I_2][H_2]$ **13.93** $NH_2NO_2 \longrightarrow N_2O + H_2O$, NaOH **13.95** 3.5×10^{-6} M/s; 3.2×10^{-6} M/s; 2.5×10^{-6} M/s **13.97** average $k = 2.5 \times 10^{-4}$/s **13.99** 8.33×10^3 s

13.101 5.50×10^3 s   **13.103** 0.74   **13.105** a 34 L/(mol·s)
b 8.2×10^{-5} M **13.107** 4.7×10^{-3} M **13.109** 2.5×10^{-4}/s
13.111 6.8×10^{-4} mol/(L·s), 3.7×10^2 s   **13.113** 114 kJ/mol;
5×10^9; 14 M/s   **13.115** Rate $= k[CH_3Cl][OH^-]$
13.117 Rate $= k[NO_2Br]$ **13.119** Rate $= k_2(k_1/k_{-1})[A]$
$[H_3O^+] = k[A][H_3O^+]$ **13.121** a Rate $= k[NO]^2[O_2]$
b 0.12 mol/L·s   **13.123** a i) decreases   ii) decreases
b i) increases   ii) decreases
13.125

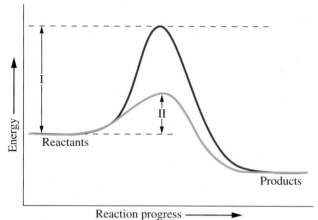

80 kJ, forward reaction will be more sensitive to a temperature change because it has the larger activation energy
13.127 a The rate is the change in the concentration of a reactant or product with time. b The rate changes because the concentration of the reactant has changed. c The rate will equal k when the reactants all have 1.00 M.   **13.147** 9.2 L
13.149 -0.210 kJ/s   **13.151** 3.06×10^{-10} mmHg/s

CHAPTER 14

14.19 $\frac{2x}{3}$ **14.21** a   **14.23** a III   b Both I and II shift right.
14.25 a shifts left   b Concentrations of A and B increase; concentration of C decreases.   **14.27** container I
14.29 2.162 mol PCl_5, 0.338 mol PCl_3, and 0.338 mol Cl_2
14.31 0.0213 mol NO, 0.0107 mol Br_2, and 0.0311 mol NOBr.

14.35 a $K_c = \dfrac{[H_2]^2[S_2]}{[H_2S]^2}$

b $K_c = \dfrac{[NO_2]^2}{[NO]^2[O_2]}$ c $K_c = \dfrac{[P(NH_2)_3][HCl]^3}{[PCl_3][NH_3]^3}$

d $K_c = \dfrac{[NO_2][NO]}{[N_2O_3]}$

14.37 $2H_2S(g) + 3O_2(g) \rightleftharpoons 2H_2O(g) + 2SO_2(g)$

14.39 $K_c = \dfrac{[NO^2][H_2O]^3}{[NH_3]^2[O_2]^{5/2}}$

14.41 $K_c = 0.543$ **14.43** 39.1   **14.45** $3.1 \times .10^3$

14.47 $1.0 \times .10^{-3}$ **14.51** 3.0×10^{-5} **14.53** 56.3

14.55 a $K_c = [Pb^{2+}][I^-]^2$ b $K_c = \dfrac{[CO]^2}{[CO_2]}$

c $K_c = \dfrac{[CO_2]}{[CO]}$ d $K_c = \dfrac{[CO_2]}{[SO_2][O_2]^{1/2}}$

14.57 a nearly complete   b not complete   **14.59** No, decomposition does not occur; 3.2×10^{-48} mol/L; yes it agrees with what is expected.   **14.61** a goes to the left   b goes to the left   c goes to the right   d equilibrium   **14.63** goes to the left   **14.65** 1.3×10^{-3} M **14.67** $[I_2] = [Br_2] = 5 \times 10^{-5}$ M, and $[IBr] = 5.1 \times 10^{-4}$ M **14.69** 0.14 mol PCl_3, 0.14 mol Cl_2, and 0.26 mol PCl_5 **14.71** $[CO] = 0.0613$ M, $[H_2] = 0.1839$ M, $[CH_4] = 0.0387$ M, $[H_2O] = 0.0387$ M **14.73** forward

14.75 a no effect   b no effect   c goes to the left   **14.77** high temperatures because an increase in temperature increases the amounts of products of an endothermic reaction.   **14.79** decrease
14.81 low temperature and low pressure   **14.83** 8.6   **14.85** 0.125
14.87 reverse   **14.89** the reverse direction (left)   **14.91** [HBr] $=$ 0.008 mol; $[H_2] = 0.0010$ mol; and $[Br_2] = 0.0010$ mol
14.93 57.6%   **14.95** CO: 0.48 mol; H_2: 2.44 mol; CH_4: 0.52 mol; and H_2O: 0.52 mol   **14.97** exothermic

14.99 $K_p = \dfrac{[NH_3]^2(RT)^2}{[N_2](RT)[H_2]^3(RT)^3} = \dfrac{[NH_3]^2}{[N_2][H_2]^3}(RT)^{-2}$

$K_p = K_c(RT)^{-2}$, or $K_c = Kp\,(RT)^2$

14.101 a 22.9   b 1.31 atm, 5.49 atm   c no effect
14.103 a $[PCl_3] = [Cl_2] = 0.031$ M, $[PCl_5] = 0.004$ M
b 0.89   c Less PCl_5 would decompose   **14.105** 0.335
14.107 a 0.062 M b 0.21   c shift the equilibrium to the left and decrease the fraction of $COCl_2$ that has decomposed
14.109 a 7.2×10^{-5} M, 1.6×10^{-4} M b hydrogen bonding occurs   c decrease   **14.111** 1.1×10^2 **14.135** 0.430
14.137 0.29 atm

CHAPTER 15

15.21 Hydroxide ion forms in the reaction $NH_3(aq) + H_2O(l) \rightleftharpoons NH_4^+(aq) + OH^-(aq)$ **15.23** $HPO_4^{2-}(aq) + OH^-(aq) \longrightarrow PO_4^{3-}(aq) + H_2O(l)$ (acid), $HPO_4^{2-}(aq) + H_3O^+(aq) \longrightarrow H_2PO_4^-(aq) + H_2O(l)$ (base)
15.25 HNO_2 **15.27** b
15.29 $HOBr(aq) + H_2O(l) \rightleftharpoons H_3O^+(aq) + OBr^-(aq)$
　　　　　acid　　　　base　　　　　acid　　　　base
15.31 a NO^{2-} b PO_4^{3-} c $HAsO_4^{2-}$ d HS^-
15.33 a $HClO$ b H_3PO_4 c $HTeO^{3-}$ d AsH^{4+}
15.35
a $HPO_4^{2-}(aq) + NH_4^+(aq) \rightleftharpoons H_2PO_4^-(aq) + NH_3(aq)$
　base　　　　　acid　　　　　　base　　　　　acid
b $Al(H_2O)_6^{3+}(aq) + H_2O(l) \rightleftharpoons$
　acid　　　　　　　　base

$Al(H_2O)_5(OH)^{2+}(aq) + H_3O^+(aq)$
　　　acid　　　　　　　　base

c $SO_3^{2-}(aq) + NH_4^+(aq) \rightleftharpoons HSO_3^-(aq) + NH_3(aq)$
　base　　　　acid　　　　　　base　　　　acid
d $HSO_4^-(aq) + NH_3(aq) \rightleftharpoons SO_4^{2-}(aq) + NH_4^+(aq)$
　acid　　　　base　　　　　acid　　　　base
15.37 $AlCl_3 + NH_3 \longrightarrow AlCl_3NH_3$; $AlCl_3$ Lewis acid, NH_3 Lewis base
15.39 a $I^- + I_2 \rightleftharpoons I_3^-$

$$\left[\ddot{\ddot{I}}\!:\right]^- + :\ddot{I}\!:\ddot{I}: \longrightarrow \left[\ddot{I} - \dot{I} - \ddot{I}:\right]^-$$
　　base　　　　acid

b $AlCl_3 + Cl^- \rightleftharpoons [AlCl_4]^-$

$$:\ddot{Cl}-Al\begin{smallmatrix}\ddot{Cl}:\\ \\ \ddot{Cl}:\end{smallmatrix} + [:\ddot{Cl}:]^- \longrightarrow \left[:\ddot{Cl}-\overset{:\ddot{Cl}:}{\underset{:\ddot{Cl}:}{Al}}-\ddot{Cl}:\right]^-$$
　　acid　　　　　　base

15.41 a Cu^+: Lewis acid, H_2O: Lewis base
b BBr_3: Lewis acid, AsH_3: Lewis base
15.43 $H_2S + HOCH_2CH_2NH_2 \longrightarrow HOCH_2CH_2NH_3^+ + HS^-$
　Lewis　　　　Lewis
　acid　　　　　base
The H^+ ion from H_2S accepts a pair of electrons from the N atom in $HOCH_2CH_2NH_2$.   **15.45** $HCN + SO_4^{2-} \longrightarrow$ $CN^- + HSO_4^-$, reaction occurs in the opposite direction

15.47 a left b right c left d right   **15.49** richloroacetic acid; the equilibrium favors the formation of the weaker acid
15.51 a H_2SO_3; acid strength increases with increasing electronegativity   b H_2S; acid strength increases with the increasing size   c HIO_4; acid strength increases with the number of oxygen   d HBr; acid strength increases with increasing electronegativity   e H_2S; acid strength decreases with increasing anion charge
15.53 a 6.45×10^{-15} M, 1.55 M   b 3.3×10^{-14} M, 0.30 M   c 0.056 M, 1.8×10^{-13} M   d 0.47 M, 2.1×10^{-14} M
15.55 $[H_3O^+]$ = 0.50 M; $[OH^-]$ = 2.0×10^{-13} M
15.57 6.4×10^{-4} M, 1.6×10^{-11} M   **15.59** a acidic
b basic c acidic d neutral   **15.61** basic
15.63 a basic b acidic c neutral d acidic
15.65 a Basic b Neutral c Acidic d Acidic
15.67 a 4.00 b 9.49 c 4.64 d 10.536   **15.69** 2.12
15.71 a 5.72 b 11.92 c 2.56 d 6.32
15.73 12.70   **15.75** 7.6×10^{-6} M   **15.77** 4.3×10^{-3} M
15.79 13.16   **15.81** 5.5 to 6.5, acidic   **15.83** a P_4O_{10} is an acid; $P_4O_{10} + 10H_2O \longrightarrow 4H_3O^+ + 4H_2PO_4^-$ b Na_2O is a base; $Na_2O + H_2O \longrightarrow 2Na^+ + 2OH^-$ c N_2H_4 is a base; $N_2H_4 + H_2O \longrightarrow N_2H_5^+ + OH^-$ d H_2Te is an acid; $H_2Te + H_2O \longrightarrow H_3O^+ + HTe^-$ **15.85** a $H_2O_2(aq) + S^{2-}(aq) \longrightarrow HO_2^-(aq) + HS^-(aq)$ b $HCO_3^-(aq) + OH^-(aq) \longrightarrow CO_3^{2-}(aq) + H_2O(l)$ c $NH_4^+(aq) + CN^-(aq) \longrightarrow NH_3(aq) + HCN(aq)$ d $H_2PO_4^-(aq) + OH^-(aq) \longrightarrow HPO_4^{2-}(aq) + H_2O(l)$

15.87 a $ClO^- + H_2O \rightleftharpoons HClO + OH^-$
      base   acid   acid   base

$$\left[:\!\ddot{C}l\!-\!\ddot{O}\!:\right]^- + H\!-\!\ddot{O}\!-\!H \longrightarrow :\!\ddot{C}l\!-\!\ddot{O}\!-\!H + \left[:\!\ddot{O}\!-\!H\right]^-$$

b $NH_4^+ + NH_2^- \rightleftharpoons 2NH_3$
   acid   base

$$\left[\begin{array}{c} H \\ | \\ H\!-\!N\!-\!H \\ | \\ H \end{array}\right]^+ + \left[\begin{array}{c} H \\ | \\ :\!N\!: \\ | \\ H \end{array}\right]^- \longrightarrow 2\begin{array}{c} H \\ | \\ :\!N\!-\!H \\ | \\ H \end{array}$$

15.89 $H_2S + CN^- \longrightarrow HS^- + HCN$ **15.91** $H_2S <$
$H_2Se < HBr$ **15.93** $[H_3O^+] = 34.0 \times 10^{-14}$ M
15.95 2.82   **15.97** 1.4×10^{-10} M
15.99 a $H_2SO_4(aq) + 2NaHCO_3(aq) \longrightarrow$
$Na_2SO_4(aq) + 2CO_2(g) + 2H_2O(l)$
$H_3O^+(aq) + HCO_3^-(aq) \longrightarrow CO_2(g) + 2H_2O(l)$ b 0.180
c 60.66%, $NaHCO_3$ **15.101** amphiprotic or amphoteric
$H_2PO_4^-(aq) + H_2O(l) \longrightarrow H_3O^+(aq) + HPO_4^{2-}(aq)$;
$H_2PO_4^-(aq) + H_2O(l) \longrightarrow H_3PO_4(aq) + OH^-(aq)$;
$H_2PO_4^-(aq) + K^+(aq) + OH^-(aq) \longrightarrow K^+(aq) +$
$HPO_4^{2-}(aq) + H_2O(l)$; $H_2PO_4^-(aq) + H^+(aq) +$
$NO_3^-(aq) \longrightarrow H_3PO_4(aq) + NO_3^-(aq)$
15.103 $CaH_2(s) + 2H_2O(l) \longrightarrow Ca(OH)_2(s) + 2H_2(g)$;
The hydride ion is a stronger base   **15.129** the structure of H_3PO_4 is $(HO)_3PO$; 0.976 g NaOH
15.131 $BF_3 + :NH_3 \longrightarrow F_3B:NH_3$; 12.5 g $F_3B:NH_3$
   Lewis   Lewis
   acid   base

CHAPTER 16

16.23 A   **16.25** a $HF(aq) + H_2O(l) \rightleftharpoons H_3O^+(aq) + F^-(aq)$
b $F^-(aq) + H_2O(l) \rightleftharpoons HF(aq) + OH^-(aq)$
c $C_6H_5NH_2(aq) + H_2O(l) \rightleftharpoons C_6H_5NH_3^+(aq) + OH^-(aq)$
d $C_6H_5NH_3^+(aq) + H_2O(l) \rightleftharpoons C_6H_5NH_2(aq) + H_3O^+(aq)$
16.27 Acidic: $RanH^+(aq) + H_2O(l) \rightleftharpoons Ran(aq) + H_3O^+(aq)$
16.29 high pH at the start and a gradual decrease in pH as the titration proceeds   **16.31** a weak acid b 8.5 c thymol blue or phenolphthalein (an indicator that changes color in the

pH range of about 7 to 10)   **16.35** 1.9×10^{-6} **16.37** 5.53;
2.0×10^{-4} **16.39** $[H_3O^+] = [C_6H_4NH_2COO^-] = 1.1 \times 10^{-3}$ M
16.41 0.048 M   **16.43** 1.3×10^{-3} M, 1.44   **16.45** 0.361 M
16.47 a 3.7×10^{-3} M b 3.9×10^{-6} M
16.49 $C_6H_5NH_2(aq) + H_2O(l) \longrightarrow C_6H_5NH_3^+(aq) + OH^-(aq)$

$$K_b = \frac{[C_6H_5NH_3^+][O^-]}{[C_6H_5NH_2]}$$

16.51 3.2×10^{-5} **16.53** 4.8×10^{-5} M; 9.68
16.55 a $NH_2NH_3^+ + H_2O \rightleftharpoons H_3O^+ + NH_2NH_2$

$$K_a = \frac{K_w}{K_b} = \frac{[H_3O^+][NH_2NH_2]}{[NH_2NH_3^+]}$$ b no hydrolysis

c no hydrolysis d $OCl^- + H_2O \rightleftharpoons HOCl + OH^-$

$$K_b = \frac{K_w}{K_a} = \frac{[HOCl][OH^-]}{[OCl^-]}$$

16.57 $Cu(H_2O)_6^{2+}(aq) + H_2O(l) \longrightarrow Cu(H_2O)_5(OH)^+(aq) + H_3O^+(aq)$ **16.59** a acidic b basic c basic d acidic
16.61 a acidic b nearly acidic   **16.63** a 2.2×10^{-11}
b 7.1×10^{-6} **16.65** 8.64, 4.4×10^{-6} M   **16.67** 5.58, 2.6×10^{-6} M   **16.69** a 0.028 b 0.0011   **16.71** 3.17
16.73 10.85   **16.75** 9.49   **16.77** 9.26, 9.09   **16.79** 2.71
16.81 10.5   **16.83** 0.34 mol   **16.85** 12.22   **16.87** 8.59
16.89 5.97   **16.91** a 11.15 b 9.12 c 5.24
d 1.90   **16.93** 9.08   **16.95** 3.6×10^{-3} **16.97** 1.1×10^{-2}
16.99 $K_b(CN^-) = 2.0 \times 10^{-5}$; $K_b(CO_3^{2-}) = 2.1 \times 10^{-4}$;
CO_3^{2-} **16.101** 2.84   **16.103** 7.20   **16.105** 11/1   **16.107** 2.4
16.109 7.47   **16.111** a 0.100 M b 0.11 M
16.113 a 2.3×10^{-11} b 4.4×10^{-4} c 10.39
16.115 a false b true c true d false e true
f false   **16.117** a 45.1 g/mol b 4.7×10^{-4}
16.119 a Initial pH = 11.13; 30% titration pH = 9.62; 50% titration pH = 9.62; 100% titration pH = 5.28
b acidic   **16.121** a $H_2PO_4^-$ and HPO_4^{2-} have a pKa value closest to a pH of 6.96 b add 36 mL and 64 mL
16.123 a 0.215 M b 3.55 c bromphenol blue
16.125 a $H_3O^+(aq) + NH_3(aq) \longrightarrow NH_4^+(aq) + H_2O(l)$
b $NH_4^+(aq) + OH^-(aq) \longrightarrow NH_3(aq) + H_2O(l)$ c moles NH_3 before/after: 0.10/0.090; moles NH_4^+ before/after: 0.10/0.11; pH 9.2 d This is a buffer system.   **16.127** a 0.59
b 7.16 c 7.55   **16.129** a $H_2A + 2H_2O \rightleftharpoons 2H_3O^+ + A^{2-}$
b $[H_2A] \gg [H_3O^+] = [HA^-] \gg [A^{2-}]$ c pH = 2.34;
$[H_2A] = 0.0205$ M d 4.6×10^{-5} M   **16.131** a 0.095 M
b 410 mL c 0.42 M   **16.151** 0.95% NH_3 **16.153** $-1.7°C$
16.155 9.1 mL

CHAPTER 17

17.17 From the K_{sp} values, these are the more soluble:
a magnesium hydroxide b silver chloride   **17.19** a
17.21 middle beaker   **17.23** no; no   **17.25** a soluble
b insoluble c insoluble d soluble   **17.27** a $K_{sp} = [Ca^{2+}]^3[AsO_4^{3-}]^2$ b $K_{sp} = [Fe^{3+}][OH^-]^3$ c $K_{sp} = [Mg^{2+}][OH^-]^2$ d $K_{sp} = [Sr^{2+}][CO_3^{2-}]$ **17.29** 8.6×10^{-5}
17.31 1.2×10^{-7} **17.33** 1.8×10^{-11} **17.35** 0.27 g/L
17.37 3.1×10^{-3} M   **17.39** 2.0×10^{-4} g/L   **17.41** 1.1×10^{-5} g/L   **17.43** 0.4 g/L   **17.45** a yes b yes   **17.47** Lead chromate will precipitate.   **17.49** No precipitation   **17.51** no precipitation   **17.53** 0.0018 mol   **17.55** 6.9×10^{-9} M
17.57 $MgCO_3(s) + 2H_3O^+(aq) \rightleftharpoons Mg^{2+}(aq) + CO_2(g) + 3H_2O(l)$ **17.59** BaF_2; F^- is the conjugate base of the weak acid HF.   **17.61** $Fe^{3+}(aq) + 6CN^-(aq) \rightleftharpoons Fe(CN)_6^{3-}(aq)$
$K_f = [Fe(CN)_6^{3-}]/[Fe^{3+}][CN^-]^6 = 9.1 \times 10^{41}$
17.63 5.5×10^{-19} M   **17.65** At equilibrium, a precipitate will form, and the solution will be saturated   **17.67** 3.0×10^{-3} M
17.69 Add HCl to precipitate only the Pb^{2+} as $PbCl_2$, leaving the others in solution. After decanting or filtering, add 0.3 M HCl and H_2S to precipitate only the CdS away from the Sr^{2+} ion.

17.71 Mn^{2+} **17.73** 1.3×10^{-26} M   **17.75** [a] $6.9 \times 10^{-7}\ M$
[b] 3.2×10^{-4} g/L   **17.77** [a] $5.2 \times 10^{-6}\ M$ [b] pOH = 4.81
17.79 26 g/L   **17.81** 1.4×10^{-24} M   **17.83** $8.0 \times 10^{-3}\ M$
17.85 1.0×10^{-5}; unsaturated   **17.87** 2.2×10^{-6} g Na_2SO_4
17.89 0.14 M **17.91** 8.2×10^{-5} M (the molar solubility of
AgI in 2.2 M NH_3)   **17.95** $5.3 \times 10^{-5}\ M$ **17.97** [a] $6.4 \times$
$10^{-7}\ M$ [b] $2.4 \times 10^{-4}\ M$ [c] yes   **17.99** [a] $6.6 \times$
$10^{-4}\ M$ [b] $1.6 \times 10^{-3}\ M$, 11%   **17.101** [a] 2.5×10^{-18}
[b] $2.1 \times 10^{-10}\ M$ [c] The common-ion effect (a Le Châtelier
stress) decreases the solubility of Be^{2+} **17.103** [a] 3.1×10^{-3}
[b] 0.044 mol; 0.044 M; 0.89 mol   **17.105** [a] $1.5 \times 10^{-4}\ M$
[b] 99%   **17.129** 0.3 M **17.131** $2.7 \times 10^{-4}\ M$
17.133 $Pb^{2+}(aq) + 2Cl^-(aq) + 2Ag^+(aq) + SO_4^{2-}(aq) \rightleftharpoons$
$PbSO_4(s) + 2AgCl(s)$; $Pb^{2+} = 2.50 \times 10^{-6}\ M$; $SO_4^{2-} =$
0.0067 M; $Cl^- = 1.4 \times 10^{-8}\ M$; $Ag^+ = 0.0133$ M

CHAPTER 18

18.23 [a] false [b] true [c] false [d] false [e] false
18.25 [a] 2.0 mol CO_2 [b] butane gas [c] CO_2 at $-80°C$
[d] bromine vapor   **18.27** [a] negative [b] negative
[c] negative [d] zero   **18.29** [a] K_c does not change.
[b] $Q > K_c$; the reaction will go to the left [c] ΔG is positive;
ΔS is negative   **18.31** 106 J/K   **18.33** 152 J/K
18.35 [a] positive [b] not predictable [c] positive
[d] negative   **18.37** [a] 313.6 J/K [b] -56.4 J/K
[c] -181.7 J/K [d] -57.9 J/K   **18.39** -266.90 J/K; entropy
decreases, as expected from the decrease in moles of gas.
18.41 -1452.8 kJ; -161.4 J/K; -1404.7 kJ
18.43 [a] $Mg(s) + \frac{1}{2}O_2(g) \longrightarrow MgO(s)$ [b] C(graphite) +
$\frac{1}{2}O_2(g) + Cl_2(g) \longrightarrow COCl_2(g)$ [c] C(graphite) +
$2F_2(g) \longrightarrow CF_4(g)$ [d] P(white) + $\frac{5}{2}Cl_2(g) \longrightarrow PCl_5(g)$
18.45 [a] -56.2 kJ [b] -1331.4 kJ   **18.47** [a] Spontaneous
reaction [b] Spontaneous reaction [c] Nonspontaneous
reaction [d] Equilibrium mixture; significant amounts of
both [e] Nonspontaneous reaction   **18.49** [a] 850.2 kJ, 838.8
kJ; endothermic reaction with mainly reactants at equilibrium
[b] -72.2 kJ, -142.0 kJ; exothermic reaction with mainly
products at equilibrium   **18.51** -474.2 kJ; 0   **18.53** -15.8 kJ
18.55 [a] $K = [Mg^{2+}][OH^-]_2$ [b] $K = [Li^+]_2[OH^-]_2 P_{H_2}$
[c] $K = K_p = \dfrac{P_{CO_2}P_{H_2}}{P_{CO}P_{H_2O}}$
18.57 -107.00 kJ; 6×10^{18} **18.59** 2×10^4
18.61 -144.39 kJ; 2×10^{25} **18.63** 1.2×10^2; yes
18.65 734.7 K   **18.67** Reaction is spontaneous because
$\Delta S°$ is large and positive.   **18.69** -184 kJ; exothermic ;
positive; both the ΔH term and the $-T\Delta S$ term are negative
18.71 -14.3 J/(K·mol)   **18.73** [a] positive [b] negative
[c] positive [d] negative   **18.75** negative   **18.77** -136.0 J/K
18.79 nonspontaneous   **18.81** $\Delta H°$ is positive, $\Delta S°$
is positive   **18.83** 57.0 kJ; 1×10^{-10} **18.85** 109.6 kJ;
136.6 J/K; 68.9 kJ; -37.0 kJ; nonspontaneous reaction at 25°C
but a spontaneous reaction at 800°C   **18.87** [a] -84 kJ
[b] -173.78 J/K [c] -31.7 kJ/mol   **18.89** 205.4 J/mol·K
18.91 [a] Since $\Delta G°$ is positive, K will be less than 1.
[b] 125 J/K [c] 917 K [d] change in entropy
18.93 [a] formic acid [b] products   **18.95** [a] products
[b] need to heat to increase rate of reaction
18.97 [a] -454 kJ/mol [b] -1177.6 kJ [c] -43.0 J/K
18.99 [a] 0.017 [b] The change in entropy is negative.
18.123 0.65   **18.125** 87%   **18.127** 31 J/K; 65 kJ

CHAPTER 19

19.25 [a] no change [b] no change [c] The Zn strip would
dissolve, and the blue color of the solution would fade, and solid
copper would precipitate out of the solution. [d] no change

19.27 The Zn is a sacrificial electrode that keeps the hull from
undergoing oxidation by the dissolved ions in seawater. Zn
works because it is more easily oxidized than Fe.
19.29 Since there is more zinc present, the oxidation-
reduction reactions in the battery will run for a longer period of
time. This assumes zinc is the limiting reactant.
19.31 Pick elements or compounds to be reduced and one to be
oxidized so that when the half-reactions are added together, the
$E°_{cell}$ is about 0.90 V. For example:

$Cd(s) \longrightarrow Cd^{2+}(aq) + 2e^-$	0.40 V
$I_2(s) + 2e^- \longrightarrow 2I^-(aq)$	0.54 V
$I_2(s) + Cd(s) \longrightarrow 2I^-(aq) + Cd^{2+}(aq)$	0.94 V

19.33 [a] brighter [b] dimmer [c] dimmer [d] off
19.35 [a] $Cr_2O_7^{2-} + 3C_2O_4^{2-} + 14H^+ \longrightarrow$
$2Cr^{3+} + 6CO_2 + 7H_2O$
[b] $3Cu + 2NO_3^- + 8H^+ \longrightarrow 3Cu^{2+} + 2NO + 4H_2O$
[c] $MnO_2 + HNO_2 + H^+ \longrightarrow Mn^{2+} + NO_3^- + H_2O$
[d] $5PbO_2 + 2Mn^{2+} + 5SO_4^{2-} + 4H^+ \longrightarrow$
$5PbSO_4 + 2MnO_4^- + 2H_2O$
[e] $3HNO_2 + Cr_2O_7^{2-} + 5H^+ \longrightarrow 2Cr^{3+} + 3NO_3^- + 4H_2O$
19.37 [a] $Mn^{2+} + H_2O_2 + 2OH^- \longrightarrow MnO_2 + 2H_2O$
[b] $2MnO_4^- + 3NO_2^- + H_2O \longrightarrow 2MnO_2 + 3NO_3^- + 2OH^-$
[c] $Mn^{2+} + 2ClO_3^- \longrightarrow MnO_2 + 2ClO_2$
[d] $MnO_4^- + 3NO_2 + 2OH^- \longrightarrow MnO_2 + 3NO_3^- + H_2O$
[e] $3Cl_2 + 6OH^- \longrightarrow 5Cl^- + ClO_3^- + 3H_2O$
19.39 [a] $8H_2S + 16NO_3^- + 16H^+ \longrightarrow S_8 + 16NO_2 + 16H_2O$
[b] $2NO_3^- + 3Cu + 8H^+ \longrightarrow 2NO + 3Cu^{2+} + 4H_2O$
[c] $2MnO_4^- + 5SO_2 + 2H_2O \longrightarrow 5SO_4^{2-} + 2Mn^{2+} + 4H^+$
[d] $2Bi(OH)_3 + 3Sn(OH)_3^- + 3OH^- \longrightarrow 3Sn(OH)_6^{2-} + 2Bi$
19.41 [a] $MnO_4^- + I^- + H_2O \longrightarrow 2MnO_2 + IO_3^- + 2OH^-$
[b] $Cr_2O_7^{2-} + 6Cl^- + 14H^+ \longrightarrow 2Cr^{3+} + 3Cl_2 + 7H_2O$
[c] $3S_8 + 32NO_3^- + 32H^+ \longrightarrow 32NO + 24SO_2 + 16H_2O$
[d] $3H_2O_2 + 2MnO_4^- \longrightarrow 3O_2 + 2MnO_2 + 2H_2O + 2OH^-$
[e] $5Zn + 2NO_3^- + 12H^+ \longrightarrow N_2 + 5Zn^{2+} + 6H_2O$
19.43

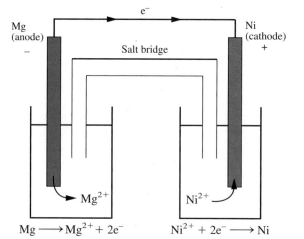

$Mg \longrightarrow Mg^{2+} + 2e^-$ $Ni^{2+} + 2e^- \longrightarrow Ni$

19.45 $Zn \longrightarrow Zn^{2+} + 2e$
$Ag^+ + e^- \longrightarrow Ag$
Electron flow in circuit is from Zn to Ag.
19.47 Anode: $Zn(s) + 2OH^-(aq) \longrightarrow Zn(OH)_2(s) + 2e^-$
Cathode: $Ag_2O(s) + H_2O(l) + 2e^- \longrightarrow 2Ag(s) + 2OH^-(aq)$
Overall: $Zn(s) + Ag_2O(s) + H_2O(l) \longrightarrow Zn(OH)_2(s) + 2Ag(s)$
19.49 $Ni(s)|Ni^{2+}(aq)||Pb^{2+}(aq)|Pb(s)$
19.51 $Ni(s)|Ni^{2+}(1\ M)||H^+(1\ M)|H_2(g)|Pt$
19.53 $Fe(s) + 2Ag^+(aq) \longrightarrow Fe^{2+}(aq) + 2Ag(s)$
19.55 $Cd(s) \longrightarrow Cd^{2+}(aq) + 2e^-$
$\underline{Ni^{2+}(aq) + 2e^- \longrightarrow Ni(s)}$
$Cd(s) + Ni^{2+}(aq) \longrightarrow Cd^{2+}(aq) + Ni(s)$

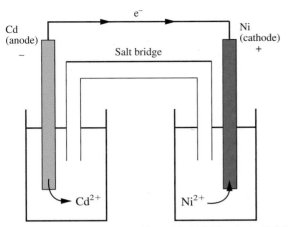

Cd (anode) −   e⁻ →   Ni (cathode) +

Salt bridge

Cd²⁺   Ni²⁺

19.57 −69 kJ **19.59** −96 kJ **19.61** $NO_3^-(aq)$, $O_2(g)$, $MnO_4^-(aq)$ **19.63** $Zn(s)$ is the strongest reducing agent, and $Cu^+(aq)$ is the weakest. **19.65** a not spontaneous b spontaneous **19.67** Cl_2 will oxidize Br^-; $Cl_2(g) + 2Br^-(aq) \longrightarrow 2Cl^-(aq) + Br_2(l)$ **19.69** 1.54 V **19.71** 1.28 V **19.73** -3.6×10^2 kJ **19.75** -1.6×10^2 kJ **19.77** 3.16 V **19.79** 1.934 V **19.81** 10^{14} **19.83** 1×10^6 **19.85** 0.57 V **19.87** 0.29 V **19.89** 0.004 M **19.91** a The cathode reaction is $Ca^{2+}(l) + 2e^- \longrightarrow Ca(l)$ The anode reaction is $S^{2-}(l) \longrightarrow S(l) + 2e^-$ b The cathode reaction is $Cs^+(l) + e^- \longrightarrow Cs(l)$ The anode reaction is $4OH^-(l) \longrightarrow O_2(g) + 2H_2O(g) + 4e^-$ **19.93** a anode: $2H_2O(l) \longrightarrow O_2(g) + 4H^+(aq) + 4e^-$ cathode: $2H_2O(l) + 2e^- \longrightarrow H_2(g) + 2OH^-(aq)$ overall reaction: $2H_2O(l) \longrightarrow 2H_2(g) + O_2(g)$ b anode: $2Br^-(aq) \longrightarrow Br_2(l) + 2e^-$ cathode: $2H_2O(l) + 2e^- \longrightarrow H_2(g) + 2OH^-(aq)$ overall reaction: $2Br^-(aq) + 2H_2O(l) \longrightarrow Br_2(l) + H_2(g) + 2OH^-(aq)$ **19.95** 3.87×10^7 C **19.97** 0.360 g Li **19.99** a $3Fe(s) + 2NO_3^-(aq) + 8H^+(aq) \longrightarrow 3Fe^{2+}(aq) + 2NO(g) + 4H_2O(l)$ b $3Fe^{2+}(aq) + NO_3^-(aq) + 4H^+(aq) \longrightarrow 3Fe^{3+}(aq) + NO(g) + 2H_2O(l)$ c $Fe(s) + NO_3^-(aq) + 4H^+(aq) \longrightarrow Fe^{3+}(aq) + NO(g) + 2H_2O(l)$ **19.101** a $16MnO_4^-(aq) + 24S_2^-(aq) + 32H_2O(l) \longrightarrow 16MnO_2(s) + 3S_8(s) + 64OH^-(aq)$ b $IO_3^-(aq) + 3HSO_3^-(aq) \longrightarrow I^- + 3SO_4^{2-}(aq) + 3H^+(aq)$ c $CrO_4^{2-}(aq) + 3Fe(OH)_2(aq) + 4H_2O(l) \longrightarrow Cr(OH)_4^-(aq) + 3Fe(OH)_3(s) + OH^-(aq)$ d $Cl_2(aq) + 2OH^-(aq) \longrightarrow Cl^-(aq) + ClO^-(aq) + H_2O(l)$ **19.103** $4Fe(OH)_2(s) + O_2(g) + 2H_2O(l) \longrightarrow 4Fe(OH)_3(s)$ **19.105** $Ca(s)|Ca^{2+}(aq)\|Cl_2(g)|Cl^-(aq)|Pt(s)$; Anode: $Ca(s) \longrightarrow Ca^{2+}(aq) + 2e^-$; Cathode: $Cl_2(g) + 2e^- \longrightarrow 2Cl^-(aq)$; 4.12 V **19.107** a spontaneous b Fe^{3+} will oxidize Sn^{2+} to Sn^{4+}. **19.109** 0.36 V **19.111** a 2.2 b 0.3 M **19.113** a 1.0 F, 9.6×10^4 C b 2.0 F, 1.9×10^5 C c 10.11 F, 1.1×10^4 C d 0.028 F, 2.7×10^3 C **19.115** 2.67×10^{-4} g As **19.117** a $Cd(s) + 2Ag^+(aq) \longrightarrow 2Ag(s) + Cd^{2+}(aq)$; 1.20 V **19.119** Anode: $2H_2O(l) \longrightarrow O_2(g) + 4H^+(aq) + 4e^-$ Cathode: $Cu^{2+}(aq) + 2e^- \longrightarrow Cu(s)$ **19.121** a 1.48×10^3 s b 0.398 g **19.123** a 0.323 F b 3+ **19.125** a 1.06 V b 8.8×10^{-3} M **19.127** a -23.2 kJ b -0.28 V; $Cd(s) + Co^{2+}(aq) \longrightarrow Cd^{2+}(aq) + Co(s)$ **19.145** $E° = 0.02$ V; $\Delta G° = -10$ kJ; $\Delta H° = -174.7$ kJ; $\Delta S° -5.5 \times 10^2$ J/K **19.147** -0.05 V; non spontaneous **19.149** -0.04 V; non spontaneous **19.151** $K_{sp} = 1 \times 10^{-7}$

CHAPTER 20

20.23 a 0.1% b If you had a large quantity of material, 0.1% still would be a significant quantity. Also, if the material were particularly toxic in addition to being radioactive, the amount would be a significant quantity. **20.25** Some of the expected mass is in the form of energy, the mass defect. **20.27** a At $t = 0$ min,

the container contains 16 spheres; at $t = 30$ min, the container contains 2 spheres. b Incorrect at $t = 5$ min; at $t = 5$ min, there should be fewer that 750 atoms. **20.29** The large, positively charged He nucleus that makes up alpha (α) radiation is unable to pass through the atoms that make up solid materials such as wood without coming into contact or being deflected by the nuclei. Gamma (γ) radiation, however, with its small wavelength and high energy, can pass through large amounts of material without interaction, just as x rays can pass through skin and other soft tissue. **20.31** Irradiation of food atoms with gamma radiation does not result in the creation of radioactive elements; therefore, the food cannot become radioactive. The addition of a radioactive element directly to the meat would make the meat radioactive. Under the right conditions, nuclear bombardment could also lead to the production of radioactive elements. **20.33** $^{87}_{37}Rb \longrightarrow ^{87}_{38}Sr + ^{0}_{-1}e$ **20.35** $^{232}_{90}Th \longrightarrow ^{228}_{88}Ra + ^{4}_{2}He$ **20.37** $^{18}_{9}F \longrightarrow ^{18}_{8}O + ^{0}_{1}e$ **20.39** $^{210}_{84}Po \longrightarrow ^{206}_{82}Pb + ^{4}_{2}He$ **20.41** a $^{122}_{51}Sb$ b $^{204}_{85}At$ c $^{80}_{37}Rb$ **20.43** a alpha emission b positron emission or electron capture c beta emission **20.45** $^{235}_{92}U$ decay series; $^{232}_{90}Th$ decay series. Alpha emission decreases the mass number by 4. Beta emission has no effect upon the mass number. **20.47** a $^{26}_{12}Mg(d, \alpha)^{24}_{11}Na$ b $^{16}_{8}O(n, p)^{16}_{7}N$ **20.49** a $^{27}_{13}Al + ^{2}_{1}H \longrightarrow ^{25}_{12}Mg + ^{4}_{2}He$ b $^{10}_{5}B + ^{4}_{2}He \longrightarrow ^{13}_{6}C + ^{1}_{1}H$ **20.51** 1.22×10^9 kJ/mol **20.53** a $^{4}_{2}He$ b α **20.55** plutonium-239 **20.57** 1.79×10^{-9} s^{-1} **20.59** 9.1×10^{-8} s^{-1} **20.61** 1.3×10^{13} y **20.63** 3.84×10^{-12}/s **20.65** 2.1×10^2 Ci **20.67** 6.0×10^{-4} g **20.69** 0.574, 3.4 μg **20.71** 4.16 s **20.73** 44.6 d **20.75** 5.3×10^3 y **20.77** 5.2 disintegrations/(min·g) **20.79** -4.37×10^{-9} g **20.81** -1.688×10^{12} J; -17.50 MeV **20.83** 3.02184 amu; 32.01 MeV; 5.336 MeV **20.85** Na-20 is expected to decay by electron capture or positron emission. Na-26 is expected to decay by beta emission. **20.87** seven alpha emissions and four beta emissions **20.89** $^{209}_{83}Bi + ^{4}_{2}He \longrightarrow ^{211}_{85}At + 2^{1}_{0}n$ **20.91** $^{246}_{98}Cf$ **20.93** 6.3 y **20.95** 2.42 pm **20.97** -6.02×10^{14} J; -1.64×10^5 kJ **20.99** a $^{47}_{20}Ca \longrightarrow ^{47}_{21}Sc + ^{0}_{-1}\beta$ b 41 μg **20.101** a $^{82}_{35}Br \longrightarrow ^{82}_{36}Kr + ^{0}_{-1}\beta$; $2H^{82}_{35}Br(g) \longrightarrow H_2 + 2^{82}_{36}Kr + 2^{0}_{-1}\beta$ b 0.0123 mol; 0.396 atm **20.123** 1.75×10^{13} nuclei/s **20.125** 0.001 L **20.127** 2.732×10^9 kJ; 4.91×10^7 L

CHAPTER 21

21.71 It would react with the oxygen by the following equation: $C(s) + O_2(g) \longrightarrow CO_2(g)$ **21.73** The metallic character decreases from left to right and increases going down a column. **21.75** carbon dioxide $2LiOH(s) + CO_2(g) \longrightarrow Li_2CO_3(s) + H_2O(l)$ $Li_2CO_3(s) + H_2O(l) + CO_2(g) \longrightarrow 2LiHCO_3(s)$ **21.77** Oxygen is a very electronegative element, and its bonding involves only the s and p orbitals, in contrast to bonding using the d orbitals in sulfur, etc. Molecular oxygen is a reactive gas, but it forms mainly compounds in which its oxidation state is -2, compared with compounds of sulfur, etc., which exhibit positive oxidation states as well as the -2 state. **21.79** no **21.81** $Fe_2O_3(s) + 3H_2(g) \longrightarrow 2Fe(s) + 3H_2O(g)$ **21.83** 36.9 kg Fe **21.85** -417.9 kJ (exothermic) **21.89** 2.48×10^{19} **21.91** $CO_2(g) + NH_3(g) + NaCl(aq) + H_2O(l) \longrightarrow NaHCO_3(s) + NH_4Cl(aq)$ $2NaHCO_3(s) \xrightarrow{\Delta} Na_2CO_3(s) + CO_2(g) + H_2O(g)$ $Na_2CO_3(aq) + Ca(OH)_2(aq) \longrightarrow 2NaOH(aq) + CaCO_3(s)$ **21.93** a $2NaOH(s) + CO_2(g) \longrightarrow Na_2CO_3(s) + H_2O(l)$ b $Li_2CO_3(aq) + 2HNO_3(aq) \longrightarrow H_2O(l) + 2LiNO_3(aq) + CO_2(g)$ c $2K(s) + Br_2(l) \longrightarrow 2KBr(s)$ d $K_2SO_4(aq) + Pb(NO_3)_2(aq) \longrightarrow PbSO_4(s) + 2KNO_3(aq)$ e $2K(s) + 2H_2O(l) \longrightarrow 2KOH(aq) + H_2(g)$

21.95 $^{227}_{89}Ac \longrightarrow ^{223}_{87}Fr + ^4_2He$ **21.97** by adding SO_4^{2-} ion, precipitating $BaSO_4$ but not K_2SO_4 **21.99** Add Na_2SO_4 to the solution. $BaSO_4(s)$ can then be filtered from the solution. The filtrate will contain Mg^{2+} ions.
21.101 $^{223}_{88}R \longrightarrow ^{219}_{86}R + ^4_2H$
21.103 a $2NaOH(aq) + MgSO_4(aq) \longrightarrow Mg(OH)_2(s) +$ $Na_2SO_4(aq)$ b $BaCO_3(s) \xrightarrow{\Delta} BaO(s) + CO_2(g)$
c $Mg(s) + NiCl_2(aq) \longrightarrow Ni(s) + MgCl_2(aq)$
d $Ba(s) + 2H_2O(l) \longrightarrow Ba(OH)_2(aq) + H_2(g)$
e $Mg(OH)_2(s) + 2HNO_3(aq) \longrightarrow 2H_2O(l) + Mg(NO_3)_2(aq)$
21.105 0.11 g   **21.107** $Al(H_2O)_6^{3+}(aq) + HCO_3^-(aq) \longrightarrow$ $Al(H_2O)_5OH^{2+}(aq) + H_2O(l) + CO_2(g)$
21.109 Test portions of solutions of each compound with each other; the results can differentiate the compounds. For example, if one solution is poured into both of the other solutions, and both give a precipitate [$\longrightarrow Al(OH)_3$] and $\longrightarrow [Mg(OH)_2 + BaSO_4]$, that solution must be $Ba(OH)_2$. Continuing the addition of the $Ba(OH)_2$ will dissolve the $Al(OH)_3$ as $Al(OH)_4^-$ but will not dissolve the $Mg(OH)_2 + BaSO_4$. Thus, all three will be identified. If, instead, $MgSO_4$ is poured into $AlCl_3$, no reaction will occur, and that will mean the third solution is $Ba(OH)_2$
21.111 $Pb(H_2O)_6^{2+}(aq) + H_2O(l) \longrightarrow Pb(H_2O)_5(OH)^+(aq) +$ $H_3O^+(aq)$ **21.113** $PbO_2 + 4HCl \longrightarrow PbCl_2 + Cl_2 + 2H_2O$
21.115 a $2Ga(OH)_3(s) \longrightarrow Ga_2O_3(s) + 3H_2O(g)$
b $8Al(s) + 3Mn_3O_4(s) \longrightarrow 9Mn(s) + 4Al_2O_3(s)$
c $Al(s) + 3AgNO_3(aq) \longrightarrow 3Ag(s) + Al(NO_3)_3(aq)$
d $Pb(NO_3)_2(aq) + 2NaI(aq) \longrightarrow PbI_2(s) + 2NaNO_3(aq)$
e $Al_2O_3(s) + 3H_2SO_4(aq) \longrightarrow$

$$Al_2(SO_4)_3(aq) + 3H_2O(l)$$

21.117 272 kJ   **21.119** a -1 b $+1$ c -4
d $+6$ **21.121** a Carbon has four valence electrons. These are directed tetrahedrally, and the orbitals should be sp^3 hybrid orbitals. Each C—Cl bond is formed by the overlap of a $3p$ orbital of a chlorine atom with one of the singly occupied sp^3 hybrid orbitals of the carbon atom.   b Carbon has four valence electrons, and nitrogen has five valence electrons. The carbon-nitrogen triple bond is a sigma bond formed by the overlap of sp hybrid orbitals and two pi bonds formed by the overlap of the unhybridized p orbitals of the C and N atoms. The C—H bond is formed by the overlap of a $1s$ orbital of a hydrogen atom with one of the singly occupied sp hybrid orbitals of the carbon atom.   c Silicon has four valence electrons. These are directed tetrahedrally, and the orbitals should be sp^3 hybrid orbitals. Each Si—F bond is formed by the overlap of a $2p$ orbital of a fluorine atom with one of the singly occupied sp^3 hybrid orbitals of the silicon atom.   d Carbon has four valence electrons. In the CH_3 group, these are directed tetrahedrally, and the orbitals should be sp^3 hybrid orbitals. Each C—H bond is formed by the overlap of a $1s$ orbital of a hydrogen atom with one of the singly occupied sp^3 hybrid orbitals of the carbon atom. The C—C bond between the CH_3 and the COOH groups is formed by the overlap of the sp^2 orbital of the middle carbon and the sp^3 orbital of the CH_3 carbon. In the COOH group, there is a C=O double bond and a C—O single bond. The C=O double bond is a sigma bond formed by the overlap of the sp^2 hybrid orbitals, and a pi bond formed by the overlap of the unhybridized p orbitals of the C and O atoms. The C—O bond is a sigma bond formed by the overlap of the C sp^2 hybrid orbital and the O sp^3 hybrid orbital. The O—H bond is similar to the C—H bond.
21.123 a 74.87 kJ   b 137.15 kJ   **21.125** a $CO_2(g) +$ $Ba(OH)_2(aq) \longrightarrow BaCO_3(s) + H_2O(l)$
b $MgCO_3(s) + 2HBr(aq) \longrightarrow CO_2(g) + H_2O(l) + MgBr_2(aq)$
21.127 $C(s) + O_2(air) \longrightarrow CO_2(g)$
$CO_2(g) + NaOH(aq) \longrightarrow NaHCO_3(aq)$
$2NaHCO_3(aq) \xrightarrow{\Delta} Na_2CO_3(s) + H_2O(l) + CO_2(g)$

21.129 Each silicon atom is tetrahedrally bonded to four other silicon atoms. Each bond is formed by the overlap of an sp^3 hybrid orbital on each silicon atom to make a sigma bond.
21.131 0.827 kg Si
21.133

The drawing represents two of the sets of tetrahedra forming a portion of a long chain. There is a net -8 charge on the fragment, which includes two sets of two SiO_3 units. Thus, the $(SiO_3)_2$ unit has a -4 charge. This would have to be balanced by one Li^+ and one Al^{3+} for every two SiO_3. Thus, the empirical formula is $LiAl(SiO_3)_2$.   **21.135** 2.53 g
21.137 $4NH_3(g) + 5O_2(g) \xrightarrow{Pt} 4NO(g) + 6H_2O(g)$
$2NO(g) + O_2(g) \longrightarrow 2NO_2(g)$
$3NO_2(g) + H_2O(l) \longrightarrow 2HNO_3(aq) + NO(g)$
$NH_3(g) + HNO_3(aq) \longrightarrow NH_4NO_3(aq)$
$NH_4NO_3(aq) \longrightarrow NH_4NO_3(s)$
$NH_4NO_3(s) \xrightarrow{Pt} N_2O(g) + 2H_2O(g)$
21.139 $3Ag(s) + NO_3^-(aq) + 4H^+(aq) \longrightarrow 3Ag^+(aq) +$ $NO(g) + 2H_2O(l)$
21.141 tetrahedral; Each bond is formed by the overlap of a $4p$ orbital from Br with an sp^3 hybrid orbital from P.
21.143 $H_3PO_3 + H_2SO_4 \longrightarrow H_3PO_4 + SO_2 + H_2O$
21.145 14.1 g H_3PO_4 **21.147** a $2Ca(s) + O_2(g) \longrightarrow$ $2CaO(s)$ b $4PH_3(g) + 8O_2(g) \longrightarrow P_4O_{10}(s) + 6H_2O(g)$
c $4HOCH_2CH_2NH_2(l) + 13O_2(g) \longrightarrow 8CO_2(g) +$ $14H_2O(g) + 2N_2(g)$ **21.149** a $+6$ b $+6$ c -2 d $+4$
21.151 $8H_2SeO_3(aq) + 16H_2S(g) \longrightarrow Se_8(s) + 2S_8(s) +$ $24H_2O(l)$ **21.153** 49.1 g
21.155 $Ba(ClO_3)_2(aq) + H_2SO_4(aq) \longrightarrow 2HClO_3(aq) + BaSO_4(s)$
21.157 $I_2 + 10HNO_3 \longrightarrow 2HIO_3 + 10NO_2 + 4H_2O$
21.159 a sp^3d; T-shaped   b sp^3; bent (angular)
c sp^3; trigonal pyramidal
21.161 a $NaBr(s) + H_3PO_4(aq) \xrightarrow{\Delta} HBr(g) + NaH_2PO_4(aq)$
b $Br_2(aq) + 2NaOH(aq) \longrightarrow$

$$H_2O(l) + NaOBr(aq) + NaBr(aq)$$

21.163 nonspontaneous at standard conditions

21.165

sp^3d^2, square planar

21.167 $XeO_3 + 6H^+ + 6I^- \longrightarrow Xe + 3H_2O + 3I_2$
21.169 0.675 metric ton Pb   **21.171** 74.40%
21.173 -850.2 kJ; -425.1 kJ/mol Fe   **21.175** 84.5%
21.177 13.91 g   **21.179** 1582 K   **21.181** $E° = 0.19$ V; -37 kJ; yes   **21.183** 238 g $KClO_3$ **21.185** 69.6%
21.187 12.6 hr   **21.189** 2.58%

CHAPTER 22

22.31 $[Co(NH_3)_4(NO_2)_2]Cl$ **22.33** "Hexacyano" means that there are six CN^- ligands bonded to the iron cation. The Roman numeral II means the oxidation state of the iron cation is $+2$, so the overall charge of the complex ion is $4-$. This requires four potassium ions to counterbalance the $4-$ charge.
22.35 geometric   **22.37** a $+2$ b $+4$ c $+2$ d $+6$
22.39 $3Fe^{2+} + NO_3^- + 4H^+ \longrightarrow 3Fe^{3+} + NO + 2H_2O$
22.41 a Six   b Four   c Six   d Four
22.43 a $+3$ b $+2$ c $+3$ d $+3$
22.45 a $+3$ b Aqua; Ammine; Hydroxo
c Six   d -3 **22.47** a Potassium hexafluoroferrate(III)

b Diamminediaquacopper(II) ion
c Ammonium aquapentafluoroferrate(III)
d Dicyanoargentate(I) ion
22.49 **a** Octacarbonyltungsten
b Diaquabis(ethylenediamine)cobalt(VI) sulfate
c Potassium octacyanomolybdate(VII)
d Tetraoxochromate(VI) ion (Chromate is the usual name.)
22.51 **a** $K_3[Mn(CN)_6]$ **b** $Na_2[Zn(CN)_4]$
c $[Co(NH_3)_4Cl_2]NO_3$ **d** $[Cr(NH_3)_6]_2[CuCl_4]_3$

22.53 **a**

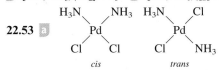

b no geometric isomerism   **c** no geometric isomerism
d

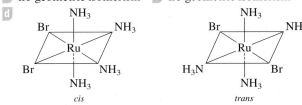

22.55 No optical isomers   **22.57** **a** two   **b** two   **c** five
22.59 **a** seven electrons in square planar field (low spin)
b five electrons in tetrahedral field (high spin)   **c** seven electrons in tetrahedral field (high spin)   **d** eight electrons in square planar field (low spin)   **22.61** purple   **22.63** yes; H_2O is a weaker-bonding ligand   **22.65** 1.76×10^5 J/mol (176 kJ/mol)   **22.67** Zn^{2+} ion has a completely filled d subshell with no openings for d electrons to be promoted to a higher energy level for which the difference in energy would correspond to that of visible light.

22.69 three geometric isomers

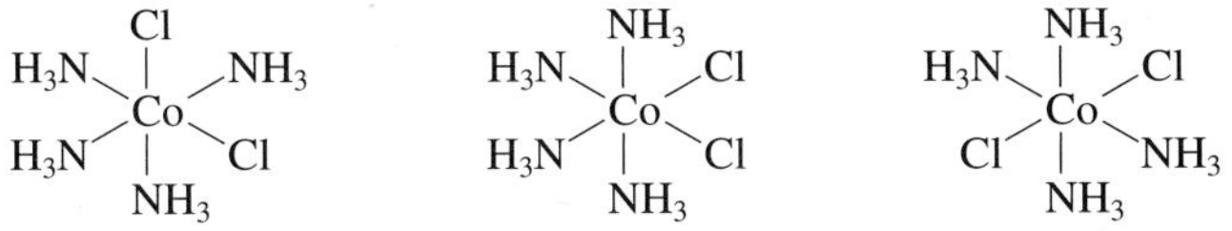

H₃N, Cl, NH₃ / Co / H₃N, Cl / NH₃ (structures shown)

22.71 0.050 M; 2.4×10^{-4} M; 0.050 M

CHAPTER 23

23.19 **a** He is incorrect. The longest chain of carbon atoms is seven, making the compound a heptane. The correct name would be 4-ethylheptane.
b

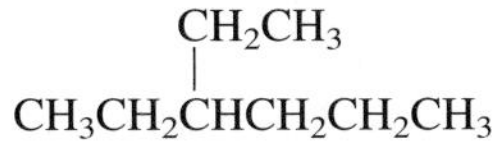

CH₂CH₃
CH₃CH₂CHCH₂CH₂CH₃

c Low-molecular-mass hydrocarbons have fewer polarizable electrons; therefore, they have weaker London forces than the longer chains and, as a result, boil at a lower temperature.
d The more compact 2,3-dimethylbutane would boil at a lower temperature.   **23.21** Since carbon would have more than four bonds in this case, CH_5 would be in violation of the octet rule.   **23.23** **a** trimethylamine, C_3H_9N; acetaldehyde, C_2H_4O; 2-propanol, C_3H_8O; and acetic acid, $C_2H_4O_2$
b trimethylamine: $N(CH_3)_3$ acetaldehyde:

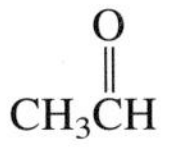

O
∥
CH₃CH

2-propanol: OH
                    |
             CH₃CHCH₃

acetic acid: O
                    ∥
             CH₃COH

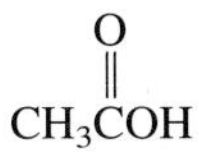

c The functional groups in the molecules are trimethylamine, amine (tertiary); acetaldehyde, aldehyde; 2-propanol, alcohol; and acetic acid, carboxylic acid.

23.25 The molecules increase regularly in molecular mass. Therefore, you expect their intermolecular forces (dispersion forces) and thus their melting points to increase.   **23.27** $CH_3CH_2CH_2CH_3$
23.31 **a** $C_2H_4 + 3O_2 \longrightarrow 2CO_2 + 2H_2O$

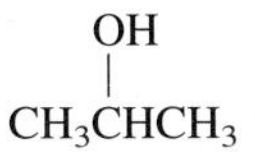

b $3CH_2{=}CH_2 + 2MnO_4^- + 4H_2O \longrightarrow$
                    OH   OH
                    | |
$3CH_2{-}CH_2 + 2MnO_2 + 2OH^-$

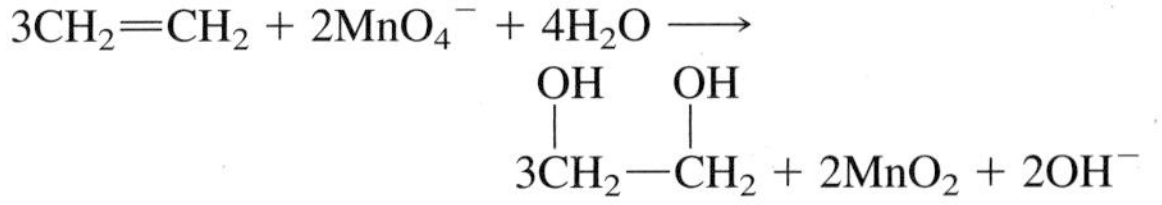

c $CH_2{=}CH_2 + Br_2 \longrightarrow$
Br   Br
| |
$CH_2{-}CH_2$

d (toluene + Br_2 $\xrightarrow{FeBr_3}$ p-bromotoluene + HBr)

e (toluene + HNO_3 $\xrightarrow{H_2SO_4}$ p-nitrotoluene + H_2O)

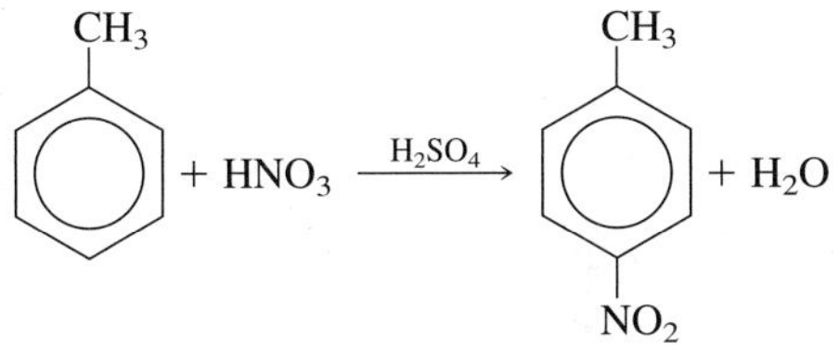

23.33 $CH_3CH_2CH_3 + Cl_2 \longrightarrow CH_3CHClCH_3 + HCl$

23.35
                    Br
                    |
$CH_3{-}C{-}CH_3$
                    |
                    CH_3

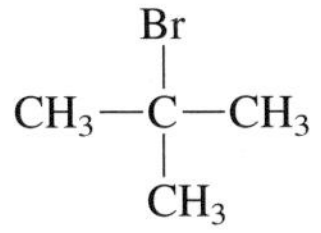

23.37 **a** 2,3,4-trimethylpentane   **b** 2,2,6,6-tetramethylheptane   **c** 4-ethyloctane   **d** 3,4-dimethyloctane

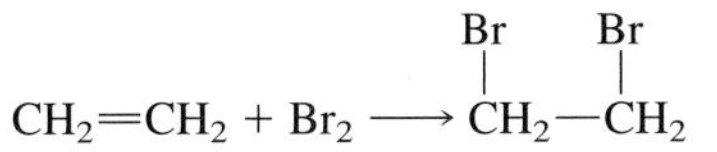

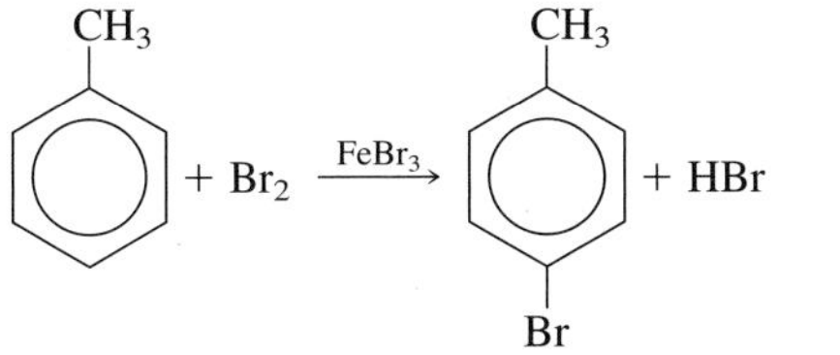

23.39 **a**
                    CH_3
                    |
$CH_3CHCH_2CHCH_2CH_2CH_3$
                    |
             CH_3CHCH_3

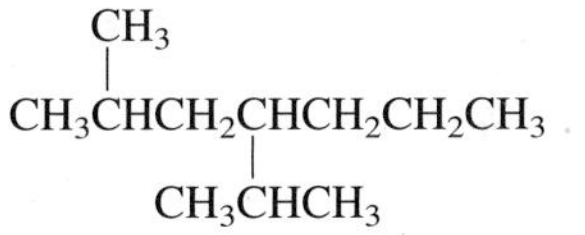

b
       CH_3 CH_3
       | |
$CH_3C{-}{-}CCH_2CH_3$
       | |
       CH_3 CH_3

c
             CH_2CH_3
             |
$CH_3CH_2CHCH_2CH_2CH_3$

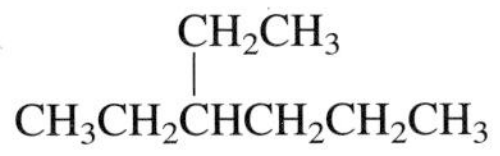

d
             CH_3
             |
$CH_3CHCHCH_2CH_2CH_3$
             |
             CH_3

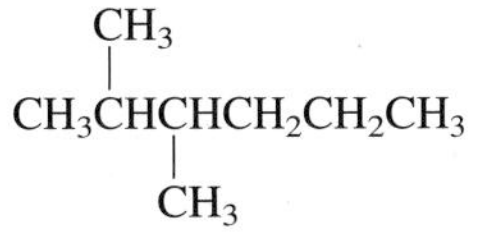

23.41 **a** 2-pentene   **b** 2,5-dimethyl-2-hexene

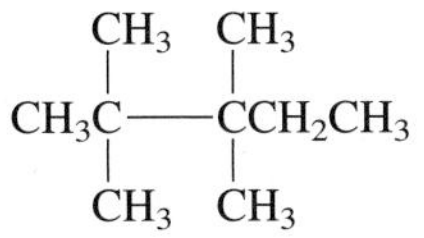

23.43 **a** $CH_3C{=}CHCHCH_2CH_3$ (CH_3, CH_2CH_3)   **b** $CH_3CH{=}CCH_2CH_3$ (CH_2CH_3)

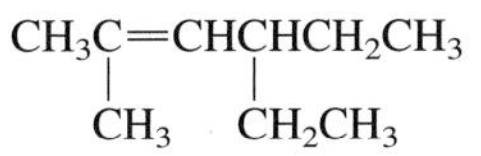

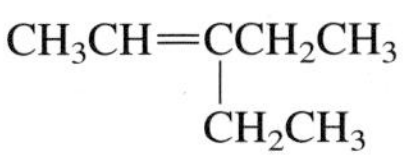

23.45 *cis*-2-butene and *trans*-2-butene   **23.47** **a** 2-butyne
b 3-methyl-1-pentyne

23.49 **a** **b**

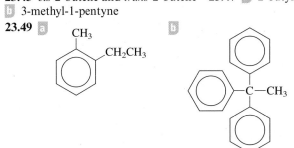

23.51 [a] $CH_3-\overset{O}{\overset{\|}{C}}-CH_2CH_2CH_3$ *ketone* [b] $CH_3-\underset{OH}{\overset{}{C}}HCH_2CH_3$ *alcohol*

[c] $HO\overset{O}{\overset{\|}{C}}-CH_2CH_3$ *carboxylic acid* [d] $H-\overset{O}{\overset{\|}{C}}-CH_2CH_3$ *aldehyde*

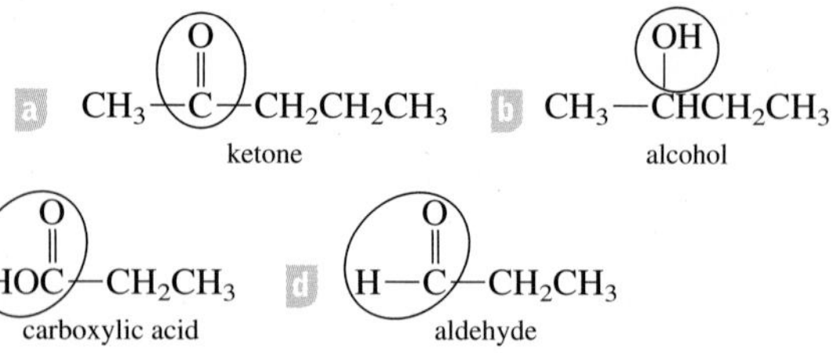

23.53 [a] 1-pentanol [b] 2-pentanol [c] 2-propyl-1-pentanol
[d] 6-methyl-4-octanol **23.55** [a] secondary alcohol
[b] secondary alcohol [c] primary alcohol [d] primary
alcohol **23.57** [a] ethyl propyl ether [b] methyl isopropyl
ether **23.59** [a] butanone [b] butanal
[c] 4,4-dimethylpentanal [d] 3-methyl-2-pentanone
23.61 [a] secondary amine [b] secondary amine
23.63 [a] 3-methylbutanoic acid [b] *trans*-5-methyl-2-hexene
[c] 2,5-dimethyl-4-heptanone [d] 4-methyl-2-pentyne

23.65 [a] $CH_3CH_2\overset{O}{\overset{\|}{C}}-O-\underset{CH_3}{\overset{CH_3}{\overset{|}{C}}}H$ [b] $CH_3-\underset{CH_3}{\overset{CH_3}{\overset{|}{\underset{|}{C}}}}-NH_2$

[c] $CH_3CH_2CH_2CH_2\underset{CH_3}{\overset{CH_3}{\overset{|}{\underset{|}{C}}}}COOH$ [d] $\underset{H}{\overset{CH_3CH_2}{C}}=\underset{H}{\overset{CH_2CH_3}{C}}$

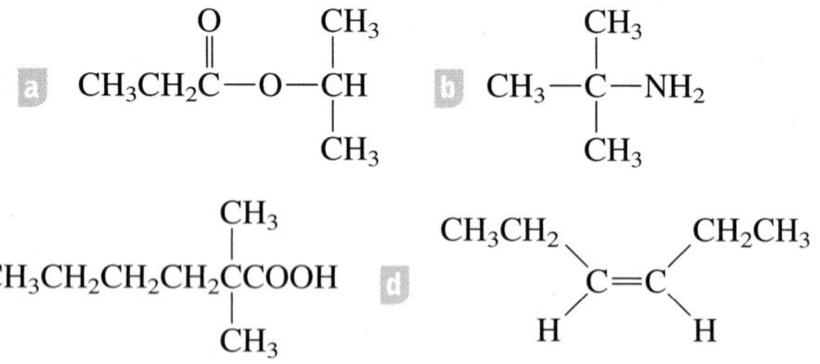

23.67 [a] Add Br_2 in CCl_4 each compound. The first compound
reacts; the color of Br_2 fades. Benzene does not react. [b] Addition
of dichromate ion in acidic solution to propionaldehyde will
cause the reagent to change from orange to green as the aldehyde is
oxidized. Under similar conditions, acetone (a ketone) would
not react. **23.69** [a] toluene [b] methanol [c] methylamine
[d] ethylene **23.71** methyl propanoate

CHAPTER 24

24.21 [a] —A—A—A—A—A—A—, —B—B—B—
B—B—B— [b] —A—B—A—B—A—B—
[c] —A—C—A—A—C—A—
24.23 gly—gly—ser, gly—ser—gly, and ser—gly—gly
24.25 AAA CCU GCU UUU UGG GAA CAU GGU UAA.
24.29 $HOCH_2CH_2OH$ $HOOCCH_2CH_2COOH$
24.31 [a] would give a polymer having alternating single and
double bonds, in which resonance might be expected to give a
conducting polymer.
24.35 Two dipeptides are possible:

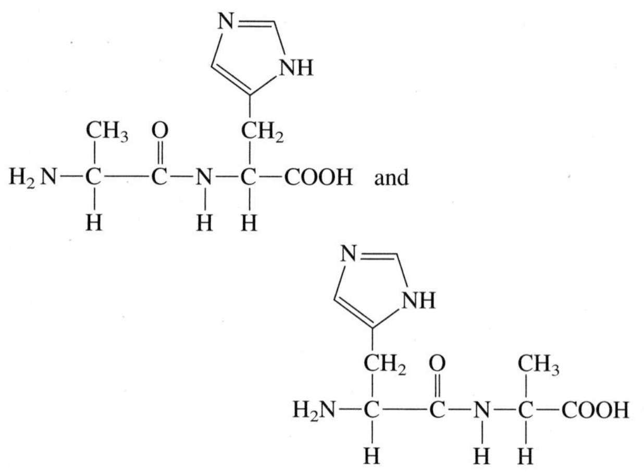

$H_2N-\underset{H}{\overset{CH_3}{\overset{|}{\underset{|}{C}}}}-\overset{O}{\overset{\|}{C}}-\underset{H}{\overset{}{N}}-\underset{\overset{|}{CH_2}}{\overset{|}{C}}H-COOH$ and

(imidazole ring: N=... NH, CH_2)

$H_2N-\underset{H}{\overset{CH_2}{\overset{|}{C}}}-\overset{O}{\overset{\|}{C}}-\underset{H}{\overset{}{N}}-\underset{H}{\overset{CH_3}{\overset{|}{\underset{|}{C}}}}-COOH$

24.37 A:T, 1:1, G:C, 1:1

24.39

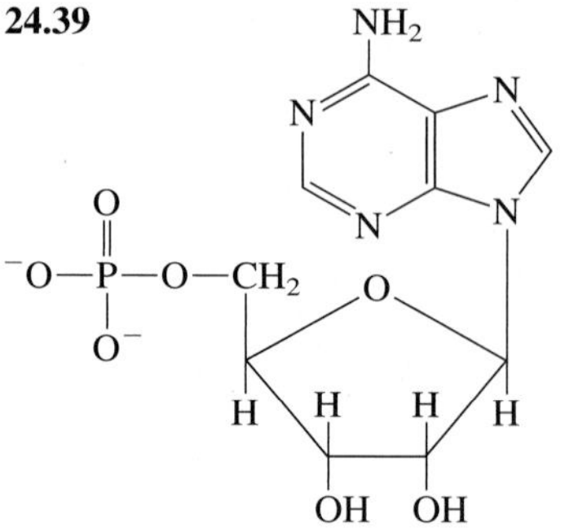

24.41 Two, hydrogen bonding in a uracil-adenine base pair is
the same as in a thymine-adenine base pair, so the strength of
the bonding should be the same **24.43** 16; no
24.45 hydrogen bonding; less readily
24.47 Ile—Gly—Ala—Arg—Ser—Asn—Glu—Pro—Ser
24.49 AAC GUG UGA CUU
24.51 UAU AUU CCU CAU CUU CAU ACU UCU UUU
AUG

24.53 $-\underset{Cl}{\overset{H}{\overset{|}{\underset{|}{C}}}}-\underset{Cl}{\overset{H}{\overset{|}{\underset{|}{C}}}}-\underset{Cl}{\overset{H}{\overset{|}{\underset{|}{C}}}}-\underset{Cl}{\overset{H}{\overset{|}{\underset{|}{C}}}}-$

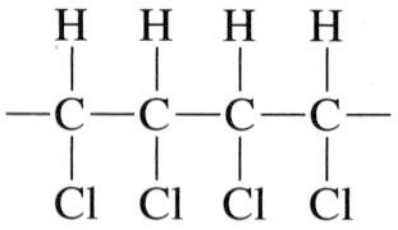

24.55 $HOOC-C_6H_4-COOH$, $NH_2-C_6H_4-NH_2$

24.57 $CH_3SCH_2CH_2\underset{\overset{|}{NH_2}}{\overset{}{C}}H-COOH$ has a nonpolar side chain

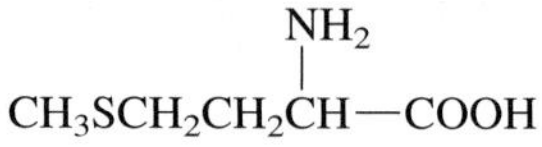

24.59 $^+H_3N-\underset{\overset{|}{CH_2}}{\overset{\overset{|}{COO^-}}{C}}-H$ (CH_2, OH)

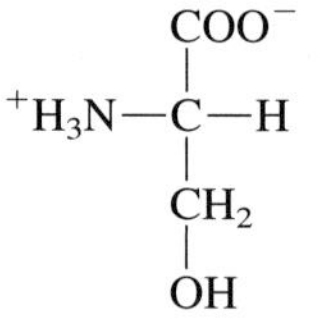

24.61

$CH_3SCH_2CH_2\underset{\overset{|}{NH_2}}{\overset{}{C}}H-\overset{O}{\overset{\|}{C}}-NH-\underset{\overset{|}{CH_2SH}}{\overset{}{C}}H-COOH$

$HSCH_2\underset{\overset{|}{NH_2}}{\overset{}{C}}H-\overset{O}{\overset{\|}{C}}-NH-\underset{\overset{|}{CH_2}}{\underset{\overset{|}{CH_2SCH_3}}{\overset{}{C}}}H-COOH$

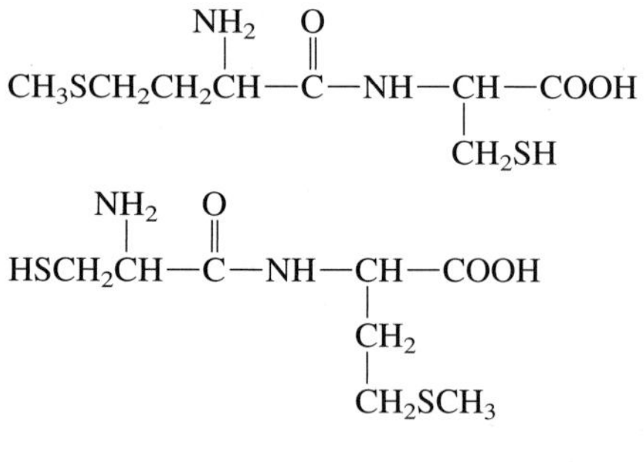

24.63 720

Glossary

The number given in blue at the end of a definition indicates the section where the term was introduced. In some cases, more than one section is indicated. In a few cases, the term is defined in an introductory passage just *before* the section indicated.

Absolute entropy see *Standard entropy.*

Absolute temperature scale a temperature scale in which the lowest temperature that can be attained theoretically is zero. (1.6)

Accuracy the closeness of a single measurement to its true value. (1.5)

Acid (Arrhenius definition) a substance that produces hydrogen ions, H^+ (hydronium ion, H_3O^+), when it dissolves in water. (4.4 and 15.1); (Brønsted–Lowry definition) the species (molecule or ion) that donates a proton to another species in a proton-transfer reaction. (4.4 and 15.2)

Acid–base indicator a dye used to distinguish between acidic and basic solutions by means of the color changes it undergoes in these solutions. (4.4)

Acid–base titration curve a plot of the pH of a solution of acid (or base) against the volume of added base (or acid). (16.7)

Acidic oxide an oxide that reacts with bases. (8.7)

Acid-ionization (or acid-dissociation) constant (K_a) the equilibrium constant for the ionization of a weak acid. (16.1)

Acid rain rain having a pH lower than that of natural rain, which has a pH of 5.6. (16.2)

Acid salt a salt that has an acidic hydrogen atom and can undergo neutralization with bases. (4.4)

Actinides elements in the last of the two rows at the bottom of the periodic table; actinium plus the 14 elements following it in the periodic table, in which the 5f subshell is filling. (22.1)

Activated complex (transition state) an unstable grouping of atoms that can break up to form products. (13.5)

Activation energy (E_a) the minimum energy of collision required for two molecules to react. (13.5)

Activity of a radioactive source the number of nuclear disintegrations per unit time occurring in a radioactive material. (20.3)

Activity series a listing of the elements in order of their ease of losing electrons during reactions in aqueous solution. (4.5)

Addition polymer a polymer formed by linking together many molecules by addition reactions. (24.1)

Addition reaction a reaction in which parts of a reactant are added to each carbon atom of a carbon–carbon double bond, which then becomes a C—C single bond. (23.3)

Adsorption the binding or attraction of molecules to a surface. (13.9) See *Chemisorption* and *Physical adsorption.*

Aerosol a colloid consisting of liquid droplets or solid particles dispersed throughout a gas. (12.9)

Alcohol a compound obtained by substituting a hydroxyl group (—OH) for an —H atom on a tetrahedral (sp^3 hybridized) carbon atom of a hydrocarbon group. (23.6)

Aldehyde a compound containing a carbonyl group with at least one H atom attached to it. (23.6)

Aliphatic hydrocarbon a saturated or unsaturated hydrocarbon; a hydrocarbon that does not contain benzene rings. (23.1)

Alkali metal a Group IA element. The Group IA elements are reactive metals. (2.5 and 8.7)

Alkaline dry cell a voltaic cell that is similar to the Leclanché dry cell but uses potassium hydroxide in place of ammonium chloride. (19.8)

Alkaline earth metal a Group IIA element. The Group IIA elements are reactive metals, though less reactive than the alkali metals. (8.7)

Alkane an acyclic saturated hydrocarbon (one without a ring of carbon atoms); a saturated hydrocarbon with the general formula C_nH_{2n+2}. (23.2)

Alkene a hydrocarbon that has the general formula C_nH_{2n} and contains a carbon–carbon double bond. (23.3)

Alkyl group an alkane less one hydrogen atom. (23.5)

Alkyne an unsaturated hydrocarbon containing a carbon–carbon triple bond. The general formula is C_nH_{2n-2}. (23.3)

Allotrope one of two or more distinct forms of an element in the same physical state. (6.8)

Alloy a material with metallic properties that is either a compound or a mixture. (21.2)

Alpha emission emission of a $_2^4He$ nucleus, or alpha particle, from an unstable nucleus. (20.1)

Amide a compound derived from the reaction of ammonia, or a primary or secondary amine, with a carboxylic acid. (23.7)

Amide bond see *Peptide bond.*

Amine a compound that is structurally derived by replacing one or more hydrogen atoms of ammonia with hydrocarbon groups. (23.7)

Amino acid a compound containing an amino group (—NH_2) and a carboxyl group (—COOH). (24.3)

Amorphous solid a solid that has a disordered structure; it lacks the well-defined arrangement of basic units (atoms, molecules, or ions) found in a crystal. (11.7)

Ampere (A) the base unit of current in the International System (SI). (19.11)

Amphiprotic species a species that can act as either an acid or a base (that is, it can either lose or gain a proton). (15.2)

Amphoteric hydroxide a metal hydroxide that reacts with both bases and acids. (17.5)

Amphoteric oxide an oxide that has both acidic and basic properties. (8.7)

Angstrom (Å) a non-SI unit of length; $1 Å = 10^{-10}$ m. (1.6)

Angular geometry see *Bent geometry.*

Angular momentum quantum number (l) also known as the azimuthal quantum number. The quantum number that distinguishes orbitals of given n having different shapes; it can have any integer value from 0 to $n - 1$. (7.5)

Anion a negatively charged ion. (2.6)

Anode the electrode at which oxidation occurs. (19.2)

Antibonding orbitals molecular orbitals having zero values in the region between two nuclei and therefore concentrated in other regions. (10.5)

Aromatic hydrocarbon a hydrocarbon that contains a benzene ring or similar structural feature. (23.1)

Arrhenius equation the mathematical equation $k = Ae^{-Ea/RT}$, which expresses the dependence of the rate constant on temperature. (13.6)

Association colloid a colloid in which the dispersed phase consists of micelles. (12.9)

Atmosphere (atm) a unit of pressure equal to exactly 760 mmHg; 1 atm $= 101.325$ kPa (exact). (5.1)

Atom an extremely small particle of matter that retains its identity during chemical reactions. (2.1)

Atomic mass as a general term, the mass of an individual atom; but usually we mean the average atomic mass for the naturally occurring element, expressed in atomic mass units. (2.4)

Atomic mass unit (amu) a mass unit equal to exactly one-twelfth the mass of a carbon-12 atom. (2.4)

Atomic number (Z) the number of protons in the nucleus of an atom. (2.3)

Atomic orbital a wave function for an electron in an atom; pictured qualitatively by describing the region of space where there is a high probability of finding the electron. (7.5)

Atomic radius as an approximate measure of the radius, the maximum in the radial distribution function of the outer shell of the atom. Also, the value for an atom in a set of covalent radii assigned to atoms in such a way that the sum of the covalent radii of atoms A and B predicts the approximate A—B bond length (often called the covalent radius). (8.6)

Atomic symbol a one- or two-letter notation used to represent an atom corresponding to a particular element. A temporary name with a three letter notation (denoting the atomic number) is given to newly discovered elements. (2.1)

Atomic theory an explanation of the structure of matter in terms of different combinations of very small particles (atoms). (2.1)

Aufbau principle see *Building-up principle.*

Autoionization a reaction in which two like molecules react to give ions. (15.6)

Avogadro's law equal volumes of any two gases at the same temperature and pressure contain the same number of molecules. (5.2)

Avogadro's number (N_A) the number of atoms in a 12-g sample of carbon-12, equal to 6.02×10^{23} to three significant figures. (3.2)

Axial direction one of two directions pointing from the center of a trigonal bipyramid along its axis. Compare *Equatorial direction.* (10.1)

Azimuthal quantum number see *Angular momentum quantum number.*

Balanced chemical equation a chemical equation in which the numbers of atoms of each element are equal on both sides of the arrow. (2.10)

Band of stability the region in which stable nuclides lie in a plot of number of protons against number of neutrons. (20.1)

Band theory molecular orbital theory of metals. (21.3)

Bar a unit of pressure equal to 1×10^5 Pa, slightly less than 1 atm. (5.1)

Barometer a device for measuring the pressure of the atmosphere. (5.1)

Base (Arrhenius definition) a substance that produces hydroxide ions, OH^-, when it dissolves in water. (4.4 and 15.1) (Brønsted–Lowry definition) the species (molecule or ion) that accepts a proton in a proton-transfer reaction. (4.4 and 15.2)

Base-ionization (or base-dissociation) constant (K_b) the equilibrium constant for the ionization of a weak base. Thus K_b for NH_3 is 1.8×10^{-5}. (16.3)

Base pairing the hydrogen bonding of complementary nucleotide bases. (24.4)

Basic oxide an oxide that reacts with acids. (8.7)

Bayer process a chemical procedure in which purified aluminum oxide, Al_2O_3, is separated from the aluminum ore bauxite. (21.2)

Bent geometry (angular geometry) nonlinear molecular geometry, in the case of a molecule of three atoms. (10.1)

Beta emission emission of a high-speed electron from an unstable nucleus. (20.1)

Bidentate ligand a ligand that bonds to a metal atom through two atoms of the ligand. (22.3)

Bimolecular reaction an elementary reaction that involves two reactant molecules. (13.7)

Binary compound a compound composed of only two elements. (2.8)

Binary hydride a compound that contains hydrogen and one other element. (21.7)

Binding energy (of a nucleus) the energy needed to break a nucleus into its individual protons and neutrons. (20.6)

Body-centered cubic unit cell a cubic unit cell in which there is a lattice point at the center of the unit cell as well as at the corners. (11.7)

Boiling point the temperature at which the vapor pressure of a liquid equals the pressure exerted on the liquid (atmospheric pressure, unless the vessel containing the liquid is closed). (11.2)

Boiling-point elevation a colligative property of a solution equal to the boiling point of the solution minus the boiling point of the pure solvent. (12.6)

Bond distance see *Bond length.*

Bond enthalpy the average enthalpy change for the breaking of a bond in a molecule in the gas phase. Also called *bond energy.* (9.11)

Bonding orbitals molecular orbitals that are concentrated in the regions between nuclei. (10.5)

Bonding pair an electron pair shared between two atoms. (9.4)

Bond length (bond distance) the distance between the nuclei in a bond. (9.10)

Bond order in a Lewis formula, the number of pairs of electrons in a bond. (9.10)   In molecular orbital theory, one-half the difference between the number of bonding electrons and the number of antibonding electrons. (10.5)

Born-Haber cycle a thermochemical cycle originated by Max Born and Fritz Haber in 1919 to obtain the lattice energy of an ionic solid. (9.1)

Boyle's law the volume of a sample of gas at a given temperature varies inversely with the applied pressure. (5.2)

Brønsted–Lowry concept a concept of acids and bases in which an acid is the species donating a proton in a proton-transfer reaction, whereas a base is the species accepting a proton in such a reaction. (4.4 and 15.2)

Buckminsterfullerene a molecular form of carbon (C_{60}) informally called "buckyball," referring to its soccer-ball shape; a fullerene. (21.8)

Buffer a solution characterized by the ability to resist changes in pH when limited amounts of acid or base are added to it. (16.6)

Building-up principle (Aufbau principle) a scheme used to reproduce the electron configurations of the ground states of atoms by successively filling subshells with electrons in a specific order (the building-up order). (8.2)

Calorie (cal) a non-SI unit of energy commonly used by chemists, originally defined as the amount of energy needed to raise the temperature of 1 g of water by 1°C; now defined as 1 cal = 4.184 J (exact). (6.1)

Calorimeter a device used to measure the heat absorbed or evolved during a physical or chemical change. (6.6)

Carboxylic acid a compound containing the carboxyl group, —COOH, whose H atom is acidic. (23.6)

Catalysis the increase in rate of a reaction as the result of the addition of a catalyst. (13.9)

Catalyst a substance that increases the rate of reaction without being consumed in the overall reaction. (2.9 and 13.1)

Catenation the covalent bonding of two or more atoms of the same element to one another. (21.8)

Cathode the electrode at which reduction occurs. (19.2)

Cathode rays the rays emitted by the cathode (negative electrode) in a gas discharge tube (tube of low-pressure gas through which an electric current is discharged). (2.2)

Cation a positively charged ion. (2.6)

Cation exchange resin a resin (organic polymer) that removes cations from solution and replaces them with hydrogen ions. (11.10)

Cell potential the maximum potential difference between the electrodes of a voltaic cell. (19.4)

Cell reaction the net reaction that occurs in a voltaic cell. (19.2)

Celsius scale the temperature scale in general scientific use. There are exactly 100 units between the freezing point and the normal boiling point of water. (1.6)

Chain reaction, nuclear a self-sustaining series of nuclear fissions caused by the absorption of neutrons released from previous nuclear fissions. (20.7)

Chalcogen a Group VIA element. (8.7)

Change of state (phase transition) a change of a substance from one state to another. (11.2)

Charles's law the volume occupied by any sample of gas at a constant pressure is directly proportional to the absolute temperature. (5.2)

Chelate a complex formed by polydentate ligands. (22.3)

Chemical bond a strong attractive force that exists between certain atoms in a substance. (Ch. 9 opener)

Chemical change see *Chemical reaction.*

Chemical equation the symbolic representation of a chemical reaction in terms of chemical formulas. (2.9)

Chemical equilibrium the state reached by a reaction mixture when the rates of forward and reverse reactions have become equal. (14.1)

Chemical formula a notation that uses atomic symbols with numerical subscripts to convey the relative proportions of atoms of the different elements in a substance. (2.6)

Chemical kinetics the study of how reaction rates change under varying conditions and of what molecular events occur during the overall reaction. (13.1)

Chemical nomenclature the systematic naming of chemical compounds. (2.8)

Chemical property a characteristic of a material involving its chemical change. (1.4)

Chemical reaction (chemical change) a change in which one or more kinds of matter are transformed into a new kind of matter or several new kinds of matter. (1.4) The rearrangement of the atoms present in the reacting substances to give new chemical combinations present in the substances formed by the reaction. (2.1)

Chemisorption the binding of a species to a surface by chemical bonding forces. (13.9)

Chiral possessing the quality of handedness. A chiral object has a mirror image that is not identical to the object. (22.5)

Chlor–alkali membrane cell a cell for the electrolysis of aqueous sodium chloride in which the anode and cathode compartments are separated by a special plastic membrane that allows only cations to pass through it. (19.10)

Chlor–alkali mercury cell a cell for the electrolysis of aqueous sodium chloride in which mercury metal is used as the cathode. (19.10)

Chromatography a name given to a group of similar separation techniques that depend on how fast a substance moves, in a stream of gas or liquid, past a stationary phase to which the substance may be slightly attracted. (1.5)

cis-trans **isomers** see *Geometric isomers.*

Clausius–Clapeyron equation an equation that expresses the relation between the vapor pressure P of a liquid and the absolute temperature T: $\ln P = -\Delta H_{vap}/RT + B$, where B is a constant. (11.2)

Claus process a method of obtaining free sulfur by the partial burning of hydrogen sulfide. (21.10)

Close-packed structure a crystal structure in which the atoms or other units are packed together as closely as possible. See *hexagonal close-packed structure* and *cubic close-packed structure.* (11.8)

Coagulation the process by which the dispersed phase of a colloid is made to aggregate and thereby separate from the continuous phase. (12.9)

Codon a sequence of three bases in a messenger RNA molecule that serves as the code for a particular amino acid. (24.4)

Colligative properties properties that depend on the concentration of solute molecules or ions in a solution but not on the chemical identity of the solute. (12.4)

Collision theory the theory that in order for reaction to occur, reactant molecules must collide with an energy greater than some minimum value and with proper orientation. (13.5)

Colloid a dispersion of particles of one substance (the dispersed phase) throughout another substance or solution (the continuous phase). (12.9)

Combination reaction a reaction in which two substances combine to form a third substance. (4.5)

Combustion reaction a reaction of a substance with oxygen, usually with the rapid release of heat to produce a flame. (4.5)

Common-ion effect the shift in an ionic equilibrium caused by the addition of a solute that provides an ion that takes part in the equilibrium. (16.5 and 17.2)

Complementary bases nucleotide bases that form strong hydrogen bonds with one another. (24.4)

Complete ionic equation a chemical equation in which strong electrolytes (such as soluble ionic compounds) are written as separate ions in the solution. (4.2)

Complex (coordination compound) a compound consisting either of complex ions and other ions of opposite charge or of a neutral complex species. (22.3)

Complex ion an ion formed from a metal ion with a Lewis base attached to it by a coordinate covalent bond. (15.3, 17.5, and 22.3)

Compound a substance composed of two or more elements chemically combined. (1.4); A type of matter composed of atoms of two or more elements chemically combined in fixed proportions. (2.1)

Concentration a general term referring to the quantity of solute in a standard quantity of solvent or solution. (4.7 and 12.4)

Condensation the change of a gas to either the liquid or the solid state. (11.2)

Condensation polymer a polymer formed by linking many molecules together by condensation reactions. (24.1)

Condensation reaction a reaction in which two molecules or ions are chemically joined by the elimination of a small molecule such as water. (21.8)

Condensed (structural) formula a structural formula in which the bonds around each carbon atom are not explicitly written. (23.2)

Conjugate acid in a conjugate acid–base pair, the species that can donate a proton. (15.2)

Conjugate acid–base pair two species in an acid–base reaction, one acid and one base, that differ by the loss or gain of a proton. (15.2)

Conjugate base in a conjugate acid–base pair, the species that can accept a proton. (15.2)

Constitutional (structural) isomers isomers that differ in how the atoms are joined together (22.5); compounds with the same molecular formula but different structural formulas. (23.2)

Contact process an industrial method for the manufacture of sulfuric acid. It consists of the reaction of sulfur dioxide with oxygen to form sulfur trioxide using a catalyst of vanadium(V) oxide, followed by the reaction of sulfur trioxide with water. (21.10)

Continuous spectrum a spectrum containing light of all wavelengths. (7.3)

Control rods cylinders composed of substances that absorb neutrons, such as boron and cadmium, and can therefore slow a nuclear chain reaction. (20.7)

Conversion factor a factor equal to 1 that converts a quantity expressed in one unit to a quantity expressed in another unit. (1.8)

Coordinate covalent bond a bond formed when both electrons of the bond are donated by one atom. (9.4)

Coordination compound a compound consisting either of complex ions and other ions of opposite charge or of a neutral complex species. (22.3)

Coordination isomers isomers consisting of complex cations and complex anions that differ in the way the ligands are distributed between the metal atoms. (22.5)

Coordination number in a crystal, the number of nearest-neighbor atoms of an atom. (11.8)   In a complex, the total number of bonds the metal atom forms with ligands. (22.3)

Copolymer a polymer consisting of two or more different monomer units. (24.1)

Coulomb (C) unit of electric charge. (2.2)

Coulomb's law the potential energy obtained in bringing two charges Q_1 and Q_2, initially far apart, up to a distance r apart is directly proportional to the product of the charges and inversely proportional to the distance between them. (9.1)

Covalent bond a chemical bond formed by the sharing of a pair of electrons between atoms. (9.4)

Covalent network solid a solid that consists of atoms held together in large networks or chains by covalent bonds. (11.6)

Covalent radius the value for an atom in a set of covalent radii assigned to atoms in such a way that the sum of the covalent radii of atoms A and B predicts the approximate A—B bond length. (8.6 and 9.10)

Critical mass the smallest mass of fissionable material in which a chain reaction can be sustained. (20.7)

Critical pressure the vapor pressure at the critical temperature; it is the minimum pressure that must be applied to a gas at the critical temperature to liquefy it. (11.3)

Critical temperature the temperature above which the liquid state of a substance no longer exists regardless of the pressure. (11.3)

Crystal a kind of solid having a regular three-dimensional arrangement of atoms, molecules, or ions. (2.6)

Crystal field splitting (Δ) the difference in energy between the two sets of d orbitals on a central metal ion that arises from the interaction of the orbitals with the electric field of the ligands. (22.7)

Crystal field theory a model of the electronic structure of transition-metal complexes that considers how the energies of the d orbitals of a metal ion are affected by the electric field of the ligands. (22.7)

Crystal lattice the geometric arrangement of lattice points of a crystal, in which we choose one lattice point at the same location within each of the basic units of the crystal. (11.7)

Crystalline solid a solid composed of one or more crystals; each crystal has a well-defined ordered structure in three dimensions. (11.7)

Crystal systems the seven basic shapes possible for unit cells; a classification of crystals. (11.7)

Cubic close-packed structure (ccp) a crystal structure composed of close-packed atoms (or other units) with the stacking ABCABCABCA···. It has a face-centered cubic unit cell. (11.8)

Curie (Ci) a unit of activity, equal to 3.700×10^{10} disintegrations per second. (20.3)

Cycloalkane a cyclic saturated hydrocarbon; that is, a saturated hydrocarbon in which the carbon atoms form a ring; the general formula is C_nH_{2n}. (23.2)

Cyclotron a type of particle accelerator consisting of two hollow, semicircular metal electrodes, called dees (because the shape resembles the letter D), in which charged particles are accelerated by stages to higher and higher kinetic energies. Ions introduced at the center of the cyclotron are accelerated in the space between the two dees. (20.2)

Dalton's law of partial pressures the sum of the partial pressures of all the different gases in a mixture is equal to the total pressure of the mixture. (5.5)

d-Block transition elements those transition elements with an unfilled d subshell in common oxidation states. (8.2)

de Broglie relation the equation $\lambda = h/mv$ relating the wavelength λ associated with a particle of mass m and speed v. (7.4)

Decomposition reaction a reaction in which a single compound reacts to give two or more substances. (4.5)

Degree of ionization the fraction of molecules that react with water to give ions. (16.1)

Delocalized bonding a type of bonding in which a bonding pair of electrons is spread over a number of atoms rather than localized between two. (9.7)

Denitrifying bacteria bacteria that use nitrate ion, NO_3^-, as a source of energy; they convert the ion to gaseous nitrogen. (21.9)

Density the mass per unit volume of a substance or solution. (1.7)

Deoxyribonucleic acid (DNA) the hereditary constituent of cells; it consists of two polymer strands of deoxyribonucleotide units. (24.4)

Deposition the change of a vapor to a solid. (11.2)

Derived unit see *SI derived unit.*

Desalinate remove ions from brackish (slightly salty) water or seawater, to make drinkable or industrially usable water. (12.7)

Deuterons nucleus of a hydrogen-2 atom. (20.2)

Dextrorotatory refers to a compound whose solution rotates the plane of polarized light to the right (when looking toward the source of light). (22.5)

Diamagnetic substance a substance that is not attracted by a magnetic field or is very slightly repelled by such a field. This property generally means that the substance has only paired electrons. (8.4)

Diffraction a property of waves in which the waves spread out when they encounter an obstruction or small hole about the size of the wavelength. The spreading waves may interfere in a way to produce a distinctive pattern, called a diffraction pattern. (7.2 and 11.10)

Diffusion the process whereby a gas spreads out through another gas to occupy the space uniformly. (5.7)

Dimensional analysis (factor-label method) the method of calculation in which one carries along the units for quantities. (1.8)

Dipole–dipole force an attractive intermolecular force resulting from the tendency of polar molecules to align themselves such that the positive end of one molecule is near the negative end of another. (11.5)

Dipole moment a quantitative measure of the degree of charge separation in a molecule and therefore an indicator of the polarity of the molecule. (10.2)

Dispersion forces see *London forces.*

Displacement reaction (single-replacement reaction) a reaction in which an element reacts with a compound, displacing another element from it. (4.5)

Dissociation constant of a complex ion (K_d) the reciprocal, or inverse, value of the formation constant. (17.5)

Distillation the process in which a liquid is vaporized then condensed; used to separate substances of different volatilities. (1.4)

Distorted tetrahedral geometry see *Seesaw geometry*.

Disulfide linkage a covalent linkage formed between two cysteine groups within a protein molecule by oxidation of the two groups. (24.3)

Double bond a covalent bond in which two pairs of electrons are shared by two atoms. (9.4)

Downs cell a commercial electrochemical cell used to obtain sodium metal by the electrolysis of molten sodium chloride. (19.9)

Dow process a commercial method for isolating magnesium from seawater. (21.2)

Dry cell see *Alkaline dry cell* and *Zinc–carbon dry cell*.

Dynamic equilibrium an equilibrium in which molecular processes are continuously occurring. (11.2)

Effective nuclear charge the positive charge that an electron experiences from the nucleus, equal to the nuclear charge but reduced by any shielding or screening from any intervening electron distribution. (8.6)

Effusion the process in which a gas flows through a small hole in a container. (5.7)

Elastomer elastic or rubbery type of polymer. (24.1)

Electrochemical cell a system consisting of electrodes that dip into an electrolyte and in which a chemical reaction either uses or generates an electric current. (19.2)

Electrolysis the process of producing a chemical change in an electrolytic cell. (19.9)

Electrolyte a substance, such as sodium chloride, that dissolves in water to give an electrically conducting solution. (4.1)

Electrolytic cell an electrochemical cell in which an electric current drives an otherwise nonspontaneous reaction. (19.1)

Electromagnetic spectrum the range of frequencies or wavelengths of electromagnetic radiation. (7.1)

Electromotive force (emf) see *Cell potential*.

Electron a very light, negatively charged particle that exists in the region around the atom's positively charged nucleus. (2.2)

Electron affinity the energy required to remove an electron from the atom's negative ion (in its ground state); the negative of the energy change obtained when the neutral atom picks up an electron. (8.6)

Electron capture the decay of an unstable nucleus by capturing, or picking up, an electron from an inner orbital of an atom. (20.1)

Electron configuration the particular distribution of electrons among available subshells. (8.1)

Electron-dot formula see *Lewis electron-dot formula*.

Electronegativity a measure of the ability of an atom in a molecule to draw bonding electrons to itself. (9.5)

Electron spin quantum number (m_s) see *Spin quantum number*.

Electron volt (eV) the quantity of energy that would have to be imparted to an electron (whose charge is 1.602×10^{-19} C) to accelerate it by one volt potential difference. (20.2)

Element a substance that cannot be decomposed by any chemical reaction into simpler substances. (1.4)
A type of matter composed of only one kind of atom, each atom of a given kind having the same properties. (2.1)
A substance whose atoms all have the same atomic number. (2.3)

Elementary reaction a single molecular event, such as a collision of molecules, resulting in a reaction. (13.7)

Empirical formula (simplest formula) the formula of a substance written with the smallest integer (whole-number) subscripts. (3.5)

Emulsion a colloid consisting of liquid droplets dispersed throughout another liquid. (12.9)

Enantiomers (optical isomers) isomers that are nonsuperimposable mirror images of one another. (22.5)

Endothermic process a chemical reaction or physical change in which heat is absorbed (q is positive). (6.3)

Energy the potential or capacity to move matter. (6.1)

Energy levels specific energy values in an atom. (7.3)

Enthalpy (H) an extensive property of a thermodynamic system defined as the internal energy, U, plus pressure, P, times volume, V: $H = U + PV$. (6.3)

Enthalpy of formation see *Standard enthalpy of formation*.

Enthalpy of reaction (ΔH) the change in enthalpy for a reaction at a given temperature and pressure; it equals the heat of reaction at constant pressure. (6.3)

Entropy (S) a thermodynamic quantity that is a measure of how dispersed the energy of a system is among the different possible ways that a system can contain energy. (18.2)

Enzyme a protein that catalyzes a biochemical reaction. (24.3)

Equatorial direction one of the three directions pointing from the center of a trigonal bipyramid to a vertex other than one on the axis. Compare *Axial direction*. (10.1)

Equilibrium see *Chemical equilibrium*.

Equilibrium constant K see *Thermodynamic equilibrium constant*.

Equilibrium constant K_c the value obtained for the equilibrium-constant expression when equilibrium concentrations are substituted. (14.2)

Equilibrium constant K_p an equilibrium constant for a gas reaction, similar to K_c, but in which concentrations of gases are replaced by partial pressures (in atm). (14.2)

Equilibrium-constant expression an expression obtained for a reaction by multiplying the concentrations of products, dividing by the concentrations of reactants, and raising each concentration term to a power equal to the coefficient in the chemical equation. (14.2)

Equivalence point the point in a titration when a stoichiometric amount of reactant has been added. (16.7)

Ester a compound formed from a carboxylic acid, RCOOH, and an alcohol, R′OH. (23.6)

Ether a compound formally obtained by replacing both H atoms of H_2O by hydrocarbon groups R and R′. (23.6)

Exact number a number that arises when you count items or sometimes when you define a unit. (1.5)

Exchange (metathesis) reaction a reaction between compounds that, when written as a molecular equation, appears to involve the exchange of parts between the two reactants. (4.3)

Excited state a quantum-mechanical state of an atom or molecule associated with any energy level except the lowest, which is the ground state. (8.2)

Exclusion principle see *Pauli exclusion principle*.

Exothermic process a chemical reaction or physical change in which heat is evolved (q is negative). (6.3)

Experiment an observation of natural phenomena carried out in a controlled manner so that the results can be duplicated and rational conclusions obtained. (1.2)

Face-centered cubic unit cell a cubic unit cell in which there are lattice points at the center of each face of the unit cell in addition to those at the corners. (11.7)

Factor-label method see *Dimensional analysis*.

Faraday constant (F) the magnitude of charge on one mole of electrons, equal to 9.6485×10^4 C. (19.4)

f-Block transition elements (inner transition elements) the elements with a partially filled f subshell in common oxidation states. (8.2)

Fibrous proteins proteins that form long coils or align themselves in parallel to form long, water-insoluble fibers. (24.3)

First ionization energy (first ionization potential) the minimum energy needed to remove the highest-energy (that is, the outermost) electron from a neutral atom in the gaseous state. (8.6)

First law of thermodynamics the change in internal energy of a system, ΔU, equals $q + w$ (heat plus work). (6.2)

Formal charge (of an atom in a Lewis formula) the hypothetical charge you obtain by assuming that bonding electrons are equally shared between bonded atoms and that the electrons of each lone pair belong completely to one atom. (9.9)

Formation constant (stability constant) of a complex ion (K_f) the equilibrium constant for the formation of a complex ion from the aqueous metal ion and the ligands. (17.5)

Formula see *Chemical formula* and *Molecular formula.*

Formula mass (FM) the sum of the atomic masses of all atoms in a formula unit of a compound. (3.1)

Formula unit the group of atoms or ions explicitly symbolized in the formula. (2.6)

Fractional (isotopic) abundance the fraction of the total number of atoms that is composed of a particular isotope. (2.4)

Fractional precipitation the technique of separating two or more ions from a solution by adding a reactant that precipitates first one ion, then another, and so forth. (17.3)

Frasch process a mining procedure in which underground deposits of solid sulfur are melted in place with superheated water, and the molten sulfur is forced upward as a froth using air under pressure. (21.10)

Free energy (G) a thermodynamic quantity defined by the equation $G = H - TS$. (18.4)

Free energy of formation see *Standard free energy of formation.*

Freezing the change of a liquid to the solid state. (11.2)

Freezing point the temperature at which a pure liquid changes to a crystalline solid, or freezes. (11.2)

Freezing-point depression a colligative property of a solution equal to the freezing point of the pure solvent minus the freezing point of the solution. (12.6)

Frequency (ν) the number of wavelengths of a wave that pass a fixed point in one unit of time (usually one second). (7.1)

Frequency factor the symbol A in the Arrhenius equation, assumed to be a constant. (13.6)

Fuel cell essentially a battery, but it differs by operating with a continuous supply of energetic reactants, or fuel. (19.8)

Fuel rods the cylinders that contain fissionable material for a nuclear reactor. (20.7)

Fullerene a family of molecules consisting of a closed cage of carbon atoms arranged in pentagons as well as hexagons. (21.8)

Functional group a reactive portion of a molecule that undergoes predictable reactions. (2.7 and 23.6)

Fusion see *Melting;* see also *Nuclear fusion.*

Galvanic cell see *Voltaic cell.*

Gamma emission emission from an excited nucleus of a gamma photon, corresponding to radiation with a wavelength of about 10^{-12} m. (20.1)

Gamma photon a particle of electromagnetic radiation of short wavelength (about 1 pm, or 10^{-12} m) and high energy. (20.1)

Gas the form of matter that is an easily compressible fluid; a given quantity of gas will fit into a container of almost any size and shape. (1.4)

Gas chromatography a chromatographic separation method in which a gaseous mixture of vaporized substances is separated into its components by passing the mixture through a column of packing material. Substances in the gaseous mixture are attracted to the packing material to different extents and thus move through the column at different rates. (1.5)

Geiger counter a kind of ionization counter used to count particles emitted by radioactive nuclei. It consists of a metal tube filled with gas, such as argon. (20.3)

Gene a sequence of nucleotides in a DNA molecule that codes for a given protein. (24.4)

Geometric isomers isomers in which the atoms are joined to one another in the same way but differ because some atoms occupy different relative positions in space. (10.4, 22.5, and 23.3)

Glass electrode a compact electrode used to determine pH by cell potential measurements. (19.7)

Globular proteins proteins in which long coils fold into compact, roughly spherical shapes. (24.3)

Goldschmidt process a method of preparing a metal by reduction of its oxide with powdered aluminum. (21.6)

Graham's law of effusion the rate of effusion of gas molecules from a particular hole is inversely proportional to the square root of the molecular weight of the gas at constant temperature and pressure. (5.7)

Gravimetric analysis a type of quantitative analysis in which the amount of a species in a material is determined by converting the species to a product that can be isolated completely and weighed. (4.9)

Ground state a quantum-mechanical state of an atom or molecule associated with the lowest energy level. States associated with higher energy levels are called *excited states.* (8.2)

Group (of the periodic table) the elements in any one column of the periodic table. (2.5)

Haber process an industrial process for the preparation of ammonia from nitrogen and hydrogen with a specially prepared catalyst, high temperature, and high pressure. (3.6 and 21.9)

Half-cell the portion of an electrochemical cell in which a half-reaction takes place. (19.2)

Half-life ($t_{1/2}$) the time it takes for the reactant concentration to decrease to one-half of its initial value. (13.4)   The time it takes for one-half of the nuclei in a sample to decay. (20.4)

Half-reaction one of two parts of an oxidation–reduction reaction, one part of which involves a loss of electrons (or increase of oxidation number) and the other a gain of electrons (or decrease of oxidation number). (4.5 and 19.1)

Hall–Héroult process the commercial method for producing aluminum by the electrolysis of a molten mixture of aluminum oxide in cryolite, Na_3AlF_6. (21.2)

Halogen a Group VIIA element. The Group VIIA elements are reactive nonmetals, except possibly element 117. (2.5 and 8.7)

Heat an energy transfer (energy flow) into or out of a thermodynamic system that results from a temperature difference between the system and its surroundings. (6.2)

Heat capacity (C) the quantity of heat needed to raise the temperature of a sample of substance one degree Celsius (or one kelvin). (6.6)

Heat of formation see *Standard enthalpy of formation.*

Heat of fusion (enthalpy of fusion) the heat needed for the melting of a solid. (11.2)

Heat of phase transition the heat absorbed or evolved during a phase transition. (11.2)

Heat of reaction the heat, q, absorbed or evolved from a reaction system to retain a fixed temperature of the system under the conditions specified for the reaction (such as a fixed pressure). (6.3)

Heat of solution the heat absorbed (or evolved) when an ionic substance dissolves in water. (12.3)

Heat of vaporization (enthalpy of vaporization) the heat needed for the vaporization of a liquid. (11.2)

Henderson–Hasselbalch equation an equation relating the pH of a buffer for different concentrations of conjugate acid and base: $pH = pK_a + \log [\text{base}]/[\text{acid}]$. (16.6)

Henry's law the solubility of a gas is directly proportional to the partial pressure of the gas above the solution. (12.3)

Hess's law of heat summation for a chemical equation that can be written as the sum of two or more steps, the enthalpy change for the overall equation equals the sum of the enthalpy changes for the individual steps. (6.7)

Heterogeneous catalysis the use of a catalyst that exists in a different phase from the reacting species, usually a solid catalyst in contact with a gaseous or liquid solution of reactants. (13.9)

Heterogeneous equilibrium an equilibrium that involves reactants and products in more than one phase. (14.3)

Heterogeneous mixture a mixture that consists of physically distinct parts, each with different properties. (1.4)

Heteronuclear diatomic molecule molecule composed of two different nuclei. (10.6)

Hexagonal close-packed structure (hcp) a crystal structure composed of close-packed atoms (or other units) with the stacking ABABABA $\cdots$; the structure has a hexagonal unit cell. (11.8)

High-spin complex ion a complex ion in which there is minimum pairing of electrons in the orbitals of the metal atom. (22.7)

Homogeneous catalysis the use of a catalyst in the same phase as the reacting species. (13.9)

Homogeneous equilibrium an equilibrium that involves reactants and products in a single phase. (14.3)

Homogeneous mixture (solution) a mixture that is uniform in its properties throughout given samples. (1.4)

Homologous series a series of compounds in which one compound differs from a preceding one by a fixed group of atoms, for example, a —CH_2— group. (23.2)

Homonuclear diatomic molecule molecule composed of two like nuclei. (10.6)

Homopolymer a polymer whose monomer units are all alike. (24.1)

Hund's rule the lowest-energy arrangement of electrons in a subshell is obtained by putting electrons into separate orbitals of the subshell with the same spin before pairing electrons. (8.4)

Hybrid orbitals orbitals used to describe bonding that are obtained by taking combinations of atomic orbitals of the isolated atoms. (10.3)

Hydrate a compound that contains water molecules weakly bound in its crystals. (2.8)

Hydration the attraction of ions for water molecules. (12.2)

Hydrocarbons compounds containing only carbon and hydrogen. (2.7 and 23.1)

Hydrogen bonding a weak to moderate attractive force that exists between a hydrogen atom covalently bonded to a very electronegative atom and a lone pair of electrons on another small, electronegative atom. It is represented in formulas by a series of dots. (11.5)

Hydrologic cycle the natural cycle of water from the oceans to fresh water sources and back to the oceans. (11.10)

Hydrolysis the reaction of an ion with water to produce the conjugate acid and hydroxide ion or the conjugate base and hydrogen ion. (16.4)

Hydronium ion the H_3O^+ ion; also called the hydrogen ion and written $H^+(aq)$. (4.4)

Hydrophilic colloid a colloid in which there is a strong attraction between the dispersed phase and the continuous phase (water). (12.9)

Hydrophobic colloid a colloid in which there is a lack of attraction between the dispersed phase and the continuous phase (water). (12.9)

Hypothesis a tentative explanation of some regularity of nature. (1.2)

Ideal gas law the equation $PV = nRT$, which combines all of the gas laws. (5.3)

Ideal solution a solution of two or more substances each of which follows Raoult's law. (12.5)

Immiscible fluids fluids that do not mix but form separate layers. (12.1)

Inert gases see *Noble gases.*

Initiator a compound that produces free radicals in a reaction for the preparation of an addition polymer. (24.1)

Inner transition elements the two rows of elements at the bottom of the periodic table (2.5); the elements with a partially filled f subshell in common oxidation states. (8.2 and 22.1)

Inorganic compounds compounds composed of elements other than carbon. A few simple compounds of carbon, including carbon monoxide, carbon dioxide, carbonates, and cyanides, are generally considered to be inorganic. (2.8)

Integrated rate law a mathematical relationship between concentration and time. (13.4)

Intermolecular forces the forces of interaction between molecules. (11.5)

Internal energy (U) the sum of the kinetic and the potential energies of the particles making up a system. (6.1)

International System of Units (SI) a particular choice of metric units that was adopted by the General Conference of Weights and Measures in 1960. (1.6)

Ion an electrically charged particle obtained from an atom or a chemically bonded group of atoms by adding or removing electrons. (2.6)

Ion exchange a process in which a water solution is passed through a column of a material that replaces one kind of ion in solution with another kind. (11.10)

Ionic bond a chemical bond formed by the electrostatic attraction between positive and negative ions. (9.1)

Ionic compound a compound composed of cations and anions. (2.6)

Ionic equation see *Complete ionic equation.*

Ionic radius a measure of the size of the spherical region around the nucleus of an ion within which the electrons are most likely to be found. (9.3)

Ionic solid a solid that consists of cations and anions held together by the electrical attraction of opposite charges (ionic bonds). (11.6)

Ionization energy the energy needed to remove an electron from an atom (or molecule). Often used to mean *first ionization energy*. (8.6)

Ion product (Q_c) the product of ion concentrations in a solution, each concentration raised to a power equal to the number of ions in the formula of the ionic compound. (17.3)

Ion-product constant for water (K_w) the equilibrium value of the ion product $[H_3O^+][OH^-]$. (15.6)

Ion-selective electrode an electrode whose cell potential depends on the concentration of a particular ion in solution. (19.7)

Isoelectronic refers to different species that have the same number and configuration of electrons. (9.3)

Isomers compounds of the same molecular formula but with different arrangements of the atoms. (10.4)

Isotope dilution a technique to determine the quantity of a substance in a mixture or the total volume of solution by adding a known amount of an isotope to it. (20.5)

Isotopes atoms whose nuclei have the same atomic number but different mass numbers; that is, the nuclei have the same number of protons but different numbers of neutrons. (2.3)

Joule (J) the SI unit of energy; $1 \text{ J} = 1 \text{ kg} \cdot \text{m}^2/\text{s}^2$. (6.1)

Kelvin (K) the SI base unit of temperature; a unit on the absolute temperature scale. (1.6)

Ketone a compound containing a carbonyl group with two hydrocarbon groups attached to it. (23.6)

Kilogram (kg) the SI base unit for mass; equal to about 2.2 pounds. (1.6)

Kinetic energy the energy associated with an object by virtue of its motion. (6.1)

Kinetic-molecular theory (kinetic theory) the theory that a gas consists of molecules in constant random motion. (5.6)

Kinetics, chemical see *Chemical kinetics*.

Lanthanides the first of the two rows of elements at the bottom of the periodic table; lanthanum plus the 14 elements following it in the periodic table, in which the 4*f* subshell is filling. (22.1)

Laser a source of an intense, highly directed beam of monochromatic light; the word *laser* is an acronym meaning *l*ight *a*mplification by *s*timulated *e*mission of *r*adiation. (7.3)

Lattice energy the change in energy that occurs when an ionic solid is separated into isolated ions in the gas phase. (9.1 and 12.2)

Law a concise statement or mathematical equation about a fundamental relationship or regularity of nature. (1.2)

Law of combining volumes a relation stating that gases at the same temperature and pressure react with one another in volume ratios of small whole numbers. (5.2)

Law of conservation of energy energy may be converted from one form to another, but the total quantity of energy remains constant. (6.1)

Law of conservation of mass the total mass remains constant during a chemical change (chemical reaction). (1.3)

Law of definite proportions (law of constant composition) a pure compound, whatever its source, always contains definite or constant proportions of the elements by mass. (1.4)

Law of effusion see *Graham's law of effusion*.

Law of heat summation see *Hess's law of heat summation*.

Law of mass action the values of the equilibrium-constant expression K_c are constant for a particular reaction at a given temperature, whatever equilibrium concentrations are substituted. (14.2)

Law of multiple proportions when two elements form more than one compound, the masses of one element in these compounds for a fixed mass of the other element are in ratios of small whole numbers. (2.1)

Law of partial pressures see *Dalton's law of partial pressures*.

Lead storage cell a voltaic cell that consists of electrodes of lead alloy grids; one electrode is packed with a spongy lead to form the anode, and the other is packed with lead dioxide to form the cathode. (19.8)

Le Châtelier's principle when a system in equilibrium is disturbed by a change of temperature, pressure, or concentration variable, the system shifts in equilibrium composition in a way that tends to counteract this change of variable. (12.3 and 14.7)

Leclanché dry cell see *Zinc–carbon dry cell*.

Levorotatory refers to a compound whose solution rotates the plane of polarized light to the left (when looking toward the source of light). (22.5)

Lewis acid a species that can form a covalent bond by accepting an electron pair from another species. (15.3)

Lewis base a species that can form a covalent bond by donating an electron pair to another species. (15.3)

Lewis electron-dot formula a formula in which dots are used to represent valence electrons. (9.4)

Lewis electron-dot symbol a symbol in which the electrons in the valence shell of an atom or ion are represented by dots placed around the letter symbol of the element. (9.1)

Ligand a Lewis base that bonds to a metal ion to form a complex ion. (17.5 and 22.3)

Limiting reactant (limiting reagent) the reactant that is entirely consumed when a reaction goes to completion. (3.8)

Linear geometry a molecular geometry in which all atoms line up along a straight line. (10.1)

Line spectrum a spectrum showing only certain colors or specific wavelengths of light. (7.3)

Linkage isomers isomers of a complex that differ in the atom of a ligand that is bonded to the metal atom. (22.5)

Lipids biological substances like fats and oils that are soluble in organic solvents, such as chloroform and carbon tetrachloride. (12.9)

Liquefaction the process in which a substance that is normally a gas changes to the liquid state. (11.2)

Liquid the form of matter that is a relatively incompressible fluid; a liquid has a fixed volume but no fixed shape. (1.4)

Liter (L) a unit of volume equal to a cubic decimeter (equal to approximately one quart). (1.7)

Lithium–iodine battery a voltaic cell in which the anode is lithium metal and the cathode is an I_2 complex. (19.8)

London forces (dispersion forces) the weak attractive forces between molecules resulting from the small, instantaneous dipoles that occur because of the varying positions of the electrons during their motion about nuclei. (11.5)

Lone pair (nonbonding pair) an electron pair that remains on one atom and is not shared. (9.4)

Low-spin complex ion a complex ion in which there is more pairing of electrons in the orbitals of the metal atom than in a corresponding high-spin complex ion. (22.7)

Magic number the number of nuclear particles in a completed shell of protons or neutrons. (20.1)

Magnetic quantum number (m_l) the quantum number that distinguishes orbitals of given n and l—that is, of given energy and shape—but having a different orientation in space; the allowed values are the integers from $-l$ to $+l$. (7.5)

Main-group element (representative element) an element in an A column of the periodic table, in which an outer *s* or *p* subshell is filling. (2.5 and 8.2)

Manometer a device that measures the pressure of a gas or liquid in a sealed vessel. (5.1)

Markownikoff's rule a generalization stating that the major product formed by the addition of an unsymmetrical reagent such as H—Cl, H—Br, or H—OH is the one obtained when the H atom of the reagent adds to the carbon atom of the multiple bond that already has the greater number of hydrogen atoms attached to it. (23.3)

Mass the quantity of matter in a material. (1.3)

Mass defect the total nucleon mass minus the atomic mass of a nucleus. (20.6)

Mass number (A) the total number of protons and neutrons in a nucleus. (2.3)

Mass percentage parts per hundred parts of the total, by mass. (3.3)

Mass percentage of solute the percentage by mass of solute contained in a solution. (12.4)

Mass spectrometer an instrument, such as one based on Thomson's principles, that measures the mass-to-charge ratios of atoms. (2.4)

Material any particular kind of matter. (1.1)

Matter all of the objects around you (1.1); whatever occupies space and can be perceived by our senses. (1.3)

Maximum work the maximum (useful) work obtained from a spontaneous process equals the change of free energy, $w_{max} = \Delta G$. (18.5)

Maxwell's distribution of molecular speeds a theoretical relationship that predicts the relative number of molecules at various speeds for a sample of gas at a particular temperature. (5.7)

Melting (fusion) the change of a solid to the liquid state. (11.2)

Melting point the temperature at which a crystalline solid changes to a liquid, or melts. (11.2)

Messenger RNA a relatively small RNA molecule that can diffuse about the cell and attach itself to a ribosome, where it serves as a pattern for protein synthesis. (24.4)

Metal a substance or mixture that has a characteristic luster or shine, is generally a good conductor of heat and electricity, and is malleable and ductile. (2.5 and 21.2)

Metallic solid a solid that consists of positive cores of atoms held together by a surrounding "sea" of electrons (metallic bonding). (11.6)

Metalloid (semimetal) an element having both metallic and nonmetallic properties. (2.5)

Metallurgy the scientific study of the production of metals from their ores and the making of alloys having various useful properties. (21.2)

Metal refining in metallurgy, the purification of a metal. (21.2)

Metaphosphoric acids acids with the general formula $(HPO_3)_n$. (21.9)

Metastable nucleus a nucleus in an excited state with a lifetime of at least one nanosecond (10^{-9} s). (20.1)

Metathesis reaction see *Exchange reaction.*

Meter (m) the SI base unit of length. (1.6)

Micelle a colloidal-sized particle formed in water by the association of molecules or ions each of which has a hydrophobic end and a hydrophilic end. (12.9)

Millimeters of mercury (mmHg) a unit of pressure also known as the *torr*. A unit of pressure equal to that exerted by a column of mercury 1 mm high at 0.00°C. (5.1)

Mineral a naturally occurring inorganic solid substance or solid solution with definite crystalline form. (21.2)

Miscible fluids fluids that mix with or dissolve in each other in all proportions. (12.1)

Mixture a material that can be separated by physical means into two or more substances. (1.4)

Moderator a substance that slows down neutrons in a nuclear fission reactor. (20.7)

Molality the moles of solute per kilogram of solvent. (12.4)

Molar concentration (molarity), *M* the moles of solute dissolved in one liter (cubic decimeter) of solution. (4.7)

Molar gas constant (*R*) the constant of proportionality relating the molar volume of a gas to *T/P*. (5.3)

Molar gas volume (*V_m*) the volume of one mole of gas. (5.2)

Molarity see *Molar concentration.*

Molar mass the mass of one mole of substance. In grams, it is numerically equal to the formula mass in atomic mass units. (3.2)

Mole (mol) the quantity of a given substance that contains as many molecules or formula units as the number of atoms in exactly 12 g of carbon-12. The amount of substance containing Avogadro's number of molecules or formula units. (3.2)

Molecular equation a chemical equation in which the reactants and products are written as if they were molecular substances, even though they may actually exist in solution as ions. (4.2)

Molecular formula a chemical formula that gives the exact number of different atoms of an element in a molecule. (2.6)

Molecular geometry the general shape of a molecule, as determined by the relative positions of the atomic nuclei. (10.1)

Molecularity the number of molecules on the reactant side of an elementary reaction. (13.7)

Molecular mass (MM) the sum of the atomic masses of all the atoms in a molecule. (3.1)

Molecular orbital theory a theory of the electronic structure of molecules in terms of molecular orbitals, which may spread over several atoms or the entire molecule. (10.5)

Molecular solid a solid that consists of atoms or molecules held together by intermolecular forces. (11.6)

Molecular substance a substance that is composed of molecules, all of which are alike. (2.6)

Molecule a definite group of atoms that are chemically bonded together—that is, tightly connected by attractive forces. (2.6)

Mole fraction the fraction of moles of a component in the total moles of a mixture. (5.5); the moles of a component substance divided by the total moles of solution. (12.4)

Mole percent percent, in terms of moles, of a component in a solution; obtained by multiplying mole fraction by 100. (12.4)

Monatomic ion an ion formed from a single atom. (2.8)

Monodentate ligand a ligand that bonds to a metal atom through one atom of the ligand. (22.3)

Monomer the small molecules that are linked together to form a polymer (2.6); a compound used to make a polymer (and from which the polymer's repeating unit arises). (24.1)

Monoprotic acid an acid that yields one acidic hydrogen per molecule. (4.4)

Monosaccharides simple sugars, each containing three to nine carbon atoms, generally all but one of which bear a hydroxyl group, the remaining one being part of a carbonyl group. (24.4)

Nernst equation an equation relating the cell potential, E_{cell}, to its standard potential cell, $E°_{cell}$, and the reaction quotient, Q. At 25°C, the equation is $E_{cell} = E°_{cell} - (0.0592/n) \log Q$. (19.7)

Net ionic equation an ionic equation from which spectator ions have been canceled. (4.2)

Neutralization reaction a reaction of an acid and a base that results in an ionic compound and possibly water. (4.4)

Neutron a particle found in the nucleus of an atom; it has a mass almost identical to that of the proton but no electric charge. (2.3)

Neutron activation analysis an analysis of elements in a sample based on the conversion of stable isotopes to radioactive isotopes by bombarding a sample with neutrons. (20.5)

Nickel–cadmium cell a voltaic cell consisting of an anode of cadmium and a cathode of hydrated nickel oxide (approximately $NiOOH$) on nickel; the electrolyte is potassium hydroxide. (19.8)

Nitrogen cycle the circulation of the element nitrogen in the biosphere, from nitrogen fixation to the release of free nitrogen by denitrifying bacteria. (21.9)

Nitrogen-fixing bacteria bacteria that can produce nitrogen compounds in the soil from atmospheric nitrogen. (21.9)

Noble-gas core an inner-shell configuration corresponding to one of the noble gases. (8.2)

Noble gases (inert gases; rare gases) the Group VIIIA elements; all are gases consisting of uncombined atoms. They are relatively unreactive elements. (8.7)

Nomenclature see *Chemical nomenclature.*

Nonbonding orbital an orbital that is neither bonding nor antibonding. (10.6)

Nonbonding pair see *Lone pair.*

Nonelectrolyte a substance, such as sucrose, or table sugar ($C_{12}H_{22}O_{11}$), that dissolves in water to give a nonconducting or very poorly conducting solution. (4.1)

Nonmetal an element that does not exhibit the characteristics of a metal. (2.5)

Nonpolar molecule a molecule having a dipole moment of zero. (10.2)

Nonstoichiometric compound a compound whose composition varies from its idealized formula. (11.7)

Nuclear bombardment reaction a nuclear reaction in which a nucleus is bombarded, or struck, by another nucleus or by a nuclear particle. (20.1)

Nuclear equation a symbolic representation of a nuclear reaction. (20.1)

Nuclear fission a nuclear reaction in which a heavy nucleus splits into lighter nuclei and energy is released. (20.6)

Nuclear fission reactor a device that permits a controlled chain reaction of nuclear fissions. (20.7)

Nuclear force a strong force of attraction between nucleons that acts only at very short distances (about 10^{-15} m). (20.1)

Nuclear fusion a nuclear reaction in which light nuclei combine to give a stabler, heavier nucleus plus possibly several neutrons, and energy is released. (20.6)

Nucleic acids polynucleotides folded or coiled into specific three-dimensional shapes. (24.4)

Nucleotides the building blocks of nucleic acids. (24.4)

Nucleus the atom's central core; it has most of the atom's mass and one or more units of positive charge. (2.2)

Nuclide a particular atom characterized by a definite atomic number and mass number. (2.3)

Nuclide symbol a symbol for a nuclide, in which the atomic number is given as a left subscript and the mass number is given as a left superscript to the symbol of the element. (2.3)

Number of significant figures the number of digits reported for the value of a measured or calculated quantity, indicating the precision of the value. (1.5)

Octahedral geometry the geometry of a molecule in which six atoms occupy the vertices of a regular octahedron (a figure with eight faces and six vertices) with the central atom at the center of the octahedron. (10.1)

Octet rule the tendency of atoms in molecules to have eight electrons in their valence shells (two for hydrogen atoms). (9.4)

Optical isomers (enantiomers) isomers that are nonsuperimposable mirror images of one another. (22.5)

Optically active having the ability to rotate the plane of light waves, either as a pure substance or in solution. (22.5)

Orbital diagram a diagram to show how the orbitals of a subshell are occupied by electrons. (8.1)

Ore a rock or mineral from which a metal or nonmetal can be economically produced. (21.2)

Organic compounds compounds that contain carbon combined with other elements, such as hydrogen, oxygen, and nitrogen. (2.7 and 23.2)

Osmosis the phenomenon of solvent flow through a semipermeable membrane to equalize the solute concentrations on both sides of the membrane. (12.7)

Osmotic pressure a colligative property of a solution equal to the pressure that, when applied to the solution, just stops osmosis. (12.7)

Ostwald process an industrial preparation of nitrogen monoxide starting from the catalytic oxidation of ammonia to nitric oxide. (21.9)

Overall order of a reaction the sum of the orders of the reactant species in the rate law. (13.3)

Oxidation the part of an oxidation–reduction reaction in which there is a loss of electrons by a species (or an increase in the oxidation number of an atom). (4.5)

Oxidation number (oxidation state) either the actual charge on an atom in a substance, if the atom exists as a monatomic ion, or a hypothetical charge assigned by simple rules. (4.5)

Oxidation potential the negative of the standard electrode potential. (19.5)

Oxidation–reduction reaction (redox reaction) a reaction in which electrons are transferred between species or in which atoms change oxidation number. (4.5)

Oxide a binary compound with oxygen in the -2 oxidation state. (21.10)

Oxidizing agent a species that oxidizes another species; it is itself reduced. (4.5)

Oxoacid an acid containing hydrogen, oxygen, and another element (often called the central element). (2.8); a substance in which O atoms (and possibly other electronegative atoms) are bonded to a central atom, with one or more H atoms usually bonded to the O atoms. (9.6)

Pairing energy (P) the energy required to put two electrons into the same orbital. (22.7)

Paramagnetic substance a substance that is weakly attracted by a magnetic field; this attraction generally results from unpaired electrons. (8.4)

Partial pressure the pressure exerted by a particular gas in a mixture. (5.5)

Particle accelerator a device used to accelerate electrons, protons, and alpha particles and other ions to very high speeds. (20.2)

Pascal (Pa) the SI unit of pressure; 1 Pa = 1 kg/(m·s^2). (5.1)

Pauli exclusion principle no two electrons in an atom can have the same four quantum numbers. It follows from this that an orbital can hold no more than two electrons and can hold two only if they have different spin quantum numbers. (8.1)

Peptide (amide) bond the C—N bond resulting from a condensation reaction between the carboxyl group of one amino acid and the amino group of a second amino acid. (24.3)

Percentage composition the mass percentages of each element in a compound. (3.3)

Percentage yield the actual yield (experimentally determined) expressed as a percentage of the theoretical yield (calculated). (3.8)

Period (of the periodic table) the elements in any one horizontal row of the periodic table. (2.5)

Periodic law when the elements are arranged by atomic number, their physical and chemical properties vary periodically. (8.6)

Periodic table a tabular arrangement of elements in rows and columns, highlighting the regular repetition of properties of the elements. (2.5)

Peroxide a compound with oxygen in the -1 oxidation state. (21.10)

pH the negative of the logarithm of the molar hydrogen-ion concentration. (15.8)

Phase one of several different homogeneous materials present in the portion of matter under study. (1.4)

Phase diagram a graphical way to summarize the conditions under which the different states of a substance are stable. (11.3)

Phase transition see *Change of state.*

Phospholipid a type of lipid that contains a phosphate group. These molecules have hydrophobic and hydrophilic ends, and they aggregate to give a layer structure. The process is similar to the formation of micelles. (12.9)

Phospholipid bilayer a part of a biological membrane consisting of two layers of phospholipid molecules. (12.9)

Photoelectric effect the ejection of electrons from the surface of a metal or other material when light shines on it. (7.2)

Photon particle of electromagnetic energy with energy E proportional to the observed frequency of light: $E = h\nu$. (7.2)

Physical adsorption adsorption in which the attraction is provided by weak intermolecular forces. (13.9)

Physical change a change in the form of matter but not in its chemical identity. (1.4)

Physical property a characteristic that can be observed for a material without changing its chemical identity. (1.4)

Pi (π) bond a bond that has an electron distribution above and below the bond axis. (10.4)

Planck's constant (h) a physical constant with the value 6.63×10^{-34} J·s. It is the proportionality constant relating the frequency of light to the energy of a photon. (7.2)

Plasma an electrically neutral gas of ions and electrons. (20.7)

Polar covalent bond a covalent bond in which the bonding electrons spend more time near one atom than near the other. (9.5)

Polar molecule a molecule having a dipole moment. (10.2)

Polyamide a polymer formed by reacting a substance containing two amine groups with a substance containing two carboxylic acid groups. (24.1)

Polyatomic ion an ion consisting of two or more atoms chemically bonded together and carrying a net electric charge. (2.8)

Polydentate ligand a ligand that can bond with two or more atoms to a metal atom. (22.3)

Polyester a polymer formed by reacting a substance containing two alcohol groups with a substance containing two carboxylic acid groups. (24.1)

Polymer a very large molecule made up of a number of smaller molecules repeatedly linked together. (2.6); a chemical species of very high molecular mass that is made up from many repeating units of low molecular mass. (24.1)

Polymerase chain reaction (PCR) a technique used to amplify the quantity of DNA in a sample. (24.4)

Polynucleotide a linear polymer of nucleotide units. (24.4)

Polypeptide a polymer formed by the linking of many amino acids by peptide bonds. (24.3)

Polyphosphoric acids acids with the general formula $H_{n+2}P_nO_{3n-1}$ formed from linear chains of P—O bonds. (21.9)

Polyprotic acid an acid that yields two or more acidic hydrogens per molecule. (4.4 and 16.2)

Positron a particle that is similar to an electron and has the same mass but has a positive charge. (20.1)

Positron emission emission of a positron from an unstable nucleus. (20.1)

Potential difference the difference in electrical potential (electrical pressure) between two points. (19.4)

Potential energy the energy an object has by virtue of its position in a field of force. (6.1)

Precipitate an insoluble solid compound formed during a chemical reaction in solution. (4.3)

Precision the closeness of the set of values obtained from identical measurements of a quantity. (1.5)

Pressure the force exerted per unit area of surface. (5.1)

Pressure-volume work work equal to the negative of the pressure times the change in volume of the system. (6.3)

Primary alcohol an alcohol in which the hydroxyl group is attached to a carbon atom that is itself bonded to only one other carbon atom. (23.6)

Primary structure (of a protein) the order, or sequence, of the amino-acid units in the protein. (24.3)

Principal quantum number (n) the quantum number on which the energy of an electron in an atom principally depends; it can have any positive integer value: 1, 2, 3, ... (7.5)

Product a substance that results from a chemical reaction. (2.9)

Protein a biological polymer of small molecules called amino acids; a polypeptide that has a biological function. (24.3)

Proton a particle found in the nucleus of the atom; it has a positive charge equal in magnitude, but opposite in sign, to that of the electron and a mass 1836 times that of the electron. (2.3)

Pseudo-noble-gas core the noble-gas core together with $(n - 1)d^{10}$ electrons. (8.2)

Qualitative analysis the determination of the identity of substances present in a mixture. (4.9 and 17.7)

Quantitative analysis the determination of the amount of a substance or species present in a material. (4.9)

Quantum (wave) mechanics the branch of physics that mathematically describes the wave properties of submicroscopic particles. (7.4)

Racemic mixture a mixture of equal amounts of optical isomers. (22.5)

Rad the dosage of radiation that deposits 1×10^{-2} J of energy per kilogram of tissue. (20.3)

Radioactive decay the process in which a nucleus spontaneously disintegrates, giving off radiation. (20.1)

Radioactive decay constant (k) rate constant for radioactive decay. (20.4)

Radioactive decay series a sequence in which one radioactive nucleus decays to a second, which then decays to a third, and so on. (20.1)

Radioactive tracer a very small amount of radioactive isotope added to a chemical, biological, or physical system to facilitate study of the system. (20.5)

Radioactivity spontaneous radiation from unstable elements. (20.1)

Raoult's law the partial pressure of solvent, P_A, over a solution equals the vapor pressure of the pure solvent, P°_A, times the mole fraction of the solvent, X_A, in solution: $P_A = P^\circ_A X_A$. (12.5)

Rare gases see *Noble gases.*

Rate constant a proportionality constant in the relationship between rate and concentrations. (13.3)

Rate-determining step the slowest step in a reaction mechanism. (13.8)

Rate law an equation that relates the rate of a reaction to the concentrations of reactants (and catalyst) raised to various powers. (13.3)

Reactant a starting substance in a chemical reaction. (2.9)

Reaction intermediate a species produced during a reaction that does not appear in the net equation because it reacts in a subsequent step in the mechanism. (13.7)

Reaction mechanism the set of elementary reactions whose overall effect is given by the net chemical equation. (13.7)

Reaction order the exponent of the concentration of a given reactant species in the rate law, as determined experimentally. (13.3) See also *Overall reaction order.*

Reaction quotient (Q_c) an expression that has the same form as the equilibrium-constant expression but whose concentration values are not necessarily those at equilibrium. (14.5)

Reaction rate the increase in molar concentration of product of a reaction per unit time or the decrease in molar concentration of reactant per unit time. (13.1)

Redox reaction see *Oxidation–reduction reaction.*

Reducing agent a species that reduces another species; it is itself oxidized. (4.5)

Reduction the part of an oxidation–reduction reaction in which there is a gain of electrons by a species (or a decrease of oxidation number of an atom). (4.5)

Reduction potential see *Standard electrode potential.*

Reference form the stablest form (physical state and allotrope) of an element under standard thermodynamic conditions. (6.8)

Refining see *Metal refining*.

Rem a unit of radiation dosage used to relate various kinds of radiation in terms of biological destruction. It equals the rad times a factor for the type of radiation, called the relative biological effectiveness (RBE): rems = rads × RBE. (20.3)

Representative element see *Main-group element*.

Resonance description a representation in which you describe the electron structure of a molecule having delocalized bonding by writing all possible electron-dot formulas. (9.7)

Reverse osmosis a process in which a solvent, such as water, is forced by a pressure greater than the osmotic pressure to flow through a semipermeable membrane from a concentrated solution to a more dilute one. (12.7)

Ribonucleic acid (RNA) a constituent of cells that is used to manufacture proteins from genetic information. It is a polymer of ribonucleotide units. (24.4)

Ribosomal RNA the RNA in a ribosome. (24.4)

Ribosomes tiny cellular particles on which protein synthesis takes place. (24.4)

Roasting the process of heating a mineral in air to obtain the oxide. (21.2)

Root-mean-square (rms) molecular speed a type of average molecular speed, equal to the speed of a molecule having the average molecular kinetic energy. It equals $\sqrt{3RT/M_m}$, where M_m is the molar mass. (5.7)

Rounding the procedure of dropping nonsignificant digits in a calculation result and adjusting the last digit reported. (1.5)

Salt an ionic compound that is a product of a neutralization reaction. (4.4)

Salt bridge a tube of an electrolyte in a gel that is connected to the two half-cells of a voltaic cell; it allows the flow of ions but prevents the mixing of the different solutions that would allow direct reaction of the cell reactants. (19.2)

Saturated hydrocarbon a hydrocarbon that has only single bonds between carbon atoms; all carbon atoms are bonded to the maximum number of hydrogen atoms (that is, the hydrocarbon is *saturated* with hydrogen). A saturated hydrocarbon molecule can be cyclic or acyclic. (23.1)

Saturated solution a solution that is in equilibrium with respect to a given dissolved substance. (12.2)

Scientific method the general process of advancing scientific knowledge through observation, the framing of laws, hypotheses, or theories, and the conducting of more experiments. (1.2)

Scientific notation the representation of a number in the form $A \times 10^n$, where A is a number with a single nonzero digit to the left of the decimal point and n is an integer, or whole number. (1.5)

Scintillation counter a device that detects nuclear radiations from flashes of light generated in a material by the radiation. (20.3)

Second (s) the SI base unit of time. (1.6)

Secondary alcohol an alcohol in which the hydroxyl group is attached to a carbon atom that is itself bonded to two carbon atoms. (23.6)

Secondary structure (of a protein) the relatively simple coiled or parallel arrangement of a protein molecule. (24.3)

Second law of thermodynamics the total entropy of a system and its surroundings always increases for a spontaneous process. Also, for a spontaneous process at a given temperature, the change in entropy of the system is greater than the heat divided by the absolute temperature. (18.2)

Seesaw geometry the geometry of a molecule having four atoms bonded to a central atom, in which two of these outer atoms occupy axial positions of a trigonal bipyramid and the other two occupy equatorial positions. (10.1)

Self-ionization see *Autoionization*.

Semimetal see *Metalloid*.

Shell model of the nucleus a nuclear model in which protons and neutrons exist in levels, or shells, analogous to the shell structure that exists for electrons in an atom. (20.1)

SI see *International System*.

SI base units the SI units of measurement from which all others can be derived. (1.6)

SI derived unit a unit derived by combining SI base units. (1.7)

SI prefix a prefix used in the International System to indicate a power of 10. (1.6)

Sigma (σ) bond a bond that has a cylindrical shape about the bond axis. (10.4)

Significant figures those digits in a measured number (or in the result of a calculation with measured numbers) that include all certain digits plus a final digit having some uncertainty. (1.5)

Silica a covalent network solid of SiO_2 in which each silicon atom is covalently bonded in tetrahedral directions to four oxygen atoms; each oxygen atom is in turn bonded to another silicon atom. (21.8)

Silicate a compound of silicon and oxygen (with one or more metals) that may be formally regarded as a derivative of silicic acid, H_4SiO_4 or $Si(OH)_4$. (21.8)

Silicone a polymer that contains chains or rings of Si—O with one or more of the bonding positions on each Si atom occupied by an organic group. (21.8)

Simple cubic unit cell a cubic unit cell in which lattice points are situated only at the corners of the unit cell. (11.7)

Simplest formula see *Empirical formula*.

Single bond a covalent bond in which a single pair of electrons is shared by two atoms. (9.4)

Single-replacement reaction see *Displacement reaction*.

Sol a colloid that consists of solid particles dispersed in a liquid. (12.9)

Solid the form of matter characterized by rigidity; a solid is relatively incompressible and has fixed shape and volume. (1.4)

Solubility the amount of a substance that dissolves in a given quantity of solvent (such as water) at a given temperature to give a saturated solution. (12.2)

Solubility product constant (K_{sp}) the equilibrium constant for the solubility equilibrium of a slightly soluble (or nearly insoluble) ionic compound. (17.1)

Solute in the case of a solution of a gas or solid dissolved in a liquid, the gas or solid; in other cases, the component in smaller amount. (12.1)

Solution see *Homogeneous mixture*.

Solvay process an industrial method for obtaining sodium carbonate from sodium chloride and limestone. (21.4)

Solvent in a solution of a gas or solid dissolved in a liquid, the liquid; in other cases, the component in greater amount. (12.1)

Specific heat capacity (specific heat) the quantity of heat required to raise the temperature of one gram of a substance by one degree Celsius (or one kelvin) at constant pressure. (6.6)

Spectator ion an ion in an ionic equation that does not take part in the reaction. (4.2)

Spectrochemical series an arrangement of ligands according to the relative magnitudes of the crystal field splittings they induce in the d orbitals of a metal ion. (22.7)

Spin quantum number (m_s) the quantum number that refers to the two possible orientations of the spin axis of an electron; its possible values are $+\frac{1}{2}$ and $-\frac{1}{2}$. (7.5)

Spontaneous fission the spontaneous decay of an unstable nucleus in which a heavy nucleus of mass number greater than 89 splits into lighter nuclei and energy is released. (20.1)

Spontaneous process a physical or chemical change that occurs by itself. (18.2)

Square planar geometry the geometry of a molecule in which a central atom is surrounded by four other atoms arranged in a square and in a plane containing the central atom. (10.1)

Square pyramidal geometry the geometry of a molecule in which a central atom is at the apex of a pyramid and four other atoms form the square base of the pyramid. (10.1)

Stability constant (of a complex) see *Formation constant.*

Standard cell potential (E°_{cell}) the potential of a voltaic cell operating under standard-state conditions (solute concentrations are 1 M, gas pressures are 1 atm, and the temperature has a specified value—usually 25°C). (19.5)

Standard electrode potential (E°) the electrode potential when the concentrations of solutes are 1 M, the gas pressures are 1 atm, and the temperature has a specified value—usually 25°C. (19.5)

Standard enthalpy of formation (standard heat of formation), ΔH°_f the enthalpy change for the formation of one mole of a substance in its standard state from its elements in their reference forms and in their standard states. (6.8)

Standard enthalpy of reaction (ΔH°) the enthalpy change for a reaction in which reactants in their standard states yield products in their standard states. (6.8)

Standard (absolute) entropy (S°) the entropy value for the standard state of a species. (18.3)

Standard free energy of formation (ΔG°_f) the free-energy change that occurs when one mole of substance is produced from its elements in their reference forms (usually the stablest states) at 1 atm and at a specified temperature (usually 25°C). (18.4)

Standard heat of formation see *Standard enthalpy of formation.*

Standard state the standard thermodynamic conditions (1 atm and usually 25°C) chosen for substances when listing or comparing thermochemical data. (6.8)

Standard temperature and pressure (STP) the reference conditions for gases, chosen by convention to be 0°C and 1 atm. (5.2)

State function a property of a system that depends only on its present state, which is determined by variables such as temperature and pressure and is independent of any previous history of the system. (6.3)

States of matter the three forms that matter can commonly assume—solid, liquid, and gas. (1.4)

Steam-reforming process an industrial preparation in which steam and hydrocarbons from natural gas or petroleum react at high temperature and pressure in the presence of a catalyst to form carbon monoxide and hydrogen. (21.7)

Stereoisomers isomers in which the atoms are bonded to each other in the same order but that differ in the precise arrangement of the atoms in space. (22.5)

Stock system a system of chemical nomenclature in which the charge on a metal atom or oxidation number of an atom is denoted by a Roman numeral in parentheses following the element name. (2.8)

Stoichiometry the calculation of the quantities of reactants and products involved in a chemical reaction. (3.6)

Stratosphere the region of the atmosphere occurring at 10 to 15 km above the ground level wherein the temperature increases with increasing altitude; it lies just above the troposphere, the lower portion of the atmosphere. (10.7)

Strong acid an acid that ionizes completely in water; it is a strong electrolyte. (4.4 and 15.1)

Strong base a base that is present in aqueous solution entirely as ions, one of which is OH⁻; it is a strong electrolyte. (4.4 and 15.1)

Strong electrolyte an electrolyte that exists in solution almost entirely as ions. (4.1)

Structural formula a chemical formula that shows how the atoms are bonded to one another in a molecule. (2.6)

Structural isomers see *Constitutional isomers.*

Sublimation the change of a solid directly to the vapor. (11.2)

Subshell the orbitals of a given shell n that have a particular l quantum number. (7.5)

Substance a kind of matter that cannot be separated into other kinds of matter by any physical process. (1.4)

Substitution reaction a reaction in which a part of the reacting molecule is substituted for an H atom on a hydrocarbon or a hydrocarbon group. (23.2)

Substrate the substance whose reaction an enzyme catalyzes. (13.9)

Superoxide a binary compound with oxygen in the $-\frac{1}{2}$ oxidation state; it contains the superoxide ion, O_2^-. (21.10)

Supersaturated solution a solution that contains more dissolved substance than does a saturated solution; the solution is not in equilibrium with the solid substance. (12.2)

Surface tension the energy required to increase the surface area of a liquid by a unit amount. (11.4)

Surroundings everything in the vicinity of a thermodynamic system. (6.2)

Synthesis gas reaction a chemical reaction in which a mixture of CO and H_2 gases is used in the industrial preparation of a number of organic compounds, including methanol, CH_3OH. (21.7)

System see *Thermodynamic system.*

Termolecular reaction an elementary reaction that involves three reactant molecules. (13.7)

Tertiary alcohol an alcohol in which the hydroxyl group is attached to a carbon atom that is itself bonded to three carbon atoms. (23.6)

Tertiary structure (of a protein) the structure associated with the way the protein coil is folded. (24.3)

Tetrahedral geometry the geometry of a molecule in which four atoms bonded to a central atom occupy the vertices of a tetrahedron with the central atom at the center of this tetrahedron. (10.1)

Theoretical yield the maximum amount of product that can be obtained by a reaction from given amounts of reactants. (3.8)

Theory a tested explanation of basic natural phenomena. (1.2)

Thermal equilibrium a state in which heat does not flow between a system and its surroundings because they are both at the same temperature. (6.2)

Thermochemical equation the chemical equation for a reaction (including phase labels) in which the equation is given a molar interpretation, and the enthalpy of reaction for these molar amounts is written directly after the equation. (6.4)

Thermochemistry the study of the quantity of heat absorbed or evolved by chemical reactions. (6.1)

Thermodynamic equilibrium constant (K) the equilibrium constant in which the concentrations of gases are expressed in partial pressures in atmospheres, whereas the concentrations of solutes in liquid solutions are expressed in molarities. (18.6)

Thermodynamics the study of the relationship between heat and other forms of energy involved in a chemical or physical process. (6.1 and Ch. 18 opener)

Thermodynamic system the substance or mixture of substances that we single out for study (perhaps including the vessel) in which a change occurs. (6.2)

Third law of thermodynamics a substance that is perfectly crystalline at 0 K has an entropy of zero. (18.3)

Titration a procedure for determining the amount of substance A by adding a carefully measured volume of a solution with known concentration of *B* until the reaction of *A* and *B* is just complete. (4.10)

Torr see *Millimeters of mercury.*

Transfer RNA (tRNA) a small RNA molecule; it binds to a particular amino acid, carries it to a ribosome, and then attaches itself (through base pairing) to a messenger RNA codon. (24.4)

Transition elements the B columns of elements in the periodic table. (2.5); the *d*-block transition elements in which a *d* subshell is being filled. (8.2); those metallic elements that have an incompletely filled *d* subshell or easily give rise to common ions that have incompletely filled *d* subshells. (22.1)

Transition state see *Activated complex.*

Transition-state theory a theory that explains the reaction resulting from the collision of two molecules in terms of an activated complex (transition state). (13.5)

Transmutation the change of one element to another by bombardment of the nucleus of the element with nuclear particles or nuclei. (20.2)

Transuranium element an element with atomic number greater than that of uranium ($Z = 92$), the naturally occurring element of greatest Z. (20.2)

Trigonal bipyramidal geometry the geometry of a molecule in which five atoms bonded to a central atom occupy the vertices of a trigonal bipyramid (formed by placing two trigonal pyramids base to base) with the central atom at the center of this trigonal bipyramid. (10.1)

Trigonal planar geometry the geometry of a molecule in which a central atom is surrounded by three other atoms arranged in a triangle and in a plane containing the central atom. (10.1)

Trigonal pyramidal geometry the geometry of a molecule in which a central atom is at the apex of a pyramid and three other atoms form the triangular base of the pyramid. (10.1)

Triple bond a covalent bond in which three pairs of electrons are shared by two atoms. (9.4)

Triple point the point on a phase diagram representing the temperature and pressure at which three phases of a substance coexist in equilibrium. (11.3)

T-shaped geometry the geometry of a molecule in which three atoms are bonded to a central atom to form a T. (10.1)

Tyndall effect the scattering of light by colloidal-size particles. (12.9)

Uncertainty principle a relation stating that the product of the uncertainty in position and the uncertainty in momentum (mass times speed) of a particle can be no smaller than Planck's constant divided by 4π. (7.4)

Unimolecular reaction an elementary reaction that involves one reactant molecule. (13.7)

Unit a fixed standard of measurement. (1.5)

Unit cell the smallest boxlike unit (each box having faces that are parallelograms) from which you can imagine constructing a crystal by stacking the units in three dimensions. (11.7)

Unsaturated hydrocarbon a hydrocarbon that has at least one double or triple bond between carbon atoms; not all carbon atoms are bonded to the maximum number of hydrogen atoms (that is, the hydrocarbon is *unsaturated* with hydrogen). (23.1)

Unsaturated solution a solution that is not in equilibrium with respect to a given dissolved substance and in which more of the substance can dissolve. (12.2)

Valence bond theory an approximate theory to explain the electron pair or covalent bond in terms of quantum mechanics. (10.3)

Valence electron an electron in an atom outside the noble-gas or pseudo-noble-gas core. (8.2)

Valence-shell electron-pair repulsion (VSEPR) model predicts the shapes of molecules and ions by assuming that the valence-shell electron pairs are arranged about each atom so that electron pairs are kept as far away from one another as possible, thus minimizing electron-pair repulsions. (10.1)

van der Waals equation an equation that is similar to the ideal gas law but includes two constants, *a* and *b*, to account for deviations from ideal behavior. (5.8)

van der Waals forces a general term for those intermolecular forces that include dipole–dipole and London forces. (11.5)

Vapor the gaseous state of any kind of matter that normally exists as a liquid or solid. (1.4)

Vaporization the change of a solid or a liquid to the vapor. (11.2)

Vapor pressure the partial pressure of the vapor over the liquid, measured at equilibrium at a given temperature. (5.5 and 11.2)

Vapor-pressure lowering a colligative property equal to the vapor pressure of the pure solvent minus the vapor pressure of the solution. (12.5)

Viscosity the resistance to flow that is exhibited by all liquids and gases. (11.4)

Volatile refers to a liquid or solid having a relatively high vapor pressure at normal temperatures. (11.2)

Volt (V) the SI unit of potential difference. (19.4)

Voltaic cell (galvanic cell) an electrochemical cell in which a spontaneous reaction generates an electric current. (19.1)

Volumetric analysis a method of analysis based on titration. (4.10)

Wavelength (λ) the distance between any two adjacent identical points of a wave. (7.1)

Wave mechanics see *Quantum mechanics.*

Weak acid an acid that is only partly ionized in water; it is a weak electrolyte. (4.4 and 15.1)

Weak base a base that is only partly ionized in water; it is a weak electrolyte. (4.4 and 15.1)

Weak electrolyte an electrolyte that dissolves in water to give a relatively small percentage of ions. (4.1)

Work an energy transfer (or energy flow) into or out of a thermodynamic system whose effect on the surroundings is equivalent to moving an object through a field of force. (6.2)

Work function the minimum energy that a photon can have and still eject an electron from a material in the photoelectric effect. It is frequently expressed in electron volts, eV. (7.2)

Zinc–carbon (Leclanché) dry cell a voltaic cell that has a zinc can as the anode; a graphite rod in the center, surrounded by a paste of manganese dioxide, ammonium and zinc chlorides, and carbon black, is the cathode. (19.8)

Zwitterion an amino acid in the doubly ionized form, in which the carboxyl group has lost an H^+ to give —COO$^-$ and the amino group has gained an H^+ to give —NH$_3^+$. (24.3)

Index

TABLE OF ATOMIC NUMBERS AND ATOMIC MASSES

Name	Symbol	Atomic Number	Atomic Mass	Name	Symbol	Atomic Number	Atomic Mass
Actinium	Ac	89	(227)	Molybdenum	Mo	42	95.96
Aluminum	Al	13	26.9815386	Neodymium	Nd	60	144.242
Americium	Am	95	(243)	Neon	Ne	10	20.1797
Antimony	Sb	51	121.760	Neptunium	Np	93	(237)
Argon	Ar	18	39.948	Nickel	Ni	28	58.6934
Arsenic	As	33	74.92160	Niobium	Nb	41	92.90638
Astatine	At	85	(210)	Nitrogen	N	7	14.0067
Barium	Ba	56	137.327	Nobelium	No	102	(259)
Berkelium	Bk	97	(247)	Osmium	Os	76	190.23
Beryllium	Be	4	9.012182	Oxygen	O	8	15.9994
Bismuth	Bi	83	208.98040	Palladium	Pd	46	106.42
Bohrium	Bh	107	(272)	Phosphorus	P	15	30.973762
Boron	B	5	10.811	Platinum	Pt	78	195.084
Bromine	Br	35	79.904	Plutonium	Pu	94	(244)
Cadmium	Cd	48	112.411	Polonium	Po	84	(209)
Calcium	Ca	20	40.078	Potassium	K	19	39.0983
Californium	Cf	98	(251)	Praseodymium	Pr	59	140.90765
Carbon	C	6	12.0107	Promethium	Pm	61	(145)
Cerium	Ce	58	140.116	Protactinium	Pa	91	231.03588
Cesium	Cs	55	132.9054519	Radium	Ra	88	(226)
Chlorine	Cl	17	35.453	Radon	Rn	86	(222)
Chromium	Cr	24	51.9961	Rhenium	Re	75	186.207
Cobalt	Co	27	58.933195	Rhodium	Rh	45	102.90550
Copernicum	Cn	112	(285)	Roentgenium	Rg	111	(280)
Copper	Cu	29	63.546	Rubidium	Rb	37	85.4678
Curium	Cm	96	(247)	Ruthenium	Ru	44	101.07
Darmstadtium	Ds	110	(281)	Rutherfordium	Rf	104	(267)
Dubnium	Db	105	(268)	Samarium	Sm	62	150.36
Dysprosium	Dy	66	162.500	Scandium	Sc	21	44.955912
Einsteinium	Es	99	(252)	Seaborgium	Sg	106	(271)
Erbium	Er	68	167.259	Selenium	Se	34	78.96
Europium	Eu	63	151.964	Silicon	Si	14	28.0855
Fermium	Fm	100	(257)	Silver	Ag	47	107.8682
Fluorine	F	9	18.9984032	Sodium	Na	11	22.98976928
Francium	Fr	87	(223)	Strontium	Sr	38	87.62
Gadolinium	Gd	64	157.25	Sulfur	S	16	32.065
Gallium	Ga	31	69.723	Tantalum	Ta	73	180.94788
Germanium	Ge	32	72.64	Technetium	Tc	43	(98)
Gold	Au	79	196.966569	Tellurium	Te	52	127.60
Hafnium	Hf	72	178.49	Terbium	Tb	65	158.92535
Hassium	Hs	108	(270)	Thallium	Tl	81	204.3833
Helium	He	2	4.002602	Thorium	Th	90	232.03806
Holmium	Ho	67	164.93032	Thulium	Tm	69	168.93421
Hydrogen	H	1	1.00794	Tin	Sn	50	118.710
Indium	In	49	114.818	Titanium	Ti	22	47.867
Iodine	I	53	126.90447	Tungsten	W	74	183.84
Iridium	Ir	77	192.217	Ununhexium	Uuh	116	(293)
Iron	Fe	26	55.845	Ununoctium	Uuo	118	(294)
Krypton	Kr	36	83.798	Ununpentium	Uup	115	(288)
Lanthanum	La	57	138.90547	Ununquadium	Uuq	114	(289)
Lawrencium	Lr	103	(262)	Ununseptium	Uus	117	(294)
Lead	Pb	82	207.2	Ununtrium	Uut	113	(284)
Lithium	Li	3	6.941	Uranium	U	92	238.02891
Lutetium	Lu	71	174.968	Vanadium	V	23	50.9415
Magnesium	Mg	12	24.3050	Xenon	Xe	54	131.293
Manganese	Mn	25	54.938045	Ytterbium	Yb	70	173.54
Meitnerium	Mt	109	(276)	Yttrium	Y	39	88.90585
Mendelevium	Md	101	(258)	Zinc	Zn	30	65.38
Mercury	Hg	80	200.59	Zirconium	Zr	40	91.224

A value in parentheses is the mass number of the isotope of longest half-life.

UNIT CONVERSIONS

LENGTH
SI base unit = meter, m

$1\ \text{Å} = 10^{-10}\ \text{m} = 100\ \text{pm}$
$1\ \text{in} = 2.54\ \text{cm}$ (exact)
$1\ \text{cm} = 0.3937\ \text{in}$
$1\ \text{yd} = 0.9144\ \text{m}$ (exact)
$1\ \text{m} = 1.0936\ \text{yd}$
$1\ \text{mi} = 1.6093\ \text{km}$
$1\ \text{km} = 0.6214\ \text{mi}$
$1\ \text{mi} = 5280\ \text{ft}$

MASS
SI base unit = kilogram, kg

$1\ \text{amu} = 1.660538921 \times 10^{-27}\ \text{kg}$
$1\ \text{lb} = 453.59237\ \text{g}$ (exact)
$1\ \text{kg} = 2.2046\ \text{lb}$
$1\ \text{lb} = 16\ \text{oz}$ (exact)
$1\ \text{ton} = 907.18474\ \text{kg}$ (exact)
$1\ \text{ton} = 2000\ \text{lb}$ (exact)
$1\ \text{metric ton} = 10^3\ \text{kg}$ (exact)
$1\ \text{metric ton} = 2204.6\ \text{lb}$

TEMPERATURE
SI base unit = kelvin, K

$0\ \text{K} = -273.15°\text{C}$
$0\ \text{K} = -459.67°\text{F}$

$$T_K = \left(t_C \times \frac{1\ \text{K}}{1°\text{C}}\right) + 273.15\ \text{K}$$

$$t_C = \frac{5°\text{C}}{9°\text{F}} \times (t_F - 32°\text{F})$$

$$t_F = \left(t_C \times \frac{9°\text{F}}{5°\text{C}}\right) + 32°\text{F}$$

VOLUME
SI derived unit = cubic meter, m³

$1\ \text{L} = 1\ \text{dm}^3$ (exact)
$1\ \text{mL} = 1\ \text{cm}^3$ (exact)
$1\ \text{L} = 1.0567\ \text{qt}$
$1\ \text{in}^3 = 16.387064\ \text{cm}^3$ (exact)
$1\ \text{gal} = 3.785411784\ \text{L}$ (exact)
$1\ \text{gal} = 4\ \text{qt}$ (exact)
$1\ \text{qt} = 16\ \text{fluid oz}$ (exact)

ENERGY
SI derived unit = joule, J

$1\ \text{J} = 1\ \text{kg·m}^2/\text{s}^2$ (exact)
$1\ \text{J} = 0.2390\ \text{cal}$
$1\ \text{J} = 1\ \text{C V} = 1\ \text{coulomb volt}$ (exact)
$1\ \text{cal} = 4.184\ \text{J}$ (exact)
$1\ \text{eV} = 1.6022 \times 10^{-19}\ \text{J}$
$1\ \text{eV/molecule} = 96.4853\ \text{kJ/mol}$
$1\ \text{J} = 9.4781 \times 10^{-4}\ \text{Btu}$

PRESSURE
SI derived unit = pascal, Pa

$1\ \text{Pa} = 1\ \text{kg/m·s}^2$ (exact)
$1\ \text{atm} = 101{,}325\ \text{Pa}$ (exact)
$1\ \text{atm} = 760\ \text{torr}$ (exact)
$1\ \text{torr} = 1\ \text{mmHg}$ (exact)
$1\ \text{bar} = 10^5\ \text{Pa}$ (exact)
$1\ \text{atm} = 14.696\ \text{lb/in}^2$

FUNDAMENTAL PHYSICAL CONSTANTS
(values from 2010 CODATA)

Atomic mass unit	$1\ \text{amu} = 1.660538921 \times 10^{-27}\ \text{kg}$	
Avogadro's number	$N_A = 6.02214129 \times 10^{23}\ \text{mol}^{-1}$	
Electron (proton) charge	$e = 1.602176565 \times 10^{-19}\ \text{C}$	
Electron mass	$m_e = 9.10938291 \times 10^{-31}\ \text{kg}$	
Faraday constant	$F = 9.64853365 \times 10^4\ \text{C/mol}$	
Molar gas constant	$R = 0.08205746\ \text{L·atm/(K·mol)}$	
	$= 8.314462\ \text{kPa·dm}^3/\text{(K·mol)}$	
	$= 8.314462\ \text{J/(K·mol)}$	
	$= 1.9872041\ \text{cal/(K·mol)}$	

Molar gas volume, STP $\quad V_m = 0.022413968\ \text{m}^3/\text{mol}$
$= 22.413968\ \text{L/mol}$

Neutron mass $\quad m_n = 1.674927351 \times 10^{-27}\ \text{kg}$

Planck's constant $\quad h = 6.62606957 \times 10^{-34}\ \text{J·s}$

Proton mass $\quad m_p = 1.672621777 \times 10^{-27}\ \text{kg}$

Speed of light (in vacuum) $\quad c = 2.99792458 \times 10^8\ \text{m/s}$ (exact)

Standard acceleration of gravity $\quad g = 9.80665\ \text{m/s}^2$ (exact)

LOCATIONS OF IMPORTANT INFORMATION